American College of Veterinary Internal Medicine (ACVIM) Forum 2013

Seattle, Washington, USA
12-15 June 2013

Editors:

Rance K. Sellon
J. Trenton McClure

ISBN: 978-1-62748-856-3

Printed from e-media with permission by:

Curran Associates, Inc.
57 Morehouse Lane
Red Hook, NY 12571

Some format issues inherent in the e-media version may also appear in this print version.

Printed by Curran Associates, Inc. (2013)

For permission requests, please contact the American College of Veterinary Internal Medicine at the address below.

American College of Veterinary Internal Medicine
1997 Wadsworth Boulevard Suite A
Lakewood CO 80214-5293

Phone: (800) 245-9081
Fax: (303) 231-0880

acvim@acvim.org

Additional copies of this publication are available from:

Curran Associates, Inc.
57 Morehouse Lane
Red Hook, NY 12571 USA
Phone: 845-758-0400
Fax: 845-758-2634
Email: curran@proceedings.com
Web: www.proceedings.com

ACVIM Specialty Symposium (Pre-Forum) Proceedings

ACVIM Cardiology

ACVIM Large Animal Internal Medicine - Equine

ACVIM Large Animal Internal Medicine - Food Animal

ACVIM Neurology

ACVIM Oncology

ACVIM Small Animal Internal Medicine

Veterinary Scientific Program Proceedings

AAEP

AAVN/ACVN (Food Animal)

ABVP

ACVCP/AAVPT

ACVIM Business Management

ACVIM Cardiology

ACVIM Large Animal Internal Medicine - Equine

ACVIM Large Animal Internal Medicine - Food Animal

Title / Author **Page#**

ACVIM Neurology

Title / Author **Page#**

ACVIM Oncology

ACVIM One Medicine

Title / Author **Page#**

ACVIM Small Animal Internal Medicine

ASVCP

AVHTM

CAPC

CGS

ISCAID

Liver Study Group

SCE

SVN/U

VCRS (Equine)

VCRS (Small Animal)

VECCS

Technician Program Proceedings

Technician

WELCOME TO THE 2013 ACVIM FORUM IN SEATTLE, WA

Welcome to Seattle! This is one of our organizations favorite settings for the Forum because it gives us a chance to couple a state-of the art meeting with a location that offers great entertainment, dining and hotel options. We have an outstanding program this year thanks to the efforts of the ACVIM Forum Program Committee, the ACVIM staff, our sponsors, and the many specialists and investigators who present the very latest advances in veterinary research to our attendees.

This year's Forum agenda is full, with a scientific program of approximately 270 lectures, more than 300 research abstracts, and the technician and business lecture series. Early morning and evening sessions also offer a variety of special interest group (SIG) sessions, workshops, case discussions, and labs that address diverse topics in veterinary medicine. Veterinary technicians will be pleased to see a scientific program and workshops designed specifically for them, and they are always welcome to attend any of the veterinary scientific sessions as well.

The Opening Thursday Night Reception, sponsored by Hill's Pet Nutrition and the ACVIM, provides a great opportunity to catch up with friends and colleagues. On Friday morning, please join us at the Keynote Address which will be delivered by Dr. Ford Bell, DVM. Dr. Bell is the current President of the American Alliance of Museums and his lecture, titled: "From Prostate Cancer to Museums: A Veterinary Journey" will inspire us with the story of his unusual and exciting life-journey as a veterinarian, researcher, educator, senatorial candidate, and advocate for non-profit organizations that promote science learning, conservation and community health. Following the Keynote address, please stay for the presentation of the first draft of the Small Animal Consensus Statement titled "the Safe Use of Chemotherapy in Veterinary Practice" or the first draft of the Large Animal Consensus Statement titled "Recommendations for Equine Athletes with Cardiovascular Abnormalities."

Research posters will be on display throughout the meeting. Take advantage of the Friday night Wine and Cheese Reception sponsored by Nestlé Purina to talk directly with the investigators. The ACVIM Resident Research awards will be announced during the reception as well.

To keep up with the latest in medical equipment and supplies, take a break and visit with our many exhibitors in the Exhibit Hall. You will be surprised by the number and diversity of companies that come to bring you the very latest in veterinary technology, pharmaceutical advances, superior nutritional products, educational materials, and career opportunities.

As you can see, there will be a lot to do while you're at the Forum. If you just can't make it every session in the scientific program that interests you, don't forget the Proceedings. While there is no substitute for the lectures, the Proceeding notes will serve to fill in the gaps.

The Forum is among the largest conferences in specialty medicine. The ACVIM is indebted to the 2013 ACVIM Forum Program Committee (led by the chair, Dr. Suzi White and the assistant chair, Dr. Frederic Gaschen) for the incredible effort they have invested to assemble a program of this quality and magnitude. We are also grateful to our many speakers, facilitators, abstract presenters, and moderators who are willing to share their knowledge, talent, and time to make this program happen. The ACVIM Forum continues to grow substantially in both size and quality due to a wide range of sponsor support and we greatly appreciate their contributions towards our efforts to provide outstanding continue education experience for the veterinary community. Roberta Herman (ACVIM CEO), Mya Sadler (ACVIM Director of Conference Management and Constituent Services) and the rest of the dedicated ACVIM staff members must also be acknowledged. They have handled all of the organizational details needed to bring this meeting to you, and are available at the registration desk to help you with any special needs or requests. Please take a few minutes to introduce yourself and to personally thank them for a job well done.

The members of the ACVIM Board of Regents would like to thank you for choosing the 2013 ACVIM Forum in Seattle as your source of professional continuing education, and we hope that our program well exceeds your expectations. Next year we'll be in Nashville, Tennessee (June 4 through 7) where we plan to provide you another outstanding continuing education opportunity in a great location! Mark your calendars and for now, we wish you a safe and rewarding visit to Seattle.

Dr. Virginia Buechner-Maxwell, DVM, MS, DACVIM (LAIM)
President, ACVIM Board of Regents

LETTER FROM THE EDITORS

We are pleased to be able to offer the Proceedings in flash drive format and online. You may purchase additional copies by contacting the ACVIM office. The unit cost is $49.00 (US) for a flash drive, which includes domestic shipping (additional charges apply with international shipping including Canada). Payment must accompany your order. To order by mail, send your check or money order (US funds only) to the ACVIM, 1997 Wadsworth Blvd., Lakewood, CO 80214-5293. To place your order using Visa, MasterCard or American Express, call the office at 800-245-9081 (US or Canada), 303-231-9933, or e-mail ACVIM@ACVIM.org.

Format

The manuscripts are arranged according to specialty group, then by session title. The sections designated in the Table of Contents correspond to the specialty headings in the grid. All programs for the Liver Study Group are listed together; all programs for the ACVIM Large Animal Internal Medicine Group are listed together; etc. The research abstracts and author index are also listed in the Table of Contents with their corresponding page numbers.

We acknowledge that not all manuscripts presented for publication conform to the guidelines distributed (e.g. outlines and paper length). We are thankful that the majority of speakers were able to keep their paper within the length and format specified.

Acknowledgments

These Proceedings represent a collaborative effort by a number of people. We greatly appreciate the time and commitment of the speakers that make these Proceedings possible, and the efforts of the moderators in reviewing submissions.

Rance K. Sellon
DACVIM (Small Animal Internal Medicine & Oncology)
Co-Editor

J Trenton McClure
DACVIM (Large Animal Internal Medicine)
Co-Editor

Ivy M. Leventhal
Publications Administrator

Fact Sheet

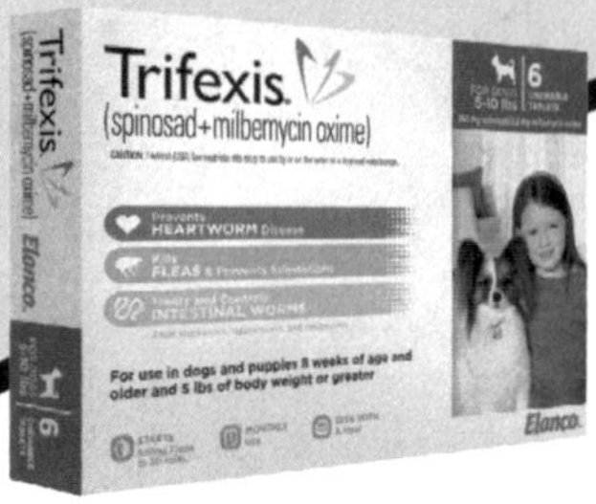

Trifexis®
(spinosad+milbemycin oxime)

Trifexis is an FDA-approved, monthly, chewable tablet indicated to kill fleas and prevent infestations, prevent heartworm disease and treat and control intestinal parasite infections (hookworm, roundworm and whipworm)

ACTIVE INGREDIENTS: Spinosad and milbemycin oxime

MODE OF ACTION: The primary target of action for spinosad is an activation of nicotinic acetylcholine receptors (nAChRs). Milbemycin oxime acts by binding to glutamate-gated chloride ion channels in invertebrate nerve and muscle cells. Increased permeability by the cell membrane to chloride ions causes hyperpolarization of affected cells and subsequent paralysis and death of the intended parasites.

ADMINISTRATION: Chewable, beef-flavored tablet administered orally once monthly for at least three months after the last exposure to mosquitoes. Available in five tablet sizes for dogs ranging from 5 to 120 lbs.[1] Give with food for maximum effectiveness.

EFFECTIVENESS: **Heartworm.** In a well-controlled laboratory study, Trifexis was 100% effective against induced heartworm infections when administered for 3 consecutive monthly doses. Two consecutive monthly doses did not provide 100% effectiveness against heartworm infection. In another well-controlled laboratory study, a single dose of Trifexis was 100% effective against induced heartworm infections.

Fleas. In a controlled laboratory study, spinosad began to kill fleas in 30 minutes and was 100% effective in 4 hours.[2] Kills fleas before they can lay eggs.[2]

Intestinal Parasites. A single dose of Trifexis was effective in removing adult roundworm, whipworm and hookworm. Spinosad did not interfere with the effectiveness of milbemycin oxime.[3]

USE: Trifexis is demonstrated safe for dogs and puppies 8 weeks of age and older and 5 lbs of body weight and over. Not for human use. Keep out of reach of children.

IMPORTANT SAFETY INFORMATION: Serious adverse reactions have been reported following concomitant extra-label use of ivermectin with spinosad alone, one of the components of Trifexis chewable tablets.

Treatment with fewer than three monthly doses after the last exposure to mosquitoes may not provide complete heartworm prevention. Prior to administration of Trifexis, dogs should be tested for existing heartworm infection. Use with caution in breeding females. The safe use of Trifexis in breeding males has not been evaluated. Use with caution in dogs with pre-existing epilepsy.

The most common adverse reactions recorded in clinical trials were vomiting, pruritus, lethargy and diarrhea. To ensure heartworm prevention, observe your dog for one hour after administration. If vomiting occurs within an hour of administration, redose with another full dose. Puppies less than 14 weeks of age may experience a higher rate of vomiting.

For product label, including complete safety information, see back.

[1] Dogs over 120 lbs. should be administered the appropriate combination of tablets
[2] Trifexis prescribing information [3] Trifexis FOI

TRIFEXIS®

(spinosad + milbemycin oxime) Chewable Tablets

Caution: Federal (USA) law restricts this drug to use by or on the order of a licensed veterinarian.

Description:
TRIFEXIS (spinosad and milbemycin oxime) is available in five sizes for oral administration to dogs and puppies according to their weight. Each chewable flavored tablet is formulated to provide a minimum spinosad dose of 13.5 mg/lb (30 mg/kg) and a minimum milbemycin oxime dose of 0.2 mg/lb (0.5 mg/kg). Spinosad is a member of the spinosyns class of insecticides, which are non-antibacterial tetracyclic macrolides. Spinosad contains two major factors, spinosyn A and spinosyn D, derived from the naturally occurring bacterium, *Saccharopolyspora spinosa*. Spinosyn A and spinosyn D have the chemical compositions $C_{41}H_{65}NO_{10}$ and $C_{42}H_{67}NO_{10}$, respectively. Milbemycin oxime is a macrocyclic lactone anthelmintic, containing two major factors, A_3 and A_4 of milbemycin oxime. The approximate ratio of A_3:A_4 is 20:80. Milbemycin A_4 5-oxime has the chemical composition of $C_{32}H_{45}NO_7$ and milbemycin A_3 5-oxime has the chemical composition of $C_{31}H_{43}NO_7$.

Indications:
TRIFEXIS is indicated for the prevention of heartworm disease (*Dirofilaria immitis*). TRIFEXIS kills fleas and is indicated for the prevention and treatment of flea infestations (*Ctenocephalides felis*), and the treatment and control of adult hookworm (*Ancylostoma caninum*), adult roundworm (*Toxocara canis* and *Toxascaris leonina*) and adult whipworm (*Trichuris vulpis*) infections in dogs and puppies 8 weeks of age or older and 5 pounds of body weight or greater.

Dosage and Administration:
TRIFEXIS is given orally, once a month at the minimum dosage of 13.5 mg/lb (30 mg/kg) spinosad and 0.2 mg/lb (0.5 mg/kg) milbemycin oxime body weight. For heartworm prevention, give once monthly for at least 3 months after exposure to mosquitoes (see **EFFECTIVENESS**).

Dosage Schedule:

Body Weight	Spinosad Per Tablet (mg)	Milbemycin oxime Per Tablet (mg)	Tablets Administered
5 to 10 lbs	140	2.3	One
10.1 to 20 lbs	270	4.5	One
20.1 to 40 lbs	560	9.3	One
40.1 to 60 lbs	810	13.5	One
60.1 to 120 lbs	1620	27	One
Over 120 lbs	Administer the appropriate combination of tablets		

Administer TRIFEXIS with food for maximum effectiveness. To ensure heartworm prevention, owners should observe the dog for one hour after dosing. If vomiting occurs within an hour of administration, redose with another full dose. If a dose is missed and a monthly interval between doses is exceeded, then immediate administration of TRIFEXIS with food and resumption of monthly dosing will minimize the opportunity for the development of adult heartworm infections and flea reinfestations.

Heartworm Prevention:
TRIFEXIS should be administered at monthly intervals beginning within 1 month of the dog's first seasonal exposure and continuing until at least 3 months after the dog's last seasonal exposure to mosquitoes (see **EFFECTIVENESS**). TRIFEXIS may be administered year round without interruption. When replacing another heartworm preventative product, the first dose of TRIFEXIS should be given within a month of the last dose of the former medication.

Flea Treatment and Prevention:
Treatment with TRIFEXIS may begin at any time of the year, preferably starting one month before fleas become active and continuing monthly through the end of flea season. In areas where fleas are common year-round, monthly treatment with TRIFEXIS should continue the entire year without interruption.

To minimize the likelihood of flea reinfestation, it is important to treat all animals within a household with an approved flea protection product.

Intestinal Nematode Treatment and Control:
TRIFEXIS also provides treatment and control of roundworms (*T. canis, T. leonina*), hookworms (*A. caninum*) and whipworms (*T. vulpis*). Dogs may be exposed to and can become infected with roundworms, whipworms and hookworms throughout the year, regardless of season or climate. Clients should be advised of measures to be taken to prevent reinfection with intestinal parasites.

Contraindications:
There are no known contraindications to the use of TRIFEXIS.

Warnings:
Not for human use. Keep this and all drugs out of the reach of children. Serious adverse reactions have been reported following concomitant extra-label use of ivermectin with spinosad alone, a component of TRIFEXIS (see **ADVERSE REACTIONS**).

Precautions:
Treatment with fewer than 3 monthly doses after the last exposure to mosquitoes may not provide complete heartworm prevention (see **EFFECTIVENESS**).

Prior to administration of TRIFEXIS, dogs should be tested for existing heartworm infection. At the discretion of the veterinarian, infected dogs should be treated with an adulticide to remove adult heartworms. TRIFEXIS is not effective against adult *D. immitis*. While the number of circulating microfilariae may decrease following treatment, TRIFEXIS is not indicated for microfilariae clearance (see **ANIMAL SAFETY**).

Mild, transient hypersensitivity reactions manifested as labored respiration, vomiting, salivation and lethargy, have been noted in some dogs treated with milbemycin oxime carrying a high number of circulating microfilariae. These reactions are presumably caused by release of protein from dead or dying microfilariae.

Use with caution in breeding females (see **ANIMAL SAFETY**). The safe use of TRIFEXIS in breeding males has not been evaluated.

Use with caution in dogs with pre-existing epilepsy (see **ADVERSE REACTIONS**).

Puppies less than 14 weeks of age may experience a higher rate of vomiting (see **ANIMAL SAFETY**).

Adverse Reactions:
In a well-controlled US field study, which included a total of 352 dogs (176 treated with TRIFEXIS and 176 treated with an active control), no serious adverse reactions were attributed to administration of TRIFEXIS. All reactions were regarded as mild.

Over the 180-day study period, all observations of potential adverse reactions were recorded. Reactions that occurred at an incidence >1% (average monthly rate) within any of the 6 months of observation are presented in the following table. The most frequently reported adverse reaction in dogs in the TRIFEXIS group was vomiting.

Average Monthly Rate (%) of Dogs With Adverse Reactions

Adverse Reaction	TRIFEXIS Chewable Tablets[a]	Active Control Tablets[a]
Vomiting	6.13	3.08
Pruritus	4.00	4.91
Lethargy	2.63	1.54
Diarrhea	2.25	1.54
Dermatitis	1.47	1.45
Skin Reddening	1.37	1.26
Decreased appetite	1.27	1.35
Pinnal Reddening	1.18	0.87

[a]n=176 dogs

In the US field study, one dog administered TRIFEXIS experienced a single mild seizure 2 ½ hours after receiving the second monthly dose. The dog remained enrolled and received four additional monthly doses after the event and completed the study without further incident.

Following concomitant extra-label use of ivermectin with spinosad alone, a component of TRIFEXIS, some dogs have experienced the following clinical signs: *trembling/twitching, salivation/drooling, seizures, ataxia, mydriasis, blindness and disorientation*. Spinosad alone has been shown to be safe when administered concurrently with heartworm preventatives at label directions.

In US and European field studies, no dogs experienced seizures when dosed with spinosad alone at the therapeutic dose range of 13.5-27.3 mg/lb (30-60 mg/kg), including 4 dogs with pre-existing epilepsy. Four epileptic dogs that received higher than the maximum recommended dose of 27.3 mg/lb (60 mg/kg) experienced at least one seizure within the week following the second dose of spinosad, but no seizures following the first and third doses. The cause of the seizures observed in the field studies could not be determined.

For technical assistance or to report a suspected adverse drug reaction, contact Elanco Animal Health at 1-888-545-5973. Additional information can be found at www.trifexis.com. Alternatively, suspected adverse drug reactions may be reported to FDA at 1-888-FDA-VETS or http://www.fda.gov/AnimalVeterinary/SafetyHealth/ReportaProblem/ucm055305.htm

Post Approval Experience (Mar 2012):
The following adverse reactions are based on post-approval adverse drug event reporting. The adverse reactions are listed in decreasing order of frequency: vomiting, depression/lethargy, pruritus, anorexia, diarrhea, trembling/shaking, ataxia, seizures, hypersalivation, and skin reddening.

Mode of Action:
The primary target of action of spinosad, a component of TRIFEXIS, is an activation of nicotinic acetylcholine receptors (nAChRs) in insects. Spinosad does not interact with known insecticidal binding sites of other nicotinic or GABAergic insecticides such as neonicotinoids, fiproles, milbemycins, avermectins and cyclodienes. Insects treated with spinosad show involuntary muscle contractions and tremors resulting from activation of motor neurons. Prolonged spinosad-induced hyperexcitation results in prostration, paralysis and flea death. The selective toxicity of spinosad between insects and vertebrates may be conferred by the differential sensitivity of the insect versus vertebrate nAChRs.

Milbemycin oxime, a component of TRIFEXIS, acts by binding to glutamate-gated chloride ion channels in invertebrate nerve and muscle cells. Increased permeability by the cell membrane to chloride ions causes hyperpolarization of affected cells and subsequent paralysis and death of the intended parasites. Milbemycin oxime may also act by disrupting the transmission of invertebrate neurotransmitters, notably gamma amino butyric acid (GABA).

Effectiveness:
Heartworm Prevention:
In a well-controlled laboratory study, TRIFEXIS was 100% effective against induced heartworm infections when administered for 3 consecutive monthly doses. Two consecutive monthly doses did not provide 100% effectiveness against heartworm infection. In another well-controlled laboratory study, a single dose of TRIFEXIS was 100% effective against induced heartworm infections.

In a well-controlled six-month US field study conducted with TRIFEXIS, no dogs were positive for heartworm infection as determined by heartworm antigen testing performed at the end of the study and again three months later.

Flea Treatment and Prevention:
In a well-controlled laboratory study, TRIFEXIS demonstrated 100% effectiveness on the first day following treatment and 100% effectiveness on Day 30. In a well-controlled laboratory study, spinosad, a component of TRIFEXIS, began to kill fleas 30 minutes after administration and demonstrated 100% effectiveness within 4 hours. Spinosad, a component of TRIFEXIS, kills fleas before they can lay eggs. If a severe environmental infestation exists, fleas may persist for a period of time after dose administration due to the emergence of adult fleas from pupae already in the environment. In field studies conducted in households with existing flea infestations of varying severity, flea reductions of 98.0% to 99.8% were observed over the course of 3 monthly treatments with spinosad alone. Dogs with signs of flea allergy dermatitis showed improvement in erythema, papules, scaling, alopecia, dermatitis/pyodermatitis and pruritus as a direct result of eliminating the fleas.

Treatment and Control of Intestinal Nematode Infections:
In well-controlled laboratory studies, TRIFEXIS was ≥ 90% effective in removing naturally and experimentally induced adult roundworm, whipworm and hookworm infections.

Palatability:
TRIFEXIS is a flavored chewable tablet. In a field study of client-owned dogs where 175 dogs were each offered TRIFEXIS once a month for 6 months, dogs voluntarily consumed 54% of the doses when offered plain as if a treat, and 33% of the doses when offered in or on food. The remaining 13% of doses were administered like other tablet medications.

Animal Safety:
TRIFEXIS was tested in pure and mixed breeds of healthy dogs in well-controlled clinical and laboratory studies. No dogs were withdrawn from the field studies due to treatment-related adverse reactions.

In a margin of safety study, TRIFEXIS was administered orally to 8-week-old Beagle puppies at doses of 1, 3, and 5 times the upper half of the therapeutic dose band, every 28 days for 6 dosing periods. Vomiting was seen in all groups including control animals with similar frequency. Adverse reactions seen during the course of the study were salivation, tremors, decreased activity, coughing and vocalization.

Body weights were similar between control and treated groups throughout the study. Treatment with TRIFEXIS was not associated with any clinically significant hematology, clinical chemistry or gross necropsy changes. One 5X dog had minimal glomerular lipidosis observed microscopically. The clinical relevance of this finding is unknown.

Plasma spinosyn A, spinosyn D, milbemycin A_3 5-oxime and milbemycin A_4 5-oxime concentrations increased throughout the study. At each dosing period, plasma spinosyn A and spinosyn D concentrations were greater than proportional across the dose range 1 to 5X. Plasma milbemycin A_4 5-oxime concentrations appeared to be dose proportional across range 1 to 5X by the end of the study. Plasma concentrations of spinosad and milbemycin oxime indicate that expected systemic exposures were achieved throughout the study.

In an avermectin-sensitive Collie dog study, TRIFEXIS was administered orally at 1, 3, and 5 times the upper half of the recommended therapeutic dose band every 28 days. No signs of avermectin sensitivity were observed after administration of TRIFEXIS during the study period to avermectin-sensitive Collie dogs. The adverse reactions observed in the treatment groups were vomiting and diarrhea. Body weights in all treatment groups were comparable to the control group. Hematology and clinical chemistry parameters showed no clinically significant changes from study start to end, and all dogs were considered healthy throughout the study.

In a heartworm positive safety study, TRIFEXIS was administered orally at 1, 3, and 5 times the upper half of the therapeutic dose band to Beagle dogs with adult heartworm infections and circulating microfilariae, every 28 days for 3 treatments. Vomiting was observed in one dog in the 1X group, in three dogs in the 3X group, and in one dog in the 5X group. All but one incident of vomiting was observed on the treatment day during the first treatment cycle. The vomiting was mild and self-limiting. Hypersensitivity reactions were not observed in any of the treatment groups. Microfilariae counts decreased with treatment.

In a reproductive safety study, TRIFEXIS was administered orally to female dogs at 1 and 3 times the upper half of the therapeutic dose band every 28 days prior to mating, during gestation and during a six-week lactation period. Dogs with confirmed fetal heartbeats on ultrasound examination were evaluated for reproductive safety. One 3X and one 1X group female did not become pregnant. No treatment-related adverse reactions or signs of avermectin toxicosis were noted for adult females. Adult females in the 3X group lost weight during the 6-week pre-mating period, while control group females gained weight during that time. The body weights of the treated groups were comparable to the control group during gestation and post-parturition phases of the study. Gestation length, litter average body weight, litter size, stillborn pups, pup survival and the proportion of pups with malformations were comparable between treated and control dam groups. Malformations in the 1X group included a pup with cleft palate and a littermate with anophthalmia, fused single nares, misshapen palate, hydrocephalus, omphalocele and malpositioned testes; a pup with a malformation of the anterior tip of the urinary bladder and umbilical blood vessel; and a pup with patent ductus arteriosus (PDA). Malformations in the 3X group included three littermates with PDA. Malformations in the control group included a pup with a malformed sternum and a pup with PDA and a malpositioned superior vena cava. Clinical findings in pups of the treated groups were comparable to the control group except for one 1X group pup that was smaller and less coordinated than its littermates and had tremors when excited. The relationship between spinosad and milbemycin oxime treatment and the 1X and 3X dogs that did not become pregnant, the specific pup malformations and the unthrifty 1X group pup are unknown. The incidence of cleft palate is not unexpected based on the historical data collected at the breeding site.

In a margin of safety study with spinosad alone, 6-week old Beagle puppies were administered average doses of 1.5, 4.4, and 7.4 times the maximum recommended dose at 28-day intervals over a 6-month period. Vomiting was observed across all treatments, including controls, and was observed at an increased rate at elevated doses. Vomiting most often occurred 1 hour following administration and decreased over time and stabilized when puppies reached 14 weeks of age.

Storage Information:
Store at 20-25°C (68-77°F), excursions permitted between 15-30°C (59-86°F).

How Supplied:
TRIFEXIS is available in five tablet sizes. Each tablet size is available in color-coded packages of 6 tablets.

5-10 lbs (140 mg spinosad and 2.3 mg milbemycin oxime)
10.1-20 lbs (270 mg spinosad and 4.5 mg milbemycin oxime)
20.1-40 lbs (560 mg spinosad and 9.3 mg milbemycin oxime)
40.1-60 lbs (810 mg spinosad and 13.5 mg milbemycin oxime)
60.1-120 lbs (1620 mg spinosad and 27 mg milbemycin oxime)

NADA 141-321, Approved by the FDA

Manufactured for Elanco Animal Health,
A Division of Eli Lilly & Company
Indianapolis, IN 46285

Elanco and TRIFEXIS are trademarks owned or licensed by Eli Lilly and Company, its subsidiaries or affiliates.

NDC 0986-4332-06
NDC 0986-4333-06
NDC 0986-4334-06
NDC 0986-4335-06
NDC 0986-4336-06
CA4332AM
CA4333AM
CA4334AM
CA4335AM
CA4336AM

PA086076AMB_Mkt1 (V03-09-2012)

KEYNOTE ADDRESS

Inspiring Events for Connecting and Learning

FRIDAY

JUNE 14

10:30 – 11:20 am

Keynote Address
From Prostate Cancer to Museums:
A Veterinary Journey

WSCC - Room 6B

Extra-Ordinary Veterinarian: Dr. Ford Bell

Ford W. Bell has more than 30 years of experience as a nonprofit executive, board chair, donor, trustee, educator and scientist. He has a unique background that spans veterinary oncology, nonprofit leadership and advocacy for a variety of causes. He received his DVM from the University of Minnesota in 1982, and he formerly served as a Clinical Assistant Professor of Oncology at the University of Minnesota's College of Veterinary Medicine. He is currently president of the American Association of Museums, which is the largest museum organization in the world. Bell's many visits to the James Ford Bell Museum of Natural History, named after his grandfather, helped him develop a lifelong love of nature and science and influenced his decision to become a veterinarian. Don't miss this fascinating keynote address.

Sponsored by: Hill's

OPEN-HEART SURGERY FOR CORRECTION OF CONGENITAL HEART DEFECTS: NIHON UNIVERSITY EXPERIENCE

Masami Uechi, DVM, Phd
Yokohama, Japan

Congenital heart disease in dogs and cats

Severe congenital heart disease has a poor prognosis. Although surgical intervention is required in these cases, open- heart surgery has not commonly been applied. Recently, cardiopulmonary bypass has been performed successfully in small-breed dogs and cats. This has led to an increasing success rate for open-heart surgery in the treatment of mitral regurgitation and congenital heart disease. We have reported several successful open heart procedures under cardiopulmonary bypass in small-breed dogs and cats: e.g., surgical techniques for reducing severe pulmonic stenosis, including open and closed patch grafting, pulmonary commissurotomy, and valvotomy; closure of a septal defect; and repair of mitral and tricuspid valves.

Septal defects

Atrial and ventricular septal defects are one of the most commonly observed congenital heart diseases. Patients with small defects are usually managed conservatively (e.g., with vasodilators) to minimize the risk of cardiac failure. For cases with large defects or with echocardiographic evidence of a significant shunt, surgical correction is recommended. In dogs, a ratio of pulmonary to systemic vascular resistance (Qp/Qs) >2.5 warrants surgical intervention; however, surgical repair for the management of a septal defect is not common, because of financial and technical issues. We performed successful open-heart surgery in dogs and cats with atrial or ventricular septal defects.

In cats, atrial or ventricular septal defect detected by echocardiography was confirmed. Under cardiopulmonary bypass, open surgical repair was achieved with an expanded polytetrafluoroethylene (e-PTFE) graft. The priming volume of the cardiopulmonary bypass circuit was minimized by cutting the cardiopulmonary bypass tubing and partially replacing the priming fluid with whole cat blood. To prevent hemodilution associated with the use of cardioprotective agents, surgery was performed on the beating heart. At 1-year echocardiographic evaluation, the repair was intact, and at 3 years, the cats were alive with no need of medication. Thus, atrial or ventricular septal defect in cats can be repaired using e-PTFE under cardiopulmonary bypass.

Endocardial cushion defect

Endocardial cushion defect is classified into partial and complete forms based on the anatomic involvement. Partial endocardial cushion defect is characterized by primum atrial septal involvement with a septal mitral leaflet cleft, whereas a complete defect is characterized by the presence of both primum atrial septal defect and inlet ventricular septal defect with a common atrioventricular valve. When complete endocardial cushion defect is present, atrioventricular valve regurgitation and left-to-right shunting may develop, which are likely to result in pulmonary hypertension and a poor prognosis. Early surgical correction in this case can improve the prognosis.

Surgical repair of a complete endocardial cushion defect was performed in 2 dogs. Echocardiographic examination revealed an ostium primum atrial septal defect, an inlet ventricular septal defect, mitral regurgitation and tricuspid regurgitation; thus, complete endocardial cushion defect was diagnosed. Surgical correction was performed under cardiopulmonary bypass via a right atriotomy. An e-PTFE patch was secured along the margin of the inlet ventricular septal defect using a simple continuous suture, then the cleft in the septal mitral leaflet was sutured. The cleft in the septal tricuspid leaflets was sutured in a similar fashion. To complete the inlet ventricular septal defect closure, the ventricular septal defect patch was secured to these sutured leaflets by a simple continuous suture. Another e-PTFE patch was used to close the ostium primum atrial septal defect. After surgery, mitral regurgitation, tricuspid regurgitation, and interventricular shunting were decreased. The dogs were alive 5 and 8 years after the surgery, with no evidence of an interventricular shunt, tricuspid regurgitation, or other clinical signs.

Pulmonic stenosis

Small dogs and cats with severe pulmonic stenosis can be appropriate candidates for surgical correction under cardiopulmonary bypass. We have performed valvotomy with or without biomembrane patch grafting in cases of severe pulmonic stenosis. The subjects comprised dogs and a cat, weight 2.1–4.8 kg, that were diagnosed with severe pulmonic stenosis and right ventricular hypertrophy, with an average preoperative pressure gradient of >80 mmHg on echocardiography. The pressure overload caused by the severe pulmonic stenosis was reduced by valvotomy: i.e., open pulmonary valve commissurotomy, with or without biomembrane patch grafting, under cardiopulmonary bypass. The reduced pressure gradient was maintained at about 50 mmHg at final follow up. Thus, open valvotomy, pulmonary valve commissurotomy, and biomembrane patch grafting were effective in reducing the obstruction in severe pulmonic stenosis in dogs and cats.

Open valvotomy may induce pulmonic regurgitation, so valve replacement may be required in such cases. We have developed autologous prosthetic implants using simple and safe in-body tissue architecture technology. An autologous valved conduit with the sinus of Valsalva (BIOVALVE), manufactured using this unique technology, was subsequently used to replace the pulmonary valves in a beagle model. A mold from the BIOVALVE organization was assembled using 2 types of specially designed silicone rods with a small, trileaflet-shaped aperture between them. The concave rods had 3 projections that resembled the protrusions of the sinus of Valsalva. The molds were placed in the dorsal subcutaneous space of beagle dogs for 4 weeks, until they were covered with autologous connective tissue. Thus, BIOVALVEs with 3 leaflets on the inner side of the conduit, with the sinus of Valsalva, were created. These valves had adequate burst strength, similar to that of native valves. Tight valvular coaptation and a sufficient open orifice area were observed in vitro. These BIOVALVEs were implanted to the main pulmonary arteries as allogenic conduit valves. Postoperative echocardiography demonstrated smooth movement of the leaflets with trivial regurgitation. Histological examination of specimens obtained at 84 days showed that the surface of the leaflet was covered by endothelial cells and neointima, including an elastin fiber network formed on the luminal surface of the conduit at the anastomosis sites.

When successfully implanted, these BIOVALVEs may be new candidate valves for pulmonary valve replacement. They could possibly also be used for aortic valve replacement.

PEARLS AND PITFALLS IN THE TREATMENT OF PULMONIC STENOSIS

Amara H. Estrada, DVM, DACVIM (Cardiology), Gainesville, FL
Sophy A, Jesty, DVM, DACVIM (Cardiology and Internal Medicine,) Knoxville, TN
N. Sydney Moise, DVM, DACVIM (Cardiology and Internal Medicine), Ithaca, NY
Romain Pariaut, DVM, DACVIM (Cardiology,) Baton Rouge, LA

INTRODUCTION

The following summarizes information from the limited published data and the collective experiences from four universities in treating pulmonic valve stenosis using balloon valvuloplasty.

SELECTION OF PATIENTS

One of the first questions regarding balloon valvuloplasty (BV) for dogs with pulmonic stenosis (PS) is when to intervene. Firm guidelines for deciding which animals need to be treated have not been developed. In pediatric cardiology, most interventionalists agree that treatment should be performed in any child with moderate stenosis or a gradient > 50 mmHg. Most veterinary cardiologists would agree that animals with gradients in the severe range should be treated and a recent study[1] showed that due to a greater risk of cardiac death, dogs with a gradient > 60 mmHg may benefit from BV. Reliance on Doppler derived estimates of gradient alone however, may not be sufficient to make a treatment decision. Several other factors should be taken into consideration when making the decision for treatment, such as the degree of RV hypertrophy, presence of clinical signs, and severity of concurrent tricuspid insufficiency.

Previously, BV was considered contraindicated in English bulldogs with a coexisting R2A-aberrant coronary artery or dogs with tetralogy of Fallot. However, recent reports have shown that treatment might be feasible. In English bulldogs with right sided congestive heart failure due to PS and an aberrant coronary artery, less aggressive dilation with a balloon that is 0.8 to 1.0 times the size of the pulmonary annulus has been shown to be safe and possibly effective.[2] In fact, the very 1st BV in a child in 1982, the safety and efficacy of the procedure was tested in 1980 in an English Bulldog with an R2A aberrant coronary artery.[3] A small balloon was used in this test procedure, and fatal disruption of the anomalous left coronary artery (CA) did not occur. Selected dogs with tetralogy of Fallot may also benefit from balloon valvuloplasty if there is truly a valvular component that is amenable to this .[4-6]

Patients with dynamic stenosis are another group which deserves special consideration and planning for intervention. Our group has found that titration of atenolol dose to achieve a heart rate of 70-80 bpm prior to intervention can be helpful for guidance of catheters into the outflow tract and also likely helps to reduce the occurrence and/or severity of acute exacerbations of a dynamic stenosis. Other steps we take in the attempt to reduce this complication include use of specific anesthetic protocols which allow for low heart rates, (the authors prefer a heart rate of 60-90 bpm) and volume loading with intravenous crystalloids pre-operatively. Our typical anesthetic protocol includes premedication with an opioid combined with a benzodiazepine intravenously (typically fentanyl and midazolam). Induction is then obtained with etomidate and anesthesia is maintained with a fentanyl CRI and lidocaine CRI in addition to very low concentrations of inhalants (isoflurane or sevoflurane). In the crash cart besides our emergency drugs we also have doses ready of esmolol, hypertonic saline and hetastarch. We have found this to be particularly helpful in small dogs with very severe stenosis. We often also perform bubble studies pre-operatively in our patients in attempt to identify any right to left shunting through a patent foramen ovale. Knowing that this exists prior to intervention can be helpful in the planning of the anesthetic protocol and for trouble shooting during intervention should significant shunting develop. Use of pressors such as phenylephrine can be helpful in reducing or balancing a right to left shunt at the atrial level.

EQUIPMENT CONSIDERATIONS WHEN PERFORMING BALLOON VALVULOPLASTY

Diameter of the dilation balloon

The diameter of the balloon is based upon the size of the annulus of the pulmonic valve. This measurement will depend on the resolution of the image, imaging plane, imaging modality, time during the cardiac cycle, heart rate, volume loading, and the definition held by the observer in identifying the hinge points of the pulmonic valve. Obviously, the higher the resolution of the image the more accurate the measurement can be and this translates into a higher repeatability. When using transthoracic echocardiography, selection of the highest frequency ultrasound probe will improve resolution. Importantly, use of the octave function, which permits the sending of lower frequency sound waves and the return of higher frequency sound waves (taking advantage of harmonics), will also improve image quality. When using echocardiography, the pulmonic valve may be seen in several different planes, including at least three views from the right parasternal position and one view from the left apical position. There is no 'best' view. The ideal view from which the measurement is obtained is that which shows the valve hinge points with the greatest clarity. Transesophageal echocardiography can also be used to view the pulmonic valve and the hinge points. Measurements of the pulmonic valve annulus were initially described from angiocardiographic images. There are several potential problems that must be considered when measurements are taken via radiography or cine-loops. First, the amount of magnification must be considered and corrected for when calculating the actual measurement. Either internal or external calibration with a marker catheter or ruler should be used to adjust the magnified image. The risk of using an external marker is that the degree of magnification between the marker and the valve may differ on the image if it is not carefully placed on the same horizontal plane. Higher frame rates during angiography will more likely permit the cardiac cycle to be considered. Usually a measurement is determined during systole. There are significant differences in the values obtained when the pulmonic annulus is measured in systole as opposed to diastole. If a slow frame rate with no timing is used, the cardiac cycle is estimated. Another potential variable in delineating the valve leaflets and hinge points may rest on whether the angiocardiogram was obtained using a power injector or contrast was injected by hand. The volume of contrast and the speed of the injection will alter the appearance of the angiogram and will therefore influence the clarity of the images to be measured. The heart rate, which affects diastolic filling time and thus diastolic volume, will influence stroke volume and thus, the systolic diameter that is measured. Usually, attempt is made to inject the contrast within one cardiac cycle, but it is not possible if the heart rate is elevated. Finally, even amongst cardiologists, the identification of the pulmonic valve hinge points is highly variable. One solution to the latter problem is that in a particular catheterization lab, a consensus on the definition of the image hinge points of the pulmonic valve is established. With this type of preparation, the results that occur can be individualized amongst that group of individuals.

The size of the balloon is calculated using the hinge point measurement of the pulmonic valve; this is known as the balloon:annulus ratio (BAR). For single ballooning, the ratio usually varies from 1.2 to 1.4. Sometimes a lower ratio is used, as discussed in dogs with aberrant coronary arteries, or a larger ratio is used when smaller ratios have been unsuccessful (but rarely greater than 1.5 BAR). In past years, balloons were manufactured in 2 mm increments (e.g. 10 mm, 12 mm, 14 mm, etc.). Today it is possible to obtain balloons in 1 mm increments, which affords more accuracy and should theoretically provide the opportunity for more success.

Length of the dilation balloon

Another consideration for selection of dilation balloons is the length. To determine the proper length of the dilation balloon, the distance between the pulmonic valve and the protrusion (cap) of the pulmonic trunk (post-stenotic dilation) is measured; this is the maximum length of the balloon to be used. A dilation balloon that is too long can lead to damage to the main pulmonary artery during inflation, particularly when the balloon is not kept centered over the valve and instead becomes trapped between the valve and main pulmonary artery during inflation. Tearing of the pulmonary artery or avulsion of the pulmonary trunk is a potentially fatal complication of a dilation balloon that is excessively long. Another potential complication with a long balloon is damage to the tricuspid valve when the inflated balloon is trapped proximal to the pulmonic valve leaflets. A dilation balloon that is too short can also be problematic. In this situation, the balloon is more difficult to pin across the pulmonic valve and it will constantly bounce back and forth with the cardiac cycle during the time of the inflation. We attempt to manipulate the guide wire during inflation such that any forward or backward motion of the balloon created during inflation is countered with an opposite push or pull of the guidewire to keep the balloon centered on the valve.

Low and high profile balloons

Although low profile balloons permit the use of smaller introducers, which is usually advantageous for small dogs, the amount of pressure that these balloons can endure is low. In some cases higher pressures are required in order to tear the valve leaflets to permit a drop in the pressure gradient.[7] If a low profile balloon is inflated too aggressively, it is possible to rupture the balloon. When hand inflation is used with a small syringe and a hard push, it is possible to rupture the low profile balloons. Because balloons are filled with diluted contrast (one part contrast media to five parts sterile saline) there is likely no danger to the patient; however, there is the expense of a dilation balloon that is now not reusable. With an inflation device, the actual atmospheres of pressure can be monitored.

Guidewires

Although the balloon catheter manufacturers suggest the size of the exchange wire, additional considerations which can have a profound effect during the procedure must be considered. Consideration of whether the exchange wire is considered stiff or superstiff is important. The guide wires should be twice the length of the catheter in order to allow for exchanging of balloon wedge pressure catheters and dilation balloons, and guide wires should be checked to ensure that these fit all the way through beforehand. Careful placement of guidewires in the appropriate location deep enough in the left pulmonary artery to provide enough support for the balloon, but careful enough placement to avoid damage to the vascular endothelium is important. Use of a super stiff guidewire may cause more arrhythmias and tensing of the tricuspid valve in small dogs so we often use regular stiff wires in these small patients. Sometimes, however, these wires do not provide enough support for allowing the balloon dilation cathether to be passed around the right ventricular chamber and into the right ventricular outflow tract and a super stiff wire is necessary. We routinely have dogs on a lidocaine constant rate infusion during passage of the guidewires and balloons with bolus injections of lidocaine given immediately before balloon inflation.

PROCEDURAL CONSIDERATIONS FOR BALLOONS VALVULOPLASTY

Large pulmonic valve annulus

Dogs with annuli of > 18 mm in diameter may be problematic. Our experience has been that larger balloons are not able to generate enough pressure to effectively dilate the valves of these larger dogs, even with inflation devices. Additionally, these larger single balloons often require large introducers, are slow to inflate and deflate, are more difficult to maintain catheter position and are more frequent to eject during cardiac systole generating avulsing forces. Therefore, we often use a double ballooning technique (from either the jugular veins or femoral veins) in Italy dogs with annuli measuring greater than 18 mm.[8] We also now routinely use pressure inflation devices for these cases in order to maximize the tearing versus simple stretching of the valve. Other benefits of this technique are the rapid inflation and deflation possible with two smaller balloons, better maintenance of cardiac output during inflation, and two smaller introducers with less concern for post-operative bleeding if using a percutaneous approach.

Small Dogs

Severe pulmonic stenosis in very small dogs (2-3 kg) represents another challenge. A major problem is positioning in the outflow tract. It is important to select a catheter with the right curve which permits passage through the outflow tract and then positioning across a very stenotic valve. We have used tip deflectors in an angled glide catheter (as small as 4 Fr, which still has an 0.038" internal diameter), shaped the tip of a catheter with a heat gun, and used very small gauge and flexible guidewires or used hydrophilic guide wires (which are 'slippery' allowing you greater flexibility in manipulation across a valve, but also easily dislodged because they are so slippery, expensive and non-reusable) to get across very stenotic valves. These very small dogs with very severe PS can desaturate quickly, especially if there is a right-to-left shunting patent foramen ovale. When across the pulmonic valve, right ventricular output drops, which increases shunting. Even without a shunt, simply being across a severely stenotic valve in a small dog can substantially drop systemic pressure. The most important thing to keep in mind is that the only way to improve the situation is to open the valve, but working efficiently and quickly is important.

Dogs with large right atria/tricuspid dysplasia

Patients with severely enlarged right atria represent another challenging situation in which actually guiding angiographic catheters, guidewires and balloon wedge pressure catheters into the right ventricular chamber and/or right ventricular outflow tract is difficult. In these situations, we have found that it is sometimes beneficial to change the direction of the approach (i.e. come from the femoral vein instead of jugular vein). When we have patients with moderately to severely enlarged right atria, we routinely prepare both femoral and jugular introducer sites just in case we need to change direction. Other 'tricks' that we have learned are the use of tip deflectors or partial inflation of the balloon dilation catheter to give more support to the catheter.

EXPECTATIONS FOR OUTCOME

The response to treatment with balloon valvuloplasty is variable. Outcome is likely dependent on multiple factors. Some of these include the following:

1. Type of pulmonic stenosis: purely valvular stenosis without leaflet thickening and a normal annulus size (type A) is likely to have a better outcome vs.a hypoplastic annulus and/or pulmonic valve with marked leaflet thickening (type B)[9]
2. The amount that the pressure gradient drops is somewhat dependent on where it began. A dog with a purely valvular stenosis and extremely high gradient will have a greater fall than a dog with a dysplastic valve or a dog with a gradient that marginally required treatment. We have arbitrarily defined success as a gradient that is less than 50 mmHg or a drop of more than 50%. Of course, it may be optimal to have a gradient less than 20 mmHg.
3. Experience of cardiologist
4. Attention to the details of measurement and selection of proper equipment
5. Intracardiac pressures should be measured during the procedure to document before and after pressure gradients. However, it is common and expected that these gradients under anesthesia and with a different methodology are quite different from those obtained with Doppler echocardiography. During the procedure, when comparisons are made before and after balloon dilation, it is vital that the systemic blood pressure be taken and comparisons made when these are similar. If the systemic pressure has fallen greatly under anesthesia at the end of the procedure the gradient across the pulmonic valve will be less. Using a ratio of pulmonary:systemic pressure is often helpful for making comparisons intra-operatively

6. With severe pulmonic valve stenosis that has resulted in marked right ventricular hypertrophy, the initial drop across the pulmonic valve may not be as great as that observed approximately 3 months after the procedure. A further drop in the gradient across the pulmonic valve is not usually apparent unless the dog is in the severe category. This further drop may be due to remodeling of the right ventricular wall and interventricular septum and resolution of secondary obstruction.
7. If the balloon dilation catheter does not develop a waist during the inflation, which is then seen to disappear at the peak of inflation, it is likely that the required damage to the valve to permit the relief of the obstruction has not occurred. In such cases, decisions must be made with regards to selecting a larger balloon, a balloon that can withstand higher inflation pressures, or an acceptance that the pathology is not amenable to this treatment.

PROBLEMS THAT CAN OCCUR

The major complications of balloon valvuloplasty include cardiac perforation, arrhythmias and damage to the right ventricular outflow tract, pulmonary artery and/or tricuspid valve apparatus during the procedure with the development of significant tricuspid insufficiency post-operatively[10-13]. Other complications that our group has experienced are endothelial tears or dissection of the pulmonary artery[14], avulsion of the pulmonary valve leaflet or main pulmonary artery, reactive infundibular obstruction also known as a 'suicide right ventricle'[15], and acute pulmonary reperfusion hemorrhage[16]. Additional rare complications that are reported in the literature that we have not experienced include traumatic pulmonary valve hematomas[17], dissecting aneurysm of the main pulmonary artery[18], and transient complete heart block[19].

REFERENCES

[1]Francis AJ, et al. **JSAP** 2011;52:282-288. [2]Fonfara S, et al. **JVIM** 2010;24:354-359.[3]Buchanan JW, et al. **JVIM** 2002; 16(1):116-117. [4]Wu ET, et al. **Card** 2006; 105(1) 52-56. [5]Robinson JD, et al. **J Thor Card Surg** 2011; 142(6): 1367-1373. [6]Sreeram N, et al. **JACC** 1991:18(1): 159-165. [7]Moquillansky D, et al. **Cong Heart Dis** 2010; 5(2): 134-140. [8]Estrada AH, et al. **JVIM** 2006; 20(4)862-872. [9]Bussadori C, et al. **JVIM** 2001; 15(6):553-558. [10]Rao PS. **J Invas Card** 2005; 17:323-325. [11]Chen CR, et al. **NEJM** 1996;335:21-25. [12]Kan JS, et al. **NEJM** 1982;307:540-542. [13]McCrindle BW, et al. **Circ** 1991;83:1915-1922. [14]Bogdan J, et al. **JASE** 2006; 19(9): 1191.e5-1191.e8. [15]Buheitel G, et al. **Z Kardiol** 1999; 88(5):347-352. [16]Cheng HI, et al. **J Chin Med Assoc** 2009; 72(11):607-610. [17]Özeren M, et al. **Anadolu Kardiyol Derg**. 2012;12(7):605-606.[18]Li Q, et al. **Circ** 2009 Feb 10;119(5):761-763. [19]Steinberg C, et al. **Pediatr Cardiol** 1992;13(3):181-183.

LEFT VENTRICULAR OUTFLOW TRACT OBSTRUCTIONS IN DOGS: WHAT WE KNOW AND DON'T KNOW

David Sisson, DVM, DACVIM (Cardiology)
Corvallis, Oregon

INTRODUCTION

Left ventricular outflow tract (LVOT) obstructions are detected commonly in dogs with congenital heart disease and many of the anatomic variants documented in people have also been identified in canine patients. It is noteworthy that coarctation and interruption of the aorta as well as supravalvular aortic stenosis occur rarely in dogs and are worthy of only passing reference in the context of this presentation. Elastin deficiency, as occurs in humans with the Williams-Beurens syndrome and other nonsyndromic conditions resulting in supravalvular aortic stensosis, has not been documented in dogs.[1] Congenital valvular aortic stenosis, most often manifested as a bicuspid or less commonly as a unicuspid aortic valve, is particularly common in humans is very uncommon in dogs.[2] Valvular aortic stenosis does occur in bull terriers and mini bull terriers but the nature of the valvular pathology is quite different from the human disorder in that the aortic leaflets of affected bull terriers are thickened and inflexible but usually are not fused with a normal tricuspid configuration. Fixed subvalvular aortic stenosis is a common disorder in dogs, comprising 20 to 25% of the congenital cardiac malformations recognized in dogs.[3,4] In contrast, fixed subvalvular aortic stenosis (SAS) comprises about 6.5% of the defects found in adults with congenital heart disease.[2] In both dogs and humans, SAS is sometimes found in association with pulmonic stenosis, mitral valve dysplasia, patent ductus arteriosus, ventricular septal and atrioventricular septal defects.[5-7] Insight into these associations is beginning to emerge as the developmental interrelationships and co-dependencies of these different anatomic constituents of the growing heart are understood. Dynamic subvalvular aortic stenosis (DSAS) also occurs in humans and dogs, but the predominance of hypertrophic cardiomyopathy as the underlying disorder has not been conclusively established in dogs.[8] Clarification of the etiology of DSAS is often confounded by the occurrence of systolic anterior motion (SAM) together with the anatomic features of fixed SAS or in association with other anatomic derangements such as mitral valve dysplasia, pulmonic stenosis, and tetralogy of Fallot.[9-13]

PATHOLOGIC FORMS OF LVOT OBSTRUCTION IN DOGS

The outflow tract lesions in humans with fixed SAS are often conveniently classified either as short- or long-segment obstructions. In the former, a discrete fibroelastic ridge or membrane is located below or at the base of the right and non-coronary cusps of the aortic valve leaflets and may extend as a horseshoe-shaped lesions onto the anterior leaflet of the mitral valve or as a complete ring encircling the entire LVOT.[14] This lesion may be present at birth but often develops or becomes more severe during adolescence, often in dramatic fashion and with severe consequences. Some of the more interesting observations made in some affected human patients include displacement of the aorta to the right of the crest of the interventricular septum, reduction in the angle of the aorta relative to the interventricular septum, and an increase in the distance between the mitral and aortic valves (mitral-aortic distance).[14, 15] These highly variable alterations are thought to cause flow disturbances in the LVOT promoting the growth of the obstructing membrane.[16] Long-segment lesions are encountered less commonly than the discrete form of SAS and consist of a variety of malformations causing tubular narrowing of the LVOT. More extensive fibromuscular obstructions, sometimes in combination with a discrete membrane, are observed in this circumstance often together with an increased mitral-aortic distance, and reduced aortic diameter.

Pyle and colleagues provided the first detailed pathologic descriptions of SAS in dogs, defining 3 grades of lesions in affected Newfoundland dogs and also noted that, at least in some instances, the lesions were not present at birth but developed over the first few months of life.[17] Mildly affected dogs had small nodules below the right and non-coronary cusps of the aortic valve (grade 1) and more extensive partial (grade 2) and complete rings (grade 3) encircled the outflow tract in more severely affected dogs. Of particular interest are the observations of remodeled intramural coronary arteries and descriptions of subendocardial myocardial fibrosis in dogs with severe outflow tract obstruction.[17] Such changes are consistent with the perfusion deficits that have been documented in dogs with SAS18 and are often cited as the probable substrate for the formation of malignant ventricular arrhythmia, the usual presumed cause of sudden death. A more diverse group of complex lesions have since been described in reports of SAS in other breeds of dogs.[19-21] Sometimes, the LVOT is narrowed by a diffuse fibromuscular obstruction rather than a discrete membrane. Reduced dimensions of the aorta and aortic annulus are sometimes observed in English bulldogs, Boxers, golden retrievers, and bull terriers. Other abnormalities that have been described include prominent muscular trabeculae extending from the LV apex to the outflow tract, accessory or abnormally positioned papillary muscles and other malformation of the mitral valve, often in association with evidence of dynamic left ventricular obstruction.18 Such changes appear to occur more often in golden retriever and Rottweiler dogs with SAS, but systematic studies correlating anatomic variations of the LVOT in relation to breed have not been reported. Given the diversity of pathologic changes observed in different breeds (and individuals), the prudent observer is forced to consider the possibility that SAS in dogs (and humans) may not be a single disorder but several different genetic disorders with different spectra of anatomic lesions leading to obstruction of the LVOT.

The pathologic lesions observed in bull terriers are clearly unique and distinct from other forms of LVOT obstruction in dogs. In this unique disorder the aortic valves are remarkably rigid and composed of dense fibrous tissue with islands of cartilaginous metaplasia. This breed often manifests congenital mitral valve stenosis and/or insufficiency in association with aortic stenosis.[22] Dogs with SAM and dynamic LVOT obstruction that do not evidence any trace of fixed obstruction in the outflow tract may constitute a disease entity totally separate from fixed SAS.[8-10] Dogs with dynamic outflow tract obstruction are often regarded as having obstructive hypertrophic cardiomyopathy, but there is rarely documented evidence of myocardial fiber disarray at necropsy, which is typical of the disorder in humans and in many affected cats. As a result, the appropriateness of this classification remains uncertain.

NORMAL DEVELOPMENT OF THE LVOT TRACT

The development of the left ventricular outflow tract is a complex and incompletely understood process involving the synchronized participation of the developing muscular interventricular septum, the proximal portion of the conotruncal cushions, and the “anterior” margin of the atrioventricular cushions.[23, 24] Some of the many regulatory genes that coordinate this complex process have been identified, but many of the participating elements remain unidentified. The subvalvular region of the LVOT is partitioned from the RVOT via a process involving growth and muscularization of the proximal portion of the conal cushions which fuse with the crest of the muscular interventricular septum and with a mound of atrioventricular cushion tissue residing at the base of the developing anterior leaflet of the mitral valve. Appropriate alignment of these participating tissues and temporal coordination of their growth are essential to normal development of the outflow tract. In addition, developmental events in the LVOT are influenced by sculpturing processes in the left ventricular myocardium and the developing mitral valve apparatus and papillary muscles. Blood flow patterns in the developing embryo not only reflect the anatomic development of the interior of the heart but also appear to influence the morphogenesis of developing structures located downstream and upstream. Hence the superior and inferior atrioventricular endocardial cushions direct blood flow from the atria to the developing ventricles and, as the heart loops and the truncus is partitioned, blood flow changes so that right-sided flow directed at the inner curvature is re-directed to the outer curvature of the developing ventricles and left-sided flow initially directed at the outer curvature is re-directed to the inner curvature and out the developing aorta. Inasmuch as partitioning of the truncus arteriosus proceeds from the distal to proximal end, malformations of the aortic arches can also influence the blood flow proximally, including the developing left ventricular outflow tract.

PROPOSED PHYSIOPATHOGENESIS OF SUBVALAVULAR AORTIC STENOSIS

The discrete form of subvalvular aortic stenosis in humans is regarded by many investigators to be a consequence of disturbed blood flow in an abnormally formed LVOT.[16, 17, 25] It is hypothesized that turbulent blood, via the local induction of growth promoters, stimulates the development of a crescent, horseshoe-shaped, or encircling ridge of fibroelastic tissue that protrudes into the left ventricular outflow tract. Persistence of endocardial cushion tissue in the left ventricular outflow tract has also been cited as a possible contributing (or essential) element for the development of the obstructing membrane in patients developing SAS, but this issue has not been resolved. The inciting regional disturbances of blood flow in the developing LVOT have been attributed to a variety of morphologic alterations including displacement of the aorta to the right side of the crest of the interventricular septum, reduction in the angle of the aorta relative to the interventricular septum, an increase in the distance between the mitral and aortic valves (mitral-aortic distance), reduced diameter of the LVOT, or a combination of these abnormalities. Importantly, the precise causes of these alterations have not been defined. Moreover, abnormal muscular trabeculae below the outflow tract, abnormal papillary muscles and aberrant ventricular bands may also induce flow disturbances in the LVOT tract with similar consequences. Flow disturbances resulting from upstream and downstream anatomic alterations also influence the development and morphology of the LVOT tract both prior to and after birth and LVOT obstructions are, in fact, sometimes present in affected human infants at birth. Interestingly, Pyle et al did not observe lesions in dogs less than 3 weeks of age.[17] There appears to be considerable variation in the postnatal progression of SAS in humans and dogs. In some affected individuals, LVOT obstruction progresses quite remarkably but in many individuals no or only modest increases in the severity of obstruction are observed. Severe obstructions rarely if ever become less severe over time but it is not currently possible to predict the eventual progression of mild to moderate SAS in dogs that present at 8 to 12 weeks of age.

HERITABILITY OF SAS

Familial associations of SAS are well established in Newfoundlands, Golden Retrievers and Rottweilers, but the precise mode(s) of inheritance and specific genetic mutations responsible for SAS have not yet been clearly ascertained.[26, 27]

Pyle et al postulated that SAS in Newfoundlands was either a polygenic trait or was inherited as an autosomal dominant trait with modifiers.[17] More recently, Reist-Marti et al reported that the occurrence of SAS in Newfoundlands best fit an autosomal codominant pattern of inheritance with lethal homozygosity with a penetrance of 1/3rd in heterozygotes.[28] Stern et al reported in 2011[29] that genome wide association analysis had implicated several regions of interest on chromosome 21 in both Rottweiler dogs and golden retrievers which differ from the candidate area(s) identified on a different chromosome in Newfoundlands (Dr. Kate Meurs, NC State, personal communication). With regards to the possible heritability of "hypertrophic cardiomyopathy" in dogs, the author investigated the heritability of dynamic LVOT obstruction in hounds over 3 generations and determined that a recessive mode of inheritance was likely and that no instances of fixed SAS were observed in the offspring.

IMAGING IN DOGS WITH LVOT OBSTRUCTION

A good deal of attention has been focused on the detection and quanitation of SAS via Doppler echocardiography as evidenced by the seminal work of Darke, Lahmkuhl, and Bonagura.[30-32] Further discussions on a threshold velocity for the detection of mild obstructions during screening examinations is pointless from the author's perspective in that such considerations will be altered when genetic testing is available as an additional guide for counseling owners and breeders. Unfortunately, little attention has been focused on the anatomy of the obstructing lesions, detailed evaluations of LV geometry, and better descriptions of coexisting abnormalities. Notable exceptions include reports evaluating the aorto-septal angle, the size of the aortic annulus and a single report of the correlation of disease severity and the cross-sectional area of the obstructive orifice and degree of LV hypertrophy.[33-35] Understanding of the diversity of SAS in dogs would certainly be aided by more thorough descriptions of the anatomy of LVOT obstructions that included relational measurements of the aortoseptal angle, the relationship of the plane of the aortic and mitral valves and measures of the mitral-aortic separation. Such evaluations may prove valuable to patient management. A review of the angiographic anatomy of dogs with SAS shows alterations of the plane of the aortic valve to the plane of the mitral valve and the long axis of the left ventricle. Recent studies indicate that multiplanar review analysis of 3D echocardiographic studies are helpful for defining the complex anatomy of LVOT obstructions, but relational anatomy is optimally displayed via contrast-enhanced computed tomography.[36, 37]

NATURAL HISTORY OF SAS IN DOGS

Kienle et al have provided the first and most detailed description of the untreated natural history of canine SAS.[38] Of 96 untreated dogs, 1/3rd had signs of illness varying from fatigue to syncope. Sudden death occurred in 22% , typically in the first 3 years of life and mainly in dogs with severe gradients (> 80 mm Hg). Just over 11% developed infective endocarditis (6.3%) or left heart failure (7.3%), typically later in life and in dogs with less severe obstructions. Progression of the severity of SAS has been reported in Boxer dogs[39] but has not been systematically evaluated in other affected breeds.

TREATMENT OF LVOT OBSTRUCTIONS

Attempts to remedy or palliate subvalvular aortic stenosis in dogs have included beta-receptor blocking drugs (atenolol), balloon valvuloplasty using traditional and cutting balloons, and surgical excision of the obstructing lesion with and without revision of the left ventricular outflow tract. At the time this manuscript was written, there were no published reports indicating that beta-blockers alter the course of SAS in dogs. Meurs et al reported no difference in median survival in dogs treated by balloon valvuloplasty (55 months) when compared to dogs treated with atenolol (56 months).[40] In this study, it should be noted that the mean pressure gradient (measured in 10 dogs) following valvuloplasty was 119 +/- 32.6 mm Hg compared to 167 +/- 40.1 mm Hg at baseline. It is difficult to surmise from this and other studies whether or not certain anatomic obstructions are more amenable to treatment with balloon valvuloplasty or whether there would be any correlation between the final gradient realized and any improvement in clinical signs, recurrence of obstruction, or long-term outcome (survival).[41] The efficacy and promise of various surgical interventions are similarly difficult to interpret. Excision of the obstructive lesion in seven dogs with SAS using cardiac bypass was first reported by Komtebedde et al.[42] All 7 dogs survived the surgery and 6/7 dogs were reported as alive and stable after a mean follow time of 15.8 months. Orton et al reported outcomes in 22 dogs with SAS undergoing cardiac bypass and surgical membranectomy with or without septal myectomy.[43, 44] The systolic pressure gradient prior to surgery (128 +/- 55 mm Hg) declined substantially to 54 +/- 27 mm Hg, but there was no difference in survival when operated dogs were compared to 22 non-operated control dogs treated with atenolol. The inference that dogs with SAS develop an anatomic substrate that leads unavoidably to sudden death, even after relieve of obstruction, is interesting but unproven. From a completely different perspective, it is interesting to speculate about the possibility of targeting the development of the developing membrane in young dogs prior to the development of severe obstruction.

REFERENCES

[1]Merla G, et al Circ Cardiovasc Genet. 2012;5(6):692. [2]Aboulhosn J and Child JS Circulation. 2006;114:2412. [3]Oliveira P, et al J Vet Intern Med. 2011 May-Jun;25(3):477. [4]Oyama et al Textbook of Vet Int Med 7th Edition 2010; p1283. [5]Radermecker M, et al Acta chir belg, 2005, 105 ;475. [6]Chetboul V, et al. J VetM ed A Physiol Pathol Clin Med 2006;53:346. [7]Fernández del Palacio MJ, et asl J Small Anim Pract. 1998;39(10):481. [8]Liu SK et al Am J Pathol. 1979;94(3):497.

[9]Thomas, WP, et al J Am Anim Hosp Assoc 1984;20:253. [10]Connolly JD and Boswood A J Small Animal Practice 2003;44:319. [11]Sisson D and Thomas WP J Am Anim Hosp Assoc 1984;20:657. [12]Loureiro J, et al J Small Animal Practice 2008;49:578. [13]De Majo M, et alVet Res Commun. 2003 Sep;27 Suppl 1:391. [14]Rosenquist GC, et al Circulation. 1979;60:7014. [15]Zielinski P, et al Circulation. 1987;75:1124. [16]Gewillig M, et al J Am Coll Curdled 1992;19:818. [17]Pyle RL, et al Am Heart J. 1976;92(3):324. 18.Borkon AM, et al J Thorac Cardiovasc Surg 1982;84(6):876. [19]Buoscio DA J Am Anim Hosp Assoc 1994;30:100. [20]McConkey MJ Can Vet J 2011;52:1248. [21]Robinson NA and Armien AG J Vet Diagn Invest 2010;22:667. [22]O'Leary CA and Wilkie I Vet Path 2009; 46(6):1149[23]Anderson RH, et al Heart 2003;89:1110. [24]Moorman AFM and Christoffels VM Physiol Rev 2003;83:1223. [25]Cilliers AM and Gewillig M Heart 2002; 88(4):335. [26]O'Grady MR, et al Can Vet J. 1989 Oct;30(10):811. [27]Sterns JA et al J Small Anim Pract. 2012;53(4):213. [28]Reist – Marti SB, et al Vet Rec. 2012 Jun 9;170(23):597. [29]Stern JA, et al J Vet Int Med 2011;25(4):971. [30]Lehmkuhl LB, et al J Am Soc Echocardiogr. 1995;8(5 Pt 1):611. [31]Belanger MC, et al J Vet Intern Med. 2001;15(5):430. [32]Bonagura JD J Vet Intern Med. 2001;15(5):427. [33]Lang RM, et al J Am Soc Echocardiogr 2005; 18(12):1440. [34]Menegazzo L, et al J Anim Sci 2012;90: 2419. [35]Quintavalla C et al Vet J. 2010;185(3):332. [36]Oyama MA and Thomas WP J Am Anim Hosp Assoc. 2002;38(3):209. [37]Barucha T et al European Journal of Echocardiography 2008;9: 614.

[38]Stolzman P et al J Thorac Cardiovasc Surg 2007;134:240. [39]Kienle RD et al J Vet Intern Med. 1994;8(6):423. [40]French A, et al J Small Anim Pract. 2000 Oct;41(10):451. [41]Meurs KM, et al J Am Vet Med Assoc. 2005 Aug 1;227(3):420. [42]DeLellis LA, et al J Vet Intern Med. 1993 May-Jun;7(3):153. [43]Komtebedde J et al Vet Surg. 1993;22(6):419. [44]Monnet E, et al J Am Vet Med Assoc. 1996;209(7):1255. [45]Orton EC, et al J Am Vet Med Assoc. 2000;216(3):364.

ACVIM Candidate Function: Review of Hemodynamic Calculations
Mark A. Oyama, DVM, DACVIM (Cardiology)
Philadelphia, PA

Various hemodynamic parameters can be calculated from data gathered during cardiac catheterization and/or echocardiography, including cardiac output and valve orifice area. These calculations yield a deeper assessment of cardiac anatomy and performance than do more simple parameters such as pressure gradient. This presentation will review several different cardiac calculations and their applicability to an example of cardiac disease.

Calculation of cardiac output using angiography or echocardiography

Cardiac output (CO) is defined as the volume of blood volume pumped by the heart over time and is commonly represented by the formula:

Cardiac output (L/min) = Stroke volume (L) x heart rate (bpm)

CO in a healthy adult human is ~5L/min. In dogs, CO varies tremendously with body size, however CO can be indexed to body surface area as cardiac index (CI). CI in dogs has been reported as 3.9 +/-0.4L/min/m^2 (Kittleson, 1998). Stroke volume can be calculated using angiography or echocardiography. The volume of the left ventricular cavity during end-diastole and end-systole is estimated using dimensions of ventricular length, width, and/or area. The most common volume formulas that have been validated in the dog include the prolate ellipse methods, cylinder-cone methods, cylinder-hemiellipsoid methods, and the modified Simpson's rule (method of discs) (Weyman, 1994). Each of these methods calculates the ventricular volume by assuming that the ventricular cavity can be defined by a particular combination of geometric shapes whose dimensions can be measured using angiography or echocardiography.

The simplest method is to assume that the ventricle is ellipsoid in nature and that its longest radius is twice the length of the two short axis radii (rz= 2 * rx).

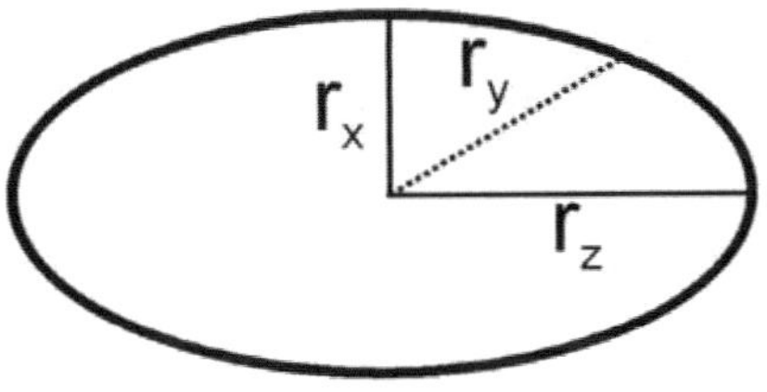

Volume of sphere

$$V=4/3 *\pi *r^3$$

Volume of prolate ellipse

$$V=4/3 *\pi *r_x*r_y*r_z$$

$$V=4/3*\pi*\frac{d_x}{2}*\frac{d_x}{2}*\frac{2d_x}{2}$$

$$V=d_x^3$$

Thus, in a case of the prolate ellipsoid, $V=D_{shortaxis}^3$. Obviously, this is a crude approximation of the LV cavity shape but is remarkably easy to calculate. For greater accuracy, the three main axes of the ventricle can be independently measured from biplane angiography or from the left apical 4 and 2 chamber echocardiographic views.

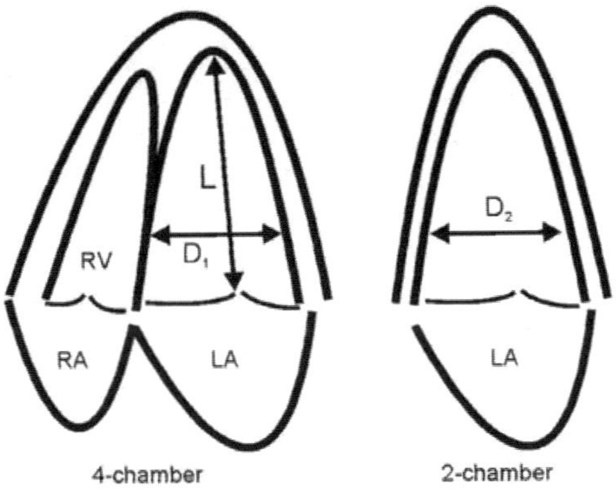

Volume of prolate ellipse

$$V=4/3*\pi*\frac{L}{2}*\frac{D_1}{2}*\frac{D_2}{2}$$

$$V=\frac{\pi*L*D_1*D_2}{6}$$

Going further, the shape of the ventricle can be approximated as a cylinder with either a hemiellipse or cone at one end. These formulas utilize the 2-dimensional short axis area of the LV as traced by planimetry and the LV length.

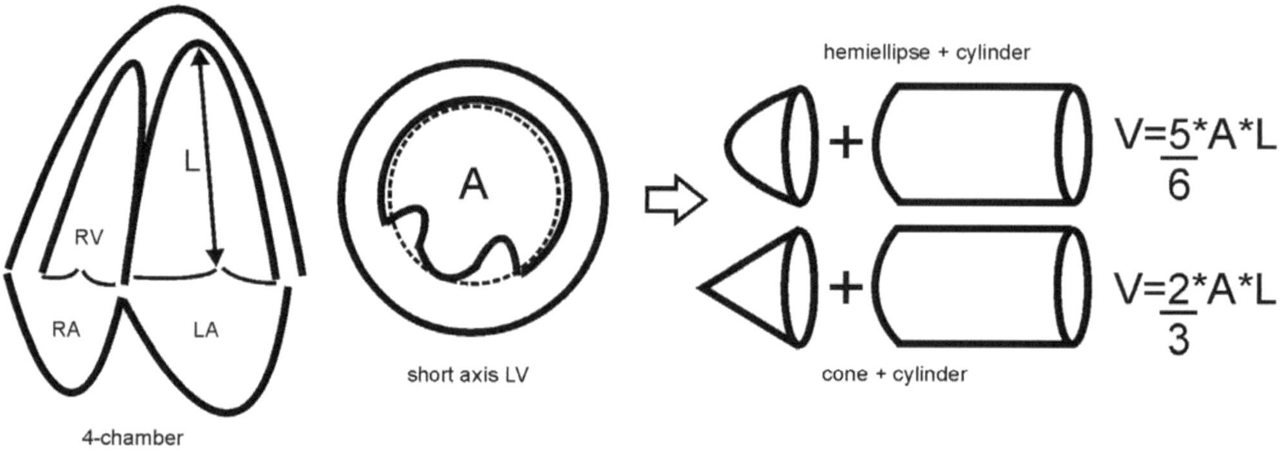

These volume calculations are performed at end-diastole and end-systole so that stroke volume can be then calculated. These techniques calculate total stroke volume of the LV and if significant mitral regurgitation is present, the forward stroke volume will be less than the total.

Calculation of valve orifice area using the Gorlin equation

Calculation of CO allows for calculation of valve orifice area. This calculation can be used to calculate the orifice area of stenotic pulmonic valves or the stenotic area of a subaortic stenosis lesion. The area of an orifice is proportional to the flow rate across the area divided by the mean pressure gradient that drives that flow. The flow rate is not equivalent to the CO but instead is the instantaneous volume of blood that moves across the orifice during the appropriate portion of the cardiac cycle (mls/s). The flow rate is the cardiac output divided by the product of the heart rate and the time during each heart beat that the flow is occurring across the valve orifice. For pulmonary artery and aortic flow, the time spent is the systolic ejection period (SEP) and for the mitral and tricuspid annulus the time spent is during the diastolic filling period (DFP)

Flow rate (mls/s) = CO (mls/min)/([SEP or DFP (sec/beat)] * HR (beats/min))

The *mean* pressure gradient (not the maximal PG) drives flow across the orifice during the entire duration of systole or diastole and is used in the denominator. The gravitational and fluid density components of the formula are captured by use of a constant of 44.3 in the denominator. In conditions of weightlessness (ie., zero gravity), this constant would not be needed. When Gorlin and Gorlin (a father and son team) compared their original calculated orifice areas to the actual areas as measured at necropsy, they found that the formula needed to be slightly modified to best approximate the necropsy gold standard. If the aortic, pulmonic, or tricuspid valves are being measured, the denominator remains unchanged by multiplying by 1, however, if the mitral valve is being measured, the denominator is multiplied by a mitral-specific constant of 0.85.

$$A_{\substack{aortic\\pulmonic}} = \frac{CO/(SEP * HR)}{1*44.3*\sqrt{PG_{mean}}} \qquad A_{mitral} = \frac{CO/(DFP * HR)}{0.85*44.3*\sqrt{PG_{mean}}}$$

$A=cm^3$
CO=mLs/min
HR=bpm
SEP or SFP=sec/beat
PG=mmHg

Practical use of the Gorlin formula

The success of pulmonic balloon valvuloplasty is typically assessed by measuring the pre- and post-procedure PG across the stenotic valve orifice. This can be done by direct pressure measurement using fluid-filled catheters or by Doppler echocardiography and the modified Bernoulli equation. The PG across the pulmonic orifice is used as a surrogate measure of severity, however PGs not only are affected by the stenotic orifice area (resistance) but also by the amount of flow through the orifice, as can be seen in Ohm's law ($\Delta P=QR$). Thus, use of PGs to estimate orifice area assumes that the flow through the orifice is the same during each measurement, which might or might not be the case. For instance, the PV PG is commonly lower during anesthesia vs. the baseline measurement in the awake dog. After balloon procedure, it is reasonable to assume that flow through the stenosis is improved and the PG might not be an accurate reflection of the residual severity.

Measurement of cardiac output and the resultant flow across the valve orifice allows the actual stenotic area to be calculated and this might be a better measure of disease severity and interventional success. Calculation of the valve orifice area requires calculation of cardiac output by either the aforementioned methods or by thermodilution. In this example, CO is calculated by angiographic methods and the mean pressure gradient is calculated by either Doppler or by direct pressure tracing.

Pre-balloon **Post-balloon**

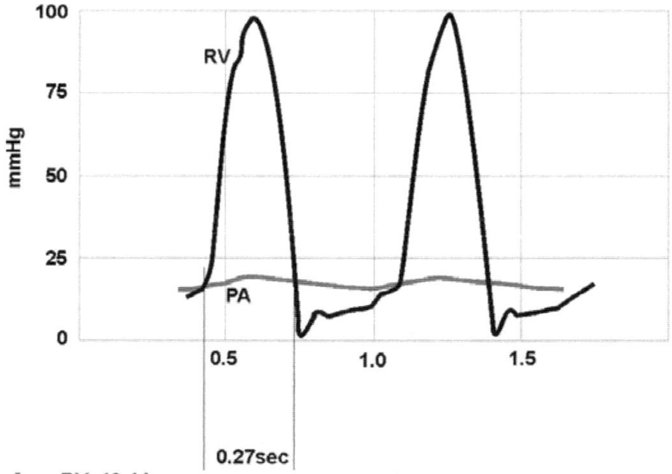

Area RV=13.44
Area PA=0.283
Area Difference=13.16
Mean Pressure Difference=13.16/0.27=48.73mmHg

CO=2140ml/min
HR=98bpm

PVA=[2140/(98*0.27)]/[44.3*SQR48]
PVA (cm^2)=0.264

Calculated PV radius=2.9mm

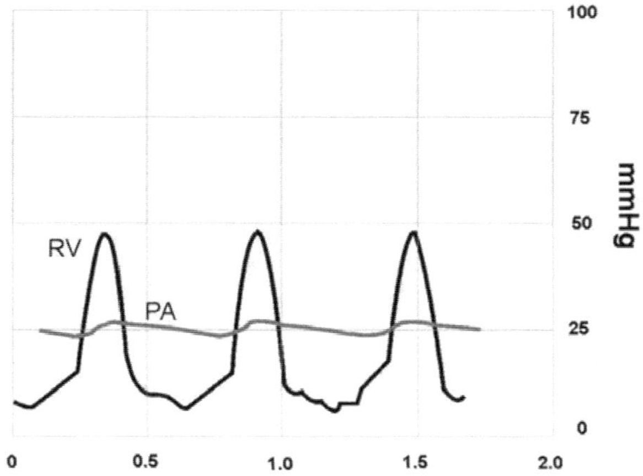

SEP=0.130sec
Area PA=0.138
Area RV=1.914
Area difference=1.776
Mean pressure difference=1.776/0.130=13.662

CO=2.38l/min
HR=114bpm

PVA=[2380/(0.130*114)]/(44.3*SQRT13.662)
PVA=0.981cm^2

Calculated PV orifice radius=5.6mm

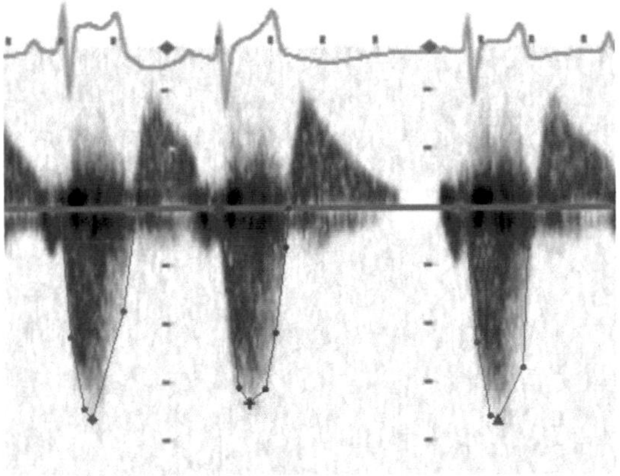

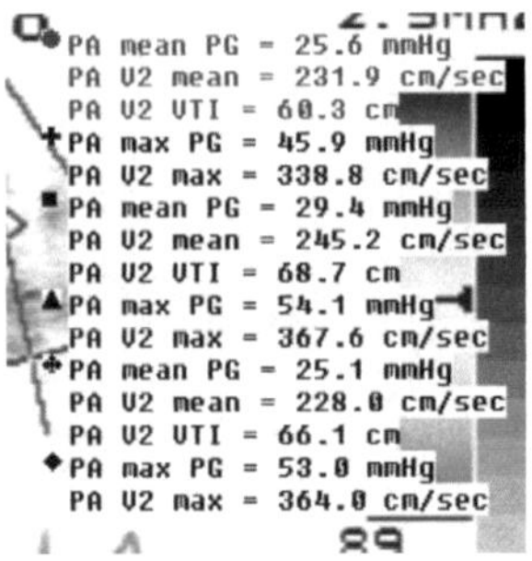

Doppler calculation of mean pressure gradient

REFERENCES

Kittleson MD, Kienle RD. Small Animal Cardiovascular Medicine. Mosby 1998. Weyman AE. Principles and Practice of Echocardiography, 2nd ed. Williams and Wilkins, 1994.

Found in Translation: Pearls of Veterinary Cardiology from the Overseas Literature (Japan)

Masami Uechi, DVM, PhD
Fujisawa, Japan

Torasemide (torsemide) is a new loop diuretic that combines the effects of furosemide and spironolactone. Its main tubular site of action is the ascending limb of the loop of Henle, where it interacts with Na, Cl, and K co-transporters localized on the luminal surface. Torasemide is a more potent and longer-lasting diuretic than furosemide in both cats and dogs. Torasemide, furosemide, and placebo were orally administered to cats and dogs. Their urine samples were obtained at baseline and at 1, 2, 3, 4, 5, 6, 8, 12, and 24 h after administration of each drug dose. The urine volume and urinary sodium and potassium levels were measured. Compared with the baseline value, short-term administration of furosemide or torasemide immediately increased urine volume significantly, while long-term administration of either drug decreased urine specific gravity significantly. In both cats and dogs, furosemide caused a dose-dependent increase in urine volume that peaked at 2–3 h after administration and the diuretic effect lasted for 6 h. On the other hand, the effect of torasemide peaked at 2–4 h after administration and persisted for 12 h. In dogs, torasemide significantly decreased urine potassium excretion for 7 days. Compared with the baseline value, the plasma aldosterone concentration increased significantly after long-term administration of either drug and was significantly higher after torasemide treatment than after furosemide treatment in dogs. In conclusion, in dogs and cats, torasemide is ten times more potent than furosemide and has a longer diuretic effect. These data suggest that torasemide is useful for treating congestive heart failure or edema in cats and dogs. However, both loop diuretics increase blood urea nitrogen and plasma creatinine concentrations. These effects may be more potent in the case of torasemide and need to be further explored.

FOUND IN TRANSLATION: PEARLS OF VETERINARY CARDIOLOGY FROM THE OVERSEAS LITERATURE (ITALY)

Roberto Santilli, DVM, PhD, DECVIM-CA (Cardiology)
Samarate, Varese, Italy

UTILITY OF HOLTER MONITORING IN DOGS WITH ARRHYTHMIC SYNCOPE[1]

Syncope is a non traumatic transient loss of consciousness (T-LOC) and postural tone induced by global cerebral hypoperfusion.[2] This event is secondary to a reduction of stroke volume, peripheral resistance or total blood volume, leading to systemic arterial blood pressure fall to values lower than 60 mmHg. A complete loss of consciousness occurs when cerebral hypoperfusion lasts for a period of 6-8 seconds.[2] According to the different causes, syncope can be classified into: neuromediated or vasovagal, arrhythmic, dysautonomic or orthostatic and cardio-respiratory. Neuromediated syncopes are not well documented in veterinary medicine and are associated with inappropriate vagal reflex and sympathetic withdrawl that induce bradycardia, vasodilation or both. [2-4] Vaso-vagal syncope can occur in both healthy patients and in patients with underlying structural heart diseases. A common type of neurally-mediated T-LOC is situational syncope, typically triggered by coughing, urination, defecation or pain[4]. Intrinsic rhythm disturbances such as high grade atrioventricular block, sinus node dysfunction and rarely supraventricular or ventricular tachycardia can induce arrhythmic syncope[2]. Primary dysautonomia or drug-induced dysautonomia have been reported as potential cause of orthostatic hypotension.[2] Finally obstruction of aortic or pulmonary outflow, myocardial contractile deficit or pulmonary arterial hypertension should be considered as possible causes of cardio-respiratory T-LOCs.[2]

The evaluation of a patient presenting with episodes of T-LOC is sometimes extremely complex, since the spectrum of differential diagnosis is truly vast, so that in humans half of the episodes remain undiagnosed. [3-6] The first step in the diagnosis of T-LOC is an electrocardiogram, since in people, 5-30 % of the syncopal events are arrhythmic. Its diagnostic yield is however influenced by the paroxysmal behavior of some arrhythmias.[5] Holter monitoring and loop recorders represent alternative diagnostic tools to study patients with suspect arrhythmia-induced T-LOC. [6-7] In veterinary medicine these devices have shown a good accuracy in establishing the cause of syncope in dogs with a diagnostic yield ranging from 34.7 and 84.4 %. [8-12]

The purpose of this study was to test the clinical utility of Holter monitoring in a large group of patients presenting with a history of T-LOC and a normal basal electrocardiogram. The records of 112 dogs were retrospectively analyzed and data regarding history, clinical findings, electrocardiogram, thoracic radiographs, echocardiography and Holter monitoring collected. Holter monitoring was recorded for 24 hours using Cardioline devices with the orthogonal bipolar system (X,Y,Z). All tracing were manually edited by one author (RAS) with particular attention to the period where a T-LOC was reported by the owner on the diary. The clinical utility was evaluated considering the test positive for arrhythmic or neutrally-mediated syncope when during the T-LOC event an arrhythmias was detected on the tracing, negative when during the event no arrhythmias could be found. Diagnostic yield was then calculated as the sum of the percentage of patients with positive results and the percentage of patients with negative results but with arrhythmias in other parts of the tracing that could have induced the syncopal event. The population of dogs was composed by 72 males and 40 females with a mean age of 7.8 ± 4.2 years and a mean body weight of 25.7 ± 13.4 kg, 29 were mongrel, 28 Boxer, 5 Labrador, 4 Cocker Spaniel, 2 Great Dane, 2 Doberman and 40 of other breeds. Seventy-two had no underlining heart diseases, while 40 had either acquired or congenital cardiac abnormalities (23 chronic valvular diseases, 7 subaortic stenosis, 6 dilated cardiomyopathy, 2 arrhythmogenic right ventricular cardiomyopathy, 1 pulmonic stenosis, and 1 heart base tumor). 22/112 (19.6%) dogs experienced a T-LOC during Holter monitoring: 11/22 bradyarrhythmias, 2 ventricular tachycardias and in 9 dogs no rhythm disturbances were present during the event. In the remaining 90 dogs that did not experience syncope during the Holter recording, the entire 24 hour tracing was analyzed and 62 (51.6 %) did not present any arrhythmias, while in 28 cases different rhythm abnormalities were noted: 14 sustained ventricular tachycardia, 7 bradyarrhythmias, 6 paroxysmal supraventricular tachycardia, and 1 pacemaker malfunctioning. The overall diagnostic yield of Holter monitoring to identify dogs with arrhythmic or neutrally-mediated syncope was 44.6 %, a value similar to the data previously reported. [8] The only discrepancy found is that in the literature there was a difference in the diagnostic yield of the Holter in the group of dogs with cardiac diseases (55%) compare to the group of normal dogs (45%), while in our study no differences were found. The prevalence of intrinsic or neutrally-mediated arrhythmias found in our sample was similar to the result present in human literature wherein almost one fourth of patients rhythm disturbances are the cause of T-LOC. [5] Our study presents some limitations: the population studied experienced frequent T-LOCs and the calculation of the diagnostic yield of the Holter included also dogs with no T-LOC during the recording, but with different arrhythmias that could be considered possible cause of the syncopal episodes.

In conclusion the study herein reported showed that Holter recording could be considered an important diagnostic tool in patients with suspected arrhythmic and/or neutrally-mediated with cardio-inhibition T-LOCs, particularly when frequent episodes are reported. Other devices should be considered to increases the diagnostic yield of prolonged rhythm recording in the diagnosis of T-LOC in dogs.

VAGALLY-MEDIATED PAROXYSMAL ATRIAL FIBRILLATION IN 7 DOGS [13]

The objectives of the study was to investigate the electrocardiographic features of vagally-mediated paroxysmal atrial fibrillation (AF). We studied seven dogs (2 boxer, 2 mongrels, 1 labrador, 1 pug and 1 beagle), 4 males and 2 female with an average age of 8.6 years and an average body weight of 20.83 kg. Three dogs suffered from cardiac diseases (1 subaortic stenosis, and 2 chronic valvular disease respectively at stage B2 and C). All dogs underwent 24-hour Holter recording for transient loss of consciousness (T-LOC) and all of them experienced an episode during the recording (2 after exercise, 2 after cough) followed by paroxysmal AF. All Holters were manually edited and were analyzed starting 10 minutes before the syncopal episodes reported in the diary until AF stopped.

Before the syncopal episodes 5 dogs presented sinus bradycardia with an average duration of 6.17 seconds and a rate of 54 bpm, while 3 dogs sinus rhythm. These rhythms preceded a sinus arrest with an average length of 22.78 seconds interrupted by junctional escape rhythm with an average duration of 24.36 seconds and a rate of 29 bpm. After the sinus arrest or the junctional escape AF started and presented an average duration of 171.97 seconds and a rate of 138 bpm.

According to our findings the T-LOCs herein reported were suspicious of neurally-mediated events.[14-17] Enhanced parasympathetic tone results in heterogeneity of atrial refractoriness, atrial action potential abbreviation and increase pulmonary vein firing, all effects that trigger and maintain AF. [18-20]

COMPARISON BETWEEN CONVENTIONAL AND COMPUTERIZED ELECTROCARDIOGRAPHY IN CATS[21]

The electrocardiogram (ECG) provides a representation of the electrical activity of the heart and is a fundamental part of a clinical evaluation. Thus, the ECG is commonly used in small animal practice to diagnose conduction disturbances and other cardiac arrhythmias. Comparing the conventional (Cv) to the computerized (Cp) ECG method, the second one had shown a higher accuracy, because of its ability to detect readings of 1 millisecond (ms), while in the Cv method, readings can only be made from 5 ms in paper speed of 100mm/sec. [22] The aim of the present study was to analyze the differences between conventional and computerized methods of recording.

Thirty healthy adult cats were housed in appropriate cages, fed a dry pellet diet twice a day, and received water ad libitum. The animals were allocated into two groups according to gender: group F, n=16 females ranging from 2.74 to 4.53 kg and group M, n=14 males ranging from 3.57 to 6.10 kg. All animals were submitted to conventional (Cv) and computerized (Cp) electrocardiography. In the Cv method, a monochannel (ECG6 – ECAFIX) electrocardiographic device, provided with thermal sensitive millimetrated paper was used. For the Cp electrocardiographic method (EGC-PC version 2.07 – TEB), an electronic circuit connected to a personal computer (Pentium, 1660MHz, 32MB) provided with specific software, able to calculate and analyze electrocardiographic data, was used. Recorded tracings were show on the computer screen. After performing the computerized method, ECG tracings of each subject were printed in order to analyze the recorded data, which was named as computerized printed method (CpP). The paper used for printing was endowed with millimetrated spaces for the recording of waves, intervals, and segments defined by the electronic program, by means of a jet printer connected to the computerized ECG. The animals were positioned on right lateral recumbency, on an appropriate table, with limbs perpendicularly positioned to the trunk. Electrodes were attached to forelimbs and hindlimbs. Speed used for all recordings was 50mm/s, with calibration of voltage of 1 centimeter for each millivolt (1mV=1cm). Both registrations were obtained with a regular sequence, first the Cv than Cp method. The ECGs were recorded in the bipolar lead II (L II) and the evaluated parameters were: heart rate, cardiac rhythm, duration (milliseconds-ms) and amplitude (millivolts-mV) of the P wave, duration (ms) of the PR, QT interval and QRS complex, amplitude (mV) of the R wave, polarity characteristics of the T wave, presence or absence of uneven ST segment, and values (in degrees) of the mean electrical axis (MEA). The electrocardiographic measurements were analyzed in accordance to described method.[23] Data recorded by Cv and Cp ECG were analyzed using paired t test in order to interpret possible effects capable to change the means of each variable regarding electrocardiographic methods. Values of $P<0.05$ were considered significant.

All animals in this study showed a normal sinus rhythm with no wandering pacemaker. The P wave duration and amplitude, QRS duration, and the PR and QT intervals were greater ($P<0.05$) in both groups by the Cp ECG. The MEA values were not different between groups ($P>0.05$). Heart rate presented similar values between groups and methods ($P<0.05$). However, the P wave duration was greater ($P<0.05$) in Group F by the CpP method compared to Cv. Group M presented greater P wave amplitude in the Cv ECG compared to CpP. No differences were found in PR duration, QRS, and QT intervals, as well as the MEA. No differences were observed in the heart rate, P and R waves amplitudes, and PR and QT intervals. Both groups presented P wave and QRS complex duration greater ($P<0.05$) by Cp method. The MEA did not differ between groups. Considering the T wave polarity, similar values of positive and biphasic waves for Group F were observed in all methods. Negative waves were not observed in Group F. Controversially, a great percentage of positive waves compared to the biphasic waves were observed in Group F. Group M presented similar values for positive and biphasic waves in Cp and CpP methods. A great percentage of positive waves were observed in this group.

Conventional ECG is a non invasive and feasible diagnostic method commonly used in veterinary medicine to access alterations in the cardiac rhythm, as well as baseline values of intervals, segments, amplitudes, and wave durations. [23-24] However, there is a requirement for more studies to obtain references values from the computerized ECG method in cats. The comparison of results from the three different methods (Cv, Cp, Cp P), highlighted that the P wave duration (ms) and QRS complex (ms), in both groups (F and M), featured significant increased values for the Cp and CpP methods with regard to Cv method. [23] Results in dogs were reported in a previous study, when the Cv, Cp, and CpP methods were compared. [27] Other variables significantly changed when comparing the three different ECG methods, notably the P wave amplitude (mV), duration of the QT interval, and polarity of the T wave, nevertheless these results did not exceed the values preconized for the Cv method, which was also seen in another study with dogs. [27] In the meantime, significant differences were not evidenced in some variables, such as duration of the PR interval (ms), R wave amplitude (mV), unlevelling of the ST segment, and MEA, in contrast with another study that reported these differences in dogs. [27] Notwithstanding, the cardiac rhythm did not vary between the various ECG methods used in this study, been characterized as 100% sinus rhythm, in accordance with other authors, which verified rhythm constancy, independently of the method used, with the exception of respiratory sinus arrhythmia in all groups of dogs . [27] Such difference may be explained due to the higher activity of the sympathetic autonomous component present in the feline species when comparing with the canine, which more often shows enhanced parasympathetic tonus. [23,25] The differences between the Cv and Cp methods seen in this study may be attributed to the higher accuracy of the computerized measurement because, according to Shiwen et al.[22], this method is able to detect readings of 1ms, whereas the Cv method, the reading can only be made from 5ms, with paper speed of 100mm/sec. Moreover, it was noticed that the Cp ECG method allows monitoring for longer period, which ultimately makes it a value alternative in intensive care units and it confers the capacity of managing and storing larges volumes of electrocardiograms in a shorter period. In addition, it enables the concomitant observation of several leads, beyond an easier and faster method to measure waves and intervals. [23] Furthermore, depending of the operational system, the informatized method shows reproducibility of its measurements, improves the control of quality, and allows comparisons between electrocardiographic tracings obtained in different periods.

In conclusion, this study ensures that there are differences among the three ECG methods herein evaluated, notably the computerized method, which showed mean values of duration of the P wave and QRS complex above the reference values preconized for the conventional method. Considering these differences, special attention should be given during the computerized electrocardiographic interpretation in both clinically healthy cats and those suffering from any cardiac disease.

REFERENCES

[1] Santilli RA, et al. **Veterinaria** 2009;4:13. [2] Moya A et al **Eur Heart J** 2009;30:2631. [3]Jhanjee R. et al. **Pacing Clin Electrophysiol** 2006;29:1169. [4]Brignole M et al. **Europace** 2000;1:66. [5]Chang-Sing P et al. **Cardiol Clin** 1991;9:641. [6]Kapoor WN et al. **Am J Med** 1987;83:700, [7] Holter NJ, **Clin Eng** 1980;8:65, [8]Petrie JP et al. **Am J Cardiol** 1990;66:214, [8] Miller RH et al. **J Vet Intern Med** 1999;13:111, [9] Bright JM et al. **J Am Vet Med Assoc** 2000;216:1114, [10] James R et al**. J Small Anim Pract** 2008;49:564, [11] Santilli RA et al. **Am J Vet Med Assoc** 2010;236:78, [12] MacKie BA et al. **J Vet Cardiol** 2010;12:25, [13]Porteiro DM et al. **VetMadrid 2013** – XXX Congreso annual Medicina y cirurgia del sistema cardiorespiratorio. 2013, [14] Moya A et al. **Eur Heart J** 2009;30:2631, 15 Campagna JA et al. **Anaesthesiology** 2003;98:1250, 16 Mehlsen J et al, **Europace** 2008;10:91, 17 Hamlin RL et al. **Am J Physiol** 1968;215:560, [18] Wang et al **Am J Physiol** 1996;271:148, [19] Liu L et al. **Am J Physiol** 1997;273:805, [20] Patterson E et al. **J Am Coll Cardiol** 2006;47:1196, [21] Camacho AA et al, **Arq Bras Med Vet Zootec** 2010;62:765, [22] Schiwen Y et al. **PACE** 1996;19:297, [23] Tilley LP 1992;270, [24] Hoffman BF et. **Am J Med** 1964;37:670, [25] Kittleson MD et al 1998;449, [26] Ware W et al, **Am J Vet Res** 1999:60:1426, [27] Wolf R et al. **Arq bras Med Vet Zootec** 2000;52:610.

THE FUNDING LANDSCAPE FOR RESEARCH IN VETERINARY MEDICINE & TACTICS FOR SUCCESSFUL GRANT WRITING

David Haworth, DVM, PhD
Denver, CO

INTRODUCTION

Like human medicine, veterinary medicine is a profession built on scientifically-validated research into the diseases and conditions relevant to the species that we care about most. It is not necessarily so. There are several examples of professions that devote very little effort or resources towards generating new knowledge. However, veterinary medicine has traditionally eschewed sales pitches and anecdotes in favor of hypothesis-driven, scientifically-based research. This is not always easy. Despite the fact that there is only 10.4 times the number of physicians practicing in the United States compared to veterinarians, there is approximately 134 times less funding for veterinary research than human-centric research. Without an understanding of where the limited money that veterinary research requires can be found, researchers feeding our profession's need for new knowledge will face difficult decisions, and forward progress in the discovery, diagnosis, treatment and prevention of veterinary-specific diseases could slow.

Where and how to get funding for research are topics that can cause dyspepsia in new and established veterinary researchers. Deciding where to apply for funding is one of the more difficult tasks faculty members, residents, house officers and institutional researchers face. The costs of poorly placed grant applications are time and lost opportunity. No one, especially investigators early in their career, can afford to waste either. Even established investigators realize that a grant application that is funded during one grant cycle may have little chance of getting funded the next round by the same institution. A holistic and complete understand is essential to success.

Often granting agencies do a poor job explaining what they consider important, and sometimes the decision-making process for funding grants can appear capricious and random. For these reasons and others, a solid framework with which to evaluate where a given idea (a grant proposal) should be submitted is essential for anyone whose career involves research. Additionally, many inherently good ideas are rejected because they are presented poorly, presented to the wrong organization, or presented in a manner that makes it easy for administrators (as opposed to qualified evaluators) to reject them. Not only are researchers disappointed in these cases, but more importantly, good science is left undone and animals go without potential cures.

The first part of this presentation focuses on creating a framework for use by researchers in veterinary-specific areas to find the best funding sources for their work. The second part of the presentation will focus on tactics and approaches to writing successful grant applications, regardless of the funding agency.

SOURCES OF FUNDING

"Public" or Government Sources

There is essentially no incentive for Federal or State funding agencies to give grants for research focused on veterinary-specific diseases. In the United States, taxes are theoretically raised for the betterment of society. To date, that does not include our veterinary species. Livestock disease is an exception, in which governments have an economic incentive to protect and promote health of these species. Clearly, however, there is no incentive to, and subsequently little government funding for, research into diseases specific to dogs, cats, exotic species and wildlife. While this is generally true throughout the world, some countries do provide public funding for veterinary-specific research. The United States does not.

This is not to imply that veterinary medicine does not receive a substantial amount of government funding for research; or that many successful and fulfilling research careers have not been built on federal or state funding. For many decades, veterinary researchers have successfully argued that spontaneous disease in veterinary species is a better model for a number of human diseases than in vitro or artificial models created in traditional laboratory species (rodents, zebra fish, non-human primates, etc.). However, this approach has two important consequences: 1) it does not allow funding for the many diseases that veterinary species suffer that have marginal or no human counterparts (GDV, laminitis, white-nose syndrome, etc.) and 2) it forces the position that veterinary species are at best secondary beneficiaries of the knowledge created by that research. As such, government funding of veterinary research is an important, but imperfect method to generate knowledge relevant to the veterinarian.

Additionally, there are good reasons to suspect that the sustained increases in federal funding for basic and applied research enjoyed since World War II are a thing of the past. Current fiscal realities at the federal, state and local levels of government in the United States, coupled with increasing costs of research and societal pressures to address more immediate needs, are resulting in fewer and fewer grant applications to government agencies being successful. There is a lot of money still available, but researchers must be clear about which agencies they should apply to and how they go about applying.

"Private" or Corporate Sources

Another important source of funds for veterinary research is corporations that have an incentive to investigate veterinary diseases. These can be companies with diagnostic or therapeutic products, or those that would otherwise benefit from a disease or condition being cured. It could also be companies that sponsor research in order to solicit the good will of investigators or institutions. There is great good that has come from research sponsored by private companies, and indeed, it would be difficult to imagine how new products could come to the market without research sponsored by them. In fact, for product-based therapies, the only way for the knowledge to result in applied good is for companies to translate it into a product we can all use.

A fundamental difference between knowledge derived from private money and knowledge derived from other sources is that knowledge created on behalf of a private company is a private good. That means that its sole purpose is to give the funding company a competitive advantage or increase shareholder value. Incredible good has come about because of this relationship, but often researchers outside corporate R&D programs resent or misunderstand the fact that the knowledge they helped create belongs to the company that funded it. By contrast, knowledge created (i.e. research funded) by public or independent sources are a public good, and should belong to the public for the sole purpose of bettering society. Misunderstandings, bad feelings and accusations of bad intent on both sides have resulted from this misunderstanding.

"Independent" or Non-Profit Sources

A third source of research funding is the "independent" or non-profit sector. This source of funding includes individual donors, private foundations with missions to help animals (either generally or specific types of animals), and public not-for-profit organizations. These individuals or organizations are over-represented in veterinary research, mostly due to the relative under-representation of government funding. There are numerous such organizations and each has slightly different models of operating, missions behind their existence, and processes by which researchers apply for funding. Some organizations have well-qualified boards of specialists that evaluate applications with strict policies to eliminate or minimize conflicts (or even the appearance of conflicts) of interest. Others have single benefactors that may or may not be qualified to evaluate proposed ideas. Most agencies fall somewhere in between. Some organizations have substantial wealth, and can fully fund researchers for significant portions of their careers, while others have precarious funding situations, able to only variably provide small grants year-to-year. Every non-profit organization in the US is required by the IRS to have a publically declared mission statement, which researchers evaluating funding sources can (and should) use to determine whether their idea or area of interest fit with a given agency.

The sheer variability within this sector can dissuade investigators from seeking funding through these sources. As government sources of research funding are reduced, this could be a mistake. Particularly with information-gathering sites like Charity Navigator ™ and GivingWell ™, what may appear daunting is actually well-worth the effort for new and established investigators. This is certainly not intended to imply that these sources are easy, convenient, or even logical. However, this source of funding has generated science that often would not have been possible with funds from government or private sources.

HOW TO APPLY FOR GRANTS

Granting agencies exist to give away money in a manner that is most effective in furthering their mission. Their entire existence should be designed to maximize the impact of the dollars they give away, regardless of whether those dollars come from taxpayers, shareholders or donors. Their ability to do this often depends on the wishes of those sources of funds. Funding research that specifically would help cats, but has no realistic chance of helping people, would likely not please a majority of taxpayers. However, a passionate donor who wants to use their personal wealth to eliminate feline cancer from the earth would likewise be displeased to find out their funds were going to help giraffes. Therefore, the most important step in a grant-writing process is to determine which agency has the best alignment with a specific idea or area of interest.

After the source is determined, the actual process is usually painstaking and time-consuming. For larger agencies, the number of grant applications is so great that rigorous rules must be in place to logistically manage the load. It is not sadism that leads to specific rules about abstract length, formatting and organization. It is survival, mixed with the fact that the supply of funds is so dramatically out-stripped by the demand of good ideas that funding agencies can afford to be more selective than Harvard admission committees.

This situation requires that prospective researchers must pay meticulous attention to the details of an application process. Having well-written, clear application packages that contain all the required pieces will maximize the chances that a specific grant will be funded. This takes time. Often it takes institutions to invest in infrastructure like Offices of Grant Management, or specific and enforced mentorship programs. Sometimes it takes driven researchers to use each other to review, criticize and edit each other's applications. Regardless of the method, successful grant applicants generate application packages that are clean, easy to review, and fit according to the standard format that reviewers are expecting.

CONCLUSION

There is money available for veterinary-specific research. It is not easy to find, nor is it easy to acquire once it is found, but it is present, and funding agencies exist to give it away. They do so in order to create new knowledge for public or private use, and each has its own unique mission, interest, and process by which grants are awarded. For researchers that are starting out in their career and for established investigators that have dozens of grants on their CV, a clear understanding of where the money is and how best to go about getting it is essential to success as a scientific researcher.

PHYSICAL DIAGNOSIS IN RUMINANTS: A REVIEW

Bradford P Smith, DVM, DACVIM (Large Animal Internal Medicine)
Davis, CA

This discussion of physical diagnosis is aimed at large animal internists who may have been trained chiefly with horses, and only see the occasional ruminant patient. It is assumed that ruminant experts already possess this level of knowledge. During this same ACVIM Forum a presentation dedicated to examination of the ruminant nervous system will be given, so most discussion of the nervous system has been omitted here. In this paper I review the efficient technique I prefer for a basic ruminant physical examination (PE), moving around the animal from left to right and looking at the head last, rather than an approach that examines system by system. Fertility exam is beyond the scope of the basic PE and will not be discussed here.

As soon as the call comes to you, you should begin thinking of the possibilities that go with the presenting complaint. If you are in your vehicle, driving time is a good time to think. If you are in your clinic you may even want to open a book or jump on your computer and open Cornell`s Consultant (www.vet.cornell.edu/consultant/consult.asp) to look at possibilities. For example if you are called to see a sick 3 year old postpartum dairy cow, think of the most common conditions: Mastitis, metritis, and milk fever, the 3 M`s. Before arrival at the animal, you want to be sure that you have the basic tools needed to ensure a complete PE: Forms for taking history and recording PE[1], pencil/pen, watch with second hand, cups for urine and feces, urine dipsticks, CMT paddle and reagent, electronic thermometer or thermometer with clip so it does not fall on the concrete, stethoscope, plexor and pleximeter (many people like a rubber hammer and tablespoon), needles and syringes, blood tubes and needles, good light (preferably an LED headlamp) with spare batteries, rectal exam gloves and lubricant, latex gloves for examining the mouth and a small towel to help grasp the tongue, dental wedge to help examine the mouth of small ruminants, and fluorescein strips to stain the cornea.

The physical examination of cattle, goats and sheep often reveals the basic problem. This is especially good news when ruminants are involved because the individual animal is likely to have a financial value that limits some diagnostic options, such as laboratory tests, biopsies, and ancillary imaging, due to their cost.. Value also dictates that the amount of time spent performing the PE may be limited. In order to assess these factors, the veterinarian should ask the owner what he/she expects to get from your exam, i.e. is this a particularly valuable individual, is this representative of a herd problem, is this likely to be a legal case, is this for a health certificate (HC) for interstate shipping, is this an insurance or pre-purchase exam? If for a HC, you are concerned mainly with parasites and infectious diseases. If this is a herd problem, try to select three representative animals to examine. A Holstein cow which is in the top tier of milk production for the herd will be treated differently by the owner than a herd-mate in the bottom tier. A registered bull will be considerably more valuable than a commercial bull. All these factors will influence your exam, as well as treatment choices.

I prefer to use forms[1] with choices for answers for the history and PE because they make keeping good records easier and quicker, ensure that my exam is complete and consistent, and help avoid extensive writing. Choices can simply be circled, rather than written out, which gives a tremendous time advantage. The form should note the date, owner or herd name, and animal ID including name or tag number, date of birth, breed, sex, date of last parturition for females, approximate weight, and any identifying characteristics. As you list the presenting complaint include such information as whether the onset was gradual or sudden, recent changes in feeding, type of environment, deworming and vaccination history, and whether this individual has had previous illnesses.

As you consider the information gathered develop a mental list of rule outs, while keeping an open mind. The signalment and history of a male pygmy goat with straining and discomfort should prompt you to rule out urolithiasis and carefully examine the urethral process. A Holstein dairy cow which calved 3 weeks ago and is eating poorly, has scant feces, and is producing little milk should clearly send a signal to rule out left displaced abomasum (LDA).

Also look at the environment as you talk to the owner. Is the housing crowded, bunk space adequate and feed available, is it safe and clean, are there hazards visible such as old batteries, paint cans, oil drums, or toxic plants? Are there 40 goats on a small overgrazed plot of irrigated grass, which are likely to have a major burden of Haemonchus? I sometimes look in a bale of hay as I pass it, and have found such undesirable contents as oleander (cardiac glycoside) or senecio vulgaris (pyrrolizidine alkaloid).

Now watch the animal before you restrain it for the exam. Is the overall body condition score (BCS) normal for use? Is the respiratory rate and effort normal, or is she open mouth breathing with expiratory dyspnea? Are there any obvious swellings or edema? Is the head carriage and mentation normal? What is the body shape and posture? Is the abdomen gaunt, normal or distended? When viewed from the side is the back arched indicating lameness or abdominal pain? When viewed from the rear is she bloated, overly distended or papple shaped? Abducted elbows often indicate thoracic pain.

As you or the owner moves the animal, note any lameness. Common causes of lameness in cattle include papillomatous digital dermatitis (foot warts), sole abscess, footrot, and mycoplasma ployarthritis if multiple joints appear affected. In small ruminants footrot is common in sheep and mycoplasma polyarthitis is frequently seen in young goats. If a cow is lying down can she rise normally? If she is down and unable to rise, is she depressed or alert? Rule outs for a recumbent bovine unable to rise will include hypocalcemia, hypokalemia, and hypomagnesemia. If alert but recumbent, examine the legs for fractures or joint swelling, and when you later palpate the pelvis, you will want to pay special attention for fractures and swellings. If all these are ruled out, nerve deficits such as calving paralysis or spinal insult must be considered if the history is compatible. Multiple recumbent cows should bring botulism to mind. A depressed pregnant recumbent sheep or goat should bring to mind pregnancy toxemia as a rule out.

Once the animal is restrained appropriately, note the hair coat and look for external parasites, scars and other visible signs. If the unpigmented areas are sunburned, rule out photosensitization and whether it is primary (plants, drugs, sprays) or hepatogenous in origin. Wool and hair color changes can be seen with chronic copper deficiency. Also palpate the animal to put a number on the BCS. Adopt a system for BCS and use it consistently, whether it is based on a 1-5 or 1-8 system or some other. Beef cattle have a 1-8 system, dairy cattle and sheep have a 1-5 system[2]. Sheep with heavy wool are particularly difficult to assess until you are able to palpate the muscles.

It is now time to restrain the cow using the minimal amount of restraint necessary to keep you and the animal safe. Ask the owner NOT to use dogs or hotshots that will get the animal upset. It is preferable to have a capable assistant to perform tail jack restraint when it is needed. Be aware of kick zones and keep your head safe at all times. If you are using a squeeze chute, never put hands and arms through small spaces between bars; instead remove a bar. Never reach under the animal without removing the bottom access panel of the chute. Small spaces can lead to fractures or your hand or arm. Also be very careful around the head of a horned animal. Unlike horses, cattle given tranquilizers often decide to lay down which makes the exam difficult and can lead to strangulation in the head gate apparatus, so drugs to calm the animal are to be avoided. Sheep should be held by an assistant using a hand under the chin and the other hand under the tail. Goats can be restrained using a rope halter.

Always perform your exam in a similar manner, so that it becomes automatic and you`re less likely to skip a step and miss something. Do not start your "hands on" PE with the head as you might with a horse, as ruminants will react adversely because they are not used to people being near the head. The heart rate (HR) and respiratory rate (RR) will go up, and the animal may struggle to escape. Save the head for last, or just before the rectal exam. First talk gently to the animal, then touch her where she is least likely to react fearfully; with dairy cattle that means around the rear or near the udder. Lifting the tail and inserting the thermometer is often a good first step. Normal rectal temperature for cattle is up to 102.5 F or 39.1 C. Remember that cattle, especially large bodied dark animals, are especially prone to having an elevated rectal temperature (up to 106 F or 41 C) when ambient temperature is above this level. Fever must be carefully evaluated in such instances. Normal temperatures for sheep and goats can be up to 103.5 F or 39.7 C. In neonates thermoregulation is poor, and fevers as high as 108 F or 42.2 C can be seen during hot weather, resulting in convulsions. Subnormal temperatures are found in animals with hypocalcemia, septic shock, cachexia, uremia, or generalized muscle weakness. In cold environments, neonates with hypovolemia and acidosis and/or hypoglycemia commonly have subnormal temperatures.

You should now move to the left thoracic area and to palpate the brisket for edema or masses and observe the jugular veins for distention and pulsation. Brisket and submandibular edema with distended jugulars is characteristic of heart failure, and brisket edema, along with submandibular edema, are typical of hypoalbumenia. A large mass at the thoracic inlet is characteristic of the thymic form of lymphosarcoma in young cattle. If edema is noted, also palpate the ventral abdomen when you have safe access to reach the area, and make a mental note to check for submandibular edema later. Also run your hands over the withers and back to check for lumps or crepitation. Rapid respiration and expiratory dyspnea with dorsal crepitation due to subcutaneous emphysema are characteristic of atypical interstitial pneumonia.

Now, auscult the heart and lungs. Take a HR and reassess by taking it again later to see if she has calmed down, as initially elevated HR is often due to excitement. Listen for strength, rhythm, and murmurs. Normal HR for cattle is 40-80 bpm, with calves 100-140. Sheep are 60-120 bpm and lambs 120-160. Goats are 70-110 bpm with kids 120-160. Many factors besides excitement cause an increased HR including fever, sepsis, pain, hypocalcemia (usually also weak heart sounds), anemia or hypovolemia. Conditions associated with decreased HR include uremia and hypokalemia. Ruminants that have not been eating and have an empty rumen also tend toward low HR. Pay attention to the strength of the heart sounds…loud pounding may be associated with anemia. Muffled sounds may be due to pericarditis, which later develops into splashing "washing machine" sounds as a gas-fluid interface develops.

The most common arrhythmia in adult cattle is atrial fibrillation, which is irregularly irregular and associated with gastrointestinal disturbances and electrolyte abnormalities. This generally does not require specific treatment. In calves the most common cause of cardiac arrhythmia is hyperkalemia (potassium above 8 meq/L), and immediate treatment is indicated. Ventricular arrhythmias can be heard in animals that have ingested oleander, azalea, or laurel leaves, which contain a potent cardiac glycoside.

Heart murmurs should be characterized by where the sound is loudest, the intensity (grade 1-6), and described in terms of systole and diastole, or continuous. The mitral valve is heard best in the left fifth intercostal space just above the point of the elbow. Pulmonic valve sounds are heard best at the third to fourth intercostal spaces under the triceps muscle just below the level of the shoulder. The aortic valve is located dorsal and caudal to the pulmonic. The tricuspid valve will be checked over the right third to fourth intercostal space when you reach the other side. Congenital septal defects can result in loud murmurs and signs of congestive heart failure that might mimic pneumonia. Cattle with foot abscesses or other purulent process may develop septic endocarditis with fever, with the murmur most audible over the valve or valves involved. Severe anemia can also produce murmurs.

The respiratory rate and effort should next be recorded. The normal respiratory rate varies greatly according to the ambient temperature and whether in sun or shade. Cattle run from 12-36 /min, calves 30-60/min, goats15-40/min, kids 40-65, sheep 12-70 (because they tend to pant in hot weather), and lambs 30-70. Also note whether the effort is excessive, and if so whether inspiratory dyspnea or expiratory dyspnea is present. As a rule of thumb inspiratory dyspnea is from a problem from the nose to the thoracic inlet such as pharyngeal trauma or laryngeal disorders, while expiratory dyspnea is associated with intra-thoracic problems. The left thorax is next ausculted for pulmonary sounds. Healed fractures discovered as lumps on the ribs should be noted. In cattle with thick chest wall, no breathing sounds may be audible. In thin dairy cattle and goats with normal lungs, one can usually hear only inspiratory sounds over the large airways. In sheep with very moist mucous filled but normal airways, both inspiratory and expiratory sounds are often heard. Breath sounds should be described as normal, harsh, wheezes, or crackles. Harsh sounds are associated with airway inflammation and debris or with pulmonary consolidation, wheezes with narrowed small to middle size airways, and crackles with interstitial emphysema or popping sounds of mucous in airways. Significant pathology can be present in ruminant lungs without auscultable changes. Loss of sounds on one side only may be associated with pleural fluid (loss of ventral sounds) or collapsed lung if dorsal sounds are absent.

If any abnormalities are noted, the chest should be percussed to determine whether any areas of dullness or hollowness are present. I prefer to utilize a rubber hammer and a tablespoon. Percussion is most effective in goats, calves, and lambs. Adult cattle are more difficult to assess, although thin dairy cows can be successfully percussed, whike sheep with a thick fleece are almost impossible. Dull sounds ventrally are associated with consolidated lung such as found in severe bronchopneumonia, or with pleural effusion. Hollow sounds dorsally are associated with collapsed lung such as might be found with atypical interstitial emphysema and pneumothorax. Before leaving this area, the left prescapular lymph node should be palpated for size. Other lymph nodes are palpated as one moves around the animal.

The ruminant abdomen is often the key to solving a problem in dairy cows. You have already noted the shape and fill viewed from both rear and side, and whether or not the back was arched. Note any surgical scars in the left flank. Auscult for normal rumen contractions with one primary contraction every 30 to 60 seconds. This contraction should be strong enough to push your hand out at the left flank at least 1 cm, preferable 2 or 3 cm. The secondary contraction which pushes gas from the cardia (eructation) occurs about once every 2 minutes. The normal rumen should feel doughy dorsally in the left paralumbar fossa, and resilient and fluid filled ventrally. A small barely active rumen is typical of an animal that has not eaten normal amounts for several days. Gas in the dorsal rumen indicates either frothy or free gas bloat, which will have a monotone boink or ping in the left flank often extending up to the hip.

A large inactive fluid filled rumen is typical of vagal indigestion. You can also succuss the rumen by listening with your stethoscope while forcefully ballotting the rumen ventrally and listening for fluid sounds. Now perform simultaneous auscultation and percussion over the left abdomen, by snapping your finger forcefully against the body wall while listening for a ping. A ping is a metallic sound sometimes described as hitting crystal, and is generated by a gas filled area under some pressure. When you hear a ping, keep your stethoscope in that location and move your snapping finger around to determine the shape and size of the gas pocket being heard. A ping is associated with a gas distended viscus, often with a fluid line, such as a left displaced abomasum (LDA). With LDA the abomasum is trapped between the left body wall and the rumen. Because the rumen is medial to the trapped abomasum but still contracting, the pressure in the LDA changes, resulting in changes in pitch of the ping you are hearing. There may also be gas bubbling sounds as gas enters the abomasum. Monotone pings can be due to gas in the rumen, sometimes LDA, or some other cause. The ping of an LDA is heard on a line between the hip and the elbow and normally only extends slightly behind the last rib in Holstein cattle, whereas the boink of gas in the rumen is not heard on that line and often extends up to the spinal column and back to the hip/pelvis. The examiner can next auscult the dorsal aspect of the ping, while vigorously pushing in and out below it, so that a fluid line in the abomasum can be located by sloshing sounds. This process is called succussion. If the rumen is large and fluid filled as in some types of vagal indigestion, as well as in grain overload, similar sounds may be heard. If you suspect grain overload make a mental note to come back and secure a needle puncture sample of rumen fluid to check pH toward the end of the exam.

In cattle check for cranial abdominal pain by pinching forcefully just caudal to the withers…a normal response is to dip the back, but an animal with pain will resist this. You can also push up on the xiphoid area forcefully if this can be done safely, and listen carefully for a grunt indicating that this causes pain. This procedure is possible in most dairy cows in a stanchion and in sheep and goats, but use caution regarding kicks from beef cattle in a squeeze chute. In calves, lambs, and goats, stand behind the animal and use two hands to push abdominal structures together and check for fetuses, masses and painful areas much as you would palpate a dog`s abdomen. If the bladder of a small ruminant has ruptured there may be a large fluid filled abdomen. Now palpate the udder and milk veins for hematoma and other problems, and examine the milk from all 4 quarters. Visually examine the teats for color, inflammation, or damage. Coliform mastitis is associated with swelling, redness, and pain in the affected quarter, plus milk that is serum-like with chunks of fibrin in it. If appropriate collect milk samples for culture. Check the ventral abdomen for edema or masses. Severe ventral swelling that is not hot may be found in bulls or steers with urethral rupture. In males palpate the prepuce and note whether there are crystals on the preputial hair tuft. Later, if you are suspicious of urolithiasis in a sheep or goat, extrude the penis and observe the urethral process for stones. It is usually safer and easier to palpate the scrotum and testicles from the rear. Before heading to the rear, palpate the left prefemoral lymph node.

Move to the rear and if not already done, take out your thermometer and record the temp. Also note the color and odor of the feces. Next examine the vulva for edema, mucous membrane color (pale, hyperemic, icteric, or cyanotic), pustules due to IBR, and check for vaginal discharges. Is the discharge considered normal lochia for this post parturient animal? Lochia should have minimal odor, be reddish brown in color, and become more viscous in the days after a normal parturition. If you have any doubt, make a mental note to also do a vaginal exam later near the end of the PE. Palpate the supramammary lymph nodes in the escutcheon. In males palpate the scrotum and testicles.

You may now elect to collect urine or wait until just before the rectal exam. Move people away from the front of the animal so stress and nervousness is minimized, and have a cup ready to catch the sample as you gently stroke the vulva in an upwards motion meant to simulate the licking of a bull. Remember that she has to relax to urinate! In males you can try tickling the preputial orifice. Check the urine for gross appearance and use a dipstick to look for ketones, blood, glucose, pH, and protein. If the urine is grossly abnormal, you should save the sample for a complete urinalysis.

Now move to the right side, note any surgical scars, ballotte the abdomen for late term fetus or masses, and also assess whether painful. A grossly enlarged liver may have a palpable caudal border behind the last rib. Palpate the prefemoral lymph node, and perform simultaneous auscultation and percussion looking for pings. Many sick cattle have a small (20-25 cm diameter) circular ping under the last rib indicative of gas in the spiral colon, but not of specific diagnostic significance. Cecal pings will be high, larger and often run from the hip to cranial of the last rib, and are diagnostic of large amounts of cecal gas associated with cecal distention or volvulus. The ping associated with a right diplaced abomasum (RDA) or abomasal volvulus (AV) can be found on a line from hip to elbow, and is usually centered cranial to the last rib, although the ping of an AV can also extend caudal and cover an area up to a meter in diameter.

Check the right thorax and right heart as you did for the left before moving to the head. Be sure the head is appropriately restrained so that you can safely approach. This is a good time to put on your headlamp if you have not yet done so, and if the animal is down or has CNS signs so that rabies is a possibility, put on latex gloves. Check the head for bilateral symmetry…is there a swollen sinus nor dental abscess or edema of one side? If so be sure to also percuss the sinuses. Post dehorning frontal sinusitis should be ruled out. Palpate the ears, larynx, pharynx, and parotid and submandibular lymph nodes for edema, pain, heat, or swelling. Note the position of the eyes in the orbit, noting whether they are sunken, or if there is any strabismus or nystagmus. Remember that head position will influence eye position. Check for a menace or blink response as a crude test of vision and eyelid function. Pinch and tent the eyelids to assess hydration. The ocular mucous membranes should be examined for color (the oral pigments in many cattle make the mouth a poor location to assess mucous membrane color) as well as petechia and inflammation. The sclera should be white and scleral vessels small, red and few in number. Numerous large dark engorged scleral vessels are indicative of sepsis or toxemia, and are the equivalent of oral mucous membrane color changes in horses. Now shine a light in the eyes and look for corneal ulcers, tumors or other abnormalities. Systemic disorders such as MCF and some cases of BVD have ocular abnormalities such as corneal clouding and conjunctivitis.

With facial nerve paralysis, the animal may be drooling rumen contents (cud) from one side of the mouth. The nose should be clean and moist with beads of moisture. A dry nose is indicative of fever, dehydration or hypocalcemia, since calcium is necessary for sweat glands to function. A dirty nose often indicates that the animal has not been eating. Shine the light into the nares and note petechia, ulcers, and discharges. IBR may cause small lesions in the nares. Be sure to smell the breath. Infected bone associated with a tooth abscess has a typical odor, and necrotic tissues in the pharynx, larynx, or lungs can be detected. Restrain the head of cattle with nose tongs and examine the mouth. Look at incisors for age, check jaw tone, note crepitation associated with fractured mandible, and then using a cloth or small towel insert your hand in the interdental space and grasp the tongue. Gently pull it out to the side, note tone, and look for lesions such as actinobacillosis. Shine your light into the mouth looking for tooth symmetry and dental pad lesions such as found with BVD in cattle, bluetongue or contagious ecthyma in sheep, or contagious ecthyma in goats. If you are contemplating passing a stomach tube, this is a good time to do so while the head is restrained.

Finally it is time for the rectal exam in cattle. Put on a long sleeved glove and lubricate generously. As you remove feces note the characteristics and put a sample in a cup if it is abnormal. Most animals that are "off feed" have scant amounts of relatively dry feces. As you get wrist deep run your hand around the pelvis for symmetry and masses. Just cranial to the pelvis and at approximately 2 and 10 o`clock are the internal iliac lymph nodes. Ventrally grasp the cervix and assess size and weight of the uterus. Retract it if it is involuted. Determine pregnancy status and estimate the fetal age. Adhesions may indicate a source of peritonitis. With peritonitis one may often get the sensation of grittiness rather than smooth moist peritoneum. If the bladder contains large quantities of urine it should be easy to identify ventrally. In cases of obstructive urolithiasis the base of the penis may be pulsating and painful. Masses caused by fat necrosis, tumors, or abscesses may be discovered as the hand sweeps the abdomen and moves cranial. Now move your hand cranial and try to touch the rumen or LDA …this is usually difficult in large cattle. Near midline lies the lobulated left kidney. The small intestines and cecum are not notable when normal, but a gas filled cecum may feel like a loaf of French bread or a balloon just right of midline. With intussusception, a painful sausage-like mass may be palpated near the midline or just to the right. One would also expect to see scant dark to bloody feces with an intussusception. In smaller cattle the operator may also be able to reach the caudal pole of the right kidney and the caudal border of a grossly enlarged right liver lobe.

If you suspect abnormal uterine contents in a postpartum animal, clean the perineal area thoroughly, and using clean glove and sterile lubricant, perform a vaginal exam. Check for retained membranes, lacerations, and other pathology, and note the odor. Normal lochia is not malodorous. If the animal is lame, it may be necessary to sedate and cast the cow at this time in order to work on the feet. After the general PE you should now proceed to a further examination of the suspect system or organs.[3]

It may be necessary to cast or sedate a lame cow or small ruminant to examine the foot in question, unless she can be appropriately restrained while standing. For recumbent cows that are unable to stand, the float tank or hip lifter can be used to further assess whether one leg in particular is the problem. As the cow is floated it may become apparent which leg is knuckling, and appropriate splints can be applied. If the cow is down but has not recently calved and does not have severe systemic illness (rule out the 3M`s), collecting a sample of spinal fluid from the LS space may be indicated.

Consider whether you need additional samples at this point, such as rumen fluid or blood for CBC or chemistry. The tail vein is used for blood collection in adult cattle, while the jugular is most accessible in calves and small ruminants. You can also consider special procedures to further assist in diagnosing the problem, such as imaging, joint tap, needle puncture of the rumen, biopsy, or laparotomy.

ADDITIONAL READING

Terra R: Chap 1 in Large Animal Int Med, 4th ed, Elsevier, 2009, pg 3. Radostits OM et al: Chap 1 in Vet Med, 9th ed, WB Saunders, 1999, pg 4. Rosenberger G, et al: WB Saunders, 2nd ed, 1977.

REFERENCES

Terra RL: In Smith BP Large Ani Int Med, 4th ed, Elsevier, 2009, pg 4. Maas J, Stratton-Phelps, M: In Smith, BP, Large Ani Int Med, 4th ed, Elsevier, 2009, pg 166. Radostits OM, et al: Vet Med 9th ed, WB Saunders, 2009, pg 25.

CLINICAL TRIAL FOR CANINE DEGENERATIVE MYELOPATHY: ANTISENSE OLIGONUCLEOTIDE THERAPY

Joan R. Coates, DVM, MS, DACVIM (Neurology)
Columbia, MO

INTRODUCTION

Mutations in the superoxide dismutase 1 (*SOD1*) gene, which encodes cytosolic Cu/Zn superoxide dismutase (SOD1), are most prevalent with the inherited form of ALS (fALS). Familial ALS (fALS) accounts for 10% of all ALS cases; the rest appear as sporadic ALS (sALS).[1] Although the etiology of the more common sALS is largely unknown, symptoms and pathology associated with *SOD1*-related ALS mirror those in sALS. Since the initial discovery that mutations in *SOD1* could cause ALS,[2] more than 160 *SOD1* mutations have been identified in ALS patients (http://alsod.iop.kcl.ac.uk/), accounting for approximately 20% of fALS cases.[3] SOD1 aggregates in motor neurons are the histopathologic hallmark of *SOD1*-mediated ALS.

Our group recently discovered that mutations in the superoxide dismutase 1 gene (*SOD1*) also underlie canine degenerative myelopathy (DM).[4,5] We first identified a *c.118G>A* transition in *SOD1* that predicts an E40K missense mutation in Pembroke Welsh Corgis (Corgis) with DM. In addition, we demonstrated that, as in ALS, cytoplasmic aggregates that bind anti-SOD1 antibodies were present in spinal cords of *SOD1:c.118A* homozygotes with DM. No similar SOD1 antigen-containing aggregates are found in control spinal cords from *SOD1:c.118G* homozygotes. *Thus, we have identified canine DM as the first naturally occurring animal disease model of human ALS that results from SOD1 mutations.* Since not all *SOD1:c.118A* homozygotes develop clinical signs, DM appears to be an incompletely penetrant autosomal recessive disease; whereas most human *SOD1* mutations cause dominant forms of ALS. However, we have identified several dogs with signs of DM that are heterozygous for the E40K mutation, suggesting that this mutation may also show dominant inheritance. Similarly, the human D90A *SOD1* mutation is recessively inherited in some families and dominant in others. The course of disease in families with recessive inheritance resembles canine DM in that initial signs are typically in the lower limbs and progress slowly.[6-8]

Transgenic rodent models have provided a research platform by which to investigate the underlying disease mechanism and to evaluate strategies and protocols for disease intervention. Importantly, these studies have determined that neurodegeneration in h*SOD1*^{m} mice and in ALS patients results from a toxic gain-of-function rather than a loss-of-function.[9] In contrast to the progressive neurodegeneration exhibited by transgenic mice, nullizygous *SOD1* knockout mice do not develop ALS-like disease.[10]

As *SOD1*-mediated ALS is a dominantly inherited disease, the mechanism responsible for this toxic gain of function still remains unknown. We argue that clinical similarities between *SOD1*-related ALS patients and sporadic ALS patients, and recent findings of SOD1 aggregates in pathology in sporadic ALS patients, suggest that further study of SOD1 still is relevant to all ALS.[11] Moreover, a recent report demonstrated that oxidative damage to wild-type SOD1 can be pathogenic in sALS and identified a SOD1-dependent mechanism common to fALS and sALS.[12] There also is increasing evidence that propagation of ALS occurs by cell-to-cell transmissibility of pathological misfolded proteins.[13,14]

SUPEROXIDE DISMUTASE 1 PATHOGENESIS

Superoxide dismutase 1 is ubiquitously expressed in the cytosol as a homodimer that converts superoxide, a toxic by-product of mitochondrial oxidative phosphorylation to molecular oxygen and hydrogen peroxide. The active SOD1 subunit is a dimer consisting of two153 amino acids polypeptides complexed with a copper ion and a zinc ion which are required for conformational stability and enzymatic activity. The folding properties of SOD1 are tightly linked to the interaction between copper and zinc within an eight-stranded antiparallel β-barrel which drives the formation of an intramolecular sulfide bond and thus the homodimerization of SOD1.[15]

Mutations of *SOD1* in ALS patients are distributed in all five exons of the gene resulting in alteration of amino acids throughout the SOD-1 protein where they are believed to destabilize protein conformation and promote misfolding. Some mutations affect the active site, whereas others are positioned at the dimer interface or inside the β-strands or connecting loops.[16-19] The canine SOD1, amino acid position 40 lies within the short "Greek key" connecting loop that stabilizes the alignment between the two sets of β-strands which comprise the β-barrel.[20,21] The codon for amino acid position 40 of human SOD-1 lies within a region that contains a cluster of missense mutations that are associated with human ALS.[16,22]

The misfolding of mutant SOD1 may induce neurodegeneration by forming insoluble aggregates that alter intracellular function or by altering the enzyme substrate specificity and generate toxic byproducts.[19] Mutant SOD1 aggregates in the neurons of ALS patients may be toxic or may alter other proteins while preventing normal intracellular functions.[23] Different mutations of *SOD1* can induce aggregation through different biophysical effects: destabilization, altered copper or zinc binding and decrease in overall net charge.[17] The SOD1 isoforms with reduced net negative charge may be prone to aggregation because of reduced repulsive Coulombic forces or because of increased interaction with anionic membrane surfaces.[16-18] The canine E40K mutation and several other ALS-associated *SOD1* mutations, reduces the net negative charge. Protein misfolding is probably an initiator of aggregate formation. Still it remains to be established whether SOD1aggregates are directly toxic or a neutral by-product of the disease process, or even have a beneficial effect by sequestering misfolded SOD1.[17]

ANTISENSE OLIGONUCLEOTIDE THERAPY

In ALS and neurodegenerative disease, multiple methods are currently being studied for their ability to decrease levels of unwanted proteins, including immunization strategies, small molecules, gene therapy, RNA interference, and antisense oligonucleotides (ASO).[24,25] Nucleic acid-based therapies offer the promise of modifying or arresting the course of neurodegenerative disease in which down-regulation of a protein is a treatment objective. RNA inference is triggered by short, double-stranded RNA, while ASOs are short, single stranded DNA-like molecules. Recognition of RNA as a regulator of gene expression came with the discovery of interfering RNAs.[26] RNA interference is a collection of small RNA directed mechanisms that result in sequence specific inhibition of gene expression. Both RNA interference and ASOs are posttranscriptional gene down-regulation approaches that destroy messenger RNA (mRNA) in a sequence specific fashion, leads to lower mRNA levels and results in decreased protein production. The use of ASOs and interfering RNA molecules to lower concentrations of mutant mRNA slowed disease progression and increased survival in the *SOD1* transgenic rodent models.[27-31] Nucleic acid-based therapies do not affect a protein that is already expressed; instead they prevent expression of the protein. Thus in ALS, such treatments would reduce synthesis rather than remove aggregated SOD1 protein.

In ASO therapy, expression of a mutant gene is down-regulated by the targeted ASO, reducing the total amount of mutant protein generated. ASOs are 13-25 nucleotides long and are specifically designed to hybridize to corresponding RNA (complementary mRNA) by Watson-Crick binding. The specificity of the hybridization makes antisense treatment an attractive strategy to selectively modulate the expression of genes in pathogenesis of disease [Figure 1].[32,33] The target of the ASO is mRNA; thus, modifies gene expression at the translational step. The mRNA function is inhibited in several ways, including inhibiting splicing, transport and translation of protein. The most important mechanism seems to be the utilization of endogenous RNase H1 enzymes by the ASOs. The RNase H1 recognizes the mRNA-ASO duplex and cleaves the mRNA strand, leaving the ASOs intact to bind to other target RNA. ASOs are perhaps the most direct therapeutic strategy to approach RNA.[33] ASOs must reach the targeted RNA inside the cytoplasm or nucleoplasm to be effective. In other words, the oligonucleotide must traverse from the cell surface, across the plasma membrane and then to the target RNA. Results suggest that ASOs are transported into cells by endosomal transport mechanisms.[33] Thus, the oligonucleotide is chemically modified for effectiveness and administered locally or systemically.

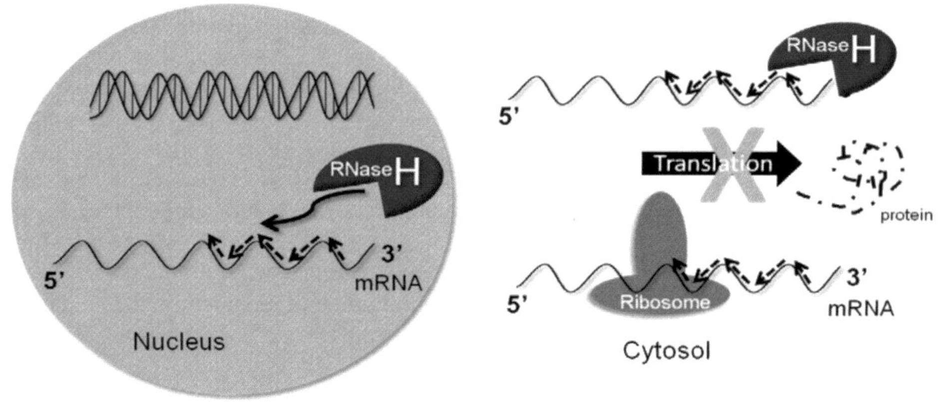

Figure 1. Mechanism of action of antisense oligonucleotide: ASOs (dotted arrows) bind by Watson-Crick base-pairing to mRNA either in nucleus or cytosol. They disrupt mRNA by activation of endogenous RNAase H. RNase H is activated by the oligo:RNA dimer and cleaves the mRNA leaving the oligo intact. Modified ASO molecules inhibit translation of mRNA by steric hindrance of ribosome translocation.

The blood-brain barrier constitutes a challenge for delivering exogenous molecules to the CNS. Parenterally administered ASOs or interfering RNAs do not cross the blood-brain barrier; thus reduced SOD1 expression within neurons and surrounding non-neuronal cells requires direct administration into the cerebrospinal fluid (CSF) via intraventricular or intrathecal routes. Since RNA-like molecules have been difficult to stabilize in biological fluids, effects have been limited with direct infusion into the systemic circulation or CSF.[34] In contrast, the current generation of ASOs is considered stable in biological fluids with half-lives of days to weeks; therefore, ideal for delivery directly into CSF. Although the safety of nucleic acid-based therapies in the human CNS is unknown, in the periphery, ASOs have been proven to be well tolerated in humans with minor injection site reactions and effects on the coagulation system.[32] One of the major benefits of ASOs and stabilized interfering RNAs is that their doses can be adjusted or treatment can be stopped. To provide continuous down-regulation of mutant mRNA, these RNA interference compounds would need to be administered directly into the CNS through an intrathecal pump from a reservoir or as a bolus injection.

A promising therapeutic strategy for ALS is direct infusion ASO into the CNS.[25] Using an ALS rat model, ASOs complementary to SOD1 mRNA appear to distribute widely in the CNS, decrease SOD1 mRNA and protein, and provide neuroprotection by slowing of disease progression.[28] Based on these data, a Phase I human trial of safety, tolerability, and pharmacokinetics of ASOs administered intrathecally to patients with SOD1 familial ALS has been completed (NCT01041222; www.clinicaltrials.gov). We believe therapies that decrease the amount of SOD1 in neurons are likely to reverse or slow the disease in *SOD1*-related human ALS as well as in canine DM.

REFERENCES

[1]Valdmanis PN, et al. **Curr Neurol Neurosci Rep** 2009;9:198. [2]Rosen DR, et al. **Nature** 1993;362(6415):59. Dion PA, et al. **Nat Rev Genet** 2009;10:769. Awano T, et al. **Proc Natl Acad Sci U S A** 2009:2794. [5]Wininger FA, et al. **J Vet Intern Med** 2011;25(5):1166. [6]Andersen PM, et al. **Brain** 1996;119:1153. [7]Khoris J, et al. **Eur J Neurol** 2000;7(2):207. [8]Robberecht W, et al. **Neurology** 1996;47:1336. [9]Bruijn LI, et al. **Annu Rev Neurosci** 2004;723. [10]Reaume A, et al. **Nat Genet** 1996;1343. [11]Forsberg K, et al. **PLoS ONE** 2010;5:1. [12]Bosco DA, et al. **Nat Neurosci** 2010;13:1396. [13]Prudencio M, et al. **Mol Neurodegen** 2011;6:77. [14]Grad LI et al. **Proc Natl Acad Sci U S A** 2011;108:16398. [15]Lindberg MJ, et al. **Proc Natl Acad Sci U S A** 2004;101:15893. [16]Sandelin E, et al. **J Biol Chem** 2007;282:21230. [17]Shaw BF, et al. **Trends Biochem** 2007;32:78. [18]Valentine JS, et al. **Ann Rev Biochem** 2005;74:563. [19]Rakhit R, et al. **Biochimica et Biophysica Acta** 2006;1762:1025. [20]Green SL, et al. **J Hered** 2002;93:119. [21]Boissinot M, et al. **EMBO J** 1997;16:2171. [22]Deng HX, et al. **Science** 1993;261:1047. [23]Bruijn LI, et al. **Science** 1998;281:1851. [24]Zinman L, et al. **The Lancet Neurol** 2011;10:481. [25]Miller TM, et al. **Arch Neurol** 2008;65:447. [26]Fire A, et al. **Nature** 391:806. [27]Raoul C, et al. **Nat Med** 2005;11:423. [28]Smith RA, et al. **J Clin Invest** 2006;116:2290. [29]Wang H, et al. **J Biol Chem** 2008;283:15845. [30] Yokota T, et al. **Arch Neurol** 2007;64:145. [31]Miller TM et al. **Ann Neurol** 2005;57:773. [32]Tamm I, et al. **Lancet** 2001;358:489. [33]Bennett CF, et al. **Annu Rev Pharmacol Toxicol** 2010;50:259. [34]Tiemann K, et al. EMBO Mol Med 2009;1:142.

CEREBELLAR DEGENERATION DUE TO AN AUTOPHAGY DEFECT IN OLD ENGLISH SHEEPDOGS AND GORDON SETTERS. NJ Olby, C Agler, G Urkasemsin, D Nielsen. North Carolina State University College of Veterinary Medicine, Raleigh, NC.

Hereditary cerebellar degenerative disorders have been described in many species, including the dog. The genetic basis of the disease in Finnish Hounds and Beagles has been described. Old English Sheepdogs suffer from juvenile onset cerebellar cortical degeneration causing progressive gait disturbance. The purpose of this study was to map the disease trait in Old English Sheepdogs and identify the underlying genetic cause.

A family of affected Old English Sheepdogs was genotyped using the Illumina SNP beadchips and the region associated with the disease was mapped using a genome-wide association study (GWAS). The region associated with the trait was sequenced using targeted sequence capture and next generation sequencing techniques, and variants were identified using GATK. Variants that segregated with phenotype were sequenced in a larger group of Old English Sheepdogs, normal dogs from different breeds, and affected dogs from other breeds.

GWAS linked a region of CFA4 with the disease phenotype. Sequencing of the region identified 6 exonic, non-synonymous SNPs that segregated with phenotype. Sequencing of a larger population of Old English Sheepdogs and normal dogs eliminated all but one SNP in a proposed autophagy gene. The same mutation was found in Gordon Setters with cerebellar degeneration. Electron microscopy of affected dogs revealed accumulation of autophagic vacuoles within the processes of Purkinje neurons. We conclude that cerebellar degeneration in Old English Sheepdogs and Gordon Setters is a result of a mutation in an autophagy gene, providing strong evidence of the importance of autophagy in neuronal health.

JOB TASK ANALYSIS UPDATE

Simon R Platt, BVM&S, DACVIM (Neurology), DECVN
Athens, GA

INTRODUCTION

A job task analysis (JTA) is performed to provide validation of the content of a training program and/or examination.

Without a formal analysis of the current job requirements for a neurology specialist, producing a blueprint for what the examination should include, examinations are inherently arbitrary. Once the content of the examination is determined (using the examination blue-print developed as a result of the JTA), it is a separate task to validate the pass fail cut off score based upon, for example, criterion referencing using a definition of a minimally qualified or competent practitioner.

In addition to the academic benefits, JTA provides some protection for the College against charges that a professional certification examination was conducted arbitrarily.

AIMS OF THE JTA TASK FORCE:

The aims of the task force when they originally formed in 2011 were to (i) develop a comprehensive survey of tasks performed by board certified neurologists and the knowledge required to perform them; (ii) develop a new blue print for the certifying examination based on the objective details of daily job requirements; (iii) better define minimally qualified practitioner / neurologist (MQP); and (iv) determine a more accurate scope of residency training.

METHODS

A master task and knowledge statement list was developed with the premise that a competent specialist in the discipline must know or be able to perform the individual items in their work. A list of 255 tasks /knowledge statements was then refined into a survey format, arranging the knowledge and skills into separate categories:

The 5 sections and subsections were:

1. Pre-clinical & Basic Training
   a. Anatomy & Physiology
   b. Pharmacology & Toxicology
   c. Genetics & Molecular Biology
2. Clinical Training: Medical
   a. Canine & Feline methodology
   b. Equine methodology
   c. Ruminant / Food Methodology
   d. Canine & Feline pathogenesis
   e. Equine pathogenesis
   f. Ruminant / Food Pathogenesis
   g. Canine & Feline diagnosis and treatment
   h. Equine diagnosis and treatment
   i. Ruminant / Food diagnosis and treatment
3. Clinical Training: Neurosurgery
   a. General
   b. Understand technique
   c. Perform technique
4. Clinical Training: Neuroradiology
5. Clinical Training: Neuropathology

The survey required that both the importance and frequency of use of each task was ranked on a scale of 0-4 based on a competent and effective performance by an 'entry level' neurologist. {0- Not performed/ 1= Not significant / 2= somewhat significant / 3 – quite significant / 4 – extremely significant.}

Tasks were excluded based upon a low percentage of respondents performing them (<50%) and or a low mean importance rating. Additionally an overall mean significance task rating of 2.00 (i.e., somewhat significant) or higher was required for task inclusion.

RESULTS

The survey was distributed to all ACVIM Neurology Diplomates on-line in the spring of 2011. Eighty-two responses were received; this was a 45% response rate of the specialty. Sixty three percent were from specialty practice and 31% were from academia. Eighty-six percent of respondents practice in the USA. Neurosurgery was performed by all 82 respondents.

Thirty-nine tasks were removed because >50% of respondents rated the task as not performed or were not significant. 216/255 tasks remained after exclusions were made final and were assigned into one of the 5 examination / subject sections.

An averaged percentage weighting was assigned to each of the 5 sections of the exam content allocation in accordance with observed working patterns. See Table 1

Table 1.

	% Weighting of each section
Preclinical & Basic Training	40
Medical Neurology	22
Surgical Neurology	12
Neuroradiology	16
Neuropathology	10
TOTAL	*100*

CONCLUSIONS

The JTA defined the scope of the job in 2011 and as such identified the entire testable content domain / exam blue print. Additionally, the division /weighting of the subject areas has been assigned. The JTA itself did not serve on its own to define the MQP, which is a separate ongoing process.

FURTHER WORK TO DO

In an attempt to integrate the results into meaningful changes in examination protocols, it is felt that several vital areas of work remain. These are:

1. Exam committee to review the entire testable content domain / exam blue print
2. Categorize every item in current exam bank to one of tasks in outline: this will serve as inventory and identify areas in need of question creation
3. Determine the MQP by:
    a. Determination of problematic areas in the content for a MQP
    b. Determination of areas which should be standard for a MQP

REFERENCES

[1]Rodgers SH. A functional job analysis technique. **Occup Med** 1992;7(4):679. [2]Holman D, et al.. **Appl Ergon**. 2002;33(3):197. [3]Nelson EC, et al. **Med Care**. 1975;13(2):104. [4]Fullerton JT, et al. **J Midwifery Womens Health**. 2000;45(2):150.

NOVEL THERAPEUTIC OPTIONS FOR BRAIN TUMORS

John H. Rossmeisl, Jr., DVM, MS, DACVIM (SAIM and Neurology)
Blacksburg, VA

Malignant gliomas (MG), and in particular glioblastoma multiforme (GBM), represent some of the most biologically aggressive and treatment refractory malignancies recognized in humans and dogs. Despite several decades of intensive research focused on MG/GBM, significant progress towards improving the survival of humans with GBM has been lacking. Currently, the average survival of a human with GBM is approximately 15 months. Only 5% of humans with GBM survive for 5 years, which is a statistic that has remained largely unchanged for several decades. Multiple independent investigators have demonstrated that the clinical manifestations, natural course of disease, pathology, molecular etiopathogenesis, and genetic aberrations of canine and human brain tumors have been shown to be similar, which makes dogs an ideal translational model for the study of brain cancer. This session will review the background and justification, pre-clinical development, and preliminary results of several novel therapeutic approaches to the treatment of brain tumors currently being investigated in the author's laboratory.

ELECTROPORATION-BASED THERAPIES

Electroporation, which results in an increase in the permeability of the cell membrane, is initiated by exposing cells or tissues to electric pulses. As a function of the induced transmembrane potential the electroporation pulse can result in: no effect on the cell membrane, reversibly permeabilize the cell after which cells can survive (reversible electroporation), or permeabilize the cell membrane in a manner that leads to cell death (irreversible electroporation [IRE]) presumably through a loss of homeostasis if not from other superimposed damage modes. This increase in transmembrane potential is dependent on a variety of conditions such as tissue type, cell size, and pulse parameters including pulse shape, duration, number, and repetition rate. However, for a specific tissue type and set of pulse conditions, the primary parameter determining the extent of electroporation is the electric field to which the tissue is exposed. Reversible electroporation has been used for years to facilitate genetic modifications of cell function via delivery of plasmid DNA, mRNA, and viruses through the reversible creation of nanopores within the cell membrane. Reversible electroporation can be used to increase permeability in order to allow intracellular delivery of otherwise impermeable cytotoxic macromolecules for targeted cell death in a process known as electrochemotherapy (ECT).

In 1993, Salford et al. first reported on the use of ECT with bleomycin in a rodent glioma model, demonstrating a doubling in survival time compared to animals treated with bleomycin alone. The rats showed no immediate adverse effects from the delivery of the pulsed electric fields. It should be noted that the effects of electric pulses alone were never investigated. Since this time, ECT has evolved as a promising new anticancer therapy for the treatment of cutaneous and other systemic tumors, but investigations of its use for the treatment of brain disorders were largely abandoned until recently. ECT has been used effectively in the European Union for the treatment of cutaneous neoplasms in veterinary patients, and there is currently a Phase I clinical trial evaluating ECT for the treatment of brain metastases in humans in Denmark (ClinicalTrials.gov Identifier NCT01322100).

The destruction of tumors and other undesirable tissue through focal ablation has become an important, minimally invasive alternative to open surgery. Mechanisms of tissue destruction include changes in local temperature resulting in coagulation necrosis (radiofrequency lesioning, LITT, and ultrasound) and cell membrane rupture (cryoablation), generation of free-radicals (phototherapy) and *in situ* tissue fixation (microwaves). The confines of the cranial vault and the precision necessary for treatment have limited the use of some of these tools for intracranial disorders. When the electric field strength is above a critical value, IRE is achieved, resulting in cell death within the treated volume. IRE has been reported in conjunction with tissue thermal damage caused by electrical Joule heating. However, it is possible at certain electric parameters to produce IRE without thermal damage to the tissue in a process referred to as non-thermal IRE.

IRE offers several advantages over more traditional surgical resection methods for selected brain tumors. The small electrode size makes the procedure minimally invasive and adaptable to virtually any neuroanatomic location with existing stereotactic guidance systems. It also offers biologic advantages (minimal heat generation during treatment and sparing of major blood vessels) that may make it attractive as a complimentary technique to tumor resection. IRE creates a sharply delineated volume of ablated tissue with sub-millimeter resolution. Moreover, outside the zone of tissue ablation, there is a penumbra of non-destructive increase in membrane permeability. The cells outside the zone of ablation but within this penumbra are more susceptible to exogenous agents such as chemotherapy. IRE also offers the advantage of speed, as typical treatment durations range from 5-15 minutes, and numerous commercial entities manufacture relatively inexpensive electrical generators capable of pulse delivery within the realm of IRE. Finally, preliminary studies also suggest that the mechanism by which IRE induces cell death is effective in neoplasms that have developed chemo- and radio-resistance.

In vitro investigations in our laboratory have demonstrated the cytotoxic effects of IRE against canine, rodent, and human glioma cell lines; identified and capitalized on the effects of various pulse paradigms to maximize anti-neoplastic field effects while simultaneously preserving major vascular structures in the treated volume and minimizing formation of vasogenic edema; and provided insight into the mechanism of cellular death induced by IRE which has implications for selecting combinatorial treatment approaches. Additional studies have indicated that tumor cytotoxicity can be enhanced (Irreversible Electroporation Enhanced Pharmacotherapy [IREEP]) within the reversible penumbra of electroporation using various chemotherapeutic agents with known efficacy against gliomas at doses 10-100 fold less than those recommended for systemic use, thus minimizing any systemic dose-related toxicosis. Recent progress in device development has also yielded advancements in electrode design and pulse generator capabilities that allow for delivery of high frequency IRE pulses (H-FIRE) that obviate the need for administration of neuromuscular blocking agents.

Additionally, we have shown *in vivo* that IRE ablation of brain tissue: 1) can be achieved safely and predictably in rodent and canine models; 2) can be monitored using routine clinical procedures such as ultrasound, CT, or MRI; 3) occurs without any contribution from thermal damage; and 4) is capable of submillimeter resolution between ablated and normal cells. Our *in vivo* work has also shown that normal and neoplastic brain tissue can be safely and controllably ablated using electrical charges lower than those typically delivered to the human brain during electroconvulsive therapy. Although an intact immune system is not required to achieve the desired effects of IRE, studies in brain and extraneural tumors suggest that IRE treatment may be associated with induction of host tumor infiltrating lymphocytes, which has been shown in some other spontaneous neoplasms to have a positive correlation with therapeutic response and survival. Preliminary work has also indicated that IRE may also have a radiosensitizing effect in the brain, which may offer additional advantages for treating intracranial neoplasms with adjunctive radiotherapy.

MOLECULARLY-TARGETED THERAPIES

Several recent human studies have demonstrated the potential value of the clinical usage of molecularly targeted cytotoxins for the treatment of GBM. Investigators in our collaborative research group have demonstrated that the majority of human GBMs express the monomeric from of the IL-13 receptor (IL-13Rα2) on their surface, which is not expressed in normal brain. This discovery led to the development of first generation IL-13 based, Pseudomonal exotoxin (PE) derived cytotoxin, cintredekin besudotox (IL13-PE38QQR). IL-13 based cytotoxins have been one of the few compounds in the last two decades to have to have been shown to have demonstrable clinical efficacy in the setting of recurrent GBM. However, these trials have also highlighted the obstacles researchers and clinicians are faced with when designing and applying molecularly targeted cytotoxins. Ultimately, the clinical utility of any molecularly targeted cytotoxin will depend on three fundamental factors: 1) the target specificity, 2) the relative percentage of the patient population that can be predicted to respond to the cytotoxin, and 3) effective and efficient delivery of the cytotoxin to the tumor.

In an effort to improve upon previous results, ongoing research in our laboratory has resulted in the identification of an additional molecular target for GBM, the tyrosine kinase receptor EphA2. Overexpression of EphA2 receptor has been demonstrated in many human GBM specimens, as well as various other types of human malignancies, including breast, ovarian, and pancreatic carcinoma. Physiologically, activated EphA2 negatively regulates integrin-mediated cellular adhesion, cell spreading, and migration. However, in most solid tumors studied to date, EphA2 is present in its inactive (non-phosphorylated) state and thus cells on which it is expressed may have tendencies to be more motile, faster-growing, and invasive, features which are characteristic of the phenotype of tumors on which EphA2 overexpression have been documented. However, most interesting is that results to date indicate that overexpression of IL-13Rα2 and/or EphA2 together can be documented in nearly 100% of human GBM, thus indicating that predictable anti-tumor efficacy should be nearly universal if a combinatorial treatment approach using both of these molecular cytotoxins is employed.

As justification for performance of a combinatorial molecularly targeted cytotoxin clinical trial in dogs with MG, we have further shown that the molecular fingerprints of canine high-grade gliomas and GBM are homologous to that of humans. Investigations by our collaborative group have demonstrated that IL-13Rα2 and EphA2 overexpression also occur in the majority of canine high-grade gliomas, but not in normal brain tissue. Thus, we provide additional evidence that the dog is an excellent model for translational brain tumor research and the molecular signature of IL-13Rα2 and Eph2A overexpression represents a unique, specific and predictable target for human and canine glioma treatment. In preparation for canine clinical trials, new generation IL-13 based cytotoxic mutants have been designed and shown to have potent and more specific anti-glioma activity than previous cytotoxins. A prototype recombinant EphA2 based cytotoxin has also been conceived, manufactured, and is currently undergoing testing in our laboratory.

Due to the inherent limitations of the blood-brain-barrier, macromolecules such as the targeted cytotoxic proteins previously discussed require local delivery into brain tissue. To accomplish this task in our clinical trial, we use a convection enhanced delivery (CED) technique. In summary, our pre-clinical work has identified specific and predictable molecular signatures within canine and human gliomas that can serve as therapeutic targets for bacterial based cytotoxins with potent activity against gliomas. In conjunction with recent improvements in CED techniques, we have initiated a clinical trial using spontaneous canine gliomas to further evaluate the *in vivo* efficacy of these molecularly targeted cytotoxic compounds.

TECHNICAL ADVANCES IN CONVECTION ENHANCED DELIVERY

CED, pioneered at the NIH and significantly refined in canine models of glioma, is essentially a low-pressure continuous infusion process that occurs over hours to days that has been shown to be capable of achieving substantial tissue concentrations of macromolecules over large areas of the brain without producing tissue injury, substantially increasing intracranial pressure, or decreasing blood flow. Compared with conventional chemotherapy, CED exposes brain tissue to high concentrations of therapeutic agents, while minimizing systemic and CNS toxicity. Although CED theoretically offers an improvement over traditional systemic chemotherapy in terms of effective drug delivery, it has yielded only minor improvements in survival in Phase III clinical trials in humans with malignant glioma. This can be attributed to the limited ability of CED to 1) uniformly distribute drug throughout the tumor and 2) broadly disseminate drug to the infiltrative MG cells residing in the primary tumor periphery (not detected by MRI) which contribute to tumor recurrence.

CED involves stereotactic insertion of small-caliber catheters into the brain parenchyma. Several previous attempts at CED techniques in the brain have been troubled by unpredictable, undesirable, or unknown delivery of agents due to technical issues with catheter design and placement. Building on the pioneering CED work done in dogs by Dickinson et al., several new innovations relating to CED catheter design and intraoperative neuroimaging have now made it possible to efficiently and effectively monitor delivery of macromolecules into the brain using CED. We have carried these studies further by using real time PET based imaging techniques to monitor catheter placement, drug delivery and distribution, while preventing shortcomings related to MRI techniques dependent on infusion of standard gadolinium based contrast agents, which can linger when injected locally into tumor beds.

The anatomical heterogeneity of the brain is also a major limiting factor in perfusion by CED. Portions of the brain are inherently difficult to saturate with drugs due to variations in permeability of white and gray matter, tumor tissue, the ventricular system, and vascular beds. The high lipid content of the brain also makes predictable movement of aqueous drug formulations problematic. Traditional catheters do not effectively perfuse drug into distant infiltrative cells. Studies have shown that controlled hyperthermia enhances the cytotoxic effects of several chemotherapeutic drugs in a process coined thermochemotherapy. This enhancement may be due to increased tumor cell membrane permeability and greater drug metabolism by the cells. Hyperthermia may also provide a means to increase regional circulatory dilation and perfusion, potentially allowing more effectively perfusion of tumor tissue. No studies, however, have investigated the effect of CED perfusion of MGs with laser-induced local hyperthermia (photothermochemotherapy; PTC).

Motivated by the limitations of current CED therapy of MGs, we have invented a fiberoptic microneedle device, FMD, which allows for simultaneous delivery of laser energy and liquid infusates. The novel features of the FMD include its arborizing ability, and capability of performing co-localized PTC. The device is comprised of a bundle of 3-7 fiberoptic microneedles (diameter 250 μm OD; 150 μm ID), which are passed through a rigid guide cannula which can then be arborized to penetrate several centimeters into tissues. Fiber tips may be individually and precisely positioned to enhance drug delivery to the desired target volume and shape. The unique arborizing, multifiber catheter feature of the FMD is well suited to deliver chemotherapeutic drugs locally to the tumor site and large surrounding margins to reach infiltrative cells. We have developed innovative microneedle fabrication techniques, demonstrated the therapeutic optical diffusive abilities of FMD devices and feasibility of mechanical penetration into brain tissue, and have evaluated the performance and efficiency of the FMD compared to traditional CED catheter designs both *in vitro* and *in vivo*. These advantageous features of the FMD devices make them highly suitable for clinical use with CED-based approaches to MG treatment.

REFERENCES

Edd JF, et al. IEEE Trans Biomed Eng 2006; 53(7): 1409-1415, Salford LG, et al. Biochem Biophys Res Comm 1993; 194(2): 938-943, Agerhol-Larsen B, et al. Cancer Res 2011; 71(11): 3753-3762, Spugnini EP, et al. J Vet Int Med 2011; 25(2) 407-411, Kodre V, et al. In Vivo 2009; 23(1): 55-62, Ellis TL, et al. J Neurosurg 2011; 144: 681-688, Garcia PA, et al. J Membrane Biology 2010; 236: 127-136, Garcia PA, et al. Technol Cancer Res Treat 2011; 10(1): 73-83, Kosoglu MA, et al. Lasers in Surgery and Medicine, 2011. 43(9): 914-920.

ADVANCED IMAGING FOR THE ASSESSMENT OF NOVEL THERAPEUTICS AND ANTITUMOR RESPONSE

David M. Vail, DVM, DACVIM (Oncology)
Madison, WI

INTRODUCTION

Advanced molecular imaging techniques, which measure biologic processes at the cellular level,[1-3] are quickly becoming commonplace in physician-based oncology and have the potential to play an important role in the tailoring of cancer therapy in veterinary patients. Advanced imaging modalities have several applications, including the early detection of cancer, more accurate staging of cancer, determination of response to therapy (including prediction of long-term prognosis) and early detection of cancer recurrence. Further, advanced imaging modalities that provide information on the functional properties of tumors, including metabolic rate, proliferative status, presence of receptors and tumor associated antigens, and vascularity are now within our grasp. Additionally, with the development and availability of advanced image-modulated radiation therapy delivery devices (e.g., tomotherapy, stereotactic radiation therapy, stereotactic radiosurgery), the co-development of advanced imaging technologies are critical in order to take advantage of these advancements. While the benefits of more accurate and extensive imaging modalities are intuitive, there exist inherent problems created by their addition into the current standard of care. That is, if we are able to diagnose tumors earlier, 'lead time bias' – the increased survival that results from early detection - can confuse comparisons of treatment and population groups from that point forward. Additionally, the ability to stage tumors more thoroughly may place patients in a more negative prognostic categories when indeed 'stage-migration' has made this artificial.[4] The advantages of imaging modalities currently available in veterinary medicine are summarized in Table 1. As most of us are familiar with standard imaging modalities such as radiographs, magnetic resonance (MR) and computed tomographic (CT) imaging, this discussion will be limited newly developed enhancements of these standard modalities, as well as discussion on positron emission tomography (PET) and PET/CT combinations, dynamic contrast-enhanced MR (DCE-MR) and nuclear scintigraphy.[5-7]

Table 1. Comparison of imaging modalities used in veterinary medicine[8]

MODALITY	COST	SENSITIVITY	SPECIFICITY	AVAILABILITY
‡Radiography	Low	Moderate-high for bone lesions only	High for bone lesions only	High
‡*Ultrasound	Moderate	High	Low-moderate	Moderate-high
‡*CT	High	Moderate-high	Moderate-high	Moderate-high
‡*MRI	High	High	Moderate-high	Moderate-low
*NM	Mod-High	Moderate-high	Low	Moderate-low
*SPECT & PET	High	High	Moderate	Low
‡*PET/CT	High	High	High	Low

Abbreviations: Computed tomography (CT), magnetic resonance imaging (MRI), nuclear medicine (NM), single photon emission computed tomography (SPECT), positron emission tomography (PET), moderate-high (Mod-High), ‡anatomic imaging, *functional/physiologic imaging

ENHANCED CT IMAGING

Computerized tomography (CT) is a commonly used diagnostic tool which provides information regarding the presence of tumors often too small to detect using conventional radiographs. Contrast agents that preferentially bind to tumors are often administered just prior to a CT scan to further enhance the chance of detecting a tumor. One of the limitations of currently used contrast agents is their short half-life, disappearing from the body within hours of administration. A contrast agent that remains in the body longer (for several days without adverse effects) would allow for the performance of multiple CT scans over time which could provide valuable information regarding the effectiveness of anticancer treatments. Novel imaging agents that have a half-life of days allow for the performance of multiple CT scans over time following a single injection. In addition, some of these agents preferentially bind to the tumor vasculature (blood vessels) or other tumor markers (e.g., antigens, receptors) and could theoretically provide greater resolution, which increases the ability to detect tumors at an earlier stage of development. This is extremely important with regard to the ability to detect metastatic lesions in patients diagnosed with cancer and for the evaluation of the effect of their cancer treatment. Examples will be shown and discussed.

PET/CT IMAGING

There is a tremendous interest in molecular imaging for the diagnosis, staging, and evaluation of response for individuals with cancer. PET/CT imaging, as implied by the name, is a marriage of accurate anatomic imaging (CT) which provides high spatial resolution and supplemental functional imaging through PET. PET is a nuclear medicine imaging technology that uses agents labeled with positron emitting radioisotopes as molecular probes in order to image and measure various biological processes. Essentially, the PET image moves beyond anatomy and characterizes tissue based on various metabolic functions such as metabolic rate, DNA replicative rate (proliferation), bone deposition, vascularity, cell receptor kinetics and many more. This is important as changes occurring at the molecular or cellular level in tumors often occur before subsequent alterations in anatomic structure are detected by standard imaging. An ability to observe these changes can further our knowledge of tumor biology, help visualize and quantify cellular receptors and pathways of tumor progression, allow evaluation of pharmacodynamic and pharmacokinetic characteristics of new drugs, and measure and predict patient response early in the course of therapeutic intervention. Several molecular markers for PET/CT functional and anatomic imaging have been applied in human medicine and are becoming important in veterinary medicine. Table 2 summarizes the commonly used markers along with their functional relevance.

Table 2: Examples of PET/CT Molecular Markers and Functional Relevance

Molecular Marker	Functional Relevance
^{18}F-Fluorodeoxyglucose (FDG)	• **A glucose analogue that is used for measuring glucose metabolic rate** • **FDG is trapped intracellularly** • **Cancer cells tend to have increased glucose utilization** • **Used for many tumor types**
^{18}F-fluorothymidine (FLT)	• **Thymidine incorporated into DNA** • **Uptake reflects cell growth/proliferation** • **Correlates well with Ki67 or PCNA indices** • **Useful for many tumor types**
^{60,62,64}Cu-methylthiosenmicarbaxone (Cu-ATSM) **^{18}F-fluoromisonidazole (FMISO)** **^{18}F-Fluoroazomycin arabinoside (FAZA)**	• **Markers of hypoxia** • **May be useful for radiation planning**
18F-sodium fluoride (NaFl)	• **Bone turnover** • **Useful for primary and metastatic tumor imaging**

PET/CT radiopharmaceuticals are currently being developed with three main goals: (1) quantifying a therapeutic target; (2) identifying resistance factors; and (3) measuring early response to therapy. Specific targets under development include markers of new blood vessel growth (angiogenesis), tyrosine kinase growth factor receptors, stromal receptors, and markers of apoptosis and metastasis. Indeed, PET/CT scans performed only a few days or weeks after initiation of therapy have been shown to predict long-term progression-free survival in people with non-Hodgkin's lymphoma, allowing clinicians to make treatment modification decisions prophylactically, prior to actual recurrence. In dogs with lymphoma, FLT uptake, monitored by PET/CT, has been shown to precede clinical or cytologic evidence of recurrence. Several examples of the current and potential utility of PET/CT in veterinary species will be presented.

ENHANCED MR IMAGING

Dynamic Contrast-Enhanced Magnetic Resonance (DCE-MR).

Angiogenesis, long recognized as a key factor in tumor growth and metastasis, has been the target of new anticancer treatment paradigms and DCE-MR has been developed as a means of characterizing changes in microvasculature. In this modality, MRI is first performed with dogs under general anesthesia in an MR instrument and after routine anatomic MR imaging, dynamic contrast-enhanced MR is done by controlled injection of gadolinium diethylenetriaminepentaacetic acid i.v. while simultaneously repeating three-dimensional spoiled gradient echo T1-weighted scans through the tumor volume over several minutes. These allow estimations of both tumor perfusion and vascular permeability. The effects of treatment, particularly therapies relying on or effecting tumor-supporting vasculature and angiogenesis can be measured. Examples will be used in lecture to illustrate.

Diffusion MR

Diffusion MR allows for the mapping of the diffusion process of molecules, mainly water, in biological tissues, in vivo and non-invasively. It evaluates the mobility of water molecules in tumors as a function of cell density and tissue architecture, with an increase in tumor diffusion indicating a positive treatment response and cell death.[9,10] This can allow assessment of the effect of therapy on the cancer without relying on measurable changes in tumor size. Therefore, it has the potential to change the way therapeutic response is delivered such that rather than waiting for results of weeks of therapy (and the associated side effects and progression that occurs with unsuccessful therapy), scanning within days of treatment initiation may predict subsequent response.

MR Spectroscopy Imaging (MRSI)

MRSI uses a strong magnetic field to obtain metabolic information (spectra) that identifies the relative concentrations of various metabolites in the cell cytoplasm and the extracellular space.[11] The use of *in vivo* MR spectroscopy for measuring malignant metabolic biomarkers, such as choline relative to normal surrounding tissue, is moving from the research realm into human clinical use for brain and prostatic cancer and also has future potential for veterinary diagnosis.[11-13] With respect to prostate cancer, malignancy is associated with proportionately lower levels of citrate and higher levels of choline and creatine than are seen in benign prostatic hyperplasia (BPH) or in normal prostate tissue. This difference can be detected by magnetic resonance spectroscopic imaging. This technique can identify metabolic differences in prostate tissue (BPH, prostate cancer, and normal prostate tissue). The anatomic and metabolic criteria used to identify and localize prostate cancer through the use of endorectal MRI/MRSI (endoMRI/MRSI) have been standardized in people, and these criteria improve the accuracy of the examination and limit interobserver variations in interpretation. Evidence is now emerging that MRSI may be helpful in assessing response to treatment, most commonly with radiation and/or androgen deprivation.

REFERENCES

1. Higgins, LJ, et al. *Semin Oncol* 2011, 38:3.
2. Smith, RA, et al. *Front Biosci* 2011, 16:393.
3.Zhu, A, et al. *Semin Oncol* 2011, 38:5.
4. Flory AB, et al. *J Vet Intern Med* 2007, 21: 1041.
5. Lawrence J, et al. *Vet Rad Ultrasound* 2009, 50:660.
6. Lawrence J, et al. *Vet Comp Oncology* 2010, 8:163.
7. Cheng HL. *Current Clin Pharmacol* 2007, 2:111.
8. Forrest LJ, et al. Imaging in oncology, In: Small Animal Clinical Oncology, 5th Edition. 2012 Eds: SJ Withrow, DM Vail & R Page. Elsevier, St. Louis, MO, pp665.
9. Yankeelov TE, et al. *Semin Oncol* 2011, 38:16..
10. Lin C, et al.. *Eur Radiol* 2010, 20:2027.
11. Carroll PR, et al. *Urology* 2006, 8:S4.
12. Kraft SL: Cancer imaging, in Gavin PR, Bagley RS (eds): Practical Small Animal MRI, Vol 1. Aimes, WIley-Blackwell, 2009, pp 333.
13. Glunde K, et al. *Semin Oncol* 2011 38:26.

VIRUSES AND CANCER FOR DIPLOMATES AND CANDIDATES

Sandra Axiak-Bechtel, DVM, DACVIM (Oncology)
Columbia, MO

INTRODUCTION

There are 2 groups of oncogenic viruses: DNA tumor viruses that have linear or circular double stranded DNA, and retroviruses which are RNA containing tumor viruses. The initial discovery that viruses induce cancer was in 1976 with the Rous sarcoma virus, which contains nucleotide sequences that could not be found in related retroviruses, but were found in normal chicken cells. This discovery demonstrated that the Rous sarcoma virus transforming gene originates from a normal cell gene in the chicken. Numerous other transforming retroviruses contain oncogenes derived from a cellular counterpart; the normal cellular gene is referred to as c-onc, and the retroviral gene counterpart is referred to as v-onc. Although most cancers are not caused by viruses this discovery allowed the study of the molecular mechanisms of cancer.

RNA VIRUSES

Retroviruses have a common life cycle. All have an RNA genome that is reverse transcribed into a double stranded DNA provirus. As an obligatory part of the normal life cycle of retroviruses, the DNA provirus is randomly integrated into the genome of the infected host cell. There are 3 major retroviral genes: *gag*, *pol*, and *env*. The *gag* gene encodes for the capsid (internal) proteins, the *pol* gene encodes for the reverse transcriptase enzyme, and the *env* gene encodes the viral envelope proteins. Gene expression of retroviruses is under the control of potent viral transcription regulators (long terminal repeats, LTR) that flank the end of the genome; these regulatory regions determine the tissues in which the virus replicates. The ability of a retrovirus to become oncogenic is made possible by several characteristics of their life cycle.1) Infection is not cytolytic, enabling cells to survive and acquire more mutations. 2) The integration of the provirus damages the integrity of the genome. 3) The integrated provirus acquires the intact or damaged cellular gene, resulting in the incorporation of portions of the cell genome. 4) And /or the function of the cellular genes is affected by the retrovirus.[1,4]

Retroviruses produce cancer in 2 ways. The first is by transduction, which is the rapid recombination of proviral DNA and the host genetic material resulting in a hybrid genome. During the incorporation of host cell genetic material, most viruses lose some of their own genome and are unable to replicate (the Rous sarcoma virus is one exception) on their own. These oncogenes are expressed at high levels because they are controlled by potent viral promoters, so infected cells are rapidly and efficiently transformed into cancer cells. Feline sarcoma virus is another example of a transducing virus – it is closely related to FeLV and contains a portion of the FeLV genome and one of several oncogenes. Each strain of feline sarcoma virus is therefore unique because each arises from a recombination of FeLV and cellular oncogenes. So, cats with FeSV-induced sarcoma are always infected with FeLV, FeSV cannot replicate independently, and horizontal transmission does not occur.[2,4]

The second mechanism by which retroviruses canproduce cancer is through pro-viral insertion. These viruses do not contain an oncogene but, during proviral insertion, they insert their viral transcription regulatory elements of the long terminal repeat (LTR) region near cellular oncogenes and activate them. These tumors develop over a long period of time compared to transduction; this is because with pro-viral insertion, integration in the genome is random and the virus must integrate into or near an area of the genome with a protooncogene. Veterinary examples are FeLV, avian leukosis virus (ALV), and bovine leukemia virus (BLV). [2,4]

FeLV is subclassified into 3 receptor interference groups, A, B, and C, which are defined by genetic sequence variation in the *env* gene (including receptor binding domain). Each subgroup uses different kinds of transporter proteins as a receptor and persistent infection is associated with many disorders. FeLV-A is considered minimally pathogenic without other subvariant infection, FeLV –c is associated with pure red cell aplasia, and FeLV-B with thymic lymphoma, non-regenerative anemia, other lymphoid neoplasms, and acute myeloid leukemia. Six common integrations sites are associated with FeLV in the cat genome, one of which is the *c-myc* protooncogene. *Myc* is intimately associated with cell cycle progression and proliferation. When viral insertion occurs near the *myc* locus, *myc* becomes controlled by the quite powerful viral promoters resulting in up-regulated myc expression. The end result is the prevention of cells from entering G0 and cell cycle progression. Although not all cats with FeLV develop lymphoma, when infected with FeLV the risk of developing lymphoma is >60 times higher than non-infected cats.[2,3]

In cattle, the enzootic leukemia/lymphoma complex is associated with BLV infection. Unlike FeLV or avian leukemia virus, BLV has a definitive cell association; it is found in the latent form only in B cells. Although FeLV, BLV, and ALV are all considered type C retroviruses based on electron microscopy, cattle infected with BLV have no free virus in blood. BLV also has a *tax* gene (in addition to the *gag*, *pol*, and *env*) which regulates the transcription of provirus, and is usually silent; however, when activated the tax protein activates other cell proteins that bind the LTR and up-regulate transcription.[1,2,5]

DNA VIRUSES

Families of oncogenic DNA viruses are herpesvirus, papovavirus (papillomaviruses and polyomavirses), adenoviruses, hepadnavirses, and pox viruses. DNA viruses transform infected cells by expressing genes of viral origin. Genes of DNA viruses involved in oncogenesis are essential for the normal viral life cycle. In the normal life cycle of a DNA virus, a permissive cell is infected and early and late viral genes are expressed. The early genes are responsible for subverting control of the cell to support viral replication. Some early genes encode for proteins that interact with host cell genes and block apoptosis or affect regulation of cell replication. Once the viral progeny mature, the infected cells lyse and viral particles are released due to expression of late genes. However, in some cases an aberrant infection of a permissive cell or infection of non-permissive cells or semi-permissive cells occurs, and only a part of the viral life cycle is completed. In these cases, no viral progeny are produced and a portion of the viral genome can become integrated into the cellular DNA. If the integrated portion contains early genes, the encoded proteins can deregulate cell growth controls, including Rb and p53. Without the expression of viral late genes, the cells do not lyse and are now at high risk of developing into malignant cells. [1,2,4]

Malignant transformation of epithelial cells is associated with certain strains of papilloma viruses with affinity for p53 and Rb. These include bovine papilloma viruses (BPVs), separated into 2 subgroups: 1) A (the fibropapilloma viruses) including BPV 1, 2, &5 and 2)B (the epitheliotrophic papillomaviruses) including BPV 3, 4, &6. BPV-2 is responsible for the cutaneous wart in cattle. High risk viruses have the early genes produced (E6 and E7) which immortalize cells and contribute to malignancy. However, further genetic insult is needed for malignancy in addition to the effects of E6 and E7. In healthy cattle, the papillomas normally regress; however, co-carcinogen exposure will increase the risk of cancer formation. BPV-2 has been associated with bladder cancer, and BPV-4 with upper alimentary tract cancer, in cattle also exposed to bracken fern. Activation of *ras* and inactivation of p53 also likely play a role in carcinogenesis. [2,5]

Canine papilloma viruses are responsible for a wide variety of clinical syndromes in the dog, including benign cutaneous and mucocutaneous papillomas that, in rare cases, transform into squamous cell carcinomas. They rarely integrate into the host cellular genome and remain episomal. Clincal signs of disease present 4-6 weeks following infection and dogs present with hyperplasia and hyperkeratinization. In the oral cavity, these lesions will resolve spontaneously in 4-6 months and, on the skin, -within 6-12 months. Antibody levels are not correlated with growth or regression of the lesions and regression is thought to be mediated through cellular immunity. Treatment is usually not indicated unless the lesions are causing a decline in quality of life; surgery (cryosurgery, laser, electrosurgery) of some lesions has reportedly induced spontaneous regression of other lesions. Papillomatosis is rare in the cat.[1]

Herpesvirus saimiri is responsible for the formation of oropharyngeal ulcerative lesions in squirrel monkeys. When this virus infects owl monkeys and New Zealand white rabbits the virus produces T cell lymphoma or lymphoid leukemia. In chickens, herpesvirus infection produces lymphoma (Marek's disease). Liver cancer risk increases significantly in squirrels and woodchucks with hepadnvirus, specifically hepatitis B virus. Interestingly, hepadnaviruses produce tumors when in the typical host in cells that are normally infected. Lastly, the poxvirus Shope fibroma virus is responsible for myxomas and fibromas in their typical host, the rabbit.[1]

CONCLUSIONS

Viruses are important in producing a wide array of malignancies in multiple different species. In general, DNA viruses down-regulate tumor suppressor genes (p53 and Rb) while retroviruses either transduce cellular genes encoding from growth factors, receptors, tyrosine kinases, G-proteins, adaptors, and transcription factors or activate these genes through insertional mutagenesis. Many viruses contain genes that inhibit apoptosis or inhibit immune recognition of the infected cell. Each viral mechanism of cancer induction represents an excellent model for further study of carcinogenesis.

REFERENCES

[1]Cullen JM et al. **Tumors in domestic animals** 4th ed 2002; 3-44. [2]Argyle DJ et al. **The Vet Journal** 2008; 177: 311-323; [3]Fujino Y et al. **Vet Imuno Immunopath** 2008; 123: 138-143. [4]Richardson CD. **The Basic Science of Oncolog**y 2005; 5th ed: 100-122. [5]Beniston RG et al. **Carcinogenesis** 2001; 22 (7): 1069-1076.

FELINE ADIPOSE-DERIVED MESENCHYMAL STEM CELLS: WHAT DO WE KNOW?

Tracy L. Webb, DVM, PhD
Fort Collins, CO

INTRODUCTION

Stem cells were discovered approximately 50 years ago, and the field has rapidly expanded in the last decade. In general, stem cells are defined by the following three characteristics: capability of dividing and renewing through cell division for long periods of time, unspecialized, and capability of giving rise to specialized cell types. Stem cells are divided into two main categories: embryonic and adult (or somatic). Adult stem cells can be divided into hematopoetic and mesenchymal types. Mesenchymal stem cells (MSC), also referred to as multipotent stromal cells, are identified mainly by their ability to differentiate into multiple mesenchymal tissues: chondrocytes, osteoblasts, and adipocytes. In fact, tri-lineage differentiation is one of the three defining characteristics of MSC in addition to plastic-adherence and self-renewal.[1] MSC can be derived from a multitude of adult tissues. Two of the most common tissues used for MSC generation are bone marrow and adipose tissue. MSC have been generated from many species, including recently from the cat. MSC are considered important for both their potential regenerative capabilities as well as their immunomodulatory properties. MSC avoid allo-recognition and therefore represent an interesting potential therapeutic for several disease processes affecting both human and veterinary patients.

FELINE MSC

There is currently very little published on feline mesenchymal stem cells. However, the available research shows that feline MSC (fMSC) are similar to MSC from other species. Feline MSC have been generated from bone marrow, adipose tissue, umbilical cord blood, and fetal fluid and membranes.[2-9] In all of these studies, the fMSC show appropriate stem cell morphology and plastic adherence as well as the ability to propagate for several passages *in vitro*. All of the tested tissue types have also shown multi-lineage differentiation potential when cultured in the appropriate induction medium, including osteogenesis, chondrogenesis, adipogenesis, and neurogenesis. Zhang et al. found that fMSC have higher plasticity for neural induction than human or mice MSC, which they postulated may be related to the moderate expression of several pluripotent and neural genes in untreated bone marrow-derived fMSC.[6]

The International Society of Cellular Therapy states that MSC should express CD73, CD90, and CD105 and lack expression of hematopoietic markers (CD45, CD14, and CD34).[1] Although full characterization of fMSC is hampered by the lack of feline-specific antibodies, all of the studies published thus far show fairly consistent surface marker expression in the generated fMSC. In 2002, Martin et al. found the bone marrow-derived fMSC to be CD9, CD44, and MHC 1 positive and CD4, CD8, CD13, CD14, CD18, CD41/61, CD45, and MHC II negative.[2] Also examining bone marrow-derived fMSC, Jin et al. found the cells to be strongly CD9 and CD44 positive while CD18 and CD45 negative.[3,8] Adipose-derived fMSC were found to be CD44, CD105, and CD90 positive while mainly CD4 and MHC II negative.[5,7] Feline MSC from fetal tissues was also CD44, CD90, and CD105 positive and CD14 negative.[7] Human CD73 does not appear to cross-react with fMSC.[9] In general, the fMSC evaluated in the studies were thought to be very similar to MSC from other species based on the available antibodies.

Direct comparisons of the potential tissues of origin of fMSC are limited. One study found that, consistent with other studies in other species, feline adipose tissue generates significantly more MSC in a shorter time period than bone marrow.[5] Clinically useful numbers of early passage adipose-derived fMSC can be generated in 7-10days. However, the frequency of cells in feline adipose tissue was not determined in that study. Martin et al. determined the frequency of bone marrow fMSC to be between 1 in 4.7×10^4 and 1 in 5.9×10^5.[2]

The majority of the available fMSC literature has been descriptive, and functional data for fMSC is lacking. MSC from other species are known to suppress T-cell function *in vitro*. Adipose-derived fMSC are similarly capable of suppressing mitogen-induced T-cell proliferation (Webb, unpublished data). Of additional importance, adipose-derived fMSC can be safely administered to allogenic recipients without generating a clinically-apparent immune response.[7,11,12] Allogeneic adipose-derived fMSC are currently being used in clinical trials to evaluate their therapeutic effects on several feline disease processes including chronic kidney disease, asthma, and inflammatory bowel disease.[7,11,12] Data from a pilot study evaluating intrarenal fMSC injections in chronic kidney disease found autologous bone marrow- and adipose-derived fMSC to be safe, and two of the four treated cats saw some improvement in measured parameters.[7] Data presented in abstract form from studies using allogeneic adipose-derived fMSC in an experimental model of feline asthma additionally demonstrate the safety of repeated intravenous fMSC injections.[11,12] Some aspects of the asthmatic immune response in fMSC-treated cats were altered in a sustained fashion in the pilot study.[11]

Of clinical significance, information concerning the ideal treatment parameters for fMSC is lacking. The ideal donor (autologous versus allogeneic, donor age, etc.), tissue of origin, number of cells, method of cell preparation, site/route of injection, and number of treatments remain unknown. These parameters may be dependent on multiple factors including the disease process being treated. On-going research and additional results from the current clinical trials should provide much needed information on these issues.

CONCLUSION

Feline mesenchymal stem cell research has been in progress for just over a decade but is currently expanding significantly as human and rodent studies have suggested several potential clinical applications of MSC for disease processes. Although fMSC thus far appear similar to MSC from other species, little published data is currently available. There is a significant lack of information on optimal details for fMSC use and function at this time. Current clinical trials have demonstrated safety and feasibility of administration of autologous and allogeneic adipose-derived fMSC, and further results from on-going *in vitro* and *in vivo* studies should provide additional information in the near future.

REFERENCES

[1]Dominici M et al. Cytotherapy 2006;8:315-317. [2]Martin DR et al. **Exp Hematol** 2002;30(8):879-86. [3]Jin G et al. **J Vet Sci** 2007;8(4):377-382. [4]Munoz JL et al. Differentiation 2012;84:214-22. [5]Webb et al. J **Fel Med Surg** 2012; 14(2):165-8. [6]Zhang Z et al. **Cell Reprogram** 2011;13(5):385-90. [7]Quimby J et.al **J Fel Med Surg** 2011;13:418. [8]Jin G et al. **J Vet Med Sci** 2008;70(7):723-726. [9]Iacono E et al. **Vet Res Commun** 2012;36(2):107-18. [10]Reinero C et al. **ACVIM Abstract** 2012. [11]Trzil JE et al. **VCRS Abstract** 2012.

STEM CELLS IN INFLAMMATORY BOWEL DISEASE

Craig B. Webb, Ph.D., DVM, DACVIM (SAM)
Fort Collins, CO

INTRODUCTION

Inflammatory bowel disease (IBD) involves disruption of gastrointestinal mucosal immunity and loss of tolerance to intestinal antigens.[1] Intestinal inflammation characterizes the disease and therapy is aimed at modulating the gastrointestinal immune response and reducing antigenic stimulation.[1,2] Dietary manipulation and glucocorticoids are the mainstays of therapy for IBD, but treatment is challenging and often unsatisfactory.

Mesenchymal stem cells (MSC) have been shown to alter immune responses and reduce inflammation through direct interactions with T-cells, NK cells, neutrophils and dendritic cells, and by increasing regulatory T cells (Tregs).[3,4] MSC can be generated from adult tissues and do not induce a clinical immune response when delivered to allogeneic recipients. In mouse models of acute colitis, administration of a single injection of adipose-derived MSC has been shown to ameliorate clinical and microscopic signs of colitis, reduce systemic and mucosal pro-inflammatory cytokine production, increase IL-10 secretion, and induce regulatory Tregs in mesenteric lymph nodes.[5,6] Experimental and clinical evidence shows that MSC therapy is safe and induces long-lasting remission in many patients with active severe Crohn's disease that is otherwise refractory to standard treatments.[7,8] Adipose-derived MSC are currently being used in phase III clinical trials for inflammatory bowel disease in humans.[9]

In Vitro

The immunomodualtory properties of MSC have been extensively characterized *in vitro.* It would appear prior activation by the inflammatory cytokines IFN- , TNF- , and IL-1 is required. This activation results in the MSC production of prostaglandin E2 (PGE2), an immune suppressant that inhibits Tcell mitogenesis and IL-2 production, induces a T helper (Th2) response, and stimulates IL-10 production by macrophages. Activated MSCs also produce IL-6 and HLA-G5, which inhibits the differentiation of monocytes to dendritic cells and suppresses T cell proliferation and cytotoxicity. Perhaps of most relevance, activated MCS decrease the production of inflammatory cytokines IL-12, IFN- , and TNF- by dendritic cells and promote the generation of CD4+CD25+FoxP3+ T regulatory cells (Tregs) and the anti-inflammatory cytokine IL-10.[10] MCS appear to exert their immunomodualtory effects through both direct contact and the production of soluble factors.

Mouse Models

A murine model of IBD (Crohn's disease) is induced by the administration of trinitrobenzene sulfonic acid. Mice are then treated with human adipose-derived mesenchymal stem cells (hASCs). Both the clinical (weight-loss, diarrhea, survival) and histopathological severity of the disease were significantly reduced by this treatment.[5] The therapeutic effect appears to be mediated by an hASC-induced down regulation of the Th1 immune and inflammatory response. Inflammatory cytokine production, including TNF- , IFN- , IL-6, IL-1 , and IL-12 was decreased while levels of the anti-inflammatory cytokine, IL-10, was increased in IBD mice following hASC treatment. Consistent with the hASC cell surface receptor profile, there appeared to be a specific homing of the administered stem cells to inflamed and lymphoid tissue; colonic inflammation led to colonic recruitment of administered stem cells.[5]

The cellular make-up of colonic tissue was shifted from Th1 differentiation and activation to proliferation of IL-10 producing T cells by hASC administration, increasing the number of CD4+CD25+Foxp3+ T cells and down-regulating the Th1 inflammatory response. Similar results were seen with hASC treatment in a dextran sulfate sodium-induced IBD murine model.[6]

Human Models

Twelve patients with Crohn's disease refractory to standard therapy received hematopoietic stem cell transplantation (HSCT) and all 12 achieved remission (CDAI < 150), with no serious or lasting complications. A large percentage of these patients remained in remission and without the need for medication up to 5 years after treatment.[11] In a separate study, 3 patients with severe refractory Crohn's disease received autologous HSCT. Stem cell therapy resulted in long-term remission in all 3 patients, with 1 patient able to discontinue immunosuppressive medications for several years. HSCT therapy requires immune ablation of the patient prior to stem cell transplantation, a significant deterrent to the use of this particular strategy in the veterinary patient population.[12]

Patients with refractory Crohn's disease were treated with bone marrow derived mesenchymal stem cells. Two IV injections, 1-2x10^6 cells/kg, 7 days apart, were administered. The Crohn's disease activity index (CDAI) was monitored and colonoscopies were performed prior to treatment and 6 weeks following stem cell administration. Three of 10 patients showed significant clinical improvement (a decrease in the CDAI) while in 3 other patients the disease progressed and required surgical intervention. Importantly, there were no serious adverse events associated with stem cell injections, supporting the safety and feasibility of this treatment modality.[7]

Fifty patients with either ulcerative colitis or Crohn's disease received IV injection of bone-marrow derived MSC ($1.5x10^{8}$ cells). A significant decrease in clinical indices of disease and tissue inflammation was seen in treated patients compared to patients who did not receive MCS therapy. Clinical remission was achieved in 40 of patients and the use of corticosteroids was discontinued in 34 patients, while MCS treatment was deemed ineffective in 10 patients.[13]

Stem cell therapy has been used extensively and effectively to ameliorated the pain, diarrhea, hemorrhage, inflammation and fistulization seen in cases of Crohn's disease refractory to established therapies, other fistulizing bowel diseases, and radiation-induced colitis. To date, over 100 human patients with these disorders have taken part in MCS clinical trials and adverse side-effects appear virtually non-existent, even with repeated treatments.[10] The positive response in these patients is attributed to a modulation of the inflammatory lymphocyte subsets from effector T cells to regulatory T cells, stimulation of angiogenesis, and decreasing fibrosis. Administered MCS appear to favor the proliferation of endogenous tissue stem cells through the release of paracrine trophic factors.[10]

Feline Inflammatory Bowel Disease

Feline inflammatory bowel disease is not subdivided into ulcerative colitis and Crohn's disease, as IBD is in human patients. The cytokine profile in cats with IBD compared to cats with non-IBD GI disease shows an increase in both immunomodulatory cytokines IL-10 and TNF- as well as the proinflammatory cytokines IL-6, IL-18, TNF- , and Il-12p40.[14] In a separate study the proinflammatory cytokines IL-1, IL-8, and IL-12 were increased in cats with IBD.[15] Clearly there is significant immune dysregulation in feline IBD, and although the cytokine profile is complex and incompletely understood, it appears consistent with a Th1 response, as seen in humans with Crohn's disease. The trophic properties along with the anti-inflammatory and immunomodulatory effects of MSC administration make it a theoretically beneficial therapeutic modality for the treatment of feline IBD. The early success reported in animal models and clinical trials with human patients suffering from Crohn's disease further suggest that the use of MSC therapy in feline IBD warrants further investigation.

CONCLUSION

Our laboratory has shown that feline adipose-derived MSC (fMSC) can be generated in large quantities to allow for clinical use, and that these fMSC are plastic-adherent, spindle-shaped cells that possess tri-lineage differentiation capabilities and suppress T-cell proliferation in vitro.[16] Allogeneic fMSC have been safely and repeatedly administered to healthy and diseased cats with no notable side effects. We are currently conducting a double-blind placebo control study to evaluate the safety, feasibility, and clinical effect of allogeneic fMSC as a treatment for feline IBD.

ACKNOWLEDGMENTS

Our current fMSC study is supported by a generous grant from the The Winn Feline Foundation (George and Phyllis Miller Trust, The San Francisco Foundation)

REFERENCES

[1]Trepanier L, et.al **J Fel Med Surg** 2009;11:32. [2]Janeczko S et.al **Vet Microbiol** 2008;128:178. [3]Panés J et.al **Gut** 2009;58:898. [4]Uccelli A et.al **Nat Rev Immunol** 2008;8:726. [5]González M et.al **Gastroenterology** 2009;136:978. [6]Gonzalez-Rey E et.al **Gut** 2009;58:929. [7]Duijvestein M et.al **Gut** 2010;59:1662. [8]Panés J et.al **Expert Rev Clin Immunol** 2010;6:597. [9]Lanzoni G et.al **Cytotherapy** 2009;11:1020. [10]Voswinkel J et al. **Clinic Rev Allerg Immunol** 2013:Epup. [11]Oyama Y et al. **Gastroenterology** 2005;128:552. [12]Hommes DW et al. **J Crohn's and Colitis** 2011;5:543. [13]Lazebnik LB et al. **Ter Arkh** 2010;82:38. [14]Van Nguyen N et al. **Vet Immunol Immunopathol** 2006;113:404. [15]Janeczko S et al. **Vet Microbiol** 2008;128:178. [16]Quimby J et.al **J Fel Med Surg** 2011;13:418.

IS THERE A ROLE FOR MESENCHYMAL STEM CELLS IN THE TREATMENT OF FELINE ASTHMA?

Carol Reinero, DVM, DACVIM, PhD
Columbia, MO

OVERVIEW

Allergic asthma is one of the most common bronchopulmonary disorders in pet cats. The mainstay of therapy relies on environmental modulation along with injectable, inhaled or oral glucocorticoids and bronchodilators. However, these treatments are only palliative, and fail to alter the underlying immune dysfunction (T helper 2-driven response to aeroallergens); thus they must be given lifelong. Importantly glucocorticoids and bronchodilators may be contraindicated in cats with heart disease, diabetes mellitus, or certain infections leaving no good treatment options. It is imperative that safe new treatments are developed which are capable of targeting the underlying immunopathophysiologic events which cause development and progression of asthma. In particular treatment addressing the three hallmark features of allergic asthma: eosinophilic airway inflammation, airway hyperresponsiveness and remodeling is desirable

ALLERGIC ASTHMA IN CATS.

Feline asthma is responsible for substantial morbidity, occasional mortality and a high cost of repeated hospitalizations. Clinical signs range from cough and wheeze to episodes of respiratory distress ("status asthmaticus"). These clinical signs are mediated by three pathologic features driven by an allergic response: eosinophilic airway inflammation, airflow limitation and architectural changes in the lung (remodeling). Inhaled aeroallergens picked up by local antigen presenting cells in the respiratory tract stimulate the development and activation of T helper 2 (Th2) immune responses. These Th2 lymphocytes orchestrate pro-inflammatory cytokine release and pulmonary recruitment of inflammatory cells. Repeated exposure to allergens leads to chronic changes in the lung. Airway remodeling in particular may over time become less responsive to therapy.

While feline asthma is clearly an important health concern, current treatments are only palliative, must be given life-long and do not address early immunologic events that trigger and perpetuate this disease. Importantly, no single therapy in cats can ameliorate all three hallmark features of asthma. Stem cell therapy has emerged as a novel treatment which may modulate the aberrant immune response in asthmatics and could attenuate all three components of asthma: inflammation, hyperreactivity and remodeling.[1]

HOW DO STEM CELLS WORK?

Embryonic and adult stem cells have emerged as novel therapeutic approaches for a variety of devastating and incurable lung diseases.[1] Stem cells were originally touted as being able to engraft into tissue and replace damaged cells, thus helping to regenerate diseased organs; however, more recent studies suggest that is unlikely[2,3] and that their beneficial role is predominantly through paracrine effects. Paracrine effects can be broken down into six major categories: immunomodulation, anti-apoptosis, angiogenesis, support of growth and differentiation of stem and progenitor cells, antifibrosis and chemoattraction.[4] Immunomodulation is achieved by production of a number of different bioactive factors which suppress effector cell development, maturation and function. It remains unclear if stem cell effects are immunosuppressive (ie, non-specifically downregulating immune responses) or if they induce/restore immune tolerance to specific antigen. Allergic asthma results from a hypersensitivity to aeroallergens in which tolerance to these allergens is lost, thus restoration of tolerance could shut off the abnormal immune cascade and perhaps promote a cure for this disease. The other paracrine effects of stem cells may serve to decrease or reverse the structural changes in the lung (termed remodeling) by release of bioactive factors that orchestrate the function of local cells.

EVIDENCE FOR A BENEFICIAL ROLE OF STEM CELLS IN RODENT MODELS OF ASTHMA

There is ample evidence in murine models of asthma that MSCs diminish airway eosinophilia, which is accompanied by decreases in Th2 cytokines such as IL-4, IL-5 and IL-13.[5-7]The hallmark antibody associated with allergic asthma, allergen-specific IgE, also decreases with MSC treatment.[6,7] Regulatory T cells, which serve to dampen allergic airway inflammation and may restore tolerance to allergen, were increased in bronchoalveolar lavage fluid in one study.[7] Airway hyperresponsiveness with bronchoprovocation using methacholine was dampened with administration of MSC.[6] Finally, airway remodeling in chronic models of murine asthma was ameliorated with MSC treatment.[5,8] Overall these results suggest that MSC impact the three major features of asthma--airway inflammation, hyperresponsiveness and remodeling—and deserve further evaluation for feline asthma.

It is important to realize there are several limitations to the murine asthma studies. First, most studies evaluated the effects of stem cells <u>prior</u> to induction of asthma. Clearly we would not be able to predict which cats might develop asthma and give them MSCs preemptively. Second, murine models are likely overly simplistic with limited clinical relevance to pet cats—i.e, it is highly unlikely in an asthmatic cat with complex genetics and complicating environmental influences that a single injection of MSCs will completely cure asthma as has been shown in some murine studies. Finally, no study to date has determined long term effects of MSCs on the asthmatic phenotype after repeated injections. This is highly clinically relevant as it is crucial to know if MSCs need to be given as a finite series or indefinitely. If MSCs could induce immunologic tolerance to allergen, it is more likely that a finite series of injections could be used therapeutically.

MESENCHYMAL STEM CELLS IN CATS AND APPLICATION TO FELINE ASTHMA

Allogeneic MSCs have been successfully collected, cultured, characterized and administered to cats.[9,10] Pilot data suggests that repeated intravenous injections of MSCs are safe in cats. Administration of MSCs by intravenous injection leads to localization within the lung prior to distribution to other tissues, which is ideal for the treatment of asthma.[5]

Critical evaluation of new treatments like stem cell therapy for feline asthma is desirable, but study is difficult in the naturally-developing disease because clinicopathologic changes wax and wane depending on time and duration of exposure to allergen and the number and types of allergens to which cats have been sensitized. Using an experimental model of feline asthma that mimics the naturally-developing disease allows a more carefully controlled evaluation of stem cell therapy.[11] The model was developed by selecting allergens implicated in pet cats presenting to a Veterinary Medical Teaching Hospital with clinical and laboratory evidence of spontaneously developing asthma. It allows us to sensitize cats with known allergen(s), co-house them to limit the effects of different environments, and challenge them with a standard dose of allergen on a weekly basis to eliminate varied allergen exposures. Preliminary results of studies assessing both the early (immediate) effects and late (over the subsequent year) effects on airway inflammation, airflow limitation and remodeling will be presented.

REFERENCES

1. Sueblingvong V, et al. *Translational Res* 2010;156:188-205.
2. Chang J, et al. *Am J Respir Cell Mol Biol* 2005;33:335-342.
3. Kotton D, et al. *Am J Respir Cell Mol Biol* 2005;33:328-334.
4. Singer N, et al. *Annu Rev Pathol Mech Dis* 2011;6:457-478.
5. Bonfield T, et al. *Am J Physiol Lung Cell Mol Physiol* 2010;299:L760-770.
6. Goodwin M, et al. *Stem Cells* 2011:doi 10.1002/stem.1656.
7. Nemeth K, et al. *PNAS* 2010;107:5652-5657.
8. Firinci F, et al. *Int Immunopharmacol* 2011;11(8):1120-6.
9. Webb TL et al. *J Fel Med Surg* 2012;14:165-168.
10. Quimby JM et al. *J Fel Med Surg* 2011;13:418-26.
11. Norris Reinero C, et al. *Int Arch Allergy Immunol* 2004;135:117-131.

SURPRISING ANTIMICROBIAL PROPERTIES OF MESENCHYMAL STEM CELLS

Valerie Johnson DVM, DACVECC, Kelly Walton DVM, Tracy Webb DVM, Ph.D., Steven Dow DVM, Ph.D., DACVIM
Fort Collins, Colorado

INTRODUCTION

The need for novel therapeutic strategies to combat increasing incidence of infections with multidrug resistant organisms is a big challenge facing the medical community today. Veterinary hospitals are experiencing similar trends as are human hospitals with increasing incidence of resistant infections resulting in potential failure of conventional pharmaceutical therapies.[1] Infections that cannot be cleared often result in chronic infections due to the inability of the host immune response to clear the invading organism. Bacterial infections result in stimulation of cellular signaling through toll-like receptors (TLRs) which initiates a cascade of events culminating in the release of inflammatory mediators. These mediators activate inflammatory cells thus eradicating the invading pathogens. Traditional antibiotics combat the microorganism directly to clear infections when the host response is inadequate. Stimulation of the immune system to enhance host mediated bacterial killing is a novel approach to management of resistant infections. This approach avoids reliance on development of new pharmaceuticals which are subject to rapid development of antimicrobial resistance. A surprising candidate in this arena is mesenchymal stem cells (MSCs). MSCs are pluripotent cells that are easily isolated from adipose tissue and have been demonstrated to possess immunomodulatory properties. These cells are generally thought of as possessing immunosuppressive properties and therefore the discovery that these cells possess antimicrobial properties and may assist in clearing chronic wound infections is somewhat counterintuitive. However, rodent models, together with *in vitro* experimental evidence reveals promising, if somewhat surprising results.

IMMUNOMODULATORY PROPERTIES OF MSC

MSCs are most well known for their immunosuppressive properties including suppression of T cell proliferation, induction of regulatory T cell (T-reg) proliferation, interference with production of antibodies by B cells and interruption of maturation of dendritic cells.[2] However, it has been observed that under different conditions, MSCs can exert a pro-inflammatory effect. This observation led to the hypothesis that MSCs, like T cells and monocytes, can change their phenotype based on the environment and therefore exert a pro-inflammatory or anti-inflammatory effect.[3] Immune cells respond to invading pathogens through cell surface receptors called Toll-like Receptors (TLRs). These receptors recognize patterns inherent in various pathogens such as peptidoglycan (TLR2), double stranded RNA (TLR3), LPS (TLR4), and other pathogen associated molecular patterns. MSCs have repeatedly been shown to express TLRs 2, 3, 4, 7 and 9, and expression of other TLRs has also been reported. TLR ligation of MSCs has been demonstrated to influence the cytokine profile and migration capacity of immune cells and thus have the potential to act in an immune-suppressive or immune-stimulating manner depending on the environment to which the MSC is exposed. [4] As such, it is perhaps not unexpected that these cells would behave differently in the environment of infection than in the environment of an autoimmune disease for example.

EVIDENCE FOR ANTIMICROBIAL ACTIVITY OF MSC

The demonstration of anti-inflammatory actions of MSCs in conditions such as asthma and acute lung injury (ALI) prompted investigators to examine the effects of MSC administration in the pro-inflammatory environment of early sepsis.[5] Using a rodent model of sepsis, Mei et al. demonstrated reduced bacterial loads, increased survival, decreased organ failure and lower levels of pro-inflammatory cytokines.[6] Other studies have confirmed this effect in several different infection models including *Escherichia Coli* induced pneumonia and *Pseudomonas aeruginosa* induced peritonitis.[7,8]

These studies prompted our lab to examine whether MSCs would exert a similar antimicrobial effect in the setting of a chronic wound infection in the absence of the overwhelming systemic inflammation displayed in sepsis. Interestingly, we discovered a dramatic difference in wound infections in mice treated with intravenous MSC compared to untreated animals. The model involves implantation of surgical mesh impregnated with *Staphyococcus aureus*. Mice infected with this implant consistently develop an abscess at the site of infection. As shown in figure 1, the surprising finding was that mice that were treated with MSC did not display any abscess formation at the site of the implant in contrast to the untreated mice. This exciting finding could have huge implications in the treatment of chronic wound infections with antibiotic resistance, contamination or foreign material or other impediments to effective treatment by conventional means.

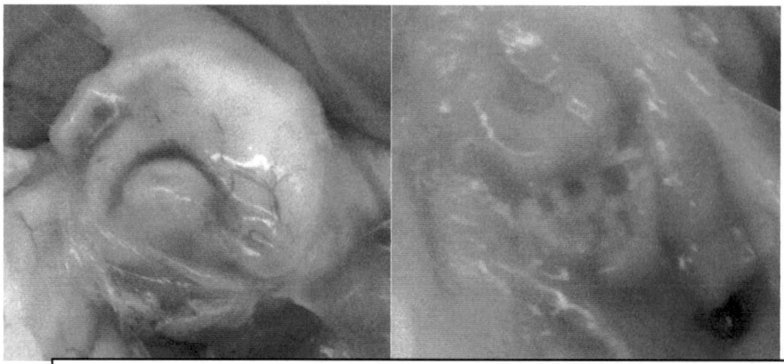

Figure 1: Infected wound in a mouse treated with antibiotics (left) and antibiotics with MSC (right).

PROPOSED MECHANISMS OF ANTIMICROBIAL ACTION

Studies implicate a variety of mechanisms involved in the observed antimicrobial effect of MSCs and it is likely that no single pathway is responsible for the full effect (Table 1). MSCs have been shown to migrate to sources of inflammation.[9] Thus it is possible that MSCs possess a direct bactericidal effect. In fact, direct killing by antimicrobial peptide LL-37 produced by MSCs has been shown to be partially responsible for the increased bactericidal activity observed in a pneumonia model of sepsis.[7] MSCs are known to exert paracrine effects as well. Bacterial stimulation of MSC through TLRs is thought to alter the responsiveness of inflammatory cells to infectious stimuli by influencing migration, activation or proliferation. Various infection models have demonstrated increased opsonization and phagocytosis by monocytes and macrophages in response to MSC treatment. Many studies have reported MSC induced increased production of the anti-inflammatory cytokine IL-10. It has been postulated that the shift towards anti-inflammatory cytokines may inhibit neutrophil influx into tissues preventing organ damage while simultaneously increasing neutrophil killing of bacteria in blood.[10] IL-17 production by T cells has also been postulated to be involved in enhancing neutrophil activity.[11] Ligation of different TLRs has been shown to influence the type of response generated by MSCs.[12,13] Ligation of TLR3 and TLR4 on MSC have been demonstrated to induce activation of NFkB with subsequent cytokine production.[4] In particular, binding of TLR3 has shown the most profound induction of NFkB, is critical for neutrophil migration and results in increased neutrophil longevity and viability.[14]

Table 1

Proposed Mediator or Action	**Cell type involved**	**Direct contact of MSC required for effect**	**Mechanism of Killing**
Antimicrobial Peptides	MSC	Yes	Membrane disruption of bacteria
Increased superoxide production and longevity	Neutrophils	No	↑ phagocytosis and killing by reactive oxygen burst
Increased phagocytosis	Macrophages	No	↑ engulfment and killing in phagolysosome
Cytokines – increased production of IL-17, IL-10	T cells, Macrophages, Monocytes, Neutrophils	No	↑neut activation, ↓ neutrophil influx into tissues, ↑ neut in blood
Increased chemokine release (IL-8, MCP-1, CXCL10)	MSC, Monocytes	no	Recruit inflammatory cells to site of insult

ACTIVATION OF MSC TO ENHANCE ANTIMICROBIAL EFFECT

The ability of MSC to modulate their activity in response to environmental signals suggests that they could be primed to function in a particular capacity. Activation of MSC by TLR agonist poly I:C has been shown to increase neutrophil activation and MSC migration as described above. To evaluate the effect of TLR3 ligation on bacterial killing by MSC in vitro, MSC were stimulated by TLR3 agonist poly I:C for 1 hour and the activated MSC tested for bactericidal efficacy against *S. aureus*. When combined with subtherapeutic doses of antibiotics, activated MSCs displayed superior bactericidal activity over unactivated MSC or antibiotic alone.

CONCLUSIONS/IMPLICATIONS

The development of new therapeutics to treat infections is of paramount importance due to the increasing incidence of resistant infections with limited pharmacologic treatment options. Stem cell based therapies are particularly attractive for several reasons including the relative ease of harvesting and propagating the cells and the ability to potentially tailor the desired immune response. Veterinary medicine is often limited in pharmacologic interventions for multidrug resistant organisms due to cost and difficulty in administering long term intravenous medications. Stem cell therapy in cats and dogs has proven beneficial for several inflammatory diseases. Use of this therapy for chronic or resistant infections shows great potential and bears further investigation.

REFERENCES

1. Pellerin et al. **Comp Immunol Microbiol Inf Dis** 1998;21(2):115
2. Gebler et al. **Trends in Molecular Medicine** 2012;18(2):1471
3. Waterman et al. **PLoS one** 2010;5(4):1
4. Tomchuck et al. **Stem Cells** 2008;26:99
5. Xu et al. **Am J Physiol Lung Cell Mol Physiol** 2007;293:L131
6. Mei et al. **Am J Respir Crit Care Med** 2010;182:1047
7. Krasnodembskaya et al. **Stem Cells** 2010;28(12):2229
8. Krasnodembskaya et al. **Am J Physiol Lung Cell Mol Physiol** 2012;302:L1003
9. Chen et al. **Biomaterials** 2011;32:3189
10. Nemeth et al. **Nature Medicine** 2009; 15(1):42
11. Hsu et al. **Immunobiology** 2013; 218:90
12. DelaRosa et al. **Mediators of Inflammation** 2010; 2010:1
13. Lei et al. **Cellular Immunology** 2011; 271:147
14. Cassatella **Stem Cell** 2011;29: 1001

DERMATOLOGICAL MANIFESTATIONS OF INTERNAL MEDICINE DISEASE

Karen L. Campbell, DVM, MS, DACVIM (SAIM), DACVD
Urbana, IL

INTRODUCTION

Cutaneous markers of internal diseases are common and are often highly indicative for a specific systemic disease; therefore the recognition of the significance of these markers will facilitate diagnosis of the underlying disease processes. The cutaneous symptoms may be the result of the same etiologic agent or pathogenic process as the internal disease—as occurs in some bacterial, protozoal, fungal, immune-mediated and neoplastic disorders, or may be the result of hormones affecting the skin, or may be due to genetic links or other factors.

COLLAGEN-VASCULAR DISEASES

Collagen-vascular diseases include a multitude of conditions characterized by lesions affecting skin, connective tissue and internal organs. Some of these are hereditary and others are classified as immune-mediated diseases.

Cutaneous Asthenia (Ehlers Danlos Syndrome, Dermatosporaxis)

Cutaneous asthenia is a hereditary disease with defective formation of connective tissue resulting in skin fragility and in some cases joint hyperlaxity, hernias and vascular complications. Several forms occur in animals and humans. Reported cases in cats have been characterized by reduced quantities of dermal connect tissue, shortened and fragmented collagen fibers.[1-3] Clinical signs include hyperextensible skin that tears with minor trauma.[1-3] Dyspnea may occur secondary to a diaphragmatic hernias and one affected cat also developed a perineal hernia.[3] A defect in procollagen peptidase was identified in one cat.[1] A mutation of ADAMTS2 gene has been identified as the cause of dermatosparaxis in White Dorper sheep.[4,5] Affected dogs have exhibited skin hyperextensibility and fragility, vessel fragility, coxofemoral joint dislocation, medial patellar luxation, subcutaneous hematomas and periodontal disease.[6] Some cattle with dermatosparaxis have a mutation in the procollagen I N-proteinase gene,[7] while others have a different, as yet unidentified, mutation.[8] Quarter horses with hereditary equine regional dermal asthenia (HERDA) have an increased incidence of mutations in cyclophilin B.[9-11] Affected rabbits have also been reported.[12]

Dermatomyositis (DM)

Dermatomyositis is an inflammatory disease affecting skin and/or muscle. In dogs, DM occurs most often in collies and Shetland sheepdogs, although other breeds have also been affected. Skin lesions vary from mild to severe and start with small focal areas of crusting, scaling and alopecia, most commonly affecting the face, lower extremities and tail. Muscle disease may result in dysphagia, reduced gag reflex, and gait abnormalities. Affected dogs may develop a megaesophagus and aspiration pneumonia. The disease is thought to be autosomal dominant with incomplete penetrance in collies.[13] A study of linkage disequilibrium (LD) mapping in shelties had evidence of LD linkage to microsatellite marker FH3570 on canine chromosome 35.[14] Gene expression profiles of 285 genes failed to detect any disease-specific autoantibodies.[15] This disorder has also been classified as an ischemic dermatopathy.[16] Recommended treatments include pentoxifylline, vitamin E, tacrolimus, tetracycline/niacinamide and corticosteroids.[16-18]

Lupus Erythematosus

Lupus erythematosus is an autoimmune, inflammatory, multisystemic disorder of complex etiologies that may have clinical manifestations restricted to the skin or may involve multiple organs.. Discoid lupus erythematosus (DLE) is generally restricted to skin lesions affecting the face, particularly the planum nasale and dorsal surface of the nose. However severely affected animals may have involvement of the nasal turbinates with epistaxis and secondary infections. Treatments include pentoxifylline, vitamin E, tacrolimus, tetracycline/niacinamide , corticosteroids and hydroxychloroquine.[19] Exfoliative cutaneous lupus erythematosus (ECLE) is a familial form of lupus affecting German shorthaired pointer dogs.[20] Skin lesions include excessive scaling and crusting. As the disease progresses the dogs develop severe lameness and renal disease. Diagnosis is made by clinical observations and skin biopsies demonstrating findings compatible with lupus—hyperkeratosis, epidermal atrophy, basal cell vacuolar degeneration, basement membrane zone thickening and mononuclear cell infiltration at dermoepidermal interface. Most affected dogs are euthanized before 4 years of age. Treatment has been attempted with cyclosporine, corticosteroids, hydroxychloroquine and adalimumab with generally poor responses.[21] A genome-wide association study identified a SNP allele on canine chromosome 18 that segregated with the disease in the 267 dogs tested.[22] Systemic lupus erythematosus (SLE) is a multisystem disease with variable clinical presentations. Skin lesions are present in 40 to 50% of cases.[23, 24] Skin lesions are usually symmetrical and vary from mild scaling and/or alopecia to severe ulcerations. Numerous other findings may be present depending upon the organ systems affected—these include fever, malaise, anemia, weight loss, lymphadenopathy, lameness, weight loss, depression and behavioral changes.

Sterile Nodular Panniculitis (SNP)

Sterile nodular panniculitis is characterized by sterile subcutaneous inflammatory nodules are generally present in the absence of any other known specific diseases or infectious agents. Pancreatitis and pancreatic tumors have been associated with panniculitis and all dogs and cats presenting with panniculitis should be evaluated for the possibility of pancreatic disease.[25-29] Other differentials for nodular panniculitis include foreign bodies, immune-mediated reactions, infections, vitamin E deficiency, injection reactions, burns and trauma. Minimum data base evaluations should include CBC, biochemical profiles, pancreatic lipase immunoreactivity screening, abdominal ultrasonography and deep biopsies for histopathology and culture/PCR evaluations. Treatment options for SNP include vitamin E, corticosteroids, tetracycline/niacinamide, azathioprine and cyclosporine.

METABOLIC DISEASES

Certain metabolic diseases may present with cutaneous lesions as the first symptoms of the disorder. Examples include xanthomas as manifestations of hyperlipidemias, periocular and pressure point dermatitis as manifestations of zinc-related disorders, and superficial necrolytic dermatitis as a manifestation of liver disease, diabetes mellitus or as a paraneoplastic sign of a glucagonoma.

Xanthomas

Xanthomas appear as yellow-white papules, nodules or plaques that are most often found on the head, limbs, feet or over bony prominences. The lesions may ulcerate and are often surrounded by erythema. The lesions are composed of foamy macrophages and giant cells often surrounding deposits of lipid in the dermis. These lesions are cutaneous manifestations of abnormalities in lipid metabolism or consumption of extremely high-fat diets. Xanthomas develop in cats with hereditary hyperlipoproteinemia and in animals with diabetes mellitus, hypercortisolism or animals that have been treated with progestational compounds or fed high-fat foods and treats. [30-33] Treatment includes feeding a low-fat diet and management of the underlying disease.

Zinc-Responsive Dermatitis

Skin lesions associated with zinc-deficiency or zinc-responsive dermatitis include areas of erythema, alopecia, crusts and scales starting on the face, mucocutaneous junctions, feet and footpads. Other findings may include depressed appetite, weight loss, impaired wound healing and increased susceptibility to secondary infections. Alaskan malamutes and Siberian huskies are most often affected and many have a decrease in intestinal absorption of zinc.[34, 35] Pharaoh hounds have also been reported with a severe form of the disease.[36] Treatment includes oral supplementation with zinc or in severely affected dogs intravenous administration of zinc sulfate at weekly intervals for 4 weeks and then monthly for maintenance.

Lethal Acral Dermatitis/Acrodermatitis

Bull terriers have an autosomal recessive condition with defective zinc and copper metabolism and altered production of 13 liver proteins. Affected puppies are lighter than normal skin pigmentation and may have difficulty in chewing and swallowing. They develop splaying of their digits, crusting and cracking of their footpads and ulcerated crusted lesions on their ear pinnae and mucocutaneous junctions. Foot lesions progress to onychodystrophy, paronychia and interdigital pyoderma. Systemic signs may include diarrhea and respiratory infections, most succumb to secondary infections before 7 months of age. Diagnosis is made based on signalment and clinical signs, serum and liver zinc levels are usually low and skin biopsies show parakeratotic hyperkeratosis.[37-39]

Superficial Necrolytic Dermatitis (SND) (Metabolic Epidermal Necrosis, MEN) (Hepatocutaneous Disease) (Diabetic Dermatopathy) (Glucagonoma Syndrome) (Necrolytic Migratory Erythema, NME)

NME in humans is a cutaneous marker for alpha$_2$-glucagon-producing pancreatic islet cell tumors. The skin lesions are due to degeneration of keratinocytes resulting in epidermal edema and necrosis. In dogs only a low proportion of dogs with classical lesions of this disease have underlying glucagonomas, the majority of canine cases are diagnosed with liver disease.[40-42] The degeneration of keratinocytes is believed to be due to hypoaminoacidemia resulting in inadequate nutrients to maintain a viable epidermis. Lesions start in areas of high cell turnover including mucocutaneous junctions, face, footpads and pressure points. Skin lesions include crusts and ulcerations with perilesional erythema. Most affected dogs also have a non-regenerative anemia, mild hyperglycemia, and elevated liver enzymes. Liver ultrasonagrams are often characterized as having a honeycomb appearance. Amino acid levels are low and dogs with pancreatic tumors usually have elevated levels of plasma glucagon.[43] The disease in cats has been associated with pancreatic tumors and chronic liver disease.[44,45] Skin biopsy findings are unique for this syndrome and consist of basal cell hyperplasia, vacuolar degeneration of the stratum spinosum and parakeratotic hyperkeratosis ("blue," "white" and "red" zones). Most affected animals are euthanized within a few months. Surgical removal of pancreatic tumors may be curative if the tumor has not metastasized, however the majority are metastatic. Somatostatin may be beneficial but is expensive. Intravenous amino acids and feeding high protein diets provide palliative therapy for dogs with hypoaminoacidemia. Cholchicine has been recommended for treatment of hepatic cirrhosis and in combination with nutritional support with high levels of protein plus essential fatty acids and zinc may lead to clinical remissions lasting several months.[46]

PARANEOPLASTIC DERMATOPATHIES

Paraneoplastic dermatopathies, when recognized, are often the first clue that the pet is suffering from an underlying neoplastic condition. Examples include SND when associated with glucagonomas, feminization and linear preputial erytema associated with testicular tumors, feline paraneoplastic alopecia associated with pancreatic adenocarcinoma and cholangiocarcinoma, nodular dermatofibromas associated with renal and uterine neoplasia, feline exfoliative dermatitis associated with thymomas, pemphigus associated with splenic tumors and lymphoma, and cutaneous manifestations of Cushing's disease associated with adrenal or pituitary tumors.

Glucagonoma

As discussed above, some cases of SND in dogs are associated with glucagonomas . Clinical lesions may resolve following excision of the tumor and may relapse if metastatic disease develops.[47]

Hyperestrogensism/Testicular Tumors

Hyperestrogenism most commonly is associated with Sertoli cell tumors but may also occur with other testicular tumors, cystic ovaries, granulose cell tumors or iatrogenically following administration of estrogen supplementation or ingestion of topical estrogen creams or patches.[48, 49] Estrogen inhibits hair growth and stimulates pigment production resulting in a sparse hair coat and diffuse or multifocal hyperpigmentation. Linear preputial erytema has been reported in male dogs with Sertoli cell tumors.[50] Other symptoms include feminization with gyncomastia and prepucial keratinization. Some affected animals also develop bone marrow suppression with aplastic anemia and/or thrombocytopenia.[51]Intact animals should be evaluated for testicular or ovarian disease, overproduction of sex hormones by the adrenal cortex may also occur in some patients. Exogenous sources of estrogen include tablets used for treatment of urinary incontinence and topical estrogens prescribed for humans.

Feline Paraneoplastic Alopecia

A rapid development of extensive alopecia in an older cat may be the first symptom of underlying pancreatic or hepatic neoplasia.[52] Hair loss often begins periocularly and rapidly progresses with alopecia of the ventral neck, abdomen and legs. Hairs epilate easily and the alopecic skin appears shiny but is not fragile. Secondary malassezia infections may result in excessive grooming. Some cats also have dry, scaly footpads. Skin biopsies show miniaturization of hair follicles and adnexa with epidermal hyperplasia. The most commonly associated tumor is a pancreatic adenocarcinoma, a few affected cats were diagnosed with bile duct carcinomas. The prognosis is grave, one cat improved following surgical removal of a pancreatic adenocarcinoma and then relapsed following the development of metastatic disease.[53]

Nodular Dermatofibromas

German shepherds and related dogs have an autosomal dominant genetic disease caused by mutations in the folliculin gene (FLCN) that results in the development of multiple cutaneous nodules and multiple cysts and cystadenocarcinomas in the kidneys.[54, 55] Female dogs may also develop uterine leimyomas and leisarcomas. The skin lesions may develop months or years before the renal lesions, although renal cysts and tumors are usually detectable via CT scans or ultrasonography by 4-5 years of age.

Feline Exfoliative Dermatitis

Skin lesions of erythema, crusting and scaling with alopecia may be the first sign of disease in cats with thymomas. Secondary infections with *Malassezia* may result in pruritus. Affected cats may later develop signs of coughing dyspnea, anorexia and weight loss. Skin biopsies show a hydropic interface dermatitis with apoptotic keratinocytes. Radiology or thoracic ultrasonography can be used to demonstrate the presence of a mediastinal mass. Surgical removal may be curative.[56,57]

Paraneoplastic Pemphigus

Paraneoplastic pemphigus has been reported in dogs in association with thymic lymphoma and splenic sarcoma. Autoantibodies in affected dogs have targeted desmoplakin, envoplakin, periplakin, desmoglein 1 and desmoglein 3.

Histopath lesions have features of pemphigus foliaceous (intraepidermal pustules with acantholysis) , pemphigus vulgaris (suprabasliar clefts) and erythema multiforme (apoptotic keratinocytes). The prognosis is grave.[58, 59]

Feline Skin Fragility

Acquired skin fragility is most often seen in cats with hypercortisolism secondary to adrenal tumors or iatrogenic administration of glucocorticoids or progestational compounds. Affected cats have thin, fragile skin that tears easily. A few cases have been reported as being associated with hepatic lipidosis and hepatic neoplasia.[60]

Hypercortisolism

Cutaneous signs of hypercortisolism include atrophy of the epidermis, dermis and adnexal units resulting in thin skin, hair loss, easily visible blood vessels, phlectasias, comedones, calcinosis cutis and secondary pyodemas and otitis externa.

ENDOCRINE DERMATOPATHIES

The classic manifestations of endocrine-related dermatoses include bilaterally symmetrical alopecia, failure of hair regrowth after clipping, seborrheic dermatitis and predisposition to cutaneous infections including bacterial pyoderma, malassezia dermatitis and otitis externa. Underlying endocrinopathies include hypothyroidism, hypercortisolism, hyperthyroidism and sex hormone imbalances. Other endocrine diseases that may present with cutaneous symptoms include pitiuitary dwarfism, acromegaly and growth-hormone responsive dermatoses.

INFECTIOUS DISEASES WITH CUTANEOUS MARKERS

Infectious diseases that may be first recognized because of cutaneous symptoms include systemic mycoses, leishmaniasis, feline herpesvirus, feline leukemia virus, feline immunodeficiency virus and canine distemper infections.

REFERENCES

[1]Holbrook KA, et al. J Invest Dermatol 1980; 74(2):100; [2]Sequeira JL, et al. Vet Pathol 1999; 36(6):603;[3]Benitah N, et al. J Am Vet Med Assoc 2004; 24(5):706; [4]Zhou H, et al. Anim Genetics 2011; 43:471; [5]Vaatstra BL, et al. New Zealand Vet J 2011; 59(5):258; [6]Paciello O, et al. Vet Clin Pathol 2003; 32(1); [7]Colige A, et al. Am J Hum Genet 1999; 6592):308; [8]Holm DE, et al. J S Afr Vet Assoc 2008; 79(1):19; [9]Tryon RC, et al. Genomics 2007; 90(1):93; [10]White SD, et al. Vet Dermatol 2011; 22(2):206; [11]White SD, et al. Vet Dermatol 2004; 15(4):207; [12]Sinke JD, et al. Vet Quarterly 1997; 19(4):18; [13]Haupt K et al. Am J Vet Res 1985; 46:1861; [14]Clark LA et al. Vet Dermatol 2005; 16:392; [15]Wahl JM et al. Vet Dermatol 2008; 19:52; [16]Morris DO. Vet Clin Small Animal 2013; 43:99; [17]Rees CA et al. Vet Ther 2003; 4:234; [18]Hollar CB et al. J Dermatol Treat 2004; 15(1):35; [19]Oberkirchner U et al. Vet Dermatol 2011; 23:65; [20]Bryden SL et al. Vet Dermatol 2005;16:239; [21]Mauldin EA et al. Vet Dermatol 2010; 21(4):373; [22]Wang P et al. Immunogenetics 2011; 63(4):197; [23]Scott DW et al. J Am Anim Hosp Assoc 1983; 19:461; [24]Chabanne L et al. Compend Contin Educ Pract Vet 1999; 21:135; [25]Moreau PM et al. J Am Vet Med Assoc 1982; 180(4):422; [26]German AJ et al. J Small Anim Pract 2003; 44(10):449; [27]Mellanby RJ et al. J Small Anim Pract 2008; 44):24; [28]Fabrinni F et al. Vet Dermatol 2005; 16(6):413; [29]Paterson S. J Small Anim Pract 1994; 35(2):116; [30]Jones BB et al. Fel Pract 186; 16:7; [31]Carpenter JL et al. Vet Dermatol 1992; 3:1987; [32]Chastain CB et al. J Am Vet Med Assoc 1978; 172:1209; [33]Kwochka KW et al. Compend Cont Educ Pract Vet 1984; 6:185; [34]Colombini S et al. J Am Vet Med Assoc 1997; 211:451; [35]White SD et al. Vet Dermatol 2001; 12:101; [36]Campbell GA et al. J Vet Diagn Invest 2010; 22:663; [37]Jezyk PF., et al. J Am Vet Med Assoc 1986; 188:833; [38]Smits B et al. Vet Dermatol 1991; 2:91; [39]McEwan NA et al. J Sm Anim Pract 2000; 41:507; [40]Miller WH et al. J Am Anim Hosp Assoc 1990; 26:573; [41]Gross TL et al. Vet Pathol 1993; 30:75; [42]Cave TA et al. J Small Anim Pract 2007; 48:522; [43]Outerbridge CA et al. Vet Dermatol 2002; 13:177; [44]Patel A et al. Vet Dermatol 1996; 7:221; [45]Godfrey DR et al. J Small Anim Pract 2000; 41:324; [46]Miller WA et al. Small Animal Dermatology 2013;7:542; [47]Torres SM et al. J Am Anim Hosp Assoc 1997; 33:313; [48]Fiorito DA. Compend Cont Educ Pract Vet 1992; 14:727; [49]Medleau L. J Am Anim Hosp Assoc 1989; 25:689; [50]Nemrek JA et al. J Am Anim Hosp Assoc 1992;28:402; [51]Morris BJ. Vet Med Small Anim Clin 1983; 78:1070; [52]Turek MM. Vet Dermatol 2003; 14:279; [53]Tasker S et al. J Small Anim Pract 1999; 40:16; [54]Moe L et al. J Small Anim Pract 1997; 38(11):498; [55]Bansdorff TB et al. Mamm Genome 2009; 20:315; [56]Scott DA et al. Fel Pract 1995; 23:8; [57]Zitz JC et al. J Am Vet Med Assoc 2008; 232(8):1186; [58]Olivry T et al. Vet Dermatol 2009; 20:313; [59]Elmore SA et al. Vet Pathol 2003; 40:609; [06]Gross TL et al. Skin Diseases of the Dog and Cat. 2005; 2:1

KINASE INHIBITORS IN VETERINARY ONCOLOGY

Cheryl London, DVM, PhD, DACVIM
Columbus, OH

INTRODUCTION

In normal cells, signals are generated that begin at the outside of the cell and transmit through the cytoplasm to the nucleus, regulating cell growth, differentiation, survival, and death. These are typically induced by growth factors, cytokines, chemokines, hormones, electrolytes, and cell-cell contact among others. The intracellular signals generated from these stimuli induce a coordinated network of protein-protein interactions/protein complex formation that tightly regulate cellular proceses. Over the past decade it has become evident that many critical components of these signal transduction pathways are dysregulated in tumor cells, contributing to the classic features of cancer including uncontrolled growth, resistance to apoptosis, and ultimately enhanced mobility resulting in metastasis. Through substaintial improvements in molecular techniques the exact events that lead to this dysregulation are being characterized, thereby providing significant opportunities for the development of novel therapeutic approaches aimed at disrupting aberrent intracellular signaling in tumor cells. This review focuses on the role of a particular group of signal transducers, called protein kinases, their function in normal cells, the mechanisms by which their dysrgulation contribute to tumorigenesis, and the use of agents designed to inhibit aberrent signaling pathways in tumor cells driven by protein kinases.

SIGNAL TRANSDUCTION IN NORMAL CELLS: KINASE SIGNALING

While there are numerous signaling pathways that have been well-characterized, those involving protein kinases are probably the best described with respect to their roles in cell biology. Protein kinases are critical players in normal cell signal transduction, acting to tightly regulate essential cell processes such as growth and differentiation, survival and death, and migrationand metabolism. These proteins work through phosphorylation; they bind adenosine triphosphate (ATP) and use it to transfer the terminal phosphate group to key residues on themselves (also known as autophosphorylation) and on other molecules. The phosphorylation events alter the structure of protein kinases, permitting the binding of other proteins/kinases, driving protein-protein interactions, and the subsequent formation of protein complexes and scaffolds that act to convey signals with in the cell. This process typically occurs in response to external signals generated by growth factors, cytokines, chemokines or other stimuli that stimulate the cascade. Protein kinases are classified as tyrosine kinases (TKs) if they phosphorylate proteins on tyrosine residues or serine/threonine kinases if they phosphorylate proteins on serine and threonine residues. In some cases the kinases perform both functions and are labeled dual function kinases. Protein kinases are found on the cell surface (receptor kinases, most of which are tyrosine kinases, RTKs) in the cytoplasm, and in the nucleus (non-receptor kinases). The human genome encodes for 518 kinases, of which 100 are TKs.(*1*)

Of the 100 identified TKs, 58 are known to be RTKs. Each contains an extracellular domain that binds a growth factor (GF), a transmembrane domain that anchors the RTK in the membrane and regulates the spatial organization of these proteins, and a cytoplasmic kinase domain that positively and negatively regulates phosphorylation of the RTK.(*2-4*) RTKs typically exist as monomers on the cell surface and are dimerized through the act of GF binding. The dimerization alters the three dimensional structure of the receptor, permitting ATP to bind and autophosphorylation to occur, resulting in generation of a downstream signal through subsequent binding of adaptor proteins and nonreceptor kinases.(*2*) Specific examples of RTKs linked to human cancer include KIT and gastrointestinal stromal tumor (GIST), epithelial growth factor receptor (EGFR) and anaplastic lymphoma kinase (ALK) and lung cancer, and HER2/Neu and breast cancer. The contribution of RTKs to cancers in dogs and cats is only beginning to be elucidated.

Kinases in the cytoplasm act as a network of protein bridges, conducting signals initiated by RTKs to the nucleus through a series of intermediates that become phosphorylated.*(5)* The cytoplasmic kinases may be directly on the inside of the cell membrane or free in the cytoplasm. There are two central pathways involving cytoplasmic kinases that are often dysregulated in cancer. The first includes members of the RAS-RAF-MEK-ERK/p38/JNK families. Several members of this pathway are known to be altered in human cancers including RAS (mutations in lung cancer, colon cancer, and several hematologic malignancies) and BRAF(mutations in cutaneous melanomas and papillary thyroid carcinomas), among others. The second cytoplasmic pathway includes phosphatidyl inositol-3 kinase (PI3K) and its associated downstream signal transducers AKT, NFkB, and mTOR, among others. Abnormalities of PI3K that result in activation of this pathway are comonly found in human cancers including mutations (breast and colorectal cancers and glioblastoma) and gene amplification (gastric, lung, ovarian cancers). Dysregulation of this pathway also occurs through loss of activity of PTEN, a phosphatase that negative regulates AKT signaling through dephoshorylation.

DYSREGULATION OF PROTEIN KINASES IN CANCER

Dysfunction of protein kinases occurs frequently in cancer and characterization of specific signaling abnormalities associtated with these aberrant kinases has been increasingly well-documented. Abnormal function of kinases occurs through a variety of mechanisms, including mutation, overexpression, fusion proteins, and/or autocrine loops. In the case of mutations, these typically induce phosphorylation of the kinase in the absence of an appropriate stimulus. These mutations typically consist of a single amino acid change through a point mutation, deletion of amino acids, or insertion of amino acids, usually in the form of an internal tandem duplication (ITD). For example, a point mutation occurs in the BRAF gene (V600E, exon 15) in approximately 60% of human cutaneous melanomas. This amino acid change causes a conformation change in B-Raf mimicking its activated form thereby inducing constitutive downstream ERK signaling and abnormal promotion of cell growth and survival. Another protein for which dysregulation has been well-characterized is the RTK KIT. In approximately 25% to 30% of canine grade 2 and 3 mast cell tumors, mutations consisting of ITDs are found in the juxtamembrane domain of KIT resulting in constitutive activation in the absence of ligand binding. KIT mutations consisting of deletions in the juxtamembrane domain are also found in approximately 50% of human patients with gastrointestinal stromal tumors (GISTs) and similar mutations are found in dog GISTs, although at a lower frequency.

Overexpression of kinases can also drive aberrent signaling. This typically involves RTKs and may generate enhanced responses of cells to normal levels of growth factor; or, if the levels are high enough, the RTK can become phosphorylated in the absence of growth factor/signal through spontaneous dimerization. In humans, the RTK HER2/Neu (also known as ErbB2, a member of the EGFR family) is overexpressed in breast and ovarian carcinomas and this often correlates with a more aggressive tumor phenotype. Fusion proteins are generated when a portion of the kinase becomes attached to another gene through chromosomal rearrangement/translocation and the normal mechanisms that control protein function are disrupted via loss of regulatory constraints induced through both tertiary structure and binding of additional proteins. One of the best characterized fusion proteins is BCR-ABL which is found in 90% of patients with chronic myelogenous leukemia (CML). . Autocrine loops of activation primarily occur when the tumor cell expresses both the RTK and the growth factor; in most cases, one or the other usually is also overexpressed, resulting in constitutive activation of the RTK In canine cancers, possible autocrine loops have been documented in osteosarcoma (OSA, MET and HGF) and hemangiosarcoma (HSA, KIT and SCF).

INHIBITION OF PROTEIN KINASE FUNCTION

With the realization that specific signaling pathways are often dysregulated in particular cancer has come the opportunity to develop therapeutic strategies that target those pathways in tumor cells. Such an approach, often termed personalized or targeted therapy, relies on the notion that the targets for treatment are differentially expressed/important in tumor versus normal cells, permitting a differential that can be exploited. This is in contrast to standard chemotherapy or radiation therapy for which cell damage does not distingush between tumor and normal cells. There have been several methods employed to target individual pahways for therapy, atlhough the two most successful approaches to date have been the use of monoclonal antibodies (mAbs) and small molecule inhibitors; the small molecule inhibitors are addressed below.

Small molecule inhibitors work primarily by blocking the ATP-binding site of kinases thereby acting as competitive inhibitors; a smaller number of these inhibitors work by preventing necessary protein-protein interactions (allosteric inhibition).*(6)* In the absence of ATP, the kinase is unable to phosphorylate itself or downstream signaling elements, thereby interrupting a survival/growth signal essential to the tumor cell, ultimately resulting in cell death. Such inhibitors typically work in a reversible manner and effectively compete for ATP binding by exhibiting a higher affinity for the ATP binding pocket. More recently, irreversibly small molecule inhibitors are being explored that bind covalently in the pocket thereby providing more extended duration of action. The small molecule inhibitors are relatively easy to synthesize in large quantities, frequently orally bioavailable, and are not restricted to cell surface receptors as they can readily enter cells to bind the intended target.

The first small molecule inhibitor to be approved for human use was imatinib (Gleevec), an orally administered drug that binds the ATP pocket of ABL, as well as the RTKs KIT and PDGFRα. BCR-ABL fusion proteins are present in 90% of human patients with CML, and as such, ABL is a good target for therapeutic intervention in this disease. Imatinib has significant biologic activity in human patients with CML and it is now approved for first-line care of affected individuals. Given its inhibition of KIT, imatinib also has activity against human GIST in which 60% to 80% of the tumors have activating point mutations or deletions in the juxtamembrane domain of KIT. Response rates of 50% to 70% have been reported with imatinib in patients with GIST, far better than the 5% response rate seen with standard chemotherapy. Several small molecule inhibitors have been approved for the treatment of human cancers that possess specific mutations in kinases known to drive tumor growth and survival including EGFR inhibitors erlotinib (Tarceva) and gefitinib (Iressa), the ALK inhibitor crizotinib (Xalkori) and the BRAF inhibitor vemurafenib (Zelboraf).

KINASE INHIBITORS: THE VETERINARY EXPERIENCE

There are now two small molecule inhibitors approved for use in veterinary medicine. The first, toceranib phosphate (Palladia, Pfizer Animal Health, Madison, NJ) is an orally bioavailable small molecule inhibitor that blocks a variety of RTKs including VEGFR2, PDGFRα/β, KIT, FLt3 and CSF1R. *(7)* A recent kinome analysis (London et al, unpublished data) supports the activity of toceranib against the split kinase RET and possibly JAK family members. The first evaluation of toceranib in dogs was a phase I clinical trial in 57 dogs with a variety of cancers.*(7)* In this study, objective responses occurred in 16 dogs consisting of six complete responses (CR) and 10 partial responses (PR) with stable disease (SD) occurring in an additional 15 dogs for an overall biological activity of 54%. Responding tumors included sarcomas, carcinomas, melanomas, myeloma, and MCTs. The highest response rate was in MCTs, with 10/11 dogs with KIT mutations exhibiting response to therapy. The maximum tolerated dose was established as 3.25 mg/kg every other day (EOD) and the adverse event profile was found to be primarily gastrointestinal (GI) in nature including loss of appetite, diarrhea, and less commonly vomiting, although these toxicities were relatively well controlled with the addition of appropriate concomitant medications. Based on the phase I trial results, a placebo controlled randomized study of toceranib was subsequently performed in dogs with grade 2 and 3 MCTs. *(8)* The overall response rate for all 145 dogs was 42.8% (21 CR, 41 PR) with an additional 16 dogs experiencing SD. Dogs whose MCT had KIT mutations were more likely to respond to toceranib than those without (69% vs. 37%). Following its approval in 2009, toceranib was used to treat a variety of canine solid tumors, often in the setting of failed primary therapy or metastatic disease. A retrospective analysis of its off-label use provided evidence of biologic activity in anal gland anal sac adenocarcinoma (AGASACA), metastatic osteosarcoma (OSA), thyroid carcinoma (thyroid CA), head and neck carcinoma, and nasal carcinoma. *(9)* Clinical benefit was observed in 63/85 (74%) dogs with these tumors (2 CR, 18 PR, 43 SD) providing preliminary evidence that toceranib may have biologic activity against certain solid tumors, although prospective studies are necessary to define its true effects.

Given the use of non-steroidal anti-inflammatory medications (NSAIDs) in veterinary oncology, there was interest in combining an NSAID with toceranib to treat solid tumors. A phase I trial was performed in tumor-bearing dogs to establish the safety of toceranib/piroxicam co-administration. *(10)* The combination of standard dosages of both drugs (toceranib, 3.25 mg/kg EOD and piroxicam at 0.3 mg/kg/day) was found to be generally safe and several anti-tumor responses were noted. To begin to determine how best to combine toceranib with chemotherapeutic agents, a phase I clinical trial was performed to identify an appropriate dosing regimen combining toceranib with vinblastine in canine MCT where both drugs have single agent activity. *(11)* The dose-limiting toxicity for the toceranib/vinblastine combination was found to be neutropenia, and the maximum tolerated dose of vinblastine was 1.6 mg/m^2 every other week with toceranib at 3.25 mg/kg EOD. Despite the reduced dose of vinblastine the overall response rate was 71% to the drug combination, superior to either drug alone suggesting additive/synergistic effects of the drug combination. In addition to chemotherapy, there is interest in combining toceranib with radiation therapy (RT). Dogs with non-resectable MCTs received prednisone, omeprazole, diphenhydramine, and toceranib for 1 week prior to starting coarse fractionated radiation therapy (6 Gy given once per week for 4 weeks).*(12)* The overall response rate was 76.4%, with 58.8% of dogs achieving CR and 17.6% PR. The overall median survival time was not reached with a median follow-up longer than one year. Importantly, there was no evidence of enhanced radiation induced toxicities in this study, suggesting that combined toceranib/RT may achieve significant clinical activity in dogs without serious side effects.

The second small molecule inhibitor approved for use in dogs is masitinib (Kinavet, AB Science, Paris, France) that blocks the activity of KIT, PDGFRα/β and the cytoplasmic kinase Lyn. A large placebo controlled clinical trial was performed in over 200 dogs with MCTs in which masitinib significantly improved time to progression compared to placebo, and outcome was improved in dogs with MCTs possessing KIT mutations.*(13)* Subsequent follow-up of patients treated masitinib for 1-2 years identified an increased number of patients with long-term disease control compared to those treated with placebo (40% versus 15% alive at 2 years). *(14)* Lastly, small studies have evaluated the efficacy of imatinib for the treatment of canine and feline MCT. *(15-17)* Imatinib was well tolerated, and objective antitumor responses were observed in dogs with both mutant and wild-type KIT. Responses have also been observed in cats with MCT that have KIT mutations as well. *(18, 19)*

RESISTANCE TO THERAPY

The response of tumour cells to small molecule inhibitors in the presence of known protein dysregulation is often dramatic, with objective response rates often exceeding 50%, far higher than that typically observed with chemotherapy alone. Unfortunately, in most instances these responses are typically not durable, lasting from 6-18 months on average before relapse. The mechanisms that drive resistance to small molecule inhibitors have been well characterized for specific therapeutics such as imatinib and erlotinib, but remain only partly understood for many others.*(20, 21)* In general, more than one cellular alteration contributes to drug resistance, complicating strategies to prevent or circumvent this issue. For patients that take imatinib, the primary cause for relapse is the development of point mutations in the kinase domain of BCR-ABL that often prevent imatinib binding. Several have been identified in the ATP binding loop, activation loop, imatinib contact sites outside the loop, and other locations. Additionally, some patients develop resistance through upregulation of BCR-ABL mRNA, resulting in protein overexpression that overwhelms the ability of imatinib to block function. Lastly, eleveated p-glycoprotein expression and enhanced multi-drug efflux, as well as activation of other growth factor pathways, have been documented in some patients. Other mechanisms of resistance to kinase inhibitors that are not as well characterized include epigenetic changes secondary to alterations in histone acetylation/deacetylation and chromatin and histone methylation.*(21)* These changes act to modify the expression of genes that regulate responsiveness to kinase inhibitors, thus promoting escape from therapy. Taken together, these observations reinforce the notion that drug resistant cell populations may be selected via multiple mechanisms, making prevention and treatment of this resistance challenging.

SUMMARY

Progress in molecular biology has permitted a greater understanding of how dysregulation of signal transduction in cancer cells contributes to uncontrolled growth and survival. This has translated into a entirely new apporach to cancer therapy through the use of small molecule inhibitors, many of which have demonstrated substantial single agent activity in a variety of human cancers. The use of such agents is just beginning to be explored in veterinary oncology and this process has been accelerated through the approval of both toceranib and masitinib. Nevertheless, significant challenges remain including determining how these therapies can be effectively combined with chemotherapy and radiation therapy to provide optimal anticancer efficacy without enhancing toxicity, and identifying strategies that are less likely to result in drug resistance.

REFERENCES

1. G. Manning, D. B. Whyte, R. Martinez, T. Hunter, S. Sudarsanam, *Science* 298, 1912 (Dec 6, 2002). 2. M. A. Lemmon, J. Schlessinger, *Cell* 141, 1117 (Jun 25, 2010). 3. S. Madhusudan, T. S. Ganesan, *Clin Biochem* 37, 618 (Jul, 2004). 4. E. Zwick, J. Bange, A. Ullrich, *Trends Mol Med* 8, 17 (Jan, 2002). 5. P. Blume-Jensen, T. Hunter, *Nature* 411, 355 (May 17, 2001). 6. J. Zhang, P. L. Yang, N. S. Gray, *Nat Rev Cancer* 9, 28 (Jan, 2009). 7. C. A. London *et al.*, *Clin Cancer Res* 9, 2755 (Jul, 2003). 8. C. A. London *et al.*, *Clin Cancer Res* 15, 3856 (Jun 1, 2009). 9. C. London *et al.*, *Vet Comp Oncol*, (Jun 1, 2011). 10. E. Chon, L. McCartan, L. N. Kubicek, D. M. Vail, *Vet Comp Oncol*, (Mar 21, 2011). 11. C. Robat *et al.*, *Vet Comp Oncol*, (Jan 31, 2011). 12. K. S. Carlsten *et al.*, *J Vet Intern Med* 26, 135 (Jan, 2012). 13. K. A. Hahn *et al.*, *J Vet Intern Med* 22, 1301 (Nov-Dec, 2008). 14. K. A. Hahn *et al.*, *Am J Vet Res* 71, 1354 (Nov, 2010). 15.M. Isotani *et al.*, *J Vet Intern Med*, (Jun 18, 2008). 16. L. Marconato *et al.*, *J Vet Intern Med*, (Jun 18, 2008). 17. O. Yamada *et al.*, *Vet Immunol Immunopathol* 142, 101 (Jul 15, 2011). 18. M. Isotani *et al.*, *Vet Immunol Immunopathol* 114, 168 (Nov 15, 2006). 19. M. Isotani *et al.*, *Br J Haematol* 148, 144 (Jan, 2009). 20. R. Barouch-Bentov, K. Sauer, *Expert Opin Investig Drugs* 20, 153 (Feb, 2011). 21. S. A. Rosenzweig, *Biochem Pharmacol* 83, 1041 (Apr 15, 2012).

METRONOMIC (CONTINUOUS LOW-DOSE) CHEMOTHERAPY - FACT OR FICTION

David M. Vail, DVM, DACVIM (Oncology)
Madison, WI

INTRODUCTION

Traditionally, cytotoxic chemotherapy has been dosed according to the established maximally tolerated dose (MTD) in a pulse fashion. The MTD is used as the dose-response curve for traditional cytotoxic drugs is linear and using the highest dose that the normal tissues can recover from is theoretically ideal. However, more recently the concept of low-dose, continuous delivery of cytotoxic chemotherapy agents, sometimes called "metronomic" chemotherapy has been introduced. Metronomic chemotherapy involves the use of chemotherapy in a non-conventional way – that is, instead of giving high MTD-based doses in a pulse setting and then allowing recovery of normal tissues to occur over 1 – 3 weeks, in the metronomic setting, very low doses of chemotherapy (well below the MTD) are given daily or every other day continuously (for months or even years) (See **Figure 1**).

Figure 1: Schematic of traditional MTD dosing versus metronomic dosing: Kerbel et al. Nature Rev Canc 2004.[3]

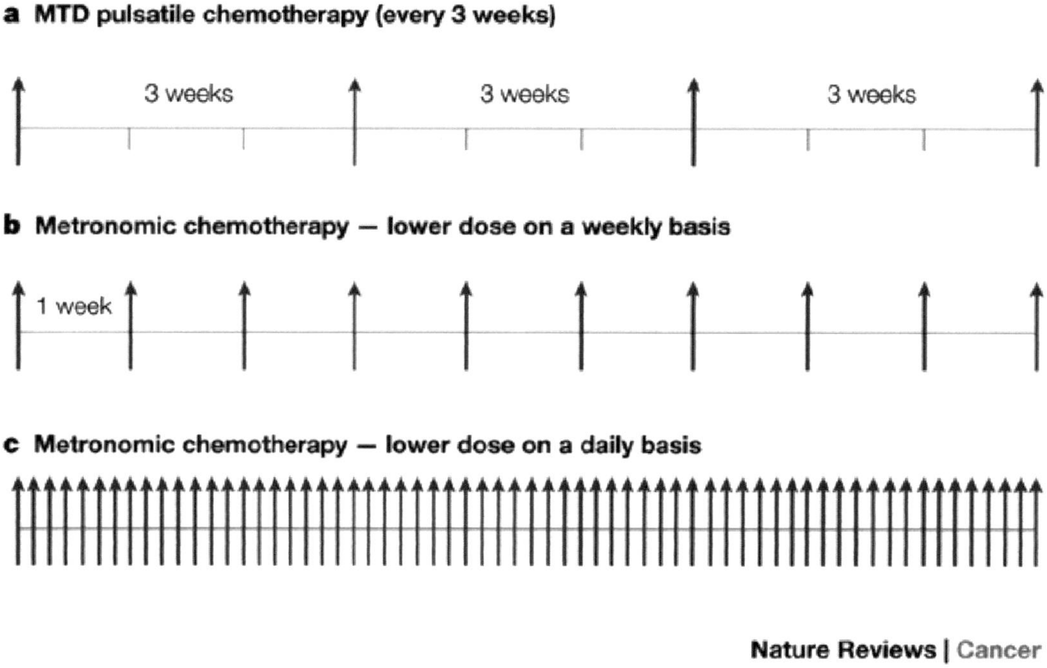

Metronomic chemotherapy has received a great deal of discussion and has been the subject of much research in the past decade.[1-3] Results have been mixed and in many cases its effectiveness is still an area of controversy. Metronomic chemotherapy is primarily reserved for cases where: 1) cancer has spread (metastasized) significantly; 2) the primary tumor is not amenable to a more curative modality (surgery or radiation), or has been declined by the caregiver) ; 3) traditional chemotherapies are unlikely to have positive activity, or; 4) the primary cancer has been effectively removed but the likelihood of future recurrence or spread is high based on the specific cancer being treated (e.g., soft tissue sarcoma excised with incomplete 'dirty' margins).

PROPOSED MECHANISMS OF ACTION OF METRONOMIC CHEMOTHERAPY

The exact mechanisms of action of metronomic chemotherapy is currently not well understood, however there exists some *in vitro* and *in vivo* (primarily in rodent tumor models and dogs) evidence to suggest that cytotoxic agents applied in the low-dose continuous fashion may act by one or several of the following possible ways:

1. Attacking existing blood vessels or preventing new blood vessels that form to provide oxygen and nutrients to a growing tumor – so called 'antiangiogenic' or 'antivasculogenesis' therapy.

2. Attacking normal host cells (so-called 'stromal cells') that feed or support the growing cancer cells

3. Producing an immunomodulatory effect whereby the patients immune system is stimulated and/or the immuno-evasion properties of tumors is inhibited leading to enhancement of antitumor immunosurveillance. In particular, metronomic chemotherapy (documented with cyclophosphamide) decreases the number of a type of immune system cell (called a 'regulatory' T-cell) that helps the cancer avoid the patient's immune system (so called immuno-evasion). This immune-modulatory effect may help the patients immune system fight the cancer more naturally

4. Possible direct anti-cancer effects against the cancer cell itself.

As metronomic therapy is at least partially thought to work by starving tumors of nutrients, support, or blood supply, it is more likely to result in stabilization of tumors (inhibiting or slowing additional growth) rather than causing actual shrinkage of tumors that are already present. However, if the tumor is in a location that is not causing problems due to presence, disease stabilization can be a positive outcome and may result in prolongation of progression-free survival.

DRUGS AND DRUG PROTOCOLS APPLIED IN THE METRONOMIC SETTING

Several different cytotoxic drugs have been used in the metronomic setting (summarised in **Table 1**). Additionally, the use of drugs that are not classical cytotoxic chemotherapeutics, such as non-steroidal antiinflammitory (NSAID) drugs or tyrosine kinase recepter inhibitors that may have antiangiogenic and immunomodulatory effects (e.g., piroxicam, toceranib) have been investigated in combination with more classic metronomic cytotoxic-class drugs. Examples of single-agent or combination metronomic agents include:

Low dose cyclophosphamide (cytoxan)

In traditional chemotherapy protocols, cyclophosphamide, an alkylating agent, is used at a high pulse MTD-based dose (250 – 300 mg/m^2) with a 3-week rest period in between. In the metronomic setting, cyclophosphamide is used at low doses (approximately 1/20th of the pulse dose) every day or every other day, continuously. Dosing schemes of 12.5 – 15 mg/m^2 daily (PO) have been documented to deplete T regulatory cells in cancer bearing dogs and to decrease microvascular density within soft tissue sarcomas of dogs.[4-6] Non-randomized data comparing dogs receiving metronomic cyclophosphomide following surgery for soft tissue sarcomas to historical controls receiving only surgery suggests a longer time to recurrence; however, randomized clinical trials have not as yet been performed to confirm this.

Low dose chlorambucil (leukeran)

This weak alkylator has also been used in the metronomic setting using a dose of 4 mg/m^2 (PO) daily or every other day.[7] While chlorambucil is in the same class of drug as cyclophosphomide, evaluation of the immunomodulatory effects (e.g., Treg depletion) of this agent in dogs have yet to be performed, so its current use for this purpose is currently theoretical. Only anecdotal use has been reported and prospective clinical trials are yet to be completed.

Low dose CCNU (lomustine)

As with cyclophosphomide and chlorambucil, this alkylating agent has also been investigated for metronomic use, using a low dose given daily, continuously.[8] However, adverse events were high in the doseage investigated and no activity data exists in the dog.

Piroxicam (feldene)

This non-steroidal anti-inflammatory (NSAID) drug has been used along with metronomic cytotoxic agents at a standard dose and its COX inhibitory effects may have antiangiogenic activity in addition to its anti-inflammatory or antiarthritic properties. Occasionally, other NSAIDs (e.g., rimadyl, derramax) are substituted for piroxicam.

Toceranib (Palladia®)

This receptor tyrosine kinase receptor inhibitor class of anticancer drug was originally developed as an antiangiogenic agent as it targets vascular endothelial growth factor receptor (VEGFR) and platelet derived growth factor receptor (PDGFR). Both of these receptors are involved in and modulate new or continued blood vessel growth. Further, toceranib has been document to also deplete Tregs in tumor bearing dogs, similar to metronomic cyclophosphomide.[6] Toceranib, while licensed for use in dogs with mast cell tumors, is currently being investigated as an antiangiogenic and immunomodulatory agent alone or in combination with cytotoxic metronomic agents. Masitanib (Kinavet®, Masivet®) does not have appreciable VEGFR activity and is theoretically less likely to have as potent an antiangiogenic effect, however, it does have PDGFR antaoginstic properties.

Side Effects (Adverse Events)

Despite their use in low doses, adverse events can still develop using metronomic alkylators (cyclophosphomide, chlorambucil, CCNU) therapy and anti-angiogenic agents are not innocuous and may have unique adverse event profiles. For example, targeting new vasculature carries the potential risk that drug effects on vessels have unwanted consequences including hypertension, edema, hemorrhage, thromboembolism, proteinuria, intestinal perforation and impaired wound healing. Further, there is some evidence in certain tumor models that antiangiogenic therapy can lead to metastatic conditioning – enhanced invasion and metastasis, which would obviously be undesirable. Through further clinical evaluation we will continue to learn the relative risks and benefits for different anti-angiogenic drugs applied in a number of tumor settings.

Specific adverse events related to specific drugs used requires carefull observation. When using low-dose cytotoxics, while the risk of clinically significant myelosuppression appears low, periodic CBC assessment (e.g., at 2 and 6 weeks after initiation and then every 8 weeks) is recommended. Additionally, while cyclophosphamide has become a popular drug for metronomic scheduling, it is still associated with the potential for causing sterile hemorrhagic cystitis. Anecdotally, it is estimated that 5 – 7% of dogs develeop sterile hemorrhagic cystitis at some point following initiation of metronomic cyclophosphomide. Client education and instructions to monitor for stranguria, polakyuria and hematuria are important considerations and periodic urine testing for hematoria is indicated. While the addition of pulse furosemide concurrent with pulse-dose cyclophosphamide generally abrogates the risk of hemorrhagic cystitis, the daily application of cyclophosphamide does not lend itself to long-term continous concommitant use of furosemide due to unavoidable polyuria and polydypsia. Many have substituted chlorambucil for cyclophosphomide if sterile hemorrhagic cystitis develops.

Chronic piroxicam use can be associated with gastrointestinal upset and renal impairment, necessitating active surveillance (periodic biochemistry profiles) for these issues. Side effects for the tyrosine kinase receptor inhibitors can include dose-dependent gastrointestinal upset, gastric bleeding, myelosuppression, azotemia, anemia, lethargy or lameness, hypertension and proteinuria.

Table 1: Examples of investigational metronomic protocols in Veterinary Medicine

Metronomic Protocol	Investigational Use
• *Cyclophosphamide 12.5 – 15 mg/m^2 every other day alternating with; • Palladia® 2.5 – 3.25 mg/kg every other day • Piroxicam 0.3 mg/kg/day orally	Investigational for a wide variety of tumor types • Osteosarcoma • hemangiosarcoma
• *Cyclophosphamide 10 – 15 mg/m^2/day orally • Piroxicam 0.3 mg/kg/day orally • Doxycycline 5 mg/kg B.I.D. orally	Investigational for a wide variety of tumor types
• *Cyclophosphamide 10 – 15 mg/m^2/day orally • Piroxicam 0.3 mg/kg/day orally	Canine soft tissue sarcoma – possible improvement in time to recurrence following incomplete surgical excision.
• Chlorambucil 4 mg/m^2 daily	Investigated for a wide variety of tumor types. Suggested activity for hemangiosarcoma in the adjuvant (post splenectomy) setting
• *Cyclophosphamide 12.5 – 25 mg/m^2/day orally alternating in 3 week cycles with etoposide 50 mg/m^2/day orally • Piroxicam 0.3 mg/kg orally daily	Canine hemangiosarcoma[9] – modest improvement in disease free and overall survival
• CCNU (lomustine) 2.84 mg/m^2 PO daily	Approximately 25% of dogs did not tolerate due to adverse events

*New evidence suggests 15 mg/m^2 cyclophosphamide is superior than lower doses for depleting Treg lymphocytes

REFERENCES

1. Potente M, et al. *Cell* 2011, 146: 873.
2. Pasquier E, et al. *Nat Rev Clin Oncol* 2010, 7: 455.
3. Kerbel et al. Nature Rev Canc 2004, 4:423.
4. Burton JH, et al *J Vet Intern Med* 2011, 25: 920.
5. Elmslie RE, et al. *J Vet Intern Med* 2008, 22: 1373.
6. Mitchel L, et al. *J Vet Intern Med* 2008, 26: 355.
7. Leach TN, *et al. Vet Comp Oncol* 2011, 10:102.
8. Tripp CD, et al. *J Vet Intern Med* 2008, 25: 278.
9. Lana S, *et al. J Vet Intern Med* 2007, 21: 764.

ACTIVE LEARNING STRATEGIES FOR THE CLASSROOM

Cynthia Ward, VMD, PhD, DACVIM (SAIM)
Athens, Georgia

Introduction

Over the years more of us involved with clinical teaching on the hospital floor have become frustrated that the students cannot think through clinical problems. We complain about them and worry about what kind of veterinarians they will become. We worry that we will not have enough time to teach them clinical problem solving skills. It has led many of us to re-examine our pre-clinical teaching methodology and ask the question as to whether our current instruction methods are designed to produce the clinicians that we want. Most veterinary faculty have no formal training in pedagogy (the science and art of education) and yet we have the enormous responsibility of training future veterinarians. In many institutions teaching is not valued in promotion and tenure decisions; therefore, there is no incentive for junior faculty to develop this area of expertise.

In addition to reflecting on how we are training our students to be clinicians, there is also a rising expectation from the students themselves. They are more demanding of their teachers and seem to require that all educational material be handed to them. Asking them to "look up" information or to read an article or textbook chapter before class seems to be an enormous burden. They are increasingly more difficult to engage in the classroom as teachers unsuccessfully compete with a constant stream of online media. Older teaching methods of information-rich lectures conflict with methods that today's students use to get data in their lives. They are used to fast pace texting, twitter, and facebook.

Engaging Students

We, as teachers, want to engage students. We want to fill them with all the passion of the subject that we love, or hopefully love since we are teaching it. We want the educational experiences to be transformative. Many teachers realize that learning is dynamic, and it consists of relating new information to what is already known. To accomplish this, students must be engaged. To really understand an idea, the student must be able to carry out practices using that idea. [1] This is the cornerstone of active learning principles: having students perform active tasks based on an idea that you want them to learn. When combined, motivation and active learning result in student engagement. The challenge for us as teachers is to be able to provide these tools to our students. This can be especially challenging when we are faced with large classes of 100 or more. Lecture-style information presentation works well in such classes, but the result is that often students do not engage and thus do not really learn the material that we desire.

What we hope to accomplish in this presentation

This presentation is designed to be interactive and hopefully to share ideas about active learning strategies employed in the classroom. We hope to stimulate discussion with others in the audience who teach pre-clinical students. Some questions to explore are:

1. How to motivate large classes of 100 students
2. How to successfully compete for attention with the computer in the classroom
3. How to present education materials in a way that will be effective for all students
4. How to get students to think outside the classroom
5. How to make learning stick with students until they make it to the clinic floor
6. How to reinforce learning through the entire veterinary curriculum

Original pedagogical studies will be presented and opened for discussion. Active learning strategies will be discussed. Our hope is to learn from each other.

REFERENCES

[1]Edgerton, R Higher education white paper, Pew Charitable Trust, 1997; Barkley, EF, Student Engagement Techniques: A handbook for college faculty,2010; Davis, BG, Tools for teaching, second edition, 2009; McKeachie, WJ and Svinicki, M, McKeachie's teaching tips: strategies, research, and theory for college and university teachers, 2006, Bloom, B.S.E et al, Taxonomy of education objectives: Handbook I: cognitive domain, 1956; Fink, LD, Creating significant learning experiences: An integrated approach to designing college courses, 2003.

TEACHING STRATEGIES FOR THE CLINIC FLOOR

Jo R. Smith, MA, VetMB, PhD, DACVIM (SAIM)
Athens, GA

Expert teachers are enthusiastic and enjoy teaching; have a genuine interest in students and foster positive and supportive relationships with learners; are able to ask questions skillfully; communicate clearly; and acknowledge their own inadequacies.[1] In addition to these personal qualities and behaviors, there are certain andragogic (adult teaching) methods and principles that can improve effective clinical teaching. The following is an explanatory summary of these teaching qualities and strategies to assist you in building learners' confidence and responsibility, performing mindful observations, providing specific feedback and eliciting reflection from learners.

DURING ORIENTATION

Be enthusiastic. The best teachers typically inspire their students. If you cannot be passionate about your chosen field, then why should it interest your students? Enthusiasm helps motivate students to learn and improve.[2] Be excited about teaching and learning.

Provide clear expectations and explain why they are established. Strive to set the learner up for success. Clarifying expectations helps orient the learners to their responsibilities. Most of us are more likely to follow instructions if we understand the logic behind them.[3] The UGA SAIM orientation materials include a syllabus and grading rubric and are made available at the start of the senior students' final clinical year. We also prime the learners with questions that help them become more effective and efficient in patient encounters; remind them of the formats and provide them with examples of discharge instructions and SOAPs; and give them advice on writing ICU orders.

Generate a team ethos. Emphasize collaboration and cooperation during orientation, these result in more comprehensive patient care and efficient completion of work. Underline that collegial interaction is expected among *all* healthcare providers, students, staff, house officers and faculty alike. Delegate responsibility to the students to organize equitable case distribution, treatment and emergency or after-hours shifts. This provides them an opportunity to practice the team work, negotiation and communication skills necessary for a success as veterinary professionals.[3]

Establish a relationship with each learner and foster a comfortable learning environment. Treat your colleagues with respect, this helps establish trust and a positive learning environment. Learners who are treated as equals will feel more comfortable asking questions and engaging their instructors.[4]

Recognize diverse learning styles and have alternate teaching/learning methods available. Being attentive to individual learning styles helps ensure student success. Adult learners prefer the use of principles, concepts and information that can be readily applied soon after learning in a problem-solving context.[5] Application of material to real-life experiences shows the relevance of the subject matter. Patient-centered teaching contextualizes information which enhances transfer of knowledge to new settings and is more powerful than abstract knowledge imparted in the pre-clinical years. This is the basis of the experiential learning that the seniors' clinical year provides.

Set goals. Evaluate learners' existing knowledge and skills to target their specific needs. Listen to them think through cases, use probing questions, and observe non-verbal cues. Adult learners want to be involved in the planning and evaluation of their instruction.[5] Ask the learners if they have specific learning goals for this rotation. Goals should be specific, measurable, attainable, relevant and achievable within a definite time period. A self-assessment with a "Day One" skills checklist may help students guide their learning.[6]

IN THE HOSPITAL

Consider a two minute exam room observation of each student.[7] Make sure that the learner knows the observation is for teaching purposes and clarify the process. The learner should explain the process to a client so that you may enter and leave the room without disrupting the encounter. Whenever in the room, you need to be outside client's visual field and refrain from participating in any manner. Make time for discussion afterwards to focus on what was done well, what can be improved and what type of independent study might be useful. (See – Provide excellent feedback.)

Assess the learner. Evaluate the learner's existing knowledge and skills to target the learner's specific needs. Listen to him or her think through cases, use probing questions, and observe non-verbal cues.

Ask better questions (Socratic instruction). Using questions as a teaching tool is one of the most important skills of a clinical teacher.[8] Mastery of questioning assumes an ability to listen, read facial expressions and adjust questions to the learner's level of understanding and thus requires exemplary communication skills. Start with clear, brief, general, open-ended questions that have more than one acceptable answer, this helps to establish rapport.[9] An alternate suggestion is to start with relatively simple, recall, close-ended that start with "What", e.g. "*What is the differential diagnosis for a dog with back pain and fever?*" Then progress to higher-order questions that require interpretation, application, synthesis or judgment with existing knowledge. These are often open-ended and start with "How" Why" or "What if", e.g. "*How would you evaluate this* *patient's back pain if she had a fever?"* These questions enable you to better assess the learner's knowledge and clinical-decision making capacity. The exceptions to this rule are, "*What do you think we should do next?"* and asking learners to consider *"What is not there, and what might we learn from that?"* when considering differential diagnoses.

The sequence of questions is important; do not erratically switch back and forth since this may confuse the learner. Follow a general question with a probing question, one that solicits clarification and supporting evidence. Do not ask *"What am I thinking?"* questions and avoid asking questions designed merely to display your own intelligence. Do not ask double-barreled questions or fail to give feedback ("You're on the right track"), and remember to allow sufficient pauses between questions.[10]

To avoid these pitfalls, try this method: (1) when you are inclined to speak, stop and craft the statement into a question, (2) when posing a question, stop and reflect, word your question clearly and consider phrasing as a higher order question - craft questions that require synthesis and application of information, (3) ask your question and wait silently for several seconds after asking the question, (4) study the learner's response and body language to create the next question.[4]

Use a similar approach to test the effectiveness of instruction. Start with a question that asks for recall of basic facts. Progress to direct application, ask the learner to apply the information in the context of a patient. Then ask the learner to apply hypothetical information to the case to test their use of the principles taught.[11]

Use teachable moments. Clinical reasoning is determined by organization of knowledge in the memory.[12] As the "expert" you should prioritize information and create a framework for learners. Identify good teaching cases - these may be relatively simple, undiagnosed cases, or seen early in their diagnostic evaluation. Extend this content to maximize teaching. Unusual, complex or advanced cases can also be used; it is not the case content that is important but how it is used for teaching. Teach general rules – ask yourself, *"What is the most important teaching point this student needs to know about this case right now?"*

Think aloud (didactic instruction). Explain your thought processes aloud – during your physical examination, when considering results or making medical decisions. Make your clinical reasoning explicit. This is invaluable after critical patient encounters when there may not be time to explain your decision-making. Thinking aloud has three primary purposes.[13] (1) Communicating the general framework for solving the problem. (2) Demonstrating evidence-based medicine and evaluating how the evidence should be applied to this specific patient's case. (3) Purposefully exposing the ambiguity and ambivalence inherent to clinical medicine by thinking "on the fly" as patients' problems arise in real time. This technique can be extended to board rounds when discussing interesting cases or soliciting advice during transfer rounds.

During student rounds. Ask the struggling student the easiest questions so that he is not stressed; do not demand clinical performance expectations that are beyond the student's current knowledge and skill levels. Make it clear that the student presenting a case will not be asked any theoretical questions. It allows this student to concentration on their presentation and encourages the other students to participate and learn from all cases on the rotation.

Decide on the instructional content. Translate the learner's needs into what should be taught. Areas include; knowledge (pathophysiology, diagnosis, treatment); communication (how the patient is presented); procedural skills; attitudes (patient advocacy) and behaviors (efficient time management). Choose one topic and teach this in detail.[11]

Teaching procedures. During a high risk procedure you are more likely to demonstrate a procedure rather than allow a student to perform a technique to preserve patient safety. Remember to give the learner a specific assignment while she is observing and give her an understanding of what is expected in terms of her participation.[7] For lower risk procedures, avoid "see one, do one, teach one" which may overestimate the learner's knowledge and leaves little room for evaluation and feedback. You should model a skill and then create a similar opportunity for the student. There are four steps to the guided inquiry and cognitive apprenticeship: (1) demonstration – the teacher performs the procedure without narration at normal speed, (2) deconstruction – the teacher demonstrates individual steps with narration, (3) comprehension – the teacher performs the procedure while the learner describes the steps, (4) performance – the learner performs the procedure and narrates.[11]

Provide excellent feedback. You want to communicate using honest, positive, supporting and constructive feedback. It should reinforce the learner's strengths, identify specific needs for improvement, and result in effective change. This is an essential skill for effective teaching.[14,15] It is important to distinguish feedback from evaluation. Feedback is formative; it is an assessment *for* learning and its aim is to modify and improve learner attainment. Evaluation is summative; it is an assessment *of* learning and involves judgment.[14]

Feedback should be: frequent – even if it is only brief; timely – ideally within 48 hours of observing behaviors; empathetic and delivered in an appropriate private location; and always labeled as feedback– most learners understand the importance of feedback, but they may not always understand that your comments actually represent feedback. Feedback should be appropriate for learner's level of competence; ensure this by making it interactive. Ask the learner for his opinion on his performance, "Please provide one example of a skill you did effectively and a skill you would do differently" and what he believes could be improved. Eliciting this information cues you into the learner's level of self-awareness and may obviate some of your comments. Then you should provide balanced and behavior- specific comments. The key point is to make your remarks specific rather than generalized, "You did really well" does not allow the learner to identify what behavior should be repeated. It should be descriptive (what a video camera would observe) rather than evaluative. Consider "You did not allow your classmate to finish her case presentation in rounds" versus "You dominated in rounds". It is hard to take offence or get annoyed with the facts. Select no more than 3-5 points to convey and "sandwich" positive, negative, and then positive components. Remember to take the "but" out; any praise that precedes "but" is overshadowed by the perceived criticism. Consider "Your tone of voice was empathetic, but you spoke quickly and had your arms folded". Ideally the action plan to remedy these issues should be mutually determined.

Be an effective role model. Remember that you are always viewed as a role model. Exhibit intellectual curiosity and practice evidence-based medicine; demonstrate that lifelong learning applies to all veterinary professionals. As formal educators we also need to epitomize non-technical competencies. Show professional integrity at all times – be positive and productive. Have exemplary communication skills. Acknowledge your inadequacies and apologize, publicly, if warranted. Model reflection to increase self-knowledge and awareness. Prepare students to cope with work- and life-related stresses by role modeling self-care and a work-life balance. Remind students that professional responsibilities can be met by a team approach.

EXAMPLES OF TEACHING FRAMEWORKS

Teaching frameworks structure learning encounters, incorporate the skills of asking questions and providing feedback, and help to assess learners. Different models provide flexibility for adapting instructions to the needs of the learner; remember to communicate expectations about what will happen and why. These techniques can also be shared with house officers to improve their teaching of students.

Five steps "microskills" model of clinical teaching[16]

(1) *Get a commitment* about what the learner thinks is going on with the patient; ask her for her diagnosis and plan. This will increase her sense of responsibility and promote collaboration in problem solving. Your questions should be directed towards diagnosing the learner rather than the patient. Asking too many questions about the patient may expose your thinking process. No commitment from the learner indicates a lack of processing, reluctance to expose a weakness, or an inability to think independently. If the learner makes a mistake in problem formulation will provide a teaching opportunity.
(2) *Probe for supportive evidence.* Elicit the learner's underlying reasoning or alternative explanations for insight into his clinical reasoning. This will help identify what the learner does and does not know. You need to avoid appearing threatening so that the learner feels comfortable thinking aloud and exposing their weaknesses.
(3) *Teach a general principle.* The initial steps may reveal the learner's limitations, thus providing a targeted opportunity to teach general rules. Alternatively, you may direct the learner towards resources to further investigate the patient's condition.
(4) *Provide positive feedback.* Reinforce what was done correctly; remember to be specific.
(5) *Correct mistakes*; remember to ask the learner for her opinion first.

SNAPPS model[17]

This model is more appropriate for experienced learners. It encourages them to do most of the work in justifying their thinking and exploring and what they do not understand.
(1) *S*ummarize the history and physical exam findings briefly. This encourages concise and accurate communication.
(2) *N*arrow down the differential to two or three possibilities.
(3) *A*nalyze the differentials by comparing and contrasting the possibilities –what is missing?
(4) *P*robe the clinical teacher by asking questions about ambiguities, challenges or alternative approaches.
(5) *P*lan the patient management.
(6) *S*elect a care-related problem for self-directed learning.

TROUBLESHOOTING TIPS[11,18]

Be flexible and anticipate a variable caseload. Encourage students to structure and use "down time" well and increase client contact and patient follow-up.

Students used as a labor force for extensive nursing care. Increase the number and types of procedures performed by students. Encourage attention to basic teaching skills in all personnel, including the ability to give feedback. Hospital and diagnostic service staff can enhance the learning environment across the hospital. The teaching of core competencies should involve more than just clinicians.

Reliance on consultations or case transfers because of specialization. This may stop a student from seeing a whole case through to completion. Emphasize the opportunity to practice communicating information accurately and concisely to a different service. Exposure to advanced techniques may educate the nascent practitioner to what is on offer for future referral cases.

Combined student and house officer rounds. Respect the medical hierarchy – never ask a student to respond to a question that a house officer has failed to answer. This will promote one-upmanship.

Students being unaware of changes in their patient's case management. This may occur if you have separate house officer rounds. Remedies include "running the board" with the entire service at the end of the day to bring the students up to speed, or having joint cage side rounds in the mornings. Promote an atmosphere of collaborative learning that encourages the student to ask the house officers why these changes occurred. Challenge the house officers to practice their teaching skills when discussing these changes with the students.

The struggling student. Identify the source of the problem: insufficient knowledge, poor clinical judgment; inefficient use of time; poor communication; unethical/professional behavior. Provide high quality feedback.

The "difficult" student - disinterested, poor work ethic or unresponsive to encouragement or feedback. Relate the topic at hand to a situation relevant to the learner. Attempt to change attitudes and perceptions with enthusiasm about the patient and potential interesting issues, emphasize elements of differential diagnoses that are diagnostically challenging. Avoid complaining about any patient encounter that may be perceived as less than interesting. Discuss permutations of the patient presentation that would make the management more difficult or complex. Use the "carrot and stick" approach – elicit from the learner what interests her the most. Explain that there will be multiple tasks during the rotation and that interesting and educational aspects of each patient will be highlighted. Commit to offering opportunities for procedures or type of patients who interest the learner the most.

The advanced student - have a more sophisticated knowledge base. Explain that you have raised your expectations of his performance. Bring new literature and appropriate references into the discussion. Ask him to model and consider peer-teaching of core competencies.

REFERENCES

[1]Sutkin G, et al. **Acad Med** 2008;83(5):452. [2]Williams G, et al. **J Gen Intern Med** 1994;9(6):327 [3]Cornell K. **J Vet Med Educ** 2008;35(1):11. [4]Beckmann T, et al. **Mayo Clin Proc** 2009;84(4):339. [5]Knowles M. **Adult Leadership** 1968;16;350. [6]Schull D, et al. . **J Vet Med Educ** 2011;38(3):251. [7]Wilkerson L, et al. **Acad Med** 1998;73(suppl 10):S67. [8]Irby D. **BMJ** 2008;336:384. [9]Penciner R. **CJEM** 2004;4:286. [10]Rowe M. **JTE** 1986;37(1):43. [11]Houghland J, et al. **Ann Emerg Med** 2010;55(5):434. [12]BordageG, et al. **Med Educ** 1987;18(6):406. [13]Reilly B. **Lancet** 2007;370:705. [14]Ende J. **JAMA** 1983;250(6):777. [15]Hewson M, et al. **J Gen Intern Med** 1998;13:111. [16]Neher J, et al. **J Am Board Fam Pract** 1992;5:419. [17]Wolpaw T, et al. **Acad Med** 2003;78:893. [18]Lane I, et al. **J Vet Med Educ** 2008;35(3):397.

BEYOND THE LECTURE IN VETERINARY EDUCATION: CREATING LIFELONG LEARNERS

Kate E. Creevy, DVM, MS, DACVIM (SAIM)
Athens, GA

Successful teaching of professional students requires far more than transfer of content knowledge. In our fast-paced society, where the rate of expansion of medical knowledge continually increases, it is incumbent upon teachers to inspire students to undertake a lifelong commitment to continual expansion of their own knowledge so as to keep pace with the available data. Beyond inspiring them to do so, we must teach them how to do so. Students must become effective and efficient discoverers, critics, interpreters and users of information and must actively pursue those behaviors throughout their careers.

In this interactive presentation, experiences from the author's institution will be used as examples of teaching strategies for literature search and critique, assessing information sources, resolving conflicting findings, decision-making, service-learning, and reflective practice. The author's examples are merely intended to initiate discussion among session attendants so that we all may benefit from the collective imagination and expertise of the room.

Example 1: SAMS 5165 "Critical Reading and Lifelong Learning"

This didactic course was developed by collaboration with student volunteers, and introduced into the curriculum of the University of Georgia College of Veterinary Medicine in 2010. Its goals are stated in the syllabus as follows.

Upon completion of this course, students shall have obtained the following.

- *Introduction to the concept of evidence-based medicine for clinical decision-making.*
- *Development of higher level cognitive skills which facilitate application of critical analysis.*
- *Enhanced understanding of learning skills for DVM degree pursuit and lifelong learning.*

Coursework includes discussion of how to use textbooks, social media, public internet searches and library search engines in the practice of veterinary medicine. Students complete solo and group projects on selection of pertinent CE meetings, identification of primary literature for specific clinical questions, and written journal article critique. Small group journal club meetings are held weekly. The instructors share their philosophy with students as follows:

> *A graduate degree involves guided learning, but also personal responsibility. You must plan, act, assess and reflect. If you identify learning goals or opportunities then you have the obligation to pursue and achieve them. We aim to inform and strengthen your ability for lifelong learning by conveying a basic understanding of effective educational strategies targeted to individual learning styles.*

Example 2: Teaching critical Decision Points in case management

A case-based e-learning module was developed by collaboration among colleagues in the Colleges of Education and Veterinary Medicine at the University of Georgia. It is designed not to emphasize correctly "solving" the case, but rather to emphasize the moments in each case where critical decisions are made (the Decision Points).[1] In the funding proposal which sponsored development of this module, the collaborators stated:

> *In order to build real-world problem solving abilities or dynamic decision-making skills that can be transferable to another situation, students must integrate three types of knowledge: situational knowledge (knowing contextual information; how to identify, interpret, and evaluate situational cues, etc.), strategic knowledge (knowing what and how to apply/modify rules and theories; monitoring problem solving process and results; how to reason, etc.), and content knowledge (knowing general facts, concepts, principles, theories, etc.) around core principles or fundamental themes of the particular field.*

As such, the module is structured around five Decision Points. At each of these, the students are asked to commit to a course of action for the patient. After this selection, they are given access to expert interviews in which the instructors outline their thought processes surrounding the decision, as well as carefully-selected, brief, didactic content pertinent to the decision at hand. Students are then permitted to revise their intended courses of action, and provide explanations of how they made their decisions. In this way, the focus moves away from content knowledge (in the above framework) to the learnable skill of decision-making within a complex problem for which several valid solutions exist.

Example 3 - Service-learning through Vets for Pets and People

A service-learning program was developed by collaboration with the UGA Veterinary Teaching Hospital and the local domestic violence shelter, Project Safe. Similar community partnership programs exist at other veterinary schools and colleges in North America.[2] The program is currently self-sustaining by donation funding; its goals as stated in the grant proposal which supported its initial development are as follows:

> *This program will partner the expertise of the Project Safe staff with the expertise and resources of the UGA CVM to accomplish complementary goals. To provide safe haven for the pets of victims of domestic violence, Project Safe staff will identify their clients who also need shelter for companion animals. The students and faculty at UGA CVM will provide shelter and wellness care for the animals, initially within the veterinary college and subsequently by foster care by veterinary students. Identities of the pet owners will be confidential.*
>
> *To inform veterinary students of the critical link between animal abuse and domestic violence, Project Safe staff will provide education including macro theories of domestic violence, forms of abuse, lethality factors, and community resources. Some students who complete specialized training may also volunteer more extensively with Project Safe, including answering calls to their hotline. Following this education, veterinary students will create informational brochures and presentations for practicing veterinarians in order to increase awareness of the link between animal abuse and domestic violence.*

By participation in this program, veterinary students take responsibility for educating themselves about an ongoing societal problem. Importantly, development of educational materials for clients and outreach materials for veterinary practitioners model the application of content knowledge about veterinary medicine and animal care to a community need. In this way, we hope to inspire students to value the flexibility of their expertise and to experience the satisfaction of applying their skills to novel circumstances.

Example 4 - Reflective practice in a fundamental skills course

Deliberate reflective practice was introduced into the fundamental surgical skills course at the University of Georgia, and described by the author and colleagues.[3] The use of reflection in education allows the learner to clarify thoughts and questions regarding the material, as well as facilitates focus on his development as an effective independent learner. Using the electronic classroom management software, students were required to respond to 3-5 questions within 24 hours following completion of each laboratory. Rather than merely asking students to review or describe the techniques performed in each laboratory, reflection questions required students to project their experiences from the laboratory to future application of the techniques and principles. Examples of reflection questions included

> *Why is a surgical scrub performed?*
>
> *What was the most difficult component of scrubbing, gowning and gloving?*
>
> *What questions do you have about what was done in lab today? What was unclear or confusing?*
>
> *Identify two take-home facts from the lab this week, things you didn't understand until you were in the laboratory.*
>
> *What do you want the surgeon who might one day perform surgery on you to be thinking about as she does a resection and anastomosis?*

Rather than emphasizing students' answers to each question, the instructors' goals were to encourage students to develop the habit of reflection, which could be pursued throughout their careers. While the value of this first step in the process of developing reflective practice questions is difficult to assess, 85% of students felt that answering the questions had benefitted their learning. Additionally, students' final grades for the surgical skills course positively correlated with the length of their responses to the reflection questions.

REFERENCES

1) Choi I, Lee K. A case-based learning environment design for real-world classroom management problem solving. *TechTrends* 2008; 52: 26-31. 2) Creevy KE, Shaver SL, Cornell KK. Domestic violence shelter partnerships and veterinary student attitudes at North American veterinary schools and colleges. *Accepted by J Vet Med Educ, pending publication.* 3) Cornell K, Creevy K, Parks A. The use of reflection in a surgical training course: student attitudes and performance outcomes. *Poster.* American Association of Veterinary Medical Colleges Conference, Alexandria, VA, Mar 9, 2012.

TOP NEPHROLOGY/UROLOGY PAPERS FROM 2012

Gregory F. Grauer, DVM, MS, DACVIM (Small Animal Internal Medicine)
Manhattan, KS

In the fall of 2012, an announcement was sent out to all members of the North American Society of Veterinary Nephrology/Urology, soliciting published manuscripts for presentation at the at the Pre-Forum track on teaching and research at the 2013 ACVIM Forum. Nominated papers were reviewed by the NASVN/U Advisory Board and the following three papers were selected.

Clinicopathological Variables Predicting Progression of Azotemia in Cats with Chronic Kidney Disease
S. Chakrabarti, H.M. Syme, and J. Elliott (J Vet Intern Med 2012;26:275–281) From the Department of Veterinary Basic Sciences, Royal Veterinary College, London, UK (Chakrabarti, Elliott); and the Department of Veterinary Clinical Sciences, Royal Veterinary College, Hatfield, UK (Syme).

Background: Chronic kidney disease (CKD) is common in geriatric cats, but often appears to be stable for long periods of time.
Objectives: To describe CKD progression and identify risk factors for progression in newly diagnosed azotemic cats.
Animals: A total of 213 cats with CKD (plasma creatinine concentration > 2 mg/dL, urine specific gravity < 1.035) were followed up until progression occurred or for at least 1 year; 132, 73, and 8 cats were in International Renal Interest Society (IRIS) stages 2, 3, and 4, respectively.
Methods: Progression was defined as a 25% increase in plasma creatinine concentration. Logistic regression was used to assess variables at diagnosis that were associated with progression within 1 year. Changes in IRIS stage during follow-up also were described. Cases that remained in stages 2 or 3, but did not have renal function assessed in the last 60 days of life, were excluded from analysis of the proportion reaching stage 4.
Results: Of the cats, 47% (101) progressed within 1 year of diagnosis. High plasma phosphate concentration and high urine protein-to-creatinine ratio (UPC) predicted progression in all cats. Low PCV and high UPC independently predicted progression in stage 2 cats, whereas higher plasma phosphate concentration predicted progression in stage 3 cats; 19% (18/94) of cats diagnosed in stage 2; and 63% (34/54) of cats diagnosed in stage 3 reached stage 4 before they died.
Conclusions: Proteinuria, anemia, and hyperphosphatemia may reflect more progressive kidney disease. Alternatively, they may be markers for mechanisms of progression such as tubular protein overload, hypoxia, and nephrocalcinosis.

Evaluation of the Efficacy and Safety of High Dose Short Duration Enrofloxacin Treatment Regimen for Uncomplicated Urinary Tract Infections in Dogs (J Vet Intern Med 2012;26:506–512) J.L. Westropp, J.E. Sykes, S. Irom, J.B. Daniels, A. Smith, D. Keil, T. Settje, Y. Wang, and D.J. Chew From the Department of Medicine and Epidemiology, University of California, Davis, CA (Westropp, Sykes); the Department of Veterinary Clinical Sciences, The Ohio State University, College of Veterinary Medicine, Columbus, OH (Irom, Daniels, Chew); and Bayer HealthCare Animal Health, Shawnee Mission, KS (Smith, Keil, Settje, Wang).

Background: Uncomplicated urinary tract infections (UTI) in dogs usually are treated with antimicrobial drugs for 10–14 days. Shorter duration antimicrobial regimens have been evaluated in human patients.
Hypothesis: A high dose short duration (HDSD) enrofloxacin protocol administered to dogs with uncomplicated UTI will not be inferior to a 14-day treatment regimen with amoxicillin-clavulanic acid.
Animals: Client-owned adult, otherwise healthy dogs with aerobic bacterial urine culture yielding ≥103 CFU/mL of bacteria after cystocentesis.
Methods: Prospective, multicenter, controlled, randomized blinded clinical trial. Enrolled dogs were randomized to group 1 (enrofloxacin 18–20 mg/kg PO q24h for 3 days) or group 2 (amoxicillin-clavulanic acid 13.75–25 mg/kg PO q12h for 14 days). Urine cultures were obtained at days 0, 10, and 21. Microbiologic and clinical cure rates were evaluated 7 days after antimicrobial treatment was discontinued. Lower urinary tract signs and adverse events also were recorded.
Results: There were 35 dogs in group 1 and 33 in group 2. The microbiologic cure rate was 77.1 and 81.2% for groups 1 and 2, respectively. The clinical cure rate was 88.6 and 87.9% for groups 1 and 2, respectively. Cure rates between groups did not differ according to the selected margin of noninferiority.
Conclusions and Clinical Importance: HDSD enrofloxacin treatment was not inferior to a conventional amoxicillinclavulanic acid protocol for the treatment of uncomplicated bacterial UTI in dogs. Further research is warranted to determine if this protocol will positively impact owner compliance and decrease the emergence of antimicrobial resistance.

Uropathogenic E. coli Promote a Paracellular Urothelial Barrier Defect Characterized by Altered Tight Junction Integrity, Epithelial Cell Sloughing and Cytokine Release (J. Comp. Path. 2012, Vol. 147, 11-19)**M. W. Wood*, E. B. Breitschwerdt*, S. K. Nordone†, K. E. Linder‡ and J. L. Gookin*** *Department of Clinical Sciences , †Department of Molecular Biomedical Sciences and ‡Department of Population Health and Pathobiology, College of Veterinary Medicine, North Carolina State University, Raleigh, NC 27607, USA

Summary

The urinary bladder is a common site of bacterial infection with a majority of cases attributed to uropathogenic Escherichia coli. Sequelae of urinary tract infections (UTIs) include the loss of urothelial barrier function and subsequent clinical morbidity secondary to the permeation of urine potassium, urea and ammonia into the subepithelium. To date there has been limited research describing the mechanism by which this urothelial permeability defect develops. The present study models acute uropathogenic E. coli infection in vitro using intact canine bladder mucosa mounted in Ussing chambers to determine whether infection induces primarily a transcellular or paracellular permeability defect. The Ussing chamber sustains tissue viability while physically separating submucosal and lumen influences, so this model is ideal for quantitative measurement of transepithelial electrical resistance (TER) to assess alterations of urothelial barrier function. Using this model, changes in both tissue ultrastructure and TER indicated that uropathogenic E. coli infection promotes a paracellular permeability defect associated with the failure of umbrella cell tight junction formation and umbrella cell sloughing. In addition, bacterial interaction with the urothelium promoted secretion of cytokines from the urinary bladder with bioactivity capable of modulating epithelial barrier function including tumour necrosis factor-a, interleukin (IL)-6 and IL-15. IL-15 secretion by the infected bladder mucosa is a novel finding and, because IL-15 plays key roles in reconstitution of tight junction function in damaged intestine, this study points to a potential role for IL-15 in UTI-induced urothelial injury.

2012 HIGHLIGHTS IN RESPIRATORY MEDICINE

Carol Reinero, DVM, DACVIM, PhD
Columbia, MO

OVERVIEW

To review the year 2012 in terms of advances made in the area of canine and feline pulmonology, this lecture will touch on relevant manuscripts in peer-reviewed journals, findings from the 30th annual Veterinary Comparative Respiratory Society Symposium and anecdotal progress in respiratory diagnostics and therapeutics at the author's institution.

ADVANCES IN DIAGNOSTICS

Imaging (computed tomography and positron emission tomography)--Computed tomography (CT) has made a considerable impact in the diagnosis and characterization of lung diseases in dogs and cats and there were several studies published in 2012 that reflect ongoing work in this area of imaging. Multi-slice CT images can markedly reduce scan times and allow for high resolution multiplanar reconstruction images. At a few institutions, CT images are being merged with positron emission tomography (PET) scans which are particularly helpful in documenting whole body detection of metastatic lesions. In dogs with cancer, ^{18}FDG-PET/CT can also be used to monitor response to therapy.[1]

Adding to the existing body of literature this past year were two studies confirming the higher sensitivity of CT compared with radiography in detecting pulmonary nodules with improved inter-observer accuracy and agreement. [2,3] In terms of airway advances, CT and radiography were compared for measurement of canine tracheal diameters with the objective of determining if the measurements from each modality would lead to the same or different tracheal stent size selection.[4] While this was a cadaveric study, results showed that tracheal measurements using CT usually led to a larger stent size selected versus radiography, suggesting that prospective studies need to be performed in dogs with tracheal collapse to compare these two imaging techniques to for optimal stent selection. Relevant to feline airway disorders, pulmonary CT in healthy cats was used to establish a normal bronchoarterial ratio (BA ratio), to be used in future studies to document feline bronchiectasis. [5] Also in cats, CT has been performed in tandem with assessment of pulmonary mechanics to simultaneously characterize airway remodeling and airway hyperresponsiveness in an experimental feline asthma model. [6] This opens the door for future studies assessing lung structure and function at the same time.

Exhaled breath analysis—The search for minimally invasive diagnostics that provide information on the health or disease state of the lung has led to investigations of volatile gases in exhaled breath (EB) and non-volatile molecules in exhaled breath condensate (EBC). A review of the application of analysis of exhaled breath in veterinary species was published this year. [7] The gaseous phase of EB can be analyzed for nitric oxide (an important biomarker in inflammatory airway diseases in humans), and carbon monoxide and hydrocarbon gases (markers of oxidative stress), among others. A collection method and analysis protocol for volatile biomarkers from canine breath using gas chromatography-mass spectrometry (GC-MS) has been recently published. [8] Of interest, there have been multiple studies in which "sniffer" dogs can detect human cancer on the basis of odor (volatile organic compounds). A comparison of detection of VOCs from humans with diagnosed cancer using GC-MS versus sniffer dogs showed a positive correlation between two VOCs identified on GC-MS with a positive trained dog response. [9]

Exhaled breath condensate reflects the non-volatile substances in the epithelial lining fluid which are aerosolized and cooled to form a liquid or frozen condensate. A variety of important molecules can be analyzed including lipid mediators, cytokines, electrolytes and many others. Even the pH of EBC may reflect lung disease. For example, feline EBC pH is more acidic in experimentally asthmatic compared with healthy airways[10], a finding which has been mirrored in human asthmatics. Metabolomics, the study of chemical fingerprints resulting from unique cellular processes, provides information of the physiology of the cell and when applied to EBC may uncover biomarkers of disease states. Nuclear magnetic resonance spectroscopy has been used for metabolomic analysis of feline EBC allowing discrimination of healthy and asthmatic cats; furthermore, successful resolution of airway inflammation with steroids led to similar biochemical fingerprints as healthy cats. The advantages of using EB and EBC for diagnostic purposes (non-invasive technique, serial evaluation possible, simple to acquire, etc) must be tempered with limitations (no standardized methodologies for collection, very dilute samples sometimes requiring sophisticated assays for analysis, variable dilution, etc.). This is surely an exciting area of research in the years to come.

ADVANCES IN THERAPEUTICS

There have been several studies this past year investigating novel applications for existing therapeutics or novel therapeutics for feline asthma. Neither short term administration of salmeterol (a long acting beta-2 agonist) nor doxycyline (given as an inhibitor of matrix metalloproteinases) dampened.airway eosinophilia or airflow obstruction in an experimental model of feline asthma. [11] The application of mesenchymal stems cells in experimental feline asthma has gained interest and is the subject of another ACVIM lecture this year. While the stem cells do not regenerate lung tissue to cure asthma, they may hold promise to blunt airway hyperresponsiveness and airway remodeling. Inhibitors of receptor and non-receptor tyrosine kinases, by blocking signaling pathways important in asthma pathogenesis, may also emerge as novel treatments. For example, masitinib reduced airway eosinophilia and improved respiratory compliance in an experimental feline asthma model. [12] However, clinically asymptomatic proteinuria (often severe) was noted in all treated cats, mandating further trials at a lower dose to determine if efficacy would be comparable. A novel non-receptor tyrosine kinase inhibitor also showed promise in experimental feline asthma with no noted toxicity. While stem cells and tyrosine kinase inhibitors may modulate asthmatic pathology, the only treatment that shows promise for a cure is allergen-specific immunotherapy (ASIT). Identification of allergens causative for allergic symptoms can be problematic with limitations to testing modalities, intermittent exposures to allergens, and concurrent therapy. To answer the question of what would happen if the incorrect allergens or if only a partial repertoire of sensitizing allergens were selected for ASIT, a study was undertaken using an experimental asthma model. [13] While both closely matched allergens and unrelated allergens to the sensitizing allergen(s) were capable of blunting airway eosinophilia, the immune mechanism responsible for this change differed (ie induction of tolerance perhaps being curative versus an active immune response requiring indefinite allergy shots). As an important follow up study, simultaneous administration of oral glucocorticoids (GC) but not inhaled GC diminished efficacy of ASIT in this model; therefore only the latter are recommended as concurrent therapy. [14]

Xylitol, a sugar alcohol that may enhance innate immune responses to promote antibacterial effects in the airways, has received attention as a possible treatment for humans with chronic bacterial airway infections (rhinosinusitis, cystic fibrosis). [15,16] To determine the safety of nebulized xylitol as a future route of administration for humans with cystic fibrosis, research beagle dogs were treated with varying concentrations for 14 days. The study concluded that xylitol was safe for clinical administration. These results may stimulate interest in future studies in dogs with chronic airway infections.

LINK BETWEEN GATROESPHAGEAL REFLUX DISEASE AND CANINE/FELINE AIRWAY DISORDERS

In humans, a strong link exists between GERD and a variety of respiratory disorders including chronic non-specific cough, reflux-related asthma, and laryngitis/laryngeal dysfunction. With one exception, there is a surprising paucity of information on the concomitant occurrence of GERD with other respiratory disorders in dogs or cats. In dogs, the major complication of GERD described in the literature is aspiration pneumonia, especially under anesthesia, with two manuscripts investigating preventative therapeutic trials in anesthetized dogs this year. In the first, neither a bolus nor continuous rate infusion of metoclopramide immediately after induction nor ranitidine iv 6 hours prior to induction diminished the incidence of GERD during anesthesia. [17] In the second study, pre-anesthetic administration of cisapride and esomeprazole decreased the total number of reflex events compared with placebo or esomeprazole alone.[18]

While GERD related anesthetic aspiration is well recognized in dogs, other respiratory complications are not. In humans, GERD is one of the top three causes of chronic cough[19,20] with GERD being clinically silent in up to 75% of patients.[21] This information is not available in dogs or cats perhaps in part because GERD may also be clinically silent and in part because we lack a good way of documenting GERD. In 2012, a report was published describing GERD inducing laryngeal dysfunction in a dog.[22] While only a single case report, the value of this publication in large part is to put the link between these two disorders on our radar and encourage future studies to try to determine how common it is and what the best treatment modalities would be. Even in humans, documentation can be tricky with laryngoscopy having poor specificity, interpretation of pH and intraluminal impedance monitoring challenging, and response to proton pump inhibitor therapy imperfect. [23] Future studies should also be performed to look for evidence of GERD as a contributing factor to chronic cough in which no primary respiratory cause can be documented.

REFERENCES

1. LeBlanc AK et al. *Vet Radiol Ultrasound* 2012;53;348-357.
2. Armbrust LJ et al. *JAVMA* 2012;240:1088-1094.
3. Alexander K, et al. *Vet Radiol Ultrasound* 2012;53:258-265.
4. Montgomery JE et al. *30th Symposium of the Veterinary Comparative Respiratory Society* 2012, Columbia MO. (abstract)
5. Reid LE, et al. *Vet Radiol Ultrasound* 2012;53:34-37.
6. Masseau I et al. *30th Symposium of the Veterinary Comparative Respiratory Society* 2012, Columbia MO. (abstract)
7. Cathcart MP et al. *The Vet J* 2012:191;282-291.
8. Dissanayake S et al. *J Breath Res* 2012; 6(4):041001. doi: 10.1088/1752-7155/6/4/041001.
9. Buszewski B et al. *Anal Bioanal Chem* 2012;404:141-6.
10. Puza N et al. *30th Symposium of the Veterinary Comparative Respiratory Society* 2012, Columbia MO. (abstract)
11. Leemans J et al. *The Vet J* 2012;192:49-56.
12. Lee-Fowler T et al. *Int Arch Allergy Immunol* 2012:158;369-374.
13. Reinero CR et al. *The Vet J* 2012;192:412-416.
14. Chang C et al. *The Vet J* 2013; in press.
15. Weissman JD et al. *Laryngoscope* 2011;121:2468-72.
16. Reed MD et al. *Inhal Toxicol* 2012;24:365-72.
17. Favarato ES, et al. *Res Vet Sci* 2011:93;466-467.

18. Zacuto AC, et al. *J Vet Intern Med* 2012:26;518-525.
19. Benich, JJ, et al. *Am Fam Physician* 2011;84;887-892.
20. Sylvester DC, et al. *Int J Otolaryngology* 2012;doi:10.1155/2012/564852.
21. Irwin RS, et al. *Chest* 1993;104:1511-1517.
22. Lux CN et al. *JAVMA* 2012;240:1100-1103.
23. Yuksel ES, et al. *Swiss Med Wkly* 2012;142:w13544.

ELECTROCARDIOGRAPHIC DIAGNOSIS: ARE WE TRYING HARD ENOUGH?

Gunther van Loon (DVM, PhD, DECEIM, Ass DECVDI)
Merelbeke, Ghent, Belgium

INTRODUCTION

Cardiac arrhythmias are commonly found in both normal and diseased horses and assessing the importance, especially in relation to performance, can be very challenging. Whenever arrhythmias are detected or suspected, an electrocardiogram (ECG) must be recorded as it is the gold standard technique for an accurate diagnosis. The clinician must be familiar with basic ECG principles, cardiac physiology and common causes of arrhythmogenesis to record and interpret the ECG. Besides ambulatory recordings, 24 hour ECG recordings are useful for rarely-occurring arrhythmias (e.g. in case of occasional syncope) or for quantification of the number of arrhythmias. Except in case of overt clinical signs (ventricular tachycardia, congestive heart failure) or obvious innocent arrhythmias (2nd degree atrioventricular block that disappears with mild stress) an exercising ECG is mandatory for assessment of the impact of an arrhythmia on performance. In selected cases, alternative ECG recordings (esophageal, intracardiac), may help to identify the origin or pathophysiology of the arrhythmia.

TECHNIQUE

Electrocardiography implies recording the cardiac electrical activity, usually from electrodes (self-adhesive rather than crocodile clamps) positioned at the skin surface. However, the surface ECG is a poor reflection of the complex 3D electrical activation patterns that occur in the heart during the cardiac cycle. In addition, movement artifacts may distort the signal and render it useless for interpretation.

Numerous studies have concentrated on optimal electrode positions and have attempted to obtain as much information from the equine ECG as from human or small animal ECGs. Einthoven's system of limb leads, adapted Einthoven's systems, semi-orthogonal systems, 12- and 11-lead recordings, have been developed for horses. However, the different Purkinje network in horses, extending throughout the myocardium, results in cancellation of many electromotive forces, reducing the information that can be obtained from the equine ECG. The range for the mean electrical axis is very wide and of limited clinical value. In addition, QRS duration is not necessarily related to ventricular size. All these factors limit the ECG of the horses largely to diagnosis of rate and rhythm. Because of the complexity and yet limited additional value of e.g. 11-lead systems[1], many clinicians use a 3 or 4 electrode system and record only one lead, a (modified) base-apex recording, which is often sufficient for accurate diagnosis of rate and rhythm. However, correct identification of an arrhythmia is not always straightforward and recording of additional leads (e.g 3 Einthoven leads) can be very useful in selected cases.

The position where the electrodes are placed depends upon equipment, the lead system used and may differ for exercising and ambulatory recordings. The purpose is to obtain a maximum amount of information and to limit artifacts. Self-adhesive electrodes provide good contact and clipping is generally not necessary. Electrodes placed along the main direction of depolarization result in the largest complex and will facilitate interpretation. The main vector of the ventricular activation is directed from ventral to dorsal and slightly from left to right while the atrial activation has a more cranial to caudal direction. This means that optimal electrode placement for atrial versus ventricular activity is different. Many clinicians use their own adapted electrode configuration for ECG recording, especially during exercise. A true base apex lead, can be recorded from lead I by placing the positive left arm electrode near the left ventricular apex and the negative right arm in the lower third of the right jugular furrow or over the right shoulder. This lead provides good quality recordings but shows important artifacts because body movement. Therefore, and especially during exercise, electrodes are often positioned more vertically with the right arm electrode placed on the left neck/shoulder or near the withers. These locations produce less artifacts and are easier for the electrodes to stick in place but the position is not optimal to record atrial activity. When using a 4 electrode system electrodes are often attached two by two to record 2 identical leads. This ensures that one lead is still recorded in case an electrode falls off during exercise. However, slightly separating both dorsally and both ventrally located electrodes allows to record slightly different leads[2,3]. As each of the leads 'looks' at the heart from a different position it can be very useful for identification of some arrhythmias (e.g. supraventricular versus ventricular).

Finally, esophageal electrocardiograms are able to better visualize atrial activity but cannot be used during exercise due to abundant artifacts[4]. Intracardiac electrograms are obtained from intracardiac catheter electrodes and can be useful in certain circumstances and for research purposes. At the same time such catheters allow to perform intracavitary pacing.

INTERPRETATION

Only good quality recordings should be evaluated in a methodical manner. However, artifacts will always be present and it is of utmost importance that the clinician is able to identify these. Heart rate can be automatically calculated by software but should be manually checked because T waves might be labeled as QRS complexes. Also the regularity of the cardiac rhythm can be assessed by software. It is important to set the threshold for RR variation rather low (around 10%) in order to spot small changes and manually track the ECG. Besides rate and rhythm, the correlation between P waves and QRS complexes, and the morphology and duration of the different complexes and intervals should be assessed. Only the depolarization or repolarization of relatively large muscle mass, such as the atrial and ventricular myocardium, produce sufficient electrical potential difference to result in a deflection on the surface ECG, while depolarization of a small group of cells, such as sinus node and atrioventricular (AV) node, remains 'silent' on the ECG. The normal cardiac rhythm starts in the sinus node (sinus rhythm) and runs subsequently through the atrial myocardium, the AV node, the Hiss and Purkinje system to the ventricles to produce the characteristic P-QRS-T waves on the ECG. Specific guidelines and flow-charts exist to define the origin and cause of an arrhythmia, based on the sequence, morphology and duration of P waves, QRS complexes and the different time intervals, and are available elsewhere[5]. However, the clinician must be aware that there are a number of pitfalls that complicate the analysis of which the most important ones will be listed below. Finally the difficulty is usually to identify artifacts and to define which waves or complexes are normal in morphology and duration and which are not.

SPECIFIC DIFFICULTIES AND PITFALLS

Artifacts

Artifact do not influence the underlying rhythm. Artifacts can have very different shapes. One should remember that conduction through the myocardium takes time and the conduction of a normal sinus beats result in the smallest QRS complex. Very sharp deflections, shorter in duration than a normal QRS complex are artifacts. In addition, artifacts don't have T waves. Typical artifacts such as those due to muscle twitching result in repetitive sharp deflections and should not be misinterpreted as atrial or ventricular tachyarrhythmias. Low-amplitude 50 or 60 Hz artifacts may occur due to interference or lack of grounding of equipment. During exercise typical movement artifacts occur and often seem more pronounced in obese animals.

Ventricular Premature Depolarizations (VPD)

Ventricular premature originate in the ventricular myocardium. Because of the different cell-to-cell conduction, the QRS complex has a different morphology and longer duration. Regarding QRS duration, one should be aware that the duration of the normal QRS shows a physiological inverse relation with HR. The VPD is easy to identify when morphology differs largely from sinus beats. However, VPDs originating near the His and Purkinje system may be very similar to the normal complexes and only show minimal changes in morphology and duration. Identifying a compensatory pause can therefore be very helpful. 'Compensatory pause' means that the duration 'normal QRS-VPD-normal QRS' is similar as twice the normal RR interval. However, on rare occasions a VPD may be interlaced: it occurs between two normal beats, not changing the underlying rhythm and not resulting in a compensatory pause. When a VPD occurs almost simultaneous with a normal beat it may produce a 'fusion beat' of which the morphology is a mixture of both normal QRS and VPD. The fusion beat can be easily overlooked as it occurs at almost the normal timing, does not interrupt the rhythm and is preceded by a P wave. Recording of additional leads can be very helpful in identifying VPDs, especially those that have a morphology only slightly different from the normal QRS complexes. As an additional lead 'looks' to the heart from a slightly different angle it might be able to better show a slightly different QRS morphology.

It has been shown during exercise in horses with atrial fibrillation that a high burden of QRS complexes with abnormal morphology can be present[6]. These can be broad isolated complexes, runs of broad QRS tachycardia or R-on-T phenomenon. Although it probably seems more likely that these abnormal complexes are supraventricular in origin and that the broadening occurs due to aberrant intraventricular conduction, a ventricular origin cannot be excluded.

Junctional Premature Depolarizations (JPD)

Junctional beats originate from the AV node or His bundle. As the ventricular depolarization follows the normal His-Purkinje pathways, QRS morphology is normal but the QRS is not associated with a preceding P wave. While in other species retrograde conduction of the junctional beat into the atria is often found resulting in an inverse P wave immediately before, during or after the QRS, this is not the case in horses.

Atrial Premature Depolarizations (APD)

The normal P wave at rest is notched or bifid but marked changes may occur due to changes in vagal tone and heart rate. The P wave of APDs can have a different morphology but often, and depending on the site of origin in the atria, the P wave of an APD may be hard or impossible to discern from the normal ones. APDs may also be buried in preceding T waves and especially during exercise it is no longer possible to identify P waves due to motion artifacts. In such circumstances it becomes increasingly important to look at QRS morphology and duration, and at the presence of a compensatory or non-compensatory pause. The non-compensatory pause that follows an APD means that the duration of 'normal QRS-APD-normal QRS' is less than twice the normal RR interval. This is due to resetting of the sinus node due to the APD. In addition, APDs may conduct slowly over the AV node or produce a longer pause making them difficult to differentiate drom sinus dysrhythmias. Regarding QRS morphology, it is important to realize that APDs may alter QRS morphology. When APDs have a coupling interval of less than 600 ms, QRS amplitude often increases[7]. An inverse proportional relation was found between coupling interval and S-wave amplitude on a base-apex ECG. Very shortly coupled APDs may not be conducted to the ventricles due to AV node or ventricular refractoriness. As the APD may be buried in preceding complexes, it is very important to look for pauses following the APD. Again, recording of different leads can be very useful to identify buried complexes as the P' wave may difficult to spot on one lead, while obvious on another lead.

Atrial Tachy-arrhythmias

Multiple successive APDs are called atrial tachycardia. A rapidly firing atrial ectopic focus may produce atrial tachycardia at high rates, which might be difficult to differentiate from atrial flutter. In human medicine differentiation is often made based upon the atrial rate whereby the higher atrial rates are due to atrial flutter. As flutter is a macro-reentry, and conduction velocity is probably similar as in humans, atrial flutter in horses has a slower rate compared to humans because of the larger atria. It has been suggested that atrial flutter in horses often varies between 160 and 250 per minute. However, especially for relatively slow atrial rates it is not possible to prove the difference between atrial tachycardia and atrial flutter. Horses with atrial flutter show P-like undulations on the ECG. When autonomic tone prevails, ventricular response is usually irregularly irregular. With stress or exercise, conduction towards the ventricles often becomes more regular because of a 3:1, 2:1 and 1:1 atrioventricular conduction[8]. During an exercise test, horses typically have a maximal heart rate that equals the atrial flutter rate. Atrial flutter or tachycardia should be differentiated from advanced 2nd degree or 3rd degree atrioventricular block where the atrial rate is much slower (usually 50 to 100 per minute). In atrial fibrillation (AF) the local atrial rate is usually between 350 and 450 per minute and ventricular response is irregularly irregular[9]. The arrhythmia is usually sustained, with the exception of paroxysmal AF in association with racing. Fibrillation waves on the ECG often change between coarse and fine f-waves in the same horse. During sudden stress or exercise many horses present QRS broadening which is caused by aberrant conduction or ventricular ectopy.

Sinus Arrhythmia

Sinus arrhythmia is often found during recovery from exercise when heart rate drops from about 150 bpm towards 50 bpm. In this period, there, is a shift from sympathetic towards a more vagal tone resulting temporarily in a cyclic increase and decrease in heart rate. The progressive and cyclic character usually makes it easy to identify this arrhythmia. The differentiation between sinus arrhythmia, sinus block and APCs may be more difficult on 24-hour ECG recordings when parasympathetic tone is high because PP intervals and P wave changes may be both physiological or pathological.

CONCLUSION

Interpretation of the ECG requires a thorough knowledge on basic ECG principles and cardiac pathophysiology. Identifying artifacts is mandatory. The clinician should be aware of the many exceptions to the general rules that make ECG interpretation challenging. In order to optimize diagnostic power of the ECG it is advisable to record more than 1 lead: often the artifact or atrial or ventricular ectopic rhythm can be better identified on certain leads than on other.

REFERENCES

1. Fregin GF. The equine electrocardiogram with standardized body and limb positions. Cornell Vet 1982;72:304-324. 2. Verheyen T, Decloedt A, De Clercq D, et al. Electrocardiography in horses - part 1: how to make a good recording. Vlaams Diergeneesk Tijdschr 2010;79:331-336. 3. van Loon G. Electrocardiography. In: Wilson D, ed. Clinical veterinary advisor: The Horse, 1st edW. B. Saunders; 2011:716-718. 4. Verheyen T, Decloedt A, De Clercq D, et al. Oesophageal electrocardiography in healthy horses. Equine Vet J 2012;44:640-645. 5. Verheyen T, Decloedt A, De Clercq D, et al. Electrocardiography in horses - part 2: how to read the equine ECG. Vlaams Diergeneesk Tijdschr 2010;79:337-344. 6. Verheyen T, Decloedt A, Van Der Vekens N, et al. Ventricular response during lunging exercise in horses with lone atrial fibrillation. Equine Vet J 2012:doi: 10.1111/j.2042-3306.2012.00653.x. [Epub ahead of print]. 7. Broux B, Decloedt A, Verheyen T, et al. Supraventricular premature beat associated QRS morphology varies with RR coupling interval: preliminary results. In: 6th Congress of the European College of Equine Internal Medicine, Le Touquet, France 2013. 8. Christmann U, van Loon G. ECG of the Month. Javma-Journal of the American Veterinary Medical Association 2013;242:165-167. 9. De Clercq D, van Loon G, Baert K, et al. Intravenous amiodarone treatment in horses with chronic atrial fibrillation Vet J 2006;172:129-134.

IS LACTATE MONITORING VALID FOR PERFORMANCE TESTING IN HORSES?

Modest Vengust, DVM, DVSc, PhD, DACVIM
Ljubljana, Slovenia
George Heigenhauser, BA, MSc, PhD
Hamilton, Ontario, Canada

Lactic acid (2-hydroxypropanoic acid) was discovered over 200 years ago, when it was associated with milk fermentation. In the early 20th century lactic acid became the focus of exercising skeletal muscle metabolism research. Oxygen insufficiency within the contracting skeletal muscle and lactic acidosis were associated with muscle fatigue[1,11]. This interpretation remains prominent in the veterinary literature today. Lactic acid has a pKa-value of 3.86; therefore, 99% of it is in the dissociated form as lactate and hydrogen ions (H^+)[15, 21].

During submaximal exercise the energy supply relies mainly on free fatty acids with a small contribution from carbohydrates (CHO). At moderate levels of exercise intensity (60% of peak oxygen consumption – peakVO_2) free fatty acids become the primary source for ATP provision, thereafter CHO become predominant.

During high intensity exercise horses require exceedingly high rates of ATP resynthesis since ATP turnover rate increases up to a 100 fold from resting value. At maximal exercise glycolysis becomes the primary source of ATP turnover[25]. In order to provide ATP at increased turnover rate degradation of phosphocreatine (PCr) and glycolysis rather than free fatty acids provide the preferred substrate. Very high rates of glycolytic ATP production results in high muscle and plasma lactate concentrations of ~43 and ~30 mmol/L, respectively[10, 21].

Plasma lactate concentration is used in equine athletes for assessment of physiological adaptations to exercise. Classical interpretation of lactic acidosis being responsible for muscle fatigue during exercise is not appropriate.

LACTATE METABOLISM

ATP turnover during exercise can be achieved by: 1) Substrate-level phosphorylation, which includes hydrolysis of PCr and glycolysis and 2) oxidative phosphorylation. The classical division of muscle energy metabolism on aerobic and anaerobic is still terminologically valid but conceptually flawed because both systems share a common central pathway and occur simultaneously[29,33]. Glycolysis and lactate production (anaerobic metabolism) during high intensity exercise generate ATP in the presence of muscle mitochondrial oxygenation levels that are well above the mitochondrial needs to support oxidative phosphorylation[23,30]. Appropriate biochemical terms for the aerobic and anaerobic metabolic pathways are oxidative and substrate-level phosphorylation, respectively[31].

Lactate production occurs during all intensity levels of exercise. When the resynthesis of ATP by oxidative phosphorylation cannot meet the demand for ATP utilization by working muscle both PCr breakdown and glycolysis increase. As exercise intensity increases the glycolysis exceeds the rate by which pyruvate can be oxidized within mitochondria by the rate limiting enzyme pyruvate dehydrogenase (PDH) reaction and pyruvate accumulates intramuscularly. Lactate is then produced by the conversion of pyruvate to lactate by the equilibrium lactate dehydrogenase (LDH) reaction[29,31]. Simultaneously, the LDH reaction provides nicotinamide adenine dinucleotide, which is essential for maintaining the substrate-level phosphorylation higher up in the glycolytic pathway. Glycolysis is a rapid mean to defend the concentration of ATP in the working muscle. It sustains glycolytic ATP regeneration at the rate that can be up to six times faster than oxidative phosphorylation alone; however, at a cost of a lower energy yield[31].

Lactate is not a waste product of contracting muscle metabolism as often portrayed in the veterinary literature. Contracting skeletal muscle both produces and utilizes lactate as fuel. Lactate is rapidly removed from contracting skeletal muscle and can be taken up by the liver for gluconeogenesis and highly oxidative tissues such us cardiac and skeletal muscle (Lactate cell-to-cell shuttle)[3,8,12,22,27]. Lactate is oxidized in the muscle with the absolute rate depending on the metabolic rate of exercising and resting muscles[16].

Lactate clearance from exercising skeletal muscles occurs primarily by a facilitated plasma membrane transporters. Some lactate moves across the sarcolemma via passive diffusion. Facilitated lactate transport is accomplished by a family of monocarboxylate transport proteins (MCTs), which is responsible for >70% of the lactate clearance in the exercising muscle. Monocarboxylate transport is bidirectional and is governed by the transmembrane lactate and H^+ gradients across the sarcolemma[15].

In horses, MCT1 and MCT4 have been identified in muscle and shown to facilitate lactate removal and uptake during exercise. MCT1 and MCT4 protein contents increase in equine gluteus medius muscle after 18 weeks of high-intensity training. The increase in the MCT1 protein content and oxidative capacity after high-intensity training were maintained for 6 weeks by moderate-intensity training, but the increase in the MCT4 protein content was not maintained after the detraining period[17]. Trained horses' exercising muscles have, therefore, increased ability to release and take up lactate.

ACIDOSIS OF EXERCISE

Acidosis of exercise is still very controversial. Three prevailing sources for interpretation of the skeletal muscular acidosis are available. Classical interpretation states that lactic acid is completely responsible for the intramuscular acidosis during exercise[1,35]. Two other interpretations suggest that lactic acid production is not exclusively responsible for acidosis during exercise but other variables also contribute to increased intramuscular $[H^+]$[9,20,29].

According to Robergs et al. (2004)[29] the reaction catalyzed by LDH produces lactate, not lactic acid. Based on stoichiometry the production of lactate in the LDH reaction consumes rather than produces H^+. This same stoichiometric approach is used in the other energy metabolic pathways to determine production of H^+. Using this approach this other metabolic energy pathways are the source of acidosis rather than lactate[29].

The third approach is based on the physicochemistry of aqueous solutions. Proponents of physicochemical approach (Heigenhauser, Jones, Lindinger) define total weak acid concentration ($[A_{tot}]$) and the strong ion difference ([SID] = [strong cations] – [strong anions]) as independent determinants for intramuscular H^+. According to the physicochemical approach the accumulation of lactate within the working skeletal muscle directly contributes to intracellular acidosis, by virtue of the fact that it is a strong acid anion that fundamentally alters the behavior of water[14,20]. It is dangerous to ignore the ionic state of the reactants because the stoichiometric interpretation of biochemical relationships yield only a shortsighted view of proton release (associated with energy metabolism in the contracting muscle)[14]. Employing physicochemical principles to understanding the acid base chemistry allows for comprehensive interpretation of intramuscular acidosis with respect to all independent determinants influencing intramuscular $[H^+]$[33].

LACTATE AND FATIGUE

Classically intramuscular acidosis due to lactic acid production has been interpreted as the cause for fatigue during high intensity exercise[1,34]. Although acidosis develops during exercise there is ample evidence that it is not causing fatigue[4]. Association between lactic acidosis and muscular fatigue was derived by correlation studies that failed to include several other factors that may influence the ability of muscle to contract (ie. decrease of ATP, increase of inorganic phosphate, increase of ADP, PCr depletion)[4]. In contrary to the common believe that lactic acidosis is detrimental, acidosis triggers several beneficial cardiovascular, respiratory and central nervous adaptations[35], which are necessary to sustain muscular exercise[2,18,21].

Beneficial effects of acidosis also occur on a cellular level. Experiments on mammalian skeletal muscle exposed to lactic acid, propionic and carbonic acids (from addition of high CO_2) showed that acidification counteracts the effects of elevated extramuscular potassium concentration, which is strongly associated with muscle fatigue [2,24].

Excitation–contraction coupling in the skeletal muscle is controlled by calcium ions (Ca^{2+}). The decline in muscle force output during muscular fatigue can be traced, in part, to a decreased release of Ca^{2+} from the muscle sarcoplasmic reticulum. Acidosis impairs Ca^{2+} release to an extent that should not be regarded to contribute substantially to impairment of the excitation–contraction coupling[6,26].

The detailed explanation of the role of acidosis in muscle function during exercise is available elsewhere [4,13,14,19,20,29].

LACTATE BASED MEASUREMENTS FOR EVALUATION OF PHYSIOLOGICAL ADAPTATIONS TO EXERCISE IN EQUINE ATHLETES

Blood lactate measurement for evaluation of the fitness of an equine athlete is widespread. Optimal determination of physiological adaptations to exercise would include 1) assessment of the cardiovascular/respiratory system (peak VO_2 determination), 2) evaluation of glycolytic capacity (blood sampling) and 3) muscle biopsy for histochemical evaluation (fiber typing, oxidative and glycolytic enzyme capacity, muscle fiber capillarization), 4) evaluation of metabolism during exercise (muscle glycogen utilization, lactate formation, muscle phosphocreatine utilization, muscle enzyme activity) and 5) genetic testing for athletic ability. Such optimal evaluation of fitness is rarely available outside the specialized laboratory setting.

Most lactate based evaluations are correlated to the percentage of the lactate threshold (LT) (also termed anaerobic threshold, ventilatory threshold, ventilatory anaerobic threshold, onset of blood lactate accumulation, and maximal lactate steady state). The LT is the point at which lactate begins to accumulate and exceeds the capacity for its clearance from the circulation. Higher LT values are seen in elite athletes[3,13,25]. Absence of methodological standardization for measurement of the percentage of LT in horses casts doubt upon the appropriateness of this measure to guide clinical/practical evaluation of fitness in horses. The LT is determined on very variable muscle and circulation lactate kinetics with pronounced individual differences; therefore, it is necessary that the interpretation of the LT is based on prospective and individualized evaluation with regards to the horse and the equestrian sport[5,32]. Differences are also present between fitness evaluations performed indoor/treadmill or on the track/field[5].

Several LT estimates had been proposed for horses that exercise at and/or near maximal power; they can be evaluated/plotted either against heart rate, work rate or VO_2 at intervals (steps) with increasing power output[5]:

1. VLac4 (Vla4, V4) = exercise velocity at which lactate concentration of 4mmol/L is mathematically derived,
2. HR4 = heart rate at the work load/velocity at which lactate concentration of 4mmol/L is mathematically derived,
3. VO2–Lac(4) = Oxygen uptake at specified blood lactate concentration (4 mmol/L).

VLac4 seems to be widely accepted measurement to assess the fitness level in horses[7].

CONCLUSION

Several aspects of muscle lactate metabolism determined almost 100 years are outdated but still present in the modern literature despite a very significant advance in understanding of the lactate muscle metabolism. Lactate production during exercise is:

1) not the product of anaerobic conditions in exercising muscles,
2) not a sole contributor to muscle fatigue, and
3) not a waste product, but a valuable energy substrate for the contracting muscle.

It is important to realize that adaptation to exercise includes more optimal production, elimination, reuptake, and oxidation of lactate in contracting skeletal muscles. It is also important to realize that lactate is not the sole factor contributing to intramuscular acidosis. It is possible that lactate production may express its significance at the finish line in competitions where milliseconds divide success from failure.

Is there any value in lactate based testing in horses? Determination of the horse fitness status is important in practice. VLac4 test is relatively easy to conduct and it seems to give good estimate of the LT. Other lactate based tests can also be valuable. What the veterinarian, owners/trainers do with the lactate based tests results is the key to success. Standardization (consensus needed) and prospective integration of lactate based tests into the individual training program are important for adequate assessment and interpretation of the equine athlete. Predetermined values and fixed formulas for determination and/or prediction of the individual equine athlete fitness should be avoided.

REFERENCES

[1]Brooks GA. **Med Sci Sports Exerc** 1991; 23(8):895, [2]Brooks GA. **J Physiol** 2001;536:1, [3]Brooks GA. **J Physiol** 2009;587:5591, [4]Cairns SP. **Sports Med** 2006;36(4):279, [5]Campbell EH. **Vet J.** 2011;190(2):199, [6]Dutka TL& Lamb GD. **Am J Physiol** 2000; 278: C517, [7]Evans DL et al. **Equine Vet J** 1995;18:422, [8]Gladden LB. **J Physiol** 2004;558:5, [9]Gladden LB. **Med Sci Sports Exerc** 2008;40(3):477, [10]Harris RC et al. **J Appl Physiol** 1987;63,12, [11]Hill AV et al. **Proc R Soc Lond B Biol Sci** 1924;16:84, [12]Holloszy JO. **J Biol Chem** 1967;242:2278, [13]Joyner MJ & Coyle EF. **J Physiol**. 2008;586(1):35, [14]Jones NL. **Clin Sci (Lond)** 1980;59:85, [15]Juel C. **Physiol Rev** 1997;77:321, [16]Kelley KM et al. **J Appl Physiol** 2002;93 :865, [17]Kitaoka Y et al. **Exp Physiol** 2011;96:348, [18]Lindinger MI et al. **J Appl Physiol** 1995;78,765, [19]Lindinger MI. **J Physiol** 2003;550:334, [20]Lindinger MI et al. **Am J Physiol** 2005;289:R891, [21]Lindinger MI. **Exp Physiol** 2011;96(3):261, [22]McCutcheon LJ et al. **Equine Exercise Physiology 2**. ICEEP Publications: 1987:348, [23]Mole PA et al. **Am J Physiol** 1999;277:R173. [24]Nielsen OB et al. **J Physiol** 2001;536(Pt 1):161, [25]Phillips SM. **Exercise Metabolism**, 2nd edition, Human Kinetics, 2006, [26]Posterino GS & Fryer MW. **J Appl Physiol** 2000;89:517, [27]Pösö AR et al. **Equine vet J** 1995; Suppl. 18:231, [28]Richter EA et al. **Am J Physiol** 1988;254:E555, [29]Robergs et al. **Am J Physiol** 2004;287(3):R502, [30]Shulman RG. **Exerc Sport Sci Rev** 2005;33(4):157, [31]Spriet LL, **Exercise Metabolism**, 2nd edition, Human Kinetics, 2006, [32]Stegmann H et al. **Int J Sports Med** 1981;2:160, [33]Stewart PA. **Can. J. Physiol. Pharmacol** 1983;61:1444. [34]Stringer W et al. **J Appl Physiol** 1992;72: 954, [35]Vengust M. **Can J Vet Res** 2012;76:275.

EQUINE CYTOLOGY: A CLINICAL WORKSHOP

Allison J. Stewart, BVSc(hons), MS, DACVIM(LAIM), DACVECC, Auburn, AL; Pete Christopherson, DVM, PhD, DACVP, Auburn AL; Katharyn Mitchell BVSc, DACVIM, Gisborne, Victoria, Australia.

INTRODUCTION

Cytologic examination by a clinical pathologist is always recommended, but with appropriate sampling, careful slide staining technique and practice, a provisional diagnosis can often be made allowing for earlier instigation of appropriate therapy.

Case 1. Acute Peritonitis in a 7 year-old Thoroughbred gelding. There was a 12 hr history of mild depression followed by acute, moderate colic. Physical examination was unremarkable other than a mild pelvic flexure impaction. PCV and TP were normal, and no fluid was obtained by abdominocentesis. Overnight the horse received oral fluids. Fever (T=102° F) was noted. Abdominal ultrasound: mild small intestinal distension, but normal motility. CBC: neutrophilia (8,197/µL), fibrinogen =500 mg/dL. Abdominocentesis was repeated and milky, pink fluid was obtained. TP =4.6 g/dL, WBC=383,000/uL: 85% non-degenerate neutrophils and 15% macrophages; some contained cellular debris. Small lymphocytes and groups of mesothelial cells were rare. Gram stain: no bacteria seen and culture negative. Diagnosis: chronic marked purulent inflammation with evidence of cytophagia. The horse responded to 7 days of IV antibiotics. Repeat abdominocentesis 2 weeks later: normal nucleated cell count and TP. There were 90% macrophages and 10% non-degenerate neutrophils and rare lymphocytes. Occasional macrophages contained large amounts of phagocytized lipid. There were several large lacy meshwork structures with entrapped cells that appeared to be a combination of fibrin and adipocytes. There had likely been some fat necrosis secondary to the inflammatory process with cellular clean-up.

Case 2. Metastatic melanoma in a 16 year-old Warmblood mare. The horse presented for radiation therapy of a large melanoma that had been surgically debulked on the crest of the neck. A second mass (15cm diameter, 6 cm depth) was apparent in the region of the left prescapular lymph node. A sample was collected from the second mass by fine needle aspirate for cytological examination. Most of the nucleated cells in the smear were neoplastic melanocytes with a mild to moderate accumulation of pigment granules. Many melanin granules were also scattered in the background. Few lymphocytes were present in the tissue to prove it was a lymph node. This was indicative of extensive metastasis to the node, essentially effacing the normal tissue.

Case 3. Uterine prolapse leading to severe metritis in a 6 year-old Trakehner mare. The mare presented for depression, severe abdominal pain and vaginal discharge 3 days after correction of a uterine prolapse. Clinical examination and blood work were suggestive of septic shock. CBC: marked leukocytosis (31,730 WBC/uL), neutrophilia (22,846/µL) with a left shift. TP =4.23 g/dL, azotemia (creatinine = 7.2 mg/dL) and hyperglycemia (360 mg/dL). Abdominocentesis: orange fluid with WBC= 13,420 /uL, TP=3.6 g/dL. The majority of the cells were moderate to severely degenerate neutrophils, but no bacteria were seen. The aerobic culture had no growth and anaerobic culture had a light growth of *Bacteroides fragilis*.

Due to the uncontrollable, persistent pain, an exploratory celiotomy was performed that revealed severe metritis. Throughout surgery the mare become progressively hypotensive despite aggressive therapy. Upon manipulation of the uterus, multiple ventricular premature contractions developed. The mare died under anesthesia.

Case 4. Cutaneous lymphoma in a 7 year-old Dutch Warmblood stallion. The horse presented to the dermatology service for intradermal skin testing. The owners had attributed several 2 cm diameter skin masses to insect bites. Multiple palpable masses were present in the testicles, and the first differential for the skin masses was cutaneous lymphoma. Fine needle aspirate of a cutaneous mass revealed predominantly intermediate sized, neoplastic lymphocytes. Histopathology of a cutaneous mass and a palpably normal mandibular lymph node indicated lymphoma. Remission of cutaneous masses was achieved with COP chemotherapy. The owners discontinued therapy, and the masses returned. The horse was euthanized 1 year later after becoming progressively ataxic.

Case 5. Sinonasal Cryptococcus in a 7 year-old Tennessee Walking Horse mare. The mare was evaluated because of bilateral mucoserosanguinous nasal discharge, intermittent right-sided epistaxis, and worsening dyspnea of 9 months duration. Multiple masses were detected via endoscopic and radiographic examinations. Cytologic examination of biopsy specimens of 1 mass revealed round yeasts with thick capsules and occasional narrow-based budding that resembled cryptococcal organisms. The diagnosis was confirmed by culture and serology. The disease was controlled with 6 weeks of fluconazole therapy but returned 2 years later. Aggressive surgical debulking and medical therapy with 14 days of IV amphotericin B and 8 months of oral fluconazole resulted in disease resolution.

Case 6. Granulomatous pneumonia in a 2 month-old draft horse cross colt. The colt presented for depression and tachypnenia. Generalized pleural irregularities were noted on thoracic ultrasound, and diffuse military opacities observed on thoracic radiographs. CBC: neutrophilia (13,695/µL) and fibrinogen = 900 mg/dL. Transtracheal wash: increased macrophages, degenerate neutrophils, intra and extracellular rod bacteria and refractile bodies. Culture: *Actinobacillus* spp. Initial response to antimicrobials, then intermittent fevers. Repeated transtracheal wash: normal differential, negative bacterial cultures, 1 *Penicillium* colony- questionable significance. Clinical deterioration. Bronchoalveolar lavage: mats of fungal hyphae and *Dictyocaulus* larvae. Fecal: *Strongyloides westeri.* Voriconazole and ivermectin administered. Hypoxia (PaO_2=62 mmHg). Increased intranasal oxygen to 9L/min. Anorexia, increased depression, continued clinical deterioration. Fibrinogen =1500 mg/dL. Lung biopsy- histopathology: foamy macrophages with engulfed neutrophils within interstitial spaces, no etiologic agents seen; culture negative. Nebulized 10% albuterol, acetylcystine and amikacin. Liquid enteral diet force fed. Continued clinical and radiographic deterioration. Discontinued NSAIDs and administered dexamethasone. Improved appetite, arterial blood gas and pulse oximetry. Continued weight loss and intermittent fevers. Profound weakness; euthanized after 3 weeks. Gross necropsy: 0.2-0.5cm slightly raised white tan foci. Tracheobronchial lymph nodes markedly enlarged- 1-5 cm. Granulomatous bronchointerstitial pneumonia, no etiological agents seen. Light growth *Penicillium* and *Aspergillus* spp.

Case 7. Pure red cell aplasia 4 year-old Arabian mare. The mare presented for anorexia, fever, anemia, tachycardia, tachypnea and ventral edema. Mucus membranes were very pale. PCV = 7%. Liver enzymes (GGT and SDH) were increased. A star-gazing pattern was observed in the liver by ultrasound. Cholangiohepatitis confirmed by liver biopsy. Hematology: severe normocytic normochromic anemia. Autoagglutination present. Coomb's test positive. Bone marrow aspirate: absence of red blood cell precursors, with only a few rubriblasts present. Myeloid:erythroid ratio 100:1. Diagnosis: pure red cell aplasia. Blood transfusion and immunosupressive doses of dexamethasone and antibiotics administered. Onset of fever. *Actinobacillus* sp. pneumonia diagnosed by transtracheal wash. Euthanized due to worsening signs of pneumonia and no improvement in bone marrow cytology.

Case 8. Disseminated large granular lymphoma in a 22 year-old Tennessee Walking Horse mare. Anorexia of 3 days duration and then decreased fecal output and mild colic. HR=46 bpm. Physical examination and thoracic and abdominal ultrasound were normal. No reflux obtained. Small colon impaction with mild diarrhea detected by transrectal palpation. CBC: leukopenia (2,400 cells/µL, with 23% neutrophils and 4% bands), fibrinogen=500 mg/dL and YP=5.2 g/dL. Abdominocentesis: clear fluid, with TP=1.2 g/dL and WBC=920 cells/µL. Cytology: large granular lymphoma (T cell).

Case 9. Peritonitis, pleuritis and pericarditis in a 7 year-old Arab gelding. The horse had a 6 month history of weight loss and reduced appetite. Body condition score was 2.5/9, but no other abnormalities were detected on physical examination. Biochemistry revealed an increased total protein of 9.1g/dL with a hyperglobulinemia of 6.1g/dL. The white cell count and fibrinogen concentration were normal. Protein electrophoresis showed an increase in α_1 and α_2 globulins, β-γ bridging due to a polyclonal increase in IgA and a moderate polyclonal increase in γ globulins, consistent with chronic inflammation and/or infection. Ultrasound revealed fibrinous peritonitis, pleuritis and pericarditis. Fluid obtained by abdominocentesis was thick, yellow and opaque. The total cell count was 750,000 cells/µL with 95% degenerate neutrophils. The total protein was 5.6 g/dL. No bacteria were seen. Direct and enriched culture of the fluid showed no growth. PCR was performed looking for bacterial 16s rRNA, and a large band was seen in the region of *Actinobacillus/Corynebacterium* spp. Further sequencing is in progress. The horse is currently receiving antimicrobial treatment and is gaining weight with an improved appetite.

Case 10. Pericarditis and congestive heart failure in a 6yo Clydesdale mare. The horse presented for evaluation of fever of unknown origin. The mare had been febrile for two weeks, with a poor appetite, diarrhea and vulval discharge. The mare had not been seen to urinate for 48hrs prior to presentation. On examination, the mare had muffled heart sounds, with a heart rate of 108 bpm. Bilateral jugular pulses and severe ventral edema were present. The respiratory rate was 24 bpm, with crackles heard around the perihilar region. CBC: WBC = 16,700 cells/µL, fibrinogen 600 mg/dL. Cardiac troponin 0.14 ng/mL, creatinine 9.4mg/dL, BUN 60 mmol/L. Ultrasound showed severe pleural, pericardial and peritoneal effusion. Pericardiocentesis drained 28L of turbid fluid with a TP 5.8g/dL and total nucleated cell count 30,800/µL. Cytology identified 100% degenerate neutrophils with very occasional rod shaped intracellular bacteria. Culture was negative. At post mortem, severe fibrinous pericarditis was evident causing constrictive cardiomyopathy and subsequent heart failure.

Case 11. Anaplasmosis in a 7yo QH gelding. The gelding was recently transported from Montana to Upstate New York with subsequent lethargy and dullness. He had developed hindlimb swelling and was reportedly 'wobbly' when ridden, with a fever of 104 °F for 3 days prior to admission. On examination, petechiae were present on the oral and nasal mucosa, the respiratory rate was 20 bpm with a tracheal rattle and cough on rebreathing bag examination. All distal limbs were warm and swollen but not painful to palpate. Mild ataxia was seen affecting all four limbs during the neurological exam. The PCV was 24%, WBC 4,800 cells/µL, platelets 40,000/µL and fibrinogen 500 mg/dL. *Anaplasma phagocytophilum* organisms were identified on the blood smear. Transtracheal wash: 83% nondegenerate neutrophils, some containing *Anaplasma* organisms. Culture negative. The horse recovered after 7 days of IV oxytetracyline.

Case 12. EIPH and RAO in a 14yo TBxWB eventer gelding. The horse presented for poor performance and heavy breathing after galloping. Physical examination: bilateral wheezes cranioventrally. Thoracic ultrasound: mild comet tails in the caudodorsal lung fields bilaterally but worse on the right side. Bronchoalveolar lavage: fluid was tinged pink with multiple floating white mucus strands. Cytology: 1% hemosiderophages, with 10% mast cells, 7% lymphocytes and 32%

neutrophils. Treatment with systemic dexamethasone, inhaled mast cell stabilizers and bronchodilators resulted in return to competition at a lower level.

Case 13. Esophageal perforation in a 9yo Friesian mare. The horse presented with an acute onset of respiratory distress, colic and fever. The rDVM had attempted to pass a NGT on the farm, but had difficulty passing the tube. No reflux was obtained, but the mare had become severely distressed. On presentation, HR=90, RR=44, T=104°F, with a full body sweat and generalized muscle fasciculations. Ultrasound revealed a large amount of pleural fluid bilaterally. 3L of foul smelling, serosanguinous fluid was obtained from the R thorax: TP=5g/dL, 80% degenerate neutrophils, with intracellular and extra cellular Gram positive and negative bacteria and plant material. Thoracic radiographs: bilateral pneumo- and hydrothorax. Esophagoscopy: 10cm tear in the esophageal mucosa just proximal to the cardia. At post mortem there was severe hypertrophy of the distal esophagus with a full thickness tear just proximal to the diaphragm.

Case 14. Ruptured gastric ulcer in an 8wk old TB filly. The foal presented with acute onset of depression and respiratory distress. On examination the foal was febrile, T=103.6°F, HR=120 bpm, RR=60 bpm, toxic mucus membranes with a delayed CRT of 4sec. PCV=55%, TP=8g/dL, WBC=2,800 cells/μL, lactate=7.5mmol/L. Abdominal ultrasound revealed a large amount of hyperechoic fluid with gas bubbles and fibrin strands. All intestinal walls were thickened and ileus was evident. Abdominocentesis revealed a large volume of yellow/orange fluid with a fermented smell. WBC=12,000/μL, RBC=20,000/μL and total protein=6.5g/dL. Cytology: 95% degenerate neutrophils with bacterial rods and cocci seen both intracellularly and extracellularly. No plant material was found. A non hemolytic streptococcus and *Candida albicans* were cultured. The foal responded initially to medical therapy before developing an adhesion 10 days later. At necropsy a small perforated gastric ulcer was found with a loop of small intestine adhered, blocking the hole in the gastric serosal surface.

Why Foals Don't Gallop In Utero: Studies in Transitions of Fetal Consciousness with Implications for Neonatal Health

J E Madigan, DVM, MS Diplomate ACVIM, Diplomate ACAW, Monica Aleman MVZ, PhD, Diplomate ACVIM, Kirstie Pickles, BVMS MSc CertEIM PhD DipECEIM MRCVS
Davis, CA

Introduction

A syndrome termed neonatal maladjustment syndrome (NMS) in the UK, and more recently as hypoxic ischemic encephalopathy (HIE) or perinatal asphyxia in the USA, is characterized by neurological and behavioral abnormalities that develop in the early neonatal period in foals. Other terms used include "dummy foals", wanderers, and barkers. The literature suggests that this is one of the most common diseases affecting newborn foals[1,2]. The cause of NMS has been attributed by some to pre-, intra- or postnatal hypoxia leading to neuronal cellular energy failure and death[3-5]. Clinical signs have been stated to be consistent with brain hypoxia and include alterations in the state of consciousness from mild obtundation to stuporous to comatose; abnormal behavior such as lack of affinity for the mare, not nursing, vocalization, and wandering; blindness; and paroxysmal activity such as paddling and seizures. Histopathological evidence of cerebral hemorrhage and hypoxia has been detected in some severely affected foals in the mid 1970's prior to the creation of the discipline of neonatal intensive critical care[6]. However, many foals in recent studies do not have histological evidence of hypoxia, edema or hemorrhage[1]. Additionally, unlike hypoxic human infants that go on to have residual neurologic deficits[7], 80% foals with NMS recover well and fully in the absence of any specific therapy other than supportive measures[8]. These disparities have led the Equine and Comparative Neurology Research group at UC Davis to question the hypoxia hypothesis and search for an alternative explanation with a focus on the transition of fetal consciousness at birth.

Neurosteroids and the Fetus

Certain steroidal compounds, predominantly 5α-reduced pregnanes such as progestagens, appear to have important neuromodulatory roles[9-11] and are thus termed neurosteroids. These neurosteroids are synthesized *de novo* in glial cells from cholesterol or blood-borne steroid precursors[11] and are potent allosteric modulators of the $GABA_A$ receptor. Low concentrations cause weak enhancement of GABA activity and high concentrations cause complete non-competitive inhibition[10]. These pregnanes are very closely related to the anesthetic alphaxalone and their intravenous infusion into rats and mice[12,13] leads to anesthesia or marked behavioral effects suggesting that these pregnanes cross the blood brain barrier and exert CNS depressant effects.

The fetal foal is subject to high levels of progesterone and other progestagens *in utero*[14], which are deemed important in providing tonic inhibition of fetal CNS activity and damping movement to prevent maternal damage[15]. Injections of progesterone or its metabolites into the ovine fetal circulation in late gestation reduce fetal electroencephalograph, electrocorticograph and electrooculograph activity, breathing movements and behavioral arousal, whilst inhibition of placental progesterone enhance these parameters[16-19]. The receptors in the fetal brain are more sensitive to these pregnanes, compared with the receptors in the adult brain[20]. The fetus must rapidly change from the quiet suppressed state *in utero* to one of arousal and attempts to rise shortly after birth. The loss of placentally derived precursors at birth and the switch to adrenal or other derived precursors causes a dramatic decline in pregnane concentrations shortly after birth in healthy neonates[21].

Evidence for a Role for Neurosteroids in NMS

Neonatal foals have high concentrations of pregnanes at birth which decrease rapidly over the first 48 hours of life[22,23]. Compared to healthy age matched neonatal foals, NMS foals show significant ongoing elevation of plasma pregnane concentrations[23]. Elevated concentrations of plasma pregnanes and a correlation between decreasing levels of pregnanes and clinical recovery have also been reported[24].

Additionally, infusion of the neurosteroid pregnane allopregnanolone to healthy neonatal foals induces obtundation, lack of affinity for the mare, and decreased response to external stimuli[25] typical of that seen in maladjusted foals. These effects were short-lasting and associated with measurable concentrations of plasma pregnanes[25]. EEG data during infusion was indicative of slow wave sleep whilst foals were standing (Aleman, unpublished data). These data suggest that these steroids can cross the blood brain barrier and exert neuromodulatory effects, which at high concentrations may have a dampening effect in the CNS with resulting alterations in states of consciousness, altered behavior, and responsiveness to stimuli, such as observed in NMS cases. The opposite may also apply; when certain pregnane concentrations are low, foals may be more alert, responsive to environmental stimuli, have affinity for the dam, and nurse.

One possible cause of persistence of fetal hormones is a failure of normal events of parturition which provide essential signaling that the foal has transitioned to outside of the mare and the desire to rise and nurse can commence. Events such as rapid passage through the birth canal or cesarean delivery might produce a failure of transition signals to allow fetal hormone reduction and the transition from the *in utero* to extra-uterine behavioral status. Another possible mechanism for elevated pregnanes is the reversion to fetal cortical status when adverse postnatal circumstances occur. The syndrome of reversion to fetal circulation is a well-known and accepted consequence of adverse birth and postnatal events which cause the neonate to revert to mechanisms which regulated the cardiovascular system *in utero*.

Novel Potential Treatment Therapies

Based on these findings the concept of reducting post birth circulating plasma pregnanes would seem to offer a potential new therapeutic option for NMS foals. We have considered physical means as well as steroidogenesis inhibitors as targets for further studies.

We wonder if the pressure associated with being squeezed in the birth canal creates negative feedback loops which causes fetal neurostroid levels, which keep the foal asleep in utero, to decline so that at birth the foal has a different level of consciousness to allow standing and nursing behavior and recognition of the dam. Encouraging the transition of the foal to extra-uterine physiologic status may provide a treatment option. In neonatal foals deep pressure lateral thoracic squeezing of healthy neonatal foals causes recumbency, non-REM sleep, muscle relaxation and increases in β-endorphins with no adverse effects[26]. We have termed this effect squeeze induced somnolence. It is believed that this technique may mimic passage through the birth canal and, as such, prime the foal for extra-uterine life, including pregnane down-regulation. Investigation of selective inhibition of the synthesis of persistent neurosteroids is also underway.

REFERENCES

1. Bernard, W.V., Reimer, J.M. and Cudd, T. (1995) Historical factors, clinicopathologic findings, clinical features, and outcome of equine neonates presenting with or developing signs of central nervous system disease. In: *Proceedings*, 41st American Association of Equine Practitioners, Lexington, KY. pp 222-224.
2. Rossdale, P.D. and Leadon, D.P. (1975) Equine neonatal disease: A review. *J Reprod Fertil* **23**, 658-661.
3. Drummond, W.H. (1988) Neonatal maladjustment syndrome: its relationship to perinatal hypoxic-ischaemic insults. *Equine Vet J* **5**, 41-43.
4. Rossdale, P.D. (1972) Modern concepts of neonatal disease in foals. *Equine Vet J* **4**, 117-128.
5. Ringger, N.C., Giguere, S., Morressey, P.R., Yang, C. and Shaw, G. (2011) Biomarkers of brain injury in foals with hypoxic-ischemic encephalopathy. *J Vet Intern Med* **25**, 132-137.
6. Palmer, A.C. and Rossdale, P.D. (1976) Neuropathological changes associated with the neonatal maladjustment syndrome in the Thoroughbred foal. *Res Vet Sci* **20**, 267-275.
7. van Handel, M., Swaab, H., de Vries, L.S. and Jongmans, M.J. (2007) Long-term cognitive and behavioral consequences of neonatal encephalopathy following perinatal asphyxia: a review. *Eur J Pediatr* **166**, 645-654.
8. Françoise Hess-Dudan and P. D. Rossdale (1996) Neonatal maladjustment syndrome and other neurological signs in the newborn foal: Part 2 *Equine Vet Educ* **8** (2) 79–83.
9. Baulieu, E.E. (1998) Neurosteroids: a novel function of the brain. *Psychoneuroendocrinology* 23, 963-987.
10. Mellon, S.H. and Griffin, L.D. (2002) Neurosteroids: biochemistry and clinical significance. *Trends Endocrinol Metab* 13, 35-43.
11. Robel, P. and Baulieu, E.E. (1994) Neurosteroids biosynthesis and function. *Trends Endocrinol Metab* 5, 1-8.
12. Naert, G., Maurice, T., Tapia-Arancibia, L. and Givalois, L. (2007) Neuroactive steroids modulate HPA axis activity and cerebral brain-derived neurotrophic factor (BDNF) protein levels in adult male rats. *Psychoneuroendocrinology* 32, 1062-1078.
13. Zhu, D., Wang, M.D., Backstrom, T. and Wahlstrom, G. (2001) Evaluation and comparison of the pharmacokinetic and pharmacodynamic properties of allopregnanolone and pregnanolone at induction of anaesthesia in the male rat. *Br J Anaesth* 86, 403-412.
14. Holtan, D.W., Houghton, E., Silver, M., Fowden, A.L., Ousey, J.C. and Rossdale, P.D. (1991) Plasma progestagens in the mare, fetus and newborn foal. *J Pediatr Psychol* 44, 517-528.
15. Mellor, D.J., Diesch, T.J., Gunn, A.J. and Bennet, L. (2005) The importance of awareness for understanding fetal pain *Brain Res Rev* 49, 455-471.
16. Crenshaw, M.C., Meschia, G. and Barron, D.H. (1966) Role of progesterone in inhibition of muscle tone and respiratory rhythm in foetal lambs. *Nature* 212, 842.
17. Crossley, K.J., Nicol, M.B., Hirst, J.J., Walker, D.W. and Thorburn, G.D. (1997) Suppression of arousal by progesterone in fetal sheep. *J. Reprod. Fertil. Dev.* 9, 767-773.
18. Nicol, M.B., Hirst, J.J., Walker, D. and Thorburn, G.D. (1997) Effect of alteration of maternal plasma progesterone concentrations on fetal behavioural state during late gestation. *J Endocrinol* 152, 379-386.
19. Nicol, M.B., Hirst, J.J. and Walker, D. (2001) Effect of finasteride on behavioral arousal and somatosensory evoked potentials in fetal sheep. *Neurosci Letters* 306, 13-16.
20. Crossley, K.J., Walker, D.W., Beart, P.M. and Hirst, J.J.(2000) Characterisation of GABA(A) receptors in fetal, neonatal and adult ovine brain: region and age related changes and the effects of allopregnanolone. *Neuropharmacology* **39,** 1514–1522.
21. Hirst, J.J., Yawno, T., Nguyen, P. and Walker, D.W. (2006) Stress in pregnancy activates neurosteroid production in the fetal brain. *Neuroendocrinol* 84, 264-274.
22. Houghton, E., Holtan, D., Grainger, L., Voller, B.E., Rossdale, P.D. and Ousey, J.C. (1991) Plasma progestagen concentrations in the normal and dysmature newborn foal. *J Reprod Fertil* **44**, 609-617.
23. Aleman, M., Haggett, E.F., Pickles, K.J., Conley, A., Stanley, S., Moeller, B., Palmero, J., Toth, B. and Madigan, J.E. (2013) Abnormal plasma neurosteroid concentrations in ill, neonatal foals presented to the neonatal intensive care unit. *Equine Vet J (in press)*.
24. Rossdale, P.D., Ousey, J.C., McGladdery, A.J., Prandi, S., Holdstock, N., Grainger, L. and Houghton, E. (1995) A retrospective study of increased plasma progestagen concentrations in compromised neonatal foals. *Reprod Fertil Dev* **7**, 567-575.
25. Madigan, J.E., Haggett, E.F., Pickles, K.J., Conley, A., Stanley, S., Moeller, B., Palmero, J., Toth, B. and Aleman, M. (2012) Allopregnanolone infusion induced neurobehavioral alterations in a neonatal foal: Is this a clue to the pathogenesis of neonatal maladjustment syndrome? *Equine Vet J* **44** S41 109-112.
26. Toth et al (2012) Evaluation of squeeze-induced somnolence in neonatal foals. *Am J Vet Res*. 73(12):1881-9

OMEGA 3 FATTY ACIDS IN CATTLE: CHANGING THE PRODUCT AND CHANGING THE BOVINE?

William S. Swecker, Jr. DVM, Ph.D, DACVN
Blacksburg, VA

"Historically, the goal of agricultural research has been to increase yield and productive efficiency, with little focus given to improving the nutrient profile of food products." [1]

INTRODUCTION

The necessity for essential fatty acids has been established for most animals for many years. The premise of essential is the fatty acid must be provided within the diet as it cannot be de novo synthesized within the animal. Within the past decades, the relative roles of the omega-3 and omega-6 fatty acids have been further defined, mostly in monogastric animals with the general theme that the omega-6 fatty acids are pro-inflammatory, the omega-3 fatty acids are anti-inflammatory and that as humans, we consume too many omega-6 fatty acids relative to omega-3 fatty acids. Humans should consume a diet that has an n-6:n-3 ratio of < 3. The role of the omega series of fatty acids in the ruminant was minimized because 1) omega-series fatty acids are polyunsaturated fatty acids (PUFAs) and the majority PUFAs (70-100%) were biohydrogenated (saturated) in the rumen environment and 2) excessive fat supplementation of the ruminant's diet, especially PUFAs, disrupted microbial activity and thus rumination. Fish oils, an omega-3 fatty acid promoted for humans, have a profound negative impact on synthesis of milk fat in the mammary gland.[2] The purpose of this manuscript is to review fatty acid nomenclature and sources; evaluate strategies to alter the omega-3 and omega-6 fatty acid content of meat and milk; and to propose potential alterations in cattle on predominant omega-3 and omega-6 rations.

FATTY ACID NOMENCLATURE

Lipids can be categorized by chain length (short, medium, long) and the number of double bonds (saturated = no double bonds, monounsaturated = one double bond, and polyunsaturated = 2 + double bonds) Description of individual fatty acids can be challenging as multiple systems can be utilized to describe a specific fatty acid which include: chemical structure, chemical name, systematic name, and trivial name. One of the naming systems describes the fatty acid structure from the methyl (CH3) end or Omega end which is represented by "n" or "ω". The number represents the location of the first double bond from the methyl end, thus a n-3 fatty acid has the first double bond 3 carbons from the methyl end, a n-6 fatty acid has the first double bond 6 carbons from the methyl end. Finally, the location of the fatty acid within the animal has physiologic relevance. Although we consider most lipid or fat in the body as a storage structure for energy, the presence of specific fatty acids within cell membranes alters physiologic function.

Chemical Name	Trivial name	Comments
C18:2n-6	Linoleic Acid (LA)	Most abundant in concentrates and oilseeds
C20:4n-4	Arachidonic Acid (AA)	Animals elongate and desaturate LA to AA, precursor to inflammatory mediators
C18:3n-3	Linolenic Acid (LNA)	Most abundant in plant leaves, stems and roots – abundant in limited number of oilseeds – Flax and canola (rape)
C20:5n-3	Eicosopentaenoic Acid (EPA)	Highest in wild raised fish. Can be elongated from LNA in animals
C22:5n-3	Docosahexaenoic Acid (DHA)	Highest in wild raised fish. Can be elongated from LNA in animals
C18:1 trans 11	Vaccenic Acid	Biohydrogenation intermediate of LA and LNA, precursor to rumenic acid in humans
C18:2 cis-9, trans 11	Rumenic Acid	The "good" CLA , higher in meat and milk of grass-fed ruminants as compared to grain fed ruminants
C18:2 trans10, cis 12		Produced by rumen hydrogenation under high grain / acidic conditions, potent antagonist of milk fat synthesis

YOU ARE WHAT YOU EAT WHEN IT COMES TO FAT

The fatty acid composition of any animal or product from that animal (meat or milk) traces back to which plant or part of a plant was consumed at the lower end of the food chain. Most concentrates (seeds) and oils extracted from those seeds that are fed to cattle tend to be high in the n-6 fatty acids and Linoleic acid (18:2 (9,12) is usually the dominant fatty acid. Conversely leaves and stems of plants tend to be high in the n-3 fatty acids with alpha-linolenic acid, 18:3 (9,12,15) as the most common fatty acid. In general, feedyard cattle and high producing dairy cows that are fed a grain-based diet will produce meat or milk that is n-6 dominant, whereas pastured cattle will produce meat or milk that is n-3 dominant. For example, Finchem et al evaluated changes in fatty acid composition of ruminal fluid, serum, and adipose tissue in yearling angus-cross steers (300 kg BW) finished for 140 days on either a corn / corn silage / soybean meal ration or grazed on naturalized cool season grasses in Virginia. Linoleic acid was the predominant fatty acid in the grain diet (58%), whereas alpha-linolenic acid was the predominant fatty acid in the grass diet (range of 62-73% over the grazing season). The n-6: n-3 ratio in adipose tissue increased from 1.3 to 5.3 over the finishing period in the grain based diet, whereas the n-6 / n-3 ration declined from 1.2 to 0.9 in the grass-based diet. Trends in ruminal fluid and serum were similar and fatty acid composition differed between the groups after 28 days. [3] Duckett et al reported that the n-6 / n-3 ratio of ribeyes of steers finished under similar conditions was 4.8 for the grain-based diet and 1.7 for the pasture based diet. [4]

The relationship of dietary lipid composition and milk lipid composition is more complex in the dairy cow. Supplementation of dairy cows with Marine oils or algae does increase the n-3 content of milk, however the severe depression in total fat production limits this option to enhance n-3 content of milk.[2] Likewise, the efficiency of transfer of dietary fish oils (EPA and DHA) is very low (<4%)[1] Zachut et al added rumen-protected flaxseed oil (source of n-3) or sunflower oil (source of n-6) to diets of mid-lactation Holstein cows. The n6:n 3 ratio in the milk was 16 in the sunflower supplemented cows and 2.4 in the flaxseed supplemented cows. Concentrations of the longer chain omega fatty acids (AA, DHA and EPA) are lower in magnitude, but similar patterns were observed. Cows supplemented with flaxseed oil had 27% lower concentrations of arachidonic acid and a 3-4 fold increase in EPA and DHA as compared to sunflower seed cows. Of interest, the plasma n 6: n3 ratio was 42 in the sunflower supplemented cows and 7 in cows supplemented with flaxseed.[5] Provision of a more desirable n 6: n3 ration in both beef and dairy production scenarios seems feasible in modern production systems either through changing the basal diet or supplementation with rumen protected forms of n-3 fatty acids.

EFFECTS OF N-6 AND N-3 FATTY ACIDS ON REPRODUCTION

Arachidonic acid is the substrate for the production of Prostaglandin $F_{2\alpha}$ ($PGF_{2\alpha}$) thus a direct connection between dietary fatty acids and reproductive function is established. Several groups have reported that production of $PGF_{2\alpha}$ and metabolites are increased when n-6 fatty acids are supplemented and decreased when n-3 fatty acids are supplemented.[6] Occurrence of post-partum disease (retained placenta, metritis and mastitis) had been decreased in some, but not all studies, when cows are supplemented with n-6 fatty acids pre-partum. Santos et al summarized that supplementation with fat sources rich in n-6 during late gestation and early lactation enhanced follicle growth, uterine PG secretion, embryo quality and pregnancy. Omega-3 fatty acid supplementation during lactation suppressed uterine PG release, improved embryo quality and maintenance of pregnancy. [6] Silvestre et al also reported a decrease in pregnancy loss after first AI in cows supplemented with Fish oil post partum as compared to cows supplemented with Palm Oil. [7] In summary, a broad physiologic basis exists for a role of the n-6 and n-3 fatty acids in bovine reproduction.

EFFECTS OF N-6 AND N-3 FATTY ACIDS ON IMMUNE FUNCTION IN CATTLE

Calder proposed that fatty acids could alter immune function through 3 broad mechanisms; membrane alterations, signal transduction and lipid mediators.[8] Dietary supplementation of n-3 fatty acids does alter the fatty acid composition of immune cell membranes[9,10] Silvestre et al utilized a model where dairy cows were fed a high n-6 diet (safflower oil) from late gestation to early lactation (d35) and then switched cows do a high n-3 ration from d 35 to d 160 of lactation. Palm oil (low n-6 and n-3) was used as a control. The acute phase proteins, haptoglobin and fibrinogen were higher in the cows fed the high n-6 diet than cows fed palm oil. Neutrophil oxidative burst and phagocytosis were higher on d 4 post partum in the high n-6 cows than the cows supplemented with palm oil.[9] In summary, the majority of data on the role of n-6 and n-3 fatty acids on immune function comes from the human and rodent literature and many of the findings are contradictory between species.[8] The role of lipid-derived inflammatory mediators can be altered through provision of different fatty acids in the diet. The potential roles of the omega series in immune function of cattle is undiscovered, yet the author does wonder why a dairy cow on a high grain diet tends to have a profound systemic reaction to metritis and mastitis while the grass-fed beef cow seems relatively unaffected by these disorders.

CONJUGATED LINOLEIC ACIDS (CLA)

In the late 70's, researchers attempted to induce cancer in mice via a combination of injection of beef fat in the skin and application of a carcinogen. To their surprise, the mice injected with the beef fat had 20% fewer tumors.[11] The protective compound(s) in the beef fat were later identified as isomers of Conjugated Linoleic Acid (CLA). Subsequent workers reinforced the anti-cancer properties of CLA and reported potential cardiovascular and metabolic benefits. The predominant source of CLA in human diets is meat and milk from ruminants and rumenic acid is the predominant CLA in those products.[1] Now, to get back to the cow, rumenic acid (C18:2 cis-9, trans 11) was classically described as an intermediate in the biohydrogenation (saturation) of LA to stearic acid. Griinari et al observed that rumenic acid can also be formed in the tissue

or udder from vaccenic acid which is a biohydrogenation product from LA or ALA. [12] This finding has relevance to human health as humans can also convert vaccenic acid to rumenic acid. [13]
The author suggests not sharing this finding with the popular media as vaccenic is a trans-fatty acid and the media has already convinced the general public that trans fats are bad.

When cows are fed grass or forages, rumenic acid is the dominant CLA. However, when grain is added to the ration, the amount of rumenic acid decreases and the amount of C18:2 trans10, cis 12 CLA increases. Bauman and colleagues reported the t10:c12 CLA is a potent inhibitor of milk fat synthesis in the mammary gland and may be a component of milk fat depression associated with high grain diets. [14] In humans, C18:2 trans10, cis 12 CLA may have effects in reducing obesity. [15] Most human supplements of CLA contain a mixture of rumenic acid and C18:2 trans10, cis 12 CLA as it appears that the chemists have not yet determined a method to make a pure source of rumenic acid.

SUMMARY

In conclusion, the amount and relative ratio of essential fatty acids consumed by humans is a fertile area of investigation. Most work to date in cattle has focused on devising methods to increase the n-3 concentration of meat an milk with limited investigation on the effects of those fatty acids on the cow producing the meat and milk. The provision of n-6 and or n-3 fatty acids to dairy cows on high grain diets may be manipulated to improve immune function. As a nutritionist, however, the author cautions the reader to consider the basal diet first before attempting to manipulate potential harmful effects of the diet with supplements.

REFERNCES

1. Lock AL, Bauman DE. Modifying milk fat composition of dairy cows to enhance fatty acids beneficial to human health. *Lipids* 2004;39:1197-1206.
2. Chilliard Y, Ferlay A, Doreau M. Effect of different types of forages, animal fat or marine oils in cow's diet on milk fat secretion and composition, especially conjugated linoleic acid (CLA) and polyunsaturated fatty acids. *Livestock Production Science* 2001;70:31-48.
3. Fincham JR, Fontenot JP, Swecker WS, et al. Fatty acid metabolism and deposition in subcutaneous adipose tissue of pasture- and feedlot-finished cattle. *J Anim Sci* 2009;87:3259-3277.
4. Duckett SK, Neel JPS, Fontenot JP, et al. Effects of winter stocker growth rate and finishing system on: III. Tissue proximate, fatty acid, vitamin, and cholesterol content. *J Anim Sci* 2009:jas.2009-1850.
5. Zachut M, Dekel I, Lehrer H, et al. Effects of dietary fats differing in n-6:n-3 ratio fed to high-yielding dairy cows on fatty acid composition of ovarian compartments, follicular status, and oocyte quality. Journal of dairy science 2010;93:529-545..
6. Santos JEP, Bilby TR, Thatcher WW, et al. Long Chain Fatty Acids of Diet as Factors Influencing Reproduction in Cattle. *Reproduction in Domestic Animals* 2008;43:23-30.
7. Silvestre FT, Carvalho TS, Francisco N, et al. Effects of differential supplementation of fatty acids during the peripartum and breeding periods of Holstein cows: I. Uterine and metabolic responses, reproduction, and lactation. *Journal of dairy science* 2011;94:189-204.
8. Calder PC. The relationship between the fatty acid composition of immune cells and their function. *Prostaglandins, leukotrienes, and essential fatty acids* 2008;79:101-108.
9. Silvestre FT, Carvalho TS, Crawford PC, et al. Effects of differential supplementation of fatty acids during the peripartum and breeding periods of Holstein cows: II. Neutrophil fatty acids and function, and acute phase proteins. *Journal of dairy science* 2011;94:2285-2301.
10. Sijben JW, Calder PC. Differential immunomodulation with long-chain n-3 PUFA in health and chronic disease. *Proc Nutr Soc* 2007;66:237-259.
11. Ha YL, Grimm NK, Pariza MW. Anticarcinogens from fried ground beef: heat-altered derivatives of linoleic acid. *Carcinogenesis* 1987;8:1881-1887.
12. Griinari JM, Corl BA, Lacy SH, et al. Conjugated linoleic acid is synthesized endogenously in lactating dairy cows by Delta(9)-desaturase. *J Nutr* 2000;130:2285-2291.
13. Mosley EE, McGuire MK, Williams JE, et al. Cis-9, trans-11 conjugated linoleic acid is synthesized from vaccenic acid in lactating women. *J Nutr* 2006;136:2297-2301.
14. Bauman DE, Griinari JM. Nutritional regulation of milk fat synthesis. *Annu Rev Nutr* 2003;23:203-227.
15. Pariza MW. Perspective on the safety and effectiveness of conjugated linoleic acid. *Am J Clin Nutr* 2004;79:1132S-1136S.

TRACE ELEMENT SUPPLEMENTATION OF CATTLE: BY NEEDLE OR BY MOUTH?

William S. Swecker, Jr. DVM, Ph.D, DACVN
Blacksburg, VA

INTRODUCTION

Trace element supplementation is commonly used to correct for deficiencies of nutrients in the feeds or overcome high concentrations of antagonists. Although trace elements are commonly added to a complete ration, supplement, or free choice mineral mix, the use of injectable products gives the practitioner and producer an alternative to provision in the feed. Supplementation during high stress periods such as weaning or in the periparturient period has been recommended, even when the basal ration may be considered adequate. This places this supplementation mode as a nutraceutical or conditionally essential nutrient that may only be beneficial during certain periods. The purpose of this manuscript is to review the functions of Copper, Manganese, Selenium and Zinc and evaluate the utility of parenteral injections of these products.

THE NEED FOR TRACE ELEMENTS IN CATTLE PRODUCTION

The National Animal Health Monitoring System (NAHMS) collected over 350 forage samples from over 325 beef cow/calf operations in 1992. Forages samples were classified at deficient, marginal or adequate in trace elements relative to their ability to meet dietary requirements. Approximately 55% of the forages were marginal or deficient in Copper (Cu), Cobalt (Co) and Selenium (Se), while 97% of the samples were marginal or deficient in Zinc (Zn). Trace element content of forages reflects the mineral concentration where the forage or concentrate was grown. Even with regional variation, the need for trace element supplementation exists across the US.

SELENIUM

The necessity of Selenium for ruminants was established by Muth et al in 1958 when they reported that dietary selenium supplementation decreased the occurrence of white muscle disease in lambs. [1] Subsequently, Se was identified as a key component of the enzyme Glutathione Peroxidase (GPX), a critical antioxidant in multiple tissues. Currently, multiple isoforms of GPX have been identified based on location within tissue or organs. Most Se in whole blood is in the form of GPX1 in the erythrocyte. This GPX1 is incorporated into the erythrocyte during formation and thus the duration of activity is related to the life span of the erythrocyte. Serum or plasma GPX3 is formed in the liver and has a much shorter half-life than erythrocyte GPX. Therefore analysis of whole blood Se or GPX1 tends to reflect long term changes (weeks to months) in Se status whereas analysis of serum Se or GPX3 reflects short term (days to weeks) changes in Se status.[2] Selenium was also identified as a key component of the iodothyronine deiodinase enzyme that converts T4 to T3, and thioredoxin reductase. Selenoprotein P is suggested to be a transport protein for selenocystine to body tissues.[2] Multiple other selenproteins have been identified and the function is still unknown on most.

Assessment of Se status is usually based on Se concentration of whole blood or plasma. Selenium is usually supplemented in the ration in the form of sodium selenite, however, yeast-based organic forms are available for dietary use.

COPPER

Copper (Cu) participates in multiple enzymes systems in the body which include: Ceruloplasmin – an acute phase protein that prepares Iron for incorporation into hemoglobin; Lysyl Oxidase – allows for the cross-linking of collagen; Tyrosinse – converts tyrosine to melanin which is important for hair color; Cytochrome oxidase – involved in electron transfer; Dopamine ß hydroxylase – involved in the conversion of dopamine to norepinephrine; and Cu-Zn Superoxide dismutase – a cellular antioxidant.[3] As copper is involved in multiple enzyme systems, several deficiency syndromes have been noted. Achromotrichia (lack of color in the hair) has been reported in cattle and enzootic ataxia (spinal cord degeneration and demyelination) has been observed in lambs. Conversely, copper toxicosis is reported commonly in sheep, thus care must be taken in Cu supplementation of sheep.

Assessment of Cu status is usually based on liver Cu concentrations in field situations, whereas serum ceruroplasmin concentrations have been used in research settings.[4] Cattle can store Cu excess Cu in the liver and then release Cu under conditions of low dietary intake, although the regulatory mechanisms are not understood. Plasma or serum copper concentrations have not been consistently reliable in assessment of Cu status other than a severe deficiency may be reflected with low serum Cu concentrations. [5,6]

ZINC

Zinc (Zn) is a component of multiple enzyme systems which include the enzymes carbonic anhydrase, lactate dehydrogenase, alkaline phosphatase and the antioxidant Cu-Zn superoxide dismutase. In addition, Zn is involved in the conversion of ß carotene to retinol in the intestinal mucosa. The classic Zn deficiency syndrome, parakeratosis, may reflect it's role in retinol metabolism. Assessment of Zn status also is based on liver concentrations in the field setting. Again, plasma Zn concentrations are of limited use other than the documentation of severe deficiency.

MANGANESE

Manganese is the least understood trace mineral. Classic deficiencies are described as bone abnormalities and poor reproductive performance. Manganese is a component of Mn superoxide dismutase and is an activator for a number of enzymes.

RATIONALE FOR PARENTERAL SUPPLEMENTATION

Parenteral supplementation is warranted when faced with a clinical deficiency syndrome, e.g neonatal calves and lambs with white muscle disease. In addition, milk is a relatively poor source of Se, thus supplementation of the lactating dam to enhance Se status of the neonate via milk may be inefficient. Conversely, cows do transfer Se transplacentally to the developing fetus to the point where the fetus may have a higher Se status than the dam; therefore oral supplementation of gestating dams is recommended to reduce the risk of white muscle disease in the next calving season.[7] Another option is the late gestation injection of Se to cows and ewes to minimize the risk of white muscle disease. Per label instructions from the manufacturer, anaphylactoid reactions have been reported in injected cattle and premature births and abortions have been reported in dairy cows injected with MU-SE®. A second indication for parenteral supplementation is when known antagonists are present in the diet; a classic example is antagonism of Cu by high dietary molybdenum. A third potential indication falls under the category of conditionally essential nutrients or nutrients that must be supplied to the body only under specific conditions such as stress or illness. This indication may apply to cattle at stressful times such as weaning or calving.

PARENTERAL SELENIUM SUPPLEMENTATION
SELENIUM AND NEONATES

Blood Se concentrations at birth of beef calves from dams injected with Se at 0.1 mg / at mid gestation did not differ from unsupplemented control dams on Se-deficient forages. Likewise, blood Se concentrations of calves given an injection of Se at 0.1 mg / kg at birth did not differ at 60 days from unsupplemented control calves.[8] Conversely, blood Se of calves from dams supplemented with a free choice mineral mix that contained 120 ppm Se was higher at birth and 60 days of age than calves from unsupplemented dams. Castellan and colleagues compared one SC injection at birth (0.05 mg Se / kg) or multiple injections at birth and d 70, 114 and 149 to unsupplemented calves. Dams consumed naturalized forages that were considered Se-deficient. Blood Se concentrations on d 160 were highest in calves with multiple injections, intermediate for the calves injected at birth and lowest for the unsupplemented controls. On d 209, blood Se concentrations were higher in the calves with multiple injections and did not differ between single injection calves and negative controls. In addition, only the calves given multiple injections reached adequate Se status.[9] Enjalbert at al reported that injection of 1.38 mg of Se at 2 and 49 days of age did not improve either erythrocyte or plasma GPX status in calves.[7] In conclusion, reliance on Se injection at birth for beef calves in Se-deficient regions provides a short term (days to weeks) improvement in Se status, but does not resolve the deficient state.

SELENIUM AND GROWING CATTLE

Maas and colleagues compared a single IM injection of Se (0.05 mg / kg) to no injection in recently weaned Herford heifers (224 kg) that were Se-deficient. Injected calves had higher blood Se concentrations than unsupplemented controls for 28 days. Blood GSH-PX was higher in Se-injected heifers than unsupplemented controls for 84 days. The authors noted, however, that blood Se concentrations associated with Se adequacy only were noted 5 hours after injection.[10] In a similar trial, Swecker and colleagues, did not find a difference in blood Se concentrations of weaned steers (237 kg) 28 and 70 days after a single injection of 0.1 mg Se / kg as compared to non-injected steers. Both groups of calves were Se-deficient at the beginning of the trial and were offered a free choice salt-mineral mix with 20 ppm Se.[11] Selenium injections (0.05 mg Se / kg) given at arrival and repeated on d 28 did increase blood and serum Se concentrations for 42 d in recently weaned steers (203-250 kg) as compared to unsupplemented steers in a series of 3 trials. The series of 2 injections, however, only increased Se status of calves to adequacy in one trial.[12] In conclusion, a single injection of Se growing calves increases Se status for days to weeks, but may not achieve Se adequacy.

PARENTERAL SUPPLEMENTATION WITH SEVERAL MINERALS

Bovine practitioners can also choose a product that provides a combination of Cu, Mn, Se, and Zn for parenteral supplementation (MULTIMIN®90). The labeled dose is 0.33 mg Cu, 1.33 mg Zn, 0.22 mg Mn and 0.11 mg Se / kg BW. Pogge and others evaluated the effect of a single injection of this product in 10 Angus and 10 Simmental steers (332 kg) on serum and liver concentrations of these minerals for 15 days post injection. Steers were fed a corn-silage based diet formulated to meet NRC recommendations for Cu, Zn, Mn, and Se. Plasma Zn and Mn concentrations were higher in injected calves than unsupplemented calves for 10 hours post injection, whereas selenium concentrations were higher in injected calves than unsupplemented calves for 24 hours post injection. A difference in plasma Cu was not detected between groups. Plasma Cu, Zn, and Se were lower in the Simmental steers than the the Angus steers. Liver concentrations of Cu, Se, and Zn were higher in injected calves than unsupplemented calves for the 15 d observation period. The authors noted a tendency (P= 0.06) for higher liver Mn in the injected calves as compared to the unsupplemented calves. [13] In a similar study, Arthington and colleagues injected a combined mineral product to weaned beef steers (316 kg) that were concurrently inoculated with a modified-live BHV-1, BVDV-1, BVDV-2, PI3, BRSV vaccine. In this trial, the calculated doses would be 0.33 mg Cu, 0.89 mg Zn, 0.22 mg Mn, and 0.11 mg Se / kg BW. Serum Zn and Se concentrations where higher on d 14 post-injection in treated calves as compared to unsupplemented controls. Serum copper increased from d 0 to d 14 in injected calves while no difference in serum Cu was detected in unsupplemented calves.[14] All serum concentrations of measured minerals were considered to reflect adequacy and the ration provided was also adequate in the measured minerals. In conclusion, parenteral administration of a combined mineral product increased plasma and liver concentrations of Cu, Zn, and Se in steers that were deemed adequate in these minerals prior to supplementation. The author notes that biological outcomes such as pregnancy rate or disease incidence may be of more interest when using this product than the provision of an essential nutrient.

SUMMARY

In conclusion, provision of trace mineral through the parenteral route does increase serum and liver concentrations of Se, Cu, Zn, and Mn for days to weeks. These injections do not, however, usually resolve the deficient or marginal state; therefore, subsequent oral supplementation is recommended for deficient cattle. Conversely, thus use of parenteral injections of trace elements to provide conditionally essential nutrients may have merit, yet double-blinded, randomized, and controlled trials should be performed across multiple sites to provide evidence for their use.

REFERENCES

1. Muth OH, et al. *Science* 1958;128:1090.
2. Brown KM, Arthur JR. *Public Health Nutr* 2001;4:593-599.
3. Brewer NR. *J Am Vet Med Assoc* 1987;190:654-658.
4. Blakley BR, Hamilton DL. *Can J Comp Med* 1985;49:405-408.
5. Lopez-Alonso M, et al. *J Vet Diagn Invest* 2006;18:71-75.
6. Tessman RK, et al. *J Am Vet Med Assoc* 2001;218:756-760.
7. Enjalbert F, et al. *J Anim Sci* 1999;77:223-229.
8. Swecker WS, Jr., et al. *Am J Vet Res* 1995;56:450-453.
9. Castellan DM, et al. *J Am Vet Med Assoc* 1999;214:816-821.
10. Maas J, et al. *J Vet Intern Med* 1993;7:342-348.
11. Swecker WS, Jr., et al. *Am J Vet Res* 1989;50:1760-1763.
12. Swecker WS, Jr., et al. *J Vet Intern Med* 2008;22:443-449.
13. Pogge DJ, et al. *J Anim Sci* 2012;90:2692-2698.
14. Arthington JD, Havenga LJ. *J Anim Sci* 2012;90:1966-1971.

IMMUNE SUPPRESSION OF CALVING: WHY? THE ROLE OF METABOLIC DISORDERS

Jesse P. Goff DVM, PhD
Ames, IA

DOES MILK PRODUCTION AFFECT IMMUNE STATUS?

Neutrophil and lymphocyte function is diminished in the periparturient period, especially in the dairy cow. The onset of milk production imposes tremendous challenges to the mechanisms responsible for energy, protein, and mineral homeostasis in the cow. Negative energy, protein, and/or mineral balance associated with the onset of lactation may be partially responsible for the immunosuppression observed in periparturient dairy cattle.

HYPOCALCEMIA AND MASTITIS SUSCEPTIBILITY

Milk fever cows are at increased risk of developing mastitis . Why?

Calcium is necessary for proper contraction of muscle. Severe hypocalcemia prevents skeletal muscle contraction to the point that the clinical syndrome known as milk fever occurs. Muscle contraction is reduced by any decrease in blood calcium, however it must be severe before we observe the "downer cow". The teat sphincter consists of smooth muscle which must contract if closure of the teat end is to occur. If low blood calcium reduces teat sphincter contraction the teat canal may remain open, inviting environmental pathogens to enter the mammary gland. We generally associate milk fever with the day of calving but we have demonstrated that many cows remain subclinically hypocalcemic for the first week of lactation. Hypocalcemic cows tend to spend more time lying down than do normocalcemic animals. Again, this could increase teat end exposure to environmental opportunists.

Hypocalcemia also acts as a stressor to the cow. Cows typically exhibit a 3-4 fold increase in plasma cortisol as part of the act of initiation of parturition. However subclinically hypocalcemic cows may have 5-7 fold increases in plasma cortisol on the day of calving and the typical milk fever cow may exhibit plasma cortisol concentrations that are 10 –15 fold higher than pre-calving plasma cortisol concentration . Cortisol is generally considered a powerful immune suppressive agent and may exacerbate the immune suppression normally observed in the periparturient period. Exacerbation is a more probable state than rather than causationbecause most studies suggest that immune suppression begins 1-2 weeks before calving (Kehrli et al., 1989), and the cortisol surge is fairly tightly confined to thefirst 1-2 days after calving..

Kimura and colleagues studied calcium fluxes inside peripheral blood mononuclear cells (PBMC) isolated from cows during the transition period. Because intracellular calcium signaling is a key early feature in immune cell activation, we have hypothesized that the increased demand for calcium in periparturient cows may adversely affect intracellular calcium stores of immune cells. This reduction in intracellular calcium stores in immune cells could blunt intracellular calcium release following an activating stimulus, contributing to the immune suppression seen in these animals. To test this hypothesis, peripheral blood mononuclear cells were obtained from 27 multiparous dairy cows spanning a period of 2 weeks before and 2 weeks after parturition. Following activation of these cells by anti-CD3 antibodies along with secondary antibodies, intracellular calcium release from intracellular stores was measured. The intracellular calcium released in response to the activation signal declined as calcium demand for lactation became more intense and recovered as plasma calcium normalized. Intracellular calcium stores in peripheral mononuclear cells, estimated by pretreating cells with pervanadate and ionomycin, significantly decreased at parturition and returned to normal levels as the cows' blood calcium returned to normal levels. Hypocalcemia, which is common in periparturient dairy cows, is associated with decreased intracellular calcium stores in peripheral mononuclear cells. Our data suggest that this is the cause of a blunted intracellular calcium release response to an immune cell activation signal. It is concluded that intracellular calcium stores decrease in peripheral blood mononuclear cells (PBMC) before parturition and development of hypocalcemia. This decrease in PBMC intracellular calcium stores before parturition and the development of hypocalcemia contributes to periparturient immune suppression.

KETOSIS AND MASTITIS SUSCEPTIBILITY

Ketosis is diagnosed whenever there are elevated levels of ketones in the blood, urine, or milk of a cow. The disease is always characterized by a decline in blood glucose as well. In lactation, the amount of energy required for maintenance of body tissues and milk production exceeds the amount of energy the cow can obtain from her diet, especially in early lactation when dry matter intake is still low. As a result, the cow must utilize body fat as a source of energy. Every good cow will utilize body reserves in early lactation to help her make milk. However, there is a limit to the amount of fatty acid that can be handled and used for energy by the liver (and to some extent the other tissues of the body). When this limit is reached, the fats are no longer burned for energy but begin to accumulate within the liver cells as triglyceride. Some of the fatty acids are converted to ketones. The appearance of these ketones in the blood, milk, and urine is diagnostic of ketosis. As fat accumulates in the liver it reduces liver function - and a major function of the liver in the dairy cow is to produce glucose.

On average, dry matter intake decreases by 20-30% 1 or 2 days before calving, and does not recover until 1 to 2 days after calving (Bertics et al., 1992). Interestingly, liver biopsies showed that liver triglycerides were increased 3-fold the day of calving. Triglyceride buildup in the liver is a much earlier phenomenon than previously assumed. Even more interesting; when cows were fitted with rumen fistulas and dry matter intake was not allowed to drop around the time of calving by forcing feed into the rumen, liver lipids and triglycerides increased only a small amount. The conclusion is that energy intake must not be compromised during the days around calving. Any factor restricting feed intake around calving (such as milk fever or retained placenta) increases fat accumulation in the liver, affecting the energy deficit of the cow and increasing the risk of fatty liver andketosis.

The fresh cow is also in negative protein balance shortly after calving. Generally this is not perceived to be as big a problem as the negative energy balance of early lactation but the typical cow will lose 37 lbs body protein during the first 2 weeks of lactation. Much of this body protein is being used to support the amino acid and glucose requirements of milk production. Therefore, in many respects, the dairy cow in early lactation is in a physiological state comparable to that of humans and rodents with prolonged protein-calorie restriction. Glutamine is the most abundant free amino acid in human muscle and plasma and is utilized at high rates by rapidly dividing cells, including leucocytes, to provide energy and optimal conditions for nucleotide biosynthesis. As such, it is considered to be essential for proper immune function. Interestingly, plasma glutamine does not seem to be depressed in the dairy cow at calving – when body fat mobilization is rapidly increasing and presumably protein catabolism would also be increasing rapidly. Plasma glutamine does decrease as the cow progresses into the early weeks of lactation (Meijer,1995).

Does the relative calorie-protein deficiency of the early lactation cow impact her immune response? In humans, protein-calorie restriction has severe effects on cell–mediated immunity. There is often widespread atrophy of lymphoid tissues and this can cause a 50% decline in the number of circulating T-cells. Surprisingly, antibody responses are intact and phagocytosis of bacteria is relatively normal. However destruction of bacteria within the phagocytes is impaired.

A few studies have been done on lymphocytes isolated directly from normal cows and cows with clinical and subclinical ketosis and then placed into culture. Leukocytes of cows with clinical symptoms and the highest concentration of ketones and free fatty acids in blood responded with the lowest levels of interferons alpha and gamma to three interferon inducers: Newcastle Disease Virus, phytohemagglutinin (PHA) and concanavalin A . Depression in PHAstimulated interferon synthesis correlated with a very low mitogenic response of blood lymphocytes. A correlation between the stage of ketosis and the level of interferon production in milk leukocytes was also observed (Kandefer-Szerszen et al., 1992).

Zerbe and colleagues examined the relationship between liver triacyl glycerol content and immunophenotypical and functional properties of neutrophils of dairy cows in the peripartum period. Increased liver TAG content, (> 40mg/g, considered the upper level of normal) was associated witha reduced expression of function-associated surface molecules on blood neutrophils. Moreover, in cows with high liver TAG levels, the antibody-independent and -dependent cellular cytotoxicity (AICC, ADCC) of blood PMN was markedly reduced. PMNs were also less capable of reactive oxygen species generation after stimulation with phorbol myristate acetate.

WHAT IS THE ENERGY COST TO MOUNT AN IMMUNE RESPONSE?

Little to no work has been done to examine this issue in cattle. However, in humans suffering from severe infection causing sepsis (various degrees of fever, increased WBC count, and acute phase protein production), the resting energy expenditure (determined by indirect calorimetry), increased progressively over the first week of the infection to around 40% above normal and was still elevated 3 weeks from the onset of illness. As an aside; over a 3-week period patients lost 13% of their total body protein (Plank and Hill, 2003).

No such measurements have been reported for cattle. However, if we are allowed to extrapolate and speculate we can go through a few calculations. Maintenance energy for a 600 kg dairy cow is approximately 9.7 Mcal Net energy / day. If the cow must also increase energy expenditure 40% to mount an inflammatory response the energy requirement increases by nearly 4 Mcal / day. This is roughly equivalent to a requirement that the cow consume an additional 2.4 kg of diet (assuming a diet that provided 1.65 Mcal Nel/kg). Can the periparturient cow, already in negative energy balance, be expected to successfully mount a rapid immune response? If she is in fact in negative protein balance as well, will her immune system produce the immunoglobulins and acute phase proteins necessary to fight an infection while it is still in the acute phase, in order to prevent it from escalating to a clinical infection?

RETAINED PLACENTA

. Gunnink's theory suggested the fetal placenta must be recognized as "foreign" tissue and rejected by the immune system after parturition to cause expulsion of the placenta. We hypothesized that impaired neutrophil function causes retained placenta. We examined the ability of neutrophils to recognize fetal cotyledon tissue as assessed by a chemotaxis assay which utilized a placental homogenate obtained from a spontaneously expelled placenta as the chemoattractant. Neutrophil killing ability was also estimated by determining myeloperoxidase activity in isolated neutrophils. Blood samples were obtained from 142 periparturient dairy cattle in 2 herds. Twenty cattle developed RP (14.1 %). Neutrophils isolated from blood of cows with RP had significantly lower neutrophil function in both assays prior to calving and this impaired function lasted for 1-2 wk after parturition. Addition of antibody directed against interleukin-8 (IL-8) to the cotyledon preparation used as a chemoattractant inhibited chemotaxis by 41%, suggesting one of the chemoattractants present in the cotyledon at parturition is IL-8. At calving, plasma IL-8 concentration was lower in RP cows (51 ± 12 pg/ml) than in cows expelling the placenta normally (134 ± 11 pg/ml) (Kimura et al., 2002). These data suggest neutrophil function determines whether or not the cow will develop RP. These data also suggest that depressed production of IL-8 may be a factor affecting neutrophil function in cows developing RP. This suppressed immune system could also explain why the same cows are more susceptible to mastitis. Retained placenta probably does not cause mastitis but is symptomatic of a depressed immune system.

REFERENCES

Bertics SJ, **Grummer** RR, et al., J Dairy Sci. 1992 Jul;75(7):1914-22.
Gunnink JW. Vet Q. 1984 Apr;6(2):52-4.
Kandefer-Szerszen M, Filar J, et al., Dtsch Tierarztl Wochenschr. 1992 Nov;99(11):440-3.
Kehrli ME Jr, Nonnecke BJ, **Roth** JA. Am J Vet Res. 1989 Feb;50(2):207-14.
Kimura K, Reinhardt TA, **Goff** JP. J Dairy Sci. 2006 Jul;89(7):2588-95.
Kimura K, **Goff** JP, Kehrli ME Jr, Reinhardt TA. J Dairy Sci. 2002 Mar;85(3):544-50.
Meijer GA, Van der Meulen J, Bakker JG, etal., J Dairy Sci. 1995 May;78(5):1131-41.
Plank LD, **Hill** GL. Proc Nutr Soc. 2003 May;62(2):545-52.
Zerbe H, Schneider N, et al., Theriogenology. 2000 Sep 15;54(5):771-86.

MACROMINERAL PROBLEMS OF RUMINANTS: LATEST ISSUES AND HOW TO FIX THEM

Jesse P. Goff, DVM, Phd
Ames, IA

PREVENTION OF MILK FEVER

Reducing Diet Cation-Anion Difference to Prevent Hypocalcemia

Reduction of the number of absorbable dietary cations and/or increasing the number of absorbable dietary anions greatly diminishes the incidence of hypocalcemia and milk fever in dairy cows. The major cations present in feeds and the charge they carry are Na (+1), K (+1), Ca (+2), and Mg (+2). The major anions and their charges found in feeds are Cl (-1), SO_4 (-2), and phosphate (assumed to be -3). In theory, all the cations and anions in a diet are capable of exerting an influence on the electrical charge and hence the pH of the blood. Cations or anions present in the diet will only alter the electrical charge of the blood if they are absorbed into the blood. Trace elements present in diets are absorbed in such small amounts that they are of negligible consequence to acid-base status. Organic acids such as the volatile fatty acids are generally absorbed in the un-dissociated neutral form. They also are rapidly metabolized within the liver so they have only a small effect on general acid-base balance under most circumstances.

The difference between the number of cation and anion particles absorbed from the diet determines the general acid-base balance of the body and therefore the pH of the blood. The cation-anion difference of a diet is commonly described in terms of mEq/kg DM (some authors prefer "mEq/100 g" diet DM) of just Na, K, Cl, and SO_4, although it must be kept in mind that Ca, Mg, and P absorbed from the diet will also influence blood pH. While DCAD equations provide a theoretical basis for dietary manipulation of acid-base status they are not necessary for formulation of mineral content of prepartum dairy cow rations. With the exception of K and Cl, the rate of inclusion of the other minerals can be "fixed".

The diet Na requirement of a late gestation cow is about 0.12%. A little salt is added to the diet to meet Na requirement and prevent pica (often manifest as a desire to drink urine from the floor). Unlimited access to NaCl in late gestation will increase the risk of udder edema.

At least two studies demonstrated that inclusion of calcium (Ca) in the diet at NRC required levels or several fold above NRC requirement does not influence the degree of hypocalcemia experienced by the cow at calving. Close-up diet Ca concentration should be between 0.85 and 1.0% Ca.

To ensure adequate concentrations of Mg in the blood of the periparturient cow the dietary Mg concentration should be 0.35-0.4%. This higher dietary Mg concentration allows the cow to take advantage of passive absorption of Mg across the rumen wall should active transport of Mg be compromised.

Dietary P concentration should be fed at a level to meet the NRC requirement for P in the late gestation cow. This is generally about 0.35 -0.4% P for most cows. A diet supplying more than 80 g P/d will block renal production of 1,25-dihydroxyvitamin D and will actually cause milk fever.

Dietary S must be kept above 0.22% to ensure adequate substrate for rumen microbial amino acid synthesis. Corn silage diets are notoriously low in sulfur. Diet S should be kept below 0.4% to avoid possible neurological problems associated with S toxicity.

Now, with the exception of K and Cl, the "variables" in the various proposed DCAD equations have become more or less "fixed". The key to milk fever prevention (at least with Holstein cows) is to now keep K as close to the NRC requirement of the dry cow as possible (about 0.85% diet K, which is very difficult to achieve). The key to reduction of subclinical hypocalcemia(i.e. not just milk fever), is to add Cl to the ration to counteract the effects of even low diet K on blood alkalinity. For formulation purposes, the concentration of Cl required in the diet to acidify the cow is approximately 0.5% less than the concentration of K in the diet. In other words, if diet K can be reduced to 1.3%, the Cl concentration of the diet should be raised to 0.8%. Then, the pH of the urine should be observed urine pH after 3-4 days. This is often a conservative approach and the final concentration of Cl needed to truly acidify the urine (pH 6.2-6.8) may have to be brought up within 0.4-0.3% of dietary K. NEVER START AN ANION SUPPLEMENTATION PROGRAM WITH HIGHER LEVELS OF CL. If you have over-acidified the cows at the onset it becomes very difficult to evaluate urine pH as feed intake will quickly be reduced. Always work the dry cow pen up to the adequate Cl dose. If dietary K cannot be reduced below 2.0% the diet Cl would need to be at least 1.5% to acidify the cow; a level of Cl in the diet likely to decrease dry matter intake independent of over-acidification. Chloride sources differ in their palatability and since achieving low dietary K can be difficult it is prudent to use a palatable source of Cl when formulating the diet. Ammonium chloride (or ammonium sulfate) can be particularly unpalatable when included in rations with a high pH. In high pH diets, a portion of the ammonium cation is converted to ammonia, which is highly irritating when smelled by the cow. Prilling the Cl (and SO_4) salts reduces the unpleasant taste of the salts. In our experience hydrochloric acid has proved the most palatable source of anions. Hydrochloric acid can be extremely dangerous to handle when it is procured as a liquid concentrate. Several North American companies now manufacture hydrochloric acid based anion supplements, which are safe to handle, effective, and reasonably palatable.

These are simply guidelines for anion supplementation used by this author and are based on inclusion of Ca, Na, S, Mg, and P at the levels outlined above. Urine pH is a good gauge of the appropriate level of anion supplementation. Urine pH on high cation diets is generally above 8.2. Limiting dietary cations will reduce urine pH only a small amount (down to ~7.8). For optimal control of subclinical hypocalcemia, the average pH of the urine of Holstein cows should be between 6.2 and 6.8, which essentially requires addition of anions to the ration. If the average urine pH is between 5.0 and 5.5, excessive anions have induced an uncompensated metabolic acidosis and the cows will suffer a decline in dry matter intake. Urine pH can be checked 48 or more hours after a ration change. Urine samples should be free of feces and made on midstream collections to avoid alkalinity from vaginal secretions. Anion-supplemented diets are generally fed for the last 3 weeks before calving, though the length of time these diets need to be fed to induce a compensated metabolic acidosis is no more than 4-5 days. When fed longer than six weeks, the urine pH of cows ready to freshen will have risen as the bone successfully buffers the acidity generated by the diet. In practice, this means when anions are fed the entire dry period the urine pH target for the final week before calving must be increased by about 0.5 units. Alternatively one can simply monitor urine pH in those cows that have been on the diet for at least one week but not longer than three weeks.

How do I interpret urine pH? Let's say I check urine pH of ten close-up cows. I am interested in the average – there is always one odd ball cow. Don't adjust the diet to accommodate the one cow!!

- Scenario 1 - average pH = 6.3 $\pm$ 0.6
  - Congratulations - you have induced a compensated metabolic acidosis.
- Scenario 2 - average pH= 7.4 $\pm$ 0.5
  - Add anions in small increments (raise Cl by 0.1%) , wait 5 days and check pH
- Scenario 3 - average pH 5.2 $\pm$ 0.5
  - You have induced an uncompensated metabolic acidosis= sick cows! Reduce anions so Cl is 0.2% less and check urine pH in 5 days
- Scenario 4 - 4 cows at pH 5.2 and 6 cows at pH 7.8
  - First – eliminate sorting & lack of bunk space (overcrowding) as possible causes. These are often the problem with rations where straw or forage is too long! If you feel certain this is not the cause of the variable urine pH then the variation is likely an indication of over-acidification. Cows at pH 5.2 this afternoon ate well this morning but will not this evening due to uncompensated metabolic acidosis. The cows at pH 7.8 this afternoon went off feed early this morning from over-acidification and the kidneys have over-compensated for the severe acidosis caused by the ingestion of the diet last night. These cows will likely eat well tonight and be over-acidifed tomorrow morning. Reduce anions to bring chloride down by 0.2%, , wait 5 days, check pH, and start increasing chloride as needed by 0.1% increments until average urine pH is in target zone.

Feeding a Ca Deficient Diet To Stimulate PTH Secretion Pre-Calving To Prevent Hypocalcemia

When cows are fed a diet that supplies less Ca than they require, the cows are in negative Ca balance. This causes a minor decline in blood Ca concentration stimulating PTH secretion, which in turn stimulates osteoclastic bone resorption and renal production of 1,25-dihydroxyvitamin D. At parturition the cow's osteoclasts are already active and in high numbers and the lactational drain of Ca is more easily replaced from bone Ca. If provided with Ca in the lactation ration, the previous stimulation of enterocytes by 1,25-dihydroxyvitamin D will allow efficient utilization of dietary Ca and the cow avoids hypocalcemia. This works even in the face of metabolic alkalosis as metabolic alkalosis reduces but does not totally eliminate tissue PTH sensitivity. **Prolonged** exposure to high PTH levels induced by the low Ca diet overcomes the reduced tissue sensitivity to PTH.

The absorbable Ca requirement of the late gestation cow is from 14 g /day in Jerseys to about 22 g in large Holsteins. A truly low Ca diet must supply considerably less absorbable Ca than required by the cow if it is to be capable of stimulating PTH secretion. For example, a 600 kg cow consuming 13 kg DM must be fed a diet that is less than 0.15% **absorbable** Ca if it is to provide less than 20 g absorbed Ca/ day.

Oral Ca Treatments at Calving To Pevent Hypocalcemia

Ca administered to the fresh cow may arguably be called a treatment rather than a preventative measure for hypocalcemia. Briefly, the concept behind oral supplementation is that the cow's ability to utilize active transport of Ca across intestinal cells is inadequate to help her maintain normal blood Ca concentrations. By dosing the animal with large amounts of very soluble Ca orally it is possible to force Ca across the intestinal tract by means of passive diffusion between, not across, intestinal epithelial cells. Best results are obtained with doses of Ca between 50 and 125 g Ca / dose. Ca chloride has been used but can be caustic. Large or repeated doses of Ca chloride can induce an uncompensated metabolic acidosis in the cow, especially if the cow is already being fed an acidogenic diet. Ca propionate is less injurious to tissues and is not acidogenic. It has the added benefit of supplying propionate, a gluconeogenic precursor. For best control of hypocalcemia a dose is given at calving and again 24 hrs later. Larger or more frequent dosing can be toxic. Toxic doses of Ca can be delivered orally – about 250 g Ca in a soluble form will kill some cows. The benefit of adding oral Ca on top of a properly formulated low DCAD program does not seem to warrant the added expense.

ASSURING ADEQUATE MAGNESIUM FOR RUMINANTS

Magnesium (Mg) is included in dairy rations for two reasons: to maintain adequate levels of Mg in the blood and as a rumen fluid alkalinizer. Maintaining adequate blood Mg concentrations in the cow is essential to calcium homeostasis as hypomagnesemia interferes with bone and kidney tissues' ability to respond to parathyroid hormone. Hypomagnesemia is still a common cause of hypocalcemia and milk fever at calving and is, in my experience, the most common cause of "mid-lactation" milk fever cows! Cows cannot mobilize body stores of Mg to maintain blood Mg levels. Mg homeostasis is dependent on daily absorption of diet Mg. For some unknown reason, adult ruminants are nearly totally dependent on the forestomach for absorption of Mg. The small intestines do not absorb Mg as they do in monogastrics. Instead it is a site of net secretion of Mg . Two mechanisms exist for absorption of Mg across the rumen wall. One is an active transport process, allowing the cow to absorb adequate Mg even when the diet is as low as 0.24% magnesium. Unfortunately, the integrity of this Mg pump is often compromised by diets high in potassium and perhaps non-protein nitrogen, rendering this means of Mg absorption unreliable. A second, passive transport system for Mg exists which is not poisoned by potassium. This transport system only works when Mg in solution in the rumen fluids is high. It contributes to Mg homeostasis when the diet Mg exceeds 0.35%. For either transport mechanism it is critical that the Mg utilized in the ration be available to the cow. The MgO must be soluble in rumen fluid. The feed industry utilizes MgO which is about 54% Mg (>58% Mg MgO is suitable only for making steel!). MgO takes up little room in the ration, costs less, and is more palatable than some other sources of Mg. Unfortunately, there is tremendous variability in MgO quality – and price does not always reflect bioavailability. The best sources are also ground to a fine powder to increase solubility at rumen fluid pH. A quick test can tell you the relative availability of MgO sources. Place 3 g of the MgO in a container and slowly add 40 ml white vinegar (5% acetic acid). Cap the container and shake vigorously for ten seconds. Let the mixture sit for 30 minutes, thencheck the pH. Vinegar will typically show a pH of 2.6-2.8! The best MgO sources will bring the pH up to 8.2. The worst sources will increase the pH to just 3.8. Since pH is a log scale, this represents >10,000 fold difference in the number of hydrogen ions buffered. Remember in lactating rations, we expect MgO to help combat rumen acidosis- and we are not getting that action from these insoluble MgO sources. For close-up diets, it is often wise to use Mg chloride or Mg sulfate to ensure adequate and available Mg for the diet. These two sources are very soluble in rumen fluid and have the advantage that the anion they containcan help reduce diet DCAD to help acidify the cow's blood.

REFERENCES

1. Goff JP. Vet J. 2008 Apr;176(1):50-7.
2. Goff JP. Vet Clin North Am Food Anim Pract. 2004 Nov;20(3):471-94,
3. Goff JP, Horst RL. Acta Vet Scand Suppl. 2003;97:51-6.

EVALUATION OF 25 OH VITAMIN D IN CANINE CANCER AND HEART DISEASE: IS THERE A LINK?

Joseph J Wakshlag, DVM, PhD, DACVN, DACVSMR

Vitamin D Metabolism

The metabolism of vitamin D holds a circuitous path in canine and feline metabolism considering in the mid 1990's it was discovered dogs and cats do not undergo similar metabolism of cholesterol into the common precursor 7-deoxycholesterol. Typically 7-deoxycholesterol is metabolized at the skin surface through UV light exposure to form the precursor Vitamin D3 or cholecalciferol, Dog and cats rather metabolize 7-deoxycholesterol readily to other sterol derivatives.[1] This explains why dogs have not shown diurnal fluctuation in serum D in polar environments like some other polar species.[2] This finding led to inclusion of cholecalciferol in the diets of dogs and cats as a requirement which is currently mandated by AAFCO to be in pet food at the concentration of 500 IU/kg of food based on a 3,500 kcals diet or 125 IU (3.75 ug)/1000 Kcals. The pet food industry typically utilized cholecalciferol since studies in cats show that cats preferentially utilize cholecalciferol (D3) over ergocalciferol (D2) in the synthesis of active vitamin D.[3]

The metabolism of vitamin D3 occurs rapidly in the liver to 25–OH D3 which can be measured easily in a handful of diagnostic laboratories in the country. As such it stands a surrogate marker for vitamin D3 sufficiency as the active form of the vitamin 1,25-OH D3 is only formed when needed through metabolism by 1 hydroxylase in the renal epithelium when calcium in the blood is insufficient and is up-regulated by parathyroid hormone activity.[4] Therefore, to truly examine active vitamin D quantification of 1,25-OH D3 is essential and rarely performed clinically. Equally as important as the activation of vitamin D 3 to make active 1,25-OH D3; is the inactivation of 1,25-OH D3, which is initiated by hepatic cytochrome P450-24 (CYP24) primarily in the liver making either 24,25-OH D3 or 1,24,25-OH D3 which is preferentially excreted though renal mechanisms. This hepatic and renal activation and deactivation and excretion is highly regulated in vitamin D homeostasis.[4]

Vitamin D and Cancer

Substantial epidemiological data indicate that low serum vitamin 25-OH D3 concentrations, is influenced in humans by things such as geographic latitude or season, and vitamin D3 is associated with an increased risk of a variety of cancers in people.[5-13] The mechanisms by which vitamin D status can alter cancer development are still being studied, but what is now known is:

1) Many neoplastic cell types contain vitamin D receptors. [14]
2) When receptors are activated by 1,25-OH D3, the most active metabolite of vitamin D, they induce differentiation and inhibit proliferation, invasiveness, angiogenesis, and metastatic potential. [14]
3) Recent murine models whereby vitamin D receptor has been knocked out of the genome and these mice develop cancer more frequently among other problems.[14]
4) Mice fed diets deficient in vitamin D deficient for extended periods of time develop cancer more frequently.[15]
5) Vitamin D cancer hypothesis has been supported by laboratory studies that show 1,25-OH D3 has potent antineoplastic activity against numerous tumor cells.[16]
6) Much of the human data now suggests similarly that those humans that are vitamin D sufficient have decreased rates of certain cancers including breast, prostate and colon cancer. [8-13]

Vitamin D and Heart Disease

There is increasing evidence that vitamin D influences the cardiovascular (CV) system at many sites and plays a role cardiac function.[17] The possibility that Vitamin D might play a role in CV disease has arisen from some compelling evidence.

1) Vitamin D receptor knockout mice develop typical signs of CHF such as cardiac hypertrophy, overstimulation of the renin-angiotensin-aldosterone system, hypertension, and increased levels of atrial natriuretic peptide.[15]
2) Most groups of CHF-patients studied to date have had insufficient concentrations of 25-OH D3 with one large cohort study in adults with CHF showing 34% lower 25-OH D3 levels than healthy subjects; while a similar large cohort of patients where each 10 ng/mL lower 25-OH D3 concentration was associated with a 25% greater relative hazard of heart disease.[18-23]
3) Finally, calcitriol, the active form of vitamin D can normalize the impaired contractility of the myocardium that is observed under experimental vitamin D deficiency in rodent models.[24]

Canine Vitamin D Sufficiency in Acute/Chronic Disease

The species differences leaves dietary intake as the primary means of obtaining vitamin D_3 for conversion to 25-OH D3 thereby is eliminating a potentially confounding variable during investigation in dogs and cats. Five recent studies have studied the idea of serum 25-OH D3 concentrations in normal dogs versus a specific disease entity. A very comprehensive study examined 25-OH D3 concentration, ionized serum calcium, PTH and 1,25-OH D3 in 4 groups of dogs including control, non-systemically ill hospitalized patients IBD patients and PLE patients. Findings from this study suggest that only PLE dogs had a decreased 25-OH D3 concentration and 1,25 OH D3 concentrations while IBD patients showed no differences and potentially even elevated 1,25-OH D3 concentrations. In addition the PLE patients have low albumin and ionized calcium and elevated PTH suggesting true secondary hyperparathyroidism.[25] Without any intake date in this study a cause cannot be definitively determined, however malabsorption is thought to be the underlying cause.

Another recent report examined circulating 25-OH D3 in a cross-section of CHF dogs compared to normal dogs. This report also performed a prospective cohort study to determine the impact of 25-OH D3 status on incident cardiovascular CV outcomes in dogs with CHF. The findings suggest that dogs with CV disease have marginally yet significantly lower serum 25-OH D3 concentrations than normal dogs; and that dogs stratified into two cohorts of CHF based on vitamin D status showed those with lower vitamin D actually have a decreased survival interval. Equally if not more interestingly is when food questionnaires assessed presumed vitamin D intake, vitamin D intake was higher for those dogs with CV disease.[26]

In canine oncology a similar study was performed. Cutaneous mast cell tumors (MCTs) occur more frequently in dogs than any other species; MCTs account for 7-21% of all canine skin and subcutaneous tumors.[27,28] Although MCTs have been reported in most breeds, the risk is increased in Labrador Retrievers.[28] Examination of one breed allowed for elimination of a potential confounder and it is known that the majority of canine MCTs express the vitamin D receptor, a likely prerequisite to respond to vitamin D or its derivatives.[29] Recent *in vitro* evidence shows that 1,25-OH D3 inhibits canine mastocytoma proliferation,[16] and that high oral dosing of 1,25-OH D3 induces tumor regression in dogs with spontaneously occurring MCTs, yet not without clinical signs of toxicity.[30] The objective of the Labrador mastocytoma study was to investigate 25-OH D3 serum concentrations as an indicator of disease accounting for individual dog vitamin D intake. Much like the CV study the serum concentration s of normal control dogs were marginally, yet significantly higher than that of the Labradors with mastocyoma and the intake between the two groups was not significantly different, however there was a trend towards enhanced intake in the mastocytoma group.[31]

Another recent study examined similar questions regarding 25-OH D3 status in dogs with experimentally induced endotoxemia. The study examined 6 dogs that were fasted during the experiments whereby either PBS or lipopolysaccharide were infused into the cepahalic vein and multiple parameters including calcium status, PTH and 25-OH D3 status were assessed. Besides a rise in PTH, a decrease in 25-OH D3 was observed within 2-4 hours of treatment suggesting an increased loss of vitamin D3. The reason for this rather dramatic decrease could not be elucidated from the study design and was an unexpected finding, yet renal or intestinal losses were the proposed mechanisms as well as a decrease in vitamin D binding protein which has been observed in sepsis.[32]

Renal disease has previously been associated with urinary losses of B vitamins, and this has been further verified with recent data examining complete vitamin and mineral status of 19 dogs with chronic renal failure versus 64 similarly aged healthy control dogs. This investigation suggested that the 25-OH D3 serum concentration in dogs with renal failure was approximately half of that in healthy controls dogs. Although the dietary intake was not part of this investigation the authors postulated that the decreased serum 25-OH D3 concentrations were due to either hypophagia or proteinuria resulting in loss of vitamin D binding protein.[33]

What does it mean?

The relevance to these finding remains clinically elusive and a few major questions need to be answered to address these findings regarding serum 25-OH D3 concentrations:

Question: Is this clinically significant?

Answer: This will depend on whether there is a disease outcome that can be ameliorated with supplementation. In humans the current thought is to only supplement deficiency; and this is well characterized in human populations as the inflection point where PTH begins to rise due to deficiency. This point is reached in humans at a much lower concentration than concentrations typically observed in canine serum. Adequate supplementation in commercial canine diets results in higher baseline serum concentration and the inflection point where PTH begins to rise has yet to be fully determined.

Question: What is it about these diseases that result in low serum 25-OH D3?

Answer: It seems fairly clear in PLE that absorption may be a significant issue, however in CV disease, renal disease and mastocytoma, the reasons are less clear, particularly since calculated intake appears to not be an issue. Yet, it must be clear that this is calculated D3 assumed intake and not based on feed analysis, therefore we do not truly know the vitamin D content of these diets. Assuming similar intakes, then the reasons may be that drug treatment or chronic inflammation may alter hepatic metabolism of vitamin D through upregulation of CYP24 and excessive renal disposal. There is little literature regarding chronic inflammation or drug therapy at induction of CYP24. One intriguing study in humans suggests that a cultural chewing of the "betel nut" made into a complex chew, similar to actions of tobacco chewing, is directly correlated to low serum 25-OH D3 in South Asians.[34] There are likely genetic mechanism at play such as vitamin D receptor, alpha 1 hydroxylase activity (CYP 1), and CYP 24 hepatic haplotypes suggesting genetic links which may be further related to the genetics behind diseases like CV disease and mastocytoma. Further investigations are needed to determine if other cancers display this serum anomaly and whether inherent or external factors are involved in diminished 25-OH D3 concentrations in dogs.

Question: Is vitamin D supplementation worthwhile or necessary?

Answer: the most current recommendations in human health suggest that supplementation is not necessary when taking in 600-800 IU of D3 meeting the recommended daily intake, particularly since D3 intervention has not proven successful in disease prevention.[35,36] In dogs and cats daily intake is usually sufficient when consuming a majority of calories as commercial pet food, and even in disease states there is no apparent reason for supplementation particularly in light of many supplements being formulated for humans on the market and the risk for over-supplementation of small breed dogs. Until there is proof that supplementation helps increase longevity or clinical signs associated with a specific disease process we must assume that sufficiency particularly in light of the severe clinical manifestations of over-supplementation of cholecalciferol.

REFERENCES

[1.] How KL, et al. **Gen Comp Endocrinol** 1994; 96: 12-8. [2.] Griffiths P and Fairney A. **Comp Biochem Physiol B**. 1988; 91: 511-6. 3.Beitz Ed. **Nutrient Requirements of Dogs and Cats.** Ch 8: Vitamins; pp194-246. 4.) Battault S et al. **Eur J Nutr**, 2012 (e-ahead pub).[5.] McCullough ML, et al. **Am J Epidem**, , 2010; 45: 1-15. [6.] Trump DL, et al. **Br J Urol Intern** 2009; 30: 1-6. [7.] Bertone-Johnson ER, et al. **Canc Epidem Biomarker Prev** 2005; 14: 1991-1997. [8.] Ng K, et al. J Clin Oncol 2008; 26: 2984-2991. [9.] Goreham ED, et al. **J Ster Biochem Molec Bio** 2005; 97: 179-194. [10.] Abbas S, et al. **Carcinogene**sis 2007; 29; 93-99. [11.] Yin L, E et al. **Aliment Pharmacol Ther** 2009; 30: 113-125. [12.] Chlebowski RT. **Breast Canc Res,** 2011; 13: 217-225. [13.] Schottker B et al. **Aging Res Rev,** 2012 (e ahead print) [14.] Deeb KK, et al. Rev Canc 2007; 7, 684-700. [15.] Welsh J (2004) **Am J Clin Nutr**, 80: 1721S-4S. [16.] Rassnick KM,et al. **Cancer Chemo Pharm** (2008) 62, 881-891. [17.] Bouillon RL et al. **Endocr Rev** 2008; 29:726-776. [18.] Dobnig H, et al. **Arch Intern Med** 2008;168:1340-1349. [19.] Pilz S, et al. **J Clin Endocrinol Metab** 2008;93:3927-3935. [20]Zittermann A, et al. **Clin Chem** 2009;55:1163-1170. [21.] Shane E, et al. **Am J Med** 1997;103:197-207. [22.] Schleithoff SS, et al. **Am J Clin Nutr** 2006;83:754-759. [23]Zittermann A, et al. **J Am Coll Cardiol** 2003; 41:105-112. [24.] Lee JH, et al. **J Am Coll Cardiol** 2008;52: 1949-1956. [25] Gow AG et al **J Small Anim Pract.** 2011; 52: 411-8. [26.] Kraus MS et al. **J Vet Intern Med**, 2012 (Under Review). [27.]Dorn CR, et al. **J Nat Canc Inst** 1968; 40, 307-318. [28.] Cohen D, et al. **Canc Res** 1974; 34, 2859-2868. [29.] Russell DS, et al. **J Comp Pathol** 2010; 143: 223-226 [30.] Malone EK, et al. **Vet Comp Oncol** 2010; 8:209-220. [31.] Wakshlag JJ, et al. **Br J Nutr** 2011;106 Suppl 1:S60-3. [32.] Holowaychuk M.K. et al. **J Vet Intern Med**, 2012; 26: 244-251. [33]Galler A et al. **Vet J**, 2012:226-231.[34.] Ogunkalode WB et al. **J Clin Endocrinol Metab**, 2008; 91: 2612-2617. [35.] Aloia JF. **J Clin Endocrinol Metab**, 2011, 96: 2987-2996.[36] Buttiglieаro C et al. **Oncologist,** 2011, 16: 1215-1227.

FUNCTIONAL FOODS: FRIEND OR FOE?
Joseph J Wakshlag DVM, PhD, DACVN, DACVSMR

There are a vast number of nutraceuticals utilized by clients for their dogs and cats, which is increasing in veterinary medicine [1], however there is a paucity of information regarding their bioavailability, and there is even less information regarding their activities in canine and feline cancer cells; the disease where these supplements are often implemented. Though these molecules may show promising for prevention or even treatment in rodent models and in-vitro systems, there is a tremendous amount of work to be done to verify their pharmacokinetics, metabolism and activity in canine and feline systems. The major nutraceuticals that are derived from functional foods where there is some examination of bioavailability in dogs and or cats include the flavonoids from green tea extract, soybeans and grape skins; as well as the carotenoid lycopene found in many red fruits and vegetables. These stand as examples of where we are at in our understanding of these molecules in veterinary medicine.

FLAVONOIDS

In foods there are 6 classes of flavonoids that are abundantly expressed that can be consumed in large enough quantities to be considered pharmacologic, hence, these ones have been the focus in the human literature of the 6000 flavonoids that are found in nature.[2-5] Human epidemiologic data suggests that flavonoid consumption is around 1 gram per day in the United States.[4] This seems robust in light of other studies in Denmark and the Netherlands suggesting flavones, flavonols and flavanone consumption to be closer to 25-30 mg per day.[4] For many of these molecules to achieve pharmacologic concentrations in the body it is usually necessary to use extracts or concentrated forms of these functional foods as nutraceuticals.

Green Tea Extracts (GTE)

The flavanols can be found in wines, teas, and cocoa, yet the extensive processing of these products lead to theaflavin and other conjugates in most of these products with the exception being green teas which are not as highly processed thereby retaining more natural flavanols.[6-8] Of particular interest is epigallocathecin-gallate (EGCG) which is likely to be the most potent of the falvan-3-ols and is most abundant in green tea.[6] Part of the anti-inflammatory effects of GTE and EGCG are through Nf-κβ inhibition of macrophage chemotactic protein -1 production, diminishing the cytokine response and COX activity in inflammatory cells. Much like other flavonoids the concentrations in cell culture, to achieve growth inhibition of neoplastic cells 10 uM of higher needs to be used, with only a handful of studies showing any growth suppression between 1-10uM, which is potentially the upper limit of physiologic concentrations. Dog studies on pharmacokinetics comparable to a human dose of green tea polyphenols (250-600 mg a day) given to 25 kg dogs at a dose of 200 mg daily of green tea extract showed similar pharmacokinetics to that observed in people, with EGCG being absorbed without modification and maximal serum concentration at around 2 hours, and rapid elimination by 6 hours post consumption, except for EGC glucuronide (hepatic conjugated form of EGC) which has a slightly longer ½ life with elimination closer to 10 hours.[9] Additionally, the ECG glucuronide reached serum concentrations of 800 nM, while EGCG reached Cmax of 400 nM. Chronic dosing schemes in dogs at 200 mg/kg and up led to toxic events in all groups including the lowest treatment group which was 200 mg/kg body weight. [10] All dogs showed lack of weight gain, some degree of anorexia, and erythroid/myeloid hypoplasia and myeloid hyperplasia in bone marrow resulting in anemia and high WBC counts. Part of this toxic effect was attributed to giving the extract in an unfed state and when follow up studies were done with the 200 mg/kg dose it proved to be far less toxic when given with a meal. [10] A follow up study using a similar product for 13 weeks at 50, 150 and 300 mg/kg of EGCG proved that the 50 mg/kg dose was safe in fasted dogs, while the 150 and 500 mg doses proved to be lethal over time with similar signs as the previous study with additional renal tubule necrosis and hepatic necrosis.[11] A recent study in obese dogs suggested that green tea extracts (not straight EGCG) can be used at 80 mg/kg body weight without some clinical effects on lipid and skeletal muscle metabolism, making around 50 mg/kg of ECGC or green tea extract the upper end for safe consumption based on present information.[12] Furthermore, at similar doses, a recent paper suggests that the myeloid hyperplasia response may be due to the up-regulation of pro-proliferative pathways in canine leukocytes based on real time PCR.[13] Hence, making the use of green tea and green tea extracts tenuous at best.

Isoflavones

Isoflavones are found almost exclusively in leguminous plants and their concentrations are highest in soybeans.[14] The two major isoflavones found in legumes are genistein and dadzein which have estrogenic activity and have been the most widely examined flavonoid in dogs and cats to date.[15-18] Genistein is considered the more bioactive of the two isoflavonoids predominantly found in the diet. Genistein has been studied extensively in cell culture systems including leukemia, lymphoma, ovarian, cervical, leiomyoma, melanoma, neuroblastoma, gastric pancreatic, prostate and breast cancer cells and has the ability to suppress cell cycle at the Go or the G2M stage in different cell types through down regulation of cyclin B or upregulation of cyclin dependent kinase inhibitors like p21, p27, and p16. [19] AKT phosphorylation, a potentiator of cell survival, is also inhibited through dampening of upstream MAP kinases or tyrosine kinase activity from receptors such as EGF, HER-2 and PDGF. [19] Additionally, genistein can cause inhibition of IGF-1 signaling though inhibition of STAT protein phosphorylation. Additional evidence shows that genistein also alters MMP-2 and MMP-9 production suggesting a role in preventing cell migration and metastatic behavior of neoplastic cells. [19] Similar studies have been performed in canine lymphoma cell lines showing this alteration in mitochondrial permeability leading to cell death.[20] The most encouraging result from these canine cell culture studies is that alteration in cell proliferation can been observed in the range of 5-10 uM, which is close to physiologically achievable concentrations. Genistein is one of the few flavonoids where extensive bioavailability work has been performed in cats and dogs due to its modest estrogenic influence, which may have ramification in satiety as a feedback mechanism to the hypothalamus, like normal estrogen.[17-18] Long term toxicity studies were performed using orally administered genistein in beagles. Other than changes in reproductive tissue weights (increased uterine weight in females and increased testicular weight in males) even at the highest concentrations (500 mg/kg body weight) daily administration caused no adverse side effects, although reproductive performance was not evaluated. [19] In cats short term toxicity and oral feeding studies post castration showed no toxicity other than changes in behavior, with the oral dosing showing a maximal serum concentration at around 4 hours at about 1 uM for the aglycone and 15 uM for conjugated genistein, and half lives of both the aglycone and glycosylated forms of around 17 hours and 9 hours respectively.[17,18]

Resveratrol

Resveratrol is considered a plant antibiotic that is made during stress and acts as a natural inhibitor of cell proliferation. It can be found in high concentrations in red wine and peanuts and is purported to be one of the benefits of red wine consumption, however consumption would have to be excessive.[4,5] It can also be found in grapes, berries, red cabbage and plums, as well as 70 plant species.[2,3] The anticarcinogenic properties of resveratrol are attributed to a number of cell functions including initiation of cell cycle arrest, suppression of tumor cell proliferation, induction of apoptosis, inhibition of metastasis, diminished inflammation and anti-angiogenesis, with much of this data being generated in cell culture systems.[21,22] The cell actions of resveratrol are numerous including: tyrosine kinase inhibition, MAP Kinase activation, Nf-κB inhibition, activation of antioxidant response elements, cytochrome p450 pre-carcinogen activation, estrogen and androgen inhibition capability, anti-inflammatory capabilities. These anti-neoplastic properties seem to differ depending on cell type examined with a majority of the studies showing a need for extremely high concentrations (above 10 uM) with only anti-estrogenic and antioxidant response element studies suggesting lower concentrations for these cellular effects (less than 10 uM). [22] Numerous clinical trial with resveratrol do suggest potential anti-athelosclerotic, anti-inflammatory and metabolic implications, as well as some ongoing studies in colon carcinogenesis showing promise.[23] Until recently the only study examining dogs showed that sub-chronic feeding of a resveratrol rich grape color extract from grape skins was fed to dogs for 90 days at either 7.5% or 15% of their overall dietary consumption with the only adverse effect being weight loss in the high dose group when compared to control dogs, however there are a range of flavonoids (anthocyanins) besides resveratrol in grape extracts.[24] More recently a study was performed examining the effects of a resveratrol supplement at concentrations from 200 mg/kg to 1200 mg/kg body weight. Dogs were supplemented for 13 weeks and maximal serum concentrations in the 200 mg/kg groups exceeded 10 uM, while the 1200 mg/kg group was around 30 uM. No adverse side effects to this chronic supplementation were observed when examining hepatic and renal parameters, with the only adverse side effect being that both male and female dogs on the 1200 mg/kg supplementation did not gain weight during the clinical trial like the other groups suggesting possible appetite suppression based on food intake data.[25]

CAROTENOIDS

Carotenoids fall into 2 major classifications; the hydrocarbon carotenoids and the xanthophylls. This distinction takes into account the oxygenation or hydroxylation of the structures, thus oxygenated or hydroxylated carotenoids are considered xanthophylls. Through there are over 600 different carotenoids only a handful are extensively studied in the scientific literature due to their prevalence in foods and potential associations with decreased risk of cancer (lycopene, β-carotene, α-carotene, β-cryptoxanthin, lutein, zeoxanthin, astaxanthin and fucoxanthin.[26]

Absorption of carotenoids within considering normal fruit and vegetable intakes range from 10-50% for most carotenoids, lending to the relatively high plasma concentrations found in people, but the absorption can be influenced by many variables including fasting, processing, cooking, dietary constituents and whether supplied as a supplement (often in oil or water soluble microparticles). Processing of foods and heating of foods breaks down cell walls and microcrystalline structure where carotenoids are stored releasing them generally leading to better absorption. However, over cooking and release also allows for oxidation, yet in most cases carotenoids become more available with cooking and processing of foods.[27,28] For example, lycopene from tomato paste is far more available than from crushed tomato or fresh tomato. The intestinal absorption is dependent on dietary fat as carotenoids are assimilated into fat micelles for enterocyte uptake and lymphatic uptake like typical dietary fats. Once absorbed as chylomicrons into the lymphatics, most carotenoids are distributed into high density and low density lipoproteins and are distributed to tissues with various carotenoids showing slightly different tissue concentrations.[27] Besides fat influencing absorption, there have been associations between phosphatidylcholine and fiber content of diets which can negatively influence overall absorption of a of carotenoids, however on a comparative basis to other natural products carotenoids are well absorbed and retained in tissue making them attractive chemopreventative and potential therapeutic modalities.[28]

Lycopene

Lycopene is the general backbone precursor from which all other carotenoids are formed and is one of the most well studied carotenoids from an epidemiologic perspective due to its extremely high content in processed tomato products.[27-29] Lycopene's mechanism of action at the cellular level is similar to other carotenoids, but in some aspects (antioxidant response element signaling and connexin upregulation) are superior to other prevalent carotenoids in the diet. Lycopene at the low micromolar concentrations have been associated with decreasing cell proliferation through increases in cyclin D transcription and translation in prostate carcinoma cells, as well decreasing Bcl and increasing Bax expression altering mitochondrial permeability causing apoptosis.[30,31] Recent work in canine osteosarcoma cells shows apoptosis induction associated with diminished AKT signaling. [32] Carotenoids in general have the ability to increase cell-cell contact through enhancing gap junctions via enhanced connexin 43 transcription and translation, which can suppress proliferation.[33] Lycopene in particular has the unique ability to enhance Nrf-2 translocation from the cytosol to the nucleus enhancing anti-oxidant response element genes like glutathione synthase improving anti-oxidant status.[34]
Epidemiologically many retrospective case controlled studies showed that serum lycopene concentrations or processed tomato intake correlated with the relative cancer risk for many cancers including prostate, breast, gastric, lung, colorectal and skin. However, longitudinal cohort studies and nested case control studies were less convincing with as many studies showing a lack of significance for all carotenoids including lycopene.[35,36] Lack of efficacious interventional studies with lycopene across many cancers types except for prostate cancer has led the FDA consensus statement to strongly negate the effects of lycopene with more liberal wording when elaborating on the benefits in prostatic disease.[36]

In an extended exposure study with a small group of dogs receiving 30 mg/kg daily for 28 days showed that serum concentrations reached just over 1 uM from day 7 onward and that even after dosing was discontinued (5 days) serum concentrations were still at 200 nM proving that lycopene is not rapidly eliminated from the body. [37] Additionally, tissue concentrations were examined showing a range of accumulation in tissues with most tissues having concentrations in the 100-1 uM/kg range while colon and jejunum were in the 1.6 uM/kg range and ileum was around 2.9 uM/kg. Lymphoid tissue contained 5-6 uM/kg, while adrenals reached 18.5 uM/kg and liver was over 50 uM/kg.[37] All of these concentrations were similar to what could decrease cell proliferation of osteosarcoma cells in cell culture.[32]

The "Antioxidant"Conundrum

Much of the evidence regarding the flavonoids and carotenoids is that they may be effective in the prevention of cancer with the most influential veterinary evidence being that Scottish Terriers that eat vegetables as part of their dietary regimen have a lower incidence of transitional cell carcinoma.[38] These functional ingredients are thought to be the reason for the decreased prevalence; many of which are termed "antioxidants", yet have a plethora of cell signaling and cell altering capabilities as well. The most important question regarding the use of these functional food components is: will these compounds hinder or help common chemotherapeutic or radiation treatments? There is little known about how survival and outcomes are effected when implementing these nutraceuticals or functional foods during traditional treatments. There is some in-vitro evidence that there may be some mild protective effects for canine cancer cells as well as some mild synergistic effects to improve cytotoxicity which appears to be dependent on the compound and the chemotherapeutic utilized.[32,39,40] Additionally, there is some evidence that commonly used nutraceuticals may actually decrease some of the cytotoxic hepatic and renal effects of commonly used chemotherapies such as CCNU and platin analogs.[41,42] These concepts regarding nutraceutical use have not been well studied, particularly in veterinary medicine, and considerable research is necessary before recommendations can be made regarding their possible utility or harm.

REFERENCES

[1] Lana SE, et al. **J Am Anim Hosp Assoc**. 2006;42:361-5 [2] Bravo L. **Nut. Rev**. 1998; 56: 317-333. [3] Ross JA et al. **Ann Rev. Nutr.** 2002, 22: 19-34. [4]Crozier A et al. **Nat. Prod. Rep**. 2009; 26: 1001-1043. [5] Khan N et al. **Antiox Redo Sign** 2008: 10: 1-36. [6] Naghma K et al. **Canc Lett.** 2008; 269: 269-280. [7] Butt MS et al. **Crit Rev Food Sci Nut**, 2009; 49: 463-473. [8] Boehm K et al **Chochrane Database Rev**, 2009, Issue 3. pp 1-25. [9] De-Lourdes Mata-Bilbao, M et al. **Br J of Nutr,** 2008; 100: 496-502. [10] Kapetanovic IM et al. **Tox**, 2009; 260: 28-36. [11]Isbrucker et al. **Food and Chem Tox**, 2006; 44: 636-650. [12] Serisier S et al. **Br J Nutr**, 2007; 99: 1-9. [13] Salas A et al. **J Nutrigenet Nutrigenomics**, 2009; 2: 43-52. [14] Jian L. **Mol Nutr Food Res**. 2009; 53: 217-226. [15] Cerundolo R et al. **Am J Vet Res**, 2009; 70: 353-359. [16] McClain RM et al. **Food and Chem Tox**, 2005; 43: 1461-1482. [17] Bell KM et al. **J Nut**, 2006; 2004-2006. [18] Cave NJ et al. **J Vet Pharmacol Ther**, 2007; 30: 227-335. [19] Banergee S et al. **Canc Lett**, 2008; 269: 226-242. [20] Jamadar-Shroff V et al. **Clin Canc Res**, 2009; 15: 1269-1274. [21] Goswami S.K. et al. **Canc. Lett**. 2009, 284; 1-6. [22] Bishayee A. **Cancer Prev Res**, 2009; 2: 409-418 [23] Wenzel E et al. **Metab Mol Nutr Food Res**, 2008, 49: 472-481. [24] Becci PJ et al. **Food Chem Tox**, 1983; 21: 75-77. [25] Johnson WD et al. **Food Chem Tox**, 2011; 49: 2219-3327. [26] Krinsky NI et al. **Mol Aspects Med**, 2005; 26: 459-516. [27] Faulks RM et al. **Biochim Biophys Acta**, 2005; 1740: 95-100 [28]Yonekura L et al. **Mol Nutr Food Res**, 2007; 51: 107-115. [29] Rao AV et al. **Pharm Res**, 2007; 55: 207-216. [30] Van Breemen RB et al. **Canc Lett,** 2008; 269: 339-351. [31] Mein JR et al. **Nutr Rev**, 2008; 66: 667-683. [32] Wakshlag JJ, Balkman CE. **Am J Vet Res**, 2010;71:1362-70. [33] Bertram JS et al. **BiochimBiophys Acta**, 2005; 1740: 170-178 [34] Ben-Dor A et al. **Mol Canc Ther**, 2005; 4: 177-186. [35] Kavanaugh CJ et al. **J Natl Cancer Inst**., 2007; 99: 1074-1085. [36] Cohen L **Exp Biol Med**, 2002; 227: 664-668. [37] Vertzoni M et al. **J Pharm. Pharmacol**, 2006; 58: 1211-1217. Korytko PJ et al **J Nutr**, 2003; 133: 2788-2792. [38] Raghaven M et al. **J Am Vet Med Assoc**, 2005; 227: 94-100. [39] Helmerick EC et al. **Vet Comp Oncol.** 2012 Dec 10. [Epub ahead of print] [40] Wakshlag JJ et al. **Am J Vet Res**, 2010, 71: 89-96. [41] Skorupski KA, et al. **J Vet Intern Med**. 2011; 25:838-45. [43] Guerrero-Beltrán CE et al. **J Nutr Biochem**. 2012; 23:494-500.

Heart Disease in Cats: Signs, Diagnosis, and Treatment

Joshua A Stern, DVM, DACVIM (Cardiology)
Raleigh, NC

INTRODUCTION

This lecture will focus on feline heart disease from a general practice perspective. It will review common presentations, clinical examination findings and the latest diagnostic and therapeutic approaches for managing cats with all stages of heart disease. While we will briefly mention congenital malformations, the bulk of the lecture focuses on feline cardiomyopathies: hypertrophic, dilated and restrictive.

SIGNS OF CARDIAC DISEASE

Heart murmurs remain an important finding in cats with heart disease. The prevalence of heart murmurs in the general cat population is startling. Depending on the level of excitement and study referenced, the percent of healthy cats with heart murmurs approaches 25%.[1,2] It is important to remember that presence of a heart murmur does not indicate a diagnosis of heart disease. Innocent heart murmurs in the cat are quite common with a prevalence of wide variation of 25-69% in the literature.[2,3] Cats with significant heart disease may remain completely asymptomatic such that presence of a heart murmur may be the only supportive finding that cardiovascular evaluation is warranted.

Exercise intolerance is admittedly difficult to ascertain in many cats. However, cats with significant heart disease are often reported to be less interactive or begin avoiding what was once normal play activity. Cats with heart disease and frequent arrhythmias may experience syncopal events that typically only last a few seconds. A hallmark of syncope is complete recovery without the postictal period characteristic of seizure disorders.

Dyspnea, tachypnea and respiratory distress are frequent presenting complaints for cats with significant heart disease. Cats with heart disease may manifest congestive heart failure as pulmonary edema, pleural effusion, pericardial effusion, abdominal effusion or any combination of these findings. Far and away the most common manifestations of heart failure in cats are those of pleural effusion and pulmonary edema.

Thromboembolic disease most commonly seen as "saddle thrombus" is another presenting complaint for cats with significant heart disease. The blood stasis, endothelial damage and left atrial enlargement that accompanies many congenital and acquired heart diseases in the cat predisposes them to clot formation and embolism.

CONGENITAL HEART DISEASE

Congenital heart disease in the cat is far less common than the acquired diseases such as hypertrophic cardiomyopathy. Congenital heart disease remains an important differential diagnosis for young cats with physical exam findings or clinical complaints that implicate cardiac disease. Historically less than 5% of cats evaluated by a cardiologist for the first time are diagnosed with congenital heart disease.[4] Although congenital heart disease may be suspected due to the signalment (typically kittens to adolescent cats with significant heart murmurs with or with possible exercise intolerance, syncope, or congestive heart failure), congenital heart disease remains a diagnosis made thorough echocardiographic imaging.

Atrioventricular valve dysplasia (mitral > tricuspid) and ventricular septal defects remain the most common congenital heart defects in cats. Many other rare defects are also reported including: mitral valve stenosis; aortic stenosis; pulmonic stenosis; pulmonary artery stenosis or atresia; atrial septal defects; atrioventricular septal defects (common AV canal / endocardial cushion defects); cor triatriatum sinister or dexter; patent ductus arteriosus; Tetralogy of Fallot, and double chamber right ventricle. Interventional and surgical therapy is still a rare treatment option in the cat, although smaller occlusion devices, improved procedural descriptions in the literature and partnering with human pediatric cardiologists have resulted in successful therapy for several of these defects. The remainder of medical therapy is aimed at palliation of cardiac signs and follows that described below for acquired heart disease in the cat.

ACQUIRED HEART DISEASE

Endocardiosis / Endocarditis

For this topic it is important to remember that cats are not like their small dog counterparts. Atrioventricular valve degeneration / myxomatous mitral valve degeneration / endocardiosis is exceedingly rare in the cat and not an important cause of heart disease in this species. Endocarditis is also a rare condition in the cat that will not be discussed during this session due to time constraints.

Hypertrophic Cardiomyopathy (HCM)

Hypertrophic cardiomyopathy is defined by left ventricular hypertrophy in the absence of systemic disease or other cardiovascular disease. It is the most common cause of heart disease in the cat. While this is a very common disease, the specific etiology is in most cases not known. Familial HCM is inherited in Maine Coone and Ragdoll cats with a causative mutation in the Myosin Binding Protein C gene that mirrors HCM causing mutations in human beings. In these breeds HCM is inherited as an autosomal dominant trait. Genetic screening tests for these breed specific mutations are available (http://www.cvm.ncsu.edu/vhc/csds/vcgl/cat-dna.html). Other cat breeds are known to have familial HCM including the Sphynx, Bengal, Norwegian Forest Cat, American Shorthair and a few others.

HCM is characterized by hypertrophy of the left ventricle (either uniformly of asymmetrically within the free wall and/or interventricular septum). This thickened left ventricle has reduced compliance, small luminal size and subsequently reduced diastolic function. The thickening can distort the mitral valve annulus and cause regurgitation or result in forces that entrap the mitral valve during systole and cause outflow tract obstruction called systolic anterior motion of the mitral valve (SAM). The resistance to diastolic filling in the LV results in increased filling pressures and subsequent left atrial pressure increase. This left atrial pressure increase can result in left atrial pressure dilation and elevated pulmonary venous pressures. This pulmonary venous pressure elevation can cause traditional left-sided congestive heart failure and pulmonary edema or translate to right heart pressure elevation and subsequent pleural effusion. Left atrial dilation and blood stasis due to diastolic dysfunction may result in thrombus formation, which may dislodge and become lodged in the systemic circulation (most often the distal aorta causing the "saddle thrombus" scenario discussed previously).

Clinical signs of cats with HCM are highly variable. Affected cats are often asymptomatic at the time of evaluation, however respiratory distress, syncope, thromboembolic disease and exercise intolerance may all be signs associated with this condition. This disease can also be silent in its presentation with the first adverse event being sudden death.

Physical exam findings may include systolic heart murmur from either left ventricular outflow tract obstruction and/or mitral valve regurgitation. A gallop rhythm may be heard as an indication of altered left ventricular filling. Cats may also present with signs of congestive heart failure such that dyspnea and tachypnea can be observed. It is important to remember that cats may manifest heart failure frequently with pleural effusion such that muffled heart sounds and/or a notable fluid line on percussion are common. Crackles as an indication of pulmonary edema are uncommon in cats.

Electrocardiogram may be within normal limits however conduction disturbances such as left anterior fascicular block and arrhythmias (ventricular or supraventricular) can be noted. Patterns of chamber enlargement may be observed but are not a reliable screening test.

Chest radiographs may be useful to evaluate for heart enlargement or evidence of congestive heart failure. Chest radiographs cannot specify the underlying type of heart disease. Much like physical examination findings, the observations from chest radiographs vary widely in cats with heart disease and radiographically silent HCM is possible.

Echocardiogram represents the single best diagnostic tool for evaluation of cats with heart disease. Concentric left ventricular hypertrophy can be quantified and described. Cats can be segregated into obstructive and non-obstructive disease varieties, which may dictate treatment decisions, and severity of the disease can be well differentiated. Diagnosis of HCM is generally accepted as a LV free wall or interventricular septum measurement that exceeds 6mm in diastole by either m-mode or 2D long axis imaging. This diagnosis is made only if hyperthyroidism and systemic hypertension have been excluded. If obstruction to outflow is observed by color-flow and/or Spectral Doppler the distinction of hypertrophic obstructive cardiomyopathy may be made. Severity of left atrial enlargement and presence of absence of cardiac thrombi are important imaging findings that can be ascertained by echocardiogram.

Blood pressure measurement and assessment of thyroid function is an important process in screening cats for hypertrophic cardiomyopathy. Both systemic hypertension and hyperthyroidism can lead to left ventricular hypertrophy and mandate a different course of treatment. HCM is an adult onset disease such that most cats with findings consistent with HCM should have routine thyroid and blood pressure assessments made prior to diagnosis and treatment.

Cardiac biomarkers have become an important aspect to screening for heart disease in the general practice setting. Brain natriuretic peptide assessed as NT-proBNP has been shown to discriminate between cats with congestive heart failure and respiratory distress of other origin reasonably well.[5] As a screening test for cats with heart murmurs the test performs moderately with a fair number of false positive and negative results. As a screening test without any additional suspicion of heart disease, the author does not recommend this test. If other portions of the diagnostic workup have already implicated heart disease, measurement of NT-proBNP is unlikely to be of clinical benefit.[6]

Treatment of hypertrophic cardiomyopathy is directed at decreasing heart rate to improve the time allotted for diastolic filling of the left ventricle, reducing the outflow tract obstruction if present, and controlling congestive heart failure when present.

Optimal therapy for asymptomatic cats is unknown however cats that are persistently tachycardic (HR >220) or have outflow tract obstruction as noted on echocardiogram should be treated. The most commonly recommended therapy is beta blockade with the drug Atenolol (B_1 Selective, ~3mg/kg orally, q12hrs). This drug is utilized to reduce heart rate and outflow tract gradient if SAM /obstruction is present. It's important to remember that beta blockers should not be administered to cats in active congestive heart failure. In these cats clinicians may prefer to use calcium channel blockers (diltiazem) to reduce obstruction and lower heart rate once their initial congestive heart failure episode is controlled.

Cats with congestive heart failure are generally treated with an ace-inhibitor and diuretic. The author's treatment of choice is benazepril at a dose of 0.25-0.5mg/kg PO q24hrs and furosemide at the lowest daily dose that controls congestion. Recent attention has been given to the use of oral pimobendan in cats with congestive heart failure although this is not FDA approved at this time.[7] If atenolol therapy was ongoing at the time congestive heart failure developed it should not be withdrawn.

Cats with evidence of thrombus formation or severe atrial enlargement are generally treated with an anti-coagulant. No clear evidence has been offered to prove a benefit of platelet inhibition by clopidogrel over standard aspirin therapy. The author's preference is the use of clopidogrel (18.75mg PO q24hrs per cat) since the generic drug formulation is now available.

The prognosis for HCM varies depending on the etiology and some cats may progress rapidly to congestive heart failure or severe disease outcomes such as sudden death or thromboembolism. Other cats may live years without incidence of heart failure. Prognosis is perhaps best understood by observation of disease progression and understanding that cats with CHF and/or thromboembolic disease have a poorer prognosis.

Dilated Cardiomyopathy

Dilated cardiomyopathy in the cat is a disease of systolic dysfunction causing progressive left ventricular dilation and subsequently elevated left ventricular diastolic pressures. These changes ultimately distort mitral valve conformation, permit mitral valve regurgitation and result in the aforementioned left atrial enlargement and congestive heart failure physiology. This disease is uncommon in cats but can be associated with Taurine deficiency (an essential feline amino acid). While most commercial diets are well supplemented with taurine, some new diets and homemade diets may be deficient. A small number of cats may have this disease with normal plasma taurine levels owing to an unknown etiology

Important differences between DCM and HCM include the need to obtain whole blood taurine levels (normal >200nmol/ml). Taurine levels in taurine deficient DCM cases are usually low (<100 nmol/ml). Taurine supplementation (250-500 mg PO q12hrs) should be administered in addition to the above-mentioned therapy for CHF (benazepril, furosemide) if needed until confirmation of normal levels is obtained. Taurine deficient DCM cats generally have a good prognosis when supplemented and can even be removed from cardiac medications in the future. DCM cats without taurine deficiency have a guarded prognosis and often progress to congestive heart failure. Pimobendan may be beneficial in these cases although not yet FDA approved for use in cats.[8]

Restrictive Cardiomyopathy

Restrictive cardiomyopathy is an uncommon myocardial disease characterized by myocardial fibrosis and impaired ventricular filling. The systolic function is variable and bi-atrial enlargement is considered a common feature. RCM may lead to biventricular congestive heart failure and severe atrial enlargement. Thrombus formation in the severely enlarged atria is a possible feature of this cardiomyopathy.

Therapy is identical to the aforementioned treatment of non-obstructive hypertrophic cardiomyopathy. If systolic function is also observed consideration for the addition of pimobendan therapy may be warranted, although this is not clearly established at this time and not FDA approved for use in cats at this time.[7,8]

CONCLUSION

It is important to remember that while the etiology and some special characteristics of cardiomyopathies may differ, the overarching principles of signs, symptoms, diagnostic approach and therapy are very similar. The single best diagnostic test for these diseases remains echocardiography; treatment of asymptomatic cats remains controversial; and prognosis prior to the onset of congestive heart failure is highly variable.

REFERENCES

[1]Cote E, et al. J Am Vet Med Asoc 2004;225:384-388. [2]Paige CF, et al. J Am Vet Med Assoc 2009;234:1298-1403. [3]Bonagura JD. J Fel Med Surg 2000;2:147-151. [4]MacDonald KA.Vet Clin No Am Sm An Pract 2006;36:503-531. [5]Fox PR, et al. J Vet Cardiol 2009;11:S51-S61. [6]Hsu A, et al. J Vet Cardiol 2009;11:S63-S70. [7]Gordon SG, et al. J Am Vet Med Assoc 2012;241(1):89-94. [8]Hambrook LE, et al. J Fel Med Surg 2012;14(4):233-239.

CANINE AND FELINE CUSHING'S DISEASE

Cynthia R, Ward, VMD, PhD, DACVIM
Athens, Georgia

Introduction

Cushing's syndrome is defined as the constellation of clinical signs that result from effects of hypercortisolism in the body. Cushing's syndrome resulting from excess release of ACTH from the pituitary gland is termed Cushing's disease. However, in small animal medicine the terms Cushing's syndrome, Cushing's disease, and hyperadrenocorticism (HAC) tend to be used interchangeably. Cushing's syndrome is common in dogs and ferrets and is becoming increasingly more recognized in cats. In dogs, most cases are true Cushing's disease with 80-85% resulting from pituitary-dependent disease. Adrenal tumors resulting in hypercortisolemia occur less commonly and rare causes of hypercortisolemia include concurrent pituitary and adrenal tumors or ectopic ACTH secretion.

Clinical Signs

Clinical signs of HAC in dogs include the 5 "p"s: polyuria, polydipsia, polyphagia, panting, and pot-bellied appearance (hepatomegaly and flank muscle atrophy). A host of dermatological signs are also commonly seen including alopecia (bilaterally symmetrical, endocrine), comedones, thin skin, hyperpigmentation, pyoderma, calcinosis cutis, and failure of hair regrowth. Urinary incontinence as a result of polyuria and urinary tract infections and insulin-resistant diabetes mellitus may occur. Rare clinical signs include pseudomyotonia, facial nerve palsy, cruciate ligament rupture, and thromboembolic disease. Reproductive failure may also occur. Neurological signs may be present if there is a pituitary macroadenoma. The presence of these clinical signs often prompts the practitioner to run screening labwork. Common findings include a stress leukogram, thrombocytosis, increased alkaline phosphatase, hypercholesterolemia, hyperglycemia, hypertriglyceridemia, low urine specific gravity, and proteinuria. Some dogs may exhibit several of the clinical and biochemical signs; and some dogs only a few. This may be due to differing severity of the disease and differing sensitivity to cortisol excesses by individual dogs. Due to the increasing recognition of HAC by practitioners, suspicion of the disease is often raised early.

Diagnosis

The diagnosis of HAC can be difficult and confusing. No one test is 100% accurate leading the practitioner to have to employ different tests or repeat tests. This can be frustrating to explain to owners. As a testimony to the difficulty of HAC testing, one of the 2012 ACVIM consensus statement sessions addressed this issue.[1] Although the final manuscript detailing the panel's recommendations has not yet been published, the practitioner is encouraged to read it upon release. Testing for any endocrine disease should be preceded by a clinical suspicion of the disease. Having appropriate clinical signs consistent with HAC preselects the testing population and increases the diagnostic accuracy of the tests. Testing should also be avoided in dogs with severe non-adrenal illness as the associated stress of chronic illness may affect testing of the endocrine axis. Commonly used screening tests for HAC include the urine cortisol to creatinine ratio (UCCR), ACTH-stimulation test, and low dose dexamethasone suppression test (LDDST). Baseline cortisol levels are not useful for HAC screening because of episodic fluctuations in the bloodstream. Of these tests, the UCCR is an excellent test to rule OUT HAC out since it has a sensitivity approaching 99% when 2 urine samples are in the reference range. However, the UCCR has a low specificity and dogs with UCCR above the reference range must be retested with another screening test in order to diagnose HAC. Of the remaining screening tests, the LDDST is the screening test of choice to diagnose HAC. It is more sensitive at 85-95%, while the reported sensitivity of the ACTH-stimulation test is: 57-83%. Additionally, the LDDST is much cheaper to run than the ACTH-stimulation test due to the expense of synthetic ACTH. Cortisol precursors, especially 17-alpha-hydroxyprogesterone have been measures in the ACTH-stimulation tests; however, cortisol has been shown to be the most sensitive and specific for diagnosis of HAC.[2,3]

After the diagnosis of HAC has been made, it is important to differentiate between pituitary (PDH) or adrenal (AT) dependent HAC in order to provide optimal treatment and prognostic information. Endogenous plasma ACTH levels may be measured and should be elevated in PDH and decreased in AT. This test can show excellent discrimination when the correct assay is used. [4] Measurement is also hampered by special sample handling which can be difficult in practice settings. Imaging studies, especially ultrasound and CT have been used to measure size of the adrenal glands in an attempt to determine whether both are enlarged (PDH) or one is enlarged or irregular and the other is small (AT). Overlap between normal dogs and dogs with PDH is considerable and there is breed and age variation as well as variation depending on the observer. [5,6] CT measurements may be more promising.[7]

Treatment

For many years, mitotane was the only effective medical treatment for HAC. It works by selective necrosis of the cortisol-secreting layers of the adrenal cortex. Mitotane treatment normalizes hypercortisolemia and, thus, resolves the clinical signs associated with HAC. The side effects of mitotane therapy can be severe and have frightened many practitioners into avoiding this drug. Trilostane is a 3-beta hydroxysteroid dehydrogenase inhibitor. It works by competitive inhibition of the 3-beta hydroxysteroid dehydrogenase enzyme that is important in the formation of mineralocorticoids, glucocorticoids, and sex hormones in the adrenal gland. It is most effective on blocking the conversion of 17 hydroxypregnenolone to 17 hydroxyprogesterone, a precursor to cortisol. Although deemed a "safer" alternative to mitotane, trilostane treatment can also have severe side effects such as causing partial to complete mineralocorticoid and/or glucocorticoid deficiencies. As with mitotane, these effects can be reversible. Trilostane can also cause adrenal necrosis. [8] Many of these effects can be prevented by carefully controlling the dosing of trilostane. Lower twice daily dosing has been shown to be more effective at controlling clinical signs while decreasing undesirable side effects. [9] Larger dogs may need lower dosing per body weight than smaller dogs. [10] Until recently, trilostane could only be obtained through compounding pharmacies in the United States. Huge variation in amount of active trilostane exists between products manufactured by different pharmacies and even between different pills from the same pharmacy which is probably responsible for some of the early frustration with trilostane use. [11] The use of a consistently manufactured product that is now available in the US, Vetoryl by Dechra, is recommended for more predictable results.

Monitoring

Monitoring of medical therapy with HAC is primarily through evaluation of an ACTH-stimulation test measuring cortisol. In mitotane-treated animals, the test should occur that day after loading and then anytime during maintenance treatment. The test measures adrenal reserve by stimulating cortisol secretion with pharmacological doses of ACTH. ACTH-stimulation testing is recommended to monitor trilostane therapy as well. This test should be done 4-6 hours post-trilostane administration. For both mitotane and trilostane therapy, the cortisols should be within the reference range. However, clinical signs in the patient are most important and may override dosing decisions just based on cortisol levels. Because of the expense of the synthetic ACTH used for the stimulation test, other tests such as the use of baseline cortisol concentrations or UCCR have been examined to determine if they could be used for trilostane monitoring. Although some animals may be monitored with baseline cortisol concentrations, the UCCR has been shown to be unhelpful. [12,13] Although expensive, ACTH-stimulation testing is the most helpful when monitoring mitotane or trilostane therapy in HAC.

REFERENCES

[1]Behrend et al, 2012 ACVIM Forum Consensus Statement, [2]Sieber-Ruckstuhl, et al., Vet Rec, 2008 162:673,[3] Monroe, et al., JVIM 2012,26:945, [4] Rodriguez Pineiro et al., JVIM, 2009, 23:850, [5] de Chalus et al, Anat Histol Embryol, 2012, 42:57, [6], Barberet et al., Vet Rad Ultra, 2010 51:656,[7], RodriguezPineiro et al., JVIM, 2011, 25:1066, [8], Chapman et al., JSAP, 2004, 45:307, [9] Vaughan et al., JAVMA, 2008, 232:1321, [10] Feldman and Kass, JVIM, 2012, 26:1078, [11] Cook et al., JAAHA, 2012, 48:228, [12] Cook and Bond, JAVMA, 2010, 23:801, [13], Galac et al, JVIM, 2009, 23:1214.

FELINE PLATELET DISORDERS

Marjory Brooks, DVM, DACVIM (Small Animal Internal Medicine)
Ithaca, NY

INTRODUCTION

Platelets play a critical role in physiologic hemostasis by controlling hemorrhage from the microvasculature and localizing clot formation to sites of vascular injury. Platelet plug assembly depends on adequate platelet numbers, platelets' response to activating stimuli, and their ability to form stable bridges with the vessel wall and other platelets. Thrombocytopenia is the most common platelet defect recognized in clinical practice[1,2], however thrombopathia (platelet dysfunction) may also cause signs of a bleeding diathesis. Since platelet defects are typically acquired, rather than hereditary, resolution requires correction of an underlying disease process and providing appropriate supportive care.

DIAGNOSTIC EVALUATION

Clinical signs of petechiae (pinpoint hemorrhage), ecchymoses (bruising), mucosal hemorrhage, and prolonged hemorrhage at sites of injury are typical manifestations of platelet defects. Performing a platelet count and coagulation panel as screening tests help streamline diagnosis of common disorders and guide selection of ancillary tests for more complex cases.

Diagnostic Algorithm

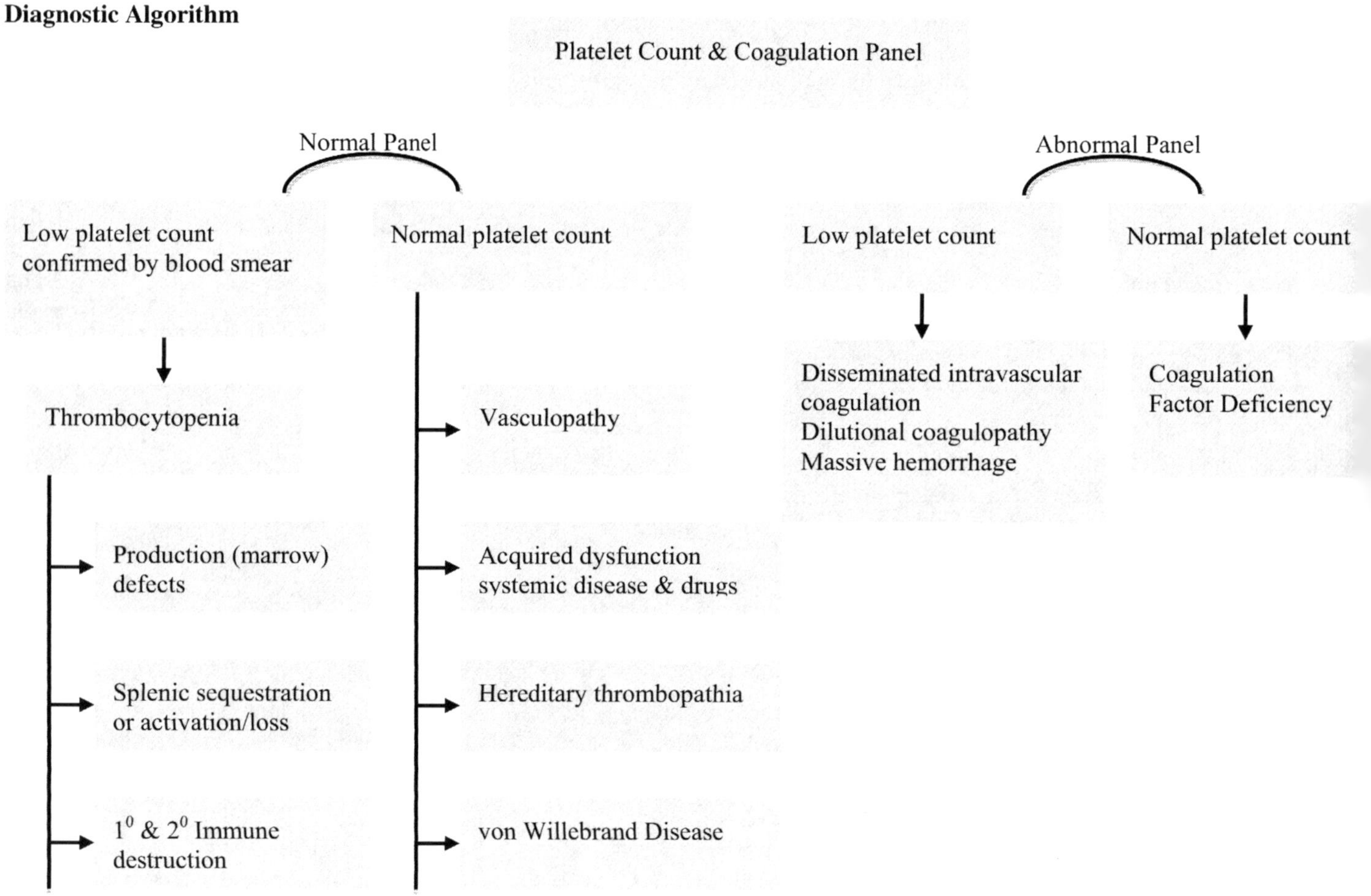

PATHOGENIC CLASSIFICATION OF PLATELET DEFECTS

Table 1.

CATEGORY	DISORDER	MECHANISMS: EXAMPLES
Quantitative Defects	Thrombocytopenia	Production defects: Bone marrow aplasia, dysplasia, infiltration
		Consumption defects: Hypersplenism, peripheral activation/thrombosis
		Immune destruction: autoimmune, infection or drug-induced
	Thrombocytosis	Reactive: blood-loss & iron deficiency, chronic inflammation
		Myelodysplasia, essential thrombocythemia
Qualitative Defects	Extrinsic dysfunction	Uremia, liver failure, hyperglobulinemia, plasma expanders
		Von Willebrand factor deficiency
	Intrinsic dysfunction	Hereditary thrombopathias, anti-platelet drugs

SPECIFIC DISORDERS

Thrombocytopenia

Feline platelets are readily activated during blood collection and processing. Spuriously low platelet counts are common artifacts and low automated platelet counts should be confirmed by examination of a blood film to detect platelet clumping.[3] The reference range for normal feline platelet count is approximately 200,000 to 600,000/ uL. Values should be interpreted in the context of laboratory established or in-house reference ranges. Thrombocytopenia rarely causes signs of abnormal hemostasis unless the platelet count falls below 50,000/uL, with spontaneous petechiae and mucosal hemorrhage most likely at counts below 20,000/uL.[4,5] The most severe thrombocytopenias are generally found in patients with production defects or immune-mediated destruction, whereas consumption and sequestration defects typically cause relatively moderate thrombocytopenia. Thrombocytopenia combined with leukopenia or anemia suggests a production defect. Table 2 displays an overview of diagnostic tests for thrombocytopenia.

Table 2.

Diagnostic Tests for Thrombocytopenia	
CBC	Detect multi-lineage cytopenias, pathogens, neoplastic cells, schistocytes
Bone marrow examination	Assess megakaryocyte number, dysplasia, neoplasia, myelofibrosis, pathogens
Splenic/lymph node aspirate	Identify neoplastic cells, pathogens,
Serology & blood culture	Antibody titers to pathogens and antinuclear antibody, pathogen detection
Polymerase chain reaction	Pathogen detection
Anti-platelet antibody	Detect platelet-bound antibody, platelet-bindable antibody
Drug history	Examples: doxorubicin, sulfonamides, chloramphenicol, griseofulvin, methimazole

Thrombocytopenia secondary to infection is often multifactorial, including bone marrow suppression and increased rates of peripheral consumption. Many different feline pathogens are associated with thrombocytopenia, such as viral agents (FeLV, panleukopenia virus, FIP-coronavirus), insect or arthropod-borne agents (eg. Hemoplasma, Anaplasma, Cytauxzoon), fungal agents (Histoplasma, Cryptococcus) and sepsis caused by gram negative and gram positive bacteria. Infectious agents usually cause signs of systemic illness in addition to thrombocytopenia. Thrombocytopenia may also develop secondary to neoplasia, inflammatory disorders such as pancreatitis, and hepatic disease. Signs of the underlying disorder, rather than spontaneous hemorrhage, typically dominate the clinical presentation. In contrast, primary immune-mediated (autoimmune) thrombocytopenia causes clinically apparent petechiae and mucosal hemorrhage in the absence of an inciting disease process.[6]

Thrombocytosis

Reactive thrombocytosis and splenectomy may cause high platelet count, but do not require treatment. Reactive thrombocytosis develops secondary to hemorrhagic and inflammatory disorders that induce upregulation of thrombopoetin and inflammatory cytokines. Essential thrombocytosis refers to a myeloproliferative disorder that results in markedly high platelet counts (> 900,000/uL). This is a rare disorder, associated with both microvascular thrombosis and hemorrhage.

Platelet Dysfunction

Many diseases and drugs cause acquired platelet function defects. A mild to moderate bleeding tendency, attributed to changes in platelet metabolism and blood flow properties, may develop in patients with anemia, uremia, liver failure, paraproteinemias, and DIC. Drugs may directly inhibit platelet function by interacting with membrane receptors, signaling pathways, or calcium mobilization. While drugs and disease processes that impair platelet function may not cause spontaneous hemorrhage, the bleeding tendency complicates disease management and may cause significant blood loss if patients undergo invasive procedures. Hereditary platelet function defects (thrombopathias) are rare, but likely underdiagnosed because of the difficulties of performing comprehensive platelet function studies.[7] Most of these analyses require referral of the patient to a testing center for platelet isolation and immediate analyses in order to maintain platelet viability. Table 3 displays a summary list of platelet function testing.

Table 3.

Diagnostic Tests of Platelet Function	
Bleeding time	In vivo test of platelet adhesion, aggregation, von Willebrand factor
Platelet function analyzer (PFA100)	Point of care instrument; test of platelet adhesion, aggregation, von Willebrand factor
Aggregometry	Measure aggregation response to specific agonist stimuli
Lumiaggregometry	Measure dense granule secretion
Flow cytometry	Detect membrane receptors, assess agonist-induced activation response
Microscopy	Assess structures related to function: dense granules, alpha granules, microtubules
VWF:Ag	Plasma von Willebrand factor concentration

Von Willebrand Disease

Von Willebrand factor (VWF) is a plasma protein that supports platelet adhesion at sites of vascular injury. Von Willebrand disease (VWD) is a hereditary bleeding disorder caused by a lack of plasma VWF. Although affected patients' platelets are functional, in the absence of VWF they are unable to bind stably to subendothelial collagen under conditions of high shear. Von Willebrand disease is a common hereditary trait in many breeds of dog, however few cases have been identified in cats. Clinical signs of VWD are similar to intrinsic platelet function defects, including epistaxis and other mucosal hemorrhage, and bleeding and bruising after surgery or trauma. Affected patients have normal coagulation screening tests, therefore the diagnosis of VWD is based on specific measurement of plasma VWF concentration (VWF:Ag).

MANAGEMENT OF PLATELET DEFECTS

Effective management of platelet defects includes control of active hemorrhage, stabilization of the patient to reverse signs of blood loss anemia or hypovolemic shock, and resolution of underlying diseases that cause or exacerbate thrombocytopenia or platelet dysfunction.

Transfusion Therapy

Feline platelet rich plasma and platelet concentrates are not widely available, therefore fresh whole blood is the major source of platelets in clinical practice. In order to enhance the viability and maximize the number of transfused platelets, whole blood should be drawn into plastic tubes or bags containing citrate-based anticoagulants, held at room temperature, and transfused as soon as possible after collection.[8] The volume to transfuse generally ranges from approximately 8 to 16 mL/kg. Patients with concomitant anemia and hypovolemia can receive relatively high volume transfusion, however circulatory overload will limit the total dose of platelets that can be administered as fresh whole blood. Nevertheless, transfusion may supply sufficient platelets to support hemostatic plug formation or temporarily stabilize the patient pending resolution of an underlying disorder. Transfusion of fresh whole blood to maintain platelet count above a threshold value is rarely feasible, rather, transfusion should be given to patients with signs of blood loss anemia, uncontrolled active hemorrhage, or central nervous system hemorrhage. The standard blood banking procedures of donor pathogen screening and confirming donor-recipient blood type compatibility prior to transfusion must be followed to minimize the risk of transfusion reaction.[9]

Non-Transfusion Therapy

Most thrombocytopenic cats, and those with acquired platelet function defects, are successfully managed by identifying and correcting the primary disease process. Drugs with antiplatelet action, surgery, and invasive procedures such as jugular catheterization should be avoided. Cage rest to minimize trauma should be considered for severely thrombocytopenic patients (< 20,000 platelets/uL). Antibiotic therapy is often initiated pending serologic or culture results for patients at risk for tick-borne disease or other infectious agents based on geographic location, access to outdoors, and results of preliminary testing. Because primary immune thrombocytopenia is uncommon in cats, steroid therapy should be reserved until after a complete work-up has ruled out an inciting disease. Prednisolone (1 to 2 mg/kg) induced remission of thrombocytopenia in 4 of 5 cats with primary immune thrombocytopenia, however gastrointestinal hemorrhage was a complication of prolonged steroid therapy.[6] The vasopressin analogue, desmopressin acetate (DDAVP) is used as an adjunct to transfusion in human beings and dogs with VWD and platelet function defects. A pre-operative dose of 1 microgram/kg SQ is given to dogs, however clinical data is lacking for its use in cats.

REFERENCES

[1]Jordan HL, et al. J Vet Intern Med 1993;7:261. [2]Peterson JL, et al. J Vet Intern Med 1995;9:298. [3]Normal EJ, et al. Vet Clin Pathol 2001;30:137. [4] Strauss RG. Curr Opin Hematol 2005;12:499. [5] Abrams-Ogg AC. Vet Clin North Am Small Anim Pract 2003;33:1401.[6] Wondratscheck C, et al. J Am Anim Hosp Assoc 2010;46:12. [7] Boudreaux MB. J Am Vet Med 2008;8:1251. [8] Callan MB, J Vet Emerg Crit Care 2009;19:1476. [9] Castellanos I, J Vet Intern Med 2004;18:529

JOINT NUTRACEUTICALS AND DIETS, CRITICAL EVALUATION OF EFFICACY
Steven C. Budsberg DVM, MS, DACVS
Athens, GA

INTRODUCTION

Nutritional and dietary supplements are often termed nutraceuticals as an amalgamation of the terms nutritional and pharmaceutical. Nutraceuticals still fall into a grey area from a regulatory point of view, yet in recent times have grown into a billion-dollar industry. These products have come a long way since entering companion animal medicine in the early 1990s. The enormous public interest in the relationship between diet supplements and OA often takes center stage when discussing OA and potential therapeutic plans with clients. Unfortunately, speculative information on nutritionally based therapies to treat OA has permeated every form of media. The paucity of well-designed scientific studies exploring these treatments leaves clinicians at a distinct disadvantage when discussing possible therapies with clients and colleagues alike. This discussion will focus on the use of evidence-based medicine (EBM) when evaluating nutraceuticals. Evidence is defined as "the data on which a judgment or conclusion may be based, or by which proof or probability may be established."[1] The pragmatic use of EBM involves the integration of the best research evidence, clinical expertise and client/patient considerations. There are many ways by which to analyze and integrate evidence into the practice of veterinary medicine, and several schemes by which to rank the strength of evidence. This presentation will be based upon the Food and Drug Administration's (FDA) evidence-based ranking system for scientific data adapted and utilized for study rating and ranking.[2-4] Additionally we will include data which is based upon a modification of this system.[5] Briefly, three separate ratings are used for study evaluation: 1) a rating for study design, 2) a rating for study quality, and 3) a group agent rating for the strength of the entire body of evidence. A final rank of the scientific evidence is decided based upon the classifications from the three rating systems. This system does not use the terms rate and rank interchangeably. These notes will use data directly from two previous publications.[2,4,5] Certainly everyone can use different criteria to assemble their data, and this is only one example of methodology and subsequent outcome. This presentation should be not considered a complete resource for all the data available to evaluate nutraceuticals, but instead an example of how to evaluate some of the products currently available.

Omega-3 Based Diets

Six trials were identified describing the use of Omega-3 based diets as the major component for improving the clinical signs associated with osteoarthritis (OA) in dogs.[6-11] All the studies were prospective, randomized, and received a Type I classification. Examination of the quality of the studies showed that they had adequately addressed issues of scientific quality relating to data collection, analysis, bias and generalizability. The studies were consistent in finding positive effects of the diets on the clinical signs associated with OA. The studies did vary in outcome measures and this created some differences in ability to assess overall strength, but an overall rating of the strength of the evidence concludes that one can have a **high level of comfort** with the results of the aforementioned studies. Given the consistency of the findings, one can also have confidence in the data that show that these diets are effective. (Appendix 1)

Compounds based upon Chondroitin sulfate, and glucosamine hydrochloride

Two trials were identified describing the use of compounds with chondroitin sulfate, and glucosamine hydrochloride as major components for improving clinical signs associated with OA in dogs. [12,13] Both study designs were prospective, randomized, and received a Type I classification. One study subjectively showed a positive effect,[13] while the other showed no positive effect.[12] Examination of the quality of the studies showed that they had adequately addressed issues of scientific quality relating to data collection, analysis, bias and generalizability. There was a low consistency rating meaning the results were inconsistent between the studies. The studies do provide information to conclude that there was some suggestion that the effect will be physiologically meaningful and achievable. An overall rating of the strength of the evidence concludes that one can have a moderate level of comfort with the results of the aforementioned studies. However confusion arises as to which study you should use to help make your decision on whether or not to use this type of product.

Green-lipped Mussel Preparation

Three trials were identified using a compound that's main ingredient was green-lipped mussel (*Perna canaliculus*) for the treatment of OA in dogs.[14-16] The studies were prospective, randomized in design and received a Type I rating. While the studies subjectively showed a positive effect, the studies get a quality rating that suggests some uncertainties exist related to the scientific quality. There was a moderate level of consistency between the studies. The studies do provide information to conclude that there was some suggestion that the effect will be physiologically meaningful and achievable. An overall rating of the strength of the evidence indicates a moderate level of comfort with the results of the aforementioned studies.

Elk Velvet Preparation

Velvet from elk antlers was tested as an additive to the food of OA dogs. One randomized, blinded, placebo-controlled, parallel group clinical trial as a treatment for osteoarthritis was evaluated.[17] This study is classified as type I . The study had positive efficacy data from objective and subjective outcome measures. Examination of the quality of the study showed that it had adequately addressed issues of scientific quality relating to data collection, analysis, bias and generalizability. The study does provide information to conclude that there was some suggestion that the effect will be physiologically meaningful and achievable. An overall rating of the strength of the evidence concludes that one can have a moderate level of comfort with the results of the aforementioned study.

***Brachystemma calycinum* D. Don**

Brachystemma calycinum D. Don was tested in one randomized, blinded, placebo-controlled, parallel group clinical trial as a treatment for osteoarthritis was evaluated.[18] This study is classified as Type I . The study had positive efficacy data from objective and subjective outcome measures. Examination of the quality of the studies showed that it had adequately addressed issues of scientific quality relating to data collection, analysis, bias and generalizability. The study does provide information to conclude that there was some suggestion that the effect will be physiologically meaningful and achievable. An overall rating of the strength of the evidence concludes that one can have a moderate level of comfort with the results of the aforementioned study.

S-adenosyl l-methionine

S-adenosyl l-methionine (SAMe) was evaluated in one randomized, blinded, placebo-controlled, parallel group clinical trial as a treatment for canine osteoarthritis.[19] This study is classified as Type I. The study had consistent data from objective and subjective outcome measures showing no positive effect. Examination of the quality of the studies showed that it had adequately addressed issues of scientific quality relating to data collection, analysis, bias and generalizability. The study does provide information to conclude that there was some suggestion that the effect will be physiologically meaningful and achievable. An overall rating of the strength of the evidence concludes that one can have a moderate level of comfort with the results of the aforementioned study.

P54FP

P54FP is an extract of *Curcuma domestica* and *Curcuma xanthorrhiza.* One randomized, blinded, placebo-controlled, parallel group clinical trial of P54FP as a treatment for osteoarthritis of the canine elbow or hip was evaluated.[20] This study is classified as Type I. The study had conflicting data from objective and subjective outcome measures. Examination of the quality of the studies showed that it had adequately addressed issues of scientific quality relating to data collection, analysis, bias and generalizability. The study does provide information to conclude that there was some suggestion that the effect will be physiologically meaningful and achievable. An overall rating of the strength of the evidence concludes that one can have a low level of comfort with the results of the aforementioned study.

Resin extract of Boswellia serrata

One trial with an herbal dietary supplement consisting of a natural resin extract of *Boswellia serrata* was conducted to evaluated effects on OA in dogs.[10] The study is classified with a Type III rating.[2] Subjective clinical improvements were identified. As for a quality rating, the study did not adequately addressing the important issues of scientific quality as defined by Aragon 2007. An overall rating of the strength of the evidence concludes that one can have a low level of comfort with the results of the aforementioned study.

It is important to note that this type of evaluation system does not rate the products or studies. What it does do is provide the reader with a sense of the reliability of the data presented to them and how comfortable they can be with the results that they are presented in each study. Furthermore, systematic reviews rarely consider the safety of any therapies and focus solely on efficacy. Weakness of systematic reviews will be discussed.

REFERENCES

1. *The American Heritage Dictionary of the English Language.* Boston: Houghton Mifflin Company, 1978.
2. Aragon CL, Hofmeister EH, Budsberg SC. A systematic review of the pharmacological and nutraceutical agents currently used in the treatment of canine osteoarthritis. J Am Vet Med Assoc 2007;230:514-521.
3. U. S. Food and Drug Administration Web site. Interim evidence-based ranking system for scientific data. Available at: www.cfsan.fda.gov/~dms/hclmgui4.html. Accessed June 12, 2008.
4. Sanderson RO, Beata C, Flipo RM Systematic review of the management of canine osteoarthritis. Vet Rec 2009;164(14):418-24.

5. Vandeweerd JM, Coisnon C, Clegg P, et al. Systematic review of efficacy of nutraceuticals to alleviate clinical signs of osteoarthritis. J Vet Intern Med. 2012;26(3):448-56.
6. Fritsch D, Allen TA, Dodd CE, et al. Dose-titration effects of fish oil in osteoarthritis dogs. J Vet Intern Med 2010;24:1020–1026.
7. Fritsch DA, Allen TA, Dodd CE, et al. A multicenter study of the effects of dietary supplementation with fish oil omega-3 fatty acids on carprofen dosage in dogs with osteoarthritis. J Am Vet Med Assoc 2010;236:535–539.
8. Roush JK, Dodd CE, Fritsch DA, et al. Multicenter veterinary practice assessment of the effects of omega-3 fatty acids on osteoarthritis in dogs. J Am Vet Med Assoc 2010;236:59–66.
9. Roush JK, Cross AR, Renberg WC, et al. Evaluation of the effects of dietary supplementation with fish oil omega-3 fatty acids on weight bearing in dogs with osteoarthritis. J Am Vet Med Assoc 2010;236:67–73.
10. Moreau M, Troncy E, Del Castillo JR, et al. Effects of feeding a high omega-3 fatty acids diet in dogs with naturally occurring osteoarthritis. J Anim Physiol Anim Nutr (Berl). 2012 Jul 14. doi: 10.1111/j.1439-0396.2012.01325.x
11. Hielm-Björkman A, Roine J, Elo K, et al. An un-commissioned randomized, placebo-controlled double-blind study to test the effect of deep sea fish oil as a pain reliever for dogs suffering from canine OA. BMC Vet Res. 2012 Sep 6;8:157. doi: 10.1186/1746-6148-8-157
12. Moreau, M., J. Dupuis, et al. (2003). "Clinical evaluation of a nutraceutical, carprofen and meloxicam for the treatment of dogs with osteoarthritis." Vet Rec **152**(11): 323-9.
13. McCarthy, G., J. O'Donovan, et al. (2007). "Randomised double-blind, positive-controlled trial to assess the efficacy of glucosamine/chondroitin sulfate for the treatment of dogs with osteoarthritis." Vet J **174**(1): 54-61.
14. Bui LM, Bierer TL. Influence of green lipped mussels (perna canaliculus) in alleviating signs of arthritis in dogs. *Vet Therapu* 2001;2:101-111.
15. Bierer TL, Bui LM. Improvement of arthritic signs in dogs fed green-lipped mussel (perna canaliculus). *Am Soc Nutr Sci* 2002;132:1634S-1636S.
16. Pollard B, Guilford WG, Ankenbauer-Perkins KL, et al. Clinical efficacy and tolerance of an extract of green-lipped mussel (Perna canaliculus) in dogs presumptively diagnosed with degenerative joint disease. N Z Vet J. 2006 Jun;54(3):114-8.
17. Moreau M, Dupuis J, Bonneau NH, et al. Clinical evaluation of a powder of quality elk velvet antler for the treatment of osteoarthrosis in dogs. Can Vet J. 2004 Feb;45(2):133-9.
18. Moreau M, Lussier B, Pelletier JP, et al.Brachystemma calycinum D. Don Effectively Reduces the Locomotor Disability in Dogs with Naturally Occurring Osteoarthritis: A Randomized Placebo-Controlled Trial. Evid Based Complement Alternat Med. 2012;646191:doi;10.1155/2012/646191.
19. Imhoff DJ, Gordon-Evans WJ, Evans RB, et al. Evaluation of S-adenosyl l-methionine in a double-blinded, randomized, placebo-controlled, clinical trial for treatment of presumptive osteoarthritis in the dog. Vet Surg. 2011 Feb;40(2):228-32
20. Innes JF, Fuller CJ, Grover ER, et al. Randomised, double-blind, placebo-controlled parallel group study of P54FP for the treatment of dogs with osteoarthritis. Vet Rec. 2003;152:457-60.
21. Reichling J, Schmokel H, Fitzi J, et al. Dietary support with Boswellia resin in canine inflammatory joint and spinal disease. Schweiz Arch Tierheilkd. 2004;146:71-9.

Appendix 1: Strength of evidence ranking

A "high level of comfort" ranking advises that qualified scientists agree that a specific claim is scientifically valid. This highest level of ranking possesses a very low level of probability that new scientific data will overturn the conclusion that the relationship in question is valid or significant. This rank is based on relevant, high quality studies of study design Types I and II with sufficient numbers of individuals resulting in a high degree of confidence that the results are relevant to the target population.

A "moderate level of comfort" ranking describes a relationship as promising but not definitive. The claim is based on relevant, high to moderate quality studies of study design Type III and higher and sufficient numbers resulting in a moderate degree of confidence that the results could be extrapolated to the target population.

A "low level of comfort" ranking possesses a low consistency. The relationship is based on moderate to low quality studies of study design Type III and has insufficient numbers of individuals tested resulting in a low degree of confidence that the results could be extrapolated. Uncertainties would also exist as to whether the proposed benefit(s) would be physiologically meaningful and achievable.

An "extremely low level of comfort" ranking has very low consistency and is based on moderate to low quality studies of design Type III and insufficient numbers resulting in a very low degree of confidence that the results could be extrapolated.

URINARY TRACT NUTRACEUTICALS, CRITICAL EVALUATION OF EFFICACY

Kate S. KuKanich, DVM, PhD, DACVIM (SAIM)
Manhattan, KS

INTRODUCTION

The term nutraceutical refers to any non-drug substance including dietary supplements, vitamins, minerals, and herbs that can be administered orally with the intention to promote good health.[1] These products can be marketed and sold without a prescription, and many consumers and pet owners perceive that they are all natural and therefore risk-free. For these reasons, many dog and cat owners have turned to nutraceuticals to help solve frustrating urinary problems such as recurrent infections and incontinence in their dogs and cats. Unfortunately, there are still limited data to support use of these nutraceuticals for these indications. Furthermore, many pet owners do not realize that nutraceuticals are not approved by the FDA, which means that their labels can be very misleading and that safety, efficacy, true contents of each pill, quality control, and interactions with other medications are poorly understood for these products. This review will discuss the many nutraceutical products on the veterinary market today which advertise a holistic strategy for managing urinary tract infections and urinary incontinence in dogs and cats, and specifically point to the literature-based evidence or lack of evidence for use of these products in companion animal medicine.

MANAGING URINARY TRACT INFECTIONS

Novel therapeutic strategies using nutraceuticals are being explored clinically for their potential benefit in the prevention and management of complicated UTIs. Although these therapies may show promise in individual animals when used in conjunction with appropriate antibiotic therapy and monitoring, there is little compelling evidence at this time supporting their efficacy in preventing or treating UTI in companion animals. The Antimicrobial Guidelines Working Group of the International Society for Companion Animal Infectious Diseases does not support use of any of the nutraceuticals listed below in the management of UTIs in companion animal medicine.[2]

Cranberry Extract

Cranberries have been associated with prevention of UTI for approximately 100 years.[3] Initially it was thought that cranberries cause a bacteriostatic urinary environment by acidifying pH; however this is not achieved with normal servings of cranberry.[4] Instead, research suggests that proanthocyanidins, which are polyphenolic compounds found in cranberries, ultimately interfere with adhesion of *E. coli* to the patient's uroepithelium by reducing the genotypic expression of fimbriae or by altering the phenotypic conformation of fimbriae.[5] Two randomized, double-blinded, placebo-controlled studies in women found positive clinical evidence that cranberry juice consumption can decrease the number of symptomatic UTIs over a 12-month period.[4,6] Therefore, cranberry juice is touted to be helpful in human healthcare to prevent UTIs in women.

In veterinary medicine, there are many cranberry products on the market that make claims of supporting a healthy urinary tract in dogs and cats, with some suggesting that use of their product will prevent the need for further use of antibiotics and thus development of antibiotic resistance. Although proanthocyanidins are considered the active ingredient in all of these cranberry products, ingestion of cranberry juice vs. cranberry extract may not produce the same effect, and each formulation of cranberry should be tested for efficacy and safety. The effect of cranberries may be different in humans vs. animals, and clinical trials using each of the various formulations of cranberries should be performed in each species for intended use. Most companies marketing cranberry products for companion animals do not provide research to support their claims of efficacy or safety in companion animals.

What evidence do we have to support cranberry extract use in veterinary medicine? In 2010, Howell et al presented an *ex vivo* study using 6 male research beagles administered the Nutramax cranberry product, *Crananidin,* for 21 days and documented anti-adhesion activity (inhibiting the attachment of p-fimbriated *E. coli* to uroepithelium) of cranberry beginning within 3 hours of administration and peaking at 7 days.[7] In 2011, Smee et al presented a study where 5 healthy client-owned female dogs received the Vetoquinol product, *Paxon*, for 30 days, and compared urine before and after for bacterial adhesion via an *in vitro* assay.[8] They found that administration of the cranberry extract product led to a 30% reduction in bacterial adhesion of *E. coli.*[8] The clinical application of the results from these 2 studies is not yet fully understood. It is also unclear whether these results would apply to other cranberry products due to differences in extraction, potency, stability, or purity. While these studies suggest that cranberry extract (these specific tested products) may decrease adherence of *E. coli* to the bladder wall of dogs, no prospective studies have been performed in dogs with clinical recurrent *E. coli* UTIs to determine if administration of cranberry extract does in fact help prevent recurrence or aid in the management of this condition. These studies are warranted to determine if a cranberry nutraceutical is worthy of implementation into clinical practice. In addition, further studies would be warranted to determine the most appropriate dose and dosing interval.

While no adverse effects were reported in these two small studies, the full safety profile and potential for interactions with antibiotics and other drugs have not been studied. Similarly, no data exist to support use of this product for prevention of UTI from other bacterial species *(Proteus, Staphylococcus, Enterococcus*, etc). Further, there are no data currently to suggest that cranberry extract is effective in treating existing UTIs in dogs. At this time, it is the consensus of the Antimicrobial Guidelines Working Group of the International Society for Companion Animal Infectious Diseases that there is not sufficient evidence to support use of cranberry extract to prevent recurrent UTIs in dogs and cats.[2]

D-Mannose

D-mannose is a sugar that competitively binds to mannose-sensitive fimbriae on certain strains of *E. coli* that possess these fimbriae. This inhibits adhesion to polysaccharide receptors on the uroepithelial surface.[9] In one study, addition of D-mannose to human epithelial cells to which *E. coli* was pre-attached caused a rapid displacement of organisms from the epithelial cells, suggesting this nutraceutical has the potential to be therapeutic in people.[9] However, no studies are available suggesting clinical efficacy in humans, no similar studies have been performed with canine or feline epithelial cells, and no clinical trials have been performed in dogs or cats. Little data exist regarding veterinary uropathogens other than *E. coli* that express these fimbriae, and thus extrapolating how other bacterial species causing veterinary UTIs may respond to this nutraceutical is not possible at this time. For example, while most human uropathogenic *Proteus* isolates do express this type of fimbriae, many are reported to be mannose-resistant which may alter their anticipated response to this nutraceutical.[10] D-mannose has been used anecdotally for the treatment and prevention of recurrent UTIs in companion animals, but no studies are available on efficacy, safety, or most appropriate dose for this indication in dogs and cats.

Forskolin

Treatment with forskolin, the main active ingredient in the Asiatic herb *Coleus forskohlii*, has been shown to increase cAMP levels and reduce the level of intracellular bacteria in the bladders of infected mice, which is thought to occur by stimulating the exocytosis of *E. coli*-containing vesicles.[11] Therefore, in theory, administration of this dried herb could be beneficial in certain cases of recurrent *E. coli* UTIs that involve a nidus of infection in the bladder wall where antimicrobial therapy may have inadequate penetration.[12] Potential side effects to consider are that components of *Coleus forskohlii* reportedly cause vasodilation, are positive inotropes, and decrease intraocular pressure.[13] Although no clinical trials are available to provide information on efficacy, dose, or safety, this herb has been used anecdotally in veterinary medicine, in combination with appropriate antimicrobial therapy, for control of recurrent UTIs.

Methenamine

Methenamine is a urinary antiseptic that is hydrolyzed to formaldehyde and ammonia in acidic environments ($pH<5.5$) and has bactericidal activity against *E. coli* and a variety of other microbes, making it useful to help treat or prevent recurrence of UTIs. It is available as a salt with either mandelic acid or hippuric acid, which act to acidify the urine; additional urine acidifiers may be needed as well. Although this nutraceutical has been used for years in veterinary medicine, studies regarding safety, efficacy, and appropriate dosing for veterinary patients are lacking. Gastrointestinal upset is a reported side effect, and methenamine should be avoided in patients with diseases associated with metabolic acidosis, such as chronic kidney disease. Methenamine is not well tolerated by cats.

Vaccination

On the horizon, one strategy to combat UTIs that may become available to clinicians is vaccination. Vaccine development has targeted *E. coli's* virulence factors, including adhesins, toxins, and lipopolysaccharide. One promising vaccine, developed by deleting an O antigen ligase gene, stimulates urothelial cytokine secretion and has prevented persistent colonization of *E. coli* in the bladders of mice upon challenge.[14] No vaccines for uropathogenic *E. coli* are available for dogs or cats at this time.

Miscellaneous

Various other nutraceuticals have been used or discussed in companion animal medicine, including acidifiers (methionine, vitamin C), glycosaminoglycans, probiotics, *Echinacea*, marshmallow root, blueberry extract, olive leaf, Oregon grape root, couchgrass root, juniper berry, and parsley root. To the author's knowledge, no efficacy studies have been performed using any of these nutraceuticals for urinary tract infections in dogs and cats. At this point in time, there is insufficient evidence to support use of these nutraceuticals for management of urinary tract infections in dogs and cats, and their use is not currently recommended by the Antimicrobial Guidelines Working Group of the International Society for Companion Animal Infectious Diseases.[8]

MANAGEMENT OF URINARY INCONTINENCE

Urinary incontinence can be a frustrating chronic condition to manage in dogs and cats; identifying an underlying cause is often the first step toward successful management. A thorough workup should be performed initially to rule-out anatomical defects, such as an ectopic ureter, which may be surgically addressed. A complete neurologic exam should be performed to determine if there is a neurologic component to the incontinence (such as disc disease). Any causes of local inflammation (such as UTI or uroliths) should be identified and treated appropriately. Clinicians should also clarify that the patient has true incontinence rather than polyuria, especially in older patients, as polyuria would require a different avenue of diagnostics and therapy. After ruling out all possible underlying causes of incontinence, many female dogs are ultimately diagnosed with sphincter incontinence, which can be congenital or acquired (often after ovariohysterectomy). A major pathophysiological mechanism of this condition is relative lack of estrogen affecting the collagen in the urethral sphincter and leading to a decreased intrinsic tone; however, sphincter incontinence is considered to have a multifactorial etiology that is not well understood. For sphincter incontinence in dogs, traditional therapy has included medical management with either phenylpropanolamine (PPA), an alpha-adrenergic receptor agonist, which stimulates contraction of the internal urethral sphincter, or to provide estrogen effects via diethylstilbestrol (DES). PPA is an effective FDA-approved drug in dogs for urinary incontinence due to urethral sphincter hypotonus.[15-17] Side effects of concern for PPA include tachycardia, hypertension, CNS excitement, and appetite suppression; clinicians should also be aware of potential drug interactions (other sympathomimetic drugs, antidepressants, etc).[18] In mice, administration of estrogen has been shown to improve periurethral blood flow and to enhance collagen formation within the sphincter which may lead to improved urinary continence.[19] Response to estrogens in dogs has ranged from 41-82% in the veterinary literature.[20-22] DES is not FDA-approved for animal use and is currently only available in the U.S. through compounding pharmacies. Although hormones are typically regulated by the FDA, compounded preparations lack FDA approval and therefore do not have guaranteed identity/composition, potency, stability, purity, quality control, bioequivalence, or efficacy.[23] Conjugated estrogens (Premarin, Pfizer Inc.) are available commercially in the U.S. and can be used extra-label in dogs as an alternative for DES. Newer to the market is an FDA-approved estriol product (Incurin, Merck Animal Health) which is labeled for oral use to control estrogen-responsive urinary incontinence in spayed female dogs. In general, the main side effect of concern for estrogen supplementation is bone marrow suppression (mainly anemia), although there may also be increased risk of pyometra and neoplasia. For refractory cases, PPA and DES can be used in combination, with the patient weaned to the lowest effective doses to minimize side effects. In the past decade, submucosal collagen injections have provided continence to some dogs that were previously refractory to medical management.[24] Additional novel therapies, such as GnRH analogues, continue to be studied to provide additional options.

There are many nutraceutical products on the veterinary market targeting urinary incontinence in dogs and cats. Most of these products contain phytoestrogens as the active ingredient. Phytoestrogens are derived from plants and are structurally and functionally similar to 17-estradiol.[25] These ingredients include (but are not limited to): *Angelica sinensis* (dong quai), *Cimicifuga racemosa* (black cohosh), *Dioscorea villosa* (wild yam root), *Glycyrrhiza uralensis* (licorice), *Psoralea corylifolia*, and *Vitex agnus-castus* (Chaste tree). Other ingredients are added to some formulations for potential antioxidant or antimicrobial contribution, such as bee pollen,[26] or anti-inflammatory activity, such as *Rehmannia*.[27] While the author is not aware of evidence-based research to support the efficacy of these products for use in companion animal medicine, nutraceuticals marketed for incontinent animals may be a preferred option in some situations if PPA is not effective or the side effects render it unacceptable or inappropriate. Although these nutraceuticals may be marketed as all-natural or safe alternatives, it should be considered possible that administration of nutraceuticals containing estrogen-like compounds could result in side effects (such as bone marrow suppression) from their estrogen-like properties. Until full safety profiles are available on the use of these extracts in dogs and cats, caution should be used. For example, it is now widely known that ingestion of grapes and raisins by dogs can cause acute renal failure, and it is possible that other potentially severe adverse reactions could occur from seemingly safe all-natural products that may be found in nutraceuticals.[28]

REFERENCES

[1]Bauer JE. JAVMA 2001;218:1755-1760. [2]Weese JS, et al. Vet Med Int 2011; doi:10.4061/2011/263768. [3]Howell AB. Mol Nutr Food Res 2007;51: 732-7.[4]Avorn J, et al: JAMA 1994;271:751-4. [5]Liu Y, et al: Biotechnol Bioeng 2006;93:297-305.[6]Kontiokari T, et al. BMJ 2001;322:1571. [7]Howell A, et al. JVIM 2010;24:660. [8]Smee et al. JVIM 2011;25:722-723 [9]Ofek I, et al. Nature 1977;265:623-5. [10]Jansen AM, et al. Infection Immun 2004;72:7294-7305. [11]Bishop BL, et al. Nat Med 2007;13:625-30. [12]Anderson GG, et al. Trends Micro 2004;12:424-430. [13]Dubey MP, et al. J Ethnopharmacol 1981;3:1-13. [14]Billips BK, et al. J Infect Dis 2009;200:263-72. [15]Richter KP, et al. JAVMA 1985;187:605-611. [16]White RA, et al. Vet Rec 1989;125:478-480. [17]Bacon NJ, et al. Vet Rec 2002;151:373-376. [18]Peterson KL, et al. JAVMA 2011;239:1463-1469. [19]Rizk DE, et al. Urol Res 2003;31:147-151. [20]Holt PE, et al. JSAP 1985;26:181-190. [21]Nendick PA, et al. Aust Vet J 1987;64:117-118. [22]Mandigers PJJ, et al. Vet Rec 2001;149:764-767. [23]AVMA, Veterinary Compounding 2011; accessed Jan 2013 at: https://ebusiness.avma.org/EBusiness50/ProductCatalog/product.aspx?ID=155. [24]Byron JK, et al. JVIM 2011;25:980-984. [25]Russel L, et al. Am J Med Sci 2002;324:185-188.[26]Fatrcova-Sramkova K, et al J Env Sci Health 2013;48:133-138. [27]Prieto JM, et al. J Pharm Pharm 2003;55:1275-1282. [28]Eubig PA et al. JVIM 2005;19:663-674.

NUTRACEUTICAL EVALUATION: MARKETING AND SURVEILLANCE

Butch KuKanich, DVM, PhD, DACVCP
Manhattan, KS

DIETARY SUPPLEMENTS AND ANIMALS

Dietary supplements are Food and Drug Administration (FDA) regulated products for use in people only [1]. There are no regulatory categories of dietary supplements for animals and as such they will fall into either a category of a food or drug. A food is defined as "articles used for food or drink for man or other animals...and articles used for components of any such article." Animal foods do not need pre-market approval by the FDA and animal foods are required to be pure and wholesome, contain no harmful or deleterious substances, and be truthfully labeled. A drug is defined as "an article intended for use in the diagnosis, cure, mitigation, treatment, or prevention of disease, or an article intended to affect the structure or function of the body other than food." If a substance affects the function of the body, other than provision of nutrients, it is considered a drug. For example, a supplement marketed to improve joint function is considered a drug. It is also important to note that a substance including food, but is intended for the treatment or prevention of disease, is also considered a drug. For example, a food marketed to decrease clinical signs associated with renal disease is considered a drug.

Dietary supplements are not required to demonstrate safety or efficacy prior to marketing since they are not approved drugs. There are also no requirements for minimum manufacturing standards, such as potency (correct amount of active ingredients), stability, purity, or batch to batch consistency as they are not approved drugs. Another substantial problem with unapproved drugs is the lack of a central database for reporting adverse drug events, including lack of efficacy. In contrast, approved veterinary drugs have a central adverse drug experience database (see below). Although some dietary supplements are included in the reports, they are not robustly reported and therefore cannot be directly compared to adverse events reported for FDA approved veterinary drugs. Manufacturers of approved drugs are required to report adverse events to the FDA. The manufacturers of unapproved drugs are not required to report adverse events as the drug is not approved. There are limitations of adverse event reporting even for approved veterinary drugs. There is no certainty of the adverse event to the administered drug. The adverse event may have been due to concurrent drug administration, an underlying disease, environmental, or coincidental factors. The total number of animals treated is unknown, therefore some drugs may have a larger number of reported adverse effects if they are administered more frequently or have been on the market longer. Additionally, common or expected adverse events are often not reported. Not all adverse events are reported due to time constraints or lack of knowledge of the system by the veterinarian or owner. The duration of treatment is not reported in the adverse drug experience reports. The accuracy of the information submitted is not known. Information on how the drugs were administered may not be accurate, for example the correct dose may not have been administered. The drug may have been administered in an extralabel manner or with concurrent drugs which increased the risk for the adverse event. The magnitude of the adverse event may not be detailed in the reports, such as an increase in ALT may be reported, but may not be clinically relevant. For example many experts consider an increase of 2 times the upper end of the reference range for AST as a clinically relevant increase, but the report may include an increase of AST by 15% over the upper end of the reference interval which may or may not be clinically relevant.

DIETARY SUPPLEMENTS AND THE FDA

Dietary supplements were placed into their own regulatory category by the FDA under the Dietary Supplement Health and Education Act of 1994 and are considered a category of food[2]. A dietary supplement is defined as a product that contains any dietary ingredient to supplement or add nutritional value to the diet. Dietary supplements may be produced in many different forms including tablets, capsules, liquids, powders, etc. The dietary ingredient may include any one or combination of:

- Vitamin
- Mineral
- Herb
- Botanical
- Amino acid
- Dietary substance to increase total dietary intake
- Concentrate, metabolite, constituent, or extract

Dietary supplements may ensure adequate intake of essential nutrients or may help reduce your risk of disease, but NOT intended to treat, diagnose, cure, or alleviate the signs of disease. It is important to note that human dietary supplements are not FDA approved and do not need to be assessed by the FDA prior to marketing. Therefore safety, efficacy and dosing studies are not needed prior to marketing. The manufacturer of the supplement has the responsibility to ensure its products are safe, claims are not false or misleading, and the products comply with the Food Drug and Cosmetics Act and other FDA regulations. However not all companies assume these responsibilities. For example, Riger Naturals S.A. is a company in Mexico that marked a product called Reumofan Plus [3]. Reumofan Plus was marketed as a "natural" dietary supplement for the treatment of conditions such as arthritis, however the FDA received numerous complaints regarding this product including liver injury, severe bleeding, corticosteroid withdrawal syndrome, adrenal suppression and stroke. The product was evaluated by FDA laboratories and found to contain the corticosteroid dexamethasone, the NSAID diclofenac, and the muscle relaxant methocarbamol. The FDA issued a safety alert for undeclared drug ingredients and product recall for Reumofan Plus. Subsequently, the same product was marketed as WOW to deceptively sell remaining supplies, which the FDA also sent a notification. This product is not the only one which the FDA has discovered to contain undeclared drug ingredients. Many "all natural sexual enhancement" supplements have contained sildenafil (Viagra) or analogues, and "all narutal" weight loss supplements have contained sibutramine, fenfluramine, ephedrine, and propranolol. Additionally, some dietary supplements have been removed from the market due to poor safety profiles including reports of liver failure, seizures and death for weight loss supplements and loss of smell for cold relief products.

Due to the lack of premarket efficacy and safety testing of dietary supplements and lack of oversight on manufacturing, it is truly "Buyer Beware" when using these products. Additionally there are liability considerations for the veterinarian that dispenses or recommends these products in lieu of FDA approved drugs.

USEFUL INFORMATION AVAILABLE ONLINE

The FDA has many resources available for food, dietary supplements and approved drugs. An email listserv is available which sends out notices of recalls of dietary supplements in addition to FDA approved drugs, over the counter drugs, and food recalls. The FDA also has a website that describes human dietary supplements.

Listserv for FDA MedWatch including product recalls, information, changes, et al:
https://public.govdelivery.com/accounts/USFDA/subscriber/new?pop=t&topic_id=USFDA_9

Update on Animal Dietary Supplements
http://www.fda.gov/AnimalVeterinary/NewsEvents/FDAVeterinarianNewsletter/ucm110415.htm

FDA Consumer Updates for Animal & Veterinary, human drugs, dietary supplements et al.
http://www.fda.gov/ForConsumers/ConsumerUpdates/default.htm

FDA 101: Dietary Supplements
http://www.fda.gov/ForConsumers/ConsumerUpdates/ucm050803.htm

Instructions for reporting an adverse drug event for veterinary drugs
http://www.fda.gov/AnimalVeterinary/SafetyHealth/ReportaProblem/ucm055305.htm

FDA Veterinary Adverse Drug Experience Reports
http://www.fda.gov/AnimalVeterinary/SafetyHealth/ProductSafetyInformation/ucm055369.htm

Pet Food Complaints
http://www.fda.gov/AnimalVeterinary/SafetyHealth/ReportaProblem/ucm182403.htm

REFERENCES

[1]Grassie, L. **FDA Veterinarian Newsletter** 2002;17(3):3. [2] Dietary Supplement Healthy and Education Act of 1994. [3] FDA MedWatch Safety Alert, December 2012: http://www.fda.gov/ForConsumers/ConsumerUpdates/ucm333188.htm (accessed 1/31/2013)

NUTRACEUTICALS AND DRUG INTERACTIONS

DM Boothe, DVM, PhD, DACVIM, DACVCP
Auburn, AL

INTRODUCTION

For the purposes of this manuscript, although the term nutraceutical will be used, the topic of discussion is not limited to nutraceuticals (eg, endogenous compounds supplemented to affect function or structure). Rather, the discussion applies to (animal) dietary supplements (ADS). Although not a legal term, ADS has been defined by the Committee on Examining the Safety of Dietary Supplements for Horses, Dogs, and Cats (National Academy of Sciences) as a substance for oral consumption by horses, dogs, or cats … intended for specific benefit to the animal by means other than nutrients recognized as essential…. Accordingly, the term incorporates compounds that be considered endogenous, or truly "natural" to the animal, such as glucosamines, sAME or other compounds, as well as biologics or herbs which are "natural" in the sense that they are not synthesized but are not natural to the patient. The use of the word "compound" intermittently throughtout this discussion is intended to remind the reader that interactions involve a wide variety of compounds, including drugs, nutrients, or endogenous compounds (whether generated by the body or supplemented exogenously), and other xenobiotics such as herbs or biologics. No one term entirely captures the broad nature of compounds that might be involved in interactions. Likewise, the site of interactions among these compounds is not limited to one class of molecules (eg, drugs versus biologics vs endogenous compounds); rather, interactions can occur among any.

DRUG INTERACTION DEFINED

An interaction occurrs between or among compounds whenever the action of one compound is modified by the presence of another, concurrently administered compound. Interactions may occur prior to absorption (most commonly reflecting a chemical interactions), often referred to as a "pharmaceutic" phase, during the "pharmacokinetic" phase, during which when compound one alters the disposition of another (most commonly due to altered absorption or metabolism) or during the "pharmacodynamics" phase if one compound alters the body's response to another. Interactions can occur between active ingredients, additives or excipients, carriers or other components of an animal dietary supplement, or with storage containers or the environment. Interactions may occur between animal dietary supplements and drugs or nutrients. Among the dietary supplements, interactions have been best described for herbs. The focus of this discussion will be pharmacokinetic interactions, although the type is not always identified. Much of the information presented here is from human medicine. Interactions that occur inside the body may lead to life-threatening adversities either due to therapeutic failure or increased side effects. Pharmacokinetic interactions may impact any phase of compound movement: absorption, distribution, metabolism, or elimination. Over 50% of recent drug withdrawals in the United States have occurred because of serious drug –drug interactions, emphasizing the increasing recognition of the role of interactions in adversity.

Table 1. Sites of interactions involving nutraceuticals (compounds) and drugs

PHARMACOKINETIC	
Absorption	
Increased splanchnic blood flow (nutrients)	Increased rate of absorption
Gastric motility	Degree of ionization dependent on compound pKa
Gastric pH	Ionization decreases rate and extent; status depends on compound pKa
Competition for transport proteins	Altered rate or extent
Inhibition or inducdtion of drug metabolizing enzym	Altered rate or extent
Distribution	
Competition for transport proteins in circulation	Altered concentration of free drug in circulartion
Competition for transport proteins in tissues	Altered concentration of free drug in tissue
Altered regional blood flow	Altered drug delivery
Metabolism	
Competition for transport proteins	Altered clearance
Induction of phase I enzymes (CYP 450)	Increased clearance of other compounds
	Increased risk of toxic metabolites
Inhibition of phase I enzymes (CYP 450)	Decreased clearance of other compounds
Increased phase II enzymes (glutathione)	Increased clearance, especially toxic metabolites
Excretion	
Competition for transport proteins	Altered renal clearance
Altered renal blood flow	Altered renal clearance
Altered urinary pH	Altered passive resorption and clearance
PHARMACODYNMIC	
Additive or synergistic interaction	Enhanced response
Competitive or irreversible inhibition	Decreased response

TRANSPORT AND DRUG METABOLIZING ENZYMES

The majority of pharmacokinetics interactions currently recognized as significant involve either transport proteins or drug (or compound) metabolizing enzymes. Both protein types are intended to protect the body from unwanted compounds. As such, these proteins are located either at portals of entry (e.g, the GI tract, skin or respiratory tract), or at barriers encountered at sanctuaries (such as the brain, CSF, eye, testicles). Transport proteins located on cell membranes act to efflux or influx drug; efflux prevents the compound from moving to an undesirable site; drug metabolizing enzymes tend to work in concert at the same site in order to removed compounds that are able to circumvent transport proteins. Both classes of proteins act on a variety of compounds, including endogenous compounds that may require manipulation, and foreign compounds that do not belong. Compounds that serve as substrates for these proteins often induce the proteins.

Because induction generally reflects formation of more protein, induction impacts most substrates. Substrates may also act as inhibitors; such inhibition is often competitive and may not be a broad in specificity as induction. Not surprisingly, some compounds simultaneously serve as substrates, inducers or inhibitors of both transport or metabolizing enzymes.

Transport proteins act alone or in concert with one another. Some are specific whereas others are broad in substrate specificity. Several families have been identified including 1) those located on the apical cell membrane (organic anion-transporting polypeptides (OATPs) of the liver, kidney, and brain and thus able to influence distribution and excretion. These proteins transport a large number of amphipathic drugs (e.g., digoxin, steroids) and endogenous compounds (e.g., steroids, thyroid hormones). (2) Proton-dependent oligopeptide transporters (POTs), driven by a proton gradient, are present in the kidney, brain, and the gastrointestinal tract, where their activity appears to increase from the duodenum to the ileum. Among the drugs influenced by these transporters are beta-lactam antimicrobials, ACE-inhibitors, and prodrugs of selected antivirals. (3) The family of ATP-binding cassette (ABC) superfamily of transporters includes the P-glycoproteins. These include multidrug resistance protein (MRP1) and a multispecific organic anion transporter (MOAT) also located in other tissues. An efflux transporter, P-glycoprotein is characterized by broad substrate specificity. Because it is abundant in enterocytes, it can profoundly decrease the oral bioavailability of target drugs. It is associated and shares substrate specificity with selected cytochrome-P450 enzymes (most notably, CYP3A), which decreases drug that is not effluxed from the cell by the transporter. Examples drug substrates for P-gp include: antimicrobial drugs (macrolides, imidazole antifungals, selected fluoroquinolones); anticancer drugs (doxorubicin, vincristine, vinblastine, mitoxantrone), anthelmintics (avermectins), cardiac drugs (digoxin, quinidine, diltiazem, verapamil),CNS active drugs (selected morphine derivatives, behavior modifying drugs, phenothiazines), steroidal hormones, and endogenous substrates such as bilirubin and bile acids. Inhibitors of P-gp include macrolides, cyclosporine, imidazole antifungals, quinidine, verapamil, and selected morphine derivatives. Inducers include dexamethasone, morphine, rifampin, phenothiazines, St. John's wort and others.

Drug Metabolizing Enzymes

Pharmacokinetic interactions frequently alter the metabolism of a concurrently administered compound. Most of the interactions result from modulation of hepatic (phase I) drug metabolizing enzymes . Although referred to as drug metabolizing enzymes, the enzymes function to metabolize any type of compound including dietary supplements. Induction of drug metabolizing enzymes is a protective mechanism that facilitates excretion of potentially toxic compounds. However, induction is a double-edged sword: although inducers generally increase the elimination of a potentially toxic xenobiotic, increased formation of toxic or carcinogenic metabolites may also occur. Most CYP enzymes are inducible, although the response to inducers varies within and among species. Human CYP known to be influenced by inducers include CYP 1A1/2, 2A6, 2C9, 2C19, 2E1, and 3A4. Inducers are substrates for the enzymes which they induce and induction generally is dose dependent. Induction is accompanied by an increase in the intracellular concentration if the induced enzyme, often as a result of increased transcription. Although inducers tend to exhibit enzyme specificity, they may induce more than one CYP. Inducers may result in significant activity of an enzyme that otherwise (constitutively) is either present in very low concentrations or altogether absent. Maximal transcription in response to an inducer generally requires 10 to 12 hrs of exposure to a compound with activity returning to baseline by 18 hrs (dependent on dose) following discontinuation of exposure to the inducer. However, the impact of the inducer may persist, depending on the rate of CYP degradation. The impact of induction on the metabolism of selected substrates varies with the extent being dose dependent. Inhibition of CYP enzymes also occurs, with some inhibitors by broad-enzyme interactions toward a number of different CYPs, and others selective for a single enzyme (Hollenberg 2002). Most inhibitory interactions are transient, resolving when therapy with the inhibitor is discontinued. Like induction, reversible inhibition appears to be dose-dependent.

NUTRACEUTICALAND DRUG INTERACTIONS

The most clinically relevant pharmacokinetic interactions that have been described involving dietary supplements reflects inhibition or induction of absorption or metabolism. In the author's opinion, an important distinction should be made between nutraceuticals, which are endogenous and thus ARE natural to the patient, versus herbs or biologics, which are not.The more foreign the compound is to the patient, the more likely induction or inhibition will take place. The latter are more likely to stimulated protective mechanisms, such as induction of proteins such that the offending compound is less likely to reach the body or sanctuaries. However, it is also important to recognize that the interactions can go both ways, that is, the dietary supplement can be impacted. The absence of information regarding the impact of drugs on dietary supplements is likely to reflect a number of factors. These include the lack of mandated safety studies, limited well-designed clinical trials (for a variety of reasons, including the variability of products used, differences in "dosing" regimens, the lack of control data, lack of knowledge regarding anticipated duration to effect, etc), the questionable quality of the studied products (including limited quality assurance considerations), the wide diversity in compounds that comprises biologics and herbs (thus requiring sophisticated analytical techniques as well as difficulties in identifying exactly which compound was causing the effect). For example, garlic contains sulphur compounds, flavonoids/isoflavanoids, prostaglandins, saponins, terpenes, and others. The variable names under which herbs or biologics are marketed confounds the identification of an adversity as a result of a supplement-drug interaction. Finally, the lack of a mandated or well recognized robust adverse event reporting system for dietary supplements in either humans or animals confounds our ability to recognized adversities and the potential role of interactions in their advent. The report of the Committee on Examining the Safety of Dietary Supplements for Horses, Dogs, and Cats (http://www.nap.edu/catalog.php?record_id=12461) addresses these issues more in depth.

Despite the lack of mandated quality assurance or adverse event surveillance, several reviews have described interactions between drugs and botanicals or herbs and foods. Most reports involving herbs or botanicals reflect case reports, case series or open, uncontrolled studies rather than randomized, controlled studies (Hu et al 2005). Mechanisms are often difficult to identify, in part because of broad substrate specificity. Those most commonly identified involve altered pharmacokinetics of the xenobiotic due to interference with either P-glycoprotein or xenobiotic metabolizing enzymes, particularly CYP3A4. Oftimes, the cause of an interaction cannot be identified; rather, changes in pharmacodynamic response might be documented, but the underlying cause may be changes in the disposition (pharmacokinetics) of the drug induced by the active ingredient of the supplement. Pharmacodynamic interactions occur when one compound directly alters the chemical or physiologic response to another. Pharmacodynamic interactions can enhance the response of a compound due to additive or synergistic effects at the same receptor, or at different sites but with the same physiologic reaction. Pharmacodynamic interactions may decrease the response of some compounds due to competitive antagonism at the same receptor site or due to antagonistic responses mediated at distant, but physiologically related sites. The risk of drug interactions in persons consuming herbs has caused the American Society of Anesthesiologists to generate a brochure for its members entitled ''What You Should Know About Your Patients' Use of Herbal Medicine (https://ecommerce.asahq.org/p-131-considerations-for-anesthesiologists-what-you-should-know-about-your-patients-use-of-herbal-medicines.aspx?).'' Examples of potential interactions involving the central nervous system (CNS) include enhanced stimulation by caffeine, ephedra, yohimbine, guarana, and ginseng as well as enhanced sedation by valerian, kava, and St. John's wort. Drug-induced hemostasis defects may be potentiated by garlic, ginger, gingko, ginseng, chamomile, feverfew, and bromelain, whereas the effects of hypoglycemia may be exacerbated by bilberry, bitter melon, dandelion, and garlic. Echinacea and astragalus may offset the immunomodulatory effects of corticosteroids or other immunosuppressants. Other examples are listed in Table 2.

Some pharmacokinetic interactions have been documented between drugs and supplements. Interactions between nutraceuticals and drugs may involve distribution. The most likely site is competition for protein binding sites (Table 1), but these are not well characterized. Further, displaced, unbound compounds are likely to be cleared more rapidly, thus balancing the increase in free, active drug. Interactions involving absorption and metabolism are probably best described and of most importance, with most of these interactions occurring at the level of the protein. Food – drug interactions have been described,

Table 2. Examples of interactions between drugs and dietary supplements				
Compound	**Reference**	**Alternative names**	**Drug**	**Effect**
Cranberry juice extract	Gardiner 2008		Warvarin	Prolonged bleeding
Ephedra	Flanagan 2001		Cardiac glycosides	Arrhythmias
Feverfew	Flanagan 2001		Coumarin derivatives, aspirin, antiplatelet drugs	Prolonged bleeding
Fish oil	Gardiner 2008		Warvarin	Prolonged bleeding
Garlic	Hu 2006	Allium sativum	Aspirin, antihemostatics	Prolonged bleeding
Ginger	Flanagan 2001		Aspirin, antihemostatics	Prolonged bleeding
Ginseng	Flanagan 2001		Oral hypoglycemics, cardiac stimulants, warfarin	Hypoglycermia, tachycardia, hypertension
Ginseng (siberian)	Hu 2006	Eleutherococcus senticosus	digoxin	Increased AUC, increased effects
Gingko	Flanagan 2001		?	Hemorrhage
Licorice	Flanagan 2001			hyperkalemia,
Milk Thistle	Hu 2006	Silymarin, silybum	antiviral drugs	decreased efficacy
Pepper	Hu 2006	Piper nigrum, P. longum linn	rifampin, phenytoin, propranolol, theophylline	Increased AUC, increased effects
Senna	Flanagan 2001		Oral drugs, diuretics	Decreased absorption, hypokalemia
St. Johns wort	Flanagan 2001	Hypericum perforatum	Anesthetics, behavior modifying drugs	Prolonged effects, ehanced side effects
	Hu 2006, Gardiner 2008		Cyclosporin, midazolam, tarcrolimus, amitryptyline and other TCA, midazolam and other benzodiazepines, methadone,	Decreased AUC, decreased efficacy
	Hu 2006		selective serotonin inhibitors (fluoxetine, paroxetine),	serotonin syndrome
	Hu 2006		(oral contraceptives)	unplanned pregnancies
	Consumerlab		Fluoroquinolones	increased risk of phototoxicity
Valerian	Flanagan 2001		Anesthetics	Prolonged effects

most commonly occur ing in the gastrointestinal tract and reflect therapeutic failure due to reduced bioavailability of the drug. For example, absorption has been impacted by a number of foods (dietary nutrient) during the pharmaceutic phase. For example, tetracyclines (but not doxycycline) and fluoroquinolones bind to cationic (calcium, magnesium) dietary constituents, precluding drug absorption. Less commonly, increased drug concentrations reflect altered appetite or gastrointestinal acidity. Absorption of drugs will be impacted by interactions that influence both efflux proteins or drug metabolizing enzymes (Table 1). Note that the diverse nature of ingredient s (including state of the ingredient, ranging from extracts to ground substrates, whole or partial plant materials) should cause concern about binding of drug to these materials, with similar impact on rate and extent of drug absorption. Although minimally described, interactions involving renal excretion, and specifically active tubular secretion, as well as metabolism should be anticipated. Examples of dietary supplements with known drug interactions include St. John's Wort (induction of CYP3A4, particularly intestinal, resulting in increasd clearance and decreased efficacy of many drugs), echinacea (induction and inhibition of intestinal CYP), ginkgo biloba (induction of CYP 219A), and grapefruit (inhibition of CYP3A4 and inhibition of P-glycoprotein). Flavenoids are inhibitors of several P-glycoprotein substrates. Characterization of the xenobiotic interactions as a cause of adverse effects is difficult because large numbers of subjects generally must be studied to identify even common interactions. Of potential interactions impacting absorption, competition among substrates for transport proteins have probably been the best described.

WEBSITES

Obtaining information regarding the potential risk of interactions between drugs and herbs can be difficult. The following (all accessed February 17, 2013) provides useful information. Among the more informative sites is www.consumerlab.com . This for profit organization tests the quality of a variety of different supplements at the manufacturers' behest. Considerations include properly labeled strength, lack of contaminants, including drugs and other criteria. Products are sent to an independent lab for testing. For an approximate $25/year subscription fee, users can obtain information on which products passed, and which products failed and why. The income generated from the subscriptions is used to study non-solicited products which also is reported (not, in general, pet products, which are not often studied, fare more poorly compared to human products). However, among the information at this web site is an encyclopedia that describe the active ingredients of the supplements, including links to clinical trials that address evidence (or lack thereof) of efficacy. Drug interactions are included in the "cautions" section for each product. HerbMedicine (http://www.hebmed.org) provides a similar format at no charge. Medscape (sponsored by the NIH [free]: http://www.nlm.nih.gov/medlineplus/druginfo/natural/1109.html) also offers a website on dietary supplements, with drug interactions delineated within the document for each erb. Dr. Oz [free] has an apparently informative website that links to clinical trial reports among other resources (http://www.doctoroz.com/videos/drug-interaction-checker).This website is dedicated to interactions and is among the easiest to use. The National Medicine Comprehensive Data Base (http://www.naturaldatabase.com) includes a subscription-based site that allows typing drugs and supplements of concern.

REFERENCES

Fennell, C.W et al. Journal of Ethnopharmacology 2004; 95:113-121. Flanagan K et al: J Perianeths Nursin 2001; 16:19-26; Gardiner P, et al: Am Fam Physician. 2008 Jan 1;77(1):73-8. Hollenberg, P.F. 2002: Drug Metabolish Reviews 34(1&2):17-35.Hu, Z et al: Drugs. 2005: 65(9):1239-1282.Huang, S.-M et al: 2004; Journal of Clinical Pharmacology 44:559-569. Kunta, 2004;Current Drug Metabolism 5:109-124.Schmidt, et al: 2002; Drugs 62(10):1481-1502. Zhang et al: 2003. The Journal of Pharmacology and Experimental Therapeutics 304(3):1258-1267.

ANTIOXIDANT SUPPLEMENTS: CRITICAL EVALUATION OF EFFICACY

Katrina R. Viviano DVM, PhD, DACVIM, DACVCP
Madison, WI

INTRODUCTION

Reactive oxygen species (ROS) are a by-product of oxidative metabolism and are highly reactive molecules that damage lipids, proteins, carbohydrates, and DNA. When ROS are produced in excess of the capacity of the endogenous antioxidant network, increased cellular oxidative products are generated leading to a state of oxidative stress.[1,2] Biomarkers of oxidative stress include endogenous antioxidants and oxidation products of cellular components. Endogenous antioxidants function independently and synergistically within a highly interconnected cellular antioxidant network to modulate or quench endogenous ROS.[3] It is the interplay between antioxidants and antioxidants with ROS that maintain cellular homeostasis and redox balance.

Increasing evidence supports the role of oxidative stress in the pathogenesis of many systemic diseases in humans,[2,4] as well as dogs and cats,[5] making antioxidant supplementation a rational therapy to block the formation of ROS, scavenge ROS, and augment endogenous antioxidants. Many clinical studies in humans have evaluated antioxidant supplementation during illness with mixed results, in part due to the lack of standardization of methodologies including the type, dose, and route of the antioxidant intervention and the biomarkers evaluated.[6,7] By comparison, much less is known about oxidative stress and antioxidant supplementation in ill veterinary patients. Most of what is known is limited to experimental studies, case reports, and small clinical studies. This review will focus on the current published evidence to support or refute the use of antioxidants in the treatment of a variety of disease states in dogs and cats.

LIVER DISEASE

Studies in humans describe decreased liver glutathione concentrations in naturally occurring liver disease and animal models of induced hepatopathies.[8-11] Similar to humans, both dogs and cats with liver disease have decreased hepatic glutathione levels.[12,13] For example, approximately 50% of dogs and cats with inflammatory liver disease have low liver glutathione concentrations.[12] Multiple mechanisms likely contribute to the low glutathione levels including limited glutathione precursors, impaired hepatic synthesis of s-adenosylmethionine, increased glutathione hepatic efflux, increased glutathione utilization, and reduced redox recycling of oxidized glutathione. Therefore glutathione supplementation is a logical approach to treating patients with liver disease.

The results of human clinical studies assessing antioxidant supplementation in liver disease patients are difficult to compare as most trials consist of a low number of patients and have a high risk of bias and heterogeneity. However, some reported clinical trials and meta-analyses support the use of glutathione precursors (S-adenosylmethionine, N-acetylcysteine (NAC) and silibinin) to replenish hepatic glutathione concentrations. Extrapolated from human studies the use of antioxidants supplementation in dogs and cats with liver disease is common despite the paucity of species specific controlled clinical trials or meta-analyses.

S-adenosylmethionine (SAMe)

In humans, supplementation with SAMe (a glutathione source) is recommended for the treatment of hepatotoxicity and cholestatic liver disease.[14] SAMe increases liver glutathione concentrations in both dogs and cats. In dogs treated with prednisone, SAMe administration increased both hepatic and erythrocyte glutathione concentrations.[15,16] In a case report, SAMe was successful in increasing glutathione levels in a dog with acetaminophen toxicosis.[17] In healthy cats, SAMe administration increased hepatic glutathione concentrations and decreased markers of systemic oxidative stress.[18]

N-acetylcysteine (NAC)

In humans NAC is used in the treatment of acute hepatoxicity,[19] non-alcoholic steatohepatitis,[20,21] and alcoholic hepatitis.[22] NAC remains the standard of care for the treatment and prevention of the oxidative damage associated with acetaminophen induced hepatoxicity dogs.[23,24]

Silibinin (milk thistle)

Silibinin is commonly used as an adjunctive therapy in humans with chronic hepatitis or liver failure secondary to Amanita phalloides toxicosis, alcoholic or viral hepatitis, non-alcoholic steatohepatitis, and cirrhosis.[25] Experimental studies in dogs support that silibinin prevents the phalloidin toxin (the toxic principle in Amanita mushroom toxicosis) from binding hepatocytes minimizing hepatotoxicity.[26] In normal cats, silibinin is well tolerated when administered orally with a bioavailability of 6-7%.[27] In healthy cats, granulocyte glutathione concentrations increased following silibinin administration.[27]

Denamarin®

Denamarin®, a combination product of silibinin and SAMe, is often used in dogs and cats with liver disease. Recently, a heterogeneous group of dogs diagnosed with neoplasia that were treated with lomustine (CCNU) were randomized to treatment with CCNU alone or in combination with Denamarin®. Only 68% of the Denamarin® treated dogs experienced increases in their liver enzymes associated with CCNU administration versus 85% of the dogs treated with CCNU alone.[28]

Vitamin E

Decreased vitamin E concentrations are associated with chronic liver disease of different etiologies in humans. The overall clinical benefits of vitamin E in the treatment of hepatobiliary disease in humans remains controversial.[29] However, a recent study supports the use of vitamin E as part of a combination protocol in the treatment of non-alcoholic steatohepatitis.[30]

Decreased vitamin E concentrations and the use of vitamin E in the treatment of spontaneous liver disease in dogs and cats is not well documented. Anesthetized dogs with experimentally induced hypoxia experience a time-dependent depletion in plasma and liver vitamin E concentrations.[31] In a small group of dogs with chronic inflammatory liver disease, dogs were randomized to treatment with or without vitamin E. The vitamin E supplemented dogs had a significant increase in serum and liver vitamin E concentrations, increased glutathione concentrations, and decreased ALTs relative to the controls.[32] A feline case report describes the therapeutic use of vitamin E and selenium in a cat suspected of a tetracycline-induced hepatotoxicity with a marked increase in ALT.[33]

KIDNEY DISEASE

Through multiple mechanisms chronic kidney disease (CKD) leads to a pro-oxidant environment and oxidative stress. Urinary markers of oxidative stress are increased in humans with CKD.[34] In dogs and cats with CKD oxidative stress likely contributes to renal interstitial fibrosis, glomerulosclerosis, glomerular hypertension, renal inflammation, and a decline in kidney function.[35] In dogs with CKD, vitamin concentrations (decreased concentrations of 25-hydroxycholecalciferol and folate and increased concentrations of ascorbate and vitamin A) are altered relative to healthy dogs.[36]

Many antioxidant trials in human CKD patients have produced equivocal results.[37] However in a few small studies antioxidant supplementation in pre-dialysis CKD patients suggests a slower progression to end-stage disease.[38] Dietary antioxidants have been the preferred supplements in dogs and cats with CKD.

Vitamins

A recent meta-analysis of clinical studies evaluating antioxidant supplementation during illness concluded that trace element and vitamin supplementation not only supports antioxidant function but may reduce mortality.[39] Vitamins A, C, E, and carotenoids act as free radical scavengers that may limit oxidative injury. Diets supplemented with vitamin A, C, and beta-carotene reduced markers of oxidative stress in cats with chronic kidney disease.[40]

Vitamin E has been reported to have some benefit in human CKD patients.[41] In dogs, vitamin E in combination with omega-3 polyunsaturated fatty acids have been used to reduce glomerular hypertension, proteinuria, and minimize pro-inflammatory mediators.[42,43] In an experimental study, dogs treated with gentamicin for 9 days and supplemented with either vitamin E or silibinin had a reduction in gentamicin-induced nephrotoxicity relative to the dogs treated with saline.[44]

CARDIOVASCULAR DISEASE

Dogs in congestive heart failure due to both chronic valvular disease and dilated cardiomyopathy have significant increases in plasma oxidized glutathione, ascorbate and isoprostane levels, providing evidence of an oxidative state associated with congestive heart failure in dogs.[45] In addition, vitamin E concentrations negatively correlate with disease severity in dogs with idiopathic dilated cardiomyopathy.[46] These altered antioxidant levels are speculated to be a compensatory adaptation in dog with cardiac dysfunction suggesting a population of dogs that may benefit from antioxidant supplementation.

Studies designed to assess the effects of antioxidant supplementation in dogs with naturally occurring cardiac disease or dogs in congestive heart failure are lacking. A few experimental studies in laboratory beagles have studied the role antioxidant supplementation may have in preventing doxorubicin-induced cardiotoxicity. Supplementation with vitamin E and selenium was not cardioprotective[47] but more recently the antioxidant lycium barbarum and free-radical scavenger edaravone were protective against doxorubicin-induced cardiomyopathy.[48]

CENTRAL NERVOUS SYSTEM DISEASE

The central nervous system, due to its high lipid content, is susceptible to lipid peroxidation and oxidative stress. Dogs with spinal cord disease, specifically intervertebral disk disease (IVDD) experience oxidative stress.[49] In a randomized, blinded, placebo-controlled clinical trial, dogs with acute IVDD were treated with a single IV infusion of NAC at the time of surgery. This systemically delivered cysteine source had no effect on urinary isoprostane concentrations or neurological outcome.[50]

The aging brain of humans and dogs accumulate oxidative damage that may contribute to neuron dysfunction.[51] In aged beagles dietary antioxidant supplementation maintained or improved cognition and reduced oxidative damage.[52] Overall suggesting that antioxidant supplementation may be a preventative therapy in aged dogs with cognitive dysfunction.

SYSTEMIC ILLNESS

Hospitalized ill dogs experience systemic oxidative stress including depletion of endogenous antioxidants (erythrocyte glutathione and serum vitamin E) and increased urinary isoprostane concentrations, a marker of lipid peroxidation.[5,53] Specific canine diseases associated with underlying oxidative stress and reduced antioxidant reserve include immune-mediated hemolytic anemia (IMHA) and lymphoma. Dogs with IMHA have increased plasma malondialdehyde and decreased vitamin E serum concentrations in comparison to healthy dogs.[54] At the time of diagnosis, dogs with lymphoma have reduced vitamin E concentrations and increased levels of oxidative markers including isoprostanes and glutathione peroxidase activity which returns to normal post chemotherapy induced disease remission.[55]

In heterogeneous group of clinically ill cats, increased ascorbate concentrations are associated with illness, suggesting a very different antioxidant profile associated with illness in cats.[5] Specific systemic diseases in cats which are associated with oxidative stress include with diabetes mellitus,[56] acute FIV,[57] and acetaminophen toxicosis.[23]

Studies in dogs and cats support the state of oxidative stress in association with various systemic diseases. Collectively, the currently available data in dogs and cats supports the need for prospective interventional studies to assess the role antioxidant supplementation may have in the treatment of systemic illness including species and disease specific studies. Currently, studies evaluating the use of antioxidant in the treatment of systemic illness have primarily focused on acetaminophen toxicosis in cats.

S-adenosylmethionine (SAMe)

Treating acetaminophen toxicosis in cats with S-adenosylmethionine has been reported to be protective in limiting Heinz body formation and erythrocyte destruction, as well as increasing hepatic glutathione concentrations.[58]

Silibinin (milk thistle)

In cats experimentally exposed to acetaminophen, prophylactic silibinin administration prevented the development of acetaminophen induced toxicosis.[59]

N-acetylcysteine (NAC)

As in dogs, NAC remains the standard of care in the treatment and prevention of the oxidative damage associated with acetaminophen toxicosis in cats, specifically methemoglobinemia.[23] In addition, cysteine combined with vitamin E has been used in an experimental study in cats to reduce the oxidative state induced by acetaminophen.[60]

A randomized placebo-controlled clinical study supplemented glutathione deficient ill dogs with NAC for 48 hours. The NAC supplemented dogs stabilized their erythrocyte glutathione concentrations compare to the control dogs which had a further decline in erythrocyte glutathione. However, NAC did not resolve the systemic oxidative state, improve serum vitamin E concentrations, or impact long-term outcome. Further studies are needed to investigate whether longer duration or combination antioxidant therapy would benefit ill dogs.

AGING

Previous reports in laboratory animals suggest that cellular oxidative damage accumulates with age supporting the free-radical theory of aging.[61,62] In geriatric humans, glutathione concentrations have been shown to decline with age suggesting impaired antioxidant defenses.[63] Based on rodent studies the reduction in glutathione with age may be due to decreased *de novo* glutathione synthesis as there is an inverse relationship between age and gamma-glutamylcysteine synthetase gene expression.[64]

In healthy research beagles, older dogs, especially the older male dogs, had reduced erythrocyte glutathione levels relative to their younger counter parts.[65] In healthy client-owned dogs a similar correlation of decreasing erythrocyte glutathione levels with age has been reported.[5] However, in a follow-up study that specifically recruited a group of older and younger dogs, no significant difference was found in the erythrocyte glutathione concentrations with age; refuting the need for routine glutathione supplementation in geriatric dogs.[66]

In healthy client-owned dogs no significant differences in plasma cysteine or ascorbate concentrations were associated with age or sex.[5] Other studies in dogs have also reported plasma ascorbate concentrations to be unaffected by age[67] which is in contrast to that reported in aging mice[68] or human.[63,69] Much less is known about aging and antioxidant status or oxidative stress in cats. In one study antioxidants, glutathione, cysteine, or ascorbate, concentrations did not correlate with age in healthy cats.[5]

REFERENCES

[1]Cowley HC, et al **Crit Care Med** 1996;24:1179. [2]Roth E, et al **Curr Opin Clin Nutr Metab Care** 2004;7:161. [3]Temple MD, et al Trends Cell Biol 2005;15:1000. [4]Nathens AB, et al. **Ann Surg** 2002;**236**:814. [5]Viviano KR, et al **J Vet Intern Med** 2009;23:250. [6]Halliwell B **Free Radic Res** 1996;25:57. [7]Hermans N, et al **Curr Med Chem** 2007;14:417. [8]Lieber CS **Alcohol Res Health** 2003;27:220. [9]Mato JM, et al **Drugs** 1990;40:58. [10]Pisi E **Drugs** 1990;40:65. [11]Siegers CP, et al **Res Commun Chem Pathol Pharmacol** 1982;37:163. [12]Center SA, et al **Am J Vet Res** 2002;63:1187. [13]Spee B, et al **Comp Hepatol** 2005;4:3. [14]Frezza EE, et al Minerva Gastroenterol Dietol 1992;38:145. [15]Center SA J **Small Anim Pract** 1999;40:106. [16].Center SA Vet Clinc North Am Small Anim Pract 2004;34:67. [17]Wallace KP, et al J Am Anim Hosp Assoc 2002;38:246. [18]Center SA, et al **J Vet Intern Med** 2005;19:303. [19]Tsai CL, et al **Clin Ther** 2005;27:336. [20]Pamuk GE, et al **J Gastroenterol Hepatol** 2003;18:1220. [21]de Oliveira CP, et al **J Am Coll Nutr** 2008;27:299. [22]Nguyen-Khac E, et al **N Engl J Med** 2009;365:1781. [23]Villar D, et al **Vet Hum Toxicol** 1998;40:156. [24]MacNaughton SM **Can Vet J** 2003;44:142. [25]Hackett ES, et al **J Vet Intern Med** 2013;27:10. [26]Saller R, et al Drugs 2001;61:2035. [27]Webb CB, et al **J Feline Med Surg** 2009;11:271. [28]Skorupski KA, et al **J Vet Intern Med** 2012;25:838. [29]Singal AK, et al **Liver Int** 2011;31:1432. [30]Pacana T, et al **Curr Opin Clin Nutr Metab Care** 2012;15:641. [31]El-Bassiouni EA, et al **Toxicology** 1998:128:25. [32]Twedt DC, et al **J Vet Intern Med** 2003:17.418. [33]Kaufman AC, et al **J Am Vet Med Assoc** 1993;202:628. [34]Agarwal R **Am J Physiol Renal Physiol** 2003;284:F863. [35]Brown SA **Vet Clin North Am Small Anim Pract** 2008;38:157. [36]Galler A, et al, **Vet J** 2012;192:226. [37]Galle J, et al **Nephrol Dial Transplant** 2003;18:1452. [38]Jun M **Cochrane Database Syst Rev** 2012;10:1. [39]Heyland DK, et al **Intensive Care Med** 2005;31:327. [40]Yu S, et al **Vet Res Commun** 2006;30:403. [41]Galle J **Nephrol Dial Tranplant** 2001;16:2135. [42]Brown SA, et al **J Lab Clin Med** 1998;131:447. [43]Brown SA, et al **J Lab Clin Med** 2000;135:275. [44]Varzi HN, et al **J Vet Pharmacol Ther** 2007;30:477. [45]Freeman LM, et al **J Vet Intern Med** 2005;19:537. [46]Freeman LM, et al J Am Vet Med Assoc 1999;215:644. [47]Van Vleet JF. et al **Am J Pathol** 1980;1:13. [48]Xin Y, et al **Biol Pharm Bull** 2011;34:1523. [49]McMichael MA **J Am Vet Med Assoc** 2007;231:714. [50]Baltzer WI, et al **Spine** 2008;33:1397. [51]Head E, et al **Vet Clin North Am Small Anim Pract** 2008;38:167. [52]Dowling et al, **Biochim Biophys Acta** 2012;822:685. [53]Viviano KR, et al **J Vet Intern Med** 2013;In press. [54]Pesillo SA, et al **Am J Vet Res** 2004;65:1621. [55]Winter JL, et al **J Vet Intern Med** 2009;23:311. [56]Webb CB, et al **Am J Vet Res** 2009;70:57. [57]Webb CB, et al **J Feline Med Surg** 2008;122:16. [58]Webb CB, et al **J Feline Med Surg** 2003;5:69. [59]Avizeh R, et al **J Vet Pharm Ther** 2010;33:95. [60]Hill AS, et al **Am J Vet Res** 2005;66:196. [61]Harman D **Proc Natl Acad Sci1981**;78:7124. [62]Boross M, et al **Z Gerontol** 1991;24:76. [63]Lenton KJ, et al **Am J Clin Nutr** 2000;71:1194. [64]Liu R, et al Free Radic Biol Med 2000;28:566. [65]Vajdovich P, et al **Vet Res Commun** 1997;21:463. [66]Moyer et al, **J Am Vet Med Assoc** 2009;234:95. [67]Stowe HD, et al **J Nutr** 2006;136:1844. [68]Taylor A, et al Exp Eye Res 1995;61:55. [69]Lykkesfeldt J, et al **Mech Ageing Dev** 2005;126:892.

CARDIAC SUPPLEMENTS: CRITICAL EVALUATION OF EFFICACY

Lisa M. Freeman, DVM, PhD, DACVN
North Grafton, MA

INTRODUCTION

Cardiac disease is one of the most common disease affecting dogs and cats (the prevalence is approximately 10% in both species). In animals with cardiac disease, nutrition is an important part of the medical therapy. In patients with cardiac disease, nutrition can not only improve the patient's quality of life but also can help to modulate the underlying disease. Although every animal with cardiac disease is an individual and requires an individualized approach to medical and nutritional management, there are five goals for the optimal care of dogs and cats with cardiac disease:

1. Maintain optimal body condition
2. Avoid nutritional deficiencies
3. Avoid nutritional excesses
4. Avoid drug-nutrient interactions
5. Consider potential benefits of nutritional pharmacology

Dietary supplements play an important role in each of these nutritional goals. For example, certain supplements may be beneficial for maintaining optimal body condition or for correcting nutritional deficiencies. However, supplements can have adverse effects by contributing to nutritional excesses (and resulting side effects) or interacting with cardiac mediations. Finally, supplements can be used for a "pharmacologic" effect in animals with cardiac disease, with some potential benefits.

To achieve these five goals in the patient with cardiac disease, a thorough diet history is required. A complete diet history includes (WSAVA Nutritional Assessment Guidelines Taskforce, 2011):

- Pet food – specific type (brand, product, flavor, and form); recipe, if feeding a home-prepared diet. Amount and frequency
- Pet treats – specific type, flavor, size, and frequency
- Table food – specific foods, amounts, and frequency
- Other foods – e.g., dental chews, rawhides
- Foods used for medication administration – this is often a hidden source of sodium and other nutrients in the diet
- Dietary supplements – two studies showed that 31% of dogs and 13% of cats were receiving dietary supplements (Freeman et al, 2003; Torin et al, 2007). However, owners usually don't provide information on dietary supplements unless specifically asked.

From the medical and diet history and a careful physical examination (which includes body weight, body condition score, and muscle condition score), clinicians can determine whether the diet is optimized. In many cases desired nutrient modifications can be achieved through diet alone (along with treats, table food, and medication administration). However, supplementation of certain nutrients may be desirable if they are not available in a particular diet or are not present in high enough levels in that diet to achieve the desired effect. Some supplements that may have potential benefits in dogs and cats with cardiac disease are discussed below.

OMEGA-3 FATTY ACIDS

Fat provides calories to the diet but also has significant immunological, inflammatory, and hemodynamic effects. The omega-3 fatty acids, eicosapentaenoic acid (EPA) and docosahexaenoic acid (DHA), are normally in low concentrations in the cell membrane but levels can be increased by a food or supplement enriched in omega-3 fatty acids. Dogs with congestive heart failure (CHF) have been shown to have decreased concentrations of EPA and DHA compared to normal dogs, and fish oil supplementation can normalize these plasma fatty acid abnormalities. There are a number of potential benefits of omega-3 fatty acids supplementation. The eicosanoids derived from omega-3 fatty acids are less potent inflammatory mediators than those from omega-6 fatty acids, and omega-3 fatty acids decrease production of TNF and IL-1. In addition, omega-3 fatty acids have antiarrhythmic effects.

Although an optimal dose of omega-3 fatty acids has not been determined, the author currently recommends a dosage of fish oil to provide 40 mg/kg EPA and 25 mg/kg DHA for animals with cardiac disease, and especially in those with anorexia or muscle loss (cachexia). Unless the diet is one of a few specially designed therapeutic diets or certain specific over-the-counter diets, supplementation will be necessary to achieve this omega-3 fatty acid dose. When recommending a supplement, it is important to know the exact amount of EPA and DHA in the specific brand fish oil since supplements vary widely. The most common formulation of fish oil, however, is one gram capsules that contain approximately 180 mg EPA and 120 mg DHA. At this concentration, fish oil can be administered at a dose of 1 capsule per 10 pounds of body weight to achieve the author's recommended EPA and DHA dosage. Fish oil supplements should always contain vitamin E as an antioxidant, but other nutrients should not be included to avoid toxicities. Cod liver oil should not be used to provide omega-3 fatty acids because it contains high levels of vitamins A and D which can result in toxicity. Inefficient hepatic elongation of plant-based omega-3 fatty acids (i.e., α-linolenic acid) to EPA and DHA in dogs (and particularly in cats) make flax seed oil a much less effective source of omega-3 fatty acids for these species. In addition, ventricular arrhythmias in dogs were not significantly reduced by flax seed oil supplementation, as they were with fish oil (Smith et al, 2007).

Omega-3 fatty acids do have potential adverse effects as a result of their effects on platelets. In addition, some animals develop diarrhea when administered omega-3 fatty acids at this dose. On a practical note, not all animals like the taste of fish oil. In these patients, if the owner cannot pill the animal (to avoid them tasting it), the potential benefits of this supplement do not outweigh the risks of decreased food intake and fish oil supplementation should be avoided. Finally, it is important to be aware that there may be breed and species differences in the relative effects of omega-3 fatty acids. In one study, Boxers and Doberman pinschers had significant differences in plasma fatty acids (Smith, et al, 2008), and had different relationships between arrhythmias and fatty acid concentrations. Cats with cardiomyopathy had *higher* plasma concentrations of omega-3 fatty acids compared to healthy controls (Hall, unpublished data). Much additional research on fatty acid metabolism in animals with cardiac disease is needed.

TAURINE

There has been a dramatic reduction in cases of feline dilated cardiomyopathy (DCM) since the late 1980's when increased dietary supplementation of taurine was instituted after the association between DCM and taurine deficiency was reported. Most of the current cases of feline DCM are not taurine deficient but taurine deficiency still should be suspected in all cases of feline DCM until proven otherwise. Cats that have been fed a poor quality, homemade, vegetarian, or otherwise unbalanced diets are at risk for taurine deficiency. Correction of taurine deficiency in cats with taurine deficiency-induced DCM can reverse the disease, if the cat's condition is not already too advanced.

Taurine deficiency is now suspected in some cases of canine DCM. Unlike cats, dogs are able to synthesize adequate amounts of taurine and so are not thought to require dietary taurine. Most dogs with DCM do not have taurine deficiency, but low taurine concentrations have been found in some dogs with DCM, especially the American Cocker Spaniel, Golden Retrievers, Labrador Retrievers, St Bernards, Newfoundland's, Portuguese Water Dogs, and English Bulldogs. Taurine deficiency in dogs may be related to dietary factors as it is thought to be more common in dogs eating high fiber or certain lamb meal and rice based diets and has been induced by feeding a low protein, low taurine diet long-term to dogs. Taurine deficiency also may be the result of increased renal or fecal loss of taurine or other metabolic defects present in certain breeds. Taurine supplementation (with or without carnitine supplementation) may be beneficial in some dogs with taurine deficiency but, even in dogs that respond, the response usually is not as dramatic as in taurine deficient cats with DCM. The exact role of taurine in canine DCM still is unclear and some of the potential benefits may be due to its positive inotropic effects or role in calcium regulation in the myocardium. Taurine has minimal side effects.

L-CARNITINE

L-Carnitine is required for normal fatty acid metabolism and energy production, and is concentrated in skeletal and cardiac muscle. Carnitine deficiency is associated with primary myocardial disease in a number of species, including a family of Boxers. To date, no blinded prospective studies have been conducted in dogs so a causative role has not been established. Nonetheless, the syndrome of CHF appears to be associated with a secondary carnitine deficiency. Therefore, even if carnitine deficiency is not the inciting cause of DCM, L-carnitine supplementation could be beneficial by improving myocardial energy metabolism in animals with CHF. In human DCM patients, most studies of L-carnitine have not been well-controlled. However, one randomized, double-blind, placebo-controlled study showed improved three-year survival in people with DCM who received L-carnitine supplementation. Carnitine has minimal side effects.

COENZYME Q10

Coenzyme Q10 is a cofactor in energy production pathways but is also acts as an antioxidant. There are many anecdotal reports of benefits to coenzyme Q10 supplementation but controlled prospective studies are necessary to accurately judge the efficacy of this product. Most human studies of coenzyme Q10 supplementation have not been well-controlled and results are conflicting. Possible reasons for the reported benefits of supplementation include correction of a deficiency, improved myocardial metabolic efficiency, or increased antioxidant protection. Coenzyme Q10 appears to have minimal side effects.

ANTIOXIDANTS

Reactive oxygen species are a normal by-product of oxygen metabolism and, in the normal animal, are balanced by endogenous antioxidants production. However, an imbalance between oxidant production and antioxidant protection could increase the risk for certain types of heart disease. In addition, in animals with cardiac disease, inadequate antioxidant protection may contribute to progression of disease. Antioxidants are produced endogenously but also can be supplied exogenously with either enzymatic antioxidants (e.g., superoxide dismutase, catalase, glutathione peroxidase) or oxidant quenchers (e.g., vitamin C, vitamin E, or glutathione). Dogs with CHF have been shown to have elevated oxidative stress and reduced vitamin E concentrations compared healthy controls. Although antioxidant supplementation has been shown to reduce oxidative stress in dogs with asymptomatic chronic valvular disease, long term benefits of antioxidant supplementation in dogs are cats with cardiac disease are not known. Depending on the antioxidant used and the dose, side effects are possible (e.g., vitamin E and coagulopathy).

POTASSIUM

Potassium and magnesium are nutrients of concern in dogs and cats with cardiac disease because depletion of these electrolytes can cause cardiac arrhythmias, decreased myocardial contractility, and muscle weakness, and can potentiate the adverse effects of cardiac medications. Deficiencies of these nutrients are most likely in animals with CHF because many of the medications used in animals with CHF, such as loop diuretics (e.g., furosemide) and thiazide diuretics (e.g., hydrochlorothiazide), can predispose a patient to hypokalemia. Inadequate dietary intake of potassium, which occurred in nearly half of dogs with heart disease in one study, also can predispose an animal to hypokalemia (Freeman et al, 2003). However, it is also important to note that ACE inhibitors and spironolactone can increase potassium retention and make hyperkalemia a potential problem in animals with CHF. Finally, some commercial pet foods are high in potassium, which can exacerbate hyperkalemia.

Serum potassium concentrations should be monitored routinely in CHF patients, particularly those receiving an ACE inhibitor or large doses of diuretics. If serum concentrations are low, the dietary content of potassium should be evaluated. Potassium concentrations can be increased either through the diet or with dietary supplementation. If potassium concentrations are elevated and the dietary content also is high, dietary modification should be recommended. It also is important to assess the amount of potassium coming from other sources (i.e., supplements, treats, table food, medication administration)

MAGNESIUM

Magnesium is critical for normal cardiovascular function, and medications, such as loop diuretics, can be associated with magnesium depletion. Therefore, animals with CHF have the potential to develop hypomagnesemia. Studies have reported highly variable prevalence rates of hypomagnesemia (from 2% to over 50%) in dogs with heart disease. Because hypomagnesemia can increase the risk of arrhythmias, decrease cardiac contractility, cause muscle weakness, contribute to renal potassium loss, and potentiate the adverse effects of certain cardiac medications, it is important to monitor magnesium concentrations in animals with CHF. The challenge, however, is that serum magnesium concentrations are a poor indicator of total body stores. In animals with hypomagnesemia, a thorough diet history may reveal contributing factors, such as an unbalanced diet or a pet food that is low in magnesium. In these animals, correction of dietary issues may correct the hypomagnesemia or supplementation may be required.

B VITAMINS

Thiamine deficiency is known to be a cause of cardiomyopathy in people. However, there has been little research on the role of B vitamins as a primary or secondary issue in dogs and cats with cardiac. In one study of cats with cardiomyopathy, vitamins B_6, B_{12}, and folate were significantly lower compared to healthy controls, an effect that was unrelated to diet or furosemide use. Anorexia and urinary loss of water soluble vitamins can contribute to low B vitamin concentrations so there could be some potential benefit of B vitamins in the CHF population. However, the specific B vitamins and doses require further research.

OTHER SUPPLEMENTS

There are a variety of other supplements that have been recommended for cardiac disease. Some of these are proposed for use in people to specifically address coronary heart disease (e.g., niacin to reduce cholesterol, red yeast rice). Others may have more general effects that could be beneficial in dogs and cats with cardiac disease, but this requires much research before any recommendations can be made. It also is important to be aware of supplements that pet owners might use because they've read about beneficial effects in people but might have adverse effects in dogs and cats (e.g., garlic). For most other supplements that have been recommended for cardiac disease (e.g., artichoke, hawthorne, gingko, ginseng, etc), there is minimal scientific evidence available for people, let alone for dogs and cats. Careful scientific evaluation of these products to determine if they have benefits (and that they have no adverse effects) is required before adopting their use in veterinary medicine.

DIETARY SUPPLEMENTS: GENERAL ISSUES

The clinician and pet owner should understand that dietary supplements can be marketed without proof of safety, efficacy, or quality. Therefore careful consideration of the type, dose, and brand may be important to avoid toxicities or a complete lack of efficacy. Potential interactions with cardiac medications also are important to consider, especially in animals with CHF that are often on numerous cardiac medications. Another important issue related to administration of dietary supplements is their usage in place of standard cardiac medications. Animals with severe CHF already may be receiving many pills each day, and it may not be clear to owners that it is more important to give the cardiac medications than the dietary supplements. In situations in which pill administration is becoming overwhelming for an owner, the veterinarian can assist the owner in determining which dietary supplements have the least potential benefits (or might have adverse effects) and should be discontinued.

REFERENCES

1. Fascetti AJ, Reed JR, Rogers QR, Backus RC. Taurine deficiency in dogs with dilated cardiomyopathy: 12 cases (1997-2001). *J Am Vet Med Assoc* 2003; 223: 1137-1141.
2. Freeman LM, Rush JE, Cahalane AK, Kaplan PM, Markwell PJ. Dietary patterns in dogs with cardiac disease. *J Am Vet Med Assoc* 2003; 223: 1301-1305
3. Freeman LM, Rush JE. Nutritional management of cardiac disease. In: Ettinger SJ (ed). *Textbook of veterinary internal medicine, 7th ed.* St Louis: Elsevier, 2010: 579-583.
4. Freeman LM. Beneficial effects of omega-3 fatty acids in cardiovascular disease. *J Small Anim Pract* 2010; 51: 462-470.
5. Kittleson MD, Keen B, Pion PD, Loyer CG. Results of a multicenter spaniel trial (MUST): Taurine- and carnitine-responsive dilated cardiomyopathy in American cocker spaniels with decreased taurine concentration. *J Vet Intern Med* 1997; 11: 204-211.
6. McMichael MA, Freeman LM, Selhub J, Rozanski EA, Brown DJ, Nadeau MR, Rush JE. Plasma homocystein, B vitamins, and amino acid concentrations in cats with cardiomopathy and arterial thromboembolism. *J Vet Intern Med* 2000; 14: 507-512.
7. Smith CE, Freeman LM, Meurs KM, Rush JE, Lamb A. Plasma fatty acid concentrations in Boxers and Doberman pinschers. *Am J Vet Res* 2008; 69: 195-198
8. Torin DS, Freeman LM, Rush JE. Dietary patterns of cats with cardiac disease. *J Am Vet Med Assoc* 2007; 230: 862-867
9. Tufts HeartSmart website [information on heart disease for pet owners, including nutrition (e.g., low sodium diets for dogs and cats, treats, omega-3 fatty acid supplements)]. Available at: www.tufts.edu/vet/heartsmart January 30, 2013.
10. WSAVA Nutritional Assessment Guidelines Taskforce: Freeman L, Becvarova I, Cave N, et al. 2011 Nutritional Assessment Guidelines. *J Small Anim Pract* 2011; 52: 385-396.

CREATING VALUE BETWEEN REFERRAL PRACTICES AND rDVM'S

Karen E. Felsted, CPA, MS, DVM, CVPM and Robin Brogdon, MA
Dallas, TX

INTRODUCTION

Veterinary medicine has changed from a recession proof profession to one that has been heavily impacted by the many economic challenges facing the United States. All areas of veterinary medicine have been struggling with one or more significant challenges and it's unclear what the future will hold.

The issues impacting the profession over the past five to ten years have been widely discussed; some of the key trends that most veterinarians are familiar with but continue to trouble us include:

- Decrease in number of cases referred to specialty hospitals
- Reduction in client spending at specialty hospitals
- A decline in veterinary visits overall, particularly feline visits
- Reduction in the percentage of pet owners in the US and the actual number of owned pets
- A softening in veterinary earnings
- A growth in the number of veterinarians and veterinary practices
- Increase in the cost of veterinary care and the client pushback against it
- Low profit margins in many veterinary hospitals
- Evidence that many pets aren't getting basic care
- A lack of understanding about the need for veterinary care by pet owners
- A perception by pet owners that their veterinary practice doesn't offer value equivalent to the price paid for the services

Are these problems going to continue into the future? And more importantly, what do practices have to do in order to be successful? Since specialty practices are so dependent on referrals from general practitioners, one of the areas that deserves focus is the value of the relationship between specialists and referring veterinarians. Not only must this value genuinely exist but it also must be clearly communicated amongst the parties.

COMPONENTS OF VALUE

Since specialists are more dependent on general practitioners than vice versa, much of the onus for creating value in the relationship falls on the specialist. Some of the key areas that create value for general practitioners are as follows:

- Quality of medical and surgical care offered by the specialist
- Satisfaction of pet owners in receiving advanced care
- Opportunity to learn from the specialist
- Pet owner satisfaction in a good quality client service experience and value for the dollars paid
- Included as part of the team (specialists and general practitioners) treating the pet
- Timely communication regarding the treatment of the pet
- Respect for the part the general practitioner plays in the pet's health and return of the case to the general practitioner when appropriate and desired by the referring veterinarian

The first area identified above is essentially a given; it is unusual for referring veterinarians to feel "bad medicine" is practiced by the specialists. Complaints by general practitioners are more common, however, in the other areas and this is where specialty practices need to focus attention on improvement. Business owners and managers are always looking for the "big new idea" that will radically transform their business or the new way to "wow" their clients (both pet owners and referring veterinarians.) There is nothing wrong with this but what generally makes businesses successful is a solid attention to basics such as those mentioned above. Most practices do some good things in some of these areas some of the time but few do it all well all of the time.

The value components above can be grouped into two categories—the client service experience the pet owner receives and the relationship with the referring veterinarians.

PET OWNER CLIENT SERVICE EXPERIENCE

An article in the Harvard Business Review ("Clueing in Customers", Feb, 2003, Berry & Bendapudi) studied the communication of value at the Mayo Clinic, probably the best-known brand in human medicine. The Mayo Clinic happens to do this extremely well while spending little money on traditional forms of advertising. The authors noted, "When a company's offerings are hard to judge, customers look for subtle indicators of quality." Translated, this means that clients can't judge the quality of the medicine but will look at the aspects of a practice they CAN understand to make a judgment about the medicine. The things they can understand are generally about the service experience and about communication. This is as true in veterinary medicine as it is in human medicine. Pet owners can't judge the quality of the medical care; they assume it is good, particularly at the specialist level. Instead they focus on things such as the the level and frequency of communication, whether the pet was seen on time, the cleanliness and attractiveness of the facility and their rapport with the team (both veterinarians and non-veterinarians.)

Unfortunately, pet owners don't always have the experience they want. While pet owners in the Bayer Veterinary Care Usage Study released in 2011 were generally happy with many aspects of their veterinary practices, one of the things they considered a weakness was the value they received for the price paid. This is not a sustainable situation; ultimately consumers of any kind will drift away from a business in which they do not find value. The situation is somewhat different in a referral hospital than in a specialty hospital in that the specialists may only see a pet owner for one problem in the pet's life and that pet owner may not need to make a future decision about seeing a specialist again. Certainly they won't make that decision as frequently as they make the decision about whether to see their general practitioner again. However, pet owners talk—they will tell their friends, neighbors, on-line buddies, work colleagues, etc. about their experience (good or bad) and that will influence whether those pet owners consider a referral experience. And, of course, they will relate their experience to the veterinarian who referred them. Too many of these bad service reviews and the referring doctor will stop referring or will find someone else to refer to. While the pet owners in the Bayer study mostly had visited general practices, there is no indication that the service experience is automatically better at a specialty practice and yet it must be, both to support the greater price and to encourage continued referrals.

The following story illustrates some of the things that can go wrong in the specialty medicine experience as perceived by a pet owner. The pet owner described here is a colleague of one of the authors of this paper; she doesn't work at a veterinary practice but has spent her career working with companies that provide services for veterinarians. She is familiar with the veterinary profession and a strong supporter and admirer of both the profession and the people she has worked with. She is also a very devoted pet owner who does what her veterinarian recommends. She had an older dog that recently was having a series of GI complaints. Her regular veterinarian worked up the case, diagnosed a mass in the abdomen and referred the dog to a board certified oncologist. The oncology practice recommended surgery and wanted to repeat the blood work that was done about a week earlier. The cost for the panel was $50 more than her general practitioner charged. When the client asked why, a technician told her that the higher cost was "because we are a specialty hospital." The pet owner elected to have the repeat blood work done at her own veterinarian but before she could do that the dog crashed and needed to go to surgery ASAP. The surgery practice operated out of the oncology facility but was a separate legal entity. The surgeons wanted to repeat the blood work immediately before surgery and quoted yet another price, one that was less than the oncology practice charged but more than the referring vet. The blood work was done; the dog went to surgery and ultimately did not survive. The pet owner knew the risks and there was no blame of the doctors involved but she found the experience to be a hard one. In addition to the price issues, she found communication to be "poor"—she wasn't really sure what was going on, people weren't available to talk to her, she didn't know which person at which practice to contact, etc. At a difficult time emotionally, none of this contributed to a good service experience for the pet owner. No doubt there were some logical reasons (at least logical to the practice) for some of what happened, but practice owners and managers often don't recognize what the perceptions of pet owners are. Consumers (of all types) look for value when they spend money—blood work done at a specialty hospital doesn't automatically have additional value that warrants a $50 surcharge, what does have value is the additional <u>benefit</u> the pet owner gets—perhaps the panel is a different one with additional tests included or its being done STAT instead of being sent out. The benefit not only has to be a real one but it has to be communicated to the client, neither of which happened here. The communication protocols (if there were any) may have been set up the way they were because of the separate legal status of the practices but to a pet owner who perceives this as being "one" practice, they created angst and she didn't get the information she wanted and needed. Pet owners who have this kind of experience complain to their referring veterinarians and friends.

Of course, in order to provide value to clients, practice owners and managers have to first know what they want and to look at the service experience through their eyes. Clients who complain give the practice one opportunity to understand what pet owner concerns are but there are other ways to get this information as well. The majority of veterinarians in the Bayer study either completely or somewhat agreed with the statements: "I would do almost anything to ensure my customers are satisfied" and "I would change how my practice operates if I knew it would increase client satisfaction." However, only 20% of practice owners agreed that they measured client satisfaction through after-service surveys. If practice owners don't know what clients want and don't know if they are meeting client needs, they won't be able to provide a level of value that supports the prices charged and encourages clients to talk about their great experience. Client surveys are a very simple activity that practices can implement in order to improve their offerings and their service.

As discussed above, one component in creating value in the specialist/referring veterinarian relationship is about creating a good client service experience for the pet owner. The other component involves the communication that goes on between the specialty practice and the rDVM. This will be discussed more in the paper for the next session "Communicating Value between Referral Practices and rDVMs."

COMMUNICATING VALUE BETWEEN REFERRAL PRACTICES AND rDVM'S

Karen E. Felsted, CPA, MS, DVM, CVPM and Robin Brogdon, MA
Dallas, TX

INTRODUCTION

Veterinary medicine has changed from a recession proof profession to one that has been heavily impacted by the many economic challenges facing the United States. All areas of veterinary medicine have been struggling with one or more significant challenges and it's unclear what the future will hold. Since specialty practices are so dependent on referrals from general practitioners, one of the areas that deserves focus is the value of the relationship between specialists and referring veterinarians. Not only must this value genuinely exist but it also must be clearly communicated amongst the parties.

COMPONENTS OF VALUE

Since specialists are more dependent on general practitioners than vice versa, much of the onus for creating value in the relationship falls on the specialist. Some of the key areas that create value for general practitioners are as follows:

- Quality of medical and surgical care offered by the specialist
- Satisfaction of pet owners in receiving advanced care
- Opportunity to learn from the specialist
- Pet owner satisfaction in a good quality client service experience and value for the dollars paid
- Included as part of the team (specialists and general practitioners) treating the pet
- Timely communication regarding the treatment of the pet
- Respect for the part the general practitioner plays in the pet's health and return of the case to the general practitioner when appropriate and desired by the referring veterinarian

The first area identified above is essentially a given; it is unusual for referring veterinarians to feel "bad medicine" is practiced by the specialists. Complaints by general practitioners are more common, however, in the other areas and this is where specialty practices need to focus attention on improvement. Business owners and managers are always looking for the "big new idea" that will radically transform their business or the new way to "wow" their clients (both pet owners and referring veterinarians.) There is nothing wrong with this but what generally makes businesses successful is a solid attention to basics such as those mentioned above. Most practices do some good things in some of these areas some of the time but few do it all well all of the time.

The value components above can be grouped into two categories—the client service experience the pet owner receives and the relationship with the referring veterinarians. The pet owner client service experience and its influence on the value proposition between the referral practice and the referring veterinarians was discussed in the previous session "Creating Value Between Referral Practices and rDVMs;" in this session, the focus will be on the direct relationship between the referral practice and general practice.

COMMUNICATION OF VALUE

Specialists will be able to communicate value to referring veterinarians better if they understand what's going on with general practices? What's happening in their world? What are their pain points? It isn't possible to provide solutions to problems unless you know specifically what they are. Therefore, its important to be involved in the GP community—attend VMA meetings, engage in one-on-one discussion, read AAHA, AVMA and other publications. Ultimately this investment of time and money will help build the relationships that cause referrals to flourish. What are some of the issues general practitioners are dealing with and how can specialists add value?

Keeping Up with New Medical Information

This is a challenge for everyone. The body of knowledge doubles every three years and much of what veterinarians learned in school has become irrelevant. Specialists are in a perfect position to help with this by putting on regular CE seminars for general practitioners in their community, holding rounds type events, visiting practices and talking about cases and sending out a regular case-oriented newsletter.

Managing Their Practices Effectively

It used to be possible to run a financially successful veterinary hospital with little effort. Those times are gone for good in both the GP and the specialty world. All the CE offered by specialists doesn't have to be medically oriented—include some management content using outside speakers or the managers in your own hospital. Share the things you've found work in your practice not only during the formal CE programs but in your newsletter and other communications as well.

Providing Value to Pet Owners

Pet owners have many options these days in choosing where (or if) to get veterinary care. The profession needs pet owners to understand why veterinary care is so important to their pets and to rely on veterinarians as their primary resource for pet health care. Everyone suffers if pet owners don't regularly visit their general practitioner for preventive care and sick pet services; GP revenue dries up, as do referrals. The pet owners surveyed in the 2011 Bayer Veterinary Care Usage Study made it very clear they thought veterinary medicine was too expensive and they didn't fine value for what they paid.

There are many things referral practices can do in order to help general practitioners provide value to pet owners. First of all, make sure the clients referred to you have a great experience. Secondly, give the referring veterinarian the tools they need to look good and provide value to the pet owner. Make it easy for them to refer—provide them with an easy to use referral form and plenty of brochures and business cards to hand out to their clients, let them know what you need from them for the visit to go smoothly, communicate with them rapidly after you've seen the case so when the pet owner talks to the GP about the experience, the GP is already in the loop. Don't undercut what they charge for drugs that are available both from generalists and specialists and don't criticize the care they provided. There will be times when a specialist thinks the care provided by a generalist was substandard but expressing that opinion to the pet owner is never a good idea; one such occurrence will cut off those referrals forever. Call the veterinarian instead and tactfully use the situation as a training moment.

Growing the Market

Every five years, the AVMA publishes the most comprehensive report available on pet populations, veterinary visits and veterinary expenditures; the most recent version released last year is the "2012 U.S. Pet Ownership and Demographics Sourcebook."

Overall, the percentage of U.S. households owning a pet declined 2.4% from 2006 to 2011. In 2011, 56% of all U.S. households owned a pet at year-end, compared to 57.4% at the end of 2006. Both the number of households owning dogs and cats also declined, 1.9% for dog households and 6.2% for cat households. The number of absolute dogs and cats owned also declined. At the end of 2011, the U.S. dog population was approximately 70 million, down from approximately 72 million in 2006; this is a 3.0% decline. Cat numbers dropped dramatically from approximately 81.7 million at the end of 2006 to 74.1 million at the end of 2011, a 9.4% drop.

The AVMA report also reviewed trends in veterinary visits and expenditures. Total veterinary expenditures for all household pets increased 14.3% from 2006 to approximately $28 billion in 2011. Dogs represented 68.1% of total veterinary expenditures and cats, 26.3%. Dogs visited a veterinary practice an estimated130.4 million times in 2011, an increase of 9.2% from 2006. Cat visits unfortunately declined 4.4% from 2006 to an estimated 60.5 million visits._Among dog-owning households, 81.3% had at least one visit to the veterinarian in 2011, down 1.7% from 2006. Cats again didn't fare so well with only 55.1% having at least one visit to the veterinarian in 2011, down 13.5% from 2006. Remember that cat visits had declined dramatically also from 2001 to 2006.

Although veterinary visits were down in 2011, total expenditures on veterinary visits for dogs and cats increased from 2006 to 2011. The mean expenditure per dog was $227 in 2011, a 13.5% increase compared to 2006. Note, however, that inflation was almost as much during this same five-year period so the "real" amount spent increased very little. The mean expenditure per cat was $90 in 2011, a $9 increase from five years ago, resulting in an 11.1% increase per household over the past five years. Again, the "real" amount spent increased very little when inflation is factored in.

These are concerning statistics but there is also all sorts of evidence to indicate that many pets don't get the basic care they need. No doubt of all those pets were seeing a veterinarian regularly, all practices would have more cases than they could handle. Pet owners need education and this can come from both generalists and specialists. Work together on campaigns to reach out to pet owners and give them the information they need whether this is through community events, a marketing campaign, jointly provided information on your websites or Facebook or other means.

THE NEED FOR SYSTEMS

A recent article on www.businessknowhow.com discussed what makes small businesses succeed. Listed below are the key components according to these authors:

- A defined business concept and current strategic business plan
- Positive, committed, persistent and patient senior management
- A structured and functional organization
- Basic, automated tracking systems to support the organization and make it efficient

What was particularly interesting about this list was the emphasis on systems and metrics. It's not just enough to have a great business concept and good intentions; a business has to be able to execute. Much of what is important to referring veterinarians is about timely communications and achieving this goal means having systems (both human and electronic) in place to achieve that.

Protocols should be in place for how rDVM communication will be handled and include specific procedures for getting information from the rDVM before the appointment (including good contact info for the rDVM), timely updates (phone, email, written referral reports) once the pet has been seen, availability of specialists for consultations and follow-up calls, handling liability concerns for pets who have not been seen ever or not recently, notifying the rDVM immediately if the pet dies, and regularly providing rDVMs with brochures, cards and referral forms. Systems need to be regularly evaluated for effectiveness through subjective means and the use of metrics.

CREATING VALUE BETWEEN REFERRAL PRACTICES AND PET OWNERS

Robin Brogdon, MA and Karen E. Felsted, CPA, MS, DVM, CVPM
Irvine, CA

INTRODUCTION

Largely dependent upon general practitioners for patient referrals, the specialty practice needs to both create and communicate value for the pet owner as well as the referring veterinarian. How well you execute on this will determine the value a customer places on your services and whether or not their experience met, fell short, or exceeded the perceived value. Value Perception (VP) is the opinion your potential and current customers have of your product or service. This perception determines the value it adds to them in line with the problems it needs to solve or aspirations they want it to fulfill. Also evaluated is your offering's relevance and importance, over and above that of your competitors.

CREATING THE RIGHT VALUE PERCEPTION FOR YOUR CUSTOMERS

So why is the customers' Value Perception so important? Irrespective of your customer's opinion being right of wrong in your mind, their opinion matters none the less, in fact, critically so. Some may suggest; "But the customer may have it wrong" In this instance we respond: whose fault is that: the customer, referring veterinarian, front office person, marketing agency or practice owner? One thing is for sure, it certainly isn't the customers fault.

In a successful service-focused company inspirational service leadership always flows from the very top of the organization all the way down to the frontline and through to your customers.

The responsibility in creating and distributing customer value is always ordered as follows:

1. Company owners
2. Marketing/brand department or agency
3. Front office supervisor/administration
4. Front office team
5. Supporting team (technicians, assistants)
6. Customer

How Your Customers Perceive Your Value Determines Your Value!

How does your practice currently create and distribute value?

- Are the practice owners genuinely attempting to connect positively with customers?
- Is the marketing and branding positively connected to the customer?
- Is the front office manager positively connected to the customer?
- Are the front office people effectively adding value to the customers?
- Is the after service support adding value to the customers?
- Do your customers get it? (do they understand your value)

Value cascades down the value delivery system into the customer. A breakdown on any level can be detrimental to a company's success. The customer's positive perception, along with an effective service process will help the customer make the appropriate buying decision or comply with recommendations. Yes, NO or even worse MAYBE, will largely be determined by the value your offering adds in relation to either solving the buyers problem or fulfilling their aspirations.

Are Your Customers Getting the Right Message?

If not a change needs to be implemented at the top of value process. We recommend a sales/service culture development process for your practice's senior leaders to determine what your organization actually does and how this adds value to your customers, in order to create the right customer value perception. This will help your practice align all communication, both internally and externally, to this value proposition, we call it a Unique Selling Proposition (USP). This statement then becomes the blueprint and catalyst to inspire positive change and align and direct all customer facing activities. Most importantly, the front office team knows how they can create, engage and develop meaningful value based relationships.

Companies that understand the power of their Value Perception are usually the quickest to thrive and indeed prosper!

It's Also About Differentiation

To help earn a perception of great value, there has to be something tangibly different about what you do or how you deliver it. Many practices offer the same or similar specialty services. But creating some sort of differentiation in the mind of your target audiences is the key to being chosen time and time again, as well as receiving the highly desired word-of-mouth (WOM) added value.

The most effective way to differentiate your practice is to create an experience that will transcend the time a customer is actually at your facility. First you must determine and define the experience you want to deliver By defining the experience, I mean making a deliberate choice on how clients will feel when they touch your practice. You can begin to differentiate yourself by defining how you want your staff to answer your telephone and handle calls. If you're trying to create a personal environment, maybe your phones are always answered by a person – not an answering device. Whatever you choose – it's defined by you – or you and your staff. And it's communicated to your staff. Then they are trained to deliver the experience you define and most importantly held accountable for delivering it.

Creating Value is Nearly Always the Result of an Exceptional Experience.

There are countless opportunities throughout the customers' time interacting with your practice to separate yourself from others and lay the foundation for a long term, ongoing relationship. Each element of the experience needs to meet this level of service however if it is to succeed. Any place along the service continuum that breaks down has the ability to derail even the best intentions and even negate other aspects of value that work well. In the specialty environment, and given the fact that your services are in most cases discretionary and more expensive than that of the general practitioner, you need to do everything within the realm of possibility to create value for your services to support the cost structure. And most importantly, because you are so dependent on receiving cases by referral, it is critical that your customers go back to their family veterinarian with messages of outstanding service and an experience that makes the general practitioner inclined to send the next case over as it enhances his or her reputation while serving the needs of the patient.

COMMUNICATING VALUE BETWEEN REFERRAL PRACTICES AND PET OWNERS

Robin Brogdon, MA and Karen E. Felsted, CPA, MS, DVM, CVPM
Irvine, CA

INTRODUCTION

According to Wikipedia, **customer service** is the provision of service to customers before, during and after a purchase. Customer service is a series of activities designed to enhance the level of customer satisfaction – that is, the feeling that a product or service has met the customer expectation.
Perceived value: A customer's opinion of a product's value to him or her. It may have little or nothing to do with the product's market price, and depends on the product's ability to satisfy his or her needs or requirements.

How do we Communicate Value?

- Show
  - Process
  - Standards
  - Consistency
  - Frequency
  - Dependability

No One Does it Better than Disney

And while no one visits the veterinarian for entertainment, a lot can be learned about exceptional service and the many details that make for a truly satisfying experience. An excerpt from their appearance standards:
"Since the Disneyland Park opened in 1955, our guests have treasured the elements that set our Disney theme parks and resorts apart – the cleanest venues, the friendliest Cast Members and the most exceptional show. A big part of that show is you, with your quick smile, your eagerness to help and your willingness to maintain the Disney Look that our guests have come to associate with our very special brand. No matter where you work or what your role is, anytime you are in a public area, you are 'on stage.' Your attitude and performance are direct reflections on the quality of our Disney show. It's those things that you don't always think of that detract from our guests' enjoyment--chewing gum, having poor posture, using a cell phone...or frowning. Like these actions, smoking and eating on stage are also strictly prohibited. All of this adds up to one of the most important aspects of your role in our show: good stage presence".

- Tell
  - Directly-verbally
  - Indirectly-in writing
  - Indirectly-on line
  - Indirectly-visually/images

Earn your SAG Card

Truth is, you are in a performance and every client deserves an encore-worthy experience. This can only be accomplished if you treat each day like 'open-house' day, where everyone is a VIP and receives the proverbial red carpet treatment. This is done by unrelenting attention to detail. Clients do not know how to evaluate your medicine, but they all make judgments about their experience, so give them every reason to believe you are outstanding and a true advocate for their pet.

The military is another place you can glean great ideas for how to address the details. Repetition and consistency are hallmarks of the military and allows for precise execution of any process or procedure, in the exact same manner. This creates credibility, trust and ultimately comfort.

There is no better way to start or enhance this process for your team then to have written documentation, but they should participate in writing the standards. It's much easier to get the team to comply if they've determined what needs to be done and therefore much easier to hold the team accountable. If you are AAHA Accredited, start there and simply add to the documentation, you do not need to start from scratch. Build processes and systems around each element of service that you and your team deem the highest priorities. For a thorough step-by-step plan for how to communicate value at every touch point, break down your process into pre, during and post visit.

Pre-Visit

- Telephone greeting
- Appointment scheduling – dissemination of information and appropriate questions to ask

- Website
- Signage
- Referral information

During Visit

Be sure the clinic is clean, the parking lot free of debris, doggy bags are stocked, lighting is in working order, door signage is visible, door mat in place and clean, landscaping is attractive, smells good.

Some tips by clinic role:

Reception

- Grooming – uniform and appearance standards (keep a change of clothes and a hygiene kit at work)
- Smile – friendly, warm, welcoming, service attitude – clients sign your paycheck
- Welcome mat (literally and figuratively)
- Greet by name, wear a name tag – use of service pins – 'since 1999'
- Introduction to include role and how they can help
- Stand up to greet – shake hands - good posture, body language, eye contact
- Use of client and pet name – get the sex right!
- Offer beverage/refreshment
- Assess client demeanor
- Explain visit – provide information to lessen anxiety and set expectations! If you ask them to fill out forms, let them know how the forms are used – explain the privacy policy.

Technician

- Use pet owner and client name when escorting them to an exam room
- Explain your role and title
- Offer information on what will happen at this stage in the visit – manage expectations

Doctor

- Explain technicians capabilities and who will follow up with them and when
- Offer appropriate 'take home' materials or resources of reputable sources

Check out

While first impressions are lasting, the last touch that a patient and client have with your practice is the one they will likely recall. That's why you need to really see how your patients and clients are being treated upon checkout, as this will complete the experience. As with the greeting area, the appearance of the checkout area is important. When you get back to your offices next week, stand where your patients and clients stand when they check out. Observe what they observe. Take note of what they see, what kind of impression is this visual picture making for them. Then ask yourself, what are we doing as a practice to make this client and patient feel appreciated and valued? Is there some semblance of privacy for them to discuss their bill?

- All questions answered?
- Need Rx refill?
- Schedule recheck appointment
- Update communication preferences for follow up
- Can I sign you up for our newsletter?

Post visit

Consider communication of lab results, recheck appointments, thank you notes, solicit feedback.

Other details to consider that reflect on the practice:

- Appearance standards outside of clinic - where can the uniform or logo be worn and visible outside the clinic?
- Social media - get your handbook up to date and make sure there is someone monitoring places where information might get posted about your practice
- Encourage wellness among the team
    - Meals - when bringing in lunch, make sure there is a healthy option, encourage walks at lunch breaks

- Offer discount gym memberships as a benefit
- Provide an incentive for annual physical
- Offer incentives to quit smoking and create a smoking policy
- Know how to identify drug or alcohol abuse and have a policy to deal with it
- Keep an eye out for compassion fatigue and provide resources to manage this type of stress

Realizing that your practice has unlimited opportunities to communicate value and enhance the experience of every customer is the first step in defining the processes for your staff to follow day in and day out.

CRITICAL APPRAISAL OF THE ROLE OF ECHOCARDIOGRAPHY IN ACQUIRED CANINE HEART DISEASE

PD Dr Gerhard Wess, DACVIM (Cardiology), DECVIM-CA (Cardiology and Internal Medicine)
Munchen, Germany

Dilated cardiomyopathy (DCM) is an important cause of cardiac morbidity and mortality in dogs and is the most common acquired cardiac disorder in large and giant breed dogs. Mitral regurgitation (MR) in dogs is most often caused by chronic degenerative valvular disease (CDVD), which is the most common cardiac disease in small breed dogs. Echocardiography plays an important role in the diagnosis and staging of both diseases. Many echocardiographic methods have been described for this purpose, but all methods have their advantages and disadvantages.

The diagnosis of overt DCM is normally not problematic, although the importance of active exclusion of other causes of the dilated, hypokinetic heart is emphasized. A challenge remains the early diagnosis in the occult stage of the disease. The ESVC published 2009 "proposed guidelines for the diagnosis of canine idiopathic dilated cardiomyopathy". The diagnosis of DCM is based on the identification of myocardial (predominantly but not solely systolic) dysfunction with the active exclusion of other acquired or congenital cardiac disease. To diagnose systolic or diastolic chamber enlargement it is recommended to use breed specific M-Mode reference values, if available. In humans, ejection fraction is quoted as the main parameter assessing systolic function and ejection fractions (calculated on echo from apical views using Simpson's rule in most studies) are said to be abnormal at values ranging from less than 40% to less than 50%. In dogs, the ESVC guidelines consider that an ejection fraction less than 40% is abnormally low. The guidelines further state that an ESV-I over 80 ml/m2 offers unequivocal evidence for systolic dysfunction, although this may be an excessively conservative value, as the normal canine ESV-I has been suggested to be <30 ml/m2. Unfortunately, there is such a large gray zone between 30 and 80 ml/m2 that this recommendation makes a use in clinical practice difficult. Also, reference values for cardiac chamber volumes are not published, with the exception of Doberman Pinschers. In man it is stated that the Teichholz method to calculate chamber volumes from M-Mode should not any longer be used, as the geometric assumptions made by this method are not correct in dilated hearts. In humans the Simpson method of disc is the preferred method and is probably superior to M-Mode measurements – with the problem of lacking reference values.

M-mode echocardiography used to be considered the gold-standard to diagnose echocardiographic changes in Doberman Pinschers with DCM. A recent study in Doberman Pinschers compared several established with new calculated M-Mode reference values using a longitudinal follow-up study. A multivariate analysis including weight, age, and gender revealed that only gender had a significant influence on M-mode values. Therefore, separate M-mode cut-off values should be used for male and female dogs. The best discriminating values to differentiate healthy dogs from DCM were for the left ventricular diastolic diameter, LVIDd >48mm for males and >46 mm for female Doberman Pinschers. The best cut-off value for the systolic left ventricular diameter, LVIDs was >36 mm and can be used for both, male and female dogs. However, M-mode is a 1-dimensional method that relies on geometric assumptions that may assess geometric changes occurring in diseases such as DCM less accurately. Accordingly, in humans it is now recommended to use the Simpson's method of disks (SMOD) for 2D volume measurements. Using SMOD, heart volume is measured as the summation of parallel cylinders, whose diameters are derived from endocardial border tracing performed on 1 or 2 orthogonal LV apical views. The principle underlying this method is that the total LV volume is calculated from the summation of a stack of elliptical disks. The SMOD method was recently also evaluated in Doberman Pinschers using a longitudinal follow-up study and compared to M-Mode measurements. The ability of the new SMOD cut-off values, normalized to body surface area (BSA), for left ventricular end-diastolic volume (LVEDV/BSA >95ml/m2) and end-systolic volume (LVESV/BSA >55ml/m2) to detect echocardiographic changes in Doberman Pinschers with DCM was compared to currently recommended M-mode values. Dogs with elevated SMOD values but normal M-mode measurements were followed-up using a prospective longitudinal study design. The study showed that the SMOD method is superior to conventional M-mode measurements in its ability to detect early echocardiographic changes in Doberman Pinschers with DCM. SMOD detected correctly early echocardiographic abnormalities in Doberman Pinschers that only later also became abnormal using M-Mode. Therefore, SMOD could be considered as a new gold-standard in the diagnosis of echocardiographic changes in Doberman Pinschers.

The ESVC guidelines also introduced the sphericity index. The authors proposed that a ratio of LV diastolic length (from right parasternal (RPS) long axis four chamber view) to the M-mode LV diastolic dimension (LVIDd) <1.65 represents increased sphericity. Although the sphericity index might be clinically useful, there are no published studies evaluating its value compared to other methods. Furthermore, the guidelines proposed a scoring system including major and minor criteria.

Proposed major criteria for the diagnosis of canine DCM

1. Left ventricular M-mode systolic or diastolic dimensions exceeding 95% confidence intervals for the individual based on regression equations or predicted reference values, or outside other established breed specific reference ranges.
2. Increased sphericity < 1.65

3.a. EITHER:

M-mode fractional shortening of <20% or 25% (depending on breed-specific reference values).

3.b. AND / OR:

Left ventricular ejection fraction less than 40%.

Proposed minor criteria for the diagnosis of canine DCM

1. The presence of an arrhythmia in a specific breed where the arrhythmia has been shown to be strongly associated with DCM (e.g. increased (for age) ventricular ectopy in Dobermans or Boxers) 2. Atrial fibrillation 3. Increased mitral valve M-mode E point to septal separation (EPSS). 4. PEP:ET ratio increased over 95% confidence intervals (e.g. over 0.4) 5. M-mode fractional shortening in equivocal range (depending on breed-specific references). 6. Left or bi-atrial enlargement.

If major criteria score 3 points and minor criteria score 1 point each, a total score of SIX or more should identify dogs with DCM. A problem using this scoring system is that there are so far no studies evaluating the value of the scoring system.

The scoring system includes systolic time intervals, EF or FS measurements, but these parameters are all preload and afterload dependent and are not necessarily a good indicator of myocardial contractility. The scoring system also includes parameters such as PEP/ET. The ratio of PEP to left ventricular ejection time (LVET) is increased in DCM secondary to slow rate of pressure rise during isovolumic contraction. However, this value is not sensitive, as PEP/LVET ratios may still be normal in dogs with overt heart signs of heart failure. Furthermore, systolic time intervals have limitations to their use since heart rate and loading conditions do affect them significantly.

Newer methods such as strain or strain rate parameters are not yet part of the standard echocardiographic screening for DCM.

Chronic degenerative valvular disease (CDVD), also called valvular endocardiosis or myxomatous valvular disease, develops slowly and increases in severity over a period of years. Worsening of CDVD leads to several adaptive changes and later on to complications such as cardiac remodeling, systolic and diastolic myocardial dysfunction, increased left ventricular filling pressure and pulmonary arterial hypertension.

Echocardiography is recommended by the "Guidelines for the Diagnosis and Treatment of Canine Chronic Valvular Heart Disease" in small breed dogs with typical murmurs to answer specific questions regarding either cardiac chamber enlargements or the cause of the murmur if those questions are not answered adequately by auscultation and thoracic radiography (Atkins et al., 2009). In the guidelines no specific echocardiographic parameters are mentioned that should be used for staging or that could be useful for prognostic purposes. Many conventional two-dimensional (2D), M-mode, and Doppler parameters are reported for staging purposes and for the initial and longitudinal assessment of dogs affected by CDVD, providing information on mitral valve anatomy, MR severity, left ventricular (LV) size and function, as well as cardiac and vascular pressures. However, in contrast to humans no guidelines for the echocardiographic assessment of CDVD exist. The Society of Echocardiography Task Force on the Assessment of Native Valvular Regurgitation in man recommends integrating several parameters to assess the severity of mitral regurgitation (Zoghbi et al, 2003). This helps eliminate error in technique and measurement and helps define which parameters may be erroneous because of technical error, physiologic reasons, and chronicity of the disease process.

A mitral regurgitation index has been devised in man that assigns a number from 0 (absent) to 3 (significant) for six different parameters used in the evaluation of mitral regurgitation in man. These parameters include PISA radius, jet expansion into the left atrium, vena contracta, pulmonary venous flow, pulmonary artery pressure, and left atrial size. The average of these scores is calculated (Thomas et al, 1999).

A recent study showed that measures of left heart size and plasma NT-proBNP concentration independently estimate risk of first-onset of CHF in dogs with CDVD (Reynolds et al, 2012). Therefore, the easiest staging system could be the assessment of LV and LA size.

Two-dimensional (2D) echocardiography allows an evaluation of the valvular structure as well as the impact of the volume overload on the cardiac chambers. Valve lesion such as thickening of the valve, rupture of chordae tendineae, motion abnormalities, prolapse or flail leaflet can be recognized.

Doppler echocardiography is the most common technique used for the detection and evaluation of severity of valvular regurgitation. Several indexes have been developed to assess the severity of regurgitation using Color Doppler, Pulsed wave (PW) and Continuous wave (CW) Doppler. MR severity can be assessed by several methods. One of the methods commonly used to assess MR severity in dogs with MVD consists of assessing the regurgitant jet size and width or by calculating the maximal ratio of the regurgitant jet area signal to LA area (ARJ/LAA ratio) using color-flow Doppler mode. The severity grading using this method is not the same in all studies, but some studies report that MR is usually considered as mild if the ARJ/LAA ratio is <20 - 30%, moderate if it is >20-30% but <70%, or severe if it is >70% (Chetboul et al., 2010). However, using the regurgitant jet size in relation to the LA is not without problems. First, if the LA is normal, it might be easier for the jet to fill the LA to a quite big proportion compared to a large LA, yet nevertheless the MR would be graded moderate or severe. But does this represent truly advanced disease? The size of the regurgitation jet by color Doppler and its temporal resolution however, are significantly affected by transducer frequency and instrument settings such as gain, output power, Nyquist limit, size and depth of the image sector. Other factors, including systemic arterial blood pressure, LA pressure and spatial orientation of the jet also play an important role. Jet area is inversely proportional to PRF, and substantial error can be introduced with use of higher or lower settings than the nominal settings to which echocardiographers have become accustomed. Regarding hemodynamic factors, eccentric, wall-impinging jets appear significantly smaller than centrally directed jets of similar hemodynamic severity, mainly because they flatten out on the wall of the receiving chamber.

Another method that is occasionally mentioned to staging is the vena contracta. The vena contracta is the narrowest portion of a jet that occurs at or just downstream from the orifice. It is characterized by high velocity, laminar flow and is slightly smaller than the anatomic regurgitant orifice due to boundary effects. Thus, the cross-sectional area of the vena contracta represents a measure of the effective regurgitant orifice area (EROA), which is the narrowest area of actual flow. The size of the vena contracta is independent of flow rate and driving pressure for a fixed orifice. This is the narrowest portion of the MR jet that occurs at or just downstream from the regurgitant orifice. However, few data are available in the dog for this method and the vena contracta is also prone to errors, particularly in case of dynamic regurgitant orifices, or if several regurgitant jets exist.

Proximal isovelocity surface area (PISA) is an area of flow acceleration and convergence proximal to the mitral valve as the regurgitant jet approaches the regurgitant orifice. This method assumes that flow accelerates as it approaches the small orifice and that it does so in concentric hemispheres on the ventricular side of the valve. The more blood volume that moves backward through the valve the larger the proximal hemisphere is. Studies have shown that this measurement can be used to determine not only the regurgitant orifice area but also regurgitant fraction and volume when used in conjunction with regurgitant flow velocity. It has been reported that this technique is repeatable and reproducible in dogs when performed by a trained observer. However, calculation of regurgitant fraction requires the measurement of three additional parameters besides the PISA: the VTI – Velocity Time Integral of mitral regurgitant flow using CW-Doppler, the aortic diameter (or cross sectional area) and VTI of aortic flow using PW-Doppler. This can be quite time consuming and as each method is prone to measurement errors, this is certainly a disadvantage. A problem using the regurgitant volume or effective orifice area is that the results can't be compared between different body sizes. For example, 10 ml regurgitant volume is much more relevant in a Chihuahua compared to a Great Dane. It is more accurate for central jets than for eccentric jets, and for regurgitation with a circular orifice. If the image resolution allows the flow convergence to be seen well, and a Nyquist limit can be chosen at which the flow convergence has a hemispheric shape, it is easy to identify the aliasing line of the hemisphere. However, it can be difficult to judge the precise location of the orifice and the flow convergence shape. Any error introduced is then squared, which can markedly affect the resulting flow rate and EROA. Although promising, further experience is needed with these methods, ideal would be a 3D PISA method.

There are several Pulsed and CW Doppler methods that give indirect clues to the significance of valvular regurgitation.

Continuous-wave Doppler studies of MR can be used to identify elevated LA pressure and LV systolic and diastolic dysfunction. Mitral regurgitation flow profile reflects LA pressure, systolic LV function, preload as well as arterial blood pressure.

The contour of the velocity profile and its density might be useful. A truncated, triangular jet contour with early peaking of the maximal velocity indicates elevated LA pressure or a prominent regurgitant pressure wave in the LA. However, in eccentric significant MR, it may be difficult to record the full envelope of the jet because of its eccentricity, while the signal intensity shows dense features.

Canine studies have demonstrated a close relationship between catheter-measured and Doppler-estimated dp/dt. One simple method for calculating this variable involves measuring the time period between the 1 and 3 m/s time points along the MR spectrum. However, reference values for dp/dt using this method in small-breed dogs with mild MR and normal LV function have not been published. Using CW Doppler, the tricuspid or pulmonic regurgitation jet should be interrogated in order to estimate pulmonary artery systolic pressure. Although the precise tricuspid regurgitant velocity defining PH in dogs is not established, velocities exceeding more than 3 m/s (in the absence of right ventricular outflow obstruction) generally indicate elevated pulmonary artery pressures. The presence of pulmonary hypertension provides another indirect clue as to MR severity and compensation to the volume overload.

Pulsed Doppler tracings at the mitral leaflet tips are commonly used to evaluate LV diastolic function. The transmitral flow profile reflects diastolic function (relaxation and compliance, volume and recoil) as well as LV filling pressure. Patients with severe MR often demonstrate a mitral inflow pattern with a dominant early filling (increased E velocity) due to increased diastolic flow across the mitral valve, with or without an increase in left atrial pressure. However, as age also influences E- and A-waves and as the method is quite load dependent, diastolic evaluation using only E- and A-waves is difficult. Tissue Doppler imaging methods such as tissue velocity, strain or strain rate are less load dependent and might provide better assessment of the diastolic function. However, the PW- and color tissue Doppler methods provide different values due to technical factors. Also, different segment and ventricular walls have different reference ranges (with often a quite wide SD), the method is age and weight dependent which makes interpretation difficult.

Other Doppler measures such as pulmonary venous flow patterns are potentially useful, but often confusing due to projection of MR into the pulmonary veins and difficult to obtain. Calculation of the isovolumic relaxation time (IVRT), defined as the time interval from the Doppler signal of aortic valve closure to the beginning of the mitral E wave and assessment of peak E/IVRT may be of interest as these variables have been shown experimentally to be correlated with LA pressure and LV end-diastolic pressure in the dog.

Assessment of systolic function. Whereas the diagnosis of CDVD is usually easy in small breeds, the differentiation between DCM and CDVD with secondary systolic dysfunction might be challenging in large breed dogs. Also, the identification of LV systolic or diastolic dysfunction in the setting of CDVD is quite challenging, as there is no totally load independent measure available (Bonagura et al., 2009).

The traditional measures of global LV systolic function are the ejection and shortening fractions. Ejection fraction (EF) is commonly measured by 2D Echo using a single-plane (long axis or apical) image of the LV. Shortening fraction (SF) is measured in most situations from the M-mode echocardiogram but also can be determined from a frozen 2D image. Systolic heart failure is classically characterized by reduced EF and SF and subjectively by reduced radial shortening (fractional area shortening); however, in CDVD these measures can be very misleading. Significant MR increases LV preload and reduces afterload. These alterations foster a normal to hyperdynamic ventricular contraction, even in the presence of impaired myocardial cell contraction. Therefore, often a normal SF or EF in a dog with advanced CDVD is considered abnormal. The ESVI index has been shown to be a relatively load-independent predictor of post-operative systolic dysfunction in humans with MR and a relatively independent variable of experimental increased preload in the dog. The ESVI can be calculated by M-Mode methods, such as the Teichholz method, which is however not anymore recommended in humans – and it should probably also not be used anymore in animals, as this method relays on geometrical assumptions that are changing in cardiac diseases. A good method to calculate volumes is the Simpson method of disc – however, a disadvantage of this method is that there are not breed specific or even body size dependent reference values published.

CAN I TRUST MY ECHOCARDIOGRAPHIC ASSESSMENTS OF CARDIAC DIMENSIONS AND VOLUMES?

Ingrid Ljungvall, DVM, PhD and Jens Häggström, DVM, PhD, DECVIM-CA (Cardiology)
Uppsala, Sweden

INTRODUCTION

Evaluation of left atrial (LA) and ventricular (LV) size and shape is one of the most common reasons for performing echocardiography in cardiac patients. A variety of echocardiographic techniques have been described for assessment of LA and LV geometry, and traditional clinical echocardiographic assessments rely on 1-dimensional (M-mode) and two-dimensional (2D) images, and more recently three-dimensional (3D) datasets. These assessments can be flawed by use of different reference points for measurements amongst cardiologists, assumptions of geometry, and by oblique image plane positioning; potentially leading to inaccuracies in measurements. These technical limitations might be pronounced when LA and LV morphology is changed due to presence of cardiac diseases.[1,2] Measurements in dogs and cats need to be normalized for body size, and several systems have been utilized for this purpose, which may create confusion. Furthermore, cutoff values between supposedly normal and abnormal measurements have in many instances not been thoroughly investigated. Accordingly, cut-off values differ amongst cardiologists. This presentation is aimed at identifying some of these issues and to comment on them from our perspective, but we do not aim at providing definite answers.

LEFT ATRIAL LINEAR DIMENSION

Left atrial size assessment allows estimation of severity of left-sided heart disease associated with volume overload. LA diameter is commonly normalized by linear aortic diameter as the aortic diameter will vary with body size and absolute size and shape of the aortic root rarely is altered with disease in dogs and cats animals. This method is dependent on that accurate measurements of aortic diameters are obtained and that the heart condition does not affect aortic size. The benefits with this method are that it is easy to do, and that differences in body condition need not be accounted for. Two methods of obtaining a linear aortic diameter from 2D right parasternal short-axis projections have been described in dogs in order to define reference intervals for normalized LA size (Left-atrial-to-aortic ratios; LA/AO) in healthy adult dogs.[3,4] Both methods of aortic measurement are also commonly used for LA size estimation in cats. However, the reference intervals specifying the upper "normal" limit differ in these two studies: LA/Ao ratios of 1.6[3] and 1.3[4], respectively. A recently performed study (Georgiev *et al.* pending publication) aimed to compare the two echocardiographic methods of aortic valve linear dimensions assessment in dogs and cats. The results showed sufficient agreement between the two methods to suggest that these methods can be used interchangeable for aortic linear dimension assessment. Accordingly, the reason for differences between the two methods in reference interval specifying "normal" LA/Ao ratio remains unsolved. Both the two mentioned studies[3,4] described similar image acquisition techniques. Still, the view each group of investigators in the two mentioned studies considered "optimal" for measurement could potentially differ; resulting in slightly different reference interval.

LEFT VENTRICULAR LINEAR DIMENSIONS

Measurements of heart dimensions must be corrected for body size before presence of LV dilatation can be assessed. Over the years, several methods have been suggested for correcting for body size. One simple way is to divide the one-dimensional measurements obtained from the M-mode of 2D mode with aortic diameter, as commonly done for LA size (see above).[5] Other methods of correcting for body size are correcting for body weight. The relationships between body weight and cardiac dimensions are not linear; it is curvilinear. Previously, heart dimensions have been corrected for body size using the body surface area (which is proportional to body weight $^{(2/3)}$).[6] However, it has been shown the relationship between dimensions and body weight is best described as a proportionality between cardiac dimension and the cubic root of body weight (body weight $^{(1/3)}$).[6]This relationship is logical because a cardiac dimension is a one-dimensional measurement, whereas body weight is indirectly a 3-dimensional measurement; body weight is highly associated with body volume, because body water is the most dominant constituent of the body. A linear relationship of the one-dimensional cardiac measurement and the cubic root of the body weight have therefore been shown in dogs with a wide range of body sizes. Today, this method of correcting for body size, first described by Cornell et al.,[6] is frequently used in veterinary medicine. The method is still limited by comparably wide normal reference ranges, and lack of breed specific reference ranges for many breeds. Finally, the drawback is, like any other linear measurement of heart size, it is used for assessing changes in a 3-dimensional structure. One-dimensional measurements will accordingly often underestimate true cardiac size with increasing underestimation with increasing cardiac size.

MEASUREMENT OF WALL THICKNESS IN CATS

Few, if any, issues in veterinary cardiology are more controversial than measurements of wall thickness, normal reference ranges and cut-off values between normal and abnormal measurements in cats. Measurements of wall thickness are an important part in diagnosing hypertrophic cardiomyopathy. The currently published normal reference ranges are derived from comparably old studies from the 80 and 90:ies including M-mode measurements in apparently normal cats in a comparably narrow range of body weight. There have been considerably improvements in the performance of the ultrasound systems since these studies were published, particularly concerning image resolution and frame rate. These technological improvements now allow better visualization of ectopic papillary muscles and false tendons: structures that previously were confused with endocardial borders. An alternative to measure wall thickness using M-mode is to measure in the frozen 2D image. Different approaches have been suggested for measuring in the 2D image, such as maximal wall thickness and mean thickness defined segments of the LV wall.[7] There is currently no consensus to which method is superior to the others. Furthermore, there is a problem applying the published normal reference ranges in the general cat population where small cats and very large cats are being scanned for the presence of heart disease. For many years, the cut-off value for normal and abnormal measurements were one for all cats, regardless of body size. However, cats are like any other mammal; heart dimensions are proportional to the cubic root of the body weight. The difference between dogs and cats is that the range in body weight is smaller in cats than in dogs, which means that the effect of body size is smaller than in dogs. Still, it is present. A cut-off value of 6 mm has been widely employed to discriminate between cats with thickened and normal wall thicknesses.[7] However, many cardiologists consider this cut-off to be too generous and there is a problem applying this cut-off for cats with different body size and breeds. Recently, data from cat screening including a large number of cats showed a major effect of body weight on wall thickness measurements with the upper 95% confidence range of septal and lateral wall thicknesses in all cats (including healthy and diseased) <5 mm in cats <6 kg.[8] Including only healthy cats, the weight range for cats having wall thickness <5 mm increased to <7kg. Furthermore, both lateral and septal diastolic wall thicknesses increased with age, were higher in males, varied between breed groups, and between screeners (median difference 0.5 mm). In conclusion, this shows that the cut-off between the normal range and abnormal values for wall thickness in cats is less than 6 mm in cats, but is influenced by many factors. Accordingly, measurements corrected for body weight, breed and gender should be established.

LEFT ATRIAL VOLUME

Estimation of LA/AO ratio is the most common method for assessing LA size, and it can be performed in either 2D parasternal short-axis or long-axis views.[3] However, left atrial dilatation may occur in all directions. [9,10] Because of this problem, other methods of assessing LA size has been used, including LA circumference, and cross-sectional area indexed to Ao circumference and cross-sectional area, respectively. However, as LA is a 3D structure, the accuracy of both diameter and area measurements of LA size in single or 2D planes may be questioned. RT3D echocardiography is expected to more accurately estimate LA size. Estimation of LA volume by RT3D echocardiography has been reported to be a major predictor of clinical outcome in human patients with severe LV dysfunction, and its clinical value has been shown to be superior to that of 2D techniques.[11] Optimal performed RT3D echocardiography is a relatively time consuming technique, and rapid on-line M-mode or 2D methods are still required in clinical practice. The question may therefore arise which of the most commonly used 2D methods for estimating LA size shows best agreement with the RT3D method. In a study by Tidholm *et al.*, assessments of LA volumes in healthy dogs and in dogs with myxomatous mitral valve disease (MMVD) were investigated using a RT3D method and four different 2D-based echocardiographic methods.[12] Good correlation could not be demonstrated between body weight-indexed RT3D LA volumes and any of the indexed 2D methods of estimating LA size. In comparison with the RT3D method, allometric scaling of the LA diameter, LA/Ao in short-axis view, and LA long axis diameter compared to Ao in short-axis view underestimated LA size at higher values of LA size, whereas the LA area/ Ao area overestimated LA size at all values. Good correlations were found between LA volume estimated by RT3D, and LA volume estimated by allometric scaling of LA long axis diameter (according to the formula: LA_{vol} (ml) = 0.46x $(LA_{lax} (cm))^{2.67}$) followed by LA short axis diameters. Once estimates of LA volume have been obtained, they must be corrected for differences in body size similar to one-dimensional measurements. The currently most common way to do this is to divide volume with body weight. Studies indicate that left atrial volumes in normal dogs are approximately 0.65 ml/kg in atrial diastole and about 0.4 ml/kg in systole, and values may increase up to 4-6 ml/kg in severe LA enlargement.[12]

LEFT VENTRICULAR VOLUMES

Three-dimensional (RT3D) echocardiography allows for superior outlining of the LV in real time. Due to the ability to align the true short- and long-axes of the LV, the problems with chamber foreshortening and oblique imaging planes can be reduced; thus providing more anatomically correct views than conventional 2D echocardiography.[13; 14]

Estimation of LV size using RT3D echocardiography and cardiac magnetic resonance imaging (MRI) has been shown highly correlated in human cardiac patients.[15] The correlation between RT3D echocardiography and a gold standard technique such as MRI has, to our knowledge, never been completely investigated in dogs. Most studies in people comparing RT3D and 2D methods show that 2D methods, but not RT3D, consistently underestimate LV volumes compared with a gold standard technique such as MRI. The potential of LV volume assessment by RT3D echocardiography has recently been demonstrated in dogs by our group.[16,17] In a study comparing LV volumes estimated by M-mode, 2D echocardiography, and RT3D

echocardiography in dogs with and without acquired heart disease, a good agreement was demonstrated between RT3D and 2D estimations of LV volumes, even though both end diastolic-and end systolic volumes estimated by RT3D echocardiography were greater than those estimated by 2D echocardiography. Teichholz method was shown to overestimate LV volumes by a factor of approximately 2. Similar to LA volumes, LV volumes need to be corrected for differences in body size for assessment of LV dilatation. The currently most common way to do this is to divide volume with body weight. Studies indicate that left atrial volumes in normal dogs are approximately 2.0 ml/kg in atrial diastole and about 1.0 ml/kg in systole, and values may increase up to 7 ml/kg in diastole in severe LV dilatation.[16,17]
Another advantage of using RT3D echocardiography is that the RT3D dataset can be further divided into regional segments by sectioning the LV from base to apex, perpendicular to the LV long-axis; thereby allowing regional volume assessment.[18] Abnormal changes in LV shape, accompanying LV dilation, can furthermore be assessed by an echocardiographically derived sphericity index.[19,20] A 3D echocardiographically derived sphericity index, which can be obtained by dividing the end-diastolic volume by the volume of a sphere, has been demonstrated an earlier and more accurate predictor of remodeling compared to other echocardiographic variables following acute myocardial infarction in human patients.[21] Sphericity index may be calculated in the 2D image by dividing LV length with LV width, which means that this method is very sensitive to LV foreshortening. Studies indicate 3D sphericity index is approximately 0.8 in normal dogs, but may increase up to 1.4-1.5 in dogs with severe LV dilatation.[16] Measurement of 3D sphericity is, accordingly, a more accurate method, because it compared the LV volume with a volume of a sphere, and image planes may be optimized before volume estimations are performed. The sphericity index is usually not included in the routine echocardiographic protocol when assessing LV remodeling in dogs, but this index might have a potential when investigating progression of LV remodeling in dogs, as previously shown by our group in dogs affected by myxomatous mitral valve disease (MMVD).[16]

REFERENCES

[1]Kupferwasser, et al. **J Am Soc Echocardiogr** 1997;10:840-852, [2] Lang, et al. **J Am Soc Echocardiogr** 2005;18:1440-1463, [3] Rishniw, et al. **J Vet Intern Med** 2000;14:429-435, [4]Hansson, et al. **Vet Radiol Ultrasound** 2002;43:568-575, [5] Hall, et al. **J Vet Cardiol** 2008;10:11-23, [6] Cornell, et al. **J Vet Intern Med** 2004;18:311-321, [7] Wagner et al. **J Vet Cardiol** 2010;12:171-182, [8] Häggström, et al. **Proc ACVIM Forum** 2011;2011, [9] De Castro, et al. **Heart** 2007;93:205-209, [10]Suh, et al. **J Am Soc Echocardiogr** 2008;21:439-445, [11] Badano, et al. **J Cardiovasc Med** 2008;9:476-484, [12] Tidholm, et al. J **Vet Intern Med** 2011;25:1320-1327, [13] Lu, et al. **Am Heart J** 2008;155:946-953, [14] Jacobs, et al. **Eur Heart J** 2006;27:460-468. [15] Sugeng, et al. **Circulation** 2006;114:654-661, [16] Ljungvall, et al. **J Vet Intern Med** 2011; 25:1320-1327, [17] Tidholm, et al. **J Vet Intern Med** 2010;24:1414-1420, [18] Cerqueira, et al. **Circulation** 2002;105:539-542, [19] Monaghan MJ. **Heart** 2006;92:131-136, [20] Di Donato, et al. **Eur J Cardiothorac Surg** 2006;29 Suppl 1:S225-230. [21] Mannaerts, et al. **Eur Heart J** 2004;25:680-687.

A REVIEW OF NEUROENDOCRINE HORMONES AS CARDIAC BIOMARKERS IN DOGS

Katja Höglund, DVM, PhD and Jens Häggström, DVM, PhD, DECVIM-CA (Cardiology)
Uppsala, Sweden

Cardiac biomarkers

A biomarker is an endogenous substance which changes character or concentration in body fluids when disease develops. Ideally, changes of a biomarker should facilitate diagnosis, staging or prognostication of a disease or be used to assess response to treatment. A clinically useful cardiac biomarker requires an upper reference limit in healthy dogs and cut-off values to identify dogs with subclinical disease, congestive heart failure (CHF), or increased risk to develop CHF or experience cardiac sudden death. However, in determining such limits one has to take into consideration the many physiological and hemodynamic factors that determine cardiac function and that the borderline between physiological and pathological levels often is not sharp. Another factor of special interest in dogs is the emerging evidence of breed differences.

Neuroendocrines as cardiac biomarkers

There is an intimate relationship between brain, heart and kidneys in regulation of fluid balance. During progression of heart disease, complex neuroendocrine activation takes place in order to maintain cardiac output and tissue perfusion. Important mechanisms include up-regulation of the sympathetic nervous system (SNS) and the renin-angiotensin-aldosterone system (RAAS). Although these compensatory mechanisms initially improve cardiovascular function, their activity leads to gradually increasing peripheral vascular resistance and fluid retention with deterioration of myocardial function and promotion of adverse myocardial remodeling (such as myocardial fibrosis). Other neuroendocrines contributing to vasoconstriction include endothelin and vasopressin. In response to the developing volume overload, natriuretic peptides (NPs) are secreted, leading to natriuresis, diuresis, vasodilation and counteraction of cardiac hypertrophy.[1,2]

Natriuretic peptides as cardiac biomarkers

Since their detection in the 1980's,[3,4] atrial natriuretic peptide (ANP) and B-type natriuretic peptide (BNP), have both gained large interest as cardiac biomarkers. The production and release of ANP and BNP are shown in Fig 1. The active C-terminal fragment of each peptide is released in equimolar quantity to the inactive N-terminal (NT) fragment, but due to the longer half-lives of inactive segments they are often used as diagnostic tools. BNP and NT-proBNP have been established as important biomarkers in the diagnosis and prognosis of CHF in people and they are also considered important in early detection of ventricular dysfunction.[5] Also in dogs, NPs have been found useful to distinguish CHF from respiratory disease[6-8] and increased plasma concentrations of NPs have been identified in dogs with dilated cardiomyopathy (DCM).[9-13] In dogs with myxomatous mitral valve disease (MMVD), concentrations of NPs have been shown to increase with disease severity,[10,14,15] and NT-proBNP has been shown predicative of outcome.[14,16] However, in several canine studies, a considerable overlap between investigated groups has been identified, and so far clear reference values are not available.

In a recent large European investigation of more than 500 healthy dogs examined within the LUPA project, highly significant breed differences were shown in plasma concentrations of midregional proANP (MR-proANP) and NT-proBNP.[17] Interestingly, the breeds with highest and lowest concentrations of each NP differed between MR-proANP and NT-proBNP.

ANP and BNP in health and heart disease

Both NPs are released in response to volume expansion and pressure overload, but their modes of production and release differ. ProANP is stored intact in mature secretory granules, while BNP is a constitutively secreted hormone with little intracellular storage. In healthy dogs, both ANP and BNP are produced in the atria, while both atria and ventricles contribute in production of ANP and BNP in dogs in CHF.[18,19]

In healthy people, NT-proBNP concentration is higher in females compared to males,[20,21] whereas MR-proANP has been shown unaffected by gender.[22] Both NPs increase with age in people,[20,22] while canine studies on influence of gender and age on NP concentration show conflicting results.

In the healthy dogs of the LUPA project, most breeds had higher concentrations of MR-proANP than NT-proBNP, which is in consistency with recent findings in healthy people.[21,22] However, the range within each dog breed was larger for NT-proBNP. A high individual weekly variability of NT-proBNP concentration has been shown in healthy dogs[23] and NT-proBNP has been shown to vary considerably in people with stable compensated CHF.[5] Furthermore, physical activity has been shown to affect NT-proBNP to a higher extent than NT-proANP.[24] Several recent human studies have shown similar diagnostic and prognostic value for MR-proANP compared to BNP and NT-proBNP.[25,26] A known limitation in people with dyspnea is the rather large gray zone of BNP or NT-proBNP concentrations, a problem which is encountered also in dogs. In people, measurement of MR-proANP concentration has been shown to provide incremental diagnostic value when BNP or NT-proBNP concentrations fall into gray zone[26] and MR-proANP has been suggested to be a more potent prognostic marker of mortality.[27] Partly, these findings could be attributed to the high biological stability shown for MR-proANP,[22] making it an interesting molecule for further studies in dogs.

The renin-angiotensin-aldosterone system, catecholamines and endothelin as cardiac biomarkers

Several studies have shown that components of RAAS, catecholamines (epinephrine and norepinephrine), and endothelin-1 are altered in dogs with clinical and subclinical heart disease and they have all been suggested as biomarkers. However, the potential of many of these markers has at best been only partially explored in veterinary patients. Components of RAAS have been shown to be increased in dogs with DCM already in the pre-clinical phase.[28] In dogs with MMVD, the situation is less clear. The RAAS is highly influenced by blood pressure, sympathetic drive and volume status. Dogs with MMVD, as opposed to DCM, do not suffer the same decline in forward stroke volume (and forward cardiac output), which could be one reason for the difference in RAAS response between diseases. Furthermore, many types of heart failure drugs affect plasma concentrations, especially diuretics such as furosemide,[29,30] complicating interpretation of test values. Nevertheless, urinary aldosterone concentration has recently been shown to increase with increasing heart dimensions in dogs with MMVD, indicating an association between aldosterone and cardiac remodeling (eccentric hypertrophy and dilatation).[31] However, data pending publication could not identify an association between degree of histopathological findings of fibrosis and arteriosclerosis, and plasma renin activity or plasma aldosterone concentration in dogs with MMVD.[32] The few veterinary studies available on the usefulness of endothelin-1 and catecholamines as cardiac biomarkers show conflicting results, although increasing concentrations of big endothelin-1 and norepinephrine might be of predictive value in dogs with DCM.[9,28] Very little attention has been made in the veterinary literature on breed differences in components of RAAS, catecholamines and endothelin-1. In fact, recent data from the LUPA project indicate comparably large breed differences also for these substances,[17,33] differences that may affect outcome of studies in which these biomarkers are investigated.

Future use of cardiac biomarkers in dogs

The comparably wide breed variations found in the LUPA project indicate that breed variation should be taken into account when designing clinical tests including these substances. Possible causes for breed differences include variation of all aspects of these substances, i.e. stimulating factors, production, storage, release, and clearance.[34-36] Likely these differences are influenced by genetic factors, since breeding has created dogs which are homogenous within breeds, but show large phenotypic variation between breeds.[37] The interactions between different substances should also be taken into account when interpreting concentrations of different neuroendocrines. By way of example, increased concentration of ANP has been shown to suppress RAAS activation in early-stage heart failure in dogs,[38] whereas activation of local and circulating angiotensin II and endothelin has been suggested involved in the increased synthesis and release of NPs in dogs with CHF.[18,19]

Several studies show the potential of NPs as cardiac biomarkers in dogs. However, further studies are needed on the many factors, physiological as well as pathological, affecting NP concentrations in dogs. Results from the LUPA project suggest that breed-specific reference ranges should be established.

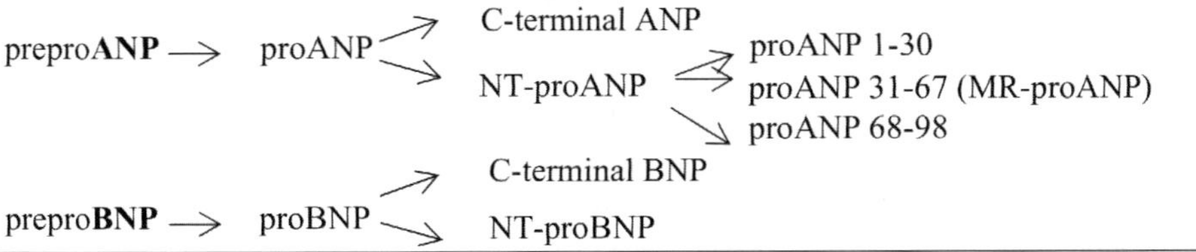

Figure 1. Production and release of natriuretic peptides. MR-proANP=midregional proANP

REFERENCES

[1]Yasue H, et al. **Circulation** 1994;90:195-203. [2]Ghosh N, et al. **Can J Physiol Pharmacol** 2011;89:587-591. [3]de Bold AJ, et al. **Life Sciences** 1981;28:89-94. [4]Sudoh T, et al. **Nature** 1988;332:78-81. [5]Maisel A, et al. **Eur J Heart Fail** 2008;10:824-839. [6]Boswood A, et al. **J Small Anim Pract** 2008;49:26-32. [7]DeFrancesco TC, et al. **J Vet Intern Med** 2007;21:243-250. [8]Prosek R, et al. **J Vet Intern Med** 2007;21:238-242. [9]O'Sullivan ML, et al. **J Vet Intern Med** 2007;21:92-99. [10]Oyama MA, et al. **J Am Vet Med Assoc** 2008;232:1496-1503. [11]Oyama MA, et al. **Am J Vet Res** 2007;68:42-47. [12]Singletary GE, et al. **J Vet Intern Med** 2012;26:1330-1336. [13]Wess G, et al. **Am J Vet Res** 2011;72:642-649. [14]Tarnow I, et al. **Vet J** 2009;180:195-201. [15]Wolf J, et al. **J Vet Cardiol** 2012;14:399-408. [16]Reynolds CA, et al. **J Vet Cardiol** 2012;14:193-202. [17]Stephenson H, et al. Proc. **ECVIM Conf**, Maastricht, Holland, 2012. [18]Luchner A, et al. **Eur J Heart Fail** 2000;2:379-386. [19]Luchner A, et al. **Am J Physiol** 1998;274:H1684-1689. [20]Balion CM, et al. **Clin Biochem** 2008;41:231-239. [21]Hess G, et al. **Clin Chim Acta** 2005;360:187-193. [22]Morgenthaler NG, et al. **Clin Chem** 2004;50:234-236. [23]Kellihan HB, et al. **J Vet Cardiol** 2009;11 Suppl 1:S93-97. [24]Wijbenga JA, et al. **Eur Heart J** 1999;20:1424-1428. [25]Dieplinger B, et al. **Heart** 2009;95:1508-1513. [26]Maisel A, et al. **J Am Coll Cardiol** 2010;55:2062-2076. [27]Moertl D, et al. **J Am Coll Cardiol** 2009;53:1783-1790. [28]Tidholm A, et al. **Am J Vet Res** 2001;62:961-967. [29]Häggström J, et al. **Am J Vet Res** 1996;57:1645-1652. [30]Lantis AC, et al. **Am J Vet Res** 2011;72:1646-1651. [31]Hezzell MJ, et al. **Am J Vet Res** 2012;73:1765-1774. [32]Falk T, et al. Proc. **ACVIM Conf**, Anaheim, CA, 2010. [33]Höglund K, et al. **J Vet Intern Med** 2012;26:1300-1308. [34]Boomsma F, et al. **Cardiovasc Res** 2001;51:442-449. [35]Koller KJ, et al. **Circulation** 1992;86:1081-1088. [36]Meyer M, et al. **J Cardiovasc Pharmacol** 2007;49:100-105. [37]Parker HG, et al. **PLoS genetics** 2005;1:e58. [38]Stevens TL, et al. **J Clin Invest** 1995;95:1101-1108.

LEFT ATRIAL SIZE, ANATOMY AND FUNCTION IN DOGS ASSESSED BY 2D- AND 3D ECHOCARDIOGRAPHY AND MDCT

David Sisson, DVM, DACVIM (Cardiology) and Kate Scollan, DVM, DACVIM (Cardiology)
Corvallis, Oregon

INTRODUCTION

The left atrium is a dynamic structure with a variable and irregular shape that complicates accurate and reproducible estimates of its volume. Inasmuch as considerable variation in the dimensions of the left atrium has been recognized in humans[1] due to differing body size, thoracic conformation, and athleticism, the challenges for accurate estimation of left atrial volume in different dog breeds with radically different body sizes and conformations can be readily appreciated even without the additional complexities contributed by different types of congenital and acquired heart disease. Nonetheless, even crude assessments of left atrial size have proven useful for estimating the likelihood of heart failure in patients with ambiguous clinical signs and predicting the risk of adverse events or death from heart disease.[2-11] In people with heart disease, measures of left atrial size and function have also been employed to estimate the risk of atrial fibrillation and to determine the likelihood of a thromboembolic event, such as stroke.[12-14] To maximize the utility of left atrium as a biomarker for adverse cardiovascular events, it would be helpful to develop consensus regarding how best to assess its size and function and to determine how much accuracy is needed to satisfy the intended purposes of the evaluation.[15]

ANATOMY OF THE LEFT ATRIUM

The left atrium of dogs consists of the following regions: (1) an inflow region containing the various protuberances or diverticula associated with the main confluences of pulmonary venous inflow, (2) the body of the left atrium, and (3) the left atrial appendage.[16-18] The left atrial appendage is an irregularly shaped, pouch-like structure with crenelated walls that extends from the cranial aspect of the left atrium as a finger like projection. Its volume normally occupies less than 10% of the volume of the entire left atrium but its proportional size increases when the left atrium enlarges. Most methods of estimating left atrial size do not include the contribution made by the left atrial appendage, but its size and function are clearly relevant to thrombus formation and the risk of systemic thromboembolism. The dorsocaudal region of the left atrium receives the confluence of the pulmonary veins from the left caudal lung lobe laterally and the typically conjoined veins draining the right caudal and the right accessory lung lobes medially. Another prominent diverticulum of the left atrium extends from the cranio-dorsal-medial aspect of the left atrium to receive the confluences of the pulmonary veins draining the right cranial and right middle lung lobes. A third, more variable, diverticulum receives venous flow from the confluence of the cranial and caudal veins draining the left cranial lung lobe. The body of the left atrium is that portion of the left atrium that remains when the left atrial appendage and diverticula are excluded. The overall shape of the left atrium, minus the left atrial appendage, has been wishfully likened to that of either a prolate ellipsoid by some investigators and to an asymmetric trapezoid by others. The shape of the left atrium has not been well characterized in dogs.

LEFT ATRIAL FUNCTION

In most of the published clinical applications, the size of the left atrium is measured at a single point in time, end-systole, when it is at its maximal volume, but there may also be considerable value in measurements made at other time points.[18, 19] Left atrial pressure and volume change over the course of the cardiac cycle, and review of the left atrial pressure-volume loop facilitates understanding of the mechanical function of the left atrium. During ventricular systole and isovolumetric relaxation, the left atrium acts as **a reservoir** where blood flowing into the LA from the pulmonary veins is stored. When the mitral valve opens with the onset of ventricular diastole, the atria empties passively and the left atrium serves as **a conduit**. When the atria actively contracts at end diastole, the **pumping function** of the left atrium propels blood into the left ventricle. Thus, the reservoir, conduit, and pumping functions of the atrium can be described via equations employing estimates of left atrial volume (1) at the end of ventricular systole when maximal LA volume is achieved ($LA\text{-}V_{max}$), (2) just prior to atrial contraction ($LA\text{-}V_{con}$ or $LA\text{-}V_{pre\ A}$), and (3) at the end of atrial systole when left atrial volume is at a minimum ($LA\text{-}V_{min}$). From these volume determinations total emptying volume, passive and active emptying volumes and emptying fractions can be calculated (Table 1). The clinical value of these determinations for elucidating a patient's hemodynamic or at-risk status is currently uncertain and the subject of ongoing clinical appraisal. Competing techniques for measuring left atrial function include Doppler estimates of pulmonary venous and transmitral flow velocities, tissue Doppler estimates of annular and atrial wall velocities, and strain and strain rate imaging techniques used to assess segmental LA passive and active emptying.

Table 1. Left Atrial Functional Measurements

Left Atrial Total Emptying Volume	Vmax-Vmin
Left Atrial Passive Emptying Volume	Vmax-VpreA
Left Atrial Active Emptying Volume	VpreA-Vmin
Left Atrial Total Emptying Fraction	(Vmax-Vmin)/Vmax
Left Atrial Passive Emptying Fraction	(Vmax-VpreA)/Vmax
Left Atrial Active Emptying Fraction	(VpreA-Vmin)/VpreA

LEFT ATRIAL IMAGING: ECHOCARDIOGRAPHY AND COMPUTED TOMOGRAPHY

Anatomic imaging of the left atrium can be accomplished via echocardiography, contrast-enhanced computed tomography, or magnetic resonance imaging (MRI), and each of these imaging modalities has particular attributes that recommend its application in certain defined circumstances.[20-27] Echocardiography has the distinct advantages of ease of application, cost-effectiveness and excellent temporal resolution. In contrast, MRI and CT offer more accurate representations of the anatomy and spatial relationships of the left atrium, its appendage, and the pulmonary veins, albeit at greater expense and with reduced temporal resolution. Moreover, inspection of LA images obtained either with CT and MRI lend particular insight to the problems that plague the approximation of LA size using M-mode and two dimensional echocardiography.[28-31] Clearly, a single M-mode- or 2D-derived linear measurement of the left atrium does not accurately reflect atrial volume because the dimensions of left atrium do not change increase symmetrically as the left atrium enlarges. In humans, the estimation of left atrial volume by two-dimensional echocardiography is best accomplished from apical two- and four-chamber views via the biplane Simpson method disks [Volume = $\pi/4 * (h) * \Sigma (D1) * (D2)$] or via biplane area length measurements using the formula for a prolate ellipse [Volume = $8/3\ \pi * (A1) * (A2) * (L)$].[32] It is likely that these 2DE methods will eventually default to volume estimates derived from 3DE imaging platforms given the large number of publications reporting the improved accuracy attainable with this modality, the ease of measurement now being incorporated into analysis software, and the anticipated improvements in the resolution of 3D images.[33]

At our institution, we have had the interesting opportunity to evaluate and compare the imaging capabilities of computed tomography with two- and three-dimensional echocardiography. Gated, multi-detector/slice computed tomography (MDCT/MSCT) is an extraordinary resource to explore the complex and dynamic anatomy of the pulmonary veins and the left atrium. In addition to its utility for identifying structural abnormalities of these structures, MSCT provides reasonably accurate life-like 4D reconstructions of the left atrium and can be employed as a reference standard to improve the evaluation of left atrial volume and function when the left atrium is evaluated by two- or three dimensional echocardiography. In our CT facility cardiac image acquisition is accomplished with a Toshiba Aquilion 64-slice CT Scanner and image processing is accomplished using a Vitrea workstation. To reduce respiratory motion artifact, dogs are anesthetized and intubated to accomplish involuntary breath holding at the time of image acquisition. Highly iodinated contrast is administered at a dose of 0.5 to 1.0 ml/kg at a flow rate varying from 0.1 to 0.5 ml/kg/second, depending on the precise application. The contrast injection is usually followed by either a saline or diluted-contrast chaser of similar volume. For gated studies, we typically use a retrospective image acquisition protocol wherein overlapping projections are continuously acquired together with the ECG signal. The volumetric data sets from different phases of the cardiac cycle are then temporally sorted and reassembled relative to the R wave of the QRS complex. For analysis, the cardiac cycle is divided into 10 sequential periods of time representing data obtained during that portion of the cardiac cycle. Hence, the temporal resolution of this method is considerably less than that which is accomplished via echocardiography.

Each of the temporal volumetric data sets can be displayed in transverse, sagittal, and coronal imaging planes or the display can be customized using tangential imaging planes defined by the user. Imaging planes similar to those used for echocardiographic imaging can be used to derive the corresponding measures of LA linear dimensions or cross-sectional area. Three-dimensional rendering using a variety of customizable formats allows visualization of the topographical features of the pulmonary veins and left atrium or of the contrast-enhanced interior. Once that is accomplished the three-dimensional, volume-rendered images of the left atrium can be displayed in a variety of video formats for review and comparison with 3D images derived from echocardiographic evaluations of the same patient. The volume of the left atrium can also be calculated at each time period using dedicated cardiac software (Vitrea) designed for evaluating the left ventricle. As a result, the estimation of left atrial volume requires the evaluator to adjust the identified margins of the left atrium to exclude the pulmonary veins and the left atrial appendage (if desired). The volume of the left atrial appendage can be determined directly or by subtraction when studies are replicated with and without appendage exclusion. Other, more recently introduced software versions include more robust algorithms specifically designed for studying the left atrium. In normal dogs, left atrial volume is at its nadir (LA-V_{min}) during the "0" time interval of CT data set which commences with the R wave signal from the gating electrocardiogram. The left atrium attains its largest size (LA-V_{max}) at the time interval corresponding to the peak of left ventricular systole, typically during the "30%" time interval. The volume of the left atrium can also be displayed as a function of time permitting comparison with echocardiographic-derived volume estimates made over the entire expanse of the cardiac cycle.

REFERENCES

[1]Patel DA, et al. The Ochsner Journal 2009;9:191. [2]Reynolds CE, et al J Vet Cardiol. 2012;14(1):193. [3]DHezel MJ, et al J Vet Cardiol 2012 14(1):269. [4]Ristow B, et al Am J Cardiol 2008;102:70. [5]Tsang TS, et al J Am Coll Cardiol 2006;47:1018. [6]Rossi A, et al Eur J Heart Fail. 2009; 11:929. [7]Lim TK, et al Am J Cardiol 2007;100(5):870. [8]Welles CC, et al J Am Coll Cardiol 2012;59:673. [9]Moller JE, et al Circulation 2003;107:2207. [10]Borgarelli M, et al J Vet Intern Med. 2012;26(1):69.
[11]Chetboul V and Tissier R J Vet Cardiol. 2012;14(1):127.[12]Tsang TS, et al Mayo Clin Proc. 2001;76:467. [13]Benjamin EJ, et al. Circulation. 1995; 92:835. [14]Gardin JM, et al Am J Cardiol. 2001; 87:1051. [15]Hoit BD Journal Am Coll Cardiol: Cardiovascular Imaging 2012;5(8):778. [16]Brewer FC, et al J of Veterinary Cardiology 2012;14, 293. [17]Holt DE, et al Anat Histol Embryol 2005;34:273. [18]Verheule S, et al Cardiovascular Research 2002;55:727. [19]Blume GG, et al Eur J Echocardiography 2011;12:421. [20]Cameli M et al Cardiovascular Ultrasound 2012;10:4. [21]Kataoka A, et al International J of Cardiology 2011;153:47. [22]Christiaens L, et al International J of Cardiology 2009;131:217. [23]Christiaens L, et al International J of Cardiology 2010;140:189. [24]Schwartzman D, et al J Am Coll Cardiol 2003;41:1349. [25]Artang R, et al Cardiovascular Ultrasound 2009;7:16. [26]Mor-Avi V, et al J Am Coll Cardiol Img 2012;5:769. [27]Hudsmith LE, et al J of Cardiovasc Mag Resonance (2005;7:775. [28]Rishniw M, et al J Vet Intern Med 2000;14:429. [29]Oyama M, Sisson D Vet Radiol & Ultrasound 2005:46(4): 331. [30]Hannson K et al Vet Radiol & Ultrasound 2002:43(6): 568. [31]Tidhom A, et al J Vet Intern Med 2011;25:1320. [32]Lang RM, et al J Am Soc Echocardiogr 2005; 18(12):1440. [33]Lang RM, et al J Am Soc Echocardiogr 2012;25:3.

LEFT VENTRICULAR VOLUME AND FUNCTION IN DOGS ASSESSED BY 3D ECHOCARDIOGRAPHY AND MDCT

Katherine F. Scollan, DVM, DACVIM (Cardiology) and D. David Sisson, DVM, DACVIM (Cardiology)
Corvallis, OR

INTRODUCTION

The quantification of left ventricular volumes and systolic function, particularly ejection fraction (EF), are common indications for cardiac evaluation and can be assessed by multiple imaging modalities. Over the past decade, advances in echocardiography and computed tomography (CT) imaging technologies have expanded the ability to assess left ventricular size and function, each with their own attributes and limitations. This lecture offers an overview of quantification of left ventricular volume and function by three-dimensional echocardiography (3DE) and cardiac-gated multi-detector CT (MDCT). In addition, current clinical and research applications of these modalities in veterinary medicine as well as future directions will be addressed.

QUANTIFICATION OF LEFT VENTRICULAR SIZE AND FUNCTION

Historically, volume measurements of the left ventricle were determined by invasive methods including single and biplane angiography. The technologic developments over the past several decades in ultrasound have allowed tremendous leaps since the first echocardiograms were performed in the late 1950s and early 1960s. Interest in non-invasive methods to estimate left ventricular size and function has spurred newer methods and models for quantification. Determination of three-dimensional left ventricular volumes from two-dimensional data is accomplished by using formulas that fit ventricular shape to primary geometric figures. Ellipse, cylinder, cone, and truncated polyhedrons have all been assessed and previously validated[1-5]. Originally, linear M-mode measurements were used to extrapolate volumetric data from a single plane using the Teichholz formula [$Volume= 7D^3/(2.4+D)$] [6]. This formula assumes an elliptical ventricular shape and because it is a one-dimensional approach, requires strict adherence to the geometric assumptions and amplifies inaccurate measurements as they are cubed within the formula. Further studies with 2D echocardiography revealed the biplane modified Simpson's rule [$Volume = \pi/4 * (h) * \Sigma (D1) * (D2)$] showed the best correlation of left ventricular volumes when compared with angiography and formalin fixed ventricles[1,3]. The principle underlying Simpson's method of discs is that the left ventricular volume is calculated from the summation of a stack of elliptical discs. The height of each disc is calculated as a fraction (usually $1/20^{th}$) of the longer LV lengths from the 2- or 4- chamber left apical views. The cross sectional area of each disc is calculated from the two measured diameters. While cumbersome early on, built in software on many ultrasound machines now make this calculation easy and rapid. While the modified Simpson's rule is recommended by the American Society of Echocardiography for LV quantification, if the apical region cannot be traced adequately due to poor endocardial definition the area-length method [$Volume = (5 \times A \times L)/6$] can be used which assumes a bullet shaped left ventricle[7].

The quantification of left ventricular volumes from three-dimensional data, including 3DE, MDCT, and cardiac magnetic resonance imaging (MRI) does not require geometrical assumptions of ventricular shape. Commercial analysis software for both 3DE and MDCT data allow semi-automated detection of the endocardial surface followed by the calculation of the volume contained within that surface in three dimensions. Volumes are calculated at end diastole and end systole to assess ejection fraction and thus global left ventricular function. These modalities that utilize three-dimensional data avoid the necessity of geometrical assumptions to quantify left ventricular volumes and therefore may be more suited to assessing volumes within distorted ventricular shapes. This may be particularly useful in veterinary medicine as left ventricular shape in a range of species may not conform to the geometric assumptions validated for use on human ventricles.

TECHNOLOGY

Three-dimensional Echocardiography (3DE)

The need for more accurate assessment of ventricular function and display of complex anatomy prompted the investigation into three-dimensional echocardiography only shortly after the introduction of two-dimensional ultrasound. The early approach to 3DE was the offline reconstruction of 2D images. These images were acquired utilizing differing technique including linear, fan-like, and rotational manual transducer motions to obtain data sets for 3D reconstruction. The development of real time 3DE required ultrasound machines to be able to acquire and process vast amounts of ultrasound data and also display that data in live action. The leap forward to 3DE was achieved by expanded used of sparse matrix array transducers (approximately 512 elements) with parallel processing, which permits the reception of several lines for each single transmitted signal. Newer matrix array transducers have as many as 2,500 elements, which are grouped together in small arrays and combine their output by micro-beam formers. The micro-beam formers greatly enhance the image quality and processing speed while minimizing the actual size and space of the hardware required. The matrix array transducer ultimately produces a pyramidal array of ultrasound to collect three-dimensional data.

The essential limitation of 3DE technology is the balance between frame rate and resolution. Most 3DE ultrasound platforms allow for several imaging options to address this limitation. Real-time, or live, 3DE yields an ultrasound array in a

30° x 60° pyramidal shape, with the angulation degree dependent on the manufacturer. The 3D zoom imaging modality overcomes the limitation of the narrow scanning sector of live 3D by allowing the sonographer to manually define a small region of interest. Full volume imaging allows for scanning of larger volumes by acquiring smaller sub-volumes in each cardiac cycle. These sub-volumes are then stitched together to create one large volume data set, reaching angles of 104° x 104°, which can encompass the entire left ventricle or the entire heart. The practical limitation of full volume acquisition are stitching artifact caused by respiratory motion or irregular cardiac rhythms. Continued advances in ultrasound technology have already engineered single-beat full volume acquisition of the left ventricle to surpass these limitations[8].

Multi-detector Computed Tomography (MDCT)

Computed tomography technology has made remarkable advances during the past three decades since the first CT scanners were introduced in 1972[9]. The introduction of slip-ring technology in the late 1980s allowed the gantry to rotate continuously, therefore enabling single slice helical scanning. Helical scanning dramatically reduced scanning times and increased temporal resolution, although was still insufficient to obtain motion free cardiac images. Further advances in the 1990s and 2000s brought the development of multi-detector rows, referred to as MDCTs. Increased numbers of detectors, increasing from 4 to 16 to 64 and now 256, as well as increased gantry rotation speeds have improved spatial and temporal resolution to allow motion free cardiac imaging. In addition to improved gantry and detector hardware, the use of cardiac gating also improves temporal resolution and minimizes motion artifacts. Two approaches to cardiac gating are generally used: prospective ECG triggering and retrospective ECG gating. Prospective ECG triggering generates X-rays during diastole when cardiac motion is minimal and minimizes the total radiation dose, although is sensitive to arrhythmias and does not allow ventricular functional analysis.

Retrospective ECG gating techniques acquire partially overlapping MDCT projections continuously with a simultaneously recorded ECG[10]. Algorithms are used to sort the data from different phases of the cardiac cycle. Following data acquisition, images are retrospectively reconstructed to produce sequences of images of the entire heart at multiple phases of the cardiac cycle, often segmenting the R-R interval into 10% increments. The end diastolic and end-systolic sets of images are then identified allowing volumetric measurements of the left ventricle and the assessment of global and regional systolic function.

CLINICAL AND RESEARCH APPLICATIONS

In the evaluation of left ventricular size and function, the use of 2 dimensional imaging methods, mainly 2DE, while widely available and noninvasive, is inherently unreliable. The required of assumptions of geometrical shape and symmetry as well as the unintentional acquisition of foreshortened views likely account for the relative inaccuracy and less reproducibility when compared to three-dimensional imaging modalities[11-13]. The current on and off-line computer software packages for 3DE allow for volumetric measurements of left ventricular size. Semi-automated border detection delineates the contours of the ventricle and calculates the volume within the three-dimensional region by the total number of voxels. These methods eliminate the necessity of geometrical assumption for volume calculations. Several imaging studies in human medicine have shown better agreement of left ventricular size and ejection fraction with 3DE compared to 2DE against cardiac MRI as the gold standard.

3DE has also been demonstrated to have superior performance over 2DE in evaluating left ventricular volumes in patients with hypertrophic cardiomyopathy[14] and in the presence of LV aneurysms[15]. This seems to substantiate the theory that volumetric data from 3D imaging would improve reproducibility and accuracy of measurements in ventricles that violate the geometrical assumptions for 2D based measurements.

3DE has been used in veterinary medicine previously to assess LV volume and shape and dogs with and without acquired cardiac disease[16]. Global LV volumes as well as regional changes to LV shape have been assessed with chronic degenerative valve disease[17]. The continued use of 3DE in veterinary medicine in both clinical and research settings may provide more accurate LV volume measurements, reveal additional information regarding ventricular geometrical changes with disease, and allow the assessment of function by strain using 3D speckle tracking.

Due to the necessity of anesthesia and the use of radiation, cardiac-gated MDCT likely has limited use in a clinical application for the measurement of left ventricular volumes and function. Since volumes measured by MDCT and cardiac MRI in human studies have shown very high measures of agreement[10, 18, 19], it would be feasible to use MDCT as a gold standard for other imaging modalities in a research setting for veterinary medicine. The fast data acquisition times of MDCT in comparison to MRI allow for minimal anesthetic time and its inherent alterations of cardiac function. In addition, outside of its use for left ventricular quantification, cardiac-gated MDCT may prove invaluable for the evaluation of complex congenital defects and the assessment of right ventricular function with pulmonary hypertension.

CONCLUSIONS AND FUTURE DIRECTIONS

Real-time 3DE and cardiac-gated MDCT have become viable imaging modalities to assess left ventricular size and function in veterinary medicine. Both modalities have the significant advantage of more accurate quantification of left ventricular volumes and function by acquiring three-dimensional anatomic data and allowing volumetric calculations rather then using mathematical models. Further advances in ultrasound technology will allow the acquisition of single-beat full volume data minimizing the limitations of heart rate regularity and respiratory motion artifacts. Real-time 3DE and MDCT each have important advantages that may be useful both in investigation and study of veterinary cardiac disease and daily clinical practice.

REFERENCES

[1]Parisi AF, et al. **Clin Cardiol** 1979;2;257. [2]Folland ED, et al. **Circulation** 1979;60(4):760. [3]Wyatt HL, et al. **Circulation** 1980;61(6):1119. [4]Dumesnil JG, et al. **Circulation** 1982;65(1):91. [5]Mercier JC, et al. **Circulation** 1982;65(5):962. [6]Teichholz LE, et al. **N Eng J Med** 1974;291:1220. [7]Lang RM, et al. **Eur J Echocardiography** 2006;7:79. [8]Macron L, et al. **Circ Cardiovasc Imaging** 2010;3:450. Desjardins B, et al. **Am J Roentgenol** 2004;182:933. [10]Belge B, et al. **Eur Radiol** 2006;16:1424. [11]Sapin PM, et al. **JACC** 1994;24(4):1054. [12]Jenkins C, et al. **JACC** 2004;44(4):878. [13]Dorosz JL, et al. **JACC** 2012;59(20):1799. [14]Bicudo LS, et al. **Echocardiography** 2008;25(7):717. [15]Qin JX, et al. **JACC** 2000;36(3):900. [16]Tidholm A, et al. **JVIM** 2010;24:1414. [17]Ljungvall I, et al. **JVIM** 2011;25:1036. [18]Mahnken AH, et al. **Eur Radiol** 2006;16:1416. [19]Guo K, et al. **Eur Radiol** 2009;19:2107.

CLINICAL CHARACTERIZATION OF 100 BOXER DOGS WITH ARVC

William A. Brown, DVM, DACVIM (Cardiology)
Birmingham, MI

BACKGROUND

In 1983 Harpster described a group of 64 Boxer dogs with an atypical form of cardiomyopathy. Severe ventricular arrhythmias of right ventricular origin were a hallmark of the disease. Affected dogs were equally divided between 3 groups. An asymptomatic group was comprised of apparently healthy Boxers in which ventricular ectopy was noted during a routine wellness exam. Boxers with clinical signs of weakness, exercise intolerance and syncope formed the second group. The final group included dogs with overt signs of congestive heart failure (ascites, pulmonary edema). A genetic cause for the cardiomyopathy was proposed based on the observation that affected dogs were often related.

Necropsy evaluation of 18 affected dogs revealed myocytolysis, mild inflammation, fibrosis and extensive fatty infiltration of nearly all sections of atrial and ventricular myocardium examined. The right ventricular free wall was most severely affected with variable involvement of the interventricular septum and left ventricular free wall.

Harpster reported that the "natural course of the disease for most dogs is a progression from mild arrhythmias to serious potentially life-threatening arrhythmias and eventually to congestive heart failure." Asymptomatic dogs were described to live for several years, while dogs with signs of congestive heart failure rarely survived more than 6 months.

Subsequent to this original report, Boxer cardiomyopathy was renamed arrhythmogenic right ventricular cardiomyopathy (ARVC) for its clinical and pathologic similarities to ARVC in humans.

GENETICS / PATHOGENESIS

Recent studies suggest that ARVC is a disease of the desmosome. Eight different genes (plakoglobin, desmoplakin, plakophilin-2, desmoglein-2, desmocollin-2, cardiac ryanodine-2 receptor, transforming growth factor beta-3 and transmembrane protein 43) and numerous mutations are all reported to cause ARVC in humans. In 2010, Meurs reported the discovery of an 8 base pair deletion in the 3' untranslated region of the striatin gene in Boxer dogs affected with ARVC. The defect resulted in a significant reduction in striatin protein. While the role of striatin in the heart is not well understood, it appears to be associated with desmosomal function. Boxers that are homozygous for the defect appear to have a more severe form of the disease based on a higher number of ventricular premature complexes (VPCs).

The desmosome plays an integral role in cell-to-cell adhesion, as well as transmission of force between the junctional complex and the intermediate filaments in the cytoskeleton. Dysfunction of the desmosome may lead to myocyte detachment and death. Because the myocardium has limited regenerative capacity, repair by fibrofatty replacement is the likely sequelae to myocyte loss. It is theorized that islands of fibrofatty tissue generate macro-reentry circuits and form the arrhythmogenic substrate for malignant ventricular arrhythmias to develop. However, emerging data suggests that ventricular arrhythmias can occur even in the absence of histologic disease. These findings suggest that desmosome dysfunction may lead to gap junction remodeling and may thereby compromise intercellular electrical coupling. A recent transmission electron microscopy study in Boxers affected with ARVC demonstrated a significant decrease in desmosomes, gap junctions and adherens junctions. A second study reported a similar disruption of the gap junctions and a decrease in connexin 43 localization at the end-end sites of ARVC affected Boxers. These data suggest that a loss of gap junctions may be responsible for the severe ventricular arrhythmias seen in this breed.

CLINICAL DIAGNOSIS

Currently, there is no antemortem gold standard that confirms the diagnosis of ARVC. Instead, the best strategy consists of combining information from several diagnostic tests to achieve a definitive diagnosis.

Because ARVC is inherited in Boxers, a family history of ARVC affected dogs must be considered a risk factor. In addition, the presence of the recently identified striatin mutation must also be considered a contributing factor. Dogs that are both heterozygous and homozygous for the mutation appear to be at risk for developing ARVC. Unfortunately, the absence of the striatin mutation does not reduce or eliminate the risk of ARVC. A significant number of Boxers affected with ARVC are negative for the striatin mutation.

Electrocardiography

The finding of increased ventricular ectopic activity on physical examination, resting ECG or ambulatory ECG monitoring in a Boxer is strongly suggestive of ARVC. The absolute number of VPCs required to establish a diagnosis of ARVC is unknown. Several ambulatory ECG studies suggest that normal dogs have very few VPCs (0-24) in a 24-hour period. In addition, normal dogs have single VPCs and rarely demonstrate complex ventricular arrhythmias. Thus, the finding of numerous single VPCs (likely >100), ventricular couplets, triplets or runs in a Boxer should be considered significant. The presence of exercise intolerance, unexplained weakness or collapse should further increase the clinical suspicion of ARVC.

Studies by Spier and Meurs suggest a high degree of day-to-day variability (80%) in the frequency of VPCs in Boxers with ARVC. This suggests that brief in-hospital ECGs may be of limited value (high specificity, low sensitivity) in evaluating Boxers suspected of having ARVC. Therefore, ambulatory ECG monitoring is recommended for screening Boxers for ARVC. Multiple day recordings may ultimately prove to be even more valuable than a single 24-hour recording.

Echocardiography

Because ARVC is a myocardial disease, echocardiography is an ideal non-invasive tool to assess cardiac chamber dimensions and myocardial function. Several studies in humans with ARVC have demonstrated both qualitative and quantitative abnormalities of the right ventricle.

The qualitative abnormalities described in human ARVC patients include hyper-reflective moderator bands, trabecular derangement, focal aneurysms and regional wall motion abnormalities. To date there are no echocardiographic studies that describe any qualitative RV changes in Boxers with ARVC.

Humans with ARVC frequently have dilation of the right ventricle. An enlarged right ventricular outflow tract was found in 100% of affected humans. While there are individual reports of ARVC Boxers with varying degrees of right ventricular enlargement, there are no large studies that document the overall frequency of RV dilation. Quantification of RV enlargement in ARVC Boxers is further limited by a lack of normal RV dimensions and appropriate cut off values for mild, moderate and severe enlargement.

Global RV dysfunction is also often noted in humans with ARVC. The complex geometry of the right ventricle and problems with complete RV visualization make the assessment of RV function difficult in many human and canine patients. The RV fractional area change (FAC) from the parasternal long axis view has been shown to be a useful correlate of RV function in humans with ARVC. In addition, the RV myocardial performance index (MPI) has been proposed as a simple Doppler derived index that combines systolic and diastolic ventricular function into a composite value. Yoerger et al. reported preliminary data demonstrating reduced RV MPI in humans with ARVC. A similar study conducted by Baumwart and Meurs did not detect any difference in RV MPI between normal Boxers and Boxers with ARVC.

Initial reports of ARVC in humans suggested that left ventricular involvement was an endstage complication, occurring after the development of right ventricular dysfunction. However, recent clinical studies have demonstrated significant left ventricular involvement as a common feature of ARVC, which may occur in the absence of right ventricular dysfunction. Left-dominant arrhythmogenic cardiomyopathy (LDAC) has now been described as 1 of 3 patterns of disease expression in human families with ARVC. In contrast to "classic" ARVC which causes right ventricular dysfunction and "biventricular" ARVC, with parallel involvement of both ventricles, LDAC is characterized by early and predominantly LV involvement.

Initial descriptions of ARVC in Boxers suggested that left ventricular myocardial failure is an uncommon finding occurring in approximately 7% of affected dogs. However, a more recent retrospective study by Palermo in Boxers with ARVC revealed left ventricular dysfunction in 74%. Surprisingly, the dogs with LV involvement tended to be younger (5.7 years) than dogs without LV dysfunction (7.6 years). In addition, 17% of the dogs with LV dysfunction exhibited signs of right heart failure (ascites), while 59% of the dogs with LV dysfunction had signs of left heart failure (labored breathing, pulmonary edema). This data suggests that similar to humans with ARVC, left ventricular involvement may not be an endstage complication of ARVC. Further, it appears that some Boxers have predominantly left ventricular disease and signs of left heart failure in the absence of gross right ventricular disease.

The echocardiographic findings of left ventricular involvement in Boxers with ARVC are similar to those described in dogs with idiopathic dilated cardiomyopathy. Both the left ventricular end-diastolic and end-systolic dimensions are increased, while the left ventricular contractility is decreased. Baumwart and Meurs have previously described the echocardiographic findings in 48 Boxers with ARVC and left ventricular systolic dysfunction.

TREATMENT

The primary goal of treatment of both humans and Boxers with ARVC is to prevent sudden death. Because ARVC is a progressive myocardial disease, it is important to recognize that a patient's risk of sudden death may increase over time. In humans with ARVC, a retrospective analysis suggested that young age, prior cardiac arrest, fast and poorly tolerated multifocal ventricular tachycardia, syncope, severe RV dysfunction, heart failure with LV involvement, and familial occurrence of juvenile sudden death are the major determinants in predicting sudden death and worse outcome. Thus far, there are no studies in Boxers with ARVC that define the risk factors associated with sudden death. There are numerous reports that suggest that Boxers with myocardial failure have a worse prognosis than those without myocardial failure.

Antiarrhythmic drug therapy is often the first choice treatment in human ARVC patients. Studies suggest that either sotalol or amiodarone (alone or in combination with a beta-blocker) are the most effective drugs with a low pro-arrhythmic risk. A study by Meurs and Spier comparing the efficacy of four different antiarrhythmic treatments in Boxers with ARVC, revealed that therapy with sotalol and the combination of mexiletine and atenolol are superior to either procainamide or atenolol alone. Both sotalol and the combination of atenolol and mexiletine decreased the total number of VPCs and the complexity of the ventricular arrhythmia. Ideally, Boxers that are treated with antiarrhythmic drugs would be assessed by comparing pre-treatment and post-treatment 24-hour ambulatory ECG recordings. A significant (>80%) decrease in the total number of VPCs, a decrease in the complexity of the ventricular arrhythmia and a decrease or resolution of clinical signs (syncope) should be considered a positive treatment response. Unfortunately, there is no evidence at this time that decreasing the total number of VPCs or the complexity of the ventricular arrhythmia decreases syncope, prevents sudden death or improves long term survival in Boxers with ARVC. Further studies will be required to assess the potential benefits of antiarrhythmic therapy in Boxers.

In human patients with ARVC, internal cardiac defibrillators (ICD) are indicated in patients with cardiac arrest, syncope or ventricular tachycardia that is poorly responsive to antiarrhythmic drug therapy. The results of clinical trials suggest an improvement in long term prognosis using ICD therapy in these high risk patients. To date ICD implantation has only been described in one Boxer with ARVC. This dog experienced recurrent ventricular tachycardia associated with weakness and lethargy. Numerous attempts to treat the dog with oral antiarrhythmic medications (mexiletine, sotalol, and atenolol) were unsuccessful. Ultimately, an ICD was successfully implanted. Unfortunately, several complications were described including oversensing T waves and double counting the heart rate, delivering inappropriate "shocks" during exercise due to an elevated heart rate (>240/min), and infection at the generator site. Based on this report, it appears that ICD implantation is a reasonable option for Boxers with ventricular arrhythmias (and clinical signs) that are refractory to antiarrhythmic drug therapy. However, more work needs to be done to limit inappropriate shocks and reduce complications.

In addition to antiarrhythmic treatment, some degree of exercise restriction in Boxers with ARVC is strongly recommended. Data in humans with ARVC suggests that vigorous exercise (football, basketball, etc.) is associated with a five-fold increase in the risk of sudden death. Current theories suggest that acute increases in the RV workload and increased sympathetic activity may be responsible for this increased arrhythmogenic risk. A similar tendency has been described in Boxers with ARVC. Our data and the original reports by Harpster, suggest that syncopal episodes in Boxers are often precipitated by heavy exercise (running) or excitement (greeting another dog or the owner returning home from work). Therefore, exercise restriction combined with antiarrhythmic drug therapy may be the best method to manage these dogs.

Boxers with myocardial failure and signs of congestive heart failure are managed using the same strategies employed in dogs with idiopathic dilated cardiomyopathy. Rapid atrial arrhythmias (atrial fib or SVT) are typically controlled using either diltiazem or a combination of diltiazem and digoxin. Fluid retention is most often treated with a blend of furosemide, an ACE inhibitor and spironolactone. In addition, thoracocentesis or abdominocentesis can be used as needed to manage dogs that have an incomplete or poor response to oral drug therapy. Finally, therapy with pimobendan (Vetmedin) is recommended in all Boxers with myocardial failure. While initial reports have suggested that Boxers with myocardial failure have a survival of 4-6 months, these reports were published prior to the introduction of Vetmedin. Therefore, new studies will be required to determine if Vetmedin improves quality of life or survival time in Boxers with ARVC and myocardial failure.

REFERENCES

Harpster N. **Curr Vet Ther VIII** 1983:329-337. Harpster N. **Vet Clin N Am** 1991;21(5):989-1003. Delmar M, et al. **Circ Res** 2010;107:700-714. Meurs K, et al. **Hum Genet** 2010;128(3):315-324. Gemayel C, et al. **J Am Coll Cardiol** 2001;38:1773-1781. Sen-Chowdhry S, et al. **J Am Coll Cardiol** 2007;50:1813-1821. Oxford E, et al. **J Vet Cardiol** 2011;13:101-113. Oxford E, et al. **Heart Rhythm** 2007;4:1196-1205. Meurs K, et al. **JAVMA** 2001;218(8):1291-1292. Stern J, et al. **JAVMA** 2010;236(4):430-433. Spier A, et al. **JAVMA** 2004;224(4):538-541. Meurs K, et al. **JAVMA** 2001;218(2):222-224. Yoerger-Sanborn D, et al. **Arrhythmogenic RV Cardiomyopathy/Dysplasia** 2007:129-134. Baumwart R, et al. **AJVR** 2008;69(8):1029-1032. Palermo V, et al. **J Vet Cardiol** 2011;13:45-55. Baumwart R, et al. **JAVMA** 2005;226(7):1102-1104. Basso C, et al. **Lancet** 2009;373:1289-1300. Corrado D, et al. **Heart** 2009;95:766-773. Meurs K, et al. **JAVMA** 2002;221(4):522-527. Nelson O, et al. **J Vet Intern Med** 2006;20:1232-1237.

EFFECT OF VAGAL MANEUVERS ON THE HEART RATE OF HEALTHY DOGS. E. Côté,[1] E. Trageser, [1] A. Trivett.[2] 1. Atlantic Veterinary College, University of Prince Edward Island, Charlottetown, PE, Canada. 2. Department of Engineering, University of Prince Edward Island, Charlottetown, PE, Canada.

The effects of vagal maneuvers (VM; ocular pressure [OP], carotid sinus massage [CSM]) are widely described but poorly quantified in dogs. The purposes of this study were to measure the changes in heart rate (HR) response to VM, assess HR post-VM, compare the changes elicited with OP to those with CSM, and characterize the duration of these changes, in healthy dogs.

Thirty-two volunteered pet dogs of both genders and various breeds were enrolled based on absence of known illness, normal physical examination, and owner consent. After acclimation in a quiet room, each dog was held gently in right lateral recumbency and was instrumented to a 3-lead electrocardiograph (ECG) to undergo 4 maneuvers. Each maneuver lasted 15 seconds and was separated from the next by 45 seconds' rest. The 4 maneuvers were OP, CSM, sham OP (hand placed near the eyes) and sham CSM (hand placed near the neck). The sequence of maneuvers was changed from dog to dog (Latin square design). The principal investigator (PI) performing the maneuvers was blinded to results during data acquisition and during ECG measurement. Beat-to-beat HR was measured and data were partitioned into 6 time periods: PRE (7.5 seconds preceding maneuver), INTRA1 and 2 (0-7.5 seconds and 7.6-15 seconds of VM, respectively), and POST1, 2, and 3 (1-5, 6-10, and 11-15 seconds following cessation of the maneuver, respectively). After all dogs were studied, the approximate force and pressure applied by the PI during OP and CSM were quantified by having the PI mimic performing the maneuvers *in vitro* on electronic pressure and force gauges.

HR PRE was no different among groups. For CSM, HR INTRA 2 was lower than HR PRE (P=0.028) and HR INTRA 1 (P<0.001) and HR remained significantly lower at POST1 compared to INTRA 1 (P=0.003), suggesting that the HR-lowering effect of CSM took >7.5 seconds to occur and persisted after cessation. Equivalent differences were not noted with sham CSM. For OP, HR during INTRA 2 was significantly lower than HR in POST1, POST2, or POST 3 (P=0.023, P=0.006, P=0.012, respectively) but not lower than OP HR PRE, suggesting that a rebound increase in HR existed after termination of OP. Equivalent differences were not noted with sham OP. INTRA1 HR was significantly higher for CSM than for OP (P=0.012), indicating a more rapid onset of HR slowing with OP, an initial increase in HR with CSM, or both. *In vitro* replications (10 repetitions each) by the PI indicated mean pressures of 1.52 +/-0.12 kPa and 0.5 +/- 0.05 kPa applied during CSM and OP, and mean forces of 6.45 +/- 0.5 N and 1.64 +/- 0.2 N exerted during CSM and OP, respectively.

These results indicate that in healthy dogs, VMs produce characteristic changes in HR. These data, and quantification of force and pressure with which maneuvers are performed, provide baseline information for clinical applications of VMs.

INTERNATIONAL COLLABORATIVE STUDY TO ASSESS CARDIOVASCULAR RISK AND EVALUATE LONG TERM HEALTH IN FELINE HYPERTROPHIC CARDIOMYOPATHY (REVEAL). PR.Fox,[1] AA.Motsinger-Reif,[2] BW.Keene[2] for the REVEAL-Investigators. [1]Animal Medical Center, NY, NY, [2]North Carolina State University, Raleigh, NC.

Hypertrophic cardiomyopathy (HCM) is the most prevalent feline myocardial disorder. REVEAL was designed to illuminate natural history, compare health outcomes of asymptomatic HCM vs. normal cats, ascertain incidence of events [heart failure (CHF), arterial thromboembolism (ATE), all-cause cardiac death (ACD), noncardiac death (NCD)], and overall survival.

1,362 cats (median;interquartile range: weight-4.8kg; 3.9-5.8kg; age-5.6yrs; 2.3-10yrs) were classified via echocardiography:HCM (n=321), HOCM (n=464), and normal (n=577). Baseline demographics were NS between cohorts.

Cox proportional hazards regression to evaluate CHF, ATE, ACD, event rates, and survival (HCM, HOCM, normal) showed differences between HCM or HOCM and normal for all outcomes ($p<0.0001$). Pair-wise, between cohort comparisons revealed NS difference for any outcome between HCM and HOCM cohorts ($p>0.05$), so hazard ratios (HR) were constructed for a combined HCM + HOCM group [H(O)CM]. For H(O)CM cats, HR's were: CHF- 49.2 (95% CI:15.7-154.1)- i.e. CHF was 50 times more likely in H(O)CM vs normal cats; ATE- 13.2 (CI:5.34 to 32.6); ACD- 32.4 (CI:13.3 78.7). These differences are reflected in the respective incidence rates (IR)/cat/year: CHF IR-4.47% for H(O)CM (i.e., 22.3% at 5yrs) vs. <0.1% per cat/year for normal; ATE IR-1.99% for H(O)CM vs. <0.2% for normal; ACD IR-4.9% for H(O)CM vs. <0.3% for normal. Median survival was 4.97yrs (2.6 - 6.2yrs) for H(O)CM vs. 5.45yrs (3.6 - 6.6yrs) for normal cats.

EFFECT OF DIET ON THE PROGRESSION OF HYPERTROPHIC CARDIOMYOPATHY (HCM) IN CATS. LM Freeman, JE Rush, SM Cunningham, BJ Bulmer. Tufts Cummings School of Veterinary Medicine, North Grafton, MA.

A wider variety of commercial cat foods with varying nutrient profiles have become available in recent years and, during this same time, the incidence of HCM in cats may have increased. The goal of this pilot study was to evaluate the effects of 3 reduced sodium diets with varying macronutrient profiles and ingredients on progression of disease in cats with HCM. Cats with asymptomatic (International Small Animal Cardiac Health Council stage 1b) HCM were enrolled. To be eligible, cats had to have typical echocardiographic features of HCM (e.g., $\geq$ moderate left ventricular hypertrophy with or without left atrial enlargement). After baseline physical examination, echocardiogram, blood pressure measurement (Doppler), and blood collection (serum biochemistry profile, insulin, cardiac troponin I, NT pro-BNP), cats were randomized to 1 of 3 diets which all were <100 mg/100 kcal sodium but varied in carbohydrate content and ingredients. After eating the assigned diet for 6 months, all baseline measurements were repeated. Owners were not blinded to the diet assignment but all cardiologists remained blinded to diet for the duration of the study. The same cardiologist performed the baseline and follow-up echocardiogram in each cat.

Median age was 4.3 yrs (range, 1.1-16.6 years) and the study included 26 castrated males and 4 spayed females. Twenty-six cats were domestic short- or long-hairs, with 1 each of Persian, Peterbald, Ragdoll, and Siamese. Age, sex, and breed were not significantly different among the 3 diet groups. Median baseline body weight and body condition score (BCS) were 4.6 kg (range, 3.5-7.2 kg) and 6.0 (4.5-9.0), respectively and were not different among the 3 diet groups.

Twenty-seven cats have completed the study: 1 cat died of congestive heart failure before the re-evaluation and 2 cats have not yet completed the study. After 6 months, only cats eating Diet C had a significant change in body weight compared to baseline (median difference = -0.4 kg [range, -0.7 to + 0.2 kg]). On echocardiography, cats eating Diet C had a significant decrease in left ventricular free wall in diastole (p=0.05), interventricular septum in systole (p=0.03), left atrial dimension (p=0.04), and the ratio of the left atrium to aorta (p=0.007). In addition, cats eating Diet C had a significant decrease in alanine aminotransferase (p=0.04) and alkaline phosphatase (p=0.03). Cats eating Diets A and B had significant increases in BUN (both p<0.01) and cholesterol (both p<0.05). Cats eating Diet B also had a significant decrease in NT pro-BNP (p=0.04). Tropinin I analyses are pending.

Although larger prospective studies are warranted, these data suggest that diet can influence biochemical parameters and echocardiographic measurements in cats with HCM.

OVERVIEW OF POTENTIAL HUMAN FACTORS CONTRIBUTIONS TO TRAINING IN VETERINARY MEDICINE. AC McLaughlin, JF Sprufera, SP Tou, JA Stern, MK Ames, TC DeFrancesco, BW Keene,

In human factors psychology the knowledge of human capabilities and limitations is applied to the design or alteration of systems to make them more safe, usable, and effective. The field has long informed human medical practice, including usability for medical devices, the acquisition of skills, and encouraging pro-health behaviors such as hand hygiene. Though veterinary medicine has commonalities with human medicine, there are important differences that necessitate separate analyses of veterinary systems and procedures.

In the current study we performed the first step for understanding human needs in a domain – task analyses of high-skill veterinary tasks, using cardiac catheterization procedures as our domain and cardiac residents as our target group for improvement. Our analysis covered three common cardiac procedures: balloon valvuloplasty, patent ductus arteriosus occlusion, and pacemaker insertion. Multiple instances of each procedure were recorded and analyzed for the knowledge demands, interaction of team members, and learning required during procedures.

General results revealed task characteristics with the potential to affect resident performance: human device/animal mismatch, limited inventory, lack of access to technicians for emergency procedures, and requirements for learning difficult perceptual and motor skills under stressful multitasking conditions. The last characteristic, difficulty of learning under stress, is well documented in the psychological literature, allowing us to suggest how task steps may be practiced in a part-task training paradigm or simulated outside live procedures to avoid overload. Abstracting the type of sub-tasks best matched to part-task practice can provide generalized guidelines for training program development beyond the three procedures here.

IDENTIFICATION OF GENETIC POLYMORPHISMS IN THE CANINE AND FELINE RENIN ANGIOTENSIN ALDOSTERONE SYSTEM GENES, ANGIOTENSIN I-CONVERTING ENZYME, ANGIOTENSINOGEN AND ANGIOTENSIN RECEPTOR. 1. KM Meurs, JA Stern, Y Reina Doreste, L Chdid. North Carolina State University, Raleigh, NC.

Genetic variability in the rennin angiotensin aldosterone system (RAAS) genes, Angiotensin I-converting enzyme (ACE), angiotensinogen (AGT) and angiotensin receptor 1 (AGTR1) in human beings has been shown to have an impact on clinical response to certain cardiac drugs as well as on clinical presentation of cardiovascular disease. Genetic variability in these genes in dogs and cats has not yet been evaluated. The purpose of this study was to evaluate these genes in the dog and cat to evaluate for genetic variation as described in human beings.

DNA samples from 15 dogs of 3 breeds (Doberman pinscher, Boxer and Cavalier King Charles Spaniel) and from 15 cats of 3 breeds of (Maine Coon, Ragdoll, Sphynx) were evaluated with PCR-based sequencing based on the known sequence of the AGTR1 and AGT genes. Only the intronic sequence was evaluated in the ACE gene since this is the area previously shown to be variable in human beings. Sequence results were evaluated for variation using the CLustalW Alignment program.

Several single base pair changes as well as deletion polymorphisms were identified in all of the genes. In some cases the polymorphism changed the amino acid produced. However, the variability was not identical to that observed in human beings. We conclude that these RAAS genes have genetic variability in dogs and cats. The clinical and functional significance of these polymorphisms deserves further study.

ASSOCIATION OF DILATED CARDIOMYOPATHY WITH THE STRIATIN MUTATION GENOTYPE IN BOXER DOGS. KM Meurs[1], JA Stern[1], DD Sisson[2], MD Kittleson[3], SM Cunningham[4], MK Ames[1], CE Atkins[1], TC DeFrancesco[1], TE Hodge[1], BW Keene, AA Motsinger-Reif[1], SP Tou[1] 1.North Carolina State University, Raleigh, NC. 2. Oregon State University, Corvalis, OR. 3. University of California-Davis, Davis, CA. 4.Tufts University, North Grafton, MA.

Dilated cardiomyopathy (DCM) in the Boxer dog is characterized by ventricular systolic dysfunction. Arrhythmogenic right ventricular cardiomyopathy (ARVC) also exists in this breed and has been shown to be associated with a deletion in the 3' untranslated region of striatin, a desomosmal protein, at least in some Boxers. In human ARVC caused by mutations in genes encoding for desmosomal proteins, a small percentage of affected individuals develop a classical form of DCM. We hypothesized that DCM in the Boxer may be a variant of ARVC and may be associated with the striatin mutation in some dogs. The objective was to determine if Boxer dogs with DCM were positive for the striatin mutation.

Thirty-three adult boxer dogs (4 female, 10 female spayed, 9 male, 10 male castrated) with a DCM phenotype as defined by an left ventricular diastolic diameter (LVDD) >4.8 cm (median 5.5, range, 4.8-7.1), left ventricular systolic diameter (LVSD) >3.3 cm (median 4.8, range 3.8-6.4) & fractional shortening (FS) < 21% (median 14, range 1.0 – 20) were evaluated by PCR and AFLP for the striatin mutation. Confidence intervals were evaluated to determine the lower and upper limits of the 95% confidence interval for the proportion of dogs with DCM that had the mutation. An association of the deletion with the phenotype was tested using a case-only design, testing for overrepresentation of the alleles in the case samples for the DCM phenotype with a Chi^2 test of association. T-tests evaluated for a difference between positive heterozygous and positive homozygous groups for age, LVDD, LVSD and FS.

Thirty (91%), (confidence intervals of 76-97%)) dogs were positive for the striatin mutation (16 positive heterozygous, 14 positive homozygous). The deletion mutation was strongly associated with the DCM phenotype (P <0.0001). There was no statistical difference between the groups for age, LVDD, LVSD, and FS.

In conclusion, the striatin mutation was frequently observed in Boxer dogs with DCM suggesting that as in people, desmosomal alterations may result in the development of DCM. The observation that 3/33 dogs developed DCM without the striatin mutation suggests that there is at least one other cause of DCM in the Boxer dog.

MICRODOSE COMPUTED TOMOGRAPHIC CARDIAC ANGIOGRAPHY IN NORMAL CATS. K. Rodriguez, S Hartman, C McReynolds, A Mulherin, M O'Brien, R O'Brien. University of Illinois at Urbana-Champaign, Urban, IL.

The purpose of this study was to determine if very low dose intravenous iodinated contrast computed tomographic (CT) angiography evaluation resulted in sufficient enhancement for characterization of cardiac chambers in lightly sedated normal cats using a novel movement limiting device, for future clinical studies on cats in congestive heart failure.

Normal cats (n=7), as determined by complete physical examination, blood count, T4 analysis, serum chemistries, Doppler blood pressure evaluation, urinalysis and echocardiography, were administered 0.3 mg/Kg body weight butorphanol intramuscularly approximately 15 minutes prior to imaging. Cats were placed in a novel motion restriction device (VetMouseTrap™) and routine scout CT and precontrast transverse images were obtained of the thorax. Thereafter the cats were administered 0.2 ml/Kg body weight of nonionic iodinated contrast medium (Iohexol, 300 mg Iodine/ml) mixed with an equal volume of saline by hand injection intravenously over 6 seconds. CT imaging began concurrently with the start of injection of contrast medium. Three contiguous CT series of the heart region were obtained with a 16-slice CT system using 250 mA and 120 kV, and a collimation/slice thickness of 1.25 mm, image reconstruction interval of 0.625 mm, pitch of 1.3, with speed of 13.75mm/sec, and tube rotation time of 0.5 s. Studies were repeated, as necessary due to patient motion, to a maximum of three imaging sessions. The timing and mean attenuation (Hounsfield Units) of the right ventricle, left ventricle, left atrium and aortic lumen were calculated from measurements using 5.0 cm diameter circular regions-of interest (ROI's) from 3 contiguous reconstructed slices on precontrast and the three phases of the post-contrast CT images. Chamber attenuation of and time of maximum enhancement were compared statistically using Student's t-test and a level of significance of $p < 0.05$.

Cats tolerated the injected dose of contrast medium and restraint within the VetMouseTrap™ without clinical detectable effects. CT image acquisition duration was approximately 30 seconds per study for the 3-phase post-contrast studies and between 10 and 20 minutes total time for each cat. Significant contrast enhancement was seen in all four luminal spaces during every study. Mean maximum contrast enhancement in the right ventricle (2.67 seconds) was significantly ($p=0.02$) earlier than the aorta and left atrium (5.65 seconds) and left ventricle (7.25 seconds). Mean maximum contrast enhancement of the left atrium (171 HU), aorta (167 HU) and left ventricle (154 HU) were not significantly different from each other, but were significantly ($p=0.01$) higher than the right ventricle (102 HU) and all chambers were significantly higher at peak enhancement than on precontrast images.

Microdose multi-slice CT angiography resulted in significant contrast enhancement of the heart chambers in lightly sedated normal cats restrained in the VetMouseTrap™.

EVALUATION OF THE BREEDING PROGRAM AIMING AT REDUCING THE PREVALENCE OF MYXOMATOUS MITRAL VALVE DISEASE IN CAVALIER KING CHARLES SPANIELS IN DENMARK. Birkegaard AC[1], Reimann MJ[1], Häggström J[2], Martinussen T[3], Pedersen HD[4], Olsen LH[1]. 1. Department of Veterinary Disease Biology, University of Copenhagen, Frederiksberg, Denmark. 2. Department of Clinical Sciences, Swedish University of Agricultural Sciences, Uppsala, Sweden. 3. Department of Biostatistics, University of Copenhagen, Denmark. 4. Novo Nordisk A/S, Maaloev, Denmark.

Cavalier King Charles spaniels (CKCS) are predisposed to myxomatous mitral valve disease (MMVD). Studies have demonstrated a strong genetic background. The objective of the study was to evaluate the effect of a breeding program involving auscultation and echocardiography aimed at reducing the prevalence of MMVD.

Purebred CKCS (n=997) were examined for the Danish Kennel Club mandatory breeding program during the period from 2002 to 2011. Each dog was evaluated 1-4 times with the total of 1380 examinations. Auscultation and echocardiography were performed to evaluate mitral regurgitation murmur (MRM) severity and degree of mitral valve prolapse (MVP).

The risk of having MRM in 2010-2011 compared to 2002-2003 was estimated using linear regression test (exact condition binary logistic regression), adjusting for repeated measurements. In 2010-2011, dogs were estimated to have 71% reduced risk of having MRM than dogs in 2002-2003 if they were a product of the breeding program (P=0.0017). A dog was defined as a product of the breeding program if both parents where approved by the breeding program. If non-products of the breeding program (for example imported dogs with parents with unknown cardiac status) were included in the analysis, reduced risk of having MRM did not reach statistical significance (OR 1.71; P=0.06).

In conclusion, the mandatory breeding program, based on auscultation and echocardiography, significantly reduced the prevalence of MMVD over the 8 to 10-year period. Therefore, such breeding programs are recommended for CKCS.

SLEEPING AND RESTING RESPIRATORY RATES IN DOGS AND CATS WITH STABLE, CONTROLLED, CONGESTIVE HEART FAILURE.

Mark Rishniw[1], Ingrid Ljungvall[2], Dan Ohad[3], Luca Ferasin[4], Francesco Porciello[5], Jens Haggstrom[2]. 1. Veterinary Information Network, Davis, CA. 2. Swedish University of Agricultural Science, Uppsala, Sweden. 3. Hebrew University of Jerusalem, Rehovot, Israel. 4. SVCC, UK. 5. University of Perugia, Italy.

Recent studies have demonstrated that at-home sleeping and resting respiratory rates (SRR and RRR) in healthy dogs and cats, and dogs and cats with subclinical cardiac disease are generally <30breaths/minute. However, little information is available about at-home SRR and RRR in dogs and cats with well-controlled congestive heart failure (CHF).

We prospectively acquired at-home SRR and RRR in a cohort of 28 dogs and 5 cats with CHF managed by cardiologists and clinicians with an interest in cardiology. All patients were diagnosed with CHF at least 6 weeks prior to data collection, and were considered "well controlled" by the clinician and the owner at the time of data collection on a minimum of 3mg/kg/day (dogs) and 1mg/kg/day (cats) furosemide. Owners collected 10 SRR and 10 RRR measurements over the following 1-2 months.

Average canine and feline furosemide doses were 4.8 and 5.1 mg/kg/day. Median time in CHF at time of data collection was 180 days in dogs and 115 days in cats. Within-dog SRR_{mean} and RRR_{mean} were 19±3 and 24±3 breaths/min. Within-cat SRR_{mean} and RRR_{mean} were 22±3 and 29±3 breaths/min.

Our data suggest that dogs and cats with well-controlled CHF have at-home SRR and RRR similar to those of dogs and cats with subclinical heart disease, and infrequently exceed 30 breaths/minute. Clinicians should strive for a stable SRR <30 breaths/minute to achieve optimal control in their CHF patients whenever possible. Persistent SRR >30 breaths/minute in a patient with CHF warrants additional evaluation of the degree of CHF control.

ATRIAL DEPOLARIZATION WAVES LOCALIZATION ON SURFACE ELECTROCARDIOGRAM IN DOGS WITH SUPRAVENTRICULAR ARRHYTHMIAS.

R.A. Santilli[1], T. Vezzosi[1], M. Perego[1] 1. Clinica Veterinaria Malpensa, Samarate, Italy.

Previous studies showed that atrial depolarization (AD) wave position may differentiate focal atrial tachycardia (FAT) from orthodromic atrioventricular reciprocating tachycardia (OAVRT), although at multivariate analysis only QRS complex alternans and P' wave axis on the frontal plane resulted independent predictors of the arrhythmia type. The aim of our study was to determine AD waves localization in a larger number of supraventricular arrhythmias (SVA) and to assess its utility to differentiate SVAs in dogs.

We selected 77 dogs of different breeds with a mean ($\pm$ SD) age of 3.7 $\pm$ 3.3 years, a mean body weight of 31.5 $\pm$ 14.5 kg, a M:F of 1.8, with different types of SVA assessed by electrophysiologic study: 8 atrial flutter (AFL), 21 FAT, 3 focal junctional tachycardia (FJT), 32 OAVRT, 2 permanent junctional reciprocating tachycardia (PJRT) and 11 sinus tachycardia (ST). For each SVA AD position on surface electrocardiogram was assessed according to the sequence of atrial activation during endocardial mapping. Different electrocardiographic measurements of AD waves, QRS complex, T wave and QT interval were manually calculated. The position of the AD during SVA along the R-R interval were grouped as following: AD on pre-R, R-AD short, AD on T, R-AD long and AD on post-R. Normalcy was tested by the Shapiro-Wilk W-test, differences in proportion were compared using a $\chi 2$ test, different variables were compared amongst the SVA with GLM, and correlation calculated by Pearson. A logistic regression model was performed having SVA type as dependent variable and interval ECG measures (QRS, R-R, AD-AD, RT-T, TR-T, T duration, AD-R, R-AD, QRS-T, AD duration, AD amplitude, AD axis, atrial rate, ventricular rate) as independent variable. Sensibility, specificity and area under the ROC curve were calculated according to SVA type. Reproducibility was calculated by Pearson correlation. AD on pre-R group included 2 FJT with 1:1 ventriculo-atrial conduction; R-AD short group included 31/32 OAVRT, 2 AFL with 1:1 AV conduction ratio and 4/21 FAT; AD on T group included 1/32 OAVRT; R-AD long group included 17/21 FAT, 2/2 PJRT and 11/11 ST; AD on post-R group included 1 case of FJT with isorhythmic AV dissociation and synchronization type II. 6/8 of AFL presented 2:1 AV conduction ratio with 2 AD per each R-R interval located respectively in the AD con pre-R and AD con post-R. Logistic regression showed that AD axis, atrial rate, QT interval duration and ventricular rate were predictor of tachycardia type ($P<0.05$). R-AD duration less than 120 msec and R-AD duration/RR duration < 50% appeared to have the best sensibility (97% and 94%) and best specificity (92 % and 82 %) to differentiate OAVRT from FAT and ST. Reproducibility of the measurements, since guided by endocardial electrograms, was very high and > 0,982 in all cases. According to the results of the present study AD position may help in the differentiation of most common SVA of the dogs. The rarity of other SVAs precludes so far the possibility to construct detailed diagnostic algorhythm, although the evaluation of AD position could suggest, also in these cases, the underlining electrogenic mechanism.

RESULTS OF THE UK CAVALIER KING CHARLES BREEDING PROGRAM: 1991 – 2010. S Swift[1] and P Cripps[2] . 1. Northwest Surgeons, Sutton Weaver, UK 2. School of Veterinary Science, University of Liverpool, UK.

This study examined the database derived from breed screening in cavalier King Charles spaniels in the UK from 1991 to 2010. We sought to determine whether a selective breeding program had had an impact on age of onset of murmurs consistent with degenerative mitral valve disease (DMVD).

Dogs were presented at shows or at private practices and assessed for the presence or absence of a murmur consistent with DMVD by cardiologists or GP vets. These results were entered on a database. Breeding advice stated that dogs could be used if they were over 5 years of age before they developed a murmur or if they were younger, both parents were over 5 before they developed a murmur.

A total of 16,887 examinations were performed on 8860 dogs. Multivariant models showed male dogs developed murmurs at a younger age than females (7.9 years v. 8.1 years) and cardiologists detected murmurs in younger dogs than GP vets (7.2 years v. 8.8 years). Comparing 4 quartile time periods, there was a significant increase in the median age at which 50% of cavaliers will be murmur free determined by GP vets (8.3 to 9.9 years, P 0.001) over 1991-1995 and 2006-2010. Although this trend was seen for cardiologists (7.0 to 7.4 years), it was not significant.

This study suggests the current UK breed program is having a significant impact increasing the age of onset of murmur consistent with DMVD in cavalier King Charles spaniels.

MUSCULOSKELETAL DISORDERS IN VETERINARY ECHOCARDIOGRAPHERS. KA MacDonald[1], DL King[2]. [1]VCA Animal Care Center, Rohnert Park, CA, [2]Ithaca College, Ithaca, NY.

A survey was submitted to the ACVIM Veterinary Cardiology List-serve to study the relationship between performing echocardiograms (ECHOS) and work-related musculoskeletal disorders (WRMSDs). 198 of 487 (40.6%) list-serve members, evenly divided between men and women, responded. 76.8% were <50 years old; 7.6% were > 60 years. 96% currently perform ECHOS. 92.1% perform ECHOS every week (w) except vacation, and most (76.1%) perform ECHOS 4-5 days/w. 87 people (44.4%) experience pain associated with ECHOS (ECHOSpain) that is mild and occasional (77%), frequent and moderate (20.7%), or frequent and severe (2.3%). W/year performing ECHOS was associated with ECHOSpain (χ^2(2,188), 5.458, p=0.065). Those performing ECHOS 31-50 w/year were 2.7 times more likely to have ECHOSpain compared to those performing ECHOS < 31w/year with an odds ratio of 5.5 (p = 0.028). 51.7% with pain reported performing ECHOS as the sole cause of pain and 31% reported ECHOs aggravating pain from a pre-existing problem, 17.2% reported unrelated pain. The most common areas of pain were neck (53.5%), shoulder (52.3%), wrist (41.9%), and back (34.9%). 78.2% had pain at two or more sites. Pain impaired job performance in 36%, required sick leave or disability in 12%, required reduction in ECHOS/day in 36%, and affected lifestyle in 22.1%. 1 of 8 people not currently performing ECHOS had to stop due to ECHOSpain. 9.9% of people have been diagnosed with WRMSD of which 21% have a permanent disability. In conclusion, ECHOSpain is common (44.4%) in veterinary echocardiographers and is associated with weeks/year performing ECHOS.

ANALYSIS OF THE FELINE ARTERIAL THROMBOEMBOLISM: CLOPIDOGREL VS. ASPIRIN TRIAL (FAT CAT). D. Hogan[1], P. Fox[2], K. Jacob[3], B. Keene[4], N. Laste[5], S. Rosenthal[3]. 1. Purdue University, West Lafayette, IN. 2. The Animal Medical Center, New York, NY. 3. Chesapeake Veterinary Cardiology Associates, Annapolis, MD. 4. North Carolina State University, Raleigh, NC. 5. Angell Memorial Animal Hospital, Boston, MA.

Cardiogenic arterial thromboembolism (CATE) is a devastating and usually fatal clinical condition in cats. Antithrombotic drugs are commonly employed for primary and secondary prevention of CATE although there have been no prospective clinical trials to support such an approach. The aim of this prospective, double-blinded, multicenter study was to determine if there was a difference in secondary recurrence of CATE between aspirin and clopidogrel (Plavix) over a 1 year study period.

Cats that experienced a CATE event and survived 1-3 months were eligible for enrollment. The cats were randomly assigned to receive either aspirin (81 mg PO q 3 d) or clopidogrel (18.75 mg PO q 24 hr). The primary endpoint was recurrent CATE; secondary endpoints included all-cause mortality, cardiac death, and study termination due to perceived adverse drug event. Cats that remained on the study after 12 months continued to be followed for primary and secondary endpoints. Kaplan-Meier survival analysis was performed to evaluate for recurrent CATE, combined recurrent CATE or cardiac death, and all cause endpoints for both the 1 year and total study period.

A total of 72 cats were enrolled in the study with 36 cats in each study group. Randomization was adequate with regard to age, weight, gender, underlying cardiac disease, and recovery from the CATE event. For the 12 month study period, median survival time for the primary endpoint of CATE was 192 d in the aspirin group and >365 d in clopidogrel group (difference of >173 d, p=0.021); for CATE or cardiac death was 128 d in the aspirin group and 346 d in the clopidogrel group (difference of 218 d, p=0.028); and for all cause endpoint was 116 d for the aspirin group and 248 d in the clopidogrel group (difference of 132 d, p=0.051). For the total study period, median survival time for CATE was 192 d in the aspirin group and 443 d in the clopidogrel group (difference of 251 d, p=0.019); for CATE or cardiac death was 128 d in the aspirin group and 346 d in the clopidogrel group (difference of 218 d, p=0.025); and for all cause endpoint was 116 d for the aspirin group and 248 d in the clopidogrel group (difference of 132 d, p=0.046). The majority of cats tolerated the study drugs although there was 1 cat from each study group that had a perceived adverse drug event necessitating removal from the study.

Clopidogrel was associated with a significant improvement in survival compared to aspirin for the 1 year and total study period. This is the first prospective thromboprophylactic trial in CATE to demonstrate a significant improvement in survival, thereby providing objective clinical guidance for antithrombotic therapy. This study can also serve as the basis for future trials evaluating additional antithrombotic drugs.

EFFECT OF ATENOLOL ON QUALITY OF LIFE, ACTIVITY AND CARDIAC BIOMARKERS IN CATS WITH APPARENTLY ASYMPTOMATIC HYPERTROPHIC CARDIOMYOPATHY. AE Coleman[2], TC DeFrancesco[1], BDX Lascelles[1], DJ Kleisch[1], J Sauer[3], EH Griffith[1], CE Atkins[1], BW Keene[1]. 1. North Carolina State University College of Veterinary Medicine, Raleigh, NC. 2. University of Georgia College of Veterinary Medicine, Athens, GA. 3. Cummings School of Veterinary Medicine at Tufts University, North Grafton, MA.

Apparently asymptomatic cats with an auscultable cardiac abnormality were screened for hypertrophic cardiomyopathy (HCM) via echocardiography at NCSU. Normotensive, euthyroid cats with concentric left ventricular hypertrophy and left atrial dilatation (LA:Ao $\geq$ 1.5) were included in a prospective, double-blinded, placebo-controlled clinical study. At baseline and following 6 months of treatment with either placebo (PL) or atenolol (6.25 mg/cat PO q12 hr), physical examination, blood pressure, echocardiography, NTproBNP and cTnI were performed, owners completed a questionnaire assessing quality of life (QOL), and activity levels were assessed via accelerometry. QOL and activity of 30 age-, body weight-, "life style"-, and gender-matched normal control cats were also evaluated.

32 HCM cats completed the study (n=16, AT; n=16, PL). 7 additional cats were withdrawn due to medication non-compliance (n=4), change in environment between visits (n=1) and development of illness (n=3). Baseline characteristics were not different between the two treatment groups. Mean ± SE atenolol dose was 1.12 ± 0.22 mg/kg q 12h. Compared to baseline, treatment with AT decreased heart rate ($P < 0.001$) and rate-pressure-product ($P = 0.015$). Overall QOL score was higher in the AT group at 6 months vs. baseline ($P = 0.034$), but not in the PL group ($P = 0.15$). No difference was identified in overall and peak activity between normal and HCM cats pre-treatment. Post-treatment peak activity trended lower in the PL group ($P = 0.09$). A treatment effect was not identified for any of the other tested variables including overall activity, biomarkers and echocardiography.

PROSPECTIVE EVALUATION OF NT-PROBNP, HIGH SENSITIVITY TROPONIN I AND PDK4 FOR THE DETECTION OF OCCULT DCM IN 225 DOBERMAN PINSCHERS

S Gordon[1], L Braz-Ruivo[2], L Drourr[3], A Estrada[4], K Meurs[5], N Morris[6], M O'Grady[7], M Oyama[8].
1. College of Veterinary Medicine and Biomedical Science, Texas A&M University, College Station, TX, 2. Dog and Cats Veterinary Referral and Emergency, Bowie, MD, 3. Sausalito, CA, 4. College of Veterinary Medicine, University of Florida, Gainesville, FL, 5. College of Veterinary Medicine, North Carolina State University, Raleigh, NC, 6. Mass Veterinary Cardiology Services Inc West Springfield MA, 7. Wasaga Beach, ONT, Canada, 8. Matthew J. Ryan Veterinary Hospital of the University of Pennsylvania School of Veterinary Medicine, University of Pennsylvania, Philadelphia.

The use of NTproBNP, Troponin I and PDK4 screening for occult DCM in the Doberman Pinscher has been previously reported. The objective of this study was to screen a large population of asymptomatic/apparently healthy Dobermans (N=225) at the Doberman National Specialty show in 2012 to evaluate the clinical utility of blood based screening using a variety of current tests for the detection of occult DCM. Eligible dogs were > 3 years of age (median age = 6). Screening included auscultation, a 3 min ECG, an echocardiogram and blood testing for NTproBNP, high sensitivity Troponin I and PDK4. NTproBNP plasma samples were tested on a commercially available NTproBNP ELISA (Cardiopet® proBNP Test) and an experimental 2nd generation NTproBNP ELISA.

Dogs were classified as normal (NL) if their LVIDd and LVIDs were < the PROTECT Study entry criteria and they had no VPCs. Dogs were considered affected (DCM) if their LVIDs was ≥ the PROTECT entry criteria with or without the presence of VPCs. ROC analysis comparing the 2 groups (NL and DCM) was done for the Cardiopet® proBNP Test, the experimental 2nd generation NTproBNP ELISA and Troponin I. The AUC was 0.964, 0.97 and 0.948 respectively with the percentage correctly classified equal to 86.6, 91.6 and 90.8 when a cut-off of 638 pmol/l, 689 pmol/l and 0.207 ng/ml were used respectively.

SUDDEN DEATH IN A FAMILY OF ENGLISH SPRINGER SPANIELS WITH PROLONGED QT INTERVALS AND A SUBSTITUTION MUTATION IN THE KCNQ1 GENE AFFECTING CARDIAC REPOLARIZING POTASSIUM CURRENT (IKs). WA Ware[1], Y Reina-Doreste[2], J Stern[2], KM Meurs[2]. 1. Iowa State University College of Veterinary Medicine, Ames, IA. 2. North Carolina State University College of Veterinary Medicine, Raleigh, NC.

Inherited Long QT syndrome (LQTS) is a cardiac ion channelopathy commonly recognized in people and associated with syncope and sudden death, often in the young. Inherited LQTS has not been previously identified in dogs. Four young dogs whose 3 littermates and dam had died suddenly during periods of exercise, and 10 extended family members, were evaluated. Clinical exam, electrocardiogram (ECG), pedigree information, and DNA samples were obtained from all. Using PCR-sequencing, the exonic, untranslated and splice site regions of the canine KCNQ1 gene were evaluated for a sequence change between affected dogs, published normal canine sequence, and controls (110 dogs of 21 breeds) .

Clinical exams were normal. Three of the 4 young dogs had QT prolongation and prominent biphasic T waves on ECG; none of the others dogs did. DNA sequencing of the KCNQ1 gene revealed a heterozygous single base pair change in codon 1 of exon 8 in all affected dogs and none of the unaffected or control dogs. The observed change altered a highly-conserved amino acid from threonine (T) to lysine (K) at position 377 in the protein transcript, within the protein's cytoplasmic region. No other mutations were identified. The mutation is predicted to have a damaging effect on the protein produced and to alter the structure of the protein's intracytoplasmic region.

One affected dog has since died suddenly. We conclude that a missense mutation in the canine KCNQ1 gene in this family of English Springer Spaniel dogs leads to QT prolongation and sudden cardiac death.

THE FIRE WITHIN: THE ROLE OF INFLAMMATION IN CARDIOVASCULAR HEALTH AND DISEASE

Suzanne M. Cunningham, DVM, DACVIM (Cardiology)
North Grafton, MA

INTRODUCTION

The molecular pathways underlying inflammatory signaling are complex, and the same pathways may exhibit both beneficial and deleterious effects on cardiac function under different circumstances. (**Figure 1**) Controlled activation of the inflammatory response plays a role in normal cardiovascular homeostasis and likely exerts salutary effects on cardiac healing and repair following an acute cardiovascular insult. However, similar to neurohumoral activation, sustained or excessive inflammatory activation is maladaptive and contributes to a vicious cycle of adverse ventricular remodeling and cardiac dysfunction.[1] There is an ever-growing body of evidence from both human and experimental animal models to suggest that activation of the innate immune response plays an important role in the etiopathogenesis of ventricular dysfunction and congestive heart failure (CHF).[2]

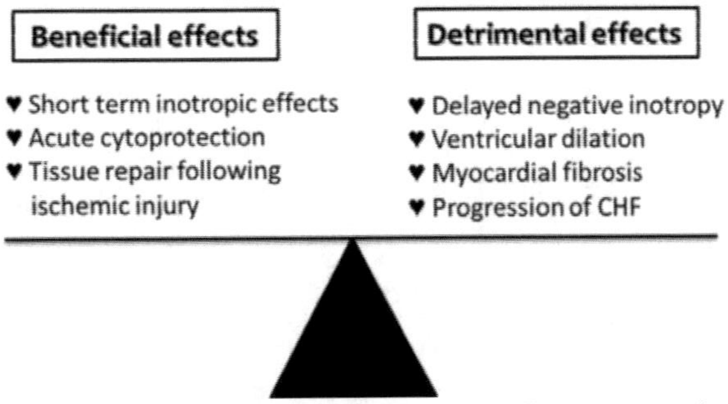

Figure 1: The dual nature of cardiac inflammatory signaling.
Modified from: Vallejo, et al. Progress Inflamm Res 2010.

CYTOKINES: FRIENDS OR FOES?

Cytokines are secreted both locally in the heart and systemically by leukocytes in a wide range of cardiac conditions, including CHF, cardiomyopathy, sepsis-induced myocardial dysfunction, and myocarditis. Cytokines exert direct effects on the myocardium and also influence the composition of the extracellular matrix. Whether those effects are beneficial or deleterious depends on a complex interplay of physiologic factors, including the level and chronicity of cytokine release, the presence of co-expressed cytokines and neurohormones, and the redox and metabolic state of the myocardium. Pro-inflammatory cytokines are thought to play a major role in the progression of CHF via deleterious effects exerted on the heart, vasculature, and peripheral tissues.[3] In people, increased cytokine levels are associated with an increased risk of developing CHF,[4] and large-scale human studies have demonstrated increasing cytokine levels in proportion to CHF severity, with increased tumor necrosis factor (TNF)-α and interleukin (IL)-6 levels serving as powerful independent predictors of mortality.[5] These findings naturally prompted consideration of the therapeutic potential of TNF-antagonists in human patients with CHF.

Paradoxically, though small preliminary trials of TNF-antagonists in people with CHF demonstrated dose-dependent improvements in left ventricular (LV) ejection fraction and quality of life, larger scale placebo-controlled clinical trials showed either neutral or adverse effects of these agents in people with moderate to severe heart failure. Though the reasons for these unfavorable outcomes are not entirely clear, these contradictory findings demonstrate the double-edged nature of the effects of inflammatory signaling within the heart. Similar to the effects of adrenergic stimulation, TNF-α has been shown to elicit a bimodal inotropic response in vivo, with conscious dogs exhibiting an immediate increase in contractile function during TNF-α infusion that is subsequently followed by a delayed and profound reduction in contractile function and the development of cardiac dilation that persists well beyond cessation of TNF administration. A better understanding of the complex molecular pathways involved in cardiac inflammatory signaling might allow for more specific targeting of the inflammatory response and improved selection of the patients most likely to benefit from anti-inflammatory therapies.

MECHANISMS OF INFLAMMATORY ACTIVATION IN HEART DISEASE

Several theories have been proposed to explain the immune activation that accompanies chronic CHF, (**Figure 2**) and the inflammatory triggers in this setting are likely multifactorial and synergistic. There is increasing evidence that toll-like receptors (TLR) play a major role in initiating and potentiating the local production of cytokines within the heart. Toll-like receptors are a class of highly conserved pattern recognition receptors that serve as a first line of defense against infectious pathogens and play a pivotal role in inciting the innate immune response and elaborating pro-inflammatory cytokines. Though TLR have been best studied in cells of the immune system, they are also widely expressed by vascular endothelial cells and cardiomyocytes and there is growing interest in the potential link between TLR-induced activation of inflammatory mediators and the pathogenesis of pathologic LV remodeling and CHF.[3,6]

TLR pathways are activated in response to either conserved pathogen-associated molecular patterns (PAMP), such as bacterial endotoxin; or danger-associated molecular patterns (DAMP) released during host cellular injury or death. Increased circulating levels of endotoxin and monocyte-derived TNF-α are found in people with decompensated CHF and have been shown to decline with resolution of congestive signs.[7] Though the source of endotoxemia in heart failure is uncertain, it has been proposed that reduced intestinal perfusion, along with acute venous congestion and bowel wall edema could result in altered intestinal permeability to endotoxin. Studies have demonstrated morphological and functional alterations of the gut in people with CHF along with increased circulating endotoxin; and positive correlations between increased colonic wall thickness and elevations in circulating levels of C-reactive protein (CRP) and peripheral leukocyte counts have also been identified.[7]

In the heart failure setting, TLR-signaling and the resultant local and systemic inflammatory response may be activated by gut-derived endotoxemia and/or release of DAMPs from local tissue injury. In people with CHF, both monocyte and myocardial expression of TLR-4 are up-regulated, and increasing TLR-4 levels are correlated with worsening LV systolic dysfunction and CHF severity.[8] Expression of TLR-2 and TLR-4 have been documented in the myocardium of normal dogs, however, to the author's knowledge, no studies to date have evaluated TLR expression or endotoxin activity in dogs with spontaneous cardiac disease.

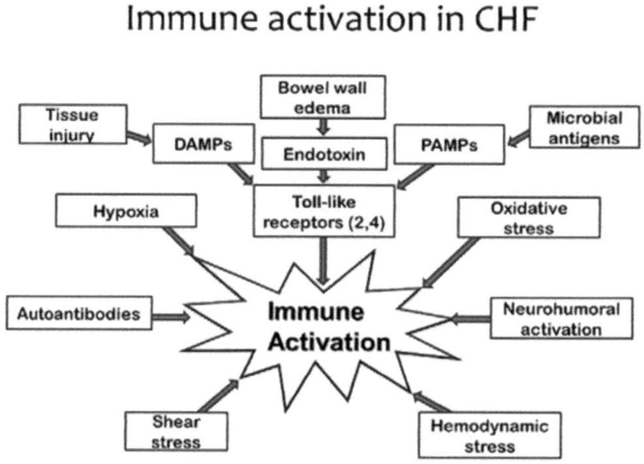

Figure 2: Factors proposed to play a role in the chronic immune activation associated with congestive heart failure (CHF). *Modified from: Yndestad A, et al. Curr Cardiol Rep 2007.*

DOES SYSTEMIC INFLAMMATION PLAY A ROLE IN VETERINARY PATIENTS WITH HEART DISEASE?

Inflammation in dogs with DMVD and CHF

Though studies of the inflammatory response in animals with naturally-occurring cardiac disease remain sparse, limited studies have identified evidence of systemic inflammation in dogs and cats with heart disease or heart failure. C-reactive protein is an acute phase protein that is both a marker and mediator of the systemic inflammatory response, serving to activate monocytes and stimulate the production of pro-inflammatory cytokines. Increased CRP levels are associated with progression of CHF from various etiologies in people and are independently associated with increased risk of cardiovascular death.[9] Studies evaluating CRP in dogs with cardiac disease have yielded somewhat divergent results. Elevated CRP concentrations have been demonstrated in dogs with pre-clinical DMVD relative to healthy controls,[10] and more recently in dogs with moderate CHF secondary to DMVD or DCM.[11] In the latter study, increasing CRP levels were associated with worsening CHF severity and impaired endothelial function, implying a role of systemic inflammation in the progression of heart failure in the dog. Two other studies evaluating CRP in dogs with DMVD did not show increased CRP concentrations in dogs with varying degrees of mitral regurgitation; however the study by Ljungvall, et al.[12] identified significant associations between concentrations of circulating CRP and cardiac troponin-I, suggesting a possible link between systemic inflammation and myocardial injury.

Monocyte chemoattractant protein-1 (MCP-1) is another emerging biomarker of systemic inflammation with important cardiovascular implications.[13] MCP-1 is a chemokine produced by a variety of cell types that stimulates monocyte mobilization and honing to sites of inflammation. In animal models, increased myocardial expression of MCP-1 has been shown to promote adverse ventricular remodeling and the development of CHF. Increased plasma levels of MCP-1 have been documented in people with heart failure, and MCP-1 levels are correlated with systolic dysfunction and increased CHF severity. (Wrigley, 2011) Increased MCP-1 levels have also recently been documented in dogs with DMVD,[14] however the initiating factors and role of MCP-1 in animals with cardiac disease remains incompletely characterized.

Immune activation in animals with primary cardiomyopathy

Though increased TNF-α concentrations have been reliably demonstrated in people with CHF, limited studies of circulating TNF-α levels in dogs and cats with heart disease have yielded inconsistent results. Increased TNF-α concentrations have been identified in some cats with CHF due to cardiomyopathy, especially in those with right-sided CHF.[15] The few studies evaluating TNF-α using various methodologies in dogs with CHF due to various etiologies have not consistently demonstrated increased circulating TNF-α concentrations in these patients. However, in one study of dogs with CHF due to dilated cardiomyopathy TNF-α concentrations were shown to correlate positively with the severity of cardiac cachexia, and fish oil supplementation was associated with improved cardiac cachexia and a reduction in circulating IL-1 concentration.[16]

In human patients with arrhythmogenic right ventricular cardiomyopathy (ARVC), recent studies have identified increased circulating concentrations of CRP and various pro-inflammatory cytokines (TNF-α, IL-1β, IL-6, IL-6R, IL-8, MCP-1, macrophage inflammatory protein 1β).[17-19] In one study, CRP levels were found to be highest in the first 24 hours following presentation for ventricular tachycardia, raising the question of whether the systemic inflammation was the antecedent or the aftereffect of malignant arrhythmia.[17] Inflammatory infiltrates are commonly found in the myocardium of people and animals with ARVC, and recent postmortem studies of human heart samples strongly support the hypothesis that inflammation plays a key role in modulating the severity of the disease.[20,21] In the latter studies the presence of multifocal inflammatory infiltrates was correlated with increased phenotypic severity and biventricular involvement in people with ARVC. Similarly, one study examining the clinicopathologic findings of ARVC in Boxers identified myocarditis affecting one or both ventricles in the majority of affected dogs and none of the controls.[22] Moreover, in this study myocarditis was associated with the more advanced fibrofatty form of the disease and was conspicuously present in all dogs that exhibited sudden death, suggesting a possible role of myocardial inflammation in modulating the electrical instability and malignant arrhythmias that characterize this disease. Further investigation into the role that inflammation plays in the pathophysiology of ARVC in the Boxer is warranted and currently ongoing in the author's clinical laboratory.

CONCLUSIONS AND FUTURE DIRECTIONS

The exact pathophysiologic mechanisms involved in the progression of myocardial disease, DMVD, and CHF remain elusive and treatment options for our patients remain limited at this time. Further research to elucidate the factors contributing to the pathogenesis and progression of these conditions could pave the way for new therapeutic interventions to improve the quality of life and survival of animals with cardiac disease. Mitigating the release or toxic effects of inflammatory mediators on the cardiovascular system may represent a viable treatment option for heart disease, in much the same way that neurohumoral antagonists are used successfully to block the harmful effects of chronic adrenergic and renin-angiotensin-system activation.

Despite the cautionary tale of prior clinical trials of specific TNF-antagonists in human heart failure patients, there is renewed enthusiasm for developing novel pharmacologic agents to target the inflammatory response in people with cardiovascular disease. HMG-CoA reductase inhibitors, or statins, have myriad cardioprotective and anti-inflammatory effects that are independent of their lipid-lowering properties.[23] Statins reduce *in vitro* monocyte TLR-4 expression[8] and cytokine concentrations in human CHF patients. They have been shown to improve cardiac and endothelial function,[23] and some studies in people have demonstrated improved CHF symptoms and survival.[24] Statins may also hold promise as novel therapeutic agents for the treatment of cardiac disease in veterinary patients. Statins have been shown to be well-tolerated in experimental canine models, and a preliminary study by the author demonstrated good tolerability of atorvastatin and a reduction in leukocyte counts in dogs with CHF.[25]

A number of specific antagonists of the TLR pathway are also under development.[6] The TLR-4 antagonist eritoran has been shown to reduce inflammatory cytokine expression and improve cardiac function in animal models of sepsis.[26] It is currently undergoing evaluation in Phase III clinical trials of sepsis in people[6] and could have additional utility for the treatment of CHF. There is also great interest in developing pharmacologic inhibitors of Myeloid differentiation factor 88 (MyD88), an adaptor protein that coordinates the downstream inflammatory response to TLR signaling. MyD88 inhibition has been shown to attenuate ventricular remodeling in animal models[27] and may represent another potential novel therapy to attenuate cardiac hypertrophy or ameliorate signs of CHF.

REFERENCES

[1]Coggins M, et al. **Circ Res** 2012; 110(1):116. [2]Mann DL, et al. **Circ Res** 2011; 108(9):1133. [3]Nakagomi A, et al **J Card Fail** 2010; 16(7):562. [4]Kalogeropoulos A, et al. **J Am Coll Cardiol** 2010; 55(19):2129. [5]Miettinen KH,et al. **Eur J Heart Fail** 2008; 10(4):396. [6]Topkara VK, et al. **J Mol Cell Cardiol** 2011; 51(4):594. [7]Sandek A, et al. **Int J Cardiol** 2010; 157(1):80. [8]Foldes G, et al. Int J Cardiol 2008; 124(1):80. [9]Araujo JP, et al. J Card Fail 2009; 15(3):256. [10]Rush JE, et al. **J Vet Intern Med** 2006; 20(3):635. [11]Cunningham SM, et al. **J Vet Intern Med** 2012; 26(3):547. [12]Ljungvall I, et al. **J Vet Intern Med** 2010; 24(1):153. [13]Wrigley B, et al. **Eur J Heart Fail** 2011; 13(11):1161. [14]Zois NE, et al. **Vet J** 2012; 192(1):106. [15]Meurs KM, et al. **Am J Vet Res** 2002; 63(5):640. [16]Freeman LM, et al. **J Vet Intern Med** 1998; 12(6):440. [17]Bonny A, et al. Cardiol Res Pract 2010; doi:pii: 919783. [18]Campian ME, et al. **Eur J Nucl Med Mol Imaging** 2010; 37(11):2079. [19]Asimaki A, et al. **Circ Arrhythm Electrophysiol** 2011; 4(5):743. [20]Tavora F, et al. **Hum Pathol** 2012; 43(4):592. [21]Campuzano O, et al. **J Clin Pathol** 2012; 65(12):1077. [22]Basso C, et al. Circulation 2004; 109(9):1180. [23]Zhou Q, et al. **Circ J** 2010; 74(5):818. [24]Lipinski MJ, et al. **Am J Cardiol** 2009; 104(12):1708. [25]Cunningham SM, et al. **J Vet Intern Med** 2008; 22(3):687. [26]Ehrentraut H, et al. **Eur J Heart Fail** 2011; 13(6):602. [27]Van Tassell BW, et al. **J Cardiovasc Pharmacol** 2010; 55(4):385.

TRANSCATHETER VALVE REPLACEMENT: THE HUMAN EXPERIENCE AND THE VETERINARY HORIZON

George A. Kramer, DVM, DACVIM (Cardiology)
Bohemia, NY

INTRODUCTION

Valvular heart disease in dogs (acquired and congenital) is a significant cause of morbidity and death. Approximately 7 million dogs in the United States are estimated to have cardiac disease; the majority of those cases are due to valvular disease. A significant portion of those animals have severe enough regurgitation to develop clinical signs and suffer chronic progression of the disease. As in humans with chronic valvular disease, despite years of clinical experience and clinical drug trials, there is no definitive medical treatment for the disease that will prevent the progression to CHF. As a result, surgery is the gold standard of treatment in people. The AHA/ACC and ESC guidelines for treatment of valvular heart disease state that mitral valve repair is the treatment of choice for patients with symptomatic severe MR. (1,2) However, many patients (estimated to be 50%) in USA and Europe do not undergo surgical repair for a variety of reasons. Over the last 10 years, interventional transcatheter valve devices and procedures have been developed to help address the unmet surgical need of those patients. The transcatheter devices tend to be constructed of some combination of metal (usually nitinol or stainless steel), various polymers and biologic materials. They are designed to modify or repair the defective valve or insert a new valve within the defective native valve.

Surgical repair in dogs has also shown to be effective in reducing or eliminating regurgitations. (3,4) However, surgical replacement or repair of cardiac valve in dogs is rarely done due to the high cost, small number of existing surgical programs and technical barriers that inhibit new surgical valve programs from becoming established. The development of successful interventional transcatheter or transapical therapies for the treatment of valvular diseases in dogs could dramatically improve our success at treating these patients.

THE HUMAN EXPERIENCE

The human experience began with animal models. The potential for interventional transcatheter valves was first demonstrated by Andersen et al. in 1992, who implanted a porcine bioprosthesis attached to a balloon expanding stent at various aortic positions into closed chest pigs.(5) In 2000 Bonhoeffer et al successfully placed bovine jugular vein valves mounted on balloon expanding stents within the pulmonic outflow tract in lambs.(6) That same year the first human percutaneous valve replacement was performed by Bonhoeffer, where a transcatheter valve made from bovine jugular vein mounted on a platinum-iridium stent was successfully placed in a stenotic right ventricle to pulmonary artery conduit.(7) This prototype was eventually developed into the Melody Transcatheter Pulmonic Valve (Medtronic) which was the first transcatheter valve device that was approved for use by the FDA. Most of the R&D work over the last 10 years has been on developing devices for the aortic and mitral valves. To date, very little attention has been paid to the tricuspid valve.

Pulmonic Valve

The Melody® Transcatheter Pulmonary Valve is made by sewing a glutaraldehyde fixed bovine jugular vein valve into a balloon expanding stent. It is approved for the use in failed RVOT conduits that had been previously placed surgically. The valve comes in sizes 18, 20, and 22mm diameters. The valve is crimped onto a balloon-in-balloon catheter delivery system and deployed into the RVOT conduit via the transfemoral venous route. In some cases, prior to implantation of the valve a nitinol stent is placed in the RVOT to act as a "landing zone" for the valve device. The valve has also been placed successfully within failed surgically implanted bioprosthetic valves in the mitral, aortic and tricuspid positions (off-label use). (8)

Aortic Valve

Sapien Valve (Edwards) and CoreValve (Medtronic) are the two commercially available devices for transcatheter aortic valve implantation (TAVI) for cases of severe aortic stenosis or severe aortic insufficiency. Over 50,000 TAVI procedures have been performed worldwide. The Sapien valve is a bovine pericardial valve fixed onto a balloon-expandable stainless steel frame, whereas the CoreValve is made from porcine pericardium and is fixed onto a self-expanding nitinol frame. Both valves can be placed via a transcatheter femoral approach. In cases where the femoral approach is not an option, the Sapien valve is usually place transapically. It has also been placed via a transaortic approach through a mini thoracotomy allowing access to the aortic arch. The CoreValve has a smaller delivery system (18 Fr) compared to the Sapien valve (24 Fr) and has been able to be placed via the subclavian approach. Both valves have good procedural results but serious complications can occur including; death, bleeding, complete AV block and stroke. At 30 days, there was no difference in the rate of all-cause mortality, cardiovascular mortality, MI, stroke, or device success, as well as no major difference in the rate of major vascular complications or life-threatening bleeding events. The CoreValve was associated with a higher rate of permanent pacemaker implantation, with 22.5% of patients requiring a pacemaker compared with 5.9% of patients treated with the Sapien valve. (9, 10, 11)

Mitral Valve

Device development for mitral interventions have focused on mitral valve repair (MVRe) or mitral valve replacement (MVR). At this point in time the only approved mitral device is the MitraClip (Abbott) which is an MVRe device. The MitraClip works as a leaflet plication device by creating a double inlet mitral valve similar to the Alfieri suture technique that is performed surgically. The device is advanced into the left atrium via a femoral vein approach and atrial septal puncture. The device is maneuvered under TEE guidance to stabilize and grab the anterior and posterior leaflets of the valve and then clip them together reducing the regurgitant flow. In some cases a second clip may be placed if there is still significant MR seen on TEE. This device is used in both degenerative MR cases and functional MR (i.e. caused by annular dilation secondary to ischemic heart disease or DCM) cases. In the Everest II Trial, the MitraClip was shown to be non-inferior to surgical repair of 3+ or 4+ MR in high-risk patients. The functional MR patients had less residual MR than the degenerative MR patients did when compared to the surgically treated patients. (12, 13, 14)

Other MVRe devices that are in development include ones that fall into certain categories: leaflet plication (e.g. MitraFlex), leaflet ablation (e.g. Thermocool), leaflet coaptation (e.g. Percu-Pro), indirect annuloplasty (e.g. Monarc, Carillon, and Viacor), direct annuloplasty (e.g. Mitralign, Accucinch), chordal implants (e.g. NeoChord, MitraFlex, and Babic), and LV remodeling (e.g. Mardil-BACE, Myocor). These devices and others are in various stages of development and many undoubtedly will never make it to market. Ultimately the most successful MVRe method may utilize a combination of repair techniques (i.e. annuloplasty, leaflet placation and chordal implants) to more closely mimic current open-heart surgical repair techniques.

The other mitral valve approach is the MVR, or valve-in-valve technique. There are several different designs in development including the Tiara (Neovasc, Inc), CardiAQ (CardiAQ Valve Technologies, Inc), Endovalve-Herrmann prosthesis (Endovalve, Inc) and Lutter Valve. These devices utilize different stent designs and bioprosthetic valves and are delivered either transapically or transseptally. The Tiara and the CardiAQ valves have both recently reached the first-in-man level of development. The other valves are still in the animal model stage of development.

Tricuspid Valve

Currently there are no approved devices for tricuspid valve repair or replacement. The MitraClip has been used off label in Europe in 5 cases successfully reducing TR flow (unpublished results). The Melody valve and the Sapien valve have been implanted (off label) into failing bioprosthetic valves that had previously surgically replaced the native tricuspid valves. There has also been a stented valve prototype that was placed in the vena cava in a human patient on compassionate grounds to treat (successfully) severe flow reversal caused by a TR jet and pulmonary hypertension.

THE VETERINARY HORIZON

Mitral Valve

The MitralSeal (Avalon Medical) is being developed in collaboration with Colorado State University and Dr. Chris Orton (co-inventor). It is the first interventional MVR device specifically designed for veterinary patients, but is not yet commercially available. It is a self-expanding bioprosthesis with an atrial cuff and a ventricular tether anchoring system that is deployed via a transapical approach. The device is a hybrid surgical-transcatheter system that utilizes collaboration between a veterinary surgeon and cardiologist to place the device. The valve portion of the device is made from porcine pericardium and is mounted on a nitinol stent. It has been placed experimentally in some clinical cases but is still in the animal trial stage of development.

The animals are heparinized prior to surgery. A small thoracotomy is performed to isolate and stabilize the left ventricular apex. A 20 Fr deployment catheter is introduced in the left ventricle through the apex and passed across the mitral annulus into the left atrium along a guidewire under fluoroscopic and TEE guidance. The device is deployed into the left atrium and pulled into position at the mitral annulus by pulling on tethers that exit the ventricular apex. Once proper positioning is confirmed by TEE the tethers are fixed in position and the chest is closed. The initial results have demonstrated good valve function resulting in marked reduction of the MR jet. (15)

Tricuspid Valve

The Tucker valve (Ultravet Medical Devices) is being developed by the author for placement in the tricuspid position in cases of severe tricuspid regurgitation (e.g. tricuspid dysplasia). The device is a leaflet coaptation device that can be placed through an 8 Fr delivery sheath. The device is anchored in the RV apex through active fixation. A small coaptation balloon is inflated within the RV cavity and extends across the tricuspid annulus. The valve is still in the prototype stage and has been placed in clinical cases for a proof of concept study. Results of that study have been submitted as an abstract presentation at this meeting. The small profile of the valve makes it feasible for use in the mitral position via a trans-septal approach.

Future Development

These first two veterinary devices need to be further tested and refined. Multiple iterations may be needed before they are ready for commercial use. Additionally, other devices should be developed in the future for veterinary use to treat some of the conditions we commonly see in veterinary cardiology. For example, a chordal implant device could be effective in correcting a flail leaflet in degenerative valve cases, a stented valve conduit might be effective in decreasing the transvalvular pressure gradient in severe subaortic stenosis, a similar device may turn out to be superior to balloon valvuloplasty in cases of severe pulmonic stenosis, a window device could be place across the atrial septum to reduce left atrial pressure and decrease pulmonary capillary wedge pressure in cases of severe chronic CHF, a percutaneous ventricular remodeling device could reduce the size of the left ventricle in cases of DCM, or a small intravascular ventricular assist device could improve cardiac output in DCM cases. We have plenty of unmet clinical needs in the veterinary space that could be addressed and treated by veterinary specific device development programs. If the human experience can act as a guidepost, we must realize that interventional device development takes time and one must expect failures along the way and learn from them. If we do that I think the veterinary horizon is bright.

References

Nishimura RA, et al. **J Am Coll Cardiol**. 2008;52(8):676-685. [2] **Eur Heart J** 2012 & **Eur J Cardio-Thoracic Surg** 2012. [3]Griffiths LG, et al. **J Am Vet Med Assoc** 2004;224:1941–1945. [4] Masami U. et al. (2012) **J Am Vet Med Assoc 240**:10, 1194-1201. [5] Andersen HR, et.al. **Eur Heart J** 1992;13:704–8. [6] Bonhoeffer P, et al. **Circ** 2000;102:813– 6. [7] Bonhoeffer P, et al. **Lancet** 2000;356:1403–5. [8]Hasan BS, **Circ Cardiovasc Interv.** 2011 Dec 1;4(6):615-20. [9] Grube E, et al. **J Am Coll Cardiol** 50 2007 69-76. [10] Moynagh AM, et al. **J Am Coll Cardiol**. 2011;57(5):634-635. Smith CR, et al **N Engl J Med** 2011; 364:2187-2198. [2]Ussia GP, **Inter J Cardio** 2012: 155(2) 194-200. [3] Feldman, T., et al. **ACC late breaking clinical trial** (2010). [4]Thomas M, et al.**Circulation**. 2011; 124: 425-433. [5]Orton C, ACVS Veterinary Symposium 2011:311-312

TOP TEN TREATMENT TIPS FOR FELINE HEART DISEASE: FEEDING AND PHARMACOLOGY

Lisa M. Freeman, DVM, PhD, DACVN and John E. Rush, DVM, MS, DACVIM (Cardiology), DACVECC
North Grafton, MA

Despite the increasingly common diagnosis of cardiac disease in cats, managing cardiac disease in this species can be particularly challenging. Successful management requires careful selection of treatment strategies, regular monitoring, and clear client communication. The authors have identified a number of nutritional and pharmacologic tips that can be helpful in the care of feline cardiac patients.

TOP TEN TIPS

1. **Body condition AND muscle condition scores should be determined for every patient with cardiac disease at every visit.** Body weight and body condition score, which primarily assesses body fat, are important because they have been associated with overall survival. Cats that are underweight and cats that are significantly overweight have the shortest survival while cats in moderate body condition (4-6/9) had the longest survival (Finn et al, 2010). However, it also is important to assess muscle condition at every visit, particularly in cats with cardiac disease. The muscle condition score is a subjective score which assesses muscle condition in the areas of the epaxial, gluteal, scapula, and temporal areas and graded as normal muscle condition or mild, moderate, or severe muscle loss. Because of increased production of inflammatory cytokines in congestive heart failure (CHF), muscle is lost preferentially to fat. Therefore, even in overweight cats, significant muscle wasting can be present. This muscle loss has important clinical implications as it negatively impacts strength, immune function, wound healing, and survival. By assessing muscle condition at every visit, muscle loss can be identified at its early stages when intervention is more likely to be successful. Optimizing diet (e.g., ensuring nutritionally balanced diet with adequate calorie and protein intake) and omega-3 fatty acid supplementation can help to reduce muscle loss.
2. **Pick the right dose of furosemide** – Furosemide is indicated once clear cut congestive signs have developed – pulmonary edema, pleural effusion or ascites that is cardiogenic in origin. It has been documented that furosemide will activate undesirable neuroendocrine responses if administered to animals prior to the onset of CHF, and most cardiologists would avoid use of furosemide as the sole drug for chronic treatment of CHF. After the onset of CHF, almost all cats will need ongoing furosemide to prevent recurrence of signs of CHF. It can be difficult to define the exact dose of diuretic required by any individual cat with CHF, although the desired dose will clear significant edema accumulations and avoid undesirable side effects such as electrolyte disturbance, dehydration and the development of pre-renal azotemia. The combined use of angiotensin-converting enzyme (ACE) inhibitors and diuretics compromises one of the kidneys' normal compensatory mechanisms (vasoconstriction of the efferent arteriole) and can lead to elevation of BUN and creatinine when 1) excessive diuretic dose is initiated or 2) significant pre-existing renal disease is present. Still, most cardiologists now concurrently use ACE inhibitors and diuretics for management of CHF, thus caution is advised to avoid excessive doses of furosemide. We usually try to use the lowest possible dose of furosemide which with control signs of CHF, and since this dose will change over time we give the owner upper and lower limits for acceptable furosemide dose, and carefully explaining to them that they should "give more for difficulty breathing or rapid respirations (> 35 breaths per minutes when not purring), and give less if the animal seems weak, lethargic, anorexic, or depressed". In cats with relatively mild CHF, a dose of 6.25 mg/cat q 24 to 48 hrs may be adequate. Cats with ascites or significant volume of pleural effusion often require more furosemide, especially those with enough pleural effusion to require thoracentesis, and a starting dose of at least 6.25 mg/cat q 12 hours is usually needed. When this dose does not control signs of CHF then diuretic resistance has likely developed and drugs such as spironolactone, pimobendan, torsemide, or eplerenone may be used in addition to the furosemide. Alternatively, going to a higher dose of ACE inhibitor, reviewing dietary sodium intake, or taking away the beta-blocker can be alternative to going to higher doses of furosemide. Mild azotemia (e.g., BUN 30-39 mg/dl, creatinine 2.0-2.4 mg/dl) commonly develops and might not necessarily be a trigger for diuretic dose reduction if the cat seems normally hydrated, is eating well, and has normal total solids. BUN concentrations > 50 mg/dl and creatinine > 3.0 mg/dl are much less likely to be tolerated on a long term basis and efforts should be made to alter cardiac medications and/or diet and see whether azotemia can be reduced.
3. **Don't make major dietary changes while the cat is hospitalized.** Most cats with acute CHF will not eat but should begin eating when the CHF is better controlled. While hospitalized, avoid diets that are high in sodium but don't make major dietary changes during hospitalization. Once the cat is home, feeling better, and stabilized on medications, a gradual change to a new and more optimal diet can be made. Forced dietary changes when the cat is sick can induce food aversions. Knowing what the cat eats at home can help to know what food preferences will be (e.g., canned vs dry, flavor, kibble shape) since some cats have very distinct preferences.

4. **Collect complete diet information at every visit to determine whether the diet is optimized.** A complete diet history includes the main cat food being provided, but also treats, table food, dietary supplements, and foods used for medication administration. There is usually not a single "best" diet for any cat. The authors typically determine several diets that would be appropriate for an individual cat based on the cat's individual clinical situation (e.g., severity of cardiac disease, clinical signs, physical examination findings, laboratory values, and individual preferences of the cat and the owner). These diets are offered as choices for the owner and for the cat. Once they determine which diet the cat prefers, this diet can be used. Be sure to make specific diet recommendations, not just "a low sodium diet" or "a senior diet." Also, even if the cat is discharged with an optimal diet plan, don't assume that the owner is following these recommendations at subsequent visits; ask exactly what the owner is feeding at each visit.

5. **The bigger the left atrium, the more an ACE inhibitor is indicated.** ACE inhibitors are commonly used in the management of CHF in cats. Blocking the action of the renin angiotensin system leads to diminished plasma levels of angiotensin II and reduced stimulation of aldosterone. As a result, fluid retention and vasoconstriction are blunted. The beneficial effects of ACE inhibition likely result from both the vasodilation and the drug's effects to reduce cardiac remodeling. In well-designed human and canine heart failure trials, ACE inhibitors resulted in improved clinical signs and a prolonged survival or the time until additional therapy is needed. There are few well conducted clinical trials in cats with which to form solid guidelines regarding clinical use of ACE inhibitors, although one study indicated that in Maine coon cats with asymptomatic HCM, no benefit was seen on cardiac enlargement or remodeling. Most veterinary cardiologists would agree that ACE inhibitors are probably effective in cats with congestive heart failure. There is also general agreement that ACE inhibitors are useful when systemic hypertension complicates feline heart disease, especially when hypertension persists despite amlodipine. ACE inhibitors are also indicated in cats with asymptomatic heart disease and concurrent renal disease that is associated with even modest proteinuria. Many veterinary cardiologists recommend an ACE inhibitor in cats with cardiomyopathy and moderate to marked left atrial enlargement, even if signs of CHF have not yet developed. Certainly there is no consensus on this topic, but the smaller the left atrium the less likely it is that a cat will benefit from ACE inhibition.

6. **The bigger the left atrium, the more likely you should think about an antithrombotic -** Arterial thromboembolism (ATE) is a common complication of feline heart disease, occurring in 30-40% of cats with cardiomyopathy. The thrombus may develop in either the left ventricle or the left atrium; however, a left atrial origin is most common. Three factors are thought to contribute to thrombus formation: altered coagulability, altered blood flow, and endothelial damage. Cats with cardiomyopathy often have dilatation of the left atrium with blood stasis, they may have endothelial disruption or reduced atrial pump function, and altered coagulation can also been seen in cats with cardiomyopathy. Once ATE develops, treatment is very difficult and often unsuccessful, therefore prevention of this devastating event becomes very important.
   - **Unfractionated heparin** is a very familiar drug to most veterinarians, and this drug is commonly used to prevent further enlargement of a thrombus in cats with active ATE. While several doses of heparin have been proposed in this setting, and the author's opinion is that heparin should be used at high doses (275-350 IU/kg q 6-8 hours) for active ATE. However, the long term use of unfractionated heparin is not usually advisable due to the need for frequent administration and the number of side effects that can develop with chronic use.
   - Low molecular weight heparins (**Dalteparin and Enoxaparin**) have been recently used to prevent thrombus formation in cats at risk of ATE. We have used dalteparin (Fragmin; 160 to 175 U/kg) subcutaneously twice a day. Research has suggested that the high doses are needed to inactivate clotting factors for the entire day, and administration of the drug q 8 hrs may be advisable in cats with active ATE. Another low molecular weight heparin option is enoxaparin, which has been studied at 1 mg/kg subcutaneously q 12 hr. Both dalteparin and enoxaparin can be expensive to use on a long-term basis and these drugs must be given by subcutaneous injections. Yet, many owners prefer injections to oral medications in cats, and the drug is currently the author's first choice for prevention of ATE in cats.
   - **Clopidogrel** is an antiplatelet drug (Plavix) and is the other drug often used to prevent ATE in cats. This drug reduces platelet activation, reduces platelet degranulation, and inhibits modification of glycoprotein IIb/IIIa receptor which leads to reduced aggregation. Clopidogrel seems to be well tolerated in many cats, although foaming at the mouth or vomiting has been seen in some cats, and increased liver enzymes and bilirubin can also be seen infrequently with the drug. The author prefers clopidogrel to aspirin as the latter has been associated with altered renal function, and this may be especially concerning in cats taking furosemide and ACE inhibitors. There is no routine monitoring of clotting times or any other blood tests for these drugs. Clopidogrel

is supplied as 75 mg tablets and the proposed dose (pending further clinical experience) is 2-4 mg/kg/day (¼ of a tablet orally once a day), although in cats less than 3-4 kg the drug should likely be compounded to avoid overdose.

7. **Warn owners of cyclical/variable appetite in cats with CHF.** In cats with CHF, appetite often becomes more variable/cyclical (i.e., the cat will eat one diet well for a few days but then stop eating it. However, they'll often eat a different food for a few days or more). It is helpful to warn the owner about appetite and to provide multiple appropriate diet options for the owner to feed (i.e., they can feed one diet for several days until the cat no longer has interest for that food and then switch to a different food. They'll often be willing to eat the original food again so owners can use a "rotation" of foods to keep the cat eating). In addition to having different food options, other tips to increase appetite include:
   - Warm the food to body temperature.
   - Feed several small frequent meals during the day, rather than just one or two meals.
   - Feed the cat from a different type of dish (e.g., a new cat food dish or a human dinner plate).
   - Try feeding in a different location in the house.
   - Add homemade chicken, beef, or fish broth to the food (even low sodium store-bought broths are too high in sodium).
   - Add a small amount (1-2 teaspoons) of cooked meat (hamburger, chicken, or fish) to the food. Be sure to instruct the owner not to use any prepared foods, such as rotisserie chicken, deli meats, or canned meats or fish due to their high sodium content.
   - Fish oil, which is high in omega-3 fatty acids, reduces inflammatory cytokines and may have modest benefits for appetite. Fish oil (rather than flax or cod liver oil) should be used and a good quality brand should be used since there is little regulation of dietary supplements. The dose used by the authors is enough fish oil to provide 40 mg/kg eicosapentaenoic acid (EPA) and 25 mg/kg docosahexaenoic acid (DHA). For cats, a one gram capsule containing 180 mg EPA and 120 mg DHA can be easily used to provide this dose when given as one capsule per cat per day. Specific brands can be found on the Tufts HeartSmart website (www.tufts.edu/vet/heartsmart)

   One note of caution: If the cat also is acting lethargic, is tachypneic, won't eat *any* foods or if the cat has been eating very well and then suddenly stops eating, this usually indicates the need for reassessment and adjustment of medications.

8. **Be sure to specifically discuss treats** with the owner since many owners give treats (either cat treats or "people food"). Most owners are unaware of treats that would be contraindicated (eg, high sodium). The author typically provides a list of foods that are appropriate and foods to avoid as treats to assist the owner in wise selection. Specifically discuss (and demonstrate, if necessary) appropriate methods for medication administration and if foods are used, be sure to provide appropriate (low sodium) choices. A final part of the diet history is dietary supplements. Ask at each visit if the owner is administering dietary supplements. If so, ensure that the supplements are safe, are not interacting with the diet or medications, and are being administered at an appropriate dose. Fewer cats than dogs with cardiac disease receive dietary supplements (13% of cats vs. 31% of dogs; Torin et al, 2007; Freeman et al, 2003) but addressing this issue with the owner is important as "Internet surfing" for alternative treatments of cardiac disease is common. In addition to safety and efficacy issues, there are significant concerns about the quality control of dietary supplements since they do not require proof of safety, efficacy, or quality control to be marketed. Therefore careful selection of type, dose, and brand is important to avoid toxicities or complete lack of efficacy.

9. **Pimobendan can be used in most cats for management of chronic CHF** – Pimobendan, a calcium sensitizing drug that is useful as a positive inotrope, also has vasodilating properties due to concurrent phosphodiesterase III inhibition. The combined positive inoptropism and vasodilation has been referred to as inodilation. Pimobendan use in cats with CHF has been reported, but the optimal dose and the optimal clinical situation has not yet been established. Pimobendan also seems to be associated with a low side effect profile in cats. The somewhat empiric dose for pimobendan in cats is 0.25 mg/kg q 12 hours, although some authors have suggested a lower dose should be considered. Several questions remain regarding pimobendan in cats, given the lack of approved drugs for treating heart disease and the retrospective nature of the studies on pimobendan to date. Most cardiologists would give pimobendan in combination with a diuretic and an ACE inhibitor to cats with CHF. Most would not endorse use in cats prior to the onset of CHF, especially in cats with asymptomatic hypertrophic cardiomyopathy. For cats with HCM and concurrent hypertrophy at the top of the interventricular septum causing obstruction the flow of blood through the left ventricular outflow tract (e.g., systolic anterior motion of the mitral valve, increased aortic outflow tract velocity, and often secondary mitral regurgitation), there are concerns that pimobendan might worsen the pathophysiology of left ventricular outflow tract obstruction. In general, cats with HCM and dynamic left ventricular

outflow tract obstruction are thought to be at risk of hypotension or worsening obstruction when given drugs that are positive inotropes or vasodilators, thus drugs like pimobendan might not be advisable. Because the drug has a positive inotropic effect, there has been a tendency for veterinarians to use pimobendan in cats after myocardial failure, evidenced by a reduced fractional shortening, is apparent. Further prospective studies are needed to better outline whether this drug has benefits in survival time or quality of life in cats with CHF, and which clinical situations are most appropriate for the use of the drug. The author currently is most likely to use pimobendan in cats under the following circumstances: 1) the cat has advanced heart disease and congestive heart failure 2) there is no left ventricular outflow tract obstruction and 3) the cat has either echocardiographic evidence of myocardial failure or signs of CHF that have not fully responded to furosemide and an ACE inhibitor <u>or</u> 4) the cat has combined heart and renal failure and there has been a deterioration in renal function (and creatinine > 2.5-3 mg/dl) following initiation of an ACE inhibitor.

10. **Maintain quality of life** - Cats with heart disease can have many factors that reduce their quality of life, including respiratory distress, weakness after arterial embolism, drug administration issues and drug side effects, trips to the veterinarian, and altered interactions with owners or other animals in the household. Identification of these factors, and determining which are most important for an individual cat and cat owner, can help limit the impact of heart disease on the quality of life. Reducing frequency of medication administration, and consider other alterations in drug administration method, can reduced disruption of the bond between cat and owner.

REFERENCES

1. Finn E, Freeman LM, Rush JE, Lee Y. The relationship between body weight, body condition, and survival in cats with heart failure. *J Vet Intern Med* 2010; 24: 1369-1374
2. Freeman LM, Rush JE, Cahalane AK, Kaplan PM, Markwell PJ. Dietary patterns in dogs with cardiac disease. *J Am Vet Med Assoc* 2003; 223: 1301-1305
3. Freeman LM, Rush JE. Nutritional management of cardiac disease. In: Ettinger SJ (ed). *Textbook of veterinary internal medicine, 7th ed.* St Louis: Elsevier, 2010: 579-583.
4. Freeman LM. Beneficial effects of omega-3 fatty acids in cardiovascular disease. *J Small Anim Pract* 2010; 51: 462-470.
5. Freeman LM. Cachexia and sarcopenia: Emerging syndromes of importance in dogs and cats. *J Vet Intern Med* 2012; 26: 3-17.
6. Torin DS, Freeman LM, Rush JE. Dietary patterns of cats with cardiac disease. *J Am Vet Med Assoc* 2007; 230: 862-867
7. Tufts HeartSmart website [information on heart disease for pet owners, including nutrition (eg, low sodium diets for dogs and cats, treats, omega-3 fatty acid supplements)]. Available at: www.tufts.edu/vet/heartsmart January 17, 2013.
8. WSAVA Nutritional Assessment Guidelines Taskforce: Freeman L, Becvarova I, Cave N, et al. 2011 Nutritional Assessment Guidelines. *J Small Anim Pract* 2011; 52: 385-396.

OBESITY IS LINKED TO HEART DISEASE IN HUMANS, WHAT ABOUT DOGS?

Denise S Schwartz, DVM, MS, PhD
São Paulo, SP, Brazil

Obesity, a result of imbalance between caloric intake and expenditure, has been considered as a global epidemic in humans[5] and it has also affected pets.[3] In 2010, the CDC reported high numbers of obesity, counting 35.7% of American adults as obese, and 17% of American children.[9] It is currently estimated that obesity affects 20-40% of the canine population.

In humans, obesity may result in cardiac structural and functions changes, even in the absence of arterial hypertension and coronary heart disease. Although heart disease is cited as a problem related to obesity in the veterinary literature, there are only few studies that evaluate the actual cardiovascular effects of obesity in pets.

Obesity has been cited as having a minor effect on blood pressure in dogs, but hypertension has been reported in 23[13] and 45%[4] of obese dogs in two separate studies, and 67% of the obese dogs presented systolic blood pressure above 160mmHg (based on current criteria)[2] in another study[1]. Body condition score (BCS) has been positively correlated with blood pressure in dogs, and the authors suggested that obesity should be considered as a risk factor for secondary hypertension in this specie.[7]

An investigation of preclinical effects of natural obesity on left ventricular function of dogs of various breeds, using pulsed TDI found changes in left myocardial systolic and diastolic function. An increased early mitral flow velocity to early diastolic myocardial velocity suggests increased left ventricular filling pressure in obese dogs, compared to controls.[6]

Increased relative wall thickness i.e., left ventricular wall to ventricular diastolic diameter (LVW/LVDD) is observed in obese dogs, and the pattern of hypertrophy appears to be different depending on the BCS. Obese dogs also showed an increased LV mass indexed to estimated ideal body weight. Although the measurements may still be within normal ranges, the results show concentric increased LVW thickness as an adaptation to overload caused by obesity in dogs, which may be related to increase in LV wall tension and diastolic dysfunction. [12]Based on these data, it seems that obese dogs show changes similar to what is described in humans[15], even in the absence of systolic arterial hypertension.

Obese human patients are shown to present lower levels of natriuretic peptides, regardless of an increased volume and sometimes, pressure load, what raises the possibility of compromised natriuretic response in these individuals.[14] There are controversial findings in obese humans and a lack of studies reported in obese dogs regarding natriuretic peptide levels. Nt-proANP concentration was decreased in obese dogs in comparison to dogs with a normal (4-5/9) body condition score (BCS) in one study.[9]

A clinical study on naturally obese dogs shows also controversial results on Nt-pro BNP levels. We would expect that in the presence of a decreased diastolic function and increased ventricular wall thickness, Nt-pro BNP levels should be increased, but there was a similarity of Nt-proBNP levels between BCS9 and BCS5 dogs. These findings strengthen the need for further studies, especially when considering obese dogs with heart disease, not included herein, which may need lower Nt-proBNP cut-off values in order to differentiate the presence of heart failure as a cause of respiratory distress. This study provides important information, which should be better explored in more prolonged studies with a larger number of dogs.[6]

A decreased heart rate variability (HRV) was observed in naturally obese dogs during 24 hour Holter monitoring, both in time and frequency domains. There was a decrease in Low frequency (LF) and High frequency (HF) power for BCS 8 and BCS 9 dogs compared to controls (BCS 5). SDNN24 was decreased as BCS increased. In this same study, obese dogs showed a greater number of arrhythmias of ventricular origin than dogs in ideal weight, although not frequent.[10]

Exercise intolerance is always a complaint of obese dogs owners, but it is difficult to quantify, because most of these dogs could be considered sedentary as they are not submitted to exercising. A six-minute walk test (6MWT) was applied to quantify the exercise intolerance in healthy obese dogs in relation to healthy lean dogs. Dogs with BCS9/9 presented exercise intolerance compared to BCS5/9 dogs, based upon the shorter distance walked during the 6MWT.[11]

Obese dogs, without overt heart disease present preclinical left ventricular diastolic dysfunction, relative ventricular hypertrophy, controversial natriuretic peptide findings, a decreased HRV and exercise intolerance, as found in humans, and not related to coronary disease. These findings may not themselves be clinically relevant, but may contribute as extra overload to changes observed in the progression of naturally occurring heart diseases, as chronic degenerative mitral valve disease. Although an obesity paradox has been raised as a protection to heart failure patients, this has still not been sufficiently explored. Attention should be taken regarding the observed changes and their impact on existent heart disease progression.

REFERENCES

1. Aptekmann, KP; Carvalho Filho, AS; Schwartz, DS. Cardiovascular evaluation and cTnI levels in obese dogs. JVIM, 20: 778, 2006. Abst 246.
2. Brown S, Atkins C, Bagley R et al. Guidelines for the identification, evaluation and management of systemic hypertension in dogs and cats. JVIM 2007; 27:542-558.
3. German AJ. The growing problem of obesity in dogs and cats. J Nutr. 2006 Jul;136 (7 Suppl):1940S-1946S.
4. Jerico, MM; Silva, MBFP; Machado, FLA. Avaliação cardiovascular em cães obesos:mensuração da pressão arterial e acahdos eletrocardiográficos. Revista Clínica Veterinaria, 61:66-71, 2006.
5. Litwin, SE. The growing problem of obesity and the heart. JACC 47:617-619, 2006.
6. Melo, PRR; Mazini, AM; De Oliveira, VMC; Goldfeder, GT.; Duarte, CN; Schwartz, DS. Plasma levels of NT-pro BNP and ventricular diastolic function assessed by pulsed tissue doppler echocardiography in obese dogs. JVIM 2012; 26: 721-722 Abstract C-41.
7. Montoya JA, Morris PJ, Bautista I, Juste MC, Suarez L, Peña C, Haskett RM, Rawlings J. Hypertension: a risk factor associated with weight status in dogs. J. Nutr, 136: 2011S-2013S, 2006.
8. National Obesity Trends, CDC NCHS, 2010, retrieved 2012-03-26
9. Schwartz DS, Melo PRR, Mazini AM, De Oliveira VMC, Goldfeder GT, Azedo MR. NT-PRO ANP in obese dogs. JVIM 2011 25: 655-656.
10. Schwartz DS, Duarte CN, Mazini AM, Melo PRR, Goldfeder, GT. Six minute walk test for assessment of exercise tolerance in obese dogs. In: Mandigers, P.J.J; German, A.J. Congress Proceedings 2011 ECVIM 8-10 Sept., Sevilla, Spain. POSTER CA-P-17. ISSN 1726-3204, p. 230.
11. Schwartz DS, Mazini AM, Melo PRR, De Oliveira VMC, Goldfeder GT, Duarte CN. Heart rate variability is decreased in obese dogs. In: Mandigers, PJJ; German, AJ. Congress Proceedings 2011 ECVIM 8-10 Sept., Sevilla, Spain. POSTER CA-P-25, P. 233.
12. Schwartz DS, Melo PRR, Mazini AM, De Oliveira VMC, Goldfeder GT, Duarte CN. Echocardiographic findings in obese dogs. JVIM 2012, 26:722. Abstract C-43.
13. Verwaerde P, Senard JM, Galilier M et al. Changes in short-term variability of blood pressue and heart rate during the development of obesity-associated hypertension in high-fat fed dogs. J hypertens 1999; 17:1135-43.
14. Wang, TJ, Larson, MG, Levy D, Benjamin, EJ, Leip EP, Wilson, PWF, Vasan, RS. Impact of obesity on plasma natriuretic peptide levels. Circulation 2004; 109:594-600.
15. Wong CY, O'Moore-Sullivan T, Leano R, Byrne N, Beller E, Marwick TH Alterations of left ventricular myocardial characteristics associated with obesity. Circulation 2004; 110:3081-87

TREATMENT OF COMMON ARRHYTHMIAS: HOW AND WHEN TO TREAT

Gunther van Loon (DVM, PhD, DECEIM, Ass DECVDI)
Merelbeke, Ghent, Belgium

INTRODUCTION

Dysrhythmias are common in horses and can be physiological or pathological. Most common physiological arrhythmias, such as 1st and 2nd degree atrioventricular (AV) bloc and sinus arrhythmia, bloc or arrest are associated with vagal tone and obviously don't need any treatment as long as they disappear with exercise. The most important pathological arrhythmias are atrial fibrillation (AF) (or flutter), atrial premature beats and atrial tachycardia and ventricular premature beats and tachycardia. Treatment aims to convert the rhythm to sinus rhythm or to decrease the burden of ectopy. One should be aware that all anti-arrhythmic drugs have pro-arrhythmic properties and, as such, they should be used with caution.

ATRIAL FIBRILLATION (ATRIAL FLUTTER)

Atrial fibrillation is the commonest cardiovascular cause of poor performance in racehorses. AF is a reentry arrhythmia that generally presents as a sustained, self-perpetuating arrhythmia. Occasionally, AF is short-lived and terminates spontaneously (usually within the first 24-48 hours after a race), which is called paroxysmal AF. This form of AF has been shown to be a cause of reduced performance in slowly finishing Thoroughbreds.

When

Atrial fibrillation may occur as primary disease ('lone AF') or develop following other underlying disease usually leading to atrial dilatation (mitral regurgitation, congestive heart failure). The choice whether or not to treat AF depends on the presence of underlying disease, the effect on performance and potentially associated risk factors. In the presence of underlying disease AF is often left untreated because treatment may be associated with a risk (eg. horse in heart failure) and the arrhythmia is likely to return because of the predisposing cardiac disease. These horses mainly benefit from supportive care with digoxin, ACE inhibitors or furosemide. Horses that have no significant underlying disease (often mild to moderate valvular regurgitation and normal to slightly dilated atria) and that present reduced performance because of AF (racehorses, jumping and dressage horses performing at high level) should be cardioverted. However, because AF does not affect the horse at rest and as treatment of AF can hold a risk, treatment is rarely indicated in horses at rest or broodmares. Opinions have changed regarding the treatment of horses that perform at lower levels (low level jumping, dressage). Although many of these horses do not show clear clinical signs it has recently been shown that many horses with AF show disproportionate tachycardia, QRS broadening and even R-on-T phenomena during sudden stress or exercise[1]. Although the exact origin of the abnormal QRS complexes is yet unknown (supraventricular with aberrant conduction or ventricular in origin) it is considered a potential risk factor. Therefore, an exercise stress ECG is needed in these patients and presence of abnormal ventricular rhythms should encourage for AF treatment.

How

Only in racehorses in which AF is discovered immediately after the race, one should delay treatment for about 48 to 72 hours as spontaneous cardioversion may occur. In other horses, AF is permanent which means that it will never terminate without treatment. Medical treatment or transvenous electrical cardioversion can be applied to treat AF and is successful in up to 90% of the lone AF cases[2-4].

Treatment with quinidine sulphate (QS) has been used for many decades and has proven its efficacy[3,5]. However, the drug becomes more expensive and is being taken off the market in many countries. Treatment must be performed in a quite environment. Venous access and continuous ECG monitoring must be available. Generally, a 22 mg/kg dose is administered via nasogastric tube every two hours until conversion or severe toxic side effects occur. Usually not more than 6 doses a day are administered. 'Normal' toxic effects include depression, nasal edema, tachycardia. More severe side-effects, such as colic, diarrhea, laminitis, ataxia, hypotension, collapse, ventricular rate exceeding 120/min, >25% QRS broadening, may require termination of treatment or increasing the dosing interval (up to 6 hours which is the half-life of QS). Supraventricular tachycardia can be treated with digoxin (2.2 µg/kg IV), and, if unsuccessful, with propranolol (0.03 mg/kg IV). If no severe signs of hypotension are present, administration of an alfa$_2$-adrenergic agonist is often very effective to slow down the supraventricular rate immediately. Further emergency treatment includes administration of bicarbonate at 1 meq/kg bodyweight. Ventricular tachycardia is treated with magnesium sulphate, followed by lignocaine IV. Hypotension can be treated with crystalloids and phenylephrine. Horses that do not cardiovert and show severe supraventricular tachycardia should be pretreated with digoxin. During QS treatment atrial activity gradually becomes more organized, resulting in P-like waves on the ECG that still lack a constant relation with QRS complexes. At cardioversion, suddenly one P is found before each QRS complex (due to the vagolytic properties of QS AV block will not occur). Occasionally, during the QS treatment, there is one P-like wave buried in each T wave: this rhythm is very easily misinterpreted as sinus rhythm but actually cardioversion has not yet occurred. Termination of treatment at this stage may result in a progressive return to initial AF status as QS levels gradually fall.

Flecainide treatment for AF has been reported to produce potentially dangerous wide-QRS tachycardia and risk for sudden death[6]. Amiodarone showed moderate efficacy and chronic administration was associated with side-effects[7,8]. Propafenone and sotalol have been sporadically used but little is known about their efficacy.

Transvenous electrical cardioversion (TVEC) can be applied under general anesthesia and is very effective to immediately restore sinus rhythm[4,9,10]. The cardioversion procedure starts with the positioning of 2 cardioversion catheters in the standing horse: one in the left branch of the pulmonary artery and one in the right atrium. A third pacing catheter can be placed in the right ventricle for backup pacing during the cardioversion procedure[11]. Catheter placement can be fully guided by ultrasound. Also pressure traces from the catheter and radiography might be used for correct placement. After induction of anesthesia catheter positions are double-checked. Both cardioversion catheters and the surface ECG are connected to the defibrillator, which is operated in synchronous mode. This mode continuously identifies the R wave from the surface ECG and only delivers a shock on the R wave. Correct detection of the R wave is vital as shock delivery on a T wave almost in evitably results in fatal ventricular fibrillation. Position of the skin electrodes must be changed when wrong R wave detection occurs. Incremental shock are applied usually starting around 125 Joules. Most horses convert between 125 and 250J. If cardioversion cannot be achieved, shock delivery after administration of an anti-arrhythmic drug, such as amiodarone, may result in successful cardioversion. The drug is also used to suppress abundant atrial ectopy in the immediate post-cardioversion period.

Aftercare of a successfully cardioverted horse is different depending on the duration of AF. Recent onset AF (eg 1 week) quickly recovers and horses usually return to full work within days. Experiment work suggests that horses with longstanding AF (more than 4 months) require several weeks to recover from AF-induced electrical and contractile remodeling. However, data from clinical cases with chronic AF suggest that even after 6 weeks of follow-up, AF cases showed larger atria, lower 2D ejection phase indices and prolonged TDI-based conduction time compared to controls (unpublished data). Recurrence of AF is around 25% in racehorses (where AF is often relatively acute) and is around 35% in warmbloods (often chronic AF). Horses that show a high burden of atrial ectopy after treatment of AF may have an increased risk for recurrence of AF. These horses may benefit from steroid treatment. Anti-arrhythmic drugs may be administered although their dose and effect is poorly known.

VENTRICULAR PREMATURE BEATS AND VENTRICULAR TACHYCARDIA

Ventricular premature depolarisations (VPD) or tachycardia (VT) originate from the ventricular myocardium. The exact etiology in horses is not well understood but similar as in other species it is associated with electrolyte disturbances, myocardial hypoxia, inflammation or necrosis, aortic regurgitation or underlying systemic disease. VPDs are a possible cause for sudden induction of ventricular tachycardia that may deteriorate to ventricular flutter or fibrillation which is fatal.

When and how to treat

A few monomorphic, isolated ventricular premature beats in the immediate post-racing period rarely indicate cardiac pathology and do not require treatment. Horses with VPDs on a resting ECG should be rested and require a full cardiac exam and serum biochemistry. Cardiac troponin I may be increased indicating active cell damage. Serum digoxin concentrations should be determined if there is suspicion of cardiac glycoside intoxication. Treatment aims to treat any underlying disorder (electrolyte, acid-base, anemia, …) and to restore sinus rhythm and suppress ventricular ectopy. Steroids (dexamethasone, followed by oral prednisolone for several weeks) may be administered in order to decrease myocardial inflammatory reaction. Occasional isolated monomorphic VPDs generally do not require aggressive anti-arrhythmic treatment and one should be aware that all anti-arrhythmic drugs have pro-arrhythmic properties, some of them being worse than the arrhythmia *per se*. Chronic oral administration of anti-arrhythmic drugs such as sotalol, phenytoin or propafenone has been performed anecdotally but limited information is available so far regarding dosing regimens, efficacy and potential side effects. Clinical signs at rest and presence of frequent VPDs and VT at high rates, especially if polymorphic, carry a risk for sudden deterioration into ventricular tachycardia and ventricular fibrillation and require immediate anti-arrhythmic treatment. Magnesium sulphate at 2-6 mg/kg/min to effect, up to 55 mg/kg total dose is often administered initially. It rarely converts the ventricular rhythm but is thought to decrease the risk of torsades de pointes during subsequent anti-arrhythmic treatments. Different drugs are available and no single drug is claimed to be the best one. Lidocaine IV is readily available, is the most rapidly acting drug and therefore often the first administered drug. Repeated slow IV doses of 0.25-0.5 mg/kg up to 2-4 mg/kg, followed by 0.05mg/kg/min CRI are given. Diazepam must be available for emergency treatment of excitement and seizures associated with the lidocaine treatment. Quinidine glucate can be administered at at 1-2.2 mg/kg IV every 10 minutes to effect or a maximal total dose of 12 mg/kg. Procainamide CRI is given at 1 mg/kg/min IV up to 20 mg/kg total dose. Amiodarone CRI is started at 5 mg/kg/h for 1 hour followed by 0.83 mg/kg/h for 23 hours (up to 48 h). Potential side effects include hind limb weakness and diarrhea. Phenytoin is administered at 5-10 mg/kg IV, followed by 1-5 mg/kg IM BID or 10-15 mg/kg PO BID. It is thought to be the treatment of choice in case of cardiac glycoside intoxication. Plasma level monitoring is readily available and doses should be decreased when signs of sedation, lip and facial twitching or gait deficits occur. Other drugs include porpafenone (1 mg/kg in 5% dextrose over 5-8 min IV), propranolol (0.03 mg/kg IV BID) or bretylium tosylate (3-5 mg/kg slow bolus IV up to a total of 10 mg/kg). Transvenous electrical cardioversion has been successfully applied in unresponsive VT but is technically challenging and requires general anesthesia in an unstable patient[12]. After successful treatment of ventricular tachycardia horses should be rested for at least 2 months and should be carefully checked (ultrasound, exercise ECG, biomarkers) before being able to gradually return to work.

ATRIAL PREMATURE BEATS AND ATRIAL TACHYCARDIA

Atrial premature depolarisations (APD) originate in the atrial myocardium. They are usually detected incidentally in an otherwise healthy horse. APDs often disappear during exercise. Thorough evaluation of electrolytes and other body systems should be performed, followed by echocardiography, cardiac biomarkers and exercise tests to assess the effect on performance. Echocardiography should rule out cardiac chamber enlargement and major structural lesions. Frequent APBs or runs of atrial tachycardia are a potential trigger for atrial fibrillation.

When and how to treat

Atrial premature beats rarely require immediate anti-arrhythmic drug treatment. Corticosteroid treatment (dexamethasone followed by oral prednisolone for 2-4 weeks) may resolve atrial ectopy in some cases and should be administered along with a rest period. Chronic oral administration of anti-arrhythmic drugs such as sotalol, phenytoin or propafenone has been performed anecdotally but limited information is available so far regarding dosing regimens, efficacy and potential side effects. These drugs have been used in an attempt to prevent recurrence of AF after successful cardioversion. Atrial tachycardia only requires further treatment if it occurs at high rates and results in high ventricular rates. Digoxin reduces AV nodal conduction and slows down ventricular response in these animals. Especially at higher rates, atrial tachycardia may be difficult to differentiate from atrial flutter. Treatment of sustained atrial tachycardia can be attempted with quinidine sulphate, similar as for treatment of atrial fibrillation.

CONCLUSION

For each dysrhythmia, the clinician should investigate if there is an underlying cause and treat that cause. Atrial fibrillation is the most important arrhythmia and fairly good guidelines exist for deciding when to treat and how to treat. Both medical and electrical treatment are efficacious. Other atrial arrhythmias rarely require emergency treatment. Increased ventricular ectopy is regarded as a potential risk for induction of life-threatening arrhythmias and require the horse to be rested and treated. Emergency anti-arrhythmic treatment is only indicated in unstable dysrhythmias as the treatment *per se* also carries a risk. Further treatments include steroids and oral anti-arrhythmic drugs in order to suppress ventricular ectopy

REFERENCES

1. Verheyen T, Decloedt A, Van Der Vekens N, et al. Ventricular response during lungeing exercise in horses with lone atrial fibrillation. Equine Vet J 2013;45:309-314. 2. McGurrin MKJ, Physick-Sheard PW, Kenney DG. Transvenous electrical cardioversion of equine atrial fibrillation: Patient factors and clinical results in 72 treatment episodes. J Vet Intern Med 2008;22:609-615. 3. Reef VB, Levitan CW, Spencer PA. Factors affecting prognosis and conversion in equine atrial fibrillation. J Vet Intern Med 1988;2:1-6. 4. De Clercq D, van Loon G, Schauvliege S, et al. Transvenous electrical cardioversion of atrial fibrillation in six horses using custom made cardioversion catheters. Vet J 2008;177:198-204. 5. Reef VB, Reimer JM, Spencer PA. Treatment of atrial fibrillation in horses: new perspectives. J Vet Intern Med 1995;9:57-67. 6. van Loon G, Blissitt KJ, Keen JA, et al. Use of intravenous flecainide in horses with naturally-occuring atrial fibrillation. Equine Vet J 2004;36:609-615. 7. De Clercq D, van Loon G, Baert K, et al. Effects of an adapted intravenous amiodarone treatment protocol in horses with atrial fibrillation. Equine Vet J 2007;39:344-349. 8. De Clercq D, van Loon G, Baert K, et al. Intravenous amiodarone treatment in horses with chronic atrial fibrillation Vet J 2006;172:129-134. 9. McGurrin MKJ, Physick-Sheard PW, Kenney DG, et al. Transvenous electrical cardioversion of equine atrial fibrillation: Technical considerations. J Vet Intern Med 2005;19:695-702. 10. McGurrin MKJ, Physick-Sheard PW, Kenney DG. How to perform transvenous electrical cardioversion in horses with atrial fibrillation. J Vet Cardiol 2005;7:109-119. 11. van Loon G, De Clercq D, Tavernier R, et al. Transient complete atrioventricular block following transvenous electrical cardioversion of atrial fibrillation in a horse. Vet J 2005;170:124-127. 12. van Loon G, De Clercq D, Schauvliege S, et al. Successful electrical cardioversion of sustained ventricular tachycardia in an adult horse. In: European Veterinary Conference, 42nd, Proceedings, Amsterdam, The Netherlands 2009;374.

EXERCISE INDUCED CHANGES IN THE EQUINE ATHLETE'S HEART – DOES THE CLINICIAN NEED TO BOTHER?

Rikke Buhl, DVM, PhD
Copenhagen, Denmark

INTRODUCTION

Prolonged physical exercise induces dramatic changes in body composition and function. Luckily for all of us, the heart has a unique capacity for adapting to altered loading conditions. With increased oxygen demands during exercise the demand for blood flow is met by increased stroke volume as explained by Starling's law of the heart as well as increased heart rate (HR), both resulting in increased cardiac output. For the short-term, the heart can easily cope with the greater volume and/or pressure loads. However, when the overload is repeated or long lasting, a second mechanism becomes activated, leading to an altered expression of cardiac genes and an ensuing increase in ventricular muscle mass. Together with other cardiac changes, this physiological phenomenon is described in both human and equine athletes as the "Athlete's heart".

The cardiovascular system of the horse is in many aspects superior compared to other mammals including humans, therefore making it extremely relevant to study this animal in translational aspect. Its resting HR is low around 30 bpm and its maximum HR is relatively high up to 240 bpm. When this large range of HR is combined with the great capacity of increasing stoke volume, this can result in a 10-fold increase in cardiac output up to 300 l/min during peak exercise. In comparison, the man or dog is capable of increasing cardiac output up to 7 fold.

This presentation will focus primarily on the structural and functional adaptive changes within the heart due to prolonged physical training, and it is discussed whether or not there is significant evidence that these changes should be taken into consideration when clinicians examine equine athletes.

ATHLETE'S HEART – HISTORICAL PERSPECTIVE

Cardiac enlargement due to exercise was demonstrated in man in the late 1800s. In 1899, a German physician examined a group of cross-country skiers and found that they had larger hearts compared to untrained control persons; he was the first to introduce the phrase "The Athlete's heart", which has been used ever since. Also among racehorses focus on Athlete's heart has existed for more than 200 years. In 1789, the legendary racehorse Eclipse was autopsied and the heart was shown to be more than twice the normal weight of an average heart from racehorses at that time. Two hundred years later, the heart of Secretariat, a record-breaking American Thoroughbred racehorse, was alleged to have had a heart that weighed over 10 kg at post-mortem examination, while normal Thoroughbred hearts weigh about 4.5 kg. For several racehorse breeders and trainers these early reports of cardiomegaly have emphasized the importance of a large heart for superior racing performance. Fortunately nowadays we don't have to euthanize the animals anymore in order to measure the heart and to examine its morphology: an ultrasound machine and an ECG will provide substantial information about the heart. And due to extensive cardiology research in several species including horses, we now know that Athlete's heart is not only related to cardiac hypertrophy, but that due to prolonged exercise also valvular function, cardiac rhythm and HR are remodelled. These effects will be discussed in the following paragraphs.

THE PHYSIOLOGICAL EFFECT OF A LARGE HEART

The importance of a large heart for racehorses is related to the capacity for aerobic energy production. Delivery of oxygen to tissues during exercise depends mainly on cardiac output and arterial oxygen content. Maximal cardiac output is determined by maximal HR multiplied by stroke volume. Maximal heart rate during exercise remains unaffected by athletic training in horses, but stroke volume increases with increasing training intensity[1,2]. Stroke volume is largely determined by cardiac dimensions and myocardial performance and stroke volume of the equine heart is believed to be an important factor for the maximal aerobic capacity. It has been shown that maximum oxygen uptake (VO_{2max}) is strongly correlated to heart size in Thoroughbreds and has been related to athletic performance in horses.[3,4] This relationship has also been documented in humans.[5]

Horses have a higher aerobic capacity than humans. In humans, the anaerobic capacity assumes a more substantial role, contributing 50% of the energy in races of short duration.[6] This is in contrast to racehorses, where the aerobic component of energy supply predominates during racing. Anaerobic energy sources might supply less than 30% of the total energy during shorter races.[7] It therefore appears that the aerobic energy production in horses is far more important than in humans. All these studies taken together thus explain the physiological background for the importance of heart size for athletic performance.

HEART RATE

In human athletes, it is well known that resting HR decreases due to training.[8,9] Whether this is true in horses is not clear, as some studies do not document decreases in HR after a training period[1,10] whereas others observe significant decreases in HR due to training.[11,12] In contrast to the resting HR, it is clear that maximal HR does not change with training. Nevertheless, maximal HR is dependent on the age of the athletes as it decreases with ageing.[13] Although maximal HR is not influenced by training, the effect of training would result generally in increased V_{200} (speed where a

HR of 200bpm is reached) or V_{HRmax} (speed where the maximal HR is reached) meaning that the trained horse would run at a given speed with a lower HR compared to the unfit athlete. This can be explained by an increased stroke volume due to cardiac hypertrophy, although both muscular and pulmonary changes may also contribute to the improvement.

HEART SIZE

During the last ten years, echocardiographic studies have documented that the heart of racehorses undergo hypertrophy due to training.[14-16] Together with the numerous studies of human athletes, we now have a detailed description of the adaptive myocardial changes following training. Two different morphological forms of the human Athlete's heart have been reported: an 'endurance-trained' heart and a 'strength-trained' heart[17]. The endurance-trained heart is characteristic of athletes involved in sports with a dynamic and aerobic component (e.g. running, cycling). Such athletes develop predominantly increased left ventricular volume with a proportional increase in thickness of the myocardial walls. This is due to volume overload caused by prolonged increase in stroke volume, which will result in an eccentric left ventricular hypertrophy with an unchanged relationship between the myocardial wall thickness and the internal diameter of the left ventricle. In contrast, athletes involved mainly in anaerobic strength-training (e.g. sprinting, weight-lifting) experience pressure overload and high systemic arterial blood pressure. Thus, these athletes will develop concentric left ventricular hypertrophy, which is characterized by an increased thickness of the myocardial walls.

Cardiac changes in Thoroughbreds and Standardbreds have demonstrated an increase in both myocardial wall thickness as well as the internal diameter of the left ventricle, similar to the morphology of the endurance-trained heart.[15,16] There has been considerable scientific debate over the years whether large hearts are associated with racing success. The rationale behind this hypothesis is that stroke volume is increased in athletes with a larger heart and maximum oxygen uptake (V_{O2max}) will increase, which is beneficial for the performance of the horse and has been documented in Thoroughbreds.[18] These results support the findings in cross sectional and longitudinal studies of Thoroughbreds and Standardbreds demonstrating a positive relationship between left ventricular heart size and racing performance.[15,16]

VALVULAR REGURGITATION

Valvular regurgitations are defined as a backflow of blood through the orifices of the heart valves due to imperfect closing of the valves. If the regurgitant is sufficient it will result in an audible murmur heard on cardiac auscultation. In addition to cardiac hypertrophy, both human and equine athletes develop valvular regurgitations due to intensive athletic training and these valvular regurgitations are also considered part of the Athlete's heart.[19-21] In this line, it has been shown that high intensively trained racing horses had significantly more tricuspid regurgitation than low intensity trained horses.[20] The physiological explanation for the development of valvular regurgitations is unclear, but it is possible that the altered geometry of the ventricle is responsible for the valvular regurgitation. If this is the case, then the valve leaflets and/or the *chordae tendineae* may become relatively too short and regurgitation of blood through the valves will subsequently occur. In general, these valvular regurgitations are small and do not increase in severity over time. Moreover, it has been shown that the majority tends to disappear after exercise.[22] Others have speculated whether increased pulmonary arterial systolic pressure result in increased right-sided valvular regurgitation in athletes.[23] Whilst the size of the heart appears to significantly influence racing performance, no studies have demonstrated any influence of physiological training induced valvular regurgitation on racing performance in horses or humans.[19-21] Therefore one should in general consider very mild murmurs (grade I-II) heard by auscultation as a relative normal finding with no significant influence on performance.

CARDIAC ARRHYTHMIAS

Exercise-induced hypertrophy may also, like in the hypertensive heart, result in ECG abnormalities with both supraventricular and ventricular arrhythmias. Studies of arrhythmias in normal performing racehorses as well as riding horses has been published recently.[24-27] In the normal performing horses, a quite high amount of supraventricular arrhythmias and also ventricular arrhythmias are observed during and after exercise. Interestingly it seems that show jumping horses develop relatively more ventricular arrhythmias during the show jumping course compared to the other disciplines.[26] An explanation to this could be the substantial HR fluctuations during the showjumping courses. Until now, only few studies have examined arrhythmias in horses with reduced performance, but no correlation between heart size and prevalence of arrhythmias has been studied.[28,29] Therefore, a more thorough knowledge and understanding of cardiac electrophysiology in athletes are essential to differentiate pathological from physiological induced arrhythmias. High performance horses are at greater risk of dying from cardiac failure while performing at maximum capacity; and horses have collapsed and died during or after high-intensity exercise. The pathophysiological changes underlying collapse or sudden cardiac death in horses are unknown, but probably similar to the situation in humans.[30,31] Horses that suddenly die have a big impact on animal welfare, rider safety, finances and public relations. A large proportion of racehorses that suddenly die do not show sufficient lesions to account for death on necropsy.[30] As described above, spontaneous arrhythmias are common in the horse, but studies coupling specific arrhythmias to sudden death are sparse.[32] Therefore it can be speculated that cardiac repolarization disorders, such as the Long QT syndrome, may underlie some of these sudden death as described in humans, but this area is very poorly described in horses. Measurements of the QT interval in horses have been reported recently and some reference values are established.[33] Future studies on ECGs from horses suffering from sudden death should enlighten the significance of Long QT syndrome in horses.

CONCLUSION

Well-trained horses will probably over time develop a larger heart, lower resting HR, mild low-grade cardiac murmurs and most likely some supraventricular arrhythmias during and after exercise – all characteristic for Athlete's Heart. According to the present knowledge there is no reason for the clinician to worry about these cardiac changes, but they should be aware that these findings are normal adaptive changes.

REFERENCES

[1]Betros CL, et.al **Equine vet J Suppl** 2002;34:100-105.
[2]Thomas DP, et.al **Am J Physiol** 1983;245:160-165.
[3]Young LE, et.al **Equine vet J Suppl** 2002;34:467-471.
[4]Harkin JD, et.al **Equine vet J** 1993;25:53-60.
[5]Osborne G, et.al **Int J Sports Med** 1992;13:219-224.
[6]Hermansen L, Medbø J **Med Sports Sci** 1984;17:56.
[7]Eaton MD, et.al **J appl Physiol** 1995;78:1564-1568.
[8]Fagard RH, **Int J Sports Med** 1996;17:140-144.
[9]Triposkiadis F, et.al **Eur J Clin Invest** 2002;32:16-23.
[10]Bayley WM, et.al **Am J Vet Res** 1983;44:544-553.
[11]Kuwahara M, et.al **Equine vet J Suppl** 1999;30:178-180.
[12]Ohmura H, et.al **Equine vet J Suppl** 2002;34:140-146.
[13]Rose RJ, Hodgson DR In: **The Athletic horse** 1994;245-257.
[14]Seder JA et.al **J EquineVet Sci** 2003;23:149-167.
[15]Young LE, et.al **J Appl Physiol** 2005;99:1278-1285.
[16]Buhl R, et.al **J Am Vet Med Ass** 2005;226 :1881-1887.
[17]Morganroth J et.al **Ann Intern Med** 1975;82:521-524.
[18]Young LE et.al **Equine Vet J** 2002;34:467-472.
[19]Douglas PS, **Am J Cardiol** 1998;64:209-212.
[20]Buhl R, et.al **J Am Vet Med Ass** 2005;227:1630-1635.
[21]Young LE, et.al **J Vet Intern Med** 2008;22:418-426.
[22]Buhl R et.al **Equine Vet J Suppl** 2006;36:178-182.
[23]Erickson BK, et.al **Equine Vet J Suppl** 1990;9:47-52.
[24]Ryan N, et.al **Equine Vet J** 2005;37:265-268.
[25]Physick-Sheard PW, McGurrin MK **J Vet Intern Med** 2010;24:1158-1166.
[26]Buhl R, et.al **Equine Vet J Suppl** 2010;42:196-201.
[27]Barbesgaard L, et.al **Equine Vet J Suppl** 2010;42:202-207.
[28]Martin BB, et.al **J Am Vet Med Assoc** 2000;216:554-558.
[29]Jose-Cunnilleras E, et.al **Equine Vet J Suppl** 2006;26:163-170.
[30]Gelberg HB, et.al. **J Am Vet Med Assoc** 1985;187:1354-1356.
[31]Lyle CH, et.al **Equine Vet J** 2011;43:324-331.
[32]Kiryu K, et.al **J Vet Med Sci** 1999;921:928.
[33]Pedersen PJ, et.al **J vet Cardiol** 2013 in press: http://dx.doi.org/10.1016/j.jvc.2012.08.002.

VALVULAR DISORDERS OF EQUINE ATHLETES

Emmanuelle Van Erck, DVM, PhD, DECEIM
Waterloo, BELGIUM

INTRODUCTION

Although cardiac problems are relatively rare in horses, they can have significant repercussions on exercise tolerance and may be dangerous for the rider. They rank as the third in problems causing poor performance in horses[1,2]. Valvular regurgitations are a common finding in athletic horses and depending on the valves concerned, they can have very different consequences on the horse's exercise capacity. As veterinarians, we are responsible for establishing a precise diagnosis and giving a prognosis for the safe pursuit of an athletic career, taking into consideration the sport, the length of the career and the age of the horse. Valvular leakage is generally associated with murmurs, audible on auscultation. In working horses, distinguishing them from physiological murmurs is important, as physiological murmurs are commonly heard. Mapping these murmurs is helpful to orientate diagnosis. In depth morphological and functional investigation using ancillary exams, such as Doppler echocardiography and exercising ECG, is warranted to better define athletic prognosis.

In the athletic population, the prevalence of valvular insufficiencies is related to age and to the type of exercise performed. Chronic pulmonary hypertension and training induced morphological changes can promote the pathogenesis of atrio-ventricular valve regurgitation in horses[3,4]. Consequently, the prevalence of heart murmurs has been found higher in athletic individuals, such as racehorses[5,6] and in older individuals engaged in longer careers[6].

Evaluating the pathological character of valvular regurgitations and their functional consequences is important in competition horses, both to determine if it can be the cause of poor performance and if the horse can be safely kept at work. Doppler echocardiography (DE) not only allows establishing a diagnosis but also is also helpful to have both a morphological and functional appreciation of the heart. Doppler is a very useful feature of echocardiography, as it can readily detect abnormal blood flows. Doppler mapping of the regurgitant jet is used to characterize the severity of the valvular insufficiency. Pulsed wave and continuous Doppler yield important functional information. Assessing the rate of progression of valvular disorders and their consequences (whatever their degree of importance) by regular echocardiographic follow-up is important if the horse is to continue work and competition.

This paper addresses acquired valvular diseases in athletic horses. Pulmonary regurgitation is a rare condition that will not be dealt with here.

MITRAL REGURGITATION

Clinical Signs

Mitral regurgitation (MR) is the valvular insufficiency most commonly associated with poor performance. In its early stages, MR can be asymptomatic in horses performing in low intensity sports (for instance dressage) or competing at lower level. Symptoms associated with early MR are often insidious and translate as a lack of energy at work, breathlessness, prolonged post-exercise cardio-respiratory recovery. The horse's behavior at work can vary from day to day according to climatic conditions. If some horses can continue to work and perform with milder degrees of MR, riders should be made aware of the possible complications associated with MR. Work should be discontinued and the horse presented for re-evaluation if the horse shows a decrease in exercise tolerance, tachypnea or cough.

Because the left heart is a high-pressure system, the risks of progression of MR can be higher in horses that are worked at high intensities, like racehorses or eventers. In horses with MR, pulmonary arterial pressures are higher both at rest and during exercise in horses with MR. Increased pulmonary hypertension can lead to exercise-induced pulmonary hemorrhage (EIPH) or pulmonary edema. More rarely, it has been associated with fatal rupture of the pulmonary artery in racehorses.

Chronic MR usually leads to progressive enlargement of the left atrium (LA) and atrial fibrillation is a potential complication. Other arrhythmias can be associated with MR, namely atrial (APB) and ventricular premature beats (VPB). These findings are sometimes only present during exercise or recovery and are indicative of myocardial dysfunction or chamber dilation.

Diagnosis

Mitral regurgitation (MR) is associated to a holo- to pansystolic murmur, with a maximal point of intensity located on the left side at the level of the cardiac apex. Mitral murmurs are usually graded III to VI/VI and radiate dorsally and cranially, towards the base of the heart. In many cases the intensity of the murmur is proportional to the severity of the regurgitation, although this is not systematic. Diagnosis of MR can be made on auscultation only, however DE is important to assess its severity and the extent of morphological and functional consequences. Because of the angle at which the probe apprehends the left atrio-ventricular flows, the mapping of the MR jets can be difficult and deceptive. In addition to identifying a systolic regurgitant jet retrograding into the LA, LA and LV volume and pulmonary artery diameter should be systematically assessed. The valvular structures and movements are noteworthy (valve thickening, prolapse, flail leaflet,...). Dosage of specific plasmatic parameters such as cardiac troponin I or CK iso-enzymes can be useful to assess myocardial damage.

Prognosis

Generally, compensated valvular regurgitations are well tolerated in performance horses and appear to have a slow progression. Factors influencing progression may be the level of work performed by the horse, the rhythm and strain of competition and concurrent diseases. Occurrences of clinical signs triggered by exercise are of poor prognosis. On DE examination, signs of volume overload, PA dilation and changes in valvular structure or motion are also of poor prognosis for sport. Repeated ultrasound examinations should be done to monitor evolution and help in providing a more accurate prognosis. However, to our current knowledge, there is a large variability in the rate of progression amongst individuals.

TRICUSPID REGURGITATION

Clinical signs

Tricuspid regurgitation (TR) is a relatively common finding in equine athletes of higher level or under intensive training[5,6]. The prevalence of TR is particularly high in Thoroughbred and Standardbred populations. Tricuspid valve incompetence seems to have fewer consequences on performance in comparison to mitral valve regurgitation. It is often found in the absence of clinical signs and seldom has repercussions on the horse's ability to perform. However, horses with any signs of right-sided heart failure (peripheral edema, exercise intolerance, possibly combined with atrial fibrillation) should be evaluated carefully for tricuspid insufficiency.

Some authors have suggested that TR could be a consequence of pulmonary hypertension, either secondary to intensive exercise or pulmonary disease. Either condition generates chronic pulmonary hypertension and could exert excessive strain on the tricuspid valve annulus (TV) and result in leakage. This mechanism seems to occur in all athletic horses, whatever their discipline. Because it can be related to other problems or less tolerated in some individuals, investigation of TR remains important in athletic horses.

Diagnosis

TR is associated with a right-sided pan or holosystolic murmur with a PMI over the TV region. As for other valvular insufficiencies, DE is the best exam to assess both morphological and functional consequences of TR. If the TR jet, retrograding in the right atrium (RA) during systole can be easily detected when present, measurement of RV or RA size is difficult to standardize. This is due to the many trabeculations that make up the internal surface of the RV. So the general cardiac morphology is subjectively assessed and the TR jet mapped and graded. The tricuspid valves seldom demonstrate structural changes.

Prognosis

For mild to moderate TR, the prognosis for sport is generally good, even in intensely exercising horses. Severe regurgitant jets will be clinically more significant in exercising horses. Changes in valve structure or motion are a negative finding (thickening or prolapse of the valve, flail cusp). Repeated DE is recommended to monitor progression. In case of chronic RA overload, atrial fibrillation can occur.

AORTIC REGURGITATION

Clinical signs

Aortic regurgitation (AR) is more frequently encountered in adult horses. This higher prevalence in mature horses has been attributed to changes in the quality of the aortic valve (AV) tissue, resulting in a progressive thickening and possible subsequent leakage of the valve.

AR murmurs are usually distinctive holodiastolic murmurs: long, decrescendo, with a PMI at the base of the heart (over the aortic valve). For grades 3 or beyond, the murmur can be very musical or even harsh, however in low grades, these murmurs are sometimes very soft and difficult to pick up if the environment is noisy or the horse fidgety. Intensity of the murmur is not proportional to severity. Palpation of the pulse is another important step in the clinical exam: because LV stroke volume is increased, the difference in diastolic and systolic pressures is increased, resulting in a stronger pulse. Weakening of the pulse is a sign of AR decompensation.

In the majority of cases, mild forms of AR are well tolerated and have little consequence on the horse's performances. When the disease progresses, heart rate (HR) during exercise can be affected. Some horses develop associated arrhythmias during exercise. Exercising ECG should be considered in all athletic horses at work diagnosed with AR.

Diagnosis

Once again DE is the technique of choice to confirm AR diagnosis, suspected on auscultation. The AV is examined from the right parasternal view (long and short axis) to check for structural remodeling or lesions. Abnormal vibrating motion of the valve can be seen during diastole. The AR flow can be detected flowing backwards in the LV during diastole and mapped to appreciate severity. Many normal adult horses will have a small regurgitant flow. The velocity of the regurgitant jet is related to the pressure gradient across the valve during diastole and can be estimated with continuous Doppler. LV size should be carefully assessed and measured as LV volume overload has direct prognostic implications.

Prognosis

If mild cases of AR can perform normally, the question of maintaining a horse with moderate AR at work will depend on the equestrian discipline in which the horse is involved and its level of performance, as well as on the presence of concurrent problems (high HR, prolonged recovery, arrhythmia, myocardial lesions,…) or MR. Aortic incompetence is more frequently associated with arrhythmias during exercise and may generate more acute clinical signs when the condition progresses.

Some horses with relative LV overload are best kept at low to moderate exercise as this can help limit eccentric dilation by maintaining higher myocardial tonicity. Simultaneous MR is a significant finding that compromises the future athletic use of the horse.

REFERENCES

[1] Martin B. et al. **J. Am. Vet. Med. Assoc.**, 2000,216, 554-558.
[2] Morris E and Seeherman H.J. **Equine Vet. J.**, 1991,23, 169-174.
[3] Young and Wood, 2000, **Equine Vet. J.**, 2000,32, 195-199.
[4] Buhl et al.**J Am Vet Med Assoc** 2005,226, 1881-1887
[5] Kriz et al., **J. Am. Vet. Med. Assoc.,** 2000,216, 1441-1445
[6] Young et al., **J. Vet. Intern. Med.**, 2008,22, 418-426.

AN EMERGING ROLE FOR THE EQUINE AND ASININE GAMMA-HERPESVIRUSES?

Philip J. Johnson, BVSc(Hons), MS, DACVIM, DECEIM, MRCVS
Columbia, MO

INTRODUCTION

Equidae may be infected with one or more of nine distinct herpesviruses, of which EHV-2, EHV-5, and EHV-7 (also known as AHV-2) are gamma-herpesviruses.[1] The gamma-herpesviruses are regarded as ubiquitous in horse (and foal) populations worldwide, but specific prevalence may differ depending on the age of the host; numerous studies support this observation, most employing PCR-based detection methods performed on peripheral blood and/or nasal swabs.[2-4] However, EHV-5 appears to be less ubiquitous in normal horse populations than EHV-2. Both EHV2 and EHV5 can be more readily identified (appear to have greater prevalence) in horses affected with airway inflammation than healthy controls.[4]

EQUID HERPESVIRUS-2 (EHV-2)

Equine herpesvirus type 2 (EHV-2, equine cytomegalovirus), a member of the Gammaherpesvirinae[5], is often isolated from tissues of apparently healthy horses.[6,7] Although this virus is clearly ubiquitous in the equine population, its significance as a pathogen has been generally regarded with uncertainty. Respiratory tract inflammation, keratoconjunctivitis superficialis, fever, pharyngitis, inappetance, immune suppression, lymphadenopathy, lymphosarcoma, ataxia, dermatitis, abortion, poor performance, and general malaise have all been attributed to or associated with infection by EHV-2.[7-16] The virus has been isolated from numerous tissue locations following natural exposure including circulating leukocytes, nasal and nasopharyngeal swabs, kidney, bone marrow, spleen, mammary gland, salivary gland, vagina, tracheal mucosa, cornea, and conjunctiva. Additionally, EHV-2 has also been detected in lymphoid and neural tissues (including the trigeminal ganglia) of naturally-infected ponies and horses.[17] In one study, EHV2 was isolated from tracheal aspirates in few healthy foals (1/20 foals) but from two-thirds of foals (20/30 foals) with clinical signs of lower respiratory disease.[18] In that same study, EHV was also identified in peripheral blood mononuclear cells from all foals (both groups). Those results further demonstrate a greater prevalence of EHV2 in lower respiratory secretions in foals with clinically apparent lower respiratory disease, but a cause and effect relationship between EHV-2 and lower respiratory disease remains to be elucidated.[18] Following treatment with dexamethasone (immune suppression), Borchers et al. (1998) inoculated (intranasal) 2 healthy ponies using EHV-2 isolates that had been obtained from naturally-occurring cases of respiratory infection and keratoconjunctivitis superficialis.[19] These experimental ponies developed respiratory tract symptoms, conjunctivitis, coughing and viral shedding accompanied by an antibody response. The authors noted difficulties associated with experimental EHV-2 infections as a result of prior exposure of experimental principals to the virus.[19] Moreover, studies by Purewal et al. (1992) suggested that EHV-2 could play a role in transactivation and, possibly, reactivation of EHV-1.[20]

Brault et al (2010) suggested that the cellular immunopathologic responses to EHV-2 may lead to an immunologically mediated disease of foals that is analogous to infectious mononucleosis caused by Epstein-Barr virus (EBV) infection in humans.[21] Keratoconjunctivitis superficialis has been attributed to and associated with infection by EHV-2.[9,22] Moreover, EHV-2 has been recovered from tissues of apparently healthy horses.[7,22] After natural exposure to infection, virus has been isolated from nasal and nasopharyngeal swabs, kidney, bone marrow, spleen, mammary gland, salivary gland, vagina, tracheal mucus membrane as well as cornea and conjunctiva.[7] Perhaps similar to the situation in human medicine in which both Kaposi's sarcoma and Burkitt's lymphoma have been associated with infection by gammaherpesviruses, Vander Werf (2011) recently reported that equine lymphosarcoma might also be associated with infections by EHV-5.[16] In that study, EHV-5 was identified in 5 of 6 tested lymphoma cases. Moreover, herpes virus particles were identified in lymphoma tissue from one of those horses.

EHV-5, AHV-5 AND EQUINE MULTINODULAR PULMONARY FIBROSIS (EMPF)

Interest in the role of gamma-herpesviruses as a cause of pulmonary fibrosis in equine species was prompted by the identification of asinine herpesvirus-4 and AHV-2 in outbreaks of interstitial pneumonia amongst donkey herds in the mid-west.[23] AHV-5 has also been identified in horses affected with poor performance and respiratory infections[4] (Fortier et al., 2009) and, in horses affected with Equine Multinodular Pulmonary Fibrosis (EMPF).[24] Equine herpesvirus-5 was first recognized in 2 quarantined horses that developed respiratory infections in Australia in the 1970s.[25] More recently, Williams et al., (2007) described a novel chronic and debilitating pulmonary condition of mature (>2 years) horses that has been named EMPF.[26] Many cases of EMPF have subsequently been reported from numerous centers, worldwide.[24,27-34] The Equid herpesvirus 5 (EHV-5; family Herpesviridae, subfamily Gammaherpesvirinae, genus Percavirus 2) has been identified in almost all reported cases of EMPF. Williams (2012) administered EHV-5 directly (via endoscopic guidance) into the lobar bronchus of the accessory lung lobe of 6 healthy horses and demonstrated both an increase in total lung collagen and the appearance of myofibroblasts (alpha-smooth muscle actin IHC), thus largely fulfilling Koch's postulates regarding EHV-5 as the cause of EMPF.[35] Although infection by EHV-5 may directly and independently cause pulmonary fibrosis, there has also been conjecture that severe clinical disease may be facilitated by co-infection when EHV-5 acts alongside other gamma-herpesviruses such as EHV-2[26,27,31] and/or AHV-5.[24] A possible synergistic effect between EHV-2 and EHV-5 is suggested by the observation that both gamma-herpesviruses have been identified in approximately one third of reported cases of EMPF.

Horses affected with EMPF generally exceed 2 years of age and are presented for veterinary examination for a variety of medical problems. Commonly-reported owner complaints include fever, lethargy, inappetance, weight loss, increased respiratory rate, cough, difficult breathing, shallow breathing, exercise intolerance, and a bilateral nasal discharge (not always present). Clinical signs identified upon veterinary examination have also included tachypnea, abnormal breath sounds (increased bronchovesicular sounds, crackles, wheezes), tracheal rattle upon tracheal auscultation, and mandibular lymphadenopathy. Other less commonly-reported clinical signs that might result from EMPF include unwillingness to move, acute-onset respiratory distress, oral cavity ulcerations, and keratoconjunctivitis.[28,31,36,] Abnormalities detected by routine clinico-pathological laboratory tests usually include hyperfibrinogenemia, leukocytosis, neutrophilia, lymphopenia, anemia, hypoxemia, and hypoalbuminemia.[24,30] Elevated circulating serum amyloid A concentrations have also been reported. Although presently uncorroborated, the viral infection has been proposed to cause immune-mediated destruction of mature cells in the peripheral circulation and/or of hematopoietic precursors in the bone marrow, leading to cytopenia.[37] Hart and others (2008) detected EHV-5 in the bone marrow of a horse with EMPF and pancytopenia.[27]

Endoscopic examination of the airways of EMPF-affected horses often demonstrates the presence of exudate. Cytological examination of bronchoalveolar lavage fluid or a trans-tracheal wash aspirate often reveals mild inflammation (predominantly characterized by segmented neutrophils). In some cases, eosinophilic to amphophilic intranuclear inclusion bodies with chromatin margination may be identified in macrophages obtained for airway cytology. Results of aerobic and anaerobic bacteriological culturing of airway fluid often yields negative results or the presence of opportunist bacterial pathogens. Thoracic radiographic changes are often striking and characterized by a diffuse interstitial pattern associated with air bronchograms and thickened peri-bronchial walls and moderately well-defined soft tissue opacities (severe nodular interstitial pattern).[28,33] The radiographic abnormalities are not pathognomonic for EMPF and other forms of interstitial pneumonia must be considered, including idiopathic granulomatous pneumonia or mycotic pneumonia, as well as lung neoplasia (Wilkins 2003).[38] Ultrasonographic abnormalities are usually detected on both or either side of the chest and may include generalized pleural roughening, comet tail artifacts, and sub-pleural hypoechoic nodules (0.5-5.0 cm, diameter).

Gross post-mortem abnormalities identified in EMPF-affected horses are restricted to the lungs and include enlarged lungs that fail to collapse (and often retain rib imprints), extensive thickening throughout the lung parenchyma by multifocal to coalescing whitish to tan, firm, well-demarcated fibrous nodules (varying in size from < 1 cm to 5 cm), and enlargement of bronchial and mediastinal lymph nodes. The pleural surfaces of affected lungs become uneven and severely engorged vessels with a distinct superficial branching pattern are evident (tortuous neovascularization). Interestingly, two distinctly different gross patterns of pulmonary lesion distribution have been described for horses affected with EMPF.[28,29] The 'disseminated nodular form' of the disease is characterized by multifocal, coalescing, white fibrotic nodules varying from <1 to 5 cm in diameter (potentially affecting up to 80% of the lung parenchyma). In this form of the disease, there is very little normal-appearing pulmonary parenchyma. A 'discrete nodular' form of the disease has also been described, albeit much less common than the former type. The 'discrete nodular' manifestation of EMPF is characterized by large discrete nodules (8-10 cm in diameter) that are separated by normal-appearing parenchyma. Histologically, affected areas of the pulmonary parenchyma are characterized by marked, diffuse interstitial fibrosis, most prominently observed in subpleural spaces that multifocally coalesce into ill-defined fibrous nodules. Alveolar septa are usually expanded by collagen, and alveoli are lined by low cuboidal epithelium resulting from Type 2 pneumocyte hyperplasia. Air spaces often contain large numbers of neutrophils within the lumen mixed with fewer alveolar macrophages, multinucleated giant cells, and eosinophilic proteinaceous material. Intranuclear inclusions may be seen in either or both macrophages and epithelial cells in some cases.[26,33,34]

Although the diagnosis of EMPF must be corroborated based on results of histopathological characterization (lung biopsy or post-mortem examination), clinical suspicion is certainly supported by results of ultrasonographic and radiographic imaging and airway cytology (especially if intranuclear inclusions are identified). Diagnosis of EMPF is further strengthened if the presence of EHV-5 is demonstrated in airway fluid (bronchoalveolar lavage) or lung tissue (biopsy or post-mortem) using molecular diagnostic techniques (PCR). It remains to be seen how important that other gammaherpesviruses (EHV-2, AHV-5) are for the development of EMPF.

EMPF-affected horses are typically treated supportively and symptomatically with broad-spectrum antimicrobials (guided by airway bacteriological culture results), anti-inflammatory agents (including corticosteroids), and oxygen (intranasal). The herpes anti-viral drug acyclovir has been used for the management of some cases of herpes myeloencephalopathy and EMPF with variable to disappointing results reported.[28,33,39] When administered orally to horses, acyclovir (10 or 20 mg/kg) has poor bioavailability and there is marked variability in serum concentrations between horses.[40,41] However, this treatment approach may result in adequate antiviral plasma concentrations in some horses.[42] If antiviral drug treatment for EMPF is eventually shown to be efficacious, the acyclovir pro-drug valacyclovir may represent a better choice. Valacyclovir has much greater bioavailability (compared with acyclovir) when administered orally.[43] It has been suggested that EMPF might – not surprisingly – be complicated by the development of pulmonary hypertension in some cases.[28] Pulmonary hypertension might be suspected on the basis of radiographic and echocardiographic findings. Pulmonary artery dilation was observed post-mortem in one EMPF-affected horse.[28] It has been suggested that pulmonary hypertension is a poor prognostic indicator in horses affected with EMPF; pulmonary artery diameter (in comparison to aortic diameter) and pulmonary artery pressure should be determined if possible. Pulmonary hypertension (& right ventricular dilation) was previously reported in a horse affected with interstitial lung disease.[44] Suspected pulmonary hypertension was managed in one EMPF-affected horse by administration of sildenafil and furosemide.[39] Other pharmaceutical agents that are sometimes recommended for the management of EMPF have included nebulized N-acetylcysteine, pentoxifylline, and doxycycline.

The gammaherpesviruses (especially EHV-5, but likely including AHV-5) are interesting in that, although they are identified in healthy horses worldwide, they have the propensity to cause severe interstitial lung disease (pulmonary fibrosis) under some circumstances in some horses. Although Williams (2012) demonstrated that EHV-5 is capable of provoking pulmonary fibrosis in the absence of other co-factors or other gammaherpesviruses, the lesions provoked by his experimental challenges were mild, leaving open the possibility that the etiology of EMPF in nature may be more complicated.[35] It is always difficult to ascribe primary and independent responsibility to the gammaherpesviruses because they can so often be identified in healthy horses. Recognition of the etiological link between EHV-5 and EMPF was facilitated by the fact that gammaherpesviruses (especially Epstein-Barr virus) may cause pulmonary fibrosis in humans and that asinine gammaherpesviruses was identified as etiological agents of interstitial pneumonia (pulmonary fibrosis) in donkeys.[23] Moreover, a murine gammaherpesvirus (MHV68) may contribute to the pathogenesis of pulmonary fibrosis in inbred mice (with propensity to develop Th-2 biased T-Helper lymphocyte responses). Current thinking suggests that gammaherpesviruses may lead to pulmonary fibrosis following infection of and injury to alveolar epithelial cells if there is a subsequent Th-2 biased T-Helper lymphocyte response.[45,46] As is the case for human patients affected with interstitial lung disease (the third leading cause of death for people in the USA), identification of etiological agents for affected equids is often evasive or impossible. The fact that EHV-5 is an important etiological consideration for interstitial pneumonia/pulmonary fibrosis in horses (and that AHV-5 is similarly important in donkeys) represents a significant improvement in our understanding of this debilitating and often fatal condition in our veterinary patients.[23] Hitherto, pulmonary interstitial fibrosis had commonly been regarded as idiopathic in horses.[47] EMPF is distinct from other fibrosing pulmonary conditions of equids in that it has a nodular pattern, there is remodeling of lung architecture and eosinophilic intranuclear viral inclusion bodies may be identified within intraluminal macrophages.[31]

FINAL COMMENTS

It has long been recognized that the equine and asinine gammaherpesviruses are often readily detected in ostensibly healthy horses. In the past decade or so, it has become clear that this interesting group of viral pathogens may play a significant role in the pathogenesis and risk of a diverse group of disease conditions.

REFERENCES

[1]Davison AJ, *et al.*, **Arch Virol** 2009;154:171. [2]Bell SA, *et al.*, **Vet Microbiol** 2006;116(4):249. [3]Wang L, *et al.*, **Vet Microbiol** 2007;121(1-2):18. [4]Fortier G, *et al.*, **Vet J** 2009;182(2):346. [5]Telford EAR, *et al.*, **Virol** 1993;195:492. [6]Harden TJ, *et al.*, **Vet J** 1974;50:483.[7]Browning GF, *et al.*, **Vet Bull** 1988;58:775. [8]Pálfi V, *et al.*, **Zentralbl Veterinarmed B** 1978;25:165. [9]Miller TR, *et al.*, **Equine Vet J Suppl** 1990;(10):15. [10]Collinson PN, *et al.*, **J Am Vet Med Assoc** 1994;205:329. [11]Borchers K, *et al.*, **Arch Virol** 1997;142:917. [12]Dunowska M., *et al.*, **N Zeal Vet J** 2002;50:132. [13]Galosi CM, *et al.*, **J Vet Diagn Invest** 2005;17(5):500. [14]Sledge DG, *et al.*, **Vet Pathol** 2006;43(4):548. [15]Richter N., *et al.*, **Pferdeheilkunde** 2009;25:38. [16]Vander Werf KA, *et al.*, **J Vet Intern Med** 2011;35(3):673. [17]Edington N, *et al.*, **Equine Vet J** 1994;26:140. [18]Murray MJ, *et al.*, **Equine Vet J** 1996;28(6):432. [19]Borchers K, *et al.*, **Virus Res** 1998;55(1):101. [20]Purewal AS, *et al.*, **J Gen Virol** 1992;73:513. [21]Brault SA, *et al.*, **Vet Immunol Immunopathol** 2010;137(1-2):136. [22]Thein P, *et al.*, **Zbl Vet Med** 1976;23:507. [23]Kleiboeker *et al.*, **J Vet Diagn Investig** 2002;14:273. [24]Back H, *et al.*, **Acta Vet Scand** 2012;54:57. [25]Studdert MJ, *et al.*, **Aust Vet J** 1970;46:83. [26]Williams KJ, *et al.*, Vet Pathol 2007;44(6):849. [27]Hart K. A., *et al.*, **Equine Vet Educ** 2008;20:469. [28]Wong DM, *et al.*, **J Am Vet Med Assoc** 2008;232(6):898. [29]Kubiski SV, *et al.*, **J Am Vet Med Assoc** 2009;235(4):381. [30]Williams K, *et al.*, **Pferdeheilkunde** 2009;25:598. [31]Niedermaier G, *et al.*, **Vet Rec** 2010;166:426. [32]Lehmbecker A, *et al.*, **Tierarztl Prax Ausg G Grosstiere Nutztiere** 20122;39(4):237. [33]Marenzoni ML, *et al.*, **J Vet Diagn Invest** 2011;23(4):802. [34]Soare T, *et al.*, **Vet Rec** 2011;169(12):313. [35]Williams KJ. Proceedings 30th **Symp Vet Comp Resp Soc** 2012:34. [36]Vengust M., *et al.*, **J Vet Diagn Invest** 2008;20:811. [37]Smirnov SV, *et al.*, **Virology** 2008;360:6. [38]Wilkins PA. **Vet Clin N Am: Equine Pract** 2003;19:101. [39]Wilkins P. **Proceedings 22nd Am Coll Vet Intern Med Forum** 2004;178. [40]Wilkins PA, *et al.*, **J Vet Emerg Crit Care** 2005;15:174. [41]Bentz BG, *et al.*, **J Vet Intern Med** 2006;20:589. [42]Wilkins P. **Proceedings 22nd Am Coll Vet Intern Med Forum** 2004;170. [43]Garre B, *et al.*, **Antimicrob Agents Chemother** 2007;51:4308. [44]Cargile J. **Proceedings 11th Am Coll Vet Intern Med Forum** 1993;598. [45]Doran P, *et al.*, **Am J Physiol Lung Cell Mol Physiol** 2005;289:L709. [46]Mora AL, *et al.*, **Am J Physiol Lung Cell Mol Physiol** 2005;289:L711. [47]Donaldson MT, *et al.*, **Equine Vet J** 1998;30:173.

THEILERIA EQUI: A REVIEW OF CLEARANCE THERAPEUTICS AND NOVEL DIAGNOSTIC TECHNIQUES

L. Nicki Wise DVM, DACVIM and Don Knowles DVM, PhD, DACVP
Pullman, WA

INTRODUCTION

Theileria equi, one of the causative agents of equine piroplasmosis, is a tick borne, global apicomplexan parasite of horses. Infection of naïve equids results in hemolytic anemia and systemic illness which can be fatal if untreated. Resolution of initial infection invariably results in life-long persistence and subsequent tick borne or iatrogenic transmission risk. The recent re-emergence of equine theileriosis within the U.S. has exposed significant limitations of current infection and disease control strategies[1]. It is estimated that only 10% of horses worldwide reside in non-endemic regions making the task of remaining free of infection and disease increasingly challenging. The horse industries of endemic nations suffer significant economic losses due to poor performance, diagnostics, treatment costs and, most profoundly, international trade restrictions with non-endemic nations. In 2005, it was estimated that the United States generated 462 million dollars in equid exportation. Therefore it is clear that reestablishment of this parasite within the United States would have a profound impact on our economy. Currently, detection of infected horses followed by mandated quarantine or elective euthanasia is being enforced but this strategy is economically unsustainable. Clearance of the organism from an infected horse is not recommended in endemic countries since life-long immunity (premunity) is assumed to be conferred with chronic, inapparent infection. However, in non-endemic regions attempting to remain free of theileriosis, treatment of infected horses with the intent of clearance (chemosterilization) is desired. Research outlined in this review demonstrates current techniques for detecting infection, achieving chemosterilization and effectively exhibiting clearance of this organism.

PATHOGENESIS & CLINICAL RESPONSE

The biological nature and life cycle of this organism contributes substantially to obstacles of infection and disease prevention. The details of the life cycle of *T. equi* have not been fully elucidated and may vary depending on the tick species involved. In general, as infected ticks feed on horses, *T. equi* sporozoites are injected into the new host with the tick's saliva. As with other theilerial species, these sporozoites first invade peripheral blood mononuclear cells (PBMCs) which can be detected 2-3 days after transmission[2]. Asexual reproduction leads to mass production of merozoites which rupture from PBMCs and enter the circulation to invade erythrocytes. Inside the erythrocyte, they replicated asexually, causing physical rupture of infected erythrocytes and release of merozoites into the circulation. The parasite is transiently free in blood without succumbing to or stimulating a pathogen eliminating immune response. This allows for invasion of new erythrocytes and continual parasite replication. Some merozoites transform into ring forms within the erythrocyte, termed gamonts. Ticks can ingest gamonts, where they grow in the tick's midgut and fuse, forming zygotes. Inside the zygote a kinete is formed which penetrates tick salivary gland cells. In tick salivary cells, sporonts, sporoblasts, and then sporozoites are formed.

The typical clinical response to acute infection is varying degrees of anemia due to parasite induced erythrolysis. Hemolysis results from the physical rupture of erythrocytes during release of merozoites (intravascualar) as well as from removal of infected erythrocytes from the circulation by the spleen (extravascular). Other clinical signs of acute infection include fever, lethargy, peripheral edema, petechiations (due to thrombocytopenia) and associated systemic sequelae from the infection. The severity of these signs varies from life threatening to quite mild. Regardless of clinical severity, after the acute phase of infection, equids become asymptomatic carriers and serve as an efficient reservoir for both natural tick and iatrogenic transmission. Critical to tick transmission is the ability of *T. equi* to persist in the individual equid host at levels of 10^3 to 10^6 infected erythrocytes per milliliter of blood[3]. Parasite levels in the clinically normal horse enable efficient vector-borne transmission and contribute to the difficulty in maintaining an infection free status. The mechanism by which this parasite persists at this high level within the immunocompetent horse without causing apparent disease are unknown.

IMMUNITY & THE EMA PROTEINS:

A better understanding of the correlates of immune control is essential to delineating how this parasite persists and to improve our detection and control measures. Both the innate and adaptive immune systems are important for control of *T. equi* infection yet the exact role each of these plays remains unclear. The spleen aids in eliminating hemoprotozoan parasites and equids with intact spleens are typically able to control *T. equi* infections and survive. Splenectomized horses infected with *T. equi* develop terminally high levels of parasitemia and inapparent carriers of *T. equi* will also develop fatal disease upon surgical removal of the spleen[4,5]. Despite its important role, the spleen and innate immune system are insufficient for protection against *T. equi* in the absence of adaptive immunity. SCID (severe combined immunodeficiency) foals, which lack mature T and B lymphocytes and are incapable of antigen-specific immune responses, have intact spleens and a competent innate immune system. SCID foals infected with *T. equi* are unable to control parasite replication and become severely anemic[6]. This demonstrates that the spleen is unable to control *T. equi* infection in the absence of parasite-specific immune responses, and that erythrolysis caused by *T. equi* does not require a specific immune response.

T. equi infected horses produce antibodies against proteins on the surface of the merozoites, termed equi merozoite antigens (EMAs). Recent completion of the *T. equi* genome revealed that 10 genes make up the *ema* family[7]. To date, the surface proteins EMA 1, 2, 3 & 4 have been documented and demonstrated to stimulate an immune response in the infected horse. High antibody titers to EMA1 and 2 are correlated with control of parasitemia[8]. During acute stages of infection with *T. equi*, high levels of IgGa (now termed IgG1) and IgGb (IgG 4&7) correlate with control of parasitemia. IgG(T) (IgG3 & 5) levels increase during the phase of chronic infection. Specific antibodies to these EMA proteins are first detected 7 to 11 days after experimental infection in ponies and reach a peak 30 to 45 days after infection. The antibody titer against an epitope conserved on EMA1/EMA2 is the basis for the competitive ELISA, the regulatory test mandated by the OIE for international horse transport[9]. Despite the fact that antibody responses to EMAs is correlated with control of parasitemia, it is unable to eliminate the parasite and the antibodies directed to the EMA1/EMA2 epitope and EMAs in general continue to be detectable after elimination of the parasite.

CLEARANCE THERAPEUTICS:

As stated above, infection of equids with *T. equi* without aggressive chemotherapeutic sterilization results in life-long persistence of the parasite. Currently, the only drug shown effective in eliminating *T. equi* infection is imidocarb diproprionate (ID). Current data indicate that imidocarb diproprionate through 4 doses of 4 mg/kg can eliminate *T. equi* infection. This was demonstrated both in experimentally and naturally infected horses within the United States[1,10]. In both instances, clearance (chemosterilization) was defined as negative nested polymerase chain reaction (PCR) results and inability to transmit the parasite to a naïve splenectomized horse via blood transfusion. Ueti and others reported that chemosterilization was achieved in 25 naturally infected horses using the aforementioned treatment regimen. One of the 25 horses required an additional treatment before clearance was accomplished. Since the publication of this paper, a total of 171 horses involved in this outbreak have been treated and successfully cleared. Additionally, Grause and others demonstrated that clearance of experimentally infected horses could also be achieved at this dose. In this study, one of the 6 treated horses was unable to be cleared after 2 courses of imidocarb therapy. A third treatment was not attempted.

The results of these studies, in which one horse required an additional treatment and another was not cleared, offers the possibility that *T. equi* resistant parasites and/or horses exist. While imidocarb is relatively safe, especially when used in conjunction with anticholinergic drugs, therapy does not come without risk. Common side effects noted after administration include sweating, signs of agitation, colic and diarrhea and horses undergoing therapy with this drug should be monitored carefully for development of complications[11, 12]. Transient azotemia and/or elevations in urinary GGT/creatinine ratios and transient elevations in liver enzymes (AST, ALT, ALP, and SDH) may be observed during treatment but typically resolve with discontinuation of the drug. ID toxicity has been reported and can be fatal. Donkeys and mules are exquisitely sensitive to ID therefore its use in these species is not recommended. In the United States, if a horse is diagnosed with *T. equi*, the owner and veterinarian can enroll the horse in the USDA controlled treatment program as to ensure appropriate quarantine, treatment and subsequent release of cleared horses.

DIAGNOSIS: CURRENT AND NOVEL TECHNIQUES

Currently the test mandated by the OIE for international horse transport is the competitive ELISA (cELISA). As mentioned previously, antibodies made by the infected horse against EMA 1 or 2 are detected to yield a positive result. Once a horse is infected it can take up to 12 weeks to reach high enough levels to be positive[10]. In the acute phase of infection, blood smears can be used to make a diagnosis or alternative serum antibody tests can also be performed. The complement fixation test (CFT), which detects antibodies that fix complement, can yield a positive result in as little as 8 days post infection[13]. However, as mentioned above, the anti-EMA antibody profile changes as the horse enters the chronic phase of infection. At this phase, the predominant antibody is IgG5, which is unable to fix complement. This explains the insensitive nature of the CFT for the chronically infected horse. The cELISA is however the most sensitive test in the chronic/inapparent carrier phase making it the ideal test for screening prior to international transport[14].

During the collection of imidocarb diproprionate clearance data, it became apparent that although the parasite was eliminated from the horse, the cELISA continued to be positive, indicating that horses continued to harbor antibodies against the EMA proteins[1]. The reason for this phenomenon remains unclear but in human literature, the role of the rogue memory B cells is well documented and often implicated in measurable antibody long after an infection has been eliminated[15]. Of the 25 cleared horses reported in the paper by Ueti et al, all of these horses eventually became cELISA negative yet the time to negative status was highly variable between horses and in some cases, remained positive for up to 24 months post clearance. These results indicate that the imidocarb therapy is working but a test other that the cELISA is needed to demonstrate clearance in a reliable timely manner. Without an alternative test, most cleared horses remain positive in the eyes of the regulatory agencies for much longer than necessary creating a burden for the owner and the horse.

PCR tests for the presence of the organism of interest by amplifying and detecting parasite specific DNA. This test is exquisitely sensitive. Three variations of PCR include real-time PCR, nested PCR and nested PCR with hybridization.

Comparison of these tests and the results from these tests is difficult given that no standardization exists between laboratories. Nested PCR for *T. equi* using the EMA-1 gene sequence has been shown to detect a positive result equivalent to a percent parasitemia of 0.000006%[16]. One report also indicated that nested PCR detects 3.6x more infections that microscopy and 2.2x more than traditional PCR[17]. Given the nature of PCR as a test, in order for it to be used as a routine diagnostic and surveillance test, a significant amount of standardization is required[18]. As it is currently performed, it is unlikely that nested PCR will ever be standardized and commercially marketed for detection of *T. equi* infection.

Western blot (WB) functions as a diagnostic test by detecting an antibody reaction to a specific antigen using gel electrophoresis and an immunoblot format. Serum samples from horses that are infected with *T. equi* demonstrate a characteristic banding pattern on WB, as the anti-EMA antibodies react with merozoite antigen. A positive result can be detected as early as 7 days post experimental infection. In horses that have been cleared with imidocarb, this characteristic banding pattern diminishes as early as 4 weeks post treatment. Although validation of this test is underway, the WB test for clearance is a reliable, timelier means of demonstrating successful chemosterilization.

CONCLUSION

As it has for decades, infection with *T. equi* continues to disrupt global equid movement. The economic impacts include interference with international completion and general movement and trade. Advances in diagnostics and chemotherapeutics are offering to greatly reduce the impact of *T. equi* infection on the global movement and trade of equids.

REFERENCES

(1)Ueti MW, et al. **PLoS One** 2012;7:e44713. (2) Mehlhorn H, et al. **Parasit Res** 1998;84:467-475. (3) Ueti MW, et al. **Infect Immun** 2008;76:3525-3529. (4) Ambawat HK, et al. **Vet Parasit** 1999;85:319-324. (5) Guimarães AM, et al. **J Eq Vet Sci** 1997;17:211-216. (6) Knowles DP, et al. **Infect Immun** 1994;62:1909-1913. (7) Kappmeyer LS, et al. **BMC Genomics** 2012;13:603. (8) Cunha CW, et al. **Clin Vaccine Immunol** 2006;13:297-300. (9) Knowles DP, et al. **Clin Microbiol** 1991;29:2056-2058. (10) Grause JF, et al. **Vet J** 2012, Nov 28 epub. (11) Belloli C, et al. **Equine Vet J** 2002;34:625-629. (12) Adams LG. **Res Vet Sci** 1981;31:54-61. (13) Brüning A. **Br Vet J** 1996;152:139-151 (14) Knowles DP, et al. **J Clin Microbiol** 1992;30:3122-3126. (15) Bernasconi NL, et al. **Science** 2002; 298(5601):2199-202. (16) Nicolaiewsky TB, et al. **Vet Parasit** 2001;101:9-21. (17) Rampersad J, et al. **Vet Parasit** 2003;114:81-87. (18) Yang S, et al. **The Lancet Inf Dz** 2004;4:337-348.

ADVANCES IN THE DIAGNOSIS AND TREATMENT OF BOTULISM

Amy L. Johnson, DVM, DACVIM (LAIM & Neurology)
Kennett Square, PA

INTRODUCTION

Botulism is a disease of people and animals exposed to the neurotoxin of *Clostridium botulinum* (BoNT), which causes progressive flaccid paresis and cranial nerve deficits. *C. botulinum* is a spore-forming, anaerobic, Gram-positive bacterium with 8 different serotypes (A-G, with C-α and C-β subgroups), each of which is distinguished by the unique properties of its toxin. These toxins are antigenically different and act specifically on various parts of the synaptic vesicle docking and fusion apparatus at the motor end plate to prevent release of acetylcholine. BoNT is widely regarded as the most potent and lethal toxin known to humankind; although it has never been used successfully as a bioterrorism agent, the potential for this use has generated a great deal of interest and research in strategies for its detection, treatment, and prevention.

EPIDEMIOLOGY

Natural intoxication has been well-documented to occur in people and animals exposed to BoNT. Botulism in horses can be categorized in two ways, either by the *C. botulinum* serotype involved or by the route of acquisition of disease. Certain serotypes are prevalent in certain geographic regions. In the United States, type B in endemic in the mid-Atlantic states and Kentucky, while type A is reported primarily in the western United States (California, Idaho, Montana, Oregon, Nebraska,Washington, and Wyoming).[1,2] Type C is seen sporadically across the United States, whereas confirmed cases of types D-G have not been reported in horses. Both type B and type A are associated with soil, while type C is associated with carrion. Equine type B cases are most common and are estimated to account for over 85% of equine cases diagnosed in the United States. Horses with botulism almost always acquire the disease in 1 of 3 ways: ingestion of preformed toxin with food ("food-borne" botulism in adults), ingestion of *C. botulinum* spores that subsequently germinate in the gastrointestinal tract and elaborate toxin ("toxicoinfectious" botulism in foals), or contamination of wounds with *C. botulinum* and subsequent bacterial growth with toxin release ("wound" botulism). Food-borne botulism in adults and toxicoinfectious botulism in foals are more common than wound botulism. Considering both serotype and route of acquisition, several trends are evident: toxicoinfectious botulism due to type B is common in foals in the mid-Atlantic region and Kentucky, food-borne botulism in adult horses most commonly results from spoiled forage contaminated with type B (in the eastern part of the country) or type A (in the western part), and food-borne botulism due to type C is occasionally seen when carrion contaminates forage.

CLINICAL DIAGNOSIS

The gold standard of botulism diagnosis is identification of BoNT in serum samples from patients with compatible signs by using the mouse bioassay test. However, diagnosis is often based on compatible history and clinical signs, since laboratory confirmation with the mouse bioassay can be difficult. Regardless of botulinum toxin type, the clinical signs of intoxication are the same: progressive flaccid paresis or paralysis and cranial nerve deficits such as dysphagia, weak tongue tone, weak eyelid tone, and slow pupillary light reflexes. Two clinical tests have been described to diagnose botulism in the horse. The first is known as the tongue tone or tongue stress test, in which the examiner gently withdraws the tongue from the horse's mouth and assesses the ability of the horse to pull it back while holding the jaw closed. The second is known as the grain test and involves feeding the horse 8 ounces of grain in a bucket while timing how long it takes for the horse to consume the feed. Normal horses finish in less than 2 minutes, whereas horses with dysphagia do not. Additionally, horses with botulism frequently drop grain admixed with saliva from their mouths or noses and may leave a trail of saliva in the bottom of the feeding pan.

Electrodiagnostic testing may be performed as an adjunct to the neurologic examination and has been described in 3 foals with botulism.[3] As with clinical examination, electrodiagnostic testing cannot provide a definitive diagnosis. However, findings can confirm a neuromuscular problem and support a diagnosis of botulism. Results from those three foals were similar to those from human patients including decreased baseline compound motor action potentials (CMAPs) and an incremental response to repetitive nerve stimulation.

LABORATORY DIAGNOSIS

Although clinical diagnosis by experienced practitioners is probably reasonably accurate in predicting botulism, and serotype may be predicted based upon known serotype prevalence in a geographic region and likely means of exposure, confirmation of serotype must be made in the diagnostic laboratory. For suspected equine cases or outbreaks, the most appropriate samples to submit include forage, gastrointestinal contents, and feces. Serum might be appropriate if the case is particularly severe and rapidly progressive. As aforementioned, the gold standard for laboratory diagnosis is the mouse bioassay, which detects BoNT.

Unfortunately, detection of BoNT in equine samples is difficult. When compared to other species, horses are highly sensitive to BoNT and develop clinical signs after exposure to smaller amounts of toxin. Any circulating toxin is quickly bound to receptors and internalized in motor endplates, becoming undetectable. Furthermore, toxin remaining in the gastrointestinal tract is degraded by microbial organisms and their enzymes. Although preformed toxin may be detected by the mouse bioassay in a small number of foal cases (approximately 20%), it is rarely detected in adult cases. Therefore, our laboratory routinely repeats the mouse bioassay on samples that have undergone culture enrichment, which allows any spores in the original sample to germinate and elaborate toxin, thus increasing the diagnostic yield of the assay. These culture-enriched samples are positive in approximately 70% of suspected foal cases and 30% of suspected adult cases. The identification of *C. botulinum* spores in the original sample is considered presumptive evidence of botulism in animals with compatible clinical signs, as this organism is not part of the normal gastrointestinal flora.

Performing the mouse bioassay on both non-enriched and enriched samples extends the lag time for results. The protocol involves an initial overnight extraction in gelatin phosphate buffer. The samples are then divided into two aliquots, one of which is used to detect preformed toxin and one of which is used to detect *C. botulinum* spores. The first aliquot is centrifuged and the supernatant, which contains any preformed toxin from the original sample, is injected into mice. The second aliquot is culture-enriched for 5-7 days prior to performing the mouse bioassay. The mouse bioassay may take up to 4 days to complete; results are obtained more quickly when more BoNT is present in the original or enriched sample. Development of clinical signs of botulism in mice injected with the non-culture-enriched sample indicates the presence of preformed BoNT in the original sample. Development of clinical signs of botulism in mice injected with the culture-enriched sample indicates the presence of *C. botulinum* spores in the original sample. If either test is positive, the mouse bioassay is repeated to determine *C. botulinum* serotype. Aliquots of the positive sample are mixed with serotype-specific monovalent antitoxins prior to inoculation, with the expectation that the only surviving mice will be those receiving the protective antitoxin.

PCR technology has been investigated to improve diagnostic accuracy and speed while reducing live animal use. This modality is particularly appropriate for equine samples, which more commonly contain *C. botulinum* spores than preformed toxin. Our laboratory has optimized and validated a quantitative real-time PCR assay for the detection of the neurotoxin gene of *C. botulinum* type B in equine diagnostic samples; results indicate that this assay is faster, more sensitive, and more economical than the mouse bioassay.[4] Following these promising results, PCR capabilities were expanded to a multiplex assay for types A and B as well as a singleplex assay for type C. Unpublished results indicate that both the A/B multiplex assay and the C singleplex assay are highly specific and more sensitive than the mouse bioassay for detecting *C. botulinum* in equine samples.

Although PCR is a very useful assay for equine samples, it detects the neurotoxin gene rather than the biologically active neurotoxin and therefore cannot completely replace the mouse bioassay. Cell-based assays, immunoassays, and spectrometric assays have been described to detect BoNT. Does this mean that despite the high sensitivity and specificity, the PPV might not be high? Meaning, not necessarily linked to disease. Unfortunately, many of these assays have only been validated for relatively simple matrices (water, milk, serum) as opposed to the complex sample materials (forage, gastrointestinal contents) of equine samples. Therefore, optimization and validation using equine samples will be required before any of these methodologies can be utilized.

TREATMENT

Antitoxin administration is considered a mainstay of therapy and should be administered as soon as possible. Currently there are 2 commercially available antitoxins: a trivalent antitoxin against types A, B, and C (Lake Immunogenics, Ontario, NY) and a monovalent antitoxin against type B (Plasvacc, Templeton, CA). What is the current recommendation: give antitoxin based on geographical location (type B if in in Mid-Atlantic states and KY) and trivalent elsewhere? Or should we always be using trivalent "just in case". How many units of antitoxin is recommended to give. Antitoxin will only bind circulating toxin, not toxin bound at motor end plates or internalized, so clinical signs can continue to progress for 24 hours after administration. Since toxin is irreversibly bound and regrowth of motor end plates is required for recovery, administration of antitoxin will not reverse clinical signs.

Supportive care is an essential component of therapy. Dysphagic horses require their hydration and nutritional requirements to be met through enteral (preferred) or parenteral administration. Horses with moderate to severe myasthenia require thick bedding and/or padding to prevent complications associated with excessive recumbency. Even horses receiving high-quality nursing care will develop necrosis of skin and underlying muscle over pressure points if recumbency is prolonged. Horses should not be forced to rise frequently or stand for prolonged periods, as depletion of acetylcholine stores will increase weakness and lead to progression of signs. Therefore, use of sling support is generally not recommended during initial treatment (first 1-3 days). However, after antitoxin treatment and stabilization of signs some severely affected horses will benefit from use of a sling to assist them to rise. Evacuation of the bladder and rectum may be necessary.

Complications will occur in severely affected, hospitalized horses. These may include decubital ulcers, cellulitis, myositis, colic, cystitis, pneumonia, and salmonellosis.

PROGNOSIS

A dataset of 30 foals treated for botulism at New Bolton Center showed an overall 96% survival rate among treated foals, with an 88% survival rate among the 8 treated foals requiring mechanical ventilation.[5,6] Based on laboratory confirmation (when available) and characteristics of the referral area, these foals likely all had type B botulism.

There are no large retrospective studies investigating the outcomes of adult horses with botulism. Based on samples received and consultations performed by the National Botulism Reference Laboratory at New Bolton Center, adult equine mortality is estimated to be about 75%. This figure is in contrast to the adult human mortality rate, which is <10%. Clearly the discrepancy is due to several factors. Early symptoms of human botulism include double or blurred vision, slurred speech, and dry mouth, which often prompt early medical treatment. Similar early signs would not be detected in horses and therefore equine patients generally are not presented for care until the disease progresses to severe paresis of pharyngeal, laryngeal, and appendicular muscles. Owners may be dissuaded from seeking appropriate treatment due to the financial cost; veterinary bills typically run from thousands to tens of thousands of dollars depending on the severity of intoxication. Finally, the inability to provide adequate supportive care, including mechanical ventilation, for the more severely affected adult equine patients lowers survival rate.

It is possible that serotype influences mortality. People with type A botulism tend to have more severe disease than those with type B, with faster onset of paralysis and increased likelihood of requiring mechanical ventilation, and a higher case fatality rate.[7] Although limited data are available, the same trend may be true for horses, with reported type A outbreaks having an overall case fatality rate of 90-100%, type B outbreaks having an overall case fatality rate of about 70%, and one type C outbreak having a case fatality rate of 80%.[1]

PREVENTION

Only one USDA-approved vaccine against equine botulism is available in the United States (BotVax® B, Neogen Corporation). This product is a killed (toxoid) vaccine directed against *C. botulinum* type B. The initial series consists of three doses administered at 4 week intervals, with subsequent yearly boosters. Pregnant mares should receive their primary series and then be revaccinated 4-6 weeks prior to foaling. Passively acquired maternal antibodies do not appear to interfere with the foal's serologic response, so foals may have their primary series started as early as 2 weeks of age, although it is more commonly started at 1-3 months of age.

Unfortunately, there are no licensed vaccines available for preventing types A or C botulism, and there is no cross-protection between types.

Horses affected by botulism may not develop protective immunity, as the amount of toxin required for intoxication is less than that required to stimulate an immune response. Therefore, affected horses should be vaccinated during or after recovery assuming that *C. botulinum* type B is the serotype of concern.

BoNT AS A PHARMACEUTICAL

Although BoNT is the most potent known toxin, it is also the most frequently used toxin in medicine.[8] In the 1820s Justinus Kerner, a German doctor and poet, was both the first to describe the symptoms of botulism and the first to recommend BoNT for medicinal purposes including St. Vitus's dance (Huntington's chorea), rabies, mania, excessive sweating, and mucous production. It was not until the mid-1900s that the mechanism of action of BoNT was identified and a stable, standardized form of BoNT/A was produced. Subsequently, animal experiments demonstrated that local injection into chicken embryos caused localized muscle paralysis and that BoNT/A could treat strabismus in monkeys. In 1980 Dr. Alan Scott treated human patients with BoNT/A to correct strabismus, and this drug was approved by the FDA in 1989 to treat strabismus, blepharospasm, and hemifacial spasms. Over the next 12 years, additional approval was received for treatment of cervical dystonia, glabellar lines, hyperhidrosis, and migraine headaches. Additionally, BoNT is widely accepted as a treatment for spasticity and muscle pain disorders, and it is used for countless off-label treatments.[8]

Equine clinicians have investigated BoNT for several indications. Injection of BoNT/B into the anal sphincter was investigated as a potential treatment for mares recovering from perineal lacerations.[9] Treatment of healthy horses resulted in decreased peak anal pressure, indicating that this treatment has therapeutic potential. However, one treated horse developed clinical signs of botulism, highlighting the narrow therapeutic window and potential for adverse effects. In another experimental study, intra-articular injection of BoNT/A was investigated as a potential treatment for synovitis.[10] Results of this pilot study suggested that BoNT-A can attenuate lameness due to experimentally-induced synovitis and may be a potential treatment for arthritis. A case series of horses with laminitis treated by chemical denervation of the deep digital flexor muscle has been published; the lack of a control treatment makes efficacy impossible to determine but no adverse effects attributed to treatment were reported.[11] Finally, BoNT/A has been used in a pilot study to treat stringhalt by reducing muscle activity in the affected leg. Although the abnormal gait was not abolished, the authors concluded that BoNT/A would be a potential treatment for stringhalt and spastic muscular conditions in the horse.[12]

SUMMARY

Equine botulism remains a relatively common problem in endemic areas and a sporadic problem in the rest of the country. Rapid clinical diagnosis and implementation of therapy is essential for a favorable outcome. The availability of PCR assays has improved the accuracy and rapidity of laboratory diagnosis and serotype identification. As additional molecular techniques are validated for equine samples, our ability to identify cases will improve. The mainstays of treatment have not changed but the availability of an affordable, commercially-available multivalent antitoxin and improved large animal slings and facilities might improve our ability to care for severely affected adult horses. As more practitioners begin to use BoNT as a therapeutic drug, clinicians should anticipate the possibility of more iatrogenic botulism cases.

REFERENCES

1. Johnson AL, et al. J Vet Diagn Invest 2010;22:165.
2. Ostrowski SR, et al. J Vet Diagn Invest 2012;24:601.
3. Aleman M, et al. J Vet Intern Med 2011;25:365.
4. Johnson AL, et al. Vet J 2012;194:118.
5. Wilkins PA, et al. J Vet Intern Med 2003;17:702.
6. Wilkins PA, et al. J Vet Intern Med 2003;17:708.
7. Horowitz BZ. Crit Care Clin 2005;21:825.
8. Pellett S, et al. Curr Opin Microbiol 2012;15:292.
9. Adam-Castrillo D, et al. Am J Vet Res 2004;65:26.
10. DePuy T, et al. Am J Phys Med Rehabil 2007;86:777.
11. Carter DW, et al. J Equine Vet Sci 2009;29:595.
12. Wijnberg ID, et al. Equine Vet J 2009;41:313.

ASSESSING CARDIOVASCULAR FUNCTION IN THE CRITICALLY ILL HORSE

Gayle D. Hallowell, MA, VetMB, PhD, CertVA, DACVIM, DACVECC, FHEA, MRCVS
Nottingham, UK

INTRODUCTION

Accurately assessing the cardiovascular system is of paramount importance when evaluating the critically ill foal and horse irregardless of the underlying cause of SIRS, this system is invariably affected. The tools available to assess the cardiovascular system keep expanding. The aims of this talk are to evaluate the value of current and future tools available to assess the cardiovascular system.

SIRS and the cardiovascular system

Although there are many causes of critical illness in horses and foals, many present with certain clinical signs in common. Most are tachycardic and hypotensive, the latter partly due to vasodilatation. These animals however often have a normal to increased cardiac output[1,2] and are in hyperdynamic hypovolaemic shock with some degree of distributive shock.

The difference between the global and microvascular tissue perfusion

Assessing global tissue perfusion can be accomplished using many techniques that are discussed further below. Each of these techniques provides slightly different information, but is mainly focused upon assessing the function of the heart, arteries and veins. These techniques do not however provide information of capillary blood flow and that to organs.

The microcirculation lies between the arteries and veins and is responsible for oxygen and nutrient transport to cells[3]. The microcirculation has been proposed to be the first organ that fails when sepsis progresses to MODS[4,5]. In addition, microvascular dysfunction in human patients with shock, particularly septic shock, occurs despite the presence of normal or supranormal global cardiovascular parameters[4-6], and this is also likely the case in our equine patients with SIRS. It has been proposed that improving microcirculatory blood flow as well as global cardiovascular parameters may improve survival in sepsis[7]. Again this may be of importance in equine patients, where despite increased standards of pre- and post-operative care for horses undergoing colic surgery for strangulating lesions, the short term mortality rate is still 16-26%[8-11]. However assessing the microcirculation is more challenging than that of the global circulation, but new techniques may allow for this to be undertaken in our patients and perhaps help improve survival and reduce morbidity.

ASSESSING GLOBAL TISSUE PERFUSION

Physical examination

Heart rate, mucous membrane colour, capillary refill time, pulse quality and jugular filling provide valuable information regarding global tissue perfusion. Additionally urine output provides information regarding both global perfusion and the microcirculation. This aspect of cardiovascular evaluation will remain the mainstay of initial and repeated assessment of the critically ill patient. However, although increases in heart rate usually are attributed to hypovolaemia, a small component can be due to stress, pain and fear. One study evaluating experimentally induced hypohydration in horses did not find a correlation between reduction in body weight and any vital signs (or PCV)[12].

Basic blood work – packed cell volume (PCV), total solids (TS) and lactate

PCV and TS can be used to assess haemoconcentration and thus directly allowing assessment of hydration status, but care is required in its interpretation. Splenic contraction can increase PCV by 25-30%, occurring secondary to catecholamine release and PCV is breed dependent so what is normal for a Thoroughbred is very abnormal in a pony or draught breed. Total solids measured on a refractometer are pointless following colloid administration as it has been shown not to change, thus over-estimating TS[13]. In addition many critically ill patients have primary GI disease and thus can be concurrently hypoproteinaemic; when this found in conjunction with hypovolaemia, TS can be within normal range.

Lactate concentrations can be very valuable for assessment of poor tissue perfusion. However it is not only tissue hypoperfusion that can cause increased lactate concentrations. Other causes include inappropriate anaerobic metabolism, renal or hepatic dysfunction, increased aerobic glycolyis and abnormalities of aerobic metabolism. In addition, when lactate is measured on initial assessment it will be increased, but will often increase further after initial fluid resuscitation, as lactate is washed from closed capillary beds that are now being perfused. Thus looking too soon and too often can be misleading.

Measurement of urine specific gravity and observation of urine output not only provides information regarding hydration status, but also organ perfusion. Frequently at presentation, these equine patients are anuric so production of any urine can provide valuable information that fluid therapy is being successful as can urine colour, although this varies from horse to horse. Serial measurements of USG will allow a more quantitative approach, where USG greater than 1.040 is suggestive of hypovolaemia and a reduction to less than 1.008 may suggest more than enough fluid has been administered in adult horses. However in critically ill foals, it is easier to measure urine output and targeted endpoints are USG <1.008 and urine production of at least 2ml/kg/hour. Creatinine concentrations can also be used to attempt to identify mild hypovolaemia and response to therapy in the absence of intrinsic renal disease.

Mixed venous oxygen concentrations

This parameter can be very valuable to assess the oxygen concentration in blood after it has passed through the capillary beds. Normal PvO2 is approximately 50mmHg. During low cardiac output states, oxygen extraction increases and thus PvO2 decreases. Other causes of decreased PvO2 include anaemia, decreased oxygen saturation and increased oxygen demand (shivering, hyperthermia, pain, seizures). Increased PvO2 includes decreased oxygen demand (anaesthesia, hypothermia and sepsis or increased oxygen supply (hyperoxia). In animals where this is less than 30-35mmHg, they are more likely to respond to additional therapy, such as fluid therapy, inotropes and pressors, whereas this is unlikely to be true when >60mmHg. Advanced use of this technique may come from the ability to easily measure it continuously.

Electrocardiography

This can be of value in horses with unexplained tachycardia or dysrhythmias. The critically ill horse may be predisposed to dysrhythmias secondary to SIRS, bacteraemia, hypoperfusion and electrolyte derangements. Obviously monitoring electrolyte concentrations, especially sodium, potassium, chloride and magnesium are important in the critically ill animal and even more so in those with dysrhythmias, but is beyond the scope of this talk.

Echocardiography

This can be valuable for interpretation of murmurs in critically ill animals to distinguish between those due to changes in blood viscosity and those due to degenerative valvular disease, or rarely to identify endocarditic lesions. However care should be taken in interpretation of cardiac appearance and dimensions in hypovoloaemic horses as internal ventricular dimensions will be reduced and free wall thicknesses increased when compared with the normovolaemic state[14].

Central venous pressure

CVP is a moderately easy measurement to acquire, but is dependent upon head and body position[15] and trends are felt to be more useful than absolute numbers. Thus measurements taken prior to anaesthesia cannot be compared with those during anaesthesia. Recently it has been shown that CVP correlates with bodyweight, TS and blood lactate concentrations, but not PCV[12]. In addition, jugular and central venous pressures have been shown to correlate in anaesthetized, laterally recumbent horses[16]. This has the advantage that it is slightly easier to perform, will still provide trends, although 'normal' reference ranges will be slightly different.

Arterial blood pressure

When considering the formula for mean arterial pressure (MAP; MAP = (CO/SVR) +CVP), it must be remembered that MAP measures global cardiovascular function, but doesn't provide information about tissue blood flow. There needs to be a 'minimum' mean pressure required to overcome resistance in tissue in order to perfuse it. However, more is not better. In fact, normal to increased blood pressure may be due to increased SVR, which in turn may reduce tissue perfusion. This is of particular relevance in the critically ill animals that secondary to SIRS may be vasodilated. This should be interpreted with caution alongside other measures of global and microcirculation. If this parameter is measured in hypovolaemic adult horses, direct blood pressure is often preferable as readings will be intermittent with indirect measurements obtained using a cuff around the tail head. Indirect blood pressure measurements are more successfully obtained in the sick foal from the distal limb. This modality is most frequently used in critically ill foals and in some critically ill adults, especially when inotropes and pressors form part of the management.

Cardiac output

In the last 15 years, there have been 10's of publications evaluating or using various methods to measure cardiac output in adult horses and foals. The gold standard for measurement of cardiac output is aortic impedance, which requires an implant being placed around the aorta and is obviously not practical for use in a clinical setting or even an experimental setting in the horse. Beyond this, all invasive techniques rely on one of two methods: Fick oxygen method or indicator dilution. Until 10 years ago, thermodilution[17-19] was by far the most commonly used method to measure cardiac output and this is an adaptation of the indicator method using thermal energy. Other invasive methods reported to have been used in the horse include lithium dilution[1, 2, 20-21], and its main advantage over thermodilution is that it does not require a catheter being placed in the pulmonary artery.

Less invasive methods for measuring cardiac output that have been evaluated in the horse include Doppler transoesophageal echocardiography[22-23], which requires anaesthesia, and also transthoracic Doppler echocardiography[24], which has wide limits of agreement, likely due to difficult alignment to the aorta and/or pulmonary artery. Newer methods using echocardiography, particularly the bullet method can be used in conscious patients[25-26] and have narrower limits of agreement when compared to previous attempts using this technique. Pulse contour analysis has also been used in the horse[2, 27] and calculates cardiac output from the area under the arterial pressure waveforms after it has been calibrated with another method. The advantage of this technique is that it allows cardiac output to be continuously reported. Another continuous method that has been used in the horse uses the Fick principle and involves partial rebreathing of carbon dioxide [28-29]. This is a very valuable measure to have available in the critically ill horse, but is a measure of global circulation and not that going to the microcirculation or ultimately the organs.

ASSESSING THE MICROCIRCULATION

Many methods available to assess the microcirculation include crude measures such as blood lactate concentrations, intestinal capnography and mixed venous oxygen concentrations[30-33]. Laser Doppler and tissue O_2 measurements satisfactorily detect global decreases in tissue perfusion but not heterogeneity of microvascular perfusion. However orthogonal polarization spectral (OPS) imaging assesses both tissue perfusion and heterogeneity of flow.

Capillaroscopy

A recently developed technique for direct assessment of the microvasculature is OPS imaging, which is a non-invasive technique that creates an image from reflected light in the tissues[34-35]. This technique has been evaluated most extensively in human patients[4, 5, 36], and experimentally in pigs[37-38] and sheep[39] with sepsis and septic shock. It has also been validated in healthy dogs and horses[40].

SUMMARY

In summary, there are many methods for assessment of global cardiovascular function in the critically ill equine adult and foal. However, with more knowledge and methods to assess the microcirculation, it may enable therapy to be better targeted and ultimately allow reduced morbidity and mortality.

REFERENCES

1.Hallowell GD and Corley KT. *J Vet Intern Med* 2006;20:980-986. 2.Hallowell GD and Corley KT. *Vet Anaesth Analg* 2005;32:201-211. 3. Klijn E et al. *Clin Chest Med* 2008;29:643-654, viii. 4.Chierego M. et al. *Minerva Anestesiol* 2006;72:199-205. 5.De Backer D et al. *Am J Respir Crit Care Med* 2002;166:98-104. 6.Sakr Y et al. *Crit Care Med* 2004;32:1825-1831. 7.Rivers E et al. *N Engl J Med* 2001;345:1368-1377. 8. Archer DC et al. *Equine Vet J*;43 Suppl 39:56-62. 9.Ellis CM et al. *Vet Surg* 2008;37:786-790. 10. Freeman DE et al. *Equine Vet J Suppl* 2000:42-51. 11. Prange T et al. *Vet Surg*;39:748-753. 12. Nolen-Walston RD et al. *J Vet Intern Med*;25:570-574 13.Bumpus SE et al *Journal of Veterinary Emergency and Critical Care* 1998;8:21-26. 14. Underwood C, Norton JL, Nolen-Walston RD, et al. Echocardiographic changes in heart size in hypohydrated horses. *J Vet Intern Med* 2011;25:563-569. 15. Norton JL et al. *J Vet Intern Med*;25:575-578. 16. Tam K et al. *Vet Anaesth Analg* 2011;38:580-583. 17. Dunlop CI et al. *Am J Vet Res* 1991;52:1893-1897. 18. Muir WW et al *Am J Vet Res* 1976;37:697-700. 19. Wilkins PA et al. *Am J Vet Res* 2005;66:878-884. 20. Corley KT et al. *Equine Vet J* 2002;34:598-601. 21. Durando MM et al.. *Am J Vet Res* 2008;69:1054-1060. 22. Linton RA, et al. *Am J Vet Res* 2000;61:731-737. 23. Young LE et al. *Br J Anaesth* 1996;77:773-780. 24. Blissitt KJ et al. *Equine Vet J* 1997;29:18-25. 25. McConachie E et al. Doppler and Volumetric Echocardiographic Methods for Cardiac Output Measurement in Standing Adult Horses. *J Vet Intern Med*. In press. 26.Shih A et al. *J Vet Emerg Crit Care (San Antonio)* 2009;19:438-443. 27.Shih AC et al. *Am J Vet Res* 2009;70:334-339. 28.Giguere S et al. *J Vet Intern Med* 2005;19:737-743. 29. Valverde A et al. *Am J Vet Res* 2007;68:141-147. 30.Fiddian-Green RG. *Chest* 1999;116:1839-1841. 31. Fiddian-Green RG and Baker S. P*Crit Care Med* 1987;15:153-156. 32. Bakker J et al. *Chest* 1991;99:956-962. 33. Bakker J et al. *Am J Surg* 1996;171:221-226. 34. Goedhart PT, et al. *Opt Express* 2007;15:15101-15114. 35. Harris AG et al. *J Vasc Res* 2000;37:469-476. 36. De Backer D et al. *Intensive Care Med*;36:1813-1825. 37. Verdant CL et al. *Crit Care Med* 2009;37:2875-2881. 38. Wester T et al. *Clin Physiol Funct Imaging*;31:151-158. 39. Salgado DR et al. *Shock*;35:542-549. 40. Silverstein DC et al.. *J Vet Emerg Crit Care (San Antonio)* 2009;19:579-587.

DYSREGULATION OF CALCIUM HOMEOSTASIS IN THE CRITICALLY ILL HORSE

Gayle D. Hallowell, MA, VetMB, PhD, CertVA, DACVIM, DACVECC, FHEA, MRCVS
Nottingham, UK

Unlike other species, horses absorb a large proportion of dietary calcium, and are very reliant up dietary calcium to meet their daily requirements. Majority of calcium absorption takes place in the proximal half of the small intestine, with some also absorbed in the distal small intestine and the dorsal colon. The cecum and ventral colon are main sites for calcium secretion, although excess calcium is primarily excreted through the renal system as calcium carbonate. Parathyroid hormone is released in response to hypocalcemia and calcitonin in response to hypercalcemia. 1,25dihydroxyvitamin D_3 is known to decrease PTH secretion, but plays little role in intestinal calcium absorption.

Hypocalcemia, is a common finding in sepsis and is the most common cause of hypocalcemia in equine patients[1,2], particularly in those with severe gastro-intestinal disease. Adult horses and foals with signs consistent with SIRS and sepsis have been shown to have hypocalcemia and inappropriately low PTH suggestive of parathyroid gland dysfunction[1,3]. The mechanisms responsible for sepsis-induced hypocalcemia are unknown, but suggested mechanisms include calcium loss through the kidney[4] or gastro-intestinal tract[5], intracellularization and secondary accumulation of calcium[6], poor calcium mobilization[7] and parathyroid gland dysfunction[1].

Calcium administration has been a commonly used therapeutic in equine colic patients. Peri-operatively it has been used as an aid to increase mean arterial blood pressure (MABP) due to its action upon vascular smooth muscle. It has been shown to cause vasoconstriction as well as directly acting upon the cardiac myocytes as a positive inotrope[8-10] in both normocalcemic as well as hypocalcemic animals. This treatment is often administered to those severely hypotensive, cardiovascularly unstable horses who are not responding appropriately to dobutamine infusions. Post-operatively, calcium infusions or fluid supplementation with calcium are used in an attempt to stimulate gastro-intestinal motility in colic patients that are normocalcemic or hypocalcemic. Anecdotally, calcium is reported to increase gastro-intestinal borborygmi[11]. The rationale behind calcium administration is that it has been shown to mediate normal contractile activity of both intestinal and mesenteric vascular smooth muscle[12]. In intestinal smooth muscle, ion movement across smooth muscle membranes has been shown to generate the slow wave or basic rhythm of the intestine[13].

There are however several studies in both experimentally induced endotoxemic animals and in septic human patients, that have shown calcium has adverse effects on intracellular function. In septic human patients, it has been shown that there is an increased risk of mortality associated with calcium administration[14]. The reason for this are thought to be multi-factorial. Although ionized calcium may be decreased in the intravascular space, it has been shown to be normal or increased intracellularly[15], secondary to changes in calcium flux[16]. Supplementation will lead to even higher intracellular concentrations, which then leads to propagation of ischemic cell injury. The increase in intracellular calcium causes digestive enzyme release, cytokine release, free radical production and inhibition of ATP synthesis[17-20]. In addition, increased intracellular calcium within red cells (RBC) causes stiffening of the outer RBC membrane and a decreased RBC half-life[21]. Stiffening of this membrane may lead to RBC dysfunction and tissue hypoxia. In endotoxemic rats, calcium has been shown to increase endotoxin concentrations either by depressing neutrophil phagocytosis or increasing bacterial translocation[22]. This did occur at a calcium infusion rate lower than would be used clinically in horses under anesthesia[23]. Other effects of calcium-linked signalling derangements in sepsis include altered skeletal muscle protein catabolism and resistance to insulin-mediated glucose use[24].

In experimentally-induced endotoxin animal models, pre-treatment with calcium antagonists has been shown to increase survival and improve cardiovascular parameters[15,25-27], but not in peritonitis models using a cecal ligation and perforation study[28]. Pretreatment with levosimendan, a novel inodilator that improves cardiac contractility by sensitizing troponin C to calcium, in pigs subjected to endotoxin shock improved cardiac output and systemic and gut oxygen delivery. In addition, pulmonary hypertension largely was attenuated without any adverse effects on gas exchange[29].

Septic foals had been shown to be hypocalcemic with increased serum PTH with nonsurviving septic foals having higher PTH concentrations than survivors. Survival has also been associated with hypocalcemia, or failure to normalize following treatment in septic dogs[25], cats[30] and horses[2] with gastro-intestinal disease. However one has to be cautious in interpreting this; the low calcium is likely to be a marker for the severity of SIRS rather than the underlying cause of death.

Currently, there have no studies conducted investigating the effects of calcium administration in horses with SIRS secondary to gastro-intestinal disease. Horses are exquisitely sensitive to the effects of endotoxin[31,32]. It has been shown that post-operative ionized calcium concentrations are decreased in surgical colic patients that have strangulating compared to non-strangulating lesions[33,34], and that small intestinal lesions have lower ionised calcium concentrations when compared to large intestinal lesions[2]. Supplementation with calcium borogluconate returned ionised calcium concentrations to within the normal reference range in a group of surgical colic patients[33]. Ionized calcium has also been shown to be decreased in horses with post-operative ileus[34], but is this cause or effect. Data will be presented assessing a group of horses with SIRS secondary to small intestinal strangulating lesions that were either treated or not treated with intravenous calcium supplementation.

Conclusions

Hypocalcemia is a common finding in the critically ill animal. It appears that this is likely to be secondary to SIRS and sepsis. Although plasma calcium concentrations may be decreased, increased intracellular calcium, that ultimately will trigger apoptotic cell death may contribute to the increased morbidity and mortality in these patients. Thus, attempting to normalize ionized plasma calcium concentrations may be detrimental rather than beneficial in these patients, when clinical signs of hypocalcemia are not evident.

REFERENCES

1.Toribio RE et al. *J Vet Intern Med* 2005;19:223-231; 2. Delesalle C et al. *Vet Q* 2005;27:157-172; 3. Hurcombe SD, et al. *J Vet Intern Med* 2009;23:335-343; 4. Zaloga GP et al. *Circ Shock* 1988;24:143-148; 5. Nakamura T et al. *Horm Metab Res* 1998;30:88-92; 6. Crouser ED and Dorinsky PM. *Clin Chest Med* 1996;17:249-261; 7. Sperber SJ et al. *Rev Infect Dis* 1990;12:736-739; 8. Gasthuys F et al. *Zentralbl Veterinarmed A* 1991;38:728-736; 9. Grubb TL et al. *Am J Vet Res* 1999;60:1430-1435. 10. Grubb TL et al. *J Vet Intern Med* 1996;10:401-404. 11. White NA, Moore J, Mair T. *The Equine acute abdomen*. 2nd ed. Jackson: Teton NewMedia, 2009; 12. Walus KM et al. *Circ Res* 1981;48:692-700; 13. Davies JV. *Vet Clin North Am Equine Pract* 1989;5:271-281; 14. Lind L et al. *Crit Care Med* 2000;28:93-99; 15. Malcolm DS et al *Crit Care Med* 1989;17:900-903; 16. Bhattacharyya J and Sayeed MM. *Indian J Physiol Pharmacol* 1997;41:344-352; 17. Kasden SE and Kirkpatrick JR. *Am Surg* 1990;56:707-714; 18. Hotchkiss RS et al. *Shock* 1995;3:337-342; 19.Sakaguchi S, et al *Microbiol Immunol* 1989;33:99-110; 20. Lau YT et al. *Circ Shock* 1992;38:238-244; 21. Todd JC, 3rd and Mollitt DL. *Crit Care Med* 1995;23:459-465; 22. Zaloga GP et al. *Circ Shock* 1992;37:226-229; 23. Hall LW, Clarke KW, Trim C. *Veterinary anaesthesia*. London ; New York: W.B. Saunders, 2001; 24. Sayeed MM. *Arch Surg* 2000;135:1432-1442; 25. Bosson S,et al. *Circ Shock* 1985;16:307-316; 26. Bosson S et al. *Circ Shock* 1986;19:69-74. 27.Dudley CM et al. *Prog Clin Biol Res* 1988;264:373-378; 28. Hotchkiss RS and Karl IE. *New Horiz* 1996;4:58-71; 29. Oldner A et al. *Crit Care Med* 2001;29:2185-2193; 30. Kellett-Gregory LM et al. *J Vet Emerg Crit Care (San Antonio)*;20:398-405; 31. Moore JN and Morris DD. *J Am Vet Med Assoc* 1992;200:1903-1914; 32. Steverink PJ et al. *Vet Q* 1994;16 Suppl 2:S117-121; 33. Dart AJ et al. *J Am Vet Med Assoc* 1992;201:1244-1248; 34. Garcia-Lopez JM et al. *Am J Vet Res* 2001;62:7-12.

RENAL REPLACEMENT THERAPY IN THE HORSE: A VIABLE CLINICAL OPTION?

Dr. David M. Wong, DACVIM, DACVECC
Ames, IA

INTRODUCTION

Renal replacement therapy is a term used to describe artificial techniques that mimic or augment renal function, typically in the clinical setting of acute renal failure (ARF), with the following goals: 1) maintaining fluid, electrolyte, acid-base, and solute balance, 2) preventing further kidney insult, 3) promoting healing and renal recovery and 4) permitting supportive measures (i.e. nutrition) to proceed without limitation.[1] These techniques include modalities such as intermittent hemodialysis (IHD), peritoneal dialysis (PD) and continuous renal replacement therapy (CRRT). A thorough discussion of each modality is beyond the scope of this review, but a brief description is provided below. Subsequently, a more thorough look into CRRT and its use in horses will follow.

IHD is a standard form of therapy accepted worldwide for the treatment of ARF. In IHD, a large volume of blood is removed from the human patient (200-250 ml/min) and passed through a dialyzer that is exposed to high countercurrent dialysate flow rates which consequently facilitates diffusion and removal of various sized molecules (solutes).[2] This method is very efficient, but can be associated with large shifts of solutes and fluid over a brief period of time (1-6 hours) resulting in hemodynamic instability in the patient. PD involves the infusion and drainage of dialysate into the abdominal cavity allowing the peritoneum to serve as the filter between the blood and dialysate. Dialysate is instilled and drained from the abdominal cavity over a set dwell time (e.g. 3 to 6 hours) and dextrose can be added to the dialysate to promote fluid removal from the patient, if desired.[3] PD can be implemented in an intermittent or continuous fashion. Disadvantages of PD include adverse effects on respiration, limitations with use in patients with peritonitis or abdominal surgery and the increased incidence of iatrogenic peritonitis.[1,3] CRRT is similar to IHD in that blood is passed through thousands of straw-like semipermeable membranes within the dialyzer, but is performed in a slower and more continuous (e.g. > 24 hour) method. CRRT is preferred over IHD by some clinicians because this modality allows more accurate treatment delivery, better control of blood flow and fluid removal, and more gradual volume and solute concentration changes; these benefits are especially necessary in critically ill patients.[4] An on-going debate exists in the literature in regard to the superiority of IHD verse CRRT. Proposed theoretical advantages of CRRT over IHD include better hemodynamic stability allowing more adequate fluid removal, better recovery of renal function (as a result of fewer hypotensive episodes) and more efficient removal of small and large molecules.[2] However, controlled trials provide equivocal results or fail to support these theoretical advantages. In comparison, IHD theoretical advantages include easier implementation of IHD (i.e. patient has dialysis-free periods), comparatively more cost-effective therapy and fewer bleeding complications.[4,5]

MECHAMISMS OF SOLUTE REMOVAL

The main mechanisms by which solutes are removed in renal replacement therapies are diffusion, convection, and/or, to a lesser extent, adhesion. Diffusion is the transfer of solute through a semipermeable membrane from a higher concentration to a lower concentration and is the main principle governing solute removal in IHD.[1] Efficiency of molecular movement is inversely related to molecular size and directly proportional to the magnitude of the concentration gradient. In addition, faster dialysate flow rates promote faster mass transfer under the circumstances of diffusion. Convection exposes the extracorporeal blood to a positive transmembrane pressure resulting in an ultrafiltrate to be pushed out of the blood and across the membrane. At the same time, other molecules are carried along with the ultrafiltrate (solvent drag) promoting solute removal. This principle is involved with various modes of CRRT. Higher ultrafiltrate rates result in more rapid convection. Adhesion plays a smaller role in solute removal and involves adherence of molecules to the membrane that are removed from the circulation.

INDICATIONS

The main indication for renal replacement therapy is acute renal failure (ARF) where patients are not responsive to conventional therapy (i.e. IV fluid therapy, correction of acid-base and electrolyte derangements, diuretics) is the main indication for renal replacement therapy.[4] The presence of volume overload, hyperkalemia, metabolic acidosis and signs of severe uremia are also common renal indications to consider renal replacement therapy in people,[4,6] whereas non-renal indications include removal of toxins and inflammatory cytokines.[7] In veterinary medicine, CRRT has been used to treat dogs with ARF, tumor lysis syndrome, heatstroke, pre- and postsurgical ureteral obstruction and various toxicities.[8] IHD and PD have been implemented in a few reported cases of ARF in horses and foals.[10-12] However, CRRT has not been thoroughly investigated as a potential therapeutic modality in horses and will be further discussed.

MODES OF CRRT OPERATION

A brief review of the modes of operation that can be employed in CRRT while be covered. CRRT was initially developed based on combined arterial and venous access (arterial pressure driving blood flow through circuit); however, current methods drive blood through the circuit by pump-driven veno-venous access. Modes of operation (illustrated at right) include slow continuous ultrafiltration (SCUF), continuous venovenous hemofiltration (CVVH), continuous venovenous hemodialysis (CVVHD) and continuous venovenous hemodiafiltration (CVVHDF).[1,8] In SCUF, blood is pumped through a hemofilter and plasma water (ultrafiltrate) is produced through convection, which is then discarded as effluent. This mode of operation is used in patients with volume overload (e.g. congestive heart failure); no dialysate or replacement fluids are used in this mode. CVVH is similar to SCUF with the addition of a balanced electrolyte solution (replacement fluid) that is administered within the circuit to replace the volume of ultrafiltrate that was removed. CVVHD is similar to IHD in that diffusion is the main principle of solute removal. In this mode, blood entering the dialyzer is bathed in dialysate solution that is flowing in the opposite direction of the blood. Solute that is in high concentration within the blood (e.g. creatinine) is able to permeate through the dialysis membrane and is transferred to the dialysate and removed with the effluent. CVVHDF is the most common mode used and combines aspects of several modalities. Both dialysate and replacement fluids are added to the circuit with convection and diffusion facilitating solute removal.

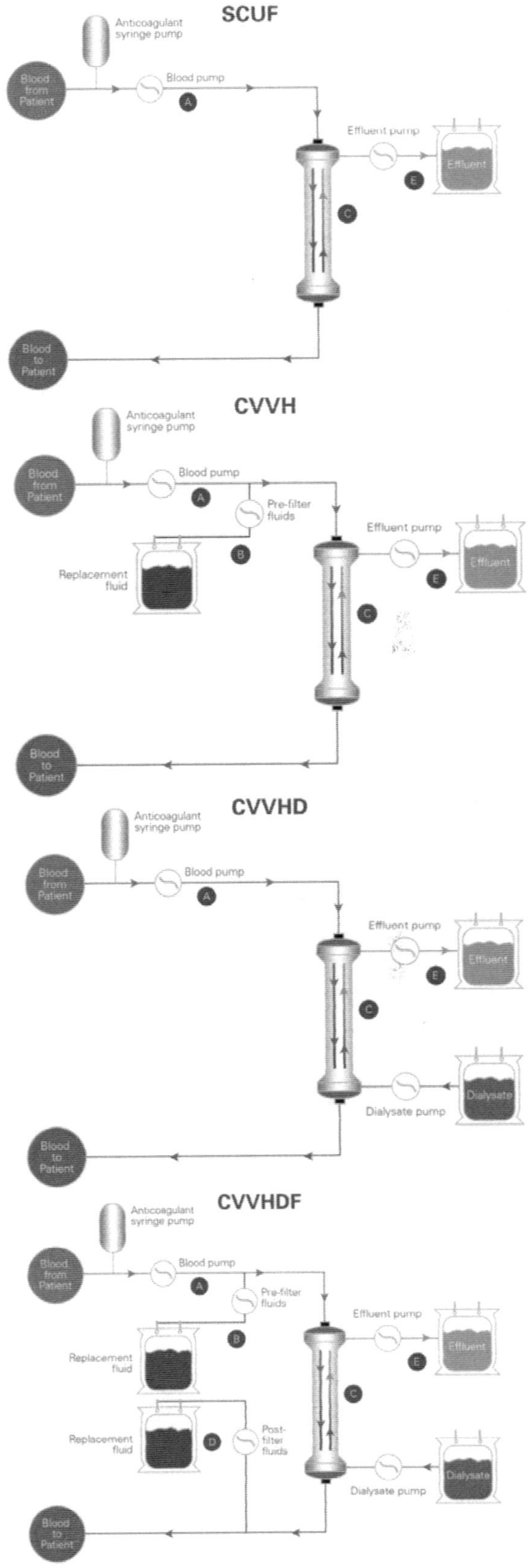

EQUIPMENT SET UP

A number of commercially available CRRT machines are available and include the Acquarius (Baxter), BM25 (Baxter), Diapact (Braun), Equa-Smart (Medica), HF 400 (Infomed), Hygeia plus (Kimal), Performer (Rand), Prisma (Gambro), Prismaflex (Gambro), and Multifiltrate (FMC).[13] As technology has advanced, CRRT machines have incorporated safety mechanisms and have become more user-friendly. Each individual machine will have specific features, but most are equipped with integrated safety alarms, pumps, fluid balancing controls, scales and pressure sensors to allow easier and safer implementation of CRRT.[13] A number of biocompatible membranes are commercially available with many of them made of polyacrylonitrile (An-69 membrane). No specific manufacturer or machine is recommended here, but an excellent review of individual features of the aforementioned CRRT machines is available.[13]

Vascular access is obviously necessary to remove the patient's blood and is one of the first steps in preparing a patient for CRRT. In the horse, the jugular vein provides the best and most accessible vascular access. No equine specific dialysis catheter is available, but numerous double-lumen dialysis catheters are commercially available for people. The largest bore catheter will allow highest flows with least resistance. Typically, the tips of the dialysis catheters are staged to minimize recirculation of blood returning from the circuit to the patient (e.g. return tip longer than intake tipt). A variety of dialysis filters exist as well; in the adult horse, the largest filter available should be used; even with the use of the largest filter, it is likely

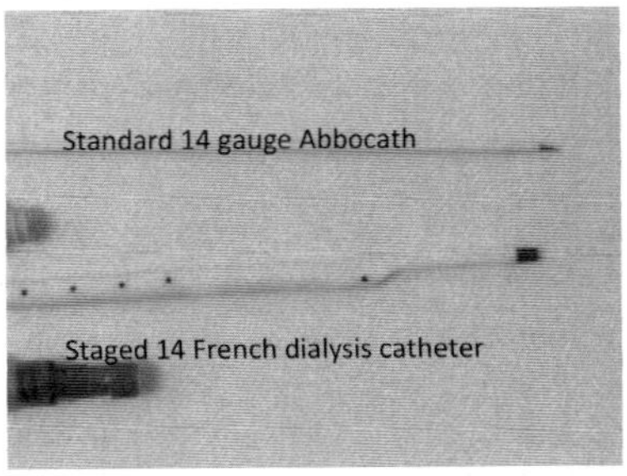

undersized for a horse. Sedation might be necessary depending on the demeanor and healthy status of the horse (e.g. horse may be lethargic from primary disease).

Due to the fact that the patient's blood will be exposed to foreign surfaces (e.g. catheters, tubing, dialysis chambers), administration of anticoagulation is typically necessary to prevent excessive clotting of blood within the dialysis filter or tubing. The most common anticoagulants are heparin, which is administered systemically, or citrate, which is administered locally.[14,15] Heparin binds to antithrombin and potentiates its activity of blocking factor Xa and thrombin and provides systemic anticoagulation. Coagulation parameters should be monitored with the use of heparin and dose of anticoagulants adjusted accordingly. When utilizing heparin, 1.5-2 x normal PTT is targeted.[14,15] Alternatively, citrate is administered as blood is removed from the body (arterial limb of circuit) which subsequently binds to calcium that is necessary for the coagulation cascade to progress. Systemic hypocalcemia is avoided by administering calcium chloride or calcium gluconate at a site away from the CRRT circuit. Thus, citrate provides regional anticoagulation within the dialysis circuit, thereby eliminating bleeding tendencies often observed with heparin administration. Serum calcium should be monitored if citrate is used. A disadvantage of citrate use arises from the fact that citrate is metabolized by the liver and skeletal muscle to bicarbonate; therefore, patients receiving citrate as an anticoagulant should be monitored for metabolic alkalosis.[14,15] In certain clinical situations (i.e. existing coagulopathy, DIC, hepatic failure), periodic saline flushes of the circuit are used in place of anticoagulants; however this practice is associated with decreased circuit life span and decreased ability to deliver CRRT.[1,3]

CRRT IN HEALTHY HORSES

Recently, CRRT was performed in 5 healthy adult horses with a mean body weight of 544 kg at Iowa State University. In this study, a 14 French double-lumen dialysis catheter was inserted into the left jugular vein to perform CRRT while 3 other over-the-needle catheters were placed to monitor direct arterial blood pressure (facial artery), administer a continuous-rate-infusion (CRI) of sedation (cephalic vein) and allow blood collection (right jugular vein). Sedation was provided by a staged CRI of detomidine (8.4 μg/kg, IV bolus, followed by 0.5 μg/kg/min for 15 minutes, 0.3 μg/kg/min for 15 minutes and then 0.15 μg/kg/min thereafter).[16] Heparin was administered using a bolus loading dose of 100 IU/kg, IV followed by a CRI of 50 IU/kg/hr, IV.[17] Each horse was then interfaced with a commercial CRRT machine (Prismaflex) utilizing a 1.5m^2 filter. The machine was set to deliver continuous venovenous hemodiafiltration (CVVHDF) with a blood flow rate of 250 mL/min, dialysate rate of 3000 mLs/hr, and a pre- and post-filter replacement fluid rate of 3000 and 2000 mLs/hr, respectively. Physical examination parameters (heart rate, respiratory rate, body temperature, blood pressure, heart rhythm) were monitored and blood was collected periodically to monitor serum electrolyte concentrations, renal enzymes and coagulation parameters.

CRRT was performed successfully for 6 consecutive hours with no major adverse effects observed. A decrease in heart rate and rectal temperature was observed at variable time points as well as occasional ventricular premature contractions (VPCs) and 2nd degree atrio-ventricular block on electrocardiogram. Coagulation parameters were prolonged and platelet count decreased over the study period. A decrease in blood urea nitrogen was observed over various time points as well. Minor complications of the CRRT procedure included hematoma formation at the catheter insertion site in a few horses.

FUTURE DIRECTIONS

To date, renal replacement therapies are infrequently utilized in equine medicine and critical care. This may relate to the availability of this modality, the labor-intensive nature of providing therapy, and perceived high cost. However, CRRT machines are becoming more readily available at relatively competitive cost to users. Future investigations should focus on applying CRRT to clinical cases of horses with ARF. Other potential uses include drug overdoses (i.e. aminoglycoside or non-steroidal anti-inflammatory toxicity) and treatment of endotoxemia or septicemia. The future of CRRT in equine medicine is unknown, but might provide an adjunct therapy to improve outcome to equine patients with ARF.

REFERENCES

1. Mehta R. Supportive therapies: Intermittent Hemodialysis, Continuous Renal Replacement Therapies, and Peritoneal Dialysis. In: Schrier RW, ed. Atlas of Kidney Diseases, Vol 1. Philadelphia, PA: Current Medicine; 1999: 301-316.
2. Bloom CA, Labato MA. Intermittent hemodialysis for small animals. Vet Clin Small Anim 2011;41:115-133.
3. Cooper RL, Labato MA. Peritoneal dialysis in veterinary medicine. Vet Clin Small Anim 2011;41:91-113.
4. Sutherland SM, Alexander SR. Continuous renal replacement therapy in children. Pediatr Nephrol 2012;27:2007-2016.
5. Vanholder R, et al. Pro/con debate: Continuous versus intermittent dialysis for acute kidney injury. Crit Care 2011;15:204
6. Thadhani R, Pascual M, Bonventre JV. Medical Progress: Acute Renal Failure. N Engl JMed 1996;334:1448-60.
7. Oda S, Sadahiro T, Hirayma Y. Non-renal indications for continuous renal replacement therapy: Current status in Japan. Contrib Nephrol 2010;166:47-53.
8. Acierno MJ. Continuous renal replacement therapy in dogs and cats. Vet Clin Smal Anim 2011;41:135-146.
9. Vivrette S, et al. Hemodialysis for treatment of oxytetracycline-induced acute renal failure in a neonatal foal. *J Am Vet Med* Assoc 1993;203: 105-107.

10. Thornhill JA, et al. Hemodialysis in a horse with acute renal failure induced by rhabdomyolysis (Abstr.) Proceedings Am Assoc Zoo Vet Ann Mtng. 1983, Pg. 16.
11. Kritchevsky JE, et al. Peritoneal dialysis for presurgical management of ruptured bladder in a foal. J Am Vet Med Assoc 1984;185:81-82.
12. Gallatin LL, et al. Use of continuous-flow peritoneal dialysis for the treatment of acute renal failure in an adult horse. J Am Vet Med Assoc 2005;226:756-759.
13. Cruz D, Bobek I, Lentini P, et al. Machines for Continuous Renal Replacement Therapy. Semin Dial 2009;22:123-132.
14. Brophy P, et al. Multi-centre evaluation of anticoagulation in patients receiving CRRT. Nephrol Dial Transplant 2005;20:1416-1421.
15. Ross S. Anticoagulation in intermittent hemodialysis: pathways, protocols, and pitfalls. Vet Clin Small Anim 2011;41:163-175.
16. Goodrich LR, Clark-Price S, Ludders J. How to attain effective and consistent sedation for standing procedures in the horse using constant rate infusion. Proceedings of the 50th Annual Convention of the American Association of Equine Practitioners, Lexington KY, December 4, 2004, Pg. 239-232.
17. Oliveira J, Palhares MS, Frossler TC, et al. Adequacy of hemodialysis system in horses. Ary Bras Med Vet Zootec 2008;60:1346-1357.

RECENT ADVANCES IN EQUINE IMMUNODEFICIENCIES

Sandra D. Taylor, DVM, PhD, DACVIM (LAIM)
West Lafayette, IN

INTRODUCTION

Immunodeficiencies can be classified as primary (genetic) or secondary (acquired) and are characterized by an incomplete or absent immune system that allows organisms to invade the host and cause disease. Primary immunodeficiencies are relatively uncommon in the horse, but present a diagnostic challenge to equine practitioners as affected animals often present with vague clinical signs and recurrent infections. Examples of primary immunodeficiencies include severe combined immunodeficiency (SCID), common variable immunodeficiency (CVID), and foal immunodeficiency syndrome (also known as Fell Pony Syndrome) (FIS). Failure of transfer of passive immunity (FTPI) in neonatal foals is the most common immunodeficiency encountered by equine practitioners and represents a secondary immunodeficiency. Selective IgM deficiency has been reported in foals and adult horses, and may be classified as primary or secondary.

PRIMARY IMMUNODEFICIENCIES

SCID

Severe combined immunodeficiency is a genetic disorder described in foals, humans, mice and dogs. Equine SCID is caused by a frameshift mutation in the gene encoding the catalytic subunit of DNA-dependent protein kinase (DNA-PKcs) [1,2] and has an autosomal recessive mode of inheritance.[3] The mutation is seen primarily in Arabian horses, but has been described in a Caspian foal.[4] Affected foals lack functional B and T lymphocytes and thus, are unable to mount an adaptive immune response against microorganisms. Foals with SCID are normal at birth but typically develop clinical signs of generalized infection at 1-3 months of age, which is associated with increased exposure to opportunistic pathogens and waning maternal antibody. Upper respiratory tract infection and pneumonia are the most common presenting complaints, with diarrhea associated with enterocolitis less commonly observed. Adenoviral pneumonia is seen in approximately two thirds of affected foals, with *Pneumocystic carinii*, *Rhodococcus equi*, *Streptococcus sp.* and *Cryptosporidium sp.* also commonly isolated.[5] Foals with SCID may initially respond to antimicrobial treatment and supportive care, but eventually succumb to overwhelming infection within 2-5 months of age.[5] Although immunologic reconstitution and "recovery" are possible through transplantation with bone marrow cells from a histocompatible donor, this is not a feasible option in most cases.[6,7]

Severe lymphopenia (< 500/μl) is highly suggestive of SCID in an Arabian foal with chronic infections. Cells counted as "lymphocytes" on a complete blood count are usually natural killer cells, as SCID foals are born with zero lymphocytes.

Definitive diagnosis of SCID requires genetic testing of blood or cheek swabs (www.vetgen.com). The mutated gene is identified through polymerase chain reaction (PCR) testing of DNA isolated from cells. It is estimated that approximately 8% of Arabian horses in the United States carry the SCID mutation (heterozygous), yielding a prevalence for SCID (homozygous) of 0.18%.[8] The estimated prevalence of SCID carriers in Morocco is similar at 7%[9], while the United Kingdom reported only a 2.8% frequency.[10] Although the prevalence is uniformly low, genetic testing of breeding Arabians is recommended, as prognosis for affected offspring is grave.

CVID

Combined variable immunodeficiency was first described in horses in 2002, and is characterized by late-onset recurrent bacterial infections, hypo- or agammaglobulinemia, progressive B cell lymphopenia or depletion, and poor response to vaccination.[11-14] Although CVID is considered a primary immunodeficiency in both humans and horses, the majority of equine cases present in adulthood.[12] Many different breeds and both sexes have been affected, and there does not appear to be a regional geographical distribution. The most common clinical signs upon presentation include recurrent fever and pneumonia, but infections of the nervous system, gastrointestinal tract, gingiva, sinuses, liver and skin have also been reported.[12] Diagnosis of CVID is based on compatible history and clinical signs, and the presence of intermittent or persistent lymphopenia (<1,500/μl). Neutrophilia and hyperfibrinogenemia may also be present, indicating chronic bacterial infection. Peripheral blood lymphocyte phenotyping typically reveals severe B cell lymphopenia, markedly reduced or absent serum IgM concentrations, and decreased serum IgG concentrations.[12] Serum IgA concentrations typically decline as the disease progresses. Although some horses have been managed for 1-5 years on continuous or intermittent antimicrobial and immunoglobulin therapies, the majority of horses diagnosed with CVID are euthanized.

In a review of 14 horses diagnosed with CVID, the average age at the time of diagnosis was 10.7 years, with a range of 2-19 years.[12] Six females and 8 males were affected, and affected breeds included Thoroughbreds (n=8), Quarter Horses (n=4), and one each of Warmblood and Pony. None of the horses were related. Clinical signs and/or diagnoses among these cases included fever (n=14), pneumonia (n=12), neurologic signs of ataxia or seizure (n=8), meningitis confirmed by CSF analysis (n=5), septicemia (n=3), hepatitis (n=2), uveitis (n=2), peritonitis (n=1), sinusitis (n=1), diarrhea (n=1), gingivitis (n=1), and skin abscesses (n=1).[12] The average time between onset of clinical signs and diagnosis was 5 months. Overall, 12 of 14 horses were euthanized, and the average time from diagnosis to euthanasia was 7 months. Necropsy was performed in 10 horses, all of which demonstrated absence of primary and secondary lymphoid follicles in the lymph nodes and spleen.[12]

Horses with CVID demonstrate impaired B cell production in the bone marrow. Since B cells are continuously generated from hematopoietic stem cells over an individual's lifetime, a study was conducted to determine the extent of ongoing B cell differentiation in CVID horses and healthy horses.[15] Expression of essential B cell genes, including transcription factors and cell receptor genes, was measured using real time RT-PCR. Results of this study suggest that B cell development is impaired at the transition between pre-pro-B cells and pro-B cells in horses affected with CVID.[15] The mechanism of B cell depletion and the initial trigger for loss are still under investigation.

FIS

Foal immunodeficiency syndrome, previously known as Fell Pony Syndrome, is a primary immunodeficiency that affects Fell and Dales pony breeds. This syndrome was first described in 1998 and is inherited as an autosomal recessive trait.[16] FIS is estimated to affect 10% of Fell and 1% of Dales foals, and is invariably fatal.[17] This disease is particularly important in Fell ponies, as the breed is already considered threatened because of the small population size worldwide. Affected foals present with severe anemia, B cell lymphopenia and clinical signs associated with opportunistic infections. Foals with FIS are generally normal at birth, but develop clinical signs that may include weakness, inappetance, poor growth, nasal discharge, diarrhea and pale mucous membranes at 2-6 weeks of age. Routine blood work at the time of illness reveals severe and progressive anemia in addition to lymphopenia and neutrophilia.[18,19] Although signalment, clinical signs and routine blood analysis may be highly suggestive of FIS, peripheral blood lymphocyte phenotyping may aid in diagnosis by identifying B cell lymphopenia. One would expect to find decreased serum immunogloblulin in affected foals, but hypoglobulinemia is not a consistent finding given the presence of maternal colostral antibodies.[20,21] Death occurs generally by 3 months of age. Postmortem examination typically reveals small numbers of erythroid precursors in the bone marrow, a small thymus, lack of secondary lymphoid follicles or germinal centers, and peripheral ganglionopathy.[16,20]

Recently, a prospective study evaluated potential FIS-affected Fell Pony foals to determine clinical, immunological, and molecular parameters at birth and up to 5 weeks of age.[22] In the first few weeks of life, affected foals were clinically normal and had normal blood work. Lymphocyte cell surface markers were evaluated using flow cytometry and revealed normal B cell and T cell subpopulation distribution. However, foals developed anemia at approximately 3 weeks of age, and bone marrow evaluation at this time revealed erythroid hyperplasia with left-shifted maturation and concurrent erythroid and myeloid dysplasia. Sequential bone marrow aspirates revealed progression of erythroid dysplasia to severe erythroid hypoplasia at 5 weeks of age. In addition, FIS-affected foals did not demonstrate an increase in serum IgM concentration with age, unlike unaffected Fell Pony foals. Immunohistochemistry at necropsy showed the absence of B cells in the bone marrow and lymph nodes, and absent germinal centers in the spleen of FIS-affected foals. Genes involved in B cell hematopoiesis and development were screened for mRNA expression in these foals, and low *PAX5* gene expression in the bone marrow was detected. This gene is essential for B cell commitment and development, as well as B cell survival in peripheral tissues.[23,24] To summarize, FIS-affected foals appear to be born with erythroid precursors and B cells in the bone marrow, suggesting productive fetal hematopoiesis. Postnatal development of progressive anemia and B cell lymphopenia coincides with erythroid and myeloid dysplasia and severe erythroid hypoplasia in the bone marrow, and is associated with decreased expression of at least one gene responsible for B cell development.

Recently, a genetic mutation was identified in foals affected with FIS.[17] The sodium/myoinositol cotransporter gene (*SLC5A3*) demonstrates a single nucleotide polymorphism that results in an amino acid substitution, which likely alters the function of SLC5A3. Although the mechanism has not been elucidated and a causal relationship between the SLC5A3 mutation and FIS has not been established, dysfunction of this protein may lead to erythropoiesis failure and myeloid hypoplasia. It is possible that other genomic mutations may be responsible for the FIS phenotype, and further investigation is ongoing.

SECONDARY IMMUNODEFICIENCIES

FTPI

The most common immunodeficiency syndrome in horses is failure of transfer of passive immunity, formerly termed "failure of passive transfer." FTPI occurs when neonatal foals fail to ingest or absorb maternal colostral antibodies. Causes of FTPI may include poor colostral quality, premature lactation, failure of the foal to ingest an adequate amount of antibodies, and insufficient intestinal absorption of ingested antibodies.[25-27] FTPI is associated with increased risk of bacteremia, sepsis and death in neonatal foals.[28-31] Serum immunoglobulin (sIgG) concentration of < 400 mg/dl is used to define complete FTPI, while sIgG of 400 – 800 mg/dl is considered partial FTPI and sIgG > 800 mg/dl is considered adequate FTPI.[25,32,33] Quantitative measurement of sIgG in healthy Thoroughbred foals is approximately 1500 mg/dl, which highlights the significance of measurements < 800 mg/dl.[34,35] The reported incidence of FTPI among neonatal foals in the United States is 13-18%.[36-38]

Maximal absorptive efficiency in the intestinal tract of neonatal foals occurs immediately after birth, decreasing to 22% efficiency at 3 hours of age and < 1% by 20 hours of age.[39,40] Therefore, testing for sIgG should be performed at 12-24 hours of age. Several diagnostic tests are available to assess transfer of passive immunity in neonatal foals and include 1) single radial immunodiffusion (SRID), 2) zinc sulfate turbidity, 3) glutaraldehyde coagulation, 4) latex agglutination, 5) immunoassays, 6) measurement of serum total protein concentration and 7) infrared spectroscopy. Most tests are accurate in diagnosing complete FTPI, but there is significant variation in the ability of these tests to detect partial FTPI. The most accurate diagnostic test for FTPI is the SRID (VMRD, Pullman, WA; Plasvacc USA, Templeton, CA, Equine IgG Test Kit, Kent Laboratories, Bellingham, WA), which is considered the reference standard. This test is based on antigen/antibody precipitation on agar gel plates and results are obtained in 24 hours. Thus, it is not practical for use in sick foals when rapid diagnosis and treatment are necessary.[41-43] The zinc sulfate turbidity test (Equi Z, Equine FPT Test Kit, VMRD, Pullman, WA) is based on zinc turbidity that is usually proportional to sIgG concentration. False elevations in sIgG concentration may occur if hemolysis is present, and poor handling or quality of reagents can influence results.[41,43] One study found that the zinc sulfate turbidity test was 81% sensitive and 57% specific in detecting sIgG < 800 mg/dl when compared to SRID.[44] The glutaraldehyde coagulation test (Gamma-Check-E, Plasvacc USA, Templeton, CA) is based on coagulation of sIgG with glutaraldehyde that ultimately forms a gel. Results from two studies that compared this test with the SRID identified a sensitivity and specificity in detecting sIgG < 800 mg/dl of 93-100% and 59%, respectively, and a sensitivity and specificity in detecting sIgG < 400 mg/dl of 95% and 80%, respectively.[44,45] The latex agglutination test (Foalcheck, Centaur, Overland Park, KS) is based on agglutination of sIgG to anti-equine IgG bound to latex particles. The sensitivity and specificity of this test are relatively low at 72% and 79% for sIgG < 800 mg/dl and 53% and 79% for sIgG < 400 mg/dl, making it a poor choice for rapid screening.[45] Available stall-side semiquantitative enzyme immunoassays include the Snap Foal IgG Test (IDEXX Laboratories, Westbrook, ME) and the Midland Plasma Foal IgG Quick Test Kits for 400 mg/dl or 800 mg/dl (Midland BioProducts Corp, Boone, IA). Using the SRID as a reference standard, the sensitivity and specificity for detecting sIgG < 800 mg/dl for the Snap Foal IgG Test has been reported as 81-95% and 52-95%, respectively. Similarly, the sensitivity and specificity for detecting sIgG < 400 mg/dl for the Snap Foal IgG Test has been reported as 90% and 79-96%, respectively.[44,46] The sensitivity for detecting sIgG < 800 mg/dl was considered too low for the Midland Plasma Foal IgG Quick Test Kit (800 mg/dl) to use as a screening kit for FTPI.[44] A quantitative colorimetric immunoassay (immunoturbidometric assay; DVM Stat, VDx Inc., Belgium, WI) has been shown to be the most sensitive and specific of the commercially available assays when compared to the reference standard (SRID).[34,44] This test can be performed on a biochemistry analyzer, and relies on agglutination of sIgG with anti-equine IgG that results in turbidity which is proportional to the sIgG concentration.[43] Sensitivity and specificity have been reported to be 81-98% and 83-86%, respectively, for sIgG < 800 mg/dl. For detection of sIgG < 400 mg/dl, sensitivity and specificity were 63-100% and 92-96%, respectively.[43,44] Serum total protein concentration is limited as a stand-alone screening test for FTPI, but may provide useful information. Using a serum total protein concentration of 4.5 g/dl as a cut-off value and comparing to SRID, the sensitivity and specificity for detecting sIgG < 800 mg/dl has been reported as 52% and 96%, respectively, and for detecting sIgG < 400 mg/dl, sensitivity and specificity were 43% and 97%, respectively.[44] Foals with a serum total protein concentration < 4.5 g/dl are likely to have sIgG < 400 mg/dl, and foals with a serum total protein concentration > 5.5 g/dl are likely to have sIgG > 800 mg/dl; results between 4.5 – 5.5 g/dl do not reliably predict sIgG concentration.[44,46] Finally, infrared spectroscopy is a relatively novel method for quantifying sIgG concentration, whereby infrared radiation is transmitted through the serum sample and the wavelength dependence of radiation absorption is read by a spectrometer. When compared to SRID, infrared spectroscopy had a sensitivity of 97% and a specificity of 93% for detecting sIgG < 400 mg/dl.[42] A recent study evaluated mare sIgG using infrared spectroscopy at the time of post-parturient foal assessment and concluded that spectra of dam serum have the potential to provide a basis for a screening method for mares at risk of having a foal susceptible to FTPI.[47] Future studies aim to assess dam spectra during parturition. In summary, the relatively low specificity of the zinc turbidity test, the glutaraldehyde coagulation test, and the Midland Plasma Foal IgG Quick immunoassay suggests that foals with adequate transfer of passive immunity may be diagnosed with FTPI when these tests are used. Conversely, relying only on serum total protein concentration may result in false negatives. Therefore, use of the Snap Foal IgG Test or the colorimetric immunoassay is recommended for screening tests for FTPI, with questionable results confirmed by SRID.

Treatment of foals with FTPI depends on the age of the foal at diagnosis and concurrent clinical signs. An affected foal < 12 hours of age with a normal gastrointestinal tract should receive colostrum through a nasogastric tube. If the foal is able to suckle, a bottle with an appropriately-sized nipple may be used, with care taken to prevent aspiration pneumonia. A plasma transfusion should be performed in an affected foal > 12 hours of age, given that "gut closure" is likely complete by this time. A study assessing the effect of plasma transfusions on survival in septic and sick non-septic foals found that administration of plasma harvested from mares hyperimmunized with endotoxin was associated with increased survival compared to conventional hyperimmune plasma.[48] Septic foal sIgG concentrations should be reassessed following plasma transfusions given that transfused antibodies may be utilized quickly in these cases, which may necessitate additional transfusions.

REFERENCES

1. Shin, E, et al., **J Immunol.** 1997;158:3565. 2. Wiler R, et al., **Proc Natl Acad Sci USA.** 1995;92:11485. 3. Perryman L, et al., **J Am Vet Med Assoc.** 1980;176:1250. 4. Larson J, et al., **J Vet Intern Med.** 2011;25:954. 5. Perryman L, et al., **Am J Vet Res.** 1978;39:1043. 6. Bue C, et al., **Transplantation**. 1986;42:14. 7. Perryman L, et al., **Vet Immunol Immunopathol.** 1987;17:495. 8. Bernoco D, et al., **Anim Genet**. 1998;29:41. 9. Piro M, et al., **Equine Vet J.** 2008;40:590. 10. Swinburne J, et al., **Vet Rec.** 1999;145:22. 11. Flaminio M, et al., **J Am Vet Med Assoc**. 2002;221:1296. 12. Flaminio M, et al., **J Clin Immunol.** 2009;29:107. 13. Pellegrini-Masini A, et al., **J Am Vet Med Assoc**. 2005;227:114. 14. Tennent-Brown B, et al., **Equine Vet Educ**. 2010;22:393. 15. Tallmadge R, et al., **Mol Immunol.** 2012;51:169. 16. Scholes S, et al., **Vet Rec**. 1998;142:128. 17. Fox-Clipsham L, et al., **Plos Genetics.** 2011;7. 18. Dixon J, et al., **Vet Clin Path.** 2000;29:84. 19. Bell S, et al., **Equine Vet J.** 2001;33:687. 20. Richards A, et al., **Equine Vet J**. 2000;32:386. 21. Thomas G, et al., **Vet Rec.** 2003;152:618. 22. Tallmadge R, et al., **Clin Vaccine Immunol**. 2012;19:1054. 23. Michaelson J, et al., **J Immunol.** 1996;156:2349. 24. Nutt S, et al., **EMBO J**. 1998;17:2319. 25. Giguere S, et al., **Vet Clin North Am Equine Pract.** 2005;21:241. 26. Sellon D, **Vet Clin North Am Equine Pract.** 2000;16:117. 27. Crisman M, et al., **Vet Clin North Am Equine Pract**. 2008;24:299. 28. McGuire T, et al., **J Am Vet Med Assoc**. 1977;170:1302. 29. Tyler-McGowan C, et al., **Aust Vet J.** 1997;75:56. 30. Raidal S, **Aust Vet J.** 1996;73:201. 31. Cohen N, **J Am Vet Med Assoc.** 1994;204:1644. 32. Stoneham S, et al., **Vet Rec.** 1991;128:416. 33. LeBlanc M, et al., **J Am Vet Med Assoc.** 1986;189:57. 34. Nath L, et al., **J Am Vet Med Assoc.** 2010;236:1085. 35. LeBlanc M, **J Reprod Fertil Suppl**. 1987;35:735. 36. LeBlanc M, et al., **J Am Vet Med Assoc.** 1992;200:179. 37. Baldwin J, et al., **J Am Vet Med Assoc**. 1991;198:423. 38. Clabough D, et al., **J Vet Intern Med**. 1991;5:335. 39. Jeffcott L, **Vet Rec**. 1971;88:340. 40. Jeffcott L, et al., **J Comp Pathol.** 1974;84:455. 41. Clabough D, et al., **J Am Vet Med Assoc.** 1989;194:1717. 42. Riley C, et al., **J Vet Intern Med**. 2007;21:828. 43. Davis D, et al., **J Vet Intern Med.** 2005;19:93. 44. Davis R, et al., **J Am Vet Med Assoc.** 2005;227:1640. 45. McClure J, et al., **Proc ACVIM Annual Forum**, May 2002. 46. Metzger N, et al., **J Vet Intern Med.** 2006;20:382. 47. Riley C, et al., **Aust Vet J.** 2012;90:387. 48. Peek S, et al., **J Vet Intern Med.** 2006;20:569.

IMMUNITY TO *SARCOSYSTIS NEURONA*

Sharon Witonsky DVM, PhD, DACVIM
Blacksburg, VA

INTRODUCTION

Sarcocystis neurona is the most common cause of Equine Protozoal Myeloencephalitis. Incidence of disease is between 0.5-1% of all horses that will get disease during their lifetime (NAHMS). National seroprevalence is >50% with some regional variation (MacKay, 1997). With high exposure rates, but low incidence of disease, it would suggest that the immune system would be critical in prevention of disease. Some studies report stressful factors or steroid treatment can increase incidence of disease in horses (Saville et al., 2001). Other studies in horses that develop EPM, both from natural and experimental exposure, have altered immune responses (Tornquist et al., 2001; Spencer et al., 2004; Yang et al., 2006; Witonsky et al., 2008). Additionally, some immunocompromised mouse models infected with *S. neurona* develop disease (Sellon et al., 2004; Witonsky et al., 2003, 2005). This talk with provide a brief review of the pathophysiology of disease in horses and mice, as well as review what is known about the protective immune response. The focus of the talk will be on the knowledge we have gained regarding the immune response to *S. neurona*, both from studies in mice and horses.

LIFE CYLE OF *S. NEURONA*

The definitive host for *S. neurona* is the opossum. Intermediate hosts are skunk, cat, raccoon, armadillo, and sea otters. *S. neurona's* life cycle includes both sexual and asexual stages. In the lumen of the gut of the opossum, sarcocysts release bradyzoites. These enter the small intestine and transform into male and female gamonts; the male gamonts divide into gametes. Fertilization by the male gamete of the female gamont causes an oocyst to be formed. Within the lamina propria, the oocysts sporulate to produce two sporocysts. Each sporocyst has four sporozoites. Opossum feces contain fully sporulated oocysts or sporocyst (Dubey et al., 2001a; Sellon and Dubey, 2007).

INFECTION

Horses ingest the sporozoites, which invade the intestinal epithelium and undergo asexual replication in many tissues. The mechanism of dissemination to the tissues is not well understood. *S. neurona* replicates in the peripheral tissues and central nervous system (CNS) by schizogony (endopolygeny). During this process, the nucleus becomes lobulated, with chromatin connecting the strands. The merozoites can undergo replication, and can also develop to become mature schizonts. The mechanism by which the merozoites enter the central nervous system (CNS) is unknown, but it may involve mononuclear cells (Ellison et al., 2004; Lindsay et al., 2006). Some studies demonstrate that *S. neurona* can infect endothelial cells as well as leukocytes (Witonsky et al., 2005). Therefore, both endothelial cell invasion and entrance via infected mononuclear cells are possible mechanisms by which *S. neurona* invades the CNS. Studies in gamma interferon knock-out mice (KO), demonstrate that *S. neurona* can infect the peripheral tissues prior to invasion of the CNS. It is not certain whether this applies in immunocompetent animals (Dubey et al., 2001a; Sellon and Dubey, 2007).

Typical lesions in the CNS, which often affect predominantly brain stem and spinal cord, grossly include multifocal hemorrhagic foci, which on histopathology, consist of hemorrhage, nonsuppurative inflammation, small areas of necrosis, and perivascular cuffing with mononuclear cells. Organisms are identified less than 50% of the time (Sellon and Dubey, 2007). A small study demonstrated an increased number of T-cells vs. B-cells present in the lesions (Scott et al., 2005).

ALTERED IMMUNE RESPONSE INCREASING RISK OF DISEASE

In initial efforts to establish a valid, reproducible, experimental model for *S. neurona* infection, several methods have been attempted, all of which involve some sort of "stress and/or immune compromise" (Saville et al., 2001). Shipping and glucocorticoids were used in some models (Cutler et al., 2001; Saville et al., 2001). In another model, immunodeficient horses were used (Sellon et al., 2004). Finally, a model has been established which involves inoculating horses using infected autologous leukocytes (Ellison et al., 2004). This method may allow for evasion of some of the host immune response early in the disease because cells are infected directly. Mononuclear cell infection may also represent the normal method of entry into the CNS. Lindsay et al., (2006) and Heid et al., (2008), both demonstrated that *S. neurona* can infect leukocytes. As a correlate for the role of the immune response in disease, Saville et al., (2000), compiled data from multiple EPM vs. non neurologic controls and determined that risk factors for disease include racing or showing as well as adverse health incidents. This study would support that "stress" may increase incidence of disease. Additionally, there are several studies in mouse models demonstrating that *S. neurona* infected immunocompetent mice are resistant to disease, whereas infected immunocompromised mice develop disease (Dubey et al., 2001; Witonsky et al., 2003a, b; 2005; Sellon et al., 2004a). Collectively, these data support a role for a dysfunctional immune response associated with disease.

PROTECTIVE IMMUNE RESPONE TO *S. NEURONA*

The protective immune response to many intracellular agents is a cell mediated CD4 T-helper-1 (CD4 Th1) with CD8 T-cytotoxic-1 (CD8 Tc1) response, with associated interferon gamma production. Data from studies in horses and mice, which will be discussed below, suggest that the protective immune response to *S. neurona* is a CD4 Th1 CD8 Tc1 response.

MOUSE MODELS OF S. NEURONA: WHAT HAVE WE LEARNED?

To better understand the pathology of *S. neurona* infection, the development of EPM, and the role of the immune response, investigators performed some studies in immunodeficient mice to determine the pathology and the role of each immune component in disease. Data in immunocompetent mice demonstrated that C57BL/6 mice infected with *S. neurona* develop early activation and memory CD8 response (Witonsky et al., 2003a). Furthermore, additional studies in CD8KO, interferon (IFN)-gamma KO, and B-cell deficient mice, all of which were on a C57BL/6 background, confirmed the most significant role for CD8 cells (Witonsky et al., 2003b, 2003c, 2005). *S.neurona* infected IFN-gamma KO mice developed an increased number of memory CD4 and CD8 lymphocytes in the spleen (Witonsky et al., 2003b). Infected CD8 KO mice develop splenomegaly and lymphadenopathy, histopathological changes including follicular hypoplasia, endothelitis and perivascular inflammation in the liver, and meningoencephalitis in the brain. Brains from infected CD8 KO mice were positive by PCR for *S. neurona* (Witonsky et al., 2005a). B-cell deficient mice also did not develop encephalitis (Witonsky et al., 2005b). *S. neurona* infected mice also developed significant CD4 and CD8 effector (CD45RBhigh) and memory (CD44high) cells in lymph nodes and spleen. These data support the role for both a CD4 and CD8 cell mediated immune response, and in particular, the role for CD8 cells (Witonsky et al., 2005a).

One of the main cytokines that both CD4 and CD8 cells produce is IFN-gamma. The studies in CD4 and CD8KO mice as well as the studies in IFN-gamma KO mice, all confirm the critical role for IFN-gamma. Besides CD4 and CD8 cells, NK cells, macrophages and dendritic cells also make IFN-gamma. In studies performed by Sellon et al., (2004a), with immunodeficient mice, severe combined immunodeficient (SCID) mice did not prevent parasitemia, but did prevent encephalitis. However, when the SCID mice were treated with antibodies to interferon-gamma, the mice developed encephalitis. Similar studies in SCID foals demonstrated that SCID foals developed parasitemia, but not neurologic disease (Sellon et al., 2004b). These data suggest that one or more IFN-gamma producing cell populations likely prevents *S. neurona* from causing encephalitis, possibly by preventing *S. neurona* from infecting specific cell types and crossing the blood brain barrier. Additional studies are warranted to investigate these mechanisms.

HORSES INFECTED WITH S. NEURONA HAVE ALTERED IMMUNE RESPONSES

Tornquist et al., (2001) was one of the first studies demonstrating a decrease in percentage of CD4 cells in EPM horses compared to asymptomatic seropositive and seronegative horses. Asymptomatic horses had increased Concanavalin A (Con A), a marker of T-cell proliferation, and antigen specific proliferation compared to symptomatic and seronegative horses. Spencer et al., also showed that there EPM affected horses had decreased Surface antigen 1 (SAG-1) mediated proliferation as well as IFN-gamma vs. IL-4 expression (Spencer et al., 2004; 2005). In contrast to Tornquist's study, Yang et al., (2006) identified a significantly increased number of CD4 cells and neutrophils in naturally occurring cases of EPM, but that non-antigen specific proliferation to mitogens phorbol myristate acetate and ionomycin (PMA/I) was decreased in cells incubated overnight. The decreased PMA/I proliferation response was confirmed in experimentally infected horses. Experimentally infected horses also demonstrated a trend towards decreased antigen specific interferon gamma responses (Witonsky et al., 2008). In a subsequent study to define the mechanism of suppressed responses, the antigen specific proliferation to merozoites was decreased compared to non-specific PMA/I induced proliferation. These data suggest that antigen presenting cell function (i.e. dendritic cells) may have decreased function (Lewis et al., 2011). In addition to the roles for cell mediated immunity for protection, Ellison and Witonsky (2009) demonstrated that there was a correlation between antibody titer and protection in SAG-1 vaccinated horses against disease. Collectively, these data suggest that while the cell mediated immune response may be critical for protection, multiple components of the immune response, innate, CD4, CD8 and B-cell, are likely needed for optimal immunity.

REFERENCES

Cutler TJ, et al. **Vet Parasitol** 2001. 95:197-210, 2001. Dubey JP, et al., **Vet Parasitol** 2001a: 95(2-4):89. Dubey JP, et al., **J Parasitol** 2001b; 87(5): 1171-3. Ellison SP, et al., **Intern J Appl Res Vet Med** 2003: 1: 318. Ellison SP, et al., **Intern J Appl Res Vet Med** 2004: 2: 79. Ellison SP, Witonsky SG: **Can J Vet Res** 2009: 73: 176. Ellison SP, et al., **Intern J Appl Res Vet Med** 2003:1: 318. Heid B., et al, **ACVIM EPM SIG,** 2008. Lewis R, et al., **ACVIM EPM SIG**, 2011. Lindsay DS, et al., **Vet Parasitol** 2006: 138:371. Long MT et al, **Exp Parasitol** 2001: 100:150. MacKay RJ: **J Am Vet Med Assoc** 1997: 210:482. National Animal Health Monitoring System: *Equine protozoal myeloencephalitis (EPM) in the U.S.* 2001, USDA:APHIS:VS, CEAH, NAHMS. Saville WJ et al., **J Am Vet Med Assoc** 2000: 217:1174. Scott et al., **J Parasitol** 2005. 91(6): 1499. Sellon DC, et al., **Clin Diagn Lab Immunol** 2004b: 11:1134. Sellon DC, et al:, **J Parasitol** 2004a: 90:782. Sellon DC and Dubey JP in. **Equine Infectious Disease**, p453. Spencer JA, et al: **J Parasitol** 2004: 90:428. Spencer et al., **Clin Diag Lab Immunol** 2005. 12(5): 644. Tornquist SJ et al., **Eq Vet J** 2001: 33: 726. Witonsky SG, et al., **J Parasitol** 2003b: 89:932. Witonsky SG, et al, **Int J Parasitol** 2005a:35:113. 116. Witonsky SG, et al., **J Parasitol** 2003a: 89:924. Witonsky SG et al., **J Parasitol** 2005b: 91(4): 830. Yang J, et al., **Vet Parasitol** 2006:138: 200. Witonsky S, et al., **J Parasitol** 2008: 94(5):1047.

IDENTIFICATION OF THE CAUSE OF SEASONAL PASTURE MYOPATHY IN HORSES

Stephanie J. Valberg, DVM, PhD, DACVIM[1]; Beatrice Sponseller Dr med vet, DABVP[2], Adrian D Hegeman PhD[1], Jennifer Earing PhD [1], Jeffery B. Bender DVM MS[1], Krishona L Martinson PhD[1], Steven E Patterson PhD[3], Lawrence Sweetman PhD[4]

St. Paul, MN[1], Ames IA[2] and Minneapolis MN[3], Dallas, TX[4]

INTRODUCTION

A highly fatal muscle disease termed Seasonal Pasture Myopathy (SPM) has been described for many decades in horses on pastures in the Midwestern USA and eastern Canada [1]. A similar disorder called Atypical Myopathy (AM) is noted with increasing frequency in the United Kingdom and Northern Europe [2]. Outbreaks of SPM and AM vary from year to year and are seasonal with most cases occurring in the fall and fewer cases occurring in the subsequent spring. In North America, only a few horses are affected on a given pasture whereas in Europe large outbreaks occur with many horses on the same premises being affected [1,2]. Horses that develop SPM and AM are usually kept on sparse pastures with an accumulation of dead leaves, dead wood and trees in or around the pastures. Until very recently the cause of SPM and AM was unknown.

CLINICAL SIGNS

Initial clinical signs include reluctance to move, muscular weakness, stiffness and fine muscle tremors, leading to increased periods of recumbency. Tachycardia, myoglobinurua and occasionally choke are additional clinical signs. Signs usually progress within 48 hours to a rapid respiratory rate and dyspnea and death within 72 h in at least 75% of cases.

CLINICAL PATHOLOGY AND PATHOLOGY

The major abnormalities present in the serum chemistry profile of SPM and AM cases include marked elevations in serum creatine kinase (CK) and aspartate transaminase (AST) activity, lactic acidemia, hyperglycemia, electrolyte derangements such as hypocalcemia, hyonatremia, hypochloremia and hyperphosphatemia, mild elevations in alkaline phosphatase and variable elevations in creatinine [1,2,3]. Myoglobinurua is consistently present. Derangements are largely the consequence of severe myonecrosis causing fluid and electrolyte shifts into damaged muscle, efflux of intramuscular phosphate into the blood stream, renal compromise and damage to respiratory and cardiac muscles.

The primary histopathologic lesion associated with SPM is acute myodegeneration primarily affecting deep postural muscle, the diaphragm and intercostals muscles. In 50% of SPM cases scattered acute myocardial necrosis is also observed. Oil red O staining of frozen muscle sections reveals a marked intracellular accumulation of lipid droplets in those muscles undergoing myonecrosis.

ABNORMALITIES IN LIPID METABOLISM

Within skeletal muscle of SPM and AM horses there is a deficiency of multiple mitochondrial dehydrogenases that catalyze β-oxidation of short, medium and long chain fatty acids, and dehydrogenases involved in amino acid and choline metabolism [4]. This acquired deficiency is called multiple acyl-CoA dehydrogenase deficiency or MADD Enzymes impaired in MADD include Disruption of enzymatic pathways in MADD leads to accumulation of enzyme specific substrates (acyl-CoAs) that are eliminated as acylcarnitines or glycine conjugates or metabolized by alternate metabolic pathways. Peroxisomal β-oxidation or microsomal ω-oxidation results in a variety of disease specific secondary metabolites in urine [4]. A diagnosis of MADD in horses can therefore be made based on characteristic elevations in serum short (C2-C5) and medium chain (C6, C8, C10) acylcarnitines and a lesser abnormal elevation in long chain acylcarnitines [3,5]. In addition, urine from SPM and AM cases has abnormal elevations in ethylmalonic, methylsuccinic and lactic acid as well as specific glycine conjugates ([iso]valeryl-, butyryl- and hexanoylglycine).

INVESTIGATION OF THE ETIOLOGY OF SPM

In 2011 a team of researchers based out of the University of Minnesota and Iowa State University investigated 12 SPM cases that occurred on 11 farms [6]. Two cases were defined as suspected cases based on the criteria; unexercised horses on pasture with acute clinical signs of muscle pain, weakness and recumbency as well as high serum creatine kinase (CK) activity [>2,000 for undiluted samples, > 45,000 when dilution performed] or pigmenturia. Ten cases were confirmed cases (10 horses) based on, in addition to the above criteria, abnormal elevations in MADD-specific urine organic acids and serum acylcarnitines or demonstration of a lipid storage myopathy in at least one of the muscles sampled. In October and November of 2011 a site visit of SPM pastures located in Minnesota and Iowa was performed. Digital images of the pasture, plants and trees were obtained and notes were made with regard to the presence of dead wood, pasture topography, standing water and over-grazed pastures. For farms outside Minnesota and Iowa where a site visit was not possible, digital images of the pastures and trees were requested. Twenty three Control farms located in Minnesota did not have a history of SPM occurrences in the preceding years.

EPIDEMIOLOGY: One specific tree, the box elder tree laden with seeds, was a consistent feature of all SPM pastures (11/11) and this tree was present in significantly fewer control pastures (61%; 14/23) [6]. All SPM horses spent 24 h/day on pasture which was significantly longer than horses on control farms. All of the SPM pastures were overgrazed compared to 44% (10/23) of control horse pastures. It was significantly less common for horses on SPM farms to receive supplemental hay or concentrate compared to controls. Thus, the common features for SPM horses were little supplemental feeding and prolonged grazing of sparse pastures that were in close proximity to seed laden box elder trees.

ANALYSIS OF BOX ELDER SEEDS: Box elder seeds were collected from a random sampling of trees on 7/11 SPM farms in the fall of 2011 [6]. A concurrent control sample of seeds from an Ash tree was also obtained. Individual seeds were dissected from the samara husks, pulverized and extracted in dilute HCl for methyl chloroformate derivatization and amino acid analysis by GC-MS. One nonproteogenic amino acid, hypoglycin A, was highly abundant in box elder seeds and not ash seeds. It was in fact the most abundant amino acid in box elder seeds. This amino acid was thought to be highly significant because in humans ingestion of hypoglycin A contained in unripe ackee fruit causes MADD (Jamaican Vomiting Disease) [7]. The mechanism behind MADD is conversion of hyoglycin A to methylenecyclopropylacetate (MCPA) and its CoA ester which acts as a suicide inhibitor of the general-acyl-CoA dehydrogenases. This causes profound disturbances of the oxidation of fatty acids, and some amino acids.

ANALYSIS OF BLOOD AND URINE METABOLITES: Serum and urine samples from horses confirmed to have SPM were analyzed for MCPA-carnitine or MCPA-glycine respectively [6]. UPLC-MS/MS was utilized to identify MCPA-carnitine and MCPA- glycine conjugates were isolated from inorganic anions and amino acids using liquid partition chromatography (LPC) and GCMS. MCPA-carnitine ranged from 4.8 nM to a remarkably high level of 102.4 nM in serum from 4 SPM horses and was below the limit of detection (approximately 0.001 nM) in serum of nine reference control horses. MCPA-glycine in urine ranged from 0.79 to 3.97 mmole/mole creatinine in 6 of the 7 SPM horses compared to one SPM and 5 reference control horses where it was below the limit of detection (<0.01 mmole/mol creatinine).

THE ETIOLOGY OF SPM IN NORTH AMERICA

Hypoglycin A, present in the seeds of box elder trees appears to be a significant cause of SPM based on (1) the abundance of hypoglycin A in box elder seeds obtained from SPM pastures (2) the presence of the toxic metabolite of hypoglycin A, MCPA, in the serum and urine of affected horses and (3) the fact that MCPA is a specific inhibitor of multiple acyl CoA dehydrogenases, the pathophysiologic mechanism responsible for SPM.

THE ETIOLOGY OF ATYPICAL MYOPATHY IN EUROPE

The box elder tree was introduced into Europe as early as 1688 and is found as a wind-break and shelter-belt tree, however, it is not widely distributed on equine pastures in Europe. Another tree within the same Sapindacae family as Ackee and box elder trees, the European sycamore maple (*Acer pseudoplatanus*), is found widely in pastures within northern Europe. Hypoglycin A previously has been identified in the seeds of the sycamore maple [8]. Thus hypoglycin A may be a common cause of both SPM and AM. This is supported by (1) common epidemiology of SPM and AM, (2) the presence of other hypoglycin A containing Acer species in European pastures (3) preliminary data to be presented. The lethal dose of hypoglycin A for horses is unknown. Studies have been performed in guinea-pigs, rabbits, dogs, cats, rats and mice, listed in order of decreasing sensitivity to hypoglycin's hypoglycemic effects.

PREVENTION

Some horses do not appear to develop SPM after years of being housed on affected pastures. This can make discussions with owners regarding prevention complicated. Young horses and those new to an affected pasture appear to be at increased risk, particularly if pastures are overgrazed, if turn out time is >12 h/day and if no supplemental hay is provided on pasture. To address potential concerns of SPM developing in horses, removal of box elder or sycamore trees from affected pastures can be one approach but may not always be feasible. Decreasing turnout time on affected pastures from October through mid December, and in the early spring are other possible approaches. Other important preventative measures could include providing additional forage if pastures are overgrazed, preventing over grazing of pastures through rotational grazing and limiting turnout to less than 12 hours per day during fall and early spring.

More research needs to be done to determine if seasonal and annual fluctuations occur in hypoglycin A content of box elder and sycamore maple trees.

REFERENCES

1. Finno,C.J. et al. **J.Am.Vet.Med.Assoc** 2006; 229:1134-1141.
2. van Galen,G. et al. **Equine Vet J** 2012;44(5):614-620.
3. Westermann,CM, et al. **Neuromuscul.Disord** 2008;18:355-364.
4. Gregersen,N. **Scand J Clin.Lab Invest Suppl** 1985; 174:1-60.
5. Sponseller,BT. **J Vet Intern Med** 2012; 4:1012-1018.
6. Valberg,SJ et al. **Equine Vet J** 2012; Sep 28 epub ahead of print
7. Fowden,L et al. **Phytochemistry** 1972;12:1677-1681
8. Sherratt HSA et al. **Trends in Pharm Sciences** 1986;7"186-191.

INHALED CORTICOSTEROIDS AND ANTIGEN AVOIDANCE IN HEAVES: A YEARLONG STUDY

Mathilde Leclere, DVM, Ph.D., DACVIM (Internal medicine), Anouk Lavoie-Lamoureux, Ph.D., Jean-Pierre Lavoie, DVM, DACVIM (Internal medicine)
Montreal, QC

INTRODUCTION

Heaves (RAO) is a chronic respiratory condition of horses characterized by airflow obstruction and airway inflammation. Clinical signs are triggered or exacerbated mainly by inhalation of antigen-rich organic dust present in hay and indoor environment such as molds, mites and endotoxin. These clinical signs include coughing, exercise intolerance, and increased respiratory efforts at rest [1]. Heaves is estimated to affect approximately 10 to 20% of adult horses in temperate climates [2]. The mechanisms by which antigen exposure causes airway obstruction and inflammation remain incompletely understood, but immune response alterations have been observed in affected horses. Medical management include reduction of exposition to indoor airborne particles, more specifically by preventing access to hay. Antigen avoidance is an effective mean of controlling clinical signs and pulmonary neutrophilia [3,4] but persistent subclinical peripheral airway obstruction can be detected even after three years on pasture [3]. Corticosteroids are potent anti-inflammatory drugs that remain the most effective pharmacological treatment available for the control of heaves. They control clinical signs and improve lung function, even with persistent antigen exposure [5,6]. Their effect on pulmonary neutrophilia is variable (from a significant decrease to no detectable effect) [7,8], but if administered in conjunction with a reduction of hay exposure, they improve bronchoalveolar lavage neutrophilia to a greater extend than when antigen avoidance is used alone [4]. To avoid the potential side effects of systemically administered corticosteroids, inhaled corticosteroids are now commonly used for the treatment of equine inflammatory lung diseases, but until recently, there was little information on their effects when used over prolonged period of time.

One of the potential contributing factors to the hyperreactivity to inhaled antigens, the worsening of clinical signs over time and the persistent airway obstruction in the absence of measurable inflammation, is the presence of airway remodeling. Airway remodeling is believed to be the consequence of chronic inflammation and is characterized by thickening of the airway wall due in most part to an increase in airway smooth muscle mass and subepithelial collagen. Airway smooth muscle remodeling has been observed in heaves [9,10] and the increase in mass appears to be maintained in a dynamic equilibrium by an elevated myocyte turnover [9], suggesting that it may be reversible with prolonged treatment. This conference describes the effects of inhaled corticosteroids, used alone or combined with antigen avoidance, and compares them to the effects of antigen avoidance alone in the long-term treatment of heaves. Results of this study have been reported elsewhere [11,12].

METHODS

Eleven adult heaves-affected horses were stabled and exposed to hay to induce clinical signs of airway obstruction (Baseline). Treatment consisted of antigen avoidance as sole therapy for 5 horses. The remaining 6 horses were treated with inhaled corticosteroids (ICS). They remained in the same environment that led to their clinical exacerbation for the first 6 months of the study (ICS with antigen exposure), and were then turned out on pasture for the second half of the study (ICS combined with antigen avoidance). Antigen avoidance consisted in keeping horses on pasture and supplementing them with a pelleted diet. In the ICS group, fluticasone propionate (Flovent® HFA, GlaxoSmithKline, Montreal, Canada) was administered at a dosage of 2000 µg q12h for a month, then dosages were adjusted to control clinical signs (2000 to 3000 µg, q12h to 24h). Pulmonary function was measured and bronchoalveolar lavage fluid and blood samples collected at Baseline, and after 1, 6, 7 and 12 months of treatment. Peripheral lung biopsies performed under thoracoscopic guidance were performed at Baseline, 6 and 12 months. Morphometric data included airway smooth muscle area and myocyte count, subepithelial collagen area and histological scoring based on the presence of endobronchial, peribronchial and interstitial inflammation.

RESULTS

All horses developed airflow limitation and airway inflammation with hay exposure (Baseline). Horses tolerated well the daily administration of fluticasone and no adverse reaction associated with the administration of ICS over a 12-month period were observed. Pulmonary resistance declined more rapidly in the ICS group than in the Antigen avoidance group, and this despite continuous antigen exposure in the former group. This improvement was faster with ICS but not complete, as an additional decrease in resistance was observed when ICS-treated/antigen exposed horses joined the Antigen avoidance group at 6 months. This was not observed in the Antigen avoidance group, in which maximal improvement in function was reached between 2 and 4 months.

In the Antigen avoidance group, neutrophil percentage in bronchoalveolar lavage fluid had returned to normal values by month 6. In the ICS group, pulmonary neutrophilia only decreased once antigen exposure ceased. Individual data observation did not show a difference between horses receiving high or low doses of fluticasone, or showing a greater improvement in function. In both groups, there was a significant decrease in IL-8 mRNA expression from bronchoalveolar lavage cells at 6, 7 and 12 months compared to Baseline. This decrease was more pronounced in the Antigen avoidance group at 6 and 7 months, compared to the ICS group.

In the Antigen avoidance group, airway smooth muscle mass decreased gradually over the study period, reaching an approximately 30% decrease at 12 months. A similar maximal improvement was also seen in the ICS-treated horses, but this 30% decline was reached at 6 months with no further significant decrease in the second half of the study. Overall, airway smooth muscle had decreased in 8 horses at 6 months, and in all but one at 12 months. There was also a small decline in subepithelial collagen area over the study period.

The following table summarizes the effects of antigen avoidance and ICS (with or without antigen avoidance) on heaves-affected horses over a 12-month period.

<table>
<tr><th rowspan="2"></th><th rowspan="2">Pulmonary resistance[12]</th><th colspan="3">Inflammation[12]</th><th colspan="2">Remodeling[12]</th><th rowspan="2">Immune system[11]</th></tr>
<tr><th>BAL fluid neutrophils</th><th>BAL cells IL-8 mRNA</th><th>Histological inflammation</th><th>ASM remodeling</th><th>Collagen remodeling</th></tr>
<tr><td>ICS + Antigen exposure (6 months)</td><td>↓ (Fast)</td><td>No effect</td><td>↓</td><td>No effect</td><td>↓</td><td rowspan="2">Marginal ↓[a]</td><td>No effect</td></tr>
<tr><td>ICS + Antigen avoidance (6 months)</td><td>Further ↓</td><td>↓↓</td><td>Further ↓</td><td>No effect</td><td>No further effect</td><td>No effect</td></tr>
<tr><td>Antigen avoidance (12 months)</td><td>↓</td><td>↓↓</td><td>↓↓</td><td>Marginal ↓</td><td>↓</td><td>Marginal ↓[a]</td><td>No effect</td></tr>
</table>

ICS: inhaled corticosteroids. BAL: bronchoalveolar lavage. IL-8: interleukin-8. ASM: airway smooth muscle.
[a] Comparing 12 months with baseline.

DISCUSSION

Inhaled corticosteroids-treated horses had a faster decline in lung function than horses treated with antigen avoidance only, and this despite continuous exposure to hay. Despite this faster decline in pulmonary resistance, a better control of inflammation was not detected in these horses, whether it was assessed by bronchoalveolar cytology, bronchoalveolar cells gene expression, or by histological inflammation scores done on peripheral lung biopsies. This apparent dissociation between the effect on function and inflammation suggests that inhaled corticosteroids primary anti-inflammatory effects may be on other cell types such as epithelial cells or lymphocytes, as suggested by others [4,8], or may affect the level of activation of neutrophils, but not their numbers. Alternative explanation would be that a non-negligible part of corticosteroids action may take place by affecting directly smooth muscle contractility and bronchospasm [13]. It should be noted that within the ICS-treated group, there was a large variation in pulmonary resistance between individuals. The response was not associated with the initial severity of the airway obstruction, or the percentage of neutrophil in the bronchoalveolar lavage. The lack of effect of ICS on bronchoalveolar lavage cytology could also be in part due to the pulmonary function-directed dosing, as higher doses may be required to control inflammation [14]. As systemically administered corticosteroids also failed to improve airway neutrophilia in several studies where horses were maintained in an antigen-rich environment [15,16], this latter explanation appears less probable.

The results indicate that chronic airway smooth muscle remodeling is partially reversible, even in animals that have experienced multiple episodes of antigen exposure and airway obstruction over a period of many years. A reduction in airway smooth muscle remodeling was hastened by inhaled corticosteroids but the maximal improvement was of the same magnitude in both groups at the end of the 12-month study period. The improvement is also modest, with mean values remaining approximately twice those of healthy controls from two previous studies [9,10]. We also observed here an apparent contradiction with the current working hypothesis that airway remodeling is a consequence of chronic inflammation while there was little evidence of a decrease in pulmonary inflammation in the inhaled corticosteroids group during the first half of the study. As mentioned earlier, corticosteroids could have direct effects on other cell types (such as structural cells like airway smooth muscle), or could affect airway smooth muscle remodeling independently from their anti-inflammatory effects, for example by decreasing myocyte proliferation [17].

CONCLUSION

From a clinical point of view, we could summarize that the benefits of using inhaled corticosteroids in the treatment of heaves include a more rapid effect on lung function and airway smooth muscle remodeling compared to antigen avoidance alone, without inducing detectable side effects with prolonged use. However, antigen avoidance leads to a better control of inflammation and it has additional beneficial effects on pulmonary function of inhaled corticosteroids-treated horses, even when clinical signs are apparently controlled by medication.

This study was funded by the Canadian Institutes for Health Research, Allergen, NCE Inc, NSERC, The Clinical Research Pfizer Fund, and le Fonds du Centenaire.

REFERENCES

1.Robinson NE. **Equine Vet J** 2001;33(1):5 2.Hotchkiss JW, et al. **Equine Vet J** 2007;39(4):301 3.Miskovic M, et al. **J Vet Intern Med** 2007;21(5):1060 4.DeLuca L, et al. **J Vet Intern Med** 2008;22(2):427 5.Rush BR, et al. **Am J Vet Res** 1998;59(8):1039 6.Leclere M, et al. **Equine Vet J** 2010;42(4):316 7.Couetil LL, et al. **J Vet Intern Med** 2006;20(2):399 8.Rush BR, et al. **Am J Vet Res** 1998;59(8):1033 9.Leclere M, et al. **Am J Respir Cell Mol Biol** 2011;45(1):181 10.Herszberg B, et al. **J Allergy Clin Immunol** 2006;118(2):382 11.Dauvillier J, et al. **J Vet Intern Med** 2011;25(3):549 12.Leclere M, et al. **Am J Respir Cell Mol Biol** 2012;47(5):589 13.Nabishah BM, et al. **Gen Pharmacol** 1991;22(2):389 14.Zitt MJ. **Allergy Asthma Proc** 2007;28(2):114 15.Lapointe JM, et al. **Am J Vet Res** 1993;54(8):1310-1316 16. Lavoie JP, et al. **Am J Vet Res** 2002;63(4):579 17.Stewart AG, et al. Br J Pharmacol 1995;116(8):3219

DO HEMATOLOGIC AND ULTRASONOGRAPHIC SCREENING METHODS PREDICT CLINICALLY APPARENT *RHODOCOCCUS EQUI* IN FOALS? M. K. Chaffin,[1] N. D. Cohen,[1] G. P. Blodgett,[2] M. Syndergaard,[2] . [1]Texas A&M University, College of Veterinary Medicine & Biomedical Sciences, Department of Large Animal Clinical Sciences, 4475 TAMU, College Station, TX, USA, 77843-4475; [2]6666 Ranch, Guthrie, TX, USA.

The objective was to estimate the sensitivity and specificity of 3 hematologic (measurements of white blood cell [WBC], neutrophil [NEUT], and fibrinogen [FIB] concentrations) and 3 ultrasonographic (measurements of pulmonary consolidations including total maximal diameter [TMD], total cross-sectional area [TCSA] and total number of consolidations [TNC]) screening methods for predicting subsequent onset of clinically-apparent *R. equi* pneumonia.

The methods used were as follows: 270 foals were studied at an endemic farm. Foals were screened biweekly from 3 to 19 weeks of age. Farm personnel were blinded to screening results. Foals were not treated with antimicrobials unless they demonstrated clinical signs of pneumonia. Tracheobronchial aspirates were obtained from all pneumonic foals.

Our results showed that 46 (17%) foals developed clinically-apparent *R. equi* pneumonia. Cumulative sensitivities for WBC (using threshold [TH] $\geq$ 13,000 cells/ul), NEUT (using TH $\geq$10,000 cells/ul), FIB (using TH $\geq$ 500 mg/dl) were 59%, 50% and 59%, respectively. Respective cumulative specificities were 37%, 55%, 33%. Cumulative sensitivities for TMD (using threshold [TH] $\geq$ 200 mm), TCSA (using TH $\geq$200 mm^2) and TNC (using TH $\geq$2) were 89%, 83%, and 78%, respectively. Respective cumulative specificities were 62%, 64%, and 64%. 216 (80%) foals developed sonographically visible pulmonary consolidations, and of those, 46 (21%) foals progressed to clinically apparent *R. equi* pneumonia.

We concluded that hematologic screening demonstrated limited performance for predicting subsequent onset of clinically-apparent *R. equi* pneumonia. Ultrasonographic screening was sensitive, but had limited specificity. In over 75% of foals with sonographic consolidations, pulmonary lesions resolved without treatment.

UMBILICAL ARTERY LACTATE VALUES IN FOALS AT BIRTH. B. Barr, M. Wendel, B. Cornman. Rood and Riddle Equine Hospital, Lexington, KY.

In human neonates umbilical cord lactate concentration is used to describe the neonate's metabolic status at birth. A higher value has been associated with a greater degree of hypoxia and stress during labor, which results in higher morbidity and mortality. Peripheral blood lactate concentration has been evaluated in neonatal foals, but umbilical cord lactate concentration at birth has not. The aim of the study was to determine umbilical cord blood lactate values in foals delivered unassisted and assisted.

Samples were collected from unassisted births at a local farm and assisted births at Rood and Riddle Equine Hospital. The samples were obtained from the umbilical artery, collected into a sodium fluoride/potassium oxalate blood tube and analyzed using the Accutrend® lactate analyzer.

A total of 81 samples were obtained: 47 unassisted vaginal deliveries, 8 assisted vaginal deliveries, 14 controlled vaginal deliveries, and 12 caesarean sections. The mean umbilical artery lactate values are as follows: unassisted vaginal delivery 3.6±4.6 mmol/l, assisted vaginal delivery 5.2±3.1mmol/l, caesarean section 7.2±4.6 mmol/l and controlled vaginal delivery 8.3±5.0 mmol/l.

Foals born unassisted had the lowest values, whereas foals born by controlled vaginal delivery had the highest values. This suggests that there is more risk of neonatal disorders in foals requiring assistance for delivery. Further studies are needed to determine the relation to neonatal disorders, outcome and hyperlactemia.

CYCLOPHILIN B AND LYSYL-HYDROXLASE-1 INTERACTION IN HOMOZYGOUS AND HETEROZYGOUS QUARTER HORSES WITH HEREDITARY EQUINE REGIONAL DERMAL ASTHENIA (HERDA) MUTATION. JE Bowser[1], S Pyott[2], PH Byers[2], A Hill[1], L Schwarz[3], M Pasquali[3], CE Swiderski[1]. 1. Departments of Clinical Sciences, College of Veterinary Medicine, Mississippi State University, Mississippi State, MS. 2. University of Washington Medical Center, Seattle, WA. 3. University of Utah School of Medicine, Salt Lake City, UT.

The goal of this investigation was to identify the molecular mechanisms by which the HERDA mutation in the gene encoding Cyclophilin-B (PPIB) causes the clinical syndrome of skin fragility in HERDA Quarter Horses. Based upon prior investigations in our laboratory[1] demonstrating decreased lysine hydroxylation and an associated increase in deoxypyridinoline in HERDA skin, we hypothesized that the HERDA PPIB mutation modifies an uncharacterized interaction between Cyclophilin B (CYPB) and Lysyl Hydroxylase-1 (LH1), interfering with hydroxylysine synthesis and its availability for collagen crosslink formation. *In-situ* interactions between CYPB and LH1 in cultured dermal fibroblasts from HERDA homozygote, heterozygote, and control horses were assessed using proximity ligation assay (PLA). Interaction between CYPB and LH1 was confirmed in all equine dermal fibroblasts but this CYPB-LH1 interaction was reduced in both HERDA homozygotes and heterozygotes when compared to controls. These results substantiate our hypothesis that a protein-protein interaction occurs between CYPB and LH1 during collagen processing. In light of our prior evidence of decreased lysine hydroxylation and altered collagen cross-linking in HERDA skin[1], this information refines the mechanistic link between the HERDA PPIB mutation and the disease phenotype of fragile hyper-extensible skin. In addition, the evidence of decreased interaction between CYPB and LH1 in cells from HERDA heterozygotes represents a quantifiable phenotype in HERDA carriers. This provides a mechanistic basis from which to investigate the link between heterozygosity for the HERDA allele and a performance advantage in the western performance discipline of cutting[2]. (References on request)

INTRAVENOUS TECHNETIUM-99M LIPOSOMES ACCUMULATE IN LAMELLAR TISSUE DURING LAMINITIS ONSET. C Underwood[1], CC Pollitt[1], JM Metselaar[3], L van Bloois[4], JM van den Hoven[5], P Laverman[2], SN Collins[1], CE Medina-Torres[1], G Storm[4], A.W. van Eps[1]. 1. University of Queensland, Gatton, QLD, Australia. 2. Radboud University Nijmegan Medical Centre, NL. 3. Department of Targeted Therapeutics, MIRA Institute, University of Twente, NL. 4. Utrecht Institute for Pharmaceutical Sciences, University of Utrecht, NL. 5. Department of Pharmacy and Pharmacology, Sloterevaart Hospital, Amsterdam, NL

Laminitis is a crippling disease of the equine foot. At present there is no pharmaceutical means of laminitis prophylaxis. Several candidate anti-laminitis drugs have been identified. Their use is precluded by expense, systemic side effects and rapid clearance after intravenous dosing. Liposomes are phospholipid nanoparticles used for targeted drug delivery. They extravasate at sites of increased vascular permeability, including inflamed tissues. The aim of this study was to determine whether intravenous (IV) liposomes accumulate in lamellar tissue during laminitis development, and hence whether they have potential for targeted lamellar drug delivery.

Ten horses were enrolled in the study. Six horses received 10g/kg oligofructose (OF) via nasogastric tube to induce laminitis. PEG-liposomes were prepared via the film hydration method and labelled with 220 mCi ^{99m}Tc-hexamethyl-propylene-amine-oxime. Horses received 1500 µmol PEG-liposomes by slow IV infusion at the time of OF administration (n=2), 12 hours after OF administration (n=2) and 18 hours after OF administration (n=2). Four control horses received water via nasogastric tube followed by IV liposome infusion. Scintigraphic imaging was performed at 6 and 12 hours post-infusion. The horses were euthanized at 12 hours post infusion (n=8) apart from the 2 horses that received liposomes at the time of OF administration (n=2). These horses were euthanized at 24 hours post injection. The percentage of the injected dose of liposomes per kilogram (%ID/kg) of tissue was calculated for multiple organs. Lamellar samples were collected for histology. Results were analysed using the Mann-Whitney U Test, with significance set at $p<0.05$.

All horses developed clinical signs of laminitis (Obel grade 2-3). The %ID/kg in lamellar tissue was higher in the OF group compared to control horses (median [interquartile range], 0.21 [0.14-0.3]% vs. 0.07 [0.06-0.11]%, p=0.02). The %ID/kg was also higher in OF vs. control skin (0.17 [0.15-0.28]% vs 0.08 [0.07-0.10]%, p=0.01), muscle (0.04 [0.03-0.06]% vs 0.02 [0.01-0.02]%, p=0.03), jejunum (0.2 [0.17-0.25]% vs. 0.11 [0.09-0.13]%, p=0.048), colon (0.27 [0.25-0.41]% vs 0.14 [0.13-0.17]%, p=0.03) and kidney (1.37[1.06-1.47]% vs 0.82 [0.74-1.03]%, p=0.04).

PEG-liposomes accumulate in lamellar tissue during OF induced laminitis development and have potential for lamellar drug delivery. Liposome accumulation also occurs in the skin, muscle, jejunum, colon and kidneys. Liposomes accumulate in inflamed tissues due to increased vascular permeability. These findings indicate that lamellar inflammation occurs early in the developmental phase of laminitis. This is the first study to demonstrate an increase in lamellar vascular permeability during laminitis development.

LAMINAR IGF-1 RECEPTOR SIGNALING IN A MODEL OF ENDOCRINOPATHIC LAMINITIS. HE Lane[1], RJ Geor[2], TA Burns[1], MR Watts[1], LJ McCutcheon[2], JK Belknap[1]. 1. Ohio State University College of Veterinary Medicine, Columbus, OH. 2. Michigan State University College of Veterinary Medicine, East Lansing, MI.

Hyperinsulinemia is a risk factor for equine metabolic syndrome (EMS)-associated laminitis (EMSAL), and results in the development of laminitis when experimentally induced in normal horses. The laminar basal epithelial cell (LBEC) layer, the cell layer which dysadheres from the underlying basement membrane leading to laminar failure in laminitis, does not have insulin receptors but expresses insulin-like growth factor-1 receptors (IGF-1Rc) which can bind insulin. The two primary pathways downstream of IGF-1Rc are the PI3K/AKT/mTOR and the Ras/ERK pathways, both of which can result in activation of ribosomal protein S6 (RPS6). RPS6 activation can occur through phosphorylation of serine moieties 235/236 and 240/244; whereas activation of the PI3K/AKT/mTOR pathway can result in phosphorylation of all 4 moieties (via p70S6K), the Ras/ERK pathway only results in phosphorylation of Ser 235/236 (via p90RSK). We hypothesized that, in EMSAL, hyperinsulinemia leads to aberrant LBEC signaling due to activation of pathways downstream of the IGF-1Rc. The purpose of this study was to assess laminar activation states of proteins central to the PI3K/AKT/mTor and Ras/ERK pathways in an equine model of EMSAL. For this study, 22 ponies stratified by body condition (lean [LN, n=11] and obese [OB, n=11]) were exposed to a low non-structural carbohydrate (NSC) diet (CON, n=5 for OB and LN groups) or a high NSC diet (CHO, n=6 for OB and LN groups) for 7 days. Serum insulin concentrations were obtained prior to the feeding protocol; insulin concentrations and laminar samples were obtained at the end of the 7 day feeding protocol. Immunoblotting was employed to assess laminar concentrations of both activated (phosphorylated) and total protein for key proteins of the PI3K/AKT/mTOR and Ras/ERKpathways. Significant correlations existed ($P<0.05$; $r> 0.5$) between the 7 day insulin concentrations and the following phospho-proteins: phospho-mTOR, phospho-p70S6K, phospho (Ser 240/244)-RPS6; no significant correlations existed with phospho-ERK or phospho (Ser 235/236)-RPS6. The only change ($P<0.05$) in activation state of the proteins of interest between groups was a marked increase in laminar phospho (Ser 240/244)-RPS6 in obese CHO-challenged ponies compared to obese CON ponies. Both the association of serum insulin concentrations with laminar activation of members of the PI3K/AKT/MTOR pathway and the distinct increase in phosphorylation of the Ser 240/244 moieties (but not the Ser 235/236 moities) of RPS6 in the obese CHO-challenged ponies indicate activation of the PI3K/AKT/mTOR pathway in the laminae in EMSAL. As RPS6 activation is associated with decreased adhesion properties of epithelial cells in human disease, further investigation of this pathway may reveal new therapeutic targets for LBEC dysadhesion in EMSAL.

EVALUATION OF AN OMEGA–3 FATTY ACID CONTAINING FEED SUPPLEMENT (*ALEIRA*™) IN THE MANAGEMENT OF HORSES WITH CHRONIC LOWER AIRWAY INFLAMMATORY DISEASES. N Nogradi, JR Burgess, JB Messick, LL Couetil. Purdue University, West Lafayette, IN.

Chronic lower airway inflammatory diseases, such as recurrent airway obstruction (RAO) and inflammatory airway disease (IAD) are common in horses. RAO and IAD present many similarities with asthma in people. Based on studies in human asthma, omega – 3 fatty acids have been proposed to improve clinical signs in horses with chronic lower airway inflammatory disease. The purpose of this study was to evaluate the clinical efficacy of an equine feed supplement (*Aleira*™) in horses with chronic lower airway disease

The study was a randomized, double blind, placebo controlled clinical trial. Client-owned horses with a history of chronic cough (>3 months) were recruited. Horses were examined twice, 2 months apart. Baseline evaluation included physical examination, clinical scoring, pulmonary function testing, and bronchoalveolar lavage. Additionally, an owner assigned visual analog scoring system (VAS) was utilized to evaluate treatment efficacy based on cough, respiratory effort and performance. Eligible horses were randomly assigned into one of 3 treatment groups fed daily for 2 months with: 30 g of the supplement, 60 g of the supplement or 30 g of placebo. Additionally, all horses in the study were maintained on a complete pelleted diet for the duration of the study with no exposure to hay. At the end of the trial, horses returned for a recheck evaluation, where testing procedures were repeated. Data was analyzed using Wilcoxon matched pairs test. Data were expressed as median [25th-75th percentiles] and $P<0.05$ was considered significant.

A total of 32 horses were included in the study, 14 were diagnosed with RAO and 18 with IAD. There was a significant improvement in clinical score in IAD cases receiving the 1x dose of the supplement (6.0 [6.0-15.0] before and 3.0[1.0-6.0] after treatment; P= 0.04), while IAD horses receiving the 2x dose had a significant improvement in VAS cough score (50 [20-70] before and 100 [95-100] after treatment; P=0.04). In horses diagnosed with RAO, there was a significant improvement in clinical score in horses receiving the 1x dose (11.0 [9.0-12.0] before and 4.0 [3.0-5.0] after treatment; P=0.04), as well as in horses receiving the 2x dose (9.0 [8.0-10.0] before and 4.0 [3.0-4.0] after treatment; P=0.04). Clinical score did not improve significantly in IAD and RAO horses receiving placebo. However, owner-reported VAS cough score improved significantly in IAD horses fed the placebo for 2 months (P=0.01). The supplementation had no significant effect on either lung function or BAL fluid cytology in both horses with IAD and RAO.

Our results indicate that daily feeding with 30 g or 60 g of omega-3 fatty acid-rich feed supplement for 2 months in addition to management practices eliminating exposure to hay in the environment can result in improvement in the clinical status of horses when compared to feeding placebo along with the low dust diet.

COMPARISON OF ASSAY RESULTS FOR MEASUREMENT OF PLASMA ADRENOCORTICOTROPIN CONCENTRATION. H Schott II[1], K Refsal[2]. 1. Department of Large Animal Clinical Sciences, and 2. Diagnostic Center for Population and Animal Health, College of Veterinary Medicine, Michigan State University College of Veterinary Medicine, East Lansing, MI.

Measurement of plasma adrenocorticotropin (ACTH) concentration is currently advocated as the endocrine test of choice for screening horses for pituitary pars intermedia dysfunction (PPID). However, assays used to measure ACTH differ between commercial laboratories and there is little published data validating that these assays specifically measure equine $ACTH_{1-39}$. In one study, incubation of adrenocortical explants with plasma collected from PPID-affected horses with high ACTH concentrations did not result in proportionally greater cortisol release into the supernatant than incubation with plasma collected from normal horses. This raises the question of whether antibodies in commercially available assay kits cross react with incomplete ACTH fragments or other pro-opiomelanocortin (POMC) derived peptides that may be present in higher concentrations in PPID-affected horses. Such peptides could cross react with assay anti-ACTH antibodies to produce either falsely elevated or lowered plasma ACTH concentrations, depending on the assay method. To assess variability between assays, ACTH concentration was measured in 39 plasma samples collected from PPID-affected and normal horses using a radioimmunoassay kit (MP Biomedical™), an immunoradiometric kit (Scantibodies™), and a chemiluminescent immunometric assay (Immulite™). Spearman rank order correlation revealed fairly good correlation between results of the three assays with correlation coefficients ranging from 0.88 to 0.93. However, the slope of the regression line between assay results ranged from 0.24 to 3.6 with one assay (Immulite™) yielding ACTH concentrations that were 3.6-fold greater than those measured with the Scantibodies™ kit while the MP Biomedical™ kit yielded intermediate values. Bland-Altman analysis revealed bias ± limits of agreement of 19.3 ± 26.5 (MP Biomedical™ - Scantibodies™), 24.2 ± 62.9 (Immulite™ - Scantibodies™), and 4.4 ± 43.0 (Immulite™ - MP Biomedical™). Further, the positive bias of the Immulite™ assay increased with increasing ACTH concentrations. In conclusion, currently available ACTH assay kits can yield markedly different ACTH concentrations for the same plasma sample. Consequently, practitioners using this test to screen equids for PPID should use laboratory specific reference ranges to interpret results. Further, researchers should clearly specify which assay kit is used in their studies along with steps taken to validate assay performance, including available information on cross reactivity of ACTH fragments and recovery of added $ACTH_{1-39}$. Further work is needed to determine which ACTH fragments and other POMC derived peptides are produced in excessive amounts in PPID-affected equids and how these peptides may interfere with measurement of $ACTH_{1-39}$.

ACTH CONCENTRATIONS IN ILL HORSES DURING HOSPITILIZATION.

Allison J. Stewart[1], Eddy Kwessi[2], Taylor J Towns[1], Heather Weaver[1], Bradly Johnson[1]. [1] College of Veterinary Medicine, [2]College of Mathematics and Statistics, Auburn University, Auburn AL.

Concentrations of ACTH are expected to rise with illness associated stress, but have never been reported in adult horses. Our purpose was to determine the effects of illness on pituitary function in adult horses.

Horses >1 year of age admitted to Auburn University with colic or systemic illness were divided according to severity of illness (mild, moderate, severe) based on physical examination findings and clinicopathologic data. Plasma ACTH concentrations were measured from frozen batched samples by radioimmunoassay on admission, and days 2, 4 and 6. ACTH concentrations over time were compared using 1-way ANOVA with $P<0.05$ considered statistically significant.

The study involved 136 horses: 48 mild, 46 moderate and 42 severely ill. At admission ACTH concentration was higher in severely ill horses (120 ± 160pg/mL; n=42) compared to mild (33 ± 48pg/mL; 45, $P<0.0001$) and moderately ill horses (30 ± 32pg/mL; 45, $P=0.0005$). By day 2, most mild (20 ± 26pg/mL; 21) and moderately (14 ± 16pg/mL; 33) ill horses had ACTH concentrations within the reference interval, while severely ill horses still had elevated ACTH concentrations (47 ± 45pg/mL; 14). On day 4, all mild (10 ± 5pg/mL; 6) and moderately (9 ± 7pg/mL; 23) ill horses and most severely (26 ± 21pg/mL; 12) ill horses had ACTH concentrations within the reference interval.

Admission ACTH concentrations were associated with illness severity, and remained high in severely ill horses for several days. ACTH concentrations returned to within reference intervals in most mild and moderately ill horses by day 2.

REGULATION OF GLUCOSE TRANSPORT DURING EQUINE HYPERINSULINAEMIA.

M. A. de Laat[1], C. K. Clement[1], C. C. Pollitt[2], C. M. McGowan[3], M. N. Sillence[4] and V. A. Lacombe[1]. 1. Center for Veterinary Health Sciences, Oklahoma State University, Stillwater, OK. 2. Australian Equine Laminitis Research Unit, University of Queensland, Gatton, Qld, Australia. 3. Faculty of Health and Life Sciences, University of Liverpool, Neston, UK. 4. School of Earth, Environmental and Biological Sciences, Queensland University of Technology, Brisbane, Qld, Australia.

Glucose uptake is mediated by glucose transporters (GLUTs) through insulin-dependent or -independent processes. During equine hyperinsulinaemia, which is a known risk factor for laminitis, GLUT trafficking may be altered. The aim of this study was to investigate the regulation of glucose transport in two models of hyperinsulinaemia. Striated muscle and lamellar samples were collected from Standardbred horses treated with a prolonged euglycaemic, hyperinsulinaemic clamp (p-EHC) or hyperglycaemic clamp (HC) and balanced electrolyte-treated controls. The p-EHC and HC induced clinical and subclinical laminitis respectively. Crude membrane and total protein extracts from striated muscles and lamellae were analyzed with western blotting for GLUT (1, 4, 8, 12) expression. In control horses, GLUT1 protein expression was greatest in lamellae, whereas GLUT4 was greatest in the heart and almost absent in lamellae. In horses treated with a p-EHC, GLUT1 content decreased in skeletal muscle whereas other GLUT isoforms did not change compared to controls. Lamellar GLUT protein expression in horses with subclinical laminitis did not differ from controls. During clinical laminitis, lamellar GLUT1 and 4 protein expression did not differ from controls whereas novel GLUT isoform expression (8, 12) was increased. A lack of GLUT4 down-regulation in skeletal muscle during hyperinsulinaemia suggests that insulin resistance is not required for the development of laminitis. Abundant lamellar GLUT1 protein expression relative to insulin-sensitive GLUT4, which was minimal, confirms prior gene expression studies. Our data further suggest that increases in novel GLUT isoforms identified in the lamellae of horses may play a role in laminitis pathophysiology.

SERUM FREE CORTISOL FRACTION AND INSULIN CONCENTRATION IN NORMAL HORSES, OBESE HORSES AND HORSES WITH PITUITARY PARS INTERMEDIA DYSFUNCTION. KA Hart[1], DM Wochele[1], D McFarlane[2], AA Wooldridge[3], N Frank[4]. 1. University of Georgia College of Veterinary Medicine, Athens, GA; 2. Oklahoma State College of Veterinary Medicine, Stillwater, OK; 3. Auburn University College of Veterinary Medicine, Auburn, AL; 4. Tufts Cummings School of Veterinary Medicine, North Grafton, MA.

Laminitis and insulin resistance (IR) are important complications of Pituitary Pars Intermedia Dysfunction (PPID) in horses, and laminitis, insulin resistance and obesity are hallmarks of Equine Metabolic Syndrome. However, the pathogenesis of laminitis and insulin resistance in horses with endocrine disease is poorly understood. Cortisol has been theorized to play a role in the development of endocrinopathic laminitis and insulin resistance, but total serum cortisol concentration is not typically increased in affected horses. Measurement of total cortisol concentration quantifies both protein-bound and free cortisol, though only free cortisol is biologically active. In people, increased free but not total cortisol is associated with insulin resistance and obesity. To the authors' knowledge, free cortisol has not been investigated in horses with endocrine disease, insulin resistance, or obesity. Thus, the objectives of this study were to compare free cortisol fraction (FCF) between horses with PPID and age-, sex- and season-matched healthy horses, and to determine if FCF and hyperinsulinemia are correlated in PPID and obese horses. Body condition scores (BCS) were determined and blood was collected in October, January, April, and July from 57 healthy adult horses. Archived frozen serum samples were obtained from 20 horses previously diagnosed with PPID and from 19 obese horses (BCS 6-9/9). Serum insulin and total cortisol concentrations and FCF were measured with previously validated radioimmunoassay, chemiluminescent immunoassay, and ultrafiltration/ligand-binding assay respectively. Hyperinsulinemia was defined as a non-fasted insulin concentration > 45 μIU/ml. Data were analyzed with linear regression and Mann Whitney tests with statistical significance set at $P < 0.05$. In healthy horses, there was a significant association of increasing BCS with increasing insulin concentration ($P < 0.0001$) and increasing FCF ($P < 0.0001$). Total cortisol concentration was not significantly different between PPID and healthy horses, but both insulin concentration and FCF were significantly higher ($P=0.011$ and $P=0.0004$ respectively) in PPID horses (insulin 46.2 ± 43.2 μIU/ml; FCF 8.9 ± 5.8%) than healthy horses (insulin 17.0 ± 8.9 μIU/ml; FCF 5.0 ± 0.9%). There was no difference in FCF between hyperinsulinemic and normoinsulinemic PPID horses ($P = 0.6341$), but in obese horses, FCF was significantly higher in hyperinsulinemic horses than normoinsulinemic horses (8.3 ± 5.2% vs. 5.7 ± 1.6%; $P = 0.037$). These data suggest that cortisol binding dynamics may be altered in horses with obesity and PPID, resulting in increased free cortisol. Further study is needed to determine if such alterations in free cortisol directly impact insulin sensitivity in horses.

EFFECT OF A SUPPLEMENT (SMARTGUT® ULTRA) ON GASTRIC ULCER SCORES AND GASTRIC JUICE PH. F Andrews[1], P Lofton[1], G Gammon[2], P. Camacho[1], J Cartmill[1], F Garza, Jr.[1], M Keowen[1], M Kearney. [1]Louisiana State University, Baton Rouge, LA, [2]Tuskegee University, Tuskegee, AL

Non-glandular (NG) gastric ulcers are common in horses. Omeprazole paste (OME, GastroGard®; Merial, LTD) is effective in treatment of gastric ulcers but recurrence is common. Also, long-term treatment is expensive, must be administered orally and increases gastric juice pH, which may affect digestion. Botanicals and other natural supplements are being used in people to improve gastric health, but effectiveness in horses is questioned. SmartGut® Ultra (SmGU; SmartPak® Equine, LLC, Plymouth, MA), a supplement containing botanicals, natural coating agents and buffering compounds, might be useful in protecting the stomach from the effects of acid injury after pharmacologic treatment. The purpose of this study was to determine the effect of SmGU on gastric lesion scores and gastric juice pH in stall-confined horses treated with OME.

A masked 2-period crossover study was performed using adult Thoroughbreds (n=8). SmGU (40 g) was mixed with grain (Omelene 100, Purina Mills, LLC, Grey Summit, MO.; 1 Kg) or grain alone (control) was fed to horses twice daily for 42 days. Horses were stratified by NG lesion score, then by sex and assigned to treatment groups (n=4). There were two 42-day periods, where horses were stall-confined. Horses received OME (4 mg/kg, PO, Q24h) from days 0-13, then were subjected to alternating feed-deprivation from days 28-35 and allowed to recover for 7 days (Day 35-42). Gastroscopy was performed on days -1, 14, 28, 35, and 42. Non-glandular lesion number (NGN) and severity (NGS) scores were assigned by a masked clinician (FMA) and gastric juice pH was measured using a handheld pH meter. Gastric lesion scores were expressed as mean (SEM) and an ANOVA for repeated measures using the Mixed Procedure in SAS (SAS Institute, Cary, NC) was used to compare ulcer scores between groups and over time. A post-hoc least square means test (LSM) was used to determine differences ($P<0.05$).

All horses (n=8) were treated for 42 days. Mean NGN and NGS scores significantly decreased in both groups by day 14, as a result of OME treatment. By day 28, 2 weeks after OME treatment was discontinued, mean NGN scores did not significantly increase in SmGU-treated horses, whereas NGN scores significantly increased in untreated control horses. Also, by day 35, after feed-deprivation, NGN scores increased in both groups, but NGN scores were significantly lower in the SmGU-treated horses, when compared to the untreated controls on the same day. Ulcer scores were lower in both groups by day 42. SmGU did not significantly increase gastric juice pH. SmGU treatment resulted in fewer ulcers two weeks after discontinuing OME treatment and blunted the increase in the number of gastric ulcers after feed-deprivation, without altering gastric juice pH. Supplementation with SmartGut® Ultra might be an affordable alternative to aid in the protection of the non-glandular stomach from the rebound acid effects once omeprazole treatment is discontinued and in stall-confined horses undergoing intermittent feeding.

ADMINISTRATION OF ONCE DAILY TRIMETHOPRIM-SULPHADIMIDINE DOES NOT IMPROVE HEALING OF GLANDULAR GASTRIC ULCERATION IN HORSES RECEIVING OMEPRAZOLE: A BLINDED, RANDOMIZED CLINICAL TRIAL. BW Sykes[1], KM Sykes[1], GD Hallowell[2]. 1. Upper Orara, NSW. 2. University of Nottingham, Sutton Bonington, UK.

The use of antimicrobials as an adjunct to acid suppressor therapy has been advocated for the treatment of equine glandular gastric ulcer syndrome (EGGUS). The purpose of this study was to evaluate whether the addition of an antimicrobial to omeprazole therapy improved outcome in EGGUS. Thoroughbred racehorses in training with grade 2-4/4 glandular ulcers were identified on gastroscopic examination. Horses were randomized to receive either 2.0 grams of enteric coated omeprazole (OM) paste, equivalent to 4 mg/kg for a 500 kg horse, with or without 15 grams trimethoprim-sulphadimidine (TMPS), equivalent to 30 mg/kg for a 500 kg horse, PO SID. The medications were administered 1–3 hours prior to morning exercise. Repeat gastroscopy was performed at 25±2 days. The data was not normally distributed and was analysed using a Fishers exact test and Mann Whitney U test. Significance was assumed at $p<0.05$.

Twenty-nine horses met the inclusion criteria and were randomised with 14 horses in the OM and 15 in the OM+TMPS group. One horse was subsequently excluded from the OM group. Eleven and 13 horses, in groups OM and OM+TMPS respectively, had been treated previously with omeprazole for 28±2 days. Two horses in each group were identified with EGGUS on the first examination of their stomach. There were no differences between the groups at enrolment, time to follow up ($p=0.51$), number of race starts ($p=0.65$) or mean glandular ulcer score at enrolment ($p=0.64$). A reduction in mean glandular ulcer score was observed over time in both groups (OM - 2.9±0.3 to 1.9±1.0; $p=0.002$ and OM+TMPS - 3.0±0.5 to 1.5±1.3; $p=0.0003$), but there was no difference between the OM and OM+TMPS groups ($p=0.21$). Improvement, defined as a decrease in ulcer score by at least 1 grade, occurred in 69% and 80% of the OM and OM+TMPS horses respectively. Healing, defined as a return to grade ≤ 1/4 occurred in 23% and 47% of the OM and OM+TMPS horses respectively. Worsening of ulcer grade was not observed in any animal. There was no difference between groups in the number of horses that showed improvement ($p=0.99$), or in those that showed healing ($p=0.41$) of glandular ulcers. In conclusion, the results of the study suggest that the addition of trimethoprim-sulphadimidine does not enhance healing over omeprazole therapy alone. Until further studies demonstrate a positive effect, antimicrobial therapy should not be recommended as a part of therapy for EGGUS.

REDUCTION OF BETA2 TOXIN GENE EXPRESSION OF *C. PERFRINGENS* BY THE CELL FREE SUPERNATANT OF FOUR *LACTOBACILLUS* STRAINS AND *BIFIDOBACTERIUM ANIMALIS* SSP *LACTIS*. A Schoster[1,2], L Guardabassi[1] B Kokotovic[3]. 1. University of Copenhagen, Faculty of Health and Medical Science, Copenhagen, Denmark. 2. University of Zurich, Vetsuisse Faculty, Zurich, Switzerland. 3. Technical University of Denmark, Copenhagen, Denmark.

Clostridium perfringens, particularly strains harboring the beta2 toxin gene (*cpb2*) are important enteropathogens in humans and animals. In horses, *C. perfringens*, particularly beta2-toxigenic *C. perfringens* has been associated with necrotizing enterocolitis. Beta2 toxin has been shown to be cytotoxic to porcine and human endothelial cells, further supporting its role in necrotizing enterocolitis and suggesting a similar pathogenesis in humans and animals. Probiotics have gained increased importance in human and veterinary medicine to treat and prevent enteric disease. The mechanisms of action are multifactorial, and aside from inhibition of pathogen growth also include interference of gene expression through quorum sensing. The objective of this study was to assess the ability of selected commercial probiotic strains to reduce toxin gene expression by *C. perfringens*.

The time point of maximum gene expression of *C. perfringens* was determined in a growth curve experiment. Reverse transcriptase real time PCR was used to assess the relative differences in toxin gene expression of *C. perfringens* normalized to a reference housekeeping gene. Addition of the cell free probiotic supernatant obtained from five probiotics (*Lactobacillus plantarum* (n=2), *L. rhamnosus* (n=2) and *Bifidobacterium animalis* ssp. *lactis*) was compared to an untreated culture at the time of maximum gene expression. Quiagen REST 2009 software was used to compare normalized toxin gene expression between treated and untreated cultures.

Expression of *tpi* (housekeeping gene) and *plc* (encoding alpha-toxin) started at 2h of growth while *cpb2* expression could first be detected after 10h of growth. Expression of all genes reached the highest levels after 18h and started to decline thereafter. Cell free probiotic supernatants from *L. plantarum* LPAL, *L. rhamnosus* LRH10, *L. rhamnosus* SP1 and *B. animalis* ssp. *lactis* were able to significantly down-regulate beta2 toxin gene expression 9-10 fold around the peak of expression (after 16h of incubation). Expression of *plc* was only reduced by the cell free supernatant of *L. plantarum* LPAL.

This study demonstrates that some probiotic strains cause a significant reduction in *C. perfringens* toxin gene expression under *in vitro* conditions. These data support the hypothesis that probiotics could be an alternative to antibiotics for treatment of *C. perfringens* enteric disease.

RAPID *SALMONELLA* DETECTION IN FECAL AND VETERINARY HOSPITAL ENVIRONMENTAL SAMPLES USING COMMERCIALLY AVAILABLE LATERAL FLOW ANTIGEN DETECTION SYSTEMS. BA Burgess, DS Bolte, DR Hyatt, DC Van Metre, PS Morley. College of Veterinary Medicine and Biomedical Sciences, Colorado State University, Fort Collins, CO.

Salmonella enterica is one of the most commonly reported organisms associated with epidemic disease at veterinary hospitals, significantly impacting animal morbidity and mortality and adversely affecting the public's perception of these hospitals. Comprehensive screening and rapid detection of *S. enterica* in relevant samples is essential for effective infection control in high-risk populations. The objective of this series of projects was to adapt commercially available rapid *Salmonella* detection systems, used in food safety microbiology, for use as point of care tests with 24-48 hr results, and to characterize test accuracy with relevant veterinary samples.

Equine fecal samples were cultured using two techniques: 1-gm in tetrathionate broth (TET) and 10-gm in buffered peptone water (BPW) and Rappaport-Vassiliadis (RV); and veterinary hospital environmental samples were cultured using four techniques: BPW-RV, BPW-TET-RV, BPW-TET, and BPW. In addition, a formal random process was used to select 112 *S. enterica* isolates, collected as part of long-term surveillance at the CSU-VTH, representing the 10 most common serotypes (6 serogroups) in this archive. One gram equine fecal samples were inoculated with 1ml of standardized inoculum (~10^2 cfu per ml) into 9ml TET and incubated at 43°C for 18 hr. Finally, clinical fecal samples were collected from cattle on a high and low-ration. One gram fecal samples were cultured in 9ml TET at 43°C for 18 hr. All samples were tested in a blinded fashion.

Immunoassay sensitivity and specificity were generally good for equine fecal and environmental samples. They had greater sensitivity with enrichment of 1-gm fecal samples in TET compared to 10-gm samples in BPW-RV which may be related to incubation times. All environmental sample techniques had similar sensitivities. However, serotype tended to vary by immunoassay with the poorest sensitivity for serotypes Cerro (Group K), Mbandaka (Group C1) and Montevideo (Group C1). In contrast, rapid test strips displayed poor specificity (i.e., had a high false positive rate) when testing fecal cultures obtained from cattle on a high-ration. Of 116 fecal samples from dairy cattle on a high-ration, 96.5% (112/116) were test strip positive, however only 25.8% were culture positive.

The findings of this project demonstrate that the utility of these commercial immunoassays varies by type of sample as well as by isolate serotype. Understanding the reactivity of these immunoassays is critically important to their implementation in clinical practice. This study demonstrates that commercially available rapid detection systems may be valuable as point-of-care diagnostic tests when managing animal facilities such as veterinary hospitals.

EQUINE POSTOPERATIVE ILEUS: ARE WE ON THE RIGHT TRACK?

Neil P.H. Hudson, MA, VetMB, PhD, DEIM, DipVetClinStud, MRCVS
Edinburgh, United Kingdom

INTRODUCTION

Disorders of the gastrointestinal tract are among the most common problems in equine veterinary medicine. Furthermore motility disorders are a significant cause of morbidity and mortality. Colic is reported by insurance companies and universities as the single greatest killer of horses.[1] The incidence of colic in the horse has been reported in the range of as many as 4 to 10 cases per 100 horses per year.[2-5] The economic impact on the equine industry is huge with an annual estimate of cost in the USA of over $115 million.[4] Furthermore, all horses undergoing a laparotomy for an acute abdominal crisis are at risk of developing postoperative ileus, the pathophysiology of which still remains unclear.[6,7]

ILEUS

There are numerous and confusing definitions of ileus and postoperative ileus. Broadly, ileus is a functional problem as opposed to motility problems due to mechanical or physical obstructions. Further terms include adynamic, postoperative and paralytic ileus. This author's opinion is that these terms can be confusing in the equine setting, especially those that suggest more prolonged time courses (e.g. paralytic). For the purposes of this discussion, the definition of ileus used will be: ileus is the syndrome of functional inhibition of propulsive bowel motility.[8] Ileus frequently occurs in the immediate period following abdominal surgery (postoperative ileus [POI]) and constitutes a serious postoperative complication in horses. Key to an understanding of these issues is an appreciation of the basic mechanisms that control gastrointestinal smooth muscle. The enteric nervous system (ENS) which is involved in all aspects of gastrointestinal function, has been shown in the horse to consist of a myenteric plexus and a two-layered submucous plexus.[9] Intimately associated with the ENS and the intestinal smooth muscle are the interstitial cells of Cajal (ICC). ICC are the pacemakers and mediators of certain forms of enteric motor neurotransmission in the gastrointestinal tract and also facilitate active propagation of electrical events.[10] Slow waves are rhythmic pacemaker currents initiated by the ICC and are the rate limiting step for peristalsis.[11-14] When action potentials (spikes) are superimposed on slow waves (Fig 1), contractile activity occurs.[15]

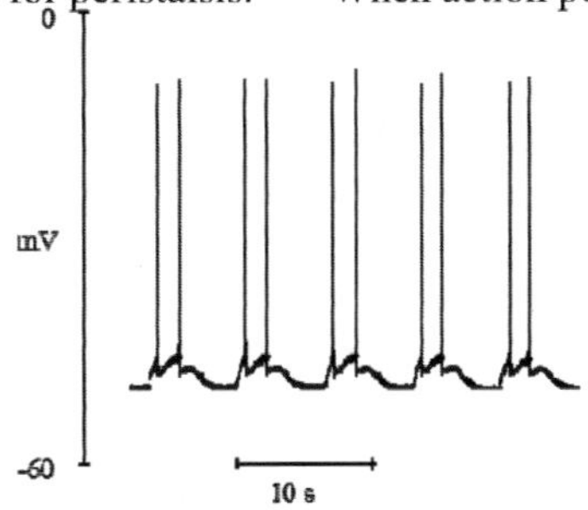

Figure 1. *In vitro* intracellular recording from the longitudinal smooth muscle of normal equine ileum showing slow waves and associated superimposed spontaneous action potentials (reproduced with permission from Hudson *et al.*[16]).

Equine ileus

Ileus is seen in both adult horses and foals, but this discussion will concentrate on syndromes seen in the adult animal. Intestinal dysmotility can be present in a variety of scenarios in the horse including large intestinal impactions (e.g. pelvic flexure and caecum), grass sickness, peritonitis. endotoxaemia, duodenitis/proximal jejunitis and of course POI. In many situations the exact cause of the ileus is unknown; this can be seen in the wider context of colic where in many cases the exact gastrointestinal dysfunction is unknown.[2] In examining the pathophysiology of equine intestinal dysmotility it can be difficult to ascertain whether a finding (such as a change in intestinal cellular profile) is a primary cause of the syndrome or an effect due to the intestinal disease process itself. If the change observed is primary, it does beg the question as to whether certain horses are predisposed to developing signs of colic and ileus because of this change. Ileus syndromes may have a link with the motility control systems, namely the interactions between the enteric nervous system, the interstitial cells of Cajal (ICC) and the intestinal smooth muscle. Horses with chronic obstructive disorders of the large colon or recurrent caecal impactions have been shown to have reduced densities of neurons in the large colon or caecum .[17,18] A reduction in ICC density has been demonstrated in the pelvic flexure of horses with large intestinal obstructive disorders, suggesting a possible involvement of ICC in equine intestinal dysmotility.[19] Recently, a reduction in enteric neuronal activity and ICC immunoreactivity in cases of equine colic has been reported.[20] The neuronal pathology of grass sickness has been well described in the autonomic ganglia and ENS, including a decrease in neurons in the ENS.[21] Additionally, a reduction in ICC in grass sickness cases[22] may also contribute to the observed motility disturbances. Ileus can also have a link with management practices. Numerous risk factors have been identified for the development of colic, including management changes, diet, stabling, exercise levels etc. Interestingly, a recent ultrasonographic study showed a reduction in large intestinal motility in stabled compared to pastured horses.[23] The role of parasites such as tapeworms and cyathastomins in affecting intestinal motility should also be considered. Inflammatory intestinal conditions, such as inflammatory bowel disease can have some clinical signs associated with a disruption to normal motility. There have been sporadic reports of equine myenteric ganglionitis where inflammatory infiltration of the myenteric region resulted in signs of intestinal pseudo-obstruction.[24-26] Of course, the classical intestinal inflammation/ileus relationship in the horse is the duodenitis/proximal jejunitis (DPJ) syndrome.

POSTOPERATIVE ILEUS

Equine postoperative ileus

Postoperative ileus (POI) constitutes a serious post-operative complication in horses. The prevalence of POI has been reported in the range of 10-50% of colic cases. [27,28] POI has been described as being responsible for between around 9% and 40% of postoperative deaths.[28,29]. Indeed, the fatality rate of POI cases sadly can be high; ranges of 13-86% are described. [30,31]

When evaluating the clinical diagnostic and treatment approaches to equine POI it is important to be aware of the varying definitions and descriptions of the condition. Much hinges on the interpretation by the clinician as to whether indeed the horse has POI or not. The presence of nasogastric reflux upon intubation is frequently an important inclusion criterion but there is variability as to how much reflux should be present; in an early study of postoperative complications no volume of reflux was specified. [29] Blikslager et al. [32] used a value of > 2 litres at any intubation; Roussel et al.[30] used > 20 litres during a 24 hour period or a volume of > 8 litres at any single sampling postoperatively. Merritt and Blikslager [31] have suggested going forward with the adoption of a consensus on the classification of clinical criteria for POI including: ≥ 4 litres on any given intubation or > 2 litres/hour on repeated intubation, heart rate persistently > 40 bpm, mild-severe colic, rectal and/or ultrasonographic evidence of multiple loops of fluid distended small intestine. Freeman [33] also adds another dimension to the debate regarding the definition of POI; perhaps some cases diagnosed and treated as POI may actually be due to mechanical obstructions or surgical complications. Also it is worth bearing in mind the proportion of horses that may have reduced faecal output with or without signs of colic after general anaesthesia not related to the gastrointestinal tract. [31,34]

Human postoperative ileus

In all species including humans, postoperative ileus (POI) remains a poorly understood clinical problem and the lack of specific and efficacious therapies means that it has a significant economic impact because of prolonged hospitalisation. In the USA costs of human POI in terms of hospitalisation and treatment are in the region of $1.46 billion per year. [35] There has been much learned in avoiding POI in humans, including avoidance of starch in gloves, optimised postoperative fluid and electrolyte therapy and early feeding, modification of pain management strategies, as well as attempts at chemical means to reduce adhesion formation. Drug therapies such as alvimopam (a peripherally acting mu-opioid antagonist) are showing promising results. [36] Strategies used in combination as part of Fast-track multimodal treatment plans for patients lead to a decrease in the time to return to normal intestinal function and shortened hospital stay. [37]

Current thinking on the pathophysiology of postoperative ileus

POI is thought to be the result of loss of normal co-ordination of intestinal contraction by the intrinsic electrical activity of the gut. The classical view of POI was that it involved an early phase that consisted of inhibition of motor activity mediated through neural reflexes. It is now clear that there is also a clinically-relevant prolonged inflammatory phase to POI. Manipulation of bowel leads to local inflammation and impaired muscle function. It has been suggested that POI is mediated through a leucocytic inflammatory response in the *muscularis externa.* [38] Jejunal manipulation in the rat causes a pan-enteric inflammation and prolonged dysmotility. [39] It is becoming increasingly evident that complex multidirectional interactions exist between intestinal immune cells (such as mast cells and macrophages) and motility effector cells (neurons, smooth muscle cells, interstitial cells of Cajal [ICC] and glial cells).[40] Recent work in rodent models of POI has strongly supported a pivotal role for intestinal macrophages in initiating and maintaining ileus.[40] The so-called gastrointestinal field effect is thought to be immune mediated, including important roles for dendritic cells, T helper type 1 cells and intestinal macrophages. [41-43]

Inflammation and equine postoperative ileus

Neutrophilic inflammation has been observed in the proximal resection margin of jejunum of colic surgery cases; in addition, serosal and neuromuscular neutrophilic inflammation was observed in tissues harvested 18 hours after laparotomy incorporating intestinal ischaemia and reperfusion. [44] Horses with obstructions or strangulating lesions of the large colon and small intestine have been shown to have high numbers of apoptotic cells (including neurons, smooth muscle and glia), possibly due to ischaemia and subsequent inflammation. [45] In this study, increased numbers of apoptotic cells were observed distant to the site of intestinal strangulation. Recently it has been shown that there is evidence of cellular stress in the smooth muscle and neurons at the outer macroscopically-viable margins of small intestinal resections from strangulating small intestinal colic cases. [46] In addition, in a separate equine study it has been shown that there is a rapid colonic inflammatory response after colonic mechanical irritation and after the manipulation of small intestinal segments.[47] This suggests, not surprisingly, that the pan-enteric inflammatory effect described by Schwarz et al.[39] may indeed also be occurring in equine surgical colic cases.

Perspectives on the treatment and management of equine postoperative ileus

The mainstays of the treatment and management of equine POI include: nasogastric decompression, replacement and maintenance of fluids and electrolytes, control of the effects of endotoxaemia, control of infection, effective pain management, prokinetic therapy and attempts to reduce inflammation. It is important to bear in mind attempts to try to reduce the incidence or severity of POI by instituting some therapies intra-operatively and early in the postoperative period. Individual clinicians have their own preferences with regard to the treatment of POI and although there is a degree of commonality there is also a variety of approaches taken. As in the human setting, further work is necessary to improve our treatment regimens for POI and to improve the evidence base for clinical decision making. Below is an attempt to highlight some areas of commonality and also debate with regard to the care of equine POI patients. Many clinicians in their fluid therapy protocols may supplement calcium, magnesium and potassium. Calcium supplementation (optimally by measuring ionised calcium levels) may be helpful in improving outcome in colic cases[48]; however there is a school of thought urging caution on the supplementation of calcium in horses with endotoxaemia due to the evidence from laboratory mammals of calcium administration causing an increase in mortality in cases of endotoxic shock.[49] With the increasing evidence for the important role of inflammation in POI, it is no surprise that anti-inflammatory treatment is of benefit to equine patients. There is significant variability in the choice and dose of NSAIDs used depending on availability and spectrum of action (e.g. non-selective NSAIDs and COX-2 selective drugs). In addition to anti-inflammatory effects, some NSAIDs have varying effects on intestinal motility and there are differences observed between normal and diseased intestine.[50] Early feeding is highlighted as a priority in human POI cases and there is an increasing move towards feeding equine patients earlier. If feeding is not feasible, even measures such as tying a hay net outside the stall appear to have some merit. Also, early hand-walking postoperatively is advocated by many clinicians. It has been suggested that the intraoperative administration of carboxymethylcellulose in small intestinal surgical cases may improve survival, potentially by reducing adhesion formation.[51] Against this argument is the finding of another group who observed that in their cases the use of carboxymethylcellulose was associated with a higher adhesion rate.[52] Another contentious area in POI cases is the decision as to whether and when to take a horse back for repeat laparotomy. On one side there is the argument that an early decision to 'go back in' and decompress and potentially correct a mechanical obstruction or a surgical complication may be warranted.[33] Counterbalancing this is the evidence that the survival rate for horses requiring relaparotomy is reduced and the complication rate high.[53] There is considerable use of prokinetic therapy in the management of equine POI cases. The most commonly chosen drug is lignocaine/lidocaine.[54] Other commonly chosen drugs include metoclopramide, erythromycin and bethanechol. This list is not exhaustive and the reader is directed to reviews on the subject.[7,55] It is worth at this point mentioning a few points with regard to lidocaine. Some of the literature and clinical perceptions of lidocaine are conflicting. There are some surgeons who are very keen to start lidocaine infusions when the horse is on the table, due to the perception that it has anti-inflammatory effects (see below). From the anaesthetic point of view there is also the benefit that the lidocaine may reduce the isoflurane requirement.[56] However some anaesthetists are more reluctant to choose to start the infusion of lidocaine intraoperatively because of its potential adverse effects on the quality of anaesthetic recovery.[57] The proposed mechanisms of lidocaine are becoming more apparent in recent years. Classically it has been thought of as a prokinetic agent but there is increasing evidence that it may be having its beneficial effects on horses via alternative mechanisms. It has been proposed that lidocaine suppresses the firing of afferent neurons and has centrally-mediated analgesic effects. Lidocaine however may play more of a role in somatic analgesia compared to visceral analgesia.[58] Lidocaine is thought to have a stimulatory effect on intestinal smooth muscle. It has been shown to increase the *in vitro* contractile activity in equine duodenum but not the pylorus or jejunum[59]; other workers however showed that there was a positive *in vitro* effect on normal jejunal smooth muscle.[60] An electrophysiological *in vivo* study of horses without gastrointestinal disease showed lignocaine had no significant effect postoperatively on the electrical activity in the proximal jejunum.[61] The authors of the latter study concluded that the use of lignocaine as a prokinetic agent cannot be supported by their study; however, they quite rightly add that this is with reference to normal horses and results may be different in clinically abnormal horses. An illustration of this point is a clinical study by Malone et al.[62] which showed that intravenous infusion of lignocaine improved the clinical course in 32 refluxing horses suggesting that it does have beneficial prokinetic actions. Torfs et al.[28] showed that prophylactic use of lidocaine (immediately postoperatively) reduced the incidence of POI and that lidocaine treatment improved survival. Potentially lidocaine may be having its effects via anti-inflammtory and protective actions. Lidocaine may inhibit granulocyte migration but there are conflicting reports of this. It has however been shown that lidocaine has anti-inflammtory effects on ischaemia-injured equine jejunum .[63] Peiro et al.[64] showed that lidocaine inhibits TNF-α activity in peritoneal fluid in horses with experimental endotoxaemia. Guschlbauer et al.[60] showed that the lidocaine-stimulated *in vitro* contractility of ischaemia-injured jejunum was more pronounced than that seen normal jejunum. In addition it has been recently shown that lidocaine and the structure-related drug mexiletine enhanced contractility of ischaemia-reperfusion injured equine jejunum and reduced creatine kinase release.[65] Perhaps therefore, lidocaine also may have interactions with smooth muscle membranes and be in some way membrane protective.

CONCLUSIONS

There is still much we have to learn about equine postoperative ileus. There is much debate in this field and it is important to move towards clarity in terms of the definitions of the POI that we are using. Research approaches in this area need to be cognisant of the important differences in the behaviour of normal and diseased tissues and horses. Inflammation appears to be a pivotal player in the development and maintenance of POI and perhaps lidocaine is having its beneficial effects via anti-inflammatory actions rather than the traditional 'prokinetic' actions. Finally, comparative gastroenterological work in different species holds promise to elucidate disease causality and pathogenesis. This has considerable potential to benefit both human and equine patients.

REFERENCES

[1]White NA. In: White, N.A. (ed): **The Equine Acute Abdomen**. Philadelphia, Lea & Febiger, pp49-64; 1990.[2]Archer DC et al. **Vet J** 2006; 172: 29. [3]Tinker, MK et al. **Equine Vet. J.** 1997; 29, 448-453. [4]Traub-Dargatz JL et al. **J. Am. Vet. Med. Assoc.** 2001; 219; 67. [5]Dukti S et al. **Vet. Clinics N. Am. Equine Pract.** 2009; 25: 217. [6]Dart AJ et al. **Aust. Vet. J.** 1998; 76: 25. [7]Doherty TJ. **Vet. Clinics N. Am. Equine Pract.**2009; 25: 351. [8]Livingston EH et al. **Dig. Dis. Sci.** 1990; 35, 121 -32. [9]Pearson GT. **Cell Tissue Res** 1994; 276: 523. [10]Sanders KM. **Ann. Rev. Physiol.** 2006; 68: 307. [11]Ward SM et al. **J. Physiol.** 1997; 505: 241.[12]Thomsen L et al. **Nature Med** 1998; 4: 848. [13]Koh SD et al. **J. Physiol.**1998; 513.1: 203. [14]Lee JCF et al. **Am. J. Physiol.** 1999; 277: G409. [15]Koenig J et al. **Can. Vet. J.** 2006; 47: 551. [16]Hudson NPH et al. **Vet. Rec.** 2001; 149: 707. [17]Schusser GF et al. **J. Am. Vet. Med. Assoc.** 1997; 210: 928. [18]Schusser GF et al. **Equine Vet. J.** Suppl. 2000; 32: 69. [19]Fintl C et al. **Equine Vet. J.** 2004; 36: 474. [20]Pavone S et al. **Res Vet Sci** 2012; 93: 350. [21]Cottrell DF et al **Neurogastroenterol Motil** 1999; 11: 79. [22]Hudson NPH et al **Autonom Neurosci: Basic and Clinical** 2001; 92: 37. [23]Williams S et al. **Equine Vet J** 2011; 43 (Suppl. 39): 93. [24]Burns GA et al. **Cornell Vet** 1990; 80: 53. [25]Chénier S et al. **Can Vet J** 2011; 52: 419. [26]Blake KR et al. **J Am Vet Med Assoc.** 2012; 240: 1494. [27]Cohen ND et al. **J. Am .Vet. Med. Assoc.** 2004; 225: 1070. [28]Torfs S et al. **J Vet Intern Med** 2009; 23: 606. [29]Hunt JM **Equine Vet J** 1986; 18: 264. [30]Roussel AJ et al. **J. Am. Vet. Med. Assoc.** 2001; 219: 72. [31]Merritt AM et al. **Equine Vet J** 2008; 40: 295. [32]Blikslager AT et al. **J Am Vet Med Assoc.** 1994; 205: 1748. [33]Freeman DE. **Equine Vet J** 2008; 40: 297. [34]Little D et al. **J Am Vet Med Assoc.** 2001; 218: 414.[35]Goldstein JL et al. **Pharm Therap** 2007; 32: 82. [36]Wang S et al. **Pharm Therap** 2012; 37: 518. [37]Story SK et al. **Dig surg** 2009; 26: 265. [38]Kalff JC et al. **Gastroenterol** 1999; 117: 378. [39]Schwarz, NT et al. **Gastroenterol** 2004; 126: 159. [40]De Winter BY et al. **World J Gastroenterol** 2010; 16: 5523. [41]Engel DR et al. **Nature Med** 2010; 16: 1407. [42]Bauer AJ. **Nature Med** 2010; 16: 1378. [43]Koscielny A et al. **Am J Physiol Gastrointest Liver Physiol** 2012; doi:10.1152/ajpgi.00161.2012. [44]Little D et al. **Equine Vet J** 2005; 37:329-335. [45]Rowe EL et al. **Am. J. Vet. Res.** 2003; 64: 982. [46]De Ceulaer K et al. **Res Vet Sci** 2011; 91: 294. [47]Hopster-Iversen C et al. **Equine Vet J** 2011; 43 (Suppl. 39): 1. [48]Delesalle C, et al. **Vet Q** 2005; 27: 157. [49]Malcolm DS et al. **Crit Care Med** 1989; 17: 900. [50]Marshall JF et al. **Equine Vet J** 2011; 43 (Suppl. 39): 140. [51]Fogle CA et al. **Vet Surg** 2008; 37: 558. [52]Klohnen A. **Vet. Clinics N. Am. Equine Pract.** 2009 25, 341.[53]Mair TS et al. **Equine Vet J** 2005; 37: 315. [54]Van Hoogmoed LM et al. **Vet Surg** 2004; 33: 279. [55]Wong DM et al. **Equine Vet Educ** 2011; 23: 88. [56]Villalba M et al. **Equine Vet J** 2011; 43: 721. [57]Valverde A et al.**Equine Vet J** 2005; 37: 559. [58]Robertson SA et al. **Equine Vet J** 2005; 37: 122. [59]Nieto JE et al. **Am. J. Vet. Res.** 2000; 61: 413. [60]Guschlbauer M et al. **Equine Vet J** 2010; 42: 53. [61]Milligan M et al.**Vet. Surg.** 2007 36, 214. [62]Malone E et al. **Vet. Surg.** 2006; 35: 60. [63]Cook VL et al. **Am J Vet Res** 2009; 70: 1259. [64]Peiró JR et al. **J Vet Int Med** 2010; 24: 940. [65]Tappenbeck K et al. **Vet J** 2013; htttp://dx.doi.org/10.1016/j.tvjl.2012.11.011

COLONIC ACIDOSIS: DIAGNOSIS, TREATMENT AND PREVENTION

Frank M. Andrews, DVM, MS, DACVIM and Rafat Al Jassim, PhD
Baton Rouge, LA, USA and Gatton, Queensland, AU

INTRODUCTION

The maintenance of equine gut health is dependent on the interaction of good quality forage (hay), cereal grains (concentrate) and the microbial (bacterial and protozoa) population present in the gastrointestinal tract. For a healthy gut, forages should make up a majority of the equine diet (1.5% of body weight/day) and high starch cereal grains should be a supplement (not to exceed 0.5% of body weight) on an as needed basis. Overfeeding cereal grains is counter-productive to equine gut health because it leads to alteration in gut microflora, which are essential for proper digestion and energy generation. Furthermore, overfeeding cereal grains can lead to gastric ulcers, colonic acidosis, laminitis, colic and obesity. This article will concentrate on the conditions of the cecum and colon that leads to a decrease in luminal pH, otherwise known as colonic acidosis.

DIGESTIVE FUNCTION

Under normal feeding conditions, horses spend approximately 10-12 hours/day eating. Such feeding patterns allow them to maintain a full stomach and continuous supply nutrients to body tissues and to the microbial community residing in the hindgut. This feeding behavior helps the horse overcome the problem of having a relatively small stomach compared with other herbivore species of a similar body size such as the cow. Feeding behavior of this kind is important with respect to health and meeting their nutritional needs, especially energy. In contrast, performance horses are fed concentrate (cereal grain) rich diets and in many cases are withheld from feed for extended periods before exercise. The microbial community of the hindgut responds negatively to high cereal grains which results in poor gut health. Combine the overfeeding of cereal grains, the reduction in saliva production due to less fiber, periods of feed deprivation and stall confinement leads to colonic acidosis, which can increase the risk of colic and laminitis. Maintaining health of horses under conditions of concentrate feeding is a real challenge facing nutritionists, veterinarians and owners. Such challenge requires better understanding of the anatomical, physiological and functional features of the horse gastrointestinal tract. Understanding the complexity of the microbial ecosystem, interaction among the vast numbers and diverse microbial community, between microbes and the horse and the dietary influences on the microbes are equally important.

Horses are best described as hindgut fermenters with a modified cecum and colon that harbors a complex microbial (bacterial and protozoa) community. The microbial community of the hindgut contributes to digestion, which enable the horse to extract energy from dietary components that otherwise would be wasted. Horses are subject to nutritional disorders and diseases common to intensively managed animals such as gastric ulcers, acidosis, colic, and laminitis. The digestive system of the horse has special anatomical and physiological features that need to be considered when deciding what to feed horses. Improved pastures that have been developed to maximize microbial protein synthesis in the rumen of cattle and sheep[1] and milk yield[2] may be deleterious to horses due to their high content of water soluble carbohydrates (WSC) and crude protein. The effect of improved pastures may be similar to that experienced by grain or concentrate feeding for performance or racehorses.[3]

The large intestine

The large intestine consists of an enlarged cecum, modified large and small colon, which harbors a complex microbial population. The conditions of the large intestine match the description of a continuous chemical reactor, where the input of nutrients is continuous, mixing is adequate, products are removed and reactants are in contact with enzymes, whether these enzymes are produced by the horse or the microbes. Mixing of contents allows control over the temperature, pH and the removal of gasses. The large intestine comprises approximately 64% of the volume of the equine GI tract, but only 30% of its total length.[4] It is a voluminous but short organ, holding approximately 95 to 112 L of liquid, approximately the size of the compartmental stomach of an adult cow, and similar to the rumen in having a fermentative microbial community inhabiting the organ. The walls of the large intestine lack villi and the epithelial cells lack microvilli. However, they contain glands for secreting mucus. The walls of the large intestine also contain longitudinal and circular muscle fibers. Unlike other herbivores the small intestine in the horse opens directly into the cecum through a muscular valve. Relatively close to this is a second valve where digesta passes from the cecum and into the right ventral colon.

The cecum is a large blind-ending, comma-shaped structure with a capacity of between 25-35 L and of approximately 1m (3 ft.) in length. It extends from the pelvic inlet to the abdominal floor within the right flank area of the horse. The base of the cecum extends forward to lie under the cover of the last few ribs, with its tip just behind the diaphragm.

The large intestine in the horse does not have a longitudinal muscle layer. Instead, it has bands of muscle that follow the length of the cecum, called tenia, which gather the cecal wall into sacculations called haustra. The haustra acts like buckets, delay the transport of chyme through the large intestine and contribute to the mixing of the colonic contents.

In the horse the ascending colon is modified into the right and left ventral and dorsal large colon.[5] The first segment of the colon, the ventral colon, can hold a volume of approximately 150 L, has a diameter of 250-300 mm and is about 2-4 meters (6 to 8 ft.) long. The transition (pelvic flexure) between the ventral colon and dorsal colon is narrow. This narrow segment will delay the transport of large particles from the ventral to the dorsal colon, increase their retention time and increase fermentability. The diameter varies significantly between regions however it is largest in the right dorsal colon, forming sacculations with a diameter of up to 500 mm. The large colon is folded together forming a double loop consisting of the right and left ventral colons and the left and right dorsal colons. The four parts of the large colon are connected by structures called the pelvic and dorsal and ventral (sternal) diaphragmatic flexures. Their significance probably lies in differences in microbial populations and functions between the segments.[6] The small or descending colon joins the dorsal and transverse colons to the rectum and is approximately 3 m (9 ft.) in length. The rectum is approximately 30 cm (1.5 ft.) long and is the continuation of the small colon, into the pelvic inlet where it ends at the anus. Due to the size, location and importance in digestion, the equine large intestine may be referred to as the hindgut.

Microbiology population of the large intestine

The microbial (bacteria and protozoa) community in the cecum and colon of horses is less understood and has received less attention than in cattle. Horses are less efficient than cattle and other ruminants in digestion of fiber and thus their bacterial flora are more susceptible to the effect of high carbohydrate feeding. Available information has dealt mainly with fermentation processes in the hindgut, especially those related to fiber digestion,[7-10] lactic acidosis,[11,12] and laminitis.[13,14] Early workers reported a higher bacterial population in the distal parts of the GI tract from the duodenum to the colon.[15] Horses, fed a grass diet, had a bacterial population of 2.9×10^6 per g wet weight in the duodenum, 29.0×10^6 in the jejunum, 38.4×10^6 in the ileum, 2.05×10^9 in the cecum and 1.26×10^9 in the colon. In their study they reported a relationship between mucosal and luminal counts. Bacteria associated with the mucosa were 73, 22, and 24% of the luminal counts in the duodenum, jejunum and ileum. Using similar culturing techniques we often obtain counts for total anaerobes around 4.3×10^8 in the stomach, 4.2×10^8 in the cecum, 1.0×10^8 in the proximal colon, 6.7×10^7 in the distal colon and 1.7×10^8 in the rectum. Early work focused on bacterial groups and relied mainly on culture dependent methods that had their limitations. However, molecular techniques have revealed the bacterial diversity within the equine large intestine.[9,12]

Many bacteria species in the hindgut produce D-Lactic acidosis in response to carbohydrates. Most of the work in the past focused on *S. bovis* and *S. equinus* due to their established role in acid build up and their involvement in the sequence of events leading to the development of laminitis. In recent work, changes in the bacterial community of the cecum of horses has been evaluated using the florescence *in situ* hybridization (FISH) technique.[14] Cecal and colonic pH dropped rapidly and laminitis was induced in all horses following oral administration of oligofructose at a relatively high rate of 10g per kg BW. All horses lost their appetite by 8 hr post oligofructose administration (POA) and developed profuse, watery diarrhea by 16h. Clinical signs of laminitis were observed in these horses (e.g. lameness and shifting of weight from one foot to another) from 20 hours POA onwards. The population of the Gram positive bacteria changed from <20% at 0 time to >70% between 4 and 8h POA and remained at high level up to 32h POA. This was confirmed by FISH results which showed the relative abundance of *S. bovis /S.equinus* to be >50% by 16h POA in all samples. These results confirm the association of *S.bovis/S.equinus* with acid build up and sequential events leading to the development of laminitis in horses.

Roughage feeding of horses

Under normal feeding conditions horses consume diets consisting mainly of cell wall polysaccharides including cellulose, hemicellulose and pectin, the main structural constituents of the plant cell wall. Such diets are often bulky and low in energy content. In order for the horses to meet their requirements for energy and other nutrients they graze continually and spend approximately 10-12 hours eating during a 24-hour period, in sessions of 0.5 to 3 hours under free range conditions.[16] Another report noted that a free range horse spend 16-18 hours or more out of 24-hours searching for and selecting feed to meet their requirements.[4] The fact that horses feed continuously, ensures they maintain a full stomach and continuous buffering from saliva protects the non-glandular stomach from acid injuries.

The digestive processes begin with acid hydrolysis in the stomach and enzymatic digestion by the host in the small intestine followed by extensive fermentation in the cecum and colon. In horses fed grass or hay, fermentation of structural carbohydrates by the action of microbial enzymes in the cecum and the colon produces VFA, which are readily absorbed and utilized as an energy source. The contribution of VFA to the energy requirements of the horse could be as high as 70-80%. Under these conditions the large intestine's environment remains buffered and supports the proliferation of major groups of the microbial community. One disadvantage of the system is that considerable amounts of fermentation products such as vitamins and amino acids are retained by the microbes, hence these products are not utilized by the horse and appear in the feces. One way for the horse to benefit from products that are retained by bacteria is to consume its own feces. The incidence of coprophagy appears to be related to the quality of the horse's diet. In the other hand, feeding starch rich diets, which is particularly a problem in intensively managed racehorses, disrupts the microbial ecosystem of the GI tract, particularly the large intestine.

Similar condition may results from feeding grasses rich in the oligosaccharides fructans. Fructans are polymers of fructose molecules and there are 3 types of fructans present in temperate grasses as follows: Inulin, a linear fructans generally linked by β (2→1) glycosidic bonds; Levan, linear fructans generally linked by β (2→6) glycosidic bonds; and Graminan, branched fructans linked by both β (2→1) and β (2→6) glycosidic bonds.[17] Accumulation of fructans in grasses is influenced by grass type, environmental conditions, and stage of maturity.[18] Although, fructans are water soluble the horse does not produce enzymes capable of digesting them and thus they pass through the small intestine and enter the hind gut of the horse where they undergo fermentation process by the hind gut microbes. Excessive intake of fructans can lead to acid build up and metabolic disorders similar to that experienced under grain overload conditions.[3] However, little is known about the interaction between the different bacterial species and the level of intake of fructans.

Concentrate feeding, fermentative acidosis and laminitis

Dietary supplementation is essential to maintain good health and to satisfy the nutritional needs for growth and performance of various equine classes. Grain and mixed feeds contain large amounts of water soluble carbohydrates (WSC) mainly as starch. The choice of cereal grain is based on its safety, palatability, energy value and cost. Oat grain is palatable and relatively safe because it is lower in energy density and higher in fiber than other grains.

Starch is not well digested in the small intestine of the horse and variable proportions of ingested starch that escape enzymatic digestion in the small intestine enter the large intestine and are rapidly fermented, increasing VFA and lactate production and lowering the pH. An acidic environment favors the rapid proliferation of lactic acid producing bacteria (LAB), resulting in increased lactic acid production and a further decline in the pH. Lactic acidosis is often accompanied by inflammation of the colon wall (colitis) and increases the risk for laminitis.

Accumulation of lactic acid in the hind gut of the horse results from the rapid fermentation of starch by resident bacteria. This occurs when the supply of starch by the diet exceeds the digestive capacity of the small intestine. The upper limit of 3.5–4.0 g starch per kg BW per meal was earlier suggested,[19] but other studies on variations in the digestion of starch suggest that the upper limit is < 3.5.[20] Differences can partly be explained by the source of starch. An increase in the population of lactic acid producing and utilizing bacteria and a decrease in concentration of cellulolytic bacteria in the cecal contents of horses were measured in horses fed high starch diet supplying 3.4 g starch per kg BW per meal.[21] The lowest pH and highest lactic acid values were measured at 5 and 6 h post-feeding in the cecum (6.45; 532 mg/L) and the colon (6.38; 582 mg/L), respectively. A recent report showed a decrease in cecal pH level from 7.04 to an average of 6.74 and increase in the concentration of lactate in horses fed Alfalfa cubes supplemented with different grains (e.g. barley, corn, naked oats, and oats).[22] Lactate concentration was notably high for the barley supplemented diet (2.82 μmol/g versus 0.96μmol/g of DM). This supports earlier results which showed a very low preileal digestibility for rolled barley starch in horses (e.g. 0.21).[23] The source of starch and the processing method of the cereal grain were also emphasized by McLean et al. (2000)[24] who observed unfavorable intra-cecal fermentation changes in ponies fed hay and rolled barley diet (50:50) that was formulated to supply 2.1 g per kg BW per meal. These unfavorable changes were not observed with micronized or extruded barley. Processing of grain to improve the extent of enzymatic hydrolysis in the small intestine should reduce the amount of starch reaching the large intestine. However, work by de Fombelle et al. (2003)[25] indicated the presence of significant numbers of lactic acid producing and utilizing bacteria in the horse stomach suggests that such grain processing might have the undesirable effect of increasing starch fermentation, and hence lactic acid production in the stomach. Such a situation could contribute to lactic academia, a condition associated with laminitis.[26-28] Fermentation in the stomach was also shown to be extensive under concentrate feeding conditions, particularly in the non-glandular region of the stomach (NG-S). Horses fed 4 kg grass hay and 4 kg dry-rolled or steam-flaked sorghum per head per day, had a lactic acid concentration in the NG-S greater than 40mmol/L.[29] Another important feature of the fermentation process in the stomach is the high proportion of the D-lactate isomer because in other anaerobic fermentations, the L-isomer is the more prominent. In addition to the two isomers of lactate, which are the main products, small amounts of VFA are also produced. The study also showed that lactic acid is absorbed from the gastrointestinal tract before the digesta reaches the cecum.

CONCLUSION

Horses have a unique gastrointestinal tract most suited for the digestion and utilization of fiber rich diets. In order to maintain health and derive adequate nutrients to meet their requirements, horses need to graze continually and feed on pastures that are not too rich in water soluble carbohydrates that ferment in the GI tract, alter conditions, and adversely affect the balanced microbial population. General feeding recommendations can be made to optimize equine gut health. Limiting the size of high starch meals, continuous feeding of good quality forage, adding non-starch sources (e.g. vegetable oils and beet pulp) to supply energy and minimizing stress on gut microflora are essential for optimizing equine gut health and preventing colonic acidosis.

REFERENCES

[1]Merry RJ et al. (2006) *JAS* **84**:3049-3060.
[2]Miller LA et al. (2001) *Grass and Forage Science* **56**: 383–394.
[3]Longland AC et al. (2006) In Proceedings of the Waltham International Nutritional Sciences Symposia: *Journal of Nutrition* 136 (7S): 2099S-2102S.
[4]Kohnke J et al. (1999) Feeding Horses. Union Offset, Canberra, ACT, Australia.
[5]Reece WO et al. (1997) *Physiology of Domestic Animals* (2nd edition) ed. Reece, W.O. pp. 270-311. Philadelphia:Lippincott Williams & Wilkins.
[6]Frape, D. (1998) Equine Nutrition and Feeding, 2nd Edition. London: Blackwell Science Ltd.
[7]Lin C and Stahl DA. (1995) *Applied and Environmental Microbiology* **61**: 1348–1351.
[8]Julliand V et al. (1999) *Applied and Environmental Microbiology* **65**: 3738–3741.
[9]Daly K et al. (2001) *FEMS Microbiology Ecology* **38**: 141-151.
[10]Al Jassim RAM et al. (2005) *Recent Advances in Animal Nutrition in Australia* **15**: 155-163.
[11]Al Jassim RAM et al. (1999) *Recent Advances in Animal Nutrition in Australia* **12**, 91-97.
[12]Al Jassim RAM et al. (2006) *FEMS Microbiology Letters* **248**:75-81.
[13]Milinovich GJ et al. (2006) *Environmental Microbiology* **8**: 885-898.
[14]Milinovich GJ et al. (2007) *Environmental Microbiology* http://www.blackwell-synergy.com/toc/emi/0/0.
[15]Mackie RI and Wilkins CA. (1988) *Applied and Environmental Microbiology* **54**: 2155-2160.
[16]Sjaastad ØV et al. (2003) *Physiology in domestic animals* ed. Sjaastad, Ø. V., Hove, K. and Sand, O. pp. 490-563. Oslo: Scandinavian Veterinary Press.
[17]Chatterton NJ and Harrison PA (1997) *New Phtologist* **136**: 3-10.
[18]Chatterton NJ et al. (2006) Proceedings of the Waltham International Nutritional Sciences Symposia: *The Journal of Nutrition* 136 (7S): 2111S-2113S.
[19]Potter GD et al. (1992) In: Europäische Konferenz über die Ernährung des Pferdes, pp. 107–111, Hannover, Germany.
[20]Kienzle, E. (1994) *Revue de Médecine Véterinaire* **145**: 199–204.
[21]Medina B et al. (2002) *Journal of Animal Science* **80**: 2600–2609.
[22]Hussein HS et al. (2004) *Journal of Animal Science* **82**: 1986–1996.
[23]Meyer H et al. (1993) In: *Proceedings of the 13th Equine Nutrition and Physiology Symposium*. pp. 92–97. Florida.
[24]Mclean BML et al. (2000) *Animal Feed Science and Technology* **85**: 79-87.
[25]de Fombelle A et al. (2003) *Animal Science* **77**: 293-304.
[26]Garner HE et al. (1977) *Journal of Animal Science* **45**: 1037- 1041.
[27]Rowe JB et al. (1994) *Journal of Nutrition* **124**: 2742S-2744S.
[28]Pollitt CC et al. (1998) *Equine Veterinary Journal*, Supplement **26**: 125-132.
[29]Al Jassim, R.A.M. (2006) *Animal Feed Science and Technology* **125**: 33-44.

NUTRITIONAL SUPPORT IN THE DYSPHAGIC HORSE

Janice Sojka Kritchevsky VMD, MS, DACVIM & Balazs Toth DVM, MS
West Lafayette, IN

INTRODUCTION

Meeting the nutritional and fluid requirements for a horse with prolonged dysphagia is one of the most difficult tasks an equine practitioner can face. Challenges include the need to supply large volumes of materials that must be made liquid enough to pass through a nasogastric tube combined with the relatively small size of the horse's stomach. Administering a liquid, low-residue diet usually results in a dramatic change in nutrient composition compared to the horse's normal intake, and this sudden dietary shift can result in alterations in colonic flora which, in turn, can result in complications such as colitis and laminitis. The longer a horse remains dysphagic, the more attention must be given to providing 100% of the animal's nutrient requirements rather than some smaller percentage. At the same time, the morbidity attendant with prolonged nasogastric tube placement or alternate administration techniques increase. Establishing a specific diagnosis for the cause of dysphagia is important as it allows a clinician to determine a likely time frame for duration as well as a prognosis for recovery.

HYPOPHAGIA VS. DYSPHAGIA

Despite often being discussed together in publications that deal with providing nutritional support, there is a large difference between a dysphagic and a hypophagic horse. Hypophagia is defined as reduced intake, either absolute or relative to increased caloric needs. A hypophagic horse is capable of ingesting and swallowing food, and will be able to meet a portion of its nutritional requirements voluntarily. Often a hypophagic horse will ingest hay or other forage, and the primary concern is to provide additional calories in the form of extremely concentrated feedstuffs. In contrast, the primary consideration in dysphagic horses is to provide a diet that meets the animal's caloric needs while being as balanced as possible. The risk of inhalation pneumonia from attempting to eat is not a consideration in hypophagic horses, and they can be allowed access to food at all times. In contrast, dysphagic horses are generally hungry, and will attempt to eat even when they cannot protect their airway.

ESTIMATION OF REQUIREMENTS

Resting Energy Expenditure (REE) is defined as the amount of calories required over 24-hours by the body during a non-active period. The formula used to calculate REE in adult horses[1] is:

$$\text{REE (kcal)} = 21 \times \text{BW (kg)} + 975$$

In order to calculate the total Mcal a horse needs daily in digestible energy, both sides of the equation should be divided by 1000, so that DE(Mcal/day) = 0.021X BW(kg) + 0.975.[2] Thus, a 500 kg horse would require 11.47 Mcal/day. This is quite a bit less than published estimates of a horse's regular maintenance requirement because the REE does not account for activity of any kind. The actual REE for any individual animal is influence by its age, body composition, and the presence of any disease state. A sedentary horse that is suffering from generalized disease such as sepsis or has experienced traumatic injury requires additional nutrients above the REE, although exactly how much is not known. It is recommended that the calculated DE needed per day be multiplied by a factor of 1.25 to 2.0 to account for increased metabolic demand.[2]

Protein requirements increase when there is a catabolic state, and increasing ingested protein as a percentage of the total diet may be indicated. Ensuring that 5 grams of protein are given for every 100 kcal total diet (i.e. 800 gm/16Mcal) has been recommended.[3] It is particularly important to ensure that the amino acid glutamine is supplemented. Glutamine is required for normal enterocyte and immune system function. Although it is made in the liver, it can be rapidly depleted during catabolic states, and should be considered a "conditionally essential" amino acid.[3] Similarly, blood vitamin C values are often decreased in compromised horses. Whether this represents a true deficiency is not known, but addition of ascorbic acid to the diets of such horses is often advocated.[2]

Average voluntary water intake in adult horses is approximately 24 L per day, which corresponds to the common value used to calculate fluid replacement of 60 ml/kg/day. Intake is predicated on metabolic body weight, rather than absolute body weight, however, and larger animals may require somewhat lower volumes per kg.[4] Ambient temperature has a large effect on water requirements, and losses via sweat must be accounted for in hot environments. The volume of water needed to produce a liquid slurry of feed usually exceeds a horse's daily requirements. If a very low volume of feed is used, however, additional water must be given to meet daily requirements.

Along with energy and protein needs, electrolyte requirements in equids suffering from systemic disease may be altered from those of healthy horses. How they are affected is not well understood, however. Generally, it is recommended that NRC requirements for macro and micro nutrients be met daily. Like water, electrolyte requirements increase during times of high ambient temperatures when excess sweating occurs. Potassium intake in horses on low forage diets may be compromised, and blood potassium concentrations should be monitored.

FEEDING STRATEGIES

There are two basic routes for providing needed calories and nutrients to a dysphagic horse; enteral or parenteral. In both instances, supplementation should begin at a low rate, 25% or less of the horse's daily needs, then slowly increased over several days as the horse adapts.

Parenteral

The typical target when using total parenteral nutrition (TPN) is to provide 34 kcal/kg/day. This amount was well tolerated and sufficient to maintain body weight in adult horses.[5] Parenteral fluids contain glucose as a carbohydrate source, amino acids for protein, vitamins calcium, magnesium, and trace minerals. In additional to glucose, non-nitrogen calories can be provided in the form of lipids. The use of lipid in TPN solutions is controversial when used on foals or animals particularly susceptible to hyperlipemia such as miniature horses, ponies, and donkeys. In order to provide the total caloric requirement for horses larger than 400 kg, however, it is not practical to use TPN fluids that do not contain them. The primary limitations to prolonged use of TPN when treating dysphagic horses is the large expense involved. Other long term problems may include septic thrombophlebitis, hyperglycemia, and hypertriglyceridemia.[6]

Enteral

A clinician has three options when providing enteral nutritional support for a dysphagic horse. Each has advantages and drawbacks, and which is chosen is dictated by the individual circumstances attendant with each case. The clinician can purchase and feed a commercially available liquid diet that is designed for horses. At the time of this writing there are two such products on the market. Alternatively, one can use a commercially available complete pelleted feed as the basis of the diet. Grinding the diet into a powder and then adding water will allow the pellets to dissolve into slurry that can be pumped into the stomach. Finally, a clinician can assemble raw ingredients and make a diet from scratch.

It has been documented that oral supplements designed for humans can be used successfully in horses although expense puts their use of out of reach in most situations.[7] At this time there are 2 equine use commercial products available. Both are designed to be used primarily as a nutrient-rich top dress, but their manufacturers claim they can also be given alone. Purina has recently released a complete enteral supplement - WellSolve® Well-Gel. Its composition is proprietary, but it claims to provide 100% of a horse's nutritional requirements when fed as directed, and it is fortified with glutamine. The major ingredients are alfalfa mean and soy protein concentrate, and it is extremely high in protein with a guaranteed analysis of 36%. Instructions for use state that it should be fed at a rate of 0.1% of a horse's body weight per day. This would result in giving a total of 10 pounds of feed per day to a 1,000 pound horse. It is easily mixed into a slurry that can pass through a nasogastric tube. Currently, there is a lack of published data, or even testimonials, on its suitability as a complete feed either in terms of ease of use or development of side effects. MD's Choice manufactures PART IV, a suspendable feed mix which is claimed to meet 100% of a horse's protein requirements and a substantial portion of its total caloric needs. When fed at the manufacturer's recommendations, which are six 1-pound feedings per day, it will meet 138% and 123% of a horse's protein and resting caloric requirements respectively. There are no ingredients listed on either the container or the manufacturer's website although they claim it can be considered half forage and half grain. Again, there are no data or testimonials available to support its use.

Several manufacturers produce 100% complete pelleted feeds that can be adapted for use in a horse with dysphagia. Their principle advantages include that they require no additional ingredients, are readily available, and are relatively low cost. Their disadvantages are the relatively large volume of feed needed to meet a horse's daily requirements and the difficulty in getting the pellets to dissolve and remain in solution long enough to be pumped through a tube. Since no more than 6 L of slurry or other material should be placed in a horse's stomach at one time, multiple feedings are needed to give a horse its daily requirements. Grinding the pellets to a powder prior to putting them in solution is the best way to ensure the entire amount measured gets into the horse without clogging up tubing, but this can be extremely labor intensive.

Dr. Jonathan Naylor developed a homemade liquid diet that has come to be called "The Naylor Diet."[8] It is made by combining alfalfa meal, dextrose, electrolytes, and either casein or dehydrated cottage cheese meal. When made as recommended, it will meet a horse's nutritional requirements when fed in 5 L of water in 6 feedings per day. While easy to mix and use once the necessary ingredients have been obtained, it can be quite difficult to locate and purchase all the constituents of the diet. Dehydrated alfalfa meal is easier to suspend in water and pass through a tube than most pelleted feeds. It is possible meet a portion of a horse's requirements over a short period of time with just alfalfa meal and electrolytes.

DELIVERY SYSTEMS

Nasogastric tubes are the standard method of providing enteral feedings. A tube with a 12.5 mm internal diameter is needed to allow passage of slurry diets, and generally the larger the nasogastric tube, the easier and quicker the feeding.[3] It must be decided whether the tube is passed for each feeding or sutured in place. The presence of a nasogastric tube in the stomach can depress gastric motility, and prolonged presence of a tube can cause irritation and ulceration of the nasal passages, pharynx, and esophagus.[9] Repeated intubations are extremely stressful to most horses, however, and many horses will not tolerate multiple intubations per day for very long. In those instances, leaving a tube in place is the only option.

Surgical cervical esophagostomy has been described as a means to provide extra oral feeding in dysphagic animals, although the procedure can be technically challenging and has a large number of associated complications.[10] For these reasons, esophagostomy is reserved for horses that will require prolonged tube feeding or that have lesions in the head or neck that preclude using a nasogastric tube.

Long-term dysphagia in dogs, cats, and man is often managed via the insertion of percutaneous endoscopic gastrostomy (PEG) tubes.[11,12] We have recently developed a technique for inserting PEG tubes into horses. The stomach is localized via ultrasound, but otherwise insertion is the same as in other species. PEG tube insertion holds promise for allowing a method of long-term feeding in the dysphagic horse without the morbidity attendant with esophagostomy.

MONITORING

In addition to assessing for the continued presence of dysphagia, horses being fed slurry-based or liquid diets should be monitored daily for dehydration, digital artery pulses, and fecal output and consistency. Body weight should be assessed every third to fifth day so that weight maintenance is assured. Periodic serum chemistries should be run to ensure normal blood electrolyte concentrations. Nasogastric tube placement must be verified before each feeding, and tubes should be replaced if they become stiff. If an esophagostomy or PEG tube is placed, daily attention to surgical site care is necessary as well.

REFERENCES

[1]PagenJD et. al **J Animal Sci** 1986;63:815 [2]Ralston SR **Vet Clin Equ Prac** 1990;6:339 [3]Geor RG http://www.ker.com/library/advances/249.pdf [4]Rose RJ **Vet Clin Equ Prac** 1990;6:281 [5]Hansen et. al **Am J Vet Res** 1988;49:122 [6]Krause JB et. al **Equ Vet J** 2007;39:74 [7]Sweeney RW et. al **JAVMA** 1990;197:1030 [8]Naylor JM **Current Therapy Equ Med 3**, Robinson NE(ed) 1992;736 [9]Hardy J et. al **JAVMA** 1992;201:483 [10]Freeman et. al **JAVMA** 1978;172:314 [11]Bright RM **Vet Clin Small Anim Pract** 1993;23:531 [12] Minar et. al **J Pediatr Gastroenterol Nutr** 2011;53:293

THE COMPENSATORY ANTI-INFLAMMATORY RESPONSE IN HORSES

Imogen C. Johns, BVSc, DACVIM, MRCVS
North Mymms, United Kingdom

SUMMARY

Following an inflammatory insult, a counter-balancing anti-inflammatory response develops. However, an overzealous compensatory response, known as the Compensatory Anti-Inflammatory Response Syndrome (CARS) can lead to a state of immunoparalysis, with associated poor outcome. This presentation will discuss CARS, and present the findings of a prospective study investigating anti-inflammatory cellular and cytokine responses in horses with GI inflammation.

THE INFLAMMATORY RESPONSE

An inflammatory insult, be it infectious or non-infectious (for example trauma), elicits a localized response with the aim of containing the insult, repairing tissue damage and ultimately restoring homeostasis. If the initial response is sufficiently severe, these changes are no longer restricted to the local environment, and instead become more widespread, with changes in distant organs, a state known as the Systemic Inflammatory Response Syndrome (SIRS).[1] Sepsis occurs when this inflammatory response is initiated by an infectious agent. The inflammatory response begins when injured cells release mediators such as tumor necrosis factor α (TNF-α) and interleukin-1 (IL-1). When microbial invasion occurs, specific bacterial components, known as Pathogen Associated Molecular Patterns, bind to receptors on cell membranes, or in the cell cytoplasm of blood monocytes and tissue macrophages. Binding to these Pattern Recognition Receptors results in macrophage production of pro-inflammatory cytokines such as TNF-α and IL-1, which in turn induce transcription of other inflammatory cytokines, mediators such as cyclo-oxygenase 2, inducible nitric oxide and endothelial adhesion molecules. The inflammatory cascade continues with production of other mediators such as platelet activating factor and leukotrienes, as well as activation of neutrophils, adhesion to endothelium and emigration to injured tissues. TNF-α and IL-1 also stimulate IL-6 production, which promotes the hepatic acute phase response and is a growth factor for B cells. The inflammatory response, whilst initially protective, may result in shock, hypoperfusion, and multiple organ dysfunction (MODS) if an excessive response occurs.[2]

SIRS and sepsis in horses

The equine SIRS is best characterized in adult horses with GI inflammation, where endotoxin, accessing the systemic circulation via a compromised GI mucosal barrier, elicits a widespread inflammatory response. Endotoxin infusion in healthy horses results in increases in serum IL-6 and TNF-α.[3] In horses with naturally occurring GI disease and associated endotoxemia, serum TNF-α and IL-6 concentrations are also increased, being higher in those with colic compared to healthy horses.[4] Increased IL-6 and TNF-α concentrations are also associated with the presence of a strangulating or inflammatory GI lesion, and with a higher mortality rate.[4,5] The inflammatory response characterized in septic neonatal foals appears to differ somewhat to adult horses although there are limited studies to date, and differences may be related to timing of sampling and sepsis definition. Expression of IL-6 and IL-1β mRNA did not differ between septic and non-septic foals in two studies, with TNF-α expression being lower in septic foals than in healthy controls.[6,7] IL-6 expression was however significantly higher in non-survivors compared to survivors.[6] When serum IL-6 concentrations were measured, a significantly lower concentration was found in septic foals than in healthy controls.[8]

THE COMPENSATORY ANTI-INFLAMMATORY RESPONSE SYNDROME

Historically, the hyper-inflammatory response suggested by high levels of pro-inflammatory cytokines was believed to be responsible for early deaths in septic human patients, and therapies directed at controlling this response were investigated.[9] However, widespread failure of therapies such as eritoran tetrasodium (TLR-4 antagonist) and other anti-inflammatory therapies has resulted in increased recognition of the importance of the anti-inflammatory response during SIRS and sepsis. Coincident with the systemic inflammatory response, initiation of a compensatory anti-inflammatory response has been described in human patients with sepsis and in laboratory animal models.[10] This regulatory response is directed towards limiting the SIRS and restoring homeostasis. It is characterized by the production of the anti-inflammatory cytokines IL-10, IL-4 and transforming growth factor beta (TGF-β), and by altered cellular functions such as expansion of regulatory T cells (Tregs), increased lymphocyte apoptosis, decreased monocyte production of inflammatory cytokines, increased production of anti-inflammatory cytokines and decreased monocyte antigen presenting capacity.[9,10] However, just as an over-zealous pro-inflammatory state can contribute to poor patient outcome, so too can an over-exuberant anti-inflammatory reaction. A state of 'immunoparalysis', the Compensatory Anti-inflammatory Response Syndrome (CARS), has been described in humans with systemic inflammation and sepsis, and is associated with poor clinical outcome, including nosocomial infections, multiple organ failure and high mortality.[11] Although initially described as a sequential process, more recent evidence suggests that inflammatory and anti-inflammatory responses are in fact highly dynamic biological processes, and should perhaps be best characterised as MARS: Mixed Anti-Inflammatory Response Syndrome.[12]

Key cytokines in CARS

IL-10, an anti-inflammatory cytokine produced predominately by CD4+ T lymphocytes, regulatory T cells and macrophages, has received much attention in human medicine as a key marker for this state of 'immunoparalysis'.[11,13] IL-10 markedly inhibits macrophage phagocytic functions and reduces the synthesis of pro-inflammatory mediators. In septic patients, serum IL-10 was found to be significantly higher at admission in non-survivors than survivors, with increased concentrations persisting after admission in non-survivors.[11,13] IL-10 concentrations were also significantly higher in patients with CARS vs SIRS (based on mHLA-DR expression, see below), with IL-10 production by LPS-stimulated monocytes being significantly higher in CARS patients than those with SIRS.[14] TGF-β, another anti-inflammatory cytokine, has been implicated in immunoparalysis, although evidence to support its importance is somewhat contradictory, with increased serum concentrations in CARS patients in some studies.[14] In other studies, no such increase was identified, although a significantly higher concentration in survivors was identified in one study.[15,16]

As SIRS/sepsis and CARS are not independent or sequential processes, ratios of inflammatory:anti-inflammatory cytokines may be more relevant in characterizing an individual patient's status as being predominately inflammatory or anti-inflammatory. The IL-10/TNF-α ratio has been shown to be a predictor of survival in human septic patients[11], and was significantly higher in CARS patients than SIRS patients.[14]

Cellular events in CARS

Monocytes are key effectors of the innate immune response, having the ability to phagocytize microorganisms, sense microbial products and respond by releasing inflammatory mediators such as the cytokines IL-1, IL-6 and TNF-α. Their capacity to present antigens to T lymphocytes, and thus induce specific activation, represents a link to the adaptive immune response. As such, monocytes play a central role in the acute phase of inflammation. However, during CARS, monocytes have a diminished capacity to produce inflammatory cytokines such as TNF-α and IL-6, have increased production of IL-10, and reduced phagocytic capacity.[14] In addition, monocyte antigen presenting capacity is decreased due to diminished cell surface expression of Human Leucocyte Antigen-DR (HLA-DR). Indeed, this decreased expression is considered a key marker for CARS-associated immunosuppression.[9,14]

During the inflammatory response, circulating neutrophils are rapidly recruited into infected or injured tissues, providing one of the first lines of defense against pathogens. Neutrophils generate proteolytic enzymes, and reactive oxygen and nitrogen species to destroy invading micro-organisms, followed by phagocytosis or neutrophil extra-cellular trap formation. CARS is characterized by alterations in both neutrophil function and apoptosis. During severe sepsis in human patients, decreased phagocytic activity of neutrophils has been identified, and has been shown to be a negative predictor for survival.[17] Neutrophil apoptosis and removal by macrophages from inflamed tissues is vital for the timely resolution of the inflammatory response. In human patients with sepsis, neutrophil apoptosis is suppressed, allowing for prolonged tissue injury and resultant poorer prognosis.[18,19]

Lymphocytes play a key role in the inflammatory response through their ability to interact with both the innate and adaptive immune responses, and to coordinate, amplify and attenuate that inflammatory response. During CARS, both lymphocyte number and function are altered.[10] Lymphocyte apoptosis is considered a key contributor to the state of immunosuppression following sepsis. In septic human patients this has been characterized by a peripheral lymphopenia and extensive apoptotic depletion of B and T cells from lymph nodes, spleen, gut, lung and other organs.[20] In a rabbit model of intra-abdominal sepsis, lymphocyte apoptosis as determined by flow cytometry of peripheral blood lymphocytes, was significantly greater in animals following cecal ligation and puncture than in sham operated controls, and this change occurred as early as 4 hours post surgery, supporting the concept of a co-incident inflammatory and anti-inflammatory response.[21] In addition, animals with the lowest levels of lymphocyte apoptosis were significantly more likely to survive, suggesting that lymphocyte apoptosis is a major contributor to the state of immunosuppression following sepsis.

Not only are the number of lymphocytes depleted during this state of immunoparalysis, but the function of the remaining cells has also shown to be compromised, with decreased mitogenic responses and altered cytokine production. Early reports suggested that CD4+ T cells shifted from a predominately Th1 phenotype to a more Th2, or anti-inflammatory, phenotype, although more recent studies support a more global downregulation of the T cell effector response, and that this response is correlated to patient mortality.[22]

Regulatory T cells are thought to be key players in this downregulation.[9,23] Tregs are a small subpopulation of lymphocytes (CD4+ cells characterized by expression of FOXP3) that are important regulators of anti-inflammatory immune responses. Tregs modulate the immune response at the site of inflammation by: 1) direct cytotoxicity via cell-to-cell contact; 2) inhibition of cytokine production by cytotoxic cells, especially IL-2 and TNF-α; and 3) secretion of immunomodulatory cytokines (TGF-β, IL-10). In septic patients with CARS characterized by decreased mHLA-DR expression, a marked elevation in Tregs as compared to healthy individuals was identified, with the increase persistent in non-survivors.[24] An increased percentage of Tregs in septic patients was also associated with a decreased lymphoproliferative response, suggesting that Tregs may play a role in lymphocyte apoptosis. Regulatory T cells have recently been characterized in horses although their role in sepsis has yet to be investigated.[26,27]

Relevance of CARS

Although the exact characteristics of this compensatory response remain debated, it is clear that sepsis is not dominated by a persistent pro-inflammatory state, but that profound immunosuppression, with correlating decreased ability to clear the primary infection, and increased susceptibility to secondary infections may develop in a subset of patients.[9] As such, rather than targeting inflammation in the treatment of septic patients, immunostimulatory therapies may be appropriate. However, because of the dynamic interplay between the inflammatory and anti-inflammatory state, characterization of an individual patient's immune status would be necessary to allow tailored therapy, either by stimulating the immune response or by containing it.

Equine CARS

Despite increasing evidence in man supporting the importance of the anti-inflammatory response in patient outcome, little is known about CARS in horses. In a study evaluating IL-10 and IL-6 concentrations in septic and healthy foals, the IL-6: IL-10 ratio was found to be significantly greater in healthy foals.[8] Expression of IL-4 mRNA was also found to be significantly higher in septic foals when compared to controls and IL-10 mRNA expression was significantly higher in non-survivors.[6] These results suggest that a similar anti-inflammatory response occurs in equine neonates, and may be associated with a poor prognosis.

The findings of a prospective study investigating the anti-inflammatory response in horses with GI disease will be discussed.

REREFERNCES

[1]Roy, M.F. **Vet Clin N Am: Equine Pract** 2004; 20: 41. [2]McKenzie H, et al. **Compendium: Equine** 2001;23(7):661. [3]MacKay R et al. **Am J Vet Res** 1999;60:68. [4]Steverink, et al. **J endotoxin Res** 1995;1:289. [5]Werners A, et al. **Equine vet J** 2005; 37:371. [6]Gold J, et al. **J Vet Intern Med** 2007; 21:482.[7]Pusterla N, et al. **Am J Vet Res** 2006;67:1045. [8] Burton A, et al.**Vet Immunol and Immunopathol** 2009;132:122. [9]Monneret G, et al. **Mol Med** 2008;14:64. [10]Bone R. **Crit Care Med** 1996;24:1125. [11]Gogos C, et al. **J Infect Dis** 2000;181:176. [12]Marcin F, et al. **J Immunol** 2012;189:4648. [13]Di Pino J, et al. **Arch Surg** 1995;130:1159. [14]Xu P-B, et al. **J Infect** 2012;65:380. [15]Monneret G, et al. **Immunol Letters** 2004;95:193. [16]Lekkou A, et al. **Clin Diagn Lab Immunol** 2004;11:161. [17]Danikas D, et al. **Clin Exp Immunol** 2008;154:321. [18]Milot E, et al. **Frontiers immunol** 2012;3:1. [19]Taneja R, et al. **Crit Care Med** 2004;32(7):1460. [20]Hotchkiss R, et al. **J Immunol** 2001;166:6952. [21]Atmatzidis S, et al. **Infect Dis** 2011;11:321. [22]Ward N, et al. **Clin Chest Med** 2008;29(4):617. [23]Monneret G, et al. **Crit Care Med** 2003;31:2068. [24]Venet F, et al. **Crit Care Med** 2003;31(7):2068. [25]Venet F, et al. **Intensive Care Med** 2009;35(4):678. [26]Hamza E, et al. **Immunology** 2011;134(3):292. [27]Robbin M, et al. **Vet Immunol Immunopathol** 2011;140:90.

THE DIAGNOSTIC POSSIBILITIES OF QUANTITATIVE NEEDLE ELECTROMYOGRAPHIC EXAMINATION.

Inge D. Wijnberg, DVM, PhD, KNMvD Specialist Equine Internal Medicine, DECEIM
Utrecht, The Netherlands

INTRODUCTION

In equine neurology, differentiation between spinal cord lesions, peripheral nerve disorders or muscular disorders can be a challenge even though progression has been made in unraveling the pathophysiology of muscular disorders. Quantitative EMG (QEMG) has been used for decades in human medicine, and due to the advanced computer techniques the clinical neurophysiology is still a fast developing area. In human medicine, the needle EMG technique has been applied since the 1950's. Nowadays, modern computer techniques have made this technique widely accessible for diagnostic centers. In veterinary medicine, clinical neurophysiology has gained interest, which has resulted in many interesting publications in this field [1-9]. Quantitative EMG in horses has been brought to the attention since 2000 [10-20], however is still applied only on a limited scale. Most institutes that are using needle EMG techniques are more familiar with evaluating pathological spontaneous activity in syndromes like myotonia, HYPP, or lower motor neuron disease [18]. Quantitative EMG has been proven to be a non-invasive diagnostic technique that is helpful in diagnosing generalized neurogenic conditions such as Equine Lower Motor Neuron Disease, Equine Grass Sickness, generalized myopathies, as well as in diagnosing the site of lesions, for example in cervical arthrosis [20-26].

EMG needle examination is based on intramuscular registration and analysis of electrical signals originating from the motor unit. Insertional activity induced by needle placement is evaluated, followed by observation for presence of pathological spontaneous activity. Then motor unit action potentials (MUP's) induced by low level muscle contractions are recorded and analyzed, followed by evaluation of the interference patterns during increasing muscle force [12,18b]. Former studies showed that quantitative EMG analysis can be performed quite easily in the horse, without restraint being more than using stocks and in an occasional impatient horse using light sedation. First of all the so-called insertional activity is evaluated in the muscle at rest; this is the activity generated by muscle fibers as a result of intramuscular needle placement. This activity is generated by muscle fibers that are damaged due to the needle placement and as a result, muscle fibers will discharge spontaneously. Insertional activity can be absent in case of fibrosis, because no fibers are present in this fibrotic tissue; or prolonged if diseased and therefore instable muscle fibers are activated by needle placement.

Secondly, the muscle at rest is evaluated for the presence of pathological spontaneous activity such as fibrillation potentials, positive sharp waves, complex repetitive discharges, (neuro) myotonia or doublets, triplets or multiplets. Pathological spontaneous activity (PSA) can be generated either from instable muscle fibers or from instable diseased nerves and should be discriminated from the normal presence of fibrillation potentials and positive sharp waves in the area of the endplate. Fibrillation potentials and positive sharp waves are examples of PSA originating from instable muscle fibers whereas neuromyotonia originates from instable peripheral nerves, doublets or triplets originate from diseased motor neuron nuclei located in the grey matter of the spinal cord. Increased presence of fibrillation potentials and positive sharp waved can be found in both neuropathies and myopathies. Neuromyotonia or multiplets are more commonly found in primary neurogenic disease. Myotonia is found in specific myogenic disorders such as the recently discovered congenital myotonia in New Forest ponies [27].

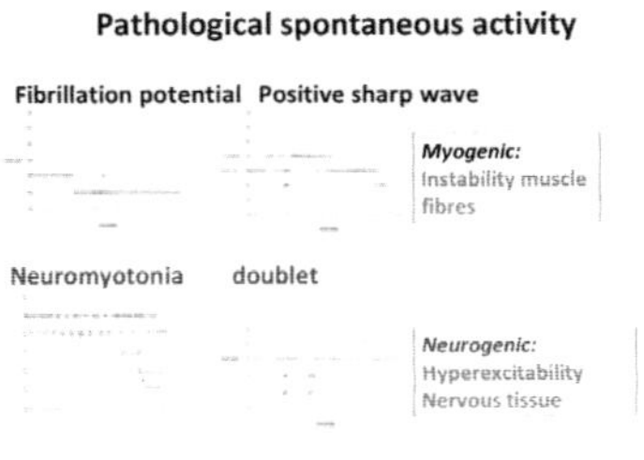

Figure1

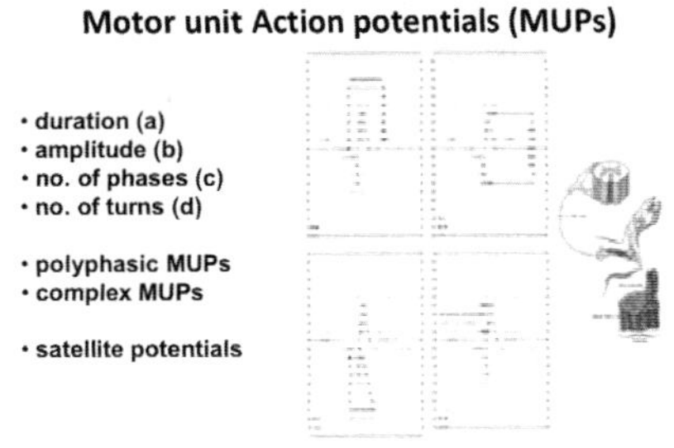

Figure2

Thirdly, MUPs generated by muscle activity at low force enables discrimination of separate potentials that can be captured and analyzed according to criteria as described in the literature. EMG enables discriminating myopathy from neuropathy, it is helpful in localizing the site of a lesion and provides information on extension or severity of the lesion. The presence of neuropathy is recognized by larger, broader, polyphasic or complex MUPs. Re-innervation is causing this increase in MUP variables such as amplitude, duration, number of phases and turns as well as Size Index of the MUP. The interference pattern can be poorly filled even at higher muscle force, indicating that recruitment of larger motor units at a certain muscle force is failing as a result of loss of functional motor units. Loss of motor units, a poor prognostic sign, is recognized by a poor - or absent interference pattern, indicating the lack of possibility to recruit more motor units if more muscle force is required. Myopathy, on the other hand, is characterized by low amplitude and short duration MUP's and low Size Index due to the loss of individual muscle fiber action potentials contributing to the MUP; the number of phases and turns might be increased, depending on the level of temporal dispersion. The interference pattern will be a low amplitude full interference pattern: at a relatively low muscle force a full interference pattern is generated. Reference values obtained from matched healthy horses for each individual muscle and for different age categories are required to enable correct interpretation, whilst temperature influences should be taken into account. Age differences need to be taken into account but do not preclude diagnosis using needle EMG. The same holds for differences of reference values as a result of muscle differences that need to be taken into account if subtle lesions are present[12, 14-26,29].

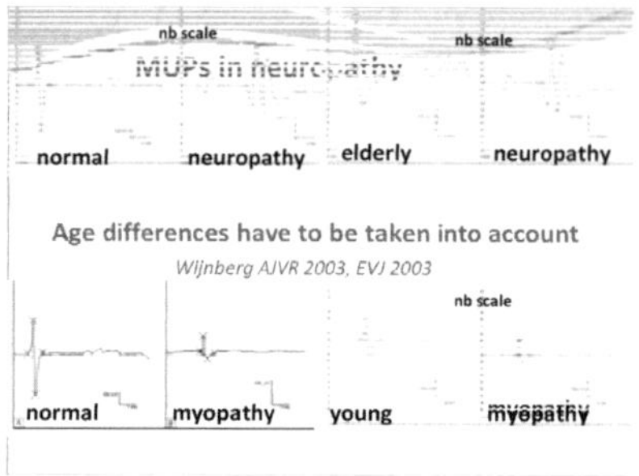

Figure 1. age differences influencing QEMG variables

Quantitative needle EMG has been proven to be useful, not only in detecting or elucidating (patho-) physiological adaptation such as pathological spontaneous activity [16], age related differences in MUP activity [17] or muscle differences [14,27,28] but also in determining and quantifying training effects [30,36]. In addition, it proved to be an important diagnostic tool in diagnosing Equine lower motor neuron disease[21,22 23-26], Equine Grass sickness [11,20], subclinical myopathies [18,19], localizing cervical lesions and explaining stringhalt [13,19,29]. The EMG evidence of lower motor neuron disease can be an explanation of the similarity in appearance (trembling, muscle wasting, tucked up abdomen, bottom narrow leg position) between EGS and EMND. The histopathological findings of lower motor neuron abnormalities seen in these two diseases is described on many occasions by several authors [20,34]. Comparing EMG results with histopathological and histobiochemical analysis of muscle biopsies and necropsies, the quantitative MUPs analysis appeared to be superior in early detection of pathological changes [12,18,35]. Histological morphometry of the muscle fibers appeared to be a useful alternative for histochemical analysis [34] and enabled earlier detection of abnormalities than histochemical analysis, but MUP analysis is also a very useful antemortem diagnostic tool in detecting early changes [12,18b, 35.] Lastly, needle EMG has played a role in optimal administration of Botox in the endplate zone in patients suffering from Stringhalt [31], or in the treatment of laminitis [32,33]. In addition, MUP analysis appeared to be more sensitive in detection of muscular causes of chronically or recurrent poor performance horses. Where blood biochemical analyses were not showing abnormalities, and muscle biopsies were normal, MUP analysis detected loss of muscle fibers contributing to the MUP. If the horses were managed as if they were suffering from a myopathy or rhabdomyolysis, the owners claimed improvement and absence of relapses[18a].

A study on the value of needle EMG in diagnosing locomotion disturbances of unknown origin[19] showed that if clinical information after a complete work-up was combined with the results of an EMG, this resulted in a shift from original determined location (local or generalized) as well as cause (neuropathic or myopathic origin) considerably. It appeared that clinicians overestimated the lumbosacral area as the primary location of the defect and underestimated the neck area as a potential location of the primary cause. In addition, myopathies were overestimated and neuropathies underestimated. This indicated that adding needle EMG in the diagnostic methods did result in changes in conclusions made. So, application of this technique did contribute in diagnosing causes of vague locomotor problems[19]. Another conclusion was, that in many cases clinicians assumed that Stringhalt like gait abnormalities were originating in the LS area, whereas needle EMG did not detect any abnormalities originating from this segment of the spinal column, but instead elucidated the presence of cervical neurogenic abnormalities in many of these cases [19]. In addition, there was a very good correlation between the pathological segments diagnosed by EMG and arthritic lesions found in these segments, with concomitant abnormalities in neurological examinations such as mild ataxia, abnormal correction reflexes and mild paresis[13].

More recent studies revealed that needle EMG was superior in detecting training adaptations than advanced biochemical muscle enzyme techniques, and might have a potential role in determining early stages of over trained or overreached horses based on the presence of more pathological spontaneous activity [30]. The training adaptations as measured by EMG indicated more muscle unit synchronization due to the training and recruitment of larger motor units at a lower stimulus level[30]. Also, EMG techniques proved useful in determining effect of the internationally discussed Low Round and Deep training methods on neuromuscular transmission in neck muscles of sport horses[36] and in evaluation of the effect of Botox for the treatment of Stringhalt [37] and laminitic horses[31,32,38].

Conclusion

Using QEMG, decision making can be facilitated regarding the presence of abnormalites, the type of abnormality (neurogenic or myogenic) and/ or its location, the severity of the lesion and sometimes information is achieved about the duration of an abnormality. In any horse in which detailed information about the origin or location of lesion, severity of the lesion or a prognosis is needed, QEMG contributes in decision making on cause of the problem, therapy required, determining a prognosis or timing for return to work.

REFERENCES

[1]Aleman M et al J **Vet Intern Med** 2008, 22(5):1196
[2]Haga HA et al **Vet Anaesth Analg** 2005;32(3):128
[3]Hawe C et al **Equine Vet J** 2001; 33(5):459
[4] Johnson CB et al **Res Vet Sci** 1994;56(3):373Beech J. et al **Muscle & Nerve** 1992; 15:932
[5]Kruluc P et al **Acta Vet Hung** 2006;54(1):35
[6]Murrell JC et al **Vet Anaesth Analg** 2005;32(4):212
[7]Otto KA et al **Vet Surg** 1996;25(3):249
[8]Whalen LR et al **Am J Vet Res** 1994;55(7):892
[9]Williams DC et al **J Vet Intern Med** 2008;22(3):630
[10]Wijnberg ID et **Tijdschr Diergeneeskd** 2000;15;125(24):743
[11]Wijnberg ID et al. **Compendium of the Finnish Veterinary Congress** ; 2001:.301
[12]Wijnberg ID. **Thesis UU** 2002. ISBN 90-393-3034-4.
[13]Wijnberg ID **Proceedings of the 41st British Equine Veterinary Association Congress**;2002:. 205
[14]Wijnberg ID et al **Equine Vet J** 2002;34(6):556
[15]Wijnberg ID et al **Am J Vet Res** 2002; 63(2):198
[16]Wijnberg ID et al **Am J Vet Res**.2002;63(6):849
[17]Wijnberg ID et al **Am J Vet Res** 2003; 64(1):70
[18a]Wijnberg ID et al. **J Vet Intern Med** 2003;17(2):185
[18b]Wijnberg ID et al **Equine Vet** J 2003;35(1):9
[19]Wijnberg ID et al. **Equine Vet J** 2004;36(8):718
[20]Wijnberg ID et al **Equine Vet J** 2006;38(3):230
[21]Benders NA et al **Tijdschr Diergeneeskd** 2001; 1;126(11):376
[22]Benders NA et al. **Am J Vet Res** 2005; 66(1):93
[23]van der Kolk JH et al **Am J Vet Res** 2005;66(2):271
[24]Swagemakers V et **al Proceedings of the International Congress Voorjaarsdagen** 2005, :318.
[25]Wijnberg ID. **Equine Vet Educ** 2005(17):123
[26]Wijnberg ID **Equine Vet Ed** 2006; 18 (3: 126-129.
[27]Wijnberg ID **J Vet Intern Med** 2011; 25 (3): 592
[28]Wijnberg ID et al. **Neurom Disord** 2012**;** 22 (4): 361
[29] E.J Cunilleras et al **JVIM** 2012, vol26,2. L03: 424.
[30]Wijnberg ID et al **J Appl Physiol** 2008;105(6):1746
[31]Wijnberg ID et al. **Equine Vet J** 2009;1(3):313
[32]Hardeman et al. **Proceedings ICEL 7** Stromsholm, Sweden 2012:153
[33]Wijnberg ID et al **Proceedings ICEL 7** Sweden;2012:154
[34]Palencia P et al **Acta Neuropathol** 2005;109(3):272
[35]van Dam, KG et al **Thesis UU** ISBN 90-393-4340-3,2006:88
[36]Wijnberg ID et al **Eq Vet J** 2010;2 (SUPPL. 38): 261
[37]Wijnberg ID et al **Equine Vet J** 2009 Mar;41(3):313
[38]Wijnberg ID et al. **Vet J** 2013; in press

PATHOPHYSIOLOGY OF EXERTIONAL RHABDOMYOLYSIS

Richard J Piercy, MA, VetMB, MS, PhD, DACVIM (LAIM), MRCVS
London, United Kingdom

INTRODUCTION

Recent years have seen improvements in the veterinary profession's understanding of the cause and pathophysiology of equine exertional myopathies; of key importance, was recognition that the syndrome of 'tying up' has several different causes. Historically, horses with signs of muscle contracture, stiffness and pain-associated elevations in the muscle-derived enzymes CK and AST, were believed to all have the same underlying form of myopathy and it has only been the advent of more routine muscle biopsy sampling and genetic testing that has enabled identification of specific entities within this syndrome. Much of this key work has been led by the group from Minnesota, including Drs McCue, Mickelson and Valberg. Despite these advances there remain many unanswered questions particularly regarding potential genetic causes in different breeds and their appropriate management. In this presentation I will focus on important historical work and new data from my own group.

Acquired causes

A variety of acquired factors underlying the development of exertional rhabdomyolysis (ER) is proposed, some of which are supported by evidence from other species. Although important as specific causes in normal horses, some apparently acquired causes may be associated with underlying genetic predisposition to ER. Unfortunately, for most proposed causes, the evidence remains speculative or hard to document with certainty due to various diagnostic constraints. Current and historical acquired causes that are proposed include overexertion or exhaustion, oxidative injury, eccentric contraction, hormonal influences and electrolyte abnormalities.

Extreme or unaccustomed exercise predisposes horses to rhabdomyolytic attacks. Muscle damage in such cases may include a combination of physical damage incurred during metabolic exhaustion, oxidative injury and excessive eccentric contractions. The latter is believed to be a major cause of the 'delayed onset muscle soreness' syndrome in humans where the eccentric contractions (contractions during muscle lengthening) are thought to predispose to sarcomeric injury. Such damage in humans is reflected by (sometimes considerable) elevations in serum muscle enzyme activities and triggers a local inflammatory response accompanied by edema. It is unknown whether horses experience this syndrome, however it seems likely. Downhill and jump exercise in particular are implicated.

Exceeding the level of training in a horse either by excessive endurance exercise or overexertion when galloping is believed to be a common cause of acute ER in horses. Although the causes are numerous and likely involve electrolyte imbalances and hyperthermia, in certain cases a presumed underlying factor is the deficiency of ATP, which results in an inability to maintain muscle ion homeostasis. In addition, free radicals are widely believed to cause post-exercise stiffness and fatigue in muscle, through several deleterious mechanisms that include the peroxidation of lipid membranes. Cell damage is normally minimized by the action of a complex cascade of free radical scavengers and antioxidants, including vitamin E and the selenium-dependent enzyme, glutathione peroxidase. Whilst evidence suggests that exercise-induced oxidative stress occurs in horses, there is no clear association with muscle damage (Chiaradia *et al.* 1998). Furthermore, antioxidant supplementation fails to attenuate exercise-induced elevations in plasma CK activity in horses, (Brady *et al.* 1978; Siciliano *et al.* 1997; White *et al.* 2001). Despite this, during conditions likely to result in significant oxidant stress, such as very strenuous or prolonged exercise, antioxidant deficiencies might exacerbate muscle damage caused by other mechanisms.

Published and anecdotal reports of improvement following correction of electrolyte clearance ratios (Harris and Colles 1988) underlie the interest in a role for whole body or muscle electrolyte defects in rhabdomyolysis. In a group of 38 ER-susceptible Thoroughbreds, about a third had potassium clearance ratios of less than 30% or low chloride clearance (Beech *et al.* 1993). These differences could potentially reflect differences in the handling of electrolytes by ER-susceptible animals, but ER-prone horses exhibit the same dietary-induced alterations to electrolyte clearance ratios as normal animals (McKenzie *et al.* 2002). In comparison with controls, normal Thoroughbreds administered frusemide and sodium bicarbonate developed lower plasma calcium, chloride, magnesium and potassium concentrations. Following exercise, serum CK activity was found to be significantly (but biologically marginally) higher in the treated group (Freestone *et al.* 1991). Some evidence therefore suggests that electrolyte imbalance might play a role in the development of certain forms of the disorder though it is not clear whether this is a primary effect or a secondary problem in animals with a genetic predisposition. Given that assessing muscle electrolyte status is notoriously difficult or unreliable in the clinical setting, testing, (such as electrolyte clearance ratios) are probably best avoided when investigating horses with ER.

Many studies report a higher incidence of ER in female horses compared with males (Harris 1991; MacLeay *et al.* 1999a; McGowan *et al.* 2002; Isgren *et al.* 2010). No correlation was found however between the stage of the estrus cycle and plasma CK activities in thoroughbreds in training (Frauenfelder *et al.* 1986), suggesting that a direct association of female sex hormones with ER is unlikely. The higher incidence of ER in females remains unexplained, although it may relate to differences in temperament.

Equine herpes virus 1 (EHV1) infection was proposed as causing an outbreak of ER in a training yard, where several horses seroconverted to the virus (Harris 1990). Equine Influenza Virus (EIV) has also been diagnosed by seroconversion as the potential cause of certain equine myopathies (Freestone and Carlson 1991). Both EIV and EHV1 infections are common in groups of young racehorses and cause and effect have not been fully established; as with other acquired causes, viral infection might modify the phenotype of genetically susceptible horses, or might simply be a coincidence.

Inherited causes

We currently recognize several forms of exercise-related myopathy in horses with underlying genetic causes. These include a condition examined extensively in a small group of Thoroughbreds in the USA that has been termed 'recurrent exertional rhabdomyolysis' (RER) (Lentz *et al.* 1999) and another condition, termed 'polysaccharide storage myopathy' (PSSM or EPSM) (Valberg *et al.* 1992; Valentine *et al.* 2000). These conditions have clinical and clinicopathological similarities and are managed similarly, though they also have key differences and breed susceptibilities (Valberg *et al.* 1999) and etiologies. In humans there are many genetic causes of rhabdomyolysis (e.g. mutations in genes encoding enzymes involved in cellular metabolism or structural proteins); it is currently unknown how many different forms of heritable rhabdomyolysis affect horses.

Although many horses with ER, of various breeds, have intermittent or repeated episodes of the syndrome, the word 'recurrent' tends to be reserved for a disorder that has been studied in detail in Thoroughbreds (Lentz *et al.* 1999, MacLeay *et al.* 1999b). Estimates of its prevalence suggest that 5-7% of Thoroughbreds worldwide are affected (Cole *et al.* 2004; MacLeay *et al.* 1999a; McGowan *et al.* 2002), although it remains unknown whether all these animals have the same disorder. Pedigree analysis of some lines of Thoroughbreds in the USA supports autosomal dominant inheritance of the trait (Dranchak *et al.* 2005; MacLeay *et al.* 1999b). Given that acquired forms of rhabdomyolysis are common in humans and other species it seems possible, or likely that some Thoroughbreds develop acquired forms of the syndrome as a result of some triggering environmental or management factor (MacLeay *et al.* 1999a). Despite this, it is not possible to rule out a common underlying genetic predisposition within the breed: it is well known that external factors modify the phenotype in many genetic conditions. Furthermore, epidemiological evidence suggests that other breeds likely have the same disorder (Standardbreds in particular) (Isgren *et al.* 2010).

A suspected abnormality in muscle cell calcium regulation has been the subject of several studies in the Thoroughbred. Researchers examined similarities between horses with ER and a condition recognised in humans and other species, known as malignant hyperthermia (MH). In particular, muscle from horses with RER and other species with MH is hypersensitive to agents (such as caffeine and halothane) that stimulate release of calcium from the muscle calcium store (sarcoplasmic reticulum) through a calcium release channel known as the ryanodine receptor (RYR1) (Bendahan *et al.* 2004; Lentz *et al.* 1999; Lentz *et al.* 2002). However, though MH has been reported in some horses following halothane anaesthesia, and indeed, though an RYR1 receptor mutation has been identified in MH-susceptible Quarter horses (Aleman *et al.* 2004), Thoroughbreds with abnormal calcium regulation do not share the same mutation and there is evidence suggesting that the RYR1 receptor is not involved in Thoroughbred ER (Dranchak *et al.* 2006). Furthermore, our recent work examining intracellular calcium responses in cultured muscle cells suggests that whilst a calcium-related defect might still exist, it appears separate from that seen in MH.

Polysaccharide storage myopathy (PSSM or EPSM) was traditionally diagnosed by muscle biopsy when pathognomonic changes include amylase-resistant inclusions detected by Periodic Acid Schiff staining. Some horses with this disorder are hypersensitive to insulin and clear glucose from the circulation more rapidly than normal animals (Annandale *et al.* 2004). This disease was first reported in detail in Quarter Horses with exertional rhabdomyolysis in the USA (Valberg *et al.* 1992). Since then, a disease with similar, if not identical histopathological features has been reported in a variety of other breeds, in particular draught horses (Baird *et al.* 2010; McCue *et al.* 2006; McCue et al. 2008a; McCue et al. 2008b; Valentine *et al.* 1997; Valentine *et al.* 2001; McGowan *et al.* 2003; Stanley *et al.* 2009). Although there are phenotypic differences in the presentations of this disease (rhabdomyolysis in athletic animals and sometimes paresis or muscle atrophy in the draught horses), many of these horses share the same underlying genotype and a specific founder mutation in the skeletal muscle glycogen synthase gene (GYS1) (McCue *et al.* 2008a; McCue et al. 2008b). Horses with this specific mutation (R309H) are regarded as having PSSM1. The pathogenesis of PSSM1 remains unclear, though the underlying defect appears to be an unregulated over-activity of glycogen synthase enzyme, one of the 2 key enzymes responsible for the normal structure of glycogen (McCue et al. 2008). An increased glycogen synthase: branching enzyme ratio appears to result in abnormally filamentous glycogen deposition, which, over time leads to insoluble, crystalline amylopectate or polysaccharide. Recent evidence supports some role for impaired structural integrity of muscle (Naylor *et al.* 2012); indeed, significant accumulations of polysaccharide and proteinaceous debris, destined for cellular degradation, increase as horses with PSSM1 age (Valentine e*t al.* 2006). Impaired energy metabolism also might play a part (Dranchak *et al.* 2007;Annandale *et al.*2005). Furthermore, genetic modifiers modify the phenotype (McCue *et al* 2009a).

Advent of simple and reliable DNA testing in affected horses, led researchers to determine that not all horses with histopathological features of PSSM carry the GYS1 mutation, revealing that additional glycogen storage disorders (currently grouped as PSSM2) likely exist in horses (Stanley *et al.* 2009; McCue *et al.* 2009b). The etiology and pathophysiology remains unclear in these cases, which unfortunately means that advice for management, prophylaxis and prognosis is, at best, speculative.

REFERENCES

Aleman, M., et al. (2004). Muscle Nerve 30, 356-365.Annandale, E.J., et al. (2004). Neuromuscul Disord 14, 666-674.Annandale EJ, et al.(2005) Am J Vet Res 66: 839–845. Baird, J.D. et al. (2010) Vet Rec. 167(20); 781-4; Beech, J., et al.(1993) Res Vet Sci 55, 43-51. Bendahan, D., et al. (2004) Acta Anaesthesiol Scand 48, 1019-1027. Brady, P.S., et al(1978). J Anim Sci 47, 492-496. Cole, F.L., et al. (2004) Vet Rec 155, 625-630. Dranchak, P.K., et al. (2005) J Am Vet Med Assoc 227, 762-767. Dranchak, P.K., et al. (2006) Am J Vet Res 67, 1395-1400. Dranchak PK, et al. (2007) Am J Vet Res 68: 1079–1084. Frauenfelder, H.C., et al. (1986) Equine Vet J 18, 371-374. Freestone, J.F. and Carlson, G.R. (1991) Equine Vet J 23, 86-90. Freestone, J.F., et al. (1991) J Vet Intern Med 5, 40-46. Harris, P. and Colles, C. (1988) Equine Vet J 20, 459-463. Harris, P.A. (1990) Vet Rec 127, 468-470. Harris, P.A. (1991) Br Vet J 147, 373-384. Isgren et al. (2010) 5(7):e11594. Lentz, L.R., et al. (1999) Am J Vet Res 60, 992-999. Lentz, L.R., et al. (2002) Am J Vet Res 63, 1724-1731. MacLeay, J.M., et al. (1999a) Am J Vet Res 60, 1562-1566. MacLeay, J.M., et al. (1999b) Am J Vet Res 60, 250-256 McCue ME, et al. Genomics. (2008);91(5):458-66. McCue et al. (2008) J Vet Intern Med. 91 (5): 458-66; McCue et al. (2009a) Neuromusc Disord. 19(1); 37-43. McCue et al. (2009); Vet Pathol 46 (6); 1281-91.; McGowan, C.M.,et al. (2002) Vet Rec 151, 623-626. McGowan, C.M., et al. (2003) Vet Rec 152, 109-112. McKenzie, E.C., et al. (2002) Am J Vet Res 63, 1053-1060. Naylor, R.J., et al. PLoS One. 2012;7(7):e42317. Siciliano, P.D., et al.(1997) J Anim Sci 75, 1553-1560. Stanley, R.L., et al. Equine Vet J. 2009. 41(6): p. 597-601. Valberg, S.J., et al. (1992) Neuromuscul Disord 2, 351-359. Valberg, S.J., et al. (1999) Equine Vet J Suppl 30, 533-538. Valentine, B.A., et al. (1997) Equine Vet J 29, 220-225. Valentine, B.A., et al. (2000) Vet Pathol 37, 193-196. Valentine, B.A., et al. (2001) J Vet Diagn Invest 13, 63-68. Valentine et al. (2006) Vet Pathol; 43 (3) 270-5; White, A., et al. (2001). Comp Biochem Physiol A Mol Integr Physiol 128, 99-104.

EQUINE COLLAPSE: CAUSES AND DIAGNOSTIC APPROACH

Neil P.H. Hudson, MA, VetMB, PhD, DEIM, DipVetClinStud, MRCVS
Edinburgh, United Kingdom

INTRODUCTION

Collapse in the horse can be a very distressing syndrome for the owner and provides a significant diagnostic challenge to the equine clinician. There are varying definitions used for this complex syndrome, but for the purpose of this discussion, collapse will be defined as an animal becoming acutely recumbent with or without the loss of consciousness. In addition, most of this discussion will centre on the adult horse. Some of the definitions and classifications of collapse can be confusing and indeed there is significant variation in the owner perception/description of the problem. It is important to apply a systematic and logical approach to the situation keeping in mind the important aspects of safety of the horse and its handlers.

APPROACH TO THE COLLAPSING HORSE

In the first instance, the practitioner may be presented with a situation that requires emergency care to stabilise the horse. Again, safety is paramount and it is vital that injury to the people around the horse is avoided. First-aid measures will include airway maintenance, breathing and circulation. Control of haemorrhage and cardiovascular support are important. Attending to acute injuries such as fractures and control of seizures with sedation or anticonvulsants may be necessary. In the acute setting, obtaining a quick history and performing a clinical examination as best you can are important. After stabilisation and initial examinations, further ancillary tests and treatments can be instituted depending on the detected problems.

In many cases of collapse, the veterinarian is not presented with the collapsed horse, but with a presenting history of intermittent collapse. In these cases, it is important to be thorough and systematic in the approach. A detailed history is indicated. Questions should include the frequency and duration of episodes, environmental details at the time, description of the events (before, during and after), state of consciousness of the horse, any changes in management, whether any other horses are affected, any possible toxic factors, drug therapies, whether the events occur at rest or at/after exercise, and whether the horse is "normal" between episodes. In many cases, the history can be vague or confusing. It may be helpful to ask the owner to keep a diary and if possible to video events. Again, safety is paramount.

Collapse can be due to problems related to a variety of body systems. Accordingly the clinical examination should be thorough, paying special attention to the cardiovascular and nervous systems. If feasible, a neurological examination is an important part of the work-up. Ancillary cardiological examinations including echocardiography and ECGs (including continuous monitoring) are important if a cardiovascular cause is suspected. Exercise testing may be indicated if considered safe. Haematological and serum biochemical analyses may be helpful if there is a systemic or metabolic background to the syndrome. Ancillary diagnostic procedures may also include modalities such as radiography and CT/MRI scanning.

CAUSES OF COLLAPSE

Frustratingly, in many cases of collapse a definitive diagnosis is not reached. The horse may or may not recover from the episode of collapse; if the horse recovers, recurrence of collapse is an important issue for the owner and the veterinarian. Collapse can be classified according to body system involvement such as cardiovascular, nervous, musculoskeletal etc. For this discussion, the broad categorisation described by Mayhew[1] will be used. This uses the categories of syncope, seizures, sleep disorders, coma, motor paralysis and generalised and metabolic disorders (the final three are grouped below as "other"). In a study of 44 horses presented for cardiological evaluation of collapse, 18% remained undiagnosed; in terms of causes of collapse, cardiac, neurological and metabolic disorders were identified in descending order of frequency.[2] In a recent retrospective study of 25 collapse cases, a definitive diagnosis was made in 11/25 cases, a presumptive diagnosis in 8/25 cases and the cause remained unknown in 6/25 cases.[3] In a *post mortem* study of 268 racehorses that acutely collapsed and died during or within an hour after exercise, a definitive cause of death was recorded in 53% (143/268) of cases, a presumptive cause of death in 25% (67/268) of cases and the cause remained unknown in 22% (58/268) of cases.[4]

Therefore, it can be seen that in a significant proportion of cases of equine collapse, determining the definitive cause remains elusive. The list of possible conditions that may be associated with equine collapse below is not exhaustive.

Syncope

Syncope involves a transient and acute loss of consciousness, with generalised weakness and sudden collapse. These "fainting" attacks are associated with insufficient cerebral blood flow. Syncope can be cardiac or extra-cardiac in origin. Syncope in the absence of cardiac failure is very rare. In human (and indeed equine) medicine it can be very difficult to delineate syncopal episodes from seizures[5] (see below) and also sleep disorders. Syncope in horses is poorly understood and many explanations are extrapolated from human conditions. Cardiac causes can include: arrhythmias (e.g. atrioventricular block – third degree more than second degree –, atrial fibrillation, ventricular premature depolarisations, ventricular tachycardia), cardiac failure (e.g. ruptured chordae tendinae, myocardial disease, severe valvular disease, endocarditis, pericardial disease). Extra-cardiac causes are difficult to validate. The vaso-vagal fainting syndrome (neurocardiogenic syncope)[6] observed in humans has been suggested to occur in horses but this is difficult to prove. A significant proportion (12/25) of the equine collapse cases in the study by Lyle et al.[3] with a diagnosis (definitive or presumptive) were described as being likely to be related to syncope.

Seizures

Seizures in the horse can be manifest as partial and generalised seizures and *status epilepticus*. Seizures represent the physical manifestation of spontaneous paroxysmal electrical discharges in the cerebrum. Repeated seizures or epilepsy can occur in foals with disease processes such as sepsis or hypoxic ischaemic encephalopathy or on occasion can have a genetic basis, for instance in Arabs. At this point in time, epilepsy in adult horses is largely considered only as acquired and as having an underlying disease process. A recent retrospective study of 104 horses presenting for seizures has helpfully improved the classification and description of equine epilepsy (2 or more recurrent seizures), applying accepted definitions from human and small animal medicine.[7] In this study, 70% (73/104) of seizure cases were described as epilepsy and of these 73 cases, 35.6% were classified as 'symptomatic' (with underlying identifiable brain pathology), 54.8% were classified as 'cryptogenic' (i.e. unknown aetiology) and 2.7% as idiopathic (suspected genetic predisposition).[7]

Sleep Disorders

In humans, narcolepsy is characterised by chronic sleepiness, a marked disorganisation of sleep/wake behaviour and pathological manifestations of rapid eye movement (REM) sleep.[8] It can be associated with or without cataplexy (flaccid paralysis). In humans there is also a syndrome of idiopathic hypersomnia, involving unwanted sleep attacks.[9] Narcolepsy with cataplexy does occur in foals such as Miniature Horses[10] and some pony breeds but it is less likely that sleep attacks in adult horses represent true narcolepsy. Defective signalling of the neuropeptide hypocretin has been implicated in certain forms of human and canine narcolepsy. Narcolepsy without cataplexy has been described in an Icelandic foal; in this case, CSF hypocretin-1 concentration was measured as normal.[11] There are limited studies of equine sleep in the literature but in one study of normal horses using electroencephalography (EEG), there were some interesting results. REM sleep was observed standing and in sternal recumbency and, in one horse, partial collapses were observed with the onset of REM sleep.[12] There is certain degree of debate regarding nomenclature but it is possible that some sleep attacks seen in adult horses (e.g. excessive drowsiness, buckling at the knees, stumbling and even total collapse) may be a form of idiopathic hypersomnia associated with chronic sleep deprivation.[13,14] Removal of the causes of this recumbent sleep deprivation with environmental modification and therapy (e.g. pain relief) has been described.[15]

Other

Coma involves recumbency, unconsciousness and unresponsiveness and can be associated with CNS disease processes such as trauma, asphyxia at birth, bacterial infection, parasites, liver disease and inadvertent intracarotid injections.

Collapse caused by loss of motor function can be associated with trauma, botulism, myasthenic syndromes and hyperkalaemic periodic paralysis (HYPP). *Generalised or metabolic causes* can include shock, hypoglycaemia (including possibly secondary to neoplasia[16]) electrolyte abnormalities (hypocalcaemia, hypokalaemia, hyperkalemia), endotoxaemia and anaphylaxis.[1]

REFERENCES

[1]Mayhew IG. In: **Large Animal Neurology** *2nd Edition. Wiley-Blackwell*, Oxford, pp 90-93; 2008. [2] Piercy RJ, et al. In: **Cardiology of the Horse**. Ed C.M. Marr, WB Saunders, London. pp 268-288; 1999. [3]Lyle CH, et al. **J Vet Intern Med** 2010; 24: 1498. [4] Lyle CH, et al. **Equine Vet J** 2010; 43 (3): 324. [5]McKeon A, et al. **Lancet Neurol** 2006; 5: 171. [6]Chen-Scarabelli C, et al. **British Med J** 2004; 329: 336. [7]Lacombe VA, et al. **Equine Vet J** 2012; 44: 646. [8]Scammell TE. **Ann Neurol** 2003; 53: 154. [9]Ohayon MM. **Sleep Medicine Reviews** 2008; 12: 129. [10]Lunn DP, et al. **Equine Vet J** 1993; 25: 483. [11]Bathen-Nöthen A, et al. **J Vet Intern Med** 2009; 23: 1299. [12]Williams DC, et al. **J Vet Intern Med** 2008; 22: 630. [13]Bertone JJ. **Proceedings 24th ACVIM Forum** 2006, *Louisville, pp 167-169.* [14]Mayhew IG. In: **Large Animal Neurology** *2nd Edition. Wiley-Blackwell*, Oxford, pp 87-89; 2008. [15]Bertone JJ. **Vet Clin N Am Equine Pract** 2006; 22: 157-62. [16]Swain JM, et al. **J Vet Intern Med** 2005; 19: 613.

VITAMIN E SUPPLEMENTATION IN HORSES WITH NEUROLOGIC DISORDERS

Carrie J. Finno, DVM, PhD, DACVIM; Stephanie J. Valberg, DVM, PhD, DACVIM; Joe D. Pagan, PhD
St. Paul, MN

INTRODUCTION

Vitamin E functions as a biological antioxidant, preventing the oxidation of unsaturated lipids within cellular and subcellular membranes by neutralizing production of free radicals. Through this mechanism, and potentially other mechanisms yet to be elucidated, vitamin E serves to maintain normal neuromuscular function. Several specific equine diseases develop in the face of vitamin E deficiency. These include nutritional myodegeneration, neuroaxonal dystrophy and equine degenerative myeloencephalopathy in young animals. Adult horses deficient in vitamin E may develop a vitamin E deficient myopathy or equine motor neuron disease. Treatment with vitamin E is usually instituted in an attempt to reverse clinical signs. However, selecting the type and amount of vitamin E to supplement can be challenging for veterinarians because the bioavailability and potency varies widely among commercial supplements. Most vitamin E supplements consist of natural or synthetic forms of α-tocopherol because it is the most biologically active and most researched isoform of vitamin E. Vitamin E, however is a complex nutrient consisting of eight closely-related fat-soluble naturally occurring compounds that form two groups; tocopherols (saturated) and tocotrienols (unsaturated). Within each group, there are four individual isoforms (α, β, γ and δ). The purpose of these proceedings is to review the equine neuromuscular disorders related to vitamin E deficiency and data on serum and tissue responses to commercial vitamin E supplements.

ASSESSMENT OF A-TOCOPHEROL DEFICIENCY

Serum or plasma is clearly the most readily available sample for determination of α-tocopherol deficiency. Blood should be chilled as soon as possible after obtaining a sample and protected from light by wrapping in tin foil. Blood samples should be centrifuged at 4°C and plasma/serum separated as soon as possible. If analysis is delayed, serum/plasma samples should be stored frozen (- 21° F, - 70° C) as vitamin E deteriorates rapidly. Reference ranges for plasma/serum concentrations of α-tocopherol in the horse are as follows:[1]

>2	μg/mL	Adequate
1.5-2	μg/mL	Marginal
<1.5	μg/mL	Deficient

Serum versus tissue α-tocopherol concentrations: Significant correlations exist between serum α-tocopherol concentrations and adipose tissue,[2] liver,[3] muscle[3] and CSF[4] in healthy horses. The normal reference ranges for muscle tissue and cerebrospinal fluid (CSF) α-tocopherol concentrations are reported at 4.2 ± 0.8 μg/g[3] and 9.5 ± 5.54 ng/mL,[5] respectively; however, the values for CSF concentrations were determined in apparently healthy, but α-tocopherol deficient horses (serum α-tocopherol <2 μg/mL). A correlation between serum/plasma and CSF α-tocopherol is also evident in deficient horses suffering from neuroaxonal dystrophy/equine degenerative myeloencephalopathy (Finno, unpublished results) and between serum α-tocopherol and nervous tissue α-tocopherol in equine motor neuron disease.[6] This correlation, however, is not consistently evident in horses with vitamin E deficient myopathy. In cases of vitamin E deficient myopathy, a discrepancy between serum and tissue levels of α-tocopherol may indicate that an abnormality exists in uptake of α-tocopherol into muscle tissues.

Equine Diseases associated with vitamin E deficiency

Vitamin E deficiency may exist in an entire herd of horses, yet only a select few develop clinical signs of neuromuscular disease. Development of neuromuscular disease appears to depend upon the age of the animal when deficiency develops, the duration of vitamin E deficiency, the genetic make-up of the individual and other concurrent dietary deficiencies or excesses. In many horses, there are no apparent ill effects of vitamin E deficiency. Therefore, diagnosis of the neuromuscular disorders described below requires not only determination of vitamin E status but also the appropriate clinical signs, supporting clinical pathology and/or muscle biopsy results and elimination of other possible diseases.

Nutritional myodegeneration (NMD)

Nutritional myodegeneration (NMD), also referred to as white muscle disease, affects skeletal or cardiac muscle of rapidly growing, active foals and is primarily due to a dietary deficiency of selenium beginning in utero.[7] In some, but not all cases, there may also be a deficiency of vitamin E.

Clinical signs: Foals with NMD often present with signs of muscle weakness, difficulty rising trembling of the limbs, stiffness, prolonged recumbency, firm painful muscles and potentially aspiration pneumonia. Depression, rapid heart rate, difficulty breathing and foamy nasal discharge or sudden death may indicate cardiac involvement.

Diagnosis: Foals with NMD have elevated serum creatine kinase (CK) and aspartate transaminase (AST) activities, whole blood selenium < 0.07 μg/ml and serum vitamin E may or may not be < 2 μg/ml.

Treatment: The cardiac form of NMD is usually not compatible with life. Foals with skeletal muscle involvement often show significant improvement within 3 to 5 days of treatment with injectable selenium products as well as general supportive care. Injectable selenium products contain 50 mg/ml (68 IU) of vitamin E as dl-α-tocopheryl acetate (all-rac-α-tocopheryl acetate) which acts a preservative for the solution and is, therefore, insufficient for vitamin E supplementation. Therefore, additional oral vitamin E is suggested to quickly increase antioxidant levels (see below for dosages). Foals with prolonged recumbency require significant rehabilitation therapy to regain normal muscle function.

Neuroaxonal Dystrophy / Equine Degenerative Myeloencephalopathy (NAD/EDM)

Equine neuroaxonal dystrophy (NAD) and equine degenerative myeloencephalopathy (EDM) are closely related disorders distinguished largely by the location of axonal or neuronal degeneration in the central nervous system. NAD/EDM occurs in multiple breeds with most cases demonstrating clinical signs by six to twelve months of age. Although the pathophysiology is not completely defined, there is strong evidence of a genetic component whose expression is highly influenced by vitamin E deficiency during the first year of life. Low serum α-tocopherol has been described in most, but not all, affected foals.

Clinical signs: Clinical signs include symmetric ataxia that is often more severe in the pelvic limbs than the thoracic limbs, abnormal base-wide stance at rest, and proprioceptive deficits. In some reports, hyporeflexia of the cervicofacial and cutaneous trunci is described in addition to an absent laryngeal adductor reflex. Horses with NAD/EDM that survive to 2-3 years of age commonly exhibit lifelong, stable neurologic deficits.[8]

Diagnosis: An antemortem diagnosis of NAD/EDM is based solely upon clinical signs, the elimination of other causes of neurologic disease, and a possible association with a low serum α-tocopherol concentration. A definitive diagnosis is only available upon histopathologic evaluation of specific tracts and nuclei within the spinal cord and brainstem tissue at post-mortem. There is no treatment for NAD/EDM and there have been no reports of spontaneous resolution.

Treatment: Suspected cases are often treated empirically with vitamin E supplementation (see below for dosages). Unfortunately, there is strong evidence that vitamin E supplementation of affected horses does not lead to neurologic improvement.[9] However, supplementation of dams with vitamin E during pregnancy may lead to a decreased incidence of NAD/EDM during the next foaling season.[10]

Vitamin E deficient myopathy (VEM)

A subset of horses with vitamin E deficiency may develop clinical signs solely related to muscle atrophy and weakness without evidence of damage to motor nerves. These horses of a wide variety of breeds have been diagnosed with a vitamin E deficient myopathy (VEM).[11] Whether VEM is an entity unto itself or a predecessor to development of equine motor neuron disease is not yet known.

Clinical signs: Horses present with loss of muscle mass, toe dragging, poor performance, weakness and muscle fasciculations.

Diagnosis: The most specific diagnosis of this disease is the histopathologic finding in fresh (not formalin-fixed) sacrocaudalis dorsalis muscle of an abnormal moth eaten mitochondrial staining pattern in the absence of neurogenic angular atrophy of myofibers. In many, but not all cases of VEM, serum α-tocopherol is low. In all affected cases assessed to date, muscle α-tocopherol concentrations have been below normal. This represents an unusual lack of linear correlation between serum and muscle α-tocopherol and may indicate deficient skeletal muscle uptake or excessive muscle consumption.

Treatment: Horses with VEM are remarkably responsive to appropriate vitamin E supplementation (see below for dosages) and show a complete recovery.

Equine Motor Neuron Disease (EMND)

Equine Motor Neuron Disease (EMND) is an acquired neurodegenerative disorder affecting motor nerves, particularly those supplying highly oxidative type 1 muscle fibers. EMND is associated with low plasma concentrations of α-tocopherol and a dietary deficiency of vitamin E of at least 18 months duration.[12]

Clinical signs: Horses with EMND show signs of generalized muscle wasting, muscle fasciculations, shifting of weight between hindlimbs, low head carriage in some cases and prolonged recumbency.

Diagnosis: Antemortem diagnosis of EMND is based upon either histopathologic evidence of degeneration of myelinated axons upon biopsy of the ventral branch of the spinal accessory nerve or the finding of neurogenic angular atrophy of predominantly type 1 muscle fibers in sacrocaudalis dorsalis medialis muscle biopsy.[13-14]
Treatment: Treatment with relatively high doses of 5000-7000 IU α-tocopherol/day is reported to produce clinical improvement in about 40% of cases within 6 weeks and some horses may appear normal within 3 months.[15] It should be noted, however, that return to performance may result in deterioration.

VITAMIN E SUPPLEMENTATION

It is important to realize that the absorption and metabolism of vitamin E in healthy horses may differ from horses with vitamin E deficiency. Thus, the choice of the type of vitamin E supplement should differ depending upon whether a maintenance supplement is being selected for a horse with normal serum α-tocopherol or whether there is a need to rapidly increase serum α-tocopherol in a horse with low serum α-tocopherol and clinical signs of vitamin E deficiency.

Vitamin E can be obtained from natural or synthetic sources, but the chemical structure of each is different. Natural vitamin E is composed of one isomer (d-α-tocopherol [RRR-α-tocopherol]), and it is the most bioactive form in animal tissues. Synthetic vitamin E is a mixture of eight isomers (dl-α-tocopherol [all-rac-α-tocopherol]), of which only one is identical to the natural isomer. These eight isomers vary greatly in relative biopotency. Furthermore, when synthetic or natural vitamin E is formulated as a feed additive, it is manufactured as an esterifed form (α-tocopheryl acetate) to prolong shelf life. In order for α-tocopheryl acetate to be utilized in the body, the ester has to be removed and the α-tocopheryl made water-dispersible by the action of bile salts (micellization). These additional steps may limit α-tocopheryl acetate absorption in the horse.

To account for differences in biopotency, the relative strengths of different forms of vitamin E are expressed as international units (IU) in which 1 mg of synthetic acetate equals 1 IU, 1 mg of natural acetate equals 1.36 IU, and 1 mg of natural alcohol equals 1.49 IU.[16] These conversion factors were developed using laboratory animal models. A recent study in horses found that relative to synthetic (dl-α-tocopheryl acetate), natural-source vitamin E acetate (d-α-tocopheryl acetate) had 1.97 and natural-source vitamin E alcohol (d-α-tocopherol) 2.52 times the potency of synthetic vitamin E.[17] To further enhance the bioavailability of α-tocopherol, water dispersible formulations of vitamin E have been developed. When tested, two of the commercially available water dispersible natural vitamin E products had a bioavailability 5.59 (Elevate WS, Kentucky Performance Products LLC, Versailles, KY) and 6.13 (Nano•E, Kentucky Equine Research, Versailles, KY) fold higher than synthetic vitamin E acetate (d-α-tocopheryl acetate).[17] The capacity of these compounds to significantly increase serum α-tocopherol concentration in horses has been studied in horses with normal baseline α-tocopherol (Table 1). Synthetic forms of α-tocopherol produced no significant increase in serum over a range of doses each fed for 14 days. In contrast, the natural acetate form caused a significant increase when fed ≥2000 IU/500 kg horse for 14 days and the water dispersible formulation of natural α-tocopherol had a significant and greater increase in serum levels at ≥ 1000 IU/500 kg horse.[18]

Table 1: Significant changes in serum concentrations of α-tocopherol when 3 formulations of α-tocopherol were each fed for 14-days to healthy unexercised horses that had normal baseline serum α-tocopherol concentrations[18]

Dosage (IU/500 kg BW)		500	1000	2000	4000	5000	8000
Type	**Formulation**						
all-rac-α-tocopheryl acetate (synthetic)	Powder	No significant effect	No significant effect	No significant effect	No significant effect	No significant effect	No significant effect
RRR-α-tocopheryl acetate (natural) (KER Eq Ester)	Powder	No significant effect	No significant effect	~1.3x increase	~1.5x increase	~1.8x increase	~2x increase
RRR-α-tocopherol (Elevate)	Liquid	No significant effect	~1.5x increase	~2x increase	~2.4x increase	~2.5x increase	~3x increase

It is important to realize when interpreting these studies that the absorption and metabolism of vitamin E in healthy horses may differ from horses with vitamin E deficiencies. In two studies using horses with a marginal/deficient α-tocopherol status but no clinical signs of deficiency, a serum α-tocopherol concentrations increased with both all-rac-α-tocopheryl acetate (synthetic) and water-dispersible RRR-α-tocopherol (natural; Emcelle, Steward Prodcuts, Bedford, TX) at 10,000 IU/500kg horse/day for 10-14 days, with levels being twice as high in the group on RRR-α-tocopherol. Cerebrospinal fluid concentrations of α-tocopherol increased from baseline (1.3 to 3.5 fold) only in horses supplemented with natural water dispersed RRR α-tocopherol.[4]

Vitamin E deficiency: Natural-source water-dispersible forms of vitamin E at 10 IU/kg body weight seem like the obvious choice for optimal treatment as they are 5-6 times more bioavailable than synthetic vitamin E acetate, and a 5000-IU dose/horse more than doubles serum vitamin E levels within 12 hr.[17] Before implementing supplementation it is important to measure serum α-tocopherol concentrations in order to determine if there is an underlying deficiency and in order to monitor the efficacy of supplementation.

Healthy horses: Current NRC daily recommendations for vitamin E in horses are 1 -2 IU/kg body weight, however, these NRC recommendations do not discriminate between natural or synthetic sources.[19] Research-based evidence for the need for additional vitamin E supplementation above 500 IU/day is lacking in healthy young and middle-aged horses receiving adequate dietary vitamin E intake. Synthetic or natural esterified forms of vitamin E are the most cost-effective means to provide vitamin E to horses with normal vitamin E status. It is important to periodically check serum α-tocopherol concentrations and, if an adequate response to supplementation does not occur, the dose or formulation of α-tocopherol should be altered. Vitamin E, unlike other fat-soluble vitamins, does not accumulate in the body to a toxic level due to protective mechanisms. Despite these mechanisms, it is important to consider that supplementation with vitamin E may alter drug metabolism and disposition since, in humans, the same cytochrome isoforms that metabolize vitamin E metabolize >50% of therapeutic drugs.[20] High levels of supplementation in humans do not enhance plasma concentrations more than 3-4 fold and excess tocopherol is sequestered in the liver, rapidly metabolized and/or excreted. The NRC has set the upper safe diet concentration at 20 IU/kg BW based on biopotency of synthetic α-tocopherol (10,000 IU/500 kg horse). Above this level, coagulopathy and impaired bone mineralization have been reported.[19] Furthermore, in healthy exercising horses, high dosage of vitamin E supplementation (10x NRC requirements) was shown to be potentially detrimental to beta-carotene absorption and thus not recommended.[21]

REFERENCES

[1]Blythe LL, et.al In: Robinson NE,ed. **Current Therapy in Equine Medicine** 1992;559-561.
[2]Steiss JE, et.al. **Equine Vet J** 1994;26:417-419.
[3]Roneus BO, et.al **Equine Vet J** 1986;18:50-58.
[4]Pusterla N, et.al **Vet Rec** 2010;166:366-368.
[5]Higgins JK, et.al **Am J Vet Res** 2008;69:785-790.
[6]Divers T, et.al **Clin Tech Equine Pract** 2006;5:24-29.
[7]Perkins G, et.al **J Vet Intern Med** 1998;12:173-177.
[8]Blythe LL, et.al **Comp Cont Educ Pract Vet** 1992;14:1215-1221
[9]Aleman M, et.al **J Am Vet Med Assoc** 2011;239:823-833.
[10]Dill SG, et.al **Am J Vet Res** 1990;51:1300-1305.
[11]Bedford H, et al. **J Am Vet Med Assoc** *in press*
[12]Divers TJ, et.al **Am J Vet Res** 2006;67:120-126.
[13]Divers TJ, et.al **42nd Annual Am Assoc Equine Pract Proc** 1992;28:215-219
[14]Jackson CA, et al. **Equine Vet J** 1996;28:215-219
[15]Divers TJ, et.al **Equine Vet J** 2011;13:63-67
[16]Weber P, et.al **Nutrition** 1997;13:450-460
[17]Pagan JD, et.al **Nordic Feed Sci Conf Proc** 2010;112-115.
[18]Pagan JD, et al. **Pferdeheilkunde** 2005;21:101-1-2
[19]NRC **National Research Council** 2007.
[20]Kliewer SA, et.al **Endocr Rev** 2002;23:587-702
[21]Williams CA, et.al **Equine Vet J Suppl** 2006;617-621

ANTIMICROBIAL USE IN FOOD ANIMALS: IMPLICATIONS FOR HUMAN AND ANIMAL HEALTH

John R. Middleton, DVM, PhD, DACVIM (LAIM)
Columbia, MO

INTRODUCTION

Antimicrobial use in food animals is not only highly regulated by government agencies, but also highly scrutinized by consumers and/or consumer advocacy groups. As such, prudent and informed use of antimicrobials in food animals is essential to the preservation of a wholesome food supply while at the same time preserving our ability to treat disease. The role of the veterinarian in ensuring appropriate use of therapeutics involves education of the animal-owning public and prescribing antimicrobials for conditions only where they are clearly warranted. These activities can only occur under a valid veterinarian-client-patient relationship.

In general, three classifications of drugs are found farms 1) over the counter (OTC) drugs, 2) prescription drugs, and 3) OTC or prescription drugs which are not used according to label directions (extra-label drug use; ELDU). Prescription drugs are those drugs which must be dispensed by or on the order of a veterinarian and will carry one of the following statements on the label: "CAUTION: Federal Law restricts this drug to use on or by the order of a licensed veterinarian" or "CAUTION: Federal Law restricts this drug to use on or by the order of a physician". Over the counter products can be purchased without a veterinary prescription and there are several with food animal disease and species label claims. However, when used in an extra-label fashion these products become prescription drugs. Extra-label drug use is often practiced due to the limited repertoire of drugs and label indications specific to food animal practice. Extra-label drug usage is governed under the Animal Medicinal Drug Use Clarification Act (AMDUCA), which will be discussed in a separate session. This session will focus on consumer perceptions about antimicrobial use in food animals, antimicrobial resistance, and evaluate whether consumer perceptions are justified.

CONSUMER PERCEPTION

In her proceedings article "Public perceptions of drug use in food animals: Why should we be concerned?", Dale Moore argues that consumer perception is a consequence of marketing.[1] She uses the example that when red meat consumption was linked to heart disease in the 1970s, the poultry industry responded by marketing their product as a healthy alternative which caused an upward trend in poultry consumption. Trends in low and non-fat milk consumption have similarly responded to news about adverse health effects of fat-laden whole milk. Upward trends in organic food consumption can also be linked to niche marketing. News that two creameries in the Northeast were moving towards marketing milk from cows not treated with rBST caused other larger fluid milk suppliers to follow suit to maintain market share. Consistent messages in the news media pushed by groups opposed to animal agriculture that highlight hand-picked specific research supporting their agenda are likely to influence buying trends especially if the perceived risk of consumption is high, regardless of whether there is a solid scientific basis for the information. Based on these assertions, frequent news articles about pathogens such as methicillin-resistant *Staphylococcus aureus* (MRSA) in the pork industry and *E. coli* O157:H7 in ground beef and raw milk increase the perception that food frequently contains bacteria that are resistant to antimicrobials. Hence, when consumer advocacy groups state that millions of pounds of antimicrobials are used annually in livestock production for non-therapeutic purposes, e.g., growth promotion, the consumer makes the "obvious" connection that antimicrobial use impacts resistance and human health whether the connection is founded or not. In response, therefore, the livestock industry including veterinarians need to market the positive attributes of using antimicrobials in food-producing animals before retailers follow the trend taken with rBST and demand food from livestock produced with minimal or no antimicrobial use. As an industry we also need to demonstrate that antimicrobials are being used prudently, and to quote Call et al.[2] use "is limited in a manner that meets animal welfare and public health needs while reducing costs for producers and consumers". Phasing out of non-therapeutic uses of antimicrobials that have significance in human health for use in livestock is already on the radar of Federal regulatory agencies and the recent FDA ban of extra-label use of cephalosporins in food-producing animals highlights a trend towards limitations on the use of antimicrobials in livestock production. It is thus important that veterinarians have an understanding of antimicrobial resistance mechanisms and the issues surrounding antimicrobial use in food-producing animals.

ANTIMICROBIAL RESISTANCE

Antimicrobial resistance (AMR) is a major concern to physicians, veterinarians, farmers and consumers worldwide because resistance can render some diseases untreatable. Whenever we treat an animal or human with an antimicrobial drug a certain selection pressure is placed on the microbial population that could ultimately select for AMR. From a public health perspective, because animal products become food, there is concern about AMR pathogens being disseminated from the livestock sector into the human population.[3] This could occur by direct contact with animals, through environmental contamination, or through the food chain. However, while these may seem like legitimate concerns, it is important to recognize that concerns do not necessarily constitute risk and the "estimated farm-to-fork risk from on-farm antibiotic use is extremely low" (White paper: Antibiotic Use in Food Animals, 2011).[4] Furthermore, "the alternative risk of sub-optimal animal health may be higher than the risk of on-farm antibiotic use" (White paper: Antibiotic Use in Food Animals, 2011).[4]

Predicting whether an infection will be likely to respond to antimicrobial therapy is most often determined in vitro using antimicrobial susceptibility testing. Susceptibility testing can be performed by either the disk diffusion or broth microdilution method where bacteria are grown in the presence of antimicrobial drugs to determine whether certain concentrations of the drugs inhibit bacterial growth. Using these methods and correlating the results with clinical outcomes, it is determined whether a bacterium will likely be susceptible or resistant to treatment when a particular drug is applied in the host. The clinical and laboratory standards institute (CLSI) publishes diagnostic breakpoints for estimating whether a drug will be effective or not based on predicted drug concentrations expected to at the site of infection. However, breakpoints for certain diseases and animal hosts are not available and susceptibility or resistance is determined by extrapolation. In the case of bovine mastitis, for example, breakpoints only exist for a few of the commonly used drugs. Because in vitro testing does not always account for drug distribution within the mammary gland, ability of the drug to penetrate microabscesses, or to kill intracellular bacteria, there can be a lack of correlation between susceptibility testing and clinical outcome. In some cases, in vitro susceptibility may overestimate our ability to treat an infection such as in the case of a chronic *Staphylococcus aureus* mastitis where microabscesses form within the mammary gland. In other cases, because intramammary infusion may achieve higher concentrations of the drug than predicted by the susceptibility testing parameters, it may underestimate treatment success. Overall, the correlation between in vitro susceptibility testing and treatment outcome for bovine mastitis pathogens is generally poor[5] and knowledge of clinical outcome of treatment of a particular pathogen with a given drug on a particular farm is the best way to predict treatment success. This paradigm can be applied to other food animal diseases such as bovine respiratory disease complex. On the other hand, if we want to monitor populations of pathogens to determine whether antimicrobial therapy is selecting increased numbers of resistant pathogens, susceptibility testing can be a useful tool.

The acquisition and dissemination of antimicrobial resistance among bacteria is a complex and multifactorial issue, and while antimicrobial use plays a role in the size of the resistant bacterial population, factors other than antimicrobial use are also known to play a role.[2] Resistance can occur via an assortment of mechanisms including the organism being intrinsically resistant to an antimicrobial, by spontaneous mutation, or by transfer of genes encoding for resistance from one bacterial host to a new bacterial host. When an antimicrobial drug is used, AMR is promoted either because there is a competitive advantage for inherently resistant bacterial strains to proliferate in the population, because use of the antimicrobial facilitates movement of resistance genes from one bacterial host to a new bacterial host, or because the antimicrobial can select for genetic mutations that allow for relatively rapid conversion to a resistant strain.[2]

Antimicrobials are used for a number of reasons in livestock production including treatment of lameness, respiratory disease, reproductive tract disorders, mastitis, and diarrhea. Additionally a large volume (estimates vary from 3M to 20M lb annually in the United States) of antimicrobials are used for non-therapeutic purposes (disease prophylaxis and growth promotion) in poultry, swine, and cattle. The long-term non-therapeutic use of antimicrobials comes under the most scrutiny because it is assumed that such use will place a selection pressure on microbial populations resulting in emergence, maintenance and transfer of AMR determinants in commensal and pathogenic bacteria.[3] However, this is a hotly debated topic with limited supportive evidence. By example, mastitis is the most common reason for antimicrobial use on dairy farms and β-lactam antimicrobials (e.g., penicillins and cephalosporins) are the most frequently used drugs to treat subclinical and clinical bovine mastitis and <u>prevent</u> infection during the dry period in the United States (NAHMS, Dairy 2007). However, despite continued prophylactic and therapeutic use of intramammary antimicrobials for the treatment and prevention of mastitis, data suggests there has been little change in the proportion of resistant pathogens and, in fact, there appears to be an overall increase in susceptibility of mastitis pathogens to antimicrobials.[6]

Oliver et al. in their review on the impact of antibiotic use in adult dairy cows on AMR of veterinary and human pathogens asked 4 poignant questions "1) Are science-based data available to demonstrate antimicrobial resistance in veterinary pathogens that cause disease in dairy cows associated with use of antibiotics in dairy cows?, 2) Are science-based data available to demonstrate that antimicrobial resistance in veterinary pathogens that cause disease in adult dairy cows impacts pathogens that cause disease in humans?, 3) Does antimicrobial resistance impact the outcome of therapy?, and 4) Are antibiotics used prudently in the dairy industry?"[3] They concluded that antimicrobial use in adult dairy cattle has not increased antimicrobial resistance of veterinary pathogens to antimicrobial drugs commonly used in dairy production, and that available scientific data do not support widespread, emerging resistance among mastitis pathogens despite several decades of routine antimicrobial use. The impact of antimicrobial resistance among pathogens found in dairy cattle on human health appears to be minimal, and the risk to human health is negligible provided milk is pasteurized. When the safeguard of pasteurization is removed by choosing to consume raw milk, the likelihood of exposure to foodborne pathogens and foodborne pathogens possessing AMR determinants increases significantly. While AMR obviously decreases the success of treatment, overall trends among mastitis pathogens suggest that treatment success has not appreciably changed over time and pathogen niches within the mammary gland, such as *S. aureus* hiding intracellularly or in microabscesses, likely have a greater impact on treatment success.

Multiple factors appear to impact shedding of AMR bacteria in food animals including stress, age, nutrition, conditions under which the animals are reared (intensive vs. extensive management), and exposure to antimicrobials.[2] These influences suggest that it is not simply antimicrobial use, but that other extraneous factors also impact the prevalence of AMR bacterial shedding. That said, it is clear that antimicrobial use in animals and humans has led to antimicrobial resistance and will continue to do so. This is not unexpected. Strategies aimed at improved animal husbandry and disease prevention that reduce antimicrobial use will be critical decreasing the emergence of resistance in microbial populations on farm.

Dissemination of resistant organisms to people via food can occur, but outbreaks are usually associated with consumption of raw, unpasteurized, or improperly cooked products. The impact of antimicrobial use and AMR in food-producing animals on public health remains uncertain with the majority of evidence suggesting the impact is small.

CONCLUSIONS

In conclusion, whenever we treat an animal with an antimicrobial drug a certain selection pressure is placed on the microbial population that will ultimately select for AMR. Interestingly, however, while AMR is generally more prevalent on conventionally managed farms that employ antimicrobial therapy than on organic farms, antimicrobial resistant bacteria can persist on organic farms after many years of antimicrobial-free management suggesting that other factors surrounding the ecology of those bacteria on the farm are also important in the size of the AMR population.[2] Prudent use of antimicrobial drugs, i.e., only applying antimicrobials where there are clearly demonstrated production and animal welfare benefits, will reduce the selection pressure for AMR. Optimizing management practices that promote animal health will be essential to reducing antimicrobial usage.

REFERENCES

[1]Moore D. **Proc Natl Mastitis Counc** 2008:3-9. [2]Call DR, et al. **Anim Health Res Rev** 2008;9(2):159. [3]Oliver SP, et al. **Foodborne Pathog Dis** 011;8(3):337. [4]White paper; Antibiotic use in food animals. www.animalagriculture.org 2011:1. [5]Owens WE, et al. **J Dairy Sci** 1997;80(2):313. [6]Erskine RJ, et al. **J Dairy Sci** 2002;85(5):1111.

COXIELLOSIS: THE DISEASE AND CLINICAL PRESENTATION

Paul J. Plummer, DVM, PhD, DACVIM (LAIM)
Ames, IA

INTRODUCTION

Coxiella burnetii is an intracellular gram-negative bacterium that is the causative agent of Coxiellosis in animals and "Q-fever" in humans. Ruminants act as reservoir and replicating hosts of the organism, although almost any species of animals can become infected and shed the organism. In the large majority of these animals the disease process is subclinical, with adverse pregnancy events being the most common clinical manifestation in animals. In recent years, the interest in *C. burnetii* has increased significantly due to the designation of the organism as a "select agent" and class "B" bioterrorism weapon. This designation for an organism that is present on upwards of 90% of the dairy cattle operations in the United States and a significant number of sheep and goat operations results in uncertainty regarding the public health risk of shedding on our livestock operations. This session will focus on the organism and its clinical presentation in both animals and humans. Subsequent sessions will provide insights to the control of this disease as well as lessons that the US can learn from a recent large-scale outbreak of Q-fever in the Netherlands.

BACTERIOLOGY AND EPIDEMIOLOGY

C. burnetii is a gram negative organism that is morphologically similar to Rickettsia and is classified in the order Legionellales. The organism occurs in two predominate variants, a large cell metabolically active form (called the large cell variant) and a small cell form that is metabolically inactive and functions similar to a spore (although it is not truly a spore). The replicative large cell variant form of the organism requires an acidified environment for growth, generally the acidified lysosome of a cell. Although traditionally considered an obligate intracellular organism recent efforts have resulted in the development of an acidified media (ph<4.5) that allows for extracellular growth. The small cell variant is highly resistant to environmental degradation and can be easily aerosolized. The aerosol route of transmission is considered to be the primary route of transmission to both animals and humans. Recent epidemiologic data from the Netherlands suggest that the organism is capable of aerosolizing up to 5 km in appropriate environmental conditions. The organism is apparently more prevalent in US ruminant populations than previously believed. Two recent studies in dairy cattle bulk tank milk (one using serology and the other using PCR) demonstrated upwards of 90% herd level prevalence in US dairy cattle operations.[1,2] Work in our laboratory and others (unpublished, personal communication) has demonstrated significant herd/flock level prevalence in sheep and goat operations although this prevalence is likely somewhat more seasonal then that of cattle due to shedding characteristics. Based on experimental infection models and observation of infected animals following abortion storm the following trends appear to occur. Dairy cattle shed the organism more in their milk then vaginal fluids (perhaps why more zoonotic risk associated with attending small ruminant abortions), however they also shed the organism in their milk for prolonged periods of time relative to small ruminants. In contrast, vaginal shedding of the organism is very high in small ruminants with lesser amounts in the milk. Shedding in small ruminants also appears to be much more prevalent in the periparturient period in comparison to cattle.

CLINICAL DISEASE

In infected herds/flocks most shedding occurs in the complete absence of clinical signs. The consequence of this finding is that the zoonotic risk associated with this organism is not confined to periods of high abortion rates. In fact, in the Netherlands outbreak the majority of source farms for human infection did not have any clinical signs of disease. When clinical disease of animals does occur it is generally associated with adverse pregnancy events (abortions, still-born, weak lambs/kids). In severe situations the abortion rate may exceed 50% of the herd. Animals that abort generally have no signs of systemic illness and the aborted fetuses/placenta are often unremarkable in their gross appearance. Live born kids may have very high loads of organism and shed this into the environment without any clinical signs of disease. Abortion storms are most common in the first year of introduction into herds/flocks when presumably the immunity of the animals is fairly low. Subsequent years often demonstrate an increase in the within-herd prevalence of shedding but a significant decrease in clinical signs (ie. lower to minimal abortions). Although the number of animals shedding is often much higher in these years the quantitative level of shedding is typically reduced when compared to the level of shedding found in abortions. This scenario is especially true for goats where the herd often remains chronic subclinical shedders. Sheep also continue to shed but drop off more rapidly than goats (ie. over a period of 4-5 years the research would suggest). In many cases the herd/flock owners may have no idea that their animals are infected.

In our experience *C. burnetti* is in the top three causes of goat abortions as a result of decreased rates of *Campylobacter* in goats compared to sheep. *C. burnetti* is also commonly associated with abortions in sheep, however *Campylobacter, Chlamydophila and* toxoplasmosis exceed it in number. The traditional wisdom concerning the disease process and abortions associated with this agent have been recently questioned in light of several large human disease outbreaks, most recently in the Netherlands associated with dairy goat operations.

DIAGNOSIS

Diagnosis of Q-fever as the cause of abortion typically relies on the identification of characteristic histopathological changes, although it should be remembered that the organism is capable of being shed in animals without histopathologic evidence. In these cases it is unlikely that the organism is the primary agent associated with the abortion however the samples should be treated with appropriate precautions to prevent zoonotic transmission. It is also important to note that our experience and that of other laboratories is that animals can also be shedding organism when the abortion is caused by another primary agent. For instance, it is not uncommon to find histopathologic evidence of toxoplasma or *Campylobacter jejuni* along with shedding of *Coxiella*. In these cases the current thought is that the organism with histopathologic evidence of disease is the most likely culprit in the abortion event. That said, the finding of the *Coxiella* shedding does confirm the presence of the organism in the farm (entire farm must be considered positive) and results in the consideration of zoonotic Q-fever. There are several commercial and non-commercial PCR test available and validated for use on vaginal fluids, milk, feces, urine and aborted tissues. These tests are highly sensitive and are likely the best option for testing shedding status. Immunohistochemistry on placenta is also worthwhile and should be performed if there is concern for *Coxiella* on the basic histopathology exam.

Serologic testing has traditionally relied on the complement fixation test, however an increasing body of evidence suggest that this test lacks sensitivity in the clinical setting. ELISA and IFA tests are rapidly surpassing the CF test as the preferred serologic assays. When discussing serology for *Coxiella* it is important to understand that the organism is able to undergo a phase shift that results in changes in the immunodominant epitopes present on the surface of the bacteria. This shift is somewhat similar to the differences between a rough and smooth variant of *E. coli*. As a result of these changes in surface antigens there is a change in serologic response and the differentiation of phase I and phase II antibody responses. In human medicine the IFA is the gold standard serologic test since it allows for determination of independent phase I and phase II responses that can assist with determining the chronicity of infection in human patients. Our laboratory has recently developed and validated an IFA assay for sheep, goats and cattle. The ELISA performs similar to the IFA assuming that an ELISA specific for both phase I and II antigens is utilized, however it is unable to distinguish between the two responses and simply provides a positive/negative result. The utility of determining the phase response with regards to clinical disease is well validated in humans but has not been rigorously tested in ruminant species. It is also critical to understand that a significant proportion (10-15%) of infected animals that are shedding the organism fail to develop a measureable serologic response, even chronically. Thus, negative serology cannot be used to rule out infection and shedding of an individual animal.

Herd/flock level testing is beneficial in many cases. Since the organism is easily spread, has an infective dose of 1-10 organisms by aerosol and can persist in the environment for prolonged periods of time, identification of a single positive sample on a farm is indicative of herd/flock infection. In dairy situations, testing of bulk milk tank samples is an appealing means of monitoring herd status. In most species the organism is shed in the milk, however the temporal dynamics of that shedding is somewhat variable between the species. Cattle shed the organism primarily in the milk and tend to shed for prolonged periods of time after parturition (albeit relatively low level shedding). These attributes along with the year-round calving that most dairy cattle farms utilize in the US, allows for a relatively consistent contamination of the bulk tank milk and shows less seasonal variation in bulk tank sampling than the small ruminants. In fact, in at least two papers (one using antibody in the milk and one using PCR) more than 90% of the dairy cattle bulk tanks tested were positive. In comparison, the data suggest that shedding in small ruminants tends to be a little shorter in duration (still measured in weeks to months) than that of cattle. Combined with the seasonality of parturition present on many of our small ruminant operations this suggests (although this has not been confirmed on farms in US) that the bulk tank testing may vary a little more than that of cattle. In all cases, increasing the number of samples submitted will increase the sensitivity for detecting an infected herd and repeated sampling around kidding/lambing season (if seasonal) will increase the sensitivity. Generally bulk tank PCR is the preferred testing option in these cases. Data does suggest that bulk tank testing is least useful in dairy sheep flocks and they are the species with the greatest variability in milk shedding. In such cases it may be beneficial to use a two-pronged approach of bulk tank testing in lambing season and some herd serology.

In non-dairy herds an alternate testing strategy is necessary. In these cases two options would exist. Testing a subpopulation of all of the animals individually by PCR or by serology. In most cases the cost of serology is much lower and may make more sense from an economic standpoint although the PCR gives direct evidence of shedding. Along those lines two important points should be remembered. 1) Shedding is very intermittent in most animals, thus a single positive or negative PCR test on an individual animal can not be interpreted as the true disease status of that animal. Likewise, culling based on shedding status is not rewarding because the disease remains in the environment and other animals are likely already infected. 2) As mentioned earlier up to 10-15% of the animals can be serologically negative and be shedding. On the herd or flock level serology is useful at determining if the population is infected (are any of them serologically positive?), however individual animal results need to be interpreted with caution and do not rule out infection or shedding. Ideally, individual serologic samples would be submitted from each animal, however in large herds/flocks sampling of a subpopulation may be appropriate. In such cases the Table 1 below gives some basic suggestions on how many animals should be sampled based on the level of certainty that one wants to achieve.

ZOONOTIC RISK

This organism poses a significant zoonotic risk to veterinarians, farmers and possibly individuals in aerosol distance from infected farms. A recent serologic survey of veterinarians attending the AVMA conference showed that roughly 25% of veterinarians were seropositive for exposure. Individuals that performed large animal veterinary services or had significant surface water exposure (ie. swim in lakes, ponds etc) were at increased risk of being seropositive. The cause for an association with surface water is unclear although the organism has been show to survive in amoebas for up to six months. In the Netherlands humans that lived within 2 km of an infected dairy goat operation (ie. bulk tank positive by PCR) were more then 30 times more likely to develop Q-fever the individuals that lived more the 5 km from those same operations. The Dutch government initially instituted a vaccination strategy along with a ban on manure spreading but the human case incidence continued to rise. Under intense pressure to address the issue in 2010 they elected to euthanize all pregnant goats on farms that had at least one positive bulk tank PCR, a decision that has been questioned by many experts. A breeding ban was placed on all non-pregnant animals from those same farms and the human case rate appears to have dropped significantly. For more information regarding the outbreak in the Netherlands please see the session devoted to that topic later in the day.

Table 1. Number of animals to randomly select and test in order to detect flock / herd seroprevalence of *Coxiella burnetii* infection as detected by a serological response. Estimated test sensitivity of 98% and test specificity of 100% with a target Type I and Type II error of 0.05%. [1] Based on calculations using FreeCalc Version 2

Flock Prevalence of *Coxiella burnetii* Serological Reactors[1]				
Flock Size	**1%**	**5%**	**10%**	**15%**
10	10	10	10	8
35	35	28	19	16
60	59	39	24	17
85	83	45	24	17
110	107	44	26	17
135	131	47	26	18
160	127	50	27	18
185	147	53	27	18
210	166	50	28	18
235	186	52	27	19
260	167	54	28	19
285	183	55	28	19
310	200	53	28	19
335	216	54	28	19
360	193	55	28	19
385	207	56	28	19
410	220	55	29	19
435	234	56	28	19
460	211	56	29	19
485	223	57	28	19

CONCLUSIONS

Coxiella burnetii is a ubiquitous bacterial agent in our livestock environments, however it should be considered a zoonotic threat. In most animals shedding is at low levels so that normal immunocompetent individuals have minimal risk of exposure. Pregnant women, immunocompromised individuals receiving chemotherapy or with HIV and individuals with preexisting heart disease are at the highest risk for becoming infected. Individuals that are handling aborted fetuses should be especially vigilant as the bacterial load in these matrices is very high.

REFERENCES

[1] McQuiston, et al. (2002). Vector Borne Zoonotic Dis. 2(3):179-91. [2]Kim SG, Emerg Infect Dis. 2005 Apr;11(4):619-21. [3]Gutteo R, Seegers H, Taurel A-F, Joly A, Beaudeau F. (2010). Vet Microbiol. 149: 1-16.

COXIELLOSIS: CONTROL AND ENVIRONMENTAL MANAGEMENT

Paul J. Plummer, DVM, PhD, DACVIM (LAIM)
Ames, IA

INTRODUCTION

Coxiella burnetii is an intracellular gram-negative bacterium that is the causative agent of Coxiellosis in animals and "Q-fever" in humans. Ruminants act as reservoir and replicating hosts of the organism, although almost any species of animals can become infected and shed the organism. In the large majority of these animals the disease process is subclinical, with adverse pregnancy events being the most common clinical manifestation in animals. In recent years, the interest in *C. burnetii* has increased significantly due to the designation of the organism as a "select agent" and class "B" bioterrorism weapon. This designation for an organism that is present on upwards of 90% of the dairy cattle operations in the United States and a significant number of sheep and goat operations results in uncertainty regarding the public health risk of shedding on our livestock operations. This session will focus on control of the organism during outbreaks and what role the environment plays in maintenance of the organism. Other sessions in this series are focused on the bacteriology and diagnosis of the disease, so readers are referred to those proceedings for further details on those aspects.

CONTROL

Controlling the spread of *Coxiella burnetii* is a complicated task in most livestock operations. Having a basic understanding of the transmission and routes of infection is critical in this regard. Infectious dose studies for this organism have primarily focused on humans, as the US military pursued development of *C. burnetii* as a bioterrorism weapon during the cold war. Those studies suggest that as low as one organism inhaled by a human can result in infection. However, the wide spread nature of the organism in our environment and the relatively low rate of human infections suggest that the normal infectious dose is likely greater than 1 organism. In any case, compared to most bacteria the organism has a very low infectious dose and a relatively low exposure can result in infection. Similar infectious dose experiments in animals have not been performed, however standard animal inoculations often are performed with 1×10^4 CFU/ml with success. Traditional wisdom suggests that ingestion of the bacteria was a significant route of transmission, and indeed this organism was one of the major reasons for the development of standardized pasteurization ordinance for milk. Pasteurized milk has minimal to no risk, however the role of raw milk consumption is more poorly defined. The organism clearly can survive in milk and consumption of raw milk is a well-described risk factor for transmission. The role of aerosol transmission has been known for a considerable time and was the primary means of delivery tested by the US military. This route of transmission also seems to be very important for the natural transmission of the disease in animals and is the most consistent means of disease induction in model infections of animals. The small cell variant (spore-like form of the organism) is very small and easily aerosolized in dust. Epidemiologic data from the outbreak of *Coxiella* in the Netherlands suggest that the organism can easily be carried 2-5 kilometers and remain infectious. Finally, vertical transmission of the organism occurs in a large percentage of offspring born on infected farms. The role and duration of this vertical transmission is not well understood and is a topic of research for our laboratory and others. Our unpublished data demonstrates that on farms that are infected, even in the absence of abortions, the majority of goat kids are born shedding the organism. This finding is not surprising given the fact that on infected farms the majority of animals will shed organism during parturition. That said, the natural course of infection in these neonates is not well understood and many of them appear to stop shedding at some point during the first year, only to become re-infected or reactivated at a later time.

The identification of the organism on a farm is sufficient to consider the entire farm infected. The small cell variant of the organism can readily survive in the environment for prolonged periods of time. It is profoundly resistant to desiccation and disinfectants and is resistant to moderate to high levels of heat (killed by pasteurization temperatures). For this reason, the environment on positive farms must be considered positive and unlikely to change in that status for a significant period of time. Although this might at first seem highly problematic it is important to realize that the organism is prevalent throughout our environment, even in areas that have no livestock – therefore, positive farms should not be condemned or deemed unsafe for livestock production. More details on environmental contamination will be discussed below.

Once a farm is identified as infected the difficult task of controlling the disease must begin. Clear evidence suggests that vaccination of animals, even animals that are already infected, is the best means of control. However, in the United States there are currently no approved vaccines for the control of *Coxiella burnetii*. Vaccines do exist in Europe, however their importation into the US has thus far been unsuccessful since they are produced in a facility overseas that producers other foreign animal disease vaccines. The vaccine currently marketed in Europe requires propagation of the organism in embryonated chicken eggs. With the select agent status of this organism production of this type of vaccine in the United States is unlikely to be cost-efficient. Canada has been successful at importing the vaccine for use in "approved" circumstance but similar efforts have failed in the US. It is important to realize that vaccination does not appear to reduce the number of animals that shed the organism. Instead vaccine appears to result in reduced levels of shedding by each animal and in some cases a reduced risk of adverse pregnancy events.

In humans, Q-fever is generally treated with a prolonged course of oral doxycycline. This is generally effective, so the use of tetracyclines has been recommended for control in animals. Several studies have evaluated the use of parenterally administered oxytetracycline in herds with previous abortions. The authors were unable to demonstrate any impact on bacterial shedding prevalence or quantitative level. [1,2] Unpublished data for the author's laboratory suggest similar results with parenterally administered tulathramycin. Given the apparent lack of efficacy following parenteral tetracycline the use of oral tetracycline is unlikely to be beneficial. The role of antimicrobial therapy during abortion storms is less clear and needs additional research. Published studies have failed to demonstrate any impact of antibiotic therapy on decreasing abortions, however these studies were all performed in flocks/herds that had abortion storms one year prior to the trial. Given that the abortion rate in these flocks in subsequent years is typically much lower than the first year of exposure the statistical power of the studies for impacting abortions was quite low.

ENVIRONMENTAL CONTAMINATION

Feces, urine, milk, vaginal fluids and placentas of infected livestock all contribute to the environmental load of the organism. Particular attention to proper disposal of fetal fluids and placentas is critical in controlling environmental load, as these matrices tend to contain large quantities of the organism. It is recommended that these matrices be collected promptly after excretion and either composted or incinerated. Likewise, aborted fetuses should be promptly collected and handled in a similar manner. Shedding of the organism in the feces of wildlife, rodents, birds and domestic pets occurs frequently. The exact role of these species in maintaining the organism in the farm environment and/or transmitting the organism between farms is generally unclear. Aborting cats are well recognized as a cause of human Q-fever and undoubtedly pose a significant risk for livestock. However, similar to livestock all of these species can shed the organism without clinical evidence of disease.

The small cell variant of *C. burnetii* is highly resistant to killing in the environment. Studies have demonstrated the ability of the organism to survive for several years in dust and feces. Aerosols of infected dust can readily carry the organism up to several kilometers. Data suggest that low environmental moisture (ie. low rainfall) and low levels of ground cover both contribute to increased levels of the organism in aerosols. Spreading contaminated manure is considered a risk factor for increased aerosol formation. Consequently, it is recommended that an effort be made to only spread manure when moist and when there is minimal wind. Composting of manure has been shown to be beneficial in decreasing the load of organism and may play a role in managing environmental contamination. Environmental disinfection with chemicals has been found to be of limited benefit and is generally not utilized.

A recent study of soil samples from different regions of the country demonstrated widespread presence of the organism. In some parts of the country up to 40% of the soil samples were positive.[3] Additionally, a significant number of samples collected from non-livestock environments such as post offices, grocery stores and banks were found to be positive. These findings demonstrate the widespread presence of the organism even when livestock are not present.

WORKER PRECAUTIONS

Because of the zoonotic nature of the organism livestock workers are at increased risk of contracting the disease. It is recommended that workers maintain dedicated barn clothing for working with animals that is removed prior to leaving the facility to minimize transmission into non-livestock areas. Use of a properly fitted (fit-tested) N95 mask provides significant protection against inhalation of the organism and should be considered in high-risk environments. Lambing/kidding facilities and periods when manure is being moved are considered the highest risk periods. During abortion storms, access to these high risk environments should also be restricted.

CONCLUSIONS

The veterinary practitioner has limited options for control of abortion storms associated with *C. burnetii.* The use of parenteral antibiotics has not been demonstrated to result in decreased shedding of the organism or abortion rate. Therefore, control of the organism must be focused on minimizing environmental load by proper disposal of highly contaminated products such as fetal fluids, placenta and aborted fetuses. Efforts to minimize aerosol formation should also be considered.

REFERENCES

[1] Astobiza I, Barandika JF, Hurtado A, Juste RA, Garcia-Perez AL. (2010) The Vet J. 184:172-175. [2] Astobiza I, Barandika JF, Juste RA, Hurtado A, García-Pérez AL. Vet J. 2012 Sep 11. [3] Kersh GJ, Wolfe TM, Fitzpatrick KA, Candee AJ, Oliver LD, Patterson NE, Self JS, Priestley RA, Loftis AD, Massung RF. Appl Environ Microbiol. 2010 Jul;76(13):4469-75.

COXIELLOSIS: LESSONS TO BE LEARNED FROM THE NETHERLANDS OUTBREAK

Paul J. Plummer, DVM, PhD, DACVIM (LAIM)
Ames, IA

INTRODUCTION

Coxiella burnetii is an intracellular gram-negative bacterium that is the causative agent of Coxiellosis in animals and "Q-fever" in humans. Ruminants act as reservoir and replicating hosts of the organism, although almost any species of animals can become infected and shed the organism. In the large majority of these animals the disease process is subclinical, with adverse pregnancy events being the most common clinical manifestation in animals. In recent years, the interest in *C. burnetii* has increased significantly due to the designation of the organism as a "select agent" and class "B" bioterrorism weapon. This designation for an organism that is present on upwards of 90% of the dairy cattle operations in the United States and a significant number of sheep and goat operations results in uncertainty regarding the public health risk of shedding on our livestock operations. The Netherlands recently experienced a significant increase in human Q-fever cases. This outbreak was epidemiologically associated with aerosols from dairy goat operations and ultimately led to the euthanasia of a large number of pregnant does. This session will focus on lessons that we can learn from this historic outbreak.

HISTORY

In 2005 two Dutch dairy goat operations experienced abortion storms in their pregnant goats. Over the following two years additional abortion storms were observed in dairy goat operations throughout the southern part of the country. This was followed by an increase in human Q-fever cases starting in 2007. Ultimately over 4000 human cases were documented. Epidemiologically there was a strong association of these cases with living in close proximity to a dairy goat operation (less than 5 kilometers). An important aspect of this disease outbreak was the very high hospitalization rate associated with this disease outbreak. The outbreak was ultimately stopped by a complete cessation of goat kidding in the impacted portion of the country.

DUTCH DAIRY GOAT INDUSTRY

Significant differences between the Dutch and US dairy goat industries exist and may be important in better understanding the disease outbreak. The Dutch dairy goat industry had undergone significant growth in the early part of the 21st century. This growth occurred mainly in the form of farm expansion by retention of additional animals for breeding. The Dutch herds are very large compared to most herds in the US and are managed in a higher biosecurity fashion than typical of our herds. This rapid growth in the industry through herd expansion likely resulted in a large population of susceptible animals. The US dairy goat industry has also undergone rapid growth over the last decade, however the farms tend to be much smaller than Dutch herds.

Additionally, Dutch dairy goat operations tend to be located in close proximity to villages. This results in a dense population of non-farm associated humans surrounding the dairy goat farms. In some parts of the country it was estimated that over 12% of the local population lived within 5 km of a dairy goat facility. In contrast, many of our dairy goat operations in the US are more isolated from dense population centers. As a consequence, the at risk population of humans that live in close proximity to farms is much smaller in the US than in the Netherlands.

DISEASE CONTROL EFFORTS

When the increase in human illness was linked to the dairy goat operations the first control measure implemented was a mandate that animals be kidded in confined quarters and that manure should not be spread, but instead composted. Despite these efforts the incidence of human illness continued to climb. At that time a vaccination strategy was adopted and a significant number of animals were vaccinated utilizing a commercially available phase I *C. burnetii* vaccine. Prior to the following kidding season the government decided that the risk of having infected animals kid was too great. Therefore, the instituted a bulk milk tank testing policy and elected to prevent kidding of does on infected farms. This was accomplished by euthanizing any pregnant does and placing a temporary breeding ban on all non-pregnant does. This action was successful in stopping the human epidemic but resulted in long lasting and permanent impacts on the dairy goat industry.

LESSONS TO BE LEARNED

The author feels that the following points may be important lessons that we can learn from this outbreak:

- The goat abortion storms primarily occurred prior to the start of the human outbreak and most farms that were involved with human illness had minimal to no ongoing abortion issues. This emphasizes the fact that the organism can be shed and pose zoonotic concerns even in the absence of abortions.

- Individuals in close proximity to infected herds are at increased risk. Fortunately, these populations are likely much smaller in the US than in the Netherlands
- The use of a manure spread banwas insufficient to stop the outbreak. This suggests that the primary source of the infectious aerosols was more likely from the abortions than the manure.
- Bulk tank milk testing many be beneficial for determining herd status but additional work needs to be done to validate the testing protocol
- Strain variations do exist and may be important in virulence.
- Perhaps most importantly, the use of vaccination did result in a quantitative decrease in pathogen shedding. The lack of an approved vaccine for *C. burnetii* in US livestock is a serious deficit and needs to be addressed.

CONCLUSIONS

Although significant differences exist between the dairy goat industries of the Netherlands and the US, we can gain insights into risk and control measures that should be considered if we were to experience a similar outbreak. This outbreak exemplifies the "One Health" role of veterinarians and the zoonotic nature of this pathogen.

REFERENCES

van der Hoek W, Morroy G, Renders NH, Wever PC, Hermans MH, Leenders AC, Schneeberger PM. Adv Exp Med Biol. 2012;984:329-64.

ABOMASAL PROBLEMS OF DAIRY CALVES: HERD-BASED APPROACH TO PROBLEM SOLVING

Sheila M. McGuirk, DVM, MS, PhD, DACVIM (Large Animal)
Madison, WI

INTRODUCTION

At birth, the abomasum of the newborn calf accounts for 70% of the volume of the forestomach compartments. By 8-weeks of age, rumen development has progressed to the point that the abomasum represents only 50% of the volume of the forestomachs. During this period, the calf is functionally a pre-ruminant and is susceptible to a number of pathologic abomasal conditions, including ulcers, bloat, abomasitis and tympany. Whether the conditions are related, represent a continuum of pathologic change, share risk factors or have a common etiology is not completely clear. What is clear is that morbidity and mortality due to abomasal problems is a serious problem on some farms, during certain seasons of the year, and under some nutritional conditions. The purpose of this paper is to develop a herd-based problem solving approach to abomasal problems of young dairy calves.

Bacteria that have been incriminated in abomasal conditions of dairy and beef calves are *Clostridium perfringens* type A, *Clostridium perfringens* type D, *Sarcina ventriculi*, *Lactobacillus sp.*, *Streptococcus sp.*, *E.coli* and *Salmonella enterica* serotype *Typhimurium*.[1-5] Abomasal tympany and abomasitis were reproduced in beef calves using *Clostridium perfringens* type A.[6] Abomasal tympany with mural emphysema was produced in calves by administering a milk replacer, corn starch and glucose mixture.[7] Necrotic enteritis was induced in calves with a non-enterotoxin, non-beta 2 toxin producing *Clostridium perfringens* type A.[8] Despite the ability to experimentally reproduce selective pathologic abomasal conditions of dairy calves, the relevance to the spontaneous disease seen on farms is not clearly established. Abomasal syndromes appear to be the result of a combination of factors, including nutritional substrate, the presence of gas producing organisms that can utilize the substrate for proliferative growth, fermentation and/or toxin elaboration along with an underlying motility disturbance.

CLINICAL SIGNS AND POST MORTEM FINDINGS

Clinical signs vary with the abomasal syndrome that is present and can range from an apparent absence of any clinical signs with abomasal ulcers to depression, anorexia, diarrhea, bloat, rapidly progressive abdominal distension with succussible fluid, colic, cold extremities, dehydration, recumbency, and death with the more severe abomasal syndromes.[9, 10] Abomasal disease can occur in calves of any age, but severe abomasitis, bloat and tympany are most common in dairy calves between 4 and 21 days of age.

Medical treatment of mild abomasal syndromes like ulcers in dairy calves on the farm may be successful if started early and administered consistently in the milk or milk replacer. Antacids or kaolin pectate are frequently chosen for this purpose. A concomitant increase in feeding frequency may be beneficial. The prognosis for calves with abomasitis syndrome is poor as many perceive that less than 50% of calves survive. When treatment is successful, early and aggressive therapy is required. Medical therapy can include initial hypertonic saline administration, followed by intravenous balanced electrolyte crystalloid fluids, anti-inflammatory drugs, and a combination of parenteral and oral penicillin. Many administer therapeutic dose of *Clostridium perfringens* Type C&D antitoxin. Needle decompression of abomasal bloat has met with limited success for calves on the farm. Surgical treatment of abomasal conditions is reserved for selective patients of individual value, but prognosis is guarded.

Post mortem examination is necessary to characterize the abomasal condition that is present and to reveal important risk factors like trichobezoars. Abomasal ulcer type and location is useful information to determine a therapeutic or preventive strategy. In more severe abomasal syndromes, post mortem usually finds an abomasum distended with gas, fluid, clotted milk or milk replacer and blood-tinged fluid. Forestomach and abomasal hyperemia, petechial and ecchymotic hemorrhage is common, along with edema, mucosal necrosis, mural emphysema and tympany and abomasal rupture.

RISK FACTORS

Many risk factors for abomasal syndromes in dairy calves have been proposed. Commonly, the risk factors are associated with feeds and feeding practices, especially those that result in the delivery of an inconsistent product. From the temperature of the liquid feed, to feed ingredients, feed additives, the amount fed, feeding frequency, size of the meal, and the type of feeding equipment, diet changes are not well tolerated by young dairy calves and pose a risk for pathologic abomasal syndromes. Large meals, high total solids content of the milk or milk replacer, high protein milk replacers, increased milk carbohydrate content, fast feeding, increased fiber content of the milk replacer and poorly digested fiber in the milk replacer have been incriminated as factors contributing to pathologic abomasal syndromes of dairy calves. Limited access to water, adverse weather conditions, calf age, and season of the year can be directly or indirectly (via stress) associated with abomasal conditions. Housing conditions may be considered a risk factor when there is over-crowding, inadequate resting space, poor bedding, and ventilation or air quality concerns. Some farms have endemic problems while others have none. A genetic predisposition has been suggested, but not proven. Motility disturbances from nutritional concerns or enteritis may be a significant risk for abomasal syndromes. Tricho- or phytobezoars have been associated with abomasal conditions of calves. Copper and selenium deficiencies have been suggested as additional risk factors, but in an experimental model of abomasitis due to *Clostridium perfringens* type A, copper deficiency was not a risk factor.

Milk replacer diets are more commonly associated with pathologic abomasal conditions of dairy calves than whole milk diets. Stressful calf handling, painful procedures, and frequent vaccinations could also be potential risk factors.

CASE PRESENTATION

A Wisconsin dairy that raises approximately 275 replacement, genomically valuable Holstein heifer calves per year experienced 6 deaths in calves between the ages of 7 and 28 days of age over a course of three weeks beginning late December, 2011. Post mortem findings were abomasitis in all calves with abomasal perforation in 2. At the time, calves were housed in individual Calf-Tel hutches and were being fed 2 quarts of a 26% crude protein,17% fat milk replacer twice daily. Two different calf feeders were employed by the farm to deliver morning and night feeding, respectively. The milk replacer concentration had recently been increased due to severe cold temperatures, but a precise mixing rate of powder to water was not known. A texturized 18% crude protein calf starter was fed once daily, beginning on day 7 of age. Water was provided to calves after the morning feeding once daily, beginning at 7-days of age. Calves were given *Clostridium perfringens* type C&D antitoxin on day 7 and a 7-way *Clostridium* bacterin toxoid at 2-weeks and again at 4-weeks of age.

Milk replacer samples from 3 different calves (first, middle and last calves in the feeding order), collected at each of 5 successive feedings were analyzed for total solids, the results of which are shown in Table 1.

Table 1. Milk replacer percent total solids determined by Brix readings.

	Calf 1	**Calf 2**	**Calf 3**
Day 1 AM	16.4	16.7	19.6
Day 1 PM	18.5	17.4	18.4
Day 2 AM	20.5	19.4	17.8
Day 2 PM	19.4	17.7	22.0
Day 3 AM	17.6	19.6	17.5

Interpretation

A reasonable milk replacer total solids target for winter feeding is 15% (~12 ounces of powder added to 2 quarts of water). The goal is to have the total solids variation between calves within a single feeding and between different feedings be 1% or less. The total solids should never be greater than 18%. These data show significant variability in percent total solids between calves and between successive feedings. In addition to unacceptable inconsistency, the total solids percent is dangerously high (greater than 18%) for 2 or 3 of the 5 feedings represented in this data set. Dietary inconsistency at this level along with high solids concentration of a high protein milk replacer presents a significant risk for abomasitis syndrome in dairy calves.

HERD PROBLEM SOLVING

In the case example above, feeding inconsistencies were evident when milk replacer sample total solids calculations were made. In other dairy herds with abomasitis, ulcers or bloat in preweaned calves, a systematic approach is needed to find the the most important risk factors.

- Use total solids, osmolality, sodium concentration, mixing and feeding temperatures, amount (volume or weight) of additives in each feed mix/batch, delivery method and feeding pattern to find any dietary inconsistencies.
- A digital refractometer can be used to determine Brix readings from sequential samples. For whole milk, add 2 to the Brix reading to predict percent total solids.[11] For milk replacer readings, make a standard curve by plotting known milk replacer solutions on the y-axis (10, 12, 14 and 16% solutions) against the Brix reading on the x-axis. Use the line correlation equation to solve for y, which is the actual total solid percent of the unknown sample. Obtain the milk or milk replacer sample as it is delivered to the calf. Sample at least 3 calves (beginning, middle and end of the feeding delivery process) for 5 successive feedings or some representative sample that includes at least 2 different feeders or 2 different mixes. When the percent total solids is greater than 18 or the variability between feedings or between calves within the same feeding is greater than 1, there is risk for abomasitis and abomasal bloat.
- Osmolality or sodium concentration determination may be particularly useful for milk or milk replacer mixes that have medications or other additives. Most laboratories with a freezing point depression osmometer can measure osmolality of milk or milk replacer samples. The osmolality of milk is approximately 300 mOsm/L. Osmolality of

milk replacers can range from 300 to 600 mOsm/L. If osmolality is inconsistent or over 600 mOsm/L, make sure water (warm temperature in winter) is delivered or available immediately after every milk or milk replacer feeding. Sodium concentration in milk or milk replacer should be less than 120 mEq/L.

- Is there a history of recent feed change, especially a change in the powder to water ratio of a milk replacer mix for winter feeding? Are all the calf feeders trained to mix the milk/milk replacer or balance total solids of a whole milk blend? Is a volume or weight measurement used for milk replacer mixing? A water weight or measurement is required for consistency amongst batches of milk replacer mixes. Consider how milk replacer bags are stored as powder can be compressed when bags are stacked, creating potential variability if volume serves as the basis of the amount of powder put in the milk replacer solution.
- Check the milk/milk replacer delivery system to determine whether the final product is agitated, heated or monitored for consistency and quality from the first calf fed to the last.
- Monitor milk pH at successive feedings as a measure of consistent milk quality fed to calves.
- Is the blended whole milk fed to calves coming from a dynamic population of fresh, treated or other non-saleable milk cows? If so, the percent total solids should be monitored at least daily for consistency and/or to assess the effect of a total solids balancer. Is water added to milk to warm it up or dilute it?
- Determine standard plate counts of successive milk/milk replacer samples to determine the level of bacterial contamination. Of particular interest is growth of *Clostridium perfringens, E.coli, Salmonella species* and other bacterial counts. Collect as-fed milk samples for standard plate count, bulk tank culture, coliform count and/or *E. coli* count. If the level is high at calf level, use a step-wise approach, moving closer and closer to the source of milk or milk replacer to elucidate the point of contamination.
- Assess the liquid feed digestibility by monitoring successive milk/milk replacer samples for sediment, color, and temperature at mixing and at delivery to the calf.
- Observe water delivery after milk feeding. Is it warm in winter? Are water buckets cleaned daily?
- Observe and test colostrum feeding program. Volume, quality, timing, cleanliness, and feeding method of colostrum are relevant when trouble shooting abomasal problems in dairy calves. The use of colostrum replacement and supplement products should be noted. Observe storage, mixing and delivery of the products and the timing of administration. Of particular importance are additives put into maternal colostrum or a colostrum replacement product. The effect of probiotic, prebiotic, nutritional supplements or electrolyte additions to colostrum, milk or milk replacer should be tested by percent total solids, sodium concentration, osmolality measurements and standard plate count.
- Is there evidence of calf stressors like aggressive calving intervention, a large number of assisted deliveries, or numerous pen changes for calves? How often are calves handled for vaccinations? Is there a health screening process that goes beyond appetite or feed consumption? Assess the bedding by evaluating nesting scores, bedding bacterial counts or bedding cultures. Comfortable calves should be sleeping at least 75% of the day.[12]

Systematic assessment of these and other risk factors for abomasal bloat, ulcers and abomasitis will clarify problem issues, resolution of which should successfully resolve most abomasal problems on dairy farms.

SUMMARY AND CONCLUSIONS

Consistency of feeding practices is essential including time, temperature, total solids, additives and delivery. Feed milk at body temperature (100-103° F). Don't cut the end of nipples. Provide water within 30 minutes of milk or milk replacer feeding. Choose a milk replacer total solids mixing rate and be consistent. It needs to be determined whether mixing milk replacer powder by weight is more consistent than mixing by volume, but it is clear that precise measurements of powder and water are needed. Carefully scrutinize the milk replacer ingredients, not only watching crude protein and fat, but fiber content and fiber source. Feed changes should be gradual and not exceed 1% difference in total solids per day. Some advocate increasing the total solids for winter feeding. Make a gradual change and monitor the mix on a regular (daily) basis. Milk storage, mixing, distribution and feeding equipment should be cleaned and sanitized daily with protocols similar to those used for milking parlor equipment. Starter grain that is texturized and at least 19% crude protein should be offered daily and fed to minimal refusal. Consider the effect of any feed additives on digestive tract flora. Support flora with prebiotics and probiotics. Intestinal immune function may be enhanced with immunoglobulins, natural immune products. Minimize stress by comfortable housing, air quality, deep bedding, adequate resting space, and manure handling. Health comes from colostrum, nutrition and environmental management with procedures and protocols for health screening that finds problems early and results in early, effective treatment.

REFERENCES

[1]Roeder BL, et. al. **J Am Vet Med Assoc** 1987;190:12.
[2]Assis RA, et. al. **Arch Med Vet** 2002;2:287.
[3]Carlson SA, et. al. **Vet Micro** 2002;85:233.
[4]Songer JG, et. al. **Anaerobe** 2005;11:290.
[5]Edwards GT, et. al. **Vet Rec** 2008;163:291.
[6]Roeder BL, et. al. **Am J Vet Res** 1988;49:201.
[7]Panciera RJ, et. al. **J Vet Diagn Invest** 2007;19:392.
[8]Morris WE, et. al. **J Vet Diagn Invest** 2011; 23:254.
[9]Welchman D deB, et. al. **Vet Rec** 1987;121:586.
[10]Shoemaker DE, et. al. **Tri-State Dairy Nutr Conf** 2008;85.
[11]Moore DA, et. al. **J Dairy Sci** 2009;92:3503.
[12]Panivivat R, et. al. **J Dairy Sci** 2004;87:3736.

METRITIS: DAIRY HERD-BASED PROBLEM SOLVING

Sheila McGuirk, DVM, MS, PHD, DACVIM (Large Animal)
Madison, WI

INTRODUCTION

Metritis is a prevalent problem in dairy herds, affecting approximately 20% of dairy cows.[1,2] Defined as an acute systemic illness due to bacterial infection of the uterus, it is characterized by a malodorous reddish-brown, watery vaginal discharge in a cow with fever and signs of systemic illness that is in the first 20 days in milk (DIM).[3] The uterine bacterial pathogens *E. coli, Trueperella pyogenes, Prevotella melaninogenicus* and *Fusobacterium necrophorum* are associated with metritis.[4] Metritis has a negative impact on the performance, profitability, and welfare of affected cows. The total metritis cost, including cost due to culling in the first 60 DIM, milk loss, losses due to reproductive issues, treatment cost and lost milk may be as high as $386/case, depending on the antibiotic selected for treatment.[5] With early detection, accurate diagnosis and effective treatment of metritis, the negative outcomes can be minimized.

METRITIS RISK FACTORS

Retained placenta (RP), the retention of fetal membranes for at least 24-hours after parturition, is an important risk factor for metritis and in most dairy herds the metritis incidence is closely related to, but slightly greater than, the incidence of RP. Not all cows with an RP develop metritis and not all cows with metritis have had an RP. The challenge for veterinarians is the herds in which metritis and RP show no relationship.

Dairy herds with increased RP and metritis share many of the same risk factors. These include mechanical factors such as an increased dystocia rate or routine provision of calving assistance, especially in heifers. Abortion, the delivery of twins, or the birth of a stillborn calf are risk factors for RP and metritis. Nutritional risk factors for RP and metritis are mineral (copper, zinc, selenium), iodine, or vitamin A and E deficiencies.[6,7] Clinical and subclinical hypocalcemia are risks for RP and metritis in cattle. Factors that decrease immune function create risk for RP and metritis.

A risk factors unique to metritis is the temperature humidity index (THI), especially for lactation 1 cows.[8] Depression in dry matter intake before calving is expected in cows and in heifers, but when it occurs the week before, rather than the day before calving, there is an increased risk of metritis.[9,10] Ketosis or post-partum negative energy balance puts cows at risk for metritis as does compromised cow and calving pen hygiene. Increased stress around calving may directly or indirectly increase the risk of metritis. Stressful events could include pen changes, overcrowding, inadequate bunk space, diet change, vaccination, noise, and calving intervention.

Risk factors for metritis have been categorized into 2 groups; 1) those that affect the bacterial contamination of the uterus (RP, poor calving pen hygiene, inadequate preparation of the cow for calving assistance, vaginal examination, intrauterine treatments) and; 2) factors that decrease immunity (hypocalcemia, mineral and vitamin deficiencies, body condition score factors, selective mycotoxins, molds and stress).[1]

METRITIS PROBLEM SOLVING IN A HERD

Dairy herds with a metritis crude incidence rate (CIR) greater than 15% may benefit from a focused herd investigation, the goal of which is define and resolve responsible risk factors. A reasonable goal for most dairies is a metritis CIR less than 10%, where CIR is equal to the number of new cases of metritis in a specified time period divided by the number of animals at risk in the same specified time period. Average herd size can be used to represent the number of animals at risk in the specified time period.

A metritis herd-based trouble shooting investigation can be broken down into stages: 1) History taking and review of records; 2) Interview phase to establish current farm definition of a metritis case and standard vaccination and treatment protocols; 3) Observation of cows, people, environment, drug inventory and calving equipment; 4) Fresh cow examinations; 5) Systematic steps to rule-out the metritis risk factors; and 6) Assessment and recommendations.

History and Records Review

A complete history should establish the onset of the metritis problem, perceived metritis prevalence, DIM at onset of metritis, changes in the herd (size, animal source, housing, nutrition, protocols, pen changes, cow groups, management, production, or health), and response to interventions. Herd records of importance are stillbirth rate, twinning rate, calving ease and/or calving assistance scores, RP rate, metritis rate, herd calving patterns, herd distribution for days carrying calf (DCC), lactational patterns for RP and metritis, herd removals by DIM, and other fresh cow health problems.

Interview Data

From on-farm interviews, determine location of far-off and close-up dry cows, calving cows and fresh cow and heifer groups. Inquire about diets, pen sizes, lengths of stay in a pen, and frequency of bedding changes (additions or removal). Note when lactation one heifers return to the farm and in which pen they commingle with lactation two and older cows for the first time. Determine the vaccination schedule for all cows and heifers from 200 DCC through 45 DIM. What criteria are used for moving cows into a calving pen and for providing calving assistance? Establish who works in the calving and fresh cow pens, the frequency of observations or interventions, what training is provided, how often performance is monitored and the role of the veterinarian on the farm. Establish the process, procedures and basis for fresh cow health screening and examinations. Are fresh cow examinations targeted by calving status, DIM, appetite, milk weights, activity monitors, or rectal temperature or are all cows examined? Is there a defined procedure for fresh cow exams and are conditions recorded? Determine the herd definition of metritis and the current treatment protocols for RP and metritis.

Observations

This phase of the herd investigation is used to validate the information obtained from records and interviews. A simple walk through to observe cows, people, feeds and the environment is necessary but targeted participation and discussions during a calving that requires assistance, in a pen after feed delivery, during the fresh cow screening, while exams are being performed, and when treatments are administered is extremely beneficial. Are protocols written in the primary language of the workers? Establish whether the metritis case definition and treatment protocol is appropriate and whether there are consistent, effective methods being used to find cows with metritis. Are there routine or unnecessary vaginal exams or uterine infusions? Determine which examination parameters or additional tests are regularly performed. Do the drug inventory, calving equipment and appearance of the cows reflect current protocols? Assess the calving, exam and treatment supplies and equipment, pens and animals for cleanliness and sanitation procedures. Evaluate metritis case recording in written and computer records. Find out whether protocols and practices that are employed to find, evaluate, treat or prevent metritis have input from or review by the farm veterinarian and whether the veterinarian ever works alongside the calving pen or fresh pen workers. Furthermore, assess whether the farm workers recognize their limits and know when to call the veterinarian for assistance or advice if outcomes are unexpected.

Fresh Cow Examinations

Not all fresh cows will need an examination, but it is important that the farm is examining all cows that had twins, dystocia, calving assistance, have a retained placenta, did not lock up, show gait abnormalities, weakness, lethargy, or muscle tremors. Cows with abnormal activity, milk deviations, or feed intake or appetite change should be added to the exam list. A view of cows from the front of the pen should find additional cows with appetite variation, abnormal nasal or ocular discharge, coughing or labored breathing that need examination. Similarly, a view of fresh cows or heifers from the back of the pen should add cows to the exam list that have abnormal abdominal contours, malodorous vaginal discharge, udder asymmetry or diarrhea. Routine exam parameters should be reviewed but include rectal temperature, fecal consistency and appearance, odor and appearance of vaginal discharge, rumen activity, udder appearance, breathing effort, coughing and nasal discharge. For a metritis herd investigation, establish that the examination parameters are sufficient to find true metritis cases. A cow in the first 20 DIM with foul smelling brownish red, watery vaginal discharge along with one or more accompanying sign from the list below is treated for metritis.

- Fever
- Milk production drop
- Reduced intake or loss of appetite
- Depression

Additional tests for a cow with metritis may be neck skin tent to determine dehydration, a ketosis test and abdominal percussion.

To complete a metritis herd investigation, it is important to establish pen dimensions, cow numbers, stall numbers and bunk space for each of the close-up, calving and fresh cow pens. Determine the prevalence of RP's, metritis, cows with elevated tails, straining, vulvar injury, or vaginal bruising. Rectal examinations may be useful to assess uterine tone in post-partum cows. It is essential to establish that metritis is not a diagnosis loosely applied to cows with a fever and no other accompanying signs or that metritis is not a consequence of routine vaginal examinations or intrauterine infusions. Additional diagnostic testing that may be useful in a metritis herd investigation include bedding bacterial counts and culture, pre-fresh cow non-esterified fatty acid (NEFA) or post-partum cow beta-hydroxy butyric acid (BHBA) concentrations, serum calcium (total or ionized) concentrations and an in-depth nutritional evaluation.

Systematic Review of Metritis Risk Factors

If there are appropriate health records, establish the relationship between RP and metritis rates. For most herds, the metritis rates is approximately 1.2 times the RP rate. In an average dairy herd with an 8% RP rate, the expected metritis rate is 9.6%. If both the RP and metritis rates are high, but the relationship is appropriate, review the risk factors shared by both conditions. If the metritis rate is elevated and much greater than the RP rate, establish the validity of the metritis case definition. Metritis may become a "default diagnosis" for cows in the fresh pen with a fever (or temperature elevation) of unknown origin. This is not unusual on dairies that obtain a rectal temperature in every fresh cow every day, even in the absence of other abnormal signs or calving status risk. If possible determine the frequency and level of calving assistance, recognizing that assistance may be disproportionate to dystocia rates and that farm records may not accurately depict dystocia or calving assistance. For cows moved to a calving pen when feet are showing, note whether cows and heifers are given time to resume labor or whether intervention is immediately provided. With accurate cow records, it may be possible to find an association between calving assistance and metritis, a finding that may implicate procedural problems, equipment sanitation lapses, poor cow preparation or the need for calving assistance training. Regardless of established training or written protocols, routine vaginal exams or intrauterine instillation of pills or medications for fresh cows may be practiced. Intrauterine treatment of metritis is not superior to systemic antibiotic therapy and may do more harm than good if it is practiced routinely.[11]

For dairy herds with increased rates for twinning (> 5%) or stillbirths (> 8%), check for an association between twinning or stillbirths and metritis. Rule out contributing metabolic problems in fresh cows by doing appropriate herd tests for hypocalcemia, hypokalemia, and ketosis. If the body condition score of 10% or more of the fresh cows or heifers is 4 or greater, metritis may be a consequence of hepatic lipidosis and subsequent reduced immune function. Rule out poor pre-fresh or fresh pen hygiene as a cause of metritis. Bacterial counts and cultures can pinpoint problems in pen environments or with calving equipment, but observation of the people working with cows, the equipment used on fresh cows, preparation of fresh cows for calving assistance, and lubrication employed during calving assistance is necessary. Finally, find potential stressors that could put cows at risk for metritis. Consider frequent pen changes, pre-fresh or fresh pen overcrowding, poor feed quality, quantity or ingredient problems as well as excessive or rough cow handling for examinations, vaccinations or other routine procedures performed in the close-up, calving or fresh pens.

Assessment and Recommendations

Specific herd findings will determine the follow up and recommendations made to the farm, but clarification of the metritis definition, with enhanced training of workers in the fresh pen will resolve many problems. Adherence to systemic antibiotic treatment as the first-line therapy for metritis is valuable in some herds with regular involvement of the veterinarian in fresh pen work or clarification of when to call the veterinarian for treatment failures. Assist herds that treat by observation only, that medicate cows with fever and no other accompanying clinical signs, and support adherence to established criteria for metritis. In addition, be engaged and responsive to protocols that need adjustments or change.

Enhance health and reduce stress in close-up and calving heifers and cows by tracking body condition scores, monitoring the diet and strategically timing vaccinations before and after calving. Close-up and calving cows should have a well-formulated diet that cannot be sorted and the feed bunk should never be empty. Bunk space at 30 inches per cow and pens with one stall per cow should be the goals. Avoid hypocalcemia by feeding anionic salts and low potassium alfalfa for dietary cationic-anionic (DCAD) control for lactation 2 and older cows. If the close-up period is short, do not add anionic salts, but routinely administer an absorbable calcium source to lactation 2 and older cows. Control bedding hygiene and comfort by targeting the addition of 25 pounds of new straw per cow per day and having a minimum of 100 sq ft of bedded pack space per cow.

Metritis treatment other than with systemic antibiotics should not be the routine farm protocol. Protocols should be regularly reviewed and have input by the veterinarian in the context of an established veterinarian-client-patient relationship.

REFERENCES

[1]Dubuc J, et.al. **J Dairy Sci** 2010;93:5764.
[2]Galvão KN, et.al. **J Dairy Sci** 2009;92:1532.
[3]Dubuc J. **WCDS Adv Dairy Technol** 2011;23:255.
[4]Sheldon IM, et. al. **Theriogenol** 2006;65:1516.
[5]Overton M, et.al. **Proc Dairy Cattle Reprod Council Conv** 2008;39.
[6]Smart ME, et. al. **Can Vet J** 1981;22:372.
[7]Laven RA, et. al. **Vet Rec** 1996;139:465.
[8]Benzaquen ME, et. al. **J Dairy Sci** 2007;90:2804.
[9]Huzzey JM, et.al. **Pacific NW Anim Nutr Conf Proc** 2006;137.
[10]Urton G, et. al. **Pacific NW Anim Nutr Conf Proc** 2003;172.
[11]Drillich M, et. al. **J Dairy Sci** 2001;84:2010.

FOOD ANIMAL VACCINOLOGY
WHY VACCINES MAY OR MAY NOT WORK

Sherrill A. Fleming, DVM, DACVIM
Mississippi State, MS

INTRODUCTION:

The definition of a vaccine is a preparation of living or inactivated pathogens used as an antigen. The act of vaccination is the administration of a living or inactivated pathogen to induce an immune response. The purpose of vaccination is to stimulate an immune response in the animal to provide protection against infectious disease. It is important to understand the formulation of the vaccine that is being used, the expected protection of the vaccinated animal and limitations of what can be accomplished by vaccination in food animals.

VACCINE LICENSING

The licensing of veterinary vaccines is regulated by the USDA APHIS at the Center for Veterinary Biologics. The Virus-Serum-Toxin Act of 1913 (last amended Dec 23, 1985) covers the preparation and sale of vaccines, importation of vaccines, regulation and inspection of production premises. Information is given for documentation necessary to support the licensing of a particular vaccine. The regulations include the demonstration of host-animal safety and a reasonable expectation of efficacy.

EXPECTED VACCINE RESPONSE

The response of the host-animal is to produce immunity that is cell-mediated and/or humoral. Cell-mediated immunity is provided by T-lymphocytes, macrophages and other leukocytes rather than antibodies. Humoral immunity is mediated by antibodies which are produced by B-lymphocytes.

TYPES OF VACCINES

Classically, there are 6 types of vaccines that may be produced. These include killed (completely inactivated organisms), modified-live (modified organism that is less virulent or attenuated), purified subunit (purified antigens from microbes or inactivated toxins), recombinant vaccines (genes of target organism are inserted into an anttenuated vector organism), DNA vaccines (pathogen DNA is inserted into a plasmid that is inserted into organism that produces translated proteins) and gene deletion (attenuates organism by removing specific genes) vaccines. There some vaccines that retain some of their virulence and resemble clinical disease such as the contagious echthyma vaccine.

IDEAL VACCINE/IDEAL DISEASE

A vaccine that would be highly desirable is one that induces immunity to an infectious disease by stimulating long-lived effector and memory cells. The characteristics of a disease pathogen that is relatively easy to effectively vaccinate the individual are stable antigens with little variation, require a larger infective dose and the infection is limited to one host species. Unfortunately, many current vaccines induce excellent humoral immunity, but we are uncertain the level of cell-mediated immunity produced due to challenges in identifying cell-mediated immunity. Many of our most challenging food animal diseases involve a reservoir host that allow for reintroduction of the disease, with Leptospirosis being an example.

The targets of vaccines can be any type of infectious disease and includes bacteria, viruses, parasites, fungi/yeasts and venoms. Historically, bacteria, viruses and to some extent venoms have been much more successfully addressed in vaccine development. Parasites and fungi/yeasts have proved to be much more challenging in developing a protective vaccine. Often experimental vaccines appear to achieve the desired response only to lose efficacy when attempting to mass produce the vaccine.

PASSIVE IMMUNIZATION

The transfer of pre-formed antibodies to an individual transfers immediate humoral immunity to the individual. However, once the antibodies are gone, there is no ability of the individual to produce more antibodies of their own. The absorption of immunoglobins from the colostrum that a neonate ingests at birth is the most common example of passive immunization. However, there are a variety of specific antibody preparations that address diseases such as tetanus, botulism, enteroxemia and pit viper venom. In general these preparations have relatively short shelf-life and maybe difficult to keep in stock.

KILLED VERSUS MODIFIED-LIVE VACCINES

The goal of producing immune memory with the ability to produce sufficient antibodies for disease protection has a two phase aspect. The initial exposure to an antigen causes the immune system to respond by selecting and multiplying a population of B-lymphocytes to produce antibodies that peaks in approximately 2 weeks. If no further antigenic stimulation occurs, this initial priming response will fall back to pre-exposure levels and may not have a strong memory response. This is particularly important in killed vaccines as a second vaccination is required within 2-4 weeks of the initial vaccination. The second booster produces a stronger anamnestic response which contributes to the development of immune memory cells which provide longer term protection. The advantages of killed vaccines include their avirulence, stability in storage and safety in immunosuppressed animals. However, they may require the addition of adjuvants to increase the time period that antigen stimulates the immune system and to provide a stronger response from the immune system. This unfortunately can also lead to an increase in the local adverse effects of the killed vaccines. Killed vaccines always require a minimum of 2 initial doses and may require annual boosters.

Modified-live vaccines introduce a vaccine that subclinically "infects" the individual. The organism is able to produce antigen and stimulates the immune system similar to the infectious disease. The initial antigen dose is lower and no booster dose is required. Therefore, both a cell-mediated immunity and a longer-lived humoral response are obtained. The disadvantages of modified-live vaccine is that they may be accidentally inactivated (disinfectants, sunlight), have a risk of reversion to virulence, more risk of contamination of unwanted viruses from cell culture and may produce disease in immunosuppressed hosts.

ADJUVANTS

Adjuvants are substances that enhance the immunogenicity of vaccines and are often included with killed vaccines or purified/subunit antigens. The purpose is to increase the duration and amount of immunostimulation to more closely resemble modified-live vaccines. As a result they cause an increased inflammatory reaction. There are a wide range of possible adjuvants. The most common are oil emulsions and metal salts such as aluminum hydroxide. As knowledge of the immune system expands the use of specific arms of the immune response such as toll-like receptor ligands and cytokines are being included as adjuvants. The negative effects of adjuvants have been excessive inflammation resulting in abscesses or masses and possible tumor induction. Adjuvants also increase the likelihood of allergic or anaphylactic responses. There is increased risk to the administrator of the vaccine from accidentally self-injecting an irritating substance.

MATERNAL INTERFERENCE WITH VACCINATION

Passive immunity obtained from the dam through the placenta or through colostrum is important in protecting the newborn from challenges in their immediate environment. This protection declines over 6-16 weeks. The challenge for the veterinarian is to introduce vaccines to the newborn inducing their own immunity before they become susceptible to disease. The gap between waning passive immunity and effective self-produced immunity is called the immunity gap. Multiple vaccinations are given over this period in an attempt to overcome this potential gap. There is some evidence that vaccination can overcome maternal immunity. For example, lambs have shown an antibody response to enterotoxemia vaccine at 4 weeks of age despite having a maternal antibody titer. Recominant vaccines are thought to allow an earlier, stronger response in neonates.

CORE VERSUS NON-CORE VACCINES

Core vaccines are vaccines that all animals within a species or group should receive. Non-core vaccines are those that should be limited to certain individuals or groups. Non-core vaccines are those that are for diseases that are limited to a geographical location, present a less severe threat to some animals, have a low benefit/risk ratio or lack sufficient scientific information to evaluate safety and/or efficacy.

ECONOMIC CONSIDERATIONS

The difficulty of efficiently providing vaccination to some groups of food animal is a large problem. Initial vaccinations of beef calves are a particular tactical problem. The risk of collecting and working cattle with young calves may outweigh the benefit of getting disease protection. The cost of administering the vaccine is much greater than the cost of the vaccine due to the labor requirement.

UNKNOWNS

For a vaccine to be licensed there is good information on dosing and expected response to vaccine. However, there is little published in the area of the optimal revaccination intervals. There is currently a push to increase the vaccination intervals in small animals. We have many multivalent products available and several vaccines are often used simultaneously. Do we have good evidence that these multitudinous antigen loads are having effective immune responses?

REFERENCES

Vaccination Principles (Oversight: COBTA; Approved by the AVMA Executive Board April 2001; revised April 2007) . http://www.avma.org/issues/vaccination/vaccination.asp. http://www.aphis.usda.gov/animal_health/vet_biologics/.

THIRD COMPARTMENT ULCERS IN CAMELIDS

Geof W Smith, DVM, PhD, DACVIM
Raleigh, NC

Third compartment (gastric) ulcers are a common cause of morbidity and mortality in New World camelids of all ages.[1-3] Although true prevalence studies are lacking, in a review of 87 camelid cases handled by the Oregon State University Veterinary Diagnostic Lab, perforating ulcers of the third compartment were identified as the cause of death in approximately 5% of the cases and were a contributing cause of death in over 20% of the cases.[2] Ulcers also appear to be present in a significant number of camelids presented for colic. A study done in 1993 by the International Llama Association examined 271 deaths of camelids where necropsy data was available. Perforating third compartment ulcers were found as the primary cause of death in 6% of the cases reported. Ulcers occur in both crias and adult animals and there is some suggestion that breeding age females have a higher occurrence than males.

Third compartment ulcers are typically limited to the distal 20% of the third compartment and are also occasionally found in the duodenum. Although the exact cause and pathogenesis of gastric ulceration in camelids is not completely understood, risk factors appear to include stress, high-grain diets, nonsteroidal anti-inflammatory drugs, the presence of other diseases, and reflux of bile from the duodenum into the stomach.[1-3] Ulcers are common in camelids with other diseases (such as orthopedic patients, dog bites, early weaned crias, recently purchased/moved animals, etc). Although the specific role of these factors remains unclear, they likely lead to a decrease in third compartment pH, gastric emptying, and mucosal protection.[1] Bile acid concentrations are also elevated in camelids with ulcers indicating some degree of duodenogastric reflux.[1] It is unclear whether bile acids serve as a cause of third compartment ulcers or if ulcers cause gastric atony resulting in elevated bile acids.

Clinical Signs and Diagnosis

Clinical signs tend to be vague, but can include colic, anorexia, decreased fecal output, bruxism and depression. The most common clinical presentation is usually depression and/or anorexia. Third compartment ulcers are difficult or impossible to diagnose ante-mortem. Contrast medium will not pass through the first compartment fast enough for radiographic studies to be useful and endoscopy of the third compartment is not possible. Abdominocentesis might be helpful and usually shows an elevated protein and leukocyte concentration. Fecal occult blood tests are almost always negative.

Treatment

Since clinical signs are vague and a definitive diagnosis is difficult, many "stressed" or "sick" camelids are given various drugs to treat and/or prevent ulcers. Accepted treatments of gastric ulcers in other species include acid neutralization, protection of the damaged mucosa, and prevention of acid secretion.[2] Pharmacological agents commonly used to suppress gastric acid production include histamine type 2 (H_2) receptor antagonists, prostaglandin analogues, and proton pump inhibitors. The H_2-receptor antagonists decrease the production of gastric acid by parietal cells and include drugs such as cimetidine, ranitidine, nizatidine, and famotidine.[4] Although they are the most commonly used drugs to decrease gastric acid production in many species, they have been shown to be ineffective in camelids.[4,5] In one study, llamas were given cimetidine at a dose of 3.3 mg/kg, IM, every 12h and third compartment pH was measured at 30 minute intervals on days 1 and 3.[5] As compared to the baseline value (1.43 $\pm$ 0.06), gastric pH was slightly increased 30 minutes after cimetidine administration, but was not increased at any subsequent time points. The study concluded that the usefulness of cimetidine for the treatment of gastric ulcers in llamas was limited. Another study measured third compartment acid production in llamas before and after ranitidine was given at a dose of 1.5 mg/kg, IV.[4] Third compartment content samples were collected at 15 minute intervals from a gastrostomy site for 2 hours preceding and 12 hours following drug administration. Ranitidine produced a small, transient (<1 hour) decline in acid production that the authors concluded would not be of therapeutic benefit for treating gastric ulcers in camelids. The plasma concentrations and half-life values reported in this study were similar to what has been reported in humans and the blood ranitidine concentrations were in the therapeutic range that has been reported for humans.[4,19] Therefore the short duration of gastric acid suppression was not thought to be caused by an inadequate dose or low plasma concentrations of ranitidine. The authors of the study suggested that camelids likely have few H_2 receptors in their third compartment or that the receptors have a low affinity for this class of drugs.[4]

The prostaglandin-E analogue, misoprostol, has been shown to effectively reduce gastric acid production and increase mucus and bicarbonate secretion in humans.[6] Although IV administration of misoprostol produced a significant (>50%) reduction in gastric acid production in llamas for greater than 6 hours, a number of adverse effects including collapse, recumbency, disorientation, and regurgitation were frequently observed.[2,4] Because of the potential for toxicity in camelids, misoprostol is not recommended for use.[2,4]

Another therapeutic approach to treating gastric ulceration is inhibition of the proton pump (H^+-K^+-ATPase enzyme) of the gastric parietal cell that is responsible for acid production in the stomach. The drug omeprazole has previously been shown to significantly reduce third compartment pH in llamas with intravenous doses of 0.4 and 0.8 mg/kg.[4] The IV administration of omeprazole has been considered the optimal treatment in camelids showing clinical signs of gastric ulcers, or in stressed animals at high risk for developing ulcers. However the intravenous formulation of omeprazole is no longer available in the United States and intravenous administration is not practical in camelids for on-farm use since it generally requires the assistance of a veterinarian. Multiple oral formulations of omeprazole are widely available in North America, however the results of studies performed in llamas indicated the bioavailability of the drug is very low, and drug concentrations were well below levels previously shown to be associated with a therapeutic reduction in third compartment acid production.[4,7] In a pharmacokinetic study, llamas were given oral omeprazole (GastroGard®) at 4 mg/kg for 6 days. Results of this study demonstrated that the bioavailability was extremely low and blood concentrations were well below levels previously associated with gastric acid suppression.[7] Even at higher doses (8 mg/kg and 12 mg/kg), the C_{max}, T_{max}, and area under the curve values for omeprazole were not increased as compared to the 4 mg/kg dose. The results of this study suggested that orally administered omeprazole is likely degraded in the first compartment of camelids. A subsequent study gave oral omeprazole to alpacas that had been surgically fitted with third compartment cannulae to allow for measurement of gastric pH.[8] Alpacas were given omeprazole at a dose of 4 mg/kg following a 24 hour fasting period and during a normal 5 day ad libitum feeding period. Third compartment pH was monitored regularly over a 24 hour period. There was no significant difference in gastric pH at any time point following the administration of oral omeprazole as compared to baseline pH.

Pantoprazole is a proton pump inhibitor that is similar in action to omeprazole but more widely available. It is an irreversible proton pump inhibitor and is marketed in more than 60 countries for therapy of reflux esophagitis and gastric ulcers in humans. Pantoprazole is similar to omeprazole in that once it is absorbed into the blood, the drug accumulates in the parietal cells of the gastric mucosa and suppresses gastric acid secretion by reacting with the H^+-K^+-ATPase (proton pump).[9] In a recent study, alpacas were fitted with a third compartment cannula for measuring gastric pH.[10] Following recovery, alpacas received 1 mg/kg pantoprazole IV, every 24h for 3 days or 2 mg/kg SC every 24h for 3 days. Alpacas received both IV and SC pantoprazole, with a minimum of 3 weeks between treatments. Third compartment pH was recorded and plasma samples were taken for pharmacokinetic analysis. The results demonstrated that pantoprazole induced a slow but sustained increase in third compartment pH when given by both the IV and SC routes. Third compartment pH was significantly increased as compared to baseline values (1.81 ± 0.7; mean $\pm$ SD) at 24 (2.47 ± 0.8), 48 (3.53 ± 1.0), and 72 hours (4.03 ± 1.3) following daily IV administration of pantoprazole. Third compartment pH increased from 1.73 ± 0.6 at baseline to 3.05 ± 1.1, 4.02 ± 1.4, and 3.61 ± 1.6 at 24, 48, and 72 hours following SC administration. Based on the results of this study, pantoprazole represents a safe and effective drug for increasing third compartment pH in camelids. Either IV or SC administration is likely an effective treatment for gastric ulcers. Although not specifically evaluated in the study noted above, based on the high bioavailability of pantoprazole as well as clinical experience, lower doses (1 to 1.5 mg/kg) SC seem to be efficacious in reducing gastric acid production and would lower the daily treatment cost (as compared to 2 mg/kg). Pantoprazole is readily available from several veterinary drug distributors and represents an effective treatment for reducing gastric acid production in camelids.

REFERENCES

1. Cebra CK, et al. Bile acids in gastric fluids from llamas and alpacas with an without ulcers. J Vet Int Med 2003;17:567-570.
2. Smith BB, et al.. Third compartment ulcers in the llama. Vet Clin Food Anim 1994;10:319-330.
3. Smith JA. Noninfectious diseases, metabolic diseases, toxicities, and neoplastic diseases of South American camelids. Vet Clin Small Anim 1989;5:101-144.
4. Drew ML, et al. Effect of flunixin meglumine and cimetidine hydrochloride on the pH in the third compartment of the stomach of llamas. J Am Vet Med Assoc 1992;201:1559-1563.
5. Christensen JM, et al. Pharmacokinetics and pharmacodynamics of antiulcer agents in llama. J Vet Pharmacol Therap 2001;24:23-33.
6. Walt RP. Misoprostol for the treatment of peptic ulcer and antiinflammatory-drug-induced gastroduodenal ulceration. N Engl J Med 1992; 327:1575-1580.
7. Poulsen KP, et al. Pharmacokinetics of oral omeprazole in llamas. J Vet Pharmacol Therap 2005;28:539-543.
8. Johnson JL, et al. The effects of oral omeprazole on third compartment pH in healthy male alpacas. *J Vet Int Med* 2007;21:584 (abstract).
9. Fitton A, Wiseman L. Pantoprazole: review of its pharmacological properties and therapeutic uses in acid-related disorders. Drugs 1996;51:460-482.
10. Smith GW, et al. Efficacy and pharmacokinetics of pantoprazole in alpacas. J Vet Int Med 2010; 24:949-955

USE OF THORACIC ULTRASONOGRAPHY FOR THE EVALUATION OF LUNG PATHOLOGY IN BEEF CATTLE WITH ACUTE BOVINE RESPIRATORY DISEASE.

R. Rademacher [1], M. Edmonds[2], H. Tripp[3], EG Johnson[2], S. Buczinski.[4] College of Veterinary Medicine, Oregon State University, Corvallis, OR. [2]Johnson Research LLC, Parma, ID. [3]Oklahoma State University, Stillwater, OK. [4]Faculté de Médecine Vétérinaire, Université de Montréal, St-Hyacinthe, QC, Canada.

Little information is currently available concerning the use of ultrasonography (US) for the evaluation of bovine respiratory disease (BRD) in feedlot cattle. We hypothesize that US may be useful to predict death in BRD affected calves and aid in individual case management. This case-control study was performed on 232 high-risk steers. The steers were then evaluated daily for BRD clinical signs and included in the study if pulled by the penchecker and using based on depression and respiratory scores. Final enrolment selection was made by temperature (>104F) and pathogen load as determined by bronchoalveolar lavage (BAL) culture (>10^5 CFU *M. haemolytica*). For every 2 BRD cases, one healthy control steer was included (depression and respiratory score 0, temp <104F, CFU <10^5).

A complete thoracic US was performed using a linear 8.5MHz probe at the day of first pull (Day 0) in cases and controls and analyzed off-line for the maximal depth of consolidation (DEPTH), the area of consolidated lung (AREA), the total consolidated area (TCA), the presence and number of comet-tail artifacts (COMET) as well as maximal depth of pleural effusion (EFF). The outcome of each case was noted as survivor (S) or non-survivor (NS) at the end of the study. Wilcoxon rank sum analysis was performed in the different US parameters and P <0.05 was considered as significant. Receiver operating characteristic (ROC) curves were drawn for each significant US parameter.

Twenty-nine cases and 15 control steers were included in the study. DEPTH, AREA and TCA were significantly higher in NS than in S. The area under ROC curve (95% CI) of DEPTH, AREA and TCA were 0.865 (0.728-0.949), 0.875 (0.740-0.955) and 0.885 (0.753-0.962). Although conducted in a research station on a small number of animals, this study demonstrated that US may be useful for BRD management in feedlot calves.

ANTIMICROBIAL RESISTANCE SURVEILLANCE IN BEEF FEEDLOT CATTLE – IS THERE ONE ANSWER? KM Benedict[1], SP Gow[2], RJ Reid-Smith[3], NR Noyes[1], CW Booker[4], TA McAllister[5], SJ Hannon[4], CL Waldner[6], SL Checkley[7], PS Morley[1]. 1. College of Veterinary Medicine and Biomedical Sciences, Colorado State University, Fort Collins, CO 2. Laboratory for Foodborne Zoonoses, Public Health Agency of Canada, University of Saskatchewan, Saskatoon, SK. 3. Laboratory of Foodborne Zoonoses, Public Health Agency of Canada, Guelph, ON 4. Feedlot Health Management Services, Okotoks, AB 5. Lethbridge Research Center, University of Lethbridge, Lethbridge, AB 6. Western College of Veterinary Medicine, University of Saskatchewan, Saskatoon, SK 7. Faculty of Veterinary Medicine, University of Calgary, Calgary, AB.

The overarching purpose of this study was to explore development of an antimicrobial surveillance system for use in commercial feedlot operations. To this end, various sampling, testing and analytical methodologies were compared with the goal of determining which methods represented the best combination of accuracy, cost-effectiveness and practicality.

5,498 cattle in 310 pens from four Alberta feedlots were randomly enrolled in the three-year longitudinal study. Individual rectal and composite pen floor fecal samples were collected multiple times from cattle and pens, respectively, and cultured for non-specific type *Escherichia coli*. Nasopharyngeal swabs were taken from cattle and cultured for *Mannheimia haemolytica*. Isolates of both organisms were tested for antimicrobial resistance (AMR) using broth microdilution (BM) and disk diffusion (DD). Antimicrobial use (AMU) data was recorded for all enrolled individuals and pens. Associations between AMU and AMR were identified using generalized estimating equations (GEE), GEE with alternating logistic regression (ALR), generalized linear mixed modeling (GLMM) and multivariate analysis. Performance of BM and DD was compared using latent class analysis.

There were substantial differences in model results when comparing AMU-AMR associations in individuals versus pens. Modeling methodology substantially impacted model results, particularly when modeling pen-level associations. Resistance patterns differed when comparing *E. coli* and *M. haemolytica* in the same population. Bayesian analysis indicated that disk diffusion and broth microdilution testing methods generally exhibited comparable accuracies, however this was somewhat dependent on the organism and drug being tested.

These results emphasize the complexities surrounding AMU-AMR research and surveillance, and suggest that there is no "one answer". Consequently, research and surveillance programs must focus on specifics – specific organisms, drugs, animal and human populations, and specific sampling, microbiological, and analytical methods. The specifics, in turn, would ideally be dictated by the concerns and circumstances of the producers, governmental agencies, researchers and public health officials involved in implementing surveillance systems.

CRYPTOCOCCOSIS: AN EMERGING DISEASE IN THE PACIFIC NORTHWEST

Rob J. Bildfell, DVM, DACVP
Corvallis, Oregon

INTRODUCTION

Cryptococcosis is an important fungal disease of humans and animals. Much of the recent notoriety of this disease has coincided with the AIDS pandemic; it has been estimated that this mycosis is responsible for over 600,000 mortalities annually, mainly affecting AIDS patients and other immunocompromised individuals.[1] In the late 1990s, veterinarians identified an unusual number of cases of cryptococcosis in animals on Vancouver Island, British Columbia, Canada and subsequent investigations determined a similar spike in human cases.[2] The majority of these cases were due to infection with *C. gattii*. These fungi were further subclassified using Multi Locus Sequence Typing (MLST), revealing most to be Variety Gattii type IIa, or less commonly, VGIIb. These were subsequently referred to as the major (VGIIa) or minor (VGIIb) strains. From a taxonomic perspective, there are four major *C. gattii* genotypes (I –IV) and each may be further subtyped with designations such as *a,b,c* assigned according to the various combinations of genes present. It should be noted that *C. gattii* has only been recognized as a species for the past ten years; prior to 2002 all of the organisms in Table 1 were merely classified as variants of *Cryptococcus neoformans*.[3]

Table 1: Nomenclature of *Cryptococcus neoformans-gattii* complex

Species and variety	**Serotype**	**Molecular type**	Subtype
C. neoformans var. *grubii*	A	VNI VNII	
C. neoformans var. *neoformans*	D	VNIV	
C. neoformans hybrid	AD	VNIII	
C. gattii	 B C	VGI VGII VGIII VGIV	 VGIIa VGIIb VGIIc

For the past dozen years this outbreak has been perceived to expand in the Pacific Northwest, with isolates of environmental, veterinary, and human origin recovered in mainland British Columbia, Washington, California and Oregon.[1] This was surprising as *C. gattii* had usually been associated with tropical and subtropical climates rather than temperate environments. Initial patient demographics from the Vancouver Island outbreak suggested that the *C. gatti* isolates had increased virulence; in contrast to the pattern seen with *C. neoformans*, many of the human patients lacked any history of predisposing disease problems or immune system compromise. There is now also experimental evidence for increased virulence of some of these Pacific Northwest *C. gattii* isolates, including a VGIIc strain that appears to unique to Oregon.[1] Public health concerns regarding the disease have been elevated to the point that both human and non-human cases of *C. gattii* infection became notifiable/reportable in Washington (Jan 2011), Oregon (2007) and BC (June 2003). The Center for Disease Control has also maintained a national *Cryptococcus* surveillance program since 2008.

Clinical disease in humans

Although both fungi are capable of inducing the same range of clinical problems, epidemiologists have noted that *C. gattii* tends to target the respiratory system more frequently, and is more prone to the formation of solitary or multiple cryptococcomas, when compared to *C. neoformans* cases.[4] Most of the BC cases initially presented with respiratory clinical signs and there is also experimental support of an increased respiratory vs CNS tropism for *C. gattii*, at least in some strains of mice.[5] Commonly reported *C. gattii*-associated symptoms include cough, headache, dyspnea, nausea, and fever with clinical work up identifying problems such as pneumonia (57% of cases), meningitis, encephalitis, and space-occupying/solid lesions in lung or brain (cryptococcomas).[6] Reports vary regarding the percentage of patients suffering concurrent immune-compromising or predisposing conditions but concur that such factors appear to be less important with *C. gattii* infections than has been recorded with *C. neoformans* cases.[4]

Clinical disease in animals

Cats remain the veterinary species most likely to develop cryptococcosis in North America and also the species most frequently affected by *C. gattii* in the Pacific Northwest. However, recent publications suggest that infections in superficial locations such as the skin and nasal region could reflect virulence attributes associated with different *Cryptococcus* lineages.[7] If this is confirmed, then isolation and genotypic identification of the fungus may prove critical in designing treatment protocols. Dogs are more likely to develop nasal/skin and CNS manifestations than pneumonia and are the second most likely species to develop *C. gattii* infection in the Pacific Northwest. For both dogs and cats, exposure to outdoor environments is a significant risk factor for development of cryptococcosis, but not a requirement.

A variety of factors have made it difficult to track trends in the incidence of small animal cryptococcosis in Oregon via diagnoses at the Oregon State University Veterinary Diagnostic Laboratory (OSUVDL) but it appears that these diagnoses are increasing. More interestingly, of the cases in which a speciation result could be achieved, the vast majority have been due to *C. gattii*.

Large animal species are less likely to develop cryptococcosis in the Pacific Northwest, although the disease is reported with some frequency in horses in Australia, often as a result of *C. gattii* infection.[8] With the exception of New World camelid species, the diagnosis remains rare in large animal specimens submitted to the OSUVDL but our records do suggest a mild increase in case frequency over the past decade. Once again, for almost all of the cases in which some degree of fungal speciation has been performed, the pathogen has been *C. gattii*.

Wildlife and other semi-domestic species in the Pacific Northwest have also developed cryptococcosis and in most instances this has proven to be due to *C. gattii*, when a speciation test was performed. Numerous Dall's and harbor porpoises, a dolphin, several elk, and a few birds (two psittacines and one rhea) are believed to have been infected with *C. gattii*. Preliminary PCR tests on archived elk (*Cervus elaphus roosevelti*) tissue at the OSUVDL indicate that this animal died of *C. gattii* infection in 1997, just before the fungus was first gaining scientific attention on Vancouver Island.

Life cycle and virulence factors of the fungus

Cryptococcal infections are unlikely to be the result of animal to animal or animal to human transmission; almost all infections are believed to be initiated by spores produced in the environment. The 10 - 60 µm diameter yeast form of *Cryptococcus* has a rather distinctive microscopic appearance and these are the round to oval organisms found in vertebrate tissues. However, this *dimorphic* fungus can assume an entirely different appearance in the environment, completely independent of a vertebrate host. It is these environmental filamentous hyphae which produce basidiospores, the 2 – 3 µm structures thought to be the infectious propagules for most clinical cases. These filamentous forms have mating types (**a** and **α**) and under the proper conditions sexual reproduction can occur (in addition to asexual budding of yeast or hyphal forms). This may enhance the exchange of genetic material (including virulence factors) and promote genotypic diversity.[9] Studies suggest that *C. gattii* prefers an environmental niche with plant/tree detritus, and nutrient- poor, acidic soil. The fungus is susceptible to freezing but can survive in both fresh water and sea water. In contrast, *C. neoformans* has traditionally been associated with sites rich in avian and bat guano.[9]

Recognized virulence factors include melanin production, temperature sensitivity, capsule formation and the formation of titan (large) cells. The first is believed to protect the organism against free radicals and other oxidative stressors.[9] This protection has implications in both the extracellular and intracellular compartments of the host. Similarly, components of the capsule appear to help the yeast resist phagocytosis as well as enhance survival if phagocytosis does occur. The VGIIa and VGIIc strains of the Pacific Northwest tend to exhibit increased proliferation rates within murine phagocytes in comparison to reference strains and are considered hypervirulent (this was further supported by live animal studies).[1] *C. gattii* strains may have capsular elements/secretions that inhibit neutrophil migration, altering the immune response and inhibiting effective Th1 immunity. [10]

Host response

The typical incubation period is estimated to be between 2 months to over a year and infective organisms have been also recovered from the nasal passages of a few asymptomatic animals. As infections are sometimes localized and the immune/inflammatory response tends to be muted, any alterations in CBC parameters are inconsistent and unreliable. Indeed, local inflammation at a site of infection may be restricted to small numbers of macrophages and a few multinucleate giant cells, with the bulk of the lesion formed by masses of yeast organisms (cryptococcoma).

Diagnosis

Options for imaging internal lesions are highly variable since the infection may localize in a variety of tissues. Once a site has been identified, a definitive diagnosis of cryptococcosis can often be obtained by cytologic or histologic examination of the affected tissue. CSF analysis is particularly rewarding for cryptococcosis with CNS involvement as there is often a mild elevation in protein, a mild pleocytosis (usually neutrophilic, sometimes eosinophilic, plus mononuclear cells) and yeast organisms are sometimes visible. Organisms which might be confused with *Cryptococcus* on microscopic examination include *Blastomyces sp.* (minimal capsule, broad based budding), *Histoplasma sp.* (more uniform size, no capsule) and *Sporothrix sp.* (oval, smaller, no capsule).

Other popular and useful tests involve the detection of cryptococcal antigen in serum, plasma or CSF samples using immunoassays or latex agglutination based tests such as the CALAS® (Cryptococcal Latex Agglutination Test System,

Meridian). An ELISA, Premier™, manufactured by Meridian and used primarily on serum, is a widely used test and the manufacturers claim 90 – 100% sensitivity and 97 -100% specificity. A newer Lateral Flow Immunoassay, CrAg LFA, by IMMY ® may have superior sensitivity. It should be noted that the CALAS® test was developed *for C. neoformans* and is somewhat less sensitive for *C. gattii* detection. False negatives may occur with cases involving cryptococcomas or localized skin/nasal lesions. False positives may occur if titers are in the low range (<1:200) and should be confirmed by other diagnostic modalities. Tracking antigen titers is also useful in following response to treatment. It should be noted that titers often rise initially, reflecting degradation of organisms killed by antifungal treatment.

Although a diagnosis of cryptococcosis can easily be achieved without isolation of the fungus, it is recommended that this test be performed; as we accumulate more knowledge about the different species and strains, treatment protocols may be tailored to more effectively target specific strains while minimizing host side effects.

Treatment

If diagnosed, then early aggressive treatment of cryptococcosis may lead to a favorable outcome in small animal patients but there is scant documentation of successful treatment regimes in large animal species. When possible, surgical excision of discrete lesions should be pursued. The American Medical Association has published treatment protocols and does not make a distinction between therapies for *C. gattii* vs *C. neoformans* cases.[11] However, research on some the isolates from the Pacific Northwest indicate that the Minimum Inhibitory Concentration (MIC) required for fluconazole is higher for most VGII isolates than for other isolates and another study also found fluconazole to be less effective than itraconazole, ketaconazole, and amphotercin.[12,13] Voriconazole and posaconazole are used in human medicine but have not been employed extensively in animals. For both human and veterinary anti-cryptococcal protocols a central tenet is long term therapy; ie. months to years of treatment. Animals that have converted to an antigen test seronegative state should be retested one month after the cessation of therapy to confirm seronegative status.

CONCLUSION

Cryptococcus gattii has likely been present in the Pacific Northwest for at least several decades, but in the past ten years it has become a more significant veterinary and public health problem. This could the consequence of the development of new, hypervirulent strains resulting from genetic recombination events and thus a reflection of a comparatively recent genetic event. Alternatively, such strains may have been endemic for a long time and their emergence as a more significant health threat attributable to a greater number and wider distribution of organisms influenced by environmental changes, or perhaps human-related dispersal mechanisms. The monitoring and characterization of these fungi is important to provide our patients with better health care, promote public health, and to better understand our changing environment.

REFERENCES

[1] Byrnes III EJ, et al. **Plos Pathog** 2010; 6(4):e100850. [2] MacDougall L, et al.**Emerg Infect Dis** 2007;13(1): 42. [3] Kwon-Chung K, et al. **Taxon** 2002;51:804. [4] Brynes III EJ et al. **Microbes Infect** 2011;13(11):895.[5] Perfect J. **mBio** 2012;3(4):e00165. [6] DeBess E, et al. **MMWR**; 2010 ;59: 865.[7] Trivedi S, et al. **J Fel Med Surg** 2011;13:163. [8]McGill S, et al. **Med Mycolog** 2009; 47(6):625. [9]Lin X et al.**Ann Rev Microbiol** 2006; 60:69. [10]Brown G. **Annu Rev Immunol** 2011;29:1. [11] Perfect J, et al.**Clin Infect Dis** 2010;20:291. [12]Chong H, et al. **J Clin Microbiol** 2010;48:4115. [32] Iqbal N,et al. **J Clin Microbiol** 2010; 48:539.

CRYPTOCOCCOSIS AND OTHER FUNGAL DISEASES IN CAMELIDS

Rob J. Bildfell DVM, DACVP
Corvallis, Oregon

INTRODUCTION

For most domestic animal species, fungal organisms lack important economic significance and are a less frequent cause of health concerns than most other classes of infectious agents. However, the recognition of diseases such as chytridiomycosis (*Batrachochytrium dendrobatidis*), white nose syndrome (*Geomyces destructans*) and hive collapse syndrome (*Nosema ceranae*) remind us that these pathogens do have the capacity to cause devastating infections. Fortunately, fungal infections are also relatively uncommon in New World Camelids (NWC), although these hosts may have an increased susceptibility to at least two mycoses: coccidioidomycosis and cryptococcosis. This presentation briefly reviews reported cases of fungal disease in NWC and also presents new findings regarding 16 cases of cryptococcosis diagnosed in Oregon, between 1999 - 2012.

Dermatophytosis

Detailed published reports of ringworm in NWC are lacking, but most cases are ascribed to *Trichophyton* and *Microsporum spp.*. Only 3 of 111respondents in a 2006 United Kingdom survey had documented fungal skin disease in their camelid stock and all cases were ringworm.[1] The lesions in NWC generally correspond to those seen in cattle (alopecic, hyperkeratotic and roughly circular foci), and are typically located in short-haired areas. Treatment regimes similar to those for cattle are effective.[2]

Candidiasis

As in other host species, these yeasts tend to be opportunistic in NWC, generally invading the tissues of immune compromised individuals or those animals that have received long term antibiotic treatment. Reports of *Candida albicans* induced disease include: multisystemic pyogranulomas in a 1 yr old male alpaca that had received long term dexamethasone treatment; renal and cerebral candidiasis in a 6 yr-old male llama that had received multiple dexamethasone injections; renal and gastric involvement in 7 yr old that had received multiple antibiotic and corticosteroid treatments. A C3 ulcer was postulated as the portal of entry for the latter animal.[3,4] Candidiasis confined to localized skin lesions has been reported in two adult llamas. The lesions appeared as crusting hyperkeratotic plaques and were successfully resolved with chlorhexidine topical ointments plus a 60 d course of nystatin.[5]

Aspergillosis

Reports of disease due to *Aspergillus sp.* include: granuloma formation in the frontal sinus, two cases of fatal embolic pneumonia associated with fungal colonization of a C3 ulcer, and a locally extensive pneumonia that included co-infection with *C.albicans*.[6,7,8,9] This last case included the unusual feature of the production of oxalate crystals by the offending *A. niger*. Disseminated aspergillosis has also been reported on at least two occasions, both involving young alpacas.[10,11]

Zygomycosis

There are two reports of infection with *Conidiobolus coronatus* involving the nasal/muzzle area of llamas. These granulomatous lesions were deformative and obstructive but did not spread to other organs or deeper tissues. One animal was treated unsuccessfully with intravenous sodium iodide and later euthanized while the other responded to radical surgical excision of the nasal masses. In neither case was a predisposing condition identified.[12,13] *Rhizopus* has been reported as a cause multiorgan disease in an adult llama, with initial clinical findings related to dysphagia and facial paralysis.[14]This infection may have been initiated as a rhinitis before extending to neural and pulmonary tissues.

Phaeohyphomycosis

An unusual infection with a dematiceous fungus (*Cladophialophora bantiana*) resulted in the unanticipated death of an 8 year old male alpaca. The granulomatous response to this pigmented fungus was limited to a solitary cerebral mass and again, no predisposing condition was identified.[15]

Histoplasmosis

A solitary report of histoplasmosis described pulmonary granulomas in a 5 year old female llama that had been imported from Bolivia seven months earlier. The pulmonary lesions were fatal with only minor involvement of other tissues. No antifungal treatment had been attempted prior to death.[16]

Coccidioidomycosis

This mycosis is recognized as an important but geographically restricted problem of NWC, as they appear to be more prone to clinical disease than many other species. The causative agents, *Coccidioides immitis* and *C. posadasii*, are dimorphic fungi, existing in specific arid environments of the southwestern US, Mexico, Central America and South America. The progressive disease that develops in camelids may initially present with problems as varied as cutaneous granulomas, neurologic signs, respiratory difficulty, or merely weight loss, owing to the wide range of organs which may be damaged.[17,18,19,20] Animals less than 2 years of age are infrequently affected. Any organ may be altered by foci of caseous necrosis and granulomatous inflammation; pneumonia, osteomyelitis, and meningoencephalitis are some of the more common lesions. Confirmation of the diagnosis is based on visualization of characteristic 10 – 80 µm spherules containing endospores in diagnostic specimens (cytology or histopathology). In camelids, a high serologic titer via Agar Gel ImmunoDiffusion test is strongly supportive of a diagnosis. Culture is generally discouraged as the mycelial form may develop and pose a health risk to laboratory personnel. There are anecdotal reports of attempted antifungal therapy but to this author's knowledge complete recovery of confirmed camelid patients with internal lesions has not been achieved.

CRYPTOCOCCOSIS

Early reports of cryptococcosis in camelids include a diagnosis in a vicuna, an alpaca, and a llama.[21,22,23] The alpaca was located in Australia and the organism cultured identified as *Cryptococcus neoformans* var *gattii*. At the time, this fungus was strongly correlated with eucalyptus-rich environments and tropical to subtropical locations. Documentation of an outbreak of cryptococcosis in animals and humans on Vancouver Island, British Columbia, Canada has challenged the perceived geographic restrictions associated with this variety of fungus. Also, *C. gattii* has now received status as a species distinct but closely related to *C. neoformans*.[24] The yeast form of *Cryptococcus*, which proliferates in mammalian tissues, is believed to have much less infectious capacity than the spores released from the filamentous form that proliferates in the environment. Thus, neither animal-animal nor animal-human transmission is believed to be a significant threat, although obviously appropriate basic medical hygiene should be maintained.

The BC outbreak has included at least two *C. gattii*-infected camelids (Dr. Stephen Raverty, pers. comm). Preliminary results of a retrospective study of cases of camelid cryptococcosis at the Oregon State University Veterinary Diagnostic Laboratory suggest that for all cases in which a speciation result has been obtained (either through culture or PCR testing) the organism has been *C. gattii type II*, not *C. neoformans*. This includes the llama diagnosed in 1999, when the BC outbreak was first being recognized, and another unpublished case, in 1997, preceding the BC outbreak. In total, the OSUVDL records contained 17cases; 8 llamas and 9 alpacas. All had a fatal outcome with cause of death or euthanasia attributed to cryptococcosis. A clinical history or detailed pathology report could not be obtained for the 1997 case so it was excluded from further analysis. Points of note regarding the signalment and presentation of the remaining cases included:

No gender bias: 7 females:9 males.

Age range from 2 – 18 years, average age = 9.4.

Course of clinical disease ranged from 1 day to 6 weeks, and one animal was merely found dead.

Most animals initially noted to have neurologic disease, or anorexia followed shortly by neurologic disease.

Only one case presented with primary respiratory disease.

Clinical signs noted in those animals showing neurologic disease included: ataxia (9/12), blindness (4/12), seizures (3/12) and difficulty swallowing (2/12). Antemortem blood work was available for review in a limited number of cases. Leukocytosis was present at a least one time point in only a few cases. Only a few biochemical profiles were available for review and these were either normal or tended toward non-specific "sick camelid" changes such as elevations in bilirubin and altered non-esterfied fatty acids levels. Antemortem tests that provided a definitive diagnosis included CSF analysis (2/2) and Cryptococcal Antigen Latex Agglutination System (CALAS)® serologic testing (1/1 with a titer of 1:16384).

Necropsy revealed involvement of multiple organs in most cases, although some submissions were of field necropsies with limited tissue sampling. Of particular note, gross or microscopic lesions were found in CNS tissues (14/14), lung (15/15), and kidney (10/14). However, other tissues were sometimes extensively affected, including intestine and mesenteric lymphatics in one alpaca and the myocardium in another. When present, gross lesions in CNS tissue included grey-white meningeal plaques and foci of yellowish gelatinous malacia of the parenchyma. Pulmonary lesions ranged from locally extensive or multifocal, firm, grey-red areas to a more diffuse change in which the lungs failed to collapse. When visible grossly, renal lesions were multifocal, firm, raised grey nodular lesions in the cortices. Microscopically, most lesions were dominated by yeast, with only small numbers of various other leukocytes present, mostly macrophages. We have also noted that *C. gattii* appears to display more size variation than the 10 -30 µm diameter typically reported for *C. neoformans*, with some cells achieving nearly twice that size.

Cultures of tissues collected at necropsy were generally successfully when Sabourad's medium was employed (13/16). At least one of the culture negative cases utilized previously frozen tissue, making successful isolation very unlikely.

Notes on cryptococcal antigen testing.

Multiple commercial detection systems are available and include classic agglutination and semi-quantitative ELISA, and Lateral Flow Immunoassay. These tests can utilize serum, plasma, or CSF. For the Premier™ ELISA and CALAS® (Meridian Inc) tests the sensitivity for serum samples is estimated at 97% and the specificity 93 – 100%. Note that these parameters were calculated for *C. neoformans* and the detection threshold for *C. gattii* may be tenfold lower. Hemolysis and lipemia adversely affect the test, as does the use of certain detergents, soaps, or anaerobic transport media during sample processing.

Recommendations for antemortem diagnosis

Cryptococcosis should be an important differential for any adult camelid in the Pacific Northwest with neurologic disease, especially if the signs include ataxia. It is also worthy of consideration for animals with respiratory disease or poorly defined problems such as weight loss and prolonged anorexia. Diagnostics can include serum testing for cryptococcal antigen, but for neurologic disease patients a CSF analysis (+/- antigen testing) may permit a definitive diagnosis. Cytology and/or biopsy should provide a rapid diagnosis if there is involvement of the upper respiratory tract or skin. *Swabs or tissues collected for fungal culture should not be frozen and cryptococcal culture should be specifically requested.*

Treatment of fungal disease in camelids (courtesy Dr. C. Cebra, OSU CVM)

With the caveat that pharmacokinetic and safety trials are completely lacking in camelids, the following would seem reasonable choices: Fluconazole (14 mg/kg, IV or PO as a loading doses, followed by 5 mg/kg, q24h for 30 days), Itraconazole (5 mg/kg, IV or PO, q24h for 30 days), or Amphotericin B (0.5 mg/kg diluted in 0.5 to 1 L of 5% dextrose IV over 1h, q24 for 30 days; pretreatment with flunixin meglumine (0.25 mg/kv, IV) may decrease adverse reactions. Uncomplicated cases of ringworm may be treated removal of debris followed by via topical application of iodine (7%) or captan (1 oz per gal) biweekly.[25]

CONCLUSION

Fungal diseases are rare in NWC but at least two, coccidioidomycosis and cryptococcosis, have considerable importance in key geographic locations. Clinicians need to have a high index of suspicion of fungal disease when dealing with camelids in these locations or when there is travel history involving these regions. It is very likely that initial colonization of the respiratory tract is a part of the pathogenesis of cryptococcosis (involved in 14/14 animals in this study) and perhaps early diagnosis and treatment of the disease, before there is involvement of the central nervous system, might permit more favorable clinical outcomes. A hurdle facing treatment of fungal diseases in camelids is a lack of scientifically established antifungal drug protocols. Long term antifungal treatment of camelids is undoubtedly expensive and difficult, but there will be owners wishing to pursue this option. Correction of this gap in the pharmacokinetic data base should be an area of research emphasis.

REFERENCES

[1]D'Alterio G, et al. **Vet Rec** 2006;158: 86. [2]Rosychuk R. **VCNA Food An Pract** 1994;10: 228. [3]Smith, S, et al. **Vet Rec** 2008;162: 485. [4]Kramer K, et al. **J. Comp Path** 2008;139:141.[5]Lamm C ,et al **JAVMA** 2009;234(8):1013. [6]Quist C, et al. **JAVMA** 1998;212(9):1438. [7]Byers S, et al. **Can Vet J** 2007;48:939. [8]Hughes K, et al. **J Vet Diagn Invest** 2008;20:672. [9]Muntz A. **Vet Pathol** 1999;36:631. [10]Severo L, et al. **J Med Vet Mycol** 1989;27:193. [11]Pickett J, et al. **JAVMA** 1986;187(11):1241.[12]Moll H, et al. **JAVMA** 1992;200(7):969. [13]French R, et al. **Vet Pathol** 1994;31:120. [14]Fowler M. Medicine and Surgery of South American Camelids, 1998: pp162. [15]Frank C, et al. **J Comp Path** 2011;145:410. [16]Woolums A, et al. **J Vet Diagn Invest** 1995;7:567.[17]Muir S, et al. **JAVMA** 1982;181(11):1334.[18]Fowler M, et al. **JAVMA** 1992;201:1609.[19]Shubitz L. **Ann NY Acad Sci** 2007;111:395.[20]Coster M, et al. **Vet Ophthal** 2010;13(1):53.[21]Griner L. Pathology of Zoo Animals. Zoological Society of San Diego 1983, pp503.[22]Goodchild L, et al. **Aust Vet J** 1996;74:428.[23]Bildfell R, et al. **J Vet Diagn Invest** 2002;14:337.[24] Kwon-Chung K, et al. **Taxon** 2002;51:804.[25]Thedford T, et al. **VCNA Food An Pract** 1989;5(1):145.

DETECTION OF SHIGA TOXIN-PRODUCING *ESCHERICHIA COLI* O157:H7 IN MEAT GOATS AT HARVEST. DM Foster, CC Balcomb, ME Jacob. NC State University College of Veterinary Medicine, Raleigh, NC.

Human infections with Shiga toxin-producing (STEC) *E. coli*, including *E. coli* O157:H7, are commonly associated with exposure to cattle or their feces or consumption of undercooked ground beef. Recent illnesses associated with goat contact at livestock exhibitions demonstrate the potential role small ruminants can play in outbreaks of STEC diarrhea and subsequent hemolytic-uremic syndrome. Yet, little work has been done to quantify the risk of STEC infection due consumption of goat meat in the U.S. The objective of our work was to determine the prevalence and relatedness of *E. coli* O157:H7 in a meat goat processing setting, ultimately aimed at better quantifying the risk to consumers of goat meat.

Fecal, hide and carcass samples were collected from 300 goats slaughtered at a USDA inspected abattoir in the southeastern United States. Samples were collected on nine days during a 12 week period from August to October. Selective enrichment and culture methods were used to isolate *E. coli* O157:H7, and isolates were confirmed to be *E. coli* O157:H7 with a six-gene multiplex PCR. Pulsed-field gel electrophoresis was used to determine the relatedness of isolates within and between each goat.

A total of 296 fecal, 300 hide, and 299 carcass samples were collected. The prevalence of *E. coli* O157:H7 was 11.1%, 2.3%, and 3.0% in the fecal, hide, and carcass samples, respectively. The overall goat level prevalence was 14.3%. There was a significant association ($p<0.001$) between the day of collection and the prevalence of culture positive fecal samples, hides, and carcasses. Fecal prevalence ranged from 0 to 52.5% on different collection days, while carcass prevalence ranged from 0 to 12.5%. Of the 43 positive goats, 28 were positive by only fecal culture, 5 by only hide culture, and 4 by only carcass culture. Six goats were culture positive at two of the three collection sites; no goats were culture positive at all three. Only 1 animal was positive on both feces and hide cultures or hide and carcass, while 4 animals were culture positive on both fecal and carcass samples. Of the 5 animals positive only on hide culture, 2 were slaughtered immediately before or after a fecal positive animal; whereas, 3 of the 4 goats positive only on carcass culture were slaughtered immediately before or after a fecal positive animal. Pulsed-field gel electrophoresis analysis to determine the genetic relatedness of these isolates is in progress.

The prevalence of *E. coli* O157:H7 in the feces of goats at slaughter is higher than previously demonstrated in goats. The fecal prevalence contributes to carcass contamination that appears to be more closely correlated to fecal shedding as opposed to hide contamination. Additional control measures to mitigate contamination of goat carcasses at slaughter with *E. coli* O157:H7 may be necessary, and further epidemiologic data is needed to determine risk factors for fecal carriage of *E. coli* O157:H7 in goats.

CONCENTRATION OF TULATHROMYCIN IN STERILE VERSUS INFECTED TISSUE CAGES IN GOATS FOLLOWING SUBCUTANEOUS OR INTRA-CHAMBER ADMINISTRATION. KE. Washburn, V. Fajt, S. Lawhon, W. Bissett. Texas A&M University, College of Veterinary Medicine, College Station, TX.

The purposes of this study were to determine the concentration of tulathromycin in sterile or *Corynebacterium pseudotuberculosis* (CPSTB) infected tissue chambers following subcutaneous or intra-chamber administration and the effect on CPSTB over time within infected chambers.

Tissue cages were surgically implanted into the right and left paralumbar fossa of 12 goats. One tissue cage was infected with *Corynebacterium pseudotuberculosis* while the other was not. Six goats were administered a single dose of tulathromycin at 2.5 mg/kg subcutaneously, while the other 6 goats received the same dose by injection directly into infected chambers. Samples of plasma and tissue chamber fluid were obtained at selected times for tulathromycin assay and bacterial culture of chamber fluid.

Concentrations of tulathromycin in infected and uninfected tissue cages were similar to those in plasma following subcutaneous administration. Following intra-chamber administration, concentrations in infected chambers were above the MIC for the CPSTB isolate (2.0 µg/mL) for the entire study (360 hr). There were no significant differences for plasma area under the curve and elimination half-lives between subcutaneous and intra-chamber administration of tulathromycin. All total, six of the 12 infected chambers had no growth of CPSTB by time 360 hr post administration.

Results of our study indicated that tulathromycin achieved concentrations inside infected tissue cages following subcutaneous administration. Further, when administered directly into infected chambers, tulathromycin displayed antimicrobial properties and reached concentrations in the uninfected chambers comparable to those following subcutaneous administration. Therefore, the treatment of goats with caseous lymphadenitis with intralesional and/or subcutaneous tulathromycin may be beneficial.

PREVALENCE OF DIARRHEA CAUSING ENTEROPATHOGENS IN NEW-BORN CALVES OF CATTLE *(BOS INDICUS)* AND WATER BUFFALO *(BUBALUS BUBALIS)* USING (FASTEST® STRIPS) RAPID TESTS IN EGYPT. Abou El-Ella G.A.[1,2], A.M. Mohamed [1,2], A.A. Aamer [1]. [1] Clinical Laboratory Diagnosis, Department of Animal Medicine, Faculty of Veterinary Medicine, Assiut University, Egypt. [2] Laboratory Medicine, Faculty of Applied Medical Sciences, Umm Al-Qura University, Saudi Arabia.

Neonatal calf diarrhea, also known as calf scours, is a common disease affecting the newborn calf characterized by diarrhea, progressive dehydration and death. Enteropathogenic infectious agents are the most common causes of calf scours.

The aim of the current study was to determine the prevalence of calf scours causing agents (cryptosporidia,rotavirus, bovine coronavirus, and *Escherichia coli*) in diarrheic calves of cattle *(Bos indicus)* and water buffalo *(Bubalus bubalis*) using rapid field test.

Fecal samples from a total of 124 cattle and water buffalo calves were collected from 2 adjacent districts (Alfateh and Sahel Seleem) of Assiut, Egypt. Calves were divided into 4 age groups (0–4 days old, 5–14 days old, 15–21 days old and >21 days old). Immuno-chromatographic rapid tests (*FAST*est® Strips) were used for the detection of bovine coronavirus, rotavirus, *Cryptosporidium parvum* and *E.coli*-K99 (F5)in fecal samples.

Of alldiarrheic calves of both species, 16.9%, 29.8%, 51.6% and 21.8% of animals were recorded positive for coronavirus, rotavirus, *C. parvum*and *E.coli*-K99 (F5) infections, respectively. In cattle calves (n = 76), *C. parvum* was the most frequently encountered pathogen among all age groups, followed by rotavirus with an overall rate of 59.2% and 35.5%, respectively. In water buffalo calves (n = 48), the highest positive test rate was recorded for *C. parvum* (39.6%) followed by both rotavirus and *E.coli*-K99 (F5) (20.8% each) in all age groups with exception of the first age group (0-4-days-old), where the highest positive test rate was recorded as *E.coli*-K99 (F5) (57.1%) followed by *C. parvum* (28.6%) and rotavirus (14.3%). Of all calves of both species for which at least one pathogen was identified in the fecal sample,62.1% were infected with a single pathogen, while 26.0% were infected with 2 or morepathogens. *C. parvum* was the most frequently encountered enteropathogen among the single-pathogen group of both cattle and water buffalo calves, while *C. parvum* and rotavirus was the most frequently identified combination of two pathogens.

In conclusion, *C. parvum* was reported as the most frequently encountered causative agent among both cattle and water buffalo calves.. The higher rate of *C. parvum* infection was recorded among the second age group (5-14 day old) of cattle calves and the forth age group (> 21 day old) of water buffalo calves, whilst the highest rate of *E.coli*-K99 (F5) infection was recorded among the first age group (0-4 days) of both cattle and water buffalo calves. In addition, cattle calves were found more likely to have mixed infections than water buffalo calves.

RISK FACTOR ANALYSIS FOR PARATUBERCULOSIS IN SWISS DAIRY AND BEEF HERDS. R. Künzler[1], P. Torgeson[2], S. Keller[2], M. Wittenbrink[2], R. Stephan[2], G. Knubben[3], B. Berchtold[1], M. Meylan[1]. 1.Vetsuisse Faculty, University of Berne, Switzerland. 2.Vetsuisse Faculty, University of Zurich, Switzerland. 3. Ludwig-Maximillian University, Munich, Germany.

Paratuberculosis is a chronic granulomatous enteritis of ruminants caused by *Mycobacterium avium* subsp. *paratuberculosis* (MAP). The clinical phase of the disease is characterized by profuse diarrhea and weight loss, leading to emaciation and death. Economic losses caused by the disease include decreased production, increased replacement rates and loss of cull value. Risk factors associated with paratuberculosis have been mainly described for large herds. The purpose of the present study was to investigate risk factors for paratuberculosis in small herds typical of traditional farming in Switzerland.

Seventeen infected herds, 10 dairy and 7 beef herds, with a history of paratuberculosis in the previous 5 years were recruited. All herds were visited, and fecal samples from all animals ≥ 2 years old were collected. In addition, a questionnaire focused on management practices considered to be potential risk factors for MAP transmission was conducted with the herd managers. The prevalence of MAP shedding was assessed by PCR and fecal culture. Infected herd status was confirmed by the presence of at least one animal shedding MAP. For each infected herd, the questionnaire was applied to 4 matched non-infected herds. Multivariate logistic regression was used to compare risk factors in infected and control herds, and risk factors associated with the prevalence level of MAP shedding in infected herds were investigated by univariate analysis.

The prevalence of MAP shedding ranged from 1.2 to 16.7% in the infected herds. The most important risk factor for infected herds vs. control herds was purchase of 2 or more animals per year (OR=7.66, p=0.005); this factor was also significant ($p \leq 0.05$) for dairy and beef herds separately (OR=7.83 and 9.75, respectively). Within infected herds, the factors significantly associated with the prevalence of MAP shedding were direct contamination of the heifer area with cows' feces (OR=1.94), the heifers being kept in the same barn as cows (OR=1.79), and the heifers being pastured on areas contaminated with feces from adult animals (OR=2.22). In dairy herds only, direct fecal contamination of the heifer area (OR=2.40), of the heifer water supply (OR=2.99), of the areas pastured by heifers (OR=2.43), and increased contamination of the calving pen (OR=2.11) were significant risk factors. In beef herds, only direct contamination of the heifer water supply was significant (OR=2.94). These results confirm the importance of direct exposition of young animals to feces from adults, whereby this analysis emphasized the importance of contamination not for young calves but rather for heifers (≥ 1 year old). Risk factors for paratuberculosis in small cattle holdings are similar to those described for larger herds.

DISEASES OF THE BRAIN AND BRAINSTEM OF FEEDLOT CALVES: FROM VIDEOS TO HISTOPATHOLOGY

Michel Lévy, DVM, DACVIM Catherine Graham DVM, MVSc, DACVP
Calgary, AB, CANADA

INTRODUCTION

In a recent study to define biomarkers of neurologic diseases of feedlot cattle, we evaluated 44 calves with a primary complaint of neurologic disease. We videotaped the animals, performed a physical and neurological examination, collected samples and euthanized the calves to perform a necropsy. Our attempt for this presentation is to combine the clinical presentation with histopathological findings in selected cases.

POLIOENCEPHALOMALACIA

Polioencephalomalacia (PEM) and Cortico-cerebral necrosis (CCN) are non-specific terms associated with softening of the gray matter of the brain. These lesions can result from several causes. In cattle (and other ruminants) these lesions can result from alterations in the thiamine metabolism, sulfur intoxication, lead poisoning, salt toxicity or water deprivation[1].

Alteration of thiamine metabolism

Thiamine is an important coenzyme in several pathways of intermediate metabolism of carbohydrates and energy. It is essential for production of ATP and control of the neuronal intracellular osmotic milieu. Pre-ruminants are dependent on ingestion of thiamine. In ruminants, thiamine is produced by ruminal bacteria and protozoan under normal environment. Production is close to daily requirements and storage of thiamine is minimal. Disruption of the normal microbial environment can reduce thiamine production either by decreasing the number of bacteria synthetizing thiamine or increasing the population of thiaminase producing bacteria, specifically thiaminase type II[2]. Altered rumen environment is most common in feedlot cattle on high carbohydrate diet associated with ruminal acidosis. Plants such as bracken fern, horsetail and Nardo fern in Australia also produce a thiaminase resembling thiaminase type I[3]. Drugs such as Amprolium may act as thiamine analogs[4]. Common clinical signs include central blindness, ataxia, muscle fasciculations, nystagmus, head pressing, seizures and progress to recumbency, coma and death[5].

Definitive diagnosis is made on histopathological findings. Presumptive diagnosis can be made on history and clinical signs. There are increases in blood pyruvate, and erythrocyte transketolase activity is reduced. Cerebrospinal fluid protein may be increased and cellularity varies from normal to mild pleocytosis (5-10 WBC/dl, Ref :< 5 WBC/dl)[5]. The mainstay of treatment is thiamine (10 mg/kg) IV or IM every 6 hours for 1 to 3 days then daily. Prognosis is fair if the animals are still standing with some improvement in 24 hours and up to a week for complete recovery.

Sulfur-induced polioencephalomalacia (S-PEM)

In case of S-PEM, alteration of the thiamin metabolism is not as clear[5]. Sulfur compounds in the rumen are metabolized in sulfide ion by sulfate reducing bacteria. In the rumen, about half of the hydrosulfide is in its gaseous form: H_2S. Hydrosulfide is eructated from the rumen and H_2S can be aspirated in the lungs and cause PEM. The most common theory is that H_2S may block the enzyme cytochrome c oxidase causing cellular energy deficiency but some recent work suggests that cytotoxicity caused by H_2S may involve reactive S species that deplete reduced glutathione and activates oxygen to form reactive oxygen species. Levels of thiamine do not appear to change significantly[6]. Sources of excess sulfur are multiple: in many areas of Canada and the United States, surface and deep waters are high in sulfate. Distiller grains with solubles as a by-product of ethanol production from corn, other co-products of corn, sugar cane and sugar beet processing as well as corn gluten feed are usually high in sulfur. Some cruciferous forages are important sources of sulfur. Some trace elements influence H_2S production by their presence or absence[7]. Two clinical forms are recognized: an acute form is characterized by blindness recumbency, seizures and frequently death. In the subacute form, animals show blindness, twitching of the ears, facial muscles and ataxia[7].

When suspecting S-PEM, both water source and feed should be tested. Ruminal hydrogen sulfide can be measured by obtaining samples of rumen gas through percutaneous needle aspiration and using a precision gas sampler.* Other methods such as quantifying thiosulfate concentrations in the urine may also be useful[8]. It is important to remember that concentration of H_2S in the rumen usually decreases when animals become anorectic. Treatment is supportive; administration of thiamine and corticosteroids may be helpful although the success rate is less than in cases of thiamine alteration. Removal of animals from sources high in sulfur is an important control measure. Regular analysis of feed and water for sulfur concentration can help making management decisions[7].

Acute Lead Toxicity

Lead has been a leading cause of accidental poisoning for a long time in ruminants [5,9]. Main sources of lead include batteries, asphalt, discarded crankcase oil, old paint and solder. Most ingested lead forms an insoluble complex which is excreted in the feces. Absorbed lead is distributed to blood (red blood cells and to a lesser extent albumin) and tissues (bone, teeth, liver, kidney, brain and spleen)[10]. Lead interferes with multiple biochemical mechanisms. It binds sulfhydryl groups and inactivates enzymes involved in heme synthesis causing red blood cell abnormalities[11]. Several mechanisms may be involved in neurotoxicity of lead: disruption of the blood brain barrier by injuring endothelial cells, epithelial and glial cells as well as causing apoptosis of neurons[12]. In the acute form of lead poisoning, animals may be found dead in pasture. Others may be staggering, showing muscle tremors with champing of the jaws, twitching of the eyelids and ears and bellowing. Blindness and head pressing are common. The affected animals may fall and show tonic-clonic convulsions, hyperesthesia to the touch and sounds and opisthotonos. In the subacute form, cattle show blindness, ataxia, circling, muscle tremors and hyperesthesia[1].

Measurement of blood lead concentrations is a common means of diagnosis. Administration of Calcium EDTA may increase urinary lead excretion. Hematologic evaluations may reveal normochromic normocytic anemia and basophilic stippling of erythrocytes, especially with chronic exposures. However, these findings are not pathognomonic. Methods that can be useful in management of lead toxicosis include elimination of lead from the gastrointestinal tract (rumenotomy, administration of cathartic), chelation therapy (calcium EDTA), general supportive care and removal of lead source from the animal's environment. Administration of thiamine seems to reduce the signs of neurologic disease.

Sodium salt poisoning and water intoxication

Cattle have been reported to show neurologic signs associated with various combinations of excess dietary sodium salt, water deprivation or restriction followed by free access to water[13]. Water can be restricted because of broken pumps, drought or frozen water supply. High salt sources include sodium salt added to the diet, incorrect dosing of electrolytes, and use of feeds such as whey and bakery by-products[2]. Excess sodium results in inhibition of anaerobic glycolysis and decrease in the rate of sodium export. In cases of excessive drinking after water deprivation, there is a sharp drop in plasma sodium concentration due to hemodilution. This leads to increase shift in water into the cells, particularly in the brain causing edema[14]. In the blood, hypotonicity leads to lysis of the cells and hemolysis causing a severe anemia and hemoglobinuria. In a group of calves, clinical signs can be manifested in all or a few animals. The condition is manifested by hemoglobinuria, diarrhea and nervous signs such as hyperesthesia, muscular tremors, head pressing, blindness, nystagmus and lethargy[15]. Some cases may progress to depression, coma and death within a day or two.

In conjunction with corroborating historical findings, supportive evidence for the diagnosis is an elevated serum osmolality or sodium concentration in salt intoxication or decreased serum osmolality or sodium concentration in water intoxication with concomitant changes in the CSF. Low doses of furosemide with IV normal or hypertonic saline are indicated in hyponatremic cases. Mannitol may be indicated. Frequent administration of small amounts of water per os and IV 5% dextrose are recommended for hypernatremic animals. As a rule, therapeutic alterations of serum osmolality should be made slowly[16].

Post mortem findings[17,18]

All of the above named diseases may result in changes in the brain which are classified as polioencephalomalacia. Grossly, this lesion is characterized by swelling of the brain, particularly of the cerebral hemispheres. The brain is often pale, soft and there may be gray discolouration at the gray-white junction. The lesion is most severe in the dorsal/caudal regions of the cerebrum and is associated with the blood supply region of the middle and caudal cerebral arteries. In the classical polio lesion, the affected regions of the brain show autofluorescence when observed under a UV light. This is due to the accumulation of lipid metabolites in the damaged areas. The histologic lesion in this group of diseases is a laminar necrosis, usually in the deeper layers of the cerebrum. There is hypertrophy of endothelial cells, and the neuropil of affected regions is pale, may be spongiotic and contains dark red shrunken neurons which are necrotic. In severe cases, there may be full thickness necrosis of the gray matter. If the animals survive for several days, grossly there may be a yellow/brown discolouration at the gray/white matter junction and separation or cavitation may be observed. The brain shows severe edema, and there may be herniation in the caudal portions of the brain. Histology is similar, but larger areas are affected and there is a marked accumulation of gitter cells and gliosis may be severe. If animals are treated and survive beyond several weeks, healing by a glial scar may be seen. If the lesion was more extensive, gray matter may be lacking over the affected regions in chronic disease.

Differentiation of the causes of laminar cortical necrosis is difficult by pathological examination alone. History, epidemiology of the herd and toxicology will be of assistance in arriving at a definitive diagnosis. In pigs which present with laminar cortical necrosis, the presence of a large number of eosinophils in the meninges of the brain is strongly suggestive of salt intoxication but this is not often present in ruminants with this condition. Most cases of lead intoxication do not show significant lesions within the CNS, or only very mild vascular lesions are noted. Some cases of lead intoxication may also show vascular changes and gliosis in the cerebellum. Chronic lesions of polioencephalomalacia or very severe cases may also have softening and/or hemorrhagic foci in the thalamus and midbrain and anecdotally these may be more commonly association with sulphur induced polioencephalomalacia.

BRAIN ABSCESS

Brain abscesses are a relatively rare condition in cattle[19]. Affected animals are typically less than 1 year of age. Abscesses may be single or multiple and they are either hematogenous in origin, result from trauma or from direct extension of a local suppurative process[20]. The clinical signs are usually not pathognomonic. The rectal temperature may be normal and the animal may be depressed. The neurologic signs depend on the location and extent of the abscess. Reported signs are often assymetrical, and include head tilt, ventral strabismus, circling or hypermetria, intention tremors and ataxia when the abscess is localized to the cerebellum.

The ante mortem diagnosis of a cerebral abscess is often presumptive. Cerebrospinal fluid analysis may support the diagnosis but the abnormal findings depend on the degree of encapsulation and amount of ependymal involvement. Reported changes include pleocytosis with increases in monocytes and lymphocytes[20]. Computed tomography or magnetic resonance imaging allows antemortem diagnosis and localization[21]. Treatment is usually not attempted.

Post mortem findings[22]

A brain abscess may form due to haematogenous spread of bacteria, and in these cases additional abscesses may be observed in other organs. Brain abscess can also result from extension of suppurative processes elsewhere in the head, such as otitis, or can occur secondary to trauma or foreign body penetration. Abscess formation as a sequela to leptomeningitis is rare.

The gross appearance of the lesion is typical of abscesses elsewhere in the body. As they are slowly developing lesions, most present in the chronic state as one or more discrete areas of purulent inflammation, surrounded by a tough fibrous tissue capsule of variable thickness. The capsular components of abscesses which form in the CNS are slightly different than elsewhere in the body. The capsule is formed both by condensation of vascular structures already present in the brain with a lesser contribution by fibroblasts. The masses are usually expansile, resulting in deformation or asymmetry of the surrounding brain. Depending on the location, they may obstruct the flow of CSF within the brain and secondary hydrocephalus is common.

Histologically, the lesions are also typical of abscesses. The center of the lesion contains abundant cell debris including many degenerate neutrophils and some bacterial colonies may be present. The capsule is composed of fibrous tissue containing histiocytes or gitter cells, and the surrounding neuropil is usually compressed. Moderate amounts of gliosis and edema may be seen in the surrounding tissue, and when near the meninges, suppurative inflammation may also be present on the surface of the brain. If secondary hydrocephalus is present, atrophy of brain matter is also present.

THROMBOEMBOLIC MENINGO ENCEPHALITIS

Histophilus somni (formerly *Haemophilus somnus*) is a gram negative bacterium that is a normal inhabitant of the bovine genital tract and nasal cavity. Exposure may occur by direct contact or aerosol, and neurological disease is often coincidental to the occurrence of respiratory disease, transportation and other stressors. *H. somni* can cause a clinical syndrome including pneumonia, reproductive failure, mastitis, myocarditis, arthritis and meningo encephalitis[23]. Feedlot cattle are more often affected than pastured or dairy cattle. Most of the clinical syndromes are result from septicemia and hematogonous spread of bacteria[24]. The relationship of *H. somni* to vascular endothelium is unclear. Recent studies suggest that *H. somni* induces apoptosis in bovine endothelial cells in vitro and that the bacteria forms biofilms in blood vessels[25]. Sudden death may be the first sign of infection. In the early stages of the neurologic form, calves can have a fever and profound depression (sleeper syndrome). Blindness of central origin may be present but that could also be due to ocular lesions such as retinal hemorrhages, edema and necrosis[26]. Some cattle may be in lateral recumbency with opisthotonos and show convulsions. Standing cattle show ataxia and weakness[27].

Etiologic diagnosis is usually confirmed at necropsy. Changes in the complete blood count are indicative of inflammation. Grossly the CSF of animals with meningoencephalitis is cloudy and cytologic evaluation is consistent with bacterial infection and hemorrhage. *H. somni* can be isolated from the CSF or other exudative body fluids. Treatment may be attempted in the early stage. Tetracycline is most commonly used, combined with non-steroidal anti-inflammatory drugs. In face of outbreaks, mass medication with antibiotics is often implemented. Immunization can reduce morbidity and mortality

Post mortem findings[28]

Peracute death, often without lesions may be seen. In this case, vessels plugged with bacteria are found throughout the body. A subacute syndrome is more common; meningoencephalitis and inflammatory lesions in a multitude of other organ systems are related to a vasculitis, with secondary thrombosis.

Grossly the meninges are cloudy and may appear multifocally thickened. Within the brain, multifocal areas of hemorrhage and necrosis are found. These are variable in size, up to several centimetres in diameter. They may appear randomly distributed, but are more commonly found in the thalamus and at the gray/white matter junction in the cerebral cortices.

Histologically the lesion is identified as inflammation of small veins, with resulting endothelial damage and thrombus formation, thrombophlebitis. The areas of hemorrhage and necrosis are secondary to this vascular damage. Definitive diagnosis is by culture of the organism.

REFERENCES

[1]Radostits OM et al **Veterinary Medicine: A textbook of the diseases of cattle, horses, sheep, pigs, and goats** 10th Ed. WB Saunders 2007; 1803. [2]Rachid MA, et al. **Asian J Anim Vet Adv** 2011; 6(2):126. [3]Gould DH **J Anim Sci** 1998;76:309. [4]Spicer EM et al **Aus Vet J** 1981;57(5):230. [5]Cebra CK, et al **Vet Clin Food Anim** 2004;20:287. [6]Drewnoski ME, et al. **J Anim Sci** 2012;90 :4478. [7]Gould DH **Vet Clin Food Anim** 2000;16:481. [8]Drewnoski ME, et al **J Vet Diagn Invest** 2012;24(4):702. [9]Blakley BR **Can Vet J** 1984;25:17. [10] Humphreys DJ **Br vet J** 1991;147:18. [11]Plumlee K **Clinical Veterinary Toxicology** 2003: [12]Liu J et al **Neurochem Res** 2010;35:473. [13]Gilchrist F **Can Vet J** 1996;37:490. [14]Njoroge EM et al **Jl S Afr vet Ass** 1999;70(4):177. [15]Maribei JM et al **Indian j Anim Sci** 1998;68(6):531. [16]Mayhew **Large Animal Neurology** 2nd Ed. Wiley-Blackwell 2008;324. [17]Maxie MG **Pathology of Domestic Animals** 5th Ed. Elsevier Ltd 2007; 351. [18] Nietfeld JC **Vet Clin Food Anim** 2012; 28: 515-534. [19]Fecteau G et al **Vet Clin Food Anim** 2004;20:363. [20]Gerros TC et al **Vet Radiol & Ultrasound** 1998;39(4):322. [21]Tsuka T et al **J Vet Med Sci** 1999;61(4):425. [22]Maxie MG **Pathology of Domestic Animals** 5th Ed. Elsevier Ltd 2007;393. [23]Orr JP **Can Vet J** 1992;33:719. [24]O'Toole et al **Proceedings ACVP/ASVCP annual meeting** 2009. [25]Sylte MJ et al **Infect Immun** 2001;69(3):1650. [26]Dukes TW **Can Vet J** 1971;12(9):180. [27]Harris et al **Can Vet J** 1989;30:816. [28]Maxie MG **Pathology of Domestic Animals** 5th Ed. Elsevier Ltd 2007:408.

* Precision gas sampler is a general term for gas analyzers. The technology used to detect H2S varies depending on the company: some will use silver nitrate; H2S will react with silver nitrate and form silver sulfide, others will use lead acetate, electrochemical technology. GC-MS can be used as well.

DISEASES OF THE CEREBELLUM AND SPINAL CORD OF FEEDLOT CALVES: FROM VIDEOS TO HISTOPATHOLOGY

Catherine Graham DVM, MVSc, DACVP and Michel Lévy, DVM, DACVIM
Lethbridge, AB

CEREBELLAR DISEASES

Cerebellar hypoplasia

Cerebellar hypoplasia is a congenital disorder that can be hereditary or due to an intrauterine viral infection. It is an inherited disorder in Shorthorn, Hereford, Angus, Ayrshire and Holstein calves. In most cases the mode of inheritance is autosomal recessive[1]. It is also commonly seen with pestiviral infections such as with Bovine Viral Diarrhea Virus. Experimentally, these congenital malformations are more likely to occur when susceptible heifers are inoculated between 100 and 170 days of gestation and, in field cases, between 75 and 150 days[2]. Severely affected calves are unable to rise and show opisthotonos and extensor rigidity of all limbs. Moderately affected calves may require assistance to stand, and show a wide base stance, ataxia, hypermetria and intention tremors. Other congenital defects such as retinal atrophy, cataract and microphthalmia are sometimes present as well and result in blindness. Calves may also become persistently infected if the infection occurs early in that period of nervous system development. If infected later (after about 125 days), the calf will have antibodies at birth. [3].

Post mortem findings

Cerebellar hypoplasia is the most common and often the most dramatic lesion resulting from intrauterine viral infection i.e. BVD Virus, but other developmental lesions in the CNS such as microencephaly, hydrocephalus, and porencephaly are also reported. In addition, retinal dysplasia, optic neuritis and microphthalmia may also be found in conjunction with CNS defects. When the fetus is exposed to BVD virus at the most susceptible time during gestation, there is direct virus-induced necrosis of the granular layer of the cerebellum and also a vasculitis is induced which results in edema, hemorrhage and tissue necrosis in utero. This results in depletion of granule cells, swollen Purkinje cell axons within the granular layer and loss or ectopia of Purkinje cells. A birth, the inflammatory lesion is resolved, and the virus is undetectable.

Hereditary syndromes of cerebellar hypoplasia show similar pathologic changes, but present without other developmental defects in the CNS. The cerebellum size may range from grossly normal, to visible only histologically. Thinning of cerebellar peduncles may be observed in some cases. Histologically, there is narrowing of the granular layer, and loss of Purkinje cells. The structure of the cerebellar cortex appears disorganized. Changes may be segmental, localized or global throughout the cerebellum.

Cerebellar Abiotrophy

Cerebellar abiotrophy is an inherited disorder of Holstein and Angus cattle. Clinical signs may appear suddenly and include intention tremors of the head, lack of menace reflex, ataxia, base wide stance and dysmetria. Animals with abiotrophy are usually clinically normal at birth and onset of signs usually starts between 3 and 24 months of age[5].

Post mortem findings [6]

Abiotrophy is defined as premature degeneration of normally formed tissues or cells, and is presumed to be due to an intrinsic metabolic defect. Gross lesions of abiotrophy may be absent, the cerebellum is most often of normal size and structure in spite of severe clinical disease observed. Histologic changes indicate ongoing degeneration of cells. Purkinje and Golgi cells may be shrunken and dark, or swollen and pale. Often there is degeneration and segmental loss of Purkinje cells resulting in 'empty baskets' at the junction between the granular and molecular layers of the cerebellum. Secondary astrogliosis and a spongy appearance in this region are common in later stages. Secondary loss of cells in the granular layer is also seen.

eta-Mannosidosis

Beta-Mannosidosis is an inherited defect of glycoprotein metabolism. The disorder is associated with a deficiency of the lysosomal acidic -Mannosidase in calves of the Salers breed, resulting in the accumulation of mannose based oligosaccharides in lysosomes in neural and extraneural tissues[7]. Some affected neonatal calves are not able to rise. Others have intention tremors and nystagmus. The head may be moderately domed and some calves exhibit mild superior brachygnathism[8]. Results of pedigree examination and breeding trials are consistent with an autosomal recessive trait. Decreased beta mannosidase activity is present in the serum or plasma of affected calves. This test can be used to detect heterozygotes

Post mortem findings [9,10]

Gross pathology lesions in the brains of calves presenting with Beta-Mannosidosis include dilation of ventricles, and pale and/or decreased amounts of white matter of the cerebrum and cerebellum. On histological examination, there is generalized fine cytoplasmic vacuolation of neurons. These vacuoles appear to be empty following normal histologic slide preparation procedures, but by ultrastructural examination they contain membrane bound lysosomes containing lucent material. Many, sometimes large axonal spheroids are present and there is hypomyelination/dysmyelination which may assist in the differentiation of this disease from other storage diseases in the bovine. Multifocally, macrophages with vacuolated cytoplasm may be found perivascularly and mineralization may be found multifocally in the cerebrum and cerebellum. Similar cellular vacuolation may be found elsewhere in the body, especially in the epithelium of kidneys and thyroid gland, and reticuloendothelial cells.

ACQUIRED SPINAL CORD DISEASES

Acquired spinal cord diseases result most commonly from infections, trauma, metabolic/nutritional or toxic causes and rarely, from neoplasia. Clinical signs depend on the location of the lesions.

Spinal cord trauma

Spinal cord trauma is often associated with vertebral fractures that compress the spinal cord. They may result from riding injury in heifers and cows, chute accidents in feedlot calves or forced delivery in neonatal calves. The lesions may be caused by disruption of the spinal cord tissue, compression from a hematoma or the fractured bone. Clinical signs appear soon after the injury as flaccid paralysis, caudal to the lesion and the animal will remain recumbent. Sensation may also be reduced caudal to the lesion[11]. Infrequently, the neurologic signs can appear later if the spinal cord compression is associated with callus formation. Careful inspection and palpation of the vertebral column may reveal evidence of trauma or elicit pain or abnormal movement localized at the affected area. Rectal examination can help detecting fractures with displacements. Diagnosis can be confirmed with radiography and/or myelography in young calves. The cerebrospinal fluid is usually normal unless there is bleeding in the CSF. Treatment consists in the administration of large doses of corticosteroids or non-steroidal anti-inflammatory drugs to limit or decrease the edema associated with the trauma. Prognosis is guarded at best.

Spinal/vertebral abscesses

Vertebral body abscesses/osteomyelitis or spinal abscesses usually occur by spread of bacteria hematogenously to the vertebrae. The origin of the infection may be the lungs, the liver, and the kidneys or associated with an omphalitis in young calves, a traumatic reticulitis in adult cattle. Bacteria most commonly isolated include *Arcanobacterium pyogenes* (mostly in calves) and *Spherophorus necrophorus* (more frequent in adults). In a study of 21 animals with vertebral abscesses, 12 were calves ranging in age from 2 weeks to seven months. Nine animals were adults from 1 to 5 years of age[12]. The abscesses involve one vertebra in most cases but sometimes up to 3 adjacent vertebrae or between vertebrae. Cervical, thoracic and lumbar vertebrae can be involved. Clinical signs vary from ataxia, to reluctance to walk, adopting a dog sitting position, weakness and inability to stand without assistance. Hematologic findings include increased serum protein levels and plasma fibrinogen. Cerebrospinal fluid can be unremarkable or show a slight increase in protein and white blood cells if the abscess extends in the epidural space. Radiographs are helpful to determine the affected vertebrae. Treatment is rarely performed but surgical drainage is reported.[13]

Post mortem findings [14]

Compressive or traumatic lesions in the spinal cord show similar pathologic changes, with minor variations related to cause. In simple traumatic injuries, uncomplicated by inflammation the initial reaction to trauma may be seen grossly as hemorrhage and edema surrounding and within the spinal cord. The spinal cord appears swollen and wet at the site of injury, more chronic lesions may show visible narrowing (if the compression was severe enough) and yellow discolouration. In compression related to abscesses and/or osteomyelitis, the meninges may appear cloudy or thickened and adhesions may be present. Areas of necrosis may be found, either as a result of the inflammatory process, or due to vascular compression or damage. This may occur directly from an acute traumatic injury or as a result of stretching /compression of vessels over bony prominences such as fracture calluses.

Histologic lesions show similar progression no matter the cause. Initial compression results in swelling of axons and dilation of myelin sheaths. The location, distribution and severity depend very much on the insult severity and duration of compression. Reaction to this damage is also found in the neuronal cell bodies, visible as central chromatolysis and clumping of Nissl substance. Distal to the injury, Wallerian degeneration can be identified as swelling and fragmentation of axons. The anatomic location of these changes follows the axonal tracts and may be in a different region of the white matter than the initial injury. Removal of axonal fragments in the spinal cord is initially performed by microglia, and later by macrophages. Secondary demyelination and myelin fragmentation can be found. Chronically, healing is by astrogliosis and glial scars may be found in very longstanding lesions.

REFERENCES

[1]Mayhew **Large Animal Neurology** 2nd Ed. Wiley-Blackwell 2008;184. [2]Brown TT et al **Vet Path** 1974;11:486. [3]Radostits OM, et al **Veterinary Medicine: A textbook of the diseases of cattle, horses, sheep, pigs, and goats** 10th Ed. WB Saunders 2007; 1258. [4] Maxie MG **Pathology of Domestic Animals** 5th Ed. Elsevier Ltd 2007; 310-311, 319-320 [5]Mitchell PJ et al **Aus Vet J** 1993;70(2):67 [6] Maxie MG **Pathology of Domestic Animals** 5th Ed. Elsevier Ltd 2007; 311-312 [7]Patterson JS et al **J Neuropathol Exp Neurol** 1991;50(5):538. [8] Abbitt B et al **J Am Vet Med Assoc** 1991;198(1):109 [9]Maxie MG **Pathology of Domestic Animals** 5th Ed. Elsevier Ltd 2007; 327 [10] Bryan L et al Vet Pathol 1993; 30: 130-139 [11]Divers T **Vet Clin Food Anim** 2004;20:231. [12]Finley GG **Can Vet J** 1975;16(4):114. [13]ZaniDD et al **Vet Surg** 2008;37:801 [14] Maxie MG **Pathology of Domestic Animals** 5th Ed. Elsevier Ltd 2007; 287, 343-345, 395-404

VAC® THERAPY: A NEW APPROACH TO WOUND CARE IN FOOD ANIMAL CASES

Suzanne G. Genova, DVM, DACVIM
Alvarado, TX

INTRODUCTION

Wound dressing and care can pose a challenge in food animal cases due to location, size, and animal temperament amongst other things. Vacuum assisted closure, also known as negative pressure wound therapy or dressing, subatmospheric pressure, topical negative pressure, and sealed surface wound suction, has been utilized in human medicine for the treatment of many types of wounds and surgical sites. Negative pressure has been employed since the 1940s with the application of drains and suction catheters. VAC® therapy was first described in the 1980s with the earliest studies using animal models in the early 1990s. The VAC® system in veterinary medicine is more often used in small animals and horses, but can be utilized for many situations encountered in our livestock species.

WOUND HEALING

All wounds no matter the cause undergo the same healing steps. These steps include inflammation, proliferation, and maturation or remodeling. These stages are complex and there is synergism between each phase. The inflammatory phase begins immediately and lasts 2-5 days. Hemostasis starts within seconds to minutes following injury. Inflammation and invasion of inflammatory cells occurs in 2-5 days. The proliferative phase characterized by granulation, contraction, and epithelialization occurs from approximately days 4-21. It is reliant upon migration of epithelial cells and structures such as hair follicles, and sebaceous glands. The final stage, remodeling begins on day 8 and lasts for one year or longer. Wound strength following injury is decreased and 3 months following injury the wound strength is only 80% of normal issue strength.[1]

VAC® MECHANISM

The exact mechanism by which negative pressure or vacuum assisted wound closure promotes wound healing is not fully understood. The VAC® system is used to promote granulation, reduce edema, increase circulation, draw wound edges together, and remove exudate and infectious material.[2] It has been found that 125 mmHg subatmospheric pressure results in a four-fold increase in blood flow which lasts 5-7 minutes. This in turn resulted in increased tissue oxygenation and 63.3% faster granulation formation when compared to wet to dry bandaging.[3] The mechanical forces placed on cells under subatmospheric pressure lead to cell proliferation and promotion of granulation tissue formation similar to remodeling of bone placed undergoing mechanical stress.[4] It has also been proposed that VAC® removes cytotoxic inhibitory factors which inhibit migration of cells to wounds. The removal of bacteria and bioburden from wounds is controversial. It has been shown that there is a decrease in the number of organisms below the level considered to be consistent with clinical infection.[3] Conflicting evidence has also found that there is no reduction in bacterial number, but suggested that the vacuum assisted closure modulated the bacterial population to one that is more favorable for healing.[5]

EVIDENCE

Vacuum assisted wound closure has been used for treating acute and chronic injuries, non-healing wounds, open fractures, open/septic abdomens, compartment syndrome, incisions, burns, and skin flaps or grafts. There are very few veterinary publications supporting the use of the VAC® system. In human medicine the first published report was in 1993 which looked at the use of negative pressure in the treatment of open fractures. It was found that this resulted in cleaning and conditioning of the wound with proliferation of granulation tissue.[6] It has also been shown in the treatment of sternal wounds that the use of the VAC resulted in a decreased number of bandage changes, decreased days between debridement and closure, and decreased soft tissue flaps.[7]

In the veterinary literature it was found that the system can be used as a treatment for distal limb injuries in dogs and can improve skin graft survival rates.[8] In a randomized controlled study looking at healing rates in antebrachial injuries in dogs it was found that negative pressure wound therapy accelerated the appearance of granulation tissue, but may prolong wound contraction and epithelialization.[9] Vacuum assisted closure has also been utilized in the management of septic peritonitis in small animals.[10, 11] The system has also been used in cats for treatment of tissue necrosis following prepubic urethrostomy,[12] post-operative feline injection site sarcoma removal,[13] and management of skin wounds.[14, 15] The VAC® has also been used in exotic species such as tigers,[16] tortoise shell injuries,[17] and rhinoceros.[18] In the horse negative pressure wound therapy has been used following skin graft on the distal limb,[19] and traumatic neck wounds.[20]

MATERIALS NEEDED

There is key equipment needed in order to use the vacuum assisted wound closure. The sterile foam or sponge is made of open cell polyurethane foam with 400-600 μm pores. This specific pore size allows for even distribution of the negative pressure over the wound bed. An adjustable suction/vacuum pump that can maintain pressures of -50 mmHg to -200 mmHG and the tubing which connects the dressing to the canister. Also needed is the T.R.A.C™ pad, a sterile plastic adhesive drape, and medical adhesive spray and/or tag cement. Items needed to prepare the area for application of the VAC® system includes sterile water-based lubricant, a surgical clipper blade, razor, sterile gloves, sterile towels, scrub, and alcohol. It is also useful to have a hair dryer or other dryer that can be used to dry the skin around the site which the system is to be applied. Depending on what suction pump is being used various connecting pieces may be needed such as needle adapters or other adapters. Different types of suction tubing may be needed as well.

SITE PREPARATION

Sterile lubricant is applied to the wound. The hair around the wound must be clipped with wide margins and then a razor should be used to provide a smooth surface taking care to not cut the patient. All of the hair, dirt, and oils need to be removed from the skin. This area is scrubbed with betadine or chlorhexadine scrub. The wound itself should be debrided and aseptically prepped. The foam is then cut to the size of the wound and placed within the wound itself. The skin margins are then wiped with alcohol and dried with a dryer and towels. Tag cement or medical spray adhesive, such as Hollister adhesive, is applied to the skin 1-3 cm from the wound margins. A sterile plastic drape is then placed over the wound making sure to smooth the drape so that no wrinkles exist. A small hole is then made in the drape and the T.R.A.C. pad applied over the hole in the drape. The T.R.A.C. pad is then connected to the suction device and suction is turned on. Suction tubing can be attached to the patient to prevent tangling.

FOOD ANIMAL CASES

The utilization of the VAC® system in food animal cases has not been described. We have used the system on many species including cattle, small ruminants, camelids, and cervids. Bovine cases included non-healing or delayed healing claw amputation sites, severe frontal sinusitis in a rodeo bull, bilateral pleural abscessation with staged rib resections, large traumatic wounds, a large flank abscess, post-operative sequestrum removal on the distal limb, and post-operative flexor tendon repair through a window cast. Small ruminant cases include pleural abscessation and traumatic head wounds. It has also been used in an alpaca following total ear canal ablation for treatment of chronic otitis. In a white tail deer fawn the VAC system was used following a traumatic wound to the medial upper hind limb with bone involvement.

Complications were few and included dermatitis where the adhesive was applied, loss of suction due to tears in the drape or leaks in the system. Other possible complications reported in other species include pain, bleeding, infection, and pressure necrosis around the tube site.

The VAC® system can be used in food animal cases for a wide variety of reasons. It can be used on large wounds on various areas of the body with few complications. While economics may limit use on all cases many livestock species and individual animals are valuable enough that the VAC® should be considered for wound management.

REFERENCES

1. Janis JE, et. al. Plast Reconstr Surg 2010; 125(6):230e-44e.
2. Hodshon RT, et. al. NAVC Clinicians Brief 2012; June 62-67.
3. Morykwas MJ, et. al. Ann Plast Surg 2001; 47(5):547-51.
4. Plikaitis CM, et. al. Expert Rev Med Devices 2006; 3(2):175-84.
5. Moues CM, et. al. Wound Repair Regen 2004;12(1):11-7.
6. Fleischmann W, et. al. Unfallchirurg 1993;96(9):488-92.
7. Song DH, et. al. Plast Reconstr Surg 2003; 111(1):92-7.
8. Ben-Amotz R, et. al. Vet Surg 2007;36(7):684-90.
9. Demaria M, et. al. Vet Surg 2011;40(6):658-69.
10. Cioffi KM, et. al. J Vet Emerg Crit Care (San Antonio) 2012; 22(5):601-9.
11. Buote NJ, et. al. J Am Anim Hosp Assoc 2012; 48(3):164-71.
12. Jordan CJ, et. al. Vet Comp Orthop Traumatol 2012;25(5):433-7.
13. Woods S, et. al. J Feline Med Surg. 2012;14(3):214-8.
14. Owen L, et. al. Vet Comp Orthop Traumatol. 2009;22(5):417-21.
15. Guille AE, et. al. J Am Vet Med Assoc. 2007;230(11):1669-73.
16. Lafortune M, et. al. J Zoo Wildl Med. 2007;38(2):341-4.
17. Adkesson MJ, et. al. J Am Vet Med Assoc. 2007;231(8):1249-54.
18. Harrison TM, et. al. J Zoo Wildl Med. 2011;42(2):317-21.
19. Jordana-Garcia M, et. al. Vlaams Diergeneeskundig Tijdschrift 2011; 80(5):343-350.
20. Gemeinhardt KD, et. al. Equine Vet Educ 2005;17(1):27-33.

DIAGNOSIS AND MANAGEMENT OF HYPERKERATOTIC SKIN LESIONS IN NEW WORLD CAMELIDS

Sandra D. Taylor, DVM, PhD, DACVIM (LAIM)
West Lafayette, IN

INTRODUCTION

Hyperkeratotic skin disorders in New World camelids include zinc-responsive dermatosis, ichthyosis, idiopathic necrolytic neutrophilic hyperkeratosis (INNH, also known as "munge"), and chronic mite infestation. Although these disorders are relatively uncommon, clinical signs may be severe and aesthetics may negatively impact the value of animals used for production. Hyperkeratotic lesions are typically alopecic with varying degrees of skin thickness. Pruritis is not usually a characteristic of hyperkeratotic skin disorders, with the exception of sarcoptic and psoroptic mite infestation. Biopsy is an important diagnostic modality, and treatment depends on etiology; therapy may include steroids, antimicrobials, zinc supplementation, and/or antiparasitic treatment. Although studies investigating the prevalence of skin disorders in New World camelids are limited, a recent retrospective analysis of 68 alpacas with skin disorders found that 10% of animals demonstrated mite infestation, 8% had zinc-responsive dermatosis, and 4.5% had ichthyosis.[1] Diagnosis of INNH was not definitively made in these cases as the disease was thought to be secondary to several other skin disorders. In a separate study, the most common hyperkeratotic disease reported in New World camelids from the United Kingdom was zinc-responsive dermatosis which was diagnosed in 35% of animals with skin lesions, followed by mite infestation in 29% of animals; INNH and ichthyosis were not diagnosed in this report.[2]

ZINC-RESPONSIVE DERMATOSIS

Zinc-responsive dermatosis may be due to a true zinc deficiency or a keratinizing disorder that is responsive to high dosages of zinc supplementation; thus, the disorder has also been termed "idiopathic hyperkeratosis." Lesions in affected animals consist of thickened skin with tightly adhering crusts that are found most commonly on hairless areas of the body (ventral abdomen, axilla, medial thighs, and inguinal area) but can also occur on the face, pinnae and neck.[1] Colored fleeced New World camelids appear to be more susceptible than white fleeced New World camelids, and young animals (1 - 2 years of age) are more frequently affected.[3,4] A herd of 48 llamas and alpacas was evaluated for nutritional status and skin lesions; 25% were found to have lesions consistent with zinc-responsive dermatosis despite being fed grass hay and commercial camelid supplemental feed. [3] The grass hay and camelid feed contained 15.8 and 51.3 mg/kg zinc, respectively. Mean serum zinc concentration were 0.17 μg/ml for all alpacas in the herd, and 0.22 μg/ml for all llamas in the herd. The proposed reference range for serum zinc concentrations for llamas is 0.30 – 0.50 μg/ml.[5] There was no significant difference in serum zinc concentrations between New World camelids with skin lesions and those without. In this particular study, only females were affected, but reports in males have also been documented.[6,7] The authors of this study made the following conclusions: 1) skin lesions are most likely to occur in young breeding females when the mineral content of feed is low; 2) dark fleeced animals are more affected because dark fleeces contains higher levels of zinc and copper than white fleeces and exert higher demands on mineral metabolism; and 3) serum zinc concentrations may not reflect total body zinc levels, but affected animals may demonstrate low serum zinc concentrations. It is important to note that zinc-responsive dermatosis often affects only one individual in a herd, even when all animals are fed the same diet.[1]

Biopsy is an important component of diagnosis. Histologic changes in affected New World camelids include epidermal and follicular orthokeratotic hyperkeratosis with mild to moderate perivascular dermatitis containing lymphocytes, macrophages, plasma cells and occasional eosinophils.[3,6-8] This is different than the typical parakeratotic hyperkeratosis reported in sheep,[9,10] goats,[11,12] cattle,[13] swine[14] and dogs[15] with zinc deficiency. Serum zinc concentrations should be tested, even though normal levels do not rule out zinc-responsive dermatosis. Blood samples for serum zinc analysis should be placed into plastic tubes with no anticoagulant using a plastic syringe that does not contain rubber. Rubber in tube stoppers may contain zinc that may falsely elevate levels. Also, erythrocyte lysis can release zinc into the serum.[7] Finally, diagnosis may be aided by response to treatment. It is recommended that affected New World camelids receive 1 - 2 g zinc sulfate or 2 - 4 g zinc methionine once daily by mouth for several weeks. Although improvement is often seen in 30 - 90 days, lesions may not resolve for up to 12 months after the start of zinc supplementation.[1,3,7,8]

ICHTHYOSIS

Ichthyosis is a congenital disorder characterized by focal or diffuse hyperkeratosis and scaling, and has been described in humans, dogs, cattle, pigs, mice and New World camelids.[16-21] The disease is caused by defects in terminal differentiation of keratinocytes and desquamation, which occur in the upper layer of the epidermis. The result of defective desquamation is increased cohesion of keratinocytes. In humans, there are several clinical forms of ichthyosis including lamellar ichthyosis (LI), which is an autosomal recessive disease considered to be the form most similar to that in animals. Lamellar ichthyosis is a severe nonepidermolytic form of disease, and has been described in golden retriever and Jack Russell terrier dogs, as well as Chianina cattle.[19,21-24] In humans, approximately 40 genes are involved in ichthyosis, mutations of which can lead to the lamellar form. Mutations within the transglutaminase 1 (*TGM1*) gene have been shown to decrease or inhibit transglutaminase activity, which results in the LI phenotype in humans and dogs.[18,25-27] Recently, an indel mutation in the *PNPLA1* gene was found to be highly associated with LI in humans and golden retriever dogs.[28] Both genes (*TGM1* and *PNPLA1*) are important in formation of the epidermal lipid envelope.[28,29] A less common form of ichthyosis in humans and cattle is ichthyosis fetalis (harlequin ichthyosis), which is the most severe form of congenital ichthyosis and is characterized by diffuse, large, diamond-shaped or plate-like scales.[23,30] Because of cracked skin in locations where normal skin would fold, bacterial dermatitis can lead to fatal infection. In New World camelids affected with ichthyosis, lesions resemble those of LI. A genetic cause has yet to be elucidated, but it is reasonable to expect that a *TGM1* mutation is associated with disease.

Diagnosis of ichthyosis in New World camelids relies on age, clinical signs, and biopsy. Lesions are present at birth or shortly thereafter.[1,16] Focal or diffuse nonpruritic, nonpainful hyperkeratotic plaques are typically observed. Histologic changes include prominent laminated orthokeratotic hyperkeratosis of the epidermis and infundibula of hair follicles with minimal epidermal hyperplasia.[1,16,17] This is similar to histology of affected dogs, cattle and humans.[18,21,23,25] The absence of inflammatory cells helps distinguish ichthyosis from zinc-responsive dermatosis. Treatment is not typically attempted in these animals, as most are otherwise healthy. In humans with ichthyosis, oral retinoids such as isotretinoin (Vitamin A derivative) have greatly improved quality of life and may be a candidate treatment for animals with the disease pending further research.

IDIOPATHIC NECROLYTIC NEUTROPHILIC HYPERKERATOSIS (INNH)

Idiopathic necrolytic neutrophilic hyperkeratosis (INNH), also referred to as "munge," is a hyperkeratotic disorder that affects alpacas and llamas. Two general forms of INNH are recognized. Focal INNH affects the perinasal and perioral regions with some extension to the periocular and periaural areas. Thick crusts may obstruct the nostrils in severe cases. Diffuse INNH is typically seen in llamas that are 1 - 2 years of age.[7] This condition is poorly understood and is likely a cutaneous reaction in the skin of New World camelids that is caused by host, environmental and pathogen factors. Disorders that may initiate the development of INNH include bacterial folliculitis, dermatophilosis, dermatophytosis, mite infestation, fly bites, viral papillomas/fibropapillomas, contact dermatitis, and zinc-responsive dermatosis.[1,7]

Histology often reveals parakeratotic and orthokeratotic hyperkeratosis with a seropurulent, palisading crust associated with epidermal hyperplasia, degenerate nuclear and hyperkeratotic debris, epidermal edema and keratinocyte necrosis. Secondary bacterial dermatitis with neutrophils may also be observed.[6,8] A positive bacterial culture and antimicrobial susceptibility of the skin lesion may guide appropriate therapy. Serum zinc concentration may reveal deficiency that can be addressed with zinc supplementation.

Because the etiology of INNH is unknown and predisposing factors are likely multifactorial, treatment often consists of "the kitchen sink." Cases have been reported to respond to topical and/or systemic antimicrobials, topical and/or systemic corticosteroids, oral zinc supplementation, or to spontaneously regress.[1,7,8] A popular treatment that is reported to be effective anecdotally includes a mixture of gentamicin, ivermectin, dimethyl sulfoxide and mineral oil.[1]

MITE INFESTATION

New World camelids are exposed to and affected by several ectoparasites, the most common of which are mites. They are susceptible to sarcoptic, psoroptic, and chorioptic mange, and can be simultaneously infested with all three.[31] However, chorioptic mange appears to be the most common mite infestation.[32]

Sarcoptic mange

Sarcoptic mange is caused by *Sarcoptes scabiei* var. *auchinae*, and has been reported in many countries in South America, Europe and New Zealand.[31,33-35] The disease is thought to be rare in the United States because of routine use of ivermectin.[4,6,8] Sarcoptic mange is a significant cause of weight loss and fiber loss. In some populations, sarcoptic mange can be found in 40% of animals and can be responsible for up to 95% of monetary losses secondary to ectoparasite infestation.[33] Affected animals present with pruritis and alopecia.[1,34,35] Early lesions most commonly affect the ventral abdomen, axillae and groin, and may extend to the medial thighs, extremities, feet and face. Initially, the infested skin is erythematous with yellow to grey crusts, but with time the skin becomes thickened, lichenified and hyperpigmented. Secondary bacterial infection may occur and lead to further morbidity. Importantly, sarcoptic mange in New World camelids is a potential zoonosis.[7,8,34]

Psoroptic mange

Psoroptic mange in New World camelids is caused by a mite that has not been specifically named, so is referred to as *Psoroptes sp.*[1,7] Skin lesions predominantly affect the head, face and pinnae, with less common sites including the shoulders, dorsum, and perineum.[2,7,8,31,36,37] If the ear canals are affected, the animal may present with ear twitching, head shaking, and head tilt. Like with sarcoptic mange, pruritis and alopecia are often observed and are associated with papules and crusts early in infestation; chronic disease is associated with thickened skin. Purulent discharge from the ears may indicate secondary bacterial infection. Psoroptic mange is not considered zoonotic.

Chorioptic mange

The most common mite infestation of New World camelids is *Chorioptes bovis*, and is recognized in many countries including the United States. Unlike other types of mange, pruritis is generally absent or mild. Animals may appear clinically normal despite heavy parasite infestation, while animals with few mites may develop severe lesions. Early in the course of infestation, lesions are typically found on the ventral abdomen, perineum, ventral tail and medial thighs, and may progress to include the axillae, face, pinnae, distal limbs and interdigital spaces.[7,8,38,39] Alopecia, scaling and crusting lesions often progress to include lichenified, thickened skin. Like psoroptic mange, chorioptic mange does not constitute a zoonotic risk.

Diagnosis

Diagnosis of mange is confirmed by the presence of mites in skin scrapings; however, the absence of visible mites does not rule out disease.[7,8] In a study evaluating the prevalence of *C. bovis* infestation in a herd of alpacas in the United Kingdom, skin scrapings were positive in 55% of clinically normal animals that had direct contact with animals bearing skin lesions, while only 28% of animals with skin lesions were positive for mites.[32] Chorioptic mites are most readily recovered by scraping the interdigital spaces of the forefeet.[32,38] Biopsy can be helpful in supporting the diagnosis of mange, with histopathology often revealing eosinophilic interstitial dermatitis, marked parakeratotic hyperkeratosis, and mites within surface crusts.[1,35] In addition to these histology findings, eosiniophilic epidermal microabscesses and pustules have been described in New World camelids with chorioptic mange.[40] Eosinophilia may be present with heavy mite infestation but normal peripheral eosinophil counts do not rule out disease.

Treatment

Avermectins in different formulations have been shown to be effective in treating mange in cattle,[41,42] sheep,[43,44] pigs,[45,46] and dogs.[47] Although its use is considered off-label, ivermectin at a dose of 0.2 - 0.4 mg/kg every 1 – 2 weeks subcutaneously for 2 – 4 injections is considered an effective means to treat mange in New World camelids; however, treatment failures with ivermectin, doramectin, and eprinomectin have been reported.[7,34,48] Topical amitraz administration has been shown to resolve clinical signs of sarcoptic mange for at least 10 months in infested alpacas that did not respond to avermectin treatment.[34] There is evidence that chorioptic mange may be more difficult to clear than sarcoptic or psoroptic mange. In one study evaluating treatment in a herd with simultaneous infestation with *Sarcoptes sp.*, *Psoroptes sp.* and *Chorioptes sp.*, ivermectin (1%) was effective in eliminating *Sarcoptes sp.* and *Psoroptes sp.* after subcutaneous administration of 0.2 mg/kg on day 0 and day 10; however, additional treatment with 50 µg/kg ivermectin as a pour-on on day 24 and day 34 was required to eliminate *Chorioptes sp.*[31] This is consistent with previous findings.[36,49,50] Topical administration of eprinomectin at a dose of 0.5 mg/kg weekly for four weeks was found to be highly effective at reducing *Chorioptic sp.* mite numbers in a herd of alpacas, but failed to completely eradicate them.[38] Similarly, the same dose of eprinomectin applied topically weekly for 10 weeks failed to eradicate chorioptic mange in a herd of llamas and alpacas.[39] The use of organophosphate dips to control mite infestation is not recommended for New World camelids as it is for small ruminants, as dipping can be stressful for the animals and there is no safety information available. Movement of animals to clean pasture following treatment and disinfection of fomites may aid in successful eradication; however, mites can survive in the environment for up to 2 months, making reinfestation possible.[51]

REFERENCES

1. Scott D, et al., **Vet Dermatol**. 2011;22:2. 2. D'Alterio G, et al., **Vet Rec**. 2006;158:86. 3. Clauss M, et al., **Vet J.** 2004;167:302. 4. Rosychuk R., **Vet Clin North Am Food Anim Pract.** 1994;10:228. 5. Johnson L., **Vet Clin North Am Food Anim Pract.** 1989;5:37. 6. Rosychuk R., **Vet Clin North Am Food Anim Pract.** 1989;5:203. 7. Foster A, et al., **In Practice**. 2007;29:216. 8. Plant J, **International Camelid Health Conference**. 2007;127. 9. Masters D, et al., **Aust J Biol Sci.** 1985;38:355. 10. Suliman H, et al., **Trop Anim Health Prod**. 1988;20:47. 11. Miller W, et al., **J Dairy Sci**. 1964;47:556. 12. Krametter-Froetscher R, et al., **Vet Dermatol**. 2005;16:269. 13. Singh A, et al., **Indian J Anim Sci**. 1994;64:35. 14. Norrdin R, et al., **Cornell Vet.** 1973;63:264. 15. White S, et al., **Vet Dermatol.** 2001;12:101. 16. Belknap E, et al., **J Am Vet Med Assoc**. 1990;197:764. 17. Charney V, et al., **J Am Vet Med Assoc**. In press. 18. Cao X, et al., **Clin Exp Dermatol.** 2009;34:904. 19. Molteni L, et al., **Vet Rec**. 2006;158:412. 20. O'Goshi K, et al., **Skin Pharmacol Appl Skin Physiol.** 2000;13:120. 21. Guaguere E, et al., **J Small Anim Pract.** 2009;50:227. 22. Mauldin E, et al., **Vet Pathol.** 2008;45:174. 23. Testoni S, et al., **Dtsch Tierarztl Wochenschr.** 2006;113:351. 24. Raoofi A, et al., **Vet Rec.** 2001;149:563. 25. Credille K, et al., **Br J Dermatol.** 2009;161:265. 26. Huber M, et al., **Science**. 1995;267:525. 27. Russell L, et al., **Nat Genet**. 1995;9:279. 28. Grall A, et al., **Nat Genet.** 2012;44:140. 29. Oji V, et al., **Eur J Dermatol.** 2006;16:349. 30. Rajpopat S, et al., **Arch Dermatol.** 2011;147:681. 31. Geurden T, et al., **Vet Rec.** 2003;153:331. 32. D'Alterio G, **Small Ruminant Res.** 2005;57:221. 33. Leguia G, **Parasitol Today**. 1991;7:54. 34. Lau P, et al., **Vet Dermatol**. 2007;18:272. 35. McKenna P, et al., **N Z Vet J.** 2005;53:213. 36. D'Alterio G, et al., **Vet Rec**. 2001;149:96. 37. Bates P, et al., **Vet Rec**. 2001;149:463. 38. D'Alterio G, et al., **Vet Parasitol**. 2005;130:267. 39. Plant J, et al., **Vet Dermatol.** 2007;18:59. 40. Petrikowski M, **Adv Vet Dermatol.** Boston: Butterworth-Heinemann, 1998;2. 41. Rehbein S, et al., **Vet Parasitol.** 2003;114:267. 42. Logan N, et al., **Vet Parasitol.** 1993;49:67. 43. Papadopoulos E, et al., **J Vet Pharm Ther.** 2000;23:111. 44. Fthenakis G, et al., **Vet Parasitol**. 2000;87:207. 45. Geurden T, et al., **Vet Rec**. 2003;153:272. 46. Jensen J, et al., **Acta Vet Scand**. 2002;43:75. 47. Curtis C, **Vet Dermatol**. 2004;15:108. 48. Borgsteede F, et al., **Tijdschr Diergeneeskd**. 2006;131:282. 49. Curtis C, et al., **Vet Rec**. 2001;149:208. 50. Johnson L, **Vet Clin North Am Food Anim Pract.** 1994;10:248. 51. Scott D, **Large Anim Dermatol**. Philadelphia: W. B. Saunders, 1988.

EMERGING ISSUES IN GASTROINTESTINAL DISORDERS OF NEONATAL CALVES

Peter Constable, BVSc (Hons), MS, PhD, DACVIM, DACVN (Honorary)
West Lafayette, IN

Calf diarrhea should be considered as a complex when diagnosing enteric diseases as there is the typical interaction between calf, agent, and environment (management).[1,2] The 6 most common causes of diarrhea in calves < 21 days of age and their most common age of occurrence are: 1) Enterotoxigenic *E. coli* (ETEC; <4 days); 2) Rotavirus (4-21 days); 3) Coronavirus (4-21 days); 4) *Cryptosporidium parvum* (5-28 days); 5) Salmonella; and 6) Nutritional. Most enteropathogenic studies of calves with diarrhea around the world, conducted using currently available diagnostic tests, indicate that rotavirus and *Cryptosporidium parvum* are the most common isolates from neonatal calves with diarrhea.[3,4]

PATHOPHYSIOLOGY OF DIARRHEA

Rotavirus

Calves are infected with rotavirus from other calves, environment, or dam feces. Rotavirus infects epithelial cells of the small intestinal villus with minimal destruction of the crypts. Infected cells are sloughed, resulting in villus atrophy and decreased villous length; however, because the crypts are left intact, the repair and regeneration of intestinal villi occurs rapidly. Rotavirus diarrhea in calves is caused by Group A rotavirus, but there are numerous field strains (G1, G3, G5, G6, G8, G10, G15) which may be different to vaccine strains (typically G6).[5,6] There is some data to suggest that G6 and G10 are the most prevalent strains in calves with diarrhea; G6 maybe more prevalent in beef herds and G10 in dairy herds, although the dominant strain can change over time in a geographic location. The importance of the strain difference in disease severity or resistance to rotaviral diarrhea in calves remains uncertain.

Historically rotavirus has been considered to be primarily a maldigestion/malabsorption diarrhea because of the decrease in absorptive surface area in the small intestine, but recent studies of rotaviral diarrhea in other species indicate that there is an important toxin mediated component due to nonstructural glycoprotein 4, NSP4. The NSP4 toxin is produced during rotaviral replication within the enterocyte and is secreted or released during enterocyte death, acting in a paracrine manner to affect other epithelial cells. NSP4 inhibits a sodium glucose transporter SGLT1 that plays an important role in absorption of sodium and free water from the intestinal lumen.

Coronavirus

Calves are infected with coronavirus from other calves, the environment, or dam feces; cows shed virus more in winter and at parturition. Once ingested, coronavirus attaches to the enterocyte via spike and hemagglutinin glycoproteins with subsequent endocytosis. Coronavirus infects the epithelial cells of both the small and large intestine and this is an important distinguishing difference from rotaviral diarrhea. In addition, the crypt epithelium is also damaged by coronavirus which leads to a longer repair time than rotaviral diarrhea (intestinal crypts are not damaged in rotavirus). Coronavirus diarrhea in calves is caused by a virus is group 2a of the Coronaviridae family; the distinguishing feature of this family is the presence of a surface glycoprotein called hemagglutinin-esterase and a spike (S) glycoprotein for the peplomers that is the major target of neutralizing antibodies.[7] A 2008 molecular analysis study indicated three main strains of coronavirus exist in calves world-wide.[8]

Historically coronavirus has been considered to be primarily a maldigestion/malabsorption diarrhea because of the decrease in absorptive surface area in the small and large intestine and because replacement cells have decreased lactase activity. These combine to produce a maldigestion/malabsorption diarrhea (osmotic/fermentative) diarrhea, but it is likely there is also a secretory component.

Salmonella enterica

Salmonella enterica serotype *dublin* (host adapted) and *Salmonella enterica s*erotype *typhimurium* are the most frequent pathogenic salmonella isolates from calves with diarrhea. Hundreds of other strains have also been isolated and we need to focus more on the pathogenicity of salmonella serotypes in bovine practice rather than lumping all serotypes into one box. This is because some salmonella serotypes have been isolated from cattle feces but have never been associated with clinical disease in cattle. The infective dose of salmonella plays an important role in the clinical outcome when calves are exposed to pathogenic seroptypes; the actual dose required to induce clinical disease depends on the serotype, age of the calf, and immune status. Salmonella causes an invasive enteritis of the ileum, cecum, and colon, resulting in maldigestion/malabsorption, loss of protein, fluid, and electrolytes into lumen, colonization of lymph nodes, bacteremia, fever, and possibly osteomyelitis.

EFFICACY OF VACCINES

Protection against rotaviral or coronaviral diarrhea is believed to exist as long as colostral Ab is in lumen of the small intestine to interact with ingested virus particles.[6,7,9] There is a low antibody concentration in milk; this is one reason that rotavirus and coronavirus usually cause diarrhea in calves > 4 days of age; age immunity in older calves has been attributed to slowed virus entry of cells, but the evidence is not strong. The first major vaccine challenge is providing sufficient protection from >3 days of age. The second major vaccine challenge is ensuring vaccine strains of rotavirus and coronavirus have sufficient antigenic cross protection against the majority of field strains. The optimal vaccination approach is to vaccinate the dam in late gestation with rotavirus and coronavirus antigens that are cross protective against all field strains, and then to optimize the transfer of passive immunity of colostral immunoglobulins. This may require the presence of more than one strain of rotavirus and coronavirus. Vaccination of the dam ensures protection of the calf while colostrum is being ingested. Continued feeding of colostrum to dairy calves is therefore beneficial, and the ideal vaccine would induce a high colostral immunoglobulin concentration against both rotavirus and coronavirus. It is currently believed that some of the absorbed immunoglobulin directed against rotavirus and coronavirus is resecreted into the small intestine.[6,9]

Vaccination against Salmonella has been shown to be most efficacious with oral avirulent strains that replicate in the intestinal tract. This is because IgA and cell-mediated immunity are thought to play an important role in protection. The major challenge is that efficacious oral vaccines are challenging to produce, handle and deliver. There are also concerns about transfer of genetic material in enteric bacteria, particularly Salmonella (some countries prohibit the administration of genetically modified organisms), and shedding of live vaccine bacteria in the feces of vaccinated calves.[10] Consequently, recent vaccine development has focused on parenteral vaccination.

Standard approaches for vaccination against Salmonella include: 1) parenteral vaccination of the dam or calf with Salmonella bacterins that are group specific, so this farm and strain specific protection (autogenous vaccine). These generally provide poor protection against other strains. 2) vaccination of the calf with mutant gram-negative bacteria, such as the Re mutant = *S typhimurium* (Endobac) or the Rc mutant = *E coli (*J5*)*. This provides gram negative protection which is directed more against endotoxin (core lipopolysaccharide). In general, core lipopolysaccharide vaccines provide poorer protection against Salmonellosis than autogenous bacterins. Adverse reactions can occur with both approaches (due to the endotoxin content in vaccine, not an anaphylactic reaction); 3) vaccination with a *Salmonella Newport* siderophore receptor and porin protein (SRP) vaccine.[11]

The author is not aware of any peer reviewed efficacy studies that document vaccine efficacy for Salmonella in calves. Until such studies are available, routine vaccination of calves against Salmonella is not currently recommended. Improved vaccines are under development but need to be cross protective. It is important to remember to feed milk replacer instead of fresh milk or colostrum in herds that are endemically infected with *S. dublin* , because it is shed in the milk of dairy cows. Other potential strategies are competitive exclusion bacteria (bacteriocin) and acetate in oral electrolyte solution (acetate inhibits Salmonella growth). A modeling paper concluded that the most effective method for controlling Salmonellosis in calves was to assign workers and equipment to specific groups by pairing workers with groups of calves or by using a separate set of clothing and boots for each group.[12]

OPTIMIZING TRANSFER OF PASSIVE IMMUNITY

Neonatal calves must ingest colostrum during the first 24 hours after birth to acquire passive immunity via the active uptake of maternal IgG across small intestinal epithelial cells. The mass of IgG absorbed from small intestine of the colostrum-fed calf depends on the IgG concentration in colostrum, the volume of colostrum administered, and the apparent efficiency of absorption (AEA) of ingested IgG. For calves < 24 hours of age the AEA ranges from 6 to 66%, with typical mean values for AEA of 30 to 35% for fresh colostrum and 20 to 30% for colostral replacement products. An AEA of 33 to 45% appears to be near the practical biologic limit when 2 L or more of colostrum is fed shortly after birth.[13]

The rate of abomasal emptying influences the rate at which colostral IgG is delivered to the site of IgG absorption which is the small intestine. We have recently reported here that the AEA of colostral IgG was linearly and negatively associated with abomasal emptying rate in dairy calves, with abomasal emptying rate explaining 22% of the variation of AEA.[13] This new finding is consistent with long-held beliefs that the rate of gastrointestinal motility and therefore the rate of abomasal emptying is an important determinant of the AEA of colostral IgG. As a consequence, there are four major factors that affect colostral IgG absorption: age at first feeding, concentration of IgG in colostrum, volume of colostrum fed (within limits), and the rate of abomasal emptying.

Inadequate absorption of colostral IgG in dairy calves increases the risk of morbidity and death. To reduce the risk of failure of transfer of passive immunity in dairy calves, producers should harvest colostrum as soon after parturition as possible, and preferably within 2 hours after calving if the cow is strong enough to permit milking.[14] This is because colostral IgG concentration decreases by 3.7% during each subsequent hour after calving because of postparturient secretion of milk by the mammary glands.[14] Holstein-Friesian calves should be fed at least 153 g of total IgG within the first 2 hours after birth, equivalent to 3 L of first milking colostrum.[15] This is because transfer of passive immunity is optimized when colostrum is fed to calves within the first 2 hours after birth.[15] In summary, current research indicates that the optimal method to transfer colostral immunoglobulin is to feed 3 L of first milking colostrum harvested within 2 hours of parturition to new born calves within 2 hours of parturition. The 3 L of colostrum should be administered by oroesophageal intubation or by allowing the calf to suckle for up to 10 minutes followed by intubation of the remaining volume.

REFERENCES

1. Radostits OM, Gay CC, Hinchcliff KW, Constable PD. *Veterinary Medicine. A textbook of the diseases of cattle, sheep, pigs, goats, and horses.* ISBN 13:978 0702 07772. 10th ed. London, England: W.B. Saunders Company; 2007, 2156 pages. 2. Constable PD. *Bovine Neonatology.* Veterinary Clinics of North America, Food Animal Practice. Philadelphia, PA: W.B. Saunders Company; 2009:25:101-120. 3. Bartels CJM et al. Prev Vet Med 2010; 93:162-169. 4. Izzo MM et al. Aust Vet J 2011;89:167-173. 5. Dhama K et al. Vet Res Commun 2009; 33:1-23. 6. Gonzalez DD et al. Vet Immunol Immunopathol 2010;137:155-160. 7. Hasoksuz M et al. Methods Mol Biol 2008;454:43-59. 8. Takiuchi E et al. Braz J Med Biol Res 2008;41:277-282. 9. Parreno V et al. Vet Immunol Immunopathol 2010;136:12-27. 10. Mizuno T et al. Vet Res 2008;39:26. 11. Mohler VL et al. *Bovine Neonatology.* Veterinary Clinics of North America, Food Animal Practice. Philadelphia, PA: W.B. Saunders Company; 2009:25:37-54. 12. Lanzas C et al. Vet Res 2008:39:61. 13. Mokhber-Dezfooli MR et al. J Dairy Sci. 2012;95:6740-6749. 14. Morin DE et al. J Am Vet Med Assoc.2010; 237(4):420-428. 15. Chigerwe M et al. Am J Vet Res 2008;69:1158–1163

RECENT ADVANCES IN ORAL REHYDRATION THERAPY AND ABOMASAL DISORDERS IN NEONATAL CALVES

Peter Constable, BVSc(Hons), MS, PhD, DACVIM, DACVN(Honorary)
West Lafayette, IN

CALF DIARRHEA AND DEHYDRATION

Diarrhea in calves results in an increased loss of electrolytes and water in the stool, decreased milk intake, and changes in the small intestinal bacterial flora. Calves with diarrhea die because of: 1) septicemia, 2) acidemia, 3) hyperkalemia, 4) prolonged malnutrition and hypoglycemia, and 5) hypothermia. The 5 main treatment goals in calf diarrhea are therefore to: 1) replace free water, electrolyte, and base deficits; 2) provide nutritional support; 3) facilitate repair of injured intestinal epithelium; 4) eliminate *Escherichia coli* bacteremia; and 5) decrease *E coli* concentrations in the proximal small intestine.

Dehydration is due to increased loss of fluid in the stool and decreased fluid intake (inappetance or intentional milk deprivation). The latter effect is generally overlooked but plays an important role in calf diarrhea. Most calves do not drink water during the first week of life, and are therefore totally dependent upon the fluid load provided to them by the dam or producer. The result of dehydration is decreased plasma volume and extracellular fluid volume, which in turn leads to decreased cardiac output, peripheral perfusion, and oxygen delivery (hypovolemic shock). Dehydration does not appear to be the important cause of death, but is associated with acid-base and electrolyte disturbances which will directly cause death. Fluid therapy therefore remains an important cornerstone of treating calves with diarrhea.

The method used for fluid administration in calves with diarrhea should be based on the presence or absence of a suckle reflex and degree of dehydration. The most accurate methods for assessing dehydration are the extent of eyeball recession into the orbit and skin tent duration in the neck region. All other methods of assessment are inferior to these 2 methods. Eye recession is measured by rolling the lower eyelid out to its normal position and measuring the distance between the cornea and lower eyelid. The recommended formula to estimate hydration status is: % dehydration = 1.7 x (eyeball recession in mm). Eyeball recession is less accurate in calves with chronic weight loss than in calves that have been on a reasonable plane of nutrition because chronic weight leads to loss of the periorbital fat and eye recession. Skin tent duration is measured by pinching a fold of skin, rotating it 90 degrees while pulling gently away from the calf, and then releasing. The skin tent should disappear within 2 seconds of release. Measuring eyelid skin tent duration on the upper or lower eyelid is not recommended because skin is this area is poorly elastic. Calves that are able to suckle and are less than 8% dehydrated (eye recessed < 4 mm into the orbit) should be administered an oral electrolyte solution; some of the fluid can be intubated using oro-esophageal intubation if needed. Calves that are not able to suckle or that are 8% or more dehydrated (eye recessed 4 or more mm into the orbit) should receive intravenous fluids. These can be administered by placement of a jugular venous catheter or auricular vein catheter.

Intestinal electrolyte loss and decreased milk intake results in a total body deficit of Na, Cl, and K. Serum Na concentration is almost always decreased in calves with diarrhea. Serum K concentration is variable in diarrheic calves and does not reflect body stores, because K is primarily an intracellular ion. Serum K concentration is also influenced by the degree of metabolic acidosis. The assessment of serum electrolyte abnormalities really requires laboratory evaluation. If unavailable, total body electrolyte deficits of sodium and potassium should always be assumed in calves with diarrhea. Bradycardia, cardiac arrhythmias, and clinical signs of profound weakness are suggestive of hyperkalemia.

ORAL ELECTROLYTE SOLUTIONS

Our knowledge of the requirements for the ideal oral electrolyte solution for diarrheic calves continues to evolve. However, much progress has been made over the last 30 years, and the critical issues in formulating the ideal oral electrolyte solution are osmolality, sodium concentration, source of the alkalinizing agent, and the energy content (which is intimately tied to osmolality). An unknown component is whether solutions should contain agents such as glutamine that may facilitate repair of damaged intestinal epithelium.

The osmolality should range from isotonic (300 mOsm/kg) to hypertonic (700 mOsm/kg). The effective osmolality at tip of the intestinal villus is approximately 600 mOsm/kg due to a countercurrent exchange mechanism. Although markedly hypertonic fluids should be avoided in animals having severe villous damage, it is impossible to predict which animals have severe villous damage on the basis of the physical examination findings and measurement of fecal pH or other body parameter. Low osmolality fluids (300 mOsm/kg) have inadequate energy content because they have insufficient glucose. For this reason, if milk is withheld, then hypertonic oral electrolyte solutions (~600 mOsm/kg) should be administered. If milk is fed, then isotonic oral electrolyte solutions (300 mOsm/kg) should be administered, because inadequate energy content is no longer an issue.

The sodium concentration should be between 90 and 130 mM/L. Adequate sodium absorption is the main determinant of successful expansion of the extracellular space. Sodium concentrations < 90 mM/L provide an inadequate sodium load, whereas sodium concentrations >130 mM/L can lead to hypernatremia and additional free water loss. The oral electrolyte solution should also contain glucose and either acetate, propionate, or glycine to facilitate Na absorption and provide energy. There are co-transport mechanisms for Na and glucose, Na and volatile fatty acids such as acetate and propionate, and Na and amino acids in the luminal membrane of villus epithelial cells. Administration of glucose, acetate, propionate, and glycine facilitates Na absorption. These transport mechanisms are unimpaired in enterotoxigenic *E.coli* and are at least partially functional in malabsorptive/maldigestive diarrheas.

The alkalinizing agent should ideally be acetate or propionate instead of bicarbonate, at a concentration range of 50 to 80 mM/L. Acetate-containing fluids are as effective as bicarbonate-containing solutions at correcting mild to moderate acidosis [acetate- = CH_3COO-] + H^+ +$2O_2 \leftrightarrow 2CO_2 + 2H_2O$. Acetate must be metabolized to be effective, and metabolism may be impaired in severely dehydrated or acidemic animals. Acetate-containing fluids can be fed with milk as acetate does not raise abomasal pH or inhibit milk clotting. Bicarbonate-containing fluids are more effective at rapidly correcting severe acidemia, since bicarbonate reacts directly with H^+ ions ($HCO_3^- + H^+ \leftrightarrow H_2CO_3 \leftrightarrow H_2O + CO_2$). The main disadvantages of bicarbonate-containing oral fluids are that the pH of the abomasum (a natural defense mechanism) is increased. We have recently found that a theoretical disadvantage that bicarbonate inhibits clotting of milk in the abomasum is not true, at least when low bicarbonate solutions are fed. In summary, oral electrolyte solutions containing acetate or propionate are preferred over solutions containing bicarbonate, because bicarbonate excessively alkalinizes the abomasum and proximal small intestine, thereby decreasing the effectiveness of the "abomasal sterilizer" in killing ingested enteric pathogens, and promoting enteroxigenic *E. coli* attachment to epithelial cells and STa enterotoxin production. Moreover, bicarbonate does not inhibit growth of Salmonella in the intestinal lumen, whereas acetate and propionate both inhibit Salmonella growth. However, the disadvantages of bicarbonate are minimized when bicarbonate is mixed with milk, because of the buffering capacity of milk.

ABOMASAL ULCERS IN CALVES

Abomasal ulcers penetrate the mucosa, and may extend through serosa. In contrast, abomasa erosions are discrete mucosal defects with no penetration of the muscularis mucosa that are usually multiple, circular, small, and 1 to 20 mm diameter. Abomasal ulcers often have a central crater with a fibrinonecrotic covering. Ulcers heal by wound contraction and granulation tissue to produce a permanent scar. Ulcers are categorized as: Type 1: non-perforating ulcer - minimal intraluminal hemorrhage; Type 2: non-perforating ulcer - with severe blood loss. The ulcer erodes a major submucosal blood vessel, leading to severe intraluminal hemorrhage. This can be fatal if not treated; Type 3: perforating ulcer with local peritonitis - the ulcer perforates leading to localized peritonitis (outside the omental sling or inside omental bursa); Type 4: perforating ulcer with diffuse peritonitis - the ulcer perforates leading to massive abdominal contamination. This form is 100% fatal.

Abomasal ulceration occurs commonly in calves with a prevalence of 5 to 76% in healthy veal calves. Abomasal ulceration is also a common cause of death in suckling beef calves aged 1 to 2 months old in the north central region of the US and along the eastern slopes of the Rocky Mountains in Canada, with cases occurring more frequently during or after a period of inclement weather. The cause for the high prevalence of abomasal ulceration amongst suckling calves remains uncertain, although diet has long been suspected to play an important role in the etiopathogenesis. In beef calves, ulcers occur in the mid-section of the fundus, usually along the greater curvature of abomasum. In milk fed veal calves, ulcers occur predominantly in the pyloric antrum. Consumption of straw in veal calves is associated with increased incidence of ulceration in the aboral 5 cm of pyloric antrum; it is believed that straw may exert a direct abrasive effect. Abomasal contents after straw feeding are more watery and contain less milk of lower pH than calves with no straw.

Four main etiologies have been proposed for abomasal ulceration in suckling calves: 1) mechanical abrasion of the pyloric antrum due to ingestion of coarse roughage such as straw or the presence of trichobezoars, 2) infection with bacteria such as *Clostridium perfringens* type A or unidentified fungi, 3) trace mineral deficiencies, and 4) concurrent stress. None of these factors or theories alone adequately explains abomasal ulcer formation in suckling calves.

Mechanical injury to the abomasal mucosa has been considered for many years to be the principle cause of ulceration in calves as the underdeveloped rumen purportedly allows passage of rough food that has not been adequately ruminated. Feeding roughage to veal calves during the fattening period increased the prevalence of abomasal ulceration by 20 to 30% in some, but not all studies. Although the high prevalence of ulceration in veal calves has been partially attributed to abrasion of the abomasal mucosa in the pyloric antrum, roughage cannot be the critical factor causing ulceration as the majority of veal calves fed only milk replacer also have ulcers. Mechanical abrasion is not considered a necessary requirement for the pathogenesis of ulceration in suckling beef calves because the presence of trichobezoars in the abomasum does not increase the risk of fatal ulceration. The abomasal luminal pH of veal calves fed straw has been reported to be lower than that of veal calves not fed straw. However, the increased prevalence of abomasal ulceration in veal calves fed straw is more likely to be due to a lower luminal pH rather than the direct effect of mechanical injury to the pyloric antrum.

Bacteria and fungi have been frequently isolated from abomasal ulcers. *Clostridium perfringens* type A appears to be a primary cause of abomasitis, and has been isolated from calves with abomasal tympany, abomasitis, and abomasal ulceration, and ruminal and abomasal tympany, abomasitis, and abomasal ulceration have been experimentally induced by intraruminal inoculation of *Clostridium perfringens* type A. In general, bacterial and fungal agents are considered post mortem invaders from the alimentary tract or opportunistic secondary infective agents that invade after injury to the abomasal mucosa.

A nutritional deficiency of copper has been associated with abomasal ulceration in beef calves in the Western US; however, these studies did not have an appropriate control group and definitive evidence of an association between copper deficiency and abomasal ulceration is lacking.

Stress has also been considered for many years to be involved in the pathogenesis, as ulcers in calves reportedly are found more frequently in calves around the time of weaning or after a period of feed deprivation due to inclement weather. Stress is hypothesized to induce ulcer formation through increased plasma cortisol concentrations, although there is little evidence to support a psychosomatic cause for abomasal ulceration. Fasting leads to sustained periods of low pH and abomasal hyperacidity is probably the major factor promoting ulcer formation. A plausible hypothesis for the development of abomasal ulcers in suckling beef calves is that a sustained period of weather-induced inappetance leads to low abomasal luminal pH thereby promoting ulcer formation. In addition, the presence of poorly ruminated roughage in the abomasum of inappetant beef calves could act to promote ulceration through mechanical damage.

PATHOPHYSIOLOGY OF ULCER DEVELOPMENT

In calves, abomasal pH is close to neutral pH (7.0) at birth and luminal pH rapidly decreases over the first 24 to 48 hours of life. A low abomasal pH is thought to be extremely important in increasing the resistance of newborn calves to enterotoxigenic *Escherichia coli* (ETEC) diarrhea (colibacillosis), as experimental induction of ETEC in calves requires suckling of sodium bicarbonate to increase abomasal pH and facilitate passage of viable *E.coli* bacteria into the small intestine.

In an attempt to develop an effective and practical medical treatment for abomasal ulceration in suckling calves, we developed a method for the continuous measurement of abomasal luminal pH and completed a series of studies examining the effect of orally administered cimetidine, ranitidine, $Al(OH)_3/Mg(OH)_2$, omeprazole, feeding frequency, and oral electrolyte solution formulation on abomasal luminal pH in milk-fed dairy calves. Collectively, these studies indicated that preprandial abomasal luminal pH in suckling dairy calves is low (mean 1.4), and that suckling milk replacer at 12% body weight per day divided into 2 feedings results in sustained periods of low luminal pH. Abomasal hyperacidity is considered to play an important role in the pathogenesis of abomasal ulceration in calves, as back diffusion of hydrogen ions from the lumen into the mucosa permits penetration of the proteolytic enzyme pepsin into the deeper layers of the abomasal mucosa, leading to further damage and ulceration. Because sustained periods of decreased luminal pH increase the likelihood of injury to the abomasal mucosa, it is likely that feeding frequency exerts a profound effect on mean 24 h luminal pH in calves fed milk replacer and thereby influences the risk of abomasal ulceration.

An abomasal luminal pH > 3.0 and > 4.0 will be accompanied by a slower activation rate of pepsinogen and prochymosin, respectively, thereby markedly decreasing the proteolytic activity of abomasal secretions. Because abomasal ulcers are most frequently observed in older calves where pepsinogen is the most important proteolytic agent, a general goal of prophylaxis should be to increase abomasal pH to > 3.0 for as long as possible. We have determined that suckling at 3 h intervals (8x) increases luminal pH >3.0 for the greatest percentage of the 24 h recording period. An important goal when preventing abomasal ulceration in milk-fed calves is presumably to increase luminal pH and thereby inhibit the proteolytic activity of pepsin and chymosin. Chymosin is the major proteinase present at birth in calves, but pepsinogen secretion occurs at or soon after birth. Bovine pepsinogen is activated when pH < 5.0 and the optimal pH for activation is 2.0 to 2.5. Bovine chymosin has an optimal pH for activation of 3.0 to 3.8.

Beef calves usually suckle 3 to 6 meals per 24 h (range 1 to 11 meals per 24 h). Increasing the suckling frequency in dairy calves above the industry standard of every 12 h will more closely reflect suckling behavior in beef calves, not adversely effect on growth rate, but may decrease the prevalence of abomasal ulceration by increasing mean 24 h luminal pH and percentage of the 24 h period that luminal pH >3.0.

REFERENCES

1. Ahmed AF et al. Effect of orally administered cimetidine and ranitidine on abomasal luminal pH in clinically normal milk-fed calves. Am J Vet Res 2001;62:1531-1538. 2. Ahmed AF et al. Effect of an orally administered antacid agent containing aluminum hydroxide and magnesium hydroxide on abomasal luminal pH in clinically normal milk-fed calves. J Am Vet Med Assoc, 2002;220:74-79. 3. Ahmed AF et al. Effect of feeding frequency and route of administration on abomasal luminal pH in dairy calves fed milk replacer. J Dairy Sci, 2002;85:1502-1508. 4. Ahmed AF et al. Effect of orally administered omeprazole on abomasal luminal pH in dairy calves fed milk replacer. J Vet Med A. 2005;52:238-243. 5. Constable PD et al. Comparative effects of two oral oral rehydration solutions on milk clotting, abomasal luminal pH, and abomasal emptying rate in suckling calves. J Dairy Sci. 2009;92:296-312. 6. Constable PD et al. Effect of suckling cow's milk or milk replacer on abomasal luminal pH in dairy calves. J Vet Intern Med. 2005;19:97-102. 7. Constable PD. The treatment of the diarrheic calf: an update. In: Kaske M, Editor. Recent developments and perspectives in bovine medicine: Keynote Lectures of the XXII World Buiatrics Congress. Hannover, Klinik fur Rinderkrankheiten. Philadelphia, PA: W.B. Saunders Company; 2002:132-143. 8. Constable PD. Fluids and Electrolytes. In: Brumbaugh GW, Guest Editor. Clinical Pharmacology. Veterinary Clinics of North America, Food Animal Practice. Philadelphia, PA: W.B. Saunders Company; 2003:19(3):1-40. 10. Constable PD, Gohar HM, Morin DE, et al (1996). Use of hypertonic saline-dextran solution to resuscitate hypovolemic calves with diarrhea. Am J Vet Res 57:97-104. 11. Constable PD et al. Comparison of two oral electrolyte solutions for the treatment of dehydrated calves with experimentally-induced diarrhoea. Vet J 2001;162:129-140. 12. Constable PD et al. Clinical and laboratory assessment of hydration status of neonatal calves with diarrhea. J Am Vet Med Assoc. 1998;212(7):991-996. 13. Constable PD et al. Use of peripheral temperature and core-peripheral temperature difference to predict cardiac output in dehydrated calves housed in a thermoneutral environment. Am J Vet Res. 1998;59(7):874-880. 14. Goodell GM et al. An alkalinizing oral rehydration solution containing lecithin-coated citrus fiber is superior to a non-alkalinizing solution in the treatment of 360 calves with naturally acquired diarrhea. J Dairy Sci. 2012;95:6677-6686. 15. Hemmingsen I. Erosiones et ulcera abomasi bovis. Nord Vet Med 1966;18:354–365. 16. Marshall TS. Abomasal ulceration and tympany of calves. Vet Clin Nth Am Food Anim Pract. 2009;25:209-220. 17. Marshall TS et al. Effect of suckling an isotonic solution of sodium acetate, sodium bicarbonate, or sodium chloride on abomasal emptying rate and luminal pH in calves. Am J Vet Res. 2008;69:824-831. 18. Nouri M, Constable PD. Comparison of two oral electrolyte solutions and route of administration on the abomasal emptying rate of Holstein-Friesian calves. J Vet Intern Med. 2006;20:620-626. 20. Sen I et al. Effect of suckling isotonic or hypertonic solutions of sodium bicarbonate or glucose on abomasal emptying rate in calves. Am J Vet Res. 2006;67:1377-1384. 21. Smith DF et al. Abomasal ulcer disease in adult dairy cattle. Cornell Vet 1983;73:213–224. 22. Smith GW et al. Effect of orally administered electrolyte solution formulation on abomasal luminal pH and emptying rate in dairy calves. J Am Vet Med Assoc, 2012;241:1075-1082. 23. Walker PG et al. Comparison of hypertonic saline-dextran solution and lactated Ringers solution for resuscitating severely dehydrated calves with diarrhea. J Am Vet Med Assoc 1998;213:113-121.

THE 5 BEST FOOD ANIMAL MEDICINE PAPERS OF 2012

Peter Constable, BVSc(Hons), MS, PhD, DACVIM, DACVN(Honorary)
West Lafayette, IN

Peer-reviewed food animal medicine articles published in 2012 were examined. English-language articles published in top tier veterinary journals and a smaller number of animal agricultural journals based on 5-year impact factor (ISI Web of Knowledge) and articles published in other journals based on recommendations from colleagues were considered. The title and abstract of eligible articles were reviewed and articles selected for further consideration. Five evidenced-based articles that dealt with important but different topics were subsequently selected for presentation and discussion. Criteria used for selection included diversity, excellence in study design and statistical analysis, and a perceived likelihood that the results of the study will impact the way we practice food animal medicine. The papers selected contained a clearly stated rationale and hypothesis, and a well described materials and methods section that will permit anyone, anywhere, to reproduce the study.

Study designs used in food animal medicine include descriptive studies (case reports, case series, surveys) and analytic studies (observational studies such as case-control, cross-sectional, and cohort studies; experimental studies such as laboratory studies and randomized clinical trials; Dohoo et al., 2003). The papers selected used an appropriate method of randomization (where appropriate) and had appropriate statistical power as well as internal validity and external validity (generalizability). In general terms, large multi-site studies have greater external validity and small controlled laboratory or experimental station studies have greater internal validity (Lean et al., 2009). Although the ideal study has excellent internal and external validity, completion of the ideal study usually requires extensive resources which are usually unavailable. Guidelines for materials and methods section for randomized controlled trials were developed in 2010 (the REFLECT statement; O'Connor et al., 2010) and should be adhered to whenever possible.

The intent of limiting the list was to focus attention and discussion on a small number of evidenced-based papers that are likely to become "game changers". A similar approach was used to select papers previously at ACVIM Forums (Constable 2009; Constable and Roussel 2010). The selected papers for discussion at the 2013 ACVIM Forum increase our understanding of the pathophysiology of disease or change the way we diagnose, treat, and control disease of food animals. The five papers are listed below in alphabetical order:

1. Bleul U, Bylang T. Effects of doxapram, prethcamide and lobeline on spirometric, blood gas and acid-base variables in healthy new-born calves. *Vet J* 2012;194:240-246.
2. Grosche A, Furll M, Wittek T. Peritoneal fluid analysis in dairy cows with left displaced abomasum and abomasal volvulus. *Vet Rec* 2012;170:413-418.
3. Hill KL, Hunsaker BD, Townsend HG, van Drunen Littel-van den Hurk S, Griebel PJ. Mucosal immune response in newborn Holstein calves that had maternally derived antibodies and were vaccinated with an intranasal multivalent modified-live virus vaccine. *J Am Vet Med Assoc* 2012;240:1231-1240.
4. Mömke S, Sickinger M, Rehage J, Doll K, Distl O. Transcription factor binding site polymorphism in the Motilin gene associated with left-sided displacement of the abomasum in German Holstein cattle. *PLoS ONE* 2012;7(4):e35562.
5. Trefz FM, Lorch A, Feist M, Sauter-Louis C, Lorenz I. Metabolic acidosis in neonatal calf diarrhea – clinical findings and theoretical assessment of a simple treatment protocol. *J Vet Intern Med* 2012;26:162-170.

REFERENCES

1. Constable PD. Top ten evidenced-based papers in food animal medicine and surgery for 2008. *Proceedings of the 27th ACVIM* Forum, June 2009. 2. Constable PD & Roussel AJ. Five of the best food animal internal medicine journal articles published in 2009. The other five best food animal internal medicine journal articles published in 2009. *Proceedings of the 28th ACVIM Forum*, June 2010. 3. Dohoo IR, Martin W, Stryhm H. *Veterinary Epidemiologic Research*, 2nd edition. AVC Inc, Charlottetown, Prince Edward Island, Canada, 2010. 4. Lean IJ, Rabiee AR, Duffield TF, Dohoo IR. Invited review: Use of meta-analysis in animal health and reproduction: Methods and applications. *J Dairy Sci* 2009; 92:3545-3565. 5. O'Connor AM, Sargeant JM, Gardner IA, Dickson JS, Torrence ME and consensus meeting participants. The REFLECT statement: methods and processes of creating reporting guidelines for randomized controlled trials for livestock and food safety. *J Vet Intern Med* 2010;24:57-64.

THE OTHER 5 BEST FOOD ANIMAL MEDICINE PAPERS OF 2012

Geof W Smith, DVM, PhD, DACVIM
Raleigh, NC

Peer-reviewed articles in ruminant medicine published in 2012 were examined. To form a list of potential articles – a search of the CAB Abstract database was performed in January of 2013. Search keywords used to scan title and abstract were cow,* cattle, bovi,* calf, calves, heifer, sheep, goat, alpaca, llama and camelid. These words were combined with medicine, surger,* treatment,* prognosis diagnos*. The * symbol at the end of some words allows the inclusion of all possible terminations of the word (ie. surgery, surgeries, surgical, etc). The initial search found 3,355 potential manuscript. The search was then limited to English language articles only. Initially, a list of about 70 papers were selected based on title and abstract. Then manuscripts were scanned in greater detail to select those the presenter felt had the highest impact or would be the most useful for presentation to an audience of internal medicine diplomates. Amongst criteria used to select manuscripts were: relevance to internal medicine, strength of study design, novelty of the data, and geographic distribution. A deliberate attempt was made to highlight research done in different countries and to illustrate a couple emerging diseases in different parts of the world (such as Schmallenberg virus). Ultimately I attempted to select manuscripts that would be educational to the audience and would impact our ability to practice food animal medicine

Study designs used in food animal medicine include descriptive studies (case reports, case series, surveys) and analytic studies (observational studies such as case-control, cross-sectional, and cohort studies; experimental studies such as laboratory studies and randomized clinical trials).[1] Guidelines for materials and methods section for randomized controlled trials were developed in 2010 (the REFLECT statement)[2] and should be adhered to whenever possible. The five papers are listed below in alphabetical order:

1. Cernicchiaro N, White BJ, Renter DG, Babcock AH, Kelly K, Slattery R. Associations between the distance traveled from sale barns to commercial feedlots in the United States and overall performance, risk of respiratory disease, and the cumulative mortality in feeder cattle from 1997 to 2009. J Anim Sci 2012; 90:1920-1939.
2. Garigliany MM, Hoffmann B, Dive M, Sartelet A, Bayrou C, Cassart D, Beer M, Desmecht D. Schmallenberg virus in calf born at term with porencephaly, Belgium. Emerg Infect Dis 2012; 18:1005-1006.
3. Kasonta R, Sauter-Louis C, Holsteg M, Duchow K, Cussler K, Bastian M. Effects of the vaccination scheme on PregSure-BVD induced alloreactivity and the incidence of bovine neonatal pancytopenia. Vaccine 2012; 30:6649-6655
4. Mokhber-Dezfooli, MR, Nouri M, Rasekh M, Constable PD. Effect of abomasal emptying rate on the apparent efficiency of colostral immunoglobulin G absorption in Holstein-Fresian calves. J Dairy Sci 2012; 95: 6740-6749
5. Ollivett TL, Nydam DV, Linden TC, Bowman DD, van Amburgh M. Effect of nutritional plane on health and performance in dairy calves after experimental infection with *Cryptosporidium parvum*. J Am Vet Med Assoc 2012; 241:1514-1520.

Table 1. Characteristics of selected best papers of 2012

First Author	Animal Studied	Journal	Country Research was Performed	Study Design
Cernicchiaro	Beef calves	Journal of Animal Science	USA	Retrospective study
Garigliany	Beef calf	Emerging Infectious Diseases	Belgium	Case report
Kasonta	Cattle	Vaccine	Germany	Experimental challenge
Mokhber-Dezfooli	Dairy calves	Journal of Dairy Science	Iran	Prospective study
Ollivett	Dairy calves	Journal of the American Medical Association	USA	Experimental challenge

REFERENCES

1. Dohoo IR, Martin W, Stryhm H. *Veterinary Epidemiologic Research*, 2nd edition. AVC Inc, Charlottetown, Prince Edward Island, Canada, 2010.
2. O'Connor AM, Sargeant JM, Gardner IA, Dickson JS, Torrence ME and consensus meeting participants. The REFLECT statement: methods and processes of creating reporting guidelines for randomized controlled trials for livestock and food safety. *J Vet Intern Med* 2010;24:57-64.

SONOGRAPHIC ASSESSMENT OF THE ABDOMEN IN CATTLE

Christian Gerspach, Dr. med. vet., MS, DACVIM
Zurich, Switzerland

INTRODUCTION

Ultrasonography is an ideal extension of the physical examination for the evaluation of abdominal disorders in cattle.
This tool is used to better locate a lesion and to confirm a clinical diagnosis. Ultrasonography has also become an important tool to make a decision in abdominal surgery in cattle. The most common indications for abdominal ultrasonography are traumatic reticuloperitonitis, intestinal obstruction, and displacement of the abomasum.[1] If blood analysis reveals suspicion, the abdomen can be screened for an abscess.

EXAMINATION TECHNIQUE

A complete examination of the abdomen is performed, following a logical procedure. Based on a tentative diagnosis, an examination of the region of interest can be performed. But important findings might be missed. Both sides of the abdomen are examined. The hair is clipped and alcohol is applied to the skin, followed by ultrasound gel. Alternatively, the hair can be soaked with water or alcohol. This takes less time, but the quality of the images is reduced.
The examination starts on the left side, ventral at the reticulum, followed by the spleen and the rumen. On the right side the small and large intestine, right kidney, liver, omasum, and the abomasum are visualized.[2] Depending on the stage of pregnancy the uterus can be seen on either side. Rectal ultrasound is useful to evaluate uterus, left kidney, ureter, urinary bladder, and the pelvic cavity.

RETICULUM

The reticulum can be visualized at the ventral aspect of the thorax, left and right to the sternum. Starting on the left side, the structure of the wall, the motility and structures nearby are evaluated. The layers of the reticular wall can only be imaged, in cattle with ascites. Foreign bodies inside the reticulum cannot be detected by ultrasound. In order to evaluate the reticular motility the reticulum is observed over a period of three minutes. Within 3 minutes the reticulum has 3 biphasic contractions, of which the first phase is incomplete. In cows that are ruminating there is a third phase. Next to the caudal aspect of the reticulum the craniodorsal blindsac of the rumen can be seen. The abomasum can be imaged between reticulum and rumen. The spleen may be located between peritoneum and reticulum.
In cattle with reticuloperitonitis free abdominal fluid and fibrin deposits are often imaged on the reticular serosa and structures nearby the reticulum. Also abscesses can be found next to the reticulum. In these cases a complete examination of the abdomen should be performed, in order to differentiate local from generalized peritonitis.
The motility may be restricted by adhesions. In cows that are stressed the number of contraction decreases and the interval between contractions is longer.[3] In cattle with vagal indigestion the reticular motility may be reduced, normal or increased.[4] Cattle with proximal functional stenosis often have reticular hypermotility. However, based on reticular motility proximal and distal functional stenosis cannot be differentiated and normal motility does not rule out vagal indigestion.[4] In conjunction with characteristic clinical signs ultrasonography can support the diagnosis of vagal indigestion.

RUMEN

The rumen can be imaged at the left abdominal wall. The wall appears echogenic and ruminal contents usually cannot be imaged. Moving the probe from dorsal to ventral, the dorsal sac, the ruminal groove and the ventral sac are seen. In cattle with peritonitis fibrin can be located in the ruminal groove. Fluid, fibrin or abscesses between the ventral ruminal sac and the left body wall are typical findings in omental bursitis.

SPLEEN

The spleen is located on the rumen, in the region of the intercostal spaces 7 – 12. The parenchyma is homogeneous and contains several anechogenic blood vessels.[5] Enlargement in infectious processes, abscesses or tumors can be imaged in the spleen.

OMASUM

The omasum is visible between intercostal spaces six to eleven, on the right side of the abdomen and the size is greatest in the ninth intercostal space.[6] The wall of the omasum appears as a thick echogenic line. The contents cannot be visualized. The attachement of the omasal laminae are only occasional visible.[6] In cows with abnormal increased fluid in the omasum the laminae are seen on ultrasonography. The omasum does not appear to have active motility in healthy cows. In cows with reticulo-omasal stenosis omasal motility has been observed.[6] Ultrasonography was found not to be useful to diagnose omasal impaction.[7]

ABOMASUM

Using ultrasonography, size, position, and content of the abomasum can be assessed.[1] The abomasum is seen caudal of the xyphoid from ventral midline and in the left and right paramedian region.[8] The pylorus usually can be imaged ventrally on the right side. The wall of the abomasum appears as a thin echogenic line. The abomasal content can be seen, as well as the abomasal folds.

In cows with left displacement of the abomasum the abomasum is seen between the left body wall and the rumen. Moving the probe from dorsal to ventral, within the last three intercostal spaces, the abomasal gas cap can be seen, followed by ingesta and abomasal folds. Sometimes the pylorus is imaged between left body wall and rumen.

In cows with right displacement of the abomasum the abomasum is seen within the last three to four intercostal ribs on the right. The displaced and dilated abomasum displaces the liver from its usual position. Abomasal displacement and abomasal volvolus cannot be differentiated by ultrasonography.

Functional and mechanic stenosis of the pylorus or intestinal obstruction results in dilation of the abomasum. The abomasum is distended, filled with liquid ingesta, but not displaced.

Type 3 abomasal ulcers can cause a local peritonitis, which appears similar to peritonitis caused by hardware. Type 4 abomasal ulcers cause a generalized peritonitis, with free abdominal fluid, fibrin and adhesions.

Tentative diagnoses of abomasal lymphoma[9] and lymphosarcoma[10] have been made by ultrsonography. In both cases the abomasum was in an abnormal position, the wall was thickened and the pylorus was dilated.

INTESTINE

For ultrasonography of the small intestine, the area from the right tuber coxae to the eighth intercostal space and from the transverse processes of the vertebrae to the linea alba on the right side is examined.[1, 11] Diameter, contents and motility are assessed. Usually the ingesta can be seen within the intestinal loops. The diameter is 2 – 4 cm. The duodenum is seen close to the liver and gallbladder. It is usually located within the tenth and eleventh intercostal space. The diameter of the cranial part of the duodenum is 0.9 – 5 cm.[11] The jejunum and the ileum cannot be differentiated. The small intestines are constantly moving.

Diameter and motility are the most important parameters to diagnose ileus. In cattle with intestinal obstruction or ileus, the motility is reduced or absent. The diameter of the loops is more than 3.5 cm. If less than 5 dilated loops are seen a proximal ileus (duodenum) is most likely. If more than 5 dilated loops are seen distal ileus (jejunum or ileum) is more likely.[12] In cows with ileus, the diameter of dilated loops does not change. Oftentimes empty loops are seen adjacent to dilated loops. Liquid intestinal contents appear anechoic. In cows with haemorrhagic bowel syndrom blood clots appear hyperechoic. Only in 19 % of cows with haemorrhagic bowel syndrome blood clots are seen on ultrasonography.[13] Intestinal intussusception sometimes results in a bull's eye appearance of the invaginated intestine.[12] This finding is also occasional imaged in cattle with intussusception. In cases of incomplete perforation of the intestine, free fluid or fibrin can be seen.

A dilated cecum can be imaged from the right flank. In a gas filled cecum only the wall towards the body wall can be imaged.

LIVER

Assessment of the liver reveals findings consistent with hepatic abscesses, fatty liver, neoplasia, fascioliasis and disease of major vessels.[14] The liver is located at the right body wall. The examination is performed from caudal to cranial, beginning caudal to the last rib, ending at the fifth intercostal space. The caudal vena cava is visualized in the twefth and eleventh intercostal spaces and has a triangular shape. The portal vein is located in the twelfth to eighth intercostal spaces.[14] The gallbladder is situated between the ninth and eleventh intercostal spaces and appears as a fluid-filled vesicle with anechoic content.

Hepatic abscesses can be found within the whole liver. The content appears anechogenic to hyperechogenic and may be homogeneous or heterogeneous. There can be single or multiple abscesses in varying size. Hepatic tumors appear as single or multiple circumscribed structures.

In cows with fatty liver the parenchyma appears white on the ultrasonogram. The contrast between tissue and vessels is decreased.

Chronic fascioliasis is associated with calcified bile ducts. The bile ducts are hyperechoic with distal shadowing. In obstructive cholestasis the bile ducts appear dilated. A proximal obstruction is associated with dilated bile ducts and a distal obstruction (duodenal papilla) is associated with a dilation of the common bile duct and the gallbladder.[14] The gallbladder can

be enlarged in anorectic cows, because there is no stimulus for emptying the gallbladder.
If the caudal vena cava is dilated it appears round. The reasons for a dilated caudal vena cava are right-sided cardiac insufficinecy, compression by a mass, and thrombosis of the vena.

KIDNEYS

The right kidney is examined over the right paralumbar fossa and in the last intercostal space.[15] The left kidney and the urinary bladder can be approached by transrectal ultrasonography. Oftentimes it is not possible to evaluate the entire kidneys.
In cows with pyelonephritis hyperechogenic debris is seen in the renal sinus and calices. Cysts appear round to oval and anechoic. They are often unilateral.[15] If the kidneys are otherwise normal, cysts are incidental findings.
In patients with suspected urolithiasis an examination of the entire urinary tract is indicated. Urinary calculi are hyperechoic and cast acoustic shadows.
Especially in pyelonephritis and urolithiasis it is important to differentiate between unilateral and bilateral kidney lesions, which significantly affect the prognosis.

REFERENCES

[1]Braun U. **Vet Clin Food Anim**. 2009;25:567. [2]Braun U. **Vet Clin Food Anim**. 2005;21:33. [3]Braun U. et al. **Vet. Rec.** 2008;163:571. [4]Braun U. et al. **Vet. Rec.** 2009;164:11. [5] Braun U. et al.**Vet J.** 2006;171(3):513. [6] Braun U. et al. **Vet. Rec.** 2006;159:812. [7]Imran S. et al. **Vet Med Int.** 2011;2011:1. [8] Braun U. et al. **Vet. Rec.** 1997;140(4):93. [9]Buczinski S. et al. **JAVMA**. 2011;238(8):1044. [10]Braun U. **BMC Veterinary Research**. 2001;7:20. [11] Braun U. et al. **Vet. Rec.** 1995;136(10):239. [12] Braun U. et al. **Vet. Rec.** 1995;137(9):209. [13] Braun U. et al. **Vet. Rec.** 2010;166:79. [14] Braun U. **Vet Clin Food Anim**. 2009;25:591. [15] Floeck M. **Vet Clin Food Anim**. 2009;25:651.

ULTRASONOGRAPHY FOR THE DIAGNOSIS OF BOVINE CARDIORESPIRATORY DISEASES

Sébastien Buczinski, DrVét, DÉS, MSc, DACVIM,
Faculté de médecine vétérinaire, St-Hyacinthe, Université de Montréal, Qc, Canada.

Bovine respiratory disease (BRD) is one of the leading cause of disease in both dairy and beef cattle.[1-2] Although less common, heart diseases have an important impact on cattle since their diagnosis most of time carry a poor prognosis.[3] Despite diagnosis of BRD is mostly made based on typical clinical signs (e.g. fever, depression and alteration of respiratory function), this diagnosis may be difficult when clinical signs are not obvious or in chronic cases of BRD.

Cardiac disease may also lead to clinical challenge since the diagnosis is generally not obvious until obvious clinical signs of heart failure are present. Moreover, the clinical setting and the size of the patients generally leads to a lesser sensitivity of cardiac auscultation to detect any cardiac anomaly.

Moving from these constatations, ancillary tests that can be performed cow-side are of utmost interest since they allow a more accurate diagnosis. Among these ancillary tests, ultrasonography is one of the more practical tool that directly allow to examine the lesions (e.g. consolidated lung for BRD ; or assessment of the various parts of cardiac anatomy as well as cardiac function). The aim of this session is therefore to present an up-to-date review of bovine cardiorespiratory disease focusing not only on the diagnosis *per se* but also on the prognostic application of ultrasound in bovine cardiorespiratory problems.

ULTRASONOGRAPHY OF THE LOWER RESPIRATORY TRACT (LRT)

The ultrasound of the LRT may be easily performed with the high frequency probes that are used for reproductive purpose since this examination mostly focus on the examination of the pleura and peripheral part of lung parenchyma.[4] The intercostal spaces can be clipped with ultrasonographic gel applied in the area of interest, but the examination can also be performed without clipping and only applying alcohol directly on the area of interest.[4,5] This is of clinical interest to perform this examination faster in a farm setting.

The scanning then begins and consists in a rapid screening of the ICS especially on the ventral part since this is the most common location of pneumonia in cattle. The typical ultrasonographic findings in healthy pleura and lung consist in hyperechoic visceral and parietal layers of the pleura sliding during the breathing movements. Due to its gaseous content, no lung parenchyma is observed and all ultrasounds are reflected back to the transducer occasionally leading to a reverberation artifact.

The various anomalies encountered during BRD consist in (1) pleural irregularity or thickening, (2) visualization of the lung parenchyma due to inflammation and exsudate deposit in the parenchyma or atelectasia and (3) various amounts of pleural fluid accumulation. Ultrasound of the LRT can therefore be used for assessing the extension of the lung consolidation and non ventilated pulmonary parenchyma.

Pleural irregularity or thickening

The pleural line is normally a thin echoic line with a smooth contour. The early lung anomalies associated with pneumonia and that can be observed with ultrasonography consist in the observation of comet tails artifacts which are numerous hyperechoic parallel lines due to reverberation artifacts associated with small reflective surfaces within the pleural space. These lesions can also be observed in cases of lung edema as well as in diffuse parenchymatous diseases. Comet-tails can also be observed in healthy animals so they should not be considered as a reliable marker of lung disease.

Pleural fluid accumulation

Pleuritis and bronchopneumonia are characterized by ventral accumulation of various amounts of hypochoic to anechoic fluid with variable quantity of echoic fibrin strands.[4] The most striking feature in these cases will be the observation of separation of both visceral and parietal pleura. Depending on the amount of pleural fluid, other observation associated can be lung parenchyma visualization despite the absence of lung infection but secondary to lung atelectasia.

Lung consolidation

Consolidated lung has a hypoechoic appearance close to the liver echotexture. The consolidation can only be observed if adjacent to the pleural line since deeper consolidated area can not be observed due to the reflection of the ultrasound beam by healthy lung tissue. The consolidated parenchyma is most of time heterogenous due to (1) the appearance of hyperechoic bronchi that could also contain residual gas or (2) secondary gas production by the bacteria responsible of the pneumonia. Cavitary lesions can also be observed with a thick capsula and heterogenic content when abcedative lesions are present.

Lung ultrasonography as a prognostic tool

Most of the available data concerning ultrasonography are only descriptive information without any information on the prognostic application of this ancillary test. To date, only a few studies mentioned this practical application.

In the feedlot side, a study was performed in Albertan feedlots using systematic screening of 3 intercostal spaces in calves with a clinical suspicion of BRD base on penchecker clinical assessment. The authors did not find any prognostic information when performing ultrasonography during the processing or when pulled to the chute with BRD suspicion.[6] On the dairy side, a study conducted nearly 100 pre-weaned dairy calves using a systematic screening of both sides of the thorax found that consolidation was negatively associated with growth rate and can also be found in calves without obvious signs and history of BRD.[5]

In a recent study on feedlot calves in Idaho with naturally occuring BRD, systematic thoracic ultrasonography showed that maximal depth of lung consolidation, maximal area of consolidated lung per site, and total area of consolidated lungs were significantly associated with survival.

These different results showed that further studies need to be performed to be able to recommend practical guidelines concerning the calf-side application of thoracic ultrasonography as a prognostic ancillary test.

TRANSTHORACIC ECHOCARDIOGRAPHY (TTE)

Transthoracic echocardiography (TTE) allows a non-invasive diagnosis of the most common heart diseases in cattle. Among these diseases pericarditis, bacterial endocarditis (and cardiac lymphoma in country where bovine leukosis virus infection is present) are the main differential diagnosis of heart diseases in adult animals. In calves, congenital heart diseases (CHD) need to be added in the differential list (especially ventricular septal defect). The TTE is generally performed using a low-frequency probe especially in adult cows due to the required depth of examination. The shape of the probe is ideally thin and could be easily manipulated in the narrow intercostal space. If available a phased array probe should be preferred. The area of interest is clipped, rinsed and copious amount of ultrasonographic gel are applied on the ventral part of the thorax from the 3rd to the 6th intercostal space. If needed the cranial limb can be pushed forward to be able to place the probe more adequately. The author personnally place his leg medial to the cranial limb and try to create a gentle abduction of the limb. This is well tolerated by most of the cows and is easy to perform with some experience.

Normal echocardiographic findings

A thorough examination of the main cardiac structures need to be performed including the pericardium, the myocardium as well as the endocardium. The assessment of the cardiac cavities should also be or calculated looking for dilation of the cavities.[7,8] The examination is performed from the right and the left side, however, from a practical perspective the more informative views are obtained using right-side TTE views. With some experience, this two-dimensional examination is performed in less than 10 minutes, and can be used in the farm in order to confirm or exclude heart disease.[9] This may potentially have an important impact on the therapeutic decision (decide to treat or to not invest money in an animal with poor chances of recovery).

The right parasternal long axis views consist in (1) the four chamber view allowing to see both right and left ventricles (RV, LV), right and left atria (RA, LA) and atrioventricular valves (tricuspid (TV) and mitral (MV) valves), (2) the left ventricular outflow view allowing one to see the aortic root and aortic valve in addition with the previous structures seen in the precedent view, (3) the right ventricular outflow tract seeing the RA, TV, RV, the pulmonary trunk, pulmonary valve and the aorta in transverse section .

Pericardial effusion and pericarditis

Pericardial fluid accumulation is observed in cases of septic pericarditis (traumatic origin or extension of pneumonia), idiopathic hemorrhagic pericarditis (IHP) and cardiac lymphoma. Depending of the amount of pericardial fluid, different signs of cardiac tamponade may be seen including collapsed right ventricle during diastole and atrial collapse during atrial diastole.

The echogenicity of pericardial fluid may help in the differential diagnosis of these conditions. Anechoic pericardial fluid is classically observed in cases of lymphoma or IHP. Heterogenic echogenicity with echogenic fibrin clots is classically seen with septic pericarditis but can also be seen with other pericardial conditions. For these reasons, pericardiocentesis should always be performed to confirm the diagnosis.

Endocarditis

Endocarditis is the most common endocardial disease in cattle. Echocardiography is a sensitive method (84.3%, 95% CI : 60.4-95.0%) to detect valvular thickening when compared with necropsic findings as recently reported in a meta-analysis.[10] The 4 cardiac valves should be systematically evaluated. The three right parasternal long axis views are required to assess the tricuspid, mitral, aortic and pulmonary valves. The typical echocardiographic findings in endocartitis in cattle consist in irregular valvular thickening with a shaggy appearance. Secondary to valvular insufficiency, turbulences may be observed using Doppler examination. Dilation of cardiac chambers may also be observed secondary to chronic regurgitation (eg distended right atrium with tricuspid endocarditis).

Cardiac lymphoma

There are two main echocardiographic findings in cardiac lymphoma including (1) massive amounts of anechoic pericardial fluid accumulation with signs of cardiac tamponade and (2) a mass effect in the right atrial wall.[11] These echocardiographic findings are not specific (differential diagnosis include pericarditis and/or endocarditis) and therefore should be confirmed using other ancillary tests (eg : evidence of neoplastic cells using body fluid analysis (pericardial fluid, paracentesis…).

Ventricular septal defect and other congenital heart diseases

Echocardiography is a useful tool to assess simple or complex congenital heart diseases (CHD). We will just describe echocardiographic findings in cases of ventricular septal defect (VSD) which is by far the most common CHD in calves. The clinical presentation is generally sufficient to raise a clinical suspicion of VSD and can be confirmed using transthoracic echocardiography. The best view to observe the defect in the membranous part of the septum is the right parasternal long-axis view of the left ventricular outflow tract.[12] The defect is classically observed just below the aortic cusp. It is important to remember that the four chamber view alone is not recommended for the diagnosis of VSD.

Other more complex CHD can also be diagnosed using 2-dimensional and Doppler ultrasonography, but require more advanced skills in echocardiography and are less frequent. The reader is refered to more specialized textbooks addressing specific echocardiographic findings for diagnosing these less commonly encountered CHD.[13]

In conclusion, thoracic ultrasonography is an interesting and practical tool that can be used by the internist and the clinician for the diagnosis of various diseases. In cases of suspected BRD, thoracic examination can be performed with units that are available for reproduction work and may help to assess lung consolidation. The prognostic information brought by this ancillary tests may be promising but data are scant and contradictory.

Echocardiography is a little more technical to perform and requires more specialized equipment (at least a low frequency probe but phased array probe are more adequate) but may be helpful to rule in or to rule out heart disease. This information is of utmost interest since it will have an impact on case management.

REFERENCES

[1]McGuirk SM. **Vet Clin North Am Food Anim Pract** 2008;24:139. [2]Miles DG. **Anim Health Res Rev** 2009;10:101. [3]Buczinski S, et al. **Vet J** 2010;184:258. [4]Babkine M et al. **Vet Clin North Am Food Anim Pract** 2009;25:633. [5]Ollivett TL, et al. **Am Assoc Bovine Pract Proc** 2001;44:148. [6]Abutarbush SM, et al. **Can J Vet Res** 2012;76:23. [7]Braun U, et al. **Vet Rec** 2001;148:47. [8]Hallowell GD, et al. **J Vet Card** 2007;9:91. [9]Buczinski S. **Vet Clin North Am Food Anim Pract** 2009;25:611. [11]Buczinski S, et al. **Vet J** 2012;193:349. [12]Buczinski S. **J Am Vet Med Assoc** 2012;241:1083. [13]Buczinski S, et al. **Can Vet J** 2006;47:246-252. [14]Boon JA. Congenital shunts and atriovalvular valve dysplasia. In :Boon JA, Ed **Veterinary Echocardiography** Wiley-Blackwell, 2nd ed, 2011, 437.

ELECTROENCEPHALOGRAPHY AND CLINICAL FINDINGS IN 37 DOGS AND 4 CATS.

WW Bush[1], GA Garcia[1], CE Weaver[1], MG Young[1], DE Cuff[1], DC Williams[2], MM Stecker[3].
[1]Bush Veterinary Neurology Service, Leesburg, VA; [2]William R. Pritchard Veterinary Medical Teaching Hospital, Davis, CA; [3]Winthrop University Hospital, Mineola, NY.

EEG is a regular part of the evaluation in humans with either altered mental status or possible seizure but is not regularly performed in clinical veterinary medicine. In our clinic, EEG studies are interpreted by a physician skilled in both the interpretation of human and animal EEG. We retrospectively evaluated all EEG from last 4 years to better understand the value of veterinary EEG studies. We hypothesized that seizure-like events could be evaluated and classified quickly using EEG. We hypothesized that a positive or abnormal EEG was associated with an abnormal MRI, abnormal CSF, more hospitalization time and cost, and higher mortality. We also hypothesized that abnormal EEG patterns could be associated with a higher chance of mortality.

Cases were included if the medical record contained an EEG report, exam findings at time of the EEG, itemized invoice, and when performed a record of the MRI and CSF findings. EEG was considered positive when there was spike, sharp wave, electrical seizure, non-convulsive status epilepticus (NCSE), generalized periodic complexes, periodic lateralized epileptiform discharges or a burst suppression pattern. Mortality was recorded if the patient died or was euthanized for brain disease. The EEG studies were recorded using a variety of sedative techniques, at least six active recording electrodes, and standard filter settings. Cases were placed into the acute group if the indication for the EEG was an active movement disorder, severe confusion and inability to walk, or coma, whereas the chronic group was normal on examination at the time of the EEG.

EEG was positive in 17/41 (41%) of cases. MRI and CSF were abnormal in 13/32 (40%) and 12/31(39%) of cases, respectively. When the EEG was positive, an abnormal MRI and abnormal CSF was noted in 9/13 (69%) and 6/11 (55%) of cases, respectively. The EEG was positive in 14/28 (50%) and 3/13 (23%) for the acute and chronic group, respectively. Mortality at 3 months post-EEG was 10/41 (24%) and 4/24 (17%) for EEG positive and EEG negative cases, respectively. Hospitalization time and costs for patients in the EEG positive group were twice that of the EEG negative group. Significant correlation was found between a positive EEG and a positive MRI and positive EEG and mortality at 3 months. Generalized periodic complexes were significantly associated with mortality. Three patients were identified with NCSE and 1 patient with convulsive activity and negative EEG improved rapidly without seizure medication.

These findings suggest that EEG was a better tool for detecting seizure than observation alone. Within this patient population a positive EEG was associated with abnormal advanced test results, longer and more expensive hospital stays and higher mortality. EEG appeared to be a valuable test and should be considered in evaluating patients with acute brain dysfunction.

PLACEMENT OF DEEP STIMULATING AND RECORDING ELECTRODES IN THE CANINE BRAIN USING THE BRAINSIGHT® FRAMELESS STEREOTAXY SYSTEM.

S Long[1], S Frey[2], D Freestone[3], J Giftakis[4], P Stypulkowsk[4], M Cook[3]. 1 Melbourne University Faculty of Veterinary Science, Werribee, Australia. 2 Rogue Research, Montreal, Canada. 3 St Vincent's Hospital, Melbourne, Australia. 4 Medtronic Inc., Minneapolis,MN.

Up to 30% of canine epileptics and a similar proportion of their human counterparts suffer from refractory epilepsy. The development of novel treatment strategies is therefore an active field of research, and one of the most promising of these is the prediction or detection of seizure activity that can be abrogated by targeted electrical stimulation within certain nuclei in the brain. The purpose of this study was to examine the feasibility of placing electrodes in the thalamus and hippocampus of normal greyhounds as part of a wider study evaluating a novel stimulating and recording device.

Four normal greyhounds were enrolled in the study. Imaging using 1.5T MRI was performed on the day prior to surgery with fiducial arrays attached to the frontal bone for trajectory planning. The targets chosen were the anterior (rostral) nucleus of the thalamus and the hippocampus on the ipsilateral side in each dog. During surgery, following registration of the subject with the Brainsight system, a rostral craniotomy approach was used to expose the calvarium. Using the Brainsight system, the target and trajectory were identified and burrholes drilled at the point of entry for each electrode. Following incision of the dura, a canula and stylette were advanced along the appropriate trajectory to the target depth, after which the stylette was removed and either a 0.6mm diameter or a 4mm diameter electrode, depending on the location, threaded through the canula to target. Following removal of the canula, electrodes were secured in place using ceramic screws and methylmethacrylate bone cement then connected via an extension lead to the stimulating and recording device which was placed within a pocket in the muscle on the left side of the neck of each dog. Postoperative CT scanning and co-registration with the preoperative MRI was then performed to confirm placement of the electrodes and calculate placement error before the dogs were recovered.

All four dogs recovered normally and showed no neurological deficits post implantation. Postoperative CT showed that in one dog both leads pulled out during closure, although intraoperative recording and impedance checks together with electrode tracks visible on CT suggested correct placement initially. In the remaining three dogs co-registration of CT and MRI showed that all leads were found to be in contact with the intended target regions.

This study showed that the Brainsight system can be successfully used to place stimulating and recording electrodes with precision and minimal morbidity in deep regions of the brain of dogs. Future trials involving epileptic dogs are planned in order to further evaluate the ability of the stimulating device to abrogate seizures in the clinical setting.

RAPID IDENTIFICATION OF MENDELIAN DISEASE VARIANTS IN THE WHOLE-GENOME SEQUENCES OF INDIVIDUAL DOGS. G.S. Johnson, D.P. O'Brien, J.R. Coates, T. Mhlanga-Mutangadura, R. Zeng, D.H. Gilliam, J.F. Taylor, and R.D. Schnabel. University of Missouri, Columbia, MO.

Recently whole-genome sequencing (WGS) from human patients or small family groups has been used to identify disease-causing mutations. As costs decrease, WGS is likely to become the preferred method for such studies. The identification of disease alleles among the numerous sequence variants, revealed by the alignment of a patient's whole genome sequence to a reference genome sequence, is facilitated by an accurate and well-annotated reference sequence and the availability of a panel of alleles in dbSNP known to occur in individuals unaffected by the targeted disease. Unsurprisingly, build 3 of the canine reference genome is neither as accurate nor as well annotated as build 37 of the human reference sequence. Furthermore, dbSNP contains only 3,328,578 canine entries compared to 63,222,716 entries for human (12/18/12). Thus, it was uncertain if the canine resources were of sufficient quality to support mutation discovery by WGS.

We describe a strategy for using WGS of individual affected dogs to rapidly identify the alleles responsible for autosomal recessive diseases. We used Illumina sequencing technology to generate whole genome sequences for 23 dogs, including 15 dogs which are (1) diagnosed with autosomal recessive diseases and (2) for which we have DNA samples from other dogs likely to be affected by the same disease. NextGENe software (SoftGenetics) was used to align the reads from each dog to the canine genome reference sequence and to identify sequence variants predicted to affect the primary amino acid sequences of the gene products. The variants were further filtered to identify those that were homozygous in the sequenced dog, but absent from the other 22 sequenced genomes. If the diseases were previously mapped, the variants were further filtered to remove all sequence variants not harbored within the mapped regions. The resulting candidate sequence variants are prioritized based on the known biological functions of the gene products and the likelihood that the variants will alter these biological functions. DNA from other dogs believed to have the same diseases can then be genotyped for the highest priority candidate sequence variants to determine if they are also homozygous for the pertinent sequence variant.

We have used this strategy to identify the sequence variants likely to cause the diseases in four dogs. Subsequent speakers will provide clinical details for three of these dogs. We continue to refine our data processing pipeline to discover causal sequence variants for the other 11 dogs and to generate whole genome sequences for additional dogs with unique recessive diseases.

IDENTIFICATION OF TWO BREED-SPECIFIC MUTATIONS ASSOCIATED WITH CANINE MULTIPLE SYSTEM DEGENERATION USING WHOLE GENOME ESEQUENCING. D. P. O'Brien, R. Zeng, R. D. Schnabel, J. F. Taylor, and G. S. Johnson. University of Missouri, Columbia MO.

Canine Multiple System Degeneration (CMSD) is a hereditary movement disorder of Kerry Blue Terriers (KBT) and Chinese Crested Dogs (CCD) characterized by degeneration of the cerebellum, caudate nucleus, putamen, and substantia nigra. This study was conducted to identify the mutation(s) responsible within the 5.2Mb locus previously mapped (O'Brien et al, 2005).

We performed whole genome resequencing of DNA from an affected dog from each breed with Illumina Genome Analyzer II and HiSeq 2000 instruments to over 20x depth of coverage. This sequence was processed using a novel analysis pipeline developed at the University of Missouri. Variants identified within the locus were screened for their potential to cause disease. We identified a different plausibly causal variant in the *SERAC1* gene in each breed: a nonsense mutation in exon 15 that produces a premature stop codon in KBT and a 4bp deletion of the exon 4 splice donor site in CCD. RT-PCR of RNA from an affected CCD and a heterozygous littermate confirmed exon-skipping. Genotyping assays were designed, and genotypes for 99 CCD and 85 KBT were concordant with their phenotypes.

We conclude that mutations in *SERAC1* cause CMSD, and a DNA test is being developed to allow breeders detect carriers. *SERAC1* mutations cause human MEGDEL syndrome, a mitochondrial encephalopathy with impaired oxidative phosphorylation and altered cholesterol & phospholipid metabolism. CMSD can serve as a model disease to understand the role of these processes in neurodegeneration. The results also illustrate the utility of whole genome sequencing for identification of disease-causing mutations.

IDENTIFICATION OF A MUTATION ASSOCIATED WITH SPINOCEREBELLAR ATAXIA, MYOKYMIA AND SEIZURES IN JACK/PARSON RUSSELL TERRIERS USING WHOLE GENOME RESEQUENCING. J.R. Coates, D. Gilliam, D.P. O'Brien, R.D. Schnabel, J.F. Taylor, and G.S. Johnson. University of Missouri, Columbia MO.

Spinocerebellar ataxia (SCA) is a hereditary disease of Jack and Parson Russell Terriers (J/PRT). In addition to cerebellar signs, affected dogs can exhibit myokymia, and seizures. To identify the responsible mutation, we performed whole genome resequencing of DNA from an affected dog with Illumina Genome Analyzer II and HiSeq 2000 instruments to over 20x depth of coverage. This sequence was processed using a novel analysis pipeline developed at the University of Missouri. Screening the detected homozygous variants identified a plausibly causal variant in *KCNJ10*. We genotyped 200 J/PRT and 75 randomly sampled dogs of 65 other breeds. All 15 J/PRT homozygous for the mutant allele showed SCA except one with no phenotype information. One dog reported as SCA without necropsy confirmation was homozygous wild-type. The mutant allele was not found in other breeds.

We conclude that the mutation in *KCNJ10* causes most SCA in J/PRT, and developed a DNA test to detect carriers. *KCNJ10* encodes a glial potassium channel that regulates neuronal excitability by buffering extracellular potassium and facilitating glutamate uptake. Mutations in human *KCNJ10* cause a syndrome of ataxia, seizures, deafness and electrolyte imbalances. Knockout mice show severe ataxia, paraparesis, deafness, and premature death, while conditional knockout in glia produces ataxia and seizures with hypomyelination, vacuolation and axonal swellings similar to SCA. Further study of SCA in J/PRT can aid in understanding the role of this channel in these disease processes. The results also illustrate the utility of whole genome sequencing for the identification of disease-causing mutations.

GENOME WIDE METHYLATION SEQUENCING TO PROFILE EPIGENETIC MODIFICATIONS IN CANINE GLIOMA. P.J. Dickinson[1], R.P. Nagarajan[2], D. Lin[2], M. Britton[1], J. Fass[1], N. Joshi[1], R.J. Higgins[1], R.A. LeCouteur[1], J.F. Costello[2]. 1. University of California Davis, Davis, CA. 2. University of California San Francisco.

Epigenetic mechanisms of gene activation and silencing, such as aberrant methylation of promoter elements and genome wide alterations in methylation status have been implicated in gliomagenesis, with or without associated genomic alterations. Methylation status in human tumors is often assessed in individual genes known to play a role in specific tumor development. Genome-wide screening can provide a more systematic and unbiased approach for the detection of novel gene targets and patterns of methylation, particularly relevant in canine gliomas where specific candidate genes are less well defined, and assumptions about commonality of underlying causes with human tumors are not always valid.

We developed a protocol utilizing the canine reference genome assembly (Broad CanFam3.1) and massively parallel sequencing methodology to provide high resolution, genome wide profiles of 5-methylcytosine in canine gliomas and normal brain. Use of methylated DNA immunoprecipitation sequencing (MeDIP-seq) combined with methylation-sensitive restriction enzyme sequencing (MRE-seq) allowed comprehensive methylome coverage at low cost.

The methodology was found to be robust and alignment of sequence data was achieved for both MRE-seq and MeDIP-seq libraries as well as non methylation dependant input libraries. Detailed genome wide analysis of a canine GBM demonstrated similarities to human GBM including hypermethylation of the HOXA10 gene and genomic deletions of the INK4a/ARF locus, NF-1, and TUSC1 genes.

These preliminary data suggest that canine gliomas may have similar genetic and epigenetic alterations to human tumors and that this genomic approach may be used in a cost-effective manner to define oncogenic pathways and therapeutic targets in canine gliomas.

STEM CELL SOURCES IN ADULT DOGS. JH Lim, NJ Olby. North Carolina State University College of Veterinary Medicine, Raleigh, NC.

Cell-based therapeutic strategies have generated great interest in the past decade as novel tools in the field of regenerative medicine. Interest is now focused on developing clinically relevant autologous stem cell sources that are readily harvested. Adipose tissue has several advantages as a source of mesenchymal stem cells. It is easily accessible with a minimally invasive harvesting procedure, and yields large numbers of adipose tissue-derived stromal cells (ADSC). The present study was conducted to determine whether autologous neural progenitor cells (NPC) generated from ADSC had comparable characteristics to NPC derived from the subventricular zone (SVC) and could be employed as a practical stem cell source for neurologic disorders.

Stem cells were isolated from dorsal cervical subcutaneous fat and the SVZ of the lateral ventricle in adult dogs. NPC were generated from ADSC (ADSC-NPC), and their characteristics compared with NPC isolated from the SVZ using mRNA expression analysis and immunocytochemistry (ICC). The capacity for neural differentiation and the karyotypic stability were also evaluated.

ADSC-NPC and SVC-NPC were successfully generated. Both sources of NPC showed similar gene expression profiles and immunocytochemical characteristics. ADSC-NPC demonstrated a capacity to produce neurons and glia, and showed a high level of chromosomal stability. We conclude that ADSC-NPC represent a powerful, safe autologous stem cell source, which is clinically relevant for cellular transplantation trials related to neurologic disease.

THE NEBULIN GENE (*NEB*) IS LIKELY CANDIDATE FOR NEMALINE ROD MYOPATHY IN A FAMILY OF AMERICAN BULLDOGS. G. D. Shelton[1], C. Rinz[2], J. Huska,[3] L. Gaitero,[3] L. Guo,[1] L.A. Clark[2]. 1. University of California San Diego, La Jolla, CA. 2. Clemson University, Clemson, SC. 3. University of Guelph, Guelph, ON, Canada.

Nemaline rod myopathy (NM) is a congenital muscle disease associated with weakness and the presence of nemaline bodies (rods) in muscle fibers. To date, recessive mutations in six genes have been implicated in human NM including tropomyosin 3 (*TPM3*), skeletal a-actin (*ACTA1*), nebulin (*NEB*), tropomyosin 2 (*TPM2*), troponin T (*TNNT1*) and cofilin 2 (*CFL2*). Single case reports of nemaline rod myopathy are in the veterinary literature but specific mutations have not yet been identified. Five month old male and female American Bulldog littermates were evaluated for non-progressive muscle weakness, exercise intolerance and tremors beginning at approximately 2 months of age. Serum CK activities were mildly elevated. Cryosections of muscle biopsies demonstrated the presence of numerous nemaline rods in multiple muscles using the modified Gomori trichrome stain. Nemaline rods were also confirmed ultrastructurally by electron microscopy. To narrow the list of candidate genes, single nucleotide polymorphism (SNP) profiles were generated for both affected littermates, an unaffected littermate and the unaffected parents. Genotypes were analyzed to determine the pattern of inheritance in chromosomal regions surrounding the six candidate genes. SNPs flanking *NEB* and *TNNT1* were consistent with an autosomal recessive mode of inheritance. Analyses of microsatellites showed homozygosity for markers proximal to *NEB*, but not *TNNT1*. These data suggest that *NEB* harbors a mutation that was inherited from a common ancestor in the affected littermates. Directed sequencing of canine *NEB* is in progress.

RODENT SCI MODELS AND THEIR APPLICATION TO DOGS

Jonathan M. Levine, DVM, DACVIM (Neurology)
College Station, TX

INTRODUCTION

The concept of spinal cord injury (SCI) models dates back 100 years, to Dr. Alfred Allen a neuropathologist working at the University of Pennsylvania.[1] Since that time, animal models – in particular rodent models – have become an accepted platform for understanding SCI. Validated rodent SCI models are easy to replicate, share pathologic features with human and canine SCI, have outcome metrics that are reliable and resemble those used in clinical SCI, and provide an environment which allows novel therapeutics to be studied.[2] Despite the benefits of a rodent-based approach to SCI research, these models do have significant limitations and the best means of applying rodent-derived data has been questioned.[3,4] The goals of this presentation are to i) review common rodent SCI models; ii) discuss the strengths and weakness of these models; and iii) provide thoughts on how data can be applied to canine SCI.

MODELS

Contusion

Contusion is believed to be an important primary mechanism underlying various forms of canine SCI and can be modeled in rodents.[5-7] Typically, animals are anesthetized, a dorsal laminectomy is performed over the mid-thoracic vertebral column, and an apparatus is used to drop a weight or impactor onto the exposed spinal cord. Over time, weight drop devices have become increasingly more sophisticated; for example, the Multicenter Spinal Cord Injury Study Impactor (MASCIS) uses optical potentiometers to precisely measure trajectory and can adjust for vertebral column movement at the time of injury.[8] Weight drop devices induce contusion by deforming the spinal cord and forcing parenchymal movement in the longitudinal plane.[8,9] This longitudinal movement is greatest in the center of the spinal cord over the impact site; there is relatively little movement at the pial surface.[9] Lesions are oriented over the impact site, have a "football shape", and acutely consist of large diameter axonal injury with regions of grey and white matter hemorrhage. Over weeks, spinal cord cross sectional area shrinks and centrally located parenchymal cavitations may develop.

Laceration/transection

Spinal cord laceration may not adequately mimic many in vivo SCIs and can be deceptively challenging to model. Traditionally, it has been performed via an open laminectomy. Recently, hemisection models have been utilized as a viable technique to mimic SCI in the cervical spinal cord, as more complete injuries may result in disabilities not compatible with survival. Caution must be used when assessing outcomes in hemisectioned animals as return of function can occur due to plasticity associated with preserved contralateral pathways.[10] Some noted limitations of laceration models include: 1) possibility for unintentional sparing of ventral funiculi axons; 2) disruption of the dura and subarachnoid space; and 3) spinal cord retraction with variable gap distance.[6] Limited standardized data exist concerning histopathologic outcomes in various laceration models, although laceration injuries seem to induce robust glial scar at lesion margins.

Crush

Like contusion models, experimental crush injuries may mimic lesions seen in vertebral column fracture/luxation or other compressive/contusive SCIs. To create these lesions rodents are anesthetized, a laminectomy is performed, and hemostats are used to cross-clamp the spinal cord for 3-5 seconds. Although data are more limited with regard to crush injury compared to contusion, advocates of this technique feel that it is reliable, easy to perform, leaves the meninges intact as is the case in human/canine SCI, and does not result in spinal cord retraction.[11] Within hours, hemorrhage appears within the spinal cord and often extends significant distances cranial and caudal to the lesion within sensory tracts. Over hours to days, areas of spinal cord necrosis develop. Following crush injury there may be complete spinal cord disruption or disruption encompassing the entire spinal cord except the ventral funciculi.[11]

Compression/Clip

Traditional contusion models generate SCI by transferring momentum to the spinal cord, which results in viscoelastic distortion of parenchyma. Unlike many in vivo SCIs, continued spinal cord compression does not occur in rodent contusion models. The clip model was developed in 1978 as a means to replicate impact and continued compression.[12] Animals are anesthetized and the spinal cord is exposed by laminectomy. An aneurysm clip is placed around the dorsal and ventral spinal cord and released with a set amount of force. The clip is often left in place for several minutes to induce a state of continued compression. The histopathologic lesions associated with the clip model include central spinal cord necrosis, sparing of the sub-pial white matter, and white matter lesions that seem to resemble focal ischemic infarction.[12] These lesions resemble those in humans and dogs with contusive-compressive SCI.

STRENGTHS

Injury in rodent models is highly stereotyped, which enhances repeatability and facilitates scientific discovery. For example, the MASCIS device protocol standardizes the age of rodents, anesthesia protocols, and injury timing. Variation in spinal cord edema and ionic content is minimal when impact velocity is held constant.[13] And, at a moderate contusion velocity, there is minimal variation in percent tissue sparing (9.9%+/-4.8%).[14] Rodent models also allow investigators to perform experimental treatments at uniform time intervals in animals with relatively homogeneous genetic background. And, genetically manipulated mice are relatively easy to obtain – these animals may facilitate identification of important injury pathways. Many valid outcome metrics exist in rodents, including ordinal behavioral injury scales, computerized gait analysis (kinematic/kinetic), systems for measuring lesion volumes, and in vivo imaging techniques.[3,7] Finally, rodents are a relatively inexpensive platform to investigate neurotrauma.

WEAKNESSES

Some of the strengths of rodent models may also be weaknesses. For example, homogenous injuries are not a clinical reality. In many human and canine SCIs primary mechanisms vary. For example, laceration may occur in fracture luxation and compression is often prolonged. Most rodent SCI models have spinal cord compression of <30 minutes total duration. Rodents also exhibit strain-strain variations in response to injury even when identical injury devices are used, which may reflect intra-strain differences in response. The spinal cord of a rodent is much smaller than that of a dog or human. Axons can bridge lesions much more readily in a small spinal cord and therefore data on regenerative or plasticity-inducing therapies in rodents need to be viewed cautiously. In mice, post-injury cavitations are uncommon, whereas in other species these are frequently identified in the chronic phases of injury and may be an important source of neuropathic pain. Questions have also been raised about bias, inter-laboratory variation in measuring outcomes, and lack of blinding in rodent SCI datasets.[15,16] Recently, the NINDS has tried to alleviate concerns regarding the quality of rodent-derived SCI data by encouraging independent replication of published studies. Replication of an experiment that had previously shown positive outcome effects associated with Nogo-66 receptor antagonism in mice generated dissimilar findings as have several other replication studies.[16]

APPLICATION

For the time being, rodent SCI models are an important tool for investigating post-injury events and validating novel therapeutics. Data from rodent models, however, need to be viewed with caution. Experiments that have been independently replicated likely have greater reliability than those performed at a single laboratory. And, findings regarding basic biology or response to therapy that have been demonstrated across 2 or more species are likely more clinically meaningful than single species data.[17]

How do clinicians apply data from rodent SCI models to naturally occurring canine SCI? 1) One of the challenges in answering this question is the realization that clinical SCIs are heterogeneous relative to primary and secondary events. For example, mechanisms underlying a vascular SCI are likely to differ (to some extent) from those that underlie a fracture/luxation. Utilizing rodent contusion data broadly to explain mechanisms underlying all forms of SCI is therefore unlikely to be appropriate. And, by extension, promising therapies developed in a contusion model may not be applicable to certain clinical forms of SCI with divergent mechanisms. 2) Mechanistic data from injury models need to be validated in dogs with naturally occurring SCI before becoming broadly accepted in this species. Several recent studies in dogs have identified secondary events that parallel those that are well recognized in rodent contusion models such as microglial activation, T cell infiltration, and metalloproteinase activation.[18,19] 3) Rodent models offer a wealth of data concerning novel therapies that can be combined with mechanistic data from dogs to develop robust canine SCI studies. Implementing therapies derived from rodent models in dogs with clinical SCI without thorough, careful investigation is of dubious merit. Therapeutic studies in dogs with SCI should follow accepted clinical trial principals including a) careful development of a biologically relevant agent; b) open label safety study in 10-20affected individuals assessing a limited number of clinical outcomes (phase I); c) randomized, blinded, placebo controlled study of the agent at a single center using a moderate number of animals with the assessment of several validated outcomes (phase II); and d) multi-center, randomized, placebo controlled study (phase III).

REFERENCES

[1] Allen AR **J Am Med Assoc** 1911;57:878. [2]Das GD. **Brain Res Bull** 1989;22:32. [3]Dobkin BH. **Dev Neurobiol** 2007;67:1147. [4]Courtine G et al **Nat Med** 2007;13:561. [5]Norenberg MD, Smith JM, A. **J Neurotrauma** 2004;21:429. [6]Steward O et al. **Exp Neurol** 1999;157:19. [7]Metz GAS et al. **J Neurotrauma** 2000;17:1. [8]Young W. **Prog Brain Res** 2002;137:231. [9]Blight AR, Decrescito V. **Neuroscience** 1986;19:321. [10]Vinit S, Kastner A. **Respir Physiol Neurobiol** 2009;169:115. [11]Guth L et al. **Exp Neurol** 1994;126:76. [12]Rivlin AS, Tator CH. **Surg Neurol** 1978;10:38. [13]Kwo S et al. **J Neurotrauma** 1989;6:13. [14]Basso DM et al. **Exp Neurol** 1996;139:244. [15]Aguilar RM, Steward O. **Exp Neurol** 2010;221:38. [16]Steward O et al. **Exp Neurol** 2008;209:446. [17]Levine JM et al. **J Neurotrauma** 2011;28:675. [18]Bock P et al. **Brain Pathol** 2013;23:82. [19]Boekhoff TM et al. **J Neurotrauma** 2012;29:1000.

THE USE OF HYPERBARIC OXYGEN THERAPY IN SMALL ANIMAL INTERNAL MEDICINE

Diane M. Levitan, VMD, DACVIM (SAIM)
Commack, New York

The value of hyperbaric oxygen has become increasingly evident in human medicine and the knowledge gained has been valuable to establish effective and rational uses for hyperbaric oxygen therapy (HBOT) in small animal internal medicine. Hyperbaric oxygen therapy in humans (HBOT) fell out of favor in the mid-1900s due to misuse and false claims as well as a lack of controlled studies to support its effectiveness in certain conditions. In the past 40 years, thousands of evidence based medicine papers have help to elucidate not only what conditions it is helpful in, but also the cellular and physiologic mechanisms for its actions.

Hyperbaric Oxygen Therapy (HBOT) is a medical treatment in which the patient is entirely enclosed in a pressure chamber breathing 100% pure oxygen (O2) at greater than one atmosphere of pressure. It is the use of oxygen under pressure as a drug. HBOT is delivered in a hyperbaric chamber. Multiplace chambers are capable of treating multiple individuals simultaneously. Monoplace chambers are the most common type used in veterinary medicine. These chambers are best suited to one person, however, up to 3 pets could potentially be treated together in a 'monoplace' chamber. The risks and benefits of this must be determined on a case by case basis. There are no attendants and no monitoring can occur inside the monoplace chamber.

Pressure terminology

At sea level, or normobaric conditions, the pressure is 1 atmosphere. The pressure on an individual is equal to the weight of the atmosphere around them. 1 atmosphere of pressure is equal to 14.7 pounds per square inch (PSI) or 760 mmHg or 760 Torr or 0 ft of sea water. The sum of pressures that something is exposed to is called atmospheres absolute (ATA). Therefore, if an animal is in a chamber at sea level (14.7 PSI or 0 feet of sea water), and the chamber pressure is brought up to 2 ATA, then the pressure is raised in the chamber an additional atmosphere to 14.7 PSI. The effect of 2 ATA of pressure equals the pressure of being surrounded by 33 feet of sea water. Most hyperbaric treatments in small animal medicine involve pressures of 1.5 to 3.0 ATA—one atmosphere from the earth's atmosphere plus an additional one half to two atmospheres exerted by pressurizing the hyperbaric chamber.

Physiologic effects of hyperbaric oxygen

Physiologic effects of hyperbaric oxygen are discussed here in detail as these effects will be sited when discussing rationale of HBOT for any internal medicine applications. There are two basic effects of HBOT on tissues—(1) effects related to the mechanical forces of the pressure itself and (2) the effects related to the increased oxygen in blood and tissues.

Mechanical effects of pressure

Pressure gets gas into solution, shrinks bubbles and decreases diffusion distances. At normal atmospheric pressure arterial oxygen tension is 100mmHg and tissue oxygen 55mmHg. In 100% Oxygen at 3 ATA, arterial oxygen tension is 2000mmHg and tissue oxygenation is 500mmHg. Instead of 3ml of oxygen per liter, there is 60 ml oxygen per liter. (3) This results in oxygen transport and tissue survival without the need for hemoglobin, as was demonstrated by Boerema in 1955. (4) (5) (6) High oxygen gradients result in diffusion of oxygen into areas of low oxygen. This provides immediate help to ischemic and compromised tissue even with marginal or no blood flow. Oxygen dissolved in plasma can be delivered from capillaries to tissues at least three times further than delivered when carried only by hemoglobin. (5) (6) HBOT decreases the volume of any gas in proportion to the increase in atmospheric pressure delivered by the chamber (Boyle's law). At elevated pressures, oxygen and nitrogen diffuse into solution.

HBOT helps scuba divers with decompression sickness by dissolving large toxic nitrogen bubbles back into solution. The same concept applies in the case of a venous or arterial gas embolism in any animal. At 2.8 atmospheres, any bubble volume reduces by almost two thirds. Hyperbaric oxygen speeds the dissolution of the inert-gas bubble by replacing the inert gas in the bubble with oxygen, which is then rapidly metabolized by the tissues. The use of hyperbaric oxygen also prevents the formation of new bubbles. (3) Compression in a chamber will decrease gas from ileus, bloat or severe intraluminal gas accumulation. HBOT is particularly useful in perioperative gastrointestinal obstruction, ileus, gas gangrene, emphysematous biliary or urinary bladder tissues, etc. due to a reduction in gas volume, nitrogen diffusion into tissues and blood and a high gradient for nitrogen removal via respiration. The decrease in pressure from gas in tissues or cavities results in less injury from vascular compromise, decreased bacterial translocation, necrosis, pain and swelling. (7) (8) (9) (10) (11)

Effects related to the increased oxygen in blood and tissues

The effects of increased oxygen on tissues

Hyperbaric Oxygen therapy decreases inflammation by many mechanisms

Vasoconstriction

High levels of oxygen cause vasoconstriction, limiting blood flow and oxygen transport via red blood cells. There are organ specific vasoconstrictive responses to hyperbaric oxygen- but the most profound vasoconstriction is in the brain, retina and skeletal muscle. For the most part, oxygen induces smooth muscle contraction in all muscularized vessels on both the arterial and venous sides of circulation. (15) This seems counterintuitive, but overall oxygen delivery to tissues is increased regardless of the vessel sizes. (16) (17) (18)

Reduction of swelling and edema

Blood and lymphatic vessel disruption, dilation and widening following damage to tissue results in decreased blood and lymphatic flow. Increase in blood flow from damaged tissue (vasogenic) coupled with increased vascular permeability from cellular injury (cytogenic) leads to tissue swelling/edema. (19)This creates increased local pressure further compromising blood flow. If this pressure approaches or exceeds that in the blood vessels, blood flow will slow or cease altogether. Swelling contributes to tissue hypoxia by increasing capillary distances within tissues. With HBOT, there is decreased filtration of fluid from the capillary to the extracellular space as a consequence of vasoconstriction while reabsorption of fluid at the capillary level is maintained. Reducing edema decreases intercapillary distances, therefore increasing oxygen perfusion. Fluid reabsorption helps inflammatory cells remove cell debris and micro-organisms that are in the vicinity. This decreases edema and is important in the treatment of burns, crush injuries and injured tissues in general. Edema reduction has been mostly attributed to oxygen induction of vasoconstriction, but is also likely due to reduction in inflammation. (16) (17) (18) (20) (21)

HBOT Enhances efficacy of leukocytes

White blood cells use oxygen-independent and oxygen-dependent mechanisms to kill microorganisms. HBOT's effects on neutrophil activity result in decreased inflammation and increased bactericidal capabilities. In tissues where oxygen tensions fall below 30 to 40 mmHg, leukocytes lose their effectiveness against gram-positive and gram-negative organisms, allowing bacteria to flourish. HBOT improves phagocytosis impaired by hypoxia. HBOT enhances granulocyte production of antimicrobial agents such as free radicals and reactive/toxic oxygen species. (22) (23) (13) Phagocytosis of pathogens results in an "oxidative burst" comprised of oxygen radicals such as hydroxyl radicals, peroxides, and superoxides to kill microorganisms. The oxidative burst increases the oxygen consumption of the leukocyte 30 times; therefore production of these oxygen radicals requires an oxygen rich environment which is restored with HBOT. (24) (25) (26) Anaerobic bacteria are very susceptible because they have poor defenses against free radicals and other toxic oxygen species created by neutrophils in that environment.

HBOT will decrease inflammation by decreasing inflammatory cytokines that result in neutrophil rolling, activation, transmigration into tissues and local adherence by impairing cGMP synthesis in activated leukocytes. (27) High oxygen concentration results in reduced expression of endothelial adhesion molecules and increases nitric oxide production which also inhibits neutrophil adhesion. (28) (29) (30) HBOT reduces pro-inflammatory cytokine induction which also dampens the inflammatory response. (31)

Macrophages, like neutrophils, are affected by tissue oxygen tension. In hypoxic conditions macrophages are unable to scavenge effectively or produce peroxides (32) Hypoxia induces macrophages to produce the inflammatory cytokines TNF-alpha, IL-1, IL-8, and intracellular adhesion molecule-1, which decrease response to infection. (33) HBOT increases oxygen to normal levels so that macrophages can function normally. (13)

Direct Antibacterial effects of HBOT

HBOT has direct bacteriostatic and bactericidal effects against gram positive, gram negative aerobic and anaerobic microorganisms equal to some antibiotics. HBOT increases the concentration of superoxides both intracelluarly and extracellularly. Increased superoxide levels lead to increased production of hydrogen peroxide and other toxic oxygen radicals all of which impair microbial metabolic reactions. Anaerobic microorganisms cannot survive the amount of reactive oxygen species/oxygen based free radicals created in hyperoxic environments because they lack the superoxide dismutase, the superoxide degrading enzyme and catalse, the hydrogen peroxide degrading enzyme. Free radicals oxidize proteins and membrane lipids, damage DNA and inhibit metabolic functions (protein and nucleic acid synthesis) needed for bacterial growth and so are lethal for microorganisms that lack protection from them (particularly anaerobes). Clostridium perfringens is an anaerobe easily killed in the hyperbaric environment. (34) (35) HBOT will suppress secretion of alpha toxin in *C. perfringens* infections, which is the major source of tissue injury and death. When HBOT is utilized in the early course of infection with appropriate antibiotics and surgical attention, many of the deleterious effects of bacterial endotoxin release are minimized. HBOT treatment of anaerobic infections caused by Clostridial perfringens has increased patient survival, reduced number of surgeries, shortened hospital stays, reduced morbidity, extension of necrosis, and systemic toxicity. (36) (35) (37)

(38) Actinomycosis is also very susceptible to HBOT. Some *Actinomyces* spp. are obligate anaerobes while others are facultative anaerobes. HBOT combined with proper surgical debridement and antibiotics greatly enhances healing and speeds clearing of infection. (39)

Leukocyte mediated killing of several aerobic bacteria

Proteus vulgaris, Salmonella typhimurium, Klebsiella pneumoniae, Serratia marcescens, Pseudomonas aeruginosa and Staphylococcus species - is significantly reduced in hypoxia. Increasing the concentration of oxygen over ambient levels has been shown to reduce infection (40) (41) (42)HBOT enhances phagocytic clearance of S. aureus, including Methicillin-resistant S. aureus and helps potentiate the bactericidal effects when used with appropriate antibiotics. (43) (44) (45) Additionally HBOT aids in the treatment of mycobacteria such as Mycobacterium ulcerans (46), Mycobacterium tuberculosis (47), and Mycobacterium fortuitum (48) Several studies have suggested effective treatment of Borrelia Burgdorferi (Lyme disease) with HBOT. The studies have not proven eradication of the disease but show excellent evidence that HBOT improves quality of life for patients who have failed all other treatment regimes. (49) (50)

Increases efficacy of antibiotics

When oxygen tension falls below 30 mmHg, bacteria quickly destroy tissue. Evidence suggests that hyperoxygenation and antibiotics combine to yield a synergistic effect. (51) (52) HBO raises oxygen tension in hypoxic tissue to level necessary for killing of bacteria by PMNs and optimal activity of certain antibiotics. HBOT aids in the penetration, efficacy and action of aminoglycosides and the antimetabolites/sulfonamides trimethoprim, sulfamethoxazole, and sulfasoxazole. Efficacy of cephalosporins, nitrofurantoin and amphotericin B are also improved with HBOT. Other antibiotics, like vancomycin and fluoroquinolones, become less efficient under hypoxic conditions but recover their usual effectiveness when normal oxygen tension is restored. Presumably their efficacy may be increased with increased oxygen tension. (53) (54) (35)

HBOT as an antifungal

HBOT has also been shown to inhibit the growth of some fungi and to potentiate the antifungal effect of amphthericin B. Antifungal effects are attributed to membrane damage, enzyme inactivation, and reduced protein synthesis caused by increased production of oxygen-based free radicals during exposure to HBOT (55) HBOT use as proven efficacious as adjunctive therapy for treating invasive rhinocerebral mucormycosis, (56) (57) aspirgillosis (58),*Candida albicans*, (59) (60) Pneumocystis carinii has been considered to be a fungus. In vitro studies show oxygen to be lethal to P.Carinii due to low levels of antioxidant enzymes. (35)

Hyperoxia facilitates healing by interfering with the inflammatory cascade

Most importantly, oxygen relieves hypoxia to damaged tissues. Studies have proven that HBOT decreases neutrophil rolling, activation, transmigration into tissues and local adherence by imparing cGMP synthesis in activated leukocytes. (27) Hyperoxia results in reduced expression of endothelial adhesion molecules and increases nitric oxide production which also inhibits neutrophil adhesion (28) (29) (30) HBOT significantly reduces proinflammatory cytokine induction which also contributes to minimizing inflammatory response (31)

Hyperoxia reduces free radical damage of reperfusion injury

Ischemia/reperfusion injury occurs in many disease states when the restoration of blood flow to tissues causes dangerous vascular and tissue damage, further impairing blood flow and healing. Ischemia/reperfusion injury occurs in many instances such as stroke, crush injury, embolic disease, myocardial infarctions, CO poisoning, and any other disease state suffering an ischemic insult followed by the restoration of blood flow. During reperfusion, the restoration of blood flow actually leads to progressive vascular damage and an expanding area of poor blood flow. (61) (62) HBOT interferes with the sequestration of neutrophils on injured endothelium and lessens the initiation of the reperfusion injury. It affects neutrophil function, endothelial cellular adhesion molecule expression, nitric oxide production, nitric oxide synthetase expression, cellular energetics, lipid peroxidation and microvascular blood flow. (63) (64) HBOT protects tissues from the harmful effects of toxic oxygen radicals or reactive oxygen species in many ways by providing enough oxygen for the reperfused tissues to generate scavengers (by increasing local endothelial surface superoxide dismutase and peroxidase activity) to detoxify the reactive oxygen species. (65) (62) HBOT antagonizes lipid peroxidation of cell membranes from toxic oxygen radicals by preventing the conversion of endothelial xanthine dehydrogenase to xanthine oxidase, the essential step required for lipid peroxidation. (65) (63)

Longer term effects (days to weeks)

Effects of HBOT are prolonged after HBOT sessions end and tissue oxygen levels return to pre-treatment values. Gas and tissue levels of O2 decrease rapidly after HBOT but are still elevated for at least 4 hours. (12) (13) Marx and Johnson demonstrated that for irradiated wounds, HBOT induces neovascularization, which becomes significant after about 14 treatments and continues for years after the HBOT treatments have ended. (14)Other important beneficial effects of HBOT are attributed to improved tissue repair by improving necrotic tissue proteolysis, fibroblast proliferation, collagen deposition, angiogenesis, migration of epithelial cells, and bone remodeling by osteoblastic/osteoclastic activity, which are all significantly diminished in hypoxic tissues (66) A number of days following injury there is migration of fibroblasts (connective tissue cells responsible for collagen production) into the areas of damage. These cells then divide and replicate producing large amounts of collagen-(connective tissue used to repair damage to tissue) that acts as the building block for the healing of tissue and wounds. (67) When the environment of fibroblast cells has an oxygen tension of less than 10mmHg, the cells can divide, but can no longer make collagen. It also cannot migrate to where it is needed for healing. When the oxygen tension is increased, the fibroblast can again carry out these wound healing functions. The collagen produced by these cells forms a fibrous matrix, and new capillaries grow into this. HBOT increases fibroblast growth, production of collagen and cellular matrix via cytokine down regulation and growth factor up-regulation. HBOT up-regulates collagen synthesis through pro-al(I) mRNA expression. (68) In ischemic flaps HBOT up-regulates fibroblast growth factor (FGF) causing an increased effect over that seen with FGF alone. (69) In situations where FGF is ineffective, HBOT can make it more effective (70) (13) Research has shown that increasing oxygen concentration in tissues above normal will increase collagen production rate above normal. (71) (66) (72)

HBOT increases osteoclast and osteoblast production, proliferation, differentiation and activity. This aids in bone regeneration and healing. Bone healing as well as resorption of necrotic bone by osteoclasts is enhanced with HBOT. (73) (74) (75)A recent study suggests that the proliferative effects of HBOT on osteoblasts may contribute to the recruitment of osteoblasts at fracture sites. The same study demonstrated that HBOT stimulated significantly increased mRNA expression of fibroblast growth factor (FGF)-2 as well as many other bone promoting proteins promoting osteoblast proliferation and cell cycle progression after 3 days of treatment. (76)

Angiogenesis

Destruction or injury of tissues results in damage of the supporting blood vessels. The healing process relies on supply of the blood containing the cells and nutrients that carry out and enable healing as well as remove damaged cells, debris and foreign microorganisms. Angiogenesis occurs in response to high oxygen concentrations (77) HBOT helps create collateral blood flow by increased angiogenesis (new vessel growth budding from existing intact vessels) and vasculogenesis (when bone marrow derived endothelial precursor cells populate the tissue and differentiate and grow into new vessels). Recent evidence shows that hematopoietic stem cell derived vessels occur much more than previously believed. (78) (79) (80)

The increased oxygen supply and increased pressure employed by hyperbaric oxygen supply are both responsible for the stimulation of new capillary growth and the repair of existing capillaries. Collateral blood vessels are a result of increased fibroblast production of collagen and are oxygen dependent and collagen is the foundation matrix for angiogenesis. Growth of new blood vessels in the micro-circulation and repair of damaged capillaries are stimulated within the damaged tissues by hyperbaric oxygen therapy, providing increased oxygen availability to these areas. (81) (82) (77) (83) (80)

Hyperbaric Oxygen Mobilizes Stem Cells

One of the most exciting effects of HBOT is the increased mobilization of stem cells in circulation of patients. This finding could explain much of what has been seen in the physiologic activity of HBOT. American Journal Physiology - Heart and Circulatory Physiology reports a single 2-hour exposure to HBOT at 2 ATA doubles circulating CD34+ progenitor stem cells (primordial cells targeted to salvage and restore damaged structures); and at approx. 40-hours of HBOT; circulating CD34+ cells increases eight fold (800%). (81)

There are pharmaceutical options for increasing stem cell levels in the body such as erythropoiesis stimulating agents (erythropoietin, darbepoetin) and colony stimulating factors such as granulocyte colony stimulating factors (G-CSF). According to Thom, et al publication in the American Journal of Physiology-Heart and Circulation Physiology, exposure to HBOT mobilizes stem/progenitor cells from the bone marrow by a nitric oxide (·NO) -dependent mechanism. The population of $CD34^+$ cells in the peripheral circulation of humans doubled in response to a single exposure to 2.0 atmospheres absolute (ATA) O_2 for 2 hours. Over a course of 20 treatments, circulating $CD34^+$ cells increased eightfold, although the overall circulating white cell count is not significantly increased without any pharmacologic intervention or underlying disease processes. This ground breaking study offers a new, safe, therapeutic option for stem cell proliferation with little risk of side effects. (84) This was demonstrated to be particularly useful in diabetic patients where increases in stem cells were directly correlated to improved refractory lower extremity neuropathic ulcers. Nitric oxide synthase activity was found to be acutely increased in patient's platelets following HBOT and remains elevated for at least 20 hours. HBOT aided healing may be due to increased vasculogenic stem cell mobilization to skin wounds. (85)

The rationale for HBOT in many different internal medicine scenarios will be discussed in this lecture. The following are disease processes that are responsive to HBOT.

Central Nervous System Brain trauma Spinal cord trauma Stroke Cerebral ischemia Compressive cord diseases GME Fibro-cartilaginous emboli Peripheral nerve injury Head trauma Acute cerebral edema Peripheral nerve damage Polyradiculononeuritis Intracranial abscess	**Inflammatory, infectious and immune mediates diseases:** Allergic reactions/anaphylaxis GME RMSF induced vasculitis/ meningitis/arthritis Immune mediated polyarthritis IMHA ITP Inflammatory bowel disease Pancreatitis Cholangitis: Osteomyelitis Septic arthritis Septicemia Endotoxemia Mold exposure Aspergillosis Mucormycosis, Candidiasis pneumocystis Lyme disease Anaerobic infections: Actinomyces, clostridium, etc Necrotizing fasciitis Intracranial and abdominal abscess	**Gastrointestinal/hepatic** Post op enterotomy or GI surgery Ileus Pancreatitis Inflammatory bowel disease Peritonitis Gastric ulceration Reperfusion Hepatic inflammatory diseases Biliary tract inflammation Biliary tract infection Hepatic necrosis Oral ulceration or wounds Esophageal inflammation	**Emergency Medicine** Carbon monoxide Cyanide toxicity Smoke inhalation Lymphangitis Burns Thromboembolic disease Gas embolism Acute traumatic ischemia Bloat Necrotizing faciitis Blood loss anemia Tetanus Septicemia Head trauma Acute cerebral edema Skin burns Fractures and acute trauma Envenomation- spider, snake	**Musculoskeletal** Athletic injuries Tendonitis Periostitis Fracture Myositis Crush injuries All wounds Refractory osteomyelitis Severe degloving Wounds Burns Delayed wound healing Acute cruciate injuries Delayed fracture healing Skin grafts
Dermatology: Infection control: deep pyoderma and MRSA Immune mediated skin dz Improved skin healing Hair regrowth Decrease otitis inflammation and infection +/- myringotomy	**Oncology:** Post-operative healing Pre and post radiation therapy Palliative care for neoplasia, pain and swelling Radiation burns/necrosis Osteoradionecrosis Soft tissue radionecrosis	**Cardiovascular** Hypotension Shock Cardiac infarction Acute anemia Thromboembolic disease	**Renal** nephrotic syndrome, ischemia reperfusion, inflammation, hypoxia	

REFERENCES

1. **Hardy, Kevin.** The physics of Hyperbaric Oxygen Therapy. [book auth.] Stephen R. Thom Tom S. Newman. *Physiology and Medicine of Hyperbaric Oxygen Therapy.* Philadelphia : Saunders, 2008, pp. 57-64. **2. Kindwall, Eric.** The Physics of Diving and hyperbaric Pressures. [book auth.] Harry T. Whelan Eric P. Kindwall. *Hyperbaric Medicine Practice.* Flagstaff : Best Publishing Company, 2004, pp. 21-36.**3. Patrick M.Tibbles, John S. Edelsberg.** Hyperbaric-Oxygen Therapy. *NEJM.* June 1996, Vol. 334, 25, pp. 1642-1648.**4. Boerema I, Meijne NG, Brummelkamp WH, et al.** *Life without blood. J Cardiovasc Surg 182: 133, 1960.*

5. Weibel, ER. Delivering oxygen to the cells. [book auth.] Weibel ER. *The Pathway for Oxygen.* Boston : Harvard University Press, 1984, pp. 175-210. **6. AL Gill, CNA Bell.** Hyperbaric oxygen: its uses, mechanisms of actions and outcomes. *Q J Med.* 2004, 97, pp. 385-395.**7.** *The striking effect of hyperbaric oxygenation therapy in the management of chronic idiopathic intestinal pseudo-obstruction.* **Yokota T, Suda T, Tsukioka S, Takahashi T, Honma T, Seki K, Matsuzawa J, Miura M, Aoyagi Y, Asakura H.** 1, Jan 2000, Vol. 95, pp. 285-8.**8.** *Hyperbaric oxygen therapy for intestinal obstruction in children: an exceptional experience in a compromised child.* **Ohno Y, Kanematsu T.** 33, Oct 1998, J Pediatr Surg., Vol. 10, pp. 1543-5.**9.** *Hyperbaric oxygen ameliorates bacterial translocation in rats with mechanical intestinal obstruction.* **Akin ML, Uluutku H, Erenoglu C, Ilicak EN, Elbuken E, Erdemoglu A, Celenk T.** 45, Jul . 2002 Jul;45(7):967-72., Dis Colon Rectum, Vol. 7, pp. 967-72.**10. LM, Ludwig.** The role of hyperbaric oxygen in current emergeny medical care. *J Emerg Nurs. 1989 May-Jun;15(3):229-37.* **11. Ambiru S, Furuyama N, Aono M, Kimura F, Shimizu H, Yoshidome H, Miyazaki M, Shimada H, Ochiai T.** Hyperbaric oxygen therapy for the treatment of postoperative paralytic ileus and adhesive intestinal obstruction associated with abdominal surgery: experience with 626 patients. *Hepatogastroenterology.* Oct-Nov 2007, Vol. 54, 79, pp. 1925-9.**12. Piantadosi, Claude A.** Pulmonary Gas Exchange, Oxygen Transport, and Tissue Oxygenation. [book auth.] TS Neuman and SR Thom. *Physiology and*

Medicine of Hyperbaric Oxygen Therapy. Philadelphia : Saunders, 2008, pp. 133-185.**13.Nylander G, Nordstrom H, Eriksson E.** Effects of hyperbaric oxygen on oedema formation after a scald burn. *Burns Incl Therm Inj.* 1984, Vol. 10, 3, pp. 193-196.**14. Shupak A, Gozal D, Ariel A, Melamed Y, Katz A.** Hyperbaric oxygenation in acute peripheral posttraumatic ischemia. *J Hyperbaric Med.* 1987 , Vol. 2, pp. 7-14.**15. JL., Phillips.** Air as medicine. [book auth.] Phillips JL. *The Bends.* New Haven : Yale University Press, 1998, pp. 197-203.**16. Strauss, Michael.** Crush Injury Compartment Syndrome and Other Acute Traumatic Peripheral Ischemias. [book auth.] Harry T Whelan Eric P Kindwall. *Hyperbaric Medicine Practice.* 2nd. Flagstaff : Best Publishing Company, 2004, 30, pp. 753-778.**17. Skyhar MJ, Hargens AR, Strauss MB, Gershuni DH, Hart GB, Akeson WH.** Hyperbaric oxygen reduces edema and necrosis of skeletal muscle in compartment syndromes associated with hemorrhagic hypotension. *J Bone Joint Surg Am.* 1986, Vol. 68, 8, pp. 1218-1224.**18. McFarlane RM, Wermuth RE.** The use of hyperbaric oxygen to prevent necrosis in experimental pedicle flaps and composite skin grafts. *Plast Reconstr Surg.* 1966, Vol. 37, 5, pp. 422-430.**19. Allen DB, Maguire JJ, Mahdavian M, Wicke C, Marcocci L, Scheuenstuhl H, Chang M, Le AX, Hopf HW, Hunt TK.** Wound hypoxia and acidosis limit neutrophil bacterial killing mechanisms. *Arch Surg.* 1997, Vol. 132, 9, pp. 991-996.**20. Knighton DR, Halliday B, Hunt TK.** Oxygen as an antibiotic. The effect of inspired oxygen on infection. *Arch Surg.* 1984, Vol. 119, 2, pp. 199-204.**21. Wright, James.** http://www.worldwidewounds.com/2001/april/Wright/HyperbaricOxygen.html#ref14. *Hyperbaric oxygen therapy for wound healing* . [Online] May 2001.**22. Chen Q, Banick PD, Thom SR.** Functional inhibition of rat polymorphonuclear leukocyte B2 integrins by hyperbaric oxygen is assoicated with impared cGMP synthesis. *J Pharmacol Exp Ther.* 1996, Vol. 276, pp. 929-933.**23. Buras JA, Reenstra WR.** Hyperbaric oxygen decreases endothelial cell E-selectin protein expression in an in-vitro model of ischemia/reperfusion. 32:S17. *Ann Emerg Med* . 1998, 32, p. S17.**24.** Banick PD, Chen Q, Xu YA, Thom SR: Nitric oxide inhibits neutrophil β β2 integrin function by inhibiting membrane-associ- ated cyclic GMP synthesis.1997, 172:12-24. *J Cell Physiol.* 1997, Vol. 172, pp. 12-24.**25. Bitterman, Haim.** Bench to Bedside review: Oxygen as a drug. *Critical Care.* 2009, Vol. 13, 1.**26. Masatomo Yamashita, Mamuro Yamashita.** Hyperbaric oxygen treatment attenuates cytokine induction after massive hemorrhage. *Am J Physiol Endocrinol Metab.* 2000, 278, pp. E811-E816.**27. Mrotek JJ, Henderson R, Kiel J, et al.** Normobaric oxygen concentrations affect cultured mouse macrophage responses. . *Fed Proc.* 1986, Vol. 45, p. A958.**28. G., Scannell.** Leukocyte responses to hypoxic/ischemic conditions. . *New Horiz.* 1996, Vol. 4, 2, pp. 179-183.**29. IA, Silver.** Cellular microenvironment in healing and non-healing wounds. [ed.] Heppenstall RB, Pines M. Edited by Hunt TK. *Soft and Hard Tissue Repair.* New York : Praeger, 1984, pp. 50-66.**30. Park, Matthew.** Effects of Hyperbaric Oxygen in Infectious Diseases: Basic Mechanisms. [book auth.] Harry T. Whelan Eric P. Kindwall. *Hyperbaric Medicine Practice.* Second. Flagstaff : Best Publishing Company, 2004, 10, pp. 205-243.

31. Korhonen K, Hirn M, Niinikoski J. Hyperbaric oxygen in the treatment of Fournier's gangrene. *Eur J Surg.* 1998, Vol. 164, 4, pp. 251-255.

32. Korhonen K, Klossner J, Hirn M, Niinikoski J. Management of clostridial gas gangrene and the role of hyperbaric oxygen. *Ann Chir Gynaecol.* 1999, Vol. 88, 2, pp. 139-142.**33. Demello FJ, Haglin JJ, Hitchcock CR.** Comparative study of experimental Clostridium perfringens infection in dogs treated with antibiotics, surgery, and hyperbaric oxygen. 1973, 73, p. 936.**34. Manheim S.D., Voleti C., Ludwig, A., and Jacobson J.H.** JAMA . *Hyperbaric oxygen in the treatment of actinomycosis.* 1969, Vol. 210, 3, pp. 552-553.**35. DC., Hohn.** Host resistance of infection: Established and emerging concepts. [ed.] Hunt TK. *Wound Healing and Wound Infection.* New York : Appleton-Century-Crofts, 1980, pp. 264-280.**36. Knighton DR, Halliday B, Hunt TK.** Oxygen as an antibiotic. A comparison of the effects of inspired oxygen concentration and antibiotic administration on in vivo bacterial clearance. *Arch Surg.* 1986, Vol. 121, 2, pp. 191-195.**37. Greif R, Akca O, Horn EP, Kurz A, Sessler DI.** Supplemental perioperative oxygen to reduce the incidence of surgical-wound infection. Outcomes Research Group. . *N Engl J Med.* 2000, Vol. 342, 3, pp. 161-167.**38. Tsuneyoshi I, Boyle WA 3rd, Kanmura Y, Fujimoto T.** Hyperbaric hyperoxia suppresses growth of Staphylococcus aureus, including methicillin-resistant strains. *J Anesth.* 2001, Vol. 115, 1, pp. 29-32.**39. Knighton DR, Fiegel VD, Halverson T, Schneider S, Brown T, Wells CL.** Oxygen as an antibiotic. The effect of inspired oxygen on bacterial clearance. *Arch Surg.* Jan;125(1):97-100. 1990, Vol. 125, 1, pp. 97-100. .**40. Jain, K.K.** HBO Therapy in Infections. . [ed.] K.K. Jain:. *Textbook of hyperbaric medicine.* 5. Cambridge : Hogrefe& Huber Publishers, Inc., 2009, 13, pp. 135-148.**41. Krieg RE, Wolcott JH, Confer A.** Treatment of Mycobacterium ulcerans infection by hyperbaric oxygenation. *Aviat Space Environ Med. 1975 Oct;46(10):1241-5.* Oct 1975, Vol. 46, 10, pp. 1241-1245.

42. Warren J, Sacksteder MR, Thuning CA. Therapeutic effect of prolonged hyperbaric oxygen in adjuvant arthritis of the rat. . *Arthritis Rheum.* Apr 334-339, 1979, Vol. 22, 4.**43. GL., Goodhart.** J Orthop Trauma. Mycobacterium fortuitum osteomyelitis following trauma. 1993, Vol. 7, 2, pp. 142-145.**44. Fife, W. P., and D. M. Freeman.** "[abstract] Treatment of Lyme disease with hyperbaric oxygen therapy.". *Undersea and Hyperbaric Medical Society, Inc. (http://www.uhms.org) http://archive.rubicon-foundation.org/6541998.* [Online] 1998. **45. Taylor RS, Simpson IN.** Review of treatment options for lyme borreliosis. *J Chemother.* 17, Sep 2005 , Vol. Suppl 2, pp. 3-16.**46. Jonsson K, Hunt TK.** Ann Surg. . *Oxygen as an isolated variable influences resistance to infection.* 1988, Vol. 208, 6, pp. 783–787.**47. Knighton D, Halliday B, Hunt TH.** Oxygen as an antibiotic: a comparison of inspired oxygen concentration and antibiotic administration on in vivo bacterial clearance. *Arch Surg. 1986 Feb;121(2):.* Feb 1986, Vol. 121, 2, pp. 191-5.**48. Gottlieb SF, Solosky JA, Aubrey R, et al.** Synergistic action of increased oxygen tensions and PABA-folic acid antagonists on bacterial growth. *Aerosp Med.* Aug 1974, Vol. 45, 8, pp. 829-33.**49. M K Park, K H Muhvich, R A Myers, and L Marzella.** Hyperoxia prolongs the aminoglycoside-induced postantibiotic effect in Pseudomonas aeruginosa. *Antimicrob Agents Chemother.* April 1991, Vol. 35, 4, pp. 691–695.**50. Bitterman, Haim.** Hyperbaric Oxygen for Invasive Fungal Infections. *Isr Med Assoc J.* May 2007, Vol. 9, 5, pp. 387-388.

51. Yohai RA, Bullock JD, Aziz AA, Markert RJ. Survival factors in rhino-orbital-cerebral mucormycosis. *Surv Ophthalmol. 1994 ;39(1):3-22.* Jul-Aug 1994, Vol. 39, 1, pp. 3-22.**52. Ferguson BJ, Mitchell TG, Moon R, Camporesi EM, Farmer J.** Adjunctive hyperbaric oxygen for treatment of rhinocerebral mucormycosis. *Rev Infect Dis.* May-Jun 1988, Vol. 10, 3, pp. 551-559.**53. García-Covarrubias L, Barratt DM, Bartlett R, Metzinger S, Van Meter K.** Invasive aspergillosis treated with adjunctive hyperbaric oxygenation: a retrospective clinical series at a single institution. *South Med J.* Apr 2002, Vol. 95, 4, pp. 450-6.**54. Gudewicz TM, Mader JT, Davis CP.** Combined effects of hyperbaric oxygen and antifungal agents on the growth of Candida albicans. *Aviat Space Environ Med. Jul;58(7):673-8.* Jul 1987, Vol. 58, 7, pp. 673-678.**55. WJ., Cairney.** Effect of hyperbaric oxygen on certain growth features of Candida albicans. *Aviat Space Environ Med.* Aug 1978 ;49(8):956-8., Vol. 49, 8, pp. 956-958.**56. J., Buras.** Basic mechanisms of hyperbaric oxygen in the treatment of ischemia-reperfusion injury. *Int Anesthesiol Clin.* Winter, 2000, Vol. 38, 1, pp. 91-109. **57. Buras JA, Reenstra WR.** Endothelial-neutrophil interactions during ischemia and reperfusion injury: basic mechanisms of hyperbaric oxygen. *Neurol Res.* Mar 2007, Vol. 29, 2, pp. 127-131.**58. SR, Thom.** Antagonism of CO-mediated brain lipid peroxidation by hyperbaric oxygen. . *Toxicol Appl Pharmacol.* 1990, Vol. 105, p. 340.**59. Yogaratnam JZ, Laden G, Guvendik L, Cowen M, Cale A, Griffin S.** Hyperbaric oxygen: a novel technology for modulating myocardial ischemia-reperfusion via a single drug. *Adv Ther.* Jul-Aug 528-33, 2006, Vol. 23, 4.**60. Kaide, Sorabh Khandelwal. Colin G.** Hyperbaric Oxygen Therapy. s.l. : McGraw-Hill Companies, Inc, 2010.**61. CH, Wells.** Tissue Gas measurements during hyperbaric oxygen exposure. *6th International Congress of Hyperbaric Medicine 1977: 118-124.* . 1977, pp. 118-124.**62. Marx RE, Johnson RP.** Problem wounds in oral and maxillofacial surgery: The role of hyperbaric oxygen. [book auth.] Hunt TK Davis JC. *Problem Wounds - the role of oxygen.* New York : Elsevier, 1988, pp. 65-123.**63. Gimbel ML, Hunt TK.** Wound healing and hyperbaric oxygenation. [ed.] Whelan HT. Kindwall EP. *In Hyperbaric Medicine Practice.* 2nd. Flagstaff : Best Publishing Company, 1999, pp. 169-204.

64. S.C., Hunt TK. In: H. Columbia:. Oxygen and wound healing. [book auth.] 8th Annual Advanced Symposium yperbaric Medicine 2000. s.l. : Palmetto Richland Memorial Hospital and the University of South Carolina School of Medicine, 2000.**65. Ishii Y, Myanaga Y, Shimojo H, Ushida T, Tateishi T.** Effects of hyperbaric oxygen on procollagen messenger RNA levels and collagen synthesis in the healing of rat tendon laceration. *Tissue Eng.* 1999, Vol. 5, pp. 279-86.

66. Bayati S, Russell RC, Roth AC. Stimulation of angiogenesis to improve the viability of prefabricated flaps. *Plast Reconstr Surg.* 1998, Vol. 101, 5, pp. 1290-1295.
67. Wu L, Pierce GF, Ladin DA, Zhao LL, Rogers D, Mustoe TA. Effects of oxygen on wound responses to growth factors: Kaposi's FGF, but not basic FGF stimulates repair in ischemic wounds. *Growth Factors.* 1995, Vol. 12, 1, pp. 29-35.**68. Asano T, Kaneko E, Shinozaki S, Imai Y, Shibayama M, Chiba T, Ai M, Kawakami A, Asaoka H, Nakayama T, Mano Y, Shimokado K.** Hyperbaric oxygen induces basic fibroblast growth factor and hepatocyte growth factor expression, and enhances blood perfusion and muscle regeneration in mouse ischemic hind limbs. *Circ J.* 71, Mar 2007, Vol. 3, pp. 405-411.**69. Kang TS, Gorti GK, Quan SY, Ho M, Koch RJ.** Effect of hyperbaric oxygen on the growth factor profile of fibroblasts. *Arch Facial Plast Surg.* Jan-Feb 2004, Vol. 6, 1, pp. 31-35.
70. Strauss MB, Malluche HH, Faugere MC. Effect of hyperbaric oxygen on bone resorption in rabbits. . *Seventh Annual Conference on Clinical Application of HBO.* June 1982 , pp. 8-18. **71. Penttinen R, Niinikoski J, Kulonen E.** Hyperbaric oxygen and fracture healing: a biomechanical study with rats. . *Acta Chir.* 1972, 138, pp. 39–44.**72. Ueng SW, Lee SS, Lin SS, Wang CR, Liu SJ, Yang HF, Tai CL, Shih CH.** Bone healing of tibial lengthening is enhanced by hyperbaric oxygen therapy therapy: a study of bone mineral density and torsional strength on rabbits. *J Trauma.* 1998, 44, pp. 676–681.**73. Hsieh CP, Chiou YL, Lin CY.** Hyperbaric oxygen-stimulated proliferation and growth of osteoblasts may be mediated through the FGF-2/MEK/ERK 1/2/NF-κB and PKC/JNK pathways. *Connect Tissue Res.* . Dec 2010, Vol. 51, 6, pp. 497-509.**74. Knighton DR, Silver IA, Hunt TK.** Regulation of wound-healing angiogenesis-effect of oxygen gradients and inspired oxygen concentration. *Surgery.* Aug 1981, Vol. 90, 2, pp. 262-270.**75. Milovanova TN, Bhopale VM, Sorokina EM, Moore JS, Hunt TK, Hauer-Jensen M, Velazquez OC, Thom SR.** Hyperbaric oxygen stimulates vasculogenic stem cell growth and differentiation in vivo. *J Appl Physiol.* Feb 2009, Vol. 106, 2, pp. 711-728.**76. OC., Velazquez.** Angiogenesis and vasculogenesis: inducing the growth of new blood vessels and wound healing by stimulation of bone marrow-derived progenitor cell mobilization and homing. *J Vasc Surg.* Jun 2007, Vol. 45, Suppl A, pp. A39-47.**77. Gallagher KA, Goldstein LJ, Thom SR, Velazquez OC.** Hyperbaric oxygen and bone marrow-derived endothelial progenitor cells in diabetic wound healing. *Vascular.* Nov-Dec 2006, Vol. 14, 6, pp. 328-37.**78. Sheikh AY, Gibson JJ, Rollins MD, et al.** Effect of hyperoxia on vascular endothelial growth factor levels in a wound model. *Arch Surg.* . Nov 2000, Vol. 135, 11, pp. 1293-7.**79.** Hunt TK, Pai MP. The effect of varying ambient oxygen tensions on wound metabolism and collagen synthesis. *Surg Gynecol Obstet.* Oct 1972 , Vol. 135, 4, pp. 561-567.**80. Sander AL, Henrich D, Muth CM, Marzi I, Barker JH, Frank JM.** In vivo effect of hyperbaric oxygen on wound angiogenesis and epithelialization. *Wound Repair Regen.* Mar-Apr 2009, Vol. 17, 2, pp. 179-84.**81. Stephen R. Thom, Veena M. Bhopale, Omaida C. Velazquez, Lee J. Goldstein, Lynne H. Thom, and Donald G. Buerk.** Stem cell mobilization by hyperbaric oxygen. *Am J Physiol Heart Circ Physiol.* 2006, 290, pp. H1378-H1386.**82. Thom SR, Milovanova TN, Yang M, Bhopale VM, Sorokina EM, Uzun G, Malay DS, Troiano MA, Hardy KR, Lambert DS, Logue CJ, Margolis DJ.** Vasculogenic stem cell mobilization and wound recruitment in diabetic patients: Increased cell number and intracellular regulatory protein content associated with hyperbaric oxygen therapy. *Wound Repair Regen.* 2011, Vol. 19, 2, pp. 149–161.**83.** Hyperbaric oxygen therapy reduces severity and improves survival in severe acute pancreatitis. *J Gastrointest Surg.* Aug 2007, Vol. 11, 8, pp. 1008-15.**84. Christophi C, Millar I, Nikfarjam M, Muralidharan V, Malcontenti-Wilson C.** Hyperbaric oxygen therapy for severe acute pancreatitis. *J Gastroenterol Hepatol.* Nov 2007 , Vol. 2211, pp. 2042-2046.**85. Cuthbertson CM, Su KH, Muralidharan V, Millar I, Malcontenti-Wilson C, Christophi C.** Hyperbaric oxygen improves capillary morphology in severe acute pancreatitis. *Pancreas.* . Jan 2008, Vol. 36, 1, pp. 70-75.**86. Yu X, Li YG, He XW, Li XR, Din BN, Gan Y, Xu M.** .Hyperbaric oxygen reduces inflammatory response in acute pancreatitis by inhibiting NF-kappaB activation. *Eur Surg Res.* . 2009, Vol. 42, 2, pp. 130-135.

FROM CATS TO CHILDREN: TREATING NIEMANN-PICK TYPE C DISEASE IN MULTIPLE SPECIES

Charles H Vite, DVM, PhD, DACVIM (Neurology)
Philadelphia, Pennsylvania

The NIH/TRND (Therapeutics for Rare and Neglected Diseases) team collaboratively set a goal to develop a safe, effective, and FDA-approved therapy for previously untreatable Niemann-Pick type C disease (NPC disease; childhood Alzheimer-like disease). Treatment for this rare disease, affecting 1/150,000 people with over 300 mutations, depended on strong private, corporate, intramural, and national funding of many labs to support the selection of potential compounds, animal studies, and bioanalytical and biomarker development. The path toward the development of therapy and the assembly of the collaborative team necessary to begin the phase 1 and phase 2 trials scheduled for this year will be discussed.

INTRODUCTION

Niemann-Pick disease type C (NPC disease) is an incurable cellular lipid trafficking disorder resulting in the accumulation of unesterified cholesterol and glycosphingolipids in the CNS and in visceral organs. Clinical features include progressive neurological dysfunction, hepatosplenomegaly, and early death. Brain histology and biochemistry reveals widespread neuronal cytoplasmic vacuolization and intracellular cholesterol and ganglioside storage, neuroaxonal dystrophy and neuronal loss most severely affecting Purkinje cells, cortical pyramidal neuron meganeurite formation and ectopic dendritogenesis, gliosis, and inflammation. The mechanisms by which deficient activity of NPC1 or NPC2 proteins result in clinical disease and neuropathology are not understood.

MIGLUSTAT

Miglustat, a small imino sugar that inhibits glucosylceramide synthase and the synthesis of all glucosylceramide-based glycosphingolipids including lactosylceramide and the gangliosides, ameliorates NPC disease in murine and feline disease models in which daily administration was found to delay the onset of neurological dysfunction, to significantly increase lifespan, to improve Purkinje cell survival, and to decrease GM2 ganglioside accumulation in the cerebellum. *Ex vivo* examination of microglia from the brains of treated cats also revealed normalization of CD1c and MHC1 expression as well as reactive oxygen species generation. Together, these results support a beneficial influence of miglustat on the clinical progression of NPC disease and suggest that prolonged Purkinje cell survival, reduced glycosphingolipid accumulation, and/or the modulation of microglial immunophenotype and function contribute to the mechanistic basis of miglustat-induced neurological improvement in treated cats.

Miglustat is the only drug presently known to have measurable positive influence on NPC disease progression in human patients. A randomized controlled study in patients of various ages was performed with therapeutic efficacy assessed primarily by quantification of horizontal saccadic eye movements, and, secondarily, by assessments of swallowing, auditory ability, ambulatory ability, and cognition. Based on outcome data from this study and subsequent studies, miglustat was approved for use in the treatment of NPC disease in several countries but remains unapproved in the US pending further proof of efficacy.

CYCLODEXTRINS

Recently, hydroxypropyl-ß-cyclodextrin (HPßCD) was shown to be endocytosed by npc1-/- cells *in vitro* and to reduce cholesterol storage by acting from inside endocytic organelles. Cyclodextrins are a family of cyclic oligosaccharides with a hydrophilic exterior and a non-polar cavity; they are used as pharmaceutical excipients that increase the solubility of poorly water-soluble drugs. HPßCD studies suggest that it can replace membrane-bound NPC1 in restoring cholesterol homeostasis by allowing lysosomal cholesterol to flow into the cytosolic ester pool. The ability of HPßCD to affect disease is evident from studies in the murine NPC disease model. Weekly intraperitoneal administration of 1500 mg/kg of HPßCD to $npc^{-/-}$ mice resulted in improvement in hepatic disease but had no effect on neurological disease or lifespan. In contrast, the administration of a single subcutaneous (s.c.) dose of 4000 mg/kg of a 20% solution of HPßCD to 7-day–old $npc^{-/-}$ mice significantly decreased neurodegeneration, improved hepatic function, and prolonged lifespan. This study also identified that although the plasma half-life of HPßCD was measured in hours, the effect on sterol homeostasis lasted at least 6 days. Notably, repeated s.c. administration of 4000 mg/kg to affected mice also significantly decreased neuronal cholesterol, ganglioside and sphingosine accumulation. High doses of HPßCD were hypothesized to be needed to ameliorate neurologic dysfunction due to an increased ability to cross the blood-brain barrier or due to increased binding of circulating sterols resulting in enhanced cholesterol egress from the CNS. In each of these animal studies, no significant toxicity was observed following the administration of HPßCD.

STUDIES IN THE CAT

In the feline model of NPC1 disease, the first clinical signs of the disease were observed at approximately 6 weeks of age. At the age of 6 weeks, intention tremors and ataxia were observed indicating onset of a progressive cerebellar and vestibular dysfunction. The disease also affected the liver and the lungs. Progression of disease was rapid and the expected average life span for diseased cats was 20.5 ± 4.8 weeks. NPC1 cats were treated beginning at 3 weeks of age at different dose levels of HPβCD via intrathecal (IT) administration at the cerebellomedullary cistern (120 mg once every 2 weeks, equivalent to ~30 mg/mL CSF or ~20 mM in 4.4 mL CSF) or s.c. (1000, 4000, or 8000 mg/kg once per week) administration. Treated cats showed improvements in the liver and neurological disease at the 2 higher s.c. dose levels (4000 and 8000 mg/kg/week). At the highest dose level, the greatest improvement in neurological disease was observed, while no improvement in the neurological disease was observed at the lowest dose level (1000 mg/kg s.c. weekly). Cats dosed with 120 mg every 2 weeks IT remained neurologically normal (coordination, physical appearance) but with a lower weight gain. The IT treated cats had an increased alanine aminotransferase (ALT) level through 24 weeks of treatment, although lower than in the untreated cats, suggesting a limited effect of the IT treatment on liver disease. The efficacy investigation of HPβCD in NPC1 cats showed a major side effect. Weekly s.c. administration of 4000 mg/kg and 8000 mg/kg HPβCD and IT administration of 120 mg HP-β-CD given once or multiple times caused partial or substantial loss of hearing in NPC1 cats due to irreversible destruction of hair cells as observed during histopathological investigation. This side effect was also noted in normal cats treated at similar dose levels and route of administration. Studies in NPC1 cats treated at 30 and 60 mg/dose (~7 and 14 mg/mL CSF respectively) IT every other week also showed a partial or substantial loss of hearing. It was observed that cats administered the high dose of 8000 mg/kg given s.c. once a week for 21 weeks showed alveolar damage; this was observed in 2 of 5 cats treated at this dose level, and was considered related to the cause of death. However, HPβCD treatment resulted in an increased life span in cats treated with 4000 mg/kg and in the surviving cats at the 8000 mg/kg s.c. dose level. Two of the 5 cats that received 8000 mg/kg HPβCD survived 20 to 30 weeks longer than untreated NPC1 cats. The 5th cat treated at 8000 mg/kg was terminated at 24 weeks for scheduled necropsy. Cats that received IT treatment survived for longer than 2 year with normal neurological exams and are still being treated with HPβCD and examined regularly. NPC1 cats that received combined s.c. (1000 mg/kg) and IT (120 mg/dose) treatment survived for greater than 2 years with normal neurology exams and ALT levels.

STUDIES NECESSARY FOR PHASE I TRIAL

What safety information was needed to support dosing HPBCD into NPC1 patients in a Phase 1 clinical trial? The safety is well established as a pharmaceutical excipient in oral and IV administered drug. Also, 5 compassionate use single-patient INDS were approved. Preclinical studies needed to support the phase I study included 1) NPC1 animal studies to determine the lowest efficacious dose, 2) pharmacokinetic/pharmacodynamic studies to support brain distribution, 3) biomarkers development to be able to measure biochemical change specific to NPC 1 disease, 4) bioanalystical method development to measures HPBCD in plasma and CSF, 5) CMC support for formulation and clinical supply of HPBCD, 6) toxicology studies to determine safe starting dose in patients, and 7) clinical endpoints to determine changes that would indicated positive signs of efficacy in patients. The Phase 1 study is a non-randomized, open-label, single-center, dose-ranging, multiple-dose study to examine the safety, tolerability, feasibility, and pharmacokinetics of ICV administration of HPβCD via an Ommaya reservoir to patients with NPC1 disease and to determine an MTD. This trial began in January 2013. The study will also evaluate the use of candidate plasma and CSF biomarkers, explore the use of potential clinical outcome measures, and evaluate the effect of HPβCD on cholesterol mobilization. This intra-patient dose escalation study will include 6 patients, each of whom will receive monthly doses of HP-β-CD ranging from 12.5 mg to 200 mg.

ACKNOWLEDGMENTS

The speaker gratefully acknowledges the assistance of the following in the preparation of this talk:
Elizabeth Ottinger, PhD, Project Manager NIH/TRND at NCATS
Forbes Porter MD, PhD, Acting Program Director PDEGEN, NICHD
Daniel Ory MD, Professor of Medicine, Cell Biology, and Physiology, Washington University St Louis
Steven Walkley, DVM, PhD, Director Rose F Kennedy Intellectual and Developmental Disabilities Research Center
Mark Kao, PhD, Lead in Preclinical Development Drug Safety Sciences at Johnson and Johnson Pharmaceutical Research and Development.

REFERENCES

Pentchev PG, et al. PNAS USA. 1985;82:8247; Patterson MC et al. Scriver CR, Beaudet AL, Sly WS, Valle D, editors. New York: McGraw-Hill Medical Publishing Division; 2001. pp. 3611–3633. The Metabolic & Molecular Bases of Inherited Disease, vol. 3, 8th edition; Walkley SU and Suzuki K. Biochim Biophys Acta. 2004;1685:48; Sarna JR et al. J Comp Neurol. 2003;456:279; Infante RE et al. J Biol Chem.2008;283:1052; Infante RE et al. PNAS USA. 2008;105:15287; Liu Y et al. Hum Mol Genet. 2000;9:1087; Lachmann RH and Platt FM. Expert Opin Investig Drugs. 2001;10:455; Zervas M et al Curr Biol. 2001;11:1283; Patterson MC et al. Lancet Neurol. 2007;6:765; Griffin LD et al. Nat Med. 2004;10:704; Liu B et al. J Lipid Res. 2008;49:663; Liu B et al. PNAS USA.2009;106:2377; Pacheco CD et al. Hum Mol Genet. 2007;16:1495; Liao G et al. Am J Pathol. 2007;171:962; Jeyakumar M et al..2003;126:974; Irie T and Uekama K. J Pharm Sci. 1997;86:147; Lloyd-Evans E et al. Nat Med. 2008;14:1247; Walkley SU. Semin Cell Dev Biol. 2004;15:433; Camargo F et al. Sci. 2001;70:131; Monnaert V et al J Pharmacol Exp Ther. 2004;310:745; Benarroch EE. Neurology. 2008;71:1368; Zhang M et alJ Neurosci Res. 2008;86:2848; Rajewski RA and Stella VJ. J Pharm Sci. 1996;85:1142; Gould S and Scott RC. Food Chem Toxicol. 2005;43:1451; Loftus SK et al. Science. 1997;277:232; Vance JE et al. Sem Cell Dev Biol 2005; 16: 193; Li H et al. J Neuropathol Exp Neurol 2005; 64: 323; Vite CH et al. Ped Res 2008; 64: 544; Stein VM et al. J Neuropath Exp Neurol 2012l 71: 434; Ward SA Ped Res 2010; 68: 52; Bagel J et al J Neuropath Exp Neurol (in press)

MAGNETIC RESONANCE IMAGING OF THE SPINE

Ronaldo C. da Costa, DMV, MSc, PhD, DACVIM (Neurology)
Columbus, Ohio, USA

Advanced imaging of the spine and mostly magnetic resonance imaging (MRI) has revolutionized the diagnosis of many spinal conditions such as fibrocartilaginous embolic myelopathy. MRI is the only modality that allows direct visualization of soft tissue structures such as the spinal cord and nerve roots, thus allowing characterization of spinal disorders well beyond those achieved with myelography and computed tomography. The applications of spinal MRI are many and are expanding progressively. In addition to routine MRI sequences, such as T2 and T1 weighted images (WI), special sequences such as diffusion tensor imaging and angiography, are beginning to be used, (1-3) and may be routinely applied in the future.

Normal Anatomy

The anatomic features of the spinal region being studied must be taken into consideration when interpreting advanced imaging studies. Each region of the vertebral column (cervical, thoracic, lumbar and sacral) has specific anatomic features. Due to the exquisite sensitivity of MRI it is important to assess the imaging findings in the light of the patient's clinical signs and examination findings. The presence of asymptomatic spinal abnormalities is widely recognized in human medicine. The presence of multiple spinal abnormalities is commonly seen in aged dogs and cats. Clinically silent spinal cord compression, intervertebral disc degeneration, intervertebral disc protrusion, nerve root compression, intervertebral foraminal stenosis, and vertebral canal stenosis have all been reported without concurrent clinical signs in dogs. (4-6)

Congenital diseases

Spinal arachnoid cysts

Arachnoid cysts consist of outpouchings of the arachnoid matter or focal dilatations in the subarachnoid space that are filled with cerebrospinal fluid. (7) The use of the term 'cyst' to describe these lesions is a misnomer, as these outpouchings are not lined by epithelium. They are most commonly located in the dorsolateral area of the cervical spine but may be seen ventrally and in other regions of the spine column. They may be bilaterally paired, having a bi-lobed appearance. (8) There is still discussion as to whether these 'pseudocysts' are congenitally acquired or possibly secondary to trauma or arachnoiditis-induced adhesions. Arachnoid adhesions resulting from chronic inflammation and microtrauma may be the instigating cause of pseudocyst development. (9) It has also been postulated that the arachnoid adhesions, rather than the subarachnoid outpouchings, are the cause of spinal cord compression and the resultant neurologic deficits. (9) As the Rottweiler breed is over-represented, a genetic or hereditary component is likely in at least some patients. (7-9) An acquired arachnoid cyst has been reported in a cat. (10) CT myelography and MRI are equally adept at identifying cyst-like lesions of the subarachnoid space. MRI appearance is of a "tear-drop" appearance on sagittal images with focal accumulation of CSF at the level of the dilation. There may also be, particularly cranially to point of greatest compression, T2W hyperintensity within the spinal cord parenchyma, consistent with edema, malacia and/or syringohydromyelia. (11) Heavy T2 weighted images (known as MR myelogram sequences) may be helpful to identify these "cysts" and are more sensitive than traditional T2 weighted sequences.

Vascular diseases

Fibrocartilaginous embolic myelopathy (FCEM)

(12)MRI is the imaging modality of choice for the diagnosis of FCEM. MRI depicts the spinal cord parenchyma allowing visualization of the ischemic spinal cord region. The appearance of FCEM on MRI is typically of a focal (although sometimes extensive) hyperintense T2-weighted lesion affecting primarily the grey matter and usually clearly asymmetric. The lesion is commonly isointense on T1-weighted images but can also be hypointense. Contrast enhancement is not commonly seen, it was observed in only 12% of cases in a study. (13) It is more likely to see contrast enhancement when MR imaging studies are performed 5 to 7 days after the ischemic event. (13) Spinal cord swelling is a common feature of FCEM. The severity of intramedullary lesions seen on MRI has been correlated with the severity of neurological signs at presentation and with outcome. (13) Interestingly, the presence of MRI changes was not associated with timing of imaging. (13) In some cases, primarily those imaged within 24 to 72 hours after the ischemic injury, no signal changes are seen on MRI. (14) These cases could be diagnosed with diffusion weighted MR imaging, although application of this technique in dogs is challenging. (14)

Inflammatory Spinal Diseases

Discospondylitis

Discospondylitis has been, by convention, a radiographic diagnosis. Clinical signs often precede radiographic evidence of vertebral endplate lysis and medullary sclerosis, typically seen with more advanced disease. CT is much more sensitive than radiographs for identifying endplate osteolysis early in the course of disease. MRI is much more sensitive for detecting soft tissue inflammation of the disc that usually precedes bony changes. Also, early marrow changes in the affected vertebra (increased hyperintensity on T2 weighted images and contrast enhancement on T1 weighted images) may be seen with MRI prior to overt osteolysis. As such, MRI may be preferred to CT in order to screen for early cases of discospondylitis where bony changes are not observed radiographically. (15, 16). Two recent studies reported the MRI findings in dogs with discospondylitis. The involved endplates and adjacent marrow were T1-hypointense or had mixed intensity, with hyperintensity in short tau inversion recovery (STIR) images in most dogs. Contrast enhancement of endplates and paravertebral tissues was also frequently seen. The intervertebral disks were fequently hyperintense in T2W and STIR images and had contrast enhancement in most cases. (17, 18)

Spinal Neoplasia

Spinal neoplasia is an important differential diagnosis for dogs presenting with either chronic or acute neurologic signs. Several classification systems are used to categorize spinal neoplasms. A common classification categorizes the tumors according to the location into: extradural, intradural-extramedullary, and intramedullary. Extradural tumors such as osteosarcoma and fibrosarcoma are the most common. (19)

In one study, myelography was more useful in differentiating between intradural-extramedullary and intramedullary tumors than MRI. (20) Careful evaluation of the images in all 3 planes (transverse, sagittal and dorsal) may assist in defining the location of the tumor. Dorsal images are particularly useful. The MRI findings of spinal meningiomas, have been well described. (21, 22) Most meningiomas are iso- to hyperintense on T1-weighted images and hyperintense on T2-weighted images. (20-22) Homogeneous, strong contrast enhancement and presence of dural tail are also consistently observed. (20-22)

MRI findings for other spinal tumors have not been well described. Osseous tumors have very variable signal intensity, ranging from iso- to hyper- or hypointense on both T1- and T2-weighted images. Contrast enhancement is also variable. It is important to use fat suppression techniques (e.g. STIR) when observing hyperintense lesions on T1- and T2-weighted images within osseous structures, as these changes may be caused by fat infiltration. Intramedullary tumors have more consistent imaging patterns. Due to the common presence of edema, hyperintensity on T2-weighted images and hypointensity on T1-weighted images, in relation to the surrounding spinal cord is commonly observed. The pattern of enhancement is variable.

Primary or secondary nerve sheath tumors (NST) are the most common cause of neurogenic lameness. Approximately 45% of the tumors are located in the nerve roots proximal to the spinal cord, while 55% are located in the plexus area or peripheral nerves. (23) This emphasizes the importance of using a larger field of view when imaging patients suspected of having nerve sheath tumors. The visualization of NST is facilitated using MRI. On MRI, nerve sheath tumors are consistently hyperintense on T2 weighted images, hypointense on T1 weighted images, and show homogenous or inhomogeneous contrast enhancement. The tumor may appear as a diffuse brachial plexus nerve thickening or a circumscribed mass. (24)

Intervertebral disc disease

MRI allows clear visualization of intervertebral disc disease. The normal nucleus pulposus of the intervertebral disc appears a hyperintense ellipsoid area on sagittal T2-weighted images. Intervertebral disc degeneration leads to a decrease in signal intensity and the degenerated disc becomes isointense to hypointense relative to the surrounding annulus fibrosus. The degree of brightness of the T2 signal of the nucleus pulposus correlates with the proteoglycan concentration but not with water or collagen concentration. (25) It is important to emphasize that disc degeneration per se does not lead to clinical signs, except in uncommon cases of discogenic spinal pain. MRI also allows visualization of the spinal cord, which facilitates comparison when multiple sites are affected. Sagittal and transverse images should be used concurrently to assess the severity and lateralization of spinal cord compression. It is common to observe lateralized disc extrusions causing nerve root compression and this may not be seen on mid-sagittal images. Parasagittal images allow visualization of lateralized disc herniations and nerve root compression. Hemorrhage associated with disc extrusion can cause a signal void on MR images. (26) Gradient echo sequences can confirm the presence of hemorrhage.

MRI also allows assessment of the spinal cord parenchyma and detection of spinal cord signal changes. In cases with multiple sites of spinal cord compression, identification of hyperintensity on T2-weighted images indicates the site with the worst compression. The spinal cord hyperintensity seen on T2-weighted images correlate with the severity of clinical signs. (27) The degree of severity of spinal cord compression, however, does not correlate with the severity of clinical signs.

(27-29) Three studies in dogs have indicated that the presence and extension of spinal cord signal changes have prognostic implications. (28, 30, 31) A study suggested that areas of hyperintensity longer than 3 times the body of L2 were associated with poor prognosis with only 20% of dogs with this signal change regaining ambulatory status. (30) The presence of hyperintensity was a more reliable prognostic indicator than absence of nociception. (30) Even in non-compressive intervertebral disc extrusions, the extent of spinal cord hyperintensity predicted the outcome. (31)

Chronic disc extrusions may be associated with inflammatory reaction surrounding the extruded disc leading to a ring enhancement pattern surrounding the disc. The image may be mistaken as a granulomatous or neoplastic lesion. It is important to be aware of this imaging feature of IVDD to avoid misdiagnosing it as other conditions. (32) Intervertebral disc herniation can cause many different types of imaging patterns and should always be considered in the differential diagnosis for unusual MRI findings. In contrast to non-enhanced CT, normal MRI findings in all three imaging planes rules out the diagnosis of IVDD.

Cervical Spondylomyelopathy

Cervical spondylomyelopathy (CSM) is characterized by static and dynamic spinal cord compressions. When planning advanced imaging studies of the cases suspected of having CSM, it is important to plan the field of view to cover the entire cervical spine up to the 3rd thoracic vertebrae, as dogs with CSM have been shown to have cranial thoracic compressions. (33)

MRI has been considered the best imaging technique for humans with cervical spondylotic myelopathy for many years. (34, 35) CT-myelography is still used in humans for equivocal cases where cervical radiculopathy secondary to foraminal stenosis is the main clinical problem. (36) The main advantage of MRI over CT-myelography is the ability to directly visualize the spinal cord. This allows detection of spinal cord signal changes that are helpful to determine the primary spinal cord lesion in cases with multiple compressions. Two recent studies indicate that multiple spinal compressions are seen in 63% of dogs with CSM. (33, 37)

A study compared MRI and myelography for dogs with CSM. It was concluded that MRI allows identification of more sites of abnormalities than cervical myelography. Although myelography could identify the location of the lesion in most patients, MRI was more accurate in predicting the site, severity and nature of the spinal cord compression. Spinal cord signal changes were seen in the majority of patients and assisted in precise lesion identification. (38) A study looked specifically at factors associated with spinal cord signal changes (SCSC) in 102 dogs with CSM. (39) The overall incidence of SCSC was 55.9%. SCSC were significantly more common in dogs with a chronic history, more severe neurological deficits, and with moderate or severe spinal cord compressions. The location and direction of the compressive lesions had no influence on the development of SCSC in dogs with CSM. (39) Another recent study compared myelography, CT-myelography and MRI in 22 dogs with disc-associated CSM. The study found moderate agreement of all modalities in identifying the most severe compressive lesions and the authors conclude that CT-myelography and MRI could be seen as complementary modalities. (40)

Lumbosacral disease

Lumbosacral (LS) disease, also known as degenerative lumbosacral stenosis (DLSS) or cauda equina syndrome is a common disease of large breeds dogs associated with nerve root compression at L6-L7 or L7-S1 vertebrae. The degree of lumbosacral compression detected using MRI has no correlation with the severity of clinical signs. (41) Fecal and urinary incontinence were observed in dogs with minimal LS compression, while other dogs had severe compression of the LS region and showed pain only without neurological deficits. (41) MRI usually reveals ventral compression caused by intervertebral disc degeneration and protrusion. Intervertebral disc protrusion can be seen with loss of the normal bright hyperintense signal of the intervertebral disc. (41) Dorsal compression can also be observed caused by joint capsule thickening, osteophyte formation and hypertrophy of the ligamentum flavum. Foraminal stenosis is an important component of the complex of the DLSS and the parasagittal and transverse images should be carefully examined. Transverse images allow assessment of the dorso-ventral diameter of the foramina, while the parasagittal images allow evaluation of the cranio-caudal diameter of the foramina. Dynamic studies of the LS spine can be performed, (32) however, criteria for testing and normal reference ranges have not been established and as such, interpretation of the results can be problematic. CT and MRI findings for LS disease showed a high agreement between both modalities, however, the correlation between CT or MRI findings with surgical findings is low. (42, 43) MRI and CT findings also had no correlation with outcome. (42) It is also important to remember that clinically normal dogs can have imaging characteristics of LS diseases without clinical signs. (6)

Spinal Trauma

Limited information is available on MRI and spinal trauma. In a study of 11 dogs with spinal fracture or subluxation, MRI identified rupture of soft tissue structures and/or fracture of 2 vertebral compartments (based on the 3 compartment theory). The majority of dogs had spinal cord changes (hyperintensity on T2 weighted images), hemorrhage, or swelling. Epidural hematoma was also identified in most dogs. (12) Another study compared the diagnostic sensitivity of survey radiographs and CT in dogs with confirmed spinal fractures and luxations. Radiographs missed approximately 25% of the lesions detected on CT. (44) CT is routinely used as the screening imaging of choice for spinal trauma in people due to its availability and rapidity of imaging.

REFERENCES

1. Westworth DR, et al. **Vet Radiol Ultrasound** 2006;47(3)265-269. 2. Kerkovsky M, et al. **Spine (Phila Pa 1976)** 2012;37(1)48-56. 3. Pease A and R. Miller. **Vet Radiol Ultrasound** 2011;52(5)492-497. 4. da Costa RC, et al. **Am J Vet Res** 2006;67(10)1601-1612. 5. Axlund TW and J. A. Hudson. **Vet Radiol Ultrasound** 2003;44(6)630-634. 6. Jones JC and K. D. Inzana. **Vet Radiol Ultrasound** 2000;41(1)19-26. 7. Jurina K and V. Grevel. **J Small Anim Pract** 2004;45(1)9-15. 8. Rylander H, et al. **J Vet Intern Med** 2002;16(6)690-696. 9. Gnirs K, et al. **Vet Radiol Ultrasound** 2003;44(4)402-408. 10. Sugiyama T and D. J. Simpson. **Aust Vet J** 2009;87(7)296-300. 11. Westworth DR and B. K. Sturges. **Vet Clin North Am Small Anim Pract** 2010;40(5)951-981. 12. De Risio L, et al. **J Vet Intern Med** 2007;21(6)1290-1298. 13. De Risio L and S. R. Platt. **Vet Clin North Am Small Anim Pract** 2010;40(5)859-869. 14. Cherubini GB, et al. **J Small Anim Pract** 2004;45(8)417-420. 15. Gonzalo-Orden JM, et al. **Vet Radiol Ultrasound** 2000;41(2)142-144. 16. Carrera I, et al. **Vet Radiol Ultrasound** 2011;52(2)125-131. 17. Harris JM, et al. **J Am Vet Med Assoc** 2013;242(3)359-365. 18. Dernell WS, et al. **J Am Anim Hosp Assoc** 2000;36(3)245-251. 19. Kippenes H, et al. **Vet Radiol Ultrasound** 1999;40(6)627-633. 20. McDonnell JJ, et al. **Vet Radiol Ultrasound** 2005;46(5)368-374. 21. Petersen SA, et al. **J Vet Intern Med** 2008;22(4)946-953. 22. Brehm DM, et al. **J Am Anim Hosp Assoc** 1995;31(4)349-359. 23. Kraft S, et al. **Vet Radiol Ultrasound** 2007;48(1)1-7. 24. Pearce RH, et al. **J Rheumatol Suppl** 1991;2742-43. 25. Tidwell AS, et al. **Vet Radiol Ultrasound** 2002;43(4)319-324. 26. Besalti O, et al. **J Am Vet Med Assoc** 2006;228(6)902-908. 27. Levine JM, et al. **J Vet Intern Med** 2009;23(6)1220-1226. 28. Penning V, et al. **J Small Anim Pract** 2006;47(11)644-650. 29. Ito D, et al. **J Am Vet Med Assoc** 2005;227(9)1454-1460. 30. De Risio L, et al. **J Am Vet Med Assoc** 2009;234(4)495-504. 31. Bagley RS, et al. Diagnosis of Spinal Disease. Practical Small Animal MRI, 2009. 32. da Costa RC, et al. **Vet Radiol Ultrasound** 2012;53(1)64-70. 33. Modic MT, et al. **Clin Orthop Relat Res** 1989;(239)(239)109-120. 34. Larsson EM, et al. **Acta Radiol** 1989;30(3)233-239. 35. Song KJ, et al. **J Spinal Disord Tech** 2009;22(5)353-357. 36. da Costa RC and J. M. Parent. **J Vet Int Med** 2009;23(3)740. 37. da Costa RC, et al. **Vet Radiol Ultrasound** 2006;47(6)523-531. 38. da Costa R. **J Vet Int Med** 2012;26(3)807-808. 39. De Decker S, et al. **J Am Vet Med Assoc** 2011;238(12)1601-1608. 40. Mayhew PD, et al. **J Am Anim Hosp Assoc** 2002;38(6)555-562. 41. Jones JC, et al. **J Am Vet Med Assoc** 2000;216(11)1769-1774. 42. Suwankong N, et al. **J Am Vet Med Assoc** 2006;229(12)1924-1929. 43. Johnson P, et al. **Vet Radiol Ultrasound** 2012;53(5)552-559. 44. Kinns J, et al. **Vet Radiol Ultrasound** 2006;47(6)563-570.

TEMPORAL EVALUATION OF PATHOLOGICAL AND ULTRASTRUCTURAL FEATURES IN CANINE CENTRONUCLEAR MYOPATHY. GL Walmsley[1], S Blot[2], K Venner[3], SC Brown[1], L Tiret[4], CA Sewry[5] and RJ Piercy[1]. 1: RVC, London, UK 2: UPR de Neurobiologie, ENVA, Paris, France 3: IoN, London, UK 4: INRA, ENVA, Paris, France 5: ICH, London, UK.

Centronuclear myopathies (CNM) comprise a genetically heterogeneous group of inherited neuromuscular diseases that affect humans and dogs. They are characterized clinically by paresis and skeletal muscle atrophy, and histologically, by abnormal centralization of myonuclei. Causal mutations have been identified for several forms of human CNM, and their functional consequences include defective membrane trafficking and excitation-contraction coupling in muscle. Autosomal recessive CNM in Labrador Retrievers is caused by a SINE insertion in the *PTPLA* gene, but the associated protein's function in normal muscle and the mutation's effect on intracellular membrane systems are poorly understood. We aimed to describe light microscope and ultrastructural features of muscle pathology in CNM-affected dogs as they age.

Biceps femoris muscle biopsy samples were taken at 11 and 30 months of age from 4 CNM-affected dogs and a heterozygote littermate. Additional samples were taken from 12 year aged dogs with CNM and a heterozygote; six normal dogs of various ages and breeds were used as controls. Fibrosis, fat infiltration, variability of myofibre size (with prominent hypotrophic fibres) and abnormal oxidative staining patterns were documented in affected dogs. The proportion of fibres with internalized ,myonuclei increased with age. Immunostaining for T-tubule (DHPR) and sarcoplasmic reticulum (SERCA and RyR1) markers was disorganized in CNM-affected dogs and most commonly featured a peripheral halo with radial remnants directed towards the sarcolemma, analogous to the appearance of serial sections stained with oxidative histochemistry. More generalized membrane abnormalities were also evident including mislocalization of dysferlin and caveolin 3 proteins (normally involved in membrane repair and which localize to the developing T-tubule system). Dystrophin staining confirmed the presence of invaginated and internalized sarcolemmal membranes. Intermediate filament (desmin) and myofibrillar (sarcomeric myosin) proteins were also disorganized. Ultrastructural abnormalities, which were most severe in the older CNM dogs, included: centralized nuclei, Z-line abnormalities such as rod formation and myofibrillar disorganization, mitochondrial aggregation and morphologically-abnormal T-tubules. Numerous tubular and vesicular membranous structures were observed in 30 month old dogs which resembled a radial alignment with mitochondria. Morphometric analyses confirmed a significant difference in T-tubule shape and size between CNM-affected and control dogs that progressed with age. No abnormalities were detected in the heterozygote or control dogs.

In conclusion, Labradors with CNM display progressive changes in internal tubuloreticular and sarcolemmal membranes, triads and mitochondria and recapitulate prominent features of related diseases in humans. These findings have important implications for our understanding of CNM pathogenesis and the role of PTPLA in membrane trafficking in skeletal muscle.

IS APOPTOSIS PRESENT IN THE SPINAL CORD OF DOGS WITH CERVICAL SPONDYLOMYELOPATHY? Ronaldo C. da Costa[1], Juliet Armstrong[1], Duncan Russell[1], Phillip Popovich[2], Dana McTigue[2]. [1]Dpt of Veterinary Clinical Sciences and Biosciences, College of Veterinary Medicine, and [2]Dpt. of Neuroscience College of Medicine, The Ohio State University, Columbus, OH.

Cervical spondylomyelopathy (CSM) is a challenging neurologic disease. In spite of a multitude of treatment options including dozens of surgical techniques, the long-term outcome and survival of affected dogs is poor. The most disseminated theory to account for this is the development of additional lesions (known as "domino" lesions), however, such mechanism is only documented in 20% of dogs. It is likely that other mechanisms play a role in the deterioration seen in dogs with CSM. We hypothesized that apoptosis is present in the spinal cord of dogs with CSM. This study aimed at investigating whether dogs with CSM had apoptotic changes in their spinal cords and to characterize the cell type (oligodendrocytes or neurons) undergoing apoptosis.

Spinal cord of 12 dogs with CSM and of 12 large breed dogs euthanized for reasons unrelated to the study were used. The control group was composed of dogs lacking gross evidence of compressive lesions at necropsy. All spinal cords were evaluated histologically using hematoxylin-eosin staining to characterize the severity of gray and white matter changes. Immunohistochemical (IHC) evaluation with caspase-3 was used to document the presence of apoptotic cells. Double labeling with terminal deoxynucleotidyl transferase dUTP nick end labeling and anti-myelin associated glycoprotein (TUNEL + MAG) or anti-neuronal nuclei markers (TUNEL + NeuN) were used to identify oligodendrocytes and neurons, respectively. For purposes of cell counts, the spinal cord was divided into white and gray matter regions and 10 specific areas from each region were consistently used for cell counts. Caspase-3, TUNEL+MAG and TUNEL + NeuN positive cells were counted in a blinded manner by one of the investigators. Comparisons between the CSM and control group were performed using a Wilcoxon rank-sum test. Significance was set at $p<0.05$.

The mean number of caspase-3 positive cells in the CSM group was 83.3 (±40.9), whereas for the control group it was 36.5 (±12.6), which was significantly different between groups ($p < 0.001$). The mean number of TUNEL + NeuN cells was 4.5 (±2.91) and 4.5 (±2.28), for the CSM and control group, respectively ($p = 0.83$). The mean number of TUNEL + MAG positive cells for the CSM group was 4.68 (±2.78), whereas for the control group it was 1.78 (±0.84), $p = 0.002$. The distribution of TUNEL + MAG positive cells was significantly higher in the dorsal, lateral and ventral funiculi of the CSM group.

The results indicate that CSM-affected dogs have apopotic changes in their spinal cord, affecting preferentially oligodendrocytes. Oligodendrocyte apoptosis may be related to the demyelinating and remyelinating process associated with chronic spinal cord compression, and may play a substantial role in the deterioration and outcome of CSM-affected dogs.

RECOVERY FROM CANINE SPINAL CORD INJURY - A TREADMILL BASED SCORING SYSTEM. NJ Olby, JH Lim, C Domaracki, K Bach, K Babb. North Carolina State University College of Veterinary Medicine, Raleigh, NC.

Outcome measures are important when performing clinical trials. The ideal measure provides objective, continuous data that quantifies a clinically relevant outcome or addresses a mechanistic question. There are ordinal scales for quantifying outcome of canine spinal cord injury; kinematic analysis generates continuous data, but requires specialized equipment. The aim of this study was to generate continuous data on hind limb function following acute spinal cord injury using a treadmill.

Normal dogs were walked on a treadmill to establish stepping patterns. Dogs with acute spinal cord injuries were grouped as non-ambulatory paraparetic, paraplegic and paraplegic without nociception and videotaped while walking on a treadmill with and without support at 48 hours, 2, 4 and 6 weeks. Dogs were also videotaped when walking on a non-slip surface and their gait was quantified using an established ordinal open field scale. 50 continuous stepping cycles were quantified by counting the number of hind limbs steps expressed as a percentage of forelimb steps (the stepping score), and as the percentage of coordinated hind limb steps (the coordination score).

Normal dogs walked with a cruciate pattern, which was defined as normal coordination. The treadmill was well tolerated by all dogs. The recovery of the three groups of dogs was significantly different at each time point using the stepping and coordination scores, while the open field scale failed to discriminate between non-ambulatory paraparetic and paraplegic dogs. We conclude that this treadmill based scoring system is a sensitive way to score outcome in dogs following spinal cord injury.

EVALUATION OF THE BICEPS REFLEX IN THE DOG. F Giebels[1], B Kohn[1], HA Volk[2], N Shihab[2], S Loderstedt[1]. [1]Free University of Berlin, Germany, [2]The Royal Veterinary College, London, UK.

In veterinary literature the biceps reflex is often described as not consistent for neurological examination in dogs. Innervated by the musculocutaneous nerve (spinal cord segments C6 to C8) the biceps brachii reflex response involves an elbow-flexion and/or movement over the biceps brachii muscle. The aim of this study was to prospectively evaluate the clinical assessment of the biceps reflex in healthy dogs.

Thirty-two dogs of different breeds, age, weight and fur length were examined and video taped. Including criteria were a normal clinical and neurological examination. The biceps reflex was examined and videotaped in lateral recumbency under standardized conditions. A randomized number anonymized each examination-sequence. The examination footage was evaluated by four different blinded observers (F.G., S.L., H.V., N.S.). The observers were asked to evaluate if reflexes were present (1) or absent (0) and the reflex activity by using the international accepted scoring system (0 = absent; 1 = reduced; 2 = normal; 3 = exaggerated; 4 = clonus). For statistical analysis of interobserver agreement κ and κw was used. Differences between dogs of different weight and age groups as well as fur length were calculated using the chi-square test.

The evaluation of the biceps reflex being present or absent (κ 0.485 - 0.852, mean 0,706) and the gradual reflex scoring (κw 0.414 - 0.595, mean 0.528) showed a good to very good interobserver agreement. There was no significant difference ($p <0,05$) in the reflex evaluation between dogs of ages <3y, 3 - 6y and > 6y, between dogs of <15kg, 16 - 30kg and >30kg or between long and short fur dogs.

In conclusion, the bicipital reflex in dogs might be a reliable parameter in the neurological examination of the dog. The good interobserver agreement implies an adequate comparability of the biceps reflex examination under clinical setting.

HEREDITARY POLYNEUROPATHY IN ALASKAN MALAMUTES – PHENOTYPE AND GENOTYPE OF HISTORICAL AND RECENT CASES. K. Hultin Jäderlund[1], M. Berendt[2], A. Espenes[1], C. Rohdin[3], Ø. Stigen[1], I. Bjerkås[1], M. Fredholm[2], L. Moe[1]. 1. Norwegian School of Veterinary Science, Oslo, Norway. 2. University of Copenhagen, Frederiksberg, Denmark. 3. Swedish University of Agricultural Sciences, Uppsala, Sweden.

Thirty years ago, a breed-specific polyneuropathy in Alaskan Malamutes was studied in Norway. The clinical picture included inspiratory stridor, exercise intolerance and tetraparesis. Pedigree studies and a test mating suggested an autosomal recessive mode of inheritance for the trait. A breeding strategy was recommended. Subsequently, cases were not diagnosed in Scandinavia for more than 20 years. However, during the last few years the disease has recurred in Scandinavia. A research project was therefore initiated, and we found the genetic basis for this disease to be a Gly98Val mutation in the N-myc downstream regulated gene 1 *(NDRG1)*. The aim of the present study was to link the historical cases with the cases of the last years, both by phenotype and genotype.

Archived material and information from historical cases as well as data from recent cases were analyzed and compared, according to clinical features, electrophysiological characteristics and pathological changes. The relationship between affected dogs was studied and the genotype of historical material was determined.

A common ancestor was found in pedigrees of all fathers and mothers of all affected dogs, and genetic testing of archived material confirmed homozygosity for the causative mutation in the *NDRG1* gene also in historical material. The phenotypic expression was considered equal in historical and recent cases, confirming these two groups of dogs to have the same degenerative nervous system disorder. In conclusion, historical Norwegian and recent Scandinavian cases of inherited polyneuropathy in Alaskan Malamutes have the same disease.

PREDNISONE AS A SOLO IMMUNOSUPPRESANT IN DOGS WITH PRESUMPTIVE MENINGOENCEPHALITIS OF UNKNOWN ETIOLOGY. H Barnes Heller[1]. M Mercier[2]. 1. University of Wisconsin School of Veterinary Medicine Madison, WI. 2. VCA Aurora Animal Hospital Aurora, IL.

The purpose of this study was to evaluate prednisone as a solo immunosuppressant in dogs diagnosed with meningoencephalitis of unknown etiology (MUE). Specific objectives were a) describe the cerebrospinal fluid (CSF) analysis after 1 month on immunosuppressive prednisone and b) to describe survival from the diagnosis of MUE.

Dogs were included if they met the following inclusion criteria: 1) normal minimum database 2) no glucocorticoid use within 5 days 3) MRI consistent with a diagnosis of meningoencephalitis/myelitis 4) negative geographically specific infectious titers 5) abnormal CSF analysis without evidence of neoplastic or infectious etiology.

Sixteen dogs met the inclusion criteria. Median CSF RBC, WBC, and protein at time of diagnosis were 65 cells/hpf, 39 cells/hpf, and 49 mg/dl respectively. Median CSF RBC, WBC and protein after 1 month on prednisone were 70 cells/hpf, 1 cell/hpf and 23 mg/dl respectively. Eight of 16 dogs (56%) had a normal CSF analysis at 1 month. Median survival for all dogs was 320 days, range 45-654 days. Median survival for responders was 373.5 days compared to 212.5 days in non-responders.

Dogs that achieved immunosuppression by 1 month had a longer survival than those that did not achieve immunosuppression by this time. Statistical analysis was limited by the study size.

REHABILITATION AND THE NEUROLOGIC PATIENT
Dr. H. Steven Steinberg

Unfortunately, rehabilitation of the neurologic patient today consists of extrapolation from the experiences in human physical therapy; application of modalities that have been tested on rodents, lagomorphs, and humans; and anecdotal conclusions. This presentation will demonstrate a structured approach to the neurologic rehab patient emphasizing neuro-anatomic localization, healing, retraining, and plasticity. Realistic expectations will also be explored.

Neuroanatomic localization even without a specific diagnosis is extremely useful when planning a rehabilitation program. Cervical spinal cord lesions require a completely different approach than the dog that has lumbo-sacral disease, although both may present with hindlimb weakness.

Intracranial rehabilitation has very limited usefulness in our patients because their ability to heal is exceptional compared to the experience in humans. The most chronic deficit that we deal with is the head tilt related to vestibular disease in cats and dogs. Middle ear disease, brainstem bleeds and strokes are the most common source of these and under most circumstances, these are not serious and as severe as they look (when the signs are acute). In almost all cases the signs will resolve and usually quickly. Finding the underlying cause is the challenge. The chronic head tilt is one that is very hard to correct. The oft-touted Eply procedure has never worked in my hands. We do work often with dogs that do not return to normal and the level of compensation can be quite remarkable but in those dogs with long-standing vestibular issues a cure is not likely.

Spinal cord disease can be segregated into segmental (LMN) and suprasegmental (UMN) disease processes and from a clinical stand point gray matter and white matter disease processes can have very different outcomes if the intumescence is involved. Many incomplete long-tract diseases will return to normal. Incomplete information getting to the central pattern generators is often adequate, in sub-primates, for a complete recovery.

At the intumescence, gray matter disease can be devastating, as motor function cannot return unless appropriate motor neurons are available. Similar difficulties are appreciated with peripheral nerve damage but many of these if treated properly will return to normal function. Contractures that are the result of permanent nerve damage (neurogenic atrophy) are impossible to relieve completely and keeping a limb healthy while waiting for a nerve to regrow can be a daunting task.

Aside from localization, compressive lesions are much more likely to heal and our experience in this area is quite extensive as at least 85% of the spinal cord injuries that we see are compressive. With advanced imaging the amount of longitudinal swelling has become quite evident. The longitudinal anatomy of the spinal cord is not given a great deal of attention but the effects of local compression/trauma can be far reaching, with swelling developing for several segments in both a cranial and caudal direction. Given time, most of these cases will demonstrate slow but steady resolution of the swelling and over several weeks, generally with restricted activity, function often returns to an acceptable level. It is hard to separate the roles that recovery, retraining and plasticity play in the clinical setting but it is reasonable to assume that all three are involved in the return to function.

As our canine population is aging polyneuropathy, with ascending clinical signs seems to be on the increase. It is our impression, that large breed male dogs are overrepresented.

Dogs with evidence of low cervical compression and lumbosacral nerve root entrapment seem to also reflect an aging canine population. Success with treating these patients with rehabilitation has been rewarding and very successful.

Retraining seems to be a primary goal for many of our neurologic post-treatment patients. Many paralyzed dogs that return to normal strength are presented because they have not regained proper mobility. Although we call this ability proprioception, the exact cause of this abnormality is poorly understood and has sensory and motor implications that go well beyond the simple "position sense" that is commonly presented to our students. Even the neuro-anatomic explanation for "knuckling" does not hold up when scrutinized.

Since many of these dogs present months after their initial recovery and our success is high, it is reasonable to classify this as retraining. Getting a dog to pay more attention to their paws and to automatically cause them to lift their limbs higher and to place them more efficiently, has born fruit for so many of my colleagues who work at these tasks every day.

It is well documented that the reflexes described by Charles Scott Sherrington are modifiers that help the coarse subliminal gaits determined by the central pattern generator(s) to respond to environmental changes, such as abrupt changes in terrain, without having to have a large diversion of attention. There is some reasonable evidence that there can be a major shift in roles where the reflexes themselves play a greater role in movement when the pattern generators are compromised.

In addition, the plasticity of the system allows for new connections to form, well documented in the spinal animal. These new connections are the result of new growth predominately arising out of the central neurons of the reflex arcs. How these new connections that take afferent reflex information and can initiate normal looking pattern generation is an area of great interest.

Lastly, one must keep in mind when developing a plan for the neurologic patient, the pet owner's goals and one's ability to reach a sensible outcome. The owner who simply wants their "couch potato " to stop dragging their nails when they are out for a fifteen minute walk twice per day, will require a completely different approach than the dog that drops stool when it walks across the floor. Likewise the attention span of the individual patient and their demeanor may demand very different exercises in different individuals even when in each case, the goal is the same.

For those of us in this business, we find no greater reward than having an owner tell us their pet went carelessly unaided up the stairs for the first time in two years.

ADVANCEMENTS IN THE DIAGNOSIS OF CEREBROVASCULAR DISEASE IN DOGS AND CATS

Laurent S. Garosi DVM, DECVN, MRCVS
Hitchin-Hertfordshire, England

INTRODUCTION

The term cerebrovascular disease is defined as any abnormality of the brain resulting from a pathological process compromising its blood supply. Stroke or cerebrovascular accident (CVA) is the most common clinical presentation of cerebrovascular disease and is characterized by the abrupt onset of a neurological deficit referable to a specific vascular territory. By convention, these signs must continue for more than 24 hours to qualify for the diagnosis of stroke, which is usually associated with permanent damage to the brain. A transient ischemic attack (TIA) is an acute neurological syndrome of vascular etiology that resolves within 24 hours and often predicts the risk of a significant incoming stroke. CVAs are caused by an abrupt disruption of blood flow to the brain owing to blockage of an artery, depriving brain tissue of oxygen and glucose (ischemic stroke), or to the rupture of a vessel, resulting in hemorrhage into or around the brain (hemorrhagic stroke).

IMAGING OF ISCHAEMIC STROKE

The primary aims when imaging patients presenting with suspected CVA (acute onset of brain signs) are to determine the stroke type (ischemic or hemorrhagic stroke), to exclude conditions that may mimic stroke (head trauma, infectious encephalitis, encephalitis of unknown etiology and decompensation from primary or metastatic brain tumor) and to evaluate the viability and reversibility of brain lesions. Computed tomography (CT), conventional and functional magnetic resonance imaging (MRI) as well as cerebral digital subtraction angiography (DSA) are discussed herein in the context of these goals.

Computer tomography (CT)

Computed tomography (CT) images are frequently normal during the acute phase of ischemia; therefore the diagnosis of ischemic stroke using CT relies upon the exclusion of mimics of stroke. Early CT signs of ischemia can be subtle and difficult to detect even by experienced readers and include parenchymal hypodensity, loss of grey-white matter differentiation, subtle effacement of the cortical sulci, and local mass effect. CT has until recently been the preferred imaging modality in man to determine the presence of hemorrhage in early stroke, with hemorrhage appearing hyperdense in the early stages. Recent developments in magnetic resonance imaging (MRI) mean this is no longer the case and CT now has no advantage over MRI in the diagnosis of ischemic stroke.

Conventional magnetic resonance imaging

Conventional MRI can be used to depict ischemic stroke within 12 to 24 hours of the onset and to distinguish hemorrhagic lesions from infarction. Although infarcts can sometimes be difficult to differentiate from other pathologic processes such as inflammatory diseases, they tend to have certain distinguishing characteristics on conventional MR images.

- A distinctive diagnostic aspect of ischemic stroke is their location and distribution, which depends on the vascular territory involved. The conformity of an ischemic infarct to a vascular territory is an important element in the diagnosis that helps in distinguishing these lesions from brain tumors, inflammation, and trauma. Ischemic infarcts are caused by occlusion of a cerebral blood vessel. They therefore occur and are limited to the region of the brain vascularized by the affected vessel with often sharp demarcation with the surrounding normal brain tissue and minimal or no mass effect.
- Ischemic/infarcts are caused by blood perfusion failure and therefore energy depletion. The consequence for the cell is failure of the Na^+/K^+ pump and accumulation of Na^+ and water within the cell, i.e. cytotoxic oedema. The MRI changes seen in ischemic parenchyma rely on an increase in tissue water content. Gradually T2-weighted or fluid-attenuated inversion recovery (FLAIR) images become more hyperintense (T2 prolongation which produces higher signal in areas of increased tissue water content) in the ischemic region particularly over the first 24 hours.
- MRI changes are best appreciated in the grey matter and are well visualized in deep grey matter structures such as the thalamus and basal nuclei due to the selected vulnerability to ischemia.
- Contrast-enhancement (associated with reperfusion) is not usually seen until at least 7 to 10 days.

T2-weighted and FLAIR images are particularly useful in ischemic stroke to give a more anatomical image of the brain and depict edema, old infarcts, microangiopathic changes, tumors and other pathology. With these sequences, ischemic infarction appears as a hyperintense lesion. Differentiation of ischemic core from penumbral tissue is however not possible. T2*-weighted gradient echo (GRE) images are use to show the presence of or exclude intracranial hemorrhage.

Functional magnetic resonance imaging

Several functional magnetic resonance imaging (fMRI) techniques have been developed in the early diagnosis of stroke and follow-up stroke treatment in human. They include diffusion and perfusion imaging and magnetic resonance angiography (MRA). Diffusion and perfusion MRI are new techniques that monitor water transport in the microenvironment at cellular or capillary levels. They provide complementary information about the pathophysiological processes following cerebral ischemia.

- Diffusion weighted images (DWI)

DWI is used commonly in humans to improve the sensitivity and specificity for the diagnosis of acute stroke making it an ideal sequence for positive identification of hyperacute stroke, excluding stroke mimics. The temporal evolution of the DWI signal also allows the discrimination of acute versus chronic lesions. Moreover, the sequence is sensitive, detecting lesion as small as 4 mm in diameter.

Magnetic resonance DWI detects the random molecular motion of water in vivo. The degree of this mobility can be quantified by a physical parameter known as the apparent diffusion coefficient (ADC). In biological systems, free diffusion is restricted by physical barriers (e.g. cell membranes) and chemical interactions (e.g. binding to macromolecules). Acute infarction leads to water trapping within the cells and causes reduced diffusion. This phenomenon of decreased diffusion and cytotoxic edema produces a regional hyperintensity on DWI. The image intensity on DWI is dependent on the ADC as well as on transverse relaxation time (T2). Because DWIs are affected by T1 and T2 contrast, stroke lesions also become progressively brighter owing to concurrent increases in brain water content, leading to the added contribution of hyperintense T2-weighted signal known as "T2-shine through". To differentiate between true restricted diffusion and "T2-shine through", bright lesions on DWI should always be confirmed with apparent diffusion coefficient maps, which exclusively measure diffusion. The ADC map helps to remove the effect of T2-weighted hyperintensity (associated with cytotoxic edema) that can contribute to diffusion-weighted hyperintensity. Diffusion weighted images can be used to improve the sensitivity and specificity for the diagnosis of acute stroke. This type of study also permits the discrimination of acute versus chronic lesions. The classic appearance of acute infarction is hyperintensity on DWI and reduced ADC. The ADC values remain below normal (reflect cellular swelling) over the first 8 to 10 days and progress to pseudonormal and supernormal values at time points beyond 10 days (due to cellular necrosis or lysis).

- Perfusion-weighted images (PWI)

In addition to diffusion-weighted imaging, MR perfusion-weighted imaging is used to depict brain regions of hypoperfusion and derives the tissue at risk by comparing the results with the findings on DWI. Perfusion-weighted imaging involves the repeated and rapid acquisition of images before and after the injection of a contrast agent using a two-dimensional gradient-echo or spin-echo EPI sequence. The shortening of the T1-relaxation rate over time is proportional to the concentration of the contrast agent and thus provides information on tissue perfusion.

Diffusion and perfusion MRI techniques have made possible the distinction between the two compartments (central core and peripheral penumbra) of ischemic tissue. With perfusion-weighted images, the blood supply of the tissue and area of hypoperfusion can be monitored, whereas DWI approximately reflect the irreversibly damaged infarcted core. The volume difference between the two, also termed PWI/DWI-mistmatch, has some correlation with the ischemic penumbra.

- Magnetic resonance angiography (MRA)

In addition to its use for tissue evaluation, MRA can non-invasively assess the intracranial vascular status of stroke patients. Two techniques can be used: time of flight (TOF) MRA and contrast-enhanced MRA. One of the main limitations of MRA is its lower resolution compared with conventional angiography. This becomes progressively worse as the luminal size decreases. In human, angiographic techniques are particularly used for screening of carotid artery stenosis, vascular malformation (such as arteriovenous malformation, venous angioma) and aneurysms. The use of MRA in dogs has been described and may allow identification of underlying vascular lesions in cases of canine stroke.

IMAGING OF HEMORRHAGIC STROKE

Computer tomography (CT)

CT is exquisitely sensitive for detection of acute hemorrhage. Acute hemorrhage is evident as a hyperdensity on CT due to hyperattenuation of X-rays by the globin portion of blood. The attenuation decreases until the hematoma is isodense at about 1 month after the onset. The periphery of the hematoma contrast enhances from 6 days to 6 weeks after the onset due to re-vascularisation.

Conventional magnetic resonance imaging

The MR signal intensity of intracranial hemorrhage is also influenced by several intrinsic (time from ictus, source, size and location of hemorrhage) and extrinsic (pulse sequence and field strength) factors. The causes of these intrinsic and extrinsic variations in hematoma intensity are difficult to evaluate with clinical studies since it is frequently impossible to precisely ascertain the interval between hemorrhage and MR imaging. Gradient echo sequences have been proven to be the most accurate of all of the MR pulse sequences, and more accurate than CT, in predicting the extent of hemorrhage on pathologic examination in a dog model. Compared to other sequences, gradient echo scans demonstrate readily detectable hypointensity regardless of the time from ictus, the source and location of hemorrhage, or the field strength. Hypointensities on GRE images is however not specific for hemorrhage and may also be seen with calcification, air, iron, foreign bodies and melanin. Air, calcification and often foreign bodies would also be normally hypointense on all the pulse sequences. In a recent study on non-traumatic intracranial hemorrhage in dogs, concurrent conditions relating to single lesions ≥5 mm were most commonly associated with A. vasorum infection and neoplasia. Whereas concurrent conditions relating multiple lesions <5 mm were most commonly associated with endocrinopathies and hypertension.

IDENTIFICATION OF UNDERLYING CAUSES OF STROKE

Causes of ischemic strokes

Ischemic strokes have been reported infrequently in veterinary publications when compared with human medical literature. Most reports have been based on postmortem results in dogs that either died or were euthanized as a result of the severity of the ischemic stroke and/or the suspected underlying cause of the stroke. This retrospective reporting may affect the prevalence and type of underlying causes identified in canine CVA, because only the most severely cases are likely to be reported. Similarly, dogs in which infarction occurred secondarily to a disease with a poor prognosis would also die or be euthanized. Suspected underlying causes identified in histopathologically confirmed cases included septic thromboemboli, atherosclerosis associated with primary hypothyroidism and Miniature Schnauzers with hypertriglyceridaemia, aberrant parasite migration (*Cuterebra* spp.) or parasitic emboli (*Dirofilaria immitis*), embolic metastatic tumor cells, intravascular lymphoma, aortic or cardiac source of embolism, and fibrocartilaginous embolism. In our MRI–based study a concurrent medical condition was detected in just over 50% of dogs affected by brain infarcts. Hypertension was documented in 30% of dogs. Of these dogs, chronic kidney disease and hyperadrenocorticism were the most commonly suspected underlying causes for hypertension. No underlying cause could be identified antemortem in nearly half of the dogs. An infarct of unknown origin is called cryptogenic. No age, sex, or breed predisposition was identified. However, cavalier King Charles spaniels and greyhounds appeared overrepresented.

Reports of ischemic strokes in cats are limited. Compared to dogs, a concurrent medical condition is more frequently found. Concurrent conditions identified include cuterebra parasitic migration, migrating heartworms, intracranial telangiectasia, cardiovascular disease, hyperthyroidism, liver disease and neoplasia elsewhere in the body.

Ancillary diagnostic tests in ischemic strokes

Ancillary diagnostic tests in ischemic stroke should focus on evaluating the animal for hypertension (and underlying causes), endocrine disease (hyperadrenocorticism, hypothyroidism, hyperthyroidism, diabetes mellitus, phaeochromocytoma), kidney disease (especially protein-losing nephropathy), heart disease (a greater risk factor in cats with cardiomyopathy), and metastatic disease. In cases of ischemic strokes, thromboelastography, D-dimer assays and/or antithrombin III evaluation should be routinely included in the screening tests for thromboembolic disease as a possible cause of ischemic stroke. D-dimer is considered to be more useful test for detecting thromboembolic disease in dogs when compared to traditional tests currently in use (platelet count and clotting times).

Causes of hemorrhagic strokes

In hemorrhagic stroke blood leaks from the vessel directly into the brain, forming a hematoma within the brain parenchyma or into the subarachnoid space. The mass of clotted blood causes physical disruption of the tissue and pressure on the surrounding brain. This alters central nervous system volume/pressure relationships with the possibility of increasing intracranial pressure (ICP) and decreasing cerebral blood flow (CBF). In contrast to high incidence in humans, intracerebral hemorrhage resulting from spontaneous rupture of vessels is considered rare in dogs. Secondary hemorrhage in dogs has been associated with rupture of congenital vascular abnormalities, primary and secondary brain tumors, inflammatory disease of the arteries and veins, intravascular lymphoma, brain infarction (hemorrhagic infarction), or impaired coagulation. Nontraumatic subarachnoid hemorrhage has been reported in dogs but remains very rare. This is in contrast to human patients in whom aneurysmal rupture of blood vessels is the most common underlying cause of hemorrhagic stroke. Parenchymal hemorrhage in cats has been associated with primary or secondary hypertension, intracranial neoplasia, cerebral amyloid angiopathy and feline infectious peritonitis.

Ancillary diagnostic tests in hemorrhagic strokes

In cases of hemorrhagic stroke, diagnostic tests should be targeted to screen the animal for a coagulation disorder (and underlying causes), hypertension (and underlying causes), and metastatic disease (particularly hemangiosarcoma): serial blood pressure measurements, complete blood count, serum biochemistry profile, buccal mucosa bleeding time, prothrombin time, activated partial thromboplastin time, thoracic radiographs, abdominal ultrasound.

References

Adams RD, et al (1997) Cerebrovascular diseases. In: Adams RD & Victor M eds. Principles of neurology. 6th ed. New York: McGraw-Hill Inc. pp 777-873. Graves MJ (1997). Magnetic resonance angiography. **Brit J Radiol** 70, 6-28. Garosi LS, et al. (2005) Brain infarct in dog and human: a comparative review. **J Small Anim Pract** 46:521-529**.** Garosi LS, et al. (2005) Results of diagnostic investigations and long-term outcome of 33 dogs with brain infarction (2000-2004). **J Vet Intern Med** 19:729-731. Garosi LS, et al. (2006) Clinical and topographical magnetic resonance characteristics of suspected brain infarctions in 40 dogs. **J Vet Intern Med**.*20:311-321.* Hakim AM (1998) Ischaemic penumbra: the therapeutic window. **Neurol** 51, S44-46. Heiland S (2003) Diffusion- and perfusion-weighted MR imaging in acute stroke: principles, methods, and applications. **Imaging Decisions in MRI** 4, 13-25. Kalimo H et al (2002) Vascular diseases. In: Graham DI & Lantos PL eds. Greenfield's neuropathology. 7th ed. London: Arnold. pp 233-280. McConnell JF, et al. (2005) MRI findings of presumed cerebellar cerebrovascular accident in twelve dogs. **Vet Radiol Ultrasound** 46, 1-10. Platt SR, Garosi L (2003) Canine cerebrovascular disease: do dogs have strokes? **J Am Anim Hosp Assoc** 39, 337-342. Lowrie M et al (2012) Concurrent medical conditions and long-term outcome in dogs with non-traumatic intracranial hemorrhage. **Vet Radiol Ultrasound** 53, 381-388. Garosi LS et al (2010) Cerebrovascular disease in dogs and cats. **Vet Clinic North Am – Small Anim Pract** 40, 65-79.

DIAGNOSIS OF CANINE AND FELINE MYOPATHIES

G. Diane Shelton, DVM, PhD, DACVIM (SAIM)
La Jolla, CA

INTRODUCTION

Neuromuscular diseases affect muscle, the neuromuscular junction (NMJ), peripheral nerves, and the ventral horn cells of the spinal cord. As such, corresponding diseases are referred to as myopathies, disorders of neuromuscular transmission (or junctionopathies), peripheral neuropathies and motor neuron diseases, respectively. Neuromuscular diseases can be acquired (such as immune-mediated myasthenia gravis and focal or generalized inflammatory myopathies) or inherited (such as in the muscular dystrophies and congenital myopathies). Although there are many neuromuscular diseases, they present with relatively few clinical signs, and in most cases, it is not possible to distinguish one from another just by visual examination. For this reason, a combined approach including attention to the history and signalment, thorough physical and neurologic examinations, appropriate laboratory testing, possibly muscle imaging, electrodiagnostic examinations if available, and the muscle biopsy are required to reach a diagnosis. Molecular based testing is available for a limited number of disorders but the list is growing. For this discussion, the focus is on diagnostic testing most useful in the dog or cat with muscle disease. The diagnosis of muscle diseases can be very challenging, but rational use of laboratory testing can result in a definitive diagnosis or direct the course of further diagnostics.

RECOGNIZE THE CLINICAL MYOPATHIC PHENOTYPE

The signalment is helpful in identifying congenital or breed specific inherited muscle diseases, and clinical signs are usually apparent at a young age. A detailed history including clinical signs of systemic disease and disease course, and vaccination and travel history can assist the clinician in prioritizing potential rule-outs. A through physical examination is critical with evaluation of all organ systems. A careful and complete neurological examination is essential to determine the correct anatomic localization to a specific portion of the neuromuscular system. Weakness, a stiff, short-strided gait, and muscle atrophy or hypertrophy are hallmarks of muscle disease.

THE NEED FOR CONFIRMATORY DIAGNOSTICS

- Some neuromuscular diseases are treatable, others are not
- An accurate prognosis depends on reaching a diagnosis
- Identification of a specific therapy depends on reaching a diagnosis
- Premature institution of corticosteroid therapy without an adequate diagnostic work-up may lead to delays in or failure to establish an accurate diagnosis
- Delay or failure to obtain an accurate diagnosis can result in irreversible damage to skeletal muscle in treatable disorders such as masticatory muscle myositis or polymyositis
- Delay or failure to obtain an accurate diagnosis can result in death in cases of myasthenia gravis due to aspiration pneumonia or respiratory failure
- An accurate diagnosis is critical to the development of DNA based testing for genetic diseases of purebred dogs and cats.

MINIMUM DATABASE FOR MUSCLE DISEASES

Routine laboratory screening including a CBC, serum chemistry profile and urinalysis can aid in the identification of a number of common systemic diseases that may result in muscle weakness. Specific abnormalities that may be identified on routine screening include anemia, hypo- and hyperglycemia, hypo- and hyperkalemia, hypo- and hypernatremia, hypo- and hypercalcemia, hypophosphatemia and hypomagnesemia. Additional testing procedures that should be part of baseline testing in any dog or cat with clinical neuromuscular disease includes measurement of serum creatine kinase (CK) activity, cardiac troponin I, plasma lactate concentration, and determination of thyroid status.

Serum Creatine Kinase Activity

Creatine kinase is a specific enzyme marker of myofiber damage and the one most commonly used in the diagnosis of muscle diseases.[1] Most notably CK is elevated in necrotizing, inflammatory and dystrophic myopathies and is usually normal or only mildly elevated in non-inflammatory muscle diseases. However, the specificity of serum CK is low resulting from the many causes of external muscle damage and the very short half-life of serum CK activity lasting only about 2 hours. The magnitude and persistence of CK elevation is an indication of severity and activity of muscle lesions. Intramuscular injections of drugs can cause an increase in serum CK with peak elevation about the fourth hour, consistent elevation for about 24 h, and return to baseline values by day three. Serum CK activity can be elevated as much as 12-fold in animals in prolonged recumbency.CK activity can also be elevated during and after surgery with the peak increase within 6-12 hours.

Electromyography can result in CK elevation. Anorexia in cats may be associated with moderate to marked elevations in serum CK activity and should not be used directly for a diagnosis of myopathy until feeding has resumed. If the serum CK activity is persistently increased and external causes can be eliminated, an underlying muscle disease should be considered.

Serum CK activity should be part of every neuromuscular minimum database, and most importantly on pre-neuter blood evaluations in young dogs. Creatine kinase activity may be normal in the presence of muscle disease, and muscle disease should not be ruled out based on a normal CK activity. Marked or persistent increases of CK may be indicative of a congenital or inherited muscle disease; even though the animal may be clinically asymptomatic at the time. The most marked increases in serum CK activity (>20,000 IU/L) are associated with necrotizing myopathies or muscular dystrophies. Generalized inflammatory myopathies usually show moderate increases in CK activity (2,000-20,000 IU/L), while the CK activity in focal inflammatory myopathies such as masticatory muscle myositis (MMM), endocrine myopathies, neuropathies and other congenital muscle diseases are normal or only mildly increased (0-2000 IU/L).

Cardiac Troponin I

Diseases that affect skeletal muscle can also affect cardiac muscle. Cardiomyopathy has been convincing demonstrated in muscular dystrophies, mitochondrial myopathies, and most recently, in a dog with severe necrotizing myopathy [2] and in horses with atypical myopathy.[3] Troponin is located primarily in myofibrils, and is the regulatory protein of contractile skeletal and cardiac muscle, regulating the interactions of actin and myosin in muscle contraction. Troponin is composed of 3 subunits: troponin C the calcium binding subunit, troponin I the inhibitory component, and troponin T the tropomyosin-binding subunit. Genes distinct from those encoding for skeletal muscle isoforms encode cardiac troponin-I, and this troponin is a marker of myocardial cellular injury. Thus, cardiac troponins are more specific for cardiac damage, compared with the specificity for lactate dehydrogenase, CK activity or myoglobin. Further, cardiac troponin levels are not generally affected by non-specific types of skeletal muscle damage including trauma, IM injections or strenuous exercise.[1] Following myocardial damage, troponin I increases in 3-6 hrs, peaks at 14-20 hrs and returns to normal in 5-7 days.

Lactate and Pyruvate Analysis

Lactate, the product of anaerobic glucose metabolism, accumulates when aerobic metabolism is impaired, causing a shift in the oxidized-to-reduced NAD+/NADH ratio within mitochondria. Elevated plasma lactate and/or pyruvate levels may occur in a wide range of conditions, and in humans, is an essential part of baseline evaluations for mitochondrial disease. Elevations may be physiological as with anaerobic exercise, may be a result of systemic diseases such as hypoxia, hypotension, shock sepsis or renal failure, or in metabolic diseases such as organic acidemias, urea cycle defects, or mitochondrial disorders. Defects of oxidative metabolism, including deficiency of pyruvate dehydrogenase or pyruvate carboxylase, and defects of the Kreb's cycle or the mitochondrial respiratory chain will result in increased blood lactate and pyruvate concentration. To date, only a small number of metabolic disorders associated with lactic acidosis and exercise intolerance have been described in dogs. Exercise intolerance and post-exercise collapse has been reported in Clumber and Sussex spaniels with markedly elevated resting and post-exercise plasma lactate and pyruvate concentrations.[4] A mutation in the pyruvate dehydrogenase phosphatase 1 gene has recently been demonstrated in dogs of these breeds and a genetic test is now available [5]. Despite the lack of specificity, an elevated plasma lactate or pyruvate level at rest and/or post-exercise can be an important marker of metabolic myopathies associated with mitochondrial disease or fatty acid oxidation defects, and exercise intolerance syndromes.

Unfortunately, the accurate interpretation of abnormal lactate and pyruvate concentrations is not always straightforward. Spurious plasma lactate elevations can occur as a result of physical activity before collection, particularly if the animal is excited or struggling during blood collection, and poor venipuncture technique. Special collection and handling are required, and specific tubes and solutions should be collected prior to blood collection. Blood for lactate should be collected into sodium fluoride/potassium oxalate tubes, and the plasma separated and stored at -20°C until further analyzed. Point of care testing has made lactate measurements more readily available in a clinical setting. For pyruvate analysis, blood should be mixed immediately following collection in a 1:1 ratio of either 8% perchloric or trichloroacetic acid for deproteinization. The mix should be placed on ice for 10 minutes then centrifuged, the supernatant collected, then stored frozen and analyzed within 30 days. Fasting samples should also be collected as elevated pyruvate levels can occur in the first few hours after a meal in normal individuals.

Alterations in the lactate to pyruvate ratios (L/P) help to localize the enzymatic defect in human beings with pathologic lactic acidemia.[6] However, previous studies demonstrated that direct correlations with alterations in the L/P in human beings with pathologic lactic acidemia are not possible in dogs.[7] For exercise intolerance syndromes, it is recommended that resting and post-exercise blood samples are collected. It is important that an age and size matched (and ideally breed matched) control dog is exercised at the same intensity and duration as the affected dog, and blood collected at the same time-points. In addition to collection of blood for lactate and pyruvate concentrations, post-exercise CK, electrolytes (particularly potassium) and acid-base status should also be evaluated.

Thyroid Evaluation

Skeletal muscle weakness, cramping, exercise intolerance and myalgia have been reported as potential sequelae of hypothyroidism in humans and dogs. Thyroid hormones play a key role in energy and lipid metabolism and exert most of their known biological effects via stimulation of gene expression, although the target genes regulating tissue-specific energy metabolism and metabolic efficiency are largely unknown. In a recent study, substantial myopathic changes indicative of altered muscle energy metabolism and depletion of skeletal muscle carnitine were described in an experimental canine model of hypothyroidism.[8] Such abnormalities may contribute to the nonspecific clinical signs, such as lethargy and exercise intolerance, often reported in hypothyroid dogs. At the time of initial diagnosis of hypothyroidism, up to 80% of humans report problems attributable to skeletal muscle dysfunction and nearly 40% have clinical evidence of muscle weakness.[9] Similarly, muscle stiffness and myalgia may be an early indicator of hypothyroidism in dogs without obvious classical clinical signs. Screening measurements should include total thyroxine (T4), free T4 by equilibrium dialysis or thyroid-stimulating hormone. Evaluation of thyroid function should also be performed in older cats as weakness can be a prominent and reversible clinical sign of hyperthyroidism.

2M Antibody

The most extensively studied inflammatory myopathy in dogs is masticatory muscle myositis (MMM) which selectively affects the muscles of the masticatory muscle group (including the temporalis, masseter, pterygoid, and rostral digastricus muscles which are innervated by the mandibular branch of the trigeminal nerve) and spares limb muscle. The masticatory muscle group contains a unique muscle fiber type (type 2M fibers) and protein composition compared to limb muscle that could partially explain the selective involvement of muscles in this group. Autoantibodies specifically target type 2M fibers in MMM, and these antibodies are currently used to diagnose the disease in dogs using an ELISA assay which is specific (100%) and sensitive (85-90%). In early studies it was shown that autoantibodies in MMM recognized masticatory muscle-specific myosin heavy and light chains plus an unknown protein. Such antibodies are not present in dogs with polymyositis, even though cellular infiltration and myofiber damage is clearly present in the masticatory muscles. Thus, the autoantibodies in MMM are disease specific and not just an epiphenomena secondary to myofiber damage. Most recently, the unknown protein was shown to be a novel myofibril-associated protein, which is a member of the myosin binding protein-C family (MyBP-C), specifically expressed in masticatory muscle type 2M fibers.[10]

Muscle Biopsies

Muscle biopsies should be collected early in the course of diagnostic evaluation of an animal with a neuromuscular disease, rather than waiting until extensive muscle damage, fiber loss and fibrosis has occurred and the chances for a successful treatment diminished. A delay in diagnosis and initiation of appropriate therapy may result in irreversible fibrosis and limb contractures. Muscle and nerve biopsies should be evaluated by laboratories with expertise in muscle diseases. Muscle biopsies should be collected by an open-biopsy technique and evaluated in frozen sections, using a standard panel of histochemical stains and reactions including fiber typing.[11] The usefulness of evaluating muscle and nerve in only paraffin sections is limited. Ultrastructural analysis is necessary for definition of structural abnormalities in selected congenital myopathies and is useful for peripheral nerve diseases. Results of tissue biopsies should guide further diagnostic testing focusing on laboratory tests that are relevant to the specific disease group, or identify a specific disease and therapy if available. Immunohistochemical staining can identify decreased or absent proteins in muscle and are particularly relevant to obtaining a specific diagnosis in the muscular dystrophies and other congenital myopathies. In addition to the diagnosis, results of muscle biopsies can also help determine a prognosis in many cases, as some neuromuscular diseases are treatable and others are not.

Genetic Testing

The breed of an affected animal is one of the most useful distinguishing diagnostic criteria. Several DNA based tests are currently available for many neuromuscular diseases. Many more DNA based tests will be established in the future. It is critical that the full armamentarium of diagnostic modalities available in veterinary medicine is used in the classification of unfamiliar disorders. Clinical diagnosis should be detailed as accurately and completely as possible, and all organ systems need to be evaluated. Exact diagnostic criteria must be established to decide which cases that look clinically similar are actually the same disease, even when they occur in animals of the same breed. For example, a poorly muscled and weak young Labrador retriever may be affected with CNM,[12] XLMTM,[13] dystrophin deficient muscular dystrophy[14] or even an inflammatory myopathy, and the diagnosis cannot be distinguished by clinical examination alone.

REFERENCES

[1] Shelton GD; **Vet Clin Pathol** 2010;39;278. [2] Wells RJ et al; **J Am Vet Med Assoc** 2009;234;1049. [3] Votion DM et al; **J Vet Intern Med** 2007;21;1380. [4] Abramson CJ et al; **J Small Anim Pract** 2004;45;162. [5] Cameron JM et al; **Mol Genet Metab** 2007;90;15. [6] Robinson BH; **Mol Genet Metab** 2006;89;3. [7] Matwichuk CL; **Am J Vet Res** 1999;60;88. [8] Rossmeisl JH et al; **Am J Vet Res** 2009;70;879. [9] Horak HA; **Neurol Clin** 2000;18;203. [10] Wu X; **J Immunol** 2007;179;4939. [11] Dickinson PJ et al; **Vet Clin North Am** 2002;9;63. [12] Pele M et al; **Hum Mol Genet** 2005;14;1417. [13] Beggs AH et al; **Proc Nat Acad Sci** 2010;107;14697. [14] Bergman RL et al; **J Am Anim Hosp Assoc** 2002;38;255.

DIAGNOSIS OF CANINE AND FELINE NEUROPATHIES

G. Diane Shelton, DVM, PhD, DACVIM (SAIM)
La Jolla, CA

INTRODUCTION

The frequency and variability of inherited or acquired neuropathies is underappreciated by clinicians. An accurate diagnosis is first based on pattern recognition of the disorder by gait analysis and neurological examination. Proper neuroanatomic localization is crucial to the direction of further diagnostic testing. The diagnosis and treatment of polyneuropathies can be challenging. The age of onset can be variable and clinical signs of the same disorder can begin at a few months of age or not until an older age. Onset of clinical signs may be insidious and not recognized for many years. The animal may be considered "clumsy" or have unexplained orthopedic injuries. What was thought to be an "acquired" disease may be unresponsive to treatment. Peripheral neuropathies are often painless and thus neurologic impairments may go unnoticed. Involvement of related animals may be incorrectly attributed to another cause. More often than not, the exact cause of a peripheral neuropathy remains elusive despite extensive biochemical, toxicological and electrophysiological, histochemical and histopathological studies of peripheral nerve biopsies. This is a least partially a result of the limited number of ways in which peripheral nerves are able to react to disease.

RECOGNIZE THE CLINICAL NEUROPATHIC PHENOTYPE

- Quality of paresis: Flaccid
- Gait: Short strided, may have tremors or bunny-hop
- Ataxia is not present unless a sensory neuropathy
- Tone: Decreased to absent
- Reflexes: Diminished to absent
- Muscle mass: Denervation atrophy in 7-10 days
- Pain: Usually painless, but painful neuropathies are described in humans. Perceives noxious stimulus but may not be able to withdraw
- Posture: Crouched stance, limbs more centered under the trunk to help support weight, ventral neck flexion

THE NEED FOR CONFIRMATORY DIAGNOSTICS

- Some peripheral nerve diseases are treatable, others are not
- Identification of a specific therapy depends on reaching a diagnosis
- Need for early recognition and diagnosis: extensive nerve fiber loss and fibrosis may not be treatable
- Premature institution of corticosteroid therapy without an adequate diagnostic work-up may lead to delays in or failure to establish an accurate diagnosis
- An accurate diagnosis is critical to the development of DNA based testing for genetic diseases of purebred dogs and cats
- Development of markers for early diagnosis

MINIMUM DATABASE FOR PERIPHERAL NERVE DISEASE

- CBC, biochemistry panel and urinalysis. These routine analyses may reveal easily diagnosed causes of neuropathy including hyperglycemia (diabetes mellitus) or hypoglycemia (insulinoma), hypercholesterolemia suggestive of hypothyroidism, or elevations of alkaline phosphatase with or without elevations of other liver enzymes, cholesterol and glucose suggestive of hyperadrenocorticism. The urinalysis may assist in the diagnosis of diabetes mellitus (glycosuria) and immune-based diseases such as SLE (proteinuria).
- Blood tests aimed at specific etiologies include serum fructosamine levels for diabetes mellitus, a combination of serum total T4 and free T4 (by equilibrium dialysis) and TSH for hypothyroidism, a corticotrophin stimulation test in combination with an endogenous corticotrophin assay for hyperadrenocorticism, and a serum *N. caninum* assay for suspected congenital neosporosis.
- Radiography, ultrasonography and special imaging of thorax and abdomen to investigate the possibility of neoplasia as a cause for peripheral neuropathy (paraneoplastic neuropathy).
- Electrophysiological assessment of peripheral nerves and nerve roots in combination with muscle and fascicular nerve biopsy, and lumbar CSF analysis. Despite the importance of these procedures to our understanding of the nature, severity, and distribution of the neuropathy, they may not provide the investigator with a specific etiology.

PERIPHERAL NERVE BIOPSIES

In conjunction with the neurological examination, electrophysiological testing and muscle biopsy evaluation, examination of an affected peripheral nerve should provide definition of an underlying neuropathic process. In most instances, nerve biopsy evaluation allows an accurate determination of the specific pathologic process involved in the disease such as axonal degeneration, demyelination, or invasion of inflammatory or neoplastic cells. Extent of nerve fiber loss, endoneurial fibrosis, and presence or absence of regeneration will aid in determination of long-term prognosis. Due to the limited number of ways nerves can react to injury, several disorders may have similar pathological changes and a precise diagnosis cannot be made in many instances. As molecular testing becomes more available, it is critical that pathological changes are correctly identified, particularly for genetic diseases.

Selection of Nerve for Biopsy

The nerve should be affected by the disease process as evidenced by neurologic abnormalities in areas innervated by the nerve (atrophy, hypotonia, hyporeflexia, paresis, sensory deficits) and by abnormal electrophysiological investigations. If the disease process is generalized, a nerve should be selected that is easily biopsied with low morbidity, has established normal electrophysiological and morphometric data available, and innervates a muscle that is routinely biopsied and for which normal histochemical and morphometric data are available. The common peroneal nerve, a mixed nerve containing both motor and sensory fibers, is easily identified and relatively easy to biopsy because of its flat anatomy and easily identified fascicles. (For technique of common peroneal nerve biopsy see reference 1). Other mixed nerves including the tibial nerve in the pelvic limb and the ulnar nerve in the thoracic limb may also be easily biopsied. When a predominantly sensory neuropathy is suspected, biopsy of cutaneous sensory nerves such as the caudal cutaneous antebrachial nerve in the thoracic limb and caudal cutaneous sural nerve in the pelvic limb are suggested. Intramuscular nerve branches are often found in muscle biopsy specimens and provide information about the integrity of the most distal nerve branches. Biopsy of nerve roots may be necessary when pathologic change is restricted to the most proximal portions of the peripheral nervous system. Nerve root biopsy is an invasive procedure requiring extensive surgical exposure and laminectomy.

A 1" segment of peripheral nerve should be collected by careful dissection. Handle the nerve carefully to prevent artifactual changes. Lay the nerve onto a tongue depressor (do not stretch, just maintain length) then immerse directly into fixative. For diagnostic purposes 10% neutral buffered formalin is adequate. For research purposes, fixation into 2.5% glutaraldehyde is optimal. Ship the nerve specimen to the laboratory that will be performing the testing within 48 hours for continued processing to ensure optimum fixation of myelin. A shorter length of unfixed nerve should be shipped under refrigeration and flash frozen in the event immunostaining is required.

Electron Microscopy, Teased Nerve Fibers, Morphometry and Immunohistochemistry

For most diagnostic purposes, resin embedded nerve evaluated in 1 μm sections is adequate for evaluation of general pathological changes such as axonal degeneration, regeneration, demyelination and abnormalities of supporting structures. Limited information can be obtained from paraffin embedded nerve, but it is optimal for evaluation of cellular infiltrates or neoplasia. However, individual cases or research projects require additional evaluations. Ultrastructural examinations are required for evaluation of unmyelinated fibers, confirm a single basal lamina around regenerating clusters, and for critical examination of organelles. Teased nerve fibers provide additional information in cases of demyelinating, hypomyelinating or hypermyelinating neuropathies. Quantitative assessment of myelinated fiber density, axonal size frequency distributions, and myelin thickness requires morphometric analysis. In my laboratory, immunostaining for phenotypic identification of cellular infiltrates or for missing or misplaced proteins is performed on frozen sections.

Genetic Testing and the Future

With the advent of new molecular testing methods, the genetic basis of inherited polyneuropathies in several canine breeds such as sensory polyneuropathy in Golden Retrievers[2], motor and sensory neuropathy in the Greyhound[3], Leonberger and St. Bernard dogs (early onset form)[4], and most recently in the Alaskan Malamute[5], have been identified. Chronic relapsing forms of polyneuropathy in young cats, most notably in the Bengal breed[6], have been characterized and research is in progress to evaluate possible risk alleles. Exciting new work has documented esophageal involvement in the polyneuropathy/laryngeal paralysis complex in older Labrador retrievers and other breeds[7], and genetic studies are in progress. It is clear that much work remains to be done, but the future holds great promise. The need for accurate clinical and pathological characterization of polyneuropathies of unknown cause remains.

REFERENCES

[1]Dickinson PJ et al; **Vet Clin North Am** 2002;9;63. [2]Baranowska I et al; **PLoS Genet** 2009;5;e1000499.doi:10.1371. [3]Drogmuller C et al; **PLoS One** 2010;5;e11258.doi:10.1371. [4]Ekenstedt KJ et al; In: **5th International Conference on Advances in Canine and Feline Genomics and Inherited Diseases** 2010;32. [5]Sichlau Brun CV et al; **PLoS One** 2012; in press. [6]Bensfield AC et al; **J Vet Intern Med** 2011;25;882. [7]Stanley BJ et al; **Vet Surg** 2010;39;139.

ASSESSING CLINICAL TRIAL DESIGN: CRITICAL EVALUATION OF REPORTED OUTCOMES

David M. Vail, DVM, DACVIM (Oncology)
Madison, WI

INTRODUCTION

Clinical trials in veterinary oncology have gained interest and focus over the last decade, with both veterinary and comparative oncology drug development leading the effort. There have also been a growing number of consortia and cooperative groups that have functioned successfully by uniting multiinstitutional efforts, advocating for veterinary clinical trials, and emphasizing the synergy between basic science and clinical progress. With support from clients who are motivated to seek advanced care for their pets and to enroll them in investigational trials that offer new therapies, clinical research in veterinary oncology is growing in scope and importance.

Oncology clinical trials attempt to answer questions and find better ways to prevent, diagnose, or treat cancer. Their model is different from trials involving infectious or even chronic diseases because the risks involved have greater morbidity and mortality. Traditional drug development follows a strict, step-wise paradigm that begins with a phase I dose-finding trial, followed by a phase II efficacy/activity trial, and concludes with a phase III "pivotal" trial that pits a novel agent against or with the current standard of care (Table 1). Although veterinary oncology trials sometimes combine these concepts, their individual descriptions serve as the framework for new drug development. Clinical designs, pertinent endpoints and analyses, the process for drug approval, and clinical trial ethics will be explored in the following sections.

CONSORTIA

One of the most exciting achievements in veterinary oncology over the last decade has been the development of successful and collaborative consortia groups that are purposed to perform multicenter clinical trials and prospective tumor biospecimen repository collections. Consortia infrastructures allow larger scale clinical trials and provide the voice for collective advocacy in veterinary and comparative oncology. Their success is an example of the growing importance of the study of tumor and clinical biology. Some of these efforts are profiled here. Examples include the Comparative Oncology Trials Consortium (COTC), the Canine Comparative Oncology and Genomics Consortium (CCOGC; www.ccogc.net) and Animal Clinical Investigation (ACI; www.animalci.com). COTC (https://ccrod.cancer.gov/confluence/display/CCRCOPWeb/Comparative+Oncology+Trials+Consortium) is an active network of 20 centers, centrally managed by the National Institutes of Health-National Cancer Institute's Comparative Oncology Program (NCI-COP), that functions to design and execute clinical trials in dogs with cancer to assess novel therapies. The goal of this effort is to answer biologic questions geared to inform the development path of these agents for use in human cancer patients. As of 2012, the COTC has completed or is currently conducting 18 clinical trials and has been successful in promoting the utility of comparative oncology modeling within the drug development community. The CCOGC is an informal collaboration of veterinary and medical oncologists, pathologists, surgeons, geneticists, and molecular and cellular biologists with a common interest in the comparative study of canine and human genomics and cancer. The goals of the CCOGC are to facilitate partnerships and collaborations focused on the problem of cancer in dogs. Early priorities included advocacy for the field of comparative oncology, the development of a mechanism to share reagents and resources in the community, and the development of a biospecimen repository. The repository houses tumor tissue, normal tissues, serum, plasma, peripheral blood mononuclear cell preparations, genomic DNA, RNA, and urine samples. As of 2011 the CCOGC has collected 1600 of 3000 anticipated samples. The repository will serve as a seminal resource for the research community to acquire tissues for target biology studies, as well as its mechanism for prospective collections of unique samples. Future efforts will focus on dissemination of these tissues. ACI is a privately organized and run specialty network of veterinary hospitals that designs, conducts, and reports clinical studies for the animal health industry. ACI trials emphasize oncology drug development but have recently expanded to include other medical conditions, including inflammatory and metabolic disease, cardiovascular disease, and arthritis. ACI provides multi-site, pivotal, or nonpivotal studies and commercialization support to help define effective novel veterinary therapeutics. It is the first example of a commercial clinical research organization (CRO) in veterinary medicine. Over 30 clinical trials have been conducted through this network over the last 10 years.

THE TRADITIONAL DRUG DEVELOPMENT FLOW

Traditionally, *first-in-species trials* start with a *Phase I* dose-finding trial, followed by a *Phase II* efficacy/activity trial and concluding with a *Phase III* "comparative" trial that pits the novel agent against or with the current standard of care (Table 1).

Phase I Trials (Dose-Finding)

Phase I trial design and statistical considerations have been reviewed.[1-5] The primary goal of Phase I trials is to determine the maximally tolerated dose (MTD) or for targetted therapies biological optimal dose (BOD) to be used in future Phase II studies, by evaluating safety, tolerance and dose-limiting toxicities (DLT) in treatment cohorts of increasing dose. Activity/efficacy is **NOT** a primary goal of Phase I trials. In fact, response rates in Phase I trials are seldom more than 10%. This is particularly important with respect to informed consent as even though clients are informed that they may be receiving a drug with nonexistent activity or at a suboptimal dose in early dosing groups, a full 50% of humans entering Phase I trials believe they will experience a response. Secondary goals of Phase I trials may include scheduling issues, response rate, pharmacokinetic information (ADME = absorption, distribution, metabolism and elimination), and effects on molecular targets or pathways.

Setting the Starting Dose. Generally, some preclinical data exists (in other than the target species) and that data is used to inform a starting dose for Phase I. If other species (e.g., rodent) toxicity data exists, 1/3rd the "no observable adverse event level" (NOAEL) or 1/10th the severe toxicity dose in the most sensitive species is used to start with. If normal laboratory dog (usually beagle dogs) data is available, we have found over the years that it is prudent to start at 50% the MTD in beagles, as they appear to be less sensitive to toxicity than tumor-bearing patient dogs. If the starting dose is too low the length of the trials are longer and there is poor utilization of resources and the number of patients exposed to sub-optimal doses is increased. Some patient advocate groups are allowing patients to pick a starting dose based on varying degrees of risk; that is, some patients are willing to risk more toxicity for a higher likelihood of activity.

Dose escalation strategies. As with starting dose, escalation strategies greatly affect the number of patients treated at a potential ineffective dose, the length of the trial and the risk of toxicity. The traditional method of escalation uses a 3 + 3 cohort design where dose escalations are made with 3 dogs per dose level and the MTD is set based on the number of patients experiencing a dose-limiting toxicity (DLT). A DLT is defined as ≥ grade III toxicity in any category (except hematologic) according to predefined adverse event categories such as those in the Veterinary Cooperative Oncology Group Common Terminology Criteria for Adverse Events (VCOG-CTCAE v1.0)[6]. Grade IV is the cut-off most preferred for the DLT for bone marrow suppression in human trials as these events are usually considered manageable and transient. The MTD is defined as the highest dose level in which no more than 1/6 dogs develops a DLT. Traditionally, a fixed-dose modified Fibonacci method of dose-escalation is used where the dose is escalated 100, 67, 50, 40 and then 33% of the previous dose as the cohorts increase. Similar to starting at too low of a dose, If the escalations are too conservative, more patients receive a sub-optimal dose; however, if the escalations are too rapid, more patients are at risk for significant toxicity and the accuracy of the MTD is poor. Alternative "accelerated titration" dose-escalation strategies have been suggested and in the end, it is always a trade-off of risk *versus* benefit; however, rapid accelerations is less likely to deny efficacious dosing to someone with a fatal disease.

TABLE 1. Goals of Phase I–III Clinical Trials

CHARACTERISTIC	PHASE OF CLINICAL TRIAL		
	PHASE I (DOSE FINDING)	PHASE II (ACTIVITY/EFFICACY)	PHASE III (PIVOTAL)
Primary goals	• Determine MTD • Define DLT • Characterize type and severity of adverse events	• Determine activity/efficacy in defined populations • Inform the decision to move to a Phase III trial	• Compare a new drug or combination to therapy currently regarded as standard of care
Secondary goals	• PK/PD issues • Scheduling issues • Target modulation effects • Preliminary efficacy data • Investigate surrogate biomarkers of response	• Estimate therapeutic index • Expand adverse event data • Evaluate additional dosing groups • Expand target modulation and biomarker data • Quality of life measures	• Quality of life comparisons • Comparative costs

MTD, Maximum tolerated dose; *DLT,* dose-limiting toxicity; *PK/PD,* pharmacokinetic/pharmacodynamic.

Phase II Trials (Activity/Efficacy Trials)

The primary goal of Phase II trials is to, utilizing the MTD established in phase I, identify the clinical or biologic activity in defined patient populations (e.g., tumors of a particular histology or tumors with a particular molecular target) and inform the decision to embark on a larger, pivotal Phase III trial.[3,7-9] The traditional Phase II design (phase IIA), the single-arm open label phase II trial is a non-randomized, non-blinded activity assessment of a novel drug/therapeutic modality that lacks a control group, or uses historical controls which are prone to bias (selection, population drift, and stage migration bias). Simplistically, a minimum of 9 patients of the same histology or molecular target are treated with the investigational drug in order to test the *null hypothesis* of insufficient efficacy. Assuming the likelihood of spontaneous regression is less than 5% and expecting at least a 25% response rate for the agent to be clinically useful, with a $p \leq 0.05$ (*type I [α] error*; false positive) and a power of 0.8 (*type II [β] error*; false negative), if no responses are observed after 9 cases the study ends. However, if a response is noted in 1 of the cases, then the accrual is increased to 31 to get an accurate response rate. If you expect a less sizable response rate, (e.g., 5 – 20%), then the initial accrual number must be increased. Several modifications of Phase II trial design will be discussed.

Endpoints of Activity/Efficacy. Since the primary goal of Phase II trials is assessment of activity/efficacy, the endpoints used to evaluate response are critical to the design. With traditional cytotoxic chemotherapeutics, response criteria are fairly straight forward as size and/or volume are used to assess response according to several published methodologies (e.g., VCOG-RECIST[9] or VCOG Lymphoma response criteria[10]). However, it is readily evident that such criteria may not be appropriate for the newer molecular targeting agents that are more likely to be cytostatic than cytotoxic and result in stabilization of disease rather than measurable regression. In such cases, temporal measures such as progression-free survival (PFS) or time to progression (TTP) would seem more appropriate end-points; however, these often take too long to mature for timely Phase II trials. Alternatively, an adequate compromise could be progression-free rate (PFR) at predetermined time-points. Other endpoints for targeted agents could be a validated surrogate biomarker or measure of a molecular effect such as dephosphorylation of a growth factor receptor, changes in microvascular density, or specific target modulation that is linked to clinical outcome. Secondary endpoints that can be evaluated in Phase II trials are quality of life assessments, comparative cost of therapy, days of hospitalization, toxicity, etc.

Phase III Trials (Comparative/Confirming Trials)

It has been suggested that if Phase II trials are "learning" trials, Phase III trials are "confirming" trials.[1,8,12] These larger, randomized, blinded controlled trials have the goal of comparing a new drug or combination to therapy regarded as the standard-of-care. They are often performed by large cooperative groups that ensure greater case accrual. They are not common in veterinary medicine due to their size and expense. Sample size and Power: The overriding function of clinical research is to provide a definitive answer to a clinical question. However, it is rarely that simple. It is possible that once complete, clinical trial conclusions may be incorrect based on chance or design error. Chance error results when an erroneous inference is drawn from a study sample group that is not actually representative of an entire patient population. Type I or α error (false positive) occurs if an investigator rejects the null hypothesis when it is actually true. This is also referred to as the study's level of significance. Type II or β error (false negative) occurs when one fails to reject the null hypothesis when it is actually incorrect. Type I and II errors are due to chance and cannot be avoided completely, although steps can be taken to reduce their potential impact by increasing sample size and augmenting study design or measurements. Power is the ultimate measure of a clinical trial's results and also must be prospectively controlled. Power is defined as 1-β, the probability of correctly rejecting the null in the sample if the actual effect in the population is equal to (or greater than) the effect size. Power is governed by sample size, with the goal being to enroll enough patients to accurately allow for a difference to be seen between groups. Power is irrelevant if the results are statistically significant, but if not, it is important to ensure the study had adequate numbers to detect a difference between groups. If a study to detect the difference between two cancer treatments is designed with an α of 0.05, then the principal investigator (PI) has set 5% as the maximum chance of incorrectly rejecting the null hypothesis if it is true. This is the level of doubt the PI is willing to accept when statistical tests are used to compare the two treatments. If β is set at 0.10, the PI is willing to accept a 10% chance of missing an association of a given effect if it exists. This represents a power of 0.90 (1-β), or a 90% chance of finding an association of that size or greater. α and β levels are determined prior to trial initiation, and their set points are based on the importance of avoiding either a false-positive or a false-negative result.

New Trial Designs

Several new designs will be discussed, including "accelerated titration" dose escalation strategies for phase I trials, phase 0 and phase IV designs, randomized phase II trials, 'adaptive' and 'continuous reassessment' trial designs (e.g., Bayesian methods), randomized discontinuation, phase combinations, and enrichment. In the end, it is always a tradeoff of risk versus benefit; however, rapid acceleration of trials are less likely to deny efficacious dosing to someone with a fatal disease.

TRIAL REPORTING

The transparent reporting of clinical trials has received considerable attention in the last decade. For randomized trials, the current gold standard guidelines for reporting of data is the Consolidated Standards of Reporting Trials (CONSORT; www.consort-statement.com). In these guidelines, logical and step-wise guides to the reporting of patient enrollment, patient allocation (between and within treatment groups), patient follow-up, and data analysis are clearly spealled out. Anyone devising, performing and indeed critically appraising clinical trials should be well-versed in these guidelines.

CONCLUSIONS

Clinical trials are an important research discipline to improve care and outcome for cancer patients in both human and veterinary oncology. Steps should be made to ensure study aims are achievable within a crafted study design and protocol. Rules governing design are prospective and involve questions of dose and schedule selection, toxicity, activity, and comparison to known effective therapies. Statistical expertise is also necessary to ensure appropriate clinical trial design. Regulatory oversight of veterinary oncology trials is increasing, and new approval of veterinary oncology agents will emphasize these processes over the next decade. Comparative oncology trials also are key to the inclusion of pet animals in the evaluation of novel anticancer therapeutics, imaging strategies, and medical devices. The field of veterinary oncology will continue to grow through the proper use and design of both traditional and novel clinical trials.

REFERENCES:

1. Teicher BA, et al. Anticancer drug development guide : preclinical screening, clinical trials, and approval. 2004, Totowa, NJ, Humana Press.
2 Potter DM.. *J Biopharm Stat* 2006 16:579.
3. Kummar S, et al. *Br J Clin Pharmacol* 2006, 62:15.
4. Le Tourneau C et al. *J Natl Cancer Inst* 2009, 101:708.
5. Bria E, et al. *J Exp Clin Cancer Res* 2009, 28:66.
6. Vail et al. Vet Comp Onc e-Pub ahead of print: 20JUL, 2011. DOI: 10.1111/j.1476-5829.2011.00283.x
7. Gray R, et al. *Clin Cancer Res* 2006, 12:1966.
8. Lee JJ, et al. *J Clin Oncol* 2005, 23:4450.
9. Brown SR, et al. *Br J Cancer* 2011, 105:194.
10. Bernard s, et al. *Vet Comp Oncology* 2013, in press.
11. Vail et al. *Vet Comp Oncology* 2010 8:28.
12. Hulley SB, et al. Designing clinical research. 2007, Lippincott Williams & Wilkins, Philadelphia.

ANTI-TUMOR ACTIVITY OF IRREVERSIBLE SELECTIVE INHIBITORS OF NUCLEAR EXPORT (SINE) IN DOGS WITH CANCER. S Shacham[1], S Barnard[3], W Kisseberth[3], M Breit [2], M Bear[3], D Ito[4], K Jensen[4], A Borgatti[4], M Henson, H Wilson[5], D McCauley[1], JF Modiano[4], M Kauffman[1], C London[2,3]. [1]Karyopharm Therapeutics, Inc, Natick, MA, [2]Department of Veterinary Biosciences, [3]Department of Veterinary Clinical Sciences, The Ohio State University, Columbus, OH, [4]Department of Veterinary Clinical Sciences and Masonic Cancer Center, University of Minnesota, MN, [5]Department of Veterinary Medicine, Texas A&M University, TX.

KPT-SINE are Selective Inhibitors of Nuclear Export that irreversibly block the activity of CRM1/XPO1, the major nuclear export protein in cells, forcing nuclear retention of key tumor suppressor and growth regulatory proteins ultimately resulting in tumor cell death. KPT-SINE: KPT-185, KPT-276 and KPT-335 demonstrate potent cytotoxic activity against human and canine tumor cells *in vitro* at nanomolar concentrations (IC50 < 500 nM) and in multiple mouse xenograft models with minimal effects on normal cells in vitro (IC50 > 5 μM). Based on these data, a Phase I study was performed to evaluate the clinical toxicities, establish the maximum tolerated dose/dosing regimen, and provide a preliminary assessment of biologic activity of KPT-276 and KPT-335 in dogs with cancer. Toxicity and oral bioavailability issues resulted in the transition exclusively to KPT-335 for the maximum tolerated dose (MTD) study. Biologic activity of KPT-335 was observed in dogs with lymphoma, including both partial responses to therapy and stable disease lasting from 4-23 months duration were observed at doses ranging from 1-1.75 mg/kg. The MTD was established as 1.75 mg/kg, with clinical toxicities consisting of primarily anorexia, vomiting and diarrhea; hepatotoxicity was noted at 2 mg/kg. A subsequent expansion study established 1.5 mg/kg given on a Monday, Wednesday, Friday basis as the regimen to be utilized in a subsequent ongoing registrational study for MUMS approval for the treatment of lymphoma in dogs. In summary, the novel oral SINE KPT-335 exhibits single agent biologic activity in dogs and is well-tolerated at biologically active doses of drug.

EVALUATION OF THE NOVEL HSP90 INHIBITOR STA-1474 IN DOGS WITH CANCER. C London[1], M Bear[1], J McCleese[1], K Selting[2], S Axiac[2], K Foley[3], R Paalangara[3], T Inoue[3], W Ying[3], J Barsoum[3], D Proia[3]. [1]Department of Veterinary Biosciences, The Ohio State University, Columbus, OH, [2]Department of Clinical Sciences, University of Missouri, Columbia, MO, [3]Synta Pharmaceuticals Corp., Lexington, KY.

STA-1474 is a highly soluble prodrug of ganetespib (formerly STA-9090), a novel resorcinol-containing small molecule inhibitor of HSP90. We demonstrated that STA-1474 kills canine tumor cell lines *in vitro* at nanomolar concentrations and inhibits the growth of murine tumor xenografts. A subsequent Phase I clinical trial was performed to investigate the safety and efficacy of STA-1474 in dogs with cancer as a prelude to future human clinical work. The maximum tolerated dose of drug was 9.5 mg/kg given over 8 hours, and biologic activity was observed in dogs with mast cell tumors, osteosarcoma, melanoma and thyroid carcinoma (25% PR [6/25], 36% CB [9/25]). Toxicities were primarily gastrointestinal in nature (anorexia, vomiting, diarrhea) and were manageable with concomitant medications. Based on additional data generated in mouse models, a 2-day in a row dosing regimen was established to circumvent upregulation of HSP90 following treatment and a subsequent Phase 2 was performed in dogs with mast cell tumors. STA-1474 (6 mg/kg/week) was given under one of four different dosing schedules: once weekly with a 1 hour infusion or 8 hour infusion, or twice weekly with a 1 hour infusion given 24 hour or 72 hour between doses. STA-1474 given on two consecutive days yielded the greatest objective responses (50%) and biologic activity of 100%. Biomarker changes from pre- and two post-treatment biopsies are currently being evaluated by reverse phase protein array. These data supported the evaluation of a 2-day in a row dosing regimen of ganetespib in people.

GLUCOSE, LACTATE AND THE METASTATIC PHENOTYPE

Douglas H. Thamm, VMD, DACVIM (Oncology)
Fort Collins, CO

THE WARBURG EFFECT

From the work of Warburg and the Coris in the early 20th century,[1-4] it has been known that most tumor cells exhibit increased glucose utilization and lactate production, implying enhanced utilization of glycolytic pathways over more bioenergetically efficient pathways such as oxidative phosphorylation (OXPHOS). While tumor hypoxia may explain decreased OXPHOS *in vivo*, normoxic tumor cells *in vitro* exhibit this same phenotype (**aerobic glycolysis**).

Despite the high glycolytic phenotype and dramatically increased glucose uptake of most tumors, respiration rate is changed little from surrounding normal tissue.[5] This disparity is best explained by the fact that rapidly growing tumors have an increased requirement for biosynthetic precursors necessary for fatty acid and nucleotide synthesis, which is best met via augmentation of glycolytic pathways.[6-8] However, a recent flux balance model of ATP production suggests that, under conditions of high glucose uptake, glycolysis may actually be a more bioenergetically efficient mode of ATP generation than OXPHOS.[9]

GLYCOLYTIC INHIBITORS AND CANCER THERAPY

The selective *in vivo* uptake of glucose by tumors has been observed since the original experiments of Warburg and the Coris,[3,4] and forms the basis for the diagnostic utility of PET-CT with ^{18}fluorodeoxyglucose, a universal cancer imaging modality with >90% sensitivity and specificity for detecting carcinoma metastasis.[10] Multiple pharmacologic inhibitors of glycolysis are in development. 2-deoxyglucose (2-DG), 3-bromopyruvate, fluoro-2-deoxyglucose, pyruvate kinase M2 inhibitors and lonidamine have all been evaluated *in vitro* and to some degree in preclinical models.[11-18]

2-Deoxy-D-Glucose: 2-DG is a competitive inhibitor of glucose transport. It is phosphorylated by hexokinase to form 2-DG-6-phosphate, which cannot be metabolized through the glycolytic or pentose-phosphate pathways. Some of 2-DG's effects may be unrelated to hexokinase inhibition, including induction of the unfolded protein response (UFPR) and autophagy, and inhibition of N-linked glycosylation. While *in vitro* growth inhibition and apoptosis have been observed in a variety of tumor types,[19] and these are accentuated under hypoxia,[17] variable single-agent efficacy has been observed *in vivo*. Combining 2-DG and other glycolysis inhibitors with chemotherapy and radiation therapy holds promise, as glycolysis inhibition may interfere with energy-dependent DNA repair pathways.[20] Interestingly, *tumor selectivity* and even protection of normal cells from radiation has been observed.[21] In contrast, both impairment in T lymphocyte function and, with chronic high-dose administration, cardiomyocyte vacuolation have been observed in 2-DG treated rodents.[22,23]

2-DG was initially studied as a cancer treatment in the 1950's, in humans with metastatic carcinomas and leukemias treated with high-dose intravenous boluses. Although reduction in leukemic cell burden was reported, diaphoresis and CNS disturbances discouraged clinical development.[24] Recently, early-phase studies have evaluated oral 2-DG, both in combination with radiation therapy in patients with glioma, and as a single agent in patients with prostatic carcinoma.[25-27] Asymptomatic QT prolongation was dose-limiting at a dose of 60 mg/kg daily in prostate cancer, but doses of 250 mg/kg/day were administered without dose-limiting toxicity to glioma patients. 2-DG exhibited linear pharmacokinetics with C_{max} values of 80-800 ug/mL, and a half-life of approximately 90 minutes. Additional studies of 2-DG in combination with chemotherapy or radiation therapy are in progress.

BIOENERGETICS AND TUMOR PROGRESSION – CLINICAL DATA

Recently, we undertook a re-analysis of the classic Ramaswamy dataset of primary and metastatic tumors,[28] in which we showed that metastatic tumors have a further increased dependency on glycolytic pathways and concomitantly down-regulate pathways involved in OXPHOS, when compared with primary tumors (**Fig. 1**).[29] This striking observation was confirmed using several other independent datasets.[29,30] These observations go beyond the original tenets of Warburg, suggesting that *further augmentation of this glycolytic preference may be a hallmark of metastasis*.

Additional data supporting these findings come from several investigators evaluating different measures of the "bioenergetic signature". For example, Cuezva *et al* have identified **β-F1-ATPase** as a key regulator of OXPHOS and mitochondrial biogenesis, and the ratio of β-F1-ATPase to **GAPDH** (as a measure of glycolytic capacity) as a prognostic factor in a number of human carcinomas.[31,32] Similarly, Coy *et al* have demonstrated tumor-restricted expression of a novel protein transketolase isozyme (TKTL1), which enhances glycolysis.[33] Upregulation of TKTL1 has been associated with poor outcome and/or enhanced invasiveness or metastatic progression in multiple human carcinomas.[34] Finally, recent studies have demonstrated that enhanced glycolysis is associated with lung metastasis from hepatocellular carcinoma in mice, and that lactate production correlates with *in vivo* metastatic capacity in human breast cancer cells.[35,36]

BIOENERGETICS AND TUMOR PROGRESSION – MECHANISTIC DATA

Whether the enhanced glycolytic phenotype is a cause or an effect of metastatic behavior remains unclear. A variety of classical oncogenic changes, including Myc, Hif-1 and Akt activation, as well as p53 inactivation, can alter glucose utilization in cells. Furthermore, increased lactate and decreased mitochondrial transcripts are observed in the stepwise multistage fibroblast carcinogenesis model described by Weinberg *et al*,[37] as cells progress to a fully transformed and malignant phenotype.[38] However, accumulating evidence suggests that *modulation of the bioenergetic state can have direct effects on tumor cell behavior*. Perhaps the most compelling clinical data demonstrating this come from studies of cancer patients receiving the antidiabetic drug metformin. Multiple studies have demonstrated reduced cancer risk in patients on metformin therapy,[39] as well as reduction in the rates of tumor progression and increased progression-free survival for some patients with established cancer.[40, 41] Metformin is a modulator of AMP kinase activity, which has effects on Akt signaling and glucose utilization.

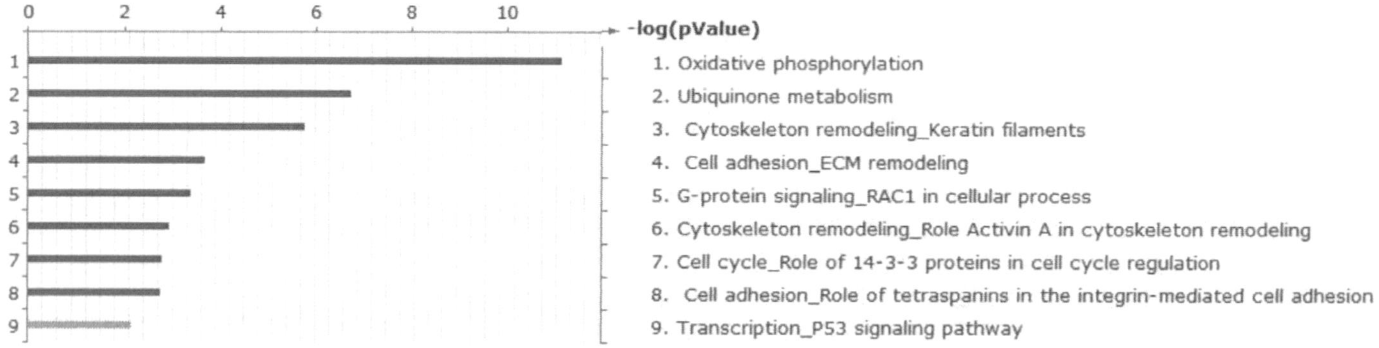

Fig. 1: *Re-evaluation of the Ramaswamy dataset, using pathway based analysis.* OXPHOS pathways are significantly down-regulated in metastatic, vs. primary tumors.

Lactate production by glycolytic tumor cells may have pro-survival and pro-metastatic effects in the tumor microenvironment. These include conditioning of the microenvironment to support tumor-associated stroma,[42] suppression of immune effector function,[43] and enhancement of tumor cell invasion, through facilitation of extracellular matrix degradation.[44] Defects in OXPHOS may also confer resistance to apoptosis through suppressed activation of the pro-apoptotic proteins Bax and Bak.[45]

Avadhani *et al.* have shown that mitochondrial depletion in A549 cells, increasing dependency on glycolysis, increases *in vitro* invasiveness and apoptosis resistance.[46, 47] Furthermore, increased expression of the cysteine protease *cathepsin L* appears partly responsible for this phenotype.[46-48] These characteristics can be reversed by allowing mitochondrial recovery.[46, 49]

We hypothesized that pharmacologic agents impairing glycolysis may force aggressive cancer cells away from the glycolytic phenotype, shifting them back toward OXPHOS and reducing metastatic potential. We demonstrated a *reversal of the invasive phenotype of aggressive and highly metastatic osteosarcoma (OS) cells upon incubation with low concentrations of 2-DG* (**Figs. 2A and B**).[50] *This was associated with a reduction of cathepsin L expression*, in keeping with the findings described above (**Fig. 2C**). A similar inhibition of migration and invasion was observed in canine, human and murine breast cancer and OS cells.[50]

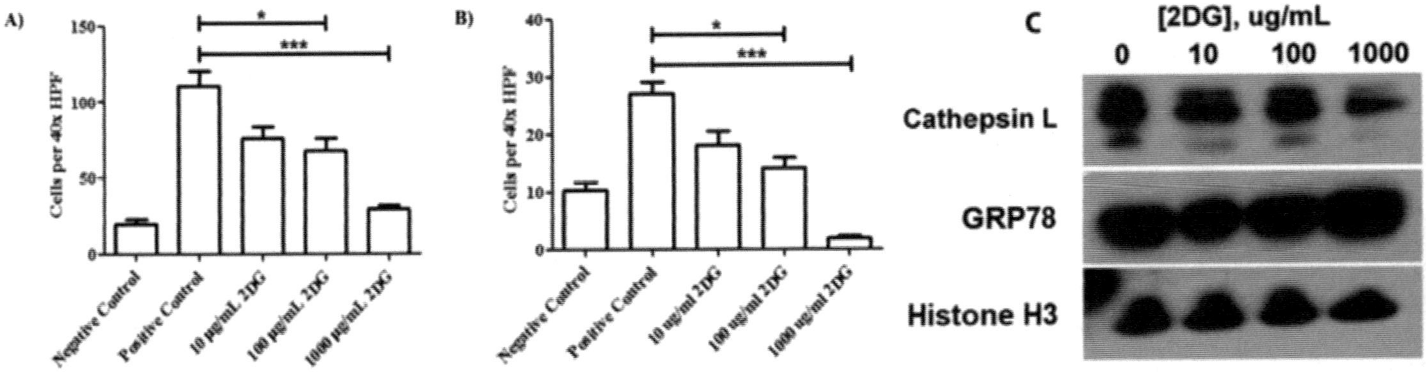

Fig. 2. *2-DG inhibits OS migration and invasion.* DLM8-luc-M1 cells were plated in migration or invasion (Matrigel) Boyden chambers for 18 hours; 2-DG was placed in the upper well. There was a significant ($p<0.05$) inhibition of migration (**A**) and invasion (**B**) in the presence of increasing 2DG concentrations. No effect on proliferation was observed. Similar results were achieved using canine and human osteosarcoma and breast cancer cells. **C.** *2-DG reduces cathepsin L expression.* DLM8 lysates were subjected to western analysis after a 24-hour incubation in 2-DG. Dose-dependent inhibition of cathepsin L expression was observed.

Although acquired mutations in mitochondrial genes can impair OXPHOS,[51, 52] accumulating evidence points to **dynamic regulation** of glucose utilization pathways. Transcription of key metabolic enzymes, including hexokinase 2, β-F1-ATPase and fructose bisphosphatase 1, can be modulated through altered promoter methylation.[53, 54] Additionally, we and others have demonstrated *reactivation of gene expression pathways related to OXPHOS following treatment of tumor cells with the HDAC inhibitor valproic acid (VPA)*.[55, 56]

These findings underscore the plasticity of bioenergetic pathways in cancer cells, and suggest that pharmacologic manipulation of these pathways could result in decreased tumor cell aggressiveness. Furthermore, therapeutic targeting of glucose metabolic processes may be a broadly applicable antimetastatic strategy since these pathways are a common theme of tumor cells, and not necessarily tumor type specific.[29, 57]

2-DG and the Metastatic Phenotype:

Indirect evidence suggests an ability to reverse the metastatic phenotype with 2-DG. Treatment of cells with 2-DG can reverse the aggressive bioenergetic phenotype associated with β-F1-ATPase depletion described by Cuezva *et al,* resulting in increased oxygen utilization and diminished lactate production.[49] In the stepwise carcinogenesis model of Weinberg,[37] fully transformed cells were the most sensitive to the antiproliferative effects of 2-DG and responded to treatment with up-regulation of transcription of key OXPHOS-associated genes.[38] The concentrations of 2-DG utilized in the experiments outlined above were very high and likely unachievable *in vivo*.

We evaluated the effect of 2-DG on the development of spontaneous metastasis

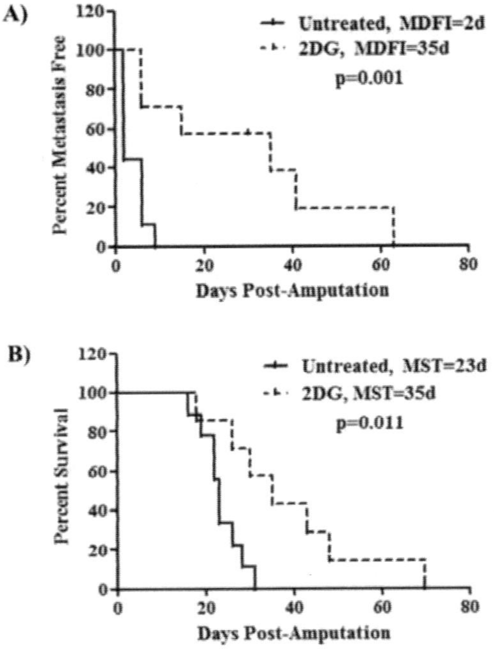

Fig. 3: *OS metastases are inhibited by 2-DG.* Mice (n=8 per group) were challenged orthotopically with DLM8-luc-M1 murine OS cells followed by amputation. 24 hours post-amputation, treatment with 500 mg/kg 2-DG IP thrice weekly for one month was initiated. Mice were imaged twice weekly for the development of luciferase positive metastases and euthanized upon showing signs of distress. **2-DG led to a significant delay in metastasis and prolongation in overall survival.**

in a luciferase-engineered, syngeneic, orthotopic, spontaneously metastatic OS model we developed.[58] *Postoperative 2-DG treatment significantly delayed metastasis and prolonged overall survival in this model* (**Fig. 3**).[50] Since cell luminescence is an ATP-dependent process, we confirmed that our observations were not an artifact of diminished luminosity of 2-DG treated cells, either *in vitro* or *in vivo* (not shown).[50]

Preliminary results of Nextgen sequencing of 2-DG treated DLM8 OS cells suggests modulation of pathways related to glucose metabolism, actin/tubulin reorganization and protein translation as possible mediators of 2-DG's antimetastatic effects.

In conclusion, the dynamic regulation of glucose utilization pathways by tumor cells appears to serve a number of purposes, including provision of building blocks for cell division, immune avoidance and regulation of migration and invasion. Further study of ways to modulate tumor metabolism holds significant promise for future cancer therapy.

REFERENCES

1. Warburg O. **Science** 1956;123:309. 2. Warburg O. **Science** 1956;124:269. 3. Warburg O, et al. **J Gen Physiol.** 1927;8:519. 4. Cori CF and Cori GT. **J Biol Chem** 1925;65:397. 5. Koppenol WH and Bounds PL. **Science** 2009;E-letter. 6. Mathupala SP, et al. **Semin Cancer Biol** 2009;19:17. 7. Vander Heiden MG, et al. **Science**. 2009;324:1029. 8. Gatenby RA and Gillies RJ. **Nat Rev Cancer.** 2004;4:891. 9. Vazquez A, et al. **BMC Syst Biol.** 2010;4:58. 10. Mankoff DA, et al. **Clin Cancer Res.** 2007;13:3460. 11. Fanciulli M, et al. **Clin Cancer Res.** 2000;6:1590. 12. Fanciulli M, et al. **Oncol Res.** 1996;8:111. 13. Floridi A, et al. **Biochem Pharmacol.** 1998;56:841. 14. Ganapathy-Kanniappan S, et al. **Curr Pharm Biotechnol.** 2010;11:510-7. 15. Kurtoglu M, et al. **Antioxid Redox Signal.** 2007;9:1383-90. 16. Cao X, et al. **Cancer Chemother Pharmacol.** 2008;62:985-94. 17. Lampidis TJ, et al. **Cancer Chemother Pharmacol.** 2006;58:725-34. 18. Spoden GA, et al. **Int J Cancer.** 2008;123:312-21. 19. Wangpaichitr M, et al. **Mol Cancer Ther.** 2008;7:1506-13. 20. Maschek G, et al. **Cancer Res.** 2004;64:31-4. 21. Prasanna VK, et al. **J Cancer Res Ther.** 2009;5 Suppl 1:S44-7. 22. Cham CM, et al. **Eur J Immunol.** 2008;38:2438-50. 23. Minor RK, et al. **Toxicol Appl Pharmacol.** 2010;243:332-9. 24. Landau BR, et al. **J Natl Cancer Inst.** 1958;21:485-94. 25. Mohanti BK, et al. **Int J Radiat Oncol Biol Phys.** 1996;35:103-11. 26. Singh D, et al. **Strahlenther Onkol.** 2005;181:507-14. 27. Stein M, et al. **Prostate**. 2010;70:1388-94. 28. Ramaswamy S, et al. **Nat Genet.** 2003;33:49-54. 29. Ptitsyn AA, et al. **BMC Bioinformatics.** 2008;9 Suppl 9:S8. 30. Ptitsyn A. **BMC Bioinformatics.** 2009;10 Suppl 11:S6. 31. Ortega AD, et al. **Carcinogenesis**. 2008;29:2053-61. 32. Lin PC, et al. **Int J Colorectal Dis.** 2008;23:1223-32. 33. Coy JF, et al. **Clin Lab.** 2005;51:257-73. 34. Zhang S, et al. **Cancer Biol Ther.** 2008;7:517-22. 35. Robey IF, et al. **Neoplasia**. 2005;7:324-30. 36. Li ZF, et al. **Rapid Commun Mass Spectrom.** 2010;24:2765-75. 37. Hahn WC, et al. **Nature**. 1999;400:464-8. 38. Ramanathan A, et al. **Proc Natl Acad Sci U S A.** 2005;102:5992-7. 39. Noto H, et al. **PloS ONE.** 2012;7:e33411. 40. Hitron A, et al. **Cancer Epidemiol.** 2012;36:e243-50. 41. Romero IL, et al. **Obstet Gynecol.** 2012;119:61-7. 42. Koukourakis MI, et al. **Cancer Res.** 2006;66:632-7. 43. Fischer K, et al. **Blood**. 2007;109:3812-9. 44. Gatenby RA and Gawlinski ET. **Cancer Res.** 1996;56:5745-53. 45. Tomiyama A, et al. **J Natl Cancer Inst.** 2006;98:1462-73. 46. Amuthan G, et al. **Oncogene**. 2002;21:7839-49. 47. Amuthan G, et al. **EMBO J.** 2001;20:1910-20. 48. Guha M, et al. **J Biol Chem.** 2007;282:14536-46. 49. Sanchez-Arago M, et al. **Carcinogenesis**. 2010;31:567-76. 50. Sottnik JL, et al. **Clin Exp Metastasis.** 2011;28:865. 51. Brandon M, et al. **Oncogene**. 2006;25:4647-62. 52. Zhou S, et al. **Proc Natl Acad Sci U S A.** 2007;104:7540-5. 53. Li RJ, et al. **Ann Oncol.** 2010;21:1506-14. 54. Liu X, et al. **Oncogene**. 2010;29:442-50. 55. Duenas-Gonzalez A, et al. **Cancer Treat Rev.** 2008;34:206-22. 56. Wittenburg LA, et al. J Cell Biochem. 2010;113:773. 57. Acebo P, et al. **Transl Oncol.** 2009;2:138. 58. Sottnik JL, et al. **Clin Exp Metastasis.** 2010;27:151-60.

TARGETING CANCER ACID ENVIRONMENT: A NOVEL ANTICANCER STRATEGY

Enrico P Spugnini, DVM, PhD, DACVIM (Oncology), DECVIM-CA (Oncology)
Rome, Italy

INTRODUCTION

Tumor cells rely on H^+ exchangers to relieve themselves from the dangerous protons byproduct of cancer metabolism that could trigger a cascade of lytic enzymes that ultimately would lead to self-digestion. Among these mechanisms, one of the most investigated are the vacuolar H^+-ATPases (V-ATPases).[1-4] V-ATPases are ATP dependant H^+ transporters that utilize the energy freed by the hydrolysis of ATP with the active transport of protons from the cytoplasm to the lumen if intracellular compartments or, if located within the cytoplasmic membrane, the extracellular compartment. Structurally speaking, the V-ATPases are composed of a peripheral domain (V_1) that carries out ATP hydrolysis and an integral domain (V_0) responsible for exchanging protons.[5] The regulatory mechanisms of V-ATPases are neutralized in malignant tumor cells that overexpress lysosomal proteins on the cell surface.[6,7] The acidic tumor environment is a consequence of anaerobic glucose metabolism with secondary production of lactates byproducts through the upregulation of hypoxia-inducible factor 1α or can be due to inadequate tumor perfusion, tumor hypoxia or enhanced transmembrane pH regulation.[8,9] These pumps, coupled with other ion exchangers, play a crucial role in the establishment and maintenance of malignant tumor environment and promote the selection of more aggressive cell phenotypes able to survive in this highly selective ambient.

ROLE OF V-ATPASES IN TUMOR SPREAD

V-ATPases play a critical role in the maintenance of an appropriate relatively neutral intracellular pH, an acidic luminal pH, and an acidic extracellular pH by actively pumping protons either through ion exchange mechanisms or by segregating H^+ within cytoplasmic organelles that are subsequently expelled.[10] It is hypotesized that the low extracellular pH of tumors might trigger proteases, leading to the dissolution of extracellular matrix. This phenomenon, as is well known, significantly contributes to tumor invasion and dissemination.[11-12] In fact, tumor invasion depends on tumor acidifying ability that ultimately leads to secretion and activation of several classes of proteases. [13-14] The most commonly triggered enzymes are MMP-2, MMP-9, cathepsin B, and cathepsin L. Low pH can also up-regulate or activate proangiogenic factors such as VEGF-A and IL-8.[15-16]

ROLE OF V-ATPASES IN CHEMORESISTANCE

The acidic microenvironment caused by changes in the pH gradient between the intracellular and the extracellular compartments as well as the pH gradient between the cytoplasm and the intracellular organelles can be significantly involved in the mechanism of drug resistance.[17,18] There are several mechanisms involved in this phenomen, including decreased uptake or neutralization of weakly basic drugs by the acidic tumor microenvironment or the sequestration of chemotherapy drugs within lysosomal vesicles.[18,19] An accelerated turnover of acidic vesicles may represent an additional tumor strategy of drug resistance based on counteracting current transportation. Several investigators developed new approaches to better characterize tumor pH in animal models mostly through imaging systems in order to identify novel targets.

RESULTS IN PRECLINICAL STUDIES

Several studies in murine models have shown that the neutralization of the H^+ transporters could result in increased tumor control and extended survival. A recent study showed that blocking V-ATPases through RNA interference led to prevention of cancer metastases in a murine model. As a result, new approaches have been developed to modulate drug efficacy within the low pH tumor milieu including the use of RNA interference, bicarbonates or the induction of metabolic alkalosis.[20,21] Finally, two recently published articles describe the chemosensitizing action of proton pump inhibitors in a murine model of orthotopic cutaneous melanoma, a well known chemo-refractory neoplasm, opening a novel field of investigation.[22,23] Meanwhile, Acridine Orange, was used in mice carrying osteosarcoma as a diagnostic and therapeutic drug coupled with phototherapy. It has various unique biological activities, and has been shown to be a useful fluorescent dye specific for DNA and RNA, a pH indicator, photosensitizer, antitumor and antimalarial drug, and a detector of bacteria and parasites, apoptosis and sperm mobility. This molecule has shown the tendency to concentrate in an acid environment such as tumors and to be more slowly metabolized by tumors than other tissues.[24] Finally, electroporation could be as well a technique to modulate tumor acid environment, thus improving drug uptake and efficacy.[25]

CURRENT CLINICAL APPLICATIONS IN HUMANS AND PETS

DIAGNOSIS/PROGNOSIS: many studies pointed out that there exists a very interesting association between oral squamous cell carcinoma in human patients and overexpression of *ATP6V1C1* 54, which seems to be the main gene involved in the regulation of V-ATPase enzymes and the acidity of solid oral tumors. [26]Moreover, in brush cytology samples it has been shown that *ATP6V1C1* levels were significantly higher in patients with OSCC than in healthy controls, with expression increasing with higher tumor stage. ROC analysis showed that the measurement of *ATP6V1C1*expression levels is a highly sensitive and specific diagnostic method. [27]

THERAPY: Surgeons are exploiting the property of Acridine Orange to be concentrated within tumors to show and remove residual disease during marginal surgeries for the therapy of sarcomas.[28,29] Several phase II trials are ongoing in humans with advanced oral carcinoma, lymphoma and melanoma that combine proton pump inhibitors (PPIs) with conventional chemotherapy. The preliminary results appear very promising, especially for chemoresistant subtypes. A phase II trial has been recently conducted to identify the best drug(s), their dose and schedule in dogs and cats with advanced tumors showing a high percentage of chemoresistance reversal.[30] Currently studies are being conducted in pets to develop novel protocols with inhibitors of acidity and chemotherapy.

FUTURE DIRECTIONS

Novel vacuolar ATP-ase inhibitors are being made and evaluated in cancer patients, in combination with different chemotherapy agents in order to maximize tumor control. Finally, besides being a potential target of anticancer drugs, it is conceivable that V-ATPases might become a predictive factor of tumor behavior and final outcome through the immunohistochemical evaluation of their expression and cellular distribution in tumor biopsies.

REFERENCES

[1]Spugnini et al. **J Exp Clin Cancer Res** 2010; 29: 44. [2]Forgac. **Nat Rev Mol Cell Biol** 2007 ; 8: 917. [3]Cipriano et al. **Biochim Biophys Acta** 2008; 1777: 599. [4]Jefferies et al. **Arch Biochem Biophys** 2008; 476: 33. [5]Arai et al. **J Biol Chem** 1988; 263: 8796. [6]Saitoh et al. **J Biol Chem** 1992; 267:5700. [7]Glunde et al. **Neoplasia** 2003; 5: 533. [8]Gatenby et al. **Nat Rev Cancer** 2004; 4: 891. [9]Fais et al. **Cancer Res** 2007; 67: 10627. [10]Nishi et al. **Nat Rev Mol Cell Biol** 2002; 3: 94. [11]Martinez-Zaguilan et al. **Am J Physiol** 1993; 265: 1015. [12]Martínez-Zaguilán et al. **Clin Exp Metastasis** 1996; 14: 176. [13]Razaq et al. **Eur Spine J** 2003 ; 12: 341. [14]Webb et al. **Novartis Found Symp** 2001; 240:169. [15]Koukourakis et al. **Br J Cancer** 2003; 89: 877. [16]Rofstad et al. **Cancer Res** 2006; 66: 6699. [17]Mahoney et al. **Biochem Pharmacol** 2003; 66: 1207. [18]Simon et al. **Proc Nat Acad Sci USA** 1993; 91: 1128. [19]Raghunand et al. **Biochem Pharmacol** 1999; 57: 1047. [20]Raghunand et al. **Biochem Pharmacol** 1999; 57: 309. [21]Raghunand N. **Methods Mol Med** 2006; 124: 347. [22]Luciani et al. **J Natl Cancer Inst** 2004; 96:1702. [23]De Milito et al. **Int J Cancer** 2010; 127: 207.[24] Kusuzaki et al. **In Vivo** 2007; 21: 205. [25] Spugnini et al. **Microsc Res Tech** 2007; 70: 1041. [26] Pérez-Sayáns et al. **Cancer Letters** 2012; 295:135. [27] Pérez-Sayáns et al. Cancer Biology & Therapy 2010,9:12, 1-8. [28]Kusuzaki et al. **Curr Pharm Des** 2012;18:1414. [29]Matsubara et al. **Clin Orthop Relat Res** Epub. [30] Spugnini et al. **J Trasl Med** 2011; 9:221.

ONCOLYTIC VIRUS THERAPY IN MODELS OF FELINE AND CANINE CANCERS

Amy L. MacNeill, DVM, PhD, DACVP (Clinical)
Urbana, IL

INTRODUCTION

The idea of using pathogens to eliminate neoplastic cells was introduced in the early 20th century with reports of remission in women with cervical carcinoma after receiving post-exposure rabies vaccinations (Dock, 1904). Since then, numerous scientific advances have been made in cancer biology, immunology, and pathogenesis of infectious diseases. In 2005, China approved a conditionally-replicative adenovirus (CRAd) for the treatment of human head and neck squamous cell carcinoma (Yu & Fang, 2007). This oncolytic viral therapy may be beneficial, but its efficacy likely can be improved. Several clinical cancer trials are ongoing to evaluate the effectiveness of genetically altered viruses in humans. Many of these oncolytic viral therapies were successful at clearing xenografts in murine tumor models, but failed to significantly extend remission in humans. To improve the efficacy of new cancer therapeutics, it is imperative that scientists begin to utilize cancer models which recapitulate the tumor microenvironment and the immunotolerance known to occur during oncogenesis. Dogs and cats with spontaneously arising cancers are an underutilized population of patients that are excellent animal models of human cancers and simultaneously could benefit from adjunctive treatment with an appropriate oncolytic virus.

Attributes of viral cancer therapy

An ideal oncolytic virus destroys tumor cells throughout the body and induces antitumor immunity without damaging healthy cells. Safety is a key concern with oncolytic viruses, particularly when used in immunosuppressed patients, so virotherapeutics should be derived from wildtype viruses that do not cause severe disease. It also is imperative that recombination of the genetically modified oncolytic virus with wildtype virus does not form a new pathogen. Preferably, viruses used as oncolytic therapeutics also lack the ability to integrate into the patients' DNA. These considerations make genetic stability an important attribute of an oncolytic virus.

Many oncolytic viruses replicate in and directly lyse infected cells. Therefore, tumor specificity enhances the safety of an oncolytic virus. Interestingly, some of the common defects found in neoplasms allow viruses to productively and selectively infect tumor cells. Several tumor cells have mutations in key components of intracellular signaling pathways which block apoptosis and/or promote proliferation making them susceptible to viral infection. Although some viruses are inherently tumor-selective, others are genetically altered to improve tumor specificity. In CRAds, the viral E1A and/or E1B gene is deleted so that the viruses can only replicate in cells with defective retinoblastoma or p53 pathways (Barker & Berk, 1987). Additionally, several oncolytic viruses encode tumor-specific promoter elements that only permit replication of the virus in tumor cells.

Oncolytic virotherapy has the potential benefit of altering the tumor microenvironment enough to break existing tumor immunotolerance. Failure of the immune system to recognize tumor cells may be due a paucity of stimulated immune cells infiltrating the tumor and/or masking of tumor antigens. Aberrant cytokine patterns in the tumor microenvironment may severely limit an antitumor immune response (Tsung et al., 1997; Zou, 2005). Additionally, regulatory T cells and myeloid-derived suppressor cells may be present in increased numbers in cancer patients. Infected tumor cells express viral antigens that trigger a fully-functional cellular immune response and alter cytokine production to promote infiltration of immune cells. The cellular immune response lyses infected tumor cells and presents viral antigens to additional immune cells. Theoretically, tumor cell lysis will also expose hidden tumor antigens that the immune system can respond to. This would lead to clearance of the virus by the immune system and concurrent amplification of an antitumor immune response. Indeed, oncolytic viruses that eliminate xenografts in murine cancer models have been shown to induce an antitumor immune memory that prevents re-engraftment of the tumor cells (Kirn et al., 2007).

One of the most powerful attributes of oncolytic viral therapy is the ability to design viruses that express exogenous proteins. Viral expression of tumor-associated antigens, cytokines, or chemokines could promote antitumor immune responses that eliminate tumor cells (Thorne et al., 2005). Intratumoral injection of an oncolytic poxvirus expressing granulocyte-monocyte-colony stimulating factor (GM-CSF) caused regression of the tumor and metastatic lesions. In this study, virus was not recovered from the metastatic lesions but T cell infiltration was documented (Mastrangelo et al., 1999). Furthermore, tumor angiogenesis can be limited and tumor growth suppressed by recombinant viruses expressing vascular endothelial growth factor (VEGF) inhibitors (Kottke et al., 2010). Additionally, genetically altered oncolytic viruses can be used to target chemotherapeutics to the tumor and minimize systemic effects of the drugs.

Oncolytic virus treatment of feline and canine cancers

Of the many viruses being tested as therapies for human cancers, only a few have been evaluated in cat and dog cancers. CRAds are the most common virus tested. Adenoviruses are non-enveloped, linear, double-stranded DNA viruses. These viruses are promising as vectors for expression of exogenous antigens, however their ability to maintain gene expression in cells is linked to their ability to evade the immune system and prevent cell death during viral replication, attributes which may make them less ideal as oncolytic virus therapies (Thorne et al., 2005). Nonetheless, a conditionally-replicating canine adenovirus 2 (CAV2) constructed with an osteocalcin promoter (to drive E1A gene expression and limit viral replication to osteoblasts) elicited cell death in canine osteosarcoma cultures and improved outcome in a murine xenograft model of canine osteosarcoma (Hemminki et al., 2003). When osteosarcoma cells were used as carriers to deliver this virus intravenously, viral expression in the liver was minimized and tumor growth in murine models of canine osteosarcoma was inhibited (Alcayaga-Miranda et al., 2010). Likewise, a CRAd that expresses canine p53, inhibited growth of canine cancer cells (Yazawa et al., 2003) and tumors in two murine xenograft models of canine osteosarcoma (Kanaya et al., 2011). A recent toxicity trial using a similar CRAd that expressed p53 showed no adverse effects on cardiovascular or respiratory systems of healthy cats injected intramuscularly with 3000× the clinical dose of virus (Su et al., 2008). The conditionally-replicating human adenovirus 5 (CRAd5) replicates in many canine cancer cell lines and primary canine cancer cell cultures, but the level of virus production is moderate in canine cells as compared to human cells (Ternovoi et al., 2005). When functional components encoded by CAV2 were inserted genetically into CRAd5, virus replication in canine osteosarcoma cells was improved and increased expression of a reporter gene encoded by the virus was observed in a murine xenograft model of canine osteosarcoma (Le et al., 2006). When this virus was administered intravenously to dogs, neutropenia was observed and virus was recovered from feces, urine, liver, and spleen at 96 h after virus inoculation (Smith et al., 2006). Recent studies have utilized CRAd5 expressing CD40 ligand to target the virus to cells expressing $\alpha_v\beta_3$ integrin. This virus was injected intratumorally into a dog with stage III oral melanoma and a dog with a melanoma on his eyelid; tumor resolution was observed in both patients (von Euler et al., 2008).

Paramyxoviruses, including measles virus and canine distemper virus, also have been tested in canine cancer cells. Paramyxoviruses are negative-strand RNA viruses that induce syncytia formation in the cells they infect. These viruses infect and deplete lymphocytes. If constructed correctly, they may be effective treatments for lymphoma. Suter et al. (2005) demonstrated that canine lymphocytes and polymorphonuclear cells express the CD150 receptor needed for measles virus and canine distemper virus entry. This receptor was absent in canine osteosarcoma and melanoma cells.

Poxviruses have many attributes that make them excellent oncolytic viruses. Unlike paramyxoviruses, poxviruses do not require a specific cellular receptor to infect cells. However, they are large viruses that naturally target neoplasms where new leaky vessels are being formed (Thorne et al., 2005). Poxviruses are enveloped, double-stranded DNA viruses with genomes that replicate with high fidelity and allow for insertion of up to 25 kb of contiguous DNA making them excellent vectors for exogenous protein expression (Kaufman et al., 2002). In addition, poxviruses have their own replication machinery and remain in the cytoplasm of infected cells, so recombination of viral DNA into the host DNA does not occur and gene transcription begins immediately upon entering the host cell. Additionally, poxviruses virions are released from infected cells before the cells are lysed, allowing for efficient spread of virus to neighboring cancer cells (Thorne et al., 2005). Importantly, poxviruses elicit a strong cell-mediated immune response and are ultimately cleared from the body by the humoral immune system, preventing poxviruses from causing latent or recurrent infection (Xu et al., 2004; Chaudhri et al., 2006). Vaccinia virus (VACV) is the poxvirus most frequently used in oncolytic virotherapy and was the vaccine used to eradicate smallpox. Wildtype VACV causes mild to moderate disease in several species including humans and mice. Genetic manipulation of VACV has produced attenuated viruses with improved targeting to malignant cells and antitumor effects (Buller et al., 1985; Gnant et al., 1999a; Gnant et al., 1999b; McCart et al., 2001; Kudo-Saito et al., 2006). Recombinant VACV have shown promising results in many murine xenografts of human tumors and in some human clinical trials. Clinical trials using poxviruses in cats with soft tissue sarcomas also had good results. Jourdier et al. (2003) reported that cats with soft tissue sarcoma given post-operative subcutaneous injections of canarypox virus expressing feline IL-2 had a recurrence rate of 28% (5/18 cats), patients treated with VACV encoding human IL-2 had a recurrence rate of 39% (7/18), while cats that did not receive viral therapy had a recurrence rate of 61% (11/18). This therapy did not cause any adverse effects. VACV with deletion of three viral genes has been shown to replicate in canine tumor cells and significantly decrease tumor volume in murine xenograft models of canine mammary tumors and soft tissue sarcoma (Gentshev et al., 2009; Gentshev et al., 2010; Gentshev et al., 2012a). However, the virus caused mild weight loss and replicated at low levels in the liver, lung, spleen, and ovary of mice (Gentshev et al., 2010). A similar recombinant VACV encoding anti-VEGF was effective in reducing tumor volume in mice with canine soft tissue sarcoma or prostate carcinoma xenografts. Weight loss was not evident in these mice, but low levels of viral replication were detected in the liver, lung, kidney, and spleen of the mice with soft tissue sarcoma xenografts (Patil et al., 2012). Another modified VACV induced cell death in canine soft tissue sarcoma, melanoma, osteosarcoma, and prostate carcinoma cells. This virus effectively reduced tumor growth of canine soft tissue sarcoma and prostate carcinoma xenografts in mice. Clinical trials are ongoing in dogs with various tumors

(Gentschev et al., 2012b). Myxoma virus (MYXV) is a poxvirus that shares most of the attributes of VACV, but does not cause disease in any animal except the rabbit. MYXV infects tumor cells from several species, including dogs and cats (MacNeill et al., 2011; Spylua et al., 2004; Wang et al., 2004; Urbasic et al., 2012) and has shown promise as an oncolytic agent in murine xenograft models of human tumors (Lun et al., 2005; Stanford et al., 2008; Lun et al., 2010; Thomas et al., 2011; Doty et al., 2013).

CONCLUSIONS

Oncolytic viruses are exciting novel therapeutic options that could benefit dogs and cats with cancer while providing better insight into which recombinant viruses will be most effective in human cancer patients. Conversely, as more oncolytic viruses enter human clinical trials, more oncolytic therapeutics may become available for use in companion animals. Prudent use of new oncolytic viruses will require an understanding of the biology of the viruses and the exogenous proteins they express.

REFERENCES

Alcayaga-Miranda, F, M Cascallo, J J Rojas, J Pastor, and R Alemany. 2010. Osteosarcoma cells as carriers to allow antitumor activity of canine oncolytic adenovirus in the presence of neutralizing antibodies. *Cancer gene therapy* 17(11): 792–802.

Barker, D. D., and A. J. Berk. 1987. Adenovirus proteins from both E1B reading frames are required for transformation of rodent cells by viral infection and DNA transfection. *Virology* 156(1): 107-121.

Buller,R.M., Smith,G.L., Cremer,K., Notkins,A.L., and Moss,B. 1985. Decreased virulence of recombinant vaccinia virus expression vectors is associated with a thymidine kinase-negative phenotype. *Nature* 317: 813-815.

Chaudhri,G., Panchanathan,V., Bluethmann,H., and Karupiah,G. 2006. Obligatory requirement for antibody in recovery from a primary poxvirus infection. *J. Virol.* 80: 6339-6344.

Dock G. 1904. Rabies virus vaccination in a patient with cervical carcinoma. *Am J Med Sci* 127:563.

Doty, Rosalinda A, Grant Mcfadden, Edward J Roy, and Amy L Macneill. 2013. Histological evaluation of intratumoral myxoma virus treatment in an immunocompetent mouse model of melanoma. *Oncolytic Virotherapy* 2: 1–17.

Gentschev, I, M Adelfinger, R Josupeit, S Rudolph, K Ehrig, U Donat, S Weibel, N G Chen, Y A Yu, Q Zhang, M Heisig, D Thamm, J Stritzker, A Macneill, and A A Szalay. 2012a. Preclinical evaluation of oncolytic vaccinia virus for therapy of canine soft tissue sarcoma. *PloS one* 7(5): e37239.

Gentschev, I, K Ehrig, U Donat, M Hess, S Rudolph, N Chen, Y A Yu, Q Zhang, J Bullerdiek, I Nolte, J Stritzker, and A A Szalay. 2010. Significant Growth Inhibition of Canine Mammary Carcinoma Xenografts following Treatment with Oncolytic Vaccinia Virus GLV-1h68. *Journal of oncology* 2010: 736907.

Genetschev, I, SS Patil, M Adelfinger, S Weibel, U Geissinger, A Frentzen, NG Chen, YA Yu, Q Zhang, G Ogilvie, AA Szalay. 2012b. Characterization and evaluation of a new oncolytic vaccinia virus strain LIVP6.1.1 for canine cancer therapy. *Bioengeneered* 4(2). Epub ahead of print..

Gentschev, I, J Stritzker, E Hofmann, S Weibel, Y A Yu, N Chen, Q Zhang, J Bullerdiek, I Nolte, and A A Szalay. 2009. Use of an oncolytic vaccinia virus for the treatment of canine breast cancer in nude mice: preclinical development of a therapeutic agent. *Cancer gene therapy* 16(4): 320–328.

Gnant,M.F., Noll,L.A., Irvine,K.R., Puhlmann,M., Terrill,R.E., Alexander,H.R., Jr., and Bartlett,D.L. (1999a). Tumor-specific gene delivery using recombinant vaccinia virus in a rabbit model of liver metastases. *J. Natl. Cancer Inst.* 91: 1744-1750.

Gnant,M.F., Puhlmann,M., Alexander,H.R., Jr., and Bartlett,D.L. 1999b. Systemic administration of a recombinant vaccinia virus expressing the cytosine deaminase gene and subsequent treatment with 5-fluorocytosine leads to tumor-specific gene expression and prolongation of survival in mice. *Cancer Res.* 59: 3396-3403.

Hemminki, Akseli, Anna Kanerva, Eric J Kremer, Gerd J Bauerschmitz, Bruce F Smith, Bin Liu, Minghui Wang, Renee A Desmond, Anne Keriel, Brian Barnett, Henry J Baker, Gene P Siegal, and David T Curiel. 2003. A canine conditionally replicating adenovirus for evaluating oncolytic virotherapy in a syngeneic animal model. *Molecular therapy : the journal of the American Society of Gene Therapy* 7(2): 163–73.

Jourdier, TM, C Moste, MC Bonnet, F Delisle, JP Tafani, P Devauchelle, J Tartaglia, P Moingeon. 2003. Local immunotherapy of spontaneous feline fibrosarcoma using recombinant poxviruses expressing interleukin 2. *Gene Therapy* 10: 2126-2132.

Kanaya, Noriko, Mitsuhiro Yazawa, Yuko Goto-Koshino, Manabu Mochizuki, Ryohei Nishimura, Koichi Ohno, Nobuo Sasaki, and Hajime Tsujimoto. 2011. Anti-tumor effect of adenoviral vector-mediated p53 gene transfer on the growth of canine osteosarcoma xenografts in nude mice. *The Journal of veterinary medical science / the Japanese Society of Veterinary Science* 73(7): 877–83.

Kaufman, H. L., K. Flanagan, C. S. Lee, D. J. Perretta, and H. Horig. 2002. Insertion of interleukin-2 (IL-2) and interleukin-12 (IL-12) genes into vaccinia virus results in effective anti-tumor responses without toxicity. *Vaccine* 20(13-14): 1862-1869.

Kirn, D. H., Y. Wang, F. Le Boeuf, J. Bell, and S. H. Thorne. 2007. Targeting of interferon-beta to produce a specific, multi-mechanistic oncolytic vaccinia virus. *PLoS Medicine* 4(12): e353.

Kottke, T., Hall, G., Pulido, J., Diaz, R. M., Thompson, J., Chong, H., Selby, P., et al. 2010. Antiangiogenic cancer therapy combined with oncolytic virotherapy leads to regression of established tumors in mice. *Journal Of Clinical Investigation*, *120*(5), 1551–1560.

Kudo-Saito,C., Hodge,J.W., Kwak,H., Kim-Schulze,S., Schlom,J., and Kaufman,H.L. (2006). 4-1BB ligand enhances tumor-specific immunity of poxvirus vaccines. *Vaccine* 24: 4975-4986.

Le, L P, a a Rivera, J N Glasgow, V V Ternovoi, H Wu, M Wang, B F Smith, G P Siegal, and D T Curiel. 2006. Infectivity enhancement for adenoviral transduction of canine osteosarcoma cells. *Gene therapy* 13(5): 389–99.

Lun, X., T. Alain, F. J. Zemp, H. Zhou, M. M. Rahman, M. G. Hamilton, G. McFadden, J. Bell, D. L. Senger, and P. A. Forsyth. 2010. Myxoma virus virotherapy for glioma in immunocompetent animal models: Optimizing administration routes and synergy with rapamycin. *Cancer Research* 70(2): 598-608.

Lun, X., W. Yang, T. Alain, Z. Q. Shi, H. Muzik, J. W. Barrett, G. McFadden, et al. 2005. Myxoma virus is a novel oncolytic virus with significant antitumor activity against experimental human gliomas. *Cancer Research* 65(21): 9982-9990.

MacNeill, A. L., T. Moldenhauer, R. Doty, and T. Mann. 2011. Myxoma virus induces apoptosis in cultured feline carcinoma cells. *Research in Veterinary Science*.

Mastrangelo, M. J., H. C. Maguire Jr, L. C. Eisenlohr, C. E. Laughlin, C. E. Monken, P. A. McCue, A. J. Kovatich, and E. C. Lattime. 1999. Intratumoral recombinant GM-CSF-encoding virus as gene therapy in patients with cutaneous melanoma. *Cancer Gene Therapy* 6(5): 409-422.

McCart,J.A., Ward,J.M., Lee,J., Hu,Y., Alexander,H.R., Libutti,S.K., Moss,B., and Bartlett,D.L. 2001. Systemic cancer therapy with a tumor-selective vaccinia virus mutant lacking thymidine kinase and vaccinia growth factor genes. *Cancer Res.* 61: 8751-8757.

Patil, SS, I Gentschev, M Adelfinger, U Donat, M Hess, S Weibel, I Nolte, A Frentzen, AA Szalay. 2012. Virotherapy of canine tumors with oncolytic vaccinia virus GLV-1h109 expressing an anti-VEGF single-chain antibody. *PloS one* 7(10): e47472.

Smith, Bruce F, David T Curiel, Vladimir V Ternovoi, Anton V Borovjagin, Henry J Baker, Nancy Cox, and Gene P Siegal. 2006. Administration of a conditionally replicative oncolytic canine adenovirus in normal dogs. *Cancer biotherapy & radiopharmaceuticals* 21(6): 601–6.

Stanford, M. M., M. Shaban, J. W. Barrett, S. J. Werden, P. A. Gilbert, J. Bondy-Denomy, L. Mackenzie, K. C. Graham, A. F. Chambers, and G. McFadden. 2008. Myxoma virus oncolysis of primary and metastatic B16F10 mouse tumors in vivo. *Molecular Therapy* 16(1): 52-59.

Stanford, M. M., S. J. Werden, and G. McFadden. 2007. Myxoma virus in the European rabbit: Interactions between the virus and its susceptible host. *Veterinary Research* 38(2): 299-318.

Stanford, M. M., J. W. Barrett, S. H. Nazarian, S. Werden, and G. McFadden. 2007b. Oncolytic virotherapy synergism with signaling inhibitors: Rapamycin increases myxoma virus tropism for human tumor cells. *Journal of Virology* 81(3): 1251-1260.

Su, Changqing, Hui Cao, Shuping Tan, Yao Huang, Xiaoyuan Jia, Lixin Jiang, Kai Wang, Ying Chen, Ju Long, Xinyuan Liu, Mengchao Wu, Xiaobing Wu, and Qijun Qian. 2008. Toxicology profiles of a novel p53-armed replication-competent oncolytic adenovirus in rodents, felids, and nonhuman primates. *Toxicological sciences* 106(1): 242–50.

Suter, Steven E, May B Chein, Veronika von Messling, Becky Yip, Roberto Cattaneo, William Vernau, Bruce R Madewell, and Cheryl a London. 2005. In vitro canine distemper virus infection of canine lymphoid cells: a prelude to oncolytic therapy for lymphoma. *Clinical cancer research* 11(4): 1579–87.

Sypula, J., F. Wang, Y. Ma, J. Bell, and G. McFadden. 2004. Myxoma virus tropism in human tumor cells. *Gene Therapy & Molecular Biology* 8 : 103-114.

Ternovoi, Vladimir V, Long P Le, Natalya Belousova, F Smith, Gene P Siegal, David T Curiel, and Bruce F Smith. 2005. Productive Replication of Human Adenovirus Type 5 in Canine Cells Productive Replication of Human Adenovirus Type 5 in Canine Cells. *Journal of Virology* 79(2): 1308-1311.

Thomas, D. L., R. Doty, V. Tosic, J. Liu, D. M. Kranz, G. McFadden, A. L. Macneill, and E. J. Roy. 2011. Myxoma virus combined with rapamycin treatment enhances adoptive T cell therapy for murine melanoma brain tumors. *Cancer Immunology, Immunotherapy : CII* 60(10): 1461-1472.

Thorne, S. H., D. L. Bartlett, and D. H. Kirn. 2005. The use of oncolytic vaccinia viruses in the treatment of cancer: *Current Gene Therapy* 5(08): 429-443.

Tsung, K., J. B. Meko, G. R. Peplinski, Y. L. Tsung, and J. A. Norton. 1997. IL-12 induces T helper 1-directed antitumor response. *J. Immunol.* 158(7): 3359-3365.

Urbasic, Ashlee S, Stacy Hynes, Amy Somrak, Stacey Contakos, Masmudur M Rahman, Jia Liu, and Amy L MacNeill. 2012. Oncolysis of canine tumor cells by myxoma virus lacking the serp2 gene. *American Journal of Veterinary Research* 73(8): 1252–1261.

Wang, F., Y. Ma, J. W. Barrett, X. Gao, J. Loh, E. Barton, H. W. Virgin, and G. McFadden. 2004. Disruption of erk-dependent type I interferon induction breaks the myxoma virus species barrier. *Nature Immunology* 5(12): 1266-1274.

Xu,R., Johnson,A.J., Liggitt,D., and Bevan,M.J. 2004. Cellular and humoral immunity against vaccinia virus infection of mice. *J. Immunol.* 172: 6265-6271.

Yazawa, M, A Setoguchi, SH Hong, R Uyama, T Nakagawa, N Kanaya, R Nishimura, N Sasaki, K Masuda, K Ohno, H Tsujimoto. 2003. Effect of an adenoviral vector that expresses the canine p53 gene on cell growth of canine osteosarcoma and mammary adenocarcinoma cell lines. *Am J Vet Res* 64(7): 880-888.

Yu, W., and H. Fang. 2007. Clinical trials with oncolytic adenovirus in china. *Current Cancer Drug Targets* 7(2): 141-148.

Von Euler, Henrik, Arian Sadeghi, Björn Carlsson, Patricio Rivera, Angelica Loskog, Thomas Segall, Olle Korsgren, and Thomas H Tötterman. 2008. Efficient adenovector CD40 ligand immunotherapy of canine malignant melanoma. *Journal of immunotherapy* 31(4): 377–84.

Zou, W. 2005. Immunosuppressive networks in the tumour environment and their therapeutic relevance. *Nature Reviews.Cancer* 5(4): 263-274.

COMPARATIVE ONCOLOGY: WHERE ARE WE NOW AND WHAT CHALLENGES LIE AHEAD? M. Paoloni, C. Mazcko, C. Khanna. Comparative Oncology Program, National Cancer Institute, Bethesda, MD.

Comparative Oncology is defined as the inclusion of naturally occurring cancers in pet animals to study cancer biology and drug development. Its' goals are to inform human oncology progress through the conduct of drug, device, and immunotherapy clinical trials. The field has grown in significance over the past decade and interest from clients, industry and academia place us on the cusp of our next series of challenges. Where do we go from here? Will we expand our reach? Can we progress from calling this a "novel" model? And has comparative oncology improved veterinary medicine?

The future of comparative oncology requires rigorous characterization of the molecular drivers of veterinary cancers to define subtypes and signatures shared and different from the human disease. This necessitates a large-scale push for gene expression, NEXTgen sequencing, proteomic and epigenetic profiling in a collaborative and synchronized way. Such molecular refining will set the stage for more advanced biology studies and identify appropriate models for targeted therapy clinical trials. From here, the opportunity to model personalized medicine prescriptive studies and advance rationally selected combinations through innovative trial designs can be realized. One of comparative oncology's greatest attributes is the freedom to work outside the traditional regulatory paradigm of phase structured drug development. Hence the challenge forward is to conduct a series of multi-arm studies in the adjuvant setting to identify agents that improve outcome up-front and early. These directed priorities will lead the way for comparative oncology's success in the next decade and promote its' maximal impact.

VARIABILITY OF T CELL FOXP3 EXPRESSION IN THE AFFECTED LYMPH NODES OF DOGS WITH MULTICENTRIC B CELL LYMPHOMA. D Pinheiro[1], T Dalessandri[1], L Davison[2], G Polton[3], OA Garden.[1] [1]The Royal Veterinary College, London, UK; [2]The Wellcome Trust Centre for Human Genetics, University of Oxford, Oxford, UK; [3]North Downs Specialist Referrals, Surrey, UK.

To explore the pathogenic role of regulatory T cells (Tregs) in canine lymphoma, the proportions of $FOXP3^+$ T cells in fine needle aspirates of affected lymph nodes of dogs presenting with B cell lymphoma (BCL) were measured by flow cytometry. Control aspirates were collected from dogs presenting with reactive lymphoid hyperplasia (RH).

The lymph nodes of dogs with BCL possessed significantly lower proportions of $FOXP3^+$ T cells (0.90% (median), range 0.05-6.01%; n=21) than dogs with RH (4.67%, range 0.3-17.3%, n=14; p=0.0006), attributed to the effacement of normal lymphoid tissue by neoplastic B cells. However, gated $CD5^+$ cells showed highly variable FOXP3 expression in both BCL (8.08%, range 0.77-18.1%) and RH (5.61%, range 0.48-20.6%; p=0.435) patients. Similarly, FOXP3 expression by peripheral blood mononuclear cells (PBMCs) showed no significant difference between the groups (BCL 1.83%, range 0.04-8.45%; RH 3.05%, range 0.03-10.6%; p=0.495).

To explore whether Tregs could be induced by neoplastic B cells *in vitro,* PBMCs were treated with concanavalin A (Con A) for 72 hours. $CD5^+$ T cells positively selected from the activated PBMCs were co-cultured with neoplastic B cell lines 1771 (T_{1771}) or CLBL1 (T_{CLBL1}) for 96 hours, and then maintained in monocultures supplemented with IL-2 for an additional 24 hours. Parallel streams of T cells cultured with either Con A (T_{CA}) or B cells pre-stimulated with lipopolysaccharide (T_{Bp}) were used as controls. FOXP3 expression showed a significant increase in both T_{CA} (15.75±0.35% (mean±SEM), n=3; p=0.043) and T_{1771} (26.20±3.00%, n=3; p=0.021), but not in T_{Bp} (6.01±0.90%, n=3; p=0.116) or T_{CLBL-1} (17.67±9.44%, n=3; p=0.490) at 24 hours. However, FOXP3 expression returned to baseline (5-7%) in all conditions examined after an additional 96 hours of monoculture, suggesting transience of FOXP3 expression characteristic of activated conventional T cells rather than Tregs.

These data suggest that $FOXP3^+$ Tregs are unlikely to represent a universal tumor immune subversion mechanism in BCL of the dog. Further work is required to elucidate both the clinical and pathological characteristics of the cases showing higher proportional FOXP3 expression, and the prevailing regulatory mechanisms in BCL cases showing low FOXP3 expression.

COMPARISON OF SERUM DOXORUBICIN LEVELS FOLLOWING DIFFERENT DEB CHEMOEMBOLIZATION TECHNIQUES AS WELL AS SYSTEMIC ADMINISTRATION IN THE SAME CANINE PATIENTS WITH NATURALLY-OCCURRING HEPATOCELLULAR CARCINOMA. C. Weisse[1], A. Allyson[1], M. Soulen[2]. 1. Interventional Radiology, Animal Medical Center, New York, NY, United States. 2. Interventional Radiology, University of Pennsylvania, Philadelphia, PA, United States.

The ideal technique for administration of DEBs has not been completely elucidated but the most recent recommendations suggest maintaining tumor perfusion rather than continuing to complete vascular stasis in order to facilitate future treatments. It is unclear how intratumoral and systemic doxorubicin levels are effected by DEB delivery performed with or without vascular stasis. The investigators hypothesized that there would be a decrease in systemic doxorubicin exposure following chemoembolization to stasis versus delivery without stasis, and intravenous administration would result in the highest exposure.

Dogs with diagnosed with naturally occurring nonresectable HCC were scheduled to receive three separate treatments ~5 weeks apart containing the same systemic dose of doxorubicin; 1) 100-300 DEB delivered without stasis, 2) 100-300 DEB delivery followed by Beadblock to stasis, and 3) intravenous doxorubicin. Serum doxorubicin levels were measured postoperatively.

Serum doxorubicin levels were detectable for a median of 180, 30, and 1080 minutes following delivery without stasis (5 dogs), with stasis (7 dogs), and intravenously (4 dogs), respectively (FIGURE 1). Median maximum serum doxorubicin levels, median area-under-the-curve systemic doxorubicin levels, and median relative systemic doxorubicin exposures are reported in TABLE 1.

Different DEB chemoembolization techniques may achieve different serum doxorubicin concentrations and therefore different intratumoral drug concentrations. When performing DEB chemoembolization in dogs with naturally occurring hepatocellular carcinoma, delivery to stasis results in lower systemic doxorubicin exposure (and presumably higher intratumoral concentrations) than when embolization to stasis in not performed. Intravenous administration results in the highest systemic exposure.

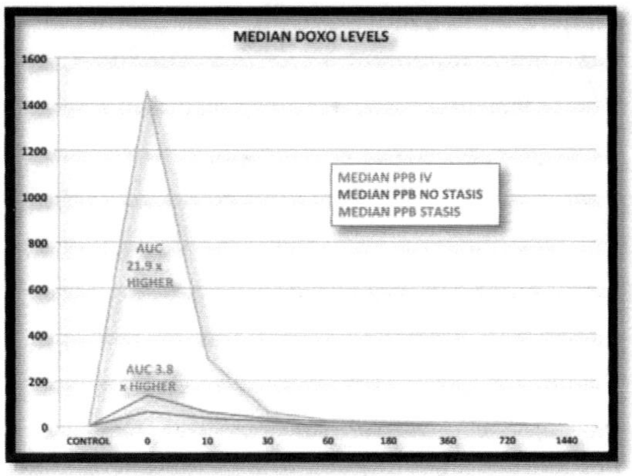

FIGURE 1

	NO STASIS	STASIS	INTRAVENOUS
Median maximum serum doxorubicin levels	91.7 ppb	63 ppb	1456.5 ppb
Median area-under-the-curve systemic doxorubicin levels (PPB)	5056 ppb	1337.5 ppb	29221.2 ppb
Median relative systemic doxorubicin exposures (PPB)	3.8 fold	BASELINE	21.9 fold

TABLE 1

LOW STORES OF 25-HYDROXYVITAMIN D IN DOGS AND ITS ASSOCIATION WITH INCREASED CANCER RISK. [1]KA Selting, [2]CR Sharp, [3]R Ringold, [1]R Backus. [1]University of Missouri, College of Veterinary Medicine, Columbia, MO. [2]Tufts University, Cummings School of Veterinary Medicine, North Grafton, MA, [3]Veterinary Diagnostic Institute, Inc., Simi Valley, CA.

The objective of this study was to evaluate the association between serum 25(OH)D concentrations ([25(OH)D]) and cancer in dogs. We hypothesized that clinically healthy dogs of at risk breeds with low [25(OH)D] would be more likely to subsequently develop cancer than those with adequate [25(OH)D]. Additionally, we hypothesized that in dogs with hemoabdomen those with splenic malignancy would have lower [25(OH)D] than dogs with benign splenic masses.

We measured [25(OH)D] in a large population of healthy dogs and defined vitD sufficiency using regression analysis to determine optimal [25(OH)D] where intact parathyroid hormone concentrations ([iPTH]) plateaued. Clinically healthy dogs were followed for 6 months to assess for the development of cancer. [25(OH)D] was also determined in dogs with hemoabdomen. [25(OH)D] in healthy dogs was then compared with [25(OH)D] in the 2 cohorts of cancer-bearing dogs.

In healthy dogs (n=320), [iPTH] correlated inversely to [25(OH)D] and plateaued at 8.2 pg/ml (range 0.9-127 pg/ml) which correlated to 100-120 ng/ml of 25(OH)D (9.5-249 ng/ml). Thirteen dogs within the at risk group developed cancer. Relative risk of developing cancer with [25(OH)D] below 85 ng/ml was 1.3 (P=0.004).

Hemoabdomen dogs (n=63, 43 malignant and 20 benign) had [25(OH)D] ranging from 19.4 to >150 ng/ml. Relative risk of having cancer with [25(OH)D] below 40, 60, 85, and 100 ng/ml was 3.5, 1.9, 1.5, 1.1 respectively (P<0.0001).

[25(OH)D]in dogs vary widely, although low [25(OH)D] is associated with an increased risk of developing cancer in at risk breeds and of being diagnosed with splenic malignancy in dogs with hemoabdomen.

ANTIPROLIFERATIVE AND PROAPOPTOTIC EFFECTS OF DUAL PI3K/mTOR INHIBITION WITH PWT32976 IN CANINE LYMPHOMA. DH Thamm[1], BJ Rose[1], R Ventura[2], DJ Matthews[2], 1. Colorado State University, Fort Collins, CO. 2. Pathway Therapeutics Inc., San Francisco, CA.

It has been well established that dysregulation of the PI3K/Akt/mTOR pathway mediates key oncogenic functions associated with tumor growth, survival, metastasis and energy utilization. Recent data suggest that inhibition of this pathway is a viable therapeutic strategy in human B-cell lymphoma, and we previously observed objective antitumor responses in dogs with B-cell lymphoma treated with ibrutinib (PCI32765), an inhibitor of the Btk kinase upstream of PI3K/Akt/mTOR (Honigberg *et al.*, PNAS 107:13075, 2010). PWT32976 is a dual inhibitor of both PI3K and mTOR with nanomolar potency. The goal of this study was to evaluate the *in vitro* effects of PWT32976 on proliferation in a panel of canine cell lines and primary lymphoma cultures, with expanded evaluation of effects on apoptosis, chemosensitivity and target inhibition in the CLBL1 canine B-cell lymphoma cell line.

A panel of 25 canine tumor-derived cell lines was incubated with PWT32976 for 72 hours, followed by determination of relative viable cell number using a bioreductive fluorometric assay. Similar methodology was used to evaluate PWT32976 against a panel of CD40 ligand (CD40L)-stimulated primary B cell lymphoma cultures. Changes in cell cycle distribution were evaluated using propidium iodide (PI) staining and flow cytometry. Induction of apoptosis was evaluated using annexin-V/PI staining with flow cytometry and an activated caspase-3,7 ELISA. Inhibition of Akt phosphorylation was evaluated using western analysis.

PWT32976 exerted dose-dependent antiproliferative effects against canine tumor cells, with calculated 50% inhibitory concentrations (IC_{50}s) ranging from >10 uM to approximately 80 nM. The canine B-cell lymphoma derived cell line CLBL1 was the most sensitive. PWT32976 induced apoptosis in CLBL1 cells in a dose- and time-dependent fashion: no other alterations in cell-cycle distribution were observed. Akt phosphorylation was reduced in a dose-dependent fashion as well. PWT32976 demonstrated more potent growth inhibition than the human clinical candidates ibrutinib (PCI32765), fostamatinib (R406) and idelalisib (CAL-101) in CLBL1 cells. PWT32976 enhanced the antiproliferative effects of doxorubicin in an additive fashion. PWT32976 inhibited CD40L-stimulated proliferation in 14/15 primary B-cell lymphoma cultures, with IC_{50}s ranging from 40 nM to approximately 1 uM.

In conclusion, the dual PI3K/mTOR inhibitor PWT32976 exerts dose-dependent antiproliferative effects against a variety of canine tumor cell lines, with preferential activity against canine B-cell lymphoma. Much of the effects on growth may be due to induction of apoptosis. These data strongly justify clinical evaluation of PWT32976 in canine lymphoma. Modulation of Akt phosphorylation may be a useful pharmacodynamic biomarker.

PEGYLATED ARGININE DEIMINASE IN THE DOG FROM BENCHTOP TO CAGESIDE. C. O. Rodriguez, Jr., Department of Veterinary Surgical and Radiological Sciences, University of California, Davis, CA.

Pegylated arginine deiminase (PEGADI) is capable of degrading plasma arginine. Many tumor types are auxotrophic for L-arginine due to very low or no specific activity of argininosuccinate synthetase (ASS1) and are thus sensitive to depravation of this single nutrient. PEGADI has entered clinical trials in humans for melanoma, hepatocellular carcinoma, and prostatic carcinoma. The purpose of this study is to identify a dose of PEGADI that safely reduces circulating plasma arginine in the dog.

Dogs (n=25) with diffuse hepatocellular carcinoma, malignant melanoma, lymphoma, or other sarcomas, or carcinomas, were given either 85 IU/m^2 (n=12) or 170 IU/m^2 (n=13) subcutaneously once. Arginine and citrulline in deproteinated plasma were quantitated with HPLC before and at 8 and 24 hours after treatment and then once weekly for the first month after PEGADI administration.

Plasma arginine was not detectable in one of twelve dogs treated with 85 IU/m^2 where as plasma arginine was not detectable in twelve of thirteen dogs treated with 170 IU/m^2. Arginine remained undetectable for 1-3 weeks and returned to pre-treatment levels in all dogs by the fourth week. Plasma citrulline was elevated in all dogs that had undetectable levels of arginine. Toxicity noted was noted but mild and self-limiting. Specifically, neutropenia (grade I; n=4) and thrombocytopenia (grade III; n=3) were the most commonly encountered toxicities.

Based on these data, 170IU/m^2 PEGADI administered subcutaneously on a monthly basis can be used safely for Phase II trials. Such a trial is underway.

DEFINING CANINE B-CELL LYMPHOMA USING HUMAN IMMUNOHISTOCHEMICAL ALGORITHMS. S Suter[1], K Richards[2], Y Fedoriw[2], L Borst[1]. 1. North Carolina State University College of Veterinary Medicine, Raleigh, NC. 2. University of North Carolina, Chapel Hill, NC.

Canine B-cell lymphoma (cBCL) is phenotypically and biologically similar to human diffuse large B-cell lymphoma (hDLBCL). hDLBCL is a clinically heterogeneous disease that can be subdivided into germinal center B-cell lymphoma (GCB, ~50%) and activated B-like (ABC, ~50%) subtypes that are histopathologically and cytologically indistinguishable. ABC lymphomas are biologically distinct, therefore, human ABC lymphoma patients have a decreased survival compared to GCB lymphoma patients receiving the same treatments.

Immunohistochemical algorithms can differentiate human GCB and ABC subtypes. The most commonly used antibodies bind the germinal center antigens CD10, GCET1, Bcl6, and LMO2 and the post-germinal center antigens Mum1/IRF4 and FoxP1. These algorithms correlate well with corresponding gene expression analysis (GEP) and demonstrate clear survival differences between the GCB and ABC groups. Our GEP analysis suggests GCB-like and ABC-like subtypes exist in cBCL, so we sought to define an immunohistochemical algorithm for cBCL.

In this study, we utilized immunohistochemial analysis of 45 cases of cBCL (36 diffuse large B-cell; 9 marginal zone) using the antibodies listed above to determine if GCB and ABC subtypes could be indentified. 15/45 (33.3%) were positive for CD10, the most important antigen used to delineate the GCB subtype in hDLBCL. In contrast, there was little staining variability using the remaining antibodies - the vast majority of cases were FoxP1/GCET1/LMO2 positive and Bcl6/Mum1/IRF4 negative. When analyzing the staining patterns using 5 hDLBCL algorithms (Choi, modified Choi, Hans, modified Hans, Tally), >95% of cBCL cases were classified as GCB. Since these results are in striking contrast to hDLBCL data and our canine GEP data, we suggest that hDLBCL immunohistochemical algorithms are not useful in defining GCB or ABC subtypes in cBCL and that other, more canine-specific germinal center and post-germinal center markers need to be identified to validate that these subtypes exist in cBCL.

EVALUATION OF TAVOCEPT, PIROXICAM AND CISPLATIN FOR THE TREATMENT OF CANINE TRANSITIONAL CELL CARCINOMA OF THE URINARY BLADDER.

CJ Henry[1], DJ Tate[1], KA Selting[1], JN Bryan[1], SM Axiak-Bechtel[1], JC Lattimer[1], FH Hausheer[2].
1. University of Missouri, Columbia, MO. 2. BioNumerik Pharmaceuticals, Inc, San Antonio, TX.

Cisplatin/piroxicam combination therapy provides superior response rates for canine bladder cancer, but is not recommended due to nephrotoxicity. The chemoprotectant, Tavocept, mitigates cisplatin toxicity in people and in normal dogs. We hypothesized that Tavocept would reduce toxicity of cisplatin/piroxicam combination therapy and facilitate a shortened saline diuresis protocol in dogs with bladder cancer.

Fourteen dogs with treatment-naïve transitional cell carcinoma (TCC) were enrolled. Baseline evaluations included CBC, UA, serum creatinine and urea nitrogen, thoracic radiographs, and abdominal ultrasound and CT. One hour after maropitant pretreatment, Tavocept (18.4 g/m^2), then cisplatin (60 mg/m^2) were infused using a 90-minute diuresis protocol with 0.9% NaCl. Four Tavocept/cisplatin treatments (q21d) and daily piroxicam (0.3 mg/kg/day PO) were planned. Renal function was assessed before and one week after each treatment. Tumor response was assessed on days 42 and 84.

Eight dogs completed the protocol; 6 required protocol alteration due to complications or disease progression. The dose-limiting toxicity was elevation in BUN, with (n= 6) or without (n= 8) concurrent creatinine increase. Two dogs had concurrent isothenuria and elevated creatinine. The shortened saline diuresis protocol was clinically well tolerated. Overall response rate was 28%, with no CR, 4 PR, and 6 SD. Seven dogs are alive (median follow-up = 141 days). Median survival time has not been reached (range = 19 to >598 days).

We conclude that Tavocept markedly reduces the time required for saline diuresis in dogs receiving cisplatin. Although survival analysis is pending, response rates are similar to those reported for other TCC protocols.

CHEMOTHERAPY FOR MALIGNANT MAMMARY CARCINOMA

Antony Moore BVSc, MVSc, DACVIM (Oncology)
Wauchope, NSW, Australia

INTRODUCTION

Mammary tumors are the most common neoplasm in intact female dogs and account for approximately half of all neoplasms in the bitch. In contrast, mammary neoplasms are rare in male dogs, accounting for less than 1% of these types of tumors. The incidence of mammary tumors is higher in dogs than any other domesticated animal and is three times the incidence in humans. Approximately half of the tumors are malignant, and half of these have metastasized by the time they are initially diagnosed.

ETIOLOGY

Sex hormones certainly play a role in development of mammary tumors in the bitch. Intact females have a seven-fold increased risk of developing mammary cancer compared to neutered females. The age at which ovariohysterectomy takes place is directly related to the risk of developing mammary cancer. Data clearly indicate the preventive role of ovariectomy prior to the second estrus. Interestingly, when only the dogs that had 2 or more estrus cycles before spaying are considered, there is still a benefit to earlier spaying. If they were spayed before 2.5 years of age, then the risk was 6% of those dogs not spayed at all; (that approximates to the 8% quoted for 1 estrus cycle). If they were older than 2.5 years, the risk was 40% of unspayed dogs, but that difference was not significant. Essentially the greatest reduction in risk comes if you spay before the first heat, but then as long as you spay before 2.5 years, the risk is very low. Counter to all current thought one group reviewed the statistics in all the literature regarding the risk of neutering and found the data weak.(1) We continue to recommend spaying at an early age.

Obesity may be a factor in mammary neoplasia in dogs. In one case-controlled study, the risk of mammary carcinoma among ovariectomized dogs was significantly reduced (by 40%) in dogs that had been thin at 9 to 12 months of age. Among intact dogs that were thin at 9 to 12 months, however, risk of mammary cancer was not reduced.(2) In contrast another study, showed obesity at 1 year of age almost trebled the incidence of mammary cancer in intact dogs.(3) In that study, dogs with a higher intake of red meat in their diet were also at higher risk for developing mammary cancer. Interestingly, feeding of a high fat diet, and obesity in the year before being diagnosed with a mammary cancer had no influence of tumor development.(2;4) Caregivers should be counseled to ovariectomize their dogs before the first estrus, and to ensure weight control is practiced in at least the first year of life. Ovariectomies performed later in life, while not as protective, should not be discouraged

STAGING AND DIAGNOSIS

The two most common sites of metastases are lungs and regional lymph nodes. Therefore, staging should include a minimum database of tumor measurements, thoracic radiographs, complete blood count (CBC), biochemical profile, urinalysis, and an evaluation of regional lymph nodes (axillary and inguinal at minimum) by palpation, fine-needle aspiration cytology, and (if indicated) biopsy.

PROGNOSTIC FACTORS

Somewhat surprisingly, the prognosis for dogs with mammary cancer is not influenced by either tumor location or number of tumors.(4;5) Other factors that are not prognostic are number of pregnancies, age at first pregnancy and occurrence of pseudopregnancies. The following are prognostic factors that have been shown in studies to predict survival or disease-free interval.

Stage

Dogs with stage 1 tumors were more likely to survive longer than dogs with any other stage tumor.(4) This effect of tumor stage was similar in other studies and is detailed in the following tables.(6)

Effect of tumor stage on survival.(6)

Stage	Percent alive 1 year after surgery	Percent alive 2 years after surgery
1	97.9	97.9
3	75.8	66.4
4	13.6	13.6

Tumor Size

This is probably one of the most important prognostic factors for a dog with a mammary mass. Dogs with mammary tumors less than 3-cm in diameter have a significantly better prognosis than dogs with larger tumors. In another study, dogs with tumors larger than 5-cm in diameter were more likely to develop metastases than dogs with smaller tumors and were 7 times more likely to die in the first 2 years after surgery.(7)

Metastasis

Metastases to regional lymph nodes has been associated with an increased risk for tumor recurrence and for decreased overall survival. The presence of distant metastases was found to be prognostically important in other studies. (6) Those dogs with no metastases were more than 3 times as likely to survive one year from diagnosis.(4)

Histopathology

Several histologic grading schemes are of prognostic significance. Important factors include histologic classification, degree of nuclear differentiation, and the presence of lymphoid accumulation. In general, the more highly differentiated the tumor, the better the prognosis. Poorly differentiated tumors are much more likely to recur than well-differentiated tumors.(8;9), The chance of recurrence for poorly differentiated canine mammary tumors is 90%; for moderately differentiated tumors, 68%; and for well-differentiated tumors, 24%. Dogs that have mammary cancer but no evidence of lymphoid cellular reactivity at the time of initial mastectomy have a threefold increased risk of developing recurrence within two years compared to those with such reactivity. Dogs with inflammatory carcinomas also have a poor prognosis.

When reviewing a histopathology report, the clinician should look for information regarding completeness of the surgical excision; invasion into lymphatics or blood vessels, and differentiation of the tumor.

Dogs with tumors that showed a high proportion of Ki-67 staining (which is an immunohistochemical marker for cellular proliferation) were more likely to develop metastases in studies.(10;11) Additionally Ki-67 staining was inversely related to survival time.(10;11)

Degree of Invasion and Ulceration

Dogs with tumors that ulcerate overlying skin have a worse prognosis (shorter overall survival times) than dogs with tumors without ulceration. Rapid and invasive growth correlates with a worse prognosis, which may be recognized as fixation of the tumor to the underlying skin. Vascular or lymphatic invasion is a poor prognostic factor; dogs with histologic evidence of invasion have a shorter median survival.(6)

Ovariectomy Status

In one study dogs that were intact at the time of surgery for a mammary carcinoma survived a shorter time (median survival 9.5 months) compared to dogs ovariectomized within the 2 years before surgery (median survival 25 months).(5) Dogs ovariectomized more than 2 years before mammary tumor surgery did not benefit to the same extent. In addition, dogs that were intact had a higher proportion of solid and anaplastic carcinomas than either group of ovariectomized dogs (80% solid carcinomas in intact dogs compared to 20% (<2 years) and 7% (> 2 years)).(5) In another study dogs that were *not* entire at the time of diagnosis were 2.5 times more likely to survive 2 years after surgical excision of the mammary malignancy.(7) In contrast ovariohysterectomy at the time of tumor removal had no effect on survival in another study with approximately 60% of dogs with malignant tumors dying within 2 years of surgery whether they were spayed at the time or not.(12)

Duration of tumor presence

Dogs that had a mammary malignancy present for longer than 6 months were more likely to develop metastases than dogs with tumors removed soon after diagnosis.(7)

TREATMENT

Extent of Surgery

The extent of surgery influences neither survival nor disease-free interval but rather the histologic completeness of surgical margins as assessed by histopathology has been shown to be prognostic for survival so the best surgery to achieve complete margins is the surgery that should be offered.

Adjuvant chemotherapy

Chemotherapy agents that have been anecdotally reported for mammary carcinoma as causing clinical tumor responses when used as single agents (the most important criterion for inclusion in a combination protocol) include doxorubicin, cisplatin, mitoxantrone and paclitaxel, cyclophosphamide and 5-fluorouracil. Laboratory evidence also exists for the efficacy of doxorubicin and carboplatin against cultured canine mammary tumor cells. Unfortunately there are few studies to compare survival statistics with and without chemotherapy in dogs. While that seems surprising, the reason is that most of the clinical research happens in the USA, where dogs are routinely spayed at an early age, so mammary cancer is rare. The other places where dogs are not spayed (such as Japan and Europe) is where the more recent information has come using surgery alone…but chemotherapy has only been recently more widely used in those countries.

The finding that normal mammary tissue has no evidence of cyclo-oxygenase-2 (cox-2) expression, and that 24% of benign and 56% of malignant mammary tumors do express cox-2 had led to speculation that non-steroidal inhibitors of cox-2 may play a role in the treatment of mammary carcinoma, and possibly the prevention of new lesions. In another study, mammary carcinomas that were more anaplastic had higher levels of cox-2 expression. In an early evaluation of piroxicam (a non-steroidal anti-inflammatory drug) in the treatment of cancer in dogs, one of 3 dogs with mammary carcinoma had a partial response to therapy. There are a couple of reports from Brazil showing improved survival after piroxicam for dogs with inflammatory mammary cancer (which thankfully Jill does not have). This implies that piroxicam may have some efficacy on the treatment of this disease. It could be considered for use alone as a "low-cost" alternative to chemotherapy, but with uncertain effectiveness; or as an adjunct to chemotherapy. In addition, nonsteroidal anti-inflammatory drugs that primarily target cox-2 may have some role in chemoprevention. The mechanism appears to be partly through cell death, but may also include anti-angiogenesis.

One small study from Greece showed that survival in dogs with stage 3 mammary cancer treated with chemotherapy was 24 months compared with 6 months if they did not get chemotherapy (they used cyclophosphamide and 5-fluorouracil (the latter drug does carry a risk of fatal neurotoxicity), but included 16 dogs total so not very convincing for owners). There were more long-term survivors in another small German study if chemotherapy was used, although the median survival rate did not change. A very recent study found that dogs receiving carboplatin and cox-2 inhibitor had longer survivals than comparable patients treated by surgery alone.(13)

Beyond that, we are left with telling owners that there are some intriguing early results, and that the efficacy of chemotherapy in women means that we most likely would help dogs with advanced stage disease by using chemotherapy.

REFERENCES

[1.] Beauvais W, etal. J Small Anim Pract 2012;53(6):314. [2.] Sonnenschein EG, etal. Am J Epidemiol 1991;133:694. [3.] Alenza DP, etal. J Vet Intern Med 1998;12:132. [4.] Philibert JC, etal. J Vet Intern Med 2003;17(1):102. [5.] Sorenmo KU, etal. J.Vet Intern Med 2000;14: 266. [6.] Yamagami T, etal. J Vet Med Sci 1996;58(11):1079. [7.] Chang SC, etal. J Am Vet Med Assoc 2005;227(10):1625. [8.] Kurzman ID, Gilbertson SR. Semin Vet Med Surg (Small Anim) 1986;1(1):25. [9.] Pena L, etal. Vet Pathol 2013;50(1):94. [10.] Pena LL, etal. J Vet Diagn Invest 1998;10(3):237. [11.] Zuccari DA, etal. Vet Clin Pathol 2004;33(1):23. [12.] Morris JS, etal. Vet Rec 1998;142(24):656. [13.] Lavalle GE, etal. In vivo 2012;26(3):375.

IMMUNODYSFUNCTION AND HOW TO MODIFY IT

Sandra Axiak-Bechtel, DVM, DACVIM (Oncology)
Columbia, MO

INTRODUCTION

The immune system plays a dual role in the development of cancer. The defensive role of the immune system is to fight and cure infectious agents that cause cancer, help resolve inflammation that can promote cancer, and lastly recognize and kill tumor cells when they form. It is therefore not surprising that underlying immune dysfunction exists in cancer patients, both human and veterinary. The immune system may also play a pro-tumor formation role, and chronic inflammation in the presence or absence of infection can promote carcinogenesis when unregulated. Furthermore, the immune system can support tumor formation by selecting cancer cells that can best survive in an immunocompenent host. Then, the tumor itself evades the immune system and utilizes the immune system to its advantage by secreting molecules that result in immunosuppression and protect it from the body's natural defenses. Based on this knowledge, a further understanding of the immune system's role in cancer is important prior to utilizing targeted therapies that can manipulate immunity back to the advantage of the host.[1]

IMMUNE DYSFUNCTION AND LOSS OF IMMUNOSURVEILLANCE

The cellular component of the innate immune system is comprised of neutrophils, macrophages, dendritic cells (DCs), and natural killer cells (NKs); these are rapidly acting and not specific. The adaptive arm of the immune system is quite specific to the insult, and responds more robustly when exposure is repeated; adaptive immunity consists of T and B lymphocytes. Both arms of the immune system are critical for defense against cancer.

Immune surveillance is the system by which the immune system recognizes cancer cells as foreign, and triggers their elimination. The proof of this concept lies in murine model studies, where mice lacking interferon-γ (IFNγ) responsiveness or adaptive immunity are more susceptible to both spontaneous and carcinogen induced cancer. This theory is further supported by findings in other species with underlying immunodysfunction and increased cancer risk. For example, dogs with immune mediated thrombocytopenia have a higher incidence of lymphoma compared to dogs without immune mediated thrombocytopenia, and cats with renal transplantation and long term cyclosporine administration had a six fold higher risk of developing lymphoma.[2,3] Dogs with cancer have reduced neutrophil oxidative burst, NK proliferation, lymphokine activated killer function, and blunted inflammatory response to bacterial pathogen associated molecular patterns compared to normal dogs.[4-6] Furthermore, dogs with cancer have increased percentages of T regulatory cells compared to normal dogs, particularly dogs with carcinomas.[7]

Tumors that form in a deficient immune host are more immunogenic than tumors that form in the immunocompetent. In fact, the immune system may help shape clonal selection and the immunogenicity of the tumor itself. For example, the immune system can promote tumor formation by selecting for cells best adapted to survive in a normal immune system environment or by creating conditions within the tumor microenvironment that facilitate tumor growth. Cancer immunoediting refers to the elimination of cancer by the immune system, the selection of less immunogenic tumor cells during the anti-tumor response, and/or evasion of the tumor from the immune system. Mechanisms of tumor evasion of the immune system are loss of tumor antigens, down-regulation of antigen presenting molecules, and tumor resistance to cytotoxic pathways including over-expression of anti-apoptotic molecules. Tumors secrete immunosuppressive molecules such as transforming growth factor –β (TGF-β) and interleukin-10 (IL-10). TGF-β binds to its receptors on lymphocytes, altering the lymphocyte phenotype and therefore its cytokine secretion profiles. These cytokine profiles are implicated in the generation of T regulatory cells that inhibit other T cells. IL-10 binds to its receptor expressed on immune cells to upregulate genes responsible for preventing maturation of DCs; furthermore, IL-10 has direct effect on CD4+ T cells and inhibits their proliferation and their production of cytokines. Further study evaluates the ability of cancer to cause immunosuppression to improve its chances of survival.[1] The function of myeloid cells (monocytes, macrophages, DCs) when exposed to soluble factors excreted from canine tumor cells was significantly deceased; phagocytic activity was blunted leading to decreased tumor antigen uptake, decreased MHC class II expression lead to decreased tumor antigen presentation, and diminished stimulation of the adaptive immune system due to decreased CD80 expression was found.[8]

THE ROLE OF INFLAMMATION IN CANCER FORMATION

Inflammation is an important physiological response against injury or infection; inflammation is also highly regulated and short lived in those with intact immune systems with the help of anti-inflammatory mediators. However, when inflammation is not well controlled and becomes chronic, it contributes to carcinogenesis.

A cell component of the chronic inflammatory response is macrophages, producing tumor necrosis factor –α (TNF-α). TNF-α will coordinate the inflammatory response by inducing a range of effector molecule release, and many of these effector molecules are proteins that perpetuate inflammation. Furthermore, leukocytes produce reactive oxygen species

(ROS) and nitric oxide (NO) to help fight infection, however, these both cause DNA damage (through the formation of peroxynitrite. So chronic inflammation, leading to increased ROS and NO production, leads to an increase in DNA damage and increased DNA mutation rates. The increase in DNA mutations rates leads to an increased risk of cancer. [1,9]

Several pro-inflammatory products, including TNF-α, interleukins, and chemokines, mediate processes that are known to be critical in tumor formation: cellular proliferation, apoptosis, angiogenesis, and metastasis. The final link between inflammation and cancer is the transcription factor NF-KB, a transcription factor that regulates the expression of genes that encodes for these pro-inflammatory products. NF-KB is induced by macrophages, target cells of inflammation, and cancer cells. It will inhibit apoptosis, promote angiogenesis, and promote metastasis. In addition to inducing anti-apopotic protein production (Bcl-x_L, c-Flip), promoting angiogenesis (vascular endothelial growth factor), and various cytokines, NF-KB activates cyclin D, which stimulates the cell to progress through the cell cycle and expression of pro-inflammatory genes, such as cytokines and COX-2. Overall, NF-KB helps maintain the inflammatory response and promotes metastasis.[1]

Cancers linked to chronic inflammation are ocular sarcoma and injection site sarcoma in the cat, urinary bladder transitional cell carcinoma, squamous cell carcinoma, and osteosarcoma in the dog, and lymphoma in the dog and cat. Cancers linked to chronic infection in the cat are lymphoma (FeLV), sarcoma (FeSV), sarcoid (bovine papilloma virus), and squamous cell carcinoma (papilloma virus). In the dog, 2 cancers are associated with chronic infection: esophageal sarcoma (Spirocirca lupi) and squamous cell carcinoma (papilloma virus). In addition, there is a suspected link between lymphoma and inflammatory bowel disease in both the dog and cat.[9]

NON-SPECIFIC IMMUNE STIMULATION

Biologic response modifiers

The goal of non-specific immune activation is to stimulate the innate and adaptive immune system to recognize tumor cells as non-self. Immune stimulators are, for the most part, innate immune stimulators as increased antigen presenting cell activity (DCs, macrophages) will lead to more effective T and B cell responses.[10] Immunotherapy must overcome many of the mechanisms of tumor escape described above, however, immunotherapy has been successfully carried out and shown efficacy, alone and in combination with other therapies. Evidence of a blunted immune system and overcoming this is described in dogs with osteosarcoma. Dogs with localized osteomyelitis following a limb-sparing procedure had nearly double the survival time of dogs with no infection following surgery.[11] A follow up study using the murine model of canine osteosarcoma revealed that bacterial osteomyelitis elicited non-specific tumor growth inhibition mediated by NKs, inflammatory monocytes, and tumor associated macrophages.[12] These results are encouraging for the development of future immunotherapies, and, in fact, systemic immune stimulation is successful in the treatment of other canine cancers.

The bacillus of Calmette and Guerin (BCG) is a modified strain of *Mycobacterium bovis*. Infusion of BCG into the bladder is an integral treatment for human non-invasive bladder cancer; unfortunately, dogs and cats develop invasive bladder cancer and BCG is ineffective. However, intralesional BCG has been used successfully in the treatment of equine sarcoids and bovine ocular squamous cell carcinoma.[10] Immunocidin™ is a formulation of mycobacterial cell wall is UDA licensed for use intra-lesionally in canine mammary cancer. A large clinical trial has not been published at the time of these proceedings, but preliminary data from the licensing study appear promising.

Liposome-encapsulated muramyl tripeptide (L-MTP-PE) is composed of mycobacterial cell wall components encapsulated in a liposome. It activates monocytes and macrophages with the result of increased production of pro-inflammatory cytokines, include IL-1, IL-6, IL-7, IL-8, IL-12, and TNF-α. When used intravenously following amputation in stage II canine osteosarcoma as the sole adjuvant treatment, survival was significantly prolonged compared to dogs not treated with adjuvant therapy. The response with combined L-MTP-PE and cisplatin chemotherapy post-op showed no benefit when administered concurrently. However, dogs treated with surgery and cisplatin/ L-MTP-PE sequentially did have a survival advantage compared to those dogs receiving adjuvant cisplatin alone. A survival advantage was also found in dogs diagnosed with splenic hemangiosarcoma – following surgery, dogs treated with doxorubicin, cyclophosphamide, and L-MTP-PE had higher serum levels of TNF-α and IL-6 and longer survival compared to dogs treated with surgery, chemotherapy, and empty liposome. Availability of this product, however, is limited.[10]

Liposomal DNA complexes are composed of bacterial DNA motifs which stimulate the activation of DCs. Antitumor activity is mediated by NKs, DCs, and the release of pro-inflammatory cytokines. These complexes can also be used for gene delivery, and when administered IV with a gene encoding for IL-2 to dogs with pulmonary metastatic osteosarcoma, potent immune activation and NK cell activity was noted in dogs, with an increase in overall survival.[13]

Oncolytic viruses are viruses that preferentially replicate within cancer cells, and result in lysis of the tumor cells. Oncolytic viruses are tumor selective and can directly kill tumor cells; additionally, the y can be used for targeted delivery of drugs, cytokines or genes. Currently, canine adenovirus and canine distemper virus are under investigation for this technique.[10]

Cytokine Therapy

Interleukin-2 activates DCs, macrophages, and B cells in addition to inducing local expansion of T cells and stimulating NK cells. Therefore, it plays an important role in immune system stimulation and is a promising therapeutic intervention. Although side effects are considerable when IL-2 is used in people, IL-2 has been used with success in the treatment of canine cancer. For example, in dogs with carcinoma of the urinary bladder or urethra, intralesional IL-2 was administered into the mass or post-operatively into the tumor bed, with some responses.[14] Intravenous gene delivery using liposome-DNA IL-2 complexes is safe in canine osteosarcoma and systemic immune activation resulted; 3 of 20 dogs had partial or completion regression of lung metastases [13]; an inhalational formulation of liposomal IL-2 was also found to be safe, with some evidence of preliminary efficacy[15].

Interleukin 12 (IL-12) also has pronounced stimulatory effects on both the innate and adaptive immune system, in addition to having an anti-angiogenic effect. IL-12 and IL-15, which stimulates NK cells and promotes proliferation of T cells, are under investigation for the treatment of canine cancer.

THE ADAPTIVE IMMUNE SYSTEM

The adaptive immune system detects tumors through tumor associated antigens, recognized by cytolytic T cells and antibodies. Tumor associated antigens can be specific for a particular tumor, arise from a mutated gene product, or they can be normal cellular antigens that are overexpressed. The ultimate goal of a tumor vaccine is an antitumor response that results in regression of the tumor. Because the adaptive immune system takes longer to induce than the innate immune system, a response to cancer vaccination may take several months. The currently available vaccine for canine cancer treatment is Oncept® for the treatment of melanoma. This particular vaccine is a xenogeneic DNA vaccine; each dose contains plasmid DNA with the gene encoding for human tyrosinase. This is administered transdermally every 2 weeks for 4 treatments, then once every 6 months. Studies have shown that this treatment is safe, and is promising for prolonging survival in dogs with loco-regional disease control (surgery and/ or radiation therapy).[16]

Other approaches to tumor vaccination are the use of whole tumor cell and tumor cell lysate vaccines (these are made directly from the patient's tumor cells or from a cell line of the same tumor type and administered with an adjuvant for immune stimulation. This has shown promise in combination with standard care in canine hemangiosarcoma and canine lymphoma. Dendritic cell vaccination is also under investigation due to their potent antigen presenting ability. Monoclonal antibodies, commonly used in the treatment of cancer in people as an adjunct to standard therapy, are not yet developed for use in veterinary medicine.[16,17]

CONCLUSIONS

The immune system is defective in patients with cancer, and both the innate and adaptive arms of the immune system are affected. Current research is further elucidating mechanisms of immune system dysfunction in dogs and cats with cancer. Promising research is ongoing to determine how we can manipulate the immune system, and, with combined our current standards of care, improve the prognosis of dogs and cats with cancer.

REFERENCES

[1]Vesely MD et al. **Annu Rev Immunol** 2001; 29: 235-71. [2]Keller **Cancer** 1992; 70: 2334-2337. [3]Schmiedt et al **Vet Comp Onc** 2009; 7(1): 45-53. [4]LeBlanc et al. **Vet Clin Path** 2010; 39 (2): 157-63. [5]Funk J et al. **Cancer Immunol Immunother** 2005; 54: 87-92. [6]Fowler BL et al. **Vet Immuno Immunopath** 2011; 144: 167-171. [7]O'Neill K et al. **J Vet Intern Med** 2009; 23: 875-881. 8Wasserman J et al. **Vet Immuno Immunopath** 2012; 145: 420-430. [9]Morrison WB. **J Vet Intern Med** 2012; 26: 18-31. 10 Biller BJ. **Vet Clin North Am Small Ani Prac** 2007; 1137-1149. [11]Duncan B et al. **Annals of Surgical Oncology** 2005; 12(12): 1073-1083. [12]Sottnik JL et al. **Cancer Immunol Immunother** 2010; 59: 367-378. [13]Dow et al **Hum Gene Ther** 2005 16(8) 937-46. **14**Konietschke U et al **In Vivo** 2012 26(6) 931-5. [15]Khanna C et al **Cancer** 79 1997 1409-1421. [16] Bergman PJ **Vet Clin Small Anim** 2010; 40: 507-518. 17 Guth AM et al **Small Animal Clinical Oncology** 5th ed 2013; 198-214.

CHEMOTHERAPY CLASH: MAXIMUM TOLERATED DOSE VS. METRONOMIC PROTOCOLS

Anthony J. Mutsaers, DVM, PhD, DACVIM (Oncology)
Guelph, ON, Canada

INTRODUCTION

Conventional cytotoxic chemotherapy drugs have been the corner stone of medical oncology practice for many decades. Exactly the same drugs are utilized in both human and veterinary oncology, as well as the same principles of administration, with the caveat that tolerance for drug-related toxicity is generally less in the veterinary community. Recently, preliminary clinical trials of low dose continuous "metronomic" chemotherapy administration protocols have been reported. Does this represent a significant departure from conventional drug dosing in veterinary oncology? Indeed, many questions remain to be answered regarding how best to utilize chemotherapy drugs in future oncology practice, whether as single agents, or in combination with other cytotoxic or targeted oncology drugs.

MTD Chemotherapy

Conventional dose and scheduling of chemotherapy has generally followed the principle of maximum tolerated dose (MTD), in which the administered dose is defined by the highest allowable toxicity to rapidly dividing normal tissues, such as cells in the bone marrow and intestinal epithelium. Much of the theory put forth to explain the potential benefit of this dosing approach was outlined by Skipper et al in 1970.[1] The experiments conducted to demonstrate a logarithmic relationship between drug dose and cancer cell kill was conducted using cell culture systems. Cancer cell lines grown in optimal conditions of logarithmic growth phase were tested after a constant drug exposure time (Figure 1). While the use of MTD protocols has been successful for the treatment of many cancers, with outright cure achieved for some, critics of the model point out that the artificial cell culture system does not take potential contributions of the tumor microenvironment into account, which is an area from which the modern concept of metronomic chemotherapy dosing was born.

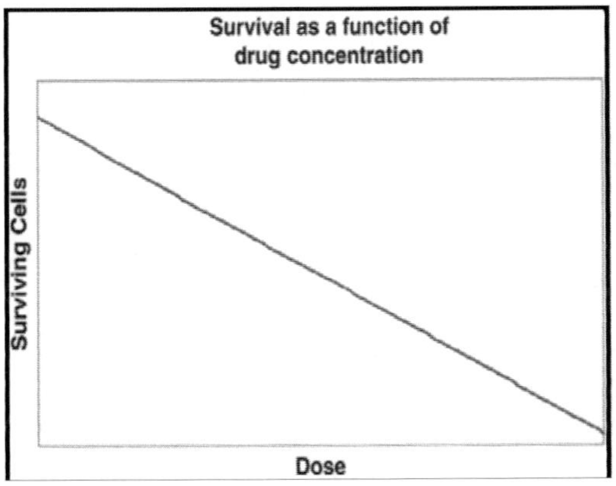

Figure 1. A representation of the log-dose survival curve as part of the basis for MTD chemotherapy dosing, originally detailed by Skipper et al. The surviving fraction (SF) of cells was defined as SF=e-kc where c equals drug exposure. Drug exposure is calculated as CxT (concentration x time) but in this model only drug concentration was considered. Time was held constant.[2]

"Metronomic" Chemotherapy

Because MTD protocols give the highest possible doses that normal tissue toxicities will allow, there is an unavoidable break period between doses to allow for repair of these cells. In veterinary oncology high-grade toxicity is not an acceptable component of treatment and doses are adjusted accordingly, but break periods are built in to protocols to allow for normal tissue repair. The principle goal of metronomic administration of chemotherapy is reduction or elimination of the break period between doses. As such, reduction of each individual dose is required to accomplish the goal of continual administration. The low cost, ease of administration, and generally well-tolerated adverse event profile has made metronomic administration of chemotherapy drugs appealing in veterinary oncology. However, to date formal clinical evaluation of this treatment strategy is in the early stages and a complete understanding of the types of cancers that are sensitive to treatment, as well as the optimal drugs and dosages to utilize requires further study in rigorous clinical trials.

MECHANISMS

As mentioned above, the goal of MTD-based chemotherapy dosing is to kill as many cancer cells as possible. This likely represents a mechanism of action for metronomic dosing as well. However, in addition to a direct tumour cell cytotoxic effect, metronomic administration of certain chemotherapy drugs has been shown to modulate specific aspects of the tumour microenvironment. The two most studied areas within the microenvironment have been a reduction in tumour angiogenesis by various mechanisms, and stimulation of anti-tumour immunity through alterations of T cell lymphocyte subset populations.

Angiogenesis

Belief in the anti-angiogenic mechanisms of low dose cyclophosphamide and other drugs is largely based on studies that demonstrate upregulation of the endogenous angiogenesis inhibitor thrombospondin-1 (TSP-1), downregulation of pro-angiogenic factors such as vascular endothelial growth factor (VEGF), and direct endothelial cell cytotoxicity at very low chemotherapy concentrations.[3] The fact that chemotherapy agents preferentially target rapidly dividing cells makes their cytotoxic effect on the growing tumor endothelium a logical anti-angiogenic mechanism. Indeed, MTD protocols have also been shown to have anti-angiogenic effects through this mechanism. The difficulty with maintaining this effect in the context of MTD protocols is the rapid regrowth of tumor blood vasculature/endothelium during the break period between doses. Similarly, the changes in growth factors such as VEGF also likely occur to some extent with MTD protocols, but are short lived as the body regenerates lost cell populations during the break period.

Immune Modulation

Immunotherapeutic mechanisms for metronomic cyclophosphamide are based on reduction in circulating T regulatory cells that suppress the immune response, and possible concomitant increases in T helper subset 17 (Th17) cells, which translates to a boost in anti-tumour immunity.[4] In addition to the role of lymphoid cells to the antitumor response of chemotherapy, another potential contribution could be made by subsets of myeloid cells, such as the myeloid-derived suppressor cell (MDSC), which is of monocytic origin. As with angiogenesis, there is a likely effect of MTD protocols on the relative balance of immune effector cells in the tumor microenvironment but it may be distinct, less consistent, and possibly less efficacious at tumor surveillance when compared to metronomic protocols.

It is worth mention that not all metronomic chemotherapy drugs and regimens are likely to be equally efficacious, despite the fact that sustained angiogenesis and immune evasion, are general hallmarks of tumour growth. Different drugs and schedules are associated with differing mechanisms of action.

CLINICAL TRIALS

While there are many reported clinical trials investigating MTD chemotherapy regimens in veterinary oncology, there have been only a few reports of metronomic chemotherapy in a single tumor type, and no randomized trials that compare the two strategies head-to-head. Of the many cytotoxic chemotherapy drugs available, metronomic administration of alkylating agents, such as cyclophosphamide, chlorambucil, and lomustine, have arguably received the most investigation, and preliminary studies of metronomic schedules with each of these drugs have been reported.[5-9] Metronomic cyclophosphamide with piroxicam has been shown to delay or prevent the regrowth of incompletely excised canine soft tissue sarcomas.[8] A metronomic "chemo-switch" regimen of cyclophosphamide and etoposide may also have activity in canine hemangiosarcoma.[9] This study included a historical comparison group of cases treated with MTD single agent doxorubicin.

The list of chemotherapy agents that may have utility when administered in a metronomic fashion likely extends beyond alkylating agents. The ability to provide frequent administration, most often necessitating oral dosing, is a key component to drug suitability. Platinum agents, taxanes, anthracyclines, antimetabolites, among other drug families, are potential candidates for exploration of metronomic scheduling, but each requires proper evaluation. Slow release formulations of platinum agents, such as the OPLA platform, could represent a form of metronomic delivery.[10] A single subcutaneous release of carboplatin has also been recently reported for the treatment of canine osteosarcoma,[11] and as an oral formulation, satraplatin could also be potentially administered in a metronomic schedule.[12]

COMBINATION THERAPY

As is often the case in oncology, combination therapy approaches may result in superior efficacy. As more clinical results are reported, we will gain an increased understanding of drug combinations and schedules that will be most beneficial to patients. Many early metronomic clinical trials have evaluated combinations of drugs, usually a chemotherapy drug coupled with a non-steroidal anti-inflammatory agent such as piroxicam. This practice may hamper the ability to properly evaluate relative efficacy if these agents/doses/schedules have not first been evaluated clinically in a single agent context. The use of low dose metronomic chemotherapy protocols does not necessarily equate to an abandonment of the principles of MTD administration, as these approaches have been applied in combination in preclinical studies and anecdotally in veterinary oncology. Metronomic chemotherapy administration may also be a promising approach to combination therapy with emerging targeted agents, such as receptor tyrosine kinase inhibitors. Both strategies utilize drugs given in a chronic fashion at optimal biologic doses that may be considerably less than the maximum tolerated dose. Finally, the break period between MTD drug administrations may represent a unique opportunity to intervene with metronomic and/or targeted agents to specifically inhibit the compensatory mechanisms induced within the tumor microenvironment, leading to a superior outcome, however this theory has yet to be tested in a clinical setting.

SIDE EFFECTS

Conventional MTD chemotherapy protocols have not necessarily been designed for, or evaluated after, chronic application. Low dose metronomic protocols have generally been well tolerated in the short term. However, when using a metronomic treatment strategy, clinicians must be mindful of side effects that may only occur after prolonged drug exposure. The most concerning side effect encountered to date has been sterile cystitis with cyclophosphamide use, although the occurrence has not been reported to be higher than that observed with conventional protocols.

BIOMARKERS FOR PATIENT SELECTION AND MONITORING

Much work is required in the area of biomarkers to predict optimal biologic dose, and cancer types or subtypes that may benefit most from both MTD and metronomic chemotherapy protocols. Candidate biomarkers that have received preliminary evaluation, both clinically and pre-clinically, include plasma VEGF levels, tumour vascular density, as well as circulating endothelial cells and their precursor cells (CEC/CEP). Veterinary evaluation of these potential biomarkers has begun, and functional imaging studies, such as dynamic contrast MRI and PET-CT may also be useful.

DRUG RESISTANCE

The challenge of multi-drug resistance is well established in the use of MTD chemotherapy protocols. Resistance to metronomic therapy is also likely to be a significant problem. This is true despite early predictions that anti-angiogenic and immunotherapy approaches should evade resistance by targeting genetically stable endothelial cell and lymphocyte populations, respectively. Use of a "chemo-switch" regimen, where exposure to one drug is alternated with another, is one strategy aimed at addressing this problem. A recent preclinical study with prostate cancer xenografts has suggested that onset of resistance to continuous low doses of cyclophosphamide does not select for resistance to the same drug administered at the maximum tolerated dose. This research is welcome news to those concerned that the chronic application of lower chemotherapy doses would easily prime tumor cells for drug resistance.[13]

REFERENCES

[1]Skipper HE, et al. Cancer Chemother Rep 1970;54(6):431. [2]Kamen B. J Pediatr Hematol Oncol, 2006;28(6):325. [3]Mutsaers AJ. Small Animal Clinical Oncology. 5[th]ed, 2013:231. [4]Burton JH, et al. J Vet Intern Med 2011;25:920. [5]Leach TN, et al. Vet Comp Oncol 2012;10:102. [6]Marchetti V, et al. Invest New Drugs 2012;30(4):1725. [7]Tripp CD, et al. J Vet Intern Med 2011;25:278. [8]Elmslie RE, et al. J Vet Intern Med 2008;22:1373. [9]Lana S, et al. J Vet Intern Med 2007;21:764. [10]Withrow SJ, et al. Ann Surg Oncol 2004;11:705. [11]Simcock JO, et al. J Am Vet Med Assoc 2012;241(5):608. [12]Selting KA, et al. J Vet Intern Med 2011;25:909. [13]Emmennegger U, et al. Neoplasia 2011;13(1):40.

CUTTING EDGE TREATMENTS (SCALPEL NOT REQUIRED)

Anthony J. Mutsaers, DVM, PhD, DACVIM Oncology
Ontario Veterinary College, University of Guelph, Canada

Receptor Tyrosine Kinase Inhibitors

The arsenal of the medical oncologist expanded significantly with the relatively recent approval of the receptor tyrosine kinase inhibitors (RTKIs) toceranib (Palladia®) and masitinib (Masivet®, Kinavet®) for veterinary use in treatment of aggressive canine mast cell tumours. These drugs are examples of "targeted" cancer therapy, which began in human oncology with clinical use of drugs such as imatinib (Gleevec®), trastuzumab (Herceptin®), and bevacizumab (Avastin®), which were approved for chronic myelogenous leukemia, breast cancer, and colon cancer, respectively. Now, over 10 years later, there are numerous other RTKI drugs in clinical use, and potentially more that could come to veterinary oncology as well.

Receptor tyrosine kinases are transmembrane proteins that receive signals from extracellular growth factors that bind to the receptor. The intracellular portions of these RTKs contain kinase domains, which phosphorylate tyrosine residues facilitating the propagation of the signal by attracting various intracellular proteins to bind. The signal ultimately culminates in the nucleus and alters gene transcription in favour of cell growth, survival, or other processes. Normal RTK function becomes altered or dysregulated in a number of cancers. In canine mast cell tumours the RTK c-kit may be mutated resulting in constitutive signalling. Drugs such as imatinib, toceranib, and masitinib are able to bind to c-kit and abolish the signal. While these drugs have been used as single agents for the treatment of aggressive recurrent canine mast cell tumours, they may also have activity in other cancer types. Toceranib has shown activity in cancers such as canine anal sac adenocarcinoma, osteosarcoma, thyroid, nasal and oral carcinomas. The mechanism(s) underlying this activity remain to be determined.

The other major category of targeted drugs for RTKs is monoclonal antibodies, such as trastuzumab and bevacizumab. The extracellular target of these drugs may be highly conserved between dogs and man, however use of antibodies as therapeutics in veterinary oncology has been hampered by the fact that animals will mount a neutralizing immune response, which will negate their anti-cancer effects. However, as biotechnology techniques advance, it is becoming possible to manufacture canine-specific versions of these antibody drugs, and clinical assessment of monoclonal antibody therapy in veterinary oncology may be on the horizon.

Cancer Immunotherapy

There has been a long history of attempting to harness the body's own immune defences to attack malignant cancer cells. Historically, veterinary studies have investigated the potential biologic and therapeutic role of immune cytokines such as interleukins, and non-specific immunostimulation from compounds such as Muramyl Tripeptide Phosphatidylethanolamine (Mifamurtide®). More recently, the focus has shifted to determining the role of T cell subsets in immune surveillance, and designing cancer vaccines based on unique tumour antigens, particularly for treatment of canine malignant melanoma.

Within the T cell family of lymphocytes exists subsets that both enhance and suppress the overall immune response. The presence of a tumour is associated with elevation in T regulatory cells that suppress immune attack, and this has been demonstrated in dogs. Interestingly, low doses of cyclophosphamide, given in a continuous/metronomic fashion have been shown to decrease the fraction of T regulatory cells, which may contribute to the mechanism of action of this form of chemotherapy.

Tumour vaccination has also become a clinical reality with the recent approval of a vaccine (Oncept®) to treat canine stage II or III oral malignant melanoma in the adjuvant setting. This vaccine uses human DNA for the melanoma-specific antigen tyrosinase and is administered using a special transdermal device. It is important that local control of the primary tumour be accomplished with surgery and/or radiation therapy as the vaccine is only effective for suppression of microscopic metastases. Anti-cancer activity has also been reported with the use of other canine melanoma vaccines, which supports the notion that melanoma in particular is an "immunogenic" cancer worthy of this therapeutic approach.

High Dose Radiation Therapy

The capabilities of external beam megavoltage radiation therapy units continue to expand with technological advances that have real clinical application. While availability is currently limited, these machines are finding their way into an increasing number of veterinary cancer centres. Linear accelerators capable of delivering very focused radiation that spares normal tissue through techniques such as tomotherapy, intensity-modulated delivery, gamma knife, and stereotactic radiosurgery permit changes to conventional fractionation protocols. Currently, frequent low doses of radiation are delivered in curative intent protocols because the dose of radiation given to "late-responding" normal tissues such as bone and nerves must be limited to prevent irreparable damage. With newer technology comes the ability to deliver the same or higher total dose in only one or a few fractions. It has been demonstrated that canine osteosarcoma and transitional cell carcinoma tumours are more sensitive to high dose radiation than conventional fractionated protocols, and preclinical studies may determine whether this is the case for other cancers as well, by calculating their alpha/beta ratio. Finally, some newer radiation therapy units contain on-board imaging CT for modifying the table position prior to each dose to compensate for patient movement, increasing treatment precision. There is no doubt that availability to the latest types of radiation therapy in veterinary medicine will continue to expand as oncology clients continue to request access to the latest treatment options for their pets.

Minimally Invasive Procedures

The emergence of techniques that utilize laparoscopy or interventional radiology has decreased patient surgical morbidity and expanded the number/type of cancer treatment options available in veterinary oncology. Laparoscopic biopsies offer a less invasive approach than surgical laparotomy while still allowing direct visualization and manipulation of the tumour, including direct control of potential haemorrhage, unlike image-guided biopsy techniques. Laparoscopy has also been utilized to perform pericardectomy for cases of right atrial hemangiosarcoma with pericardial effusion.

The use of stents for management of neoplastic occlusion of the urethra, ureters, colon, and esophagus has also become available recently. This treatment can prevent life-threatening occlusion, but provides palliation only and does not treat the tumour directly. Finally, image guidance, such as fluoroscopy-guided tumour vascularity studies can facilitate the accurate delivery of chemotherapy at high doses or embolization agents directly into the tumour via feeder arteries. Chemoembolization has been utilized in liver cancer, which is a tumour that is very resistant to conventional chemotherapy in both humans and dogs.

Selected References

London CA. Tyrosine kinase inhibitors in veterinary medicine. *Vet Clin North Am Small Anim Pract* 2009; 24(3):106-12.

London CA, Malpas PB, Wood-Follis SL, et al. Multi-center, placebo-controlled, double-blind, randomized study of oral toceranib phosphate (SU11654), a receptor tyrosine kinase inhibitor, for the treatment of dogs with recurrent (either local or distant) mast cell tumor following surgical excision. *Clin Cancer Res* 2009;15(11):3856-65.

Hahn KA, Ogilvie G, Rusk T, et al. Masitinib is safe and effective for the treatment of canine mast cell tumors. *J Vet Intern Med* 2008;22:1301-9.

Bergman PJ. Cancer Immunotherapy. *Vet Clin North Am Small Anim Pract* 2010; 40(3):507-18.

Biller BJ, Guth A, Burton JH, Dow SW. Decreased ratio of CD8+ T cells to regulatory T cells associated with decreased survival in dogs with osteosarcoma. *J Vet Intern Med* 2010;24(5):1118-23.

Lawrence JA, Forrest LJ. Intensity-modulated radiation therapy and helical tomotherapy: its origin, benefits, and potential applications in veterinary medicine. *Vet Clin North Am Small Anim Pract* 2007; 37(6):1151-65.

Farese JP, Milner R, Thompson MS, et al. Stereotactic radiosurgery for treatment of osteosarcomas involving the distal portions of the limbs in dogs. *J Am Vet Med Assoc* 2004;225(10):1567-72.

Berent AC, Weisse C, Beal MW, et al. Use of indwelling, double-pigtail stents for treatment of malignant ureteral obstruction in dogs: 12 cases (2006-2009). *J Am Vet Med Assoc* 2011;238(8):1017-25.

Weisse C, Berent A, Todd K, et al. Evaluation of palliative stenting for management of malignant urethral obstructions in dogs. *J Am Vet Med Assoc* 2006;229(2):226-34.

Wiesse C. Hepatic chemoembolization: a novel regional therapy. *Vet Clin North Am Small Anim Pract* 2009; 39(3):627-30.

BLADDER MATTERS.
TREATMENT OPTIONS FOR TRANSITIONAL CELL CARCINOMA

Anthony J. Mutsaers, DVM, PhD, DACVIM Oncology
Ontario Veterinary College, University of Guelph, Canada

While bladder cancer is not common in the dog, transitional cell carcinoma is by far the most common form of this disease, and clinicians are detecting cases at earlier stages than in the past. Risk factors for development of TCC include female gender, obesity, topical insecticide exposure, and certainly breed. In North America many terrier breeds are overrepresented including Scottish, West Highland White, and Wire Haired Fox. Beagles and Shetland sheepdogs also appear to be predisposed. Stranguria, hematuria and pollakiuria are common clinical signs. Transient or lack of response to appropriate antibiotic therapy for treatment of chronic urinary tract infections is also a frequent historical finding. Cystocentesis may result in seeding tumour cells, so urine is generally collected by catheterization (may be difficult with the presence of an obstructive mass) or free catch when TCC is suspected. Urinalysis may reveal neoplastic cells in the sediment of approximately 30% of dogs with TCC. Bladder tumour antigen tests have been evaluated in dogs, with low specificity for TCC in the presence of other urologic diseases. However, the sensitivity was considered acceptable, and therefore potential screening capabilities exist for highly TCC-susceptible breeds.

Staging and Treatment

The most common location within the bladder for TCC is the trigone, which presents difficulty for complete surgical removal, and predisposes to neoplastic obstruction of the urethra and/or ureters. Conventional imaging studies to assess tumour burden include double contrast cystography and urethrography, however trans-abdominal ultrasound is now more frequently utilized. Standardization of bladder size for sequential ultrasound measurements to assess tumor response to medical treatment is challenging, however techniques have been developed and used for TCC chemotherapy clinical trials. Ultrasound is also useful for assessment of tumour involvement of sublumbar lymph nodes and/or abdominal organs. While the majority of TCC cases do not present with obvious metastatic extension to local lymph nodes, lungs or bone, regional and/or distant metastasis will develop in approximately half of cases, and this rate is likely higher for cases that achieve good local tumour control.

Despite the localized nature of most tumours at the time of diagnosis, TCC has typically been treated medically. Surgical removal has generally been considered not feasible due to the common trigone and urethral tumour locations. However, recently veterinary oncologic surgeons have become more interested in attempting total cystectomy with transposition of ureters to the remaining urethra in select cases. Radiation therapy for TCC treatment has also not been a conventional choice due to the side effect profile to normal tissues in the field, such as urethra and colon. These side effects would include colitis and urethritis during and shortly after treatment, with the possibility of stricture in the longer term. As with surgery, radiation therapy is now being utilized more commonly for local control, in both definitive and palliative protocols, likely due to improvements in sparing of normal tissues that surround the bladder, recognition of optimal treatment protocols, and the technology to deliver them. Finally, obstruction of the urethra has conventionally been managed quite successfully with placement of a cystostomy tube. However, recently the placement of urethral or ureteral stents to treat obstruction through an interventional minimally invasive approach is becoming more widely available.

Medical treatment of TCC has included chemotherapy and non-steroidal anti-inflammatory drugs (NSAIDs), such as piroxicam and now other, more cyclooxygenase-2 (COX-2) selective versions, such as deracoxib. At the moment it is not clear how important or relevant COX selectivity may be to the likelihood of treatment response, nor the relative superiority/inferiority of "newer" NSAIDs for TCC treatment. The exact role of COX enzymes in the biology of TCC is an area of ongoing investigation, but previous studies have implicated these enzymes in TCC growth, angiogenesis and apoptosis. The most common chemotherapy protocols have been mitoxantrone, carboplatin, vinblastine, and a few others, including metronomic chlorambucil, often in combination with an NSAID. Nephrotoxicity from platinum drugs in combination with piroxicam limits their application to TCC, despite early studies of high efficacy.
In addition to potential nephrotoxicity from NSAID treatment and chemotherapeutics such as carboplatin, ascending pyelonephritis and obstructive hydronephrosis also pose risks to kidney function during TCC management.

Selected References

Mutsaers AJ, Widmer WR, Knapp DW. Canine transitional Cell Carcinoma. *J Vet Intern Med* 2003;17(2):136-44.

McMillan SK, Boria P, Moore GE, et al. Antitumor effects of deracoxib treatment in 26 dogs with transitional cell carcinoma of the urinary bladder. *J Am Vet Med Assoc* 2011;239(8):1084-9.

Arnold EJ, Childress MO, Fourez LM, et al. Clinical trial of vinblastine in dogs with transitional cell carcinoma of the urinary bladder. *J Vet Intern Med* 2011;25(6):1385-90.

Henry CJ, McCaw DL, Turnquist SE, et al. Clinical evaluation of mitoxantrone and piroxicam in a canine model of human invasive urinary bladder carcinoma. *Clin Cancer Res* 2003;9(2):906-11.

Nolan MW, Kogan L, Griffin LR, et al. Intensity-modulated and image-guided radiation therapy for treatment of genitourinary carcinomas in dogs. *J Vet Intern Med* 2012;26(4):987-95.

SOFT AND HARD TISSUE SARCOMAS

Anthony J. Mutsaers, DVM, PhD, DACVIM Oncology
Ontario Veterinary College, University of Guelph, Canada

Canine Soft Tissue Sarcomas

Despite a multitude of names, based on their presumed cells of origin, these tumours are generally considered collectively with regard to prognosis and treatment options (with the notable exception of hemangiosarcoma). These are tumours of mesenchymal tissues, including muscle, adipose, fascial, fibrous, and neurovascular origin. It is sometimes difficult for pathologists to determine the exact histogenesis of the tumour, and immunohistochemical procedures using antibodies to various tissue markers are utilized in this situation.

Treatment of canine soft tissue sarcomas is primarily by surgical removal that requires extensive margins for cure. These tumours often have poorly defined margins, and are very locally invasive. Pre-surgical planning with advanced imaging such as MRI or CT is recommended to assess the degree of tissue infiltration prior to surgery. These tumours generally have a low metastatic rate (collectively seen as less than 20% risk, but increased likelihood with higher grade), however as our ability to manage local control continues to improve, the long-term metastatic rate may begin to increase.

In addition to surgery, there is a key role for radiation therapy in the adjuvant setting for cases with incomplete surgical margins. In cases with very large tumours, radiation therapy has also been applied in the neo-adjuvant setting (prior to surgery) to shrink the margins and increase the likelihood of surgical success. Soft tissue sarcomas tumours are generally considered much more resistant to radiation applied to gross disease, and as a single form of treatment, radiation is considered only palliative. The success of radiation decreases dramatically to approximately 50% response rates when applied in the gross disease setting. The role of chemotherapy for local control is not established, and likely to be of minimal benefit when delivered in conventional protocols. However, there is recent interest in treatment with metronomic cyclophosphamide and piroxicam for prevention of tumour regrowth in dogs with dirty surgical margins and no other ancillary therapy.

The accurate prediction of late metastasis is facilitated by tumour grade, and as in other tumours, mitotic index appears to play a role in prognosis. It is worthy to note that there may occasionally be discrepancy between the grade assigned at presurgical biopsy and review of the surgical specimen, so it is worth cautioning clients about this possibility if treatment decisions were based on a presurgical biopsy report. Overall, if assessed and treated appropriately, the prognosis for canine soft tissue sarcomas is good. Management of tumours that recur after the first surgery is more difficult than the original primary tumour.

Feline Soft Tissue and Injection Site Sarcomas

These two types of sarcoma, while of the same histologic origins, are very different clinical entities. Feline soft tissue sarcomas are similar to their canine counterparts in terms of growth pattern, metastatic rate, and response to treatment (surgery and radiation). However, sarcomas that are linked to injection site reactions have a more aggressive behaviour and are typically of a higher histologic grade. Many critical growth factors have been documented to be associated with the characteristic inflammatory infiltrate, fibroblasts, and myofibroblasts associated with these tumours. The vaccine associated feline sarcoma task force recommends removal of post vaccinal masses still growing after 4 weeks, still evident after 3 months, or > 2 cm in diameter. In order to have the best chance of local control of these tumours a combination of both wide surgical resection and curative intent radiation therapy are required. Removal of bone, such as dorsal spinous processes and/or scapulae, is commonly recommended for interscapular locations. Recently, the recommended minimum 2-3 cm surgical margin has been challenged by a study of 100 cats treated with 5 cm margins and 2 fascial planes deep, with good outcome.

Canine Hard Tissue Sarcomas

Osteosarcoma is by far the most common primary bone tumour in dogs, but chondrosarcoma, hemangiosarcoma, and fibrosarcoma, and perhaps also lymphoma can exist as more rare primary tumours of bone. Although there is a small early peak age incidence at 1.5-2 years, most dogs are middle age to older and the disease manifests in the appendicular skeleton of large and giant breeds. Small dogs (< 15 kg) are more likely to have osteosarcoma arise in the axial skeleton. The metaphyseal region is the most common location in the bone, and front limbs are affected approximately twice as often as hind limbs. Although not clinically detectable in approximately 85% of cases at diagnosis, over 90% of canine osteosarcoma cases will ultimately die of metastatic disease, with lung being the most common metastatic site. Metastasis to other bone sites is not uncommon as well. Mandibular osteosarcoma has been reported to have a less aggressive biologic behaviour than other anatomic locations. In addition to certain anatomic site-related prognostic factors, elevated serum alkaline phosphatase values have been associated with a shorter disease-free interval and survival in canine osteosarcoma.

Diagnosis of osteosarcoma is frequently suggested based on consistent signalment, history, physical examination, and imaging results. Therefore it is not unusual for primary surgical tumour removal, most often by amputation, to provide the definitive histologic confirmation of the diagnosis. Limb sparing techniques for tumours of the distal radius have become more widely available from specialized surgical oncology centres. Provided the tumour fits criteria for complete removal, limb spare cases have a similar prognosis to amputation cases. Limb spare surgery has been attempted for other anatomic sites, however results showed a high complication rate. Interestingly, limb spare cases that develop post-operative surgical site infection have a better outcome. Radiation with external beam therapy has been used for limb sparing. While the role of radiation therapy is typically palliative for analgesia, the possibility exists that stereotactic delivery of very precise doses may lead to a role for radiation as a definitive treatment choice for select lesions. Another form of radiation therapy using the bone seeking isotope ^{153}samarium has been shown to provide palliation and delay progression of disease. Palliation of bone pain has also been accomplished with systemic bisphosphonate therapy, which deceases osteoclast activity and bone resorption.

Adjuvant cytotoxic chemotherapy is the mainstay of treatment to suppress microscopic metastases following surgical treatment of the primary tumour. Many protocols have been reported with similar efficacies. Carboplatin and doxorubicin are the most commonly utilized chemotherapy agents, alone or in combination. Recently, a study of slow release carboplatin from a subcutaneously placed infusion pump demonstrated similar survival times after 1 treatment as conventional multidose treatment protocols. This form of treatment may represent "metronomic" administration of chemotherapy. Immunotherapy has also shown promise in the past treatment of canine osteosarcoma, using liposome encapsulated muramyl tripeptide-phosphatidylethanolamine (L-MTP-PE). Finally, the molecular targeted agent toceranib (Palladia®), which targets a number of receptor tyrosine kinases that may be active or altered in osteosarcoma, has shown activity in some cases with metastatic disease, which is generally an extremely treatment-resistant clinical scenario.

Feline hard tissue sarcomas

These tumours are rare but usually malignant osteosarcomas. Feline osteosarcomas are reported to have a lower metastatic rate and may be cured by surgery alone. Metastatic disease, while unusual, has been reported. There are no proven efficacious adjuvant chemotherapy protocols for this disease.

Selected References

Perry JA, Culp WT, Dailey DD, et al. Diagnostic accuracy of pre-treatment biopsy for grading soft tissue sarcomas in dogs. *Vet Comp Oncol* 2012 *epub* May 22.

Phelps HA, Kuntz CA, Milner RJ, et al. Radical excision with five-centimeter margins for treatment of feline injection-site sarcomas: 91 cases (1998-2002). *J Am Vet Med Assoc* 2011;239(1):97-106.

Simcock JO, Withers SS, Prpich CY, Kuntz CA, Rutland BE. Evaluation of a single subcutaneous infusion of carboplatin as adjuvant chemotherapy for dogs with osteosarcoma: 17 cases (2006-2010). *J Am Vet Med Assoc* 2012;241(5):608-14.

CANINE OSTEOSARCOMA OUTCOMES WITH SEQUENTIAL DOXORUBICIN - CARBOPLATIN CHEMOTHERAPY

Angela E. Frimberger, VMD, MANZCVS, Diplomate ACVIM (Oncology)
Wauchope NSW Australia. www.vetoncologyconsults.com

INTRODUCTION

Osteosarcoma is the most common primary bone tumor of dogs, with a high incidence rate of approximately 6000 cases per year. 1 This review will focus on canine appendicular osteosarcoma.

PRESENTATION AND NATURAL BEHAVIOR

The natural behavior of this disease has been fairly well defined. [1-6] The primary tumor usually occurs at the metaphysis of long bones, particularly the proximal humerus, distal radius, distal femur and proximal tibia, in that order. It affects primarily large and giant dogs, and usually presents initially with lameness and a painful bone swelling at the site of the primarily tumor. A sudden severe worsening of pain often indicates pathologic fracture.

Epidemiology

Giant and large breed dogs are more likely to develop osteosarcoma than dogs that weigh less than 10 kg. Rapid early growth and increase in stress on weight-bearing limbs may explain why the risk is particularly associated with limb length.[7] Osteosarcoma can affect dogs of any age, although most are middle-aged to elderly animals with a median age of 8 years.

STAGING AND DIAGNOSIS

Dogs with a suspected primary bone tumor should have high-detail radiographs of the lesion evaluated. The tumor has a typical radiographic appearance that is practically diagnostic, but definitive diagnosis depends on histopathology.

Although pulmonary metastases are rarely detectable at the time of diagnosis, three-view thoracic radiographs still should be taken; right and left lateral views and a dorsoventral view should be obtained because occasionally lateralizing metastases are seen on only one lateral view but not the other. The presence of radiographically detectable pulmonary metastases at diagnosis imparts a poor prognosis, with or without treatment.

In addition, routine blood work and urinalysis should be performed.

Many clinicians delay definitive histological diagnosis until after amputation to expedite amputation for palliative reasons; and because the entire tumor is a more diagnostic specimen, and other diseases in dogs that could be represented by the same radiographic appearance are all appropriately treated by amputation.

Cytologically, osteosarcoma can be difficult to distinguish from other sarcomas but alkaline phosphatase staining has excellent sensitivity and good specificity in making this distinction. [8, 9] Histologically, osteosarcomas frequently include areas of cartilage and fibrous tissue as well as osteoid and are often surrounded by new bone. For these reasons, a biopsy diagnosis of chondrosarcoma, fibrosarcoma, or "reactive" bone should be considered suspect and further samples obtained at definitive surgery should be submitted for histopathologic examination.

Prognostic Factors

Serum ALP

Several studies have shown an association between elevated serum alkaline phosphatase levels at diagnosis and worse survival after amputation for dogs with appendicular osteosarcoma. Both total serum ALP and bone ALP are independently predictive. [10, 11] For example, in one study, 11 median survival times for dogs with normal or increased total ALP activities before treatment were 12.5 and 5.5 months, respectively; and median survival times for dogs with normal or increased bone ALP activities before treatment were 16.6 and 9.5 months, respectively. A recent meta-analysis of prognostic factors for osteosarcoma identified serum ALP level as one of the strongest prognostic factors for dogs with appendicular osteosarcoma. [12] This means that while serum ALP does not absolutely determine the chance of long-term survival, it has a great impact.

Tumor grade

Histologically high-grade tumors are associated with a worse prognosis for appendicular osteosarcomas. In one study, dogs with grade 3 tumors accounted for more than 75% of tumors; those with low grade tumors had a better prognosis even with less aggressive treatment. [13]

Lymph node metastasis

A recent study found that the presence of nodal metastasis is rare but imparts a poor prognosis , with median survival time of 59 days. [14]

Monocyte and lymphocyte counts

A recent study found that higher numbers of circulating monocytes (>0.4 x 10^9/L) and lymphocytes (>1.0 x 10^9/L) before treatment were significantly associated with shorter disease-free interval. [12] The impact of the monocyte count was much stronger than for lymphocyte count. Higher numbers of circulating monocytes (>0.4) before treatment was associated with shorter time to metastasis (6.6 months vs. 15.3 months if monocytes <0.4). [15]

TREATMENT

Surgery

The most pressing clinical problem for dogs with osteosarcoma is pain control, and this can be very difficult without surgery as this tumor is often very refractory to medical pain management. The quickest and single best pain relief measure for dogs with appendicular osteosarcoma is amputation; and generally dogs feel better within a few days of amputation than they did before. For these reasons, most clinicians perform surgery as early as possible for palliation. Dogs do not have the same psychological and functional considerations that humans do, and most dogs recover extremely quickly and have excellent function following amputation. The surgery eliminates the primary tumor with little to no reduction in mobility and quality of life for most dogs. Although surgical treatment of osteosarcoma by amputation dramatically improves quality of life, it only increases survival time slightly; and therefore chemotherapy should be strongly considered (below). The median survival time for dogs treated by amputation alone is about 4 months, and only 10% of dogs are alive one year after surgery. Dogs usually die as a result of pulmonary metastases. [16]

Limb-sparing surgery is important in human patients, for whom cosmetic appearance and function are impaired by amputation. In dogs, limb-sparing surgery may be an option for relatively small lesions [17-22]

Radiation Therapy

If an animal is not considered an eligible candidate for amputation or limb sparing, consideration may be given to palliation with radiation therapy. Coarse fractionation of radiation delivered in weekly fractions of 8 to 10 Gy has been reported to palliate pain for dogs with osteosarcoma. Improved limb function was seen in approximately 75% of the dogs treated with either 10 Gy on days 0, 7 and 21 of a 3-week protocol, or 8 Gy on days 0 and 7. Improvement lasted for a median of 2 months regardless of the protocol and toxicities were rare.[23-26] A potential risk of pain relief is increased use of the limb leading to a stress-induced pathologic fracture.

As an alternative to external beam radiotherapy, samarium-153 provide similar palliation. [27-29] The main disadvantages are the need for isolation and the potential for myelosuppression meaning it cannot be used concurrently with chemotherapy.

Chemotherapy

For dogs that can achieve pain control either through surgical or nonsurgical means, metastasis becomes the life-threatening issue.

Survival of dogs with osteosarcoma can be significantly prolonged by adjuvant chemotherapy. The best time after surgery to start chemotherapy is a compromise between allowing time for surgical recovery and not allowing time for micrometastases to grow. In one study, chemotherapy was delayed for either 2 days or 10 days after amputation. No difference in survival was seen regardless of the delay, but increased toxicity was seen in the dogs treated 2 days after surgery. [30] Most oncologists agree a delay of 10 days to 2 weeks between surgery and starting chemotherapy is acceptable. Chemotherapeutic approaches that have been reported include:

- Cisplatin monotherapy. [31-35] Cisplatin markedly improves median survival times to between approximately 6 and 13 months and 1-year survival rates to between 30% and 62%; 2-year survival rates are between 7% and 21%. The main disadvantages of cisplatin is the need for diuresis to prevent nephrotoxicity and its emetogenicity.
- Carboplatin monotherapy. [36-38] Carboplatin given adjunctively after amputation is associated with median survival times of between approximately 10 and 11 months. Carboplatin seems to be as effective as cisplatin in treating canine appendicular osteosarcoma, but this drug is not renal toxic and therefore does not require fluid diuresis, which makes carboplatin an easier drug to administer for many private practitioners. Subcutaneous infusion of carboplatin was recently reported to yield outcomes similar to those previously reported for intravenous infusion of carboplatin [39]
- Lobaplatin monotherapy has been reported to be associated with one-year survival fraction of 31.8%, similar to that seen with other platinum monotherapy, and also did not require pretreatment diuresis. [40]

- Doxorubicin monotherapy. Doxorubicin prolonged survival in dogs treated by amputation and five bi-weekly doses with similar results to those achieved with cisplatin. [41] If chemotherapy with doxorubicin is being considered, a baseline echocardiogram should be obtained. Decreased cardiac contractility, even in a dog without clinical signs of heart disease, should prompt the clinician to avoid doxorubicin. Cardiac screening is particularly important in breeds of dogs that are prone to congenital cardiomyopathy (e.g. Doberman pinschers, boxers). Some clinicians elect not to treat these "at-risk" breeds with doxorubicin at all.
- Doxorubicin and cisplatin combination. Combinations of cisplatin and doxorubicin, alternating every 21 days for two cycles, gave a median survival comparable to cisplatin chemotherapy alone. [42 - 44] 47% of 102 dogs treated with cisplatin and doxorubicin given on the same day survived 1 year after surgery, 28% were alive 2 years after surgery, and 16% were alive 3 years after amputation and chemotherapy. [30] The intensity of scheduling appears to play a role in the outcomes in this combination. (See tablet 1.)
- Doxorubicin and carboplatin combination. [45 – 48] Combinations of carboplatin and doxorubicin, have been reported to yield median survival times of approximately 8 to 11 months.
- Carboplatin and gemcitabine combination has been reported to be associate with a median survival time of approximately 9 months. [49]

Taken together, these results show that chemotherapy dramatically improves survival times over surgery alone for dogs with appendicular osteosarcoma. However most reported chemotherapeutic approaches yield similar median survival times, so the choice or protocol for the individual patient is based primarily on the best fit of the toxicity profile and practical factors such as cost and time commitment. Moreover, there appears to be a "ceiling" beyond which present conventional chemotherapy does not appear to be able to improve median survival times. On the other hand, while median survival times are similar between all these published protocols, there are some more subtle differences in outcomes when results are compared in terms of 2-year (and beyond) survival fraction, or in other words, the proportion actually cured. Table 1 makes a comparison of selected published chemotherapy in terms of the relative dose intensity (D.I.) and tumor control outcomes and suggests that the intensity of scheduling, while not able to break the aforementioned "ceiling", can help- to maximize the proportion cured within these apparent limitations.

We have sought to intensify the scheduling of a doxorubicin – carboplatin protocol using a sequential combination schedule and outcomes from this treatment will be presented.

Table 1. Comparison of dose intensity and outcomes for selected published chemotherapy trials for canine appendicular osteosarcoma

Protocol	MG/M^2	INTERVAL WEEKS	D.I.	Protocol DI	N dogs	Median (months)	1 year (%)	2 year (%)	3 year (%)
CISPLATIN	70	3	1	1	100	10.5	46	11	
CARBOPLATIN	300	3	1	1	203	10.1	37	19	11
DOXORUBICIN	30	3	1	1	303	9.5	40	21	10
Kent CARBO	300	6	0.50	1	32	10.5	48	18	
Kent DOX	30	6	0.50						
Berg CIS	60	3	0.86	1.25	102	11	48	28	19
Berg DOX	17.5	3	0.39						

SUMMARY

Combinations of surgery and chemotherapy make osteosarcoma one of the most treatable cancers in dogs. Advances in chemotherapy in the last 25 years have led to tremendous improvement in survival rates one year after amputation when compared to amputation alone. However permanent cure still is only achieved in approximately 1 out of 5 dogs, so further refinements in chemotherapy agents and scheduling and combination of presently available agents is needed.

REFERENCES

1 Hahn KA, et al. in vivo 8:133-144;1994. 2. Tjalma RA. J Natl Cancer Inst 36:1137-1150;1966. 3. Owen LN. Proc Royal Soc Med 60:1309-1310;1967. 4. Brodey RS, Riser WH. Clin Orthop Rel Res 62:54-64;1969. 5. Misdorp W, Hart AA. J Natl Cancer Inst 62:537-545;1979. 6. Misdorp W. Am J Pathol 98:285-288;1980. 7. Ru G et al: Vet J. 1998;156:31. 8. Barger A, et al. 2012 Sep;41(3):391-5. 9. Ryseff JK, Bohn AA. Vet Clin Pathol. 2012 Sep;41(3):391-5. 10. Ehrhart N et al. J Am Vet Med Assoc. 1998;213:1002. 11. Garzotto CK et al. J Vet Int Med. 2000;14:587. 12. Boerman I, et al. BMC Vet Res. 2012 May 15;8:56. 13. Kirpensteijn J et al. Vet Pathol. 2002;39:240. 14. Hillers KR, et al. J Am Vet Med Assoc. 2005 Apr 15;226(8):1364-7. 15. Sottnik JL et al. J Vet Int Med 2010;24:1439. 16. Spodnick GJ et al. JAVMA 200:995-999, 1992. 17. Straw RC et al. VCNA 26:135-143, 1996. 18. Kuntz CA et al. Vet Surg. 1998;27:417. 19. Degna MT et al. Vet Comp Orthop Traumatol 2000;13:18. 20. Morello E, et al. Vet Surg. 2003 Nov-Dec;32(6):539-44. 21. Lascelles BD, et al. Ann Surg Oncol. 2005 Dec;12(12):1073-83. 22. Liptak JM, et al. Vet Surg. 2006 Aug;35(6):518-33. 23. McEntee MC et al. Vet Radiol Ultrasound 1993;34:367. 24. Ramirez O 3rd, et al. Vet Radiol Ultrasound. 1999 Sep-Oct;40(5):517-22. 25. Walter CU, et al. Vet Comp Oncol. 2005 Mar;3(1):1-7. 26. Mueller F, et al. In Vivo. 2005 Jul-Aug;19(4):713-6. 27. Lattimer JC, et al. J Nucl Med. 1990 Aug;31(8):1316-25. 28. Milner RJ, et al. J S Afr Vet Assoc. 1998 Mar;69(1):12-7. 29. Barnard SM, et al. J Am Vet Med Assoc. 2007 Jun 15;230(12):1877-81. 30. Berg J et al. Cancer 1997;79:1343. 31. Shapiro W, et al. J Am Vet Med Assoc. 1988 Feb 15;192(4):507-11. 32. Straw RC, et al. J Vet Intern Med. 1991 Jul-Aug;5(4):205-10. 33. Kraegel SA, J Am Vet Med Assoc. 1991 Oct 15;199(8):1057-9. 34. Thompson JP, Fugent MJ. J Am Vet Med Assoc. 1992 Feb 15;200(4):531-335. Berg J, J Am Vet Med Assoc. 1992 Jun 15;200(12):2005-8. 36. Bergman PJ et al. J Vet Int Med. 1996;10:76. 37. Phillips B, et al. J Am Anim Hosp Assoc. 2009 Jan-Feb;45(1):33-8. 38. Saam DE, et al. J Am Vet Med Assoc. 2011 Jan 15;238(2):195-206. 39. Simcock JO, et al. J Am Vet Med Assoc. 2012 Sep 1;241(5):608-14. 40. Kirpensteijn J, A et al. Anticancer Res. 2002 Sep-Oct;22(5):2765-70. 41. Moore AS et al. J Vet Int Med 2007;21:783. 42. Mauldin GN, et al. J Vet Intern Med. 1988 Oct-Dec;2(4):177-80. 43. Chun R, et al. J Vet Intern Med. 2000 Sep-Oct;14(5):495-8. 44. Chun R, et al. J Am Anim Hosp Assoc. 2005 Nov-Dec;41(6):382-7. 45. Bailey D, et al. J Vet Intern Med. 2003 Mar-Apr;17(2):199-205. 46. Kent MS, et al. J Vet Intern Med. 2004 Jul-Aug;18(4):540-4. 47. Bacon NJ, et al. J Am Vet Med Assoc. 2008 May 15;232(10):1504-10. 48. Lane A, et al. Aust Vet J. 2012 Mar;90(3):69-74. 49. McMahon M, et al. J Vet Intern Med. 2011 May-Jun;25(3):511-7.

TREATMENT OF HEAD AND NECK TUMORS USING NEW RADIATION THERAPY MODALITIES

Dr. Susan M. LaRue
Fort Collins, Colorado

Canine and feline head and neck tumors present a therapeutic challenge. It is an anatomically dense region. Surgery is often not an option, and if it is, it can come with loss of optimal function or cosmetics. Traditional fractionated radiation therapy has been a viable treatment option; however acute radiation effects are often so severe that many owners will select a more palliative approach. Chemotherapy and palliative radiation therapy can improve clinical signs, but duration of response is limited. Recent technological advances in the field of radiation oncology have provided promising new options for the treatment and management of these cancers. This lecture will focus on the application of specific technologies, such as intensity modulated radiation therapy (IMRT), and stereotactic radiation therapy (SRT) for treatment of tumors of the head and neck region.

Nasosinal tumors

Tumors of the paranasal sinuses and nasal cavity make up 1% of all canine cancers. Carcinomas, including adeno-, solid- and squamous cell, represent 60% of these tumors. Sarcomas, including fibro-, chondro-, and osteosarcoma make up approximately 30%. Other tumor types, such as lymphoma, mast cell tumor, transmissible venereal tumor, hemangiosarcoma, neuroendocrine tumors and histiocytic sarcomas are seen less commonly. Most nasal tumors are locally invasive and generally have not metastasized at the time of diagnosis. These tumors occur most commonly in middle to older dogs, and dolichocephalic breeds may be over represented. Clinical signs include intermittent and progressive history of unilateral or bilateral mucopurulent discharge or epistaxis. Neurological signs may occur in cases with lysis of the cribiform plate and tumor extension into the cranial vault. Cross-sectional imaging with MR or CT can be pathognomonic for the presence of a tumor, however histological confirmation, generally using endoscopy or transnasal techniques, is necessary for tumor histological identification. Most of these tumors are not surgically resectable, and traditional fractionated radiation therapy has long been the primary treatment modality used for these tumors. Reported survival times range from 8-20 months, although tumors causing cribiform lysis with extension into brain are associated with a poor outcome. Over the last 2 decades most radiation therapy centers have treated these tumors with a technique called 3 dimensional conformal radiation therapy (3D-CRT). This technique results in a high level of certainty that the tumor receives the prescribed dose. However, there is limited ability to minimize dose to surrounding normal tissues such as the eyes, skin, tongue, and mucous membranes of the oral cavity. Therefore, the technique is associated with severe acute radiation effects. Although these effects will resolve with time and supportive care, the patients are often extremely uncomfortable and rigorous pain management is required. Despite the efficacy of the treatment, many owners bypass radiation therapy in favor of more palliative techniques.

Two technological advances in radiation oncology have vastly improved the ability to decrease radiation acute effects while still being able to administer clinically effective doses of radiation. **Intensity Modulated Radiation Therapy (IMRT)** controls the intensity of each beam through motorized collimator leaves (MLC) that move during treatment, allowing the dose to be sculpted around important normal structures. This technique requires a specific treatment planning system that uses ***inverse planning.*** Inverse planning requires that the various tumor structures as well as critical normal tissue structures be identified. The gross tumor volume (GTV) is the macroscopic portion of the tumor. The clinical target volume is the area where microscopic disease may be present. The planning target volume (PTV) accounts for potential error in patient positioning. For inverse treatment planning these structures of interest are identified by the radiation oncologist and outlined (contoured) on each slice of the CT scan, which has been downloaded into the treatment planning computer. Dose objectives for each structure are entered, and a sophisticated algorithm attempts to meet all objectives. The 2 major benefits associated with IMRT are: 1) Dose to adjacent normal tissue structures can be minimized, dramatically reducing acute effects and 2) Radiation dose to the tumor can be increased, improving the probability of tumor control. Although most IMRT is administered using a linear accelerator based system, it can also be administered using a technology called tomotherapy. IMRT is now available at a number of veterinary radiation therapy centers. Recent publications have verified a dramatic decrease in radiation effects and survival times comparable to 3D-CRT(1;2). IMRT is now available at a number of veterinary radiation oncology centers.

Stereotactic radiation therapy (SRT) is a new technology that has been used for the treatment of nasal tumors. SRT involves the use of high doses per fraction but overcomes the radiobiological limitations with stereotactically verified positioning and treatment delivery techniques that leave a minimal volume of normal tissue in the high dose area. SRT, by definition, requires: 1) a tumor for targeting (not microscopic disease); 2) treatment planning and administration that will provide a dramatic dose drop off between the tumor and the surrounding normal tissue structures; and 3) A method of stereotactically verifying patient positioning. The result is that normal late responding tissue structures are spared through dose avoidance rather than by administering small doses per fraction. This technique is appealing because only 1-5 fractions of radiation are required and treatments are generally well tolerated by the patient. A number of different systems have been developed for SRT including the Body Gamma Knife™, Cyberknife™ and linear accelerator based systems, such as the Varian Trilogy™. Like IMRT, acute radiation effects are minimized using this technique. In a series of 29 dogs treated at Colorado State University with three fractions of SRT, the median survival time exceeded 400 days. Long term studies to evaluate tumor control and late effects are ongoing.

Oral tumors

Oral tumors make up approximately 4% of canine tumors. Malignant melanomas (30-40%), squamous cell carcinomas (17-25%), fibrosarcomas (8-25%) and anantomatous ameloblastomas (5%) are the most common. In cats, squamous cell carcinomas account for over 70%, followed by fibrosacoma (15%). Surgery is often the treatment of choice, and a number of sophisticated surgical approaches have been developed. However, radiation alone, or combined with surgery, is an important part of the treatment arsenal.

IMRT can markedly reduce radiation acute effects; however, there will be mucositis in regions directly overlying tumor. In our experience, keeping the mucositis geographically restricted with IMRT makes pain more easily managed than with traditional 3D-CRT, where much of the mouth can be affected. An additional bonus of accelerator based IMRT is that regional nodes can be easily treated while minimizing dose to normal tissue structures (like the base of the tongue) that may be located between the primary tumor and the nodal bed. IMRT can be used alone, or can follow surgical excision where adequate margins have not been obtained. SRT can also be used for the treatment of oral tumors, but case selection is important. Tumors that have eroded the mucous membranes may not be candidates for treatment. SRT appears to be beneficial for the treatment of large tumors that have no surgical option. While long term tumor control data is immature, clinical response is often positive and acute radiation effects are minimal.

Other tumors

IMRT has been used on a limited basis for tumors of the pharynx, ear, and thyroid on a limited basis. Long term efficacy and toxicity are being evaluated; however acute radiation effects are minimal. SRT has also been used on thyroid tumors and salivary gland tumors.

Conclusions

The new renaissance in radiation oncology has introduced exciting technologies that may benefit canine and feline patients with tumors in the head and neck region. While decreasing acute effects is an important component of these technologies there is also the potential to increase dose resulting in better local tumor control.

REFERENCES

1. Lawrence JA, Forrest LJ, Turek MM et al. Proof of principle of ocular sparing in dogs with sinonasal tumors treated with intensity-modulated radiation therapy. *Vet.Radiol.Ultrasound* 2010;51:561-70.
2. Hunley DW, Mauldin GN, Shiomitsu K, Mauldin GE. Clinical outcome in dogs with nasal tumors treated with intensity-modulated radiation therapy. *Can.Vet.J.* 2010;51:293-300.

ADVANCES IN RADIOTHERAPY TREATMENT OF NEUROLOGICAL TUMORS

Dr. Susan M. LaRue
Fort Collins, Colorado

Canine and feline intracranial tumors present a diagnostic and therapeutic challenge. Obtaining histological confirmation can be expensive and has associated morbidity. While surgery is often the preferred method of treatment, many tumors are in a location where there are no surgical options. Traditional fractionated radiation therapy, alone or combined with surgery, has been a viable treatment option for many years. Recent technological advances in the field of radiation oncology are providing promising new options for the treatment and management of these cancers. This lecture will focus on the application of specific technologies, such as intensity modulated radiation therapy (IMRT), and stereotactic radiation therapy (SRT) for treatment of intracranial tumors in dogs and cats.

Meningiomas

Meningiomas are the most common intracranial tumor in dogs. Surgery is considered the standard of care; however, many tumors are not found in a location where surgical excision is possible. In addition, some of these tumors are friable and difficult to completely remove, so radiation therapy is often advocated following excision. Until recently, fractionated radiation therapy in veterinary medicine was administered using a technique called 3D-CRT. For many brain tumors this technology, even though it is older, is still very appropriate. Reported median survival times for surgery followed by 3D-CRT range from 540-900 days (1-3). Radiation alone is associated with tumor control of up to 699 days(4).

A number of advances in radiation oncology have vastly improved the ability to focus radiation dose on the tumor while decreasing dose to surrounding tissues. **Intensity Modulated Radiation Therapy (IMRT)** controls the intensity of each beam through motorized collimator leaves (MLC) that move during treatment, allowing the dose to be sculpted around important normal structures. This technique requires a specific treatment planning system that uses ***inverse planning.*** Inverse planning requires that the various tumor structures as well as critical normal tissue structures be identified. The gross tumor volume (GTV) is the macroscopic portion of the tumor. The clinical target volume is the area where microscopic disease may be present. The planning target volume (PTV) accounts for potential error in patient positioning. For inverse treatment planning these structures of interest are identified by the radiation oncologist and outlined (contoured) on each slice of the CT scan, which has been downloaded into the treatment planning computer. Dose objectives for each structure are entered, and a sophisticated algorithm attempts to meet all objectives.

The 2 major benefits associated with IMRT are: 1) Dose to adjacent normal tissue structures can be minimized, dramatically reducing acute effects and 2) Radiation dose to the tumor can be increased, improving the probability of tumor control. Although most IMRT is administered using a linear accelerator, it can also be administered using a technology called tomotherapy. IMRT is now available at a number of veterinary radiation therapy centers.

Interestingly, for brain tumors IMRT may not always be clinical advantage over 3D-CRT. That is because, depending on location, many brain tumors do not have adjacent acutely responding normal tissues, like skin, that would benefit from decreased dose. Dogs with tumors near the eyes (olfactory tumors) or near the auditory apparatus experience fewer acute effects with IMRT. Perhaps more critical is fact that IMRT can decrease the dose to normal brain in the region. This is important because the dose that can be delivered to the tumor is constrained by the dose delivered to normal surrounding brain. If higher doses could be delivered to the tumor (dose escalation) using IMRT, while maintaining appropriate dose constraints to normal tissues, then greater tumor control could be achieved. This has been demonstrated in a number of tumors in people treated with IMRT, including prostate tumors and some head and neck tumors.

The development of immobilization devices, which allow greater confidence in treatment reproducibility, has also contributed to reduced radiation dose to normal tissues(5).

Stereotactic radiation therapy (SRT) is a new technology that has been used for the treatment of brain tumors in dogs(6). SRT involves the use of high doses per fraction but overcomes the radiobiological limitations with stereotactically verified positioning and treatment delivery techniques that leave a minimal volume of normal tissue in the high dose area.

Stereotactic radiation therapy, by definition, requires: 1) a tumor for targeting (not microscopic disease); 2) treatment planning and administration that will provide a dramatic dose drop off between the tumor and the surrounding normal tissue structures; and 3) A method of stereotactically verifying patient positioning. The result is that normal late responding tissue structures are spared through dose avoidance rather than by administering small doses per fraction. This technique is appealing because only 1-5 fractions of radiation are required and treatments are generally well tolerated by the patient. A number of different systems have been developed for SRT including the Body Gamma Knife™, Cyberknife™ and linear accelerator based systems, such as the Varian Trilogy™. Like IMRT, acute radiation effects are minimized using this technique. A series of 25 dogs with presumed meningiomas were treated at Colorado State University using SRT. The median survival time exceeded 500 days. Long term studies to evaluate tumor control and late effects are ongoing.

Pituitary Somatotroph Tumors (Feline Acromegally)

This is a disease of older cats resulting from excessive growth hormone secretion from the pituitary gland. Although originally thought to be a rare condition, recent finding suggest that it is often associated with, but not recognized in, cats with insulin resistant diabetes. Other clinical manifestations include broadening of the head, increased spacing between teeth, and large feet. Organomegaly and cardiac arrhythmias have also been observed with this condition. In humans this condition is primarily treated with surgery or radiation therapy. In veterinary medicine, the availability of this surgery is limited and efficacy has not been established. A recent publication showed that fractionated radiation therapy resulted in improved control of diabetes in 13 of 14 cats with this condition(7). The human literature suggests that for the although neurological responses are comparable, that endocrine response is faster and more durable in patients treated with SRT, compared to those receiving fractionated radiation therapy. At Colorado State University we have treated over 20 acromegalic cats using SRT. The treatment was well tolerated and almost all cats required a substantial decrease in insulin dose.

Canine Pituitary Tumors and Pituitary Dependent Hypercortisolism (PDH)

Radiation therapy has long been the standard of care for dogs with macroadenomas. Fractionated radiation therapy provides long term control of neurological impairment, but the associated endocrinopathy has generally not resolved. A series of dogs with PDH were treated at Colorado State University with SRT. Treatment was well tolerated. Long term efficacy and toxicity is currently being evaluated.

Other Tumors

IMRT and SRT have been used to treat other brain tumors including glial tumors, choroid plexus tumors and trigeminal nerve sheath tumors. Efficacy and toxicity is being evaluated. These technologies have also been used on a limited basis for vertebral tumors, such as osteosarcoma, that can compromise the spinal cord.

REFERENCES

1. Axlund TW, McGlasson ML, Smith AN. Surgery alone or in combination with radiation therapy for treatment of intracranial meningiomas in dogs: 31 cases (1989-2002). *Journal of the American Veterinary Medical Association* 2002;221:1597-600.
2. Uriarte A, Moissonnier P, Thibaud JL, Reyes-Gomez E, Devauchelle P, Blot S. Surgical treatment and radiation therapy of frontal lobe meningiomas in 7 dogs. *Can.Vet.J.* 2011;52:748-52.
3. Theon AP, LeCouteur RA, Carr EA, Griffey SM. Influence of tumor cell proliferation and sex-hormone receptors on effectiveness of radiation therapy for dogs with incompletely resected meningiomas. *Journal of the American Veterinary Medical Association* 684;216:701-7.
4. Bley CR, Sumova A, Roos M, Kaser-Hotz B. Irradiation of brain tumors in dogs with neurologic disease. *J.Vet.Intern.Med.* 2005;6:849-54.
5. Harmon J, Van UD, LaRue S. Assessment of a radiotherapy patient cranial immobilization device using daily on-board kilovoltage imaging. *Vet.Radiol.Ultrasound* 2009;50:230-4.
6. Lester NV, Hopkins AL, Bova FJ et al. Radiosurgery using a stereotactic headframe system for irradiation of brain tumors in dogs. *Journal of the American Veterinary Medical Association* 2001;219:1562-7.
7. Dunning MD, Lowrie CS, Bexfield NH, Dobson JM, Herrtage ME. Exogenous insulin treatment after hypofractionated radiotherapy in cats with diabetes mellitus and acromegaly. *J.Vet.Intern.Med.* 2009;23:243-9.

TCC LASER ABLATION – THE TAMU EXPERIENCE

Audrey Cook, BVM&S, MRCVS, DACVIM (SAIM), DECVIM-CA
College Station, TX

INTRODUCTION

Transitional cell carcinoma (TCC) is the most common malignancy reported in the canine lower urinary tract, and accounts for up to 2% of all cancers in dogs.[1] Affected animals are usually middle aged or geriatric. Females are predisposed, and strong breed predispositions are reported, with Scottish Terriers, Beagles, and Shetland sheepdogs disproportionally affected. Environmental triggers have also been implicated, including exposure to certain drugs and insecticides.

Clinical signs reported in dogs with TCC include hematuria, pollakiuria, stranguria and urinary tract obstruction. Fecal tenesmus and obstruction can be seen in male dogs with prostatic involvement. In addition, signs related to renal failure can occur if ureteral obstruction occurs. Signs related to metastatic disease are relatively uncommon, but include straining to defecate due to rectal compression from sublumbar lymphadenopathy, spinal pain +/- neurologic deficits due to invasion of the lumbar vertebrae, and coughing or dyspnea from pulmonary metastatic disease.

Most patients are euthanized due to complications from the primary tumor, rather than distant disease. Survival times are markedly influenced by the location of the tumor within the urinary system. Patients with disease at the trigone or involving the urethra have much shorter survival times than those with a tumor in the apical region.

Complete surgical cure for TCC is extremely unlikely, although surgery may have a palliative role in some individuals. Current treatment strategies include non-steroidal anti-inflammatory agents such as piroxicam, systemic or intravesicular chemotherapy, radiation therapy, and urethral stenting. Multimodal therapy is commonly used. Patient response to treatment is unpredictable, and it is often difficult to provide a reliable prognosis. Despite our best efforts with these strategies, patients are expected to die of this disease. The decision to euthanize is generally driven by poor quality of life due to severe dysuria or urinary obstruction.

Laser debulking of trigonal and proximal urethral TCC was first reported in 2006.[2] A carbon dioxide laser was used via cystotomy in 8 dogs. Although an improved survival time could not be confirmed, clinical signs were mitigated substantially by the procedure. In 2010, ultrasound-guided endoscopic laser ablation (UGELAB) was described.[3] By using a cystoscope to introduce a diode laser, the need for a surgical approach was eliminated, thereby minimizing the morbidity associated with a laparotomy, including seeding of the abdomen. Thirty eight patients with TCC were described in this report. Most received some form of adjunctive therapy, and survival times ranged from 11-1906 days, with a median of 380 days. A diode laser has some substantial advantages, related to the physics of the laser itself. These include the near infrared wavelength of the diode system, which makes for greater efficacy in a fluid-filled environment. Also, highly vascular tissue such as TCC is more vulnerable to the diode wavelength.

A team at Texas A&M University began to perform UGELAB in April 2010, using a method modeled from the one reported. This lecture will describe specific details of the procedure itself, the personnel and equipment needed, short term complications, and patient outcomes.

PATIENT SELECTION

Based on our experience, UGELAB is an appropriate and beneficial palliative option for patients with a trigonal or urethral TCC. It is unlikely to provide a substantial benefit to a dog with a mass situated in the body of the bladder or in the apical region. Although such a mass can be effectively debulked, masses in these areas cause little clinical morbidity and it is unlikely that ablation will prolong survival. Most dogs with trigonal or urethral masses experience a variable degree of dysuria, and the majority are ultimately likely to obstruct. UGELAB can mitigate clinical signs and maintain an adequate outflow in these patients.

The procedure has limited applicability to male dogs (see below) and may be technically challenging in very small patients.

PERSONNEL

A team of at least four people is essential: cystoscopist, ultrasonographer, technician to assist with the laser and supplies, and anesthesia technician. The ultrasonographer is key to success. This person must be able to maintain a view of the cystoscope and the laser fiber and make confident judgments about the integrity of the bladder and urethral walls. The cystoscopist and the ultrasonographer need to clarify directional terms (dorsal, ventral, up, down, left, right) before starting, to prevent any confusion while the laser is active.

EQUIPMENT

Cystoscope

A rigid cystoscopy system is required. It is advantageous to have a range of sizes, so that more patients can be effectively treated. A 0° or 12° scope is ideal as a greater angle than this makes it harder to guide and visualize the laser fiber. Our 'go-to' scope is an Olympus 0° human pediatric rigid cystoscope, 3 mm x 20 cm. This has a working sheath to accommodate the laser, making for a maximal diameter of 14 French and an insertion length of 16 cm. If this scope is not long enough, we use a 12 ° human adult rigid cystoscope. We occasionally use a pediatric scope with an integrated working channel. This gives us access in extremely small dogs or patients with marked narrowing of the urethral lumen. However, the integrated working channel also serves as the fluid port. With the laser in place, the in-flow of saline is restricted and it can be hard to maintain an adequate view, particularly if hemorrhage occurs.

We have tested a few fiberoptic scopes (human ureteral scopes) and had problems with poor flexibility with the laser in place, as well as limited directional capability. Adequate fluid inflow is also an issue as the laser fiber shares the in-flow access. To date, we have not managed to successfully laser a tumor using a flexible scope.

Ultrasound system

Most of the imaging is fairly shallow, so a high frequency probe is needed. We use a Seimens Acuson Antares system, with a 4-9 MHz probe (VFX9-4)

Laser system

The laser used for this purpose is a diode laser. The device comes in various sizes, defined by the wattage. We use a 30 watt device, although we have never exceeded the 20 watt level. The laser fibers also come in various sizes, with the size inversely proportional to the burn power. We use a 600 μm fiber. Each fiber can be used numerous times, and is sterilized with a low temperature hydrogen peroxide gas plasma system. Sterile scissors and a sterile fiber stripper are needed to adjust the tip of the laser during use.

Safety equipment

Most institutions will require safety training. The laser used is a class 4, as it is capable of causing severe damage to the retina and can burn tissue, even without direct contact. The laser should only be used in a secured room, with entry limited to trained and protected personnel. A laser-safe black out curtain must be drawn across any windows (including those in a door) to prevent accidental exposure to passersby. Goggles appropriate for a class 4 laser must be worn by everyone in the room.

Other considerations

Prepare to use 3-5 liters of saline for the average patient. Wear protective clothing and plastic booties to minimize damage to clothes. A suction mat system or a V-bag suction system will also reduce the amount of water on the floor.

As the procedure regularly takes 2-3 hours, make sure that the team is comfortable. Our cystoscopist uses an ophthalmology chair, as the height is easily adjustable and the arm rests reduce strain. Try to position the scope screen and the ultrasound screen where both are easily viewed by the cystoscopist.

PRE-PROCEDURE EVALUTION

A pre-procedure ultrasound scan is essential to evaluate the location and extent of the disease. As the intrapelvic urethra is not easily visualized with an abdominal window, scanning via the perineum may provide additional useful information. Evidence of ureteral involvement +/- hydronephrosis has important prognostic impact. Other imaging modalities such as a contrast study are rarely indicated, unless the patient is potentially a candidate for urethral stenting. A digital rectal and vaginal examination should be performed.

In most patients, a definitive diagnosis of TCC can be made with a cytological assessment of a voided urine sample or via traumatic catheterization. It is unwise to collect a biopsy during UGELAB, as hemorrhage may limit the cystoscopist's view.

MALE DOGS

As we have not been able to successfully laser via a flexible scope, male patients have required a temporary urethrotomy so that we can access the urethra with a rigid cystoscope. A red rubber catheter is placed into the urinary bladder to outline the location of the urethra, and a small incision is made over the urethra ventral to the anus, in the mid-perineal region. To reduce the risk of post-operative stricture, this opening is subsequently left to heal without surgical closure. Due to severe post-operative hemorrhage in the first male patient, a modified procedure with partial apposition of the tissues was subsequently used. We have not documented seeding of this site with neoplastic cells, but this is a major concern.

PROCEDURE

The first ten minutes

We routinely perform a urine culture on a sample collected through the cystoscope at the start of the procedure. This avoids any potential tumor seeding or leakage from a cystocentesis. A dose of cefazolin (22 mg/kg) is then given intravenously. This is repeated every 90 minutes, with a final dose administered at extubation.

Getting started

The cystoscope should be positioned cranial to the trigone in a mid-sized bladder. The laser fiber is then introduced into the working channel. The fiber can be very hard to see unless it is in the ready mode, as this makes the tip glow. Advance the fiber out in to the lumen of the bladder, and make sure that the ultrasonographer can easily identify the tip. As the laser fiber could perforate the urethra or bladder wall, we always come back to the center of the bladder before exteriorizing it, and then reverse or redirect as needed.

Setting the wattage

We usually do a test burn on a prominent area, starting at 7 watts. If the tissue appears unaffected, the wattage is slowly increased until an obvious blanching or burn is noted. The amount of power required to see an effect is variable, and seems to depend on the density of the tumor. Some are very solid where as others are very 'frondlike' and more difficult to contact effectively. When working in the urethra, we are particularly careful to keep an eye on the power level. It is unusual for us to exceed 16 watts.

Using the laser

The effect of the laser extends beyond the tissue in direct contact. Blanhing or burning may be clearly evident, but tissue adjacent to the laser is also impacted. Some tissue is essentially vaporized and simply disappears. This can be seen on the ultrasound, with a hyperechoic appearance to the devitalized tissue. If the mass is fairly solid, the laser can be slowly advanced through the mass with the tip of the fiber monitored closely by ultrasound to prevent penetration of the bladder wall. We usually burn for 5-15 seconds, depending on how firm the tissue is and how it looks on the ultrasound image. As the procedure progresses, it can become more difficult to assess the damage to the tumor, as hyperechoic debris confuses the image.

Things to watch for

In small dogs, the bladder can quickly fill, with limited space for fluid to exit around the cystoscope. The ultrasonographer should assess bladder size periodically, so that fluid can be drained before problems occur.

POST-PROCEDURE

Bladder lavage

Prior to waking the patient up, we pass a red rubber catheter and lavage the bladder several times with warm saline. Variable amounts of debris are flushed out. The goal is to reduce the amount of material (tissue or blood clot) which may occlude the urethra.

Urinary catheter placement

We do not routinely place an indwelling urinary catheter, as obstruction following the procedure has not been an issue for us. In addition, these patients are at substantial risk of acquiring a resistant hospital-acquired urinary tract infection. We do place a catheter if there is any concern about compromise to the bladder wall or urethra, and leave it in place for 2-3 days.

Fluid therapy

We often continue fluid therapy overnight, to be sure that the patient is urinating adequately before discharge. If the procedure was fast and uncomplicated, fluids are discontinued at extubation.

Medical therapy

Opioid analgesia is routinely administered at the time of extubation, to counter discomfort associated with the cystoscopy.

The majority of patients are receiving a non-steroidal anti-inflammatory as part of their medical plan. This is given as soon as the dog is willing to eat. If there has been extensive ablation of urethral tissue, we often administer oral diazepam, in an effort to prevent urethral spasm. This is continued for 2-3 days in most cases and seems to be of some benefit.

COMPLICATIONS

Overt rupture of the urinary bladder or urethra may occur. Although a small perforation may be successfully managed with a urinary catheter, a sizeable defect is unlikely to heal spontaneously. Clients must be warned of this eventuality and a plan should be discussed ahead of time.

Transitional cell carcinoma is notorious for seeding on contact. It seems likely that iatrogenic trauma during the procedure may implant cells in previously healthy tissue. Many of our patients have developed multiple 'new' masses following UGELAB, but it is unclear if these are seeded lesions or if the tumor is simply spreading across the mucosa.

Post-operative infection has not been a major issue in our patients, although these dogs do develop urinary tract infections in association with their underlying disease.

The vast majority of patients appear comfortable with immediate improvement in their ability to void urine. Post-procedure urethral spasm appears to be transient, if it occurs at all.

COMPARISON TO OTHER OPTIONS

It is very difficult to compare the results of UGELAB to other treatments, as the location and severity of disease is so variable. The most analogous treatment to UGELAB is probably urethral stenting. This has emerged as a useful option for dogs with obstructive disease, with median survival times of 78-251 days reported in two recent studies. Again, multimodal therapies were used in these patients, so direct comparisons of outcome are difficult.[4,5] Potential advantages of UGELAB compared to stenting include maintenance of continence, no risk of implant migration, and minimal post-procedure discomfort. In our institution, a single UGELAB costs less than placement of a stent, and the equipment required to perform UGELAB is substantially less expensive.

SUMMARY

Laser ablation of TCC has a role in the management of dogs with this condition. Patients with trigonal or urethral disease are most likely to benefit from this therapy. Adjunctive medical therapies should be considered in patients undergoing UGELAB to prolong survival times and maintain patient comfort.

REFERENCES

[1]Mutsaers AJ, et al. **J Vet Int Med** 2003;17:136. [2]Upton ML, et al. **J Am Vet Med Assoc** 2006;28:549. [3]Cerf DJ, et al. **J Am Vet Med Assoc** 2012;240:51. [4]McMillan SK, et al. **J Am Vet Med Assoc** 2012;241:1627. [5]Blackburn AL, et al. **J Am Vet Med Assoc** 2012;242:59.

TUMOR ABLATION WITH IRREVERSIBLE ELECTROPORATION

John H. Rossmeisl, Jr., DVM, MS, DACVIM (SAIM and Neurology)
Blacksburg, VA, USA

In collaboration with Dr. Rafael Davalos, a co-inventor of irreversible electroporation, one of the major research foci of my laboratory is the systematic investigation of the feasibility and efficacy of irreversible electroporation (IRE) for the treatment of solid tumors. Using small animals with spontaneous tumors and experimental animal models, we have demonstrated the safety and therapeutic potential of IRE for the treatment of a variety of tumor types. Currently, IRE is being evaluated as a treatment for human prostate, hepatic, renal, pancreatic and lung cancers in over 15 clinical trials across the globe.

ELECTROPRATION THEORY AND HISTORY

Electroporation is a biophysical process in which biological cells are transiently or permanently permeabilized by exposing them to a series of low-energy pulsed electric fields. These external pulses induce an increase in the resting transmembrane potential (TMP) of the cells. The cell membrane is a complex biological system that separates the interior of a cell to the extracellular environment. The membrane is composed primarily of a thin phospholipid bilayer with embedded proteins that regulate the movement of substances in and out of the cell. The permeability of the membrane to molecules is dependent on the electric charge, size, and conformation of the molecules. The cell membrane is basically a non-conductive barrier that protects and controls molecular transport in and out of the cell through ion channels and the membrane's resting potential. Electroporation is a molecular tool and an electromechanical method used to introduce polar molecules into a cell through the cell membrane via intense and short electrical fields.

When cells are exposed to pulsed electric fields that are much longer than their charging time, the induced TMP reaches steady-state within microseconds of the exposure. In this general case, the membranes act as electrical insulators that prevent the current from entering the cell. The local distortions of the electric field around the cells and its vicinity can be attributed to the discrepancy between the membrane's much lower electric conductivity compared to both the intracellular and extracellular environments. Since the electric conductivity is several orders of magnitude smaller than the intracellular and extracellular compartments, most of the electric field in the cell is concentrated within the membrane itself. The magnitude of the TMP is proportional to the amplitude of the applied electric field, and with a sufficiently strong field, there is a significant increase in membrane permeability. The induced increase in the TMP is dependent on the electric pulse parameters (i.e. strength, duration, repetition rate, and pulse number) and physical geometry of the electrodes used to deliver the pulses. Depending on the magnitude of the induced TMP the electric pulses can have no effect, transiently increase membrane permeability to exogenous agents (reversible electroporation) or cause cell death due to loss of cell homeostasis (irreversible electroporation; IRE). The molecular and biophysical phenomena responsible for creation of electroporation-induced defects in cellular membranes remain unknown. However, the putative hypothesis is that there is a rearrangement of the lipid bilayer in the plasma membrane that creates water-filled "pores", which provide a pathway for ions and molecules that are not normally permeable to the cell membrane. Electroporation aqueous pore dynamics are considered a four-step process: pore induction, expansion, stabilization and resealing.

It is difficult to discern when IRE was first observed as a discrete phenomenon. The literature suggests that the initial studies could have been as early as the 18th century. However, it was not until 1967 that Sale and Hamilton demonstrated the non-thermal and lethal effect of high electrical fields on organisms. They concluded that the damage to the cell membrane occurs when TMP of around 1V are achieved. This threshold is based on the seminal theoretical electric potential distribution of an insulating sphere in a conducting medium for the field of electroporation. For decades, IRE has been studied extensively within *in vitro* cellular systems, particularly in industry for sterilization and preprocessing of food. IRE has also been considered an effective means to destroy bacteriaand amoebae with regards to water decontamination. Another context in which IRE has been studied is in the delayed cell damage in high-voltage accidents and the post-electric-shock arrhythmias during defibrillation. It had also been observed in medical applications involving thermal ablation that electrosurgical instruments also induce electroporation and that tissue coagulation may be due to electrofusion of cellular membranes. Reversible electroporation has been also been used for years to facilitate genetic modifications of cell function via delivery of plasmid DNA, mRNA, and viruses through the reversible creation of nanopores within the cell membrane. Reversible electroporation can be used to increase permeability in order to allow intracellular delivery of otherwise impermeable cytotoxic macromolecules for targeted cell death in a process known as electrochemotherapy (ECT). ECT has been used effectively for years in the European Union for the treatment of cutaneous neoplasms in veterinary patients.

DEVELOPMENT AND EVOLUTION OF IRE AS A TISSUE ABLATION MODALITY

In the traditional field of electroporation, which was mainly investigating the reversible realm, IRE was viewed seen as a potential negative effect since the purpose objective of reversible electroporation is to introduce a macromolecule through the membrane without necessarily affecting cellular viability. It was not until 2005 when Davalos, Mir and Rubinsky postulated that IRE could be induced *in vivo* to destroy substantial volumes of targeted tissue prior to the onset of thermal damage that the beneficial effects and biomedical applications of IRE were considered. Their hypothesis that IRE could be used as an independent modality for tissue ablation was later confirmed in small animal models in the liver, and on rodent tumors, as well as in large animal models in the liver, the prostate, and the heart. These initial *in vivo* studies yielded a wealth of information pertaining to the additional benefits of IRE ablation procedures, and served as the seminal studies for current field of irreversible electroporation.

IRE offers several advantages over more traditional surgical resection methods for solid tumors in veterinary patients. The small electrode size makes the procedure minimally invasive and adaptable to virtually any anatomic location, and the procedure can be guided using imaging modalities commonly employed in veterinary practice, such as ultrasound, CT, or MRI. It also offers biologic advantages (minimal heat generation during treatment and sparing of major blood vessels and the extracellular matrix) that may make it attractive as a complimentary technique to tumor resection. IRE creates a sharply delineated volume of ablated tissue with sub-millimeter resolution. Moreover, outside the zone of tissue ablation, there is a penumbra of non-destructive increase in membrane permeability. The cells outside the zone of ablation but within this penumbra are more susceptible to exogenous agents such as chemotherapy. IRE also offers the advantage of speed, as typical treatment durations range from 5-15 minutes, and numerous commercial entities manufacture relatively inexpensive electrical generators capable of pulse delivery within the realm of IRE. Although an intact immune system is not required to achieve the desired effects of IRE, studies in normal and tumor bearing animals suggest that IRE treatment may be associated with induction of host tumor infiltrating lymphocytes, which has been shown in some other spontaneous neoplasms to have a positive correlation with therapeutic response and survival. Finally, preliminary studies also suggest that the mechanism by which IRE induces cell death is effective in neoplasms that have developed chemo- and radioresistance.

IREVERSIBLE ELECTROPORATION IN VETERINARY CLINICAL ONCOLOGY

At our institution, IRE treatment is performed with either a commercially available electroporation system (Nanoknife®, Angiodynamics, Inc.), or a high-frequency (H-FIRE) pulse generator custom designed in our laboratory (INSPIRE, VoltMed). The Nanoknife® generator currently has FDA 510K approval for the surgical ablation of soft-tissue. Our team uses a three-step approach to execution and evaluation of the IRE treatments: therapeutic planning and target tissue characterization, real-time confirmation of pulse delivery, and post-treatment therapeutic assessments. The treatment process will be illustrated using canine, feline, and equine patients with spontaneous tumors such as glioma, soft-tissue sarcomas, squamous cell carcinomas, and melanoma.

Although the Nanoknife® system contains a software algorithm that allows for some degree of customization and individualization of IRE, the current system has several limitations. Thus, we routinely perform pre-treatment planning for each patient's unique tumor type, geometry, and anatomic location using the patient's medical images and computational modeling software. In tissue, there are a number of conditions that determine the extent of electroporation, such as tissue type, heterogeneity, and temperature, as well as a number of pulse parameters, including duration, number, shape, and repetition rate. However, for a given set of conditions, the primary parameter affecting the degree of electroporation is the local electric field strength. Therefore, in order to design protocols for an IRE procedure, the electric field distribution, which is dependent on the procedure's specific electrode-tissue geometry, pulse amplitude, and tissue conductivity distribution, must be determined. Furthermore, to verify that a specific protocol does not induce thermal effects, the temperature distribution can be calculated from the electric field distribution, the electric pulse parameters, and tissue electrical and thermal properties. Knowledge of the electric field and temperature distribution enables surgeons to reliably predict the results of an IRE procedure. This insight allows for planning and optimization the electrode geometry and voltage parameters for varying types of tissue and heterogeneities in order to: insure treatment of the entire and appropriate region, especially when multiple applications are necessary; minimize applied voltages in order to reduce charge delivered; visualize where potential thermal damage may occur to surrounding tissues; and reduce treatment time, invasiveness, and number of procedures.

The actual delivery of IRE treatment involves insertion of electrodes into or around the target tissue prior to pulse delivery. For IRE treatments, patients are anesthetized and a neuromuscular blocking agent administered to prevent unintended pulse-induced muscular contractions. The electrode and pulse parameter configurations are determined during the pre-treatment planning process. The delivery of pulses is also synchronized to the patient's ECG to prevent cardiac dysrhythmias. Prior to therapeutic pulse delivery, a test pulse train is delivered to assess actual tissue conductivity and compare with the theoretical conductivities used in treatment planning analysis. During IRE treatment, current delivery is monitored and recorded in real time, and diagnostic imaging is used to confirm electrode placement and pulse delivery. IRE induced lesions evolve rapidly and can be visualized using ultrasound, CT, or MRI. All tumors are serially evaluated using either manual or imaging quantification of tumor response. For readily accessible tumors, post-treatment therapeutic efficacy evaluation is also assessed with serial biopsy. Based on the existing literature and our clinical experience, IRE induced-tumor death occurs by both apoptosis and necrosis.

HIGH FREQUENCY IRREVERSIBLE ELECTROPORATION (H-FIRE)

In heterogeneous tissues such as tumors, information on multiple tissue types, including their electrical properties and intricate geometrical architecture, complicates treatment planning. For example, subcutaneous tumors treated non-invasively with IRE through the skin using plate electrodes may experience sub-lethal electric fields. This can be attributed to the low conductivity of the skin, which results in a large potential drop across it that increases the likelihood of thermal damage. Traditionally, IRE pulses have been unipolar, meaning that they have a strong DC frequency component (0 Hz). We have theoretically shown that by delivering high-frequency, bipolar bursts, the impedance barrier of the skin or other epithelial layers can be overcome. Specifically, squarewave bursts with a center frequency around 500 kHz (duration of single polarity equal to one microsecond) produce more homogenous, predictable electric field distributions in a heterogeneous geometry and reduce the potential for thermal damage. This is due to the fact that tissue contains both resistive and capacitive components, resulting in a bulk electrical impedance decrease as the frequency of the applied field increases. We have also investigated the *in vivo* potential of high-frequency (H-FIRE), bipolar bursts to destroy tissue through IRE. As opposed to traditional IRE that uses unipolar pulses delivered at a repetition rate of one pulse per second, we demonstrated that IRE can be achieved with a series of high-frequency, bipolar bursts delivered at the same repetition rate. In initial studies, we investigated H-FIRE lesion development in a rat brain model at center frequencies up to 500 kHz, which maintain the potential to overcome impedance barriers posed by epithelial layers.

In addition to its ability to more uniformly ablate tissue, H-FIRE has other benefits that are desirable for clinical applications. H-FIRE has the potential for reducing nerve stimulation. Currently, clinical applications of IRE require the administration of a general anesthesia and paralytic agents in order to mitigate movement and pain associated with muscle contractions during each pulse. In some cases, even with a neuromuscular blockade, muscular contractions are still visible. In addition to the associated pain, muscle contractions may affect the location of implanted needle electrodes, which may invalidate treatment planning algorithms and prove harmful in treatments near vital structures. The threshold for nerve stimulation increases as the center frequency of bipolar waveforms is increased, and above around 100 kHz, muscle stimulation stops. Our results have indicated that H-FIRE can non-thermally ablate tissue without causing muscle contractions, and we are actively investigating the use of H-FIRE for outpatient, minimally-invasive tumor ablation in veterinary patients using local or regional anesthetics.

REFERENCES

Davalos RV, et al. Ann Biomed Eng, 2005; 33(2):223-31, Hamilton WA, et al. Biochimica et Biophysica Acta (BBA), 1967; 163:37-43, Kodre V, et al. In Vivo 2009; 23(1): 55-62, Ellis TL, et al. J Neurosurg 2011; 144: 681-688, Garcia PA, et al. J Membrane Biology 2010; 236: 127-136, Garcia PA, et al. PLoS ONE 2012; 7(11): e50482, Neal RE,et al. J Clin Oncol 2011;29(13): e372-e377, Garcia PA, et al. Technol Cancer Res Treat 2011; 10(1): 73-83, Arena C, et al. Biomedical Engineeering OnLine 2011; 1:102.

MEDICAL, SURGICAL & RADIATION ONCOLOGY: PRACTICAL APPROACH TO DIAGNOSIS AND MANAGEMENT OF MAST CELL TUMOURS IN DOGS & CATS

J. Paul Woods, DVM, MS, DACVIM (Small Animal Internal Medicine, Oncology), Ontario Veterinary College, University of Guelph
Sarah Boston, DVM, DVSc, DACVS, College of Veterinary Medicine, University of Florida
Michelle Turek, DVM, DACVIM (Oncology), DACVR (Radiation Oncology), CVM, University of Georgia

INTRODUCTION

Mast cell tumours (MCT) (histiocytic mastocytoma, mast cell sarcoma, mastocytosis) are the most common cutaneous tumour in the dog representing 16-21% of all canine skin tumours. In the cat MCT represent 15% of all tumours (second most common cutaneous tumour). No sex predilection has been found in the dog, but in the cat the male:female ratio is 2:1. MCT occur most commonly in middle-aged dogs and cats with a mean age of about 8 years. Siamese cats, boxers, Boston terriers and other breeds of bulldog descendants are at higher risk to develop MCT.[1-4]

Mast cell tumours may exist in cutaneous or extracutaneous sites. In dogs the most common sites are the skin of the trunk and perineal region (50%). the skin of the extremities (40%), and the cutaneous head and neck (10%). MCT can arise in multiple cutaneous locations (11%). In cats most MCT are found in the head and neck region. Occasionally MCT are located only in the spleen in cats.[1-4]

The biologic behavior of MCTs is extremely variable. Histologic grading can help determine the likelihood of recurrence or metastasis, but it will not predict the biologic behavior of an individual tumour. Mast cell tumours have a high metastatic potential and many dogs will present with regional lymph node metastasis or rarely, splenic metastasis with splenomegaly. Dogs with advanced disease may have gastric or duodenal ulcers (related to histamine levels with secondary excess of hydrochloric acid secretion in the stomach), and present with vomiting, melena, and anemia. Occasionally mechanical manipulation during examination will result in degranulation of mast cells which results in erythema and wheal formation (Darier sign).[1-4]

Diagnosis of MCT can often be made by a fine needle aspirate (FNA) cytology but excisional biopsy is required for accurate histologic grading. Establishing a diagnosis prior to surgical treatment can be very helpful because the treatment should be aggressive and precaution may need to be taken to prevent or control heparin and histamine release. Diagnostic work-up may consist of staging with a complete blood count (CBC) (eosinophilia, basophilia, mast cells), buffy coat smear, bone marrow aspirate, FNA of the lesion and regional lymph node, abdominal radiographs or ultrasound (splenomegaly). In advanced canine MCT, a coagulogram (PT, PTT, FDP) may be performed.[1-4] The presence of the occasional mast cell in buffy coat or draining lymph node must be interpreted with caution because non neoplastic diseases can also result in mast cells.[5] The tumour biopsy may be submitted for additional tests, such as immunohistochemistry (IHC) for markers of Ki67, c-kit mutations (CD117), and proliferation markers such as AgNOR staining and mitotic index.[6-8]

THERAPY

The therapeutic approach for MCT is based on the clinical stage and histologic grade of the tumour.

Surgery remains the cornerstone of surgical treatment for mast cell tumours. Wide excision with 2-3cm margins and one fascial plane deep is recommended. This may require advanced imaging for surgical planning, depending on the tumour size and location. Lymph nodes should be evaluated by fine needles aspirate prior to surgery and draining lymph nodes should be removed if accessible for diagnostic and therapeutic purposes.

A common question is whether or not mast cell tumours should be downstaged prior to surgical excision, using either corticosteroids or chemotherapy. There is no definitive evidence to determine whether or not this should be done. However, the use of corticosteroids can decrease the size of the tumour by decreasing inflammation. Corticosteroids are not cytotoxic to the neoplastic mast cell tumours and therefore, it is possible that corticosteroids will decrease inflammation, but that neoplastic mast cell tumours will remain at the periphery of the original tumour mass, making it difficult to plan complete excision. Corticosteroid use will facilitate surgery, but carries the risk of a lesser resection than what is necessary to achieve clean margins of excision. This approach is appropriate if resection involves a planned positive margin and radiation is part of the local control plan. The use of chemotherapeutics that are cytotoxic to mast cells may result in successful downstaging because they do not just decrease inflammation, but they will result in death of tumour cells and consolidation of the tumour itself. In general, if a complete excision is possible without downstaging, this approach is recommended. If it is not possible, downstaging can be performed, but caution should be used in interpreting the histological margins for completeness.

Chemotherapy is less effective than surgery or radiation. However, multiple tumours or advanced local disease can only be managed by chemotherapy and although remissions may be achieved, cure is unlikely. In dogs, glucocorticoid therapy (oral prednisone [0.5-1.0 mg/kg PO or 20-40 mg/m^2 PO q24 hr] or triamcinolone [1 mg/cm diameter of tumour intralesionally q 2 weeks]) frequently results in partial remission (PR) or occasionally complete remission (CR); however feline MCT appears to be less responsive to glucocorticoids. Remission times are usually 10-20 weeks; however, dogs that are tumour free after 6 months have a low incidence of recurrence and therefore therapy is usually discontinued at that time. Multiagent protocols (e.g. prednisone/vinblastine, lomustine/vinblastine) may achieve a higher response rate than single agent therapy.[1-4,9-20] Recently, tyrosine kinase inhibitors (TKI) (toceranib [Palladia], masitinib [Kinovet], imatinib [Gleevec]) have become more available in veterinary medicine.[21-29] Kit, a tyrosine kinase, has been found to be mutated in 9-30% of high grade MCT in dogs.[22] In future, TKIs in combination with other chemotherapy may result in higher response rates.

Radiotherapy has played an important role in the management and treatment of canine cutaneous MCT. As with most tumour types, radiation therapy is used for the purpose of achieving local and regional control of the tumour. If excision of a cutaneous mast cell tumour is incomplete, a second and wider surgery should be considered. If, due to the location and/or size of the scar, re-excision is not possible, radiotherapy is the treatment of choice. Curative-intent radiation schedules of daily small-dose fractions for 3 to 4 weeks (total dose approximately 40-52 Gray) have proven to be effective for local control of incompletely excised cutaneous MCT. [1] Two-year control rates of 85-95% are reported when low or intermediate grade MCT are treated in this manner.[1] It is important to note that the reported recurrence rate of incompletely excised grade II MCT is approximately 23%, suggesting that a minority of incompletely excised MCT recur and that adjuvant therapies may not always be necessary.[30] However, local recurrence has been shown to negatively affect the survival of affected dogs,[30] and adjuvant radiotherapy remains the treatment of choice to minimize the risk of local recurrence if complete surgical excision is not possible. When regional lymph node metastasis is present, extirpation of the affected node is recommended along with excision of the primary tumour. It has been reported that dogs with low to intermediate grade MCT with metastasis limited to the local regional lymph node may have a favorable prognosis (median disease free interval of 41 months) when treated with a combination of surgery and definitive (curative-intent) radiation therapy.[31] These findings suggest that systemic therapy may not be necessary for some dogs with low and intermediate grade tumours that might intuitively be considered "high risk", such as those with lymph node metastasis. Additional studies are needed to confirm this conclusion. In the meantime, adjuvant systemic therapy should be considered in dogs with negative prognostic factors including high-grade tumours, lymph node metastasis, recurrent tumours, tumour ulceration, GI signs, tumours in unfavorable locations, rapid tumour growth and large tumours.[1] For incompletely excised grade III or high grade MCT, numerous radiotherapy schedules have been proposed, including standardly fractionated (daily, small-dose fractions for 3 to 4 weeks) and coarse-fractioned protocols (weekly, large-dose fractions for 3 to 4 weeks), but these have not been compared.[1] Systemic adjuvant therapy is recommended for grade III or high-grade tumours due to the higher metastatic potential. Interestingly, a report described positive outcomes in dogs with grade III MCT treated with surgery and radiation alone, including the regional lymph node (median remission and survival times of 28 months).[32] Despite these interesting findings, the majority of veterinary oncologists feel that local therapy alone is insufficient for optimal control of poorly differentiated or high-grade MCT.

Radiotherapy can also serve as a means to achieve palliation of clinical signs associated with non-resectable tumours. Although irradiation of bulky MCTs is generally less effective than treatment of microscopic disease, excellent tumour responses may be observed, including complete responses. Coarse-fraction protocols (weekly, large-dose fractions for 3 to 4 weeks) are commonly used for palliation. Tumour response rates greater than 70-80% have been reported when MCTs were irradiated in the gross disease setting in combination with Palladia[33] and/or glucocorticosteroids[33,34]. Degranulation of tumours in response to irradiation is possible, but rare. This may be of more concern when large tumours are treated. Ancillary medications such as those suggested below and glucocorticoids should be used when MCTs are irradiated in the gross disease setting.

Ancillary drug therapy is important with canine MCT. Dogs with mastocytosis, palpable MCT, or evidence of gastrointestinal bleeding should receive H_2 antagonists (i.e. famotidine, ranitidine) to reduce gastric acid secretion and prevent gastrointestinal ulceration. Sucralfate (0.5-1 g PO q 8h) may be given in cases with evidence of gastrointestinal ulceration and bleeding. H_1 antagonists (e.g. diphenhydramine 2-4 mg/kg PO q 12 h) are used along with famotidine prior to and following surgical removal of canine MCT to help prevent the negative effects of local histamine release on fibroplasia wound healing.[1-4]

PROGNOSIS

Prognosis is based on species, breed, tumour location, growth rate, extent of disease, and the histologic grading of the tumours. The grading is based upon cellular differentiation, cellular pleomorphism, cytoplasmic granules, mitotic figures, and depth of invasion. In general, cutaneous MCT carry a more guarded prognosis in the dog than in the cat. MCT in the boxer are usually of a lower histologic grade than those found in other breeds. MCT in Siamese are of the less malignant histiocytic type. Tumours located in the perineal or preputial area are likely to metastasize both locally and to deep lymph nodes. Growth rate but not tumour size is also an important prognostic indicator. Dogs with MCT growth rates of greater than 1 cm/week have only a 25% chance of surviving an additional 30 weeks. The more undifferentiated tumour then the higher the grade and the poorer the prognosis.[1-4,35]

Regardless of local therapy chosen, even dogs with low to intermediate grade tumours should be evaluated regularly for local recurrence and possible systemic spread.

CHEMOTHERAPEUTIC PROTOCOLS FOR CANINE MAST CELL TUMOUR[36]

Prednisone/Vinblastine[10-12]
Prednisone 2 mg/kg PO q 24 h;
Vinblastine 2 mg/m^2 IV q weekly for 4 weeks followed by 4 treatments every 2 weeks
Continue Vinblastine 2 mg/m^2 IV every 2 weeks for as long as appears to have an objective response. Taper and discontinue prednisone over 12-26 weeks.

CCNU (lomustine)[13]
CCNU (lomustine) 70-90 mg/m^2 PO

Vinblastine/CCNU (lomustine)[14]
Vinblastine 2 mg/m^2 IV week 1 then every 4 weeks
CCNU (lomustine) 60 mg/m^2 PO week 3 then every 4 weeks

CVP (lomustine/vinblastine)[3,15]
CCNU (lomustine)/**V**inblastine /**P**rednisone
[15]CCNU (lomustine) 70-90 mg/m^2 PO week 1 then every 4 weeks
Vinblastine 3.5 mg/m^2 IV week 3 then every 4 weeks
Prednisone 2 mg/kg PO q 24 h for 2 weeks, then 1 mg/kg PO q 24 h for 2 weeks, then 1 mg/kg PO q 48 h until week 24 then taper and discontinue prednisone over 4 weeks.
or
[3]Vinblastine 2 mg/m^2 IV week 1 then every 4 weeks
CCNU (lomustine) 70 mg/m^2 PO week 3 then every 4 weeks
Prednisone 0.5 mg/kg PO q 48 h

Chlorambucil/prednisolone[16]
chlorambucil 5 mg/m^2 PO q 48h
Prednisolone 40 mg/m^2 PO q 24 h for 14 days, then 20 mg/m^2 PO q 48 h

CVP (cyclophosphamide/vinblastine)[17]
(**C**yclophosphamide/**V**inblastine/**P**rednisone)
Cyclophosphamide 50 mg/m^2 PO q 48 h or for 4 days per week; Vinblastine 2 mg/m^2 IV once a week; Prednisone 20-40 mg/m^2 PO q 48h
or
Day 1 Vinblastine 2-3 mg/m^2 IV (start at 2 mg/m^2 and increase by 10-30% with each dose if tolerated); Days 8,9,10,11 Cyclophosphamide 50-75 mg/m^2 PO ; Daily Prednisone 1 mg/kg PO q 24 h
Repeat cycle at Day 21 for 6 months. Prednisone tapering began at 4 months and discontinued by 7 months.

Paclitaxel (Paccal Vet®)[18,19]
150 mg/m^2 IV q 3 weeks

Hydroxyurea[20]
60 mg/kg PO q 24 h for 2 weeks then 30 mg/kg PO q 24h

Tyrosine Kinase Inhibitors (TKI)
Toceranib (Palladia®) [21-25,29]
(3.25 mg/kg PO q 48 h)
2.5-2.8 mg/kg PO Monday, Wednesday, Friday London

Masitinib (Masivet®, Kinavet®)[26]
12.5 mg/kg PO q 24 h

Imatinib (Gleevec)[23,27,28]
5 to 10 mg/kg PO q 24 h

REFERENCES

[1]London CA, et al, in **Withrow and MacEwen's Small Animal Clinical Oncology**, ed 5, 2013:335, [2]Fox LE, in **Cancer in Dogs and Cats Medical and Surgical Management**, ed 2, 2002;451, [3] Welle MM, et al, **Vet Derm** 2008;**19**:321, [4]Blackwood L, et al **Vet Comp Onc** 2012;10: e1, [5]Bookbinder PF, et al, **J Am Vet Med Assoc** 1992;200:1648, [6]Scase TJ, et al, **J Vet Intern Med** 2006;20:151, [7]Romansik EM, et al, **Vet Pathol** 2007;44:355, [8]Kiupel M, et al, **Vet Pathol** 2011;48(1):147, [9]Davies DR, et al, **J Am Anim Hosp Assoc** 2004;40:124, [10]Thamm DH, et al, **J Vet Intern Med** 1999;13:491, [11]Thamm DH, et al, **J Vet Med Sci** 2006; 68: 581–587, [12]Vickery KR, et al, **Vet Comp Onc** 2008; **6**: 111–119, [13]Rassnick KM, et al, **J Vet Intern Med** 1999;13:601, [14]Cooper M, et al, **Vet Comp Onc** 2009;**7**:196, [15]Rassnick KM, et al, **Vet Comp Onc** 2010; **8**:138, [16]Taylor F, et al, J Small Anim Pract 2009;50(6) 284, [17]Elmslie R, **Vet Cancer Soc Newsl** 1997;20:6, [18]Rivera P, et al, **J Small Anim Pract** 2013;54(1):20, [19]Vail DM, et al, **J Vet Intern Med** 2012;26(3):598, [20]Rassnick, K M, **Vet Comp Onc** 2010;8(2):103, [21]London CA, et al. **Clin Cancer Res** 2003;9:2755, [22]Downing S, et al, **Am J Vet Res** 2002;63:1718, [23]London CA, **Top in Comp An** 2009;24(3):106, [24]Pryer NK, et al, **Clin Cancer Res** 2003;9:5729,
[25]London CA, et al, **Clin Cancer Res** 2009;15(11):3856, [26]Hahn KA, et al, **J Vet Intern Med** 2008;22:1301, [27]Isotani M, et al, **J Vet Intern Med** 2008;22(4):985, [28]Lachowicz JL, et al, **J Vet Intern Med** 2005;19:860, [29]London C, et al, **Vet Comp Onc** 2011;10(3):194, [30]Seguin B, et al, J Vet Intern Med 2006;20(4):933, [31]Chaffin K, et al, **Vet Radiol Ultrasound** 2002;43(4):392, [32]Hahn KA, et al, **J Am Vet Med Assoc** 2004;224(1):79, [33]Dobson J, et al, **Vet Comp Onc** 2004;2(3):132, [34]Carlsten KS, et al, **J Vet Intern Med** 2012;26:135, [35]Bostock DE, **J Small Anim Pract** 1973;14:27, [36]Woods JP, In: **Handbook of Veterinary Drugs**, ed 3, 2005;72,

MEDICAL, SURGICAL & RADIATION ONCOLOGY: PRACTICAL APPROACH TO DIAGNOSIS AND MANAGEMENT OF HISTIOCYTIC SARCOMAS IN DOGS & CATS

J. Paul Woods, DVM, MS, DACVIM (Small Animal Internal Medicine, Oncology), Ontario Veterinary College, University of Guelph
Sarah Boston, DVM, DVSc, DACVS, College of Veterinary Medicine, University of Florida
Michelle Turek, DVM, DACVIM (Oncology), DACVR (Radiation Oncology), CVM, University of Georgia

INTRODUCTION

Histiocytic proliferative disorders in dogs are a spectrum of disease from reactive to neoplastic.[1,2] They can be difficult to differentiate from granulomatous, reactive inflammatory, or lymphoproliferative disorders.[3]

Cutaneous Histiocytoma:	-single lesion, young dog (< 3 years), regress, -epidermal Langerhans cells: CD1a, CD11c, E-cadherin, MHC class II
Cutaneous Histiocytosis:	-single or multiple lesions wax & wane, limited to skin and subcutis -interstitial dendritic cells: CD1a, CD11c, MHC class II, Thy-1, CD4 (E-cadherin negative)
Systemic Histiocytosis:	-skin lesions +/- mucous membrane, lymph node and organ involvement (spleen, liver, lung), 3-9 years old -BMD, golden retrievers, Irish wolfhound -interstitial dendritic cells: CD1a, CD11c, MHC class II, Thy-1, CD4
Histiocytic Sarcoma:	-local or systemic lesions -variable cell origin: dendritic cells CD1, CD11c, CD18, MHC II (E-cadherin negative), hemophagocytic HS histiocytes express CD11d instead of CD11c

Histiocytic sarcomas (HS) are aggressive cancers occurring in predisposed breeds such as Bernese mountain dogs (BMD), rottweilers, flat-coated retrievers, golden retrievers and also sporadically in many other breeds.[4-6] Affected dogs are commonly middle-aged to older although young dogs (3 years) have been reported. Lesions have been reported in spleen, lymph nodes, lung, stomach, pancreas, mediastinum, bone marrow, skin and subcutis, skeletal muscle, brain, eyes, nasal cavity, and periarticular tissue of appendicular joints.[3]

Presenting history and clinical signs will vary depending on the site of tumour involvement (e.g. mass, lymphadenopathy, lameness, cough, vomiting). Nonspecific signs (lethargy, inappetence, weight loss) are common. Clinical signs may be due to marked anemia or thrombocytopenia (particularly with hemophagocytic HS).

Diagnosis may be obtained via cytology or histology of tumour mass. However, definitive diagnosis may be difficult and may require immunocytochemistry and/or immunohistochemistry to rule out lymphoma and poorly differentiated mast cell tumours. Complete staging is recommended consisting of compete blood count (CBC), biochemical profile, urinalysis, three-view thoracic radiographs, abdominal ultrasound, +/- advanced imaging, +/- bone marrow aspiration cytology.

THERAPY

The biological behaviour of localized HS is usually progressive whereas disseminated HS is rapidly fatal without treatment. From a surgical standpoint, histiocytic sarcoma should be managed similarly to other soft tissue sarcomas (STS) and considerable effort should be focused on obtaining local control of the tumour. An incisional biopsy is necessary to achieve a histological diagnosis prior to definitive surgery. Often immunohistochemistry is necessary for a definitive diagnosis and this is especially important for periarticular tumours, as synovial cell sarcomas and histiocytic sarcomas can have a similar appearance histologically. When possible, wide excision of histiocytic sarcoma using 3 cm margins and one fascial plane deep is recommended for adequate local control. If this is not possible, a marginal excision followed by a course of radiation is recommended.

Presently, CCNU (lomustine) has been used to treat HS. A retrospective study of 59 dogs were treated with CCNU (60-90 mg/m^2 PO q 3-4 weeks) resulted in an overall response rate of 46% in 56 dogs with gross measurable disease.[7] Three dogs with minimal residual disease experienced tumour relapse but lived 433 days or more after starting CCNU. The median survival of all 59 dogs was 106 days. Another retrospective study of localized HS in 16 dogs treated with curative-intent surgical excision followed by adjuvant CCNU (60-76 mg/m^2 PO q 3 weeks) reported a median disease-free interval of 243 days.[8] Two dogs had local recurrence and 8 dogs developed metastatic disease with a median time to relapse of 201 days in these 10 dogs. The median survival time for all 16 dogs was 568 days.

A prospective study with 21 dogs (14 with disseminated HS and 7 with localized HS) treated with CCNU (90 mg/m^2 q 4 weeks) had an overall response rate of 29% with a median of 96 days.[9] Three dogs (1 disseminated, 2 localized) had complete responses (CR) lasting for 54–269 days and 3 dogs (2 disseminated, 1 localized) had partial responses (PR) lasting for 78–112 days.

A prospective study of metronomic chlorambucil chemotherapy (4 mg/m^2 q 24h) had 2 HS of which I dog had a partial remission of 39 weeks duration (and the other dog had progressive disease).[10]

Other medical therapies have included liposomal doxorubicin[11], paclitaxel[12], and liposomal clodronate. Liposome-encapsulated clodronate (0.5 ml/kg IV over 60 minutes q 2 weeks) was given in a pilot study to 5 dogs with HS which had failed lomustine therapy. Tumour regression occurred in 2 of the 5 dogs.[13]

The role of radiotherapy in the treatment of HS is not well defined. Based on limited information and anecdotal reports, these tumours appear to be radiosensitive.[8,15] In the gross disease setting, it is one of the author's (MT) experience that course fractionated protocols can result in rapid palliation of clinical signs associated with localized HS. This includes alleviation of pain and ulceration, as well as a reduction in tumour size. The reduction in size of irradiated tumours is often marked; this is corroborated by the high complete and partial response rates described anecdotally in the literature.[8,15] Common palliative protocols include three to four weekly 8 Gray (Gy) fractions, or five weekly fractions of 6 Gy. For stage 0 localized HS, the optimal adjuvant approach for eradication of residual microscopic disease following surgery is not known and awaits further investigation. Standardly fractionated definitive (curative-intent) protocols (e.g. 12-19 daily fractions of 3-4 Gy for a total dose of 48-57 Gy) have been anecdotally described for HS of the skin or subcutaneous tissues in this setting.[8,15] These protocols are extrapolated from experience with canine subcutaneous soft tissue sarcoma, and have not been evaluated or optimized in HS. Given the high risk of metastasis associated with many histiocytic tumours, chemotherapy using CCNU is often recommended in combination with radiation to address the potential for systemic spread. A recent study of 16 dogs with localized HS treated with CCNU and local therapy reported a median survival time of 1.5 years (568 days).[8] Therefore, as an adjuvant therapy, radiotherapy may have a role in long-term tumour control of some dogs with localized HS, and appears to provide palliation for non-resectable tumours.

PROGNOSIS

The ability to definitively diagnose the histiocytic diseases and differentiate reactive from malignant conditions has been greatly enhanced by the recent immunohistochemical distinctions determined amongst these disorders. In addition, biomarkers may allow for earlier detection of the disease particularly in genetically predisposed breeds.[14] However, chemotherapy at present has been disappointing with many dogs being euthanized at the time of diagnosis because of poor clinical condition and poor prognosis. Therefore more effective chemotherapy is needed. As mentioned above, dogs with localized HS may benefit from local therapy (surgery +/- radiation) in combination with chemotherapy.[8] The relative importance of each treatment modality in the outcome of dogs with localized HS is not defined. Coarse fractionated radiotherapy can result in palliation of clinical signs associated with localized, non-resectable tumours.

REFERENCES

[1]Affolter VK, **Vet Pathol** 2002;39:74, [2]Moore PF, et al, **Vet Pathol** 2006;43:632, [3]Clifford CA, et al, in **Withrow and MacEwen's Small Animal Clinical Oncology**, ed 5, 2013:706, [4]Hedan B, et al, **BMC Cancer** 2011:11(1): 201, [5]Abadie A, et al, **J Hered** 2009;100 Suppl 1:S19, [6]Ostrander EA, **N Engl J Med** 2012:367(7):636, [7]Skorupski KA, et al, **J Vet Intern Med** 2007;21(1):12, [8]Skorupski KA, et al, **Vet Comp Oncol** 2009;7(2):139, [9]Rassnick KM, et al, **J Vet Intern Med** 2010;24(6):1528, [10]Leach TN, et al, Vet Comp Oncology 2012;10(2):102, [11]Vail DM, et al, **Cancer Chemother Pharmacol** 1997;39(5):410, [12]Poirier VJ, et al, **J Vet Intern Med** 2004;18(2):219, [13]Hafeman S, et al, **Cancer Immunology, Immunotherapy** 2010;59(3):441, [14]Nielsen LN, et al. **Vet Comp Oncology** 2012:10:12, [15]Fidel J, et al. **Vet Comp Oncol** 2006;4(2):63

MEDICAL, SURGICAL & RADIATION ONCOLOGY:
PRACTICAL APPROACH TO DIAGNOSIS AND MANAGEMENT OF NASAL TUMOURS IN DOGS & CATS

J. Paul Woods, DVM, MS, DACVIM (Small Animal Internal Medicine, Oncology), Ontario Veterinary College, University of Guelph
Sarah Boston, DVM, DVSc, DACVS, College of Veterinary Medicine, University of Florida
Michelle Turek, DVM, DACVIM (Oncology), DACVR (Radiation Oncology), CVM, University of Georgia

INTRODUCTION

Tumours of the nasal cavity and paranasal sinuses represent only 1 to 2% of canine neoplasia,[1] yet they are one of the most commonly irradiated cancers in veterinary medicine. Approximately two thirds of intranasal tumours are epithelial in origin (carcinoma and adenocarcinoma), and the bulk of the remainder are sarcomas (fibrosarcoma, chrondrosarcoma, osteosarcoma, etc.). The biologic behaviour, treatment and prognosis of these tumours are similar and will be discussed together. Intranasal tumours are locally invasive and can cause bone erosion with tumour extension into the surrounding subcutaneous tissues, calvarium, oral cavity, or retro-orbital space. They are slow to metastasize, and are usually confined to the nasal cavity and surrounding tissues at the time of diagnosis. The classic presentation is that of a middle to older aged dog with unilateral or bilateral nasal signs including nasal mucopurulent discharge, epistaxis, congestion and/or, in some cases, facial deformity. Rarely, dogs may present solely with neurologic signs associated with tumour extension through the cribriform plate. Common differential diagnoses include fungal or bacterial rhinitis, lymphoplasmacytic rhinitis, foreign body, bleeding disorders, and hypertension. Rare tumours affecting the nasal cavity include lymphoma, mast cell tumour, hemangiosarcoma, melanoma and others; these will not be considered here.

A definitive diagnosis of an intranasal tumour requires a tissue biopsy. This is usually obtained by transnostril blind biopsy using cup forceps, by fiberoptic-guided biopsy during rhinoscopy, or by fine-needle aspiration of facial deformities. Coagulation profiles should be verified prior to tissue sampling, and care should be taken to avoid the cribriform plate during transnostril biopsy techniques. Conventional radiography can be useful to detect the presence of a soft-tissue opacity within the nasal passages, however advanced imaging (CT or MRI) has the greatest imaging value and allows for determination of the local extent of the tumour (staging).[1] Staging of the primary tumour is important for treatment planning as well as for prognostication. Tumours invading through the cribriform plate have been associated with a worse prognosis following radiotherapy.[2] Complete tumour staging includes cytology of enlarged regional lymph nodes (mandibular and/or retropharyngeal nodes) as well as thoracic radiographs to rule out the presence of lung metastasis. Metastasis occurs in approximately 10-20% of cases at the time of diagnosis.[1] When present, it suggests a worse prognosis and drives treatment recommendations towards a more palliative approach.

In cats, the most common tumour of the nasal cavity and paranasal sinuses is lymphoma, followed by epithelial neoplasms (carcinoma, adenocarcinoma).[1] Sarcomas and other histologies occur less frequently. Clinical signs and diagnostic principles are similar to those in the dog, with the exception that cats with nasal lymphoma should be fully staged, including abdominal imaging, +/- bone marrow aspirate, due to the potential of systemic disease. Differential diagnoses include chronic or infectious rhinitis, polyp or foreign body.

THERAPY

Generally tumours that are within the nasal cavity are not primarily treated with surgery. However, there are some cases that will benefit from surgery. This will include cases that are diagnostic challenges, where a ventral rhinotomy may be performed for diagnostic and therapeutic purposes. In these cases, follow up with radiation will be required with a definitive diagnosis of neoplasia. Cases with prolonged epistaxis, discharge and odor after radiation are also likely to benefit from ventral rhinotomy. As well, in late stage cases with prolonged epistaxis, chemoembolization may also be considered as a method for short-term relief of epistaxis.

Another group of nasal tumours that can be treated with curative intent with surgery include nasal planum and septum squamous cell carcinoma. Evaluation of these tumours with an incisional biopsy and CT is recommended to determine the definitive diagnosis and determine the extent of disease. If the mass is confined to the nasal septum, the septum can be resected without removal of the nasal planum. If the planum is involved, a planectomy is required. Although in the past, this involved a surgery that was functional, but less than cosmetic, a new technique has been recently described to reconstruct the nasal planum using the upper lip and this is a much more cosmetically appealing technique.

Radiotherapy as the sole treatment modality is the therapy of choice for canine intranasal tumours. Treatment objectives include 1) maximizing local tumour control (definitive approach or curative intent), or 2) short-term palliation of clinical signs. Unlike surgery, radiotherapy has the advantage of being able to 'reach' all components of the disease in the nasal cavity. The therapeutic challenge lies in the proximity of critical normal structures such as the eyes and central nervous system whose tolerance to radiation limits the radiation dose that can be delivered to the tumour. Recent advances in radiation delivery systems and treatment planning software have resulted in improved dose distributions to tumours and greater sparing of normal tissues.[1,3] Currently, 3D computerized radiation treatment planning is standardly available in veterinary medicine. 3D radiotherapy requires a CT scan to be performed with the patient in treatment position for computer planning purposes. Linear accelerators equipped with multi-leaf collimators have field shaping flexibility that allows for better dose distributions. Furthermore, the recent and increasing availability of intensity-modulated radiotherapy (IMRT) in veterinary medicine has resulted in far superior radiation delivery with steep dose gradients between tumour and normal tissues. Radiation can now be delivered with high precision to the tumour while sparing surrounding critical normal tissues. IMRT requires a linear accelerator with a multileaf collimator (for field shaping) and specialized computer software. Modern IMRT linear accelerators (such as Varian's Triology, Accuray's Tomotherapy) have even faster delivery techniques and on-board imaging to ensure accurate patient positioning prior to each treatment. Nasal tumours are excellent models to benefit from IMRT. IMRT allows for sparing of the eyes and brain, thus reducing treatment-related morbidities while ensuring dose homogeneity to the tumour.

Definitive radiation protocols include 10-15 daily fractions, and total doses in the range of 42-54 Gray (Gy).[1] The author's (MT) preferred protocol is 4.2 Gy for 10 daily fractions.[3-5] Median survival times are reported between 14-19 months.[3-5] The risk of side effects is directly related to the radiation dose received by normal tissues. This will vary depending on the location and size of the tumour and the degree of conformal dose distribution. Acute radiation side effects associated with definitive protocols can include oral mucositis, +/- moist desquamation, conjunctivitis, keratoconjunctivitis sicca and potentially corneal ulceration if KCS is not managed aggressively. The risk of life-threatening late radiation effects is believed to be 5% or less. The lens is has a low tolerance for radiation and cataract formation is a common late complication associated with ocular irradiation; blindness can also result from retinal changes that develop after exposure to high total doses of radiation. With conformal radiation delivery (IMRT), critical structures including the eyes and brain can be spared without compromising dose delivery to the tumour. This significantly reduces the morbidity associated with definitive radiotherapy for intranasal tumours. Tumour dose escalation and stereotactic radiosurgery using IMRT techniques are awaiting further investigation. There is no evidence that adjuvant chemotherapy improves tumour response to radiation or patient outcomes. To date, the longest survival time reported in dogs with intranasal tumours was described in a small study evaluating the combination of definitive radiation (4.2 Gy X 10 fractions) followed by nasal cavity exenteration at the time of tumour progression.[4] The median survival time in this cohort of 13 dogs was 47 months.

Radiotherapy also offers the opportunity for palliation of clinical signs using less intensive protocols.[6-10] Rates of clinical sign resolution with palliative protocols are high, on the order of 65-100%, but survival times and tumour control rates are not as favorable as with definitive therapy. Most palliative protocols involve coarse fractionated treatments (6 to 9 Gy) delivered weekly or biweekly.[6-9] Palliative radiation can also be delivered as a daily protocol of 4 Gy X 5 fractions.[10] The reported median duration of control of clinical signs ranges from 120 to 300 days. Acute toxicities are generally mild and short lived.

When clinical signs associated with a previously irradiated intranasal tumour recur and are attributable to progressive disease, further treatment options include re-irradiation (ideally with IMRT for the sake of sparing normal tissues as much as possible), or ligation of the carotid artery (unilateral or bilateral) to control epistaxis.

As in the dog, radiotherapy is the local therapy of choice for intranasal tumours in cats. Reports on efficacy are few and results are variable.[1] However, the inclusion of radiotherapy appears to enhance overall survival. Based on limited information, the median survival of cats with intranasal tumours treated with radiation is approximately one year or longer.[1] The relative benefit of chemotherapy and/or radiation in cases of localized feline nasal lymphoma is not yet defined.

PROGNOSIS

Without treatment, the median survival time of dogs with intranasal carcinoma is less than 4 months.[11] For both dogs and cats with intranasal tumours, radiotherapy enhances overall survival and causes improvement of clinical signs. Median survival times associated with various treatment approaches are discussed above. Negative prognostic indicators include invasion through the cribriform plate and presence of metastasis. As radiation planning and delivery technologies advance, greater strides in conformal dose distributions and normal tissue sparing will be achievable and patient outcomes may improve.

REFERENCES

[1]Turek M, et al, in **Withrow and MacEwen's Small Animal Clinical Oncology**, ed 5, 2013:435, [2]Adams WM, et al, **Vet Radiol Ultrasound** 2009;50:330, [3]Lawrence JA, et al, **Vet Radiol Ultrasound** 2010;51:561, [4]Adams WM, et al, **J Am Vet Med Assoc** 2005;227(6):936, [5]Adams WM, et al, **Vet Radiol Ultrasound** 1998;39(5):475, [6]Gieger T, et al, **J Am Anim Hosp Assoc** 2008;44:116, [7]Buchholz J, et al, **Vet Radiol Ultrasound** 2009:50(6),679, [8]Belshaw Z, et al, **Vet Comp Oncol** 2010;9(2):141, [9]Maruo T, et al, **J Vet Med Sci** 2011;73(2):193, [10]Tan-Coleman, et al, **Vet Radiol Ultrasound** 2013;54(1):89, [11]Rassnick KM, et al, **J Am Vet Med Assoc** 2006;229:401.

MEDICAL, SURGICAL & RADIATION ONCOLOGY:
PRACTICAL APPROACH TO DIAGNOSIS AND MANAGEMENT OF MELANOMA IN DOGS & CATS

J. Paul Woods, DVM, MS, DACVIM (Small Internal Medicine, Oncology), Ontario Veterinary College, University of Guelph
Sarah Boston, DVM, DVSc, DACVS, College of Veterinary Medicine, University of Florida
Michelle Turek, DVM, DACVIM (Oncology), DACVR (Radiation Oncology), CVM, University of Georgia

INTRODUCTION

Melanomas are relatively common tumour in dogs occurring in the oral cavity (30-40% of canine oral tumours), digits (nailbed, footpad), and skin (5-7% of canine skin tumours).[1] Melanomas can also occur in the eye, gastrointestinal tract, and mucocutaneous junctions. Metastatic sites can include draining lymph nodes, lungs, liver, meninges, adrenals, and other sites. Oral melanoma occurs in older dogs (mean age 11-12 years) with smaller breeds (e.g. Scottish terriers, daschunds, cocker spaniels) and also golden retrievers and poodles overrepresented. Location in the mouth (in decreasing frequency) is gingival (42-63%), lips (15-33%), hard and soft palate (10-16%) and tongue (1.5-3.3%). Oral melanomas are typically brown-to-black but may also be unpigmented (33%).

Melanomas exhibit an extremely diverse biologic behaviour ranging from rapidly growing, locally invasive, highly metastatic malignant oral melanomas to usually benign skin melanomas with nailbed/footpad melanomas in between with approximately 40-50% metastatic. Prior to prognosis; however, an accurate diagnosis must be established. Therefore some tumours (particularly anaplastic amelanotic melanomas versus soft tissue sarcomas) require a panel of immunohistochemical stains. Stains include Melan A, S100 and now PNL2 and tyrosine reactive proteins 1 (TRP-1) and 2 (TRP-2) and perhaps in future integrin alpha-3 integrin binding ligands (+/- actin-bundling protein fascin-1).[2,3]

Biologic behaviour has been found to vary based on anatomic site, size, stage, and histologic characteristics.[4,5] Generally, oral melanomas have a poorer prognosis than cutaneous (with lip or digit locations worse than other cutaneous sites). However, location alone cannot predict prognosis for an individual (i.e. exceptions). For all sites, distant metastasis was indicative of a poor prognosis. Histologic evaluation of nuclear atypia has prognostic value (minimal atypia associated with more favourable prognosis); however, assessment was subject to interobserver variation.[6] The mitotic index (MI) determined as number of mitoses per 10 hpf in bleached tissue samples and Ki67 were determined to be more objective prognostic indicators. Dogs with cutaneous melanomas with MI ≥ 3 per 10 hpf and oral or lip melanomas with MI ≥ 4 per 10 hpf had shorter survival times.[6]

For oral melanoma, primary tumour size has been found to be prognostic. Staging involves history, physical exam, complete blood count (CBC), biochemical profile, urinalysis, three-view thoracic radiographs, local draining lymph node aspiration and cytology, abdominal ultrasound.

World Health Organization (WHO) Tumour, Node, Metastasis staging scheme for dogs with oral melanoma

T:	Primary tumour
T1	Tumour $\leq$2 cm diameter
T2	Tumour 2–4 cm diameter
T3	Tumour >4 cm diameter
N:	Regional lymph nodes
N0	No evidence of regional node involvement
N1	Histological/cytologic evidence of regional node involvement
N2	Fixed nodes
M:	Evidence of distant metastasis
M0	No evidence of distant metastasis
M1	Evidence of distant metastasis

Stage I	T1 N0 M0
Stage II	T2 N0 M0
Stage III	T2 N1 M0 or T3 N0 M0
Stage IV	Any T, any N, and M1

THERAPY

The role of surgery for the control of oral melanoma depends on the size of the tumour and the extent of disease. For tumours that are detected early without evidence of metastastic spread beyond the local lymph nodes, wide excision with maxillectomy or mandibulectomy is recommended. Dogs tolerate hemimandiculectomy and maxillectomy well and are usually able to eat within 24-48 hours of surgery. Caudal maxillectomy can be achieved using a combined approach, with a skin incision under the eye and an incision within the oral cavity. When biopsying any oral mass, consideration should be given to the method of surgical excision. Generally speaking, the skin overlying the maxilla and mandible can be preserved with maxillectomy and mandibulectomy, so the incisional biopsy should be performed directly over the tumour from within the mouth. The biopsy tract should not go through the skin overlying the oral tumour, as this entire biopsy tract will need to be resected with definitive surgery and this will result in a less cosmetic result. For cases where the oral mass is not resectable and/or where there is widespread metastatic disease, a marginal excision may still be appropriate to improve patient comfort.

The efficacy of chemotherapy concurrent with radiation or in an adjuvant setting has not been established. Chemotherapy has included cisplatinum, carboplatinum, dacarbazine, melphalan, doxorubicin, toceranib (SU11654), and nonsteroidal anti-inflammatories (NSAIDs).[7] Carboplatin was reported in 27 dogs to have a response rate of 28% (for a median of 5.5 months) with gross disease. Smaller dogs receiving higher doses on an mg/kg basis were more likely to respond but also experienced gastrointestinal side effects.[8] A recent retrospective study of 63 dogs with oral (30), digital (9), or cutaneous (24) malignant melanoma compared the effect of carboplatin chemotherapy (300 mg/m^2 IV q 3 weeks for 4-6 treatments) on the survival after loco-regional control or as a sole therapy.[9] The study found the addition of chemotherapy to surgery did not confer a survival benefit in the loco-regional control setting and the addition of chemotherapy to local treatments at oral, digital and cutaneous sites did not lead to a significant increase in survival times. The overall median survival times for patients with oral melanoma was 389 days and for dogs with gross disease (8) there was no difference between survival of patients treated with chemotherapy and palliative intent therapy. However, there was an improvement in survival in the 3 dogs that responded to chemotherapy (978 days; p=0.04) compared to the 8 non-responders (147 days). In the phase I dose-escalating study of SU11654 (toceranib) 2 of 3 cases of melanoma had stable disease for 12 weeks.[10] KIT expression had been demonstrated in 51% (20/39) canine oral melanomas; however, no association was found with stage nor overall survival.[11]

Melanomas have demonstrated expression of cyclooxygenase (COX)-1 and COX-2.[12] Almost all tumours (9 oral, 2 ocular, 20 cutaneous) expressed COX-1, but COX-2 expression was restricted to the malignant tumours (i.e. 11 of the 20 cutaneous malignant melanomas, all oral malignant melanomas and in 1of 2 ocular malignant melanomas).[13] COX-1 expression did not differ significantly between benign and malignant skin lesions, but COX-2 expression was significantly greater in cutaneous malignant melanoma compared with melanocytoma (P=0.05). COX-2 labeling was particularly intense in the more highly malignant oral tumours. The results of the study suggested that NSAIDs, (particularly COX-2 inhibitors) may be useful in the treatment of canine malignant melanoma. A xenogeneic human tyrosinase DNA vaccine has proven to be safe and well tolerated. Survival times compared to historical controls was favourable.[14-17]

Radiotherapy plays an important role in the management of oral melanoma for many dogs. It is indicated as a primary treatment for non-resectable tumours and as an adjuvant therapy to control residual microscopic disease for incompletely excised (microscopic) tumours. Most protocols used in dogs are hypofractionated (i.e. high dose per fraction).[1] Common radiation protocols include three to four weekly fractions of 8 Gray (Gy), or six weekly fractions of 6 Gy.[19,20] The weekly interval between radiation treatments results in a reduced risk of acute side effects such as oral mucositis compared to standardly fractionated protocols. For tumours treated in the gross disease setting, radiation offers the opportunity for palliation of clinical signs (halitosis, pain, hemorrhage, dysphagia) and can result in a significant reduction in tumour size. The tumour response rate to radiation is 70-90%, including a complete response rate of 30-50%.[1] Regional lymph nodes, including tonsils, mandibular lymph nodes and retropharyngeal lymph nodes are often included in the radiation field, regardless of their disease status, due to the high risk of metastasis to these sites. If the primary oral tumour is resectable, then extirpation of metastatic lymph nodes is recommended prior to radiotherapy. With local therapy alone (surgery and/or radiation), most dogs with oral melanoma succumb to progressive disease (usually distant metastatic disease) within a few to several months of diagnosis.[1] The development of the therapeutic DNA vaccine (Oncept™) mentioned above offers hope that tumour progression can be delayed. In a study of 58 dogs treated with local therapy (surgery +/- radiation) and the Oncept™ vaccine, the median survival time was significantly improved compared to historical controls that received local therapy alone.[14] It has been proposed that the best outcomes associated with immunotherapy occur in dogs in which there is local control of the primary tumour by surgery and/or radiation. Additional studies are needed to confirm the efficacy and safety of combination therapies. Prognostic indicators potentially associated with a more favorable outcome for dogs with oral melanoma include small tumour size, low mitotic index, lip location, and lack of bone lysis.[1] For dogs that present with distant metastasis whose quality of life is compromised only by the presence of the oral tumour, coarse fraction radiation is a good option for palliation of clinical signs. Median survival time in this setting is about 1-3 months.

PROGNOSIS

Accurate prognosis becomes more important as more treatment options are available for treatment of melanoma. In future, prognostication could determine which dogs require additional therapy and which are likely to be cured by surgery alone. New biomarkers may be discovered which can be used to develop novel targeted imaging and therapeutic agents to aid in early diagnosis, detection of metastasis and treatment of melanoma.[18] Prognosis associated with radiotherapy in dogs with oral melanoma is discussed above.

REFERENCES

[1]Bergman PJ, et al, in **Withrow and MacEwen's Small Animal Clinical Oncology**, ed 5, 2013:321, [2]Aina OH, et al **Vet Immunol Immunop** 2011;143(1-2):11, [3]Yamada N, et al, **Vet Comp Oncology** 2012;10(4):303, [4]Bostock DE, **Vet Path** 1979;16:32, [5]Spangler WL, et al, **Vet Pathol** 2006;43(2):136, [6]Smedley RC, et al, **Vet Path** 2011;48(1):54, [7]Boria PA, et al, **J Am Vet Med Assoc** 2004;224:388, [8]Rassnick KM, et al, **J Am Vet Med Assoc** 2001;218:1444, [9]Brockley, LK, et al, **New Zeal Vet J** 2013;61(1):25, [10]London CA, et al, **Clin Cancer Res** 2003;9:2755, [11]Murakami A, et al, **Vet Comp Oncology** 2011;9(3):214, [12]Mohammed SI, et al, **Prostag Leutr Ess** 2004;70:479, [13]Pires I, et al, **J Comp Pathol** 2010;143(2-3):142, [14]Grosenbaugh DA, et al, **Am J Vet Res** 2011;72:1631, [15]Gyorffy S, et al, **J Vet Int Med** 2005;19:56, [16]Bergman PJ, et al, **Clin Cancer Res,** 2003;9:1284, [17]Bergman PJ, et al, **Vaccine** 2006;24:4582, [18]Mayayo SL, et al, **Vet J** 2011;190(2):e26, [19]Bateman KE, et al, **J Vet Intern Med** 1994;8(4):267, [20]Freeman KP, et al, **J Vet Intern Med** 2003;17:96.

ONE MEDICINE IN DISASTERS: HAITI EARTHQUAKE

Dr. Kelly Crowdis
Port Au Prince, Haiti

Veterinarians are uniquely trained in multiple species and multiple disciplines to prevent and treat injuries and disease. Because we have an "all-around" training, we are well equipped to deal with whatever comes our way, even when disaster strikes.

On January 12, 2010 a 7.0 magnitude earthquake struck Haiti, killing over 300,000 and injuring nearly 500,000. The enormity of the loss and destruction was evident all too quickly. In the aftermath of the earthquake, I came to understand how my veterinary training had uniquely qualified me to respond to this disaster.

The Haitian medical system is typical of third world countries in that they lack advanced training to deal with trauma, have minimal diagnostic capabilities, have little in the way of supplies, and are understaffed and overworked. When the earthquake hit the country was in a state of suspended animation when it came to treating the multitude of injuries in the first days following the earthquake. The hospitals were overrun within hours, the parking lots full of injured waiting for a chance to get inside. Doctors worked with what they had but their supplies were exhausted long before foreign aid arrived.

As Haitian culture doesn't allow for animals to be inside, there were few animal injuries during the earthquake, so as a veterinarian my focus – like the rest of the world's – was on the people. While the hospitals were quickly running out of supplies, I was uniquely equipped to respond to the need because I had supplies to treat both small and large animals. I had suture from 4-0 to 2, bandage supplies, splints, giggly wire, surgery packs, gallons of disinfectant, 250ml bottles of lidocaine, and oral and injectable antibiotics. Before first light on the 13th of January I was in the field. Literally in a field with my trunks of supplies and my tailgate open ready to start working. The medical training kicked in. A wound is a wound no matter the species, and the people just kept coming. I will admit to being hesitant when dealing with closed traumas with no diagnostics but I did what any good general practitioner does – refer!

What started out to be helping a family friend with a laceration became by the end of the first day a fully-staffed, functional field hospital with a Haitian doctor, several medical students, nurses, crowd control personnel, and one veterinarian. We continued on with primary wound care for the first week and then started to treat the infections that set in from improper primary wound closure and lack of antibiotics. The hospital remained open for 9 months providing primary care by Haitian nurses and doctors, preventative care, psychological counseling and obstetrical services. After the first 2 weeks, I moved back to the world of 4-legged patients and provided the hospital supplies and encouragement.

There were lots of questions that arose in my mind as journalists came and asked which hospital I was associated with. Should I say, "No I'm not an MD, I'm a veterinarian who just amputated this young lady's finger"? What is our role as a veterinarian in a disaster or medical emergency? What are we legally able to do? A literature search did not yield specific results, however Bernard E. Rollin, PhD in a 2003 Canadian Veterinary Journal article provided an informed opinion regarding "Good Samaritan Laws" which would indicate that – provided a veterinarian was skilled and comfortable with the level of care he/she could provide – they have a moral and ethical obligation to help treat people in an emergency situation.

REFERENCES

Rolin B.Can Vet J. 2003 March; 44(3): 189–191

CLINICAL PRACTICE GUIDELINES FOR MILITARY WORKING DOG EMERGENCY CARE BY HUMAN HEALTH CARE PROVIDERS

LTC Matt S. Takara DVM, MSpVM
Camp Pendleton, CA
LTC (Ret) Michael S. Lagutchik, DVM, MS, DACVECC

INTRODUCTION

The primary goal for the Military Working Dog (MWD) Clinical Practice Guidelines (CPGs) are to: provide non-veterinary healthcare providers (HCPs) with general information concerning handling of MWDs and provide – as appendices to the base document – clinical medical guidance for management of seriously ill or injured MWDs to assist in recognition and initial resuscitation and stabilization of life- and limb threatening conditions that warrant HCP intervention in the absence of veterinary personnel.

Appendices

APPENDIX A: NORMAL CLINICAL PARAMETERS FOR MILITARY WORKING DOGS
APPENDIX B: EMERGENCY AIRWAY MANAGEMENT IN MILITARY WORKING DOGS
APPENDIX C: MANAGEMENT OF PENETRATING CHEST WOUNDS AND RESPIRATORY DISTRESS IN MILITARY WORKING DOGS
APPENDIX D: CARDIOPULMONARY RESUSCITATION (CPR) OF MILITARY WORKING DOGS
APPENDIX E: MANAGEMENT OF SHOCK IN MILITARY WORKING DOGS
APPENDIX F: MANAGEMENT OF ABDOMINAL TRAUMA IN MILITARY WORKING DOGS
APPENDIX G: MANAGEMENT OF GASTRIC DILATATION-VOLVULUS SYNDROME IN MILITARY WORKING DOGS APPENDIX H: MANAGEMENT OF ENVIRONMENTAL INJURIES IN MILITARY WORKING DOGS
APPENDIX I: MANAGEMENT OF LONG BONE FRACTURES IN MILITARY WORKING DOGS
APPENDIX J: WOUND MANAGEMENT IN MILITARY WORKING DOGS
APPENDIX K: MANAGEMENT OF OCULAR INJURIES IN MILITARY WORKING DOGS
APPENDIX L: ANALGESIA AND ANESTHESIA FOR MILITARY WORKING DOGS
APPENDIX M: MANAGEMENT OF TRAUMATIC BRAIN INJURY (TBI) AND ACUTE SPINAL CORD INJURY (ASCI) IN MILITARY WORKING DOGS
APPENDIX N: MANAGEMENT OF CANINE POST TRAUMATIC STRESS DISORDER-LIKE SYNDROME
APPENDIX O: MANAGEMENT OF TRAINING AID TOXICOSES IN MILITARY WORKING DOGS
APPENDIX P: EUTHANASIA OF MILITARY WORKING DOGS
APPENDIX Q: AFTER-ACTION REVIEW OF MILITARY WORKING DOG EMERGENT CARE
APPENDIX R: ADDITIONAL INFORMATION REGARDING OFF-LABEL USES IN CPGs

Background

MWDs are critical assets for military police, special operations units, and others operating in today's combat environment. Expectations are that injured working dogs will receive a high level of resuscitative care as far forward as possible, often where the presence of trained Veterinarians or Animal Care Specialists is uncommon.

The ideal medical providers for MWDs, especially during emergent situations, are military Veterinarians and Animal Care Specialists, supporting the assigned dog handler. However, HCPs may be the only medical personnel available to MWDs that are gravely ill or injured. Generally, provision of medical care by HCPs is limited to circumstances in which the dog is too unstable to transport to supporting veterinary facilities or medical evacuation is not possible due to weather or mission constraints, immediate care is necessary to preserve life, limb, or eyesight, and veterinary personnel are not available. HCPs should only perform medical or surgical procedures necessary to manage problems that immediately threaten life, limb, or eyesight, and to prepare the dog for evacuation to definitive veterinary care. Under no circumstances should routine medical, dental, or surgical care be provided by non-veterinary personnel unless this care has been coordinated with and approved by supporting veterinary personnel.

Veterinary care is established at multiple locations throughout theater. The veterinary health care team will remain the MWD's primary provider, and every effort should be made to contact the supporting veterinary team for guidance immediately upon presentation of an ill or injured MWD. Nothing in this CPG or its appendices authorize HCPs to provide medical care except as outlined above.

Information concerning MWDs (e.g., unit of assignment, operating location, illness or injury data, functional status) should be considered confidential and treated as such.

Working dogs operated by government-contracted personnel may be presented for medical care by HCPs. These contract working dogs (CWDs) are authorized medical care on a reimbursable basis or as stipulated in contracts reviewed by the Theater MWD/CWD Program Manager; however, routine medical care will not be provided to these dogs by HCPs. In emergent situations and as outlined in this CPG for MWDs, HCPs are authorized to provide initial resuscitation and stabilization for these dogs as outlined for MWDs.

Working dogs operated by allied military forces in theater may likewise be presented for emergent care. Established agreements permit US military HCPs to provide necessary emergency resuscitative and stabilization care for allied military dogs, just as if they were US MWDs.

General Handling and Management of MWDs.

MWDs are often unpredictable and potentially dangerous animals, especially when ill, injured, or stressed, and especially when not under the control of their assigned handler. The dog handler is the best person to control the MWD. Handlers also have the most accurate information about past medical problems and current situation. All MWD handlers have received basic emergency first aid training and can assist in limited emergent care; experience levels will vary. If the assigned handler is not present, other handlers are skilled at MWD restraint and handling and should be utilized if available.

Safety of HCPs and bystanders is paramount. For safety reasons, never examine a MWD without a handler or knowledgeable person present, especially if the patient is conscious. MWDs should be muzzled whenever being handled, unless medical issues prevent muzzling. Standard muzzles issued to handlers are ideal; however, roll gauze can be used for temporary control, looped tightly around the muzzle twice, and tied behind the head. Remove the muzzle when not actively handling the dog, if the dog is sedated or anesthetized, if the dog is having breathing difficulty, or if temperature extremes prevent cooling by panting.

MWDs must be controlled and supervised at all times, especially if sedated or anesthetized; someone must be dedicated to this task if the handler is not available. A handler should be present or immediately (on site) available 24 hours a day during periods of MWD hospitalization or treatment. MWDs must never be transported by any means without a handler, or someone to control the dog if the handler is not available. If available, portable kennels are the best means of transport for stable MWDs. If a portable kennel is not available, MWDs transported in aircraft should be muzzled if their medical condition permits.

Evaluation and Treatment

Appendices A through P guide HCPs in management of specific scenarios involving life- or limb-threatening problems with MWDs. As noted in specific appendices, there may be instances in which emergent surgical management of injured MWDs is necessary by HCPs to afford a chance at patient survival. Providers must note that emergent surgical management should be considered only if 1) the provider has the necessary advanced surgical training and experience, 2) the provider feels there is a reasonable likelihood of success, and 3) the provider has the necessary support staff, facilities, and monitoring and intensive care facilities to manage the post-operative MWD without compromising human patient care. Thus, emergent surgical management should be considered only in Level 2 or higher medical facilities and by trained surgical specialists with adequate staff. Direct communication with a US military veterinarian is essential before considering surgical management, and during and after surgery, to optimize outcome.

Performance Improvement Monitoring

HCPs manage MWDs safely in emergent situations to provide life-saving measures to resuscitate and stabilize injured dogs until evacuation is coordinated to a veterinary facility. HCPs do not perform medical procedures for non-emergent reasons or if veterinary personnel are available. HCPs promptly coordinate veterinary support and provide timely feedback in instances in which care is provided to MWDs.

Performance/Adherence Measures

All encounters between HCPs and MWDs will undergo a compliance review by supporting veterinary personnel of key data sources. Each instance will have face-to-face discussions between supporting veterinary personnel and HCPs providing care.

Data Sources

Patient record generated by HCPs during care, Appendix Q (After Action Report) and system reporting & frequency will be utilized to generate clinical information and data required to review and improve current CPG.

The above constitutes the minimum criteria for monitoring of this CPG. System reporting will be performed biannually; additional monitoring and system reporting may be performed as needed. The system review and data analysis will be performed by the Joint Theater Trauma System (JTTS) Director, JTTS Program Manager, and the Joint Trauma System (JTS) Performance Improvement Branch. It is the trauma team leader‘s responsibility to ensure familiarity, appropriate compliance and monitoring at the local level with this CPG.

HCPs will notify veterinary personnel immediately upon notification of an inbound MWD or on the arrival of a MWD in any medical facility, especially in emergent situations. HCPs will complete APPENDIX Q any time emergent care is provided and forward to the supporting veterinary facility, in an effort to evaluate the utility of this and other MWD-related CPGs and to improve allocation of veterinary resources. Veterinary personnel will remain integrally involved in the decision chain for all MWD care issues.

The process for updating CPGs is described in the CENTCOM JTTS CPG Development, Approval, Implementation, and Monitoring Process document posted to the JTS website (http://usaisr.amedd.army.mil/cpgs.html). The Directors determined that one MWD CPG containing multiple appendices was the best way to ensure that the MWD CPG will not be confused with or lost within the human medical CPGs also posted on the JTS website. The in-theater proponents for updates are the JTTS Director and the TF MED/MED BDE Staff Veterinarian or a Medical Detachment (Veterinary Service Support) Commander. Theater recommendations are presented to JTTS POCs, reviewed and passed to the JTS Directorate POCs for final review and approval. The Director, DoD Military Working Dog Veterinary Service has the responsibility to update the MWD CPG in coordination with the JTTS and JTS Directors and is the final veterinary approval authority as the MWD representative of the DOD Executive Agent for Veterinary Services, IAW DODD 5101.1, DoD Executive Agent, November 21, 2003; DODD 5200.31E, DoD Military Working Dog (MWD) Program, August 10, 2011; DODD 6400.4, DoD Veterinary Services Program, November 24, 2003; and, Memorandum, Surgeon General, 9 December 2008, Subject: Delegating Roles and Responsibilities for the DoD Veterinary Service Activity (DODVSA).

REFERENCES

Bernhard M, Gries A, Kremer P, et al. Spinal cord injury (SCI)—prehospital management. Resuscitation 2005;66:127-139. Bouchama A, Dehbi M, Chaves-Carballo E. Cooling and hemodynamic management in heatstroke: practical recommendations. Critical Care [serial online]. 05/12/07;11 (issue 3):R54. Available at: www.ccforum.com/content/11/3/R54. Accessed May 15, 2010. Bright PA. Military working dog training aids: toxicity and treatment. Air Force Occupational and Environmental Health Laboratory (AFSC), Human Systems Division, Brooks Air Force Base, TX, 1989;1-32. Bruchim Y, Klement E, Saragusty J, et al. Heat stroke in dogs: a retrospective study of 54 cases (1999–2004) and analysis of risk factors for death. J Vet Intern Med 2008;1:38-46. Bonczynski JJ, Ludwig LL, Barton LJ, et al. Comparison of peritoneal fluid and peripheral blood pH, bicarbonate, glucose, and lactate concentrations as a diagnostic tool for septic peritonitis in dogs and cats. Vet Surg 2003;32:161-166. Casa DJ. Exercise in heat. I. Fundamentals of thermal physiology, performance implications, and dehydration. J Athlet Train 1999;34:246-252.

7. Centers for Disease Control and Prevention. After an earthquake: management of crush injury and crush syndrome. Accessed online 12 April 2011: http://www.bt.cdc.gov/disasters/earthquakes/crush.asp. Centers for Disease Control and Prevention. Explosions and blast injuries: a primer for clinicians. Accessed online 01 April 2011: http://www.bt.cdc.gov/masscasulaties/explosions.asp. Cole SG, Otto CM, Hughes D. Cardiopulmonary-cerebral resuscitation in small animals–a clinical practice review, Part II. J Vet Emerg Crit Care, 2003;13:1323. Crowe DT, Jr., Devey JJ. Assessment and management of the hemorrhaging patient. Vet Clin N Am (Sm Anim Pract), 1994;24:1095-1122. Dewey CW. Brain Trauma. In: The Veterinary ICU Book. Teton NewMedia, Jackson, WY, 2003.

12. Drobatz K. Triage and initial assessment. In: Manual of Canine and Feline Emergency and Critical Care. King L and Hammond R, eds. British Small Animal Veterinary Association, 1999, pp 1-7. Drobatz KJ. Heat stroke. In: Silverstein DC, Hopper K, eds. Small Animal Critical Care Medicine. St. Louis: Saunders/Elsevier, 2009;723-726. Emergency War Surgery, 3rdUS revision. US Government Printing Office, Washington, DC, 2004. Fletcher DJ, Syring R. Traumatic Brain Injury. In: Small Animal Critical Care Medicine, Silverstein DC and Hopper K, eds. Saunders/Elsevier, St Louis, MO, 2009.. Flournoy WS, Macintire DK, Wohl JS. Heat stroke in dogs: clinical signs, treatment, prognosis, and prevention. Comp Cont Edu Pract Vet 2003;25:422-431. Gall TT and Monnet E. Evaluation of fluid pressures of common wound-flushing techniques. Am J Vet Res 2010;71:1384-1386. Garzotto CK. Thermal burn injuries. In: Silverstein DC, Hopper K, eds. Small Animal Critical Care Medicine. St. Louis: Saunders/Elsevier, 2009;683-686. Garzotto CK. Wound management. In: Silverstein DC, Hopper K, eds. Small Animal Critical Care Medicine. St. Louis: Saunders/Elsevier, 2009;676-682. Gonzalez D. Crush syndrome. Crit Care Med, 2005;33(Suppl 1):S34-S41. Halling K. Wounds and open fractures. In: Mathews K, ed. Veterinary Emergency and Critical Care Manual. Guelph, Ontario, Canada: Lifelearn, Inc., 2006;702-708. Harker J, Gibson P. Heat-stroke: a review of rapid cooling techniques. Intensive Crit Care Nursing 1995;11:198-202. Hoedemaekers CW, Ezzahti M, Gerritsen M, et al. Comparison of cooling methods to induce and maintain normo- and hypothermia in intensive care unit patients: a prospective intervention study. Critical Care [serial online]. 08/24/07;11 (issue 4):R91. Available at: www.ccforum.com/content/11/R91. Accessed May 15, 2010. Levin GM, Bonczynski JJ, Ludwig LL, et al. Lactate as a diagnostic test for septic peritoneal effusions in dogs and cats. J Am Anim Hosp Assoc 2004;40:364-371. Kovacic JP. Management of life-threatening trauma. Vet Clin N Am (Sm Anim Pract) 1994;24:1057-1094. Lisciandro GR, Lagutchik MS, Mann KA, et al. Evaluation of an abdominal fluid scoring (AFS) system determined using abdominal focused assessment with sonography for trauma (AFAST) in 101 dogs with motor vehicle trauma. J Vet Emerg Crit Care, 2009;19:426-437. Lisciandro GR, Lagutchik MS, Mann KA, et al. Evaluation of a thoracic focused assessment with sonography for trauma (TFAST) protocol to detect pneumothorax and concurrent thoracic injury in 145 traumatized dogs. J Vet Emerg Crit Care 2008;18:258-269. Mathews K. Accidental hypothermia. In: Mathews K, ed. Veterinary Emergency and Critical Care Manual. Guelph, Ontario, Canada: Lifelearn, Inc., 2006;291-296. Mathews K. Burn injury and smoke inhalation. In: Mathews K, ed. Veterinary Emergency and Critical Care Manual. Guelph, Ontario, Canada: Lifelearn, Inc., 2006;682-689. Mathews K. Hyperthermia, heat stroke, malignant hyperthermia. In: Mathews K, ed. Veterinary Emergency and Critical Care Manual. Guelph, Ontario, Canada: Lifelearn, Inc., 2006;297-303. McCurnin DM, Bassert JM, eds. McCurnin's Clinical Textbook for Veterinary Technicians, 7th ed. Saunders, St. Louis, MO, 2009. McMichael M. Heatstroke. In: Cann CC, Hunsberger S, eds. Handbook of Veterinary Emergency Protocols: Dog and Cat. Jackson, WY: Teton NewMedia, 2008;228-230. NATO STANAG 2534 MED (Edition 1). Welfare, Care and Veterinary Support for Deployed Military Working Animals. NATO Standardization Agency, 02 Dec 2005. NATO STANAG 2538. Animal Care and Welfare and Veterinary Support During All Phases of Military Deployments. NATO Standardization Agency, 19 Nov 2009. Oncken AK, Kirby R, Rudloff E. Hypothermia in critically ill dogs and cats. Comp Cont Edu Pract Vet 2001;23:506-521. Pennardt A. Blast injuries, updated 28 May 2010. Accessed online 01 April 2010: http://emedicine.medscape.com/article/822587. Platt SR. Coma Scales. In: Small Animal Critical Care Medicine, Silverstein DC and Hopper K, eds. Saunders/Elsevier, St Louis, MO, 2009. Shawver D, Battaglia A. Mechanical ventilation. In: Battaglia A, ed. Small Animal Emergency and Critical Care: A Manual for the Veterinary Technician. Ithaca, NY: WB Saunders Co, 2000;100-107. Simpson SA, Syring RS, Otto CM. Severe blunt trauma in dogs: 235 cases (1997-2003). J Vet Emerg Crit Care 2009;19:588-602. Smarick SD. Heatstroke: keeping them alive. In: Proceedings 14th Int Vet Emerg Crit Care Symp. Chicago, IL. 2009;167-170. Soldier's Manual and Trainer's Guide, MOS 68T Animal Care Specialist, STP 8-91T14-SM-TG. Headquarters,

Department of the Army, 20 Aug 2002. Tabor B. Heatstroke in dogs. Vet Tech 2007;28. Available at: www.vettechjournal.com. Accessed May 31, 2010. TB MED 283. Veterinary Necropsy Protocol for Military Working Dogs and Pathology Specimen Submission Guidelines. Headquarters, Department of the Army, 24 May 2001. The Handbook of Veterinary Care and Management of the Military Working Dog. DoD Military Working Dog Veterinary Services, Lackland AFB, TX, 2004. Thulesius O. Thermal reactions of blood vessels in vascular stroke and heatstroke. Med Principles Practice 2006;15:316-321. Tillson MD. Open fracture management. Vet Clin North Am Small Anim Pract 1995;1093-1110. Todd J, Powell LL. Hypothermia. In: Silverstein DC, Hopper K, eds. Small Animal Critical Care Medicine. St. Louis: Saunders/Elsevier, 2009;720-722. Vogelsang R. Care of the Military Working Dog by Medical Providers. J Special Operations Med 2007;7:33-47. Wingfield WE, Palmer SB. In: Veterinary Disaster Response. Wiley-Blackwell, Ames, IA, 2009. Wingfield WE, Nash SL, Palmer SB, et al. In: Veterinary Disaster Medicine: Working Animals. Wiley-Blackwell, Ames, IA, 2009. Wingfield WE, Raffe MR, eds. The Veterinary ICU Book. Teton NewMedia, Jackson, WY, 2003.

ONE HEALTH IN THE EXAM ROOM

Linda Kidd DVM, Ph.D., DACVIM
Pomona, CA

INTRODUCTION

The definition of the one health initiative is as follows: "The One Health Initiative is a movement to forge co-equal, all inclusive collaborations between physicians, osteopaths, veterinarians, dentists, nurses and other scientific-health and environmentally related disciplines, including the American Medical Association, American Veterinary Medical Association, American Academy of Pediatrics, American Nurses Association, American Association of Public Health Physicians, the American Society of Tropical Medicine and Hygiene, the Centers for Disease Control and Prevention (CDC), the United States Department of Agriculture (USDA), and the U.S. National Environmental Health Association (NEHA)……"[1] The study and prevention of cross-species disease transmission is a major goal of the One Health Initiative.

In an ideal world we could screen for and eradicate every infectious agent that might affect the health of cats and dogs and their human companions. Unfortunately that is not possible. There are many agents of potential zoonotic importance that dogs and cats may play a role in transmitting to people. Testing for each one on every patient is cost prohibitive. Furthermore, depending on the organism in question, treatment can be expensive, have side effects and does not always eliminate infection. Today comprehensive diagnostic panels are available that simultaneously test for many organisms that cause a particular constellation of clinical signs. However, the sensitivity and specificity of commonly used diagnostic tests such as serology and PCR varies with the pathophysiology of the organism and other factors. Therefore, more than one modality may be necessary to confidently rule out infection in an individual patient. Furthermore, for some agents, specialized testing and particular sampling methods are required to detect infection. In addition, it is necessary to have a high index of suspicion of zoonotic disease BEFORE leaving the examination room in order to protect hospital and laboratory personnel. Therefore, familiarity with case contexts that make a particular infectious disease more likely is required to determine which organisms should be included in testing, whether more than one testing modality or repeated testing is warranted, and to use appropriate precautions to prevent in hospital disease transmission.

In this session, problems and facts identified in the signalment, history and physical examination that should trigger the consideration of selected infectious diseases of zoonotic importance will be reviewed. This is not a comprehensive list. Only major clinical signs are considered for some organisms of zoonotic importance. More comprehensive information regarding the clinical findings associated with zoonotic and other infectious disease may be found in the resources listed below. Of note, the term "zoonotic" as it is used here will include infectious disease that may be transmitted from dogs and cats to people and vice versa, and will also refer to infectious diseases where dogs may act as sentinels but they are not known to be reservoirs or transmit the organisms directly to people.

THE SIGNALMENT

Consideration of the age, breed and sex can help increase the index of suspicion for certain infectious diseases in the context of a case presentation. Age may impact the prevalence of infection and severity of clinical signs. A familiar example is *Toxocara* infections in neonates. Another example is that puppies and kittens are more susceptible to *Salmonella* infection. Rocky Mountain Spotted Fever (RMSF) has been reported to be more frequent young adult animals in some studies. The prevalence of *Giardia* infection is higher in puppies than other age groups.[2] Many zoonotic diseases that cause subclinical infection in adults are more likely to manifest clinical signs in young animals due to immune status, the pathophysiology of the organism, the environment (eg. overcrowding) and other variables. Examples include giardiasis, salmonellosis, campylobacteriosis, and leptospirosis.

Some agents are more common in certain breeds, presumably due to breed related activities such as hunting, or possibly due to immune status and other factors. Hunting breeds are overrepresented in prevalence studies for a number of vector borne diseases including RMSF and environmentally acquired infections such as histoplasmosis. Foxhounds have a high prevalence of leishmaniasis in the United States. The prevalence of *Giardia* infection is higher in purebred dogs.[2] The prevalence of some infectious diseases is also affected by sex. Possible explanations include differences in activities or housing practices that affect exposure, differences in hormone levels that change host susceptibility, or the pathophysiology of the organisms themselves. For example, organisms such as *Sporothrix shenkii* that are introduced by skin wounds are more common in male cats. Intact dogs may have increased exposure to *Brucella* through breeding and whelping. Interestingly, *Giardia* was found to be more prevalent in intact dogs in a recent large nationwide study.[2] The authors hypothesized that hormonal differences or differences in husbandry might play a role.

THE HISTORY

Facts about geographic locale and travel history impact zoonotic disease exposure. Therefore it is important for veterinarians to be cognizant of the typical geographic distribution of zoonotic diseases. For example, due to differences in geographic distribution of tick vectors, a veterinarian living in Oklahoma may consider diagnostic testing to rule out infection with *Ehrlichia ewingii* in a dog presenting with neutrophilic polyarthritis whereas a veterinarian living in Wisconsin would pursue testing for *Anaplasma phagocytophilum* and *Borrelia burgdorferi*. Similarly, a veterinarian living in Texas may perform more extensive diagnostic testing to rule out *Trypanosoma cruzi* infection in a dog with dilated cardiomyopathy than a veterinarian practicing in Minnesota. That said, clinicians should be aware that the geographic distribution of many diseases is shifting due to increases in travel and climate change. For example, an outbreak of RMSF in people in Arizona occurred in 2005.[3] Prior to that time the disease was exceedingly rare in that region of the United States. The epidemic was caused by a novel tick vector *Rhipicephilus sanguineus*, and it was subsequently found that infection was present in the dog population prior to infection in people. Therefore, knowledge of typical geographic distributions for zoonotic disease is important but the possibility of expanding geographic distributions for these agents should always be kept in mind.

Facts about health, occupation and lifestyle of owners and pets will help raise the index of suspicion or concern for screening for certain zoonotic infectious diseases in a canine or feline patient. For example, if an owner or member of the household is immunosuppressed, increased vigilance for the presence of zoonotic disease is necessary and protocols for screening even healthy pets should be followed.[4] Immunosupression in pets due to concurrent illness and/or medication may potentially result in the reactivation and shedding of zoonotic agents such as toxoplasmosis. The occupation of the owner can also impact the likelihood of various zoonotic diseases being present. For example, concern about *Ecchinococcus granulosa* might be increased if an owner is a sheep farmer. "Occupation" of the pet should also be considered. For example, a recent study showed that dogs that visited health care facilities were more likely to acquire MRSA and *Clostridium difficile* than other animal assisted intervention dogs.[5] Furthermore, dogs used for hunting are at increased risk for exposure to a number of agents acquired through direct environmental exposure and vector borne diseases. That said, it should be kept in mind that many zoonotic diseases including environmentally acquired and vector borne diseases occur in urban and suburban environments as well. Pets that roam freely in rural or urban environments have the opportunity for increased exposure. Unsupervised roaming gives cats and dogs the opportunity to ingest rodents, birds and other prey. This exposes them to many very important human pathogens such as *Yersinia pestis*, *Echinococcus multilocularis*, *Franciscella tularensis, Toxoplasma gondii*, and *Salmonella typhimurium* (songbird fever).

It is also very important to solicit information regarding frequency of routine fecal examinations, anthelmentic administration for ascarids, hookworms and cestodes, vaccination history, flea and tick control, and heartworm testing and prevention to assess risk and likel hood of infection. Recommendations for anthelmentic prophylaxis and fecal examinations are available from the Centers for Disease Control and Prevention.[6]

Information about diet is important to assess nutritional status and to assess potential exposure to pathogens. Although cooked commercial foods are not risk free with regard to contamination with pathogens, raw food or non-commercial sources have increased potential to expose pets and thus human companions to many zoonotic diseases of importance. The link between feeding raw diets and salmonellosis has been shown in several studies. Depending on the meat source, other infectious agents can be transmitted in raw diets as well, for example, the risk of hydatidosis in people was higher if dogs ingested uncooked viscera or carrion in one study.[7] Cats fed raw food or undercooked viscera and or meat had a significantly higher seroprevalence documenting exposure to *Toxoplasma gondii* than those fed commercial canned or dry food in another recent study.[8]

Clinical signs of coughing, sneezing vomiting are relatively nonspecific with regard to consideration of a specific zoonotic agent. Although the sign of diarrhea is nonspecific, it is a major clinical sign associated with many common important zoonotic pathogens that should be considered. These include but are not limited to *Toxocara sp.*, *Ancylostoma sp.*, *Giardia sp.*, *Cryptosporidium*, *Salmonella* sp., and *Campylobacter sp.* Leptospirosis should always be considered as a differential for PUPD localized to renal insufficiency in a dog. Infectious disease may be associated with weight loss due to a variety of mechanisms. Of interest, rapid and dramatic weight loss has been associated with RMSF in dogs and weight loss and emaciation is a prominent clinical sign associated with visceral leishmaniasis and histoplasmosis. Zoonotic agents of particular importance that cause the clinical signs of stillbirth and abortion include *Coxiella burnetii* and *Brucella canis*. Finally, it is important to note that many infectious agents of zoonotic importance cause subclinical infection in the majority of pets, so a lack of reported problems is not equal to a lack of potential for disease transmission.

PHYSICAL EXAMINATION

The index of suspicion for a given zoonotic disease usually requires the presence of a constellation or pattern of clinical abnormalities in the history physical examination and laboratory testing. However, some physical examination findings should trigger consideration of certain zoonotic diseases. The following is a list by affected body system of selected physical examination abnormalities and associated zoonotic diseases and/or agents that might be considered. The list of findings only includes zoonotic diseases and it is not complete. Further information can be found in the references and resources listed below.

Oral
Normal flora: Bites and licks may transmit *Pasturella* and other organisms to immunocompromised individuals.
Petechiae of mucous membranes: Ehrlichiosis, RMSF, leptospirosis.
Mucocutaneous ulceration: Leishmaniasis, sporothrichosis

Integument
Ectoparasites:

Fleas: *Ctenocephalides felis* are vectors for important human pathogens including *Bartonella* species, *Rickettsia typhi*, and *Rickettsia felis*.

Ticks: Ixodid (hard bodied ticks) are vectors for many diseases that affect humans and/or dogs and cats. In some instances dogs may act as reservoirs of infection but more commonly dogs serve as sentinels for tick borne diseases in people and may increase human contact with ticks that carry infectious agents. Examples of tick borne infectious agents infecting both small animals and people include *Ehrlichia sp.*, *Anaplasma phagocytophilum*, Spotted fever group *Rickettsia*, *Francisella tularensis* and *Borrelia* species. Ticks may also harbor infectious agents that are not known to cause disease in small animals but infect humans. For example, *Ixodes scapularis*, the vector of *A. phagocytophilum* and *B. burgdorferi* also carries *Babesia microti*, which is not unknown to cause disease in dogs and cats but does so in people. Owners of pets found to be infested with ticks should be educated in proper tick removal [9] and prevention and told to contact their physician if they have any signs of illness.

Hairloss/alopecia/scaling/ pustules: Leishmaniasis, dermatophytosis, scabies, *Malassezia pachydermatis*.
Draining tracts/nodules/abscesses: Sporotrichosis, mycotic infections, mycobacteriosis, tularemia, pasteurellosis, bartonellosis.
Petechiation: Leptospirosis, RMSF, ehrlichiosis, anaplasmosis.
Edema/vasculitis: Leptospirosis, RMSF.

Ocular
Uveitis/chorioretinitis/intraocular infections: Brucellosis, RMSF, leptospirosis, ehrlichiosis, anaplasmosis, mycobacteriosis, systemic mycoses, toxoplasmosis, leishmaniasis, bartonellosis.
Conjunctivitis: *Chlamydophila felis*, leptospirosis, leishmaniasis.

Lymphatic: Lymphadenomegaly:Bartonellosis, tularemia, sporothrichosis, leishmaniasis, acute Chagas Disease, systemic mycoses, ricketttsial disease, Plague (draining abscessed LN). Splenomegaly: Rickettsial disease, bartonellosis, histoplasmosis, tularemia, mycobacteriosis.

Nasal: Nasal discharge/rhinitis: Bordetellosis, leishmaniasis, Plague, tularemia, bartonellosis, mycoses.
Epistaxis: Infectious diseases of varied zoonotic importance can cause epistaxis by nasal turbinate destruction or by affecting primary hemostasis. For example, *Crypococcus* is a common cause of nasal turbinate destruction and epistaxis in cats, whereas the thrombocytopenia, thrombocytopathia and vasculitis may contribute to the clinical signs of epistaxis in dogs with ehrlichiosis.

Cardiac: Chagas disease, bartonellosis, ehrlichiosis.

Trachea/Bronchi: Tracheitis/bronchitis: Bordetellosis.

Pulmonary: Blastomycosis, coccidiomycosis, cryptococcosis, histoplasmosis, bordetellosis, toxoplasmosis, mycobacteriosi , *Mycoplasma*, leptospirosis, RMSF, Pasteurellosis, Pneumonic plague, tularemia, dirofilariasis.

Gastrointestinal: Intussusception: Leptospirosis. Pancreatitis: Toxoplasmosis, cryptococcosis. Liver: Icterus and hepatomegaly: leptospirosis, histoplasmosis, others.

Renal: Signs of pain or renomegaly suggesting acute renal faiure: Leptospirosis.

Central Nervous System Abnormalities: Rabies, bartonellosis,RMSF, ehrlichiosis systemic mycoses, protozoal infection, borreliosis, anaplasmosis.

Musculoskeletal: Joint effusion: Infective arthritis and/or immune mediated polyarthritis, *Bartonella spp. E. ewingii, E chaffeensis, A. phagocytophylim, Leishmania, Mycoplasma, Borrelia burgdorferi* and *Rickettsia rickettsii.* Discospondylitis: Brucellosis, mycotic infection. Muscle: Polymyositis/local myositis/myalgia: Trypanosomiasis, leishmaniasis, toxoplasmosis, leptospirosis, borrieliosis, sporothrichosis, RMSF.

Orchitis/epididymitis: RMSF, brucellosis, mycoses.

SUMMARY

Prior to leaving the exam room and performing diagnostic testing, findings in the signalment, history and physical examination can help clinicians consider relevant infectious diseases of zoonotic importance in every clinical encounter, thereby promoting the health and wellbeing of animals and their human companions.

REFERENCES

1. http://www.onehealthinitiative.com/index.php. 2. Mohomad et al Vet Parasit 2013 Jan. epub ahead of print. 3. Nicholson W.L. Ann NY Acad Sci 2006 1078:519. 4. Glaser C. A. et al In Greene C. E. Infectious Diseases of the Dog and Cat 4th ed. 2012 1141. 5. Lefebvre et al JAVMA 2009 234(11):1404. 6. http://www.cdc.gov/parasites/zoonotichookworm/resources/prevention.pdf. 7. Campos-Bueno Am J Trop Med Hyg 2000 62(3) 329. 8. Lopes et al. Vet Parasit 2008 155(3-4): 184. 9. http://www.cdc.gov/ticks/removing_a_tick.html.

ADDITIONAL RESOURCES

1. Greene C.E, Infectious Disease of the Dog and Cat 4th Ed. Elsevier 2012. 2. Sykes J. E., Canine and Feline Infectious Diseases. 1st ed. Elsevier Saunders 2013 In Press. 3. Rabinowitz P. M. Human Animal Medicine Saunders 2010.

MECHANISMS AND PATHOGENESIS OF GASTROINTESTINAL MOTILITY DISORDERS

Robert J. Washabau, VMD, PhD, DACVIM
St. Paul, Minnesota

Canine Idiopathic Megaesophagus

Etiology - Idiopathic megaesophagus is the most common cause of regurgitation in the dog. The disorder is characterized by esophageal hypomotility and dilation, progressive regurgitation, and loss of body condition. Several forms of the syndrome have been described, including congenital idiopathic, acquired secondary, and acquired idiopathic megaesophagus.

Congenital idiopathic megaesophagus is a generalized hypomotility and dilation of the esophagus causing regurgitation and failure to thrive in puppies shortly after weaning. An increased breed incidence has been reported in the Irish setter, Great Dane, German shepherd, Labrador retriever, Chinese Shar-Pei, and Newfoundland breeds, and autosomal dominant inheritance has been demonstrated in the Miniature Schnauzer and Fox terrier breeds. The pathogenesis of the congenital form is incompletely understood, although several studies have pointed to a defect in the vagal afferent innervation of the esophagus. Congenital idiopathic megaesophagus has been reported in several cats, and in one group of cats secondary to pyloric dysfunction. Acquired secondary megaesophagus may develop in association with a number of other conditions. Myasthenia gravis accounts for 25-30% of the secondary cases. In some cases of myasthenia gravis, regurgitation and weight loss may be the only presenting signs of the disease, whereas in most other cases of acquired secondary megaesophagus regurgitation is but one of many clinical signs including peripheral muscle weakness. Acquired secondary megaesophagus has also been associated with hypoadrenocorticism, lead poisoning, lupus myositis, and severe forms of esophagitis. Hypothyroidism has been suggested as a secondary cause of idiopathic megaesophagus but retrospective risk factor analysis has not identified it as an important cause.

The syndrome occurs spontaneously in adult dogs between 7 to 15 years of age without sex or breed predilection. The disorder has been compared erroneously to esophageal achalasia in humans. Achalasia is a failure of relaxation of the lower esophageal sphincter and ineffective peristalsis of the esophageal body. A similar disorder has never been rigorously documented in the dog. Several important differences between idiopathic megaesophagus in the dog and achalasia in humans have been documented. Although the etiology(ies) has not been identified, some studies have suggested a defect in the afferent neural response to esophageal distension similar to what has been reported in congenital megaesophagus.

Clinical Examination

Survey radiographs will be diagnostic for most cases of megaesophagus. Contrast radiographs may be necessary in some cases to confirm the diagnosis, evaluate motility, and exclude foreign bodies or obstruction as the cause of the megaesophagus. Endoscopy will confirm the diagnosis and may further reveal esophagitis, a frequent finding in canine idiopathic megaesophagus.

If acquired secondary megaesophagus is suspected, additional diagnostic tests should be considered, for example: serology for nicotinic acetylcholine receptor antibody, ACTH stimulation, serology for antinuclear antibody, serum creatine phosphokinase activity, electromyography and nerve conduction velocity, and muscle and nerve biopsy. Additional medical investigation will be dependent upon the individual case presentation. Hypothyroidism has been cited as an important cause of idiopathic megaesophagus in the dog, although risk factor analysis has not revealed a clear association.

Treatment

Animals with secondary acquired megaesophagus should be appropriately differentiated from other esophageal disorders and treated. Dogs affected with myasthenia gravis should be treated with pyridostigmine (1.0-3.0 mg/kg PO BID) and/or corticosteroids (prednisone 1.0-2.0 mg/kg PO or SQ BID), dogs affected with hypothyroidism should be treated with levothyroxine (22 µg/kg PO BID), and dogs affected with polymyositis should be treated with prednisone (1.0-2.0 mg/kg PO BID). If secondary disease can be excluded, therapy for the congenital or acquired idiopathic megaesophagus patient should be directed at nutritional management and treatment of aspiration pneumonia. Affected animals should be fed a high-calorie diet, in small frequent feedings, from an elevated or upright position to take advantage of gravity drainage through a non-peristaltic esophagus. Dietary consistency should be formulated to produce the fewest clinical signs. Some animals handle liquid diets quite well, while others do better with solid meals. Animals that cannot maintain adequate nutritional balance with oral intake should be fed by temporary or permanent tube gastrostomy.

Smooth muscle prokinetic (e.g., metoclopramide or cisapride) therapy has been advocated for stimulating esophageal peristalsis in affected animals, however metoclopramide and cisapride will not likely have much of an effect on the striated muscle of the canine esophageal body. Bethanechol has been shown to stimulate esophageal propagating contractions in some affected dogs and is therefore a more appropriate prokinetic agent for the therapy of this disorder. Because of the high incidence of esophagitis in canine idiopathic megaesophagus, affected animals should also be medicated with oral sucralfate suspensions (1 g q8h for large dogs 0.5 g q 8h for smaller dogs 0.25 to 0.5 g q8h to q12h for cats).

Gastric Emptying Disorders

Gastric emptying disorders are fairly common in dogs and cats. They result from disease processes that alter normal gastric functions, i.e. storage of ingesta, mixing and dispersion of food particles, and timely emptying of gastric contents into the small intestine. Disorders of gastric emptying arise from mechanical obstruction, or from defective propulsion. Anatomic lesions (e.g. malignancy, hyperplasia, foreign bodies) cause delayed gastric emptying because of mechanical obstruction. Diagnosis and management of mechanical obstruction is usually straight-forward. Disorders of defective propulsion, on the other hand, cause delayed gastric emptying because of abnormalities in myenteric neuronal or gastric smooth muscle function, or because of abnormalities in antropyloroduodenal coordination. A number of primary conditions have been associated with these functional disorders, including infectious or inflammatory disease, ulcer, and post-surgical gastroparesis. Delayed gastric emptying has also been associated with a number of secondary conditions, including electrolyte disturbances, metabolic disorders, concurrent drug usage (cholinergic antagonists, adrenergic agonists, opioid agonists), acute stress, and acute abdominal inflammation. Recovery from gastric dilation/volvulus is almost always associated with significant myoelectrical and motor abnormalities in the dog. Diagnosis and management of the delayed gastric emptying disorders may not be so straight-forward. Nutritional and medical management, including smooth muscle prokinetic agents (e.g., 5-HT_4 drugs, erythromycin, and ranitidine), are important components of therapy.

Small Intestinal Transit Disorders

A number of small intestinal transit disorders have been described in dogs and cats, including enteritis, post-surgical pseudo-obstruction, nematode infection, intestinal sclerosis, and radiation enteritis. Vomiting and diarrhea are the most important clinical signs associated with these disorders. Overgrowth of small intestinal bacteria, a common sequela to disordered motility, contributes to these clinical signs. Transit disorders associated with mechanical obstruction should always be differentiated and treated appropriately. Delayed transit associated with functional disorders should be managed with dietary modification (low fat diets) and prokinetic agents (cisapride, tegaserod, or metoclopramide). Tegaserod has been shown to normalize intestinal transit in opioid-induced bowel dysfunction in dogs, but the drug was removed from the American market in 2007.

Colonic Motility Disorders

History

Constipation, obstipation, and megacolon may be observed in cats of any age, sex, or breed, however, most cases are observed in middle aged (mean = 5.8 years), male cats (70% male, 30% female) of Domestic Shorthair (46%), Domestic Longhair (15%), or Siamese (12%) breeding. Affected cats are usually presented for reduced, absent, or painful defecation for a period of time ranging from days to weeks or months. Some cats are observed making multiple, unproductive attempts to defecate in the litter box, while other cats may sit in the litter box for prolonged periods of time without assuming a defecation posture. Dry, hardened feces are observed inside and outside of the litter box.

Physical Examination

Colonic impaction is a consistent physical examination finding in affected cats. Other findings will depend upon the severity and pathogenesis of constipation. Dehydration, weight loss, debilitation, abdominal pain, and mild to moderate mesenteric lymphadenopathy may be observed in cats with severe idiopathic megacolon. Colonic impaction may be so severe in such cases as to render it difficult to differentiate impaction from colonic, mesenteric, or other abdominal neoplasia. Cats with constipation due to dysautonomia may have other signs of autonomic nervous system failure, such as urinary and fecal incontinence, regurgitation due to megaesophagus, mydriasis, decreased lacrimation, prolapse of the nictitating membrane, and bradycardia. Digital rectal examination should be carefully performed with sedation or anesthesia especially in those cats with recurring bouts of constipation. Pelvic fracture malunion may be detected on rectal examination in cats with pelvic trauma. Rectal examination might also identify other unusual causes of constipation, such as foreign bodies, rectal diverticula, stricture, inflammation, or neoplasia.

Differential Diagnoses

Several authors have emphasized the importance of considering an extensive list of differential diagnoses (e.g. neuromuscular, mechanical, inflammatory, metabolic/endocrine, pharmacologic, environmental, and behavioral causes) for the obstipated cat. A review of published cases, however, suggests that 96% of cases of obstipation are accounted for by idiopathic megacolon (62%), pelvic canal stenosis (23%), nerve injury (6%), or Manx sacral spinal cord deformity (5%). A smaller number of cases are accounted for by complications of colopexy (1%) and colonic neoplasia (1%); colonic hypo- or aganglionosis was suspected, but not proved, in another 2% of cases. Inflammatory, pharmacologic, and environmental/behavioral causes were not cited as predisposing factors in any of the original case reports. Endocrine factors (obesity, n=5; hypothyroidism, n=1) were cited in several cases, but were not necessarily impugned as part of the pathogenesis of megacolon.

Therapeutic Plan

The specific therapeutic plan will depend upon the severity of constipation and the underlying cause. Medical therapy may not be necessary with first episodes of constipation. First episodes are often transient and resolve without therapy. Affected animals should always be re-hydrated if dehydration has contributed to the onset of clinical signs. Mild to moderate or recurrent episodes of constipation usually require some medical intervention. These cases may be managed, often on an outpatient basis, with dietary modification, water enemas, oral or suppository laxatives, and/or colonic prokinetic agents. Severe cases of constipation usually require brief periods of hospitalization to correct metabolic abnormalities and to evacuate impacted feces using water enemas, manual extraction of retained feces, or both. Followup therapy in such cases is directed at correcting predisposing factors and preventing recurrence. Subtotal colectomy will become necessary in cats suffering from obstipation or idiopathic dilated megacolon. These cats, by definition, are unresponsive to medical therapy. Pelvic osteotomy without colectomy may be sufficient for some cats suffering from pelvic canal stenosis and hypertrophic megacolon.

REFERENCES

Washabau R. Oropharyngeal motility disorders. In, Washabau and Day, Canine and Feline Gastroenterology, 2013, Elsevier, St. Louis, MO, pp 551-553.

Washabau R. Esophageal motility disorders. In, Washabau and Day, Canine and Feline Gastroenterology, 2013, Elsevier, St. Louis, MO, pp 591-595.

Washabau R. Gastric motility disorders. In, Washabau and Day, Canine and Feline Gastroenterology, 2013, Elsevier, St. Louis, MO, pp 630-634.

Washabau R. Intestinal motility disorders. In, Washabau and Day, Canine and Feline Gastroenterology, 2013, Elsevier, St. Louis, MO, pp 706-710

Washabau R. Colonic motility disorders. In, Washabau and Day, Canine and Feline Gastroenterology, 2013, Elsevier, St. Louis, MO, pp 757-764.

Washabau R. Anorectal motility disorders. In, Washabau and Day, Canine and Feline Gastroenterology, 2013, Elsevier, St. Louis, MO, pp 791-792

METHODS OF ASSESSING GASTROINTESTINAL MOTILITY
Robert J. Washabau, VMD, PhD, DACVIM
St. Paul, Minnesota

Disorders of Gastrointestinal Motility

Disorders of gastrointestinal motility represent a diagnostic and therapeutic challenge. Gastrointestinal motility disorders may result in delayed transit, accelerated transit, impaired relaxation, or inappropriate relaxation.[1] The delayed transit disorders are the most important motility disorders of companion animals and may involve the esophagus (e.g., idiopathic megaesophagus), stomach (e.g., delayed gastric emptying), small intestine (e.g., ileus or pseudo-obstruction), or colon (e.g., constipation) independently, or as a more generalized and diffuse gastrointestinal motility disorder (e.g., dysautonomia).[2]

Idiopathic megaesophagus – Idiopathic megaesophagus is the most common form of regurgitation in the dog. The disorder is characterized by esophageal hypomotility and dilation, progressive regurgitation, and loss of body condition. Several forms of the syndrome have been characterized, including congenital, acquired secondary, and acquired idiopathic megaesophagus.[2,3,4]

Gastric Emptying Disorders – Gastric emptying disorders are fairly common in dogs and cats and are an important cause of nausea and vomiting. Primary conditions that have been associated with delayed gastric emptying include infectious and inflammatory disease (e.g., IBD), ulcer, and post-surgical gastroparesis, while secondary conditions include electrolyte disturbances, metabolic disorders, concurrent drug usage (cholinergic antagonists, adrenergic agonists, opioid agonists), acute stress, and acute abdominal inflammation.[5] Recovery from gastric dilatation/volvulus (GDV) is almost always associated with significant myoelectrical and motor abnormalities in the dog.[6]

Small Intestinal Transit Disorders – Several small intestinal transit disorders have been described in the dog and cat, including inflammatory bowel disease, post-surgical pseudo-obstruction, nematode infection, intestinal sclerosis, and radiation enteritis.[2] Vomiting and diarrhea are the most important clinical signs associated with these disorders. Overgrowth of small intestinal bacteria, a common sequela to disordered motility, may contribute to these clinical signs.

Colonic Motility Disorders – Constipation, obstipation, and megacolon are primarily a disorder of the domestic cat.[2,7,8] An extensive list of differential diagnoses (e.g., neuromuscular, mechanical, inflammatory, metabolic, endocrine, pharmacologic, environmental, and behavioral causes) have been proposed, but most (> 96%) cases are accounted for by idiopathic megacolon (62%), pelvic canal stenosis (23%), nerve injury (6%), or Manx sacral spinal cord deformity (5%).[7]

Dysautonomia – Dysautonomia is a generalized autonomic neuropathy that was originally reported in cats in the United Kingdom, but that has now been documented in dogs and cats throughout western Europe and the United States. The clinical signs reflect a generalized autonomic dysfunction, including megaesophagus and esophageal hypomotility, gastric dilatation and delayed gastric emptying, ileus and intestinal pseudo-obstruction, and megacolon and obstipation.[4,9]

Methods for Measuring Gastrointestinal Motility

Methods available for the evaluation of gastrointestinal motility include (1) radiography – survey, barium contrast, and radiopaque indigestible solids (e.g., barium-impregnated polyethylene spheres or BIPS), (2) quantitative videofluoroscopy, (3) ultrasonography, (4) nuclear scintigraphic imaging, (5) tracer studies, (6) manometry, and (7) functional MRI .

Survey Radiography - Survey abdominal radiography provides very little information about G.I. motility, but it is the imaging technique of choice in the initial assessment of any gastrointestinal disorder. Survey radiographs are useful in providing information about gastrointestinal tract position and content that may help to delineate mechanical obstruction from functional motility disorders. Survey radiographs are also helpful in determining the size and shape of other abdominal organs (e.g,. spleen, liver, biliary tract, urogenital tract) and their relationship to the gastrointestinal tract.

Contrast Radiography - Liquid Barium - Barium contrast radiography is often used in clinical practice to detect gross abnormalities of esophageal peristalsis, gastric emptying , intestinal transit, and colonic motility, but the technique does have some distinct limitations.[10,11] In gastric emptying studies, for example, gastric emptying of a radionuclide was markedly delayed in a group of dogs with pyloric hypertrophy although liquid emptying of barium was thought to be normal.[12] Similarly, the barium swallow technique used to assess esophageal peristalsis provides only a qualitative assessment unless it can be coupled to quantitative videofluoroscopy.[13,14] The latter technique requires sophisticated equipment and computer software that are generally not available in most clinical practices. Barium enema is now rarely performed in clinical practice, and has been superseded by other imaging techniques. In general, liquid barium studies will be useful only in documenting gross abnormalities of gastrointestinal motility.

Contrast Radiography – Barium Meal - Esophageal peristalsis, gastric emptying, and intestinal transit are affected by the physical properties of the meal (solids vs. liquids), size of the ingested particles (large vs. small), and chemical composition (lipids vs. proteins vs. carbohydrates),[5,15] therefore barium mixed with food is thought to be a better contrast agent for the determination of gastrointestinal transit. Despite this, barium can dissociate from the food and re-distribute into the liquid phase of the ingested meal, and this likely accounts for the wide variability in reported transit times. For example, the gastric emptying time of ground kibble (8 g/kg) mixed with barium sulfate suspension (5-7 ml/kg) was reported in the range of 5-10 hours in mature Beagle dogs,[16] while total gastric emptying time ranged from 7 to 15 hours in another study.[17] As with liquid barium studies, G.I. motility disorders can be diagnosed only if the transit/emptying times are markedly prolonged.

Contrast Radiography – BIPS - Small, indigestible radiopaque markers such as barium-impregnated polyethylene spheres (BIPS) have been used to quantitate gastric emptying and intestinal transit times in dogs and cats.[18,19] BIPS are administered in food as recommended in the manufacturer's package insert, and two to four abdominal radiographs are taken at convenient intervals over the next 13 to 24 hours.[18,19] The percentage of BIPS that have transited the stomach and intestine is calculated and compared with standard emptying and transit curves (provided in the manufacturer's package insert). Unfortunately, interpretation of BIPS emptying and transit data has some of the same limitations as liquid barium and barium meal studies. However, because of the widespread availability of radiographic equipment and practitioner expertise, radiographic methods employing liquid barium, barium meal, or BIPS will continue to be the methods of choice for most practitioners.

Ultrasonography - Ultrasonographic equipment is now more widely available in veterinary practice, and recent studies suggest that ultrasound may be a useful non-invasive method for quantitative assessment of gastric emptying in dogs and cats.[15,20] In healthy dogs fed a solid meal labeled with 13C-octanoic acid, there was a strong correlation between the rate of solid-phase gastric emptying assessed by use of gastric emptying ultrasonography and the 13C-OBT (carbon 13-labeled octanoic acid breath test) in dogs.[21] Further research will be necessary to validate this method against nuclear scintigraphic imaging and to describe reference ranges.

Nuclear Scintigraphy - Nuclear scintigraphic imaging is a very effective means of evaluating gastrointestinal motility and is now considered to be the standard method of assessment.[12,15,21,22,23] ^{99m}technetium (bound to sulfur or albumin colloid, disofenin, or mebrofenin) and ^{111}indium (bound to diethylene triamine pentacetic acid [DPTA]) are the radioisotopes most widely used because they are safe, simple to use, and non-absorbable. Two radionuclide markers can be tracked simultaneously, which allows solid and liquid emptying to be assessed during the same test period. Animals are fasted for 12-24 hours after which a test meal is fed incorporating one or two radioisotopes. Left lateral, right lateral, and ventral images are acquired with a gamma-camera and integrated nuclear medicine computer system. Gastric, intestinal, and/or colonic regions of interest are identified, and the radioactive counts in these regions are recorded, usually at regular intervals for 6-9 hours (gastric emptying), 12-24 hours (intestinal transit), or 24-36 hours (colonic transit). The expense, limited availability, and radiation hazards associated with this method have limited its widespread clinical application in dogs and cats.

Tracer Studies - Several kinds of tracer studies, including gastric, plasma, and breath tracers, have been developed for the assessment of gastric emptying and/or intestinal transit. Gastric tracer studies involve the serial aspiration of gastric contents after administration in food or by gastric intubation of a known concentration of a non-absorbable marker substance.[15] Chromium oxide, polyethylene glycol, and phenol red have all been used to assess solid (chromium oxide) or liquid phase (polyethylene glycol, phenol red) gastric emptying. The invasive nature of this method precludes its use in anything other than the research setting. Plasma tracer studies take advantage of the site-specific absorption of orally administered drugs following gastric emptying (acetaminophen) or orocecal transit (sulfasalazine). Acetaminophen is poorly absorbed in the stomach and rapidly absorbed in the duodenum, and the appearance of acetaminophen in plasma therefore reflects the gastric emptying time of acetaminophen.[15] Sulfasalazine is a compound molecule of sulfapyridine and 5-aminosalicylate linked in an azochemical bond. After oral dosing, most of the sulfasalazine is transported un-metabolized to the distal G.I. tract where cecal and colonic bacteria metabolize the drug to its component parts. Sulfapyridine is largely absorbed by the colonic mucosa but much of the 5-aminosalicylate remains in the colonic lumen where it inhibits mucosal cyclooxygenase and the inflammatory cascade. The appearance of sulfapyridine in plasma therefore reflects the orocecal transit time of sulfasalazine. Acetaminophen and sulfasalazine plasma tracer studies have been validated in the dog,[15,24,25] but there are no published studies comparing animals in health vs. disease. Breath tracer studies take advantage of the site-specific absorption of orally administered compounds following gastric emptying (^{13}C-octanoic acid), or of the site-specific fermentation (molecular hydrogen [H_2] generation) of orally ingested food or carbohydrate following orocecal transit. Both can be detected in expired breath, one reflecting gastric emptying time (^{13}C), the other representing orocecal transit time (H_2). The ^{13}C-octanoic breath test (^{13}C-OBT) has been validated as a measure of solid phase gastric emptying in the dog, but there are no published studies comparing animals in health vs. disease.[21] The H_2 breath test has been validated as a measure of orocecal transit time in both dogs and cats.[25]

Manometry – Manometry will have limited application in the diagnosis of cricoesophageal and gastroesophageal achalasia, gastroesophageal reflux, and aganglionic megacolon (Hirschsprung's disease),[26] but this technique is currently only performed at major referral centers and university teaching hospitals.

Functional MRI – Functional MRI has been used to quantitate gastric emptying in humans,[27] but this technique has not yet been validated in the dog or cat. Future MRI usage will likely be limited by expense and necessity for access to specialized equipment.

REFERENCES

1. Washabau RJ, Holt DE. Pathophysiology of gastrointestinal disease. In: Textbook of Veterinary Surgery. 3rd edition. Ed. D. Slatter, WB Saunders Company, Philadelphia, 2003, pp. 1142-1153.
2. Washabau RJ. Gastrointestinal motility disorders and gastrointestinal prokinetic therapy. In: Vet. Clin. N. Amer. 2003; 33: 1007-1028.
3. Gaynor A, Shofer F, Washabau RJ. Risk factors associated with the development of canine acquired megaesophagus. J. Am. Vet. Med. Assoc. 1997; 211: 1406-1412.
4. Washabau RJ. Disorders of the pharynx and esophagus. In: Simpson JW, Hall EJ, Williams DA. ed. Manual of Canine and Feline Gastroenterology. London: British Small Animal Veterinary Association, 2005, pp. 133-150.
5. Hall JA, Washabau RJ. Diagnosis and treatment of gastric motility disorders. In: Vet. Clin. N. Amer. 1999; 29: 377-395.
6. Hall JA, Solie TN, Seim HB, et al. Gastric myoelectric and motor activity in dogs with gastric dilatation-volvulus. Am. J. Physiol. 1993; 265: G646-G253.
7. Washabau RJ, Hasler A. Constipation, obstipation, and megacolon. In: August JR, ed. Consultations in Feline Internal Medicine. 3rd ed. Philadelphia: WB Saunders Co., 1996, pp 104-113.
8. Washabau RJ, Holt DE. Diseases of the large intestine. In: Ettinger SJ and Feldman EC, eds. Textbook of Veterinary Internal Medicine. 6th ed. Philadelphia: WB Saunders Co., 2005, pp 1378-1408.
9. O'Brien DP and Johnson GC. Dysautonomia and autonomic neuropathies. Vet. Clin. N. Amer. 2002; 32: 251-265.
10. Lamb CR. Recent developments in diagnostic imaging of the gastrointestinal tract of the dog and cat. In: Vet. Clin. N. Amer. 1999; 29: 307-342.
11. Miyabayashi T, Morgan JP, Atilola MAO, et al. Small intestinal emptying time in normal beagle dogs: a contrast radiographic study. Vet. Radiology 1986; 27: 164-168,
12. Hornof WJ, Koblik PD, Strombeck DR, et al. Scintigraphic evaluation of solid-phase gastric emptying in the dog. Vet. Radiol. 1989; 30: 242-248.
13. Pollard RE, Marks SL, Davidson A, et al. Quantitative videofluoroscopic evaluation of pharyngeal function in the dog. Vet. Radiol. & Ultrasound 2000: 41: 409-412.
14. Davidson AP, Pollard RE, Bannasch DL, et al. Inheritance of cricopharyngeal dysfunction in Golden Retrievers. Amer. J. Vet. Res. 2004; 65: 344-349.
15. Wyse CA, McLellan J, Dickie AM, et al. A review of methods for assessment of the rate of gastric emptying . J Vet Intern Med. 2003; 17: 609-621.
16. Miyabayashi T, Morgan JP. Gastric emptying in the normal dog: a contrast radiographic technique. Vet. Radiol. 1984; 25: 187-193.
17. Burns J, Fox SM. The use of a barium meal to evaluate total gastric emptying time in the dog. Vet. Radiol. 1986; 27: 169-172.
18. Guilford WG, Lawoko CR, Allan FJ. Accuracy of localizing radiopaque markers by abdominal radiography and correlation between their gastric emptying rate and that of a canned food in dogs. Am. J. Vet. Res. 1997; 58: 1369-1363.
19. Chandler ML, Guilford WG, Lawoko CR, et al. Gastric emptying and intestinal transit times of radiopaque markers in cats fed a high fiber diet with and without low-dose intravenous diazepam. Vet. Radiol. Ultrasound 1999; 40:3-8.
20. Chalmers AF, Kirton R, Wyse CA, et al. Ultrasonographic assessment of the rate of solid-phase gastric emptying in dogs. Vet Rec. 2005; 157:649-652.
21. McLellan J, Wyse CA, Dickie A, et al. Comparison of the carbon 13-labeled octanoic acid breath test and ultrasonography for assessment of gastric emptying of a semisolid meal in dogs. Am J Vet Res. 2004; 65: 1557-1562.
22. Iwanaga Y, Wen J, Thollander MS, et al. Scintigraphic measurement of regional gastrointestinal transit in the dog. Am. J. Physiol. 1998; 275: G904-G910.
23. Theodorakis MC. External scintigraphy in measuring rate of gastric emptying in beagles. Am. J. Physiol. 1980; 239: 1285-1291.
24. Mizuta H, Kawazoe Y, Ogawa K. Effects of meals on gastric emptying and small intestinal transit times of a suspension in the beagle dog assessed using acetaminophen and salicylazosulfapyridine as markers. Chem. Pharm. Bull. 1990; 38: 2224-2227.
25. Papasouliotis K, Gruffydd-Jones TJ, Sparkes AH, et al. A comparison of orocaecal transit times assessed by the breath hydrogen test and sulphasalazine/sulphapyridine method in healthy beagle dogs. Res. Vet. Sci. 1995; 58: 263-267.
26. Lang IM, Dantas RO, Cook IJ, et al. Videoradiographic, manometric and electromyographic assessment of upper esophageal sphincter function in the dog. Am. J. Physiol. 1991; 260: G911-G919.
27. Feinle C, Kunz P, Boesiger P, et al. Scintigraphic validation of a magnetic resonance imaging method to study gastric emptying of a solid meal in humans. Gut 1999; 44: 106-111.
28. Orihata M, Sarna SK. Contractle mechanisms of action of gastroprokinetic agents: cisapride, metoclopramide, and domperidone. Am. J. Physiol. 1994; 266: G665-G676.
29. Gullikson GW, Virina MA, Loeffler R, et al. Alpha-2 adrenergic model of gastroparesis: validation with renzapride, a stimulator of motility. Am. J. Physiol. 1991; 261: G426-G432.
30. van den Brom WE, Happe RP. Gastric emptying of a radionucle-labelled test meal in healthy dogs: a new mathematical analysis and reference values. Am. J. Vet. Res. 1985; 47: 2170-2174.

31. Steyn PF, Twedt DF, Toombs W. The scintigraphic evaluation of solid phase gastric emptying in normal cats. Vet. Radiol. Ultrasound 1995; 36: 327-331.
32. Weber MP, Stambouli F, Martin LJ, et al. Influence of age and body size on gastrointestinal transit time of radiopaque markes in healthy dogs. Am. J. Vet. Res. 2002; 63: 677-682.
33. Allan FJ, Guilford GW, Robertson ID, et al. Gastric emptying of solid radiopaque markers in healthy dogs. Vet. Radiol. Ultrasound 1996; 37: 336-344.
34. Lester NV, Roberts GD, Newell SM, et al. Assessment of barium impregnated polyethylene spheres (BIPS) as a measure of solid phase gastric emptying in normal dogs – comparison to scintigraphy. Vet. Radiol. Ultrasound 1999; 40: 465-471.
35. Nelson OL, Jergens AE, Miles KG, et al. Gastric emptying as assessed by barium-impregnated polyethylene spheres in healthy dogs consuming a commercial kibble ration. J. Am. Anim. Hosp. Assoc. 2001; 37: 444-452.
36. Hall JA, Willer RL, Seim HB, et al. Gastric emptying of non-digestible radiopaque markers after circumcostal gastropexy in clinically normal dogs and dogs with gastric dilatation-volvulus. Am. J. Vet. Res. 1992; 53: 1961-1965,
37. Papageorges M, Breton L, Bonneau NH. Gastric drainage procedures: effects on normal dogs. II. Clinical observations and gastric emptying. Vet. Surg. 1987; 16: 332-340.
38. Sparkes AH, Papasouliotis K, Barr FJ, et al. Reference ranges for gastrointestinal transit of barium-impregnated polyethylene spheres in healthy cats. J. Small Anim. Pract. 1997; 38: 340-343.
39. Wyse CA, Preston T, Morrison DJ, et al. The C-octanoic acid breath test for assessment of solid phase gastric emptying in dogs. Am. J. Vet. Res. 2001; 62: 1939-1944.
40. Chaudhuri TK. Use of 99mTc-DTPA for measuring gastric emptying time. J. Nuc. Med. 1974; 15: 391-395.
41. Choi M, Seo M, Jung J, et al. Evaluation of canine gastric motility with ultrasonography. J Vet Med Sci. 2002; 64: 17-21.
42. Leib MS, Wingfield WE, Twedt DC, et al. Gastric emptying of liquids in the dog: serial test meal and modified emptying-time techniques. Am. J. Vet. Res. 1985; 46: 1876-1880.
43. Russell J, Bass P. Canine gastric emptying of fiber meals: influence of meal viscosity and antroduodenal motility. Am. J. Physiol. 1985; 249: G662-667.
44. Scrivani PV, Bednarski RM, Meyer CW. Effects of acepromazine and butorphanol on positive contrast upper gastrointestinal examination in dogs. Am. J. Vet. Res. 1998; 59: 1227-1233.
45. Weber MP, Martin LJ, Biourge VC, et al. Influence of age and body size on orocecal transit time as assessed by use of the sulfsalazine method in healthy dogs. Am. J. Vet. Res. 2003; 64: 1105-1109.
46. Papasouliotis K, Muir P, Gruffydd-Jones TJ, et al. Decreased orocecal transit time, as measured by the exhalation of hydrogen, in hyperthyroid cats. Res. Vet. Sci. 1993; 55: 115-118.
47. Chandler ML, Guilford WG, Lawoko CRO. Radiopaque markers to evaluate gastric emptying and intestinal transit times in healthy cats. J. Vet. Intern. Med. 1997: 11: 361-364.

NEW THERAPIES FOR GASTROINTESTINAL MOTILITY DISORDERS

Robert J. Washabau, VMD, PhD, DACVIM
St. Paul, Minnesota

INTRODUCTION

Gastrointestinal motility disorders represent a diagnostic and therapeutic challenge. Disorders of gastrointestinal motility may result in delayed transit, accelerated transit, impaired relaxation, or inappropriate relaxation. The delayed transit disorders are the most important motility disorders of companion animals and may involve the esophagus (hypomotility and megaesophagus), stomach (delayed gastric emptying), small intestine (post-operative ileus and intestinal pseudo-obstruction), or colon (constipation and megacolon).

DOPAMINERGIC D_2 ANTAGONIST DRUGS

Metoclopramide and Domperidone - The dopaminergic D_2 antagonists are a group of drugs with gastrointestinal prokinetic and antiemetic effects at peripheral (prokinetic) or central (antiemetic) dopamine D_2 receptors.[1] The best representatives in this classification, metoclopramide and domperidone, reverse gastric relaxation induced by dopamine infusion in dogs, and they abolish vomiting associated with apomorphine therapy. Although the role of dopamine receptors in chemoreceptor trigger zone-induced vomiting is fairly well established, there is no definitive evidence that inhibitory dopaminergic neurons regulate gastrointestinal motility. The prokinetic effects of metoclopramide and domperidone thus may not be readily or exclusively explained by dopamine receptor antagonism. Some dopaminergic antagonists (e.g., metoclopramide) have other pharmacologic properties, e.g., 5-HT_3 receptor antagonism and 5-HT_4 receptor agonism. Domperidone also has α_2- and β_2-adrenergic receptor antagonistic effects. The characterization of these drugs as dopaminergic antagonists is convenient but may not properly describe their overall in vivo effects.

Gastroesophageal Sphincter Disorders

The gastroesophageal sphincter prevents reflux of gastrointestinal contents into the esophageal body.[2,3] Reflux of gastric H^+ and pepsins, and of duodenal bicarbonate, bile salts, and proteases, induces chemical injury and inflammation of the esophageal mucosa.[3] Gastroesophageal sphincter tone appears to be under the regulation of dopaminergic neurons as both metoclopramide and domperidone increase sphincter tone. The gastric prokinetic effect of metoclopramide, although moderate, may also help to reduce the frequency, severity, and duration of reflux episodes.

Gastric Emptying Disorders

Metoclopramide increases the amplitude and frequency of antral contractions; inhibits fundic receptive relaxation; and coordinates gastric, pyloric, and duodenal motility, all of which result in moderate acceleration of gastric emptying.[4,5] Metoclopramide appears to have continuing clinical application as a gastric prokinetic agent in the dog and cat, although the serotonergic agonists are clearly more potent. Domperidone appears to be less effective as a gastric prokinetic agent. Although effective in humans, domperidone actually decreases the frequency of corporeal, pyloric, and duodenal contractions and deteriorates antropyloroduodenal coordination in the dog by decreasing the frequency of contractions spreading from the antrum or pylorus to the duodenum.[5]

Small Intestinal Transit Disorders

Metoclopramide and domperidone are generally considered less effective in the management of the small intestinal transit disorders. Metoclopramide enhances antropyloroduodenal coordination in the dog and may be effective when delayed gastric emptying is due to poor antropyloroduodenal coordination.[6] Domperidone has no documented effects on small intestinal transit.

Chemoreceptor trigger zone-induced emesis

SEROTONERGIC 5-HT_4 AGONIST DRUGS

Drugs acting on gastrointestinal 5-hydroxytryptamine (5-HT or serotonin) receptors have potent motility effects.[8] Serotonergic drugs that bind 5-HT_4 receptors on enteric cholinergic neurons induce depolarization and contraction of gastrointestinal smooth muscle. These drugs are not entirely selective for the 5-HT_4 receptor, however. Some of the putative 5-HT_4 receptor agonists also have 5-HT_1 and 5-HT_3 antagonistic effects on enteric cholinergic neurons, and direct non-cholinergic (perhaps 5-HT_{2a}) effects on colonic smooth muscle. Cisapride, mosapride, prucalopride, and tegaserod are the best examples in this classification. Cisapride and tegaserod have been withdrawn from several international markets because of their effect(s) on myocardial Q-T internal prolongation, although both remain available through compounding pharmacies. Mosapride is available in Japan (Pronamid - DS Pharma), and prucalopride is available in Europe (Resolor - Movetis).

Cisapride - Cisapride was widely used in the management of canine and feline gastric emptying, intestinal transit, and colonic motility disorders throughout most of the 1990's.[9,10] Cisapride was withdrawn from the American, Canadian and certain Western European in July of 2000 following reports of untoward cardiac side effects in human patients. Cisapride causes QT interval prolongation and slowing of cardiac repolarization via blockade of the rapid component of the delayed rectifier potassium channel (I_{Kr}).[11] This effect may result in a fatal ventricular arrhythmia referred to as torsades de pointes.[11] Similar effects have been characterized in canine cardiac Purkinje fibers, but in vivo effects have not been reported in dogs or cats. The withdrawal of cisapride has created a clear need for new G.I. prokinetic agents although cisapride continues to be available from compounding pharmacies. Cisapride's effect on disorders of the gastroesophgeal sphincter, stomach, intestine, and colon may be found in references [9-20].

Mosapride - Mosapride citrate, a substituted benzamide, is a novel 5-HT_4 receptor agonist that increases gastric emptying in rats and dogs, and increases electrically-evoked contractions in the isolated guinea pig ileum.[21-22] Mosapride stimulates acetylcholine release from the myenteric plexus via activation of 5-HT_4 receptors, but has no real affinity for D_2 dopamine, 5-HT_1, 5-HT_2 receptors, or α_1-adrenoceptors. Mosapride restores gastric motility in dogs with vincristine-induced gastric hypomotility,[23] and therefore may be clinically useful in other gastric emptying disorders. Mosapride is apparently without effect on distal gastrointestinal tract motility. Mosapride is marketed as Pronamid by DS Pharma Animal Health in Japan.

Prucalopride - Prucalopride is a potent partial benzamide agonist at 5-HT_4 receptors, but is without effect on other 5-HT receptors or cholinesterase enzyme activity. Prucalopride dose-dependently (0.02-1.25 mg/kg) stimulates giant migrating contractions (GMC's) and defecation in the dog.[24-26] The prucalopride effect is observed most prominently in the first hour after administration, suggesting that the prucalopride effect is a direct effect on the colon rather than on total gut transit time. Oral and intravenous doses appear to be equipotent again implying a high oral bioavailability. Prucalopride also enhances defecation frequency in healthy cats.[27] Cats treated with prucalopride at a dose of 0.64 mg/kg experience increased defecation within the first hour of administration. Fecal consistency is not altered by prucalopride at this dosage. Prucalopride also appears to stimulate gastric emptying in the dog.[28] In lidamidine-induced delayed gastric emptying in dogs, prucalopride (0.01-0.16 mg/kg) dose-dependently accelerates gastric emptying of dextrose solutions. The prucalopride effect is equipotent following oral and intravenous administration suggesting that prucalopride may have a high oral bioavailability. Prucalopride is marketed as Resolor by Movetis in Europe.

Tegaserod - Tegaserod is a potent partial non-benzamide agonist at 5-HT_4 receptors and a weak agonist at 5-HT_{1D} receptors.[29,32]
Tegaserod was approved by the U.S. Food and Drug Administration in September 2002, but later removed from the American market in 2007 because of reports of prolongation of the Q-T interval and delayed cardiac repolarization. Tegaserod is marketed as Zelnorm by Novartis in Europe.

Future Research in 5-HT Receptor Pharmacology

The 5-HT_4 receptor appears to hold the most interest and promise for future drug development. 5-HT_4 receptor activation can cause relaxation or contraction depending on the region, cell type, and animal species. In the dog, the effects of selective 5-HT_4 receptor agonists suggest that these receptors are present on jejunal mucosa, ileal mucosa, gastric cholinergic neurons, and circular colonic smooth muscle cells. Development of 5-HT_4 ligands is somewhat constrained by the effects these drugs have on cardiac 5-HT_4 receptors and the delayed rectifier potassium channel (I_{Kr}). Some, but not all, 5-HT_4 agonists prolong the QT interval and delay cardiac repolarization. Molecular biology experiments have revealed differences in the carboxyl terminus of smooth muscle and cardiac muscle 5-HT_4 receptors, but these amino acids differences are distant from the receptor binding site. Thus, receptor sub-types may exist but they may not be important from a functional or therapeutic standpoint.

MOTILIN-LIKE DRUGS

Erythromycin - The antibiotic properties of erythromycin and other macrolides were discovered in the early 1950's. Since that time, erythromycin has been widely used in treating patients with gram-positive and gram-negative bacterial and mycoplasmal infections. Physicians and veterinarians noted that erythromycin therapy was accompanied by frequent gastrointestinal side effects including nausea and vomiting. This occurrence suggested to researchers that erythromycin might have effects on gastrointestinal motility. It was subsequently demonstrated that microbially-effective doses of erythromycin stimulate retrograde peristalsis and vomiting in dogs, and that lower microbially-ineffective doses of erythromycin stimulate migrating motility complexes and antegrade peristalsis similar to that induced by the endogenous gastrointestinal hormone, motilin.[32,33,34]

Gastroesophageal Sphincter Disorders

Motilin, erythromycin, and erythromycin analogues (e.g., LY-267108) increase gastroesophageal sphincter pressure in cats and dogs.[35] Erythromycin also increases gastroesophageal sphincter pressure in cats in which the basal pressure has been lowered experimentally by esophageal acid perfusion or following intravenous isoproterenol administration.[35] These studies suggest that erythromycin should be useful in treating cats, and perhaps dogs, wth gastroesophageal reflux and reflux esophagitis.

Gastric Emptying Disorders

Erythromycin accelerates gastric emptying by inducing antral contractions similar to phase III of the interdigestive state.[36-39] The strong contractions associated with phase III normally occur only during the fasted state when they clear the stomach of large indigestible solids. After meals, intravenous or oral erythromycin accelerates gastric emptying of solid meals in dogs. EM574 is 250 times more potent than erythromycin in inducing phase III contractions in dogs and it has no anti-bacterial activity.[37,38] EM574 is as effective as cisapride in normalizing gastric contractility and emptying in dogs with clonidine-induced gastroparesis in dogs.[37,38]

Colonic Motility Disorders

Erythromycin accelerates regional colonic transit in the dog.[40] Erythromycin has been shown to stimulate canine but not feline colonic smooth muscle contraction *in vitro*.[41] These results suggest that erythyromycin may be useful in the treatment of canine colonic motility disorders.

ACETYLCHOLINESTERASE INHIBITORS AND CHOLINOMIMETIC AGENTS

Ranitidine and Nizatidine - Ranitidine and nizatidine, classic histamine H_2 receptor antagonists, stimulate gastrointestinal motility by inhibiting acetylcholinesterase activity.[42-44] As parasympathetic potentiating agents, ranitidine and nizatidine stimulate gastric emptying and small intestinal and colonic motility. The prokinetic effects of ranitidine and nizatidine appear to be more prominent in the proximal gastrointestinal tract (i.e., gastric emptying). Other members of this classification, e.g., cimetidine and famotidine, apparently have no effect on gastrointestinal motility.

Bethanechol - Bethanechol is a cholinomimetic agent that binds muscarinic cholinergic receptors and stimulates motility throughout the gastrointestinal tract.

Esophageal Motility Disorders

Smooth muscle prokinetic agents, e.g., cisapride and metoclopramide, have limited clinical application in the treatment of canine idiopathic megaesophagus. Cisapride and metoclopramide stimulate distal esophageal peristalsis in those animal species (e.g., cat, human being, guinea pig) in which the distal esophageal muscularis is composed of smooth muscle. The domestic dog has evolved with a purely striated muscle esophageal muscularis which has a different regulation and therapeutic responsiveness.[13] Bethanechol has been shown to stimulate esophageal propagating contractions in some dogs affected with idiopathic megaesophagus and is therefore a more appropriate prokinetic agent for the therapy of this disorder.[45]

Gastric Emptying Disorders

Ranitidine and nizatidine stimulate gastric antral contractions at gastric anti-secretory dosages (ranitidine 1.0-2.0 mg/kg PO BID; nizatidine 2.5-5.0 mg/kg PO BID) and may be useful as gastric prokinetic agents in dogs and cats.[43,44,46]

Colonic Motility Disorders

Infusions of ranitidine at doses of 3.0 mg/kg induce canine colonic propagating contractions *in vivo*.[46] Ranitidine and nizatidine have also been shown to stimulate feline colonic smooth muscle contraction *in vitro*.[47] These data suggest that ranitidine and nizatidine may be useful in the treatment of feline or canine colonic motility disorders.[42]

NITRIC OXIDE DONORS

Delayed gastric emptying is recognized as an important cause of upper gastrointestinal tract pathology (e.g., anorexia and vomiting) in companion animals.[14] Delayed gastric emptying has been reported in infectious and inflammatory gastric diseases, diabetes mellitus, and radiation injury in the dog.[48,49] Delayed gastric emptying has also been associated with several secondary conditions, including electrolyte disturbances (e.g., hypokalemia, hypocalcemia), metabolic disorders (e.g., hypoadrenocorticism, hypergastrinemia, uremia), concurrent drug usage (e.g., cholinergic antagonists, β-adrenergic agonists, opiates), acute stress (e.g., sympathetic stimulation, spinal cord injury), and acute abdominal inflammation.[48,49] Diabetes mellitus is the most common endocrinopathy of the domestic dog.[50] Long-standing un-diagnosed or un-treated diabetes mellitus is associated with significant gastroparesis in the dog,[51,52] just as it in humans. The pathogenesis of gastroparesis in diabetes mellitus is complex and probably multi-factorial, involving one or more of the cellular elements (neurons, smooth muscle cells, interstitial cells of Cajal) regulating gastric motility.[53] An important pathophysiologic mechanism appears to be the loss of neuronal nitric oxide synthase, the enzyme responsible for the production of nitric oxide, an inhibitory neurotransmitter that is required for relaxation of smooth muscle and therefore a critical component of normal gastrointestinal motility. In the absence of nitric oxide, the stomach cannot relax, resulting in bloating, satiety, nausea, and vomiting.[54]

Cisapride, metoclopramide, and erythromycin have all been used with variable effect in diabetic gastroparesis. Therapy aimed instead at restoring nitrergic neurotransmission could have intrinsic beneficial effects in canine diabetic gastroparesis. AMU-301, a nitric oxide (NO) donor, is recognized as an effective treatment for diabetic gastroparesis in streptozotocin-induced (STZ) diabetic rat models of delayed gastric emptying,[55] and may eventually prove useful in spontaneous canine diabetes mellitus.

REFERENCES

1. Hall JA, Washabau RJ. Gastrointestinal prokinetic therapy: dopaminergic antagonist drugs. Compend Contin Educ Pract Vet 19(2): 214-221, 1997.
2. Biancani P, Barwick K, Selling J, et al: Effects of acute experimental esophagitis in mechanical properties of lower esophageal sphincter function. Gastroenterology 87:8-16, 1984.
3. Eastwood G, Castell DO, Higgs RH: Experimental esophagitis in cats impairs lower esophageal sphincter function in the cat. Gastroenterology 69:146-153, 1975.
4. Hall JA, Solie TN, Seim HB, et al. Effect of metoclopramide on fed-state gastric myoelectric and motor activity in dogs. Am J Vet Res 57: 1616-1622, 1996.
5. Orihata M, Sarna SK. Contractile mechanisms of action of gastroprokinetic agents: cisapride, metoclopramide, and domperidone. Amer J Physiol 266: G665-G676, 1994.
6. Summers RW, Yanda R, Prihoda M, et al.: A comparative study of the effects of four motor-stimulating agents on canine jejunal spike bursts. Scand J Gastroenterol 23: 1173-1181, 1988.
7. Washabau RJ and Elie M. Anti-Emetic Therapies. *In*, Kirk R and Bonagura J, eds. Current Veterinary Therapy XII. 12th ed. Philadelphia: WB Saunders Co., 1995, 679-684.
8. Janssen P, Prins NH, Meulemans AL, et al. Pharmacological characterization of the 5-HT receptors mediating contraction and relaxation of canine isolated proximal stomach smooth muscle. Brit J Pharm 136: 321-329, 2002.
9. Washabau RJ, Hall JA. Clinical pharmacology of cisapride. JAVMA 207: 1285-1288, 1995.
10. Washabau RJ, Hall JA. Gastrointestinal prokinetic therapy: serotonergic drugs. Compend Contin Educ Pract Vet 19(4): 473-480, 1997.
11. Drici MD, Ebert SN, Wang WX, et al. Comparison of tegaserod and its main metabolite with cisapride and erythromycin on cardiac repolarization in the isolated rabbit heart. J Cardiovasc Pharmacol 34: 82-88, 1999.
12. Clark S et al.: Comparison of potential cytoprotective action of sucralfate and cimetidine - studies with experimental feline esophagitis. Amer J Med 83: 56, 1987.
13. Washabau RJ. Diseases of the esophagus. *In*, Ettinger SJ and Feldman EC, eds. Textbook of Veterinary Internal Medicine. 5th ed. Philadelphia: WB Saunders Co., 2000, pp. 1142-1153.
14. Washabau RJ, Holt DE. Pathophysiology of Gastrointestinal Disease. *In*, Slatter D, ed. Textbook of Veterinary Surgery. 3rd edition. Philadelphia: WB Saunders Co., 2003, pp 530-552.
15. Cohen ML, Susemichel AD, Bloomquist W, et al.: 5-HT_4 receptors in rat but not rabbit, guinea pig, or dog esophageal muscle. Gen Pharmacol 25: 1143-1148, 1994.
16. Edelbroek M, Schuurkes JAJ, de Ridder WJE, et al.: Effect of cisapride on myoelectrical and motor responses of antropyloroduodenal lipid and antral tachygastria in conscious dogs. Dig Dis Sci 40: 901-911, 1995.
17. Schemann M, Ehrlein HJ: 5-hydroxytryptophan and cisapride stimulate propulsive jejunal motility and transit of chyme in dogs. Digestion 34: 229-235, 1986.
18. Hasler AH, Washabau RJ. Cisapride stimulates contraction in feline idiopathic megacolonic smooth muscle. J Vet Intern Med 11(6): 313-318, 1997.
19. Washabau RJ and Sammarco J. Effect of cisapride on colonic smooth muscle function. Am J Vet Res 57: 541-546, 1996.
20. Aboumarzouk OM, Agarwal T, Antakia R, et al. Cisapride for intestinal constipation (review). The Cochrane Collaboration, 2011; http://www.thecochranelibrary.com
21. Curran MP, Robinsoni DM. Mosapride - Use in gastrointestinal disorders. Drugs 68(7):981-991, 2008.
22. Mine T, Yoshikawa Y, Oku S, et al. Comparison of effect of mosapride citrate and existing 5-HT_4 receptor agonists on gastrointestinal motility, in vivo and in vitro. J. Pharm. Exp. Ther. 283: 1000-1008, 1997.
23. Tsukamoto A, Ohno K, Tsukagoshi T, et al. Ultrasonographic evaluation of vincristine-induced gastric hypomotility and the prokinetic effect of mosapride citrate in dogs. J. Vet. Intern. Med. 24(3): 721, 2010.
24. Briejer MR, Van Daele P, Bosmans J-P, et al. Dose-dependent effects after oral and intravenous administration of R093877 on colonic motility in conscious dogs. Gastroenterology 112: A704, 1997a.
25. Briejer MR, Ghoos E, Eelen J, et al. Serotonin 5-HT_4 receptors mediate the R093877-induced changes in contractile patterns in the canine colon. Gastroenterology 112: A705, 1997d.
26. Prins NH, Van Haselen JF, Lefebvre RA, et al. Pharmacological characterization of 5-HT receptors mediating relaxation of canine isolated rectum circular smooth muscle. Brit J Pharmacol 127(6): 1431-1437, 1999.
27. Briejer MR, Engelen M, Jacobs J, et al. R093877 enhances defecation frequency in conscious cats. Gastroenterology 112: A705, 1997c.
28. Briejer MR, Meulemans AL, Wellens A, et al. R09387 dose dependently accelerates delayed gastric emptying in conscious dogs. Gastroenterology 112: A705, 1997b.
29. Nguyen A, Camilleri M, Kost LJ. SDZ HTF 919 (tegaserod) stimulates canine colonic motility and transit in vivo. J Pharmacol Exper Therap 280: 1270-1276, 1997.
30. Schikowski A, Thewissen M, Mathis C, et al. Serotonin type-4 receptors modulate the sensitivity of intramural mechanoreceptive afferents of the cat rectum. Neurogastroenterol Mot 14: 221-227, 2002.
31. Weber E, Braun E, Forgiarini P, et al. Tegaserod normalizes opioid-induced bowel dysfunction in dogs. Gastroenterology 124; A571, 2003.
32. Hall JA, Washabau RJ. Gastrointestinal prokinetic therapy: motilin-like drugs. Comp Contin Educ Pract Vet 19(3): 281-288, 1997.
33. Itoh Z, Suzuki T, Nakaya M, et al: Gastrointestinal motor-stimulationg activity of macrolide antibiotics and analysis of their side effects on the canine gut. Antimicrob Agents Chemother 26(6): 863-869, 1984
34. Sarna S, Gonzalez A, Ryan R. Enteric locus of action of prokinetics: ABT-229, motilin, and erythromycin. Amer J Physiol 278: G744-G752, 2000.
35. Greenwood B, Kieckman D, Kirst HA, et al: Effects of LY267108, an erythromycin analogue derivative, on lower esophageal sphincter function in the cat. Gastroenterology 106:624-628, 1994.
36. Sako F, Marui S, Inatomi N, et al. EM574, an erythromycin derivative, improves delayed gastric emptying of semi-solid meals in conscious dogs. Eur J Pharm 395: 165, 2000.

37. Tanaka T, Mizumoto A, Mochiki E, et al.: Effects of EM574 and cisapride on gastric contractile and emptying activity in normal and drug-induced gastroparesis in dogs. J Pharmacol Exp Therap 287: 712-719, 1998.
38. Tanaka T, Mizumoto A, Mochiki E, et al. Effect of EM574 on postprandial pancreaticobiliary secretion, gastric motor activity, and emptying in conscious dogs. Dig Dis Sci 44: 1100-1106, 1999.
39. Tsukamoto K, Tagi Y, Nakazawa T, et al. Gastroprokinetic effect and mechanism of SK-896, a new motilin analogue, during the interdigestive period in conscious dogs. Pharmacology 63: 95-102, 2001.
40. Chiba T, Thomforde GM, Kost LJ, et al. Motilides accelerate regional gastrointestinal transit in the dog. Aliment Pharmacol Ther 14: 955-960, 2000.
41. Melgarejo LT, Simon DA, Washabau RJ. Erythromycin stimlulates canine, but not feline, longitudinal colonic muscle contraction. J Vet Intern Med 15: 333, 2001.
42. Hall JA, Washabau RJ. Gastrointestinal prokinetic therapy: acetylcholinesterase inhibitors. Compend. Contin. Educ. Pract. Vet. 19(5): 615-621, 1997.
43. Bertaccini G, Poli E, Adami M, et al: Effect of some new H_s-receptor antagonists on gastrointestinal motility. Agents Actions 13: 157-162, 1985.
44. Mizumoto A, Fukimura M, Iwanaga Y, et al. Anticholinesterase activity of histamine H_2 receptor antagonists in the dog: their possible role in gastric motor activity. J Gastrointestinal Motil 2: 273-280, 1990.
45. Diamant N, Szcepanski M, Mui H: Idiopathic megaesophagus in the dog: Reasons for spontaneous improvement and a possible method of medical therapy. Can Vet J 15: 66-71, 1974.
46. Kishibayashi N, Tomaru A, Ichikawa S, et al: Enhancement by KW-5092, a novel gastroprokinetic agent, of the gastrointestinal motor activity in dogs. Japn J Pharmacol 65: 131-142, 1994.
47. Washabau RJ, Pitts MM, Hasler A. Nizatidine and ranitidine, but not cimetidine, stimulate feline colonic smooth muscle contraction. J Vet Intern Med 10: 157, 1996.
48. Hall JE, Washabau RJ. Diagnosis and Treatment of Gastric Motility Disorders. In, *Veterinary Clinics of North America*, 1999; vol. 29: 377-395.
49. Washabau RJ. Gastrointestinal Motility Disorders and G.I. Prokinetic Therapy. In, Veterinary Clinics of North America, Willard MD, ed. Philadelphia: WB Saunders Co., 2003, 1007-1028.
50. Nelson RW. Diabetes mellitus. In, Textbook of Veterinary Internal Medicine, Ettinger SJ and Feldman EC, eds. 6th edition. WB Saunders, Philadelphia, 2005, 1563-1591.
51. Takeda M, Mizutani Y, Tsukamoto K, et al. Gastric emptying in diabetic gastroparetic dogs: effects of DK-951, a novel prokinetic agent. Pharmacology 2001; 62: 23-28.
52. Koizumi F, Kawamura T, Ishimori A. Correlation between gastric emptying time and both plasma gastrin and pancreatic polypeptide in streptozotocin diabetic dogs. Japanese Journal of Gastroenterology 1989; 86: 1037-1043.
53. Camilleri M. Diabetic gastroparesis. New England Journal of Medicine 2007; 356: 820-829.
54. Pasricha PJ. The riddle, mystery, and enigma of gastroparesis. Journal of Supportive Oncology 2007; 5: 368-370.
55. Amulet Pharmaceuticals AMU-301 Fact Sheet. New Chemical Entity (NCE) for Diabetic Gastroparesis. http://www.amuletpharma.com/AMU-301_FACT_SHEET.pdf

BIOMARKERS FOR SEPSIS: ARE WE ANY CLOSER?

Amy E. DeClue, DVM, MS, DACVIM (SAIM)
Columbia, MO

INTRODUCTION

A biomarker, or biological marker, is defined as "a characteristic that is objectively measured and evaluated as an indicator of normal biological processes, pathogenic processes, or pharmacologic responses to a therapeutic intervention".[1] Another, perhaps more simple, definition is a "quantifiable measurement(s) of biological homeostasis that define(s) what is 'normal', therefore providing a frame of reference for predicting or detecting what is 'abnormal'".[2] Biomarkers include physiologic indicators (e.g., body temperature, respiratory rate, heart rate) or endogenously produced biochemical compounds found in various body fluids or tissues (e.g., TNF, IL-6, CNP). Molecular biomarkers also include nucleic acid-based biomarkers such as gene mutations or polymorphisms; and quantitative analysis of the gene expression of peptides, proteins, lipids metabolites, and other small molecules.

Sepsis, defined as the systemic inflammatory response to infection, is associated with substantial morbidity and mortality in dogs. Identifying biomarkers that would allow the rapid diagnosis of sepsis facilitating early intervention, particularly antimicrobial therapy[3], would undoubtedly benefit patients with sepsis. In addition to their role as a diagnostic tool, biomarkers may allow for temporal categorization of systemic inflammation such that tailored therapeutic interventions can be implemented. Biomarkers are also being investigated to facilitate the identification of potential therapeutic targets for immunomodulation and predict patient response to novel treatment strategies. Additionally, biomarkers that correlate with disease severity and predict morbidity and mortality could assist with client education and patient stratification in clinical trials.

Determining the most ideal biomarkers for the diagnosis, staging, and prognostication of sepsis is complex. There are many potential biomarkers for sepsis; more than 100 distinct molecules have been proposed.[4] This discussion will review biomarkers that have been evaluated in clinical canine and feline sepsis. It is important to be cognizant of the inconsistencies in the literature, including discordant conclusions regarding biomarker relevance in certain settings. Some of the sources of inconsistency include variable patient populations, assay methodology, reference standards, site of sampling, requirements for activation; and the presence of soluble receptors, circulating inhibitors, and carrier proteins which may interfere with measuring the biomarker of interest.

BIOMARKERS

Clinical evaluation – Rectal temperature, heart rate, respiratory rate and leukocyte count have been used as markers for the systemic inflammatory response syndrome in dogs. In one study of 500 dogs with a variety of illnesses, not just sepsis, presenting to a veterinary teaching hospital, mortality was significantly associated with the number of SIRS criteria fulfilled.[5] While these parameters are often used as part of the clinical definition of sepsis, it is important to note that these criteria are not necessarily specific for sepsis (i.e., infection). The sensitivity and specificity of SIRS criteria for the diagnosis of sepsis is variable based on the study and the cut off criteria used but ranges from a sensitivity of 77-97% and specificity of 64-77% when 2 of the 4 criteria were met.[6]

Multiple organ dysfunction syndrome (MODS) has recently been evaluated in a retrospective study of 114 dogs with sepsis secondary to GI leakage (defined as infection based on cytologic examination, positive bacterial culture, or gross evidence of gastrointestinal tract leakage).[7] Approximately 60% of dogs had coagulation dysfunction, 49% had hepatic dysfunction, 31% had respiratory dysfunction, 17% had cardiovascular dysfunction and 12% had renal dysfunction.[7] Multiple organ dysfunction syndrome (dysfunction of ≥ 2 organ systems) was diagnosed in 50% of the population.[7] Dogs with MODS had a mortality rate of 70% compared to 25% for dogs without MODS.[7] There was a significant association between the presence of organ dysfunction and death. Other biomarkers of organ function, particularly cardiovascular function, have been evaluated as well. In a prospective study of dogs with pyometra- induced sepsis, the need for vasoactive drugs, larger volume of crystalloid fluid resuscitation, the central venous oxygen saturation, higher base deficit and blood lactate concentration at the time of hospital admission were all associated with outcome.[8]

Endotoxin Activity (EA)- Measurement of EA in people at the time of ICU admission has a high negative predictive value (NPV) for the diagnosis of Gram-negative infection and sepsis in people; such that it may have a role in the exclusion of sepsis (Se = 85.3%, Sp = 44%, NPV 98.6%).[9] EA is not useful in people for predicting disease severity. A small number of veterinary studies have evaluated EA. The first of the canine studies evaluated 22 healthy dogs, 11 sick dogs without evidence of SIRS, and 8 dogs with suspected sepsis (defined by the authors as SIRS with toxic neutrophils on blood smear evaluation).[10] EA was significantly higher in dogs with sepsis compared to those without SIRS and healthy dogs. The authors thus concluded that the EAA is useful for endotoxin measurement in dogs and proposed a clinical cut off value of 0.346, since only one of the healthy dogs had an EA above that value. The second study evaluated the ability of the EAA to identify infection and predict outcome in 100 dogs consecutively admitted to a university teaching hospital ICU.[11] In contrast to the first study however EAA did not correlate with the presence of infection ($p>0.1$) but was useful as a part of a multivariate prognostic model to predict survival.

Tumor Necrosis Factor (TNF) -TNF has been found to have prognostic significance in a variety of human studies; with higher TNF concentrations correlating with mortality and persistent TNF production being associated with a poorer prognosis.[12-14] A variety of investigators have studied the potential role of TNF as a biomarker in veterinary patients with sepsis. TNF activity has been demonstrated in hospitalized dogs with canine parvoviral (CPV) enteritis and there was significant association between increasing TNF activity over time and morality in this population.[15] TNF concentrations have also been used as an indicator of response to treatment in clinical CPV,[16] and experimental endotoxemia in dogs.[17] In 23 dogs with septic peritonitis, there was no detectable serum TNF (measured by ELISA) in 17 dogs on day 1; whereas 16 dogs had detectable TNF in their abdominal fluid. [18] The median abdominal fluid TNF concentration was 0.69 ng/mL for surviving dogs, compared to 1.16 ng/mL for non-survivors, although this difference was not statistically significant.[18] Plasma TNF bioactivity at hospital presentation in dogs with sepsis, non-infectious SIRS and healthy controls has been evaluated. Significantly more dogs in the sepsis and SIRS groups had measurable TNF bioactivity in their plasma compared to healthy controls, but there was no difference between the sepsis and SIRS groups.[19] Additionally, plasma TNF bioactivity was not significantly correlated with mortality.[19] In cats, Plasma TNF bioactivity is significantly greater during sepsis or non-infectious SIRS compared to health.[20] However, there is no difference in plasma TNF bioactivity between infectious and non-infectious SIRS and there is no association between plasma TNF bioactivity and outcome for cats with sepsis.[20]

Interleukin (IL)-6- Several studies in human medicine have demonstrated a positive correlation between plasma IL-6 concentrations both at admission and over time with subsequent morbidity and mortality. One such study revealed a direct correlation between plasma IL-6 and the risk of death in patients with intra-abdominal sepsis; with IL-6 predicting outcome with >80% accuracy.[21] Persistent production of IL-6 seems to be more important than initial or peak concentrations in regards to patient outcome, although IL-6 may simply serve as a marker for disease severity in the absence of a causal relationship. [22-24] There are 2 studies in veterinary medicine that have evaluated plasma IL-6 concentrations in dogs with sepsis and SIRS. In the first study, plasma IL-6 concentrations at presentation predicted outcome in canine SIRS and sepsis.[25] In a study conducted at the MU CIML, plasma IL-6 concentrations were evaluated at hospital presentation in dogs with sepsis, non-infectious SIRS and healthy controls. Significantly more dogs in the sepsis or SIRS groups had IL-6 in their plasma compared to healthy controls, but there was no difference between sepsis and SIRS groups. Additionally, plasma IL-6 concentration was not significantly correlated with mortality.[19] It should be noted that these 2 studies used different assay methodologies to evaluate IL-6, the first used a bioassay and the second an ELISA technique. Cats with sepsis are more likely to have detectable concentrations of IL-6 in their plasma compared to cats with non-infectious causes of systemic inflammation or healthy cats but there is no correlation with outcome.[20]

Nitric Oxide (NO) - Nitric oxide plays an important role in the pathophysiology of sepsis. Pathogen associated molecular patterns and pro-inflammatory mediators induce iNOS in the endothelium, vascular smooth muscle, macrophages and various parenchymal calls.[26] NO is thought to be a key mediator of the myocardial dysfunction, hypotension and vascular hyporeactivity associated with sepsis, severe sepsis and septic shock.[26, 27] Nitric oxide also has profound immunomodulatory functions during sepsis.[26, 28, 29] Inhibiting NO production in experimental animal models of sepsis and human clinical trials has been shown to have beneficial effects on the hemodynamic stability.[24, 30-32] Nitric oxide has been evaluated in dogs with critical illness previously. In this study, critically ill dogs had significantly higher plasma nitrate and nitrite concentrations (stable markers of NO production) than non-critically ill or healthy dogs. [33] Additionally, plasma nitrate/nitrite correlated with disease severity and survival.[33] Investigators at the MU CIML have evaluated NO in dogs with naturally developing sepsis.[34] We found dogs with sepsis had significantly higher plasma NO concentrations than healthy control dogs; however there was no difference between septic dogs and dogs with non-infectious causes of SIRS. Nor was there a difference in plasma NO concentrations between the survivors and non-survivors.[34] Jacobson et al had similar findings in a cohort of dogs with babesiosis.[35]

IL-10 - Interleukin 10 is a potent anti-inflammatory cytokine which down regulates the inflammatory response through various mechanisms.[36] IL-10 acts to inhibit synthesis of many pro-inflammatory mediators (IFN-γ, IL-3, IL-2 and TNF) as well as enhancing production of IL-1RA by monocytes. IL-10 also blocks IFN-γ, binding at a key step in its signal transduction pathway.[36, 37] Diseases associated with increased IL-10 mRNA expression in dogs include Leishmania infantum[38] and canine distemper virus[39]; however little investigation has been carried out in clinical cases of bacterial sepsis. In a study conducted at the MU CIML, plasma IL-10 concentrations were evaluated in dogs with sepsis, non-infectious SIRS and healthy controls.[19] There was no difference in the number of dogs with measurable plasma IL-10 concentrations among groups nor were plasma IL-10 concentrations significantly correlated with survival.[19]

Procalcitonin (PCT) - Procalcitonin is perhaps the single most useful biomarker for sepsis in adult human patients. PCT is the precursor peptide of the hormone calcitonin; it is typically expressed in the C cells of the thyroid gland, but also secreted by non-thyroidal tissues. In people and experimental models it appears that circulating concentrations of PCT begin to increase within 2-6hrs of induction of bacteremia and/or endotoxemia, with a circulating half-life of approximately 24 hours.[40-42] PCT concentration will also be transiently increased with trauma, such as surgery; but return to baseline within 24 hours in the absence of infection or sepsis. While the exact role of PCT is still being determined in human medicine it appears to be useful in the diagnosis of sepsis,[43, 44] to guide antibiotic therapy (both commencement, cessation and de-escalation of antibiotics)[45, 46] and for prognostication[47]. Another interesting feature of PCT as a biomarker is that it is useful in the setting of chemotherapy induced sepsis, since it is a tissue-based rather than leukocyte-based biomarker. Unfortunately the ELISA for PCT in humans cannot be used for dogs due to differences in the carboxy terminus of the PCT molecule, such that the only study evaluating PCT in dogs investigated PCT mRNA expression by qRT-PCR. [48] This study demonstrated greater PCT mRNA in sick dogs (n=56; categorized into groups with infectious, inflammatory/non-infectious, neoplastic and other 'diseases) compared to healthy controls (n=11) ($p<0.001$), but no there was no difference in PCT mRNA expression based on the primary disease process.[48] Interestingly in five dogs from each group, PCT mRNA expression was determined at two time points; and in each dog the changes in its expression were in agreement with the clinical evaluation of patient improvement or deterioration, suggestive of a possible prognostic role.[48]

Markers of Coagulation - Altered coagulation is directly correlated to disease severity and mortality in human patients with sepsis. Hypercoagulability is also a common finding in dogs with naturally occurring sepsis as evidenced by prolongation of prothrombin time, partial thromboplastin time, increased D-dimer and fibrinogen degradation products concentrations, and depletion of antithrombin (AT)and activated protein C (aPC).[7, 49, 50] It appears that aPC may act as a useful prognostic indicator in dogs with sepsis.[49] Evaluation of hemostasis in dogs with CPV enteritis also documented decreased AT activity and hypercoagulability in affected dogs.[50] Von Willebrand factor antigen concentration has been evaluated as a marker of endothelial activation and is an independent predictor of survival in humans with sepsis. In a study of 14 dogs with sepsis and 11 control dogs, Von Willebrand factor antigen concentration was significantly higher in the sepsis group compared to control ($p<0.005$) but there was no difference between the survivors and non-survivors in the sepsis group.[51]

Natriuretic Peptides (NP) - The NP system is an endocrine mechanism that regulates salt and body fluid balance. In mammals it consists of three structurally related members; atrial natriuretic peptide (ANP), brain or B-type natriuretic peptide (BNP) and C-type natriuretic peptide (CNP). ANP and BNP are made primarily in the heart, whereas CNP is made in many organs, predominately vascular endothelial cells. NPs are synthesized as prepropeptides and then cleaved into an inactive N-terminal (NT) peptide and a biologically active C-terminal mature peptide. In many cases, the NT-peptide has greater plasma half-life and stability compared to the biologically active c-terminal peptide and thus is a preferred assay target. ANP, BNP and CNP are promising biomarkers for sepsis in humans[52-54] To the author's knowledge, ANP has not been evaluated in dogs with sepsis. In pilot study performed at MU CIML, NT-pBNP was evaluated in a small number of dogs with sepsis and healthy controls. There was no significant difference in serum NT-pBNP concentrations between the populations. We have also investigated NT-pro-CNP as a biomarker for sepsis in dogs. Serum NT-pCNP concentration was significantly greater in the sepsis group compared to the NSIRS and healthy control groups ($p<0.001$). Using a cut-off value of 10.1 pmol/l, AUC, sensitivity and specificity of NT-pCNP for differentiating dogs with sepsis from dogs with NSIRS or healthy control dogs were 0.71 (95% CI, 0.58-0.85), 65.5% (45.7-82.1%) and 89.2% (80.4-94.9%). Dogs with peritoneal forms of sepsis had significantly lower serum concentrations of NT-pCNP than dogs with other forms of sepsis ($p\leq0.04$). Of the dogs with a peritoneal source of sepsis, 9/12 had a false negative test result. The AUC, sensitivity and specificity were 0.92 (0.81-1.0), 94% (71-100%) and 89% (80-95%) when evaluating dogs with non-peritoneal sources of sepsis. Serum NT-pCNP concentration did not significantly correlate with survival in the sepsis group.[55]

C-reactive protein (CRP) - CRP is an acute phase protein that has been evaluated extensively as a biomarker both in human and veterinary medicine. In human medicine its use has been surpassed given the development of more sensitive and specific sepsis biomarkers such as PCT. A CRP assay has been validated for use in dogs[56] and increased CRP has been found in dogs associated with surgery, immune mediated disease,[57, 58] pancreatitis,[59] gastrointestinal disease and a variety of infectious insults including sepsis.[60, 61] CRP concentrations appear to be higher in critically ill dogs compared with controls.[62] The change in CRP over time has also been suggested to predict survival in dogs with sepsis.[63]

Ionized Calcium – There are multiple proposed mechanisms to explain ionized hypocalcemia in dogs and cats with sepsis. Parathyroid gland dysfunction, cytokine-mediated suppression of parathyroid hormone release, vitamin D deficiency hypomagnesemia, calcium chelation or tissue accumulation of calcium could all contribute to hypocalcemia of sepsis. Ionized hypocalcemia is associated with a more severe clinical course of disease and poorer outcome in people with sepsis. Ionized hypocalcemia has been identified in dogs and cats with critical illness and sepsis.[64-67] Ionized hypocalcemia is associated with a longer duration of hospitalization and poorer outcome in dogs with sepsis.[65] In cats, the dynamic change overtime appears to have a stronger association with outcome than evaluation of ionized calcium at a single time point.[67]

CONCLUSION

Strong criteria for the rapid diagnosis and prognostication of sepsis in dogs are needed. Large scale, multi-institutional studies to critically evaluate scoring systems for sepsis, including both clinical evaluation and endogenously produced biochemical compounds, will help establish sepsis biomarkers with the potential to play an important role in rational patient management and patient stratification for clinical trials.

ACKNOWLEDGMENT

The author thankS Dr. Claire Sharp at Tufts University for her contribution to these proceedings.

REFERENCES

[1]Group BDW. **Clin Pharmacol Ther** 2001;69:89.[2]Dalton WS, et al. **Science** 2006;312(5777):1165.[3]Garnacho-Montero J, et al. **Crit Care** 2006;10(4):R111.[4]Marshall JC, et al. **Crit Care Med** 2003;31(5):1560.[5]Okano S, et al. **Vet Rec** 2002;150:245.[6]Hauptmann J, et al. **Vet Surg** 1997;26:393.[7]Kenney EM, et al. **J Am Vet Med Assoc**;236(1):83.[8]Conti-Patara A, et al. **J Vet Emerg Crit Care (San Antonio)**.[9]Marshall JC, et al. **J Infect Dis** 2004;190(3):527.[10]Kjelgaard-Hansen M, et al. **Comp Immunol Microbiol Infect Dis** 2008;31(6):477. [11]Dickinson AE, et al., paper presented at the 15th IVECCS, Chicago, IL, 2009.[12]Martinez M, et al. **Crit Care Med** 1999;27:1303. [13]Oberhoffer M, et al. **Clin Chem Lab Med** 1999;37(3):363. [14]Gogos CA, et al. **J Infect Dis** 2000;181(1):176. [15]Otto C, et al. **JVIM** 1997;11:65. [16]Senturk S. **J Vet Pharmacol Ther** 2005;28(1):57. [17]DeClue AE, et al. **Am J Vet Res** 2008;69(2):228. [18]Humm KR, et al., in *14th IVECCS.* (Phoenix, AZ, 2008), 819. [19]DeClue AE, et al. **J Vet Intern Med**;26(3):624. [20]DeClue AE, et al. **J Am Vet Med Assoc** 2011;238(7):890. [21]Patel RT, et al. **Br J Surg** 1994;81(9):1306. [22]Casey LC, et al. **Ann Intern Med** 1993;119(8):771. [23]Barriere SL, et al. **Crit Care Med** 1995;23(2):376. [24]Tanigawa K, et al. **J Crit Care** 1999;14(4):186. [25]Rau S, et al. **Vet Clin Pathol** 2007;36(3):253. [26]Kirkeboen KA, et al. **Acta Anaesthesiol Scand** 1999;43(3):275.[27]Abraham E, et al. **Crit Care Med** 2007. [28]Clancy RM, et al. **J Clin Invest** 1992;90(3):1116. [29]Kubes P, et al. **Proc Natl Acad Sci U S A** 1991;88(11):4651. [30]Watson D, et al. **Crit Care Med** 2004;32(1):13. [31]el-Dwairi Q, et al. **Am J Physiol** 1998;274(3 Pt 1):C770. [32]Chan D, paper presented at the International Veterinary Emergency and Critical Care, Atlanta, GA, 2005. [33]Jones J, et al. **Journal of Veterinary Emergency and Critical Care** 1999;9(4):195. [34]Osterbur K, et al. **Vet Rec**;169(21):554. [35]Jacobson L, et al. **Vet Parasitol** 2002;104:27. [36]Gosain A, et al. **J Burn Care Rehabil** 2005;26(1):7. [37]Rozanski E, paper presented at IVECCS, San Antonio, Texas, 2006. [38]Boggiatto PM, et al. **Clin Vaccine Immunol** 2009. [39]Beineke A, et al. **Viral Immunol** 2008;21(4):401. [40]Meisner M, et al. **Shock** 2003;19(2):187. [41]Dandona P, et al. **J Clin Endocrinol Metab** 1994;79(6):1605. [42]Nijsten MW, et al. **Crit Care Med** 2000;28(2):458. [43]Assicot M, et al. **Lancet** 1993;341(8844):515.[44]Simon L, et al. **Clin Infect Dis** 2004;39(2):206. [45]Christ-Crain M, et al. **Lancet** 2004;363(9409):600.[46]Stocker M, et al. **Neonatology** 2009;97(2):165.[47]Jensen JU, et al. **Crit Care Med** 2006;34(10):2596. [48]Kuzi S, et al. **J Vet Diagn Invest** 2008;20:629. [49]de Laforcade A, et al. **J Vet Intern Med** 2003;17:674. [50]Otto CM, et al. **J Am Vet Med Assoc** 2000;217(10):1500. [51]Rogers CL, et al. **J Vet Intern Med** 2009. [52]Morgenthaler NG, et al. **Crit Care** 2005;9(1):R37. [53]Chen Y, et al. **Am J Emerg Med** 2009;27(6):701. [54]Hama N, et al. **Biochem Biophys Res Commun** 1994;198(3):1177. [55]DeClue AE, et al. **J Vet Intern Med** 2011;25(3):453. [56]Kjelgaard-Hansen M, et al. **J Small Anim Pract** 2008;49(6):282. [57]Griebsch C, et al. **Vet Clin Pathol** 2009. [58]Lowrie M, et al. **J Vet Intern Med** 2009;23(4):862. [59]Mansfield CS, et al. **J Am Vet Med Assoc** 2008;233(6):936. [60]Dabrowski R, et al. **Theriogenology** 2009;72(4):471. [61]Nakamura M, et al. **J Vet Med Sci** 2008;70(2):127. [62]Chan DL, et al. **J Vet Intern Med** 2009;23(3):559. [63]Gebhardt C, et al. **J Vet Emerg Crit Care (San Antonio)** 2009;19(5):450. [64]Holowaychuk MK, et al. **J Vet Intern Med** 2009;23(3):509. [65]Luschini MA, et al. **J Vet Emerg Crit Care (San Antonio)**;20(4):406. [66]Holowaychuk MK, et al. **J Vet Emerg Crit Care (San Antonio)**;21(5):521. [67]Kellett-Gregory LM, et al. **J Vet Emerg Crit Care (San Antonio)**;20(4):398.

IDENTIFICATION AND MANAGEMENT OF SEPSIS IN THE CAT

Amy E. DeClue, DVM, MS, DACVIM (SAIM)
Columbia, MO

INTRODUCTION

Sepsis is a Greek work meaning "decomposition of animal or vegetable organic matter in the presence of bacteria".[1] In the clinic, sepsis has come to mean infection with concurrent clinical evidence of the systemic inflammatory response (i.e., SIRS because of infection). At one time, sepsis and bacteremia were considered synonymous terms. However, as our knowledge of sepsis has grown we have realized that the clinical manifestations of sepsis relate to wide-spread inflammation and not necessarily wide-spread infection.

CLINICAL FINDINGS

Sepsis has been associated with many conditions in the cat including septic peritonitis, hepatic abscessation, pyothorax, bacteremia, pneumonia, endocarditis, pyelonephritis, pyometra, viral, fungal protozoal, and parasitic infections.[2-6] Accurately describing the clinical findings associated with sepsis in cats is challenging since diagnostic criteria for sepsis in cats have not been prospectively evaluated. The following data, which focuses on bacterial sepsis, has been compiled from a literature search using combinations of the following terms: feline, sepsis, septic peritonitis, bacteremia, abscess, pyometra, and pyothorax. The terms bacteremia, abscess, pyometra and pyothorax were selected because cats with those conditions are at risk for developing a systemic inflammatory response to infection. These data should be interpreted with caution as there are several important limitations to this analysis including that inclusion criteria for these studies were often focused on the source or type of infection, criteria to demonstrate systemic inflammation were not always used and most data were retrospectively collected. Thus, this information should be viewed as clinical findings in cats sepsis or cats with infections that are likely to result in sepsis.

There were ten studies identified that contained clinical and/or clinicopathologic data.[2-4, 7-13] A total of 336 cats were included in this review. Sex was reported in 294 cats; approximately 60% of cases were male and 40% were female. Age was reported for 257 cats with an overall mean age of 4.58 years. Duration of hospitalization was reported for 135 cats with an overall mean of 3.14 days. Survival statistics were reported for 214 cats with an overall survival rate of 57.9%. The location of infection was identified as pleural space (n=131), peritoneum (95), blood (71), liver (14), systemic (7), endocardium (5), lung (5), kidney (3), subcutaneous space (2), bone (1), uterus (1) and CNS (1). The majority of cats had bacterial infections (328); 8 cats had other types of infection (FeLV (2), FIP (2), *Cytauxzoon felis* (2), FIV, *Toxoplasma*). Aerobic cultures were performed in 237/324 cats (73.1%). Some studies did not indicate if anaerobic cultures were also performed. For studies that reported growth of anerobic organisms, *Clostridium* (30), mixed anaerobes (25), *Bacteroides* (17), *Fusobacterium* (15) and *Actinomyces* (13) were the most commonly reported results. The most common organisms identified with aerobic bacterial culture are listed in table 1. The frequently reported clinical signs were generally non-specific in nature and included anorexia (141/198; 71%), lethargy (14/200; 71%), GI signs (38/120; 31%), and weight loss (32/133; 24%). The most commonly reported physical examination findings included tachypnea (72/94; 77%), pale MM (87/115; 76%), abdominal pain (86/153; 56%), fever (47/136; 35%), hypothermia (52/189; 28%), and bradycardia (32/135; 24%). The most commonly reported clinicopatholgic abnormalities are listed in table 2.

Table 1: Organism	Number of positive cultures	Percent
E. coli	55	23
Pasteurella	39	16
Streptococcus	17	7
Enterococcus	17	7
Staphylococcus	11	5
Enterobacter	7	3
Pseudomonas	4	2
Corynebacterium	4	2
Salmonella	3	1
Micrococcaceae	2	1
Acinetobacter	2	1

Table 2: Clinicopathologic parameters.

Parameter	# affected	# evaluated	Percent
Hypoalbuminemia	74	116	64
Acidemia	21	34	62
Band neutrophils	93	158	59
Hyperlactatemia	30	52	58
Hyponatremia	27	50	54
Hypocalcemia	36	73	49
Hyperbilirubinemia	56	121	46
Anemia	39	91	43
Leukocytosis	48	114	42
Increased BUN	31	77	40
Hyperglycemia	55	156	35
Increased ALT	28	116	24
Hypokalemia	12	50	24
Creatinine	17	77	22
Leukopenia	13	70	19
Hypoglycemia	25	137	18
Increased ALP	13	79	16
Hyperkalemia	6	50	12
Hypernatremia	1	50	2

DIAGNOSIS

The diagnosis of sepsis is made based on identification of systemic inflammation secondary to infection. There are several diagnostic criteria that have been proposed ranging from simply the recognition of systemic consequences of infection to more elaborate diagnostic schemes requiring fulfillment of 2/4 SIRS criteria and identification of infection. The most commonly used SIRS criteria (rectal temperature>103.5°F or <100°F; heart rate >224 or <140bpm; respiratory rate >40 bpm; WBC count >19,500/ul or < 5,000/ul or >5% bands) are based on a retrospective study of clinical and necropsy findings in 29 cats with sepsis.[4] While this was an important study and certainly added to our understanding of the clinical aspects of sepsis, caution should be used when applying criteria from a study with a 100% mortality rate. It is possible that these criteria exclude cats with less severe forms of sepsis. It is important to remember that there are many noninfectious causes of SIRS. Examples of noninfectious causes of SIRS include, but are not limited to, reaction to medications or vaccinations, envenomization, pancreatitis, heat stroke, trauma, burns, autoimmune disease and neoplasia. Ruling out any potential noninfectious cause of SIRS will help strengthen your clinical suspicion of infection.

Diagnostic investigation to identify the infectious organism(s) starts with a complete history and physical examination. For some cases, identifying the inciting cause will be simple (i.e., finding an abscess on a limb in a cat with a history of a recent cat bite), while other patients will require in-depth diagnostic testing. Based on historical and physical examination findings, an anatomic region can be usually be identified as the origin of sepsis (i.e., pyothorax in the cat with dyspnea, fever and pleural effusion). Thoracic and abdominal imaging are also helpful in identifying the potential source of infection and should be routinely performed in any cat with suspected sepsis. Use clinical clues to help guide sample collection (table 6). When in doubt, collect samples from any possible source of infection. Samples should be evaluated via cytology/histopathology and with culture and sensitivity. Even if bacteria are not identified on cytology, culture and sensitivity should always be performed so that antibiotic selection can be tailored to the particular organism. Samples should be collected as quickly as possible and preferably prior to administration of antimicrobials. Aseptic technique should be used for sample collection to minimize sample contamination. For patients with suspected bacterial peritonitis, glucose concentration of the blood can be compared to the glucose concentration of the peritoneal fluid to achieve a rapid diagnosis. A difference of > 20 mg/dL between blood and peritoneal fluid glucose concentrations (blood > peritoneal fluid) is diagnostic for septic peritoneal effusion. Remember that infectious organisms that do not lend themselves well to culture may be the cause of sepsis; serology and/or PCR for vector borne disease, viral infection and fungal infection etc. should be considered.

Table 6: Sample collection	**Clinical clues to guide sampling**	**Evidence of infection** (Note: a lack of these findings does not necessarily indicate a lack of infection in the sampled area)
Blood culture	Any (perhaps all) patient with suspected sepsis, newly discovered heart murmur	Positive culture
Bronchoalveolar lavage, endotracheal wash	Cough, dyspnea, interstitial or alveolar infiltrates on thoracic radiographs consistent with infectious pneumonia; note that bacterial pneumonia is relatively rare in the cat	Positive culture, degenerative neutrophils, intracellular bacteria or evidence of fungal organisms
Abdominocentesis	GI signs, abdominal distention, abdominal effusion, abdominal pain, lack of abdominal detail on radiographs or evidence of abdominal fluid on ultrasound	Positive culture, intracellular bacteria or evidence of fungal organisms, > 20 mg/dl difference between blood and fluid glucose concentrations
Thoracocentesis	Inspiratory or mixed inspiratory and expiratory dyspnea, muffled lung sounds in the dependent lung fields, evidence of pleural effusion on imaging	Positive culture, intracellular bacteria or evidence of fungal organisms
Arthrocentesis	Evidence of joint pain or effusion	Positive culture, intracellular bacteria or evidence of fungal organisms
Aspiration of subcutaneous masses	Any mass that has recently developed or changed in size, any draining or ulcerated mass, history of trauma in the region	Positive culture, intracellular bacteria or evidence of fungal organisms
Urine culture	Any patient with sepsis of unknown origin, azotemia, pyuria, bacteriuria, evidence of pyelonephritis on ultrasound	Positive culture, bacteriuria, pyuria. Note a lack of bacteriuria or pyuria does not rule out urinary tract infection or pyelonephritis
Pericardiocentesis	Muffled heart sounds, evidence of a globoid heart on thoracic radiographs, evidence of pericardial effusion on ultrasound	Positive culture, intracellular bacteria or evidence of fungal organisms
Aspiration of abdominal organs	Identification of abnormal organ architecture or abscess on ultrasound	Positive culture, intracellular bacteria or evidence of fungal organisms
Diagnostic peritoneal lavage	Suspected abdominal source of infection but little or no abdominal effusion; any patient with sepsis of unknown origin	Positive culture, intracellular bacteria or evidence of fungal organisms

A complete patient evaluation, including CBC, serum biochemical profile, urinalysis, blood gas analysis, coagulation profile and appropriate diagnostic imaging, should be performed in any critically ill patient suspected of having sepsis. Some patients with sepsis will have multiple clinicopathologic abnormalities and others with have only a few. There is no one particular clinicopathologic finding or combination of findings that is specific for the diagnosis of sepsis. Possible serum biochemical abnormalities in include hyper- or hypoglycemia, hypoalbuminemia, azotemia, hyperbilirubinemia, electrolyte abnormalities and increased ALT and/or ALP. In general, dogs tend to be more likely to have increased liver enzymes than cats during sepsis. Leukocytosis, neutrophilia with a left shift or a leucopenia, anemia and thrombocytopenia are abnormalities that may be recognized on the CBC. Coagulation abnormalities are common. Urinalysis findings in can include isosthenuria, proteinuria and possibly pyuria and bacteriuria (if inflammation/infection of the urinary tract is present). It is important to consider the volume status of your patient when interpreting renal parameters. Most patients will present with pre-renal azotemia often in the face of isosthenuria because bacterial infection can cause secondary nephrogenic polyuria; this also contributes to dehydration and hypovolemia. Prior to diagnosis of renal failure (as opposed to pre-renal azotemia), rehydration and volume expansion should be accomplished.

Additional diagnostic evaluation should include assessment of the patient's acid base status, oxygenation and blood pressure. Many patients will develop a metabolic acidosis secondary to hypovolemia and poor tissue perfusion. Some patients with respiratory compromise will be hypoxemic or will have evidence of a respiratory acidosis on blood gas analysis. In the event that arterial blood gas analysis is not available, pulse oximetry can be used to assess oxygenation. Many patients will have hypotension. Blood pressure can rapidly change in patients with sepsis so it is advisable to frequently monitor blood pressure (and other vital parameters).

TREATMENT

The severity of bacterial sepsis varies from patient to patient. Some patients will require advanced life support measures to survive while others will require only basic supportive care. The key aspects of sepsis treatment include 1.) alleviating the inciting cause and 2.) aggressive supportive care.

Alleviate the cause

The most important aspect of treating bacterial sepsis centers on the identification and eradication of the inciting cause. Although stringent effort should be made to identify the cause of sepsis, early antimicrobial treatment is critical for survival and should not be withheld pending culture results. In humans, early antibiotic administration has been shown to decrease mortality. Broad spectrum, bactericidal antimicrobial agents (e.g., fluoroquinolone + penicillin) administered IV should be instituted as quickly as possible in patients with suspected sepsis. Antibiotic therapy should be selected based on the most likely type of organism given the site of infection. Depending upon the source of infection, surgical debridement may be necessary for infection control. Once culture specific antibiotic sensitivity is determined, the antibiotic with the narrowest spectrum of activity should be chosen and administered until there is complete clinical resolution. The remainder of therapy centers on maintenance of tissue perfusion, management of organ failure and aggressive supportive care.

Supportive care

Hemodynamic support -Most patients with sepsis will have a relative or absolute hypovolemia, hypotension and/or poor tissue perfusion. Boluses of isotonic crystalloids (10ml/kg, IV to effect) or colloids (2 ml/kg to effect) can be used for rapid initial intravascular volume resuscitation and administered based on evaluation of end points (e.g., heart rate, pulse quality, mentation, membrane color, blood pressure, central venous pressure). Then, ongoing fluid therapy should be tailored to the needs of the patient to provide for maintenance requirements, correct interstitial hydration deficits and replace ongoing losses. Patients with sepsis will have a propensity to develop interstitial edema, including subcutaneous and pulmonary edema, because of increased vascular permeability and decreased blood colloid osmotic pressure. Administration of colloid fluids (e.g., hetastarch) may help prevent interstitial edema and should be considered for volume resuscitation and during maintenance of sepsis patients, in addition to crystalloid fluid support.

Despite aggressive volume resuscitation, some patients will require additional support to maintain normal blood pressure and tissue perfusion. The safest and most efficacious treatment for septic shock has not been determined in cats. Prior to administration of any sympathomimetic drug, care should be taken to ensure that the patient is not hypovolemic. Since the goal of treating septic shock is to maintain tissue perfusion, medications that cause vasoconstriction should be used only if absolutely necessary. Positive inotropic drugs (e.g., dobutamine) may be a good initial management choice for septic shock since they help combat decreased cardiac output caused by myocardial dysfunction without inducing peripheral vasoconstriction. However, care should be taken to avoid positive inotropic drugs in cats with evidence of HCM. If volume resuscitation and positive inotropic support has failed to restore adequate blood pressure, a vasopressor agent (e.g., dopamine, norepinephrine, epinephrine or vasopressin) could be added. Although there are no clinical trials evaluating these drugs, epinephrine is generally considered the least ideal vasopresor for sepsis in cats. Relative adrenal insufficiency has been documented in cats with sepsis and refractory hypotension and should be considered in any cat requiring vasopressor therapy during sepsis.

GI/Nutrition- Bacterial translocation from the GI tract can contribute to systemic inflammation during sepsis. Early placement of a feeding tube and initiation of enteral or, in patients with vomiting, parenteral nutrition will help maintain GI barrier function. Additionally, medication aimed at maintaining normal GI protective mechanisms (e.g., omeprazole, famotidine, sucralfate) could be considered. Hyperglycemia can be a complication of nutritional therapy, especially parenteral nutrition. Hyperglycemia has been associated with increased inflammation and a poorer prognosis in people with sepsis and SIRS. Although the importance of glucose homeostasis in cats with sepsis is unknown, iatrogenic hyperglycemia should be avoided.

Cellular oxygenation- Oxygenation, acid-base status, PCV or hemoglobin concentration and organ function should be closely monitored. Along with maintaining good tissue perfusion, maximizing cellular oxygenation will help maintain tissue viability and avoid multiple organ dysfunction. This may be accomplished by ensuring good oxygen carrying capacity, providing supplemental oxygen when indicated and preventing microthrombi formation. For the majority of patients, acid-base abnormalities are related to lactic acidosis secondary to poor tissue perfusion. Typically these abnormalities will resolve with resolution of hypotension and normalization of perfusion. Therefore, bicarbonate administration is rarely needed and in fact may be contraindicated.

Nursing care/Pain management- Care should be taken to ensure adequate patient comfort including management of pain, careful catheter maintenance, bladder/bowel care and frequent patient turning/movement to prevent decubital ulcers. Almost all patients with sepsis will require analgesic administration for pain management. Although many analgesics like ketamine and buprenorphine may offer specific anti-inflammatory advantages during endotoxemia, some may be detrimental. Morphine, for instance, augments the inflammatory response to endotoxin, has a detrimental effect on mean arterial pressure and increases mortality in endotoxemic rats. While it is not known if morphine is detrimental during feline sepsis its routine use in septic patients should be avoided where possible.

REFERENCES

[1]Geroulanos S, et al. **Intensive Care Med** 2006;32:2077. [2]Sergeeff J, et al. **J Vet Intern Med** 2004;18:295. [3]Costello M, et al. **J Am Vet Med Assoc** 2004;225:897. [4]Brady C, et al. **J Am Vet Med Assoc** 2000;217(531-5). [5]Waddell L, et al. **JAVMA** 2002;221:819. [6]Barrs V, et al. **J Fel Med Surg** 2005;7:211. [7]Waddell L, et al. **J Am Vet Med Assoc** 2002;221:819. [8]DeClue A. **J Am Vet Med Assoc** 2011;238:890. [9]Barrs V, et al. **J Feline Med Surg** 2005;7:211. [10]Ruthrauff CM, et al. **J Am Anim Hosp Assoc** 2009;45(6):268. [11]Greiner M, et al. **J Feline Med Surg** 2007;9(5):404. [12]Demetriou JL, et al. **J Small Anim Pract** 2002;43(9):388. [13]Parsons KJ, et al. **J Small Anim Pract** 2009;50(10):518.

AN EXPERIMENTAL JANUS KINASE (JAK) INHIBITOR SUPPRESSES EOSINOPHILIC AIRWAY INFLAMMATION IN FELINE ASTHMA. CH Chang, JR Dodam, LA Cohn, J Trzil, H Liu, CR Reinero. Comparative Internal Medicine Laboratory, Department of Veterinary Medicine and Surgery, University of Missouri, 900 East Campus Dr. Columbia, MO 65211, USA

Janus kinase (JAK) enzymes are involved in immune cell signaling pathways for several cytokines believed important in mediating pathology in allergic asthma. We hypothesized that a novel JAK inhibitor would effective for treating experimental feline asthma; an additional objective was to assess safety.

Experimental asthma was induced according to a previously established protocol using Bermuda grass allergen in 24 cats. Eight cats/group were randomly assigned to receive placebo or positive control (a compound provided by Pfizer, Inc. with known *in vitro* and *in vivo* activity against JAK) at either 0.5 or 1 mg/kg twice daily for 28 days. Safety evaluation was performed by serial examinations and complete blood count, serum biochemical profile, and urinalysis. Bronchoalveolar lavage fluid % eosinophils and ventilator-acquired pulmonary mechanics with methacholine (MCh) challenge were used to determine drug effects on airway inflammation and airway hyperresponsiveness.

No cat displayed any adverse clinical signs or clinically important laboratory abnormality with treatment. At day 28, both low and high doses of the positive control compound significantly suppressed eosinophilic airway inflammation versus placebo ($p=0.022$ and $p=0.003$, respectively), with no significant difference between doses ($p=0.512$). There was marked variability in pulmonary mechanics responses between cats with no significant difference between treatments in the effective concentration of MCh that induced a 200% increase over baseline airway resistance.

In conclusion, these data suggest the experimental JAK inhibitor could be a safe and effective treatment for reduction of airway inflammation in experimental feline asthma. Prospective trials in pet cats with allergic asthma are warranted.

FLUTICASONE OR PREDNISOLONE FOR TREATMENT OF CATS WITH NATURALLY-OCCURRING LOWER AIRWAY DISEASE. E Rozanski, C Sharp, Oura T, Egan A, Bain P, Knoll J Tufts Cummings School of Veterinary Medicine, North Grafton, MA.

Feline lower airway disease (FLAD) is a common inflammatory/allergic disease in cats. Clinical signs include coughing, wheezing, and episodic respiratory distress. If no underlying cause/allergen is identified, life-long treatment is advised to limit airway inflammation and subsequently remodeling. Glucorticoids are considered the most effective therapy, and may be administered orally. Inhaled glucocorticoids (IG) have also been advocated as a therapy as a potential method of limiting systemic glucocorticoids side effects, yet effectively treating clinical signs and airway eosinophilia. The purpose of this study was to compare treatment with oral prednisolone versus fluticasone for the treatment of FLAD.

Cats were recruited from the medicine and emergency services with a history and thoracic radiographs consistent with lower airway disease. A CBC/chemistry profile, occult heartworm antibody/antigen testing and fructosamine were performed. Cats were anesthetized and lung function testing (LFT) was performed using Puritan Bennett 840, using a standard volume history and standardized ventilator settings. LFT included static resistance and compliance and the peak and plateau airway pressures. A tracheal wash was performed for cytological examination and confirmation of airway eosinophilia. Cats were then randomized to receive either 5 mg of prednisolone orally twice daily for two weeks, and then once a day, or prednisolone 5 mg twice daily for one week, then once daily for one week, and fluticasone 110 mcg twice daily using an AEROKAT chamber. Cats were rechecked after at least 8 weeks of therapy, with repeated physical examination, thoracic radiographs, lung function testing, airway cytology and fructosamine/glucose.

Fourteen cats were enrolled, and 9 cats successfully completed the study. Five cats received IG and four cats ORAL All cats were clinically normal after treatment and either form of therapy were well-tolerated by cats and owners. There was no difference in fructosamine or glucose after treatment of between groups. The median airway eosinophil count fell significantly from 48% to 6% in the ORAL group, and 51% to 19% in the IG. Pre and post therapy static compliance was variable, increasing in 2/5 cats treated with IG but in all 4 cats treated with ORAL. Airway resistance decreased in 4/5 cats treated with IC by a median of 13%, and increased by 91% in one cat. In the four cats treated with ORAL airway resistance decreased by 59%.

In this 8 week clinical trial, treatment with either IG and ORAL glucocorticoids was effective in eliminating clinical signs, and reducing airway eosinophilia. IG were well tolerated by all treated cats. ORAL glucocorticoids were associated with more robust improvement in airway resistance and compliance than fluticasone at 110 mcg twice a day in cats with naturally-occurring lower airway disease.

WHEN TWO ARE NOT BETTER: CANINE CHRONIC KIDNEY DISEASE AND NEPHROURETEROLITHIASIS

Joe Bartges, DVM, PhD, DACVIM, DACVN
Knoxville, TN

Upper urinary tract urolithiasis (nephroureterolithiasis) is being recognized more frequently with a 10-fold increase in occurrence in cats during a 20-year interval in; similar data are not available in dogs. There appears to be an association of nephroureteroliths with acute or chronic renal failure. Many patients are asymptomatic and nephroureteroliths are discovered when lower urinary tract uroliths are creating clinical signs or when abdominal imaging is performed for other reasons. Nephroureteroliths can lead to obstruction to urine flow; deterioration of renal function; serve as a nidus for bacterial urinary tract infection; or cause hematuria or pain. Clinical signs, when present, may include systemic signs associated with renal failure or pyelonephritis, vague signs (abdominal pain, arched back, vomiting, or anorexia) or hematuria. Sometimes animals present for an abrupt change in behavior or an acute onset of abdominal pain or vomiting when an acute ureteral obstruction occurs; this may be mistaken for intervertebral disk disease. Small, multiple uroliths are common and usually found in both kidneys and/or ureters. Small ureteroliths appear to migrate over time and can move retrograde in the ureter. Some patients with upper urinary tract uroliths have lower urinary tract uroliths. When lower tract uroliths occur in association with upper tract uroliths, it is unknown whether they formed in the lower urinary tract or whether they passed through the ureter and into the urinary bladder. A common scenario is to find upper tract uroliths in a cat with chronic renal failure. When there is unilateral nephrolithiasis, often they occur in the larger and more functional kidney.

ETIOLOGY AND PATHOPHYSIOLOGY

Calcium oxalate accounts for most (>70%) nephroureteroliths. Struvite, particularly infection-induced, and other mineral types do occur. Risk factors for calcium oxalate nephroureteroliths are the same as for calcium oxalate urocystourethroliths. Breeds considered at higher risk for forming calcium oxalate uroliths include Miniature schnauzers, Lhasa apsos, Shih tzu, toy Poodles, and Bichons Frises. They occur more commonly in male than in female dogs. Nephroureteroliths usually occur in middle-aged or older dogs. Approximately 5% of dogs with calcium oxalate uroliths have hypercalcemia, usually due to primary hyperparathyroidism. In these dogs, serum total and blood ionized calcium concentrations are increased and parathyroid hormone concentration is high or "abnormally" normal. Struvite uroliths occur usually due to a bacterial infection with a urease-producing microbial organism (infection-induced struvite). *Staphylococcus spp* is the most common urease-producing microbial organism associated with infection-induced struvite uroliths; however, other urease-producing organisms include *Proteus spp, Enterococcus spp, Streptococcus spp,* and uncommonly *Klebsiella spp*. Occasionally, other minerals such as urate and cystine form uroliths in the upper urinary tract.

THERAPY

Not all nephroureteroliths require treatment especially by surgery. In one study, presence of nephroliths without ureteral obstruction was not associated with progression of CKD in cats. If uroliths are composed of struvite or urate, medical dissolution can be attempted. Indications for treatment of nephroureteroliths include obstruction to urine flow, increase in urolith size and/or number despite appropriate therapy, if uroliths are composed of a mineral amenable to medical dissolution therapy, recurrent or persistent urinary tract infections, compromise of renal function due to uroliths, recurrent or persistent clinical signs associated with renal pain (e.g. vomiting or abdominal pain), or severe hematuria. Many dogs with nephroureteroliths already have renal compromise. Removal of nephroureteroliths in cats with CKD, especially if ureteroliths, improves survival; however, there is an associated morbidity and mortality associated with surgery.

Medical management

At present, there is no means to medically dissolve calcium oxalate uroliths. Infection-induced struvite uroliths can be dissolved by feeding a low magnesium and phosphorous diet that induces aciduria and administering an appropriate antimicrobial agent. Average dissolution time for infection-induced struvite uroliths is 8 weeks; however, nephroliths often take longer to dissolve. Sterile struvite uroliths can be dissolved by feeding a low magnesium and phosphorous diet that induces a diuresis; administering an antimicrobial agent is unnecessary. Average dissolution time is 2-4 weeks; however, nephroliths often take longer to dissolve. Urate uroliths can be dissolved by feeding an ultra-low protein diet that induces alkaluria and administering allopurinol. This protocol is effective in approximately 40% of dogs with urate uroliths; however, is not effective in dogs with portovascular anomalies. In most cases, uroliths dissolve in 4 weeks. Dissolution time is variable and often several months. Because urate uroliths are radiolucent typically, monitoring dissolution requires ultrasonography, CT, or contrast excretory urography In human beings, alpha blocking agents have been shown to promote movement of ureteroliths into the bladder (called medical expulsion therapy) especially if the uroliths are in the distal 1/3 of the ureter; however, the effectiveness in dogs has not been evaluated. Glucagon has also been used to facilitate passage of ureteroliths after fragmentation by lithotripsy; however, was not effective in cats with ureteroliths. Alpha-blocking agents (prazosin: 0.5 mg PO q12h) and anti-inflammatory drugs (prednisone: 0.5 mg/kg PO q12h, dexamethasone: 0.05 m/kg IV q12h) may facilitate passage with fluid therapy. Amitriptyline (0.5 – 2.0 mg/kg PO q24h) has been shown to relax ureteral smooth muscle and facilitate expulsion of ureteroliths in humans. Spontaneous passage of ureteroliths has been shown to occur. Ureteroliths have also been reported to move retrograde up to 4 cm in the ureter even back into the renal pelvis resulting in improvement of urine flow and azotemia.

Dialysis

Dialysis is the process of removing solutes and/or water from one solution (plasma) to another solution (dialysate) by an osmotic gradient through a semi-permeable membrane. The 3 types of dialysis are peritoneal dialysis, hemodialysis, and enteric dialysis using an oral sorbent. Acute renal failure is the prevailing indication for dialysis, although hemodialysis may be used to manage CKD. Conditions where dialysis is indicated include: acute renal failure, pre-surgical treatment of uremia (e.g. uroabdomen or renal transplantation), electrolyte abnormalities unresponsive to medical therapy (e.g. hypernatremia and hyperkalemia), metabolic and acid-base disturbances (e.g. hepatic encephalopathy and severe metabolic acidosis), severe fluid overload, and toxicities where the toxin is dialyzable (e.g. ethylene glycol, acetaminophen or salicylate toxicity, barbiturate toxicity, and grape/raisin toxicity). Dialysis involves the use of one type of semi-permeable membrane that allows the passage of small molecules while preventing the transfer and loss of large molecules. In peritoneal dialysis, the peritoneal membrane, interstitial tissue, and capillary endothelium act as the membrane. In hemodialysis, blood is circulated outside of the body through a dialyzer.

Surgical management

Surgical removal is not necessary for all upper urinary tract uroliths. Indications for surgery are similar to those listed previously. Surgery is not indicated for relatively small, non-obstructive upper urinary tract uroliths. Small upper urinary uroliths may pass into the urinary bladder. It is unknown what a safe time period to allow for ureteroliths passing without resulting in irreversible renal damage. Renal function does not recover n dogs with unilateral ureteral ligation for 40 days. If a patient is ill due to ureteral obstruction, it is not probably appropriate to wait for the ureteroliths to pass or to medically manage them for a prolonged period of time if renal function is to be preserved. If a urinary tract infection is present, removal of the upper urinary tract urolith may be necessary to control the infection; however, it is not possible to determine prior to surgery if the uroliths are the source of infection. If bilateral upper urinary tract surgery is required, the procedures should be staged with an approximate separation of 4 weeks in order to re-evaluate renal function and to allow recovery from the first procedure. In general, the side with the most renal function should be operated first in order to preserve as much renal function as possible. Nephroliths can be removed either nephrotomy or by pyelolithotomy; however with either technique, it is not always possible to retrieve all uroliths. Nephrotomy should be avoided, if possible, because it causes renal damage and loss of nephrons. Pyelolithotomy is preferred because renal parenchymal tissue is not incised and so loss of renal function is less likely to occur when compared with nephrotomy; however, pyelolithotomy cannot be performed unless the renal pelvis is dilated. Often it requires microsurgical technique Ureterotomy is technically difficult to perform because of the diameter of the ureter, especially in cats and stricture formation often occurs Nephrectomy and ureterectomy are indicated for a severely hydronephrotic, non-functional kidney.

Comparison of medical vs surgical management

In a retrospective analysis of medically versus surgically managed patients with ureteroliths, surgically managed patients tended to do better over a longer time. There was a recurrence rate of 40%.

Lithotripsy

Extracorporeal shock wave lithotripsy (ESWL) is a standard of care for many human patients with upper urinary tract uroliths, and has been performed successfully in dogs. Struvite is most easily fragmented, followed by calcium oxalate, then urate, then cystine. When upper urinary tract uroliths fragment, the small fragments pass into the urinary bladder over several weeks; their passage is not instantaneous. Animals that have undergone ESWL should have periodic examinations including serum biochemical analysis, urinalysis, and radiographs to monitor for successful passage of fragments and to detect any compromise of renal function due to the lithotripsy procedure. Acceptable candidates are those with nephroliths smaller than 2-3 cm in their greatest dimension or ureteroliths. With bilateral nephroliths, both kidneys are treated at the same time, unless there is concern about compromising renal function further. Successful fragmentation has been reported to occur in the majority of canine patients. Although renal function may be normal in otherwise healthy dogs undergoing lithotripsy, lithotripsy of clinical cases of nephroureterolithiasis suggests that some, particularly those with pre-existing renal disease, experience renal function compromise or worsening of their renal failure. While lithotripsy is considered safer and less invasive than surgical removal of upper urinary tract uroliths, there are risks. Abdominal pain, hemorrhage, and bruising of the kidneys occur, and hematuria may be observed immediately after the procedure. More significant hemorrhage within or around the kidney may occur in some cases. Residual stone fragments often take several weeks to move from the kidney into the urinary bladder. Transient or permanent ureteral obstruction can occur. If permanent and progressive ureteral obstruction occurs, it requires re-treatment by lithotripsy or surgical intervention. Uncommon complications include pancreatitis, bowel irritation, hemolysis, and systemic hypertension.

Percutaneous nephrolithotomy

A rigid endoscope can be inserted through the renal parenchyma into the dilated renal pelvis. The nephrolith is visualized and either retrieved or fragmented using laser lithotripsy and fragments retrieved.

Ureteral stent

In patients where nephroureteroliths cannot be managed surgically, urinary diversion may be accomplished by placing a ureteral stent. A double pig-tailed stent is placed surgically, fluoroscopically, or via cystoscopy. One of the pig-tails is placed so that it is within the dilated renal pelvis and the other pig-tail is placed so that it is within the urinary bladder. The body of the stent connects the 2 pigtails and provides diversion of urine flow around the obstructive ureteroliths

Subcutaneous urinary bypass device

More recently, a subcutaneous urinary bypass device (SUB) has been used in dogs and cats with ureteral obstruction. The device involves surgical placement of a pig-tailed stent through the renal parenchyma into the dilated renal pelvis with nephropexy that is connected to tubing that is tunneled subcutaneously and connected to a metallic port. The port is connected by tubing to a catheter that is inserted into the apex of the urinary bladder with cystopexy performed. The result is diversion of urine from the obstructed kidney through the SUB and into the urinary bladder.

Renal transplantation

A retrospective study of 19 cats with renal failure associated with calcium oxalate urolithiasis has been published There were 13 females and 7 males All cats were azotemic and 17 were anemic. Hypercalcemia was present in 7 cats. Mean duration of survival in all cats was 605 days. Eight cats were alive 282-1,005 days (median = 1,305 days),11 cats died 2-1,197 days (median = 300 days), 5 cats formed uroliths in their allograft kidneys, and 2 were hypercalcemic.

MANAGEMENT OF CHRONIC KIDNEY DISEASE

Many patients with nephroureteroliths also have CKD, which must also be managed. CKD may be staged using IRIS guidelines (http://www.IRIS-kidney.com) with initial staging based on serum creatinine concentration and substaging based on presence of proteinuria and presence and degree of systemic arterial hypertension. The goal of management is to minimize excesses and deficits induced by CKD in order to improve quality and quantity of patient's life. The goal of nutrition is to maintain adequate to optimum body condition and adequate muscle condition (lean body mass). Uremic gastroenteritis occurs commonly with CKD and may be treated by modifying diet and feeding patterns and pharmacologically with histamine-2 receptor blockers, gastroprotectants, and appetite stimulants. A feeding tube may also be utilized. Electrolyte imbalances occur uncommonly in dogs with CKD. Hypokalemia may result in muscle weakness, anorexia, and worsening azotemia and is treated by feeding a potassium replete diet or by supplementing with potassium (gluconate or citrate). Metabolic acidosis occurs commonly and is usually a high anion gap acidosis. It is treated by feeding a low protein diet and by administering an alkalinizing agent (e.g. potassium citrate). Proteinuria with CKD has been associated with decreased survival and with azotemic CKD is treated when the urine protein-creatinine ratio (UPC) is > 0.4 in cats and 0.5 in dogs. Treatment involves feeding a restricted protein diet, administering an angiotensin converting enzyme inhibitor (and/or an angiotensin receptor blocker), and supplementing with omega-3 fatty acids. Maintenance of hydration is vitally important for renal perfusion and can be accomplished by feeding a canned food, moistening a dry food, or by supplementing water by SQ route or through a feeding tube. As renal function declines, compounds that should be excreted are not resulting in retention of "wastes" such as nitrogen containing compounds, blood urea nitrogen (BUN) and creatinine. Feeding a reduced protein diet is associated with decreasing levels of BUN but does not influence creatinine concentration to a great extent. Renal diets may contain fiber sources (prebiotics) that select for intestinal bacteria that may result in nitrogen excretion via feces; likewise, probiotics may have the same effect. A study of a probiotic in cats with CKD, however, failed to demonstrate a benefit. In patients with CKD, avoid administering medications that are potentially nephrotoxic or that may result in dehydration (e.g. aminoglycoside antimicrobials). In patients with CKD, three neuroendocrine abnormalities may occur: renal hyperparathyroidism, hypoproliferative anemia, and systemic arterial hypertension. Renal hyperparathyroidism is associated with hyperphosphatemia and increased parathyroid hormone levels and possibly decreased vitamin D3 levels. The goals of treatment are to keep serum phosphorous concentrations < 5.0-6.0, which is accomplished by feeding a low phosphorous diet and administering phosphate binders if necessary (e.g. aluminum hydroxide, sevelamer, lanthanum, and chitosan). Alternatively, administering oral vitamin D3 has been shown to benefit dogs with CKD in IRIS stage III or IV.

Hypoproliferative is treated by maintaining a good nutritional status, minimizing uremic gastroenteritis, and hormone replacement therapy. Recombinant human erythropoietin may be used; however, darbepoetin, a synthetic analog, appears safer and is effective at once a week administration. Patients should be supplemented with iron when using darbepoetin or erythropoitin. Systemic hypertension may occur in 65-75% of patients with CKD and risk of hypertension-related disease increases as systolic blood pressure is > 160 mmHg. Goal of treatment is a systolic blood pressure < 150 mmHg. It is treated with calcium channel blockers (amlodipine), which decrease systolic blood pressure by an average of 50 mmHg and ACE-I, which decrease systolic blood pressure by an average of 10 mmHg. Because CKD is a dynamic and progressive disease, serial monitoring and adjustment of treatment is necessary.

MANAGEMENT STRATEGIES

- Most nephroureteroliths are composed of calcium oxalate abut other minerals exist that may be amenable to medical dissolution
- Dogs with chronic kidney disease should undergo periodic abdominal radiography particularly if there is an abrupt decline in renal function because of the association of chronic kidney disease with nephroureterolithiasis
- The presence of nephroliths does not necessitate aggressive management. Nephroliths are not associated with progression of chronic kidney disease; however, ureteroliths are associated with decreased survival time.
- Because often nephroureteroliths are associated with chronic kidney disease and it is unknown how long they have been present, aggressive therapy may not result in improvement in azotemia
- In patients with ureteroliths, surgical management by ureterectomy and neoureterocystostomy is an effective means if the ureter is dilated and there is at least 1 cm of proximal ureter available.
- In dogs with ureteroliths where surgical intervention is not possible, ureteral stenting or use of a subcutaneous urinary bypass device is a reasonable option

REFERENCES

Lekcharoensuk C, Osborne CA, Lulich JP, et al. *J Am Anim Hosp Assoc* 2005;41:39-46; Ross SJ, Osborne CA, Lekcharoensuk C, et al. *J Am Vet Med Assoc* 2007;230:1854-1859; Dalby AM, Adams LG, Salisbury SK, et al. *J Am Vet Med Assoc* 2006;229:1118-1121; Lulich JP, Osborne CA, Albasan H. In: Bartges J,Polzin DJ, eds. *Nephrology and Urology of Small Animals*. West Sussex UK: Wiley-Blackwell, 2011;687-706; Bartges JW. In: Birchard SJ,Sherding RG, eds. *Saunders Manual of Small Animal Practice*. 2nd ed. Philadelphia: WB Saunders, 2000;943-957; Singh SK, Agarwal MM, Sharma S. *BJU Int* 2011;107:356-368; Forman MA, Francey T, Fischer JR, et al. *J Vet Intern Med* 2001;18:390-391; Kyles AE, Hardie EM, Wooden BG, et al. *J Am Vet Med Assoc* 2005;226:937-944; Achar E, Achar RA, Paiva TB, et al. *Kidney Int* 2003;64:1356-1364; Langston C. In: Bartges J,Polzin DJ, eds. *Nephrology and Urology of Small Animals*. West Sussex UK: Wiley-Blackwell, 2011;255-285; Acierno MJ. In: Bartges J,Polzin DJ, eds. *Nephrology and Urology of Small Animals*. West Sussex UK: Wiley-Blackwell, 2011;286-292; Labato MA. In: Bartges J,Polzin DJ, eds. *Nephrology and Urology of Small Animals*. West Sussex UK: Wiley-Blackwell, 2011;293-305; Cooper RL, Labato MA. *J Vet Intern Med* 2011;25:14-19; Tobias K. In: Bartges J,Polzin DJ, eds. *Nephrology and Urology of Small Animals*. West Sussex UK: Wiley-Blackwell, 2011;596-616; Adams LG, Goldman CK. In: Bartges J,Polzin DJ, eds. *Nephrology and Urology of Small Animals*. West Sussex UK: Wiley-Blackwell, 2011;340-348; Lane IF. *Vet Clin North Am Small Anim Pract* 2004;34:1011-1025; Weisse C, Berent A, Todd K, et al. *J Am Vet Med Assoc* 2006;229:226-234; Adin CA. In: Bartges J,Polzin DJ, eds. *Nephrology and Urology of Small Animals*. West Sussex UK: Wiley-Blackwell, 2011;306-328; Aronson LR, Kyles AE, Preston A, et al. *J Am Vet Med Assoc* 2006;228:743-749.

TWO ARE NOT BETTER: CANINE CHRONIC KIDNEY DISEASE AND HEART DISEASE

Ashley B. Saunders, DVM, DACVIM (CARDIOLOGY)
College Station, TX

INTRODUCTION

Chronic kidney disease (CKD) and heart disease are both common in dogs, often with an increased incidence in the aging population. Both are typically irreversible and progressive with highly variable survival rates. The interactions between the heart and kidneys are referred to as the "cardiorenal syndrome" in humans. The pathophysiology is complex and involves reduced cardiac output, renal hypoperfusion, activation of the renin angiotensin aldosterone system (RAAS) and sympathetic nervous system. In dogs, treatment is often complicated by systemic hypertension, electrolyte abnormalities, anemia, concurrent diseases (diabetes mellitus, hyperadrenocorticism, arthritis/orthopedic) and polypharmacy situations. Poor perfusion to the kidneys activates the RAAS system resulting in sodium and water retention and increased circulating blood volume contributing to overload of the cardiovascular system. What is good for one system is counterproductive for the other. Is it possible to strike a balance? How do we manage heart failure in dogs with kidney disease?

SYSTEMIC HYPERTENSION

The International Renal Interest Society (IRIS) treatment recommendations for CKD include discontinuing nephrotoxic medications, identifying and treating pre-renal abnormalities, and measuring urine protein to creatinine ratio (UP/C) and blood pressure.[1] Systolic blood pressure >160mmHg is considered abnormal and >180mmHg increases the risk of end organ damage to the eyes, brain, kidneys, and cardiovascular system. Hypertension occurs in dogs with CKD, hyperadrenocorticism, diabetes mellitus and pheochromocytoma.[2] Persistent elevations in systemic blood pressure and/or evidence of end organ damage warrants vasodilator therapy.[2] Angiotensin converting enzyme (ACE) inhibition results in dilation of the renal efferent arterioles but the effects are often too weak to significantly reduce blood pressure alone.[3] The combined use of amlodipine and an ACE inhibitor is recommended in cases where blood pressure remains above 160mmHg or UP/C is indicative of proteinuria.[1]

HEART FAILURE THERAPY

Very little information is currently available to guide management of dogs with heart failure and kidney disease. In humans with heart disease, ACE inhibitors are generally considered safe and prolong survival while kidney dysfunction and proteinuria are associated with shorter survival times.[4] Approximately 1/3 of humans with heart disease have concurrent kidney disease[4] and in some cases, initial increases in creatinine following the initiation of an ACE inhibitor will resolve without a reduction or discontinuation of the medication.[5] In dogs, studies have demonstrated the safety of ACE inhibitors in heart disease and improved survival with their use in heart failure.[6,7] A marked increase in creatinine following initiation of an ACE inhibitor could suggest an adverse drug effect.[1] Current standard therapy for congestive heart failure typically consists of furosemide, an ACE inhibitor, and pimobendan.[8] Positive inotropes, including pimobendan, increase left ventricular stroke volume which could in theory increase kidney perfusion.[9] Many medications used to manage heart failure affect the RAAS system, the kidneys and alter electrolytes (ie. potassium wasting or sparing diuretics). Important complications that can be encountered while managing heart failure and contribute to decreased kidney perfusion include reduced cardiac output, systemic hypotension and dehydration/volume contraction associated with diuretic therapy. Dogs receiving diuretic therapy have dilute urine and can have pre-renal azotemia without a characteristic urine specific gravity making it important to use surrogates for assessing hydration. Therefore monitoring is important.

Changes in kidney function can result in drug toxicity by affecting the clearance of medications. In particular, dogs with kidney dysfunction are susceptible to digoxin toxicity, used for managing supraventricular arrhythmias. Polypharmacy can include the use of NSAIDS, a potential cause of acutely worsening kidney disease when receiving an ACE inhibitor and diuretic therapy.[10] Efferent arteriole vasodilation induced with an ACE inhibitor combined with NSAID induced cyclooxygenase inhibition creates vasoconstriction of the afferent arteriole which can result in a reduction in glomerular capillary pressure and glomerular filtration rate.

CLINICAL EVALUATION

Recheck evaluations in heart failure patients typically consist of at least blood pressure, thoracic radiographs and serum biochemistry panel including electrolytes. Although an echocardiogram is require to diagnose the type of cardiac disease present, once the diagnosis has been established, echocardiography is often reserved for patients that are not responding to therapy, to assess ventricular function, or to diagnose pulmonary hypertension. N-terminal pro-B-type natriuretic peptide (NTproBNP) is a diagnostic biomarker elevated in heart failure, however concentrations can be artificially elevated in dogs with azotemia.[11]

The goal is to maintain appetite, body condition, address the clinical signs, and provide a good quality of life. Mild azotemia, in the absence of clinical signs of CKD, may be acceptable and well tolerated.[12] Supportive therapy may include the use of H2 blockers, anti-nausea medications, or omega-3 fatty acids and optimally, identifying a palatable diet with lower sodium and protein content.[13,14]

A FEW IMPORTANT POINTS

Things to consider when making recommendations:

1. Can the owner give multiple medications- three times a day if necessary?
2. Which medications interact and what are the adverse effects?
3. What is causing the dog's clinical signs – heart disease or kidney disease?
4. Blood work and thoracic radiographs go hand in hand in heart failure. If azotemia is present, is there evidence of heart failure? If congestion is present (pulmonary edema), is there azotemia?
5. Is blood pressure accurate and repeatable?
6. Be careful with fluids. Fluid therapy for CKD can easily overload a diseased cardiovascular system. Subcutaneous fluids are not well absorbed and can still result in volume overload.
7. Encourage drinking and eating.
8. In dogs with kidney disease, ACE Inhibitors may be used at a lower dose with careful monitoring.

This discussion will elaborate on the challenges of addressing opposing goals when managing combined CKD and heart disease in dogs using case examples.

REFERENCES

[1]Iris-kidney.com [2]Brown S, et al. J Vet Intern Med 2007;21:542. [3]Lefebvre HP, et al. Curr Pharmaceut Design 2007;13:1347. [4]Shlipak MG. Circ 2004;110:1514. [5]Ljungman S, et al. Am J Cardiol 1992;70:479. [6]Kvart C, et al. J Vet Intern Med 2002;16:80. [7]Atkins CE, et al. J Am Vet Med Assoc 2007;231:1061. [8]Atkins CE, et al. J Vet Intern Med 2009;23:1142. [9]Nakayama T, et al. J Vet Intern Med 2007;21:445. [10]Lefebvre HP, et al. J Vet Pharmacol Therap. 2004;27:265. [11]Schmidt MK, et al. J Vet Cardiol 2009;11:S81. [12]Atkins CE, et al. J Vet Cardiol 2012;14:165. [13]Bartges JW. Vet Clin North Am Small Anim Pract 2012;42:669. [14] Polzin DJ. Vet Clin Small Anim 2011;41:15.

ACUTE KIDNEY INJURY: HUMAN AND VETERINARY KNOWLEDGE

Marie E. Kerl DVM, MPH, DACVIM (SAIM), DACVECC
Columbia, MO

ACUTE RENAL FAILURE OR ACUTE KIDNEY INJURY?

There has been a shift in terminology in the human medical field within the last 10 years to establish a more concise and clinically relevant definition of acute renal failure (ARF). By the traditional definition, ARF is defined as a rapid decline in glomerular filtration, causing an abrupt, sustained increase in urea and creatinine and disruption of salt and water homeostasis. This definition cannot be quantified, and comparing clinical studies is challenging. In order to provide a more concise definition for acute kidney failure, physicians have proposed a change in terminology to "acute kidney injury" (AKI). While there are over 30 discrete definitions for AKI, common themes include reduction in kidney function within 48 hours defined by at least one of the following criteria:

1. Absolute increase in serum creatinine by at least 0.3 mg/dL (26.4 mmol/L)
2. Increase of serum creatinine by more than 50% of baseline
3. Oliguria of < 0.5 mL.kg hourly for > 6 hours.[1]

In 2004, the Acute Quality Dialysis Initiative developed and published the Risk-Injury-Failure-Loss-End-stage disease [RIFLE] criteria to provide a consensus definition for AKI (Table 1).[1] Key features of the RIFLE criteria included a proportional increase in serum creatinine or decrease in glomerular filtration rate (GFR), and a proportional decrease in urine output that was sustained over a number of hours, to categorize patients according to severity. The consensus definition provided by the RIFLE criteria became widely recognized, and has been widely published in research studies to investigate AKI. Following this first step, problems identified with the RIFLE criteria to include uncertainty with classification if an initial creatinine was unknown, misalignment of creatinine and GFR criteria, and equal weighting of urine output criteria despite poor agreement with serum biochemical parameters and frequent use of diuretic therapy.[2] Therefore, in 2005, the Acute Kidney Injury Network (AKIN) convened to refine the RIFLE criteria. Recommended changes included a 3-level scheme to mirror the "R I F" parts of the RIFLE criteria, lower requirement for absolute increase of serum creatinine, omission of the GFR criteria, and assignment of all patients receiving renal replacement therapy to stage 3. This classification scheme is known as the Acute Kidney Injury Network [AKIN] criteria.[3] Although there are differences between each scheme and classification is not always identical, multiple retrospective and prospective studies using each of these schemes have proven a significant association between increasing severity of AKI and mortality, even with minimal increases of creatinine that were previously not considered important.[4-6]

TABLE 1: COMPARISON OF ACUTE KIDNEY INJURY NETWORK (AKIN) AND RISK-INJURY-FAILURE-LOSS-END-STAGE (RIFLE) CRITERIA FOR EVALUATING HOSPITAL ACQUIRED ACUTE KIDNEY INJURY

		Creatinine		GFR decrease	Urine output
		Increase from baseline	**Absolute increase**		
AKIN	Stage 1	≥150-200%	≥0.3 mg/dL		<0.5ml/kg/h ≥6 h
	Stage 2	>200-299%			<0.5 ml/kg/h ≥12 h
	Stage 3	≥300%	≥4.0 mg/dL + acute increase ≥ 0.5 mg/dl		<0.3 ml/kg/h ≥24 h or anuria ≥12 h
RIFLE	Risk (R)	≥1.5-fold		≥25%	<0.5ml/kg/h ≥6 h
	Injury (I)	≥2.0-fold		≥50%	<0.5 ml/kg/h ≥12 h
	Failure (F)	≥3.0-fold	≥4.0mg/dL + acute increase ≥ 0.5 mg/dL	≥75%	<0.3 ml/kg/h ≥24 h or anuria ≥12 h

HUMAN HOSPITAL-ACQUIRED ACUTE KIDNEY INJURY

In human medicine, AKI is primarily recognized as hospital-acquired rather than community-acquired. The term hospital-acquired acute kidney injury (HA-AKI) is used to identify this event. As compared to veterinary patients that have potential exposure to a wide variety of etiologies causing AKI, people are less likely to experience these in a community setting. In contrast, hospitalized people are more likely to be exposed to medications or diagnostic agents (e.g. radiocontrast agents) that predispose to AKI. The causes of decreased renal perfusion included volume contraction, congestive heart failure, hypotension, cardiac arrest, better blood pressure control, third space losses, and arrhythmia. Other causes of HA-

AKI included nephrotoxic medications, radiographic contrast media, postoperative, and sepsis.[7, 8] Sepsis and the systemic inflammatory response (SIRS) result in multiple organ dysfunction (MOD) in the severely ill. Aged adults often have reduced renal function compared to younger individuals. In a recent report comparing HA-AKI to community-acquired(CA) AKI in a veteran's hospital setting, volume depletion was significantly more common as a cause of CA-HAI, and drug-induced AKI was significantly more common in HA-AKI.[9]

VETERINARY HOSPITAL-ACQUIRED ACUTE KIDNEY INJURY

Both AKI and HA-AKI have been investigated in the veterinary literature. In one retrospective review of 164 dogs admitted to the ICU, 14.6% developed AKI, as defined by an increase in creatinine of at least 150% above baseline or an absolute increase of ≥ 0.3 mg/dl during the hospital stay.[10] In another retrospective review, 15.2-19.3% of dogs with at least two creatinine measurements within 2, 3, or 7 days (n = 401, 477, and 646, respectively) developed AKI, as defined by an increase in serum creatinine ≥ 0.3 mg/dl.[11] In both studies, only cases with serum creatinine < 1.6 mg/dl on the initial sample were included. There are no published prospective studies of the incidence of HA-AKI in dogs admitted to ICU, but preliminary data suggests an incidence of around 10-17%. In cats, 20-21.3% with at least two creatinine measurements within 2, 3, or 7 days (n = 128, 165, and 221, respectively) developed AKI, as defined by an increase in serum creatinine ≥ 0.3 mg/dl.[11]

Commonly recognized risk factors for HA-AKI include preexisting renal disease, dehydration and volume depletion, decreased cardiac output, advanced age, fever, hypotension, hypertension, hypoalbuminemia, sepsis, electrolyte imbalances, acidosis, hyperviscosity syndromes, liver disease, administration of nephrotoxic drugs, radiocontrast media, pancreatitis, diabetes mellitus, anesthesia, and surgery.[12] In a case series of 29 dogs, dogs developing HA-AKI had been exposed to a nephrotoxicant (72%), were > 7 years old (69%), had chronic heart disease (41%), preexisting renal disease (35%), neoplastic condition (31%), fever (28%), or had undergone anesthesia (14%).[13] The nephrotoxicant drugs included aminoglycosides alone or in combination, cardiac drug combinations, cisplatin, or nonsteroidal anti-inflammatory drugs.[13]

OUTCOME OF HA-AKI

The development of HA-AKI has a negative impact on outcome. Survival to hospital discharge was 38% in one canine study.[13] In another study, 46% of dogs survived to discharge, compared to 84% of dogs that did not develop AKI. This survival effect was significant for both mild and severe changes in creatinine.[10] In a different study, 30-day survival was 56% in dogs developing AKI within 2 days, compared to 75% survival in dogs not developing AKI; 90-day survival of these dogs was 43%, compared to 71% in those dogs not developing AKI.[11] In that study, mild changes in renal function (serum creatinine < 1.6 mg/dl despite a rise of ≥ 0.3 mg/dl) did not have a significant impact on survival at 30 days compared to those not developing AKI, but there was a significant impact on 90-day survival.

The 30-day survival rate for cats with developing AKI within 2 days was 48%, compared to 85% survival in those not developing AKI. The survival difference was not significant for cats developing mild AKI, but it was significant for cats developing more severe AKI (serum creatinine > 1.6 mg/dl and a rise of ≥ 0.3 mg/dl).[14]

A retrospective study presented the results of application of modified RIFLE criteria in 853 dogs presented to a veterinary teaching hospital from 2000-2006. [15] The authors reported increased 30-day mortality for dogs in each of 3 categories of "Risk, Injury, and Failure". Dogs were enrolled based upon creatinine value at the time of admission, and were reported to have normal kidney function 14 days prior. The authors applied modified RIFLE criteria by utilizing an estimation of glomerular filtration rate (GFR) based on creatinine and applying to the original RIFLE criteria.[15]

REFERENCES

[1]Bellomo R, et al. **Crit Care** 2004;8(4):R204. [2]Bagshaw SM. **Nat rev: Nephrol** 2010;6(2):71. [3]Mehta RL, et al. **Crit Care** 2007;11(2):R31. [4]Joannidis M, et al. **Intensive care medicine** 2009;35(10):1692. [5]Hoste EA, et al. **Crit Care** 2006;10(3):R73. [6]Chang CH, et al. **Shock** 2010;33(3):247. [7]Shusterman N, et al. **Am J Med** 1987;83(1):65. [8]Nolan CR, et al. **JASN** 1998;9(4):710. [9]Schissler MM, et al. **Nephrol** 2013: doi: 10.1111/nep.12036. [10]Thoen ME, et al. **J Vet Emerg Crit Care** 2011;21(6):648. [11]Harison E, et al. **J Vet Inten Med** 2012;26(5):1093. [12]Ross L. **Vet Clin No Amer Small Anim Pract** 2011;41(1):1. [13]Behrend EN, et al. **J Am Vet Med Assoc** 1996;208(4):537. [14]Eatroff AE, et al. **J Am Vet Med Assoc** 2012;241(11):1471. [15]Lee YJ, et al. **Vet Rec** 2011;168(10):264.

TREATMENT OF CUSHING"S DISEASE IN THE DOG: ARGUMENTS FOR AND AGAINST

Rhett Nichols, DVM, ACVIM (Small Animal Internal Medicine)
Smithtown, NY

The decision to treat any disorder is often based on the adverse effects and cost of treatment, the degree of illness, quality of life, and survival time. In addition, the quality of life of the pet owner is also a factor in the decision to treat. For example, if a severely polyuric dog is having daily urinary accidents, a frustrated owner may consider euthanasia unless the problem can be successfully resolved. Also, there is confusion as to when to start treatment. Should dogs with suspected hyperadrenocorticism (HAC) be treated if they are asymptomatic? Or should treatment only be initiated if the pet has moderate or severe clinical signs?

SURVIVAL TIME

Some veterinarians believe that the average life span of a dog with pituitary-dependent hyperadrenocorticism (PDH) is the same whether the dog is treated or not. Is this true? Based on a number of survival studies using different modes of therapy it would appear that the type of treatment as well as early treatment can greatly influence survival times.

Selective adrenocorticolysis with mitotane: In one study, 200 dogs with PDH were treated with 50 mg/kg of mitotane and low-dose glucocorticoid replacement for 5-14 days and transitioned to a maintenance dosage of 50 mg/kg weekly (1). The mean survival time was 2.2 years. Thirty-four dogs were euthanized or died during the initial 6 months; 18 of these dogs died from complications of HAC. In contrast, dogs usually did not die of complications of HAC after the first few months of treatment when they were in clinical remission. The percentage of dogs still alive after 1, 2, 4, 6, and 8 years was 72%, 42%, 27%, 8% and 1.4 % respectively. In another study, over 1500 dogs with PDH were treated with a similar protocol but without daily glucocorticoid replacement (2). The average survival time was 32 months. In both studies, some dogs lived only a few days after starting treatment while others lived over 10 years.

Nonselective adrenocorticolysis: A popular mitotane treatment protocol in Europe revolves around the central theme that it is easier to treat Addison's disease long-term than Cushing's syndrome. Treatment is aimed at complete destruction of the adrenal cortex. In a study of planned medical adrenalectomy, 129 dogs with PDH were treated with 50-75 mg/kg mitotane for 25 days divided into 3 or 4 administrations daily with food (3). On day 3, dogs were started on hormone therapy: cortisone 2mg/kg/day and fludrocortisone 0.125 mg/kg/day and NaCl (0.1 g/kg). The 1, 2, 3 year survival fraction of 111 dogs completing the protocol was 87%, 77%, and 69% percent respectively.

Dopamine agonist therapy with cabergoline (Cbg): The pars intermedia of the pituitary gland is negatively neuro-regulated by dopamine and the corticotroph cells of the pars distalis may also be inhibited by dopamine. Moreover, functional dopamine receptors (specifically a subtype called D2 receptors) are present in 60% of pituitary cells that produce ACTH. Forty dogs with PDH that were treated with Cbg, a D2 dopamine agonist, at 0.07 mg/kg/week divided into 3 doses given every 48 hours (4). Out of the 40 dogs, 17 responded to Cbg (42.5%); the nonresponders were treated with ketoconazole. A year after the treatment, there was a significant decrease in endogenous ACTH levels, the urinary cortisol: creatinine ratio, and of the tumor size evaluated by nuclear magnetic resonance. Dogs responding to Cbg lived significantly longer than those in the control group treated with ketoconazole. The average survival time of the Cbg treated dogs was 3 years versus 2 years for the ketoconazole treated group and the 4 year survival fraction of the Cbg treated dogs was 41%.

Transsphenoidal hypophysectomy: The long-term survival, disease-free fractions, and complications of hypophysectomy in 150 dogs with PDH were examined in a prospective study (5). The 1, 2 , 3, and 4 year estimated survival rates were 84%, 76%, 72%, , and 68% respectively. The 1, 2, 3, and 4 year relapse-free fractions were similar to the survival rates. Common post-operative complications included decreased tear production which was often reversible and central diabetes insipidus (CDI). The survival and relapse-free fractions after hypophysectomy decreased and the incidence of CDI increased with increasing pituitary size. Transsphenoidal hypophysectomy coupled with early diagnosis of PDH is expected to have the best outcome as a primary treatment for dogs with nonenlarged or moderately enlarged pituitary glands.

IS THE TREATMENT WORSE THAN THE DISEASE? EXPERIENCE WITH MITOTANE AND TRILOSTANE

Mitotane: Mitotane and trilostane are the most commonly used drugs used to treat Cushing's syndrome but neither drug is without its problems. In two large selective adrenocorticolysis studies with mitotane, the prevalence of adverse effects associated with the loading phase of mitotane with and without glucocorticoid replacement was 25% and 5% respectively (1,2). Adverse effects during maintenance phase were greater than 30%. Complete unplanned adrenal cortical necrosis (Addison's disease) occurred in only 2% to 5% of the cases. Compared to selective adrenocorticolysis, nonselective adrenocorticolysis is fraught with problems. In the study mentioned above in the section on survival times, the adverse effects (anorexia, vomiting, depression, weakness, and diarrhea) resulted in temporary or complete cessation of mitotane treatment or death in 40% of the dogs (3). Furthermore, fifteen of 129 dogs died during treatment and 7 of these dogs died from hypoadrenocorticism. Despite the fact that the average survival time in dogs using this approach exceeds that of dogs treated with selective adrenocorticolysis, this form of treatment is not universally recommended because of the frequency and severity of adverse effects.

Trilostane: The results of several published studies have shown that the prevalence of adverse signs range from 4% to 17% depending on the dose and frequency of administration. In general, trilostane is well tolerated by most dogs. If the numbers of dogs treated with trilostane in the initial 6 published clinical studies are combined, 39 out of the 244 dogs (16%) developed adverse signs (6-11). More recent studies have showed that administering trilostane twice daily at low dosages (0.5 to 1 mg/kg) reduced the prevalence of adverse effects to a very low 4% (11, 12). The most serious side effect of trilostane is acute adrenal necrosis which appears to be very rare. This has been documented in two case reports, one fatal and the other requiring permanent hormone replacement therapy. Interestingly, many dogs appear to develop slow-onset adrenal necrosis over time (months to years) and as a result, the drug can sometimes be discontinued indefinitely (13).

SHOULD DOGS WITH ASYMPTOMATIC CUSHING'S SYNDROME BE TREATED?

One argument against treatment is how do you improve quality of life if there are no clinical signs? Another major argument against treating an asymptomatic animal is how do you know that they actually have Cushing's disease? Hyperadrenocorticism is often viewed as a clinical syndrome and therefore, the best test to diagnose this disorder is a thorough history and physical examination. If HAC is still suspected after the history and physical exam, routine blood work, urinalysis, and specific screening tests may be recommended. Often a diagnosis of Cushing's is pursued in a clinically normal animal with an elevated serum alkaline phosphatase concentration alone – this practice is often a formula for misdiagnosis since false positive screening test results are common, especially if these tests are applied to a population of animals where the prevalence of disease is low (e.g., no clinical signs of HAC). As a result of not weighing the clinical picture more heavily than laboratory abnormalities, it has been suggested that more dogs are diagnosed with Cushing's syndrome than actually have the disorder. Conversely, early treatment of HAC during the "asymptomatic phase" may prevent or mute some of the many complications associated with this disorder.

COMPLICATIONS ASSOCIATED WITH CUSHING'S SYNDROME

Some serious and not so serious complications include chronic urinary tract infections, thromboembolic disease, neurologic impairment from an enlarging pituitary tumor, hypertension, proteinuria, and renal tubular disease, chronic pyoderma and calcinosis cutis, calcium uroliths, blindness, diabetes mellitus, and biliary mucoceles 1,2,14-18). Only a few of these complications that have received recent attention in the veterinary literature will be addressed.

Acute blindness: SARDS (sudden acute retinal degeneration syndrome) is a retinal disorder of unknown cause that produces sudden and permanent blindness in adult dogs. The syndrome is characterized by a noninflammatory degeneration and loss of retinal photoreceptors. An association has been suggested between SARDS and HAC. In a series of 44 dogs diagnosed with SARDS, 16 of these dogs had at least one abnormality in the history, physical examination and laboratory findings consistent with Cushing's syndrome (2). Of these 16 dogs, 8 dogs (18% of 44 total) had at least one endocrine test consistent with HAC. In another study, 70 dogs with PDH were evaluated (19). At diagnosis 17% (12/70) of the dogs were already blind. Of the 12 blind dogs, the cause was optic chiasm (OC) compression from the pituitary tumor in 2 dogs while the remaining 10 blind dogs were diagnosed with SARDS. In the large group of dogs that were visual, the owners reported that the clinical signs of Cushing's syndrome (PU/PD, polyphagia, weight gain) appeared progressively over time and the estimated course of the disease was 1-2 years prior to diagnosis. In the blind group of dogs without OC compression, the owners reported that the onset of clinical signs of HAC was rapid, accompanied by loss of vision anywhere from 72 hours to 20 days later. In the two dogs with OC compression, clinical signs appeared in a similar way to the dogs that remained visual. The question is often asked could SARDS be prevented by treating sighted dogs with Cushing's syndrome? Currently the answer to that question is unclear. However, if most cases of SARDS are associated with a rapid onset of clinical signs of Cushing's syndrome, the likelihood of preventing SARDS would seem to be quite remote.

Diabetes mellitus: Diabetes is commonly associated with PDH although the prevalence is only about 5% (1, 2). The most likely cause and effect relationship is that hypercortisolemia causes insulin resistance, which over time leads to B-cell exhaustion. Over the years, endocrinologists have mainly focused on the difficulties involved with treating a dog with Cushing's syndrome and concurrent diabetes *after* the dog has become an overt diabetic. In other words, diabetes prevention strategies in animals with HAC have often been ignored. In a recent retrospective study of 152 dogs with PDH, 10.5% of dogs developed diabetes mellitus within a 3 month period after diagnosing and initiating medical therapy (20). As result of this study, investigators asked the question what criteria could be used to identify prediabetic dogs? And if specific criteria could be established, what therapeutic strategy might slowdown or prevent progression to diabetes in these animals? Twenty-six PDH dogs were divided into 3 groups and treated medically: one group with fasting blood glucose (BG) levels <105 mg/dl and two groups with glycemia >105 mg/dl and <170 mg/dl, one of which received insulin detemir for 4 months at 0.1 U/kg once daily. Dogs with BG concentrations <105 mg/dl and those with BG >105 mg/dl which received insulin did not develop diabetes. In the noninsulin group 6/7 dogs developed diabetes after the third month. Overall, there was a 13-fold greater risk of diabetes in dogs with BG levels >105 mg/dl and no insulin treatment. So, what is the bottom line? While this was a small study, these results suggest that administering insulin in dogs with PDH and glycemia >105 mg/dl could prevent progression to diabetes.

Biliary mucoceles: An association between gallbladder mucoceles and HAC has been described in dogs (14-18). Interestingly, there is little mention of biliary mucoceles prior to 1990. Is this really a new disorder, possibly diet related, or is this an old disorder uncovered by the increasing use of ultrasound as a diagnostic tool? One study suggests that biliary mucoceles are 20 to 30 times more prevalent in dogs with HAC than in dogs that do not have Cushing's disease (18). However, the prevalence of mucoceles in dogs being evaluated for HAC that do not abdominal issues (so-called healthy Cushing's dogs) appears to be quite low. For example, in the large Cushing's survival studies, biliary mucoceles are not mentioned as a complication or cause of death (1, 2, 3). Despite the apparently low prevalence of gall bladder mucoceles in the healthy HAC population, it is recommended to ultrasound the abdomen in all dogs with suspected Cushing's disease, especially in dogs that are predisposed to mucocele formation (e.g., the Cocker Spaniel, Shetland Sheepdog, and Miniature Schnauzer). Conversely, any animal diagnosed with a biliary mucocele should be evaluated for underlying HAC. Recent evidence suggests that hypercortisolemia changes the composition of the bile salts to a more "toxic" form of bile that may injure the gall bladder epithelium and result in mucocele formation (21). Other predisposing causes of biliary mucoceles include hypothyroidism, diabetes mellitus, pancreatitis, and dyslipidemia (18). Whether treatment of Cushing's disease prevents gall bladder mucocele formation or aids in the dissolution process of a mucocele is unclear at this time.

Hypertension and proteinuria: In the workup and management of dogs with Cushing's syndrome, it is important to remember that hypertension and proteinuria are extremely common complications with this disorder (22-25). The prevalence of hypertension has been reported to range from 37% to 86% in dogs with untreated HAC (22-25). The mechanisms underlying the hypertension potentially include excessive secretion of renin, activation of the renin–angiotensin system, enhanced vascular sensitivity to catecholamines and a reduction of vasodilator prostaglandins (26, 27, 28). In most dogs with Cushing's syndrome, the degree of hypertension is mild to moderate and not associated with overt clinical signs of blindness (i.e., retinal detachment or hemorrhage is extremely rare). However, this moderate hypertension can exacerbate left ventricular hypertrophy and congestive heart failure and may lead to glomerular damage and proteinuria. In general, hypertension does not resolve following treatment of HAC (22, 24, 26, 27, 28). The explanation for why hypertension persists in many dogs despite successful treatment of their Cushing's disease is not clear. About half of dogs with untreated Cushing's syndrome also have proteinuria, defined as a urine protein: creatinine ratio > 0.5-1.0, in the absence of UTI (22, 26, 27, 29). However, as with hypertension, proteinuria sometimes fails to completely resolve despite treatment of HAC (22, 26, 27, 29).

Glomerular filtration rate (GFR): In a recent study of dogs with Cushing's syndrome over half the dogs had high values for GFR prior to treatment and the GFR decreased significantly after successful treatment (25). The clinical relevance of GFR changes in these dogs with HAC is currently unclear. However, the high pretreatment GFR may be the result of increased renal blood flow and decreased renal vascular resistance, as described in dogs treated with exogenous glucocorticoids (30). The new finding of a falling GFR after treatment is consistent with the fact that underlying renal disease is present in many of these dogs. Therefore, it is prudent to continue to monitor blood pressure and urinary protein excretion in all dogs with Cushing's syndrome. If hypertension or proteinuria (or both) persist, one should institute treatment with angiotensin-converting enzyme (ACE) inhibitors (e.g., enalapril or benazepril) and add amlodipine, if needed (31-35) and monitor renal function.

SUMMARY

The body of evidence would suggest that early treatment, depending on the treatment modality, may prolong survival time, and prevent some but not all the complications associated with Cushing's syndrome - especially those complications where the exact cause and effect mechanism is unclear (e.g., SARDS and biliary mucoceles) and improve the quality of life for the dog as well as the owner. Not treating because the treatment is perceived as worse than the disease may be a strong argument in the case of nonselective adrenocorticolyis but this argument loses its strength if the dog is treated with low-dose twice daily trilostane. Lastly, the argument about not treating an asymptomatic dog is quite sensible, especially if the diagnosis is in doubt. However, if the diagnosis has a strong degree of certainly and the dog has silent abnormalities such as hypertension and proteinuria, treatment of the underlying HAC could improve or slow down the progression of these problems.

REFERENCES

1. Kintzer PP and Peterson ME. J Vet Intern Med 1991;5(3):18 2. Feldman EC, Nelson RW. Canine and feline endocrinology and reproduction. 3rd ed. St. Louis (MO): Saunders; 2004 3. Den Hertog E et al. Vet Rec 1999; 144: 12 4. Castillo VA et al. Res Vet Sci 2008; 85:26 5.Hanson JM et al. J Vet Intern Med 2005; 19(5):687 6. Neiger R et al. Vet Rec 2002;150(26):799 7. Ruckstuhl NS et al. Am J Vet Res 2002; 63(4):506 8. Braddock JA et al. Aus Vet J 2003; 81(10):600 9.Alenza DP et al. J Am Anim Hosp Assoc 2006 ;42(4):269 10. Vaughan MA et al. J Am Vet Med Assoc 2008; 232(9):1321 11. Galac S et al. Vet J 2010;183(1):75. 12. Feldman EC. J Am Vet Med Assoc 2011;238(11):1441 13. Peterson ME, Nichols R: unpublished observations 14. Besso JG et al. Vet Radiology and Ultrasound 2000;41:261 15. Pike et al. J Am Vet Med Assoc 2004; 224:1615 16. Melian C et al. J Appled Ve t Anim Res 2006;30:117 17. Aguirre AL et al. J Am Vet Med Assoc 2007;231:79 18. Mesich MLL et al. J Small Anim Prac 2009;50:630 19.Cabrera Blatter MF. Research in Veterinary Science 2012:92: 387 20.Miceli DD et al. Res Vet Sci 2012;93(1):114 21. Kook PH et al. Vet Journal 2012;191:225 22.Ortega TM, Feldman EC, Nelson RW, et al. J AmVet Med Assoc 1996;209:17 23. Novellas R, de Gopegui RR, Espada Y.Veterinary Record 2008;163:592-596. 24. Lien YH, Hsiang TY, Huang HP. Acta Veterinaria Scandinavica 2010;52:61. 25.Smets PM, Lefebvre HP, Meij BP, et al. J of Vet Intern Med 2012;26:565 26.Melián C, M. Pérez-Alenza, D, Peterson ME. Hyperadrenocorticism in dogs, In: Ettinger SJ (ed): Textbook of Veterinary Internal Medicine: Diseases of the Dog and Cat (Seventh Edition). Philadelphia, Saunders Elsevier, 2010;1816-1840. 27. Herrtage ME, Ramsey IK. Canine hyperadrenocorticism. In: Mooney CT, Peterson ME, eds. BSAVA Manual of Canine and Feline Endocrinology. 4th ed. Quedgeley, Gloucester: British Small Animal Veterinary Association; 2012:167-189. 28. Goy-Thollot I, Pechereau D, Keroack S, et al. Journal of Small Animal Practice 2002;43:489. 29. Hurley KJ, Vaden SL J Am Vet Med Assoc 1998;212:369–373. 30. Hall JE, Morse CL, Smith MJ, et al. Hypertension 1980;2:139. 31. Brown S, Atkins C, Bagley R, et al. J Vet Intern Med 2007;21:542. 32.Lefebvre HP, Toutain PL. Journal of Veterinary Pharmacology and Therapeutics 2004;27:265. 33. Lees GE, Brown SA, Elliott J,et al. J Vet Intern Med 2005;19:377-385. 34. Lefebvre HP, Brown SA, Chetboul V, et al. Current Pharmaceutical Design 2007;13:1347-1361. 35. Chew DJ, DiBartola SP, Schenck PA. Chronic renal failure. In: Canine and Feline Nehrology and Urology. Second Ed. Elsevier Saunders, St. Louis. 2011;145-196

ELEVATED CALCIUM AND POTASSIUM LEVELS IN DOGS AND CATS: PURSUE OR IGNORE?

Rhett Nichols, DVM, ACVIM (Small Animal Internal Medicine)
Smithtown, NY

AN OVERVIEW OF THE CAUSES OF HYPERKALEMIA

The many causes of hyperkalemia include factitious (laboratory error and pseudohyperkalemia), acute and chronic renal failure, Addison's disease, primary hypoaldosteronism and hyporeninemic hypoaldosteronism and drugs that impair the renin-angiotensin-aldosterone axis such as spironolactone, ACE inhibitors, and angiotensin receptor blockers. Unusual and uncommon causes of hyperkalemia would include pseudo-Addison's disease due to Tricuris vulpis infestation and chylous effusions. However, in the well animal, despite the long list of rule-outs for hyperkalemia, an elevated potassium concentration on a routine biochemical profile is usually due to spurious or pseudohyperkalemia.

FACTITIOUS CAUSES OF HYPERKALEMIA

Pseudohyperkalemia is defined as a rise in serum potassium > 0.4 mEq/L above the concurrent plasma potassium concentration (1, 2, 3). This phenomenon was first described in human medicine over 50 years ago. It is correlated to thrombocytosis due to the release of potassium from platelet granules during coagulation. Pseudohyperkalemia can also be associated normal platelet counts if platelets are "activated", hemolysis, and occasionally extremely high WBC counts (e.g., chronic lymphocytic leukemia). In addition, Asian breeds such as the Akita, Sharpei, Sheba Inu, and Jindo have high intracellular potassium levels in their RBCs which causes pseudohyperkalemia when the blood clots or sits in contact with the plasma for long periods of time (4). Pseudohyperkalemia can be easily confirmed by collecting a heparinized blood sample to prevent clotting and determining a plasma potassium concentration. A serum potassium greater than 0.4 mEq/L above the plasma potassium concentration is consistent with pseudohyperkalemia.

Spurious hyperkalemia due to EDTA contamination: more common than you think?

There is little data in the veterinary literature related to the prevalence of sample contamination with EDTA. However, there are several studies that have looked at this issue in people (3). For example, as part of a service evaluation at a human hospital in the UK, EDTA was measured in hyperkalemic samples to identify EDTA contamination. These samples were further analyzed for calcium and magnesium to study whether these analytes could act as surrogate markers to reliably identify EDTA contamination. Over a one-month period, concentrations of EDTA, calcium and magnesium were measured in hyperkalemic serum samples (potassium > 6 mEq/L). In the one-month period studied, out of 28,471 serum samples there were 117 samples with a serum potassium > 6.0 mEq/L. Of these, 28 had an EDTA concentration of > 0.1 mEq/L. Eighteen samples were immediately identified and reported as possible EDTA contamination by laboratory staff prior to EDTA measurement, whereas 10 were initially reported as hyperkalemia and were subsequently identified as contaminated with EDTA. Twenty-seven of these patients were re-tested and all had repeat serum potassium concentrations within the reference range. Twenty-two (79%) and six (21%) of contaminated samples originated from inpatients and patients from general practice, respectively. Overall in this study, spurious hyperkalemia due to kEDTA contamination was common, accounting for about 25% of all hyperkalemic samples. In the 27 patients retested, serum potassium was within the reference range confirming EDTA contamination. In vitro kEDTA contamination usually occurs due to syringe contamination when delivering blood into EDTA sample tubes. Grossly abnormal results or subtle changes in potassium as well as lower calcium and magnesium may result. In summary, spurious hyperkalaemia due to kEDTA contamination is relatively common. Gross contamination is usually evident by finding unexpectedly high potassium and low calcium concentrations. Hyperkalemia due to low concentrations of kEDTA contamination may be more difficult to identify and can only be confidently detected by measurement of serum EDTA.

DISORDERS AND DRUGS ASSOCIATED WITH HYPERKALEMIA

Primary hypoadrenocorticism (Addison's disease)

Primary hypoadrenocorticism (Addison's disease) is a well described disorder in dogs characterized by a deficiency of glucocorticoid (cortisol) and mineralocorticoid (aldosterone) secretion from the adrenal cortex (5, 6, 7). Hyperkalemia and hyponatremia with a sodium: potassium ratio <23:1 are considered characteristic features of this disorder. In various studies evaluating diseases with these electrolyte abnormalities, hypoadrenocorticm was the most common explanation for the presence of concurrent hyperkalemia and hyponatremia (8). As a result of these studies, many veterinarians place great emphasis on the Na: K ratio as a useful test to aid in the diagnosis hypoadrenocorticism. Unfortunately, the Na: K ratio lacks sensitivity and specificity and the results can be very misleading. For example, in one study only 28% of the dogs with a Na: K ratio < 25:1 had hypoadrenocorticism. In the same study when the Na: K ratio was lowered to <20:1 nearly 1/3 of the dogs did not have the disorder (9). The true gold standard for confirming the diagnosis of hypoadrenocorticism is the ACTH response test. Dogs with hypoadrenocorticism have an absent or blunted response to ACTH administration. Basal and post-

ACTH serum cortisol concentrations are less than 1 μg/dl in over 75% of dogs and less than 2 μg/dl in virtually all dogs with primary hypoadrenocorticism. In situations where the diagnosis of hypoadrenocorticism is in question, and the client has money concerns, a resting cortisol is a practical and inexpensive screening test to help determine if an ACTH response test is indicated. A resting cortisol concentration > 2 ug/dl rules out the disorder in the vast majority of cases, while a resting cortisol < 2 ug/dl is not diagnostic and an ACTH stimulation test is recommended if hypoadrenocorticism is suspected (6,7). In approximately 20% of cases of primary hypoadrenocorticism, the serum electrolytes are normal despite low to low-normal resting and post-ACTH aldosterone concentrations. This version of primary hypoadrenocorticism is sometimes referred to as atypical hypoadrenocorticism.

Chronic renal failure

Hyperkalemia is rarely reported as a clinical feature of dogs with chronic kidney disease (CKD), even though it is commonly associated with acute renal injury. In a recent study, 152 dogs with CKD were presented for clinical evaluation (10). One hundred nine (72%) of these dogs had at least one documented episode of hyperkalemia in which the potassium concentration was above the upper limit of the reference range at either the initial evaluation or during the follow-up examination period (>5.3 mEq/L). Twenty-four dogs (16%) had at least one documented episode where the potassium concentration was > 6.5 mEq/L. Twenty-six of these dogs, which were all being fed a commercial renal diet, had a nutrition consultation and a prescription formulated by a veterinary nutritionist for a potassium restricted home prepared diet to manage the hyperkalemia. Using the IRIS classification for CKD nearly all the dogs were in stage III or stage IV CKD, with the majority in Stage III CKD. All but one dog showed normalization of serum potassium concentration within a two week period of starting the new diet. This study suggested that in a subset of dogs with CKD, hyperkalemia can be associated with commercial renal diets, and could restrict use of these diets. Appropriately formulated, potassium reduced diets are an effective alternative to correcting the hyperkalemia.

Hyporeninemic hypoaldosteronism

Recently a case report of a dog with hyporeninemic hypoaldosteronism (HH) was described (11). The dog, a 13 year old female spayed Yorkshire terrier, was referred for evaluation of hyperkalemia (7.5 mEg/L). Active medical problems included mild stable azotemia of 11 months duration (IRIS Stage II CKD) and no medication known to affect the RAAS was being administered. A completes workup conducted in stages included an ACTH response test to rule out Addison's disease, a serum aldosterone concentration, and a plasma renin activity (PRA). The post-ACTH cortisol levels were normal, but the resting aldosterone and PRA were low confirming the diagnosis of HH. In people, HH is a relatively common disorder accounting for 50-75% of all cases of initially unexplained persistent, asymptomatic hyperkalemia (12, 13, 14). HH is most often found in older adults with mild to moderate renal insufficiency, diabetic nephropathy, or chronic tubulointerstitial disease. Chronic volume expansion and structural damage to the juxtaglomerlar apparatus are considered likely causes of HH in these syndromes. Because most patients are asymptomatic, hyperkalemia is discovered on routine blood screening or during studies for an unassociated illness. The remainder of the patients seek medical attention because of symptoms referable to hyperkalemia such as muscle weakness or cardiac arrhythmias. If renal disease is present, the hyperkalemia is often disproportionate to the degree of renal dysfunction. Other common clinical findings include mild asymptomatic hyperchloremic acidosis, systemic hypertension, and congestive heart failure secondary to volume overload, Depending on the severity of the hyperkalemia and clinical signs, treatment is quite variable – everything from no treatment to potassium restricted diets, diuretics such as furosamide, and fludrocortisone (12).

Primary hypoaldosteronism

The first reported case of primary hypoaldosteronism (hyperreninemic hypoaldosteronism) was described in 1998 (15). The dog was evaluated for clinical signs and characteristic electrolyte abnormalities compatible with Addison's disease. Pre- and post-ACTH cortisol concentrations were normal but subsequent evaluation of aldosterone levels confirmed aldosterone deficiency, and renin activity levels were elevated. The dog clinical responded to mineralocorticoid replacement. At Antech Diagnostics, a few primary hypoaldosterone cases are diagnosed every year. Recently, hypoaldosteronism preceding glucocorticoid deficiency was diagnosed in a dog with primary hypoadrenocorticism (16).

Drug related hyperkalemia

A number of commonly used drugs affect the renin-angiotensin-aldosterone system (RAAS) and cause hyperkalemia. These include the ACEi, angiotensin receptor blockers (ARBs), spironolactone, NSAIDs, and cyclosporine. In general, if potassium levels exceed 6 mEg/L and factitious causes of potassium elevations can be ruled out, reducing or stopping these drugs is often recommended, or a homemade potassium-restricted diet can be considered. However, to date, there is no consensus among veterinary nephrologists as to what serum potassium concentration is considered excessive if the patient has no apparent adverse effects from hyperkalemia. Serious side effects related to hyperkalemia would include muscle weakness and cardiac arrhythmias. Other drugs that either cause or are associated with hyperkalemia include trilostane and mitotane. In dogs treated with trilostane that are clinically doing well and have post-ACTH stimulated cortisol in the therapeutic range (e.g., 2-7mcg/dl) it is not uncommon to find evidence of hyperkalemia during recheck appointments. This is probably due to both transient lowering of aldosterone concentrations via trilostane inhibition of synthetic pathways in the adrenal cortex and/or a "burst" of intracellular potassium from a small area of adrenal necrosis (17). In this situation the hyperkalemia is often transient and not clinically important. However, a much more serious but less common issue is that both mitotane and trilostane can initiate an Addisonian crisis secondary to adrenal rupture or rapid adrenal necrosis (7, 17, 18, 19). The frequency of this severe side effect is between 2% and 5% with mitotane treated dogs and < 1% of the dogs treated with trilostane (7, 18, and 19).

The "Florinef Creep"

Fludrocortisone (Florinef) is used to treat dogs with Addison's disease, hyporeninemic hypoaldosteronism, and the rare case of primary hypoaldosteronism. The average dose is 0.2 mg/kg divided BID. Often during the electrolyte monitoring process the electrolytes continue to be mild to moderately abnormal (low Na and elevated K), yet the owners report that their dogs are doing clinically well at home. Many veterinarians continue to raise the dose of fludrocortisone in an effort to improve the abnormal electrolyte concentrations– this phenomenon is referred to as the "Florinef Creep" – which refers to the observation that increasing dosages are needed to maintain normal electrolyte concentrations. Often the only way to put an end to the Florinef Creep and the expense involved is to transition the dog to the injectable mineralocorticoid Percortin.

ELEVATED CALCIUM CONCENTRATIONS

The calcium is elevated, what now?

An elevated serum calcium is often first noted when total calcium (tCa) is measured as part of a biochemical profile. Abnormalities in tCa should never be ignored and warrant further diagnostic investigation. Step number one is to verify that the abnormal result is repeatable. Because lipemia may interfere with calcium determination, the animal should be fasted for a minimum of 12 hours before the repeat sample is obtained. If the abnormality is repeatable, ionized calcium (iCa) should be measured for an accurate assessment of calcium status. Prediction of iCa status from tCa measurement is not accurate, and iCa needs to be specifically measured. Adverse effects of circulating calcium only occur when increased iCa is the cause of increased tCa. What's the bottom line? Total or adjusted tCa formulas are not reliable indicators of calcium status because of the high degree of diagnostic discordance between total, adjusted, and ionized calcium measurements. (20, 21) Ionized calcium should be used as the gold standard to determine if tCa is elevated or not and if a workup for hypercalcemia is warranted.

Preparing a blood sample for iCa determination

Accurate determination of iCa concentration requires that samples be collected and processed correctly. Some laboratories require an anaerobic prepared serum sample. When attempting anaerobic serum collection, exposure to air in partially filled serum tubes affects iCa concentration, and may result in a 5% decrease in iCa concentration (22). Anaerobic measurement of iCa requires meticulous separation techniques. Serum should be collected directly into a serum Vacutainer tube (Becton-Dickinson, Franklin Lakes, New Jersey). Silicone separator tubes are not recommended, because serum iCa concentration may increase as a result of release of calcium from the silicone gel (23). Blood should be allowed to clot and then be separated by centrifugation. To remove serum from the tube, a spinal needle attached to a non–air-containing syringe should be used to puncture the red-topped stopper without exposing the sample to air. iCa in canine or feline serum is stable for 72 hours at 23C or 4C and for 7 days at minus 10C (24). Accurate aerobic methods for iCa measurement have also been developed. The iCa concentration of an aerobically collected serum sample at any given pH can be mathematically corrected to the theoretical iCa concentration at pH 7.4 (25, 26).These correction formulas are based on the predictable linear change in iCa concentration with changes in pH and are species specific.

What are the common causes of hypercalcemia?

There are many potential causes of hypercalcemia (e.g., there at least 15 rule-outs). Cancer-associated hypercalcemia has traditionally been the leading cause of elevated serum calcium in dogs and cats. However, a reappraisal of this time honored observation would suggest that in cats, idiopathic hypercalcemia (IHC) is the most prominent cause of elevated serum calcium, followed by renal failure, and malignancy in first opinion clinics (27-31). In hypercalcemic dogs, neoplasia is the most common diagnosis, followed by hypoadrenocorticism, primary hyperparathyroidism, and chronic renal failure (27-31). Approximately 70% of hypercalcemic dogs are also azotemic, with azotemia being uncommon in dogs with primary hyperparathyroidism (32). In hypercalcemic cats, IHC and renal failure are more commonly associated with azotemia than neoplasia (31).

What if the calcium is elevated and there are no or minimal clinical signs?

Many animals with hypercalcemia have minimal or no observed clinical signs and a normal physical examination – the top rule-outs in this situation are IHC in the cat and primary hyperparathyroidism in the dog. Often the magnitude of the hypercalcemia is minimal to moderate and stable in these instances, so urgent treatment to correct the hypercalcemia is not warranted. Instead the trend in serum calcium is observed and further diagnostics are discussed with the owner. If the hypercalcemia is determined to be ionized, the recommendation is to restore normocalcemia even in the absence of obvious clinical signs. It is not uncommon for some patients to develop urolithiasis, renal mineralization and chronic kidney injury despite low-grade hypercalcemia. Animals that with vague GI or "ADR-like" signs, or hard signs such as vomiting, constipation, weight loss or polyuria and polydipsia will often clinically improve once the hypercalcemia is successfully managed. **The patient has renal failure and hypercalcemia - where are you going with the case?** About 20% of dogs and cats with CKD have hypercalcemia based on serum tCa (20, 21). However, when iCa is measured in the same animals, 10% of dogs and 30% of cats are found to be hypercalcemic (20, 21). In other words, tCa overestimates iCa in dogs and underestimates iCa in cats. The discrepancy between tCa and iCa in dogs with CKD is due to the increased complexed fraction of calcium by circulating organic anions (33). The high number of cats with increased iCa and CKD is thought to be associated with concurrent IHC. So, again, what's the bottom line? Ionized calcium should be measured in all cases with CKD to determine if the hypercalcemia based on the tCa is clinically important (e.g., the iCa is indeed elevated), or to determine if hypercalcemia is a problem despite a normal tCa. Dogs and cats with CKD and increased iCa are at increased risk for accelerated progression of their renal disease whereas animals with elevated tCa and normal to low iCa levels are not at the same risk for further kidney damage.

REFERENCES

1. Hadley GG,Weiss SP. Am J Clin Pathol 1955;25:1090 2. Imafuku Y, Meguro S, Kanno K, et al. Clin Chim Acta 2002;325:105 3. Cornes MP et al. Ann Clin Biochem 2008; 45:601 4.Degen M. J Am Vet Med Assoc 1987;190 (5):541 5. Peterson ME, Kintzer PP, J Am Vet Med Assoc 1996; 208(1):85 6. Kintzer PP, Peterson ME. J Vet Intern Med 1997; 11(2):43 7. Feldman EC, Nelson RW. Canine and feline endocrinology and reproduction. 3rd ed. St. Louis (MO): Saunders; 2004.

8. Roth L and Tyler R D. J Vet Diagn Invest 1999 11:60 9. Neiger R and Gunderson HC. BSAVA Congress Scientific Proceedings 2003:538 10. Segev G at al. J Vet Intern Med 2010;24:546 11. Kreissler JJ and Langston CE. J Vet Intern Med 2011; 25(4):955 12.De Fronzo RA. Kidney International 1980;17: 118 **13**. Singer GG and Brenner BM. Fluid and electrolyte disturbances. In: Kaspar DL et al, eds. Harrison's Principles of Internal Medicine 16th Ed. New York: McGraw-Hill;2005:25214. Rose BD. In: Rose BD and Post DW eds. Clinical Physiology and Acid-Base and Electrolyte Disorders 5th ED. New York: McGraw-Hill;2001:888 15. Lobetti RG.. J S Afr Vet Assoc 1998;69(1):33 16. McGonigle KM et al. J Am Anim Hosp Assoc 2013; 49 17. Reusch CE. Vet Rec 2007; 160: 219 18. Kintzer PP and Peterson ME. J Vet Intern Med 1991; 5(3):182 19. Ramsey I. Vet Clin North Am 2010; 40:260 20. Schenck PA and Chew DJ. Am J Vet Res 2005;66 (8):1330 21. Schenck PA and Chew DJ. Can J Vet Res 2010; 74 (3):209 22. Unterer S, Lutz H, Gerber B, et al. Am J Vet Res 2004;65(2):183 23. LarssonL,Ohman S. Clin Chem 1985;31(1):169 24.Schenck PA, Chew DJ, Brooks CL Am J Vet Res 1995;56(3):304 25. Lincoln SD, Lane VM.J Am Vet Med Assoc 1990;197(11):14 26. Nachreiner RF, Refsal KR. Proc Am Coll Vet Intern Med Forum, 1990 27. Cho T el al. J Vet Clinics 2007; 24(4):479 28. Elliot J et al. J Sm Anim Pract 2003; 32(11):1991 29. Messinger et al. J Vet Intern Med 2009;23(3):524 30. Midkif AM et al. J Vet Intern Med 2000;14(6):619 31. Savary KC et al. J Vet Intern Med 2000; 14(2):184 32. Feldman EC et al. J Am Vet Med Assoc 2995; 227(5):756 33. Schenck PA and Chew DJ. Am J Vet Res 2003:64(9):1181

APPLYING PHARMACOKINETIC PRINCIPLES TO INTERNAL MEDICINE CASES

Lauren A. Trepanier, DVM, PhD, DACVIM, DACVCP
Madison, WI

Pharmacokinetic data are available for most drugs approved for veterinary use, and are often available (in dogs) for drugs developed for human patients. Parameters such as bioavailability, Cmax, volume of distribution, clearance, and elimination half-life can be directly applied in the clinic setting to tailor drug therapy and clinical monitoring for individual patients.

BIOAVAILABILTY

Bioavailability after intravenous (IV) dosing is, by definition, 100%, and other routes are often compared to the IV route. Oral bioavailability is affected by drug size, lipophilicity, charge, and binding to dietary constituents. For example, enrofloxacin has decreased oral bioavailability in nursing kittens compared to chow-fed kittens. [1] Conversely, mitotane, which is quite lipophilic, is better absorbed with canned food or corn oil than in the fasted state in dogs. [2] Oral bioavailability is also affected by patient factors, to include drug degradation by stomach acid, duodenal bacteria, or brush border enzymes; drug efflux by transporter pumps, and drug biotransformation by both the intestine and liver, which together comprise the "first pass effect." The first pass effect can also lead to *increased* drug action if a drug or pro-drug is converted to an active metabolite; for example, codeine to morphine; enalapril to enalaprilat; and prednisone to prednisolone. The first pass effect is bypassed by IV, IM, SC, or transdermal drug administration.

Bioavailability can be used to calculate drug dosages for different routes of administration. Oral dose = IV dose (reference route) / oral bioavailability. For a drug with 50% oral bioavailability, the oral dose will be 2 times that of the IV dose.

CMAX AND TMAX

Cmax is the highest (peak) plasma drug concentration that is reached after a given dose. Cmax is typically proportional to the dose, but may increase non-linearly at higher doses (for example, with hydralazine dosing in dogs). [3] This phenomenon is typically due to saturation of clearance pathways such as efflux transporters or cytochrome P450s. Cmax also reflects the rate and extent of absorption, especially relative to the rate of elimination; for example, a drug with very slow absorption and a moderate rate of elimination will not accumulate and may have minimal pharmacodynamic effects (e.g. Ultralente insulin in the setting of ketoacidosis). Cmax determines the magnitude of side effects for some drugs, i.e. seizures after an IV bolus of lidocaine.

The ratio of drug Cmax to bacterial MIC is used to optimize efficacy for some concentration-dependent antimicrobials. For example, a
Cmax:MIC ratio of 10 or higher is targeted for aminoglycosides in human patients. [4]

Tmax is the time at which the peak plasma drug concentrations are reached. Tmax is affected primarily by the rate of drug absorption (relative to the rate of elimination). Both Tmax and Cmax can be affected by food. For example, food given concomitantly with L-thyroxine delays Tmax in dogs, and decreases Cmax and overall bioavailability. [5] Therefore, L-thyroxine should ideally be given on an empty stomach, or at least in a consistent relationship to food, in dogs; serum drug concentrations should also be monitored under the same feeding conditions as daily administration. Tmax tells you when to expect many dose-dependent side effects; however, some peak drug effects can lag behind peak plasma concentrations (for example, with transmucosal buprenorphine in cats) [6] Tmax also tells you when to draw a blood sample for monitoring peak drug concentrations. This is particularly important for drugs that are eliminated quickly, such as gabapentin in dogs (elimination half-life about 3.5 h), [7] since peak concentrations will be quite transient. For drugs that are eliminated slowly, such as KBr, timing is much less important, since plasma drug concentrations will be fairly stable throughout a dosing interval.

THERAPEUTIC RANGE

The therapeutic range (or therapeutic window) is the range of plasma concentrations in which most patients have the desired drug effect with minimal toxicity. The therapeutic range is often extrapolated from human studies and may not be well established in dogs or cats. The therapeutic index is the ratio between the upper (toxic) and lower (minimally effective) ends of the therapeutic range. If the therapeutic index is wide, this indicates a relatively safe drug. Conversely, a drug with a narrow therapeutic index has a small safety margin; for example, digoxin, with a therapeutic range of 0.8-1.2 ng/ml. For drugs with a narrow therapeutic range, therapeutic drug monitoring can be used to optimize dose. Therapeutic drug monitoring is also indicated for drugs that have variable bioavailability or clearance rates among individuals.

Table 1: Drugs for which therapeutic drug monitoring is indicated

Drug	Timing	Reason for monitoring
Thyroxine	Peak: 4 to 6 hours post pill Want *high normal* levels	Variable concentrations at standard dosages; [8] due to differences in bioavailability and/or clearance
Phenobarbital	Timing not critical in most dogs [9] Long half-life relative to dosing interval	Variable hepatic biotransformation (degree of P450 induction)
Cyclosporine	Trough level before next dose	Variable bioavailability (CYP3A, p-glycoprotein)
Bromide	Timing not critical Long half-life relative to dosing interval	Clearance varies with renal function and dietary chloride intake
Digoxin	6 to 8 hours after dosing	Narrow therapeutic index; Variable bioavailability (CYP3A, p-glycoprotein) and renal clearance
Theophylline	Extended release, dogs: Tmax 4 to 6 hours post pill [10] Extended release, cats: Tmax 8 hours post pill [11]	Variable bioavailability of extended release capsules
Gentamicin, amikacin	Peak: just after IV dosing, 30-40 minutes after SC dosing Trough: 24 hours after dosing	Narrow therapeutic index Cmax correlates with bacterial kill Trough (Cmin) correlates with nephrotoxicity

AREA UNDER THE CURVE (AUC)

AUC is a measure of drug exposure, and is affected by dose, absorption, and clearance. AUC is used to calculate drug bioavailability (oral bioavailability = AUC po / AUC iv). AUC (along with Cmax) is also used to determine bioequivalence between a generic formulation and an established product with demonstrated efficacy. In addition, the ratio of drug AUC to bacterial MIC can be used to predict antimicrobial drug efficacy. For fluoroquinolone antibiotics, an AUC:MIC ratio of > 100 has been recommended for gram-negative infections, including *Pseudomonas aeruginosa* pneumonia, in human patients. [12] This threshold is based on animal infection models and human clinical data; similar *in vivo* efficacy data have not been evaluated in dogs or cats.

VOLUME OF DISTRIBUTION

Volume of distribution (Vd) is the ratio between the total amount of drug administered and the resulting plasma drug concentrations. Volume of distribution can be thought of as a "patient beaker" into which you are pouring the drug. A low Vd suggests that drug distribution may be limited to the plasma space, as with highly protein-bound drugs such as Convenia® (Vd ~ 0.1 L/kg in dogs and cats). In this case, bound drug remains in the plasma space and acts as a drug depot, while free drug can still reach target sites of infection. Conversely, a high Vd indicates extensive distribution out of the plasma. This is desirable if it reflects distribution to target sites in the interstitial and intracellular spaces. However, a large Vd does not necessarily mean that the drug is well distributed to intended sites. Very high Vd estimates (> 1.0 L/kg) can result from drug binding to off-target tissues, such as bone or skeletal muscle. An example of this is digoxin, which is extensively bound to skeletal muscle and has a Vd of > 10 L /kg in dogs. [13] Vd can be used to determine the loading dose of a drug: Loading dose in mg/kg = Desired plasma concentration (mg/L) x Vd in L/kg. For a phenobarbital target concentration of 30 ug/ml (= 30 mg/L), with a Vd of 0.75 L/kg, [14] the calculated IV loading dose is 22.5 mg/kg.

CLEARANCE

Clearance reflects all routes of drug elimination, to include hepatic biotransformation, biliary excretion, glomerular filtration, and renal tubular secretion. Decreased hepatic or renal function can lead to impaired drug clearance and the need for drug dose adjustment. In human patients with significant hepatocellular dysfunction, drugs that are cleared by the liver and have relatively narrow therapeutic ranges merit dosage reduction. These same adjustments are likely to apply to dogs and cats with portosystemic shunts, acute hepatocellular necrosis, severe hepatic lipidosis, or acquired cirrhosis, although few studies have been performed in veterinary species.

Table 2: Suggested drug dosage adjustments for patients in hepatic failure

Chloramphenicol	Avoid if possible. If used, monitor CBC for bone marrow suppression Consider reducing dosage, but accurate guidelines not available
Diazepam or midazolam	Use with continual monitoring. Consider using 25-50% of standard dosages
Furosemide	Use with conservatively if at all. Hypokalemia may precipitate hepatic encephalopathy Consider substituting spironolactone/hydrochlothiazide when treating ascites
Lidocaine	Avoid or reduce dosage by 50%
Metronidazole	Reduce total daily dose to one third of standard anti-anaerobic dosage, and dose once daily
Phenobarbital	Avoid or discontinue. If using for seizures from congenital portosystemic shunting, consider startir at 25-50% of standard dosages, or choose potassium bromide
Propranolol	Decrease daily dosage by 50% or more. Monitor heart rate and blood pressure
Theophylline	Reduce oral dose (e.g. by ~50%). Measure serum theophylline concentrations

Adapted in part from recommendations for humans by the World Health Organization, WHO Model Formulary, 2008; www.who.int/archives/en/).

In human patients with renal insufficiency, dosage reductions of renally cleared drugs are often made when creatinine clearance values are less than ~ 0.7 ml/min/kg (depending on the drug). While this is equivalent to a serum creatinine of 3.5 mg/dl in cats, [15,16] minimal data are available in dogs, and we lack equivalent dose adjustment studies in both species. Until better studies are available, we can only extrapolate from human data.

Table 3: Empirical recommendations for drug dosage adjustment in dogs and cats with renal failure (adapted from human guidelines). [a]

Drug	Standard dosage	CL creatinine of 0.4-0.7 ml/kg/min (30-50 mls/min in humans). Roughly a serum creatinine of 2.3-4.0 mg/dl in cats [15-17]	CL creatinine of 0.15-0.4 ml/kg/min (10-30 mls/min in humans). Roughly a serum creatinine of 4.9-5.5 mg/dl in cats [15,18]	CL creatinine of < 0.15 ml/kg/min (<10 mls/min in humans) Roughly a serum creatinine of 8.5 mg/dl in cats [15]
Amikacin [19]	15 mg/kg q. 24h	q. 36-48 h Avoid if possible	q. 48 h Avoid if possible	Not recommended
Amphotericin B	1 mg/kg IV three times weekly	Use liposomal formulation (AmBisome; less nephrotoxic) [20]	Use liposomal formulation; only with careful monitoring	Avoid if possible
Atenolol	0.25 mg/kg q. 12h	No change to 50% of total daily dose	50% of total daily dose	25% of total daily dose
Benazepril	0.5 mg/kg q. 12h	No change	Start at 50% of total daily dose	Start at 50% of total daily dose
Cefazolin	20 mg/kg q. 8h	q. 12 h	q. 12 h	q. 24 h
Cefotetan [21]	30 mg/kg SC q. 12h	No change	q. 24h	q. 48h
Enalapril	0.5 mg/kg q. 12h	No change recommended in humans [b]	Start at 50% of total daily dose	Start at 50% of total daily dose
Enrofloxacin	5 mg/kg q. 24h	Presumption is to extend dosing interval (as for cipro in humans); substitute less retinotoxic fluoroquinolones in cats	Not recommended in cats due to risk of retinotoxicity	Not recommended in cats due to risk of retinotoxicity
Famotidine	1 mg/kg q. 12h	q 24h	q 24h	q 24h
Fluconazole	5 mg/kg q. 12h	50% of total daily dose	50% of total daily dose	50% of total daily dose
Fluoxetine	1-2 mg/kg q. 24h	No change	No change	50% of total daily dose
Gabapentin	10 mg/kg q. 8h (starting dose)	6 mg/kg q. 12h	6 mg/kg q 24h	3-4 mg/kg q. 24h
Gentamicin [19]	6-8 mg/kg q. 24h	q. 36-48 h Avoid if possible	q. 48 h Avoid if possible	Not recommended
Levetiracetam	20 mg/kg q 8h	50% of total daily dose	30-50% of total daily dose	30-50% of total daily dose
Metoclopramide	1-2 mg/kg/day CRI	50% of total daily dose	50% of total daily dose	25-50% of total daily dose
Spironolactone	1.0 mg/kg q. 12h	No change	No change	Not recommended
Trimethoprim/ sulfamethoxazole	15 mg/kg q. 12h	No change Avoid sulfadiazine due to higher risk of crystalluria	50% dose reduction q 12 h Avoid sulfadiazine	Not recommended

[a] Source: University of Wisconsin Hospital and Clinics Clinical Directive for Renal Function-Based Dose Adjustments in Adults (www.UWHealth.org; 2009), unless otherwise cited. [b] Studies in dogs with renal impairment (Clcr of 1.7 ml/kg/min) showed an increase in AUC by 80% for active enalaprilat. [22]

ELIMINATION HALF-LIFE

Elimination half-life is important for determining the time to reach **steady state** (stable or equilibrium) plasma concentrations. Time to steady state takes approximately 5 elimination half-lives for any drug. For example: the elimination half-life of modified cyclosporine (i.e. the microemulsion formulation found in Atopica™ and Neoral®) is 6-12 hours in dogs; [23,24] therefore, steady state should be reached in 5 x 12 = 60 hours, or 2.5-3 days, when whole blood levels can be drawn.

Elimination half-life is also important for determining dosing frequency, which is also based on the therapeutic range. For example, the elimination half-life of levetiracetam in dogs is about 4 hours; [25] with a proposed therapeutic range of 5-30 □g/ml in humans (ARUP Laboratories), this requires dosing every 8 hours. However, concurrent administration of phenobarbital shortens the half-life of levetiracetam to about 2 hours. [25] This could lead to sub-therapeutic levetiracetam concentrations during an 8-hour dosing interval, and needs further evaluation. Finally, elimination half-life determines the time when a drug or toxin will be eliminated after administration or exposure (also 5 elimination half-lives). For example, in order to switch from clomipramine to fluoxetine in a dog with behavioral anxiety, clomipramine, since this drug is contraindicated in combination with selective serotonin reuptake inhibitors. [26] With a t ½ of up to 16 h for clomipramine and 3 h for the active metabolite desmethylclomipramine in dogs, [27] washout would be 5 x 16h, or between 3 and 4 days before fluoxetine could be started.

REFERENCES

1. Seguin MA, et al. *Am J Vet Res* 2004;65:350-356. 2. Watson AD, et al. *Res Vet Sci* 1987;43:160-165. 3. Semple HA, et al. *Pharm Res* 1990;7:274-279. 4. Rea RS, et al. *Semin Respir Crit Care Med* 2007;28:596-603. 5. Le Traon G, et al. *J Vet Pharmacol Ther* 2008;31:95-101. 6. Robertson SA, et al. *J Vet Pharmacol Ther* 2005;28:453-460. 7. Kukanich B, et al. *Vet J* 2011;187:133-135. 8. Nachreiner RF, et al. *J Am Vet Med Assoc* 1992;201:623-629. 9. Levitski RE, et al. *J Am Vet Med Assoc* 2000;217:200-204. 10. Bach JE, et al. *J Am Vet Med Assoc* 2004;224:1113-1119. 11. Dye JA, et al. *J Vet Pharmacol Ther* 1989;12:133-140. 12. Wispelwey B. *Clin Infect Dis* 2005;41 Suppl 2:S127-135. 13. Button C, et al. *Am J Vet Res* 1980;41:1230-1237. 14. Plumb D. *Veterinary Drug Handbook.* 7th ed. ed. Ames IA: Iowa State University Press, 2011. 15. Miyamoto K. *J Feline Med Surg* 2001;3:143-147. 16. Sox EM, et al. *J Feline Med Surg* 2010;12:738-745. 17. Haller M, et al. *J Feline Med Surg* 2003;5:175-181. 18. Heiene R, et al. *Am J Vet Res* 2009;70:176-185. 19. Nayak-Rao S. *Indian J Nephrol*;20:121-124. 20. Bekersky I, et al. *Pharm Res* 1999;16:1694-1701. 21. Smith BR, et al. *Antimicrob Agents Chemother* 1986;29:887-893. 22. Lefebvre HP, et al. *J Vet Intern Med* 1999;13:21-27. 23. Allenspach K, et al. *J Vet Intern Med* 2006;20:239-244. 24. Radwanski NE, et al. *Am J Vet Res*;72:687-693. 25. Moore SA, et al. *J Vet Pharmacol Ther*;34:31-34.
26. Gillman PK. *Br J Pharmacol* 2007;151:737-748. 27. Hewson CJ, et al. *J Vet Pharmacol Ther* 1998;21:214-222.

IDIOSYNCRATIC DRUG TOXICITIES IN DOGS AND CATS

Lauren A. Trepanier, DVM, PhD, DACVIM, DACVCP
Madison, WI

Idiosyncratic drug reactions occur in only a small proportion of patients at therapeutic dosages, and are often difficult to predict. Idiosyncratic toxicity does not increase with dose in the general population (therefore they are not considered "dose-dependent"), but toxicity may increase with dose in susceptible individuals. Idiosyncratic drug toxicity is often caused by reactive metabolites, which may be variably generated among individuals. These reactive metabolites typically cause oxidative stress and/or lead to haptens that trigger a humoral or T-cell-mediated immunologic response. Although idiosyncratic drug reactions are sometimes called drug hypersensitivity reactions, they may or may not involve an adaptive immune response. Idiosyncratic drug reactions usually require discontinuation of the suspect drug, and structurally related drugs may cause a similar reaction.

The liver, skin, and bone marrow and circulating blood cells are common targets of idiosyncratic drug toxicity. The liver is susceptible because of high blood flow and a high concentration of cytochrome P450's and other enzymes that can generate reactive metabolites. Blood cells have a high tissue mass of circulating or dividing cells, which express cytochrome P450s, myeloperoxidases, and cyclooxygenases, each of which can bioactivate certain drugs. Finally, the skin has a large surface area, with a high number of antigen-presenting (Langerhans) cells, and some cytochrome P450 and cyclooxygenase activity in keratinocytes.

Potentiated sulfonamide antibiotics are quite old, but are still important for the treatment of fluoroquinolone-resistant gram-negative infections and methicillin-resistant Staph. However, potentiated sulfonamides can lead to a drug hypersensitivity syndrome in dogs, consisting of fever, hepatopathy (hepatocellular or cholestatic), blood dyscrasias (immune thrombocytopenia, IMHA, or transient neutropenia), skin eruptions (vasculitis, pemphigus foliaceus, or bullous skin diseases), polyarthropathy, proteinuria, or uveitis, typically after 5 to 14 days of treatment. KCS may also be seen, but this occurs in about 15% of treated dogs, [1] often after a month or more of administration, and does not fit a classic idiosyncratic pattern.

Sulfonamide hypersensitivity reactions are caused by an oxidized hydroxylamine metabolite that is generated by cytochrome P450s and myeloperoxidases. This hydroxylamine circulates in the plasma, and spontaneously oxidizes to a reactive nitroso metabolite that covalently binds to proteins and acts as a hapten. [2] Drug specific T-cells, some of which have been shown to mediate toxicity against keratinocytes, have been found in human patients, [3] and drug specific T-cells may also occur in dogs. In addition, dogs with sulfonamide hypersensitivity develop anti-drug antibodies; these antibodies cross-react across sulfamethoxazole, sulfadiazine, and sulfadimethoxine in about 30% of dogs. [4] Dogs with thrombocytopenia from sulfonamide antimicrobials develop anti-platelet antibodies that recognize non-covalent drug-platelet complexes. [5] Some of these antibodies require continuous presence of sulfonamide drug in order to bind to platelets. This may explain why thrombocytopenia may resolve rapidly in some dogs after sulfonamide discontinuation, without the need for glucocorticoids.

Risk factors for sulfonamide hypersensitivity in dogs are not entirely clear. Brand name and generic formulations have each been implicated. [6] Doberman pinschers appear to be over-represented among dogs with the combination of polyarthropathy, thrombocytopenia, and proteinuria. [7,8] In humans, there is no clear evidence of cross-reactivity between sulfonamide antibiotics and non-arylamine drugs containing a sulfonamide group, such as furosemide, acetazolamide, and hydrochlorthiazide. [9] Despite the term "sulfonamide hypersensitivity," it is the arylamine moiety, and not the sulfonamide part of the drug, that generates haptens.

When sulfonamide antibiotics need to be prescribed, a baseline STT should be obtained, and rechecked every 10 days or so during treatment. Owners should be vigilant for signs of lethargy, fever, lameness, or changes in the appearance of skin, eyes, mucous membranes, or urine. Early recognition and drug discontinuation can be life saving. Since the sulfonamide haptens can be reversed *in vitro* with antioxidants (unpublished observations, Lavergne and Trepanier 2006), consider treatment with ascorbate 90 mg/kg/day IV (empirical dosage) and N-acetylcysteine (140 mg/kg loading IV, then 70 mg/kg every 6 hours for 7 treatments, as for acetaminophen toxicosis), followed by SAMe on an outpatient basis until signs resolve. Consider short courses of prednisone (2-3 months) if there is persistent thrombocytopenia or auto-agglutinating hemolytic anemia. There has also been anecdotal success with intravenous immunoglobulin for sulfonamide-associated bullous skin eruptions in a dog (IV IgG at 0.5 gram/kg). [10]

Methimazole in cats is also associated with idiosyncratic toxicity affecting the liver, skin, and blood cells. Similar reactions have also been reported for carbimazole, a pro-drug of methimazole. Methimazole hepatotoxicity shows a mixed hepatocellular and cholestatic pattern in cats, and typically occurs in the first month of treatment, with a 1-2% incidence of jaundice. [11] Drug-induced increases in liver enzymes resolve within several weeks of methimazole discontinuation, but clinical improvement is more rapid. Rechallenge has led to recrudescence of hepatotoxicity. [11] These changes are distinct from "innocent" increases in ALT and SAP seen in cats with hyperthyroidism, which resolve once euthyroid and do not worsen during anti-thyroid drug treatment. Methimazole hepatotoxicity also occurs in about 9% of human patients, with a mean onset of 28 days after starting the drug.[12] Hepatotoxicity has been observed in a rodent model, and is associated with an N-methylthiourea metabolite that is generated by flavin mono-oxygenases.[13,14] Glutathione depletion is an experimental risk factor for methimazole hepatotoxicity,[13] and either N-acetylcysteine or taurine attenuate toxicity hepatocyte toxicity *in vitro*.[15,16] These findings may have relevance for methimazole hepatotoxicity in cats. Although we recently found no association between glutathione depletion and methimazole toxicity in pet cats,[17] cats were tested after their adverse reactions instead of prior to methimazole administration, and blood, not hepatic, glutathione was measured.

Blood dyscrasias are seen in about 4% of cats treated with methimazole, and include thrombocytopenia and neutropenia or agranulocytosis.[11] These reactions typically occur in the first 1-2 months of treatment, and rechallenge has led to recurrence.[11] Monitoring for blood dyscrasias is important, since theses reactions can progress to bleeding or secondary infections. In humans, methimazole- and carbimazole-induced blood dyscrasias include thrombocytopenia, hemolytic anemia, neutropenia, and agranulocytosis. These reactions appear to be immune-mediated, but can occur after years of therapy.[18] Drug-dependent antibodies targeting neutrophils (FcγRIIIb) and erythrocytes (Rh complex proteins) have been demonstrated, as well as antibodies that bind to the adhesion molecule PECAM-1 in the presence of drug.[18] PECAM-1 (a.k.a CD31) is expressed on platelets, neutrophils, and monocytes, and antibodies against this molecule may explain the involvement of multiple cell lineages in these blood dyscrasias. In addition, the requirement of drug for antibody binding *in vitro* is consistent with the development of transient protein-drug neoantigens. Such a mechanism would explain the clinical observation of relatively rapid recovery of cell counts after drug discontinuation in cats, although methimazole or carbimazole-dependent antibodies in cats have yet to be evaluated.

Pruritus and facial excoriation develop in 2-3% of cats treated with methimazole, most commonly in the first 3 weeks of treatment.[11] Lesions are typically due to self-trauma of the neck and the face anterior to the pinnae, but have not been characterized histologically. Pruritus and urticaria are also seen in about 8% of human pediatric patients[19] and up to 20% of adults[12] treated with methimazole, typically in the first 4 weeks of treatment. More severe bullous skin eruptions such as Stevens-Johnson syndrome have also been reported.[19] The mechanisms for these reactions have not been explored.

Clinical and biochemical monitoring in the first weeks of treatment can detect methimazole reactions before they become severe. If cats develop lethargy, vomiting, inappetance, jaundice, or pruritus, methimazole should be discontinued and importantly, the cat should be physically evaluated as soon as possible to distinguish between blood dyscrasias, hepatopathy, renal decompensation (not idiosyncratic), or simple GI upset. A serum T4 should also be evaluated if the cat has gotten methimazole in the last 24 hours and has been on the drug at least 10-14 days. Liver enzymes should always be compared to pre-methimazole, hyperthyroid values. If the diagnosis is simple GI upset, this should respond to a dose reduction or a switch to transdermal methimazole, which has a lower incidence of GI upset than oral methimazole.[20] If idiosyncratic blood dyscrasias, hepatopathy, or facial excoriation are diagnosed, radioiodine is indicated, or if not available, thyroidectomy with atenolol pretreatment. The efficacy of glutathione precursors or glucocorticoids for these methimazole-associated adverse reactions has not been evaluated.

Carprofen is commonly thought to have a higher risk of hepatotoxicity than other NSAIDs, but this has not been evaluated in a comparative study. The veterinary labels for meloxicam, etodolac, deracoxib, firocoxib, and robenacoxib also report liver toxicity in dogs. However, idiosyncratic carprofen hepatotoxicity has been described in the greatest detail, with acute hepatic necrosis with marked increases in ALT seen in approximately 0.05% of dogs, according to the manufacturer. Most dogs are affected within 14 to 30 days;[21] Labrador retrievers were over-represented in the initial published report, but the manufacturer could not reproduce this syndrome in Labrador dogs (personal communication). Therefore, this is unlikely to be a true breed risk. It is important to note that no reported cases of carprofen hepatotoxicity have shown increases in SAP without large accompanying increases in ALT.[21]

Idiosyncratic NSAID hepatotoxicity also occurs in human patients,[22] with diclofenac implicated most commonly. Three-fold or higher increases in ALT occur in about 5% of patients in the first 6 months of diclofenac treatment, with hepatocellular necrosis seen much less commonly. Diclofenac metabolites cause mitochondrial injury *in vitro*,[23] which could explain mild transaminase increases in some patients. The progression to hepatocellular necrosis could involve an amplification of this cytotoxicity by an immune response to drug haptens, in particular those formed by reactive acylglucuronide and quinone imine diclofenac metabolites.[24] Carprofen shares some structural similarities with diclofenac, so a similar scenario could apply to carprofen as well.

NSAID-associated blood dyscrasias and skin eruptions have been reported in isolated case reports in veterinary patients, to include thrombocytopenia and hemolytic anemia, (1 case) and bone marrow necrosis (3 cases) during carprofen treatment,[25,26] and vasculitis during meloxicam or carprofen treatment (1 case each).[25,27] Although hepatotoxicity, cytopenias, and vasculitis from NSAIDs can be severe, these idiosyncratic reactions are quite rare, and are not efficiently detected by routine biochemical monitoring. Clinical monitoring by owners is key, and every client should be advised to monitor for new vomiting, diarrhea, lethargy, or change in urine color. Dose-dependent gastric ulceration and new azotemia are much more common, and deserve routine clinical monitoring. Middle aged to older dogs are more likely to have pre-existing organ dysfunction, and should have baseline blood work performed to evaluate HCT, BUN, creatinine, albumin, and

ALT before starting any NSAID. Abnormal clinical signs in any dog during NSAID treatment should prompt drug discontinuation; a physical exam, CBC, and biochemical panel can distinguish between simple GI upset, gastric ulceration, renal compensation, and idiosyncratic reactions.

Zonisamide was recently associated with acute hepatotoxicity in two epileptic dogs.[28,29] In both cases, clinical signs developed in the first 3 weeks of treatment, one with a hepatocellular pattern and one with a mixed hepatocellular and cholestatic pattern. Liver failure progressed in one dog; necropsy showed massive hepatic necrosis with marked periportal microvesicular steatosis. Further clinical experience is needed before the incidence of zonisamide hepatotoxicity is clear; however, dog owners should be informed of this potential adverse drug reaction when zonisamide is prescribed.

Diazepam has led to idiosyncratic acute hepatic necrosis in cats.[30,31] Affected cats have been otherwise healthy and treated for behavioral problems or urethrospasm; clinical signs were typically seen after 8 to 9 days of treatment. Both generic and brand name oral diazepam have been implicated, although injectable diazepam has not.[30] The time frame for toxicity suggests an immune component (humoral or T cell-mediated), although this has not been explored. Reactions are fulminant and often fatal, but aggressive supportive care can be successful.[32] Acute idiosyncratic liver injury from benzodiazepines is rarely seen in human patients,[33,34] so potential mechanisms have not been explored.

Phenobarbital hepatotoxicity typically occurs in dogs after a year or more of phenobarbital treatment, and can present with asymptomatic increases in bile acids, clinical signs of hepatic disease, or overt liver failure.[35] **The clinical presentation does not follow a classic idiosyncratic or dose-dependent pattern**; it is seen late in the course of treatment, and can resolve with dose reduction; however, it has not been reported with loading doses and or been reproduced in experimental studies.[36,37] Phenobarbital hepatotoxicity is probably best described as dose/duration-dependent with individual modifiers, but individual risk factors are not clear. Hepatotoxicity could be related to cytochrome P450 induction by phenobarbital, with secondary bioactivation of dietary or environmental compounds. Primidone and phenytoin, which are also cytochrome P450 inducers, lead to a similar pattern of delayed hepatotoxicity.[38] Conversely, phenobarbital in cats leads to neither cytochrome P450 induction[39,40] nor hepatotoxicity. Given the popularity of phenobarbital as an anticonvulsant in dogs, this adverse reaction deserves further study.

In order to minimize the risk of phenobarbital-induced hepatic dysfunction, minimal effective dosages of phenobarbital should be used; this can often be achieved by combination therapy with bromide or other anticonvulsants. In addition, serum bile acids should be evaluated every 6 months during phenobarbital treatment; the degree of SAP induction alone is not necessarily predictive. In addition to typical signs of hepatic disease, dogs should also be monitored for increased sedation without a change in phenobarbital dosage, which can signal decreased hepatic clearance of phenobarbital. If hepatopathy develops, plan phenobarbital discontinuation over 1-6 weeks, depending on severity of liver dysfunction and prior seizures. Consider transitioning to KBr monotherapy (40-60 mg/kg/day); if seizure control has been brittle, use a KBr loading dose of 400-600 mg/kg divided into 4 doses. Do not load KBr if hepatic encephalopathy is present.

In addition to liver toxicity, phenobarbital has been associated rarely with blood dyscrasias in dogs, to include anemia, thrombocytopenia, and neutropenia, with bone marrow necrosis or myelofibrosis noted on bone marrow evaluation.[26,41,42] These reactions respond to drug discontinuation and supportive care, unless advanced myelofibrosis has developed. Finally, phenobarbital has been associated with superficial necrolytic dermatitis in dogs. In fact, about 45% of cases of superficial necrolytic dermatitis (also called necrolytic migratory erythema or hepatocutaneous syndrome) in dogs are associated with phenobarbital administration.[43] Skin lesions are characterized by hyperkeratosis, crusting, and erythema of the footpads, mucocutaneous junctions, and nasal planum. These lesions develop only after prolonged treatment (a median of 7 years) and are associated with hepatopathy; therefore, the pattern is not consistent with idiosyncratic drug toxicity. The presentation is distinct from the skin eruptions seen in human patients treated with phenobarbital, which occur early in treatment (within the first 2 months) and are not related to any accompanying liver toxicity.[44]

Table 1: Other drugs associated with idiosyncratic toxicity in dogs or cats

Drugs in dogs	Toxicity	Clinical features, mechanism(s)	Monitoring and management
Phenylbutazone	Aplastic anemia in dogs [45]	Oxidation of phenylbutazone to reactive metabolites by blood cell peroxidases [46]	Phenylbutazone not recommended in dogs or cats
Fenbendazole	Pancytopenia, bone marrow necrosis in dogs [26,47]	Not established; onset within 2 weeks of starting fenbendazole	Very rare; routine monitoring of CBC not indicated
Flucytosine	Skin eruptions with depigmentation and ulceration [48]	Median onset 2 to 6 weeks after starting flucytosine. Nasal planum, scrotum, mucocutaneous junctions	Clinical vigilance and drug discontinuation
Mitotane	Hepatopathy (single case). [49] Bone marrow necrosis [26]	Hepatopathy seen after one month of treatment [49]	Apparently quite rare
Griseofulvin	Neutropenia, pancytopenia in cats. [50,51] Pancytopenia in a dog [52]	Neutropenia reported in FIV+ cats; recurred with rechallenge. [51] Not reproduced in cats at high doses.[53]	Alternative drugs recommended
Albendazole	Bone marrow suppression in a cat [54]	One case report, erythroid and megakaryocytic lines affected, mechanism unknown.	Drug discontinuation and supportive care
Meloxicam	Vasculitis with ulcers, vesicles, and erosions in a dog [27]	Noted within 2 days of starting drug, which is atypical for delayed drug hypersensitivity	Apparently quite rare

REFERENCES

1. Berger S, et al. *J Am An Hosp Assoc* 1995;31:236-241. 2. Naisbitt D, et al. *Br J Pharmacol* 2001;133:295-305. 3. Schnyder B, et al. *Clin Exper Immunol* 1998;28:1412-1417. 4. Lavergne SN, et al. *Clin Exp Allergy* 2006;36:907-915. 5. Lavergne SN, et al. *Platelets* 2007;18:595-604. 6. Trepanier L, et al. *J Vet Intern Med* 2003;17:647-652. 7. Giger U, et al. *Journal of the American Veterinary Medical Association* 1985;186:479-484. 8. Vasilopulos RJ, et al. *J Small Anim Pract* 2005; 46:232-236. 9. Strom BL, et al. *The New England journal of medicine* 2003;349:1628-1635. 10. Nuttall TJ, et al. *J Small Anim Pract* 2004;45:357-361. 11. Peterson ME, et al. *J Vet Intern Med* 1988;2:150-157. 12. Otsuka F, et al. *Clin Endocrinol* 2012;77:310-315. 13. Mizutani T, et al. *J Appl Toxicol* 1999;19:193-198. 14. Mizutani T, et al. *Chem Res Toxicol* 2000;13:170-176. 15. Heidari R, et al. *Drug Chem Toxicol* 2012 epub. 16. Heidari R, et al. *Sci Pharm* 2012 epub;80:987-999. 17. Branter E, et al. *J Vet Intern Med* 2012;26:582-588. 18. Bux J, et al. *Br J Haematol* 2000;109:243-247. 19. Rivkees SA, et al. *Int J Pediatr Endocrinol*; 2010: 176970. 20. Sartor LL, et al. *J Vet Intern Med* 2004;18:651-655. 21. MacPhail CM, et al. *J Am Vet Med Assoc* 1998;212:1895-1901. 22. Rostom A, et al. *Clin Gastroenterol Hepatol* 2005;3:489-498. 23. O'Connor N, et al. *QJM* 2003;96:787-791. 24. Aithal GP, et al. *Clin Liver Dis* 2007;11:563-575, vi-vii. 25. Mellor PJ, et al. *J Small Anim Pract* 2005;46:237-242. 26. Weiss DJ. *J Am Vet Med Assoc* 2005;227:263-267. 27. Niza MM, et al. *Vet Dermatol* 2007;18:45-49. 28. Miller ML, et al. *J Vet Intern Med* 2011;25:1156-1160. 29. Schwartz M, et al. *J Vet Med Sci* 2011. 30. Center SA, et al. *J Am Vet Med Assoc* 1996;209:618-625. 31. Levy J, et al. *J Am Vet Med Assoc* 1994;205:156-157. 32. Park FM. *J Feline Med Surg*;14:158-160. 33. Sabate M, et al. *Aliment Pharmacol Ther* 2007;25:1401-1409. 34. Tedesco FJ, et al. *Dig Dis Sci* 1982;27:470-472. 35. Dayrell-Hart B, et al. *J Am Vet Med Assoc* 1991;199:1060-1066. 36. Aitken MM, et al. *Vet Rec* 2003;153:13-16. 37. Chauvet AE, et al. *J Am Vet Med Assoc* 1995;207:1305-1307. 38. Bunch SE, et al. *J Am Vet Med Assoc* 1984;184:444-448. 39. Maugras M, et al. *Comp Biochem Physiol B* 1979;64:125-127. 40. Truhaut R, et al. *C R Acad Sci Hebd Seances Acad Sci D* 1978;286:371-373. 41. Jacobs G, et al. *J Am Vet Med Assoc* 1998;212:681-684. 42. Weiss DJ, et al. *J Vet Intern Med* 2002;16:174-178. 43. March PA, et al. *J Vet Intern Med* 2004;18:65-74. 44. Newell BD, et al. *Pediatr Dermatol* 2009;26:536-546. 45. Weiss DJ, et al. *J Am Vet Med Assoc* 1990;196:472-475. 46. Uetrecht J. *Crit Rev Toxicol* 1990;20:213-235. 47. Gary AT, et al. *J Am Anim Hosp Assoc* 2004;40:224-229. 48. Malik R, et al. *Aust Vet J* 1996;74:285-28 49. Webb CB, et al. *J Am Anim Hosp Assoc* 2006;42:298-301. 50. Rottman JB, et al. *J Am Vet Med Assoc* 1991;198:429-431. 51. Shelton GH, et al. *J Vet Intern Med* 1990;4:317-319. 52. Brazzell JL, et al. *Vet Clin Pathol* 2006;35:413-417. 53. Kunkle GA, et al. *J Am Vet Med Assoc* 1987;191:322-323. 54. Stokol T, et al. *J Am Vet Med Assoc* 1997;210:1753-1756.

POSTNATAL DEVELOPMENT OF THE CANINE FECAL MICROBIOME. JS Suchodolski[1], CE Stephenson[1], A Buono[1], GE Lees[2], SE Dowd[3], JM Steiner[1]. [1]Gastrointestinal Laboratory, [2]Department of Small Animal Clinical Sciences, Texas A&M University, College Station, TX; [3]Molecular Research DNA Laboratory, Shallowater, TX.

The gastrointestinal (GI) microbiome plays a crucial role in host health, as intestinal microbes stimulate the immune system, aid in the competitive exclusion of enteropathogens, and provide nutritional benefits to the host. Recent studies have revealed a complex GI microbiome in adult dogs, but limited data are available about the postnatal development of the canine GI microbiome. The GI tract of a fetus is sterile, but during the natural birthing process, a life-long colonization by foreign microorganisms is initiated. This microbial establishment is believed to be crucial for the development of the host's immune system, and alterations in the microbial composition have been, for example, associated with immune-mediated diseases (e.g., atopy) in children. The aim of this study was to investigate the postnatal development of the microbiome in healthy puppies.

Twelve colony-housed research puppies from three different litters were included in this study. Fecal samples were collected from each puppy at five different time points during periods of nursing (N; <2 days, 5-7 days, and 19-21 days after birth), weaning (W; 30-35 days after birth), and post-weaning (PW; 9 to 13 weeks after birth). During the W period, puppies were nursed and received supplemental gruel, while in the PW period the puppies were fed a puppy maintenance diet. The composition of the intestinal microbiota was analyzed using 16S rRNA gene 454-pyrosequencing and quantitative PCR assays. Differences in bacterial proportions among time points were evaluated using Friedman tests. The p-values were adjusted for multiple comparisons based on the Benjamini & Hochberg False discovery rate, and significance was set at $p<0.05$.

Within 2 days of birth all puppies had established a complex microbiome, which was not significantly different in the abundance of total bacteria ($p=0.196$), but differed in composition compared to the later time points. In general, aerobic and facultative anaerobic bacteria (e.g., Proteobacteria) were most abundant during the N period, while obligate anaerobic bacteria were more abundant in the PW period. *Escherichia* spp. were significantly increased in the N period compared to PW (median % of sequences; N=20.0%; PW=0.4%; $p<0.001$). Bacteroidia (N=0.5%; PW=12.8%; $p=0.024$) and Firmicutes (N=44.7%; PW=70.8%; $p=0.002$) were significantly more abundant in the PW period. Important genera such as *Faecalibacterium*, *Lactobacillus*, and *Bifidobacterium* were also significantly increased during PW compared to N ($p<0.001$ for all).

In conclusion, the canine intestinal microbiome undergoes profound changes during the postnatal period. These early transitions may indicate periods of microbiome instability, in which puppies are more susceptible to environmental influences (e.g., enteropathogens). These results warrant further investigations to evaluate if therapeutic interventions, designed to modulate the gastrointestinal microbiome (e.g., probiotics) very early in life may have long-term benefits on the microbiome and immune system of the host.

METAGENOMIC ANALYSIS OF THE EFFECT OF WITHHOLDING FOOD ON THE SMALL INTESTINAL MICROBIOME OF HEALTHY DOGS. JS Suchodolski[1], SE Dowd[2], E Westermarck[2], JM Steiner[1], T Spillmann[3], JA Harmoinen[3]. [1]GI Laboratory, Texas A&M University, College Station, TX, USA; [2]Molecular Research Lab, Shallowater, TX, USA; [3]Department of Equine and Small Animal Medicine, Helsinki University, Helsinki, Finland.

The small intestinal microbiome has a significant impact on the health status of the host. Withholding food is commonly recommended in dogs with acute gastrointestinal disturbances, although recent studies suggest potential benefits of shortening the period of withholding food. No data are available about the dynamics of the canine small intestinal microbiome during the state of withholding food. Thus, the aim of this study was to evaluate the small intestinal microbiome during a period where food was withheld from dogs.

Five healthy laboratory Beagle dogs with a permanent jejunal fistula located 60 cm distally from the pylorus were included. Jejunal wall brush samples were collected daily for 15 days via fistula. The first 5 days served as baseline period (days 1-5; "before"). Food was withheld from the dogs for 5 consecutive days (days 6-10, "during"), which was followed by a period of daily feeding with their original maintenance diet (days 11-15, "after"). Normal tap water was given *ad libitum* during the entire study. Clinical examination was performed and serum biochemistry and hematological parameters were evaluated daily. Fecal microbial communities were characterized by quantitative PCR assays and 454-pyrosequencing of the 16S rRNA gene. The Unifrac metric followed by analysis of similarity (ANOSIM) was used to assess differences in microbial communities among the time periods (before, during, and after withholding food). Differences in bacterial groups were tested using Friedman tests. The p-values were adjusted for multiple comparisons based on the Benjamini & Hochberg False discovery rate (significance set at $p<0.05$).

No clinical, biochemical, or hematological abnormalities were observed. No significant changes in total bacterial abundance were found over the study period ($p=0.34$). However, Unifrac analysis revealed significant differences in microbial communities starting within 2 days of withholding food (ANOSIM; $p<0.005$) and these altered communities remained stable during the withholding food period. Microbial communities returned to the "before" state within 2 days of refeeding. Significant changes were observed for various bacterial phyla. Firmicutes decreased ($p=0.02$) during withholding food (median % of sequences: before 85%, during 7%, after 77%). In contrast, significant increases during withholding food were observed for Bacteroidetes (before 9%, during 48%, after 3%; $p=0.03$) and Proteobacteria (before <1%, during 4%, after <1%; $p=0.03$).

In conclusion, the small intestinal microbiota during a period of withholding food differs significantly from the microbiota during periods when dogs are fed. We hypothesize that the abundance of mucin-degrading bacteria increases during periods of withholding food. Investigating the effect of these altered bacterial populations on the immune system and the metabolic profile of the host warrants further studies.

A YEAR-LONG PROSPECTIVE, RANDOMIZED, DOUBLE-MASKED STUDY OF NUTRITION ON FELINE IDIOPATHIC CYSTITIS (FIC). J. Kruger[1], J. Lulich[2], J. Merrills[1], J. MacLeay[3], I. Paetau-Robinson[3], S. Steck[1], S. Davidson[2], C. Osborne[2]. [1]Michigan State University, E. Lansing, MI., [2]University of Minnesota, St. Paul, MN., [3]Hills Pet Nutrition, Inc., Topeka, KS.

Feline Idiopathic Cystitis (FIC) is currently managed with pharmaceutical, environmental and nutritional interventions despite lack of clear evidence for the efficacy of any particular therapy. The purpose of this study was to examine the effect of nutrition on episodes of lower urinary tract signs (LUTS) in cats with FIC.

The year-long, randomized, double-masked and controlled study examined 2 foods for long-term management of acute FIC. Test and Control foods differed in anti-oxidant and fatty acid profile, but not in protein, fat, carbohydrate or caloric density. Foods met minimum AAFCO requirements for adult cats and protocols were ACUC approved. Owners were given the choice of wet or dry foods based on cat preference and then cats were randomized into Test or Control groups. Clinical signs of periuria, stranguria, hematuria and pollakiuria were documented daily by the owner for a period of one year. Episodes of LUTS were defined as an initial day with at least 2 clinical signs and all subsequent days exhibiting signs with no more than 24 hours free of signs. After 48 hours without signs, that episode was considered resolved.

Thirty-one cats ranging in age from 1-9 years were enrolled; 9 (4m, 5f) Test dry and 6 (3m, 3f) Test wet and 11 (9m, 2f) Control dry and 5 (4m, 1f) Control wet. Four Test and 2 Control food cats had records that were determined to be unreliable. The remaining 11 (5m, 6f) Test and 14 (11m, 3f) Control cats were included in the analysis. In total 3,904 cat-days were analyzed for the Test group and 4,215 for the Control group. Data were analyzed as a binomial proportion of the number of days that an event occurred or the number of episodes of LUTS out of the total number of days a cat was in the study for a factorial arrangement of 2 diets and 2 formulations.

There was no statistical difference between wet and dry formulations on LUTS. Cats consuming the Test food had a significantly lower proportion of days with ≥ 2 clinical signs ($p < 0.05$) with 4/11 (36%) Test cats and 9/14 (64%) Control cats exhibiting ≥ 2 clinical signs during the study. At least 2 clinical signs were observed on any particular day 13 times in the Test group and 152 times in the Control group. Test food fed cats also had significantly fewer episodes of LUTS ($p < 0.05$). The number of episodes of LUTS was 5/3904 (1.28/1000) cat-days in the Test group and 43/4125 (10.28/1000) cat-days in the Control group.

This is the first study to definitively show that foods of different nutritional profiles impact the expression of FIC in cats.

ENDOSCOPIC NEPHROLITHOTOMY FOR COMPLICATED NEPHROLITHIASIS IN 9 DOGS AND 1 CAT. A Berent[1], E Branter[1]_C Weisse[1], D Bagley[2] 1. Animal Medical Center, NY, NY. 2. Thomas Jefferson University, Philadelphia, PA.

Percutaneous nephrolithotomy (PCNL) is considered the standard -of-care for removal of nephroliths >1.5 cm in people, minimizing morbidity and preserving renal function. Success rates are reported to be 90-100%. Most veterinary nephroliths remain clinically silent and removal is only recommended for complicated stones. Morbidity of nephrotomy can be severe.

To describe endoscopic-guided nephrolithotomy (ENL) in canine and feline patients and report clinical outcomes, hypothesizing it is safe and effective.

Nine dogs and 1 cat (12 renal units)

Patients that had either PCNL or surgically-assisted endoscopic nephrolithotomy (SENL) were retrospectively evaluated. A renal puncture needle and balloon-dilation-sheath combination was used for tract formation. A nephroscope provided visualization for intracorporeal lithotripsy (Ultrasound, Hol:YAG, or Electrohydraulic). Stone fragments were removed and a ureteral stent was placed.

Four had PCNL and 8 SENL. Indications included recurrent UTIs (4), worsening azotemia (4), and ureteral-outflow obstructions (2). Median weight was 8.2 kg (3.1-26.9). Stone composition was calcium oxalate (6), mixed struvite-calcium phosphate (3), urate (1), and cystine (2). Median stone size was 2 cm (0.7-5). Median pre- and 3 month post-operative creatinine was 1.3 (0.8-9.1) and 1.1 mg/dL (0.6-6.1), respectively. The median procedure time was 165 minutes. Successful removal of all stones were documented in 11/12 (92%). Procedure-related complications occurred in 3 units, all were easily managed. Median follow-up time was 360 days (4-2220). Four patients are still alive. No patient died from the ENL procedure.

ENL can be safely performed in dogs and cats, yielding similar success rates to people. Advanced endourologic experience is recommended.

DEVELOPMENT OF CUT-OFF EXO-IOHEXOL, CREATININE, ENDO-IOHEXOL CONCENTRATIONS AT THREE TIME POINTS AFTER INTRAVENOUS IOHEXOL AND CREATININE INJECTION TO PREDICT LOW GLOMERULAR FILTRATION RATE IN CATS. D.Paepe[1], H.P.Lefebvre[2], D.Concordet[2], I.vanHoek[1], S.Croubels[1], S.Daminet[1]. 1.Ghent University, Ghent, Belgium. 2. National Veterinary School of Toulouse, Toulouse, France.

We aimed to determine cut-off clearance marker concentrations to predict low glomerular filtration rate (GFR) in cats at three time points after intravenous marker injection.

Combined plasma exogenous creatinine-iohexol clearance tests of 73 cats (16 healthy, 19 diabetic, 16 hyperthyroid, 20 chronic kidney disease, 2 doubtful renal status) were used. Sensitivities, specificities and odds ratio's (OR) to predict low GFR (<1.7mL/min/kg for exo-iohexol; <1.9mL/min/kg for endo-iohexol or creatinine clearance) were calculated for exo-iohexol, creatinine, and endo-iohexol concentrations at three time points after marker injection. Cut-off values were identified.

Exo-iohexol concentrations ≥143µg/mL, ≥84µg/mL, and ≥50µg/mL 60, 120, and 180 minutes after marker injection predict low GFR with sensitivities, specificities and OR's of 93.3%, 83.7%, 72; 96.7%, 93%, 386.7; and 96.7%, 95.1%, 565.5 respectively. Creatinine concentrations ≥505µmol/L, ≥320µmol/L, and ≥250µmol/L 60, 120, and 180 minutes after marker injection predict low GFR with sensitivities, specificities and OR's of 90%, 81%, 38.3; 93.3%, 83.7%, 72; and 96.7%, 90.7%, 282.8. Endo-iohexol concentrations ≥27µg/mL, ≥15µg/mL, and ≥10µg/mL 60, 120, and 180 minutes after marker injection predict low GFR with sensitivities, specificities and OR's of 88%, 87.5%, 51.3; 96%, 91.7%, 264; and 92, 95.7%, 253.

In conclusion, cut-off marker concentrations 60, 120, and 180 minutes after iohexol and/or creatinine injection reliably predict low feline GFR.

STEM CELL THERAPY BLUNTS COMPUTED TOMOGRAPHIC MEASURES OF AIRWAY REMODELING IN EXPERIMENTAL FELINE ASTHMA. I Masseau [1], J. Trzil[1,2], CH Chang[1], T Webb[3], J Dodam[1,2], H Liu[2], S Dow[3], C Reinero[1,2]. [1]Department of Veterinary Medicine and Surgery and [2]Comparative Internal Medicine Laboratory, College of Veterinary Medicine, University of Missouri, Columbia MO. [3]Center for Immune and Regenerative Medicine, Colorado State University, Fort Collins, CO.

Feline asthma is a chronic respiratory disorder characterized by airway inflammation, hyperresponsiveness and remodeling. Glucocorticoids and bronchodilators address the first two features, but to date, there is no effective treatment to blunt airway remodeling. Stem cells have emerged as a novel therapy for asthma. Studies on longitudinal effects of repeated infusions have yet to be performed in the cat. We hypothesized that intravenous infusion of allogeneic mesenchymal stem cells (MSCs) given after experimental asthma induction in cats would dampen airway remodeling evaluated by computed tomography (CT) nine months later.

Asthma was induced in 10 cats using Bermuda grass allergen (BGA) at the University of Missouri (MU); 6 research cats were used as controls. Asthmatic cats received BGA aerosol challenges weekly to bimonthly throughout the study. Four asthmatic cats were treated with 5 i.v. infusions of allogeneic MSCs (days 0, 14, 28, 82 and 114). The MSC infusions ranged from 2X10E6-10E7 MSCs. A Toshiba Aquilion 64 multi-slice CT scanner was used to obtain images in anesthetized cats 9 months after the first MSC infusion. Breathholds were used to obtain inspiratory and expiratory CT scans after nebulization of saline followed by three doses of methacholine (MCh 0.625, 2, and 32 mg/mL). The study terminated at the dose of MCh doubling baseline resistance (EC200Raw). Lung attenuation in Hounsfield units (HU) was measured in the dorsal lung lobes at baseline and EC200Raw on three consecutive slices taken at the level of the caudal and accessory lung lobes. A two way ANOVA on repeated measures was used with $p<0.05$ considered significant.

On inspiration, untreated asthmatic cats had significantly increased lung attenuation (-816±23 HU) compared to MSC-treated (-923±25 HU) and healthy cats (-923±21 HU; p=0.01). On exhalation at the EC200Raw, untreated asthmatic cats had increased attenuation (-740±24 HU) compared to saline baseline (-775±24 HU; p=0.001) while in healthy cats, lung attenuation was similar to baseline (-867±22 HU vs -872±22 HU; p=0.42). MSC-treated cats also had increased lung attenuation at EC200Raw (-825±27 HU) compared to baseline (-843±27 HU; p=0.02). At the EC200Raw between groups, there was a significantly increased attenuation in untreated asthmatic cats compared to healthy cats (p=0.03); while MSC-treated cats were intermediate with attenuation not as high as untreated asthmatic cats.

In conclusion, MSCs appear to attenuate chronic remodeling changes based on CT imaging in experimentally asthmatic cats.

TRANSMISSION OF *BARTONELLA HENSELAE* TO PUPPIES BY EXPOSURE TO *CTENOCEPHALIDES FELIS* ALLOWED TO FEED ON *BARTONELLA HENSELAE*-INFECTED CATS. MR Lappin,[1] N Balakrishnan,[2] J. Bradley,[2] E Breitschwerdt.[2] 1. Department of Clinical Sciences, Colorado State University, Fort Collins, CO. 2. Department of Clinical Sciences, College of Veterinary Medicine, North Carolina State University, Raleigh, NC.

Bartonella henselae is transmitted amongst cats by *Ctenocephalides felis*. Client-owned dogs have been shown to be infected by *B. henselae* but the role of *C. felis* in transmission of this bacterium to dogs has not been evaluated. The purpose of this study was to determine if exposure of puppies to *C. felis* that fed on *B. henselae*-infected cats results in infection.

In Experiment 1, specific pathogen free, mixed sex kittens (n = 4) and beagle puppies (n = 16) were housed in three adjoining enclosures that were separated by mesh to allow *C. felis* to pass among groups, yet prevent the animals from contacting one another. The 4 cats were inoculated intravenously with *B. henselae* and after infection was confirmed based on positive polymerase chain reaction (PCR) assay results, the cats were housed in the middle enclosure flanked by the 2 groups of 8 puppies. *Ctenocephalides felis* (50 males and 50 females) that were raised in an insectary were placed on each of the 4 cats monthly for 13 applications. Blood was collected from all animals weekly for *Bartonella* spp. 16- 23S ITS region PCR, serology, and culture. In Experiment 2, 4 puppies that were negative for *B. henselae* antibodies and negative by both culture and PCR throughout Experiment 1 were inoculated once intranasally with 2 doses (2 puppies each dose) of *B. henselae* infected flea frass collected from *C. felis* that were allowed to feed on bacteremic cats.

In Experiment 1, 10 of 16 puppies were positive for *B. henselae* antibodies by IFA between weeks 6 and 14; all puppies were seronegative for the remainder of the 26 week study. *Bartonella henselae* DNA was amplified from EDTA blood from 5 puppies between weeks 20 and 26 after enrichment in *Bartonella* alpha proteobacteria growth medium (BAPGM); each of the puppies were seronegative at that time. Fleas or flea frass were only noted on 4 of the 10 puppies with serological evidence of exposure to *B. henselae*. In Experiment 2, one of four puppies inoculated intranasally with *B. henselae* infected flea frass developed detectable bacteremia.

The results suggest that dogs can be infected with *B. henselae* after exposure to *C. felis* that have fed on *B. henselae*-infected cats, dogs can be infected by nasal exposure to infected flea frass, and serology, culture, and PCR assay results do not always correlate as previously reported for naturally-infected dogs.

A LIPL32 BASED ELISA FOR THE DETECTION OF *LEPTOSPIRA* SPECIFIC ANTIBODY IN CANINE SERUM. RE Goldstein[1], KM Curtis[2], SE Winzelberg[1], SM Tasse[2], BA Stillman[2], MJ Beall[2], AP Potter[2], MR Lappin[3], R Chandrashekar[2]. 1. Animal Medical Center, New York, NY. 2. IDEXX Laboratories, Westbrook, ME. 3. Colorado State University, Fort Collins, CO.

The purpose of this study was to evaluate the performance of a non-serovar specific ELISA for the detection of pathogenic *Leptospira* antibody in canine serum. The target used in the assay is LipL32, an abundant, highly conserved outer membrane associated protein found only in pathogenic *Leptospira* serovars. LipL32 was expressed from a recombinant plasmid in *E. coli* and the purified LipL32 was used in both the solid and liquid reagents of a direct format ELISA. Enzyme-labeled LipL32 was mixed with the serum sample and added to the solid phase of the assay. Following incubation and washing, results were obtained with the addition of appropriate substrate. The reference method used in this study was the microscopic agglutination test (MAT).

Testing was performed on six distinct populations of dogs. The first population was comprised of dogs experimentally infected with *L. interrogans* serovar Pomona and tested weekly. On day 7, 4/4 dogs were positive on both ELISA and MAT. The second population consisted of samples (n=2316) submitted for MAT testing at the IDEXX Reference Labs. Of the 678 MAT positive samples tested, 63.3% (257/406) of samples at $\leq$ 1:400 were ELISA positive and 82.0% (223/272) of samples at $\geq$ 1:800 were ELISA positive. Of 1638 MAT negative samples, 76.6% (1255/1638) were also negative by ELISA. The third population consisted of samples from 195 MAT negative, healthy dogs with unknown travel history from a non-endemic area (Alaska) of which 93.8% (183/195) were ELISA negative. The fourth population was 30 Lyme positive (Lyme Quant $C_6^{®}$), MAT negative dogs, none of which were positive by ELISA. The fifth population consisted of 22 dogs vaccinated with 1 of 4 commercially available four-serovar vaccines and followed for 1 year. Nineteen (86.4%) were positive by ELISA and all (100%) were positive by MAT on at least one time point during the year. Finally, of 20 dogs with a clinical diagnosis of leptospirosis, 17 tested positive by ELISA. All three ELISA negative dogs were also MAT negative at the time of sampling.

Results from this study demonstrate the feasibility of the LipL32-based ELISA for detection of *Leptospira* specific antibody in canine serum, with comparable performance to the MAT reference method. Additional studies using well characterized clinical samples are needed to further assess the performance of this assay.

COGNITIVE BIASES IN MEDICAL DECISION MAKING

Jo R. Smith, MA, VetMB, PhD, DACVIM (SAIM)
Athens, GA

INTRODUCTION

Public awareness of medical errors was raised after publication of the US Institute of Medicine 1999 report "To Err is Human". It was estimated that the annual death toll of hospitalized patients resulting from medical errors was 45,000 to 98,000. Thus, medical error was the eighth leading cause of mortality that year and surpassed deaths from motor vehicle accidents. More recent reports have suggested medical errors may occur in 10-25% hospitalized cases.[1,2] The frequency of veterinary medical errors is unknown. A survey of recent UK veterinary graduates showed that 78% of them had made a mistake "resulting in a less than optimal or potentially adverse outcome for a patient" within the last 15 months.[3]

TYPES OF MEDICAL ERRORS

No fault errors

These include situations where the patient is either asymptomatic, has atypical or inconsistent symptoms; the patient is uncooperative and provides a confusing history; or due to current limitations of medical knowledge; or failure of the normative process – no definitive diagnosis is reached and choice of the most common differential diagnosis is incorrect.[4] Advances in medical knowledge and testing may reduce "no fault" errors, but it is unlikely these types of errors can ever be fully eliminated.

System errors

These include technical failures (faulty equipment or lack of appropriate testing facilities) and organizational failures.[4] The latter encompass failure to follow-up with patients; non-adherent patients; failure to adequately orient, train or supervise individuals; failure to co-ordinate care; or unavailability of required expertise. Morbidity and Mortality rounds aim to reduce the recurrence of these errors. Organizational structure, culture, policies, communication and performance distractors (e.g. excessive workload) are all scrutinized in an attempt to reduce system errors. It is unlikely that these errors will ever be eliminated because system improvements may introduce new problems or alternatively, degrade over time. System repairs also inevitably lag behind system changes.

Cognitive errors

These errors encompass inadequate or faulty knowledge, data gathering, information processing or metacognition. Cognitive errors result in greater morbidity than any other type of medical error and reports suggest that they may contribute to 75-96 % of all medical errors.[4,5,6]

So why do cognitive errors occur? One theory behind decision making proposes a duel model of reasoning.[7,8] System 1 is intuitive. It uses heuristics or short cuts and depends on inductive logic. Heuristics are experience-based techniques for problem solving. These include the rule of thumb, first impression, educated guess and pattern recognition, and are thus highly context bound. This type of thought process is unconscious and is a tradeoff between efficiency and accuracy. "*Our propensity for certain types of error is the price we pay for the brain's remarkable capacity to think and act intuitively*" – J. Reason, 1990. System 2 thinking involves critical, logical thought and normative, deductive and abstract reasoning. It leads to robust decision-making but requires conscious, deliberate and purposeful thinking. Consider the unconscious thought process of an experienced driver when driving a car (system 1) versus the deliberate thought process of learner driver (system 2). System 1 thinking appeals if you are tired since system 2 thinking requires effort and energy. The characteristics of the two types of cognitive operations are presented in **Table 1**.

Table 1: Characteristics of the two types of cognition in the dual model of reasoning.

Characteristic	**System 1**	**System 2**
Awareness	Low	High
Action	Reflexive, skilled	Deliberate, rule-based
Speed	Fast	Slow
Context importance	High	Low
Effort	Minimal	Considerable
Vulnerability to bias	**Yes**	Less
Reliability	**Low, variable**	High, consistent
Errors	**Common**	Few
Scientific rigor	Low	High

There are over 90 different cognitive biases.[9] The more common ones impacting medical diagnosis include[10]:

- *Premature closure* – counting your chickens before they hatch. This is the most common error that leads to missed diagnoses in internal medicine cases.[4] The decision-making process ends prematurely before the diagnosis is fully verified or "when the diagnosis is made the thinking stops".
- *Search satisfying* – Occam's razor or "keep it simple". The propensity to call off the diagnostic search once something is found. It is similar to *premature closure* and the risk is not finding concurrent disorders. The search may have been conducted in the wrong place and will be called off once there appears to be nothing to be found.
- *Anchoring* – jumping to conclusions. The fixation on a specific feature of the patient's presentation too early in the thought process. There is a tendency to ignore later information that would alter the earlier impression.
- *Confirmation bias* – only seeking data to confirm, rather than refute, an initial diagnosis. This tends to compound *anchoring* and can be impacted by *affective error*, the tendency to value evidence that we feel good about or want to see more highly, than information that we do not want.
- *Availability* – common things are common. Disorders that come easily to mind are more likely to be considered; recent experience with a disease will inflate its chances of being diagnosed.
- *Non-availability* – out of sight, out of mind. The tendency to underdiagnose a disease that has not been seen for a long time.
- *Base rate neglect* – chasing zebras. Ignoring the true prevalence of a disease and considering it more or less likely to occur. This reflects the difficulty of applying "general" statistics to an individual patient and tends to be influenced by experience. It is used to justify ruling out the worst case scenario or pursuing a rare or esoteric disease when it is not justified.
- *Representativeness* – if it walks like a duck, quacks like a duck ... The over reliance on pattern recognition and identifying a prototypical case of a disease. The actual frequency of the disease is disregarded and the quality of contrary evidence is ignored. This increases the risk of missing atypical variants of a disease, e.g. atypical hypoadrenocorticism or focal myasthenia gravis.
- *Overconfidence bias* – we think we know more than we do, especially experts! Tendency to act on incomplete information because of intuition or hunches. Can be re-enforced by a good outcome or a lucky guess. Risks are increased when coupled with *commission bias* – the tendency to intervene rather than not do so.

COGNTIVE BIASES AT KEY STAGES OF PATIENT MANAGEMENT

A recent study from the UK evaluated 21 human medical cases of misdiagnosis for likely biases in cognitive reasoning in temporal relation to case progression at a single teaching hospital.[11] The following information has been adapted to veterinary examples. The two most common cognitive biases for each key stage of patient management are listed first (there may be some duplication with the previous examples).

Initial assessment of a new medical condition (usually undertaken by a junior trainee)

- The clinician may accept the *framing* provided, e.g. the patient is triaged as a pancreatitis when the cause of vomiting may be atypical hypoadrenocorticism.
- This is considerably affected by *cognitive availability* – the clinician goes with what they know and often applies *Sutton's law* (choosing the obvious option).
- Tendency to rely on pattern recognition or *representative restraint*, e.g. wheezing is seen as bronchial asthma without considering less common causes.
- May be *judgmental – stereotyping*, e.g. a dog with a known portosystemic shunt has a seizure and is diagnosed with hepatic encephalopathy without excluding other causes. Signalment is an important issue in stereotyping.

Closing in on the diagnosis (steered by a more experienced trainee)

- *Confirmation bias* – the clinician seeks to confirm rather than consider.
- *Search satisfied with premature closure* – the clinician stops looking once a possibility appears to be supported by tests.
- *Vertical line failure* – the clinician has seen this pattern many times before and *thinks in silos* or in a predictable way.
- *Omission bias* – based on non-maleficence, the clinician may not push hard enough, and is comforting rather than investigating.
- *Value bias* – minimizing the serious ("the difficult") especially if *multiple alternatives* and evading the *unpacking principle,* i.e. detailed analysis.

Approving the diagnosis (undertaken by the attending clinician)

- The clinician may fail to put a brake on *diagnostic momentum*.
- Faced with the *order effect of a structured presentation* the clinician may just hear the opening statement and the conclusion and "accept".
- The authority effect or *blind obedience* may inhibit trainees from voicing a different opinion.
- The clinician may leave a case ambiguous and wait and see, rather than actively seeking an answer.
- The clinician may not consider *sunk costs* – the more clinicians invest in a particular diagnosis the less likely they are to consider alternatives. This may also reflect the pride of trainees in showing an ability to cope.
- The clinician may not seek to counteract *feedback restraint* for follow-up on previous episodes of clinical management. They may encourage *hindsight*; when knowing the outcome profoundly influences the perception of past events and leads to a false interpretation of the decision maker's abilities.

GENERAL AVOIDANCE STRATEGIES

General avoidance strategies[10,11,12]

- Develop an awareness of cognitive biases. They become easier to recognize; it is difficult to guard against what you do not know and so this permits enhanced metacognition.
- Decrease reliance on memory – use guidelines, clinical algorithms and mnemonics.
- Make access to information easier to decrease the tendency towards system 1 thinking when tired, stressed or busy.
- Encourage the change from intuitive (system 1) to analytical (system 2) thinking by tabulating clinical problems in order of perceived importance, together with actions to be taken. This may decrease over-reliance on the simple perceptive approach which may forestall analysis.
- Use second opinions and/or consult with specialists.
- Undergo specific training - consider a refresher course on epidemiology and read epidemiologic reviews, e.g. JAVMA.
- Minimize time pressures – quality decision-making requires adequate time.
- Establish accountability – hold people accountable for their actions and errors; ensure follow up is performed to discover any errors.
- Give feedback – this should be timely, detailed and specific. This allows clinicians to re-calibrate their decision-making process.

USING CHECKLISTS TO REDUCE DIAGNOSTIC ERRORS

General[13]

- Get a <u>complete</u> history.
- Perform a comprehensive and <u>purposeful</u> exam.
- Use a systematic approach when generating a list of differential diagnoses.
- Take time to pause and reflect (metacognition). Answer these questions (1) Was I comprehensive? (2) What else could it be? (3) What <u>doesn't</u> fit with my working diagnosis? (4) Is there more than one disease? (5) Was my judgment affected by any other bias? (6) Do I need to make the diagnosis now, or can I wait? (7) What is the worst case scenario?
- Execute the plan, but acknowledge uncertainty and ensure follow up. Determining a diagnosis is a longitudinal process, clinical signs develop and clues emerge over time. However, you should define a follow-up plan and then <u>stick to it</u>. Closing the feedback loop increases the correlation between the clinician's perceived and actual diagnostic accuracy.

Differential diagnosis checklist

Truly practice problem-oriented medicine and do not rely on pattern-recognition. Consult problem-specific lists of differential diagnoses; there are multiple texts and resources available. Comprehensive lists ensure relevant and/or rare diseases are considered and guards against availability bias.

Cognitive forcing checklist

These are focused on a specific problem or disease and promote forced consideration of alternate possibilities, e.g. "must do" special exams or tests. These outline important issues that may have been overlooked or missed altogether thus minimizing errors of omission and ensuring the worst case scenario is ruled out.

Summary

To reiterate: pause and reflect – the most common reason for diagnostic error is failure to consider the correct diagnosis. Ask standard questions – what else could it be, what doesn't fit, what is the worst case scenario? Use checklists.

REFERENCES

[1]Hogan H, et al. **BMJ Qual Saf** 2012:21(9):737. [2]Landrigan C, et al. **N Engl J Med** 2010;363(22):2124. [3]Mellanby R, et al. **Vet Rec** 2004;155(24):761. [4]Garber M, et al. **Arch Intern Med** 2005;165:1493. [5]Shojania K, et al. **JAMA** 2003;165:1493. [6]Zwaan L et al. **Arch Intern Med** 2010;170:1015. [7]Croskerry P. **Adv in Health Sci Educ** 2009;14:27. [8]Norman G. **Adv in Health Sci Educ** 2009;14:37. [9]http://en.wikipedia.org/wiki/Cognitive_bias. [10]Croskerry P. **Acad Med** 2003;78(8):775. [11]Neale G, et al. **Clin Med** 2011;11(4):317. [12]Garber M, et al. **BMJ Qual Saf** 2012;21:535. [13]Ely J, et al. **Acad Med** 2011;86(3):307.

THE POWER OF AN APOLOGY: DISCLOSING MEDICAL ERRORS*

Jo R. Smith, MA, VetMB, PhD, DACVIM (SAIM)
Athens, GA

"Like any other procedure, communication in the form of disclosure can be learned."[1]

An adverse event or outcome describes an unanticipated harm that results from the medical treatment rather than from the disease itself. It may result from unreasonable client expectations or biologic variability (e.g. idiosyncratic drug reaction) or system, procedural or individual errors. Clinical judgments that are wrong and yet within the "standard of care" (that which would be offered by similarly trained veterinarians) are also considered adverse outcomes. It is estimated approximately one third of adverse outcomes involve a medical error.

Types of medical errors

A veterinary error is a failure to complete a plan as intended or use of the wrong plan to achieve a clinical result. The legal definition is that peers with similar training and practicing the standard of care would have recognized the error at the time it occurred. The frequency of veterinary medical errors is unknown. A survey of recent UK veterinary graduates showed that 78% of them had made a mistake "resulting in a less than optimal or potentially adverse outcome for a patient" within the last 15 months.[2]

No fault errors include situations where the patient is either asymptomatic, has atypical or inconsistent symptoms; the patient has a confusing history; or due to current limitations of medical knowledge; or failure of the normative process – no definitive diagnosis is reached and choice of the most common differential diagnosis is incorrect.[3]

System errors include technical failures (faulty equipment or lack of appropriate testing facilities) and organizational failures.[3] The latter encompass failure to follow-up with patients; non-adherent patients; failure to adequately orient, train or supervise individuals; failure to co-ordinate care; or unavailability of required expertise.

Cognitive errors encompass inadequate or faulty knowledge, data gathering, information processing or metacognition. Cognitive errors result in greater morbidity than any other type of medical error and reports suggest that they may contribute to 75% to 96% of all human medical errors.[3,4,5] A retrospective study comparing veterinary clinical diagnoses with postmortem diagnoses found a discrepancy rate of ~15%.[6] Similar work in humans has suggested a median error rate of ~ 25% (range 4% to 50%).[7]

Human patients want to be informed when a medical error occurs. By extension, it is not unreasonable to assume that most veterinary clients expect an open disclosure policy from us. One study reported that 98% of patients wanted to know even if the error was minor; the more severe the occurrence, the information was desired.[8] Another study showed that 92% of patients, but only 60% of physicians, believed that patients should always be informed of complications.[9] This disconnect could be misinterpreted by patients as a lack of care or concern.

Disclosure of human medical errors

It is worth considering historical perspective of disclosure by physicians in human medicine, since it informs our clients' expectations of veterinarians. Previously, the medical profession operated under a culture of infallibility. Public awareness of medical errors was raised after publication of the US Institute of Medicine 1999 report (To Err is Human). It proved a watershed in patient safety. In response to the publicity generated, professional organizations, consumer advocacy groups, medical ethicists, and malpractice insurance carriers combined in a national movement to encourage physicians and healthcare organizations to be more open about adverse outcomes and medical errors.[10] The move towards a culture of patient safety involved a move away from apportioning blame to individuals and towards searching and trying to resolve system errors.

The Joint Commission issued the first nationwide disclosure standard in 2001 (Healthcare at the Crossroads: Strategies for Improving Medical Liability System and Preventing Patient Injury). It advocated the disclosure of adverse events, although not whether an error had occurred. Reporting of adverse outcomes then became mandatory because it was linked to accreditation. In 2006 a National Quality Forum evidence-based report (Safe Practices for Better Healthcare) recommended that disclosure of adverse events include whether an error had occurred and apologizing to the patient. It proposed that institutions integrate disclosure into patient safety and risk management strategies and establish a support system for physicians including education and coaching. The same year the Harvard Hospitals released a consensus statement (When Things Go Wrong: Responding to Adverse Events). This was a comprehensive review of the current leading practices in communication after an adverse event, and also proposed education, training and support in the disclosure of medical errors. Since then state governments have also mandated that institutions disclose serious unanticipated outcomes to patients. To counteract concerns about litigation exposure, "apology laws" prohibit the use of such communications, typically apologies or others expressions of regret, as evidence of liability for the disclosed event.[11]

The ethical imperative for disclosing medical errors

Disclosure of medical errors has been identified as an ethical obligation in human medicine by the American Medical Association, the American College of Physicians and the National Patient Safety Foundation. It fulfills the ethical principle of autonomy; the client has a right to receive all information necessary to make medical decisions and thus should be fully informed. The principles of beneficence and justice are also met by disclosure, whereas failure to disclose contravenes the principle of non-maleficence, it may lead to further harm. The veterinary profession also acknowledges the ethical imperative for disclosure since we also operate under the same ethical principles; however, it is on an individual basis that we fail. In the study examining the incidence of veterinary medical errors in recent UK graduates, only 60% of these errors were disclosed to the client.[2]

Barriers to disclosing medical errors

There are many perceived barriers that must be overcome to achieve transparency. Historically many healthcare providers and organizations have been reluctant to provide full disclosure because of the fear of increased litigation. However, poor communication is a bigger risk. Research into the effect of disclosure on litigation of physicians suggest that it may actually reduce the likelihood of being sued.[12,13,14]

The Veterans Affairs Hospital in Lexington, Kentucky published a report in 1999 that showed an open-disclosure program did not cause a significant change in the volume of claims or the size of payouts.[15] The University of Michigan showed that the cost and frequency of litigation decreased significantly in the five years after implementation of their open-disclosure program. Annual litigation expenses reduced from $3 million to $1 million, and the number of claims decreased by 50%.[16,17]. The best known liability insurer disclosure program is the "3Rs" program operated since 2000 by COPIC, the largest insurer for physicians in Colorado. It is a non-fault based system designed to resolve medical errors less adversarially than by traditional litigation. Other liability insurers have also adopted similar initiatives and, in addition to hospitals, now offer workshops on how to improve communication skills for when an adverse outcome occurs.[18]

The AVMA Professional Liability Insurance Trust (PLIT) also supports open-disclosure. They maintain more situations are worked out directly between the veterinarian, the client and insurance carrier, without stimulating legal action of formal complaints to licensing boards.[19] Rebuilding the rapport and trust and resolving disagreements after an adverse outcome may turn initial client disappointment into a stronger relationship than before the event. Physicians who do not disclose may use the argument of attempting to protect the patient and preserve trust (therapeutic privilege), or not wanting to add to the patient's emotional distress.[20] However, this is usually perceived as self-serving; physicians are actually putting their own interests above that of the patient.[21,22] Research supports that a lack of disclosure threatens the inherent trust in the physician–patient relationship and can lead to alienation.[8] When clinicians fail to disclose and resolve an adverse outcome properly, patients have reported two types of disappointment. Firstly, that the medical error occurred and secondly, in the way the clinicians behaved after the event. Research suggests that individuals are more forgiving of the medical error than the lack of disclosure.[13,15,23] The demonstration that an open-disclosure policy can reduce litigation costs reinforces the view that maximum compensation is frequently not the main objective for patients in the wake of a medical error.[24] When the hospital and the insurance carrier initiate the discussion of fair settlements with patients who have legitimately been affected by medical errors, the dollar amounts tend to be easier to negotiate and more reasonable than those obtained through legal action. Patient bitterness against perceived duplicity is minimized and dollars amounts are focused on reasonable compensation rather than punishment.

Veterinarians may also be reluctant to disclose a medical error because of shame or the fear of damage to their reputation within the eyes of colleagues and/or their community. The healthcare professional has been termed "the second victim" because of their vulnerability at the sharp end of the medical error.[25] It is now acknowledged that the healthcare provider implicated in a medical error should receive assistance and strategies have been developed to support medical healthcare professionals.[25,26] Admitting a mistake is certainly not easy at a time when you may feel self-doubt, anxiety or guilt.[27] It requires courage, composure and excellent communication skills.[10] However, in addition to helping re-establish trust in the veterinary-patient-client relationship, disclosing the error can help to preserve the individual's professional integrity and result in increased assertiveness, confidence and improved working relationships.[27] Another potential reason for lack of open-disclosure is because of a lack of training. New initiatives in communication training within the veterinary curriculum and in continuing education specifically include instruction in the area of disclosure. These proceedings are derived from one such educational module.[28]

In summary, the benefits of an open disclosure policy include: (1) fulfilling the medical ethical imperative and preserving professional integrity; (2) rebuilding trust; (3) helping retain clients; (4) assisting negotiation of fair settlements and (5) reducing malpractice suits and state licensing board complaints.

Approach to the disclosure of medical errors[10,19]

Research has demonstrated that disclosure of a medical error should contain the following: (1) what happened; (2) how did it occur; (3) what happens next for my pet; (4) an apology; (5) an explanation of how recurrence will be prevented; and (5) an offer of reparation.[8,29,30] The following will outline a generalized step-by-step process of disclosure in a situation where the medical error is recognized at the time it occurs and the client is available for a face-to-face meeting.

Before the disclosure

The first obvious step is to tend to the patient's immediate clinical care. If this involves consultation or referral, these services are typically not billed. Then take time to compose yourself; your emotions are likely to be heightened on this occasion. Recognize your own emotions and needs; you will need to manage your reactions to proceed effectively. You must be emotionally stable enough to communicate effectively with the client; otherwise it may be difficult to focus on the needs of the client and patient. You may fail to show compassion if you are worried about repercussions, or fail to provide adequate and direct answers in a timely manner if you are guilt-ridden. Talking to a trusted colleague may clarify your thinking and help put things into perspective. Next, investigate the details of the event and what could be done to prevent a recurrence in the future; the client is entitled to the most accurate understanding of what happened. It may take some time to clarify the details; however, so some questions may be left unanswered initially. Once you have developed some understanding of the sequence of events then it is time to call your liability provider. Discuss the legal and financial implications and if it will be possible waive relevant charges.

Next establish who should be present when making the disclosure. An empathetic, non-defensive demeanor can be as important as the actual words that are used in re-building rapport after an adverse outcome.[31] Therefore it is imperative that the key liaison has mastery of the four core communication skills: empathy, non-verbal cues, reflective listening and open-ended questions. Other factors which influence the choice of spokesperson include who has the most information and who has the best relationship with the client. Typically the veterinarian who has primary responsibility for the patient should take the lead; even if the adverse event was primarily caused by another staff member's actions. It is recommended to have a second person from the hospital sit in on the meeting. When preparing to talk to the client, try to anticipate their reactions, what questions they might have, and what they will want to know including further treatments, costs and the prognosis. It may be helpful to rehearse the actual words and phrases you will use to ensure they address these issues adequately.

Disclosing the medical error

Schedule the discussion as soon as feasible after the event; ask the client if he or she would like to include additional family members or friends. Pick a quiet location and allocate sufficient time; ensure there will be no interruptions. Pay attention to your non-verbal cues, your body language, and tone of voice, etc. Sit at the client's level, but not behind a desk; make good eye contact. Open the conversation with a warning and expression of sympathy, *"I have some difficult news to share with you. I am very sorry to have to tell you ..."* Ask permission to continue, *"Would it help if I explained what we believed happened?"* Explanations are best received by those who are ready to hear them. The discussion may be framed towards mutual understanding rather than sounding defensive by *asking* if an explanation would help. When you explain what occurred be honest, any attempt to spin the facts undermines trust. If you are unsure, be prepared to explain why these details are not yet known and what is being done to clarify the situation. Your stress level may decline as the conversation progresses; however, your client's will invariably rise. Avoid controversy by making empathetic statements; you can attempt to keep it conversational by asking the client for their reaction to the information, *"I imagine you have a number of questions."* Empathize and normalize their response, *"It is natural to feel shocked and angry ..."* Empathy is essential to re-establish rapport while normalizing shows acceptance, rather than dismissal, of the client's reaction. Listen and reflect, *"I imagine you are feeling so many things right now"*. Defensive comments will only heighten client frustration.

Then apologize appropriately. The proper type of apology can have a powerful effect on the client, making him or her less angry or suspicious. When a medical error has occurred, you should issue an apology of responsibility, *"I am so sorry we did this to you and [pet's name]."* If an adverse outcome has occurred without a medical error, then an apology of sympathy is appropriate, *Í am so sorry this has happened to you and [pet's name]."* Do not confuse the two; an apology of sympathy issued when an apology of responsibility is due may sound insincere and half-hearted, and may decrease trust.[32] Be accountable, do not blame others. Next, share your plan of care for the animal. Patient management is familiar territory for veterinarians; however, this is the first time the client has heard their pet's future discussed, so take your time. Clients may not yet be able to comprehend the gravity (or even the limited impact) of the error so include both the anticipated potential short- and long-term outcomes of their pet.

Discuss methods of reparation. The possibility of waiving fees related to this event should have been previously discussed with your liability provider. Although it's an understandable concern, you are not buying clients off; paying for errors is the ethical thing to do[33] Invite and answer all questions as honestly as possible. Remember to ask, *"Is there more we should be doing to resolve this with you?"* Above all else <u>listen</u> to the client.

Most clients will also want to know that some good will come of their experience. It is unacceptable to them that failure reflect on the incident means others may suffer a similar fate. Explain what changes you will make to reduce the likelihood of recurrence in the future. Convince them that you have learnt everything that can be learnt from this. Use clear, non-defensive, simple language to explain any technical aspects. Manage to resolution; full delivery of disclosure may require several conversations. Therefore finish your initial meeting with a plan for the next contact. You will need to document any communication (conversations, telephone calls, emails etc.) accurately in the medical record. Disclosure is all about being credible, honest, pro-active and open and so your documentation should reflect your actions.

In summary, when a medical error occurs: (1) care for the patient; (2) compose yourself and investigate details of the event; (3) contact your liability provider; (4) disclose to the client what occurred and apologize appropriately; (5) discuss with the client the plan of care for the patient; (6) be accountable and discuss methods of reparation; (7) share how you plan to keep this from happening in the future;(8) ensure documentation is accurate in the medical record.

A proactive stance is to establish a practice policy on disclosure. Points of discussion include: (1) how will your practice handle an error; (2) who will discuss it with the client; (3) who will be present during the discussion; (4) will the individuals involved be identified; (5) what time frame is acceptable; (6) where is the contact information for our liability provider; (7) when and how will we discuss this with staff members.

Despite our best efforts, medical errors are inevitable. We all know the adage, *"It's not if, but when, a medical error happens"*. We hope these guidelines provide you with a strategy for how to effectively disclose a veterinary medical error, for the benefit of the patient, client and veterinarian alike, and to facilitate restoring client trust in the aftermath of such event.

REFERENCES

[1]Weiss P, et al. **Obstet Gynecol Clin N Am** 2008 35:53. [2]Mellanby R, et al. **Vet Rec** 2004;155(24):761. [3]Garber M, et al. **Arch Intern Med** 2005;165:1493. [4]Shojania K, et al. **JAMA** 2003;165:1493. [5]Zwaan L et al. **Arch Intern Med** 2010;170:1015. [6]Dank G, et al. **J Small Animal Prac** 2012;53:679. [7]Shojania K, et al. **JAMA** 2003;2889:2849. [8]Witman A, et al. **Arch Intern Med** 1996;156:2565. [9]Hingorani M, et al. **BMJ** 1999;318:640. [10]Boyle D, et al. **Crit Care Med** 2006;34(5):1532. [11]Gallagher T, et al. **N Engl J Med** 2007;356(26):2713. [12]Hickson G, et al. **JAMA** 1992;267:1359. [13]Vincent d. et al. **Lancet** 1994;343:1609. [14]Levison W, et al. **JAMA** 1997;277:553. [15]Kraman S, et al. **Ann Intern Med** 1999;131:963. [16]Boothman R. **Physican Exec** 2006;32(7):7. [17]Clinton H, et al. **N EngJ Med** 2006;354:2205. [18]West Virginia Mutual Insurance C.A.R.E. skills development workshop: http://wvmic.com/riskmanagement/care-skills-workshop.php. [19]Bonvicini K, et al. **Compendium** 2009;31(3):105. [20]Matlow A, et al. **Pediatr Clin N Am** 2006;53:1091. [21]Herbert P, et al. **CMAJ** 2001;164:509. [22]Gallagher T et al. **Arch Int Med** 2005;165(16):1819. [23]Beckman H, et al. **Arch Intern Med** 1994;154:1365. [24]Bismark M, et al. **CMAJ** 2006;175:889. [25]Wu A. **BMJ** 2000;320:726. [26]Pratt S, et al. **Jt Comm J Qual Patient Saf** 2012;38(5):235. [27]Sirriyeh R, et al. **Qual Saf Health Care** 2010;19(6):e43. [28]The Bayer Animal Health Communication Project: https://www.veterinarycommunication.org/homepage.php. [29]Lazarre A. **JAMA** 2006;296(11):1401. [30]Blendon R, et al. **N Engl J Med** 2002;347(24):1933. [31]Larson E, et al. **JAMA** 2005;293:1100. [32]Mazor K, et al. **Arch Intern Med** 2004;164:1690. [33]Sorry Works! **Coalition Newsletter** 2006;Dec 4.

***This is a general overview of veterinary disclosure; please contact your insurer for specific advice for a particular case.**

NONVERBAL COMMUNICATION: TRICKS OF THE TRADE

Laura D. Garrett, DVM, DACVIM (Oncology)
Urbana, IL

INTRODUCTION

Communication is a critical part of the therapeutic success of a veterinarian's efforts, and it is an essential component of the veterinarian's role. It is the veterinarian's responsibility, and it cannot be delegated. Thinking of communication as a procedure will help to understand its importance and place emphasis on the fact that good communication CAN be learned. Like any procedure, mastery requires practice and experience. Too often the ability to talk to people effectively is considered an inherent skill – you either have it or you don't. However, many studies in the field of human medicine have shown that the skills involved in communication are able to be defined and, indeed, learned.[1-6] This talk will focus on nonverbal communication, one of the four core communication skills (nonverbals, empathy, open-ended questions, and reflective listening). Nonverbals can be broken down into 4 categories: kinesics (body position, gestures, touch, etc), proxemics (spacial relations, barriers), paralanguage (voice tone, speed, etc), and autonomic responses (flushing, eyes dilating, etc).

WHY WORRY ABOUT NONVERBAL COMMUNICATION?

Roughly 80% of all communication between individuals is nonverbal and generally unintentional.[7] Thus, information that cannot be hidden is being exchanged at all times – both from the client to the clinician, and vice versa. The nonverbal aspects of communication are like a poker player's "tell" – they give away what a client may be thinking or feeling, possibly contrary to what they are saying verbally. Thus, these unconscious nonverbal messages from a client more accurately reflect their feelings. Emotions are much more often communicated non-verbally than verbally. For example, we know when someone is mad because they raise the volume of their voice, have a different tone, change their posture, etc, not because they say "I'm so mad".

Most humans are outstanding at detecting emotion very quickly via observing nonverbal behavior. In one paper, people were found to be fairly accurate at identifying emotions via nonverbal cues in only 375 miliseconds.[8] A more recent study evaluating touch to communicate emotions found that anger, fear, disgust, love, gratitude, sympathy, happiness, and sadness were identified by an unacquainted partner based on touch alone (no visuals).[9] While this is not stated to encourage veterinarians to begin communicating with their clients by excessive touching, it highlights how powerful nonverbal communication is, how humans are programmed to interpret such communication rapidly, and how touch may play a role in such interactions.

Mixed messages occur when the verbal and nonverbal messages disagree. In these situations, the nonverbal message is more accurate and needs to be detected and addressed. A common experience is asking a client if they understand, and the client hesitantly says "yes", often without meeting the clinician's eyes. The nonverbals clearly say the client does NOT understand, and paying attention to that fact and trying to clarify things at that time will often save misunderstanding and upset in the future. A nonverbal expression of lack of understanding can be investigated in many ways, for example: "I have shared a lot of information with you, what are points that I can clarify further?" or "It sense some hesitation in your "yes" – I gave you a lot of information which can be overwhelming; what I can help you to understand?".

Another benefit to becoming aware of and following nonverbal clues is that they take up no extra time during an interaction. They are going on constantly during all interactions, and one needs to only consciously note them (and perhaps comment on them) to help guide the discussion in a beneficial manner. Pausing in an explanation if a client looks overwhelmed is an example of a nonverbal response to a nonverbal sign. Alternatively, one can verbally comment on a client's nonverbals; by doing this, empathy (another core skill) is shown. It is important that the client knows they are seen as a person when it comes to the medical issues, and that their emotions about these issues are noticed. A good way to show this by verbally stating what is seen: "you seem worried", "you look very scared". While these statements seem awkward or obvious at first, this technique is greatly appreciated by clients. Firstly, clients will appreciate the empathy and bond more with the clinician that shows it. Secondly, by bringing the emotion to the forefront, it can be dealt with and thus allow the client to feel it more and move on in the discussion about their pet. A statement as simple as 'I can see this is difficult for you; it can be very hard to hear that someone so close to you has cancer,' can be grounding for a client and may allow them to refocus on the medical conversation. Also, non-verbal means to show a client they are seen is to use appropriate speaking tone and rate (e.g. quiet slower speaking in times of sadness) or posture in response to the client. Handing a box of tissues to a crying client is a wonderful (and useful) means of showing empathy in response to their nonverbal actions.

PROOF THAT IT CHANGES OUTCOMES

Numerous studies show that in the physician–patient interaction, the ability of the physician to read the patient's nonverbal behavior impacts the relationship. Physicians who are good at this have patients who are more satisfied with their care, and are more likely to return for their next appointment (increased adherence).[10, 11] Another study showed that patient were more satisfied with physicians who spent less time looking at the medical chart, leaned forward, nodded and gestured more, used better eye contact and had closer interpersonal distance.[12] A more recent study had similar findings, with greater patient satisfaction when the doctor smiled more, had good eye contact, leaned forward, was expressive in tone and facially, and used increased gestures.[13] A meta-analysis of 26 studies assessing patients' and clinicians' nonverbal communication during real clinical interactions and clinically relevant outcomes found clinician warmth and listening were associated with greater patient satisfaction ($p<0.001$ both), while nurse negativity was associated with less patient satisfaction ($p<0.001$).[14] In a study of 452 families seeing 48 pediatricians in an asthma clinic, the doctors' communication behaviors, including careful listening and nonverbal attention, were related to positive perception of quality of care by the families and positive patient outcomes. Importantly, the techniques used did not increase the length of the visit.[15]

CLINICIAN'S APPEARANCE

The literature has many studies showing the effect of MDs' appearance on patients' (and others') perceptions. Several examples follow; while these studies do not involve veterinarians, many of the concepts likely cross over into the veterinary-client realm. In one study in the southeastern United States investigating physicians with facial piercings, nose and lip piercings were felt to be appropriate for a physician or medical student by only 24% and 22% of emergency room patients, respectively. Perceived competency and trustworthiness of doctors with these types of piercings were negatively affected.[16] Neckties are also a matter of ongoing debate. Several studies have shown that neckties, rarely if ever washed, are a source of potential pathogens.[17-19] Fortunately, the public attitude about whether doctors should wear ties is shifting, from 67% believing they should in 1991[20] to 76% against ties in 2008[21]. A prospective study of 316 discharged ER patients assessed the impression made by whether or not the ER doctor wore a tie. Almost 30% of patients erroneously remembered their doctor wearing a tie when no tie had been worn. No difference was found in the patients' impression of their physician or of the care they received between the tie and non-tie groups.[22] (men who hate ties, rejoice!)

Differences have been found between patients' expectations for and assessment of interactions based on whether the physician is male or female. One study looked at how specific physician verbal and nonverbal behaviors were related to perceived dominance by female and male physicians. Female physicians who spoke more, talked more while doing something else, spoke with louder voices, modulated their voices more, were oriented more toward the patients, sat at a smaller interpersonal distance, were more expansive, and had a more open arm position were perceived as more dominant, and this perception was significantly greater for female vs. male physicians. Thus, patients interpret nonverbal cues differently based on the gender of the doctor. For optimum interactions, doctors need to be aware of these patient expectations in modulating their nonverbal behavior.[23] A separate study looked to identify how specific physician nonverbal behaviors predicted patient satisfaction depending on physician and patient gender. No effect was found based on patient gender; however, the physician gender did reveal differences. Patients were most satisfied with female physicians who behaved consistent with the "female gender role", such as more eye contact, more leaning forward, and use of a softer voice, while still looking professional (wearing a lab coat, being in a medical-appearing exam room). Male physicians obtained higher satisfaction for a broad range of behaviors, including louder voice and more distance from the patient. Thus, expected nonverbal behavior of the doctor for client satisfaction may vary between the sexes for the physicians.[24]

DIFFICULT DISCUSSIONS

Nonverbal communication has been studied in relation to breaking bad news as well as disclosing medical errors. In a study evaluating bad news and neutral discussions with cancer patients, recordings were made and then modified to make the words unintelligible but maintain the paralanguage. All but one care provider slowed their speaking rate during bad news, and most decreased their pitch. Later, listeners perceived a significant difference between the nonverbal characteristics between the neutral vs bad news discussions, with speech of a reduced rate and lower pitch noted as more caring and sympathetic.[25] Remembering to control speech speed, tone, and volume during difficult discussions will nonverbally show the client you care. In regards to disclosing medical errors, a study involving 318 patients who watched videos of error discloser that varied in nonverbal competency found that with the same verbal message, nonverbal behaviors were significantly associated with higher patient ratings of closeness, trust, empathy, satisfaction, and forgiveness, and with lower ratings of emotional distress and avoidance. These associations were not affected by patient variables such as sex, ethnicity, religion and previous experiences with medical errors. Further research and training in nonverbal skills useful in disclosing medical errors are indicated.[26] Another study found that the surgeons' speech and tone of voice during a routine office visit recorded in just 40 second bites was significantly associated with malpractice claims history, with higher dominance and lower concern/anxiety expression identifying surgeons who had previous claims over those that had no claims.[27] Thus, it is not just what is said, but how it is said that can have significant effects.

MATCHING AND LEADING

Matching an individual's body language is a very powerful tool. We are social animals and desire to be amongst those that we trust and with whom we share commonalities, especially in difficult situations. Clients can become more comfortable, on a subconscious level, when clinicians mirror (without mimicking!) body language, posture, gestures, etc. For clients that are very tense, matching a body position (eg, client has hands clenched together resting on the table, the clinician places hands together also (but not clenched)) can provide comfort. Additionally, once matched and aligned, the clinician can then lead the client into a more relaxed (eg hands down at sides) posture. Matching has been shown to increase an individual's willingness to interact with and help out others, and may make a client more willing to become a partner-in-care with the veterinarian with no verbal effort!

REFERENCES

[1.]Baile WF, et al., **Curr Opin Oncol** 2005; 17(4): 331.[2.]Kurtz SM, **Can J Neurol Sci** 2002; 29 Suppl 2: S23.[3.]Lukman H, et al., **Med J Malaysia** 2009; 64(1): 51. [4.]Shama ME, et al., **J Egypt Public Health Assoc** 2009; 84(3-4): 261.[5.]Stiefel F, et al., **Ann Oncol** 2010; 21(2): 204.[6.]Nikendei C, et al., **Patient Educ Couns** 2010; 82(1): 94.[7.]Carson CA, **Vet Clin North Am Small Anim Pract** 2007; 37(1): 49.[8.]N A, et al., **Psychol Bull** 1992; 111(2): 256.[9.]Hertenstein MJ, et al., **Emotion** 2009; 9(4): 566.[10.]DiMatteo MR, et al., **Health Psychol** 1986; 5(6): 581.[11.]DiMatteo MR, et al., **Med Care** 1980; 18(4): 376.[12.]Hall J, et al., **Appl Prev Psychol** 1995; 4: 21. [13.]Griffith CH, 3rd, et al., **J Gen Intern Med** 2003; 18(3): 170.[14.]Henry SG, et al., **Patient Educ Couns** 2012; 86(3): 297.[15.]Clark NM, et al., **Clin Pediatr (Phila)** 2008; 47(1): 49.[16.]Newman AW, et al., **J Gen Intern Med** 2005; 20(3): 213.[17.]McGovern B, et al., **J Hosp Infect** 2010; 75(2): 138.[18.]Lopez PJ, et al., **Am J Infect Control** 2009; 37(1): 79.[19.]Ditchburn I, **J Hosp Infect** 2006; 63(2): 227.[20.]McKinstry B, et al., **Br J Gen Pract** 1991; 41(348): 270.[21.]Hathorn IF, et al., **Clin Otolaryngol** 2008; 33(5): 505.[22.]Pronchik DJ, et al., **J Emerg Med** 1998; 16(4): 541.[23.]Schmid Mast M, et al., **Patient Educ Couns** 2011; 83(2): 174.[24.]Mast MS, et al., **Med Care** 2008; 46(12): 1212.[25.]McHenry M, et al., **Support Care Cancer** 2012; 20(5): 1073.[26.]Hannawa AF, **Swiss Med Wkly** 2012; 142: w13576.[27.]Ambady N, et al., **Surgery** 2002; 132(1): 5.

DEALING WITH HIGH MAINTENANCE OWNERS

Laura D. Garrett, DVM, DACVIM (Oncology)
Urbana, IL

INTRODUCTION

Speaking with pet owners is something veterinarians do constantly; as such, it is the most common "procedure" performed in veterinary medicine. Yet very few veterinarians receive any guidance in this critical area during veterinary school or after. Does communication matter? Studies in human medicine have shown that the consumer perception of health care quality is highly dependent on the quality of interactions with doctor.[1] Also, the consumer satisfaction levels impact the clinical outcomes (people who liked their doctors did better!).[2,3] Other studies have shown that effective physician-patient communication has been correlated with at least five health outcomes including diagnostic accuracy, improved health status (BP, blood glucose, stress levels, pain management, etc.), improved adherence rates, better patient and physician satisfaction, and decreased malpractice risk.[4,5]

Good communication allows one to practice better medicine. Improved assessments & diagnosis will occur from acquiring accurate information, understanding the client's concerns and thoughts, actively involving the client in the treatment process, and enhancing the generation of solutions. Improved adherence to your recommendations will occur from better understanding of your client (beliefs, motivation, etc) and empowering the client to be part of treatment team. This will lead to increased preventative care and improved animal health and welfare. Estimates from human medicine on adherence showed that 50% of patients could not recall instructions. Of those that did, only 70% followed them.[6] Veterinarians overestimate adherence, with estimates of adherence for heartworm prevention, dental prophylaxis, and therapeutic diets of 70, 54, & 75% respectively, while actual adherence was 48, 35, and 19%.[7]

Client satisfaction/retention is increased with good communication. The client feels respected and valued; this will lead to increased referrals and decreased liability. One survey showed that out of a multitude of factors pet owners desire in their veterinarian, the top factor was that the veterinarian was kind and gentle, number two was respectful and informative, and third was that there was a reputation for high quality care. Price was selected as number nine in importance![8] Showing kindness, respectfulness, and providing information are all aspects of communication, aspects that can be improved upon with attention and practice.

For these multitudes of reasons, communication matters. As such a critical part of the therapeutic success of a veterinarian's efforts, communication is an essential component of the veterinarian's role. It is the veterinarian's responsibility, and it cannot be delegated. Thinking of communication as a procedure will help to understand its importance and place emphasis on the fact that good communication CAN be learned. Like any procedure, mastery requires practice and experience. Too often the ability to talk to people effectively is considered an inherent skill – you either have it or you don't. However, enumerable studies in the field of human medicine, from all over the world, have shown that the skills involved in communication are able to be defined and, indeed, learned.[9-14] There are several constructs upon which to outline and describe techniques in communication. This article will focus on the four core communication skills: non-verbals, empathy, open-ended questions, and reflective listening.

Non-verbal communication

Roughly 80% of all communication between individuals is nonverbal and generally unintentional.[15] Thus, information that cannot be hidden is being exchanged at all times – both from the client to the clinician, and vice versa. The nonverbal aspects of communication are like a poker player's "tell" – they give away what a client may be thinking or feeling, possibly contrary to what they are saying verbally. Thus, these unconscious nonverbal messages from a client more accurately reflect their feelings. Emotions are much more often communicated non-verbally than verbally. For example, we know when someone is mad because they raise the volume of their voice, have a different tone, change their posture, etc, not because they say "I'm so mad".

Mixed messages occur when the verbal and nonverbal messages disagree. In these situations, the nonverbal message is more accurate and needs to be detected and addressed. A common experience is asking a client if they understand, and the client hesitantly says "yes", often without meeting the clinician's eyes. The non-verbals clearly say the client does NOT understand, and paying attention to that fact and trying to clarify things at that time will often save misunderstanding and upset in the future.

Another benefit to becoming aware of and following nonverbal clues is that they take up no extra time during an interaction. They are going on constantly during all interactions, and one needs to only consciously note them (and perhaps comment on them, see below) to help guide the discussion in a beneficial manner.

Empathy

Empathy does NOT mean one feels what the client feels, but rather that one can try to imagine what the client has or is experiencing. It is the proverbial "walking in someone else's shoes", and a key point is not only to imagine that experience but to SHOW the client that understanding. Showing empathy is a key part in building rapport with a client. Just feeling empathy is not enough; the client will not know you feel it if you do not show it. Empathy can be shown in both verbal and non-verbal ways. Three key points of empathy are having a client know that they are being seen, being heard, and being accepted.

Being seen means not only physically seen, but seen as an individual person and not just a "client". Ways to show back to the client that they are seen are to comment on a unique aspect of that individual *unrelated* to their pet's medical concerns, such as their clothing (for example, if they are wearing a sport team's logo), their pet's collar, etc. It is also important that the client knows they are seen as a person when it comes to the medical issues, and that their emotions about these issues are noticed. A good way to show this by verbally stating what is seen: "you seem worried", "you look very scared". While these statements seem awkward or obvious at first, this technique is greatly appreciated by clients. Firstly, clients will appreciate the empathy and bond more with the clinician that shows it. Secondly, by bringing the emotion to the forefront, it can be dealt with and thus allow the client to feel it more and move on in the discussion about their pet. A statement as simple as 'I can see this is difficult for you; it can be very hard to hear that someone so close to you has cancer,' can be grounding for a client and may allow them to refocus on the medical conversation. Other non-verbal means to show a client they are seen is to use appropriate speaking tone and rate (e.g. quiet slower speaking in times of sadness) or posture in response to the client. Handing a box of tissues to a crying client is a wonderful (and useful!) means of showing empathy.

Being heard is a similar concept to being seen; the veterinarian not only needs to listen to the client, but needs to SHOW them that they are heard, and ideally understood. This process is closely tied to one of the other core skills, reflective listening (see below). One way to view this process is as an "Inhale and Exhale". Listen and take in what the client is saying (inhale) and then show appreciation and understanding of their experience (exhale).[16] This showing to the client can be accomplished both verbally and non-verbally. Reflective listening (below) describes the verbal showing. Non-verbal showing would include appropriate facial expressions, nodding, leaning in towards the client, etc.

Being accepted is an inherent desire in people, and this acceptance will help an owner to be open and honest with the clinician and help them also to voice concerns or questions that they may be embarrassed to mention otherwise. It also helps clients to work through difficult problems and see the veterinarian as supportive as opposed to adversarial. There are several categories of statements that one can make to help a client feel accepted, and intentionally using these statements will take conscious effort but can be of great benefit in improving the client-clinician bond. Three types of statements include 1.non-judgmental comment: "you were in a very difficult situation" 2. normalizing comment: "it is so common for people to miss these masses until they get very large" and 3.self-disclosing comment: "my cat has behavioral issues too".

These techniques for expressing empathy become even more critical with clients for whom it is difficult to feel empathy; verbally showing acceptance via the above phrases can help the client-veterinarian bond even for clients with whom it is very hard to bond. Note that showing a client that they are seen, heard, and accepted is also tied to reflective listening (see below).

Open Ended Questions

Open ended questions are very important in obtaining an accurate history, as well as during discussions regarding diagnostics and treatment choices. Additionally, they greatly increase the client-veterinarian bond by allowing a client to tell their story – a key part to feeling understood and valued as a person. This "story telling" does not need to be very long or unguided; with specific techniques a veterinarian can guide the answers in situations where they may be too long. Very often, open ended questions SAVE time in an interview, as they allow the veterinarian to understand the true nature of the client's concerns without misleading results that the use of only closed-ended questions may provide.

Closed ended questions begin with words such as "when, is, did, who, where". These questions result in very short answers, often times "yes" or "no" or other one word answers. Closed-ended questions create an atmosphere that is more like an interrogation than an interview, and they are very much centered on what the *clinician* wants to know. These questions certainly play a role in an interview – they are helpful in emergency-type situations as well as in clarifying information. They are best used AFTER open-ended questions.

Open ended questions often begin with "what or how". Avoiding "why" is a good idea, as that word seems to carry judgment ("why did you do that?"). Open ended questions allow the client to talk and express their thoughts and concerns. While the client is "telling their story" they will use their own natural vocabulary; this allows the clinician to ascertain what type of vocabulary to use in return when speaking to the client for greatest understanding. A very helpful way to start a visit with a client is to ask a big, broad question such as "What has been going on?". If you already know the complaint (e.g., technician took a brief history; the patient is hematuric), an open ended question can still be asked while including the information you have: "So, Fluffy is having blood in her urine. Can you tell me what has been going on?". This technique both allows the client to know that you know their concern, and also allows them to share more and feel that their thoughts are worthwhile. Open ended questions are a powerful tool in improving a bond with the client, versus a rapid fire, just-the-facts interrogation style: "When did you notice the blood? Is she drinking more? Any straining?" etc.

In regards to conversations after the initial visit, open-ended questions continue to be very helpful. Even with ongoing cases and familiar owners, don't assume an understanding of what the client is thinking. Open ended questions can avoid misunderstanding in difficult discussions. For example, rather than "Are you thinking about euthanasia?" and having an owner get hysterical because they were not, ask "what are your thoughts about options from here?". Also, "ask rather than tell" is a great rule of thumb. When a new diagnosis has been made, asking a client what they know about the disease, rather than jumping into a description of the problem for the client, can save time in a discussion and avoid unnecessary client education. This asking not only allows the clinician to begin the discussion at an appropriate level, but also importantly shows the client that their knowledge is valued – this leads to a better client/veterinarian bond as the client feels treated as a unique and important individual.

Reflective Listening

Reflective listening is the technique, also called active listening, by which an individual "reflects" back what has been said, thus showing interest in and *understanding of* the meaning of what the speaker is saying. With this technique, the listener repeats back to the client what was said or even implied, often via paraphrasing. This technique is a key component of showing empathy, as stated previously. Also, reflective listening is an excellent tool for checking understanding by the veterinarian. Reflective statements not only allow the client to know they are heard and seen, to "see" how they seem to the veterinarian, but they also allow the client to correct any misconceptions and add to the veterinarian's summary if needed.

A few simple tricks of reflective listening include non-verbals stated above, such as nods, smiles, mm-hmm's, and hand gestures. Also, when unsure what to say, or wanting an owner to expound without your input, a simple verbal repeat of the last word of phrase said can be helpful: Client : "Momo has just been acting crazy." Clinician: "Crazy?"

Reflective statements can seem awkward at first; having a "stem" with which to begin is very helpful. The classic stem is "what I hear you saying is…" Other stems may feel more natural (less like a therapist!) and include "so, you are saying…" "it sounds like…" "you are...". A complete statement could be something like: "It sounds like you are worried that Turner is having a poor quality of life right now." The client can then correct if that is not their main concern, or can feel a sense of relief that the clinician understands them if that is the concern. Another example of a reflective statement in action is: "So you were concerned about the cost of the treatment?" This may elicit a "Yes, it seems very expensive" or maybe "No, the cost isn't the problem, it is the time involved". Thus, the true concern can be addressed. This paraphrasing is one of the most effective listening tools available when one wants to convey to someone that what they are saying is important, and it is also an excellent way to verify correct understanding of what the client meant to express, which leads to improved diagnostic and therapeutic patient management.

Summary

By actively employing a few communication techniques in daily interactions with clients (or staff, family or friends!), difficult situations may not only be improved but completely avoided. Employment of these techniques takes concentration – intentional communication is a phrase used to describe this effort. While at first the effort may seem artificial and uncomfortable, the skills (as all learned skills) will grow over time and become a more natural part of one's communication with others.

References

[1]Clark PA. **J Ambul Care Manage** 2003;26:110-123 [2]Stewart M et al. **J Fam Pract** 2000;49:796-804 [3]Teutsch C. **Med Clin North Am** 2003;87:1115-1145 [4]Bayer Animal Health Communication Project: **Institute for Healthcare Communication** 2007 [5]Baile WF et al. **Curr Opin Oncol** 2005;17:331-335 [6]Kessels RP.**J R Soc Med** 2003;96:219-222 [7]AAHA Survey, **JAAHA** 2003. [8]Brown JP et al. **J Am Vet Med Assoc** 1999;215:161-183 [9]Fallowfield L et al.**Br J Cancer** 2003;89:1445-1449 [10]Gordon GH **Educ Couns** 2003;50:95-98 [11]Shama ME et al.**J Egypt Public Health Assoc** 2009;84:261-283. [12]Lukman H et al. **Med J Malaysia** 2009;64:51-55 [13]Stiefel F et al **Ann Oncol** 2010;21:204-207 [14]Nikendei C et al. **Patient Educ Couns** 2010;82:94-99 [15]Carson CA. **Vet Clin North Am Small Anim Pract** 2007;37:49-63[16.] Bayer Animal Health Communication Project, **Institute for Healthcare Communication** 2007

FEEDING THE CAT WITH ENDOCRINE DISEASE

Mark E. Peterson, DVM, DACVIM
New York, NY

INTRODUCTION

When treating cats with endocrine disease, most of us concentrate on the latest or best drug that can be used to cure or manage the disease. However, it is also important to realize that proper nutritional support also plays an important and integral role in management of these diseases. Whenever considering feline nutrition, it is paramount to remember that cats are strict carnivores that rely on nutrients in animal tissues to meet their specific and unique nutritional requirements. In their natural habitat, cats consume small prey high in protein (50-70%) with moderate amounts of fat (30-50%) but very minimal amounts of carbohydrates (1-2%).[1,2] Evolution has shaped the cat's core metabolism to adapt to this carnivorous diet of animal tissues over a period of millions of years, so it is not very "natural" for cats to eat many of the commercial diets that contain moderate to high amounts of carbohydrates (35-50%) and much lower amount of protein, especially animal protein. In this review, I will cover the 3 most common endocrine problems of cats seen in clinical practice (i.e., hyperthyroidism, diabetes, and idiopathic hypercalcemia) and discuss the way I integrate nutrition into the management of these common feline diseases. We want to use the cat's diet to support their overall body and metabolism and therefore best manage their underlying endocrine disease.

HYPERTHYROIDISM IN CATS

Hyperthyroidism is the most common endocrine disorder of cats, and is one of the most common medical problems seen in small animal practice. Surprisingly, despite the fact that nutritional factors and cat food has been proposed to have a role in the etiopathogenesis of this disease, there are only limited published recommendations about what to feed these cats. The question, "What's the best diet to feed my hyperthyroid cat?" is an extremely common one that we all get from concerned cat owners.

The Many Metabolic Problems Facing the Hyperthyroid Cat

When secreted in excess, thyroid hormones have profound metabolic effects on the whole body, and dysfunction of multiple organ systems (CNS, cardiac, gastrointestinal, hepatic, and renal) is common in hyperthyroid cats.

Weight loss and muscle wasting: Weight loss, despite a normal to increased appetite, is the classic and most common signs seen in cats with hyperthyroidism. These cats lose weight because their hyperthyroidism accelerates their metabolic rate and body's energy expenditure; they are burning up their food calories faster than they can consume their daily meals.

It's important to realize that hyperthyroidism is a catabolic state. The progressive weight loss and muscle wasting that is so characteristic of feline disease is caused by increased protein catabolism leading to a negative nitrogen balance. When hyperthyroid cats first lose weight, the case can usually be first noticed as a loss of muscle mass in the cat's lumbar paravertebral area. Despite this loss of muscle mass, most mildly hyperthyroid cats retain their "belly" during the initial stages of their thyroid disease and may even have a higher than ideal body condition score. With time, severe muscle wasting, emaciation, cachexia, and death from starvation can occur if the cat's hyperthyroidism is left untreated. In hyperthyroidism, the cat's body consumes its own muscle tissue to get the protein it needs to sustain its carnivorous life. Even with treatment of hyperthyroidism, recovery of muscle mass and function may be prolonged, lasting several weeks to months. This is especially true if these cats are not provided with enough dietary protein to rebuild and maintain their lost muscle mass.

Hyperglycemia, glucose Intolerance, insulin resistance, and overt diabetes: Hyperthyroid cats commonly develop profound changes in glucose and insulin metabolism. Mild to moderate hyperglycemia is common in hyperthyroid cats, which is generally attributed to a "stress" reaction. However, the underlying metabolic changes are actually much more complicated: hyperthyroidism frequently causes moderate to severe "endogenous" insulin resistance, as demonstrated by high resting serum insulin concentrations and an exaggerated insulin response during an IV glucose tolerance test.[3,4] This insulin resistance is associated with a decreased glucose clearance (impaired glucose tolerance), which is indicative of a prediabetic state. Occasionally, an untreated hyperthyroid cat will develop overt diabetes mellitus. Many of these diabetic cats will develop moderate resistance to the injected insulin, with poor diabetic control. Surprisingly, the insulin resistance and prediabetic state so common in hyperthyroid cats does not always improve and may even worsen despite successful treatment of hyperthyroidism.[4] This indicates that hyperthyroid cats may have long-lasting alterations of glucose tolerance and insulin secretion that cannot always be reversed by treatment. In accord with that, some of these hyperthyroid cats (not diabetic at diagnosis) will go on to develop overt diabetes in the months after treatment of hyperthyroidism. Sarcopenia of Aging: In addition to loss of muscle mass from the catabolic effects of thyroid hormone excess, cats also tend to lose muscle mass they age, independent of their thyroid status. This phenomenon, referred to as sarcopenia of aging, is also common in elderly human beings.[5] The term age-related sarcopenia is derived from Greek (meaning "poverty of flesh") and is characterized by a degenerative loss of skeletal muscle mass and strength, as well as increased muscle fatigability.

In adult cats, maintenance energy requirements decrease by about 3% per year up until the age of 11 years, and then actually start to increase again.[6] This contributes to a tendency of senior cats to lose muscle mass if their energy needs are not met. Lean body mass of aging cats drops dramatically after 12 years of age, and by age 15, cats may have a mean lean tissue mass that is a third less than cats aged 7 years or less.[6] Body fat also tends to progressively decrease in cats after the age of 12 years; this combination of reduced lean mass and body fat contributes to weight loss experienced by many elderly cats.

The ability to digest protein is also compromised in many geriatric cats. After the age of 14 years, one-fifth of geriatric cats have reduced ability to digest protein.[6,7] Reduced protein digestibility in geriatric cats seems to occur in parallel with reduction of lean tissue and it might predispose them to negative nitrogen balance. Although moderation of calorie intake might be suitable for some mature cats, it does not appear to match the needs of most geriatric cats.

Diet Recommendations for Hyperthyroid Cats

High dietary protein: Obligate carnivores, such as the cat, are unique in their need for large amounts of dietary protein (specifically, dispensable nitrogen) that separates them from omnivores and herbivore species.[1,2] This absolute requirement for dietary protein intake in cats is critically important when formulating a diet for hyperthyroid cats, in which protein catabolism and muscle wasting is universally present. Protein is the primary macronutrient responsible for maintenance of muscle mass. Restoring and preserving any remaining muscle tissue in cats treated for hyperthyroidism depends upon the cat consuming a diet with sufficient amounts of high-quality protein (40-60% calories as protein). This recommendation for higher amounts of dietary protein does not change once euthyroidism has been restored. The dogma that all older cats should be fed reduced energy "senior" diets must be questioned based on what is now known about the increasing energy requirements and nutritional needs of older cats. [6-8] In most geriatric cats, logic dictates the use of highly digestible, energy-dense food to mitigate the decline in body weight and lean body tissue and to avoid protein:calorie malnutrition. Reducing protein intake in geriatric cats, at a time when lean tissue has been lost, is contraindicated. Geriatric cats seem to have nutritional requirements closer to kittens than to mature adult cats. Remember that when deprived of protein, carnivores will continue to break down muscle tissue to create the energy they need.[1,2] By feeding only high-quality protein diets, we will help restore the cat's muscle mass and improve strength and agility.

Low Dietary Carbohydrates: Since most of these cats also have subclinical diabetes —as evidenced by their mild hyperglycemia, glucose intolerance, and insulin resistance— feeding a low carbohydrate diet (<10% of total calories) also is strongly recommended.[9] Feeding a low carbohydrate diet will lessen postprandial hyperglycemia, improve insulin sensitivity, reduce the need for exogenous insulin, and help stabilize glucose metabolism in these cats.[9,10] This may prevent the development of overt diabetes and control long-term obesity in these cats after successful control of the hyperthyroidism.

Check out this website (www.catinfo.org), which gives you a breakdown of the nutritional composition of the various prescription and over-the-counter diets (click on link entitled "Protein/Fat/Carbs Chart" on the right side-bar of home page). It turns out that many of the over-the-counter canned diets have an acceptable composition of protein and carbohydrates. However, the composition of almost all dry cat diets are much too high in carbohydrates and most are too low in protein content. That is why I believe it's best to limit the amount of dry food that is fed to hyperthyroid cats, or even better, not feed dry food at all.

DIABETES MELLITUS

Evolutionary events shaped the cat's core metabolism such that their systems are uniquely set up to metabolize a diet which is high in moisture, high in protein, and very low in carbohydrates. Because this is the diet they have relied upon for tens of thousands of years, they do not have the ability to process carbohydrates very efficiently.[1,2] This becomes extremely important when selecting a diet for cats with diabetes (see below).

Normal Glucose Metabolism and Postprandial Glycemia in Cats

So what are these specific feline adaptive mechanisms that have developed to meet the requirements of a carnivorous diet? First, cat's gluconeogenic pathway provides an almost continuous source of carbon skeletons for glucose or energy production.[1,2] Secondly, glucokinase concentrations are markedly reduced or absent, whereas hexokinase activity is increased. This is in marked contrast to the liver of omnivores (dogs, man), which contains both glycolytic enzymes that act to catalyze the phosphorylation of glucose during glycolysis. In addition, cats have reduced disaccarharide activity in the small intestine, reduced and delayed insulin secretion, and delayed gastric emptying. As a result of these differences, plasma glucose clearance rates are longer in cats compared to dogs or humans after feeding a moderate to high carbohydrate meal — in other words, even normal cats have much more prolonged postprandial period of hyperglycemia than might be expected. In healthy humans and dogs, postprandial hyperglycemia normally persists for 2 to 5 hours. In contrast, a recent study of healthy cats found that both serum glucose and insulin concentrations remained significantly increased for a median time of

12 hours following ingestion of a moderate carbohydrate meal (25% ME), and that both glucose and insulin concentrations remained above baseline values for 24 hours in approximately 20% of the cats.[10] Most feline diets contain ever higher amounts of carbohydrate and therefore would be expected to result in more severe postprandial hyperglycemia and a longer time to return to baseline.

Dietary Management of Cats with Diabetes Mellitus: Key Part of Treatment!

In cats with diabetes, a primary goal of therapy is to minimize the degree of hyperglycemia that develops after feeding in order to lessen the subsequent demand on beta cells to secrete insulin. By doing this, we decrease the effect of "glucose toxicity" and allow the pancreatic islet cell to hopefully recover. Feeding a low carbohydrate diet will improve insulin sensitivity, reduce or eliminate the need for exogenous insulin, and help stabilize glucose metabolism in these cats. Again, one major way a low-carb diet improves the diabetic state is by helping to prevent severe and prolonged postprandial hyperglycemia.

The sooner one starts the diabetic cat on a low-carbohydrate diet, the better. By "low-carb," I mean a diet that provides <10% of the calories as carbohydrate. Some cats will do fine on a slightly higher-carb diet (12-14%) whereas others do best on a diet containing <7% carbohydrate. To achieve these low carbohydrate levels, we must formulate a homemade diet or feed a canned food (go to www.catinfo.org and review the "Protein/Fat/Carbs Chart" to select an appropriate low-carb diet). Only a few available dry cat foods are low enough in carbohydrates, most are too low in protein, and all are too limited in water content for a diabetic. It's best to limit the amount of dry food that is fed to diabetic cats, or even better, not feed dry food at all.

When we reduce the content of carb in a cat food, we must raise the content of either protein or fat, or both. I like to do both, feeding my diabetic cats a diet that mimics the composition of their prey in the wild: about 40-60% of protein with the remaining amount in fat. This higher than average protein level also helps restore and maintain lost muscle mass, since many diabetic cats will develop "sarcopenia" as they age.[6-8] Another plus for feeding low-carb, hi-protein diets in cats with relatively early diabetes is that this diet composition (together with insulin treatment) greatly improves the diabetic remission rate.[11,12] If the diabetic cat goes into remission (no more insulin needed to maintain euglycemia), we recommend maintaining a restricted carbohydrate diet for life to help prevent relapse of the diabetic state.

IDIOPATHIC HYPERCALCEMIA IN CATS

Over the last two decades, a syndrome of idiopathic hypercalcemia in cats has emerged and appears to be increasing in frequency.[13-15] Although the cause of idiopathic hypercalcemia remains elusive, this is now considered the most common type of hypercalcemia in cats. In this syndrome, the serum concentrations of both total and ionized calcium are high, whereas serum PTH concentrations are low-normal to low. It has been suggested that the diet fed may predispose cats to development of idiopathic hypercalcemia, as well as formation of oxalate calculi. One plausible hypothesis is that excessive dietary vitamin D content in some cat foods contributes to this syndrome, at least in some cats.[15] The feline daily requirements for vitamin D are not really known. In any case, because the amount of vitamin D supplied in most cat diets is not listed on the label, it is difficult to choose a diet that is low in vitamin D. Others believe that feeding of acidifying, magnesium-restricted diets predisposes cats to idiopathic hypercalcemia. In support of this hypothesis is the fact that both calcium oxalate stones and hypercalcemia first became prevalent in the 1990's, shortly after the introduction of feline acidifying diets designed for prevention of struvite crystals.[13-15] Three of 5 cats in one series[13] and all 14 cats for which diet history was available in another[14] had been fed acidifying diets designed to minimize struvite crystalluria and urolithiasis. These acidifying diets can lead to a state of mild systemic acidosis,[16] which, in turn, can result in an increase in calcium resorption from bone, mild ionized hypercalcemia, and hypercalciuria. In almost all affected cats, hypercalcemia is slowly progressive and relatively longstanding. Clinical signs are usually relatively mild, at least initially. Therefore, most cats can be treated as outpatients with either dietary therapy, alone or in combination with drug therapy (glucocorticoids, bisphosphonates).[15,17]

Dietary Management of Cats with Idiopathic Hypercalcemia

I generally start with diet modification as a first-line treatment. If an acidifying diet was being fed, it should be discontinued. No matter what type of diet is chosen, it is best to fed a wet-only diet to promote urinary dilution and lessen the chance for calcium oxalate stones. To this end, we have 5 different types of cat food diets that may help lower calcium in cats with idiopathic hypercalcemia.

1. High fiber diets (e.g., Hill's w/d) may restore normocalcemia in some cats with idiopathic hypercalcemia and calcium oxalate urolithiasis.[13,15] The effects of fiber on intestinal absorption are complex and depend on the type and amount of fiber in the diet, as well as the interactions with other nutrients in the diet. However, these "high fiber diets" are usually supplemented with extra calcium; therefore, calcium content does not explain why these diets are sometimes successful in treating idiopathic hypercalcemia. Overall, I do not find high-fiber diets to be helpful in most cats and do not recommend these diets.

2. Prescription kidney diets also may result in normocalcemia in some cats with idiopathic hypercalcemia. Most renal diets are less acidifying than maintenance or high-fiber diets and are low in both calcium and phosphorus. The decreased consumption of calcium leads to a decrease in the amount of calcium absorbed, and the effects of alkalinization may decrease the release of calcium from bone. In man, acidifying diets will increase PTH secretion, leading to calcium release from bone and increased urinary calcium excretion. Remember, however, that veterinary renal diets are also restricted in phosphorus, which may lead to increased calcitriol (active vitamin D) synthesis by the kidney; the action of this increased serum calcitriol could offset the advantage of the decreased calcium absorption in cats with idiopathic hypercalcemia.[15] Because these diets are low in protein, renal diets are not my first choice — with time, these cats can become muscle wasted on these diets.
3. Diets developed to prevent calcium oxalate urolithiasis may be beneficial in the treatment of cats with idiopathic hypercalcemia. These diets are restricted in calcium and have less urinary acidification (neutral pH urine production). Some are also restricted in oxalic acid and sodium, and most have increased moisture in the canned formulation.
4. Over-the-counter diets with a composition similar to what cats would eat in the wild (i.e., 40-60% protein, 30-50% fat, and <10% carbohydrates) have also been recommended. Feeding a more natural will help normalize serum calcium concentrations in some of these cats, particularly those with mild forms of idiopathic hypercalcemia. However, magnesium levels should not be reduced and acidifiers should not be added to these diets.
5. Home-prepared diets that are both vitamin D and calcium-restricted can also be fed to help lower the serum calcium concentrations in these cats with idiopathic hypercalcemia. These diets must be specially formulated for the hypercalcemic cat, as none of the currently available, commercial cat food diets are low in calcium or vitamin D. Ideally, this diet should be formulated with the help of a veterinary nutritionist to ensure that it is nutritionally adequate for the cat.

REFERENCES

[1]Eisert R. *J Comp Physiol B* 2011;181:1. [2]MacDonald et al. *Annu Rev Nutr* 1984;4:521. [3]Hoenig M, et al. *J Endocrinol* 1989;121:249. [4]Hoenig M, et al. *Res Vet Sci* 1992;53:338. [5]Fujita S, et al. *Nutr Res Rev* 2004;17:69. [6]Perez-Camargo G: *Compend Contin Educ Vet* 2004;26(2A):5. [7]Patil AR, et al. *Compend Contin Educ Vet* 2004;26(2A):72. [8]Sparkes AH. *Top Companion Anim Med* 2011;26:37. [9]Rucinsky R, et al. *J Am Anim Hosp Assoc* 2010;46:215. [10]Farrow H, et al. *J Feline Med Surg* 2012;14:706. [11]Bennett N, et al. *J Feline Med Surg* 2006;8:73. [12]Mazzaferro EM, et al. *J Feline Med Surg* 2003;5:183. [13]McClain HM, et al. *J Am Anim Hosp Assoc* 1999;35:297. [14]Midkiff AM, et al. *J Vet Intern Med* 2000;14,619. [15]Chew DJ, et al. *Kirk's CVT XIV*. 2009;236. [16]Ching SV, et al. *J Nutr* 1989;119:902.

NEW DEVELOPMENTS IN DIAGNOSTIC TESTING FOR THYROID DISEASE IN CATS

Mark E. Peterson, DVM, DACVIM
New York, NY

INTRODUCTION

The diagnosis of hyperthyroidism and hypothyroidism in cats is primarily based on a constellation of history and clinical clinical signs, and physical examination findings (eg, palpation of thyroid nodule). Since many nonthyroidal diseases can mimic the signs of thyroid disease in cats, a complete database (e.g., CBC, serum chemistry profile, urinalysis) must always be evaluated to help exclude other illness. After reviewing this data, the next step is to then use thyroid function tests to confirm the diagnosis of thyroid dysfunction.

Over the past 2 decades, many advances have been made in the availability of tests that can be used to diagnose feline thyroid disease. However, all of the commonly used tests have limitations, especially when evaluating cats suffering from nonthyroidal illness. Therefore, it remains overwhelmingly clear that we still do not have the perfect thyroid test capable of both confirming the diagnosis of thyroid disease in all affected cats and completely excluding it in cats that do not have thyroid disease.

This overview will review the common thyroid function tests currently recommended to diagnose feline hyper- and hypothyroidism. Since definitive diagnosis of these disorders is not always straightforward, I will also concentrate on protocols used in the management of problem cases, in which misdiagnosis is not uncommon

DIAGNOSING CATS WITH HYPERTHYROIDISM

The diagnosis of hyperthyroidism is usually straightforward, as 90% of hyperthyroid cats will have a serum total T_4 concentration that is clearly high.[1-3] Serum total T_3 is not recommended as a sole screening test for hyperthyroidism because 25-30% of hyperthyroid cats have values that remain within normal, reference range limits (ie, not diagnostic for hyperthyroidism).[2-4]

Practitioners should be aware of the assay techniques being used by their commercial or in-house laboratories. In general, serum T_4 can now be measured by 4 different assay techniques:[22] 1) radioimmunoassay (RIA), long considered to be the gold standard technique, but the regulations regarding radioactivity has resulted in a search for alternative methods.[5-8] 2) chemiluminescent enzyme immunoassay (CEIA) (e.g., Immulite®) utilize the same type of antibody testing as RIA, but instead of measuring a radioactive isotope bound to the protein in question, they use a photomultiplier tube that counts light emissions have also been validated for feline serum.[5,6] 3) a homogenous enzyme immunoassay (EIA) method (DRI® thyroxine assay, Microgenics Corporation) is also now being used by many laboratories. This technique has the advantage for the laboratory of being fully automated, which provides the advantage for the practitioner and patient that results can be available sooner;[7] and finally, 4) an enzyme-linked immunosorbent assay (ELISA) test kit[c] for in-house use.[8] It is important to realize that all of these assays can provide serum T4 values that are falsely high or falsely low; no assay will provide perfect test sensitivity and specificity.

Although the correlation of serum T_4 concentrations between all of these assay methods is generally good, any of these different assays can occasionally provide serum T_4 values that are falsely high or falsely low; no assay will provide perfect test sensitivity and specificity 100% of the time. Compared to RIA, CEIA has been shown to provide very similar test results, whereas the ELISA test kit sometimes appears to underestimate the T_4 value. Although most T_4 results correlate well with the cat's thyroid state, in my experience the EIA method shows the highest false-positive results (i.e., falsely high T_4 values) in euthyroid cats. If a high serum T_4 value is found in a cat that lacks clinical signs of hyperthyroidism, especially if no thyroid nodule is palpated, we should never hesitate to repeat the serum T_4 test using a different technique, with RIA or CEIA being preferred in such cases.

Difficulties arise in the routine diagnosis of hyperthyroidism in three general scenarios: 1) when a cat is hyperthyroid but the serum T_4 concentration is within normal, reference range limits; 2) when a cat has a palpably enlarged thyroid nodule (s) but does not have other clinical or biochemical evidence of hyperthyroidism; and finally, 3) when a euthyroid cat is screened and is misdiagnosed as hyperthyroid based on a falsely-high serum T_4 (or free T_4) concentration.

Hyperthyroid Cat but Normal Serum T_4 Concentration

Although serum total T_4 is preferable as a screening test for hyperthyroidism, approximately 10% of all hyperthyroid cats (and 40% of cats with early or mild hyperthyroidism) have serum T_4 within the reference range limits.[2,3] Such T_4 values are usually within the mid- to high-end of the reference range. Thus, finding a single reference range T_4 value does not preclude such a diagnosis.

In general, there are 2 possible reasons why a hyperthyroid cat could have a normal T_4 value. (1) In early or mildly affected cats, serum total T_4 concentrations can fluctuate in and out of the reference range.[9] Such fluctuation of thyroid hormones occurs in all hyperthyroid cats but the degree of fluctuation is of little diagnostic significance in cats with markedly elevated T_4 concentrations. In early or mildly hyperthyroid cats (no concurrent illnesses), serum total T_4 concentrations will eventually increase into the diagnostic thyrotoxic range upon retesting a few weeks later. (2) Severe non-thyroidal illness is also capable of suppressing serum total T_4 concentrations to below the reference range in euthyroid cats.[10,11] Similarly, marginally elevated serum total T_4 concentrations may be suppressed to the mid- to high-end of the reference range in cats with concurrent mild hyperthyroidism and concurrent moderate to severe non-thyroidal disease.

If hyperthyroidism is suspected based upon history, clinical signs (weight loss despite a good appetite), and examination findings (tachycardia, palpable thyroid nodule) but the serum T_4 remains with the upper half of the normal range, hyperthyroidism is still possible. There is no definitive approach to diagnose hyperthyroidism in this scenario but we have several options:

(1) **Repeat total T_4:** In cats in which overt, manageable underlying disease is identified, such concurrent disease should be managed first, before proceeding with further thyroid testing. Once concurrent disease is resolved, most hyperthyroid cats will develop a clearly high T_4, confirming the diagnosis. On the other hand, in cats without overt underlying disease, simply repeating the serum T_4 concentration in 2 weeks may be diagnostic if the T_4 is fluctuating in and out of the reference range.[9] Again, the veterinarian should never hesitate to repeat the serum T_4 test using a different technique, with RIA or CEIA being preferred in such cats.

(2) **Scintigraphy:** Thyroid scintigraphy is a nuclear medicine procedure that produces a visual display of functional thyroid tissue based on the selective uptake of various radionuclides by thyroid tissue.[12] Thyroid scintigraphy is able to identify thyroid disease and define the degree of that thyroid disease relatively unaffected by the presence of concurrent nonthyroidal illness. Because thyroid scintigraphy directly visualizes functional thyroid tissue, thyroid imaging can diagnose hyperthyroidism before laboratory tests are abnormal. Thyroid scanning can also prevent misdiagnosis of hyperthyroidism in cats with falsely high serum T_4 values.

With thyroid imaging (scintigraphy), hyperthyroid cats usually exhibit increased thyroidal uptake of radioisotope administered, either radioactive iodine (^{123}I or ^{131}I) or technetium-99M as pertechnetate ($^{99m}TcO_4^-$). Percentage uptake or increased thyroid:salivary ratio may be calculated and both are strongly correlated with circulating T_4 concentration and provide a sensitive means of diagnosing hyperthyroidism.[12,13] However, apart from expense and the difficulties in dealing with radioisotopes, few veterinarians have access to the equipment needed to obtain thyroid images or perform thyroid uptake determinations.

3) **Free T_4 concentrations**. In cats with mild hyperthyroidism and normal T_4 values, free T_4 concentrations can aid in diagnosis. In support of that, serum free T_4 concentrations measured by equilibrium dialysis were more consistently elevated in hyperthyroid cats (over 98% of cases).[2] Although the free T_4 was more sensitive than the total T_4 for diagnosing hyperthyroidism, the test specificity for free T_4 was poor, with up to 20% of sick euthyroid cats having false-positive free T_4 results.[2,14] These sick cats generally have corresponding total T_4 values in the lower half or below the reference range.

Caution is therefore advised in using serum free T_4 measurements by equilibrium dialysis as the sole diagnostic test for hyperthyroidism. It is more reliable if interpreted with a corresponding total T_4 measurement. A T_4 value within the mid- to high-normal reference range combined with a high free T_4 concentration is consistent with hyperthyroidism, whereas a low or low-normal T_4 with a high free T_4 is usually associated with non-thyroidal illness. One must always combine these lab results with the cat's clinical features and the presence of a palpable thyroid nodule to make the diagnosis.

To complicate the free T4 assay situation further, unpublished results indicate that the present free T_4 by dialysis assay is not the same assay (originally by Nichols Diagnostics) that was used in all of the published studies evaluating the free T_4 assay in cats with thyroid disease. Nichols Diagnostics sold the distribution rights for the assay a few years ago, and results of a recent study in cats[15] indicate a much higher prevalence of false-positive test results than reported over a decade ago. Therefore, this current assay used to measure free T_4 by dialysis test needs to be reevaluated in a large number of normal, hyperthyroid, and sick cats to verify the accuracy of the assay as a diagnostic test.

In addition, many commercial diagnostic labs are now using many other free T_4 assays, most of which do not have the equilibrium dialysis step that we believed was important. At this time, given the high prevalence of false-positive results, it is unclear how much additional information is truly gained by present free T_4 assays over the use total T_4 estimations alone.

Many veterinarians mistakenly believe that the finding of a high free T_4 in a cat is completely diagnostic for hyperthyroidism. However, since up to 25% of these cats turn out to be euthyroid, it is clear that the free T_4 test can *never* be considered to be a "gold standard" diagnostic test. Use of free T_4 testing can lead to more confusion than clarity in some cats, and can certainly lead to misdiagnosis of hyperthyroidism in euthyroid cats.

4) **Thyroid stimulating hormone (Thyrotropin or TSH).** In human patients, measurement of circulating thyroid stimulating hormone (TSH) concentration is usually used as a first-line discriminatory test of thyroid function. A species-specific feline TSH assay has not yet been developed; however, assays for measuring canine TSH (cTSH) are widely available, and it has been suggested that its measurement may provide some diagnostic information in cats with suspected hyperthyroidism.[16-18] As in people TSH levels will be suppressed in early stages of hyperthyroidism before T_4 is elevated.

However, caution is advised in over interpreting undetectable TSH values in cats, since the current cTSH assay has poor assay sensitivity and cannot accurately measure low TSH values (i.e., the assays cannot distinguish a low-normal TSH concentration from an undetectable value).

The current canine assays only detect approximately 35% of recombinant feline TSH,[19] making it difficult to distinguish normal values from the suppressed values expected in cats with hyperthyroidism. At this time, I cannot recommend using routine cTSH determinations to help diagnose mild hyperthyroidism in cats. Perhaps the only use for such TSH measurement would be to exclude early or occult hyperthyroidism, i.e., finding a normal rather than suppressed cTSH value. If a normal serum TSH level is measured, a cat is very unlikely to have hyperthyroidism.

5) **Dynamic testing.** In the majority of hyperthyroid cats with normal total T_4 concentrations, identification of concurrent disease, repeat total T_4 analysis, or simultaneous measurement of free T_4 allows confirmation of the diagnosis. Further diagnostic tests are rarely required. Dynamic thyroid function tests (T_3 Suppression, TRH Stimulation) have been recommended in the past as helpful in confirming a diagnosis of hyperthyroidism.[1,3] Nowadays, use of these tests are only considered when repeated total T_4 concentration remains within reference range, when the free T_4 concentration is equivocal, or when thyroid scintigraphy is unavailable.

Palpable Thyroid Gland (Goiter) but Not Hyperthyroid

Non-functional enlargement of thyroid glands (goiter) has been recognized since the 1960's but has taken on new significance since functional hyperthyroidism arose as an entity in the late 1970's. Non-functional goiter was 're-recognized' about a decade ago.[20] Many believe that clinical hyperthyroidism has a prodromal period (also called 'subclinical hyperthyroidism' or 'pre-hyperthyroidism'). However, it is not clear whether all goiters are indicative that the cat will develop hyperthyroidism. Surgical removal of non-functional goiters has been proposed as a preventative measure, but there is no evidence to support this approach.

Misdiagnosed as Hyperthyroid based on Falsely-High Serum T4 Concentration

A high circulating total T_4 concentration is the biochemical hallmark of hyperthyroidism and is extremely specific for its diagnosis. False positive results (ie, a high T_4 in a cat without hyperthyroidism) are rare but are being seen with increasing frequency, especially with the automated enzyme immunoassays (EIA). However, about 25% of cats presenting with a border-high total T_4 values, sometimes together with high free T_4 concentrations, turn out to be euthyroid. The reason for this high incidence of misdiagnosis is unclear but may be related to the increasing use of screening T4s as part of the cat's annual "wellness" program.

If a high total T_4 concentration is measured in a cat without the characteristic signs of hyperthyroidism, one should always repeat the cervical palpation and verify the T_4 concentration using a different technique, with RIA or CEIA being preferred in such cats. Again, if we have any doubt about the diagnosis, thyroid scintigraphy should be considered. It's extremely important to remember that hyperthyroidism is a clinical diagnosis and should never be based on a serum T_4, free T_4, or TSH concentration alone. One MUST combine these lab results with the cat's clinical features <u>and</u> the presence of a palpable thyroid nodule to make the diagnosis.

DIAGNOSING CATS WITH HYPOTHYROIDISM

Diagnosing feline hypothyroidism can be challenging, regardless of its etiology. This syndrome rarely develops spontaneously. Most commonly, feline hypothyroidism is iatrogenic, secondary to overtreatment of hyperthyroidism. A presumptive diagnosis is based on a combination of the history, clinical signs, physical examination findings, and routine laboratory tests.[21-23]

Although early reports suggested that clinical signs associated with severe iatrogenic hypothyroidism in cats were uncommon and that most cats did not require treatment, it is now realized that milder degrees of iatrogenic hypothyroidism are relatively common and that these cats may benefit from thyroid replacement therapy. Many of these cats with mild or subclinical hypothyroidism fail to develop any overt clinical features that are noticeable to either the owner or veterinarian

However, a number of factors can make the diagnosis of hypothyroidism difficult to confirm. First, concurrent diseases can result in an "euthyroid sick syndrome," characterized by falsely-low serum thyroid hormone concentrations.[10,11] Second, one should expect weight gain and a decreasing activity level after successfully treating a cat with hyperthyroidism. Therefore, the clinical signs of iatrogenic hypothyroidism and the expected return to a euthyroid status can overlap. Finally, many cats develop a marked transient decrease in total T_4, within the first month of therapy with ^{131}I. This transient hypothyroid state is followed by a return to euthyroidism over the next 3 to 6 months as the remaining normal thyroid tissue recovers and starts to function once again.[23,24]

In most cats, we recommend waiting 3 months before making a definitive diagnosis of iatrogenic hypothyroidism after ^{131}I treatment, especially if the cat is not presenting with typical clinical features of hypothyroidism. However, the veterinarian should diagnose or exclude hypothyroidism as soon as possible in cats with renal disease because hypothyroidism, treatment for hyperthyroidism, and chronic renal disease all lower the glomerular filtration rate (GFR).[25,26] The combined effect of these 3 factors can lead to severe azotemia or even total renal failure. Treating the hypothyroidism can raise the GFR to an acceptable level.[27-29]

Serum T_4 concentration

By definition, cats with hypothyroidism should have deficient thyroid hormone secretion. Therefore, finding a low serum T_4 concentration is key in diagnosing feline hypothyroidism. However, it is possible for a cat with mild or subclinical hypothyroidism cat to maintain a *borderline* serum T_4 concentration in the low end of the reference range (e.g., 0.8-1.4 µg/dl; 10-18 nmol/L).

While important in establishing a diagnosis, a subnormal T_4 concentration alone is not definitive for hypothyroidism. First of all, the serum T_4 concentration may be falsely low or borderline-low as a result of assay error. If a low serum T_4 value is found in a cat that lacks clinical signs of hypothyroidism, especially if there is no history of treatment for hyperthyroidism or obvious nonthyroidal illness, one should never hesitate to repeat the serum T_4 test using a different technique, with RIA or CEIA being the preferred assay techniques in such cases. Secondly, the serum T_4 concentration can also be low in cats with nonthyroidal illness.[2,10,11] In general, the severity of the illness correlates inversely with the serum T_4 concentration (i.e., sicker cats have lower serum T_4 concentrations). Because multiple diseases and other factors can falsely lower the serum T_4 concentration in cats, the veterinarian must first rule out nonthyroidal disease before considering a diagnosis of hypothyroidism.

Cats with suspected hypothyroidism and a low T_4 concentration still require additional testing before the veterinarian can make a definitive diagnosis. Further tests such as serum free T_4 concentration, cTSH concentration, or thyroid scintigraphy are indicated to confirm the disease.

Serum T_3 concentrations

In most cats with hypothyroidism, serum total T_3 concentrations are low to subnormal, correlating fairly well with the low T_4 values. This is in contrast to the situation in dogs with hypothyroidism, in which serum T_3 generally remains within the reference range limits.[30] As thyroid function abates, it appears that cats cannot compensate —as dogs and humans initially do— by increasing the thyroidal secretion of T_3 so circulating concentrations fall. Like total T_4, however, T_3 concentrations can also be suppressed in euthyroid cats suffering from a variety of nonthyroidal illnesses.[1,2,23]

Serum free T_4 concentration

Free T_4 is the non-protein bound fraction of circulating T_4 that can enter cells, producing the biologic effect of thyroid hormone and regulating the pituitary feedback mechanism. Free T_4 accounts for less than 1% of circulating T_4. Because only the free T_4 is biologically active, measuring free T_4 should be (theoretically) a more sensitive test for diagnosing hypothyroidism. In addition, nonthyroidal illness influences free T_4 less than it influences the total T_4.[2] Therefore, free T_4 should be better at distinguishing a euthyroid cat with nonthyroidal disease from a hypothyroid cat.

Although measuring free T_4 concentration by equilibrium dialysis is generally more accurate stand-alone test than total T_4 concentration, free T_4 is far from perfect for confirming feline hypothyroidism for the following reasons: 1) the performance and accuracy of the current free T4 assays is variable, and false-positive and false-negative test results are not uncommon; 2) Like total T4, it is possible for a cat with mild or subclinical hypothyroidism cat to maintain a *borderline* serum FT_4 concentration in the lower end of the reference range (e.g., 0.7-1.2 ng/dl; 10-15 pmol/L); 3) Moderate to severe nonthyroidal illness can falsely <u>lower</u> the free T_4 concentration, although to a lesser degree than seen with total T_4[2] and 4) Up to 20% of cats with non-thyroidal disease develop a falsely <u>high</u> free T_4 concentration, further confusing the interpretation.[2,14]

Serum TSH concentrations

In dogs with hypothyroidism, a high serum TSH concentration confirms that the disease is primary (located within the thyroid gland). A specific assay for feline TSH is not yet available, but the commercially available cTSH assay cross-reacts with feline TSH enough to enable its use as a diagnostic test for hypothyroid cats. In one of the reported adult cats with spontaneous hypothyroidism, the serum TSH concentration was high when measured with the cTSH assay.[31] Similarly, many cats with iatrogenic hypothyroidism will also develop high serum TSH concentration as measured by the cTSH assay.[17,22,32]

Normal cats and cats with nonthyroidal illness generally maintain normal values for serum TSH. Therefore, the finding of a low total or free T_4 in combination with a high TSH concentration greatly improves the diagnostic sensitivity for hypothyroidism.

Serum thyroid hormone panels (T4, free T4, T3, TSH)

Differentiating hypothyroid from euthyroid cats can be challenging and may require evaluation of more than a single thyroid function test. The use of thyroid panels, which include serum total T_4, free T_4, total T_3, and cTSH, are now commonly offered by many commercial laboratories. Evaluation of these thyroid panels that measure multiple thyroid hormones can be very helpful in improving diagnostic sensitivity for hypothyroidism.

Although complete thyroid panels are helpful in diagnosis of hypothyroidism, discordant results are very common. For example, as noted above, it is not uncommon to see high cTSH concentrations with completely normal T_4 and free T_4 values in cats that have been treated for hyperthyroidism. Most of those cats are not clinical for hypothyroidism and it is unclear what the clinical significance of the high TSH concentration really is in this subgroup of cats. In general, however, if serum TSH is clearly high and at least 2 of the 3 thyroid hormones (total T_4, free T_4, or T_3) are low, that can be considered diagnostic for hypothyroidism.

Thyroid scintigraphy

Thyroid scintigraphy is considered the best imaging technique for dogs with suspected hypothyroidism because it can distinguish between hypothyroid dogs and dogs with a falsely low serum T_4 concentration.[33,34] In hypothyroid dogs, thyroid imaging typically reveals decreased or even absent radionuclide uptake (thyroid gland is not visible on the scan). In contrast, dogs with a falsely low serum thyroid hormone concentration secondary to illness or drug therapy will have a normal thyroid image. Thyroid scintigraphy is also a powerful tool for diagnosing adult hypothyroid cats (both spontaneous and iatrogenic). As in hypothyroid dogs, thyroid imaging shows minimal to no uptake of the radionuclide.[12,31]

REFERENCES

[1]Graves TK, et al. *Vet Clin North Am Small Anim Pract* 1994;24:567. [2]Peterson ME, et al. *J Am Vet Med Assoc* 2001;218:529. [3]Peterson ME. *Clin Tech Small Anim Pract* 2006;21:2. [4]Broussard JD, et al. *J Am Vet Med Assoc* 1995; 206:302. [5] Kemppainen RJ, et al. *Am J Vet Res* 2006; 67:259. [6]Singh A, et al. *J Vet Diag Invest* 1997;9:261. [7]Horney BS, et al. *Vet Clin Pathol* 1999; 28:20. [8]Lurye JC, et al. *J Am Vet Med Assoc* 2002; 221:243. [9]Peterson ME, et al. *J Vet Intern Med* 1987;1:142. [10]Peterson ME, et al. *J Am Vet Med Assoc* 1990;197:1203. [11]McLoughlin MA, et al. *J Am Anim Hosp Assoc* 1993;29:227. [12]Broome MR. *Clin Tech Small Anim Pract* 2006; 21:10, 2006. [13]Daniel GB, et al. *Vet Radiol Ultrasound* 2002; 43:374. [14]Mooney CT, et al. *J Am Vet Med Assoc* 1996;208:2004. [15]Peterson ME, et al. *Proceedings of the 21st Annual ECVIM Congress, Sevilla, Spain,* 2011.[16]Wakeling J, et al. *J Small Anim Pract* 2008; 49:287. [17]Wakeling J. *Can Vet J* 51:33, 2010 [18]Wakeling J, et al. *J Vet Intern Med* 2011;25:1057. [19]Ferguson DC, et al. *J Endocrinol* 2007;194:267. [20]Norsworthy G, et al. *J Fel Med Surg* 2002;4:145. [21]Peterson ME. *CVT X.* 1989,1000. [22]Daminet S. *BSAVA Manual of Canine and Feline Endocrinology*, 2012;111. [23]Baral RM, Peterson ME. *The Cat: Clinical Medicine and Management*, 2012;571.[24]Peterson ME, et al. *J Am Vet Med Assoc* 1995;207:1422. [25]van Hoek I, et al. *Domest Anim Endocrinol* 2009;36:45. [26]Williams TL, et al. *J Vet Intern Med* 2010;24:863. [27]van Hoek IM, et al. *J Feline Med Surg* 2010;12:117. [28]Gommeren K, et al. *J Vet Intern Med* 23:844, 2009. [29]Panciera DL, et al. *J Vet Intern Med* 2009;23:1045.[30]Peterson ME, et al. *J Am Vet Med Assoc* 1997;211:1396-1402. [31]Blois SL, et al. *J Feline Med Surg* 2010;12:156. [32]Williams TL, et al. *J Vet Intern Med* 2010;24:1086. [33]Diaz Espineira MM, et al. *J Vet Intern Med* 2007;21:25. [34]Shiel RE, et al. *J Small Anim Pract* 2012;53:278

NEW DEVELOPMENTS IN THE USE OF INSULIN MIXTURES AND ANALOGUES FOR THE PROBLEM DIABETIC

Mark E. Peterson, DVM, DACVIM
New York, NY

INTRODUCTION

Despite the fact that diabetes mellitus is one of the most common endocrine disorders in dogs and cats, a poor or inadequate response to insulin is a frequent complaint among many veterinarians and their owners. Clinical signs suggestive of a poor or inadequate response to insulin therapy include recurrence or persistence of clinical signs of diabetes, disorientation or seizures due to hypoglycemia, an insulin dose higher than 2 U/Kg/dose in the dog or > 6 U/dose in the cat, or recurrent ketoacidosis. Common problems that may lead to a poor response to insulin include problems with owner administration, inappropriate insulin dose or formulation, insulin-induced hypoglycemia, rapid metabolism of insulin, and insulin resistance. In many of these problem diabetics, part of the change in therapy will include a change in insulin type used; in many cases, the solution may be to switch to one of the newer human insulin analogues or insulin mixtures. My goal in this presentation is to present an overview of the commonly used insulin preparation in veterinary practice, as well as to review the newer insulin analogues and mixtures developed for use in human diabetics.

OVERVIEW OF STANDARD AND NEW INSULIN PREPARATIONS

During the past few years, new insulin products have been introduced to the USA, and more insulin products have gained FDA approval and will soon be commercially available. As these new insulins gain popularity, production of old and familiar insulins is being discontinued. Many questions have arisen regarding new insulin products.

The current trend in human medicine is towards long-acting peakless insulin analogue preparations that aim to provide continuous basal insulin supplementation; these long-acting insulin preparations are typically combined with a rapid-onset, short-acting insulin preparation, administered at meal times.[1,2] This has resulted in many changes in the market for short- and intermediate-acting insulin products and their pre-mixed combinations. It is these products that are most likely to be useful for management of diabetes in dogs and cats, especially if the veterinary insulin preparations prove ineffective or are not available.

STANDARD INSULIN PREPARATIONS AND INSULIN MIXTURES

Porcine Lente Zinc suspension

The only Lente insulin preparation now is **Vetsulin** (trade name in the USA) or **Caninsulin** (trade name elsewhere in the world), manufactured by Merck Animal Health. This is a 40 U/ml porcine insulin zinc suspension specifically registered for veterinary use. There is good evidence that this insulin product is effective for the treatment of diabetes mellitus in dogs and cats.[3-5] Lente insulin has an intermediate duration of action and is a combination of 30% short-acting Semilente insulin and 70% longer-acting Ultralente insulin. It is particularly suited to a twice-daily dosing regimen in diabetic dogs because the meal can be fed at the time of the insulin injection. This more dilute 40 U/ml veterinary preparation is advantageous for small dogs and cats, which might require a total insulin dose of only 1 or 2 U. Dosing is more simple and accurate if specific U-40 syringes are prescribed with this product.

For some time now, availability of Vetsulin in the USA has recently been severely limited by a FDA recall, as a result of their findings that the product may contain varying amounts of crystalline zinc insulin. However, despite the fact that Caninsulin is exactly the same insulin as Vetsulin, Caninsulin was never recalled or withdrawn from the market in any of the countries where it is sold, and it has maintained it's reliability during the entire Vetsulin recall period.

The good news is that Vetsulin should be back on the market in this Country soon. Merck Animal Health is working diligently with the FDA to re-introduce Vetsulin to the U.S. market as soon as possible in 2013.

Regular insulin

Products containing 100% regular crystalline insulin (recombinant human insulin) are still available (100 U/ml) and have long been recommended for the management of diabetic ketoacidosis and ketosis in dogs. These include **Humulin R** (Eli Lilly) and **ActRapid** (Novo Nordisk). If these products are withdrawn from the human market, it will be necessary for veterinarians to switch to the newer short-acting insulin analogue preparations, such as insulin lispro or insulin aspart (see below).

Isophane (NPH) insulin

Humulin N (Eli Lilly) and **Novolin N** (Novo Nordisk) are both recombinant human NPH insulin (100 U/ml) that have an intermediate duration of action. The human NPH insulins are effective for the treatment of diabetes in dogs when administered twice-daily.[6] These recombinant human NPH insulins do not work well in most cats and are NOT recommended.

Combinations of 70% isophane (NPH) and 30% regular insulins

Humulin 70/30 (Eli Lilly) is still available in most countries. This insulin preparation is available as a 100 U/ml pre-mixed combination of 30% short-acting and 70% intermediate-acting insulin. Because it has a similar duration/action curve to Lente insulin (Vetsulin), this insulin is well suited to a twice-daily dosing regimen in diabetic dogs where meals are fed at the same time as the insulin injections. In dogs that do not respond well to Humulin N alone, such combination of NPH and regular insulin may work well.

Recombinant human protamine zinc insulin

Protamine zinc insulin is a recombinant human insulin preparation FDA approved for use in cats. The efficacy of ProZinc (Boehringer Ingelheim) for treatment of diabetic cats is comparable to that of the discontinued beef/pork PZI-VET product.[7,8] ProZinc is an FDA-approved insulin and is available as a U-40 insulin concentration.

Protamine zinc insulin is not recommended as a first choice therapy for diabetic dogs. In one study evaluating the use of PZI in dogs,[9] ProZinc insulin was effective in improving or maintaining control of the diabetic state in all the 11 diabetic dogs of that study. However, the median PZI insulin dosage required (0.9 U/kg, bid) was much higher than reported for other insulin preparations in dogs with diabetes. Based on the results of that study, use of ProZinc at a starting dose of 0.5 U/kg, bid, should be considered in diabetic dogs that are poorly controlled because of the short duration of NPH or Vetsulin. The biggest disadvantage of ProZinc for use in dogs (especially larger dogs) is the relatively high cost of injecting the higher doses of insulin needed.

INSULIN ANALOGUES AND ANALOGUE MIXTURES

Rapid-acting Insulin Analogues: Lispro, Aspart, and Glulisine

The "big 3" insulin manufactures of human insulin preparations (Eli Lilly, Novo Nordisk, Sanofi-Aventis) all make their own version of a short-acting insulin analog.

Lispro insulin (Humalog, Eli Lilly) has been shown to be a safe and effective alternative to regular insulin for the treatment of diabetic ketoacidosis in dogs.[10] Engineered through recombinant DNA technology, the penultimate lysine and proline residues on the C-terminal end of the B-chain are reversed. This modification does not alter receptor binding, but blocks the formation of insulin dimers and hexamers. This allows larger amounts of active monomeric insulin to be immediately available for postprandial injections.[1,2]

Insulin aspart (NovoLog, Novo Nordisk) and **Insulin glulisine** (Apidra, Sanofi-Aventis) are other short acting insulin analogs that could be used as an alternative to regular insulin.[1,2] Like insulin lispro, these analogues have slight modifications of the insulin structure needed to prevent the formation of hexamers; this leads to larger amounts of active insulin to be available to create a faster-acting insulin. Either insulin aspart or glulisine could potentially be used as an alternative to regular insulin. However, studies in dogs and cats have yet to been reported.

Finally, although not marketed as a short-acting insulin analog, insulin glargine (Lantus) has been reported to have an action profile similar to regular insulin in cats when administered intramuscularly (see below).

Insulin Glargine

Insulin glargine (Lantus, Sanofi-Aventis) is a long-acting synthetic insulin analogue developed for the human diabetic market to provide continuous basal insulin concentrations that inhibit hepatic glucose production.[1,2] The therapeutic aim is to mimic the physiological pattern of insulin secretion of healthy subjects, and the basal insulin levels provided by these products are supplemented at meal times by administration of short-acting insulin preparations that act during the postprandial period.

Glargine insulin differs from standard insulins because it has been genetically modified, by replacing an asparagine with glycine and adding 2 arginine amino acids to the c-terminal end of the molecule. This alters the pH solubility (isoelectric point) of the molecule, so that it is soluble at a pH 4 but insoluble at neutral pH (e.g., in the body). When glargine insulin is injected into the skin, it precipitates because of the pH change and forms insoluble microcrystals, which are slowly absorbed into the circulation. This constant release of small amounts of insulin prevents major peaks and troughs from developing. The formation of micro-precipitates and slow absorption are dependant on the acidity of glargine. Therefore, glargine cannot be mixed or diluted.

Insulin glargine is efficacious for the long-term treatment of diabetes in cats and in considered the initial insulin of choice by many veterinarians.[11,12] If improved diabetic control and remission is the goal, insulin glargine must be administered twice daily.

In cats that present with diabetic ketoacidosis, glargine can be used in place of a short-acting insulin such as regular insulin. When glargine is administered intramuscularly or intravenously, it has a similar action profile to regular insulin and can be used in the treatment of cats with diabetic ketoacidosis.[13] Giving insulin glargine by both the IM and SC route twice daily (70% of the dose given SC and 30% of the dose given IM) can also be used to improve control of the diabetic state in cats with insulin resistance.

Information regarding the use of insulin glargine in diabetic dogs is limited. Initial pharmacological assessment in healthy dogs found that glargine administration resulted in an unpredictable serum insulin concentration response and failed to produce a significant glucose-lowering effect in 2 of 9 dogs after a single injection.[14]

However, a recent study found that insulin glargine administered subcutaneously twice daily is a valid method of treatment for naturally occurring diabetes in dogs, and that it is an alternative to other insulin preparations that have been shown to be effective in the treatment of canine diabetes.[15] However, in that study, only 58% of dogs in the present study obtained good control of hyperglycemia, which is less than the reported 75% of dogs with good control in a similar study where a porcine insulin zinc suspension (lente insulin) was used.[3] Therefore, although insulin glargine may be an acceptable insulin choice for some dogs with diabetes mellitus, the disease control success rate appears to be lower with glargine than with NPH or lente insulins. Certainly, if a more potent insulin preparation is needed (i.e., in a dog with insulin resistance), detemir insulin would be the insulin of choice, since determir appears to be at least 4-times more potent than the other insulin preparations which have been evaluated in the dog.[16]

Insulin Detemir

Insulin detemir (Levemir, Novo Nordisk) is another synthetic insulin analogue with long duration of action that is produced using recombinant DNA technology.[1,2] In contrast to glargine, detemir's long duration of action is made possible through modification of the insulin molecule via addition of an acylated fatty acid chain. This modification facilitates reversible binding to plasma proteins, particularly albumin, from where it is released slowly into plasma. The modification also prolongs self-association in the injection depot, which prolongs absorption from subcutaneous tissue at the injection site and contributes to the long duration of action.

In cats, detemir results in a similar action/duration profile and potency to insulin glargine.[17,18] The remission rates and time to remission for both detemir and glargine are also comparable. A dosing protocol similar to glargine is used, but some cats will require less detemir than glargine (the median dose of detemir was 30% less than glargine in one study of diabetic cats).[18]

In dogs, insulin detemir and glargine are in no way comparable.[14-16] Insulin detemir is a very potent, long-acting insulin preparation in dogs, and works much better than glargine in most problem diabetic dogs. In accord with that, we have had success with insulin detemir in dogs with brittle diabetes that were unresponsive to lente or NPH insulins or Humulin 70/30 insulins. In dogs, detemir is dosed initially at 0.1 U/kg BID, administered at time of feeding. This dose is approximately ¼ of that of NPH or Vetsulin.

Insulin Degludec

Insulin degludec (Tresiba, Novo Nordisk) is a new-generation, ultra-long-acting basal insulin analogue. This latest insulin analogue differs from other long-acting insulin preparations in having a longer half-life, flat time-action profile (less likely to cause hypoglycemia) and less day-to-day variability (less glycemic variability).[19-21]

Degludec is a neutral, soluble ultra-long-acting insulin that forms large soluble multihexamers at the subcutaneous injection site. Its molecular structure is similar to the human insulin amino acid sequence, apart from deletion of Threonine at position B30 and the addition of a 16-carbon fatty diacid attached to Lysine at position B29 via a glutamic acid spacer.

After SC administration, degludec results in the formation of a subcutaneous depot of soluble multihexamers that results in the slow release of insulin monomers into the systemic circulation. Insulin degludec has an onset of action of 30-90 minutes (similar to insulin glargine and insulin detemir). There is no peak in activity, due to the slow release into systemic circulation. The duration of action of insulin degludec is over 42 hours, unlike the 18 to 26 hours provided by current marketed long-acting insulins such as glargine and detemir. The large molecular size of the degludec multihexamers allows for continuous slow release of insulin with less pharmacodynamic variability and within-subject variability than is seen with the currently available insulin analogs.[19-21]

In studies in people with type 1 and type 2 diabetes, degludec was demonstrated to provide similar improvements in blood glucose control as glargine or detemir while also reducing the rate of overall and nocturnal hypoglycemia. One study found that degludec's long duration of action makes it possible to dose the drug at different times each day (i.e., in the morning one day, the evening the next day, the morning the day after that) without sacrificing effectiveness and safety.

Although dosing at roughly the same time each day would make sense for most patients, having the flexibility to vary the time of day when needed could be positive for some patients. In addition, the daily injection time of degludec can be varied to extreme intervals of 24-40 hrs without compromising glycemic control or increasing the risk of hypoglycemia. Therefore, degludec represents an improvement over glargine or detemir that require dosing at the same time each day to maintain sufficient insulin levels in the body and to avoid hypoglycemia.

If all goes according to plan, both degludec and degludecPlus (see below) should be approved in both the US and Europe in the second half of 2012. Obviously, no clinical trials have yet been performed on diabetic cats or dogs. Insulin degludec could potentially provide us with a reliable once-a-day or once-every-other-day therapy for animals with diabetes.

Mixtures of Short-acting and Long-acting Insulin Analogues

The latest insulin preparations released for the human market are pre-mixed combinations of a short-acting insulin analogue with a longer-acting or ultra-long-acting insulin analogue. **Humalog Mix 75/25 (Eli Lilly)** is composed of 70% insulin lispro protamine suspension and 30% insulin lispro injection. This analogue mixture is marketed for human diabetics as an improved product that can be used twice daily in replacement of Humulin 70/30. Compared with Humulin 70/30, Humalog Mix 75/25 in human subjects has a more rapid and predictable onset of glucose-lowering activity, with greater reduction in postprandial glycemia when administered with a meal, and a similar duration of action. **Humalog Mix 50/50** is also available, with half of the insulin being rapid-acting and half being long-acting analogue.

Novolog Mix 70/30 (Novo Nordisk) is composed of 70% insulin aspart protamine suspension and 30% insulin aspart injection. Its onset of action and duration are similar to the Humalog Mix 75/25 described above. **Insulin degludecPlus (Novo Nordisk)** is composed of 70% degludec and 30% insulin Aspart. Although not yet approved for use in the US or Europe, the ultra-long-acting insulin degludec will also available as a fix-dosed combination of a basal and mealtime insulin.

There is currently no published information on the use of any of these new pre-mixed combinations for the management of diabetes in dogs or cats. These synthetic insulin analogue mixtures appear to have time action/duration characteristics most similar to Humulin 70/30. At least in diabetic dogs, these insulins might be effective when administered twice daily at the same times as meals.

REFERENCES

[1]Hirsch IB. *New Engl J Med* 2005;352:174. [2]Gilor C et al. *Vet Clin North Am Small Animal Pract* 2010;40:297. [3]Monroe WE, et al. *J Vet Intern Med* 2005;19:675. [4]Fleeman LM, et al. *Vet Rec* 2009;164:232. [5]Michiels L, et al. *J Feline Med Surg* 2008;10:439. [6]Palm CA, et al. *J Vet Intern Med* 2009;23:50. [7]Norsworthy GD, et al. *J Vet Intern Med* 2008;22:729. [8]Nelson RW, et al. *J Vet Intern Med* 2008;22:730. [9]Maggiore AD, et al. *J Vet Intern Med* 2012;26:109. [10]Sears KW, et al. J Vet Emerg Crit Care 2012;22:211. [11]Marshall RD, et al. *J Feline Med Surg* 2009;11:683. [12]Roomp K, et al. *J Feline Med Surg* 2009;11:668. [13]Marshall RD, et al. *J Vet Intern Med* 2010;24:686. [14]Mori A, et al. *Vet Res Commun* 2008;32:563. [15]Fracassi F, et al. *Vet Record* 2012;170:52. [16]Sako T, et al. *Res Vet Sci* 2011; 90:396. [17]Roomp PKet al. *J Vet Intern Med* 2009;23:697. [18]Gilor C, et al. *J Vet Intern Med* 2010;24:870. [19]Wang F, et al. *Diabetes Metab Syndr Obes* 2012;5:191. [20]Gough SC, et al. *Diabetes Obes Metab* 2012. [21]Wakil A, et al. *Ther Adv Endocrinol Metab* 2012;3:55.

PITFALLS IN THE MANAGEMENT OF IDIOPATHIC HYPERCALCEMIA IN CATS

Mark E. Peterson, DVM, DACVIM
New York, NY

INTRODUCTION

Over the last two decades, a syndrome of idiopathic hypercalcemia in cats has emerged and appears to be increasing in frequency.[1-3] Although the cause of idiopathic hypercalcemia remains elusive, this is now considered the most common type of hypercalcemia in cats.

Cats with idiopathic hypercalcemia range in age from <1 year to up to 20 years.[1-5] There is no obvious breed or sex predilection, but longhaired cats appear to be overrepresented. About half of affected cats show no clinical signs at diagnosis. Signs reported in the other cats include weight loss, anorexia, or gastrointestinal complains. Most cats with idiopathic hypercalcemia are not azotemic at initial diagnosis, but many develop azotemia at a later date as the disease progresses. Uroliths (often oxalates) are observed in 10% to 15% of affected cats.

DIAGNOSIS OF IDIOPATHIC HYPERCALCEMIA

There are many potential causes of hypercalcemia in the cat. A differential diagnostic list ensures that all possibilities for the development of hypercalcemia have been considered. Hypercalcemia may be parathyroid-independent, or parathyroid-dependent (primary hyperparathyroidism). In parathyroid-independent hypercalcemia, the elevation of ionized calcium (iCa) results in suppression of parathyroid hormone (PTH) production.

The most common diagnoses in cats with persistent elevations in iCa are idiopathic hypercalcemia and malignancy-associated. Renal failure will often be accompanied by elevations in serum total calcium values, but iCa is rarely elevated in this condition. In many cases the diagnosis will be obvious upon analysis of history and physical examination. In others, the cause may not be obvious, and further work-up including hematology, serum biochemistry, body cavity imaging, cytology, and histopathology will be necessary.

The magnitude of elevation of serum tCa cannot be used to make a diagnosis, as there is considerable overlap in the degree of hypercalcemia in cats with idiopathic hypercalcemia or other conditions. Cats with idiopathic hypercalcemia may present with mild increases in tCa concentration (11-12 mg/dL), whereas others may have tCa concentrations >15 mg/dL.

The serum concentrations of both total and ionized calcium are high in cats with idiopathic hypercalcemia, whereas serum PTH concentrations are low-normal to low.[3-5] When measured, serum magnesium and 25(OH)-vitamin D concentrations remain normal. Serum concentrations of PTH-related polypeptide, secreted by some malignant tumors, are not detectable in these cats.

A diagnosis of idiopathic hypercalcemia is made when all other causes of parathyroid-independent hypercalcemia are excluded. Though idiopathic hypercalcemia is the most frequent diagnosis in cats with hypercalcemia, it is an exceedingly uncommon finding in dogs following adequate diagnostic workup.

TREATMENT OF IDIOPATHIC HYPERCALCEMIA

In almost all affected cats, hypercalcemia is slowly progressive and relatively longstanding. Clinical signs are usually relatively mild, at least initially. Therefore, most cats can be treated as outpatients with either dietary therapy, alone or in combination with drug therapy.

Dietary Therapy

I generally start with diet modification as a first-line treatment. No matter what type of diet is chosen, it is best to fed a wet-only diet to promote urinary dilution and lessen the chance for calcium oxalate stones. To this end, we have five different types of cat food diets that may help lower calcium in cats with idiopathic hypercalcemia.

(1) **High fiber diets** (e.g., Purina OM Overweight Management, Royal Canin Gastrointestinal Fiber Response HF, Iams Intestinal Plus Low-Residue, Hill's w/d) will restore normocalcemia in some cats with idiopathic hypercalcemia and calcium oxalate urolithiasis. The effects of fiber on intestinal absorption are complex and depend on the type and amount of fiber in the diet, as well as the interactions with other nutrients in the diet. However, these "high fiber diets" are usually supplemented with extra calcium, and therefore calcium content does not explain why these diets are sometimes successful in treating idiopathic hypercalcemia.

(2) **Prescription kidney diets** (e.g., Purina NF Kidney Function, Royal Canin Renal LP Modified, Iams Renal Plus, Hill's k/d) also may result in normocalcemia in some cats with idiopathic hypercalcemia. Most renal diets are less acidifying than maintenance or high fiber diets and are low in both calcium and phosphorus. The decreased consumption of calcium leads to a decrease in the amount of calcium absorbed, and the effects of alkalinization may decrease the release of calcium from bone. In one study in man, acidogenic diets lead to an increase in PTH secretion, leading to calcium release from bone and increased urinary calcium excretion.[6]

Remember, however that veterinary renal diets are also restricted in phosphorus, which may lead to increased calcitriol (active vitamin D) synthesis by the kidney; the action of this increased serum calcitriol could offset the advantage of the decreased calcium absorption in cats with idiopathic hypercalcemia. Because these diets are lower in protein, renal diets are not my first choice — with time, many cats with idiopathic hypercalcemia will become muscle wasted on these diets.

(3) **"Stone" diets**: Diets developed to prevent calcium oxalate urolithiasis (e.g., Royal Canin Urinary SO, Purina UR Urinary St/Ox, Iams Urinary-O Plus Moderate pH/O, Hill's c/d) may be beneficial in the treatment of cats with idiopathic hypercalcemia. These diets are restricted in calcium and have less urinary acidification (neutral pH urine production). Some are also restricted in oxalic acid and sodium, and have increased moisture in the canned formulation.

(4) **Calcium-restricted, home-prepared diets** can also be used to help lower the serum calcium concentrations in these cats with idiopathic hypercalcemia.[7] These diets must be specially formulated for the hypercalcemic cat, as none of the currently available, commercial cat food diets are calcium restricted. Use of a formulated, home-prepared diet also allows one to restrict the diet's vitamin D content. Ideally, this diet should be formulated with the help of a veterinary nutritionist (e.g., www.balanceit.com).

(5) **Natural species-appropriate diets:** Finally, there is some anecdotal evidence that feeding a more "natural" low carbohydrate, higher protein/fat canned or raw diet may help normalize serum calcium concentrations in some of these cats, particularly those with a mild form of idiopathic hypercalcemia.[8] Studies of cats fed "natural, species-appropriate diets" have not been reported to determine if these diets would be efficacious in controlling hypercalcemia. However, such studies seem warranted, especially since many of the dietary recommendations generally fail to fully normalize the hyperkalemia in these cats.

Unfortunately, in most cats with idiopathic hypercalcemia, dietary therapy will be of minimal long-term benefit, but this management is very unlikely to be harmful. In addition, because many cats with idiopathic hypercalcemia do not show clinical signs for months to years, it is appropriate to try dietary therapy as the first line of therapy at least in cats with mild ionized hypercalcemia.

Glucocorticoids

If normocalcemia has not been restored after a dietary feeding trial of 6 to 8 weeks, treatment with glucocorticosteroids should be considered.[3,5] Administration of glucocorticoids can decrease serum calcium concentration by decreased intestinal absorption of calcium, decreased renal tubular calcium reabsorption, and decreased skeletal mobilization of calcium.

Cats usually do not exhibit serious side effects from glucocorticosteroid treatment, as do dogs (e.g., polyuria, polydipsia). Prednisone is given orally at 5 mg/cat/day for one month before re-evaluation. If the serum ionized calcium concentration is normal, this dose is continued for several months. If ionized calcium value is still increased, the dose is increased to 10 mg/cat/day. Some cats may require as much as 15 to 20 mg prednisone/day to restore normocalcemia. Approximately 50% of cats with idiopathic hypercalcemia become normocalcemic with 5 or 10 mg prednisone/day, but some may require increasing doses to remain normocalcemic over.[3,5]

One disadvantage of prednisolone is that approximately a third of cats will develop secondary diabetes, especially with long-term, high-dose steroid therapy.[9] When dietary modification and treatment with prednisolone have been unsuccessful, treatment with oral bisphosphonates such as alendronate can be considered.

Bisphosphonates

Recently, I have been bypassing prednisolone completely and going right to oral alendronate, starting at 10 mg orally per week.[3,5,10] This drug is a bisphosphonate, which act to lower calcium by reducing the number and action of osteoclasts. Compared to glucocorticoids, alendronate is more successful in lowering calcium concentration without the diabetogenic effects.

It is extremely important to administer alendronate after a 12-hour fast, as food significantly reduces drug absorption. Because the medication may cause esophagitis, we give 6-ml of water orally immediately after administration of the drug to enhance its passage to the stomach. We also dab a small amount of butter on the cat's lips to increase licking and salivation, which will further promote the transit of the pill to the stomach. The cat's fast should be continued for at least 2 hours after alendronate administration.

Serum ionized levels should be checked at 2 weeks, 1 month, 2-3 months, and then every 4-6 months as long as the ionized calcium concentration remains within normal limits.

Most cats seem to have at least a partial calcium-lowering response to alendronate. The drug dose can be raised to up to 30 mg per week, as needed. In the future, oral bisphosphonates will likely replace prednisolone as the drug of choice for treatment of idiopathic hypercalcemia in cats.

Miscellaneous Treatments

Fluid therapy is used as an adjunct treatment in cats with idiopathic hypercalcemia, especially those that have developed renal failure. The administration of subcutaneous fluids on a daily or every other day basis may help expand the extracellular fluid, and promote calciuresis. Diuretics such as furosemide have been used effectively to decrease serum iCa during acute rescue protocols for hypercalcemia, usually in combination with IV fluids. Little is known about the effects of chronic furosemide administration regarding calcium status and the development of dehydration. It is a concern that cats receiving chronic diuretics will undergo diuresis but will not increase their water intake, leading to dehydration.

A calcimimetics (cinaclcet) is a new class of drug to be considered as a potential therapy for hypercalcemia that has emerged in human medicine over the last decade.[11] Calcimimetics interact with the calcium receptor directly, and have been proven effective in lowering iCa, phosphorus, and PTH concentrations in human patients. The potential future use of calcimimetics in the treatment of idiopathic hypercalcemia is an interesting possibility but the high cost of these agents is a limiting factor for long-term treatment of our cats.

REFERENCES

[1]Savary KC, et al. *J Vet Intern Med* 2000;14:184. [2]Midkiff AM, et al: *J Vet Intern Med* 2000;14:619. [3]Schenck PA, et al. *Waltham Focus* 2005;15:20. [4]Schenck PA, et al. *J Vet Intern Med* 2004;18:442. [5]Schenck PA et al. *BSAVA Manual of Canine and Feline Endocrinology, Fourth ed.* 2012:221. [6]Jajoo R, et al. *J Am College Nutr* 2006;25:224. [7]Fascetti AJ, et al. *Applied Veterinary Clinical Nutrition.* 2012:289. [8]Feline Idiopathic Hypercalcemia (website). *www.felineihc.org.* [9]Lowe AD, et al. *J Am Anim Hosp Assoc* 2009;45:215. [10]Whitney JL, et al. *J Fel Med Surg* 2011;13:129. [11]Herbert SC. *Annu Rev Med* 2006;57:349.

DIAGNOSIS IMHA: THE COOMBS' TEST

Urs Giger, DACVIM & DECVIM (Medicine), DECVCP, Ladina Caviezel and Karthik Raj
Philadelphia, Pennsylvania

Immune-mediated hemolytic anemia (IMHA) is one of the most common and serious hemolytic anemias in dogs, but occurs rarely in other animal species. In IMHA an immune response, including anti-erythrocytic antibodies, complement and macrophages, targets directly or indirectly erythrocytes and a hemolytic anemia ensues. There are many triggers for IMHA such as infections, drugs, and cancer leading to secondary IMHA, but in many dogs no cause is identified (so-called idiopathic, autoimmune or primary IMHA) or a genetic predisposition has been proposed (Cocker spaniels). Furthermore, alloimmune hemolytic anemias, such as hemolytic transfusion reactions, both acute and delayed, and neonatal isoerythrolysis (only litters from transfused bitches), are caused by specific anti-erythrocytic alloantibodies. In contrast to other species, dogs with IMHA also develop an often overwhelming inflammatory response resulting in thrombosis and necrosis of various organs. And while the anemia can be corrected with transfusions, these complications in dogs are causing severe morbidity and mortality despite aggressive immunosuppression and antithrombotic interventions. This session will review the diagnostic approach to IMHA with special emphasis on the Coombs' test including a comparative prospective study of different Coombs' tests.

Regardless of the underlying cause, IMHA results from a breakdown in immune self-tolerance or from a deficit in the control mechanism that regulates B and T lymphocyte activity as well as macrophage reactivity. Immune destruction of erythrocytes is initiated by the binding of IgG or IgM antibodies to the surface of erythrocytes. Under most clinical circumstances, immune destruction is an extravascular process that depends on recognition of erythrocytes opsonized with IgG, IgM and/or complement by specific receptors on reticuloendothelial cells. Macrophages with engulfed erythrocytes may be noted on cytological examination of blood and tissue aspirates as erythrophagocytosis, but this is not definitive proof of an immune-mediated process. Antibody-coated erythrocytes may also be lysed by complement fixation and the membrane attack complex, which is clinically noted as intravascular hemolysis.

A diagnosis of IMHA must demonstrate accelerated immune destruction of erythrocytes. Evidence of a hemolytic anemia is suggested clinically by icterus and a regenerative anemia with hyperbilirubinuria, and hemoglobinemia and hemoglobinuria refers to an intravascular process. However, the erythroid response in the bone marrow may be blunted by the immune and inflammatory process or the underlying disease thereby leading to non-regenerative anemias. Besides documenting a hemolytic anemia, one or more of the following three hallmarks must be present to support a diagnosis of immune-mediated hemolysis: persistent autoagglutination, marked spherocytosis and a positive direct Coombs' test result. As in human medicine, the Coombs' test should be considered the best test to definitively diagnose IMHA, although marked spherocytosis and persistent/true autoagglutination (after 3x washing with saline) are other important parameters indicating immune-destruction of erythrocytes.

AUTOAGGLUTINATION

Anti-erythrocytic IgM and in large quantities IgG antibodies may cause direct erythrocyte autoagglutination. The autoagglutination may be seen by naked eye in an EDTA tube or on a glass slide or may become apparent as small clumps of erythrocytes on blood smears. For yet unexplained reasons, canine erythrocytes have a tendency to unspecifically agglutinate in the presence of plasma and colder temperatures as well as possibly with excessive EDTA anticoagulant. Mixing blood with one drop of saline may break up rouleaux formation but not other forms of unspecific red cell agglutination. It is, therefore, important to determine whether the agglutination persists after "saline washing", which has been coined persistent or true autoagglutination. This is accomplished by adding physiologic saline to the tube containing a small amount of EDTA-anticoagulated blood, mixing, centrifuging and removing the supernatant including the plasma and repeating this saline washing 3 times. True autoagglutination is indicative of an immune process, but precludes the performance of most Coombs' test or blood typing and crossmatching procedures. If the agglutination breaks up after washing, the Coombs' test is expected to be positive, if it is a case of IMHA.

SPHEROCYTOSIS

If erythrocytes are only partially phagocytized or lysed by complement in circulation, erythrocytes with reduced surface area to volume ratio, known as spherocytes, are formed. They appear spherical and microcytic with no central pallor and are considered fragile. Large numbers of spherocytes are nearly diagnostic for IMHA, whereas small numbers may be seen with other conditions including DIC, endotoxemia and zinc intoxication. In our experience all dogs with marked spherocytosis and suspected to have IMHA had a positive Coombs' test. Hereditary spherocytosis due to a genetic membrane defect has rarely been seen in dogs but should be considered as a differential diagnosis in dogs with negative Coombs' test results.

Because of the difficulties with the Coombs' test (see below), Slappendale had proposed to use the erythrocytic osmotic fragility test at specific saline concentrations as a mean to diagnose IMHA and this test is currently used in various clinics in Europe. However, there are many other reasons for increased fragility of erythrocytes beside IMHA including hereditary red cell defects. This test is not used in human medicine and has not been shown to be superior to determination of marked spherocytosis and a positive Coombs' test in dogs with IMHA.

POSITIVE DIRECT COOMBS' TEST RESULT

The direct Coombs' test is also known as direct antiglobulin test and is used to detect antibodies and complement on the surface of erythrocytes when the anti-erythrocyte antibody strength or concentration is too low to cause spontaneous agglutination (subagglutinating titer). Separate canine-specific IgG, IgM, and C3b antibodies as well as polyvalent antiglobulin reagents are available. They are added at various concentrations after washing the patient's erythrocyte free of plasma (3x) and mixtures are generally incubated at room temperature or 37°C (cold agglutinins appear to be rarely of clinical importance and rarely cause hemolysis). The strength of the Coombs' reaction does not necessarily predict the severity of hemolysis but changes are useful in monitoring the disease.

Typically a tube or microtiter method has been used in the laboratory, but a flow cytometric method has also been introduced in a couple of places. A standardized, sensitive, and simple gel column method was available by DiaMed (Switzerland), but unfortunately the company was sold to another company which decided to not pursue the veterinary market. A novel standardized antiglobulin test method is being developed by Alvedia (France) similar to the immunochromatographic strip technique for blood typing. Although many commercial laboratories offer Coombs' testing for dogs, clinicians have questioned the tests sensitivity and specificity and often forgo the test and/or use response to therapy as a diagnostic. However, negative Coombs' test results may be seen because of technical reasons, insufficient quantities of bound antibodies, the presence of weakly bound antibodies, or the disease in remission. The Coombs' test stays positive for days to months after initiating treatment. A few days of immunosuppressive therapy will likely not reverse the Coombs' test result, as unlikely a transfusion would cause a positive Coombs' test result. Thus, dogs with negative Coombs' test results should be reevaluated for other causes of hemolytic anemia.

In a recent prospective study of anemic and non-anemic dogs we compared various direct and indirect Coombs' test methods including microtiter plate assays, capillary, and immunochromatographic techniques using polyvalent antiglobulins in a laboratory setting and found excellent correlations between tests and with spherocytosis and without noticeable interference by immunosuppressive or transfusion therapy in anemic dogs.

In conclusion, a diagnosis of IMHA requires the documentation of red blood cell destruction and an immune process. While regenerative anemia, icterus, and hyperbilirubinuria are suggesting a hemolytic anemia, evidence of true autoagglutination, spherocytosis, and/or a positive direct Coombs' test are required to document immune destruction. The authors also recommend monitoring IMHA patients for the disappearance of these immunological parameters to adjust and taper therapy.

References are available from authors.

ASPIRIN RESISTANCE: WHAT DO WE REALLY KNOW?

John M. Thomason, DVM, MS, DACVIM
Starkville, MS

INTRODUCTION

Normal blood hemostasis is a complicated process that involves platelets, vascular endothelium, numerous coagulation factors, and several interacting cells. Hemostasis is typically broken down into three distinct phases: first, the formation of a platelet plug through the interactions of platelets and the vessel endothelium (primary hemostasis), second, the interaction of coagulation factors to form a blood clot that strengthens the platelet plug with a cross-linked fibrin meshwork (secondary hemostasis), and finally, the activation of plasmin to break down fibrin and remove the blood clot (fibrinolysis). Defects in any of these phases can result in ineffective hemostasis, leading to either excessive hemorrhage or clot formation. The formation of a temporary platelet plug during primary hemostasis is one of the first steps in coagulation, but is only stable for a short period of time and, without coagulation factors, the plug will weaken and a blood clot may not form. On the other hand, continual platelet activation, especially in hypercoagulable states, will support excessive platelet plug and blood clot formation. The use of a safe, long term, affordable oral anti-platelet medication for the prevention of thromboembolism becomes a critical component of therapy in patients with in a hypercoagulable state. Aspirin has been used for many years for its anti-inflammatory and analgesia properties, but is also the most commonly used anti-platelet medication.

PLATELET PHYSIOLOGY

When exposed to injured or damaged endothelium, either due to normal cell turnover or vessel trauma, platelets will adhere via multiple adhesive ligands and receptors. Under high shear conditions, such as small and medium sized arteries, the adherence of platelets to the exposed subendothelial is primarily regulated receptors that bind collagen and vWF, glycoprotein (GP) Ib-IX-V and GP VI.[1,2] In the small/medium sized arteries, vWF is responsible for mediating platelet aggregation. Under low shear, in large arteries and veins, platelet adhesion to the subendothelium is regulated via collagen and fibronectin, and fibrinogen is the main ligand of thrombus growth.[1]

Following platelet adhesion, rapid signal transduction leads to platelet activation. The combined effects of platelet activation is to cause vasoconstriction, decrease blood flow, and increase blood stasis; ultimately increasing contact time between activated platelets, coagulation factors and damaged vascular subendothelium to initiate thrombus formation. Platelet activation typically consists of: platelet shape change, surface receptors expression, intra-platelet granule release, and eicosanoid synthesize. The conformational shape change during platelet activation will lead to the expression of several supportive surface intergrins necessary for adhesion and aggregation. Intra-platelet granules, contain multiple proteins that enhance platelet adhesion, promote cell-to-cell interactions, and stimulate vascular repair.[3,4] Substances found in α-granules include adhesion molecules (von Willebrand factor and fibronectin), pro and anti-angiogenic factors, fibrinogen, and coagulation factors; while dense (delta) granules contain adenosine diphosphate, epinephrine, and serotonin.[4] These biologically active molecules can be selectively packaged within different α-granule subsets in the bone marrow and during circulation.[3-5] Once released from membrane phospholipids, arachidonic acid is converted to several biologically active eicosanoids that influence platelet function.

Platelet activation does not only occur from exposure to negatively charged phospholipids present in the damaged vasculature, but can be initiated by the actions of platelet agonists, such as thrombin, epinephrine, ADP and thromboxane A_2. These platelet agonists help generate the conformational change of the platelet surface glycoprotein integrin $\alpha IIb\beta 3$ (formerly called the GPIIb/IIIa complex), enabling the interaction among platelets, fibrinogen, and vWF to stabilize the platelet plug.[2] To ensure the activation of the integrin $\alpha IIb\beta 3$, certain secondary/amplification pathways will help ensure platelet plug development. One of these amplification pathways involves thromboxane A_2 production via the cyclooxygenase enzyme pathway. A second amplification pathway involves the ADP-driven activation of the G protein-coupled receptors (especially $P2Y_{12}$) located on the platelet surface. These pathways of platelet activation, along with other platelet receptors and agonists, have become targets for medications that inhibit platelet function.

CYCLOOXYGENASE

Cells, including platelets, do not store thromboxane, but the prostaglandin is synthesized from arachidonic acid that has been released from phospholipid membranes through the actions of several phospholipases that include phospholipase A_2, prostaglandin H synthase, and thromboxane synthase.[6,7] Once mobilized, arachidonic acid is converted into prostaglandin H_2 by the cyclooxygenase (COX) enzyme. At least three COX isoforms have been described in dogs. COX-1 is constitutively expressed in most body systems, and functions to maintain normal cellular activities including platelet thromboxane A_2 production, maintenance of gastric mucosal protection, and preservation of renal blood flow.[7,8] Relative to COX-1, COX-2 has a far more limited tissue distribution, and expression is often at much lower levels. COX-2 expression occurs in the brain, kidney, thymus, circulating monocytes, tissue macrophages, and the vascular endothelium.[8] COX-2 expression can be induced by various endogenous and exogenous mediators including inflammatory cytokines, growth factors, and endotoxins.[7] The COX-3 isoform is transcribed from the COX-1 gene as a splice variant and has similar catalytic features, but only about 20% of the activity of COX-1.[9,10]

Both COX enzymes play key roles in normal hemostasis, and lead to the production of several biologically active prostaglandins, mainly thromboxane A_2 and prostacyclin.[11] Unlike thromboxane A_2, which is mainly produced by platelets to enhance aggregation, prostacyclin (prostaglandin I_2) originates from the vascular endothelium and has the opposite physiologic effect on primary hemostasis (vasodilation and inhibiting platelet aggregation). These two prostaglandins work in concert to maintain a balanced, normal hemostatic system.

Since mature platelets do not contain a nucleus and are therefore incapable of mRNA synthesis, it was long believed that the constitutively expressed COX-1 was the only COX isoform expressed in circulating platelets, and that COX induction could happen only at the level of the bone marrow. As a result, it was originally believed that COX-1 was the primary mediator of platelet thromboxane A_2 production, while COX-2 was the main isoform that controlled endothelial prostacyclin production.[6,12] Recently, using flow cytometry, COX-2 expression was identified in both human and canine platelets.[12,13] In humans, platelet COX-2 expression is greater during times of increased platelet turnover and higher levels of reticulated platelets.[11,14] These recent discoveries suggest that platelet COX-2 could in fact play a role in platelet function.[8,12]

ASPIRIN

Aspirin (acetylsalicylic acid) inhibits platelet function by irreversibly acetylating a hydroxyl group on the COX enzyme, preventing arachidonic acid from reaching the active site of the enzyme.[7,11] By acetylating the COX enzyme, aspirin prevents platelet thromboxane A_2 synthesis, and inhibits platelet activation and function.[7,15] Since platelets do not contain a nucleus, and are unable to synthesize new and functioning COX enzymes, aspirin-induced platelet dysfunction will persist for the entire lifespan of the platelet, and function will not return until the drug is discontinued and new platelets, not exposed to aspirin, are released from the bone marrow. The actions of COX-2 at the vascular endothelium, which is less sensitive than COX-1 to inhibition by aspirin at anti-platelet doses, allows for continued COX-2 dependent PGI_2 synthesis of anti-platelet doses, in spite of the early suppression of COX-1.[16]

The aspirin-induced inhibition of the COX enzyme, not only inhibits the prostaglandins needed for platelet function, but also inhibits all prostaglandins generated through the arachidonic acid cascade. These prostaglandins play key roles in mediating inflammation, tissue homeostasis, renal blood flow, and gastrointestinal wall protection. Aspirin's ability to inhibit the COX enzyme will also inhibit pro-inflammatory prostaglandin synthesis, making this medication an effective anti-inflammatory therapy. Unfortunately, by inhibiting the COX enzyme and decreasing prostaglandin synthesis, aspirin can cause several life-threatening side effects that include gastrointestinal hemorrhage and renal failure.[17] An anti-inflammatory dose, or high dose, of aspirin (10 mg/kg twice daily), is very effective at consistently inhibiting platelet function, but can lead to multiple gastrointestinal and renal side effects, which makes this aspirin dose an undesirable preventative therapy.

A high dose of aspirin permanently inhibits COX function in all cells that express this enzyme, including vascular endothelial cells. However, unlike the higher, anti-inflammatory dose, a low dose of aspirin (0.5-1 mg/kg once daily) will inhibit platelet thromboxane synthesis and function without permanently inhibiting COX function of the vascular endothelium, allowing prostacyclin synthesis and vasodilation to continue.[18] Lower doses will also eliminate the potentially devastating side effects that occur with high dose aspirin. Even when administered with concurrent high doses of prednisone, there does not appear to be a significant increase in the incidence of gastrointestinal lesions during low dose aspirin therapy. However, the concurrent administration of prednisone and low-dose aspirin may increase the frequency of mild, self-limiting diarrhea.[19] Low dose aspirin, both in humans and dogs, is a beneficial treatment for the prevention of thromboembolic complications. In fact, the administration of low doses of aspirin has been associated with an improved survival rate in dogs with immune-mediated hemolytic anemia (IMHA).[20]

In people, orally dosed aspirin is absorbed rapidly from the stomach and upper intestine reaching peak plasma levels approximately 30 to 40 minutes after dosing. The bioavailability of ingested aspirin is 40 to 50% over a very wide range of doses.[21] In people, the platelet COX-1 enzyme is acetylated in the pre-systemic circulation prior to metabolism in the liver and therefore the antiplatelet effect is independent of systemic bioavailability.[21] There remains considerable debate around several issues regarding aspirin therapy in people. Questions still remain as to the optimal dose of aspirin, possible antithrombotic effects of aspirin unrelated to the inhibition of platelet TXA_2 production and the possibility that some patients are "resistant" to the anti-platelet effects of aspirin.

ASPIRIN RESISTANCE

Unfortunately, despite preventative therapy, some dogs treated with aspirin will continue to develop thromboembolic complications. A similar phenomenon occurs in humans that are receiving low doses of aspirin, indicating that these patients are poorly responsive to the anti-platelet effects of aspirin ('aspirin resistant'). The true incidence of aspirin resistance in humans is unknown, but the published incidence ranges from 8-45% of human patients.[15,22] Recently, it has been shown that only 33% of healthy dogs treated with low dose aspirin (1 mg/kg PO q24 hours) consistently had impaired platelet function using closure times of the PFA-100® system.[23] Since canine clinic patients appear to inconsistently respond to low dose aspirin, it is possible that, similar to humans, dogs also experience aspirin resistance. The true incidence of aspirin resistance in dogs is, however, unknown.

The exact mechanism for aspirin resistance in both humans and dogs is also unknown; however, there have been many proposed explanations. Previous studies have suggested that about 70% of canine platelets are insensitive to thromboxane stimulation. The thromboxane insensitivity observed in some canine platelets may be due to impaired platelet thromboxane A_2 receptor linked G proteins.[24] In those dogs that have platelets that are insensitive to thromboxane-induced activation, other agonists such as ADP or serotonin may play a more important role in platelet activation, and such dogs may therefore be relatively resistant to the inhibitory effects of low dose aspirin. A recent human study suggests that resistance to aspirin may be due to a COX independent pathway and that, in affected people, inhibition of platelet function is more readily achieved using an ADP antagonist.[22]

Another proposed mechanism is the presence of an alternate source of thromboxane.[11] Since aspirin is highly effective at inhibiting COX-1 production of thromboxane, an additional thromboxane source would require a COX-1 independent pathway. One of the proposed alternate pathways of thromboxane production is the COX-2 isoform, either from the vascular endothelium or circulating platelets.[11,12] In humans, the inhibition of the COX-2 enzyme requires a much higher dose of aspirin than COX-1. For example, in humans, an aspirin dose of 81 – 325 mg per day is adequate to inhibit COX-1 function, while >325 mg per day of aspirin is need to inhibit COX-2 function.[11] Low doses of aspirin, administered to dogs at a dose of 0.5 mg/kg per day, may be sufficient to inhibit COX-1 synthesis of thromboxane, but insufficient to inhibit COX-2 production of thromboxane.

A third proposed mechanism of aspirin resistance includes genetic differences in the polymorphism of both COX-1 and COX-2, making these enzymes less sensitive to aspirin. In humans, polymorphism of the gene that codes for platelet glycoprotein IIIa, PLA2, has been associated with aspirin resistance.[25,26] Other proposed mechanisms of aspirin resistance include an increased production of new platelets that have not been exposed to aspirin, and increased production of isoprostanes, specifically 8-iso-$PGF_{2\alpha}$, which induce vasoconstriction and interact with TXA_2/PGH_2 receptors.[27] Finally, a deficiency in either owner compliance and drug absorption could lead to a perceived resistance to the anti-platelet effects of low-dose aspirin.

CATS: ASPIRIN THERAPY AND RESISTANCE

The pharmacokinetics of aspirin in cats is altered in cats due to their relative deficiency of glucuronate, which is responsible for the prolonged elimination half-life of approximately 38 hours in the cat. This difference in pharmacokinetics may prolong the exposure of cat platelets to the antiplatelet effects of aspirin prior to hepatic metabolism, although no formal studies have been performed evaluating the specific effects of feline aspirin metabolism on antiplatelet effects. The most common use of aspirin in cats is for the prophylactic therapy in patients with arterial thromboembolism. Unfortunately, there are no prospective, placebo-controlled study has confirmed the use of this medication as an effective therapy to reduce the incidence of arterial thromboembolism. However, several retrospective studies have suggested that the use of aspirin in cardiac patients that have previously developed an arterial thromboembolism may experience a reduced risk of thromboembolic complications.[28,29] Smith et al. demonstrated that there was no difference in the recurrence rate of arterial thromboemboli in patients that receive a high dose of aspirin (>40 mg/cat PO q24–72 hours) versus a low dose (5 mg/cat PO q72 hours).[29] However, when compared to the high dose, the cats receiving the lower dose of aspirin experienced considerably fewer gastrointestinal side effects, suggesting that the use of a low dose of aspirin may provide a safe and effective thromboprophylactic therapy in cats that have previously developed an arterial thromboembolism.[29]

Despite the apparent beneficial effects in cardiac patients, the anti-platelet effects of aspirin in cats are still questionable. Rawlings evaluated the use of aspirin to prevent platelet adhesion to damaged arterial surfaces in cats with heartworm disease. Even with individually adjusting aspirin doses, there was limited inhibition to platelet function. Additionally, these adjusted doses of aspirin in these cats were near the toxic dosages for cats, suggesting that aspirin may provide minimal prevention of arterial disease and thromboembolism in cats with heartworm disease.[30] Recently, Cathcart et al. demonstrated that despite a decrease in serum thromboxane B_2 (metabolite of thromboxane A_2), aspirin (5 mg/kg PO q48 hours) had no effect on whole blood aggregation using ADP and collagen as agonists.[31] The inability of low-dose aspirin to inhibit platelet aggregation with concurrent inhibition of thromboxane synthesis, may suggest, as in other species, cats also experience aspirin resistance. At this time, there are no studies that have investigated the presence or incidence of aspirin resistance in cats. Due to the limited number of studies in cats treated with aspirin, additional studies are necessary to better understand the effects of this anti-platelet therapy on feline platelet function and better determine a safe and effective dose.

REFERENCES

1. McMichael M. *J Vet Emerg Crit Care* 2005;15(1):1. 2. Rivera J. *Haematologica* 2009;94:700. 3. Italiano J, et al. *Blood* 2008;111(3):1227. 4. Sehgal S, et al. *J Thromb Haemost* 2007;5:2009. 5. Curotto S, et al. *J Comp Path* 2012(147):499. 6. McAdam B, et al. *Pro Natl Acad Sci USA* 1999;96:272. 7. Bjorkman D. *Am J Med* 1998;105(1B):8S. 8. Rocca B, et al. *Proc Natl Acad Sci USA* 2002;99:7634. 9. Munoz J, et al. *Inflammopharmacol* 2010;18:65. 10. Chandrasekharan N, et al. *Proc Natl Acad Sci USA* 2002;99(21):13926. 11. Guthikonda S, et al. *J Thromb Haemost* 2007;5:490. 12.Weber A, et al. *Br J Haematol* 2002;117:424. 13. Thomason J, et al. *J Vet Intern Med* 2011;25:1106. 14. Tanaka N, et al. *Arterioscler Thromb Vasc Biol* 2004;24:607. 15. Patrono C, et al. *Chest* 2001;119:39S. 16. Clarke R, et al. *N Engl J Med* 1991;325:1137. 17. Catella F, et al. *Proc Natl Acad Sci USA* 1986;83:5861. 18. Rackear D, et al. *J Am Anim Hosp Assoc* 1988;24(1):23. 19. Graham, et al. *J Vet Intern Med* 2009;23:482. 20. Weinkle T, et al. *J Am Vet Med Assoc* 2005; 226:1869. 21. Pedersen A, et al. *N Engl J Med* 1984;311:1206. 22. Frelinger et al. *Circulation* 2006;113:2888. 23. Dudley A, et al. *J Vet Intern Med* 2013;27:141. 24. Johnson G, et al. *J Clin Invest* 1993;92:2469. 25. Cooke G, et al. *Lancet* 1998;351:1253. 26. Macchi L, et al. *J Am Coll Cardiol* 2003;42:1115. 27. Patrignoni P. *Thromb Res* 2003;110(5-6):281. 28. Schoemann J. *J Feline Med Surg* 1999;1:221. 29. Smith S, et al. *J Vet Intern Med* 2003;17:73. 30. Rawlings C. *J Vet Intern Med* 1990;4:285. 31. Cathart C, et al. *J Vet Emerg Crit Care* 2012;22:99.

TRANSFUSION-RELATED BABESIOSIS: IS IT A REAL ISSUE?

Andrew Mackin, BSc, BVMS, MVS, DVSc, FACVSc, DACVIM
Starkville, MS

BABESIOSIS

Babesiosis (piroplasmosis) is caused by a protozoal organism that infects the red cells of many different mammalian species. More specifically, babesiosis is caused by organisms from the genus *Babesia* within the family Babesiidae in the order Piroplasmida. To date, there are over 100 different *Babesia* species recognized. The organism is named after Victor Babeș, a Romanian biologist who over a century ago first recognized the protozoan in the RBCs of cattle. Because *Babesia* organisms are often 'pear-shaped' (pyriform, from the Medieval Latin *pyrum,* for pear), babesiosis is also often referred to as piroplasmosis. In general, the various *Babesia* species can be visualized within RBCs in either a small form ('ring' shape) or a large form ('pear' or 'tear-drop' shape). *Babesia* typically has both a mammalian vertebrate host and an invertebrate host, with the invertebrate host invariably being a tick.

Babesiosis has probably been around for many millennia: the typical 'red water fever' caused by *Babesia bovis* was actually described in cattle in ancient Egypt. In dogs and humans, however, babesiosis has recently been described as an 'emerging disease', despite the fact that the organism has been seen in the RBCs of both species for well over half a century. In the US, human babesiosis is increasingly recognized as an important emerging disease, especially in infants and older individuals, and immunosuppressed or splenectomized patients. Within patients within these particular demographic groups, transfusion has emerged as one of the major risk factors for the development of subsequent babesiosis. In fact, over the past few decades, many hundreds of transfusion-transmitted babesiosis cases have been reported in human patients in the US, with a case mortality rate of well over 10%[1]. Transfusion-transmitted babesiosis now even has its own well-recognized acronym, TTB. The incidence of transfusion-transmitted babesiosis has steadily risen over the past few decades, and TTB is now one of the top five causes of transfusion-related deaths in people.

In contrast to human medicine, where hundreds of instances of TTB have been described[1,2], only a handful of publications have described transfusion-related babesiosis in dogs, and these have mostly been single case reports[3]. Although TTB in the dog is rarely published, this does not necessarily mean that the condition itself is rare. Babesiosis is well-recognized by the veterinary transfusion medicine community as a potential complication of transfusion and, in the individual dog affected by this condition, the consequences can be serious or even fatal. Certainly, to the individual infected dog, TTB *is* important. Furthermore, canine TTB is a complication that, with some foresight and planning, is probably avoidable in most dogs. In understanding transfusion-related babesiosis in the dog, we can borrow heavily from the human experience with TTB, where the condition has been much more comprehensively studied.

BABESIA IN THE DOG

A number of different *Babesia* species and subspecies are known to infect the RBCs of dogs. Determining the species and subspecies of *Babesia* infecting an individual canine patient is important because of species differences in virulence, prognosis, and response to treatment. Interestingly, not all dogs infected with *Babesia* will show clinical signs of infection, or anemia: a sub-clinical carrier state is common, particularly in those dog breeds in which babesiosis appears to be endemic.

Important Canine *Babesia* Species

Babesia canis is an example of a 'large' *Babesia*. Organisms measure 2 to 5 microns in size, and are typically pyriform. At least three subtypes of *B. canis* are recognized. *B. canis* subtype *vogeli* is endemic in the southeastern US, where it appears to be entrenched in greyhounds[4], although it is uncommon in non-greyhound breeds. Many infected greyhounds seem to be completely sub-clinical for the disease. This subtype is transmitted by the brown dog tick. *B. canis* subtype *canis* is endemic in Europe and Asia, and is typically a little more pathogenic than subtype *vogeli*. *B. canis* subtype *rossi* is endemic in Africa, and is by far the most pathogenic of the three *B. canis* subtypes. One case of *B. canis rossi* has been described in the US, in a dog recently imported from South Africa[5].

Babesia gibsoni is an example of a 'small' *Babesia*. Organisms measure 1 to 3 microns in size, and are often ring-shaped. In the US, the suspected vector for *B. gibsoni* is also the brown dog tick. *B. gibsoni* can also be directly transmitted from one dog to another by dog bites. Dogs infected by direct transmission can develop babesiosis, even if the source of infection is subclinical. The most common *B. gibsoni* genotype in the US appears to have originated in Asia, and the vast majority of infected dogs are pit bulls or related breeds[6]. The Asian genotype *of B. gibsoni* is considered to be more pathogenic than the other common US form of babesiosis, *B. canis vogeli*, although subclinical carriers are still common.

Generally, although babesiosis in people is considered to be a zoonotic disease, dogs are not considered to be an important source of human babesiosis. In people in the US, the most important forms of babesiosis are *Babesia microti*, in which the reservoir vertebrate host appears to be rodents and, to a lesser extent, *B. divergens*, in which cattle and deer are the reservoir hosts[7]. Although *B. microti*-like organisms have been reported in the RBCs of dogs in the US, this finding to date has no known zoonotic ramifications.

Transmission

Transmission of babesiosis in dogs can occur via many different mechanisms. *Babesia* species are commonly transmitted by bites from infected ticks. In kennel situations, *Babesia* infection rates are highest in kennels that have poor tick control. Infection occurs when feeding ticks inoculate infective sporozoites. *S*porozoites then invade RBCs where they multiply asexually by binary fission to form merozoites. The merozoites then rupture the cell and invade more RBCs. The cycle is perpetuated when a new tick feeds.

Babesia is also thought to be vertically transmissible from an infected dam to her puppies, probably via transplacental infection[6,8]. There is strong epidemiological evidence that *B. gibsoni*, in particular, can also be transmitted via dog bites, most likely through intermingling of blood associated with fighting. Iatrogenic *Babesia* transmission is thought to be possible via inoculation with contaminated needles or surgical instruments. Furthermore, all of the important *Babesia* species appear to be transmissible via blood transfusion. Dogs infected via transfusion often suffer from clinically severe babesiosis. In fact, reported cases of TTB in dogs clinically resemble the most extreme form of canine babesiosis, that caused by the African *B. canis rossi,* with shock and involvement of multiple organs, despite the fact that the transfused organisms are the typically more benign *B. canis vogeli* or *B. gibsoni*[3,9].

Clinical Findings

The most pathogenic form of *Babesia*, the African *B. canis rossi*, can lead to severe illness and multiple-organ dysfunction syndrome. However, in the US, signs of *Babesia* infection are more typically associated with hemolysis. Hemolysis can be due to direct parasite damage to the cell, but host responses to the parasite (immune-mediated RBC destruction) seem to be more important. Signs of canine babesiosis can range from peracute to acute to chronic to completely subclinical. Peracute babesiosis, however, is typically limited to the African version of the disease, with the unlucky exception of the occasional case of TTB[3,9].

Acute babesiosis is characterized by severe hemolytic anemia, moderate to severe thrombocytopenia, splenomegaly, lymphadenopathy, icterus, fever, lethargy and anorexia. Acute babesiosis is most common with *B. gibsoni*, particularly in puppies, but can also follow iatrogenic infection through blood transfusion of any *Babesia* species.

Chronic, low-grade manifestations of babesiosis are common with *B. canis*, and can also be seen with *B. gibsoni*. Signs can be subtle and difficult to characterize, and may include mild to moderate anemia, mild to moderate thrombocytopenia, low-grade fever, splenomegaly, lymphadenopathy, weight loss, poor appetite, failure to thrive and, in greyhounds, poor racing performance. In breeds where the organism is endemic, such as greyhounds and pit bulls, infection is often subclinical. Subclinical or occult infections may, however, progress to clinically significant disease if affected dogs undergo splenectomy or immunosuppressive therapy.

Diagnosis

Babesiosis should be included in the differential list in any dog with hemolytic anemia and thrombocytopenia, particularly if accompanied by fever, splenomegaly or lymphadenopathy, and particularly in a dog of a predisposed breed (greyhound or pit bull) or with a history of recent dog bites, splenectomy, immunosuppressive therapy or transfusion. Specific diagnostic tests include blood smear examination, flow cytometry, serology and polymerase chain reaction (PCR).

A diagnosis of babesiosis is easily confirmed if organisms are seen on blood smear examination. In many infected dogs, however, parasitemia can be mild or intermittent, and smears are therefore often negative. The chances of making a diagnosis are increased by a meticulous examination of the smear and may be enhanced by evaluating serial blood samples over several days, by examining a buffy coat smear, and by collecting samples from capillary beds (pinna or nail bed). Examination of stained RBCs by flow cytometry is a sensitive method for detecting low numbers of infected RBCs[10], but can be non-specific: reticulocytes, for example, can cause false positive results. Flow cytometry for the detection of *Babesia* organisms is, however, not routinely available except in experimental settings. Serology (ELISA or, most commonly, IFA) is available for both *B. canis* and *B. gibsoni*. High titers are suggestive of either active disease or past infection, although false positive results have been reported. False negatives are also seen, especially in puppies, immunosuppressed animals and in patients early in the course of acute infection[2]. Sensitive PCR-based tests are available for detecting both *B. canis* and *B. gibsoni*, and are the best means of detecting mild or subclinical infections with low numbers of organisms[11]. PCR-based methods are is the only reliable means of differentiating between the different *Babesia* species, because the standard serological tests can cross-react between species, and because the species can sometimes be hard to differentiate based on microscopic appearance alone (small ring forms, for example, can sometimes be seen with the large *Babesia* species). Since known infected but subclinical dogs can harbor very low numbers of organisms, sometimes even an extremely sensitive methodology such as PCR can yield false negatives. Repeated sampling in such dogs will sometimes later provide a positive PCR result.

Since the hemolysis associated with *Babesia* has an immune-mediated component, dogs with babesiosis may have positive slide agglutination or Coomb's tests. Babesiosis may be misdiagnosed as primary IMHA unless clinicians have a high index of suspicion in dogs belonging to at-risk breeds and dogs with a history of recent dog bites, splenectomy, immunosuppressive therapy or transfusion.

Treatment

In the US, the approved drug for the treatment of *B. canis* in dogs is imidocarb diproprionate, which is often very effective at clearing the organism. A combination of atovaquone and azithromycin has been shown to effectively clear *B. gibsoni* in most dogs[12]. Unfortunately, atovaquone can be prohibitively expensive for some owners. Alternatively, imidocarb alone or a combination of imidocarb and diminazene can be used to treat *B. gibsoni*, but these drugs are not as effective as atovaquone and azithromycin. Transfusion with packed red cells or whole blood is indicated in severely anemic dogs. Because the RBC destruction in dogs with babesiosis often has an immune-mediated component, glucocorticoids may arguably be indicated as a last resort in severe cases.

TRANSFUSION-TRANSMITTED BABESIOSIS IN PEOPLE

Babesiosis is now well-recognized as an important emerging disease in people in the US, and TTB is one prominent manifestation of this disease. Babesiosis has gradually become more common in North America (particularly in the north-east US) and, since the acute condition is often a straightforward diagnosis, the increase in disease frequency is probably real, and not just 'acute discovery of an old disease'. As babesiosis in general has become more common, TTB has also become more common, especially since transfused patients often already have multiple risk factors for the disease, including immunosuppressive therapy, splenectomy, and extremes of age (babesiosis is most severe in the very young and very old)[1,7]. One recent CDC study reported 162 cases of TTB in people over the past 30 years (159 cases of *B. microti* and 3 of *B. duncani*), with the vast majority occurring in the last decade[1]. The mortality rate in infected patients was almost 20% although, obviously, some patients had serious underlying conditions that may have contributed to their demise[1]. As expected, almost all cases were associated with RBC transfusions, as opposed to platelet or plasma transfusions.

For most important transfusion-transmitted infections in people, such as HIV and viral hepatitis, rigorous screening protocols have been established in order to minimize the risk of infection. Potential donors with a higher risk of carrying specific infectious agents are identified by careful pre-donation questioning, and individual bags are screened by nucleic acid testing (NAT). In contrast, for TTB, such standard measures have yet to be adopted, and not just because blood banks are reacting slowly to a new syndrome. Sub-clinical babesiosis in donors in the US is still very regional (mostly in the north-east), and to date questionnaires have not done a good job at identifying potentially infected donors[7]. Screening every blood unit in the US for *Babesia* would be cost-prohibitive for a disease that is uncommon in most areas, and the invariable small proportion of false positives associated with such testing would lead to loss of some perfectly sound donors. Blood banks have been slow to institute screening for *Babesia* despite that fact that screening has been shown to work. In one recent study, all bags of blood through one blood center over a one year period (over 2,000 units) were screened for *B. microti* by IFA and PCR, and 26 positive units were identified and eliminated[2]. Prior to institution of screening, TTB had occurred at a rate of about one case per 1,000 units of blood and, after screening began, no more cases of TTB were reported. However, despite such promising findings, rigorous screening programs still tend to be limited to *Babesia* endemic areas of the US[2].

TRANSFUSION-TRANSMITTED BABESIOSIS IN DOGS

What can we learn from recent experiences with human TTB that can be translated to dogs? While, unlike human medicine, we don't have studies detailing the extent of the problem, we know from single case reports and from patients that we have encountered that, in the individual dog, TTB can be a serious condition. Moreover, with attention to detail, it should be mostly preventable.

Donor Selection

In human blood banks, pre-donation epidemiologic screening is used to exclude high risk donors: intravenous drug use, for example, because it is associated with an increased risk of HIV and hepatitis, is an established indication for donor exclusion. Using the same approach, it would seem sensible to exclude canine donors in at-risk breeds (pit bulls, greyhounds), and dogs with a history of transfusions, immunosuppressive therapy, splenectomy or recent dog bites[13]. In most instances, this approach would not exclude an inordinate number of dogs from the donor pool. In my opinion, American Pit Bulls, American Staffordshire Terriers and pit bull crosses should never be used as donors: subclinical infection with *B. gibsoni* is common in these breeds, and the organism can be difficult to find on smears. Additionally, low numbers of organisms are believed to persist for years in dogs with *B. gibsoni* infections, perhaps even for life. In contrast, with a meticulous pre-donation screening program, I believe that greyhounds can still be used as donors. Exclusion of greyhounds would remove a large number of otherwise desirable donors from the donor pool, particularly since it is believed (though not proven with certainty) that most dogs infected with *B. canis* eventually clear the organism from their system.

In Europe, an established vaccine for *Babesia canis* subtype *canis* has been available for many years. The vaccine appears to confer an incomplete protective immunity and, when infection does occur, it is less severe in vaccinated dogs[14,15]. Perhaps, in the future, a vaccination program for *B. canis vogeli* and *B. gibsoni* in potential canine donors could be used to lessen the risk of TTB.

Donor Testing

In my opinion, it is safest to test all potential donors for *Babesia*. At the very least, greyhounds and pit bulls should be tested, as recommended by the 2005 ACVIM Consensus Statement on donor screening for infectious disease[16]. Initial screening should consist of serotesting (IFA) for *B. canis vogeli* and *B. gibsoni*. Since a positive titer suggests either active or past infection, and it is not possible to be absolutely sure that a dog has cleared a past infection, all positive dogs should be excluded. Seronegative dogs should then be followed up with PCR. In at-risk breeds, PCR should ideally be repeated several times before assuming a dog is a safe donor. In people, individual units are screened for common infections. Most canine blood banks, however, do not find this approach feasible. Rather, dogs are screened for infection and, if negative, it is presumed they will remain clear of infection for subsequent donations. For this assumption to be correct for *Babesia*, tick control programs must be rigorous, and dog bites must be avoided. Repeat testing should be performed annually, and perhaps more often in volunteer donor programs with less control over donor management.

Despite our best efforts, given the endemic and occult nature of babesiosis, it is still possible that an individual infected donor may lead to inadvertent TTB. In the past, infected canine donors have been identified by performing splenectomy, and this approach is still listed as an indication for splenectomy in recent journal articles and surgery text books[17]. Splenectomy, however, puts infected dogs at risk of overwhelming infection with *Babesia*[13,18], and for ethical reasons cannot be recommended. We recently attempted to identify donors with occult infection by administering liposomal clodronate, a drug that transiently knocks out macrophage function[19] (in essence, a temporary medical splenectomy), but were not successful at unveiling hidden *Babesia*. A novel approach that might be considered would be to attempt to infect ticks or lesser vertebrates with donor blood, and then test the sentinel species. Gerbils and hamsters, for example, typically develop severe babesiosis when given blood from people infected with *B. microti*. In the future, it is also possible that pathogen reduction techniques such as photosensitization could be used to sterilize units of infected blood[20].

REFERENCES

[1]Herwaldt BL, et.al. Ann Intern Med 2011;155:509. [2]Young C, et.al. Transfusion 2012;52:1523. [3]Freeman MJ, et.al. JAVMA 1994;204:94. [4]Taboada J, et.al. JAVMA 1992;200:47. [5]Allison RW, et.al. Vet Clin Path 2011;40:345. [6]Birkenheuer AJ, et.al. JAAHA 2003;39:551. [7]Leiby DA. Clin Micro Rev 2011;24:14. [8]Harvey JW, et.al. JAVMA 1988;192:1751. [9]Stegeman JR, et.al. JAVMA 2003;222:959. [10]Fukata T, et.al. J Parasitol 1996;82:641. [11]Birkenheuer AJ, et.al. J Clin Micro 2003; 41:4172. [12]Birkenheuer AJ, et.al. JVIM 2004; 18:494. [13]Sikorski LE, et.al. JVIM 2010;24:127. [14]Schetters TPM, et.al. Vet Parasitol 1997;73:35. [15]Schetters TPM, et.al. Vet Parasitol 2006;138:140. [16]Wardrop KJ, et.al. JVIM 2005;19:135. [17]Bakhtiari J, et.al. Int J Vet Res 2011;5:9. [18]Camacho AT, et.al. Bulletin Soc Path Exot 2002;95:17. [19]Mathes M, et.al. Exp Hematol 2006;34:1393. [20]Wagner SJ. Vox Sanguinis 2011;100:112.

A JOINT EFFORT: COMPLEMENTARY AND ALTERNATIVE MANAGEMENT OF OSTEOARTHRITIS

Cheryl A. Cross, DVM, DACVP (Anatomic)
Knoxville, TN

INTRODUCTION

Osteoarthritis (OA) is a common presenting complaint in pets. As is the trend in human medicine, many clients are seeking out complementary and alternative medicine (CAM) for their pets. The goal of this presentation is to serve as a basic introduction to two well-studied CAM management methods (acupuncture and therapeutic laser) and lesser understood and researched therapies (massage, chiropractic or veterinary orthopedic manipulation, and pulsed electromagnetic fields) to discuss their place in multimodal management of OA in animals.

INTRODUCTION TO ACUPUNCTURE

Acupuncture is a modality that has a long history of use in many pain syndromes including OA. The initial difficulty discussing acupuncture research is that there are multiple styles and point-selection strategies for any condition or presentation. In its original form, acupuncture was based on Chinese medicine and records of its use debatably date back to 10,000 B.C. It is impossible to adequately summarize the principles of historic acupuncture theory and its variations with brevity; volumes have been written in an attempt to clarify these systems. Traditional Chinese Medicine/Traditional Chinese Veterinary Medicine (TCM/TCVM) and Medical Acupuncture (MA) are the most common treatment styles taught and employed in human and veterinary acupuncture, although other styles exist. TCM/TCVM is based on a complex pattern diagnosis based on writings that predate modern diagnostic tools, describes strategies for unblocking energy flow, and is not based on a conventional medical diagnosis. The more modern medical acupuncture (MA) or structural acupuncture is based on contemporary, evidence-based information about specific acupuncture points and their neurophysiologic functions in health and disease. [1]

ROLE OF ACUPUNCTURE IN THE TREATMENT OF OA – MECHANISMS OF ACTION

The complete mechanisms of action of acupuncture for musculoskeletal pain including OA are still unknown. The neurophysiologic effects of acupuncture are at least mediated by a cascade release of endorphins and monoamines, it activates the "gate theory" of modulating nociception transmission, releases fascial planes, activates inhibitory descending pain pathways, and results ultimately in neuromodulation and immunomodulation. [2] Often acupuncture points are at the origin, insertion, or midbelly of skeletal muscles and thus some portion of acupuncture may rely on release of trigger points which can occur in these positions.[3] Overall, acupuncture has a strong role in mediating local pain, dampening central windup, and decreasing sympathetic tone in OA patients.[4]

Acupuncture traditionally involves the insertion of fine filiform needles into specific points on the body surface. Electroacupuncture, aquapuncture (injection into acupuncture points), and various forms of manual stimulation (acupressure, massage, etc.) are often employed alongside traditional needle acupuncture.

ACUPUNCTURE – CHALLENGES TO STUDY DESIGN AND STANDARDIZATION

Problems arise in acupuncture study review, as they not only face the routine challenges of adequate patient numbers, quality of research design, appropriate randomization, lack of bias, clear outcome measures, and appropriate control selection.[5] "No treatment" controls and "sham" acupuncture are typical, but sham acupuncture's definition varies from study to study. Sham treatment can include needles placed in acupuncture points that are not standard for the patient diagnosis or presentation, the needle is not fully inserted into the skin, or needles inserted in non-acupuncture points. Acupuncture point selection is complex, lacks standardization, and varies heavily depending on training. Acupuncture points chosen for very specific diseases (e.g. "OA of the stifle") can vary from study to study, and in practice the same points may not be used repeatedly in the same patient. This type of treatment variability is challenging to correct and study design typically does not mirror private practice, where treatments are often individualized based on patient presentation the day of treatment. In addition, it is important to consider that osteoarthritis is a chronic disease characterized by recurring pain over a period of many months or longer. Longterm follow-up is critical to determine the effectiveness of the treatment.

In human and veterinary medicine, research generally reports that acupuncture reduces the pain of OA.[6,7] However, although the majority of studies conclude that acupuncture is significantly better than no treatment at all, it is found often to be no more or slightly more effective than "sham" acupuncture,[8] even when owners could discern correctly when their dogs received sham or verum acupuncture.[9] Analysis of human and veterinary acupuncture OA studies indicate that the better the study design (adequate treatment numbers, randomized clinical trials, practitioner expertise and blinding, etc.), the more significant the success of acupuncture for reduction or resolution of OA pain. In addition, the more "individualized" the treatment to the patient (not only using standard set points for every treatment, but rather adding or removing points based on palpation of tender areas and/or presenting signs at the time of the acupuncture appointment), the more significant the acupuncture success when compared to sham needling or standard "fixed" points.

MANUAL THERAPIES IN THE TREATMENT OF OA

Both massage and chiropractic therapy (spinal or orthopedic manipulation) are popular methods for approaching OA in pets, but research data in animals is lacking. They face study challenges as mentioned in acupuncture (randomization, blinding, control selection, standardizing therapies that are typically very individualized). Human studies have shown massage therapy to significantly decrease pain and increase functionality in knee OA patients and resulted in lower anxiety, reduced stress hormones, and improved clinical outcomes in infants.[10-12] The biochemical and physiological processes following massage and chiropractic are largely inferred and are presumed to decrease pain, inflammation, and stiffness of OA via blockade of nociception via similar routes as acupuncture.[13] Other benefits attributed to these modalities are increased oxygenation of tissues, decreased fascial tension, removal of excess fascia, reduction and release of trigger points, and increased range of motion through passive stretch.[14]

INTRODUCTION TO LASER THERAPY

Laser therapy is a growing area in veterinary medicine, offering therapeutic treatment for a host of maladies including OA. For practitioners, laser administration is safe and easy to learn and institute into practice. It provides clients an "alternative" modality without necessitating the practitioner devote long periods of time to training, and allows the veterinarian to oversee the treatment while an assistant or technician administer the therapy.

LASER THERAPY – CHALLENGES TO STUDY DESIGN

While a bit less complicated than evaluating a variety of acupuncture study approaches to OA, laser therapy outcomes in the literature can vary based on study design Variations in wavelength, fluence, power density, pulse structure, and exposure time are all oft found in laser studies, leading to meta-analyses conclusions that variation in these parameters can result in reduced treatment effectiveness or even a negative outcome.[16] This considerable level of complexity and the biphasic nature of laser therapy have meant that the choice of parameters has often depended on the experimenter's or the practitioner's personal preference or experience rather than on a consensus statement by an authoritative body. The World Association of Laser Therapy (WALT) has attempted to provide dosage guidelines for human patients (http://www.walt.nu/dosage-recommendations.html).

ROLE OF LASER THERAPY IN THE TREATMENT OF OA – MECHANISMS OF ACTION

Laser therapy involves exposing cells or tissue to low levels of red and near infrared (NIR) light, and can be referred to as "low level" or "cold" laser because the energy densities applied are low and generate less heat compared to surgical, coagulative, or ablative lasers. The terms "photobiomodulation" or "photodynamic" are also used with laser therapy to reference the cellular mechanisms of action of laser.

The precise biochemical mechanism underlying the therapeutic effects of laser therapy are not yet well-established, but current observation suggests that it has a wide range of effects at the molecular, cellular, and tissue levels. Evidence points to a mitochondrial structure known as a chromophore within the mitochondria being the initial laser target. When the chromophore absorbs wavelength-specific laser light, mitochondria are activated, adenosine triphosphate (ATP) production is increased, there is modulation of reactive oxygen species (ROS), and the induction of transcription factors occurs. Several transcription factors are regulated by changes in cellular redox states which then cause protein synthesis that triggers further effects downstream. Cell proliferation and migration, and modulation in the levels of cytokines, growth factors, and inflammatory mediators, and increased tissue oxygenation have been documented.[17] Several animal studies suggest laser therapy has an inhibitory effect on nociceptors. On the clinical side, laser therapy in veterinary patients with OA can result in increased comfort, improved range of motion, decreased stiffness, and increased chondrocyte replacement in treated joints without increasing risks for side effects including the potentiation or exacerbation of neoplasia.[18]

INTRODUCTION TO PULSED ELECTROMAGNETIC FIELD THERAPY AND ITS ROLE IN OA THERAPY

Pulsed electromagnetic fields (PEMF) are extremely low frequency fields generated by external magnetic coils that induce electric fields in the body's conductive tissue through inductive coupling, thereby producing eddy currents. They do not require the placement of electrodes on the skin or the implantation of electrodes, and many PEMF devices are designed as wraps, mats, or beds and thus easy to use at home. While available reviews concur that the biophysical interactions between the PEMF signals and biological tissues are not well understood, observed experimental data in cell and animal models suggest that PEMF actions are anti-inflammatory and chondroprotective via upregulation of adenosine cell receptors or an increase in growth factor synthesis leading to enhanced chondrocyte proliferation and bone formation. On a cellular level, the most accepted mechanism of action suggests external magnetic stimuli interact with cells either via transmembrane receptors or ion channels, thus initiating one or more signal transduction cascades or cell functions.[19]

Despite a lack of proven mechanism, PEMF devices have grown in popularity for home use by clients with the most frequent indication for use being musculoskeletal disorders including OA.

In review of PEMF therapy studies, the same issue described previously also occurs: there is a marked variation in magnetic field intensities, frequencies, and other signal characteristics reported, making result analysis challenging. In addition, the number and duration of treatment sessions as well as the observation periods differ substantially, and accordingly outcomes vary from supportive of use in OA to finding no difference with PEMF therapy in parameters examined.[20] PEMF therapy devices are FDA-approved for non-healing, non-union fractures in humans, and animal-specific research supports its use in inflammatory OA, non-healing fractures, osteochondral lesions, and skin or wound healing. Devices are intended for frequent (often daily) use for several days, weeks, or longer. Anecdotal reports from clients and single case studies are often very supportive for PEMF application in pets for decreasing the discomfort and stiffness associated with OA. While adverse effects of short- or long-term PEMF therapy have not been reported, anecdotal data from performance horses and small animals using PEMF therapy daily for years have not brought forth concerns for safety; however, health consequences of long-term PEMF exposure (particularly whole-body) have not been studied.[21] Contraindications to PEMF therapy include cardiac pacemakers and automatic implantable cardioverter defibrillators.

CONCLUSIONS

Both acupuncture and laser therapies are becoming more utilized in veterinary patients and more popular with both clients and practitioners to address OA. When properly administered by adequately trained professionals, both modalities are incredibly safe to patient and practitioner, afford patients additional ways to treat a complex disease process, and integrate well into veterinary practice. As is oft concluded, in order to ultimately prove the success of therapies, better study design and research into the complete mechanisms of action of the therapies is needed. In the interim, both modalities are safe to recommend to clients and have a basis for support in the literature. Both modalities typically require patients to return for therapy on a weekly basis for a prescribed number of visits, and often in our integrative practice, we find that adding in these modalities enables lower doses of conventional pain medications, more frequent patient monitoring and follow-up, and clients feel more proactive leading to greater client compliance and satisfaction.

REFERENCES

[1]Kaptchuk T **Ann Intern Med** 2002;136(5):374. [2]Leung L J **Acup and Meridian Studies** 2012;5:261. [3]Grace S **J Austral Trad-Med Soc** 2011;17:30. [4]Pyne D, et al. **Rheumatol** 2008;47:1132. [5]Shuai P, et.al **Stats in Med** 2012;31(7):606. [6]Bjordal J M, et.al **BMC Musculoskeletal Disorders** 2007. [7]Marcellin-Little D J **Comp-Cont Educ for Veterinarians** 2009;31(3A):6. [8]Vickers A J, et.al **Trials** 2010;11(90). [9]Kapatkin A S, et.al **JAVMA** 2006;228(9):1350. [10]Perlman A I, et.al **Plos One** 2012;7(2). [11]Furlan A D, et.al **Spine** 2009;34(16):1669. [12]Field T **Jour of Dev and Behav Pediat** 1995;16(2):105. [13]Rapaport M H, et.al **J of Alt and Complem Med** 2012;18(8):789. [14]Posadski P **Pain Med** 2012;13(6):754. [15] Brosseau L, et.al **J of Rheum** 2000;27(8):1961. [16]Tuner J, et.al **J of Clin Laser Med & Surg** 1998;16(5):245. [17]Chung H, et.al **Ann of Biomed Eng** 2012;40(2):516. [18]Renno A, et.al **Photomed and Las Surg** 2007;25(4):275. [19]Selvam R, et.al **Life Sci** 2007;80(26):2403. [20]Trock D H **Rheum Disease Clin of N Amer** 2000;26(1):51. [21]Hug K, et.al **Bioelectromagnetics** 2012;33(2):95.

A JOINT EFFORT: OSTEOARTHRITIS AND NUTRACETICALS

Donna Raditic, DVM, CVA
Knoxville, TN

Osteoarthritis (OA) is the most common form of arthritis ranking as the 6th leading cause of disability globally in the human population. In veterinary medicine OA is the most common orthopedic disease in dogs and the prevalence of musculoskeletal disorders in all dogs has been reported as one in four at multicenter referral practices. Strategies to modify the inflammatory environment present within joints can include use of "nutraceuticals." A nutraceutical is defined by the AVMA as "…micronutrient, macronutrient and other nutritional supplements" used as therapeutic agents. The AVMA Guidelines for Complementary and Alternative Veterinary Medicine states: "claims for safety and effectiveness ultimately should be proven by scientific method. Circumstances commonly require that veterinarians extrapolate information when formulating a course of therapy. Veterinarians should exercise caution in such circumstances." The AVMA definition and guidelines are the platform for a discussion of nutraceuticals evaluated in dogs or *in vitro* using canine cells. Osteoarthritis occurs commonly in cats; however, there is a paucity of research in the role of nutrition or nutritional compounds in management.

Chondromodulating agents

A review of nonsurgical management of hip dysplasia using an evidence classification proposed by Aragon and Budsberg found fourteen articles meeting their criteria with three nutraceutical studies, two providing Level II evidence and one providing Level III evidence. The nutraceuticals in the studies are classified as antioxidants and chondromodulating agents that are purported to slow or alter the progression of osteoarthritis. Chondromodulating agents can be further divided into agents approved by the US Food and Drug Administration, such as polysulfated glycosaminoglycan, which can have label claims of clinical effect, or nutritional supplements, which are not regulated, and legally cannot claim any medical benefit. These oral nutraceuticals include glucosamine and chondroitin sulfate. Glucosamine is an amino sugar and precursor for biochemical synthesis of glycosylated proteins and sugars. Glucosamine -6-phosphate synthesized in the hexosamine pathway results in the production of UDP-N-acetylglucosamine, which is used for making glycosaminoglycans, proteoglycans, and glycolipids. Glycosaminoglycans are a major component of joint cartilage; therefore, it is believed that supplemental glucosamine may rebuild cartilage. *In vitro* studies support this claim although these studies use concentrations not achieved in serum or plasma after oral administration.

Results of studies of chondromodulants both in veterinary and human literature are mixed due to product used (glucosamine alone, glucosamine/chondroitin sulfate), dose administration, subjective and objective measurements of outcome, and length of trials. In a randomized, double blind, placebo controlled clinical trial comparing glucosamine/chondroitin sulfate to carprofen in 35 dogs with OA, carprofen treated dogs had improvement in 5 subjective measures while dogs treated with the glucosamine/chondroitin sulfate improved in 3 of 5 measures, but only after final assessment at 70 days. A 60-day, prospective, randomized, double blind, placebo controlled trial of 71 dogs comparing carprofen, meloxicam, glucosamine/chondroitin, and placebo demonstrated significant improvement in objective measurements with carprofen and meloxicam but not with the nutraceutical or placebo. Finally a large multicenter clinical trial compared an NSAID with glucosamine alone, chondroitin sulfate alone, and a glucosamine/chondroitin sulfate combination in patients stifle OA. Patients with moderate to severe stifle OA showed significant improvement in clinical function using the combination of glucosamine/chondroitin sulfate. Based on these studies and others the clinical evidence of these chondromodulants seems weak and will most likely be dependent on content, amount of disease, dosing, and length of treatment.

Another natural source of glycosaminoglycans, the New Zealand green lipped mussel (*Perna canaliculus*), is marketed for its chondromodulating effects. Numerous publications reporting the results of uncontrolled studies have been published with promising results. One randomized controlled clinical study added green lipped mussel powder to diet (0.3%) and the dogs receiving the supplemented diet vs. control had significant improvement in subjective arthritic scores. Another randomized double blinded clinical study with 45 dogs showed improvement in mobility compared to placebo, but not as effective as carprofen. An uncontrolled study of 85 dogs with presumptive OA, using green-lipped mussel supplemental diet for 50 days showed a reduction in a composite arthritic score when compared with baseline scores on various diets the dogs were consuming. In reviewing the studies of green lipped mussel extract as beneficial in treating canine OA, results seem promising, but there are uncertainties with the scientific quality of the data published.

Omega 3 (n-3) fatty acids

The potential of modifying the inflammatory components of osteoarthritis using nutritionals is another approach of nutraceutical therapy. Arachidonic acid, an n-6 fatty acid, is incorporated into cell membranes and when metabolized, yields the 2 and 4 series of prostaglandins, leukotrienes and thromboxanes. These proinflammatory pathways are where conventional drugs are used to control OA inflammation. Substituting n-3 fatty acids into cell membranes may decrease inflammation with biosynthesis of eicosanoids of the 3 and 5 series, which are less proinflammatory. In addition to modulating cytokines, n-3 fatty acids have been shown to reduce expression of cyclooxgenase 2- lipoxygenase-5, aggrecanase, matrix metalloproteinase 3 and 13, interleukin 1α and 1β, and tumor necrosis factor α. Interestingly n-3 fatty acids, eicosapentaenoic acid (EPA) and docosahexaenoic acid (DHA) have been shown to decrease inflammatory exudates in tissues through production of oxygenated products called Resolvins (resolution phase interaction products) and docosatrienes. Human studies of rheumatoid arthritis have shown beneficial effects of n-3 incorporation into diet. There is growing data showing the positive effects of n-3 fatty acids supplementation on cartilage metabolism with degradative enzymes and reducing inflammatory responses elicited by chondrocytes and joint matrix during the progression of OA. An unpublished study was performed in dogs evaluating omega 3 fatty acids and experimentally induced stifle OA. Dogs were randomly assigned to isocalcoric diet groups containing 21.4% fat (dry matter basis, DMB) differing only in fatty acid composition: a diet with an n-6 to n-3 ratio of 28:1, 8.7:1(control diet) and a diet with an n-6 to n-3 ratio of 0.7:1. Dogs began the new diet 3 months prior to surgical transection of the left cruciate ligament, were continued on the diet for 6 more months prior to surgical repair, and maintained for 12 months following repair. When compared to the high n-6 diet and control diet, the high n-3 diet was associated with lower serum cholesterol, triglycerides, and phospholipids, lower synovial concentration of prostaglandin E2, better ground reaction forces, and less radiographic changes of OA. A study of 127 dogs with OA fed a diet with a 31-fold increase in total n-3 fatty acid content and 34- fold decrease in n-6 to n-3 ratio had improved ability to rise from a resting position and play for 6 months when compared to dogs fed a control diet. Peak vertical force values and subjective lameness parameters improved with a randomized, double blinded controlled clinical trial of 38 dogs with OA fed a high (3.5%) n-3fatty acids for 90 days. A randomized, controlled clinical trial with 131 dogs with stable OA treated with carprofen over 12 weeks fed a diet supplemented with n-3 fatty acids provided a significantly faster carprofen dose reduction compared to a control diet. A recent study using a veterinary therapeutic diet (VTD) rich in n- 3 fatty acids (total n-3 fatty acid 1.47% DMB) compared to a control diet (CTR, total n-3 fatty acid 0.18% DMB) over 13 weeks with 30 dogs with naturally occurring OA evaluated peak of the vertical oriented ground reaction force (PVF) using a force platform and activity scores provided by owners. The VTD fed dogs showed significant improvement in PVF at both week 7 and week 13 compared to CTD fed dogs. These studies provide a rationale for n-3 fatty acid supplementation or feeding diets with high n-3 fatty acid content. Recommendations for n-3 supplementation for canine OA has been reported as 310mg/ ($kg^{0.75}$) or 1745 mg EPA + DHA per 10kg body weight for the dog.

Zeel®

Zeel® is an over-the-counter preparation that has been evaluated in 2 controlled studies in dogs. This is a highly diluted proprietary formulation of herbs, metabolites, minerals, and antioxidants. In a study of 68 dogs greater than one year of age diagnosed with OA, it was compared with carprofen in a multicenter, prospective, observational open-label cohort study in 12 German veterinary clinics. In another study in dogs (n=44), greater than one year of age diagnosed with OA, it was compared with carprofen and a placebo. Clinical signs and several measures of OA improved significantly with treatment in both studies; however, in one study, it was not as effective as carprofen. The composition of the products and the dosage of Zeel® differed between the two studies, which confounds interpretation of results.

Boswellia serrata

Boswellia, also known as Boswellin or Indian frankincense, comes from the Indian *Boswellia serrata* tree. Resin from the bark of this tree is purported to have anti-inflammatory properties derived primarily from 3-O-acetyl-11-keto-β-boswellic acid (AKBA), which inhibits 5-lipoxygenase and matrix metalloproteinases, and decreases tumor necrosis factor α and interleukin 1β. Boswellia resin has been shown to improve clinical signs and pain in humans in controlled studies. Boswellia resin has been evaluated in 24 dogs in an open multi-center study. Improvement in clinical signs, lameness, and pain was found in 17 of 24 dogs.

Avocado/Soybean Unsaponifiables

Avocado/soybean unsaponifiables (ASU) are composed of the unsaponifiable fractions of avocado and soybean oils in a 1/3 to 2/3 proportion. ASU has anti-OA properties by inhibiting interlukin-1 and stimulating collagen synthesis in cartilage cultures. Human clinical trials have shown some beneficial effects of ASU on clinical symptoms of OA, but conflicting data in other studies found no long term benefits. In one study of dogs, OA was induced by anterior cruciate ligament transection. Dogs then received placebo or ASU (10 mg/kg/24h). The size of macroscopic lesions of the tibial plateau, severity of cartilage lesions, synovial cellular infiltration, and inducible nitric oxide synthase were decreased significantly and there was reduced loss of subchondral bone volume and calcified cartilage thickness in the group receiving ASU.

Using nutraceuticals

In clinical practice using nutraceuticals can be part of multimodality management of patients who have had musculoskeletal trauma, orthopedic surgery for DOD, patients at risk for DOD, and geriatric patients with OA. Understanding the physiological and pathophysiological mechanisms involved may aid in selecting individual nutraceuticals that will optimize patient management. The concept of "cell" niche defines the unique attributes of a cell's microenvironment that influences its development. It means understanding all the factors that the cell is exposed to or will interact with via cell to cell interactions, cell matrix interactions, autocrine and paracrine hormonal influences, growth factors, and cytokines. For OA we must consider the stem cell niche with functional and anatomical properties that permit cells to switch between resting and activated state. Progenitor cells, skeletal stem cells, osteoblasts, osteoclasts, osteocytes, chondrocytes, synoviocytes, extra cellular matrix, and bioactive factors including inflammatory cytokines, chemokines, growth factors, and hormonal influences are the individual components that comprise this unique microenvironment. Each disease state will have a specific balance of influences required to perpetuate the course of the disease, creating a unique signature or niche. Traditionally OA has been regarded primarily as a degenerative process and most often a consequence of aging. Treatments focused solely on symptomatic management rather than halting disease progression will fail as research revisits the role of inflammation in OA. The same inflammatory cytokines that drive rheumatoid arthritis have been shown in OA including IL-1, IL6, and TNFα. Studies suggest IL-1β and TNFα occupy a key role in the development of OA, with elevated levels of these cytokines or their gene expression reported in synovial tissues, synovial fluid and/or plasma. Investigating the role of an individual nutraceutical in maintaining the balance in the OA niche and attendant inflammatory milieu can be used as a regenerative strategy for bone or cartilage in OA. This requires individual patient evaluation with a complete history, current activity level, nutrition, environmental factors, laboratory test, radiographs and/or other diagnostics. This evaluation is needed to develop a working knowledge of the patient to assess the degree of inflammation and what component(s) of the OA niche can be addressed with selected therapies.

Using therapeutic diets for OA

Therapeutic diets can be a safe starting point for using nutraceuticals to treat OA in veterinary patients. A recent review of the nutraceuticals and OA in horses, dogs, and cats found the evidence of efficacy poor with the exception of diets supplemented with n-3 fatty acids. Knowing the exact formulations and specific nutrients of concern in these diets allows for better application of these diets. When nutraceuticals are used with these diets, the total intake of at least n-3 fatty acids should be determined. Potential adverse effects of n-3 fatty acids includes altered platelet function, gastrointestinal adverse effects, altered immune function, nutrient-drug interactions, lipid peroxidation and weight gain. Adverse effects are less likely with glucosamine intake, but the potential for coagulopathies may exist.

REFERENCES

Aragon CL, Budsberg SC. *Vet Surg* 2005; 34:93-98;Bartges JW, Budsberg SC, Pazak HE, et al. Orthopedic Research Society 47th Annual Meeting 2001;Phitak T, Pothacharoen P, et al. *BMC Musculoskelet Disord* 2010;11:162;McCarthy G, O'Donovan J, et al. *Vet J* 2007;174:54-61; NIH funded study: *Arth Rheum* 2005: 9 (suppl), 622; Bui LM, Bierer TL. *Vet Ther* 2003;4:397-407;Hielm-Bjorkman A, Tulamo RM, et al.*Evid Based Complement Alternat Med* 2009;6:365-373; Roush JK, Dodd C, Fritsch DA, et al. *J Am Vet Med Assoc* 2010;236:59-66; Fritsch DA, Allen TA, et al. *J Am Vet Med Assoc* 2010;236:535-539;Moreau M, Troncy E, et al, *J of Animal physiol and Anim Nutr* 2012; epub;Bauer J. *J Am Vet Med Assoc* 2011; 239: 1441-1451;Neumann S, Stolt P, et al. *J Am Anim Hosp Assoc* 2011;47:12-20; Hielm-Bjorkman A, Tulamo RM,et al. *Evid Based Complement Alternat Med* 2009;6:465-471;Sengupta K, Kolla J, et al. *Mol Cell Biochem* 2011;Sander O, Herborn G, Rau R. *Z Rheumatol* 1998;57:11-16; Kimmatkar N, Thawani V, et al. *Phytomedicine* 2003;10:3-7; Reichling J, Schmokel H, et al. *Schweiz Arch Tierheilkd* 2004;146:71-79; Dinubile NA. *Phys Sportsmed* 2010;38:71-81;Ernst E. *Clin Rheumatol* 2003;22:285-288; Soeken KL. *Clin J Pain* 2004;20:13-18; Boileau C, Martel-Pelletier J, et al. *Arthritis Res Ther* 2009;11:R41;Williams E, Christopher J, et al. *Regen Med* 2012; 7:551-570; Bauer J, Lenox C. *J Vet Intern Med* 2013; epub;Vanderweerd J, Coisnon P, et al. *J Vet Intern Med* 2012; 26; 448-456

A JOINT EFFORT: OSTEOARTHRITIS AND NUTRITION

Joe Bartges, DVM, PhD, DACVIM, DACVN
Knoxville, TN

Osteoarthritis (OA) is a common syndrome having multiple etiologies and characterized by pathologic change of the synovial or diarthrodial joint accompanied by clinical signs of pain and disability. There have been few controlled studies evaluating nutrition and arthritis in dogs although nutrition plays a role in prevention of and management of dogs and cats with OA. In order to integrate appropriate nutritional recommendations for patient management, a two-step process is undertaken: a nutritional assessment phase (based on the "Circle of Nutrition" by the American College of Veterinary Nutrition (http://www.acvn.org)) and an analysis, interpretation, and action phase. In the assessment phase four sets of factors are assessed: animal-specific factors, diet-specific factors, and feeding management and environmental factors. Animal-specific factors include age, physiological status, disease processes, and activity of the patient. Diet-specific factors include safety and appropriateness of diet fed and nutrients of interest. Feeding factors include frequency, timing, location, and method of feeding, and environmental factors include living space and quality of the patient's surroundings. An important part of the assessment phase is obtaining body weight, body condition score, and muscle condition score. Body condition score evaluates body fat. There are 2 commonly used body condition scoring systems, a 5-point and a 9-point scale. In either scale, the middle number represents "optimal condition" with a body fat content of 15-25%; higher numbers on the scale represent various degrees of over-condition and body fat content > 25% and lower numbers on the scale represent various degrees of under-condition and body fat content < 15%. The goal for most patients is a body condition of 2.5-3.0 out of 5 or a 4-5 out of 9, which is associated with decreased health problems including musculoskeletal conditions. Muscle condition score evaluates muscle mass, which can be independent from body fat content assessed by body condition score. Evaluation of muscle mass includes visual examination and palpation over temporal bones, scapulae, lumbar vertebrae, and pelvic bones. The scale is from 1 to 3 with 3 being well-muscled and 1 corresponding to severe sarcopenia. After collecting information from the assessment phase, information is analyzed and interpreted and an action plan is implemented;, the patient undergoes repeated assessment and adjustment of the plan.

ROLE OF NUTRITION IN THE PREVENTION OF MUSCULOSKELETAL DISEASE

Prevalence of musculoskeletal disorders for all dogs at multicenter referral practices has been reported to be approximately one in four. In dogs less than one year of age, prevalence is 22%, with 20% of these possibly having a nutrition-related cause.

Developmental Orthopedic Disease (DOD)

DOD refers to a group of skeletal abnormalities (e.g. osteochondrosis dissecans, hip dysplasia, wobblers' syndrome, ligamentous laxity, and hypertrophic osteodystrophy) that affect primarily rapidly growing, large- and giant-breed dogs. Nutrient excess (calcium and energy) and rapid growth (overfeeding and excess energy) are known risk factors for DOD in dogs that have genetic risk. Restricting food intake during growth slows growth rate without significantly reducing adult body size and is associated with decreased incidence of DOD. Dietary calcium greater than 3% on a dry matter basis is associated with increased risk despite an appropriate calcium-to-phosphorous ratio. Even if the diet contains less calcium than this, excess calcium intake can occur if owners provide supplemental calcium. Limited meal-feeding and limiting rate of growth also decrease DOD. For large- and giant-breed dogs during growth, recommended dietary composition includes: Energy: 3.2-4.1 kcal/kg of diet (lower end of range if at risk for DOD or if clients use free-choice feeding); Crude fat: 8.5-11% on a dry matter basis; Docosahexaenoic acid: ≥ 0.02% on a dry matter basis; Protein: ≥ 27% on a dry matter basis; Calcium: 0.8-1.2% (≤ 3% on a dry matter basis) with a calcium-to-phosphorous ratio of 1.1-2.0:1.0

Obesity

Obesity can be defined as accumulation of body fat in excess of what is necessary to maintain optimum condition and health. Quantitatively, obesity is defined generally as exceeding ideal body weight by 15-20% or more. One technique that is useful in the management of clinical patients is a body condition score and muscle condition score as discussed previously. Obesity may contribute to the development and progression of OA because of excess forces placed on joints and articular cartilage, which may lead to inactivity and further development of obesity. Additionally, adipose tissue is recognized as being metabolically active and pro-inflammatory; therefore, obesity may contribute to inflammation. Several studies have demonstrated a relationship between obesity and OA; however, a cause and effect has not been determined. A long term study was performed of 48 Labrador retrievers from 7 litters divided into 2 dietary groups. One group was fed an adult maintenance dog food and the second group was fed the same diet at 75% of the amount. Restricted fed dogs lived on average 1.8 years longer, weighed less, had better body condition scores and had longer delay to treatment of chronic disease including OA. Maintaining optimal or slightly lean body condition may be associated with lower risk of developing OA, development of less severe OA if it occurs, and delay of onset of clinical signs of OA in dogs.

ROLE OF NUTRITION IN THE TREATMENT OF MUSCULOSKELETAL DISEASE

Nutrition and Surgery

Non-traumatized surgical patients

Patients undergoing surgery that are healthy and in optimal body and muscle condition require no further intervention nutritionally other than to continue feeding the current diet at the same amount and frequency. Patients are not fed 8-12 hours prior to anesthesia and surgery, but can resume normal feeding pattern once recovered from the procedure. It may be necessary to add flavoring agents or feed a different diet to stimulate appetite in the post-operative period. Patients should be assessed daily following surgery to insure they are eating adequately and that no adverse events occur either from the procedure or from post-operative medications such as analgesics and sedatives. As the patient becomes more active or as a rehabilitation program is initiated, energy requirements will increase. In many instances, surgery in these patients is elective; therefore, depending on the urgency of the surgical procedure, consideration should be given to weight reduction prior to surgery in patients that are obese. Reducing body weight to the optimal weight decreases anesthetic risk, improves immune function, and allows patients to more easily undergo post-surgical physical rehabilitation.

Traumatized surgical patients

Patients with surgical musculoskeletal disease associated with trauma may or may not have increased nutritional requirements depending on the degree of trauma. If the trauma is less severe, then energy requirements are similar to non-traumatized surgical patients, and energy requirements approximate resting energy requirements. Severely injured patients or those with closed head injuries typically have energy requirements exceeding resting energy requirements due to the extent of the injuries or to activation of the sympathetic nervous system, in part. In these patients, energy requirements may exceed two times the estimated resting energy requirements. Likewise, protein requirements may exceed that required for maintenance in order to facilitate recovery, immune function, and wound healing. With trauma or strenuous exercise, protein requirements may exceed these by 50-100%. Some patients may require facilitated nutritional support enterally or parenterally. Nutritional support in these patients should be individualized and intensive monitoring is required in order to maximize recovery and response to supportive care. Convalescent diets that are formulated to contain higher levels of protein and fat, are more calorically dense, highly palatable, and have a homogeneous texture suitable for feeding tubes are available.

Nutrition and Rehabilitation.

Research into nutritional needs of patients undergoing various rehabilitation programs is non-existent. Individuals developing rehabilitation programs should have knowledge of nutrition. Rehabilitation can include simple strengthening to extensive long-term physical rehabilitation; therefore, the nutritional plan for a patient must adjunctively meet the goals set during therapy and after goals are met. Nutritional plans should be included as part of the rehabilitation therapy. Water is the most essential nutrient. During rehabilitation, excitement, fear, pain, and other stress factors affect the simplest of nutritional needs, water and electrolytes. Water intake for healthy dogs approximates 50-60 ml/kg/day, but may be more with rehabilitation. Energy is a primary nutrient for maintenance, life-stage, and the therapy program. Energy requirements correlate to work intensity, duration, and frequency, with intensity being the primary determinant. Sprinters, such as greyhounds that are running with maximum intensity, have low energy requirements because of the short race duration and low frequency of training. These dogs use mainly carbohydrates to meet energy requirements. Studies of endurance sled dog with long durations of work are at the other extreme, requiring high energy intake. High fat diets are fed to meet these high-energy requirements. Exercise intensity has not been quantified in rehabilitation, unlike performance and working dogs. Therapy is likely an intermediate level of exercise, which has not been well defined. Intensity, duration, and frequency of exercise are highly variable; therefore, it is difficult to predict energy needs of these patients. Consideration of historical exercise level combined with current patient and diet assessment may provide some insight into energy needs. In a study of well-conditioned beagles running on a treadmill, time to exhaustion correlated with fat intake, diet digestibility, and energy concentration of the diet. Increasing fat intake improves performance of dogs with extreme energy requirements. In rehabilitation therapy, the level of exercise is unlikely to equate to the extreme energy requirements of sled dogs. Fat intake should still be considered when a patient is in therapy. If we use the example of a sprinting Greyhound, dogs in therapy may have similar fat requirements with fat providing 20 to 30% of the total energy requirement. This equates to a diet with a fat content of 8-10% on a dry matter basis. Protein is required for structure and function, and to a lesser extent, energy. Current recommendations with exercise are to increase protein 5 to 15% regardless of the level of exercise, with dietary protein providing approximately 25% of the energy needs. Dietary protein is the most important nutrient with regard to muscle protein and exercise increases protein requirements. Many of the goals of canine physical rehabilitation require increased muscle tissue. Exercise increases protein synthesis and breakdown with the net result of decreased muscle protein. Ingestion of amino acids stimulates muscle protein synthesis, exceeding the rate of breakdown, and muscle mass increases. Exercise combined with amino acid intake produces a greater response regarding muscle synthesis than response to individual treatments. This suggests exercise can prime the muscle response to the anabolic effects of amino acids or exercise activates muscle synthesis, but adequate amino acid precursors are needed to increase muscle mass. Studies of protein nutrition often evaluate total dietary protein content instead of amino acid composition. Investigation to optimize amino acid profiles for specific benefits has identified 'slow' and 'fast' proteins. This classification is determined by rate a protein exits the stomach, is hydrolyzed into peptides and amino acids, and absorbed. Plasma amino acid profiles of fast and slow proteins have shown that 'blends' produce maximal amino acid concentration for optimal protein synthesis. Identifying and using protein blends for maximal muscle synthesis and strength during physical rehabilitation may be a reasonable goal with these current studies.

Obesity

Three uncontrolled clinical studies of obese dogs with OA demonstrated improvement in mobility. In one study of nine dogs, with body condition score of 5 out of 5 and coxofemoral OA, obesity management resulted in loss of 11-18% of body weight, a decrease in body condition to optimal, and improvement in severity of subjective hind limb lameness scores. In the second study of 16 dogs with coxofemoral OA, weight loss of 13-29% of body weight and decrease of body condition to optimal resulted in improvement of ground reactive forces and improvement in subjective mobility and clinical signs of OA. In the third study, 14 client-owned dogs with clinical and radiographic signs of OA participated in an open prospective clinical trial. Results indicate that body weight reduction causes a significant decrease in lameness with a weight loss of 6.10% or more. Kinetic gait analysis supported the results with a body weight reduction of 8.85% or more. These results confirm that weight loss should be presented as an important treatment modality to owners of obese dogs with OA and that noticeable improvement may be seen after modest weight loss of 6.10 - 8.85% body weight. Weight loss is beneficial when used in combination with rehabilitation and physical rehabilitation in dogs. In a non-blinded prospective randomized clinical trial, 29 adult dogs that were overweight or obese (body condition score of 4/5 or 5/5) having clinical and radiographic signs of OA were evaluated. All dogs underwent weight loss. One group received caloric restriction and a home-based physical rehabilitation program and the other group received an identical dietetic protocol and an intensive physical rehabilitation program including transcutaneous electrical nerve stimulation. Significant weight loss and improved mobility were achieved in both groups; however, greater weight reduction and better mobility was obtained in the group receiving intensive physical rehabilitation.

DIETS FOR OSTEOARTHRITIS IN DOGS AND CATS

There are several commercially available diets formulated for management of dogs with OA. These diets contain higher levels of omega-3 fatty acids and may contain chondromodulators and antioxidants. A controlled experimental study in dogs where the cruciate ligament was transected demonstrated a beneficial effect of pre-feeding a diet containing high omega-3 fatty acids with less severe radiographic and functional joint disease. Available "joint" diets also contain antioxidants that may be beneficial in decreasing free radical induced injury with OA. Chondromodulating agents are also often included in "joint" diets. The amount of these compounds in commercial diets is less than recommended for treatment of OA; therefore, they may be beneficial in prevention or management of early disease but probably less effective in more advanced disease. OA occurs commonly in cats; however, there is a paucity of information regarding nutrition or nutritional compounds in management. In a prospective, blinded, parallel group study evaluating a diet high in n-3 fatty acids and supplemented with green-lipped mussel extract and glucosamine/chondroitin fed to cats with OA over 9 weeks, there was an improvement in mobility of cats fed the supplemented diet and a decline in cats fed the control diet. Although other nutritional compounds are used in cats with OA, none have been evaluated in a controlled manner.

REFERENCES

Baldwin K, Bartges J, Buffington T, et al. *J Am Anim Hosp Assoc* 2010;46:285-296; Laflamme D. *Canine Practice* 1997;22:10-15; Laflamme D. *Feline Practice* 1997;25:13-18; Lund EM, Armstrong PJ, Kirk CA, et al. *Int J Appl Res Vet Med* 2006;4:177-186; Lund EM, Armstrong PJ, Kirk CA, et al. *Int J Appl Res Vet Med* 2005;3:88-96; Michel KE, Anderson W, Cupp C, et al. *J Anim Physiol Anim Nutr (Berl)* 2009;93:806; Zoran DL. *Vet Clin North Am Small Anim Pract* 2010;40:221-239; Kealy RD, Lawler DF, Ballam JM, et al. *J Am Vet Med Assoc* 2000;217:1678-1680; Bartges JW, Kirk CA, Lauten S. *Vet Med* 2004;99:632; Hill RC. *J Nutr* 1998;128:2686S-2690S; Impellizeri JA, Tetrick MA, Muir P. *J Am Vet Med Assoc* 2000;216:1089-1091; Burkholder WJ, Taylor L, Hulse DA. *Compen Cotin Educ Pract Vet* 2000;23:74; Mlacnik E, Bockstahler BA, Muller M, et al. *J Am Vet Med Assoc* 2006;229:1756-1760; Lascelles BD, Depuy V, Thomson A, et al. *J Vet Intern Med* 2010

UNDERSTANDING BACTERIAL BIOFILMS

J Scott Weese, DVM, DVSc, DACVIM
Guelph, Ont, Canada

INTRODUCTION

A bacterial biofilm is a complex, sessile community of bacteria embedded within a self-produced matrix of carbohydrates, proteins and DNA referred to as extrapolymeric substance (EPS).[1] This EPS forms the scaffold for the 3-dimensional structure of biofilms, allows it to adheres to surfaces, maintains cohesion of the biofilm, retains water, forms a physically protective barrier, allows the accumulation of nutrients and facilitates horizontal gene transfer between cells, among other complex and important functions. By immobilizing cells, EPS keeps them in close proximity, facilitating cell-cell communications (which further support biofilm development) and often bacteria represent less than 10% of the dry matter of biofilms,[1] being dominated by the EPS matrix.

Within the biofilm, bacteria are immobilized and they different from their free-living counterparts by more than just their sessile nature. They have markedly altered metabolism and protein expression, and enhanced cell-cell communication.[1,2] They are able to evade the host immune response, the effects of antimicrobials and environmental stressors through the altered metabolism along with physical and chemical protection of the biofilm matrix. While the concept of bacterial biofilms has been recognized for some time, only recently has there been much study of the role of biofilm in infections in veterinary medicine.

ROLE OF BIOFILMS IN DISEASE

Biofilm formation can be an important virulence factor and play an important role in many types of disease. While poorly defined in veterinary medicine, biofilm presumably plays a key role in implant-associated infections, infections of invasive devises (e.g. IV catheters, urinary catheters), dental disease (since plaque is a bacterial biofilm), valvular endocarditis, otitis externa and probably a range of disorders involving focal bacterial infections. Biofilm production is common amongst many pathogenic bacteria of importance in domestic animals, including *Staphylococcus pseudintermedius* from dogs and cats,[3](AM Singh, unpublished data) and *Pseudomonas aeruginosa* from canine otitis.(C Pye, unpublished data) and has been reported in diverse microorganisms such as *Malassezia pachydermatis*.[4] Human data also indicate a role of biofilm formation in adherent-invasive *E. coli* infection,[5] something that may be of relevance for histiocytic colitis in dogs.

ANTIMICROBIAL SUSCEPTIBILITY OF BIOFILM-EMBEDDED BACTERIA

Bacteria within biofilms do not necessarily have the same susceptibility to antimicrobials as their planktonic (free-living) counterparts. It has been clearly shown that there can be differences (sometimes profound) in the minimum inhibitory concentrations (MICs) of biofilm-embedded vs planktonic bacteria,[6-8] and this may play a critical role in evasion of antimicrobial therapy. The biofilm matrix may physically protect bacteria by inhibiting diffusion of antimicrobials, particularly large polar molecules like aminoglycosides. The alteration in metabolism that occurs in biofilms may also be protective, particularly against antimicrobials that are dependent on active cellular replication. It also may facilitate horizontal transfer of antimicrobial resistance genes in polymicrobial biofilms or allow for accumulation of antimicrobial-degrading enzymes.

PREVENTION OF BIOFILM FORMATION

Prevention of establishment of biofilms will obviously negate any impact of biofilms on the ability of the immune system or antimicrobials to eliminate infections. Unfortunately, this is currently difficult. Obviously, prevention of infections from developing will negate any relevance of antimicrobials, re-inforcing the importance of routine infection control measures, particularly in high risk situations such as surgical procedures involving implants. Drug-based approaches to reduce or prevent biofilm formation when infection is underway are of interest, and there is evidence that some antimicrobials can inhibit biofilm formation by some bacteria, independent of their antimicrobial effect. For example, clarithromycin has been shown to reduce biofilm formation by *S. aureus*,[9,10] probably as a result of interference in bacterial cell-cell communication, an important component of biofilm population establishment.*)* However, no influence of clarithromycin was identified in a study of *S. pseudintermedius.*[11] In contrast, aminoglycosides were shown to induce biofilm formation by *Pseudomonas aeruginosa* by virtue of an aminoglycoside response regulatory gene that encoded for production of a bacterial second messenger that regulates surface adhesion.[12] More study of this area is required.

Another area of interest is the use of implant materials that are less amenable to biofilm formation, or the use of coatings or surface modification to reduce biofilm attachment.[13] Use of antimicrobial-coated materials for invasive devices such as intravenous or urinary catheters[14] may also help prevent biofilms through prevention of establishment of early infection.

TREATMENT OF BIOFILM-ASSOCIATED INFECTIONS

While a desirable goal, prevention of all biofilm-associated infections from occurring is not practical. Therefore, measures to facilitate treatment of biofilm-associated infections are required. Traditional antimicrobial therapy, while limited, is not necessarily futile and longterm therapy can result in elimination of some infections. However, this may require prolonged therapy and the prognosis is often guarded. That is particularly true when implants are involved, such as the commonly-encountered TPLO-associated infections. As more information is obtained regarding differential susceptibility of biofilm versus planktonic bacteria, this may help guide treatment options, allowing for use of antimicrobials that tend to have better activity against biofilm-embedded bacteria. MIC testing of biofilm-embedded bacteria is possible and is available commercially; however, clinical application of these data are currently challenging since there is no in vivo validation and the true clinical relevance of 'minimum biofilm eradication concentration (MBEC)[15] data from laboratories is unclear. As more data become available, it is possible that MBEC data might be useful clinically.[15] Certain antimicrobials may have good effects in biofilm, either alone or through synergistic action with another antimicrobial. For example, fosfomycin has been shown to enhance the effect of certain antimicrobials in biofilms *in vitro*.[16]

Similarly. 'biofilm inhibitory concentration' data can be provided, but use of these is even more questionable since we are uncommonly in a situation where we are able to treat in advance of biofilm formation. However, such an approach might be effective if it can define concentrations of drugs

Elimination of biofilm is a potentially desirable approach and there may be methods to do so in the near future. In particular, DispersinB™, an enzyme produced by *Aggregatibacter actinomycetemcomitans*, can alter biofilm integrity and result in biofilm degradation.[17,18] This is achieved through hydrolysis of linkages of poly-N-acetylglucosamine (PNAG), a required component of polysaccharide adhesins needed for biofilm formation by some bacteria. It has been shown to both reduce biofilm formation by *S. pseudintermedius* and eradicate established biofilm *in vitro*.(JS Weese, unpublished data) While not effective against biofilm produced by all bacteria, only those which involve PNAG, this may be a potentially useful tool in conjunction with antimicrobial therapy and/or other approaches.

Ultimately, elimination of an established biofilm-associated infection may be difficult and removal of any foreign material such as surgical implants may be the critical factor for successful elimination of infection.

OTHER CONSIDERATIONS

Most attention has been directed at biofilms as playing a factor in disease. However, biofilm production may also be an important factor in colonization of both commensal bacteria and opportunistic pathogens. Indeed, if biofilm production facilitates colonization of beneficial bacterial at various host sites such as the intestinal tract,[19] bacterial biofilm formation could be a beneficial property to the animal. The role of biofilms in colonization of commensal bacteria and development of the normal microbiota has received little attention but deserves consideration. Biofilm formation on hospital surfaces may also be of relevance from a cleaning and disinfection standpoint.[20] Clearly, our understanding of the relevance of biofilms is in its infancy but it is certain that biofilm production is an important bacterial property.

REFERENCES

[1] Flemming HC, et al. ***Nat Rev Micro*** 2010;8:623-633.
[2] Crowley RC, et al. ***Applied and Environmental Microbiology*** 2011;77:382-384.
[3] Osland AM, et al. ***J Antimicrob Chemother*** 2012.
[4] Figueredo LA, et al. ***Veterinary Microbiology*** 2012:1-20.
[5] Martinez-Medina M, et al. ***BMC Microbiol*** 2009;9:202.
[6] Frank KL, et al. ***Antimicrob Agents Chemother*** 2007;51:888-895.
[7] Raad I, et al. ***Antimicrob Agents Chemother*** 2007;51:1656-1660.
[8] Labthavikul P, et al. ***Antimicrob Agents Chemother*** 2003;47:3967-3969.
[9] Fujimura S, et al. ***J Orthop Sci*** 2009;14:658-661.
[10] Fujimura S, et al. ***Int J Antimicrob Agents*** 2008;32:481-484.
[11] Dicicco M, et al. ***BMC Vet Res*** 2012;8:225.
[12] Hoffman LR, et al. ***Nature*** 2005;436:1171-1175.
[13] Ensing GT, et al. ***Clin Orthop Relat Res*** 2008;466:1492-1498.
[14] Djeribi R, et al. ***Am J Infect Control*** 2012:1-6.
[15] Girard LP, et al. ***Perit Dial Int*** 2010;30:652-656.
[16] Tang H-J, et al. ***Journal of Antimicrobial Chemotherapy*** 2012:1-7.
[17] Darouiche RO, et al. ***Journal of Antimicrobial Chemotherapy*** 2009;64:88-93.
[18] Brindle ER, et al. ***Biotechnol Bioeng*** 2011;108:2968-2977.
[19] Sproule-Willoughby KM, et al. ***Journal of Microbiological Methods*** 2010;83:296-301.
[20] Yezli S, et al. ***The Journal of hospital infection*** 2012.

ESTABLISHING AN INFECTION CONTROL CULTURE: WHY AND HOW?

Maureen E.C. Anderson, DVM, DVSc, DACVIM
Guelph, ON, Canada

WHY AN INFECTION CONTROL "CULTURE"?

Compared to infection control in human medicine, infection control in veterinary medicine has remained an underdeveloped field. Whether this is a result of a lack of understanding or avoidance of infectious disease issues, inadequate training in this area among veterinary personnel or other reasons is unclear. However, with ongoing advances in veterinary care resulting in more invasive treatment modalities, animals living longer with more debilitating diseases and in more immunocompromised states, and treatment of infections becoming more challenging due to increasing bacterial antimicrobial resistance, prevention and control of infections in veterinary hospitals has become increasingly important.

One of the biggest obstacles to effective infectious disease control, be it in a human or veterinary medical setting, is achieving compliance with new and established protocols. Infection control practices often seem like an added encumbrance - no one can deny that additional time and effort are required to perform them, and they may be perceived as a hindrance during very busy periods in particular. Add to this the fact that infection control provides only a negative benefit: when it is done well, infections *don't* happen and life carries on seemingly "as it should." Particularly in the absence of a disease outbreak, or when insufficient surveillance is performed to detect an outbreak, rising endemic disease or pathogen carriage (as is often the case in small primary care facilities), there is little direct motivation for staff to put in the extra effort, and these practices may fall by the wayside. Caught in such an unprepared and unpracticed state, the clinic is then perfectly ripe for an outbreak to occur, after which there is a major push for better infection control - for a time, and then the cycle repeats.

How then does one keep staff motivated to remain diligent about their infection control practices so that outbreaks can be avoided, their impact minimized when they do occur, and to strive to protect patients, owners and themselves from infectious diseases on a daily basis? Education is often the first intervention considered to improve compliance, but, while important or even essential, improving knowledge alone is often insufficient. Achieving behavioural change can be very challenging because of the many complex and interactive factors that affect human behaviour. According to the theory of planned behaviour, behavioural achievement depends on both an individual's *ability* to perform the behaviour and their *intention* (motivation).[1] With regard to infection control, most routine practices are relatively simple (e.g. hand washing, cleaning and disinfection) and therefore a person's ability to perform them is typically not an issue. Intention, on the other hand, is determined by three factors: attitude toward the action (i.e. what does the individual believe are the pros and cons, and how likely does the individual believe various outcomes of the action are to occur), subjective norms (i.e. what does the individual believe others would have him/her do, and how important is it for the individual to comply with what others think), and perceived behavioural control (i.e. does the individual believe he/she has the skills, knowledge and opportunity to perform the action, which is likely to vary by situation). Based on this, it is easy to see why education or any single intervention alone may not be enough to move a person to action. With regard to improving hand hygiene compliance in human hospitals, multimodal interventions are currently recommended, which may include different combinations of educational or interactive campaigns in various forms for staff and patients, introduction of alternative or additional hand hygiene products, increased monitoring and feedback systems, as well as traditional poster campaigns.[2] This kind of multi-pronged approach addresses behavioural intention at several different levels, and does not rely on a "on-size-fits-all" means of communication, resulting in a (hopefully) longer-lasting effect on a wider range of individuals.

The ultimate sustained multimodal intervention is the creation of a culture which includes the desired behaviour(s) - in this case an infection control culture within a veterinary clinic, and ultimately within the veterinary profession. Although there is little published work on this topic in veterinary medicine specifically, it is quite comparable to the promotion of food safety culture and safe food handling in the food service industry (including promotion of hand hygiene and appropriate cleaning and disinfection in an often high-pressure busy environment), which has been studied in some detail.[3-5]

ESTABLISHING AN INFECTION CONTROL CULTURE

In many ways, culture - be it at the national, regional, local or even clinic level - is characterized (if not defined) by what individuals read, hear, see and talk about on a regular basis.[3] Furthermore, the most important components of a given culture are generally seen and talked about more often, even though they may not necessarily dominate the media or people's conversations every day. This is the key to establishing and maintaining an infection control culture within a clinic as well. Social marketing techniques, whereby infection control practices are promoted for the "greater good," often play an important role in this process,[6] as they address both individual attitudes and subjective norms, and can also have an impact on perceived behavioural control. The same type of process can theoretically be used to incorporate any major principle into clinic culture, such as emphasis on specific aspects of client service or patient care.

What you read

Infection control needs to be made visible in written material from and around the clinic. For example, clinic bulletins, whether directed toward staff or clients, should include a visible segment on infection control. It may tie-in with other topics in the bulletin (e.g. if discussing an enteropathogen, include a section on cleaning and disinfection after fecal accidents), or it could be its own separate "quick fact" (e.g. if your dog has a cough, don't take it to the dog park, and call ahead if you're coming to the clinic so we can make sure your dog gets to go straight to an exam room). Even a simple reminder to wash hands after doing a "stoop & scoop," though it may seem like an obvious thing to do for some, emphasizes the importance of the "little" things to clients and staff alike.

A written infection control manual can also contribute to infection control culture, as the manual is physical evidence of the clinic's commitment to protecting patients, staff and clients through sound infection control practices. Such a manual also happens to be a key component of any infection control program, and is important for several reasons. From a legal standpoint, it is often said that if it's not written down, it doesn't exist, so having evidence of written policies can help protect the clinic from litigation. The manual also serves as a reference for staff members, including new staff who are being trained and are not yet familiar with certain protocols, and for more senior staff who may need a "refresher," particularly for protocols that are not used on a frequent basis. This helps to ensure protocols remain consistent over time, and that the details do not get worn away as they are passed on from one person to the next. A staff member may also be more inclined to look up a protocol on their own if they feel that asking someone else might make it look like he/she does not know how to do his/her job. The infection control manual does not necessarily need to be lengthy, and there are several resources available that provide a framework for an infection control program and manual that can be tailored to a specific clinic,[7-9] but the more complete the manual is the more useful it will be. The manual can be printed as a hardcopy (which should be regularly updated), or it can be stored as a "virtual" manual on the clinic computer network (if applicable), but it should be made readily visible and accessible to all clinic personnel, from veterinarians to kennel staff to volunteers, at all times. Furthermore, every staff member needs to be made aware of the manual *and* encouraged to use it appropriately - the very best infection control manual will not do anyone any good if it only sits on a shelf collecting dust.

What you hear

Similar to incorporating infection control into written communications, it should constitute at least a "sound bite" during all staff meetings and rounds. This not only can improve infection control culture, but simultaneously improve practices by providing regular updates to staff regarding cases in the clinic that may warrant extra infection control precautions, and keeping staff up-to-date on any changes in protocols or relevant products being used in the clinic and the reasoning behind them. This also gives staff the opportunity to provide feedback to management, which is critical for ensuring that individuals take ownership of the infection control program and helps to address barriers. Generating conversation on the topic among peers also helps to incorporate behaviours and principles into the clinic culture.[3]

What you see

Posters are one of the simplest and most visible ways to "market" infection control practices within a clinic, but if used on their own their effect, if any, may be short lived. Components of posters and poster campaigns that can help to increase their effectiveness include using a very specific message to target a desired behavior (e.g. hand hygiene), placing the posters in the area where the behaviour should occur (so immediate action can be taken), use of both text and pictures to convey the message clearly, and keeping the final message as simple as possible. It is also helpful to change or rotate posters on a regular basis so they don't "fade into the background."[3] Posters or brief infosheets that describe clinical scenarios can also be motivating, and even help to generate conversation, particularly if they use a story-telling narration and real-world examples that emphasize the audience's role in infection control and potential consequences.[4]

Behaviours observed in others serve as strong indicators of clinic culture and what is expected of an individual (thus affecting the subjective norms), which can be a double-edged sword: if someone sees others performing infection control tasks, that person is more likely to adopt similar behaviours, but if someone who would typically observe these practices sees them being abbreviated, flaunted or ignored by others, that person is less likely to continue with them. Nonetheless, even an individual's behaviour can still have a beneficial effect on clinic culture through "positive deviance," whereby the individual is able to show and teach others who have been finding adoption of a particular behaviour (e.g. hand hygiene) difficult how to successfully incorporate the behaviour into their routines.[10]

What is supported

A chain is only as strong as its weakest link, and the same can be said for any infection control program. In order to be effective, infection control measures need to be practiced by every member of the clinic staff, from owners and veterinarians to kennel staff and volunteers. In human healthcare physicians have been reported to have poorer hand hygiene compliance than nurses,[11] but veterinarians and senior staff in particular need to serve as examples to other support staff by taking the time to perform hand hygiene and use other infection control measures. It is also important to avoid putting excessive time pressure on staff that may force them to rush and cut corners when it comes to infection control - giving staff adequate time to perform hand hygiene, proper cleaning and disinfection, changing of personal protective equipment and other tasks demonstrates the importance that is placed on these practices, and thereby contributes to the infection control culture. Several studies have suggested that involvement and support of upper-level management and administration are necessary for effective implementation of hand hygiene protocols in human health care facilities, [12-15] and it is likely that the same is true in veterinary clinics, for both hand hygiene and other routine infection control measures. Without the support of owners and management, the entire infection control culture is undermined, and the infection control program is bound to fail.

CONCLUSION

A critical point to remember is that establishing an infection control culture does not mean infection control topics need to take over clinic communications, be they in written, oral or other forms. The key is to make infection control a pervasive, but not an overwhelming, component of all these things, such that it serves as a reminder without interfering with or overpowering other messages. It is also important to realize that establishing such a culture does not happen overnight, especially when starting from scratch. Attempting to effect dramatic changes in multiple behaviours of multiple people all at once will often be met with resistance, and ultimately result in exhaustion of the person(s) at the heart of the attempt. Start small, target one or just a few factors at first, and go from there. As time passes and more individuals come on board with the ideals and everyday routine of practicing better infection control, the effect will snowball, and other/new staff will more easily accept and incorporate the infection control culture into their own. Hopefully the same phenomenon will ultimately take effect across the veterinary field. Furthermore, the promotion of an infection control culture is not something that should be limited to the backroom and back office areas - it should be readily apparent to clients as well. Emphasis on infection control should not be seen as sign that a clinic has more problems with infectious diseases than average - in reality it shows a commitment to protecting animals and people from infection, and to offering a higher standard of care overall.

REFERENCES

[1]Ajzen I **Organ Behav Hum Decis Process** 1991;50:179–211. [2]Boyce J, et al **Am J Infect Control** 2002;30:S1–46. [3]Yiannas F **Food Safety Culture** Springer, 2009 (doi:10.1007/978-0-387-72867-4_5). [4]Chapman B, et al **J Food Prot** 2010;73:1101–1107. [5]Mitchell R, et al **Int J Environ Health Res** 2007;17:9–24. [6]Andreasen A. **Marketing Social Change** Jossey-Bass, 1995. [7]Scheftel JM, et al **J Am Vet Med Assoc** 2010;237:1403–1422. [8]Canadian Committee on Antibiotic Resistance **Infection Prevention and Control Best Practices for Small Animal Veterinary Clinics** CCAR, 2008. [9]**Australian Veterinary Association Guidelines for Veterinary Personal Biosecurity** AVA, 2011 (http://www.ava.com.au/biosecurity-guidelines). [10]Marra AR, et al **Infect Control Hosp Epidemiol** 2010;31:12–20. [11]Pittet D **Emerg Infect Dis** 2001;7:234–240. [12]Whitby M, et al. **J Hosp Infect** 2007;65:1–8. [13]Kretzer EK, et al **Am J Infect Control** 1998;26:245–253. [14]Larson EL, et al **J Hosp Infect** 1995;30 Suppl:88–106. [15]Pittet D, et al **Lancet** 2000;356:1307–1312.

GETTING THE UPPER HAND: HAND HYGIENE IN VETERINARY CLINICS

Maureen E.C. Anderson, DVM, DVSc, DACVIM
Guelph, ON, Canada

INTRODUCTION

Hand hygiene is considered to be the single most important measure for control of hospital-associated (HA) infections in human healthcare facilities.[1] It is a similarly important measure for controlling the spread of pathogens between animal patients within veterinary hospitals, as well for preventing the spread of zoonotic pathogens from animals to humans and vice versa,[2-4] yet until recently hand hygiene has received very little attention in this context. Part of the reason for this is that the issue of HA infections in veterinary medicine is poorly quantified and likely under-recognized, but it is nonetheless an important problem. Results of a survey of biosecurity personnel at 38 AVMA-accredited veterinary teaching hospitals showed that 82% of these facilities reported at least one outbreak of HA disease in the last five years, and 32% had been forced to close areas of the hospital in order to contain the spread of the pathogen.[5] The rate of HA infections is more difficult to assess in most primary-care companion animal veterinary clinics because of the relatively small at-risk population, and most veterinarians in such clinics are ill-equipped to perform adequate surveillance to assess endemic rates and detect outbreaks. Nonetheless there are several reports of veterinary HA outbreaks in the literature, including outbreaks caused by *Salmonella* Typhimurium, multidrug-resistant *Salmonella* and methicillin-resistant *Staphylococcus aureus* (MRSA), involving spread between animals and between animals and people.[6-8] Infection control measures, including hand hygiene, are crucial for limiting disease spread under endemic and outbreak conditions, particularly as the prevalence of high-risk individuals (both human and animal) in the population continues to grow as medical treatment advances are made.

The goal of hand hygiene in the context of modern healthcare is to reduce or eliminate the transient microflora of the hands as atraumatically as possible and ideally to prevent rebound growth, in order to decrease the likelihood of transfer of pathogenic or opportunistic microbes from the hands to other surfaces or individuals/tissues.[9,10] This may be accomplished by physical removal of the microbes from the skin (i.e. using soap and water), or use of antimicrobial/antiseptic agents that kill or neutralize the microbes (e.g. using antimicrobial soap or alcohol-based hand sanitizer (AHS)). The resident microflora of the skin is more difficult to remove but is also less likely to include pathogens.[1,11] Maintaining the integrity of the skin even after repeated hand hygiene attempts is crucial, as skin damage provides niches for opportunistic pathogens to cause infection, resulting in pain and discomfort as well as increased risk of transmission due to the larger number of pathogenic bacteria present on the skin.[9,12] Recommendations regarding hand hygiene, including when, how and how often it should be performed, appear in guidelines from multiple healthcare-associated organizations [1,11,13-16] and in various veterinary infection control guidelines.[2-4]

EVIDENCE OF EFFICACY

Many studies in human medicine have looked at the effectiveness of hand hygiene for reducing rates of hospital- or healthcare-associated infections. Larson published two systematic reviews of the topic, covering literature from 1879 to 1986[17] and 1977 to 1998.[12] Both reviews supported an association between hand hygiene and reduced infection rates, although there were very few prospective clinical trials. Larson pointed out that this was likely due to the dramatic effects of using hand hygiene and aseptic technique on morbidity and mortality demonstrated by scientists like Semmelweis in the 19th century, resulting in these practices being widely adopted in human medicine by the turn of the 20th century. The evidence for causality provided by research studies published since the 1970s was often weak,[12,18] however there are numerous studies that support causal criteria including a temporal relationship, strength of association, consistency of association and "dose" response for hand hygiene and HA infections.[17] Backman et al.[18] reviewed 31 original studies looking at hand hygiene interventions published between 1996-2006. Of these, 20 studies were given a quality score of 2 or higher based on not containing any fatal flaws (as per Larson[19]), and 16/20 (80%) of these studies reported a significant reduction in HA infections associated with a hand hygiene intervention.

Hand hygiene studies tend to suffer from one or more significant limitations due to poor methodology, including but not limited to simultaneous implementation of other measures that may affect pathogen transmission and infection rates (which can artificially inflate or mask the effect of hand hygiene), lack of a control group or use of historical controls (as a negative control group is very difficult to justify on ethical grounds[17]), failure to report statistical significance, and the Hawthorne effect (whereby study subjects may alter their behaviour when being observed).[18] Use of historical controls can be particularly problematic when studying disease outbreaks, as it is difficult or impossible to separate the potential effect of an infection control intervention from the natural cycle and extinction of the outbreak itself.[18,20] Studies with negative findings may suffer from the same limitations as well as inadequate sample size.[18] Although the limitations of these studies must be taken into consideration when interpreting the results, this does not mean that the benefits of hand hygiene described in these studies should be dismissed. It is also important to differentiate a failure to achieve sufficient compliance with a hand hygiene protocol from failure of a hand hygiene protocol to have its desired effect on disease transmission. However, it is equally true that the effectiveness of any protocol is irrelevant if it is not sufficiently practical that personnel will comply with it and still be able to perform their required duties.

HAND HYGIENE COMPLIANCE

Compliance with hand hygiene protocols is the most challenging component of ensuring efficacy. Numerous studies have looked at hand hygiene compliance in hospitals.[1,21] In general, compliance is poor among healthcare workers (<50%), and physicians tend to have lower compliance with hand hygiene protocols than nurses.[10,22] Major barriers to compliance in many situations include skin irritation (irritant contact dermatitis), lack of accessibility to hand hygiene stations, time constraints (i.e. too busy), and lack of perceived importance (i.e. that it is generally unimportant or more likely that it is less important than other tasks/procedures which therefore take precedence over hand hygiene).[1,23] In some cases people may also simply forget to perform hand hygiene;[1] to prevent this, hand hygiene needs to become an automatic habit for healthcare personnel, like covering a sneeze or a cough.

HAND HYGIENE IN VETERINARY MEDICINE

Literature regarding hand hygiene in veterinary medicine specifically is limited. Traub-Dargatz et al.[24] examined the effectiveness of traditional hand washing versus AHS following basic physical examination of a horse, and found the AHS to be more effective at reducing bacterial counts on the hands. However, it was pointed out that the hands of veterinary personnel working in equine practice may become more heavily soiled on a regular basis than personnel working in a small animal or human hospital, which may require hand washing to remove gross contamination, rather than use of AHS alone for hand hygiene. Anderson et al.[25] found that hand washing following infectious cases and between farms had a protective effect against colonization with MRSA in veterinary personnel who work with horses. With regard to hand hygiene compliance, in two studies of companion animal veterinary personnel, self-reported compliance (i.e. always perform hand hygiene between patients) was 42-48%,[26,27] while observed compliance rates in one veterinary teaching hospital were 21-42%.[28]. In the study by Wright et al.,[27] self-reported compliance among large animal and equine practitioners was even lower (<20%). Only slightly more than half of small animal practitioners in the same survey (590/1069) reported always washing their hands before eating, drinking or smoking while at work, and the same was true for less than a third of large animal and equine practitioners. Nakamura et al.[26] reported the most commonly cited reason for not performing hand hygiene among veterinary support staff was being too busy (72%).

IMPROVING HAND HYGIENE COMPLIANCE

Providing better access to hand hygiene stations and supplies is often the first step to improving hand hygiene compliance, as it aids in minimizing the time required to comply with protocols, and acts as a visual reminder to perform hand hygiene. Because renovating most facilities to improve the location of sinks for hand washing is usually not feasible, improving access typically involves introducing AHS, introducing a different kind of AHS that is better accepted by personnel (e.g. based on less of a drying effect or better smell), or changing the location or number of AHS dispensers.[1] Increasing the use of AHS can help to address several barriers to hand hygiene, as compared to soap and water, use of such products takes less time, causes less skin damage, and dispensers can be easily placed in any area or even carried in pocket for ready access at any time.[9] This is also why AHS is the product primarily recommended for routine hand hygiene in hospital settings, unless involvement of an alcohol-resistant pathogen is known or suspected, or hands are visibly soiled.[1,13,14] Monitoring and feedback systems can also have a significant impact on compliance.[29,30] Feedback may be from patients (or patient owners in the case of veterinary clinics), researchers or infection control personnel. Making people aware of how often they neglect to perform hand hygiene, and/or giving them positive feedback when their compliance improves (the "stick and carrot" approach), seems to provide additional incentive to further improve compliance. Convincing people of the importance and utility of hand hygiene in curbing the spread of infectious agents is crucial to achieving compliance.[1] The perception that hand hygiene is relatively unimportant is likely more problematic in veterinary medicine compared to human medicine. Because there are fewer diseases that can be transmitted from animal-to-person compared to person-to-person, some veterinary personnel have a relatively cavalier attitude toward contamination of hands/clothing/equipment with animal blood, body fluids and excreta. However, hand hygiene is also a critical means of preventing potential indirect transmission of non-zoonotic pathogens from animal to animal, thus protecting veterinary patients. The potential of a zoonotic pathogen, brought into a clinic by a single person or animal, to spread first through the clinic and then into the community is significant. Examples of this kind of event have occurred,[6,7] and many more likely go unreported or simply unrecognized.

Ideally hand hygiene should be something that everyone wants to do, rather than something everyone must be asked to do, but education alone is frequently insufficient to change behaviour. Development of an infection control culture, whereby infection control practices and principles become integrated into the group mentality through ongoing communication, conversation, training and strong support from team leaders/management, is the ultimate means of improving compliance with hand hygiene and other infection control measures. A study by Murphy et al.[31], that looked primarily at environmental disinfection practices, management of infectious patients and antimicrobial use in clean surgical procedures in veterinary clinics, identified that none of the 101 participating clinics had a formal infection control program. Wright et al.[27] found that only 247/797 (31%) small animal practices, and even fewer large animal and equine practices, had written infection control policies. Infection control practices, including hand hygiene, are more likely to be underutilized under these circumstances, and their effectiveness cannot be evaluated without established policies and a formal infection control program, even if the program is very basic. Murphy et al.[31] also showed that veterinarians and technicians generally consider certain aspects of infection control, such as environmental cleaning and disinfection, important measures, yet 40-60% did not know what products were used in their own clinics for disinfecting areas contaminated with infectious body fluids, nor how to properly prepare the products for use. Wright et al.[27] found similar discrepancies in practitioner perceptions of zoonotic disease risk versus personal protective equipment (PPE) use when examining animals with potentially zoonotic infections. For example, less than 30% of small animal practitioners who were concerned about the diseases wore appropriate PPE when examining animals potentially infected with rabies, gastrointestinal parasites, gastrointestinal bacteria or dermatophytes. It is likely that the same shortfalls hold true for hand hygiene practices – even among veterinary personnel who acknowledge the importance of hand hygiene, few likely perform it how and when they should. These findings illustrate the lack of an infection control culture in veterinary clinics, and in veterinary medicine in general; this issue must be addressed if sustained improvement in hand hygiene and other infection control practices is to be achieved.

CONCLUSION

Without appropriate attention to infection control efforts, veterinary clinics can become amplifiers of both animal diseases and zoonotic diseases in the community, due to the high flow-through of out-patient animals and their caretakers, the mixing of both sick and susceptible animals (i.e. very young, very old or immunocompromised), the wide variety of patients seen at a clinic (e.g. medical, surgical, oncological, regular health check-ups), and the close interaction of animals, clinic personnel and members of the general public that occurs there. Improving hand hygiene practices in veterinary clinics has the potential to curb the spread of recognized pathogens within and outside these facilities, and could help to prevent widespread transmission of other emerging pathogens in the future before they are even recognized.

REFERENCES

[1]Boyce JM, et al **Am J Infect Control** 2002;30:S1–46. [2]Scheftel JM, et al **J Am Vet Med Assoc** 2010;237:1403–1422. [3]Australian Veterinary Association **Guidelines for Veterinary Personal Biosecurity** AVA, 2011 (http://www.ava.com.au/biosecurity-guidelines). [4]Canadian Committee on Antibiotic Resistance **Infection Prevention and Control Best Practices for Small Animal Veterinary Clinics** CCAR, 2008. [5]Benedict KM, et al **J Am Vet Med Assoc** 2008;233:767–773. [6]Weese JS, et al **Vet Microbiol** 2006;114:160–164. [7]Wright JG, et al **Emerg Infect Dis** 2005;11:1235–1241. [8]Cherry B, et al **Emerg Infect Dis** 2004;10:2249–2251. [9]Kampf G, et al **Clin Microbiol Rev** 2004;17:863–93. [10]Pittet D **Emerg Infect Dis** 2001;7:234–240. [11]Larson EL **Am J Infect Control** 1995;23:251–269. [12]Larson EL **Clin Infect Dis** 1999;29:1287–1294. [13]Muto CA, et al **Infect Control Hosp Epidemiol** 2003;24:362–386. [14]Pittet D, et al **Infect Control Hosp Epidemiol** 2009;30:611–622. [15]Siegel JD, et al **Am J Infect Control** 2007;35:S165–93. [16]Vearncombe M, et al **Best Practices Document for the Management of Clostridium difficile in all health care settings** PIDAC, 2007. [17]Larson EL **Infect Control Hosp Epidemiol** 1988;9:28–36. [18]Backman C, et al **Am J Infect Control** 2008;36:333–348. [19]Larson EL **Am J Infect Control** 2005;33:548–557. [20]Larson EL, et al **Behav Med** 2000;26:14–22. [21]Larson EL, et al **Am J Infect Control** 2007;35:666–675. [22]Pittet D, et al **Ann Intern Med** 1999;130:126–130. [23]Pittet D **Infect Control Hosp Epidemiol** 2000;21:381–386. [24]Traub-Dargatz JL, et al **Can Vet J** 2006;47:671–676. [25]Anderson M, et al **Vet Microbiol** 2008;129:410–417. [26]Nakamura RK, et al **J Small Anim Pract** 2012;53:155–160. [27]Wright JG, et al **J Am Vet Med Assoc** 2008;232:1863–1872. [28]Shea A, et al **J Am Vet Med Assoc** 2012;240:61–64. [29]Erasmus V, et al **Infect Control Hosp Epidemiol** 2010;31:283–294. [30]Sax H, et al **Am J Infect Control** 2009;37:827–834. [31]Murphy CP, et al **Zoonoses Public Health** 2010;57(6):429–438.

HEMATEMESIS: NOT AS EASY AS FIRST THOUGHT

Michael D. Willard, DVM, DACVIM
College Station, TX

Hematemesis necessitates a slightly different approach than we take with other vomiting cases because some rule-outs become more likely while others become much less likely. We will be including upper gastrointestinal bleeding of any cause in this discussion. For starters, we will not be discussing vomiting that produces "flecks" of blood because this can be seen in any dog (and perhaps cat) with vigorous vomiting in which the gastric mucosa is traumatized by the physical act of vomiting.

Most of the time, hematemesis is denoted by a "coffee-grounds"-like material that most clients (and some veterinarians) do not recognize as blood. A common mistake is being concerned over "dark stools". Noting that a patient has dark stools is generally useless. Lots of dogs have dark stools and no problems or GI blood loss at all. The color of the stool is not an issue until the stool is pitch-tar-coal-asphalt black. Then it may be melena (if it is not due to Bismuth or a lot of green bile giving it a near-black appearance). If in doubt, just place some fresh feces on absorbent white paper and see if a reddish color diffuses out from the feces, confirming that there is blood present. Melena is only seen if there is an acute loss of a lot of blood into the upper GI tract. Most dogs losing blood in the upper GI tract do not have any important changes in the color of the feces. Rather, you might see anemia and hypoalbuminemia. Also remember, you may or may not see hypoglobulinemia; it all depends upon what the serum globulin concentration was before you started losing blood. Sometimes the BUN is higher than expected based upon the serum creatinine, but again this is only expected if there is a lot of blood being lost in a short period of time. Fecal occult blood tests are seldom that helpful or necessary, but can occasionally be informative in confusing cases. However, you need to use a test for which the laboratory has substantial experience in dogs so that the results can be meaningfully interpreted.

When there is a substantial amount of blood being ejected from the mouth, there tend to be 3 major reasons: coagulopathy, swallowing blood from elsewhere and gastrointestinal ulceration/erosion (GUE).

Coagulopathies

Most coagulopathies cause concomitant bleeding from the nose or accumulation of blood in body cavities or petechia. However some patients with a systemic coagulopathies only have GI bleedingi. Therefore, it is always appropriate to check the platelet count and some measure of clotting factor adequacy in animals with hematemesis or GI blood loss. While coagulopathies are a relatively uncommon cause of GI bleeding compared to ulceration/erosion, they can have devastating consequences if not diagnosed promptly. In particular, remember that some cats with intestinal disease will malabsorb vitamin K to the point of having vitamin K-responsive coagulopathies. Also remember that just because the patient did well during a surgical procedure a few days before the current bleeding episodes does not mean that a coagulopathy is impossible; sometimes the prior surgical procedure apparently depletes the limited amounts of coagulation factors in patients with subclinical coagulation defects, causing them to become clinical after the procedure.

Ulcers and erosions

Gastritis due to any number of causes typically has some degree of GUE present. We will not discuss this cause too much since most of these cases quickly and spontaneously resolve. Renal disease is often named as a cause of gastric ulceration, but this is actually very rare.

The most common causes of chronic, unresolving GUE that are also the easiest to check for are mast cell tumor, drug administration and "stress".

Drugs are still a very important cause of GUE in the dog, despite all the newer cox-2 NSAIDs. High doses of dexamethasone also have substantial potential for significant GUE. Prednisolone by itself is generally not ulcerogenic unless it is used in very high doses (e.g., > 4 mg/kg/day) or is administered to a patient with other "ulcerogenic" risk factors (e.g., hypoxia, poor perfusion). Even then prednisolone is not particularly bad. Combining steroids and non-steroidal drugs can be devastating, but you can generally use ultra-low dose aspirin (0.5 mg/kg once daily) safely when treating IMHA dogs with steroids.

All NSAIDs have the potential to cause devastating GUE, and some of these non-steroidal drugs are renowned for their toxic effects (i.e., indomethacin, naproxen, flunixin). Ibuprofen is also particularly ulcerogenic in the dog because it undergoes an enterohepatic circulation. Flunixin is a particularly dangerous drug from the standpoint of causing GUE. It is extremely potent and can be devastating if combined with steroids like dexamethasone. While able to cause significant ulceration and bleeding all by themselves, the ulcerogenic potential of NSAIDs is particularly augmented by other factors, especially concurrent administration of another NSAID or a corticosteroid, and hypoperfusion of the alimentary tract. Even though many dogs seemingly tolerate such combination therapy, you need to realize that you are "walking on thin ice".

Many to most of the dogs treated with NSAIDs have endoscopically visible erosions, hemorrhages and/or ulcers, depending upon the drug used and the dose administered. It is important to note that most dogs with GUE due to NSAIDs may be completely asymptomatic. Finally, there is tremendous between-dog variation regarding the alimentary tract response to NSAID's; some dogs may almost bleed to death because of a small dose of aspirin while most dogs would tolerate a much larger dose with relative impunity.

While the newer Cox-2 NSAIDs (e.g., carprofen, etogesic, deracoxib, meloxicam, etc) have much less potential for causing GUE than the older NSAIDs, you can still see GUE (and even perforation) due to these drugs. Part of the problem is that these drugs are being used so extensively and casually. The problem often revolves around using inappropriately high doses, using the drug at the wrong time (e.g., when the patient is experiencing shock or has poor perfusion to the alimentary tract), and using the drug in combination with another NSAID.

Stress, when mentioned as a cause of ulcers, specifically refers to substantial decrease in visceral perfusion (e.g., hypovolemic shock, neurogenic shock, Systemic Inflammatory Response Syndrome) that is typically obvious from history and/or physical examination; or, it can refer to extreme exertion (e.g., sled dog, military working dogs).

Mast cell tumors may look like any skin lesion. In particular, they can perfectly mimic the appearance and feel of lipomas, such that the only way to distinguish them from lipomas is by aspirate cytology. When these tumors degranulate, they release histamine which if of sufficient magnitude can cause gastric acid hypersecretion. This can result in severe ulceration, especially just inside or just beyond the pylorus.

Hepatic failure seems to be another important cause of GUE in the dog. Anytime a dog with hepatic disease suddenly becomes clinically worse (especially if it becomes obviously encephalopathic), you should consider the possibility of GUE. Bleeding into the intestine counts as a high protein meal and predisposes to hepatic encephalopathy in these patients. In ill patients that cannot be scoped, aggressive therapy with famotidine and/or carafate is reasonable. Hepatic disease may cause disseminated intravascular coagulopathy which can cloud the picture when trying to determine the cause of hematemesis.

Gastric tumors may cause bleeding. The leiomyoma and leiomyosarcoma in particular may cause especially dramatic bleeding due to their propensity to ulcerate. This is especially important because this tumor is often curable with surgery, as opposed to lymphomas and carcinomas that are more common, have less dramatic signs, seldom cause GI blood loss and yet have a much worse prognosis. Unfortunately, it can be hard to adequately image the stomach with ultrasound, and these masses can be missed if there is blood, ingesta and/or air in the stomach.

Foreign objects get a lot of press as causes of GUE, but in fact they are relatively uncommon causes. However, they are particularly important in patients that have GUE because even the most innocuous of GI foreign objects (e.g., paper, small piece of soft cloth) can sometimes prevent a pre-existing ulcer from healing. They typically need to be removed in patients with GUE of any cause.

Ingesting blood

This possiblilty is typically forgotten. However, it is surprisingly easy to have bleeding pulmonary lesions in which the blood is coughed up, swallowed, and later vomited. In most of these cases, the client does not report coughing (perhaps because the hematemesis "wipes" all else from their mind). In like manner, we have seen cases in which blood was trickling from the choana into the pharynx and being swallowed, and yet the patient had no history of sneezing or coughing or nasal discharge.

Cats are rarely diagnosed with gastric ulceration/erosion. When a cat is found to have gastric ulceration/erosion, it is usually caused by lymphoma. However, we many times never determine the cause of GUE in cats.

Clinical approach to the patient with hematemesis or GI bleeding

There is often something in the history that is suggestive of the cause of the bleeding (e.g., NSAIDs, shock, etc). If that is the case, then it is often reasonable to begin appropriate therapy after requesting basic laboratory testing (e.g., CBC, serum chemistry panel) to determine the severity of the bleeding and if there are other diseases (e.g., hepatic disease, renal disease) present. Imaging (especially ultrasound) is typically appropriate but not necessarily imperative at this time. If the cause of the GI bleeding or hematemesis is not obvious, if the patient has not responded to 5-7 days of appropriate therapy, or if the bleeding is severe, then additional diagnostics are important and should be performed promptly.

Diagnostic approach: First, as stated above, it is wise to first eliminate coagulopathy with a platelet count and some measure of clotting factor adequacy. A mucosal bleeding time is a very useful screening test in these patients. Sometimes there is both a mucosal defect and a coagulopathy. In particular, if ehrlichiosis is possible, one must consider the possibility that the patient has what would normally be an insignificant mucosal defect but which is bleeding because of the effects of

Ehrlichia spp. upon platelet numbers and their function. After coagulopathy has been eliminated, then imaging should be done if it has not already occurred, and ultrasound is especially important as it may reveal masses that can be aspirated percutaneously, thus avoiding the need for endoscopy/surgery.

If these tests have not revealed the diagnosis, then gastroduodenoscopy is generally performed next. The specific reasons to do gastroduodenoscopy in a patient with GI bleeding are to:

a) determine if the cause can be surgically removed (i.e., there are only one to two ulcers which can be approached surgically -- this is for cases that are bleeding and have not responded to medical management or cases that are bleeding so badly that one cannot wait on medical management). In these cases, it is important to be sure that bleeding is not due to widespread erosions that cannot be cured surgically. There is no relationship between the size of the mucosal defect and the amount of bleeding; patients with lots of small erosions often bleed as bad or worse than patients with ulcers. It is also important to determine the number and location of such ulcers as they may be hard to find during a gastrotomy.

b) determine if there is a gastric tumor or some other infiltrate in a patient with GUE that is non-responsive to appropriate therapy.

c) determine the cause in a patient with GUE and no apparent cause on the history, physical examination, or routine blood work.

d) look for a cause of bleeding in a patient with GI blood loss of unknown cause.

It is important to note that endoscopy will not generally allow one to determine if an ulcer will or will not respond to medical management. In most cases, only by treating and observing the patient will you know if the ulcer will respond to medical management.

It is important to not give carafate within 24-36 hr before endoscopy because it can cover the lesions and make evaluation more difficult. It is best if food has been withheld for at least 24 hours. Avoid prokinetics (e.g., metoclopramide, cisapride, ranitidine). Endoscopy of these patients may be difficult is there is substantial blood present in the stomach. Patience is necessary to flush in water and then aspirate it and the blood over and over again. It is important to be able to view as much of the gastric mucosa as possible. In some cases, there are large blood clots that cannot be removed endoscopically, in which case surgery might be necessary. It can be especially easy to miss ulcers that are in the pylorus. If they pylorus is infiltrated and inflamed, it can be difficult to pass the tip of the scope through that area; therefore, the endoscopist often does not obtain a good view of this area. One may have to go in and out of the pylorus multiple times to be sure that there is or is not a lesion present. If a cause of upper GI bleeding cannot be found in the stomach or duodenum, strong consideration should be given to examining the trachea, bronchi, and choana while the patient is anesthetized. Patients with hematochezia may benefit from colonoileoscopy, but patients with hematemesis or melena rarely benefit from lower endoscopy.

If there is substantial upper GI blood loss and these tests do not allow diagnosis, then exploratory surgery is the next step. However, it is very easy to not be able to find the cause of the bleeding in these patients, so warn the owner before doing the procedure.

Medical management

If the patient is not exsanguinating, the cause is known or strongly suspected, and the patient has not had 5-7 days of appropriate medical therapy, then medical therapy is often reasonable as opposed to doing a major diagnostic work up. In distinction, if the patient is exsanguinating or if the patient has not shown any appreciable response to 5-7 days of appropriate medical therapy for the ulceration, then it is usually reasonable to surgically resect the ulcerated area. Note – when I say "response", I am not referring to the patient being cured; I am referring to clear evidence of improvement. If surgery will be considered, it is usually very wise to perform gastroduodenoscopy before the surgery to be sure that you find all of the sites of ulceration. It is very easy to fail to detect an ulcer at surgery, and endoscopy usually allows one to easily find all areas of ulceration. Sometimes intraoperative endoscopy is necessary to help the surgeon find the ulcer(s).

If medical management is elected, be sure to remove the cause of the GUE, if possible. If the cause is not removed, medical management tends to be less successful. Next, be sure that the patient is well hydrated; healing of the gut requires or is at least benefitted by adequate perfusion. If there is significant gastroduodenal reflux of bile, metoclopramide or cisapride may be helpful in preventing bile from entering and/or staying in the stomach and augmenting the ulcerogenic process.

<u>H-2 receptor antagonists</u> are commonly used. Cimetidine (5-10 mg/kg) needs to be given 3-4 times per day if you are really serious about decreasing gastric acid secretion. Famotidine (0.5-1.0 mg/kg) only needs to be given once or twice daily. The over the counter (OTC) preparations are all oral, and some patients are vomiting so vigorously that they must receive parenteral medication. Ranitidine (2.2 mg/kg) is usually effective if given twice daily, but may cause vomiting if given as an IV bolus. Side-effects of the H-2 receptor antagonists are rare but may include diarrhea, drug eruption, hyper-pyrexia, thrombocytopenia, granulocytopenia, and CNS problems including seizures. I think I have seen a couple of dogs in which ranitidine was responsible for large bowel diarrhea or seizures. The primary value of the H-2 receptor antagonists is

in treating existing ulcers and erosions. They can be helpful in preventing some types of ulcers, but this is not true with all types of ulcers (e.g., they are not effective in preventing ulcers due to NSAIDs or due to steroids). There is a newer drug, lafutidine, that works on capsacian receptors as well as histamine-2 receptors which holds considerable promise for increased efficacy.[1]

Proton pump inhibitors are the most effective antacid drugs we have[2-4]. Omeprazole, lanosprazole, esomeprazole, pantoprazole, dexlansoprazole are the most effective inhibitors of gastric acid secretion we currently have available. Omeprazole is available OTC as Prilosec®. The dose of omeprazole is 0.7-1.5 mg/kg qd, although I have often used it at up to 2 mg/kg bid in patients with severe reflux esophagitis or gastrinomas. The dose of lanosprazole (Previcid), pantoprazole (Protonix), and esomeprazole (Nexium) is 1 mg/kg IV (not approved for use in dogs). It generally takes 2-5 days for a PPI to have maximal efficacy; but, the immediate effects on gastric acid secretion are often better than that obtained by high dose H-2 receptor blockade. Dexlasoprazole is a newer formulation that lasts much longer than other proton pump inhibitors.[5]

Misoprostol (Cytotec®) is a prostaglandin E analog which was primarily designed to be a prophylactic drug used to prevent GUE due to NSAIDs.[6] It is also useful in treating existing ulcers, but its higher cost and more plentiful side effects usually make it undesirable as a first line therapy for GUE. It is typically used at a dose of 2-5 ug/kg, 3-4 times daily. It can cause abdominal cramping and diarrhea, but the drug seems relatively safe in dogs. The main disadvantage is that it must be given orally, which is not possible in some vomiting animals. Because it is a prostaglandin analog, it should not be used in pregnant females for fear of causing abortion or miscarriage. It is the best drug available that can be used to prevent NSAID-induced ulceration, but it is not uniformly effective in dogs. The main indications to use it appear to be a) the patient that must have NSAID's to function, but which evidences side-effects from them (e.g., anorexia, vomiting) and b) the patient that seemingly needs to receive NSAID's that have substantial potential for such side-effects (e.g., piroxicam).

Sucralfate seems to be effective in protecting those areas which are already ulcerated and helping them heal.[6] The only common side-effect is constipation. There is minimal absorption from the intestines, but it does have the capacity for adsorbing other drugs (e.g., enrofloxacin). While carafate is effective in treating ulcers, it is not always effective in preventing ulceration. My major problem with this drug is that it must be given orally, which does not always work in vomiting dogs. Sometimes you may dissolve the table in water or buy the suspension and have less problem with that being vomited.

REFERENCES

[1]Yamagishi H, et al; Stonger inhibition of gastric acid secretion by lafutidine, a novel H_2 receptor antagonist, than by the proton pump inhibitor lansoprazole, **World J Gastroenterol** 2008; 14:2406. [2]Bersenas A, et al. Effects of ranitidine, famotidine, pantoprazole, and omeprazole on intragastric pH in dogs. **Am J Vet Res** 2005; 66:425. [3]Tolbert K, et al, Efficacy of oral famotidine and 2 omeprazole formulations for the control of intragastric pH in dogs, **J Vet Intern Med** 2011; 25:47. [4]Williamson KK, et al: Efficacy of omeprazole versus high-dose famotidine for prevention of exercise-induced gastritis in racing alaskan sled dogs, **J Vet Intern Med** 2010; 24:285. [5]Hershcovici T, et al, Dexlansoprazole MR – A review, **Ann Med** 2011; 43:366. [6]Ward DM, et al. The effect of dosing interval on the efficacy of misoprostol in the prevention of aspirin-induced gastric injury. **J Vet Int Med** 2003; 17:282. [7]Dallwig B: Sucralfate, **J Exotic Pet Med** 2010; 19:101.

PROTEIN-LOSING ENTEROPATHIES: NOT WHAT YOU MIGHT EXPECT

Michael D. Willard, DVM, DACVIM
College Station, TX

HYPOALBUMINEMIA

When concerned with low protein from any cause, one should measure serum albumin concentration, not the serum total protein concentration. Do not use human clinical pathology laboratories; their technology sometimes does not detect canine albumin with the result that they routinely report serum albumin concentrations of < 1.5 gm/dl in clinically normal dogs. If the patient has substantial hypoalbuminemia, the first step is to examine the skin for obvious lesions which can cause protein loss. Cutaneous lesions sufficient cause such hypoalbuminemia will be obvious; you should be able to just look at the patient and know if this is the problem. Next, hepatic function testing (e.g., resting and post-prandial serum bile acid concentrations) and a urinalysis should be requested. If there is any doubt about the amount of protein being lost in the urine, then a urine protein:creatinine ratio will quantify the magnitude of urinary protein loss. Severe hypoalbuminemia (i.e., < 2 gm/dl) in an animal with diarrhea suggests a protein-losing enteropathy (PLE); however, diarrhea (even when severe) in no way is sufficient to eliminate hepatic disease as the cause of the hypoalbuminemia. Furthermore, a very substantial number of dogs and cats with PLE do not have vomiting or diarrhea. Some only have ascites, and some are found fortuitously on routine blood work. This may be especially true of dogs with primary intestinal lymphangiectasia.

In general, once severe, exudative cutaneous disease, protein-losing nephropathy, and hepatic insufficiency are eliminated, then PLE becomes a diagnosis of exclusion in patients with a serum albumin < 2.0 gm/dl. Fecal concentrations of alpha-1 protease inhibitor can be used as a means of confirming PLE if there is confusion because of concurrent hepatic or renal disease. The major use for this test in clinical medicine seems to be the hypoalbuminemic patient in which you strongly suspect PLE (e.g., based upon it having severe diarrhea or having hypocholesterolemia), but which also has PLN and/or hepatic disease. However, there are several nuances to this test, especially in sample collection which makes it potentially difficult to interpret. We seldom need this test in clinical practice. Finally, it is important to note that PLE may be associated with a low, normal or increased serum globulin concentration.

Perhaps the most important point of this discussion is that while hypoalbuminemia has repeatedly been reported to be a poor prognostic sign in patients with chronic GI disease, there are some patients that will respond well to appropriate therapy.[1] Therefore, diagnosing PLE is not necessarily a cause for despair. That being said, because many dogs and cats with PLE have severe alimentary tract disease that needs to be diagnosed promptly to maximize the chance for successful therapy, aggressive diagnostics are typically appropriate. Although therapeutic trials can be chosen in place of classic diagnostic tests in many of the more common alimentary tract diseases (e.g., dietary-responsive diarrhea, antibiotic-responsive enteropathy, parasites), such an approach is generally ill-advised if the serum albumin concentration is less than 2.0 g/dl. This is true because it may be necessary to perform an antibiotic and/or dietary therapeutic trial for 3-6 weeks in order to ascertain if it is being effective, and a patient with severe PLE can become markedly worse in that time, especially if the serum albumin concentration is falling rapidly.

Any GI disease can cause PLE if it is severe enough. Many acute GI diseases cause PLE (e.g., parvoviral enteritis); however, these diseases are comparatively easier to treat than chronic GI diseases causing PLE. Therefore, the focus in this lecture is PLE in animals with chronic GI disease. The major causes of PLE in adult dogs tend to be intestinal lymphangiectasia, inflammatory bowel disease (IBD), alimentary tract lymphoma (LSA), and fungal infections (i.e., histoplasmosis and pythiosis). Other causes include alimentary tract ulceration/erosion, severe disease of intestinal crypts, antibiotic-responsive enteropathy, and parasites. The major causes of PLE in juvenile dogs tend to be parasites and chronic intussusception. Cats with PLE usually have IBD or alimentary tract lymphoma.

Many dogs with PLE have hypocholesterolemia. Pets with protein-losing nephropathies usually have hypercholesterolemia, while those with hepatic insufficiency often have hypocholesterolemia. Fecal examinations for parasites are appropriate. Although parasites are an uncommon cause of PLE in adult animals, pets in select environments (e.g., confined areas where patients can reinfect themselves) may incur substantial parasitic loads.

DIAGNOSTIC APPROACH TO THE PLE PATIENT

Once PLE has been diagnosed, intestinal biopsy is usually the ultimate means of establishing a diagnosis. Biopsy can be done via laparotomy, laparoscopy, or endoscopy. Feeding a small, fatty meal (use canned food, not dry, and add in cream or corn oil) the night before the procedure might make it easier to diagnose lymphangiectasia. Flexible endoscopy, when done by someone who is trained in how to take diagnostic tissue samples and submit them is usually more than adequate to obtain diagnostic samples. However, if endoscopy will be used to biopsy the small intestines, one should first ultrasound the abdomen to make sure that there are no focal infiltrates that are out of reach of the endoscope, or which might be more easily diagnosed by ultrasound-guided fine needle aspiration. Furthermore, there are ultrasonographic changes that can be found in the small intestine (e.g., streaks in the submucosa) that can be diagnostic. Radiographs and barium series are seldom as sensitive as ultrasound. If flexible endoscopy will be done, one should biopsy both the duodenum and ileum. There have been numerous cases in which lymphangiectasia, IBD or LSA were obvious in the ileum but not in the duodenum. It is not necessary to enter the ileum with the endoscope to obtain a good tissue sample of the ileal mucosa.

Laparotomy and laparoscopy are good means of obtaining diagnostic samples, but it is surprisingly easy to procure non-diagnostic samples with these techniques (i.e., "full-thickness sample" is not synonymous with "diagnostic sample"). Endoscopy does have the advantage of allowing one to visualize mucosal lesions that are "invisible" when looking at the serosa. In some cases, the diagnosis can only be obtained by biopsying these focal lesions. If full-thickness biopsies are obtained in severely hypoalbuminemic animals, then serosal patch grafting will minimize the risk of suture line leakage. A nonabsorbable or a poorly absorbable suture (PDS) should also be used.

INTESTINAL LYMPHANGIECTASIA

Intestinal lymphangiectasia seems particularly common in Yorkshire terriers[2] and Soft-Coated Wheaten terriers, but may occur in any breed. Sometimes these dogs have distinct ultrasonographic findings: "steaks" in the mucosa that represent dilated lymphatics.[3,4] While histopathology is obviously the desired means of diagnosis, one can sometimes make a definitive diagnosis based upon grossly visible endoscopic findings (i.e., numerous, erratic, grossly engorged lacteals seen as large white blebs on the mucosa).[1] These lesions are "fragile" and apparently may be destroyed by biopsying them (both endoscopically and surgically) if the endoscopist or surgeon is not careful. It is important to note that lymphangiectasia can be a relatively localized disease in the intestines, being present in only the ileum or only the jejunum or only the duodenum; therefore, it is important to biopsy as much of the intestinal tract as possible. Furthermore, if one biopsies the intestines and cannot find a cause of PLE, sometimes lymphangiectasia can be tentatively diagnosed by elimination (i.e., by eliminating IBD, lymphoma, parasites, intussusception, fungal infections, etc).

Diagnosis by means of endoscopic biopsy is certainly possible if the endoscopist is trained in taking high quality tissue samples. However, recent work has demonstrated that poor quality mucosal biopsies (e.g., primarily villus tips or substantial "squash" artifact) makes is much more difficult or even impossible to find the lesions. If one is taking high quality tissue samples (i.e., total length of the villi plus subvillus mucosa down to the border of the mucosa and muscularis mucosa), it typically takes about 6-7 tissue samples to have 90-99% confidence in finding lymphangiectasia. However, it can take 5-7 times as many tissue samples to have the same assurance if you are obtaining poor quality tissue samples that primarily consist of villus tips.[5]

When doing endoscopy, it is important that ileal biopsies be taken in addition to the typical duodenal biopsies. We are finding that ileal biopsies often reveal lesions not found on duodenal biopsies. This is true for lymphangiectasia as well as lymphoma and other lesions. With basic training, an endoscopist should be able to obtain ileal biopsies endoscopically at least 85%+ of the time. Typically, ileal biopsies are often of higher quality than duodenal biopsies.

Therapy for intestinal lymphangiectasia revolves around an ultra-low fat diet, preferably with anti-inflammatory therapy designed to alleviate the lipogranuloma formation that commonly occurs within the intestinal wall and/or mesentery. Supplementation with medium chain triglyceride oil (MCT) used to be recommended because MCT oil supposedly bypasses intestinal lymphatics thus preventing further rupturing of the lacteals. Pancreatic enzymes were often added to the diet to ensure digestion of the medium chain triglyceride oil. MCT oil is seldom used anymore, probably because appropriate dietary therapy is usually more than sufficient. Feeding homemade diets that are highly digestible and ultra-low in fat (e.g., white turkey meat plus potato or rice) or feeding commercial diets is often very helpful in these patients. Commercial low fat diets can be used very successfully, but they need to have the lowest possible fat content. Such a diet can be so successful that it might occasionally be appropriate to use it as a therapeutic trial. Dogs with lymphangiectasia often show a marked increase in serum albumin concentration within 7-14 days of starting such a diet.

The important of lipogranulomas in the intestinal wall and mesentery is uncertain.[6] However, it is hypothesized that some patients fail to respond to appropriate dietary therapy because of formation of very large or excessive numbers of lipogranulomas that so completely obstruct the intestinal lymphatics that even an ultra-low fat diet cannot prevent lacteal rupture. Therefore, once a diagnosis of lymphangiectasia is made (either by histology, grossly at endoscopy, or tentatively by response to an ultra-low fat diet), it seems to be appropriate to use anti-inflammatory therapy designed to prevent granuloma formation/enlargement. Prednisolone, azathioprine, and/or cyclosporin are commonly used for this purpose. I do not like prednisolone, simply because of all the side effects it has in these patients. Be aware that if you use cyclosporine, it is critical that you measure blood levels of the drug. Not only is there a major difference between patients in how much they absorb, but the bioavailability of the same product may change as the intestine heals.

If the serum albumin is very low (e.g., $\leq$ 1.3 gm/dl), one is often tempted to administer a plasma transfusion while waiting to see what effect the diet will have. However, it is exceedingly difficult to increase the serum albumin concentration by transfusing PLE patients with plasma because so much of the albumin is quickly lost out the gut. You would probably have to give at least two units of plasma to a 15 lb dog in order to raise the serum albumin from 1.0 gm/dl to 1.8 gm/dl, and sometimes you would have to give 3 or 4 units. If it is critical to raise the plasma oncotic pressure, then administering hetastarch may be preferred because it costs less than plasma and it stays in the intravascular compartment longer than albumin.

These patients may be a increased risk for hypomagnesemia which may potentiate the problem of hypocalcemia.[7] At this time, we do not know how important it is to supplement magnesium to patients, but severe hypomagnesemia can be resolved by a constant rate infusion of magnesium sulfate.

Anecdotally, octreotide might help some patients that do not respond to more conventional therapy.[8,9]

CRYPT LESIONS

Lesions of the intestinal crypts have been recognized as being associated with PLE in dogs. We have identified two different lesions of the small intestinal crypts that can cause PLE. One type is characterized by crypts (usually duodenal) that are filled and somewhat distended with proteinaceous fluid and necrotic inflammatory cells.[10] While such dilated crypts can be found in many animals, including clinically normal dogs, finding large numbers of them in multiple tissue samples seems to be consistently associated with PLE. We do not know if this is a cause-and-effect relationship, or if the dilated crypts are simply a marker for some other process but are not causing the protein loss themselves. Several of these patients have responded to therapy with elemental diets, total parenteral nutrition, prednisolone, azathioprine, and/or metronidazole. We have seen this lesion associated with IBD as well as lymphangiectasia (especially in Yorkshire terriers).

A second type of crypt lesion that appears to be less common than the first, if characterized by focal accumulations of mucus causing massive distention of the intestinal crypts.[11] This has been reported once before, and we have seen two such cases. The most important aspect of diagnosis seems to be the fact that the lesion may be very focal, almost appearing as ulcers when looking at the intestinal mucosa through an endoscope. Therapy similar to that used on animals with the other form of crypt lesion may be helpful. We have used cyclosporin, but do not know if it is helpful, or if the clinical response is due to the other drugs that the patient is receiving.

These lesions have not been commonly reported. Recent work has shown that these lesions are extremely easy to miss if poor quality endoscopic biopsies are performed. While 7-12 high quality tissue samples (i.e., full length of villi plus subvillus mucosa down to the level of the muscularis mucosa) will find these lesions 90-99% of the time, about 7 times as many tissue samples will be needed if poor quality samples primarily consisting of villus tips are submitted.[5]

ANTIBIOTIC-RESPONSIVE ENTEROPATHY (ARE)

We believe that we are starting to recognize ARE as a cause of PLE in dogs. We now have several patients that appeared to have marked increases in their serum albumin concentration associated with antibiotic therapy. However, because dietary change is often performed simultaneously with antibiotic therapy in these patients, cause-and-effect is not clearly established.

CHRONIC INTUSSUSCEPTION

Chronic intussusception is a relatively important, and often missed cause of PLE in juvenile animals. The classic history is one of acute enteritis (e.g., parvoviral enteritis) which does not resolve as expected. The patient feels somewhat better, but continues to have diarrhea, and the serum albumin concentration gradually diminishes. It can be very hard to palpate an ileo-colic intussusception; abdominal ultrasound is clearly the preferred way to diagnose intussusception. Therapy is surgical.

PARASITES

Although uncommon, nematodes may cause PLE in adult animals if there are large numbers of them. Whipworms and hookworms in particular may occasionally be responsible for PLE in older dogs. However, giardiasis has been reported to cause PLE in people.

REFERENCES

[1] Peterson PB, et al. Protein-losing Enteropathies. **Vet Clin N Am** 2003;33:1061. [2]Kimmel SE, et al. Hypomagnesemia and hypocalcemia associated with protein-losing enteropathy in Yorkshire terriers: five cases (1992-1998). **J Am Vet Med Assoc** 2000;217:703. [3] Sutherland-Smith J, et al. Ultrasonographic intestinal hyperechoic mucosal striations in dogs are associated with lacteal dilation. **Vet Radiol Ultra** 2007;48:51. [4]Kull PA, et al. Clinical, clinicopathologic, radiographic, and ultrasonographic characteristics of intestinal lymphangiectasia in dogs: 17 cases (1996-1998). **J Am Vet Med Assoc** 2001;219:197. [5]Willard MD, et al. Effect of sample quality on the sensitivity of endoscopic biopsy for detecting gastric and duodenal lesions in dogs and cats. **J Vet Int Med** 2008;22:1084. [6]Romanucci M, et al. Xanthogranulomatous inflammation of the small bowel in a dog. **Vet Path** 2008;45:207. [7]Bush WW, et al. Secondary hypoparathyroidism attributed to hypomagnesemia in a dog with protein-losing enteropathy. **J Am Vet Med Assoc** 2001;219:1732. [8]Markham KM, et al. Octreotide in the treatment of thoracic duct injuries. **Amer Surg** 2000;66:1165. [9]Filik L. et al. A case of intestinal lymphangiectasia successfully treated with slow-release octreotide. **Dig Liver Dis** 2004;36:687. [10]Willard MD, et al. Intestinal crypt lesions associated with protein-losing enteropathy in the dog. **J Vet Int Med** 2000;14:298. [11]Willard MD, et al. Protein-losing enteropathy associated with cystic mucoid changes in the intestinal crypts of two dogs. **J Am Anim Hosp Assoc** 2003;39:187.

LABORATORY DIAGNOSIS OF PANCREATITIS IN DOGS AND CATS

Panagiotis Xenoulis, DVM, Dr. med. vet., PhD
Athens, Greece

INTRODUCTION

Pancreatitis is the most common disorder of the exocrine pancreas in dogs and cats. The term pancreatitis refers to inflammation (i.e., infiltration with inflammatory cells) of the exocrine pancreas, although it is often used to describe diseases of the exocrine pancreas characterized by necrosis, with or without an inflammatory component. Pancreatitis is generally divided into acute and chronic forms based on the absence or presence of permanent histopathologic lesions, respectively, such as pancreatic fibrosis and/or atrophy.[1] The categorization of pancreatitis into acute and chronic has potential diagnostic, therapeutic, and prognostic implications.

Despite recent advances in diagnostics, it is increasingly recognized that accurate clinical diagnosis of pancreatitis can be challenging. Several serum tests have been developed and evaluated over the past decades, but most have shown no or only limited usefulness for the diagnosis of pancreatitis in dogs and cats. Although there has been substantial progress in the development of laboratory testing for pancreatitis, a definitive diagnosis cannot always be reached using laboratory testing alone. Proper use and correct interpretation of results of multiple diagnostic modalities is crucial for a correct diagnosis.

ROUTINE CLINICAL PATHOLOGY

The results of complete blood count (CBC), serum biochemistry profile and urinalysis are often within normal limits in dogs and cats with pancreatitis. When present, clinicopathologic abnormalities are variable and unpredictable, and may include hematologic abnormalities, increased liver enzymes, increased creatinine and BUN, hypoalbuminemia, hypertriglyceridemia, hypercholesterolemia, hyperglycemia or hypoglycemia, and electrolyte imbalances.[2-7] While results of CBC, serum biochemistry profile and urinalysis in dogs and cats with pancreatitis are nonspecific and therefore non diagnostic, these tests should always be performed in animals with suspected pancreatitis because they are useful for the diagnosis or exclusion of other diseases, and also give important information about the general condition of the patient.

SERUM AMYLASE AND LIPASE ACTIVITIES

Serum amylase and lipase activities have long been considered markers for pancreatitis in dogs.[8,9] Although serum activities of these two enzymes increase during experimental canine pancreatitis, several studies have shown that these markers are not useful and should not be used for the diagnosis of spontaneous canine pancreatitis due to their low sensitivity and specificity.[8-14] Serum lipase activity increases and serum amylase activity decreases in experimentally induced acute pancreatitis in cats.[15-17] Although well designed clinical studies are lacking, both serum lipase and amylase activities do not appear to be of any clinical value in the diagnosis of spontaneous feline pancreatitis and their use is not recommended.[3,18,19]

TRYPSIN-LIKE IMMUNOREACTIVITY (TLI)

TLI assays are immunoassays that measure trypsinogen and trypsin concentrations in serum and have been shown to be of limited usefulness for the diagnosis of canine and feline pancreatitis. Serum canine TLI (cTLI) concentrations increase after experimental induction of pancreatitis in dogs, but decrease to concentrations within the reference interval as soon as 3 days after induction of pancreatitis in some dogs.[12] The sensitivity of serum cTLI for the diagnosis of spontaneous pancreatitis is low (36%-47%), possibly due to the short half-life of trypsin in serum.[20,21,14] In addition, although there is strong evidence that trypsinogen is exclusively of pancreatic origin, it is believed that it is cleared by glomerular filtration, and serum cTLI concentration can be increased in dogs with renal failure.[13,12,20] This clearly affects the specificity of the test and complicates the interpretation of increased cTLI concentrations in dogs with azotemia (which is not uncommon in dogs with pancreatitis). A clearly increased serum cTLI concentration in a dog that is not azotemic is indicative of pancreatitis. However, pancreatitis cannot be excluded on the basis of a normal serum cTLI concentration.

In cats with experimentally induced pancreatitis, feline TLI (fTLI) concentration increases sharply after induction of pancreatitis, but returns below the cut-off value for pancreatitis within 48 hours.[17] Feline TLI has been evaluated for the diagnosis of spontaneous pancreatitis in cats and several cut-off values have been suggested.[22-24] When cut-off values allowing adequate specificity of the assay are used (i.e., 100 μg/L), the sensitivity of fTLI for the diagnosis of pancreatitis in cats is suboptimal (28% - 64%). In addition, the specificity of fTLI has been questioned, because mildly increased serum fTLI concentrations have been reported in cats with no demonstrable pancreatic disease that had other gastrointestinal disorders (e.g., IBD or gastrointestinal lymphoma) or azotemia.[22,24,25] Similar to dogs, a clearly increased serum fTLI concentration in a cat that is not azotemic is indicative of pancreatitis. However, pancreatitis cannot be excluded on the basis of a normal serum fTLI concentration.

PANCREATIC LIPASE IMMUNOREACTIVITY (PLI) ASSAYS

The PLI assays are currently considered the most sensitive and specific serum tests for the diagnosis of pancreatitis in both dogs and cats. The advantages of the PLI assays over the traditional lipase activity assays lay basically on two facts: a) pancreatic lipase is exclusively of pancreatic origin and b) in contrast to the traditional activity assays for lipase, which indiscriminately measure the activity of lipases of any origin, immunoassays used to measure pancreatic lipase concentration in serum (i.e., the PLI assays) quantify exclusively lipase of pancreatic origin based on its unique structure.[26-30] Therefore, the PLI assays have inherent advantages over the traditional serum lipase activity assays that make them more suitable for specific evaluation of the exocrine pancreas in dogs and cats.

The originally developed and analytically validated immunoassays for the specific measurement of serum pancreatic lipase in dogs (canine PLI) and cats (feline PLI) have been replaced by more widely available immunoassays (Spec cPL® for dogs and Spec fPL® for cats) that show similar clinical performance with the original PLI assays.[14,31-34] The reference interval for Spec cPL is 0 to 200 µg/L and for Spec fPL 0 to 3.5 µg/L. Both assays incorporate a gray zone in the interpretation of the results (201 to 399 µg/L for Spec cPL and 3.6 to 5.3 µg/L for Spec fPL); values in the gray zone are non-diagnostic and further testing or retesting is recommended. Concentrations ≥ 400 µg/L (Spec cPL) or ≥ 5.4 µg/L (Spec fPL) are considered diagnostic for pancreatitis.

Canine PLI

Both clinical and histopathological studies on the sensitivity of cPLI for canine pancreatitis have been published and all generally agree that serum cPLI is the most sensitive serum marker for pancreatitis in dogs.[14,35-37] There is a wide overall range of sensitivities reported for cPLI (21% to 78%) which is also seen with other markers for pancreatitis and reflects the inherent differences in study design, methodology, and dog population used in each study. However, studies indicate that cPLI showed consistently the best performance (sensitivity and specificity) compared to other serum markers evaluated.[14,35-37]

Based on some evidence from recent studies and the fact that histopathologic lesions associated with chronic pancreatitis such as pancreatic fibrosis and atrophy are not expected to be associated with leakage of pancreatic enzymes, the sensitivity of cPLI is believed to be lower for chronic pancreatitis than for acute pancreatitis.[14,30,36,37]

Similar to its sensitivity, serum cPLI is considered to have the highest specificity for pancreatitis compared to any other serum test currently available, with specificities ranging between 81 and 100%.[8,12,20,35,38-41] Pancreatic sampling by ultrasound-guided fine-needle aspiration or surgical biopsy did not cause any increases in serum cPLI concentrations in healthy dogs.[42] It remains to be determined whether pancreatic sampling of a diseased pancreas would also not have any effect and it is currently recommended that determination of serum cPLI concentration is best performed before sampling of pancreatic tissue. Finally, there have been anecdotal concerns that high cPLI concentrations might occasionally be associated with pancreatic inflammation that is not clinically important. Therefore, it remains to be determined whether serum cPLI concentration can detect histopathologically mild pancreatitis that might be of minor clinical importance. In either case, a final diagnosis of pancreatitis should ideally not be based solely on the results of PLI, but rather, on a careful consideration of the general clinical and clinicopathological picture of the animal and the results of ultrasonography, pancreatic cytology or histopathology.

Canine SNAP PL

SNAP cPL, a rapid, in-clinic, semi-quantitative, visually-read test for the estimation of canine pancreatic lipase in serum has recently been developed.[42] This test incorporates a reference spot that corresponds to the upper limit of the reference interval and a sample spot that is visually compared to the reference spot.[42] Therefore, the results of this rapid assay are considered either normal (less intense than the reference spot) or abnormal (equally to or more intense than the reference spot). In the latter case, the actual PLI concentration may be in the gray zone (200 µg/L to 400 µg/L) or consistent with the diagnosis of pancreatitis (>400 µg/L for Spec cPL).[42] This cannot be determined with certainty, however, using the in-clinic assay and further testing using the quantitative reference method is recommended.

A recent validation study showed that there is a 90% to 100% agreement between the SNAP cPL and the reference Spec cPL.[42] A recent multi-institutional study reported that SNAP cPL has a sensitivity between 91% and 94% and a specificity between 71% and 78% for pancreatitis.[35] However, the main use of this diagnostic tool is the rule-out of pancreatitis (i.e., a normal result makes diagnosis of pancreatitis very unlikely) and therefore, the sensitivity of this test is more important than its specificity. Abnormal results could be seen in a dog with a Spec cPL concentration in the gray zone or consistent with the diagnosis of pancreatitis and further testing using the quantitative Spec cPL assay would be necessary in such cases. Diagnosis of pancreatitis cannot be based on the result of the SNAP cPL alone.

Feline PLI

Studies in cats with both experimental and spontaneous pancreatitis have repeatedly shown that serum fPLI concentration is the most sensitive and specific serum marker for feline pancreatitis currently available.[17,19,22-24,43,44] The sensitivity of serum Spec fPL concentration reported ranges from 54% for histopathologically mild pancreatitis to 100% for moderate to severe.[43,44]

Similar to dogs, histopathologic lesions associated with chronic pancreatitis such as pancreatic fibrosis and atrophy are not expected to be associated with leakage of pancreatic enzymes, and therefore, the sensitivity of fPLI is believed to be lower for chronic pancreatitis (without a concurrent acute pancreatitis) than for acute pancreatitis. Therefore, as in dogs, false negative results cannot be excluded especially in cats with chronic or mild pancreatitis. The clinical significance of mild chronic pancreatitis remains to be determined.

Similar to its sensitivity, the specificity of serum fPLI concentration for feline pancreatitis is very high, ranging between 67% and 100%.[43,44] Similar to dogs, laparoscopic pancreatic biopsy in healthy cats did not have any significant effect on serum fPLI concentrations.[45] It is unknown whether pancreatic biopsy of an inflamed pancreas would lead to an increase of serum fPLI concentration.

Feline SNAP PL

The SNAP fPL test has recently been released and is based on the same principles as the canine SNAP cPL test (see above). Although studies on the validation and clinical performance characteristics for the diagnosis of pancreatitis of the SNAP fPL have not been reported in the literature yet, the manufacturer indicates that this test has a 82% to 92% agreement with the Spec fPL assay. Therefore, the sensitivity of the SNAP fPL should theoretically be high and similar to the one reported for Spec fPL (see above). Consequently, the main use of this diagnostic tool is the rule-out of pancreatitis (i.e., with a normal test result). Abnormal results could be in the gray zone or consistent with the diagnosis of pancreatitis and further testing using the quantitative Spec fPL assay is necessary. Diagnosis of pancreatitis cannot be based on the results of the SNAP fPL alone.

OTHER DIAGNOSTIC MARKERS

Several other diagnostic markers for pancreatitis have been developed and studied, but none of those can currently be recommended for the routine diagnosis of canine and feline pancreatitis in clinical practice, either because their diagnostic performance has not been sufficiently evaluated clinically, or because they have been shown to have a low sensitivity and/or specificity. In addition, the availability of most of these diagnostic tests is currently limited. Such tests include serum concentrations of pancreatic elastase-1, phospholipase A2, trypsin-α_1-antitrypsin complexes, α_2-macroglobulin, plasma and urine concentrations of trypsinogen activation peptide (TAP), and lipase activity in peritoneal fluid.[14,24,46-51] Of these markers, serum pancreatic elastase-1 and TAP concentrations seem to hold some promise and might prove helpful for the diagnosis or the assessment of severity of pancreatitis in the future.

REFERENCES

[1]Xenoulis P, et. al **Compend Contin Educ Pract Vet** 2008;30:166. [2]Akol KG, et. al **J Vet Intern Med** 2003;7: 205. [3]Hill RC, et. al **J Vet Intern Med** 1993;7:25. [4]Hess RS, et. al **J Am Vet Med Assoc** 1998;213:665. [5]Kimmel SE, et. al **J Am Vet Med Assoc** 2001;219:1105. [6]Ferreri JA, et. al **J Am Vet Med Assoc** 2003;223:469. [7]Son TT, et. al **J Vet Emerg Crit Care** 2010;20:426. [8]Strombeck DR, et. al **Am J Vet Res** 1981;42:1966. [9]Jacobs RM, et. al **J Am Anim Hosp Assoc** 1985;21:795. [10]Brobst D, et. al **J Am Vet Med Assoc** 1970;157:1697. [11]Mia AS, et. al **Am J Vet Res** 1978;39:965. [12]Simpson KW, et. al **Am J Vet Res** 1989;50:629. [13]Simpson KW, et. al **Res Vet Sci** 1991;51:78. [14]Steiner JM, et. al **Vet Therap** 2008;9:263. [15]Kitchell BE, et. al **Am J Vet Res** 1986;47:1170. [16]Karanjia ND, et. al **J Surg Res** 1990;48:440. [17]Zavros NS, et. al **Europ J Inflam** 2008;6:105. [18]Simpson KW, et. al **J Sm Anim Pract** 1994;35:93. [19]Parent C, et. al **J Vet Intern Med** 1995;9:194. [20]Mansfield CS, et. al **Aust Vet J** 2000;78:416. [21]Steiner JM, et. al **J Vet Int Med** 2001;15:274. [22]Swift NC, et. al **J Am Vet Med Assoc** 2000;217:37. [23]Gerhardt A, et. al **J Vet Intern Med** 2001;15:329. [24]Allen HS, et al **Can J Vet Res** 2006;70:313. [25]Simpson KW, et. al **J Vet Intern Med** 2001;15:26. [26]Steiner JM **PhD Thesis** Texas A&M University 2000 [27]Steiner JM, et. al **Am J Vet Res** 2002;63:722. [28]Steiner JM, et. al **Am J Vet Res** 2006;67:84. [29]Hoffman WE **Clinical biochemistry of domestic animals**, 6th ed., Academic Press, Burlington 2008;351. [30]Neilson-Carley SC, et. al **Am J Vet Res** 2011;72:302. [31]Steiner JM, et. al **Can J Vet Res** 2003;67:175. [32]Steiner JM, et. al **Am J Vet Res** 2003;64:1237. [33]Steiner JM, et. al **Can J Vet Res** 2004;68:309. [34]Huth SP, et. al **Vet Clin Pathol** 2010;39:346. [35]McCord K, et. al **J Vet Intern Med** 2012;26:888. [36]Watson PJ, et. al **Vet Rec** 2010;167:968. [37]Trivedi S, et. al **J Vet Intern Med** 2011;25:1241. [38]Steiner JM, et. al **J Vet Int Med** 201;15:311. [39]Steiner JM, et. al **Am J Vet Res** 2009;70:1001. [40]Mansfield CS, et. al **J Vet Diagn Invest** 2012;24:312. [41]Cordner AP, et. al **J Vet Diagn Invest** 2010;22:702. [42]Beall MJ, et. al **J Vet Diagn Invest** 2011;23:115. [43]Forman MA, et. al **J Vet Intern Med** 2004;18:807. [44]Forman MA, et. al **J Vet Intern Med** 2009;23:733. [45]Cosford KL, et. al **J Vet Intern Med** 2009;24:104. [46]Mansfield CS, et. al **J Vet Diagn Invest** 2011;23:691. [47]Westermarck E, et. al **Acta Veterinaria Scandinavica** 1983;24:477. [48]Suchodolski JS, et. al **J Vet Intern Med** 2001;15:311. [49]Ruaux CG, et. al **Res Vet Sci** 1999;66:185. [50]Mansfield CS, et. al **J Vet Intern Med** 2000;14:346. [51]DeArespacochaga AG, et. al **J Vet Med Series A-Physiol Path Clin Med** 2006;53:119.

CANINE MAJOR ACUTE PHASE PROTEINS: BIOCHEMISTRY AND PHYSIOLOGY

Mads Kjelgaard-Hansen, DVM, Ph.D. (Veterinary Clinical Pathology)
Frederiksberg, Denmark

INTRODUCTION

When the organism of a dog is challenged by an inflammatory stimulus, a local response is evoked by the innate immune cells, especially macrophages and to lesser extent neutrophils. The local process is an orchestrated event of cellular activity and release of soluble factors (cytokines) to contain, defeat, heal and restore normal local homeostasis. This can turn into a systemic event if local containment of the inflammatory process is unsuccessful, which does not necessarily mean a dissemination of the inflammatory cause, e.g. an infectious agent, but often merely an activation of systemic inflammation because of a high local production of pro-inflammatory cytokines spilling-over to the systemic circulation activating distant cells and organs. Thus, cytokines shifting from autocrine/paracrine activity on self or neighbouring cells to an endocrine function. This systemic reaction is named the acute phase reaction (APR).

An APR can be evoked by a wide range of stimuli, like viral and bacterial infections, immune-mediated disease, neoplasia, aseptic tissue injury (trauma, surgery), necrosis and burns, and is thus not specific for any kind of inflammation.

The presence of increased concentrations of pro-inflammatory cytokines in the systemic circulation activates a wide range of organs involved in the systemic homeostasis, such as the central (CNS) and autonomic (ANS) nervous system, the liver, the bone marrow and the adrenal gland, primarily to put the organism in an alert state and establish a rapid and intense protective response. The responses from these organs are the main source of three of the hallmarks of an APR; fever, change in leukocyte counts (WBC) and a significant change in the protein composition of plasma (the acute phase proteins [APP]). Fever and WBC changes are familiar markers used in the clinical situation to suspect and/or finally establish the presence of systemic inflammation in a canine patient. The routine use of specific proteins as markers of inflammation in dogs is also getting established. Detailed knowledge of the mechanisms behind these APR markers will facilitate the clinical detection of APR in its the various stages and how APP measurements can complement traditional inflammatory markers to establish diagnosis and prognosis and for monitoring disease progress during treatment – a knowledge also useful for avoiding possible pitfalls in the interpretation.

Fever and changes in WBC

Fever is mainly viewed as a result of cytokine-induced change of prostaglandin production, especially PGE2 formation, which has a direct influence on the hypothalamus. The influence of prostaglandins is evidenced by the ability of inhibitors of cyclooxygense-2 to attenuate fever in dogs. Fever is generally not viewed as a quantitative marker of inflammation, as an APR can both result in hyperthermia or hypothermia. When using fever as a marker for APR, it is important to remember that the diagnostic specificity is compromised as hyperthermia can also be induced by anxiety, exercise, hot climate and stress as well as by any treatment inhibiting the endogenous production of prostaglandins.

Changes in WBC induced by an APR are usually biphasic, where the first wave of mature neutrophils emerges from the marginalised pool residing in the capillary network during normal homeostasis. This release is evoked by a cytokine-induced increased production of cortisol through the hypothalamic-hypophyseal-adrenal axis. The second wave is also mediated by pro-inflammatory cytokines through increased growth-factor induction in the bone marrow and increased release of leukocytes. The concentration of circulating leukocytes is a product of the release-rate from the bone marrow and the rate of leukocytes attracted to the site of inflammation by chemoattractants. This balance can result in either leukopenia or leukocytosis during an APR and makes the use of leukocyte counts as a quantitative marker for inflammatory activity difficult, as even a massive release of neutrophils from the bone marrow can result in a near-normal concentration of neutrophils in circulation if production is balanced by recruitment to site of inflammation. This will also limit the sensitivity as a marker for the APR. Specificity can be also compromised for some of the same reasons as for fever, leukocytosis being possibly induced by e.g. stress and exercise, but also during corticosteroid treatment, as this will activate the marginalised pool.

No doubt that both WBC and core temperature are useful clinical markers of systemic inflammatory activity, however complementary use of major acute phase proteins will strengthen the otherwise compromised diagnostic accuracy in certain situations as well as enable quantification of the inflammatory activity.

CANINE MAJOR ACUTE PHASE PROTEINS

Acute phase proteins (APP) can be divided into negative and positive acute APPs dependent on whether they decrease or increase in concentration during an APR[1]. The most well known negative APP is albumin. For the detection of an APR, the positive APPs are the most specific and most frequently used, and the major APPs, increasing more than ten-fold during an APR are preferred for their increased sensitivity. The increase in plasma concentration is mainly due to hepatic release upon

stimulation of pro-inflammatory cytokines, and although some extra-hepatic tissues are capable of producing APP's, their influence on the plasma concentration is believed to be negligible.
A major APP is characterised by a more than ten-fold increase of plasma concentration during an acute phase reaction of the patient and the kinetic features of a major APPs are[2]: 1) a short lag time from onset of inflammation to a measurable increase in blood concentration, 2) continued increased release from the liver as long as stimulated by increased plasma concentration of pro-inflammatory cytokines is present, 3) a short half-life causing plasma concentrations to decrease soon after inflammation has subsided, and 4) low plasma concentration in the healthy animal. These features make this group of proteins particularly well suited for detecting and monitoring systemic inflammatory activity. In dogs, the relevant major APPs are C-reactive protein and Serum Amyloid A:

C-reactive protein (CRP)

Canine CRP is a major APP in dogs with low serum concentration during normal homeostasis and well-described kinetics during and after an inflammatory stimulus: - Fast induction (increases within 4 hours) after inflammatory stimulus; - Maximum response after approximately 24 hours, continuously elevated during whole APR; - and a fast decrease when inflammatory stimulation is no longer present (elimination half-time believed to be close to 6 hours). Current knowledge is that CRP release is specifically induced by pro-inflammatory cytokines only, especially IL-6, and is produced *de novo* with no storage, i.e. with no risk of false release due to cellular damage. Thus, CRP possess characteristics to be a more specific inflammatory marker in comparison to more traditional markers, such as rectal temperature and WBC which are labile to other intrinsic factors, such as long- and short-term stress, and to be more close to a 'real-time' marker than slower reacting inflammatory markers like fibrinogen.
The specific induction by pro-inflammatory cytokines also enables the use of CRP as an inflammatory marker during steroidal and non-steroidal anti-inflammatory treatment regimens where traditional markers are known to be biased by the drugs, as the effect of these are mainly mediated through modulation of prostaglandins not affecting the induction of CRP. Reports demonstrate that neither endogenous nor iatrogenic glucocorticoids will induce CRP release[3, 4] that could impair diagnostic specificity, and nor will the presence of glucocorticoids modulate the CRP response[3, 5] in a manner which could impair diagnostic sensitivity. Suppressing inflammatory activity *en toto* with immunosuppressive dosages will, of course, result in reduced CRP also[6], as this will impair the organism to mobilise an inflammatory reaction at all, and thus limit production of pro-inflammatory cytokines.
Additionally, the CRP response is a quantitative marker of inflammatory activity[7] correlating well to indices of disease activity in various diseases. However, true relevance for routine clinical purposes also depends on cost, random accessibility and a reasonable turnover time of a test system. The identification, validation and implementation of an automated human CRP turbidimetric assay for measuring canine CRP[8, 9], with low costs, high throughput, random access and results within 15 minutes, has given us extensive experience of routine clinical application of CRP in dogs, running CRP as an integral part of the standard serum biochemistry panel for the minimum database for over eight years.
Point-of-care test systems for canine CRP have emerged, especially on the European market, which will undoubtedly further facilitate gathering experience of clinical applicability. As a note: Japanese practitioners have gathered experience since year 2000 with a species-specific canine CRP patient-near test available, but unfortunately only for the Japanese market. Please consult other proceeding by same speaker for review of evidence for specific clinical application of canine C-reactive protein: "CANINE MAJOR ACUTE PHASE PROTEINS: CLINICAL APPLICATIONS".

Serum Amyloid A

Serum Amyloid A (SAA), like CRP, is a major APP in dogs. It is a small apolipoprotein of 10-15 kDa tightly associated with high-density lipoprotein when released into plasma; it has amphiphatic properties and is insoluble in plasma in its native conformation. Serum amyloid A is produced and released mainly by the liver, but virtually all mammalian cells investigated has demonstrated the ability to synthesize SAA upon stimulation. Serum amyloid A is a family of isoforms, with very similar molecular weight, but different acid-base properties due to amino-acid substitutions, and it seems that extra-hepatic SAA response may be of a different isoform pattern than the hepatic release, possibly providing clinically useful information if test-systems can be further developed.[10] In circulation, this makes little difference, as a marker of systemic inflammatory activity, as the hepatic release dominates plasma SAA concentration during an APR by magnitudes.
As a major APP in dogs, SAA possess many of the same characteristics beneficial for clinical use; - Fast induction (increases within 4-6 hours) after inflammatory stimulus; - Maximum response after approximately 24-36 hours, continuously elevated during whole APR; - and a fast decrease when inflammatory stimulation is no longer present (half-life believed to be a bit shorter than for canine CRP). It is also induced by pro-inflammatory cytokines only, and not by excess corticosteroid administration[4], thus also not prone to be biased by stress and/or treatment. Investigations demonstrate comparable diagnostic accuracy in detecting systemic inflammatory activity between SAA and CRP[11], but in general less detail is known about

canine SAA compared to canine CRP as a clinical biomarker of systemic inflammatory activity. This is mainly because of a lack of clinically relevant methodologies until now and consequently the lack of clinical experience. However, the recent development and validation of a methodology relevant for clinical routine application may change this.[12] Future investigations will have to determine whether complementary aspects of the two major APPs are present justifying the use of both markers in parallel or whether one is superior to the other in clinical use.
Please consult other proceeding by same speaker for review of evidence for specific clinical application of canine serum amyloid A: "CANINE MAJOR ACUTE PHASE PROTEINS: CLINICAL APPLICATIONS".

REFERENCES

1 Ceron JJ, et al. **Vet Clin Pathol** 2005;34:85. 2 Kjelgaard-Hansen M, et al. **Clin Lab Med** 2011;31:51. 3 Caldin M, et al. **Vet Clin Pathol** 2009;38:63. 4 Martinez-Subiela S, et al. **Vet Rec** 2004;154:814. 5 Yamamoto S, et al. **Vet Res Commun** 1994;18:347. 6 Kjelgaard-Hansen M, et al. **Acta Vet Scand** 2006;48:9. 7 Kjelgaard-Hansen M, et al. **Vet Clin Pathol** 2013;In Press. 8 Kjelgaard-Hansen M, et al. **Vet Clin Pathol** 2003;32:81. 9 Kjelgaard-Hansen M, et al. **Vet Clin Pathol** 2004;33:139. 10 Kjelgaard-Hansen M, et al. **Vet Immunol Immunopathol** 2007;117:296. 11 Christensen M, et al. **J Vet Med Sci** 2013;In Press. 12 Christensen M, et al. **Vet J** 2012;194:332.

CANINE MAJOR ACUTE PHASE PROTEINS: CLINICAL APPLICATIONS

Mads Kjelgaard-Hansen, DVM, Ph.D. (Veterinary Clinical Pathology)
Frederiksberg, Denmark

INTRODUCTION

The use of the major acute phase protein C-reactive protein (CRP) as an inflammatory marker in patient management has been an integral part of human medicine for many decades; likewise in dogs, over thirty years of international research on canine CRP and over eight years of practical experience of routine clinical application at our university companion animal hospital, along with experiences from other veterinary facilities around the world applying CRP routinely within the last years, suggest that canine CRP is likely to be just as beneficial to canine patient management as CRP has proven to be in human medicine. Experience and evidence are gathered and published at a continuously growing rate and a picture on where CRP has the largest positive impact in the diagnostic process is gradually depicted. The dissemination of routine use was initiated by the discovery and validation of commercially available assays practically applicable for routine clinical use[1-4], further facilitated by the emergence of relevant point-of-care test systems, and has now reached global dissemination as a valuable marker of systemic inflammatory activity in dogs. Additionally, a relevant methodology for routine clinical application has been developed for Serum amyloid A (SAA)[5] another major canine acute phase protein (APP), which should also facilitate disseminated use of this marker. Please consult other proceeding by same speaker for review of the biochemical and physiological background for using the major canine APPs as clinical markers of systemic inflammatory activity: "CANINE MAJOR ACUTE PHASE PROTEINS: BIOCHEMISTRY AND PHYSIOLOGY".
This short review on the benefits of using the canine major APPs routinely in clinical settings will outline the beneficial characteristics of canine CRP and SAA and describe where the current understanding of these inflammatory markers allows them to make the largest impact on patient management. A peer-reviewed review by the speaker was recently published.[6]

Clinical application of canine major acute phase proteins

A frequent and important question asked in a diagnostic process of systemically affected patients, is whether inflammatory activity is a significant part of the pathology. During such differential-diagnostic considerations, an objective and unbiased marker with the ability to confirm or rule out inflammatory activity is of high value. In the case of a confirmed inflammatory activity, quantification of this activity would also be of great interest, as a possible measure of disease severity and to enable monitoring the efficacy of the subsequent treatment.
These are the demands that major acute phase proteins, especially CRP, have met worldwide in human medicine for decades, and seemingly also have the potential to meet in canine medicine, and have done so for eight years now at our University companion animal hospital and other institutions applying this marker in recent years.
With the known characteristics of canine CRP and SAA as major APPs (see other proceeding by same speaker) they can potentially influence patient management in several clinical situations, and have, in our experience, had a large impact, (a) as a part of the first-line tests at admission helping ruling in/out inflammatory activity at an early time-point and (b) as enabling monitoring systemic inflammatory activity during and after treatment. Recent studies including large numbers of dogs conclude that CRP enabled secure detection of systemic inflammatory activity that could not be detected by CBC[7] and other traditional markers, addressing specific disease processes, like detection of infections concurrent to hyperadrenocorticism,[8] diagnostication of autoimmune haemolytic anemia (IMHA)[9] and steroid-responsive meningitis-arteritis[10] leaving little doubt that canine CRP is a useful inflammatory marker during diagnostic work-up. However, enabling the monitoring of inflammatory activity during and after both medical and surgical treatment of inflammatory diseases might be even more valuable, as traditional markers are often severely biased by treatment and/or not reflecting inflammatory activity in a quantitative manner, as demonstrated for major canine APPs, e.g.[11].
Several reports demonstrate the ability of monitoring the efficacy of initial treatment of inflammatory diseases by means of canine major APPs[9, 12]. Our experience is that CRP will typically decrease within 24 hours if the treatment of an active inflammatory disease process is being successful. Serial assessment, e.g. during tapering of anti-inflammatory/immunosuppressive treatment, can detect disease relapse[13, 14], and post-operative inflammatory complications after surgical treatment[15]. These abilities have made canine CRP and SAA frequently requested parameter and part of the minimum database serum profile of systemically affected dogs at our university hospital.
With regard to major APPs providing prognostic information through its quantitative abilities, several clinical studies report that CRP/SAA concentration at admission, as stand-alone parameters, does not correlate to outcome in a wide range of critical illnesses, e.g. IMHA[9], critically ill inflamed dogs in general[16, 17], canine monocytic ehrlichiosis[18] or gastric dilatation-volvulus,[19] which is also our experience, whereas the magnitude of the change in CRP/SAA concentration over the initial

days of treatment seemingly hold valuable outcome information[16, 20]. One way to interpret this is that outcome is related to initial treatment efficacy, where CRP act as a valuable objective marker guiding early treatment choices and changes of treatment in critically ill inflamed dogs. However, well-designed trials are needed to answer that question in full.

Canine major acute phase proteins are markers of systemic inflammatory activity – not local inflammation

The major canine APPs detected and measured in circulation as markers of inflammatory activity are hepatically derived upon hepatic stimulation by pro-inflammatory cytokines released from the site of inflammation. This means that the APPs are only produced upon increased concentration of circulating pro-inflammatory cytokines, which again means that there will be no increased concentrations of APP if the inflammatory process is contained as a local process. This is important to remember when applying canine CRP and SAA in the clinical setting, as it has several implications: A locally contained process will not evoke an increase in major APPs, which has been demonstrated experimentally[21], and supported by observations, e.g. the majority of dogs with simple rhinitis were reported to have normal circulating concentrations of CRP[22] and simple bacterial cystitis not evoking systemic activity (own unpublished data). However, on the other hand the diagnostic sensitivity towards systemic inflammatory activity (marked and fast increase in concentration) will allow the clinician to detect systemic activity in processes, otherwise expected to be locally contained by means of these major APPs, like in massive experimentally induced bacterial cystitis[23]. Likewise, canine CRP being a quantitative marker of systemic inflammatory activity, normalisation of circulating CRP concentration will indicate that the inducing process has either resolved or have been locally contained.

CONCLUSION

In conclusion, where implemented, the use of canine major acute phase proteins as markers of systemic inflammatory activity has a major impact on patient management complementing traditional markers of inflammation. The effect is especially evident in the diagnostic work-up, increasing diagnostic sensitivity and specificity. The APPs need to be part of the minimum database to have full impact, and must be used as well to monitor treatment efficacy and progress of inflammatory disease activity. The major APPs are the only markers providing this type of information in a reliable manner.

REFERENCES

1 Kjelgaard-Hansen M, et al. **Vet Clin Pathol** 2003;32:81. 2 Kjelgaard-Hansen M, et al. **Vet Clin Pathol** 2004;33:139. 3 Kjelgaard-Hansen M. **Vet Clin Pathol** 2010;39:402. 4 Klenner S, et al. **J Vet Diagn Invest** 2010;22:544. 5 Christensen MB, et al. **J Vet Med Sci** 2012;In Press. 6 Kjelgaard-Hansen M, et al. **Clin Lab Med** 2011;31:51. 7 Nakamura M, et al. **J Vet Med Sci** 2008;70:127. 8 Caldin M, et al. **Vet Clin Pathol** 2009;38:63. 9 Mitchell KD, et al. **J Vet Intern Med** 2009;23:585. 10 Lowrie M, et al. **J Vet Intern Med** 2009;23:862. 11 Kjelgaard-Hansen M, et al. **Vet Clin Pathol** 2013;In Press. 12 Jergens AE, et al. **J Vet Intern Med** 2010;24:269. 13 Kjelgaard-Hansen M, et al. **Acta Vet Scand** 2006;48:9. 14 Yuki M, et al. **J Am Vet Med Assoc** 2010;236:767. 15 Dabrowski R, et al. **Theriogenology** 2009;72:471. 16 Gebhardt C, et al. **J Vet Emerg Crit Care** 2009;19:450. 17 Chan DL, et al. **J Vet Intern Med** 2009;23:559. 18 Mylonakis ME, et al. **J Vet Intern Med** 2011;25:811. 19 Israeli I, et al. **J Vet Intern Med** 2012;26:920. 20 Ohno K, et al. **J Vet Med Sci** 2006;68:1275. 21 Bauer N, et al. **Res Vet Sci** 2012;93:321. 22 Sheahan D, et al. **Vet Rec** 2010;167:895. 23 Seo KW, et al. **J Vet Sci** 2012;13:179.

LIVER CYTOLOGY: A CRITICAL APPROACH

Leslie Sharkey DVM, PhD, DACVP (Clinical Pathology)
St Paul, MN

INTRODUCTION

Examination of cytologic samples from the liver can be an important component in the diagnostic evaluation of patients with suspected disease in this organ. The frequency of samples from the liver has increased with the more routine availability of advanced imaging for the evaluation of patients with abdominal disease, which reveals lesions that are not appreciable with abdominal palpation or survey radiographs. Routine evaluation of the liver for metastatic disease in patients with neoplasia is another common indication for cytology. Optimal application of cytology for hepatic disease relies on an understanding of the diagnostic performance for specific conditions. Given the reliance of imaging data to guide additional testing, including the collection of pathologic samples, it is critical to be aware that multiple investigators have describe limited predictive value of imaging for specific types of pathology, including both diffuse infiltrative and focal liver lesions (Feeney et al 2008, Warren-Smith et al 2012), although the combination of imaging data and other clinical information may be informative (Murakami et al 2012).

DIAGNOSTIC PERFORMANCE OF CYTOLOGY FOR HEPATIC NEOPLASIA

Indications for hepatic cytology in general include hepatomegaly, the detection of mass lesions, altered echogenicity, and clinical suspicion of malignancies that have the potential to metastasize to the liver. These may or may not be associated with liver enzyme abnormalities or laboratory evidence of decreased liver function, although typically there is some indication of hepatic pathology on the serum biochemical profile. It is important to note that many changes may be observed in the liver secondary to the presence of neoplasia, including necrosis, various types of inflammation, extramedullary hematopoiesis, vacuolar change, and reactive hyperplasia of hepatocytes or biliary epithelial cells. These may be observed whether or not malignant cells are present in the cytologic samples collected.

Several studies have examined cytology:histopathology correlation for the diagnosis of hepatic neoplasia and found relatively low sensitivity but high specificity. Wang et al found that only 22% of histologically identified liver tumors were detected by cytology, however the false positive rate was only 2% (Wang et al. 2004). One study of 34 ultrasonographically detected focal liver lesions found neoplasia in some but not all lesions sampled, and identified neoplastic cells spreading beyond the focal lesion in some cases (Faverazani 2006). The authors note that the pattern of echogenecity was not consistently related to any particular tumor type. A more recent larger retrospective cytology:histopathology correlation study of 140 ultrasound detected focal liver lesions in dogs found a sensitivity of 52% and a positive predictive value of 87% for neoplasia (Bahr et al, in press). Within those general parameters of diagnostic performance, there was variation in the ability of cytology to detect tumor subtypes. For example, cytology was most sensitive for round cell tumors (60%) and non-hepatocellular carcinomas (55%), but was lower for hepatocellular carcinomas (35%) and sarcomas (17%), although the number of sarcomas in that study was very small. The positive predictive value of cytology for hepatocellular carcinoma was excellent (100%) and also high for non-hepatocellular carcinoma (86%) and round cell neoplasia (75%). All of these studies were retrospective, therefore selection bias could have influenced the diagnostic performance of cytology that was reported. Bahr et al used the STARD criteria to characterize the bias by evaluating patients that met all study criteria compared with those meeting all study criteria except no histology was performed. The data suggest that patients with a cytologic diagnosis of inflammation and necrosis of the liver were more likely to undergo biopsy than patients with other cytologic diagnoses (neoplasia, hyperplasia, extramedullary hematopoiesis, cholestasis, vacuolar change, normal).

Primary hepatic malignancies include hepatocellular carcinoma, cholangiocellular neoplasia, and rarely, hepatic carcinoid. Hepatocellular carcinomas can be reliably identified when there are marked criteria of malignancy, however well differentiated tumors may be impossible to distinguish from benign hepatomas or reactive hepatocellular hyperplasia without histopathology. One small study attempted to identify cytologic features of well differentiated hepatocellular carcinoma in dogs (Masserdotti and Drigo 2012). The authors suggest that compared with normal liver, well differentiated hepatocellular carcinomas were associated with dissociation of hepatocytes, acinar or palisading hepatocytes, naked nuclei, capillaries, multinuclearity, and increased N:C ratios. Unfortunately, the study did not incorporate the major differential diagnoses for these tumors, including hepatomas and hyperplastic nodules, thus limiting the clinical applicability of the findings. (Cholangiocellular tumors and hepatic carcinoids are often difficult to distinguish from metastatic lesions (see below).

Metastatic neoplasms include hematopoetic origin tumors (lymphoma, histiocytic malignancies, mast cell tumor, infiltration by leukemic cells), carcinomas (often pancreatic or intestinal), and less frequently sarcomas, with hemangiosarcoma being prominent. Round cell tumors can often be specifically identified based on cytology provided there are sufficient numbers of cells. In some low cellularity samples, histopathology or sampling of other involved organs may be required to support the diagnosis, and rarely special stains or markers may be required to define the cell line of origin. It is expected that increasingly, molecular techniques may augment cytologic analysis, such as the use of PARR or flow cytometry (Joetzke et al 2012). Occasionally, extramedullary hematopoiesis can imitate round cell neoplasia. It is often not possible to definitively determine the cell of origin of non-hepatocellular origin carcinomas, so imaging is important to characterize the scope of involvement and in some cases may help with identification of a primary tissue. Likewise, sarcomas often lack specific identifying features for subclassification

The use of cytology for the diagnosis of hepatic metastasis in canine cutaneous mast cell tumors has been studied in more detail. Independent investigators have shown that cytologic involvement of the liver is a negative prognostic indicator for survival (Stefanello et al 2009, Book et al 2011). Suggested cytologic criteria to distinguish neoplastic cell involvement from non-neoplastic mastocytosis included 1: clustering of well differentiated mast cells, 2: large numbers of well differentiated mast cells, and 3: mast cells with atypical morphology, including pleomorphic and poorly granulated cells. Of critical importance is that both studies identified patients with metastatic mast cell involvement of the liver in the absence of ultrasonographic abnormalities, so a normal abdominal ultrasound study does not exclude hepatic metastases associated with cutaneous mast cell tumors in dogs.

NON-NEOPLASTIC HEPATIC PATHOLOGY

A variety of non-neoplastic lesions can be observed on cytologic examination of liver samples. Both lipid and non-lipid vacuolar change are common lesions that are easily identified in cytology preparations with good sensitivity and specificity, particularly for hepatic lipidosis in cats. It is our clinical impression that lipid vacuolar change can be appreciated in some poorly controlled diabetic dogs. Non-lipid vacuolar change can be supportive of a diagnosis of hyperadrenocorticism or may reflect corticosteroid therapy, however this is also a common nonspecific finding of limited clinical relevance in some dogs. Cytology:histopathology correlation studies suggest that the diagnostic performance of cytology is relatively low for inflammatory liver diseases, although lesions such as abscesses can be accurately evaluated. (Wang et al 2004). Furthermore, in the absence of tissue architecture, clinically relevant distinctions between different inflammatory liver diseases cannot be made reliably. Rarely, cytologic findings can suggest other pathologic processes such as tissue degeneration and necrosis, amyloidosis, fibrosis, and others, however it is our clinical impression that the sensitivity and specificity is marginal and large cytology:histopathology correlation studies are lacking to more fully evaluate the diagnostic performance of cytology for these processes.

SUMMARY

Liver cytology can be a valuable diagnostic tool for hepatic neoplasia and vacuolar change, however the diagnostic performance for other pathologic processes is likely less reliable. While imaging findings may be useful in evaluating patients with liver disease, several large studies suggest significant limitations of imaging in the detection and categorization of pathology, so sampling for microscopic evaluation should be considered even in the absence of imaging abnormalities.

REFERENCES

Bahr K et al **J Am Anim Hosp Assoc**, In press, Book AP et al **Vet Rad Ultrasound** 2011;52:548-554., Faverzani S et al **Vet Res Comm** 2006;30(S):293-296, Feeney DA et al. Am J Vet Res 2008;69:212-221, Joetzke AE et al **Am J Vet Res** 2012;73:884-893, Masserdotti C, Drigo M **Vet Clin Pathol** 2012;41(3):382-390, Murakami T et al. **Am J Vet Res** 2012;73:821-829, Stefanello D et al. **J Vet Intern Med**. 2009;23:1051-1057, Wang KY et al **J Am Vet Med Assoc** 2004;224:75-78, Warren-Smith CMR et al **J Sm Anim Pract** 2012;53:168-173.

PANCREATIC CYTOLOGY: AN UPDATE

Leslie Sharkey DVM, PhD, DACVP (Clinical Pathology)
St Paul, MN

INTRODUCTION

Very little information is available on the cytology of the canine pancreas due to lack of robust cytology:histopathology correlation studies. Concerns are expressed about the safety of aspiration cytology and biopsy procedures involving the pancreas in dogs and cats that limit the use of cytology and biopsy, however studies are underway at the University of Minnesota, where pancreatic cytology samples are collected fairly routinely and anecdotal evidence and preliminary data from ongoing studies in dogs and cats suggest few complications. A prospective study at our institution demonstrated that ultrasound guided aspiration and clam shell biopsy of the pancreas in healthy dogs failed to increase the serum pancreatic lipase activity, however sampling did result in slightly increased trypsin like immunoreactivity levels in serum and mild peracute necrosis, inflammation, hemorrhage, and fibrin deposition at the biopsy sites (Cordner et al 2010). In this study of normal dogs, no cytologic samples collected by ultrasound guidance were of diagnostic cellularity, however 65-75% of intraoperative aspirates or touch impressions from biopsy samples were of diagnostic quality. Another study of endoscopic ultrasound-guided fine needle aspirates also failed to demonstrate any significant clinical complications to the procedure up to 48 hours afterwards (Kook et al 2012). They described adequate cytologic samples in 8/12 dogs, and noted that in human medicine, availability of a pathologist for rapid screening of samples to determine if repeat sampling is necessary is recommended. Our clinical experience suggests that diseased tissue is more exfoliative than the normal pancreatic tissue, which is encased in a fibrovascular stroma. Preliminary unpublished data from a retrospective study of canine pancreatic cytology at the University of Minnesota suggests that in sick dogs, only approximately 25% of ultrasonographically collected pancreatic cytology samples were nondiagnostic, with the remainder being of adequate to excellent quality (samples from neoplastic lesions were not analyzed separately). Preliminary data from 73 cats at the University of Minnesota suggests that as in dogs, complications of pancreatic fine needle aspiration are relatively uncommon and approximately 70-80% of the cytology samples collected were of acceptable diagnostic quality.

INDICATIONS FOR PANCREATIC CYTOLOGY

At the University of Minnesota, indications for pancreatic cytology are similar to those for the liver and include ultrasonographic changes including altered echogenicity and the identification of mass lesions.

MAJOR PATHOLOGIC PROCESSES

Histology studies of the pancreas suggest that like the liver, nodular hyperplasia is a common nonspecific change that will have to be distinguished from well differentiated acinar tumors (Newman et al 2005). Like hepatocellular tumors, our experience is that pancreatic acinar tumors exhibiting marked cytologic pleomorphism are relatively easy to identify, but moderate to well-differentiated acinar tumors will likely require histopathology for confirmation of the diagnosis. The veterinary medical literature describing the cytology of pancreatic neoplasia in the dog is largely limited to case reports and small case series, and more comprehensive studies are needed. Preliminary unpublished data from a series of sick dogs that had pancreatic cytology performed at the University of Minnesota identified inflammation, neoplasia, and hyperplasia as common lesions. For samples with a diagnosis of neoplasia, there was a predominance of carcinomas, occasional neuroendocrine neoplasia, and a single case of lymphoma involving the pancreas. In a very limited number of pancreatic cytology cases, histology was eventually performed by surgical biopsy or at necropsy; for all cases in which a cytologic diagnosis was available, it matched the ultimate histologic diagnosis. For cats, both inflammation and neoplasia made up about half the diagnoses, with a predominance of carcinomas and rare lymphomas. In a very limited subset of cats in which biopsy was available, the correlation with cytology was considered fair overall, but better specifically for neoplasia. These findings are not surprising based on more extensive correlations studies in other parenchymal organs.

The limited available information on pancreatic inflammation suggests a patchy rather than diffuse distribution, suggesting that multiple samples may be required to maximize diagnostic sensitivity (Newman et al 2004, Watson et al 2007). We commonly observe evidence of tissue degeneration and necrosis, mineralization, and suppurative inflammation in dogs and cats with moderate to severe clinically diagnosed pancreatitis. In our experience, identification of bacteria is rare.

SUMMARY

Preliminary studies and anecdotal evidence suggest that pancreatic cytology has the potential to be a safe and useful diagnostic procedure. More information will become available as current descriptive retrospective studies are completed. There is still a significant need for more comprehensive prospective cytology:histopathology correlation studies that will better define optimal techniques and safety precautions, the diagnostic performance of pancreatic cytology for specific disease processes, and more refined criteria for the cytologic diagnosis of pancreatic neoplasia. During the session, we will review images representative of the various pathologic features of pancreatic cytology specimens submitted to our laboratory.

REFERENCES

Cordner AP et al, **J Vet Diagn Invest** 2010;22:702-707, Kook PH et al, **J Vet Intern Med** 2012;26:513-517, Newman S et al. **J Vet Intern Med** 2004;18:488-493, Newman SJ et al. **Vet Pathol** 2005:42:510-513, Watson PJ et al. J Sm Anim Pract 2007;48:609-618

THROMBIN GENERATION: PRINCIPLES

Christian Bédard, DVM, DACVP (Clinical Pathology)
Saint-Hyacinthe, Québec

INTRODUCTION

Thrombin is a serine proteinase of the clotting system that plays a central role in hemostasis and thrombosis. It is involved in the process of promoting and inhibiting clotting as well as in fibrinolysis, platelet activation and inflammation.[1,2] The importance of thrombin in the clotting process has been recognized since the discovery of coagulation and hemostasis. In the early 1900s, Morawitz concluded that coagulation takes place in the presence of calcium and thromboplastin resulting in prothrombin converted into thrombin and fibrinogen into fibrin.[3] The first four coagulation factors were then recognized and subsequently numbered: fibrinogen (factor I), prothrombin (factor II), thromboplastin (factor III) and calcium (factor IV). This first scheme of coagulation known as the "Classic Theory of Blood Coagulation" remained unchanged until the early 1940s and was used by Quick to develop the prothrombin time assay. In 1953, three pioneering scientists revolutionized the laboratory diagnosis and therapeutic monitoring of patients suffering from hemophilia.[4] By reducing the tissue factor activity in the original thromboplastin used by Quick, they developed a test sensitive for hemophilia. The partial thromboplastin time (PTT) was invented and later on modified by the use of an activator and this test remains largely in use today. It permitted the discovery of the coagulation factors involved in the intrinsic pathway, leading to the publication of the "Waterfall sequence for intrinsic blood clotting". The two classical pathways of coagulation are now linked in a more and more complex model with numerous positive and negative feedback systems to produce and regulate thrombin generation for fibrin formation.

WHY MEASURE THROMBIN GENERATION?

The large number of components and cross-links makes the coagulation system difficult to evaluate when only isolated parts (ex.: intrinsic pathway or factor activities) are evaluated separately. For example, the incidence of bleeding in human neonates with hemophilia is surprisingly low despite low factor VIII activity and prolonged aPTTs.[5] An explanation for this observation resides in the peculiarities of the neonatal hemostatic system characterized by low levels of antithrombin and tissue factor pathway inhibitor (TFPI) at birth, resulting in similar thrombin generation profiles in neonates with hemophilia as well as in normal neonates. As synthesis of AT and TFPI rise, thrombin generation decreases and the balance between the procoagulants and the anticoagulants shifts towards a bleeding phenotype. This illustrates the difficulty to predict a clinical phenotype with accuracy when the coagulation process is only evaluated with conventional clotting assays, examining only parts of this complex process. Although useful for the diagnosis of marked hemostatic defects, these assays are usually considered insensitive to mild bleeding disorders or hypercoagulable states and thrombosis. Clotting assays such as aPTT or PT are performed using contact activation or an excess of tissue factor that are physiologically irrelevant. The end-point of these assays corresponds to the formation of the fibrin clot, when only a very small amount of prothrombin has been converted to thrombin.

With better understanding of the mechanisms of hemostasis and in order to better correlate with the clinical phenotype, recent interest has focused on general function tests evaluating the contribution of all plasmatic elements to the formation of the clot in a more physiologic setting. Among these new tests, measurement of thrombin generation (TG) is considered to be a promising assay for the study of hemostasis as well as for the detection of hyper and hypocoagulability and the monitoring of anticoagulant therapy. [6]

PROCESS OF THROMBIN GENERATION

In order to describe the process of thrombin generation, it is useful to divide the events taking place during coagulation into an initiation phase, a propagation phase and a regulation phase. During the initiation phase of clotting, the coagulation is triggered by the exposure of membrane-bound tissue factor (TF) to circulating blood.[7] When formed, FVIIa/TF complex will catalyze the activation of small quantities of factor IX and factor X. This extrinsic tenase complex will initially forme factor Xa . When bound to membranes, factor Xa will slowly convert small amounts of prothrombin (factor II) to thrombin (factor IIa). This small amount of thrombin is responsible for initial fibrin formation but is unable to sustain the coagulation process efficiently.[8] The initial thrombin is essential to initiate the amplification and propagation phase of coagulation. Factor V and FVIII will be activated by newly formed thrombin acting as a positive feedback activator. Thrombin will also activate platelets, exposing receptors and binding sites for activated clotting factors. Prothrombinase complex formed by FXa and FVa will then amplify the generation of thrombin. Once FVIIIa and FIXa are formed, intrinsic tenase complexes are assembled on activated platelet membranes catalyzing massive FX activation. Formation of prothrombinase complex will then result in thrombin generation 300 000-fold more than with factor Xa alone.[8]

It is now recognized that the vast majority of thrombin (>95%) is generated during the propagation phase after the clotting process and fibrin formation have already taken place.[9] The importance of thrombin generation during the propagation phase has been clearly demonstrated in hemorrhagic as well as thrombotic diseases.[9]

This process is under tight control by several regulators acting on the conversion of prothrombin or on the inactivation of thrombin. The principal regulator of the initiation phase is TFPI whose role is to inhibit the extrinsic tenase complex formed by TF and factor VIIa, preventing initial activation of factor IX and factor Xa. Antithrombin is an efficient inhibitor of all serine proteases, including thrombin, factor Xa, factor IXa and factor XIIa. Antithrombin will regulate thrombin generation and thrombin activity. Thrombin generation will also be under the control of thrombin itself through the activation of the protein C system. Thrombomodulin (TM) located on the endothelium will bind to thrombin and form a complex that will activate protein C (PC). Activated protein C (APC) can bind factor VIIIa and factor Va and decrease the assemblage of the prothrombinase complex and the intrinsic tenase complex.

HOW TO MEASURE THROMBIN GENERATION?

Measuring thrombin generation is not a new idea. In 1953, two papers were published on the measurement of thrombin generation and its clinical application.[10,11] It was used to identify hypocoagulable states in people with haemophilia and a bleeding phenotype despite normal clotting times as well as to study the effect of replacement therapy. The test was performed in citrated platelet rich plasma by simply adding calcium. At regular intervals, a subsample of the mixture was collected and added to a test tube containing a fibrinogen solution. The thrombin present in the subsamples would then convert fibrinogen to fibrin and result in clotting. The times for the fibrinogen to clot were recorded and converted into units of thrombin. This methodology was laborious and time consuming but would explained the bleeding phenotypes of patients with haemophilia and thrombocytopenia with normal clotting times.

A new approach, avoiding the subsampling techniques, was developed by the Hemker group, the Calibrated Automated Thrombogram (CAT, Thrombinoscope bv/Stago, Maastricht, The Netherlands).[12] This assay uses a fluorogenic substrate cleave by tthombin to allow direct measurement of thrombin in clotting platelet poor plasma (PPP) or platelet rich plasma (PRP) after addition of tissue factor for initiation. During clotting, a curve of the concentration of thrombin over time is generated (figure 1). From this curve (thrombrogram), lagtime, endogenous thrombin generation (ETP), peak and time to peak are calculated. Lagtime represents the delay in minutes before the initial generation of thrombin and corresponds to clotting time. ETP is area under the curve and represents the total amount of thrombin generation expressed in nM*min. Peak represents the maximal thrombin concentration or the maximal height on the thrombogram and is expressed in nM. Finally, time to peak is the time when the highest concentration of thrombin is reached and is expressed in minutes.

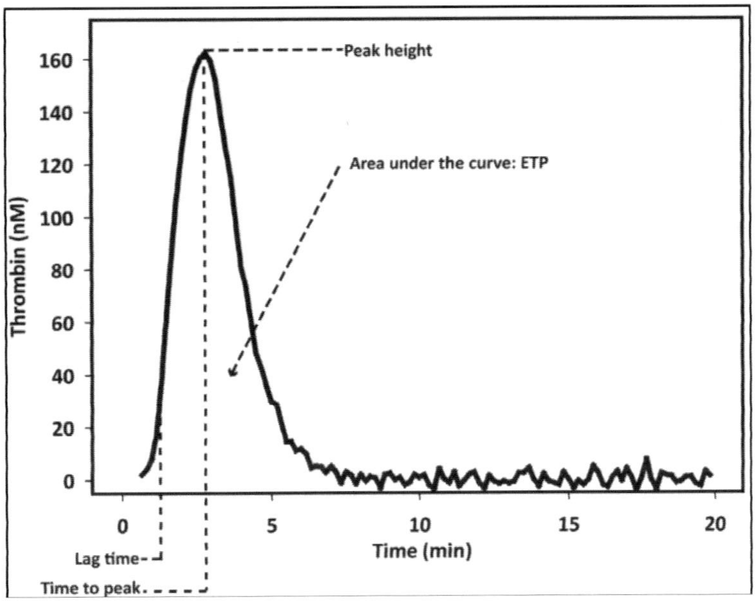

Figure 1. Normal thrombogram obtained in a beagle dog. Clotting was initiated in platelet poor plasma using a final concentration of 5ρM of TF and 4μM of phospholipids.

The CAT assay uses a fluorogenic thrombin substrate and measures thrombin generation by comparing the fluorescence in plasma during clotting to the fluorescence of a reference thrombin calibrator. Each experiment needs two sets of readings, one from wells in which thrombin generation (TG) takes place and one from wells in which the thrombin calibrator (Cal) is added. In each well, PPP or PRP is added to a buffer solution containing the trigger solution or the thrombin calibrator. The trigger solution contains a mixture of phospholipids and different concentrations of human recombinant TF. The calibrator is used to control for the appearance of the samples, aging of the lamp and filter that would both affect the fluorescence. It will also control for the inner filter effect and for the activity of alpha2-macroglobulin. Fluorescence is measured by an automated plate reader fluorometer (Fluoroskan Ascent, Thermolabsystems, Helsinki, Finland), using specific fluorescence plates and a dedicated software (Thrombinoscope bv). Analyses are usually performed in triplicate as recommended by the manufacturer. The manufacturer does not provide control material for this assay. An aliquot of pooled plasma can be included in each run as an internal control.

The trigger solutions commercially available contain different concentrations of phospholipids and human recombinant tissue factor (rhTF). For PPP, the trigger solution will add in the final mixture 4μM of phospholipids and 1, 5 or 20 ρM of rhTF. For PRP, the trigger solution will result in a final concentration of only 1 ρM of rhTF. A trigger solution containing only phospholipids for a final concentration of 4μM can be used to detect TF-bearing microparticules in plasma. It is also possible to use other sources of tissue factor and phospholipids in order to customize the assay to specific purposes. When low concentrations of tissue factor are used (<1 ρM for example), it may be necessary to add a contact phase inhibitor (ex.: corn trypsin inhibitor) to prevent unpredictable activation of the intrinsic pathway that may overpass the activation through the more physiological extrinsic pathway.[13]

CONCLUSION

Measurement of thrombin generation using the CAT assay is a global function test that evaluates all the components of the coagulation process. This assay may be useful for the study of coagulation in veterinary medicine. More studies are still needed to evaluate the usefulness of this approach.

REFERENCES

[1]Di Cera E. **J Thromb Haemost** 2007;5 Suppl 1:196-202. [2]Esmon CT. **J Thromb Haemost** 2003;1(7):1343-1348. [3]Nichols WL, el al. *A history of blood coagulation*. Rochester, Minnesota: Mayo Foundation for Medical Education and Research; 2001. [4]White GC, 2nd. **J Thromb Haemos** 2003;1(11):2267-2270. [5]Fritsch P, et al. **J Thromb Haemos** 2006;4(5):1071-1077. [6]Hemker HC, et al. **Pathophysiol Haemost Thromb** 2002;32(5-6):249-253. [7]Hoffman M. **J Thromb Thrombolysis** 2003;16(1-2):17-20. [8]Mann KG, et al. **Arterioscler Thromb Vasc Biol** 2003;23(1):17-25. [9]Mann KG, et al. **J Thromb Haemost** 2003;1(7):1504-1514. [10]Pitney WR, et al. **J Clin Pathol** 1953;6(1):9-14. [11]Macfarlane RG, et al. **J Clin Pathol** 1953;6(1):3-8. [12]Hemker HC, et al. **Pathophysiol Haemost Thromb** 2003;33(1):4-15. [13]Luddington et al. **J Thromb Haemost** 2004;2(11):1954-1959.

THROMBIN GENERATION: CLINICAL APPLICATIONS IN VETERINARY MEDICINE

Christian Bédard, DVM, DACVP (Clinical Pathology)
Saint-Hyacinthe, Québec

INTRODUCTION

Hemostasis is a complex physiological system that requires the contribution of plasma proteins as well as cellular elements including endothelial cells, platelets, leucocytes, erythrocytes and cellular-derived microparticles. A balance between the procoagulant and antithrombotic components of this system is absolutely necessary to prevent bleeding or thrombosis. In the clinical setting, it is extremely important to evaluate this balance in sick animals in order to address a potential risk for complications and eventually propose therapy. However, current conventional clotting assays using clot-based endpoints are insensitive to mild bleeding disorders or hypercoagulability. These tests, such as the activated partial thromboplastin time (aPTT) or the prothrombin time (PT), evaluate only a limited portion of the hemostasis process and often lack correlation with the phenotype of the clotting system.[1] The endpoint of these assays is the formation of a fibrin clot, when only a minimal amount of thrombin has been generated.

With a better understanding of the mechanisms of hemostasis and in order to better correlate with the clinical phenotype, recent interest has focused on global function tests evaluating the contribution of all components of the hemostatic system to clot formation in a more physiologic setting.[2,3] Among these tests, whole blood thrombelastography (TEG) is certainly a promising assay which has been used to evaluate hemostasis in healthy as well as in sick animals.[4-6] This assay measures changes in the viscoelastic properties of blood after activation. Because it is performed in whole blood, it is believed to provide a global assessment of hemostasis. Based on the TEG tracing patterns, different hemostatic abnormalities can be detected, including coagulation factor deficiency, abnormal platelet function, excessive fibrinolysis and hypercoagulability.[7] However, TEG may be affected by many factors, including the hematocrit, platelet count and fibrinogen. More recently, the Calibrated Automated Thrombogram (CAT, Thrombinoscope bv, Maastricht, The Netherlands) assay has been used to assess the coagulation status in dogs.[8] CAT is an assay that allows direct measurement of thrombin generation (TG) in platelet poor plasma (PPP) but also in platelet rich plasma (PRP).[9] CAT measures TG during clotting following activation by minimal tissue factor. During clotting, a curve of the concentration of thrombin over time is generated. From this curve (thrombrogram), lagtime (min), endogenous thrombin potential (ETP, nM*min), peak (nM) and time to peak (min) are the common calculated parameters. Measurement of thrombin generation is not only independent of fibrinogen concentration and quality but also independent of fibrin polymerization, cross-linking and fibrinolysis. It reflects the balance of the coagulation system without the contribution of its final substrate. Therefore, this assay could be a useful tool in studying hemostatic disorders as well as anticoagulant therapy in animals.

THROMBIN GENERATION IN THE PLASMA OF HEALTHY ANIMALS

The purpose of this study was to measure TG in healthy beagles and to evaluate the potential of the CAT assay in dogs.[8] Individual platelet poor plasma and plasma pooled from 20 healthy adult beagles were used. The assay was performed with the addition of recombinant human tissue factor (rhTF) and phospholipids at a final concentration of 5ρM and 4μM respectively. In all cases, analyses were completed within 8 to 9 minutes after initiation. Mean ± sd for ETP, peak, lagtime and time to peak were respectively 340.0±63nM*min, 140.0±19.8 nM. 1.17±0.22 min and 2.70±0.30 min. ETP ranged from 238.7 to 414.0 nM*min, illustrating the variation in TG between animals (figure 1). ETP intra-assay and inter-assay variation were 7.1% and 12.9% respectively. We determined that the CAT assay can be used to measure TG in canine clotting plasma. This assay can be performed on stored frozen platelet poor plasma with good analytical reproducibility under our experimental conditions.

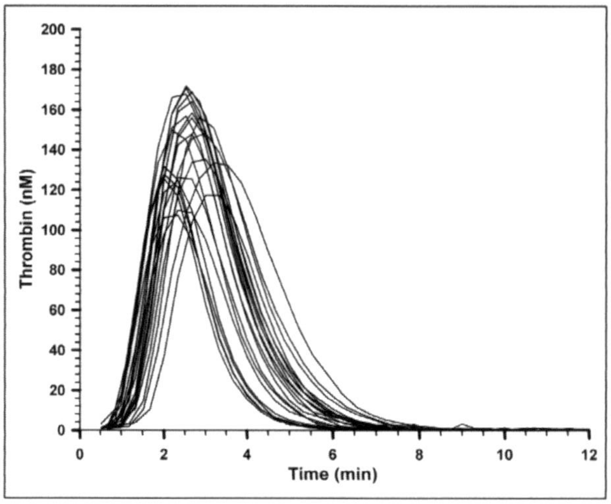

Figure 1. Thrombin generation curves of 20 adult healthy beagles obtained after recalcification and coagulation activation of platelet poor plasma by addition of recombinant tissue factor (5ρM) and phospholipds (4μM).

A similar study was performed in healthy monkeys (*Macaca fascicularis*).[10] TG was measured in the plasma of 6 healthy monkeys. Clotting was initiated using a trigger solution containing a mixture of phospholipids and rhTF at a final concentration of 4μM and 5ρM respectively with the addition of thrombomodulin (TM). Lagtime, time to peak, ETP and peak were (mean±sd) 1.8±0.1 min, 4.8±0.4 min, 1555.2 ±211.3 nM*min and 295.3±41.5 nM respectively. TG was also measured after 15 and 60 days of storage at -80°C. After 15 days of storage, only a significant increase in peak (108.75%) and a decrease in time to Peak (94.17%) were found when compared to baseline values (100%). ETP intra-assay and inter-assay variations were 7.0% and 7.8% respectively.

THROMBIN GENERATION FOR THE MONITORING OF ANTICOAGULANT THERAPY IN DOGS

Most tests used to monitor anticoagulant therapy, such as aPPT or anti-FXa activity, are not functional assay and often cannot be used to adjust therapy in cases of bleeding or thrombosis recurrence. Since the goal of anticoagulation therapy is to decrease thrombin activity, we proposed that measurement of TG could be a useful approach to assess anticoagulation with unfractionated heparin (UFH) and low-molecular weight heparin (LMWH) in dogs. A study was conducted in order to evaluate the anticoagulant therapy of unfractionated heparin (UFH) in healthy dogs.[8] In vitro, serial dilutions of UFH (0.05, 0.1, 0.2, 0.3, 0.4, 0.6, 0.8, 1.0, 1.5 and 2.0 U/ml) were used to establish a heparin sensitivity curve. Clotting was initiated using rhTF and phospholipids at a final concentration of 5pM and 4 mM. Heparin concentrations of 0.4 U/ml and above resulted in marked to almost complete thrombin activity (figure 1). The pharmacodynamics of UFH was studied on 6 control and 6 treated beagle dogs. TG was monitored for 9 hours after a single injection of UFH (75U/kg). Changes in ETP were observed after 30 minutes post injection in some dogs. Maximum effect of heparin on ETP was observed at 170 ± 36.33 minutes. At peak effect on TG, ETP was decreased by 38.5± 7.8%. Returns to baseline ETP were observed between 210 and 420 minutes for 5 dogs. In 2 dogs, ETP did not reach its baseline value at the end of the experiment. Interestingly, ETP at the end of the experiment was increased in 3 dogs when compared to baseline values. Differences in ETP between control dogs and treated dogs were significant only at 180 minutes. This study demonstrates that measurement of TG is a global method that can be used in dogs to assess the degree of anticoagulation provided by UFH.

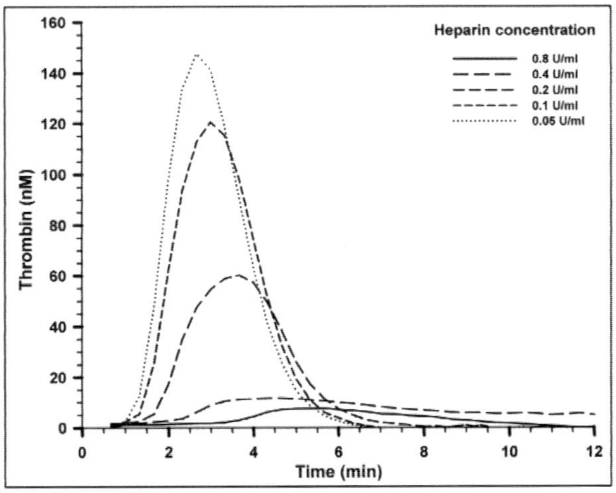

Figure 2. Thrombin generation curves obtained for serial dilutions of UFH in pooled plasma.

A similar study was performed with LMWH (Dalteparin) in healthy dogs.[11] A total of 24 dogs were randomly assigned to 4 groups of equal size and treated with a single subcutaneous injection of either 0 U/kg (saline), 50U/kg, 100U/kg or 150U/kg of dalteparin. TG was measured at baseline and overtime for 24 hrs (figure 3).

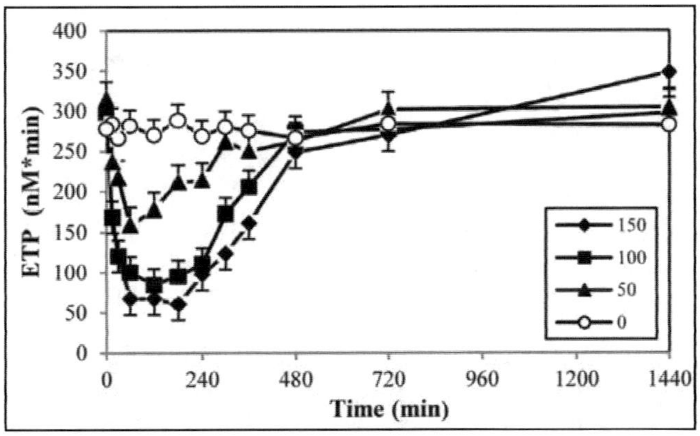

Figure 3. Endogenous thrombin potential over time after a single subcutaneous injection of 0, 50, 100 and 150 U/kg of dalteparin.

A rapid decrease in ETP was observed after only 15 minutes post injection in all dogs treated with dalteparin. The intensity of the effect and the duration of anticoagulation on TG increased with the dose of dalteparin. At peak effect on TG, ETP was decreased by an average of 54.21% ±10.6 (50 U/kg), 73.66% ±7.22 (100 U/kg) and 81.27%±4.2 (150 U/kg) when compared with the baseline. Dogs in each group demonstrated great inter-individual variability of TG (figure 4). Administration of 100U/kg showed the most homogeneous response and return to baseline amongst treated groups. In this group 2 of 6 dogs presented a plateau effect of saturation, meaning that there was a constant inhibition during a period of time. All dogs in the 150U/kg dosing group presented a plateau effect, and the maximal saturation was around 40 nM*min. In high LMWH concentration, increasing anti-FXa activity resulted in minimal decrease in ETP. Return to baseline ETP was observed at 424±173 minutes, 603±241 minutes and 894±387 minute for 50 U/kg, 100 U/kg and 150U/kg respectively.

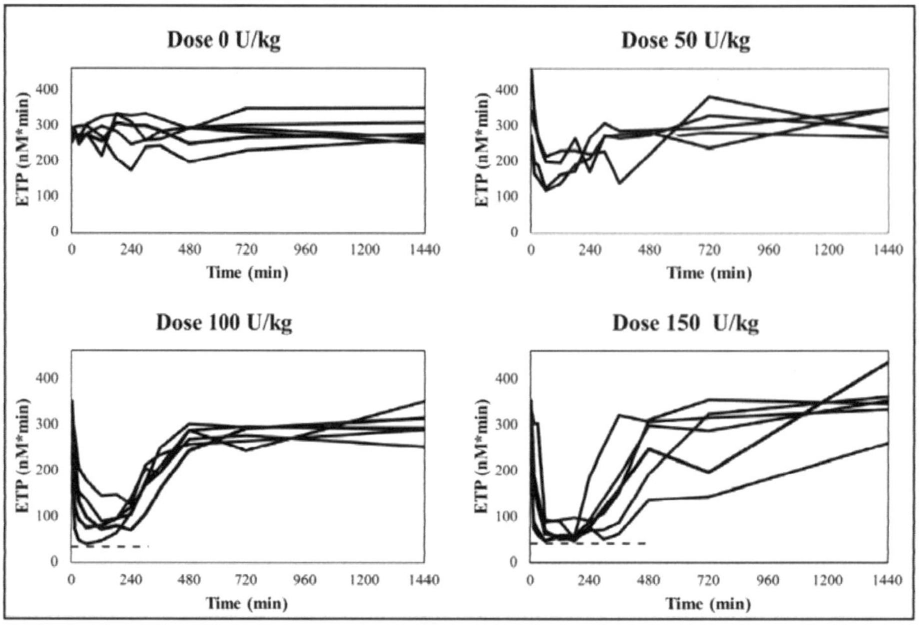

Figure 4. Individual entogenous thrombin potential measured over time after a single subcutaneous injection of 0, 50, 100 and 150 U/kg of dalteparin. The dashed lines represent the maximum thrombin generation inhibition (around 40 nM*min).

THROMBIN GENERATION AND HYPERCOAGULABILIY IN ANIMALS

In this preliminary study, TG was measured in the plasma of 13 clinically healthy dogs and in 17 dogs admitted to the intensive care unit (ICU) of the veterinary teaching hospital with various conditions (unpublished data). The assay was performed with final plasma concentrations of rhTF and phospholipids of 5pM and 4 mM respectively. ETP was significantly elevated in ICU dogs (614 nM*min) when compared to clinically healthy dogs (385 nM*min). Based on these results, measurement of TG appears applicable to canine samples and promising for detecting hypercoagulability. More studies are required to evaluate the association of a hypercoagulable profile, based on TG results, with the risk of thrombosis and the potential benefit of anticoagulant therapy in dogs.

More recently, a study using monkeys (*Macaca fascicularis*) was performed in order to evaluate the effect of a model of acute inflammation on TG.[10] The objectives were to measure TG in healthy monkeys and in monkeys treated with anti-CD3 monoclonal antibodies. TG was measured at baseline and at 24 hrs and 48 hrs post-treatment. In treated monkeys, CAT parameters were increased ($p<0.05$) at 24 hrs and 48 hrs post-treatment except for peak which was increased only at 48 hrs. Median ETP was 1203 nM*min at Baseline, 1827.5 nM*min at 24 hrs and 1491.5 nM*min at 48 hrs. According to these results, administration of anti-CD3 antibodies resulted in a hypercogulable state based on the TG tracings.

CONCLUSION

Limited data are actually available on TG in animals. However, measurement of ETP using the CAT assay appears to be a promising tool for the study of hemostasis in veterinary medicine. Its greatest utility may prove to be in the detection of hypercoagulable states and in the monitoring of anticoagulant therapy.

REFERENCES

[1]Hemker HC, et al. **Thromb Haemost** 2000;84(5):747-751. [2]Barrowcliffe TW, et al **Haemophilia** 2006;12 Suppl 3:76-81. [3]Dargaud Y, et al *Haemophilia* 2012;18 Suppl 4:81-88. [4]Leclere M, et al. **Vet Clin Path** 2009;38(4):462-466. [5]Macieira S, et al. **Vet Clin Path** 2007;36(2):204-208. [6]Wagg CR, et al **Vet Clin Pathol** 2009;38(4):453-461. [7]Donahue SM, et al **J Vet Emerg Crit Care** 2005;15(1):9-16. [8]Allegret V, et al **Vet Clin Pathol** 40(1):24-31. [9]Hemker HC, et al **Pathophysiol Haemost Thromb** 2003;33(1):4-15. [10]Bédard C, et al. **ACVP 62th /ASVCP 46th Annual Meeting** Nashville, TN 2011. [11]Gara-Boivin C, et al. **J Vet Intern Med** 2011;25:632-767(ACVIM forum abstract).

THROMBOELASTOGRAPHY - CAN IT REPLACE TRADITIONAL HEMOSTASIS TESTING?

R. Darren Wood, DVM, DVSc, Diplomate ACVP (Clinical Pathology)
Guelph, Ontario, Canada

INTRODUCTION

Thromboelastography (TEG) is a global assay of hemostasis, permitting assessment of the entire clotting process with a single whole blood sample. This seminar will review TEG pre-analytical issues, methodology, comparison to the traditional assays of hemostasis, and highlight its potential benefits and limitations. The focus will be on TEG as determined by the instrument marketed by Haemonetics®, recognizing that there are other methods available.

A TEG tracing is produced from the continuous detection of a developing clot in a sample cup into which a pin on a torsion wire is suspended.[1] The resulting changes in viscoelasticity of the sample as fibrin develops and then lyses are transmitted and converted by software into a graphical representation that can serve as both a numerical and visual snapshot of the hemostatic process (see Figure 1).

PRE-ANALYTICAL CONSIDERATIONS

Proper collection of any hemostasis sample is required to ensure valid results. Blood samples for TEG can be collected directly into a dry syringe if the instrument is cage-side and can be performed within minutes, before the sample clots. This is uncommonly performed in animals. Instead, blood is usually collected into evacuated tubes containing 3.2% sodium citrate for anticoagulation. It is important to maintain a 9:1 ratio of blood to anticoagulant, as both over- or under-citration may impact the results. Blood should be collected atraumatically from a large vein such as the jugular or cephalic. The amount of trauma encountered during venipuncture may have impact on the results.[2]

Consideration needs to be given to the timing of blood collection, since the sample needs to rest for a standardized period of time (often 30 minutes) at room temperature prior to onset of analysis. This resting period should be consistent for all samples, and established for each individual lab. The longer a sample rests (i.e. after 30 minutes), the more hypercoagulable the TEG tracing appears. Extremes of temperature[3,4] and hemolysis of the sample, should be avoided. Hemolysis has been shown to result in decreased reaction (R) time and maximum amplitude (MA).[5]

ANALYTICAL CONSIDERATIONS

Coagulation is initiated when blood is added to a cup in which calcium chloride has previously been added. In veterinary medicine, the sample for TEG is usually either performed without an additional activator ('plain' or 'native' sample), or activated with either kaolin or tissue factor. The activator is added to the blood sample prior to recalcification. The use of an activator may minimize excessive intra-assay variation that can be observed with nonactivated samples.[6] However, use of different activators produces different results, which must be factored into the interpretation of the results.[7]

Haemonetics® provides a vial containing kaolin to which a volume of whole blood is added, providing some standardization for this method. They also provide a rapid TEG reagent which contains a mixture of kaolin and tissue factor; this reagent appears unsuitable for use in dogs. Tissue factor as a sole reagent for activation is not currently provided commercially by the manufacturer, but is used by many researchers and laboratories using a separate source. The most commonly used reagent is recombinant human tissue factor[8] although the concentration used can vary substantially. The dilution used needs to be adjusted so that a sufficiently long enough *R* time develops. Tissue factor can induce very rapid onset of clotting in canine samples if too much is used, which mitigates the ability to detect a shortened *R* time.

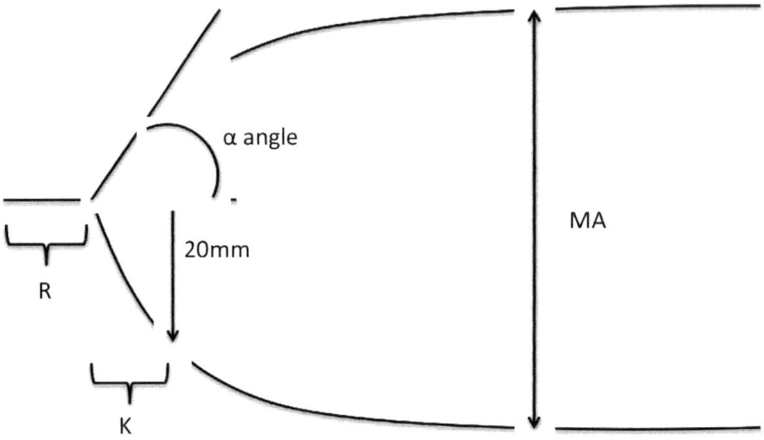

Figure 1. TEG tracing showing commonly measured variables R time, K time, α angle and MA.

MEASURED VARIABLES

Numerous variables can be derived from the TEG tracing (Figure 1), but only a few are commonly used for animals. There are four standard variables that relate to clot formation. The reaction time or *R* time, measures the time in minutes from initiation of the test to first formation of fibrin and amplitude of 2mm. This is the usual endpoint of the traditional plasma based tests, and is prolonged when there is a decrease in clotting factor activity. The *K* time measures the time in minutes from amplitude of 2mm to 20mm, and is a representation of speed of clot formation. Alpha (commonly α or angle) is a measure in degrees of the angle tangential to the developing curve as *K* is reached. Finally, the maximum amplitude, or MA, is a measure in millimeters of the maximal amplitude reached and is an assessment of maximal clot strength, as well as the overall stability of the clot. Additionally, G can be used as an assessment of clot strength, and is a derivative of MA in dynes/cm^2. Clot lysis variables LY30 and LY60 can be determined if the tracing is allowed to run long enough, and indicate percent lysis at 30 and 60 minutes respectively, after determination of MA.

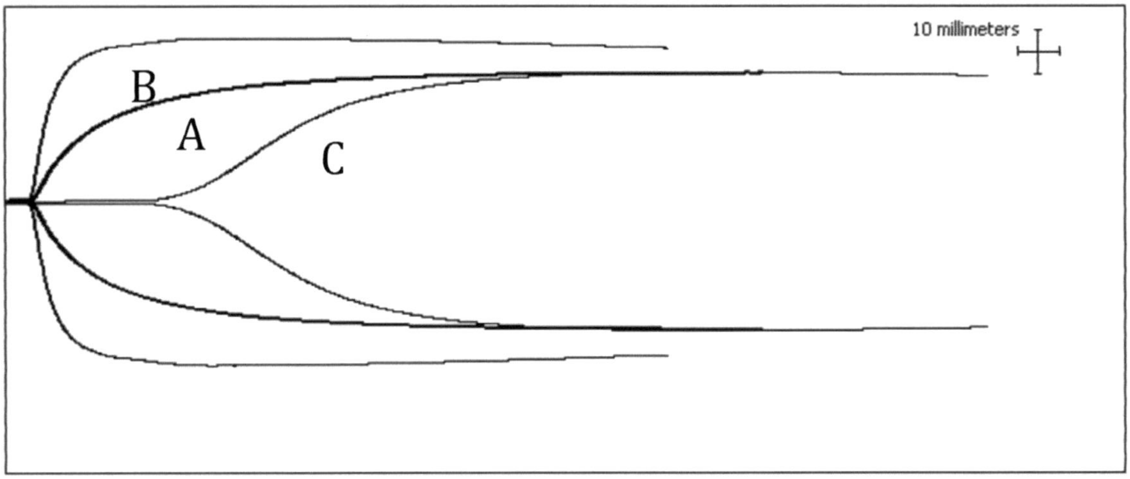

Figure 2. Examples of three different TEG tracings. A. Normal tracing from a healthy dog. B. Hypercoagulable tracing from a dog with hyperadrenocorticism. C. Hypocoagulable tracing from a dog with rodenticide toxicity.

ADVANTAGES AND LIMITATIONS

Traditional hemostasis assays tend to be plasma-based, with addition of reagents that perform the function needed for clot formation in the absence of cells. Since TEG uses whole blood, it is thought to be more representative of the *in vivo* physiological environment.[1,6] This has become more of a consideration since the adoption of the cell-based model of hemostasis, which recognizes the contribution of platelets and even leukocytes to the clotting process. Additionally, the traditional tests are not sensitive to detection of shortened clotting times, while TEG is. TEG technology can therefore allow for rapid visualization of both hypo- and hyper-coagulable states (Figure 2).

Once the sample is collected, after the 30-minute waiting period, most of the commonly used variables are determined in less than an hour, so results can be obtained for clinical use quite quickly and with minimal sample processing. Only a small volume of blood is required (less than 0.5 mL) so testing can be easily accomplished in small patients. TEG instrumentation and software are very easy to use and maintain, and there are available quality control materials to ensure testing conditions are optimal.

Since TEG is a point-of-care test using whole blood, unless a practice has the instrument, it does not replace the traditional plasma-based assays as a send-in test to a central diagnostic laboratory. This limits the use of the technology to centers that have a unit on site. There is also poor standardization between laboratories regarding specific TEG methods, and there is no uniform agreement whether an activator, and which activator, should be used. Furthermore, if tissue factor is used as activator, the source and dilution are additional variables. The methodology is not sensitive to mild clotting factor deficiencies, and platelet inhibition cannot be detected with standard TEG. There has been an effort recently by the PROVETS (Partnership on Rotational Viscoelastic Test Standardisation) group to critique the TEG literature and come up with some recommendations for better-standardized methodology and usage.

Reference values should ideally be established for each laboratory, which can be time consuming and expensive, but critical for proper interpretation of the results. Separate reference values for nonactivated and activated sample analysis should be established.[7]

CORRELATION WITH TRADITIONAL TESTS

There are several studies that have examined the relationship of TEG with traditional coagulation tests. The R time, which begins with recalcification and ends when initial fibrin is formed, should in theory approximate the activity of the proteins of the clotting cascade.[1] This has been historically determined by the prothrombin time (PT) and activated partial thromboplastin time (aPTT). Multiple components factor in to development of K and alpha angle, including platelet quantity and function, ongoing clotting factor activity and thrombin and fibrinogen concentrations.[1] MA is dictated by platelet quantity and function, FXIII activity, and fibrinogen concentration. [1]

Despite these supposed intuitive correlations, there is some conflicting information in the literature for animals, especially for R time. Wagg et al. identified correlation between PT and R time in dogs being admitted to ICU.[9] However, both fibrinogen concentration and platelet count do tend to be highly correlated with both K and MA in dogs, which was also noted in the Wagg study.

There have been only few studies to date demonstrating correlation between the TEG lysis variables and traditional tests of fibrinolysis; fibrin/fibrinogen degradation products (FDPs) and d-dimers. Wiinberg et al. did not find correlation between d-dimer concentration and TEG lysis variables in a cohort of dogs with disseminated intravascular coagulation (DIC).[10]

ADDITIONAL INTERPRETIVE CONSIDERATIONS

Red blood cell mass of the sample has an impact on some TEG variables. Compared to those with normal hematocrits, anemic blood samples tend to have higher MA and G values, while samples with erythrocytosis have lower values.[1] There is debate about whether changes in red blood cell mass have similar impact on clot formation *in vivo*, or if this is an *in vitro* artifact. The same amount of blood is used for testing regardless of hematocrit, so there can be more or less plasma in the sample, which may provide more or less clotting factor activity and impact the outcome of the tracing.[6] Results from plasma-based thrombin generation tests do not produce a similar result, however there may be a contribution from cells in the whole blood TEG sample that is lacking in thrombin generation assays.[1]

The influence of certain therapeutic drugs being used in the patient needs to be considered to properly interpret TEG tracings. Corticosteroids such as prednisone are commonly used for many reasons. It is recognized that these drugs can produce increased K, alpha angle and MA, resulting in an interpretation of hypercoagulability.[11] The same phenomenon occurs in dogs with hyperadrenocorticism[12] (Figure 2). Non-steroidal anti-inflammatory drugs such as acetylsalicylic acid and clopidogrel predominantly alter platelet function and do not appreciably impact standard TEG tracings.[13,14] A separate PlateletMapping® kit can be used to assess platelet inhibition by these drugs. Heparin at therapeutic dosages has been shown to significantly prolong R and decrease MA in dogs[15], and it has been proposed that TEG can be used to monitor dogs on heparin therapy.[16] The effect of many other drugs on TEG is limited or unknown.

Breed variation is possible in dogs, particularly in Greyhounds.[17] Citrated nonactivated TEG performed on samples from this breed had significantly slower clot kinetics and lesser over all clot strength compared with non-Greyhound dogs. This breed does tend to have higher red blood cell mass and therefore less plasma volume in the sample cup, resulting in a similar situation as was described earlier for patients with erythrocytosis. The breed can however exhibit a bleeding tendency so *in vivo* hypocoagulability may occur.

SUMMARY

Thromboelastography is an emerging diagnostic tool that can describe a patient's hemostatic ability with a single whole blood sample with a rapid turnaround time. Although standardization is an issue, there has been broad investment in the instrumentation, and much literature has been published in recent years. TEG is likely to continue to be a complementary tool in the arsenal of available assessments of the hemostatic system.

REFERENCES

[1]Kol A, et. al **Vet Clin Path** 2010;39:405. [2]Garcia-Pereira B, et. al **J Vet Emerg Crit Care** 2012;22:225. [3]Taggart R, et. al **J Vet Emerg Crit Care** 2012;22:219. [4]Ralph A, et. al **J Vet Emerg Crit Care** 2012;22:320. [5]Bauer N, et. al **Vet Clin Path** 2010;39:180. [6]McMichael M, et. al **Vet Clin Path** 2011;40:140. [7]Flint S, et. al **Vet Clin Path** 2012;41:249. [8]Wiinberg B, et. al **Vet Clin Path** 2005;34:389. [9]Wagg C, et. al **Vet Clin Path** 2009;38:453. [10]Wiinberg B, et. al **J Vet Intern Med** 2008;22:357. [11]Rose L, et. al **Vet Clin Path** 2011;40:426. [12]Rose L, et. al **J Vet Intern Med** 2013;27:207. [13]O'Kell A, et. al **Am J Vet Res** 2012;73:1569. [14]Flint S, et. al **Am J Vet Res** 2011;72:1325. [15]Pittman J, et. al **J Vet Emerg Crit Care** 2010;20:216. [16]Brainard B, et. al **Am J Vet Res** 2012;73:1577.[17]Vilar S, et. al **J Vet Intern Med** 2008;22:374.

CLINICAL APPLICATION OF THROMBOELASTOGRAPHY IN DOGS AND CATS

R. Darren Wood, DVM, DVSc, Diplomate ACVP (Clinical Pathology)
Guelph, Ontario, Canada

INTRODUCTION

Assessment of hemostasis using thromboelastography (TEG) has received much attention recently, with adoption as a diagnostic tool in many referral and institutional centers. This seminar will review the application of this test for canine and feline patients, and present a review of the current literature on the topic.

TEG VARIABLES

Numerous variables can be derived from the TEG tracing (Figure 1), but only a few are commonly used routinely. There are four standard variables that relate to clot formation. The reaction time or *R* time, measures the time in minutes from initiation of the test to first formation of fibrin and amplitude of 2mm. This is the usual endpoint of traditional plasma based tests, and becomes prolonged when there is a decrease in clotting factor activity. The *K* time measures the time in minutes from amplitude of 2mm to 20mm, and is a representation of speed of clot formation. Alpha (commonly α or angle) is a measure in degrees of the angle tangential to the developing curve as *K* is reached. Finally, the maximum amplitude, or MA, is a measure in millimeters of the maximal amplitude reached and is an assessment of maximal clot strength, and overall stability of the clot. Additionally, G can be used as an assessment of clot strength, and is a derivative of MA in dynes/cm^2. Clot lysis variables LY30 and LY60 can be determined if the tracing is allowed to run long enough, and are percent lysis at 30 and 60 minutes respectively, after determination of MA.

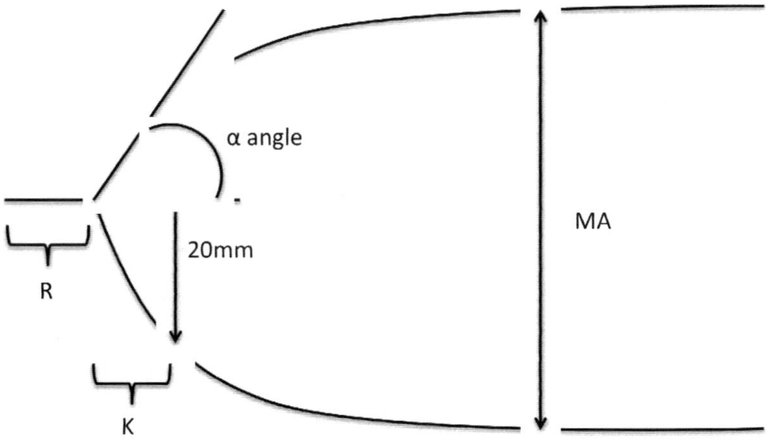

Figure 1. TEG tracing showing commonly measured variables R time, K time, α angle and MA.

CLINICAL APPLICATION IN DOGS

TEG has been evaluated as a diagnostic, monitoring, or prognostic tool in many diseases of dogs over the last several years. These are important studies, necessary to delineate the clinical usefulness of the assay. Below is a summary of the results of many of the original articles that have utilized TEG, grouped by disease process or clinical setting (see summary in Table 1).

Table 1. Diseases and settings in which TEG has been evaluated in dogs.

Inflammatory Disease	Hemostatic Derangements
Neoplasia	Fluid Therapy Monitoring
Hyperadrenocorticism	Immune-mediated Hemolytic Anemia
Peri-operative Monitoring	Anticoagulant Drug Monitoring
Disseminated Intravascular Coagulation	Protein-losing Diseases

Infectious and Inflammatory Disease

It is understood that activation of the innate immune system and development of an inflammatory response prompts concurrent activation of hemostasis. Exuberant procoagulant effects can be observed with severe systemic inflammation, from both infectious and non-infectious causes. This hypercoagulable state can contribute to tendency for either localized and/or systemic thrombosis, which could be established, and then monitored, using TEG.

In a recent study using nonactivated TEG, MA and alpha angle were significantly greater preoperatively in surviving vs non-surviving dogs with peritonitis, along with decreases in protein C and antithrombin activity.[1] MA was the most specific predictor of survival in the study.

Hypercoagulability was suspected in dogs with parvovirus infection, based on nonactivated TEG exhibiting increased MA and decreased antithrombin activity compared with control dogs.[2] Half of the dogs had evidence of thrombosis on necropsy. No significant differences were noted, however this study was hampered by a limited number of cases (n=9).

Endocrine Disease

Hyperadrenocorticism (HAC) resulting in chronic hypercortisolemia may promote a prothrombotic state in dogs. In a recent study, 12/15 dogs with HAC had evidence of hypercoagulability on tissue factor-activated TEG (one or more of decreased R and K, increased alpha angle and MA), while only 4 of those appeared hypercoagulable by thrombin generation assay.[3] Mean values for all four common variables were significantly different from control dogs. No dog exhibited clinical signs consistent with thromboembolism during the study period. Small sample size, lack of correlating hematologic data, and a mix of pituitary and adrenal dependent disease limited this study.

Another study[4] failed to demonstrate hypercoagulability in their cohort of dogs with HAC using nonactivated TEG. There was no significant difference in any of the variables between newly diagnosed HAC patients, those being treated for HAC and those that were suspicious for HAC but did not have the disease after specific testing. The actual risk of thromboembolism in dogs with HAC remains unclear and warrants further study.

Peri-opertive

Plasma samples from dogs having ovariohysterectomy or ovariectomy were hypercoagulable 4 hours after surgery as determined by tissue factor-activated TEG.[5] Both MA and G were increased at 24 hours after the surgery. TEG is not often performed on plasma samples and this does remove the impact blood cells would have on the final tracing. Results may be different using whole blood samples. However, an increase in clot strength as reported in this study, would be an expected post-surgical hemostatic response.

Eight Greyhounds classified as 'bleeders' according to a scoring system had decreased alpha angle and MA on nonactivated TEG after gonadectomy surgery, compared to 13 'non-bleeders', indicative of slower clot kinetics.[6] It was suggested that TEG could be used peri-operatively in Greyhounds to monitor for onset of bleeding complications so that appropriate therapeutic decisions can be made.

The impact of pre-anesthetic, maintenance, and non-steroidal drugs on TEG was mentioned as possible confounding variables in both studies. The effects of most of these drugs on TEG are unknown.

Neoplasia

Patients with cancer may be at increased risk for thromboembolic complications due to hypercoagulability that can accompany the disease. This was demonstrated in a cohort of dogs with various tumour types using tissue factor-activated TEG, and was particularly noted with non-mammary carcinomas.[7] Seventy percent of dogs enrolled in the study had abnormal TEG tracings, and most (94%) were hypercoagulable. There were only 3 hypocoagulable dogs, all with evidence of metastatic disease. No definitive evidence of thromboembolic disease was noted in any patient, although two of the dogs with distant metastasis had disseminated intravascular coagulation (DIC).

The TEG thrombus generation (TG) measurement, which is part of the velocity curve, was elevated in dogs with carcinomas in another study, and also indicated hypercoagulability.[8] TEG was performed without an activator. The findings were attributed to hyperfibrinogenemia in most of the cases.

In a third study of 24 dogs with malignant neoplasia, 18 were hypercoagulable (based on MA and G) with tissue factor activated TEG, while 6 were hypocoagulable.[9] Interestingly, similar to the first study mentioned above, all of the latter dogs had metastatic disease, and all exhibited hemorrhage.

Hemostatic Derangements

Nonactivated TEG was used to diagnose and monitor overt DIC in a dog, and was of particular value in documenting the hyperfibrinolytic phase of the syndrome.[10] An earlier study found that hypercoagulability was the most common conclusion from tracings of dogs with DIC (22/50 dogs enrolled) using tissue factor activated TEG.[11] Eleven out of 50 of the dogs had hypocoagulable results, and twice as many of these died (64%) versus those that were hypercoagulable

(32%). Normocoagulable tracings from an animal with DIC may represent transition from the early non-overt DIC phase, to the later overt hypocoagulable phase. This study highlighted that TEG results may be used as a prognostic indicator for patients with DIC. More studies need to be performed to verify and expand upon these results.

TEG has been used as an adjunct to study the pathogenesis and effect of new therapeutics in canine models of hemophilia.[12,13]

Using G as an indicator of hypocoagulability, Wiinberg et al. found that tissue-factor activated TEG correlated well with clinical signs of bleeding, and was superior to the routine coagulation profile with respect to both positive and negative predictive value for detecting hemorrhage.[14] It is critical to future TEG usage as a diagnostic tool that further studies such as this are carried out to determine correlations with clinical signs of hemorrhage and thrombosis.

Immune-Mediated Disease

In 30 dogs with immune-mediated hemolytic anemia (IMHA), a hypercoagulable tracing at diagnosis (predominantly based on MA using nonactivated TEG) was a positive prognostic indicator of 30-day survival.[15] In dogs that responded to therapy, these changes tended to resolve over time. Relative hypocoagulability documented at presentation usually predicted death.

Other studies have documented similar findings. All 11 dogs enrolled in one study had hypercoagulability documented using kaolin-activated TEG, as evidenced by shorter R, increased alpha angle and increased MA compared with control dogs.[16] In another study of 20 dogs with IMHA, 80% were hypercoagulable using nonactivated TEG, and all those that weren't, died.[17] A further study found that 33/39 dogs with IMHA had hypercoagulable TEG tracings, while the remaining 6 dogs had no abnormalities.[18] The latter group had increased mortality (100%) compared with the hypercoagulable group. Anticoagulant drugs are often used in these patients, and it was suggested TEG could be used to guide therapy (see below).

Interpretation of TEG in IMHA patients is challenging, as there are many potential hemostatic abnormalities that can occur, and the presence of anemia itself can induce a hypercoagulable TEG tracing. All clinical signs, drugs used, and results of other diagnostic tests (such as fibrinogen concentration, hematocrit and platelet concentration), must be in incorporated into the interpretation. It is possible that the reason some dogs with IHMA and normal TEG tracings have increased mortality is because hypercoagulability from anemia is masking an underlying hypocoagulable state.

Fluid Therapy

Administration of intravenous fluid in large volumes has been documented to impact hemostasis in people. To date, this has been examined minimally in animals. Dextrose infusion didn't impact TEG variables in one study.[19] In another study, compared with saline resuscitation, intravenous infusion of the synthetic colloid tetrastarch resulted in a more hypocoagulable tracing in dogs given lipopolysaccharide in a model of systemic inflammation.[20]

Additional prospective studies documenting how fliud products influence hemostasis are needed, for proper incorporation of their use into the interpretation of TEG tracings.

Anticoagulant Drug Monitoring

TEG has been used to monitor the effects of anticoagulant drugs such as the low molecular weight heparin dalteparin,[21,22] as well as unfractionated heparin.[23] Brainard and others,[21] using nonactivated TEG, assessed serial samples in healthy dogs collected hourly after a single injection. The Jessen study[22] added the drug to blood in vitro, and activated samples with both kaolin and tissue factor using heparinase cups. Tissue factor but not kaolin activation significantly altered all TEG variables with increasing doses of drug. Kaolin activation only prolonged R time. Pittman and co-authors[23] evaluated serial nonactivated TEG samples after injection of unfractionated heparin in 6 dogs. Peak changes (no clot formation) were noted on the tracings by 3-5 hours after administration, with return to baseline by 12 hours. Prolongation of R time appeared prior to appreciable increase in anti-Xa activity.

Protein-Losing Diseases

Dogs with protein-losing enteropathy were hypercoagulable using nonactivated TEG, with decreased R and K, and increased alpha angle and MA.[24] Antithrombin activity was not sufficiently decreased to explain hypercoagulability, so additional mechanisms may be present.

In a pilot study, Hilling and co-authors evaluated kaolin-activated TEG in 10 dogs with protein-losing nephropathy.[25] All were classified as hypercoagulable, when compared to institutional reference values. Three of 10 dogs had evidence of thromboembolic disease, but no TEG variables were significantly different from the other 7 dogs. However, this was a small study, and additional studies incorporating more cases should be performed.

CLINICAL APPLICATION IN CATS

TEG is a reproducible and valid assay for use in healthy cats according to one study.[26] However, as determined in the study, an activator such as kaolin or tissue factor should be used to minimize variability.

Alwood and co-authors used nonactivated TEG to determine the effects of ultra-fractionated (UFH) and low molecular weight (LMWH) heparins on hemostasis in healthy cats.[27] In this study, they attempted to develop reference values from 40 cats, however 15 had to be excluded due to poor reproducibility. UFH administration resulted in prolonged or absent clot formation, while a hypocoagulable effect was noted with LMWH usage. Only with UFH administration was there a correlation with anti-Xa activity. TEG may have use as a tool for monitoring anticoagulant therapy in cats.

To date, there are no published articles demonstrating the use of TEG in disease or management of feline patients, but this should be an area of developing research.

REFERENCES

[1]Bentley A, et. al **J Vet Emerg Crit Care** 2013;Jan 14:doi10.1111/vec.12013. [2]Otto C, et. al **J Am Vet Med Assoc** 2000;217:1500. [3]Rose L, et. al **J Vet Intern Med** 2013;27:207. [4]Klose T, et. al **J Vet Emerg Crit Care** 2011;21:625. [5]Moldal E, et. al **Am J Vet Res** 2012;73:1469. [6]Saavedra P, et. al **Vet Rec** 2011;169:99. [7]Andreasen E, et. al **Acta Vet Scand** 2012;54:3. [8]Vilar-Saavedra P, et. al **Vet J** 2011;190:e78. [9]Kristensen A, et. al **J Vet Intern Med** 2008;22:140. [10]Vilar-Saavedra P, et. al **J Small Anim Pract** 2011;52:656. [11]Wiinberg B, et. al **J Vet Intern Med** 2008;22:357. [12]Fischer T, et. al **J Trauma** 2011;71(Suppl 1):S171. [13]Knudsen T, et. al **Haemophilia** 2011;17:962. [14]Wiinberg B, et. al **Vet J** 2009;179:121. [15]Goggs R, et. al **Vet J** 2012;191:347. [16]Fenty R, et. al **J Am Vet Med Assoc** 2011;238:463. [17]Flint S, et. al **J Vet Intern Med** 2010;24:681. [18]Sinnott V, et. al **J Vet Emerg Crit Care** 2009;19:484. [19]Gonzales J, et. al **Am J Vet Res** 2011;72:562. [20]Gauthier V, et. al **J Vet Emerg Crit Care** 2012;22:s5. [21]Brainard B, et. al **Am J Vet Res** 2012;73:1577. [22]Jessen L, et. al **Vet Clin Pathol** 2008;37:363. [23]Pittman J, et. al **J Vet Emerg Crit Care** 2010;20:216. [24]Goodwin L, et. al **J Vet Intern Med** 2011;25:273. [25]Hilling K, et. al **J Vet Intern Med** 2009;23:690. [26]Marschner C, et. al **Acta Vet Scand** 2010;52:38. [27]Alwood A, et. al **J Vet Intern Med** 2007;21:378.

BLOOD TYPING AND CROSSMATCHING PRIOR TO TRANSFUSION

Anthony C.G. Abrams-Ogg, DVM, DVSc, DACVIM (SAIM)
Guelph, ON, Canada

INTRODUCTION

Hemolytic transfusion reactions occur when there is alloantibody directed against RBC surface antigens which define the blood groups of animals.[1] The genes, structure, and physiology of the antigens and their polymorphisms are variably understood,[2-5] but their pathophysiology is well-characterized.[2,6-7] Two key factors determining severity of a hemolytic reaction are antibody titre and type. A hemolytic reaction may occur as an acute intravascular crisis or as a delayed event, where a transfusion is "lost" over several days or weeks.

Dog and cat blood groups have been reviewed.[1] Nomenclature for dogs is not standardized; for this discussion the DEA system and *Dal* will be used.[8] There are no clinically relevant natural dog alloantibodies, but clinically relevant sensitization may occur after transfusion. The earliest reported time for sensitization of dogs after a single large transfusion is 9 days,[9] but sensitization in other species has been seen within 3 days.[10-12] The main blood group of cats is the AB system, for which there are clinically relevant natural antibodies. Natural anti-bodies have also been reported to the *Mik* antigen.[13] Based on crossmatch incompatibility and delayed transfusion reactions, there is evi-dence that other natural antibodies and sensitization to unidentified antigens in cats may occur.[14] With gastrointestinal disorders and neo-plasia in humans, antigen expression may change and/or, analogous to acquiring autoantibodies, alloantibodies may be acquired.[14-18] There are no reports in dogs or cats, but a transfusion-naïve type A, *Mik* +ve, cat with lymphoma at the Ontario Veterinary College (OVC) had an acute hemolytic reaction and was incompatible to numerous donors, but a novel blood group antigen was not ruled-out. There are three ap-proaches to detect blood groups, which in increasing order of precision are crossmatching, antibody screening, and blood typing. Serologic tests are the core of immunohematology. Molecular methods are advancing, but serology still will remain necessary for some time.

CROSSMATCHING

The crossmatch is an in vitro simulation of what will happen in vivo if a transfusion proceeds. It is the most comprehensive test. The first methods were the tube and rapid slide methods, and the former is most commonly used in veterinary reference laboratories. In a major crossmatch, recipient serum is mixed with donor RBCs; in a minor crossmatch donor serum is mixed with recipient RBCs. The mixtures are incubated, centrifuged and then observed for agglutination and/or hemolysis within the tubes. A major difficulty during crossmatching is rouleaux caused by proteins and colloids. Rouleaux may cause both false +ve results (confused with aggregates) and false -ve results (small aggregates not seen). A crossmatch is classically performed with serum to reduce rouleaux due to fibrinogen and other proteins unique to plasma, but crossmatching with either serum or plasma is performed in some laboratories. A crossmatch is typically performed with a 3-5% suspension of washed RBCs in order to reduce the concentration of serum proteins and non-specific RBC binding. Methodo-logy may be found in references,[1,6] and on veterinary blood bank and hematology laboratory websites. At OVC, reference laboratory tube crossmatching in both dogs and cats uses an immediate spin crossmatch, incubation at 22 and 37 °C, and an indirect Coombs test, but not complement. Rapid slide crossmatching is used for point-of-care (POC) testing, and this is preferred to no crossmatch. Detecting aggluti-nation/hemolysis at any step indicates incompatibility, but the quicker/stronger the test reaction, the worse the transfusion reaction. If a laboratory reports incompatibility, the point in the crossmatch at which it was detected and its strength should be clarified.

Problems with tube crossmatch include labor-intensiveness, poor technique, rouleaux, lack of standardization, and weak false +ve results. To address these, the gel column method was developed in the late 1980s. Agglutination occurs in a gel within a special microtube. The gel acts as a sieve during centrifugation, such that unagglutinated cells pellet at the bottom of the tube, while agglutinated cells are trapped in the gel forming a layer. The gel systems are easy to use, require small samples, are more sensitive, easier to grade and help standardize results. The only disadvantage is increased expense. An excellent system was produced by DiaMed AG in the previous decade and used in dogs and cats,[13,20] but veterinary support discontinued following corporate acquisition in 2007, and other manufacturers of human gel tests do not provide veterinary support. DMS Laboratories produces a gel column crossmatch kit for dogs. Anecdotes include definitive and equivocal results, which may be anticipated with any crossmatch system.

A crossmatch is indicated whenever a dog has had a previous transfusion beyond 3 days or transfusion history is not known. Previous pregnancy is not an indication. A crossmatch with pre-transfusion blood should be performed as part of the investigation of a reaction. Undoubtedly many previously transfused dogs have been re-transfused without crossmatching, and yet prevalence of DEA-based transfusion reactions is low (none at OVC over 20 years).[21,22] There is only one clinical report of a reaction to DEA 1.1.[23] This likely reflects low repeat transfusion, use of DEA 1.1 -ve donors and underreporting. However, consequences of not detecting a major incompatibility may be severe (the first *Dal* –ve dog at OVC was found by crossmatch), and vigilance is encouraged. This includes crossmatching in IMHA, which may identify a donor with a greater chance for RBC survival, but incompatibility should not be used to predict blood type. The above refers to major crossmatch. In dogs the consequences of minor incompatibilities are minor, especially using packed RBCs. Routine minor crossmatching in dogs has not been used at OVC for at least two decades. Minor crossmatching is recommended in investigation of a transfusion reaction and where any inflammatory stimulus may be very detrimental. Crossmatching in addition to typing of cats had not been routinely performed in either naïve or previously transfused cats. However, the index *Mik* case and the OVC case occurred as acute events in naïve recipients, and non-AB natural antibodies may have also caused delayed hemolysis in naïve cats.[14] Delayed transfusion reactions and/or development of crossmatch incompatibility have been seen in cats receiv-ing multiple transfusions at OVC and elsewhere, although the role of diseases is not known.[24] Crossmatching of cats is advocated.

ANTIBODY SCREENING

Antibody screening is a modified major crossmatch using RBCs of a known blood group thus screening for recipient antibodies against that group. It is extensively performed in humans and has reduced actual patient-donor crossmatching. Screening for DEA antibodies (offered by Animal Blood Resources International [ABRI]) is more expedient than crossmatching for some clinics, but does not eliminate risk for incompatibility from non-DEA/untested DEA antigens. Canine antibody screening is not performed at OVC as it is more expedient and thorough to crossmatch against banked blood. In cats antibody screening is done as back-typing: if mixing serum with type A RBCs causes agglutination, the serum is from a type B cat. If mixing serum with *Mik* +ve RBCs causes agglutination, the serum is from a *Mik* -ve cat.[13]

BLOOD TYPING

Serological blood-typing is a modified minor crossmatch, where antiserum specific to the antigen in question is used. Canine polyclonal antiserum is procured from dogs inadvertently or intentionally sensitized by single large volume or repeated small volume transfusions. Consistent with its clinical transfusion significance, DEA 1.1 antiserum can be reliably produced. Antiserum produced by alloimmunizing a DEA 1.1/1.2/1.3 –ve dog with DEA 1.1 +ve blood will agglutinate DEA1.1 +ve RBCs and cross-react weakly with DEA 1.2 or 1.3 +ve RBCs. To increase the reaction to DEA 1.2, DEA 1.2 +ve RBCs may be injected along with DEA 1.1 +ve RBCs during sensitiza-tion to produce nonspecific DEA 1.X antiserum. Several laboratories, including at OVC, provided blood-typing with such antiserum for >20 years. Adding Coombs reagent facilitates agglutination and adding complement facilitates hemolysis. A dog with a -ve reaction is classified as DEA 1.1/1.2/1.3 -ve ("A-negative"). A +ve reaction cannot be sub-classified as due to DEA 1.1, 1.2 or 1.3, although 3-4+ reactions are likely due to DEA 1.1. DEA 1.1 specific antiserum is produced by sensitizing a DEA 1.2 +ve dog with DEA 1.1 +ve RBCs; DEA 1.2 specific antiserum is produced by adsorbing DEA1.1 antibodies from DEA 1.X antiserum using 1.1 +ve RBCs. Tube methods in reference laboratories have been the historical standards against which to compare novel methods. However, discordant results are occasionally encountered between reference laboratories performing tube typing, which, may at least in part be due to differences in strength of antisera, variable expression of RBC antigens, methods of examining and grading agglutination, use of Coomb's reagent, and other procedural differences. Currently only ABRI provides commercial DEA 1.1 and 1.2 tube testing.

A major advancement was production of a monoclonal antibody (MoAb) to DEA 1.1, as the use of a MoAb provides superior consistency and stability compared to polyclonal antiserum.[25] This MoAb became commercially available in 1995 in the POC Rapid Vet-H Canine DEA 1.1 typing card from DMS Laboratories. The anti-DEA 1.1 MoAb, lyophilized on the card, forms anti-DEA 1.1 "antiserum" when mixed with PBS and agglutinates DEA 1.1 +ve RBCs when whole blood is added to the card. Similar to a rapid slide crossmatch, the agglutination reaction may be graded from 0 to 4+. Compared against reference methods, the typing cards relia-bly detect DEA 1.1 –ve and +ve healthy dogs (most 2-4+ reactions, but some 1+), but weak agglutination reactions (1+) were noted in two studies that may be due DEA 1.2,[26,27] though not in a third study.[28] In the latter, the cards accurately distinguished between DEA 1.1 +ve and –ve diseased dogs, except when autoagglutination was present that could not be eliminated with cell washing.[28] In a more recent larger study, diseased dogs, including dogs with IMHA, had an increased risk for a false +ve reaction, with some reactions ≥ 2+.[29] In that study some healthy dogs also had false +ve reactions ≥ 2+. Because of the risk for sensitizing a DEA 1.1 –ve dog typed as DEA 1.1 +ve, some clinics restrict use of the cards for selecting DEA 1.1 –ve donors and confirm DEA 1.1 +ve by another method and/or when the disease has resolved. Because dogs with IMHA may have false +ve results or cannot by typed because of autoagglu-tination, the transfusion of DEA 1.1 –ve blood is recommended. Despite manufacturer caveat, anemia did not give equivocal results.

The tube typing method suffers from the same problems as the tube crossmatch method, in addition to the variation in strength and lack of standardization of antisera. Incorporating a MoAb into gel column methodology solved these problems, and the ID-Gel Test Canine DEA 1.1 (DiaMed) system available in the previous decade gave excellent results and quickly became reference standard for other methods.[29-32] It replaced the use of antiserum at OVC for reference laboratory typing until it became unavailable.

The third typing method uses immunochromatographic (IC) lateral flow technology, which is used in many POC tests. Anti-DEA 1.1 MoAb are incorporated in the strip. The first commercial test was the DME VET DEA 1.1 (Alvedia), with others in development. This IC card has excellent performance, but false –ves may occur, with the risk higher in anemic dogs because the reaction line is hard to see.[29] Concentration of RBCs prior to performing the IC test may be beneficial. False –ve reactions are not a concern for DEA 1.1 +ve recipients if using typing to match RBCs for transfusion, but may lead to elimination of potential donors if only DEA 1.1 –ve dogs are used. In a recent study at OVC of 30 healthy and 57 diseased dogs (14 with IMHA), there was 98.8% agreement between the IC and DiaMed gel tests.[32] There were no false –ve IC results, however the +ve IC result of one dog with IMHA was discordant with the gel (-ve). IC is less affected by autoagglutination as larger aggregates cannot migrate along the strip. RBC washing was not needed to get results in any dog.[32]

The fourth typing method using anti-DEA 1.1 MoAb is the first automated device using lab-on-a-chip technology, where agglutination is detected by light transmission. It was not available in North America at the time of writing. Compared to the DiaMed gel column test, sensitivity for DEA 1.1 was 100%, but, some dogs –ve on gel test were +ve by cartridge, for a specificity of 92%.[31]

Flow cytometry may also be used to identification of DEA 1.1 by detecting antibody binding to RBCs after incubation in antisera, and demonstrates that there is variation in DEA 1.1 antigen expression on the RBC surface.[32,33] This appears to be the most accurate and precise method for detection of 1.1 but is only available in some research laboratories. Flow cytometry for DEA 7 was unsuccessful.

Although DEA 1.1 tests are good, inherent, operator and clerical errors in typing emphasize crossmatching prior to repeat transfusion. Given the increased risk for inconclusive/false results in IMHA, only DEA 1.1 –ve blood is recommended for such patients. If resources dictate use of DEA 1.1 +ve blood, then re-typing after resolution is recommended, and crossmatching before future transfusion.

DEA 1.2 tube typing is only commercially available at ABRI. A POC MoAb DEA 1.1 and 1.2 typing kit is produced by the Korea Animal Blood Bank, but the author did not have reports at the time of writing. If DEA 1.2 typing is not performed, crossmatching is the only method to detect incompatibility. Similarly, although DEA 1.3 MoAb have been produced, typing is not available, and crossmatch is needed. Serotyping for DEA 3, 4, 5, and 7 is only available at ABRI. Gel column technology was adapted for DEA 3, 4, 7 and *Dal* (but not DEA 1.2), but is not available.[34] DEA 3 and 5 antisera are more difficult to produce and exhaustion of supply may occur. Anti-DEA 3 and three other MoAb to canine RBC antigens have been produced.[35-36] These were used in a tube test by Shigeta Animal Pharmaceuticals Inc,[30,37] but seem to be no longer available. Antiserum to DEA 4 is easy to produce, but sensitizable DEA 4 –ve dogs are uncommon.[38]

In cats, AB blood typing in was initially performed by rapid slide or tube method, and, in the hands of experienced personnel, these remain the reference standards.[39-42] Microplate methodology is used to facilitate using smaller volumes, conducting large surveys, and measuring titres.[40,43] At first high-titre polyclonal anti-A and anti-B sera were used to detect agglutination;[39-41] a type A cat's natural anti-B titre could be enhanced by transfusing type B blood.[44] As with current back-typing, anti-A titres were typically high enough to cause immediate gross agglutination, but anti-B titres caused only microscopic or weak gross agglutination. Serum was heated to inactivate complement, but a standard amount of complement could then be added to detect hemolysis caused by anti-B antibodies.[40-41] Coombs reagent does not appear to be have been used to enhance the anti-B reaction. Wheat germ lectin was then reported to strongly agglutinate type B cells in 1991,[46] and has subsequently been used in rapid slide and tube methods. Lyophylized polyclonal anti-A antibodies and lectin became incorporated into a typing card in 1995 (Rapid Vet-H Feline, DMS) which greatly facilitated POC typing. In 2002 the format of the card was changed and a lyophilized murine anti-A MoAb replaced polyclonal antibodies for the anti-A reaction. The cards were evaluated against other methods in several studies, and consistently accurately identified the A antigen in healthy type A cats and B antigen in healthy type B cats, typically with strong (2-4+) reactions.[42,46-48] However, neither the A nor B antigen were always identified in healthy type AB cats.[42,26,48] In some cases, the typically weak anti-A reaction in type AB cats could lead to their misidentification as type B.[42,26,48] In other cases rouleaux formation mimicked A or B weak reactions.[49] Type A Abyssinian and Somali cats are at increased risk of misclassification as type AB.[49] At OVC, misclassification of cats as type AB has been seen in some cases of feline IMHA, with both B > A and A > B reactions; presumably undetected subtle or microscopic autoagglutination was enhanced by the typing reagents. Other type AB misclas-sifications at OVC and elsewhere include AB typing of some type A FeLV +ve cats and cats with severe inflammation, and B typing of an AB cat with inflammation.[42] Misclassifying a type A cat as type AB is not that serious as it is likely to be transfused with type A blood, and in the unlikely event of having a type AB donor, the anti-B reaction will not be severe. Misclassifying a type AB cat as type B is more concerning, because of the potentially severe anti-A reaction following transfusion of type B blood. Type AB is rare, and for this reason alone any such card typing result should be viewed with suspicion and the results confirmed. Depending upon breed, type B may also be uncommon, and the consequences of mistyping serious, so confirmation is also recommended. This may include washing RBCs and repeating the test (necessary if autoagglutination is noted), performing rapid slide or tube back typing, or, submission to a reference laboratory. Barring operator error, the author is not aware of a type A cat being typed as type B, or vice versa, with the card. The card is an excellent POC test, but for any weak reaction (1+), autoagglutination, type B result (especially in a low prevalence breed), or type AB result, the test should be repeated with washed RBCs, and confirmation of results and further investigation of discordant results obtained.

A gel column technique (DiaMed-Vet ID card A+B) using anti-A serum and lectins, and later anti-A and anti-B MoAb, was developed in the last decade.[42,46,48] As with the DEA 1.1 test, it had excellent performance, became another reference standard, and is no longer available. A commercial tube test with anti-A and anti-B MoAb[36] was in development by Shigeta, but also appears to be unavailable.

Two IC tests are now available, DME VET Quick Test A + B (Alvedia), and more recently, RapidVet H IC (DMS). Both use anti-A and anti-B MoAb and are easy to use.[42] IC methods should be unaffected by rouleaux. In a recent report the DME VET test showed even better performance that the typing card, but some erroneous typing of type A as AB, and type AB as A or B, still occurred. Preliminary results with the RapidVet H IC test are similar. The IC tests appear to be excellent POC tests, and it is anticipated they will replace agglutination cards. However, the same caveats apply as with the agglutination cards, namely that inconclusive or weak reactions, and type AB and type B results, should be confirmed by another method. Flow cytometry may also be used for AB blood-typing, but given the ease of other methods this has only been used in research laboratories.[50] *Mik* antigen testing is available from the University of Pennsylvania (Hemato-logy and Transfusion Laboratory, PennGen). The above serotype tests are all tests of phenotype. The AB genes have been cloned,[51] and PCR testing for genotype is available in North America from the UC Davis Veterinary Genetics Laboratory and from other laboratories worldwide. Typing may be performed from a buccal mucosal swab. The DNA test may be used to identify cats that will have B sero-type (homozygous for the **b** allele), and to identify B carriers (one copy of the **b** allele), which will have A or AB serotypes, but cannot distinguish between type A and AB. The DNA test can thus be used to confirm serotype B from and to investigate if an AB cat has been serotyped as B, and vice versa. Feline typing methods have excellent performance, but errors may be devastating. It is not inap-propriate to confirm any serotyping – this is may be simultaneously accomplished by a crossmatch to rule-out non-AB incompatibili-ties. At minimum confirmatory testing is needed for inconclusive, and type AB and B results, at the time and after disease resolution.

REFERENCES

[1] Weiss DJ, Wardrop KJ, eds. **Schalms Veterinary Hematology, 6th ed.** Blackwell, 2010. [2]Zimring JC, et al. **Blood** 2005;106:1105. [3]Corato A, et al. **Vet Immunol Immunopathol** 1997;59:213. [4]Daniels G. **ISBT Science Series** 2006;1:3. [5]Lin Y, et al. **Transfus Med Rev** 2009;23:292. [6]Gibson G, Abrams-Ogg A. In: Day M, Kohn B. **BSAVA Manual of Canine and Feline Haematology and Transfusion Medicine, 2nd ed.** BSAVA, 2012:289. [7]Kohn B, Weingart C. In: Day M, Kohn B. **BSAVA Manual of Canine and Feline Haematology and Transfusion Medicine, 2nd ed.** BSAVA, 2012:308. [8]Blais MC, et al. **J Vet Intern Med** 2007;21:281. [9] Young LE, et al. **Am J Vet Res** 1952;13:207. [10]Bell K. In: Agar AS, Board PG, eds. **Red Blood Cells of Domestic mammals.** Elsevier 1983:133. [11] Smith JE. **Adv Vet Sci Comp Med** 1991;36:9. [12]Schonewille H, et al. **Transfusion** 2006;46:250. [13]Weinstein NM, et al. **J Vet Intern Med** 2007;21:287. [14]Weltman J, et al. **J Vet Emerg Crit Care** 2012;22:S11. [15]Matsushita S, et al. **Jpn J Surg** 1983;13:540. [16]Mannessier L, et al. **Vox Sang** 1986;50:240. [17]Murata K, et al. **Am J Clin Pathol** 1992;98:67. [18] Matthes T, et al. **Transfusion** 2002;42:1448. [19]Yazer MH, et al. **Transfusion** 2006;46:1537. [20]Kessler RJ, et al. **Vet Clin Pathol** 2010;39:306. [21]Kerl ME, Hohenhaus AE. **J AM Vet Med Assoc**1993;202:1495. [22]Callan MB, et al. **J Am Anim Hosp Assoc** 1996;32:303. [23]Giger U, et al. **J Am Vet Med Assoc** 1995;206:1358. [24] Specht AJ, et al. **J Vet Emerg Crit Care** 2007;17:S10. [25]Andrews GA, et al. **J AM Vet Med Assoc** 1992;201:1549. [26]Giger U, et al. **Am J Vet Res** 2005;66:1386. [27]Moritz A, et al. **J Vet Intern Med** 1998;12:226). [28]Kohn B, et al. **Kleinterpraxis** 1998;43:65. [29]Seth M, et al. **Am J Vet Res** 2012;73:213. [30] Giger U, et al. **Am J Vet Res** 2005;66:1386. [31]Kohn B, et al. **J Vet Diag Invest** 2012;24:539.[32]Blois S, et al. **Vet Clin Pathol** 2013, in press. [33]Lucidi CA, et al. **Vet Clin Pathol** 2011;40:435. [34][Kesler RC, et al. **Vet Clin Pathol** 2010;39:306. [35]Hara Y, et al. **J Vet Med Sci** 1991;53:1105. [36] Kaoru A, Kyo K. **Provet** 2001:10:12. [37]Laurent O. **Bulletin USAMV-CN** 2007;64:216.[38]Callan MB, et al. **J Vet Intern Med** 1995;9:277.[39]Holmes R **J Exp Bio** 1953;30:350. [40]Auer L, Bell K. **Anim Blood Groups Biochem Genet** 1981;12:287. [41]Giger U, et al. **J Am Vet Med Assoc** 1989;195:1230.[42]Seth M, et al. **Am J Vet Res** 2011;72:203. [43]Knottenbelt, CM, et al. **J Small Anima Pract** 1999;40:365. [44]Andrews GA, et al. **Blood** 1992;79:2485. [45]Butler M, et al. **Comparative Haematoly International** 1991;1:217. [46]Stieger K, et al. **Am J Vet Res** 2005;66:1393. [47]Knottenbelt CM, et al. **J Small Anim Pract** 1999;40:115. [48]Proverbio D, et al. **Vet Clin Pathol** 2011;40:32. [49]Barrs VR, et al. **Aust Vet J** 2009;87:39. [50] Griot-Wenk ME, et al. **Am J Vet Res** 1996;57:1438. [51]Bighignoli B, et al. **BMC genetics** 2007;8:27.

STORAGE-RELATED CHANGES IN PACKED RED BLOOD CELLS

Jennifer M. Herring, DVM, MS (Resident, Emergency and Critical Care)
Urbana, IL

INTRODUCTION

Blood transfusions are essential and often life-saving therapy for medical and surgical patients, including the most critically ill. Advances in storage procedures have extended the shelf-life of pRBC to up to 42 days (dependent on the additive used) but as numerous adverse effects have been correlated with administration of stored blood, improving quality has become a priority. It has been increasingly recognized that storage itself can contribute to serious transfusion-related complications that may alter the morbidity or mortality of the recipient. In particular, it has been documented that transfusion of aged red blood cells (>2 weeks) has deleterious effects on clinical outcome compared with the transfusion of fresh red blood cells.[15] As erythrocytes are stored and as they age, there are progressive biochemical and biomechanical alterations that occur in the structure and function of the cells. These changes are collectively referred to as the red blood cell storage lesion (see table 1).[7] The interrelationship between the activities involved in the generation of these storage changes are well understood, however their physiologic significance remains unclear. Potential effects of antigenically compatible erythrocyte transfusion include febrile non-hemolytic reactions, immunosuppression, thrombocytopenia, urticaria, transfusion related acute lung injury and transfusion related immunomodulation.[9] The role of the pRBC storage lesion in these events is under active investigation.

Table 1. Storage-Related Changes in pRBC

Metabolic effects
Shape change and loss of RBC deformability
Membrane loss
Rheologic changes
Membrane carbohydrate losses
Oxidative injury to lipids and proteins
Changes in oxygen affinity and delivery
Increased adhesion of RBC to endothelial cells
Reduced RBC lifespan
Accumulation of concentrations of potassium, lactate, free hemoglobin
Shedding of active proteins, lipids, microparticles
Bacterial contamination
Coagulation effects[1]

RED BLOOD CELL STORAGE LESION

The purpose of erythrocyte transfusions is to improve oxygen delivery to the tissues by increasing the hemoglobin concentration and promoting efficient oxygen-hemoglobin dissociation. The deformability of the erythrocyte is critical in maintaining adequate tissue oxygenation as the cell must appropriately navigate the microcirculation for oxygen delivery. Factors impacting deformability include surface area to volume ratio, membrane elasticity, and intracellular viscosity.[5] Changes to shape and deformability lead to production of erythrocyte-derived microparticles (MPs) in the microvasculature and in stored products.[7] In vivo, abnormal cells are removed by the spleen, but they accumulate ex vivo over time in stored erythrocyte concentrates. Erythrocytes develop rigid membranes during storage-related shape changes and MPs are generated. Vesiculation and loss of membrane, protein oxidation, and lipid peroxidation occur leading to a loss of deformability and decreased surface area.[15] Further increased rigidity adversely impacts the survival of transfused cells. Erythrocyte-derived MPs scavenge NO, inhibiting in vitro NO availability and signaling. Other accompanying alterations include an increase in the concentrations of supernatant lipids, potassium, lactate, and free hemoglobin, and reduction in pH, glucose, 2,3-diphosphoglycerate (DPG), sodium and adenosine 5'-triphosphate (ATP).

Metabolic Effects[10,14]

Glycolysis is the only energy pathway for RBCs, in which they use glucose to make ATP and produce lactic acid and protons. Proton accumulation occurs during storage, acidifying the storage solution and RBCs, and slowing glycolysis by inhibiting phosphofructokinase and limiting ATP production over time. Glycolysis is further slowed by the standard storage of pRBC at 1-6°C. Additionally, at these storage temperatures, the Na^+-K^+ ATPase pump is nonfunctional, resulting in progressive potassium leakage from the erythrocytes and increasing in the extracellular compartment without its return to the RBC. Arrhythmias and fatal cardiac arrest have been documented in human pediatric patients, those receiving large volume transfusions, or high flow rates into central lines, owing to the high potassium levels in the product. The pH of the pRBCs (average pH = 7.1), favors the breakdown of 2,3-DPG and although this can provide some initial ATP production, depletion results in a shift of the O_2-dissociation curve to the left, increasing oxygen affinity and impairing its delivery to the tissues when the product is transfused. Impaired unloading of oxygen from hemoglobin can affect systemic perfusion.

Shape Changes/Rheologic Effects[5,7,14]

The physiology of red blood cell aging is different in vivo than it is in storage, largely due to environmental conditions such as temperature and the nature of the storage medium. Human erythrocyte concentrates undergo significant modifications and degradation over time resulting in the expression of storage lesions. Red blood cells predictably change morphology, degrade, and lose efficacy during storage. The corpuscular changes in the cells deform them from a biconcave disk to echinocytes and eventual spheroechinocytes. Much of these shape changes are reversible by warming the cells in a nutrient mixture at a neutral pH prior to transfusion in a process termed rejuvenation. This will also occur in tandem with rising ATP concentrations and 2,3-DPG. However, beyond the spheroechinocyte stage, further membrane rigidity and loss occurs, leading to MP formation through the disruption of the phospholipid asymmetry. Microvesicles arise from the tips of the echinocyte spines in an irreversible process. Membrane loss through MPs is part of the natural maturation process of the RBCs, normally controlled by high concentrations of ATP and GTP. High ATP concentrations maintain the aminophospholipid translocase activity and thus the movement of the negatively charged phospholipids, in particular, phosphatidylserine, from the luminal to the cytosolic membrane leaflet. With the loss of ATP and GTP during storage, phospholipid asymmetry reorganization results in ongoing MP production. Shape change of RBCs during storage is also associated with rheologic effects, in particular increased viscosity and diminished blood flow in the capillaries. When RBCs lose their deformability with increased duration of storage, their passage through the microcirculation is limited, which is further impaired by RBC aggregation and adhesion to the endothelium.

Microparticles[4,6]

MPs are submicron (< 1 μm diameter) membrane-derived exocytic vesicles that are released into the circulation *in vivo* and generated in stored blood products ex vivo. Platelets, endothelial cells, erythrocytes, polymorphonuclear leukocytes, lymphocytes and monocytes all produce MPs in a tightly controlled process prompted by stimuli such as shear stress, complement activation, pro-apoptotic stimulation, cellular damage or agonist interaction with cell surface receptors. Physiologic and pathologic processes are responsible for microvesicle formation in both healthy and diseased individuals and the presence of these vesicles has proven detrimental for blood transfusion recipients. MPs typically express antigens, and contain cell surface proteins, cytoplasmic contents, and nuclear components from their cell of origin. These biomolecules determine their composition, characterization and transfer of biological information. MPs have been associated with profound pro-inflammatory and pro-coagulant effects and have been implicated as an essential part of both physiological and pathologic coagulation. They also contribute to systemic inflammation and cardiovascular, hematologic, and oncologic disease states. The study of MPs in human medicine is rapidly advancing, and extends into the physiology of health, the pathophysiology of disease, and the role of MPs in transfusion medicine. In veterinary medicine, published work on MPs has been limited to the area of inherited disorders, blood storage and leukoreduction (LR).

A recent study in veterinary medicine[6] found that stored canine erythrocyte concentrates accumulate phosphatidylserine expressing MPs over time. Concentrations increased with storage time in both nonleukoreduced (NLR) and LR units, but the increase was attenuated with pre-storage LR. Transfused MPs may have a role in transfusion reactions in humans, as they are known to activate inflammation, in particular via the complement cascade and leukocytes. MPs have specifically been implicated in transfusion-associated acute lung injury, thrombotic complications in the critically ill, and post-operative cognitive dysfunction due to ischemic brain injury. Leukocytes and platelets in transfused products have been implicated in creating adverse consequences to the recipient. Transfusion induced inflammation secondary to these cells has been shown to be attenuated with LR in both humans and dogs.[9] By eliminating leukocytes and platelets through LR, there are less of these cells to generate MPs over time. It follows that if fewer MPs are generated, the adverse effects associated with transfusion of MPs could be minimized.

Oxidative Injury[2,3,11]

During storage, oxidative injury leads to erythrocyte MP generation and release of bioactive lipids from its membrane. These MPs contain CD47 on their membrane surface and can inhibit macrophages. These lipids can create immune dysfunction as in the case of transfusion associated immunomodulation and can be pro-inflammatory and increase the risk of acute lung injury. Oxidative damage occurs to lipids and proteins during the storage period and aggravates RBC injury. Oxygen saturation of hemoglobin during the pRBC collection process can off load into solution and bind other hemoglobin molecules. However, occasionally oxygen can take an electron, forming a superoxide radical and ferric methemoglobin. Normally, the RBC has superoxide dismutase and methemoglobin reductase to repair the damage but as storage occurs, glycolysis slows and glutathione concentration decreases. This can ultimately result in hydroxyl radical formation that damages lipids and proteins. Albeit a slow process over time due to the low storage temperature, total oxidative damage that accumulates and is transfused at once or with mass transfusion can have deleterious effects on recipient morbidity. In human medicine, oxidative injury to RBCs during storage leads to membrane damage and cell lysis. Specifically, values of plasma hemoglobin, potassium, lactate dehydrogenase, and markers of oxidative injury (malondialdehyde and hemoglobin oxidation) were noted to increase over a 28 day storage period.[2]

Nitric Oxide (NO)[3,4,5,8]

NO, produced by endothelial cells and induced by erythrocytes, regulates blood flow in the microcirculation. Low environmental oxygen tension results in a hemoglobin sensed NO dose adjusted vasodilation to improve capillary blood flow. Increasing hemolysis and cell-free plasma hemoglobin occur during pRBC storage time. Additionally, MPs can contain a significant amount of hemoglobin as they fragment from the erythrocytes with age. Further MP formation and hemolysis occur in vivo post-transfusion and are likely due to extravascular RBC clearance. Free hemoglobin can react with NO, resulting in its consumption. MPs can also accelerate NO scavenging at rates that exceed intact erythrocytes. In fact, using a rat model, it has been shown that free hemoglobin and microparticles react with NO approximately 1000 times faster than intact erythrocytes, resulting in potent vasoconstriction when infused into the circulation by transfusion.[4] The vasoconstrictive effects are related to the magnitude of storage-related hemolysis. Further studies have demonstrated that hemolysis also decreases NO bioavailability by releasing red cell arginase-1, an enzyme that converts arginine to ornithine. This reduces the available arginine, an important substrate for NO synthase. Disruption of NO bioavailability contributes to endothelial dysfunction, adhesion molecule expression, platelet and hemostatic activation, and reactive oxygen species generation.

Current Research

RBC storage lesion studies in veterinary medicine are emerging, especially in the areas of microparticles and cytokines in stored canine erythrocyte concentrates as well as with fresh vs. stored autologous canine transfusions.[12,13] Current investigations into markers of inflammation and coagulation in the transfusion recipients have demonstrated a greater inflammatory response with stored pRBC transfusions compared to fresh, in addition to more significant hemolysis and more marked activation of hemostatic systems. This lends support for transfusion of fresh pRBC as well as leukoreduction to minimize transfusion related concerns in our patients. Additional research is currently underway. In human medicine, four large randomized clinical trials in North America are investigating the age of pRBCs and its effects on the recipient.[11] These are summarized in Table 2.

Table 2. Ongoing Randomized Controlled Trials on RBC Storage Age

Randomized Controlled Trials	RECESS (Red Cell Storage Duration Study)	Red Cell Storage Duration and Outcomes in Cardiac Surgery	ABLE (Age of Blood Evaluation Study)	ARIPI (Age of Red Blood Cells in Premature Infants)
No. of patients	1832	2800	2510	450
Type of patients	Cardiac surgery	Cardiac surgery	ICU	Premature infants <1250g
Randomized	≤10 vs ≥21d	≤13 vs ≥21d	≤7d vs std issue	<7d vs std issue
Primary end point	ΔMODS 7d post-op	16d all-cause mortality	90d all-cause mortality	Composite end point up to 90d post-randomization

CONCLUSION

There is overwhelming evidence in support of the red blood cell storage lesion in the literature. However, as this is an in vitro phenomenon, future research to identify the clinical implications associated with these observations is needed. The areas of storage-related pRBC changes, microparticles, leukoreduction and age of pRBCs on patient hemodynamics are in its infancy in veterinary medicine. Future investigations will be instrumental in helping to improve the safety and quality of pRBC transfusions for the critically ill.

Figure 1. Potential Consequences of pRBC Storage Lesion on the Transfusion Recipient

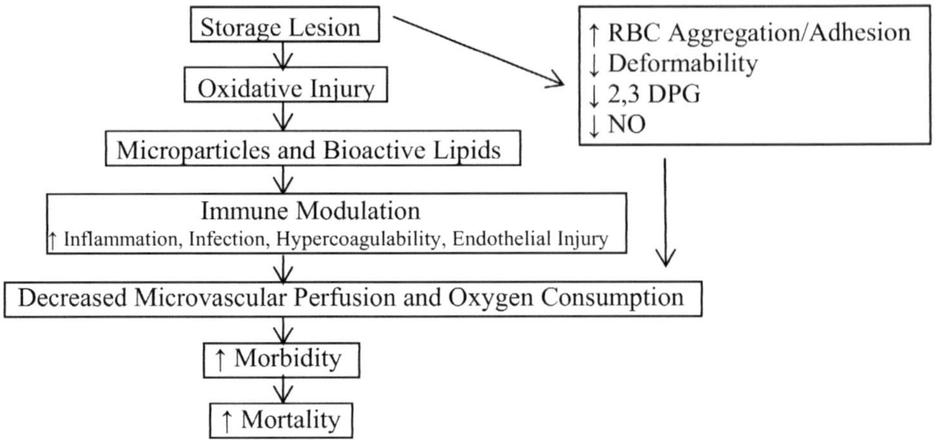

REFERENCES

[1]Aucar JA. Surgery 2012;152:697. [2]Chaudhary R, et al. Blood Transfus 2012;10:59. [3]Doctor A, et al. Semin Perinatol 2012;36:248. [4]Donadee C, et al. Circulation 2011;124:465. [5]Gladwin MT, et al. Curr Opin Hematol 2009;16:515. [6]Herring JM, et al. JVECC 2011;21:S1-S26. [7]Hess JR. J Proteomics 2010;73:368. [8]Kim-Shapiro DB, et al. Transfusion 2011;51:844. [9]McMichael MA, et al. J Vet Intern Med 2010;24:1131. [10]Patel RP. Crit Care Med 2011;39:2573. [11]Pavenski K, et al. Transf Med Review 2012;26:68. [12]Smith SA, et al. J Vet Int Med 2012;26:778. [13]Smith SA, et al. JVIM 2010;24:660. [14]Van de Watering L. Vox Sanguinis 2011;100:36. [15]Weinberg JA, et al. J Trauma 2008;65:279.

MASSIVE TRANSFUSION

Maureen McMichael, DVM, DACVECC
Urbana, Illinois

INTRODUCTION

Transfusion medicine has undergone significant changes in the past century with the development of citrate storage solutions by Lewisohn in 1914 and further work on storage solutions by Rous in 1916. Before the advent of component therapy whole blood was generally the sole product transfused. The shift from whole blood to components (RBCs, platelets, plasma, cryoprecipitate), called blood fractionation, was developed to improve access at distant remote sites, often called austere environments. The original purpose in developing and advancing component therapy was mainly driven by military requirements and has revolutionized the practice of transfusion medicine. Whole blood fractionation was developed to allow more efficient use of the individual components of blood, allowing for replacement of specific deficiencies while simultaneously decreasing waste. Component therapy grew in popularity between 1940 and 1980 essentially eliminating the demand for whole blood. Today there are very few indications for whole blood therapy. This discussion will cover the definition of massive transfusion (MT), complications of MT, current established and new MT protocols and recent research from on MT using stored blood and warm fresh blood.

DEFINITION OF MASSIVE TRANSFUSION

There is no consensus definition of MT and the most commonly used definition is the transfusion of 10 or more RBC units within 24 hours which is equivalent to one or more blood volumes. Other definitions are half a blood volume in 3 hours, 10 units within 6 hours, and infusion of blood products at 1.5 ml/kg/min over 20 minutes. Recent research in humans demonstrates a significant difference in 24 hour mortality in patients depending on the MT definition.[1] In patients receiving 10 or more RBC units before ICU admission (within 2-4 hours) mortality at 24 hours was 33% compared with 9% for the group that got $\geq$ 10 units RBCs within 24 hours. The authors concluded that the time frame of 24 hours for MT is problematic, likely evolved out of convenience, and a shorter time frame should be considered. Concern was mentioned over dilution of the study cohort with a much lower mortality rate (the 24 hour group). Current recommendations are for the 10 unit RBC threshold to occur during initial hemorrhage control. If this definition is accepted it will streamline comparison of protocols among institutions.

In veterinary medicine a single article on MT (15 cases) used the definition of MT as transfusion of blood products in excess of the patients estimated blood volume in 24 hours or half the blood volume in 3 hours.[2] Most cases were presented for hemoabdomen (neoplasia, traumatic) and GI hemorrhage. No consensus definition of MT exists for veterinary medicine. Transfusion reactions occurred in 6/15 dogs and only 4/15 survived to discharge.

COMPLICATIONS OF MASSIVE TRANSFUSION

Blood transfusion is an essential component of the treatment for many critical illnesses in companion animals, however there are multiple complications associated with the administration of blood products. Although reported clinical transfusion reactions in dogs are not common with incidence rates ranging from 3.0% to 13%, it is likely that in critically ill patients these reactions are underestimated with complications being attributed to the primary disease process. In addition it is possible that a clinically silent inflammatory response to blood transfusion might be overlooked, since many transfusion recipients have ongoing inflammation associated with the underlying disease process.

The most common adverse consequences of appropriately typed and immunologically compatible blood transfusions are febrile non-hemolytic reactions, immune suppression, decreased platelet counts, acute lung injury, and urticaria. In humans and dogs blood transfusions are associated with an inflammatory response. White blood cells (WBCs) in transfused products cause febrile reactions due to WBC cytokine production in stored products, immune suppression by decreases in natural killer cell function, phagocytosis, and decreased helper to suppressor cell ratios, decreased platelet counts due to alloimmunization, and acute lung injury from WBC aggregates in the pulmonary circulation. Leukocyte lysis during storage releases immunomodulators such as histamine, myeloperoxidase, plasminogen activator inhibitor-1, and eosinophilic cationic protein. In addition leukocyte activation during storage is associated with increased cytokines in the units and the increase appears increase with storage time. Leukoreduction attenuates and may even eliminate the inflammatory response to blood transfusion in humans. Our laboratory has documented significant inflammatory response to autologous RBC transfusion in canines,[3] to increased microparticle concentrations in stored units,[4] and to significant attenuation of these adverse effects with leukoreduction of the units.[3,4]

Complications of MT include a subset of different clinical parameters that occur due to the massive amount of blood transfused. These include hypocalcemia, hypomagnesemia, hyperkalemia, acid base disturbances, hypothermia, ARDS, dilutional coagulopathy and thrombocytopenia. The hypocalcemia and hypomagnesemia are due to citrate in the transfused products binding to these cations in vivo. These tend to develop in humans with impaired liver function or when the blood is delivered rapidly. Hypocalcemia is generally treated when hypotension or tetany develop. In the Jutkowitz study ionized calcium was low in all dogs measured (10/10) after transfusion, as was ionized magnesium. Hyperkalemia can be seen as storage time of blood products increases. Potassium tends to leak out of RBCs with prolonged storage time but hyperkalemia is an uncommon complication of MT and tends to develop more frequently when the blood is given rapidly. Acid base disturbances in the recipient arise from increasing lactic acid release with storage time in the RBC units. Acidosis can also develop from the excess citrate in the stored products. Hypothermia is a paradoxical issue in MT since several studies with severe shock (cardiac arrest, massive head trauma) report that hypothermia portends a better outcome. On the other hand hypothermia is also an indicator of hypoperfusion and would, therefore, suggest a worse prognosis. In a study of hypothermia[5] in MT, as the body temperature decreased, shock parameters, early coagulopathy, injury and severity scores and blood transfusion requirements all increased significantly. A temperature of <34C was the greatest independent risk factor for mortality after controlling for differences in injury, etc.

ARDS is a devastating complication of component therapy and is thought to be due to WBCs in the storage unit. ARDS may be reduced with transfusion of blood and plasma that has been leukoreduced to remove the WBCs along with their associated "cytokine soup." Dilutional coagulopathy and thrombocytopenia both appear to occur due to a combination of inappropriate ratio of RBCs to FFP and platelets as well as to the large amount of anticoagulants and additive solutions present in the units. Dilutional coagulopathy is such a devastating complication of MT that an entirely new approach to treatment of MT has evolved to address this. Termed "hemostatic resuscitation," this involves use of component products in a similar ratio to whole blood for patients with hemorrhage requiring MT. Also called the 1:1:1 approach, various transfusion protocols have been created to streamline the use of these products in the appropriate ratios.

MASSIVE TRANSFUSION PROTOCOLS

Multiple studies have recorded a survival benefit of early and balanced (RBC, plasma, platelet) transfusion to humans with uncontrolled hemorrhage.[6] Time to initiation of the resuscitation protocol is thought to be a major determinant in decreasing mortality rate. Anticipating a coagulopathy (with plasma and platelet transfusions) appears to be safer than trying to treat one that has already occurred. Prediction of the need for MT along with clear, easy to follow guidelines, would ideally optimize treatment and may improve mortality. Hence several centers have created MT protocols. The prospective observational multicenter major trauma for transfusion (PROMMTT) study was conducted at 10 level 1 trauma centers in the UA.[7] subsequently, data from this study has been analyzed to assess outcome associated with various MT protocols. MT protocols have several parameters in common (BP, HR, Temperature) and some are unique specific protocols (Focused abdominal sonography for trauma; FAST, etc.). A "trigger" is reached when the stated parameters meet some pre-specified point. Examples include systolic BP<90mmHg, HR ≥120bpm, Hg<11g/dl, Temperature <35.5C, INR>1.5, base deficit ≥6, and a positive FAST. Different protocols have varying numbers of triggers that must be met before the MT code is called. In a subgroup analysis of MT patients from the PROMMTT study the following triggers were significant independent predictors of a requirement for MT; systolic BP<90mmHg, Hg<11g/dl, INR>1.5, and BD ≥6. There was a significant (>80%) negative predictive value for these triggers as well.[6] Other parameters of FAST, HR≥120bpm, and penetrating injury also had a significant ability to discriminate between those requiring and those not requiring MT. If a patient had less than 2 positive triggers (MT score <2) they were highly unlikely to require MT by 6 hours.[6] The majority of the parameters used (HR, Temperature, BP) can be collected in most veterinary clinics.

Protocols identifying triggers evolved into ones outlining specific ratios of blood products to allow rapid delivery of balanced components before dilutional coagulopathy occurred. By preventing a delay in delivery due to either calculating ratios or due to sequential, repeated ordering/thawing/delivery of each component these protocols proactively streamline treatment for MT. Figure 1

Massive Transfusion Triggers
Systolic BP<90mmHg
Hg<11g/dl
Base deficit ≥6
HR ≥120 bpm
FAST +
INR >1.5
Temperature <35.5C

STORED COMPONENTS AND FRESH WHOLE BLOOD

Eighty percent of preventable deaths on the battlefield are due to hemorrhage.[8] Fresh whole blood (FWB) was the primary resuscitation fluid for hemorrhagic shock form WWI through the Vietnam War. Replacement by component therapy was established for convenience – not for outcome. Due to significant limitations in availability of fresh or stored whole blood component therapy became the treatment of choice. Component therapy also had the advantage of targeted treatment and less waste.

Due to significant limitations in availability and storage of component therapy in austere environments the military often resorts to FWB transfusions in MT casualties. These donations often occur at the scene from pre-selected donors (the buddy system) or when FWB is donated and then delivered to a scene within 24 hours.[8] FWB is a balanced product that delivers the appropriate components in the correct ratio. FWB is also less dilutional than the equivalent amount of components delivered separately. For example one unit of RBC (PCV 55%) plus one unit of FFP (80% active coagulation factor activity) and one unit of platelets (5.5 X 10^{10} platelets/ul)equals 660 ml of cold fluid with a PCV of 29%, platelets at 88,000/ul, and 65% coagulation factor activity. In comparison a 500 ml unit of FWB in 70 ml citrate delivers a product with a PCV of 43%, 130,000 to 350,000 platelets/ul and 86% coagulation factor activity.[9] The component treatment ratio results in a final product that is diluted due to additive and anticoagulant solutions in each of the individual components compared to the FWB.

Recent studies indicate improved outcome with FWB compared with component therapy.[10,11] Donors are ideally re-screened and since all military personnel are infectious disease screened at entry the risk of transfusion transmitted diseases (TTD) are minimized. If an eligible donor is nearby they are rapidly screened for anemia (copper sulfate test), ideally crossmatched, and 400-500 ml of FWB is collected from them into CPDA and transfused immediately to the victim. The entire process can be complete in 20-30 minutes. If crossmatching is not available transfusion is done as per blood type. FWB defined as either fresh or whole blood stored for less than 48 hours is associated with an improved resuscitation and hemostasis compared with component therapy.[9] Retrospective studies on 500 patients requiring MT revealed that the amount of FWB was associated with increased 48 hour and 30 day survival.[9]

Risks associated with FWB transfusion include infectious disease, WBC associated risks (ARDS, graft vs- host disease), and microchimerism. Current research is focused on performing leukoreduction in the field and pathogen inactivation procedures (exposing units to UV light after addition of riboflavin to the product).

CONCLUSION

Although military operations currently utilize FWB it is not hard to envision a mass casualty event that exhausts the stored component supplies of a city. U.S. government agencies are working on plans to use FWB from prescreened donors in case such a situation was to occur.

Future research into rapid infectious disease diagnosis, leukoreduction in the field, and rapid inactivation of WBCs, bacteria and other infectious agents will likely benefit veterinary medicine as well.

Figure 1. Massive Transfusion Protocol

MTS ≥2 plus suspect active hemorrhage

↓

Baseline blood samples

CBC, Biochemistry panel, coagulation panel, ABG

↓

Transfuse universal donor pack

6 units RBC, 4 units FFP

↙ ↘

Hemorrhage controlled | Continued bleeding

↓

Transfuse 6 units RBC, 4 units FFP, 1 unit platelets, 2 units cryoprecipitate

REFERENCES

1.Sharpe JP, et al J Traum Acute Care Surg 2012;73:1553,2. Jutkowitz LA, et al JAVMA 2002;220:1664, 3. McMichael MA, et al J Vet Intern Med 2010;24:1131, 4. Herring JM, et al. JVECC 2011;21:S1-S26, 5. Reynolds BR, et al J Trauma Acute Care Surg 2012;73:486, 6. Callcut RA, et al J Trauma Acute Care Surg 2013;74:59, 7. Rahbar MH, et al Resuscitation 2012;83:459, 8.Holcomb J, et al US Army Med Dep J 2007;Jan-Mar:24, 9.Spinella PC Crit Care Med 2008;36:S340, 10.Nessen SC, et al Transf 2013;53:107S, 11.Spinella PC, et al J Trauma 2009;66:S69-76,

DOG AND CAT BLOOD PARASITES

Dwight D. Bowman, MS, PhD, Hon. DACVM Parasitology
Ithaca, NY

INTRODUCTION

Dogs and cats can be hosts to several parasites that are considered parasites of blood. The two major forms of infection are the tick transmitted parasites of red and white blood cells which are members of the Apicomplexan group of protists and the trypanosome Excavates that are transmitted by biting flies. These infections have received greater attention in the last few years because modern molecular methods have allowed a greater ease of detection of specific types and the ability to tease out how these types are related to the observed clinical signs and induced disease. This has also produced a growing awareness that these infections are more common than previously recognized.

PARASITES OF DOGS

Dogs are hosts to species of *Babesia*, *Hepatozoon*, and *Trypanosoma*. The species or subspecies routinely encountered in North America are frequently not the same as might be encountered in other parts of the world, and some of the forms encountered in the United States are less and some are more pathogenic than forms found elsewhere.

American Triatomin-bug transmitted trypanosomes

Trypanosoma cruzi, the etiologic agent of American trypanosomiasis (Chagas' disease) is transmitted by triatomin bugs of the genera *Triatoma, Rhodnius*, and *Panstrongylus* in South and Central America and in Texas, Arizona, New Mexico, California, and Oklahoma.[1] Opossums, armadillos, rats, guinea pigs, cats, raccoons, and monkeys serve as reservoirs of infection in the wild. *T. cruzi* has been observed in hunting dogs in central Virginia that had lymphadenopathy but did not yet have clinical signs of cardiomyopathy.[2] Autochthonous cases of *T. cruzi* do occur in the United States in dogs.[3]

Trypanosomes enter the body by going through the oral, nasal, and conjunctival mucosae or by infectious bug feces being rubbed into abrasions in the skin. Infection can also occur through the placenta or by blood transfusion, and accidental self-injection presents a potential hazard of infection to persons handling blood samples from infected animals, even those specimens in which trypomastigotes cannot be demonstrated in blood films. Trypomastigotes are difficult to demonstrate in the blood of long-term carriers, and one must turn to serology, culture, polymerase chain reaction (PCR), or xenodiagnosis for recourse. In dogs, acute disease is characterized by lymphadenopathy and clinical signs associated with acute myocarditis: pale mucous membranes, lethargy, ascites, hepatomegaly, splenomegaly, and tachyrhythmia.[4] The signs during the chronic stage of the disease are related to congestive myocardial failure.[5] Megaesophagus and other megasyndromes described in humans with chronic Chagas' disease have not been reported in dogs.

Exotic Dipteran-transmitted trypanosomes

Trypanosoma evansi infects dogs and can cause both chronic and acute infections, and these infections are often fatal. A dog with an infection that was brought to Germany from Thailand presented with chronic lethargy, weight loss, unilateral anterior uveitis, pancytopenia, hyperglobulinemia, and proteinuria, and although there was an initial response to treatment with suramin the dog later relapsed and died with central nervous system disease about three months after initial presentation.[6] Other treatments for canine infections with *T. evansi* include diminazene aceturate and melarsomine dihydrochloride.

Babesia species

Babesia canis is the larger form found in canine red blood cells; with pear-shaped trophozoites, 4 to 5 µm long, typically being found in pairs in the erythrocytes. *B. canis* has been divided into three subspecies: *B. canis vogeli* of North America and northern Africa, transmitted by *R. sanguineus; B. canis canis* of Europe, transmitted by *Dermacentor reticulatus;* and *B. canis rossi* of southern Africa, transmitted by *Haemaphysalis leachi.* A survey of greyhounds in Florida has shown that large numbers (46% of 383 greyhounds) have antibodies to *B. canis vogeli*,[7] but most do not have clinical signs. The disease in dogs in the United States typically manifests with depression, anemia, anorexia, lethargy and splenomegaly. The strain of *B. canis* in Florida appears mainly to cause disease in puppies, for which the major diagnostic feature is anemia. Treatment of *B. canis* is typically with imidocarb dipropionate that is approved for this use in the United States.[8]

Babesia gibsoni is a smaller *Babesia* of the dog, being 3 µm long, and usually round to oval in shape. It is now fairly well accepted that most transmission of *B. gibsoni* in the United States is directly between dogs by wounds received when fighting.[9] The disease ranges from subclinical to fatal, with the acute disease including hemolytic anemia, thrombocytopenia, splenomegaly, lymphadenomegaly, anorexia, lethargy, pyrexia, and vomiting. Dogs often recover with treatment but remain subclinically infected. Treatment with atovaquone (13.5 mg/kg orally three times daily) and azithromycin (10 mg/kg orally every 24 hours) will reduce or eliminate *B. gibsoni* parasitemias.

Babesia conradae is a second small form of *Babesia* found in dogs that occurs in the southern California. This form seems to be related to forms found in people and wildlife in the same area. *B. conradae* is more pathogenic in dogs than *B. gibsoni*.[10] Treatment with a 10 day course of atovaquone and azithromycin removes clinical signs and clears the agent from the blood of infected dogs.[11]

Babesia microti is the species that routinely infects rodents and is transmitted by the tick *Ixodes scapularis*, and this is the species of which many clients are aware because it is a human pathogen transmitted by the Lyme disease tick. This species has not been found in dogs in the areas where human babesiosis occurs in the United States; however, Japanese workers have inoculated dogs with B. microti with a laboratory strain from rodents and showed that although they did not develop disease they did develop low parasitemias that were first evident about a month after inoculation and highest at about two months post infection.[12] The concern is that dogs might potentially serve as another reservoir of infection with this agent in these areas.

Hepatozoon americanum

The species causing disease in dogs in the United States is *Hepatozoon americanum*.[13] Throughout the rest of the world the disease is caused by *Hepatozoon canis*.[14] The vector of *H. americanum* in the United States is *Amblyomma maculatum*,[15] and coyotes serve as reservoir hosts.[16] Unlike *Babesia* species infections that are transmitted by the bite of ticks, dogs are infected with *H. americanum* either through the ingestion of ticks that have fed on another infected host and have infective oocysts in their body cavities or mammalian paratenic hosts containing cystozoites. Infected rabbits (including the European, North American swamp rabbits and cottontail rabbits), cotton rats, and mice will have persistent cystozoites of *H. americanum* within their tissues, including the spleen and myocardium.[17,18] Dogs that ingest tissues from infected rabbits will have gamonts appear in their blood within 5 weeks of having eaten the infected rabbits. Thus, dogs that hunt can get infected by eating a prey animal without ever ingesting a tick, so in these cases, tick protection will afford very little protection to the dog.

With *H. americanum*, there is typically severe disease with marked neutrophilic leukocytosis. Dogs with *H. americanum* often have significant joint pain associated with myositis and periosteal bone proliferation on the diaphysis of the more proximal long bones of the limbs; but flat and irregular bones are frequently also involved.[19] The earliest observed periosteal lesions in experimentally infected dogs were observed 32 days after infection with hypertrophy and hyperplasia of osteoprogenitor cells, and osteoblasts appearing in the cellular zone of the periosteum. The osseous lesions are similar to those of hypertrophic osteopathy in domestic dogs and other mammalian species. In cases of American hepatozoonosis, there are often large cystic forms of the organism that occurs in skeletal muscles that are not present in *H. canis* cases.

Diagnosis of *H. americanum* infection typically requires the examination of muscle tissue collected at biopsy or during necropsy to reveal the muscle forms. There are a number of laboratories now that perform molecular analysis on blood or biopsy material to make specific diagnoses.

Treatment with toltrazuril failed to prevent relapse in most of 11 treated dogs; treatment of three dogs with a combination of trimethoprim sulfate, pyrimethamine, and clindamycin also failed to prevent relapse.[20] Of 22 treated dogs, seven were humanely killed because of chronic wasting, six died of the disease, three were lost to follow-up, and six survived at least to the time of the report with three being free of signs and three having chronic wasting disease with intermittent periods of remission and relapse. Treatment to prevent remission after clinical remission is the administration of decoquinate twice a day for up to two years. It is often necessary to provide supportive therapy with non-steroidal anti-inflammatory agents to relieve pain and fever.

Dogs that travel do appear in the United States with infections with the foreign *Hepatozoon canis*, but it now appears that autochthonous cases do happen in the United States.[21] In the case of *H. canis*, dogs typically acquire their infections by ingesting an infected tick, *Rhipicephalus sanguineus*, but in Brazil it is also transmitted by *Amblyomma ovale*.[22] *H. canis* is treated with imidocarb dipropionate until gamonts are no longer present in blood smears, but treatment may frequently not clear the animals of all organisms.[22] It has been suggested that primaquine phosphate has proven efficacious in treating dogs infected with *H. canis* in Africa.[20] If there are few circulating gamonts in a dog, the prognosis is good, but must remain guarded in dogs that present with high numbers of circulating gamonts in their blood.

PARASITES OF CATS

Cats, like dogs, are host to *Trypanosoma cruzi* and to *Trypanosoma evansi*. Cats are also infected with their own forms of *Babesia* and *Hepatozoon*. Cats in the United States can also be infected *Cytauxzoon felis*, a tick-bite transmitted blood pathogen that can unfortunately often cause fulminant and fatal disease in cats.

American Triatomin-bug transmitted trypanosomes

We know that cats are infected with *T. cruzi* in Mexico and South America, and it can be assumed that they are also infected in the United States in areas where dogs are infected. For both these infections, most of the work has concentrated on cats and dogs as reservoirs of infection, and we actually have very little information on the long-term effects on the health of these hosts.

Exotic Dipteran-transmitted trypanosomes

Trypanosoma evansi infects cats, and work has shown that these infections are associated with anemia anorexia, diarrhea, loss of weight, pale mucosa, lethargy, sunken eyes, hyperthermia, and incoordination and are potentially fatal.[23] Cats have been treated successfully for this infection with diminazene aceturate and melarsomine dihydrochloride.

Exotic *Babesia* species

Fortunately domestic cats in the United States do not seem to be infected with *Babesia* species. In other parts of the world, domestic cats can be infected and develop disease with several small forms of *Babesia* (*B. felis*, *B. cati*, *B. leo*, and a *B. microti*-like form in Spain and Portugal), and several large-form *Babesia* (*B. herpailuri*, *B. canis presentii*, and again in Spain and Portugal *B. canis canis*).[24] The disease caused by these species is anemia. Treatment for the small forms is typically primaquine phosphate, and for the larger forms diminazene aceturate has been the treatment of choice.

Cytauxzoon felis

This is a sporadic but rapidly and often fatal tick-bite transmitted disease of domestic cats occurring predominantly in the south central and southeastern United States that is believed to have moved into the domestic cat population from infected bobcats, *Lynx rufus*.[25,26] Although for a number of years, the major vector was considered to be *Dermacentor variabilis*, *Amblyomma americanum* has been shown to be a highly competent vector of *C. felis*.[27] Also, and in several concurrent trials, *C. felis* was not transmitted to domestic cats by *D. variabilis*.[27]

Clinical signs when present consist of pancytopenia, icterus, pyrexia, anemia, and dehydration; with death occurring within a few days. Stained blood smears may reveal 1- to 2-μm organisms with light blue cytoplasm and dark red nucleus in the erythrocytes. Late in the course of cytauxzoonosis, enormous reticuloendothelial cells packed with schizonts appear in the peripheral blood. Histologically, parasitized reticuloendothelial cells nearly occlude the lumens of small- and medium-sized veins in the lungs, spleen, and lymph nodes. In a report on 34 cases from the mid-Atlantic states, 32 infected cats succumbed to the infection.[28] Cats can survive natural infections with *Cytauxzoon felis*.[29] Four cats without clinical signs from northwestern Arkansas and northeastern Oklahoma were identified by finding the organisms in red blood cells, and the parasitemia persisted for up to 154 days,[30] and it has been shown that cats that have survived clinical cytauxzoonosis and become carriers to serve as reservoirs for tick transmitted transfer of the infection and fatal disease to other cats.[31]

Cats, if allowed outside in areas where cytauxzoonosis is present must be protected against tick bites. The current treatment of choice for clinical cytauxzoonosis is a 10-day course of atovaquone and azithromycin.[32] Cats with a lower parasitemia, lower total bilirubin, and higher white blood cell counts were more likely to survive.

***Hepatozoon* in cats.**

Hepatozoon infections, again as with the canine form being acquired by ingestion of a tick or very likely by the ingestion of paratenic hosts, are found sometimes in cats around the world.[33] It is not known which species of ticks are important in the transmission of the feline diseases, and also, at this time, it is not known if the species in cats is the same as that in dogs. In Thailand, where both canine and feline hepatozoonosis are present, 32.3% of 300 cats were positive by PCR.[34] A *Hepatozoon* species found in a cat from Spain appeared based on molecular genetic comparisons to be more closely aligned with that of bobcats in Georgia than to that of *H. americanum*.[35] Thus, it may be that cats may be found to be infected with their own species or collection of species of *Hepatozoon*. In feline hepatozoonosis, besides gamonts circulating in around 1% of neutrophils, meronts have been identified in the myocardium and skeletal muscles of domestic cats along with elevated activities of creatine kinase.[33]

CONCLUDING REMARKS

In the United States, we unfortunately have two native species, *Cytauxzoon felis* and *Hepatozoon americanum*, that are truly dreadful disease agents relative to cats and dogs, respectively. On the other hand, we are fortunate not have the horrible *Babesia canis rossi* that is so devastating to dogs in South Africa nor its vector *Haemaphysalis leachi* in the United States, and we can only hope if it does make it here it will not find a suitable vector or that the vector does not somehow get a foothold in the United States. There is every reason to believe that with modern molecular methods and increased interest in protozoan disease of dogs and cats around the world that we will learn a great more about these diseases and the role of *T. cruzi* on the health and well-being of our canine and feline companions.

REFERENCES

[1]Kjos SA, et al **Vector Borne Zoonotic Dis** 2009; 9:41. [2]Barr SC, et al **Am J Vet Res** 1995;56:1037. [3]Patel JM, et al **Vet Parasitol** 2012;87:394. [4]Barr SC **Compend Cont Educ Pract Vet** 1991;13:745. [5]Pascon JP, et al **Pesq Vet Bras** 2010;30:115. [6]Defontis M, et al Vet Clin Pathol 2012;41(3):369. [7]Taboada J, et al **J Am Vet Med Assoc** 1992;200:47. [8]Irwin PJ: **Vet Clin N Am, Sm Anim Pract** 2010;40:1141 [9]Yeagley TJ, et al **J Am Vet Med Assoc** 2009;235:535. [10]Kjemtrup AM, Conrad PA **Vet Parasitol** 2006;138:112. [11]Cicco MF di, et al **Vet Parasitol** 2012,187:23. [12]Ohmori T, et al J Protozool Res 2011;21(2)78. [13]Vincent-Johnson NA, et al **J Parasitol** 1998;83:1165. [14]Smith TG **J Parasitol** 1996;82:565. [15]Mathew JS, et al **J Parasitol** 1999;85:1023. [16]Garrett JJ, et al **J Wildl Dis** 2005;41:588. [17]Allen KE, et al **J Parasitol** 2011;97:648. [18]Johnson EM, et al **Vet Parasitol** 2009;164:162. [19]Panciera RJ, et al **Vet Pathol** 2000;37:225. [20]Macintire DK, et al **J Am Vet Med Assoc** 1997;210:916. [21]Little L, Baneth G **J Vet Diagn Invest** 2011;23:585. [22]Baneth G **Vet Parasitol** 2011;181:3. [23]da Silva AS, et al Comp Clin Pathol 2010;19(1):85. [24]Ayoog AL, et al J Vet Emerg Crit Care 2010;20(1):90. [25]Jackson CB, Fisher T **Vet Parasitol** 2006;139:192. [26]Shock BC, et al **Vet Parasitol** 2011;175:325. [27]Reichard MV, et al **J Med Entomol** 2010;47:890. [28]Birkenheuer AJ, et al **J Am Vet Med Assoc** 2006;228:568. [29]Brown HM, et al **Vet Parasitol** 2010;172:311. [30]Meinkoth J, et al **J Vet Intern Med** 2000;14:521. [31]Reichard MV, et al **Vet Parasitol** 2009;161:110. [32]Cohn LA, et al **J Am Vet Intern Med** 2011;25:55. [33]Baneth G **Vet Parasitol** 2011;181:3. [34]Jittapalapong S, et al **Ann N Y Acad Sci** 2006;1081:479. [35]Allen KE, et al **J Parasitol** 2011;97:648.

LARVAL CESTODE DISEASE OF DOGS AND CATS

Dwight D. Bowman, MS, PhD, DACVM Parasitology (hon)
Ithaca, NY

INTRODUCTION

For the most part, we usually consider dogs and cats as the final hosts of tapeworms, i.e., the hosts with tapeworms (cestodes) in their small intestines. However, on occasion they can act as the intermediate or paratenic host of different cestode life cycles. Out of the 18 or so orders of tapeworms, only two orders actually are found as internal parasites of dogs and cats – these two orders are the Diphyllobothriidea and the Cyclophyllidea that have adult stages in mammals and utilize mammalian paratenic or intermediate hosts. The Diphyllobothriidae includes the better known genera Diphyllobothrium and Spirometra. The Cyclophyllidea includes the tapeworms of terrestrial hosts and of the 15 families within this order, larval disease in dogs and cat, and most other mammals, are caused only by member of two of the families – the Mesocestoididae (containing the genus *Mesocestoides*) and the Taeniidae (containing the genera *Taenia* and *Echinococcus*). Fortunately, infections of dogs and cats with larval cestodes are a rather rare occurrence; unfortunately, when these infections occur, the disease induced can be very difficult to diagnose and treat.

PARASITES OF DOGS

Dogs have been found to be infected with larval *Spirometra*, Mesocestoides, *Taenia* species (particularly, *Taenia crassiceps*, *Taenia solium*, *Taenia pisiformis*, and *Echinococcus multilocularis*.

Larval *Spirometra*

The plerocercoid larval stage of *Spirometra* that is found in people and animals often goes by the name sparganum (plural = spargana). It is not clear just how many species exist within this genus nor is it clear whether all species cause the same form of disease as larvae in infected hosts. In some cases, the larvae remain as single discrete white larvae that grow mainly by elongation, while in other cases, there is a form that is often called sparganum proliferum where the larval tapeworm buds and branches.

Infection with a larval *Spirometra* can occur in two different ways.[1-9] It can occur via the ingestion of the first intermediate host, a freshwater copepod that contains a larval form that has developed after the copepod ate the stage that hatched from the egg of the tapeworm. Or, hosts can be infected with the larvae by paratenesis, i.e., the ingestion of other forms that have eaten the copepod. Thus, a dog might develop a larval *Spirometra* infection by drinking water containing copepods or by eating a frog or mouse that contains a sparganum. It is not know if this has any effect on disease course. Also, it still remains unclear as to whether the proliferative form of larval development is due solely to the species of *Spirometra* involved or to deficiencies in the host's ability to contain the infection for various reasons.

Simple non-proliferative larval Spirometra infections are easy to treat through the surgical extraction of the elongate white cestode larva from the affected tissues or peritoneum. Proliferative disease often involves organ involvement and displacement, and if in the peritoneal cavity is often treated by removing the sometimes massive numbers of small larval tapeworms, sterile lavage, and repeated treatment with anti-cestocidal drugs. The disease often recurs. Unfortunately, when intestinal walls, liver, or the lungs are involved with the massive larval tapeworms that do not respond to the various cestocidal therapies, the infections are often fatal.

Larval *Mesocestoides*

In the case of Mesocestoides, the host that becomes infected with the eggs passed from the adult cestode is not known; however, due to numerous attempts to infect different animals with the eggs, it appears that it is not as simple as going straight into a mouse or a bird. Thus, when theses infections, which are much more common than those of larval spirometriosis, occur in dogs, it is unknown as to whether the dog was infected by eating one of the vertebrate paratenic hosts, that include amphibians, reptiles, birds, and mammals, or eating the unknown first intermediate host. Also, it is not known whether the horrible overgrowth of the larva stage that is called a tetrathyridium in the abdominal cavity of dogs can be caused by the ingestion of the paratenic host or only by the ingestion of the first intermediate host. However, the disease when the larvae proliferate is horrendous and very often untreatable.

Survivability of dogs with this disease has been examined as to survivability in 60 dogs undergoing treatment for this infection.[10] In this analysis, survival for 1 year was 60.5%, and dogs that were treated aggressively with surgery and peritoneal lavage along with high doses of fenbendazole were 5 times less likely to die than dogs not treated aggressively. These dogs presented with clinical signs that were ascites (60%), anorexia/weight loss (42%), vomiting (23%), diarrhea (9%) and tachypnea (9%); subclinical infections (22%) were incidentally detected, typically during ovariohysterectomy or neuter. In some cases of peritoneal mesocestodiasis, treatment with praziquantel has proven more efficacious than fenbendazole.[11]

Larval *Taenia crassiceps*

The larvae of this parasite are cysticerci which are represented by a bladder and a single scolex. The life cycle of this parasite is typically one where there are larvae in rodents and adults in foxes and other canids, including the domestic dog. The larval cysticercus undergoes asexual multiplication by budding and can develop to massive numbers in the infected host.[12] Although dogs typically serve as the final hosts, they also can on occasion be infected with the larval stage. These infections usually present as subcutaneous infections, but the budding mass can be invasive and destroy associated tissue and be found in ectopic sites as well. There is some thought that his infection occurs in dogs that have atypical immune responses.

Larval *Taenia solium*

The cysticerci of *Taenia solium* do appear in dogs that have eaten something contaminated with the eggs of this parasite with adults occur in the human small intestine. Although pigs are the typical host in which the cysticerci develop, they can develop similarly in dogs. Most of the cases that have been reported have occurred in countries where the disease in prevalent in people, but cases have been observed in the United States. In one case of a dog with neurocysticercosis in Texas, the history revealed that the dog had completed a 5-month stay in Mexico where the disease remains endemic in some areas some 4 months before the onset of signs.[13]

Larval *Taenia pisiformis*

There have been three reports, two from the United States, of dogs being infected with the cysticercal form of T. pisiformis that usually cycles between canids and rabbits.[14] In the two cases from the United States, the cysticerci were in the liver, and the hooks on the cysticerci were measured and matched in number per scolex and size those of T. pisiformis. In the third case, the cysticercus was found in the brain of the dog, and the hooks were not measured and the diagnosis presumptive at best.

Larval *Echinococcus multilocularis*

Infections with the alveolar hydatid of E. multilocularis usually occur in rodents with the adult forms occurring in foxes. It Europe there were reports of larval alveolar hydatid disease occurring on occasion in dogs; most of the reports were from Switzerland, but there were also reports from Germany and Belgium. There has now been a report of alveolar hydatid disease in a dog from British Columbia.[15] This case was treated surgically by the removal of the majority of the hepatic mass and therapy with long-term albendazole. The dog was followed for over 150 days, and although there were some side effects from the long-term albendazole therapy, it recovered.

PARASITES OF CATS

The larval tapeworms that have been reported from the cat include those of Spirometr*a*, *Mesocestoides*, *Taenia crassiceps*, *Taenia solium*, and *Taenia serialis*.

Larval *Spirometra*

Cats have been found infected with the larval form of Spirometra. In most cases cats had simple plerocercoid infections;.[16-18] although some cats in Japan had as many as 166 spargana in their tissues (inguinal and cervical regions, in fatty tissue around the kidneys, and on the chest and in the femoral regions of the body. A case of proliferative sparganosis has been reported from a cat with an abdominal mass.[19] In this case the spleen and liver were studded with numerous yellowish granules and there was a large node also on the serosal surface of the stomach.

Larval *Mesocestoides*

The tetrathyridia of *Mesocestoides* have been reported from the peritoneal cavity of cats, but they do not seem to develop the massive proliferative form described from dogs. In most cases they stay as single 1 to 8 cm long tetrathyridia in the peritoneal cavity.[20] In a report in Poland, of 100 cats that underwent abdominal surgery, 2 were found to have tetrathyridia within the abdominal cavity, one cat had 17 tetrathyridia and the other 36).[21] There is a report from more than 100 years ago, where it was stated that more than 1,000 larvae of Mesocestoides was removed from the abdominal cavity of a French cat,[22] but this is not accepted as verifiable by other workers.[23] In studies on the biology of *Mesocestoides*, some cats had very small forms without scolices that were considered to represent Mesocestoides.[24]

Larval *Taenia crassiceps*

Although as stated above, there have been numerous reports of larval T. crassiceps infections in dogs, there is only a single case in a domestic cat.[25] This case occurred in Minnesota and the cat presented due to a sudden onset of anorexia, ataxia, and lethargy along with increased incidents of vomiting. In spite of care and various treatments the cat progressed to circling, progressive ataxia, and weakness. The cat ultimate became comatose and was euthanized, and at necropsy, the left lateral ventricle was dilated by numerous cysticerci that were identified by hook size and shape as being T. crassiceps.

Larval *Taenia solium*

The have been very few reports of cats in Mexico being infected with larval *T. solium*; and a recent report has been made in South Africa.[26] In this case, the cat initially presented with ataxia. Three months later the cat had a relapse, and after being vaccinated and treated with a dewormer containing praziquantel. The cat then got progressively worse with convulsions and blindness, rigid forelimb, and hyper-reflexive hind limbs. The cat was euthanatized, and upon necropsy, there were numerous 0.5 cm cysticerci throughout the cerebral hemispheres.

Larval *Taenia serialis*

There have been a half-dozen or so cases of feline neurologic coenurosis due to *T. serialis*, with the most recent being described in a cat from the United Kingdom that had never traveled outside the country.[27] The cat presented with tetraparesis, the MRI showed two large intra-axial cystic structures, the cat recovered poorly from anesthesia, and euthanasia was deemed the appropriate course. Using DNA extraction and gene amplification, the cestode was identified as *T. serialis*.

CONCLUDING REMARKS

There have been a number of cases of larval tapeworms that have been highly fatal in dogs and cats. It is expected that with the marked improvements in imaging, that many of these animals will be saved in the future by surgical intervention.

REFERENCES

1. Koch A, Pfeil I. **Kleintierpraxis** 2012;57(5):245. 2. Simpson C, et al **Mol Cell Probes** 2012;26(1):60. 3. Ballweber LR **Vet Med** 2011;106(7):334. 4. Stief B, Enge A **Vet Pathol** 2011;48(4):911. 5. Drake DA et al **J Am Vet Med Assoc** 2008;233(11):1756. 6. Naveen K, et al **Vet Practitioner** 2004;5(1)55. 7. Bannerjee PS, et **al J Vet Parasitol** 2001;15(2):159. 8. Little S, Ambrose D **Comp Cont ed Pract Vet** 2000;22(4):299. 9. Beveridge I , et al **Austral Vet J** 1998;76(11):757. 10. Boyce WM, et al **Vet Parasitol** 2011;180(3/4):256. 11. Papini R, et al **Vet Parasitol** 2010;170(1/2):158. 12. Ballweber LR. **Vet Rec** 2009;165(23):693. 13. Buback JL, et al **J Am Vet Med Assoc** 1996;208(11):1846. 14. Ivens V, et al **Am J Vet Res** 1969;30(11):2017. 15. Peregrine As, et al **Can Vet J** 2012;53(8):870. 16. Tang CC. **Pek Natl Hist Bull** 1935;10:223. 17. Uga SGM, et al **Jap J Parasitol** 1986;35:153. 18. Schmidt RE, et al **J Small Anim Pract** 1968;9:551. 19. Buergelt CD, et al **J Parasitol** 1984;70(1):121. 20. Bowman DD, et al **Feline Clinical Parasitology**, 2002;Iowa State Univ Press. 21. Gundach JL, et al **Ann Univ Mariae Curie-Skod DD Med Vet**, 2004;59:117. 22. Neumann LG **Mém Soc Zool Fr** 1896;9:171. 23. Conn DB **J Parasitol** 1991;76:454. 24. Witenberg G **Arch Zool Ital** 1932;20:467. 25. Wünschmann A, et al **J Vet Diagn Invest** 2003;15:484. 26. Schwan EV, et al J S **Afr Vet Assoc** 2002;73(4)219. 27. Jull P, et al 2012 **J Fel Med Surg** 2012;14:646.

PROTEIN-LOSING ENTEROPATHY IN HUMANS
John K. DiBaise, MD, FACG
Scottsdale, Arizona

INTRODUCTION

In healthy individuals, loss of protein through the gut epithelium plays only a minor role in total protein metabolism; daily enteric loss of serum proteins accounts for approximately 1-2% of the entire serum protein pool and enteric loss of albumin accounts for less than 10% of total albumin (1). Indeed, most endogenous proteins found within the gut are from secretions and sloughed enterocytes. Protein-losing enteropathy (PLE) is characterized by a loss of serum protein into the gastrointestinal (GI) tract resulting in hypoproteinemia, which can be complicated by edema, ascites, pleural and pericardial effusions and malnutrition. GI protein loss in PLE has been reported to involve up to 60% of the total albumin pool. This excessive protein loss across the gut epithelium can be due to either mucosal injury, resulting in increased mucosal permeability, or to lymphatic obstruction resulting in direct leakage of protein-rich lymph. The serum protein levels most affected by this process are those with limited ability to rapidly respond to such losses and generally have longer half-lives such as albumin, most immunoglobulins and ceruloplasmin. Hepatic protein synthesis is maintained or slightly increased in patients with PLE and this is reflected by normal or increased levels of rapid-turnover proteins such as prealbumin, also known as transthyretin (2). Lower serum concentrations of other substances such as lipids, iron and other trace elements are also occasionally encountered in PLE as is lymphopenia, particularly in the setting of lymphatic obstruction.

Due to a lack of systematic screening and a wide variety of causes of hypoalbuminemia, the prevalence of this condition is poorly understood. Although thought to be rare, there are many causes of PLE (3). The primary causes, however, can be divided into erosive gastrointestinal disorders, nonerosive gastrointestinal disorders and disorders involving increased central venous pressure or mesenteric lymphatic obstruction (Table). In this presentation, cases illustrating a variety of clinical presentations and etiologies of PLE occurring in humans will be presented and the diagnostic approach and treatment of this clinically challenging condition will be reviewed.

CARDIAC ETIOLOGIES OF PLE

PLE has also been described in the setting of a number of cardiac conditions, including congestive heart failure, constrictive pericarditis, congenital heart disease and valvular disease. The pathophysiology of congestive heart failure-associated PLE appears to be secondary to lymphatic obstruction and rupture of lacteals in the gut due to an increase in central venous pressure (4). The presence of PLE does not appear to influence the treatment or prognosis of the heart failure (5). Although rare, one of the most frequently described cardiac causes of PLE occurs in the patient who has undergone the Fontan procedure, usually in the remote past. The Fontan procedure is performed in patients born with a functional single ventricle and marked underdevelopment of the left side of the heart and aims to restructure the heart so that the right side can pump enough blood to supply the body (6,7). PLE is said to occur in approximately 3% to 15% of patients after the Fontan operation and carries a high risk of morbidity and mortality (6,8). The pathophysiology of post-Fontan PLE remains incompletely understood.

GASTROINTESTINAL ETIOLOGIES OF PLE

There are two categories of gastrointestinal diseases that lead to PLE – those with and those without mucosal erosions. Even without mucosal erosions, conditions that lead to surface epithelial cell loss or malabsorption can cause excess protein loss. The most common non-erosive gastrointestinal condition that may be complicated by the development of PLE is Menetrier's disease, also known as giant hypertrophic gastropathy. This is a rare, acquired disorder characterized grossly by giant rugal folds in the stomach and histologically by foveolar hyperplasia, cystic dilation/pits and reduced numbers of parietal and chief cells. An increase in intracellular permeability and wider tight junctions between cells leads to protein loss and the presence of protein-rich exudates in the stomach (9). Recent evidence implicates heparin sulfate and increased signaling of epidermal growth factor receptor (EGFR) by transforming growth factor-alpha in the pathogenesis (10-12). Patients with Menetrier's commonly present with symptoms of abdominal pain, nausea, vomiting and weight loss. Menetrier's may be suspected during upper endoscopy because of the finding of abnormally large gastric folds although this finding is not specific. Endoscopic mucosal resection or full-thickness biopsy may be needed to confirm the diagnosis. Other non-erosive gastrointestinal processes that have been reported to be complicated by PLE include eosinophilic gastroenteritis, microscopic colitis, small intestinal bacterial overgrowth, celiac disease and *H. pylori* infection.

Both ulcerative colitis and Crohn's disease are examples of erosive gastrointestinal disease leading to PLE. The mechanism of protein loss in these conditions is thought to relate to enhanced leakage of protein-rich fluids across the eroded epithelium. The degree of mucosal involvement typically correlates with the severity of protein loss (13). Indeed, gastrointestinal protein loss has been suggested to be a marker of the severity of activity in IBD (14). Granulomatous involvement of the lymph nodes in Crohn's disease has also been demonstrated to lead to dilation of the lymphatics with leakage of lymph across the epithelium, thus, resulting in additional protein loss.

Finally, while most PLE cases due to lymphatic obstruction or increased lymphatic pressure are the result of cardiac etiologies, a rare disorder of the gut, primary intestinal lymphangiectasia (Waldmann's disease), may also be responsible. This disorder is characterized by the presence of tortuous, dilated lacteals in the mucosa and submucosa, which result in loss of protein-rich lymph into the small bowel lumen. Patients with this disorder usually present in childhood with edema, weight loss, diarrhea and lymphopenia.

CLINICAL FEATURES AND DIAGNOSIS OF PLE

The clinical manifestations of PLE are highly variable, being related to the underlying cause, but tend to mainly consist of signs and symptoms of fluid retention such as edema, pleural effusion and ascites. Diarrhea and other gastrointestinal symptoms are frequently not present. The main laboratory findings are reduced serum concentrations of albumin, total protein, gamma globulins, fibrinogen, transferrin, and ceruloplasmin. While the hypoalbuminemia contributes to the primary clinical manifestation of edema due to diminished plasma oncotic pressure, the reductions of other serum proteins rarely cause clinically significant problems. Of note, PLE may also be associated with fat malabsorption and fat-soluble vitamin deficiencies caused by the primary disease.

The initial step in the evaluation of the patient with hypoproteinemia and/or hypoalbuminemia is to exclude other more common etiologies such as malnutrition, liver and renal diseases. Importantly, malnutrition may also occur as a feature of PLE, further complicating the diagnosis. When no other cause has been identified and there is concern of possible PLE, at present, the most commonly used method to determine enteric protein loss is to determine the clearance of alpha-1 antitrypsin (A1AT) from plasma. A1AT is a protein synthesized in the liver that is neither actively secreted nor absorbed, has a molecular weight similar to albumin, and does not undergo proteolysis or degradation in the gut, thereby allowing for its intact elimination and detection in feces. The measurement of A1AT clearance requires both a blood sample to determine plasma A1AT level and a 24-hour stool collection to determine stool volume and stool A1AT level. The analysis of a random spot fecal A1AT concentration is not a reliable substitute for measurement of A1AT clearance (15). Normal A1AT clearance is $\leq$ 27 mL/24 hours. Diarrhea from any cause can increase the clearance of A1AT. Therefore, in the setting of diarrhea, the normal A1AT clearance increases to $\leq$ 56 mL/24 hours. In addition, this test does not distinguish between intestinal and gastric sources of protein loss. Because A1AT is degraded in environments with a pH $<$ 3.5, if a gastric source of protein loss is suspected or an acid hypersecretory state is known to be present, it is recommended that the test be performed while the individual is receiving acid suppressive therapy (16).

Nuclear radiology testing has also been used to quantitate enteric protein loss. The use of technetium-99m-labeled albumin to scintigraphically document protein loss in the gut has been shown to be useful not only for diagnosis but also in monitoring response to treatment (17,18). Magnetic resonance imaging has also been studied as a diagnostic test for PLE, primarily in children with lymphangiectasia, and was able to detect abnormalities in both mesenteric and intestinal edema and document the presence of dilated/prominent lymphatics. It was suggested that these findings in the setting of an appropriate clinical presentation could be used to support the diagnosis of PLE without requiring invasive testing (19).

TREATMENT OF PLE

Following confirmation of PLE, an evaluation is necessary to identify the underlying cause. As there is no specific treatment of PLE, its treatment should be directed at the underlying condition; however, careful monitoring and treatment of malnutrition and micronutrient deficiencies if present is also necessary. Fortunately, most causes of PLE can be readily diagnosed and treated. Treatment options may include dietary, pharmacological or surgical intervention, or a combination of these.

PLE associated with cardiac diseases can generally be improved with medical and surgical management of the underlying cardiac condition. Similarly, protein loss from the stomach, small bowel or colon should be treated according to the underlying disease present. For example, in cases of hypertrophic gastropathy, evidence of an infection with *H. pylori* or cytomegalovirus (CMV) should be sought and treated if present. Investigational treatment with cetuximab or gastrectomy should be considered in cases of Menetrier's disease (11). Acquired intestinal lymphangiectasia should be treated by correction of the underlying cause (20). Small intestinal bacterial overgrowth, Whipple's disease and other infectious causes of PLE should be treated with appropriate antibiotic therapy, while cancer-induced PLE requires therapy directed against the cancer. Immunosuppressive therapies may be useful in the management of PLE caused by inflammatory diseases such as Crohn's disease, ulcerative colitis, collagenous colitis, gut graft-versus-host disease, or systemic lupus erythematosus (21). Surgery may be needed in refractory inflammatory cases. Despite a lack of quality evidence supporting their efficacy in PLE, anecdotal reports suggest that other therapies such as intravenous heparin (22) and octreotide (23) may be useful to decrease fluid secretion and protein exudation from the bowel. The use of intermittent intravenous infusions of albumin is generally unhelpful over the long-term.

A high-protein diet is recommended. Normal protein requirements are generally 0.6 to 0.8 g/kg/d. In PLE, this value may increase to 2.0 to 3.0 g/kg/d to achieve positive protein balance. High protein intake within this range can usually be achieved via changes in the diet and use of commercially-available protein supplements. The ability to correct hypoalbuminemia is dependent upon the type and severity of the underlying disease. In some circumstances, a low-fat diet may also be recommended usually along with medium-chain triglyceride supplementation to offset at least some of the reduction in calories caused by the decrease in fat. Congenital intestinal lymphangiectasia is one such condition that may benefit from such dietary changes. These dietary modifications may beneficially effect albumin metabolism and diminish the need for lymphatic transport of fatty acids, thus decreasing pressure on the lymphatic system and decreasing lymph leakage. While it is always preferable to use the gut for feeding, in certain conditions such as severe erosive disease or intestinal dysmotility, parenteral nutrition support may be necessary.

Finally, supportive care to avoid complications resulting from peripheral edema with support stockings, vigilant monitoring for skin breakdown and adequate ambulation to prevent deep vein thrombosis is recommended. In contrast, diuretics are not generally helpful as the edema relates to a reduction in plasma oncotic pressure.

CONCLUSION

Protein-losing enteropathy is a syndrome of gut protein loss that may complicate a variety of diseases, most commonly cardiac or gastrointestinal conditions. The diagnosis of PLE should be considered in patients with hypoproteinemia after other causes, such as malnutrition, proteinuria, and impaired protein synthesis due to cirrhosis have been excluded. Diagnosis is most commonly based upon the determination of fecal alpha-1 antitrypsin clearance. Both the prognosis and treatment of PLE depends upon the underlying condition responsible. Treatment also includes dietary modification, supportive care and maintenance of nutritional status.

REFERENCES

1. Schmidt PN, et al. Scand J Clin Lab Invest 1995;55:35.
2. Takeda H, et al. Hepato-gastroenterol 2003;50:1963.
3. Umar SB, DiBaise JK. Am J Gastroenterol 2009;105:43.
4. Chan FK, et al. Hepatogastroenterol 1999;46:1816.
5. Berkowitz I, Segal I. Am J Gastroenterol 1990;85:154.
6. Johnson JN, et al. Nutr Clin Pract 2012;27:375.
7. Ostrow AM, et al. Ann Thorac Surg 2006;82:695.
8. Silvilairat S, et al. Congenit Heart Dis 2008;3:262.
9. Toubia N, Schubert ML. Curr Treat Options Gastroenterol 2008;11:103.
10. Bode L, et al. J Biol Chem 2006;281:7809.
11. Settle SH, et al. Clin Gastroenterol Hepatol 2005;3:654.
12. Sferra TJ, et al. J Pediatr 1996;128:213.
13. Ferrante M, et al. Acta Gastro-enterol Belg 2006;69:384.
14. Becker K, et al. Hepato-gastroenterol 1999;46:2309.
15. Strygler B, et al. Gastroenterology 1990;99:1380.
16. Takeda H, et al. Dig Dis Sci 1999;44:2313.
17. Wang SJ, et al. Clin Nucl Med 2000;25:197.
18. Uzuner O, Ziessman HA. Pediatr Radiol 2008;38:1122.
19. Liu NF, et al. Lymphol 2008;41:111.
20. Vardy PA, et al. Pediatrics 1975;55:842.
21. Landzberg BR, Pochapin MB. Curr Treat Options Gastroenterol 2001;4:39.
22. Bendayan I, et al. Pediatr Cardiol 2000;21:267.
23. Lee HL, et al. J Korean Med Sci 2004;19:466.

TABLE. CAUSES OF PROTEIN-LOSING ENTEROPATHY

Erosive Gastrointestinal Disease	**Increased Interstitial Pressure**
Inflammatory bowel disease	Intestinal lymphangiectasia
Gut malignancy	Congestive heart failure
Nonsteroidal anti-inflammatory drug enteropathy	Constrictive pericarditis
Erosive gastropathy	Congenital heart disease
Acute graft versus host disease	Fontan procedure for single ventricle
Pseudomembranous enterocolitis	Portal hypertensive gastropathy
Ulcerative jejuno-ileitis	Hepatic venous outflow obstruction
Intestinal lymphoma	Enteric-lymphatic fistula
Sarcoidosis	Mesenteric venous thrombosis
Nonerosive Gastrointestinal Disease	Sclerosing mesenteritis
Celiac disease	Mesenteric tuberculosis or sarcoidosis

Hypertrophic gastropathies
Eosinophilic gastroenteritis
Lymphocytic gastritis
Connective tissue disorders
Small intestinal bacterial overgrowth
Amyloidosis
Microscopic colitis
Tropical sprue
Whipple's disease
Parasitic diseases
Viral gastroenteritis

Neoplasia involving mesenteric lymph nodes or lymphatics
Chronic pancreatitis with pseudocyst
Congenital malformation of lymphatics
Retroperitoneal fibrosis

DIAGNOSIS OF CANINE PROTEIN LOSING ENTEROPATHY

Steve L Hill, DVM, MS, DACVIM (SAIM)
San Diego, CA

INTRODUCTION

Protein losing enteropathy (PLE) is a common clinical problem that occurs when there is excessive loss of protein into the gastrointestinal (GI) lumen.[1] When loss of albumin into the GI lumen exceeds albumin synthesis by the liver, hypoalbuminemia results. Loss of protein across the gastrointestinal mucosa is independent of molecular weight, therefore panhypoproteinemia (hypoalbuminemia and hypoglobulinemia) is expected, although in some cases only hypoalbuminemia is observed. Serum albumin contributes approximately 80% to the colloid oncotic pressure. A clinically relevant decrease in colloid oncotic pressure results when serum albumin becomes <2 g/dL. Ascites, pleural effusion (pure transudate) and peripheral edema usually do not develop until serum albumin decreases to ≤ 1.5 g/dL. With a systematic diagnostic approach it can be determined if the cause of the hypoalbuminemia is due to a PLE, and the etiology of the PLE can subsequently be determined.

PLE is classified according to the main GI mucosal abnormality with: 1) mucosal disease without ulceration and increased epithelial permeability, resulting in leakage of protein into the intestinal lumen; 2) mucosal erosion/ulceration with exudation of protein into the intestinal lumen; and 3) lymphatic dysfunction resulting in leakage of protein-rich lymph into the intestinal lumen. The most common etiologies for PLE in dogs are idiopathic inflammatory bowel disease (IBD), intestinal lymphangiectasia (IL) that can be primary or secondary, and gastrointestinal neoplasia (especially lymphoma (LSA), but also adenocarcinoma).[1,2] Other etiologies for PLE include intestinal crypt disease, regional enteritis (granulomatous, eosinophilic), infectious diseases (histoplasmosis, pythiosis, parvoviral enteritis), endoparasitism (hookworms or whipworms), chronic intussusception, GI ulceration, GI blood loss and possibly antibiotic responsive enteropathy.[1,2] PLE is uncommon in cats and when it does occur the cause is most commonly IBD or LSA.

The clinical signs and severity of disease in dogs with PLE are variable and depend on the underlying disease, severity of the hypoalbuminemia, and the presence of complications. It is important to consider the signalment as several breeds, including the Yorkshire terrier, Soft-coated Wheaten terrier, German Shepherd dog, Norwegian Lundenhund, Rottweiler, and Basenji have all been reported to be predisposed to PLE.[1] In one study, Yorkshire terriers were found to be 10.1 times more likely to develop PLE when compared to other breeds.[3] A heritable form of PLE with or without concurrent protein-losing nephropathy (PLN) is well described in Soft-coated Wheaton terriers with 10% to 15% of this breed estimated to be affected with this syndrome.[4]

Dogs with chronic enteropathies resulting in PLE most commonly present with small bowel pattern diarrhea, intermittent vomiting, weight loss, and partial or complete anorexia. It is important to recognize that some dogs may not present with classic GI signs, and the first indication of PLE may be ascites, pleural effusion, or peripheral edema. In some cases hypoalbuminemia or panhypoproteinemia may be detected on routine laboratory analysis without obvious clinical signs. Dogs presenting on an emergency basis for respiratory distress due to pleural effusion or for thromboembolic complications are uncommon.

DIAGNOSTIC APPROACH

When a patient is suspected of having PLE, the first step is to rule out non-GI disorders as the cause of hypoalbuminemia or panhypoproteinemia. Non-GI causes of hypoalbuminemia are PLN, liver failure, exudative skin disease, pleuritis or peritonitis (third spacing of protein), vasculitis, and hypoadrenocorticism. Once these conditions are ruled out, PLE is diagnosed by exclusion.

Laboratory Tests

A CBC, serum biochemical profile, and urinalysis are performed for all cases. Although panhypoproteinemia is expected with PLE there are numerous non-GI and GI causes (infectious, inflammatory, and neoplastic disorders) that will increase globulin production, resulting in a normal or even an elevated serum globulin concentration. Other biomarkers suggestive of PLE include lymphopenia, hypocholesterolemia, hypocalcemia, hypomagnesemia and hypocobalaminemia. Yorkshire terriers with PLE are at increased risk for hypocalcemia with decreased ionized calcium concentration and hypomagnesemia with the pathogenesis involving intestinal loss, intestinal malabsorption, and abnormalities in vitamin D and parathyroid hormone metabolism.[3]

PLN can be eliminated from the differential list if the urinalysis is negative for protein. If positive for protein, a urine protein to creatinine ratio (UP-C) is determined. A UP-C is only requested if the urine sediment is inactive as lower urinary tract disease, especially a urinary tract infection, can result in an elevated UP-C. Glucocorticoid therapy can also result in proteinuria, although these patients are rarely hypoalbuminemic.

Liver failure must be ruled out as the cause of hypoalbuminemia. Dogs with PLE, as those with liver failure, may present with a decrease in BUN, hypocholesterolemia, and elevated liver enzymes. Pre- and post-prandial serum bile acids concentrations are a sensitive test of liver function and normal serum bile acid concentrations eliminate liver failure as the cause of the hypoalbuminemia in most cases. A reactive hepatopathy may occur in dogs with PLE, although serum bile acids are usually normal or only mildly elevated. Obtaining liver biopsies to differentiate primary liver disease from a reactive hepatopathy is rarely necessary.

A routine flotation for ova and parasites and *Giardia* antigen testing are performed to evaluate for intestinal parasitism as a cause of PLE. If hypoadrenocorticism is suspected a baseline cortisol concentration (a value >2 μg/dL rules out hypoadrenocorticism) or an ACTH stimulation test are obtained.

Serum cobalamin (vitamin B12) and folate are indicators of intestinal function and they should be measured in all chronic enteropathy and PLE patients. Cobalamin absorption is a complex process with receptor-mediated absorption of cobalamin complexed to intrinsic factor occurring in the ileum. Hypocobalaminemia is common with PLE, is a risk factor for a negative outcome, and may be an indicator of refractoriness to treatment.[5] Dietary folate is absorbed in the proximal small intestine by specific receptors in the brush border, and decreased absorption may occur with proximal small intestinal disease. Intestinal dysbiosis (formerly known as small intestinal bacterial overgrowth (SIBO)) may also result in hypocobalaminemia with an increased serum folate concentration, although this finding is neither sensitive nor specific for dysbiosis.

Fecal α1-proteinase inhibitor (α1-PI) is a test specifically for PLE.[6] α1-PI is a plasma protein that is approximately the same size as albumin. With PLE albumin is lost into the intestinal lumen along with α1-PI, but unlike albumin, α1-PI is not digested by intestinal proteases and it is passed in the feces where it can be measured. Measurement of fecal α1-PI is not necessary for most PLE cases. Fecal α1-PI is most helpful in patients without clinical signs of GI disease in which other causes of hypoalbuminemia have been ruled out. It can also be used to detect PLE in patients with concurrent PLN or liver failure. For more information contact the Gastrointestinal Laboratory at Texas A&M University (www.vetmed.tamu.edu/gilab).

Diagnostic Imaging

Abdominal ultrasonography is recommended for most patients with PLE. Ultrasonography allows for identification of focal or regional GI lesions and for assessment of the GI mucosa. Additionally, other abdominal organs (lymph nodes, pancreas, liver, and kidneys) may are evaluated. Mucosal changes, especially hyperechoic intestinal mucosal striations secondary to villous lacteal dilation, are suggestive of PLE. Hyperechoic mucosal striations have a sensitivity of 75% and a specificity of 96% for dogs with PLE.[7] In this report hyperechoic speckles were non-specific for diagnosing IBD. In a recent report evaluating dogs with chronic diarrhea, abdominal ultrasonography was vital or beneficial in obtaining a diagnosis or identifying other factors important in case management in only approximately 25% of the cases.[8] In this study ultrasonography was most beneficial in dogs with weight loss, a palpable abdominal or rectal mass, and those with a final diagnosis of GI neoplasia. Dogs with PLE were not specifically evaluated. Fine needle-aspiration or needle-biopsy of abdominal organs or masses and collection of peritoneal fluid can be performed with ultrasound guidance when indicated. Screening the thorax for pleural fluid is also easily performed with ultrasonography. When abdominal ultrasonography is negative or is consistent with diffuse intestinal disease, the next diagnostic step is collection of GI biopsies.

Histopathologic Evaluation

Hypoalbuminemia (albumin < 2 g/dL) is a risk-factor for a negative outcome in dogs with chronic enteropathies.[5] Obtaining a definitive diagnosis based on histopathologic evaluation of small intestinal biopsies is important with PLE, whereas dietary, antibiotic or anthelmintic trials may be appropriate with chronic enteropathies without PLE. Intestinal biopsy samples can be obtained endoscopically, surgically, or laparoscopically.

Flexible GI endoscopy to obtain intestinal mucosal biopsies is preferred in most cases. PLE is most commonly caused by diffuse mucosal disease (IBD or lymphangiectasia), which can usually be diagnosed with high quality endoscopically obtained mucosal biopsies. Advantages of GI endoscopy are decreased invasiveness, decreased patient risk and decreased expense, while having direct magnified visualization of the GI mucosa. It is imperative that the small intestine be examined and sampled. Ideally, lower GI endoscopy should also be performed to obtain ileal biopsies because in some cases lesions might be more readily detected in ileal biopsies.[9,10] Endoscopic visualization of the duodenal mucosa is reported to be predictive of IL in dogs and humans. Although the presence of white villous tips has been associated with IL they can be present with other intestinal diseases, such as IBD, intestinal neoplasia, infectious diseases, and even clinically healthy individuals A recent study concluded that the duodenal mucosal appearance alone lacks specificity and has only a moderate sensitivity for the diagnosis of IL.[11] When biomarkers of PLE were considered, the sensitivity was improved. However the poor specificity supports the need for histopathologic confirmation of IL. Disadvantages of flexible endoscopy include an inability to access the entire gastrointestinal tract and the potential for not obtaining adequate tissue samples (villous tips only) for a diagnosis. Muscularis mucosa and submucosa are not reliably obtained from biopsies collected with endoscopic forceps. Quality of endoscopic biopsies affects the interpretation with larger samples providing more tissue for evaluation. An additional consideration is the significant inter-individual variability between pathologists evaluating GI biopsies.[12] Using pathologists with expertise in the interpretation of GI biopsies and the use of gastrointestinal histopathology standards (modified World Small Animal Veterinary Association criteria) for consistency is optimal.

Surgical or laparoscopic full thickness GI biopsies may be preferable in some cases. Advantages include identification of submucosal lesions that may be present with IL or intestinal neoplasia, and obtaining biopsies of segmental lesions that might not be accessible endoscopically. Obtaining full thickness biopsies is more invasive, usually more expensive, and in severely hypoalbuminemic patients there may be greater risk for surgical complications.

REFERENCES

[1]Petersen PB, et al. **Vet Clin North Am Small Anim Pract** 2003;33(5):1061.
[2] Dossin O, et al. **Vet Clin North Am Small Anim Pract** 2011;41(2):399-418.
[3]Kimmel SE, et al. **J Am Vet Med Assoc** 2000;48(1):51.
[4]Littman MP, et al. **J Vet Intern Med** 2000;14(1):68.
[5]Allenspach K, et al. **J Vet Intern Med** 2007;21(4):700.
[6]Melgarejo T, et al. **Am J Vet Res** 1998;59(3):329.
[7]Gaschen L, et al. **Vet Radiol Ultrasound** 2008;49(1):56.
[8]Leib MS, et al. **J Vet Intern Med** 2012;26:1288.
[9]Casamian-Sorrosal D, et al. **J Vet Intern Med** 2010;24(1):80.
[10]Procoli, PF, et al. **J Vet Intern Med** 2013;27:268.
[11]Larson RN, et al. **J Vet Intern Med** 2012;26:1087.
[12]Willard MD. **J Am Vet Med Assoc** 2002;22(8):1177.

CRITICAL CARE ASPECTS OF CANINE PLE

Daniel L. Chan, DVM, DACVECC, DACVN, North Mymms, Hertfordshire, UK

INTRODUCTION

Protein-losing enteropathy (PLE) refers to chronic diseases affecting the intestines leading to excessive loss of protein and is typified by sometimes severe hypoproteinemia. Intestinal lymphangiectasia, lymphoma and inflammatory bowel disease are common etiologies leading to PLE.[1] From a critical care point of view, the main challenges encountered in managing dogs with PLE stem from the severe hypoproteinemia and the associated morbidity. Managing fluid imbalances, fluid accumulations (e.g., edema, effusions) present particular challenges, especially in dehydrated edematous patients. Significant electrolyte imbalances (e.g., hypocalcemia, hypomagnesemia) can also be encountered.[2] Additional problems and complications relate to their malnourished state, issues related to gastrointestinal dysfunction (e.g., malabsorption, dysmotility) and their hypercoagulable state. A discussion of the pathophysiology of these problems as we understand them and challenges in managing complications encountered are the focus points of this talk. Potential avenues of further research in managing complications in this patient population are also highlighted.

HYPOALBUMENIA-RELATED PROBLEMS

Many of the challenges encountered in the management of dogs with PLE center on hypoalbuminemia and its associated problems. Circulating albumin generates approximately 80% of the plasma colloid osmotic pressure (COP) and is largely believed to be the key component in maintaining intravascular fluid balance. With severe albumin losses, the plasma COP decreases and leaves the patient at increased risk for developing peripheral edema. Other problems following from decreased COP may include tissue edema (e.g., gastrointestinal wall) and cavitary effusions (e.g., ascites, pleural effusion). The development of tissue edema has been postulated to relate to impaired gas exchange in the lungs, increase tissue oxygen debt, increase risk of bacterial translocation and negatively impact on wound healing.[3] The risk for developing these complications is perhaps accentuated in circumstances where the patient requires on-going fluid therapy. Because patients with PLE frequently present with significant fluid deficits (e.g., vomiting, diarrhea, dehydration, hypovolemia), the need to administer fluid therapy is often met with hesitation because of concern for decreasing COP further and perpetuating extravasation of fluid into the interstitium (i.e., edema) and into cavities (e.g., pleural space, peritoneum). Although a decrease in COP alone is blamed for the development of edema, this is an oversimplification of the problem as other factors such as impairment of tissue safety factors, alterations of hormones that regulate sodium and water retention, and failure of lymphatic clearance (particularly important in patients with lymphangiectasia) are involved.[4] The use of crystalloid fluid therapy in such patients certainly will decrease COP even further, however, there is no evidence that doing so is contraindicated. The use of colloids solely or in combination with limited quantities of crystalloids has been advocated but it is uncertain whether such measures actually reduce the risk for edema formation. Whilst the use of colloids will result in greater fluid retention within the intravascular compartment, it will also increase intravascular hydrostatic pressure and in the context of "leaky" endothelial surfaces, which may still promote efflux of fluid into interstitium and body cavities. One of the less understood functions of serum albumin appear to relate to its protective influence on capillary membrane integrity.[5] Replacing fluid deficits of a patient with colloid alone is not consistent with current recommendations for fluid therapy but there is very little data on this particular patient population so this remains an unanswered question. Correction or supplementation of serum albumin with fresh frozen plasma is considered futile and cost prohibitive. To increase serum albumin by 1.0 g/dL in at-risk patients (i.e., serum albumin < 1.5 g/dL), an approximate dose of 1.5 g/kg of albumin is required. To put this into a clinical context, a typical unit (i.e., 400 mL) of FFP only contains about 7 g of albumin. Therefore, the use of FFP to correct hypoalbuminemia in a patient that has ongoing albumin losses is considered impractical and wasteful. With the increased use of human concentrated albumin as a means to increase COP and albumin concentration in dogs,[6] the ability to correct albumin deficits became a possibility. Concentrated human albumin clearly is more effective at increasing COP from the typical 10-12 mm Hg seen in severely hypoalbuminemic animals closer to the normal of 18 mm Hg. In contrast, the use of synthetic colloids only modestly increase COP by 2-4 mm Hg.[7] Because of the myriad of functions attributed to albumin, namely its role at maintaining COP, binding and transporting hormones and drugs, serving as an antioxidant does make correction of albumin concentrations an attractive target for therapy. However, there are potential serious sequelae of administering human albumin including anaphylaxis, anaphylactoid reactions and delayed hypersensitivity reactions, and interest in using human albumin in dogs has perhaps waned.[8] Anecdotally, the rapid correction of COP has been reported to improve peripheral and intestinal edema and improve clinical response. However, proper investigations have not been carried out in animals. With the advent of canine specific albumin as a commercial product, the possibility of correcting severe albumin deficits in dogs exists, however, this has only been reported in the context of septic peritonitis and availability of the product is uncertain.[9] Equally controversial are recommendations to use synthetic colloids as the main type of fluid to be used in these patients. Synthetic colloids such as hetastarches, pentastarches and now tetrastarches have the advantage of having higher COP than crystalloids and therefore would appear to be better choices compared with crystalloids. However, the use of synthetic colloids do not translate in better patient outcomes. Even more problematic is that the now most common type of synthetic colloid is tetrastarch, which was developed as a ‘safer” colloid compared with hetastarches. However, this increased “safety” profile is related to decreased size of the average colloid molecule, which although it increases the COP of the solution (37 mm Hg in Voluven vs. 32 mm Hg hetastarch),[10] the intravascular half-life is much shorter so that the increase in plasma COP is short-lived and unlikely to change the patient’s COP significantly or for a clinical relevant period of time. Given, these issues, the optimal way to manage fluid deficits in dogs with PLE remains unresolved.

Closely related to problems with tissue edema, is the frequent finding of cavitary effusions, which can comprise breathing. Pleural effusion presents obvious compromise of pulmonary function, but excessive ascites can also compromise respiratory function by increased pressure against the diaphragm. Drainage of fluid accumulations is certainly more effective at alleviating respiratory compromise than administration of diuretics and attempts to correct COP. However, it is important to account for large volume drainage from body cavities as this will lead to overall fluid deficits and could even elicit hypovolemia. Ascites is also believed to contribute to intestinal motility problems, that can manifest as ileus, abdominal discomfort, vomiting and regurgitation. Compounding this problem is the prospect that bowel edema, also related to hypoalbuminemia, exacerbates gastrointestinal dysfunction. There are reports that the use of concentrated human albumin (which has a COP in excess of 200 mm Hg) can help alleviate these signs, however, these are only anecdotal reports.

MALNUTRITION

The excessive protein loss, the chronic inflammatory state and gastrointestinal dysfunction seen in dogs with PLE result in pronounced malnutrition. Lean body wasting is a major feature of this disease and the major contributor to the debilitated state in some patients. Because of disrupted gastrointerstinal function and potentially reduced absorptive ability, patients become both energy- and protein-malnourished. This is also closely related to quality of life aspects of the disease and may lead to euthanasia. Dietary recommendations to use ultralow fat diets, reduced protein (as a way to address possible IBD or protein hypersensitivity) certainly does not help the malnourished state in the short term unless this helps achieve remission of the disease. Potential serious sequelae of malnutrition includes poor wound and surgical healing and increased risk of infectious complications such as pneumonia. Patients euthanized for PLE often have developed complications such as pneumonia, dehiscence of surgical sites. The optimal nutritional management of patients with PLE remains speculative and is largely based on empirical experience and expert opinion.

HYPERCOAGULABILITY AND RISK OF THROMBOEMBOLISM

The excessive loss of plasma proteins seen in PLE was thought not to confer a hypercoagulable state as is seen with protein losing nephropathies because of the belief that anticoagulant proteins such as antithrombin (AT) was lost in equal proportion to pro-coagulant factors. This notion was supported by the fact that reports of thromboembolic disease in animals had greater number of animals with protein losing nephropathy. However, reports of thromboembolic complications in animals with protein losing enteropathies can be found in the literature and was the recent subject of one of our studies, which we identified a hypercoagulable state in dogs with PLE via various methods of coagulation assessment including via thromboelastography (TEG) and determination of AT activity.[11] The TEG profile of dogs with PLE indicated shortening of R time, decreased K time, increase angle and increase MA. This suggests that enzymatic hypercoagulability is a major facet of the hypercoagulability seen in dogs with PLE. We also found that decreased AT activity was not consistent in this patient population and that it was not correlated with the hypercoagulability demonstrated by TEG, indicating that the hypercoagulable state in these dogs cannot be explained by a deficiency in AT alone.[11] Management or modulation of this hypercoagulability is an area as yet unexplored. Recommendations for the use of thromboprophylaxis in the forms of anti-platelet drugs such as aspirin, is based on the perceived safety profile rather than on any evidence. Since the major concern revolves the development of pulmonary thromboembolism (PTE),[12] the reliance on anti-platelet drugs to reduce risk may be questionable since venous thrombi are typically platelet-poor. Based on the nature of venous clots, anticoagulation with heparins or Coumadin would seem more appropriate but this remains to be determined. Thus far, no studies have been conducted in dogs to better characterize the actual incidence, and risk of thromboembolic complications and no interventions (e.g., aspirin, clopidogrel, heparin) have been tested to determine risk reduction. Nevertheless, there could be a role for the use of platelet-blockers as albumin is known to have an anti-thrombotic influence on platelets[5] and therefore there may be scope of its use in reducing risk in certain patients. However, the incidence of arterial thrombosis (typically attributed to platelet-rich clots) in dogs with PLE is very low. However, assessment of platelet hyperactivity in dogs with PLE should be investigated.

CONCLUSIONS

Although PLE is considered a chronic enteropathy in dogs, there are various aspects of this disorder that present particular challenges to clinicians and may lead to serious complications related to severe hypoalbuminemia and hypercoagulability. From a critical care point of view, many of the interventions suggested for dealing with complications in this patient population require careful consideration and further evaluation as solutions are not as clear cut as once suggested.

REFERENCES

[1]Allenspach K, et al; JVIM, 2007;21:700; [2]. Kimmel SE, et al. JAVMA 2000;217(5):703; [3]Soreide E, et al. Injury 2005;36(9):1001; [4]Taylor AE. Lymphology 1990;23(3):111; [5] Neel DR et al. Best Pract Res Clin Anaesthesiol 2011;25(3):395; [6] Trow AV, et al. JAVMA 2008;233(4):607;[7] Moore Le, et al. JVIM; 1996;10:300; [8] Cohn LA, et al. AJVR 2007;68:657; [9] Craft EM et al. JVECC 2012;22(6):631; [10]Chan DL. Vet Clin No Am Small Anim Pract 2008; [11]Goodwin LV, et al. JVIM 2011;25:273; [12]Goggs RA, et al JVECC 2009;19(1):30

MANAGEMENT OF CANINE PROTEIN LOSING ENTEROPATHY: INTERNIST PERSPECTIVE

Steve L Hill, DVM, MS, DACVIM (SAIM)
San Diego, CA

Protein losing enteropathy (PLE) is a common clinical problem that occurs when there is excessive loss of protein into the gastrointestinal (GI) lumen.[1] The most common etiologies for PLE are inflammatory bowel disease (IBD), primary or secondary intestinal lymphangiectasia (IL) and gastrointestinal lymphoma (LSA).[1,2] PLE occurs infrequently in cats and is most commonly due to IBD or LSA. PLE patients are often severely affected clinically, which poses a variety of management concerns. Successful management of PLE requires nutritional support, treatment of the primary intestinal disease, and in many cases oncotic support and addressing complications.

Nutritional Support

Dogs with PLE experience negative energy and protein balance and they have specific nutritional needs. The nutritional recommendations for most dogs with PLE are below 10% to 15% fat, above 25 to 30% protein, and less than 5% crude fiber on a dry matter basis.[3] Hill's Prescription Diet® i/d® Low Fat GI Restore (canned: protein: 25.1%, fat: 8.5%, carbohydrates: 58.3%, fiber: 2.3%, 351 kcal/per 13 oz can; dry: protein 25.9%, fat: 7.4%, carbohydrate: 59.3%, fiber: 1.7%, 331 kcal/per 8 oz cup) is palatable and is not restricted in calories. Another good choice is Royal Canin® Gastrointestinal™ Low Fat™. A palatable ultra-low fat and high protein diet can also be home prepared by mixing 4 oz non-fat cottage cheese, 2 cups of rice or potatoes, a 6 oz can of tuna in water, and 2 tablespoons of non-fat powdered milk. This amount is fed per 25 lbs of body weight per day with a daily multiple vitamin and mineral supplement. I find these to be good initial diets for canine PLE patients and also good long-term diets for IL patients.

The most common diagnosis in dogs with PLE is IBD, with or without secondary IL. Increased mucosal permeability may be an important factor perpetuating inflammation in IBD patients and dietary proteins are among the most immunogenic antigens to which the gut is exposed to.[1] When diagnosed with IBD, a hypoallergenic diet (novel single protein source or hydrolyzed protein) with as low a fat content as possible is indicated. Commonly used novel protein source diets are Royal Canin® Hypoallergenic Selected Protein diets (with venison, rabbit, whitefish and duck available) and Hill's Prescription Diet® d/d (with venison, duck, and salmon available). Hydrolyzed protein diets contain low molecular weight peptides that are highly digestible and have a low antigenic potential. Commonly used hydrolyzed diets are Royal Canin® Hypoallergenic HP (dry and canned), Hill's Prescription Diet® z/d® ULTRA Allergen-Free (dry and canned), and Purina HA Hypoallergenic® (dry only).

In more severely affected patients an elemental diet may be necessary. Elemental diets are easily absorbed by the diseased intestine, cause less mucosal inflammation, and may attenuate fluid and albumin loss into the intestinal lumen.[1] Vivonex® T.E.N. (www.nestle-nutrition.com/products/fam) is a low fat readily available elemental diet, which is 1 kcal/ml and can be tube fed or given orally. In severely hypoalbuminemic patients, Vivonex® can be supplemented with 8.5% Aminosyn™ with electrolytes.[1] Patients with intractable vomiting or diarrhea may require total parenteral nutrition.

Oncotic Support

In patients with severe hypoalbuminemia, short-term oncotic support is important, especially prior to anesthesia for obtaining intestinal biopsies. Hydroxyethyl starch (HES; Hetastarch®, VetStarch™), plasma, 25% human albumin, or canine albumin (www.abrint.net) are options to increase colloid oncotic pressure. HES is administered at 20-30 mL/kg/day. The limiting factor of plasma transfusions is the large volume required to increase the serum albumin concentration. It takes approximately 45 ml/kg of plasma to increase serum albumin concentration by 1 g/dL, which increases the risk for volume overload. Concentrated human albumin will effectively raise serum albumin with a low volume infusion (2.5-5 ml/kg or 0.625-1.25 g/kg). Severe and rarely fatal adverse reactions have been reported in dogs receiving human albumin.[4] Canine albumin has recently become available and may be a good option with a lower potential for allergic reactions, although, there is limited information available for this product at this time. It is important to remember that treatment of the primary intestinal disease is necessary to prevent continued protein loss and maintain oncotic pressure long-term.

Addressing Complications

PLE patients are occasionally presented for respiratory distress due to pleural effusion and require therapeutic thoracocentesis. Diuretics are not indicated for transudative effusions due to hypoalbuminemia. Colloid support and treatment of the primary disease condition are the only effective ways of preventing effusion.

Hypocobalaminemia is common in dogs with PLE.[5] Cobalamin (vitamin B12) functions as a co-factor for many processes involving cellular turnover, and hypocobalaminemia can result in villous atrophy and reduced GI function. Low serum cobalamin (<200 ng/L) has been associated with a negative outcome in dogs with chronic enteropathies.[5] Additionally, there was a highly significant association between hypoalbuminemia and hypocobalaminemia.[5] With cobalamin deficiency resulting from intestinal malabsorption, cobalamin needs to be supplemented parenterally (www.vetmed.tamu.edu/gilab). Proximal small intestinal disease can also result in a low serum folate concentration. Whether low folate concentration is only a marker of proximal small intestinal malabsorption, or a contributor to disease has not been determined and the role of folic acid supplementation has not been fully established.

Dogs with PLE are commonly hypercoagulable and are considered at risk for thromboembolic complications. The prevalence and pathogenesis of hypercoagulability in dogs with PLE is not completely understood. Hypercoagulability due to PLE is considered to be multifactorial, cannot be solely attributed to loss of antithrombin, and has been shown to persist even after treatment of the PLE.[6] The prevalence of clinically evident TE disease is low, and it is reasonable to conclude that *in-vitro* hypercoagulability does not inevitably lead to a TE event in cases with PLE. The role of thromboprophylaxis with PLE is unknown. In severely affected cases or when thrombosis is documented, anticoagulation (enoxaparin at 1 mg/kg SQ q 8-12 hrs) and/or antithrombotic therapy (low-dose aspirin at 0.5-1.0 mg/kg PO q 24 hrs or clopidogrel at 1-3 mg/kg PO q 24 hrs) is indicated.

With severe hypocalcemia or hypomagnesemia, intravenous supplementation of calcium or magnesium, respectively is required. If hypocalcemia persists, there may be intestinal malabsorption, and treatment with parenteral vitamin D (calcitriol) may be necessary. Oral calcium and magnesium supplementation may also be necessary.

Treatment the Underlying Gastrointestinal Disease

Successful long-term management of PLE requires treating the primary disease based on intestinal histopathology. IBD with or without secondary IL, is the most common diagnosis in dogs with PLE. Although the pathogenesis of IBD is not completely understood, evidence suggests that genetic, immune, inflammatory, and environmental factors (principally the intestinal microbiota and dietary antigens) play a role in its development. In addition to dietary management of IBD, treatment is aimed at modifying the pathogenic mechanisms.

With IBD, anti-inflammatory and/or immunosuppressive therapy is indicated to reduce mucosal inflammation. Prednisone is most commonly prescribed at a dose of 1-2 mg/kg/day PO for 14 to 21 days, then at a tapering dosage by 25 to 50% at 2 to 3 week intervals, based on the clinical response. With PLE most cases are started on a second immunosuppressive drug (azathioprine 2 mg/kg PO q 24 hrs or 50 mg/m^2 PO q 24 hrs for 14 to 21 days, then reduced to q48h or modified cyclosporine at 5 mg/kg PO q 12-24 hrs) is started in combination with glucocorticoids. Modified cyclosporine has been reported as an effective treatment for steroid refractory IBD when administered at a dosage of 5 mg/kg PO q 24 hrs.[7] Recently, the combination of chlorambucil and prednisolone versus azathioprine and prednisolone for the treatment of PLE was evaluated in a retrospective study.[8] The chlorambucil dosage was 4-6 mg/m^2 q 24 h for 7-14 days, which was then tapered based on response to treatment. For dogs in the chlorambucil treatment group serum albumin and weight gain were significantly greater, and there was a survival advantage. Budesonide, structurally related to 16-hydroxyprednisolone, is an alternate glucocorticoid that can be used for the treatment of IBD. Budesonide has high topical anti-inflammatory activity and low systemic activity because of its high affinity for intestinal steroid receptors and its high first pass effect. Budesonide is dosed at 3-4.5 mg/m^2 PO q 24 hrs. Budesonide is more commonly used for refractory IBD or when there are concurrent conditions such as diabetes mellitus or hyperadrenocorticism making the treatment with systemic glucocorticoids less desirable.

Modulation of the microbiota with antibiotics and/or probiotics may decrease inflammation by reducing harmful immune responses to enteric bacteria. Dysbiosis (formerly called SIBO) occurs commonly with chronic enteropathies, and dysbiosis has been reported as a cause for PLE in human medicine.[9] Metronidazole and tylosin are commonly prescribed to address dysbiosis. Modulation of the intestinal microbiota to a more favorable composition has been suggested as an approach to managing IBD in humans. Probiotics have emerged as an innovative therapeutic concept since they target mucosa-associated bacteria and the harmful immune responses that they evoke.[10] A veterinary clinical trial evaluating a probiotic for the treatment of IBD is currently ongoing.

Octreotide has been successfully used to treat IL in humans, but its use has not yet been reported in the veterinary literature. Anecdotal reports suggest that octreotide may be beneficial in some dogs with PLE, but further studies are needed before its routine use in dogs with PLE can be recommended. The mechanism of action involves reduction of intestinal blood flow, inhibition of triglyceride absorption, and reduction of lymph flow.

Canine gastrointestinal LSA is a severe disease that is poorly responsive to treatment and carries a poor to grave prognosis.[11] Feline lymphocytic gastrointestinal LSA has a much better prognosis with chemotherapy (combination therapy with prednisolone and chlorambucil) than feline lymphoblastic gastrointestinal LSA.[12] Gastrointestinal lymphocytic LSA is an uncommon diagnosis in dogs. The author recently treated a dog with intestinal lymphocytic LSA (confirmed histologically and with PCR for antigen receptor rearrangements) with PLE for 29 months with a combination of prednisone and chlorambucil. The patient was euthanized due to disseminated high grade lymphoblastic LSA, which was histologically confirmed.

Prognosis

Management decisions are based primarily on the clinical response to treatment and normalization of the serum albumin concentration, which reflects successful treatment of the underlying intestinal disease. Initially, rechecks at 1-2 week intervals are recommended for monitoring. Treatment responses are variable and relapses are common. Therefore, the prognosis is generally guarded, although there is a subset of patients that will respond very well to treatment. With chronic enteropathies, hypoalbuminemia, hypocobalaminemia and ascites are reported risk factors for a negative outcome.[7, 13] The Canine Chronic Enteropathy Clinical Activity Index (CCECAI) includes hypoalbuminemia, ascites, and peripheral edema in the scoring and has been shown to be 91% sensitive and 83% specific with a cutoff value of 12 for predicting refractoriness to treatment and euthanasia within three years after diagnosis.[7] A recent study showed a statistically significant correlation between clinical severity of disease (CCECAI score) and the presence of hypoalbuminemia, hypocobalaminemia or ascites.[14]

REFERENCES

[1]Petersen PB, et al. **Vet Clin North Am Small Anim Pract** 2003;33(5):1061.
[2]Dossin O, et al. **Vet Clin North Am Small Anim Pract** 2011;41(2):399-418.
[3]Davenport DJ, et al. In Hand MS, et al, editors. **Small Animal Clinical Nutrition 5th edition** 2010:1077.
[4]Mathews KA. **Vet Clin North Am Small Anim Pract** 2008;38(3):595.
[5]Allenspach K, et al. **J Vet Intern Med** 2007;21(4):700.
[6]Goodwin LV, et al. **J Vet Intern Med** 2011;25:273.
[7]Allenspach K, et al. **J Vet Intern Med** 2006;20(2):239.
[8]Dandrieux RS, et al. **J Am Vet Med** In press 2013..
[9]Su, et al. **Dig Dis Sci** 1998;43(3):1998.
[10]Sartor, RB. **Gastroenterology** 2004;126(6):1620.
[11]Frank JD, et al. **J Am Anim Hosp Assoc** 2007;43:313.
[12]Fondaraco JV, et al. **Eur J Comp Gastroentero** 1999;4(2):5.
[13]Craven M, et al. **J Small Anim Pract** 2004;45(7):336.
[14]Procoli, PF, et al. **J Vet Intern Med** 2013;27:268.

BEST OF THE RECENT LITERATURE ON PROTEIN LOSING ENTEROPATHIES

Julien R. S. Dandrieux, Dr. med. vet., DACVIM (SAIM)
Wetherby, UK

INTRODUCTION

Protein losing enteropathy (PLE) is characterized by loss of proteins through the intestinal tract. Primary intestinal disease causing PLE mainly consists of lymphangiectesia, inflammatory bowel disease, crypt lesions and intestinal mucosal ulcerative damage. Gastro-intestinal disease accompanied by hypoalbuminemia has been shown to have a guarded prognosis[1,2], but there is little published information on survival or optimal treatment. Furthermore, although many studies have been published on inflammatory bowel disease, including subsets of dogs suffering from PLE, there is very little literature available on large groups of dogs with PLE.

For these reasons, the most recent literature available on protein losing enteropathies in dogs will be reviewed, but recently published abstracts will also be included.

BREEDS

Several breeds have been described to be at increased risk for PLE including the Norwegian Lundenhund, Yorkshire terrier, Shar pei, Basenji, Soft-coated Wheaten terrier, and German Shepherd dog[3]. Recently, changes found in a large number of Rottweilers (18 cases) have also been described[4].

DIAGNOSIS

The diagnosis of PLE is achieved by ruling out other causes of protein loss. Histology is also needed to characterize the underlying pathology. Endoscopy is often used to assess the intestinal mucosa and to obtain biopsies. The importance of assessing not only the mucosal appearance, but also microscopic changes has been emphasized in a recent publication, which showed that the sensitivity of endoscopy alone to diagnose lymphangiectesia was only 68%, with a specificity of 42%[5]. The number of biopsies needed depends on the quality of the samples as well as on the type of lesions[6]. Crypt lesions require the highest number of biopsies to diagnose (up to 13 good quality samples are needed to reach 99% confidence of detection). It is very important to get enough good quality biopsies to be able to achieve a diagnosis, especially when assessing for crypt lesions, which have been found in up to 68% of the cases in one publication.[7]

BIOLOGICAL MARKERS

Biological markers can be useful to confirm a diagnosis, can be used as early indicator of disease (prior to overt clinical signs), assist in determining prognosis and help assess response to treatment. Perinuclear anti-neutrophilic cytoplasmic antibodies (pANCA) have been examined in Soft-coated Weathen terriers[8]. Most dogs affected with PLE or protein losing nephropathy in this study developed positive pANCA before clinicopathologic changes. Hence, this marker has potential as an early test in this breed. In an abstract on 15 dogs from different breeds with chronic enteropathies (including 7 dogs with PLE), no significant association between pANCA and canine inflammatory bowel disease activity index (CIBDAI) score was found[9]. However, in the same study, fecal canine calprotectin (cCP) had a significant association with pre-treatment CIBDAI and there was a significant decrease in fecal cCP and serum C-reactive protein following successful treatment.

A recent study further assessed serum cCP in dogs with IBD[10]. Levels were higher in dogs with chronic diarrhoea compared to control dogs. No decrease was observed within 3 weeks of treatment despite clinical improvement and cCP levels were not correlated with C-reactive protein or severity of histopathologic changes. Serum alpha1-proteinase inhibitor was found to be significantly decreased in dogs suffering from PLE compared to food responsive diarrhoea (FRD)[11], which might be useful in determining which dogs are more likely to respond to a food trial. Other markers, which have been shown to be significantly increased in dogs suffering from chronic enteropathies include fecal cCP[12], S100A12[13], and urinary leukotriene E4[14]. None of these markers have been specifically looked at in dogs with PLE.

SURVIVAL

Although hypoalbuminemia has been recognized as a negative prognostic factor in dogs with chronic enteropathies, there are very few published studies looking specifically at survival of dogs suffering from PLE. In one study looking on hypercoagulability in dogs of different breeds with PLE, 66% of dogs died within the first 5 months[15]. In another study with only Rottweilers, the one-year survival was 47%[3].

Two recent abstracts have reported a mean survival time of 90 days in one study[11] and a median survival time of 701 days in the other[16]. However, specific details on treatment were not available in these two studies. Except for the last publication, this data is consistent with a poor prognosis for dogs diagnosed with PLE.

Other parameters, which have been associated with a poor outcome include hypocobalaminemia in dogs with chronic enteropathies[2] and the presence of crypt abscesses in dogs with PLE[17].

TREATMENT

Typically, dogs diagnosed with PLE will be treated aggressively with immunosuppressive drugs due to the guarded prognosis. Different regimes have been described including prednisolone as a single agent or in combination with azathioprine, cyclosporine, and methotrexate.

To the best of the author's knowledge, there is only one abstract that has compared two types of treatment for dogs with PLE, namely a combination of prednisolone and azathioprine versus prednisolone and chlorambucil[18]. This was a retrospective study with animals recruited at the start of the study being treated with azathioprine and animals in the second part of the study with chlorambucil. No significant differences were found between the two groups when comparing signalment, weight, clinicopathological parameters and histopathology. Following therapy, dogs treated with chlorambucil showed a significantly greater increase in albumin concentration and weight gain. Median survival time for dogs treated with azathioprine was 30 days (confidence interval 15-45 days), but was not reached for dogs treated with chlorambucil in 500 days. Survival was positively associated with the use of chlorambucil. This study suggests that chlorambucil might be more beneficial than azathioprine. However, no comparison was done with other treatments (for example cyclosporine) and the results need to be confirmed with a prospective study.

A novel treatment was assessed in Soft-coated Wheaten terriers and Norwegian Lundenhunds suffering from PLE[19]. Larazotide acetate was used in an attempt to decrease intestinal permeability by acting on tight junctions. There was a reduction in permeability following treatment but not every dog showed a clinical response and there was no difference in pre- and post-treatment albumin or cobalamin concentrations. Hence, larazotide does show a potential to decrease intestinal permeability, but this is not sufficient in every dog to control clinical signs and its use has not been assessed in other breeds suffering from PLE.

CONCLUSION

There is little literature looking at dogs suffering from PLE and often the reports are based on a limited number of animals. Several mucosal changes can lead to PLE (i.e. primary versus secondary lymphangiectesia, IBD, crypt disease,...), which represent either different disease entities or different stages of disease. There is at this time a lack of information on which treatment is best in each of these cases and more studies are clearly needed to gain more knowledge. One retrospective study described a better outcome when using chlorambucil rather than azathioprine combined with prednisolone, but this needs to be confirmed with a prospective study and this treatment was not compared to treatment with cyclosporine.

Biological markers will most likely be useful to assess prognosis and treatment response in the future, but there is only limited information in dogs suffering from IBD and even less in dogs with PLE.

REFERENCES

[1]Craven M, et al. **J Small Anim Pract** 2004;45:336. [2]Allenspach K, et al **J Vet Intern Med** 2007;21:700. [2]Dossin A, et al **Vet Clin Small Anim** 2011;41:399. [3]Dijkstra M, et al. **Tijdschr Diergeneeskd** 2010;135:406. [4]Larson R, et al. **J Vet Intern Med** 2012;26:1087. [5]Willard M, et al **J Vet Intern Med** 2008;22:1084. [6]Lecoindre P, et al. **Schweiz Arch Tierheilkd** 2010;152:141. [8]Allenspach K, et al. **Am J Vet Res** 2008;69:1301. [9]Otoni, C, et al. **J Vet Intern Med** 2012:26:768. [10]|Heilmann, R, et al. **Am J Vet Res** 2012;73:1900. [11]Equilino M, et al. **J Vet Intern Med** 2011:25:691. [12]G rellet R, et al. **J Vet Intern Med** 2011;25:1486. [13]Heilmann R, et al. **J Vet Intern Med** 2011;25:1486. [14]Im Hof M, et al. **J Vet Intern Med** 2012;26:269. [15]Goodwin L, et al. **J Vet Intern Med** 2011;2:273. [16]Owens S, et al. **J Vet Intern Med** 2011:25:692. [17]Stroda K, et al. **J Vet Intern Med** 2012;26:767. [18]Dandrieux J, et al. **J Vet Intern Med** 2011;25:1484. [19]Berghoff N, et al. **J Vet Intern Med** 2011;25:691..

NUTRITIONAL ASPECTS OF CANINE PLE: THE CRITICALIST PERSPECTIVE

Daniel L. Chan, DVM, DACVECC, DACVN, North Mymms, Hertfordshire, UK

INTRODUCTION

Protein-losing enteropathy (PLE) refers to chronic diseases affecting the intestines leading to excessive losses of protein and is typified by sometimes severe hypoproteinemia. Intestinal lymphangiectasia, lymphoma and inflammatory bowel disease are common etiologies leading to PLE. A common feature of this disease is the debilitating nature of the disease, which may result in profound weight loss, muscle wasting, gastrointestinal dysfunction in both absorptive capacity and motility issues and protein- and energy-malnutrition.[1] Despite the fact that PLE should be considered the end result of various conditions with a similar pathway rather than a discreet disease, the ensuing malnourished state is often a key reason for euthanasia. Therefore, effective management of this condition may be closely related to our ability to improve nutritional status of the patient. From a criticalist's point of view, minimizing complications during the work up of such patients (e.g., following biopsy acquisition) and during hospitalization are the main concern. Ensuring adequate nutritional intake in this patient population can be quite challenging and may involve surgical placement of feeding tubes, which can be associated with some serious morbidity.

NUTRITIONAL STATUS

The nutritional status of dogs with PLE is believed to play an important role in the pathogenesis of PLE and certainly has important implications for the assessment treatment success and ultimately for survival (i.e., animals that continue to be malnourished may be euthanized due to quality of life issues). Patients with PLE are commonly described as having lost considerable weight, and having poor body condition. Lean muscle wasting, which in itself could be prognostic, may result from a combination of poor food intake, poor assimilation and excessive protein loss. When the body is subjected to both energy and protein-malnutrition, the body catabolizes protein stores in lean muscle to support basic life functions. Compounding this is the activation of muscle proteolysis via inflammation and oxidative stress, both believed to be features of PLE. The depletion of albumin, one of the most important endogenous anti-oxidant agents, is believed to confer a degree of oxidative stress, however, this has not been quantified in dogs. As the main causes of PLE include chronic inflammatory conditions, the increase in cytokine production, particularly TNF-α and IL-1 accelerate lean muscle loss. Hypoalbuminemia, the main feature of PLE is considered to be an indicator of nutritional status in people, however, it is not considered to be a good indicator of malnutrition in dogs mainly due to lack of studies validating the utility of serum albumin to discriminate malnutrition and the relative slower turn-over in albumin synthesis in dogs, which make albumin a less sensitive candidate marker of nutritional status.[2] As such, there is no universally accepted or validated measure of nutritional status in dogs. Published studies of canine PLE have incomplete assessment of body condition and have no standardized system to quantify or qualify malnutrition. Despite the lack of definitive data, modulation of malnutrition is considered an important aspect of patient management. Given the accepted notion that nutritional status is important, development of an easily applied measure of nutritional status in dogs with PLE is warranted and could facilitate interpretation of treatment success. The most widely used tool for assessing nutritional status in animals is body condition scoring (BCS). Unfortunately, there are various proposed systems for dogs. The author prefers the 1-9 BCS as there is good evidence substantiating its repeatability. However, it should be recognized that BCS were designed (and validated) to measure body fat content. In conditions typified by catabolism and lean body mass loss, a muscle condition score has been proposed and endorsed by certain organizations such as WSAVA and ACVN, however, further research is required to validate its use in quantifying the degree of severity and correlate scores to disease progression.[3]

NUTRITIONAL SUPPORT MEASURES

Given the degree of malnutrition frequently encountered in this patient population, strategies to facilitate effective nutritional support are required. Restoration of sufficient albumin concentrations is dependent on good disease control and measures to improve voluntary food intake, such as use of anti-emetics (e.g., maropitant), adequate control of abdominal pain, prokinetic therapy (e.g., metoclopramide) should be considered. The utility of antacids such as famotidine or proton-pump inhibitors are commonly recommended, although the evidence supporting their use in the absence of gastric ulceration is lacking. Albumin has a role in helping to maintain mucosal integrity and therefore severe hypoalbuminemia is believed to perhaps increase the risk of gut bacterial translocation,[4] however, this has not be substantiated in dogs with PLE. In fact, sepsis other than as a result of intestinal dehiscence following intestinal surgical biopsies has not been described in dogs with PLE. It is worth mentioning however, that there are no large scale veterinary studies in dogs with PLE. Provision of nutritional support in severely affected dogs does present certain dilemmas. The advanced malnourished state found in some PLE dogs should prompt urgent and effective nutritional support. However, implementation of effective nutritional support measures may be complicated. Because the appetite is often affected, ensuring adequate caloric intake is difficult. Determination of protein requirements of these patients have not been evaluated other than to say that they are increased compared with a normal requirement. Recommendations to increase protein intake above 6-7g/100 kcal of food consumed is speculative as is the target of energy intake. How much above resting energy requirements should these animals be fed? No data is available in this regard. To facilitate nutritional intake, the use of feeding tubes is usually recommended and commonly used by criticalists to manage other critically ill populations. However, there may be problems associated with placement of feeding tubes in dogs with PLE. Since gastric and duodenal biopsies are commonly obtained, the possibility of placing percutaneous endoscopically-guided gastrostomy tubes is often available. However, in patients with ascites, the placement of percutaneous gastrostomy tubes is not recommended. The risk for development of septic peritonitis is believed to be significantly increased given that wound healing may be compromised in catabolic, hypoproteinemic patients and ascites itself may impair wound healing. From a criticalist's point of view, the risk of developing septic complications is too great to recommend PEG tube placement. Esophagostomy feeding tubes offer several advantages over PEG tubes, however, the main problem is the availability of suitable diets that can be fed via esophagostomy tubes. Diets recommended for patients with PLE center on ultralow fat, novel protein and even protein reduced diets. Such diets are usually completely unsuitable for esophagostomy tube feeding. Liquid or semi-liquid veterinary diets are typically high in fat, which are unlikely to be tolerated in dogs with PLE. Use of human enteral products mostly comprising of carbohydrates (i.e., maltodextrin) with amino acids and very low fat (e.g., Vivonex T.E.N. Nestle Health Science, Florham Park, New Jersey) has been anecdotally reported in patients with PLE. However, most human products are unbalanced and incomplete and therefore should only be used for short term. There may even be challenges finding suitable veterinary diets that could be used even in larger PEG tubes. In these patients, there is a temptation to consider parenteral nutrition (PN), at least temporarily until there is a clinical response and evidence of return of gastrointestinal function. There may be good reasons to consider providing PN support when significant gastrointestinal dysfunction and severe malnutrition exists, and the author has employed this mode of nutritional support. However, there are also dilemmas in the use of PN. Firstly, this usually requires central venous access and specialized equipment for compounding PN solutions. Ready-made PN solutions designed for peripheral use can also be administered, however, it is difficult to provide significant amounts of calories or protein with these solutions.[5] The other concern is that there's recent evidence outlining a hypercoagulable state in dogs with PLE.[6] As such, central catheters and PN administration could be associated with increased risk of thromboembolic complications. Despite these concerns, there is reasonable evidence that peri-operative nutritional support of debilitated patients may confer a beneficial effect, at least in people. So it follows that if faced with a debilitated dog with PLE in poor nutritional state, nutritional support either enterally or parenterally (or combination of both) should be considered.

ADDITIONAL CONSIDERATIONS

Given the serious consequences of hypoalbuminemia seen in dogs with PLE, there should be some consideration at exploring other therapeutic options. As severe hypoalbuminemia promotes the formation of tissue edema due to decreased colloid osmotic pressure, the possibility that bowel edema contributes to the gastrointestinal dysfunction, both in motility and absorptive capacity has led to speculation that hyperoncotic therapy, in the form of a concentrated albumin solution may be beneficial. The use of concentrated human albumin has been used in dogs with severe hypoalbuminemia, however, this practice is controversial given the risk of anaphylaxis, anaphylactoid and delayed hypersensitivity reactions. The advent of canine specific albumin solutions[7] has opened the possibility of a less risky treatment regimen for treatment of bowel edema and restoration of normal gut function. However, only anecdotal reports are available supporting this approach. In respect to modulation of inflammation, oxidative stress, nitrogen balance and muscle loss, novel approaches merit further investigation. One candidate option is the use of omega-3 fatty acids therapy. These polyunsaturated fatty acids have anti-inflammatory properties, may preserve lean body mass and modulate nitrogen balance and oxidative stress. However, because of the lipid malassimilation present in dogs with PLE, it is unknown how omega-3 fatty acid therapy would be tolerated. Finally, glutamine, as a preferred energy source of enterocytes, may offer some additional benefits. Glutamine is recognized as being a key immunomodulatory nutrient, and may also help to modulate nitrogen balance, preserve lean muscle mass and modulate oxidative stress. There is a recent study, which demonstrated that glutamine restored gut motility in a canine model of surgery-related ileus and therefore may have a use in dogs with motility disorders.[8]

CONCLUSIONS

From a criticalist's point of view, minimizing the risk of complications in debilitated patients is paramount. Knowing that effective nutritional support may rely on interventions such as placement of feeding tubes, or placement of central catheters for PN administration require vigilant monitoring following interventions. Given the importance of restoration of nutritional status in diseases such as PLE, consideration of novel approaches such as use of omega-3 fatty acids and glutamine warrant further investigation

REFERENCES

[1]Allenspach K, et al; JVIM, 2007;21:700; [2]Michel KE JVECC 1993;3(2):96; [3]Freeman LM, et al. JSAP 2011;52(7):385; [4]Soreide E, et al. Injury 2005;36(9):1001; [5]Gajanayake I, et al JVECC 2013; 23(3):330; [6]Goodwin LV, et al. JVIM 2011;25:273; [7]Craft EM et al. JVECC 2012;22(6):631; [8]Ohno T, et al Dig Dis Sci 2009;54:1208

TESTING FOR TOXOPLASMOSIS AND NEOSPOROSIS

Michael R. Lappin, DVM, PhD, DACVIM
Fort Collins, Colorado

Neospora caninum

This coccidian was previously confused with *T. gondii* because of similar morphology. The sexual cycle is completed in the gastrointestinal tract of canids and results in the passage of oocysts in feces. Oocyst shedding can continue for several months in some dogs (1). Sporozoites develop in oocysts within 24 hours of passage. Tachyzoites (rapidly dividing stage) and tissue cysts containing hundreds of bradyzoites (slowly dividing stage) are the other two life stages. In one study, dogs fed sporulated oocysts were infected and seroconverted but did not shed oocysts (2). Dogs are infected by ingestion of bradyzoites but not tachyzoites. Infection has been documented after ingestion of a number of different infected bovine tissues. Dogs can also become infected from ingesting intermediate hosts such as white-tailed deer and the organism has been detected in the tissues of free ranging chickens (3,4). Thus, free-roaming dogs may be at increased risk of infection. Transplacental infection has been well documented; dams that give birth to infected offspring can have repeated transplacental infection during subsequent pregnancies. Because repeated transplacental infections occur, puppies from a bitch that previously whelped infected puppies are at an increased risk. Canine neosporosis has been reported in many countries around the world. Seroprevalence of infection has varied from 0% to 100% depending on the country and lifestyle of the dog (5). Oocysts are rarely reported in fecal surveys; in one study of 24,677 dog samples, oocysts consistent with *N. caninum* were detected in 0.3% (6). The pathogenesis of the disease is primarily related to the intracellular replication of tachyzoites. Although organism replication occurs in many tissues, including the lungs, in dogs, clinical illness is primarily neuromuscular. Administration of glucocorticoids may activate bradyzoites in tissue cysts, resulting in clinical illness.

Ascending paralysis with hyperextension of the hindlimbs in congenitally infected puppies is the most common clinical manifestation of the disease. Muscle atrophy occurs in many cases. Polymyositis and multifocal CNS disease can occur alone or in combination. Cerebral ataxia and atrophy is a recently recognized clinical syndrome associated with neosporosis (7). Clinical signs can be evident soon after birth or may be delayed for several weeks. Neonatal death is common. Although disease tends to be most severe in congenitally infected puppies, dogs as old as 15 years have been clinically affected. In one dog presented primarily for respiratory disease, cough was the principal sign. Myocarditis, dysphagia, ulcerative dermatitis, pneumonia, and hepatitis occur in some dogs. Whether clinical disease in older dogs is from acute, primary infection or exacerbation of chronic infection is unknown. Administration of glucocorticoids with or without cyclosporine may activate bradyzoites in tissue cysts, resulting in clinical illness. Disease is caused by intracellular replication of *Neospora caninum* tachyzoites. Infection of CNS structures generally causes mononuclear cell infiltrates, which suggests an immune-mediated component to the pathogenesis of disease. Intact tissue cysts in neural structures are generally not associated with inflammation, but ruptured tissue cysts induce inflammation. Untreated disease generally results in death.

Hematologic and biochemical findings are nonspecific. Myositis commonly results in increased creatine kinase and aspartate aminotransferase activities. CSF abnormalities include increased protein concentration (20 to 50 mg/dL) and a mild, mixed pleocytosis (10 to 50 cells/microliter) consisting of monocytes, lymphocytes, neutrophils and, rarely, eosinophils. Interstitial and alveolar patterns can be noted on thoracic radiographs. Magnetic resonance imaging of 7 dogs with cerebellar disease showed marked bilaterally symmetrical cerebellar atrophy surrounded by an area T2-weighted hyperintense and T1-weighted hypointense signal (7).

Definitive diagnosis is based on demonstration of the organism in CSF or tissues. Tachyzoites are rarely identified on cytologic examination of CSF, imprints of dermatologic lesions, and bronchoalveolar lavage. Mixed inflammation with neutrophils, lymphocytes, eosinophils, plasma cells, macrophages, and tachyzoites was noted on transthoracic aspirate of one dog with lung disease. *Neospora caninum* tissue cysts have a wall thicker than 1 micron; *T. gondii* tissue cysts have a wall thinner than 1 micron. Oocysts can be detected in feces by microscopic examination after flotation or by PCR. The organism can be differentiated from *T. gondii* by electron microscopy, immunohistochemistry, and PCR. A multiplex PCR assay that detects both *T. gondii* and *N. caninum* for use with tissues or CSF has been reported (8). In a group of dogs with cerebellar disease related to *N. caninum* infection, four of five dogs tested were positive for DNA of the organism by PCR assay performed on CSF.

A presumptive diagnosis of neosporosis can be made by combining appropriate clinical signs of disease and positive serology or presence of antibodies in CSF with the exclusion of other etiologies inducing similar clinical syndromes, particularly *T. gondii*. Serologic cross-reactivity between *T. gondii* and *N. caninum* exist in some assays. IgG antibody titers of at least 1:200 have been detected in most dogs with clinical neosporosis; minimal serologic cross-reactivity occurs with *T. gondii* at titers of 1 :50 or higher when using the immunofluorescent assay test.

Toxoplasma gondii

This coccidian is one of the most prevalent parasites infecting warm-blooded vertebrates around the world (9). Only cats complete the sexual phase in the gastrointestinal tract and pass environmentally resistant oocysts in feces. Sporozoites develop in oocysts after 1 to 5 days of exposure to oxygen and appropriate environmental temperature and humidity. Sporozoites can penetrate the intestinal tract of cats or intermediate hosts and disseminate in blood or lymph as tachyzoites during active infection. *Toxoplasma gondii* can penetrate most mammalian cells and will replicate asexually within infected cells until the cell is destroyed. If an appropriate immune response occurs, replication of tachyzoites is attenuated and slowly dividing bradyzoites develop that persist in within cysts in extra-intestinal tissues. Tissue cysts form readily in the CNS, muscles, and visceral organs. Live bradyzoites may persist in tissues for the life of some hosts. Fatal toxoplasmosis in cats and dogs occurs most frequently from liver, lung, or CNS disease and is most common in transplacentally infected animals or those that are otherwise immunosuppressed. *Toxoplasma gondii* infection should be on the list of differential diagnoses for cats with anterior or posterior uveitis, cutaneous lesions, fever, muscle hyperesthesia, myocarditis with arrhythmias, weight loss, anorexia, seizures, ataxia, icterus, diarrhea, or pancreatitis. Respiratory, gastrointestinal, or neuromuscular infection resulting in fever, vomiting, diarrhea, dyspnea, and icterus occurs most commonly in dogs with generalized toxoplasmosis.

Cats with clinical toxoplasmosis can have a variety of clinicopathologic abnormalities, but none are pathognomonic for the disease. The organism should be on the differential list for cats with nonregenerative anemia, neutrophilic leukocytosis, lymphocytosis, monocytosis, neutropenia, eosinophilia, proteinuria, bilirubinuria, as well as increases in serum protein and bilirubin concentrations, and creatinine kinase, alanine aminotransferase, alkaline phosphatase, and lipase activities. Similar abnormalities occur in dogs with toxoplasmosis. Pulmonary toxoplasmosis most commonly causes diffuse interstitial to alveolar patterns; pleural effusion has rarely been documented as well. Mass lesions may be detected on CT or MRI examinations. Cerebrospinal fluid (CSF) protein concentrations and cell counts are often higher than normal with the predominant white blood cells in CSF being small mononuclear cells. However, increased neutrophils also are commonly found in cats with acute central nervous system disease. As in cats, clinicopathologic and imaging abnormalities are not specific. Hyperglobulinemia can occur in dogs with chronic infection (10). Increased protein concentrations and mixed inflammatory cell infiltrates occur in dogs with CNS toxoplasmosis.

The antemortem definitive diagnosis of feline toxoplasmosis can be made if the organism is demonstrated; however, this is uncommon, particularly in association with sublethal disease. Bradyzoites or tachyzoites are rarely detected in tissues, effusions, bronchoalveolar lavage fluids, aqueous humor, or CSF. *Toxoplasma gondii* oocysts are 10 X 12 µm oocysts and when found in feces of cats with diarrhea suggests toxoplasmosis. However, *Besnoitia* and *Hammondia* infections of cats produce morphologically similar oocysts. Recently, polymerase chain reaction has been used to document *T. gondii* DNA in feces and can be used to differentiate *T. gondii* from other organisms (11). *Toxoplasma gondii* DNA can be amplified from the blood of healthy cats and so positive PCR results do not correlate to clinical disease (12). Thus, PCR assays are used most frequently with tissues to document that the organisms seen were *T. gondii* or with aqueous humor or CSF (www.dlab.colostate.edu). Results of serum antigen assays or immune complex assays have also been assessed but are generally only used in research since results do not correlate with clinical illness (9).

Detection of *T. gondii* antibodies in serum is used most frequently in the diagnosis of clinical toxoplasmosis. A multitude of different techniques have been assessed including ELISA, immunofluorescent antibody (IFA), western blot immunoassay, and a variety of agglutination tests (13-17). A latex agglutination assay and an indirect hemagglutination assay are available commercially. These assays can be used with serum from multiple species and theoretically detect all classes of immunoglobulin directed against *T. gondii*. However, these assays rarely detect antibody in feline serum samples positive only for IgM (16). ELISA, IFA, and western blot immunoassay have been adapted to detect IgM, IgG and IgA antibody responses by using heavy chain specific secondary antibodies. *Toxoplasma gondii* specific serum IgA antibody responses are similar to IgG antibody responses and this antibody class is usually only measured in research studies. Several commercial laboratories in the United States offer *T. gondii* IgM and IgG testing by ELISA (www.dlab.colostate.edu). The following are common findings concerning *T. gondii* IgM and IgG antibody test results in cats. While less information is available for dogs, the findings appear to be similar to those in the cat.

***Toxoplasma gondii* IgM titers**

Using ELISA, approximately 80% of healthy, experimentally infected cats have detectable *T. gondii*-specific IgM in serum within 2-4 weeks following inoculation with *T. gondii*; these titers generally are negative within 16 weeks after infection. As described in some healthy women, persistent IgM titers (> 16 weeks) have been documented commonly in cats coinfected with FIV and in cats with ocular toxoplasmosis (18). Because of these findings and the finding that some cats never have a detectable IgM response, IgM titers cannot accurately be used to predict when a cat shed oocysts. If a clinician concerned that an individual cat is shedding *T. gondii* oocysts, a fecal flotation should be performed. In one study of cats with clinical toxoplasmosis, *T. gondii* IgM titers were detected in the serum of 93.3% of the cats while *T. gondii* IgG titers were detected in 60% of the cats. Thus, IgM antibodies have a higher positive predictive value than IgG for clinical feline toxoplasmosis (19). Some cats with chronic *T. gondii* infections that have gone from IgM positive to IgM negative, can have IgM titers detected again after repeat inoculation with *T. gondii*, primary inoculation with the Petaluma isolate of FIV, and administration of glucocorticoids without detection of clinical signs of toxoplasmosis (20). Thus, presence of IgM antibodies in feline serum does not always prove clinical toxoplasmosis in cats.

***Toxoplasma gondii* IgG titers**

Toxoplasma gondii-specific IgG can be detected by ELISA in serum in the majority of healthy experimentally inoculated cats within 3-4 weeks after infection. By the time IgG antibodies are detected in feline sera, the oocyst shedding period has usually been completed. Thus, IgG seropositive cats are of minimal public health risk. *Toxoplasma gondii* IgG antibody titers can be detected for at least 6 years after infection in experimentally inoculated cats; since the organism probably persists in tissues for life, IgG antibodies probably do as well (21). Single, high IgG titers do not necessarily suggest recent or active *T. gondii* infection; healthy cats commonly have titers > 10,000 six years after experimental induction of toxoplasmosis. Some cats with low *T. gondii* antibody titers can become seronegative based on the cutoff value used in an individual assay even though *T. gondii* is still within tissues. Based on an approximate 10% interassay variation in the ELISA technique, some cats with low positive IgG titers (1:64) can be positive for IgG antibodies on one analysis and negative on a subsequent analysis or vice versa. The demonstration of an increasing *T. gondii* IgG titer can document recent or active infection but in experimentally infected cats, the time span from the first detectable positive IgG titer to the maximal IgG titer is approximately 2-3 weeks. Thus, some cats with clinical toxoplasmosis will have reached their maximal IgG titer by the time they are evaluated serologically. Rising *T. gondii* IgG antibody titers occur in healthy infected cats as well as cats with clinical toxoplasmosis and so when assessed alone do not prove clinical toxoplasmosis. In humans and cats with reactivation of chronic toxoplasmosis from immune suppression, IgG titers only rarely increase.

For the reasons discussed above, antibody test results alone cannot be used to make a diagnosis of toxoplasmosis. However, the following combination can be used to make a presumptive antemortem diagnosis: Demonstration of antibodies in serum which suggests infection by *T. gondii;* demonstration of an IgM titer >1:64 or a four-fold or greater increase in IgG titer which suggests recent or active infection; clinical signs of disease referable to toxoplasmosis; exclusion of other common causes of the clinical syndrome; and positive response to appropriate treatment.

Because the organism cannot be cleared from the body, most cats or dogs will be antibody-positive for life, so there is little reason to repeat serum antibody titers after the clinical disease has resolved or to administer drugs with *T. gondii* activity to cats or dogs without clinical signs of toxoplasmosis.

The combination of aqueous humor or CSF *T. gondii*– specific antibody detection and organism DNA detection by PCR is the most accurate way to diagnose ocular or CNS toxoplasmosis (www.dlab.colostate.edu). Whereas *T. gondii*–specific IgA, IgG, and organism DNA can be detected in aqueous humor and CSF of both normal and clinically ill cats, *T. gondii*–specific IgM has only been detected in the aqueous humor or CSF of clinically ill cats and so may be the best indicator of clinical disease. In dogs, *N. caninum* infection should be excluded by serologic testing or PCR assay on CSF.

REFERENCES

[1]McGarry JW et al. **J Parasitol** 2003;89:628.; [2]Bandini LA, et al. **Vet Parasitol.** 2011;176:151-156.; [3]Gondim LF et al. **J Parasitol** 2004;90:1361.; [4]Gonçalves IN, et al. **Vet Parasitol**. 2012;190:74-79.; [5]Dubey JP et al. **Clin Microbiol Rev** 2007;20:323.; [6]Barutzki D, Schaper R. **Parasitol Res**. 2011;1:S45-60.; [7]Garosi L et al. **J Vet Intern Med**. 2010;24:571-578. [8]Schatzerg SJ et al. **Am J Vet Res** 2003;64:1507. [9]Lappin MR. **Top Companion Anim Med**. 2010;25:136-141.; [10]Yarim GF, et al. **Parasitol Res**. 2007;101:1197-1202.; [11]Salant H, et al. **Am J Trop Med Hyg** 2010;82:865-870.; [12]Burney DP, et al. **J Parasitol** 1999;5:947-951.; [13]Lappin MR. **Semin Vet Med Surg** 1996;11:154.; [14]Lappin MR, et al. **Am J Vet Res** 1989;50:1580-1585.; [15]Lappin MR, et al. **J Parasitol** 1994;80:73-80.; [16]Lappin MR, Powell CC. **J Vet Int Med** 1991;5:299-301.; [17]Dabritz HA, et al: **J Parasitol.** 2007;93:806-816.; [18]Lappin MR, et al. **J Parasitol** 1996;82:733-742.; [19]Lappin MR, et al. **J Vet Int Med** 1989;3:139-143.; [20]Lappin MR, et al. **J Am Anim Hosp Assoc** 1992;27:625-632, 1992.; [21]Dubey JP. **J Parasitol** 1995.

MYCOPLASMAS AND NOVEL VIRAL PATHOGENS IN CANINE INFECTIOUS RESPIRATORY DISEASE

Joe Brownlie CBE BVSc PhD DSc DSc(Hons) DipECVP FRCPath FRAgS FRASE FRCVS, Judy A Mitchell BSc PhD,
Caray A Walker BSc MSc PhD, Kerstin Erles DrMedVet PhD
Royal Veterinary College, Herts, United Kingdom

INTRODUCTION

Respiratory disease is a major hazard for both human and animal health; in the UK Government's Foresight program on 'The detection and identification of infectious diseases", respiratory diseases were identified as a most important future risk[1,2]. Aerosol transmission permits rapid transfer of infection, particularly within crowded populations eg humans on planes and dogs in kennels. If this, and the scale of global movement of people and animals, is considered, it is clear that respiratory diseases have huge potential of epidemic proportions (influenza & SARS). Moreover, respiratory infections are the most common human illness worldwide. Effective diagnostic and prophylactic vaccines are key to control.

Of all the infectious diseases of dogs, respiratory disease is arguably the most widespread and most intractable to treatment and prevention. For clinicians, it is a cause of disappointment and frustration when their patients return home following successful surgery and thereafter develop signs of respiratory disease. For training kennels and re-homing kennels, the problem can be more severe and more sustained, sometimes continuing to be a problem over many years. Possibly the most widespread respiratory problem arises within vaccational kennels, in spite of the requirement for appropriate vaccination. So the question remains whether our understanding and our presently available vaccines are adequate for today's challenges. In this paper, I will review our own studies on canine infectious respiratory disease (CIRD) over the past 10 years, review other relevant studies and thereafter present an updated hypothesis for the complex pathogenesis of this widespread syndrome.

MULTIPLE AETIOLOGY OF CIRD

Many respiratory microbes survive well in the upper airways of their hosts, rarely moving to lower airways which, to all extent, remain sterile. The innate and adaptive immune systems maintain this crucial microbial barrier. Some microbes can overcome these barriers, penetrate to the lower airways and, in the process, enable by-stander access of other microbes. This dual infection, most commonly a viral infection with subsequent bacterial superinfection, is the critical pathway but its detailed pathogenesis is poorly understood. Respiratory corona viruses[3,4] and respiratory mycoplasmas are notable invaders[5]. Few *in vivo* animal models or *in vitro* cell culture systems have been reported for dual pathogen systems.

In 1999, our laboratory was invited to investigate the underlying causes of a severe epidemic of respiratory disease in a dog re-homing kennel. Over the past years, we have discovered a novel corona virus[6] and a clinical association with canine mycoplasmas[7] which was hitherto unrecognised. We are presently pursuing vaccine development for both Canine respiratory coronavirus (CRCoV) and *Mycoplasma cynos*. However, the fundamental mechanisms of corona virus and mycoplasma pathogenesis at both the cellular and molecular levels remain unresolved. Moreover, the importance of both corona viruses and mycoplasmas in the wider context of respiratory disease of human and animals make our need for understanding these dual interactions of greater significance.

We propose that there is a mechanism by which respiratory corona viruses can potentiate bacterial superinfection with mycoplasmas. We are currently undertaking studies with the virus and the mycoplasma separately before investigating the dual sequential infections that are evidently, from our clinical studies, a cornerstone to the progression to serious respiratory disease.

Canine respiratory coronavirus (CRCoV)

Since its initial discovery in 2003[7]; canine respiratory coronavirus (CRCoV) has been detected worldwide[8-11]; and is now considered to be a significant CIRD pathogen, most frequently detected in dogs with mild respiratory clinical signs during the early stages of CIRD onset[7]. It is postulated that CRCoV plays an important role during the early stages of CIRD by predisposing the dog to more severe clinical disease from secondary infections. Through the use of an *in vitro* tracheal explant culture system, a moderate reduction in ciliary function and a down-regulation of pro-inflammatory cytokine mRNA levels (TNF-α, IL-6 and the chemokine IL-8) was observed in response to CRCoV exposure[12]. Such alterations in the mucociliary and innate immune systems could be linked to increased susceptibility to secondary infection and is consistent with the proposed role for CRCoV in CIRD. Recently published findings from an *in vivo* challenge study demonstrated that dogs were susceptible to experimental infection with different CRCoV isolates originating from geographically distinct locations. Infection gave rise to clinical signs of respiratory disease consistent with naturally occurring infection and spread rapidly to sentinel dogs which also displayed clinical signs of disease. The presence of CRCoV was associated with marked histopathological changes in the nares and trachea, with loss and damage to tracheal cilia, accompanied by inflammation. Viral shedding was readily detected from the oropharynx up to 10 days post infection, but there was little or no evidence of rectal shedding. The successful re-isolation of CRCoV from a wide range of respiratory and mucosal associated lymphoid tissues, and lung lavage fluids demonstrates a clear tropism of CRCoV for respiratory tissues and fulfils the final requirement for Koch's postulates. By 14 days post infection dogs had seroconverted to CRCoV and the antibodies raised were neutralising against both homologous and heterologous strains of CRCoV *in vitro.* Defining the role that CRCoV and other agents play in CIRD is a considerable, but important, challenge if the disease is to be managed, treated and prevented more successfully.

Canine pneumovirus (CnPnv)

A novel virus, canine pneumovirus (CnPnV), was recently identified in a retrospective study of respiratory disease in dogs from two animal shelters located in the north-eastern United States[13]. Since the initial report, CnPnV has also been detected in dogs with respiratory disease in eight other states across the USA[14] and in the UK (Mitchell et al – manuscript in preparation). Genome analysis places CnPnV in the family *Paramyxoviridae*, sub-family *Pneumovirinae* genus *Pneumovirus* together with human respiratory syncytial virus (hRSV), Bovine RSV and murine pneumovirus (MPV)[14]. The relationship between CnPnV and these important pneumoviruses suggests there is a strong pathogenic potential for CnPnV in dogs. This is supported by a recent publication in which the successful experimental challenge of mice with CnPnV was reported, resulting in viral replication within the lung which could result in severe respiratory sequelae in the infected mice[15].

Canine hepacivirus (CHepV)

Recently, a novel virus, tentatively named canine hepacivirus (CHepV) was identified in respiratory samples of dogs housed in shelters in the USA. CHV is the first non-primate hepacivirus to be described, and comparative phylogenetic analysis confirmed it as the closest genetic relative of Hepatitis C Virus to date[16]. A role for this novel virus in canine disease remains to be determined, its discovery in dogs with respiratory disease is a significant finding and it will be important to determine whether CHV causes severe respiratory illness in dogs, and contributes to CIRD pathogenesis, or whether it can alternatively be linked to other illness.

Canine bocavirus (CBoV)

Recently, a novel virus named canine bocavirus (CBoV) was identified in canine respiratory samples from dogs in the USA[17]. Genetic analysis confirmed that CBoVs are genetically distinct from minute virus of canines, the other canine bocavirus which can cause severe respiratory infections in newborn puppies, but predominantly subclinical infections in adult dogs. The study showed that CBoVs are a diverse group of viruses which can be detected in both healthy and diseased dogs; and therefore the pathogenicity of this novel group of viruses remains to be determined. However, one genotype in particular was significantly more prevalent in animals with respiratory disease compared with healthy animals. Since the initial study CBoV has also been reported in dogs in Hong Kong[18].

Mycoplasma cynos

Mycoplasma cynos is a member of the genus *Mycoplasma* within the class Mollicutes. Mycoplasmas are unique among the prokaryotes owing to the lack of a cell wall. They are the smallest living organisms capable of independent existence and are thought to have evolved from the Gram positive *Firmicutes* taxon by considerable genome reduction. *Mycoplasma cynos* was first isolated from the lungs of a dog with pneumonia in 1972[19]. In one of the more recent investigations into canine mycoplasmas Chalker *et al* (2004) isolated a diverse range of *Mycoplasma* species from throughout the respiratory tracts of both healthy and diseased dogs[20]. Interestingly, these authors report that only *M. cynos* was significantly associated with disease.

Since its initial discovery there has been intermittent research interest in *M. cynos* and, as such, there is limited knowledge available regarding the prevalence, pathogenesis and type of immune response associated with *M. cynos* infection. *M. cynos* is a fastidious organism which requires complex growth media when cultivated *in vitro*. Consequently, mycoplasma infections in dogs are often overlooked, since few laboratories routinely test for presence of the bacteria unless requested by the veterinarian[21]. To add to this problem, many of the canine mycoplasmas are identical at the colony level when grown *in vitro*. As a result, many mycoplasma infections of dogs with respiratory disease are under diagnosed; this may well have contributed to the lack of research that has gone into *M. cynos* over the years.

Mycoplasmas are naturally resistant to β-lactam antibiotics, the mode of action of these antibiotics is to interfere with cell wall biosynthesis, and since mycoplasmas do not have a cell wall the antibiotic is of no value. Treatment is usually in the form of alternative antibiotics such as doxycycline.

At present it is not known whether *M. cynos* is a primary or secondary pathogen in the CIRD disease complex. Initial work in our laboratory suggests it is capable of both; further work is required to fully understand the pathogenesis of *M. cynos* in dogs.

SUMMARY

In my lecture, I will address some of the supporting experimental and field studies that we have undertaken to support our hypothesis that canine respiratory corona virus is an important initiating cause of the complex pathogenesis of infectious respiratory disease. Furthermore, I will present some of our understanding of bacterial superinfection, particularly the mycoplasmas. Of interest will be our report of the position of Streptococcus Zooepidemicus. I will also examine the data for existing and newly-discovered canine pathogens that may contribute to this important clinical syndrome.

REFERENCES

[1]King DA, et al **Science** 2006;303:1392-1393. [2]Brownlie J, et al **Foresight OSI London** 2006:p1-83. [3]Perlman S, Dandekar AA **Nature Reviews** 2005;5:917-927. [4]Erles K, Brownlie J, **Vet Clin Small Animal** 2008;38:815-825. 5Razin S, Hayflick **Biologicals** 2010;38:183-190. [6]Erles et al **Virology** 2003;310(2):216-223. [7]Chalker V J, Brownlie **J Int J of Sys & Evol Micro**;100:77-82. [8]Priestnall SL et al **Vet Immunol Immunopathol** 2009;127:38-46. [9]Anderton TL **Microbiol** 2004;150:2843-2855. [10]Frieman M, Baric R **Microbiol Mol Biol Rev** 2008;72:672-685. [11]St-Jean JR, et al **J Virol** 2006;80:3670-3674. [12]Krishnakumar R, et al **Appl Environ Microbiol** 2010;76:5297-5299. [13]Hunt ME, Brown DR **Vet Micro** 2007;121:73-82. [14]Hudson P, et al **Infect immun** 2006;74:931-939. [15]Chopra-Dewasthaly R, et al **Int J Med Microbiol** 2005;294:447-453. [16]Reddy SP, et al **FEMS immunol med microbiol** 1996;15:199-211. [17]Zimmerman CU, Herrman R **FEMS Microbiol Lett** 2005; 253:315-321. [18]de Groot RJ, **Glyconj J** 2006;23:59-72. [19]Rosendal S **Acta veterinaria Scandinavica** 1972;13:137-139. [20]Chalker VJ et al **Microbio** 2004:150:3491-3497. [21]Chalker VJ **Res Vet Sci** 2005;79:1-8

FIP, PATHOGENESIS AND TESTING CONTROVERSIES

Anja Kipar, Dr.med.vet.habil., DiplECVP, FTA Pathologie, MRCVS
Liverpool, UK and Helsinki, Finland

INTRODUCTION

Feline infectious peritonitis (FIP) is currently the most relevant fatal infectious disease of cats. It is globally distributed among virtually all cat populations. Since its first description in the 1960s, the disease has been of interest for clinicians, pathologists and virologists, in particular due to its enigmatic pathogenesis which is reflected also by the fact that infections with Feline Coronavirus (FCoV) have a very high prevalence among cats worldwide, but the disease itself, which is caused by FIP virus (FIPV), the virulent biotype of FCoV, occurs more or less sporadically among FCoV-infected cats. Much progress has been made in recent years in the specific virology and pathogenesis of FCoV infection and FIP, which are generally followed by new diagnostic approaches.

AETIOLOGY

Feline Coronaviruses

FCoV are enveloped, single positive-stranded RNA viruses with an almost 30kb genome and 11 putative open reading frames (ORF) that belong to the family *Coronaviridae*, order *Nidovirales*. The 5' end of the genome encodes for 16 non-structural functional proteins that are mainly involved in the synthesis of the viral RNA. These are transcribed as polypeptides which are subsequently cleaved. The four structural proteins (spike [S], nucleocapsid [N], membrane [M], envelope [E]) and five group-specific, accessory proteins (3a-c, 7a and b) are expressed individually from a nested set of subgenomic mRNAs generated by discontinuous transcription mechanism from the 3' end of the genome[1,2].

FCoV form two antigenetically distinct serotypes, type I FCoV which are difficult to grow in culture, and type II which are the consequence of a double recombination between type I FCoV and canine coronaviruses[3]. Serological and molecular studies confirmed that type I FCoV dominate by far in the cat population worldwide, with a prevalence of up to 98% [4-8].

FCoV exhibit *in vivo* genetic diversity, as shown by the frequent occurrence of viral quasispecies in infected animals, i.e. both individual animals and cats from the same household[9-12]. In cats that had been infected with fecal matter from naturally Feline Enteric Coronavirus (FECV) infected, healthy cats, quasispecies occurred over time in the large intestine of individual animals[10] (Kiss et al., 2000). Also, a study on naturally infected animals showed more extensive viral quasispecies formation in cats with FIP than in healthy animals, suggesting that a higher viral mutation rate is relevant for the generation of virulent mutants[11].

However, phylogenetic analysis also indicates that the observed genetic diversity mainly applies to type I FCoV, whereas type II viruses are relatively homogenous[8,12].

Feline Enteric Coronavirus (FECV) *versus* Feline Infectious Peritontis Virus (FIPV)

FCoV occur as two pathotypes, Feline Enteric Coronavirus (FECV), defined as the "ubiquitous enteric biotype", and Feline Infectious Peritonitis Virus (FIPV), the "virulent biotype that causes FIP in individual cats"[13]. FECV and FIPV cannot be distinguished serologically or morphologically, but for many years attempts have been made to identify virulence markers that allow the identification of FIPV in infected cats.

In the past, it was assumed that the main difference between FECV and FIPV is that FECV infect the intestinal epithelium but, unlike FIPV, cannot infect and replicate in monocytes/macrophages. However, once sensitive molecular methods became available, it was shown that also FECV can infect blood monocytes and that FCoV generally spread from the initial site of infection, the intestine, via monocyte-associated viremia[14-17]. Indeed, approximately 80% of cats, healthy or with FIP, in households with endemic FCoV infection were found to harbour FCoV RNA in their blood monocytes, and healthy cats remained viremic over the 12 month test period[14].

It has since been considered most likely that FIPV are characterised by their capacity to replicate in monocytes, and protocols were developed to demonstrate viral replication in the blood through the specific detection of viral M protein mRNA[18]. These showed that FECV can replicate within monocytes in healthy cats, but generally appear to do this only intermittently , if at all, and then at far lower levels than FIPV in animals with the disease[18,19].

Initial molecular studies on FECV and FIPV isolates identified deletions in the 3c, 7a and 7b genes in FIPV[20,21]. These were then followed by more systematic studies that aimed to identify virulence markers and sequenced both the structural and accessory genes of FCoV from cats with FIP. In a proportion of cases, significant mutations were found in the 3c gene of these FIPV, resulting in variable truncation of the 3c protein in FIP lesions, whereas FCoV in the feces generally exhibited an intact 3c gene. Unfortunately, this feature was not consistently observed, but nonetheless indicated a role of 3c gene deletions in the switch of FCoV from FECV to FIPV[13,20].

Subsequent larger studies further investigated the 3c gene as a virulence marker, comparing the 3c gene of FCoV from healthy cats (FECV) and cats with FIP (FIPV). In FECV, the 3c gene was intact or only carried insignificant mutations[22,23]. After oronasal inoculation of cats with such FECV isolates, animals became infected and shed the virus with the feces[23]. In contrast, the majority (71%) of FIPV exhibited mutations, i.e. variably sized deletions and insertions, some of which were associated with severe truncation of 3c and therefore loss of 3c function. Interestingly, when FCoV was identified in the feces of cats with FIP, it generally exhibited an intact FECV 3c gene, suggesting superinfection[22]. Nonetheless, both studies identified a substantial proportion of cats with FIP (29% and 40% respectively) that harboured FIPV with an intact 3c gene in diseased tissues, confirming the results of an earlier study[20,22,23]. These FIPV isolates induced FIP in cats both after oronasal and intraperitoneal inoculation and were shed with the feces, but the fecal virus was not infectious to cats[23]. These findings led to the following conclusion: FCoV need to carry an intact 3c gene to be able to sustainably replicate in the intestinal epithelium and be infective to other cats. However, an intact/non-truncated C3 gene does not prevent a FCoV from inducing FIP[20,22,23].

A further, non-targeted approach has recently been taken in an attempt to identify further mutations that might be responsible for the change in virulence. Randomly selected FECV (from healthy cats) and FIPV (from cats with FIP confirmed by post mortem examination) were compared based on full genome sequencing. Differences were found scattered along the entire genome, but two more substantial mutations were identified in the S gene. Both mutations occur in the supposed fusion peptide of the S protein, but there is no evidence of any potential functional consequences of this alteration[24]. Considering the mutation rate of RNA viruses and thereby the expected rate of occurrence of the observed mutations as well as the relative rarity of FIP, the authors considered it as unlikely that the identified S gene mutations are solely responsible for the FECV-FIPV virulence switch[24]. Another approach has been taken by a group that has worked extensively on antibody mediated enhancement in FIP. By growing the type II FIPV 79-1146 strain in culture with monoclonal antibodies against virus neutralizing epitopes, the group has established viruses that are resistant to virus neutralizing monoclonal antibodies. These so-called mar-mutant viruses all exhibited mutations in several amino acids in the S1 region[25]. When orally administered to cats, those mar-mutant viruses that carried only the mutations in the S1 region induced FIP, whereas viruses that also showed deletions in the 7b gene exhibited reduced virulence[26].

In vitro studies complement the *in vivo* approaches and confirm previous studies indicating that the factor determining whether a FCoV belongs to the FECV or FIPV pathotype is its ability to productively and sustainably infect feline monocytes[27]. Indeed, a FIPV (79-1146) was shown to establish sustainable replication in feline monocytes, whereas a FECV (79-1683) could replicate in monocytes, but was not able to sustain the replication[27]. Interestingly, the same study found strong evidence that the monocytes of some cats can be resistant to FCoV infection in general or FECV infection alone[27]. Similar results were obtained when in bone marrow-derived macrophages were infected with a closely related FECV and FIPV. While the FIPV effectively infected and subsequently replicated in the cells, the FECV did infect less cells and appeared unable to spread the infection[28].

More conflicting results were obtained in studies to identify the viral proteins responsible for the variable replication capacity of FECV and FIPV. One group showed that the S protein, i.e. the membrane-proximal domain which is involved in virus mediated membrane fusion, determined the macrophage tropism[28]. Others showed that the deletions in the S1 gene that mar mutant FIPV carry does not affect the viral capacity to productively infect feline monocytes, unless the viruses also show a deletion in the 7b gene[26]. Subsequently, another study using FIPV isolates and recombinant viruses, provided evidence that the presence of intact 3abc proteins in type II FCoV inhibits their productive infection of feline monocytes[29]. Most recently, a further study thoroughly investigated the relevance of ORF3 and 7 proteins for the replication of FCoV in monocytes. Wildtype FIPV 79-1149 and genetically modified viruses with an ORF3abc deletion (FIPV-Δ3) established sustainable replication in isolated feline monocytes, with a lower replication capacity of FIPV-Δ3. In contrast, viruses lacking ORF7ab (FIPV-Δ7 and FIPV-Δ3Δ7) could only undergo one replication cycle, which provides further evidence that it is indeed the 7b protein that is most relevant for viral replication in monocytes[30].

PATHOGENESIS

The morphological hallmark of FIP and likely its early lesion is a granulomatous phlebitis and periphlebitis that is mediated by activated monocytes, most likely during a phase of high-level monocyte-mediated viremia in association with substantial viral replication[31]. Activation of the monocytes in association with the development of phlebitis is represented by their strong expression of cytokines, such as TNF-α and IL-1β, adhesion molecules, such as CD18, and enzymes, such as MMP-9 which leads to the destruction of the vascular basement membrane at sites of monocyte attachment and emigration[31]. A recent study also demonstrated increased VEGF transcription in monocytes of cats with FIP and increased VEGF levels in their blood, which could explain the development of effusions in FIP since VEGF is a very strong mediator of vascular permeability[32,33]. Similar cytokine patterns have been observed in peritoneal exudate cells of cats with FIP[34].

Recent *in vitro* studies examined the effect of FIPV on isolated feline monocytes/macrophages. FIPV rapidly induces activation of the p38 mitogen-activated protein kinase (MAPK), which directly regulates the expression of proinflammatory cytokines, in feline PBMCs, likely both early during virus entry and later, when virus is being produced[35]. This is associated with the induction of TNF-α and IL-1β as demonstrated in the PBMC supernatant at 24hpi[35]. In alveolar macrophages, TNF-α up-regulation was seen as a consequence of viral replication after 2 and 3 days, and vascular endothelial growth factor, G-CSF and GM-CSF were found to be up-regulated at 3 days; all upregulation levels became significant when virus inoculation was combined with an antibody against FCoV S protein that is known to induce antibody dependent enhancement[36].

FIP is not only characterized by the specific granulomatous lesions and effusions in body cavities, but is also associated with B and T cell depletion including T cell apoptosis, a feature that might be mediated by monocyte/macrophage TNF-α expression and might explain the observed drop in IFN-γ that is seen in the blood of cats with FIP[34,37,38,39]. Furthermore, there is generalised monocyte/macrophage proliferation[39]. Nonetheless, the lymphatic tissues were shown to exhibit, for example, reduced IL-12 mRNA levels, indicating impairment of the cellular immune response against the virus[37,40].

DIAGNOSIS OF FIP

Research on FIP has in particular in recent years generated a wealth of data that have elucidated many aspects of the specific aetiology and pathogenesis of FIP. However, due to their often controversial nature, they also add confusion to this topic and indicate how difficult it is to establish reliable and predictive *intra vitam* diagnostic tools that can replace the well established diagnostic panel. At present, none of the available molecular diagnostic options can be regarded as infallible and results need to be interpreted with care. A recent novel methodological approach for the quantitation of replicating virus in blood leukocytes has been reported. This detected high levels of viral replication in cats suffering from FIP[41]. However, at this stage it requires screening of larger animal cohorts to provide reliable information as to whether the quantitation of FCoV viremia would be a diagnostic tool for FIP. A similar approach should be taken in all diagnostic approaches to identify consistent genome mutations, to avoid over, or even false, interpretation.

REFERENCES

[1]Dye C and Siddell SG. **J Gen Virol** 2005;86(Pt8):2249. [2]Enjuanes L, et al. **Adv Exp Med Biol** 2006;581:13. [3]Herrewegh AA, et al. **J Virol** 1998;72(5):4508. [4]Pedersen NC, et al. **Adv Exp Med Biol** 1984;172:365. [5]Addie DD, et al. **J Gen Virol** 2003;84(Pt10):2735. [6]Kummrow M, et al. **Clin Diagn Lab Immunol** 2005;12(10):1209. [7]Shiba N, et al. **Vet Microbiol** 2007;124(3-4):348. [8]Lin CN, et al. **Vet Microbiol** 2009;136(3-4):233. [9]Gunn-Moore DA, et al. **Vet Microbiol** 1999;69(1-2):127. [10]Kiss I, et al. **Vet J** 2000;159(1):64. [11]Battilani M, et al. **FEMS Immunol Med Microbiology** 2003;39)2):141-147. [12]Duarte A, et al. **Vet Microbiol** 2009;138(1-2):163. [13]Pedersen NC. **J Fel Med Surg** 2009;11(4):225. [14]Gunn-Moore DA, et al. **Vet Microbiol** 1998;62(3):193. [15]Kipar A, et al. **Vet Microbiol** 1999;69(1-2):131. [16]Meli M, et al. **J Fel Med Surg** 2004;6(2):69. [17]Kipar A, et al. **J Fel Med Surg** 2006;8(1):69. [18]Simons FA, et al. **J Virol Meth** 2005;124(1-2):111. [19]Vogel L, et al. **Vet Res** 2010;41(5):71. [20]Vennema H, et al. **Virology** 1998;243(1):150. [21]Kennedy M, et al. **Vet Microbiol** 2001;81(3):227. [22]Chang HW, et al. **J Gen Virol** 2010;91(Pt2):415. [23]Pedersen NC, et al. **Virus Res** 2012;165(1):17. [24]Chang HW, et al. **Emerg Inf Dis** 2012;18(7):1089. [25]Kida K, et al. **J Vet Med Sci** 1999;61(8): 935. [26]Takano T, et al. **Virus Res** 2011;156(1-2):72. [27]Dewerchin HI, et al. **Arch Virol** 2005;150(12):2483. [28]Rottier PJ, et al. **J Virol** 2005;79(22):14122. [29]Balint A, et al. **J Virol** 2012;86(11):6258. [30]Dedeurwaerder A, et al. **Vet Microbiol** 2013;162(2-4):447. [31]Kipar A, et al. **Vet Pathol** 2005;42(3):321. [32]Dvorak HF, et al. **Curr Top Microbiol Immunol** 1999;237:97. [33]Takano T, et al. **Virus Res** 2011;158(1-2):161. [34]Takano T, et al. **Vet Microbiol** 2007;119(2-4):121. [35]Regan AD, et al. **Virology** 2009;384(1):135. [36]Takano T, et al. **Arch Virol** 2009;154(5):775. [37]Kipar A, et al. **Vet Immunol Immunopathol** 2006;112(3-4):141. [38]De Groot-Mijnes JDF, et al. **J Virol** 2005;79(2):1036. [39]Gelain ME, et al. **J Fel Med Surg** 2006;8(6):389. [40]Dean G, et al. **Vet Microbiol** 1993;97(102):1. [41]Hornyák A, et al. **J Virol Meth** 2012;181(2):155.

REGRESSIVE FELINE LEUKEMIA VIRUS INFECTIONS – WHAT DO THEY MEAN?

Katrin Hartmann, Prof., Dr. med. vet., Dr. Habil., DECVIM
Munich, Germany

INTRODUCTION

Since the invention of highly sensitive polymerase chain reaction (PCR) assays, cats with regressive feline leukemia virus (FeLV) infection (that are FeLV antigen-negative and FeLV PCR-positive) are commonly detected, but their role is not well understood. It is currently discussed, how common these regressive infections are, how often they are reactivated, and which clinical signs they can cause. Although FeLV infection is less commonly diagnosed than in the previous 20 years and prevalence of FeLV antigen-positive cats has been decreasing in most countries, the importance of FeLV as a pathogen is likely underestimated as it has been shown that regressively infected cats (that are negative in routinely used antigen FeLV tests) also can develop clinical signs.

WHAT IS REGRESSIVE FELV INFECTION?

Regressively FeLV-infected cats are FeLV antigen-negative, but FeLV provirus-positive. They are frequently detected, but their clinical relevance and role in FeLV epidemiology is still not fully understood. FeLV infection has different stages. Recently, novel diagnostic tools, including very sensitive PCR methods providing new data on the course of FeLV infection, have questioned the traditional understanding of FeLV pathogenesis. Cats believed to be immune to FeLV after infection were found to remain provirus-positive. Antigen-negative, provirus-positive cats are considered FeLV carriers. Following reactivation, they can act as an infection source. As FeLV provirus is integrated into the cat's genome, it is unlikely to be fully cleared over time. Antigen-negative, provirus-positive cats do not shed the virus, but reactivation with reoccurring virus shedding is possible.[1,2] Based on this information, a new classification has been proposed, in which the stages of FeLV infection are defined as abortive infection (comparable to the former "regressor cats"), progressive infection (comparable to the former "persistent viremia"), focal or atypical infection, and finally regressive infection (comparable to the former "transient viremia" followed by "latent infection"), (Table 1).[2,3,4,5,6]

Abortive infection can occur in some immunocompetent cats (formerly called "regressor cats). In these abortively infected cats, viral replication is terminated by an effective humoral and cell-mediated immune response after initial replication in the local lymphoid tissue in the oropharyngeal area. These cats never become viremic, but develop high levels of neutralizing antibodies. Neither FeLV antigen nor viral RNA or proviral DNA can be detected in the blood at any time.[6] Abortive infection is likely caused when a cat is exposed to low doses of FeLV.[7] On the contrary, in cats with progressive FeLV infection, infection is not contained. Thus, extensive virus replication occurs, first in lymphoid tissues, followed by bone marrow and mucosal and glandular epithelial tissues. Progressively infected cats remain persistently FeLV antigen-positive. They are infectious to other cats for the remainder of their life. This condition has been called "persistent viremia". Cats with progressive infection develop FeLV-associated diseases, and most of them die within a few years. A rare stage of so-called focal infections or atypical infections has been reported in up to 10% of experimentally infected cats, but is not common in the field. Focal infections are characterized by a persistent atypical local viral replication (e.g., in mammary glands, bladder, eyes). This replication can lead to intermittent or low-grade production of antigen, and therefore, these cats can have weakly positive or discordant results in antigen tests, or positive and negative results may alternate.[6]

Regressive FeLV infection develops following an effective immune response of the cat to the virus that, however, is not able to clear infection like in abortive infection. In regressive infection, virus replication and viremia are contained prior to or shortly after bone marrow infection. After initial infection, replicating FeLV spreads systemically through infected mononuclear cells (lymphocytes and monocytes). During this stage, cats have positive results on tests that detect free FeLV antigen in plasma (e.g., ELISA). They shed virus, mainly with saliva. In cats with regressive infection, this viremia, however, is terminated within weeks or months (therefore formerly called "transient viremia"). In some cats, viremia may persist longer than three weeks. After about three weeks of viremia, bone marrow cells become infected, and infected hematopoietic precursor cells develop into infected granulocytes and platelets that circulate in the body. Even if bone marrow cells become infected, a certain percentage of cats are able to clear viremia. However, they cannot completely eliminate the virus from the body, even if they terminate viremia because the information for virus replication (proviral DNA) stays conserved in bone marrow stem cells. This condition has been called "latent infection" (and is now a part of the regressive infection). The molecular basis of latency is the integration of a copy of the viral genome (provirus) into cellular chromosomal DNA. Although proviral DNA remains present within the cellular genome, no virus is actively produced. Thus, cats with regressive infection have negative results in all tests that detect FeLV antigen. During cell division, proviral DNA is replicated and the information given to the daughter cells. Thus, complete cell lineages can contain FeLV proviral DNA. However, proviral DNA is not translated into proteins, and no infectious virus particles are produced. Therefore, regressively infected cats do not shed FeLV and are not infectious to others.

Regressive and progressive infections can be distinguished by repeated testing for viral antigen in peripheral blood;[3] regressively infected cats will turn FeLV antigen-negative at latest 16 weeks after infection, while progressively infected cats will remain FeLV antigen-positive. Initially both, regressive and progressive infections are accompanied by persistence of FeLV proviral DNA in the blood detected by PCR but later are associated with different FeLV loads when measured by quantitative PCR; regressive infection is associated with low-, progressive infection with high- virus load.[3,8] Subsequently, the infection outcome is associated with different FeLV loads, and not only the overall loads but rather those of specific leukocyte subsets influence the infection outcome.[8]

Table 1. Stages of feline leukemia virus (FeLV) infection.

Stages of FeLV infection	FeLV p27 antigen in blood	Virus blood culture	Viral RNA in blood	Viral DNA in blood	Viral tissue culture	Viral shedding	FeLV-associated disease
Abortive	Negative	Negative	Negative	Negative	Negative	Negative	Not present
Progressive	Positive	Positive	Positive	Positive	Positive	Positive	Common
Focal	Variable	Variable	Variable	Variable	Positive	Variable	Unlikely
Regressive	Negative	Negative	Negative	Positive	Negative	Negative	Possible

HOW COMMON IS REGRESSIVE FELV INFECTION?

Prevalence of FeLV infection is decreasing in many countries. However, with few exceptions, FeLV prevalence studies are uniquely based on detection of FeLV antigen in blood using ELISA or similar immunochromatographic assays. However, FeLV antigen can only be detected in a part of all infected animals. Regressively infected cats are negative in antigen testing and only harbor provirus in their bone marrow cells after overcoming antigenemia.[9] Thus, antigen testing underestimates the true prevalence of infection. According to a new study, in every cat that had contact to FeLV, provirus can be detected by PCR at some stages of infection.[5] As the proviral load may be near the detection limit of the PCR, these cats can be PCR-negative intermittently but later on can turn positive again.[5]

In a study in Switzerland it was shown that in addition to 6.9% antigen-positive, provirus-positive (progressively infected) cats, 10.0% of the cat population that were negative for p27 antigen were positive for proviral DNA in blood (regressively infected).[10] In a German study, 9/495 (1.8%) cats were progressively infected and in addition 6/495 (1.2 %) were regressively infected.[11] The difference in prevalence of regressive infection could be explained by the time span between the Swiss and the German study. The Swiss study was performed in 1999 and 2000, the German study in 2011 and 2012, and in the meantime FeLV prevalence could have further decreased. Alternatively, FeLV infection rate in Switzerland might be truly higher than in Germany.

HOW OFTEN IS REGRESSIVE FELV INFECTION REACTIVATED?

Regressive FeLV infection can be reactivated because the information for producing complete viral particles is present and can potentially be reused when antibody production decreases (e.g., after immunosuppression). This usually occurs after stress and can be experimentally induced in cats by administration of high doses of glucocorticoids.[9] Reactivation is more likely the earlier the stress factor occurs after the viremic phase. In the first weeks after viremia, viral replication can be experimentally reactivated in most cats. As the time passes, regressive infections become more difficult to reactivate, even with high doses of glucocorticoids. After one year and more of infection, reactivation is considered uncommon. This can be explained by genetic code-reading mistakes that can occur if the information is frequently reproduced in these fast-dividing cells. Thus, information to produce infectious viral particles gets lost, and reactivation becomes more and more unlikely over time. However, cats can reverse to a viremic state even many years after viremia if they become immunosuppressed. It has been demonstrated that the proportion of experimentally infected cats that had regressive FeLV infections in their bone marrow decreased with time following disappearance of viremia.[12] In the first three months after recovery from viremia, integrated virus could be isolated from the bone marrow in approximately 50% of experimentally infected cats. A pronounced decrease in the incidence of regressive infections occurred 190 days after the viremia.[12,13] More than one year later, only 5/19 previously challenge-exposed cats that were FeLV antigen-negative still had FeLV detectable in several tissues (e.g., bone marrow, spleen, lymph node, small intestine).[14] At three years post viremia, only about 8% of cats still harbored FeLV infections in bone marrow, myelomonocytic cells, and stromal fibroblast cells.[12,13,15]

CAN CATS WITH REGRESSIVE FELV INFECTION DEVELOP CLINICAL SIGNS?

Most cats with regressive infections are clinically healthy because for the majority of pathogenic mechanisms by which FeLV causes clinical signs, active virus replication is necessary. This is not the case in regressive FeLV infections, in which the virus is harbored in a "dormant" and non-productive form. However, besides the potential risk of reactivation, regressive infections can have some impact on a cat's health. FeLV provirus can be inserted at many different sites in the host's genome, carrying potent regulatory signals. In the development of tumors or myelosuppressive disorders, integrated FeLV provirus may interrupt or inactivate cellular genes in the infected cells, or regulatory features of viral DNA can alter expression of neighboring genes. In addition, FeLV not only contributes its genes to the host, it also has been shown to appropriate cellular genes. Several such transducted genes that are also present in regressively infected cells have been implicated in viral oncogenesis.[16,17,18] Furthermore, because bone marrow microenvironment cells (e.g., myelomonocytic progenitor cells and stromal fibroblasts) provide a reservoir of regressive FeLV infections, it seems possible that the integrated provirus could alter cellular functions and contribute to the pathogenesis of myelosuppressive disorders. Additionally, FeLV provirus could cause bone marrow disorders by inducing the expression of antigens on the cell surface, resulting in an immune-mediated destruction of the cell.

Some studies demonstrated involvement of regressive FeLV infection in **tumors**. Prevalence of lymphomas caused by FeLV seems to be higher than indicated by conventional antigen testing of blood.[19] Cats from FeLV cluster households had a 40-fold higher rate of development of FeLV-negative lymphoma than did those from the general population. FeLV-negative lymphomas have also occurred in laboratory cats known to have been infected previously with FeLV.[18] FeLV proviral DNA was detected in lymphomas of cats that tested negative for FeLV antigen, suggesting that the virus is associated with a larger proportion of lymphomas than previously thought. FeLV has been shown to incorporate cellular genes; several such transducted genes that are also present in regressively infected cells have been implicated in viral oncogenesis.[16,17,18] Several studies showed involvement of FeLV provirus in lymphomas of FeLV antigen-negative cats.[19,20] It is still unclear, how common regressive FeLV infection is responsible for FeLV-associated tumors in the field as study results have been controversial. Proviral DNA was detected in formalin-fixed, paraffin-embedded tumor tissue in 7/11 FeLV-negative cats with lymphoma.[19] However, other groups found evidence of provirus in only 1/22[17] and in 0/50 FeLV antigen-negative lymphomas.[21]

The most important mechanism by which FeLV causes malignancy is by insertion of the FeLV genome into the cellular genome near a cellular oncogene (most commonly *myc*), resulting in activation and over-expression of that gene. These effects lead to uncontrolled proliferation of these cells (clone). A malignancy results in the absence of an appropriate immune response. FeLV may also incorporate the oncogene to form a recombinant virus (e.g., FeLV-B, FeSV) containing cellular oncogene sequences that are then rearranged and activated. When they enter a new cell, these recombinant viruses are oncogenic. In a study of 119 cats with lymphomas, transduction or insertion of the *myc* locus had occurred in 38 cats (32%).[22] Thus, FeLV-induced neoplasms are caused, at least in part, by somatically acquired insertional mutagenesis in which the integrated provirus may activate a proto-oncogene or disrupt a tumor suppressor gene. A recent study suggested that the *U3-LTR* region of FeLV transactivates cancer-related signaling pathways through production of a non-coding 104 base RNA transcript that activates *NF kappaB*.[23] Twelve common integration sites for FeLV associated with lymphoma development have been identified in six loci: *c-myc*, *flvi-1*, *flvi-2* (contains *bmi-1*), *fit-1*, *pim-1*, and *flit-1*. Oncogenic association of the loci is based on the fact that *c-myc* is known as a proto-oncogene, *bmi-1* and *pim-1* have been recognized as *myc*-collaborators, *fit-1* appears to be closely linked to *myb*, and *flit-1* insertion was shown to be associated with over-expression of cellular genes, e.g., *activin-A receptor type II-like 1* (*ACVRL1*).[24] *Flit-1* seems to have an important role in the development of lymphomas and appears to represent a common novel FeLV proviral integration domain that may influence lymphomagenesis by insertional mutagenesis. Among 35 FeLV-related tumors, 5/25 thymic lymphomas, 5/5 multicentric lymphomas, and 1/1 T-lymphoid leukemia demonstrated proviral insertion within *flit-1* locus, whereas 0/4 alimentary lymphomas examined had rearrangements in this region. Expression of *ACVRL1* mRNA was detected in the two thymic lymphomas with *flit-1* rearrangement, whereas normal thymuses and seven lymphoid tumors without *flit-1* rearrangement had no detectable *ACVRL1* mRNA expression.[25]

Hematologic disorders described in association with FeLV include anemia (non-regenerative or regenerative), persistent, transient, or cyclic neutropenia, panleukopenia-like syndrome, platelet abnormalities (thrombocytopenia and platelet function abnormalities), and aplastic anemia (pancytopenia). For the majority of pathogenic mechanisms in which FeLV causes bone marrow suppression, active virus replication is required. However, it has been demonstrated that in some FeLV antigen-negative cats, regressive FeLV infection can be responsible for bone marrow suppression. In a recent study including 37 FeLV antigen-negative cats with myelosuppression, 2/37 cats were found regressively infected with FeLV (both had non-regenerative anemia).[26] Bone marrow suppression could develop in theses cats, because FeLV provirus can interrupt or inactivate cellular genes in the infected cells, or regulatory features of viral DNA can alter expression of neighboring genes. Additionally, cell function of provirus-containing myelomonocytic progenitor and stromal fibroblasts that provide bone marrow microenvironment might be altered. Alternatively, FeLV provirus could cause bone marrow disorders by inducing the expression of antigens on the cell surface, resulting in an immune-mediated destruction of the cell.

Most FeLV-associated anemias are non-regenerative and are caused by the bone marrow suppressive effect of the virus resulting from primary infection of hematopoietic stem cells and infection of stroma cells that constitute the supporting environment for hematopoietic cells. *In vitro* exposure of normal feline bone marrow to some strains of FeLV causes suppression of erythrogenesis. FeLV infection can cause decreased platelet counts. It also can be responsible for platelet function deficits, and the lifespan of platelets is shortened in some FeLV-infected cats. Thrombocytopenia can occur secondary to decreased platelet production from FeLV-induced bone marrow suppression or leukemic infiltration. Platelets harbor FeLV, and megakaryocytes are frequent targets of FeLV infection. Neutropenia is also common in FeLV-infected cats and generally occurs alone or in conjunction with other cytopenias. In some cases, myeloid hypoplasia of all granulocytic stages is observed, suggesting infection of neutrophil precursors. In some neutropenic FeLV-infected cats, an arrest in bone marrow maturation can occur at the myelocyte and metamyelocyte stages. It has been hypothesized that an immune-mediated mechanism is responsible in cases in which neutrophil counts recover with glucocorticoid treatment ("glucocorticoid-responsive neutropenia"). Hematopoietic neoplasia ("myeloprolifertaive disorders"), including leukemia, can also cause bone marrow suppression syndromes by crowding out. Myelodysplastic syndrome (MDS), characterized by peripheral blood cytopenias and dysplastic changes in the bone marrow, is a pre-stage of acute myeloic leukemia. It was found that changes of the *LTR* region of the FeLV genome (presence of three tandem direct 47-bp repeats in the upstream region of the enhancer (*URE*)) are strongly associated with the induction of MDS.[27]

REFERENCES

[1]Torres AN, et al. **Vet Immunol Immunopathol** 2008;123:81. [2]Cattori V, et al. **Mol Cell Probes** 2006;20:172. [3]Torres AN, et al. **Virology** 2005;332:272. [4]Hofmann-Lehmann R, et al. **Vaccine** 2007;25:5531. [5]Hofmann-Lehmann R, et al. **Vet Immunol Immunopathol** 2008;123:119. [6]Levy J, et al. **J Feline Med Surg** 2008;10:300. [7]Major A, et al. **Vet Res** 2010;41:17. [8]Pepin AC, et al. **Virus Res** 2007;127:9. [9]Rojko JL, et al. **Nature** 1982;298:385. [10]Hofmann-Lehmann R, et al. **J Gen Virol** 2001;82:1589. [11]Englert T, et al. **J Feline Med Surg** 2012;14:392. [12]Pedersen NC, et al. **Feline Pract** 1984;14:32. [13]Pacitti AM, Jarrett O. **Vet Rec** 1985;117:472. [14]Hayes KA, et al. **Am J Vet Res** 1992;53:604. [15]Madewell BR, Jarrett O. **Vet Rec** 1983;112:339. [16]Rezanka LJ, et al. **Cancer Invest** 1992;10:371. [17]Sheets RL, et al. **J Virol** 1993;67:3118. [18]Rohn JL, et al. **J Virol** 1994;68:2458. [19]Jackson ML, et al. **Can J Vet Res** 1993;57:269. [20]George JW, et al. **Am J Vet Res** 2001;63:1172. [21]Stuetzer B, et al. **J Feline Med Surg** 2011;13:81. [22]Tsatsanis C, et al. **J Virol** 1994;68:8296. [23]Forman LW, et al. **FEBS Lett** 2009;583:1386. [24]Fujino Y, et al. **Vet Immunol Immunopathol** 2008;123:138. [25]Fujino Y, et al. **Virology** 2009;386:16. [26]Stützer B, et al. **J Vet Intern Med** 2010;24:192. [27]Hisasue M, et al. **Int J Cancer** 2009;124: 1133.

FELINE HERPESVIRUS-1: ANTIVIRAL DRUGS FROM BENCH-TOP TO CLINIC

David J. Maggs, BVSc, DACVO
Davis, CA

INTRODUCTION

Because no antiviral drug has been developed for cats infected with feline herpesvirus type 1 (FHV-1), veterinarians frequently "borrow" drugs developed for treating herpesviruses of humans. This session summarizes investigations regarding the in vitro efficacy of antiherpetic drugs against FHV-1 and their safety and efficacy when used in cats. The goal is to enable the reader to be better placed to make evidence-based decisions regarding antiviral therapy in FHV-1-infected cats. Although numerous antiviral agents have been used to treat cats infected with FHV-1, and each has some specific features, some general comments regarding these agents are also possible. We will start with those.

GENERAL COMMENTS RE ANTIVIRAL DRUGS USED FOR CATS INFECTED WITH FHV-1

- No antiviral agent has been developed specifically for FHV-1, although many have been tested for in vitro efficacy against this virus. Agents effective against human herpesviruses are not predictably effective against FHV-1. They should be tested in vitro before they are administered to cats. Because in vitro potency is described as the drug concentration at which viral replication is suppressed by 50% (or IC_{50}), the more potent the drug, the lower its IC_{50}. Table 1 reports antiviral efficacy (IC_{50}) against FHV-1 and human herpes simplex virus type 1 (HSV-1) for a number of antiviral drugs in common veterinary use.

IC_{50} (μM)	TFU	GCV	IDU	CDV	PCV	VDB	ACV	Foscarnet
FHV-1	1-19[1,2]	5[3]	4-7[1,3]	8-168[3-5]	1-130[3-5]	21[1]	16-222[1-6]	233[3]
HSV-1	1.7	0.8	1	9	2	4	1	74

Table 1. Comparative antiviral efficacy (IC_{50}) of various antiviral drugs against FHV-1 and herpes simplex virus type *1* (TFU = trifluridine; GCV = ganciclovir; IDU = idoxuridine; CDV = cidofovir; PCV = penciclovir; VDB = Vidarabine; ACV = acyclovir; μM = micromolar. Superscript numbers cite studies from which data were retrieved.)

- To date, no antiviral agent has been developed for cats, and agents with a reasonable safety profile in humans are not predictably non-toxic when administered to cats. All antiviral drugs should be tested for safety in cats; to date few have.
- Almost all antiviral agents require viral and sometimes host metabolism to become activated. As with other drugs, cats sometimes metabolize antiviral drugs differently from other species. Therefore, pharmacokinetic studies in cats are required and data from other species should be very cautiously extrapolated if at all.
- Antiviral agents tend to be more toxic than do antibacterial agents and most are too toxic to use systemically. For those which can be administered systemically, a relatively narrow safety margin often exists and special considerations should always be given to patients with reduced hepatic or renal function.
- All antiviral agents currently used for cats infected with FHV-1 are virostatic. This has a series of clinical implications:
    - They will have no effect against latent virus
    - They require frequent application
    - They only *assist* in (don't guarantee) viral clearing

SPECIFIC COMMENTS RE ANTIVIRAL DRUGS USED FOR CATS INFECTED WITH FHV-1

With those general comments in mind for all antiviral drugs, some additional comments specific to many commonly used drugs follow:

Trifluridine

Trifluridine is a nucleoside analogue and is too toxic to be administered systemically. Although trifluridine is the most potent antiviral drug for FHV-1 in vitro, and despite it having excellent corneal penetration in in vivo studies, many cats appear to find it irritating when applied topically as a 1% ophthalmic solution. In addition, it is expensive, even as a generic. It should be applied to the affected eye 5-6 times daily.

Idoxuridine

Idoxuridine is a nucleoside analogue that competes with thymidine for incorporation into viral DNA rendering the resultant virus incapable of replication. However, it apparently does this less effectively in FHV-1 than in HSV-1 (**see table 1**). It was available commercially as an ophthalmic 0.1% solution or 0.5% ointment, but now must be compounded in these formulations. It is generally well tolerated by cats and is inexpensive. It should be applied to the affected eye 5-6 times daily.

Vidarabine

Vidarabine appears to affect viral DNA synthesis by interfering with DNA polymerase. Because it affects a viral replication step different from that targeted by idoxuridine, vidarabine may be effective in patients whose disease seems resistant to idoxuridine. It is no longer available commercially in the USA, but can be compounded as a 3% ophthalmic ointment. Cats seem to tolerate vidarabine well. As for most of the antiviral drugs, it should be applied to the affected eye 5-6 times daily.

Cidofovir

Cidofovir is commercially available in the USA only as an injectable drug for humans; however, it has been studied as a 0.5% solution compounded in artificial tears and applied topically twice daily to cats experimentally infected with FHV-1.[7] Its use in these cats was associated with reduced viral shedding and reduced clinical disease.[7] Its efficacy at only twice daily (despite being virostatic) is believed to be due to the long tissue half-lives of the metabolites of this drug.[8] There are reports of its experimental topical use in humans and rabbits being associated with stenosis of the nasolacrimal drainage system components and, as yet, it is not commercially available as an ophthalmic agent in humans.[9,10] Although many use it for short term therapy in cats, and it appears very effective in many of these patients, at this stage there are insufficient data to support its long term safety as a topical agent. This is the only compounded antiviral drug mentioned to date for which there are some data on effects of storage on antiviral efficacy. In an in vitro study with the drug made up in saline (i.e., differently from that used in vivo), the drug retained antiviral efficacy under a number of storage conditions, the most rigorous of which was 180 days at 4^0C (in plastic or in glass).[11] Note that this study did not asses safety following storage; only antiviral efficacy.

Acyclovir

Acyclovir has very low in vitro antiviral efficacy against FHV1 (relative to HSV), has poor bioavailability in cats, and can cause bone marrow suppression in some when administered systemically.[12] Oral administration of 50 mg/kg acyclovir to cats was associated with peak plasma levels of approximately only one third the IC_{50} for this virus (33 µM).[12] For all of these reasons and because famciclovir appears to have a better safety and efficacy profile, I no longer use acyclovir systemically. However, acyclovir is available as a 3% ophthalmic ointment in some regions. In one study in which a 0.5% ointment was used 5 times daily, the median time to resolution of clinical signs was 10 days.[2] Cats treated only 3 times daily failed to resolve until therapy was increased to 5 times daily, and then took approximately twice as long to do so. Taken together, these data suggest that very frequent topical application of acyclovir may produce concentrations at the corneal surface that do exceed the reported IC_{50} for this virus without noted toxicity.

Valacyclovir

Valacyclovir is a prodrug of acyclovir that, in humans and cats, is more efficiently absorbed from the gastrointestinal tract compared with acyclovir, and is then converted to acyclovir once absorbed. Its safety and efficacy have been studied in cats.[13] Although it does elevate plasma concentrations of acyclovir in excess of the IC_{50}, it induces fatal hepatic and renal necrosis, along with bone marrow suppression (without reducing viral shedding or clinical disease severity).[13] Therefore, despite its superior pharmacokinetics, valacyclovir should never be used in cats. Note that it is likely that the toxic factor was the enhanced acyclovir concentrations achieved; not the valacyclovir per se. Thus these data reinforce the apparent toxicity of acyclovir in cats.

Famciclovir

Famciclovir is a prodrug of penciclovir, which is highly effective against FHV-1 in vitro.[3-5] However metabolism of famciclovir to penciclovir is complex and requires di-deacetylation, predominantly in the blood to BRL 42359, and subsequent oxidation to penciclovir by a hepatic aldehyde oxidase. (Neither famciclovir nor BRL 42359 have any inherent antiviral activity.) Unfortunately, the activity of this essential hepatic aldehyde oxidase is negligible in cats,[14] and thus, famciclovir administration in this species achieves lower than expected conversion rates of famciclovir to penciclovir. This necessitates cautious extrapolation to cats of data generated in humans. Indeed data to date in normal and experimentally infected cats suggest that the pharmacokinetics of this drug are extremely complex in cats, and differ markedly in comparison to other species such as humans, dogs, and rats.[15-18] For example, famciclovir appears to be very poorly bioavailable in cats relative to humans. Bioavailability in humans administered a single oral dose of approximately 7 mg of famciclovir/kg is 77%[19] whereas in cats it is only 12.5% or 7% following oral administration of 40 or 90 mg of famciclovir/kg, respectively.[15] Once absorbed, the maximum plasma penciclovir concentration (C_{max}) achieved in cats administered a single dose of 90 mg of famciclovir/kg is 1.3 μg/mL, which is approximately one-third of that in dogs administered a single dose of only 25 mg of famciclovir/kg (4.4 μg/mL).[15,17] Furthermore, comparison of data from studies utilizing famciclovir at different dose rates in cats provides evidence of nonlinear pharmacokinetics in this species. A 2.7-fold increase in famciclovir dose (from 15 to 40 mg/kg) resulted in a 4-fold increase in penciclovir C_{max} in cats; no further increase in C_{max} was observed when the dose was increased from 40 to 90 mg/kg.[15,16] Taken together, these data suggest that famciclovir is variably absorbed in cats,[15] that its metabolism to penciclovir becomes saturated at increasing doses,[15] and that the highest attainable plasma penciclovir C_{max} in cats is lower than in other species.[15-18] Despite this, there is mounting evidence that suggests famciclovir is very effective in some cats with experimentally induced or suspected spontaneous herpetic disease.[20-24] Two case series describing client-owned cats receiving a range of famciclovir doses demonstrated that this treatment is sometimes associated with improved keratitis, conjunctivitis, dermatitis, or rhinosinusitis attributed to FHV-1.[25,26] Likewise, cats experimentally inoculated with FHV-1 and treated with 90 mg of famciclovir/kg orally 3 times daily (TID) showed significant improvement in numerous systemic, ophthalmic, clinicopathologic, virologic, and histologic parameters in comparison with results for these parameters in placebo-treated cats.[27] Most recently, we have also shown that tear penciclovir concentrations likely to be efficacious against FHV-1 were achieved in a majority of client-owned cats receiving 40 mg famciclovir/kg PO TID.[28] This suggests that a topical antiviral may not be needed in cats receiving systemically administered famciclovir. However, other topical drugs such as an antibacterial agent or mucinomimetic artificial tear replacement may be critical.[24]

REFERENCES

1. Nasisse MP, et al. In vitro susceptibility of feline herpesvirus-1 to vidarabine, idoxuridine, trifluridine, acyclovir, or bromovinyldeoxyuridine. *Am J Vet Res* 1989;50:158-160.
2. Williams DL, et al. Efficacy of topical aciclovir for the treatment of feline herpetic keratitis: results of a prospective clinical trial and data from in vitro investigations. *Vet Rec* 2005;157:254-257.
3. Maggs DJ, Clarke HE. In vitro efficacy of ganciclovir, cidofovir, penciclovir, foscarnet, idoxuridine, and acyclovir against feline herpesvirus type-1. *Am J Vet Res* 2004;65:399-403.
4. Hussein IT, et al. Penciclovir is a potent inhibitor of feline herpesvirus-1 with susceptibility determined at the level of virus-encoded thymidine kinase. *Antiviral Res* 2008;78:268-274.
5. Hussein IT, Field HJ. Development of a quantitative real-time TaqMan PCR assay for testing the susceptibility of feline herpesvirus-1 to antiviral compounds. *J Virol Methods* 2008;152:85-90.
6. Collins P. The spectrum of antiviral activities of acyclovir in vitro and in vivo. *J AntimicrobChemother* 1983;12 Suppl B:19-27.
7. Fontenelle JP, et al. Effect of topical ophthalmic application of cidofovir on experimentally induced primary ocular feline herpesvirus-1 infection in cats. *Am J Vet Res* 2008;69:289-293.
8. Neyts J, et al. Particular characteristics of the anti-human cytomegalovirus activity of (S)-1-(3-hydroxy-2-phosphonylmethoxypropyl)cytosine (HPMPC) in vitro. *Antiviral Res* 1991;16:41-52.
9. Sherman MD, et al. Treatment of conjunctival squamous cell carcinoma with topical cidofovir. *Am J Ophthalmol* 2002;134:432-433.
10. Inoue H, et al. Clinical evaluation of local ocular toxicity in candidate anti-adenoviral agents in vivo. *Ophthalmologica* 2009;223:233-238.
11. Stiles J, et al. Stability of 0.5% cidofovir stored under various conditions for up to 6 months. *Vet Ophthalmol* 2010;13:275-277.
12. Owens JG, et al. Pharmacokinetics of acyclovir in the cat. *J Vet Pharmacol Ther* 1996;19:488-490.
13. Nasisse MP, et al. Effects of valacyclovir in cats infected with feline herpesvirus 1. *Am J Vet Res* 1997;58:1141-1144.
14. Dick RA, et al. Identification of aldehyde oxidase as the neonicotinoid nitroreductase. *Chem Res Toxicol* 2005;18:317-323.
15. Thomasy SM, et al. Pharmacokinetics of famciclovir and penciclovir following single-dose oral administration of famciclovir to cats. *Vet Ophthalmol* 2009;12:402.
16. Thomasy SM, et al. Pharmacokinetics and safety of penciclovir following oral administration of famciclovir to cats. *Am J Vet Res* 2007;68:1252-1258.
17. Filer CW, et al. Metabolic and pharmacokinetic studies following oral administration of famciclovir to the rat and dog. *Xenobiotica* 1995;25:477-490.
18. Pue MA, et al. Linear pharmacokinetics of penciclovir following administration of single oral doses of famciclovir 125, 250, 500 and 750 mg to healthy volunteers. *J Antimicrob Chemother* 1994;33:119-127.
19. Pue MA, Benet LZ. Pharmacokinetics of famciclovir in man. *Antiviral Chemical Chemotherapy* 1993;4(Suppl 1):47-55.
20. Thomasy SM, et al. Safety and efficacy of famciclovir in cats infected with feline herpesvirus 1 *Veterinary Ophthalmology* 2007;10:418-418.
21. Malik R, et al. Treatment of feline herpesvirus-1 associated disease in cats with famciclovir and related drugs. *J Feline Med Surg* 2009;11:40-48.
22. Thomasy SM, Maggs DJ. Treatment of ocular nasal, and dermatologic disease attributable to feline herpesvirus 1 with famciclovir. *Vet Ophthalmol* 2008;11:418.
23. Thomasy SM, et al. Evaluation of orally administered famciclovir in cats experimentally infected with feline herpesvirus 1. *Am J Vet Res*. 2011;72:85-95.
24. Lim CC, et al. Effects of famciclovir on tear film parameters in cats experimentally infected with feline herpesvirus 1. *Vet Ophthalmol* 2008;11:425.

25. Malik R, et al. Treatment of feline herpesvirus-1 associated disease in cats with famciclovir and related drugs. *J Feline Med Surg* 2009;11:40-48.
26. Thomasy SM, Maggs DJ. Treatment of ocular nasal, and dermatologic disease attributable to feline herpesvirus 1 with famciclovir. *Vet Ophthalmol* 2008;11:418.
27. Thomasy SM, et al. Evaluation of orally administered famciclovir in cats experimentally infected with feline herpesvirus type-1. *Am J Vet Res* 2011;72:85-95.
28. Thomasy SM, et al. Pharmacokinetics of famciclovir and penciclovir in tears following oral administration of famciclovir to cats: a pilot study. *Vet Ophthalmol* 2012.

CYTAUXZOONOSIS: A TREATBLE DISEASE!!!

Adam Birkenheuer, DVM, PhD, DACVIM
Raleigh, NC

HISTORY

Cytauxzoonosis is a tick-transmitted protozoan disease of cats caused by *Cytauxzoon felis* that was first reported in Missouri in 1976. Over the next thirty years the disease was only recognized in the southcentral and southeastern United States. Recently the geographic range of the organism has been recognized to extend east and north. The states in which *C. felis* has been reported are shown below.

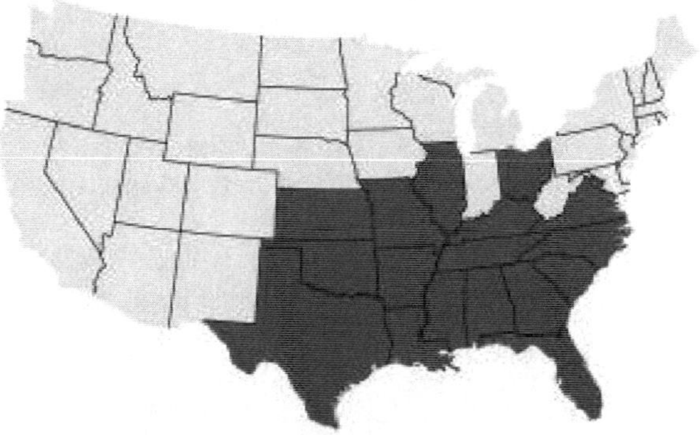

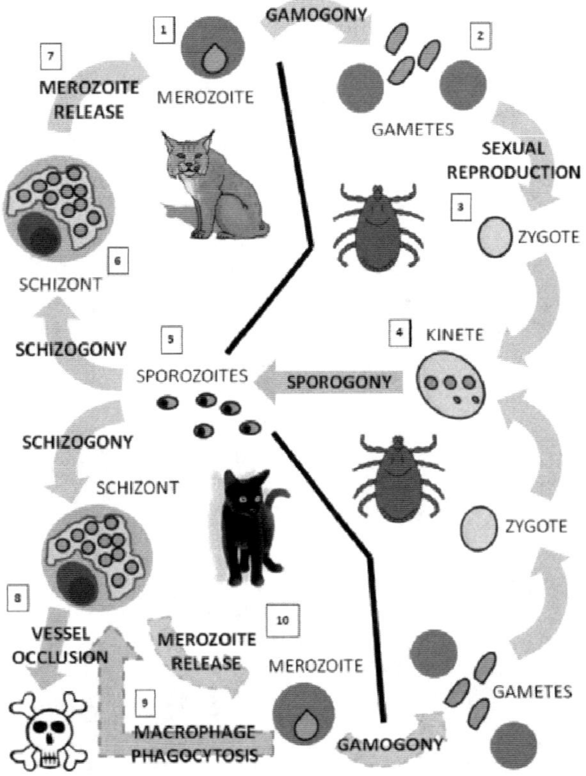

CLASSIFICATION AND LIFE-CYCLE

Based on its DNA sequence and life-cycle, *C. felis* is closely related to the *Theileria* spp. Unlike *Babesia* spp. which have only been found to infect red blood cells, *Cytauxzoon* and *Theileria* first enter a white blood cell after tick-transmission to the vertebrate host. Below is a cartoon depicting the proposed life-cycle of *C. felis* showing Bobcats (*Lynx rufus*) as the most likely reservoir hosts. *Dermacentor variabilis* and *Amblyomma americanum* have been demonstrated to transmit *C. felis* to domestic cats under laboratory conditions. The disease typically occurs between the months of April and September which correlates with the peak tick activity in most regions. Approximately 13-16 days after being bitten by an infected tick the cat develops non-specific signs of illness, typically characterized by anorexia, lethargy, pyrexia, icterus, and pallor. The initial infection of macrophages, or schizogenous phase, is responsible for nearly all the pathology associated with cytauxzoonosis. Bobcats do not appear to undergo a profound schizogenous phase of infection and therefore do not develop severe clinical disease. The erythrocytic phase of the organism alone does not induce severe disease.

CLINICAL SIGNS AND LABORATORY ABNORMALITIES

It is typically characterized by an acute febrile illness that is associated with decreases in one or more cell lines. Physical exam findings are often non-specific. Cats usually have a profound fever, but hypothermia may be identified in moribund cats. Lymphadenopathy and splenomegaly are common but not uniform findings. Cats are typically depressed and vocalization (death yowl) is a common finding in advanced cases. Some cats will have dyspnea as a prominent clinical feature (in our experience this is frequently due to pleural effusion and can be addressed easily with thoracocentesis). Laboratory abnormalities include non-regenerative anemia, leukopenia, thrombocytopenia, hyperbilirubinemia and bilirubinuria, elevations in liver enzymes (often not as severe as would be expected considering the degree of hyperbilirubinemia), hyperglycemia and hypoalbuminemia. In our experience thrombocytopenia and neutropenia are the most common CBC findings. Elevations in aPTT and non-overt DIC seem to be common. Many of the typical laboratory abnormalities become more pronounced as the disease progresses and may be absent or mild if the cat is presented for care early in the course of illness. The course of disease is short, and without treatment most cats succumb within 5 days of the onset of clinical signs.

RISK FACTORS

Cytauxzoonosis is most frequently identified in young outdoor cats. They often have a history of tick attachment. There appear to be "hot spots" of infection, and often more than one cat in a household or neighborhood will become infected. In some households where cytauxzoonosis has been diagnosed, the prevalence in the domestic cats may be >30%.

DIAGNOSTIC TESTING

Microscopy: Organisms in red blood cells are most easily identified on 1000X. Schizont infected macrophages can be seen at low power and can look like platelet clumps on the feathered edge of a peripheral blood smear. Fine needle aspiration and cytology of liver, spleen and lymphnodes can facilitate a rapid diagnosis in suspect cases. In house quick stains are usually adequate.

Polymerase Chain Reaction: PCR testing can be sensitive and specific.

TREATMENT

Aggressive supportive care is a mainstay of therapy for cytauxzoonosis. This includes IV fluid therapy and anti-coagulants (unfractionated heparin 100-300U/kg SQ TID or 300-900U/kg IV as a CRI). Corticosteroids do not appear to have a negative effect on outcome but also do not seem to provide a benefit.

Atovaquone and Azithromycin: An atovaquone (15 mg/kg PO TID with a fatty meal) and azithromycin (10mg/kg PO Q24) drug combination is the current treatment of choice for cytauxzoonosis. This combination should be administered ASAP. If you do not have this in your hospital refer to a facility that can give the meds today! A nasoesophageal feeding tube should be placed at admission to facilitate administration of medications. A randomized clinical trial demonstrated survival rates of 60% for atovaquone and azithromycin compared to 27% for imidocarb dipropionate. Treatment should be initiated within hours of admission and should be started empirically for suspect cases in endemic areas even if a definitive diagnosis is not made easily.

Imidocarb diproprionate: Imidocarb diproprionate (Imizol®, Schering-Plough) 2-4mg/kg IM once. Repeat dose in 2 weeks. Side-effects include pain at injection site and cholinergic effects (SLUD). Pre-treatment with atropine reduces cholinergic effects. Success rates have been reported to range from 0-50%.

Diminazene aceturate: Has been used with some success, but is not FDA approved or widely available in the US.

FOLLOW-UP

Time to respond: The response to therapy is NOT immediate. Most cats begin to improve within 4-7 days.

Recovery: In our experience, cats that survive acute cytauxzoonosis have complete resolution of all clinical signs (following some cats for >4 years). However, they may have persistent infection with the red blood cell form of the parasites. It is unknown whether or not these cats can serve as a reservoir for infection. It is strongly recommended that all cats are kept indoors.

Follow-up testing: Serial blood counts are recommended until PCV, WBC, and platelet count return to within normal range. In our experience, all biochemical abnormalities have returned to normal within 2-3 weeks. Parasites may remain detectable by microscopy and PCR for years.

REFERENCES

Cohn, L et. al. JVIM 25: 55-60, Birkeneheuer, A et. al. JAVMA 228(4): 568-71, Reichard, M et. al. Vetpar 161(1-2): 110-5

NUTRITIONAL CONSIDERATION OF CANINE PATIENTS WITH HEPATOBILIARY DISEASE

Dorothy P. Laflamme, MS, DVM, PhD, DACVN
Floyd, VA

Medical and dietary management of hepatic diseases should focus on treating the underlying cause and providing supportive care to allow the liver to regenerate, when possible, and slowing the progression of chronic disease. Maintaining appropriate nutrition helps support hepatic regeneration and aids symptomatic relief from clinical signs. Correction of malnutrition can improve liver function and enhance the liver's functional reserve while encouraging regeneration. However, due to the myriad of liver functions that may or may not be affected by a given disease, there is no single diet or protocol for feeding all liver patients.

PROTEIN AND ENERGY METABOLISM

Consumption of sufficient calories to maintain an optimum body weight is a critical aspect of managing patients with liver disease. In numerous studies, hepatic encephalopathy (HE) developed in dogs only after significant weight loss and loss of lean body mass. This can be explained by the fact that, in the absence of a functional liver, the skeletal muscle serves a primary role in detoxifying ammonia.[1,2] In addition to energy intake, protein plays a critical role in maintaining muscle and lean body mass.

Restriction of protein intake has become a standard part of the management of HE as a way to minimize ammonia absorption. However, protein restriction is inappropriate for most patients with liver disease, including some with HE. There is evidence that the rate of protein catabolism is increased in hepatic failure,[3] and that maintenance of positive nitrogen balance is beneficial in the face of hepatic failure.[4,5] Excess protein restriction leads to catabolism of endogenous proteins and increased ammonia production, loss of skeletal mass with decreased capacity for detoxification of ammonia, and increased potential for HE. Many serum proteins including albumin, numerous transport proteins and others normally produced by the liver, are decreased in hepatic failure. These may be further decreased by inadequate protein intake, as may occur if protein is restricted to control HE. Protein should only be restricted as much as needed to control HE. If protein restriction is necessary, a minimum intake of at least 2 g protein/kg body weight/day is recommended for dogs and 4 g/kg for cats. However, it is important that protein adequacy (e. g. serum albumin) be monitored to assure that protein depletion does not occur.

Dietary meat and blood (including endogenous blood from gastrointestinal bleeding) are poorly tolerated, perhaps because of high ammoniagenic potential from hemoglobin and myoglobin. On the other hand, protein from soybeans and dairy products are well tolerated by patients with liver failure.[6-9] For example, dogs with liver failure secondary to portosystemic shunts (PSS) that were fed a diet containing 18% isolated soy protein gained weight and remained neurologically normal for the duration of the year long study.[10] In another study, soy protein resulted in lower plasma ammonia concentrations in dogs with PSS compared to chicken protein.[11] High digestibility of dietary protein is also important. Protein that is not digested and absorbed in the small intestine can be metabolized by colonic bacteria to produce ammonia, which can subsequently be absorbed. Of the several dietary studies that seemed to "prevent" HE in PSS dogs, high digestibility was common among diets.

An amino acid imbalance is reported to occur in HE, with a relative increase in aromatic amino acids and a decrease in branched chain amino acids (BCAA) in the blood and cerebrospinal fluid.[12,13] This led to the theory that correcting the plasma amino acid profile should reverse or prevent HE. However, data regarding the use of BCAA in managing HE have been equivocal and BCAA supplemented diets do not reduce the incidence nor severity of HE in dogs with liver failure secondary to PSS.[10,14] Thus, priority should be placed on providing adequate amounts of high quality, highly digestible protein, especially those from soy and dairy.

CARBOHYDRATES AND FIBER

Glycogen storage and glucose synthesis often decrease in advanced hepatic disease, contributing to fasting hypoglycemia. Depleted glycogen stores result in premature protein catabolism to supply amino acids for gluconeogenesis. The longer a patient with hepatic failure spends in the fasting state, the greater the loss of muscle tissue and protein depletion. Nitrogen balance and nutritional status can be improved by the consumption of several small meals daily, including a bed time snack.[8] Slowly digested carbohydrates, such as from whole grains, can help provide a consistent supply of glucose and reduce the need for gluconeogenesis.

Dietary fiber and indigestible oligosaccharides

Both soluble and insoluble fibers, in moderate amounts, can be of benefit in patients with liver disease. Insoluble fibers tend to adsorb water and can dilute and bind toxins to decrease absorption. They also can contribute to firmer stools. Soluble fibers, e. g. , gums, pectins, beet pulp, fruit pomace, etc., which tend to be fermentable, can contribute to a more acidic colonic environment. In addition to other benefits, the acidity converts ammonia in the colon to ammonium, which is less absorbable, an important benefit in patients at risk for HE. Lactulose is a soluble fiber often used for this benefit. Lactulose is dosed to effect to stimulate passage of 2 soft stools daily.

FAT DIGESTION AND TOLERANCE

In some patients, hepatic dysfunction can interfere with the digestion and absorption of dietary fats and fat soluble compounds.[15] Bile acids from the liver are important for the normal digestion and absorption of long chain triglycerides (LCT), the normal lipids found in most diets. Many dogs with liver disease, including those with PSS, tolerate high fat diets without problems. For these dogs, fat restriction is not required. However, even when fat malabsorption is not clinically apparent, fat digestion may be somewhat reduced. In dogs with PSS, fat digestion decreased from 92% to 85%.[16] Colonic microflora can ferment these undigested fats, producing hydroxy-fatty acids which can cause diarrhea and enhance problems with deficiencies of fat soluble vitamins and essential fatty acids.

In patients with fat malabsorption diarrhea, restriction of LCTs is recommended. However, adequate essential fatty acids must be provided (i.e. dietary LCTs between 5 to 10% of diet dry matter). To avoid weight loss with low fat diets, alternate energy sources must be provided and may include digestible carbohydrates or medium chain triglycerides (MCT). MCTs can be readily absorbed despite a lack of bile acids and are readily oxidized as an energy source. Both dietary and parenteral MCTs have been successfully used to enhance the nutritional status in patients with liver disease.[3,15] Diets containing MCTs at levels not to exceed 50% of the dietary fat may be beneficial for dogs with liver failure.

Long chain omega-3 fatty acids from fish oil, specifically eicosapentaenoic acid (EPA) and docosahexaenoic acid (DHA) may be beneficial for reducing inflammation in certain types of liver disease. In certain models, EPA and DHA were effective at reducing production of inflammatory cytokines, reducing hepatic steatosis, and reducing hepatic fibrosis.[17-19] However, conflicting data have been reported regarding beneficial versus detrimental effects in cholestatic disease,[20,21] and there is no data regarding the use of fish oil for dogs with inflammatory liver diseases.

VITAMINS AND MINERALS

Few vitamin deficiencies are recognized in veterinary patients with hepatic disease, with the exception of Vitamin K (V_K).[9,22] Coagulopathies secondary to V_K deficiency should be treated with injectable V_K. Vitamin C (V_C), not usually recognized as an essential nutrient for dogs or cats due to endogenous synthesis, may be deficient in some patients with liver failure. Oral supplementation with 1.5 mg V_C /kg body weight/day and may be appropriate for dogs or cats with liver disease.

Zinc (Zn) deficiency is common in patients with hepatic disease, and can contribute to HE via altered nitrogen metabolism. Correction of Zn deficiency enhanced hepatic nitrogen clearance and psychomotor scores, and reduced lipid peroxidation in human cirrhotic HE patients.[23] Zinc deficiency can lead to increased iron and copper (Cu) accumulation in the liver,[24] while supplementation may help reduce hepatic fibrosis. Dietary intake should provide at least 3.5 mg zinc/ 100 Kcal metabolizable energy.

Cu storage diseases are well known causes of hepatic failure in genetically predisposed dogs. Liver Cu often reaches 10,000 ppm with primary Cu storage disease, compared to normal levels of less than 400 ppm. Copper may also accumulate in liver secondary to other hepatobiliary diseases, especially cholestatic disorders, however, secondary Cu accumulation rarely exceeds 2000 ppm.[5] Hepatic excretion can be enhanced by Cu-chelating agents or zinc acetate.[5] Dietary restriction of copper probably plays a minor role in reducing hepatic copper concentrations in diseased dogs,[9,19] although a recent study did show a reduction in hepatic Cu in dogs fed a low Cu diet.[26] Excessive chelation or excess dietary restriction can lead to Cu deficiency[27], so regular monitoring is essential.

Hypokalemia is a common precipitating cause of HE in human patients with hepatic disease. It is also linked to glucose intolerance, which is common in hepatobiliary disease. Hypokalemia can result from profuse vomiting or diarrhea, poor intake, or from excessive use of loop diuretics used in the management of ascites. Patients should be monitored for evidence of hypokalemia or acid/base imbalances.

SUPPLEMENTS AND NUTRACEUTICALS

Pre- and Pro-biotics

Prebiotics are a select group of soluble fibers that promote the growth of beneficial bacteria, such a Bifidobacteria spp and Lactobacillus spp. Probiotics are live microorganisms, including Lactobacillus spp, *Enterococcus faecium* SF68, etc, which impart a beneficial effect to the host. Human patients with chronic liver disease frequently have disturbed GI microflora.[28] Recent studies have documented a benefit of E. faecium SF68 for reducing HE in human patients and for Lactobacillus spp for reducing liver injuries following colonic inflammation in rats.[28,29] Probiotics have also proven helpful at reducing bacterial translocation in cirrhotic patients.[28] There currently is no data regarding the use of probiotics in canine or feline patients with chronic liver diseases.

Antioxidants

Many forms of hepatic disease are associated with increased oxidative stress.[30] As previously noted, supplemental V_C (1.5mg/kg body weight PO) may be appropriate for patients with various hepatopathies. At one time, there was hesitation to use V_C in patients with Cu-associated hepatopathy due to a perceived pro-oxidant effect. However, newer research does support the use of V_C as an antioxidant in these patients.[31-33] Other helpful antioxidants include Vitamin E (V_E) and silymarin from milk thistle. A recommended daily oral dose of V_E for most patients is 10 - 15IU/kg body weight. Malabsorption of V_E may occur in the face of severe fat malabsorption, which could lead to deficiency. In these patients, a parenteral form of V_E may be required.[9]

Other substances with antioxidant properties that may be useful in patients with hepatobiliary diseases include S-adenosylmethionine (SAMe: 20mg/kg/d PO), N-acetyl-cysteine (IV or PO) or Silymarin (20 – 50mg/kg/d divided and given PO Q 6 – 8 hr).[30] Silymarin, the active ingredient in milk thistle, is one of a limited number of herbal compounds that has undergone scientific testing for efficacy in liver diseases.[34,35] While not all studies showed positive effects, no adverse or toxic effects have been noted. Silymarin functions as an antioxidant and an antifibrotic agent, and it has been shown to promote hepatic regeneration. Although recommended doses are available for all of these substances, futher studies are needed to evaluate efficacy in veterinary patients with hepatic diseases.[30,35]

CLINICAL INDICATIONS FOR DIETARY THERAPY

No single diet will suit the nutritional and clinical needs of all patients with liver disease. Selection of an appropriate product depends on the specific clinical signs observed. A number of pharmaceutic agents used in the management of hepatobiliary diseases are associated with side effects that may impact dietary needs. Glucocorticoids used in the management of chronic progressive hepatitis may cause or aggravate ascites, HE, glucose intolerance and gastrointestinal ulceration. Colchicine, used to inhibit fibrosis, has been associated with nausea, vomiting and hemorrhagic diarrhea. Anorexia and vomiting may also accompany therapy with D-penicillamine or other Cu chelators. Diarrhea may result from excessive lactulose administration. These effects should be taken into consideration when planning total patient management. Introduction of a new diet while a patient is nauseated may lead to a learned aversion to that food. Patients should be stabilized before introducing a food intended for long term feeding.

Hepatic Encephalopthy

It is important to correct acid/base and electrolyte disorders, and to maintain body weight in patients with liver failure in order to reduce the risk of HE. In the absence of normal liver function, the skeletal muscle serves a primary role in detoxifying ammonia. To help maintain muscle mass, protein should be restricted only as needed to prevent HE. Use of lactulose with or without a probiotic, and supplementation with highly digestible, non-meat proteins may allow an increase in protein intake in HE patients. Vegetable proteins such as from soybeans, and milk proteins such as casein, cottage cheese and whey, are superior sources of protein for patients susceptible to HE. Monitor serum proteins regularly, and provide supplemental protein if necessary.

Feeding frequent, small meals can improve nutritional status and decrease catabolism in cirrhotic patients. It may also enhance intake in partly anorectic patients. Diet digestibility appears to be important. Of the several dietary studies that seemed to "prevent" HE in dogs with portosystemic shunts, high digestibility was common among diets.

If a previously stable patient suddenly develops HE, it should be evaluated for GI bleeding. Endogenous blood proteins can be a major source of encephalopathic toxins.

CONCLUSIONS

No single diet will suit the nutritional and clinical needs of all patients with liver disease. Selection of an appropriate diet depends on the specific clinical signs observed. Maintaining adequate calorie and protein intake are critical to support hepatic regeneration and aid symptomatic relief from clinical signs. Feeding multiple small meals per day, use of dairy or vegetable proteins, use of lactulose or other fermentable fiber to acidify the colonic contents or the use probiotics can all help increase tolerance for dietary protein. Many patients with hepatobiliary diseases may benefit from supplemental antioxidants or diets rich in these nutrients. Use of supplements or nutraceuticals should be tailored to the individual patient and monitored for safe efficacy.

REFERENCES:

1) Bémeur C, et al. J Nutr Metab 2010;doi: 10.1155/2010/489823; 2) Cabral CM, Burns DL. Nutr Clin Pract 2011;26:155-159 ; 3) Druml W, et al. Am J Clin Nutr 1995;61:812-817; 4) Bunout D, et al. Eur J Clin Nutr 1989;43: 615-621; 5) Rolfe DS, Twedt DC. Vet Clin N Am: Sm Anim Pract 1995;25:399-417; 6) Fenton JCB, et al. Lancet 1966;1:164-166; 7) Rosen HM, et al. Metabolism 1978;27: 393-404; 8) Swart GR, et al. Clin Sci 1988;75:101-107; 9) Center SA. J Nutr 1998;128:2733-2746S; 10) Meyer HP, et al. Met Brain Dis 1999;14:103-115; 11) Proot S, et al. J Vet Intern Med 2009;23:794-800; 12) Smith AR, et al. Ann Surg 1978;187: 343-350; 13) Strombeck DR, Rogers QR. JAVMA 1978;173: 93-96; 14) Laflamme DP, et al. Vet Q 1994;16:34-35S; 15) Kelly DA. Ind J Pediatr 1995;62:533-544; 16) Laflamme DP, et al. Am J Vet Res 1993;54:719-723; 17)Hao W, et al. J Pediatric Surg 2010;45:2412-2418; 18) Kajikawa S, et al. Dig Dis Sci 2011;56:1065-1074; 19)Shapiro H, et al. Clin Nutr 2011;30:6-19; 20) Chen CC, et al Dig Dis Sci 2012;doi: 10.1007/s10620-012-2489-5; 21) Chen WY, et al. J Nutr Biochem 2012;23:252-264; 22) Michel KE. Vet Clin N Am: Sm Anim Pract 1995;25:485-501; 23) Marchesini G, et al. Hepatol 1996;23:1084-1092; 24) Levinson MJ. Gastroenterologist 1995;3:234-240; 25) Marks SL et al., Compendium 1994;16:1287-1295; 26) Hoffmann G, et al.JVIM 2009;23:957-963; 27) Seguin MA, Bunch SE. JAVMA 2001;218:1593-1597; 28) Lata J, et al. World J Gastroentol 2011;17:2890-2896; 29) Hakansson A, et al. PLoS ONE 2011;7:e33510 doi:10.137/journalpone.0033510; 30) Webster CRL, Cooper J. Vet Clin N Am SAP 2009;39:631-652; 31) Gaetke LM, Chow CK. Toxicol 2003;189:147-163; 32) Valko M, et al. Curr Med Chem 2005;12:1161-1208; 33) Jomova K, Valko M. Toxicol 2011;283:65-87; 34) Ghaffari AR, et al. Hepat Mon 2011;11:464-468; 35) Hackett ES, et al. J Vet Intern Med 2012;doi:10.1111/jvim.12002.

DIETARY MANAGEMENT OF SELECT HEPATOBILIARY CONDITIONS

Dr.Vincent C. Biourge, DVM PhD DACVN, DECVCN
Aimargues, France
Catherine A Outerbridge DVM, MVSc, DACVIM, DACVD
Davis, CA

FELINE HEPATIC LIPIDOSIS

Nutritional support is the first consideration in the treatment of FHL.[1,2] The prognosis is directly related to the ability of the clinician to aggressively meet the energy and nutrient requirements via enteral feeding. In surviving cats, tube-feeding will completely reverse the condition with a return to normal liver histology within a few weeks of nutritional management.[3] The first week of treatment is critical as many cats appear sicker initially, probably due severe nutrient, electrolyte and fluid shifts associated with re-feeding.[1,3] Within a couple of weeks, a significant reduction in serum liver enzyme activities and total bilirubin should be observed.

Many highly digestible, complete and balanced feline diets have been used successfully in cats with hepatic lipidosis.[1,2] Cats with hepatic lipidosis will usually tolerate 35 to 45 % of their calories as protein, a level that is found in most commercial cat foods. Indeed, protein deficiency appears to be important in the etiology of the syndrome.[4] Protein restriction (20 to 25 % of the calories, i.e.: feline renal diets) should only be considered if the patient shows signs of encephalopathy, an uncommon presentation in this syndrome.

During fasting, the animal metabolism is geared towards using lipids as the main source of energy and to spare glucose and protein. This explains that as long as their amino acid requirements are fulfilled, FHL cats will tolerate high levels of fat (30-60 % of the calories as fat) in their diets.[1,2] Glucose intolerance was a common presentation in FHL cats before clinicians realized that prolonged fasting depresses pancreas responsiveness to carbohydrates.[5] Cats should not be fed more than 25 to 35 % of their calories as carbohydrates. Diets high in glucose or simple sugars, such as most human liquid diets should be avoided.

Cats with hepatic lipidosis are usually anorectic and tube feeding is required.[1,2] Forced feeding (i.e.: spoon or syringe feeding cats) will not allow sufficient energy intake, stress the animal and contribute to food aversion. Nasogastric (5-8 French size), esophageal (10-12 French) feeding and gastrotomy (20 French) tubes have been used successfully. Small bore nasogastric tubes can be placed with local anesthesia but require the use of liquid formulas such as feline liquid diets. Some canned diets mixed with the same volume of water and blenderized will be liquid enough to pass through nasogastric tubes. Most blenderized cat food (can mixed with 1 volume of water and dry mixed with 3 volume of warm water) will pass through larger bore gastrotomy and oesophageal feeding tubes. After 1 or 2 weeks of stabilization at the hospital, tube feeding can be managed by most owners at home. In the author's experience, although oesophageal and gastrotomy tubes are better tolerated, nasogastric tubes are well accepted, easy to place and to re-place, and associated with minor complications, but they will require an Elizabethan collar. Oesophageal and gastrotomy tube placement do require full anesthesia.

Appetite stimulants (i.e.: diazepam, oxazepam, cyproheptadine) have not been found useful in the management of FHL cats.[1] Moreover, liver dysfunction might compromise the metabolism of those agents.

Food should be introduced progressively over a 4 to 7 day period, the goal being to feed 60 to 80 kcal ME /kg/d (30 to 40 Kcal/lb/d). Initially cats will tolerate well bolus of 25-30 mL of diet QID. Increasing meal size by 5 mL each day up to meeting total daily energy requirement has been successful in the author's experience.[2] If the treatment is well tolerated, meal frequency can then be reduced to 3 meals/day. Common complications include epiphora (watery eyes with naso-gastric tubes), vomiting/diarrhea, tube removal, hypokalemia and hyperglycemia. Less common complications include hypophosphatemia, aspiration pneumonia and peritonitis. Cats should be carefully monitored for signs of infection (fever, tachycardia, abnormal breathing, abnormal auscultation). The epiphora will resolve after removal of the nasogastric tube. Vomiting can usually be controlled by giving metaclopramide parenterally (0.1-0.2 mg/lb or 0.2 - 0.4 mg/kg SQ) 10 to 15 min before the meal. Diarrhea often occurs with high carbohydrate human liquid diets or when the food is introduced too rapidly. Elizabethan collars are highly recommended to minimize the risk of tube removal. Regular blood work is advised initially to monitor serum electrolytes, glucose and phosphorus. Hyperglycemia is common with high carbohydrate human liquid diet (> 40 % of calories from glucose see above). In case of hypokalemia and/or hypophosphatemia (serum $P < 2.0$ mg/dl, $P < 0.65$ mmol/L), potassium (5-10 mEq per os/cat/day with the meal) and/or phosphate (0.01-0.03 mmol potassium phosphate /kg/hr in fluids, monitor q 3-6 hrs and stop infusion when $P > 2$ mg/dl) should be supplemented in the diet or given with the fluids.[2] Vitamin K1 (1.5 mg /kg IM, 3 treatments at 12 hour intervals) might be helpful in cats with bleeding problems. Increased bleeding time linked to Vitamin K deficiency has been observed in 50 % of FHL cats.[1,2]

L-carnitine (250-500 mg/cat) supplementation has been recommended in the management of FHL, especially in more severe cases but diets devoid of carnitine have been used successfully to control the disease.[1,2]

B-vitamins (2 x the normal daily dosage given by the manufacturer), vitamin C (30 mg/kg), vitamin E (100-400 mg/cat), taurine (250 mg), zinc (7-8 mg elemental) supplementations have also been recommended on an empirical basis in cats with hepatic lipidosis.[1,2] Controlled studies to substantiate the benefits of such supplementation above what is already in commercial cat food are not available.

Food aversion may be a complicating factor in cats with hepatic lipidosis.[2] Cats that refuse to eat a diet that they associated with nausea may continue to avoid that diet even after full recovery. Exposing cats with hepatic lipidosis to different commercial diets before tube-feeding, might predispose them to develop an aversion to all those diets. This could explain why some lipidotic cats take much longer than others before resuming voluntary intake even though all laboratory parameters appear within normal limits. It is this author's opinion that anorexia should be demonstrated with a minimum of diets and tube-feeding started immediately in cats with hepatic lipidosis. No food should be offered per mouth for the first 10 days of nutritional support. Cats expressing an interest to eat can then be presented small amounts of novel food until normal appetite resume.

Cats often look worse during the first few days of treatment than when first presented. Clinicians should not get discouraged too early. Tube feeding is the only treatment that will reverse hepatic lipidosis and those cats must be fed. Increasing the number of meals, continuous tube feeding using a peristaltic pump or total parental nutrition should be considered if the animal cannot keep the food down. A majority of cats with idiopathic hepatic lipidosis will show clinical and laboratory signs of improvement within a week. In the experience of the author, nearly all cats that survive the first week of treatment do recover completely. FHL complicated by other disease (renal disease, IBD, …) have a more guarded prognosis and require to also manage the primary disease.

Although our understanding of FHL remain partial, the recognition that long-term tube feeding with a high-fat, high protein diet will completely reverse the condition in a majority of cases (at least in the idiopathic form) should motivate clinicians to treat those patients more aggressively.

HEPATOCUTANEOUS SYNDROME

Superficial necrolytic dermatitis (SND) or hepatocutaneous syndrome (HCS) is an uncommon skin disease associated with systemic metabolic disease(s). Affected dogs most commonly have a characteristic concurrent hepatopathy, thus the popular use of the term hepatocutaneous syndrome (HCS). As different disease processes (glucagonoma, vacuolar hepatopathy, phenobarbital administration, or intestinal disease have been reported to cause similar histologic skin lesions it might be more correct to refer to the skin pathology as SND or metabolic epidermal necrolysis (MEN) or the human corollary necrolytic migratory erythema(NME).[1] The disease is typically diagnosed in older dogs. The mean age of all reported cases is 10 years of age, with a range of 4 to 16 years[1]. Male dogs comprise 64% of all reported cases[1]. Shetland sheepdogs, West Highland white terriers, Cocker spaniels and Scottish terriers may have a predisposition to develop HCS as they appear to be over represented in the literature[1].

Clinically, footpads of affected dogs develop marked crusting, fissuring and ulcerations. Erythema, crusting, exudation, ulceration and alopecia can also involve the peri-ocular or peri-oral regions, anal-genital regions, and pressure points on the limbs. Secondary cutaneous infections of lesions with bacteria, yeast (Malassezia, Candida) or dermatophytes are common. Lameness secondary to footpad lesions, inappetance and weight loss can also affect dogs with SND. Polydipsia and polyuria may be present when there is concurrent diabetes mellitus or if significant concurrent liver dysfunction is present. Diabetes mellitus has been reported to occur in a third of dogs with the hepatic form of SND[1].

Diagnosis of SND is based on obtaining skin biopsies with the typical histopathologic changes of a marked parakeratotic epidermis with striking inter- and intracellular edema from necrolysis in the upper epidermis and hyperplastic basal cells, creating the “red, white and blue” lesion that is diagnostic for this disease. Abdominal ultrasound can provide further support for the diagnosis if the pathognomonic ‘honeycomb’ pattern consisting of variable-sized hypoechoic regions surrounded by hyperechoic borders is documented[3,4]. If this ultrasonographic pattern to the liver is not visualized in a dog with a confirmed histologic diagnosis of SND on skin biopsy, evaluation for a possible pancreatic tumor is warranted. Pancreatic tumors may not be readily visible with an abdominal ultrasound examination so measurement of plasma glucagon can be pursued. In the hepatic form of SND the liver histologically has a characteristic vacuolar appearance with hepatocellular collapse. Elevation of liver enzymes and hypoalbuminemia were the most common clinicopathologic changes seen in one study[2]. Glucosuria and hyperglycemia may be documented if diabetes mellitus is present. Plasma amino acids, if measured, should document a characteristic severe hypoaminoacidemia.

The fact that severe hypoaminoacidemia is documented to occur in all cases of SND, in which plasma amino acids have been measured, regardless of associated disease, makes it likely that this metabolic derangement is directly contributing to the cutaneous lesions seen in affected dogs. The liver plays a critical role in amino acid (AA) balance. In dogs with either chronic or acute hepatitis, compromised hepatic metabolism results in increased concentrations of many plasma AA but this is not seen in dogs with SND, in which the majority of individual plasma AA concentrations are less than 60 % of normal and total AA concentrations are 30% of normal[2]. These differences suggest that the pathogenesis of hypoaminoacidemia in dogs with the hepatic form of SND cannot be explained by compromised liver function seen in hepatitis dogs. The hypoaminoacidemia is so profound that neither anorexia nor protein malnutrition could result in the same degree of hypoaminoacidemia. It seems probable that an increase in hepatic catabolism of amino acids might account for the severity of the hypoaminoacidemia documented in dogs with SND. Hyperglucagonemia in dogs with pancreatic tumors could provide a hormonal signal to initiate this hepatic catabolism but in the more common hepatic associated form of SND the signal for the catabolism remains unexplained.

The most effective symptomatic or palliative therapy for dogs with the hepatic form of SND appears to be the administration of intravenous (IV) AAs. For those patients that respond to this therapy parenteral delivery of amino acid likely results in sufficient delivery of amino acids to replete tissues prior to any liver catabolism by bypassing the first pass effect of direct delivery to the portal circulation that orally administered amino acids would face. A number of crystalline AA solutions are commercially available that vary in their concentration and the inclusion of electrolytes. Although there are minor differences in the amounts of essential and nonessential amino acids between manufacturers there is no data to suggest that one product is more efficacious than another. Solutions without additional electrolytes are preferred. These hypertonic AA solutions should ideally be administered via a central vein to diminish the chance of thrombophlebitis. Inducing a hyperosmolar state is possible if administration is too aggressive. Dogs should be watched for neurologic signs and the infusion discontinued if these occur. If compromised hepatic or renal function is present the administration of intravenous amino acids may exacerbate hepatic encephalopathy or augment increases in BUN. Such dogs warrant close monitoring with serial measurements of ammonia, BUN and osmolality during IV AA administration. Some dogs show dramatic improvement in attitude and resolution of skin lesions after receiving AA infusions while sadly others have only transient improvement in attitude or no response at all. There are no defined protocols for the administration of AA infusions in SND dogs and if of benefit repeat infusions are performed bi-monthly, monthly or when clinical signs return.

The best oral nutritional approach is not known but it seems reasonable that a high quality protein diet that can be additionally supplemented with an AA powder should be fed. Unless significant hepatic dysfunction with hyperammonemia is documented, suggesting some other concurrent liver disease and necessitating a low protein diet, most dogs with SND can not be fed enough protein to overcome the hypoaminoacidemia occurring in this disease. Zinc and essential fatty acid (EFA) supplementation and feeding egg yolks are recommended in the literature[5,6]. Zinc and EFA supplementation have been reported to benefit humans with NME[7,8]. These nutrients are carried in the plasma by albumin which is often diminished in people with NME and in dogs with SND, and could create a deficiency in these nutrients. Zinc has a myriad of physiologic roles and is vital for skin health including as a cofactor for delta 6-desaturase of linoleic acid in EFA synthesis. The skin pathology seen in NME in humans is likely a consequence of AA, zinc and EFA deficiencies, arising from a variety of causes that have direct tissue effects and influences inflammatory mediators[8]. A similar multifactorial model may be occurring in dogs with SND. The prognosis for dogs with SND is generally poor and the majority of dogs have survival times of less than 6 months. However, 20 % of the dogs in one study were maintained for 12 months or more with oral protein hyperalimentation and 50 % had periodic parenteral IV AA infusions[2]

Secondary infections should be treated with appropriate antibiotic and anti-fungal therapy with careful consideration of those drugs that may be hepatotoxic or require hepatic metabolism. Topical therapy with antibacterial and anti-yeast products can help to manage secondary infections. Although corticosteroid anti-inflammatory therapy for the skin lesions may be helpful to diminish inflammation and improve comfort, the risk of precipitating or exacerbating diabetes mellitus in these dogs makes the use of systemic glucocorticoids contraindicated. Diabetes mellitus if present requires appropriate management. Removal of a glucagonoma has been reported to result in resolution of skin lesions in one dog[9]. Octreotide (somatostatin analogue) has been of benefit in some human NME cases and serial treatments with subcutaneously administered octreotide in a dog with metastatic glucagonoma associated SND proved efficacious for 6 weeks [10]

REFERENCES

Feline Hepatic Lipidosis

[1]Center SA. Vet Clin North Am Small Anim Pract 2005, 35:225-269. [2] Biourge VC. Seminars in Vet Med and Surg (Small animal) 1997; 12:34-44. [3]Biourge VC, Pion P, Lewis J et al. J Vet Int Med 1993;7:194-197. [4]Biourge VC, Massat B, Groff JM et al. Am J Vet Res 1994;55:1406-1413. [5] Biourge VC, Nelson RW, Feldman EC et al. J Vet Int Med. 1997; 11:86-91.

Hepatocutaneous Syndrome

[1]Outerbridge CA. In Ettinger SJ, Feldman EC, editors. Textbook of Veterinary Internal Medicine. 7th edition (volume1) 2010.; 112, [2]Outerbridge CA, et al. Vet Dermatol 2002; 13(4):177, [3]Jacobson et al J Vet Intern Med 1995; 9(6) 399, [4]Nyland et al Vet Radiol Ultrasound 1996;37:200, [5]Gross TL et al. Vet Pathol 1993;30(1):75, , [6]Byrne KP.Vet Clin North Am Small Anim Pract 1999;29(6):1337., [7]Alexander EK et al Clin Endocrinol 2002;57(6):827, [8]Tierney EP et al MedGenMed 2004;6(3):4 [9]Torres, S.M.F et al JAAHA1997;33(4): 13–319., [10]Oberkirchner U, et al Vet Der

THE METABOLIC SYNDROME: COMPONENTS AND CONSEQUENCES IN 4 SPECIES

Charlotte R. Bjornvad, DVM, PhD, Dipl. ECVCN, Copenhagen, Denmark; John D Brunzell MD, Seattle WA; Nicholas Frank, DVM, PhD, DACVIM, North Grafton, MA and Sutton Bonington, Leicestershire, UK; Kurt R Verkest BVSc, MANZCVS, PhD, Sai Kung, Hong Kong.

INTRODUCTION

The prevalence of obesity is increasing in humans and companion animal species, including dogs, cats, and horses. Obesity, an increase in body adipose tissue mass, has consequences in tissues other than adipose tissue, including resistance to the glucose-lowering effects of insulin. The metabolic syndrome in humans comprises a set of risk factors, and the term "metabolic syndrome" has been proposed or applied to companion animal species. This set of lectures discusses the components and consequences of the human metabolic syndrome, and whether and how they are manifested in dogs, cats, and horses.

THE METABOLIC SYNDROME: COMPONENTS AND CONSEQUENCES IN HUMANS (Kurt Verkest and John Brunzell)

Diabetes mellitus and the distinction between type 1 and type 2 diabetes have been recognised by physicians since antiquity[1]. Research into the causes and risk factors for the development of type 2 diabetes since then has identified obesity, hyperglycemia, insulin resistance, hyperlipidemia, and hypertension (among others) as important risk factors for the development of diabetes, coronary heart disease, and stroke. The term "metabolic syndrome was devised to note that these sub-components occur together more often that would be expected. PubMed lists 42,371 scientific publications related to metabolic syndrome, 9116 of these in 2012, as of January, 2013.

The metabolic syndrome has been developed as a pathophysiological entity in human medicine since its clinical importance was highlighted by G. Reaven in 1988[2]. Reaven developed the hypothesis that insulin resistance *per se* was the driver of the pathophysiology of the metabolic syndrome, type 2 diabetes, and cardiovascular disease[3]. At the same time others highlighted the importance of central obesity[4-6]with increased intra-abdominal adipose tissue[7]. The metabolic syndrome is a cluster of clinical and physiological findings that include obesity (especially visceral obesity), high blood pressure, hyperglycemia, and hyperlipidemia[8,9]. Although various definitions of the metabolic syndrome have been devised, these definitions all contained the same essential components and reflected the same understanding of the pathophysiology of the metabolic syndrome – that obesity, especially visceral obesity, contributes to resistance to the glucose-lowering effects of insulin, high blood pressure, and increased circulating triglycerides and low HDL cholesterol[8,9]. Recently, consensus statements have been published to reflect this common understanding, and containing revised definitions of the metabolic syndrome[8-10].

The metabolic syndrome itself has no clinical symptoms, and by definition is not a disease but rather a syndrome[11]. It is a pre-morbid condition whose significance in humans is that it comprises a constellation of several important risk factors for several diseases that do cause substantial morbidity and mortality[11]. These diseases include coronary heart disease, stroke, some familial dyslipidemias and type 2 diabetes. Type 2 diabetes is associated with macrovascular disease of coronary, cerebrovascular and peripheral arteries; and with microvascular disease of the eye, kidneys and extracranial nervous system. These result in coronary heart disease, stroke, retinopathy, kidney disease and amputation. Coronary heart disease and stroke account for a large fraction of human deaths worldwide[11]. Metabolic syndrome is recognised as being part of the spectrum of type 2 diabetes mellitus[12-14], and so contributes to the risks of morbidity and mortality associated with type 2 diabetes, as well as to the risks of mortality and morbidity associated with coronary heart disease and stroke.

The fact that the metabolic syndrome is not associated with morbidity, independent of its subcomponents, led to considerable scepticism about whether the metabolic syndrome is a useful clinical entity[15,16]. One key criticism has been that diagnosing the metabolic syndrome in a patient does not change how that patient should be treated – patients with hypertension need to have their blood pressure controlled regardless of whether they have the metabolic syndrome or not, for example. This criticism was included in the arguments made by a panel of metabolic syndrome experts who agreed that it is not useful to diagnose the metabolic syndrome in clinical practice, because it does not change recommendations or outcomes. Rather, it should be regarded as a useful pathophysiological construct that helps us understand part of the development of type 2 diabetes, familial dyslipidemias, and coronary heart disease[11]. This includes proinflammatory, atherosclerotic and thrombotic abnormalities that are also seen[17].

It is well recognised that many obese human beings do not develop insulin resistance, hypertension, hyperlipidemia [3,11] due to differences in body fat distribution. These individuals are resistant to the development of the metabolic syndrome and its consequences. It is proposed that they deal with the issues of chronic nutrient excess that drive obesity and the development of diabetes by increasing insulin secretion, and adequately partitioning the excess energy into subcutaneous adipose tissue instead of visceral fat. As a result, they do not experience the substantial hyperglycemia and hyperlipidemia, and so avoid the harmful effects of these on their organs. Their livers, beta cells, muscle, and blood vessels do not develop fatty infiltration, and other harmful effects of excess glucose and lipid exposure[11]. In contrast, other humans are susceptible to the development of the metabolic syndrome due to the accumulation of visceral adipose tissue. One hypothesis is that these susceptible humans, when they are exposed to chronic excess of calories, are unable to adequately partition the excess energy, and nutrient spill-over occurs[13,18]. Nutrient spill-over in this context means that excess glucose and lipids circulate at increased concentrations, and enter non-adipose tissues where they have toxic effects, including glucose toxicity in beta cells and blood vessels[19-21], and fatty infiltration of the liver and muscle[18] An alternative concept of the metabolic syndrome was the hypothesis that visceral obesity was the key component leading to insulin resistance followed by the development of hyperlipidemia, hyperglycemia and hypertension[7].

It is also useful to consider how the metabolic syndrome was originally developed. Unlike clinically recognisable diseases such as diabetes, in which clinical problems such as polyuria, polydipsia and weight loss are demonstrable, the metabolic syndrome was originally defined as a set of risk factors for OTHER diseases[15].These risk factors tended to cluster together within individuals more often than by chance alone. This sets the metabolic syndrome apart from diseases such as cancer, heart failure, or diabetes because there are no recognisable clinical signs of the metabolic syndrome and it is not a disease itself. It was defined based on cut-off values of obesity, plasma glucose concentration, plasma HDL cholesterol concentration, plasma triglyceride concentration, and blood pressure that were associated with the subsequent development of adverse effects such as coronary artery disease or stroke. Obesity, plasma glucose concentration, plasma HDL cholesterol concentration, plasma triglyceride concentration, and blood pressure are continuous variables and there is good evidence that the increase of risk does not occur in an all-or-nothing way when these thresholds are exceeded.

Using the "harmonizing" definition of the metabolic syndrome[8] the presence of 3 of the 5 components is adequate for the diagnosis of the syndrome: plasma triglyceride >150 mg/dL (>1.7 mmol/L), HDL cholesterol < 40 mg/dL (<1.03 mmol/L; men) or 50 mg/dL (<1.29 mmol/L; women), blood pressure > 135/85 mmHg, plasma glucose >100 mg/dL (5.6 mmol/L), and gender and ethnic excess for waist circumference. Using these values almost 50% of the North American population over 50 years of age meet the criteria for the metabolic syndrome. Patients with type 2 diabetes account for 20-30% of premature coronary artery disease, while those with familial combined hyperlipidemia account for 10-20%[22]. Both diabetes and familial combined hyperlipidemia have a strong genetic basis[22,23]. Both diseases are associated with visceral obesity and insulin resistance; diabetes has a second hit in insulin secretion, familial combined hyperlipidemia has second hit with increased hepatic apolipoprotein B secretion[22,23]. Identical twin studies[24] and adoption studies[25] have demonstrated a strong genetic basis for weight. In addition, each component of the metabolic syndrome is associated with many common gene variants by GWAS with small effects.

Care needs to be taken with attempts to define the metabolic syndrome in novel situations (such as in non-human species), because the relationships between the components that comprise the human metabolic syndrome and their potential consequences in each species might be very different from those in human beings, or might not exist at all. Since genetic variation affects each of the individual components of the metabolic syndrome, this might lead to marked species differences. Recently, the term "metabolic syndrome" has been used with reference to other species, including horses[26] and dogs[27]. Investigators examining cats, the only domestic species in which type 2 diabetes has been documented, have conspicuously avoided applying the term to cats[28-30]. The remainder of this article explores whether other species exhibit an entity that is similar to the human metabolic syndrome, the known or likely clinical consequences of such an entity, and whether any such entity is worth looking for in clinical practice.

THE METABOLIC SYNDROME: COMPONENTS AND CONSEQUENCES IN DOGS (Kurt Verkest)

COMPONENTS OF HUMAN METABOLIC SYNDROME: EVIDENCE IN DOGS

There are two ways of looking at the metabolic syndrome – as a set of COMPONENTS, or as an entity with a set of CONSEQUENCES. Dogs develop all the key components of the metabolic syndrome – obesity[1], insulin resistance[2-4], hypertension[5], and hyperlipidemia[6]. Dogs have been used as models for the development of the metabolic syndrome and type 2 diabetes[4]. It is tempting to combine the components of the metabolic syndrome in dogs as they have been combined in humans. However, the metabolic syndrome in humans is important because it is part of the pathophysiological pathways that lead to adverse consequences – especially type 2 diabetes mellitus, heart attacks, and stroke. Dogs do not exhibit the pathophysiology and clinical consequences of the metabolic syndrome.

Insulin resistance exists in overweight and obese dogs[2,3], as it does in humans. Obesity in dogs involves visceral obesity[7], visceral fat in dogs is insulin resistant[8], and visceral fat contributes to whole-body insulin resistance in fat-fed dog models of obesity[4]. Omentectomy in lean dogs increases insulin sensitivity[9]. Insulin resistance in dogs with induced obesity includes both hepatic[10] and peripheral components[11]. In these ways, overweight and obese dogs are similar to human beings with the metabolic syndrome.

Dogs with naturally occurring[6,12] and induced[13] obesity exhibit hypertriglyceridemia and hypercholesterolemia. Although these changes mirror those in humans with the metabolic syndrome, there are important differences. Cholesterol circulates as part of lipoprotein complexes including low density lipoprotein (LDL, which carries cholesterol from the liver to peripheral tissues) and high density lipoprotein (HDL, which carries cholesterol from peripheral tissues to the liver). Of these, LDL contributes to atherosclerosis whereas HDL protects against atherosclerosis[14,15] – atherosclerosis in turn leads to heart attack and stroke. All definitions of the metabolic syndrome in humans define the lipid component of the metabolic syndrome as an increase in fasting triglycerides and a decrease in HDL cholesterol[16,17]. Dogs have most of their cholesterol present in circulation as HDL cholesterol[13,14,18], and the proportion of cholesterol circulating as HDL does not change substantially with the onset of obesity[13,14,18]. More importantly, concentrations of HDL cholesterol do not decrease in dogs with natural[6,18] or induced[13] obesity, and LDL cholesterol does not increase disproportionately. High fat diets, which can contribute to high cholesterol concentrations in humans and to obesity in dogs, generally do not cause substantial increases of atherogenic cholesterol fractions in dogs[19]. Triglycerides are included in the definition of the metabolic syndrome in humans. Triglycerides are markers of the presence of atherogenic cholesterol-enriched lipoprotein remnant particles but do not contribute to atherosclerosis [15]. Total serum triglycerides are increased in association with obesity in dogs[12,13], but there are no studies about whether this increase is associated with an increase in atherogenic lipoprotein remnants. In summary, obesity in dogs is associated with higher circulating total cholesterol and triglycerides, but is not associated with the more important atherogenic changes that lead to heart disease and stroke in humans. The lipid changes observed in dogs are not equivalent to those in the human metabolic syndrome.

Hypertension is more common in overweight and obese than lean humans, and is a major risk factor for heart attack (coronary heart disease) and stroke. Hypertension in dogs can damage target organs, eg the kidneys (glomerulopathy), retina (choriopathy), and brain (oedema of the white matter)[20]. Increased blood pressure has been documented in three studies in overweight and obese dogs[5,21,22], but was not found in another study[23]. The effect of obesity appeared to be small (5 mmHg) in the largest study of blood pressure in dogs to date[5], whereas the clinical consequences of hypertension in dogs are reported in dogs with much larger increases in blood pressure. Another case-control study of 19 obese and 19 matched lean dogs showed a 20 mmHg difference in systolic blood pressure between obese and lean dogs[22]. By way of comparison, humans who gained 10-20Kg of body weight had average increases of 10 to 30 mmHg systolic blood pressure in one large longitudinal study[24]. Obese dogs in the case-control study had left ventricular hypertrophy similar to that seen in hypertensive humans[22]. Similar echocardiographic changes in two other studies were reversible with weight loss in some or all of the dogs[25,26], adding to the evidence that obesity is associated with substantial hypertension in dogs and that hypertension results in some structural changes to the myocardium.

In summary, dogs develop some components of the metabolic syndrome (visceral obesity and insulin resistance, hypertriglyceridaemia, and hypertension), but not others (low HDL cholesterol) in association with overweight and obesity.

CONSEQUENCES OF HUMAN METABOLIC SYNDROME: EVIDENCE IN DOGS

In humans, the metabolic syndrome is important because it is a risk factor for the development of three important diseases[27] – type 2 diabetes, coronary heart disease, and stroke. The application of the term "metabolic syndrome" to dogs suggests that dogs will suffer similar consequences in terms of morbidity and mortality to those that occur in humans with the metabolic syndrome. However, none of the diseases associated with the metabolic syndrome in humans is common in dogs, none have been shown to be associated with obesity, and one of them (type 2 diabetes) probably does not exist at all in dogs.

Diabetes mellitus is a relatively common endocrinopathy in dogs. Diabetes in dogs can be classified with a system similar to that used in humans into type 1 (autoimmune destruction of beta cells), type 2 (failure to compensate to insulin resistance that is usually caused by obesity), gestational (first diagnosed during pregnancy or dioestrus), and other specific types (secondary to other endocrine diseases, diseases of the pancreas, etc)[28]. In dogs, a form of diabetes equivalent to type 2 diabetes has never been convincingly described[3,29]. Humans with type 2 diabetes exhibit islet amyloidosis in the islets of Langerhans. Diabetic dogs do not exhibit such accumulations[30], suggesting that type 2 diabetes is rare or does not exist in dogs. One study showed an increased prevalence of diabetes mellitus in overweight and obese dogs[1]. However, the dogs in this study did not have their diabetes classified, so that it is unclear whether obesity directly caused the development of diabetes (ie type 2 diabetes) or whether obesity contributed to another cause of diabetes such as pancreatitis and thereby increased the prevalence of diabetes of other specific types. It is also possible that diabetes was more prevalent in overweight and obese dogs because the insulin resistance caused by obesity increases insulin requirements and so highlights dogs with impaired insulin secretion attributable to other causes that would otherwise be subclinical. This would be similar to gestational diabetes in humans, wherein insulin resistance caused by pregnancy can cause previously subclinical damage to beta cells to become clinically obvious[31]. The only study to date that seemed to show type 2 diabetes in obese dogs did not rule out other causes of diabetes, eg pancreatitis and acromegaly/gestational diabetes[32]. All but one of the obese diabetic dogs in that study were older females, so these dogs might have had dioestrus- or gestational diabetes[33]. Sensitive tests for pancreatitis were not available at the time, so that chronic pancreatitis was also not ruled out as cause of diabetes in these dogs. Although we cannot be certain that type 2 diabetes is absent in dogs, its existence has not been proven.

Key steps in the development of type 2 diabetes in humans seem to be absent in dogs. In humans, type 2 diabetes occurs when insulin resistance is not compensated for by insulin secretion[34]. Early failure of compensatory insulin secretion includes loss of first phase insulin secretion during a GTT[35] and postprandial hyperglycemia[36]. In obese dogs with insulin resistance, first phase insulin secretion is retained [3]. Some obese dogs exhibit increased postprandial glycemia, but none were reported by their owners to develop clinical signs of diabetes mellitus after 2.5 years of followup[12]. Overweight, insulin resistant dogs followed for their lifetime exhibited increased, not decreased, first-phase insulin secretion[37] (one intact female overweight dog developed diabetes mellitus but the classification of the diabetes was not reported). These observations indicate that dogs compensate adequately for obesity-induced insulin resistance, and that type 2 diabetes is rare or non-existent in association with obesity in dogs. It appears that dogs are resistant to developing type 2 diabetes, an important consequence of the metabolic syndrome in humans.

Although atherosclerosis (including coronary artery involvement[38]) occurs in dogs in association with extremely high cholesterol ($\approx$ 20 mmol/L) and triglycerides[39,40], and in dogs with hypothyroidism or diabetes mellitus[38], it has not been reported in association with obesity. Obese dogs exhibit mild (mean increase $\approx$2 mmol/L) increases in total cholesterol, which are mainly accounted for by increases in HDL (1.9 mmol/L) rather than LDL (0.3 mmol/L) cholesterol[13]. In both lean and obese dogs, HDL cholesterol accounts for $\approx$90% of total cholesterol – HDL cholesterol is regarded as protective against atherosclerosis[13]. The paucity of reports of coronary heart disease in obese dogs, the strong association of cases of coronary heart disease with endocrine or other causes of extreme hypercholesterolemia, and the cholesterol profile of obese dogs suggests that coronary heart disease is a rare or non-existent complication of obesity in dogs.

Cerebrovascular accidents (strokes) have been rarely reported in dogs. The largest published case series of dogs with ischemic stroke (which is the type associated with the metabolic syndrome) included only 33 dogs, drawn from the entire caseload of three referral centres over four years, suggesting that this is a rare cause of clinical illness in dogs[41]. This study identified a range of underlying causes of stroke, including neoplasia, intravascular parasites, and cardiac disease. Hypertension (the leading cause of stroke in humans) was present in $\approx$30% of dogs, but all cases of hypertension had a cause such as chronic renal failure or hyperadrenocorticism. Obesity-related hypertension and atherosclerosis associated with obesity were not listed in that study as potential risk factors[41]. It appears that although stroke exists as an uncommon to rare disease in dogs, strokes are rarely if ever associated with obesity in dogs.

In summary, the metabolic syndrome is important because it is a set of risk factors for diseases that cause morbidity and mortality – type 2 diabetes, coronary heart disease/heart attack, and stroke. Dogs either do not develop these diseases at all, or develop them very rarely. There is no evidence that obesity or obesity-associated insulin resistance increases the risk of these diseases in dogs.

CONCLUSIONS (DOGS)

The most important consequences of the metabolic syndrome in humans either do not exist or are very rare in dogs and are not associated with obesity. This means that the metabolic syndrome, if it exists at all in dogs, is of doubtful clinical importance because it is not, on current evidence, followed by relevant morbidity or mortality. I propose that the term "metabolic syndrome" be avoided in relation to dogs unless evidence can be adduced that it is associated with meaningful morbidity or mortality attributable to pathophysiology that is similar to that observed in humans with metabolic syndrome. Although dogs have been used as models of the development of some elements of the pathophysiology of the human metabolic syndrome[4], dogs do not closely model the development of type 2 diabetes. They might instead be good models of humans who develop obesity and insulin resistance, but do not progress to clinical diabetes mellitus. The disease model approach assumes that dogs share the pathophysiology of type 2 diabetes in humans. A comparative disease approach which instead compares the physiology, pathophysiology, and disease biology amongst species and explores reasons for both the similarities and differences observed[42] might be more fruitful. The advantage of the comparative approach is that differences can be as informative as similarities. For example, the observation that overweight and obese dogs, like humans, develop insulin resistance but unlike humans do not develop a disease that is similar to type 2 diabetes implies that dogs have a protective mechanism that humans lack, or that dogs lack a crucial element of the pathophysiology of type 2 diabetes that humans have.

One key question raised by proposals to define the metabolic syndrome in dogs is whether treatment is required for this putative syndrome. The components of the metabolic syndrome are worth treating in humans because each of them independently is a risk factor for subsequent clinical disease and mortality associated with those diseases. Treatment of human hypertension, insulin resistance, and hyperlipidemia is the same regardless of whether the metabolic syndrome is diagnosed or not. In dogs, the components of the metabolic syndrome also exist and the prevalence of these is increased with overweight and obesity in dogs as it is in humans. However, treatment of these components does not appear to be warranted because there is no evidence that they are associated with clinical consequences in overweight and obese dogs. The clinical consequences of the metabolic syndrome in humans either do not exist naturally in dogs (type 2 diabetes[28]), or are rare in dogs (atherosclerosis[38], coronary heart disease[40], and stroke[41]). The deleterious consequences of obesity in dogs do not correspond to those of obese humans with the metabolic syndrome. There does not yet appear to be any benefit to dogs of attempting to define an equivalent of the human metabolic syndrome. The consequences of obesity in dogs are distinct from those in overweight humans with the metabolic syndrome, and this distinction should be marked by avoiding the use of the term "the metabolic syndrome" in dogs.

THE METABOLIC SYNDROME: COMPONENTS AND CONSEQUENCES IN CATS (Charlotte R. Bjornvad)

COMPONENTS OF HUMAN METABOLIC SYNDROME: EVIDENCE IN CATS

As for dogs, cats may develop all of the key components of the metabolic syndrome. Obesity, insulin resistance, hyperlipidemia/dyslipidemia and hypertension are all commonly observed in the pet cat population. Unlike dogs, selected components may actually predispose to disease such as feline diabetes and therefore there are clinical consequences to some, though not completely similar to humans. However, based on the current knowledge, we are still not able to define specific biomarkers and cut-off values relating to weight or body fat mass/deposition to define specific criteria for a metabolic syndrome in cats.

Overweight and obesity in cats predispose to insulin resistance and further, lean insulin resistant cats develop more severe insulin resistance when gaining weight compared with insulin sensitive cats[1]. The distribution of fat depots in cats may however not correlate with insulin sensitivity[2]. In humans, visceral fat is more metabolic active and has a closer correlation with insulin resistance than subcutaneous fat [3]. This has not been investigated closely in cats, but a study in cats found insulin sensitivity to be negatively correlated with the size of both abdominal subcutaneous and intra-abdominal fat depots with no difference between the two. Additionally the investigators found no difference in fat distribution between lean and obese cats, indicating that fat is deposited equally between subcutaneous and intra- abdominal fat depots when cats become obese [2]. Insulin resistance in cats with induced obesity primarily affects peripheral tissue (muscle and fat) while the regulation of hepatic gluconeogenesis seems to be better preserved in obese cats compared with obese humans[4]. This may partly explain why obese cats seldom have an increased fasting glucose level, which is a key parameter for evaluating insulin resistance in humans.

Obese cats exhibit hypertriglyceridemia, increased levels of very-low-density-lipoproteins (VLDL) and non-esterified fatty acids (NEFAs), but not hypercholesterolemia. Furthermore, newly obese cats had actually higher levels of HDL cholesterol compared with lean cats[5] though long-term obese cats had unchanged HDL_3 and lowered HDL_2[6]. However, not only the total concentration of a lipoprotein but also the subclass concentration and particle size within the lipoprotein fractions seems to determine the effect on the cardiovascular system and insulin resistance. Large and medium size VLDL particles[7], especially in combination with small size LDL and HDL particles has been associated with more severe insulin resistance and cardiovascular disease in humans[8]. In one investigation of dyslipidemia in long-term obese cats, obese cats had higher levels of large particle VLDL and small particle LDL and HDL cholesterol compared with lean cats[6]. Insulin resistance was not evaluated in this study, but because obese cats have been shown to be insulin resistant in several studies, the described alterations are likely to be correlated with insulin resistance in obese cats. In the same study, the effect of diet (high-protein diet compared with two high-carbohydrate diets including either saturated fatty acids or n-3 fatty acids) on lipoprotein concentrations and fractions were minimal[6]. Although obese cats share the characteristics of dyslipidemia in humans, atherosclerotic changes are uncommon in cats and have not yet been associated with obesity or dyslipidemia. In summary, feline obesity seems to result in insulin resistance and dyslipidemia comparable with components of the human metabolic syndrome.

Feline systemic hypertension is a disease seen in older cats[9]. It is primarily associated with chronic renal disease and hyperthyroidism, while it has not been associated with obesity and seems not to be common in feline diabetes[10].

In summary, cats develop many components of the metabolic syndrome including obesity, insulin resistance, hypertriglyceridaemia and dyslipidemia, but do not develop obesity-associated hypertension.

CONSEQUENCES OF HUMAN METABOLIC SYNDROME: EVIDENCE IN CATS

In humans, the metabolic syndrome is important because it is a risk factor for the development of three important diseases[11] – type 2 diabetes, coronary heart disease, and stroke. The application of the term "metabolic syndrome" to cats suggests that cats will suffer similar consequences in terms of morbidity and mortality to those that occur in humans with the metabolic syndrome. However, it seems that only selected components should be part of a possible future feline metabolic syndrome and further, only diabetes and not coronary heart disease or stroke is currently a relevant consequence to such a syndrome.

Feline diabetes mellitus shares several characteristics with human type 2 diabetes. It is seen in middle aged to older cats, obesity and physical inactivity are predisposing factors[12], it develops due to a combination of peripheral insulin resistance and inadequate insulin production, it is the only non-primate species to develop islet amyloidosis[13] and clinical complications such as peripheral polyneuropathy, glomerulopathy and retinopathy may all develop in cats with diabetes. Furthermore, in cats, first phase insulin secretion during a glucose tolerance test is significantly lower in obese versus lean cats[14].

Heart failure has been reported to occur commonly in cats with diabetes mellitus[15] and coronary arteriosclerotic lesions are described in cats[16]. However, atherosclerosis is reported to be an uncommon disease in cats and a direct relationship between the lesions and obesity has not yet been described. Despite that long-term obese cats may have alterations in their lipoprotein profiles similar to humans with metabolic syndrome[6] it does not appear to increase their risk of atherosclerosis.

Feline cerebrovascular disease was recently reviewed in a case series of 16 cats[17]. Seven cats were diagnosed with ischemic infarctions, five cats with haemorrhagic infarctions and four cats had intracranial haemorrhage. Infarctions were primarily related to pulmonary disease and hyperthyroidism while haemorrhage was associated with severe liver pathology. Obesity was not identified as a risk factor for either. Obese cats have been shown to have a faster rate of fibrin formation measured using tissue factor activated thromboelastography (TF-TEG), while conventional coagulation parameters and all other TF-TEG parameters were similar between lean and obese cats[18]. Based on the current literature, there may be a slightly faster coagulation in obese cats compared with lean, but it has not been shown to correlate with increasing risk of stroke. Stroke is a relatively uncommon diagnosis in cats and would be expected to be more prevalent if obesity predisposed to its development because approximately 35% of the pet cat population is reported to be overweight or obese[19].

CONCLUSIONS (CATS)

In summary, cats develop feline diabetes mellitus that is comparable with human type 2 diabetes, there are common risk factors and pathophysiological alterations between the two species though there are also differences. This makes it interesting to investigate further and to try to define a feline metabolic syndrome. However, cats do not share the increased risk of developing the cardiovascular diseases, atherosclerosis and stroke, with increasing adiposity and therefore a possible feline metabolic syndrome definition may only apply to the consequence of developing feline diabetes.

Because the prevalence of obesity and inactivity is already high in the feline pet population, defining a feline metabolic syndrome that could identify cats at increased risk of developing diabetes mellitus may be of clinical relevance. If an attempt should be made to define a metabolic syndrome in cats, the components defining the risk in humans cannot be directly translated but should be modified based on future research on the topic. Components such as fasting glucose and blood pressure are probably not useful in cats and a measure equivalent to visceral adiposity has not yet been identified in cats. Possible future markers for risk of developing diabetes in cats could include body condition score, levels of triglycerides and HDL cholesterol and while glucose tolerance curves are probably too laborious and strenuous for the cats the usefulness of postprandial glucose concentration or the fasting glucose/insulin ratio should be evaluated. However, it should be kept in mind that cats are sensitive to stress-induced hyperglycemia and that single glucose measurements are not always reliable. Based on the finding that cats gain weight equally between intra-abdominal and abdominal subcutaneous fat pads, body condition score[20], girth or other physical descriptors may prove useful. The role of adipokines such as leptin and adiponectin in feline obesity is currently being investigated and may also provide good future candidates.

Cats share important risk factors and pathophysiological processes with humans regarding the development of (type 2) diabetes mellitus. Similarities and differences should be further investigated to gain additional insight and hopefully this will result in better preventive and treatment strategies for both cats and humans.

The deleterious consequences of obesity in cats only corresponds with risk of developing feline diabetes mellitus and does not seem to apply to risk of atherosclerosis and stroke. A possible metabolic syndrome in cats may be defined in the future but it will not be possible to translate the components of human metabolic syndrome directly to cats.

THE METABOLIC SYNDROME: COMPONENTS AND CONSEQUENCES IN HORSES (Nicholas Frank)
INTRODUCTION

An equine metabolic syndrome (EMS) was first proposed in 2002 to encompass a collection of clinical findings associated with the development of laminitis in horses, ponies, and donkeys.[1] Parallels exist between EMS and metabolic syndrome (MetS) in people,[2] which was proposed after Reaven[3] described an insulin resistance (IR) syndrome in 1998 and hypothesized that IR was the cause of glucose intolerance, hyperinsulinemia, dyslipidemia, and hypertension. There has been an explosion of information about MetS since Reaven's paper was published and our understanding of EMS has expanded exponentially in the past 10 years. A large research study funded by the United States Department of Agriculture has also been performed to examine the genetic basis of EMS in horses and ponies, and results are eagerly awaited. In contrast to dogs and cats, EMS has an easily recognizable negative outcome in the form of laminitis, and this condition induces pain, limits performance, and can even result in euthanasia.

METABOLIC SYNDROME

Criteria used to define MetS vary according to the organization setting guidelines.[4] The World Health Organization defines MetS by obesity (body mass index > 30 kg/m^2) or visceral adiposity (weight-to-height ratio > 0.9 in males or > 0.85 in females), anti-hypertensive treatment or arterial pressure ≥ 140/90 mmHg, triglyceride concentration > 1.70 mmol/L (150 mg/dL) or high-density lipoprotein cholesterol (HDL-C) concentration < 0.90 μmol/L (males) or < 1.00 μmol/ l (females), and microalbuminuria ≥ 20 μg/mL.[5] The American Heart Association focuses on diagnostic tests that can be easily performed in clinical practice and states that patients are affected by MetS when they meet 3 of the following 5 criteria: Waist circumference > 102 cm (40 inches) in men and > 88 cm (35 inches) in women, triglyceride concentration > 1.70 mmol/L (150 mg/dL), HDL-C concentration < 40 mg/dL (1.03 mmol/L) in men and 50 mg/dL (1.3 mmol/L) in women, blood pressure > 30 mm Hg systolic or > 85 mm Hg diastolic or antihypertensive drug treatment, and elevated fasting glucose concentration > 100 mg/dL.[4]

It should be noted that insulin measurements are excluded from both definitions of MetS, even though the syndrome originated with the concept of an IR syndrome. One reason for this omission is that direct measurements of insulin sensitivity such as the euglycemic-hyperinsulinemic clamp are too complicated to perform in the primary care setting. Diagnostic evaluations do, however, include fasting insulin concentrations and proxy measures of insulin sensitivity, including 1/(fasting insulin), the quantitative insulin sensitivity check index (QUICKI) and homeostatic model assessment-insulin resistance (HOMA-IR).[6,7] Glucose tolerance is also assessed by performing an oral glucose tolerance test (OGTT) with the patient fasting overnight and then drinking a solution containing 75g dextrose.[6] Blood glucose concentrations are measured at 0, 30, 60, and 120 minutes. Insulin concentrations can ~~also~~ be measured to calculate the increment in plasma insulin concentration (delta I) to increment in plasma glucose concentration (delta G) ratio at 30 min, which assesses pancreatic insulin secretion.[8] Total and high-molecular weight (HMW) adiponectin concentrations are decreased and leptin concentrations are increased in human obesity and MetS, and these measures have been incorporated into diagnostic evaluations as independent variables or the adiponectin/leptin ratio.[9,10]

COMPONENTS OF EQUINE METABOLIC SYNDROME

In the consensus statement issued by the American College of Veterinary Internal Medicine in 2010, the three principal components of EMS listed were increased adiposity, IR characterized by hyperinsulinemia or abnormal glycemic and insulinemic responses to oral or IV glucose and/or insulin challenges, and a predisposition towards laminitis.[11] Other potential components listed included hypertriglyceridemia or dyslipidemia, hyperleptinemia, seasonal arterial hypertension, and increased systemic markers of inflammation. Components of EMS will be reviewed and compared across species.

Adiposity

Obesity has been associated with hyperinsulinemia, IR, hyperleptinemia, hypertriglyceridemia, and lower HMW adiponectin concentrations in equids, and yet obese animals vary markedly in their endocrine profiles.[12-16] Some horses and ponies have profound alterations in endocrine status, whereas others appear tolerant of obesity. The endocrine abnormalities of EMS have also been detected in ponies with normal body mass,[17,18] indicating that generalized obesity is not an essential component of the syndrome, and should be considered an exacerbating factor instead. One explanation for this finding is that genetic differences influence the relationship between obesity and endocrine abnormalities and this is supported by breed differences in the prevalence of EMS, with the syndrome being more common in ponies, Morgans, Paso Finos, and Arabians.

Regional adiposity, in the form an enlarged neck crest or fat pads close to the tail head, is commonly observed in horses and ponies when EMS is first diagnosed.[19] Neck crest measurements are positively correlated with insulin concentrations in horses and ponies, and regional adiposity has been associated with the development of laminitis.[13,15] Burns et al.[20] also demonstrated that nuchal adipose tissue has higher proinflammatory cytokine mRNA expression than other adipose tissues in the horse, although insulin-resistant horses did not differ significantly from insulin-sensitive controls. It is therefore accepted that horses and ponies exhibiting regional adiposity should be tested for EMS. However, hyperinsulinemia and IR can also develop in the absence of regional adiposity, so screening for hyperinsulinemia should be considered for all adult horses and ponies. Just as people are screened for hypercholesterolemia to determine their risk of coronary artery disease, screening for hyperinsulinemia should be performed to assess the risk of laminitis.

Hyperinsulinemia

As we discuss insulin concentrations in equids, EMS diverges from related conditions in humans, dogs, and cats. Hyperinsulinemia is a key component of EMS and higher than normal insulin concentrations persist for years in affected horses and ponies. In contrast to cats and humans, beta cell insufficiency/exhaustion is an uncommon occurrence in equids, even with marked hyperinsulinemia and accompanying obesity. Islet amyloidosis, an important feature of diabetes mellitus and cause of pancreatic insufficiency in cats and humans,[21] has not been reported in horses. Wider use of oral glucose tolerance testing in horses has also raised the question of whether postprandial versus fasting hyperinsulinemia is the first indication of insulin dysregulation, and if this problem occurs without IR in milder cases. Use of the oral sugar test (OST) and in-feed oral glucose test (OGT)[18] has led to the identification of horses and ponies with normal fasting insulin concentrations (< 20 μU/mL by radioimmunoassay) and high postprandial insulin concentrations. The OST or OGT is therefore recommended to screen horses for hyperinsulinemia. If concerns are expressed about conducting one of these tests, owners can be reassured that only transient hyperinsulinemia is induced. One major advantage of performing the OST instead of only assessing fasting insulin concentrations, as previously recommended in the consensus statement,[11] is that postprandial insulin responses can be investigated. Horses kept on pasture are eating almost continuously and non-structural carbohydrates and amino acids from the grass consumed are stimulating insulin secretion. This type of testing provides the best assessment of glucose and insulin responses to ingested sugars.

It has previously been assumed that hyperinsulinemia results from IR in horses and the concept of compensated IR is widely accepted.[22] Pancreatic insulin secretion increases to compensate for the reduced effectiveness of insulin at the tissue level and to maintain blood glucose concentrations within euglycemic range. In contrast, uncompensated IR refers to decreased tissue insulin sensitivity, inadequate insulin secretion and high-normal blood glucose concentrations. Diabetes mellitus occurs in equids, but is uncommon in comparison with cats, and often associated with pituitary pars intermedia dysfunction (PPID).[23] Transient diabetes is sometimes detected in aged horses with PPID that are stressed as a result of hospitalization and systemic disease, and catecholamine-mediated inhibition of insulin secretion via alpha 2 receptors may be a contributing factor. Many affected horses respond well to exogenous insulin and return to euglycemia once the primary problem resolves and PPID is controlled.

Since pancreatic insufficiency is uncommon in horses and postprandial hyperinsulinemia appears to be the first manifestation of insulin dysfunction in equids, a new theory is taking shape to explain the unique features of EMS. It is now postulated that EMS begins with hyperinsulinemia in genetically predisposed horses and ponies, and then IR develops and exacerbates the problem. Hyperinsulinemia might even contribute to the development of IR through the process of homologous desensitization, in which tissue insulin sensitivity decreases as blood hormone concentrations rise.[24] This is the mechanism by which insulinomas induce IR and is also referred to as the Somogyi effect when higher doses of exogenous insulin fail to lower blood glucose concentrations. Obesity acts as an exacerbating factor and insulin concentrations rise even further as tissues become insulin resistant and insulin clearance decreases. It is estimated that approximately 70% of insulin secreted by pancreatic beta cells is cleared from the portal blood by the equine liver,[25] so lipid infiltration and other effects of obesity on hepatic function would also be expected to raise insulin concentrations. Connecting (C) peptide to insulin ratios are lower in horses with IR, compared with healthy controls and this provides evidence of decreased insulin clearance.[25] A genetic basis for hyperinsulinemia would also explain why insulin concentrations respond differently to obesity in individual horses.[26,27]

Hyperinsulinemia potentially contributes to the development of obesity through the anabolic effects of insulin on lipid metabolism,[24] and a genetic basis for hyperinsulinemia would explain the observed differences among breeds with respect to the development of obesity. Horses and ponies with EMS are described as "easy keepers" and this phenotype might develop as a result of excessive insulin responses to ingested sugars and proteins. In a recent study of knockout mice, hyperinsulinemia was identified as the key factor altering lipid metabolism and increasing body fat mass in a diet-induced obesity model.[28] Expression of insulin genes was manipulated within the brain and pancreas to generate mice that were genetically incapable of diet-induced hyperinsulinemia, and these mice remained free of obesity and hepatic lipid infiltration. A vicious cycle might exist if obesity causes lipid infiltration within the liver, and this slows insulin clearance and raises insulin concentrations.

Insulin secretion is also stimulated by feeding when two incretin hormones, glucagon-like peptide 1 (GLP-1) and glucose-dependent insulinotropic polypeptide (GIP), are secreted from intestinal K and L cells, respectively.[29] Both incretin hormones are then degraded by the enzyme dipeptidyl transferase 4 (DPP4), which is present in the blood. Glucagon-like peptide 1 agonists and DPP4 inhibitors have been developed as drugs for the treatment of diabetes in humans, with the goal of increasing insulin concentrations. When administered to healthy cats, the GLP-1 agonist exenatide increased insulin secretion in a glucose-dependent manner,[30] and administration of a DPP4 inhibitor to cats in another study raised insulin concentrations during intravenous glucose tolerance testing.[31] Lower GLP-1 concentrations during an OGTT have also been detected in obese cats and may contribute to the development of feline diabetes mellitus.[32] Incretin hormones have been measured in equine plasma and active GLP-1 concentrations decreased when horses with EMS were placed on a weight gain diet.[33,34] It is intriguing to consider the idea of incretin overproduction or impaired degradation contributing to hyperinsulinemia in equids. Since beta cell mass increases with incretin hormone stimulation,[35] this theory might also explain how hyperinsulinemia can persist for years in horses without leading to beta cell exhaustion. As GLP-1 agonists and DPP4 inhibitors are considered for the treatment of diabetes mellitus in cats, they are not likely to be appropriate therapies for EMS. Laminitis has been experimentally induced in healthy ponies and horses by infusing insulin intravenously and creating hyperinsulinemia for 48 to 72 hours.[36,37]

Insulin resistance

Insulin resistance is associated with physiological adaptations to stress and pregnancy, and pathological conditions including obesity, systemic inflammation, and PPID. The euglycemic-hyperinsulinemic clamp procedure can be performed to measure tissue insulin sensitivity directly and this method was used by Vick et al.[38] to establish an association between obesity and IR in horses. Other tests involving administration of exogenous insulin can be used to assess insulin sensitivity in horses, and are reviewed elsewhere.[39] The OST or OGT is recommended for EMS because the sum effect of incretin hormones, pancreatic beta cell function, and insulin sensitivity can be assessed at once. If insulin sensitivity is the main concern, the combined glucose-insulin test or two-step insulin tolerance test can be performed in practice.[40]

Adipokines

Hyperleptinemia is associated with obesity and increased insulin concentrations in horses,[41,42] and leptin concentrations are measured for diagnostic purposes by the Cornell Animal Health Diagnostic Laboratory (Ithaca, NY). Higher than normal leptin concentrations (> 4 ng/mL) are detected in horses with EMS and these values provide additional information to aid in the interpretation of insulin results. When horses of approximately the same body condition score were compared, horses with high leptin concentrations (> 12 ng/mL) had significantly lower insulin sensitivity than those with normal leptin concentrations (< 2 ng/ml).[14] Total adiponectin concentrations have also been measured in horses and negatively correlated body fat mass.[43] Adiponectin-to-leptin ratios were calculated in the same study and found to be negatively correlated with percent fat ($r = 0.621$; $P < 0.001$; $n = 35$). It has been difficult to identify a total adiponectin assay that performs consistently with equine plasma, but Wooldridge et al.[16] recently validated an ELISA for HMW adiponectin and reported concentrations of 3.6 ± 3.9 µg/mL (mean ± SD) for obese horses and 8.0 ± 4.6 µg/mL for lean horses. There was overlap in the ranges reported, but mean values differed significantly.

Dyslipidemia and hypertension

Very low-density lipoprotein and HDL-C concentrations were significantly higher in obese insulin-resistant horses, when compared with nonobese controls.[15] Hypertriglyceridemia has also been identified as a predictor of laminitis risk with cut-off values of 57 and 94 mg/dL established from two studies of the same population of ponies.[12,13] These cut-off values have high specificity, but low sensitivity, and lower values are likely to be selected for horses.

Hypertension has been detected in laminitis-prone ponies housed on pasture and a seasonal pattern was identified.[17] No differences were detected between groups in the winter, but mean blood pressure was significantly higher in laminitis-prone ponies when they were tested in the summer. Median (interquartile range) blood pressure was 89.6 mm Hg (78.3 to 96.9 mm Hg) in laminitis-prone ponies, compared with 76.8 mm Hg (69.4 to 85.2 mm Hg) in control ponies.

Laminitis

Recent studies have demonstrated that hyperinsulinemia is detrimental to laminar tissues in horses.[36,37] The mechanisms responsible for laminitis development are incompletely understood, but may include increased glucose delivery to the hoof and localized glucotoxicity as a result of insulin-induced vasodilation and capillary recruitment.[44,45] Mitogenic stimulation by insulin and abnormal laminar tissue proliferation has also been suggested,[46,47] and insulin-like growth factor-1 receptors were recently detected on hoof epithelial cells.[48] Mean serum insulin concentrations in experimental models exceed 1,000 µU/mL and this is a higher value than commonly detected in cases; although moderate hyperinsulinemia might have the same effect with prolonged exposure. Histopathological evidence of subclinical laminitis has been detected in healthy horses after 48-hour intravenous glucose infusion that induced serum insulin concentrations between 100 and 300 µU/mL.[49] These findings support a role for hyperinsulinemia, alone or in combination with hyperglycemia, in the development of laminitis.

Other markers

Potential new markers for EMS include serum amyloid A and ferritin. Higher serum amyloid A concentrations were detected in insulin-resistant light breed horses, compared with insulin-sensitive horses[50] and in another study, a positive correlation existed between area under the OGTT insulin curve and serum ferritin concentrations.[51]

CONCLUSIONS (OVERALL)

Although humans, dogs, cats and horses all develop obesity, obesity-associated insulin resistance, and associated metabolic consequences, there are substantial differences between species in relation to the effects of obesity. Dogs, cats and horses do not uniformly express the components of the human metabolic syndrome. The clinical consequences of obesity differ among the domestic species. A disease similar to type 2 diabetes has been documented in cats but not in dogs or horses. Horses exhibit clinical conditions associated with obesity such as laminitis that are not observed in humans. Coronary heart disease and stroke are rare and do not appear to be associated with obesity in dogs, cats, or horses. Although the term metabolic syndrome has been proposed or applied to these species, it remains to be shown that the pathophysiology of obesity-associated diseases are equivalent between humans and companion animal species.

REFERENCES (Human)

1.Gorden P. ***Annal Acad Med Singapore*** 1997;26:326. 2.Reaven GM. ***Diabetes*** 1988;37:1595. 3.Reaven GM. ***Med Clin North Am*** 2011;95:875. 4.Larsson B, et al. ***BMJ*** 1984;288:1401. 5.Lapidus L, et al. ***BMJ*** 1984;289:1257. 6.Ducimetiere P, et al. ***Int J Obes*** 1986;10:229. 7.Després JP. ***Circulation*** 2012;126:1301. 8.Alberti K, et al. ***Circulation*** 2009;120:1640. 9.Alberti K, et al. ***Diabet Med*** 2006;23:469. 10.Berglund L, et al. ***J Clin Endocrinol Metab*** 2012;97:2969. 11.Simmons R, et al. ***Diabetologia*** 2010;53:600. 12.Defronzo RA. ***Med Clin North Am*** 2004;88:787. 13.Nolan CJ, et al. ***Lancet*** 2011;378:169. 14.Prentki M, Nolan CJ. ***J Clin Inv*** 2006;116:1802. 15.Gale EaM. ***Diabetologia*** 2005;48:1679. 16.Kahn R, et al. ***Diab Care*** 2005;28:2289. 17.Grundy SM, et al. ***Curr Opin Cardiol*** 2006;21:1. 18.Unger RH, Scherer PE. ***Trends in Endocrino Metabol*** 2010;21:345. 19.Giaccari A, et al. ***Nutr, metabo, cardiovascular diseases: NMCD*** 2009;19:365. 20.Lorenzi M. ***Diabetes Metab Rev*** 2009;8:85. 21.Kawahito S, et al. ***World j gastroenterol*** 2009;15:4137. 22.Carr MC, Brunzell JD. ***J Clin Endocrinol Metab*** 2004;89:2601. 23.Brouwers MCGJ, et al. ***Nature Reviews Endocrinol*** 2012. 24.Bouchard C, Tremblay A. ***J Nutr*** 1997;127:943S. 25.Sørensen T, et al. ***Int obesity metabolic disorders*** 1992;16:705. 26.Frank N, et al. ***J Vet Intern Med*** 2010;24:467. 27.Tvarijonaviciute A, et al. ***BMC Vet Res*** 2012;8:147. 28.Hoenig M, et al. ***Am J Physio - Reg, Integr Compar Physiol*** 2007;292:R227. 29.Hoenig M. ***Curr Opin Clin Nutr Metab Care*** 2006;9:584. 30.Appleton DJ, et al. ***J Feline Med Surg*** 2005;7:183.

REFERENCES (Dogs)

1.Lund EM, et al. ***Int J Appl Res Vet Med*** 2006;4:177. 2.Verkest KR, et al. ***Domest Anim Endocrinol*** 2011;41:24. 3.Verkest KR, et al. ***Am J Vet Res*** 2011;72:357. 4.Bergman RN, et al. ***Obesity*** 2006;14:16S. 5.Bodey AR, Michell AR. ***J Small Anim Pract*** 1996;37:116. 6.Chikamune T, et al. ***J Vet Med Sci*** 1995;57:595. 7.Ishioka K, et al. ***Vet Radiol Ultrasound*** 2005;46:49-53. 8.Mittelman SD, et al. ***Diabetes*** 2002;51:755. 9.Lottati M, et al. ***Obesity*** 2012;17:674. 10.Kim SP, et al. ***Diabetes*** 2003;52:2453. 11.Kolka CM, et al. ***Diabetes*** 2010;59:619. 12.Verkest K, et al. ***Domest Anim Endocrinol*** 2012;42:103. 13.Jeusette IC, et al. ***Am J Vet Res*** 2005;66:81. 14.Bauer JE. ***JAVMA*** 2004;224:668. 15.Expert Panel on Detection E, And Treatment of High Blood Cholesterol in Adults. ***Circulation*** 2002;106:3143. 16.Alberti K, et al. ***Circulation*** 2009;120:1640. 17.Alberti K, et al. ***Diabet Med*** 2006;23:469. 18.Barrie J, et al. ***J Small Anim Pract*** 2008;34:507. 19.Moses C. ***Circ Res*** 1954;2:243. 20.Brown S, et al. ***JVIM*** 2007;21:542. 21.Montoya JA, et al. ***J Nutr*** 2006;136:2011S. 22.Mehlman E, et al. ***JVIM***2012. 23.Remillard R, et al. ***Am J Vet Res*** 1991;52:561. 24.Kannel WB, et al. ***Ann Intern Med*** 1967;67:48. 25.Neto GBP, et al. ***Pesquisa Veterinária Brasileira*** 2010;30:167. 26.Pelosi A, et al. ***Vet Rec*** 2013, 27.Simmons R, et al. ***Diabetologia*** 2010;53:600. 28.Fleeman LM, Rand JS. **ACVIM Forum** 2007. 29.Rand JS, et al. ***J Nutr*** 2004;134:2072S. 30.Cotton R, et al. ***JAVMA*** 1971;159:863. 31.Buchanan TA, et al. ***Diabetes Care*** 2007;30:S105. 32.Mattheeuws D, et al. ***Am J Vet Res*** 1984;45:98. 33.Fall T, et al. ***JVIM*** 2008;22:1296. 34.Prentki M, Nolan CJ. ***J Clin Invest*** 2006;116:1802. 35.Gerich JE. ***Diabetes*** 2002;51:S117. 36.Gredal C, et al. ***Scand J Clin Lab Inv*** 2007;67:327. 37.Larson BT, et al. ***J Nutr*** 2003;133:2887. 38.Hess RS, et al. ***JVIM*** 2003;17:489. 39.Kagawa Y, et al. ***J Comp Pathol*** 1998;118:195. 40.Liu S, et al. ***JAVMA*** 1986;189:227. 41.Garosi L, et al. ***JVIM*** 2005;19:725. 42.Farber E. *Canad Vet J* 1985;26:121.

REFERENCES (Cats)

1.Appleton DJ, et al. ***J Feline Med Surg*** 2001;3:211. 2.Hoenig M, et al. ***Am J Physiol: Reg, Integr Compar Physiol*** 2007;292:R227. 3.Bergman RN, et al. ***Obesity*** 2006;14 Suppl 1:16S. 4.Kley S, et al. ***Am J Physiol: Reg, Integr Compar Physiol*** 2009;296:R936. 5.Hoenig M, et al. ***Am J Vet Res*** 2003;64:299. 6.Jordan E, et al. ***Domest Anim Endocrinol*** 2008;35:290. 7.Lucero D, et al. ***Clin Biochem*** 2012;45:293. 8.Frazier-Wood AC, et al. ***Lipids In Health And Disease*** 2011;10:237. 9.Stepien RL. ***J Feline Med Surg*** 2011;13:35. 10.Sander C, et al. ***Tierarztliche Praxis Ausgabe K, Kleintiere/Heimtiere*** 1998;26:110. 11.Simmons R, et al. ***Diabetologia*** 2010;53:600. 12.Slingerland LI, et al. ***Vet J*** 2009;179:247. 13.Lutz TA, et al. ***Aust Vet J*** 1994;71:254. 14.Hoenig M, et al. ***JVIM*** 2002;16:529. 15.Little CJ, Gettinby G. ***JSAP*** 2008;49:17. 16.Sostaric-Zuckermann IC, et al. ***BMC research notes*** 2011;4:350. 17.Altay UM, et al. ***JAAHA*** 2011;47:89. 18.Bjornvad CR, et al. ***J Animal Physiol Animal Nutr*** 2012;96:834. 19.Lund EM, et al. ***JAVMA*** 1999;214:1336. 20.Bjornvad CR, et al. ***AJVR*** 2011;72:433.

REFERENCES (Horses)

1. Johnson PJ. **Vet Clin North Am Equine Pract** 2002;18:271. 2. Nikolopoulou A, Kadoglou NP. **Expert Rev Cardiovasc Ther** 2012;10:933. 3. Reaven GM. **Journal of basic and clinical physiology and pharmacology** 1998;9:387. 4. Grundy SM, et al. **Critical pathways in cardiology** 2005;4:198. 5. Marchesini G, et al. **Diabet Med** 2004;21:383. 6. Muniyappa R, et al. **Am J Physiol Endocrinol Metab** 2008;294:E15. 7. Wallace TM, et al. **Diabetes Care** 2004;27:1487. 8. Reaven GM. **Journal of diabetes** 2009;1:142. 9. Hirose H, et al. **J Atheroscler Thromb** 2010;17:1201. 10. Mirza S, et al. **Clinical and investigative medicine Medecine clinique et experimentale** 2011;34:E290. 11. Frank N, et al. **J Vet Intern Med** 2010;24:467. 12. Treiber KH, et al. **J Am Vet Med Assoc** 2006;228:1538. 13. Carter RA, et al. **Equine Vet J** 2009;41:171. 14. Caltabilota TJ, et al. **J Anim Sci** 2010;88:2940.15. Frank N, et al. **J Am Vet Med Assoc** 2006;228:1383. 16. Wooldridge AA, et al. **Am J Vet Res** 2012;73:1230. 17. Bailey SR, et al. **Am J Vet Res** 2008;69:122. 18. Borer KE, et al. **J Anim Sci** 2012;90:3003. 19. Carter RA, et al. **Vet J** 2009;179:204. 20. Burns TA, et al. **JVIM** 2010;24:932. 21. O'Brien TD. **Mol Cell Endocrinol** 2002;197:213. 22. Treiber KH, et al. **Am J Vet Res** 2005;66:2114. 23. Durham AE, et al. **Equine Vet J** 2009;41:924. 24. Shanik MH, et al. **Diabetes Care** 2008;31 Suppl 2:S262. 25. Toth F, et al. **Equine Vet J** 2010;42:149. 26. Carter RA, et al. **Am J Vet Res** 2009;70:1250. 27. Carter RA, et al. **Am J Vet Res** 2010;71:314. 28. Mehran AE, et al. **Cell Metab** 2012;16:723. 29. Phillips LK, Prins JB. **Ann N Y Acad Sci** 2011;1243:E55. 30. Gilor C, et al. **Domest Anim Endocrinol** 2011;41:42. 31. Furrer D, et al. **Vet J** 2010;183:355. 32. Hoenig M, et al. **Domest Anim Endocrinol** 2010;38:95. 33. Chameroy K, et al. **JVIM** 2011;25:766 (abstract). 34. Duhlmeier R, et al. **Comparative biochemistry and physiology Part A, Molecular & integrative physiology** 2001;129:563. 35. Mudaliar S, Henry RR. **Am J Med** 2010;123:S19. 36. de Laat MA, et al. **Equine Vet J** 2010;42:129. 37. Asplin KE, et al. **Vet J** 2007;174:530. 38. Vick MM, et al. **J Anim Sci** 2007;85:1144. 39. Firshman AM, Valberg SJ. **Equine Vet J** 2007;39:567. 40. Bertin FR, Sojka-Kritchevsky JE. **Domest Anim Endocrinol** 2013;44:19. 41. Cartmill JA, et al. **J Anim Sci** 2003;81:2311. 42. Cartmill JA, et al. **J Anim Sci** 2005;83:1875. 43. Kearns CF, et al. **Vet J** 2006;172:460. 44. de Laat MA, et al. **Vet Clin North Am Equine Pract** 2010;26:257. 45. Cersosimo E, DeFronzo RA. **Diabetes Metab Res Rev** 2006;22:423. 46. Nourian AR, et al. **Equine Vet J** 2009;41:671. 47. Asplin KE, et al. **Equine Vet J** 2010;42:700. 48. Burns TA, et al. **Equine Vet J** 2012. 49. de Laat MA, et al. **Vet J** 2012;191:317. 50. Suagee JK, et al. **JVIM** 2012. 51. Nielsen BD, et al. **J Zoo Wildl Med** 2012;43:S61.

SELECTIVE RENAL INTRA-ARTERIAL (IA) AND NON-SELECTIVE INTRAVENOUS (IV) DELIVERY OF AUTOLOGOUS MESENCHYMAL-DERIVED STEM CELLS (MSC) IN CANINE AND FELINE PATIENTS WITH ACUTE AND CHRONIC KIDNEY DISEASE

Allyson Berent, DVM, DACVIM (SAIM)
New York, NY

INTRODUCTION

Chronic Kidney Disease (CKD) is characterized by progressive loss of renal function, and culminates with total organ failure. CKD, or chronic interstitial nephritis (CIN), is among the most common diagnosis in older feline patients, affecting up to 35% of older cats and up to 20% of all cats. The data on the long-term survival in cats with CIN suggests a mean survival time of 1-3 years, depending on the stage at which they are diagnosed. Currently there are no known therapies to aid in reversal of this condition and the focus, to this point, has been in supportive care including dietary changes, improving hydration, management of proteinuria and hypertension, chelation of excessive phosphorus and controlling renal secondary hyperparathyroidism. Renal transplantation has been the only available treatment option to improve renal function for a patient, which is expensive and met with many immunologic and technical difficulties. Most cats are diagnosed early in the stage of their disease and progressive decline in renal function is witnessed.

Protein losing kidney disease (PLN) is the leading cause of renal disease in dogs. In a randomly selected group of dogs the incidence of glomerular lesions was documented to be as high has 43-90%. This is most commonly associated with one of 2 forms: membranproliferative glomerulonephritis (MPGN) and membranous nephropathy (MN). Immune complexes that are deposited in the glomerulus can initiate glomerular damage. Cell-mediated immune mechanisms take part in the pathogenesis and once the damage has been initiated, activation of complement, influx of inflammatory cells, release of proteolytic enzymes, synthesis of cytokines and growth factors, and generation of pro-inflammatory mediators occurs, resulting in a highly inflammatory environment and severe cell damage. There are some immunosuppressive treatments aimed at minimizing immune-complex deposition and inflammatory infiltrative causes of PLN, but up to this point, overall outcomes in dogs with PLN is considered extremely poor. Traditionally, dogs that are azotemic are expected to survive less than 3 months. In one study the median survival of 53 dogs with kidney disease was 28 days. A recently reported study in 2011 compared 234 dogs with PLN and found the median survival times (MST) for dogs with a creatinine level over 1.5 mg/dL was only 13 days versus a normal creatinine with PLN was 407 days. For dogs with PLN and nephrotic syndrome (hypoalbuminemia, hypercholesterolemia, proteinuria and 3rd space effusion) the MST was only 12.5 days versus 104.5 days in dogs without nephrotic syndrome. Therefore, the investigation of alternative approaches, such as mesenchymal stem cell (MSC) therapy, in which the potential for slowing disease progression exists, is of great clinical importance. One of the largest challenges facing veterinary nephrology is improving the outcome of kidney disease in cats and dogs. The therapeutic strategies should be aimed at preventing disease progression and treating the underlying condition, but this is not currently possible in most cases. In veterinary medicine, we often treat the clinical signs while progression ensues.

Mesenchymal stem cells (MSC) are multipotent cells that reside in various parts of the body like fat and bone marrow. These cells are capable of differentiating in vitro and in vivo into different cell types. They are considered MSC based on their ability to adhere to plastic surfaces and to adopt a fibroblast-like morphology, termed colony-forming unit-fibroblasts. They are attractive candidates for renal repair because nephrons are of mesenchymal origin and stromal cells are of crucial importance for signaling, which can lead to the differentiation of both nephrons and collecting ducts. MSC express specific cell surface markers (such as CD29, 44, 73, 90, and 105). In animal models, stem cell therapy has been shown to have many potential benefits in tissue recovery, and regeneration and differentiation of damaged cells. This occurs by various mechanisms including the release of chemical mediators for the trophic effects (paracrine effects of growth factors and cytokines), angiogenic effects, anti-apoptosis, antifibrotic and anti-inflammatory effects. Previous studies in animal models have shown MSC have the potential to enhance recovery from acute renal tubular injury and repair in both interstitial and glomerular diseases. The ultimate goal is to create an environment to impede glomerular and tubular inflammation and the progression of glomerular and interstitial fibrosis. In 2006 Kunter et al reported on the use of intra-arterial MSC injection into a rat model with mesagnioproliferative glomerulonephritis. In this study 20-50% of glomeruli had fluorescently-labelled MSC detected within them. The authors reported a 50% reduction of mesangiolysis, a 3-4 fold higher intraglomerular cell proliferation, reduction in proteinuria by 28%, and increased glomerular cell proliferation. In a rat model, acute renal failure was ameliorated by MSC injection after disease induction. This study also documented MSC secreted high amounts of vascular endothelial growth factor and TGF-B1, which have anti-inflammatory properties. Overall, MSC were shown to markedly accelerate glomerular recovery, most likely related to the paracrine growth factor release.

STEM CELL DELIVERY

Because of the homing capacity of MSC in the body, once injected, these cells will ultimately be trapped peripherally in any area of diseased tissue. Older animals often have a relatively high incidence of other concurrent maladies such as inflammatory bowel disease, chronic liver disease, endocrine disease (hyperthyroidism, hypothyroidism), heart disease and osteoarthritis. With intravenous (IV) systemic delivery, the MSC's will not be delivered directly to the renal tissue, but instead will home to all areas of the body with concurrent inflammation. Stem cells that have been delivered intravenously have first pass delivery to the pulmonary capillary beds before being distributed to the rest of the body. If the cells were directly delivered into the renal artery, the first capillary bed to engraft the stem cells would be the glomeruli and vast blood supply of the diseased nephrons. This could potentially result in more dramatic effects. In rodent models, it has been shown for both myocardial disease and acute cerebral injury , intra-arterial (IA) stem cell infusion resulted in significantly higher colonization of stem cells into the diseased area. In one study where sheep had IA infusion of MSC directly into the renal artery there was no evidence of renal damage that occurred during or after the injection, confirming the safety and reliability of the IA procedure in this species. These patients were also shown to have higher engraftment of MSC's in the glomeruli, which is rarely reported in studies after IV systemic delivery. The speculation was that MSC were stopped in the glomeruli by a trapping phenomenon within the microcapillary vessels at the end of the glomerular tuft, which would not occur after first-pass elimination in the pulmonary capillary beds.

Intravenous injection is the least invasive method for delivery but may lead to low numbers of engrafted cells inside the diseased kidneys, and major trapping in the lung or other peripheral sites of concurrent injury (e.g. osteoarthritis, inflammatory bowel disease, inflammatory liver disease, hyperthyroidism). Intraarterial injection, by the use of minimally-invasive image guided interventional radiologic techniques, may be used to bypass the initial uptake of the stem cells by the non-renal systemic organs and deliver larger number of cells directly to the diseased renal tissue. Anecdotally, both canine and feline patients have been injected with autologous MSC IV for various conditions without any known ill effects (osteoarthritis, inflammatory bowel disease, inflammatory liver disease, and chronic renal disease), but evidence of long-term benefit for most of these conditions is not currently available.

Because stem cells are thought to improve, repair, and aid in the growth of damaged tissue as well as decrease fibrosis, improving the health and environment of the cells that remain could improve the overall function of the kidneys and ultimately improve the survival times and quality of life in these patients. A majority of the studies using stem cell therapy for CKD and glomerulonephritis were performed in normal animal models (rats, sheep, pigs,) where various forms of kidney disease were artificially created. Using a natural model of disease, various delivery options, and long-term follow-up on clinical patients may be a beneficial model for humans.

VETERINARY EXPERIENCE

A few studies have been done in veterinary medicine using MSC therapy in the area of osteoarthritis (OA). Intra-articular injections were done for elbow OA and Hip OA and improved lameness was concluded. One study by Quimby et al. they investigated intra-renal parenchymal injections of MSC in cats with IRIS stage 2-4 CKD and found adipose derived MSC was superior to bone marrow derived MSC but the modest improvement in creatinine and GFR were not significant.

For the past 2.5 years we have been investigating the use of regenerative medicine through adipose derived MSC therapy to aid in nephron repair, improve nephron function, decrease renal fibrosis, and ultimately slow the progression of CKD in cats, and PLN in dogs. At the Animal Medical Center we are conducting 2 separate randomized, placebo-controlled studies. The aims of our feline study are two-fold: 1) to initially perform a pilot study on 6 cats with IRIS stage III CKD using IA autologous MSC delivery into the renal artery to assess short term safety, and 2) if phase 1 proves safe over a 3 month period, we will investigate and compare the effect of autologous MSC injected via either the renal artery or cephalic (or saphenous) vein on CKD progression in a randomized placebo controlled study. The aims of our canine PLN study are 1) investigate if the use of MSC in azotemic canine patients with PLN-CKD and see if they improve renal function and protein loss over a 1 year period; 2) investigate if the delivery of the MSC selectively (IA) vs. non-selectively (IV) vs. IV placebo will improve clinical outcome in canine patients over a 1 year period. The pilot study in dogs and cats was completed. It was shown to be safe in both dogs and cats with various forms of renal failure and IA delivery was met with few complications. The Phase II trials are ongoing. The details of the procedure and some preliminary data will be presented in this lecture.

The authors hypothesize that MSC therapy coupled with conventional therapy will 1) improve renal function based on biochemical data (e.g. GFR studies, creatinine levels, proteinuria, serum phosphorus concentration), 2) reduce the mortality from renal failure, 3) reduce all cause mortality, 4) is safe and 5) will improve quality of life compared with animals treated only by conventional therapy. In addition, 6) MSC therapy delivered via renal arterial injection will prolong survival compared with animals receiving MSC through peripheral venous injection.

RECENT PUBLISHED RESULTS

In a recent abstract data on 19 cats (41 treatments) and 15 dogs (51 treatments), totaling 92 treatments (49 IA [13 feline, 36 canine], 43 IV [28 feline, 15 canine]) autologous MSCs were delivered for treatment of various causes of renal failure: PLN (5), renal dysplasia with glomerulosclerosis (5), canine CKD (2), feline CIN (19) with obstructive nephropathy (13/19), CKD of unknown histopathology (2), and AKI (1).

The presenting creatinine was 3.3 (range, 2.9-18.2) in cats and 3.8 (range, 0.8-13.1) in dogs. The presenting UP:C was 0.21 (range, 0.09-0.34) in cats and 2.5 (range, 0.75-8) in dogs. The last recorded creatinine was 3.3 mg/dL in cats and 5.3 in dogs, at a median of 179 and 213 days, respectively. The last median UP:C was 0.17 in cats and 1.8 in dogs.

For the 49 IA treatments 16 were treated unilaterally and 33 bilaterally, totaling 82 renal units treated. The median IA procedure time was 37.5 min (range, 25-75). The median hospitalization time was 0.75 days (range, 0.25-2.5). There were 3 minor complications. Two feline renal arteries had evidence of partial embolization at the time of IA cell delivery (2/13 treatments), neither of which had any clinical consequence. One feline patient had excessive subcutaneous bruising at the fat isolation site during harvesting. No patient had a known negative affect of stem cell therapy (IA or IV) at any time point after stem cell delivery (range 3-26 months) and all renal blood flow was either similar or improved to pre-IA delivery. Two cats vomited after IV delivery with no further consequence. No patient had progressive azotemia within 2 weeks of any anesthetic event. The median survival time in dogs with CKD was > 245 days (2/2 still alive), with PLN > 227 days (3/5 still alive, range 14-592 d), with renal dysplasia >198 days (5/5 dogs still alive; range, 80-773 d); and in cats with CIN> 179 days (13/19 [68%] still alive, range 60-439 d). Cats with obstructive nephropathy had a MST of > 179 days (7/12 still alive) and those with IRIS stage 3 CKD alone > 213 days (6/7 still alive).

The data we have at this time is still preliminary, and because most patients are still alive, survival times are not accurate. The goal of this talk is to discuss the potential of MSC therapy as an option for veterinary patients with CKD and AKI, and over the next few years, as placebo controlled data becomes available, the efficacy will hopefully be determined.

CONCLUSION

Overall, MSC delivery for various causes of renal failure in dogs and cats is safe with both IV and IA delivery. The need for microvascular equipment for IA delivery in feline patients in necessary and care should be taken during renal arterial catheterization, as it is more difficult than dogs. Training is encouraged prior to considering IA delivery in any patient. Efficacy for MSC therapy, both IA and IV, is currently being evaluated in a randomized placebo controlled manner over 1-3 years. The data on dogs, not in the study, would suggest that we are seeing a slower progression of disease than that which has been historically reported.

REFERENCES

1. Walczak P, Zhang J, Assaf A, et al. Dual-modality monitoring of targeted intra-arterial delivery of MSC after transient ischemia. Stroke. 2008; 39(5): 1569-1574.
2. Kunter U, et al. Transplanted MSC acclerate glomerular healing in experimental glomerulonephritis. J Am Soc Nephrol. 2006; 17:2202-2212.
3. Barbash IM, et al. Systemic delivery of bone marrow-derived MSC to the infarcted myocardium. Circulation. 2003;108(7): 863-868.
4. Newsome P. Yet another role for MSCs? Transplantation 2008; 85: 1548-1549.
5. Behr L, et al. Intra renal arterial injection of autologous mesenchymal stem cells in an ovine model in the post-ischemic kidney. Nephron Physiol. 2007; 107(3): 65-76.
6. Choi S, et al. The role of MSC in the functional improvement of chronic renal failure. Stem Cells and Develop. 2009; 18(3): 521-529.
7. Ittrich H, et al. In vivo magnetic resonance imaging of iron oxide-labeled, arterially injected MSC in kidneys of rats with acute ischemic kidney injury: detection and monitoring at 3T. J of Mag Res Imag. 2007; 25: 1179-1191.
8. Semedo P, et al. MSC attenuate renal fibrosis through immune modulation and remodeling properties in a rate remnant kidney model. Stem Cells. 2009; 27: 3063-3073.
9. Kirpatovskii VI, et al. Experimental intravenous cell therapy of acute and chronic renal failure. Bulletin of Experiment Biol and Med. 2007; 143(1): 160-1
10. Quimby J, et al. Evaluation of intrarenal MSC injection for treatment of chronic kidney disease in cats: a pilot study. JFMS 2011; 13:418-426.
11. McTaggart SJ, et al. Mesenchyal stem cells: immunobiology and therapeutic potential in kidney disease. Nephrology 2007; 12: 44-52.
12. Black LL, et al. Effect of intraarticular injection of autologous adipose-derived mesenchymal stem and regenerative cells on clinical signs of chronic osteoarthritis of the elbow joint in dogs. Vet Ther. 2008; 9(3): 192-200.
13. Boyd LM, et al. Survival in cats with naturally occurring chronic kidney disease (2000-2002). J Vet Intern Med. 2006; 22: 1111-1117.
14. Klosterman ES, et al. Comparison of signalment, clinicopathologic findings, histologic diagnosis, and prognosis in dogs with glomerular disease with or without nephrotic syndrome. JVIM 2001; 25:206-214.

TELOMERE LENGTH AND CELLULAR SENESCENCE IN FELINE CHRONIC KIDNEY DISEASE

Jessica Quimby, DVM, PhD, DACVIM
Fort Collins, CO

INTRODUCTION

Chronic kidney disease (CKD) is a common medical condition in elderly cats, yet an etiology for this disease process has yet to be elucidated. In human beings there is progressive loss of renal parenchyma and function with age.[1] Whether CKD in cats results from a similar aging process that predisposes the organ to failure or, if other factors are involved remains to be determined. A review of current literature regarding the relationship between renal disease and aging, cellular senescence and telomere length will be discussed as well as current studies assessing telomere length and cellular senescence in cats with CKD.

KIDNEYS, AGING AND SENESCENCE

The phenotype of renal aging is described not only as the loss of function and renal mass, but as the loss of an appropriate response toward injury.[2] In human beings there is progressive loss of renal parenchyma and function with age, and the kidney is one of the organs that ages the fastest.[1] In the aging human population kidney injury is more common, clinically more severe, and kidney function is less likely to recover.[3] In addition, age is the most important predictor of renal transplant long-term success; older donor kidneys are more likely to develop allograft nephropathy.[4-6] Because of these factors, the study of renal aging has garnered much interest. The exact molecular mechanisms of renal aging have not been elucidated, but theories of aging include oxidative damage, genomic instability (including telomere loss), genetic programming (such as senescence) and cell death.[7] Senescence is a fundamental cellular program similar to apoptosis and results in an organ's inability to repair and replicate, particularly in response to stresses. Senescence is associated with telomere shortening, in which case it is usually referred to as replicative senescence, but this is not the only reason it happens. Other factors may include DNA damage, oxidative stress and response to other stresses. Senescence is mediated by p16 and p53 signaling pathways.[8]

Several studies have examined aging kidneys and their association with senescence. $P16^{INKA4a}$ is an irreversible cell cycle inhibitor, which is considered to be one of the best markers for senescence.[7] When human biopsy specimens over a large age range were examined for genes associated with senescence, P16 expression, particularly in the renal cortex, was found to be most consistently correlated with age as well as histologic changes and inversely correlated with cell replication.[7] Renal senescence does appear to involve more than telomere shortening as was demonstrated by one study where telomere length was not found to shorten in the kidney with age even though increased markers of senescence, $P16^{INKa4a}$ and B gal, were shown.[9] In another study, diseased kidneys (affected by glomerulopathies) were confirmed to have evidence of increased senescence; when human biopsy specimens were examined, diseased kidneys expressed more $P16^{INK4a}$ than expected for their age.[6]

Senescence in renal age and disease has a number of critical effects on the function of the kidney. This results in a diminished capacity of aged kidney to cope with or withstand normal and abnormal stresses. The capacity for renal epithelial cell proliferation declines with aging, both baseline cell proliferation as well as capacity for bursts of proliferation after injury.[10] Susceptibility to apoptosis increases with age as well in both baseline and crisis conditions. Additionally it has been demonstrated that function of stem cells and progenitor cells decrease with age.[10] The kidney's ability to provide vital growth factors is also diminished with age and disease. Senescent cells secrete altered levels of TGF-B, EGF, IGF-1 and VEGF. VEGF is critical for vascular health and repair and reduced expression in aged and diseased kidneys likely results in the progressive rarefaction of peritubular capillaries and hypoxia that has been observed.[10,11] Reaching a better understanding of the mechanisms of renal aging and disease is a critical first step on the pathway to developing mechanism specific therapies for amelioration of age and disease related effects.

TELOMERES AND AGING

Telomeres are protective structures on the end of chromosomes. They cap the chromosome and prevent the end from being seen as a DNA strand break. DNA strand breaks lead to cell cycle arrest and repair, or apoptosis if the damage is severe. Telomeres prevent such events and are critical for chromosome stability.[12] Unfortunately telomeres shorten with each replication cycle due to "end replication problem"; DNA polymerase cannot fully replicate to the very end of the DNA and with each cycle some is lost. Additional factors can also lead to telomere shortening (deletions, oxidative damage).[12,13] Telomerase is an enzyme that can compensate for this and create new telomeres de novo, but it is generally not active in somatic cells after embryogenesis. Telomerase can be reactivated in cancer.[12]

When telomeres reach a critically short state, they are no longer able to perform the capping function and are themselves seen as a DNA strand break. At this point, p53 and p21 may be activated, cell cycle arrest occurs, and ultimately cell death.[14]. Most tissues in human beings have been documented to experience telomere shortening as a matter of age; including peripheral blood mononuclear cells (PBMC), kidney, endothelial cells, hepatocytes, intestinal epithelium, lung, muscle etc.[12] Telomere length in relation to age in cats was investigated in one study and a significant decrease in telomere length in PBMC was found with age in cats.[15] Telomeres have also been shown to be shortened in several disease states, particularly those of inflammatory nature.[13]

Telomeres are acutely sensitive to oxidative stress, in part due to their high content of guanines but also because reactive oxygen species produce single strand breaks, and telomeric DNA is thought to be deficient in the repair of single strand breaks.[13] Additionally repair of oxidative lesions appears to be less efficient than in the rest of the genome.[13] Several studies provide evidence that reduction of oxidative stress via antioxidant treatment may reduce telomere attrition and delay cellular senescence.[16,17] There appears to be an association between inflammation, oxidative stress and telomere shortening and thus investigation of telomere length is pertinent in any inflammatory disease condition. Feline CKD is a logical candidate for such studies, as it is characterized by tubulointerstitial inflammation,[18] and oxidative stress has been documented in these patients.[19]

TELOMERES AND THE KIDNEY

Several studies have assessed the association between renal age and renal disease and telomere length in human beings and rats. One study explored the relationship between age and telomere length in human surgical biopsy samples.[20] Telomere loss in the cortex occurred with age at a rate of 0.25% per year. Telomere loss in the medulla was not significant. It was concluded that telomeres shorten in an age-dependent manner, particularly in the renal cortex.[20] In another study age-dependent telomere length attrition rate was measured in the outer renal cortex and inner renal medulla of Caucasians and African Americans.[21] Telomere length attrition rate was slower in the inner medulla than in the outer renal cortex in both groups. In the outer renal cortex, telomere length attrition rate was significantly slower in African Americans than Caucasians. The proximal tubule is the most abundant nephron structure in the renal cortex and it was concluded these findings might represent a difference in proliferative growth between racial groups.[21]

The importance of functional telomeres to kidney repair has also been explored. In an intriguing study, renal ischemia-reperfusion injury led to greater impairment of renal function and increased acute and chronic histopathologic damage in fourth generation telomerase-deficient mice compared with wild-type and first generation mice.[2] Critically short telomeres, increased expression of cell cycle inhibitor p21 and more apoptotic renal cells were associated with the increased damage in fourth generation mice. These mice also had reduced proliferative capacity, highlighting the detrimental effect telomere shortening has on replication and repair.[2]

It has also been shown that other factors, such as proteinuria and hypertension, which are common in CKD, have a detrimental effect on telomere length. In rats with genetic spontaneous hypertension, telomere length was found to be shorter in all ages in comparison to normal type rats.[22] It was concluded that hypertension resulted in increased cell turnover and therefore accelerated cell aging.[22] In a study assessing differences between longevity in male vs. female proteinuric rats, it was found that increased proteinuria in male rats was associated with shortened renal telomeres, increased expression of senescence markers p21 and p53, and reduced oxidative damage protective mechanisms.[23] More specifically telomere shortening occurred with age in rats, but was more significant in male rats and was also accelerated in the renal cortex of both sexes. It was concluded that decreased antioxidant capacity and increased proteinuria in male rats might induce accelerated telomere shortening leading to senescence, renal failure and decreased longevity.[23]

Although it has been demonstrated that many factors negatively affect telomere length, there is some indication that therapeutic intervention can provide a protective effect. In an *in vitro* study the angiotensin II receptor blocker losartan was demonstrated to have a protective effect on cultured glomerular mesangial cells exposed to angiotensin II.[14] Cells which were exposed to angiotensin II + losartan had longer telomere lengths, decrease senescence as measured by B-galactosidase staining and expression of p21 and p53 than cells exposed to angiotensin II alone.[14] Although yet to be demonstrated *in vivo,* this type of therapeutic manipulation of telomere length provides encouragement that other potential therapies are conceivable.

CELLULAR SENESCENCE AND TELOMERE SHORTENING IN CKD CATS

Telomere length and cellular senescence has not been previously assessed in the feline kidney or in feline CKD. A recent preliminary study explored the relationship between telomere length and cellular senescence in cats with CKD.[24] The results of this study indicated a significant association between telomere shortening, cellular senescence and feline CKD does exist, and even though a cause and effect relationship cannot be demonstrated, the results support contribution of telomere dysfunction to predisposition for development of CKD. Additional investigation is warranted and will be necessary to further explore such possibilities and relationships. Importantly, such research also suggests novel intervention strategies and therapeutic targets for treatment of CKD; e.g., amelioration of oxidative stress and/or selective telomerase activation may represent advantageous approaches to combating critical renal telomere shortening, thereby reducing cellular senescence and slowing kidney degeneration.

REFERENCES

[1]Perico N et al. **Curr Opin Nephrol Hypertens** 2011;20:31. [2]Westhoff JH et al. **J Am Soc Neph** 2010;21:327. [3]Schmitt R et al. **Am J Kidney Dis** 2008; 52:262. [4]de Fijter JW. **J Am Soc Nephrol** 2001;12:1538. [5]Ferlicot S et al. **Hum Pathol** 2003; 34:924. [6]Melk A et al. **Am J Transplant** 2005;5:1375. [7]Melk A et al. **Kidney Int** 2004;65:510. [8]Yang H et al. **J Am Soc Nephrol** 2010; 21:1436. [9]Melk A et al. **Kidney Int** 2003;63:2134. [10]Schmitt R. **Am J Physiol Renal Physiol** 2008;294:F1265. [11]Mayer G. **Nephrol Dial Transplant** 2011;26:1132. [12]Jiang H et al. **Z Gerontol Geriatr** 2007;40:214. [13]Houben JM et al. **Free Radic Biol Med** 2008;44:235. [14]Feng X et al. **Mol Med Report** 2011;4:55. [15]McKevitt TP et al. **Am J Vet Res** 2003;64:1496. [16]Furumoto K et al. **Life Sci** 1998; 63:935. [17]Kashino G et al. **Free Radic Biol Med** 2003;35:438. [18]DiBartola et al. **J Am Vet Med Assoc** 1987; 190:1196. [19]Keegan and Webb. **J Vet Intern Med** 2010;24:514. [20]Melk A et al. **J Am Soc Nephrol** 2000;11:444. [21]Tchakmakjian L et al. **Nephron Exp Nephrol** 2004;98:e82. [22]Hamet P et al. **Hypertension** 2001;37:760. [23]Tarry-Adkins JL et al. **Am J Physiol Renal Physiol** 2006;290:F509. [24]Quimby JM et al. **Am J Physiol Renal Physiol**.2013; In revision.

VAGINAL FLORA AND UTI IN DOGS

Shelly Vaden, DVM, PhD, DACVIM (SAIM)
Raleigh, NC

INTRODUCTION

Approximately 14% of female dogs will develop a urinary tract infection (UTI) within their lifetime. While the management of dogs with acute, uncomplicated UTI is fairly straightforward, the management of dogs with chronic UTI, due to either recurrence or persistence, can be anything but straightforward. Recurrent UTI, defined as >2 UTI in a 6 month period or >3 UTI in a year, only occurs in about 0.3% of canine patients but can be difficult and expensive to eradicate.[1,2]

Complicating the entire picture is the emergence of resistant pathogens. Of particular concern are the uropathogenic *Escherichia coli* (UPEC) and *Enterococcus sp.* UPEC is the most common etiologic agent for UTI in most mammalian species and presents a particular challenge because of intracellular invasion and biofilm formation in host uroepithelial cells by UPEC strains, allowing the UPEC to avoid normal host defense mechanisms.

Antibiotic therapy is standard treatment for UTI but overuse may lead to changes in the normal host microbiome and selection for resistant bacterial strains. Although severe adverse effects of antibiotic therapy for UTI may be infrequent, mild adverse effects are probably not and chronic antimicrobial administration is expensive for the client. There has been a shift of focus to the prevention of UTI in people.

Prevention of UTI can be achieved through strengthening of the normal host defense mechanisms: improving systemic and mucosal immunity and maintaining normal urine composition, facilitating normal micturition and ensuring normal urogenital anatomy. While prophylactic antibiotic therapy has long been advocated in the prevention of UTI, a long-term benefit in people or dogs has not been demonstrated. In people, it has been shown that antibiotic prophylaxis is associated with more adverse drug effects and increased antimicrobial resistance. Novel therapeutic approaches have been explored in recent years. One such approach is the use of probiotics to reestablish a healthy vaginal microbiome in women with UTI. The purpose of this presentation is to discuss what is known about the possible association between the vaginal microbiome and UTI, providing information about what is known in women and dogs.

LACTIC ACID BACTERIA AND PROBIOTICS

Probiotics are live microbial supplements that are given to beneficially affect the host animal, generally by improving the intestinal microbiome and modulating mucosal and systemic immunity.[3] Probiotics generally contain lactic acid bacteria (LAB), which are bacteria that produce lactic acid as an end product of fermentation. The most commonly referred to LAB is *Lactobacillus* but many others exist, including *Enterococcus sp.* Benefits attributed to LAB include stimulation of vitamin and enzyme production, stabilization of gastrointestinal microbiome via competitive exclusion of pathogens, production of hydrogen peroxide and bacteriocins that antagonize growth of pathogens, reduction in serum cholesterol, decrease in risk of colon cancer and suppression of tumor growth.[3] With respect to the genitourinary tract, one of the most important properties of an LAB is the ability to maintain an acidic environment, generally maintaining a vaginal pH of <4.5 in women. In the genitourinary tract, lactobacilli also produce biosurfactants that inhibit adhesion of microorganisms.

The ability of LAB to reduce the colonization of urogenital pathogens in the urogenital tract has been studied extensively in people. The ideal LAB used in this capacity would adhere to the cells, persist and reproduce in the environment, reduce pathogen adherence, exert antimicrobial effects, maintain the acidic environment and be safe with few side effects.

HUMAN VAGINAL MICROBIOME

Most healthy women have a lactobacillus predominated vaginal microbiome. In some women, lactobacillus is not the dominant organism but other LAB become established, helping to maintain the acidic environment that helps prevent bacterial overgrowth. A non-lactobacillus predominated vaginal microbiome is not necessarily pathogenic, but instead pathogenicity is influenced by several factors, including genetic variability and the innate immune system.[4]

Bacterial vaginosis is a condition where there is a shift from the lactobacillus predominated vaginal microbiome to one that is largely characterized by pathogenic bacteria. Women with recurrent UTI often have bacterial vaginosis with increased rates of colonization of *E. coli*. However, not all women with bacterial vaginosis appear to be at risk for recurrent UTI. Furthermore, women with a normal, lactobacillus dominated, vaginal microbiome, have a reduction of episodes of recurrent UTI.[5]

Some patients with bacterial vaginosis can be returned to a more normal vaginal microbiome by administration of oral lactobacillus containing probiotics, consumption of *L. acidophilus*-containing yogurt, or vaginally inserted lactobacillus containing suppositories. For an oral formulation to be effective, the lactobacillus must maintain their structural integrity during passage through the gastrointestinal tract, be delivered to the rectal areal and subsequently colonize the vaginal tract. The different lactobacilli have varying abilities to do these things.

Published studies have varying results when lactobacillus probiotics are given to women with recurrent UTI. Although results of many of the early studies lacked power, in general the results showed promise. A recently published study used intravaginal administration of a strain of *Lactobacillus crispatus* that was isolated from a healthy woman's vagina.[5] While the risk of recurrent UTI was low in *L. crispatus* and placebo treated groups (15% vs. 27% respectively), higher vaginal colonization with *L. crispatus* was associated with a significant reduction in recurrent UTI.

CANINE VAGINAL MICROBIOME

The normal canine vaginal microbiome consists of a mixed population of anaerobic and aerobic bacteria, with greater numbers of bacteria present in the vestibule than in the cranial vagina. Because the vaginal flora originates in the gastrointestinal tract and skin, common isolates include *E. coli*, staphylococci, enterococci, *Bacteroides sp.* and mycoplasma. Less is know about LAB in the canine vaginal microbiome because many published studies did not include culture enrichment for LAB species.

In one study the vaginal microbiome of 42 healthy and ill female dogs, presumed to be intact, from Uruguay, LAB were isolated from 41 of 42 dogs (98%).[6] Dogs ranged in age from 0.25-14 years of age and 67% were had one of a variety of diseases. The only dog that did not have vaginal LAB was uremic; none of the dogs were reported to have UTI. *Enterococcus sp.* and *Lactobacillus* were the LAB isolated. These isolates inhibited the adhesion of common uropathogens (*Proteus mirabilis, Staphylococcus aureus and E. coli)* to cultured vaginal epithelial cells. Of interest was that even though LAB were isolated from these dogs, the vaginal pH was 6-7.5, much higher than the vaginal pH of < 4.5 generally found in women. Conversely, in a study evaluating healthy estrous dogs, healthy spayed dogs, and spayed dogs with recurrent UTI, LAB were isolated from fewer than 22% of dogs and did not differ between groups.[7]

Before considering the use of probiotics in the treatment of canine recurrent UTI, it would seem reasonable to determine if bacterial vaginosis occurs in dogs and, if so, to determine if it is a risk factor for recurrent UTI in dogs. The vaginal flora of 21 spayed female dogs with a history of recurrent UTI and 23 spayed female dogs without recurrent UTI was evaluated.[8] The vaginal flora was not significantly different between the two groups. *E. coli* and *S. psuedintermedius* were the most common pathogens isolated. Interestingly, *Enterococcus canintestini* was the most common LAB, isolated from 5 normal and 2 recurrent UTI dogs.

CANINE PROBIOTIC STUDIES

It is also necessary to determine if the canine vaginal microbiome can be altered in dogs as it can be in humans. Normal dogs without a history of recurrent UTI were given an oral lactobacillus containing probiotic for either 14 (n=23 dogs) or 28 (n= 12 dogs) days.[8] Vaginal swabs taken before and after probiotic administration were submitted for LAB culture. Although LAB were isolated from 8 dogs, only 1 of these dogs did not have LAB before probiotic administration but had LAB isolated following probiotic administration. Furthermore, LAB were not found in the vaginal cultures of the 2 dogs that were fed diets that contained probiotics as ingredients. It is possible that lactobacillus were not found in vaginal cultures after probiotic administration because the administered lactobacillus failed to maintain structural integrity during passage through the gastrointestinal tract or this species of lactobacillus was not able to adhere to the vaginal epithelial cells. Because *Enterococcus canintestini* was the most common LAB isolated from vaginal swabs in one study, it might be interesting to repeat this study using a probiotic containing this isolate.

It was theorized that one could achieve a higher level of LAB colonization if LAB was administered intravaginally. When 6 normal dogs were given lactobacillus-containing suppository intravaginally for 6 days, lactobacillus was isolated from vaginal cultures from 5 dogs on day 7 but only 1 dog on day 14.[9] One dog dropped out of the study because the owners found administration to be objectionable.

Much work is still needed to determine if: (1) Bacterial vaginosis is associated with recurrent UTI in dogs; (2) If probiotics can be given to alter the vaginal microbiome to become LAB dominated in dogs; and (3) To determine if altering the microbiome is associated with a reduction in the frequency of recurrent UTI in dogs.

REFERENCES

1. Weese JS, et. al. Vet Med Int 2011;2011:263768. 2. Norris CR, et. al. J Am Anim Hosp Assoc 2000;36:484. 3. Naidu AS, et. al. Crit Rev Food Sci Nutr 1999;38:13. 4. Witkin SS, et. al. Best Pract Res Clin Obst Gyn 2007; 21:347. 5. Stapleton AE, et. al. Clin Infect Dis 2011;52:1212. 6. Delucchi L, et. al. Can Vet J 2008;49:99. 7. C.S. Bailey, et. al.. Clin Therio 2012;4:405 8. Hutchins, unpublished data. 9. Ottman unpublished data.

UPDATE ON URATE UROLITHIASIS

Jennifer A. Larsen, DVM, PhD, DACVN
Jodi L. Westropp, DVM, PhD, DACVIM
Davis, CA

INTRODUCTION

There are few data supporting any effective therapy to reduce recurrence of urate urolithiasis in dogs and cats. The genetics and pathophysiology of this disease are factors that may influence the clinical manifestation as well as the efficacy of treatment of urate urolithiasis in cats and dogs.

ETIOLOGY

Urate-containing uroliths comprise approximately 25% of submissions for dogs and 10% for cats through our laboratory. Urate urolithiasis is a common disease in Dalmatians, with a prevalence of clinical disease in male Dalmatians of up to 34%. This breed, as well as dogs of an increasing number of other breeds, is affected due to a defect in uric acid metabolism. Urate urolithiasis occurs when urine is supersaturated with uric acid and related compounds, which are less soluble than the normal end-product of purine metabolism, allantoin. Uric acid is converted to allantoin by the hepatic enzyme uricase. Dalmatians are at risk for the development of urate uroliths because of a genetic defect in the hepatic and renal uric acid transporter (SLC2A9 transporter), which excludes uric acid from accessing the hepatic intracellular conversion pathway and results in reduced reabsortion of uric acid by the renal proximal tubules.[1] This results in hyperuricemia, hyperuricosuria, and an increased risk of urate urolithiasis. All Dalmatians excrete relatively high amounts of uric acid (400-600 mg/d as compared with 10-60 mg/d in non- Dalmatian dogs). Uric acid excreted in the urine may crystallize with various cations, such as ammonium, to form urate salts that precipitate to form urate uroliths (such as ammonium urate). The defect is a simple autosomal recessive trait for which all Dalmatian dogs are homozygous. Dogs of other unrelated breeds (including English Bulldogs, Black Russian Terriers, and Australian Shepherds among many others) have also been recently reported to be homozygous for the same mutation in SLC2A9.[2] There is a DNA test available to confirm the defect and guide breeding decisions as well as diagnostic plans for dogs presenting with urate uroliths (http://www.vgl.ucdavis.edu/services/Hyperuricosuria.php).

Other dogs may develop urate-containing uroliths as a result of liver disease, namely portosystemic shunts (PSS).[3] These dogs have a reduced ability to convert uric acid to allantoin, plus a reduced ability to convert ammonia to urea. Dogs with urate stones with no known genetic predisposition for urate stones should be evaluated for underlying liver abnormalities. Surgical correction of the underlying problem and reducing hyperuricosuria can prevent urate stone formation in some dogs with portal vascular anomalies. In contrast to those with genetic urate urolithiasis, allopurinol is generally not recommended in dogs with portovascular anomalies, because of alterations in metabolism of this drug.

Portovascular anomalies appear to be an uncommon cause of urate stone formation in the cat, and the incidence appears to be static.[4,5] One study showed that clinicopathological abnormalities were rare and did not reveal any associated specific disease processes.[4] In most cases, the cause is unknown; however, genetics may be involved, since a few breeds such as the Siamese and Egyptian Mau are reportedly predisposed to urate uroliths.[6] Reported risk factors for cats include neutering and age between 4 and 7 years[5], while cats with PSS developed urate uroliths at 2 years of age compared to 7 years of age for cats with no evidence of PSS.[4]

Hyperuricosuria may also have an important role as a promoter of calcium oxalate urolithiasis in human beings, and the mechanism has been reasonably well established to be a 'salting-out' effect that enhances nucleation, growth, and aggregation of calcium oxalate crystals.[7] It is not known if this relationship may influence the development of uroliths of mixed composition. Evaluating the occurrence of this in dogs and cats is challenging due to the reporting of urolith stone types in retrospective reports as “mixed” only when more than 30% of the stone is another component, as well as lack of information for the composition of “mixed” uroliths.

In people, urate stones carry a lifetime risk of up to 25% in some parts of the world; studies in Taiwan and Japan have showed that both young and older females are overrepresented, with common comorbidities being chronic kidney disease, urinary tract infection, irritable bowel syndrome, and gout.[8,9] In addition to the influence of geography, genetic risk factors, dietary habits, and systemic metabolic disease may all play a role in the apparent increasing incidence of urate urolithiasis in the human population. The link between overweight/obesity and urolithiasis in people is well established, with a specific increased prevalence of both urate and calcium oxalate uroliths.[10] When calcium oxalate and urate urolith formers are compared, those with urate uroliths have higher body weight and body mass index and lower urine pH.[11] Further, among human stone formers, obesity is inversely related to urine pH.[12] The link between body condition and altered urinary environment appears clear; however, how this impacts specific risk factors for dogs and cats remains unknown. Other systemic diseases have also been the focus of recent research in humans, with studies supporting a link between insulin resistance and urate urolithiasis; the mechanism has yet to be elucidated but may involve renal tubule dysfunction.[13] Research is needed to establish similar links between systemic disease and the impact on the local urinary environment to determine if these might be risk factors for predisposed dogs and cats.

MANAGEMENT

Prevention strategies for the management of urate stones of genetic etiology have been suggested and include low protein diets, bicarbonate supplementation, xanthine oxidase inhibitors, and increased water intake. Protein restricted diets are often prescribed for dogs with urate stones to reduce the intake of purines and decrease uric acid excretion as well as encourage renal medullary washout and urine dilution. Reduction of the amount of uric acid excreted in the urine can also be achieved by using protein sources that are low in purines (certain vegetable proteins, eggs, or dairy products), which allows for higher protein levels in the diet. Purines are found in very high concentrations in organ meats and seafood (such as anchovies, mackerel, mussels, oysters, salmon, shrimp, and tuna), moderately high concentrations in meats (skeletal muscle), legumes, lentils, and some vegetables (such as spinach and asparagus), and in negligible concentrations in breads, oils/fats, eggs, dairy, fruits, and most vegetables. Therefore, foods formulated with dairy and vegetarian protein sources are preferred to those formulated with meat and fish protein sources for the management of canine patients with urate stones. The xanthine oxidase inhibitor, allopurinol, can also be prescribed to reduce uric acid production. If used, a purine restricted diet is essential to reduce the risk of forming xanthine stones. Metabolism if allopurinol is variable; ideally the dosage should be titrated based on 24 hour uric acid excretion results.

Although nutritional therapy is the cornerstone of management of patients with urate urolithiasis, there is little data evaluating specific dietary strategies or to assess the efficacy of individual diets in prevention or reducing the recurrence rate in dogs and none in cats. Additionally, concerns regarding potential adverse effects of protein restriction in some cases have been raised. Dilated cardiomyopathy was documented in 9 Dalmatians consuming a low protein diet for the management of urate stones.[14] Although not clearly identified at the time, subsequent research suggests that a likely cause was taurine deficiency secondary to the consumption of a diet deplete in sulfur amino acid precursors for taurine production.[15] As a result, veterinary therapeutic low protein diets that are marketed for the management of urate urolithiasis have been supplemented with sulfur amino acids for many years. In addition, a higher protein option is now available, and this diet appears to be effective in maintaining low urinary uric acid concentrations, no lower urinary tract signs, and normal clinicopathologic parameters in a small number of male dogs with previous urate uroliths; data analysis is ongoing to objectively evaluate the adequacy of protein intake among other parameters.[16]

To prevent recurrence of urate stones in otherwise healthy cats, a diet that is lower in protein compared to the previous diet is often recommended. Often veterinary therapeutic diets formulated for liver disease or kidney disease are prescribed in addition to hydrolyzed soy protein diets; evidence of the effectiveness of this approach has not been published for either cats or dogs. Consideration of a thorough diet history is advisable to avoid unnecessary protein restriction.

Other dietary management strategies to minimize urate urolithiasis recurrence are increased moisture intake and urine alkalinization. Increased moisture in the diet will increase frequency of urination and decrease the concentration of stone precursors in the bladder, which will help minimize the risk of stone formation. Increasing the level of dietary sodium is a strategy used in some veterinary therapeutic diets to promote diuresis and increased water intake; canned diets and water added to dry diets is also effective. The formation of dilute urine (specific gravity <1.020 for dogs and <1.025 for cats) is considered to be critical in order to decrease the likelihood of supersaturation of the urine with urolith precursors. Additionally, alkaline urine contains lower concentrations of ammonia and ammonium ions, therefore decreasing the risk of ammonium urate urolith formation. Urine pH should be approximately 7.5 to increase the solubility of ammonium urate precursors. While dietary protein restriction (as well as the feeding of vegetable source protein) tends to promote alkaline urine, the use of alkalinizers such as sodium bicarbonate and potassium citrate are sometimes recommended if protein restriction alone does not achieve the desired urine pH.

RECURRENCE

When urate uroliths recur, ensure compliance. Review the diet history, and monitor urine specific gravity and pH. In dogs without underlying hepatic disease that have recurrent urolithiasis despite adequate urine dilution, urine pH of 7.5 or higher, and low purine intake, allopurinol should be considered. The recommended starting dose is 5-10 mg/kg PO BID. Ideally, the dose should be titrated based on determinations of 24-hour uric acid excretion, since inappropriately high doses of allopurinol can lead to production of xanthine stone formation. Clinical experience suggests if the uric acid excretion is <200 mg/day, the dose of allopurinol should be reduced.

Regular monitoring of specific gravity and presence of crystals is indicated in both dogs and cats. Periodic ultrasounds are also advisable, since removal of small cystic calculi may be accomplished with voiding urohydropropulsion. Always submit any new stones for quantitative crystallographic analysis. This is particularly important for dogs receiving allopurinol as xanthine stones can occur. Urate stones may be amenable to dissolution with nutritional and medical management. Although not reliably effective, dissolution protocols have been described for dogs, but have not been reported in cats.

REFERENCES

1.Bannasch D, et al. PLoS Genet 2008;4:e1000246. 2. Karmi N, et al. J Vet Intern Med 2010;24(6):1337-42. 3. Marretta SM, et al. J Am Vet Med Assoc 1981;178(2):133-7. 4. Dear JD, et al. J Fel Med Surg 2011;13:725-732. 5. Albasan H, et al. J Am Vet Med Assoc 2012;240:842-847. 6. Appel SL, et al. Can Vet J 2010;51:493-6. 7. Grover PK, etal. Chem Biol 2003;10:271-278. 8. Chou YH, et al. Kaohsiung J Med Sci 2012;28(5):259-64. 9. Kuruma H, et al. Int J Urol 2006;13(5):498-501. 10. Chou YH, et al. Urol Res 2011;39(4):283-7. 11. Negri AL, et al. Urol Res 2007;35(5):247-51. 12. Li WM, et al. Urol Res 2009;37(4):193-6. 13. Maalouf NM et al. Curr Opin Neph Hyperten 2004;13(2)181-189. 14. Freeman LM, et al. J Am Vet Med Assoc 1996;209:1592-1596. 15. Sanderson SL, et al. Am J Vet Res 2001;62:1616-1623. 16. Westropp JL, et al. ECVIM-CA Congress 2012 (abs).

CT AS A DIAGNOSTIC MODALITY FOR PULMONARY DISEASE

Kara M Lascola, DVM, MS, DACVIM
Urbana, IL

INTRODUCTION

Lung imaging plays a critical role in the evaluation of patients with respiratory disease. In human patients, computed tomography (CT) has become the preferred imaging modality for a variety of pulmonary diseases such as emphysema, acute respiratory distress syndrome, airway disease and neoplastic conditions.[1-5] Current CT imaging technology allows for whole lung imaging within seconds[2] and provides very high resolution allowing for more accurate anatomic and morphologic characterization of all components of the lung including airways, vessels and parenchyma. In addition, readily available image-analysis software has expanded the use of CT to perform analysis of lung mechanics[5,6,7] and lung perfusion.[8] Of particular importance has been the use of quantitative image analysis of the lung to improve timeliness of disease recognition, to better characterize specific parenchymal diseases, and to facilitate monitoring of disease progression and response to disease intervention.[5]

In veterinary medicine, the application of CT for the diagnosis of thoracic disease has been well described in dogs and cats.[9-11] In equine neonates pulmonary disease is a significant contributor to patient morbidity and mortality.[12-15] Although primary lung disease such as pneumonia is common in hospitalized neonates, a large percentage of neonates acquire pulmonary disease secondary to systemic conditions such as sepsis or systemic inflammatory response syndrome (SIRS), both common problems in hospitalized neonates.[12,15] In addition, progression of any form of pulmonary disease to equine neonatal acute respiratory distress syndrome or acute lung injury (EqNARDS/EqNALI) is a significant risk and carries a poorer prognosis.[15] Characterization of respiratory disease in equine neonates has largely relied on radiographic analysis[12] and published reports describing thoracic CT have been limited to a single case report.[16] Although extrapolation from other species has contributed towards our understanding of respiratory disease, maturational changes in this high risk population[15,17] highlight the need for species- and age-specific characterization of normal lung CT morphology in addition to disease characterization.

PRINCIPLES OF QUANTITATIVE CT ANALYSIS

Quantitative CT analysis of the lung is based on the objective measurement of lung tissue densities and the calculation of lung gas and tissue volumes[2,3,5] and was developed from empirical observation of regional increases or decreases in lung density that occur in pulmonary disease. The physical principles of CT enable measurement of density and volume for a particular tissue that is imaged. The CT tissue density is linearly correlated to the attenuation of the CT x-ray beam as it passes through that tissue and is expressed in Hounsfield units (HU). In the lung CT density is equivalent to aeration[2,3,6,18] with poorly or nonaerated regions (i.e., atelectatic lung) having greater tissue densities as compared to normal or over-aerated regions (i.e., emphysematous lung).[2-4,18] Based on the HU assigned to a given region within the lung, the percentage of air or tissue can thus be estimated. Volume measurements include the measurement of both gas and tissue volumes for the whole lung or particular regions of interest and are calculated based on voxel volume and the density of that particular voxel.[2,3] In humans, quantitative CT analysis has been used to improve the characterization, diagnosis, and treatment of several pulmonary diseases and of particular importance has been its contribution towards our understanding of the pathophysiology of acute respiratory distress syndrome (ARDS) and acute lung injury (ALI).[1-6,18-20]

Sophisticated software programs have recently introduced sensitive methods of tissue characterization to describe parenchymal pathology associated with specific pulmonary diseases.[5] However, some of the most common applications of quantitative pulmonary CT have included the use of histogram-based analysis of lung density and volumes, the evaluation of axial and vertical density gradients across the lung and the characterization of regional changes in density distribution in individuals with pulmonary disease.[2,5,19] Histogram-based analysis quantitates the distribution of lung density values within CT images by plotting the voxel frequency against the CT number[5,19] while identification of regional patterns of density changes enables pulmonary disease to be more accurately characterized as heterogeneous versus homogeneous, patchy versus diffuse, or central versus peripheral. Although radiographic images are similarly described, the marked increase in resolution of CT images provides much more sensitive differentiation of pathologic processes.[3]

Quantitative CT in Foals

In veterinary species quantitative CT analysis has been used extensively in experimental models of pulmonary disease,[8,19,21-24] in the evaluation of anesthesia-associated atelectasis,[25-27] and as a means to evaluate ventilatory strategies to improve CT image acquisition in anesthetized cats.[28] A recent pilot study performed by our laboratory utilized both quantitative and qualitative CT image analysis to describe normal lung CT morphology in healthy, sedated and spontaneously breathing neonatal foals (< 14 days of age).[29] Sedated foals were treated with butorphanol (0.05 mg/kg) and midazolam (0.1 mg/kg), and if further sedation was required an additional dose of butorphanol and midazolam or propofol (4 mg/kg administered to effect) was administered. In all foals a patchy alveolar radiographic pattern consistent with atelectasis was identified in the ventral lung fields. As predicted this corresponded to increases in lung density and a reduction in the volume of normally aerated lung in those regions. Although these were consistent findings in all foals, in foals younger than 7 days of age lung density was greater ($p=0.007$) and the volume of normally aerated lung was smaller ($p = 0.001$) compared to foals older than 7 days. Furthermore, the alveolar pattern was graded as more severe in the younger foals. Given these findings, use of CT to evaluate changes in lung density and volumes in individual foals over the first two weeks of life could provide more specific information regarding potential age-related changes.

DIAGNOSTIC CONSIDERATIONS OF CT

We have demonstrated that CT imaging is well tolerated and can be performed rapidly in healthy and hospitalized, spontaneously breathing foals using sedation or minimal intravenous anesthesia.[29] The major limitations to diagnostic utility of this protocol include atelectasis and the inability to control in which phase of respiration the image is acquired. Techniques used to address these issues include controlling ventilation (mechanically or manually) and performing ventilatory recruitment maneuvers to open atelectatic regions of lung.[2,3,7,20,21,28,30,31-33]

Atelectasis

Atelectasis of the lung can be readily detected by CT,[21,22,25-28] and along with patient motion interferes with accurate evaluation of CT images and characterization of disease distribution and severity. Sedation and general anesthesia are primary factors contributing to atelectasis in patients undergoing CT imaging. Anesthesia-associated atelectasis occurs with intravenous and inhalant anesthesia and with spontaneously breathing and ventilated patients.[30] Mechanisms contributing to atelectasis include compression of the dependent lung regions, increases in absorption of alveolar gas, and reductions in pulmonary surfactant.[25-28, 30] The onset of atelectasis is rapid and the severity increases with increased duration of anesthesia. In comparison to adults, anesthetized human pediatric and juvenile patients develop atelectasis more rapidly in the dependent lung regions. This has been attributed to differences in functional residual capacity (FRC) of the lung and in compliance of the lung and chest wall.[30]

Neonatal foals appear also to be susceptible to atelectasis in the dependent lung regions.[29] Mechanisms contributing to atelectasis in neonatal foals sedated for CT likely include compression of the dependent (ventral) lung regions by the overlying lung, age-related changes in chest wall compliance (C_w), and the potential impact of medications used for intravenous sedation on ventilation. Drugs routinely used in hospitalized neonatal foals for sedation include butorphanol, midazolam, and propofol. Propofol, butorphanol, and to a lesser extent, midazolam have dose- and rate-dependent respiratory depressant properties; propofol in particular can even induce apnea.[34, 35] This can reduce a patient's normal ventilatory response to pCO_2.[34,35] Propofol and midazolam also have skeletal muscle relaxant properties[34,35] and it is plausible that this may contribute to atelectasis through an increase in C_w. In neonates of many species, C_w is increased when compared to that in more mature individuals, and can be associated with reductions in functional residual capacity (FRC).[36,37] The increase in C_w in neonatal foals relative to more mature horses is not as extreme when compared to other species,[36] but a reduction in compliance is noted over the first 2 weeks of life.[37] Furthermore, there is some evidence that specific lung compliance is reduced in very young foals as compared to foals older than 7 days of age.[37] A further increase in C_w due to sedation could exacerbate the reduction in FRC and thus promote atelectasis.[30,37] It is possible that a reduction in dosage of sedatives may be possible when imaging critically ill patients, but the significance of different anesthetic or sedation protocols on CT image quality in neonatal foals remains to be evaluated.

Because of patient compliance anesthesia and sedation can often be avoided in humans, thus minimizing anesthesia-associated atelectasis or other potential anesthesia-associated morbidities. Furthermore, human patients respond to verbal instructions for breathing thus minimizing motion artifacts secondary to respiration and facilitating acquisition of images in specific phases of respiration. The ability to perform CT in conscious veterinary patients is limited. A recently developed specially designed restraint device has greatly facilitated CT imaging in awake or minimally sedated cats and small dogs.[9,38] Larger animals, particularly neonatal foals, cannot be imaged when awake. In the population of sick equine neonates, such as those with respiratory disease, the risks associated with general anesthesia are often perceived as greater. Thus, identification of CT imaging protocols that minimize anesthesia, as well as atelectasis and motion-associated artifacts, is essential.

Recruitment Maneuvers

Recruitment maneuvers involve the delivery of varying increases in pressure or volume over time. Common examples include intermittent delivery of a single breath hold with peak inspiratory pressures (PIP) ranging from 10-40 cmH_2O, the application of varying levels of peak end expiratory pressures (PEEP), sustained delivery of high pressure inflations at varying PEEP and PIP, and sequential stepwise increases in PIP over a series of breaths.[2,3,7,20,28,31,33] Quantitative analysis of CT images is used in clinical and experimental studies to evaluate recruitment based on measurable changes in lung densities and volumes.[2,3,6-8,20-22,27,28]

Safe performance of recruitment maneuvers has been described in critically ill pediatric patients, including human neonates.[31,32] Neonates are more susceptible to ventilatory-associated volutrauma and barotrauma when compared to adults[39] so care must be taken when introducing increases in pressure or volume, particularly during recruitment. Of the above described recruitment maneuvers the single breath hold maneuver is advantageous for many veterinary patients including neonatal foals because it does not require mechanical ventilation, can be rapidly performed in anesthetized or heavily sedated animals, and allows control over pressure and/or volume delivered to the animal. Single breath holds are routinely used in small animal patients undergoing CT and significant reductions in atelectasis and improvements in lung density and aerated volumes have been found at PIP of 20 cmH_2O.[28]

Recruitment of atelectatic lung for CT imaging has not been evaluated in neonatal foals and the minimum PIP required is unknown. Pressures of 20 cmH2O are routinely delivered to anesthetized patients undergoing surgical procedures and have been safely delivered to foals in an experimental study evaluating pulmonary function.[40] In human pediatric ARDS patients pressures >25cmH2O were delivered safely during lung recruitment.[32]

REFERENCES

1. Grenier P, et al. **Eur Radiol**, 2002; 12:1022; 2. Rouby JJ, et al. **Crit care Med**, 2003;31(4 Suppl):S285; 3. Gattinoni L et al. **Clin Chest Med**, 2006;27(4):559; 4. Reske A, et al. **Intensive care Med**, 2008; 34:2044; 5. Hoffman E, et al. **Acad radiol**, 2003; 10:1104; 6. Simon A, et al. **Proc Amer Thorac Soc**, 2005; 2:517; 7. Caironi P, et al. **Amer J Resp Crit care Med**, 2010; 181:578; 8. Fernandez-Bustamante A, et al. **Crit Care Med**, 2009; 37:2402; 9. Oliveira CR, et al. **Vet Radiol Ultrasound**, 2011; 52:268; 10. Cipone M, et al. **Vet Res Commun** 2008; 27:381; 11. Prather AB, et al. **Vet Radiol Ultrasound**, 2005; 46: 114; 12. Bedenice D, et al. **J Vet Intern Med**, 2003;17(6):868; 13. Beech J. **Vet Clin North Am Equine Pract**, 1985; Apr 1:131; 14. Freeman L, et al. **Vet Med**, 1992; 87:921; 15. Wilkins PA, et al. J Vet Emerg Crit Care, 2007;17(4):333; 16. Wion L, et al. Equine Vet J, 2001; 33:523; 17. Gillespie JR. **J Reprod Fertil Suppl**, 1975; 23:667; 18. Caironi P, et al. **Semin Respir Crit Care Med**, 2006; 27(4):404; 19. McEvoy F, et al. **Veterinary Radiology & Ultrasound**, 2009; 50:47; 20. Puybasset L, et al. **Intensive Care Med**, 2000;26:857 ; 21. Ryu HG, et al. **Anesth Intensive care**, 2008; 36:792; 22. Lee HJ, et al. **Radiology**, 2005; 234:151; 23. Grasso S, et al. **Am J Respir Crit Care Med**, 2009; 180:415; 24. Morandi F, et al. **Am J Vet Res**, 2004; 65:1114; 25. Staffieri F, et al. **Veterinary Anaesthesia and Analgesia**, 2010; 37:483; 26. Staffieri F, et al. **Am J Vet Res**, 2007; 68:925; 27. Staffieri F, et al. **Am J Vet Res**, 2010; 71:867; 28. Henao-Guerrero N, et al. **Am J Vet Res**, 2012; 73:646; 29. Lascola KM, et al. **Proc of the 29th Symposium of the Veterinary Comparative Respiratory Society**, 2011; 30. Magnusson L, et al. Ne. **Br J Anaesth**, 2003; 91:61; 31. Duff JP, et al. **Intensive care med**, 2007; 33: 1778; 32. Boriosi JP, et al. **Pediatr Crit care Med**, 2011; 12: 431; 33. Boriosi JP, et al. **Pediatr Pulmonol**, 2012; 47: 771; 34. Muir WW. In: Muir WW, Hubbell JAE, eds. **Equine anesthesia: monitoring and emergency therapy**. 2nd ed. St. Louis: Saunders, 2009; 185; 35. Muir WW. In: Muir WW, Hubbell JAE, eds. **Equine anesthesia: monitoring and emergency therapy**. 2nd ed. St. Lois: Saunders, 2009; 243; 36. Mortola, JP. **Am Rev Respir Dis** 1983; 128:569; 37. Koterba AM, et al. **Respir Physiol**, 1994; 95:21; 38. Oliveira CR, et al. **Vet Radiol Ultrasound**, 2011; 52:41; 39. Miller JD et al. **Clin Perinatol**, 2008; 35: 273; 40. Hoffman AM, et al. **Am J Vet Res**, 1997; 58: 1463

ENDOTOXIN AND THE PULMONARY SYSTEM: WHERE ARE WE NOW?

Kara M Lascola, DVM, MS, DACVIM
Urbana, IL

INTRODUCTION

Endotoxemia caused by increased systemic release of Gram-negative bacterial lipopolysaccharide (LPS, endotoxin) is a contributor to the development of Acute Lung Injury (ALI) and Acute Respiratory Distress Syndrome (ARDS) in a variety of species. The administration of exogenous LPS is a common model used in the investigation of the pathogenesis of ALI and ARDS. Despite the clinical significance of endotoxemia and the exquisite sensitivity of horses to endotoxin, its role in the development of lung injury has been incompletely defined for this species. The following discussion will review contribution of LPS to development of lung injury and what is known specifically about endotoxin-mediated lung injury in the horse.

ACUTE LUNG INJURY/ACUTE RESPIRATORY DISTRESS SYNDROME

Acute lung injury (ALI) and the more severe acute respiratory distress syndrome (ARDS) are clinical syndromes involving the acute onset of pulmonary disease in response to a variety of exogenous or endogenous inflammatory stimuli. These syndromes are defined by the presence of bilateral pulmonary infiltrates, noncardiogenic pulmonary edema, and severe hypoxemia (PaO2/FiO2 ratio of less than 200 mmHg, ARDS; PaO2/FiO2 ratio of less than 300 mmHg, ALI).[1] The pathogenesis of ALI/ARDS has been extensively reviewed and represents an imbalance between pro- and anti-inflammatory processes within the lung.[2-5] Direct (intrapulmonary) or indirect (extrapulmonary) insults trigger upregulation of local humoral and cellular mediated inflammation. Early lung injury occurs within minutes to hours of the insult and is characterized by increased alveolar-capillary permeability edema (a hallmark of ALI/ARDS), infiltration and activation of inflammatory cells, atelectasis and pulmonary hypertension.[2-5] Clinically this is represented by dyspnea, profound hypoxemia secondary to V/Q mismatch and reduced lung compliance.[2-5]

Pathogenesis of ALI/ARDS

Pulmonary edema develops in response to damage to both the pulmonary microvascular endothelium and the alveolar epithelium. Endothelial activation and injury after insult involves altered vasomotor tone as well as destruction and obstruction of the vascular bed. Loss of alveolar epithelial integrity not only disrupts the fluid transport mechanisms of the alveoli, but may also impair normal surfactant production. The combination of endothelial dysfunction, internal and external vascular occlusion, and increased vascular tone contribute to the pulmonary hypertension that is characteristic of ALI/ARDS.[6] Neutrophil migration, activation and sequestration within the alveoli and microvasculature, along with upregulation of vasoactive (e.g., ET-1), pro-inflammatory (e.g. TNF-alpha, IL-1, IL-8) and pro-coagulant (e.g., tissue factor) mediators are noted early and are central to the pathogenesis of ALI/ARDS.[2-5]

ALI/ARDS in Horses

ALI and ARDS have been described in a several veterinary species (VetALI/VetARDS), including adult and juvenile horses, and the diagnostic criteria are similar to those in humans with an added recommended requirement for evidence of diffuse pulmonary inflammation.[7] A specific subcategory also exists to define these conditions in neonatal foals, EqNARDS and EqNALI, to account for the relative hypoxemia of normal neonatal foals compared to adults during the first week of life.[7] Identified risk factors in veterinary species are similar to those in humans and include sepsis, SIRS, local or systemic infection or inflammation, severe trauma, smoke inhalation, near-drowning, multiple transfusions, and exposure to certain drugs or toxins. Published documentation of ALI/ARDS among adult and juvenile horses is limited. Among foals, published case series of ALI/ARDS usually attribute the conditions to an initial bacterial or viral insult.[8,9] Documentation of ALI/ARDS in adult horses is even more limited and includes isolated reports of near-drowning, smoke inhalation injury and experimental endotoxin exposure.[10-12]

ENDOTOXEMIA AND LPS-MEDIATED LUNG INJURY

Endotoxemia

In humans, sepsis is one of the most common predisposing factors for ALI/ARDS with an estimated 79% of ALI cases attributed to underlying sepsis.[2,4] Exposure to Gram-negative endotoxin is considered a contributing factor to the development of ALI/ARDS in humans and thus serves as a major experimental model in studies examining the pathogenesis of ALI/ARDS. Endotoxemia is a significant clinical concern in horses and is most commonly associated with gastrointestinal disease in adult horses and Gram-negative sepsis in neonatal foals.[13] In addition to causing increases in morbidity and mortality it also causes substantial economic losses associated with treatment and management.[13-19] Clinical studies have identified endotoxin in peritoneal fluid and blood in both adult horses and foals,[15,20-23] and estimate that 25-41% of adult horses presenting with colic and 50% of foals presenting for sepsis have detectable levels of LPS in their blood.[15,20,23,24] Furthermore, in horses both clinically and in experimental *in vivo* studies, increased serum TNF-α concentrations correlate with the development of characteristic clinical signs of endotoxemia and have been linked with decreased prognosis for survival.[17,23] Experimental administration of endotoxin has been associated with pulmonary inflammation and cardiopulmonary distress in horses, but clinical reports of endotoxin-mediated acute lung injury do not exist.

Endotoxemia causes a severe systemic inflammatory response through the release of pro-inflammatory cytokines and acute phase proteins, and may ultimately result in laminitis, cardiopulmonary dysfunction, coagulopathy, organ failure and ultimately death. Endotoxin forms the outer leaflet of the outer cell membrane in Gram-negative bacteria such as *Escherichia coli* and *Salmonella spp.*[25] The structural domains of the LPS molecule include the Lipid A moiety, the core oligosaccharide and the O-antigen polysaccharide.[13,25] Lipid A, the toxic component of LPS, is a potent immunostimulatory molecule.[13,25] Once released into the body LPS is recognized by the innate immune system where it has the potential to trigger a profound inflammatory response, primarily through the activation of monocytes and macrophages and the production of inflammatory mediators. The most common, and most potent, LPS-mediated pathway involves binding of lipid A to LPS-binding protein (LBP) in the circulation and transport of the LPS-LBP complex to the cell surface where it is first bound to CD14 and then to the TLR4/MD2 transmembrane complex. Binding of LPS to the TLR4/MD2 complex activates intracellular signaling cascades.[13,26] Once bound to LPS, TLR4/MD2 interacts with one of two intracellular adaptor molecules: MyD88 or TRIF.[13,27] Endotoxemia in horses appears to be mediated through the MyD88 pathway as opposed to the TRIF pathway.[27] The MyD88-dependent intracellular signaling pathway results in rapid activation of NFκβ or MAPK and the expression of cytokines such as TNF-α, IL-1β, IL-6, and IL-10. TNF-α stimulates the release of other pro-inflammatory mediators including interleukins, tissue factor, eicosanoids, and acute phase proteins.[13,23,28]

LPS-mediated lung injury

LPS-induced lung injury includes direct effects of endotoxin on the pulmonary microvascular endothelium as well as the systemic release of inflammatory mediators.[29-32] Pathologic changes in the lungs occur within a few hours to a few days depending on the dose of LPS administered, the route of delivery, and species susceptibility to LPS.[29,31] The capillary endothelium is the initial site of LPS-mediated injury.[29] TLR4/CD14 mediated uptake of LPS by the endothelial cells may directly initiate endothelial cell apoptosis and activation as well as activation of the complement cascade via C5a and C3a and subsequent neutrophil-mediated endothelial damage.[29-32] Alveolar epithelial damage is not considered a significant component of experimental LPS-induced lung injury.[29]

LPS-mediated lung injury in the horse

As previously mentioned, horses are susceptible to endotoxin-induced cardiopulmonary disturbances and, compared to other animal species, respond to a much lower lethal dose of endotoxin.[32,33] Experimental studies evaluating endotoxin-induced lung injury in horses are not limited to those involving the systemic administration of LPS. Several studies have examined the role of inhaled LPS with respect to its contribution towards airway inflammation in chronic respiratory conditions such recurrent airway obstruction.[34] With respect to systemic endotoxin-mediated lung injury, several studies in both horses and ponies have used varying dosages of *E. coli* derived LPS. These studies have primarily focused on the pathophysiology of the early stages of lung injury and inflammation[35,36] and on changes in cardiopulmonary vascular dynamics.[37-39]

Earlier studies in ponies administered LPS with dosages ranging from 20 - 80 μg/kg as an intraperitoneal injection found evidence of pulmonary artery endothelial damage, including endothelial denudement, as well as leukocyte migration.[35] Evidence for the development of pulmonary edema in horses with experimentally-induced endotoxemia has not been definitively identified. Anesthetized ponies treated with 20 μg/kg LPS IV did not demonstrate impairment of gas exchange, evidence of pulmonary edema or changes in alveolar-capillary permeability as measured by increases in extravascular lung water and increased albumin content in bronchoalveolar lavage fluid (BALF).[37] LPS administration has been reported to cause increases in pulmonary vascular resistance (PVR) and mean pulmonary artery pressure (PAP).[38,39] Increases in PAP occur early and are typically transient. It is unclear whether the increases in PAP and PVR are sufficient to cause pulmonary edema. In healthy exercising horses similar increases in PAP did not correspond to increased interstitial lung water accumulation.[40] Corresponding increases in systemic and pulmonary arterial ET-1 have not been identified[39] in horses with concurrent LPS administration, though ET-1 has been shown to be a potent pulmonary vascular vasoconstrictor in horses and is considered a mediator in ALI in humans and other species.[41-43] Increased ET-1 in BALF of horses with chronic airway inflammation has been observed[44] but changes in ET-1 have not been investigated in horses in conjunction with acute pulmonary inflammation associated with endotoxin exposure.

More recent studies have investigated the role of pulmonary intravascular macrophages (PIMs) and Toll-like Receptor (TLR) expression in the lung in relation to the pathophysiology of LPS-mediated lung injury.[32,36,45,46] Experimental evidence suggests that PIM containing species (horses, cattle, cats, sheep, pigs) are more prone to lung inflammation because

PIMs may localize LPS and other inflammatory molecules to the lung.[29,31,32,47] Depletion or inactivation of PIMs is associated with decreases in endotoxin-induced lung injury in sheep[48] as well as with reductions in endotoxin-mediated pulmonary inflammation and increases in PAP in horses[36]. TLR4 expression in equine lungs is identified on PIMs as well as endothelial cells and alveolar macrophages,[45] and experimental depletion of PIMs corresponds with marked reductions in TLR4 expression.[45] LPS administration initiates an increase in TLR4 expression and co-localization of LPS and TLR4 to PIMs has been identified.[45] Interestingly, administration of LPS to PIM-depleted horses also results in a significant increase in TLR4 expression, suggesting upregulation by other cells in the lung. Similar to macrophages elsewhere, PIMs play a central role in the production of inflammatory and vasoactive mediators associated with endotoxemia. This suggests a potential role for PIMs in systemic endotoxin-mediated inflammation in addition to any localized contribution to lung injury.[32]

TLR2 and TLR9 expression has been investigated in the lungs of healthy horses and horses administered LPS.[45,46] TLR2 responds to Gram-positive bacterial lipoteichoic acid and TLR9 is a receptor for unmethylated bacterial DNA. TLR2 was not identified in the lungs of healthy horses but was identified in horses administered LPS with and without depletion of PIMs.[45] TLR9 expression is widespread within the lungs of healthy horses and is found in PIMs, alveolar macrophages, and type-II epithelial cells.[46] Administration of LPS corresponded to increased TLR9 expression in several different cell types, including endothelial and inflammatory cells.[46] The significance of the expression of these receptors with respect to the pathogenesis of endotoxin-mediated pulmonary or systemic inflammation remains to be investigated.

REFERENCES

1. Bernard GR, et al. **Am J Respir Crit care Med**, 1994; 149:818; 2. Ware LB **Sem Respiratory Crit care Med**, 2006; 27: 337; 3. Bhatia M, et al. **J Path** 2004; 202:145; 4. Wheeler AP, et al. **Lancet** 2007; 369:1553; 5.Cehovic GA, et al. **Int Anesthesiology Clinics**, 2009; 47:83; 6. Price LC, et al **Am J Physiol Lung cell mol physiol** 2012; 302: L803; 7. Wilkins PA, et al. **J Vet Emerg Crit Care**, 2007; 17:333; 8. Dunkel B. **Clin tech Equine Pract** 2006; 5:127; 9. Dunkel B, et al. **Equine Vet. J** 2005; 37:435; 10. Sembrat R, et al. **Am J Vet Res** 1978; 39: 1147; 11. Humber KA. **J Am Vet Med** 1988; 192:377; 12. Kemper T, et al. **J Am Vet Med** 1993; 202:91; 13. Werners AH, et al. **Equine Vet J** 2005 37: 371; 14. Thoefner MB, et al. **Prev. Vet Med** 2001; 48: 201; 15. Senior JM, et al. **Equine Vet. J.** 2011; 43: 585; 16. Lohmann KL, et al. **J Endotoxin Res.** 2003; **9**: 33; 17. Morris DD, et al. **Am J. Vet. Res.** 1990; 51: 1935; 18. Morris DD, et al. **J. Am. Vet. Med.** 1991; 199: 1584; 19. MacKay RJ **Am J. Vet. Res.** 1992; 53, 748-752; 20. Fessler JF, et al. **Equine Vet. J. Suppl**. 1989; 7: 24-28; 21. Barton MH, et al. **J. Vet Int Med.** 1998; 12:26-35; 22. Peek SF, et al. **Vet. Clin. Pathol** 2004; 33: 29; 23. Steverink P, et al. **J Endotoxin Res.** 1995; 1: 289; 24. Barton MH, et al. **Equine Vet J** 1996; 28: 382; 25. Trent SM, et al. **J. Endotoxin Res**. 2006; 12: 205; 26. Fijihara M, et al. **Pharmacol ther**. 2003; 100: 171; 27. Figueiredo MD, et al. **Vet Immunol Immunopathol** 2009; 127:125; 28. Frellstedt L, et al. **Vet Immunol Immunopathol** 2012; 149: 97; 29. Maute-Bello G, et al. **Am J Physiol Lung cell mol physiol** 2008; 295: L379; 30. Wang HM, et al. **Eur Surg Res** 2008; 40:305; 31. Reiss LK, et al. **Eur J Cell Biol** 2012; 91: 590; 32. Ahronson-Raz K, et al. **Can J Vet Res** 2010; 74:45; 33. Burrows GE. **Can J Comp Med** 1981; 45: 207; 34. Pirie RS, et al. **Equine Vet J** 2001; 33:311; 35. Turek JJ, et al. **Am J Vet Res** 1985; 46:591; 36. Parbhakar OP, et al. **Vet Res** 2005; 36: 557; 37. Olson NC. **Am J Vet Res** 1985; 46:2288; 38. Pantaleon LG, et al. **J. Vet Int Med** 2006; 20:1422; 39. Tetens J, et al. **Am J Vet Res** 2004; 65: 225; 40. Wilkins PA, et al. **J Appl Physiol** 2001; 91:2442; 41. Benamou AE, et al. **Equine Vet J** 2001; 33: 337; 42. Rossi P, et al. **Intensive care med** 2008; 34: 873; 43. Jesmin S, et al. **Biomedical Research** 2011; 32: 9; 44. Benamou AE, et al. **Pulm Pharmacol Ther** 1998; 11:231; 45. Suri SS, et al. **Vet Res** 2006; 37:541; 46. Schneberger D, et al. **Anat Rec** 2009; 292: 1068; 47. Schneberger D, et al. **Am J Physiol lung cell mol physiol** 2012; 302: 498; 48. Sone Y, et al. **J Appl Physiol** 1999; 87: 1354

AIR QUALITY: WHAT OUR PATIENTS WISH WE KNEW!

Janice A. Dye, DVM, MS, PhD, DACVIM (SA-IM)
Research Triangle Park, NC

INTRODUCTION

In this session we will review health effects associated with commonly encountered outdoor (ambient) pollutants as well as indoor (residential) contaminants (e.g., irritants, allergens, and even toxins). Several ambient air pollutants readily permeate indoor spaces too. Because we spend most of our lives indoors, as do most of our small animal companions, we will also discuss how related home conditions (e.g., excessive dampness, poor ventilation) can further worsen indoor air quality. Exposures may be acute, with one or more pets (and owners) experiencing signs within a relatively short period. More often, however, exposures are episodic or chronic, occurring over prolonged periods, making it difficult to establish connections between respiratory signs and poor indoor air quality. This is in part because of the non-specificity of associated signs, and, in part, because multiple toxicants may be involved. Furthermore, one's perception of air quality is quite subjective, making inferences to pets even more problematic. Underlying respiratory conditions further complicate the clinical picture, as such individuals are often more sensitive to (and affected by) lower concentrations than are seemingly problematic for healthy housemates. In the sections that follow, agents will be broken down by class and discussed briefly. In all, we will focus on every day insults and their potential to cause and/or exacerbate airway disease. Further information may be obtained in a recent review (Dye 2013) and other references and websites listed below.

BIOLOGIC CONTAMINANTS

Allergens

Data linking respiratory disease in companion animals to allergen exposure is quite limited. Supportive evidence comes from clinical improvement after allergen avoidance trials. Furthermore, cats with bronchial disease appear to have significantly greater positive intradermal skin test (IDST) reactions to the common indoor aeroallergen — house dust mite — than do healthy cats Moriella 2009). Likewise, serum allergen-specific IgE in cats with bronchial disease were strongly reactive to dust mite allergen and select regional outdoor aeroallergens (Norris Reinero 2004).

Dampness and mold

Over the last 40 years, the average home has become more and more air-tight in an effort to make heating and cooling more economical. Modern homes recirculate more of their indoor air, again for efficiency of heating and cooling. In so doing however, the overall volume of outdoor air entering the home is significantly reduced, and can lead to stagnant air and increased humidity. So called *tight building syndrome* is especially problematic if the home's heating ventilation and air conditioning (HVAC) system is not functioning optimally. Such effects are further compounded by water intrusion (e.g., flooding, appliance malfunction, or roof leaks). Excess moisture promotes growth of mold, fungi, and bacteria — which emit spores, hyphae fragments, microbial volatile organic compounds (VOCs), and endotoxin into the indoor air (Cabral 2010).

To date, the ability to definitely link exposure to specific biocontaminants to defined respiratory or neurologic syndromes remains elusive. It is, however, an area of intense debate and active investigation. Microbes can release VOCs, which are known for their ocular and upper respiratory tract irritant properties. Some biocontaminants are known to cause respiratory or systemic infection (i.e., "Legionnaire's Disease"). Increasingly, biocontaminants are implicated in hypersensitivity responses (e.g., pneumonitis, chronic rhinosinusitis), as well as exacerbation of allergies / asthma, and poorly defined but serious chronic neurologic conditions (Thrasher 2009). Cats, not unlike young children, may be at greater risk for exposure to such agents because they typically inhale plus ingest greater quantities of microbe-ladened house dust (cats due to extensive grooming and kids due to increased floor contact time and "mouthing" behaviors). A recent case report describes two cats that developed acute pulmonary hemorrhage and died. Satrotoxin G adducts were detected in archived serum samples from the cats, consistent with exposure to *Stachybotrys chartarum* or the so-called toxic black mold (Mader 2007). The clinical course of each cat's disease was remarkably like that of infants in a series of reports from the Cleveland area who developed pulmonary hemorrhage after living in moldy indoor environments. In the cats (and infants), concurrent exposure to other respiratory irritants (e.g., cigarette smoke and inhalant anesthesia) may have contributed to the overall lung insult. Dedicated scientists are working towards improving our ability to define and quantify indoor biocontaminants (Parks 2008). Improved methods to assess exposure (e.g., detection of mycotoxins in urine) (Straus 2011) are also in progress. Much remains to be established as to how damp indoor environments are associated with adverse health outcomes.

CHEMICAL TOXICANTS

Combustion-derived products

Second-hand smoke (SHS) is the smoke from another person's cigarette. It contains gases and respirable particles. In poorly ventilated air spaces, it can cause ocular and respiratory tract irritation, especially in children. The Surgeon General concluded that there is no risk-free level of exposure to SHS. A smoke-free environment is considered the only way to fully protect nonsmokers from its hazardous effects (US Department of HHS 2010). House pets are similarly exposed to SHS. Predisposition to similar conditions is likely, and, by extension, pets with pre-existing upper or lower respiratory disease may be especially sensitive to irritating effects of SHS. Based on limited studies, in dogs SHS exposure is associated with decreased lung functional reserve, increased risk of lung cancer (in brachy- and mesocephalic dogs) and nasal cancer (in dolichocephalic breeds) (Reif 1998). In cats, SHS exposure is associated with increased risk of malignant lymphoma and oral squamous cell carcinoma (Snyder 2004), likely owing to inhalation plus oral (via grooming) exposure routes. Nicotine toxicity can occur in pets ingesting cigarette butts, and may prove fatal in birds and small dogs.

Air pollutants

Since the middle of the last century, several catastrophic air pollution episodes have occurred. They were infamous for their toll on human life (e.g., the London Fog of 1952), but they also caused considerable morbidity and mortality in animals. In North American, due to industrial accidents in combination with thermal weather inversions, key disasters included Donora, PA (1948) where effluents from steel, zinc, and sulfuric acid plants were trapped in a valley inversion, killing 20 people, 10 dogs, 3 cats, and 40% of poultry in four nearby flocks; and in Poza Rica, Mexico (1950) where release of hydrogen sulfide during a thermal inversion led to the death of 22 people and vast numbers of dogs, swine, cattle, and nearly 100% of the canaries in the area. Owing to modern day pollution control technologies and regulations, contemporary air pollution levels are greatly reduced, but nevertheless are associated with significant economic toll owing to health care costs and lost work and school productivity. Individual air pollutants and associated health effects will be discussed in brief below. It should be noted the most of the health data is inferred from epidemiologic associations in humans and data derived from relatively short-term, high-dose experimental exposures in laboratory animals.

Nitrogen dioxide is a poisonous, corrosive, brown gas. It is consider both a major indoor and outdoor air pollutant. The primary indoor sources relate to poorly ventilated biomass-burning. In pets, nitrogen dioxide exposure may result in increased frequency and severity of respiratory infection. It may also contribute to chronic bronchial and lung pathologic changes.

Carbon monoxide (CO) is a colorless, odorless gas. Unlike most pollutants, a single (likely to be encountered) exposure to CO may prove lethal. Significant exposure can occur during use of charcoal grills or unvented kerosene heaters in relatively closed air spaces (garages), or when chimneys or venting ducts of central heating systems are obstructed. Birds are especially sensitive to CO, as evidenced by their use as sentinels in coal mine shafts. CO poisoning is also an important cause of the morbidity and mortality in pets rescued from burning buildings. Having preset clinic protocols in place for managing animal victims of smoke inhalation can help to ensure a successful outcome (Fitzgerald 2006).

Ozone is a very potent oxidant gas and is one of the most directly injurious of the outdoor air pollutants. Exposure can induce respiratory tract irritation, injury, and inflammation. In some instances, ozone is emitted directly indoors by (e.g., electric generators, photocopiers and laser printers). Ironically, ionizing *air-purifiers* intentionally generate ozone to react with and remove odorants from the indoor air. Their use, however, can result in indoor ozone concentrations that exceed levels found in smog.

Sulfur dioxide is another ambient air pollutant, and is the major precursor to so-called *acid rain*. Veterinary concerns regarding this pollution primarily relate to aquatic and wildlife health. One important exception is that of unvented kerosene space heaters, which when used in closed-space situations, can lead to exorbitant total airborne particle concentrations and associated acid aerosols.

Off-gassing emissions

Formaldehyde is an overpowering, highly water soluble, reactive gas. It can cause ocular and upper respiratory tract irritation. Due to off-gassing from resins in particleboard, wood paneling, foam insulation, carpeting, upholstery, and draperies, indoor levels can be quite high, especially in new homes and mobile homes (which are made of many of these materials). Excessive acute exposure can be associated with headache, nausea, dizziness, and disorientation.

Volatile organic compounds (VOCs) can easily become vapors or gases. They are present in a wide variety of household products and aerosol sprays (e.g., butane, propane), fragrances (e.g., limonene, pinene, isoprene), and personal care products (e.g., hair spray, deodorants). To name but a few sources, VOCs (and other irritating chemicals) derive from paints, paint strippers (methylene chloride), dry-cleaned clothes (trichloroethylenes), cleaning compounds [e.g., stain removers, bleach (sodium hypochlorite), tile cleaners (ammonia hydroxide)], anti-static sprays, pesticide sprays, and fumigants (e.g., insect, roach, and flea control products). VOCs can act as sensory irritants. Not uncommonly, owners fail to use these products as per warning label instructions, making VOC exposures in pets problematic. For pets presenting with acute onset of eye, skin, and/or respiratory signs, one should consider recent exposure to VOCs. Greater, prolonged, or combined exposures may further cause dizziness, headache, nausea, and even allergic-like reactions. Avian veterinarians should know that these products are especially problematic in birds, because like CO, pet birds are very sensitive to aerosolized VOCs. Fumes from cooking pans, self-cleaning ovens, or heat lamp bulbs lined with polytetrafluoroethylene (PTFE; Teflon® or Silverstone®) have also been associated with acute and fatal toxicity in birds (Lightfoot 2008).

Particles, House Dust, and Fibers

Ambient *particulate matter (PM)* is a major environmental health concern in people. Fine ($PM_{2.5}$) and coarse (PM_{10}) particles are complex mixtures of microscopic carbon-based particles, with adsorbed materials that can be inhaled deep into the lungs. Health associations include: increased respiratory signs, decreased lung function, asthma and COPD exacerbation, arrhythmias, cardiac ischemic events, and premature death, especially in people with pre-existing heart or lung disease. Indoor PM represents both intruded outdoor PM, plus particles from indoor SHS, cooking, heating (fireplaces) and so forth. Indoor cats are further exposed to particles in litter (e.g., silicates or dusty clay), which anecdotally may be associated with respiratory distress in cats with feline bronchial disease.

House dust is a mixture of lint, particles, and adsorbed substances from household products. Dust may be laden with metals, pesticides, semivolatile persistent chemicals (e.g., polybrominated diphenyl ethers, PBDEs, a type of flame retardant), and so forth — with pets continually exposed to these contaminants through dust exposure (Dye 2007).

Asbestos fibers are naturally-occurring mineral silicates that were used extensively in friction products and as a noncombustible insulation. Asbestos fibers are small, readily inhaled, and difficult to clear from the lungs. Asbestos-associated diseases have very long latency periods (up to 30 years in humans), and include interstitial lung fibrosis, carcinoma, and mesothelioma.

IMPROVING INDOOR AIR QUALITY

Indoor air can be considerably more polluted than outdoor air (http://www.epa.gov/iaq/pubs/). The EPA lists poor indoor air quality as a major environmental concern. Nevertheless, measures can be taken to help minimize exposure for owners and their pets. Key elements of improved air quality are source control, adequate ventilation, maintaining relative humidity between 30-50%, changing air filters on a regular basis, and air duct cleaning as needed (Institute of Medicine 2004). CO monitors should be installed in the home. All gas appliances should be proper functioning, and if in question, should be inspected and adequately cleaned or repaired. In addition, gas stoves and heaters should be vented to the outside of the home. Owners should be reminded that gas ranges should *never* be used as a heat source.

Pet owners should be cautioned about so-called air purifiers that use ionization as a means of "clearing the air", but actually increase indoor ozone concentrations, at times to levels well in excess of that considered safe. Kerosene heaters should be used only as indicated by the manufacturer's instructions, and should be refueled outdoors using specially manufactured low-sulfur fuel. Cars, lawn mowers, etc., should never be left running inside a garage or shed, especially if the space is ever used to house a pet.

Pet owners should similarly be cautioned to follow the manufacturer's directions during use of common household cleaners, especially the aerosol-type products. Similar to recommendations for using flea or pest control products, the owner should ensure that there is adequate ventilation. One common mistake is for an owner to mix bleach (sodium hypochlorite) containing products with ammonia-based cleaning products. This is a recipe for serious problems because together, these compounds can generate toxic (e.g., chloramine) fumes. Pet owners, especially of birds, should minimize use of all VOC-containing products. Encourage use of unscented and non-aerosol alternatives. Many newer and so-called "green" substitutes are designed to have much lower content of known irritants (e.g., low-VOC paints). Despite these precautions, if the owner perceives that use of any cleaning product, paint, new carpeting, and like agent is associated with ocular irritation, sore throat, headache or an offensive odor, the owner should ensure that neither they nor their pet maintain prolonged exposure in the area until odors dissipate. If this is not possible, efforts to improve ventilation are vital.

Health risk from exposure to biocontaminants within damp indoor environments is recognized as a significant public health problem requiring attention and mediation (WHO 2009). It is mandatory that all water leaks and conditions contributing to excess moisture levels are repaired and eliminated. Because spores remain viable long after water intrusion has been corrected, it is also important to clean areas extensively once all structural problems have been remedied. Increasing ventilation can help not only to decrease humidity, it is thought to decrease accumulation of any and all gases or chemicals tending to "off-gas" from household materials into the indoor environment. For people and pets with known respiratory or heart conditions, the above precautions are especially important due to increased risk of exacerbating their underlying conditions owing to exposure to these respiratory toxicants. Likewise, young puppies or kittens, and geriatric pets may also be at greater risk.

Air Quality Index

Owners can be informed about the state and federal air quality index (AQI) warnings. During episodes of worsening air quality, warnings are issued in the form of "colors" similar to stop-and-go lights. During low pollution days, the air quality is "green for go outside" safely. However, on days with relatively higher ozone or PM levels, yellow, orange, or even code "red" days are broadcast for many urban areas. People and owners of pets with existing respiratory conditions can use this information to avoid exercising during peak pollutant periods (http://airnow.gov and click on *Local forecasts and conditions*).

Lastly, the one change that may be most beneficial for the pet, is for the owner(s) to stop all smoking inside the home. Again, the Surgeon General has concluded that only by eliminating cigarette smoking in indoor spaces, can one fully protect nonsmokers from SHS exposure. Relatedly, a recent pilot study on owner attitudes toward smoking showed that by informing pet owners of the risks of SHS for their pets, the owner may be more motivated to quit smoking altogether (Milberger 2009).

Disclaimer

This information has been reviewed by the National Health and Environmental Effects Research Laboratory, U.S. EPA, and approved for publication. Approval does not signify that the contents necessarily reflect the policies of the agency, nor does the mention of commercial products constitute endorsement.

REFERENCES

Bertone ER et al: Environmental tobacco smoke and risk of malignant lymphoma in pet cats, *Am J Epidemiol* 156:5, 2002.

Cabral JP: Can we use indoor fungi as bioindicators of indoor air quality? Historical perspectives and open questions, *Sci Total Environ* 408:4285-95, 2010.

Dye JA: Respiratory Toxicants of Interest to Pet Owners. *In:* Bonagura JD & Twedt D eds., *Kirk's Current Veterinary Therapy XV*, Elsevier Science, (*in press, 2013*).

Dye JA et al: Elevated PBDE levels in pet cats: sentinels for humans? *Environ Sci Technol* 41:6350-6, 2007.

Fitzgerald KT & Flood AA: Smoke inhalation, *Clin Tech Small Anim Pract* 21:205-214, 2006.

Institute of Medicine: Damp Indoor Spaces and Health, *Washington DC: National Academy of Sciences*: 2004.

Lightfoot TL and Yeager JM: Pet bird toxicity and related environmental concerns, *Vet Clin North Am Exot Anim Pract* 11:229-59, 2008.

Mader DR et al: Acute pulmonary hemorrhage during isoflurane anesthesia in two cats exposed to toxic black mold, *J Am Vet Med Assoc* 231:731-5, 2007.

Milberger SM et al: Pet owners' attitudes and behaviors related to smoking and second-hand smoke: a pilot study, *Tob Control* 18:156-8, 2009.

Moriello KA et al: Pilot study: prevalence of positive aeroallergen reactions in 10 cats with airway disease, *Vet Dermatol* 18:94-100, 2007.

Norris Reinero CR et al: An experimental model of allergic asthma in cats sensitized to house dust mite, *Int Arch Allergy Immunol* 135:117-31, 2004.

Park JH et al: Hydrophilic fungi and ergosterol associated with respiratory illness in a water-damaged building, *Environ Health Perspect* 116:45-50, 2008.

Straus DC: The possible role of fungal contamination in sick building syndrome, *Front Biosci (Elite Ed)* 3:562-80, 2011.

Thrasher JD, Crawley S: The biocontaminants and complexity of damp indoor spaces: more than what meets the eyes. *Toxicol Industrial Health* 25:583-615, 2009.

US Department of HHS: *A Report of the Surgeon General: How tobacco smoke causes disease: What it means to you*, 2010.

WHO guidelines for indoor air quality: dampness and mould, 2009.

CANINE AND FELINE RHINITIS: THEORY AND PRACTICE

Lynelle R. Johnson DVM, MS, PhD, DACVIM
Davis, CA

INTRODUCTION

Rhinitis is defined as inflammation of the mucous membranes within the nasal cavity. Inflammation can occur from a specific cause such as a foreign body, infection (fungal, viral or secondary bacteria), dental-related nasal disease, or neoplasia. In some cases, nasal regurgitation of gastrointestinal contents or lower respiratory secretions can result in secondary nasal discharge. Eosinophilic bronchopneumopathy is a disorder that can affect the epithelial surface of both the upper and lower respiratory tract. In dogs and cats with one of these defined etiologies, specific therapy might result in cure or at least control of clinical signs. The more challenging cases are those in which no specific etiology can be determined. Chronic rhinosinusitis in the cat and canine lymphoplasmacytic rhinitis represent two such diseases.

CLINICAL FINDINGS AND DIAGNOSTIC TESTING

Clinical signs are common for all etiologies of nasal discharge and include sneezing, nasal discharge (mucoid, purulent, or hemorrhagic), and possibly facial pain or asymmetry. Nasal foreign bodies, which usually result from vomiting or regurgitation of oral foreign bodies into the nasopharynx, are often recognized by an acute onset of clinical signs however animals can also present with chronic complaints. Foreign bodies usually result in unilateral disease, neoplastic of fungal disorders are often unilateral initially and then become bilateral, and inflammatory nasal disease can be unilateral or bilateral. Physical examination features that can help narrow the differential list include assessment of nasal airflow, determination of ocular retropulsion, palpation of the soft palate and regional lymph nodes, and examination for abnormalities in facial structures or the nasal planum. In dogs, fungal infection due to aspergillosis typically results in increased or preserved nasal airflow, nasal depigmentation is seen in approximately 40% of cases, ipsilateral lympadenomegaly may be seen and facial pain may be noted. Feline aspergillosis, canine or feline nasal cryptococcosis, and nasal tumors usually lead to loss of nasal airflow. Asymmetric ocular retropulsion, bulging of the soft palate and lymphadenomegaly can also be seen. Cryptococcosis might be associated with signs of disease in other organs, such as the skin, central nervous system, or the fundus. Dogs or cats with idiopathic rhinitis usually have copious mucoid nasal discharge but no systemic signs. Presence of a cough or tracheal sensitivity warrants investigation of the lower airways.

While physical examination findings can be helpful in ranking differential diagnoses for a given patient, advanced imaging and collection of samples is required for an accurate diagnosis. Occasionally, standard laboratory work can be useful. Most specifically, the LCAT or CALAS (a latex agglutination titer for cryptococcosis) is both sensitive and specific for diagnosis of cryptococcosis in dogs and cats. False positive tests are rare but have been reported in human medicine when there is interference with the test by soaps or disinfectants and it is possible that false positives also exist in veterinary medicine. False negative tests might be encountered with very localized disease (e.g. nasal or ocular involvement only).

There are a number of serologic tests for canine nasal aspergillosis, and the techniques employed in these tests include agar gel double diffusion (AGDD), agar gel immunodiffusion (AGID), counter immunoelectrophoresis, standard ELISA, and sandwich ELISA. AGID using an *Aspergillus* antigen prepared from cultures of *A. fumigatus*, *A. niger,* and *A. flavus* (the most common fungal species to infect the nose) has proven somewhat useful in confirming the clinical suspicion of aspergillosis. A recent study in dogs revealed that the sensitivity of the AGID for documenting canine aspergillosis was 68% and specificity was 98% (only one false positive). Positive and negative predictive values were 94% and 84%. Therefore, a positive serology can be taken as evidence that *Aspergillus* is the likely cause of nasal discharge, however a negative test does not rule it out. Follow-up titers unfortunately are not valuable for determining the presence or absence of nasal infection in dogs. Serum galactomannan and DNA detection in blood are of no value in the diagnosis of nasal aspergillosis in dogs or cats.

A nasal swab sample or impression of nasal discharge might occasionally allow cytologic diagnosis of cryptococcosis but is of virtually no value in diagnosing clinically significant *Aspergillus* infection. In dogs with nasal aspergillosis, fungal hyphae were observed in smears of nasal exudates or in blind swabs of the nasal cavity in only 13-20% of cases. Collection of a deep nasal flush sample might be more valuable. This can be collected by inserting a red rubber catheter into the nasal cavity to approximately the level of the medial canthus of the eye, and then injecting and aspirating 3mls of sterile fluid while compressing the soft palate to prevent loss of the fluid out the nasopharynx. Alternately, the oropharynx can be packed with a lap pad and antegrade flushing can be used to expel material onto the lap pad for microscopic examination. However, visualized nasal biopsy specimens are likely to have the highest yield in the diagnosis of fungal infections, including cryptococcosis and aspergillosis, as well as in animals with neoplastic disease. Particularly, when a suspected fungal plaque lesion is sampled, *Aspergillus* hyphae can be observed in up to 100% of cases.

Bacterial culture in animals with nasal disease is controversial. It is important to realize that almost all bacterial infections in the nose are secondary rather than primary and a culture of nasal discharge is of no value. Fungal or bacterial culture of a swab sample is likely to reveal only contaminants rather than a clinically significant organism. With this in mind, bacterial culture from a deep nasal swab can provide some direction for antibiotic therapy, particularly in cats with chronic rhinosinusitis that have been on multiple antibiotics. A deep nasal flush or nasal brush sample is useful for collection of a sample. Some clinicians recommend collection of a sample from the nasopharynx or the frontal sinus to achieve the best possible results. Fungal cultures are also controversial as there is some concern that fungal infection can complicate neoplastic disease. However, a recent study on the utility of fungal cultures in the diagnosis of canine nasal aspergillosis reported moderate sensitivity (77%) but high specificity (100%). It's important to note that in this study, the material submitted for culture was from a visualized fungal plaque and therefore, a low number of false positive values would be expected. Sampling of nasal discharge has been associated with lower sensitivity and specificity and evaluation of fungal infection in animals with nasal neoplasia has not been addressed recently. One concern might be that neoplasia and fungal infection can exist concurrently in an animal.

Extensive diagnostic testing is generally required to determine a specific etiology for nasal discharge, including imaging (skull radiographs or computed tomography) and rhinoscopy with biopsy +/- culture, followed by dental probing. Thoracic radiographs (and possibly bronchoscopy) are indicated in some animals, particularly those with cough. Skull radiogaphs are inferior to computed tomography but may be the only imaging option available. A full skull series includes lateral views, an open mouth or intra-oral view, and the frog-eye view that highlights the frontal sinuses. The most useful view is the open mouth view because it allows characterization of bony destruction and turbinate lysis on each side of the nasal cavity and comparison of symmetry between the 2 nasal cavities. Radiographic changes seen in nasal neoplasia include increased soft tissue density in the nasal cavity, lysis of the nasal turbinates, and lysis or deviation of the vomer bone. However, nasal radiography has low sensitivity in differentiating inflammatory rhinitis from neoplasia or mycotic rhinitis, because soft tissue opacification, turbinate destruction, and frontal sinus disease can be seen with all three conditions.

It is unquestionable that computed tomography provides more complete information on the nature and extent of disease within the nasal cavity. Tumors lead to turbinate destruction that can involve one or both nasal cavities, soft tissue masses that are either unilateral or bilateral, and sinus disease due to mass effect or obstruction of fluid drainage. Importantly, CT allows evaluation of the cribriform plate. Tumor-related destruction of this bony barrier to the central nervous system warrants a guarded prognosis. Finally, CT scans are recommended in the staging of nasal tumors in order to define tumor boundaries and to plan radiation therapy. CT is also critical in optimal evaluation of patients with aspergillosis because these animals can also have involvement of the cribriform plate, which impacts therapy and prognosis. With aspergillosis, CT usually reveals dramatic turbinate loss in the nasal cavity, with variable sinus involvement. In cats, mass lesions may predominate and sino-orbital disease will be evident. LPR and chronic rhinosinusitis will have imaging features that mimic both fungal and neoplastic disease but in a slightly less aggressive manner. Rhinoscopy and biopsy are critical to the final diagnosis.

Rhinoscopy follows any imaging procedure. Biopsies of mass lesions, abnormal appearing areas, or fungal plaques should be obtained with direct visualization with rhinoscopy whenever possible. Saline flush and suction can improve visualization when mucus or blood obscures the view. When sampling within the nasal cavity, the biopsy instrument should not be extended beyond the medial canthus of the eye in order to avoid penetrating the central nervous system.

RECENT ADVANCES

Molecular detection of pathogens is possible through application of the polymerase chain reaction to freshly obtained or paraffin imbedded histologic specimens. In cats with chronic rhinosinusitis, herpesvirus was rarely detected using end point PCR and even when a positive reaction is detected, determining the importance of viral presence to the disease state remains difficult to prove. Use of real time PCR may be of more benefit by providing assessment of mRNA presence and by quantitating viral loads. This might allow better determination of the role of virus in the feline condition. Real time PCR showed that the presence of FHV-1 mRNA was associated with upregulation of many pro-inflammatory cytokines in nasal tissue of cats. Shelter study by burns….

In dogs with idiopathic rhinitis, PCR has suggested that specific pathogens such as *Chlamydophila*, *Mycoplasma*, and *Bartonella* are not involved in the etiology of disease and serologic studies have suggested that *Bartonella* is not involved in canine or feline rhinitis. However, in dogs, increased exposure to fungal organisms may play a role in disease. Whether specific fungi are causing an infectious disease or triggering a hypersensitivity response remains to be investigated. In dogs with *Aspergillus*, PCR of nasal tissue is not generally important in identifying the source of infection however it has proven useful in evaluating the immune response and underlying pathogenesis. Upregulation of mRNA coding for interferon-gamma, IL-6, ILp35, IL-12p40, IL-18 and tumor necrosis factor-alpha has been demonstrated in the nasal mucosa in dogs with aspergillosis, suggesting the generation of a Th1 local immune response as well as activation of a pro-inflammatory cytokine profile. The strong Th1 response would be expected to help clear the infection and prevent systemic spread, however upregulation of IL-10 has also been shown in the nose of affected dogs. This cytokine may play a role in failure to resolve infection by modulating the inflammatory response. Nasal cytokine profiles in dogs with nasal aspergillosis differ from those with LPR in which a partial Th1 response with overexpression of IL-5 has been identified. These studies would suggest that LPR is a distinct entity rather than a different expression of fungal rhinitis. Investigation of pattern recognition receptors (Toll-like receptors and NOD-like receptors) responsible for identifying structures on microbes and signaling induction of the immune response from antigen presenting cells has revealed differences between tissues from dogs with aspergillosis and LPR, further supporting a different etiopathogenesis of disease.

REFERENCES

Barrs et al, Vet J 2012; 191: 58, Berryessa NA, et al. J Am Vet Med Assoc 2008; 233(7): 108. Hawkins EC, et al. J Am Vet Med Assoc 2008; 233(4): 597. Mercier E, et al. Vet Immunol Immunopathol. 2012; 145(3-4): 618. Johnson LR, et al. J Am Vet Med Assoc 2005; 227(4): 579. Peeters D, et al. Vet Immunol Immunopathol. 2007; 117(1-2):95. Pomrantz et al. J Am Vet Med Assoc 2007; 230: 1319, Vanherberghen M, et al. Vet Microbiol. 2012; 157(1-2):143. Windsor RC, et al. J Vet Int Med 2006; 20(2): 250.

FELINE HYPOTENSION: MAXIMIZING CARDIAC OUTPUT AND SYSTEMIC VASCULAR RESISTANCE

Christopher G. Byers, DVM, DACVECC, DACVIM (SAIM)
Omaha, NE

INTRODUCTION

Hypotension is commonly recognized in the clinical setting, and in cats has been typically defined as a systolic blood pressure <80-90 mmHg. Intermittent or prolonged hypotension may cause hypoperfusion and decreased oxygen delivery that may ultimately lead to organ damage or failure. Previously Silverstein *et al* reported hypotensive cats with lower rectal temperatures and lower PCVs had significantly lower survival rates (64% vs. 32%). So certainly identifying the underlying cause(s) is vital to reducing morbidity and mortality.

A simplistic way to view hypotension is consider it the manifestation as one of three causes: hypovolemia, decreased systemic vascular resistance (SVR), and cardiac dysfunction resulting in decreased cardiac output. Obviously there is a long list of issues may contribute to one or more of the above 3, most notably:

- Hypothermia
- Bradycardia
- Anemia
- Low ionized calcium
- Acidosis

CORRECT HYPOTHERMIA & ELECTROLYTE ABNORMALITIES

Therapeutic efforts are aimed at rapidly rewarming patients during fluid resuscitation as well as reducing additional heat loss. Resuscitative efforts should not contribute to the hypothermia. It has been recommended to warm the animal by at least 1-2 degrees Celsius per hour; however, faster rates may be necessary. Moderate intravascular volume support is recommended during active rewarming in hypovolemic shock. This will support mean arterial pressure (MAP) and resolve most cases of hypothermia-induced hypotension, bradycardia, hypoventilation, and coagulopathy while avoiding volume overload.

Rewarming hypothermic animals may be accomplished by several different methods, including passive surface (e.g., blankets), active surface (e.g., Bair hugger), and active core (e.g., peritoneal and pleural lavage) re-warming. Surface re-warming should always accompany active core re-warming to reduce core-to-peripheral temperature gradients. During external heating, care must always be taken to prevent skin burns by controlling the temperature of the external heating devices or placing a barrier between the heat source and the patients.

Electrolytes (particularly ionized calcium), acid-base status, and electrocardiography should be monitored and alterations addressed, particularly in hypothermic patients.

FOCUS ON CARDIAC OUTPUT

Provide adequate preload

The first step of volume resuscitation should always be to ensure adequate intravascular volume, and hence adequate preload to the heart to maintain appropriate cardiac output. As such one needs to provide adequate intravascular volume and stop any ongoing blood loss. Common controversies for resuscitative fluid therapy include type of fluid (colloid versus crystalloid and in case of colloid, synthetic versus natural), rate of administration, and end-points of resuscitation. Currently there is not "the one" recipe of fluid therapy for feline hypotension (or for any type of hypotension for that matter). Therefore initial fluid of choice depends on a variety of factors, including underlying disease process, colloid osmotic pressure, electrolyte and acid-based status, coagulation and major organ function. The most commonly used fluid initially is an intravenous bolus (5-10 mL/kg) of an isotonic crystalloid – exercise caution with normal saline given its acidifying nature. Boluses of hypertonic saline (2-4 mL/kg over 20-30 min) and/or synthetic colloid (2-5 mL/kg IV over 20-30 min) may also be needed depending on the factors listed above. If there is a confirmed or suspect secondary coagulopathy, fresh frozen plasma (FFP) and/or Vitamin K1 are indicated.

Given the guidelines of early goal-directed therapy (EGDT), the author prefers to maintain PCV >25%, and thus transfuse with packed red blood cells (pRBCs) or whole blood (WB) depending on the patient's protein status. Resuscitation should be performed until several endpoints are achieved including normalization of capillary refill time (CRT), restoration of normal heart rate and pulse pressure, establishment of normothermia, urinary output >1 mL/kg/hr, and normal lactate and base excess. Once the desired endpoints have been reached, initiate a crystalloid and/or colloid IV CRI based on dehydration, daily requirements (45-60 mL/kg/day), ongoing losses and COP.

Ensure Proper Contractility

Pro-inflammatory cytokines readily contribute to myocardial depression. Echocardiography frequently demonstrates poor chamber contraction contributing to reduce cardiac output. Dopamine, dobutamine, phenylephrine and epinephrine were recently evaluated for inotropic effects in the isoflurane-anesthetized cats. A positive inotropic effect and improvement in cardiac output were detected with all treatments. Among the four inotropes, dopamine had the most beneficial effects. Phenylephrine increased systemic vascular resistance index with minimal effect on heart rate but dobutamine decreased it and dopamine and epinephrine had no effect. Epinephrine was as effective as dopamine but increased lactic acid production and base deficit significantly.

FOCUS ON SYSTEMIC VASCULAR RESISTANCE

Maintain Appropriate Afterload

Afterload is determined largely by peripheral vascular tone. Clinical signs of vasodilation include hyperemic mucous membranes, brisk CRTs (<1.5 sec), and bounding pulses. Patients that have been adequately volume resuscitated but remain hypotensive and have clinical signs of vasodilation may benefit from vasopressor therapy (e.g., norepinephrine, vasopressin, or dopamine infusions). Currently one of the big current controversies in critical care medicine is selecting which vasopressor should be used first? Furthermore, some patients require use of multiple vasopressors.

Dopamine

Dopamine enhances renal and mesenteric perfusion at a dose of 2.5-5 mcg/kg/min IV CRI. At higher doses (10-15 mcg/kg/min IV CRI), dopamine exerts its effects on $\alpha 1$ and β receptors, resulting in vasoconstriction and increased myocardial contractility. At doses of 15-20 mcg/kg/min IV CRI, dopamine will cause constriction of the renal arteries, thereby reducing renal perfusion and causing potential irreversible renal damage.

Dobutamine

Dobutamine (2.5-10 mcg/kg/min IV CRI), a beta-1 agonist, increases cardiac output by enhancing myocardial contraction. One should use this medication with caution, as there are several potential serious adverse reactions, including generalized seizure activity and dysrhythmias.

Norepinephrine

Norepinephrine is a potent alpha-1 mediated vasoconstrictor and is usually reserved for severe, unresponsive hypotension at doses of 0.05-1 mcg/kg/min IV CRI. In humans, norepinephrine is frequently the vasopressor of choice in patients with septic shock.

Vasopressin

Vasopressin (0.01-0.05 units/kg/hr IV CRI) is a potent vasoconstrictor, mediated by V1a receptors on the vascular endothelium. As catecholamine receptors are reportedly dysfunctional and/or down-regulated in prolonged acidemia, use of vasopressin in patients whose metabolic acidosis is still present and/or severe may be more effective. Studies in humans have shown lower levels of endogenous vasopressin in patients with septic shock, and therefore may be chosen for blood pressure support in these specific patients. Dermal necrolysis has been reported.

Epinephrine

Epinephrine has a high affinity for all splanchnic adrenergic receptors but seems to exert greater effects on alpha-1 and beta receptors; it is the primary physiologic agonist for the pre-synaptic beta-2 receptors. Benefits include increasing cardiac output, heart rate and mean arterial blood pressure, but as a consequence may increase myocardial oxygen consumption and cause severe vasoconstriction. A commonly used IV CRI dosage is 0.005 to 0.05 ug/kg/min.

REFERENCES

[1]Wiesse A, et al **Am J Vet Res** 2012;73(6):908.
[2]Mazzaferro E, et al **Compend Contin Educ Pract** Vet 2001;23(8):728.
[3]Silverstein D, et al **J Am Vet Med Assoc** 2008;232(6):893.
[4]Clark K, et al **Cardiovasc Pharmacol** 1991;17(2):267.
[5]Pasco PJ, et al **Am J Vet Res** 2006;67(9):1491.
[6]Rivers E, et al **N Engl J Med** 2001;345(19):1368.
[7]Branson K, et al **Vet Surg** 1997;26(4):347.
[8]Sparkes A, et al **J Vet Intern Med** 1999;13(4):314.

RISK FACTORS AND MEASUREMENT FOR INTRA-ABDOMINAL HYPERTENSION IN DOGS AND CATS

Christopher G. Byers, DVM, DACVECC, DACVIM (SAIM)
Omaha, NE

INTRODUCTION

Intra-abdominal hypertension (IAH) and abdominal compartment syndrome (ACS) are associated with significant morbidity and mortality in humans. While animals have been utilized in models of IAH, prevalence studies have not been performed in critically ill veterinary patients; nevertheless IAH likely exists and is presumptively overlooked by veterinarians.

RISK FACTORS

Abdominal compartment syndrome occurs when IAH results in progressive intra-abdominal organ, cardiovascular, respiratory and/or central nervous system dysfunction. Risk factors for the development of IAH and ACS include both primary (injury or disease process within the abdominal compartment) and secondary (extra-abdominal injury or disease) causes. IAH may occur when there are increased abdominal contents, increased intra-luminal contents, reduced abdominal wall compliance and any condition capable of causing capillary leakage. Common examples include mechanical or paralytic ileus, pancreatitis, peritoneal effusion, abdominal masses, sepsis and placement of an abdominal wrap.

CONSEQUENCES OF IAH AND ACS

IAH and ACS may affect all major organ systems. Gut wall edema may develop due to decreased lymphatic flow, and capillary leakage; there is also potential for bacterial translocation and increased inflammatory mediators. Mesenteric lymph flow may be markedly decreased, and hepatic blood flow may be severely reduced, possibly triggering acute liver failure. Oliguria and anuria may occur as a result of compression of urinary collecting ducts and renal vessels. IAH may compress the abdominal caudal vena cava, thereby reducing preload (PL) and cardiac output (CO). IAP may also be transmitted trans-diaphragmatically to increase intra-thoracic pressure; this may subsequently yield decreased left ventricular compliance and reduced diaphragmatic excursions. Decreased CO and reduced cerebral venous outflow may result in reduced cerebral perfusion pressure (CPP).

TECHNIQUES FOR MEASURING INTRA-ABDOMINAL PRESSURE (IAP)

The abdomen is a closed cavity with relatively non-compressible contents with fluid characteristics; as such they behave according to Pascal's law that states that pressure exerted anywhere in a confined fluid is transmitted equally in all directions throughout the fluid. Accordingly IAP should be measurable in every location of the abdominal cavity such as the peritoneal space, urinary bladder, stomach, vessels such as the caudal vena cava, and rectum. Furthermore, IAP can be measured by direct and indirect methods and may be performed intermittently or continuously. If using an intermittent method, measurements should be made every four (4) hours.

The gold standard for IAP measurement in humans is direct measurement of peritoneal pressure. This procedure is performed by attachment of a catheter placed into the peritoneal cavity to a saline manometer or a pressure transducer. Multiple devices have been described including the piezoresistive catheter, the air-capsule catheter, the water-capsule catheter, and the Stryker intra-compartmental pressure monitor. This modality is rarely applicable in critically ill, non-surgical patients due to invasiveness, cost, and potential complications.

More commonly in people and animals, IAP is measured via placement of a urinary catheter. Specifically an indwelling urinary catheter with a balloon tip is sterilely placed, and the urinary bladder is completely emptied. Ultimately the urinary catheter is connected to a sterile closed urinary collection system with two (2) three-way stopcocks within the system. A standard water manometer is placed at the level of the first stopcock and is secured level with the pubic symphysis. A 60 mL syringe and 250 mL bag of 0.9% sodium chloride are attached to the second stopcock, and a volume of 0.5 mL/kg of 0.9% sodium chloride is infused into the urinary bladder. The manometer is then filled with 0.9% sodium chloride, and IAP is measured as the difference between the zero level and the meniscus of fluid in the manometer.

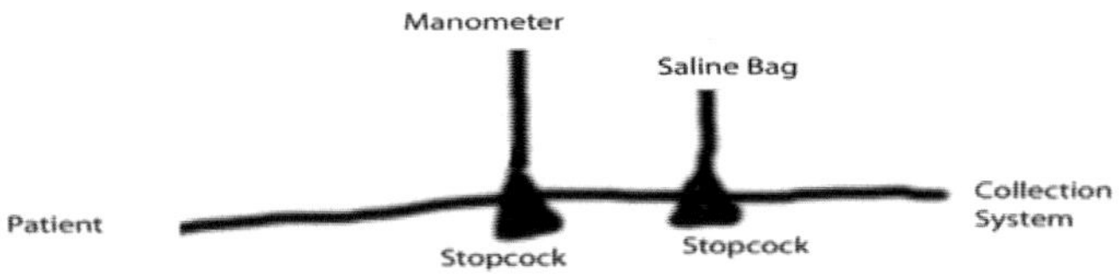

Figure 1.Schematic of a typical set-up for measuring IAP. A urinary catheter is connected to a closed collection system with 2 three-way stopcocks in series. A manometer is connected to the 1st stopcock while a saline bag and syringe are attached to the 2nd one.

GUIDELINES

The World Society of the Abdominal Compartment Syndrome (WSACS) has established guidelines to define IAP, IAH and ACS in people. IAH is defined as IAP greater than 12 mmHg, while ACS occurs when IAP is sustained at values greater than 20 mmHg with evidence of new organ dysfunction.

Guidelines for defining normal IAP and IAH have been proposed for dogs. An IAP of 0-7.4 mmHg should be considered normal; an IAP of 7.4-14.7 mmHg represents mild IAH while that of 14.7-25.7 mmHg constitutes moderate-to-severe IAP and >25.7 mmHg defines severe IAH. Some have attempted to establish a normal reference range for cats; one study identified a normal range of 0-7.7 mmHg in sedated cats and 0-13.7 mmHg in awake cats. A grading system for IAH has not yet been established for cats, but anecdotally many believe cat values of IAH mirror those proposed for dogs.

Table 1. Proposed grading system for IAH in dogs

Pressure in mmHg	Grade
0-7.4 mmHg	Normal
7.4-14.7 mmHg	Mild IAH
14.7-25.7 mmHg	Moderate-to-Severe IAH
>25.7 mmHg	Severe IAH

FACTOR AFFECTING IAP

Body Positioning

The weight of intra-abdominal organs places pressure on the urinary bladder when body position changes. Lateral recumbency has been most commonly used in dogs due to patient comfort and ease. Sternal and lateral recumbencies have been used in studies measuring IAP cats; specifically, IAP was shown to be significantly higher in right lateral recumbency compared to sternal recumbency, although both measurements were within the study's reference range.

Reference Point

Limited trials in dogs and cats have not yet established a definitive zero reference point. Studies have used the level of neck of urinary bladder, the level of the vulva, mid-way through the pelvis and the pubic symphysis.

Reproducibility

Technical inconsistencies with measuring techniques and patient issues may affect the reproducibility of consistent IAP measurements. Patient position and coughing may alter IAP measurements. Studies have also documented a more than 2 mmHg minute-to-minute variation when IAP is continuously measured, underscoring the importance of follows trends when monitoring IAP.

Infusate Volume

The optimal volume of infusate to instill into the urinary bladder of dogs and cats has not yet been determined. Over-distension of the urinary bladder may result in IAH secondary to detrusor muscle activity. Infusate volumes of 0.5-1.0 mL/kg have been used in dogs. In cats a volume of 1.0 mL/kg results in significantly higher IAP measurements compared to 0.5 mL/kg volumes. Currently, an infusate volume of 0.5 mL/kg is recommended for dogs and cats.

Obesity

Human patients with elevated body mass indices (BMI) have higher baseline IAP values. Furthermore overweight and obesity comorbidities may result in chronic IAH in people. A previous veterinary study documented progressively elevated IAP with increasing body condition scores (BCS) in cats. Given this documented trend, clinicians should consider intervening at lower magnitudes of IAH in overweight and obese patients due to the potential for pre-existing organ dysfunction.

Abdominal Muscle Contractions

Abdominal wall contraction reduces the compliance of the abdominal wall and subsequently elevated IAP. In veterinary medicine, a common cause of abdominal wall contraction is patient discomfort; patient movement due to anxiety is also a frequent cause of elevated IAP secondary to abdominal wall contracture; a recent study documented significantly higher IAP values in cats who struggled compared to those who did not. Furthermore, conscious cats had higher IAP readings compared to sedated cats. Thus when measuring IAP in critical patient, clinicians should provide adequate analgesia and anxiolytic therapy prior to IAP measurement to reduce overestimation secondary to struggling, coughing, vomiting, panting and/or whining.

REFERENCES

[1]Ivatury R, et al **Surg Clin North Am** 1997;77(4):783.
[2]Fetner M, et al **J Vet Emerg Crit Care** 2012;22(2):230.
[3]Conzemius M, et al **Vet Surg** 1995;24(3):195.
[4]Fusco M, et al **J Trauma** 2001;50(2):297.
[5]Lambert D, et al **Obes Surg** 2005;15(9):1225.
[6]Am G, et al **Crit Care Clin** 2007;36(4):1304.
[7]Parsak C, et al **World J Surg** 2008;32(1):13.
[8]Malbrain M. **Intensive Care Med** 2004;30(3):357.
[9]Rader R, et al **J Vet Emerg Crit Care** 2010;20(4):386.
[10]Schachtrupp A, et al **Intensive Care Med** 2003;29(9):1605.
[11]Serrano S **Proceedings of the 16th Annual International Veterinary Emergency and Critical Care Symposium** 2010;445.
[12]Drellich S **Compend Contin Educ Pract Vet** 200;22(8):764.
13De Waele J, et al **Intensive Care Med** 2008;34(7):1299.

COLLOID THERAPY AND CONCENTRATED ALBUMIN TRANSFUSIONS

Lisa L. Powell, DVM, DACVECC
St. Paul, Minnesota

Overview

Fluid therapy is a cornerstone of treatment for critically ill dogs and cats. Many critical illnesses cause a state of hypoperfusion, leading to organ dysfunction, which can then progress to organ failure and death. There are many different kinds of fluids to choose from, and the clinician should consider the contents of the specific fluid prior to administration. Additionally, the clinician needs to be aware of the potential effects of different fluids on varying disease states. Colloidal fluids can be advantageous in patients that are hypovolemic or hypoalbuminemic. In the past few years, a concentrated canine albumin solution has been developed for the treatment of hypovolemic and hypoalbuminemic dogs.

Synthetic Colloids

Synthetic colloids, such as dextrans and hydroxyethyl starches, are fluids that contain high molecular weight particles, enabling these fluids to increase plasma osmotic pressure (COP), holding fluid within the intravascular space. Colloids are advantageous in the treatment of patients with hypoalbuminemia, as albumin is the main contributor to COP. The hydroxyethyl starches (HES) (e.g., hetastarch, pentastarch, and tetrastarch) are most commonly used, as the dextrans were found to induce renal disease in humans. HES are the most commonly used synthetic colloid in veterinary medicine.

Hydroxyethyl starches are esterified amylopectin-containing starches that remain in the intravascular space following administration due to its high molecular weight. The differences between hetastarch, pentastarch, and tetrastarch are the average molecular weight of the particles and the degree of substitution of glucose units on the starch particle with a hydroxyethyl group. Hetastarch has the highest average molecular weight (450 kDa), with pentastarch (260 kDa) and tetrastarch (130 kDa) having lower average molecular weights. The molecular weight of the product and the degree of substitution (HES 0.5, pentastarch 0.45, tetrastarch 0.4) determines the exerted COP of the fluid and the degradation time. The higher the substitution with hydroxyethyl, the longer the fluid will persist in the intravascular space. Therefore, HES lasts about 24 hours after administration, while pentastarch and tetrastarch lasts about 12 hours. Serum α-amylase degrades the hydroxyethyl starches, and elimination occurs through the kidneys. Therefore, measured serum amylase levels will increase in patients receiving artificial colloid solutions.

When describing HES solutions, three numbers are used: the concentration of the HES solution, the mean molecular weight (MW), and the degree of substitution of hydroxyethyl groups/glucose unit.

1. The concentration of HES solutions is commonly 6%, which is iso-oncotic.
2. The higher the average MW of the HES solution, the longer the solution lasts, as larger molecules are more slowly degraded.
3. The degree of substitution of hydroxyethyl groups per glucose molecule is reported as a decimal percent. For example, Hetastarch has a degree of substitution between 0.6 – 0.75, meaning that 60-75% of the glucose molecules contain a hydroxyethyl group at either the carbon-2 or carbon-6 position (C2 or C6). The higher the degree of substitution, the longer the colloidal effects last, as the molecules are metabolized more slowly.

The synthetic colloids have been associated with side effects, including affecting coagulation and increasing the potential for volume overload due to the efficacy of expanding intravascular volume.

Concentrated Serum Albumin Solutions

Concentrated albumin solutions, including 25% Human Serum Albumin (HSA) and Canine Serum Albumin (CSA), have been used in critically ill canine and feline patients to help support blood pressure and to aid in the treatment of significant hypoalbuminemia. Studies have been published assessing the utility of HSA in the treatment of critically ill, hypoalbuminemic patients. These studies are descriptive and retrospective; there has not been a published prospective, comparative study on the use of these products in veterinary medicine. Studies assessing the affects of HSA on healthy dogs have also been published, revealing the occurrence of hypersensitivity reactions in some dogs. In addition, a study evaluating dogs that had received HSA revealed the presence of HSA autoantibodies in all dogs, increasing the risk for Type III hypersensitivity reactions. In the studies that have been published describing critically ill dogs and cats receiving HSA, there were few reported hypersensitivity reactions, and in those studies, patients with higher albumins were more likely to survive.

CSA is produced and distributed by Animal Blood Resources International. The product is lyophilized and sold in 5 gram bottles, at a cost of $125 USD/bottle. The published dose, based on company safety studies in healthy beagles, is 800-844 mg/kg over 6 hours. A prospective, blinded, comparative study was performed using CSA in dogs with septic peritonitis. This study showed CSA was safe in this population of dogs, and caused an initial increase in serum albumin that was significantly different than untreated dogs. However, at the time of discharge or death, there was no significant difference in serum albumin levels between the two groups. This study only evaluated19 dogs, so the number of patients was low; in addition, all dogs receiving CSA were only administered one dose following surgical intervention for septic abdomen. It is unknown if multiple dosing, if indicated, would improve albumin levels for a longer period of time. In addition, the study was not powered to assess outcome, so it is still not known if there is a survival benefit or decreased hospitalization when dogs with septic peritonitis are administered CSA.

WHEN TO USE COLLOIDS AND ALBUMIN TRANSFUSIONS

Traumatic Brain Injury

Dogs and cats presenting with traumatic brain injury (TBI) often require fluid resuscitation due to decreased cerebral blood flow, but are more prone to cerebral edema from cerebral vascular injury. Therefore fluids should be administered judiciously, and the patient must be monitored for worsening CNS signs. Hypertonic crystalloids in combination with a colloid is very effective in resuscitating patients with TBI using lower volumes of fluid, and increasing the COP to try and decrease the redistribution of fluid into the cerebral interstitial space. Patients with TBI should be monitored for increased intracranial pressure, evidenced by the presence of the Cushing's Reflex or worsening of mentation. The Cushing's Reflex is defined as systemic hypertension with reflex bradycardia, as the body is attempting to increase cerebral perfusion against the increase in intracranial pressure. In these cases, mannitol is usually indicated to relieve the cerebral edema and improve CNS function.

Pulmonary Disease

Diseases affecting the pulmonary system may present a challenge when fluid therapy is needed. When a patient presents with pulmonary edema, often a state of hypervolemia exists, and fluid therapy is contraindicated. Patients with pneumonia benefit from crystalloid therapy, as hydration of the inflammatory exudate helps with expectoration of the fluid. Pulmonary hemorrhage is often treated with blood products to maintain adequate coagulation and red blood cell mass, with judicious crystalloid or colloid therapy if indicated. Non-cardiogenic pulmonary edema presents a challenging situation in which a continuous rate infusion of diuretics, oxygen therapy, and, potentially, colloids are necessary.

Hypoalbuminemia

As discussed earlier, albumin is the primary contributor to colloid oncotic pressure. When hypoalbuminemia occurs, COP decreases, and fluid loss from the intravascular space into the interstitium may occur. Synthetic colloids are indicated in these patients, as these fluids will increase COP and prevent loss of intravascular fluid. In recent years human albumin has been used in dogs and cats as a colloidal therapy. Transfusion of albumin has the advantage of providing not only colloidal support, but transport of essential endogenous substances and administered protein-bound drugs. Many studies of human albumin have been published in the veterinary literature, and include the presence of allergic reactions, some of which proved fatal. A few studies describing the use of human albumin in critically ill patients concluded human albumin is likely safe in very sick patients, but the risk of devastating Type III hypersensitivity reactions still exists. Recently, Animal Blood Bank International has developed a species-specific concentrated albumin product for dogs, canine serum albumin (CSA). An ongoing study has shown CSA is safe to administer in dogs; the efficacy for improving clinical status remains to be investigated.

SUMMARY

Fluid therapy plays a vital role in the treatment of critical illness in veterinary patients. There are many different types of fluids, and the treating clinician must decide what specific fluid would be most advantageous for the patient depending on disease state, clinical status, and response to therapy. Concentrated albumin solutions have been used, although care must be taken when considering the use of the human albumin product. Canine serum albumin is safe, although its efficacy in comparison to traditional fluid administration has not been studied. Crystalloids and colloids remain the mainstay of therapy for critically ill patients, monitored by constant reassessment of perfusion parameters and clinical status of the patient.

Fluid Type	Dose Range	Frequency	Route	Indications
Crystalloid Fluids	30 ml/kg	bolus, repeat as needed based on perfusion parameters	Intravenous	hypovolemic shock sepsis/SIRS
Synthetic Colloids	5-10 ml/kg bolus 20 ml/kg/day CRI	bolus or CRI	Intravenous	low COP hypotension
Packed RBC	10-20 ml/kg	over 4-6 hrs	Intravenous	low PCV
Frozen Plasma	10-20 ml/kg	over 4-6 hrs	Intravenous	anticoagulant rodenticide toxicity
Fresh Frozen Plasma	10-20 ml/kg	over 4-6 hrs	Intravenous	any coagulopathy
Albumin Products human serum albumin (HSA) canine serum albumin (CSA)	Depends on the concentration of the fluid (see references)	CRI until albumin reaches 2.0 g/dl	Intravenous	low COP low albumin **Severe Type III hypersensitivity reactions have been reported with HSA**

REFERENCES

1. Westphal M, James MFM, Kozek-Langenencker S, Stocker R et al. Hydroxyethyl Starches. *Anesthesiology*, 2009;111:187-202
2. Muir WW and Wiese AJ. Comparison of lactated Ringer's solution and a physiologically balanced 6% hetastarch plasma expander for the treatment of hypotension induced via blood withdrawal in isoflurane-anesthetized dogs. *Am J Vet Res* 2004;65:1189-1194
3. Craft EM and Powell LL. The use of canine-specific albumin in dogs with septic peritonitis. *J Vet Emerg Crit Care* 2012;22(6):631-639
4. Trow AV, Rozanski EA, Delaforcade AM, Chan DL. Evaluation of the use of human albumin in critically ill dogs: 73 cases (2003-2006). *J Am Vet Med Assoc* 2008;233(4):607-612
5. Vigano F, Perissinotto L, Bosco VR. Administration of 5% human serum albumin in critically ill small animal patients with hypoalbuminemia: 418 dogs and 170 cats (1994-2008). *J Vet Emerg Crit Care* 2010;20(2):237-243

CANINE HYPOTENSION: ASSESSMENT AND THERAPEUTIC INTERVENTIONS

Lisa L. Powell, DVM, DACVECC
St. Paul, Minnesota

INTRODUCTION

Hypotension is defined as a systolic areterial pressure (SAP) of less than 90 mmHg, or a mean arterial pressure (MAP) of less than 60 mmHg. Blood pressure (BP) is dependent on cardiac output (heart rate x stroke volume) and systemic vascular resistance (SVR). Therefore, BP = CO x SVR.

Measurement of blood pressure in dogs can be achieved by direct or indirect methods. For direct arterial measurements, an arterial catheter must be placed and maintained. The catheter is most often placed in the dorsal pedal artery, and attached to a pressure transducer and monitor. An arterial catheter can be difficult to place and maintained, especially in smaller dogs and those that are ambulatory. Therefore, indirect BP methods are most often utilized to measure BP in dogs. Two methods used include oscillometric sphygmomanometry and Doppler ultrasound. The oscillometric method utilizes the fluctuations in the arterial wall that occurs during blood flow. A blood pressure cuff is inflated over the artery, then slowly deflated; the difference in the oscillations between the SAP, MAP, and DAP are recorded and displayed. Doppler ultrasound uses a piezoelectric crystal to detect return of blood flow through an artery after being occluded with a BP cuff. A sphygmometer is attached to the cuff, and estimates systolic blood pressure when flow becomes audible. (There is evidence that Doppler ultrasonography may closer estimate MAP in cats vs. SAP).

CAUSES OF HYPOTENSION

Hypotension can be the result of cardiogenic causes, hypovolemia, vasoplegia, or a combination of these. Myocardial dysfunction causing decreased inotropic activity results in decreased cardiac output and hypotension. This most commonly occurs in dogs with dilated cardiomyopathy; however, other cardiogenic causes, such as severe arrhythmias, end-stage valvular disease, and effects of systemic inflammation (cytokines) can affect myocardial contractility and decrease cardiac output.

Hypovolemia caused by circulatory shock, severe hemorrhage, decreased fluid intake, and increased fluid losses can contribute to hypotension. In these cases, adequate fluid resuscitation, with intravenous fluids, colloids, or blood products as indicated, restores circulating volume and normalizes BP.

Vasoplegia occurs when vessels become resistant to the effects of sympathetic compensatory mechanisms, causing persistent vasodilation that does not respond to fluid resuscitation. The most common causes of vasoplegia are anaphylaxis and septic shock. Often patients with vasoplegia require vasopressors therapy to improve BP and restore adequate tissue perfusion.

Relative adrenal insufficiency (RAI) or critical illness-related corticosteroid insufficiency (CIRCI), is a syndrome reported in humans and dogs. RAI/CIRCI results from inadequate corticosteroid release in response to severe illness, causing unresponsive hypotension, which can respond to physiologic doses of hydrocortisone.

TREATMENT OF HYPOTENSION

Treatment of hypotension should be directed toward the underlying disease process resulting in a low BP. Hypovolemia is a common cause of hypotension; therefore, intravenous fluid therapy should be considered once primary heart disease has been excluded as a cause.

Intravenous crystalloid and/or synthetic colloids should initially be given as a bolus in hypotensive patients, with dose being directed towards cardiovascular endpoints such as normalization of heart rate, respiratory rate, and blood pressure. A guideline for a dose of bolus fluids includes 20-30 ml/kg of crystalloids, 5-20 ml/kg of synthetic colloids, and 2-4 ml/kg of hypertonic saline.

A recent retrospective study by Silverstein, et al. evaluated the effectiveness of IV fluid resuscitation in dogs presenting hypotensive to an emergency room. 35 dogs were included. The study concluded dogs that were unresponsive to fluid resuscitation were more likely to be euthanized. Survival to discharge in this group of dogs was only 26%. There was no difference in the type or dose of fluids administered (crystalloids only, colloids only, or a combination of crystalloids and synthetic colloids) in dogs that responded to a fluid bolus to those that did not respond to a fluid bolus.

If hypotension is the result of whole blood loss, resuscitation with blood products, including packed red blood cells, fresh frozen plasma, or whole blood, should be considered.

If intravascular volume is adequate as evidenced by normal central venous pressure (between 5-10 cmH20) and urine output (2-4 ml/kg/hr) but blood pressure remains low, administration of positive inotropes or vasopressors should be considered. *Dobutamine* (2.5 – 10 micrograms/kg/min, use with caution in cats), a β-1 agonist, increases cardiac output by enhancing myocardial contraction. In dogs, *dopamine* enhances renal and mesenteric perfusion at a dose of 2.5 micrograms/kg/min. At higher doses (5 – 10 micrograms/kg/min), dopamine exerts its effects on α-1 and β-receptors, resulting in vasoconstriction and increased myocardial contractility. At doses of 10-20 micrograms/kg/min, dopamine can cause constriction of the renal arteries, thereby reducing kidney perfusion and causing potential irreversible renal damage. *Norepinephrine* is a potent α-1 mediated vasoconstrictor and is usually reserved for severe, unresponsive hypotension at doses of 0.05 – 1 micrograms/kg/min.

Arginine vasopressin (AVP), or anti-diuretic hormone, is produced in the pituitary gland, and released in response to hypovolemia. AVP has been found in both humans and veterinary patients to be decreased in cases of septic shock (vasoplegia), and may be the vasopressors of choice when treating dogs with septic shock-related hypotension. One veterinary study evaluated the use of vasopressin to treat sepsis-related hypotension in 5 dogs with dopamine-resistant hypotension. MAP significantly increased in all 5 dogs, with a mean dose administered of 2.1 micro-Units/kg/min.

CONCLUSION

Hypotension in dogs should be aggressively treated, as a prolonged decrease in tissue perfusion can result in increased morbidity and mortality due to organ dysfunction and failure. Fluid resuscitation should be accomplished, and if the patient remains hypotensive, vasopressors and/or positive inotropes should be considered. One retrospective study showed an increased mortality in dogs presenting with hypotension that were unresponsive to fluid therapy. Once hypotension is identified, early and aggressive therapy directed towards reversing hypovolemia and treating the underlying disease is key to improving clinical status, and likely, survival.

REFERENCES

1. "Hypotension" and "Vasopressin". In: Silverstein DC and Hopper K. Small Animal Critical Care Medicine. Saunders/Elsevier, 2009.
2. Silverstein DC, Waddell LS, Drobatz KJ, King LG. Vasopressin therapy in dogs with dopamine-resistant hypotension and vasodilatory shock. *J Vet Emerg Crit Care* 2007;17(4):399-408
3. Burkitt JM, Haskins SC, Nelson RW, Kass PH. Relative adrenal insufficiency in dogs with sepsis. *J Vet Intern Med* 2007;21(2):226-31
4. Silverstein DC, Kleiner J, Drobatz KJ. Effectiveness of intravenous fluid resuscitation in the emergency room for treatment of hypotension in dogs: 35 cases. *J Vet Emerg Crit Care* 2012;22(6):666-673

ASSESSING CEREBRAL AND PERIPHERAL TISSUE OXYGENATION WITH NEAR INFRARED SPECTROSCOPY

Søren Boysen, DVM, DACVECC
Calgary, Alberta, Canada

INTRODUCTION

Assessing perfusion in critically ill patients is difficult and often relies on indirect measures such as mental status, heart rate, pulse pressure, and blood pressure. Although clinically helpful, these indirect assessments of tissue perfusion are not specific and do not always correlate well with actual tissue perfusion. In addition these parameters are often a reflection of systemic effects and may overlook regional abnormalities in perfusion. An ideal device to assess perfusion would be non-invasive, user friendly, clinically applicable, and would provide objective information that measures oxygenation at the tissue level. Tissue oxygen saturation (StO_2) is a relatively new technique that reflects tissue perfusion in a real time continuous and non-invasive fashion. This lecture will briefly introduce the technology behind StO_2, discuss some of the applications of this technology in human and veterinary medicine, as well as highlight some of the obstacles and pitfalls associated with implementing this technique into the clinical setting.

NEAR-INFRARED SPECTROSCOPY (NIRS)

Near-infrared spectroscopy (NIRS) is a noninvasive continuous point of care diagnostic test that has received considerable attention in the human profession in the past decade. Two of its main clinical applications in the intensive care unit (ICU) setting, studied extensively in human and animal models, include evaluation of regional cerebral oxygen saturation (rSO_2 or cSO_2) and peripheral somatic (skeletal) oxygen saturation (StO_2).

The Technology

Near-infrared spectroscopy works similar to pulse oximetry, except instead of measuring only arterial hemoglobin saturation (pulsatile blood flow), NIRS measures total tissue hemoglobin saturation without regard to systole or diastole.[1] Because NIRS measures regional flow at the tissue level over which it is placed, and only 20% of the blood volume at the tissue level is intra-arterial, NIRS measurements predominantly reflect venous tissue oxygen saturation levels.[1] The exact technology varies between different devices (there are at least five different monitors evaluated to date), however, all are based on the Beer Lambert law: light transmission through a solution with a dissolved solute decreases exponentially as the concentration of the solute increases. It is important to remember that tissues/substances absorb and reflect differing wavelengths of light and that in mammalian tissue, only cytochrome aa_3, myoglobin and hemoglobin change their spectra when oxygenated vs. deoxygenated.[1] Therefore, by varying the wavelength of light passed through a tissue, the concentration of a certain substance (in this case hemoglobin) within a tissue can be determined by the amount of a specific wavelength of light that is absorbed or reflected. The absorbance spectrum of hemoglobin varies depending on its redox state, with oxygenated hemoglobin absorbing more near-infrared light in the range of 800-1000 nm and deoxygenated hemoglobin absorbing more light in the 600-800 nm range. The actual wavelength used varies between the various NIRS monitors. (Note that pulse oximeters work on wavelengths of 940 nm and 660 nm to detect oxygenated and deoxygenated hemoglobin respectively). These wavelengths pass through and are not absorbed by bone, skin, muscle and soft tissue.[1] Clinicians infer metabolic changes in oxygen supply and demand at the site of the StO_2 measurement based on the StO2 measurement obtained.

Inspectra and INVOS Monitors

The Inspectra™ StO_2 Tissue Oxygenation Monitor (Hutchinson Technology, Inc., Hutchinson, Minn), one of the more commonly used skeletal tissue oxygen saturation monitors, uses four specific calibrated wavelengths of near infrared light to noninvasively illuminate an area of tissue below a sensor placed on the skin. The depth of tissue penetration is typically 15 mm (although 25 mm probes are available). As such, the probe needs to be placed in an area where skeletal muscle is likely to lie within 15 mm of the skin surface to provide a reliable somatic reading. The Inspectra, through the use of complex algorithms, produces a ratio of oxygenated hemoglobin to total hemoglobin, expressed as percent StO_2. Furthermore, the NIRS signal is limited to vessels that have a diameter less than 1 mm (arterioles, capillaries, and venules), and therefore tends to reflect the microcirculation.[2]

The INVOS Cerebral/Somatic Oximeter (Somanetics, Troy, Mich) uses wavelengths of 730 nm (deoxyhemoglobin), and 810 nm (oxygenated Hb) and can be used noninvasively to measure both cerebral oxygen saturation and somatic oxygen saturation. It has four probes that can be placed in various locations, which subtract out surface data from the skin and skull to produce tissue oxygen, or regional oxygen saturation values. It has a field of depth to approximately 2 cm.

Somatic tissue oxygen saturation (StO_2)

Microcirculatory perfusion and tissue oxygen utilization are affected by various forms of shock, hypoperfusion and sepsis.[1,3] NIRS, through its measurement of oxygenation status of tissue hemoglobin, is affected by changes in oxygen delivery, local metabolic rates and venous outflow from tissues. As such, three general applications of NIRS have been investigated clinically: its use in hypovolemic shock, its use in septic shock, and its use in compartment syndrome of the muscles.

Decreases in StO_2 have shown strong correlation with hypoperfusion, oxygen delivery (DO_2), base deficit, lactate, prognosis, and transfusion requirements and have been used as end points to help guide resuscitative efforts, particularly in cases of *hypovolemic* shock.[2] Experimental studies have also shown that StO_2 is able to differentiate between resuscitatable and nonresuscitatable patients with *hemorrhagic* shock.[3] Predictors of multiple organ failure (MOF), blood transfusion requirements and death are desirable measures in patients presenting with shock. Base deficit and lactate have probably shown the most promise at predicting MOF, mortality and blood transfusion requirements in both human and veterinary *trauma* studies.[3,4] Recent studies have shown StO_2 to be equally effective as base deficit and lactate at predicting MOF in *trauma* patients presenting in shock to human emergency departments and more sensitive than base deficit and lactate at predicting death.[3]Sepsis, however, is more complex and the discriminatory value of StO_2 is less clear in these patients. Despite some studies demonstrating that StO_2 is lower in non-survivors than survivors after early goal-directed resuscitation of septic shock, other studies have failed to support these findings.[5] The discrepancy in results may be explained by the anatomic site of StO_2 readings. Additional studies have shown that regional-specific changes in StO_2 are often missed with central venous oxygen saturation (SvO_2), pulse oximetry (SaO_2) and blood pressure monitoring alone.[5] The use of a vascular occlusion test (VOT) allows a dynamic assessment of regional perfusion via assessment of four phases: baseline, ischemia, reperfusion and hyperemia. VOT involves continuous monitoring of dynamic changes in StO_2 in response to the application of an ischemic event on the limb proximal to the NIRS probe (i.e. application of a blood pressure cuff inflated to 30 mmHg greater than systolic blood pressure for 3-5 minutes or until StO_2 falls to 40%-50%, followed by rapid release of the occlusive pressure). During the four phases, results can be graphed which allow measurement of various parameters including the rate of deoxygenation (DeO_2 – a decrease in StO_2 during the time occlusion is applied to the limb which generally reflects local muscle metabolism and mitochondrial oxygen consumption) and rate of reoxygenation (ReO_2 – measured following release of the occlusive pressure which equates to the wash out of stagnant blood and generally reflects the reactivity (perfusion and recruitment) of the microcirculation).[6] There is evidence to suggest the VOT has greater diagnostic utility in some patients, particularly those with microcirculatory derangements associated with sepsis.[6] Several studies have demonstrated that patients with profound alteration in microvascular reactivity(determined via VOT with NIRS) and severe sepsis have a high risk of organ dysfunction and death. The VOT, however, is more a test of microcirculatory reserve than of microcirculatory perfusion.

NIRS has also been studied in relation to trauma and non-trauma induced compartment syndrome. StO_2 values in patients with compartment syndrome are lower than control groups and normalize after fasciotomy.[1,5]

Cerebral tissue oxygen saturation (cSO2 or rSO2)

Cerebral oxygen delivery is vital for normal brain function. Any disruption of oxygen delivery to the brain, particularly in the face of cerebral ischemia or injury, can have a significant impact on brain function and prognosis. With traumatic brain injury, concurrent systemic injuries are not uncommon and may result in altered systemic oxygenation (lung injury), hemoglobin content (hemorrhage), and cerebral blood flow (hypotension), or a combination of these three factors, which may significantly affect cerebral oxygenation and subsequently outcome. Monitoring cerebral oxygenation, which may be affected by any of the three factors listed above, may provide early indications of cerebral oxygen saturation derangements. The application of NIRS to continuously follow cerebral oxygenation may allow oxygen delivery and utilization to be monitored in a non-invasive and continuous fashion.

NIRS has been used successfully to confirm and localize both acute and delayed intracranial hematomas in head injury. Studies show it has 90% sensitivity at prospectively identifying hematomas, using a hand held device, when detecting changes in light wavelength absorption at 760 nm.[5]

Experimental models in cats have validated the use of NIRS to assess cerebral perfusion in the face of increased intracranial pressures, while other studies have used NIRS as an adjunct to monitor regional hypoxia following traumatic brain injury when intracranial pressures are elevated.[2,5] In experimental animal models and limited human trials, increases in intracranial pressures have been shown to correlate well with decreases in NIRS values.[2,4] There are also reports of NIRS being used to detect a loss of cerebral blood flow autoregulation due to hypotension, and detecting rapid unexplained hyperoxegenation prior to seizures.[5] Although further study is needed, the non-invasive application of cerebral tissue oxygen monitoring holds promise, especially where other modalities to assess brain injury such as MRI and CT are not available. Published guidelines in humans include mean cerebral cSO_2 values of 70% in healthy children and young adults and that cSO_2 less than 40% to 50% or a change in baseline of more than 20%, is associated with hypoxic ischemic neural injury.[5]

Limitations with NIRS

There are several limitations to the use of NIRS that have been reported in the human literature. Limitations include inaccurate readings due to placement of the probe over hematoma (note this relates to StO_2 as cSO_2 can be used to detect intracranial hematomas), fat, or bony prominences, the presence of indocyanine green and methylene blue dyes in the blood, bilirubin, melanin methemoglobin, carboxyhemoglobin and other dyshemoglobins in the blood, and placement of the probe over skin obscured by ink, pen, tattoos etc.[1-3,5,6] In addition, excessive movement has been shown to cause large rapid changes in StO_2 values. It should be pointed out that NIRS does not measure microcirculatory blood flow directly, which makes interpretation of StO_2 values challenging at times; it is difficult to distinguish altered oxygen delivery from altered tissue oxygen consumption such that a change in oxygen delivery may offset a change in tissue metabolism, or visa versa, resulting in the same StO_2 value. It should also be pointed out that standardization of this technology is lacking in human and animal model studies, showing different baseline reference values depending on the technologies used (different wavelengths of light in the different devices may not give the same reference ranges) and there is considerable variation in values obtained at different sites of the body.[6]

Application in Veterinary Medicine

To date there are limited studies investigating the application of NIRS in veterinary patients. There are two published studies that have looked at establishing normal somatic StO_2 values in dogs and its application in post ovariohysterectomy incisional wound healing.[7,8] A recent abstract has also been published, which shows that StO_2 correlates well with DO_2 and other cardiovascular variables in a hemorrhagic shock model in dogs.[9] There are no published StO_2 studies in cats. There are also very few experimental studies investigating cSO_2 in dogs and cats and no clinical studies to date. Anatomic differences between humans and veterinary species may limit the application of cSO_2 in the latter species.

In summary, NIRS measures saturation of hemoglobin in skeletal muscle and cerebral tissue and serves as an index of regional perfusion to these tissues.

REFERENCES

1. Cohn S, Crookes B, Proctor K. Near-infrared spectroscopy in resuscitation. J Trauma2003;54:S199-S202.
2. Lipcsey M, Woinarski NCZZ, Bellomo R. Near intrared spectroscopy (NIRS) of the thenar eminence in anesthesia and intensive care. Annal of Intensive Care 2012;2:11
3. Santora RJ, Morre FA. Monitoring trauma and intensive care unit resuscitation with tissue hemoglobin oxygen saturation. Crit Care 2009;13(Suppl 5):S10
4. Stillion JR, Fletcher DJ. Admission base excess as a predictor of transfusion requirement and mortality in dogs with blunt trauma: 52 cases (2007-2009). JVECC 2012;5:598-594.
5. Drayna PC, Abrama TJ, Estrada C. Near-Intrared spectroscopy in the critical setting. Ped Emer Care 2011;27:432-442.
6. Gomez H, Mesquida J, Simon P, et al. Characterization of tissue oxygen saturation and the vascular occlusion test: influence of measurement sites, probe sizs and deflation thresholds. 2009;13(Suppl 5):S3
7. Sullivan L, Campbell V, Radecki S et al. Comparison of tissue oxygen saturation in ovariohysterectomized dogs recovering on room air versus nasal oxygen insufflation. JVECC 2011; 21(6):633-38.
8. Hall K, Powell L, Beilman G, et al. Measurement of tissue oxygen saturation levels using portable near-infrared spectroscopy in clinically healthy dogs. JVECCS 2008;18(6): 594-600
9. Pavlisko N, Killos M, Ricco C, Guerrero-Henoa N. Evaluation of tissue hemoglobign saturation (stO2) via Near-Infrared Spectroscopy in a canine model of acute hemorrhagic shock (IVECCS abstract 2012)

ASSESSING TISSUE PERFUSION IN SHOCK AND SEPSIS WITH DARK FIELD SIDE-STREAM MICROSCOPY

Søren Boysen, DVM, DACVECC
Calgary, Alberta, Canada

INTRODUCTION

The microcirculation, comprised of arterioles, capillaries and venules ≤ 100 um in diameter, is an important but often overlooked entity in critical illness. Macrovascular parameters (heart rate, pulse quality, blood pressure, lactate etc.) are often used to assess tissue perfusion, however it is well established that macrovascular parameters do not always equate with local microvascular changes. With states of shock, systemic inflammatory response syndrome, and sepsis, despite adequate blood flow through the macrovascular system, blood flow through the capillaries can become heterogeneous. Due to dysregulation of arterioles governing flow at the arteriole end of capillaries, microthrombosis, increased leukocyte adhesion, and increased permeability some capillaries can have good blood flow while neighboring capillaries may have poor to no blood flow. Therefore, methodologies such as orthogonal polarization spectral (OPS) imaging and its successor, side-stream dark field microscopy (SDF), have found a niche in research settings to assess microcirculation in patients with sepsis and shock, and evaluate response to various therapeutic interventions. With further advancements in technology and ongoing research, it seems very plausible that microcirculatory assessment as a point of care test in clinical settings is not far off.

The Technology

SDF microscopy is based on the principal that some tissues reflect light while others absorb it. If light is passed through tissue such that it is reflected back by deeper tissues, the more superficial tissue is illuminated. If cells in the superficial tissue absorb light they will appear dark and will not be illuminated. SDF uses a select wavelength of 530 nm (green light), which is absorbed by hemoglobin within red blood cells (both oxygenated and deoxygenated forms) causing them to appear grey to black on the video images captured.[1] It is important to note that vessels that do not contain red blood cells cannot be visualized. It is the passage of red cells through the vessel that allows the location of vessels to be determined. If red blood cells are located outside vessels they will obscure interpretation of microvessels within the image.

By imaging the capillaries within tissues, SDF microscopy can be used to evaluate several dynamic factors of the microcirculation. These factors include perfusion parameters (microcirculatory flow index; MFI), density of vessels (total vessel density; TVD), and density of perfused vessels (perfused vessel density; PVD). The ratio of PVD/TVD is used to express the proportion of perfused vessels (PPV). When vessels < 20 *u*m are the only vessels included in the analysis, the PVD represents the functional capillary density (FCD), which is considered the major determinant of microcirculatory blood supply.[2]

To obtain consistent results it is recommended that 3-5 microcirculatory sites be imaged with a minimal recording time of 20 seconds per site.[2]

APPLICATIONS OF SDF

Evaluation of microvascular blood flow with SDF is associated with prognosis in several disease states, including several forms of shock (hemorrhagic, septic, cardiogenic), CPR, and sepsis.[3] SDF has been used to evaluate the microvascular response to several therapeutic interventions, including fluid therapy, red blood cell transfusion, vasodilators and vasopressors.[3] However, its most widely studied application is in regards to sepsis. The pathophysiology of sepsis at the microvascular level is very complex and heterogeneous, so the use of SDF to assess tissue perfusion in septic patients may help guide prognosis and therapy.

Correlation of SDF Findings and Prognosis

Microcirculatory failure has been shown to have prognostic value in human patients with sepsis, and derangements in the microcirculation are associated with increased morbidity and mortality.[3] Septic patients that have greater heterogeneity at the microcirculatory level and patients that have persistent derangements of the microcirculation following resuscitation have been shown to have a higher risk for organ failure and death.[3] There is also evidence that protocol-directed resuscitation at the microcirculatory level is associated with improved organ severity scores. In addition to having prognostic implications in septic patients, SDF has been shown to have prognostic implications in patients with acute severe heart failure and cardiogenic shock. There is also evidence to suggest that impaired microvascular flow is associated with the development of post-operative complications in patients undergoing major abdominal surgery.[3]

Applications in monitoring response to therapy

Fluid resuscitation is the cornerstone of therapy in many forms of shock. The microcirculation has been investigated following fluid challenges in patients with sepsis. Studies have shown that macrovascular responses to fluid therapy in shock patients (heart rate, cardiac output, blood pressure etc) are not always correlated with microvascular changes and that fluid loading is an effective first step in the resuscitation of the microcirculation.[3,4]

SDF microscopy has been used to investigate the impact of red blood cell (RBC) transfusion at the microcirculatory level in several subgroups of patients. Studies in cardiac surgery patients have demonstrated that although capillary flow does not increase following transfusion of RBC's, the total number of "filled" capillaries does.[3] In contrast, similar studies in patients with sepsis have failed to show an increase in microcirculatory density following RBC transfusions. It has been hypothesized that patients with sepsis may have hemorheological alterations and damage to the microcirculation that could diminish the efficacy of RBC transfusions at the microcirculatory level.[3]

Vasoactive agents are often used in patients with persistent shock following fluid resuscitation to improve blood pressure and restore normal hemodynamics. SDF has been used to assess the impact of vasopressors at the microcirculatory level following fluid resuscitation. Vasoporessors consistently increase blood pressure but can have variable effects on the microcirculation. Studies have shown that in patients with decreased microvascular function and hypotension, norepinephrine appears to improve microcirculation. However, in patients with hypotension and normal microcirculatory flow, norepinephrine tends to decrease microcirculatory flow

The microvascular effects of vasodilatory agents, such as acetylcholine (topically) and nitroglycerine (intravenously), have been investigated. Initial studies demonstrated improved microvascular perfusion with topical acetylcholine in septic patients. Intravenous nitroglycerine was then investigated in septic patients.[3] Results are conflicting, with some studies showing septic patients have improved microcirculatory perfusion following intravenous nitroglycerine with others showing no benefit.[2] In patients with cardiogenic shock, the administration of intravenous nitroglycerine showed dose dependent improvement in microvascular perfusion.[3] Finally, the administration of activated protein C (APC) to patients with severe sepsis has been investigated using SDF. APC has been shown to improve the proportion of perfused microvessels in severely septic patients when given as a continuous rate infusion. Microvascular perfusion transiently decreased when APC was discontinued.[3]

Current international multicenter trials are underway looking at the prevalence of microcirculatory alterations in all intensive care patients (microSOAP trial). Following results of these trials, it has been proposed that the microcirculation be targeted as an end point of resuscitation and be incorporated into early goal directed therapies in septic patients.

LIMITATIONS

Unfortunately, despite its promise to illuminate some of the underlying abnormalities in microvascular flow, SDF microscopy has several limitations from a clinical standpoint that still need to be addressed. Once video has been obtained, the microcirculatory flow index (MFI), perfused vessel density (PVD), and proportion of perfused vessels (PPV) of various sizes can be determined.[2] This is the rate-limiting step in SDF microscopy, which makes its utility as a point of care test cumbersome and somewhat impractical at present. Once images are obtained, the necessary steps (outline of vessels for analysis) to be able to measure the variable of PVD and PPV requires an hour or more of manual labor.

Due to the amount of time needed to perform full evaluation of the obtained video, assessment of microvascular perfusion at the bedside using more subjective analysis of microcirculatory flow (calculation of only MFI) within a few minutes has been investigated in people with some success.[3] However, this procedure omits a lot of vital information that can be obtained with SDF microscopy. Until newer software to decrease the time to analyze images can be developed or further studies performed that can validate the use of MFI (or similar point of care applications) to help determine prognosis and guide therapeutic interventions, SDF microscopy is likely to remain impractical as a point of care diagnostic test due to the cumbersome analysis currently required to evaluate the images.

Additional concerns with SDF microscopy include pressure artifact (applying too much pressure with the probe will collapse vessels below the scanner and result in the impression of decreased blood flow), movement artifact (which makes interpretation of the images impossible), excessive saliva (which impedes visualization of the microvasculature), and improper contrast and focus settings which can render obtained video images unusable.[2] To address the concerns of pressure artifact there has been an attempt to use image stabilizers with the SDF microscopy, which essentially apply a constant negative pressure to the tissue to remove erroneous pressure artifacts. The image stabilizers have allowed 42 second video clips to be obtained, compared to 8 second clips when free hand techniques without the image stabilizer are used.[3] The use of image stabilizers in locations other than the lingual frenula has not been assessed. Alternatively, it may be possible to identify pressure artifact by assessing flow through larger vessels (> 30 um) in the final video image; significant reductions in flow of larger vessels is likely reflective of pressure artifact.[2] To minimize salivary interference, it has been suggested that the tongue be wiped with gauze prior to recording the image.[2]

To overcome the time commitment required to evaluate the image clips obtained with SDF microscopy, automated image analysis programs (Automated Vascular Analysis) were developed. These early programs are still cumbersome and newer automated software, incorporated into the Autmated Vascular Analysis software, are reportedly able to allow clip analysis in under 30 seconds (as opposed to 1 hour).[3] However, the image quality obtained with the current second generation SDF devices is a limitation to the new software. A third generation SDF with higher spatial (14 megapixels versus 1.3 megapixels) and temporal (60 vs 25 frames per second) resolution is currently under consideration.[3] Further study is required but the newer software and third generation devices seem very promising and may make SDF a clinical point of care reality in assessing and therapeutically targeting the microcirculation.

Veterinary Studies

There are very few veterinary studies investigating SDF in canines or felines. A seminal article by Silverstein et al validated the technique used to assess the microcirculation in healthy dogs undergoing anesthesia. The study showed that SDF could be successfully applied at the mucogingival junction over the canine tooth.[5] Another study investigated microcirculatory changes in dogs subject to hemorrhagic shock. This study demonstrated SDF could be used for real-time imaging of the microcirculation in dogs with hemorrhagic shock, and showed there was good correlation between SDF parameters and several macrovascular variables during hemorrhage and resuscitation.[6] Two recent abstracts have evaluated a bedside scoring system to reduce the amount of time required to obtain and analyze acquired images in dogs, , although further study is required to determine the value of these systems.[7,8] There is also an abstract looking at the microcirculation of healthy cats undergoing ovariohysterectomy, which demonstrated the technology could be applied sublingually in cats.[9]

Conclusions

The use of point of care devices to non-invasively and continuously assess microvascular perfusion has become a reality. Although a lot of research still needs to be completed in order to transition this technology into the veterinary profession, initial studies are underway. With continued research, the application of such technology has the potential to greatly impact diagnostic, prognostic, and therapeutic options in many disease states faced in veterinary ICU's. In addition, the application of such devices may prove complimentary in many disease states, particularly sepsis, as these devices evaluate different aspects of the microvasulature.

REFERENCES

1. Goedhart PT, Khalilzada M, Bezemer R et al. Sidestream dark field imaging: a novel stroboscopic LED ring-based imaging modality for clinical assessment of the microcirculation. Optical Society of America 2007; 15(23):15101-15114
2. De Backer D, Hollenberg S, Boerma C et al. How to evaluate the microcirculation: report of a round table conference. Critical Care 2007;11(5):
3. Bezemar r, Bartels S, Baker J, Ince C. Clinical review: Clinical imaging of the sublingual microcirculation in the critically ill – where do we stand? Critical Care 2012;16:224.
4. Pranskunas A, Kiipmans M, Koetsier et al. Microcirculatory blood flow as a tool to select IC patients eligible for fluid therapy. Intesive Care Med. 2012;
5. Silverstein D, Pruett-Saratan A, Drobatz KJ. Measurements of microvascular perfusion in healthy anasthetized dos using orthogonal polarization spectral imaging. JVECCS 2009; 18(6):579-587
6. Peruski AM, Cooper ES. Assessment of microcirculatory changes by use of sidestream dark field microscopy during hemorrhagic shock in dogs. AJVR 2011; 72(4):438-445.
7. Allerton et al. Semiquantitative bedside evaluation of the microcirculation via sidestream dark field imaging in dogs 22nd ECIM-CA congress, 2012
8. Gommeren et la. Qualitative bedside evaluation of the microcirculation via sidestream dark field imaging in dogs. 22nd ECIM-CA congress, 2012
9. Goodnight M, Cooper E, Butler A. Assessment of microcirculatory perfusion in healthy anesthetized cats and changes associated with ovariohysterectomy using sidestream darkfield microscopy. IVECCS 2011 abstract

HOW TO USE PLEURAL, LUNG AND PULMONARY VASCULAR PRESSURES CLINICALLY

Elizabeth Rozanski, DVM, DACVIM (SAIM), DACVECC
North Grafton, MA

Pressure determinations are potentially useful in the ICU and ER, and may be considered as point-of-care evaluation of physiologic responses to critical illness or injury. Most critical care units have easy access to some sort of electronic pressure transducer for measurement of direct arterial pressure, or at least a water manometer. Measurements of any body cavity pressure may be obtained via manipulations of these devices. The utility of the measurement of any pressure specific gradient varies, but in most cases significant amount of knowledge can be gained by monitoring pressures. Additionally, for hospitals involved in teaching students or house officers, a more physiological understanding may help further cement knowledge of a specific fact. For example, visually observing a large volume pleural effusion drain spontaneously from the chest cavity will reinforce the idea that pleural pressure is positive to atmosphere when the effusion is large. Alternatively, measurement of airway pressures during ventilation is mandatory, and subsequent calculation of lung compliance may help determine if the lung function is improving separate from simple measurements of PaO2:FiO2 ratio. In fact, decreasing compliance over days 1-6 has been independently associated with death in people with ARDS. (Seeley et al). The oxygenation index, defined as = (FiO2 X Mean Airway Pressure)/PaO2 may also be used in people but has not been well described in dogs. Pulmonary pressures are more commonly estimated by use of Doppler echocardiography, but vascular pressures in the jugular vein (central venous pressure) or by placement of a pulmonary arterial catheter is also useful to provide further information as to the cardiovascular status of the patient.

PLEURAL PRESSURE

Pleural pressure may be monitoring during thoracocentesis for pleural effusion or pneumothorax or continuously if a thoracostomy tube is in place. As a brief review, pleural pressure in a healthy dog or cat is negative in relation to the atmosphere. This negative pressure gradient helps promote/maintain lung inflation, and facilitate a lower work of breathing. A small amount (< 5 ml) of effusion is present in the pleural space. However, the pleural space is dynamic with extravasated fluid being rapidly reabsorbed via lymphatic and capillaries. Pleural space disease may be acute or chronic, with most effusions being somewhat chronic. As pleural effusion forms, there is collapse of the lung parenchyma, and also an *increase* in intra-thoracic pressure. As respiratory distress develops, the intra-pleural pressure less negative or positive to the atmosphere. While typically as clinicians we assume that removal of isolated pleural effusion will result in immediate improvement in lung function, this is not always the case. In people, two separate categories of non-recruitable lung associated with pleural effusion are recognized, the first termed lung entrapment, which is a condition, which develops in associated with active pleural inflammation or neoplasia. Immature fibrin and overlying inflammation prevent re-expansion and contribute to the failure to recruit lung after thoracocentesis. During longer standing effusions, there is thickening and constriction of the visceral pleural and thickening of the parietal pleura. This may lead to the development of "trapped lung" or lung that is tightly constricted by overlying visceral pleura, and can not adequately re-expand, even in the presence of negative intra-thoracic pressure. Entrapped lung is likely more common in veterinary patients, as most effusions are associated with more active inflammation. However, the non-recruitable lung has been much less evaluated in dogs and cats than people.

Pleural manometry was designed to monitor the pleural cavity for signs of non-recruitable lung, as evidenced by large decreases in intra-pleural pressures which may be subsequently associated with the development of pain or spontaneous pneumothorax. Spontaneous pneumothorax is of particular interest as in veterinary medicine; iatrogenic pneumothorax associated with thoracocentesis is most commonly though to be associated with inadvertent laceration of the lung. The concept of non-recruitable lung is compelling, as larger increases (eg. more negative) pleural pressures might be associated with spontaneous pneumothorax as well. Pleural manometry may easily be performed with an over-damped electronic manometer. A pilot evaluation of pleural manometry in our hospital, performed by Kendra LaFaunci, a veterinary student, showed findings as below. Interesting, in the CAT 2B, a spontaneous pneumothorax developed after thoracocentesis, potentially associated with trapped lung, and negative intra-thoracic pressure. Additionally, the use of pleural manometry is also able to answer the question as if the needle or catheter is still in the thoracic cavity following either re-positioning, patient movement, or apparent completion of the procedure.

PULMONARY (LUNG) PRESSURES

During spontaneous breathing, inspiration occurs due to a negative inspiratory pressure with an open glottis. Lung compliance determines how easy or how hard it is for the lungs to inflate, with lower compliance being associated with higher pressures for the same tidal volume. Compliance can be calculated by looking at the slope of the PV curve, or by measuring changes in volume /changes in pressure. A PV curve looks as below.

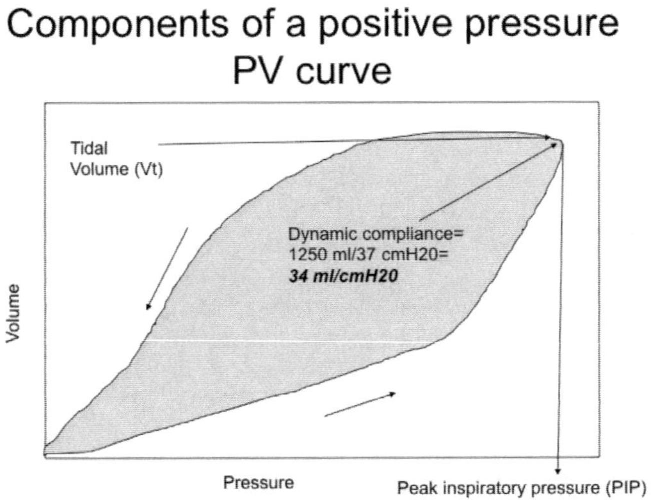

Thus when the lung start to become stiffer, the curve becomes flatter. On newer critical care ventilators, algorithms built in the software can calculate compliance and resistance, and help to better define improvement or decline of the clinical condition. Without a ventilator, King and colleagues described a simple technique for determining static lung compliance with an anesthesia machine with a pressure gauge, and a hand held spirometer. Decreasing compliance is associated with progressive lung disease or atelectasis. A sigh, or deep breathing can help to improve compliance but recruiting de-recruited airways.

Other pressures that may be evaluated during positive pressure ventilation include peak and plateau pressures. Peak airway pressure is the highest pressure that the lungs are exposed to during inspiration, and larger reflects airway flow rate and is a marker of airway resistance (larger airways) Plateau pressure, is more reflective of pressure to the smaller airways and alveoli, and requires a moment of no flow.

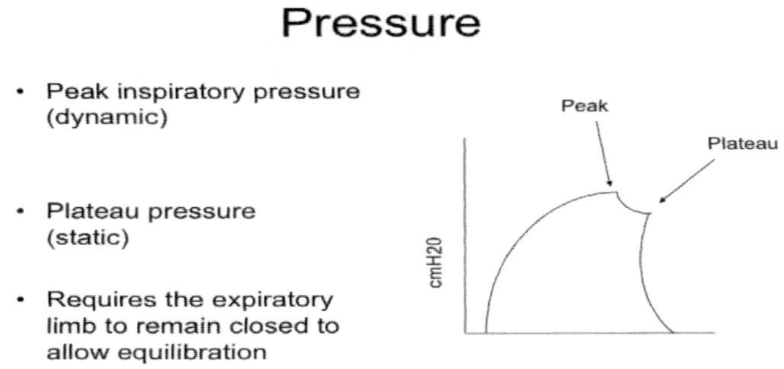

To better discuss airway pressure, it is important to recognize that mechanical ventilators work by delivering a breath to either a set volume (aka volume limited ventilation) or to a set pressure (aka pressure limited ventilation). The speed that the set volume or set pressure is reached is dependent on the inspiratory flow rate and the inspiratory time. For example, if your set tidal volume is 500 ml, and your inspiratory flow rate is 1000 ml/sec, it will take 0.5 seconds to reach 500 ml. However, if you set your patient's inspiratory flow at 5000 ml/sec, it will only take 0.1 second to reach 500 ml. In this case, the peak inspiratory pressure will be MUCH higher in the patient the higher inspiratory flow rate, but plateau pressures will be similar. Pressure changes will also be seeing if the cuff is leaking, tube is kinked or if there is a pneumothorax.

PULMONARY ARTERIAL PRESSURES

Pulmonary arterial pressure may be measured to document pulmonary hypertension, or to identify elevated capillary wedge pressure (consistent with volume overload-congestive heart failure). However, due to increased availability of echocardiography and evidence in people of harm associated with routine placement of pulmonary arterial catheters, these measurement are the least commonly pursued in the ER and ICU patient. Pulmonary hypertension is recognized with increasing frequency in dogs in recent years. As a review, the pulmonary and systemic circulation are in series, with the cardiac output of the left and right heart (in the absence of cardiac shunt) being equal. The pressure in the pulmonary vasculature however, is low in relation to the systemic circulation, with the mean pulmonary arterial pressure averaging less than 15 mmHg and a systolic pressure of < 25 mmHg at rest and less the 30 mmHg during exercise. Elevated pulmonary pressures result in syncope, weakness and fatigue, and may ultimately cause right heart failure.

NORMAL CARDIAC AND VASCULAR PRESSURES

Strain on the right heart is assessed by measurement of the pulmonary vascular resistance (PVR), which is defined as the [mean pulmonary arterial pressure-mean left atrial pressure (or pulmonary capillary wedge pressure)]/cardiac output. (dyn·s/cm5) Increasing PVR is associated with increased work for the right ventricle. If Right heart pressure exceed left heart pressure in the presence of an intra-cardiac shunt, there will be right to left shunting (Eg. Eisenmenger physiology). Increased LA pressure is hard to directly measure, but by inflating a balloon on the Swan –Ganz catheter in branch of a pulmonary artery, the LA pressure may be reasonably estimated.

WHO CLASSIFICATION OF PULMONARY HYPERTENSION

1) Primary pulmonary vascular disease 2) pulmonary venous hypertension 3) COPD/Interstitial lung disease 4) Thrombotic disease 5) Miscellaneous. Pulmonary hypertension is a global term for anything that results in increases in the pulmonary arterial pressures. Specific diseases that may result in include primary pulmonary hypertension (eg. Class I in the WHO classification scheme), which is pulmonary hypertension without an underlying cause identified. It is thought to be most common in small breed elderly female dogs. Clinical signs include shortness of breath, exercise intolerance, syncope and ascites. Cor Pulmonale is defined as the developed of right sided heart failure due to pulmonary hypertension due to lung disease. Common lung diseases that are associated with cor pulmonale include chronic bronchitis/'COPD" and pulmonary fibrosis. Eisenmeger syndrome (or physiology) is a specific form of pulmonary vascular disease associated with an intracardiac shunt, often a VSD or large ASD. In this case, the initial shunting is left to right, but over time, the volume overload to the pulmonary vasculature hypertrophies, and as the pressure increases above that in the LV, right to left shunting ensures. This is a relatively common complication of large left to right shunting defects. Dogs and cats will tend to present at 1-2 years or older with signs often attributable to neurological sequelae of either hyperviscosity due to polycythemia or stroke. Two relatively rare conditions that may exist in animals are portopulmonary hypertension and hepatopulmonary syndrome; these are considered the major pulmonary vascular complication of hepatic disease. Other sources of respiratory distress, including large volume ascites or pleural effusion, pneumonia, congestive heart failure or chronic obstructive pulmonary disease (COPD) are also common in people with hepatic disease.

Evaluating pulmonary vascular dysfunction

A variety of tests can provide useful information regarding the pulmonary vasculature, with pulmonary arterial catheterization considered the gold standard. However, since the most important question to answer is "is the right heart functionally well", the most practical test to use in the echocardiogram. In animals with significant pulmonary vascular disease of any sort, subjective and objective echocardiograph data is useful. Subjective information includes recognition that as the normal LV pressure is much higher than the normal RV pressure, is there is septal flattening, this will suggest that there is increased RV pressures, and limitation of volume delivery to the left heart. The right heart may look hypertrophied. It is important to exclude pulmonary stenosis as a cause of right heart changes prior to incriminating the pulmonary vasculature, but this is typically easily done. If available, and if there is a tricuspid regurgitant jet, Doppler may be use to estimate systolic pulmonary artery pressures using the Bernoulli equation, with the estimated pressures = **4 x velocity^2** Other advanced imaging, include CT thoracic scanning may be advised for lung disease, and may be coupled with angiography.

SUMMARY

Use pleural pressure to

1) Ensure placement of needle/catheter in the thoracic cavity
2) Watch for excessively negative pressure that could cause spontaneous pneumothorax

Use pulmonary pressures

1) To monitor for lung injury
2) As surrogate markers of resistance and compliance.
3) To adjust ventilator settings
4) To identify air leak (pneumothorax or cuff leak).

Use pulmonary arterial pressure

1) To document pulmonary hypertension
2) To evaluate (via angiography) the potential for pulmonary thromboembolism.
3) To assess pulmonary capillary wedge pressure as a surrogate marker of Left atrial pressure.

REFERENCES

Huggins JT et a. .F1000 Med Rep. 2010 Oct 21 * The unexpandable lung. King et al. AJVR1991 Oct;52(10):1597-601, Feller-Kopman D et al. Assessment of pleural pressure in the evaluation of pleural effusions.Chest. 2009 Jan;135(1):201-9. Huggins JT et al . Characteristics of trapped lung: **pleural** fluid analysis, **manometry**, and air-contrast chest CT.Chest. 2007 Jan;131(1):206-13. Koo et al. Pulmonary artery **catheters**: evolving rates and reasons for use.Crit Care Med. 2011 Jul;39(7):1613-8.

USEFUL BIOMARKERS IN THE ER

Elizabeth Rozanski, DVM, DACVIM (SAIM), DACVECC
North Grafton, MA

INTRODUCTION

A biomarker is most simply defined as a biological variable that reflects the presence or severity of a disease. Biomarkers include commonly measured units such as creatinine, glucose and hematocrit. In the critical care medicine, which includes both the ER and ICU, there is great enthusiasm for biomarkers that may prognosticate severity of illness, length of stay, and/or prognosis. The focus of this talk is on biomarkers, and we will focus primarily on biomarkers that are currently available for point of care testing in the ICU, but will also address biomarkers that are in the pipeline or available with relatively rapid turnaround times from regional references laboratories.

As with all diagnostic test menus, the clinician is encouraged to investigate current availability and utility of all tests as the biomarker field is dynamic. Additionally, some biomarkers used in people can be readily adapted to use in dogs and cats, while in other biomarkers are species specific, and not able to be adapted for use in cats or dogs.

HYPOPERFUSION

Hypoperfusion, whether recognized or occult, and may lead to tissue hypoxemia, and irreversible shock. Clinical recognition of shock is classically based upon documentation of an elevated heart rate, with or without accompanying hypotension. The clinician is urged to recall that hypotension is a relatively late finding, and animals may be hypoperfused despite a "normal" blood pressure, due to compensatory mechanisms. Heart rate is however, useful in at least preliminary assessment of a patient with suspected hypoperfusion. The *shock index* (SI), defined as the heart rate/systolic blood pressure, may be used to identify more subtle changes in cardiovascular status than might be appreciated upon individual evaluation of either parameter. In dogs, due to their higher resting heart rates, a cut-off of 0.9 has been established for normal versus affected dogs, compared with a cut-off of 0.7 which is used in people. (Porter et al, *in press,* JVECC*)*

Lactate is the most commonly used biomarker for perfusion the ICU and ER. Lactate may be easily measured using hand-held devices such as the Lactate Pro, by POC units such as the Istat, or by larger blood gas/electrolyte machines (eg. Nova Biomedical, Waltham MA). In normal animals, a lactate value of < 2.0 mmol/L is considered normal. Dogs that are excessively wiggly, or post-ictal may also have an elevated lactate value. Lactate measurement is outstanding at prompting assessment in the patient with a confusing physical examination or if assessment count be clouded by another route (eg. Sedation, presence of a beta-blocker, or AVB). All animals with signs of hypoperfusion or discordant physical and historical findings should have a blood lactate measured. Lactate measurement is also useful for improving patient assessment by more junior clinicians and technicians, particularly when a previously "stable" pet is thought to have decompensated.

Lactate clearance may also be useful in prognosis and refers simply to the rate of decrease in lactate over time. Not unexpectedly, people who have failure to improve lactate levels over time in response to therapy have a poor outcome. Lactate clearance has been evaluated in dogs with GDV, and has been demonstrated to be similarly useful. (Zacher et al) Lactate measurements ideally are repeated after a resuscitative effort in order to establish clearance has occurred. In people, lactate at presentation and 2 hours are considered the most useful. (Scott et al) A multi-center study is currently underway in dogs to determine if similar guidelines exist for dogs. In people, lactate clearance is also useful for re-assuring that the treatment plan is adequate.

Other biomarkers that have been evaluated in hypoperfusion include mixed venous and central venous saturation, and the arterial to venous carbon dioxide differences. As a refresher, mixed venous samples are those obtain after mixing behind two cardiac valves, or blood from the pulmonary artery. This is practice is obviously difficult, as it requires placement of a pulmonary arterial catheter which is not commonly performed in veterinary medicine. Central venous samples can be obtained more easily from the cranial or caudal cava, simple jugular sample collection is unlikely adequate. In one recent human study (Bracht H et al) of unplanned admissions to an human intensive care unit identified that low SCVO2 (<60%) was associated with an increased mortality rate. In group of severely septic bitches with pyometra (Conti-Patara A et al) treated with volume resuscitation and ovariohysterectomy and low ScvO2, as well as high lactate, and high base deficit were related to death, supporting the potential clinical utility of measurement of SCVO2 in dogs. Interestingly, in people, another recent article identified that high ScvO2 (~ 85%) was also associated with death in septic shock, likely due to impaired oxygen use. (Textoris J et al). The ideal ScvO2 in dogs being resuscitated remains to be determined, but is likely in the 70-80% range.

The gradient between the arterial and venous carbon dioxide has also been evaluated as a prognostic agent in shock. In an healthy patient, the gradient is approximately 5 mmHg. In 45 consecutively admitted patients with septic shock were divided into low P(cv-a)CO2 versus high P (cv-a)CO2,with low being < 6 mm Hg. All patients were treated with early goal directed therapy. Thirty patients were in the low group and 15 patients were in the high group. While in this pilot study, there was no difference in 28 day mortality, there were improved tissue perfusion and organ function the low group, supporting further investigation. (Zhao et al) On a more practical note, there is clear need for the criticalist and team to recognize that in a critical care setting, such as a cardiopulmonary arrest, there is no need to more enthusiastically ventilate a ventilated dog with a high venous CO2…

For hypoperfusion, other "sent out" biomarkers are not useful, as even a few hour delay will limit clinical utility. As in human medicine, initial efforts may require the collection and batch analysis of samples, with the ultimate goal of a bedside test. However despite this limitation, on-going investigation of a variety of point-of care markers in veterinary medicine is warranted. Due to the relative paucity of cases, most efforts would need to be multi-center, with strict inclusion criteria.

CARDIOPULMONARY

Several biomarkers and point of care testing options are available for evaluation of the patient with cardiopulmonary disease. Arterial blood gas analysis and pulse oximetry may be utilized to obtain an objection assessment of lung function. The A-a gradient may also be used to assess oxygenation. Hypoxemia, or decreased oxygen content of the blood, is confirmed by establishing a low partial pressure of oxygen dissolved in plasma (PaO_2), typically < 80 mmHg at sea level). The partial pressure of carbon dioxide ($PaCO_2$) on an ABG or VBG may be used to assess adequacy of ventilation. A $PaCO_2$ greater than 40 mmHg or $PVCO_2$ greater than 45 mmHg is supportive of hypoventilation. Historically, blood gas analysis has only been available at large referral or university hospitals, but in the last decade a variety of portable point-of care (POC) machines have become affordable and available in practices. The I-STAT (Abbott Laboratories, Abbott Park, IL) model has been widely used and is generally considered accurate and user-friendly; the IRMA (ITC, Edison, NJ) is also popular. These devices (and other similar products) are able to measure blood gas values and other blood chemistries on small volumes of blood. Disposable cartridges make each test relatively economical and the turn-around time is excellent. Larger veterinary hospitals often used the Nova Biomedical Analyzer (Nova Biomedical, Waltham, MA).

Collecting an ABG may be challenging to obtain technically, particularly in a dyspneic patient or small patient. For example, obtaining an ABG in a distressed, open-mouth breathing cat is not a viable option. Therapy should not be delayed while attempting an ABG and a patient showing respiratory distress should not be removed from supplemental oxygen for assessment of oxygen levels (pulse oximetry or arterial blood gas analysis) while breathing room air as the treatment plan should supplemental oxygen.

As mentioned previously, blood gas analysis is useful in evaluating oxygenation and ventilation. Oxygen levels should always be interpreted in light of the PCO_2 level and the inspired oxygen concentration (FiO_2). It is also important to remember that even though an arterial blood gas analysis may look "normal," it does not always imply normal oxygen delivery to tissues. The PaO_2 is an indicator of arterial oxygen tension and measured the dissolved oxygen in the blood (or whatever fluid is analyzed). Oxygen content of blood is defined as (1.34 x hemoglobin concentration x % saturation) + 0.003 PO_2 Adequate oxygen delivery is dependent on normal cardiac function, normal hemoglobin concentrations and the affinity which exists between oxygen and hemoglobin. Methemoglobin, and carboxyhemoglobin are two examples of altered hemoglobins which are markedly less effective in oxygen delivery.

It is crucial that the clinician does not just look at the PaO_2 values in interpreting blood gases, but rather to look as well at the PCO_2 values. Low partial pressure of CO_2 supports that the patient is hyperventilating in order to maintain oxygen values, while high partial pressure of CO_2 supports respiratory fatigue or neuromuscular failure. The clinician should recall that providing supplemental oxygen to a patient with respiratory failure (High $PaCO_2$) will improve oxygen level, but often worse hypercarbia.

When an ABG sample is obtained, an alveolar-arterial (A-a) gradient should be calculated to assess if pulmonary dysfunction exists. For example, a dog may be able to maintain a near normal oxygen concentration (PaO_2 80 mm Hg) by hyperventilating (as detected by a low PCO_2 of 20 mm Hg), but have moderate to severe pulmonary dysfunction, A-A gradient equal to 45. Similarly, a dog with laryngeal paralysis may have severe hypoventilation (PCO_2 of 70 mm Hg) and appear to be hypoxemic (PaO_2 of 60 mm Hg), which in actuality, his A-a gradient is normal (3), indicating normal lung function. The P/F ratio (PaO2:FiO2) is a classic marker of acute lung injury (< 300) and ARDS (<200). More recently, an abstract at IVECCS 2012 described the use of the SpO2:FiO2 ratio as potential surrogate due to potential difficulty in obtaining an ABG in distressed patient.

The dead space fraction may also be calculated in dogs or cats where concern exists for a pulmonary thromboembolism, by measuring the End tidal CO2 and arterial CO2 , and by the formula dead space= (PaCO2-ETC02)/PaCO2

Another useful screening tool to differentiate primary respiratory disease versus cardiac causes for respiratory distress is the NT pro-BNP. Chronic pressure or volume overload results in increased synthesis of pre-proBNP (a precursor molecule of BNP) by the ventricles, which is then processed to active BNP and inactive NT-proBNP. Multiple studies have evaluated the of clinical utility of NT-proBNP and found that it can be used to detect occult cardiomyopathy in cats and to distinguish cardiac and non-cardiac causes for respiratory signs in dogs.(Fox et al, Ettinger et al) Species-specific assays must be used for NT-proBNP evaluation, due to variation in the protein structure between species. In North America, NT-pro BNP is available as send out test from IDEXX, and a similar test BNP is available for dogs from ANTECH. If these tests were to convert to point of care, they could be quite useful for ER clinician.

Lastly, troponin is a biomarker that is released from ischemic myocardium. Troponin elevations have been appreciated with a variety of cardiovascular diseases, including hypertrophic cardiomyopathy and pericardial effusion. However, troponin increases are unlikely to be helpful in evaluation of the patient with respiratory distress.(Prosek et al) Troponin is well-preserved across species, human analyzers are acceptable for dogs and cats.

KIDNEY

While classically, kidney disease has been evaluating using BUN and Creatinine, there is increasing evidence in a critical care setting that kidney damage can exist without clear elevation in either parameter. Acute kidney injury (AKI) is associated with high morbidity and mortality, so early recognition and therapy may be beneficial to preventing the progression of disease. neutrophil gelatinase-associated lipocalin (NGAL) has been introduced for use in people as an early marker of kidney injury. Point of care testing is available which makes the test potentially very appealing. A recent article described the use in dogs, and further clinical trials are underway. (Lee et al 2012)

REFERENCES

Bracht H et al. **Crit care** 2007 (11) 1: R2, Conti-Patera et al. **J Vet Emerg Crit** Care 2012 22(4) 409-418, Ettinger SJ et al**. J Am Vet Med Assoc** 2012;240:171-180**,** Fox et al. **J Vet Intern Med** 2011;25:1010-1016, Lee et al. **BMC Vet Research** 2012 8:248Prosek R et al**. J Vet Intern Med**. 2007 Mar-Apr;21(2):238-42.Scott S et al. **Crit Care Research and Practice** 2010, Article ID 917053 (on-line), Textoris J et al. **Crit Care** 2011 15(4) 176. Zacher et al. **J Am Vet Med Assoc** 2010 236(8) 892-897.

A TECHNICIAN'S GUIDE TO FUNGAL INFECTIONS IN HORSES

Allison J. Stewart, BVSc(hons), MS, DACVIM(LAIM), DACVECC
Auburn, AL

INTRODUCTION

Fungal infection of the respiratory tract, such as conidiobolomycosis, cryptococcosis, pseudalleschiosis, aspergillosis, coccidioidomycosis, blastomycosis, histoplasmosis, paecilomycosis and candidiasis, can present diagnostic challenges and therapeutic dilemmas. Treatment of these fungal conditions is often unsuccessful as lesions may be surgically inaccessible, and proliferation of fibroblasts within the lesions hinders penetration of topically or systemically administered antifungal drugs. Topical/intra-lesional therapy with amphotericin B, fluconazole, enilconazole and miconazole and systemic therapy with voriconazole, itraconazole, fluconazole, iodides and immunotherapy will be reviewed.

CONIDIOBOLOMYCOSIS

Conidiobolus coronatus is a saprophytic fungus that causes granulomatous lesions of the upper respiratory tract in horses in the southern states. Single to multiple granulomatous lesions in the nasal passages, trachea or soft palate are observed endoscopically. Histologic appearance of conidiobolomycosis is similar to pythiosis and basidiobolomycosis. Hyphae of *C. coronatus* are thin-walled, highly septate, and irregularly branched (5-13μm). The lesions typically have large numbers of eosinophils (Splendore-Hoeppli phenomenon) and fewer macrophages, neutrophils, plasma cells and lymphocytes surrounding the hyphae. Definitive diagnosis is based on microbiological culture, immunodiffusion, or PCR.

Conidiobolomycosis lesions can be treated with surgical excision, cryotherapy or long-term of administration iodides or antifungals. Treatment with oral fluconazole (14mg/kg loading followed by 5 mg/kg PO q 12 hrs) for 6- 16 weeks is the treatment of choice.[1] The use of systemic iodides may be helpful, but is probably not required. The decision to discontinue therapy is based on reexamination by endoscope. Six-monthly rechecks are advised to ensure no recrudescence of disease. Surgical debulking followed by periodic intralesional injection of 20 mg amphotericin B is an alternative or adjunctive therapy.

CRYPTOCOCCOSIS

Cryptococcosis is commonly caused by *C. neoformans* (var *neoformans* and var *gattii*), a ubiqitous, saprophytic, round, basidiomycetous yeast-like fungus (3.5- 7μm) with a large heteropolysaccharide capsule (1-30 μm) that does not take up common cytologic stains. A relatively high frequency of equine cryptococcosis occurs in Western Australia, and there is an epidemiological relationship between the organism and the Australian river red gum tree (*Eucalyptus camaldulensis*) and bird (pigeon) excreta. Cytologic or histopathologic identification is most reliable for diagnosis. Serologic testing with latex agglutination to identify cryptococcal capsular antigen is useful; resolution of lesions is correlated with declining serum titers.

Cryptococcosis in horses is primarily associated with pneumonia, rhinitis, meningitis and abortion. A pony with multiple pulmonary cryptococcomas, from which *Cryptococcus gattii* was cultured from both transtracheal washings and lung mass aspirates was successfully treated with daily IV infusions of amphotericin over a 1 month period. One year after cessation of treatment, clinical signs had resolved and the cryptococcal antigen titer decreased from 4096 to 256.[2] We have successfully treated a mare that presented with a 9-month history of intermittent epistaxis, mucosanguinous nasal discharge and dyspnea due to bilateral nasopharangeal cryptococcal granulomas and an ethmoid hematoma. The mare was treated with oral fluconazole (5 mg/kg) daily for 6 weeks, potassium iodide (30 mg/kg PO q 24 hr) for 4 weeks and periodic intralesional injections of fluconazole (100 mg), formalin and then amphotericin B (50 mg) using an endoscopic delivery. The mare was disease free for 2 years, but the infection reoccurred. Extensive surgical debridement was performed through a frontochoncal bone flap and the mare was treated with amphotericin B IV for 2 weeks then 8 months of oral fluconazole. She has been disease free for 3 years.[3]

PSEUDALLESCHERIOSIS

Pseudallescheria boydii is a saprophytic ascomycete. Hyphae within tissue can not be differentiated from *Fusarium sp* or *Aspergillus sp* unless cultured. *P. boydii* cultured from the nasal cavity and sinus of a horse with chronic, malodorous nasal discharge, was susceptible *in vitro* to miconazole, ketaconazole, natamycin, and clotrimazole. After debriding and flushing the plaque, 5 grams of 2% miconazole cream was infused twice daily for 4 weeks through lavage tubing that had been passed into the nasal passage through a hole drilled in the frontal bone and sinuses. Sodium iodide (30 mg/kg IV for 4 days), then potassium iodide (0.06 mg/kg PO q 12 hrs for 14 days) was administered. The nasal cavity was free of infection at the 30 and 60 day rechecks.[4]

ASPERGILLOSIS

Aspergillosis in horses has been reviewed.[5,6] *Aspergillus* sp. characteristically have broad, septate hyphae with acute branching, parallel sides and a propensity for vascular invasion. *Aspergillus sp.* pneumonia is almost uniformly fatal, often with no or mild respiratory signs. Immune compromise (Cushings's disease, enteritis, neoplasia, surgery, EPM) is a common predisposing cause.[6] Serologic diagnosis of aspergillosis may not be reliable. Unfortunately aspergillus is not very susceptible to fluconazole. Oral itraconazole or voraconazole or IV amphotericin B is recommended. Mycotic rhinitis was successfully treated with itraconazole (3mg/kg PO q 24 hrs for 120 days) in 2 horses. Serum concentrations achieved were in excess of the mean inhibitory concentration (MIC) for the organism.[7] Topical natamycin (25 mg in 100 ml saline) was flushed via an endoscope for 3-4 days and nystatin powder was insufflated up the affected nostril in 2 horses with *Aspergillus sp.* rhinitis. A third horse was treated with natamycin solution via an indwelling facial catheter.[8] Topical enilconazole has also been used to treat *Aspergillus* nasal plaques.[9] A foal with *Aspergillus* pneumonia was treated with lung lobe resection and oral voriconazole.[10]

GUTTURAL POUCH MYCOSIS

Surgical arterial occlusion by balloon catheter technique or transarterial coil is the treatment of choice, with little evidence for the necessity of adjunctive medical therapy. Topical medications can be applied via an indwelling catheter advanced through the guttural pouch opening or directly sprayed on the lesion through an endoscope. The dorsal location of the lesions makes treatment difficult and there is always the risk of fatal hemorrhage. Successful medical treatment of guttural pouch mycosis with oral itraconazole (5mg/kg PO for 3 weeks) and topical enilconazole (60 ml) sprayed directly onto the mycotic lesion under endoscopic guidance for 2 weeks. Dysphagia resolved after 6 weeks.[11] Four other cases of guttural pouch mycosis (*Rhizopus sp*, *Aspergillus sp*, *Penicillium sp*, *Mucor sp*, *Chrysosporium sp,* and *Alternaria sp)* were successfully treated by daily local irrigation using a 0.9% or 1.7% enilconazole.[12,13]

COCCIDIOMYCOSIS

Coccidiomycosis is a systemic fungal infection caused by *Coccidioides immitis*, a soil saprophyte that grows in areas with sandy, alkaline soils and semiarid conditions. Horses present with weight loss, fever, abdominal pain, and signs of respiratory disease. Localized, recurring nasal granulomas are also reported. Diffuse infections, with granulomas in the lungs, liver, kidney or spleen have a grave prognosis.[14] Serology is very useful; a decreasing titer is associated with clinical improvement.[14] Przewalskii horses may be more susceptible. Fluconazole has been used with success,[15] and based on price would be the treatment of choice.

BLASTOMYCOSIS AND HISTOMYCOSIS

Blastomycosis is caused by inhalation of conidiae of the thermally dimorphic saprophytic fungus *Blastomyces dermatitidis. Blastomyces* yeasts can be identified on cytological examination, often within multinucleated giant cells. They are spherical, 15-17μm, with basophilic protoplasm and unstained, uniformly shaped refractile walls. Unilateral, broad-based budding is characteristic.

Blastomycosis was reported to cause pulmonary abscessation, pyogranulomatous pneumonia, pleuritis, peritonitis and abscesses in a 5-year-old horse. *B. dermatitidis* was confirmed by use of a DNA probe from transtracheal wash fluid and serology was strongly positive. Treatment with itraconazole or ketaconazole was declined and the horse was euthanized.[16] Disseminated blastomycosis was post mortem in a miniature horse with subcutaneous infections associated with a chronic pectoral wound.[17]

Histoplasmosis is caused by the saprophytic, dimorphic fungus *H. capsulatum*, and is most prevalent in moist soil containing bird or bat waste. Organisms are 2-4 μm, with a thin clear halo surrounding a round or crescent- shaped basophilic cytoplasm.

THERAPEUTICS

Amphotericin B

Intravenous amphotericin B is a very effective antifungal drug but it can be nephrotoxic. Reported doses range from 0.3-0.9 mg/kg in saline IV over 1 hr. Successful treatment of pulmonary cryptococcosis used daily infusions of amphotericin B (initially 0.35 mg/kg, increasing over 8 days to 0.5 mg/kg in 1 L 5% dextrose in water over 1 h IV for 1 month). Premedication with flunixin (0.25mg/kg IV) was administered before each treatment.[2] Topical amphotericin B was successful in the treatment of nasopharyngeal *C. coronatus.*

Azoles

Benzimidazole derivatives in the class *azoles*, such as clotrimazole, econazole, miconazole, enilconazole, ketaconazole, isoconazole, itraconazole, fluconazole and voriconazole destroy fungi by inhibition of ergosterol biosynthesis in the fungal cell membrane. Intranasal 2% miconazole was part of successful multimodal therapy against nasopharyngeal *P. boydii*.[4] Enilconazole is not commercially available in the US, but has been used topically to successfully treat guttural pouch mycosis and Aspergillus nasal palaques.[9] Aerosolization of 1.2 mg/kg enilconazole q12 hrs in 125 ml of saline resolved *Scopulariopsis* pneumonia.[18] Ketoconazole is poorly absorbed in the non-acidified form, but 30 mg/kg NGT q 12hrs mixed with 0.2 N HCl has been used. [18] Itraconazole (Sporanox solution) is well absorbed orally (bioavailability 60%). A dose of 5mg/kg PO q 24 hrs maintains concentrations above MIC for susceptible yeasts (*Histoplasma* and *Blastomyces* sp) and *Aspergillus sp* for > 24hrs, with no detectable side effects.[19] The use of compounded itraconazole is not recommended, as it is very unstable (requires low pH) and due to it's lipophilic nature is difficult to formulate in aqueous solution (Jennifer L. Davis). A loading dose of oral fluconazole (14 mg/kg) followed by 5 mg/kg every 24 hrs yields plasma, CSF, synovial fluid, aqueous humor, and urine fluconazole concentrations above the MIC reported for several equine fungal pathogens.[20] Fluconazole, however, reportedly has minimal activity against filamentous fungi (*Aspergillus sp* and *Fusarium sp*), (Jennifer L. Davis). Generic fluconazole costs less than $300 per month. Voriconazole at 4.0 mg/kg/day PO is an excellent antifungal, especially for aspergillus and fusarium infections, but is currently very expensive.[21,22]

Systemic Iodide Therapy

Treatment is inexpensive, but iodide toxicity characterized by excessive lacrimation, nonproductive cough, increased respiratory secretions, and dermatitis can occur. The recommended dose of 20% sodium iodide is 20-40 mg/kg/day IV for 7 to 10 days. Orally administered iodine is available in two forms. Inorganic potassium iodide (10 to 40 mg/kg/day) is available only as a chemical grade and is unstable in the presence of light, heat and excessive humidity. Organic ethylenediamine dihydriodide (20 to 40 mg/kg/day) is commercially available. Administration of iodine in the diet of pregnant mares may cause congenital hypothyroidism in foals.

Table 1. Characteristic Morphologic Features and Serologic tests for Fungal Organisms Infecting Horses

Agent	Cytologic Appearance	Serologic Test	
	Tissue Form	**Method**	**Veterinary Laboratory**
Cryptococcus neoformans	Round, thin-walled, yeast-like fungus (5-10 μm) with a large heteropolysaccharide capsule (1-30 μm in diameter) that does not take up common cytologic stains. Capsule is best stained using Mayer's mucicarmine stain. Organisms show narrow based budding and lacks endospores.	Capsular antigen ELISA or Latex agglutination (Antigen)	$33.00 UT $21.50 CU $21.00 CSU $16.00 UGA $18.00 NMDA
Coccidioides immitis	Spherules have a double-contoured wall, are variable in size, but can be large (20-80 μm, up to 200 μm). The mature spherules (sporangia) contain endospores (sporangiospores) 2-5 μm in diameter. In the environment, mycelium are thick-walled with barrel shaped arthroconidia.	AGID (antibody) for IgM and IgG MVista® Quantitative EIA (Antigen)	$30.00 UT $16.00 CSU $16.00 CU $14.00 NMDA $50.00 MDL
Aspergillus Spp.	*B*road (2-4 μm diameter) septate hyphae with parallel sides and acute right angled branching.	Platelia® Aspergillus Galactomannan EIA (Sandwich immunoassay) AGID (Antibody) *A. Fumigatus* only	$50.00 MDL $20.00 UT $16.00 CSU $16.00 CU $14.00 NMDA
Blastomyces dermatitidis	Yeasts are spherical (15-17μm in diameter) with basophilic protoplasm and unstained, uniformly shaped refractile walls. Unilateral, broad-based budding is characteristic. Yeasts are often located within multinucleated giant cells.	AGID (Antibody) MVista® Quantitative EIA (Antigen)	$20.00 UT $16.00 CSU $16.00 CU $14.00 NMDA $50.00 MDL
Histoplasma capsulatum	Yeasts (2-4 μm in diameter) have a thin clear halo surrounding a round or crescent- shaped basophilic cytoplasm.	AGID (Antibody) MVista® Quantitative EIA (Antigen)	$20.00 UT $16.00 CSU $16.00 CU $14.00 NMDA $50.00 MDL
Fungal serology panel	Antibody to *Histoplasma, Blastomyces, Coccidioides, Aspergillus* (AGID) The CU panal also includes *Cryptococcus* antigen		$62.75 CU $39.00 NMDA

ELISA =enzyme linked immunosorbant assay, AGID = Agar gel immunodiffusion, CF = Complement fixation *Ag (EIA)= Sandwich Enzyme Immmunoassay* UT = University of Tennessee, College of Veterinary Medicine. P (865) 974-5639 (mycology) or -5643 (serology), http://www.vet.utk.edu/diagnostic/index.php UGA = University of Georgia, College of Veterinary Medicine. P (706) 542-5812, F (706) 583-0843 http://www.vet.uga.edu/idl/tests CSU = Colorado State University. P (970) 297-1281, F (970) 297-0320 http://csu-cvmbs.colostate.edu/vdl/Pages/default.aspx CU = Cornell University, Animal Health Diagnostic Center. P (607) 253-3900, F (607) 253 3943 http://ahdc.vet.cornell.edu/ UTHSCSA = The University of Texas Health Science Center at San Antonio, Center for Medical Mycology. P (210) 567-4131, F (210) 567-4076 http://www.sacmm.org/index.html NMDA = New Mexico Department of Agriculture Veterinary Diagnostic Services. P (505) 383- 9299 http://www.nmda.nmsu.edu/vds/
MDL = Miravista Diagnostics Laboratory, Indianapolis, IL. P (317) 856-2681 F (317)-856-3685 www.miravistalabs.com

REFERENCES

[1]Taintor J, et.al. *Compend Contin* 2003;25(11):872. [2]Begg LM, et.al. *Aust Vet J* 2004;82(11):686. 3.Stewart AJ, et. al. *J Am Vet Med Assoc* 2009:15;235(6):723. [4]Davis PR, et.al. *JAVMA* 2000;217(5):707. [5]Blomme E, et.al. *Eq Vet Educ* 1998;10(2):86. [6] Sweeney CR, et.al. *JAVMA* 1999;214(6):808. [7]Korenek NL, et.al. *JVIM* 1994;8:224. [8]Greet TRC. *Vet Rec* 1981;109:487. [9]Kendall A, et al. *J Vet Intern Med*; 2008;22:5,1239. [10]Hilton H, et al. *JVIM*. 2009;23(2):375. [11]Davis EW, et.al.1994. 8(4):304. [12]Vannieuwstadt RA, et.al. *Tijd V Dierg* 1994;119(1):3. [13]Carmalt JL, et.al. *Pferdeheilku* 2004;20:542. [14]Ziemer EL, et.al. *JAVMA* 1992;201(6): 910. [15] Higgins JC, et al. *JJIM* 2005;19:482. [16]Toribio RE, et.al. *JAVMA* 1999;214(9):1357. [17]Dolente BA, et.al. *Eq Vet Edu* 2003;15(3):139. [18]Nappert G, et.al. *Eq Vet J* 1996;28(5):421.[19]Davis JL,et al. *JVIM* 2004;18:458. [20]Latimer FG, et al. *AJVR* 2001;62:1606. [21]Davis JL, et al. *AJVR* 2006;67:170. [22]Passler NI, et al. *J Pharmacol Therap Vet Therap* 2010: 33(1):35-41.

UNDERSTANDING TESTING FOR EQUINE ENDOCRINOPATHIES

Allison J. Stewart, BVSc(hons), MS, DACVIM(LAIM), DACVECC
Auburn, AL

INTRODUCTION

Reference ranges for thyroid hormones, adrenocorticotropic hormone (ACTH), cortisol and insulin concentrations vary by laboratory due to differences in assay procedures, units of measurement and populations of horses used to establish these values. It is important to verify that the laboratory has validated their assays and established their own reference ranges for horses. Seasonal reference ranges are also needed for endogenous ACTH concentrations and ACTH concentrations after administration of thyroid-releasing hormone (TRH).

PITUITARY PARS INTERMEDIA DYSFUNCTION

Earlier detection and reliable treatments mean the outlook for an older horse with this progressive endocrine condition has never been better. All horses greater than 16 years should have an annual test for PPID. Treatment can then begin prior to the onset of devastating clinical signs such as laminitis, long hair coat, loss of muscle mass, weight loss, liver dysfunction and depressed demeanor. Other than expense, is there any reason to wait until these signs develop before diagnosing and commencing treatment for PPID while it is still subclinical? Owners who value their older horses as valuable competition animals or much loved pets would probably prefer to prevent clinical signs of PPID from developing.

Tier 1 testing Dexamethasone suppression test (see below) or **Endogenous ACTH.** Normal horses have low resting ACTH, while PPID horses usually have ACTH concentrations higher than the seasonal reference range. Stress, transport and pain can falsely increase ACTH concentrations. Therefore testing a painful horse with laminitis may cause a false positive.

Advantages- single blood sample and no risk to patient. Disadvantages – may miss some early cases
Recommended for horses with hirsutism and 6 monthly monitoring of horses previously diagnosed with PPID

Tier 2 testing: Thyroid-releasing hormone (TRH) stimulation test - There are TRH receptors located in both the pars distalis and pars intermedia in horses. After TRH administration normal horses have a slight increase in ACTH concentration; subclinical PPID horses (ie pituitary hyperplasia at necropsy, but no clinical signs) have a moderate increase in ACTH; while clinical PPID horses have a dramatic increase in ACTH concentrations after TRH administration. Seasonal reference ranges have been published.[1]

i. T= 0 collect blood for baseline ACTH in a purple top EDTA tube (centrifuge quickly and put the plasma in plastic vials then ship on ice or freeze)
ii. Give 1 mg TRH IV (0.5 mg for ponies).
iii. T=30 min. Collect plasma for ACTH concentration (some prefer a 10 minute sample)

Administration of TRH can also be used to test for hypothyroidism. Collect serum at T = 0, 2 and 4 hrs. (note: true hypothyroidism is VERY rare in horses, so many owners can be talked out of testing and "treatment" for "hypothyroidism") If you collect plasma samples (purple top tube) for ACTH at 0 and 30 min and serum samples (red top tube) for T3 and T4 at 0, 2 and 4 hrs then this is a **very good** method of simultaneously testing for PPID and ruling out "hypothyroidism":

- Normal Horse: A doubling of T4 after 4 hrs and ~ tripling of T3 after 2 hrs. A normal horse will have a small increase in ACTH concentration after 30 minutes (refer to seasonal reference ranges below)
- PPID horse: T3 and T4 will increase the same as a normal horse, and the ACTH at T = 0 will be elevated and at 30 min ACTH will be markedly elevated.
- Hypothyroid horse: No marked increase in T3 or T4. (Resting T3 and T4 concentrations are not very useful) Please let us know- this would be a very exciting diagnosis! ACTH within reference ranges.

This is an **EXCELLENT test for PPID and hypothyroidism**, especially if worried about laminitis.

Endogenous Cortisol - single or repeated measurements of cortisol are of <u>no value</u>; cortisol concentration is usually within a normal range or even low in PPID horses.

Dexamethasone suppression test - dexamethasone given to normal horses causes a subsequent drop in plasma cortisol concentrations (<30 nmol/L, <10 ng/ml or 1 μg/dl) which is maintained at a low concentration for at least 24 hrs. Expect cortisol concentration to remain high in a PPID horse because the tumor in the pituitary gland is not under negative feedback control by glucocorticoids. (Pre dex serum sample (red top tube) between 5-7 pm; give dexamethasone at 40 μg/kg (0.04 mg/kg) intramuscularly (20 mg or 10 ml of Azium for average horse); ideally then take two serum samples at 15 and 19 hrs, but a single post-dex serum sample 19 hours later is sufficient.

-Normal horse - the cortisol will decrease < 30 nmol/L, (<10 ng/ml or 1 μg/dl) at 15-19 hrs

-Early PPID horse - the cortisol may drop to < 30 nmol/L, (<10 ng/ml or 1 μg/dl) at 15 hrs, but will remain suppressed for a short time only and will be > 30 nmol/L, (>10 ng/ml or 1 μg/dl) at 19 hr.

-PPID horse - the cortisol will remain > 30 nmol/L, (>10 ng/ml or 1 μg/dl) at 15-19 hrs

There are rare reports of laminitis after administration of dexamethasone. The dose of dexamethasone used in this test is small, but horses with PPID and metabolic syndrome are also more likely to develop spontaneous laminitis and may be more prone to laminitis after dexamethasone. This test has historically been the most commonly used ante-mortem diagnostic test for PPID; however, recent literature suggests that the reliability of this test has been overstated. It has been shown to be equal to the endogenous ACTH test to diagnose PPID. Both of these tests are excellent at diagnosing horses with marked clinical signs of PPID or those with large pituitary glands confirmed by necropsy. It is likely that the measurement of ACTH before and after TRH administration will soon be proven to be more accurate at diagnosing early subclinical cases of PPID.

Geriatric Horse Screening

PPID does not occur in horses < 8 years and is uncommon < 12 years. We recommend that ALL horses greater than 16 years be tested annually for PPID. It is best to diagnose and treat early subclinical PPID prior to the onset of laminitis, loss of muscle mass, hirsutism or weight loss. To test for subclinical PPID the TRH stimulation test would be ideal.

Monitoring PPID

PPID will progress with age and the dose of pergolide required to control the disease often needs to be increased each year. Treatment with appropriate dosages of pergolide can make a horse look and feel 5 years younger. After diagnosis of PPID commence pergolide treatment at 1 mg/horse per day. Retest after 3 months, then every 6 months as the dose of pergolide may need to be increased to abate clinical signs and decrease ACTH concentrations. Use of the FDA approved pergolide mesolate (Prascend™ http://www.prascend.com/) by Boehringer Ingelheim is recommended. Several compounded products have been found to have problems with stability and storage. Compounded liquid products should be kept refrigerated in dark containers and used within 30 days of manufacture.

TESTING FOR INSULIN RESISTANCE

Horses with a cresty neck, regional adiposity and a history of laminitis (characteristics of Metabolic Syndrome) as well as all horses with PPID should be tested for insulin resistance. It is best that horses are kept off feed for 6-8 hrs before testing. A single **fasting serum insulin concentration** (red top tube) can be used as a screening test. A better test is the **oral glucose challenge test**. Collect T=0 sample (red top tube for serum insulin) and then give Karosyrup (15mls per 100kg = 75mls per 1000 lb horse orally) and collect a second sample for serum insulin at 75 minutes. A **combined insulin-glucose tolerance test** can be used as a more sensitive test. This takes 4 hours and should be performed after a 6-8 hr fast.

Collect baseline serum (red top tube) for glucose and insulin concentrations

Rapidly infuse 50% dextrose solution at 150 mg/kg within 1 minute (followed immediately by 0.1 units/kg of regular insulin (equivalent to 150 mls of 500 mg/ml (50%) dextrose and 0.50 ml of 100 units/ml insulin for 500 kg horse).

Collect blood at 5, 15, 25, 35, 45, 60, 75, 90, 105, 120, 135 and 150 minutes for glucose and at 45 minutes for insulin. Insulin resistance is defined as blood glucose concentrations remaining above baseline for more than 35 minutes, or detection of serum insulin concentration above 100 μU/ml at 45 minutes. If signs of hypoglycemia occur (weakness, collapse) administer glucose as a 10% solation to effect. Note- if red top tubes are used then serum should be separated within 1 hour. Otherwise fluroacetate (grey top) tubes should be used as red blood cells will consume glucose.

Horses with insulin resistance should be fed a strict low glycemic diet. Horses >12yrs with insulin resistance and/or a history of laminitis should also be tested for PPID.

HYPOTHYROIDISM

True hypothyroidism is very rare in adult horses and is frequently and incorrectly diagnosed based on endogenous concentrations. Endogenous T3 and T4 samples may give a misleading picture of true thyroid status, because thyroid hormone metabolism, protein binding, and transport can be altered temporarily by certain physiologic or pathophysiologic states and certain drugs. For example, thyroid hormone concentrations are low in horses that have not eaten for several days. Phenylbutazone decreases serum concentrations of T_4 and T_3 by competition for binding to carrier proteins and by decreased T_4 production by the thyroid gland. Systemic illness also can result in decreased circulating concentrations of thyroid hormones (non-thyroidal illness syndrome). If single "point-in-time" measurement of thyroid hormones is the only option available for evaluation of thyroid status, measurement of free fractions of thyroid hormones in more useful than measurement of total amounts of thyroid hormones alone. However, to diagnose hypothyroidism a TRH stimulation test is required (see above).

Neonatal foals have very high circulating thyroid hormone concentrations, which slowly decrease to adult concentrations over the first weeks of life. Therefore it is important to use age-matched reference ranges. The following reference ranges have been reported for neonatal foals < 12 hours of age: total T4 = 282 ± 28.2 nmol/L, free T4 = 80 ± 20.3 pmol/L, (free T4 by dialysis = 110 ± 29.5 pmol/L), total T3 = 7.4 ± 2.3 nmol/L, free T3 = 9.1 ± 5.0 pol/L, TSH 0.31 ± 0.20 ng/ml.[2]

ADRENAL INSUFFICIENCY

Cortisol responses to ACTH stimulation in adult horses show a 2.5- to 5-fold increase in cortisol concentration 30 minutes after administration of a 0.1 to 1.0 µg/kg IV dose of exogenous ACTH (cosyntropin).[3,4] Administration of higher doses (2–10 µg/kg IV) results in a sustained and greater (5–10 fold) increase in cortisol concentration by 90 to 180 minutes after ACTH administration.[3] In neonatal foals the cortisol response to ACTH stimulation after administration of both a low (10 µg IV) and a high (100 µg IV) dose of ACTH was approximately half than that reported in adult horses.[4-6]

Reference ranges for Auburn University Endocrine Diagnostic Laboratory:

Endogenous serum insulin concentrations should be < 25µU/mL between December and July. Some normal horses have elevated concentrations in the fall, but may still benefit from low glycemic feeds as they MAY be at risk of laminitis.

Insulin after 15mls/kg of Karosyrup < 58 µU/mL

Endogenous plasma ACTH concentrations should be below 28 pg/mL between August and October and < 17 pg/mL from November through July.

Endogenous serum cortisol concentrations should be between 40 and 180 nmol/L but should not be used to diagnose PPID.

After administration of TRH, all ACTH concentrations at T=0, 10 and 30 minutes should all be < 35 pg/mL, with even lower concentrations expected between December and June.

Note: Canine ACTH is not very stable and use of the preservative aprotinin is recommended. Equine ACTH is more stable and preservative is not as critical. To minimize degradation of ACTH the blood should be collected in cold EDTA tubes (ideally with added aprotinin already added), promptly centrifuged, and the plasma placed in plastic tubes and shipped overnight on ice or stored in the freezer at –20 °C until analysis.

REFERENCES

[1] Funk FA, et al. *J Vet Intern Med.* 2011:25:579-58529. [2]Breuhaus BA. *Vet Clin North Am-Equine Pract* 2011;27:49. [3]Bousquet-Melou A, et al. *Endocrinol* 2006;147:4281-4291. 38. [4]Stewart AJ, *J Am Vet Med Assoc.* 2011:239(2):834-841 [5]Hart KA, et al. *J Vet Intern Med* 2009;23:344-351. [6]Stewart AJ, et al J Am Vet Med Assoc in press

EQUINE CLOSTRIDIAL DISEASES

Sandra D. Taylor, DVM, PhD, DACVIM (LAIM)
West Lafayette, IN

INTRODUCTION

Over 300 species of the *Clostridium* genus have been identified, and represent a diverse bacterial population of anaerobic, gram positive, spore-forming rods. Clostridial organisms are ubiquitous in the environment and are common components of the gastrointestinal flora in animals and humans. Although relatively few *Clostridium* species cause disease, they can account for significant illness and mortality in horses and other large animals. Important clostridial diseases that affect equids include enterocolitis (*C. perfringens, C. difficile*), myonecrosis (*C. perfringens, C. septicum*), botulism (*C. botulinum*), tetanus (*C. tetani*), and Tyzzer's disease.

ENTEROCOLITIS

Clostridial enterocolitis is an important syndrome in horses that can result in significant morbidity and mortality, and presents a diagnostic and therapeutic challenge as the pathogenesis of disease is incompletely understood. *Clostridium perfringens* and *Clostridium difficile* can cause severe illness in horses, but can be found in the gastrointestinal tract of normal horses as well.

Clostridium perfringens

C. perfringens is a ubiquitous clostridium that produces several toxins which contribute to pathogenesis of disease, including alpha, beta, epsilon, iota, enterotoxin (CPE) and β2 toxins. Five biotypes of *C. perfringens* have been identified and are designated A-E, based on toxin production. The alpha toxin is produced by all five types and is a lecithinase (phospholipase C) and has sphingomyelinase activity. The beta toxin is made by Type B, C and D, and is a necrotizing exotoxin that forms pores in susceptible cell membranes. The epsilon toxin is made by Type B and Type D, and the iota toxin is produced by Type E only. Enterotoxin (CPE) and β2 toxin are only produced by Type A (Table 1). The most important biotypes affecting horses are Type A and Type C. *C. perfringens* is rarely found in the gastrointestinal tract of healthy horses, with the exception of Type A in foals and broodmares. In fact, *C. perfringens* Type A may be part of commensal flora in neonatal foals, as 90% of normal 3-day old foals have been found to shed the organism.[1] Although Type A has been isolated in normal foals and broodmares, it has been shown to be associated with typhlocolitis in adult horses and enterocolitis in foals.[2-4] The mechanism of diarrhea is typically due to CPE or β2 toxin production in these cases.[3,5,6] The role of Type A strains that only produce alpha toxin is questionable, but vast genetic diversity of the gene encoding the alpha toxin among strains may elucidate its role in virulence potential.[7] *C. perfringens* Type C typically causes more severe enterocolitis than Type A, especially in neonatal foals. Although adult horses are occasionally affected, infection is more common in foals because beta toxin is highly susceptible to trypsin, inhibitors of which are abundant in colostrum.[8] Unlike Type A, Type C is never considered normal flora.[1] This is because Type C produces large quantities of beta toxin, which is considered responsible for intestinal necrosis and systemic alterations after toxin translocation. Postmortem findings consist of multifocal to segmental hemorrhage and thickened intestinal wall.[9] The small intestine is primarily involved, but the colon and cecum can also be affected. Case fatality rates for a referral population of foals with Type C infection have been reported to be as high as 83%.[4] *C. perfringens* Type B and Type D each produce alpha, beta, and epsilon toxins, while Type E produces alpha and iota toxins. These three biotypes are rare and of questionable concern in equids.

Table 1. ***Clostridium perfringens*** **Biotypes and Associated Toxins**

	Alpha	Beta	Epsilon	Iota	CPE	B2
Type A	X				X	X
Type B	X	X	X			
Type C	X	X				
Type D	X	X	X			
Type E	X			X		

Clostridium difficile

C. difficile is a ubiquitous, sporulated anaerobe that confers disease through production of several virulence factors, the most important of which are exotoxins A (enterotoxin) and B (cytotoxin). Several strains (ribotypes) have been identified in horse feces, many of which have been previously found in humans and food animals.[10-13] Some strains do not produce toxins or disease, while most toxigenic isolates produce both Toxin A and Toxin B that act synergistically to induce disease.[14-16] *C. difficile* can be found in normal horse feces as well as in diarrheic horse feces as a primary cause of enterocolitis. Factors that trigger development of colonization versus disease are not well understood. Previous studies report *C. difficile* prevalence of 0-17% in feces from normal adult horses,[14,17-20] while up to 30% of normal neonatal foals may shed the organism.[17] Individual healthy horses may shed multiple strains of *C. difficile* over time, suggesting frequent exposure but transient colonization.[19] Although colonization without disease is not uncommon, *C. difficile* is clearly an important cause of diarrhea in horses, including neonatal foals.[5] It is responsible for many cases of antibiotic-associated colitis, and has recently been implicated as a cause of duodenitis-proximal jejunitis.[17,21] Toxin A is often associated with antibiotic-associated colitis, as one study found that 97% of diarrheic horses with Toxin A in their feces had received antibiotics prior to the onset of disease.[22] The association among the presence of different strains, toxin production, and clinical outcome is unclear. One study found that horses with colitis that were positive for Toxin A demonstrated more severe disease and a higher mortality rate compared to horses with colitis that were negative for Toxin A.[22] However, Toxin B production was not evaluated in that study. Unlike observations in adult horses, antibiotic use is not a clear risk factor for development of *C. difficile*-induced diarrhea in foals, suggesting that it may be a primary pathogen in neonates.[15,23] *C. difficile* primarily affects the small intestine in foals, while adults typically have cecal and colonic lesions.[9] Most foals develop *C. difficile* enterocolitis between 7-10 days of age.[15] In a referred population, 82% of foals diagnosed and treated for *C. difficile* enterocolitis survived.[15]

Diagnosis

Diagnosis of clostridial enterocolitis is challenging, and involves anaerobic fecal culture followed by multiplex PCR on cultured isolates to identify genes encoding each toxin. However, isolation of clostridia in a horse with diarrhea is not regarded as evidence that clostridia are the primary etiology, because the organism can proliferate following disruption of intestinal microflora regardless of the initial cause. Moreover, existence of a toxin gene does not necessarily imply production of toxins themselves. Therefore, identification of fecal toxins using ELISA is prudent.[8]

Treatment

Most cases of clostridial enterocolitis require aggressive treatment, regardless of clostridial species. Intravenous crystalloid solutions are necessary to address volume loss secondary to diarrhea, and colloids may be required to maintain vascular oncotic pressure. Although antibiotic use in adult horses with diarrhea is controversial, targeted therapy is often indicated, with metronidazole as the most commonly used antibiotic in both adults and foals with clostridial enterocolitis. Metronidazole resistance is of critical importance with regard to treatment and emergence of resistant pathogens, and highlights the importance of antibiotic sensitivity testing of feces from diarrheic horses.[15,24] Di-tri-octahedral smectite is a clay that has been shown to bind clostridial exotoxins *in vitro*, without disrupting metronidazole activity.[25,26] This drug is often administered to horses with clostridial enterocolitis that can tolerate oral medications. Similarly, *Saccharomyces boulardii* is a nonpathogenic yeast that produces a protease that specifically degrades *C. difficile* toxins, and has been shown to decrease severity and duration of enterocolitis in horses.[27] Horses with acute diarrhea should be isolated and handled with strict biosecurity protocols to limit cross-contamination and reduce the risk of spread to other horses.

CLOSTRIDIAL MYONECROSIS

Clostridial myonecrosis is a syndrome of severe, necrotizing muscle infection caused by *Clostridium sp.* Affected horses often develop swelling and subcutaneous emphysema in the region associated with a wound or injection within hours of the initial insult.[28-30] Intramuscular injections of flunixin meglumine, ivermectin, or antihistamines have been particularly associated with development of clostridial myonecrosis,[28-31] although disease following injection of vaccines and other medications has also been documented.[28-30,32] Approximately 70% of cases are caused by *C. perfringens* alone, while *C. septicum* is isolated in only 15% of cases; mixed clostridial infections are identified in approximately 10% of cases.[28,33] The vast majority of cases follow intramuscular injection in the cervical region, suggesting that locally irritating compounds without labeled directions for intravenous use should be administered deep into the well-vascularized semimembranosus/semitendinosus muscles.[28] Aspiration of the swollen and emphysematous region typically yields grossly apparent gas release and demonstration of numerous gram positive rods on a gram stain. Anaerobic culture or PCR may provide a definitive diagnosis but are often not pursued due to the peracute nature of disease. Treatment typically involves immediate fasciotomy/myotomy to debride necrotic tissue and allow oxygen into the region, in addition to systemic +/- local antimicrobial therapy. Intravenous penicillin G is typically effective against these clostridial organisms and is the most common antimicrobial administered to horses with clostridal myonecrosis.[28] However, *in vitro* susceptibility testing suggests that tetracyclines, chloramphenicol and metronidazole are also efficacious.[34,35] Aggressive treatment can be associated with survival rates of up to 80% if myonecrosis is caused by *C. perfringens* alone.[28]

BOTULISM

Botulism is an uncommon but potentially devastating paralytic disease caused by *Clostridium botulinum* neurotoxins. In horses, disease is typically associated with neurotoxin/serotype B, although Types A and C have also been reported.[36-43] In the United States, cases of equine botulism in the Midwestern and Mid-Atlantic states are usually due to Type B, while cases west of the Mississippi River are often due to Type A.[37,39,43] Adult horses typically acquire disease through ingestion of preformed toxin or from contamination of a wound with *C. botulinum* spores, while foals usually ingest spores that then germinate in the intestinal tract.[44-46] Botulism neurotoxins bind to varying SNARE (Soluble NSF Attachment Protein Receptor) proteins at the neuromuscular junction to prevent release of acetylcholine, which leads to paralysis, autonomic nervous system deficits, and occasionally respiratory failure.[47] The most common clinical signs include dysphagia, ptosis, decreased tongue, tail and anal tone, ileus, progressive muscle weakness and hypoventilation.[37,39] Antemortem diagnosis of botulism is challenging. Although the mouse inoculation test is considered the reference standard for detection of botulinum neurotoxins, it is often negative in horses with compatible clinical signs.[40,41] Immediate administration of botulinum antiserum is critical for neutralization of neurotoxins prior to acetylcholine vesicle binding. The currently available trivalent antiserum is more affordable that in previous years, and provides protection against Types A, B and C. Supportive care often comprises the majority of treatment costs and labor, and is highly dependent on the severity of disease. Important consequences may include recumbency and respiratory failure, both of which are difficult to manage, especially in adult horses. With regard to prevention, a *C. botulinum* Type B toxoid is available for use in horses. This vaccine is considered highly efficacious and there are no reports of botulism in fully vaccinated horses.

TETANUS

Tetanus is caused by *Clostridium tetani*, which is ubiquitous in the environment and is a commensal of the gastrointestinal tract of domestic animals. Horses typically acquire tetanus through deep, penetrating wounds that result in anaerobic conditions, although castration sites can also become infected.[48,49] *C. tetani* is then able to germinate and produce two important exotoxins that are responsible for clinical signs. The *C. tetani* neurotoxin tetanospasmin is a lipoprotein exotoxin that is transported along peripheral nerve axons to bind Renshaw cells (interneurons) in the spinal cord. This toxin inhibits release of the inhibitory neurotransmitters glycine and GABA, resulting in hypertonia and muscular spasms. The exotoxin tetanolysin is responsible for spread of infection by causing local tissue necrosis. Clinical signs include hyperesthesia, "flashing" or intermittent prolapse of the third eyelid, extension of the head/neck, trismus, facial grimacing, demonstration of a sawhorse stance, elevation of the tail head and recumbency. Diagnosis is based on compatible history and clinical signs, although culture of *C. tetani* from the suspected site of entry may be helpful. Treatment includes providing muscle relaxation, eliminating the clostridial infection, neutralizing unbound neurotoxins and providing supportive hydration and nutritional care. The goal of treatment with tetanus antitoxin is to bind circulating toxin that has not yet bound to axons or interneurons. However, treatment with antitoxin has not been shown to increase survival rates among treated horses.[48-50] In order to establish an active immune response against tetanus toxins, horses should be immunized with tetanus toxoid at the time of treatment and again 1-2 months later. Reported survival rates among horses presented to referral hospitals with tetanus varies between 25-46%.[48-50]

TYZZER'S DISEASE

The etiologic agent of Tyzzer's disease is *Clostridium piliforme*, which causes a rapidly progressive and highly fatal hepatitis in foals 1-6 weeks of age. Although pathogenesis is not completely clear, it is suspected that ingestion of spore-containing feces from carrier horses leads to colonization in the foal intestine with subsequent infection of the liver.[51,52] Clinical signs are typically nonspecific and include fever, icterus, weakness, diarrhea, dehydration, seizures, recumbency and death. Clinical pathologic abnormalities may include hypoglycemia, increased liver enzyme activity, hyperbilirubinemia, and metabolic acidosis.[53, 54] Antemortem diagnosis is often presumptive, with transabdominal ultrasound often revealing hepatomegaly with increased liver echogenicity. Blood cultures are typically negative, but PCR may be used to identify *C. piliforme* DNA from liver obtained by biopsy. [53]Treatment involves supportive care that includes fluid resuscitation, correction of acid-base or electrolyte disturbances, caloric support with dextrose and parenteral nutrition, and anti-seizure medication as needed. Beta-lactam antimicrobials such as penicillins and ampicillin are also indicated for treatment of *C. piliforme*, as well as an aminoglycoside for gram negative coverage. To date, only one foal with confirmed Tyzzer's disease has survived, reflecting the fatality of the disease.[53]

REFERENCES

1. Tillotson K, et al., **J Am Vet Med Assoc**. 2002;220:342. 2. Herholz C, et al., **J Clin Microbiol**. 1999;37:358. 3. Waters M, et al., **J Clin Microbiol**. 2005;43:4002. 4. East LM, et al., **J Am Vet Med Assoc**. 1998;212:1751. 5. Weese J, et al., **Equine Vet J**. 2001;33:403. 6. Garmory H, et al., **Epidemiol Infect**. 2000;124:61. 7. Siqueira F, et al., **Vet Microbiol**. 2012. 8. Diab S, et al., **Vet Pathol**. 2012;49:255. 9. Uzal F, et al., **Vet Microbiol**. 2012;156:395. 10. Keel K, et al., **J Clin Microbiol**. 2007;45:1963. 11. Rupnik M, et al., **J Clin Microbiol**. 2008;46:2146. 12. Songer J, et al., **J Vet Diagn Invest**. 2009;21:377. 13. Martin H, et al., **J Clin Microbiol**. 2008;46:2999. 14. Thean S, et al., **J Med Microbiol**. 2011;60:1188. 15. Magdesian K, et al., **J Am Vet Med Assoc**. 2002;220:67. 16. Taha S, et al., **Comp Immunol Microbiol Infect Dis.** 2007;30:163. 17. Baverud V, et al., **Equine Vet J**. 2003;35:465. 18. Medina-Torres C, et al., **Vet Microbiol.** 2011;152:212. 19. Schoster A, et al., **Vet Microbiol.** 2012. 20. Ossiprandi M, et al., **Comp Immunol Microbiol Infect Dis**. 2010;33:e25. 21. Arroyo L, et al., **J Med Microbiol**. 2006;55:605. 22. Ruby R, et al., **J Am Vet Med Assoc**. 2009;234:777. 23. Jones R, et al., **J Am Vet Med Assoc.** 1988;193:76. 24. Jang S, et al., **Clin Infect Dis**. 1997;25 Suppl 2:S266. 25. Weese J, et al., **Equine Vet J**. 2003;35:638. 26. Lawler J, et al., **Am J Vet Res.** 2008;69:233. 27. Desrochers A, et al., **J Am Vet Med Assoc**. 2005;227:954. 28. Peek S, et al., **Equine Vet J**. 2003;35:86. 29. Rebhun W, et al., **J Am Vet Med Assoc**. 1985;187:732. 30. Valberg S, et al., **Can Vet J.** 1984;25:67. 31. Nation P, et al., **Can Vet J**. 1983;24:295. 32. Westman C, et al., **J Am Vet Med Assoc.** 1979;174:725. 33. Jeanes L, et al., **Comp Cont Educ Pract Vet**. 2001;23:577. 34. Marrie T, et al., **Antimicrob Agents Chemother**. 1981;19:51. 35. Stevens D, et al., **J Infect Dis**. 1987;155:220. 36. Swerczek T, **J Am Vet Med Assoc**. 1980;176:217. 37. Johnson A, et al., **J Vet Diagn Invest**. 2010;22:165. 38. Kinde H, et al., **J Am Vet Med Assoc.** 1991;199:742. 39. Ostrowski S, et al., **J Vet Diagn Invest.** 2012;24:601. 40. Ricketts S, et al., **Equine Vet J**. 1984;16:515. 41. Schoenbaum M, et al., **J Am Vet Med Assoc**. 2000;217:365-368, 340. 42. Steinman A, et al., **Equine Vet J.** 2007;39:232. 43. Whitlock R, et al., **Clin Tech Equine Pract.** 2006;5:37. 44. Swerczek T, **Am J Vet Res**. 1980;41:348. 45. Wilkins P, et al., **J Vet Intern Med**. 2003;17:702. 46. Vaala W, **Proc 9th Annual Vet Med Forum**. 1991. 47. Simpson L, et al., **Annu Rev Pharmacol Toxicol**. 2004;44:167. 48. Kay G, et al., **Equine Vet Educ.** 2007;19:107. 49. van Galen G, et al., **Tijdschr Diergeneeskd.** 2008;133:512. 50. Green S, et al., **J Vet Intern Med.** 1994;8:128. 51. Trent A, et al. **Equine Pract**. 1983;5:8. 52. Swerczek T, et al., **Mod Vet Pract**. 1973;54:66. 53. Borchers A, et al., **J Vet Intern Med**. 2006;20:1212. 54. Carrigan M, et al., **Aust Vet J**. 1984;61:199.

HYPOCALCEMIA AND ABOMASAL EMPTYING DEFECT OF RUMINANTS

Melissa G. Ackerman, DVM, MS
Pullman, WA

INTRODUCTION

Hypocalcemia

In farm animal practice, there are a number of diseases that are commonly seen. One of these is acute hypocalcemia, also referred to as "milk fever" and parturient paresis. This disease is often seen in dairy cows, but may also be seen in beef cattle, sheep, and goats. Acute hypocalcemia occurs when the animal loses more calcium to milk production or fetal skeletal demands than it is able to take in through diet and also to resorb from bone. There are also subclinical and chronic forms of hypocalcemia which will not be discussed during this presentation.

PATHOPHYSIOLOGY

Calcium is essential for proper nerve and muscle function. It is required to initiate smooth muscle, skeletal muscle, and cardiac contraction. It plays an important role in hormone release, blood coagulation, enzyme activation, and is a second messenger in many intracellular pathways. Calcium is a major mineral of bone and is also a component of milk. Calcium concentration in adult cattle is maintained at 8.5-10 mg/dL. The parathyroid gland regulates blood calcium by secreting parathyroid hormone (PTH) when calcium levels decrease. PTH then acts on kidneys to increase calcium reabsorption and to produce 1,25-dihydroxyvitamin D, which promotes calcium absorption in the gut. PTH also acts on bone to resorb calcium, and can weaken bone and cause disease if low blood calcium is prolonged.

CLINICAL SIGNS

Clinical signs may vary depending on the stage of disease and are related to calcium's physiologic role. The skeletal muscles are unable to effectively contract, which manifests as weakness, recumbency, ataxia, and muscle fasciculations. The heart is unable to contract normally, and it often sounds muffled on auscultation. Because the heart is not pumping blood effectively to the body due to decreased ventricular ejection volumes, the heart rate is increased in an attempt to compensate. The animal loses its ability to thermoregulate and may have cold ears and distal limbs. Animals are often mentally depressed. Because smooth muscle contraction is also affected by hypocalcemia, ruminal atony, stasis, and bloat may be present. Severely affected animals can present in lateral recumbency, laying with their neck in an S-shaped curve. If not treated, an animal may die within hours.

DAIRY CATTLE

An typical adult Holstein dairy cow has about 8.5 kilograms of calcium in her skeleton. Dairy cows secrete between 20 to 30 grams of calcium into their milk on a daily basis early in lactation. This increase in calcium demand comes soon after gestation, when the fetal skeletal calcium demand is already high (can be 8-10g/fetal calf/day). If calcium in the diet does not meet the cow's increased calcium output, and if the cow does not absorb sufficient calcium through her gastrointestinal tract, she can become critically ill when calcium levels drop. Most often this occurs soon after calving, but also can occur late in gestation. In large animals such as cattle, the physical pressure of recumbency will result in ischemic necrosis of the muscles and nerves, also referred to as "compartment syndrome" or "downer cow syndrome".

BEEF CATTLE, SHEEP, AND GOATS

In beef cattle, ewes, and does, especially those carrying multiple fetuses, clinical hypocalcemic disorders often occur in late gestation due to fetal skeletal demand. Lactation tends to be less of a problem because milk production is relatively lower than with dairy cattle. This disease will often occur at times of stress with sheep and goats; for example, if transported, held off feed, or forced into sudden exercise especially at times of handling. In these cases, acute hypocalcemia can present as outbreaks in a flock or herd. They can be prevented by increasing dietary calcium and/or magnesium, provided the animal is eating normally. Oral calcium formulations are also available for this purpose.

TREATMENT

Treatment should be implemented as soon as possible. Some dairy cows require several treatments that can be given based on clinical signs. The quickest way to increase blood calcium levels is to use calcium salts intravenously (IV). Calcium borogluconate is commonly used for this purpose, but commercial formulations may have different calcium concentrations. A dosage of 2 grams of calcium per 100 kilograms of body weight can be used. Because fatal arrhythmias are caused when calcium is administered too quickly, care should be taken to give calcium at a rate of 1 gram per minute.

Perfusion to the periphery is poor with hypocalcemia; therefore subcutaneous (SC) calcium salts may be less effective and not readily absorbed. Additionally, the amount injected SC into a single site should be limited to about 1 gram of calcium, so it will require multiple injections. Calcium salts designed for intramuscular injection are also available.

INTRODUCTION

Abomasal Emptying Defect

Another disease commonly seen in food animal practice is abomasal emptying defect (AED), or abomasal dilation and emptying defect. It is a disease of unknown cause that occurs predominantly in both sexes of adult Suffolk sheep. It is also reported to occur in Hampshire, Dorset, and Texel sheep, as well as Toggenburg goats. This condition is characterized by abomasal dilatation and mechanical transport failure. No hereditary association has been identified at this time and there is no lesion that has been consistently identifiable in all affected animals. Some cases have been associated with winter lambing and feeding of concentrates.

CLINICAL SIGNS

Clinical signs vary depending on the individual and stage of disease. An owner or herd manager may notice an unexplained chronic and progressive weight loss in one animal, though several animals in a flock or herd may be affected at the same time. Anorexia and mental dullness or depression may be present. As the disease progresses, the animal may develop bilateral abdominal distension with or without ruminal tympany. Feces can vary from watery diarrhea to scant pellets to normal. Rumen contractions may be decreased to absent, normal, or increased. A palpable, firm, heavy mass may be present in the right ventral abdomen. Animals may be tachycardic and have abdominal pain. Elevated ruminal chloride levels are commonly reported (normal range is 8-15 mEq/L), suggesting reflux of hydrochloric acid from the abomasum into the rumen. Diagnosis can be difficult and confirmation may require exploratory surgery or necropsy.

PATHOPHYSIOLOGY

The clinical signs described above are due to abnormal motility of the abomasum. Over time, the stagnant feed and fluid back up and the abomasum becomes enlarged. The rumen may also distend as the feed is unable to move normally through the gastrointestinal tract. Elevated rumen chloride levels occur with an obstruction or failure of normal flow of abomasal contents into the small intestine. The cause of AED remains unidentified and the prognosis is grave.

TREATMENT

Medical treatment with cathartics and laxatives are usually ineffective. Drug therapies that are targeted to improve abomasal emptying have been shown to have no benefit, including neostigmine and calcium gluconate. Surgical abomasotomy and evacuation of abomasal contents may lead to death from surgical complications or can result in re-impaction. Abomasotomy in conjunction with metoclopramide has been shown to have mixed results. Treatment with erythromycin may be effective in cattle to improve abomasal emptying rate but the effects in sheep are unknown.

NECROPSY FINDINGS

In animals that die from AED, the abomasum is greatly distended with dry or liquid ingesta that appears similar to rumen contents. The pyloris is patent, and the contents of the rest of the intestinal tract can appear normal. Histologically, reports can vary and may include thin abomasal wall layers, mononuclear cell infiltration within abomasal wall musculature, and chromatolytic change and necrosis in neurons associated with the celiacomesenteric ganglion, and changes in the vagal nerve.

CASE REPORT

This disease is well documented in sheep (Suffolk especially), but goat owners and producers should be aware of the clinical signs. In fact, the Washington State University Veterinary Teaching Hospital has diagnosed several cases of AED in goats. A case report example follows.

A six-year old Alpine/Saanen cross wether presented in September, 2011 with a complaint of a several week history of lethargy and intermittent inappetance. The owner noted that he had lost some weight and had been chewing cud less often than normal. He had been with this owner his entire life, and was used on hiking trips as a pack goat. On the most recent hiking trip several days earlier, he went off feed and became lethargic and weak. He was treated by a referring veterinarian with anti-inflammatory drugs and mineral oil given orally, with little relief.

On physical exam, he was slightly thin with a body condition score of 2/5. He was dull but responsive and mildly dehydrated. His rumen contractions varied from normal to increased at 2-4 per minute. His abdomen appeared distended ventrally on both sides. Deep palpation of the rumen to assess the feed contents indicated a lack of normal layering; it was uniformly firm rather than having a normal, doughy fiber mat above with fluid below. Deep palpation on the right side of the abdomen was abnormal; there was a large, firm, heavy mass felt on the right lower side in the area of the abomasum. Normally this side of the abdomen would feel unremarkable and relatively soft. Scant, dry feces were detected on rectal exam. Passage of a tube from the mouth to the rumen several times resulted in collection of thick, viscous rumen material that was inappropriate for rumen chloride analysis. Complete blood count was unremarkable, and serum chemistry panel showed several non-specific, minor abnormalities such as mildly elevated liver and muscle enzymes and lymphopenia. A diagnosis of abomasal emptying defect was made based on history and physical exam findings. After several treatments of magnesium hydroxide (a cathartic), and transfaunation (rumen fluid from a donor cow) with no improvement, the goat was humanely euthanized. On post-mortem exam, the rumen was mildly distended with coarse feed material. The abomasum was severely distended with long-fiber, impacted feed material, confirming the diagnosis.

REFERENCES

[1]Radostits, O.M., et. al **Veterinary Medicine: A textbook of the diseases of cattle, horses, sheep, pigs, and goats**, 2007, 10th ed., Elsevier, Edinburgh. [2]Smith, B.P. **Large Animal Internal Medicine**, 2009, 4th ed., Elsevier, MO, 863-864, 1369-1373.
[3]http://www.vivo.colostate.edu/hbooks/pathphys/endocrine/thyroid/calcium.html visited on 1/30/13. [4] Ruegg, P.L., et al. **JAVMA** 1988; 193(12): 1534. [5] Smith, R.M., et al. **Vet Record** 1992; 130 (21): 468. [5] Gabb, K., et al. **Vet Record** 1992; 131 (6): 127. [6] Rings, D.M., et al. **JAVMA** 1984; 185 (12): 1520. [7] Edwards, G.T., et al. **Vet Record** 2008; 162: 418. [7] Pruden, S.J., et al. **Vet Path** 2004; 41 (2): 164. [8] Kopcha, M. **JAVMA** 1988; 192 (6): 783.

CANINE URINARY INCONTINENCE LET'S FIX THE LEAK

Amanda J. Callens, BS, LVMT
Knoxville, TN

URINARY INCONTINENCE

Micturition is the physiologic process of storing and voiding urine and consists of a urine storage phase and a urine voiding phase. During normal micturition, urine passes from the kidneys through the ureters and into the urinary bladder, which slowly fills and relaxes. While this filling occurs, the urethra remains closed, holding urine in the bladder. Once the bladder is full, signals are sent through the nervous system, and voluntary bladder contraction causes urine to be expelled through a relaxed urethra.

During the filling phase of micturition, the sympathetic nervous system dominates, causing relaxation of the bladder and contraction of the internal urethral sphincter, preventing urine leakage. As the urinary bladder fills, afferent information is sent via the pelvic nerve. The hypogastic nerve is stimulated when the sensation of overdistention is experienced; this follows stimulation of stretch receptors in the detrusor muscle. The voiding phase is initiated when the bladder reaches its threshold volume. The parasympathetic nervous system becomes dominant and detrusor contraction and urethral relaxation occurs causing expulsion of urine.[1]

Urinary incontinence is the involuntary passing of urine. Causes of urinary incontinence include functional, neurologic and congenital defects. Although there are numerous causes of urinary incontinence, the most common causes are urethral sphincter mechanism incompetence and ectopic ureter. This paper will focus on these two disease processes, including causes and diagnostic and treatment options.

URETHRAL SPHINCTER MECHANISM INCOMPETENCE

The most common cause of urinary incontinence is urethral sphincter mechanism incompetence (USMI). In USMI, the outlet mechanism in the urethra is weak. The typical pattern of incontinence in dogs with USMI is urine leakage while lying down at rest or asleep. The typical signalment for a dog with USMI is a middle aged, large breed, spayed female dog. Up to 20% of spayed female, large breed dogs are affected by USMI.[2] In these cases, USMI is thought to be linked to changes in sex hormones after spay.

There are two urethral sphincters, the internal urethral sphincter and the external urethral sphincter. The internal urethral sphincter consists of smooth muscle and is under unconscious control. So, while at rest, urine is retained in the bladder by contraction of the internal urethral sphincter without the need for conscious control. The external urethral sphincter consists of striated muscle and is under conscious control. Once the bladder is full, signals are released and urine is consciously retained by the external urethral sphincter until the voiding phase is initiated. If the internal urethral sphincter is dysfunctional, USMI occurs and urine leaks while the dog is at rest.

DIAGNOSTICS

Urinalysis

In patients where urinary disorders are suspected, urinalysis is an important part of the diagnostic evaluation. Urinary incontinence is sometimes associated with urinary tract infection. Urinary tract infection diagnosis via urine culture and treatment with appropriate antibiotic as determined by sensitivity testing can improve urinary incontinence. Urine concentration testing in patients with urinary incontinence is important to rule out polyuric disease. Urine specific gravity is the most useful test when looking for urine concentration ability. Patients with USMI typically have the ability to appropriately concentrate urine.

Physical Exam

In dogs with no indication of urinary inflammation or infection, with well concentrated urine, and a normal physical and neurological exam and typical USMI incontinence pattern, USMI is the most likely diagnosis. However, if further investigation is warranted, the following diagnostic methods may be used.

Urodynamic Studies

A urethral pressure profile is a tool used to measure urethral pressures. It is a graphic record of intraluminal pressures exerted by the urethral wall from the bladder to the distal urethral orifice. It may be useful in evaluating patients with suspected USMI, but results are affected by the level of sedation of the patient and the level of striated muscle activity of an unsedated patient. So, results may vary.[3]

Imaging

Radiographic and cystoscopic imaging techniques can be performed to rule out other causes of urinary incontinence. These tests are useful in ruling out physical abnormalities, including congenital malformations and other findings such as urinary calculi. In patients with USMI, all radiographic and cystoscopic exam findings are likely to be normal. The normal findings on advanced imaging can support the diagnosis of USMI.

TREATMENT

Medical

Pharmacologic management of these patients is often utilized. Estrogens are reasonably effective and are often dosed using a loading phase and tapering to the lowest dose possible, tailored for each patient. By increasing alpha adrenergic receptor response, these medications improve urethral resistance. They also improve the vascularity and mucosal characteristics of the urethra.[1]

Sympathomimetics such as phenylpropanolamine are also commonly used in patients with USMI. Sympathomimetics have alpha adrenergic effects and increase the sensitivity within the urethra. Estrogens and sympathomimetics can be used in conjunction and are thought to have a synergistic effect.[1] Unfortunately, over time, some patients become refractory to medical treatment, and other options must be considered.

Surgical

A hydraulic urethral occluder is an inflatable silicone cuff with actuating tubing that attaches to a titanium inflation injection port with a silicone diaphragm. The cuff is surgically placed around the proximal urethra, and the actuating tubing is passed through the body wall. The tubing is connected to the injection port which is secured subcutaneously to the abdominal wall. The occluder is a permanent implant and can be adjusted through the injection port at any time if urinary incontinence recurs. In one study the urethral occluder was found to be an effective treatment with all cases reporting improved urinary continence scores and all owners reporting they would have the same treatment preformed again if needed.[4] The large advantage of the urethral occluder is that it is a permanent implant the can be adjusted at any time. So, in patients where urinary incontinence recurs, this is advantageous, as USMI is a lifelong disease.

Culposuspension is the most common surgical treatment described for USMI. The goal of this procedure is to reposition the urinary outflow tract from the pelvic region to the intra-abdominal region with the goal of placing the urethral sphincter in a higher pressure zone. However, most studies have found this surgical method to have a less than desirable efficacy.[5]

Interventional

Urethral bulking is a minimally invasive treatment option for urinary incontinence. During urethral bulking, a cystoscope is passed transurethrally, and a urethral bulking agent is injected submucosally while inside the proximal urethral lumen. This bulks the area, increasing urethral pressure and promoting scar tissue formation and blood vessel growth. There have been numerous products used for urethral bulking, including collagen and silicone-based products. Long-term effects of collagen injection in one study was determined to be satisfactory, where 68% of dogs were continent 1-64 months after collagen injection.[6] Urethral bulking appears to be a reasonable treatment option for patients that are non-responsive to medical treatment or whose owners prefer not to medicate.

PROGNOSIS

In dogs, USMI is the most common disease causing urinary incontinence. Noninvasive medical treatment is quite effective, and minimally-invasive treatment options are available. Hydraulic urethral occluders are fairly new to veterinary medicine and have been well tolerated and had positive results in canine patients.

ECTOPIC URETER

Congenital defects are another cause of urinary incontinence. These abnormalities include ectopic ureter and urethral hypoplasia. The most common congenital defect causing urinary incontinence is an ectopic ureter. A normal ureter travels from the kidney to the dorsolateral caudal surface of the bladder, has a short intramural course, and then empties into the trigone. An ectopic ureter is displaced and empties distal to the urethral sphincter. This causes constant urinary incontinence, and these patients tend to leak urine continuously when awake and asleep. Patients with ectopic ureters are born with the defect, so the incontinence is present at birth.

Ectopic ureters can be unilateral or bilateral and may be intramural or extramural. An intramural or tunnelling ectopic ureter exits the kidney and enters the bladder at the appropriate location, the bladder trigone, but does not terminate at the bladder. Instead, the ureter tunnels down the urethra through redundant ureteral tissue and empties distal to the bladder and urethral sphincter. This is by far the most common type of ectopic ureter seen.[7]

An extramural ectopic ureter exits the kidney and bypasses the bladder altogether. This ureter enters into an inappropriate location such as the urethra, vagina, or vestibule, bypassing the bladder and urethral sphincter.

Clinical signs of ectopic ureter include intermittent or continuous urinary incontinence since birth. Normal urine voiding may be observed, and the degree of urinary incontinence varies between patients. Some patients may have more severe incontinence when in certain positions, such as lateral recumbence. Ectopic ureters are seen most commonly in large breed female dogs, with a greater frequency in Labrador retrievers, Golden retrievers, and Siberian huskies.[8]

DIAGNOSTICS

Physical Exam

Physical exam in patients with ectopic ureters is often unremarkable, with the exception of moist fur or urine staining in the perivulvar area. Secondary pyoderma in this region due to urine scald may be noted. A thorough neurologic exam must be performed to rule out any neurologic cause for the incontinence.

Urinalysis

A urinalysis is part of the minimum database on all patients with urinary disorders. Many patients with incontinence have concurrent urinary tract infections, so urinalysis and urine culture and sensitivity testing should be performed. If a urinary tract infection is present, it should be treated with antibiotics deemed appropriate by the sensitivity report. Urine specific gravity should be assessed to look for patient ability to concentrate urine. With decreased urine concentration, the patient should be further assessed to look for polyuric disease.

Radiographic Techniques

Ultrasonography and excretory urography, such as an intravenous pyelogram, may help provide information regarding location and size of kidneys and ureters, as ectopic ureters are often dilated. Excretory urography contrast procedures may be helpful in revealing other abnormalities of the urogenital tract such as hypoplastic urethra or a caudally displaced bladder. With this diagnostic method, however, it can be difficult to delineate the terminal segment of the ureters; furthermore, this method is not definitive to rule out ectopic ureters.

Contrast computed tomography (CT) is a very sensitive imaging technique to determine the presence of ectopic ureter. A CT scan gives transverse images of the abdomen and allows for tracking of the ureters from the kidneys into the lower urogenital tract. Also, if an intramural ectopic ureter is present, CT provides information on how far the ectopic ureter tunnels through its intramural course before terminating, and therefore gives a guideline for how far to ablate the ureter back without danger of perforating the bladder and exposing the abdomen to urine excretion. A CT scan also provides information regarding the rest of the urogenital tract, looking at kidney integrity, bladder location, and urethral length.

Cystoscopy

Direct visualization of the lower urogenital tract using rigid or flexible cystoscopy dramatically improves the ability to diagnose ectopic ureters and other abnormalities in this area.[8] Cystoscopy provides magnified images of the internal structures of the distended lower urogenital tract. During cystoscopy, the entire lower urogenital tract is inspected and the ureteral orifices identified. Normally, the ureteral orifices have a C-shaped appearance in the bladder trigone, and urine can be seen occasionally pulsating from the orifices. The lack of either ureter in this area is diagnostic for ureteral ectopia. In this case, the entire urethra is examined by slowly withdrawing the cystoscopy from the bladder down the urethral lumen to look for the termination of a ureter. Oftentimes, with intramural ectopic ureters, the ureter terminates within the urethra and can be visualized with the cystoscope. If a ureter is not seen terminating in the urethral lumen, then the vestibule and vagina should be carefully examined for the presence of a ureteral opening. If a ureteral opening is seen in either the vestibule or vagina, with no evidence of tunnelling, it is diagnostic of an extramural ectopic ureter.

TREATMENT

Interventional

Cystoscopic-guided laser ablation of intramural ectopic ureters is an effective, safe, and minimally invasive alternative to surgery. This procedure is performed transurethrally via cystoscopy using either a diode or holmium: yttrium aluminum garnet laser.[9] The excess tissue tunnelling down the urethra is ablated to the bladder neck, proximal to the urethral sphincter. During this procedure a laser fiber is passed through the cystoscope, and a red rubber catheter is inserted into the ectopic ureter lumen to protect the outer urethral wall from laser damage. The laser fiber is positioned appropriately against the inner wall of the ectopic ureter, and the wall is ablated. Care must be taken to ensure that the ureteral wall is not ablated back too far as this can result in laser dissection of the bladder wall and extravasation of urine into the abdominal cavity.[9] This is where having additional information on intramural ureteral length via contrast CT is helpful.

Surgical

Extramural ectopic ureters require surgical repair. During this procedure, the extramural ectopic ureter is ligated and the distal end is implanted into the urinary bladder between the apex and the trigone. This procedure is effective in some cases, but persistent urinary incontinence is the most common complication after surgical repair.[5]

PROGNOSIS

Concurrent urinary bladder and urethral function abnormalities have been noted in patients with ectopic ureter. These abnormalities include urinary tract infection, USMI, hydroureter, hydronephrosis, and aplastic urethra.[8] In patients with multiple urinary defects, correction of the ectopic ureter may not be fully curative. Additional treatment such as medication or episioplasty for chronic urinary tract infections, and medication, urethral bulking and urethral occluder placement for USMI may be necessary.

Cystoscopic-guided laser ablation is an effective method in treating intramural ectopic ureters.[9] This procedure allows the clinician to provide a minimally-invasive treatment option and is met with few complications. Surgical procedures aimed at repositioning extramural ectopic ureters are also available and effective in some patients.[8]

REFERENCES

[1]Bartges JW, et al. Nephrology and urology of small animals. In: Fischer J, et al. eds. *Mictrition Disorders*: Wiley-Blackwell, 2011. [2]Thrusfield MV, et al. *Vet Rec* 1985;116:695. [3]Bartges JW, et al. Nephrology and urology of small animals. In: Lane I, et al. eds. *Urodynamic Studies*: Wiley-Blackwell, 2011. [4]Delisser PJ, et al. *J Small Anim Pract* 2012;53:338. [5]Rawlings C, et al. *J Am Vet Med Assoc* 2001;219:770. [6]Barth A, et al. *J Am Vet Med Assoc* 2005;226:73.[7] Bartges JW, et al. Nephrology and urology of small animals. In: Segev G, et al. eds. *Diseases of the Ureter*: Wiley-Blackwell, 2011. [8]McLoughlin MA, et al. *Clin Tech Small Anim Pract* 2000;15:17. [9]Berent AC, et al. *J Am Vet Med Assoc* 2012;240:716.

ENDOCRINOPATHIES: "HAPPENING HORMONES"

Chastity Ketchum, CVT VTS (SAIM)
Sarasota, FL

The common clinical signs of an endocrinopathy include: polydipsia, polyuria, polyphagia, weight loss/gain, appetite changes, hypotrichosis, and pyoderma.

The most recognizable clinical sign that owners will pick up on are polyuria and polydipsia. Water consumption and urine production are controlled by many factors, of them plasma osmolarity, the kidneys, pituitary gland, and hypothalamus are among the most influential. Diseases with in any of these organs can cause a disruption in the body's normal water balance. When obtaining a thorough history from an owner it is important to ask an open-ended question pertaining to water consumption and urine production.

Diabetes mellitus

Possibly all dogs at the time of diagnosis have insulin-dependent diabetes mellitus (IDDM) or type II diabetes. The cause for diabetes in dogs is likely multifactorial and the end result is the same; the permanent loss of pancreatic B-cell function, hypoinsulinemia. I feel that it is important for all technicians to understand that diabetes (especially in dogs) is a "not enough insulin" problem, versus excessive glucose.

Cats are unique in the etiology of diabetes, treatment and management of diabetes. Cats can have both IDDM and non-insulin dependent diabetes mellitus (NIDDM). Some patients will even wax and wane between both IDDM and NIDDM. A suspected 20 % of cats with IDDM become "transiently diabetic". It is very important to try to determine if there is an underlying condition (s) that can be causing (endogenous and exogenous) insulin resistance.

Technicians play an important key-role in the diagnosis, treatment and management of diabetes in dogs and cats.

Hyperthyroidism

The thyroid is a two-lobed gland located in the neck of people, dogs, cats, and other animals. One lobe is on each side of the trachea (windpipe). The thyroid produces thyroid hormone, a substance that is transported via the blood to all cells in the body. Hyperthyroidism is a common disease in middle-aged to older cats. Common clinical signs of hyperthyroidism are increased hunger, weight-loss (with an increased appetite), poor hair coat, behavioral changes, racing heart-rate and elevated liver enzymes (on routine blood-work). Hyperthyroidism is typically easily diagnosed with a physical exam, and a total T4. About 15% of hyperthyroid cats have a single thyroid tumor in one of the two lobes, called an *adenoma* (not a cancerous tumor) that produces too much thyroid hormone. About 80% of hyperthyroid cats have abnormalities in both thyroid lobes with both lobes producing excess thyroid hormone. These cats also have a benign (noncancerous) condition called *adenomatous hyperplasia*. Only about 3% of hyperthyroid cats have a malignant (cancerous) thyroid tumor. The veterinary technician plays an important role in diagnosing this common disease. Particularly with a good history and patient assessment the veterinary technician can become a significant part of discovering this disease in patients.

Primary Hypothyroidism

Hypothyroidism commonly affects middle-aged (2-6 years) dogs. Clinical signs for hypothyroidism in dogs are variable but may include dermatological signs, neuromuscular signs, reproductive, and uncommonly gastrointestinal, cardiovascular, and/or clotting abnormalities. The veterinary technician has the opportunity to ask specific questions regarding changes in these organ systems, appropriately perform thyroid function test, and participate in the management of hypothyroidism in patients.

Hyperadrenocorticism (Cushing disease)

Pituitary-dependent hyperadrenocorticism (PDH) is the most common cause of hyperadrenocorticism, responsible for about 80-85 % of diagnosed cases. There are also two other types of hyperadrenocorticism: 1) adrenocortical tumors (15-20 % of cases) and 2) iatrogenic hyperadrenocorticism (< 5 %). Clinical signs for hyperadrenocorticism include polyuria, polydipsia, inappropriate urination, polyphagia, panting, weight-gain, lethargy, generalized muscle atropy, and symmetrical alopecia or generalized hypotrichosis (thinning hair coat). The diagnosis and treatment for these patients can be complicated in some and easily accomplished in others. Veterinary technicians who communicate with the owners of these patients effectively help to ensure successful treatment/management of these patients.

Hypoadrenocorticism (Addison's disease)

There are two classifications of hypoadrenocorticism, primary and secondary adrenocortical insufficiency. The most common form is primary adrenocortical insuffiency. Primary adrenocortical insufficiency affects both mineral and glucorticoid secretion. Hypoadrenocorticism usually affects young to middle aged female dogs, and is very rare in cats. It is important to note that hypoadrenocorticism is diagnosed in older dogs and cats. There are many diverse clinical signs of hypoadrenocorticism. The most common clinical signs are waxing and waning gastrointestinal upset, lethargy, weight-loss, and weakness. An adrenocorticotropic hormone (ACTH) stimulation test is used to confirm and diagnosis hypoadrenocorticism. A veterinary technician's ability to appropriately recognize electrocardiogram (ECG) changes, clinical signs, signalment and thorough history taking can make a huge impact on diagnosis and treatment of hypoadrenocorticism.

DIABETES INSIPIDUS

Colleen Ruderman, RVT, VTS (SAIM)
Atlanta,Ga

INTRODUCTION

Diabetes insipidus is a disorder characterized by a lack in the production or response to antidiuretic hormone (ADH), also known as arginine vasopressin (AVP).ADH is responsible for urine concentration. This lack of ADH or response causes a decrease in tubular reaborption of water causing a lack of urine concentration and subsequent chronic polydipsia, polyuria and hyposthenuria. Diabetes insipidus must be distinguished from other disorders that also cause similar symptoms of polyuria and polydipsia. Hypoadrenocorticism, hyperadrenocorticism, chronic or acute renal disease, psychogenic water consumption, liver disease, urinary tract infection, hypercalcemia, and diabetes mellitus are the most common.

There are two types of diabetes insipidus, central (CDI) and nephrogenic (NDI), which can then be further classified as acquired (secondary) and congenital (primary). CDI is poor secretion or production of ADH and can be complete or partial. Complete CDI creates consistent hyposthenuria (USPG <1.008), where as partial CDI can have urine concentrations ranging from 1.008-1.020. Congenital CDI and NDI are extremely rare. Acquired CDI can be caused by trauma (permanent or temporary), neoplasia (primary or metastatic), malformations, or can be considered idiopathic. Acquired NDI is a result of other, primary diseases processes which causes a reduced number of receptors to ADH and interferes with normal mechanisms.

PATHOPHYSIOLOGY

Antidiuretic hormone helps regular water intake and excretion. It is synthesized within the hypothalamus and released by the pituitary gland. ADH then reacts with receptors on the surface of the renal tubules which then creates aquaporins, which are channels involved in water transport. Aquaporins form on the surface of the renal collecting tubule and permit renal uptake of free water and subsequent concentration of urine. Once the aquaporins are formed, water further moves into the renal interstitium via concentration gradients. ADH and this mechanism control water reabsorption, urine production and concentration, and subsequently water balance within the body. Excretion of ADH is caused by several factors. The first being a change in serum osmolality and can be as small as a 1% increase (normal 280-310 mOsm/kg). ADH secretion can also be induced by pain, fear or anxiety, in addition to a decrease in circulating blood volume.

SIGNALMENT

Congenital CDI and NDI are typically noted in patients before 1 year of age and are extremely rare. Congenital NDI has not been reported in cats. Acquired NDI and CDI can present at any age depending on the cause if identified, however patients with acquired CDI related to neoplasia are typically older. There is no obvious breed or sex disposition however there are cases of an acquired NDI associated with pyometra. In these cases bacterial endotoxins compete with the ADH receptor sites on the renal tubular membrane.

CLINICAL SYMPTOMS

The most significant clinical symptoms are severe, chronic polyuria and polydipsia. Neurologic changes can be noted in cases of acquired CDI in relation to trauma or neoplasia. These can include mental dullness, behavior changes, and seizures or stupor. When dealing with acquired CDI or NDI any sign of illness can be present, such as weight loss, poor or increased appetite, and vomiting or diarrhea. Normal water intake for dogs and cats is 20-70 mls/kg/day, with a normal urinary output between 20-45 ml/kg/day. True polydipsia is defined as a water intake greater than 80-100 ml/kg/day, and polyuria is a urinary output greater than 50 ml/kg/day. While it is important to know the ranges for PU/PD, it is often difficult to determine due to multiple pet households and the living environment of the patient. In addition water intake and urinary output measured in a hospital setting may not accurately represent the patient in its normal environment.

INITIAL DIAGNOSTICS

Initial diagnostics are not typically used to diagnose DI, but are used to rule out any causes of PU/PD other than diabetes insipidus. The start of the workup should include a detailed history, including duration of polyuria/polydipsia, as well evaluation of home life to determine if the PU/PD is occurring in response to physical stress, pain, or anxiety. In these cases clinical symptoms may be attributed to certain times of day or during certain activities. A complete medical history should also be obtained to evaluate the patient for any history of trauma which could indicate possible CDI.

Bloodwork including complete blood count, serum biochemical analysis, and urinalysis can be unremarkable with the exception of hyposthenuria. In cases of acquired NDI secondary to liver, renal, or infectious causes, azotemia, proteinuria, leukocytosis, or other abnormalities can be noted. A persistent hyposthenuria should also be proved by serial sample testing. A urine culture should also be performed to rule out urinary tract infection. This may be accomplished by having the owner collect samples at home during different periods of the day and week. In cases where owners have been restricting water intake (which should be discouraged) general bloodwork may show dehydration (erythrocytosis, hyperproteinemia, azotemia, and electrolyte abnormalities). Other testing that is often performed to rule out common primary disorders includes ACTH stimulation tests, low dose dexamethasone suppression test, and thyroid levels. Hyperadrenocorticism and hyperthyroidism are common causes for hyposthenuria in dogs and cats.

Imaging in patients with congenital CDI or NDI is often unrewarding, however in acquired cases may be helpful in determining a primary cause. Thoracic and abdominal radiographs may indicate existing diseases such as neoplasia, renal disease, and pyometra. Advanced imaging including computed tomography (CT) and magnetic resonance imaging (MRI) can also revealed central lesions or trauma potentially causing forms of CDI.

DIAGNOSTIC TESTING

Serum osmolality level may be performed. In patients with CDI serum osmolality is typically high normal to elevated (280-320 mOsm/kg) and in psychogenic polydipsic patients levels are typically low normal to decreased (<275 mOsm/kg), however this is not specific for diabetes insipidus.

Exogenous AVP administration can help differentiate between CDI and NDI, is easily performed, and does not carry the risks associated with other testing such as the modified water deprivation test. This uses a synthetic AVP (desmopressin/DDAVP) that is administered either orally (0.1mg orally TID per 20kg of body weight) or as a nasal spray administered into the conjunctival sac as drops (1 to 4 per day). An increase in urine specific gravity (USPG) of 50% or consistently above 1.030 occurs in patients with CDI, in addition to improved clinical symptoms. A poor increase is seen in patients with congenital NDI as they do not respond to exogenous AVP. However, this response test can also cause improved urine concentration in other disorders including psychogenic polydipsia and hyperadrenocorticism.

The modified water deprivation test can be performed once all other causes of polyuria/polydipsia have been ruled out. This test will help determine if ADH is released in response to dehydration and if it is being received by the kidneys appropriately. This test is not often performed and is contraindicated in patients with renal disease. It requires close monitoring for evidence of dehydration and clinical illness. It typically lasts 6-12 hours, however a 3-5 day limitation of oral water intake should be performed until a 100ml/kg/24 hours is reached to reverse medullary washout. This may be done gradually at home with client being educated on the signs of dehydration, electrolyte derangement, and clinical illness (ataxia, mental dullness, and seizures/coma). Once this water intake is reached a 12 hour fast of food and water should be performed monitoring for the same complications of dehydration.

Following a 12 hour fast the bladder should be completely emptied. This may be done in dogs with walking, or a urinary catheter may be placed. In feline patients a urinary catheter must be placed. In patients that urinate freely an empty bladder may be confirmed with ultrasound or radiographs. A baseline body weight, USPG, BUN, electrolytes, and samples for urine/serum osmolality are taken. The bladder is them emptied every 1-2 hours in addition to body weight (performed on the same scale each time), BUN, serum electrolytes, USPG, and physical examination to evaluate for clinical dehydration and hypernatremia. At each time samples may also be collected and labels for urine/serum osmolality. This test should continue until 5% of the patient's body weight has been lost (typically occurs within 3-10 hours in patients with DI), or the USPG is greater than 1.030. If the patient becomes clinically dehydrated or ill the test should be stopped immediately. Once dehydration occurs serum is obtained for vasopressin levels.

The next stop is to monitor the response to exogenously administered AVP at 2-5u given IM. Subcutaneous administration is not recommended due to variable results. Desmopressin (DDAVP) has been used however has not been completely evaluated for efficacy in this test. The water/food fast is continued and the urinary output is monitored every 30 minutes for 1-2 hours in addition to USPG, BUN, electrolyte, urine/serum osmolality, and physical examination. Following the completion of the test 10-20mls/kg of water is offered every 30 minutes for 2 hours. If no adverse issues are noted water may be given ad lib.

In normal patients a 5% body weight loss associated with dehydration will typically take 24 hours or more to achieve. Once the test is complete diagnosis is made by evaluation of urine concentration and osmolality. Normal patients will have a urine osmolality >1100 mOsm/kg and a USPG greater than 1.030 after dehydration. Patients with CDI (complete) cannot concentrate urine to these levels even with dehydration. If a dehydrated patient has a significant increase in urine concentration after AVP is given, ADH production or release is considered insufficient. This test can help determine partial or complete CDI. Partial CDI patients will have a further 10-50% increase in urine osmolality (typically >300 mOsm/kg) following dehydration and the administration of vasopressin. Patients with primary NDI cannot concentrate urine or improve osmolality in the face of dehydration or following vasopressin administration.

TREATMENT

For cases of congenital or acquired CDI desmopressin (DDAVP) may be administered either orally or using nasal spray delivered into the conjunctival sac. The oral medication is given at 0.1mg orally TID per 20kg. Oral tables have a low bioavailability and higher doses are typically needed, which can increase cost. The nasal spray is administered at 1-4 drops into the conjunctival sac once to twice a day. With both the oral and nasal forms of treatment the doses are tapered so the lowest dose possible that controls the clinical symptoms is achieved. This treatment is often costly, for this reason it may be used intermittently to limit symptoms during the most inopportune times or during times of severe PU/PD.

Chlorpropamide is a human medication used to control type II diabetes mellitus. This medication may also be helpful in treating CDI and may increase ADH stimulation of the kidney. This medication may help treat mild cases of DI. This medication does have an adverse effect of hypoglycemia so regular feeding should be stressed. Thiazide diuretics (hydrochlorothiazide, chlorothiazide) diuretics may reduce polyuria in both CDI and NDI cases. If using thiazide diuretic monitoring of serum electrolytes is recommended as iatrogenic derangements can occur. A low sodium diet (0.6g/1000 kcal ME) has also been used to reduce urinary output in DI patients. Treatment for DI is not required as long as the patient has unlimited access to water and can be kept in an environment that allows for free and frequent urination. .

Significant client education must take place in cases of diabetes insipidus. Dehydration is a significant risk in these patients and can occur with water deprivation in as little as 1-2 hours. For this reason water restriction should never take place. In circumstances where the animal is ill and anorexic or vomiting, severe dehydration can occur and immediate medical treatment should be provided. Due to dehydration and hypotension in these patients anesthetic events should be monitored closely and in patients with DI that require fasting prior to procedures intravenous fluid therapy should be administered.

Prognosis in patients with congenital CDI that are treated appropriately is good, however due to cost of treatment euthanasia is often chosen depending on the owners' ability to treat or deal with polyuria/polydipsia in their home. Untreated patients are more at risk for dehydration and typically have a more guarded prognosis. Animals with acquired CDI or primary NDI typically have a poorer prognosis due to the underlying disease or lack of treatment options.

REFERENCES

1. Kerl, Marie et. al, Standards of Care, March 2008, vol 10, no 2
2. Rossie, Tracey et. all, Compendium, Jan 2008, vol 30, no 1
3. Ettinger, Stephen, Textbook of Veterinary Internal Medicine, 5th edition, pp1374-1379
4. Feldmen, et. al, Canine and Feline Endocrinology and Reproduciton, 3rd edition, pp 2-44
5. Merrill, Linda, Small Animal Internal Medicine for Veterinary Technicians, pp13-14

PROTEIN LOSING ENTEROPATHY

Colleen Ruderman, RVT, VTS (SAIM)
Atlanta, GA

INTRODUCTION

Protein losing enteropathy (PLE) is defined as the excessive loss of plasma proteins in the small intestine. The most common associated conditions of PLE include inflammatory bowel disease (IBD), most commonly lymphocytic plasmacytic enteritis (LPE) and lymphangiectasia. Fungal, viral, parasitic and anatomical problems can also cause a protein losing enteropathy. Examples of these are histoplasmosis, parvovirus, erosion, damage/trauma and neoplasia. While PLE can affect cats and is most often associated with lymphoma, this is primarily a canine disease. During protein losing enteropathy the intestinal mucosa is affected by increased permeability through inflammation or lymphangiectasia, which causes the loss of chyle and components from dilated or ruptured lacteals causing a panhypoproteinemia.

PATHOPHYSIOLOGY

Intestinal lymphangiectasia (IL) can be further divided into two types, primary (congenital) or secondary (acquired). It is typically considered idiopathic when a primary cause cannot be determined. Congenital lymphangiectasia is less common. Either type of IL is typically characterized by fat malabsorption. Other acquired forms of IL are associated with lymph obstruction or venous hypertension and can be a result of right sided heart failure/constrictive pericarditis and neoplasia. Fat absorbed from the intestinal lumen accumulates in dilated lacteals. Increased hydrostatic pressure within the lacteals eventually leads to rupture and loss of lymph into the GI tract. The fluid accumulation in the GI tract affects nutrient absorption leading to panhypoproteinemia and clinical symptoms of PLE. With IL a loss of lymphocytes into the lumen of the GI tract can also lead to lymphopenia.

Protein losing enteropathies can affect any canine patient however Yorkshire Terriers, Wheatons, and Lundehunds are breeds that are commonly affected by this disease. Lundehund Syndrome is a specific PLE that affects Norwegian Lundehunds and includes lymphangiectasia, inflammatory bowel, and malabsorption. Many Lundehunds that do not exhibit clinical symptoms of gastrointestinal disease often have detectable lesions/abnormalities on necropsy.

CLINICAL SYMPTOMS

Clinical symptoms of PLE include chronic intermittent small bowel diarrhea, weight loss, poor appetite, peripheral edema, and abdominal distention/dyspnea secondary to effusions which may occur in patients with albumin levels less than 2.0 g/dl. Some patients may display a generalized lack of clinical symptoms and can have only weight loss as a result of poor nutrient absorption. Hypocalcemia related symptoms have also been noted (tentany, seizures) but are generally uncommon. Severely cachexic and untreated patients can present in hypovolemic and/or septic shock.

DIAGNOSTICS

Biochemical serum analysis will typically show a panhypoproteinemia, with hypoalbuminemia being the most common laboratory abnormality with PLE. It is possible to have hypoalbuminemia with normal or elevated globulins in some cases. Hypocalcemia, hypomagnesemia, hypocholesterolemia (secondary to cholesterol lost in lymphatic fluid), and occasionally elevated liver enzymes can also occur with PLE. A complete blood count often reveals a lymphopenia, and in the face of possible parasitic infection or eosinophilic enteritis, eosinophilia can also be seen. A urinalysis is typically unremarkable; however proteinuria should be quantified using a urine protein creatinine ratio (UPC) to rule out other causes for hypoproteinemia such as protein losing nephropathy, which can be seen in conjunction with PLE in some patients. Abdominal and thoracic radiographs may reveal effusions, and abdominal ultrasound will typically show enlarged gastrointestinal/mesenteric lymph nodes and GI thickening.

Additional testing includes fecal a-PI testing can further show the protein level in the feces and can help identify patients with protein loss before clinical disease is present. This can be helpful in determining if early PLE is occurring with mild gastrointestinal signs of predisposed breeds. B12 (cobalamine) levels can also be performed and are typically a marker for ileal disease. Dietary B12 is bound to animal protein within the food and absorbed within receptors located in the ileum. Ileal diseases such as PLE can destroy B12 receptors causing malabsorption of dietary B12. Serum cobalamin levels may be helpful in determining if further supplementation is needed. It is important to note that serum B12 levels may not accurately represent levels on the cellular level. Pancreatic lipase immunoreactivity (PLI) testing may also be helpful in determining other causes for gastrointestinal symptoms. Increased concentrations in dogs with IBD have been associated with a poorer response to treatment. In addition, a full fluid analysis on effusions will typically yield a pure transudate.

The gold standard in diagnosing PLE is gastrointestinal biopsy. This is typically done with endoscopy however laparoscopic and laparotomy may also be performed to obtain full thickness intestinal biopsies. Endoscopic biopsies are often preferred due to shortened anesthetic time and a decrease in postoperative complications associated with hypoproteinemia. It is important to note that hypotension is common in hypoproteinemic patients and perioperative colloidal support is recommended. Protein bound medications should be avoided or reduced, in addition to dose reduction of anesthetic agents in the face of patients with poor body mass. While fresh frozen plasma offers no long term help with hypoproteinemia it may also be used to help support oncotic pressure.

TREATMENT

Once a diagnosis is obtained, treatment is aimed at the primary cause if identified, such as removal of specific lesions and treatment of infectious causes, and often leads to a quick resolution of PLE symptoms. When a primary cause cannot be identified the treatment is multimodal and includes nutritional, medical, supportive, and supplementary treatments.

Nutritional treatment is an integral part of treatment of PLE, with novel protein, hydrolyzed protein, or low fat diet. Dietary management of IBD may include the transition to a diet with a highly bioavailable novel protein source or a hydrolyzed protein diet. A novel protein source should be determined after a complete history of dietary exposure has been obtained. A hydrolyzed diet typically contains a familiar protein source but has undergone hydrolysis to provide low molecular weight proteins which are less antigenic. In addition hydrolyzed diets can have increased absorption due to their small molecule size. When using a novel protein diet it is strongly recommended to choose prescription based diets as these are highly regulated and minute particles of non-novel protein sources may be present in over the counter limited ingredient diets. When using novel protein diets it is recommended to change them periodically if sensitivity to the new protein occurs. Some references recommend a change every 6-8 weeks after successful dietary therapy.

Some patients may benefit from a highly digestible, low fat content and high bioavailability of protein. Long chain triglycerides in dietary fat increase lymph protein and flow and worsen protein loss in the gut. Fat content for patients with IL should be kept to <10% dry matter base (DMB) for dogs and <15% for cats, and protein levels should exceed 25% DMB for dogs and 35% for cats. Severe cases may require fat content as low as 5-7% DMB. If additional protein is needed cooked egg white may be added to commercial diets as it is the protein source with the highest biologic value. It is important to note that low fat diets are not calorie dense and the patient should be monitored closely for continued weight loss or poor weight gain after treatment has been started. Diets that are high in insoluble fiber such as weight loss diets have a low energy density that may also impede weight gain. In addition insoluble fiber binds digestive enzymes and reduces pancreatic enzyme activity and can further inhibit the absorption of nutrients. With either a low fat, highly digestible diet, a novel protein diet, or a hydrolyzed diet it is important to calculate resting energy requirements based on the patients ideal body weight, not their current body weight.

Initially patients may be inappetant or anorexic, so a short term home cooked low fat diet of a lean protein source (chicken/turkey) and a highly digestible source of starch, such as white rice or potato, and can be used until the appetite returns, most often after initial medical treatment with immunosuppressive medications. If a long term home cooked diet is deisred, the recipe used should be formulated with the guidance of a veterinary nutritionist to prevent nutrient imbalances. Long term home cooked diets are often vitamin and mineral deficient and imbalanced and should be fed for no longer than three weeks. Severely cachexic patients with PLE may require parenteral nutrition. Calories can be given peripherally by using an isomolar 20% lipid solution piggy backed with standard fluid therapy. This may be helpful prior to invasive diagnostics such as endoscopy or laparotomy.

Dietary therapy alone may not yield success in severe gastrointestinal diseases such as PLE/lymphangiectasia, but may be beneficial in early forms of the disease. Medical therapy will typically begin with anti-inflammatory and immunosuppressive doses of prednisone (1-3 mg/kg/day) with a tapering dose based on clinical symptoms, resolution of effusion and normalization of protein levels. Prednisone usage for PLE is typically longer that with other forms of IBD and side effects can be severe. Longer term use can cause muscle wasting and a potbellied appearance that is not always associated with abdominal effusion. Azathioprine (1-2 mg/kg q24-48) can also be used as a second line therapy but requires laboratory monitoring for bone marrow toxicosis and hepatotoxicity. Budesonide is a glucocorticoid that can yield a similar response as prednisone with fewer side effects, however is more likely to cause iatrogenic hypoadrenocorticism with a quick withdrawal. Immunosuppression with cyclosporine may also be an acceptable second or third line treatment, however side effects are primarily gastrointestinal and it can also be cost prohibitive for larger patients. Metronidazole (10-15mg/kg PO BID) can also be helpful in treating clinical symptom and has immunomodulating and anti-diarrheal properties.

Additional supplementation may also aid in weight gain and resolution of clinical symptoms of PLE. Cobalamin deficiency is often seen with gastrointestinal disease and serum B12 levels may not accurately represent the amount of cobalimin on a cellular level. For this reason for parenteral supplementation in patient with a low-normal cobalamine level is recommended. However, there is no evidence that oversupplementaiton of B12 causes clinical disease, so supplementation in most patients with PLE is considered safe. In addition poor absorption of dietary B12 in the face of gastrointestinal disease makes parenteral cobalimin ideal. Fat soluble vitamins (D, A, K, E) may also be lost as PLE progresses and may compound other problems. Vitamin K may be administered orally and D, A, and E can be given as an injection. Although controversial, medium chain triglycerides (MCT) can also be used to increase the calorie content of low fat diets in patients that are unable to maintain or gain weight. It is typically given as an oil at 1-2 ml/kg/day however palatability problems have been noted in addition to excessive MCT being excreted in diarrhea. MCT oil has been associated with hepatic lipidosis in cats. Supplementation of probiotics and prebiotics may also be helpful in the resolution of absorption and clinical symptoms and are easily given. It should be noted that when using novel protein diets some probiotics/prebiotics may contain non-novel proteins and ingredients should be evaluated closely.

In addition to dietary, medical, and supplemental care supportive medical care is often needed, especially in cases where significant pleural or abdominal effusion is present. Therapeutic thoracocentesis can stabilize dyspneic patients, and therapeutic abdominocentesis can relieve discomfort and pressure on the diaphragm. Many patients may experience improved appetite following therapeutic centesis of effusions. In addition removal of excessive fluid accumulation can lessen anesthetic risks associated with poor oxygenation/perfusion. Use of diuretics such as furosemide and spironolactone may be used to minimize the reaccumulation of effusions, but results are typically minimal.

Patients with PLE will often lose antithrombin III due to its similar weight to albumin. This can create a hypercoaguable state and patients may experience thromboemboli. There are no current guidelines for prophylactic treatment in cases of PLE, however antithrombin II levels may be monitored and anticoagulant therapy used if needed. Low dose aspirin, Plavix, and low molecular weight heparin have been used for prophylactic anticoagulant therapy in cases of PLE.

Once a treatment protocol has been initiated detailed client education and discussion of the treatment should take place. Initial weekly reevaluation of dietary intake, weight gain, and protein levels should occur until protein levels and clinical symptoms improve. Reevaluation of medical, nutritional, and supplemental therapies should be discussed in the face of a poor response to treatment.

REFERENCES

1. Berghoff, Nora et.al, Vet Clin Small Anim, 41 (2011), 311-328
2. Simpson, Kenneth, Western Veterinary Conference 2012
3. Kittrell, Dianee et.al, Compendium, April 2012, vol 34, no 4
4. Stiener, JM, World Small Animal Veterinary Assoc. World Congress Proceedings, 2003
5. Melzer, Katja, et. al., Compendium, Dec. 2002, vol 24, no 12
6. Raux, Craig et.al, ACVIM Forum, 2012
7. Steiner, JM, ACVIM Forum 2009
8. Gonzalez, Adam et. al, Standards of Care, Nov 2002, vol 4, no 10
9. Merrill, Linda, Small Animal Internal Medicine for Veterinary Techncians, pp 216-217
10. Ettinger, Stephen et. al, Texbook of Veterinary Internal Medicine, 5th edition, pp1229-1230
11. Wynn, Susan, Nutritional Management of Lymphangiectasia, VIN, Oct 12 2010

PULMONARY HYPERTENSION REVIEW

Ashley B. Saunders, DVM, DACVIM (Cardiology)
College Station, TX

INTRODUCTION

Pulmonary hypertension (PH) is defined as an increase in pulmonary arterial systolic pressure > 30mmHg. Many factors are typically involved when PH develops in a patient. Physiologically, PH is characterized by hypoxia induced vasoconstriction, endothelial dysfunction, and vascular remodeling of the pulmonary arteries. It contributes to patient morbidity, affects quality of life, and can be life threatening.

The World Health Organization's Classification of Pulmonary Hypertension:

1. Pulmonary arterial hypertension
2. Pulmonary venous hypertension
3. PH associated with disorders of the respiratory system or hypoxemia
4. PH due to pulmonary thromboembolism
5. PH due to other causes

PH can develop in dogs with primary pulmonary disease (ie. chronic bronchitis, interstitial lung disease, chronic exposure to high altitude, heartworm disease), cardiovascular diseases with chronic elevations in left heart pressure (ie. degenerative mitral valve disease, mitral valve stenosis, and left to right shunting congenital heart disease such as patent ductus arteriosus), or pulmonary thromboembolism from a coagulopathy, protein losing disease, hyperadrenocorticism, or heartworm disease.[1-5] Cats can develop PH for similar reasons, although the frequency of reported cases is much less than in dogs.

HISTORY AND CLINICAL SIGNS

The true prevalence of PH is unknown. Because many dogs with mild PH are asymptomatic, mild PH likely goes undiagnosed. Dogs with clinical PH often present with a non-specific history of exercise intolerance, cough, breathing difficulty, and syncope/collapse. Syncope, occurring in nearly a quarter of dogs in one retrospective study of canine PH,[1] is typically attributed to hypoxemia or low cardiac output. Sudden respiratory distress may occur with pulmonary thromboembolic events. Abnormal physical examination findings may include a systolic right-sided heart murmur from tricuspid regurgitation, extra heart sounds (split S2, gallop rhythm), abnormal lung sounds, ascites, jugular distension, and rarely peripheral edema. The history and physical exam often reflect the underlying disease (i.e. chronic cough with airway disease or polyuria with endocrine disease).

DIAGNOSIS

Diagnostic tests are used to identify and determine the severity of PH and to identify any underlying causes. Routine testing may include thoracic radiography, electrocardiography, indirect blood pressure measurement, echocardiography, hematology, serum biochemistry, urinalysis, and heartworm antigen testing. Additional tests may be required when diagnosing the underlying cause and are selected based on abnormal history and physical exam findings (ie. coagulation panel, urinalysis, urine protein:creatinine ratio, abdominal imaging, fluoroscopy of the trachea).

Thoracic radiographs are useful for assessing heart size and may show abnormalities suggestive of PH, pulmonary artery enlargement for instance. Additionally, radiographs are helpful for assessing patients with chronic airway disease and left-sided heart disease, such as degenerative mitral valve disease. Dogs with PH may have electrocardiographic abnormalities indicative of right-sided heart enlargement including tall P waves, a right axis shift in the frontal plane, or S waves in leads I, II, III, and aVF. However, an ECG is neither sensitive nor specific as a screening test for PH. Echocardiography is a useful noninvasive test for diagnosing PH. Pulmonary artery pressures can be estimated indirectly with Doppler echocardiography by measuring tricuspid and/or pulmonary regurgitation velocities and converting them to a pressure gradient with the modified Bernoulli equation (Pressure=4 x Volume^2).[6] In addition to estimating pulmonary pressures, echocardiography provides information about heart enlargement and may aid in the identification of underlying cardiac disease or the presence of heartworms or thrombi.

THERAPY

Treatment goals are to improve clinical signs and quality of life as well as identify and definitively treat any underlying causes of PH (ie. airway disease, protein losing disease, coagulopathies, heartworm disease, etc).[2]

Therapeutic strategies typically address three main areas:

1. Managing the underlying disease contributing to pulmonary hypertension
2. Managing pulmonary hypertension with pulmonary arterial vasodilators
3. Managing right-sided heart failure when present

1. Managing the underlying disease contributing to pulmonary hypertension – For example, dogs with primary pulmonary disease and heartworm disease typically benefit from anti-inflammatory corticosteroid therapy as well as from bronchodilators including theophylline. Theophylline is a methylxanthine that has, in addition to bronchodilating properties, the ability to vasodilate through non-specific phosphodiesterase inhibition leading to a reduction in mean pulmonary artery pressure. Suspected or confirmed coagulation abnormalities can be addressed with anti-thrombotic medications.

2. Managing pulmonary hypertension with pulmonary arterial vasodilators - Phosphodiesterase is released from the lungs and endothelin is released from damaged vascular endothelium, which provides therapeutic targets in the form of phosphodiesterase inhibitors and endothelin antagonists. Prostacyclin and nitric oxide are vasodilators, which provide additional therapeutic targets in humans. In dogs, phosphodiesterase inhibitors have been shown to be beneficial for managing pulmonary hypertension, with sildenafil being the most utilized thus far. Caution should be used when administering any vasodilator in PH, especially in patients with clinical signs suggestive of systemic hypotension or with documented systemic hypotension.

3. Managing right-sided heart failure when present – Dogs with right-sided heart failure secondary to PH can be managed with standard therapy including the use of pimobendan, an ACE inhibitor and furosemide. Spironolactone may be a useful addition for dogs with ascites from right-sided heart failure. Pimobendan (Vetmedin) may have an important advantage in managing dogs with PH as a result of phosphodiesterase inhibition induced combined positive inotropic and balanced vasodilating properties.

Prognosis varies depending on the underlying cause and response to therapy.

This session will use case examples to highlight making the diagnosis and managing a series of patients with pulmonary hypertension.

REFERENCES

[1]Pyle RL, et al. Intern J Appl Res Vet Med 2004;2,99. [2]Kellihan HB, Stepien RL. Vet Clin Small Anim 2010;40:623. [3]Serres FJ, et al. J Vet Intern Med 2007;21:1280. [4]Schober KE, Baade H. J Vet Intern Med 2006;2;0:912. [5]Atkinson KJ, et al. J Vet Intern Med 209;23:1190. [6]Johnson L, et al. J Vet Intern Med 1999;13:440.

Cardiac Auscultation – How to Interpret What You Hear

William Rausch DVM, DACVIM (Cardiology)
Portland, OR

THE FIRST HEART SOUND

Cardiac sounds in the dog and cat can be characterized as either transient heart sounds or murmurs. Transient heart sounds are of relatively short duration and include the first and second heart sounds. The first heart sound (S_1) is initiated at the onset of ventricular systole and under normal circumstances is louder, longer and lower pitched than the second heart sound (S_2). S_1 is temporally associated with closure of mitral and tricuspid valves and is heard best in the region of these valves. However, the sound is not created by atrioventricular (AV) valve leaflets meeting per se. Rather, it is thought to be primarily due to the sudden acceleration and deceleration of blood flow and tensing of the AV valvular apparatus associated with valve closure. There is a correlation between the intensity of the first heart sound and the degree of sympathetic tone, so any disease process which increases sympathetic tone – or even simply anxiety or fear, can increase the intensity of S_1. First heart sounds can also be louder in animals who are thin, anemic or tachycardic. The first heart sound is often of diminished intensity in animals who have pathology between the heart and stethoscope placement. This could include obesity, pleural effusion, any pericardial disease or severe pulmonary disease. Other conditions sometimes associated with a softer S_1 sound include dilated cardiomyopathy, shock and first-degree AV block.

THE SECOND HEART SOUND

The second heart sound is of slightly shorter duration and higher frequency relative to the first heart sound. With a normal heart rate and rhythm, there will be a longer pause after the second heart sound than there is between the first and second sound and this relatively longer period can help properly identify the two sounds. During sinus tachycardia however, the duration of diastole shortens to a greater degree than systole and the pause between the sounds becomes more similar. In this case, it helps to remember that S_2 is the higher pitched sound and that S_1 is timed more closely with a palpable arterial pulse.

The second heart sound is generally heard best in the region of the aortic and pulmonic valves (left hemithorax in the 3rd and 4th ICS). S_2 occurs at the end of ventricular systole, near the end of the T wave on the ECG and is temporally associated with closure of the pulmonic and aortic valves. Under normal circumstances aortic valve closure precedes pulmonic valve closure by a time interval too small to be perceived by the ear, and thus S_2 is heard as a single sound.

SPLIT HEART SOUNDS

If physiologic or pathologic conditions allow the mitral and tricuspid valves to close asynchronously, S_1 can be divided into two distinct component sounds. Generally the time between AV valve closures must be at least 20 milliseconds in order for the distinction to be readily audible, but a phonocardiograph can be more sensitive for this finding. It is often an incidental finding in healthy large breed dogs and can be a rate dependent phenomenon more easily heard at lower heart rates. Split first heart sounds can also be caused by pathologic conditions that cause asynchronous AV valve closure such as bundle branch block, certain arrhythmias and congenital AV valve stenosis. In practice, a split first heart sound in a healthy large breed adult dog with no murmur and a normal ECG is usually not of clinical significance. It must be differentiated in its timing from a gallop sound or systolic click and examples will be given in lecture.

Splitting of the second heart sound can be recognized if the normal time interval between semilunar valve closures is accentuated by conditions that further delay pulmonic valve closure or markedly delay aortic valve closure. Rarely, a split S_2 can be recognized as a normal finding in healthy large breed dogs during inspiration at low heart rates. Intrathoracic pressure becomes more negative during inspiration which promotes right ventricular filling. The resultant increase in RV filling leads to a slightly longer RV ejection period and thus delayed pulmonic valve closure. Pathologic splitting of S_2 can be caused by certain arrhythmias or congenital heart diseases such as atrial septal defects, right-to-left shunting PDAs and aortic stenosis. The most common and clinically important cause for a split S_2 sound is delayed closure of the pulmonic valve as a result of pulmonary hypertension. The pulmonary hypertension can be caused by heartworm disease, chronic pulmonary disease or less commonly congenital heart disease. A split S_2 is generally the only auscultation abnormality noted in a young dog with a right-to-left shunting patent ductus arteriosus and this should be considered a strong differential in a young dog with a split S_2 with exercise intolerance or pelvic limb weakness.

GALLOP SOUNDS

The third and fourth heart sounds are not normally auscultable in dogs and cats. The third heart sound (S_3) is a low frequency sound thought to be associated with abnormally sudden deceleration of ventricular inflow during early diastole. The fourth heart sound (S_4) is associated with atrial systole and is thus heard late in diastole just prior to S_1 when present. An auscultable S_4 in a dog or cat generally indicates that an atrium is contracting to fill a ventricle that is already distended or otherwise non-compliant. As S_3 and S_4 are both low frequency sounds, they are best appreciated using the bell of the stethoscope with only gentle pressure against the skin. The "gallop rhythm" that is created when these extra sounds are present can be described as protodiastolic (S_3 only), presystolic (S_4 only) or a summation of S_3 and S_4. The distinction between these three types of gallop rhythms can be difficult at the high heart rates that these patients commonly present with. However, it is far more important that a gallop rhythm is recognized than categorized, as gallop sounds are often the first or only finding suggesting myocardial dysfunction.

Gallop sounds in canine patients are relatively uncommon, but when heard are most often associated with severe myocardial dysfunction and excessive ventricular volume – most commonly as a result of dilated cardiomyopathy, end-stage AV valvular disease or left-to-right shunting congenital heart disease. They need to be differentiated from split heart sounds, which tend to occur in different canine patient populations as described above. Gallop sounds are not uncommon in feline patients with heart disease. They are often heard in conditions associated with diastolic dysfunction including hypertrophic cardiomyopathy, hyperthyroidism, systemic hypertension and restrictive and dilated cardiomyopathy. Ventricular compliance decreases with age in people with normal hearts, and occasionally geriatric cats will have new gallop sounds in the absence of remarkable echocardiographic changes. The presence of a new gallop sound is an indication for further evaluation by echocardiography, blood pressure, laboratory work and/or radiographs as appropriate. New gallop sounds are also occasionally heard in cats after they receive IV fluid treatment for an unrelated disease. These cats may have had some degree of occult diastolic dysfunction that was revealed by the sudden increase in vascular volume.

HEART MURMURS

Heart murmurs are sustained sounds associated with abnormal turbulent blood flow within the heart or proximal great vessels. This turbulence can be caused by benign anatomic or physiologic variation, valvular insufficiencies, outflow tract disturbances, abnormal communication between chambers, anemia or neoplasia. The murmurs caused by anemia are due in large part to the drop in blood viscosity associated with anemia and thus are more likely to occur with more severe anemia. Many of these same patients will have increased flow velocities associated with the increase in cardiac output triggered by the anemia. The propensity for these changes in fluid viscosity and velocity to cause non-laminar flow in a tube was described by Osborne Reynolds in 1883. The Reynold's number (R) is a dimensionless number that can help to predict the likelihood of turbulent flow under given conditions of flow velocity, blood viscosity and vessel size.

$$R = \frac{\text{Radius x velocity x density}}{\text{Viscosity}}$$

Cardiac murmurs can be described by characteristics that allow for initial clinical deduction and clear communication between clinicians. The timing of the murmur is described relative to its place within the cardiac cycle. Systolic murmurs most often begin with the first heart sound (S_1), but can begin any time before the second heart sound (S_2). Systolic murmurs end with or prior to S_2. Diastolic murmurs begin with or following S2 and end with or prior to S_1. Therefore it is first necessary to identify S_1 and S_2 in order to place the timing of a murmur. Continuous murmurs can be heard throughout the entirety of the cardiac cycle. The timing of a murmur can be further described with modifiers such as holo- (as in holosystolic) and pan- or in their relative timing by prefixes such as pre-, early-, mid- or late. The intensity of a murmur is most often described using the six grades scale. As we will discuss, although there is a loose correlation between grade of murmur and disease severity in some conditions, this is far from true in many cases and other modalities including imaging studies are required to answer questions of disease type and severity with adequate specificity and sensitivity.

Other descriptive characteristics of murmurs include location, radiation and shape. The location, or point of maximum intensity (PMI), is simply the point where it can be heard the loudest with the stethoscope. If there is a palpable thrill associated the murmur, the PMI is usually at the location of the thrill. If a loud left-basilar murmur can be heard with similar character but slightly diminished intensity at the right cardiac base, it is said to radiate to the right base. Location and radiation help in constructing a list of differentials for the murmur in a particular case. The shape of a murmur refers to any variation in the intensity of the murmur during the cardiac cycle. These are more easily recognized on a phonocardiogram and examples will be shown, but with practice and patience these can be appreciated on exam. Plateau murmurs are of nearly equal intensity throughout the murmur. Although there are numerous exceptions, this is often the sound quality created by AV valve insufficiencies. Crescendo-decrescendo murmurs rise to a peak intensity and then soften during the course of a murmur. Again, exceptions abound, but this sound quality can often be heard in congenital outflow tract disturbances such as sub-aortic stenosis and pulmonic stenosis.

With this information as background the lecture will use case examples to focus on how various abnormal heart sounds might be interpreted in a variety of common clinical scenarios. Case examples will be used with a focus on acquired and congenital heart disease as well as "innocent" ausculatory abnormalities.

BODY CAVITY CENTESIS: REVIEW OF TECHNIQUES FOR THE PLEURAL, PERICARDIAL AND ABDOMINAL CAVITIES

H. Edward Durham Jr., CVT, LATG
Columbia, MO

INTRODUCTION

Centesis of body cavities is a very important clinical skill. Although generally performed by a veterinarian, knowledge of these skills is important for veterinary technicians as well. Understanding the procedure allows the veterinary technician to properly prepare for, assist with, and monitor patients that have under gone pericardiocentesis, thoracocentesis or abdominocentesis. The equipment and procedure for each is similar. Pericardiocentesis is perhaps the most challenging due to the proximity to the heart. Abdominocentesis and thoracocentesis are fairly straightforward but careful attention will avoid unwanted complications. There are multiple methods of performing these procedures regarding the equipment used or the exact position of the patient, but the fundamental concepts remain true. There are many reasons for effusions and a complete review of them is beyond the scope of this presentation. This presentation will present those fundamental concepts, list of equipment that can be used, and important details for avoiding complications. It can be stipulated from the beginning that all body cavity centesis procedures should be done with aseptic technique, involving proper hair removal, and disinfecting of the skin with chlorhexidine or povidone-iodine scrub before performing the centesis. Failure to do so can lead to serious infections. It can also be stipulated that all fluid removed from body cavities should be submitted for fluid analysis and histopathological examination to determine the source of the fluid.

THORACOCENTESIS

Indications: Indications for thoracocentesis include pleural effusion, chylothorax, hemothorax, pyothorax, and pneumothorax. The first four impair respiration by creating a space occupying fluid in the pleural space. Dyspnea is the primary reason for performing a thoracocentesis. By removing the fluid from the pleural space, breathing is easier for the patient. In the case of pneumothorax, air trapped in the pleural space obliterates the normal negative pressure of the thorax making it difficult to expand the lungs. The negative pressure is recreated by evacuating the trapped air with thoracocentesis.

Equipment:

1. needle or catheter of suitable length to penetrate the chest wall
2. 4 way stopcock
3. IV extension tubing
4. syringe 10-60ml depending on the size of the patient and volume of effusion expected
5. EDTA tube and plain red top tube for sample collection
6. Sterile gloves suitable sized to the operator
7. Basin for evacuating the effusion into.

Procedure: Thoracocentesis is often needed on both sides because of compartmentalization or a complete mediastinum to completely remove any effusion. Thoracocentesis is best performed in sternal recumbency. Both sides of the chest should be clipped from 5th to 11th intercostal space and from at least the costochondral junction dorsal to about 65% the way up the chest wall. After the thoracic wall is properly prepped, the 7-9 intercostal space is palpated while wearing sterile gloves. Lidocaine may be used for a local block at the puncture site. If available, ultrasound may be used to identify the optimal location for the centesis by covering the probe with a sterile probe cover. The needle or catheter is attached to the extension tubing, the stopcock is attached to the syringe and the extension tubing is connected to the stopcock, while maintaining sterility. The syringe can then be handed to a second operator who does not need to be sterile. This person will apply negative pressure to the syringe while the operator advances and manually secures the needle once evacuation has commenced. The first operator can now position the needle between the ribs in the 7-8th or the 8-9th intercostal space. The needle should be advanced on the cranial aspect of the rib to avoid the intercostal vessels. If effusion is suspected then the needle is positioned ventral on the chest and if pneumothorax is suspected then the needle is placed dorsal. As the needle is advanced into the chest, the second operator keeps a small amount of suction in the system. This will allow for immediate aspiration of any effusion and help avoid inducing a pneumothorax during the procedure. Once fluid begins flowing into the extension tubing the first operator will maintain a stable hold on the needle; the second operator will gently aspirate the effusion into the syringe. Samples for histopathology can be collected by turning the stopcock "off" to the patient and expressing a sample in to the collection tubes. Excess fluid is expressed into the basin for disposal. The second operator will continue aspiration until fluid stops flowing and they feel negative pressure, emptying the syringe as needed into the basin by turning the stopcock to "off" toward the patient. Turning the stopcock back to "off" toward the outside will allow resumption of aspiration. During the process, the second operator should never apply more than 5cc of negative pressure in aspirating the thorax. Excessive suction may induce a pneumothorax by "pulling" air through the lung tissue. Once the second operator only feels negative pressure, the first operator may try repositioning the needle by changing the angle to determine if any further fluid may be removed in that spot. If the first operator is satisfied that no more fluid is available, they may begin withdrawing the needle. The second operator should once again maintain slight negative pressure during withdrawal. The procedure is then repeated on the other side using new supplies. Ultrasound is very useful in assessing the amount of any remaining fluid, or thoracic radiographs can be used to assess the results. If the thoracocentesis is for pneumothorax, evacuation continues until negative pressure is once again achieved.

Complications: After thoracocentesis the patient should be monitored for any sign of dyspnea. Signs of hemothorax, pneumothorax and in the longer term infection should be monitored. Until the cause of the effusion is corrected a return of the effusion is always possible. Provided aseptic technique was maintained, post procedural antibiotics are not necessary.

ABDOMINOCENTESIS

Indications: When abdominal distension is present in a patient, a useful memory device is to consider the "5Fs" of abdominal distension: Fluid, Flatulence, Fat, Fetus, Feces. When the cause is determined to be fluid or ascites, abdominocentesis may be necessary. Ascites of sufficient quantity to cause difficulty breathing and diagnostic samples collection are the primary indications for abdominocentesis. Generally, unless the ascites is significant enough to cause dyspnea, then therapeutic abdominocentesis is not necessary. Treating the cause will often resolve the problem. Diagnostic sample collection will allow for the determination of the cause of the ascites, such as effusion, uroabdomen or hemoabdomen. For collecting a small amount of sample, a 21 or 20 gauge needle and a 3-5ml syringe are all that is necessary. After prepping the abdomen, simply pass the needle into the abdomen approximately 1-2 cm caudal to the umbilicus and aspirate. The collected sample is now sterile and available for analysis. For therapeutic abdominocentesis more supplies are required. There are multiple methods for performing abdominocentesis, and the described method presents the one that has been the most effective for the author.

Equipment:

1. Sterile gloves suitablly sized to the operator
2. 18-14 gauge over the needle catheter
3. Extension set
4. Basin of suitable size to collect the ascites

Procedure: The patient is positioned in lateral recumbency; either right or left is acceptable. The ventral abdomen is ascetically prepped. If ultrasound is available, the probe may be covered with a sterile cover and used to locate the largest pocket of fluid. Alternatively, the section of abdomen 1-2 cm caudal to the umbilicus is typically a good location. Lidocaine may be used for a local block at the puncture site. Insertion of the catheter directly into the *linea alba* seems to provide for the least seeping of fluid after the procedure and avoiding soft abdominal organs. The components of the system are not connected before inserting the catheter into the abdomen. The catheter and needle assembly are inserted into the abdomen 1-2 cm caudal to the umbilicus and on the *linea alba* until fluid begins to appear in the needle hub, aka, the "flash" is seen. The catheter is then advanced off the needle into the abdomen and the needle removed. The extension tubing is then attached to the catheter and the distal end is placed inside the basin, which is positioned lower than the patient. The positive pressure of the abdomen and gravity will allow the fluid to flow from the abdomen to the basin. Systems utilizing suction may be employed, but in the author's experience these methods are more likely to aspirate omentum into the tip of the catheter, thereby necessitating chronic manipulation and potentially increasing the chance of infection, catheter dislodgement or kinking. The fluid is allowed to flow passively out of the abdomen until it ceases. Gentle pressure on the up surface of the abdomen may evacuate more fluid. Removing as much fluid as possible is acceptable, but since the goal of therapy is to assist with respiration it is not necessary to remove all of the fluid. With the patient more able to breath the causes of the ascites can be addressed. Once the fluid is removed, the catheter is withdrawn and dry gauze is held over the puncture site. Seeping may continue for several minutes. If necessary a light belly wrap may be placed to absorb any seepage of abdominal fluid.

It should be stated that if the abdominal distension is due to hemoabdomen from trauma or hemorrhage of a soft tissue tumor, then abdominocentesis is NOT always the best treatment. Emergency surgery is indicated in these cases to stop the hemorrhage, and the ascites may in fact slow the bleeding until surgery is performed by creating increased abdominal pressure.

Complications: Complications include infection from breaks in aseptic technique, hemorrhage from organ puncture with the catheter needle and insufficient removal of fluid. Insufficient fluid removal may be from omentum occluding the catheter tip, compartmentalization of the fluid or by not passing the needle completely through the abdominal wall. Since the needle tip extends beyond the catheter tip, a flash may be seen from the needle tip, while the catheter tip is still in the tissue of the abdominal wall. In this case the catheter tip may track along the abdominal wall between its layers and not pass into the abdominal cavity. This problem is avoided by advancing the catheter and needle together another 1-2 mm after the flash is seen.

PERICARDIOCENTESIS

Indications: Any abnormal effusion in the pericardium that is significant enough to cause cardiac tamponade warrants pericardiocentesis. Causes of pericardial effusion are neoplastic, idiopathic, anti-thrombotic toxicity, left atrial tears, infectious and in cats cardiomyopathy leading to congestive heart failure. Cardiac tamponade describes a condition in which the pericardial effusion creates pressure in the pericardium greater than that of the right atrium (RA). The increased pressure causes the RA wall to collapse during diastole limiting filling in the right heart, and therefore cardiac output to the whole body. In cases where cardiac tamponade is not present and cardiac output is not compromised then pericardiocentesis may not be indicated. An echocardiogram is needed to diagnosis cardiac tamponade, but clinical signs of acute pericardial effusion such as collapse, muffled heart sounds and/or tachycardia are sufficient if an echocardiogram is not available. It should be noted that furosemide is contraindicated in these cases.

Many types of catheters are available for pericardiocentesis. The method described is effective, and reasonably priced. Special pericardiocentesis catheters are available with pigtail configuration to help maintain placement in the pericardium.

Equipment:

1. #11 surgical blade
2. 4-way stop cock
3. Sterile probe cover/sterile gel/elastic band as needed for ultrasound
4. Sample Tubes
    a. EDTA tube
    b. Plain tube
    c. Activated Clotting Time tube (grey top) warmed to body temperature
5. 14G X 3.25" catheter or 14G X 2" over the needle catheter (length suitable for patient size)
6. 60 cc syringe or 30 cc syringe
7. Sterile gloves suitablly sized to the operator
8. ET-30LC extension set
9. Sterile 24 x 24" fenestrated drape
10. Sterile lube packets
11. 1 -1.5 cc 2% lidocaine for skin block
12. Basin suitably sized to collect the effusion

Procedure: Placement of a peripheral indwelling catheter is recommended. Pericardiocentesis is performed from the right hemithorax to avoid hitting the left coronary arteries. The patient is placed in left lateral recumbency, and sedated as necessary to ensure they do not move during the pericardiocentesis. Continuous electrocardiography monitoring is recommended to observe for ventricular arrhythmias during the procedure. After proper aseptic scrub, if ultrasound is available, a sterile probe cover is used over the probe to locate the optimal location for the procedure. Typically this will be located in the 3-5th intercostal space, approximately 1-2 cm dorsal to the costochondral junction. The lidocaine is administered at the insertion site and injected into the skin, intercostal muscles, and to the pleura. After final aseptic scrub prep the patient is draped and a small "stab" incision is made with the #11 blade. The syringe and stopcock can be handed to a second operator who does not need to be sterile. The 14 gauge catheter of appropriate length is selected. The catheter may be fenestrated or not. The catheter/needle assembly is advanced at a 90 degree angle to the skin into the thoracic wall on the cranial edge of the rib, and through the pleura to the pericardium. Pericardial effusion is predominately blood-like in appearance and will flow freely from the catheter/needle assembly. Once the effusion has begun to flow from the catheter/needle assembly the catheter is advanced over the needle into the pericardium and the extension set attached. The extension set is previously attached to the 4 way stopcock and syringe. The second operator begins aspirating while the first operator maintains the catheter position. Samples are collected for fluid analysis (EDTA and Plain tubes) and for coagulation survey into the ACT tube (grey top). Samples for histopathology can be collected by turning the stopcock "off" to the patient and expressing a sample into the collection tubes. Pericardial effusion is devoid of clotting factors and will not clot when removed from the body. The ACT tube allows for rapid assessment of proper placement of the catheter in the pericardium and not within the lumen of the right ventricle (RV). If the RV has been catheterized, the sample in the ACT tube will clot within 2 minutes. The ACT tube must be warmed to body temperature for proper use. The sample is placed inside the tube, mixed and kept at body temperature; after 2 minutes if the sample has not formed a clot, then aspiration resumes until all available effusion has been removed. Excess fluid is expressed into the basin for disposal. The second operator will continue aspiration until fluid stops flowing and they feel negative pressure, emptying the syringe as needed into the basin by turning the stopcock to "off" toward the patient. Turning the stopcock back to "off" toward the outside will allow resumption of aspiration. It is common during the procedure for the first operator to feel the heart beating against the catheter and for ventricular extrasystoles to occur as the volume in the pericardium diminishes. Once all available effusion has been removed, the first operator slowly withdraws the catheter while the second operator maintains gentle negative pressure. Intravenous fluids may be indicated to aid in volume expansion and improve cardiac output.

Complications:

During the procedure, ventricular arrhythmias may be life threatening and treated by removing the catheter and/or lidocaine injections. The catheter may become dislodged from the pericardium. This may necessitate a second puncture, or the pericardial fluid may leak out into pleural space relieving the clinical signs. Usually in this case the pericardial effusion is of insufficient quantity to require thoracocentesis, but it may be performed as needed. Over advancement of the catheter into the heart can be a serious complication. Watching for ventricular arrhythmias, palpation of the "bouncing" catheter, and coagulation of the aspirate are all signs of a catheter placed in the heart. After the procedure, signs of hemothorax, pneumothorax and in the longer term infection should be monitored. Provided aseptic technique was maintained, post procedural antibiotics are not necessary. Pericardial effusion may recur within 24 hours and a second procedure may be necessary. Chronic effusion should be treated with surgical removal of the pericardium to prevent constrictive pericarditis resulting from frequent pericardiocentesis.

CONCLUSION

Body cavity centesis is a valuable and effective treatment for removal of large effusions. The veterinary technician needs to be proficient in understanding the procedures for centesis of various body cavities. Time is often an important factor in the overall outcome of the patient that needs a centesis procedure. By anticipating the needs of the veterinarian preforming the procedure, the veterinary technician can save valuable time for the patient and improve outcomes. Some small steps the veterinary technician can take to aid the process are: having the equipment necessary for body cavity centesis collected and prepackaged together to save time in searching for the necessary components, understanding the indications and clinical signs for conditions that would benefit from a centesis, and being able to monitor the patient post-procedure will all benefit the patient outcomes. Although veterinary technicians do not perform these procedures, they are a critical part of their smooth application.

VETERINARY TECHNOLOGY: NOT JUST JOB; MORE THAN AN ADVENTURE

H. Edward Durham Jr. CVT, LATG, VTS (Cardiology)
Columbia, MO

INTRODUCTION

The world of veterinary technology has come a long way since its inception. A review of history shows us the first persons identified as something like veterinary technicians were the graduates of the Canine Nurses Institute started in England around 1908. This program was the first to systematically train veterinary assistants. In the United States, it was not until 1967 that the American Veterinary Medical Association (AVMA) established criteria for the training of veterinary assistants. In the preface of their textbook, *Animal Hospital Technology* (1971,American Veterinary Pub Wheaton, IL) Drs. Catcott and Smithcors state "During the last 10 years there has been an unmistakable trend in the veterinary profession to rely more and more on capable lay employees to assist in providing services which the public increasingly demands." This text was the first practical textbook dedicated solely to the education of "Veterinary Assistants." It was not until 1972 that the term "veterinary technician" was officially adopted by the AVMA and a year later the first education programs were accredited by the AVMA. Today there are approximately 205 AVMA accredited veterinary technology campus based programs in the US, Canada, and Puerto Rico and 9 online distance learning programs. Veterinary technicians are recognized by the AVMA as an integral part of the profession as true paraprofessionals: "The AVMA recognizes the value of veterinary technicians as an integral component of veterinary medicine and urges full utilization of veterinary technicians. The veterinary profession is enhanced through efficient utilization of each member of the veterinary health care team by appropriate delegation of tasks and responsibilities to support staff."[1] Also there are more opportunities for veterinary technicians than ever before such as clinical practice, clinical specialization, education, research opportunities, and industry sales representative. Initially, training was an Applied Associate of Science (specialized vocational) or an Associate's of Science (2 year degree), We now have Bachelor's of Science degree level education available in veterinary technology and a Master's degree level program in the possible future.

Veterinary technicians can now become specialists in their field of choice. In 1994 the National Association of Veterinary Technicians in America (NAVTA) formed their Committee on Veterinary Technician Specialties (CVTS) opening the way for groups of technicians interested in specific areas of veterinary medicine to organize veterinary technician academies. To quote the NAVTA website, "The CVTS provides guidelines to veterinary technician organizations to facilitate the formation of a specialty organization. Academies develop advanced pathways, which a candidate must follow and complete in order to be awarded the designation of VTS (Veterinary Technician Specialist) in their specific discipline."[2] As of this writing, NAVTA recognizes 11 different academies in the disciplines of: Anesthesia, Dentistry, Internal Medicine (which consists of large and small animal internal medicine separately, cardiology, oncology and neurology), Emergency and Critical Care, Zoo Medicine, Equine Nursing, Surgery, Clinical Practice, Nutrition, and Clinical Pathology. To clarify terminology, a "college" is the title for the specialty organization of DVMs, such as American College of Veterinary Surgeons, or American College of Veterinary Internal Medicine. An "academy" is the title of a specialty organization of veterinary technicians. A "society" is a group made up of all persons interested in a particular specialty and may include doctors, technicians and lay staff. Societies do not certify candidates as specialized, but they serve more as an information dissemination system.

Within the field of research, the American Association of Laboratory Animal Science (AALAS) recognizes three levels of technical training: the Assistant Laboratory Animal Technician (ALAT), the Laboratory Animal Technician (LAT) and the Laboratory Animal Technologist (LATG). There is even a management certification now in this specialty: Certified Manager Animal Resources (CMAR).Opportunities in industry are vast as well. Many veterinary technicians are hired in the medical industry for research, as sales representative or technical representatives. Veterinary technicians are employed by all veterinary colleges and often have a direct and profound role in the education of young veterinarians, bringing decades of practical experience to the veterinary teaching hospital. Most veterinary technician schools also employ technicians to do the majority of the instruction.

GETTING STARTED

In the majority of the United States, potential veterinary technicians must graduate from an AVMA accredited program to qualify for state licensure. In 1973 Michigan State University and Nebraska Technical College were the first two programs to receive accreditation from the AVMA. During the 1970's and 1980's, many states still allowed veterinary technicians to be credentialed based on their work experience and passing a comprehensive state examination. This was largely due to the overall lack of AVMA accredited programs in the nation. States wishing to provide their veterinary practitioners with qualified staff developed examinations with questions written by their stake holders to certify and credential staff as veterinary technicians. As more AVMA accredited programs became available the credentialing of veterinary technicians has become more standardized. Each state sets its own standards for testing andcredentialing of veterinary technicians, but almost all require the applicant to pass the Veterinary Technician National Examination (VTNE), which requires the applicant to have graduated from an AVMA accredited program. This examination has supplanted the "comprehensive" examination that most state veterinary boards used to administer. A few states still administer a jurisprudence examination regarding that state's veterinary laws.

According to the Bureau of Labor Standards[3] there were approximately 80,000 veterinary technicians in the United States with a median pay of $29,710 per year. Certainly the more experience and education or specialization one has the pay can increase significantly. The projected job growth for the profession was ranked at 52% which is considered "much faster than average." In contrast human Health Care technicians are projected for only 26% job growth in the next ten years.

WHERE DO I GO NEXT

Once a veterinary technician credential has been achieved, a technician may begin looking toward becoming a VTS. The Academy of Veterinary Emergency and Critical Care Technicians was the first group to organize and receive recognition by NAVTA, blazing the trail for all other academies. Although every veterinary technician specialty academy has its own requirements, many of the required features extend across all academies. All academies require 3-5 years minimum on the job experience in the specialty of their choice before an applicant can apply. There are other documents required which may include, but are not limited to, case logs, case reports, skills list, knowledge lists, and letters of recommendation from senior people such as a VTS or Diplomats of the veterinary specialty college most closely associated with their academy. Some academies require candidates to submit questions that could appear on future examinations. Once an application is made and the required documents are approved by the academy, the candidate then must pass an examination hosted by the academy. Typically these examinations are held annually at a national meeting of that group and consist of several hundred multiple choice questions. Careful review of an academy's website information will provide the best source of the details of each academy.

There are many other professional opportunities available for motivated veterinary technicians. Several peer reviewed journals now publish articles written by veterinary technicians. The articles are usually review papers covering some topic within the technician's sphere of interest. Veterinary technicians also hold positions on editorial boards of various publications. Opportunities also exist in novel research as well. Typically this involves collaboration with veterinarians to produce some research finding that is in the general benefit of veterinary medicine. Providing the technician makes a substantial contribution to the research, many investigators will include the veterinary technician as an author. The day is fast approaching when veterinary technicians will develop their own research, affording them the coveted first authorship position on the paper.

Many veterinary technicians are accomplished public speakers. The large national veterinary conferences and most state association meetings now offer a special technician schedule. Traditionally, the technician schedule has been populated by veterinarians sharing information they felt was important to the continuing education of their staff. Increasingly; however, this role is filled by veterinary technicians. This is one situation where the designation as a VTS gives the speaker preference. The rewards for lecturing at conferences are more than simply monetary; there is a great sense of satisfaction one achieves in feeling that they have possibly made some difference in the education of other budding veterinary technicians or in the recognition that comes from one's peers.

MY GENERATION

Veterinary medicine in the United States today is multigenerational. There is often a mix of generations in a single veterinary practice. If we understand that the members of different generations bring with them a wide range of experience and cultural references, they can be used as an asset for the veterinary technician and the veterinarian alike. Traditionally, we think of the veterinarian as the most seasoned person in the practice and the technician as someone younger trained, in many cases, by the veterinarian to assist him/her. This scenario is not the case anymore. A typical multi-doctor practice will likely have doctors who are very experienced, some who have been in practice for 5-10 years and some who are recent graduates. Likewise, the paraprofessional staff may also be a mix of technicians who have more years of experience than the doctors they work with. The relative clinical experience often translates into very different age ranges. These different ages represent different generations that we can look at to understand how people see and react to the world around them.

The recognized generations that mostly come into play today are the: The Matures (aka Greatest Generation), The Baby Boomers, Generation X, and The Millennial's. Each of these groups are broad categories with world and cultural events that effected there development. Based on the foundational work in generations and how each is different by William Strauss and Neil Howe; the following table will give some idea of the formative experience that shaped each generation.[9]

	The Matures born before 1946	Baby Boomer born 1946-1964	Generation X born 1965-1980	Millennials born 1981-2000
Experiences	WWII & Pearl Harbor FDR Administration Rationing Atomic Bomb Korean War	Move to Suburbs Cold War Civil Rights Vietnam Space Race Women's liberation	Single Parents/Latch-key Watergate Iran hostage crisis MTV/Computers Challenger disaster	Technology boom Oklahoma City bombing 9/11 & Columbine Globalization Social Media
Tendencies	Dedicated Loyal & Patient Hard-working Sacrifice for common good Adhere to rules	Optimistic Idealistic Educated Driven Work to live Personal growth	Self-reliant Live for today Distrustful of authority Technology comfortable Multi-takers Skeptical & cynical	Innovative Dependent Self-confident Socially conscious Image Driven/Impatient Patriotic
Workplace	Duty before pleasure Formality & Structure Honest, but tactful Top-down leadership Change may be difficult Conservative spenders	Positive/Over confident Service oriented Good w/ relationships Bad w/ conflict May be seen a judging Not budget minded	Work to live Versatility = Security Structure = Bureaucracy Seek autonomy/flexible Result oriented Challenge status quo	Question everything Poor relationship skills Menu-driven thinking Depend on technology "Free agent" mindset Great at networking

This summary is not intended to paint a picture of EVERY individual born in the time frame noted, but simply to provides very broad definition of the formative experiences most people in that range MAY have experienced. Each individual person may have seen these events differently, depending on their own personal environment. Looking carefully at this table one can see how a person in the mature generation may have difficulty understanding how a Millennial may see the world. Those people affected by WWII remember a time of leanness in the US economy and consequently tend to save money carefully, where those raised during the 1990's remember a time of economic plenty, so saving for the future is not part of their "programming". Neither group is right or wrong, simply different. These differences play a part of problem solving skills, personal interactions, job performance, and perhaps most importantly where in life each person gets personal satisfaction.

TECHNOLOGY

Technology has literally changed the face of the veterinary practice today. Examples of this are seen everywhere. Not just new diagnostic equipment such as digital radiography and ultrasound, but in the way a veterinary practice interacts with the public. Many practices have a Facebook page, medical records are becoming electronic, doctors and technicians can be seen searching the web for information regarding medical conditions, virtually all veterinary medical journals now offer a digital version, indeed some are dispensing with a print copy. All of this technology has a tremendous potential to increase productivity and improve medical care. Every opportunity to use technology to one's advantage should be pursued, yet remembering in the end it is the human interaction that clients will remember. Although it relies heavily on technology and science at the end of the day, veterinary medicine is still an art which requires experience, clinical judgment and personal interactions.

THE NEXT STEP-TAKE A STEP

As a VTS, I hear many young technicians say something to the effect of, "Why should I bother to work so hard to get a VTS when my boss will not pay me more?" The answer is simple, because it makes you a better you. Why should we settle for only doing what we need to do to get by? Why don't we challenge ourselves for the sole reason of living up to our potential?

How many times in life have we heard people ask, "What do you want?" Usually they are asking about something like your dinner preference, but the deeper meaning of course is: What do you want out of life? How far do you wish to go with your career, your business, your family? Each of us defines what we want differently; therefore, what we define as success is different. Success is achieving something that makes us feel self-gratified or happy. What is happiness? According to Life Coach Tony Robbins[7] "Happiness is having our life condition match our blueprint for what we expect in life." So for some our blueprint may be simply a provision of security from financial disaster. For others it may be feeling loved unconditionally, no matter how poor. Whatever you define as success is within you, and no matter what that might be, we all have one thing in common we can strive for. We can all strive to be a better version of ourselves tomorrow than we are today.

Let's narrow the focus to career only. What are we looking for out of work? What life condition do we hope for in our jobs? Everyone wants good pay, but most of us want more. Most of us want is to be respected, to be seen as equals, or at the least have an important role on the team. Now we move into the concept of TEAM LIFE. Most people like feeling they are making a difference. One of the best ways to do that is to be part of a team. There are VERY few people in history who accomplished great things alone. As part of a team we can accomplish more in a more efficient manner. Ants are the perfect example of this; by working as a group they can accomplish MUCH more than any individual could. So, what is Team Life? I suggest this definition put forth by former Navy SEAL Dave Rutherford[6] "a group of qualified people, who have trained together, been trained well, in an operational capacity in which they have a common goal; with the flexibility and a certain level of autonomy that will enable them to produce without much oversight" That is exactly what the veterinary team is; a receptionist to greet clients, a veterinary technician to acquire basic information, a veterinarian to lead diagnostic and treatments with the aid of the veterinary technician to accomplish the goal of providing quality medical care for someone dear pet. That is our team mission and to achieve it we need a strong team. Mr. Rutherford goes on to elucidate 4 concepts for an effective team: Commitment, Training, Communication, and Leadership. A failure in any of these can lead to mission failure.

What makes you valuable to a team? It is being the best possible "You" every day. It is continually working to improve yourself on a personal level, not for money, but because you want that for you. There are hundreds of gurus and self-help instructors out there. It is not a bad idea to seek one that "speaks your language" and learn from him/her. Thanks to the Internet, you can often do that for free. First, you must decide to work on yourself and then start **doing** it. Ninety percent of success is showing up. If you never start you will certainly not be successful. Next do things like getting proper instruction, finding a mentor and doing what s/he says, and having an overwhelming desire to accomplish your goal. Establish your WHY; what is driving you to succeed? Mr. Eric Thomas[5] is a person who has a very inspiring message. This man was kicked out of his home at 16 by his mother telling him, "You're going to have to grow up.". He dropped out of school and lived homeless in Detroit. Today, he has a Ph.D., has published a book, and speaks internationally to students, professional sports teams, the military, and businesses about how to succeed. His "why" is remembering being homeless and how his 'now' wife helped him out. He recognizes his mother did not abort him as an unwed teen mother and he wants to honor her choice by being useful. He chose to complete school and dared himself to live up to his own potential. We all have a "why" and spending time in contemplation will help drive you. Find your "why."

You need to set goals for yourself. Clear goals help bring order to the chaos of your mind. Make a "to do" list of very specific steps. Don't write 'get Master's degree.' Instead list all the steps necessary to get the Master's degree. Establish your own self-confidence. Mr. Rutherford outlines 8 steps of Self Confidence: 1- Positive Attitude 2- Physical Training 3- Motivation 4-Earn Respect 5-Set Goals 6-Integrity 7-Mentoring 8-Have Fun. Visualize yourself as successful. Take some time to contemplate how it will look when you succeed. Then, work hard. You must be able at any moment to sacrifice what you are for what you want to become. That might mean staying up after everyone else is asleep, or getting up early. It will likely be hard, but remember pain is temporary; since you are already in pain, get a reward from it. At the end of pain is success. "Pain is temporary. It may last a minute, or an hour, or a day, or a year, but eventually it will subside and something else will take its place. If I quit, however, it lasts forever (Lance Armstrong)[4]." Don't quit, and don't be afraid to fail in your attempts. There are many clichés about failing, but I think some inspiration is best. Wayne Gretzky said "You miss 100% of the shots you do not take."[8] Michael Jordan missed 26 game winning shots and more than 9000 career shots. You may fail along the way, but if you never give up then you have only run into obstacles; if you quit then you have indeed failed.

Lastly, for true success in life, I feel you must give back. You must help others achieve their dreams. Be a mentor.

REFERENCES

1-https://www.avma.org/KB/Policies/Pages/AVMA-Policy-on-Veterinary-Technology.aspx

2- https://www.navta.net/specialties/specialties

3- http://www.bls.gov/ooh/healthcare/veterinary-technologists-and-technicians.htm

4- http://www.brainyquote.com/quotes/quotes/l/lancearmst193138.html#e2mAsqOPemDfQFh0.99

5- Eric Thomas http://etinspires.com/

6- Dave Rutherford http://www.teamfroglogic.com/

7- Anthony Robbins http://www.tonyrobbins.com/

8- http://www.brainyquote.com/quotes/quotes/w/waynegretz378694.html

9- Strauss, W. Howe, N. *Generations: The History of America's Future, 1584 to 2069* New York: Morrow, 1991.

DKA: ADVANCED NURSING CARE

Marie E. Kerl DVM, MPH, DACVIM (SAIM), DACVECC
Columbia, MO

PATHOPHYSIOLOGY OF DIABETES MELLITUS AND DIABETIC KETOACIDOSIS

Diabetes mellitus (DM) occurs when there is insufficient quantity or activity of the hormone insulin. In addition to hypoinsulinemia, glucagon and the "stress" hormones cortisol, epinephrine, and growth hormone exacerbate hyperglycemia and cause ketogenesis leading to ketoacidosis. Persistent hyperglycemia and hyperketonemia exacerbate dehydration by osmotic diuresis. Multiple electrolyte abnormalities occur due to osmolar shifting from hyperglycemia, cellular shifting from metabolic acidosis, renal wasting from osmotic diuresis, and poor oral intake.

DKA typically occurs in middle-aged to older dogs and cats. Clinical signs of DM (e.g. polyuria, polydipsia, polyphagia, weight loss) precede DKA; however clients may not recognize signs. Clinical signs for DKA include anorexia, vomiting, and lethargy. Dehydration is a common physical examination finding with DKA. Volatile ketoacids impart a characteristic "fruity" odor to the breath that some individuals can sense. Weakness can occur from severe hypokalemia, dehydration, or acidemia. Since DKA is caused by a progression of DM brought on by release of glucagon and stress hormones, the patient should be evaluated for underlying illnesses.[1-4]

DKA: MULTIPLE BIOCHEMICAL ABNORMALITIES

Diabetic ketoacidosis can be diagnosed in the emergency setting by documenting hyperglycemia, glucosuria, ketonemia or ketonuria, and metabolic acidosis.[1, 5-9] Initial diagnostic testing should include complete blood count, serum biochemical analysis, urinalysis, and blood gas evaluation. Hypokalemia, hyponatremia, hypochloremia, and hypomagnesemia should be monitored frequently throughout treatment to adjust therapy appropriately.[5,10] Identification of urine ketones on urine dipstick analysis has been a mainstay of diagnosis for DKA. If urine is not available at the time of initial presentation, serum ketones can be identified using urine test strips which measure the ketoacid acetoacetate. More recently, plasma ketone meters which measure the ketoacid 3-beta-hydroxybutyrate have been evaluated for use in dogs and cats, and have been found to improve diagnostic accuracy in dogs and cats with diabetic ketoacidosis than urine dipstick methods.[7, 9, 11-13] Some animals will develop ketosis prior to developing severe illness and metabolic acidosis. If physical examination and laboratory data reveal mild illness, less intensive therapy may be successful.[1]

Intravenous (IV) fluid therapy provides volume resuscitation, rehydration, correction of hyponatremia and hypochloremia, dilution of blood glucose, and maintenance support. Treatment should be initiated with replacement crystalloid solutions, and dosed to include maintenance, dehydration, and ongoing losses. Monitor hydration carefully, and change fluid dose accordingly. Keep in mind that initial correction of dehydration with IV fluids will reduce blood glucose by dilution.[10]

Hyponatremia is anticipated in DM. As glucose levels rise, glucose plays an increasingly important role in maintaining intravascular osmolarity. Free water shifts into the vascular space in response, diluting sodium and chloride. Normal serum sodium in the face of hyperglycemia is inappropriate, and represents excess free water loss and hyperosmolality.[10]

Hypokalemia is the most common electrolyte abnormality associated with DKA. Clinical signs include muscle weakness, cervical ventroflexion in cats, cardiac arrhythmias, and respiratory muscle failure in severely affected patients. Potassium chloride can be added to the IV fluids to correspond with the patient's serum potassium using the following scale (not to exceed a delivery rate of 0.5 mmol/kg/hr) (Table 1)[10]

Table 1: Dosing plan for potassium replacement therapy (do not exceed 0.5 mmol/kg/hr rate)

Serum K# (mmol/l or meq/l)	Add # mmol KCl/ liter to isotonic crystalloid fluid
3.5-5	20
3.0-3.4	30
2.5-2.9	40
2.0-2.4	60
< 2.0	80

Hypophosphatemia may be recognized initially or during hospitalization, and severe hypophosphatemia results in acute onset hemolytic anemia. Hypophosphatemia treatment consists of supplementing with potassium phosphate (KPO_4) via constant rate infusion (CRI). Initial dose of KPO4 is 0.01-0.12 mmol/kg/hr CRI in 0.9%NaCl for the first 6 hours, with reevaluation of serum phosphorous and adjustment of dose as needed. Alternatively, phosphorous may be supplemented by administering one-half of the potassium replacement as KPO_4.[2, 10]

Metabolic acidosis is common in DKA. Clinical manifestations include lethargy, vomiting, hyperventilation, decreased myocardial contractility, peripheral vasodilation, stupor, and coma. Administration of insulin to reverse ketogenesis, and IV fluid therapy to correct dehydration, will improve metabolic acidosis in most patients. In patients with severe acidemia, with blood pH less than 7.10-7.15, bicarbonate (HCO_3^-) values less than 8mmol/l, and clinical signs referable to metabolic acidosis, bicarbonate dose can be calculated by the formula: **Bicarbonate deficit = (Base Deficit) (BW Kg) (0.3)**

Hypomagnesemia occurs due to chronic loss and poor intake. Symptoms are difficult to characterize, but include resistance to correction of other electrolyte abnormalities. Mild hypomagnesemia may be treated by administering isotonic replacement crystalloid solutions that contain magnesium (Plasma-Lyte 148®). For severe hypomagnesemia (serum magnesium <0.4 mmol/L [1.2mg/dl]), an infusion of magnesium sulfate can be made by adding to 5% dextrose in water at an initial dose of 0.3-0.5 mmol/L/day (0.75-1mEq/kg/day) CRI in for the first 24 hours, then reducing the dose by 50% for the next 3-5 days.[10]

INSULIN THERAPY AND DKA

Insulin reduces blood glucose by promoting cellular glucose uptake, and decreasing hepatic glucose production. Ketogenesis is reversed by reducing available substrate for ketone formation, limiting hepatic production by reducing glucagon levels, and increasing peripheral ketone use. In addition to these beneficial effects, insulin can contribute to complications such as hypoglycemia, hypokalemia, and sudden changes in osmolarity. Regular insulin has traditionally been used in the initial treatment of DKA, due to rapid onset and short duration.[10] Lispro, a short-acting human synthetic insulin analog, has also been described for continuous-rate infusion in dogs with promising results.[14]

Three insulin protocols for initial DKA management are described. All can reduce blood glucose and reverse ketogenesis. Each protocol has advantages and disadvantages; therefore, selection of protocol should be made based upon the preference of the clinician. Similarities include the use of regular insulin to gradually lower blood glucose, administration of 2.5-5% dextrose when blood glucose falls below 250 mg/dl (14.2 mmol/L), frequent assessment of patient response to therapy, and change to intermediate-acting insulin for long term management once ketogenesis has resolved.[10]

IV CRI (IVCRI) provides insulin continuously to gradually lower blood glucose and reverse ketogenesis. For IVCRI, a starting dose of 2.2U/kg/day regular insulin for dogs or 1.1U/kg/day for cats is added to 0.9%NaCl and administered by CRI to achieve an initial infusion rate of 0.09U/kg/hr for dogs, and 0.045U/kg/hr for cats. A lower dose is recommended for cats to reduce risk of sudden osmolar shifts causing cerebral edema. Insulin binds to plastic and glass, therefore 50ml of insulin-containing fluid solution should be allowed to run through the line prior to administration.[14, 15] Blood glucose should be monitored at 1 to 2-hour intervals. Insulin administration rate should be increased or decreased to achieve a rate of decline of blood glucose of 50-70 mg/dl/hr (2.8 – 3.1 mmol/L/hr). Blood glucose levels should remain above 250 mg/dl (14.2 mmol/L) for the first 4-6 hours of treatment, to avoid risk of cerebral edema from rapid osmolar shifts. When blood glucose reaches 200-250 mg/dl (11.1 - 14.2 mmol/L), 2.5-5% dextrose solution should be administered with the IV fluids to avoid hypoglycemia while continuing to promote ketone metabolism through insulin administration. When ketosis has resolved and the patient is eating and drinking without vomiting, intermediate-acting insulin should begin. This protocol can safely and effectively reduce blood glucose, and reverse ketogenesis, and is less likely to stimulate glucagon and cortisol release than intermittent injections. Disadvantages include more intensive nursing care with frequent glucose monitoring times and requirement for CRI administration. Risk of excessively rapid reduction of blood glucose and electrolytes exist if the patient is not monitored closely.[10, 15]

Hourly intramuscular protocols (Q1H) prescribe regular insulin at an initial dose of 0.2U/kg IM, then 0.1U/kg hourly thereafter. Regular U-100 insulin may be diluted 1:10 with the appropriate pH-adjusted diluent or isotonic saline to achieve low doses. Insulin diluted with isotonic saline should be formulated daily to insure insulin activity. Blood glucose should be monitored hourly, with the goal being a gradual decline of 50-100 mg/dl/hr (2.8 – 5.6 mmol/L/hr) to avoid sudden osmolar shifts. Once the blood glucose concentration is below 300 mg/dl (16.7 mmol/l), 5% dextrose solution should be added to the IV fluids, and dosing frequency reduced to 4-6 hour intervals IM or SQ to keep blood glucose in the 200-300 mg/dl (11.1--16.7 mmol/l) range. Once the patient is eating and drinking without vomiting, changing to twice-daily intermediate acting insulin is recommended. The Q1H protocol can be effective with appropriate monitoring and use. This protocol has the disadvantage of requiring hourly blood glucose measurements, and intensive management by trained professional staff.[10]

Intermittent intramuscular insulin protocols (QIDIM) prescribe 0.25 U/kg regular insulin as a test dose, with subsequent doses based upon response to initial treatment. In obese animals, initial dose should be based upon estimated lean body weight to avoid overdosage and hypoglycemia. Recheck blood glucose at 4-6 hour intervals, with the glucose reduction goal of 50-70 mg/dl/hr (2.8 – 3.1 mmol/L/hr). If this goal is surpassed, the next insulin dose should be reduced by 25-50%. If this goal is not met, the next dose should be increased by 25-50%. If the blood glucose reaches 200-250 mg/dl (11.1 – 14.2 mmol/L), add 2.5-5% dextrose to the IV fluids. Advantages of the QIDIM protocol include treatment via one IV catheter, less frequent sampling times, gentle reduction of blood glucose, and resolution of ketogenesis. One disadvantage is that ketogenesis may take longer to resolve with intermittent insulin injections since regular insulin has a variable duration of activity, and may not be present for the full 6 hours between administrations.[10]

Diabetic ketoacidosis is a serious metabolic illness with a guarded prognosis. In two retrospective studies outlining response to treatment for dogs and cats diagnosed with DM or DKA, 30% (dogs) and 26% (cats) of the patients died or were euthanatized following initial diagnosis. While dogs diagnosed with DKA remain diabetic, some cats that are successfully treated for DKA may experience diabetic remission with appropriate diabetic management. [16] Patients with DKA frequently have concurrent disease processes that decrease the chances of successful therapy. [3] Some clients may elect euthanasia at the outset due to the serious nature of DKA, and long-term time and financial commitment required for DM. [3, 10]

REFERENCES

[1]Kerl ME. **Compend Cont Ed Small Anim Vet** 2001;23(3):220. [2]O'Brien MA. **Vet Clin No Amer Small Anim Pract** 2010;40(2):317. [3]Hume DZ, et al. **J Vet Intern Med** 2006;20(3):547. [4]Sallander M, et al. **Acta Vet Scand** 2012;54(1):61. [5]Durocher LL, et al. **J Amer Vet Med Assoc** 2008;232(9):1310. [6]Zeugswetter F, et al. **J Feline Med Surg** 2010;12(4):300. [7]Zeugswetter F, et al. **J Small Anim Pract** 2009;50(1):4. [8]Duarte R, et al. **J Vet Intern Med** 2002;16(4):411. [9]Di Tommaso M, et al. **J Vet Intern Med** 2009;23(3):466. [10]Kerl ME. **Compend Cont Ed Small Anim Vet** 2001;23(4):330. [11]Stojanovic V, et al. **Can Vet J** 2011;52(4):426. [12]Weingart C, et al. **J Vet Diag Inves** 2012;24(2):295. [13]Zeugswetter FK, et al. **J Small Anim Pract** 2012;53(6):328. [14]Sears KW, et al. **J Vet Emerg Crit Care** 2012;22(2):211. [15]Claus MA, et al. **J Vet Emerg Crit Care** 2010;20(5):509. [16]Sieber-Ruckstuhl NS, et al. **J Vet Intern Med** 2008;22(6):1326.

COAGULOPATHIES INVOLVING PRIMARY HEMOSTASIS

Katy W Waddell, RVT, VTS (ECC, Anesthesia)
College Station, TX

Key points

- Primary hemostasis review
- Decreased platelet production
- Platelet destruction
- Platelet consumption
- Platelet sequestering/removal from circulation

Hemostasis is divided into three separate categories, primary hemostasis, secondary hemostasis and tertiary hemostasis (also called fibrinolysis). We may define primary hemostasis as the formation of the platelet plug. Primary hemostasis is initiated when any damage to damage to small blood vessels or capillaries occur. The immediate response to the damage is a reflex that promotes vasoconstriction which in itself can diminish or decrease blood loss. Disruption or damage to the vessel exposes the endothelial lining of the vessel and. Collagen, which a supportive protein, is exposed and will provide a surface for the platelets to adhere to. Von Willebrand factor (vWf) binds to a glycoprotein in the platelet membrane. Platelets, when coming into contact with collagen fibers in the vascular wall or damaged endothelial cell, begin to swell and assume irregular forms. These irregular forms have many irradiating (star shaped) processes which protrude from their surfaces. These contractile proteins contract with force and cause the release of granules that contain multiple active factors. They become quite sticky and adhere to the collagen fibers. If the hole in the blood vessel is small, it is managed by the formation of a platelet plug rather than what we think of as a blood clot.

The function of platelets is to maintain capillary integrity by filling the spaces between the endothelial cells of the capillary bed, the initiation of coagulation and finally to retract clots. On the surface of platelets, there are many proteins which allow them not only to "stick" to disrupted areas of the blood vessel but also allow them to adhere to each other. They contain granules which have an excretory ability. Some of these proteins which are excreted during degranulation are histamine and serotonin. As platelets adhere to the damaged area of the vessel, degranulation occurs. The combined action of the platelets ability to adhere to the site of injury, to each other and the degranulation process which produces ADP and thromboxane A2 in a normal patient ensures the formation of a platelet plug. Platelets are very active structures; they have a half life in the blood of 8 to 12 days. They are then removed from the circulation mainly by the tissue macrophage system. More than half of the platelets are removed by macrophages in the spleen.

Adenosine diphosphate (ADP) functions as a platelet agonist. An agonist is simply a chemical that binds to a cell receptor and is responsible for triggering a response from the cell. ADP can/may allow platelets to change shape as well as allow aggregation. It is also responsible for the generation of thromboxane A2 which is another platelet agonist.

Thromboxane A2 is the product of activated platelets. It has prothrombotic properties. It is responsible for increasing platelet aggregation and also stimulates the activation of new platelets. Thromboxane is named for its role in clot formation (thrombosis)

What is von Willebrand factor? It is not an enzyme so it has no catalytic activity. Von Willebrand factor (Vwf) is a large glycoprotein produced in the endothelium, megakaryocytes (α- granules of the platelets) and the subendothelial connective tissue.

Why is it important? The primary function of vWF is to bind to other proteins, most particularly Factor VIII. In human medicine, this is often referred to as the antihemophilia factor. Factor VIII is a cofactor in activating factor X. We rely upon vWF to assist in platelet adhesion by binding to the collagen exposed by damaged endothelial cells within the blood vessel. This factor plays a dual role in the arrest of bleeding. It helps platelets to plug the injured blood vessel walls, and it is also a carrier for factor VIII, which is vital in the clotting cascade. In the absence of VWF, blood coagulation and the platelet plug could be abnormal. This would affect both primary and secondary hemostatis.

There is a sequence of events which occurs at the site of vascular injury. First, the platelet is attracted to the exposed sub-endothelial layer of collagen and adheres to it. To ensure this, the platelet undergoes a shape change. Secondly, the platelets release intrinsic adenosine diphosphate (ADP), among other substances. The released ADP stimulates other platelets to stick together at the wound site, and, thirdly, aggregation occurs. In this process, platelets adhere to each other to form a beginning plug. Finally, coagulation occurs and fibrin forms around the platelet aggregate to initiate repair.

We have the first two steps of primary hemostasis. The first step is vascular constriction; the stimulus of the traumatized vessel causes the wall to contract, reducing the blood flow from the vessel. The second step is the formation of the platelet plug. Platelet repair of vascular openings is based on the many abilities of the platelet itself ranging from the ability to assume irregular shapes, the ability to adhere and the release of granules many of which have excretory ability and are responsible for ADP and thromboxane A2.

Platelets can adhere directly to the exposed subendothelial mix of collagen, fibronectin and other glycoproteins. When von Willebrand's factor is released by the locally activated endothelium, more effective adhesion occurs as it "coats" subendothelial collagen to create a bridge between collagen and the glycoproteins.

As the platelets aggregate, they release the contents of their granules (ADP) which allows further bridging to occur. The formation of the platelet plug is rather like building a bridge over a mountain pass. Should the break in the vessel be small enough, the platelet plug may be sufficient to resolve the damage. But should it be of sufficient size, the exposed collagen and aggregated platelet phospholipids will promote the activation of the secondary hemostasis process.

Bone marrow disease or failure from any cause can lead to a non-regenerative anemia and pancytopenia. When diffuse marrow involvement occurs, granulocytes are the first cells affected, platelets are the second cell type affected and finally the red blood cell production is affected. It is easy to see that thrombocytopenia may well go hand in hand as part of any syndrome affecting the erythropoietin system.

Bone marrow failure, seen as aplastic anemia, typically presents as a severe anemia and thrombocytopenia resulting in the clinical symptoms of malaise and lethargy. Clinical signs may include: shock due to decreased perfusion/oxygenation, hypotension, tachycardia, and tachypnea. Bone marrow failure may be caused by infectious diseases, drug intoxication and seldom, especially in the feline patient, classified as idiopathic aplastic anemia.

Infectious agents that may cause aplastic anemia or aplastic pancytopenia for both the feline and canine patients would include: ehrlichiosis, and parvovirus. Our feline patients may also be impacted by feline leukemia virus (FeLV) and feline immunodeficiency virus (FIV). Aplastic anemia may occur in acute and chronic forms.

Ehrlichia, one of the rickettsial diseases occurring in our companion animals, may cause a decrease of production of all of the bone marrow precursor cells. This is more common in the severe acute phase of the disease. Dependent on the stage of the infection, Ehrlichia may be responsible for all three of the basic causes of thrombocytopenia that is production failure, decreased production as well as increased consumption. The acute phase of the infection sees the organism multiplying within mononuclear cells and being transported via circulation to the rest of the body systems. Infected cells adhere to the vascular endothelium causing a vasculitis. This simple act results in a subendothelial tissue infection. Remember from week one that we learned that any disruption of the endothelium/subendothelium exposes collagen providing a surface for the platelets to adhere to. Thrombocytopenia occurs due to platelet adhesion not just to one limited area as might be seen with a traumatized vessel but throughout the circulatory system. The platelets are sequestered in the affected vascular tissue and consumed making them unavailable. This process will occur with any type of vasculitis regardless of etiology. You might consider collagen to be the docking station for platelets to start their job of adhesion, swelling, assuming irregular shapes and degranulation. Very high gamma globulin concentrations seen with some cases of Ehrlichiosis can interfere with platelet function and be responsible for severe hemorrhage.

Anaplsam platys is a bacterium that was formerly classified as one of the Ehrlichia species. It is commonly referred to as infectious canine cyclic thrombocytopenia or ICCT. As with Ehrlichia, it is an acquired thrombocytopenia that results from the reduced production and increased consumption of platelets.

Parvovirus once introduced into the body, has a high affinity for rapidly dividing cells targeting those in the lymph nodes, intestinal crypts and the bone marrow. The virus can cause a panleukopenia which of course would include platelets. This would be an example of bone marrow failure or decreased production due to the destruction of precursor cells. Also bear in mind that another primary target of the virus is the intestinal crypts which have a network of capillaries and lymphatic vessels in the endothelium. Disruption of the endothelium from the virus would again cause platelets to be removed from the circulatory system by adhering to the damaged endothelium exposing collagen.

Feline leukemia virus (FeLV) has three different subtypes that may affect the feline patient with each creating its own set of signs and symptoms. An ineffective immune response on the part of the patient allows the virus to be transported to the spleen, lymph nodes, epithelial cells of the bladder and intestine, salivary glands and the bone marrow. If the patient has a partially effective immune response, the virus may become latent in the bone marrow and the virus will not replicate. In this situation, even though the patient is not viremic, a hematopoietic malignancy may occur. These nonviremic patients may be diagnosed as FeLV negative based on serology. The primary malignancy we are concerned with in this class is the Myelodysplastic syndrome. A frequent outcome of subtype C, it is characterized by anemia, leukopenia with or without thrombocytopenia. The FeLV is also capably of infecting megakaryocytes. Damage or decreased production of these megakaryocytes will cause thrombocytopenia.

Neoplasia such as lymphoma may replace normal bone marrow constituents with cancer cells. In Neoplastic disease there is a wide variety of mechanisms that may lead to thrombocytopenia. The platelets may be sequestered in the spleen, liver or in the tumor itself. Platelet consumption may be increased, for example by DIC. Platelets may be destroyed by immune reactions or platelet production may be reduced by bone marrow involvement. Very high gamma globulin concentrations associated with lymphoid or plasma cell neoplasia can interfere with platelet function. The monoclonal protein coats the platelet surface which adversely affects platelet aggregation, and the ability for

Hemangiosarcoma (HAS) is probably a malignancy that all of us are familiar with. It is a malignant neoplasia of vascular endothelial origin. The typical target organ is the spleen and it is the most common primary tumor of the right atrium in the canine patient. 30-75% of our canine patients with HAS will present with thrombocytopenia. The thrombocytopenia associated with HAS may be caused by several reasons, an immune-mediated process, platelets being sequestered, severe hemorrhage or the development of Disseminated intravascular coagulation (DIC). Laboratory testing may show thrombocytopenia, prolongation of activated partial thromboplastin time, prolonged thrombin clotting time, decreased anti-thrombin III and an increase in fibrin degradation products and or D-dimers. All in all, a wicked coagulation panel. Thirty percent of platelets are normally sequestered in the spleen. As the spleen is the primary target organ, enlargement of the spleen further removes platelets from the circulation lowering the platelet count. So we would consider hemagniosarcoma to be associated with consumptive thrombocytopenia.

With lymphomas, thrombocytopenia is another abnormality commonly seen with neoplasia (up to 40% of the dogs) mainly in dogs with lymphoma. The possible/probable causes for this thrombocytopenia are the decreased life span of the platelets due to an accelerated rate of removal from the circulation. This appears to be related to the microaggregation stimulated by the tumor or even the possibility of the coating of platelets by tumor proteins.

Immune mediated thrombocytopenia (IMT or ITP) represents the most common of causes for spontaneous bleeding seen in canines. It is rare in our feline patients. Studies have shown that it affects primarily middle-aged female canines.

Patients may present with clinical signs indicative of a primary hemostatic defect. Simply, petechiae, ecchymoses, epistaxis, gingival bleeding and/or melena may be seen. Collapse may occur if the bleeding is extreme or pronounced. These patients may also present with cavitary bleeding causing one to think of rodenticide toxicity or the presences of a compromised Hemangiosarcoma tumor based on which cavity is seen to be involved. In most of our canine patients, IMT is acute or peracute. Approximately 20% of patients with ITP will exhibit no clinical signs of bleeding despite a low platelet count and typically bleeding will not be evident until the platelet count is below 40,000 to 60,000.

ITP is often seen concurrently with Immune Mediated Hemolytic Anemia (IMHA). The concurrent disease process is referred to as Evans syndrome. The CBC in patients with IMT is characterized by thrombocytopenia with or without anemia. This may be a reflection of the degree of spontaneous bleeding or concurrent IMHA. Bone marrow examination typically reveals megakaryocyte hyperplasia. In rare instances, megakaryocyte hypoplasia with free megakaryocyte nuclei may be identified. There are different classifications of IMHA based on the causative process. Both IMHA and ITP are the result of a hypersensitivity reaction. Autoantibody, a type of protein manufactured by the immune system, attach to a cell. This triggers cytotoxicity and leads to the destruction of the cell. Researchers have not completely defined the causes of antibody productions as they appear to be quite varied. The current thought is that some autoantibody production may be due to a genetic disposition while being combined with certain triggers. These may be viruses, or prolonged exposure to certain toxic chemicals. At this time, there does not appear to be a direct genetic link.

von Willebrand's disease is a common bleeding disorder seen with certain breeds. This dysfunction affects both the primary and secondary coagulation. The release of vWF is stimulated by other substances such as histamine, fibrin and estrogen. When disruption of the endothelium occurs, vWF binds to the exposed collagen of the subendothelium. In those areas with low blood flow rate, platelets may independently adhere to the vessel walls. In those areas with a high blood flow rate, we rely upon vWF to assist platelet adherence to the subendothelium. Remember that while some vWF is located within the subendothelium, additional vWF must be released from platelets or the endothelial cell granules to provide stable platelet adherence in these high blood flow areas.

It is also a carrier for factor VIII, which is vital in the clotting cascade. In the absence of VWF, blood coagulation and the platelet plug could be abnormal. This would affect both primary and secondary hemostatis.

In von Willebrand's patients, the platelet count and platelet morphology are seen as normal. Buccal mucosal bleeding time (BMBT) in those patients with mild to moderate von Willebrand's disease (vWD) will have been between 5 to 10 minutes. With severe vWD, the bleeding time may extend to 12 minutes or longer. Remember that BMBT evaluates the primary hemostatic function of platelets and is prolonged when platelet function is abnormal.

REFERENCES

1. Small Animal Emergency and Critical Care for Veterinary Technicians, 2nd edition, Andrea M. Battaglia editor, Saunders Elsevier publishers, 2007. Chapter 15, Hematologic Emergencies
2. The Veterinary ICU Book, W. W. Wingfield and M. R. Raffe editors, Teton New Media publishers, 2002. Chapter 49, Disorders of Hemostasis
3. Emergency Procedures for the Small Animal Veterinarian, 2nd edition, S. J. Plunkett editor, W. B. Saunders publisher, 2000, Chapter 6, Hematologic Emergencies
4. North American Veterinary Conference Proceedings, 2007, NAVC editors and publisher, "Why is Fluffy bleeding? Secondary Hemostatic Disorders" L. DeWilde
5. Proceedings of the SCIVAC Congress, Rimini, Italy, 2007. "Clinical Approach to the Bleeding Patient" C. Guillermo Couto
6. Textbook of Medical Physiology, W. B. Saunders publisher, A.C. Guyton author

PETS AND IMMUNOCOMPROMISED OWNERS

J Scott Weese DVM DVSc DACVIM
Guelph, Ontario, Canada

INTRODUCTION

The role that pets play in many peoples' lives and the close contact that is inherent with this relationship can be highly beneficial, but is also associated with some risk of infectious disease transmission. Few would argue that pets are part of the family socially and emotionally. At the same time, it is becoming increasingly clear that pets are part of the family microbiologically, and that transmission of microorganisms between people and pets is probably a common event. The potential for transmission of infections from pets to people has been recognized for centuries. Zoonotic diseases are an inherent risk of any contact with animals and are generally considered an acceptable risk by most pet owners, considering the low incidence of serious zoonoses and the positive social and emotional benefits of pet contact. However, there are certain situations where people may be at higher risk for zoonotic diseases or be more prone to serious outcomes. One such group is people with compromised immune systems. The advances of modern medicine have led to a tremendous population of people that are alive but living with suboptimal immune systems. These include people with acquired immunodeficiencies (i.e. HIV/AIDS) and individuals whose immune systems are being purposefully suppressed (i.e. some cancer patients, some new biologic agents, transplant patients). Advances in medicine have been able to sustain lives of such individuals but leave them in a more susceptible state for infectious diseases, including zoonotic diseases. Others who may be at increased risk include infants, pregnant women and the elderly.

Based on numbers of organ transplant recipients and individuals infected with HIV, it is estimated that 3.6% of the United States human population has an acquired immunodeficiency.[1,2] When additional groups are added, such as the elderly and pregnant women, up to 20% of the population of the United States has some degree of immunosuppression.[3] Similar statistics probably apply to Canada and it is likely this proportion will increase in future years due to advances in life-prolonging treatments, such as chemotherapeutic agents and immunosuppressive drugs.

A large number of immunocompromised individuals own pets, as is appears that pet ownership rates among immunocompromised individuals are similar to that of the general public.[4,5] One study reported that 46% (187/408) of HIV-infected adults were living or had lived with a companion animal in the past five years (dog-38%, cat-15%, fish-8%, bird-8%, reptile-3%, rodent-2%), with 25% of pet owners having more than one pet.[4] Further, many people that do not own pets themselves have periodic contact with other peoples' pets, and demonstrated by a study in Ontario, Canada, where 37% of households that did not own pets had regular contact with pets outside of the home.[6] The large number of immunocompromised people in the population and the commonness of pet ownership clearly indicate the importance of this area, and that concerns are not limited to a small, select group of individuals with rare diseases.

A 'knee-jerk' reaction sometimes occurs when issues regarding zoonotic diseases and immunocompromised individuals are discussed. The simplest way to reduce the risk of zoonotic disease exposure from pets is to avoid all contact with pets. However, this is generally not an acceptable course of action, nor is it commonly (if ever) required. The importance of the human-animal bond and its benefits are also documented in individuals who are immunocompromised, and this fact must not be lost amongst infectious disease concerns. Such individuals may spend considerable time alone, and thus are especially vulnerable to mental and physical illness.[5,7,8] Among individuals infected with HIV/AIDS, domestic animals are often perceived as family members, serve as sources of support and affection, and protect against loneliness.[4,7,8] Siegel, et al. reported that among HIV-infected persons, those with AIDS who owned pets reported less depression than persons with AIDS who did not own pets.[5] Thus, it is clear that while there must be consideration of potential health risks associated with pet contact, the emphasis should be on determining the cost-benefit of pet ownership and implementation of methods that may be able to reduce (but never eliminate) the risk of zoonotic disease transmission.

IMMUNOCOMPROMISED INDIVIDUALS

While all people with some degree of compromise of their immune system are often lumped into the generic 'immunocompromised' category, it is critical to recognize that there are marked differences between different conditions and even between individuals with the same condition. Acquired immunodeficiencies are the most common causes of immune dysfunction. These can include numerous biologic disorders, normal physiologic states (e.g., pregnancy), and drug-induced effects on the immune system. Different conditions have different effects on the immune system, both in terms of severity of compromise and the component of the immune system that is affected. Both of these are critical factors since certain components of the immune system are critical for preventing or eliminating certain infections. A person with a drug-associated immunosuppression may have markedly different risks compared to someone without a functional spleen, despite the fact that both are at increase risk of infection in broad terms.

PATHOGENS OF CONCERN

There are various types of pathogens that are of concern in immunocompromised individuals. These include organisms that are also of concern in immunocompetent individuals, those that do not typically cause disease in immunocompetent individuals but can cause disease in immunocompromised individuals, and pathogens where disease is potentially much more severe in immunocompromised individuals. There are not always clear delineations between these groups.

Organisms that are primarily pathogens only in immunocompromised individuals

People are constantly exposed to various bacteria, viruses and fungi. Some are primary pathogens that typically cause disease when they are encountered. More are opportunistic pathogens that can cause disease under appropriate circumstances. Others are organisms that rarely cause disease in otherwise healthy people but can cause disease in individuals with compromised immune systems. For some organisms, diagnosis of infection essentially indicates that immunocompromise is present, even if it has not been previously diagnosed. Zoonotic pathogens in this category include *Capnocytophaga* spp, *Rhodococcus equi* and *Toxoplasma gondii.*

Organisms that are more likely to cause disease and/or to cause more severe disease in immunocompromised individuals

Some infectious agents may cause disease in immunocompetent people, however disease tends to be mild and either self-limiting or readily treated. The same pathogens may cause severe or life-threatening disease in immunocompromised individuals. These include *Cryptosporidium* spp, *Bartonella* spp, *Pasteurella* spp., *Chlamydophila felis, Bordetella bronchiseptica, Staphylococcus aureus, Salmonella* spp, *Leptospira* spp, *Giardia* spp, *Campylobacter* spp, *Brucella* spp, *Mycobacterium avium* complex, *Mycobacterium marinum* and *Cryptococcus* spp.

OTHER POTENTIAL ISSUES

Diet

The main concern regarding diet is feeding of diets that are at increased risk for being contaminated with pathogens such as *Salmonella*. The main concern is with feeding of raw meat diets, which have been clearly demonstrated to result in elevated risk of pets shedding of *Salmonella* and multidrug-resistant *E. coli*. Raw meat or eggs should not be fed to pets of imunocompromised individuals because of the risk of exposure to pathogens from feces (or fecal-contaminated surfaces), food bowls, water bowls or from the mouth of the animal after eating. Raw animal product based pet treats such as rawhides and pig ears have been implicated as causes of outbreaks of salmonellosis in people. While improvements in manufacturing of these products have been made, immunocompromised people should avoid any contact with raw animal based treats unless they have been treated to remove pathogens (i.e. irradiated). These treats should never be purchased from bulk bins, where contamination is more likely to occur. While the risks of *Salmonella* exposure are much lower from commercial dry foods, *Salmonella* contamination can occur and it is prudent for immunocompromised individuals to wash their hands after any contact with pet food. Pets should not be fed in the kitchen.

Veterinary Examination and Care

Regular examination is critical to detect problems that could put the owner at risk. A thorough veterinary examination should be performed at least yearly, if not more frequently. A proper preventive medicine program is important. While most of the diseases that pets are vaccinated against are not zoonotic, overall maintenance of health is important because any illness could make the pet more likely to acquire and shed potentially zoonotic organisms. A deworming program should be structured based on the risks of the parasite exposure and shedding. Flea control is also important.

Client counseling

Detailed discussion of potential risks and infection control practices is critical. The use of practical techniques can presumably greatly reduce (but never eliminate) the risk of disease transmission. Foremost of these is hand hygiene. Owners should wash their hands (or use an alcohol-based hand sanitizer) after having contact with their pets, food bowls or stool. Good stool handling practices should be used. Ideally, immunocompromised individuals should not clean litterboxes. If they must, they should clean them daily, wear gloves and wash their hands. Direct contact with the mouth of pets should be avoided. Licking, especially of the face, should be discouraged. Owners should be discouraged from sleeping in the same bed with their pets.

Immunocompromised individuals may ask for guidance in obtaining a new pet. Certain factors should be considered to reduce any risks. Young animals and animals from shelters or similar facilities are more likely to shed various infectious agents, so they are not the best choice of a pet. Further, younger animals are more likely to bite and scratch, and it is more difficult to assess their temperament. Older animals that are already in households are ideal. Certain animal species are higher risk for pathogen carriage and disease transmission. While these can be good pets in some households, it is prudent for high-risk individuals to avoid such species, including all reptiles and most exotic animal species where there is minimal information about pathogen carriage and transmission. Species such as rabbits where there is moderate information about infectious disease risks and management are relatively low risk. Birds are a somewhat controversial area. Rates of shedding of some potential pathogens can be high, including serious pathogens such as *Chlamydophila psittaci*. There are also concerns about shedding of organisms such as *Cryptococcus* spp, although the risk various geographically. There has been little objective evaluation of risks associated with pet bird ownership and some degree of risk must be considered, although it can probably be minimized by safe management practices, particularly those involving avoidance of fecal material.

It should be strongly recommended that cats remain indoors because of the higher rates of shedding of certain pathogens (i.e. *Bartonella* spp, *Salmonella, Toxoplasma gondii*) in cats with outdoor access.

Owners should be made aware of certain syndromes that might indicate an infectious process. These include diarrhea, vomiting, coughing and skin lesions. In households with immunocompromised owners, earlier or more comprehensive diagnostic testing may be indicated to allow for prompt identification of potentially zoonotic pathogens.

CONCLUSION

Virtually ever veterinarian has patients owned by immunocompromised individuals, whether they know it or not. As more immunocompromised people are present in the community, it is certain that this will continue. Therefore, veterinarians need to be aware of the potential issues. Good communication with clients is critical so that veterinarians become aware of any issues and properly counsel them. Generic information about immunocompromised owners and pets (or directions how to access them) should be readily available in clinics so that clients who do not wish to discuss their health can still be made aware of any issues and confidentially access resources. Veterinarians need to be prepared for discussions with physicians, as part of the family's healthcare team. While veterinarians rightly must refrain from direct involvement in human health issues, they must recognize the potential role of pets in human health and their role in maintenance of health of all individuals in the household. This is a complex and often daunting area, but increased awareness of zoonotic diseases, infection control and increased communication between human and veterinary medicine are creating an environment that is amenable to increasing veterinary involvement. Accordingly, veterinarians must ensure that they are adequately versed in the areas discussed above and have (or can recommend) good resources for clients, physicians and public health personnel.

References

[1] Taylor LH, et al. ***Philos Trans R Soc Lond, B, Biol Sci*** 2001;356:983-989.
[2] Kemper AR, et al. ***Eff Clin Pract*** 2002;5:84-90.
[3] Trevejo RT, et al. ***Vet Res*** 2005;36:493-506.
[4] Conti L, et al. ***American Journal of Public Health*** 1995;85:1559-1561.
[5] Siegel JM, et al. ***AIDS Care*** 1999;11:157-170.
[6] Stull JW, et al. ***BMC public health*** 2012;12:553.
[7] Castelli P, et al. ***Psychol Rep*** 2001;89:177-187.
[8] Carmack BJ. ***Holist Nurs Pract*** 1991;5:24-31.

APPROACH TO THE RESPIRATORY RESPIRATORY EMERGENCY

Steven L. Marks BVSc, MS, MRCVS, DACVIM (SAIM)
Raleigh, NC

INTRODUCTION

When a patient is presented demonstrating signs of respiratory distress or dyspnea, the veterinary clinician/technician must work quickly and efficiently. It is imperative that the entire veterinary staff is familiar with the classic signs associated with an animal having respiratory difficulty. This will enable the staff and clinicians to work as a team to benefit the patient. Although there are many causes of respiratory distress in the dog and cat, the initial assessment and therapy is very similar. Using a triage system will allow the veterinary team to prioritize problems and patients.

HISTORY

An abbreviated history relative to the respiratory system should be obtained. Specific information should include past medical history, current therapy, onset of clinical signs and progression of clinical signs. If the patient is unstable a complete history can be obtained following hospitalization.

PHYSICAL EXAMINATION

The initial physical examination should focus on assessing the stability of the patient. This will include evaluation of breathing pattern, thoracic auscultation and assessment of perfusion parameters. Observation of the breathing pattern may provide valuable information on the location of the disease and the etiology of the respiratory distress. Upper airway obstruction often leads to loud sounds (stridor) and a prolonged inspiratory phase. A prolonged expiratory phase or expiratory effort is commonly observed in lower airway diseases.

Rapid shallow breathing, often referred to as a restrictive breathing pattern is seen with pleural space disease and thoracic wall injuries. Diseases affecting the pulmonary parenchyma often result in changes in the inspiratory and expiratory phases of respiration. Thorough thoracic auscultation may be useful in identifying the underlying pathophysiology. The identification of crackles, wheezes, decreased lung sounds, murmurs or arrhythmias may help the clinician/technician narrow the differential list.

INITIAL INTERVENTION

The cornerstone of therapy for the patient in respiratory distress is oxygen supplementation and sedation. Sedation with an opioid is safest. Butorphanol is the most common anxiolytic used in our ER. The caveat with sedation of emergency room patients is the clinical team must be prepared to intubate if the patient's condition deteriorates.

Initial imaging depends on the stability of the patient. If pleural space disease is suspected, diagnostic thoracentesis should be considered. The use of FAST ultrasound will also be valuable if this is available. Radiographs should be performed only when the patient is relatively stable.

EMERGENCY OXYGEN ADMINISTRATION

Face Mask

The face mask is perhaps the easiest and least invasive technique for oxygen supplementation. Masks designed for people are not ideal and purpose made veterinary masks should be used. These masks come in an assortment of sizes that range from pediatric size to large dog sizes and are designed to fit over the muzzle. It is often awkward to administer oxygen via mask to some brachiocephalic breeds. The mask can be connected via tubing to a 100 % oxygen source and if used for long periods of time, the oxygen should be humidified. The advantages of a mask are related to ease of administration and that no specialized equipment is required. In addition, during oxygen administration the animal can be closely monitored. Disadvantages of this technique include lack of tolerance by some patients and that the mask may require constant manual application unless the patient is moribund. When using a face mask, it is estimated that 40% is the fraction of inspired oxygen.

Flow-by

This technique is a modification of the face mask technique. The mask is disconnected and the oxygen line is held close to the patient's mouth and nose. The advantages and disadvantages are similar to the face mask. However there is no way to know the fraction of inspired oxygen and it will certainly be much less than that attained with a tight fitting face mask.

Nasal insufflation

This technique is often underutilized. There are various types of cannulas available for nasal catheterization including feeding tubes and urinary catheters. In our hospital we use small bore, flexible, feeding tubes with fenestrations to avoid mucosal jet lesions. The largest tube size possible should be passed. These sizes vary from 3.5 F-10 F depending on the size of the patient. These tubes can be placed in the conscious patient but local anesthesia is generally required. Topical 2 % lidocaine, lidocaine gel or proparacaine can be used on the nasal mucosa. The animals head should be tilted with the muzzle pointing toward the ceiling. A few drops of local anesthetic can be placed in the external naris to desensitize the nasal mucosa. The tube should be pre-measured to from the external naris to the vertical ramus of the mandible and advanced to this level. This will insure that the tube is placed into the ventral nasal meatus. Once placed the tube can be secured as close to the nostril as possible with suture and or cyanoacrylate adhesive (Superglue). The tube should also be attached over the nasal bones and frontal sinus. These animals should be fitted with an Elizabethan collar to prevent removal. The nasal catheter can be adapted using a catheter (Christmas tree) adapter and syringe barrel to fit into a humidified oxygen line. The advantages of this technique include ease of application, no special equipment required, versatility in different sized animals and animals can be closely monitored. The disadvantages include lack of patient tolerance, variable flow rates required to provide oxygen, nasal mucosal irritation and gastric distension at high flow rates. The estimated fraction of inspired oxygen with this technique is 40%. There is information suggesting the benefit of utilizing bilateral nasal cannulae.

Oxygen canopy

An oxygen canopy can be created by using an Elizabethan collar and a humidified oxygen line. The patient is fitted with an Elizabethan collar and 50%-75% of the collar diameter is covered with cellophane or plastic material. The oxygen line is run inside the E-collar and provides oxygen into the cellophane covered area. Basically the patients head is within an oxygen tent. The advantages of this technique include ease of application and no special equipment requirements. The major disadvantages are patient tolerance, increased humidity within the canopy, increased temperature within the canopy and no carbon dioxide scavenging ability. If flow rates are maintained appropriately or at higher levels this may reduce heat, humidity and carbon dioxide. The estimated fraction of inspired oxygen is 40%.

Oxygen cage

This equipment usually involves a chamber that can be flooded with oxygen to provide an oxygen enriched environment. The cages can be programmed for temperature and humidity control. A carbon dioxide scavenging system is also incorporated. There are currently cages available for most sized animals. They have working ports for fluid line and patient access. The advantages of this equipment are the controlled environment provided. The disadvantages include substantial financial investment, poor access to patient, difficulty in monitoring the patient and inadvertent uncontrolled environmental conditions.

Mechanical ventilation

Mechanical ventilation can be a considered option for supplying oxygen to selected patients. Ventilator therapy has some very specific indications which are beyond the scope of this discussion. For the most part ventilator therapy is reserved for patients that demonstrate signs of hypoventilation (hypercapnia), are refractory to oxygen supplementation by the previously described routes or have respiratory muscle fatigue. Ventilator therapy requires specialized equipment as well as heavy sedation or general anesthesia in most patients. Nursing care and close monitoring are the cornerstones of patient survival.

COMMON RESPIRATORY EMERGENCIES

AIRWAY OBSTRUCTION

Larynx

Inspiratory dyspnea and respiratory stridor are classic findings with laryngeal disease. The underlying disease may be laryngeal paralysis, neoplasia, trauma, inflammation or foreign body. Cyanosis is often present on physical examination. Oxygen supplementation is beneficial during assessment of these cases. Sedation is required in most patients with significant laryngeal obstruction. If possible the obstruction is removed. Many patients with laryngeal paralysis will respond to oxygen and sedation. When the patient is not responding to initial therapy and the obstruction is unable to be relieved, an emergency tracheostomy is required. This procedure involves incising the trachea between the tracheal rings and using a purpose made tracheostomy tube or preparing a tube by modifying an endotracheal tube. Alternative techniques include needle or catheter tracheostomy.

Trachea

Tracheal obstruction can be secondary to collapse, trauma, neoplasia, inflammation and foreign bodies. The clinical signs may be difficult to differentiate from laryngeal disease. If the obstruction is located in the cervical trachea the patient usually has inspiratory dyspnea. Expiratory dyspnea is generally seen with intrathoracic tracheal disease. Oxygen therapy should be provided immediately in these cases. Sedation may be beneficial in these cases and may facilitate imaging such as radiographs. Emergency tracheostomy or bronchoscopy may be required in these cases. Nasal tracheal oxygen supplementation may provide an alternative technique.

THORACIC DISEASE

Penetrating Injuries

Penetrating thoracic injuries are often seen as the result of trauma in the dog and cat. The most common penetrating injuries are bite wounds and ballistic trauma. In all penetrating thoracic injuries the major concern is the open pneumothorax that may be present. It is imperative to seal the pleural space with a sterile occlusive bandage and eliminate the pneumothorax by thoracocentesis and if required thoracostomy tube. Stabilization of the respiratory system is the number one priority followed by correction of any hemodynamic abnormalities that may exist. It is important to remember that concurrent trauma such as hemothorax and pulmonary contusions may also be present. The use of oxygen supplementation and analgesia is important and antimicrobial agents are warranted in this situation.

Pneumothorax

Pneumothorax or air in the pleural space is most commonly seen in dogs and cats secondary to trauma. Physical examination of these animals usually indicates a restrictive breathing pattern, which is characterized by rapid shallow breathing. Thoracic auscultation reveals dull or distant lung sounds. Thoracic percussion may be hyperresonant confirming the diagnosis. Thoracocentesis should be performed on these animals immediately if dyspneic. Oxygen therapy may be used in these patients prior to and during thoracocentesis. Thoracocentesis may be performed with the pet standing or in sternal or lateral recumbency. Thoracic radiographs may be useful but should not be performed until the patient is stabilized. If thoracocentesis continues to be productive, after several attempts, thoracic drain placement may be required. As with other thoracic injuries, remember that concurrent disease may be present. Oxygen supplementation is often beneficial.

Flail Chest

Flail chest is a specific type of injury that involves multiple rib fractures. The sequel of these fractures is a free-floating section of the thoracic wall. With this degree of thoracic trauma these pets are generally dyspneic. The dyspnea may be due to concurrent injury such as pneumothorax, hemothorax or pulmonary contusions and the pain associated with the trauma. These pets will benefit from the administration of oxygen and analgesics. Local analgesia such as infiltration with lidocaine and bupivacaine may be useful in these patients. Ventilatory support may also be required. Rib fracture not associated with flail chest should be treated in a similar manner. Restrictive bandages placed around the thorax are contraindicated in patients with rib fractures or flail chest.

Pleural Effusion

Pleural effusion refers to some type of fluid in the pleural space. This is similar to another type of pleural space disease; pneumothorax which has already been discussed. The types of pleural effusion include hemothorax, hydrothorax, pyothorax, and chylothorax. The presence of fluid in the pleural space should be able to be detected on thoracic auscultation and percussion. Similar to pneumothorax, thoracocentesis is an important part of diagnosis and therapy. Oxygen therapy will be useful in the initial management of these cases. Fluid analysis should be performed as soon as possible to confirm the diagnosis. Imaging such as thoracic radiographs and ultrasound should be delayed until the patient is stable.

Pulmonary Contusions

Pulmonary contusions are a common finding in animals with thoracic trauma. This injury occurs after hemorrhage into the pulmonary parenchyma and alveoli. There may be concurrent disease such as pneumothorax and hemothorax. The diagnosis is often made after thorough physical examination, thoracic auscultation and review of thoracic radiographs. Negative thoracocentesis in a dyspnea pet may increase the clinical suspicion of pulmonary contusions. Thoracic radiographs may not reveal extent of injury during the acute assessment of the patient. Rest and oxygen supplementation are the treatments of choice. The use of antimicrobials and corticosteroids are controversial.

Diaphragmatic Hernia

Diaphragmatic hernia is another common traumatic injury in the dog and cat. Diagnosis is most commonly made after survey thoracic or abdominal radiographs. These animals are often not stable due to shock, pulmonary contusions and pain associated with the traumatic incident. The concurrent injuries should be treated and the patient stabilized prior to surgical intervention for repair of the diaphragm. Oxygen therapy may benefit these patients and thoracocentesis may be performed if pleural effusion is present.

Bronchopneumonia

Bronchopneumonia may be difficult to diagnose during the initial physical examination. The findings of pyrexia and coughing along with respiratory distress may provide clues. These animals generally have both inspiratory and expiratory dyspnea. Initially treatment with oxygen is ideal until further diagnostics such as radiography can be performed to confirm the diagnosis. Additional diagnostics may include transtracheal wash, bronchoscopy and bronchioalveolar lavage. Once the diagnosis is made therapy may include antimicrobial agents, nebulization and coupage.

Pulmonary Edema

The most common form of pulmonary edema in the dog and cat is cardiogenic pulmonary edema. This results from left-sided congestive heart failure. In the dog this is primarily due to mitral valve insufficiency or dilated cardiomyopathy. In the cat the most common etiology is hypertrophic, restrictive or dilated cardiomyopathy. It is often difficult in these cases to differentiate between primary respiratory disease and primary cardiac disease. As with the other causes of respiratory distress we have discussed, oxygen therapy is always a good choice. Thorough physical examination may reveal clues such as heart murmurs or arrhythmias that suggest cardiac disease as the etiology. Another valuable clue may be heart rate. If a patient presenting with acute respiratory distress has a slow or normal heart rate, the underlying disease is most likely respiratory in nature rather than cardiac. Additional diagnostics such as thoracic radiographs, ECG and echocardiography may be required for definitive diagnosis. Patients with pulmonary edema often have inspiratory and expiratory dyspnea. If cardiogenic pulmonary edema is diagnosed the therapy should include furosemide, nitroglycerine and oxygen. The use of morphine for sedation and venodilation may also be useful in dogs.

Non-cardiogenic pulmonary edema is a less common form of pulmonary edema. Again initial therapy may include oxygen supplementation. The pathophysiology of this type of edema can be fairly complex but it can be seen associated with upper airway obstruction, electrical cord injuries and seizures. Acute respiratory distress syndrome or ARDS is characterized by non-cardiogenic pulmonary edema and has been reported in some critically ill pets. The therapy for non-cardiogenic pulmonary edema is directed toward the underlying disease but initially is similar to treatment for cardiogenic pulmonary edema.

Pulmonary thromboembolism

Pulmonary thromboembolism is a serious hemodynamic event resulting from a thrombus or clot in the pulmonary arterial system. It has been recognized in critically ill animals secondary to heart disease, inflammatory disease, glomerular disease, neoplasia, endocrine disease and immune mediated hemolytic anemia. These pets will generally present in severe respiratory distress and must be stabilized as soon as possible. The diagnosis may be suspected based on history, clinical signs, thoracic radiographs and blood gas analysis. These animals should be treated with oxygen and evaluated for underlying disease.

MONITORING

Monitoring is an important component of the medical management of any critically ill pet. It is of particular importance in the management of the patient with respiratory distress or disease. Although initial therapy may be similar in many cases of respiratory distress, the response to therapy and the continued response to therapy may vary. Some cases that respond to therapy during the emergency phases may deteriorate during the course of hospitalization. Other cases may not respond to initial therapy and more aggressive intervention may be required. Many of these decisions will be based on patient monitoring. In addition to thorough physical examination, pulse oximetry, blood gas analysis, thoracic radiographs, electrocardiography, central venous pressure, and arterial blood gas are parameters that may be useful. Additional parameters such as A-a gradient and PaO_2/FIO_2 can be calculated and provide valuable information in the hypoxemic patient.

CONCLUSIONS

As can be seen from this discussion, the causes of respiratory distress in the dog and cat can be variable. It will often be difficult to identify the underlying disease. It is therefore imperative that the veterinary clinician and technician be able to recognize the clinical signs associated with respiratory distress and be able to provide initial therapy until a definitive diagnosis can be made and appropriate therapy provided.

REFERENCES

Camps-Palau MA et al. Compen for Cont Edu 1999 21(7):587. King LG. (ed) Respiratory Disease in Dogs and Cats. WB Saunders, Philadelphia, 2004.

GOING WITH THE FLOW: AN IN-DEPTH LOOK AT CSF

Julie A. Nettifee-Osborne, RVT, BS, VTS (Neurology)
Karen R. Muñana, DVM, MS, DACVIM (Neurology)
Raleigh, NC

INTRODUCTION

Cerebrospinal fluid (CSF) analysis is a standard part of the diagnostic workup for any animal presenting with central nervous system disease. The close association between the CSF and nervous tissue allows for changes in the brain or spinal cord to be reflected in CSF abnormalities. The degree of CSF abnormality is dependent on the location and extent of the disease within the CNS as well as the proximity of CSF collection site to the lesion. CSF findings provide information on the likelihood of certain differential diagnoses, and in rare instances, can provide a definitive diagnosis. CSF is most useful in the diagnosis of inflammatory CNS disease, whether infectious or noninfectious, as different inflammatory diseases lead to an accumulation of different types of nucleated cells in the CSF. Brain lesions due to degenerative, traumatic and neoplastic causes can also alter CSF, as can myelopathies and radiculopathies. This presentation will provide an overview of the physiology of CSF, discuss guidelines for CSF collection and laboratory analysis, and utilize case examples to illustrate the role of CSF analysis in the diagnostic evaluation of an animal with neurological disease.

OVERVIEW OF CSF PRODUCTION AND FLOW

CSF is an ultrafiltrate of plasma that is produced primarily by the choroid plexus of the lateral, third and fourth ventricles, but also by brain capillaries, parenchymal cells and ependymal cells. CSF surrounds and bathes the brain and spinal cord; its primary function is physical support and buoyancy of the brain, but it also plays a role in the transport of biologically active substances and excretion of wastes within the CNS, and is involved in autoregulatory mechanisms that function to maintain cerebral blood flow during periods of increased intracranial pressure. CSF flows caudally through the ventricular system into the subarachnoid space, where it is resorbed into the venous system via the arachnoid villi. The normal rate of CSF production ranges from 47-66 ml/minute in dogs and 20-22 ml/minute in cats.[1]

CONSIDERATIONS FOR CSF COLLECTION

The primary sites for CSF collection in companion animals are atlantooccipital (AO) and caudal lumbar. Collection from the AO site is less difficult, usually results in a larger sample volume and typically is associated with less iatrogenic blood contamination. CSF samples area more likely to be abnormal or representative of the underlying CNS pathology when they are collected caudal to, and as close to the suspected lesion, as possible.

Materials needed for CSF collection include the following:

- Clippers and surgical scrub
- Sterile gloves
- Spinal needle (22 or 20 gauge; 1.5 inch for AO site; 2.5-3.5 inch for lumbar site
- Sterile collection container: either a 3ml syringe, or plain tube (free of anticoagulant)

The skin over the collection site should be clipped and surgically prepped. For AO CSF collection, the patient is placed in lateral recumbency and the neck is flexed and held by an assistant such that the head is at a right angle to the spine. The animal's nose must be kept parallel to the table, and a non-collapsing endotracheal tube should be used to avoid occlusion of airflow during the procedure. Lumbar CSF is most frequently collected from the spinal canal at the level of the L5-L6 intervertebral disk space. The animal is placed in lateral recumbency and the lumbar spine is gently flexed by bringing the pelvic limbs forward. A twitching of the tail or pelvic limbs can be seen as the spinal needle enters the spinal canal and irritates the cauda equina or caudal spinal cord, and can be used as a signal that the needle is in the desired location.

The total amount of CSF that can be collected from each animal is one ml per 5 kg of body weight. Typically, one ml is sufficient to perform routine analysis. Additional fluid should be collected if ancillary testing, such as culture, titers, PCR, flow cytometry or electrophoresis, is to be performed. Risks of performing a CSF tap include that associated with general anesthesia, along with the potential for hemorrhage, damage to neural tissue or introduction of infection during the collection procedure. CSF collection is contraindicated in animals that have elevated intracranial pressure due to mass lesions or inflammatory disease, due to the increased risk of brain herniation (and subsequent death) that can occur from the pressure gradient that is created when the needle is inserted into the subarachnoid space. Because of this risk, brain imaging should always be performed prior to CSF collection in patients with suspected inflammatory or mass diseases. Mannitol, corticosteroids and hyperventilation can be used to decrease the risk of herniation in cases where the collection of CSF is deemed necessary. Other contraindications for CSF collection include thrombocytopenia (platelet count $\leq$ 50,000/ µl), or disorders that result in spinal instability such that positioning for the procedure can result in damage to the nervous tissue (such as with collection of CSF from the AO site in an animal with atlanto-axial subluxation).

LABORATORY EVALUATION OF CSF

CSF contains very little protein; normal CSF protein is measured in mg/dl, while serum proteins are measured in g/dl. The low protein environment causes the leukocytes that are present to be unstable and deteriorate rapidly. Consequently, CSF should ideally be analyzed or preserved for analysis within 30-60 minutes of collection to minimize cellular degradation that can hinder accurate cytologic interpretation. However, it has been demonstrated that analysis of canine CSF within 4-8 hours of collection is unlikely to alter diagnostic interpretation, especially if the sample has a protein concentration $\geq$ 50 mg/dl.[2] Samples that are not immediately analyzed should be kept refrigerated. If a CSF sample is not able to be evaluated within one hour of collection, it can be preserved by the addition of 20% fetal calf serum[2] or 10% autologous serum.[3] The sample should be divided into 2 aliquots, and the preservative added to one of the aliquots. The unaltered aliquot is used for total nucleated cell count (TNCC) and protein determination, while the preserved aliquot is used for cytologic evaluation. If the volume of fluid is insufficient to aliquot, hetastarch at a ratio of 1:1 can be added as a preservative, and the complete analysis performed on this sample.[2]

Normal CSF is clear and colorless. Routine CSF analysis includes determination of red blood cell (RBC) count, TNCC, protein concentration and cytologic evaluation. The RBC count and TNCC are performed with a hemocytometer. Normal CSF has no RBCs present, and a total NCC typically <6 cells/ µl in dogs and <3 cells/ µl in cats. The protein content of CSF is measured using a microprotein technique. The normal total protein concentration in lumbar CSF is <45 mg/dl, while fluid collected from the AO site normally has a protein concentration <30 mg/dl.[4] This difference in protein concentration between sites is believed to be due to either differences in permeability to proteins along the neuroaxis or slower circulation of CSF in the lumbar region that leads to accumulation of protein.[5] A sample of CSF should be concentrated onto a slide by a cytocentrifugation, membrane filtration or sedimentation technique, to allow for cytologic evaluation of the fluid.

CSF INTERPRETATION

Red Blood Cell Count

Elevations in RBC count can be due to hemorrhage within the subarachnoid space, but is most commonly associated with iatrogenic hemorrhage during the collection procedure. The amount of blood within the CSF often varies over the course of the collection with traumatic taps, while it remains consistent with subarachnoid hemorrhage. The presence of erythrophagocytosis and xanthrochromia can also help differentiate between hemorrhage and iatrogenic blood contamination. If pathologic CNS bleeding has occurred, macrophages will be mobilized to phagocytize the RBCs, and evidence of large, activated macrophages containing one or more RBC is often apparent, and can be seen within 4 hours.[5] Xanthrochromia is a term used to denote a pink or yellow color to the CSF, and is due to the presence of pigments associated with the breakdown of erythrocytes. Xanthrochromia can be detected within 12 hours of a bleed, and reaches maximal levels within 2-3 days.[5]

Iatrogenic blood contamination can make interpretation of CSF results more difficult, as the TNCC and protein levels can be falsely elevated. As a general rule, to correct for blood contamination, the TNCC should be decreased by 1 cell/µl and the protein concentration should be decreased by 0.5 mg/dl for every 500 RBC/µl present. However, studies suggest that iatrogenic blood contamination of CSF does not significantly alter CSF nucleated cell count or protein concentrations at low levels of contamination (RBC count <13,200/ µl).[6]

Total Nucleated Cell Count

An elevated CSF TNCC, or pleocytosis, is most commonly associated with inflammatory conditions of the CNS. The magnitude of the TNCC elevation is a reflection of the degree of meningeal and ependymal cell involvement, regardless of the cause of the inflammation. Inflammatory disease processes that are confined to the parenchyma and have little meningeal or ependymal involvement can have normal CSF TNCCs.

Alternatively, disease processes that are considered non-inflammatory can cause elevations in CSF TNCC. This includes neoplastic conditions, cerebrovascular disease, fibrocartilagenous emboli, and intervertebral disk disease. This is presumably due to inflammation or necrosis that results from the primary disease process.

Protein Concentration

An increase in protein concentration can be due to breakdown of the blood brain barrier and transudation of serum proteins into the CSF, interruption of normal CSF flow, or intrathecal antibody production. An increase in TNCC often accompanies the increase in total protein concentration. Albuminocytologic dissociation is a term utilized to describe an elevation in total protein concentration with a normal TNCC. The presence of albuminocytologic dissociation is most commonly seen with degenerative, vascular and neoplastic conditions of the CNS.

Cytologic Evaluation

Microscopic evaluation of a concentrated sample of CSF is performed to determine a differential leukocyte count and to observe for the presence of bacteria, fungal organisms, or neoplastic cells. Mononuclear cells predominate in normal CSF, with the majority of cells being small, well-differentiated lymphocytes and fewer numbers of minimally vacuolated, large mononuclear cells. Mature, non-degenerate neutrophils and eosinophils are not commonly seen in CSF samples that are free of blood contamination. Cytologic evaluation is important even in samples with normal TNCCs, as if is not unusual to see abnormalities in cell distribution with normal TNCC in certain disease processes.[7]

A neutrophilic pleocytosis is characteristically seen with bacterial or fungal CNS infections, steroid responsive meningitis, and CNS infections due to Rocky Mountain spotted fever and feline infectious peritonitis. It can also be seen with conditions that cause necrosis of the CNS tissue, such as trauma, vascular disease and neoplasia.

Mononuclear pleocytosis is most common in viral and Ehrlichial infections, as well as with granulomatous meningoencephalomyelitis (GME) and necrotizing encephalitis. CNS lymphoma should be considered in an animal with a lymphocytic pleocytosis, regardless of whether the cells appear well differentiated or atypical. Flow cytometry or PARR (PCR for Antigen Receptor Rearrangements) can be performed on CSF to look for a homogeneous population of lymphocytes, suggestive of neoplasia.

Eosinophilic pleocytosis can be recognized with parasitic CNS infections, protozoal disease, fungal infections (Cryptococcosis, protothecosis), and CNS lymphoma. In addition, a steroid responsive meningoencephalitis characterized by eosinophilic inflammation has been described in dogs, and is seen most commonly in Golden Retrievers.

CSF pleocytosis can also consist of a mixed population of small and large mononuclear cells, neutrophils, and possibly eosinophils and plasma cells. This pattern is typically seen with fungal, rickettsial, or protozoal infections, GME, and conditions that result in tissue necrosis.

Fluid should be closely evaluated for the presence of bacteria or fungal elements. Neoplastic cells are not frequently identified in CSF, but can be seen when the neoplastic condition involves the meninges or ependymal cells. Neoplastic cells are most common found in the CSF with round cell tumors affecting the CNS.

REFERENCES

[1]Dewey CW (ed) **A Practical Guide to Canine and Feline Neurology**, 2003, p57. [2]Fry MM et. al **Vet Clin Pathol** 2006;35:72. [3]Bienzle D et. al **J Am Vet Med Assoc** 2000;216:1761. [4]Bailey CS et. al **Am J Vet Res** 1985;46:1162. Fishman RA **Cerebrospinal Fluid in Diseases of the Nervous System**, 2nd ed, 1992, p183. [6] Hurtt AE et al **J Am Vet Med Assoc** 1997; 211:866). [7]Christopher MM et. al **J Am Vet Med Assoc** 1988;192:1726.

NEUROLOGICAL DISORDERS OF THE GERIATRIC DOG

Karen R. Muñana, DVM, MS, DACVIM (Neurology)
Raleigh, NC 27606

INTRODUCTION

Among the general causes of neurological disease in dogs, degenerative conditions, neoplasia and vascular disease are diagnosed more frequently in older animals. Specific disorders of the brain or cranial nerves seen in geriatric dogs include cognitive dysfunction, idiopathic peripheral vestibular disease, cerebrovascular disease and neoplasia. Spinal conditions diagnosed commonly in older dogs include degenerative myelopathy and neoplasia. This presentation will utilize a case based approach to discuss the historical and clinial findings, diagnosis, and treatment of the more common neurological conditions encountered in geriatric dogs, with emphasis on information that is most relevant to the veterinary technician's role in patient management.

CANINE COGNITIVE DYSFUNCTION SYNDROME

Canine cognitive dysfunction syndrome is a clinical disorder characterized by age-related behavioral deficits that can not be attributed to any other medical condition present in the dog. Four categories of behavioral changes associated with cognitive dysfunction have been recognized: 1) disorientation, 2) reduced social interactions, 3) changes in sleep/wake cycle or activity level, and 4) loss of housetraining.[1] A study on the prevalence of canine cognitive dysfunction demonstrated characteristic behavioral changes in 28% of dogs 11-12 years of age and 68% of dogs 15-16 years of age.[2] Morphologically, the disorder is associated with cortical atrophy and accumulation of beta amyloid in the brain, similar to that seen with Alzheimer's disease in humans.[3]

Differential diagnoses for a dog with age-related behavioral changes include vision loss, hearing loss, metabolic encephalopathy, brain neoplasia, cerebrovascular disease and encephalitis. A definitive diagnosis of cognitive dysfunction syndrome can only be made after other potential causes of the clinical signs have been excluded. A thorough evaluation should be performed to assess for systemic disease, hypertension, hearing loss, vision loss and neurological deficits. Brain imaging and cerebrospinal fluid (CSF) analysis are necessary to rule out a neoplastic, vascular or inflammatory condition of the brain. It is common to identify other conditions in geriatric dogs - such as hearing or vision loss, or metabolic, endocrine cardiovascular or degenerative joint disease - that might mimic signs of cognitive deficits. Consequently, any identified conditions should be treated and the animal monitored for improvement in clinical signs before definitive treatment for cognitive dysfunction is instituted.

Two methods of treatment for canine cognitive dysfunction are available: pharmacological and dietary. Pharmacologic treatment consists of the administration of L-deprenyl (selegiline, Anipryl). L-deprenyl is an inhibitor of monoamine oxidase, the enzyme that breaks down the neurotransmitter dopamine, and the drug is believed to function by modulating the dopaminergic system in the brain. The other treatment consists of feeding a diet enriched with antioxidants, including vitamins E and C, docasahexanoic and eicosapentaenoic acid, lipoic acid and carnitine (Hill's Prescription Diet Canine b/d). It is postulated that the diet decreases the production and consequence of free radicals, that have been shown to contribute to age associated changes in cognition. Both pharmacologic and dietary therapy have been reported to improve cognitive behavior in aged dogs in controlled studies.[4,5] However, dogs enrolled in these studies were not required to undergo a full neurodiagnostic evaluation to provide more definitive evidence that behavioral signs were due to cognitive dysfunction syndrome, and this limits the studies' conclusions. Other nutritional supplements have been advocated in the treatment of canine cognitive dysfunction, including medium chain triglycerides, phopsphalitdylserine, and S-adenosyl-L-methionine (SAMe).

CEREBROVASCULAR DISEASE

Until recently, the dogma in veterinary medicine was that dogs and cats do not get strokes. However, the routine use of magnetic resonance imaging (MRI) in veterinary neurology has uncovered the fallacy of this belief, and cerebrovascular disease is now a frequently recognized cause of brain disease in dogs and cats. The term cerebrovascular disease refers to any abnormality of the brain resulting from a disruption of its blood supply. A stroke, or cerebrovascular accident, is the most common clinical presentation of cerebrovascular disease and is characterized by a sudden onset of focal neurological signs that persist at least 24 hours. Strokes can be classified as either ischemic or hemorrhagic. In both dogs and humans, ischemic strokes are more common than hemorrhagic strokes. Brain ischemia results from occlusion of a blood vessel with a thrombus or emboli that deprives the brain of oxygen and energy. Ischemic stroke in dogs can be seen with hypertension, atherosclerosis associated with hypothyroidism, hyperlipoproteinemia in Miniature Schnauzers, septic thromboemboli, neoplastic emboli, coagulopathy causing thromboembolism, and parasitic emboli (*Dirofilaria immitis*).[6] Cerebrovascular disease due to brain hemorrhage can be spontaneous, although this is rare in dogs, or can be associated with a concurrent medical condition such as a vascular malformation, coagulopathy, neoplasia, septicemia, hypertension or parasitic infection with *Angiostrongylus vasorum*.[7]

The incidence of cerebrovascular disease is greatest in older dogs, with a median age at onset ranging from 8 to 13 years.[7-9] Neurological signs tend to be acute and nonprogressive. Deficits are often asymmetric and reflect the location of the lesion within the brain. Cerebellar involvement is most commonly recognized, but forebrain and brainstem signs can also be seen. It has recently been suggested that episodic or transient (<24 hours in duration) vestibular dysfunction in dogs can result from systemic hypertension, analogous to the transient ischemic attacks recognized in humans.[10]

Dogs presenting with signs compatible with cerebrovascular disease should undergo serial blood pressure measurements and a thorough ophthalmic examination, as animals with hypertensive encephalopathy will frequently have associated retinal hemorrhage or detachment. Treatment with antihypertensive medication often results in improvement of neurological signs in animals with hypertensive encephalopathy. Diagnosis of cerebrovascular disease is based on imaging findings. Ischemia is best visualized with MRI. Hemorrhage can be visualized on MRI or computerized tomography (CT). Analysis of CSF is frequently normal or findings are non-specific. Underlying causes for cerebrovascular disease should be investigated, including endocrine, renal and cardiac disease, metastatic neoplasia and coagulopathy. Treatment is limited to supportive care and management of any identified underlying systemic disease. Prognosis is generally favorable, but is dependent on the size and location of the infarct. Clinical signs usually begin to improve after 24-72 hours, with recovery occurring over several weeks.

CANINE GERIATRIC VESTIBULAR DISEASE

Canine geriatric vestibular disease is a common cause of peripheral vestibular signs in dogs. The etiology is unknown. The disorder typically affects middle aged or older dogs, and manifests as acute onset of vestibular signs with no evidence of limb weakness, depression of consciousness or other cranial nerve deficits that would suggest central disease. A presumptive diagnosis is based on the presence of compatible history and examination findings. Definite diagnosis can only be made by excluding other causes of peripheral vestibular disease, such as otitis media-interna and middle ear neoplasia.

No specific therapy is available. Supportive care should be provided until the condition resolves. Antiemetics such as meclizine or maropitant can be helpful in decreasing anorexia and vomiting associated with the disequilibrium. Prognosis is very good as signs resolve spontaneously in 2-4 weeks, with improvement typically noted within 24 hours. A residual head tilt can persist in some cases.

NEOPLASIA

Neoplasia is a common cause of brain dysfunction in geriatric dogs. The incidence of brain tumors has been reported to approach 3% in the dog population.[11] The median age of affected dogs is 9 years,[11-13] with 95% of dogs in one study being 5 years of age or older.[12] Primary brain tumors arise from cells normally present in or associated with the brain, and include meningiomas, glial cell tumors, and choroid plexus tumors. Meningiomas are the most common primary brain tumor diagnosed in dogs.[11] Secondary tumors originate from surrounding tissues and locally extend into the brain (such as nasal tumors), or arise from hematogenous metastasis of primary tumors in other tissues. Golden Retrievers and Boxers have been reported to be at increased risk for the development of brain tumors.[11] Breed predispositions have also been described based on tumor type; brachycephalics, especially boxers and Boston terriers are predisposed to glial cell tumors and pituitary tumors, while doliocephalic dogs are prone to developing meningiomas.

Tumor growth results in compression and invasion of normal brain parenchyma. Indirect effects of tumors include damage to the blood-brain barrier with resultant edema, obstruction of CSF flow causing secondary hydrocephalus, hemorrhage, and the potential for brain herniation due to increased intracranial pressure (ICP) from an enlarging space-occupying mass within the rigid cranial cavity. These indirect effects are often more significant that the direct effects of compression and invasion. Clinical findings reflect the location of the tumor within the brain. Seizures are the most common presenting complaint in dogs with brain tumors, as most tumors are located in the telencephalon. Other neurological signs, when present, can be acute or insidious in onset, depending on the location, rate of growth and indirect effects of the tumor.

Diagnostic evaluation of a dog with a suspected brain tumor should include minimum database (hemogram, chemistry profile, urinalysis) to evaluate for any evidence of systemic disease, as well as thoracic radiographs and possibly abdominal ultrasound to screen for metastasis. Although primary brain tumors typically do not metastasize, secondary tumors that spread to the brain can also involve the lungs and/or abdominal organs. Magnetic resonance imaging is the most useful diagnostic tool. Images are evaluated for the presence of mass effect, contrast enhancement, compression of normal brain structures and secondary hydrocephalus. Analysis of CSF frequently reveals albuminocytologic dissociation. However, CSF analysis is often not performed in cases where mass lesions are identified on imaging, due to an increased risk of brain herniation.

Symptomatic treatment is aimed at decreasing peritumoral edema with prednisone (anti-inflammatory dose administered twice daily, then slowly tapering to every other day) and treatment of seizures with antiepileptic medication. A recent study reported a median survival of 69 days with palliative therapy alone.[14] However, a significant difference in survivial was noted based on tumor location; supratentorial tumors had a median survival of 178 days compared to 28 days for infratentorial tumors.[14] Options for definitive tumor treatment include surgery, chemotherapy and radiation therapy.

Surgery allows for tumor resection or cytoreduction, and also provides a histologic diagnosis but is often reserved for tumors that are superficial and easily approachable. Reported median survival times for dogs with meningiomas treated with surgery alone range from 7 months[15] to 1254 days.[16] Radiation therapy, either alone or in combination with surgery, has been assoiciated with median survival times ranging from 4.9 - 8.3 months[13,17] and 16.5 months,[15] respectively for meningiomas. Chemotherapeutic agents such as lomustine and hydroxyurea have been used to treat canine brain tumors, but little information is currently available with respect to their efficacy in such cases. Overall prognosis is fair to guarded, and dependent on location and tumor type. In most instances, the goal of therapy is remission of signs and increased survival time rather than cure.

Spinal tumors are also more commonly diagnosed in older dogs, with a reported median age at diagnosis of 10 years.[18] An exception to this is nephroblastoma, a neoplasia typically seen in young dogs. The majority of tumors originate extradurally, including osteosarcoma, chondrosarcoma, fibrosarcoma, hemangiosarcoma and plasma cell tumors. Intradural-extramedullary tumors include meningiomas and nerve sheath tumors. Intramedullary tumors are the least common, and include round cell tumors, glial cell tumors and ependymomas. Spinal tumors tend to cause chronic progressive neurological deficits, but the clinical course can vary considerably. Pain is a common presenting sign with extradural or intradural-extramedullary lesions. Initial diagnostic evaluation should include minimum data base, thoracic radiographs and spinal radiographs. Survey radiographs can be evaluated for vertebral lysis or proliferation. MRI is the most sensitive diagnostic imaging modality. Treatment options are similar to those for brain tumors, and include palliative therapy, surgery, radiation therapy and chemotherapy. Prognosis is dependent on tumor type and location.

DEGENERATIVE MYELOPATHY

Degenerative myelopathy is a degenerative spinal cord condition recognized in middle aged to older dogs of various breeds, but is more commonly thought of as a disease of large breed dogs, particularly German Shepherd dogs. The mean age of onset is 9 years.[19] Pembroke Welsh Corgis have been shown to also have a predilection for the disease, with a reported mean age of onset of 11 years.[20] Dogs with degenerative myelopathy show signs of a slowly progressive ataxia and paraparesis, with no evidence of spinal pain. This condition was poorly understood for many years, and a presumptive antemortem diagnosis was based on the exclusion of other likely causes of myelopathy, such as neoplasia and Type II intervertebral disk disease. Recent work has identified a specific mutation in superoxide dismutase that is associated with the disease trait.[21] Genotyping of numerous dogs of many different breeds has demonstrated that the same mutation is associated with degenerative myelopathy in German Shepherd dogs, Pembroke Welsh Corgis, Boxers, Chesapeake Bay Retrievers and Rhodesian Ridgebacks. Furthermore, the disease has been demonstrated to have genetic and clinical similarities with the human condition, amyotrophic lateral sclerosis (ALS or Lou Gehrig disease). Genetic testing for degenerative myelopathy is now available. Dogs that are homozygous for the mutation are at risk for developing the disease, but many such dogs do not develop clinical signs, and age appears to affect expression of the disorder. Dogs that are heterozygous for the mutation (one mutant gene and one normal gene) are unlikely to develop degenerative myelopathy, but can pass the mutation on to their offspring. The genetic test can be used by breeders to help reduce the incidence of degenerative myelopathy in a line of dogs, and can be used to support a clinical diagnosis of degenerative myelopathy in an older dog with compatible clinical signs if other causes of myelopathy have been ruled out. Results from one study suggest that among dogs with degenerative myelopathy, those that receive physiotherapy have a better prognosis than dogs that do not.[22] No other proven definitive treatment exists and overall prognosis is poor.

REFERENCES

[1]Ruehl WW et. al **Prog Brain Res** 1995;106:217. [2]Neilson JC et. al **J Am Vet Med Assoc** 2001;218:1787. [3]Cummings BJ et. al **Neurobiol Aging** 1996;17:259. [4]Anipryl Freedom of Information Summary, Pfizer Animal Health. [5]Dodd CE et. al **Vet Med** 2003;98:396. [6]Garosi LS **Vet Clin Small Anim** 2010;40:65. [7]Lowrie M et. al **Vet Radiol Ultrasound** 2012;53:381. [8]Joseph RJ et. al **J Amer Anim Hosp Assoc** 1988;24:569. [9]Garosi L et. al **J Vet Intern Med** 2006;20:311. [10]Bentley RT et. al **J Am Vet Med Assoc** 2011;239:652. [11]Snyder JM et. al **J Vet Intern Med** 2006;20:669. [12]Bagley RS et. al **J Am Vet Med Assoc** 1999;215:818. [13]Heidner GL et. al **J Vet Intern Med** 1991;5:219. [14]Rossmeisl JH et. al **J Am Vet Med Assoc** 2013;242:193. [15]Axlund TW et. al **J Am Vet Med Assoc** 2002;221:1597. [16]Greco JJ et. al **J Am Vet Med Assoc** 2006;229:394. [17]Spugnini EP et. al **Vet Radiol Ultrasound** 2000;41:377. [18]Levy MS et. al **J Am Anim Hosp Assoc** 1997;33:307. [19]Coates JR et. al **Vet Clin Small Anim** 2010;40:929. [20]Coates JR et. al **J Vet Intern Med** 2007;21:1323. [21]Awano T et. al **Proc Natl Acad Sci USA** 2009;106:2794. [22]Kathmann I et. al **J Vet Intern Med** 2006;20:927.

UPDATES ON INFLAMMATORY CNS DISEASE, DIAGNOSIS AND TREATMENT

Christopher L. Mariani, DVM, PhD, DACVIM (Neurology)
Raleigh, NC

DEFINITIONS

- Meningitis: inflammation of the meninges (the pia, arachnoid and dura mater; protective membranes covering the brain and spinal cord).
- Encephalitis: inflammation of the brain.
- Myelitis: inflammation of the spinal cord.
- Meningoencephalitis: inflammation of the meninges and brain.
- Meningomyelitis: inflammation of the meninges and spinal cord.

INTRODUCTION

Meningoencephalitis and meningomyelitis are common problems in small animal medicine. They may affect any breed of dog or cat, and animals of multiple ages, although certain disease processes are more common in specific breeds and age ranges (see below). Small and toy breed dogs in particular are frequently afflicted with inflammatory central nervous system (CNS) disorders.

CLINICAL SIGNS

A variety of clinical signs may be seen with inflammatory CNS disease, depending on what region of the nervous system is affected. Involvement of the forebrain (cerebral cortex and thalamus) may result in seizures, altered mentation (stupor, coma, delirium, head pressing), vision abnormalities, and compulsive pacing or circling. Involvement of the brainstem may result in altered mentation, cranial nerve deficits (e.g., vestibular dysfunction [head tilt, nystagmus], facial nerve paralysis, swallowing difficulties, tongue paralysis), ataxia, and proprioceptive deficits. Spinal cord involvement leads to ataxia, paresis, and possibly segmental spinal reflex deficits. Most inflammatory CNS conditions result in pain due to involvement of the meninges. A hallmark of these conditions is the presence of multifocal neurologic signs; that is, involvement of multiple areas of the CNS (e.g., forebrain, brainstem and spinal cord). Contrary to what many believe, fever is quite uncommon in most meningoencephalitis patients.

DIAGNOSIS

The diagnosis of inflammatory CNS diseases in small animal patients involves consideration of the signalment, reported and observed clinical signs, and results of diagnostic testing. The most useful tests are advanced diagnostic imaging (particularly magnetic resonance imaging [MRI]), and cerebrospinal fluid (CSF) analysis. Although computed tomography (CT) is of some use in demonstrating brain lesions, MRI is the gold standard for brain imaging, and is much more sensitive in showing inflammatory lesions of the brain and spinal cord. Inflammatory lesions are typically hyperintense on T2-weighted, STIR, and FLAIR images, hypointense or isointense on T1-weighted images, and enhance to varying degrees after intravenous contrast administration (gadolinium compounds). Analysis of CSF in these patients typically shows elevated white blood cell numbers (pleocytosis) and protein. Cytology should be performed to determine the types of inflammatory cells present. Electrodiagnostic testing is less frequently used, but may provide useful functional information to complement the imaging and CSF results. Electroencephalography (EEG) and brainstem auditory evoked response (BAER) provide such information about the cerebral cortex and brainstem pathways respectively. Finally, testing for infectious disease agents may involve quantification of antibody titers or polymerase chain reaction (PCR)-based testing on blood and CSF samples.

THERAPY

The therapy of meningoencephalitis and meningomyelitis should be tailored to the specific etiology or suspected syndrome, while taking into account the severity of the clinical signs, comorbid disease processes, and financial constraints of the owner. Patients with identified infectious etiologies should be treated with appropriate antimicrobial medications, which may include antibiotics, antifungals, or antiprotozoals. Whenever possible, medications with good penetration of the blood-brain barrier should be selected for therapy. Such choices often include doxycycline, sulfonamide drugs, and fluconazole.

Corticosteroids are a mainstay of therapy for inflammatory CNS conditions, regardless of etiology. Although immunosuppression and long-term treatment with corticosteroids is usually contraindicated for infectious processes, short-term therapy can dramatically reduce the inflammatory response, and can be life-saving. In addition, corticosteroids form the cornerstone of therapy for the inflammatory conditions without an underlying identifiable etiology. Although many animals benefit from anti-inflammatory doses of these medications, immunosuppressive doses are often required for disease remission. The doses of corticosteroids are generally reduced over an extended period of time to the lowest dose that controls clinical signs.

Many clinicians add additional immunomodulatory or cytotoxic medications to the therapeutic regimen of patients with meningoencephalitis of unknown etiology (MUE). Such drugs include cytosine arabinoside (cytarabine, Ara-C), cyclosporine A, leflunomide, mycophenolate mofetil, lomustine (CCNU), and procarbazine.[1] These medications are useful to provide additional immunosuppression through alternate mechanisms of action, and also to reduce the side effects associated with glucocorticoid therapy, which include polyuria, polydipsia, polyphagia, weight gain, panting, gastrointestinal ulceration, coagulopathies, pancreatitis, and steroid hepatopathy.

ETIOLOGIES AND SPECIFIC SYNDROMES

A variety of infectious agents may lead to nervous system dysfunction, including viruses (canine distemper virus, FIP, FeLV, Rabies, Pseudorabies, West Nile Virus), bacteria (*Streptococcus spp., Bartonella spp., Borrellia burgdorferi., Leptospira spp.),* fungal organisms (*Cryptococcus neoformans, Coccidiodes immitis, Blastomyces dermatidis, Aspergillus spp., Cladophilophora bantiana*), rickettsiae (*Ehrlichia canis, Rickettsia rickettsia, Anaplasma phagocytophilum),* protozoa (*Neospora caninum, Toxoplasma gondii, Entamoeba histolytica)* and parasites (*Cuterebra spp., Dirofiliaria immitis, Baylisascaris procyonis, Angiostrongylus spp., Prototheca spp.*). Despite the number of potential infectious etiologies, these are rarely identified as the underlying cause of meningoencephalitis or meningomyelitis in small animals. Although cats are often presumptively diagnosed with an infectious etiology, dogs are much more likely to fit into one of the syndromes described below that have no identifiable infectious cause.

Steroid responsive meningitis-arteritis

As its name implies, this is a condition that involves the blood vessels of the meninges, leading to vasculitis and meningitis. It typically affects young (6-18 months), large breed dogs, particularly Boxers, Bernese Mountain dogs, Weimaraners, and Nova Scotia Duck Tolling Retrievers.[2,3] Beagles are also affected by this condition, which is also known as "Beagle Pain Syndrome". Affected animals usually present for moderate to severe neck pain, with a low head carriage and reluctance to move. Neurologic deficits are uncommon with this disorder, although chronically affected animals may develop ataxia or paresis if left untreated.[3] Diagnostic imaging is often unremarkable but may show meningeal enhancement.[3] Cerebrospinal fluid evaluation typically shows a profound neutrophilic pleocytosis (cell counts often exceed 1000 cells/μl). Dogs presenting in the acute stages usually respond completely to immunosuppression with corticosteroids, and additional immunosuppressive medications are rarely required.[2,4] However, more chronically affected animals may require additional immunosuppression and may have a more guarded prognosis.[3]

Generalized tremor (little white shaker) syndrome

This condition is commonly referred to as "little white shakers" or "shaker dog syndrome" due to its tendency to affect young toy breed dogs. However, it may affect dogs of any color or breed.[5] Encephalitis mainly involving the cerebellum is thought to result in the predominant clinical signs. Dogs show an obvious, constant tremor, which may worsen with movement. Additional neurologic signs are not usually found, although other cranial nerve deficits (e.g., vestibular signs), paresis, and visual deficits have been described.[6] This syndrome is often diagnosed based on clinical signs and response to therapy, and few descriptions of diagnostic imaging are available. Cerebrospinal fluid analysis usually shows a mild to moderate mononuclear pleocytosis, but may be normal.[5,6] Affected animals typically respond completely to corticosteroids and diazepam, which can usually be discontinued after a period of time.[5,6] However, recurrence of signs is possible.

Greyhound meningoencephalitis

Although originally described in greyhounds in Ireland in 2002,[7] several newer publications have documented various aspects of this unique disease entity. Affected dogs are quite young (5-18 months) and show progressive clinical signs primarily referable to forebrain dysfunction. These signs include altered mentation, abnormal behavior, circling, and postural reaction deficits, although ataxia, vestibular dysfunction and other signs of brainstem dysfunction were also seen, and a multifocal localization was common.[7,8] Interestingly, multiple dogs were often affected within the same litter. The etiology remains undetermined, and tests for infectious disease agents have been negative to date.[8,9] Cerebrospinal fluid analysis may be normal, but often shows a mild to moderate mononuclear pleocytosis.[8] A recent publication described the unique distribution of lesions visible with MRI, documenting involvement of the olfactory bulbs, cerebral cortex of the frontal lobe and caudate nuclei.[10] The prognosis is considered to be poor, as response to medical management has not been documented.

Granulomatous meningoencephalitis (GME)

This syndrome is named after the profound granulomatous inflammation of the brain of affected patients, characterized by perivascular infiltrates of macrophages, lymphocytes and neutrophils.[11,12] The brainstem is usually the area most severely affected, and dogs typically present with signs of ataxia, paresis, cervical pain, and vestibular dysfunction. Other signs such as visual dysfunction, facial nerve deficits, dysphagia, pupillary abnormalities and seizures may also be seen.[11] Clinical signs may vary from relatively mild to very severe with rapid decompensation and death. Certain breeds appear to be predisposed to the development of GME, including the toy poodle and terrier breeds. However, any breed of dog may be affected, and dogs of all ages may be seen.

Lesions noted on MRI are hyperintense on T2-weighted and FLAIR images and enhance to varying degrees after contrast administration.[11] Lesions are often found in the brainstem, forebrain, and cervical spinal cord, while meningeal involvement is rarely noted on diagnostic imaging.[13] Evaluation of CSF typically shows a moderate to severe mononuclear pleocytosis, although neutrophils can occasionally dominate the cytology.[14] In rare cases the CSF is normal. The majority of patients will show improvement with corticosteroid therapy, although this improvement is often incomplete, and most animals require persistent long-term therapy. Relapses are frequent, particularly with attempts to reduce glucocorticoid doses. Many clinicians add additional immunosuppressive medications to improve control of clinical signs and reduce some of the side effects of the corticosteroids.[11,15-19] Radiation therapy may be beneficial for some patients with focal inflammatory lesions.[20] The long-term prognosis for these patients is guarded, as many relapse and eventually succumb to the disease.[20] However, responses to treatment are quite variable, and in the author's opinion, this syndrome likely represents several different disease processes with varying etiologies and outcomes.

Necrotizing meningoencephalitis (NME)

This condition was first reported in pug dogs, and is still often known as pug encephalitis.[21] However, other young, small and toy breed dogs are commonly affected with NME, including Maltese terriers, Chihuahuas, Shih Tzus, Boston terriers, Pekingese, Papillions, Pomeranians, and others.[11,22,23] The forebrain is primarily involved with infiltration of lymphocytes, plasma cells, and macrophages at the gray-white matter junction. Chronically affected animals develop necrosis of brain tissue with cavitation.[11] Seizures are a prominent clinical sign. Other signs include altered mentation, circling, visual dysfunction, proprioceptive deficits, cranial nerve deficits, and cervical pain. Brain imaging with MRI can show a variety of lesions including T2-weighted and FLAIR hyperintensities with variable contrast enhancement, loss of the distinction between gray and white matter, cavitated regions with CSF filling, and meningeal enhancement, predominantly in the forebrain.[11,24,25] The lesions are usually asymmetric, with evidence of mass effect and occasional brain herniation seen. Cerebrospinal fluid evaluation usually reveals a mild to moderate mononuclear pleocytosis, although cell counts are occasionally within normal limits.[21,23] The response to therapy is less clear with this syndrome, although some dogs appear to respond to therapies similar to those utilized for GME (see above). These include immunosuppressive corticosteroids as well as other immunosuppressive or cytotoxic therapies.[11] Prognosis is generally poor, although some dogs may survive for a year or longer before succumbing to the disease.

A number of studies have been performed, primarily in pugs, to try and pinpoint an underlying genetic cause for NME. These studies have shown an association of disease with the dog leukocyte antigen (DLA or MHC) II region on chromosome 12, which is consistent with an autoimmune process and similar to other autoimmune diseases including multiple sclerosis in humans.[26-28]

Necrotizing leukoencephalitis (NLE)

This syndrome is similar to NME, but predominantly affects the white matter of the forebrain and brainstem.[11] This is primarily a disease of Yorkshire terriers, although other breeds have been reported, including the French bulldog.[29,30] Affected dogs are typically young (6 months to 5 years of age). Clinical signs include seizures, altered mentation, visual deficits, circling, ataxia, and cranial nerve deficits. Magnetic resonance imaging usually reveals lesions within the white matter of the cerebrum and brainstem that are hyperintense on T2-weighted and FLAIR images, and hypointense on T1-weighted images with variable contrast enhancement.[31,32] Computed tomography may show areas of decreased density within the forebrain and brainstem.[33] CSF shows a mild mononuclear pleocytosis or may be normal. Therapy is similar to NME, as is the prognosis.

Eosinophilic meningoencephalitis and meningomyelitis

Inflammatory CNS disease with a substantial eosinophilic component has been reported with several infectious etiologies including fungal and protozoal meningocephalomyelitis. However, an idiopathic syndrome resulting in similar clinicopathologic findings but without an identifiable infectious etiology has also been described.[34-36] Young to middle-aged large breed dogs are predominantly affected and Golden retrievers and Rottweilers were overrepresented in some of these reports. Clinical signs may chiefly involve the brain, spinal cord or both regions. Peripheral eosinophilia was not found in the majority of cases described. Imaging of the CNS may show focal or multifocal lesions or may be normal.[36] Cerebrospinal shows a mild to marked pleocytosis with a substantial proportion of eosinophils (>20%). Most dogs respond dramatically and often completely to glucocorticoid therapy, which can often be discontinued after a period of time.[35,36] The long-term prognosis for the idiopathic condition is good in most cases.

The hunt for infectious disease in dogs with SRMA, GME, NME, NLE and MUE

A number of studies have now been published which attempt to document the presence of infectious agents in dogs with either histologically confirmed GME, NME and/or NLE or a clinical diagnosis of SRMA or MUE. Although isolated cases have shown evidence of infectious organisms,[37] for the most part, these studies have produced negative results.[38-41] One intriguing exception has been the identification of *Mycoplasma canis* nucleic acids within brain tissue of some dogs with GME and NME, which was corroborated with positive cultures.[41] Though interesting, these results are preliminary, and will require confirmation with further work before more specific diagnostic or therapeutic recommendations can be made.

REFERENCES

1. Granger N, et al. Vet J 2010;184:290. 2. Behr S, et al. J Am Anim Hosp Assoc 2006;42:277. 3. Tipold A, et al. J Small Anim Pract. 4. Cizinauskas S, et al. J Small Anim Pract 2000;41:295. 5. Wagner SO, et al. J Am Vet Med Assoc 1997;211:731. 6. Bagley RS, et al. J Am Anim Hosp Assoc 1993;29:141. 7. Callanan JJ, et al. Vet Pathol 2002;39:56. 8. Shiel RE, et al. Vet Rec 2010;167:333. 9. Daly P, et al. Vet Microbiol 2006;118:189. 10. Terzo E, et al. Veterinary radiol ultrasound 2012;53:636. 11. Schatzberg SJ. Vet Clin North Am Small Anim Pract 2010;40:101. 12. Braund KG. J Am Vet Med Assoc 1985;186:138. 13. Cherubini GB, et al. Vet Rec 2006;159:110. 14. Bailey CS, et al. J Am Vet Med Assoc 1986;188:418. 15. Adamo FP, et al. J Am Vet Med Assoc 2004;225:1211. 16. Gnirs K. J Small Anim Pract 2006;47:201. 17. Zarfoss M, et al. J Small Anim Pract 2006;47:588. 18. Coates JR, et al. J Vet Intern Med 2007;21:100. 19. Pakozdy A, et al. Vet Rec 2009;164:89. 20. Munana KR, et al. J Am Vet Med Assoc 1998;212:1902. 21. Cordy DR, et al. Vet Pathol 1989;26:191. 22. Higgins RJ, et al. Vet Pathol 2008;45:336. 23. Stalis IH, et al. Vet Pathol 1995;32:230. 24. Flegel T, et al. Vet Radiol Ultrasound 2008;49:419. 25. Young BD, et al. J Vet Intern Med 2009;23:527. 26. Barber RM, et al. J Hered 2011;102 Suppl 1:S40. 27. Greer KA, et al. Res Vet Sci 2009;86:438. 28. Greer KA, et al. Tissue antigens 2010;76:110. 29. Timmann D, et al. J Small Anim Pract 2007;48:339. 30. Spitzbarth I, et al. J Comp Pathol 2010;142:235. 31. Lotti D, et al. Vet Radiol Ultrasound 1999;40:622. 32. von Praun F, et al. Vet Radiol Ultrasound 2006;47:260. 33. Ducote JM, et al. Vet Radiol Ultrasound 1999;40:617. 34. Bennett PF, et al. Aust Vet J 1997;75:786. 35. Smith-Maxie LL, et al. J Vet Intern Med 1989;3:167. 36. Windsor RC, et al. J Vet Intern Med 2009;23:275. 37. Schatzberg SJ, et al. J Vet Diagn Invest 2009;21:844. 38. Barber RM, et al. J Vet Intern Med 2010. 39. Jaderlund KH, et al. J Vet Intern Med 2009;23:669. 40. Schatzberg SJ, et al. J Vet Intern Med 2005;19:553. 41. Barber RM, et al. J Vet Intern Med 2012;26:962.

DIAGNOSTIC TESTING AND ITS APPLICATIONS

Melissa G. Ackerman DVM, MS
Pullman, WA

INTRODUCTION

Diagnostic testing can be a valuable tool in veterinary medicine. Tests are used to confirm or rule out the presence of disease, identify sub-clinical cases of disease, guide decision-making with therapy and treatment, monitor the course of clinical disease, and help with prognosis. It is essential that the veterinary professional understand that diagnostic tests have unique limitations. Each test provides the user with information, which must be interpreted in the context of the patient and/or population and the test. Evaluation of diagnostic testing in reality is more complex than is summarized below, but a basic understanding of sensitivity, specificity, and predictive values are important when interpreting test results.

EPIDEMIOLOGY REVIEW

Normal distributions

Test results for an individual patient are based on results in a population. Most populations from which our test results are derived have a normal or bell-shaped or Gaussian curve. This means that the majority (about 68%) of animals fall within one standard deviation from the midpoint, making up most of the middle of the bell. About 95% of animals will fall within two standard deviations from the midpoint, making up the majority of the bell. It is the small percentage of animals that fall within the thin tails of the curve that may be most susceptible to error in diagnostic testing.

Sensitivity and Specificity

For most diagnostic tests, there are two important characteristics that the user should be aware of in order to effectively interpret the data that is generated. The first is the test's sensitivity, or its ability to correctly identify diseased animals. The second is its specificity, or its ability to correctly identify disease-free animals. Several testing methods may exist for one disease, and each test has a different known sensitivity and specificity. These help the user choose the test that will be of most value given the situation.

Sensitivity is a proportion (calculated by the number of animals the test correctly identified as disease positive divided by the total number of animals that are disease positive). It is the probability of testing positive if the animal is truly diseased. Therefore, if a test has high sensitivity, there will be few false negatives. Specificity is the number of animals the test correctly identified as disease negative divided by the total number of animals that are disease negative; it is the probability of testing negative when the animal is truly negative. A test with high specificity means that there will be few false positives. Ideally, a test would have very high sensitivity and specificity. However, there is a trade-off; if sensitivity increases, specificity will decrease. This is because the frequency distribution curves of two populations of animals (disease-free and diseased) are two normal curves that will overlap. The chosen cut-off point (this point determines the classification of diseased or not diseased) will often incorporate this overlap area, so that moving the cut-off point either to the left or the right can only increase one and decrease the other.

Depending on the context of the test, the veterinary professional's desire for high specificity or sensitivity may be very different. For example, a person in charge of maintaining the disease-free status of an island would likely require disease testing on imported animals using a test with very high sensitivity, so that the probability of a false negative is very low. This would allow for the highest probability that a diseased animal would not enter the disease-free population. Alternatively, a veterinary professional that has the intention of ruling out a disease in a sick animal would choose a test with high specificity, so that the probability of a false positive is very low.

Predictive Values

In addition to sensitivity and specificity, veterinary professionals should understand the predictive value of the diagnostic test of interest, or the ability of the test to reveal the true disease status of the animal. A predictive value will vary based upon factors such as test performance and the prevalence of disease in a population. Positive predictive value is the probability that the test positive animal truly has disease. This is calculated by taking the number of animals that tested positive for the disease that truly are diseased divided by the number of total animals that tested positive for disease. Negative predictive value is the probability that the test negative animal truly does not have disease. This is calculated by taking the number of animals that tested negative for the disease that truly are disease negative and dividing by the total number of animals that tested negative.

The important concept to understand from the above calculations is that prevalence of disease impacts the predictive value of a test, which has direct bearing on the usefulness of the test. Lower prevalence of a disease decreases the predictive value of positive tests, meaning that with low disease prevalence there is less confidence in a positive test result, but more confidence in a negative result. The converse is also true. Higher prevalence of disease increases the predictive value of a positive test, meaning that with higher disease prevalence it is more likely that the animal testing positive truly is positive.

COMMON VETERINARY DIAGNOSTIC TESTS

Serology/Immunodiagnostics

Blood, tissue, and bodily fluids are often used to identify the presence of antibodies. Animals make antibodies that are specific to antigens (harmful agents) that help fight off disease. Test results reported as antibody titers (concentration of antibodies in the blood), are expressed in a ratio. If the vaccination history of the animal is unknown, it is difficult to interpret whether the presence of antibodies is due to vaccination or exposure to disease. Additionally, titers wax and wane naturally, so one serum sample reflects titers at one point in time. For this reason, if an animal is sick, acute and convalescent titers can help identify the disease. Enzyme-linked immunosorbent assay, or ELISA tests look for antigen or antibody in the sample and result in a color change if present. Similarly, antigens can be detected with fluorescent-labeled antibodies with direct or indirect fluorescent antibody tests (FA or IFA).

Immunohistochemistry

In this method, an antibody specific to a chemical agent or cellular marker is used to detect disease in blood, tissue, and bodily fluids. The antibody is marked so that it is visible. This test helps localize the affected cells within a sample.

Bacteriology

The samples listed above are also used to look for specific bacteria. Samples are used to grow colonies of bacteria, and then various methods are employed to classify the species. Because bacteria are ubiquitous, sampling handling is important when collecting a sample for bacteriology. Some bacteria need special media or temperatures in order to live outside the host and/or grow in culture, like some species of *Mycoplasma* and *Tritrichomonas*. The diagnostic lab that handles your samples is the best to advise you about sample collection details.

Often reported with the bacterial cultures is the minimum inhibitory concentration (MIC), or sensitivity of the bacteria to commonly used antibiotics. This helps identify potentially resistant infections so that therapies may be adjusted accordingly. With MIC, it is important to remember that these tests are done in vitro, and that the antimicrobial-microbe interaction within the body may be different.

Molecular Diagnostics

These tests use polymerase-chain-reaction (PCR) to detect deoxyribonucleic acid (DNA) within a sample. There are several different types of PCR-based tests that can accomplish this, and some with very little sample material. Genetic material like DNA and RNA can degrade in tissues depending on temperature, time, and storage, so it is best to check with your diagnostic lab when collecting samples for these tests.

Cytology

Cytology is the microscopic examination of cells, often used to study urine, blood, feces, and other bodily fluids. For fecal exams, the presence or the concentration of a parasite or egg may be of interest. This often differs based on the patient's age and health status. In urine, the presence of bacteria, crystals, or abnormal cells may indicate disease. In blood samples, the types and amount of cellular components are often used to indicate presence or severity of disease. These types of test results are also relatively subjective depending on the interpreter.

Clinical Pathology

Complete blood count (CBC) comprises a hemogram, leukogram, and a protein analysis. The hemogram breaks down red cell parameters that may help identify anemia and erythrocyte disorders. The leukogram analyzes white blood cells and helps evaluate inflammation, infection, and certain leukemias. The protein analysis can help with diagnosing inflammation and protein disorders. Some of these analyses may be done manually and others are automated, therefore operator error and machine error contribute to the accuracy of these tests. Some results are also open to objective interpretation. This remains true for a serum chemistry panel, which analyzes concentrations of certain electrolytes and enzymes within the serum.

CONCLUSION

There are numerous other diagnostic tests in veterinary medicine. Some results are highly subjective in interpretation, for example, assessing a patient's hydration status based on skin elasticity and feeling mucous membranes for degree of tackiness. Other test results can be a straightforward present/not present result, such as the detection of *Leptospira* spirochetes in urine. Whatever the diagnostic test, a patient's best interest may be met if results are interpreted in the context of the health status of the patient or herd and the limitations of the test.

REFERENCES

[1] http://www.vetmed.wsu.edu/depts_waddl/userguide.aspx visited 1/31/13. [2]Thrushfield, M. **Veterinary Epidemiology** 3rd ed. 2005, Blackwell Science, Oxford, UK. [3]Smith, R.D. **Veterinary Clinical Epidemiology** 3rd ed. 2006, CRC Press, FL. [4]Dohoo, I., et al. **Veterinary Epidemiologic Research** 2003, AVC Inc. Canada.

HANDLING AND SUBMISSION OF DIAGNOSTIC SAMPLES

Tamara Gull, DVM, PhD, DACVIM (LAIM), DACVPM, DACVM
Stillwater, OK

INTRODUCTION

The quality and usefulness of a diagnostic test depends in large part on the sample submitted. Diagnostic labs frequently receive samples that are inappropriately collected, stored, or transported and are subsequently of little use to the submitter. Technicians are frequently tasked with specimen handling and can be invaluable in ensuring that a valid result is obtained. This presentation will discuss where, when and how much material to collect for various types of samples, specimen transport systems and containers and how to select appropriate sample materials.

SPECIMEN COLLECTION

Proper specimen collection is the most important step in obtaining a valid and useful diagnostic result. Improperly or poorly collected specimens may lead to failure to isolate the causative agent of an infection, recovery of misleading contaminants, sample degradation or missing the actual lesion which can lead to incorrect diagnosis and therapy. There are a number of important factors to keep in mind:

Origin of specimen material

The sample should be representative of the actual site of infection with a minimal amount of contamination from the surrounding sites. Clip and scrub sampling sites as appropriate! Take deep samples rather than surface samples from wounds and draining tracts, and aspirate closed lesions rather than sampling draining lesions when available. Sample the edges of lesions, as that is where the most actively growing organisms or most active disease process will be found. Consider submitting or banking a serum sample from the animal in the event titers are needed; this could save you a trip back to the farm later.

Optimal time period

The optimal collection time period for recovery of an infectious agent varies according to the nature of the disease. It is important to collect specimens from an animal that is acutely affected rather than chronically affected if the initiating or primary etiologic agent is to be found. When culturing a chronically affected animal secondary infections may mask the presence of the primary agent. In deceased animals, anaerobic overgrowth from the GI tract occurs within 4 hours of death. Anaerobic samples must be taken as soon after death as possible. In noninfectious disease, fresh or acute lesions are more diagnostic than older lesions.

Quantity of specimens

A sufficient volume of material must be obtained to perform the tests requested. More is better! One swab is not sufficient sample for a full set of cultures (e.g. bacterial, fungal, viral). Even if there is not sufficient sample to collect measurable fluid, you can always send more than one swab from the same lesion. For tissue, send at least 4 cubic centimeters of fresh tissue and include the lesion, if visible. For enteric cases, send at least 20ml of fresh ileal or colonic content (depending on suspect organisms). If you cannot ship the entire specimen (e.g. a deceased animal), obtain large chunks of all significant organs and ship each both fresh and formalin-fixed. Don't forget fetal stomach contents and placenta (with cotyledons) in abortion cases!

Collection prior to drug treatment for suspected infectious diseases

Whenever possible, collection should precede administration of antimicrobials. If this is not possible, the causative agent may still be recovered because many antibiotics are bacteriostatic, not bactericidal, and recovery may occur when the specimen is transferred to fresh culture medium. In addition, the concentration of the antibiotic may be low enough that the minimal inhibitory concentration (MIC) for the organism was not reached. For fluid cultures, products are available to neutralize antibiotic activity.

Proper labeling

Ensure that your sample is properly labeled in a way that will not be lost or smeared. Make sure the labeling includes animal ID, site of sampling and date obtained. If the sample container contains some additive not listed on the container (e.g. formalin, saline), make sure that information is included on the label. Label the container portion of lidded sample containers, not the lid (lids can be mixed up between containers). Papers sent with the sample should be sent in a separate sealed Ziploc bag taped to the sample container; this way if the sample leaks the papers will still be legible.

Storage and shipment

Inoculate samples into the appropriate medium (see below). Tissue samples should be sealed into a screw-top vial. Small tissue samples at risk of desiccation may be wrapped in sterile saline-soaked gauze or stored in sterile saline. Samples that cannot be immediately shipped should be stored at 4°C (refrigerator temperature). Do not freeze! The exceptions are anaerobic samples: anaerobic samples should be stored/shipped at room temperature. Ship by the most expeditious means possible, and ensure temperature control during shipment with cold packs and insulated boxes. All your vaccines will come with cold packs; be sure to save a few for specimen shipment. All sample containers should be sealed within a second leak-proof container (e.g. Ziploc bag or Whirl-Pak) and shipped with absorbent material in the event of leaks. Shipping companies will refuse or discard leaky packages!

Communicate!

Whenever possible, call your diagnostic lab before you take your sample, and tell them what you suspect. They can advise you on the best sampling, storage and transport methods for your sample so that your chances of a meaningful result are maximized. They may also be able to send you specialized transport media for rare or fastidious isolates (e.g. venereal *Campylobacter*, Trich pouches, *Chlamydophila* or *Mycoplasma* medium). In some cases they may need to special order or make specific media, so a heads up is always appreciated.

TRANSPORT SYSTEMS AND CONTAINERS

The type of specimen transport system and type of container and/or device utilized for sampling and transport varies according to the type of sample needed and disease process involved. All containers and devices employed for bacterial or viral isolation must be sterile; other containers must be clean and uncontaminated. The containers should also have tight fitting caps or lids to prevent leakage or contamination during transport.

Swab Transport Culture Systems

Swabs are most often used for suspected infectious diseases. They are designed to maintain microbial flora during transport so that the organisms isolated are a true representation of the microbial population present at the sampling site. They are used to obtain specimens from mucosal and epithelial surfaces (conjunctiva and ears) and from deep within wounds or soft tissue infections. Swabs should not be used unless aspirated or tissue (biopsy) specimens are not obtainable. Micro-swabs are available for tight spaces.

Disadvantages of swabs:
They are easily contaminated with commensal surface bacteria.
It is possible to collect only small volume of sample.
They may inhibit growth of anaerobic bacteria because of oxygen trapped within fibers.

Types of swabs:
Ampule swabs contain a liquid transport medium which is located at the base of the swab system. The medium is released to the swab when the ampule is crushed.
Rubber diaphragm swabs contain a semisolid medium which is separated from the base of the swab system by the diaphragm. Pressure on the diaphragm releases the semisolid medium to the culture swab.
Guarded swabs are designed such that the swab is located inside a plastic sheath which allows selective sampling through areas that contain normal flora.
Anaerobe swabs contain an oxygen-displacing (H2 or CO2-generating) cartridge and seal tightly.

Transport medium:
Swabs are available with a choice of several media. Consult your diagnostic lab to see if a specific type of swab is necessary for the organism you suspect (e.g. Amies with charcoal for *Taylorella* suspects)
Stuart's medium: General-purpose bacterial media.
Amies medium: General-purpose bacterial media.
Amies with charcoal: Supports fastidious bacteria by absorbing toxic metabolites.
Viral or *Chlamydia* transport medium: Supports viral growth and contains antibiotics to control contaminant bacteria.
Mycoplasma medium: Contains antibiotics to control faster-growing bacteria that would obscure the target organism.

Swab material may affect the isolation of certain organisms and quality of the sample. Common swab materials include:
Cotton: Residual fatty acids on cotton fibers may be inhibitory to some fastidious bacteria. OK for cytology.

Calcium alginate: Can be dissolved in solution, releasing all collected bacteria. Good for microbiology culture samples.
Dacron/Polyester/Rayon (spun): microscopically abrasive and good for collecting mucosal cells for cytology and slide prep.
Flocked: Better for microbiology than cotton, Rayon, polyester or Dacron spun swabs; better organism release.
Polyurethane foam: Suitable for microbiology. Poor cell recovery when used for cytology.
Brush-tip: Designed to collect cells for cytology; less suitable for microbiology.

Bagged Specimen Containers

Bag-type containers prevent drying of the sample and allow individual packaging of samples from multiple tissues. They are used most often for tissue samples that have either been collected at surgery or necropsy and are also useful for fecal samples, intestinal content and milk samples if a leak-proof bag type is used. They are space-saving and trap less air/oxygen when compared to rigid containers.

Types: Several types and sizes of sterile disposable plastic bags with sealable tops are available for specimens (e.g. Whirl Pak or heat-sealed bags). Ziploc bags may leak, so are not recommended for wet samples but are perfect for paperwork or dry samples such as feed or forage. Gloves and examination sleeves should not be used for transporting samples of any type. They consistently leak causing contamination problems both to specimens and handlers of specimens.

Cup Containers

Cup-type containers may be used for solid or liquid samples. They may be obtained "clean" or sterile, depending on their intended purpose. The wide mouth of cup or jar-type containers makes them easy to use as compared to bags, but there is a considerable amount of "dead space" that can enhance drying of specimens and decrease recovery of organisms. These types of containers are most often used for tissue or fecal samples, but may be used for larger volumes of fluid (milk, large aspirates).

Types: These come in a variety of sizes and shapes. Most often they are made of clear plastic and have screw cap lids. Screw cap lids are preferable to snap-on lids. If a snap-on lid container is used, secure the lid with strong tape or parafilm. Be sure to use sterile cups for tissue or urine specimens. Sterile cups are not necessary for fecal samples.

Tube/ Vial Containers

This type of container is used most often for small-volume liquid specimens such as milk, urine, blood, CSF, exudates, and transudates. Multiple tubes with different additives may be necessary if several tests are required (e.g. both cytology and culture of an aspirate).

Types: A variety of sizes and additives are available. Vials are made of either glass or plastic and are screw-capped, snap-capped, or rubber-topped. Vacutainer tubes are one commonly used example. Purple top (EDTA) tubes should not be used for culture specimens because the EDTA may inhibit growth of some agents; however, EDTA is the preferred additive for cytologic samples. Screw-top or snap-cap vials are preferred, as it is easy to contaminate rubber stoppers when extracting them. Ensure they are sterile.

Other Devices used for Sample Containers

Diagnostic labs receive samples in an enormous variety of containers. Some are appropriate, some not. Certain specific sample types are best submitted in very specific containers to prevent deterioration or overgrowth of unwanted organisms.

Syringes: These are occasionally used for transporting aspirates and fluids. This method should no longer be practiced due to the risk of accidentally sticking your finger when recapping the needle. Syringes may be sent if the needle is removed and a syringe plug applied, but if the plunger is accidentally depressed during handling your sample may be lost. Vials, tubes and cups are more secure. Syringes sent with needles may be rejected by your diagnostic lab!

Paper envelope: This is the preferred container for hair and skin crust specimens for dermatophyte culture, or other skin/hair samples.

Petri dishes: These may be used for small non-drippy samples (e.g. hair clumps for *Dermatophilus* impressions; skin flakes or crusts for fungal culture) or organisms you have cultured at your clinic and wish to send for ID or sensitivity (e.g. on blood agar or Sabourad's dextrose agar). Be sure to seal the edges of the dish with Parafilm or similar. Desiccation may occur.

Blood culture vials: While optimized for blood, these can also be used for any fluid sample. Use only sterilely-obtained samples, as if contaminants are present overgrowth will occur! (blood culture vials are expressly designed to obtain viable cultures from extremely small bacterial numbers).

Slides: If an infectious etiology is suspected, air-dried unstained slides may be submitted with your culture sample. These slides may be useful for fluorescent antibody testing. Slides of any exudate or aspirate may also be submitted for clinicopathologic evaluation. Unstained slides should always be sent; stained slides may be sent, but should never replace unstained slides. Do not submit a swab for culture that you have used to make slides; obtain a fresh swab.

Formalin-fixed samples: These are great for histopathology, but are useless for microbial culture. If you are asking the lab for cultures, be sure that you are sending fresh as well as fixed tissue! (it's always a good idea to send both anyway) Fixed tissue *may* be usable for PCR for certain infectious agents. Formalin: tissue should be 10:1 ratio.

REFERENCES

Gilor S et al. Common laboratory artifacts caused by inappropriate sample collection and transport: how to get the most out of a sample. **Topics in Companion An Med** 2011; 26(2):109. Hodgson JL et al. Diagnosis of bacterial infections. part 1: principles of sample collection and transportation. **Eq Vet Ed** 2008; 20(11):608. Mason GL and Madden DJ. Performing the field necropsy examination. **Vet Clin N Am: Food An Pract** 2007; 23(3):503. Meinkoth JH and Allison RW. Sample collection and handling: getting accurate results. **Vet Clin N Am: Sm An Pract** 2007; 37(2):203. Murray PR et al. Manual of Clinical Microbiology, 10th ed., ASM Press, 2012. Quinn PJ et al. *Clinical Veterinary Microbiology*. Mosby, 1994.

ACID BASE STATUS FROM THE METABOLIC ASPECT

Katy W Waddell, RVT, VTS (ECC, Anesthesia)
College Station, TX

Key points

- Regulation of the pH is done by a system of buffers
- metabolic acidosis may be divided into
- Exogenous vs. endogenous
- High anion gap vs. normal anion gap

Intracellular fluid is actually not one compartment but has at least 10 separate tiny cellular packages. So we may think of the ICF compartment as a virtual compartment. The collection of the tiny cellular packages has unifying similarities which enables us to lump them into the concept of one compartment. Intracellular fluids are high in potassium and magnesium and low in sodium and chloride ions.

The intracellular fluid makes up 2/3 of the total body water. The cell membranes and cellular metabolism maintain control of the constituents of intracellular fluid. Intracellular fluid is not homogeneous in the body. It is composed of fluids from all the different cell and not only contains potassium – 160mEq/L, magnesium 26 mEq/L, sodium 10 mEq/L calcium 2 mEq/L and chloride ions but also organic anions, proteins and more.

Extracellular fluid is divided into several smaller compartments, these being plasma, interstitial fluid, fluid of bone and dense connective tissues and transcellular fluid. Their location and kinetics make them distinguishable. Interesting, this includes all of the body fluid lying outside the cells including special cavity area such as synovial fluid, urine, aqueous humor of the eye. These fluids are considered transcellular fluids which make up part of the ECF. Extracellular fluid is composed of higher concentrations of sodium and chloride and lower concentrations of potassium and magnesium.

Plasma and interstitial fluid are identical in ionic composition. The difference between plasma and interstitial fluid is protein content. Plasma contains a large amount of protein, while the interstitial fluid contains less
A buffer can either accept or give up hydrogen ions, depending on the pH of the solution. If for instance, the bicarbonate buffer system finds itself in a solution with a lot of free hydrogen ions (acid); they will act as bases and "suck" up the excess hydrogen ion. In essence, they turn a strong acid into a weak acid. Should the situation be reversed and the solution is lacking free hydrogen ions, they will "donate" some of their hydrogen ions to turn the strong base solution into a weak base solution.

Buffering is accomplished by converting a strong acid to a weak acid and/or converting a strong base into a weak base. The urinary system buffers the greatest quantity and requires hours or days to exert the effect while the respiratory system has a small effect but the effect is exerted within minutes.

The interstitial fluid contributes bicarbonate for metabolic acids. The kidney is responsible for being the gate keeper of the amount of bicarbonate but the interstitial fluid acts as one the carriers. Everyone must work together to maintain homeostasis. In research done in metabolic acidosis, it has been shown that 57% of buffering occurs intracellular and 43% occurs extracellular. Intracellular – the buffer occurs by protein, phosphate and bicarbonate buffers due to the entry of H^+ by Na^+- H^+ exchange (36%), K^+ - H^+ exchange (15%) and the every present "other" claiming 6%.

Another buffering mechanism occurs within hemoglobin. Hemoglobin is a primary buffer against carbonic acid changes In addition to binding oxygen, transporting it from the lungs to the cells which need it; hemoglobin also binds CO_2, which is released by those same cells to transport it to the lungs. If this CO_2 were not bound to hemoglobin, it would form carbonic acid in the blood, causing the blood in the veins and the pulmonary arteries to have a lower pH than the blood coming from the lungs.

What role does the kidney play in maintaining the acid base balance? Kidneys are responsible for maintaining hydrogen secretion, bicarbonate absorption and excretion of hydrogen ions. Simply stated, to manage the pH of the blood, the kidney excretes H^+ in the urine and reabsorbs bicarbonate into the blood plasma. By excreting the hydrogen ions (cation = acid) they are removed from the circulating blood. Bicarbonate (anion = base) is reabsorbed by the tubules in the kidney to maintain the balance.

Bicarbonate ions are formed by the hydration of CO_2 ($CO_2 + HOH = H_2CO_3$, and dissociation of $H_2CO_3 = H^+$ plus HCO_3^-). This combination occurs in the blood. Bicarbonate diffuses from the red blood cell into the plasma in venous blood and enters the red blood cells from the plasma in arterial blood.

What is base excess/deficit? This is another part of the checks and balances system of the body to guard homeostasis. The terms base deficit and base excess are often used interchangeably. The primary difference is that the base deficit is indicated by a positive number while the base excess is indicated by a negative number. This neutralization by the body is achieved by the buffering of HCO3 (about 50%) and the remainder from plasma, red blood cells and phosphates. If a patient were to have a base deficit of 8, this would tell me that 8mEq/L of buffer has been used by the body to neutralize metabolic acids.

Some common causes of metabolic acidosis would include diabetic ketoacidosis or starvation ketosis. Ketone bodies formed when the liver stores of glycogen are depleted are acidic. The acidity is a result of the dissociation of the H+ ion. Because of the "uncontrolled" nature of ketosis, the normal buffering systems in place are overwhelmed and unable to compensate for the change in pH. This is one reason you will notice DKA patients having a higher than normal respiratory rate. This is the body's effort to modulate pH by decreasing PCO2.

Lactic acidosis is an acidotic state which occurs during anaerobic metabolism. When cells are forced to metabolize glucose anaerobic ally due to a decrease in oxygen availability ATP production/synthesis is impaired and further glycolysis is required and the vicious cycle continues until perfusion with adequate oxygen exchange is restored.

Chronic renal failure is a common cause of metabolic acidosis due to the accumulation of sulfates, phosphates, and uric acid. The kidneys inability to filter and excrete acids and maintain bicarbonate reabsorption will lead to altered enzyme activity. The altered enzyme activity further leads to the promotion of hyperkalemia.

With metabolic acid base alterations resulting in an acidotic state, the causes must relate to the gain of a strong acid and/or a loss of base. Metabolic acidosis may be divided into two group of clarification/classification. We may refer to them as being endogenous vs. exogenous or classify the acidosis by anion gap results. The anion gap is calculated by subtracting the Cl- and HCO3- from the plasma Na concentration. By obtaining the anion gap we can determine if the metabolic acidosis is considered a "high anion gap metabolic acidosis" or a "normal anion gap metabolic acidosis". This information would lead us to further pursue the acidosis as to the determining cause.

High anion gap related causes would include:

- diabetic ketoacidosis,
- starvation ketoacidosis, and
- lactic acidosis either from impaired perfusion or by impaired carbohydrate metabolism,
- renal failure (uremic acidosis or acidosis associated with acute renal failure /acute kidney injury) or
- toxins – ethylene glycol, methanol or salicylates.

When acid is added to the body, the [H+] increases and the [HCO3-] decreases. In addition, the concentration of the anion, which is associated with the acid, increases. This change in the anion concentration provides a convenient way to analyze and help determine the cause of a metabolic acidosis by calculating what is termed the anion gap. The anion gap is estimated by subtracting the sum of Cl- and HCO3- concentrations from the plasma Na concentration.

Na + Unmeasured cations = Cl- + HCO3- + Unmeasured anions. Anion gap = [Na] – ([Cl-] + [HCO3-])

The major unmeasured cations are calcium, magnesium, gamma globulins and potassium. The major unmeasured anions are negatively charged plasma proteins (albumin), sulphate, phosphates, lactate and other organic anions. The anion gap is defined as the quantity of anions not balanced by cations.

Example

Labs : Serum chemistry: Na 132, K 5.0, Cl 104, HCO3- 16 , BUN 25, Cr 1.3, Glu 75

ABG: pH 7.30, PCO2 29, HCO3- 16, PO2 92

Serum albumin 1.0

Anion gap = (Na-(Cl +HCO3-) = 132 - (104 +16) = 12

Note that the anion gap appears to be normal, and thus lactic acidosis appears unlikely. However, note also that the patient is severely hypoalbuminemic with a serum albumin of 1.0. Because the anion gap is primarily determined by negatively charged plasma proteins such as albumin, we must adjust the normal value of the anion gap to more accurately reflect the albumin deficiency. The approximate correction is a reduction in the normal anion gap of 2.5 mEq/l for every 1g/dl decline in the plasma albumin concentration (normal value = 4 g/dl)

The normal anion gap should be:

Decline in albumin = 4 - 1 = 3 g/dl

Reduction in normal anion gap = 3 × 2.5 = 7.5

Adjusted anion gap = 12 - 7.5 = 4.5.

A calculated anion gap of 12 is high when compared to the adjusted anion gap of 4.5. .

REFERENCES

1. www.globalrph.com/anion
   Handy on line calculator to determine metabolic acid/base derangements
2. Cunningham JG: Textbook of Veterinary Physiology, 3rd ed. W. B. Saunders Company, Philadelphia, 2002, pp.459-466.
3. Dibartola SP: Fluid, Electrolyte and Acid-Base Disorders in Small Animal Practice, 3rd ed. Elsevier, City, 2006, pp. 91-121, pp. 229-251, pp. 251-283, pp. 283-296, and pp. 296-309.
4. www.acid-base.com
   Grogono's Acid-base tutorial
5. fitsweb.uchc.edu/student/selective/TimurGraham/Welcome.html
6. Acid base tutorial University of Connecticut Health Center

ELECTROPORATION IN VETERINARY MEDICINE: RIDING THE ELECTRICAL WAVEFORM

Enrico P Spugnini, DVM, Ph. D., DACVIM (Oncology), DECVIM-CA (Oncology)
Rome, Italy

INTRODUCTION

Chemotherapy drugs are effective provided that they reach their site of action. The primary obstacle for chemotherapy agents is the cytoplasmic membrane, especially for lipophobic agents such as bleomycin. Electrochemotherapy (ECT) couples the administration of an anticancer agent (usually bleomycin or cisplatin), to the delivery of electric pulses having appropriate waveforms. The application of permeabilizing pulses leads to perturbation of the cell membrane, thus resulting in increased uptake of chemotherapeutic drugs, ultimately leading to cell apoptotic death.[1] Clinical trials evaluating humans treated with ECT for cutaneous neoplasms were first reported in the early 90s, and have been followed by many trials that confirm its potential for the treatment or palliation of dermatological tumors.[2-4] Recently, a standard operating procedure has been proposed.[5]

ELECTROCHEMOTHERAPY EQUIPMENTS

Different waveforms have been adopted by investigators: exponentially decaying, square, rectangular or biphasic. In general the number of pulses delivered is set in 8 single pulses applied per cm of tumor area at a voltage of 1300 V/cm (800 V/cm for intraoperative use), with a duration of 100 μs and a frequency of 1 Hz. Treatments are repeated until the whole tumor area is covered.

RESULTS IN VETERINARY MEDICINE USING SQUARE PULSES BASED ECT

The first clinical trial in veterinary patients involved the treatment of feline sarcomas recurring after radiation therapy using bleomycin and Interleukin 2 producing cells coupled with square pulses.[6] The authors did not report any clinical response but claimed prolonged control for cats undergoing ECT compared to untreated patients. More recently, CDDP or bleomycin based ECT has been reported to treat cutanous neoplasms in dogs and cats as well as sarcoids in horses.[7-9]

EVALUATION OF BIPHASIC PULSES BASED ECT IN PETS

Experimental design

Our group investigated in the past 8 years the feasibility of electrochemotherapy in companion animals affected by different neoplasms. We adopted trains of biphasic pulses delivered as a burst instead of single impulses. The electric waveform were generated by a *Chemopulse* equipment. This device is built up by a toroidal core transformer generating a roughly rectangular pulse which is split in two halves that are sequentially driven to obtain a biphasic pulse. Pulses are not singularly produced but are created in bursts. The equipment allows choosing electric field intensity, number of pulses and pulse duration. The standard wave train used for this experiment was set to 8 pulses of 50 + 50 μs (biphasic pulse), and the electric field intensity was 1250 V/cm. The pulse repetition frequency is 1 Hz while the frequency of burst repetition is 1 kHz, resulting in a total burst duration of 7.1 ms. Bleomycin was locally injected at the tumor site and in the tumor's surrounding tissues. Five minutes after the injection, trains of biphasic pulses were administered through caliper electrodes as above described. Adherence of the caliper electrodes to the lesions was maximized using an electrophoresis gel. The treatment was repeated after 1 week.

RESULTS

In the first trial, ECT was used to directly attack neoplasms.[10] Overall response rate was 80%, with 40% long lasting remissions (in excess of 1 year). This phase I/II study led to the development of several electrodes specific for the different body districts.[11] On the basis of the preliminary studies, cohorts of dogs and cats affected by melanoma, soft tissue sarcoma mast cell tumor and nasal squamous cell carcinoma were enrolled in phase II studies.

Melanoma Ten patients with malignant melanoma (MM) of the oral cavity were enrolled in a phase II study. The overall response rate was 80% (median time to recurrence 6 months), with a 50% of patients in remission after 1 year.[12] Of interest, most of the long term responders experienced a vitiligo like discoloration at the treatment site, potentially suggestive of recruitment of the immune system after uncovering of deep antigens.

Soft tissue sarcoma A total of 72 cats affected by soft tissue sarcoma were assigned to three different groups: 1) surgery alone, 2) surgery coupled with intraoperative ECT, 3) surgery coupled with post operative ECT.[13] The median control times were respectively 4, 12 and 19 months. In this study were identified some prognostic factors such as previous treatment and tumor size. A cat with a recurring hemangiopericytoma was also referred at the time of the study, and is still in remission 2.5 years after the completion of the ECT treatment.[14] The only side effects were tumor dehiscence in 2 patients and focal inflammation in 2 patients. Systemic side effects were not reported among the enrolled patients.

In another study, adjuvant ECT was adopted to increase local control in a group of 22 dogs, obtaining a mean time to recurrence of 730 days, with 50% of the dogs still desease free at the time of writing.[15] In one dog with a large, unresectable sarcoma, ECT has been used to decrease tumor size and allow surgical excision.[16] More recently, ECT has been exploited to reintroduce potentially toxic drugs in the oncology protocols. Specifically, a cohort of 64 cats with incompletely excised soft tissue sarcoma have been enrolled in a study involving the use of cisplatin as ECT agent and matched with a control cohort of 14 cats treated with surgery. All the cats tolerated the treated very well, despite the high reported toxicity of the drug when administered systemically. In terms of tumor control mean time to recurrence was 666 days and 180 days, respectively.[18]

Mast cell tumor Adjuvant bleomycin-based ECT was used in 28 dogs with incompletely excised mast cell tumors. Overall response rate was 85%. The median time to recurrence was not reached. Estimated time to recurrence was 52.76±6.5 months.[17] One dog had wound dehiscence. Another study, enrolling 37 dogs, evaluated the efficacy of cisplatin for the treatment of incompletely excised canine mast cell tumors obtaining a mean disease-free interval of 1218 days with 78% of the patients experiencing local control at different times.[19]

Squamous cell carcinoma a small cohort of 9 cats with sun induced squamous cell carcinoma of the nasal planum were treated with intralesional bleomycin coupled with ECT with a rate of complete responses of 77%.[20]

CURRENT PROTOCOLS

Electrochemotherapy is mostly used as a adjuvant therapy, locally injecting the chemotherapy drugs bleomycin or cisplatin and then following with the application of permeabilizing pulses. For larger tumors drugs can be administered locally[16] or systemically[21], mostly on the basis of tumor volume, firmness or resistance to insertional electrodes. Caution should be exerted when using electrochemotherapy in patients previously treated with radiation therapy due to the risk of radiation recall.[22]

ELECTROCHEMOTHERAPY IN OTHER SPECIES

ECT has been used successfully in horse to treat sarcoids using intralesional bleomycin and to treat melanomas using cisplatin as chemotherapy agent, either directly or combined with surgery. The treatment has been well tolerated and resulted in high percentage and durable responses.[23-25]

CONCLUSION

Electrochemotherapy is a safe and efficacious approach to veterinary soft tissue and cutaneous neoplasms. Its low cost and ease of administration make it a valuable addition to the currently available oncological therapies. One advantage of this technique is the possibility of repeated treatments in case of local recurrence and the recruitment of the patients' immune system, as suggested by the low incidence of metastatic disease among our treated patients and by the tumor selection observed among some pets with recurring disease.

FUTURE DIRECTIONS

Ongoing studies are evaluating the efficacy of ECT when other drugs than bleomycin are used, in particular anthracyclines and their analogues (mitoxantrone) and platinum compounds such as CDDP and carboplatin. Studies are being performed to characterize the membrane alterations induced by trains of biphasic pulses in order to improve the efficacy of the technique.[26] Novel electrochemotherapy equipments adopting different electrical parameters are currently being developed. Finally, our group is assessing the potentials for electrogenetherapy using electric pulses to drive chemosensitizing plasmids within tumor cells.

REFERENCES

[1]Gothelf et al. **Cancer Treat Rev** 2003; 29: 371. [2]Belehradek et a. **Cancer** 1993; 72: 3694. [3]Daskalov et al.**IEEE Eng Med Biol** 1999 ; 18 : 62. [4]Gehl et al. **Melanoma Res** 2000; 10: 585. [5]Mir et al. **Eur J Cancer** 2006 ; 4(S) : 14. [6]Mir et al. **Br J Cancer** 1997 ; 76 : 16. [7]Tozon et al. **Anticancer Res** 2001; 21: 2483. [8]Rols et al. **Bioelectrochem** 2002 ; 55 : 101. [9]Tozon et al. **Anticancer Res** 2005 ; 25 : 839. [10]Spugnini et al. **J Exp Clin Cancer Res** 2003 ; 22 : 571. [11]Spugnini et al. **J Exp Clin Cancer Res** 2005; 24 : 245. [12]Spugnini et al. **Melanoma Res** 2006; 16 : 23. [13]Spugnini et al. **Cancer Chemoter Pharmacol**; 2007; 59 : 375. [14]Spugnini et al. **Vet Rec** 2006; 159 : 598. [15]Spugnini et al. **In Vivo** 2007; 21: 819.[16]Spugnini et al. **Vet Rec** 2008; 162: 186. [17] Spugnini et al. **J Transl Med** 2011; 9; 152. [18]Spugnini et al. **Anticancer Res** 2006; 26: 4585.[19] Spugnini et al. **J Vet Intern Med** 2011; 25: 407.[20] Spugnini et al. **Vet J** 2009; 179: 117.[21] Spugnini et al. **In Vivo** 2008; 22: 51. [22] Spugnini et al. **In Vivo** 2008; 22: 751. [23]Tamzali et al. **Equine Vet J** 2012; 44:214. [24]Spugnini et al. **J Equine Vet Sci** 2011; 31: 430: [25] Scacco L et al. **J Equine Vet Sci** In press.[26]Spugnini et al. **Microsc Res Tech** 2007; 70: 1041.

THE PERSISTENT AND PESKY PROBLEM OF FELINE INJECTION SITE SARCOMAS

Kelly R. Hume DVM, DACVIM (Oncology)
Ithaca, NY

BACKGROUND AND EPIDEMIOLOGY

Feline injection site sarcomas (FISS) are mesenchymal neoplasms that arise at the site of injections, most commonly vaccinations. They were first described in the early 1990s. These tumors have a marked inflammatory component that may include macrophages containing a gray-brown foreign substance, identified as aluminum by electron probe microanalysis.[1] Identification of these tumors coincided with legal changes requiring rabies vaccinations of cats and changes within the vaccine industry, including licensing of the first killed rabies vaccine for subcutaneous administration and a killed vaccine for FeLV.[2] Epidemiological studies have shown an increased risk for sarcoma development at vaccination sites in comparison to sarcomas at other sites, with an odds ratio of 5.49 for development of a sarcoma at a FeLV vaccination site and an odds ratio of 1.99 for development of a sarcoma at a rabies vaccination site.[3] There is also an increased risk for sarcoma development if multiple vaccinations are administered at the same site simultaneously. The Vaccine-Associated Feline Sarcoma Task Force (VAFSTF) distributed recommendations in 1996 suggesting that specific vaccines always be given in specific sites: rabies – distal right rear, FELV – distal left rear, and FVRCP – right shoulder.[4] Subsequent to these recommendations, there was a shift in location of diagnosed FISS (see Table 1).[5] An association between a particular vaccine product and sarcoma development was not identified in early epidemiological studies. Sarcomas at injection sites of other substances have also been documented: long-acting antibiotics, steroids, non-absorbable suture, microchip, and, more recently, meloxicam and cisplatin.[6-10]

Table 1. Change in location frequency of ISS after VAFSTF recommendations[5]

Location	*% of ISS from 1990-1996*	*% of ISS from 1997-2006*
Interscapular	53	40
Right thorax	10	4
Left thorax	9	1
Right thoracic limb	1	10
Right pelvic limb/abdomen	13	25
Left pelvic limb/abdomen	11	14

Cats with sarcomas at vaccination sites are significantly younger, have larger tumors, and have tumors that recur rapidly and more frequently than cats with sarcomas at other sites.[11] Time to development of FISS can vary from 3 months to 3-10 years after injection. Although the latency period can be long, these tumors are characterized by rapid and invasive growth. The true incidence of FISS is not known. Literature reports vary from 1/1000 to as low as 0.63/10,000 vaccinated cats – average incidence may be 8/10,000 cats.[12] Explanations for variability include statistical chance, possible differences in vaccine manufacturers between studies, regional variations in the epidemiology of the disease, short follow-up periods for a disease that can have a long latency, and the interactions of multiple vaccinations over a cat's lifetime.

Relatively little is known about the molecular pathogenesis of FISS. It is hypothesized that growth factors elucidated at the site of inflammation interact with mesenchymal cells to result in oncogenesis. Aberrant p53 staining[13] and allelic deletion[14] have been associated with negative outcomes.

DIAGNOSIS AND HISTOPATHOLOGIC CHARACTERISTICS

Early recognition is extremely important; all injection sites should be monitored by owners and veterinary staff. The inflammatory component complicates cytologic evaluation and cytology may only be diagnostic in 50% of cases.[15] Cytology is useful to rule out other lesions and tumor types such as abscess, basal cell tumor, mast cell tumor, and lymphoma. Incisional biopsy is the recommended technique for definitive (or confirmatory) diagnosis. Attempts at simple excision usually result in local recurrence and can lead to a more difficult second surgery. It is important to ensure that the entire biopsy tract can be excised during definitive surgical treatment. The VAFSTF made the following recommendations, the so called "1-2-3 rule", for when to biopsy a mass at an injection site: 1) if the mass is increasing in size *1* month after injection, 2) if the mass is greater than *2* cm in diameter, and 3) if the mass persists for more than *3* months.[4]

Inflammation is a component of these tumors, with lymphocytes and macrophages frequently noted at the periphery. These tumors will be vimentin positive, but may also stain for α-smooth muscle actin,[16] desmin,[17] and KIT.[18] These tumors are frequently necrotic and have an increased mitotic index compared to non-injection site sarcomas. The cells are quite pleomorphic. Tumor stroma may contain collagen, osteoid, chrondroid, or myxomatous matrix.[19] Histologic subtype is most commonly fibrosarcoma, but can also be malignant fibrous histiocytoma, undifferentiated sarcoma, osteosarcoma, rhabdomyosarcoma, liposarcoma, and chrondosarcoma.[17,20]

Various authors[18,21] have tried to apply a grading scheme to FISS that is similar to the commonly used Kuntz scheme[22] for canine soft tissue sarcomas. There is not yet conclusive evidence that grade can be used to identify cats with increased risk of recurrence or metastasis; therefore, current treatment recommendations are NOT based on grade. *All ISS should be considered aggressive.*

STAGING

The metastatic rate of FISS is ~ 25%, with the lung as the most common site. Other reported sites include lymph node, pelvis, kidney, spleen, intestine, and skin. Three view thoracic radiographs are recommended for all cats diagnosed with FISS. Abdominal imaging is recommended if the tumor is caudal to the diaphragm or if physical exam or laboratory findings are consistent with abdominal disease. Complete blood count, serum biochemistry, and urinalysis are recommended to assess overall systemic well being. There is no known association with retroviruses and FISS; testing is recommended for cats with risk of exposure as part of a general health screen. Given the expense and time associated with definitive treatment, it is reasonable to consider both abdominal imaging and retroviral testing for all cats diagnosed with FISS. Advanced pre-surgical imaging with CT or MRI is recommended to define the true extent of the local disease and allow for proper surgical planning. Volume of the tumor based on CT images is typically twice that detected via physical examination.[20]

TREATMENT RECOMMENDATIONS AND PROGNOSIS

It is important to note that the management and prognosis of FISS is very different from that of canine soft tissue sarcoma. Complete surgical resection is essential, but is often inadequate as sole therapy. Reported recurrence rates of FISS treated with surgery alone vary from 30 – 70%. Tumors with "dirty" margins are 10 times more likely to recur than tumors with "clean" margins.[23] Even with histopathologically "clean" margins, the recurrence rate may be as high as 50%.[20] Median tumor-free interval when clean margins were obtained was 16 months versus only 4 months when excision was incomplete.[24] In one study of cats treated with surgery alone, only 14% of cats survived greater than 2 years.[25] Factors that may delay the time to tumor recurrence include surgery at a referral institution and a radical first excision.[25] Surgical margins of 3-5 cm of macroscopically healthy tissue and one fascial plane deep have been recommended. Radical surgical resection for tumors in the interscapular region may require removal of vertebral spinous processes and partial scapulectomy. In a more recent study, radical surgical excision with 5 cm margins including a deep margin of 5 cm of subcutaneous tissue or two muscle planes or bone deep to the tumor was associated with a median survival time of 901 days and a local recurrence rate of 14%.[26] Major complications occurred in 11% of cats.

Due to the likelihood of local recurrence, adjuvant radiation therapy (usually 16 – 19 fractions x 3 Gy) is typically always recommended. Possible exceptions include very small tumors treated with wide surgical resection that have clean margins histologically and tumors on the distal limb that are treated with amputation. Consultation with a radiation oncologist is recommended prior to surgical resection – preferences vary regarding pre-op vs. post-op radiation therapy. For cats with complete surgical margins receiving adjuvant radiation therapy, median progression free intervals of ~ 3 years have been reported, yet recurrence rates can still reach 40%.[27,28] The exact impact of adjuvant chemotherapy is unknown. Issues in reported studies include bias and lack of power.[15,27]

For tumors initially deemed non-resectable, consultation with a board-certified surgeon may be helpful. If surgical options do not exist, palliative radiation therapy may provide symptomatic relief and/or disease stabilization. Palliative protocols (5 x 4 Gy or 4 x 8 Gy) have been associated with an overall response rate of 50-70%, with median progression-free intervals and survival times ranging from 4 – 10 months.[28,30] The response rate of macroscopic FISS to cytotoxic chemotherapy varies from 25 – 50%, with median duration of response ranging from 42 – 125 days.[31-34] Drugs to consider include doxorubicin, cyclophosphamide, carboplatin, lomustine, and ifosfamide. There is in vitro evidence that these tumors may be sensitive to tyrosine kinase inhibitors,[35,36] however the duration of response in a small number of cats treated with imatinib was only 2 months.[37]

For cats with macroscopic tumors, symptomatic therapy with analgesics and anti-inflammatory drugs is warranted. Cats may also suffer from cytokine-induced nausea and secondary infections. Appropriate supportive care can often improve quality of life even in the face of tumor progression.

STRATEGIES FOR CLIENT COUNSELING

Vaccination is a medical procedure that has risks and benefits, and it should be tailored to the individual patient. It is important to know the law in your state regarding rabies vaccination. Consult your public health officials. Know what you are responsible for, and know the responsibility born by the owner. Liability is not always straightforward. Informed consent is key. Do not perpetuate unwarranted fear or provide misinformation. An accurate discussion of legal regulations and consequences of violations should occur.

Decisions to vaccinate should be made based on risk. Panleukopenia, herpes, calici, and rabies are all considered core vaccines. Non-core vaccines include FeLV, FIV, FIP, Chlamydophila, and Bordetella. When choosing to administer vaccines you should also consider what is and isn't known about duration of immunity of a given product (know what evidence does or does not back up the label claim). Use of single dose vials is recommended. Recent studies indicate that nonadjuvanted vaccines lead to less tissue inflammation than adjuvanted vaccines,[38] and a decreased risk for sarcoma formation was seen in cats receiving recombinant vaccines compared to cats receiving inactivated vaccines.[39]

Maintain thorough and accurate records and report adverse events to both the vaccine manufacturer and the USDA Center for Veterinary Biologics (1-800-752-6255; http://www.aphis.usda.gov/animal_health/vet_biologics/vb_adverse_event.shtml). Additional resources include the AAFP 2006 Feline Vaccination Guidelines (http://www.catvets.com/professionals/guidelines/publications/?Id=176). The VAFSTF has been disbanded but their recommendations are still available online (https://www.avma.org/About/AlliedOrganizations/Pages/vafstf.aspx).

REFERENCES

[1]Hendrick MJ. **Cancer Res** 52: 5391-4, 1992. [2]Hauck M. **Vet Clin North Am Small Anim Pract** 33: 553-7, vii, 2003. [3]Kass PH, et al. **J Am Vet Med Assoc** 203: 396-405, 1993. [4]Morrison WB and Starr RM. **J Am Vet Med Assoc** 218: 697-702, 2001. [5]Shaw SC, et al. **J Am Vet Med Assoc** 234: 376-80, 2009. [6]Buracco P, et al. **Vet J** 2002;163:105-7. [7]Kass PH, et al. **J Am Vet Med Assoc** 2003;223:1283-92. [8]Daly MK, et al. **J Fel Med Surg** 2008;10:202-5. [9]Munday JS, et al. **J Feline Med Surg** 2011;13:988-91. [10]Martano M, et al. **J Feline Med Surg** 2012;14:751-4. [11]Hendrick MJ, et al. **J Am Vet Med Assoc** 205: 1425-9, 1994. [12]Martano M, et al. **Vet J** 188: 136-41, 2011. [13]Hershey AE, et al. **Vet Pathol** 2005;42:805-11. [14]Banerji N and Kanjilal S. **Am J Vet Res** 2006;67:1766-72. [15]Martano M, et al. **Vet J** 170: 84-90, 2005. [16]Couto SS, et al. **Vet Pathol** 39: 33-41, 2002. [17]Hendrick MJ, et al. **J Natl Cancer Inst** 86: 341-3, 1994. [18]Smith AJ, et al. **Vet Pathol** 46: 934-9, 2009. [19]Doddy FD, et al. **J Comp Pathol** 114: 165-74, 1996. [20]McEntee MC and Page RL. **J Vet Intern Med** 15: 176-82, 2001. [21]Romanelli G, et al. **J Am Vet Med Assoc** 232: 1193-9, 2008. [22]Kuntz CA, et al. **J Am Vet Med Assoc** 1997;21:1147-1151. [23]Giudice C, et al. **Vet J** 186: 84-8, 2010. [24]Davidson EB, et al. **Vet Surg** 26: 265-9, 1997. [25]Hershey AE, et al. **J Am Vet Med Assoc** 216: 58-61, 2000. [26]Phelps HA, et al. **J Am Vet Med Assoc** 2011;239:97-106. [27]Kobayashi T, et al. **Vet Radiol Ultrasound** 43: 473-9, 2002. [28]Eckstein C, et al. **Vet Comp Oncol** 7: 54-68, 2009. [30]Kleiter M, et al. **Vet Radiol Ultrasound** 51: 349-55, 2010. [31]Barber LG, et al. **J Am Anim Hosp Assoc** 36: 416-21, 2000. [32]Kisseberth WC, et al. **J Vet Intern Med** 22: 83-8, 2008. [33]Fan TM, et al. **J Am Anim Hosp Assoc** 38: 357-63, 2002. [34]Rassnick KM, et al. **Am J Vet Res** 67: 517-23, 2006. [35]Katayama R, et al. **Cancer Chemother Pharmacol** 2004;54:25-33. [36]Lawrence J, et al. **Vet Comp Oncol** 2012;10:143-54. [37]Lachowicz JL, et al. **J Vet Intern Med** 2005;19:860-4. [38]Day MJ, et al. **Vaccine** 2007;25:4073-84. [39]Srivastav A, et al. **J Am Vet Med Assoc** 2012;241:595-602.

URINARY TRACT CANCER DIAGNOSTIC AND TREATMENT OPTIONS

Amanda J. Callens, BS, LVMT
Knoxville, TN

INTRODUCTION

Urinary Tract Cancer

Urinary tract neoplasia includes cancer of the kidneys, ureters, bladder, and urethra. The bladder is the most common site and transitional cell carcinoma the most frequently diagnosed type of neoplasm. Primary renal neoplasms are uncommon. Other tumor types include squamous cell carcinoma, adenocarcinoma, and undifferentiated carcinoma.

Tumors of the urinary tract account for 2% of all canine malignant neoplasia. Of these tumors, 90% are transitional cell carcinoma.[1] The most common site for transitional cell carcinoma in dogs is in the trigone of the urinary bladder. The trigone is the triangular region of the internal urinary bladder and is made up of the right and left ureteral orifices and the urethral orifice. In cats, the bladder wall distal to the trigone is affected as often as is the trigone itself.[2] Most tumors are solitary, but may cover a large portion of the urinary bladder mucosa. They tend to form with the mass bulk growing in toward the bladder lumen. Transitional cell carcinoma is one of the most malignant neoplasms in domestic animals, with the most common sites for metastases being the lungs, lymph nodes, urethra, prostate, and bone.[3]

The causes of transitional cell carcinoma are unknown but are likely multifactorial. Causes have been linked to herbicide and pesticide exposure. The disease is seen more commonly in female dogs and dogs that are obese.[4] Breeds predisposed to transitional cell carcinoma include Scottish terriers and Shetland sheepdogs.

Clinical signs associated with urinary tract neoplasia tend to be consistent with lower urinary tract disease. These include hematuria, stranguria, pollakiuria, inappropriate urination, urinary incontinence, and dysuria. Diagnostics should be performed to definitively diagnose urinary bladder cancer. A small percentage of patients present with clinical signs related to metastatic disease, such as dyspnea from pulmonary metastases or lameness due to bone metastases.[1]

Growth of the primary bladder tumor is often problematic and can seriously affect the patient's quality of life. It is not uncommon for a bladder tumor to spread down the urethra, infiltrating the urethral lumen. In addition, as these tumors often sit in the trigone at the point of ureteral implantation, ureteral obstruction can occur secondary to tumor growth. Blockage of urine outflow from the ureters or the urethra will cause serious disease and is an emergency situation.

DIAGNOSTICS

Urinalysis

In patients where urinary disorders are suspected, urinalysis is an important part of the diagnostic evaluation. A large percentage of patients with bladder or urethral tumors have one or more abnormalities noted on urinalysis. These abnormalities include hematuria, pyuria, proteinuria, and bacteriuria. About 30% of dogs with transitional cell carcinoma have malignant cells detectable in the urine.[1] However, because reactive, non-neoplastic transitional cells can look similar to these malignant cells, cytologic analysis of the urine should be accompanied by other findings and histopathologic confirmation when possible.

An ELISA bladder tumor antigen test designed to aid in testing for lower urinary tract malignancies in dogs has been proven to be sensitive for distinguishing dogs with disease of the urinary tract from dogs without urinary tract disease. However, it is not specific for differentiating dogs with malignant disease from dogs with non-malignant urinary tract disease, such as urinary tract infection. Therefore, it is not recommended as a definitive diagnostic tool for detection of urinary tract malignancies in dogs.[5]

Hematology

Changes in blood chemistries and complete blood counts can be seen with bladder neoplasia. On blood counts, neutrophilia and anemia are possible. Anemia can be due to chronic inflammation or blood loss in the urine. Blood chemistry changes include azotemia, hypercalcemia and elevated liver enzymes.[1] Azotemia is likely post renal and due to some degree of ureteral or urethral obstruction, resulting in lack of urine outflow from the kidneys or bladder.[3] Increased alkaline phosphatase activity may occur with transitional cell carcinoma, likely associated with bone or liver metastasis.

Radiographic Techniques

Radiographic techniques can be useful in these cases, but contrast is often necessary, as these masses are of soft tissue density. When using contrast, filling defects are noted with space-occupying masses. When transitional cell carcinoma is suspected, three-view chest radiography is important to look for possible metastases. Other evidence of metastatic disease can be detected on radiography, including prostatomegaly, sublumbar lymphadenopathy, and pelvic or vertebral bone metastases.[3]

Ultrasound is also a useful imaging modality with transitional cell carcinoma. The advantage of ultrasound is its ability to evaluate internal luminal structures such as the kidneys, ureters, urethra, and urinary bladder without the use of contrast media.[6] Ultrasonography is used to visualize, quantify, and aid in fine needle aspiration of a mass for cytologic diagnosis. Ultrasound-guided transabdominal aspiration and cytology can be useful in diagnosing and differentiating bladder neoplasia. Some precaution should be taken, as there is a low risk of needle tract tumor implantation with this method.[7] Ultrasound also allows for visualization of the kidneys and ureters to rule out hydronephrosis and hydroureter secondary to obstruction of urine outflow. One limitation of ultrasonography is the inability to see the intrapelvic urethra, which is obscured by the pelvis.

With computed tomography, contrast is used to acquire images of the urinary tract, providing superior detail and contrast resolution. With this imaging modality, specificity and sensitivity of lesion detection and diagnosis are often improved compared to other imaging methods.[6] Computed tomography can be used to image urethral masses that would be difficult to characterize by other imaging modalities. Also, imaging of additional abdominal anatomy, such as lymph nodes, is useful to rule out metastatic disease.

Cystoscopy

Transurethral cystoscopy provides magnified images of the entire lower urogenital tract. In cases of lower urinary tract masses, information is collected regarding location, size, shape, and friability of the mass. Location is especially important because urethral or ureteral obstruction causes increased morbidity. Biopsies are obtained using cup biopsy instruments passed through the cystoscope's operating channel. In female dogs, tissue sampling is especially effective because of their larger urethral diameter, facilitating larger scopes and biopsy instruments, resulting in larger tissue samples. Male dogs are more of a challenge because of narrower, u-shaped urethral anatomy, limiting the size of instruments used and therefore smaller tissue samples. Impression smears of the biopsied tissue can be prepared for cytologic examination. Tissue sample submission for histopathologic exam is crucial for definitive diagnosis of transitional cell carcinoma.

Traumatic urethral catheterization or prostatic wash can be useful in differentiating urethral neoplasia.[1] During traumatic catheterization, a urinary catheter with side holes is passed transurethrally, and a syringe is attached to the catheter. Negative pressure is applied to the syringe while the catheter holes are in the area of the tumor. During a successful traumatic catheterization, tissue is aspirated through the catheter and into the syringe and submitted for histopathologic examination.

Cytology and Biopsy

Cytologic diagnosis can be obtained using urine samples, urethral catheterization and wash samples, tissue aspirates or impression smears of tissue obtained via biopsy. The changes seen in these samples include numerous, large, aplastic epithelial cells in clusters or individually with little to no inflammation. These cells are very large with marked cytologic and nuclear variability. Some cells may contain large cytoplasmic vacuoles. The more numerous these abnormal cells and the less inflammation seen increases the likelihood of neoplasia.[3] Histopathologic examination of larger biopsy pieces provides more definitive data.

TREATMENT

Medical

The mainstay of treatment for transitional cell carcinoma is medical therapy, which usually consists of cyclooxygenase (COX) inhibitors and possibly chemotherapy.[8] Most medical therapies are well tolerated and can lead to palliation of disease. Piroxicam is a COX inhibitor used as a palliative treatment for transitional cell carcinoma and has proven antitumor activity.[9] Mitoxantrone, carboplatin, and vinblastine are chemotherapeutic agents that have shown efficacy in palliating transitional cell carcinoma.[10-12] These chemotherapeutic medications are often used in conjunction with a COX inhibitor.

Surgical

Because of the typical location of these masses within the bladder trigone, surgical removal can be difficult. The trigone of the urinary bladder contains the ureteral openings and the urethral outflow tract, making surgical resection difficult or impossible. The decision of whether to pursue surgery should be made based on the location of the tumor, its histology, and its invasiveness. With transitional cell carcinoma, even grossly normal tissue often contains neoplastic tissue, so these tumors tend to recur shortly after resection.[13]

Radiation

Radiation therapy works by damaging genetic material within cancer cells, limiting their ability to successfully reproduce. Radiation therapy is often used in conjunction with medical therapy, using a multimodal approach to therapy. Image-guided radiation therapy has been shown to be generally well tolerated in dogs and in some studies has shown superior overall survival times.[14]

Interventional

Because of the risk of urine outflow obstruction in these patients, urinary diversion is used as a palliative treatment. Ureteral obstruction secondary to tumor growth in the bladder trigone over the ureteral orifice or into the ureter is seen in patients with transitional cell carcinoma. A ureteral stent can be placed to facilitate urine outflow from the kidney into the bladder. Another option for patients with ureteral obstruction is the placement of a Subcutaneous Ureteral System (SUB, Norfolk Medical). The SUB consists of a locking loop pigtail catheter that is placed in the kidney and a cuffed catheter that is placed in the bladder with a titanium shunting access port that connects the two catheters. Once this system is surgically placed, it facilitates urine flow from the kidney to the bladder, bypassing the affected ureter.

Urethral obstruction due to primary tumor growth down the urethra is a risk with transitional cell carcinoma. There are several options for urinary diversion in patients with urethral involvement. The placement of a urethral stent is a quick and effective way to achieve urethral patency. These self-expanding metallic stents are placed transurethrally under fluoroscopic guidance, avoiding a surgical incision. A urethral stent works by pushing the invading tumor tissue out of the urethral lumen. The stents are available in various sizes, so that only the affected portion of the urethra needs be stented. Depending on the urethral length stented, some patients are incontinent after stent placement. Alternatively, an indwelling urethral catheter can be placed and managed at home but has risk of predisposing patients to urinary tract infections. Surgically, a cystostomy tube can be placed, facilitating urine removal from the bladder. Cystostomy tubes and urinary catheters require emptying of the device by the client at home.

PROGNOSIS

Unfortunately, overall prognosis for patients with transitional cell carcinoma is poor. Most treatments are palliative and complete remission is unlikely. Survival time may not the best measure of success, but rather resolution of clinical signs. There are multiple modalities for diagnosis and treatment: traditional and modern, interventional and minimally invasive.

REFERENCES

[1]Norris A, et al. *J Vet Intern Med* 1992;6:145. [2]Wilson HM, et al. *J Am Vet Med Assoc* 2007;231:101. [3]Meuten D. Tumors in Domestic Animals. 4th ed. Ames, Iowa: Iowa State Press, 2002. [4]Glickman LT, et al. *J Toxicol Environ Health* 1989;28:407. [5]Billet JP, et al. *Am J Vet Res* 2002;63:370. [6]Bartges J, Nephrology and urology of small animals. In: MacLeod AG, eds. *Computed Tomography and MagneticResonance Imaging of the Urinary Tract*: Wiley-Blackwell, 2011. [7]Nyland TG, et al. *Vet Radiol Ultrasound* 2002;43:50. [8]Withrow S. Small animal clinical oncology. In: Knapp DW, ed. *Tumors of the Urinary System*: Saunders, 2012. [9]Knapp DW, et al. *J Vet Intern Med* 1994;8:273. [10]Henry CJ, et al. *Clin Cancer Res* 2003;9:906. [11]Boria PA, et al. *Vet Comp Oncol* 2005;3:73. [12]Arnold EJ, et al. *J Vet Intern Med* 2011;25:1385. [13]Henry CJ. Cancer management in small animalpractice. In: Henry CJ, ed. *Bladder and UrethralTumors*: Elsevier, 2010. [14]Nolan MW, et al. *J Vet Intern Med* 2012;26:987.

OSTEOSARCOMA: HOW TO SPARE THE LIMB

Sandra Axiak-Bechtel, DVM, DACVIM (Oncology)
Columbia, MO

INTRODUCTION

Canine osteosarcoma (OSA) is the most common primary bone tumor in the dog, and represents most cancers in the skeleton. The median age at diagnosis is 7 years, although a bimodal distribution exists with an increase in incidence at 18-24 months. Large to giant breeds are most commonly affected and higher risk breeds are the St Bernard, Great Dane, Irish Setter, Doberman pinscher, Rottweiler, German Shepherd, and golden retriever. Most OSAs occur in the appendicular skeleton; more specifically in the metaphyseal region of long bones. The distal radius and proximal humerus are the 2 most common sites, with distal femur and distal & proximal tibia being the most common sites in the hindlimb; less than 10% of OSAs are multicentric at the time of diagnosis[1].

The etiology of OSA has been studied, and many different factors have been found that may influence the incidence of this disease. Physical factors, such as major weight bearing bones, adjacent to late closing physes in heavy dogs, may have multiple minor trauma and subsequent injury in cells leading to mitogenic signals, increasing chance of mutant cells. An association with metallic implants, chronic osteomyelitis, and even fractures without internal fixation has also been found. Other causes are exposure to ionizing radiation (plutonium, radium), or a rare, late complication of ionizing radiation: 3% of dogs with OSA were treated for a soft tissue sarcoma with radiation therapy in that region. Molecular and genetic factors involved in OSA development are p53 (a tumor suppresser gene) that was found to be mutated and non-functional, alterations in growth factors, cytokines, and hormone signaling systems, alterations in matrix metalloproteinases , which are matrix degradative proteins that may allow disease progression and metastasis (MMPs 2 and 9 present in high concentrations in OSA), blood vessel density greater in those presenting with metastasis and increased plasma vascular endothelial growth factor (alterations in blood vessel formation), and telomerase reverse transcriptase gene upregulated in some OSA cell lines allowing cell immortalization.[1,2]

OSA is a malignant mesenchymal tumor of primitive bone cells that produces extracellular matrix of osteoid, which is the basis of histological diagnosis. Subclassifications, although not definitively prognostic, include osteoblastic, chondroblastic, fibroblastic, telangiectatic, and poorly differentiated. Radiographically, these are aggressive locally, causing bone production and bone lysis; soft tissue swelling common; and pathologic fracture can occur. It is rare for an OSA to cross the joint surface secondary to collagenase inhibitors limiting progression through the synovium. Differential diagnoses for neoplasia that does cross the joint surface would include synovial cell sarcoma and histiocytic sarcoma. OSA is aggressive systemically as well, with metastasis being common and the cause of loss of life in those dogs with adequate control of local disease and pain. Although less than 15% of dogs have radiographically detectable metastasis at diagnosis, 90% will die of metastatic disease. As a sarcoma, OSA moves through the hematogenous route and movement through lymphatics to local lymph nodes is rare; when lymphatic movement does occur the prognosis declines dramatically. The most common sites of metastatic disease are lung, other bones, and other soft tissue sites. Clinical signs when metastasis occurs may be vague and hypercalcemia is very rare with osteosarcoma.[1]

DIAGNOSIS AND STAGING

Presentation and differential diagnoses

Most dogs present with a history of lameness, sometimes following trauma, and swelling at primary site. Pain is caused microfractures and disruption of periosteum induced by osteolysis of cortical bone with tumor extension from the medullary canal. Diagnosis is first suspected on physical examination and radiographs of the lesion. Common radiographic findings are cortical lysis, soft tissue extension, soft tissue swelling, and new bone (tumor or reactive bone). Codman's triangle is a finding often seen with OSA; as tumor invades the cortex, the periosteum is elevated and new bone is laid down leading to a triangular appearing deposition of dense new bone on the cortex, at the periphery of the lesion. Please note that this finding is consistent with but not definitive for OSA. Differential diagnoses for these findings are other primary bone tumors (chondrosarcoma, fibrosarcoma, hemangiosarcoma), metastatic cancer to bone (transitional cell carcinoma, prostate carcinoma, mammary carcinoma), multiple myeloma or lymphoma of bone, systemic mycosis with bone localization, bacterial osteomyelitis, or bone cysts.

Diagnosis and staging

Diagnosis of OSA is confirmed by histopathology, although cytology is less invasive and can support the diagnosis. Unlike tumors in locations other than bone, it is important, regardless of method chosen, to sample the center of the lesion to obtain the best diagnostic sample. At the lesion periphery, reactive bone is present and may hinder confirmation of OSA. Staging is an important consideration for prognostic information, and may also determine treatment course. Although lymph node metastasis is rare, local draining lymph nodes should be palpated and aspirated if palpable. Chest radiographs, three views, are recommended to detect pulmonary nodules indicative of metastasis. A bone scan is a highly sensitive tool used to detect areas of bone metastasis; because a bone scan is so sensitive, radiographs of any suspicious area is recommended for confirmation. Any region of osteoblastic activity, including tumor, arthritis, or infection, will be identified on a bone scan. The staging scheme used for osteosarcoma is the following: Stage I: low grade lesion, no metastasis (rare); Stage II: high grade lesion, no metastasis; Stage III: regional or distant metastasis.

TREATMENT

When discussing potential treatment options with owners whose pets have been diagnosed with osteosarcoma, it is important to separate the treatments of location and systemic disease. Treatment of local disease is often the most pressing, as uncontrolled pain causes a decline in the quality of life. The first and foremost treatment option still discussed with owners is amputation, which takes away the major source of the dog's pain. Many dogs do quite well following surgery, and a significant improvement is often reported post-operatively, as dogs become pain free and adjust to life with 3 limbs. Complete forequarter amputation or coxofemoral disarticulation amputation is recommended for complete tumor excision.

ALTERNATIVE TREATMENT OPTIONS FOR LOCAL DISEASE

For some dogs, amputation is not a possibility due to owner reluctance or concurrent orthopedic or neurologic disease. However, many treatment options exist with the goal of palliation of pain and improvement in quality of life.

Surgical limb spare

Limb sparing surgery is one such option, if the lesion is of the distal radius or ulna. Limb spare surgery in other locations is not recommended as arthrodesis of the associated joint is required, and arthrodesis of the scapulohumeral, coxofemoral, stifle, or tarsal joints results poor function and a high complication rate. Additional criteria include OSA clinically and radiographically confined to the leg, where the primary tumor affects <50% of bone, no pathologic fracture, less than 360 degree involvement of soft tissues, and a firm/ definable mass (not edematous). Surgical options for replacement of the bone, following tumor excision, include frozen allograft (bone harvested into a bone bank), metal endoprosthesis , pasteurized tumor autograft(excision of the tumor, followed by pasteurization at 65C for 40 minutes, followed by reimplantation), longitudinal bone transport osteogenesis, or ulnar transposition limb sparing of a distal radial OSA[2]. Potential complications of limb sparing techniques include infection, implant failure, and tumor recurrence.

External Beam Radiation Therapy

Should limb spare not be an option, radiation therapy has been used with success in palliation of pain for the treatment of OSA. There are no site restrictions, and radiation therapy can be used to treat one or multiple sites in the palliative setting. A bone scan can be recommended prior to treatment to complete staging and to treat multiple sites if indicated. Protocols reported include accelerated, 2-, 3-, and 4-fraction protocols. Overall, palliative protocols provide 50-93% response rates for a median duration of analgesia of 53-180 days.[3]

Furthermore, radiation therapy has been used in a definitive treatment setting. The goal of definitive radiation therapy is complete local tumor control. A median survival time of 209 days for fractionated high dose radiation and chemotherapy for systemic therapy was reported.[3] Stereotactic radiation therapy has also used to treat canine osteosarcoma. The advantage to stereotactic radiosurgery is the ability to deliver very high doses of radiation therapy to the tumor, while sparing normal tissues. In this report, acute side effects (hair loss, irritation) were mild to moderate, and in a small group of dogs the median survival time of stereotactic radiation therapy with follow up chemotherapy was 363 days. The main complication was pathologic fracture of the treated site.[3]

Samarium (Quadramet®)

Samarium-153-ethylenediamine-tetramethylene-phosphonic acid (^{153}Sm-EDTMP) is a bone seeking radioisotope administered intravenously. This treatment is commonly used in people for palliation of metastatic cancer to bone. It is used with success in institutions if the ratio of tumor dose to surrounding tissues is favorable – determined using a bone scan. This particular treatment provides pain relief, and in some cases, tumor growth delay. Administered intravenously, it is particularly helpful in cases of multiple bone site involvement, or in cases of other tumors metastasizing to bone. One important side effect to note is transient bone marrow depression following treatment, which will delay chemotherapy. This particular treatment also requires isolation in a radiation facility for 5-7 days before release. [4]

Bisphosphonates

One final treatment option for localized disease and pain control is pamidronate, which is a bisphosphonate that inhibits bone resorption by binding to hydroxyapatite crystals, inhibiting further calcium and phosphorous dissolution. Furthermore, it blocks osteoclastic activity and induces apoptosis of osteoclasts. The result is inhibition of bone resorption.

This particular treatment can also target multiple OSA sites and can be used in combination with the above mentioned treatments. Pamidronate is renally excreted and administered over 2 hours as an intravenous infusion; renal parameters should be checked prior to each monthly infusion.[5]

Pain control palliation

The last and least aggressive option for treatment of local disease in dogs diagnosed with OSA is pain control alone with a combination of non-steroidal anti-inflammatory drugs and opioids. These can be used alone for those pet owners not wanting to pursue more aggressive therapy or in combination with the other treatments discussed above.

Pathologic Fractures

When any of the above limb sparing options fail or disease progresses, amputation is the treatment of choice. However, in some dogs, this is still not a viable option. A recent study reported internal and external fracture repair in dogs with a pathologic fracture due to osteosarcoma or undifferentiated sarcoma of the appendicular skeleton. Some dogs experiences pain relief associated with fracture fixation, and, should amputation be ruled out due to concurrent disease or owner's wishes, this may provide an alternative for improvement in quality of life.[11]

CHEMOTHERAPY FOR MICROMETASTATIC DISEASE

Chemotherapy is used with the goal of delaying the onset to OSA metastasis. Although not effective at controlling local disease, chemotherapy has been shown to extend median survival time in both amputees and limb spares. Doxorubicin, carboplatin, cisplatin, or combinations of doxorubicin with a platinum agent are those most commonly recommended in practice at this time. With local disease control using amputation or limb spare followed by a course of chemotherapy using a platinum containing drug, MSTs range from 8-10 months. In contrast, the prognosis with amputation alone is MST 4-5 months.[6-10]

Prognostic factors

Many prognostic factors are helpful to discuss with owners in regards to potential and expected outcomes following suspicion of, or confirmation of osteosarcoma. The first prognostic indicator is location, and a list of locations and expected outcomes are listed below; MST is median survival time.

- Mandible: mandibulectomy alone median survival time (MST) 1year
- Maxilla: maxillectomy alone MST 5 months
- Orbital: complete excision, long term survival
- Rib: surgical excision only MST 3 months, surgery + chemotherapy MST 8 months
- Scapula: poorer prognosis
- Distal to carpus/ tarsocrural joint: Surgical excision followed by chemotherapy: MST 466 days
- Soft tissue OSA: surgery alone MST 1 month, surgery with chemotherapy MST 5 months
- Mammary OSA: surgery alone MST 3 months
- Vertebral: surgery, radiation therapy, and chemotherapy: MST 4 months

Additional negative prognostic factors are stage III disease (metastatic disease), and dogs with lymph node metastasis have a MST of 59 days with surgery and chemotherapy compared to a MST of 318 days in dogs without lymph node metastasis. Elevation in alkaline phosphatase (ALP) is a consistent poor prognostic factor in many different studies. A pre-operative elevation in total ALP (>110 U/L) or the bone isoenzyme (>23 U/L) revealed a shorter disease free interval and MST; furthermore, dogs with increased ALP pre-operatively that did not return to normal within 40 days post-operatively failed earlier than dogs with normal ALP due to metastatic disease. In dogs with appendicular OSA, those dogs less than 5 years of age had a shorter survival than dogs older than 5 years when treated with amputation alone. And in OSA of the flat bones, small dogs with complete excision have a better prognosis than larger dogs.

CONCLUSIONS

Presentation and radiographs can lead to an increased index of suspicion for OSA, and cytology or histopathology can be used to confirm suspected diagnosis. Overall, important considerations to discuss with dog owners when faced with treatment decisions are the facets of localized disease control and systemic disease control. Local disease control to provide pain relief and quality of life is the most important aspect of treatment. Although amputation is still our current standard many different treatment options are available for pain and tumor control. These options include limb spare (if a distal radial or ulnar lesion), external beam radiation therapy, Samarium, and Pamidronate. In addition, systemic disease control using chemotherapy should also be discussed to maximize quality of life time. The prognostic factors discussed above can be very useful in aiding pet owners in the decision making process.

REFERENCES

[1]Chun R et al **Vet Clin Small Anim** 2003; 33:491-516. [2]Selvarajah GT et al. **The Vet Journal** 2010: 185: 28-35. [3]Coomer A et al.**Vet Comp Onc** 2009; 7: 15-27. [4]Barnard SM et al. **J Am Vet Med Assoc** 2007, 230: 1877-1881. [5]Fan TM et al. **J Vet Intern Med** 2007, 21: 431-439. [6]Spodnick GJ et al. **J Am Vet Med Assoc** 1992; 200 (7) 995-999. [7]Bacon NL et al. **J Vet Am Med Assoc** 2008, 232: 1504-1510. [8]Chun R et al. **J Am Ani Hosp Assoc** 2005: 41(6); 382-7. [9]Saam DE et al. **J Am Vet Med Assoc** 2011, 238: 195-206. [10]Berg J et al. **J Am Vet Med Assoc** 1992, 200: 2005-2008. 11Boston SE et al. **Veterinary Surgery** 2011: 40: 431-7.

Cachexia in the chronically ill patient

Ann Wortinger BIS, LVT, VTS (ECC, SAIM)
Dearborn, MI

INTRODUCTION

Cachexia is a multifactorial syndrome characterized by progressive weight loss that is often, but not always, accompanied by anorexia (Tisdale 2009). Unlike weight loss seen with starvation or anorexia, cachexia is distinguished by a loss of adipose tissue with accompanying loss of lean body mass, primarily muscle. Non-muscle protein as found in the organs is preserved in cachexia but not in starvation. In dogs with illness or injury, amino acids from lean body mass are the primary source of energy as opposed to glucose from carbohydrates during normal digestion (Freeman 2006). Loss of adipose tissue in cachexia is due to increased lipolysis, increasing the overall loss of body fat. Significant loss of mineral content in the bones can also be seen contributing to the overall weakness found in many patients (Tisdale 2009). The beginning stages of cachexia can be very subtle, even in obese animals and many clients may perceive these changes as simply 'growing old' by their pets (Freeman, Rush 2012). When evaluating an animal for cachexia it is important to remember than cachexia is a process, the loss of lean body mass, and not an end-stage syndrome (Freeman 2012). A similar syndrome known as sarcopenia is also commonly seen in dogs and cats- it is differentiated from cachexia in that sarcopenia is the loss of lean body mass associated with aging in the *absence* of disease (Freeman 2012).

This loss of lean body mass has a deleterious effect on strength, immune function and overall survival (Freeman, Rush 2012). Weight loss provides an important prognostic indicator in overall survival of the patient. Increased weight loss is inversely proportional to survival time (Tisdale 2009, Case 2011). The loss of muscle mass can first be noticed over the epaxial, gluteal, scapular and temporal muscles (Freeman 2006) and is easily detected during a routine physical exam. There does not appear to be a cause and effect relationship between anorexia and cachexia, with weight loss often exceeding that expected with simply a decrease in caloric intake (Tisdale 2009).

Cachexia is a profound state of general illness, malnutrition and disability that can affect animals with cancers, cardiac disease, renal disease and other significant illnesses and injuries (Saker 2010). Loss of 25-50% of the lean body mass compromises the immune system and affects muscle strength, with death resulting from infections, pulmonary failure or both (Saker 2010).

CAUSES

The primary underlying cause for cachexia is alterations in the body's metabolism of carbohydrates, fats and proteins (Case 2011). These metabolic alterations can, over time, lead to anorexia, fatigue, weight loss, impaired immune function and malnutrition (Case 2011). As cachexia is often due to various cancers, the tumors can actually be inducing many of these changes occurring within the patient's body.

In the case of cancers, most cancer cells use anaerobic glycolysis to generate ATP for energy at a higher rate than normal tissues use (Tisdale 2011). For the most part these cancer cells are unable to obtain significant amounts of energy from either aerobic glycolysis or fat oxidation (Case 2011). The end result of this process is a gain of energy by the tumor cells and a net loss of energy for the host, and generation of large amounts of lactic acid that the host must still convert to glucose (Case 2011).

ASSESSMENT OF CACHEXIA

The presence of cachexia increases the morbidity and can adversely affect the quality of life a patient experiences (Freeman 2012). Being able to identify the early stages and begin interventions sooner can be important in the overall success of treatment and happiness of the patient and their owner.

PHASES OF CACHEXIA

Three phases of cachexia have been identified in humans; we presume that dogs and cats also follow the same pattern. During the first phase, the patient does not exhibit any clinical signs, but the biochemical changes are already occurring. These include elevations seen in lactate levels through the glycolysis process, an increase in insulin levels, causing peripheral insulin resistance, and alterations in amino acid and lipid profiles (Case 2011).

During the second phase of cachexia, clinical signs can become evident presenting as anorexia, weight loss, and depression. Many owners attribute these early signs as their pet 'getting old' and do not recognize the clinical significance. The final phase is characterized by marked loss of body fat and protein stores, severe debilitation, weakness and biochemical evidence of negative nitrogen balance (Case 2011). If left untreated cachexia can be the ultimate cause of death.

BIOCHEMICAL CHANGES

The biochemical changes caused by the cancer or disease process lead to inefficient energy utilization by the patient and enhanced energy use by the tumor or disease process (Case 2011). The prevalence of glucose intolerance and insulin insensitivity mandate the need to limit and carefully select the type of carbohydrates used in foods for these animals. With both the patient and the tumors having obligatory protein requirements, negative nitrogen balances are often seen in cachexic patients. If this nitrogen imbalance is not corrected, it will eventually lead to increased loss of skeletal muscle, hypoalbuminemia, compromised immunity, impaired gastrointestinal function and delayed wound healing as would be seen in any condition that caused impaired protein utilization by the body. With cancer, several tumor-derived factors and cytokines have been identified that influence this increased protein breakdown (Case 2011).

Loss of lean body muscle accounts for the majority of weight loss seen in animals and humans with cancer cachexia. Though reduced food intake is a significant contributor to this loss, there is also decreased nutrient absorption and increased metabolic alterations that trigger muscle loss rather than fat loss (Freeman 2012). One of these metabolic alterations is also a decrease in leptin production; this exerts an inverse relationship related to the intensity of the inflammatory response created by the body. As leptin levels decrease, the inflammatory mediators increase causing increased production of the inflammatory cytokines (Tisdale 2009). These cytokines contribute to the anorexia seen in the patient, increase energy requirements and increase catabolism of lean body mass (Freeman, Rush 2012). The cytokines most commonly identified include proteolysis induction factor, tumor necrosis factor-alpha and interleukin-1-beta (Case 2011).

Tumors obtain energy primarily through anaerobic metabolism of glucose, this process results in an excess production of lactate in the animal (Case 2011). This lactate must then be recycled at additional energy cost by the host, resulting in a net energy loss. This process is not the only energy wasting process that a cachexic animal undergoes. Additional energy loss can included cytokine-induced increases in glucose recycling, protein degradation and overall energy expenditure (Case 2011). Since cachexia develops in stages as previously outlined, and biochemical alterations typically precede clinical signs, animals that are in the preclinical stage would be expected to have normal energy requirements. Those that are in stage 2, active untreated cachexia, would be expected to have elevated energy expenditures, while those in the final stages may actually be hypometabolic (Case 2011).

DIETARY MANAGEMENT OF CACHEXIA

While it is not possible to reverse the wasting process with nutritional supplements alone, our patients do need to be fed, and through manipulation of nutrients benefits can be seen in some of the changes the body is undergoing (Tisdale 2009). Adequate nutrition can be a key to management of cachexia with provision of calories, protein, fat and modulation of cytokine production (Freeman, Rush 2012). Specific dietary recommendation should consider the stage of disease, the patient's energy needs, current and past nutritional status and ability or willingness to eat (Case 2011).

The caloric distribution in the food should emphasize calories obtained from fats and proteins, rather than carbohydrates since we know that glucose is the preferred fuel for tumor cells, and fatty acids and amino acids are not (Case 2011). Our goal is to feed the patient, and starve the tumor cells. Ideally a food should contain 50-60% of the calories from fat, 30-50% of the calories from protein and the remaining portion from carbohydrates (Case 2011).

Studies using animal models have shown that supplementation with EPA and DHA can help to prevent cachexia and metastatic disease processes (Case 2011). The fatty acids obtained from the omega-3 class of fats produce less potent inflammatory mediators than do the omega-6 class of fatty acids. When the body uses the omega-3 class of fatty acids to produce the cytokines, the inflammatory response produced is decreased in proportion to the level of omega-3 fatty acids in the diet. Dr. Freeman recommends a dose of 40 mg/kg of EPA, and 25 mg/kg of DHA. Using the common formulation of most fish oil capsules, to achieve this concentration you would need to provide ~ 1 gram capsule/10# body weight (Freeman, Rush 2012). Many recovery diets already have this incorporated into the diet and additional supplementation is not recommended.

Protein levels in the diets should be at the upper limits recommended by AAFCO, and of a high quality to be easily digestible and utilizable by the body. Protein levels in the diet of 30-50% DM would be recommended, with dogs at the lower end, and cats at the higher end (Case 2011). Looking at protein levels from an energy standpoint, the minimum recommend intake is 5.14 grams/100 kcal with 6-7 grams/100 kcal preferred (Freeman, Rush 2012).

As stated previously, tumor cells preferentially use glucose for energy. Selecting a carbohydrate with a lower glycemic index would provide a slower release of carbohydrate generated glucose into the blood stream than would those with higher glycemic indexes. Rice is one of the carbohydrates with a highest glycemic index, with barley, sorghum and corn having much lower glycemic indexes (Case 2011). Moderately soluble fibers can also be used to slow the entrance of glucose into the blood stream with the added benefit of providing additional short fatty acids specifically for the enterocytes.

FEEDING METHODS

Enteral feeding is always the preferred delivery method, provided the patient has a functional gastrointestinal tract. Oral feeding of a canned or dry pet food would be the first choice with cachexia patients (Case 2011). When a patient is unwilling or unable to consume the desired amount of food orally, various feeding tubes can be used. The specific tube selected will vary based on the patient, the desired length of use and the owner's willingness to feed at home. Diet selection would also be based on the route selected for feeding. Energy density of a diet being fed through a feeding tube can directly affect the success of feeding- a food that has a low energy density, may actually contribute to vomiting or osmotic diarrhea in the patient due to high feeding volumes or poor digestibility of the diet (Case 2011). Feeding multiple, smaller meals frequently is easier on the patient, though this may not be easier on the client. Smaller meals decrease the incidence of nausea related to feeding, increases the transit time of food out of the stomach and ensures that the nutrition is actually usable by the animal (Case 2011)

Preparation of home-cooked diets can be used short term to tempt the patient to eat, but should only be used long-term after consultation with a nutritionist. Home-made diets are frequently unbalanced, and this is to the detriment of the animals. Being able to calculate the energy density, nutrient levels and micronutrient amounts is also difficult if not impossible for the average clinician, technician or client. A Board certified veterinary nutritionist can help a client to formulate a diet that the animal will accept that will also provide the best possible nutrition. There are also a number of excellent therapeutic diets available that fit the recommended diet profile and can be fed both orally and via tube feeding.

CONCLUSION

While cachexia is a fairly obvious sign that there are biochemical imbalances within the body, these changes have been occurring for a while by the time we see the physical changes. We need to continue to feed our patients through disease and cancer treatment. By recognizing the changes these processes have caused within the body, our diet recommendation can be used to help the patient rather than the tumor or disease process resulting in more positive outcomes and happier patients and clients.

REFERENCES

Case L, Daristotle L, Hayek M, Raasch M. (2011) Nutritional care of cancer patients in Canine and Feline Nutrition 3rd ed. Mosby Elsevier, Maryland Heights, MO. pp 479-486

Freeman LM, Rush JE (2006) Cardiovascular diseases: nutritional modulation in Encyclopedia of Canine Clinical Nutrition (eds Pibot P, Biourge V, Elliott D). Royal Canin, Aimargues France EU. pp 321- 336

Freeman LM, Rush JE. (2012) Nutritional Management of Cardiovascular Diseases in Applied Veterinary Clinical Nutrition (eds Fascetti AJ, Delaney SJ). Wiley-Blackwell Ames, IO. pp 304-307

Freeman LM. (2012) Cachexia and Sarcopenia: Emerging syndromes of importance in dogs and cats in JVIM 2012:26:3-17

Saker K, Remillard RL. (2010) Critical Care Nutrition and enteral-assisted Feeding in Small animal Clinical Nutrition 5th ed (eds Hand MS, Thatcher CD, Remillard RL, Roudebush P, Novotny BJ). Mark Morris Institute Topeka KS. pp 441- 442

Tisdale M. (2009) Mechanisms of Cancer Cachexia. Physiology Review April 2009 vol 89 no. 2 pg 381-410

BENEFITS OF FISH OILS AND FATTY ACIDS IN CARDIAC PATIENTS WITH VALVULAR DISEASE

Kara M. Burns, MS, MEd, LVT, VTS (Nutrition)
Wamego, KS

Cardiovascular disease is a common disorder in dogs and cats, with 11% of canines and up to 20% of feline populations affected by cardiac disease.[1,2] Chronic valvular disease has been found to be the most common acquired heart disease in dogs with an overall incidence greater than 40%. [3] Chronic mitral valvular disease is the most common acquired cardiac abnormality in dogs, affecting more than one-third of patients over 10 years of age. [4] In approximately 30% of cases the tricuspid valve is involved. However, disease of the tricuspid valve is usually less severe. Valvular disease is more prevalent in small breed dogs. Acquired valvular disease in cats is rare. Following the discovery that taurine deficiency was the principal cause of dilated cardiomyopathy in cats in 1987, the prevalence of this disease has decreased significantly. [3,4] Now hypertrophic and restrictive cardiomyopathies are more prevalent causes of myocardial failure in cats.

In human medicine, there is substantial evidence supporting the positive effects of omega-3 fatty acids in primary and secondary prevention of cardiac disease. Recent evidence in veterinary medicine has been supportive of the use of Omega-3 fatty acids in dogs with cardiac disease.
The benefits of omega-3 fatty acids include anti-inflammatory effects and anti-arrhythmic effects which help in managing the patient's lean body mass and arrhythmias, two problems associated with cardiac disease.

Dietary fat is important for energy, as a carrier of fat soluble vitamins and is a source of essential fatty acids. Although the term 'fat' in general offers negative connotations in today's society, dietary fat is essential for life. Dietary fat is the most concentrated form of energy in pet foods, with fat providing 2.25 times (9 kcal/gram) the metabolizable energy of proteins and carbohydrates. Fats affect immune function, inflammation, and hemodynamics. Fat, especially omega-3 fatty acids, has also been shown to affect the cardiovascular system. [5,6]

Fatty Acids

The structure of lipids ranges from simple to complex. Hydrocarbon molecules are basic subunits of lipids and are linked by covalent bonds in various manners to themselves and to other molecules, thus resulting in the numerous functions and structures occurring in the animal.[7]

Fatty acids in the omega-3 fatty acid family have the first double bond between the third and fourth carbon; Omega-6 fatty acids have the first double bond between the sixth and seventh carbon; and the omega-9 fatty acids have the first double bond between the ninth and tenth carbon.[5-7] The omega-3 and omega-6 fatty acid families are essential fatty acids (EFA) because they cannot be synthesized in animals. Members of the omega-6 family include linoleic acid (18:2n-6), g-linolenic acid (18:3n-6) and arachidonic acid (20:4n-6). Dogs are able to synthesize linoleic acid to form arachidonic acid. Consequently, linoleic acid is customarily listed as an EFA for dogs. However in cats, both linoleic acid and arachidonic acid are EFAs. Omega-6 fatty acids are necessary for growth, reproduction and precursors of eicosanoid and prostaglandin synthesis. Omega-3 fatty acids include a-linolenic (18:3n-3), eicosapentaenoic (20:5n-3) (EPA) and docosahexaenoic (22:6n-3) (DHA) acids. Omega-3 fatty acids are needed for proper brain and retinal function. Omega-3 and Omega -6 fatty acids are responsible for cell membrane fluidity and skin health.

The make-up of the diets for dogs, cats, and humans is primarily omega-6 fatty acids. Nevertheless, dietary adaptation or supplementation of omega-3 fatty acids can significantly increase omega-3 fatty acid concentrations in the blood and tissues as omega-3 and omega-6 fatty acids compete with one another for enzymes required for their metabolism. As a result, the higher the amount of omega-3 fatty acids in the diet, the more omega-3 fatty acids will be utilized and incorporated into cells.[5,7]

Nutrition is an important part of management of cardiac disease and each cardiac patient should have a nutritional assessment and specific dietary recommendation by the veterinarian and the veterinary technician as part of their overall management. The goals of nutritional management of cardiac patients include; 1.) maintaining the pets' ideal body condition, 2.) maintaining appropriate balance of nutrients, and 3.) receiving benefits from pharmacologic doses of certain nutrients. It is exciting to note that omega-3 fatty acids play a role in each nutritional goal. [5] Although cardiac cachexia is prevalent in cardiac disease, this discussion will review the role of omega-3 fatty acids in cardiac disease with respect to arrhythmias, valvular disease, etc. Cardiac cachexia and the role of nutrition and fatty acids specifically deserve its own presentation and discussion.

Fatty Acid Deficiencies and Cardiac Disease

It has been noted that when compared to normal dogs, dogs diagnosed with heart failure had a deficiency of plasma EPA and DHA.[5,8] It was also shown that in dogs with heart failure secondary to dilated cardiomyopathy (DCM), the supplementation of fish oil for eight weeks stabilized plasma fatty acid abnormalities. [5,8] Although the use of omega-3 fatty acids helps to improve myocardial energy metabolism and mitochondrial function is in its infancy, the results thus far have been promising. Therefore in an attempt to correct deficiencies in cardiac patients and to benefit energy metabolism, omega-3 fatty acids may be supplemented. [5,6]

Arrhythmia

To date, studies in humans have correlated the benefits of fish consumption with the reduction in risks for developing coronary heart disease and subsequent death from myocardial infarction. [5,6] Human studies historically have looked at fatty acids and ventricular arrhythmias. However, recent studies have looked at and have shown benefits of omega-3 fatty acids on atrial fibrillation. Results suggest that omega-3 fatty acids reduce atrial fibrillation. Human populations which eat higher amounts of fish or have higher omega-3 fatty acid concentrations were linked to a lower incidence of atrial fibrillation

There has only been one study reported to date that looked at the effect of omega-3 fatty acids on naturally occurring disease in dogs. In this study, the number of ventricular arrhythmias in Boxers was reduced. This reduction was noted after six (6) weeks of fish oil supplementation as compared to the control - sunflower oil. Obviously, more research will need to be done as this is only one study and with only one breed of dog. Therefore, although the results of omega-3 fatty acids on arrhythmias has been positive, at this point, omega-3 fatty acids should only considered as an addition to medical therapy for dogs with significant arrhythmias.

Omega-3 fatty acids and the overall effect on the cardiac patient

Omega- 3 fatty acids have been shown to have a significant effect on survival times when used in dogs diagnosed with DCM or chronic valvular disease (CVD).[9] The effect of the omega-3 fatty acids may be attributed to: anti-inflammatory effects, cachexia prevention, improved appetite, or anti-arrhythmic effects. The veterinary healthcare team should also be aware of further effects of omega-3 fatty acids on the patient. The healthcare team must be mindful of the fact that omega-3 fatty acids have the potential to alter immune function. This alteration in immune function may contribute to the cardiovascular effects of omega-3 fatty acids. Also, omega-3 fatty acids reduce platelet aggregation resulting from the production of thromboxane B5. The reduction in platelet aggregation might be of benefit in cats with cardiac disease and at risk for thrombus formation. However, this effect is also important to be mindful of when using omega-3 fatty acids in animals with coagulopathies.

The discussion has only breached the tip of the research of the effects of fatty acids and cardiac disease. Although more studies and discussion is needed in the long term, it is believed that dogs and cats with cardiac disease may benefit from omega-3 fatty acid supplementation. However, the healthcare team must consider a number of factors: 1) dose, 2) timing, and 3) omega-3 fatty acid form.

At this point in time, no optimal dose of omega-3 fatty acids has been established for humans, cats, or dogs. The current recommendation from nutritionists studying fatty acids and cardiac disease is a dose of 40 mg/kg EPA and 25 mg/kg DHA for both dogs and cats.[5]

Timing also needs to be taken into consideration when supplementing omega-3 fatty acids. The healthcare team should remember and educate owners that the majority of omega-3 fatty acids benefits occur after peak plasma and tissue concentrations have been achieved. Although plasma concentrations may increase significantly in the first week of omega-3 fatty acid supplementation, typically 4 to 6 weeks are required to reach peak plasma concentrations.

EPA and DHA can be provided via the diet or as a dietary supplement. There are a few therapeutic pet foods with high levels of EPA and DHA, but the majority of foods manufactured today do not achieve the recommended level of EPA and DHA. The current recommended dose is 40 mg/kg EPA + 25 mg/kg DHA. Therefore a manufactured food would need to contain between 80 and 150 mg/100 kcal EPA + DHA.[5] Other factors that would need to be taken into consideration would be the size of the pet and the amount of food consumed. If the pet is not prescribed one of the high fatty acid foods, a recommendation of fish oil supplementation would be necessitated. However, caution must be given when making a supplement recommendation, as fish oil supplements vary widely in the amount of EPA and DHA they contain. The healthcare team should be familiar with various brands of fish oil supplements and plan to make a recommendation based on a specific brand with which the concentration of EPA and DHA have been researched and confirmed.

Fatty acids – especially omega-3 fatty acids – have been shown to provide benefit to dogs and cats suffering with cardiac disease. Although more research is needed, omega-3 fatty acids should and have been recommended – not only for dogs and cats, but to a large degree in humans with cardiac disease as well. Following the correct dose and timing of omega-3 fatty acids, and being cognizant of the risks (such as the potential reduction in platelet aggregation) is beneficial to overall patient care.

REFERENCES

1. Buchanan JW. Prevalence of cardiovascular disorders. In: *Textbook of Canine and Feline Cardiology. 2nd ed.* Eds P. R. Fox, D. Sisson and N. S. Moise. 1999. Saunders, Philadelphia, PA, USA. pp 457-470.
2. Paige C F, Abbott JA, Elvinger F, Pyle R L. Prevalence of cardiomyopathy in apparently healthy cats. *Journal of American Veterinary Medical Association*, 2009, 234, pp.1398-1403.
3. Rush JE. Chronic Valvular Disease in Dogs. In *Kirk's Current Therapy XIV, 14th ed.* Bongura JD, Twedt DC, eds. 2009. Suanders Elsevier, St. Louis, MO
4. Roudebush P, Keene BW. Cardiovascular Disease. In *Small Animal Clinical Nutrition, 5th ed.* Hand M, Thatcher C, Remillard R, Roudebush P, Novotny B., eds. 2010. Mark Morris Institute, Topeka, KS.
5. Freeman LM, Beneficial effects of omega-3 fatty acids in cardiovascular disease. *Journal of Small Animal Practice.* Vol 51. September 2010.pp. 462 – 470.
6. Lunn J, and Theobald HE. The health effects of dietary unsaturated fatty acids. 2006 *British Nutrition Foundation Nutrition Bulletin*, 31 , pp. 178–224.
7. Gross KL, Yamka RM, Khoo C., et al. Macronutrients. In *Small Animal Clinical Nutrition, 5th ed.* Hand M, Thatcher C, Remillard R, Roudebush P, Novotny B., eds. 2010. Mark Morris Institute, Topeka, KS.
8. Freeman L M, Rush J E, Kehayias J J, Ross, J N JR, et al. Nutritional alterations and the effect of fish oil supplementation in dogs with heart failure. *Journal of Veterinary Internal Medicine*, 1998, 12, pp.440-448.
9. Slupe J L, Freeman LM, Rush JE. Association of body weight and body condition with survival in dogs with heart failure. 2008. *Journal of Veterinary Internal Medicine,* 22,pp. 561-565

WEIGHT & OBESITY: IMPACTS ON DISEASE

Vicky Ograin, MBA, RVT
Topeka, KS

INTRODUCTION:

Obesity is a disease and is increasing in pets at an alarming rate. Obesity is thought to be the number one cause of malnutrition in dogs and cats.[1] An estimated 54% or 93 million dogs and cats in the United States are overweight or obese[2]. Dogs and cats become overweight when they consume too many calories, in excess of their caloric expenditure. Pets are overweight if they are 10-20% above their ideal weight; 20-30% over ideal weight is considered obese.

Fat is the largest endocrine organ and is recognized as the most abundant source of hormones in the body. Overweight and obese pets are in a chronic state of inflammation. Adipose tissues expand rapidly and adipocytes enlarge when a pet is overweight. When a pet is obese, adipocytes will release cytokines, adipokines and free fatty acids. The inflammatory state is increased locally and systemically within adipose tissue, liver, muscle, and it causes insulin resistance.[3] Obesity can lead to many life threatening diseases.

There are many tools to determine overweight and obesity levels in pets. A new tool, morphometric measurements, is available to accurately determine a pet's ideal weight; this will allow an accurate determination of the amount of food a pet should receive to achieve weight loss. Feeding the correct amount will lead to greater weight loss success. There are many weight loss food options to help pets reach their ideal weight. Technicians play a vital role in communicating with the pet owner and assisting with their pet's weight loss.

RISK FACTORS:

Taking in too many calories is a risk for obesity. Owners contribute to excess calories by free- feeding; also giving too many treats and table food could contribute calories, and in excess may lead to an unbalanced diet. The activity level (or lack thereof) can contribute to weight gain. Most owners feel their dogs to be moderately or very active when, in fact, they may not be very active. Most cats are kept indoors and have limited activity; also pets that are unable to exercise due to a physical condition are at increased risk of gaining weight. Pets on long term central nervous system depressants or corticosteroids are at increased risk for weight gain. Age can be a risk factor for gaining weight because as pets age, many become less active.[1]

Pets are at risk for genetic and environmental causes of obesity. Certain breeds are predisposed to weight gain: Labrador retrievers, golden retrievers, cairn terriers, cocker spaniels, long- haired dachshunds, Shetland sheepdogs, basset hounds, cavalier King Charles spaniels, pugs, Dalmatians, beagles and cats of mixed breeding.[1]

Spayed female dogs are about twice as likely to be overweight than are intact female dogs. Similar trends have been seen in castrated male dogs. Male cats are predisposed to being overweight; neutering further increases the risk of obesity by decreasing the metabolic rate by at least 25%. Removal of estrogens may also increase food consumption, independent of the decreased metabolic rate. This may also be accompanied by an increased appetite following de-sexing surgery[3].

HEALTH RISK OF OBESITY:

Overweight and obese pets are at greater risk of disease, and life expectancy is decreased. The chronic low grade inflammation may affect many body systems. Cytokines and oxidative stress are a common thread to many diseases related to obesity. Cytokines are a product of adipocytes secretions; these are called fat-derived peptides. Obesity amplifies long term oxidative stress on DNA, proteins and cell membranes causing many diseases: cancer, urinary tract disease, heart disease, diabetes mellitus and liver disease. Obese dogs are at increased risk of high triglycerides, cholesterol and altered lipoprotein; these decreased when the dogs lose weight. Congestive heart failure has been related to obesity. High blood pressure has been demonstrated in experimental conditions, as well as structural changes in the heart have been demonstrated after weight gain over only nine weeks in dogs with obesity related hypertension. Anorectic obese cats are prone to hepatic lipidosis. Obese dogs and cats have a decreased heat tolerance and stamina. Other conditions thought to be related to obesity include dystocia, dyspnea, dermatological concerns, lowered immune response, oral disease and increased anesthetic risk. Overweight dogs are twice as likely to form bladder stones. Overweight cats are three times as likely to form calcium oxalates.[1]

Overweight dogs also may be at risk for renal pathology and pancreatitis. Dogs and cats that are overweight have been shown to be at increased risk of certain cancers: transitional cell carcinoma of the urinary bladder[4] and mammary tumors.[3]

A lifelong study on paired dogs showed the lean dogs lived a median life span of 13 years compared to 11.2 years for the moderately overweight dogs[1]. The dogs in the lean group had decreased risk of osteoarthritis and hip dysplasia as well as improved glucose tolerance, compared to the overweight dogs[2].

Overweight cats have a fourfold risk of developing diabetes. Abnormal glucose tolerance and insulin secretion have been reported in both overweight dogs and cats. Dogs and cats have developed glucose intolerance and hyperinsulinemia when they have gained weight, but it resolves when they lose the excess weight. While dogs are most often affected with type I diabetes, cats develop type II diabetes, which is associated with increased weight.[4]

Orthopedic disorders are increased in obese pets. Obese dogs required long-term treatment for osteoarthritis 50% sooner when the dog was obese. The extra weight causes stress on the joints and produces many inflammatory mediators that could possibly cause osteoarthritis.[1] Osteoarthritis is improved when weight is reduced. Preventing or treating obesity may delay and/or prevent many of these obesity-related diseases.

NEW TOOLS

There are many ways to determine if a pet is overweight: breed standards, body condition score (BCS), body fat index (BFI), dual x-ray absorptiometry (DEXA) and morphometric measurements. These tools are used to determine how overweight a pet is, as well as to determine how much to feed to reach the pet's ideal weight. Most weight loss programs fail because food dose calculations are based on inaccurate estimations of ideal body weight. Current methods of estimating ideal body weight (body condition score/ guessing) are most likely to underestimate % body fat and therefore overestimate ideal weight, which leads to overfeeding. Pets are being overdosed on food because of inaccurate estimations of ideal body weight.

Breed standards

Breed standard charts are not very accurate; most charts provide a range of weights. For example, according to the American Kennel Club, a Labrador retriever male could weigh 65-80 pounds and a female could weigh 55-70 pounds, a difference of about 25%,[3] which could lead to obesity if feed incorrectly. Breed standards do not give any weight recommendation for mixed breed, especially the very popular designer breeds, such as puggles and labradoodles.

Body condition score

Currently the most practical and effective method of assessing body composition in clinical practice is body condition score which suffers from major limitations. Precision and accuracy are highly dependent on the training and skill of the individual doing the assessment. The current body condition scoring scales (5 and 9 point scale) do not differentiate between levels of obesity. For example, using the 5 point scale all pets >35% body fat fall into a single score of 5. The result is a pet with 60% body fat and one with 36% body fat both receive the same score. This leads to over-estimation of ideal body weight and feeding amounts for the fatter animal resulting in slow or ineffective weight loss. There is no direct connection between the BCS score and the pet's ideal weight.[3]

Body fat index

A pet's body fat index simply refers to its current body fat percent. This new science was developed in partnership with the University of Tennessee and Hill's® Pet Nutrition and is a more effective way of connecting a pet's body fat percentage to a meaningful number that can be shared with pet owners, similar to body mass index (BMI) in humans. The BFI chart addresses all levels of pet obesity up to 65% fat.

DEXA

Dual-energy X-ray absorptiometry utilizes X-rays to differentiate the type and amount of each tissue in the part of the body being scanned. It has been used to estimate body composition and is a noninvasive method for estimating body fat content.[5] This is a very accurate technique, but DEXA machines are generally not found in general practice; DEXA machines can be found in Universities and some pet food companies with a research facility.

Morphometric measurements

The University of Tennessee School of Veterinary Medicine has developed a new technique to accurately determine a pet's ideal weight by doing a set of measurements. This method uses simple body measurements (morphometric measurements) which are entered into an online website or mobile app to calculate ideal weight. These simple body frame (morphometric) measurements were developed and validated by investigators at the University of Tennessee. They are an objective, quantitative method of evaluating the severity of hyperadiposity (body fat) in overweight pets. Using a tailor's tape, the evaluator records four simple measurements for dogs and six measurements for cats. The measurements take about 1 ½ to 2 minutes to complete and only need to be done once.[3]

NUTRITIONAL OPTIONS

There are many therapeutic food options available. These foods have been formulated to provide the proper nutrients with controlled calories or formulated to alter metabolism, so the pet will lose weight. Using over-the-counter wellness foods is not advised for weight loss. They have not been formulated for weight loss, so while calories can be restricted; nutrients will also be restricted, which could result in deficiencies in important nutrients, such as protein, minerals or vitamins.

There are many options for weight loss foods. There are restricted calories and fat foods; typically calories are replaced with fiber, water or air to give a feeling of fullness. The pet can eat the same amount, but receive fewer calories. There are weight loss foods that alter metabolism. High protein foods have been used successfully to alter metabolism to burn fat. Research has been completed using neutrogenomic technology to alter metabolism in dogs and cats.[3]

A weight loss food needs to have an appropriate amount of protein to prevent loss of lean body mass. Besides an appropriate amount of protein, the protein quality is also important. A weight loss food should have the right amino acid makeup. A study showed a food fed to dogs with an optimized amino acid ratio, (a higher lysine-to-calorie content), lost more weight than dogs eating the control food.

Dogs should get ≥ 25% of protein for weight loss and ≥ 18% for weight maintenance dry matter basis (DMB), with at least ≥ 1.7% lysine DMB. Cats on a calorie- restricted food should get ≥ 35% protein for weight loss and maintenance. A metabolic- control cat food should have 47% to no more than 55% protein DMB.[1]

Energy will be restricted in some weight loss and weight management foods. Canines eating a calorie restricted food for weight loss or maintenance should eat a food with ≤ 3.4kcal DMB metabolizable energy (ME)/g. Fat should be ≤ 9% DMB for weight loss and ≤ 14% DMB for weight maintenance. Felines eating a calorie restricted food should eat a food with ≤ 3.4kcal metabolizable energy (ME)/g DMB and ≤ 3.8 kcal ME/g DMB for weight maintenance. Fat should be ≤ 10% DMB for weight loss and ≤ 18% DMB for weight maintenance. For cats on a metabolic-control food the fat should be ≤ 25% DMB.[1]

L-Carnitine has been shown to assist in weight loss by performing a variety of functions, including fat metabolism and energy production. Canines should receive ≥ 300 ppm DMB and felines should receive ≥ 500 ppm DMB for weight loss and maintenance.[1]

Satiety can be accomplished using fiber, which dilutes calories. Foods should contain 12-25% fiber DMB for weight loss and 10-20% DMB for weight maintenance for canines and 15-20% DMB for weight loss and 6-15% DMB for weight maintenance for felines. Water is used in canned food and air is used in dry foods to decrease caloric density of a food.[1] The extra moisture in canned foods allows a pet to eat more volume while consuming fewer calories, since about 78% of the food is water.

Carbohydrates are used as a source of energy. Look for a food with ≤ 40% DMB carbohydrate for weight loss and ≤ 55% DMB for weight maintenance in caloric restricted foods for canines. For felines, carbohydrates for weight loss food should be ≤ 35% DMB and ≤ 40% DMB for calorie restricted food and ≤ 20% DMB for metabolic-control foods.[1]

COMMUNICATION:

Technicians can assist the pet owner with a weight loss program and help with successful weight loss. Calculating how much to feed and communicating the amount is important; also discuss treats and exercise. Make sure you advise the owner that the amount you are recommending is a starting point. The amount may need to be adjusted based on how the pet responds to the feeding amount. If available, use a brochure that discusses the new food. If the pet food company does not provide brochures, make your own.

It is important to discuss a transition from the old to new food, over at least 7 days. Sometimes a transition should be longer, especially for cats. A transition is vital to the success of the pet accepting the new food. There are two reasons to do a transition: 1) occasionally a pet will have a GI upset when switched to a new diet, 2) a pet will accept a new food better when a transition is done, to allow the pet to get use to the new texture and flavor. It is important to give a measuring cup (if the pet food companies supply them), or instruction to purchase the correct 8 oz. measuring cup. Sometimes using a food scale to weigh out grams of food is more successful, because the error in measurement is smaller.

It is important to follow up with the pet owner to make sure everything is going well. The technician should call in 2 -3 days, 2 weeks, and monthly to see how the pet is doing. The first follow up is the most important; call in 2-3 days. This will be during the transition and this is when most people will give up on the new food. It is an opportunity to be the owner's support and cheerleader and help with any issues that have come up. Check in at 2weeks and then monthly, until the pet reaches its goal weight. Encourage weighing every 2-4 weeks until the pet reaches its goal weight, then every 6 months. It is important when a pet is on a therapeutic food to stay on it uninterrupted, a technician can be invaluable in assuring the pet stays on the food consistently.

CONCLUSION:

Weight loss in pets can be a challenge for the pet owner; new tools allow accurate calculation of ideal weight, allowing accurate calculation of feeding amounts. With an accurate ideal weight and a specifically formulated weight loss food, a pet has a better chance of reaching its ideal weight and maintaining the weight loss. Weight loss may help manage many weight related diseases, such as diabetes. Technicians can play an integral role in providing support for the pet owner and help through the process of weight loss and maintenance.

REFERENCES

1. Hand et al., Small Animal Clinical Nutrition, 5th edition, 2010; 501-512, 2. http://www.petobesityprevention.com/pet-obesity-fact-risks/, assessed January 5, 2013, 3. Todd T.L., **Practical Weight Management in Dogs and Cats**, 2011; 3-97. 4. Fascetti A.J., **Applied Veterinary Clinical Nutrition**, 2012; 110, 5. Mawby, DI, et all., **Comparison of Various Methods for Estimating Body Fat in Dogs** J Am Anim Hosp Assoc 2004;40:109–114.

Nutritional Management of Patients with Vomiting and Diarrhea

Ann Wortinger BIS, LVT, VTS (ECC, SAIM)
Dearborn, MI

INTRODUCTION

One of the most important things to remember when dealing with a patient with vomiting or diarrhea, is that neither of these are in themselves diseases, but rather signs of disease. There are few clinical studies that have evaluated specific dietary manipulation on either the prevention or management of gastrointestinal disease in dogs and cats (Cave 2012). To date we have no physical test finding, lab test results or historical facts that are predictive of which feeding method will be most effective in managing signs in a particular patient, even with a definitive diagnosis individual feeding trials and routine follow-up visits with the veterinary team are essential in managing these signs ((Buffington 2004). If there is no improvement in clinical signs over a 2-3 day period for acute vomiting or diarrhea or if the patient continues to deteriorate despite treatment, reassessment is necessary (German 2006).

CAUSES OF VOMITING

Vomiting can be causes by any number of things, not all of which are related to the stomach. Vomiting can be induced with motion sickness, ingestion of emetogenic substances, gastrointestinal tract obstruction, abdominal infection or inflammation or extra-gastrointestinal diseases that stimulate the chemoreceptor trigger zone (Willard 2009). The most important clinical sign of chronic gastric disease is vomiting (German 2006).

The dietary behavior and lifestyles of carnivores, such as cats, are such that occasional vomiting is not considered clinically significant especially when the vomitus is mixed with hair. In cats specifically, vomiting can be seen as a sign of a more distal intestinal disorder, even in the absence of other clinical signs (Zentek 2008).

CAUSES OF DIARRHEA

Diarrhea is a non-specific sign of many diseases, and effective management will depend on identification of the underlying cause, and resolution or removal of this cause if possible (Buffington 2004). Acute diarrhea is most typically caused by dietary indiscretion or intolerance, intestinal parasites and infectious diseases such as *Salmonella* or *E.coli* (Willard 2009). Chronic diarrhea is most commonly caused by intestinal disease, though other systemic diseases may affect intestinal function and can induce secondary hypersecretion or intestinal malabsorption (Zentek 2008). The most common cause of chronic vomiting and diarrhea in dogs and cats is caused by the constellation of diseases found under the heading of inflammatory bowel disease (Buffington 2004).

Acute diarrhea can often be treated symptomatically and will resolve within 2-3 days with minimal treatment. Chronic diarrhea is rarely self-limiting, with treatment being based on definitive diagnosis (German 2006).

TO WITHHOLD FOOD OR NOT

For years our standard treatment for acute vomiting or diarrhea has been to withhold food for 12-48 hours. The belief has been that by removing food from the already upset GI tract we could allow it to clear itself of luminal contents that may be causing the problem, to help prevent mucosal cell abrasion by the GI contents, deprive the opportunistic luminal bacteria of nutrients preventing proliferation, preventing absorption of dietary antigens that may be present due to maldigestion and to permit restabilization of the brush border enzyme function (Buffington 2004, Case 2011).The problem with this approach is that in almost all cases of enteritis, regardless of the cause, there is a decreased motility with delayed gastric emptying and reduced segmental contractions. Therefore fasting would not immediately provide any physical rest for clearing for the GI tract (Cave 2012).

Studies in human patients have shown that early reintroduction of feeding resulted in shorter time for the return of gut sounds and a trend towards shorter hospital stays. Evidence also suggests that the prokinetic effect of feeding may decrease the vomiting response in some patients (Cave 2012). We know that oral fasting alone can induce an intestinal insult, with mucosal atrophy occurring within 12 hours of a fasting state (Cave 2012, Zentek 2008). While it may be easier to convince clients to continue to feed a patient with diarrhea, it is unlikely a client will continue to feed an animal with vomiting.

TREATMENT OPTIONS

The primary goal for management of vomiting or diarrhea is to maintain delivery of nutrition to the GI tract, prevent nutrient deficiency and malnutrition (Case 2011). The most common first step is introduction of a highly digestible, low fat diet (German 2006, Buffington 2004). Many times we will recommend that the patient be started on small amounts of a bland diet. There is little evidence to support this recommendation, and the term 'bland' is poorly defined and vague. With any inflammatory process in the GI tract, the type of diet offered is more important than the 'blandness' of the ingredients (Case 2011). Diet selection should be based on the specific disease being treated, the area of the GI tract that is affected and the ability of the diet to promote tissue healing and maintain remission of clinical signs (Case 2011).

Our long-term goal in management of vomiting and diarrhea is to help repair the damaged GI tract lining, restore normal GI bacterial populations, to promote normal GI motility and function, to support the immune system and decrease GI tract inflammation (Case 2011). An animal with chronic vomiting and diarrhea with subsequent loss of nutrients and energy may present with weight loss, poor hair coat, and dry flakey skin. These patients will respond well to a food with increased nutrient density in addition to the other nutritional modifications (German 2006). As always, it is important to remind clients that when we provide nutrition to the debilitated patient, the body has the final decision on where those nutrients and where that energy is distributed to. We cannot decide that we want increased muscle mass when the body wants to replace intracellular protein and fat stores. Some clients may find this frustrating, but the body makes all the final decisions not the veterinary team. Our job is to provide the building blocks to rebuild the damage so the patient can achieve full recovery.

ADVERSE FOOD REACTIONS AND FOOD RESPONSIVE DIARRHEAS

Food allergies in dogs and cats is seen as being uncommon, though some clinicians believe that they are more likely to occur with diets that are consumed during periods of acute gastritis as larger pieces of peptides are allowed to cross the intestinal barrier during these times potentially sensitizing the immune system (Buffington 2004). If an adverse food reaction is suspected, the current recommendation is to feed a highly digestible, single source novel protein diet. The thought being that a highly digestible diet has lower antigenicity because fewer intact proteins are absorbed across the inflamed intestinal barrier (Case 2011). The ideal diagnostic and long-term management of suspected cases is based primary on using novel proteins or protein hydrolysates and monitoring patient response (Cave 2012). The initial use of a sacrificial protein may be recommended. This sacrificial protein may still develop some allergic response, but will allow the intestinal tract to heal. Once healing has been achieved, a long-term maintenance diet utilizing a different protein can be introduced.

Dietary management of these conditions can be difficult in practice because they are slow to respond, and the animals remain at risk for relapse if they gain access to the problem ingredient. A minimum trial of 3-4 weeks is recommended, though some clinicians will go for 3-4 months before making any diet changes (German 2006).

Adverse reactions to food are more common than actual food allergies, and have been reported to be as high as 29% of all cases of chronic gastrointestinal disease in cats (Cave 2012). Adverse reactions can be seen with lactose intolerance in adult animals as well as loss of digestibility to certain nutrients and development of flatulence with digestions of some fibers.

Gluten, specifically wheat gluten has received a lot of press lately with the advent of new, 'grain-free' and 'gluten-free' diets for pets. Gluten is a plant protein that provides an excellent source of amino acids. On average, gluten contains 80-82% protein, with low dietary fiber and a digestibility of approximately 99%. Gluten is also an important source of the amino acid, glutamine which has been shown to be important in maintaining digestive integrity and conserving lean muscle mass during intensive activity (German 2006). Irish setters are the only breed of dog or cat in which gluten sensitivity has been documented. For these dogs a gluten-free diet would be recommended. Rice and corn do not contain gluten, where wheat, rye, barley and oats all contain gluten (German 2006).

PROTEIN

A protein source that is of high quality, easily digested and assimilated, and contains all the essential amino acids in the correct proportions would be recommended. Quality of a protein is measured by digestibility, as determined by feeding trials and by the biologic value of the protein. Biologic value (BV) is a determinant of the availability of the essential amino acids within a specific protein (Case 2011). The protein source with the highest biologic value is eggs with a BV of 100. Protein digestibility can vary between animals that are healthy as opposed to those who have disease, and can be affected by food processing methods (Cave 2012). Dietary proteins also present the largest source of dietary antigens, and are recognized and responded to by the immune system (Cave 2012).

Proteins in the diet are essential for secretion of hormones and enzymes required for digestion. This includes the release of the pancreatic enzymes as well as the hormones insulin, gastrin, cholecystokinin (Zentek 2008, Cave 2012).

Insufficient caloric availability in the diet, intestinal protein loss, increased catabolism and decreased absorption can all result in lower protein levels in the body (Cave 2012). Minimum total diet digestibility of 85-88% dry matter (DM) are recommended with a protein digestibility of > 92% (Zentek 2008).

FATS

While fats are the most calorically dense nutrient and inclusion in the diets would benefit those patients who are malnourished, absorption of fats through the intestinal lymphatic's may be impaired and could contribute to a postprandial increase of fluid influx into the intestinal tract further contributing to the development of secretory diarrhea (Cave 2012, Case 2011). In cats, higher fat diets may be beneficial in slowing gastric emptying especially with diarrhea (Buffington 2004). The content is recommended to be ~ 11-15% DM or less of the diet (Case 2011).

The presence of the Ω-3 class of fatty acids has been found to be helpful in management of inflammatory responses in the body, including those occurring in the GI tract (Case 2011). Both Ω-3 and Ω-6 fatty acids are essential and therefore required in the diet, by manipulating the ratio of these two fatty acids; we can decrease the inflammatory response.

CARBOHYDRATES AND FIBERS

Similar to proteins, carbohydrates should be easily digested and assimilated by the intestinal tract. The carbohydrate with the highest digestibility is cooked and blended white rice (Case 2011). Alternatives to rice would include potato, tapioca and corn. All of these should be blended or ground before feeding to break down the cellulose outer shell allowing to body to reach the nutrients within the cells. Many times we will recommend cooked rice in a home-made diet, but the whole rice grain may pass through the entire GI tract relatively intact due to impaired ability to access the contents through the fiber shell. Blended rice or baby food flaked rice are more digestible than is plain cooked rice.

Fibers are useful in the management of GI disease, as well as maintenance of GI health. The type of fiber used as well as the level found in the diet is important (Case 2011). Those fibers that result in production of the short-chain fatty acid (SCFA) butyrate are preferred. Butyrate is the preferred energy source for the colonocytes, obtaining ~ 70% of their energy from luminally derived SCFA (Case 2011). Fibers can also increase peristaltic contractions, and may inhibit colonic contractions. SCFAs have been shown to protect intestinal tissue and promote restoration of normal intestinal function (Case 2011). The addition of highly fermentable fibers that produce high levels of SCFAs can cause diarrhea and flatulence and may interfere with the digestion and absorption of other nutrients (Case 2011).

Both soluble and insoluble fibers may be beneficial in symptomatic treatment of large bowel diarrhea by allowing modulation of water retention (Zentek 2008). Which fiber will be most beneficial may depend greatly on the patient's response to fiber inclusion in the diet, unfortunately there is no one fiber fits all. Including highly fermentable fibers such as gar gum, lactulose and pectin, may actually increase the incidence of diarrhea, flatulence and may interfere with nutrient digestion and absorption (Case 2011)

PREBIOTICS

Prebiotics are specific short-chain carbohydrates usually classified as fibers based on their digestibility by the GI tract. Some of the more commonly available prebiotics include inulin, galactooligosaccharides (GOS), lactulose, fructooligosaccharides (FOS) and manooligosaccharides (MOS) (Case 2011). The benefit to the intestinal tract of these specific fibers is their ability to be broken down by the intestinal bacteria with production of SCFAs and their ability to help modulate GI motility. Butyric acid or butyrate is one of the SCFAs produced that provides energy for the colonocytes and has been shown to have anti-inflammatory properties on the surrounding tissue (German 2006).

The fibers used can also help modify the composition and metabolic activity of the GI bacteria, and allow the non-pathogenic bacteria to out compete the pathogenic bacteria allowing readjustment of small intestinal bacterial overgrowth (German 2006). This occurs through changes in pH caused by the production of the SCFAs.

PROBIOTICS

Probiotics are actual bacteria that are introduced into the GI tract with the intention of reestablishing a population of good bacteria and outcompeting the bad bacteria that may be contributing to the disruption of the GI tract (Case 2011). The most important thing with including probiotics in the diet is ensuring that they actually survive the stomach acid, the bile acids and are able to make it alive into the duodenum. Not all probiotics are created the same, and if they don't survive to arrive at the small intestines, you have just included a really expensive protein source in the diet. Look for actual feeding trials documenting the survivability of the specific strains of bacteria included in the product. Explaining to clients the advantages that veterinary specific strains of bacteria provided over what they can get OTC is also important.

CONCLUSION

Diet selection will be ultimately determined by the cause of the signs as well as individual patient response. There are many commercially available therapeutic diets that offer options for management of intestinal diseases including low fat, novel protein and modified fiber diets. As there are no physical test finding, lab test results or historical facts that are predictive of which feeding method will be most successful, individual response and follow up with the veterinary staff is necessary for optimal success.

REFERENCES

Buffington CA, Holloway C, Abood SK. (2004) Clinical Dietetics in Manual of Veterinary Dietetics. Elsevier Saunders, St. Louis MO. pp 82-89

Case L, Daristotle L, Hayek M, Raasch M. (2011) Nutritional management of gastrointestinal disease in Canine and Feline Nutrition 3rd ed. Mosby Elsevier, Maryland Heights MO. pp 455-472

Cave N. (2012) Nutritional Management of Gastrointestinal Diseases in Applied Veterinary Clinical Nutrition (eds Fascetti AJ, Delaney SJ). Wiley-Blackwell Ames, IO. pp 175-204

German A, Zentek J. (2006) The most common digestive diseases: the role of nutrition in Encyclopedia of Canine Clinical Nutrition (eds Pibot P, Biourge V, Elliott D). Royal Canin, Aimargues France EU. pp 103-123

Saker K, Remillard RL. (2010) Critical Care Nutrition and enteral-assisted Feeding in Small animal Clinical Nutrition 5th ed (eds Hand MS, Thatcher CD, Remillard RL, Roudebush P, Novotny BJ). Mark Morris Institute Topeka KS. pp 441- 442

Willard, MD. (2009) Clinical Manifestations of Gastrointestinal Disorders in Small Animal Internal Medicine 4th ed (eds Nelson RW, Couto CG). Mosby Elsevier, Maryland Heights MO. pp 355-361

Zentek J, Freiche V. (2008) Digestive diseases in cats: the role of nutrition in Encyclopedia of Feline Clinical Nutrition (eds Pibot P, Biourge V, Elliott D). Royal Canin, Aimargues France EU. pp 85-92

MAKING THE MOST OF CYTOLOGY IN YOUR PRACTICE

Laura D. Garrett, DVM, DACVIM (Oncology)
Urbana, IL

INTRODUCTION

Veterinary technicians can play a critical role in sample acquisition, slide preparation, and screening cytology. Cytology is a critical tool not only for the oncologist, but also for veterinarians in general practice. Cytology is easy, safe, fast, relatively inexpensive, and is often diagnostic. Cytology can be used to evaluate lymph nodes and masses (external or internal), and also fluids such as effusions and discharges. Results may change further diagnostic plans, treatment recommendations, or what a client chooses to do. For example, studies comparing evaluation of lymph nodes for metastatic disease via physical examination (PE) versus cytology showed 20-40% of patients felt to have normal or slightly enlarged lymph nodes on PE had metastatic disease found with cytology. Conversely, only 50-60% of patients with moderately to severely enlarged lymph nodes on PE had metastatic disease. Thus, PE only of lymph nodes has poor sensitivity and specificity for metastatic evaluation. Cytology is of great benefit in these cases.

ANY LUMP OR BUMP SHOULD BE ASPIRATED!! Mast cell tumors, for example, can feel just like lipomas. We routinely "map" all of our patients. This procedure includes measuring and aspirating all masses. The masses are numbered on a body chart, and the fine needle aspiration results are recorded. Once a mapping has been accomplished, it stays in the animal's chart and can be referenced if an owner thinks a mass is growing or has identified a new mass. Thus, veterinarians unfamiliar with the case will not need to go through testing that has already been performed.

OBTAINING A SAMPLE

Sample collection and preparation for cytology is simple and requires little equipment. Needles (25, 22, 20, or 19 gauge), 6 cc syringes, clean glass slides, and occasionally a scalpel blade for scraping are all that is needed. In vivo collection technique can be performed with a needle alone or with a syringe attached to the needle. Before obtaining the sample, make sure the slides are set out (at least 4-6) and the syringe is ready. Stabilize the mass in your non-dominant hand. For the needle alone technique, direct the needle into the mass; withdraw (yet stay within skin) and push back into the mass several times rapidly. Draw several cc air into the syringe, connect it to needle and expel needle contents onto a slide (make sure the bevel is down and the contents come out near the frosted end). For the needle plus syringe procedure, connect the needle and syringe, direct the needle into the mass, draw back plunger and release several times, with pressure released withdraw the needle from mass, disconnect the needle and draw several cc air into syringe, reconnect and expel needle contents onto the slide.

Cytology can also be of benefit even when a sample is being obtained and sent for histopathology. In some tumor types, particularly round cell tumors, cytology can provide a more precise diagnosis that histopathology. Thus, one may want to collect a cytologic sample in vitro, or after tissue has been removed from the patient. This can be done via impression smears or scrapes. Impression smears are best for round cell or epithelial tissues – tissues that exfoliate more easily. Without disturbing surgical margins, use a blade to cut a fresh section of tissue (maximum 1x1 cm). Gently blot the cut face on gauze or paper to rid it of excess moisture and blood. Press the blotted area onto the slide. Make several rows of impressions. If cellular yield is low with this approach, the scrape technique can be used. This may be best for mesenchymal tissues. Begin in the same manner as with the impression technique, but after blotting the sample, use the blade to scrape cells from the cut surface. Gently smear the collected cells from the blade across the slide in one direction.

PREPARING SLIDES

Slide preparation is key for good cytology results. The samples must be smeared out quickly to prevent clotting and need to be in a thin layer for best cellular evaluation. Even if only a small dot of sample is expelled onto the slide, smearing it is important. Cells that dry in a little round "puddle" of fluid maintain more of a 3-dimensional appearance, do not spread out their cytoplasm, and are difficult to assess. After the contents are expelled onto a slide, place a second slide onto the first cross-wise to the first slide) and pick up a SMALL amount of material. Pick up a third slide, gently place the second slide across the third slide to squash the sample flat and smear the sample by pulling the second slide away from the frosted edge of the third slide while maintaining slide contact with gentle pressure. The second slide needs to stay FLAT on the third slide (no edge to catch cells and drag them). Now, the second slide can be used again to pick up another small bit of sample from slide 1, and can be smeared onto a 4th slide, etc. When just a small amount of sample remains on slide 1, use slide 2 to smear it as well. Most aspirates will yield 3-6 good slides, allowing for in-house staining and evaluation of one slide while leaving several unstained for laboratory evaluation. Also, by making multiple smears, the samples are more thinly spread out which improves diagnostic ability. This is particularly helpful with very cellular aspirates.

Preparation and staining of the slides is as follows. Air dry the slides. Standard hematologic stains are preferred for most purposes (Wright's, Wright's-Giemsa, Diff Quick, etc.) If submitting slides to an outside laboratory, submit them air-dried and unfixed, or fixed and stained. Do not submit them fixed and unstained or cell quality will diminish. Do not have cytology slides near formalin. Even the fumes can distort cytologic quality. Smears should be made and kept away from biopsy formalin jars.

EVALUATING THE SLIDES

Once the slide is stained it is ready for microscopic examination. Scan the slide on low power (10x) first to get an overall feel for the cellularity of the sample and then go to higher power to evaluate the cell types and individual cytologic features of the cells. The two most important questions as a practitioner to answer when looking at a slide are 1. Did I hit the intended tissue? (salivary gland vs. lymph node, blood vs. bone marrow) and 2. Is it a good sample? (cells not all ruptured, enough cells for evaluation). After assessing these points, evaluate if the cells are from normal tissue. Note if inflammation (see cell types below) is present. If so, caution must be used in interpreting other cells on the slide. Inflammation can induce cells to look "malignant" when they are actually just reactive. Next ask if there are cells in abnormal places (non-lymphoid cells in lymph node, non-hepatic cells in liver). Finally, note the predominant cell type (as follows).

Inflammation:

Inflammatory processes usually have a mixed population of cells including inflammatory and tissue cells. The inflammatory process is characterized by the cell types present. When neutrophils predominate (>85%), this is considered a suppurative response. The character of the neutrophils should be closely evaluated. If there is a bacterial infection, the neutrophils will often undergo karyolysis and the nucleus will appear pale and swollen. These cells are often called degenerative neutrophils. Also the cells should be closely examined for intracytoplasmic microorganisms. A pyogranulomatous response consists of a mixture of neutrophils, epitheliod macrophages and multinucleated giant cells. This type of inflammatory response is commonly observed with fungal infections or responses to foreign material. Granulomatous inflammation consists primarily of macrophages with multinucleated giant cells and lesser numbers of other inflammatory cells. This type of inflammatory response does not always exfoliate well and may be of low cellularity. Eosinophilic inflammation is more subtle. Cytologic specimens containing greater than 12% eosinophils are considered eosinophilic inflammatory lesions. Common causes include allergic or hypersensitivity response, parasitic infestation or paraneoplastic response.

No Inflammation?: Tissue cells are classified as epithelial, mesenchymal (connective tissue) or round cells.

Epithelial:

Epithelial/glandular cells are found in the linings of tracts (respiratory, digestive/oral, urogenital), in skin and mucous membranes, and in glands (pancreas, thyroid, adrenal, prostate, mammary, etc.). Benign epithelial lesions are adenomas, while malignant versions are carcinomas. The cells are generally round and the sample is often highly cellular. The major defining feature of this cell population is clustering, or clumping, of cells. This feature can be observed on low power evaluation. Note that any very cellular aspirate can give the appearance that the cells of the sample are clustered together, even with mesenchymal masses. Assess less cellular regions to ascertain if in these areas cells are still bound together with other cells or if they do appear individually, as mesenchymal cells will. Round cells tend to be pressed together, as they exfoliate very well and thus provide cellular samples, but they do not tightly adhere to each other or swirl together to form 3-dimensional structures. Small spaces can often be seen between non-epithelial cells.

Neuroendocrine tissues such as thyroid and adrenal organs, while in the epithelial category, do have a few distinctive features. These cells tend to rupture very easily. Thus, samples from these tissues may have fields of naked nuclei in a smooth bluish background representative of the lost cytoplasm. Additionally, while most neuroendocrine tumors are malignant (thyroid carcinomas, insulinomas, etc), the cytologic appearance of these tumors is often very uniform with few features of malignancy. In these cases, knowing the site of origin of the mass and the fact that the tissue appears neuroendocrine is more important for predicting biologic behavior than are the malignant features of the cells.

Mesenchymal:

Mesenchymal cells are found in connective tissues such as cartilage, bone, smooth and striated muscle, blood vessels, fibroblasts, etc. When benign, the masses are categorized as the cell of origin and then "oma" – e.g. fibroma. When malignant, the masses are termed sarcomas. Mesenchymal cells may not exfoliate as well as epithelial tissue; however, most mesenchymal masses will provide enough cells for cytologic evaluation, and some exfoliate very well. In benign lesions, it is not uncommon to see only 1-2 cells on a slide. The cells are commonly spindle shaped with wispy cytoplasmic borders. The nuclei are often oval in shape. Their cell shape, indistinct cytoplasmic borders, and the fact that they occur as single cells in the preparations characterize mesenchymal cells. At times mesenchymal masses have a round cell appearance, most notably with osteosarcoma. Malignant osteoblasts are notable for their "plasmacytoid" appearance in some cases.

Round cell:

Round cells are called such based on their cytologic appearance: discrete, round cells with distinct cytoplasmic margins. This category includes lymphoid tissue, plasma cells, mast cells, histiocytomas, transmissible venereal tumors (TVT), and melanoma (which can have a round or mesenchymal appearance). Often the specific cell type can be cytologically determined in round cell tumors, unlike in epithelial or mesenchymal masses. The round cells will have round to bean-shaped nuclei, lacy to smudged chromatin and are generally small to medium sized cells. Individual features of the listed round cell types can help to provide a diagnosis. Lymphoblasts have a high nuclear:cytoplasmic (N:C) ratio, are as large or larger than neutrophils, have basophilic cytoplasm and open chromatin, and often have multiple nucleoli of varying sizes. Plasma cells have round, eccentrically located nuclei with ropy or clumped chromatin. The cytoplasm is more abundant than in lymphocytes and has a dark blue/grey hue. A perinuclear clear zone is often apparent. The most notable feature of mast cells is the presence of pink/purple-staining cytoplasmic granules. The granules may be scant or so abundant that the cell nuclei are completely obscured. Often granules litter the background of the slide and may be mistaken for cocci. Mast cells are large round cells with a centrally located round nucleus ("fried egg" appearance). Mast cells secrete chemoattractant factors for eosinophils and occasionally fibroblasts. Some mast cell populations have granules which fail to take up Diff Quick stains; these granules will often stain with other stains. If agranular cells with a "fried egg" appearance and eosinophils are noted on a mass aspirate, unstained slides should be sent to a laboratory that will use a Wright's stain to assess for granules. Histiocytomas are large cells with a moderate amount of blue-grey cytoplasm with clearing at the periphery of the cytoplasm. The cells can show extreme pleomorphism despite the tumor's benign behavior. Histiocytomas can be difficult to distinguish from some plasma cell tumors cytologically. The clinical appearance of the lesion is supportive of the diagnosis in these situations. TVTs also have a large amount of cytoplasm, usually grey-blue to blue in color, and can be quite pleomorphic. The defining feature of the TVT (other than the clinical presentation!) are unevenly distributed vacuoles in the cytoplasm. Melanoma can have a round cell appearance – assess for fine black/blue/green cytoplasmic granules. Amelanotic melanoma can be difficult to diagnose, yet even amelanotic masses will often have a few cells with pigment. Malignant melanomas are highly pleomorphic tumors with large cells that contain prominent and bizarre nucleoli.

Once the cell type has been characterized, the next step is to determine if the process is hyperplastic or neoplastic and if it is neoplastic, is it benign or malignant. It is important to evaluate for features of malignancy in epithelial or mesenchymal cells. The biologic behavior of the round cell tumors is based on cell type and site of the mass rather than specific cytologic features. Malignant criteria include: prominent and multiple nucleoli, anisocytosis, anisokaryosis, multiple nuclei, mitotic figures and nuclear molding. There are others, but these are the most common. When these criteria are present it is fairly easy to identify a malignant tumor; however, not all malignancies express these criteria. Therefore, cytology cannot always rule out a malignant process. Histopathology may be needed for diagnosis of malignancy. For example, soft tissue sarcomas are locally very aggressive yet may have a benign cytologic appearance. If ≥ 3 features of malignancy are present, then the lesion is most likely malignant. If malignant features are lacking, the lesion may still have a malignant behavior and a biopsy is indicated.

The final step in cytologic evaluation is examination of the material in the background. Proteinaceous material is eosinophilic and granular and is usually observed evenly distributed throughout the sample. Granules from ruptured mast cells or melanocytes must be distinguished from bacteria. Most cytologic samples contain some level of peripheral blood contamination. With the red blood cells come white blood cells. It is important to remember this when trying to distinguish inflammation from peripheral blood contamination. It may be necessary to do a CBC on the patient to determine the WBC in situations where increased numbers of WBC are noted on a cytology slide.

REFERENCES

[1] Leblanc CJ et.al **Vet Clin Pathol** 2009 [2]Ghisleni G et al. **Vet Clin Pathol** 2006. [3]Williams LE et al **J Am Vet Med Assoc** 2003 [4]Ballegeer EA et al. **J Am Vet Med Assoc** 2007 [5]Britt T et al. **J Small Anim Pract** 2007

Understandings ECGs
Karen Wiley, CVT, VTS (cardiology)

We have all heard the letters ECG or EKG when referring to cardiology. The initials stand for electrocardiogram (or if you are German electrokardiogram). An electrocardiogram (ECG) is a recording of the electrical activity of the heart and one must realize that this is different from an echocardiogram which is an ultrasound of the heart.

Indications for using an ECG in veterinary practice include recording an accurate heart rate, detecting an arrhythmia, monitoring anesthesia and detecting electrolyte abnormalities (specifically K+).

Normal conduction of electricity in the heart begins in the Sino-atrial (SA) node. The impulse depolarizes the atria and passes through the Atrio-ventricular (AV), down the bundle branches and then depolarizes the ventricular muscle.

Normal parts of an ECG include the P wave, the P-R interval, the QRS complex, the S-T segment and the T wave. The P wave represents depolarization of the atria. The QRS complex represents depolarization of the ventricles. The T wave represents the repolarization of the ventricles. The PR interval represents the impulse passing through the AV node. The ST segment is part of the repolarization of the ventricle.

There are several different ways of measuring an ECG. Single lead rhythm strips are used in anesthesia monitoring and in the ER. Multiple lead (6 or 12 leads) ECG, or "diagnostic" ECG are used in cardiology/specialty practices. Telemetry is used in the ICU setting for hospitalized patients, and ambulatory ECG (in the form of Holter or Event monitors) can also be used to evaluate rhythm disturbances/electrical activity at home.

To perform a "proper" ECG the patient should be placed in right lateral recumbency. Typically four electrodes are placed, one on each foot. By convention, black is placed on the left forelimb, white on the right forelimb, red on the left hindlimb and green on the right hindlimb. Alcohol or gel is applied to the skin to ensure good contact. To avoid artifact on an ECG use lots of alcohol, reduce motion/shaking of the patient and choose a quiet comfortable environment.

ECG recording are displayed as a series of 'leads'. Lead one on an ECG represents the difference in electrical potential between the right and left forelimb. Lead II represents the electrical potential between the right forelimb and the left hindlimb. Lead III represents the electrical potential between the left forelimb and left hindlimb.

Interpreting an ECG

From an ECG one can reliably determine heart rate and rhythm. One can also gain information regarding chamber enlargement via axis shifts and changes in waveforms. However, these assessments are less reliable, and it is important to recognize that a normal ECG does not mean that he heart is normal. When determining the heart rate one must ask if the rate is fast, slow or reasonable. When answering the question one must consider the species (dog, cat, other), and the breed. The next step to interpreting and ECG is to determine if the rhythm is sinus in origin, which is typically assessed by evaluating if there is a P wave for every QRS complex and a QRS complex for every P wave, and if the P wave and QRS complexes normally or abnormally conducted? The regularity of the rhythm and the presence of any pauses or premature beats is also an important feature in the interpretation of an ECG.

Normal rhythms seen on ECG include normal sinus rhythm, sinus tachycardia, sinus bradycardia, sinus arrhythmia (+/- wandering pacemaker). Abnormal rhythms include atrial (supraventricular) arrhythmias, ventricular arrhythmias and Atrio-ventricular (AV) block.

The goal of the presentation is to familiarize the audience with the theory, acquisition and interpretation of the electrocardiogram. Examples of normal and abnormal rhythms will be provided for illustrative purposes.

Cardiovascular Diagnostics to Differentiate Between Cardiac and Non-Cardiac Emergencies

John M. MacGregor, DVM, DACVIM (Cardiology)
Portsmouth, NH

INTRODUCTION

Emergencies due to cardiac causes and non-cardiac causes can be difficult to distinguish and the correct treatment for a non-cardiac condition could be potentially life threatening when administered for a cardiac condition. Diagnostic testing is crucial in making this distinction. Evaluation of testing methods will include the use of physical examination, blood tests, radiography, ultrasound and electrocardiography in an emergency setting. The broadest categories of emergencies that are likely to be cardiac in nature are collapse and breathing difficulty and a systematic evaluation of the above testing methods will allow successful differentiation of the cause for these emergencies. In such cases, collapse will not be defined as a collapse that occurred at an earlier time point but rather a patient that is collapsed currently.

DIAGNOSTIC TESTING

Physical examination

A thorough but expedient physical examination can often provide strong evidence for the cause of an emergent situation. At a minimum, it should provide the examiner with a direction for further diagnostic testing. Heart rate, respiratory rate and effort, characteristics of the mucus membranes and mentation should be assessed at initial triage. A more complete physical examination should follow, including heart rate and rhythm, respiratory rate, body temperature; assessment of pulse quality and synchrony, assessments of heart and respiratory sounds and assessment of the abdominal cavity.

Blood testing

Emergency blood testing should always include a blood glucose level, a packed cell volume and total solids. While these can help assess very basic causes of emergencies, a more comprehensive testing protocol should include electrolytes, renal values and blood lactate levels. This battery of tests can determine a great deal of information about the cause of an emergency very quickly. Less emergent testing can include blood gas levels and NT-proBNP or BNP testing for patients with dyspnea. Coagulation studies may be helpful in determining whether a pulmonary thromboembolism is the cause of dyspnea.

Radiographs

From the perspective of differentiating cardiac from non-cardiac emergencies, well-positioned radiographs are useful for assessing cavities for fluid, assessing organs in very broad terms, and assessing the pulmonary parenchyma.

ECG

An ECG should be used as a diagnostic tool in emergency situations if significant arrhythmia has been determined during physical examination. In these patients, significant arrhythmia that has potential to cause a collapse is defined as severe bradycardia or even more severe tachycardia. Single premature beats, while potentially diagnostic clues meriting further investigation, are not the cause of collapse in the patient that is currently collapsed.

Ultrasound

In emergent situations, ultrasound should be used to identify fluid in abnormal places. (FAST) These can include abdominal effusion, pericardial effusion or pleural effusion. In patients with dyspnea, when radiographs are not conclusive as the cause of dsypnea, echocardiograms can assess whether the heart is likely to be the cause of the dyspnea. In these cases, a significantly enlarged left atrium almost always exists. Echocardiography can determine whether there is pulmonary hypertension by measuring either tricuspid regurgitation velocity, or pulmonic insufficiency velocity or by using more advanced Doppler echocardiographic techniques. In some cases, pulmonary thromboembolism can be identified.

Computed Tomography

While not thought of as a primary emergency tool, CT scans can determine whether acute lung injury (ALI) or a pulmonary thromboembolism is present. Contrast CT can provide information about potential pulmonary thromboembolism.

SPECIFIC EMERGENCIES

Collapse

Collapse can be caused by a variety of clinical conditions. Collapses could be due to neurologic disease, compromised cardiac pump function (pericardial effusion, pulmonary hypertension, abnormality of heart rate/rhythm, external or internal blood loss, or metabolic disease (Addisonian crisis among others). Using the physical examination in a logical and systematic fashion, supplemented by diagnostic testing, can allow the cause of the collapse to be determined.

Difficulty Breathing

Difficulty breathing can also be caused by a variety of clinical conditions. These include diseases of the upper airway (laryngeal paralysis, nasopharyngeal polyps, elongated soft palate, collapsing trachea), thoracic cavity (pleural effusion, pleural fibrosis, pneumothorax), pulmonary parenchyma (pulmonary fibrosis, pneumonia, ARDS, ALI, pulmonary edema, lower airway disease, asthma) air exchange issues and the heart (pulmonary edema in the dog or cat and pleural effusion in the cat). Physical examination can help determine which tests to prioritize, but identification of the cause/source of dyspnea generally requires additional diagnostic testing. Diagnostic radiography and ultrasonography become critical for making this diagnosis. Recently introduced blood tests, while having a high degree of sensitivity and specificity for distinguishing respiratory disease from cardiac disease, are not useful in an emergent setting as these tests are not bedside. Submission is still recommended if possible for confirmation of diagnosis.

Summary

A thoughtful physical examination followed by appropriate diagnostic testing will help determine whether the cause of an emergency is cardiac or non-cardiac. It is important to keep an open mind and be prepared to modify differential diagnoses based on new information as it is obtained.

REFERENCES

1.Oyama, M et al. J Am Vet Med Assoc .2009; 235(11):1319. 2. Fine D et al J Am Vet Med Assoc. 2008;232(11):1674. 3.Kellihan HB J Vet Cardiol. 2009; Suppl 1:S93. 4. Raffan E J Vet Intern Med 2009; 23(6):1184. 5. Rush et al ECVIM (abs), 2010. 6. Connolly DJ et al. J Vet Cardiol. 2009; Suppl 1:S41. 7. Fox et al J Vet Cardiol. 2009; Suppl 1:S51. 8. Oyama et al Vet Clin North Am Small Anim Pract. 2010 40(4):545. 10. Connolly DJ Vet Clin North Am Small Anim Pract. 2010;40(4):559. 11. Boswood A J Vet Cardiol. 2009;Suppl 1:S23. 11. Porcello F et al Aust Vet J. 2008;86(10):390. 12. Hansson K J et al Small Anim Pract. 2009; Suppl 1:44. 13. Boswood A et al J Small Anim Pract. 2003;44(3):104. 14. Schober K et al. J Vet Intern Med. 2010;24(6):1358. 15. Boysen et al J Am Vet Med Assoc .2009; 235(11):1319. 16, Lisciandro GR. J Vet Emerg Crit Care 2011;21(2):104-22.17. Kelmer E et al Can Vet J 2012;53:167–173. 18. Scillia P, et al Crit Rev Comput Tomogr. 2004;45(5-6):293-307.

Fluoroscopy for Cardiac Patients: Preparation, Monitoring and Diagnoses

John M. MacGregor, DVM, DACVIM (Cardiology)
Portsmouth, NH

INTRODUCTION

Fluoroscopy is an imaging technique that uses x-rays to obtain real time, moving images of internal organs. Fluoroscopic technology has improved dramatically since the advent of powerful computers and digital image storage. Fluoroscopy is a vital part of a complete cardiology practice and is the basis for all cardiac catheterization procedures, both diagnostic and therapeutic. Fluoroscopy, when utilized with injectable radio opaque dye, provides vital information about the flow of blood and intra-cardiac anatomy. Although echocardiography has supplanted angiography and ventriculography as a primary diagnostic tool for many common cardiac conditions, fluoroscopic contrast studies are still necessary for almost all interventional cardiology procedures, crucial for determining anatomy of coronary arteries and very helpful in diagnosis of certain congenital heart diseases. Since cardiac patients tend to have significant cardiac and need to be anesthetized for fluoroscopy, patient preparation and monitoring are crucial to a safe procedure.

Preparation

Proper planning prevents potential perils. Patients under going any sort of cardiac catheterization tend to be higher risk patients than many other anesthetic patients. A comprehensive anesthetic plan should be made beforehand and the proper doses of all ~~E~~emergency medications should be calculated prior to anesthesia. A modern defibrillator should be present at induction and throughout the procedure and all surgical preparation that can be done prior to anesthesia should be performed. Broad-spectrum peri-operative antibiotics are indicated for all catheterization procedures. Longer-term antibiotics are indicated when devices such as pacemakers or coils are implanted. Cardiac patients with rhythm disturbances should have temporary pacing capacity in place prior to induction of anesthesia. In general, most cardiac catheterization procedures utilize an anesthetist, a non-sterile technician and one sterile assistant in addition to the cardiologist performing the procedure. Since personnel involved in cardiac catheterization procedures are potentially exposed to much higher levels of radiation than during standard plain film radiography, proper protective equipment is essential for their safety.

Monitoring and Anesthesia

Anesthesia for cardiac procedures has been described by experienced technicians as long periods of mind numbing boredom interspersed with short bursts of white knuckled terror. Patient monitoring is critical for a successful cardiac catheterization. Monitoring should include, at a minimum, blood pressure, ECG, and pulse oximetry. End tidal CO2 is useful for longer procedures. Mechanical ventilation is generally not utilized as it can make procedures difficult. Communication with the cardiologist is vital to anticipate potential trouble points including induction of arrhythmias and the occlusion of blood flow through the heart.

Radiographic anatomy/Fluoroscopic Anatomy

Understanding angiographic anatomy is also important for cardiac catheterization procedures. Important anatomic features to be able to identify are the position of the left and right ventricles, left and right atria, cranial and caudal vena cavae, aorta and pulmonary artery. Other important features are coronary artery anatomy, the position of various malformations, the position of various aortic arch remnants and the position of a patent ductus arteriosus.

Disease States

Prior to the advent of higher definition echocardiography, all congenital cardiac abnormalities required cardiac catheterization with fluoroscopic contrast angiography for definitive diagnosis. The vast majority of congenital disease can now be diagnosed noninvasively, but fluoroscopy needed for diagnosis but any catheter-based intervention utilizes fluoroscopy. Broadly speaking for cardiac congenital abnormalities, catheter-based interventions either deals with holes are not supposed to be there (ventricular septal defects, atrial septal defects, patent ductus arteriosus) or abnormal narrowings (pulmonic stenosis, subaortic stenosis). Another category of diseases where fluoroscopy is a vital part of the procedure is bradycardia requiring a pacemaker (generally advanced AV block, sick sinus syndrome or atrial standstill).

Procedures

The two types of cardiac catheterization procedures, pacemaker implantation and amelioration of congenital are very different in general flow of the procedure. Pacemaker placement tends to occur in older animals that are inherently unstable because they have a very low heart rate. In these patients, the pre-procedural preparation and the ability to capture the rhythm prior to pacemaker placement are the most crucial parts. A secondary critical moment occurs when the final pacemaker positioning and testing occurs. In congenital conditions, the most critical part of the procedure generally occurs when correction of the condition occurs. For some procedures, this involves occluding blood flow through the either sided of the heart for 8-12 seconds.

SUMMARY

Use of Fluoroscopy is critical to cardiac catheterization procedures. These procedures generally have a higher risk of patient mortality than almost any other non-emergent surgery. However, with a good understanding of the procedures and risks by all the members of the catheterization team, risks can be minimized.

REFERENCES

1. The Heart. In Owens JM, Biery DN. Radiographic interpretation for Small Animal Clinicians. 1999. Williams and Wilkins, Baltimore MD. 185-216. 2. Sleeper MM. Special Diagnostic Techniques for Evaluation of Cardiac Disease. In Tilley LP, Smith FK, Oyama MA, Sleeper MM, eds. Manual of Canine and Feline Cardiology. 2008. Saunders Elsevier, St. Louis, MO. 99-110. 3. Kealy JK. The Cardiovascular System in Kealy JK, Diagnostic Radiology of the Dog and Cat, W.B.Saunders, Philadelphia, PA. 213-253. 4. Root CR, Bahr RJ. The Heart and Great Vessels in Textbook of Diagnostic Radiology, Thrall DE, ed. 1994. WB Saunders, Philadelphia, PA 304-315. 5. Losonsky, JM The Pulmonary Vasculature in Thrall DE, ed. Textbook of Diagnostic Radiology, 1994. WB Saunders, Philadelphia, PA 319-330. 6. Thomas WP, Sisson D. Cardiac Catheterization and Angiocardiography in Fox PR, Sisson D, Moise NS, eds Textbook of Canine and Feline Cardiology: Principles and Clinical Practice 1999. WB Saunders, Philadelphia, PA 173-193. 7. Mason DE, Hubbell JA Anesthesia and the Heart in Fox PR, Sisson D, Moise NS, eds Textbook of Canine and Feline Cardiology: Principles and Clinical Practice 1999. WB Saunders, Philadelphia, PA 853-866. 8. Johnson MS, Martin M, Edwards D, French A, Henley W Pulmonic stenosis in dogs: balloon dilation improves clinical outcome.. J Vet Intern Med. 2004 Sep-Oct;18(5):656-62. 9. Kleman ME, Estrada AH, Maisenbacher HW 3rd, Prošek R, Pogue B, Shih A, Paolillo JA. How to perform combined cutting balloon and high pressure balloon valvuloplasty for dogs with subaortic stenosis.J Vet Cardiol. 2012;14(2):351-61. 10, Estrada A, Moïse NS, Erb HN, McDonough SP, Renaud-Farrell S.Prospective evaluation of the balloon-to-annulus ratio for valvuloplasty in the treatment of pulmonic stenosis in the dog.J Vet Intern Med. 2006 Jul-Aug;20(4):862-72. 11. DeFrancesco TC, Hansen BD, Atkins CE, Sidley JA, Keene BW. Noninvasive transthoracic temporary cardiac pacing in dogs. J Vet Intern Med. 2003 Sep-Oct;17(5):663-7. 12. Sanders RA, Green HW 3rd, Hogan DF, Trafney D, Batra AS. Efficacy of transesophageal and transgastric cardiac pacing in the dog. J Vet Cardiol. 2010 Apr;12(1):49-52. 13. Genovese DW, Estrada AH, Maisenbacher HW, Heatwole BA, Powell MA. Procedure times, complication rates, and survival times associated with single-chamber versus dual-chamber pacemaker implantation in dogs with clinical signs of bradyarrhythmia: 54 cases (2004-2009). J Am Vet Med Assoc. 2013 Jan 15;242(2):230-6. 14. Wess G, Thomas WP, Berger DM, Kittleson MD.Applications, complications, and outcomes of transvenous pacemaker implantation in 105 dogs (1997-2002). J Vet Intern Med. 2006 Jul-Aug;20(4):877-84. 15. Bulmer BJ, Sisson DD, Oyama MA, Solter PF, Grimm KA, Lamont L. Physiologic VDD versus nonphysiologic VVI pacing in canine 3rd-degree atrioventricular block. J Vet Intern Med. 2006 Mar-Apr;20(2):257-71.

PROTECTING OUR TEAM FROM CHEMOTHERAPY TOXICITIES

J.A. Impellizeri DVM, DACVIM (Oncology)
Wappingers Falls, New York

Proper dosing, possible complications, drug interaction and safety to staff are all factors that must be considered each and every time a potent, potentially dangerous chemotherapy drug is administered. Technicians who devote their careers to treating our animal cancer patients have been surprisingly lax in protecting themselves from workplace exposures that increase their cancer risk. More than 30 years after it became known that oncology nurses treating human beings are exposed to toxic drugs on the job, many still resist protective measures that could minimize their exposure. This is also the case in veterinary oncology.

Human studies with the National Institute for Occupational Safety and Health (NIOSH) in 1999 revealed widespread antioneoplastic contamination on countertops, carts, trays, and other surfaces where IV bags were placed. A similar study by this group in 2010 showed results that were nearly the same.

Why we handle chemotherapy drugs in human and veterinary medicine without formal protective methods stems from ambiguity in occupational law. Workers in other industries are required to follow protective measures while working with toxic materials. If they do not, the Occupational Safety and Health Administration (OSHA) can take action against their employer. NIOSH has issued voluntary guidelines for safe handling of toxic drugs since 1986, but compliance is not required.

A lack of impetus to change the law is attributed to the latency of the effects of chronic exposure. Although several population studies in human medicine have linked antineoplastic exposure in the workplace to infertility, miscarriage, and cancer, connecting such outcomes to exposures in individuals is nearly impossible. Major splashes from a broken IV bag, for example, are known to cause quick reactions including headache, nausea, and vomiting, but such exposures are rare. The most common cause of long-term harm is thought to be minute, everyday drips and spills.

A significant 2010 study in human oncology documented chromosomal abnormalities in healthy nurses and pharmacists who worked with chemotherapy agents. The 63 study volunteers came from three university-based cancer centers that endorse safe-handling practices. More of the characteristic chromosomal damage was found in exposed workers than in controls, and the incidence of abnormalities increased with the number of drug-handling events.

Without clear, legal enforcement, it is up to veterinary technicians and oncologists to promote safe drug handling. The first step is education. Ongoing training for every member of the oncology team responsible for administering or holding a patient for chemotherapy is a must.

Our Center utilizes an agreement that all new staff must sign after the safety lecture. It affirms that they understand the risks of handling antineoplastics and the need for protection. The agreement also promises that if employees are concerned about chemotherapy exposure because they are pregnant, breastfeeding, or trying to conceive, that they can have a temporary reprieve from handling antineoplastics on the job.

The second step is protective equipment. Gowns and gloves are just the beginning and should be considered only basic protection. We all hear the common complaint "they take longer to put on and take off than to administer the chemotherapy and can be hot." Eye protection (goggles for those who do not wear glasses) is commonly neglected. To minimize the chance of drips, the most progressive oncology centers in human and veterinary medicine use closed-transfer systems which prevent leaks from escaping when drugs are injected into an IV line. They also reduce the risk of aerosolization during "drawing up" of the drugs and administration. Biologic safety hoods are still required, even with closed-transfer systems.

According to a recent questionnaire completed by the technicians at the 2011 VCS meeting, many were using closed-transfer systems, but others were using perceived closed-transfer systems that were not truly closed. In addition, some specialty centers were only using a biologic safety hood.

The federal government is calling on human health care providers to improve drug-handling safety. In April 2011, the directors of OSHA, NIOSH, and the Joint Commission issued a letter to hospital administrators warning about the risks and promoting the guidelines of safe drug handling. The establishment of a culture of safety needs to be a priority in every hospital that administers chemotherapy.

CHEMOTHERAPY: MECHANISMS AND TOXICITIES

Sandra Axiak-Bechtel, DVM, DACVIM (Oncology)
Columbia, MO

INTRODUCTION

Chemotherapy is commonly used in veterinary practices to treat a variety of canine and feline cancers. Adjuvant therapy is chemotherapy treatment following surgical excision of the primary tumor. The goal of adjuvant therapy is to treat micrometastatic disease and a very common indication is chemotherapy following amputation or limb spare in dogs with osteosarcoma. In this case, adjuvant chemotherapy is quite successful in delaying metastatic disease onset, almost doubling the median survival time compared to dogs treated with amputation alone. Primary or neoadjuvant chemotherapy is using chemotherapy prior to surgical intervention, with the goal of decreasing tumor size. An example is the use of vinblastine and prednisone in a dog with a mast cell tumor; this combination has an ~40% overall response rate in the face of bulky disease and, in some cases, is used with the intention to follow with surgical excision. Induction chemotherapy is the initial drug treatment phase in systemic or "liquid" cancers with the intention of inducing remission; induction chemotherapy is commonly used in dogs and cats with lymphoma, followed by consolidation therapy to sustain this remission. Rescue therapy is chemotherapy used after the patient has failed induction or consolidation therapy, or has relapsed. Palliative chemotherapy is the use of chemotherapy to decrease clinical signs associated with bulky tumors.

MYYELOSUPPRESSION

Mechanisms

In general, most chemotherapeutics target rapidly dividing cells, both cancer and normal cells. Myelosuppression is commonly seen with many chemotherapeutics. Chemotherapeutics used in dogs and cats that very commonly cause myelosuppression are lomustine (CCNU), cyclophosphamide, vinblastine, carboplatin, and doxorubicin. Treatment affects rapidly proliferating precursor cells with minimal effect on the mature, non-proliferating, circulating cells. Therefore, blood counts (specifically neutrophils and platelets) may remain within the normal range immediately following treatment as the more mature cells will continue to differentiate, however, the numbers will fall rapidly when the earlier, now depleted cells should have completed maturation. Neutrophils have a circulating half-life of 6-10 hours, and platelets 5-10 days. Neutropenia, therefore, occurs 5-10 days after treatment, when the more mature cells are depleted, and platelet decreases may occur at similar time points or later. There is variation amongst drugs between time of treatment and fall of neutrophil count – this likely reflects different effects of each drug on the rate of cell maturation. Acute anemia caused by chemotherapy administration is rare since red blood cells have a lifespan of 100 days in the dog and 70 days in the cat. Therefore, if acute significant anemia is noted following chemotherapy, other sources of blood loss (such as gastrointestinal loss) or destruction (such as immune mediated hemolytic anemia) should be investigated.[1-3]

Intervention

The severity of neutropenia and clinical condition of the pet determine if invention is required. Infection following chemotherapy can originate from a number of sources; endogenous sources such as normal bacterial flora (skin, oral cavity, respiratory tract, gastrointestinal tract) and exogenous sources. In general, with a neutrophil count of above 1500/μl without evidence of anorexia or fever, a recheck complete blood count is recommended in 1 week. If the neutrophil count is between 1000 and 1500/μl, without evidence of anorexia or fever, a recheck complete blood count is recommended in 1 week, and the owners are instructed to monitor their pet's temperature and bring in their pet immediately if fever, lethargy, or anorexia is noted. Oral antibiotics are recommended in the afebrile neutropenic patient with a neutrophil count between 500 and 1000/μl, and a CBC is repeated in 2-3 days. A repeat CBC in 24 hours with oral antibiotics is recommended in the afebrile patient with a neutrophil count of <500/μl; granulocyte –colony stimulating factor or granulocyte-monocyte colony stimulating factor can also be considered, although most patients will recover without this intervention. If the patient is febrile with any grade neutropenia, intravenous fluids and intravenous antibiotics are required for appropriate supportive care until the fever resolves, at which time oral antibiotics can be continued until the neutropenia resolves. Anemia due to chemotherapy administration tends to be mild, normocytic, and normochromic (without evidence of regeneration), not requiring intervention.[1-3]

GASTROINTESTINAL UPSET

Mechanisms

Mature cells that live on the tips of the intestinal villi are sloughed into the intestinal lumen and replaced by rapidly dividing cells from the crypts of Lieberkuhn. Chemotherapy can lead to damage to the intestinal mucosa when the production of cells in the crypts is interrupted by chemotherapy killing these cells. The end result is ulceration of the intestines and diarrhea in addition stimulation nausea. Nausea can also be acute in nature – the best drug example of this side effect is cisplatin in dogs – via direct stimulation of the chemoreceptor trigger zone (vomiting center).[1]

Intervention

Appropriate intervention is an important component to supportive care. Diarrhea that is mild to moderate can often be managed with a bland diet; metronidazole or loperamide are also commonly used. Severe hemorrhagic gastroenteritis requires hospitalization with intravenous fluids and antibiotics. Acute nausea due to cisplatin chemotherapy can be prevented in many dogs with the use of maropitant prior to treatment. Maropitant is also effective in post-chemotherapy nausea and vomiting following other chemotherapy treatments. Other anti-emetics to consider are metoclopramide, ondansetron, and dolansetron.

THE CASE OF THE MUTATED MDR GENE

P-glycoprotein is a pump that transports drugs across the cell membrane; this pump is normally found in the intestines, kidneys, gallbladder, and brain with the function of driving drugs from within the cell out; it is postulated to play a protective role for these organs. Many drugs (see Table 1) are substrates for these pumps. A mutation in this pump has been found in many breeds: Collie, Australian shepherd, English shepherd, McNab, Old English sheepdog, German shepherd, Border collie, Longhaired whippet, and silken windhound. Clinically, the presence of a mutation in the gene encoding for this pump (MDR1-1Δ) can dramatically affect the chemotherapy choice and dosage for mutant patients. Heterozygotes and homozygotes with this genetic mutation can experience increased myelosuppression and gastrointestinal toxicity from p-glycoprotein substrate chemotherapeutics. Genetic testing (cheek swab) can be submitted to Washington State University's College of Veterinary Medicine for mutation status prior to chemotherapy treatment. In dogs with lymphoma, L-asparaginase and cyclophosphamide can be used without increased risk of toxicity while awaiting test results and dose reductions are recommended for doxorubicin and vincristine if a mutation is found.[4]

Table 1. Summary of expected adverse drug reactions, with commonly used drugs in cancer patients, with MDR1-1Δ mutation

Drug	Adverse effect
Doxorubicin	Bone marrow, GI
Vincristine/ vinblastine/vinorelbine	Bone marrow, GI
Paclitaxel/ docetaxel	Bone marrow, GI
Loperamide	CNS depression
Dexamethasone	Greater hypothalamic pituitary adrenal axis suppression
Ondansetron	CNS depression
Morphine	CNS depression
Butorphanol	CNS depression

UNIQUE DRUG TOXICITIES

Several chemotherapeutics have additional and/or unique toxicities that may warrant preventative measures, observation, or intervention.

Cyclophosphamide

Cyclophosphamide is an alkylating agent used intravenously or orally (divided over 3-4 days); this is metabolized by the liver into its active form; however, an additional metabolite is acrolein. Acrolein is excreted in the urine and is irritating to the bladder. Necrosis of the smooth muscle and chronic cystitis, with the end result of bladder fibrosis, can occur with long term use. Clinical signs of sterile hemorrhagic cystitis are hematuria, stranguria, and pollakuria. A free catch sample should be analyzed for the presence of hematuria, and full urinalysis and urine culture should be performed in all dogs and cats to rule out concurrent urinary tract infection. Once this particular side effect has occurred, the use cyclophosphamide is discontinued; chlorambucil may be used as a substitute in some protocols. Sterile hemorrhagic cystitis may be prevented by the some of the following measures: encouraging fluid intake, giving the medication in the morning, walking frequently throughout the day, and using prednisone to facilitate polyuria and polydipsia.[1-3]

Lomustine (CCNU)

Lomustine is an orally administered alkylating agent that is lipid soluble and crosses the blood brain barrier. Lomustine may cause a severe, delayed, and prolonged thrombocytopenia; additionally, prolonged neutropenia also occurs. The myelosuppression caused by lomustine in cats can be particularly troublesome as it may not occur for 1-6 weeks following treatment. Furthermore, lomustine is hepatotoxic in dogs; concurrent administration of Denamarin® may reduce the frequency of lomustine associated hepatotoxicity in dogs. Both the complete blood count and a liver panel should be closely monitored prior to and following lomustine administration.[2,3]

Doxorubicin

Doxorubicin is an anthracycline glycoside that irreversibly binds DNA and prevents both RNA and DNA synthesis. Iron mediated free radical formation is an additional mechanism of tissue damage. This free radical formation is the cause of its cardiotoxicity; dexrazoxane is an iron chelating agent that reduces cardiotoxicity in people. Cardiotoxicity is dependent on the cumulative dose administered and occurs at total doses of 180-240mg/m^2 over a lifetime. Lifetime dose is therefore limited to 6-8 treatments in dogs; an echocardiogram can be used to measure systolic function before and during the treatment course.

Doxorubicin is a severe tissue vesicant. Tissue necrosis occurs 1-2 weeks following extravasation and, if severe, may necessitate amputation of the affected limb. Therefore, precautions must be taken prior to and during administration, including the use of a clean, first-stick intravenous catheter and direct monitoring of the catheter during administration. This will minimize the amount of extravasated doxorubicin and the severity of phlebitis and necrosis. Dexrazoxane can also be used following extravasation; benefit may be seen if administered IV at 10 times the doxorubicin dose within 3 hours of extravasation, and again at 24 and 48 hours. However, dexrazoxane is expensive and can be difficult to obtain. Doxorubicin is renal toxic in cats (it can cause renal tubular damage); therefore, renal function must be closely monitored throughout treatment. Finally, doxorubicin can cause systemic histamine release and an anaphylactic type reaction - therefore, doxorubicin should be administered diluted over 20-30 minutes. [1-3]

Cisplatin

Cisplatin is fatal to cats, causing pulmonary edema, dyspnea, and death with 48-96 hours after administration. Treatment of cats with cisplatin is contraindicated. Cisplatin also has many unique toxicities in the dog. The first is acute nausea and vomiting during and immediately after administration due to direct stimulation of the vomiting center. The use of anti-emetics, particularly maropitant, prior to administration will greatly decrease the occurrence of this side effect. Cisplatin is nephrotoxic (renal tubular damage) and must be used with a pre-and post-treatment diuresis protocol to decrease the risk of nephrotoxicity. Cisplatin is excreted through the kidneys and into the urine, so personal protective equipment is necessary to prevent inadvertent personnel exposure to this drug.[2,3]

L-asparaginase

L-asparaginase is an enzyme derived from *E. coli* that depletes the body of asparagine. While normal cells are able to synthesize asparagine and avoid toxicity associated with this drug, cancer cells – specifically lymphoblastic lymphoma and lymphoblastic leukemia, depend on extracellular asparagine for survival. This causes a rapid response to treatment, although resistance can develop quite quickly. Asparaginase is a foreign protein, and, as such, can cause a severe allergic reaction, especially with repeated administration. The drug should be administered intramuscularly, as intravenous use will greatly increase the risk of an anaphylactic reaction. Although anaphylaxis is rare in the dog following L-asparaginase administration, it can be life threatening. Anaphylaxis can result in facial edema, bronchospasm, hypotension, urticaria, erythema, pruritis, pharyngeal/ laryngeal edema, arrhythmias, acute vomiting, and collapse. If a reaction is noted, immediate intervention with intravenous fluids, corticosteroids, and, in some cases, epinephrine is indicated. Rarely, L-asparaginase can also cause an acute necrotizing pancreatitis.[2,3]

Vinca alkaloids

The Vinca alkaloids – vincristine, vinblastine, and vinorelbine are tissue vesicants and extravasation causes localized inflammation and significant pain. Therefore, a clean first stick butterfly needle is used (a conventional catheter may be placed as well), and drug is administered as a bolus; saline flush following drug administration ensures no drug is remaining in the catheter or on the needle tip that may cause tissue irritation. Vincristine is the most likely of the vinca alkaloids to cause a sensory and motor neuropathy, with the cat potentially being more sensitive than the dog. When neuropathy occurs, severe nerve fiber degeneration and focal axonal swellings with secondary demyelination of peripheral nerves is seen. Treatment is discontinuation of vincristine and initiation of supportive care.[2,3]

CONCLUSIONS

Many of the chemotherapeutics commonly used in veterinary medicine have distinct side effects that must be monitored for. A working knowledge of these side effects will facilitate client education and patient care.

REFERENCES

[1]Boyer J et al. **The Basic Science of Oncology** 2005; 4th ed; 349-375. [2]Gustafson DL et al. **Small Animal Clinical Oncology** 2013; 5th ed: 157-179. [3]Wilson H et al. **Small Animal Clinical Pharm & Ther** 2012; 2nd ed: 1210-1235. [4]Mealey KL **Compendium** 2006; 23-33.

CHEMOTHERAPY OPTIONS FOR CARCINOMAS AND SARCOMAS IN PETS

Antony Moore BVSc, MVSc, DACVIM (Oncology)
Wauchope, NSW, Australia

INTRODUCTION

Technicians/nurses are often asked by owners for guidance in deciding about the benefits or otherwise of chemotherapy. It is important to recognize that although they are dealing with a rarely-curable disease, their pet can be treated in such a way that a high quality of life is achieved. In this respect, it is similar to treating heart or kidney disease, which can also be fatal eventually, but which can be managed for some time with a good quality of life. Perhaps the most important factors influencing the patient's quality of life and remission time are the care of the pet owner and veterinary cancer treatment team.

One of the very important roles for veterinary technicians/nurses in treating cancer with chemotherapy is emphasizing to owners that delaying starting treatment until their pet "needs it" is not a good strategy. In various studies of dogs with various cancers (such as lymphoma) that are unwell at the start of therapy, median remission times were approximately half of those reported for dogs who were well at the start of therapy. Even when a patient's history indicates they have been well, it is important to start therapy before such symptoms arise. It is important to stress to owners that delaying starting treatment until Symptoms of illness are seen decreases treatment success and may increase the risk of side effects.

Similarly when treating dogs for solid tumors, such as carcinomas or sarcomas, the effectiveness of chemotherapy is always going to be greater against the least number of cells; therefore, waiting to start chemotherapy until macroscopic evidence of metastasis is seen very likely to result in failure.

EXAMPLES OF TUMORS TREATED WITH CHEMOTHERAPY

Osteosarcoma (OSA)

A good example of the ability of chemotherapy to prevent metastasis is osteosarcoma in dogs. On the average, because microscopic lung metastases continue to grow after surgery alone, dogs live 5 months after surgery (amputation or limb salvage), and approximately 10% of dogs live over 1 year. Nevertheless, we still recommend amputation as the best method for relieving pain. Because amputation does not prevent growth of the metastases, the most important advance in treatment of osteosarcoma has been the finding that chemotherapy following surgery can slow the growth of the metastases, dramatically improve life expectancy, and in some cases result in a cure. Past experience has shown that three chemotherapeutic medications: cisplatin, carboplatin and doxorubicin, are effective in treating dogs with osteosarcoma. Overall (not accounting for elevations in serum ALP) approximately 45% of dogs live over 1 year when any of these drugs is given alone. Many oncologists recommend, for otherwise healthy dogs , a chemotherapy protocol that combines doxorubicin and carboplatin. For dogs completing this protocol, the overall chances of living 2 years beyond amputation are about 25%; approximately 20% of dogs live 3 years or more.

In contrast, the effectiveness of chemotherapy in treating visible osteosarcoma metastases is very poor. In one report, 45 dogs with osteosarcoma metastases were treated with cisplatin, doxorubicin, mitoxantrone, or sequential combinations of these drugs every 3 weeks. Only one dog experienced a partial remission for 21 days with doxorubicin.[1]

It is important to delay starting chemotherapy after major surgery. In a study that compared starting chemotherapy for dogs with osteosarcoma at 2 days or at 10 days post amputation, there was no difference in survival. However, dogs that were treated 2 days after surgery were more likely to show bone marrow suppression, presumably because the bone marrow had been stimulated by post-operative inflammation.[2] It is recommended that chemotherapy be delayed until wound healing has occurred, usually 10 to 14 days after surgery. This means that the best outcome for any patient comes from adjunctive therapy, delivered soon after surgery, but not immediately.

Transitional Cell Carcinoma (TCC)

Signs referable to the urinary tract are common with tumors of the lower urinary tract, the most common being TCC. Most presenting signs are identical to those in animals with infections or inflammatory conditions of the bladder or urethra, which may occur simultaneously with a tumor. Dogs with TCC may show clinical improvement when treated with antibiotics, but relapse with clinical signs. Chemotherapy for gross disease in TCC has limited efficacy, but has a reasonable chance of at least providing significant palliation. Being able to debulk a bladder tumor prior to chemotherapy appears to be advantageous for patients. The ability of chemotherapy to cause a complete remission in TCC is fairly low. However, any reduction in tumor volume usually improves clinical signs dramatically and TCC may take a long time to regrow. Many dogs that have a minimal (<50%) response, still survive a year.

Piroxicam has the best single agent efficacy with a reduction in tumor size of more than 50% in about 30% of dogs with no surgical intervention.

The best published chemotherapy outcomes are with mitoxantrone and piroxicam in combination. Response data were available for 48 dogs and included one complete response, 16 partial responses, 22 with disease stabilization, and 9 with progressive disease for an overall 35% measurable response rate. Most importantly, subjective improvement in clinical signs occurred in 75% of treated dogs. The median survival time in this study was 10 months, compared with 6 months with piroxicam alone.

Some veterinary oncologists have also had success using carboplatin chemotherapy for dogs with tumors resistant to piroxicam and mitoxantrone.[3] In a recent study, vinblastine caused a partial response in 36% of patients and stable disease in 50%,[4] which has led to some oncologists recommending a combination protocol. Other studies reported doxorubicin (and piroxicam) with a partial response rate of 8%, and stable disease in 62%; and chlorambucil which had a 3% PR rate (1 dog) and 67% stable disease; implying neither of the last two treatment are as effective as mitoxantrone, carboplatin and vinblastine.

Anal sac adenocarcinomas (ASAC)

ASAC usually occur in older dogs. Their typical behaviour is malignant, with both local infiltration and metastasis presenting important clinical problems. Metastatic disease at presentation is common. One study evaluated all factors in dogs with ASAC and developed a prognostic "grading" scheme. Dogs with only local tumor are in stage 1 or 2 (depending on the size of the cancer), while iliac lymph node enlargement elevates patients to stage 3, and the cutoff between 3a and 3b is node size of 4.5 cm.[5] If pulmonary metastases are present the median survival was found to be very short. As for other tumor types, removal of the primary tumor and any accessible lymph node metastases improves the chances for survival, and probably improves the chances for chemotherapy to be effective.

Chemotherapy for metastatic ASAC has not been evaluated in a controlled clinical trial; however, when metastatic disease is found, chemotherapy is the only treatment that has the potential to improve survival. One study showed that approximately one-third of dogs receiving platinum chemotherapy experienced a > 50% decrease in tumor size, so chemotherapy alone has approximately a 30% chance of causing a response. Efficacy is usually considered to be better if only microscopic disease is present (after surgery). We have used chemotherapy in an attempt to reduce the size of a perianal tumor and nodes prior to further surgery (occasionally with success, probably about the 30% reported above). In another study, dogs that received chemotherapy alone (usually because tumors were too extensive to approach surgically) had 67% survival at 6 months, with median survival of 7 months. Chemotherapy is always more effective when used against microscopic disease, so surgery first would be preferred.

Based on the responses seen in the paper above, most veterinary oncologists believe that the best chemotherapy drugs are platinum derived chemotherapy, but with other drugs such as doxorubicin, melphalan and mitoxantrone showing efficacy particularly in combination.

OTHER FACTORS IN TREATING WITH CHEMOTHERAPY

We recommend a course of prophylactic broad-spectrum antibiotics (trimethoprim-sulfa or enrofloxacin) each time a new myelosuppressive agent is administered, (until the patient's bone marrow response to each drug is known) – note that the nadir may vary between patients and drugs – for at least 3 weeks after the first carboplatin and at least 2 weeks after the doxorubicin.

Many pet owners wonder whether they should feed their pet anything special during chemotherapy. There is some evidence that higher protein and fat and lower carbohydrate content is beneficial to dogs with lymphoma, and possibly to dogs with other cancer types. A commercially available tinned food for dogs that addresses this need and is supplemented with some other nutrients that may benefit dogs with lymphoma is Hill's Prescription diet n/d, but home-made diets may also provide a similar benefit. Again, quality of life is the most important criterion for pets undergoing treatment for cancer and it is important that the diet chosen is one that the patient enjoys.

MOST COMMONLY USED CHEMOTHERAPY DRUGS

The chemotherapy drugs used most commonly in the treatment of carcinomas and sarcomas (not just the examples above) are provided in table form below.

REFERENCES

[1] Ogilvie GK, etal. JAVMA 1993;202:304. [2] Berg J, etal. Cancer 1997;79:1343. [3] Henry CJ, et al. Clinical Cancer Research 2003;9:906. [4] Arnold EJ, et al. JVIM 2011;25(6):1385. [5] Polton GA, Brearley MJ. JVIM 2007;21(2):274.

Table 1. Selected chemotherapy drugs used to treat carcinomas and sarcomas in pets.

Carboplatin

Available:	Available as: 450 mg in 45mL, or 150mg in 15mL
Route:	10mg/mL diluted to a total dose volume of 25mL (cats) or 50mL (dogs) with 5% dextrose and give slow IV push over 15-20 minutes OR undiluted, slowly inject into the injection port in the line of a running infusion of 5% dextrose over 15-20 minutes
Dosage:	According to protocol, usually every 3 (dogs) to 4 (cats) weeks.
Storage:	Refrigerate vials.
Monitoring:	Do not administer if neutrophil count is less than 3,000/uL

Cisplatin

Available:	50mg vials, reconstituted to 1mg/mL.
Route:	IV with catheter, or intracavitary given according to protocol.
Dose:	DO NOT USE IN CATS. Dogs: every 3-4 weeks, but dose varies with weight and according to specific protocol. Precede with dolasetron (Anzemet) or maropitant (Cerenia).
Administration:	Cisplatin must be preceded and followed by a diuresis period. Needles or intravenous sets containing aluminum parts that may come in contact with solution should not be used for preparation or administration. Aluminum reacts with cisplatin, causing precipitate formation and loss of potency.
Storage:	Store drug refrigerated.
Monitoring:	Renal function (serum creatinine and urine specific gravity) before each treatment). Do not administer if renal insufficiency.

Doxorubicin

Available:	10mg, 20mg, 50mg vials at concentration of 2mg/mL
Route:	Give intravenously with over-the-needle catheter. Administer into the injection port of a running saline infusion over 10 - 20 minutes (rate of about 1mL per minute).
Administration:	Rapid infusion may be associated with histamine release and anaphylactoid reaction (restless, urticaria, vomiting, even collapse); slow infusion avoids this. Use plain normal saline for flushing – heparin may cause a precipitate to form. Vesicant when given perivascularly.
Dosage:	Dosage usually every three weeks according to protocol
Storage:	Store in refrigerator and protect from light.
Monitor:	Do not administer if neutrophil count is less than 3,000/uL

Mitoxantrone

Available:	10mg, 20mg, 25mg vials at concentration of 2 mg/mL
Route:	Intravenously with over-the-needle catheter. Administer into the injection port of a running saline infusion over 5 - 10 minutes.
Administration:	Use normal saline for flushing - no heparin - may form a precipitate.
Dosage:	According to protocol, but usually every 2 to 3 weeks
Storage:	Refrigerate.

Vinblastine

Available:	10mg, vials at concentration of 1 mg/mL
Route:	IV with catheter as a bolus injection.
Administration:	Vesicant when given perivascularly.
Dosage:	According to protocol, but usually every 1 to 2 weeks
Storage:	Refrigerate. Reconstituted solution is stable for 30 days refrigerated and if protected from light
Monitor:	Do not administer if neutrophil count is less than 3,000/uL

THE NUTS AND BOLTS OF PACING THERAPY IN SMALL ANIMAL PRACTICE

Amara H. Estrada, DVM, DACVIM (Cardiology)
Gainesville, FL

WHICH PATIENTS AND HOW?

The first fully implantable pacemaker was placed in a person in 1958[1]. The first artificial pacemaker implantation in a dog was described more than 40 years ago[2]. Since that time, the history of artificial cardiac pacing has seen a long line of innovation resulting in a constantly evolving field of medicine. In veterinary medicine, the most common reasons to place an artificial cardiac pacemaker are atrioventricular (AV) block (2nd degree and complete) and sinus node dysfunction/sick sinus syndrome[3]. However, pacing therapy is not a 'one size fits all' practice and some options and considerations for pacemaker type, lead location and programming are important for patients with different types of bradyarrhythmias which require artificial pacing.

ATRIOVENTRICULAR (AV) BLOCK

Dogs versus cats

Atrioventricular (AV) block is the most common indication for permanent pacemaker implantation. Dogs and cats differ in terms of their escape rhythm rates, with cats typically having a faster ventricular escape rhythm (90-120 bpm) such that they do not frequently require artificial cardiac pacing[4]. Occasionally, cats with cardiomyopathy will develop bradyarrhythmias with slow enough escape rhythms where pacemaker therapy is required to either keep them out of congestive heart failure or improve quality of life if the rate is slow enough to be affecting their activity. There have been reports of cranial vena caval obstruction and resultant development of chylothorax in cats with transvenous pacing[5], and thus most cardiologists choose to place an epicardial lead on the left ventricular apex through an abdominal/diaphragmatic approach in cats if pacing is required. There are however, small diameter leads (4.1 French) that may be suitable to use in cats but thus far the experience with this is minimal.

Single versus dual chamber pacing

Most dogs affected by AV block (whether 2nd degree or complete) have been treated by single chamber ventricular based pacemaker (VVI) implantation. These systems consist of one right ventricular apical lead and therefore do not maintain AV synchrony or the normal ventricular activation sequence. Numerous studies in humans and animals have shown negative consequences of such non-physiologic pacing[6-11]. The term pacemaker syndrome is used in human medicine to describe adverse clinical and hemodynamic consequences of non-physiologic pacing. Compared with right ventricular pacing, studies have shown improved hemodynamics and quality of life benefits in people with dual chamber pacing, in which atrioventricular synchrony and the contribution of atrial contraction to ventricular filling is preserved[12-14]. Additionally, pacemakers with ventricular pacing sites other than or in addition to the right ventricular apex (interventricular septum, left ventricular free wall via the coronary sinus) allow for a more physiologic ventricular activation sequence and have also shown to provide improved hemodynamics in dogs[15-19] and humans[20,21] and quality of life benefits in humans[22]. At the University of Florida, we have been evaluating pacing site in dogs with complete AV block for both short and long term benefit or complications associated with different lead configurations and pacing sites. Thus far, we have been able to show that there is dyssynchrony created with right ventricular apical pacing and that biventricular (via RV apex and coronary sinus leads), but not LV pacing (via coronary sinus lead) appears to preserve LV synchronization in dogs with complete AV block treated with artificial cardiac pacing[15]. The significance and long term functional effects of biventricular pacing in dogs with complete AV block is unknown but currently under evaluation by our group.

It has been previously reported that dual chamber pacing systems, particularly those with 2 leads, are more difficult and more time consuming to place in dogs because of smaller vessels and the requirement for 2 leads[3]. Additionally, dual chamber pacemakers require more complex programming. Thus, it has been assumed that implantation of dual chamber pacemakers would have higher complication rates both initially and in the long-term. Our institution has recently published data that shows that while dual chamber and multiple leads does prolong anesthesia and procedural time, dogs do not have any increased incidence of either short term or long term complications associated with multiple leads or longer anesthesia/procedural time[23]. Another type of pacing, atrial synchronous pacing (via VDD implantation), allows for dual chamber pacing with only a single lead. For atrial synchronous pacing, 1 lead is implanted in the right ventricle, but unlike typical VVIs this lead has a floating electrode in the right atrium that is capable of sensing inherent atrial electrical activity and stimulating ventricular contraction after an appropriate delay, therefore allowing for atrioventricular synchrony. Although an atrial synchronous pacing system has been used in veterinary medicine, the set distance between the floating atrial electrode and ventricular pacing tip often prevents its use in small dogs. Studies comparing complication rates of single versus dual chamber pacemakers in the human literature have yielded conflicting results. In veterinary medicine, owner satisfaction with single chamber pacemakers has been high, thereby discouraging veterinary cardiologists from change cardiac pacing methods.

SINUS NODE DYSFUNCTION/SICK SINUS SYNDROME (SND/SSS)

Atrial pacing versus dual chamber pacing versus ventricular back up pacing

A recent survey of cardiologists presented at ACVIM in 2011 showed that the majority of interventional cardiologists use single chamber ventricular (VVI) systems in dogs with SND, regardless of their ability to naturally conduct through the AV node. It is important to keep in mind that the majority of dogs with SND retain the ability to conduct through the AV node. This is important not only because this may be used for conduction of antegrade pacing from the atrial myocardium during atrial based pacing, but also because retrograde conduction of impulses through the AV node is also possible in dogs paced with ventricular based pacing systems. In a pilot study performed at our institution, we have looked at dogs with SND that were paced with VVI systems and found that a large portion of these patients have retrograde conduction and echo beats which occur following ventricular pacing. It is therefore important that if using VVI pacing in dogs with SND, it is used only as 'back up' for periods of bradycardia in order to retain as much AV nodal conduction as possible.

Another recent retrospective trial in veterinary medicine failed to show improved survival in dogs paced with dual chamber pacing systems in comparison to VVI[23]. The lack of benefit is thought to be attributable to the untoward effects associated with RV apical pacing. In humans, RV apical pacing results in negative cardiac remodeling, decreased myocardial perfusion, systolic and diastolic left ventricular dysfunction, increased catecholamine levels, and ultimately, heart failure[24-26]. Similar findings have been reported in dogs. Abnormal activation of the ventricles results in histologic myocardial abnormalities, whereas dogs paced with normal activation of their ventricles have a normal myocardial appearance histologically[27]. Additionally, an acute decrease in systolic function has been shown with RV apical pacing in dogs[28]. Finally, new or worsened mitral regurgitation has been documented in humans both acutely and with chronic RV apical pacing as a direct result of the abnormal activation sequence of the ventricles[29-32]. This may be particularly important in the canine population as many dogs with SND have concurrent degenerative valve disease with mitral regurgitation prior to pacemaker implantation.

Thus, the ideal pacemaker for these dogs would be one that alleviates their symptoms associated with bradycardia, but also maintains AV synchrony and normal ventricular activation; atrial-based pacemaker (AAI) would achieve these goals. AAI pacing is currently considered the most physiologic method of pacing. Several large human trials have investigated atrial pacing. The Danish pacing trial[33,34] compared AAI to VVI pacing in patients with SND. The AAI group had a lower incidence of atrial fibrillation and thromboembolic events, whereas the VVI group showed progression of heart failure signs and had a significantly higher mortality rate. More recently, the SAVE PACE trial[35] compared conventional dual chamber pacing systems to newer pacing systems that employed algorithms to allow for minimal ventricular pacing and predominantly AAI pacing in SND patients. The group with minimal ventricular pacing showed a 40% lower incidence of persistent atrial fibrillation. Some human cardiologists avoid atrial pacing due to concerns for patients developing AV block over time, particularly those with a Wenckebach point of less than 120 bpm. The Wenckebach point is the heart rate at which a patient develops physiologic heart block at the level of the AV node to prevent significant tachycardia that could be hemodynamically compromising. While the risk for such patients to develop AV block appears to be significantly less than originally proposed[36], assessment of Wenckebach point remains standard of care prior to atrial pacemaker implantation in humans. In a group of 16 dogs with SND and atrial based pacing, only one developed AV block following the procedure and this patient had no assessment of Wenckebach point prior to atrial lead implantation[37]. In another group of normal canine patients in a pilot study run at the University of Florida, the Wenckebach point has been tested under only isoflurane anesthesia and dogs were found to be able to conduct normally through the AV node with the highest atrial pacing of 220 ppm.

Atrial pacing site

Conventionally, the atria have been paced from the right atrial appendage, however several complications have been reported when this site was attempted in small breed dogs with naturally occurring SND, including partial and complete lead dislodgement, pneumothorax and lead perforation[37]. While atrial-based pacing is preferable in such patients, the right auricular appendage does not appear to be an ideal site for these small dogs (7-10kg). Preformed atrial "J" leads are designed for adult human atrial appendages and these have proven unsuitable for these small patients. In the human literature, alternative atrial pacing sites have been proposed including the interatrial septum, right atrial freewall and left atrial pacing via the coronary sinus. Studies have shown significant reduced interatrial conduction time and reduced P wave duration with interatrial septum pacing, leading to a more synchronous atrial contraction[38,39]. This may result in a near-normal atrial contribution to ventricular filling, which has been shown to contribute up to 30% of the total ventricular filling volume in dogs. We have been evaluating this pacing site in small breed dogs and while it does appear that it is possible to implant leads in the inter-atrial septum for permanent pacing in small breed dogs with SND, there is a learning curve for placement of leads in this location which is likely a similar learning curve experienced with ventricular pacing for dogs with AV block.

REFERENCES

[1]Elmqvist R, et al. **Medical Electronics**; 1960: 253. [2]Buchanan JW, et al. **JAVMA** 1968;152(8):1099-1109. [3]Johnson MS, et al. **JSAP**;2007:4-11. [4]Kellum HB, et al; **JVIM** 2006:20(1):97-103. [5]Ferasin L, et al. **JSAP** 2002;43(3)124-128. [6]Nishimura RA, et al. **Pac Clin Elect** 1982;5(6):903-910. [7] Rediker RE, et al. **Am J Card** 1988;61(4):323-329. [8]Karpawich PP, et al. **Pac Clin Elect** 1999;22(9):1372-1377. [9]Karpawich PP, et al. **Am Heart J** 1990;119(5):1077-1083 [10]Sweeney MO, et al. **Circ** 2003;107:2932-37. [11]Lee MA, et al. **JACC** 1994;24(1):225-232. [12]Linde-Edelstam C,at al. **Pac Clin Elect**, 1992;15(10 pt 1):1467-1476. [13]Lamas GA, et al. **N Engl J Med** 1998 ;338(16):1097-1104. [14]Dretzke J, et al. **Coch Data Sys Rev** 2004;(2):CD003710. [15]Estrada AH, et al. **JVC**, 2009;11(2):79-88. [16]Vanagt WY, et al. **Pac Clin Elect**, 2004;27(6 pt 2):837-843. [17]Prinzen FW, et al. **Pac Clin Elect**,1998;21(11 pt 2):2256-2260. [18]Fei L, et al. **J Card Elect**, 1999;10(7):935-946. [19]Peschar M, et al. **JACC**, 2003;41(7):1218-1226. [20]Mills RW, et al. **Circ Arrhythm Elect**, 2009;2(5);571-579. [21]Lieberman R, et al. **JACC**, 2006;48(8):1634-41. [22]Krahn AD, et al. **Can J Cardiol**, 2002:18(4);380-387. [23]Genovese D, et al. **JAVMA** 2013; 242(2) 230-236. [24]Occhetta E. **World J Cardiol**, 2011;3(1):32.[25]Tse HF, et al. **JACC**, 1997;29(4):744–749. [26]Lee MA, et al. **JACC**,1994;24(1):225–232. [27]Karpawich PP, et al. **Am Heart J**,1991;121(3 Pt 1):827–833. [28]Maisenbacher H III, et al. **AJVR** 2009;70(4):455-63. [29]Alizadeh A, et al. **Europace**; 2011;13(11):1587–1590. [30]Lewicka-Nowak E, et al. **Kardiologia Polska**, 2006;64(10):1082–93. [31]Maurer G, et al. **JACC** 1984;3(4):986–991. [32]Sassone B, et al. **Ital Heart J**, 2001;2(6):441–448. [33]Andersen HR, et al. **Lancet**,1997;350(9086):1210–1216. [34]Andersen HR, et al. **Circ**, 1998;98(13):1315–1321. [35]Sweeney MO, et al. **NEJM**, 2007;357(10):1000–1008. [36]Adachi M, et al. **Europace**, 2008;10(2):134–137. [37]Estrada AH, et al. **JVIM**, 2012; 26(3):558-564. [38]Jamidar Het al. **Pac Clin Elect**, 1993;16(5 Pt 1):959–963. [39]Dabrowska-Kugacka A, et al. **Circ**, 2010;74(7):1308–1315.

SHAKE, RATTLE AND ROLL: RECOGNITION OF MOVEMENT DISORDERS

Theresa E. Pancotto, DVM, MS, DACVIM (Neurology)
Blacksburg, VA

INTRODUCTION

Movement disorders are characterized by involuntary muscle contractions in conscious patients. The integrity of the conscious mind is what makes these phenomena unique from seizures and sleep disorders. Movement disorders may arise from alterations within the skeletal muscle, peripheral nerves, or central nervous system. The simplest way to classify movement disorders is based on neuroanatomic origin, however they may also be classified based on phenotypic appearance e.g. tetany/tetanus, tremor, etc. Because of variability in the clinical appearance of movement disorders, phenotypic classification may be difficult. Many of the movement disorders identified in the veterinary field to date are breed-related, suggesting a genetic basis for these conditions (1-3). Although genomic analysis is advancing rapidly, some movement disorders have yet to be genotypically defined. Once recognized, electrophysiology is useful in further characterizing the movement disorder and determining the neuroanatomic origin.

The following is a summary of clinical appearance and neuroanatomic location of some of the more frequently seen movement disorders, most of which occur in dogs. The presentation will focus on videos demonstrating the various phenotypes of movement disorders.

CAUSED BY DISEASE OF THE SKELETAL MUSCLE

Myotonia congenita

Myotonia is persistent, sustained, muscle contraction with delayed relaxation. Congenital myotonia presents in the first few months of life and is most commonly recognized in Miniature Schnauzers and Chow Chows. The appearance of myotonia is that of generalized stiffness that may improve with exercise. Affected animals may also develop hypertrophy of the proximal limb muscles and tongue. In miniature Schnauzers a DNA test is available to detect genetic carriers (1). The mutation prevents repolarization of the muscle via effects on the voltage-gated chloride channels such that the muscle cell remains in prolonged contraction.

Myotonic discharges on EMG evaluation are diagnostic for the disease and are characterized by high-frequency, waxing and waning discharges. The sound has been dubbed the "dive-bomber."

Interestingly the most well-known congenital myotonia is that of the "fainting goats." These animals also have a defect in the chloride-channel of the muscle preventing normal repolarization. They experience sudden onset of diffuse extensor muscle tone after being startled. The rigidity may lead them to fall over or "faint" (consciousness is normal in these animals throughout the episode). Unlike dogs, the goats fully recover, without stiffness, between episodes.

Pseudomyotonia

Pseudomyotonia clinically appears similarly to myotonia (above) but animals with pseudomyotonia have acquired the disease, usually as a complication of spontaneous or iatrogenic hyperadrenocorticism (Cushing's disease). The EMG has a similar high frequency discharge but does not wax and wane.

Dancing Doberman

Affected Dobermans of either sex or any age (6 months- 7 years) are presented for intermittent flexion of one pelvic limb while standing. Over time the other pelvic limb becomes affected resulted in a dancing movement. Atrophy of the gastrocnemius (calf muscle) may occur over time along with neurologic deficits. The disease is thought to result of from changes within the muscle although some animals undergoing necropsy evaluation have had disease in the peripheral nerves and/or spinal cord. EMG changes are not specific but progress in severity over time (4).

CAUSED BY DISEASE OF THE PERIPHERAL NERVOUS SYSTEM

Myokimia

Myokimia results in persistent, undulating muscle contractions that spread across the surface of the muscle in the limbs or face. It has been described as looking like worms crawling under the skin. The contractions can persist through sleep but resolve under anesthesia. They have a characteristic burst pattern on an awake EMG.

Neuromyotonia

Clinically neuromyotonia is characterized by persistent muscle contracture that may result in transient deformities of the distal limb or face (mild to moderate) or generalized muscle contracture and collapse episodes (severe). Severe episodes are commonly associated with hyperthermia as a result of the ongoing muscle activity. It's important to recognize in animals that develop neuromyotonic collapse, that the witnessed event is not a seizure. Neuromyotonic discharges have a higher intraburst frequency on EMG than myokimic discharges.

Both myokimia and neuromyotonia are a result of the same underlying pathology, which typically affects potassium channels of the peripheral nerve. Previously myokima/neuromyotonia has also been referred to as continuous muscle fiber activity (CMFA). Although the disease appears to be common in terrier breeds (especially Jack Russel Terrier) a genetic mutation has not been identified. Although a large number of Jack Russel Terriers with myokimia have developed signs consistent with hereditary ataxia, this disease is due to a different underlying pathologic (degenerative) change (5). In people, in addition to genetic defects, toxins and paraneoplastic syndromes have resulted in myokimia/neuromyotonia.

Constant Repetitive Myoclonus

Myoclonus refers to a sudden contraction of a group of muscles followed immediately by relaxation. Constant repetitive myoclonus usually affects a single limb or the jaw and leads to rhythmic contractions that are more pronounced at rest but also occur during action and sleep. This is most commonly seen in association with canine distemper encephalomyelitis but has also been reported in a dog with lead intoxication. Constant repetitive myoclonus is thought to result from a functional disturbance in the microenvironment of the lower motor neurons (peripheral motor nerves).

Essential Postural Repetitive Tremor of the Head (Head-bobbers)

Tremors may be best thought of clinically as rapidly occurring myoclonus. Head bobbing (essential postural tremor of the head) is commonly mistaken for continuous seizure activity and animals are frequently rushed to the emergency clinic. In fact this disorder is quite benign. Affected dogs will have rapid movement of the head and neck in a "yes" direction or a "no" direction. They often resemble the once and newly popularized bobble-head dolls. Head bobbing has never been demonstrated to progress, the events may wax and wane over the course of an animal's life and the events are easily interrupted by distracting the pet. Breeds most commonly affected are the brachycephalics (Boxers and bulldogs) as well as the Doberman. The underlying etiology remains to be determined but a may be a result of a defective stretch-receptor mechanism in the cervical musculature (6).

CAUSED BY DISEASE OF THE CENTRAL NERVOUS SYSTEM

Resting Myoclonus/Resting Tremor

In humans a resting tremor occurs due to degeneration of the substantia nigra from Parkinson's disease. A similar entity has not been recognized in domestic animals. Interestingly, horses that ingest the toxin present in yellow star thistle develop degeneration of the substantia nigra and globus pallidus (basal nuclei of the brain) but do not show resting tremors, nor do Kerry blue terriers with abiotrophy of the cerebellum and subsequent substantia nigra degeneration (7).

Tetany/Tetanus

Tetany is a temporary state of sustained muscle contraction usually affecting the extensor muscle groups. Tetanus is when this extension becomes sustained and occurs without periods of relaxation. Many diseases that cause tetanic contractions are a result of inhibiting inhibitory-interneurons, thus leading to hyperexcitability. This is usually a result of loss of the influence of the inhibitory neutrotransmitters (i.e. glycine, GABA) on the normal tonic inhibition of motor neurons to the extensor muscles.

Tetany/tetanus may develop as a result of an inherited defect, metabolic defect (hypocalcemia, hypomagnesemia), intoxication, or traumatic insult.

Familial Reflex Myoclonus

The name is misleading for this disease as it more typically resembles a tetanic state. Clinical signs include tetanic spasms of the muscles of the limbs and sometimes face resulting in generalized extensor rigidity and grimacing. The disorder occurs in Labrador puppies and is thought to be associated with a heritable deficiency of glycine receptors in the spinal cord.

Tetanus: *Clostridium tetani* intoxication

Wounds infected with the anaerobic bacterium *Clostridium tetani* may develop tetanus as a result of production of tetanospasmin by the bacteria. This neurotoxin travels up the peripheral nerves to the central nervous system to come to rest in the interneurons of the brain and spinal cord. Interneurons normally provide inhibitory tone via release of glycine and GABA within the central nervous system; the toxin tetanospasmin prevents this by binding to the neurotransmitter vesicle and preventing release of neurotransmitters. A characteristic facial expression develops a result of tetanospasmins actions on interneurons in the brain. This expression is frequently called "risus sardonicus" and affected animals have their ears pulled taut and high, the eyelids elevated, and the commissures of the lips retracted. These animals may appear to have repetitive ocular retraction as a result of spasms affecting the extraocular muscles.

Stiff Person Syndrome (SPS)

SPS is a rare condition of people that develop anti-bodies against the enzyme GAD (glutamic acid decarboxylase). GAD is needed to make the inhibitory neurotransmitter GABA. Without GAD there is a shortage of GABA and inability to provide inhibitory tone, leading to a generalized hyperexcitability syndrome with tetanic contractions of the axial musculature.

Action-related Tremors

This type of tremor, when confined to the head and neck, is seen classically and commonly in association with cerebellar diseases. It is frequently referred to as an intention tremor. A fine tremor of the head and neck occurs in association with movement of the head and neck, such as trying to prehend food or water. The tremor is either absent or less pronounced at rest. Affected animals usually have other signs indicative of cerebellar dysfunction such as a dysmetric gait, truncal sway, or vestibular dysfunction. Many different primary diseases including congenital (cerebellar hypoplasia), degenerative, inflammatory, neoplastic, and vascular insults can result in this clinical sign.

Steroid responsive tremor syndrome is an auto-immune syndrome resulting in the acute-onset of *diffuse* action related repetitive tremors. The syndrome has also previously been known as "white shaker disease/little white shakers" since it has a tendency to affect small, white dogs and causes generalized "shaking." However, since its original description, steroid responsive tremor syndrome has been reported in many breeds of many coat colors. Affected animals are usually young adults and the clinical signs develop as a result of non-suppurative inflammation of the cerebellum secondary to immune dysfunction. Unlike other inflammatory disorders, this syndrome responds quite well to steroids alone.

In animals that develop diffuse action related tremors at or before the time of ambulation, disorders of myelin formation should be considered. These disorders may include a deficiency of myelin production (hypomyelination) or abnormally produced myelin (dysmyelination). Other symptoms of diffuse CNS dysfunction (e.g. seizures) may occur. In some instances (Chow chow, Weimaraner breeds with inherited hypomyelinoses), the disease will resolve spontaneously (8). These animals will frequently bunny-hop or have a pronounced bouncing component to their gait.

Many neurotoxins incite a diffuse whole-body action tremor as the first clinical sign of intoxication. Toxins include metaldehyde (snail bait), lead, chlorinated hydrocarbons, and mycotoxins among others. Some toxins may have a peripheral site of action such as pyrethrins, carbamates, and organophosphates. Intoxications may be preceeded by GI signs and could progress to seizures and coma.

Orthostatic Tremors

Orthostatic indicates that something is related to a particular posture. In the case of orthostatic tremors, a tremor typically of the limbs develops when affected animals are in the standing or erect posture. This condition has been recognized in large and giant breed dogs including Great Danes and mastiffs. The tremor is absent at rest, walking, running, or off-loading weight from the limbs. It is progressive over time, though periods of clinical remission may occur. The EMG reveals a firing rate that is more rapid that most other forms of tremor. The EMG tremor pattern is present in conscious animals at rest. Previous investigations are suggestive that orthostatic tremor is driven by a supraspinal generator and all the limbs and motor nuclei are affected similarly (2).

A clinically similar entity occurs in older dogs of any breed and is of unknown significance and undetermined origin. This postural tremor is known as essential tremor or benign geriatric tremor and appears to occur almost exclusively in the pelvic limbs. It may be more similar physiologically to the head-bobbers (above) as it appears to require a certain degree of muscle tension to be present (7).

Scottie Cramp

This is a syndrome of episodic hypertonicity elicited by stress, excitement, or exercise. It results from a deficiency of the inhibitory effects of serotonin. It has been reported in young Scottish Terriers (6 weeks – 18 months) and is inherited (9).

REFERENCES

1Bhalerao DP, et al. **Am J Vet Res** 2002;63(10):1443-1447. 2 Garosi LS, et al. **J Vet Int Med** 2005;19(4):606-609. 3 March P, et al. **Prog Vet Neuro** 1994;4(1):19-24. 4 Chrisman CL. **Prog Vet Neuro** 1990:1(1):83-90. 5 Vanhaesebrouck AE, et al. **J Vet Int Med** 2010;24(4):882-889. 6 Dewey CW. **A Practical Guide to Canine and Feline Neurology** 2008:316. 7 de Lahunta A, et al. **Compendium** 2006;28(7):516-529. 8 Millan Y, et al. **J Sm Anim Pract** 2010;51:594-598. 9 Shelton GD. **Vet Clin Small Anim** 2004;34:1483-1496.

ELECTROENCEPHALOGRAPHY IN THE NEUROLOGIC DIAGNOSTIC TOOLBOX

Christine Weaver, LVT, VTS (Neurology)
Leesburg, VA

INTRODUCTION

Electroencephalography (EEG) is the practice of graphically recording the spontaneous electrical activity of the cerebral cortex, interpreting that recording, and the science related to the electrical activity of the brain. EEG is a regular part of the evaluation in human ICU medicine for patients with central nervous system (CNS) disease[1] (either altered mental status or possible seizure) but is not regularly performed in clinical veterinary medicine. In our clinic, EEG studies are interpreted by a physician skilled in both the interpretation of human and animal EEG. Consequently, EEG has become a regular part of the diagnostic evaluation in dogs and cats suspected of having seizures as evidenced by severe confusion and an inability to walk, coma and semi-comatose states, and ongoing or active movement disorders. EEG has proven itself better than observation alone in distinguishing seizures from seizure-like events.

HISTORY

Human EEG was discovered and developed by German scientist Hans Berger. In 1892 he was nearly killed in a military training exercise, and many miles away, his sister was suddenly felt he was in grave danger. After he retired from military service, Berger was determined to find a physiologic basis for "psychic energy" and strove to define a correlation between objective brain activity and the subjective "psychic phenomenon." He studied the research of Richard Caton, a British scientist who used a sensitive galvanometer on the surfaces of living brain of dogs and apes. Caton found that there were variations in the electrical current that changed with sleep, onset of death and after death. Berger recognized the importance of the findings of electrical activity of the brain and in 1924 made the first EEG recording and in 1929 discovered the alpha rhythm.

INDICATIONS

Prior to the development of advanced imaging techniques a common use of EEG was to detect the location of diseased brain tissue. EEG may be useful in the detection of hydrocephalus but the advent of MRI and CT has replaced the use of EEG for this diagnosis. Current indications for use of EEG includes the confirmation and characterization of seizure disorders, detection of non-convulsive status epilepticus (NCSE)[2], assisting in determining end-point of treatment in NCSE[2], diagnosis of REM sleep disorder[9], and evaluation of comatose patients. In humans EEG can be a predictor of mortality, morbidity, and costs of hospitalization[3]. Pilot data from our clinic suggest that this may also be true in veterinary patients.

PROCEDRE

One of the major challenges to obtaining a diagnostic EEG in a veterinary patient is the artifacts created by muscle and movement during the study. Various strategies to overcome this dilemma have been explored and include sedation, anesthesia, topical anesthetics and placement of subdural electrodes[4]. Most patients will require sedation for short term scalp recording. Recording should be performed in a dimly lit, quiet room. A standardized sedation protocol using the fewest medications with the least influence of the natural EEG is recommended. The goal of sedation is to minimize muscle artifact as well as induce slow wave sleep (SWS), a state which may improve the chances of recording epilpetiform discharges. Ideally the patient is sedated with an Alpha-2 agonist, such as dexmedetomidine, which produces SWS without altering the natural EEG in people, dogs, cats, and horses[8]. A disadvantage of dexmedetomidine induction of slow wave sleep is that it makes it more difficult to identify pathologic slowing of the EEG.

Due to convenience of use, the subdermal needle electrode is the most commonly used electrode in veterinary medicine EEG[1]. Surface electrodes are primarily used in human medicine. Subdural cortical electrodes are also available but require anesthesia and surgery increasing risk to the patient[4]. The subdermal electrodes are placed in specific locations on the scalp to correspond with locations in the cerebral cortex. Generally, more electrodes improves localization of a lesion. With the transition from paper to digital EEG, the same recording can be reformatted to the EEG can be displayed in numerous ways. The EEG recording is performed while the patient is sedated and continues through the transition from sleep to alert levels of awareness. Interpretation of the EEG consists of discriminating artifact from pathology against the backdrop of different drugs and states of awareness. Interpretation of the EEG is challenging and requires an experienced expert.

WHAT THE EEG SHOWS

When reviewing an EEG recording, the clinician is looking at the background rhythm, and paroxysmal (rapid onset and offset) activity. The background rhythm is the persistent, or dominant, part of the EEG that reflects state of awareness. The waveform of the background rhythm is described in amplitude and frequency. The amplitude is measured in microvolts (μV) from the peak of positivity to the peak of negativity and the frequency is measured in cycles/second or Hertz (Hz). An alert state of awareness is recorded as high frequency and low amplitude. A slow wave sleep state of awareness is recorded as low frequency and high amplitude. This sleep state is most frequently associated with epileptiform activity[5]. REM, or paradoxical sleep, is noted with low amplitude, high frequency waveforms and rapid eye movements.

Paroxysmal activity can be from a seizure focus, sleep associated or from artifact. An artifact is an activity recorded on EEG that does not originate from the brain. Artifacts are patient-origin artifacts, physiologic artifacts, external non-physiologic artifacts[6]. Paroxysmal activity can be physiologic or pathologic. Difficulty in interpreting the EEG recording is encountered when trying to determine precisely what specifically each artifact or transient activity represents. Examples of external non-physiologic artifacts include 60 cycle interference secondary to cell phones and running electrical units, infusion pumps causing drip artifacts, movement of the electrodes or dirty electrodes. Physiologic artifacts can be caused by muscle, ear, eye, swallowing, tongue and respiratory movements, ECG and gross patient movement. Filtering electrical noise, high quality differential amplifiers, and recording during sleep or sedation to reduce muscle artifact will facilitate fewer artifacts as well as use of video-EEG or meticulously documented records for movement of and around the patient.

Physiologic and pathologic transients are the activities of greatest interest to the clinician. Sleep associated physiologic transients can be the most difficult to discriminate from a seizure focus and include sleep beta bursts, sleep spindles, vertex sharp waves, which when seen together are called a K complex. Pathologic transients indicating a possible seizure focus include spike, spike with slow wave or sharp wave. Periodic epileptifom complexes or discharges are recognized as spike, spike and slow wave or sharp waves noted a regular interval. When these periodic complexes speed up or evolve within the record instead of firing at a regular interval and are noted at greater than 2 cycles per second then an electrical seizure or status epilepticus (SE) is noted.

THE VALUE OF EEG

Though regarded as an antiquated technology, recent research has shown that EEG in human and veterinary medicine provides valuable prognostic information as well as provide the clinician with objective assessment of patients with suspected brain disease. Studies in the human medical industry shows that patients with abnormal EEG have greater cost of healthcare, significantly higher mortality and more re-examinations[3]. Cost of healthcare for patients with abnormal EEG were 3.7 times greater than that of patients with normal EEG[3]. A recent retrospective study in veterinary neurology suggests that abnormal EEG findings in patients that also had abnormal MRI and CSF findings had higher mortality, higher hospitalization costs and longer hospitalization stays.

Determining a definitive diagnosis of seizure, SE, NCSE or movement disorder allows the clinician to best treat a patient and give owners a more accurate prognosis. Seizure, the manifestation of hypersynchronous neuronal activity, is often diagnosed by subjective observation of the veterinary clinician. The only definitive means of diagnosing a seizure event in a patient is by use of EEG. NCSE, the most common type of SE in people, may not be recognized and can go untreated, leaving the patient vulnerable to irreversible neuronal damage as a consequence of SE[7]. Conversely, patients that appear to be having seizures, may be treated with anti-convulsant therapy with little to no adverse effects if they are suffering from a movement disorder and not a seizure disorder.

REFERENCES

[1]James F., et al. Investigation of the use of three electroencephalographic electrodes for long-term electroencephalographic recording in awake and sedated dogs. AJVR, Vol. 72, No. 3, March 2011, pp 384-390. [2]Walker M. Diagnosis and Treatment of Nonconvulsive Status Epilepticus. CNS Drugs 2001: 15 (12) pp 931-939. [3] Stecker MM. The EEG as an independent indicator of mortality and healthcare utilization. Clin Neurophys 120 (200)(1777-1781). [4] Davis K. et al. A Novel Implanted Device to Wirelessly Record and Analyze Continuous Intracranial Canine EEG. Epilepsy Research (2011) 96, pp 116-122. [5]Wrzosek M, et al. Visual and quantitative EEG analysis of healthy young and adult cats under medetomidine sedation. The Veterinary Journal 180 (2009) 221-230; [6]James F., et al. Investigation of the use of three electroencephalographic electrodes for long-term electroencephalographic recording in awake and sedated dogs. AJVR, Vol. 72, No. 3, March 2011, pp 384-390. [7]Raith K, et al. Continuous electroencephalographic monitoring of status epilepticus in dogs and cats: 10 patients (2004-2005) [8]Fink,et al. Guidelines for the use of EEG methodology in diagnosis of epilepsy. Acta Neurol Scand 2002:106:1-7. [9]Bush WW et al. Diagnosis of rapid eye movement sleep disorder with electroencephalography and treatment with tricyclic antidepressants in a dog. Journal of American Animal Hospital Assoc. 2004 Nov-Dec;40(6):495-500.

A CRITICAL REVIEW OF ZOMBIES IN THE ARTS: NEUROANATOMY, NEUROPHYSIOLOGY, AND NEUROPATHOLOGY

Charles H Vite DVM, PhD, DACVIM (Neurology) [1], Adam Moeser DVM[1], Carrie Jurney DVM, DACVIM (Neurology) [2], Evelyn Galban DVM, MS, DACVIM (Neurology) [1], Peter Early DVM, DACVIM (Neurology) [3], Sheldon A. Steinberg VMD, DMSc, DACVIM (Neurology), Dipl ECVN[1]

[1]Philadelphia, Pennsylvania, [2]San Francisco, CA, [3]Raleigh, NC

"The Haitian peasant recognizes two types of zombies. One is a spirit, a part of the Vodoun soul which has been sold to or been captured by a bokor and which, if released, will be "doomed to wander the earth until its destined time arrives to return to God" (Herskovits 1975). These spirit zombies, once captured, however, are usually carefully stored in jars that they may later be transmuted magically into insects, animals, or humans in order to accomplish the particular work of the bokor (Davis 1985). The other type of zombie is the more familiar living dead: innocent victims raised in a comatose trance from their graves by malevolent sorcerers, led away under cover of night to distant farms or villages where they must toil indefinitely as slaves. These are known as zombi jardin (Dewisnle 1957) or zombi corps cadaz're (Davis 1983a, 1983b, 1985). According to folk belief, these sad individuals may be recognized by their **docile natures, their glassy, empty eyes, the nasal twang to their voices (not unlike that of the Guede spirits of the dead), and by the evident absence of will, memory, and emotion** (Simpson 1945). **They are said to dwell in the continuous present, where the past is dead and the future consists of fear and impossible desires**. Their existence lies solely in the "misty zone which divides life from death" (Vetraux 1972)…" "Regarding those zombies; **they can't lift their heads. Zombies cannot look up**"… " It must be emphasized that the fear in Haiti is not of zombies, but rather of becoming a zombie."…"Significantly, all of the reputed zombies were pariahs within their communities at the time of their demise."

-Wade Davis, Passage of Darkness: The Ethnobiology of the Haitian Zombie, Univ NC Press, 1988

The zombie entered US consciousness between 1915 and 1934, as a result of the US occupation of Haiti resulting in the films "White Zombie" and "I Walked with a Zombie". The common image of the zombie is of "**a corpse in tattered rags, trailing remnants of necrotic flesh as it rises from the cemetery in a state of trance-like animation, entirely subservient and beholden to the nefarious authority of some unknown master**." Modern cinema has taken this image a step farther. Zombies now may be hyperanimated, virally-infected humans which may never have died but rather have been "highjacked" by pathogens. One quickly notes that their folklore docility has been replaced with a more Romero-esque overriding drive to feed. The modern zombie is no longer the result of an extraordinary form of social punishment, but a disease (Ataxic Neurodegenerative Satiety Deficiency Syndrome (ANSD) or Consciousness Deficit Hypoactivity Disorder (CDHD)).

We have evaluated the literature and cinema in an attempt to document the variously described characteristics of the zombie. Then, being neurologists, we have attempted to localize the dysfunction in the nervous system (abnormal gait, amnesia, strong drive to bite/feed, maintenance of continence, hyperosmia, aphasia, loss of libido, aquaphobia, aggression, absence of will, diminished pain perception, inability to lift the head); since we are also veterinarians, our localization may differ somewhat from the localizations performed by the noted Harvard psychiatrist, Dr. Steven C. Schlozman as we also consider companion animal zombies. So be it… our goal is not to be the ultimate source for knowledge on zombie neurology (that remains to be written) but to entertain you with our not uncommon interest in the undead.

Finally, we will review our newest interest… "resuscitating the dead in practice", by reviewing some of the work of Dr. Lance Becker in resuscitation science.

REFERENCES

Below is a partial list of movies, television shows, books, and journal articles that we will discuss which you are also welcomed to review prior to the talk:

Movies: Dawn Of The Dead, Braindead, Night Of The Living Dead, Return of the Living Dead, Shaun of the Dead, Zombie, Day Of The Dead, Zombieland, Cemetery Man, Reanimator, Fido, I Walked With A Zombie, Rec, The Crazies, 28 Days Later, 28 Weeks Later, Resident Evil, I am Legend, Last Man on Earth

Television: The Walking Dead

Books: Serpent And The Rainbow, Passage of Darkness, The Zombie Autopsies, Frankenstein, I am Legend

Lampe JW, Becker LB: sAnn Rev Med 2011, Vol. 62: 79-93

INFLAMMATORY BOWEL DISEASE (IBD) IN THE CANINE AND FELINE PATIENT

Rachel V. Poulin, RVT, Coral Springs, FL

INTRODUCTION

Inflammatory bowel disease (IBD) is the result of the accumulation of inflammatory cells within the lining of the stomach, small intestines and/or large intestines[1]. IBD is distinguished by persistent, recurrent gastrointestinal (GI) signs and histological evidence of intestinal inflammation[2]. Although the most common clinical signs are chronic vomiting and diarrhea[3], presenting complaints can include changes in appetite (hyporexia, anorexia or even polyphagia), weight loss, lethargy, abdominal distension, polyuria and polydipsia (PU/PD).

The two most common forms of IBD are lymphocytic-plasmacytic enteritis (LPE) and eosinophilic enteritis (EE). LPE is characterized by infiltration of lymphocytes and/or plasma cells into the intestines. EE is characterized by eosinophilic infiltration of any region of the intestinal mucosa. The exact etiology of IBD remains largely unknown. Environmental and genetic factors as well as abnormal immunologic responses are thought to play a large role however, at this time, IBD is thought to be idiopathic.

CLINICAL SIGNS

Vomiting and diarrhea are among the most common clinical signs of IBD. Other signs can include weight loss, appetite changes, poor haircoat, mucoid feces, tenesmus, abdominal pain, PU/PD, increased frequency of defecation, and hematochezia. Clinical signs may wax and wane, with or without a precipitating factor[4].

Primarily clinical signs are associated with the region affected. That being said, one would associate large volumes of soft stools, weight loss, and/or melena with small intestinal inflammation while associating the presence of hematochezia, mucus in the stool, dyschezia, tenesmus, and an increased frequency of defecation with large intestinal inflammation.

Table 1.
CLINICAL DIFFERENTIATION BETWEEN CHRONIC SMALL INTESTINAL FROM LARGE INTESTINAL DIARRHEA[21]

Sign	Small Intestinal Disease	Large Intestinal Disease
Weight Loss	To Be Expected	Rare
Polyphagia	Sometimes	Rare to absent
Frequency of Bowel Movements	Often near normal	Sometimes very increased
Soft stool	Expected	Sometimes
Volume of feces	Often increased	Sometimes decreased due to increased frequency
Blood in feces	Melena (rare)	Hematocheziea (sometimes)
Mucus in feces	Uncommon	Sometimes
Tenesmus	Uncommon	Sometimes
Vomiting	May be seen	May be seen
Dyschezia	Uncommon	Sometimes

It is important to note that watery diarrhea can be a component of both small and large bowel inflammation. Large amounts of soft stool without tenesmus, dyschezia and/or mucus is characteristic of small intestinal disease. In addition, failure to lose weight is the most reliable indication that the animal has large bowel disease[22].

In severe untreated cases it is possible for a patient to have developed protein losing enteropothy (PLE), and/or lymphangiectasia. These cases can present with severe weight loss as well as bicavitary effusion in addition to other edematous states.

DIAGNOSTICS AND LAB WORK

Along with the presence of histological findings, which appropriately explain the patient's clinical signs, the diagnosis of IBD is one of exclusion (i.e. ruling out all other possibilities). Diagnostic biopsies are necessary because there are no pathognomonic historical, physical examination, or clinical pathological findings for this disorder[5].

Diagnosis starts by exclusion of other causes of gastrointestinal inflammation such as parasitic pathologies, diet responsive and antibacterial responsive conditions, pancreatitis, neoplasia, endocrine or metabolic pathologies, foreign bodies and/or extra-intestinal infections.

Food allergies and intolerances must be ruled out. Except for very mild cases of IBD, this disease does not tend to respond to dietary manipulation alone. If a food trial is permissible, a 4 to 6 week trial is recommended, although shorter trials have been adequate[9]. Food trials should consist of a novel protein/hypoallergenic diet. Client education is essential in order to increase chances of compliance and increase likelihood of a successful outcome.

Table 2.

DIAGNOSITIC FINDINGS MAY INCLUDE:	
CBC	Usually within normal limits although occasional neutrophilia, eosinophilia, lymphopenia, anemia, or thrombocytopenia may be noted
Chemistry Panel	Usually within normal limits but in severe cases of PLE, panhypoproteinemia (hypoalbuminemia and hypoglobulinemia) along with hypocholesterolemia are present
Fecal Floatation Giardia Snap Testing Fecal PCRs	Necessary to rule out parasites such as hookworms, whipworms, and giardia PCRs can also detect chronic bacterial infections like Campylobacteriosis amongst others
Urinalysis	Abnormalities may include a USG <1.010, and an increased protein level may be present

Cobalamin and Folate

Cobalamin (vitamin B12), the largest of the water-soluble B vitamins, is absorbed in the ileum. Folate (vitamins B10 and B11[6]) is absorbed in the proximal small intestine, specifically the duodenum. Animals with IBD may have serum concentrations consistent with active inflammation. Serum concentration levels of these vitamins can thus allow for localization of the inflammation. If both serum levels are decreased it can indicate malabsorption and thus be a marker for inflammation. While either hypocobalaminemia or hypofolatasemia are consistent with distal or proximal small intestinal disease respectively, hyperfolatasemia should raise the suspicion for the presence of intestinal dysbiosis. Fasting vitamin levels allow for localization of intestinal pathology which may be most likely inflammatory or neoplastic in nature.

Table 3.

RESULTS	FOLATE	COBALAMIN
DECREASED	Proximal small intestinal disease Low dietary intake Some medications can cause a decrease in folate absorption	Distal small intestinal (Ileum) disease Pancreatic insufficiency Hereditary malabsorption Low dietary intake
INCREASED	Bacterial overgrowth Vitamin oversupplementation High dietary intake	Vitamin oversupplementation High dietary intake

Ultrasound

Ultrasonography is an important diagnostic tool. Thickening of the large and small intestinal walls as well as enlarged lymph nodes are common ultrasonographic findings They are characteristic of inflammatory disease processes but are not specific to IBD. Ultrasonography can provide information on the location and severity of lesions, suggesting the location were biopsies may be warranted[8].

Biopsy

Histopathologic findings are required for the proper diagnosis of IBD. Biopsy samples can be obtained either endoscopically or surgically. Endoscopic sampling is less invasive and tends to be less expensive, however endoscopy does have its limitations. Endoscopic sampling allows for only the sampling of the mucosa and submucosa thus many times being unable to sample the muscularis or the serosal layers of the intestinal tract. Endoscopic samples are also limited to the gastric, duodenal, proximal jejunal (in some cases) ileal and colonic areas. Full thickness surgical biopsies are the ideal diagnostic samples especially if deep intestinal layer involvement or multi-organ involvement are suspected. Surgical biopsies are more invasive and require a longer recovery period therefore not being an ideal option older patients, patients with concurrent diseases, or a patient with advanced disease and decreased healing abilities. Full-thickness and extra intestinal biopsies are important diagnostic tools as if they are not obtained, a diagnosis may be missed or a misdiagnosis can be made (e.g., IBD can be easily confused with lymphoma). Although surgical and laparoscopic biopsies are complex, they are estimated to be the gold standard for the diagnosis of IBD[10].

TREATMENT

The treatment for IBD is aimed at reducing inflammation and in many instances suppressing an overactive immune system. The way to achieve this is multimodal. Case-specific nutrition as well as nutritional supplementation is necessary to help aid and reduce inflammation to the gastrointestinal tract and help the body absorb essential nutrients. Gastroprotectants are commonly used as the primary disease and some of the medications used to treat it can cause gastrointestinal upset. Immunosuppression is often necessary to further aid in controlling inflammation while antimicrobial therapy is used to control intestinal dysbiosis.

Nutrition and Supplementation

As previously discussed, IBD classically presents with main gastrointestinal clinical signs, such as vomiting, diarrhea, flatulence, and decreased appetite. Part of these clinical signs can due to a concurrent food intolerance and/or allergy. The goal of dietary therapy is to eliminate adverse food reactions[12]. In order to accomplish this the prescribed diet should meet the patient's nutritional needs, be highly digestible, be low in antigens, and in cases of lymphangiectasia or PLE, be fat restricted. A single novel protein source is ideal. It is important to remember that there is no perfect food option for IBD, each patient will have different dietary needs and disease process restrictions. There are many options available, in the event that no commercial diet can be found a custom made balanced home-made diet should be considered.

Hydrolyzed Diets

A hydrolyzed protein diet consists of a single protein source that is specially processed to break the structure of the protein down into multiple, tiny particles that the immune system will not recognize as an allergen. It is a chemical process in which a molecule is cleaved into two parts by the addition of a molecule of water. One fragment of the parent molecule gains a hydrogen ion (H^+) from the additional water molecule. The other group collects the remaining hydroxyl group (OH^-). This method has been used for several years in infant formula to decrease food hypersensitivity. Soy and chicken are common protein sources used in making these diets. Soy may be preferable since fewer dogs seem to have been fed soy based diets previously so it is less likely that they have been sensitized to it. It has been found that hydrolyzed soy is extremely digestible and well absorbed from the gut[23].

Novel Protein Diets

A novel protein diet consists of using a single protein source which the patient has not been exposed to and thus have not been sensitized to in the past. Novel protein diets have more options in terms of potential protein sources as well as canned varieties compared to the hydrolyzed diets. These diets are typically more calorie dense than the hydrolyzed diets so portion control must be followed closely to prevent weight gain. One thing to keep in mind when choosing a diet is that pets with beef allergies may react to venison and pets with chicken allergies may react to duck protein sources[36].

Fat Restricted Diets

In the cases of secondary PLE and lymphangiectasia an ultra-low fat diet is indicated. A restricted low fat diet helps to prevent further intestinal lacteal engorgement and subsequent protein loss[37].

Table 4.
Common veterinary diets used for elimination trials and in the treatment of IBD include:

	HILL'S®	ROYAL CANIN®	PURINA®	RAYNE CLINICAL NUTRITION®
Cat	ZD/DD/ID	Hypoallergenic Selected Protein Adult Duck, Venison, and Whitefish	HA, EN, OM	Low Fat Kangaroo Rabbit & Squash Kangaroo & Sweet Potato
Dog	ZD/DD/ID lowfat	Gastrointestinal Low Fat Hypoallergenic Selected Protein Adult Duck, Venison, and Whitefish Soy and Potato	HA, EN, OM	Low Fat Kangaroo Rabbit & Squash Kangaroo & Sweet Potato

Cobalamin and Folate

Folate and cobalamin levels are frequently abnormal in patients with IBD. Supplementing these vitamins will aid in proper nutrient absorption and digestion and in some cases increase the patients' appetite. In animals with reduced cobalamin absorption, regardless of the cause, it is reasonable to expect that eventual depletion of body cobalamin stores will occur and cobalamin deficiency will ensue. As all cells in the body require cobalamin for single carbon metabolism, it has been hypothesized that cobalamin deficiency may contribute to the clinical signs and manifestations of gastrointestinal disease in some patients[11]. Supplementation of cobalalmine is thus an important part of treatment. This can be done with a subcutaneous injection given weekly.

Cobalamin is essential in many biochemical processes involving single carbon transfers. Cobalamin deficiencies have been very well described in dogs and cats having gastrointestinal pathologies. Cobalamin deficiency has been associated with worse clinical outcomes in dogs with chronic enteropathies. Cobalamin supplementation has been found to improve clinical signs in cats with chronic enteropathies.

Cobalamin can be measured in fasted serum samples. Supplementation is recommended when the cobalamin level is subnormal or at the low normal range (<300ng/L)[17].
Supplementation recommendations per the Gastrointestinal Laboratory at Texas A&M have changed as stated below:

Table 5[11]

SPECIES- WEIGHT	DOSE OF COBALAMIN (μG) GIVEN SUBCUTANEOUSLY/WEEK
Cats	250 μg
Dogs: <10 lbs	250 μg
Dogs: 10-20 lbs	400 μg
Dogs: 20-40 lbs	600 μg
Dogs: 40-60 lbs	800 μg
Dogs: 60-80 lbs	1000 μg
Dogs: 80-100 lbs	1200 μg
Dogs: >100 lbs	1500 μg

It is recommended to give the subcutaneous or IM dose every 7 days for 6 weeks, then a single dose 30 days later, while retesting 30 days after the last dose. If the underlying disease process has been well controlled and the cobalamin stores are replenished, the cobalamin levels should be supranormal. If the cobalamin concentrations are within the normal range, at least monthly supplementation should be continued. If the cobalamin level is still subnormal, then further evaluation of the underlying process needs to be pursued and the cobalamin supplementation should be instituted weekly or bi-weekly[18]

Omega 3 (n-3) Fish Oils

Omega 3 fish oils have been used more recently in IBD therapy because of their anti-inflammatory properties. N-3 fatty acids can be administered orally in an effort to reduce pro inflammatory cytokines synthesized from n-6 fatty acids. Their efficacy has been demonstrated in a number of species and inflammatory conditions, and they are commonly utilized as adjunctive therapy for canine atopy. Preliminary data on n-3 fatty acid use in rodent IBD models and in people is encouraging, although they have yet to be evaluated in dogs and cats for the ancillary treatment of IBD[19].

Prebiotics and Probiotics

Prebiotics and Probiotics can be utilized in inflammatory bowel disease to restore gastrointestinal health. The normal intestinal flora of the immune system is made up of beneficial bacteria that keeps the animal healthy and maintains the health of the gastrointestinal system. Prebiotics are a type of carbohydrates called oligosaccharides that encourage the growth of beneficial bacteria and suppress the growth of pathogenic bacteria[26]. Probioitics are actual bacteria that enhance intestinal health. They survive the acid and bile in the GI tract, they exclude and reduce pathogenic bacterial adherence, and coaggregate to help achieve a normal and balanced microflora. They need to be safe, noninvasive and nonpathogenic, and non carcinogenic to the host[27]. Care should be taken when using these nutraceuticals as it is not always appropriate to introduce bacteria into the GI tract. If ulcers or hematochezia is present, probiotics can be seen as potentially dangerous. In addition timing of antibiotics must be noted so that that the patients are receiving the benefits of both medications.

Antimicrobials

Metronidazole is a broad-spectrum antibacterial and antiprotozoal agent. It is usually the first antimicrobial of choice for IBD patients due to its ability to balance intestinal microflora, reduce obligate anaerobe load and reduce inflammation[13]. Metronidazole (10mg/kg PO BID) is recommended. Special care should be taken in animals that are on long-term metronidazole therapy because of potential risks of neurotoxicosis and hepatotoxicosis[14]. Other antimicrobials that may be used include tylosin (10-20mg/kg PO BID). There is currently little evidence to support the use of antibiotics in combination with glucocorticoid therapy. The mechanism by which the antibiotics exert their effect is poorly understood but may involve decreased bacterial adhesion, altered intestinal flora populations, and anti-inflammatory effects. Lower doses can be used chronically for both medications if the clinical signs, especially diarrhea, flatulence and borborygmi, return as the antibiotic therapy is discontinued[15].

Immunosuppression

Corticosteroids are the most commonly used immunosuppressive agents for the treatment of IBD. Commonly used oral corticosteroids include prednisone, prednisolone and budesonide. An initial dose for immunosuppression is 1-2mg/kg PO BID ideally given for 2-4 weeks and then slowly tapered over the next several months. Ideally patients should be weaned off corticoid therapy completely once clinical signs have resolved. It is important for the veterinary team to reassess these patients every 2-3 weeks as they slowly wean this therapy. It is likely that many patients need to be on long-term low-dose corticosteroid therapy. It is up to the medical team to find the lowest most effective dose of glucocorticoids. A CBC and chemistry should be performed 2 weeks after starting therapy to monitor for any acute reactions. After that testing a CBC every 2-4 weeks is important in order to monitor for bone marrow suppression. Iatrogenic hyperadrenocorticism is a possibility, therefore a chemistry should be performed every 2-3 months in order to screen patients elevated cholesterol and triglyceride levels which would let us know about developing hyperadrenocorticism. In addition proteinuria is a possibility on chronic prednisone therapy so a urinalysis should be performed with the chemistry. Feline patients have 50% the glucocorticoid receptors as dogs and therefore do not suffer side effects to the same level as dogs however, still need to be carefully monitored. Cats can have an increase in insulin resistance as a result of glucocorticoid therapy thus making them more vulnerable to developing diabetes. Patient on chronic glucocorticoid therapies are also prone to urinary tract infections and thus a urinalysis and a urine culture should be performed at least once to twice yearly. Due to their side effects and long term iatrogenic risks it is important to monitor these patients closely and reduce/discontinue therapy when it is deemed to be safe. If discontinuation of high doses of steroid therapy is not possible then concurrent therapy with other non-steroid immunosuppressive agents may be indicated in order to maintain control of the disease while minimizing the potential life-threatening side effects of glucocorticoids.

Budesonide is also a glucocorticoid that can be a good option for patients that have severe side effects while on prednisone or prednisolone. Budesonide is relatively new on the market and was originally developed to help humans with certain types of IBD that were suffering the side effects of other corticosteroids. Budesonide differs from other glucocorticoids because it reportedly has reportedly has potent topical anti-inflammatory activity on the GI mucosa and minimal systemic side effects[28]. While budesonide is a good option for patients suffering from side effects, the medication can still pose the same threat of iatrogenic hyperdrenocorticism and adrenal atrophy. Because of this clients will still need to be educated well on potential risks, side effects and a titration schedule will need to be implemented once clinical signs are under control. Budesonide (dogs: 2 mg/dog/day, PO; cats: 1 mg/cat/day, PO)[31]

Azathioprine is an option for dogs that are already on glucocorticoids and need additional suppression. It can also replace a glucocorticoid if the patient did not respond well to the original therapy. Azathioprine takes 2-3 weeks to become fully effective, therefore, if the goal is to wean off of glucocorticoids with the addition of azathioprine the patient should not begin tapering until three weeks into therapy[29]. Azathioprine can cause bone marrow suppression, therefore complete blood counts (CBC's) should be performed every 4-6 weeks. A biochemical profile should also be done due to azathioprine's association with hepatotoxicity and pancreatitis. Cats should not receive azathioprine because of rapid, lethal bone marrow suppression. Azathioprine (dogs: 2.2mg/kg PO once daily)[31]

Cyclosporine is heavily used in human medicine for transplant and immune-mediated patients. It has proven to be beneficial in dogs and cats with immune-mediated disease. Cyclosporine works by inhibiting cellular immunity; it is metabolized by the liver and excreted into the bile. When dosing cyclosporine it is important to make sure the client is getting the recommended product. Neoral/Atopica® [30] is much more effective in small animals because of a higher bioavailabilty then Sandimmune® [30]. Cyclosporine should be administered on an empty stomach and can cause GI side effects such as vomiting, anorexia, and diarrhea, these signs can usually be treated with gastrointestinal protectants such as famotidine, metoclopramide, and maropitant. Cyclosporine is one of the most commonly used second-line immunosuppressants. Cyclosporine (dogs: 5-10mg/kg/day PO; cats: 1-5mg/kg/day divided BID)[31]

Mycophenolate mofetil is an immunosuppressant that is increasingly used in dogs. While it is used commonly in human medicine it is fairly new to veterinary medicine. Mycophenolate suppresses the immune system by suppressing lymphocyte proliferation and decreasing antibody production Mycophenolate is not the first medication of choice because we have limited information on long-term effects. In addition mycophenolate should be used with caution in patients with IBD because it can be the cause of inflammatory diarrhea.[32] Serial monitoring of CBC's and chemistries and electrolytes should be performed to detect cell suppression, bone marrow suppression, acute liver and kidney reactions. Mycophenolate mofetil (dogs: 12-39 mg/kg/day)[33].

Chlorambucil is an anti-cancer agent with effective immunosuppressive properties. It has a 2-4 week delay of action therefore any decreasing of glucocorticoids should begin 3-4 weeks after starting chlorambucil. Chlorambucil is one of the most commonly used second line immunosuppressants. Serial monitoring of CBC's in necessary in order to monitor for myelosuppression that can manifest in anemia, leukopenia, thrombocytopenia, and gastrointestinal toxicity[34]. Chlorambucil (dogs: 0.1-0.2mg/kg once daily PO; cats: 0.25-0.5mg/kg q48-72 hours)[34]

Supportive Agents

Many of the medications used in the treatment of IBD can have gastrointestinal side effects such as, lack of appetite, borborygmus, vomiting, diarrhea, and GI ulcerations. Because of this supportive agents like H_2 blockers, H pump inhibitors, and motility modifiers should be considered as adjunct therapy.

Table 6[38].

Medication:	Function:	Canine Dose:	Feline Dose:	Warnings:
Famotdine (pepcid®)	H_2 receptor antagonist used to reduce GI acid production	0.5 -1 mg/kg PO SID-BID	0.5mg/kg PO once daily	H_2 have been demonstrated to be relatively safe and exhibit minimal adverse effects.
Metoclopramide (reglan®)	Stimulates upper GI motility and has antiemetic properties	0.2-0.4mg/kg PO BID-TID	0.2-0.4mg/kg PO BID-TID	Behavior changes; Changes in mentition, behavior, disorienta and constipation.
Maropitant (cerenia®)	neurokinin (NK1) receptor antagonist that blocks the pharmacological action of substance P in the central nervous system (CNS)[35]	2mg/kg PO once daily for four days	Off label for cats	has not been evaluated in dogs used for breeding, pregnant or lactating bitches, dogs with gastrointestinal obstruction, or dogs that have ingested toxins. Use with caution in dogs with hepatic dysfunction.
Ondansetron (zofran®)	5-HT_3 receptor antagonist for severe vomiting	0.1-1mg/kg PO SID-BID	0.22mg/kg BID-TID	Potential of constipation, extrapyramidal symptoms, arrhythm and hypotension
Sulcralfate (carafate®)	Locally acting treatment for GI ulcers	0.5-1 gram PO BID-QID	0.25-0.5 gram PO BID-TID	Constipation possible Should be administered on an empty stomach
Omemprazole (prilosec®)	Proton pump inhibitor used for (Ulcers and erosion	0.5-1 mg/kg PO SID	0.7 mg/kg PO SID	Hypersensitivity is possible Caution should be taken in patient with hepatic or renal disease

PROGNOSIS

The prognosis for IBD is usually good. Contributing factors in prognosis include how far along in the disease process the patient is, how compliant the client is, and the patient's response to treatment. In addition concurrent disease processes or secondary diseases must be taken into consideration. Patients with secondary pancreatitis, lymphangiectasia or protein losing enteropathy tend to have a poorer prognosis and a different response to treatment.

CLIENT EDUCATION

The role that a technician plays in educating the pet owner about IBD is an extremely important part of the overall therapeutic plan. The veterinary technician should be able to answer questions for the client spanning from the initial diagnostics to specific medical treatments and side effects. Nutritional counseling and the importance of owner compliance in regards to diet should be able to be communicated by the technician. The technician can play a very important role in helping the owner understand and preparing them for what lays ahead as the treatment develops. With excellent communication with the veterinary staff, positive patient response and owner compliance this is a disease that can have a favorable prognosis.

REFERENCES

1,9,10,11,13) Dunbar, K. (2011) *Inflammatory Bowel Disease in Cats.*[online]. Veterinary Technician, (volume 32, No 10). Available from: http://www.vetlearn.com/veterinary-technician/inflammatory-bowel-disease-in-cats 2) Summers, A. (2007) Disease of the Digestive System, In: Teri Merchant (ed.) *Common Diseases of Companion Animals 2nd edition.* St. Louis, Missouri, USA, Mosby. pp 50-51. 3,4,5,8,13,30,31, 34) Merril, L. (2012) Gastrointestinal, In: Linda Merrill (ed.) *Small Animal Internal Medicine for Veterinary Technicians and Nurses 1st edition.* Ames, Iowa, USA, Wiley-Blackwell. pp 220-222,240,413. 6,22,23,26) Nelson, R. Couto C. (1998) Disorders of The Intestinal Tract, *Small Animal Internal Medicine 2nd edition.* USA, Mosby. pp 356, 451,455 7) Wortinger, A. (2007) Vitamins, In: Ann Wortinger (ed.) *Nutrition for Veterinary Technicians and Nurses 1st edition.* Ames, Iowa, USA, Blackwell Publishing Professional. p 43 11,18,19) Steiner, J, Suchodolski, J. (n.d) Cobalamin: *Diagnostic Use and Therapeutic Consderations* [online]. Available from: http://vetmed.tamu.edu/gilab/research/cobalamin-information#GIDisease 15) Purcell, S. Cooks, K. (2010) *How to Manage Feline Chronic Diarrhea, Part II: Treatment.* [online]. Available from http://veterinarymedicine.dvm360.com/vetmed/article/articleDetail.jsp?id=679007&sk=&date=&pageID0. 16,20) Bissett, S. Boyle, T.(2007) *Idiopathic Inflammatory Bowel Disease.*[online] Standards of Care, (volume 9, No2). Available from:http://www.vetlearn.com/standards-of-care/idiopathic-inflammatory-bowel-disease. 17) Krecic, M. (2001) *Feline Inflammatory Bowel Disease Treatments, prognosis, and new developments.* [online]. Compendium, (Volume 23, No.11). Available from: http://www.eadveterinaria.com.br/material/270/1972/doenca_intestinal_inflamatoria_felina_2.pdf 21,25,26) Wortinger, A. (2010) *Nutrition know how: Prebiotics and Probiotics: What they can do for dogs and cats* [online]. Veterinary Technician (Volume 31, No 3). Available from: http://www.vetlearn.com/veterinary-technician/prebiotics-and-probiotics-what-they-can-do-for-dogs-and-cats. 29,33) Fogle, J. Bissett, S. (2007) *Mucosal Immunity and Chronic Idiopathic Enteropathies in Dogs.* [online]. Compendium (Vol 29, No.5). Available from: http://www.vetlearn.com/compendium/mucosal-immunity-and-chronic-idiopathic-enteropathies-in-dogs 32,34,38) Plumb, D. (2011) *Veterinary Drug Handbook 7th edition.*In: Gigi Davidson (ed.) Stockholm, Wisconsin, USA, Wiley Blackwell 35) Cerenia® [package insert] NY, NY: Pfizer 2012. 36,37) (2008) *Hydrolyzed Diet Vs. Novel Protein Diet.* [online]. Veterinary Forum (Voumel.25, No. 7) Available from: http://www.vetlearn.com/veterinary-forum/what-to-feed-hydrolyzed-diet-vs-novel-protein-diet

B-HARMONY: ARE YOU THE RIGHT TYPE FOR ME? BLOOD COMPONENT COMPATIBILITY

Kenichiro Yagi, BS, RVT, VTS (ECC)
Los Altos, CA

INTRODUCTION

Blood components transfusions are regularly utilized as a method of treatment for various ailments. Red blood cells (RBCs) are utilized for anemia to supplement oxygen carrying capacity; plasma and its contents are used to provide coagulation factors in coagulopathies; platelets, when available, are used to aid in hemostasis in patients with life threatening hemorrhaging due to thrombocytopenia; and supportive evidence is emerging for canine specific albumin to positively affect the outcome in hypoalbuminemic patients. Less commonly, intravenous immunoglobulin (IVIG) is administered for its immunomodulatory effect in immune-mediated hemolytic anemia (IMHA), immune-mediated thrombocytopenia (ITP), and sudden acquired retinal degeneration syndrome; in addition, specific immunoglobulins are utilized as antitoxins for snake envenomation and tetanus. Complication rates are variable depending on the component being transfused, and can be immunologic or non-immunologic in nature.

IMMUNOLOGIC RESPONSE AND ANTIGENIC BLOOD COMPONENTS

Immunologic complications are a result of the body's immune system responding to the foreign blood components transfused. Hypersensitivity reactions result from exposure to antigenic proteins in the transfused blood component. While a hypersensitivity response can occur at first exposure, the response may be even more severe with repeated exposures to the antigens due to the body's ability to "remember" past exposures and elicit a secondary ("anamnestic") immune response. Hypersensitivity reactions prompted by blood component transfusions are classified as Type I (Allergic), Type II (Cytotoxic), and Type III (Immune Complex) hypersensitivity reactions in a traditional classification scheme.

Type I Hypersensitivity

Type I hypersensitivity reactions, commonly known as allergic reactions, are mediated by immunoglobulin E (IgE) resulting in the production and release of inflammatory mediators. When IgE expressed on the surface of mast cells are cross-linked with an introduced antigen, the mast cell is triggered to release its granules (degranulation). These granules contain inflammatory mediators, enzymes, and cytokines which stimulate the production of more inflammatory mediators. The release of these mediators such as histamine, serotonin, prostaglandins, and leukotrienes results in an acute inflammatory reaction.

Clinical signs can arise in seconds to minutes, giving type I hypersensitivity its other name of "immediate hypersensitivity". Symptoms of type I hypersensitivity are usually mild, and include vomiting, diarrhea, fevers, hives (urticaria), pruritus, and facial swelling. In severe cases, a type I hypersensitivity reaction may result in anaphylaxis. Anaphylaxis in dogs typically manifests as hemodynamic collapse due to the occlusion of the hepatic veins leading to a lowered venous return, cardiac output, and arterial blood pressure. These dogs can show signs of weakness or collapse, become comatose, convulse, or die in a very short time. In cats, anaphylaxis can lead to salivation, vomiting, dyspnea, collapse, and may also lead to death. Evidence of bronchoconstriction, pulmonary hemorrhage, and edema in the respiratory system are seen on necropsy of these feline patients.

These reactions can be triggered by any blood product but seem most commonly triggered by antigens on donor white blood cells and platelets. In addition, RBCs that have been stored longer have a higher tendency of causing type I hypersensitivity. Cytokines produced by leukocytes having a role in the hypersensitivity reaction may be the reason for this observation. Any signs of an acute hypersensitivity reaction should prompt the staff to immediately stop the transfusion. In the case of mild reactions an H1 antihistamine such as diphenhydramine may be used for treatment, and cautious administration may be possible after the signs subside. In the case of anaphylaxis, administration of epinephrine, a glucocorticoid such as dexamethasone, and aggressive fluid resuscitation is recommended.

Type II Hypersensitivity

Type II hypersensitivity reactions can occur in response to genetically different RBCs being transfused into a recipient. Surface glycoproteins and glycolipids, which function as cell membrane components such as transport channels, serve as antigens for antibody response and complement activation, leading to intravascular hemolysis. The antibodies also serve as opsonins, which allow for phagocytosis and destruction of the RBCs. This leads to extravascular hemolysis through the mononuclear phagocyte system. The combination of this intravascular and extravascular hemolysis is known as type II hypersensitivity, or cytotoxic hypersensitivity. The time to onset depends on the existence of antibodies against the erythrocyte antigens being introduced. Presence of preexisting antibodies can result in agglutination and hemolysis of the transfused cells. When type II hypersensitivity is triggered due to preexisting antibodies, it is known as an acute hemolytic transfusion reaction (AHTR) due to its rapid onset. If there are no preexisting antibodies to the novel erythrocyte antigens, a delayed hemolytic transfusion reaction (DHTR) may still occur as the transfused cells circulate within the body while antibodies are produced against them, leading to elimination of these red cells. Any subsequent exposure to these antigens will result in AHTR.

Clinical signs of type II hypersensitivity vary in severity depending on the amount of incompatible red cells transfused, and the antigenic property of the erythrocyte antigen. The mildest sign is a febrile response. Most severe cases of AHTR can be seen in incompatible transfusions to a previously sensitized patient. These transfusions result in hemolysis of the transfused cells, leading to hemoglobinemia and hemoglobinuria. A large amount of intravascular hemolysis can trigger the coagulation cascade and result in disseminated intravascular coagulation (DIC). Activation of the complement system results in degranulation of mast cells, leading to systemic release of cytokines and inflammatory mediators, ultimately resulting in circulatory consequences of hypotension, bradycardia, and resulting shock. Salivation, vomiting, and diarrhea may also be seen as a result of sympathetic response. DHTR can be symptomatic or asymptomatic. The most common sign is a significant drop in PCV or Hb after the transfusion. Less common signs are fever, hyperbilirubinemia, and reduced urine output. These signs can manifest 3-21 days post transfusion.

Type II hypersensitivity reactions are of concern primarily when RBCs are transfused (whole blood, PRBC). However, a potential for red cell contamination of plasma products exist, and while rare, should not be ignored as a possibility. When a reaction is seen, the transfusion should be stopped immediately. Treatment for AHTR involves IV fluids, vasopressor and/or inotropic therapy to combat shock, glucocorticoid administration to suppress the immune response, and supportive care, especially to offset the nephrotoxic effects of hemoglobin, until the foreign red cells are eliminated from circulation.

Type III Hypersensitivity

Generalized type III hypersensitivity reactions occur when antigens are introduced intravenously. Serum antibodies form immune complexes with the antigens, and these complexes are normally removed from circulation by binding to erythrocytes and platelets, or eliminated by the mononuclear phagocyte system. In the face of a large antigenic load overwhelming clearance mechanisms, complexes are deposited on vessel walls, leading to vasculitis or arteritis; sites of high blood flow resulting in glomerulonephritis or synovitis; and on blood cells themselves, causing anemia, leukopenia, or thrombocytopenia.

Clinical signs of type III hypersensitivity include fever, erythema, edema, urticaria, neutropenia, lymph node enlargement, joint swelling, and proteinuria, and can be seen 1-3 weeks after exposure. The signs seem to be more severe in healthier patients, possibly because of a stronger immune response.

Because type III hypersensitivity reactions are caused by formation of immune complexes, blood products containing unbound antigens are potential triggers. Type III reactions have been reported in dogs receiving human serum albumin, and are postulated to occur with IVIG, snake antivenom, tetanus antitoxin, and incompatible plasma transfusions. Patients with symptoms are treated with supportive care.

Transfusion-Related Acute Lung Injury (TRALI)

A clinical syndrome characterized by an acute onset of dyspnea and hypoxemia with diffuse pulmonary infiltrates without signs of circulatory overload has been described in human transfusion recipients. This syndrome, called transfusion-related acute lung injury (TRALI), is seen within 6 hours of the start of transfusions, and can include signs such as fever, hypotension, dyspnea and tachypnea. The pathophysiology of TRALI is not well understood, but is thought to include at least two simultaneously occurring factors. The first factor involves the patient's underlying disease causing endothelial activation and adherence of neutrophils to the pulmonary endhothelium. The second factor involves transfused biologically active lipids and cytokines activating the adhered neutrophils to cause endothelial damage, leading to pulmonary edema. In addition, donor leukocyte antibodies reacting with recipient leukocyte antigens causing an immune reaction leading to lung injury may be involved. The endothelial damage is believed to cause an increase in permeability in the pulmonary circulation, leading to protein containing fluid entering the alveoli and interstitium. Plasma containing blood products are usually the cause, though TRALI can be seen with red cell transfusions.

Management of a patient with TRALI is largely supportive, involving oxygen supplementation. In severe cases, mechanical ventilation may be required. Glucocorticoid administration is thought to be beneficial, though definitive evidence in its efficacy has not been established.

COMPATIBILITY TESTING

While there are a many possible immunologic responses to blood products, immunologic complications are relatively rare. This observation may partially be due to a low natural occurrence of incompatibilities or a depressed immune response in critically ill patients. However, our ability to test for compatibility of blood components and select appropriate blood products is the largest factor in minimizing the chances of complications.

Blood Typing

Since type II hypersensitivity reactions are triggered by RBC surface antigens, blood typing of the donor and recipient is essential for a safe transfusion and is the first step in testing for compatibility. Matching donor and recipient blood types will minimize the chances of a transfusion reaction.

The DEA (dog erythrocyte antigen) blood group system has been described for canines, where DEA 1.1, 1.2, 3, 4, 5, 6, 7, and 8 are recognized, with 5 additional antigens described. An additional blood group antigen called *dal* has also been identified as lacking in some Dalmatians. Type II hypersensitivity due to DEA 1.1 and 1.2 has been known to cause severe AHTR, but because of the lack of naturally occurring antibodies to these antigens in dogs, an acute reaction is not seen when DEA 1.1 or 1.2 positive blood is transfused to a transfusion naïve, negative patient. However, a first transfusion of positive blood results in sensitization to the DEA, which leads to severe AHTR on repeated exposure. Naturally occurring antibodies for DEA 3 and 5 (10% of population) and DEA 7 (20-50%) are seen in dogs typed negative for these DEAs, but as they occur at a low titer level, clinically significant reactions do not occur. While full DEA typing is only available through commercial blood banks and university laboratories, in-house typing kits for DEA 1.1 are available. Typing every recipient for DEA 1.1 prior to transfusion will allow use of both DEA 1.1 negative and positive donor blood for transfusion, increasing the donor pool, and avoid sensitization. DEA 1.1 negative or "universal" blood (though the notion of "universal" blood is questionable due to the existence of antigens we are unable to test for) can be used without prior typing in patient certain to never have been transfused, or emergency situations when time is of the essence.

Cats are known to have the AB blood group system, where 75-95% have the blood type A, 5-25% have type B, and a very small (~1%) number have type AB. About 35% of type A cats have naturally occurring anti-type B antibodies at a low titer level. When these type A cats are transfused type B red cells, a mild form of type II hypersensitivity occurs. Type B cats have naturally occurring anti-A antibodies at a high titer level, such that even transfusions of a small amount (~1mL) of type A blood will cause severe, hemolytic reactions. Plasma transfusions of the incorrect type can also cause a hemolytic reaction due to the anti-A and anti-B antibodies present. Type AB cats lack both anti-A and anti-B antibodies, making them a universal recipient for pRBCs. However, hemolytic transfusion reactions can be seen in both type A and B plasma or whole blood transfusions to AB cats due to the antibodies present in the plasma. The Mik antigen is a blood group identified in cats distinct from the AB group, for which there are currently no methods to type in clinical settings. Because mismatched transfusions in the AB group system will result in mild to severe reactions, feline donors and recipients should be typed prior to both red cell and plasma transfusions.

In-house agglutination based card tests are available for both canine DEA 1.1 and feline A/B. The DEA 1.1 test kit has a DEA 1.1 positive control, negative control, and the test well. The feline test kit has patient control, type A test, and type B test wells. The test well contains murine monoclonal antibodies against erythrocyte antigens to be tested. Agglutination occurs with antibody-antigen complex formation, indicating that the patient is positive for the blood type under evaluation. The control wells can serve to identify any presence of auto-agglutination due to auto-immune disease. However, because the test relies on agglutination for blood type detection, interpretation of blood samples obtained from a patient with auto-agglutination is unreliable.

Another type of available in-house blood typing test utilizes immunochromatography to determine blood types. Immunochromatography uses a porous strip impregnated with antibodies in two locations. In the initial sample area, red cells with the target antigens form immune complexes with antibodies that are labeled with a chromatographic substance such as colloidal gold or selenium. The red cells then pass through the detection area with antibodies fixed in place, which stops the migration of the red cells by attaching to them. This results in a colored band as the indicator for the blood type if positive, and a lack of a band if negative. Immunochromatographic tests have the advantage that they can filter agglutinated cells, allowing for blood type determination even when auto-agglutination is present.

A comparison study of agglutination card, immunochromatographic cartridge, and gel column (not readily available in clinical settings) blood typing kits for DEA 1.1 found all methods to be highly accurate (89-91%, 93%, and 100% respectively), making both cage-side test types viable options for blood typing. The immunochromatographic cartridges have the advantages of removing subjectivity on interpretation and allowing typing even with auto-agglutination present as advantages.

Cross-Matching

Cross-matching is the act of exposing donor red cells to recipient plasma (major cross-match) and recipient red cells to donor plasma (minor cross-match), with agglutination or hemolysis indicating incompatibility. While the exact protocol for cross-matching is slightly variable depending on the practice, it involves suspending RBCs in saline to achieve a 3-5% solution, washing the red cells 3-4 times, and preparing four different combination mixtures of recipient red cells and plasma with donor red cells and plasma (major and minor cross match, recipient and donor control). The tubes are then incubated, centrifuged, and graded for the level of agglutination and hemolysis. Some protocols call for microscopic evaluation for agglutination in the absence of macroscopic agglutination, or for comparison of the degree of agglutination in the case of recipient auto-agglutination or when samples from all potential donors show a positive cross-match. This allows for determination of the least reactive match, which would indicate selection of blood least likely to cause harm, and most likely to survive the longest with the recipient. Commercial cross-matching kits are also available, which may help standardize the process, though adding expense.

For first time transfusions in canines, the necessity for a cross-match is often debated due to the lack of naturally occurring antibodies for DEA 1.1 and 1.2 and the unlikelihood of an obvious incompatibility reaction. However, if there is any uncertainty in the transfusion history, or if there is a definite history of a previous transfusion, a cross-match would be in order. However, there are good reasons for feline transfusions to be cross-matched despite our ability to match the blood types through AB type testing. There have been reports of AHTR occurring even with an AB system match, likely due to the Mik positive blood being transfused to Mik negative cats. Some evidence also exists for antigens not belonging to the AB and Mik antigen groups. Because of these observations, a cross-match prior to all feline transfusions is indicated, in addition to mandatory typing.

BEYOND THE COMPATIBILITY TESTING

While the above mentioned methods are quite effective in minimizing transfusion reactions caused by erythrocyte antigens and alloantibodies contained within plasma products, the compatibility tests available to us cannot screen for every possibility of type II hypersensitivity reactions and reduction of red cell life span. In addition, there are no readily available laboratory methods to test for compatibility of plasma component products. This leads us to think about methods to minimize transfusion-related complications in conjunction with compatibility testing. One of these approaches is appropriate blood product selection through application of concepts in blood component therapy. Reducing the transfusion of unnecessary components or the use of alternatives when available will only help minimize occurrences of immunologic complications. In addition, proper administration protocols and the technical staff's role in the monitoring of patients receiving transfusions is invaluable in early detection and treatment of complications, should they arise.

REFERENCE

[1]Craft E, Powell L, **J Vet Emerg Crit Care** 2012;22(6):631-639. [2]Creedon J and Davis H, **Advanced Monitoring and Procedures for Small Animal Emergency and Critical Care** 2012;Ch.59:p759-775. [3]Tizard I, **Veterinary Immunology** 2013;Ch.28-30:p326-364. [4]Seth M et al, **Am J Vet Res** 2012;73(2):213-219. [5]Tocci LJ, **J Vet Emerg Crit Care** 2009;19(1):66-73. [5]Weiss DJ and Wardrop KJ, **Schalm's Veterinary Hematology** 2010; Ch 92:711-724.

Author Index

A

B

Cryptococcosis and Other Fungal Diseases in Camelids

Biourge, Vincent, DVM, PhD
Dietary Management of Select Hepatobillary Conditions in Dogs

Birkenheuer, Adam, DVM, DACVIM (SAIM), PhD
Cytauzoonosis

Bjornvad, Charlotte, DVM, PhD
Feline Metabolic Syndrome

Boothe, Dawn, DVM, PhD, DACVIM (SAIM), DACVCP
Herbal/Nutraceutical Drug Interactions

Boston, Sarah, DVM, DVSc, DACVS
Medical, Surgical & Radiation Oncology: Diagnosis and Management of Histocytic Sarcomas

Boston, Sarah, DVM, DVSc, Dipl ACVS
Medical, Surgical & Radiation Oncology: Diagnosis and Management of Mast Cell Tumors
Medical, Surgical & Radiation Oncology: Diagnosis and Management of Nasal Tumors
Medical, Surgical & Radiation Oncology: Diagnosis and Management of Melanoma

Bowman, Dwight, MS, PhD
Dog and Cat Blood Parasites
Larval Cestode Disease in Dogs and Cats

Bowser, Jacquelin
Cyclophilin B and Lysyl-Hydroxlase-1 Interaction in Homo- and Heterozygous Quarter Horses with Hereditary Equine Regional Dermal Asthenia (HERDA) Mutation

Boysen, Soren, DVM, DACVECC
Assessing Cerebral and Peripheral Tissue Oxygenation with Near Infrared Spectroscopy
Assessing Tissue Perfusion in Shock and Sepsis with Dark Field Side-Stream Microscopy

Brogdon, Robin, MA
Creating Value Between Referral Practices and rDVM's
Communicating Value Between Referral Practices and rDVM's
Creating Value Between Referral Practices and Pet Owners
Communicating Value Between Referral Practices and Pet Owners

Brooks, Marjory, BS, DVM
Feline Platelet Disorders

Brown, William, DVM, DACVIM (Cardiology)
Arrhythmogenic Cardiomyopathy in 100 Boxer Dogs

Brownlie, Joe, BVSc, PhD, DSc, ARAgS, DipECVP, FRCPath, FRCVS
Mycoplasmas and Novel Viral Pathogens in Canine Infectious Respiratory Disease

Brunzell, John, MD
Human Metabolic Syndrome

Buczinski, Sebastien, DMV, MSc, DACVIM (LAIM)

Bovine Respiratory Disease

Ultrasonography for Diagnosis of Bovine Cardiorespiratory Disease

Budsberg, Steve, DVM, BS, MS, DACVS

Joint Nutraceuticals and Diets: Critical Evaluation of Efficacy

Buhl, Rikke, DVM, PhD

Exercise Induced Changes in the Equine Athlete's Heart

Burgess, Brandy, DVM, MSc, DACVIM (LAIM)

Rapid *Salmonella* Detection in Fecal and Veterinary Hospital Environmental Samples Using Commercially Available Lateral Flow Antigen Detection Systems

Burns, Kara, MS, MEd, LVT

Benefits of Fish Oils and Fatty Acids in Cardiac Patients with Valvular Disease

Bush, Bill, VMD, DACVIM (Neurology)

Electroencephalography Findings in 41 Cases

Byers, Christopher, DVM, DACVECC, DACVIM (SAIM)

Feline Hypotension: Improving Cardiac Output and Systemic Vascular Resistance

Risk Factors and Measurement for Intra-Abdominal Hypertension in Dogs and Cats

C

Callens, Amanda, BS, LVMT

Canine Urinary Incontinence: Let's Fix the Leak

Urinary Tract Cancer: Diagnostic and Treatment Options

Campbell, Karen, DVM, DACVIM (SAIM)

Dermatological Manifestations of Internal Medicine Disease

Chaffin, Morgan, DVM, MS, DACVIM

Do Hematologic and Ultrasonographic Screening Methods Predict Clinically Apparent *Rhodococcus Equi* in Foals?

Chan, Daniel, DVM, DACVECC, DACVN, MRCVS

Critical Care Aspects of Canine PLE

Nutritional Aspects of Canine PLE: The Criticalist and Internist Perspective

Claire Underwood

Intravenous Technetium-99M Liposomes Accumuolate in Lamellar Tissue During Laminitis Onset

Coates, Joan, DVM, MS, DACVIM (Neurology)

Identification of a Mutation Associated with Spinocerebellar Ataxia, Myokymia and Seizures in Jack/Parson Russell Terriers Using Whole Genome Sequencing

Coates, Joan, DVM, MS, DACVIM (Neurology

Clinical Trial for Canine Degenerative Myelopathy

Constable, Peter, BVSc (Hons), MS, PhD, DACVIM (LAIM)

Emerging Issues in Gastrointestinal Disorders of Neonatal Calves

The 5 Best Food Animal Medicine Papers of 2012

Cook, Audrey, BVM&S, MRCVS, DACVIM (SAIM), DECVIM-CA
TCC Laser Ablation: The TAMU Experience

Coté, Etienne, DVM, DACVIM (SAIM & Cardiology)
Effect of Vagal Maneuvers on the Heart Rate of Healthy Dogs

Creevy, Kate, DVM, MS, DACVIM (SAIM)
Beyond the Lecture in Veterinary Education: Creating Lifelong Learners

Cross, Cheryl, DVM, DACVP
A Joint Effort: Complementary and Alternative Management of Osteoarthritis

Crowdis, Kelly, DVM
One Medicine in Disasters: Haiti Earthquake

Cunningham, Suzanne, DVM, DACVIM (Cardiology)
The Fire Within: The Role of Inflammation in Cardiovascular Health and Disease

d

da Costa, Ronaldo, DMV, MSc, PhD, DACVIM (Neurology)
Is Apoptosis Present in the Spinal Cord of Dogs with Cervical Spondylomyelopathy?
Spinal MRI - Clinical and Research Applications

Dandrieux, Julien
Best of the Recent PLE Literature

DeClue, Amy, DVM, MS, DACVIM (SAIM)
Biomarkers for Sepsis: Are We Any Closer?
Identification and Management of Sepsis in the Cat

DeFrancesco, Teresa, DACVIM (Cardiology)
Effect of Atenolol on Quality of Life, Activiy and Cardiac Biomarkers in Cats with Apparently Asymptomatic Hypertrophic Cardiomyopathy

DiBaise, John, MD
Protein Losing Enteropathy in Humans

Dickinson, Peter, BVSc, PhD, DACVIM (Neurology)
Genome Wide Methylation Sequencing to Profile Epigenetic Modifications in Canine Gilioma

Dominique Paepe
Development of Cut-Off Exo-lohexol, Creatinine, Endo-lohexol Concentrations at 3 Time Points After Intravenous Isohexal and Creatine Injection to Predict Low Glomerular Filtration Rate in Cats

Durham, H. Edward, CVT, LATG
Body Cavity Centesis: Review of Centesis Techniques for the Pleural, Pericardial, and Abdominal Cavities
Not Just a Job and More Than an Adventure

Air Quality: What Our Patients Wish We Knew!

E

Erck, Emmanuelle van
Valvular Disorders of Equine Athletes

Estrada, Amara, DVM, DACVIM (Cardiology)
The Pearls and Pitfalls of Pulmonic Stenosis
The Nuts and Bolts of Pacing Therapy in Small Animal Practice

F

Felsted, Karen, CPA, MS, DVM, CVPM
Creating Value Between Referral Practices and rDVM's
Communicating Value Between Referral Practices and rDVM's
Creating Value Between Referral Practices and Pet Owners
Communicating Value Between Referral Practices and Pet Owners

Finno, Carrie, BSc, DVM, DACVIM (LAIM)
Vitamin E Supplementation in Horses with Neuromuscular Disorders

Fleming, Sherrill, DVM, DACVIM (LAIM), DABVP (Food Animal)
Vaccinology for Food Animals, Why Vaccines May or May Not Work

Foster, Derek, DVM, MS, DACVIM (LAIM)
Detection of Shiga Toxin-Producing *Escherichia Coli* O157:H7 in Meat Goats at Harvest

Fox, Philip, DVM, MS, DACVIM (Cardiology), DECVIM (Cardiology), DACVECC
International Collaborative Study to Assess Cardiovascular Risk and Evaluate Long Term Health in Feline Hypertrophic Cardiomyopathy (Reveal)

Frank, Nicholas, DVM, PhD, DACVIM (LAIM)
Equine Metabolic Syndrome

Freeman, Lisa, DVM, PhD, DACVN
Effect of Diet on the Progression of Hypertrophic Cardiomyopathy (HCM) in Cats
Cardiac Supplements: Critical Evaluation of Efficacy
Top Ten Treatment Tips for Feline Heart Disease: Feeding and Pharmacology

Frimberger, Angela, VMD, MACVSc, DACVIM (Oncology)
Canine Osteosarcoma Outcomes with Sequential Doxorubicin - Carboplatin Chemotherapy

G

Garden, Oliver, DVM, DACVIM (Oncology)
Variability of T Cell FOXP3 Expression in the Affected Lymph Nodes of Dogs with Multicentric B

Garosi, Laurent, DVM, DECVN
Advancements in the Diagnosis of Cerebrovascular Disease in Dogs and Cats

Garrett, Laura, DVM, DACVIM (Oncology)
Nonverbal Communication: Tricks of the Trade
Dealing with High Maintenance Owners
Making the Most of Cytology in Your Practice

Genova, Suzanne, DVM, DACVIM
VAC® Therapy: A New Approach to Wound Care in Food Animal Cases

Gerspach, Christian, DVM, MS, DACVIM
Sonographic Assessment of the Abdomen in Cattle

Giger, Urs, PhD, DVM, MS, FVH, DACVIM (SAIM), ECVIM, ECVCP
Diagnosis IMHA: the Coombs' Test

Goff, Jesse, DVM, PhD
Immune Suppression of Calving: Why? The Role of Metabolic Disorders
Macromineral Problems of Ruminants: Latest Issues and How to Fix Them

Goldstein, Richard, DVM, DECVIM-CA, DACVIM (SAIM)
A LIPL32 Based ELISA for the Detection of *Leptospira* Specific Antibody in Canine Serum

Gordon, Sonya, DVSc, DACVIM (Cardiology), DVM
Prospective Evaluation of NT-PROBNP, High Sensitivity Troponin I and PDK4 for the Detection of Occult DCM in 225 Doberman Pinschers

Graham, Catherine, DVM
Brain and Brainstem Diseases of Feedlot Calves: From Videos to Histopathology
Diseases of the Cerebellum and Spinal Cord of Feedlot Calves

Grauer, Gregory, DVM, MS, DACVIM (SAIM)
Top Nephrology/Urology Papers from 2012

Gull, Tamara, DVM, PhD, DACVPM, DACVIM (LAIM)
Handling and Submission of Diagnostic Samples

H

Haggstrom, Jens, DVM
Can I Trust My Echocardiographic Assessments of Cardiac Dimension and Volumes?
A Critical Review of Neuroendocrine Hormones as Cardiac Biomarkers in Dogs

Hallowell, Gayle, MA, VetMB, PhD, CertVA, DACVIM (LAIM), DACVECC, FHEAMRCVS
Assessing Cardiovascular Function in the Critically Ill Horse
Dysregulation of Calcium Homeostatis in the Critically Ill Horse

Hart, Kelsey, DVM, DACVIM (LAIM)
Serum Free Cortisol Fraction and Insulin Concentration in Normal Horses, Obese Horses and Horses with Pituitary Pars Intermedia Dysfunction

Regressive Feline Leukemia Virus Infections: What Do They Mean?

Haworth, David, DVM, PhD
ACVIM Candidate Function: 1) The Funding Landscape for Research in Veterinary Medicine; 2) Tactics for Successful Grant Writing

Heller, Heidi Barnes, DACVIM (Neurology)
Prednisone as a Solo Immunosuppressant in Dogs with Presumptive Meningoencephalitis of Unknown Etiology

Henry, Carolyn, DVM, MS, DACVIM (Oncology)
Evaluation of Tavocept, Prioxicam and Cisplatin for the Treatment of Canine Transitional Cell Carcinoma of the Urinary Bladder

Herring, Jennifer, DVM
Storage-Related Changes in Packed Red Blood Cells

Hill, Steve, DVM, MS, DACVIM (SAIM)
Diagnosis of PLE
Management of Canine PLE

Hogan, Daniel, DVM, DACVIM (Cardiology)
Analysis of the Feline Arterial Thromboembolism: Clopidogrel vs. Aspirin Trial (Fat Cat)

Hoglund, Katja, DVM
A Critical Review of Neuroendocrine Hormones as Cardiac Biomarkers in Dogs

Hudson, Neil, DVM
Equine Postoperative Ileus: Are We on the Right Track?
Equine Collapse: Causes and Diagnostic Approach

Hume, Kelly, DVM, DACVIM (Oncology)
The Persistent and Pesky Problem of Feline Injection Site Sarcomas

I

Impellizeri, Joseph, DVM, DACVIM (Oncology)
Protecting the Team: Chemotherapy Safety in Veterinary Medicine

J

Jäderlund, Karin Hultin, DVM, PhD, DECVN
Hereditary Polyneuropathy in Alaskan Malamutes - Phenotype and Genotype of Historical and Recent Cases

Jesty, Sophie, DVM, DACVIM (LAIM & Cardiology
The Pearls and Pitfalls of Pulmonic Stenosis

Johns, Imogen, BVSc, DACVIM (LAIM), MRCVS
The Compensatory Anti-inflammatory Response in Horses

Advances in the Diagnosis and Treatment of Botulism

Johnson, Gary
Novel Gene Sequencing Technique

Johnson, Lynelle, DVM, PhD, DACVIM (SAIM)
Canine and Feline Rhinitis: Theory and Practice

Johnson, Philip, BVSc, MS, MRCVS, DACVIM (LAIM)
An Emerging Role for the Equine and Asinine Gamma-herpes Viruses?

Johnson, Valerie, DVM
Surprising Antimicrobial Properties of Mesenchymal Stem Cells

K

Kerl, Marie, DVM, MPH, DACVIM, DACVECC
Acute Kidney Injury: Human and Veterinary Knowledge
DKA Advanced Nursing Care

Ketchum, Chastity, VTS
Endocrinopathies: Happening Hormones

Kidd, Linda, DVM, PhD, DACVIM (SAIM)
One Medicine in the Exam Room: Antibiotic Resistant Colonization and Infection in Humans and Animals

Kipar, Anja, DECVP, MRCVS, PD
FIP, Pathogenesis and Testing Controversies

Kjelgaard-Hansen, Mads, DVM, PhD
Canine Major Acute Phase Proteins: Biochemistry and Physiology
Canine Major Acute Phase Proteins: Clinical Applications

Kramer, George, DVM, DACVIM (Cardiology)
Transcatheter Valve Replacement: The Human Experience and the Veterinary Horizon

Kritchevsky, Janice Sojka, VMD, MS, DACVIM (LAIM)
Nutritional Support in the Dysphagic Horse

Kruger, John, DVM, DACVIM (SAIM)
A Year-Long Prospective, Randomized, Double-Masked Study of Nutrition on FIC

KuKanich, Butch, DVM, PhD, DACVCP
Nutraceutical Evaluation: Marketing and Surveillance

KuKanich, Kate, DVM, PhD, DACVIM (SAIM)
Urinary Tract Nutraceutics: Critical Evaluation of Efficacy

L

Regulation of Glucose Transport During Equine Hyperinsulinamia

Laflamme, Dorothy, DVM, PhD, DACVN
Nutritional Consideration of Canine Patients with Hepatobillary Disease

Lappin, Michael, DVM, PhD, DACVIM (SAIM)
Testing for Toxoplasmosis and Neosporosis
Transmission of *Bartonella Henselae* to Puppies by Exposure to *Ctenocephalides Felis* Allowed to Feed on *Bartonella Henselae*-Infected Cats

Larsen, Jennifer, DVM, PhD, DACVN
Update on Urate Urolithiasis

LaRue, Susan, DVM, MS, PhD
Treatment of Head and Neck Tumors Using New Radiation Therapy Modalities
Advances in Radiotherapy Treatment of Neurological Tumors

Lascola, Kara, DVM. DACVIM (LAIM)
CT as a Diagnostic Modality for Pulmonary Disease
Endotoxin and the Pulmonary System: Where are we Now?

Leclere, Mathilde
Inhaled Corticosteroids and Antigen Avoidance in Heaves: A Yearlong Study

Levine, Jonathan, DVM, DACVIM (Neurology)
Rodent SCI Models and their Application to Dogs

Levitan, Diane, DVM, DACVIM (SAIM)
Use of Hyperbaric Oxygen Therapy in Small Animal Medicine

Levy, Michel, DVM, PhD
Brain and Brainstem Diseases of Feedlot Calves: From Videos to Histopathology
Diseases of the Cerebellum and Spinal Cord of Feedlot Calves

Lim, Jihey
Stem Cell Sources in Adult Dogs

Ljungvall, Ingrid, DVM
Can I Trust My Echocardiographic Assessments of Cardiac Dimension and Volumes?

Loderstedt, Shenja
Biceps Reflex in the Dog

London, Cheryl, DVM, PhD, DACVIM (Oncology)
Anti-Tumor Activity of Irreversible Selective Inhibitors of Nuclear Export (SINE) in Dogs with Cancer
Evaluation of the Novel HSP90 Inhibitor STA-1474 in Dogs with Cancer
Tyrosine Kinase Inhibitors in Chemotherapy

Long, Sam, BVSc, PhD, DECVN
Placement of Deep Stimulating and Recording Electrodes in the Canine Brain Using the Brainsight® Frameless Stereotaxy System

M

MacDonald, Kristin, DACVIM (Cardiology)
Musculoskeletal Disorders in Veterinary Echocardiographers

MacGregor, John, DVM, DACVIM (Cardiology)
Cardiovascular Diagnostics to Differentiate Between Cardiac and Non-Cardiac Emergencies
Fluoroscopy for Cardiac Patients: Preparation, Monitoring, and Diagnoses

Mackin, Andrew, BSc, BVMS, MVS, DVSc, FACVSc, DACVIM (SAIM)
Transfusion Related Babesiosis: Is It a Real Issue?

MacNeill, Amy, DVM, PhD, DACVP
Oncolytic Virus Therapy in Models of Feline and Canine Cancers

Madigan, John, DVM, DACVIM (LAIM)
Why Foals Don't Gallop In Utero

Maggs, David, BVSc (hons), DACVO
Feline Herpesvirus-1: Antiviral Drugs from Bench-top to Clinic

Mariani, Christopher, DVM, PhD, DACVIM (Neurology)
Updates on Inflammatory CNS Disease, Diagnosis and Treatment

Marks, Steven, BVSc, MS, MRCVS, DACVIM (SAIM)
The Approach to the Respiratory Emergency

Masseau, Isabelle
Stem Cell Therapy Blunts CT Measures of Airway Remodeling in Experimental Feline Asthma

McGuirk, Sheila, DVM, MS, PhD, DACVIM (LAIM)
Abomasal Problems of Dairy Calves: Herd-Based Approach to Problem Solving
Metritis: Dairy Herd-Based Problem Solving

McLaughlin, Anne, DVM
Overview of Potential Human Factor Contributions to Training in Veterinary Medicine

McMichael, Maureen, DVM
Massive Transfusions

Meurs, Kathryn, DVM, DACVIM (Cardiology)
Identification of Genetic Polymorphisms in the Canine and Feline Renin Angiotensin Aldosterone System Genes, Angiotensin I-Converting Enzyme, Angiotensinogen and Angiotensin Receptor 1
Association of Dilated Cardiomyopathy with the Striatin Mutation Genotype in Boxer Dogs

Meylan, Mireille, Dr.med.vet., FVH, MS, PhD, Dr.habil, DACVIM (LAIM)
Risk Factor Analysis for Paratuberculosis in Swiss Dairy and Beef Herds

Middleton, John, DVM, PhD, DACVIM (LAIM)
Antimicrobial Resistance and Food Animals: Implications for Human and Animal Health

Mitchell, Katharyn, DVM, DACVIM (LAIM)
Equine Cytology: A Clinical Workshop

Prevalence of Diarrhea Causing Enteropathogens in New-Born Calves of Cattle (Bos indicus) and Water Buffalo (Bubalus bubalis) Using (FASTest® Strips) Rapid Tests in Egypt

Moise, Nancy, DVM, MS, DACVIM (Cardiology & SAIM)
The Pearls and Pitfalls of Pulmonic Stenosis

Moore, Antony, BVSc, MVSc, MACVSc, DACVIM (Oncology)
Chemotherapy for Malignant Mammary Carcinoma
Chemotherapy Options for Carcinomas and Sarcomas in Pets

Morley, Paul, DVM, PhD, DACVIM (LAIM)
Antimicrobial Resistance Surveillance in Beef Feedlot Cattle - Is There One Answer?

Munana, Karen, DVM, MS, DACVIM (Neurology)
Going with the Flow: An In-Depth Look at CSF
Neurological Disorders of the Geriatric Dog

Mutsaers, Anthony, DVM, DACVIM (Oncology)
Chemotherapy Clash: Maximum Tolerated Dose vs. Metronomic Protocols

N

Nichols, Rhett, DVM, DACVIM (SAIM)
Treatment of Cushing's Syndrome in Dogs: Arguments For and Against
Elevated Calcium or Potassium in Asymptomatic Patients: Pursue or Ignore?

Nogradi, Nora, DACVIM (LAIM)
Evaluation of an Omega-3 Fatty Acit Containing Feed Supplement (Aleira) in the Management of Horses with Chronic Lower Airway Inflammatory Diseases

O

O'Brien, Dennis, DVM, PhD, DACVIM (Neurology)
Identification of Two Breed-Specific Mutations Associated with Canine Multiple System Degeneration Using Whole Genome Resequencing

O'Brien, Robert, DVM, MS, DACVR
Microdose Computed Tomographic Cardiac Angiography in Normal Cats

Ograin, Vicky, MBA, RVT
Weight & Obesity: Impacts on Disease

Olby, Natasha, VetMB, PhD, DACVIM (Neurology)
Recovery From Spinal Cord Injury: A Treadmill Based Scoring System
Cerebellar Degeneration Due to an Autophagy Defect in Old English Sheepdogs and Gordon Setters

Olsen, Lisbeth, DVM, DrVetSci
Evaluation of the Breeding Program Aiming at Reducing the Prevalence of Myxomatous Mitral Valve Disease in Cavalier King Charles Spaniels

Going with the Flow: An In-Depth Look at CSF

Outerbridge, Catherine, MVSc, DVM
Dietary Management of Select Hepatobillary Conditions in Dogs

Oyama, Mark, DVM, DACVIM (Cardiology)
ACVIM Candidate Function: Hemodynamic Calculations

P

Pancotto, Theresa, DVM, MS, DACVIM (Neurology)
Shake, Rattle and Roll: Recognition of Movement Disorders

Paoloni, Melissa, DVM, DACVIM (Oncology)
Comparative Oncology: Where Are We Now and What Challenges Lie Ahead?

Pariaut, Romain, DVM, DACVIM (Cardiology), DECVIM-CA (Cardiology)
The Pearls and Pitfalls of Pulmonic Stenosis

Parnell, Nolie, DVM, DACVIM (SAIM)
Nutritional Aspects of Canine PLE: The Criticalist and Internist Perspective

Peterson, Mark, DVM, DACVIM (SAIM)
Feeding the Cat with Endocrine Disease
New Developments in Diagnostic Testing for Adrenal Disease in Cats
New Developments in the Use of Insulin Mixtures and Analogues for the Problem Diabetic
Pitfalls in the Management of Idiopathic Hypercalcemia in Cats

Piercy, Richard, MA, VetMB, MS, PhD, DACVIM (LAIM), MRCVS
Pathophysiology of Exertional Rhabdomyolysis

Platt, Simon, BVM&S, MRCVS, DACVIM (Neurology), DECVN
Job Task Analysis Update

Plummer, Paul, DVM, PhD, DACVIM (LAIM)
Coxiellosis: The Disease and Clinical Presentation
Coxiellosis: Control and Environmental Management
Coxiellosis: Lessons to be Learned from the Netherlands Outbreak

Poulin, Rachel, RVT
Inflammatory Bowel Disease: Diagnostics and Diagnosis

Powell, Lisa, DVM, BS, DACVECC
Colloid Therapy and Concentrated Albumin Transfusions
Canine Hypotension: Assessment and Therapeutic Interventions

Q

Quimby, Jessica, DVM, DACVIM (SAIM)
Telomere Length and Cellular Senescence in Feline CKD

R

S

Lactobacillus Strains and *Bifidobacterium Animalis* SSP *Lactis*

Schott, Harold, II, DVM, PhD, DACVIM (LAIM)
Comparison of Assay Results for Measurement of Plasma Adrenocorticotropin Concentration

Schwartz, Denise, FMVZ
Obesity is Linked to Heart Abnormalities in Humans. What About Dogs?

Scollan, Katherine, DVM, DACVIM (Cardiology)
Left Atrial Size and Function in Dogs Using Echocardiography and MDCT
Left Ventricular Volume and Function in Dogs Assessed by 3D Echocardiography

Sharkey, Leslie, DVM, PhD, DACVP (Clinical Pathology)
Liver Cytology: Critical Approach
Pancreatic Cytology

Sharp, Claire, DVM, BVMS, DACVECC
Low Stores of 25-Hydroxyvitamin D in Dogs and Its Association with Increased Cancer Risk

Shelton, G. Diane, DVM, PhD, DACVIM (SAIM)
The Nebulin Gene (NEB) is Likely Candidate for Nemaline Rod Myopathy in a Family of American Bulldogs
Diagnosis of Feline and Canine Myopathies
Diagnosis of Feline and Canine Neuropathies

Sisson, David, DVM, DACVIM (Cardiology)
Left Ventricular Outflow Tract Obstructions in Dogs: What We Know and Don't Know
Left Atrial Size and Function in Dogs Using Echocardiography and MDCT

Smith, Bradford, DVM, DACVIM
Physical Diagnosis in Ruminants: A Review

Smith, Geof, DVM, PhD, DACVIM (LAIM)
3rd Compartment Ulcers in Camelids
The Other 5 Best Food Animal Medicine Papers of 2012

Smith, Joanne, MA, VetMB, PhD, DACVIM (SAIM)
Active Learning Strategies for the Classroom
Teaching Strategies for the Clinic Floor
Cognitive Biases in Ethical Decision Making
The Power of an Apology: Disclosing Medical Errors

Spugnini, Enrico, DVM, PhD, DACVIM (Oncology), DECVIM-CA (Oncology)
Targeting Cancer Acid Environment: A Novel Anticancer Strategy
Electroporation in Veterinary Medicine: Riding the Electrical Waveform

Stern, Joshua, DVM, DACVIM (Cardiology)
Heart Disease in Cats: Signs, Diagnosis, and Treatment

Steven Steinberg
Rehabilitation and the Neurologic Patient

Stewart, Allison, BVSc (Hons), MS, DACVIM (LAIM), DACVECC

Equine Cytology: A Clinical Workshop

A Technician's Guide to Equine Fungal Infections

Understanding Testing for Equine Endocriopathies

Suchodolski, Jan

Postnatal Development of the Canine Fecal Microbiome

Metagenomic Analysis of the Effect of Withholding Food on the Small Intestinal Microbiome of Healthy Dogs

Suter, Steven, MS,VMD,PhD, DACVIM (Oncology)

Defining Canine B-Cell Lymphoma Using Human Immunohistochemical Algorithms

Swecker, William Terry, DVM, PhD, DACVN

Omega 3 Fatty Acids in Cattle: Changing the Product and Changing the Bovine?

Trace Element Supplementation of Cattle: By Needle or By Mouth?

Swift, Simon, MA, VetMB, CertSAC, DECVIM-CA (Cardiology), MRCVS European and RCVS Recognised Spclst in Vet. Cardio

Results of the UK Cavalier King Charles Breeding Program: 1991 - 2010

Sykes, Benjamin, BSc, BVMS, MS, DACVIM(LAIM), MBA

Administration of Once Daily Trimethoprim-Sulphadimidine Does Not Improve Healing of Glandular Gastric Ulceration in Horses Receiving Omeprazole: A Blinded, Randomized Clinical Trial

T

Takara, Matt

One Medicine on the Battlefield: Clinical Practice Guidelines for Military Working Dog Emergency Care by Human Health Care Provicers

Taylor, Sandra, DVM, PhD, DACVIM (LAIM)

Recent Advances in Equine Immunodeficiencies

Diagnosis and Management of Hyperkeratotic Skin Lesions in Camelids

Equine Clostridial Diseases

Thamm, Douglas, VMD, DACVIM (Oncology)

Antiproliferative and Propoptotic Effects of Dual PI3K/mTOR Inhibition with PWT32976 in Canine Lymphoma

Glucose, Lactate and the Metastatic Phenotype

Thomason, John, DVM, MS, DACVIM (SAIM)

Aspirin Resistance: What do we Really Know?

Toth, Balazs, DVM, MS

Nutritional Support in the Dysphagic Horse

Trepanier, Lauren, DVM, PhD, DACVIM (SAIM), DACVCP

Practical Approach to Applying Pharmacokinetic Principles to Internal Medicine Cases

Idiosyncratic Drug Toxicities in Dogs and Cats

Medical, Surgical & Radiation Oncology: Diagnosis and Management of Mast Cell Tumors
Medical, Surgical & Radiation Oncology: Diagnosis and Management of Histocytic Sarcomas
Medical, Surgical & Radiation Oncology: Diagnosis and Management of Nasal Tumors
Medical, Surgical & Radiation Oncology: Diagnosis and Management of Melanoma

U

Uechi, Masami, DVM, PhD
Open-heart Surgery for Correction of Congenital Heart Defects: Nihon University Experience
Found in Translation: Pearls of Veterinary Cardiology from the Overseas Literature (Japan)

V

Vaden, Shelly, DVM, PhD, DACVIM (SAIM)
Vaginal Flora and UTI in Dogs

Vail, David, DVM, DACVIM (Oncology)
Advanced Imaging for the Assessment of Novel Therapeutics and Anti-Tumor Response
Metronomic (low-dose continuous) Chemotherapy: Principles, Rationale, Protocols
Assessing Clinical Trial Design: Critical Evaluation of Reported Outcomes

Valberg, Stephanie, DVM, PhD, DACVIM (LAIM), ACVSMR
Ingestion of Hypoglycin A in Seeds of the Box Elder Tree Causes Seasonal Pasture Myopathy

van Loon, Gunther, DVM, PhD, DECEIM
Electrocardiographic Diagnosis: Are We Trying Hard Enough?
Treatment of Common Arrhythmias: How and When to Treat

Vengust, Modest, DVM, DACVIM (LAIM)
Is Lactate Monitoring Valid for Performance Testing in Horses?

Verkest, Kurt, DVM, ACVS
Canine Metabolic Syndrome

Vite, Charles, DVM, PhD, DACVIM (Neurology)
From Cats to Children: Treating Nieman-Pick Type C in Multiple Species
A Critical Review of Zombies in the Arts: Neuroanatomy, Neurophysiology, and Neuropathology

Viviano, Katrina, DVM, PhD, DACVIM (SAIM)
Antioxidant Supplements: Critical Evaluation of Efficacy

W

Waddell, Katy, LVT
Coagulopathies Involving Primary Hemostasis
Acid Base Status from the Metabolic Aspect

Does Vitamin D Play a Role in Canine Heart Disease and Cancer?

Nutraceutical Use in Canine Cancer: Friend or Foe?

Walmsley, Gemma, MA, VetMB, MRCVS

Temporal Evaluation of Pathological and Ultrastructural Features in Canine Centronuclear Myopathy

Ward, Cynthia, VMD, PhD, DACVIM (SAIM)

Active Learning Strategies for the Classroom

Teaching Strategies for the Clinic Floor

Canine and Feline Cushing's Disease

Ware, Wendy, DVM, MS, DACVIM (Cardiology)

Sudden Death in a Family of English Springer Spaniels with Prolonged QT Intervals and a Substitution Mutation in the KCNQ1 Gene Affecting Cardiac Repolarizng Potassium Current (IKs)

Washabau, Robert, VMD, PhD, DACVIM (SAIM)

Mechanisms and Pathogenesis of Gastrointestinal Motility

Methods of Assessing Gastrointestinal Motility

New Therapies for Gastrointestinal Motility

Washburn, Kevin, DVM, DABVP, DACVIM (LAIM)

Concentration of Tulathromycin in Sterile Vs Infected Tissue Cages in Goats Following Subcutaneous or Intra-Chamber Administration

Weaver, Christine, LVT, VTS (Neurology)

Electroencephalography in the Neurologic Diagnostic Toolbox

Webb, Craig, DVM, PhD, DACVIM (SAIM)

Stem Cells in Inflammatory Bowel Disease

Webb, Tracy, DVM

Feline Adipose-derived Mesenchymal Stem Cells: What Do We Know?

Weese, J. Scott, DVM, DVSc, DACVIM (LAIM)

A Joint Effort: Nutritional Management of Osteoarthritis

Understanding Bacterial Biofilms

Pets and Immunocompromisd Owners

Weisse, Chick, VMD, DACVS

Comparison of Serum Doxorubicin Levels Following Different DEB Chemoembolization Techniques as well as Systemic Administration in the Same Canine Patients with Naturally-Occurring Hepatocellular Carcinoma

Wess, Gerhard, DVM, Dr. Med Vet., PhD, DACVIM (Cardiology), DECVIM (Internal Medicine & Cardiology)

Clinical Appraisal of the Role of Echocardiography in Acquired Canine Heart Disease

Wijnberg, Inge, DVM, PhD, DECEIM

The Diagnostic Possibilities of Quantitative Needle Electromyographic Examination

Wiley, Karen, LVT

PQRST - What Does It Mean to You and Me?

Hematemesis: Not as Easy as First Thought

Protein-Losing Enteropathies: Not What You Might Expect

Wise, L. Nicki, DVM, MS, DACVIM (LAIM)

Theiler Equi: A Review of Clearance Therapeutics and Novel Diagnostic Techniques

Witonsky, Sharon, DVM, DACVIM (LAIM)

Immunity to *Sarcocystis Neurona*

Wong, David, DVM, MS, DACVIM (LAIM), DACVECC

Renal Replacement Therapy in Horses: A Viable Clinical Option?

Wood, Darren, DVM

Thromboelastography: Can it Replace Traditional Hemostasis Testing

Clinical Application of TEG in Dogs and Cats

Woods, J. Paul, DVM, MS, DACVIM (Oncology)

Medical, Surgical & Radiation Oncology: Diagnosis and Management of Mast Cell Tumors

Medical, Surgical & Radiation Oncology: Diagnosis and Management of Histocytic Sarcomas

Medical, Surgical & Radiation Oncology: Diagnosis and Management of Nasal Tumors

Medical, Surgical & Radiation Oncology: Diagnosis and Management of Melanoma

Wortinger, Ann, LVT

Cachexia in the Chronically Ill Patient

Nutritional Management of Patients with Vomiting and Diarrhea

X

Xenoulis, Panagiotis

Laboratory Diagnosis of Pancreatitis in Dogs

Y

Yagi, Kenichiro, BS, RVT, VTS(ECC)

B-Harmony: Are You the Right Type for Me? Blood Component Compatibility